THE ENCYCLOPEDIA OF
Molecular Biology

THE ENCYCLOPEDIA OF
Molecular Biology

EDITOR IN CHIEF
SIR JOHN KENDREW

EXECUTIVE EDITOR
ELEANOR LAWRENCE

**Blackwell
Science**

© 1994 by
Blackwell Science Ltd
Editorial Offices:
Osney Mead, Oxford OX2 0EL
25 John Street, London WC1N 2BL
23 Ainslie Place, Edinburgh EH3 6AJ
238 Main Street, Cambridge,
  Massachusetts 02142, USA
54 University Street, Carlton,
  Victoria 3053, Australia

Other Editorial Offices:
Arnette Blackwell SA
  1, rue de Lille, 75007 Paris
  France

Blackwell Wissenschafts-Verlag GmbH
  Kurfürstendamm 57
  10707 Berlin, Germany

  Feldgasse 13, A-1238 Wien
  Austria

First published 1994
Reprinted as a paperback 1995

Set by Semantic Graphics, Singapore
Printed in Great Britain at the Alden Press Limited,
Oxford and Northampton
and bound by Hartnolls Ltd, Bodmin, Cornwall

DISTRIBUTORS

Marston Book Services Ltd
PO Box 87
Oxford OX2 0DT
(*Orders*: Tel: 01865 791155
            Fax: 01865 791927
            Telex: 837515)

USA
  Blackwell Science, Inc.
  238 Main Street
  Cambridge, MA 02142
  (*Orders*: Tel: 800 215-1000
             617 876-7000
             Fax: 617 492-5263)

Canada
  Oxford University Press
  70 Wynford Drive
  Don Mills
  Ontario M3C 1J9
  (*Orders*: Tel: 416 441-2941)

Australia
  Blackwell Science Pty Ltd
  54 University Street
  Carlton, Victoria 3053
  (*Orders*: Tel: 03 347-0300
             Fax: 03 349-3016)

A catalogue record for this title
is available from both the British Library
and the Library of Congress

ISBN 0-632-02182-9
      0-86542-621-X (pbk)

Contents

Long Entries
Listed Alphabetically

Long Entries
Listed by Subject Group

General

Molecular biology *J.C. Kendrew*

Structural biology *(see also Cell biology; Immunology; Molecular genetics)*

Chromatography *H.B.F. Dixon*
Circular dichroism *M. Suzuki*
Electrophoresis *H.B.F. Dixon*
ESR/EPR spectroscopy *R. Cammack*
EXAFS (extended X-ray absorption fine structure) *G. Diakun*
Glycans *R.W. Stoddart*
Glycolipids *R.W. Stoddart*
Haemoglobin(s) *J. Tame*
Immunoglobulin structure *A.M. Lesk & A. Tramontano*
Mechanisms of enzyme catalysis *W.V. Shaw*
Molecular chaperones *R.J. Ellis*
Neutron scattering and diffraction *J.P. Baldwin, C.D. Reynolds & S.J. Lambert*
NMR spectroscopy *J. Martin & J. Waltho*
Nucleic acid structure *M. Churchill*
Post-translational modification of proteins *R.B. Freedman*
Protein engineering *J.J. Holbrook*
Protein folding *P.A. Evans*
Protein structure *A.M. Lesk*
Protein–nucleic acid interactions *J.W.R. Schwabe & D. Rhodes*
Serpins *R.W. Carrell*
Transfer ribonucleic acid *D. Moras & A. Poterszman*
X-ray crystallography *A.C. Bloomer*

Molecular genetics *(see also Bacteria and bacteriophages; Developmental biology; Immunology; Molecular medicine; Plant molecular biology)*

Adenovirus *R.T. Hay*
Animal virus(es) *N.J. Dimmock*
Antigenic variation in African trypanosomes *P. Borst*
Chemical carcinogens and carcinogenesis *S. Neidle*
Chromatin *D. Clark*
Chromosome structure: molecular aspects *H. Cooke*
Databases and information handling in molecular biology *P. Stoehr & R. Fuchs*
DNA (deoxyribonucleic acid) *L.H. Naylor*
DNA methylation *A.P. Bird*
DNA repair *R.D. Wood*
DNA replication *R. Laskey*
DNA sequencing *A.R. Coulson & R. Staden*

DNA typing *P. Gill*
DNA typing: applications in conversation and population genetics *D.A. Galbraith*
DNA typing: forensic applications *P. Gill*
Drosophila M. Ashburner
Epstein–Barr virus *P.R. Smith*
Eukaryotic gene expression *D.S. Latchman*
Eukaryotic gene structure *A. Flavell*
Gene *M.F. Tuite*
The genetic code *D. Moras & A. Poterszman*
Genetic engineering *T.A. Brown*
Genome organization *M.J. Carden*
Heat shock *P.W. Piper*
Hybridization *A.J.P. Brown*
In vitro translation *D. Colthurst*
Mendelian inheritance *J.R.S. Fincham*
Mitochondrial genomes: animal *H.T. Jacobs*
Mitochondrial genomes: lower eukaryote *H.T. Jacobs*
Mutation *J.M. Parry*
Papovaviruses *N.S. Krauzewicz*
Parental genomic imprinting *W. Reik*
Polymerase chain reaction *K.M. Sullivan*
Protein synthesis *C.G. Proud*
Recombination *K.A. Stacey*
Restriction enzymes *S.E. Halford*
Retroviruses *R. Vile*
Ribozymes *N.K. Tanner*
RNA (ribonucleic acid) *A.J.P. Brown*
RNA editing *J.M. Grienenberger*
RNA splicing *E.J. Mellor*
Saccharomyces cerevisiae M.F. Tuite
Transcription *E.J. Mellor*
Transcription factors *D.S. Latchman*
Transfection *L.H. Naylor*
Transgenic technologies *M. Evans*
Transposable genetic elements *D.J. Finnegan*
X-chromosome inactivation *M.F. Lyon*
Yeast mating-type locus *R. Egel*

Bacteria and bacteriophages *(see also Molecular genetics; Plant molecular biology)*

Bacterial chemotaxis *G.L. Hazelbauer*
Bacterial envelopes *T.R. Hirst*
Bacterial gene expression *J.E.G. McCarthy*
Bacterial gene organization *S.J.W. Busby*
Bacterial pathogenicity: molecular aspects *J.R. Saunders*
Bacterial protein export *T.R. Hirst*

List of Contributors

A. BETZ *MRC Laboratory of Molecular Biology, Hills Road, Cambridge CB2 2QH, UK*

A.P. BIRD *Institute of Cell and Molecular Biology, University of Edinburgh, King's Buildings, Mayfield Road, Edinburgh EN9 3JR, UK*

N.J.M. BIRDSALL *National Institute for Medical Research, Mill Hill, London NW7 1AA, UK*

A.C. BLOOMER *MRC Laboratory of Molecular Biology, Hills Road, Cambridge CB2 2QH, UK*

T. BOEHM *Klinikum der Albert-Ludwigs-Universität, Abt Innere Medizin I, Hugstetterstrasse 55, D7800 Freiburg, Germany*

P. BORST *Division of Molecular Biology, Netherlands Cancer Institute, Plesmanlaan 121, 1066 CX Amsterdam, The Netherlands*

P. BOSSIER *Laboratorio de Genetica Molecular, Instituto Gulbenkian de Ciencia, Rue de Quinta Grande, Apartado 14, 2781 Oeiras Codex, Portugal*

D.J. BOWLES *Centre for Plant Biochemistry and Biotechnology, University of Leeds, Leeds LS2 9JT, UK*

M.D. BRAND *Department of Biochemistry, University of Cambridge, Tennis Court Road, Cambridge CB2 1QW, UK*

G. BRICOGNE *MRC Laboratory of Molecular Biology, Hills Road, Cambridge CB2 2QH, UK*

A.J.P. BROWN *Department of Molecular & Cell Biology, University of Aberdeen, Marischal College, Aberdeen AB9 1AS, UK*

T.A. BROWN *Department of Biochemistry & Applied Molecular Biology, UMIST, PO Box 88, Manchester M60 1QD, UK*

G. BURNSTOCK *Department of Anatomy & Developmental Biology, University College London, Gower Street, London WC1E 6BT, UK*

S.J.W. BUSBY *School of Biochemistry, University of Birmingham, PO Box 363, Birmingham B15 2TT, UK*

V.S. BUTT *Department of Plant Sciences, University of Oxford, Parks Road, Oxford OX1 3PF, UK*

B. BYTH *Department of Genetics, Hospital for Sick Children, 555 University Avenue, Toronto, Ontario, Canada M5G 1X8*

M. CAMBRAY-DEAKIN *School of Biological Sciences, University of Sheffield, Sheffield S10 2TN, UK*

R. CAMMACK *Division of Life Sciences, King's College, University of London, Campden Hill Road, London W8 7AH, UK*

M.J. CARDEN *Biological Laboratory, University of Kent, Canterbury, Kent CT2 7NJ, UK*

R.W. CARRELL *Department of Haematology, University of Cambridge Clinical School, MRC Centre, Hills Road, Cambridge CB2 2QH, UK*

V.K.K. CHATTERJEE *Department of Medicine, University of Cambridge, Addenbrooke's Hospital, Hills Road, Cambridge CB2 2QQ, UK*

M. CHURCHILL *Department of Cell and Structural Biology, 506 Morrill Hall, 505 S. Goodwin Avenue, Urbana, IL 61801, USA*

D. CLARK *Bldg 5, Room 204, Laboratory of Molecular Biology, NIDDK, NIH, 9000 Rockville Pike, Bethesda, MD 20892, USA*

F.A.L. CLOWES *Department of Plant Sciences, University of Oxford, South Parks Road, Oxford OX1 3PF, UK*

S.P. COBBOLD *Sir William Dunn School of Pathology, University of Oxford, South Parks Road, Oxford, OX1 3RE, UK*

P.T.W. COHEN *MRC Protein Phosphorylation Unit, Department of Biochemistry, The University, Dundee DD1 4HN, UK*

G. L. COLLINGRIDGE *Department of Pharmacology, The Medical School, University of Birmingham, Edgbaston, Birmingham B15 2TT, UK*

A. COLMAN *Pharmaceutical Proteins Ltd, Roslin, Midlothian EH25 9PP, UK*

D. COLTHURST *Biological Laboratory, University of Kent, Canterbury, Kent CT2 7NJ, UK*

H. COOKE *MRC Human Genetics Unit, Western General Hospital, Crewe Road, Edinburgh EH4 2XU, UK*

A.R. COULSON *MRC Laboratory of Molecular Biology, Hills Road, Cambridge CB2 2QH, UK*

G.L. DANIELS *MRC Blood Group Unit, Wolfson House, University College London, 4 Stephenson Way, London NW1 2HE, UK*

K.E. DAVIES *Molecular Genetics Group, Institute of Molecular Medicine, John Radcliffe Hospital, Oxford OX3 9DU, UK*

S.D. DAVIES *Division of Neurobiology, MRC Laboratory of Molecular Biology, Hills Road, Cambridge CB2 2QH, UK*

P. DE MEYTS *Hagedorn Research Laboratory, Niels Steensens Vej 6, DK-2820 Gentofte, Denmark*

T.M. DEXTER *Kay Kendall Laboratory, Paterson Institute for Cancer Research, Christie Hospital, Wilmslow Road, Manchester M20 9BX, UK*

G. DIAKUN *SERC Daresbury Laboratory, Daresbury, Warrington, Cheshire WA4 4AD, UK*

N.J. DIMMOCK *Department of Biological Sciences, University of Warwick, Coventry CV4 7AL, UK*

C. DINGWALL *Wellcome/CRC Institute, Tennis Court Road, Cambridge CB2 1QR, UK*

H.B.F. DIXON *Department of Biochemistry, University of Cambridge, Tennis Court Road, Cambridge CB2 1QW, UK*

M.B.A. DJAMGOZ *Neurobiology Group, Department of Biology, Imperial College of Science, Technology and Medicine, Prince Consort Road, London SW7 2BB, UK*

J.E.G. DOWNING *Neurobiology Group, Department of Biology, Imperial College of Science, Technology and Medicine, Prince Consort Road, London SW7 2BB, UK*

R.R. EADY *Nitrogen Fixation Laboratory, AFRC Institute of Plant Science Research, University of Sussex, Brighton BN1 9RQ, UK*

F. EBLING *Department of Anatomy, University of Cambridge, Downing Street, Cambridge CB2 3DY, UK*

R. EGEL *Department of Genetics, Institute of Molecular Biology, University of Copenhagen, Øster Farimagsgade 2A, DK-1353 Copenhagen K, Denmark*

R.J. ELLIS *Department of Biological Sciences, University of Warwick, Coventry CV4 7AL, UK*

H. EPSTEIN *Department of Anatomy, University of Cambridge, Downing Street, Cambridge CB2 3DY, UK*

E.M. EVANS *HUGO Europe, 1 Park Square West, London NW1 4LJ, UK*

M. EVANS *Wellcome/CRC Institute, Tennis Court Road, Cambridge CB2 1QR, UK*

P.A. EVANS *Department of Biochemistry, University of Cambridge, Tennis Court Road, Cambridge CB2 1QW, UK*

R. FARQUHAR *Discovery Biology, Pfizer Central Research, Sandwich, Kent CT13 9NJ, UK*

A.R. FARUQI *MRC Laboratory of Molecular Biology, Hills Road, Cambridge CB2 2QH, UK*

M. FERGUSON *Department of Biochemistry, The University, Dundee DD1 4HN, UK*

C.K. FFRENCH-CONSTANT *Wellcome/CRC Institute, Tennis Court Road, Cambridge CB2 1QR, UK*

J.R.S. FINCHAM *20 Greenbank Road, Edinburgh EH10 5RY, UK*

D.J. FINNEGAN *Institute of Cell & Molecular Biology, University of Edinburgh, King's Buildings, Mayfield Road, Edinburgh EH9 3JR, UK*

A. FLAVELL *Department of Biochemistry, University of Dundee, Dundee DD1 4HN, UK*

T.P. FLEMING *Department of Biology, Biomedical Sciences Building, University of Southampton, Southampton SO9 3TU, UK*

S.P. FRASER *Neurobiology Group, Department of Biology, Imperial College of Science, Technology and Medicine, Prince Consort Road, London SW7 2BB, UK*

R.B. FREEDMAN *Biological Laboratory, University of Kent, Canterbury, Kent CT2 7NJ, UK*

R. FUCHS *EMBL, Meyerhofstrasse 1, D-6900 Heidelberg, Germany*

D.A. GALBRAITH *The Durrell Institute of Conservation & Ecology, University of Kent, Canterbury, Kent CT2 7NX, UK*

J. GAMBLE *Biological Laboratory, University of Kent, Canterbury, Kent CT2 7NJ, UK*

R. GIBBONS *Institute of Molecular Medicine, John Radcliffe Hospital, Oxford OX3 9DU, UK*

P. GILL *Central Research and Support Establishment, Home Office Forensic Science Service, Aldermaston, Reading RG7 4PN, UK*

B.D. GOMPERTS *Department of Physiology, University College London, Rockefeller Building, University Street, London WC1E 6JJ, UK*

N.M. GREEN *Laboratory of Protein Structure, National Institute for Medical Research, Mill Hill, London NW7 1AA, UK*

J.M. GRIENENBERGER *Institut de Biologie Moléculaire des Plantes du CNRS, 12 rue du Général Zimmer, 67084 Strasbourg Cedex, France*

K. GULL *Department of Biochemistry & Molecular Biology, University of Manchester, Stopford Building, Oxford Road, Manchester M13 9PT, UK*

W.J. GULLICK *ICRF Molecular Oncology Group, Hammersmith Hospital, Du Cane Road, London W12 0HS, UK*

S.E. HALFORD *Department of Biochemistry, University of Bristol Medical School, University Walk, Bristol BS8 1TD, UK*

A. HALL *Chester Beatty Laboratories, Institute of Cancer Research, Fulham Road, London SW3 6JB, UK*

D.G. HARDIE *Department of Biochemistry, University of Dundee, Dundee DD1 4HN, UK*

K.G. HARDY *Glaxo Institute of Molecular Biology SA, 46 route des Acacias, 1227 Carouge, Geneva, Switzerland*

A.D. HARTLEY *Section of Cell Growth, Regulation and Oncogenesis, Box 3686, Duke Comprehensive Cancer Center, Durham, NC 27710, USA*

D.A. HARTLEY *Department of Biochemistry, Imperial College of Science, Technology and Medicine, London SW7 2AZ, UK*

A.J. HARWOOD *MRC Laboratory of Molecular Biology, Hills Road, Cambridge CB2 2QH, UK*

R.T. HAY *Department of Biochemistry & Microbiology, University of St Andrews, St Andrews, Fife K16 9AL, UK*

G.L. HAZELBAUER *Department of Biochemistry & Biophysics, Washington State University, Pullman, WA 99164–4660, USA*

M. HAZELWOOD *Department of Immunology, The Medical School, University of Birmingham, Birmingham B15 2TT, UK*

J.K. HEATH *Department of Biochemistry, University of Oxford, South Parks Road, Oxford OX1 3QU, UK*

P. HEDDEN *Department of Agricultural Science, University of Bristol, AFRC Institute of Arable Crops Research, Long Ashton Research Station, Bristol BS18 9AF, UK*

S. HEKIMI *Department of Biology, McGill University, 1205 Dr Penfield Avenue, Montreal PQ, Canada H3A 1B1*

J.M. HENLEY *Department of Pharmacology, University of Birmingham, Birmingham B15 2TT, UK*

A.G. HEPBURN *PO Box 580, Verplanck, NY 10596, USA*

M. HIRST *Molecular Genetics Group, Institute of Molecular Medicine, John Radcliffe Hospital, Oxford OX3 9DU, UK*

T.R. HIRST *Biological Laboratory, University of Kent, Canterbury, Kent CT2 7NJ, UK*

M. HOBART *Mechanisms in Immunopathology Unit, MRC Centre, Hills Road, Cambridge CB2 2QH, UK*

J.J. HOLBROOK *Department of Biochemistry, University of Bristol Medical School, University Walk, Bristol BS8 1TD, UK*

N. HOLDER *Developmental Biology Research Centre, Randall Institute, Division of Biomolecular Sciences, King's College London, 26–29 Drury Lane, London WC2B 5RL, UK*

P.W.H. HOLLAND *Department of Zoology, University of Oxford, South Parks Road, Oxford OX1 3PS, UK*

J. HOPE *Institute for Animal Health, AFRC & MRC Neuropathogenesis Unit, Ogston Building, West Mains Road, Edinburgh EH9 3JF, UK*

N.D. HOPWOOD *Wellcome/CRC Institute, Cambridge CB2 1QR *Present address Cambridge Wellcome Unit for the History of Medicine, University of Cambridge, Free School Lane, Cambridge CB2 3RH, UK*

M.A. HORTON *ICRF Haemopoiesis Research Group, St Bartholomew's Hospital, London EC1A 7BE, UK*

C.J. HOWE *Biochemistry Department, University of Cambridge, Cambridge CB2 1QW, UK*

C.H.V. HOYLE *Department of Anatomy & Developmental Biology, University College London, Gower Street, London WC1E 6BT, UK*

R. HULL *John Innes Institute, Colney Lane, Norwich NR4 7UH, UK*

S.P. HUNT *Division of Neurobiology, MRC Laboratory of Molecular Biology, Hills Road, Cambridge CB2 3QH, UK*

T. HUNT *ICRF Clare Hall Laboratories, South Mimms, Herts EN6 3LD, UK*

J.C. HUNTER *Parke-Davis Neuroscience Research Centre, Addenbrooke's Hospital, Cambridge CB2 2QB, UK*

D.A. JACKSON *Sir William Dunn School of Pathology, University of Oxford, South Parks Road, Oxford OX1 3RE, UK*

M.B. JACKSON *Department of Agricultural Science, University of Bristol, AFRC Institute of Arable Crops Research, Long Ashton Research Station, Bristol BS18 9AF, UK*

H.T. JACOBS *Institute of Genetics, University of Glasgow, Church Street, Glasgow G11 5JS, UK*

R. JEFFERIS *Department of Immunology, The Medical School, University of Birmingham, Birmingham B15 2TT, UK*

P. JEFFS *Department of Anatomy, University of Cambridge, Downing Street, Cambridge CB2 3DY, UK*

R. JENKINS *Division of Neurobiology, MRC Laboratory of Molecular Biology, Hills Road, Cambridge CB2 2QH, UK*

A. JOHNSON *Department of Anatomy, University of Cambridge, Downing Street, Cambridge CB2 3DY, UK*

M.H. JOHNSON *Department of Anatomy, University of Cambridge, Downing Street, Cambridge CB2 3DY, UK*

A.W.B. JOHNSTON *School of Biological Sciences, University of East Anglia, Norwich NR4 7TJ, UK*

SIR JOHN KENDREW *The Guildhall, 4 Church Lane, Linton, Cambridge CB1 6JX, UK*

J. KENDRICK-JONES *MRC Laboratory of Molecular Biology, Hills Road, Cambridge CB2 2QH, UK*

R. KEYNES *Department of Anatomy, University of Cambridge, Downing Street, Cambridge CB2 3DY, UK*

A.E. KNIGHT *MRC Laboratory of Molecular Biology, Hills Road, Cambridge CB2 2QH, UK*

N.S. KRAUZEWICZ *Department of Virology, Royal Postgraduate Medical School, Hammersmith Hospital, Du Cane Road, London W12 0HS, UK*

J.M. LACKIE *Yamanouchi Research Institute, Littlemore Hospital, Sandford Road, Littlemore, Oxford OX4 4XN, UK*

R.A. LAKE *St Mary's Hospital Medical School, Department of Immunology, Wright–Fleming Institute, London W2 1PG, UK*

S.J. LAMBERT *Liverpool John Moore's University, School of Biomolecular Sciences, Byrom Street, Liverpool L3 3AF, UK*

E.M. LASATER *Departments of Physiology & Ophthalmology, University of Utah, Salt Lake City, UT 84108, USA*

R. LASKEY *Wellcome/CRC Institute, Tennis Court Road, Cambridge CB2 1QR, UK*

D.S. LATCHMAN *Division of Molecular Pathology, University College & Middlesex School of Medicine, University College London, The Windeyer Building, Cleveland Street, London W1P 6DB, UK*

M.-P. LEFRANC *Laboratoire d'immunogénétique Moléculaire, URA CNRS 1191, Université Montpellier II, Place Eugénie Bataillon, CPO12, 34095 Montpellier, Cedex 5, France*

A.M. LESK *Department of Haematology, University of Cambridge Clinical School, MRC Centre, Hills Road, Cambridge CB2 2QH, UK*

J.H. LEWIS *ICRF Developmental Biology Unit, Department of Zoology, South Parks Road, Oxford OX1 3PS, UK*

J. LUCOCQ *Anatomisches Institut, Universität Bern, CH-3000 Bern 9, Switzerland*

A.G.S. LUMSDEN *Division of Anatomy and Cell Biology, UMDS, Guy's Hospital, London Bridge, London SE1 9RT, UK*

J.P. LUZIO *Department of Clinical Biochemistry, University of Cambridge, Addenbrooke's Hospital, Hills Road, Cambridge CB2 2QR, UK*

M.F. LYON *MRC Radiobiology Unit, Chilton, Didcot OX11 0RD, UK*

J.F. McBLANE *Building 2, Room 304, NIDDKD, NIH, Bethesda, MD 20892, USA*

J.E.G. McCARTHY *Department of Gene Expression, Gesellschaft für Biotechnologie Forschung mbH, Mascheroder Weg 1, W-3300 Braunschweig, Germany*

J.R. McINTOSH *Department of Cellular and Developmental Biology, University of Colorado, Boulder, CO 80309-0347, USA*

S.K. MACIVER *MRC Laboratory of Molecular Biology, Hills Road, Cambridge CB2 2QH, UK*

I. MacLENNAN *Department of Immunology, University Medical School, Edgbaston, Birmingham B15 2TT, UK*

P.A. McNAUGHTON *Department of Physiology, King's College London, The Strand, London WC2R 2LS, UK*

A.I. MAGEE *Laboratory of Eukaryotic Molecular Genetics, National Institute for Medical Research, Mill Hill, London NW7 1AA, UK*

M. MARSH *MRC Laboratory for Molecular Cell Biology, University College London, Gower Street, London WC1E 6BT, UK*

C. MARTIN *The John Innes Institute, Norwich Research Park, Colney, Norwich NR4 7UH, UK*

J. MARTIN *Department of Molecular Biology and Biotechnology, University of Sheffield, PO Box 594, Sheffield S10 2UH, UK*

A. MARTINEZ ARIAS *Department of Zoology, University of Cambridge, Downing Street, Cambridge CB2 3EJ, UK*

E.J. MELLOR *Department of Biochemistry, University of Oxford, South Parks Road, Oxford OX1 3RA, UK*

L. MENGLE-GAW *Monsanto Company, 700 Chesterfield Village Parkway, Chesterfield, MO 63198, USA*

M.J. MERRICK *Nitrogen Fixation Laboratory, AFRC Institute of Plant Science Research, University of Sussex, Brighton BN1 9RQ, UK*

R.H. MICHELL *School of Biochemistry, University of Birmingham, Edgbaston, Birmingham B15 2TT, UK*

G. MOLINEUX *Kay Kendall Laboratory, Paterson Institute for Cancer Research, Christie Hospital, Wilmslow Road, Manchester M20 9BX, UK*

P. MORANDINI *MRC Laboratory of Molecular Biology, Hills Road, Cambridge CB2 2QH, UK*

D. MORAS *Institut de Biologie Moléculaire et Cellulaire du CNRS, 15 rue René Descartes, 67084 Strasbourg Cedex, France*

N.E. MURRAY *Institute of Cell & Molecular Biology, University of Edinburgh, King's Buildings, Mayfield Road, Edinburgh EH9 3JR, UK*

L.H. NAYLOR *Biological Laboratory, University of Kent, Canterbury, Kent CT2 7NJ, UK*

S. NEIDLE *Institute of Cancer Research, CRC Biomolecular Structure Unit, Cotswold Road, Sutton, Surrey SM2 5NG, UK*

M.S. NEUBERGER *MRC Laboratory of Molecular Biology, Hills Road, Cambridge CB2 2QH, UK*

D.H. NORTHCOTE *The Master's Lodge, Sidney Sussex College, Cambridge CB2 3HU, UK*

M. OSMOND *Department of Anatomy, University of Cambridge, Downing Street, Cambridge CB2 3DY, UK*

P.J. PARKER *ICRF, PO Box 123, Lincoln's Inn Fields, London WC2A 3PX, UK*

J.M. PARRY *School of Biological Sciences, University College Swansea, Singleton Park, Swansea SA2 8PP, UK*

D. PENNY *Molecular Genetics Unit, Massey University, Private Bag, Palmerston North, New Zealand*

S. PICKERING *Department of Anatomy, University of Cambridge, Downing Street, Cambridge CB2 3DY, UK*

P.W. PIPER *Department of Biochemistry and Molecular Biology, University College London, Gower Street, London WC1E 6BT, UK*

D.J. PORTEOUS *Molecular Genetics Group, MRC Human Genetics Unit, Western General Hospital, Crewe Road, Edinburgh EH4 2XU, UK*

A. POTERSZMAN *Institut Biologie Moléculaire et Cellulaire du CNRS, 15 rue René Descartes, 67084 Strasbourg Cedex, France*

S. POVEY *MRC Human Biochemical Genetics Unit, University College London, Wolfson House, 4 Stephenson Way, London NW1 2HE, UK*

D. POYNER *Pharmaceutical Sciences, Aston University, Aston Triangle, Birmingham B4 7ET, UK*

H. PRATT *Department of Anatomy, University of Cambridge, Downing Street, Cambridge CB2 3DY, UK*

C.G. PROUD *Department of Biochemistry, University of Bristol Medical School, University Walk, Bristol BS8 1TD, UK*

R. QUINLAN *Department of Biochemistry, University of Dundee, Dundee DD1 4HN, UK*

P.H. RABBITTS *Clinical Oncology & Radiotherapeutics Unit, MRC Centre, Hills Road, Cambridge CB2 2QH, UK*

P.B. RAINEY *Department of Plant Sciences, University of Oxford, South Parks Road, Oxford OX1 3RB, UK*

G.A. REID *Institute of Cell & Molecular Biology, University of Edinburgh, Mayfield Road, Edinburgh EH9 3JR, UK*

W. REIK *Department of Molecular Embryology, AFRC Institute of Animal Physiology & Genetics Research, Babraham, Cambridge CB2 4AT, UK*

C.D. REYNOLDS *School of Science and Technology, Liverpool John Moore's University, Byrom Street, Liverpool L3 3AF, UK*

G.P. REYNOLDS *Department of Biomedical Science, University of Sheffield, Sheffield S10 2TN, UK*

D. RHODES *MRC Laboratory of Molecular Biology, Hills Road, Cambridge CB2 2QH, UK*

A.J. RIVETT *Department of Biochemistry, University of Leicester, Leicester LE1 7RH, UK*

M.S. ROBINSON *Department of Clinical Biochemistry, University of Cambridge, Addenbrooke's Hospital, Hills Road, Cambridge CB2 2QR, UK*

J.H. ROGERS *Physiological Laboratory, Downing Street, Cambridge CB2 3EG, UK*

J.H. ROTHMAN *Department of Biochemistry, University of Wisconsin–Madison, 420 Henry Mall, Madison, WI 53706-1569, USA*

T. ROWE *MRC Laboratory of Molecular Biology, Hills Road, Cambridge CB2 2QH, UK*

N.J.P. RYBA *NIDR, National Institutes of Health, 9000 Rockville Pike, Bethesda, MD 20892, USA*

H. SAKANO *Department of Molecular and Cell Biology, Division of Immunology, University of California, Berkeley, CA 94720, USA*

J.R. SAUNDERS *Department of Genetics & Microbiology, The University, PO Box 147, Liverpool L69 3BX, UK*

P.J. SCAMBLER *Molecular Medicine Unit, Institute of Child Health, 30 Guildford Street, London WC1N 1EH, UK*

J.W.R. SCHWABE *MRC Laboratory of Molecular Biology, Hills Road, Cambridge CB2 2QH, UK*

S. SCOTT-DREW *Wellcome/CRC Institute, Tennis Court Road, Cambridge CB2 1QR, UK*

W.V. SHAW *Department of Biochemistry, University of Leicester, Leicester LE1 7RH, UK*

J. SLACK *ICRF Developmental Biology Unit, Department of Zoology, University of Oxford, South Parks Road, Oxford OX1 3PS, UK*

H. SMITH *Department of Botany, Adrian Building, University of Leicester, Leicester LE1 7RH, UK*

J.C. SMITH *Laboratory of Developmental Biology, National Institute for Medical Research, Mill Hill, London NW7 1AA, UK*

P.R. SMITH *Department of Virology, Royal Postgraduate Medical School, Hammersmith Hospital, Du Cane Road, London W12 0HS, UK*

A.K. SOUTAR *MRC Lipoprotein Team, Royal Postgraduate Medical School, Hammersmith Hospital, Du Cane Road, London W12 0HS, UK*

K.A. STACEY *Biological Laboratory, University of Kent, Canterbury, Kent CT2 7NJ, UK*

R. STADEN *MRC Laboratory of Molecular Biology, Hills Road, Cambridge CB2 2QH, UK*

I. STANSFIELD *Biological Laboratory, University of Kent, Canterbury, Kent CT2 7NJ, UK*

F.A. STEPHENSON *School of Pharmacy, 29–39 Brunswick Square, London WC1N 1AX, UK*

C.D. STERN *Department of Human Anatomy, University of Oxford, South Parks Road, Oxford OX1 3QX, UK*

R.W. STODDART *Department of Pathological Sciences, University of Manchester, Stopford Building, Oxford Road, Manchester M13 9PT, UK*

P. STOEHR *EMBL, Meyerhofstrasse 1, D-6900 Heidelberg, Germany*

K.M. SULLIVAN *Home Office Forensic Service Research and Support Establishment, Aldermaston, Reading RG7 4PN, UK*

M. SUZUKI *MRC Laboratory of Molecular Biology, Hills Road, Cambridge CB2 2QH, UK*

J. TAME *Department of Chemistry, University of York, Heslington, York YO1 5DD, UK*

D. TANNAHILL *Department of Anatomy, University of Cambridge, Downing Street, Cambridge CB2 3DY, UK*

N.K. TANNER *GENSET, 1 Passage Etienne Delaunay, 75011 Paris, France*

P.E.R. TATHAM *Department of Physiology, University College London, Rockefeller Building, University Street, London WC1E 6JJ, UK*

E. TERZAGHI *Molecular Genetics Unit, Massey University, Private Bag, Palmerston North, New Zealand*

J. THOMAS *Department of Biochemistry, University of Dundee, Dundee DD1 4HN, UK*

P. TIPPETT *MRC Blood Group Unit, Wolfson House, University College London, 4 Stephenson Way, London NW1 2HE, UK*

A. TRAMONTANO *Istituto di Richerche di Biologia Molecolare, Via Pontina Km 30 600, 00400 Pomezia, Rome, Italy*

M.F. TUITE *Biological Laboratory, University of Kent, Canterbury, Kent CT2 7NJ, UK*

H. VAN DEN BERGHE *Centre for Human Genetics, University of Leuven, Campus Gasthuisberg, Herestraat, B-3000 Leuven, Belgium*

V. VAN HEYNINGEN *MRC Human Genetics Unit, Western General Hospital, Crewe Road, Edinburgh EH4 2XU, UK*

R. VILE *ICRF, PO Box 123, Lincoln's Inn Fields, London WC2A 3PX, UK*

A. WALMSLEY *Department of Molecular Biology & Biotechnology, University of Sheffield, PO Box 594, Sheffield S10 2UH, UK*

F.S. WALSH *Department of Experimental Pathology, UMDS, Guy's Hospital, London Bridge, London SE1 9RT, UK*

J. WALTHO *Department of Molecular Biology and Biotechnology, University of Sheffield, PO Box 594, Sheffield S10 2UH, UK*

G. WARREN *Cell Biology Laboratories, ICRF, PO Box 123, Lincoln's Inn Fields, London WC2A 3PX, UK*

C. WATTS *Department of Biochemistry, University of Dundee, Dundee DD1 4HN, UK*

SIR DAVID J. WEATHERALL *Nuffield Department of Clinical Medicine, John Radcliffe Hospital, Headington, Oxford OX3 9DU, UK*

M. WEST *Department of Biochemistry, University of Dundee, Dundee DD1 4HN, UK*

R.A.H. WHITE *Department of Anatomy, University of Cambridge, Downing Street, Cambridge CB2 3DY, UK*

The late M. WILCOX *MRC Laboratory of Molecular Biology, Hills Road, Cambridge CB2 2QH, UK*

A. WILKIE *Institute of Molecular Medicine, John Radcliffe Hospital, Oxford OX3 9DU, UK*

G.G. WILSON *Department of Pharmacology, University of Leeds, Woodhouse Lane, Leeds LS2 9JT, UK*

S.W. WILSON *Developmental Biology Research Centre, Randall Institute, Division of Biomedical Sciences, King's College London, 26–29 Drury Lane, London WC2B 5RL, UK*

R.D. WOOD *ICRF, Clare Hall Laboratories, South Mimms, Herts EN6 3LD, UK*

D. WRAY *Department of Pharmacology, University of Leeds, Woodhouse Lane, Leeds LS2 9JT, UK*

C.C. WYLIE *Wellcome/CRC Institute, Tennis Court Road, Cambridge CB2 1QR, UK*

K.M. YAMADA *Laboratory of Developmental Biology, NIDR, NIH, Bethesda, MD 20892, USA*

Preface

MOLECULAR biology had its beginnings in the 1930s, but when the major breakthroughs were achieved in the 1950s it still seemed to be a subject remote from practical applicability. However, with remarkable speed those same breakthroughs — in genetics, in the molecular mechanisms of development and of cell function, and in the structures of macromolecules such as DNA and the proteins — have become increasingly relevant to applied fields such as medicine and pharmaceuticals, agriculture, and industrial biotechnology. They have led to dramatic advances of great utility and value: gene therapy, drug design, genetically improved crops, and the synthesis of important chemicals, to mention only a few. These practical outcomes, which would have surprised the pioneers who thought of their research simply as a fascinating exploration of the unknown, have involved medical practitioners, agriculturalists and industrialists more and more in the applications of molecular biology, and demand an understanding of its principles and achievements and the capacity to read its literature. This wider circle, not brought up in the field, needs help with the jargon, and guidance in the principles, of molecular biology. And indeed, as in other proliferating branches of science, even within the field itself professional molecular biologists on occasion find they are unable to understand the jargon they encounter in reading the literature of areas other than their own.

In the nineteenth century it was possible for scientists in most subjects to write of their researches in language that would be understood by anyone capable of reading the daily newspapers. Alas! it is no longer so; what proportion of the latest issue of *Nature* or *Science* can any one individual, whether scientist or lay person, understand? Every active science has invented its own private jargon, and molecular biology is no exception. Indeed, jargon is exported and imported as well as being invented. 'Clone', imported into molecular biology from classical botany, has now been re-exported and has become a fashionable term used by the general public, not always appropriately. And in the contrary direction, 'downstream', 'upstream', 'chaperone' and even 'induction' have meaning to the molecular biologist quite different from their usage in ordinary life.

The purpose of this book is to help those both inside and outside the field to understand the methods, concepts and findings of molecular biology, even if they are initially ignorant of the jargon of parts or the whole of the subject. It is an encyclopedia and not a dictionary; that is to say, its authors have not been content simply to list dictionary-type definitions of words — though such definitions will be found in great numbers in its pages — but they have striven also, and more importantly, to provide a conspectus of major topics like protein synthesis or membrane behaviour or genetic manipulation, in language such that those outside the field can gain an understanding of topics that interest them or that are relevant to their own professional activity. It is for this reason that the text is made up of a mixture of long entries (more encyclopedic) and short entries (more dictionary-like). There are liberal cross-references, and the text includes a large number of diagrams, to assist navigation through and between different topics. All the many contributors are active research workers who have made important contributions to their own fields, and so have a first-hand knowledge of the topics they address. The editors hope that these features of the book will make it valuable in the work of all those who need an understanding of the concepts of molecular biology.

JOHN KENDREW
Cambridge, 1994

xix

Introduction

THE AIM of the *Encyclopedia of Molecular Biology* is to provide a comprehensive and concise reference source to the core topics of molecular biology itself and also to include general background material for those fields to which molecular biology now contributes greatly. It thus provides, as far as is possible in a single volume, a self-contained work of reference, accessible to all professional biologists and students of biology, and professionals and students in related disciplines such as medicine. Its wide coverage also makes it a convenient source for non-biologists seeking information on the current state of molecular biology.

The encyclopedia contains more than 5500 entries of varying length, arranged in alphabetical order. Numbers, Greek letters, and configurational letters at the beginning of names are ignored in the alphabetic ordering, so that, for example, β-galactosidase is listed under G, *N*-acetylgalactosamine under A. The entries include 217 longer reviews on selected topics, supplemented by more than 4000 shorter definitions of terms in current usage. The remainder of the keywords comprise common abbreviations, synonyms and cross-references directing the reader to the appropriate entries. A list of standard abbreviations used without definition throughout the encyclopedia is given on pp. xxv–xxvi.

Contents lists of the review-type entries, arranged both alphabetically and by subject area, are given on pp. vi–xi.

Within each entry, cross-references to other entries providing additional or background material appear in SMALL CAPITALS. Additional cross-references will be found in small capitals at the end of the entry.

All longer entries and many shorter entries are followed by references, which are intended as an entrance into the literature and not as comprehensive bibliographies.

ELEANOR LAWRENCE

Acknowledgements

We are grateful to authors and publishers for permission to reproduce the following copyright material. Full bibliographic details will be found in the entry accompanying the figure.

Fig. A15: from Watson et al. *The Molecular Biology of the Gene*. © 1987 by the Benjamin/Cummings Publishing Company, Inc.

Figs B16, B19, B26, G49, M3, M17: from *Nature*. © Macmillan Magazines, Ltd.

Fig. C4: Fig. 18 from White, Southgate, Thomson & Brenner (1986) The structure of the nervous system of the nematode Caenorhabditis elegans. *The Philosophical Transactions of the Royal Society of London*. **B314**, 1–340, by permission of the Royal Society of London.

Figs A35, C5, P60: from *Science*. © The American Association for the Advancement of Science.

Figs C43*a,b*, N36*b,c,e,f*: reproduced from the *Journal of Cell Biology* by copyright permission of the Rockefeller University Press.

Fig. C7: from *FEBS Letters*, by permission of Elsevier Science. © The Federation of European Biochemical Societies.

Figs C48, M28, S10, X2: from *Trends in Genetics*. by permission of Elsevier Science.

Fig. R29: from *Trends in Biochemical Science*, by permission of Elsevier Science.

Figs C43, D37, H12: from *Cell*. © Cell Press.

Figs N24, N25, N26, N27: from *Biochemistry*. © American Chemical Society.

Figs H4, N23, P62, S8: from *Journal of Molecular Biology*. © Academic Press Ltd.

Fig. C31: from Cook & Stoddart (1973) *Surface Carbohydrates of the Eukaryotic Cell*. © 1973 by Academic Press Ltd.

Figs M41, M42: from Huxley, A.H. (1983) *Molecular Basis of Contraction in Cross-striated Muscles and Relevance to Motile Mechanisms. Muscle and Nonmuscle Motility*. © 1983 by Academic Press, Inc.

Fig. Z3: (drawn by S Kimmel) from Kimmel et al. (1991) *Cell–cell Interactions in Early Development* (Wiley-Liss). © Wiley-Liss.

Figs B7, B15: from (Neidhart, F.C., Ed.) *Escherichia coli and Salmonella typhimurium: Cellular and Molecular Biology*. © 1987 American Society for Microbiology.

Fig. N17: from *Neuron*. © 1989 by Cell Press.

Figs C9, L6: from *Annual Review of Biochemistry* vols 55 & 58. © 1986 & 1989 Annual Reviews Inc.

Figs A49, N38, R20: from *The EMBO Journal* by permission of Oxford University Press. © EMBO/Oxford University Press.

Fig. X8: from Levitan & Kaczmarek (1991) *The Neuron: Cell and Molecular Biology* (Oxford University Press) by permission of Oxford University Press.

Figs B24, N39: from Saenger (1984) *Principles of Nucleic Acid Structure* (Springer-Verlag, New York). © Springer-Verlag.

Figs A55, D36, M50, P5, V4, Plate 8*e*: from *Development*. © Company of Biologists Ltd.

Figs C27, H13, M24: from *BioEssays*. © ICSU Press.

Fig. C26: from *Journal of Cell Science, Supplement*. © 1986 Company of Biologists Ltd.

Fig. C39: from *Molecular and General Genetics*, Springer International. © Springer-Verlag Berlin Heidelberg.

Figs P12, P14: from Alberts et al. (1989) *Molecular Biology of the Cell* 2nd edn (Garland Publishing Company, Inc). © Alberts et al.

Fig. P28: from *Photochemistry & Photobiology*. © American Society for Photobiology.

Fig. B14: from *Current Biology*. © 1991 Current Biology.

Fig. P62*d*: from *Current Opinion in Structural Biology*. © Current Opinion in Structural Biology.

Figs L8, V6: from *Current Opinion in Neurobiology*. © Current Opinion in Neurobiology.

Figs G10, G23, R35: from Singer & Berg (1991) *Genes and Genomes* (University Science Books). © 1991 University Science Books.

Fig. V5: from *Journal of Neurobiology*. © 1990 by John Wiley & Sons Inc.

Figs C1, C2: from *Developmental Biology*. © 1977, 1983 Academic Press, Inc.

Figs M27, N14, R20*a*: from *Nucleic Acids Research* (IRL Press), by permission of Oxford University Press.

Figs M9, M10, M11: from Evans & Graham (1989) *Membrane Structure & Function* (IRL Press) by permission of Oxford University Press, 1989.

Fig. B4: from *Canadian Journal of Microbiology*. © National Research Council of Canada 1988. All rights reserved.

Fig. N9: from *Neuroscience*. © 1988 IBRO.

Fig. N10: from *Developmental Brain Research*, by permission of Elsevier Science.

Figs D25, D27, D30. © British Crown Copyright, 1991.

Fig. A43: from *The New Biologist*. © 1992 W.B. Saunders, Philadelphia.

Fig. H3: from Perutz (1990) *Mechanism of cooperativity and allosteric regulation in proteins* (Cambridge University Press). © Miss Vivien Perutz and Dr R.N. Perutz, 1989, 1990.

Fig. H5: from Fermi & Perutz (1981) *Atlas of Molecular Structures in Biology: Haemoglobin & Myoglobin* (Clarendon Press), by permission of Oxford University Press 1981.

Fig. I36: from *Diabetes*. © American Diabetes Association.

Fig. E13: from *Genetics*. © 1964 by the Genetics Society of America.

Fig. M49: from Lash & Whittaker (Eds) (1974) *Concepts of Development*. © 1974 Sinauer Associates, Inc., Sunderland, MA.

Fig. A37: from *Control of Virus Diseases*, Society of General Microbiology Symposium **45**, 1990, by permission of the Society for General Microbiology.

Fig. M26: from *Gene*, by permission of Elsevier Science.

Fig. N13: from *European Journal of Biochemistry*, by permission of Elsevier Science.

Fig. H7: from Branden & Tooze (1991) *Introduction to Protein Structure*. © Garland Publishing Company, Inc.

Fig. R20c: from *Journal of Biological Chemistry*, © American Society for Biochemistry and Molecular Biology, Inc. 1990.

Plate 7 (a,b,c,d) © Scripps Clinic and Research Foundation.

Figs C10, C50, C81, C82, E2, E8, F2, H6, K5, L1, L5, M51, P63, S6, S35, T49, U1 and Plates 3, 4, 5, 6(b & c) courtesy of A.M. Lesk.

Units and Conversions

Quantity	SI name	SI abbreviation
SI base units		
length	metre	m
mass	kilogram	kg
time	second	s
electric current	ampere	A
temperature	T	°C*
luminous intensity	candela	cd
amount of substance	mole	mol
SI derived units		
force	newton	$N = m\,kg\,s^{-2}$
pressure	pascal	$Pa = N\,m^{-2}$
energy	joule	$J = N\,m$
power	watt	$W = J\,s^{-1}$
electrical potential	volt	$V = W/A, J/C$
electric field	volt m^{-1}	$V\,m^{-1}$
electric charge	coulomb	$C = A\,s$
capacitance	farad	$F = C\,V^{-1}$
resistance	ohm	$\Omega = V\,A^{-1}$
conductance	siemen	$S = \Omega^{-1}$
radioactivity	bequerel	$Bq = disintegrations\,s^{-1}$
absorbed radiation	gray	$Gy = J\,kg^{-1}$
dose equivalent	sievert	$Sv = J\,kg^{-1}$
frequency	hertz	Hz
luminous flux	lumen	$lm = cd\,sr$
illuminance	lux	$lx = lm\,m^{-2}$

Quantity	SI name	SI abbreviation	
Non-SI units			
length	Angstrom unit	Å	$10^{-10}\,m$
pressure	atmosphere, standard	atm	101 325 Pa
	bar		$10^5\,Pa$
mass	dalton	D,	1 atomic mass unit
		dal	$= 1.660\,54 \times 10^{-27}\,kg$
energy	calorie, thermochemical	cal	4.184 J
	electronvolt	eV	$1.602\,18 \times 10^{-19}\,J$
temperature	degree Celsius	°C	
volume	litre	l	$1 \times 10^{-3}\,m^3$
concentration	molar	M	$l\,mol^{-1}$
length (of a polynucleotide)	kilobase	kb	1000 bases
radioactivity	curie	Ci	$3.7 \times 10^{10}\,Bq$

*The SI unit of temperature is strictly the degree Kelvin (K), K = 273.15 + °C.

Prefixes for SI units

Factor	Prefix	Symbol	Factor	Prefix	Symbol
10^{-1}	deci	d	10^{1}	deca	da
10^{-2}	centi	c	10^{2}	hecto	h
10^{-3}	milli	m	10^{3}	kilo	k
10^{-6}	micro	μ	10^{6}	mega	M
10^{-9}	nano	n	10^{9}	giga	G
10^{-12}	pico	p	10^{12}	tera	T
10^{-15}	femto	f	10^{15}	peta	P
10^{-18}	atto	a	10^{18}	exa	E

List of Standard Abbreviations

A	adenine, alanine	Hb	haemoglobin
ADP	adenosine diphosphate	His	histidine
Ala	alanine	I	isoleucine
AMP	adenosine monophosphate	Ig	immunoglobulin
cAMP	cyclic AMP	Ile	isoleucine
cGMP	cyclic GMP	K	lysine
Ara	arabinose	K_d	dissociation constant
Arg	arginine	L	leucine
Asn	asparagine	Leu	leucine
Asp	aspartic acid	Lys	lysine
ATP	adenosine triphosphate	M	methionine
ATPase	adenosinetriphosphatase	M_r	relative molecular mass
C	cytosine, cysteine	Man	mannose
cDNA	complementary DNA	Met	methionine
CDP	cytidine diphosphate	mRNA	messenger RNA
CMP	cytidine monophosphate	Mur	muramic acid
CTP	cytidine triphosphate	MurNAc	N-acetylmuramic acid
cyclic AMP	adenosine 3′,5′-cyclic monophosphate	N	asparagine
cyclic GMP	guanosine 3′,5′-cyclic monophosphate	NAD^+	nicotinamide adenine dinucleotide (oxidized)
Cys	cysteine	NADH	nicotinamide adenine dinucleotide (reduced)
Cyt	cytochrome	$NADP^+$	nicotinamide adenine dinucleotide phosphate (oxidized)
D	aspartic acid		
d	2′-deoxyribo-	NADPH	nicotinamide adenine dinucleotide phosphate (reduced)
DNA	deoxyribonucleic acid		
DNase	deoxyribonuclease	NeuAc	N-acetylneuraminic acid (sialic acid)
E	glutamic acid	P	proline
EC	Enzyme Commission (Nomenclature Committee of the International Union of Biochemistry)	Phe	phenylalanine
		P_i	inorganic orthophosphate
		PP_i	inorganic pyrophosphate
F	phenylalanine	Pro	proline
$FAD(H_2)$	flavin adenine dinucleotide (reduced)	Q	glutamine
fMet	formylmethionine	R	arginine
$FMN(H_2)$	flavin mononucleotide (reduced)	Rib	ribose
Fuc	fucose	RNA	ribonucleic acid
G	guanine, glycine	rRNA	ribosomal RNA
GABA	γ-aminobutyric acid	snRNA	small nuclear RNA
Gal	galactose	tRNA	transfer RNA
GalNAc	N-acetylgalactosamine	RNase	ribonuclease
Glc	glucose	RNP	ribonucleoprotein
GlcA	glucuronic acid	Ru	ribulose
GlcNAc	N-acetylglucosamine	S	serine
Gln	glutamine	Ser	serine
Glu	glutamate	SV40	simian virus 40
Gly	glycine	T	thymine, threonine
GDP	guanosine diphosphate	TF	transcription factor
GMP	guanosine monophosphate	Thr	threonine
GTP	guanosine triphosphate	Trp	tryptophan
GTPase	guanosinetriphosphatase	TTP	thymidine triphosphate
H	histidine		

Tyr	tyrosine	V	valine
U	uracil	Val	valine
UDP	uridine diphosphate	W	tryptophan
UMP	uridine monophosphate	Xyl	xylose
UTP	uridine triphosphate	Y	tyrosine

A

A (1) ADENINE.
(2) The AMINO ACID alanine.

Å Angstrom unit, a measure of distance (equal to 0.1 nm) often used for convenience even though it is a nonstandard unit.

A antigen *See*: ABO BLOOD GROUP SYSTEM.

A-DNA *See*: DNA; NUCLEIC ACID STRUCTURE.

A-form *See*: NUCLEIC ACID STRUCTURE.

A-site *See*: PROTEIN SYNTHESIS.

A subfibre Component of axoneme of eukaryotic cilia and flagella. *See*: MICROTUBULES.

A23187 A monocarboxylic acid IONOPHORE isolated from *Streptomyces chartreusensis* (Fig. A1). It has a high specificity for divalent rather than monovalent cations and catalyses the exchange of one Ca^{2+} or Mg^{2+} for two H^+. It is used in experimental cell biology to raise the intracellular level of free Ca^{2+}.

AB antigen *See*: ABO BLOOD GROUP SYSTEM.

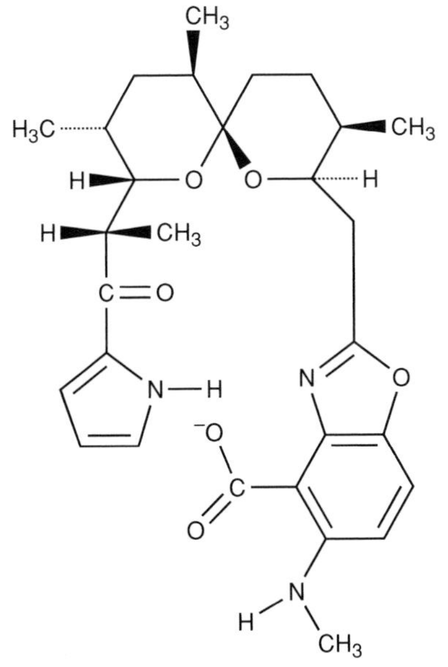

Fig. A1 Structure of A23187.

ABA Abscisic acid. *See*: PLANT HORMONES.

ABC proteins ATP-BINDING CASSETTE PROTEINS.

abdominal histoblasts In *Drosophila*, the precursors of the adult abdominal epidermis. These cells are arranged in segmentally repeated groups in the abdomen of the larva. They undergo division in the pupa and, at METAMORPHOSIS, replace the larval epidermis. *See*: DROSOPHILA DEVELOPMENT; IMAGINAL DISKS.

abdominal segments *See*: DROSOPHILA DEVELOPMENT.

Abelson murine leukaemia virus (AbMLV) Acutely transforming RETROVIRUS which carries the ONCOGENE *abl*.

abl, ABL The cellular homologue of the viral ONCOGENE v-*abl* from the acutely transforming RETROVIRUS, Abelson murine leukaemia virus. In patients with chronic myeloid leukaemia, the characteristic t(9;22)(q34;q11) translocation generates a composite gene comprised of exons from the *BCR* locus on chromosome 22 and the *ABL* gene on chromosome 9. The composite *BCR-ABL* gene spans the chromosomal junction of t(9;22)(q34;q11) and encodes a fusion protein with distinct biochemical properties. *See also*: CHROMOSOME ABERRATIONS.

ablation The selective physical destruction or removal of a cell or group of cells in the developing embryo, by surgery or laser treatment for example, in order to determine the effect of their loss on development. The term is also now used to describe the selective removal or destruction of a gene, as, for example, in a transgenic animal.

AbMLV Abelson murine leukaemia virus. *See*: ONCOGENES; RETROVIRUSES.

ABO blood group system

THE ABO histo-blood group system is a polymorphic system of antigens carried on red blood cells and other tissues [1]. It was the first BLOOD GROUP system — the first genetic polymorphism — to be identified by immunology, by Landsteiner in 1900. Compatibility for the ABO system is the most important consideration in transfusion practice because of the regular presence of the reciprocal ANTIBODIES. Recognition of red cell antigen by the reciprocal antibody in incompatible blood leads to haemolysis.

The presence or absence of A and B antigens on the red cell

Table A1 The ABO blood group system

Phenotype	Red cell antigens	Antibodies in serum	Genotype
O	—	Anti-A and anti-B	OO
A	A	Anti-B	AA or AO
B	B	Anti-A	BB or BO
AB	A and B	None	AB

surface and the occurrence of anti-A and anti-B in the serum of people who lack the relevant antigens define the four common phenotypes: A, B, O, AB (Table A1). Anti-A and anti-B are usually IgM antibodies (*see* ANTIBODIES) and cause severe and often fatal haemolytic transfusion reactions. They may also be IgG, which can cross the placenta and cause haemolytic disease of the newborn, although this is generally mild.

ABO antigens are carbohydrate structures, oligosaccharide chains with characteristic immunodominant sugars (*see* GLY-CANS). The immunodominant sugars are *N*-acetylgalactosamine and D-galactose for A and B antigens respectively. ABO antigens occur as GLYCOPROTEINS and GLYCOLIPIDS on red cells and in plasma, and as glycoproteins in secretions. ABO antigens are intrinsic to red cells and to certain other tissues, and comprise a histo-blood group system relevant to graft rejection. The wide tissue distribution appears early in embryonic life but the antigens may be modified or lost in proliferative diseases of bone marrow and in malignant cells. Secreted ABO antigens may be adsorbed onto cells (e.g. lymphocytes) that lack the intrinsic antigens. ABO antigens are found in species other than humans and are widely distributed in nature; they are even found in some bacteria.

The A and B antigens are synthesized by addition of terminal sugars to a precursor structure, the H antigen. There are several

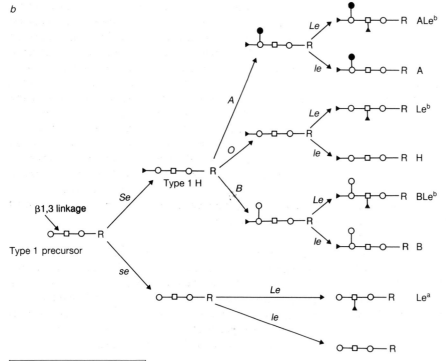

Fig. A2 *a*, Biosynthesis of H, A, and B antigens on Type 2 precursor oligosaccharide chains. *H, A, O,* and *B* represent the genes that control the pathways and which, with the exception of *O,* encode glycosyltransferases (see key). *O* represents the case where no active enzyme is produced and the H antigen remains unmodified. All these antigens are membrane bound. *b*, Biosynthesis of A, B, and H antigen, and of the Lewis antigens Lea and Leb, on a Type 1 precursor. The involvement of the secretor gene (*Se*), closely linked to the H locus on chromosome 19, will lead to secreted ABO and Lewis antigens. *Se* encodes a fucosyltransferase; *se* is an inactive allele. Individuals who are homozygous for *se* have no ABO or Lewis antigens in their secretions although they do produce the ABO antigens in membrane-bound form on their red blood cells. *Le* and *le* represent genes at the Le locus which control the production of Lewis antigens. *Le* encodes a fucosyltransferase; *le* is an inactive allele. Note that the same oligosaccharide chain can carry both A and Leb (ALeb), or both B and Leb (BLeb) antigenic determinants.

Key:
- ○ Galactose
- □ *N*-acetylglucosamine
- ● *N*-acetylgalactosamine
- ► Fucose in α1,2 linkage
- ▲ Fucose in α1,4 linkage

closely related precursor oligosaccharides for the H antigen. Figure A2 shows the synthesis of the A and B antigens from the Type 2 precursor (Fig. A2*a*) and the A and B antigens together with the closely related LEWIS BLOOD GROUP antigens Le[a] and Le[b] from the Type 1 precursor (Fig. A2*b*). ABO antigen expression is controlled by genes at the ABO locus on human chromosome 9 (Table A1) and the H locus on chromosome 19. The precursor oligosaccharide is modified by an L-fucosyltransferase encoded by the *H* gene to give an H antigen. The products of the alleles *A* and *B* at the ABO locus are glycosyltransferases, N-acetyl-galactosaminyltransferase and D-galactosyltransferase respectively, enzymes that transfer specific sugars from a nucleotide donor to an acceptor precursor oligosaccharide chain, the H antigen. The *ABO* genes have been cloned and sequenced [2]: *A* and *B* differ by four nucleotides whereas the *O* gene has a single nucleotide deletion causing a FRAMESHIFT and consequently no active enzyme is formed. On cells of group O people the H antigen remains unaltered.

The H antigen is produced by fucosylation of a precursor oligosaccharide chain. At least four types of precursor are known (Types 1–4); only Type 1 and Type 2 are shown in Fig. A2. Red cell Type 2 H antigen is produced by α2-L-fucosyltransferase encoded by the *H* (or *FUT1*) gene on chromosome 19. In the absence of *H* gene product, no H antigen, and consequently no A or B antigens, can be made. This results in the rare red cell phenotype, O_h or Bombay. Type 1 H is not intrinsic to red cells and is found in secretions of individuals with an *Se* gene. The polymorphic locus (*SE* or *FUT2*) is closely linked to *H* on chromosome 19. The *Se* gene product is probably another α2-L-fucosyltransferase.

P. TIPPETT
G.L. DANIELS

1 Clausen, H. & Hakomori, S. (1989) ABH and related histo-blood group antigens; immunochemical differences in carrier isotypes and their distribution. *Vox Sang.* **56**, 1–20.
2 Yamamoto, F. et al. (1990) Molecular genetic basis of the histo-blood group ABO system. *Nature* **345**, 229–233.

ABP ACTIN-BINDING PROTEIN; AUXIN-BINDING PROTEIN.

abscisic acid (ABA) *S*-(+)-5-[hydroxy-2,6,6-trimethyl-4-oxocyclohex-2-en-1-yl]-3-methyl-2Z,4E-pentadienoic acid (Fig. A3). M_r 264.3. The only member of its class of PLANT HORMONES. Its structure was confirmed by synthesis. Although originally isolated as an apparent inducer of leaf abscission and bud dormancy, its major functions are in adaptation of plants to environmental stress. ABA induces stomatal closure in reponse to water deficit or flooding, which is due to a reversible reduction in turgor of the stomatal guard cells as a result of an efflux of K^+ ions. Other responses include changes in the rate of gene expression; it induces the synthesis of proteins in leaves and developing seeds that protect against the effects of desiccation, and of defence proteins, such as proteinase inhibitors, in response to wounding. In germinating cereal grains, ABA counteracts the induction of hydrolytic enzymes by GIBBERELLIN and may thus protect seeds from precocious germination. Only the *S*-(+)-enantiomer is found naturally, although the *R*-(–)-enantiomer has comparable activity in some bioassays, such as inhibition of immature barley embryo germination, but none in others, such as stomatal closure. The (–) form is taken up and metabolized by plant cells much less effectively than is the natural isomer. ABA is produced in leaves and roots, particularly in response to water stress when its level may increase up to 100-fold. It is transported in xylem sap from roots to shoots of droughted plants. It is found in high concentrations in seeds in the late stages of development and in dormant buds. It is also produced by the fungi CERCOSPORA ROSICOLA and *C. cruenta* at concentrations up to $60 \, mg \, l^{-1}$ culture medium, although by a different biosynthetic route than that in higher plants.

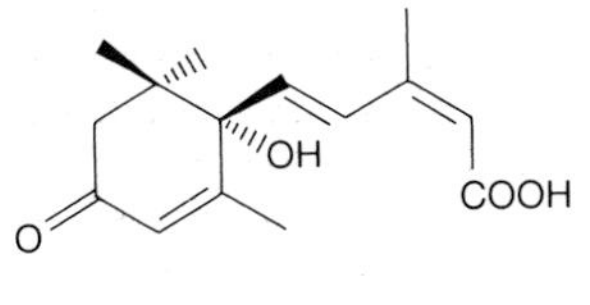

Cornforth, J.W. et al. (1965) *Nature* **206**, 715.

absolute configuration The handedness (i.e. left or right) of a molecule lacking a centre of SYMMETRY. For example, most AMINO ACIDS and all monosaccharides (*see* SUGARS) can occur as enantiomers. The two mirror-image molecular configurations around a given centre of asymmetry in such chiral molecules are conventionally designated D- and L-. PROTEINS are comprised entirely of L-isomers of amino acids; natural polysaccharides are comprised predominantly of D-sugars; and the ribose and deoxyribose in NUCLEIC ACIDS is in the D-form. Structural units such as an α-helix or a twisted β-sheet in proteins or the double-helical form of DNA thus each have a handedness which is almost invariant. Therefore, the absolute configuration of a molecule can be established at lower RESOLUTION than the highest level of atomic detail. For crystals studied by X-RAY DIFFRACTION it can be determined uniquely in the presence of ANOMALOUS SCATTERING, but in other cases an arbitrary initial choice may be made. If this appears to give, for example, left-handed α-helices, then all coordinates (*x,y,z*) are inverted to (*–x,–y,–z*) to give the correct absolute configuration.

abundance (of mRNA) The average number of molecules of a particular mRNA in a given cell. The mRNAs in many types fall into two classes: the abundant class which typically consists of up to 100 different mRNA species present in 1000–10 000 copies per cell, and the scarce or complex class which consists of up to 10 000 different species of mRNA each of which is only represented by a few (usually 1–10) copies per cell.

abzyme Engineered protein with both ANTIBODY and enzymatic activity. *See*: PROTEIN ENGINEERING.

Ac/Ds (Activator/Dissociation) *See*: TRANSPOSABLE GENETIC ELEMENTS: PLANTS.

Fig. A3 Structure of abscisic acid (ABA).

Acanthamoeba Genus of soil-dwelling amoebae which can be grown in culture and which has been used extensively in studies of CELL MOTILITY.

acatalasia, acatalassaemia AUTOSOMAL RECESSIVE condition associated with the lack of the enzyme CATALASE. It is a heterogeneous disorder with more than five subtypes. In the tissues catalase normally acts to protect against oxidation by converting the powerful oxidant H_2O_2 (produced by local bacteria) to O_2 and H_2O. In the absence of catalase the clinical manifestations of the condition range from small ulcers in the mouth to extensive tissue and bone destruction.

ACC (1-aminocyclopropane-1-carboxylic acid) Immediate precursor of the gaseous plant hormone ETHYLENE in vascular plants except club mosses, horsetails and some ferns (Fig. A4). M_r 101.1. Synthesized from S-ADENOSYLMETHIONINE (often abbreviated to SAM or AdoMet) by the enzyme ACC SYNTHASE. ACC applied exogenously to plants or excised parts (e.g. at 1 mol m^{-3}), rapidly forms ethylene, and thus can be used to increase the internal ethylene titre. Its transport in xylem sap is important in root/shoot coordination. Internal concentrations of ACC are assayed directly by gas chromatography-mass spectroscopy or, more usually, indirectly in acidic plant extracts as ethylene after decomposition in alkaline hypochlorite solution. *See also*: PLANT HORMONES.

Adams, D.O. & Yang, S.F. (1979) *Proc. Natl. Acad. Sci. USA* **76**, 170–174.

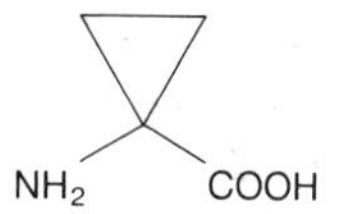

NH$_2$ COOH

Fig. A4 Structure of ACC (1-aminocyclopropane-1-carboxylic acid).

ACC oxidase Also known as the ethylene forming enzyme (EFE). A labile enzyme that oxidizes ACC (1-aminocyclopropane-1-carboxylic acid) to the gaseous plant hormone ETHYLENE with hydrogen cyanide as a secondary product that is detoxified by β-cyanoalanine synthase. Activity is inhibited by anoxia and cobalt ions, and is stereospecific. The enzyme has been prepared *in vitro* from melon and avocado and shown to be soluble and to require Fe^{2+} and ascorbate. It has been cloned from tomato fruit (M_r ~35 000) and shown to belong to a class of non-haem Fe-containing dioxygenases. *See also*: PLANT HORMONES.

McGarvey, D.J. & Christoffersen, R.E. (1992) *J. Biol. Chem.* **267**, 5964–5967.

ACC synthase (ACC methylthioadenosine lyase) Enzyme (EC 4.1.1.14) catalysing the rate limiting step in the formation of the gaseous plant hormone ETHYLENE by converting S-ADENOSYL-METHIONINE to ACC (1-aminocyclopropane-1-carboxylic acid). The enzyme requires pyridoxal phosphate as a cofactor and is possibly a dimer of identical subunits of M_r 65 000 with a short half life (0.5–2 h). *De novo* synthesis is stimulated by environmental stress, wounding, fruit ripening, and hormones, especially AUXINS, and can be either promoted or suppressed by ethylene. It is encoded by a MULTIGENE FAMILY, members of which are differentially regulated in a tissue- or inducer-specific manner. Enzyme action is substrate inhibited and can be competitively inhibited *in situ* by AVG (aminoethoxyvinylglycine) and amino-oxyacetic acid (AOA). AVG and AOA are widely used to help establish ethylene involvement in physiological processes. *See also*: PLANT HORMONES.

acceptor splice site The sequence at the border of the 3′ exon–intron junction which is cleaved during intron splicing (*see* RNA SPLICING). Also called 3′ splice site or downstream splice site. *Cf.* DONOR SPLICE SITE.

acceptor stem Double-stranded region of a TRANSFER RNA (tRNA) molecule which contains both the 5′ and 3′ ends of the tRNA. It usually involves the extreme 5′ seven nucleotides base-paired to seven nucleotides adjacent to the four-nucleotide single-stranded 3′ end (5′ XCCA 3′).

accessory cells In mammals, the accessory cells form the primordial follicle enveloping the OOCYTE. In the primordial follicle the oocyte is surrounded by a single layer of epithelial granulosa cells and mesenchymal thecal cells. During follicular maturation the granulosa cells proliferate until the mature follicle is enveloped by several granulosa cell layers. Granulosa cells play an important part in the progression of the menstrual cycle; in response to follicle-stimulating hormone and luteinizing hormone they secrete oestrogen.

acellular slime moulds A group of simple eukaryotic soil microorganisms (the Myxomycota), the so-called true slime moulds, which are considered as a division of the Fungi. The vegetative stage is a motile multinucleate plasmodium from which sporangia are produced. The acellular slime moulds (e.g. *Physarum*) are good material for studies in CELL MOTILITY. *Cf.* CELLULAR SLIME MOULDS.

acentric (1) Chromosome or chromosome fragment lacking a CENTROMERE, and which does not therefore segregate at MITOSIS or MEIOSIS.
(2) Applied to REFLECTIONS within a set of DIFFRACTION intensities which have a complex (i.e. real and imaginary) STRUCTURE FACTOR, in consequence of the crystal lacking a centre of SYMMETRY. *See*: X-RAY CRYSTALLOGRAPHY.

Acetabularia mediterranea Large unicellular alga which is differentiated into a cap and a stalk region and which has been used to study development in a unicellular eukaryote. Experiments on *Acetabularia* in the 1930s demonstrated the nuclear control of differentiation and morphogenesis.

acetosyringone 3′,5′-dimethoxy-4′-hydroxyacetophenone (Fig. A5). One of a group of small phenolic molecules synthesized by wounded or metabolically active plant cells. It is a chemoattractant of virulent *Agrobacterium* strains and is also an inducer of the virulence operons (*vir*) on the TI PLASMID.

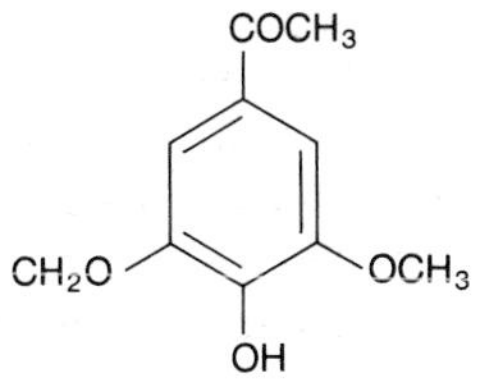

Fig. A5 Acetosyringone.

N-acetyl-D-galactosamine (GalNAc) A monosaccharide component of GLYCOLIPIDS, GLYCOPROTEINS and GLYCOSAMINOGLYCANS (Fig. A6). Present always in mucin-type oligosaccharides (a type of O-linked oligosaccharide), but seldom in N-linked oligosaccharides. *See also*: GLYCANS.

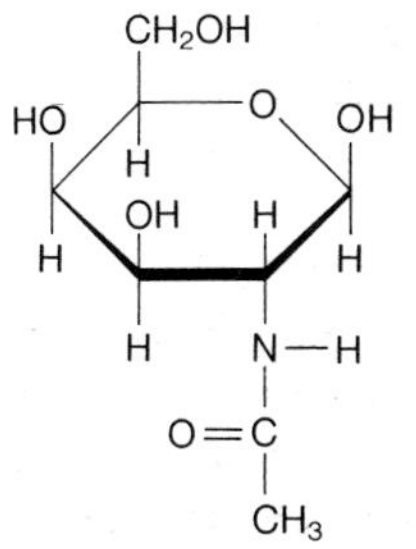

Fig. A6 *N*-acetyl-D-galactosamine.

N-acetyl-D-glucosamine (GlcNAc) A monosaccharide component of GLYCOLIPIDS, GLYCOPROTEINS and GLYCOSAMINOGLYCANS (Fig. A7). Present always in eukaryotic N-linked oligosaccharides, and often in O-linked oligosaccharides and glycolipids. *See also*: GLYCANS.

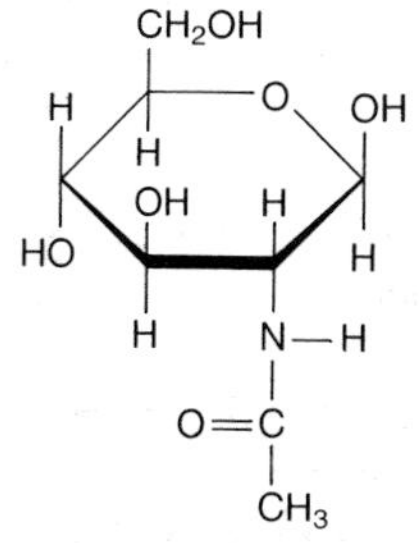

Fig. A7 *N*-acetyl-D-glucosamine.

acetylation The transfer of an acetyl (—COCH$_3$) group. Some proteins undergo N-terminal acetylation of the α-amino group as a POST-TRANSLATIONAL MODIFICATION. Histones are acetylated post-translationally on lysine side chains (*see also* CHROMATIN). Acetylation reactions are catalysed by acetyltransferases (EC 2.3.1) with acetyl-CoA as the usual donor.

acetylcholine (ACh) NEUROTRANSMITTER active at cholinergic synapses and MOTOR END PLATES (Fig. A8). It interacts with two types of receptor: G protein-coupled MUSCARINIC RECEPTORS in the central and peripheral nervous system, and ligand-gated ion

Fig. A8 Acetylcholine.

channel NICOTINIC RECEPTORS in the central nervous system and at motor end plates. Acetylcholine is removed from the synaptic cleft by hydrolysis to choline and acetic acid in a reaction catalysed by acetylcholinesterase. *See also*: G PROTEIN-COUPLED RECEPTORS.

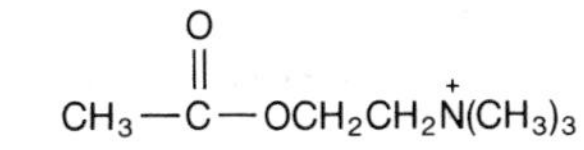

acetylcholine receptors *See*: MUSCARINIC RECEPTORS; NICOTINIC RECEPTORS.

acetylcholinesterase Enzyme (EC 3.1.1.8) (also known as cholinesterase) which is found in the synaptic cleft and which hydrolyses the neurotransmitter acetylcholine to choline and acetic acid. The rapid destruction of acetylcholine which starts as soon as it is released from the presynaptic terminal produces a sharp time-limited signal. *See*: SYNAPTIC TRANSMISSION.

N-acetylneuraminic acid (NAM, NANA, NeuAc, sialic acid) The 5-*N*-acetyl derivative of NEURAMINIC ACID. It is a common component of GANGLIOSIDES and GLYCOPROTEINS, where it is usually present as a terminal residue in α(2,3) or α(2,6) glycosidic linkage to galactose. As a prominent feature of the surfaces of glycoproteins and cells, *N*-acetylneuraminic acid (Fig. A9) has roles in molecular and cellular recognition. For example, it prevents the clearance of mammalian serum (circulating) glycoproteins by masking galactose residues and thereby preventing binding to hepatic lectins (*see* ASIALOGLYCOPROTEIN). It is also the ligand that is recognized by INFLUENZA VIRUS receptors during attachment of the virus to the host cell surface. *See also*: GLYCANS; GLYCOLIPIDS.

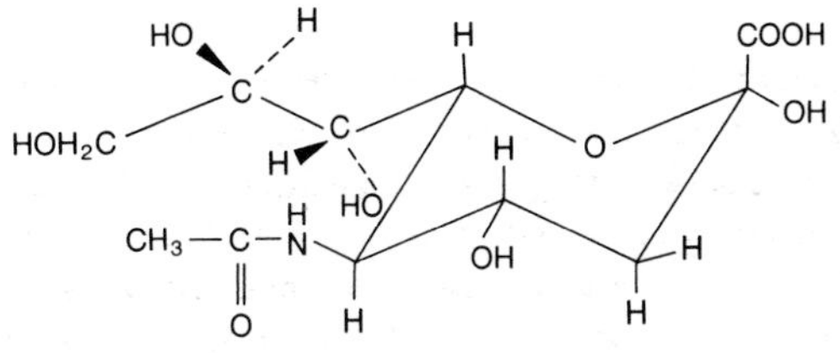

Fig. A9 *N*-acetylneuraminic acid.

achaete-scute complex A complex locus in *Drosophila* which has an important role in the generation of the nervous system in the embryo and in bristle formation in the adult. The complex contains four TRANSCRIPTION UNITS encoding TRANSCRIPTION FACTORS of the helix-loop-helix class. Lack of the entire complex results in elimination of most of the embryonic nervous system. Mutations in individual transcription units may only affect the development of specific adult bristles. Both neurogenesis and bristle formation occur in a two-step process; first a rough cluster of proneural cells is defined and then a single cell within the cluster differentiates into the NEUROBLAST (or into the bristle mother cell). The genes of the achaete-scute complex are expressed in the proneural clusters and also appear to be involved in the process of selection of the cell destined to differentiate into the neuroblast, as expression is highest in this cell.

achondroplasia The condition of short-limbed dwarfism, with an incidence of around 1 : 26 000 births. The characteristic features are shortening of the limbs, depressed nasal bridge, vertebral changes and occasional hydrocephalus. Life span and IQ are normal. It is associated with a failure of epiphyseal growth cartilage. It is inherited in an AUTOSOMAL DOMINANT fashion. Eighty per cent of cases are sporadic, that is the result of new mutation. The MUTATION RATE is around 1 : 100 000, increasing with paternal age.

acid hydrolases Enzymes which catalyse reactions of the type:

$$A—B + H_2O \rightarrow A—H + B—OH$$

and have acidic pH optima. Many are found in LYSOSOMES where they degrade biological macromolecules including nucleic acids, proteins, polysaccharides, and lipids to low molecular weight products.

acid phosphatase An ACID HYDROLASE that catalyses the hydrolysis of an orthophosphoric monoester to an alcohol and an orthophosphate (EC 3.1.3.2). Typically, there is wide substrate specificity as well as an acidic pH optimum. Lysosomal acid phosphatase is often used as a marker enzyme for LYSOSOMES in biochemical or cytochemical assays. Lysosomal acid phosphatase is synthesized on the rough ENDOPLASMIC RETICULUM as an integral membrane protein and reaches the lysosome by a pathway independent of the MANNOSE 6-PHOSPHATE RECEPTORS probably by targeting first to the cell surface and then via ENDOCYTOSIS by COATED PITS. Proteolytic cleavage then produces the soluble mature form of the enzyme.

acid protease, acid proteinase PROTEINASES that act in relatively acidic conditions. *See also*: CARBOXYL PROTEINASE; LYSOSOMAL ENZYMES.

acidic activation domain Protein domain rich in acidic amino-acid residues which occurs in certain TRANSCRIPTION FACTORS and which is involved in activation of TRANSCRIPTION.

acinar cells Epithelial secretory cells which are arranged in grape-like clusters round the lumen of a gland, for example pancreatic acinar cells which secrete digestive enzymes via the pancreatic duct into the duodenum.

ACPD *See*: EXCITATORY AMINO ACID RECEPTORS; TRANS-ACPD.

ACPD receptor *See*: EXCITATORY AMINO ACID RECEPTORS.

acquired immune deficiency syndrome *See*: AIDS; IMMUNODEFICIENCY VIRUSES.

acridine dye A chemical MUTAGEN which binds to DNA and intercalates between adjacent base pairs. This insertion distorts

the DNA strand so that during DNA REPLICATION an extra base is either added or deleted, generating a FRAMESHIFT mutation. Phage and mitochondrial genomes are particularly susceptible to the mutagenic action of acridine dyes. Examples of acridine dyes include acridine orange, acriflavin, ETHIDIUM BROMIDE, and proflavin.

acriflavin(e) ACRIDINE DYE which induces FRAMESHIFT mutations.

acrocentric Chromosome in which the CENTROMERE is very near one end.

acrosin A SERINE PROTEINASE, whose ZYMOGEN form is known as proacrosin, localized in the acrosome of mammalian spermatozoa (*see* ACROSOME REACTION). The proteins and their genes have been characterized in several species. During FERTILIZATION proacrosin autoactivates to acrosin by modification at both N-terminal and C-terminal ends, and the enzyme has long been presumed to aid penetration of spermatozoa through the ZONA PELLUCIDA by proteolytic digestion. Recent evidence also suggests that proacrosin/acrosin binds nonenzymatically to polysulphate groups of certain zona pellucida GLYCOPROTEINS. Mammalian proacrosin has properties similar to the egg adhesion protein BINDIN, present in the acrosomal granule of sea urchin spermatozoa. Spermatozoa are also thought to bear other egg-binding proteins.

acrosome reaction A change in sperm morphology which occurs during FERTILIZATION after sperm have bound to the egg via species-specific receptors on the egg coats (ZONA PELLUCIDA in mammals). The reaction consists of a membrane fusion event in which the contents of the acrosome (a large lysosome-like vesicle overlying the sperm nucleus) are exposed following vesiculation of the anterior region of the sperm plasma membrane and outer acrosomal membrane. Acrosomal enzymes (such as proacrosin/acrosin and bindin) are involved in binding to and penetration of the egg coats and the membranes exposed may assist in binding the sperm membrane to that of the egg and assist their fusion (fertilization).

ACTH ADRENOCORTICOTROPIN.

actin Protein (monomer M_r 42 000) present in almost all eukaryotic cells where it forms the MICROFILAMENT components of the cytoskeleton. Actin monomers (G-actin) polymerize end to end to form actin filaments (F-actin) ubiquitous in the cytoplasm of most eukaryotic cells and which also make up the thin filaments of muscle. In muscle (and nonmuscle cells) actin filaments form a contractile complex with MYOSIN. *See*: ACTIN-BINDING PROTEINS; CELL MOTILITY; MICROFILAMENTS for details of actin structure; MUSCLE; NEURONAL CYTOSKELETON.

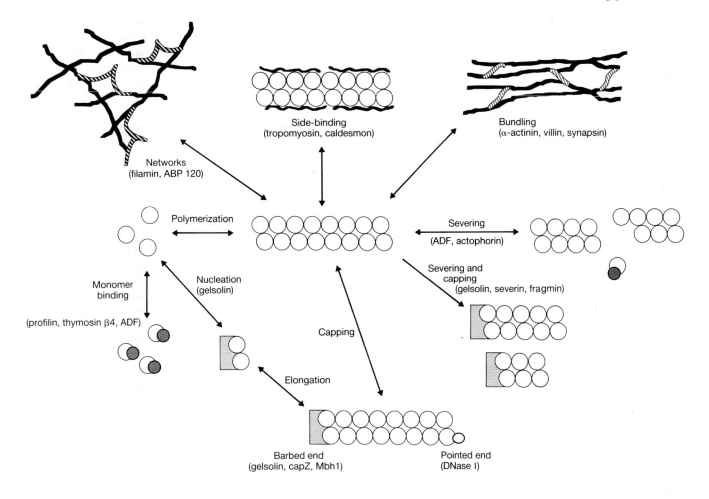

Side-binding
(tropomyosin, caldesmon)

Bundling
(α-actinin, villin, synapsin)

Networks
(filamin, ABP 120)

Polymerization

Severing
(ADF, actophorin)

Severing and
capping
(gelsolin, severin, fragmin)

Monomer
binding

Nucleation
(gelsolin)

Capping

(profilin, thymosin β4, ADF)

Elongation

Barbed end
(gelsolin, capZ, Mbh1)

Pointed end
(DNase I)

Fig. A10 The different roles of actin-binding proteins. Adapted from Way, M. & Weeds, A. (1990) *Nature* **344**, 292.

Actin-binding proteins

ASSOCIATED with the actin filaments of the cytoskeleton of eukaryotic cells (*see* MICROFILAMENTS) are a great variety of actin-binding proteins. These variously bind to actin monomers, cross-link actin filaments into bundles (bundling proteins) or gels (gelating proteins), or sever the actin filament and cap its growing ends (severing and capping proteins) (Fig. A10). The list of proteins that bind actin continues to increase but it is becoming apparent that not only can most be placed into families which share similar properties (e.g. bundling, severing), but that the proteins themselves are in many cases composed of domains shared by other actin-binding proteins in the same family and domains from other families. These domains may be responsible for actin binding itself or for various controlling and structural functions (e.g. the EF hands binding calcium in α-actinin, phosphorylation sites in MARCKS, and the structural repeated domains of dystrophin and **spectrin**). Actin-binding proteins are listed alphabetically below, with synonyms. Only proteins which are known to bind directly to actin are discussed; the number of

'cytoskeletal associated proteins' is very much larger. Proteins known by numerical nomenclature (e.g. 30kd protein) are listed toward the end. Cross-references in bold type are to proteins discussed here, those in small capitals will be found elsewhere in the encyclopedia.

ABP1p (actin-binding protein 85kd). A very hydrophilic, proline-rich protein (M_r 65 500, by sequence data) from the yeast SACCHAROMYCES CEREVISIAE. It contains an SH3 DOMAIN found in a number of membrane-associated proteins. The protein is localized to actin structures in cells; overexpression leads to deformations in actin structures.

ABP-50. *See*: **EF-1a.**

ABP-67. A homologue of **fimbrin** from *S. cerevisiae*, encoded by the *SAC6* gene.

ABP-85. *See*: **ABP1p.**

ABP-120 (mini-filamin). A calcium-insensitive gelation protein from *Dictyostelium discoideum* belonging to the same broad group as **filamin**/ABP from vertebrates. It exists as an antiparallel homodimer of M_r of 240 000 (2 × 120 000). The actin-binding region is close to the N terminus, followed by six motifs which appear to be cross β-sheet structures. A short 27 amino-acid region has been identified as necessary for actin binding.

Bresnick, A.R. et al. (1990) *J. Biol. Chem.* **265**, 9236–9240.

ABP-240. A gelating protein from *D. discoideum*. It exists as a homodimer of length 140 nm with a monomer length of half this value, indicating that the self-associating domain is close to the end. ABP-240 is moderately abundant, present at a ratio of 1 molecule per 88 actin molecules. No sequence data are yet available, but the physical properties suggest that it is related to **filamin** and **ABP-120.**

Hock, R.S. & Condeelis, J.S. (1987) *J. Biol. Chem.* **262**, 394.

ABP-280. Alternative name for **filamin.**

Actin binding protein (ABP). Alternative name for **filamin.**

Actin-gelation protein. Previous name for **filamin.**

Actobindin. An 88 amino-acid polypeptide from *Acanthamoeba castellanii* composed of two very similar repeated units. Actobindin binds one or two actin monomers depending on the mixing ratio. Two identical actin-binding sites (LKHAET), in actobindin have been identified, which are similar to regions of other actin-binding proteins.

Lambooy, P.K. & Korn, E.D. (1986) *J. Biol. Chem.* **261**, 17150; Vandekerckhove, K. & Vancompernolle, K. (1992) *Curr. Opinion Cell Biol.* **4**, 36.

Actophorin. A low molecular weight severing protein from *Acanthamoeba castellanii* which also binds to actin monomers. It is similar in sequence to **cofilin**, **ADF/destrin** and **depactin**, but has some distinct properties. Actophorin is not sensitive to calcium, but is inhibited by inorganic phosphate, TROPOMYOSIN, and phalloidin. It interacts more strongly with the ADP form of actin than with the ATP form. Severing activity depends on filament length, longer filaments being more rapidly severed.

Actin depolymerizing factor (ADF, destrin). The name now given to a low molecular weight (M_r 19 000) severing protein, originally isolated from chick brain, although the same name was first given to **gelsolin**. ADF also binds to monomeric actin; both severing and monomer binding activities are calcium insensitive. Human ADF is identical in sequence to **destrin** from pig and very similar to **cofilin** from a variety of vertebrates, but seems to be biochemically distinct from cofilin. ADF is most similar to **actophorin** in function. It binds phosphatidylinositol 4,5-bisphosphate *in vitro*.

Adams, M.E. et al. (1990) *Biochemistry* **29**, 7414; Moriyama, K. et al. (1990) *J. Biol. Chem.* **265**, 5768.

α-actinin. An actin filament cross-linking and bundling protein of M_r 120 000, localized to the MUSCLE Z line, focal adhesions, intermediate junctions (*see*: CELL JUNCTIONS), and along MICRO-FILAMENTS. It is typically calcium sensitive in nonmuscle cells and calcium insensitive in muscle cells. Many of the proteins which figure in early reports of actin-gelating or bundling activity are now known to belong to the α-actinin family. Their early

discovery is probably due to the fact that they are found in large amounts in a variety of tissues and organisms. α-Actinin is composed of two identical antiparallel peptides, with the actin-binding domain close to the N terminus, followed by four SPECTRIN-like repeats and terminating in two EF hands (calcium-binding motifs; *see* CALCIUM-BINDING PROTEINS). In calcium-insensitive isoforms these EF hands are not active. It is thought that calcium sensitivity of actin-binding arises from the proximity of the EF hand at the tail of one molecule abutting on the actin-binding head of its partner. α-Actinin contains a large structural domain responsible for self association. This domain is similar to those in both spectrin and DYSTROPHIN. In addition to being controlled by calcium, α-actinins may require phosphatidyl-inositol 4,5-bisphosphate (PIP$_2$) for some functions. Quite unexpectedly it has been discovered that PIP$_2$ is actually located in the Z-disc in muscle in association with α-actinin. PIP$_2$ greatly increases the viscosity of actin and α-actinin. As well as binding actin filaments, α-actinin makes specific interactions with VINCU-LIN and INTEGRINS (integral membrane proteins enriched at focal contacts).

Fukami, K. et al. (1992) *Nature* **359**, 150–152.

β-actinin. Identical to **CapZ.**

Actinogelin. An alternative (older) name for α-ACTININ, no longer in popular use.

Acumentin. Originally described as capping the 'pointed' end of actin; it now seems likely that **DNase I** is the only protein which binds the pointed end of actin filaments.

Adducin. A heterodimeric protein associated with the MEMBRANE SKELETON, which promotes the binding of **spectrin** to actin in a calcium/calmodulin-dependent manner. Adducin is composed of two similar subunits, α-adducin (M_r 103 000) and β-adducin (M_r 97 000) which together form higher order structures (dimers and tetramers). Each subunit is composed of two distinct domains — a globular, protease-resistant head domain and a helical tail domain. The head domain contains a sequence which is similar to the actin-binding domain of the ABP-120, DYSTROPHIN, **α-actinin** group. However, as attempts to bind this domain to F-actin have failed, the significance of this is uncertain. The tail domain contains the putative calmodulin-binding sites and the sites of phosphorylation by PROTEIN KINASE A. PROTEIN KINASE C sites are thought to be close to the junction of the head and tail domains. The dimers are thought to bind in an antiparallel fashion, which would be consistent with the reported bundling activity. Although this protein was first isolated from erythrocyte membrane it has since been found in other tissues such as kidney, brain, and liver.

Gardner, K. & Bennett, V. (1987) *Nature* **328**, 359–362.

Adseverin. A severing and barbed-end filament-capping protein (M_r 74 000) from adrenal medulla. Some properties, domain structure, and sequence homology indicate a relationship with **gelsolin**, but, because of its size, it is likely to be most similar in function and structure to **severin** and **fragmin**. It is probably synonymous with **scinderin**.

Maekawa, S. et al. (1989) *J. Biol. Chem.* **264**, 7458.

Aginactin. An agonist-regulated actin-capping protein (M_r 70 000) from *D. discoideum*. Aginactin is of interest as its capping action is regulated by a biochemical cascade initiated by the extracellular binding of CYCLIC AMP (a chemoattractant for *D. discoideum*; *see* DICTYOSTELIUM DEVELOPMENT), and so may provide a mechanism for agonist-stimulated actin polymerization in this and other cell types. Aginactin neither severs nor nucleates actin filaments and is calcium insensitive. Aginactin is 72% identical to the HEAT SHOCK protein **HSP70** from various species.

Sauterer, R.A. et al. (1991) *J. Biol. Chem.* **266**, 24533.

Aldolase. A glycolytic enzyme (EC 4.1.2.13, fructose 1,6 biphosphate aldolase) which like a few others (e.g. PHOSPHOFRUCTO-KINASE) binds actin filaments, probably in order to increase efficiency of enzyme action by locally concentrating enzyme and substrate.

Annexins. Some ANNEXINS have actin filament binding activity.
1 Annexin I. Also widely known as lipocortin I. Binds F-actin, but does not bundle, probably because the protein does not exist as a dimer. Annexin I has a motif similar to the putative annexin II actin-binding domain.
2 Annexin II. Widely known as calpactin I. Exists as a dimer apparently being held together by a 'light chain'. Binds and bundles F-actin in a calcium-dependent manner; this bundling activity is inhibited by the peptide VLIRIMVSR, corresponding to a putative actin-binding domain in the protein. Annexin II is phosphorylated by the transforming tyrosine PROTEIN KINASE from the Rous sarcoma virus (the *src* oncoprotein, *see* ONCO-GENES).
3 Annexin IV. Also known (*inter alia*) as endonexin I and thromboplastin inhibitor. A protein (M_r 35 000) which binds to F-actin in a calcium-sensitive manner. High concentrations of calcium are required and so the physiological significance of this activity has been questioned.
4 Annexin VI (lipocortin VI, protein III, p70, p68, 73k, 67k, calelectrin, chromobindin 20, calphobindin II). A protein (M_r 68 000) whose actin binding is positively regulated by calcium (other actin-binding proteins are typically negatively regulated by calcium). Annexin VI may bind G-actin as well as actin filaments and also binds lipids. Annexin VI has been localized to stress fibres, membrane ruffles, microspikes and focal contacts. On stress fibres annexin VI is periodic and coincident with **α-actinin**, consistent with the fact that there is no competition between the two proteins. Annexin VI contains eight annexin repeats.
See also: CELL JUNCTIONS; CELL MOTILITY.

ASP-56. An actin monomer sequestering protein (M_r 56 000) from platelets which has ~60% homology to adenylate cyclase-associated protein (CAP) from yeast.

Gieselmann, R. & Mann, K. (1992) *FEBS Lett.* **298**, 149.

Band 4.1. Not itself believed to bind actin, band 4.1 is included as the archetype of a large group of proteins, some of which — for example **radixin** and TALIN — do bind actin. Band 4.1 binds **spectrin**. *See also*: MEMBRANE SKELETON.

Band 4.9. *See*: **Dematin.**

Brevin. An old name for plasma **gelsolin**.

Caldesmon. A protein of M_r 87 000 which binds F-actin in the absence of calcium/calmodulin, found in smooth muscle and nonmuscle cells. Caldesmon binds actin filaments decorated with TROPOMYOSIN and interacts directly with tropomyosin. Caldesmon and tropomyosin are thought to regulate myosin in smooth and nonmuscle tissues. Caldesmon is phosphorylated during MITOSIS by the cell-cycle dependent kinase p34^{cdc2} kinase which dissociates caldesmon from the filaments. Both smooth and nonmuscle caldesmon can quickly bind to F-actin structures in nonmuscle cells but are not incorporated into the thin filaments of skeletal muscle cells. *See also*: MUSCLE.

Sobue, K. et al. (1988) *J. Cell. Biochem.* **37**, 317.

Calpactin. An older name for some members of the **annexin** group. Calpactin I = annexin II; calpactin II = annexin I = lipocortin I.

Calponin. A smooth muscle protein which binds G- and F-actin. Binding of calcium/calmodulin and phosphorylation inhibits actin binding, but the concentration of calmodulin required for inhibition seems to be too high to be of physiological relevance. Calponin binds **tropomyosin**, another actin-binding protein.

Winder, S.J. & Walsh, M.P. (1993) *Current Topics in Cellular Regulation* **34**.

Capping protein. *See*: **CapZ.**

CapZ. An actin filament barbed-end capping protein from the Z-disc of skeletal MUSCLE, composed of two subunits α and β. Similar proteins from a variety of phyla show extensive homology in both subunits. The α subunits range between 32 000 and 36 000 M_r while the β subunits range between 28 000 and 32 000 M_r. Both subunits are required for capping activity, but actin binding has so far only been ascribed to the β subunit. Localization studies in nonmuscle cells have not always produced a consistent picture; one member displays a nuclear distribution, while chicken CapZ is concentrated in CELL JUNCTION complexes of epithelial cells. Yeast capping proteins are found at the membrane in regions rich in actin.

Weeds, A.G. Maciver, S.K. (1993) *Curr. Opinion Cell Biol.* **5**, 63.

Cap42(a). A capping protein from *Physarum polycephalum* which is now known to be **fragmin**.

Cap42(b). A capping protein from *P. polycephalum* which is now known to be ACTIN. The apparent capping activity of this actin may be due to the kinase which phosphorylates actin when complexed to **fragmin**.

Cap100. A calcium-insensitive, filament capping protein (M_r 100 000) from *D. discoideum* which is regulated by phosphatidylinositol 4,5-bisphosphate.

Cofilin. A pH-dependent, calcium-insensitive, filament binding and severing protein which also binds monomeric actin. This protein shares many properties with **ADF**, a slightly smaller protein **actophorin**, and **depactin**. Cofilin becomes localized in the nucleus of heat-shocked cells, a movement coincident with dephosphorylation. In the nucleus it is present in peculiar bundled actin filaments which appear to be helical. Cofilin (and ADF) has a putative nuclear localization sequence which may be responsible for nuclear accumulation. When bound to monomeric actin, cofilin decreases the rate at which the actin exchanges its bound nucleotide. Actin binding has been attributed to two motifs in the sequence, WAPECAPLKSKM and DAIKKKL, which are also similar in ADF and actophorin.

Ohta, Y. et al. (1989) *J. Biol. Chem.* **264**, 16143.

Connectin. The name 'connectin' has been given to two very different proteins. One is an intermembrane protein which binds laminin (*see* EXTRACELLULAR MATRIX MOLECULES) on the extracellular side and actin intracellularly. The other is more commonly known as **titin**, a huge elastic protein from MUSCLE.

Brown, S.S. et al. (1983) *Proc. Natl. Acad. Sci. USA* **80**, 5927.

Coronin. An F-actin binding protein (M_r 55 000) from *D. discoideum* with sequence similarity to the β subunit of heteromeric G proteins (*see* GTP-BINDING PROTEINS). Coronin is associated with 'crown-shaped' cell surface projections. On stimulation by CYCLIC AMP (a chemoattractant for *D. discoideum*), coronin is localized to the leading edge of the cell. No homology between coronin and other actin-binding proteins has yet been recognized.

de Hostos, E.L. et al. (1991) *EMBO J.* **10**, 4097.

Dematin (Band 4.9). A trimeric protein from the erythrocyte MEMBRANE SKELETON with an M_r of 48 000. Dematin bundles actin filaments in a time-dependent fashion, bundles being initially loose but later forming cross-links in register 36 nm apart. Bundling is abolished by phosphorylation by CYCLIC AMP-DEPENDENT PROTEIN KINASE.

Siegel, D.L. & Branton, D. (1985) *J. Cell Biol.* **100**, 775.

Depactin. An actin-binding protein from echinoderm oocytes with properties similar to **ADF**, **cofilin**, and **actophorin**.

Mabuchi, I. (1983) *J. Cell Biol.*, **97**, 1612.

Destrin. An alternative name for **ADF**.

DNase I. A deoxyribonuclease that binds actin. This is the only protein which is currently believed to bind at the pointed end of actin filaments. The physiological importance of the very tight association of DNase I with actin is not certain as the two proteins would not normally be expected to meet; however DNase I has proved a valuable tool in the field as it can be used in techniques to measure the G-actin content of actin solutions. These use the fact that when DNase I is bound by actin it no longer has the capacity to cleave DNA. Its use to prevent actin polymerization at high protein concentrations has enabled the crystallographic solution of the actin monomer structure (*see* MICROFILAMENTS).

DYSTROPHIN.

DRP (dystrophin related protein; utrophin). A slightly smaller (M_r 395 000) homologue of the huge DYSTROPHIN molecule, very widely expressed in vertebrate tissues and with extensive homology to dystrophin throughout the sequence, including the actin-binding domain.

Tinsley, J.M. et al. (1992) *Nature* **360**, 591–593.

EF-1a. ELONGATION FACTOR 1a (M_r 50 000) from the cellular slime mould *D. discoideum* which also bundles actin filaments. The bundles are very tightly packed (∼5 nm between filaments). EF-1a from other species also bind and bundle actin.

Yang, F. et al. (1990) *Nature* **347**, 494–496.

Epidermal growth factor receptor (EGF receptor). A transmembrane glycoprotein (M_r 170 000) which has a cytoplasmic protein tyrosine kinase domain (*see*: GROWTH FACTOR RECEPTORS). On binding epidermal growth factor (EGF) (*see*: GROWTH FACTORS) the receptor dimerizes and phosphorylates a number of proteins including the receptor itself. The EGF receptor can bind actin via a region similar to *Acanthamoeba* **profilin**. This particular region in profilin can be cross-linked to actin.

Fascin. A monomeric actin filament bundling protein (M_r 58 000) originally isolated from sea urchin eggs. Fascin is spaced at 11 nm intervals along the filament.

Filamin (ABP, ABP-280, gyronemin). A large (M_r 240 000–280 000), calcium insensitive, dimeric actin cross-linking protein from a variety of sources. It cross-links microfilaments into an elastic and rigid orthogonal gel with a limited tendency to produce bundles. These rigid gels approximate covalently cross-linked filaments. Cells lacking filamin exhibit an unstable cytoskeleton and impaired locomotion. Filamin is enriched in stress fibres, indicating that bundles might be formed in association with other actin-binding proteins in cells. One particular isoform of filamin has been localized at the ends of stress fibres, possibly as a result of a different specificity of glycoprotein binding. Filamin has a glycoprotein-binding domain close to the self-associating C terminus. In the platelet this domain binds the glycoproteins, 1b(a,b) and IX. Like **ABP-120**, filamin has an actin-binding headpiece close to the N terminus followed by 24 cross β-sheet structures. The protein gyronemin, associated with INTERMEDIATE FILAMENTS is an isoform of filamin.

Janmey, P.A. (1990) *Nature*, **345**, 89–92.

Fimbrin (plastin, SAC6). A monomeric actin filament bundling protein (M_r 67 000) which has two **ABP-120**-like actin-binding motifs but has no rod-like domain. Fimbrin has two EF-hands (domains which typically bind calcium) but as no calcium sensitivity has been demonstrated, this is of uncertain significance. Fimbrin can be phosphorylated on serine *in vivo*.

de Arruda, M.V. et al. (1990) *J. Cell Biol.* **111**, 1069–1079.

Fodrin. An alternative name for **spectrin** found in tissues other than erythrocytes. Also known as calspectin.

Fragmin. An actin filament severing and capping protein (M_r 42 000) from *P. polycephalum* which shares many properties with **severin** and **gelsolin** but also has some rather surprising features. Actin in complex with fragmin is phosphorylated on threonines 202 and 203. These residues are involved in contacts with **DNase I** and actin itself, so this is likely to result in inhibition of polymerization. Also, the PROTEIN KINASE responsible for the phosphorylation binds very tightly to the fragmin–actin complex. A fragmin-like protein (M_r 60 000) has been isolated from *Physarum*.

Hinssen, H. (1981) *Eur. J. Cell Biol.* **23**, 225.

gCap39 (MCP, Mbh1). A capping protein (M_r 39 000) from mouse which is 93% identical to **macrophage capping protein** (MCP) from rabbit. Both proteins belong to the **gelsolin** group, but gCap39 has no severing activity, probably because it lacks an F-actin binding domain.

Dabiri, G.A. et al. (1992) *J. Biol. Chem.* **267**, 16545.

Gelactin. An alternative (older) name for **α-actinin**.

Gelactins I–IV. A series of poorly characterized filament gelation proteins of low molecular weight isolated from *Acanthamoeba castellanii*. Gelactin I is a monomer of M_r 23 000; II, III, IV are dimers of M_r 28 000, 32 000, and 38 000. The later two may be related to a 36 000 M_r dimeric protein from *Physarum*. None of these is calcium sensitive.

Gelsolin. A calcium- and phosphoinositide-regulated protein found in most vertebrate tissues and serum, encoded in humans by a single gene. Gelsolin is an actin capping and severing protein with three distinct actin-binding sites distributed throughout the molecule, which is composed of six rather similar domains. The domain arrangement is representative of a number of capping/severing and capping proteins. **Severin** and **fragmin** from lower eukaryotes are half the size of gelsolin and are analogous to the first three gelsolin domains. **Villin** is also similar to gelsolin but has an additional F-actin binding domain. Gelsolin in serum is identical to cytoplasmic gelsolin apart from a short SIGNAL SEQUENCE at the N terminus. Mutations in the gelsolin gene are responsible for a type of familial AMYLOIDOSIS, in which gelsolin is fragmented and forms the core of amyloid plaques.

Weeds, A.G. & Maciver, S.K. (1993) *Curr. Opinion Cell Biol.* **5**, 63.

Heat shock proteins (HSP, Hsp). A number of proteins whose expression is upregulated on HEAT SHOCK bind actin. HSPs have many and varied effects and properties, from entering mitochondria to association with membrane receptor proteins. These properties as well as their actin-binding properties remain enigmatic at present. Actin-binding HSPs are listed below in order of increasing molecular weight.

1 HSP27 (also called HSP28, HSP25 depending on mammalian species). Curiously HSP27 is structurally related to α-crystallin. HSP27 is phosphorylated *in vivo* on serines 15 and 68.

2 HSP70. A CHAPERONIN (a protein which aids the correct folding of other proteins) homologous to **aginactin**, an agonist-regulated actin-capping protein from *D. discoideum*. HSP70 has folding topology structures similar to those of actin despite a lack of sequence similarity.

3 HSP90. A calmodulin-regulated, actin-binding protein related to **HSP100**.

4 HSP100. A calmodulin-regulated, actin-binding protein with strong homology to HSP90. HSP100 is nearly identical to an ENDOPLASMIC RETICULUM protein ERp99 and the glucose-regulated GRP94 from yeast.

Hisactophilin. A histidine-rich (31 histidines out of 118 amino acids) actin-binding protein (M_r 17 000) from *D. discoideum*. It is concentrated at the cell cortex, generally coinciding with F-actin. Probably because of the high histidine content the protein binds to actin in a pH-sensitive manner (like **cofilin**). Hisactophilin induces actin polymerization even in the absence of Mg^{2+} or K^+ at low pH but this effect is less at higher pH within the physiological range.

Insertin. An actin filament capping protein (M_r 30 000) from chick gizzard which inhibits actin polymerization in an unusual manner, only partially blocking assembly at the barbed ends to which it binds. A large block of sequence is shared between insertin and **tensin** but the relationship between the two proteins is not clear.

Weigt, C. et al. (1992) *J. Mol. Biol.* **227**, 593.

Lipocortins. A group of **annexins**.

LSP1 (lymphocyte-specific protein). A 330 amino-acid phosphoprotein associated with the cytoplasmic face of the plasma membrane. LSP1 binds F-actin but not G-actin. The F-actin binding domain is contained within the basic C terminus which shares significant homology with the M_r 20 000 actin-binding region of **caldesmon**.

Jongstra-Bilen, J. et al. (1992) *J. Cell Biol.* **118**, 1443.

MAPs (microtubule-associated proteins). Some MAPs can cross-link actin MICROFILAMENTS to MICROTUBULES and also cross-link microfilaments to each other. The interaction between MAP2 and actin filaments is inhibited by phosphorylation by calmodulin-dependent PROTEIN KINASE II.

Macrophage capping protein (MCP). A filament capping protein (M_r 41 000) originally isolated from rabbit macrophages, and with sequence similarities to the **gelsolin** group. MCP seems to be expressed only in macrophage-related cell types and is very similar to **gCap39** and **Mbh1**.

Dabiri, G.A. et al. (1992) *J. Biol. Chem.* **267**, 16545.

MARCKS.

Mbh1. A **gelsolin**-related protein (M_r 45 000) from mouse. Mbh1 is partially localized to the nucleus and is very similar in sequence (94% identity) to **gCap39**.

Metavinculin. A high molecular weight muscle-specific isoform of **vinculin**, which in porcine tissues differs from vinculin by an extra 68 amino acids which are inserted close to the C terminus just after the proline-rich motif.

MYOSIN.

Nebulin. A very large (M_r 600 000) actin-binding protein from vertebrate MUSCLE, where it has been postulated to form a 'molecular ruler' regulating the length of the thin filament, and from brush border and other tissues. Nebulin also binds to α-actinin and MYOSIN.

Labeit, S. et al. (1991) *FEBS Lett.* **282**, 313.

Physarum 210-kDa protein. An actin bundling protein from the plasmodia of *P. polycephalum*, where is is localized to large bundles. Its activity is regulated by calcium/calmodulin. It is a long molecule (97 nm) with a bend in the middle giving it a 'V' or a 'Y' shape.

Plastin. *See*: **Fimbrin.**

Ponticulin. An abundant integral membrane F-actin binding glycoprotein from *D. discoideum*, accounting for between 0.4 and 1.0% of total membrane protein. Ponticulin both binds F-actin and nucleates actin assembly.

Profilactin. A complex of **profilin** and monomeric actin from *Thyone* (sea cucumber) sperm. The complex is apparently dissociated upon activation to permit the almost explosive polymerization of actin to form the acrosomal process of this species.

Profilin. A low molecular weight (M_r ~15 000), actin-binding protein found in all eukaryotic species examined, including plants. It binds to actin monomers. Profilin is unusual in its ability to bind poly-L-proline, and although the physiological significance of this is unknown, it has provided an easy way to purify the protein. Profilins from different phyla are not highly conserved, but they all bind actin, polyproline, and the phospho-inositides.

Tanaka, M. & Shibata, H. (1985) *Eur. J. Biochem.* **151**, 291.

Radixin. An actin filament barbed-end capping protein (M_r 68 500) localized to CELL JUNCTIONS and the cleavage furrow. Radixin belongs to the **band 4.1** family.

SAC6. ABP 67, a yeast homologue of **fimbrin**.

Scruin. An actin-binding protein (M_r 102 000) from the acrosomal process of horseshoe crab (*Limulus polyphemus*) sperm. On stimulation by a small change in the twist in the individual actin filaments the acrosomal process becomes elongated. Scruin, present at a 1 : 1 ratio to actin in the acrosomal process, is thought to be responsible for this change in twist.

Scinderin. A cytosolic protein (M_r 80 000) from the adrenal medulla which is thought to be involved in cytoskeletal rearrangements during secretion. Upon nicotinic stimulation of CHROMAFFIN CELLS, scinderin redistributes from the cytosol to submembranous patches which stain for F-actin. It has been suggested that this redistribution is brought about by increases in Ca^{2+} and concurrent increases in pH mediated by PROTEIN KINASE C. These changes in the intracellular milieu result in an increase in scinderin's affinity for phosphatidylserine and phosphatidylinositol 4,5-bisphosphate. It is not yet certain whether scinderin is the same protein as **adseverin**, another capping protein found in the same tissue.

Severin. A calcium-sensitive actin filament severing and capping protein (M_r 40 000) from *D. discoideum*. Severin is homologous to **fragmin**, a protein with similar properties from *P. polycephalum* and also to **gelsolin** and **villin**. Like these, severin is also controlled by phosphatidylinositide lipids. *Dictyostelium* amoebae lacking severin are apparently normal and able to undergo the developmental cycle (*see* DICTYOSTELIUM DEVELOPMENT). The domain structure of severin is broadly comparable to the N-terminal half of gelsolin.

Spectrin. Originally isolated from the erythrocyte membrane, the spectrins have been found in most vertebrate tissues and even in protozoans. Spectrins are elongated, apparently flexible heterodimers of α and β subunits ranging in M_r from 220 000–265 000. Both subunits are related and both contain multiple internal repeats. The β subunit contains an actin-binding domain similar to the domain found in **α-ACTININ**, DYSTROPHIN, **ABP-120**, **filamin**, and ADDUCIN. The α subunit lacks such a domain and so the spectrin heterodimer alone cannot cross-link actin filaments. However, the αβ heterodimer is able to form higher order structures such as a tetramer by association at their ends, which is able to cross-link filaments. Cross-linking of actin filaments is not likely to be the primary role of the spectrins in most cell types; their purpose is rather to form a MEMBRANE SKELETON in concert with a number of other proteins. Immunofluorescence microscopy in the majority of cell types indicates that spectrin is localized solely to the plasma membrane, possibly as a result of spectrin's own weak affinity for lipids but more probably perhaps because of the many interactions made with the cytoplasmic domains of membrane proteins.

Bennet, V. (1990) *Physiol. Rev.* **70**, 1029–1065.

β-Spectrin. Together with α-spectrin it forms the **spectrin** heterodimer. β-Spectrin has an ABP-120/α-actinin-like actin-binding domain.

Synapsins. Actin-binding proteins originally identified in association with SYNAPSES. The two major types of synapsin —

synapsins I and II — are present in quite high concentration in bertebrate brain. Synapsin I is localized to the cytoplasmic surface of synaptic vesicles and believed to have a role in the regulation of meurotransmitter release from the visicles. Synapsins are composed of a number of recognizable domains some of which are common to both. All synapsins contain an N-terminal A domain, followed by a B domain, followed by a C domain, after which considerable diversity is found. Synapsin I can bind and bundle actin in a phosphorylation-dependent manner, and also has actin nucleating activity (i.e. it enhances spontaneous actin polymerization by the formation of 'seed' nuclei). Domain C has been implicated in actin binding as has domain D, which would explain the bundling activity of synapsin I (synapsin I has domain D, synapsin II does not). The current status of synapsin II as an actin-binding protein is unclear. Two types each of synapsins I and II exist — synapsins Ia and Ib, and synapsins IIa and IIb. The a and b forms in each case are derived from the same gene by differential RNA SPLICING. No homology between the synapsins and other actin-binding proteins is now believed to exist (former reports were based on erroneous sequence data). On SDS-PAGE, synapsin Ia is the largest synapsin with an M_r ~86 000, followed by synapsin Ib (80 000), synapsin IIa (74 000) and synapsin IIb (55 000). Synapsin IIb has been implicated in synaptogenesis.

De Camilli, P. et al. (1990) *Annu. Rev. Cell Biol.* **6**, 433–460.

Talin. A vinculin-, integrin-, and actin-binding protein (M_r 25 000) which, with other proteins, provides a link between actin MICROFILAMENT bundles and the plasma membrane at focal contacts (*see* CELL JUNCTIONS). Talin also nucleates actin assembly. It is an elongated molecule (60 nm) with a globular head, which is thought to mediate binding to the membrane, and a flexible tail which binds **vinculin**. As well as binding to vinculin, which itself associates with INTEGRINS, talin has been reported to bind lipids directly. Talin is localized to the focal contact and other adhesive contacts in tissue culture cells.

Rees, D.J.G. et al. (1990) *Nature* **347**, 685–688.

Tensin. Originally identified as a contaminant of the standard **vinculin** preparation, tensin is found in two forms, M_r 150 000 and 200 000. It is localized to the focal contact in tissue culture cells (*see*: CELL JUNCTIONS) and to the Z-line of skeletal MUSCLE. Tensin shares a large portion of its sequence with **insertin**, an actin filament capping protein.

Davis, S. et al. (1991) *Science* **252**, 712.

Thymosin β4. A peptide (M_r 5000) with actin monomer sequestering activity. In platelets, it is thought to account for most of the monomeric actin pool. It binds monomeric actin with a K_d between 0.4 and 0.7 μM. Thymosin β4 strongly inhibits nucleotide exchange in actin while bound, an opposite effect to that of **profilin** which competes for actin binding. A model involving thymosin β4 and profilin for the regulation of actin polymerization in cells has been suggested.

Safer, D. et al. (1991) *J. Biol. Chem.* **226**, 4029.

Titin. A huge protein from MUSCLE with an M_r of ~3×10^6 and a length of ~1 μm. Titin is associated at one end with the M-line in striated muscle where it forms a 'molecular ruler' for the myosin filament. The molecule extends from the other end of the filament to form an elastic connection with the Z-line. Titin has also confusingly been named connectin. Together with **nebulin**, another huge protein, titin forms a largely insoluble framework comprising some 13% of the myofibril protein. The binding of titin to actin filaments has been demonstrated by sedimentation measurements and by direct electron microscopic evidence.

Maruyama, K. et al. (1987) *J. Biochem.* **101**, 1339.

TROPOMYOSIN. A family of dimeric coiled coil proteins of variable molecular length which bind along actin filaments. MUSCLE tropomyosins bind seven actin monomers, while those of platelets bind six. Binding by tropomyosins stabilizes the MICROFILAMENT against spontaneous fragmentation, and against severing by ADF and gelsolin. Tropomyosin may interact directly with gelsolin. Tropomyosins are known to interact with a variety of proteins to control actin–myosin interactions in muscle and nonmuscle cells. In striated muscle, tropomyosin binds the TROPONIN complex (troponins C, T, and I). Together this complex 'hides' the myosin head binding site on the actin filament in the absence of calcium. Given micromolar concentrations of calcium the tropomyosin–troponin complex moves so as to permit productive actin–myosin interactions. In smooth muscle and nonmuscle cells, tropomyosin interacts with other actin filament binding proteins, CALDESMON and calponin.

Cote, G.P. (1983) *Mol. Cell. Biochem.* **37**, 127–146.

Utrophin (dystrophin-related protein, DMDL protein, DRP). A nonmuscle homologue of muscle DYSTROPHIN. The protein is widely expressed and is membrane-associated in a number of cell types. *See also*: **DRP**.

Villin. A severing/capping/bundling protein from the microvilli of the brush border. It is similar to **gelsolin** in stucture and function but has an additional actin-binding domain at the C terminus accounting for its bundling activity. In the absence of calcium, villin bundles, whereas at calcium levels in the micromolar range it acts as a severing/capping protein. Fibroblastic cell types transfected with villin cDNAs produce microvillus-like structures.

Vinculin. A protein (M_r 117 000) enriched at the focal contact and adhesive CELL JUNCTIONS in vertebrate cells, especially fibroblasts. The status of vinculin as an actin-binding protein has been uncertain, as previous reports of its ability to bind actin seemed to be caused by low abundance contaminants in the purification procedure. However, recombinant protein fragments have demonstrated an actin-binding capacity. Vinculin associates with a number of other actin-binding proteins including **talin** and **α-actinin**. *See also*: **Metavinculin**.

Burridge, K. et al. (1988) *Annu. Rev. Cell Biol.* **4**, 487.

Vitamin D binding protein (VDBP, group-specific component, γc-globulin). A dual-function glycoprotein (M_r 53 000) from vertebrate serum present at about 10 μM, with sequence similarities to serum albumin and α-FOETOPROTEIN. As well as binding vitamin D and its metabolites, VDBP binds G-actin extremely tightly (K_d 1 nM), at a site towards the C terminus (the sterol-binding domain is close to the N terminus). The physiological function of its actin binding is thought to be the inactivation of actin which leaks into the serum as a result of tissue damage, and the prevention of clot formation by actin polymerization. VDBP is produced mainly by the liver.

25-kD inhibitor of actin polymerization. *See*: HSP27.

30kDa-actin bundling protein (p30b). A calcium-insensitive bundling protein (M_r 30 000) from *D. discoideum*. Distinct from **p30a**.

34-Kilodalton actin bundling protein (p30a). A bundling protein (M_r 34 000) from *D. discoideum* distinct from **p30b**. It is monomeric, calcium regulated, is selectively present in filopodia and can inhibit actin filament depolymerization.

120-kD gelation factor. *See*: ABP-120.

S.K. MACIVER

See also: CALCIUM-BINDING PROTEINS.

---

actin-bundling proteins Proteins that cross-link actin filaments into bundles. *See*: ACTIN-BINDING PROTEINS.

actin filament *See*: MICROFILAMENTS.

actin–myosin interaction *See*: MICROFILAMENTS; MUSCLE; MYOSIN.

actinidin Thiol proteinase isolated from the plant *Actinidia chinensis* (EC 3.4.22.14). *See*: Fig. M33 in MOLECULAR EVOLUTION: SEQUENCES AND STRUCTURES.

α-actinin *See*: ACTIN-BINDING PROTEINS.

β-actinin *See*: ACTIN-BINDING PROTEINS; CALCIUM-BINDING PROTEINS.

actinogelin *See*: ACTIN-BINDING PROTEINS.

actinomycin D A cyclic peptide antibiotic produced by *Streptomyces chrysomallus* which consists of coplanar heterocyclic rings carrying two cyclic oligopeptide side groups (Fig. A11). It is a potent inhibitor of RNA synthesis (*see* TRANSCRIPTION), in both prokaryotes and eukaryotes. It intercalates into DNA thereby preventing the movement of the DNA-dependent RNA polymerase along the template by steric hindrance. Ribosomal RNA synthesis is particularly sensitive to this antibiotic. Actinomycin D will also inhibit DNA synthesis but to a much lesser extent than RNA synthesis.

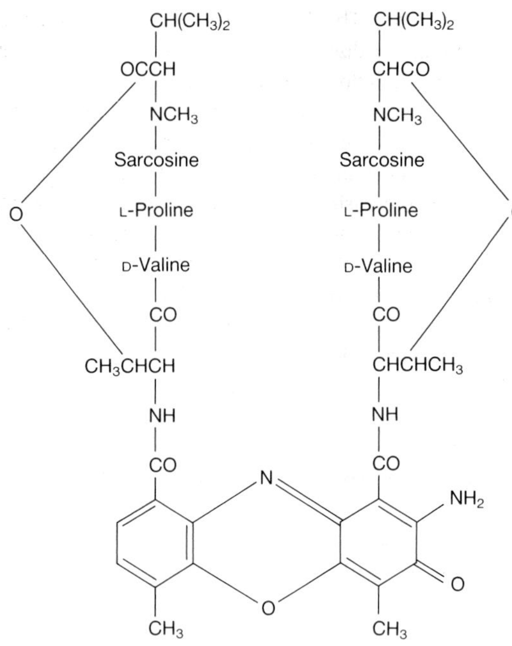

action potential Rapid, transient, self-propagating electrical exci-

Fig. A11 Actinomycin D.

tation of the membrane of electrically excitable cells, also known as a spike or nerve impulse (in neurons). Action potentials occur in nerve, muscle, and neuroendocrine cells and are generated and propagated through the actions of VOLTAGE-GATED ION CHANNELS. An action potential is essentially a brief (~1 ms long), regenerative change in the MEMBRANE POTENTIAL which occurs in electrically excitable cells. The most commonly found action potential is due to influx of Na$^+$ resulting from an initial DEPOLARIZATION of the membrane (e.g. an EXCITATORY POSTSYNAPTIC POTENTIAL). The event is regenerative and all-or-none, owing to the positive feedback between depolarization and Na$^+$ entry through voltage-gated sodium channels:

depolarization → increase in Na$^+$ CONDUCTANCE → Na$^+$ influx → depolarization.

Consequently, this type of electrical activity is nondecremental (oscillating between the resting membrane potential and the Na$^+$ EQUILIBRIUM POTENTIAL) and is used by nerve fibres for communication over any distance in the body. Action potentials are essentially digital signals which encode information as a temporal sequence. This type of activity is suppressed by voltage-dependent Na$^+$ channel blockers (e.g. TETRODOTOXIN). Some neurons generate Ca^{2+}-dependent action potentials, especially in early development. The latter are somewhat longer in duration (several milliseconds or more) and used by cells when absorption of Ca^{2+} from the extracellular medium is needed. Ca^{2+} action potentials are suppressed by Ca^{2+} channel blockers, such as Co^{2+} and Cd^{2+}.

Aidley, D.J. (1989) *The Physiology of Excitable Cells*, 3rd edn (Cambridge University Press, Cambridge).

activation (of eggs) The initiation of the developmental programme seen in fertilized eggs, involving an increase in egg metabolism and initiation of DNA synthesis, and cleavage. This sequence is normally initiated by the fertilizing sperm but other nonspecific physical and chemical stimuli can activate the egg (parthenogenetic activation). In the sea urchin, the egg membrane becomes increasingly permeable to Na^+, generating a transient membrane DEPOLARIZATION, and cytoplasmic Ca^{2+} is released (which is involved in the block to POLYSPERMY). An increase in cytoplasmic pH due to an efflux of H^+ coupled with an influx of Na^+ also occurs. These ion fluxes are thought to initiate the later activation events including increases in protein synthesis using maternal mRNA and ribosomes.

activation energy In chemical reactions the Gibbs energy of activation symbolized by $\Delta G^\dagger$, which is equal to the difference in free
energy between the transition state and the ground state. *See*: MECHANISMS OF ENZYME CATALYSIS.

Activator (Ac) *See*: TRANSPOSABLE GENETIC ELEMENTS: PLANTS.

active site *See*: MECHANISMS OF ENZYME CATALYSIS.

active transport *See*: MEMBRANE TRANSPORT SYSTEMS.

activin *See*: GROWTH FACTORS.

actobindin *See*: ACTIN-BINDING PROTEINS.

actophorin *See*: ACTIN-BINDING PROTEINS.

acute lymphoblastic leukaemia (ALL) An acute malignant disease in which the normal haematopoietic tissue in BONE MARROW is substituted by immature BLAST CELLS belonging to the lymphoid system.

acute phase proteins Proteins synthesized by liver cells and present in blood of healthy individuals in trace amounts but whose levels are greatly increased after injurious stimuli such as infection, burns and trauma, in certain chronic inflammatory conditions, and in some cancers. Their synthesis by hepatocytes is stimulated by the LYMPHOKINES interleukin-6 (IL-6) and interleukin-1 (IL-1). They include the PENTRAXIN C-reactive protein, α2-macroglobulin and other proteinase inhibitors, the animal lectin MANNOSE-BINDING PROTEIN, and the blood clotting protein fibrinogen (*see* FIBRIN). C-reactive protein, the proteinase inhibitors, and mannose-binding protein are thought to be involved in nonspecific (innate) immunity to bacterial infection.

acyclovir (acycloguanosine, Zovirax) An antiviral compound specific for herpesviruses 1 and 2. Acyclovir is selectively phosphorylated by HERPESVIRUS-induced THYMIDINE KINASE and in this form inhibits DNA polymerase and thus herpesvirus replication.

N-acylation The post-translational addition of a long chain FATTY ACID such as myristic acid to a protein which serves to anchor some cytoplasmic proteins or the cytoplasmic tails of some membrane proteins, to the inner face of the plasma membrane.

acylcyclohexanedione A group of PLANT GROWTH RETARDANTS that inhibit late steps (2β- and 3β-hydroxylations) in the biosynthesis of the GIBBERELLIN hormones. Commercial examples are prohexadione (Fig. A12) and cimectacarb. They are general inhibitors of 2-oxoglutarate-dependent dioxygenases, with K_i $\sim 10^{-6}$ M with respect to 2-oxoglutaric acid, the enzyme cosubstrate. *See also*: PLANT HORMONES.

Griggs, D.L. et al. (1991) *Phytochemistry* **30**, 2513–2517.

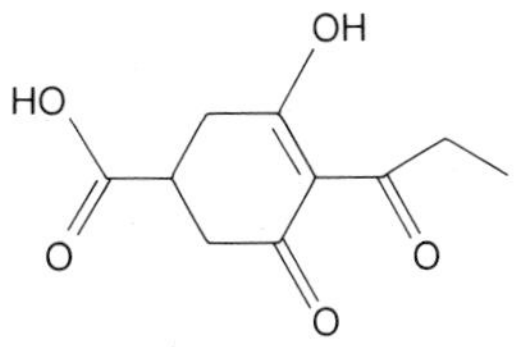

Fig. A12 Structure of prohexadione.

ADA The enzyme adenosine deaminase. *See*: ADENOSINE DEAMINASE DEFICIENCY.

adaptin Protein of M_r $\sim 100\,000$ associated with clathrin-coated vesicles (*see* COATED PITS AND VESICLES). The adaptins are components of the heterotetrameric protein complexes called adaptors, which are believed to link the clathrin lattice to the cytoplasmic domains of selected transmembrane proteins. Different adaptors are associated with the plasma membrane and the *trans* Golgi network (TGN) (*see* GOLGI APPARATUS). There are three classes of adaptins: α, β, and γ. Each adaptor contains one copy of a β-adaptin, one copy of an adaptor-specific adaptin (α for the plasma membrane and γ for the TGN) (Fig. A13), and one copy each of two smaller proteins with M_r of 47 000–50 000 and 17 000–20 000.

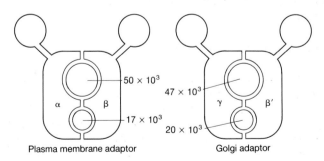

Fig. A13 Models of the two adaptor complexes.

adaptive response (in bacteria) *See*: DNA REPAIR.

adducin *See*: ACTIN-BINDING PROTEINS.

adenine (A) 6-Amino purine, one of the four nitrogenous BASES in DNA and RNA and which is also the base in the adenine nucleotides ATP, ADP, and AMP and other metabolic cofactors such as FAD and NAD (*see* NUCLEOSIDES AND NUCLEOTIDES). In DNA it pairs with thymine and in RNA with uracil (*see* BASE PAIR).

adenocarcinoma A CARCINOMA of glandular epithelium.

adenosine The nucleoside derivative of adenine (*see* NUCLEO-SIDES AND NUCLEOTIDES). *See also*: PERIPHERAL NERVOUS SYSTEM; PURINERGIC SYSTEMS.

adenosine deaminase deficiency A rare RECESSIVE inherited defect in the gene for the enzyme adenosine deaminase (ADA) (EC 3.5.4.4) located on chromosome 2 which results in a deficiency of the enzyme. ADA is a hydrolytic enzyme which catalyses the reaction:

$$\text{adenosine} + H_2O = \text{inosine} + NH_3.$$

In its absence deoxyadenosine and deoxyadenosine triphosphate accumulate and ATP is depleted. dATP is highly and selectively toxic for dividing lymphocytes especially T CELLS (it blocks DNA synthesis by inhibiting ribonucleotide reductase and transmethylation reactions). ADA deficiency thus results in a profound deficit of mature T cells and B CELLS, resulting in SEVERE COMBINED IMMUNODEFICIENCY (SCID). ADA deficiency comprises ~20% of all cases of genetically caused SCID. Sufferers rarely survive childhood untreated. ADA deficiency can be alleviated to some extent by injections of ADA, and can be corrected by transplantation of compatible normal bone marrow. It was the first genetic deficiency in humans to be treated by somatic GENE THERAPY (in 1990).

adenosine monophosphate, diphosphate, triphosphate (AMP, ADP, ATP) *See*: NUCLEOSIDES AND NUCLEOTIDES.

adenosine triphosphatase *See*: ATPASES.

***S*-adenosylmethionine (SAM)** A coenzyme which acts as a donor of methyl groups in many reactions (Fig. A14).

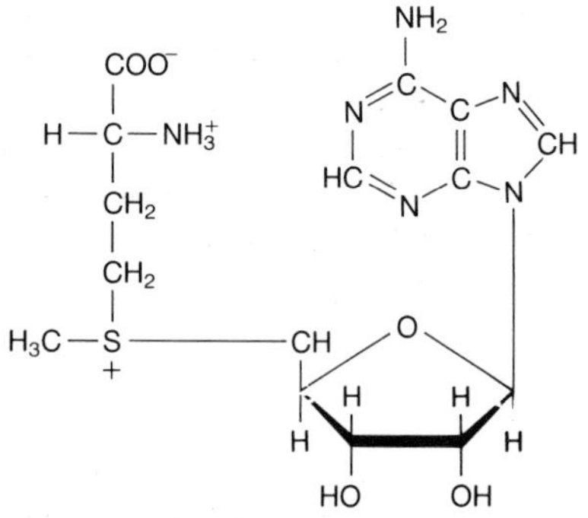

Fig. A14 *S*-adenosylmethionine.

Adenovirus

INVESTIGATIONS into the molecular biology of adenoviruses have led to discoveries with implications far beyond the biology of the virus itself. RNA SPLICING was first described in adenovirus-infected cells and the first eukaryotic *in vitro* systems for DNA REPLICATION and TRANSCRIPTION were also a result of adenovirus studies. The observation that some adenovirus serotypes had oncogenic potential and could induce tumours in rodents led to the identification of adenovirus ONCOGENES. Further studies on their mode of action led to the identification of TUMOUR SUPPRESSOR GENES and shed light on the genesis of human cancers.

Structure and properties of the virion

Adenoviruses are double-stranded (ds) DNA viruses belonging to the family Adenoviridae (*see* ANIMAL VIRUSES). The human adenoviruses (Ads) belong to the genus Mastadenovirus and, to date, 47 distinct serotypes have been identified. These are divided into six subgenera (A to F) on the basis of shared immunological and biochemical properties, and are associated with a variety of acute infections, primarily respiratory, ocular, and gastrointestinal. Serotypes 40 and 41 can be isolated in high yield from faeces of young children with acute gastroenteritis and are second only to rotaviruses as the major cause of infantile viral diarrhoea. The adenovirus virion is a nonenveloped icosahedral capsid surrounding a linear double-stranded DNA genome, which in Ad2 is 35 937 base pairs long [1]. Virions are 80–110 nm in diameter with 252 capsomers each 8–10 nm in diameter. The 240 non-vertex capsomers (hexons) are formed by the interaction of three identical polypeptides. An atomic structure of hexon has been determined by X-RAY CRYSTALLOGRAPHY [2] which indicates that hexon consists of two distinct parts: a triangular top with three towers exposed to the environment and bearing type-specific antigenic determinants [3], and a pseudo hexagonal base with a central cavity. Penton, which forms the 12 vertices of the icosahedron is a pentamer of penton base tightly associated with a trimer of fibre projecting from the surface. Polypeptides VII and IX are associated with hexon, while polypeptides IIIa and VI link the capsid to the nucleoprotein core (viral genome plus polypeptides V, VII, and X).

Gene expression

The adenovirus genome contains more than 50 OPEN READING FRAMES that could potentially code for proteins of more than 100 amino acids long. Transcription of these open reading frames is accomplished by the host RNA polymerase II and a complex pattern of pre-mRNA splicing dramatically increases the number of unique polypeptide species that are produced from the genome.

Temporal control is exerted on transcription of the viral genome with 'early' genes being transcribed before viral DNA replication and 'late' genes being transcribed after [1]. Viral cytoplasmic mRNAs are, like the bulk of cellular mRNAs, capped and polyadenylated. At early times after infection six blocks of

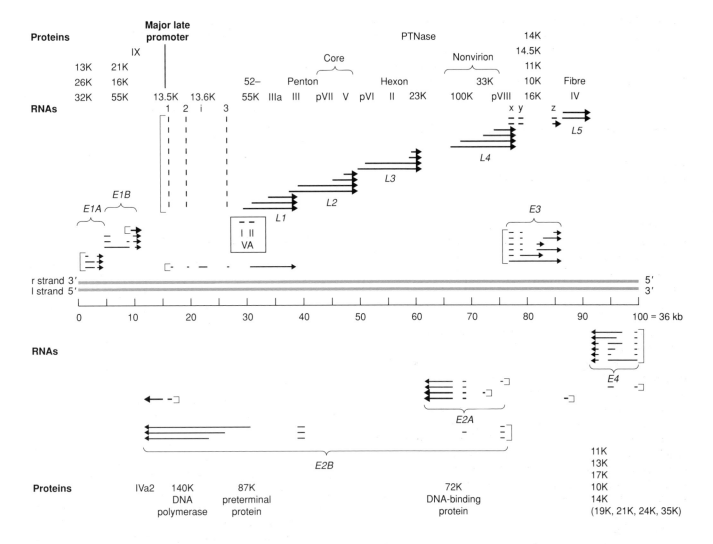

Fig. A15 The human adenovirus genome and its RNA transcripts. Transcripts sharing the same promoter are indicated by the bracket at the 5′ end of each family. The proteins encoded by each are listed above and below, named by function or molecular weight. The roman numerals indicate virion proteins. Large arrow heads indicate late transcripts, small arrow heads early transcripts. From [10].

genes are transcribed — E1a, E1b, E2a, E2b, E3, and E4 (Fig. A15). E1a encodes proteins with the properties of transcriptional regulation whereas E1b gene products are involved in both host cell shut-off and mRNA transport. Both E1a and E1b are involved in oncogenic TRANSFORMATION. Proteins encoded by the E2a and E2b regions—DNA polymerase (pol), preterminal protein (pT-P),and DNA-binding protein (DBP)—are directly involved in viral DNA replication. E3 gene products are dispensable for growth in tissue culture but are thought to be involved in modifying the infected cell such that it can escape immune surveillance by the host. Like the E1 region, the products of the E4 region are thought to be transcriptional activators, although their mode of action is less well defined.

At late times after infection, when the DNA templates have been replicated, transcription mainly initiates from the major late promoter to generate five families of 3′ co-terminal transcripts (Fig. A15). All late mRNAs have at their 5′ end the same tripartite leader sequence which increases the efficiency with which late viral mRNAs are translated [4]. Protein products of the late genes are predominantly components of the virion and proteins involved in the assembly process, although additional proteins such as the virus-coded protease are involved in proteolytic processing of viral proteins.

The VA genes of adenovirus are transcribed by the host RNA polymerase III and although they do not encode protein they dramatically influence late gene expression. It is thought that VA RNAs function by blocking the action of the cellular dsRNA-activated protein kinase (*see* RNA-ACTIVATED PROTEIN KINASE). In the absence of VA RNA, the dsRNA-activated protein kinase would be activated in the late phase of adenovirus infection, when dsDNA accumulates, thus leading to shut-down of all protein synthesis.

Transformation

When injected into newborn rodents certain adenoviruses are oncogenic. Subgroup A viruses (e.g. Ad12) are highly oncogenic when injected into newborn rodents, causing tumours with high frequency and short latency period. Subgroup B viruses (e.g. Ad3) are described as weakly oncogenic, as they induce tumours at low frequency and only after a long latent period. Although subgroups C, D, E, and F are by the above criteria nononcogenic it should be noted that all adenovirus subgroups have the ability to transform primary cultures of rodent cells *in vitro*. Analysis of tumour cell DNA from rodents infected with oncogenic adenovirus reveals that the 'leftmost' 11% of the genome is invariably retained in an integrated form. This region codes for the adenovirus oncogenes — the E1a and E1b genes — the products of which are expressed in virally transformed cells [5].

Although RNA splicing can generate at least ten different mRNAs from the E1a and E1b region it is clear that two polypeptide species from each gene block predominate. In the case of E1a, 13S and 12S mRNAs encode polypeptides of 289 and 243 residues that differ by only 46 internal amino acids. In contrast, the E1b region codes for polypeptides of M_r 19 000 (p19) and 55 000 (p55) which are unrelated in amino-acid sequence.

The E1a region alone from the highly oncogenic Ad12 can fully transform primary rodent cells in culture although the additional presence of E1b increases the frequency of transformation and reduces the time taken to achieve the transformed phenotype. It is clear that in PERMISSIVE CELLS the function of the E1a and E1b regions is to push the cells into the CELL CYCLE so that they will constitute the optimal environment for viral replication. This effect has recently been shown to involve interaction of the E1a and E1b gene products with cellular regulatory proteins and it is this property which may account for the transforming potential of adenoviruses. The 289-residue E1a protein contains three conserved domains, two of which are also present in the 243-residue protein. Conserved regions 1 and 2, which are common to both proteins, are required for transcriptional repression, transformation, and induction of cellular DNA synthesis, whereas conserved region 3, which is unique to the 289-residue protein, is involved in transcriptional activation.

In transformed cells, E1a gene products are associated with a number of cellular proteins which include polypeptides of M_r 300 000, 107 000, 105 000, and 60 000. The 105K protein is the product of the retinoblastoma susceptibility gene (RB) which is known to act as a tumour suppressor. The regions in the E1a protein that are required for transformation (conserved regions 1 and 2) coincide with the regions that are required to form the complex with RB [6]. In normal cells RB negatively regulates the activity of cellular TRANSCRIPTION FACTORS such as E2F which are involved in coordinate activation of genes required for chromosome replication. It is thought that E1a may function in transformation by disrupting the RB–E2F complex such that E2F is freed from RB and unscheduled DNA synthesis takes place.

Like E1a, the products of the E1b genes are also known to interact with cellular proteins, one of which is the tumour suppressor gene p53. p53 is a transcription factor that is inactivated in a wide range of human tumours and it is thought that its function may be to respond to mutagenic DNA damage by activating transcription of genes that are involved in DNA repair. It is therefore possible that the p55 product of E1b, when bound to p53, inactivates the protein allowing cells to accumulate

mutations that lead to transformation [7]. Although required for transformation, the role of the E1b p19 protein has yet to be determined.

DNA replication

Adenovirus is an excellent system in which to study DNA replication as the viral genome can be replicated *in vitro* by the action of three viral proteins — DNA-binding protein (DBP), preterminal protein (pTP), and DNA polymerase (pol) — and two cellular proteins — nuclear factor I (NFI) and nuclear factor III (NFIII). The adenovirus genome is a linear dsDNA of 35–36 kb with inverted terminal repeats (ITRs) of about 100 bp, the exact size depending upon serotype. DNA synthesis is initiated at either of the termini by transfer of dCMP, the terminal nucleotide, onto pTP (M_r 80 000) in a template-dependent reaction. The 3′OH of the pTP–dCMP complex serves as a primer for synthesis of the nascent strand by the viral DNA polymerase. Concomitant displacement of the nontemplate strand generates a single-stranded molecule which then acts as a template for a second round of DNA synthesis.

Adenovirus origins of replication are located at the ends of the genome within the inverted terminal repeats. Covalently attached to each 5′ end of the DNA is a terminal protein (TP) which is likely to be a *cis*-acting protein component of the replication origin. Although removal of TP reduces the efficiency of Ad2 and Ad4 DNA replication *in vitro* it does not abolish replication, and 'plasmid' templates, provided that the origin has been exposed by restriction enzyme cleavage, have 25% of the activity of protein-linked genomes. Using plasmid templates, extensive mutational analysis has revealed that four regions within the terminal 51 bp of the Ad2 genome influence origin activity *in vitro* and *in vivo*. The origin of replication of Ad2 consists of a core domain comprising the terminal 18 bp of the genome which alone is only capable of supporting a low level of initiation, and an auxiliary region encompassing nucleotides 25–50 which contains recognition sequences for the sequence-specific DNA-binding proteins NFI (also known as CTF) and NFIII (also known as Oct-1). Occupancy of the recognition site by NFI increases the frequency of initiation of viral DNA replication both *in vivo* and *in vitro*. Separating the core and the NFI binding site is a region of DNA where sequence changes are tolerated, but insertions or deletions are not. Immediately adjacent to the NFI recognition site is the binding site for NFIII. The mechanisms by which these host factors increase the efficiency of DNA replication has yet to be fully established but in each case it has been shown that the DNA-binding domain of the protein is sufficient to stimulate DNA synthesis. Genes for the three viral replicative proteins pTP, pol, and DBP have all been identified and the proteins expressed in a variety of heterologous systems, thus facilitating large-scale purification and analysis. DBP (M_r 68 000) is expressed at high levels in infected cells and functions during elongation of nascent DNA chains by increasing the processivity of the polymerase and by coating displaced single strands. On the basis of the available data it is possible to formulate a model for the formation of a preinitiation complex at the adenovirus origin of DNA replication (Fig. A16). Viral genomes coated with DBP are bound by NFI which

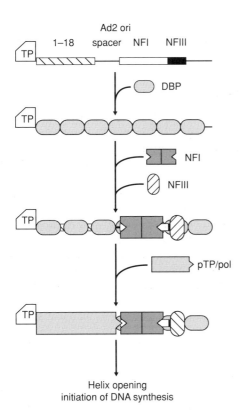

Fig. A16 Model for the formation of a preinitiation complex at the adenovirus 2 (Ad2) origin of DNA replication. Details are as described in the text.

through interactions with DNA polymerase directs the pTP/pol heterodimer to the 1–18 bp minimal origin sequence. This complex is further stabilized by interactions between the pTP/pol heterodimer and specific DNA sequences within the origin region. Unwinding of the DNA in this complex would then be required to expose the single stranded DNA template for the initiation reaction [8,9].

Transfection assays carried out with Ad4 showed that in contrast to Ad2 only the terminal 18 bp of the genome, which are identical in both serotypes and in Ad2 constitute the core origin, were required for efficient DNA replication *in vivo*. This is also the case *in vitro* where it was shown that linearized plasmid containing only the terminal 18 bp of the ITR could support initiation of DNA replication as effectively as a template containing a complete Ad4 ITR. The protein requirement for DNA replication differs markedly between Ad2 and Ad4 in that Ad4 seems to have circumvented the need for the host factors NFI and NFIII. It does not possess an NFI recognition site and although it does have a binding site for NFIII neither factor is required for DNA replication *in vivo* or is capable of stimulating DNA replication *in vitro*.

R.T. HAY

See also: EUKARYOTIC GENE EXPRESSION.

1 Akusjarvi, G. et al. (1986) Structure and function of the adenovirus genome. In *Adenovirus DNA* (Doerfler, W., Ed.) (Martinus Nijhoff Publishing, Boston).
2 Roberts, M.M. et al. (1986) Three-dimensional structure of the adenovirus coat protein hexon. *Science* **232**, 1148–1151.
3 Toogood, C.I.A. et al. (1992) Antipeptide sera define neutralizing epitopes on the adenovirus hexon. *J. Gen. Virol.* **73**, 1429–1435.
4 Logan, J. & Shenk, T. (1984) Adenovirus tripartite leader sequence enhances translation of mRNAs late after infection. *Proc. Natl. Acad. Sci. USA* **81**, 3655–3659.
5 Boulanger, P.A. & Blair, G.E. (1991) Expression and interactions of human adenovirus oncoproteins. *Biochem. J.* **275**, 281–299.
6 Whyte, P. et al. (1988) Association between an oncogene and an antioncogene: the adenovirus E1a proteins bind to the retinoblastoma gene product. *Nature* **334**, 124–129.
7 Lane, D.P. (1992) p53, guardian of the genome. *Nature* **358**, 15–16.
8 Mul, Y.M. & van der Vliet, P.C. (1992) Nuclear Factor-I enhances adenovirus DNA replication by increasing the stability of a preinitiation complex. *EMBO J.* **11**, 751–760.
9 Temperley, S.M. & Hay, R.T. (1992) Recognition of the adenovirus type-2 origin of DNA replication by the virally encoded DNA-polymerase and preterminal proteins. *EMBO J.* **11**, 761–768.
10 Tooze, J. (Ed.) (1980, 1981) *Molecular Biology of Tumor Viruses: DNA Tumor Viruses* (Cold Spring Harbor Laboratory, New York).

adenylate cyclase (adenylyl cyclase) Membrane-associated enzyme found in bacterial, fungal, and animal cells (EC 4.6.1.1) which eliminates pyrophosphate from ATP, simultaneously transferring the 5′-phosphate to the 3′-hydroxyl of the ribose ring to form adenosine 3′,5′-cyclic monophosphate (CYCLIC AMP). It was discovered by Earl Sutherland as the first receptor-controlled enzyme to form an intracellular second messenger. In animal cells, it is best understood as an enzyme controlled by the receptor-activated α subunit of the G protein G_s, but cloning has revealed multiple forms of adenylate cyclase, some of which are controlled (at least in part) by intracellular free CALCIUM concentrations or by the βγ subunits of G proteins. *See:* GTP-BINDING PROTEINS; SECOND MESSENGER PATHWAYS.

ADF *See:* ACTIN-BINDING PROTEINS.

Adh ALCOHOL DEHYDROGENASE GENE.

ADH ALCOHOL DEHYDROGENASE. *See also:* NAD$^+$-ALCOHOL DEHYDROGENASE GENE; NAD-DEPENDENT DEHYDROGENASES; PROTEIN STRUCTURE.

adh(a)erens junction Intermediate junction. *See:* CELL JUNCTIONS.

adhesin *See:* BACTERIAL PATHOGENICITY: MOLECULAR ASPECTS.

adhesion molecules/adhesion receptors Cell-surface proteins involved in mediating the recognition and adhesion of cells to their substrate and to other cells. *See:* CELL ADHESION MOLECULES (for NCAM and related molecules of the immunoglobulin superfamily, and the cadherins); CELL–CELL INTERACTIONS; CELL–MATRIX INTERACTIONS; EXTRACELLULAR MATRIX MOLECULES; INTEGRINS; LECTINS; SELECTINS.

adhesion plaque Focal contact. *See:* CELL JUNCTIONS.

adipocyte (fat cell) A CONNECTIVE TISSUE cell specialized for the storage of fat droplets. Adipose tissue is found in the subcutaneous layer and in tissue that cushions internal organs such as the kidneys.

adjuvant Any compound which when injected together with an antigen nonspecifically enhances the immune response to that antigen. Adjuvants are insoluble and undegradable substances (e.g. inorganic gels such as aluminium hydroxide, or water-in-oil emulsions such as incomplete Freund's adjuvant). They retard the destruction of antigen and allow the persistence of low but effective levels of antigen in the tissues and also nonspecifically activate the lymphoid system by provoking an inflammatory response. One of the most effective adjuvants is Freund's complete aduvant in which mycobacteria are suspended in a water in oil emulsion but the intense inflammmatory reponse it provokes precludes its clinical use.

ADP Adenosine 5′-diphosphate. *See* NUCLEOSIDES AND NUCLEOTIDES.

ADP-ribosylation The transfer of an adenosine diphosphate ribose moiety from NAD to another molecule, usually a protein. ADP-ribosylation is involved in the mechanism of action of certain toxins. The A domain of DIPHTHERIA TOXIN catalyses the ADP-ribosylation of the eukaryotic PROTEIN SYNTHESIS elongation factor-2 (which is necessary for translocation of mRNA on ribosomes), thus leading to a general inhibition of protein synthesis. The A subunit of CHOLERA TOXIN and related toxins of enterobacteria catalyse the ADP-ribosylation of an arginine side chain in the α subunit of the GTP-BINDING PROTEIN G_s. This blocks the GTPase activity of G_s rendering it permanently active. Its subsequent prolonged stimulation of adenylate cyclase and the resulting increase in CYCLIC AMP second messenger activity (*see* SECOND MESSENGER PATHWAYS) leads to disruption of ion flow into and out of the cell. PERTUSSIS TOXIN (from the causal agent of whooping cough, *Bordetella pertussis*) catalyses the ADP-ribosylation of a cysteine side chain of G_i, a GTP-binding protein that normally inhibits adenylate cyclase. ADP-ribosylation locks G_i in an inactive form which cannot carry out its normal inhibi-

Fig. A17 Adrenaline.

tory role. Nearly all known G proteins are ADP-ribosylated by cholera or pertussis toxins.

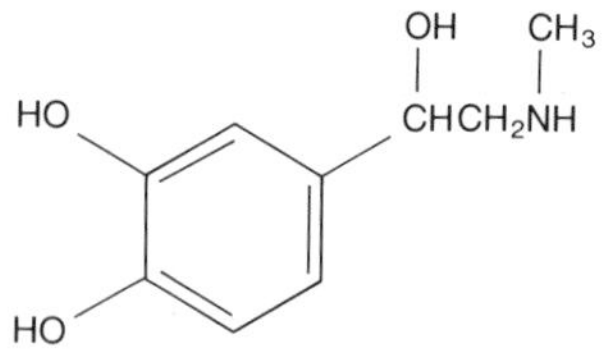

adrenal cell Adrenal cortical cells secrete a large number of STEROID HORMONES (corticosteroids) under the influence of ADRENOCORTICOTROPIN. Corticosteroids can be divided into two major functional groups — mineralocorticoids, which influence electrolyte balance, and glucocorticoids, which affect carbohydrate metabolism. Cells of the adrenal medulla secrete ADRENALINE and various OPIOID PEPTIDES.

adrenaline (epinephrine) Catecholamine hormone and neurotransmitter (Fig. A17) which is the originally identified CATECHOLAMINE NEUROTRANSMITTER in the frog. In mammals, however, its role as a neurotransmitter is mainly taken by NORADRENALINE. A small number of neurons containing adrenaline, identified by the presence of the biosynthetic enzyme phenylethanolamine-N-methyltransferase, do occur in the mammalian brain stem, some of which innervate the hypothalamus. As a hormone, adrenaline is synthesized and released from the medulla of the adrenal gland. It acts at a number of different ADRENERGIC RECEPTORS in a wide variety of tissues. Particularly well studied are its stimulation of glycogen breakdown in skeletal muscle and liver, mediated by the second messenger CYCLIC AMP (*see* SECOND MESSENGER PATHWAYS), and its action on heart muscle to speed up the heartbeat.

adrenergic receptors The receptors for noradrenaline and adrenaline, of which numerous subtypes exist (Table A2), are

Table A2 Adrenergic receptor subtypes

Mammalian subtype	Peptide length (amino acids)	Homology in transmembrane domains (%)	Introns	Chomosomal location (human)	Tissue distribution (rat)	Major G protein	Effector system
β_1	477	71 (β_2)	0	10q24–q26	Heart, pineal		↑adenylate cyclase
β_2	413	—	0	5q31–q33	Lung, prostate	G_s	↑Ca^{2+} channel
β_3	402	63 (β_2)	0	—	Adipose tissue		
α_{2A}	450	42 (β_2)	0	10q24–q26	Aorta, brain		↓adenylate cyclase
α_{2B}	450	36 (β_2), 74 (α_{2A})	0	2	Liver, kidney	G_i	↑K^+, ↓Ca^{2+} channels
α_{2C}	461	39 (β_2), 75 (α_{2A})	—	4	Brain		↑Na^+–H^+ channel
							↑phospholipase C, A_2
α_{1A}	—	—	—	—	Vas deferens, brain		↑phospholipase C
α_{1B}	515	42 (β_2)	1	5q32–q34	Liver, brain		↑Ca^{2+} channel
α_{1C}	466	42 (β_2), 72 (α_{1B})	1	8	Olfactory bulb	G_q	↑phospholipase A_2
α_{1D}	560	43 (β_2), 76 (α_{1B})	—	20p13	Vas deferens, brain		↑phospholipase D

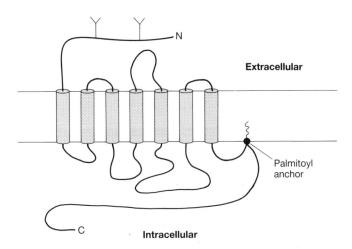

Fig. A18 The β₂-adrenergic receptor.

grouped into three families — the β receptors (β₁ and β₂), the α₁ receptors, and the α₂ receptors. They are all seven-span G PROTEIN-COUPLED RECEPTORS linked variously to the adenylate cyclase and phosphoinositidase SECOND MESSENGER PATHWAYS. The β₂-adrenergic receptor (Fig. A18) was one of the first seven-span receptors to be sequenced; its gene is unusual amongst eukaryotic genes in lacking introns.

Trends Pharmacol. Sci. (1993) Receptor supplement (Elsevier Science Publishers, Oxford).

adrenoceptors *See*: ADRENERGIC RECEPTORS.

adrenocorticotropin (adrenocorticotropic hormone, corticotropin, ACTH) Polypeptide hormone of 39 amino acids synthesized by the anterior pituitary and released under the influence of corticotropin-releasing factors (CRF). It is synthesized as part of the pro-opiomelanocortin molecule (*see* POLYPROTEINS, and Fig. O4 in OPIOID PEPTIDES AND THEIR RECEPTORS) from which it is released by proteolytic processing. It stimulates the growth of the adrenal cortex and the secretion of corticosteroids such as cortisol, and also stimulates triacylglycerol hydrolysis in fat cells, so leading to enhanced release of fatty acids.

adseverin *See*: ACTIN-BINDING PROTEINS.

adsorptive endocytosis ENDOCYTOSIS of ligands following binding to the cell surface. As with receptor-mediated endocytosis, binding to the cell surface concentrates ligands at the plasma membrane and increases the efficiency of uptake compared to nonadsorptive fluid phase endocytosis. By contrast to receptor-mediated endocytosis, adsorptive endocytosis refers to ligands which bind to multiple, different cell surface components (e.g. cationized proteins, LECTINS, ANTIBODIES and viruses). Adsorptive endocytosis occurs through the constitutive formation of clathrin-coated vesicles and results in delivery of the ligand to ENDOSOMES and LYSOSOMES. *See also*: COATED PITS AND VESICLES.

Marsh, M. & Helenius, A. (1980) *J. Mol. Biol.* **142**, 439–454.

aequorin Luminescent protein isolated from certain jellyfish (*Aequorea*). It emits light in the presence of Ca²⁺ and responds to changes in the Ca²⁺ concentration over the range 0.5 to 10 μM. It is used for monitoring calcium flux and distribution in living cells (*see* CALCIUM).

AER APICAL ECTODERMAL RIDGE.

aerenchyma Tissue composed of PARENCHYMA cells and which contains large gas-filled spaces. Found in roots and stems and often prominent in the cortex of aquatic plants. Spaces arise by either separation or breakdown of cells and may form complex channels which provide aeration, buoyancy and mechanical strength.

AEV Avian erythroblastosis virus, a RETROVIRUS.

Affective disorders

THE affective disorders are severe disturbances of mood that can be subdivided into depression and manic-depressive illness. As psychiatric diseases they share with SCHIZOPHRENIA the difficulties of objective diagnosis without an unequivocal physiological marker. Clinical depression is more than just a severely depressed mood; patients manifest a range of other symptoms that may include insomnia, suicidal thoughts, and losses of appetite and weight. Manic-depressive illness, often referred to as bipolar affective illness, is usually distinguished from depression (unipolar affective illness) by a cyclic alternation between depressed and elated mood, and in this latter manic state the patient can be overactive, wildly garrulous and generally excessive in behaviour.

Affective disorders occur with a lifetime risk of approximately 8%. They are twice as common in women than men, although this difference in incidence diminishes with increasing age. Unipolar disorder accounts for the majority of this morbidity; bipolar illness, in which there is equal sex incidence, occurs in less than 1% of the population.

Neurochemistry and pharmacology

Unlike schizophrenia, the affective disorders have not yielded any consistent evidence pointing to an underlying neuropathology, although this may well change in the future with the application of modern imaging techniques and quantitative histology. Nevertheless, abnormalities of certain NEUROTRANSMITTER systems have long been implicated in depression and mania [1]. The monoamine hypotheses of depression developed from the observation that compounds such as reserpine, which deplete MONOAMINE NEUROTRANSMITTERS, can induce depression, whereas the drugs that are often effective in treating the disorder tend to increase the effects of NORADRENALINE and 5-HYDROXYTRYPTAMINE (serotonin, 5HT). Thus the first ANTIDEPRESSANT DRUGS [2] were found to act by decreasing removal of transmitter either by inhibiting breakdown via MONOAMINE OXIDASE (as does tranylcypromine) or by blocking neuronal reuptake (Table A3). For example, the tricyclic antidepressant desipramine inhibits reuptake of noradrenaline, while clomipramine and amitriptyline inhibit both noradrenaline and 5HT reuptake. Nevertheless, these drugs also act at other receptor sites in the central nervous system and so the involvement of different mechanisms in their antidepressantaction cannot be excluded.

Both the noradrenaline and 5HT hypotheses have had their proponents, although the more recently introduced antidepressant drugs seem to have more selective actions on 5HT neurons. Thus fluvoxamine is a specific 5HT uptake inhibitor, while other antidepressants (e.g. mianserin) act postsynaptically as antagonists of 5HT RECEPTORS. The discrepancy between the rapid primary effect of most of these drugs and the period of days to weeks before a clinical improvement is apparent has led to the suggestion that a secondary effect on, perhaps, receptor numbers may mediate an antidepressant action. Several receptor systems are affected by antidepressants; one action common to all these drugs is an ability to decrease the number of $5HT_2$ receptors after chronic administration.

A nonpharmacological therapeutic strategy effective in the majority of depressed patients is that of electroconvulsive treatment (ECT). Again, the antidepressant effect of this treatment is still not understood although animal models of ECT have sug-

Table A3 The major antidepressant drugs and their pharmacological effects

Tricyclic antidepressants that inhibit reuptake of noradrenaline and, in some cases, 5HT*	Desipramine
	Doxepin
	Imipramine
	Clomipramine
	Amitriptyline
Specific 5HT reuptake inhibitors	Paroxetine
	Fluoxetine
	Fluvoxamine
$5HT_2$ receptor antagonists*	Mianserin
	Trazodone
	Amitriptyline
Irreversible nonspecific monoamine oxidase inhibitors	Phenelzine
	Tranylcypromine
Reversible monoamine oxidase type A inhibitor	Moclobemide
Increased 5HT synthesis	Tryptophan

This is not a comprehensive listing but provides examples of drugs with the indicated effects. Other sites of action (particularly for the groups indicated with an asterisk) may well be important, and different primary pharmacological effects may lead to common secondary antidepressant mechanisms. The most widely used antidepressants are the 'tricyclic' compounds, named after the chemical structure of the original members of this group. These drugs block neuronal reuptake of noradrenaline, and in some cases also 5HT, thereby diminishing the rate of removal of neurotransmitter from the synapse. However other actions, such as blockade of α_2-adrenergic receptors, have been suggested to be important for some of these compounds. More recently introduced antidepressants include reversible inhibitors of monoamine oxidase type A, selective inhibitors of 5HT neuronal reuptake and antagonists of various 5HT receptors. Tryptophan too has an antidepressant action, presumably by increasing the synthesis of 5HT.

gested a variety of possible mechanisms. These include increased noradrenaline release, decreased β-ADRENERGIC RECEPTOR density, various regulatory changes in 5HT receptor subtypes and effects on SECOND MESSENGER function; however, several of these actions are likely to be interrelated.

Mania will respond fairly rapidly to ANTIPSYCHOTIC DRUGS although, for chronic and prophylactic treatment of bipolar affective illness, lithium is particularly effective. As well as a specific action on the manic phases of bipolar disorder, lithium has an antidepressant effect in this disease. Its mode of action is not understood, although it presumably acts by stabilizing the system(s) mediating the mood swings. Effects on both the synthesis and release of noradrenaline and/or 5HT have been proposed to be important in this respect, as have effects on second messenger and receptor function.

Neurochemical studies have not provided very strong support for an involvement of one or other monoamine transmitter in affective disorders. However, low cerebrospinal fluid concentrations of 5-hydroxyindoleacetic acid (5HIAA, the major 5HT metabolite) have occasionally, if inconsistently, been reported in depressed patients, while one of the most interesting of these studies is the particular association of low 5HIAA with violent suicide attempts. Post-mortem studies have failed to identify the source of these changes within the brain, despite occasional reports of 5HT or 5HIAA being reduced below control levels in depressed (usually suicidal) subjects. Investigations of various 5HT receptors in these brains have also been unable to identify consistent abnormalities, although in some studies cortical $5HT_2$ receptor density is increased. Various neuroendocrine markers measured in blood have been found to be abnormal in depression, some of which have been used, with limited success, as potential diagnostic tools. However, many of these peripheral measures have been found to be state dependent and, although they may be under the control of central neurotransmitter activity, have not been particularly useful in understanding the neuronal basis of affective disorders.

Aetiology and genetics

Depression has in the past been classified as reactive or endogenous, implying differences in the relative contribution of exogenous environmental factors to the aetiology of the disease. However it has long been apparent that there is a strong genetic contribution to both unipolar and bipolar affective disorders [3]. This is illustrated by the high CONCORDANCE of affective disorder in monozygotic twins of 50–70%, with bipolar subjects generally demonstrating higher concordance than unipolar depressives. Other family studies also suggest that bipolar illness has a higher genetic component. Although the type of disorder tends to be similar within families, it is not uncommon for unipolar patients to be related to bipolars, indicating that these are not two genetically distinct subtypes of affective disorder. Studies of adoptees have generally shown that heredity and not family environment is primarily responsible for the familial occurrence of affective disorder.

The observation of apparent linkage to COLOUR BLINDNESS in some families provided the first genetic marker for affective disorder. Subsequent studies showed that this indication of an X-chromosome site for manic-depressive illness is true in some pedigrees but could be excluded in others. Clearly this provides evidence for an aetiological heterogeneity in the disease. Investigation of a large pedigree in the North American Amish community initially provided strong evidence for linkage of bipolar illness to chromosome 11, although more rigorous reinvestigation subsequently excluded linkage to chromosome 11 markers in this, and several other, pedigrees. Continued interest in chromosome 11 remains, owing to the presence of the 'candidate' gene for tyrosine hydroxylase, the enzyme controlling the synthesis of the CATECHOLAMINE NEUROTRANSMITTERS, and the positive results for schizophrenia which some suggest may exhibit diagnostic and aetiological overlap with bipolar affective illness. The recent cloning of various neurotransmitter receptors and related proteins provides the means to investigate further candidate genes in affective disorders.

G.P. REYNOLDS

1 Deakin, J.F.W. (1986) *The Biology of Depression* (Royal College of Psychiatrists, London).
2 Leonard, B.E. (1992) *Fundamentals of Psychopharmacology*, 55–86 (Wiley, Chichester).
3 Ciaranello, R.D. & Ciaranello, A.L. (1991) Genetics of major psychiatric disorders. *Annu. Rev. Med.* **42**, 151–158.

affinity Term widely used to describe the strength of binding of an ANTIBODY for its cognate ANTIGEN, or of any receptor for its ligand. The affinity can be measured as the ratio of receptor–ligand complex to free reactants at equilibrium, and the affinity constant is equivalent to the ASSOCIATION CONSTANT of the binding of a monovalent ligand to one binding site on the antibody. For antibody–antigen interactions it should therefore be distinguished from the AVIDITY of an antibody for its antigen, which is the measure of overall strength of binding of an antigen to antibody taking into account the increased strength of binding when the antigen and antibody are multivalent. The affinity of an antibody for a monovalent hapten can be determined from a SCATCHARD PLOT of equilibrium binding experiments. For monoclonal antibodies where only one class of binding site is present, the slope of a linear Scatchard plot is the negative of the affinity constant, which is the reciprocal of K_d, the equilibrium binding (dissociation) constant for the antigen–antibody interaction. For a population of antibodies with slightly different affinities for their cognate antigen, a curved Scatchard plot is obatined and the average affinity is calculated.

affinity chromatography *See*: CHROMATOGRAPHY.

affinity maturation The increase in the average AFFINITY for the immunizing antigen (protein antigens only) of the ANTIBODIES produced by an individual as an immune response proceeds, and after subsequent immunization with that antigen. It arises from the HYPERMUTATION of the DNA encoding the V region in individual antigen-stimulated B cells, leading to production of immunoglobulin with higher affinity for the immunizing antigen. Clones of B cells bearing these high-affinity antigen receptors will subsequently be selected during the progression of the immune response or on subsequent immunization and develop into PLASMA CELLS producing high-affinity antibodies. *See also*: B CELL DEVELOPMENT; MEMORY CELLS.

afibrinogenaemia An AUTOSOMAL RECESSIVE condition characterized by the lack — total or virtually so — of the clottable protein FIBRIN in plasma. It is associated with severe bleeding from birth. The GENE FREQUENCY is 1 : 1000. *See*: BLOOD COAGULATION AND ITS DISORDERS.

aflatoxins Toxic secondary polyketide metabolites of strains of the fungus *Aspergillus flavus* which are produced when growing on groundnuts (peanuts) and cereals. They cause aflatoxicosis in poultry and cattle and are also carcinogenic (causing liver cancer) in animals and humans. *See*: CHEMICAL CARCINOGENS AND CARCINOGENESIS.

Heathcote, J.G. & Hibbert, J.R. (1978) *Aflatoxins: Chemical and Biological Aspects* (Elsevier Scientific, Amsterdam).

agammaglobulinaemia Very low serum immunoglobulin levels, a clinical state of severe IMMUNODEFICIENCY characterized by repeated pyogenic bacterial infections. The term agammaglobulinaemia is commonly used in cases where the serum GAMMA-GLOBULIN level is <1 g l^{-1}. *See also*: DYSGAMMAGLOBULINAEMIA; HYPOGAMMAGLOBULINAEMIA; X-LINKED AGAMMAGLOBULINAEMIA.

agarose A linear polysaccharide polymer based on the repeat unit (-3DGal(p)β1,4LGal(p)α1-). It is extracted from agar which comes from a red seaweed and is used to prepare gels for ELECTROPHORESIS of nucleic acids. When a solution of agarose is heated to boiling point and then allowed to cool it hardens, forming a matrix whose density is determined by the agarose concentration. DNA fragments in the range 0.2–50 kb can be separated by agarose gel electrophoresis. Chemically modified forms of agarose that are structurally weakened and so melt and gel at low temperatures are available and are used in preparative electrophoresis of DNA fragments.

ageing *See*: MITOCHONDRIAL GENOMES: LOWER EUKARYOTE; MITOCHONDRIOPATHIES.

agglutinin A general term for an agent such as an ANTIBODY, a LECTIN, or other bi- or polyvalent reagent which can cause the aggregation and adhesion of particles or cells. *See also*: HAEMAGGLUTININ.

aggrecan A major cartilage PROTEOGLYCAN. *See*: EXTRACELLULAR MATRIX MOLECULES.

aggregation chimaera An organism produced by the aggregation *in vitro* of PLURIPOTENT embryonic cells of different origin (e.g. of different GENOTYPES or even of different species). The cells are usually obtained from early embryos and the cells of one genotype carry a genetic marker allowing later identification. Aggregation chimaeras are used in the study of vertebrate development, and are particularly useful for FATE MAPPING studies or for studies of lethal genotypes, as cells from organisms which might otherwise die may contribute to some tissues in a chimaera. *See also*: ALLOPHENIC MICE.

aginactin *See*: ACTIN-BINDING PROTEINS.

agnathans A group of jawless vertebrates which display primitive feeding mechanisms (filter feeding or parasitic sucking) before the evolutionary development of jaws. Most are extinct; living representatives include lamprey and hagfish.

agnogene/agnoprotein *See*: PAPOVAVIRUSES.

agonist Any compound that activates a receptor.

agouti (A) A mouse COAT COLOUR GENE which acts within the hair follicle to make each hair black with a band of yellow near the top. The gene controls the balance between pheomelanin and eumelanin synthesis by the melanocytes; grafting experiments have shown, however, that it is required in the epidermal or dermal cells of the follicle, rather than in the melanocytes. Certain ALLELES at the *agouti* locus also affect embryonic development; for example the *yellow* allele (A^y) has dominant pleiotropic effects including obesity and diabetes when heterozygous, and is embryonic lethal when homozygous.

AGR AGROPINE.

Agrobacterium A genus of Gram-negative, coliform, soil-borne plant pathogenic bacteria, belonging to the Rhizobiaceae. The genus contains four species, only three of which cause known plant diseases. *Agrobacterium tumefaciens* is a plant pathogen, the causative agent of the plant tumour CROWN GALL. Tumorigenic strains carry a TI PLASMID, part of which is transferred into the plant cell genome during infection and is responsible for producing and maintaining the tumour phenotype. The Ti plasmid is widely used as a vector in PLANT GENETIC ENGINEERING. *Agrobacterium rhizogenes* is a plant pathogen, the causative agent of the tumorous disease HAIRY ROOT, in which a multitude of tiny rootlets are induced. Tumorigenic strains carry an RI PLASMID, part of which is transferred into the plant cell genome during infection and is responsible for producing and maintaining the tumour phenotype. *Agrobacterium rubi* is a plant pathogen, the causative agent of the disease cane gall. Tumorigenic strains carry a plasmid similar to the Ti plasmid, part of which is transferred into the plant cell genome during infection and is responsible for producing and maintaining the tumour phenotype. *Agrobacterium radiobacter* is nonpathogenic.

agrocin 84 Antibacterial agent produced by *Agrobacterium tumefaciens* strain K84. Active against NOPALINE STRAINS of *Agrobacterium* which catabolize agrocinopines C and D.

agrocinopine Family of OPINES in which each member comprises two sugar moieties linked by a phosphodiester bond. Agrocinopines A and B induce conjugal transfer of the TI PLASMID in NOPALINE STRAINS of *Agrobacterium tumefaciens*. Agrocinopines C

and D are associated with strains which produce neither octopine nor nopaline opines (sometimes called agropine strains).

agropine (AGR) An OPINE, a lactonized derivative of MANNOPINE. The gene (*ags*) involved in its synthesis is found in the T_R-region of TI PLASMIDS of OCTOPINE STRAINS of *Agrobacterium tumefaciens*, in a group of strains which cannot catabolize either octopine or nopaline (originally null strains; now agropine strains) and in the T_R-region of some RI PLASMIDS of *A. rhizogenes*.

ags gene Agropine synthase gene. Found in the T_R-region of TI PLASMIDS of OCTOPINE STRAINS of *Agrobacterium tumefaciens* and in the T_R-region of some RI PLASMIDS of *A. rhizogenes*.

AHG *See*: BLOOD COAGULATION AND ITS DISORDERS.

Aic glycosylated haemoglobin *See*: HAEMOGLOBIN AND ITS DISORDERS.

AIDS Acquired immune deficiency syndrome. A disease caused by the human IMMUNODEFICIENCY VIRUSES (HIV). First identified in the USA and France in 1983, the virus is transmitted in the population by an exchange of cells and body fluids (blood or semen) and infection leads to profound immunosuppression and lymphadenopathy and increased susceptibility to opportunistic pathogens. This syndrome is due to infection and functional destruction of T CELLS of the CD4 class, in which the cell-surface CD antigen CD4 is the receptor for the virus. Infection causes T cells to fuse to form large syncytia that eventually lyse. Progression to full-blown AIDS is invariably fatal and there is currently no cure and no vaccine.

alanine (Ala, A) A small hydrophobic AMINO ACID with a methyl side chain.

albinism A heterogeneous group of AUTOSOMAL RECESSIVE disorders caused by mutations of the gene directing synthesis of tyrosinase, an enzyme involved in the synthesis of the pigment melanin. The absence of melanin leads to deficiency of ocular and cutaneous pigmentation.

albino (c) locus Locus on mouse chromosome 7 encoding the gene for the enzyme tyrosinase (monophenol monooxygenase EC 1.14.18.1), which acts in the biochemical pathway that converts tyrosine to melanin. Animals homozygous for a null allele at the *albino* locus are white with red eyes, owing to a total absence of melanin in the coat and retina. The optic nerves are often crossed in these mutants, causing visual defects. Albino individuals of many other vertebrate species including frogs, chickens, and humans, are known. *See also*: ALBINISM; COAT COLOUR GENES.

alcohol dehydrogenase (ADH) Usually refers to the zinc-containing enzyme (EC 1.1.1.1) from a wide variety of sources, which catalyses the reaction

$$\text{alcohol} + \text{NAD}^+ = \text{aldehyde or ketone} + \text{NADH} + \text{H}^+$$

Horse liver alcohol dehydrogenase has a subunit M_r of 40 000.

Active enzyme is composed of two subunits; isoenzymes EE, ES, or SS may be found. Nonmetalloenzyme alcohol dehydrogenases (short-chain dehydrogenases) have also been isolated from several sources. *See also*: NAD⁺-ALCOHOL DEHYDROGENASE GENE; NAD-DEPENDENT DEHYDROGENASES; PROTEIN STRUCTURE.

alcohol dehydrogenase gene (*Adh*) Possibly the best characterized gene in the fruit fly *Drosophila melanogaster*, encoding a short chain alcohol dehydrogenase, a protein of 256 amino acids which catalyses the reduction of the primary alcohols present in the fermenting fruits favoured by these flies to aldehydes and ketones. The gene is of developmental interest because its expression is controlled by two distinct PROMOTERS. The proximal promoter lying adjacent to the INITIATION CODON controls gene expression in larvae, whereas the adult (distal) promoter, 700 bp upstream, controls gene expression in the adult. Experiments involving the genetic transformation of flies with a bacterial *lacZ* REPORTER GENE fused to parts of the *Adh* promoter support the hypothesis that specific promoter DNA sequences are responsible for correct tissue-specific and developmental expression. Altering the structure of the promoter *in vitro* results in the gene taking on a different developmental expression profile.

aldolase Fructose-1,6-bisphosphate aldolase (EC 4.1.2.13), a widely distributed metabolic enzyme catalysing the conversion of D-fructose 1,6-bisphosphate to dihydroxyacetone phosphate and D-glyceraldehyde 3-phosphate. It also has actin-binding activity of unknown function.

aldosterone The major mineralocorticoid hormone secreted by the adrenal cortex (*see* STEROID HORMONES). It is involved in the regulation of electrolyte balance (especially that of Na⁺ and K⁺) in extracellular fluids. *See also*: STEROID HORMONE RECEPTORS.

alfalfa mosaic virus group Named after the type and only member of the group which is a MULTICOMPONENT VIRUS with bacilliform particles of the same diameter (18 nm) but at least four different lengths (58, 48, 36, and 28 nm). The four (+)-strand linear RNA species (Fig. A19) are encapsidated in separate components. Bottom component contains RNA 1 which, together with RNA 2 encapsidated in middle component, are monocistronic mRNAs encoding subunits of the RNA-dependent RNA polymerase used to replicate viral RNA. Top b component contains RNA 3 which is bicistronic, encoding at the 5′ end the cell-to-cell movement protein and at the 3′ end the viral coat protein; the coat protein is expressed from the subgenomic RNA 4 which is also encapsidated in top a component. As with ILARVIRUSES, RNAs 1, 2 and 3, together with coat protein molecules or RNA 4 are required for infection. *See also*: PLANT VIRUSES.

Francki, R.I.B. (1985) In *The Plant Viruses*, Vol. 1, 1–18 (Plenum, New York).

alginic acids Linear copolymers of L-guluronic and D-mannuronic acids found in brown algae (e.g. *Laminaria* spp.) in which there are roughly equimolar amounts of the two sugars, mostly arranged in short domains. The polymer is synthesized initially as

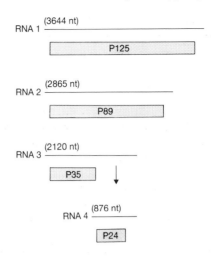

Fig. A19 Genome organization of alfalfa mosaic virus (AlMV). The lines represent the RNA species and the boxes the proteins (P) with M_r given as × 10⁻³; nt, nucleotides.

a mannuronan, probably by means of a single glycosyl transferase and GDP-ManA; the L-guluronosyl residues are generated after polymer formation by a 5-epimerase (*cf.* GLYCOSAMINOGLYCANS). Nothing is known of chain initiation, termination, or the specification of the site of epimerase action. Alginic acids are important food texturing agents and are used in dental moulding materials.

alignment (of sequences) Comparisons of two nucleotide or amino-acid sequences for phylogenetic purposes, or to detect HOMOLOGY between them, depend on finding the appropriate parts to align. Two sequences are said to 'match' when they share the same nucleotide (for nucleic acids) or the same amino acid (for proteins) at a position. A weaker match can be counted when nucleic acids both have a purine (or pyrimidine) at the comparable position or when proteins have an amino acid with similar chemical properties.

In the following example parts of two sequences show 8 matches out of a possible 25, only marginally above the 6 matches expected by chance:

```
Sequence A    . . TGAACTCCACCAAAAAGGAAGACTG . .
Sequence B    . . TAGGCTCACGCACCAACATCATACT . .
              *   ***   ** **              8/25
```

but in many pairs of sequences the number of matches can be increased significantly. In this example the number of matches can be increased to 15 by inserting two small gaps, one in each sequence:

```
Sequence A    TGAACTC---CACCAAAAAGGAAGACTG
Sequence B    TAGGCTCACGCACCAACAT--CATACT
              *   ***   ****** *   *  *** 15/25
```

Given just two sequences one cannot decide whether the gaps represent insertions in one sequence or deletions in the other.

The term indel is a neutral expression which avoids indicating the direction of change. However, as shown below, it is possible to just keep adding gaps to the first sequence, thus increasing the proportion of matches for this sequence to 7/8 for the first part of sequence A.

```
Sequence A   T-G----AA-C-------TC---C-ACCAAAAAGGAAGACT
Sequence B   TAGGCTCACGCACCAACATCATACT
             *  *   * *       **   * 7/8
```

To avoid the absurdity of any sequence showing a high similarity to every other sequence a gap penalty is introduced and is subtracted from the number of matches. It is often a two-parameter penalty where a gap spans more than one nucleotide; there is a higher weighting for introducing a gap and a lesser penalty for extending a gap. As yet, the weights given the gap penalty are arbitrary in that values are estimated, rather than calculated from first principles. It is probably possible to calculate the expected value for a gap penalty but impractical in practice, so values are usually estimated from simulations. In this approach, many pairs of random sequences could be generated. For each pair of random sequences the maximum number of matches would be found and the increase in number of matches counted as gaps (indels) are added.

The calculation becomes even more complicated for 'multiple alignment' where three or more sequences are compared. The optimum alignment is dependent on the tree which relates the sequences to each other because it improves the score if a single indel can account for changes to two or more sequences.

Other information useful for sequence alignment includes finding similar two-dimensional folding, particularly in RNA sequences; three-dimensional similarities, particularly in proteins; and similar positions of intron/exon boundaries. Alignment remains a major computational problem but is a vital part of a molecular study. Conserved regions are, as predicted by the neutral theory (*see* MOLECULAR EVOLUTION), the important functional parts of macromolecules so sequence alignment (the phylogenetic method) is an important aspect of any study of a new gene sequence.

Waterman, M.S. et al. (1991) In *Phylogenetic Analysis of DNA Sequences* (Miyamoto, M.M. & Cracraft, J., Eds) 59–72 (Oxford University Press, Oxford).

alkaline phosphatase Hydrolytic enzyme (EC 3.1.3.1) used to catalyse the removal of 5′-phosphate residues from nucleic acids. It is a dimeric GLYCOPROTEIN consisting of two identical subunits (subunit M_r 69 000) and four atoms of zinc per molecule. Usually obtained from either *Escherichia coli* (BAP) or calf intestine (CIP), it has two major applications in molecular biology:

1 To remove 5′-phosphate residues from a linearized double-stranded DNA to prevent subsequent self-ligation.

2 To remove the 5′-phosphate residue from a single-stranded DNA or RNA molecule to allow subsequent labelling of the 5′-end using T4 polynucleotide kinase (*see* END-LABELLING).

It is also used as a component of a conjugated antibody immunodetection system.

alkaptonuria An AUTOSOMAL RECESSIVE inborn error of metabolism resulting from a deficiency of homogentisic acid oxidase which catalyses the oxidation of homogentisic acid to maleylacetoacetic acid. The homogentisic acid slowly polymerizes and the resulting brown-black product may be seen deposited in cartilage, tendons, and ligaments. Ochronotic arthritis may develop. The GENE FREQUENCY is 1 : 1000.

alkylating agent Any of a large group of mutagens and carcinogens which generally act by covalent modification of DNA to generate cytotoxic and mutagenic adducts. Their action typically involves methylation or ethylation of nitrogen and oxygen atoms in the polynucleotide. Some environmental mutagens such as nitrosamines require metabolic activation by mixed function oxidases (CYTOCHROME P450) to be converted into reactive species. Certain anticancer drugs such as cyclophosphamide become alkylating aents by activation in this way. Other agents in clinical use (nitrosoureas, nitrogen mustards) are used directly as reactive compounds. The short-lived alkylating intermediate itself arises when an alkylating agent decomposes spontaneously in aqueous solution to produce an alkyl carbonium cation that reacts with DNA. *See*: CHEMICAL CARCINOGENS AND CARCINOGENESIS; DNA REPAIR.

ALL ACUTE LYMPHOBLASTIC LEUKAEMIA.

allele A variant form of a given GENE. The term was first coined by Gregor Mendel to describe the alternative forms of the inheritable 'factors' that encoded a particular characteristic of the pea plant. For example, yellow and green peas carry different alleles of the gene determining pea colour (*see* MENDELIAN INHERITANCE). A large number of genes are present in two or more allelic forms in the population and some genes, notably those encoding MHC MOLECULES, have numerous alleles. A diploid organism carrying two different alleles of a gene is said to be heterozygous for that gene, whereas a homozygote carries two copies of the same allele.

The geneticist H.J. Muller coined the terms neomorphic allele (neomorph), hypomorphic allele (hypomorph), hypermorphic allele (hypermorph), amorphic allele (amorph), and antimorphic allele (antimorph), to classify different mutations of a given gene. The effects of the test mutation when homozygous, and when heterozygous with a complete deletion of the gene, were compared with the effects of a homozygous deletion of the gene: that is, *mutation/mutation* is compared with *mutation*/deletion and with deletion/deletion. The results of the combinations are interpreted as:

1 Amorphic allele: when *mutation*/deletion shows an effect identical to a homozygous deletion of a gene. This indicates that the mutant allele has no gene activity.

2 Antimorphic allele: when *mutation/mutation* has a more severe phenotype than *mutation*/deletion. This indicates that the mutant allele possesses activity opposite to that of wild type.

3 Hypermorphic allele: these alleles possess a greater than wild-type activity, but are difficult to detect because most are fully recessive and would produce a phenotype when homozygous which would be indistinguishable from wild type.

4 Hypomorphic allele: when *mutation*/deletion has a phenotype which is similar to but more extreme than *mutation*/*mutation*. This indicates an allele with reduced activity compared with wild type.
5 Neomorphic allele: possesses qualitatively different activity from wild type. Such alleles are almost always dominant to both wild-type and deficiency alleles.

allelic Produced by or occurring in alleles of the same gene; applied to different forms of a protein, to mutations, etc.

allelic exclusion The situation in which only one of the pair of ALLELES at a genetic LOCUS is expressed in a diploid cell. It is characteristic of the expression of the immunoglobulin loci and T cell receptor loci in B CELLS and T CELLS respectively (*see* B CELL DEVELOPMENT; IMMUNOGLOBULIN GENES; T CELL DEVELOPMENT; T CELL RECEPTOR GENES). In B cells, for example, there will be a productive rearrangement of the V_H, D, and J_H gene segments leading to the synthesis of a heavy chain polypeptide at only one of the two immunoglobulin heavy chain alleles (*see* V-(D)-J RECOMBINATION). The heavy chain allele on the other chromosome of the homologous pair will either have its V_H, D, and J_H segments in the unrearranged germ-line configuration, or will carry some form of incomplete or otherwise abortive rearrangement such that this second allele does not yield a heavy chain polypeptide that can be expressed on the cell surface. The choice of parental allele for expression is apparently random. Allelic exclusion occurs in a similar fashion at the immunoglobulin light chain loci and at the T cell receptor loci, and ensures that each B and T lymphocyte expresses antigen receptors of a single antigen specificity. *See also*: ISOTYPIC EXCLUSION.

allelic rescue A method for directly cloning a mutant ALLELE of a given gene from the genome of SACCHAROMYCES CEREVISIAE. It involves TRANSFORMATION of *S. cerevisiae* with a plasmid carrying a wild-type allele of the target gene linearized by deletion of a DNA sequence internal to the target gene. Upon transformation into a strain of *S. cerevisiae* carrying the desired mutant allele, the gap in the transforming plasmid is spontaneously repaired by the host cell's DNA synthesis machinery using the chromosomally located mutant gene as a template (gap repair). The plasmid now contains a copy of the mutant allele and can be recovered from the culture.

Orr-Weaver, T.L. et al. (1982) *Methods Enzymol.* **101**, 228–245.

alloantigens Self molecules that may be recognized as foreign between members of the same species. Such molecules usually have the same or similar function but are products of different ALLELES of the same gene, for example, T CELL RECEPTORS recognizing ALLO-MHC molecules. Here the alloantigen is in many cases specified by a particular peptide bound in the peptide-binding site.

alloantisera ANTIBODIES raised in one member of a species that specifically recognize molecules of a genetically distinct member of the same species. Human alloantisera are produced by multiparous women who have been exposed to paternal allotypes from the foetus, multiply transfused individuals and immunized volunteers.

alloenzymes Variant forms of an enzyme found in different individuals of the same species and resulting from the existence of multiple alleles within the population.

allogeneic, allogenic Describes genetic differences within species, that is, differences between individuals of the same species. Allogeneic cells are therefore simply those from another individual of the same species, but the term is used particularly in immunology to describe differences in MHC MOLECULES (*see* MAJOR HISTOCOMPATIBILITY COMPLEX) on lymphocytes from genetically nonidentical individuals (*see* ALLO-MHC). Allogeneic cells have been used to investigate mechanisms of tolerance and the specificity of the immune response. Thus, injecting semi-allogeneic parental cells into an F1 hybrid mouse might be expected to cause a one-way GRAFT-VERSUS-HOST response.

allograft A graft of tissue between two genetically nonidentical individuals of the same species. Allografts are generally rejected in vertebrates because of the immune response mounted against the foreign tissue (*see*: ALLO-MHC; MAJOR HISTOCOMPATIBILITY COMPLEX), but are generally accepted in invertebrates.

allo-MHC Allogeneic MHC, which describes a situation of genetic nonidentity at the loci of the MAJOR HISTOCOMPATIBILITY COMPLEX (MHC) between individuals of the same species. Humans, other than identical twins, are related allogeneically, as are outbred mice and mice from different inbred strains. Allogeneic individuals display different allelic forms of at least some MHC antigens (MHC MOLECULES) on their cell surfaces, so that tissue grafts exchanged between such individuals are rejected. Immune cells from different individuals also cannot normally cooperate to produce an MHC-restricted immune response.

allophenic mice CHIMAERIC or MOSAIC mice generated by aggregating two or more genetically distinct cleavage-stage mouse embryos and allowing them to develop to term in a foster mother. This experimental system was developed in the 1960s for investigating CELL LINEAGE, cell allocation, cell fusions and CELL–CELL INTERACTIONS during development.

allophycocyanin PHYCOBILIPROTEIN pigment that forms the core of the PHYCOBILISOMES of red algae and CYANOBACTERIA. The chromophore is phycocyanobilin, an open-chain tetrapyrrole, which is covalently linked to the apoprotein through thioether linkages. Like other phycobiliproteins, the protein is composed of an αβ monomer which forms trimeric and hexameric aggregates. Its absorption maximum at 650 nm (670 nm in allophycocyanin B) lies at lower energy (longer wavelength) than those of PHYCOERYTHRIN and PHYCOCYANIN, so facilitating excitation energy transfer from the latter pigments to allophycocyanin and hence, via the linker protein, to the reaction centre of PHOTOSYSTEM II. *See also*: PHOTOSYNTHESIS.

alloreactivity The immune reactivity shown by an individual to tissues from another genetically nonidentical individual of the same species.

allosteric effect, allostery The phenomenon in which a molecule (an allosteric effector) binding at one site on a protein causes a conformational change in the protein such that the activity of another site on the protein is altered.

The term was first introduced by Monod, Changeux, and Jacob in 1963 to explain competitive inhibition of enzymes by compounds bearing little resemblance to the natural substrate, hence the term allosteric, from the Greek *allos*, other, *stereos*, solid or space. They suggested that these allosteric inhibitors had binding sites that were distinct from, and probably distant from, the active site, and that they exerted their competitive inhibitory effects by inducing a conformational change in the protein such that the binding of inhibitor prevented substrate binding and vice versa.

The original concept of allostery has since been extended to include the cooperative binding of substrate shown by many oligomeric enzymes (*see* COOPERATIVITY); the cooperative oxygen binding by mammalian haemoglobin; the behaviour of ligand-gated ion channels, in which binding of ligand induces opening or closing of the channel; and the control of the binding of regulatory proteins to RNA and DNA (*see e.g.* LAC REPRESSOR). The term homotropic is often used to describe cooperative interactions between identical ligand molecules (e.g. substrate), whereas the interaction of an allosteric effector (in the original sense) and a substrate is termed heterotropic. Allosteric effectors may be activators or inhibitors. Many allosteric proteins show both homotropic and heterotropic interactions. Enzymes subject to allosteric control include those that control key points in metabolism, such as PHOSPHOFRUCTOKINASE, GLYCOGEN PHOSPHORYLASE, and fructose-1,6-bisphosphatase.

The most widely used models for allosteric interactions in oligomeric proteins are the Monod–Wyman–Changeux (MWC) concerted model and the Koshland–Némethy–Filmer (KNF) sequential model. These are described under COOPERATIVITY. The MWC model introduced the idea of two conformational states for allosteric proteins, an R-state (high activity) and a T-state (low activity), which differ in the constraints between subunits, the T-state being more constrained. Binding of substrate or allosteric activators favours the R-state; binding of allosteric inhibitors favours the T-state, thus making the transition to the R-state more difficult.

The structural changes that occur in allosteric interactions have been studied in greatest detail in haemoglobin and in the classical allosteric enzymes such as aspartate transcarbamylase and phosphofructokinase where crystal structures of the different conformational states are available. Structural studies show that the allosteric transition from one functional conformation to another (R ↔ T) involves movement of subunits relative to each other, and results in changes in both quaternary and tertiary structures although the overall symmetry of the protein is preserved. The detailed structural changes, and the mechanisms by which the affinity of active sites for substrate and regulatory sites for their ligands are altered in allosteric interactions, vary from protein to protein.

Allosteric control of enzyme activity is of two extreme types which are non-exclusive. In K systems, the binding of the regulatory allosteric effector alters the apparent K_m of the active site for substrate and vice versa; the R and T states have different affinities for both regulator and substrate. In V systems, there is no difference in affinity for substrate or regulator between the R and T states, but one state has a much higher catalytic rate than the other.

Monod, J. et al. (1963) *J. Mol. Biol.* **6**, 306–329.
Perutz, M. (1990) *Mechanisms of Cooperativity and Allosteric Regulation in Proteins* (Cambridge University Press, Cambridge).
Evans, P.R. (1991) Structural aspects of allostery. *Curr. Opinion Struct. Biol.* **1**, 773–779.

allotopic gene expression Introduction (and expression) of a suitably altered gene into an organelle in which it is not normally found. The gene is previously modified so that its product is targeted into the correct organelle. Thus a chloroplast gene can be inserted into the nucleus, and if the gene is modified so that the coding sequence is fused to a sequence encoding a chloroplast TRANSIT PEPTIDE, the protein produced is targeted to the chloroplast. This therefore avoids the need for manipulation of the chloroplast genome. The same approach can be taken for mitochondrial genes.

allotypes Protein products of allelic genes. The term is commonly applied to serologically defined variants of immunoglobulin polypeptide chains (*see* ANTIBODIES). The accepted nomenclature specifies the immunoglobulin class, subclass and the genetic marker (allele) in parentheses e.g. G1m(a), A2m(1). For the human IgG allotypes there is an accepted alphameric and a numeric notation in common usage.

de Lange, G.G. (1989) *Exp. Immunogenet.* **6**, 7–17.

allozymes ALLOENZYMES.

alpha Headwords with the prefix alpha or α- are indexed under the initial letter of the headword itself.

αβ receptors *See*: T CELL RECEPTORS.

alphoid DNA *See*: CHROMOSOME STRUCTURE.

ALS AMYOTROPHIC LATERAL SCLEROSIS.

alternative pathway *See*: COMPLEMENT.

***Alu* sequence** DNA sequence of 300 bp, so-called because it contains a single *Alu*I restriction site. It is a major component of the human GENOME. Around 1 million copies of the *Alu* sequence are evenly distributed throughout the genome and constitute between 3 and 6% of the total DNA in a human cell. The *Alu* sequence shows significant homology at both its 5′ and 3′ ends to an abundant small RNA called 7SL RNA (*see* SIGNAL RECOGNITION PARTICLE), but few if any of the *Alu* sequences are transcribed *in vivo*. The structure of the *Alu* sequence (Fig. A20) is characteristic of a PROCESSED PSEUDOGENE and suggests that it may have spread through an RNA-mediated process involving REVERSE TRANSCRIPTION.

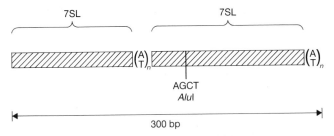

Fig. A20. The *Alu* sequence. The regions showing homology with the 7SL RNA sequence are indicated as is the location of the poly(AT)$_n$ regions and the position of the single *Alu*I restriction enzyme site diagnostic of this repetitive DNA element.

ALV Avian leukosis virus. *See:* RETROVIRUSES.

Alzheimer's disease A type of senile dementia that affects around 5 per cent of people over 65 years of age and ~10 per cent of those over 80, and which is characterized by memory loss and personality changes, eventually leading to dementia and senility. A rare familial form (familial Alzheimer's disease, FAD) with earlier onset also occurs, and has been found to be genetically heterogeneous. Alzheimer's disease is a NEURODEGENERATIVE DISORDER characterized histopathologically by neuronal degeneration in certain areas of the brain; deficits of neurotransmitters, their synthetic enzymes, and receptors; and the presence of senile plaques and neurofibrillary tangles in the brain.

Neurofibrillary tangles are intracellular and are composed of pairs of 10-nm diameter helical filaments twisted about each other. They are found in the cell bodies of neurons and in dendrites and axonal processes, and impair or destroy the functioning of the affected neuron. Protein components of the helical filaments include a fragment of the microtubule-associated protein, tau (*see* MICROTUBULES; NEURONAL CYTOSKELETON; UBIQUITIN).

Senile plaques are areas of disorganized neuropil up to 150 μm across that often have a core of extracellular AMYLOID-containing fibrils of 4–8 nm diameter. The protein in senile plaque amyloid (amyloid β-protein or A4 peptide, around 42 amino acids long) has been identified as a part of a transmembrane GLYCOPROTEIN abundant in normal brains — β-amyloid precursor (APP, ~695 amino acids, M_r ~ 92 000) — which is encoded by a gene on chromosome 21. Similar plaques of abnormal β-amyloid are found in the brains of older DOWN'S SYNDROME patients. The function of the β-amyloid precursor is unknown. The amyloid precursor mRNA is variously spliced and contains a domain with homology to certain proteinase inhibitors.

Both neurofibrillary tangles and senile plaques are thought to be the result of some as yet unidentified underlying pathogenesis. The gene for APP has been found mutated in some families with the familial form of the disease but not in others.

amalgam *See:* IMMUNOGLOBULIN SUPERFAMILY.

α-amanitin A cyclic octapeptide isolated from the poisonous 'death cap' mushroom (*Amanita phalloides*) (Fig. A21). It inhibits RNA POLYMERASE II at a concentration of 10 PM, thereby inhibiting TRANSCRIPTION. It will also inhibit RNA POLYMERASE III at 1 μM but RNA POLYMERASE I is insensitive to this toxin.

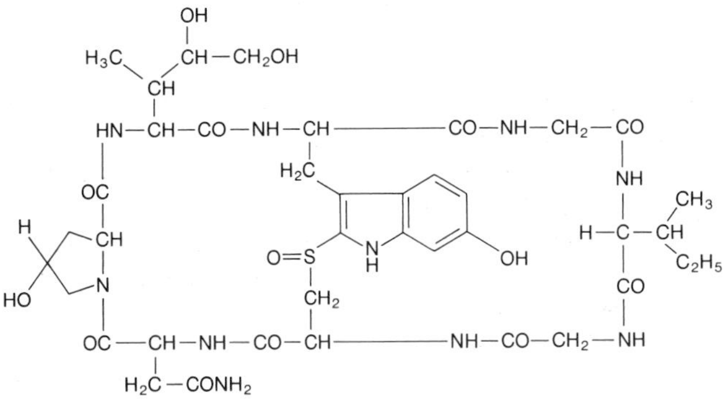

Fig. A21 α-Amanitin.

amber codon The TERMINATION CODON UAG. *See* GENETIC CODE.

amber mutation A NONSENSE MUTATION caused by a single base-pair substitution which results in a codon specifying an amino acid (e.g. AAG) becoming the TERMINATION CODON UAG (the amber codon) and thus resulting in premature termination of TRANSLATION.

amber suppressor A mutation which eliminates the phenotypic effects of an AMBER MUTATION. An amber suppressor mutation usually lies in a gene unlinked to the amber mutation and which encodes a TRANSFER RNA with a mutated ANTICODON that allows the tRNA to translate the UAG (amber) TERMINATION CODON as an amino acid (*see also* NONSENSE SUPPRESSOR).

Ames test A very sensitive test for identifying chemicals that are MUTAGENS and potential CARCINOGENS. The original test made use of a histidine-requiring mutant of the bacterium SALMONELLA TYPHIMURIUM. When mutant strains are grown in the presence of the chemical to be tested an increased incidence of wild-type REVERTANTS (i.e. not requiring histidine) indicates that the chemical is mutagenic. Later versions of the test incorporated liver microsomes to carry out the conversion of some non-mutagens into mutagens (and thus possible carcinogens) that occurs *in vivo* (*see* CHEMICAL CARCINOGENS AND CARCINOGENESIS). The test was developed and refined by Dr Bruce N. Ames.

McCann, J. et al. (1975) *Proc. Natl. Acad. Sci. USA* **72**, 5135.

amino acid(s) α-Amino acids are the fundamental structural units of proteins (*see* PROTEIN STRUCTURE; PROTEIN SYNTHESIS). They have the general structural formula

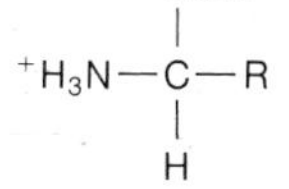

$$^+H_3N - \overset{\displaystyle COO^-}{\underset{\displaystyle H}{\overset{|}{\underset{|}{C}}}} - R$$

where R is a side chain of varying structure (*see* Fig. A22). α-Amino acids (except glycine) are chiral, occurring as L- and

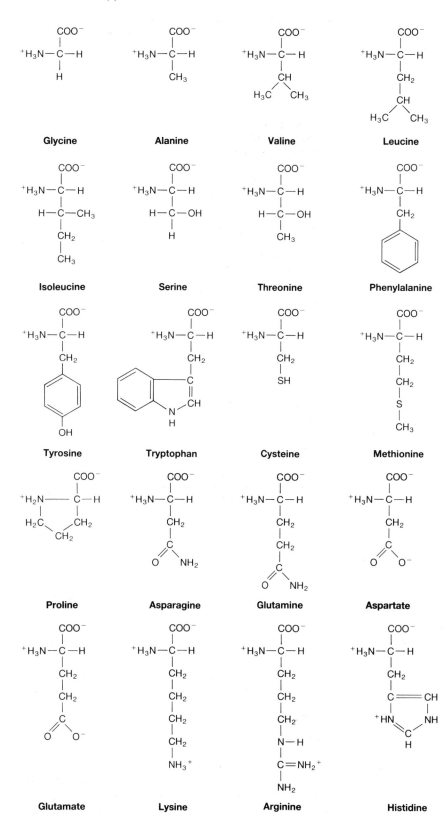

Glycine

Alanine

Valine

Leucine

Isoleucine

Serine

Threonine

Phenylalanine

Tyrosine

Tryptophan

Cysteine

Methionine

Proline

Asparagine

Glutamine

Aspartate

Glutamate

Lysine

Arginine

Histidine

Fig. A22 The amino acids found in proteins.

Table A4 The amino acids commonly found in proteins and their abbreviations

Amino acid	Three-letter abbreviation	Single-letter notation
Alanine	Ala	A
Arginine	Arg	R
Asparagine	Asn	N
Aspartic acid	Asp	D
Asparagine or aspartic acid	Asx	B
Cysteine	Cys	C
Glutamine	Gln	Q
Glutamic acid	Glu	E
Glutamine or glutamic acid	Glx	Z
Glycine	Gly	G
Histidine	His	H
Isoleucine	Ile	I
Leucine	Leu	L
Lysine	Lys	K
Methionine	Met	M
Phenylalanine	Phe	F
Proline	Pro	P
Serine	Ser	S
Threonine	Thr	T
Tryptophan	Trp	W
Tyrosine	Tyr	Y
Valine	Val	V

Table A5 Properties of amino acids in proteins

Amino acids with neutral side chains
  Aliphatic
    Glycine
    Alanine
    Valine: hydrophobic
    Leucine: hydrophobic
    Isoleucine: hydrophobic
    Serine: hydrophilic, polar
    Threonine: hydrophilic, polar

  Aromatic
    Phenylalanine: hydrophobic
    Tyrosine: somewhat hydrophobic, polar at neutral pH
    Tryptophan: hydrophobic, polar

  Sulphur-containing amino acids
    Cysteine (neutral as a component of disulphide bonds, a small proportion of the sulphydryl amino acid is negatively charged and reactive).
    Methionine: hydrophobic

  Amino acid with a secondary amino group
    Proline

  Amides of dicarboxylic amino acids
    Asparagine, polar
    Glutamine, polar

Charged amino acids (hydrophilic)
  Acidic
    Aspartic acid (aspartate)
    Glutamic acid (glutamate)

  Basic
    Lysine
    Arginine
    Histidine (weakly basic), polar

See Fig. A22 for structures of amino acids.

D-ISOMERS, but only the L-isomers are found in proteins. D-amino acids are rarely found in biological macromolecules, being mostly confined to bacterial cell walls (D-alanine). Other than the 20 α-amino acids commonly found in proteins (Table A4, Fig. A22) there is a small number of other types of amino acid with important biological roles, for example the neurotransmitter γ-aminobutyric acid (GABA) and β-alanine in the vitamin pantothenic acid. Other α-amino acids, for example L-ornithine, L-dihydroxyphenylalanine, L-DOPA, and homoserine are important metabolites.

The α-amino acids found in proteins can be classified according to the chemical and physical properties of their side chains, which are important in determining the properties of the proteins and protein domains they make up. Amino acids may be variously grouped according to whether their side chains are charged or uncharged, acidic or basic, aliphatic or aromatic, hydrophobic or hydrophilic, polar or nonpolar (Table A5). Consideration of these properties is important in analysing the amino-acid sequences of proteins in order to determine, for example, the probable topography of a transmembrane spanning region by identifying stretches of sequence rich in nonpolar amino acids which can insert into the lipid membrane.

Unusual amino acids are found in some proteins, generally as a result of POST-TRANSLATIONAL MODIFICATION, such as HYDROXYLYSINE and HYDROXYPROLINE in collagen, and selenomethionine and selenocysteine in some bacterial proteins.

amino acid neurotransmitters The AMINO ACIDS glutamate, glycine, and GABA (γ-aminobutyric acid) have been identified as neurotransmitters in the central nervous system. *See*: EXCITATORY AMINO ACID RECEPTORS; GABA AND GLYCINE RECEPTORS; LONG-TERM POTENTIATION.

amino-acid notation The amino acids found in proteins are denoted either by three-letter abbreviations or by a single-letter notation (*see* Table A4).

amino-acid sequence The order of amino acid residues in a polypeptide chain. It can be determined directly by PROTEIN SEQUENCING if pure protein is available, but nowadays is more generally determined indirectly from the nucleotide sequence of the corresponding gene (*see* DNA SEQUENCING).

amino sugars *See*: SUGARS.

amino-terminus The end of a polypeptide chain that bears the free α-amino group. It is often abbreviated to N terminus (the abbreviation used in this book). Protein biosynthesis proceeds from the N terminus to the C terminus. *See*: AMINO ACIDS; PROTEIN STRUCTURE; PROTEIN SYNTHESIS.

α-amino-3-hydroxy-5-methylisoxazole-4-propionic acid *See*: AMPA.

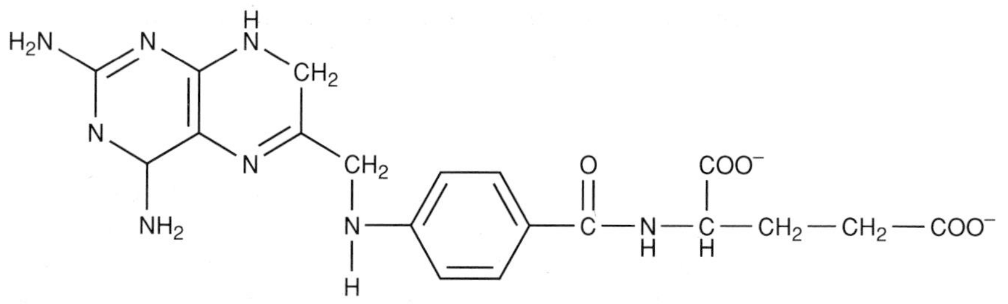

Fig. A23 Aminopterin.

aminoacyl-tRNA A TRANSFER RNA (tRNA) charged with the appropriate amino acid, which is esterified to the 3-OH of the 3′-terminal adenosine residue. *See*: GENETIC CODE; PROTEIN SYNTHESIS.

aminoacyl-tRNA synthetases Enzymes that attach the appropriate amino acid to the 3′ end of a TRANSFER RNA (tRNA). Each enzyme is specific for a particular tRNA. They are ligases catalysing the reaction

ATP + amino acid + tRNA = AMP + pyrophosphate
+ aminoacyl-tRNA.

The recognition of a tRNA by its cognate aminoacyl-tRNA synthetase is not yet fully understood (*see* GENETIC CODE; PROTEIN–NUCLEIC ACID INTERACTIONS; PROTEIN SYNTHESIS).

γ-aminobutyric acid *See*: GABA; GABA AND GLYCINE RECEPTORS.

aminopeptidase Proteolytic enzyme (EC 3.4.11) which hydrolyses the peptide bond linking the terminal amino acid at the free N-terminal end of a POLYPEPTIDE CHAIN, thus splitting off an N-terminal residue. Some aminopeptidases are of broad substrate specificity, others are limited to splitting off particular amino acids. *Cf.* CARBOXYPEPTIDASE; ENDOPEPTIDASE.

aminopterin An analogue of DIHYDROFOLATE and a potent competitive inhibitor of the enzyme DIHYDROFOLATE REDUCTASE (DHFR) (Fig. A23). It is a component of HAT MEDIUM, in which it is used to inhibit the biosynthesis of dTMP by inhibiting the regeneration of tetrahydrofolate from dihydrofolate (a tetrahydrofolate derivative acts as the methyl donor for the synthesis of dTMP from dUMP).

2-aminopurine A BASE ANALOGUE of the BASE adenine, which is mutagenic when incorporated into DNA, causing AT → GC transitions. Adenine normally pairs with thymine but 2-aminopurine can form a pair with cytosine.

amniocentesis The removal of amniotic fluid for diagnostic purposes, particularly PRENATAL DIAGNOSIS.

amnion *See*: EXTRAEMBRYONIC MEMBRANES.

AMO-1618 2′-Isopropyl-4′-(trimethylammonium chloride)-5′-methylphenyl piperidine-1-carboxylate (Fig. A24). M_r 368.5. An experimental PLANT GROWTH RETARDANT. It inhibits the cyclization of geranylgeranyl pyrophosphate to copalyl pyrophosphate (activity A of *ent*-kaurene synthetase) in the GIBBERELLIN biosynthetic pathway with I_{50} at ~10^{-6} M for the enzyme partially purified from *Marah macrocarpus* endosperm. It may, in addition, inhibit certain steps in the phytosterol biosynthetic pathway when present at high concentrations (>10^{-4} M).

Frost, R.G. & West, C.A. (1977) *Plant Physiol.* **59**, 22–59.

Amoeba Genus of free-living motile unicellular protozoa characterized by cells of continually changing shape, which move by pushing out temporary cytoplasmic projections called pseudopodia, and which are used in studies of CELL MOTILITY. Pseudopodia are used in feeding; two or more extend and engulf particulate matter forming a food vacuole into which digestive enzymes are released. By analogy, unicellular phases of other eukaryotic organisms which move in a similar way, such as the unicellular phase of cellular slime moulds (*see* DICTYOSTELIUM DEVELOPMENT), are also sometimes termed amoebae.

amorphic allele (amorph) A term in the classification of allele type devised by the geneticist H.J. Muller (*see* ALLELE for a full explanation of this classification). It denotes an allele with no gene activity, that is, a null allele.

AMP Adenosine 5′-monophosphate. *See*: NUCLEOSIDES AND NUCLEOTIDES.

AMP-activated protein kinase The central component of a PROTEIN KINASE CASCADE which is activated five- to tenfold by 5′-AMP and a further tenfold by phosphorylation catalysed by a distinct kinase kinase. It phosphorylates and inactivates several biosynthetic enzymes (e.g. HMG-CoA reductase, acetyl-CoA carboxylase). As well as causing direct activation, AMP also promotes phosphorylation by the kinase kinase, so a rise in AMP in cells can activate the system 50-fold. As AMP is a sensitive

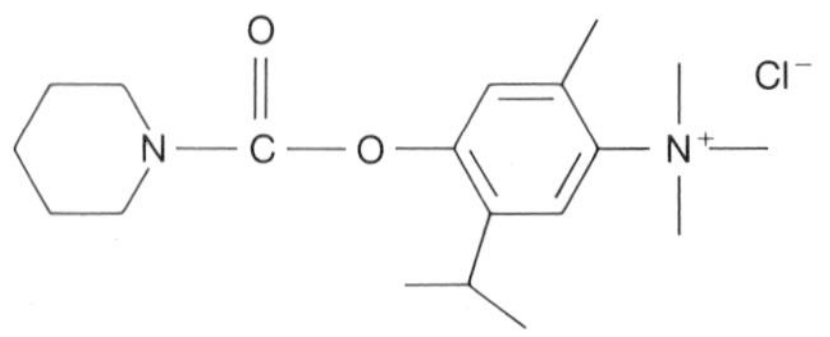

Fig. A24 Structure of AMO-1618.

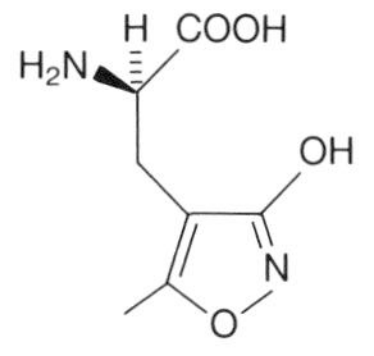

Fig. A25 Structure of AMPA.

indicator of the energy state of the cell, this may represent a protective mechanism to switch off biosynthesis whenever ATP levels become depleted, for example during HEAT SHOCK or other types of cellular stress.

Hardie, D.G. (1992) *Biochim. Biophys. Acta* **1123**, 211–238.

AMPA (α-amino-3-hydroxy-5-methylisoxazole-4-propionic acid)

The prototypic agonist for one of the subtypes of glutamate receptor (Fig. A25) (*see* EXCITATORY AMINO ACID RECEPTORS). The receptor subtype activated by AMPA is therefore referred to as the AMPA receptor. Activation of this receptor results in the opening of associated Na^+ channels and hence the generation of excitatory postsynaptic potentials (*see* SYNAPTIC TRANSMISSION). The EAA receptor subtype formerly known as the quisqualate receptor has since been renamed the AMPA receptor because of the higher specificity and affinity of this ligand.

Amphibian development

WITHIN the enormous number of amphibian species a wide variety of developmental strategies are used. Among these, a certain type of development, shown by some anurans (frogs and toads) and urodeles (newts and salamanders), permits the direct manipulation of early development. In these species the eggs are laid outside the maternal body in large numbers (often >1000 per laying). They develop at room temperature in simple hypotonic salt solutions. In isotonic buffered salt solutions they can be surgically manipulated, and small pieces of the embryos as well as individual cells can be cultured for several days, during which untreated embryos will complete early development to the swimming larval stage. Although much of their volume is yolk, the eggs are large enough (1–2 mm in diameter) and laid in sufficient quantity for routine biochemistry. This combination of properties has allowed major advances in the experimental analysis of early development. The first manipulations of embryos were carried out on amphibians, and they have continued to be one of the major animal groups studied.

Early experiments were carried out on common European and American species of anurans and urodeles. Although still used, the disadvantage of these 'classical' species is that they are seasonal. In contrast, African frogs of the genus *Xenopus* will lay eggs at any time of the year when injected with mammalian gonadotrophic hormones (from the pituitary or placenta). Originally the basis of a pregnancy test, this capacity has now made *Xenopus laevis* the most commonly used amphibian species in developmental studies. This account will therefore relate to *Xenopus* embryos, unless otherwise stated.

The egg

The egg is large, about 1.2–1.4 mm in diameter, and has an obvious animal–vegetal axis. The animal half is pigmented, whereas the vegetal half is not. There are several other known asymmetries. The nucleus is in the animal half whereas the vegetal cortex (the most superficial cytoplasm of the vegetal half) contains visible cytoplasmic masses that are inherited by the germ-line cells, and are hence called 'germ plasm'. There is also at least one species of RNA, known as Vg1, localized to the vegetal cortex, as well as others that are localized in the animal hemisphere. These asymmetries develop during oogenesis, and are thought to be mediated by the cytoskeleton. However, few details are known. The large size of the egg permits simple microinjection of a wide variety of reagents directly into the egg cytoplasm. Antibodies, inhibitors of RNA and protein synthesis, antisense constructs, competing peptides, dominant negative constructs, and functional mRNAs have all been used. mRNAs injected into the egg will continue to be translated until post-neurula stages, thus allowing the functional analysis of individual proteins during early development.

Cleavage

The fertilized egg divides by a relatively long first CELL CYCLE (about 1.5 hours) and then by 11 very rapid (~20 min each) and synchronous cycles. The rate of these cycles, and the ability of egg extracts to support most components of the cell cycle *in vitro*, has made this a particularly useful model system for studying cell cycle control. Several molecules that control the passage of cells through the cell cycle have been identified in *Xenopus*. During the first cell cycle cytoplasmic movements take place that establish the dorso-ventral axis of the embryo. Application of small spots of vital dye to the embryo have revealed a rotational movement of the cortical cytoplasm with respect to the deeper layers, which is required for normal formation of dorsal axial structures. It is not known how this movement establishes the dorsal axis. Thus by the end of the first cell cycle the embryo has animal–vegetal, and presumptive dorso-ventral axes. As the first division is in the mid-sagittal plane of the embryo, it also divides it into left and right halves (Fig. A26).

The mid-blastula transition

After 12 synchronous divisions, transcription of the zygotic genes starts. The trigger for this is unknown. However, the microinjection of large amounts of viral DNA pre-fixes the onset of transcription, suggesting that titration of an inhibitor against the increasing amount of nuclear DNA may underlie this event. Several other events start at this time. Cells of the embryo become motile, the mitotic rate drops and becomes asynchronous. This embryo-wide change in behaviour is called the mid-blastula transition (MBT). The transcriptional silence before the MBT means that all synthesis during this period is controlled by

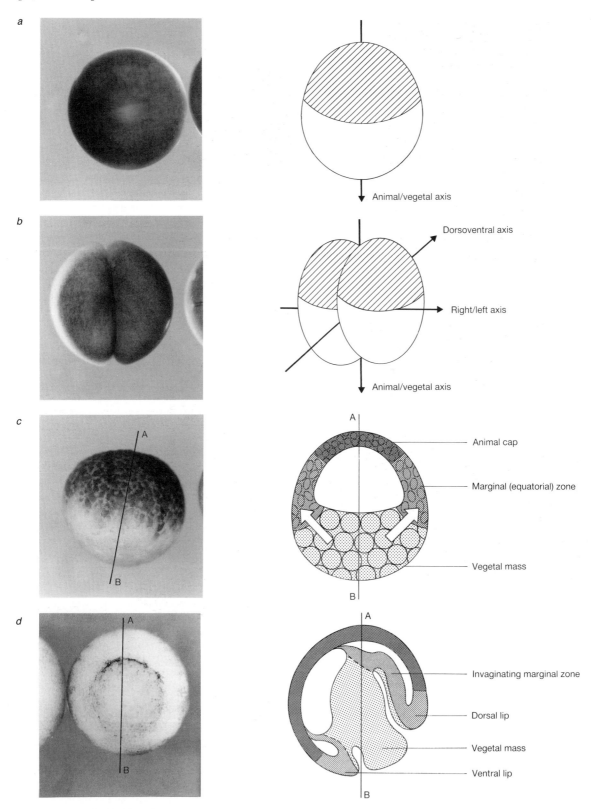

Fig. A26 Some representative stages of early *Xenopus* embryos. Left, Microphotographs of developing embryos. Right, Diagrams of cross-sections of the developing embryo (along section indicated A-B on photographs). *a*, Fertilized egg; *b*, 2-cell stage; *c*, blastula; *d*, mid-gastrula.

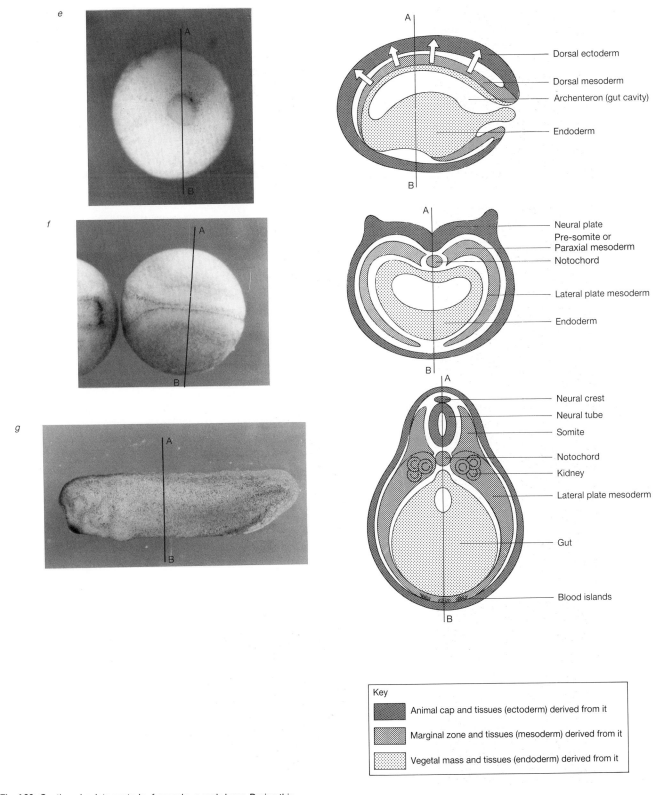

e

f

g

Dorsal ectoderm
Dorsal mesoderm
Archenteron (gut cavity)
Endoderm

Neural plate
Pre-somite or
Paraxial mesoderm
Notochord
Lateral plate mesoderm
Endoderm

Neural crest
Neural tube
Somite
Notochord
Kidney
Lateral plate mesoderm
Gut
Blood islands

Key

Animal cap and tissues (ectoderm) derived from it

Marginal zone and tissues (mesoderm) derived from it

Vegetal mass and tissues (endoderm) derived from it

Fig. A26 *Continued. e*, late gastrula; *f*, neurula; *g*, early larva. During this entire period of development, the biomass of the embryo does not change. Any small amount of growth that does occur is due to water influx.

maternal RNA and protein inherited from the oocyte. As many of the key events in axis and GERM LAYER formation occur during this period, it is of some importance to analyse the maternal molecules controlling these, by manipulation of the oocyte.

The full-grown *Xenopus* oocyte is an enormous single cell. The follicle cells that normally surround it can be easily removed, and the oocyte cultured for several days. Injected mRNAs are translated and their products assembled, so this has become a useful model system for the study of various cell surface molecules such as neurotransmitter and hormone receptors (*see* XENOPUS OOCYTE EXPRESSION SYSTEM). There has been comparatively little manipulation of the oocyte followed by fertilization. However, methods have recently been described that allow the culture of full-grown oocytes, followed by fertilization. This should allow the identification and functional characterization of developmentally important molecules in the oocyte.

The blastula

Until the embryo becomes a swimming larva, there is no significant increase in biomass; thus with every cell division the cells of the embryo get progressively smaller. Because of this, cell-autonomous vital dyes injected into the egg, or into individual blastomeres (the early cells produced by cleavage), are inherited by all the descendant cells with little dilution. This has allowed particularly accurate FATE MAPS to be constructed in *Xenopus*. Fate maps constructed at the blastula stage are not exact, as there is some cell mixing during development. However they remain an essential prerequisite for the interpretation of experiments designed to change cell fate. Vital dyes have also allowed 'potency mapping' experiments, where individual labelled cells, or groups of cells, are transplanted to new locations in the embryo, and the differentiation of their descendants followed.

During the blastula stage (Fig. A26), which starts at around the 32-cell stage (about 3 hours post-fertilization) and ends with the onset of gastrulation (about 10 hours post-fertilization), a fluid-filled space (the blastocoel) appears in the animal half of the embryo. The blastula is conventionally divided into three regions defined by their normal fate; the animal cap, marginal (or equatorial) region, and vegetal mass. After the cell movements of GASTRULATION, these regions become the three primary germ layers — the ECTODERM, MESODERM and ENDODERM, respectively.

Induction

Microdissection and culture experiments have revealed a great deal about the mechanisms whereby blastomeres become directed into one of these three pathways. We know most about the mesoderm. Cells of the marginal zone are directed into the mesoderm pathway by a combination of GROWTH FACTOR-mediated signals which originate from the vegetal blastomeres, and are collectively called mesodermal induction (*see* INDUCTION). As well as specifying mesoderm, these signals also specify the dorsoventral patterning of the mesoderm. The most dorsal mesoderm, as well as differentiating into notochord, has the property of initiating neural differentiation in the overlying ectoderm by a second induction event called 'neural induction'. Thus the dorso-

ventral pattern established during the first cell cycle becomes obvious at this time.

Single-cell transplantation experiments have revealed that individual blastomeres become committed to their appropriate germ layer during the blastula stage. When early blastula vegetal blastomeres are transplanted to the blastocoel, their descendants are found in all germ layers. However, by the end of the blastula stage, only a few hours later, individual vegetal blastomeres give rise only to endodermal cells when transplanted in the same way. Cell sorting experiments also show that during the blastula stage, changes in the cell surface occur which lead to regional differences in the late blastula. These foreshadow the physical segregation of these into the three primary germ layers during gastrulation.

Gastrulation

The earliest visible cell movements in amphibian embryos are those of gastrulation. During gastrulation, the marginal cells and vegetal cells of the blastula are invaginated inside the animal cells, which are left on the outside of the embryo as the ectoderm. The surface area of the ectoderm increases during this process by a stretching and flattening of individual cells, known as epiboly. In fact epiboly can be seen to start at the late blastula stage. However, the onset of gastrulation is conventionally judged to be the first appearance of the blastopore, so these are often called pregastrulation movements. The motor driving the invagination of the marginal zone is thought to be its deep cells, which perform a complex series of movements called convergence–extension. These movements on their own would elongate the embryo to make a tube-like structure. However, they are accompanied by an invagination of the marginal zone, starting in the vegetal hemisphere, which results in part of the marginal zone being pushed inside the sphere, being turned inside out as it does so. The invagination first appears as a pigmented depression on the prospective dorsal side of the embryo, called the dorsal lip of the blastopore. Invagination movements spread around the lower edge of the marginal zone, so that eventually the blastopore becomes a ring of invaginating tissue. Invagination is much more extensive on the dorsal side of the embryo, so that the dorsal marginal zone cells reach further into the embryo than the ventral or lateral ones. These movements are also thought to draw the vegetal mass inside the embryo through the blastopore. Once this has happened, the blastopore closes. During these movements, the three regions of the blastula become segregated from each other, to form the concentrically arranged primary germ layers, the ectoderm, mesoderm, and endoderm, respectively, from the outside in. Each layer gives rise to a characteristic set of tissues in the larval and adult animal.

The large size of the amphibian gastrula, and the ease of manipulation, have led to some important studies of the cell movements involved in gastrulation. As the marginal zone invaginates into the embryo, its leading edge migrates over the inner aspect of the ectoderm, which forms the wall of the blastocoel. Microinjection of antibodies and competing peptides into the blastocoel have revealed that CELL–MATRIX INTERACTIONS are essential for this migration to occur correctly. These experiments were done in *Triturus* embryos, and have not been repeated in *Xenopus*. The convergence–extension movements have also been reproduced in culture, using fragments of the dorsal marginal zone.

Further development

Following gastrulation, the first major organ rudiment appears, that of the nervous system. The ectoderm covering the most dorsal surface of the embryo becomes thickened to form the neural plate. This then rolls up into folds, whose lateral edges meet and fuse, and then sink below the surface to form a tube inside the embryo. This tube (the neural tube), composed initially of a single layer of cells, then differentiates into the central nervous system. As the tube forms, a layer of ectoderm-derived cells remains as a crest (the neural crest) on its dorsal axis. These cells migrate away from the central nervous system to form a number of different tissues, including pigment cells, neurons, cartilage, bone and endocrine cells. This sequence of events is initiated by a second induction, mediated by the most dorsal mesoderm upon the dorsal ectoderm. This property of the dorsal mesoderm provides one piece of evidence that the blueprint for the dorsal axis of the embryo is laid down in the marginal zone of the blastula during mesodermal induction. More evidence for this is provided by culture experiments, where different growth factors, or different concentrations of the same growth factor, can induce mesoderm characteristic of different parts of the dorso-ventral axis in isolated animal caps.

After neural induction, other organ rudiments appear. The mesoderm becomes differentiated into the notochord (a skeletal rod that is the precursor of the vertebral column), somites (segmental precursors of skeletal muscle), kidney, and lateral plate mesoderm, arranged from dorsal to ventral. The endoderm differentiates into the lining of the gut, and further regionalization of the nervous system takes place. Just four days from fertilization, the embryo becomes a larva, capable of active swimming movements, and with most of the functional organ systems characteristic of vertebrates. The elaboration of these from three concentrically arranged primary germ layers is an extraordinary example of patterning. Many genes with sequence similarities to pattern-forming genes in *Drosophila* have been identified in *Xenopus* using antibodies and sequence similarities. The distribution of the products of these genes, and effects of their overexpression by injection of mRNA into individual blastomeres, have revealed that these play pattern-forming roles in *Xenopus* also. The most recent evidence for this comes from homeobox-containing genes, and from *Xenopus* homologues of the *Drosophila wingless* gene.

Genetics

The biggest drawback of *Xenopus* and other amphibian embryos, is that their enormous genomes (*see* GENOME ORGANIZATION) and long life cycles preclude genetic experiments. *X. laevis* also suffers from two other major drawbacks from the genetic point of view. First, most loci have been duplicated during its evolution, so that most genes that are single copy in other species have two nonallelic forms in *Xenopus*. Second, attempts to make TRANS-

GENIC embryos by introducing exogenous DNA have been disappointing. Expression of DNA injected into eggs is variable, and in a MOSAIC pattern, so that only a few cells in a particular tissue make the expected protein. The reasons for this are not clear; it does, however, make experiments selectively blocking the action of particular genes by ANTISENSE constructs or by gene deletion extremely difficult. The routine introduction of foreign DNA into the germ line is currently precluded by these properties of *Xenopus*. However, despite this problem, we know more about the general mechanisms that convert the egg into the body plan in *Xenopus* than in any other vertebrate, and probably than in any other animal group.

C.C. WYLIE

See also: AVIAN DEVELOPMENT; CAENORHABDITIS DEVELOPMENT; CAENORHABDITIS NEURAL DEVELOPMENT; DROSOPHILA DEVELOPMENT; INSECT NEURAL DEVELOPMENT; MAMMALIAN DEVELOPMENT; VERTEBRATE NEURAL DEVELOPMENT; ZEBRAFISH DEVELOPMENT.

1 Slack, J.M.W. (1991) *From Egg to Embryo*, 2nd edn (Cambridge University Press, Cambridge).

---

amphipathic Containing both hydrophobic and hydrophilic moieties, such as membrane phospholipids which are composed of hydrophobic fatty acid tails and a polar headgroup. The term is also used to describe an α-helix (*see* PROTEIN STRUCTURE) which is hydrophobic on one side and hydrophilic on the other.

amphiregulin *See*: GROWTH FACTORS.

ampicillin A β-lactam antibiotic (Fig. A27). Ampicillin is widely used to select for *Escherichia coli* cells transformed with a PLASMID carrying the *bla* gene (encoding β-lactamase) from the transposon Tn*1*. It arrests growing *E. coli* cells by inhibiting a number of enzymes in the cell membrane and thereby disrupting cell wall synthesis.

amplicon A region of the mammalian genome that becomes amplified into multiple repeated copies after exposure to a compound that inhibits the function of the gene it contains (e.g. methotrexate, an inhibitor of the enzyme dihydrofolate reductase (DHFR), causes the amplification of the DHFR gene). The unit of amplification often contains sequences in addition to the gene whose product is actually inhibited by the compound. *See*: GENE AMPLIFICATION.

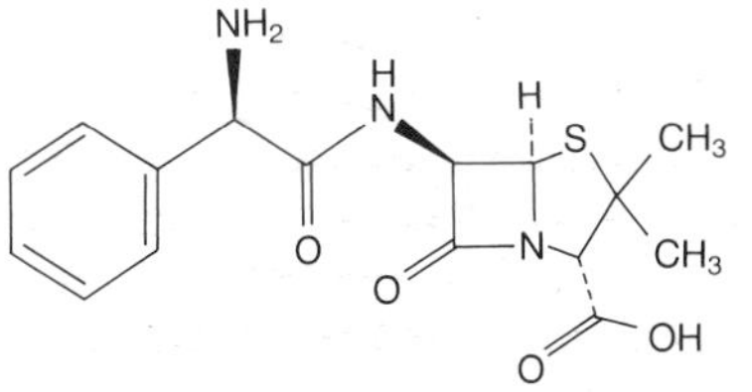

Fig. A27 Ampicillin.

amplification *See*: AMPLICON; CHLORAMPHENICOL AMPLIFICATION; CHORION GENE AMPLIFICATION; GENE AMPLIFICATION; RRNA GENE AMPLIFICATION.

amylase Any glycoside hydrolase that will hydrolyse starch, glycogen and related glucans. α-Amylases (1,4α-D-glucan glucanohydrolases, EC 3.2.1.1) hydrolyse α1,4 glucan links in trisaccharides or larger glycans. β-Amylase is a confusing term for 1,4α-D-glucan maltohydrolase (EC 3.2.1.2) which releases maltose units successively from nonreducing termini of α1,4 glucan chains, as β-maltose (i.e. with an inversion).

amyloid (1) Plant amyloids are a family of polysaccharides from seeds in which a β1,4 glucan core is substituted at C6 by α-D-xylopyranosyl residues, to which other sugars (e.g. D-galactopyranose at C2) may be attached. These GLYCANS cannot assume a cellulose-like pattern of hydrogen bonding and are highly water-soluble components of the hemicellulose fraction of plant cell walls (*see* PLANT CELL WALL MACROMOLECULES).
(2) Amyloid of animal tissues is a complex material composed partly of protein fibrils, and is deposited in various disease states. The protein varies from disease to disease. A diagnostic test for animal amyloid is the green birefringence seen under polarized light after staining with the aromatic dye congo red. *See*: ALZHEIMER'S DISEASE; AMYLOIDOSIS; NEURODEGENERATIVE DISORDERS; TRANSMISSIBLE SPONGIFORM ENCEPHALOPATHIES.

amyloid P component *See*: EXTRACELLULAR MATRIX MOLECULES; SERUM AMYLOID P COMPONENT.

amyloidosis The deposition of AMYLOID in tissues, which occurs in various diseases. Amyloid deposition in the brain is a feature of ALZHEIMER'S DISEASE. Similar amyloid deposits also occur in the brains of those with DOWN'S SYNDROME. In systemic amyloidoses amyloid deposits occur throughout the viscera. The rare inherited non-neuropathic systemic amyloidoses may be caused by mutations in various genes. Mutated apolipoprotein I is found in amyloid in some families, whereas mutated lysozyme has been found in others. Other amyloidogenic proteins include TRANSTHYRETIN (prealbumin) and β_2-MICROGLOBULIN.

amylopectin A branching α-GLUCAN of STARCH, with $\sim 10^4$–10^5 residues per molecule. Chains are of α1,4 D-glucopyranosyl residues with α1,6 branches at about every 23 residues on average. *See also*: AMYLOPECTIN; GLYCOGEN; PHYTOGLYCOGEN.

amyloplast Colourless PLASTID containing one or more starch grains.

amylose Small α1,4-GLUCAN of STARCH, having almost unbranched chains of up to about 10^3 residues of D-glucopyranose. Branches, if present, are α1,6. *See also*: AMYLOPECTIN; GLYCOGEN; PHYTOGLYCOGEN; STARCH.

amyotrophic lateral sclerosis (ALS) Also known as Lou Gehrig's disease or motor neuron disease, a late onset degeneration of motor neurons in cortex, brainstem, and spinal cord, leading to progressive paralysis. An AUTOSOMAL DOMINANT familial variant has been linked to defects in cytosolic SUPEROXIDE DISMUTASE (*SOD1* gene on chromosome 21q). One hypothesis suggests that damage is caused by overproduction of glutamate which, as an excitatory transmitter, provokes the synthesis of the free radical of NITRIC OXIDE, which, in the absence of superoxide dismutase combines with the superoxide anion $O_2^{\cdot-}$ to form peroxynitrate ($ONOO^-$) which is neurotoxic.

Rosen, R.R. et al. (1993) *Nature* **362**, 59–62.

Anabaena Genus of CYANOBACTERIA within Section IV of the classification of Rippka et al. Members are filamentous and unbranched. A proportion of cells will differentiate in the absence of combined nitrogen to form heterocysts. Some strains should be regarded as belonging to the genus *Nostoc*.

Rippka, R. et al. (1979) *J. Gen. Microbiol.* **111**, 1–61.

Anacystis nidulans Unicellular CYANOBACTERIA within Section I of the classification of Rippka et al. Division is by binary fission. They do not fix nitrogen. Strains are also known as *Synechococcus*, with *A. nidulans* UTEX 625, *A. nidulans* 602 and *A. nidulans* R2 designated *Synechococcus* PCC6301 (also *Synechococcus leopoliensis*), PCC7943 and PCC7942 respectively. *A. nidulans* R2 is readily transformable.

Rippka, R. et al. (1979) *J. Gen. Microbiol.* **111**, 1–61.

anaemia *See*: FANCONI'S ANAEMIA; HAEMOGLOBIN AND ITS DISORDERS (for sickle-cell anaemia and thalassaemia); HAEMOLYTIC ANAEMIA; NONSPHEROCYTIC HAEMOLYTIC ANAEMIA.

anaphase The stage of MITOSIS or MEIOSIS during which each already duplicated CHROMOSOME is separated into its two genetically identical parts, and the two resulting sets of identical chromosomes are segregated to distinct parts of the cell. The chromosome movements of anaphase are brought about by the action of the MITOTIC SPINDLE. They comprise anaphase A, during which the chromosomes move to the poles of the spindle, and anaphase B, during which the spindle poles move apart. The speed of chromosome movement during anaphase A is usually quite constant, but varies from one cell type to another: 0.5–3 μm min^{-1}. During anaphase B chromosome speed is around one-half that of anaphase A.

anaphylaxis Hypersensitivity reaction to intravascular antigens such as the venom toxins of biting or stinging insects or proteins introduced by passive immunization. Anaphylaxis results from the production, on first encounter with antigen, of specific ANTIBODIES of the IgE class which attach via their constant regions to the FC RECEPTORS on MAST CELLS. Upon a second encounter with the antigen, these antibodies will bind the antigen. This leads to the release of mast cell chemical mediators such as HISTAMINE. A biphasic response of vasoconstriction followed by peripheral vessel dilation results. Pooling of blood in the periphery and a concomitant drop in blood pressure may lead to a fatal anaphylactic shock. Thus, individuals may die from a simple bee sting.

anastral spindle A MITOTIC SPINDLE that lacks an ASTER. Such spindles are found in all higher plants and in some lower plants and protozoa.

anatoxin a A toxin from CYANOBACTERIA.

anchorin *See*: EXTRACELLULAR MATRIX MOLECULES.

ancymidol α-Cyclopropyl-4-methoxy-α-(pyrimidin-5-yl)benzyl alcohol (Fig. A28). M_r 256. A PLANT GROWTH RETARDANT. It is a noncompetitive inhibitor ($K_i \sim 2 \times 10^{-9}$ M) of the oxidation of *ent*-kaurene to *ent*-kaurenoic acid in the biosynthetic pathway to the GIBBERELLINS. It has been used mainly to restrict the growth of ornamental plants.

Coolbaugh, R.C. et al. (1978) *Plant Physiol.* **62**, 571–576.

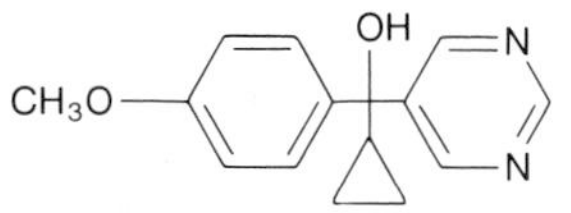

Fig. A28 Structure of ancymidol.

androgen Steroid male sex hormones (e.g. androsterone and testosterone) concerned with the development of the male reproductive system and production and maintenance of secondary sexual characteristics. They are secreted chiefly by the testis. *See*: STEROID HORMONES; STEROID HORMONE RECEPTORS.

androgenetic (androgenic) embryo Embryos possessing two paternal sets of chromosomes. Very occasionally androgenetic embryos arise spontaneously in nature when a fertilized egg loses its female pronucleus, but they are almost never viable. In humans, this condition results in a uterine tumour called a hydatiform mole. Androgenetic embryos can also be produced in the laboratory using the technique of pronuclear transplantation. In this technique, a fine glass needle attached to a micromanipulator is used to remove the male PRONUCLEUS from one fertilized egg and inject it into another egg. The female pronucleus is then removed from the recipient egg so that the ZYGOTE now contains two paternally derived sets of chromosomes. Such embryos are useful for studies of the imprinting of maternal and paternal genomes. *See also*: PARENTAL GENOMIC IMPRINTING.

anergy The functional inactivation of lymphocytes on encounter with an antigen, which may be one mechanism of maintaining tolerance to self-antigens.

aneuploidy The increase or decrease of the chromosome complement of a cell by one or more chromosomes.

Angelman syndrome Syndrome in which severe mental retardation is associated with ataxic movements and inappropriate laughter and which is associated with the absence of maternal chromosome region 15q11q13. *See*: PARENTAL GENOMIC IMPRINTING; GABA AND GLYCINE RECEPTORS.

angiogenesis Formation of blood vessels. The term is generally used to include only the final stages of blood vessel development in vertebrates, the coalescence of cells derived from the LATERAL PLATE MESODERM to form a network of hollow tubes that give rise to the adult circulatory system. It is also applied to the formation of blood vessels during wound repair, regeneration, and the infiltration of tumours with capillaries.

angiokeratoma *See*: FABRY'S DISEASE.

angiosperms Major division of the plant kingdom, commonly called the flowering plants. Reproductive organs (stamens and ovary) are carried in flowers, and after fertilization the closed ovary containing the seeds develops into a fruit. In some classifications angiosperms are called the Magnoliophyta. Virtually all cultivated plants (apart from conifers) belong to the angiosperms.

angiotensinogen Protein of M_r 61 000 acting as a peptide donor, the precursor of the vasoactive octapeptide angiotensin II. Inhibition of this activation process is a target for anti-hypertensive drugs. Angiotensinogen is a noninhibitory member of the SERPIN family, which is cleaved by the acid protease renin to release inactive N-terminal decapeptide angiotensin I. The two C-terminal residues of this are removed by angiotensin-converting enzyme to leave the active form angiotensin II (Fig. A29).

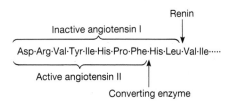

Fig. A29 Human angiotensinogen N-terminal sequence, and the formation of angiotensins I and II.

animal cell Animal cells are enormously diverse in size and structure. The common features found in most animal cells are illustrated in Fig. A30. *See also*: NEURON.

animal mitochondrial genomes *See*: MITOCHONDRIAL GENOMES: ANIMAL.

animal pole The area of an oocyte, egg, or zygote containing the female nucleus or pronucleus. It is often marked by the site of extrusion of the polar body and contains pigment granules and other organelles (especially mitochondria). Generally opposite the VEGETAL POLE, forming an animal–vegetal axis which may determine the anterior–posterior axis of the embryo. *See also*: AMPHIBIAN DEVELOPMENT.

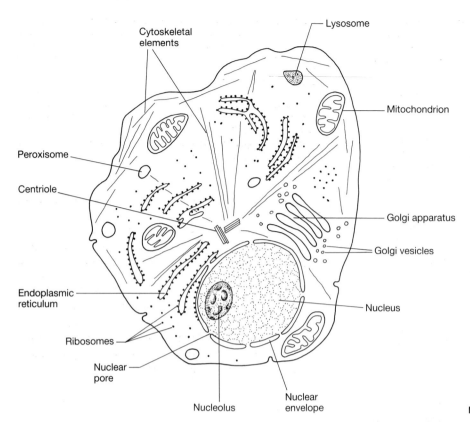

Fig. A30 Schematic diagram of a typical animal cell.

Animal virus(es)

VIRUSES are obligate intracellular parasites which have no protein-synthesizing or energy-generating mechanism [1,2]. They consist of an RNA or DNA genome (not both) contained within a protein coat (capsid) forming a nucleoprotein structure, or a nucleoprotein (the nucleocapsid) enveloped in a lipid membrane (envelope) containing further viral proteins. Viral lipids and carbohydrate are host encoded but most virions (virus particles) contain no host-encoded protein. Depending on their strategy of replication and/or other requirements, some virions contain enzymes; others have none. This article deals with the classification, structure and multiplication of animal viruses; diseases are covered separately (*see* ANIMAL VIRUS DISEASE).

Classification

Animal viruses are classified into 26 families according to their host, genome and morphology (Fig. A31). The number of virus families infecting different groups of organisms is shown in Table A6. Viruses that infect bacteria (BACTERIOPHAGES) or algae, fungi or protozoa are morphologically distinct, but two families of animal viruses (Rhabdoviridae and Bunyaviridae) also infect plants (*see* PLANT VIRUSES). Of the 26 virus families in the Animal Kingdom, four are unique to invertebrates and ten to vertebrates [3]. Thus although most major biological groups have a unique collection of viruses, there is considerable overlap between vertebrates and invertebrates. The overlap with plants may have arisen through invertebrates which feed on and transmit the virus to plants.

Some animal viruses have a complex life cycle involving both vertebrates and invertebrates; some replicate in both, and some only replicate in the vertebrate host and are carried passively by the invertebrate. This classification system takes no account of the diseases that viruses cause; these show no taxonomic pattern. Although the system is basically linnaean, viruses have not, until 1992, been designated as species.

New families of viruses are still being discovered. Investigation

Table A6 Distribution of virus families

	Bacteria	Algae, fungi, protozoa	Plants	Animals Invertebrates	Animals Vertebrates
Bacteria	**15**				
Algae, fungi, protozoa		**4**			
Plants			**34**	(2)*	(2)*
Invertebrates				**14**	(10)
Vertebrates					**22**

The total in any one group is shown in bold; those virus families which have representatives in other groups are in parentheses: thus there are 15 families of viruses unique to bacteria; the Animal Kingdom has 26 families of viruses.
* Rhabdoviridae and Bunyaviridae.

Table A7 Classification of viruses by type of nucleic acid in the genome

Class I	All viruses with a double-stranded DNA (dsDNA) genome. In this class the designation of '+' and '−' is not meaningful as different mRNA species may come from either strand
Class IIa	Viruses with a single-stranded DNA (ssDNA) genome of the same sense as the mRNA
Class IIb	Viruses with a ssDNA genome complementary to the mRNA. (These do not have a class to themselves as they (the autonomous parvoviruses) were discovered after this classification was devised.) Before the synthesis of mRNA can proceed, the DNA must be converted to a double-stranded form
Class III	Viruses with a double-stranded RNA (dsRNA) genome. All known viruses of this type have SEGMENTED GENOMES but mRNA is only synthesized on one strand of each segment
Class IV	Viruses with a single-stranded RNA (ssRNA) genome of the same sense as mRNA (positive-strand viruses). Synthesis of a complementary strand precedes synthesis of mRNA
Class V	Viruses with a ssRNA genome complementary in base sequence to the mRNA (negative-strand RNA viruses)

This classification is extremely useful in unifying the biochemistry of viruses but takes no account of their biology as, for example, viruses of humans (variola/smallpox) and bacteria (T2 phage) are both members of class I.

has concentrated on viruses of humans and farm animals, other economically important animals including fish, animal pests (as a control measure) or animal pests of plants (usually various sorts of biting arthropods) and domestic pets. It seems likely that every species can be infected by a spectrum of viruses characteristic of the vertebrate or invertebrate subkingdom. However individual viruses have a limited host distribution and are predominantly species specific.

An alternative scheme of classification named after its proposer, David Baltimore, is based entirely on how viral messenger RNA (designated as (+) (positive) strand RNA, or RNA$^+$) is produced. Using this convention, six groups of viruses can be distinguished (Table A7).

Structure

Diagrammatic representations of virus structures are shown in Fig. A31. Protein subunits are arranged either isometrically or helically with respect to the nucleic acid. The lipid envelope (where present) is derived from cellular membranes by budding. Nucleoprotein structures are built up of many copies of a few polypeptides varying in number from three (Parvoviridae) to more than 100 (Poxviridae), whereas some plant virus particles have only one. Spherical virus particles assemble spontaneously from their individual polypeptides which interact with precision to allow the exact geometry of the particle to form. The small number of repeated subunits allows an economy of genetic information. Spherical particles are all icosahedra, solids built of 20 identical equilateral triangles. The problem is to arrange nonsymmetric polypeptides to form a symmetric particle. This is done by placing them in a regular manner (Fig. A32). Thus each icosahedron is composed of 60 units. With animal viruses that

Table A8 Types of nucleic acid found in the 26 families of animal viruses

	ssRNA*	dsRNA	ssDNA	dsDNA
No. of families:				
Nonenveloped	4	2	1	3
Enveloped	11†	0	0	5‡

* May be messenger sense RNA or negative-sense RNA; some genomes are segmented (see Table T35).
† One family (Retroviridae) replicates via a DNA intermediate and ‡ one via an RNA intermediate (Hepadnaviridae).

have more than 60 polypeptides, of the same type or a few different types, the icosahedral faces can be envisaged as subdivided into smaller triangles each with three units.

The crystal structure of four members of the Picornaviridae — poliovirus, rhinovirus, mengovirus and foot-and-mouth disease virus — has been determined. A virion contains 60 copies of each of four different polypeptides — VP1, VP2, VP3 and VP4. VP4 is entirely internal and can be disregarded here. There is one molecule of each of VP1, VP2 and VP3 in each triangle and these are arranged as shown in Fig. A33, so that vertices are composed either of five units of VP1 or six units of alternating VP2 (three molecules) and VP3 (three molecules) [4].

Genomes

The distribution of types of nucleic acid in animal viruses is shown in Fig. A31 and Table A8. There are both single-stranded (ss) and double-stranded (ds) RNA and DNA viruses, but ssDNA only occurs in the Parvoviridae. Some genomes are segmented (i.e. contained in two or more unique nucleic acid molecules) (Table A9) which enables them to recombine at high frequency through exchange of segments with closely related viruses; this has important evolutionary consequences (e.g. antigenic shift in type A INFLUENZA VIRUSES). RNA genomes range in size from about 7 kilobases (kb) (Retroviridae) to 16 kb (Paramyxoviridae) and DNA genomes from about 5 kb (Parvoviridae) to 375 kbp (Poxviridae). Members of three families have circular DNA genomes; circular RNA genomes occur only in hepatitis D virus, a newly discovered virus which has not yet been classified, and in VIROIDS. The Papovaviridae and Polydnaviridae both have circular dsDNA, the former having one molecule per virion and

Table A9 Viruses with segmented genomes

	No. of segments per virion		
	ssRNA	dsRNA	dsDNA
Arenaviridae	2		
Nodaviridae	2		
Bunyaviridae	3		
Orthomyxoviridae	7 or 8		
Birnaviridae		2	
Reoviridae		10–12	
Polydnaviridae			Many*

* Circular genomes; the rest in the table are linear.

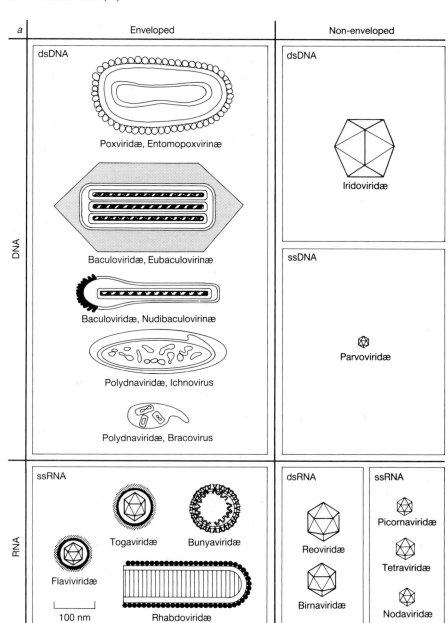

Fig. A31 Classification of animal viruses is based on having an invertebrate (*a*) or vertebrate (*b*) host and having a DNA or RNA genome, which may be double stranded (ds) or single stranded (ss). Those virions not shown solely as a geometric structure have one (usually) or more lipid membranes containing additional proteins. From [3].

the latter multiple molecules per virion; the exact situation in the Polydnaviridae is complicated as the different DNA molecules share some sequences. The DNA of Hepadnaviridae forms a circle composed of a complete minus strand whose 5′ and 3′ termini are not covalently joined. There is an incomplete plus strand, which overlaps the termini of the minus strand forming a circle which is partially single stranded. Other types of linear genomes are circularized by a terminal protein: for example one is covalently linked to the 5′ end of the dsDNA of Adenoviridae and is joined noncovalently to the 3′ end.

Multiplication

This is defined as the production of progeny virions and occurs only inside living cells. It can be divided into attachment, entry, uncoating, transport, replication, protein synthesis, assembly, and release of progeny virions.

Attachment

This occurs by specific interaction between the virus attachment site on one of the surface proteins and a cell surface receptor molecule (virus receptor) [5]. This determines absolutely the susceptibility of a cell to natural infection, although there are also intracellular restrictions on replication. The virus receptor is usually a protein but may be a carbohydrate (sialic or *N*-acetyl-neuraminic acid for Orthomyxoviridae, Paramyxoviridae and members of some other families) or (rarely) lipid. Cellular recep-

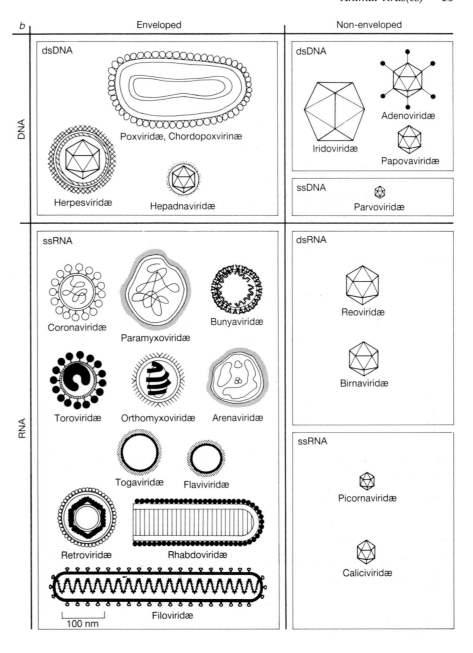

Fig. A31 *Continued*

tors for viruses have only recently been identified. They are mostly constitutively expressed proteins, some of which have a role in the immune system such as: CD4 (human and simian immunodeficiency viruses); MHC class I (adenovirus, human T cell leukaemia virus type 1, simian virus 40); MHC class II (lactate dehydrogenase elevating virus); CD21 (Epstein–Barr virus); an unidentified member of the IMMUNOGLOBULIN SUPERFAMILY (poliovirus types 1, 2 and 3); the CELL ADHESION MOLECULE ICAM-1 (human rhinoviruses); INTEGRINS (foot-and-mouth disease viruses); β-ADRENERGIC RECEPTOR (reovirus type 3); receptor for polymeric IgA (hepatitis B virus); acetylcholine receptor (rabies virus) [5].

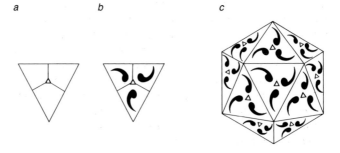

Fig. A32 The formation of an icosahedron from asymmetric units. Each of the 20 triangular faces is divided into three (*a*) and into each division a unit is placed symmetrically (*b*). These are then oriented on the icosahedron by fivefold or twofold axes of symmetry (*c*).

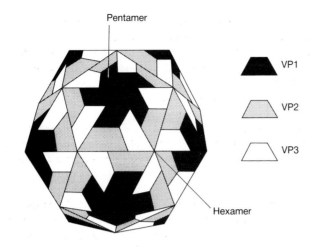

Pentamer

VP1

VP2

VP3

Hexamer

Fig. A33 Schematic diagram of a picornavirus.

Entry and uncoating

The virus receptors can move laterally through the plasma membrane and bind to virions in number. This stimulates the cell to take up many types of nonenveloped and enveloped viruses by ENDOCYTOSIS. The endocytic vesicle forms an ENDOSOME by fusion with other intracellular vesicles and the vesicle is acidified by a cellular proton pump to ~pH 5.5–6.5. For some enveloped viruses this causes conformational rearrangements of virion coat proteins such that fusion of viral and vesicle membranes ensues (acid-dependent fusion) and the nucleocapsid enters the cytoplasm. Nonenveloped viruses are rendered transcriptionally active by this process but how they escape from the vesicle is not known. Other enveloped viruses enter cells by fusing with the plasma membrane at neutral pH. There are probably further rearrangements of nucleocapsid structures before genomes can be expressed.

Transport

Transport of the uncoated subvirion particle to its site of replication, which may be cytoplasmic or nuclear, is presumably determined by specific amino-acid sequences of the sort known to target nascent viral proteins within the cells (*see also* PROTEIN TARGETING). Most DNA viruses replicate in the nucleus, but the Poxviridae replicate in the cytoplasm, setting up 'mini-nuclei' on their own. Most RNA viruses replicate in the cytoplasm but some (notably the influenza viruses: Orthomyxoviridae) replicate in the nucleus.

Replication

Strategies of replication depend on the nature and strandedness of the viral nucleic acid. Cells have no enzymes that can replicate RNA and so positive-strand RNA viruses (e.g. Picornaviridae: poliovirus) initially translate their single-stranded messenger sense genome to provide the required REPLICASE. Such enzymes probably complex with cellular TRANSCRIPTION FACTORS to form the HOLOENZYME and this provides an element of cell specificity.

Replication first produces a full length transcript (negative-sense RNA) which then acts as a template—the replicative intermediate—for the simultaneous formation of multiple new molecules of positive-sense RNA. The double-stranded RNA has no role in replication. In the cell, positive-sense viral RNA is around 100 times more abundant than the negative-sense RNA. Positive-sense RNA acts both as message and genome for progeny virus.

All negative-strand RNA viruses (e.g. Paramyxoviridae: measles virus) carry a transcriptase in the virion as they are unable to synthesize any protein from their antisense genome. Usually their positive-sense RNA is synthesized in two size classes: subgenomic length positive-sense mRNAs, and full length positive-sense RNA which is the template for the synthesis of new genomes. The segmented negative-strand RNA viruses (e.g. influenza virus types A, B and C) multiply along the same lines. Each segment in segmented viruses encodes one or more mRNAs and one template strand.

Double-stranded RNA viruses (e.g. Reoviridae: reovirus or rotavirus) also carry a particle-associated transcriptase. Replication is conservative and parental RNA is retained in viral cores in the cytoplasm. New positive-sense RNA molecules serve both as mRNAs and, after encapsidation into nascent virions, as templates for the synthesis of a complementary negative RNA strand to form the mature double-stranded genome.

DNA viruses are transcribed conventionally in the nucleus by cellular DNA-dependent RNA polymerase II (*see* TRANSCRIPTION). Poxviridae are the exception as multiplication is cytoplasmic. DNA viral genomes are replicated either by cell- or virus-encoded DNA POLYMERASES.

The Retroviridae are RNA viruses that replicate via a DNA intermediate. The positive-sense viral ssRNA genome is converted by the REVERSE TRANSCRIPTASE contained in the virion into a DNA copy which is then integrated into the cellular genome. This then behaves like a cellular gene and is transcribed into mRNAs and a complete positive-sense RNA genome. The integration of the retroviral genome, particularly the site of integration and its ability to transduce cellular proto-oncogenes, is the key to the transforming and cancer-inducing properties of some retroviruses (*see* ONCOGENES; RETROVIRUSES). However, of the three retroviral subfamilies (Oncovirinae, Lentivirinae and Spumavirinae) only the Oncovirinae and a few of the Lentivirinae are oncogenic. The double-stranded DNA of the Hepadnaviridae has been called a temporally permuted version of the retrovirus genome as it replicates via an RNA intermediate. Integration of viral DNA into the host genome is not required for replication, but hepatitis B virus, the cause of serum hepatitis, can progress through a chronic infection to primary hepatocellular (liver) carcinoma, at which stage hepatocytes are found to have integrated viral DNA sequences.

Protein synthesis

Viral proteins are synthesized using the host cell's protein synthesizing machinery. What varies from virus to virus is the ability to affect, usually to downregulate, host protein synthesis. Some viruses have no effect whereas others can virtually abolish this activity. Although this is deleterious to the infected cell in the long run, the process is probably too slow to be responsible for its death (see below). For the most part mechanisms of downregulation have not been elucidated, but poliovirus is known to inactivate the cap-binding factors required for translation from capped mRNA (*see* PROTEIN SYNTHESIS; RNA PROCESSING). Poliovirus mRNA is not capped and so escapes inhibition. The virion RNA has the viral protein VPg covalently attached at the 5′ terminus but in all other ways resembles the message; VPg is presumably detached before the infecting virion RNA is translated.

RNA viruses. RNA animal viruses have diverse strategies of GENE EXPRESSION ranging from that of segmented single-stranded (influenza) and double-stranded (reo) RNA genomes which have segments that express just one polypeptide, to that of nonsegmented RNA viruses like poliovirus, which synthesize a single POLYPROTEIN which is cleaved post-translationally to form the functional viral proteins (Fig. A34). Virus-encoded proteins are either destined to form part of the virus particle (structural proteins) or are concerned in other aspects of multiplication (nonstructural proteins).

There are numerous strategies of gene expression intermediate between the extremes noted above: in the nonsegmented positive-sense RNA Alphaviridae, nonstructural proteins are synthesized from a full (virion) length mRNA but the 3′ terminal part of the genome encoding the structural proteins is not used. Structural proteins are synthesized on a separate shorter mRNA molecule. A more extreme version of this strategy is seen in other nonsegmented positive-sense ssRNA viruses (Coronaviridae), which produce seven MONOCISTRONIC mRNAs. Each consists of a short 5′ terminal sequence from the virion RNA which is spliced to a DOWNSTREAM gene and continues to the 3′ terminus of the virion RNA. This seems wasteful, as the RNA downstream of the expressed gene is not used. A different strategy, although with the same end result, is found in the nonsegmented negative-strand ssRNA viruses of the Rhabdoviridae. All mRNAs are initiated from a single PROMOTER and can terminate after any one of the five genes. The frequency of mRNAs declines as their distance from the promoter increases, with readthrough becoming progressively less common towards the 5′ end of the template. The Arenaviridae are classified as having two segments of negative-strand ssRNA. However, the 5′ half of the smaller segment is positive-sense RNA. This is termed an ambisense RNA. Each half encodes a discrete mRNA.

The Orthomyxoviridae are unusual in using 5′ capped cellular mRNAs to prime the synthesis of their own messengers. The viral polymerase complex includes a cap-recognition function, an ENDONUCLEASE which cleaves after 10–13 nucleotides and a transcriptase. Thus viral mRNAs commence with a cap (*see* RNA PROCESSING) followed by a heterogeneous sequence derived from cellular mRNA. A similar situation exists in the Bunyaviridae and in protozoa (TRYPANOSOMES).

There is alternative splicing, and/or the use of different READING FRAMES in some but not all RNA viruses, and some RNA viruses exhibit temporal control, with early and late expressed genes.

DNA viruses. The synthesis of mRNAs from DNA viruses is essentially similar to that of cellular mRNA as, with the exception of the Poxviridae, viral DNA templates are located in the nucleus and most transcription is by cellular DNA-dependent RNA polymerase II (*see* EUKARYOTIC GENE EXPRESSION). Some but not all DNA viruses utilize RNA SPLICING in the synthesis of most of these mRNAs (*see* ADENOVIRUS). There is strong temporal control in expression of most DNA virus genomes: generally the transition from the early to the late phase of mRNA production is punctuated by genomic DNA synthesis. The Herpesviridae have three phases; first, α mRNAs are initiated by the presence of one of the virion polypeptides in conjunction with cellular transcription factor(s) and polymerase II. The α gene products permit transcription of β mRNAs and viral DNA synthesis, and the β gene products switch on the final phase of γ mRNAs which encode the structural proteins.

Maturation of progeny virions

Nonenveloped virions are believed to be formed by a SELF-ASSEMBLY process, but there is little detailed information on the processes involved. Within the cell small subvirion elements are found which, with time, are found in larger particles. The formation of the poliovirion for example progresses through 5S, 14S, 73S and 125S subvirion particles to the mature 160S virion. RNA is not associated with the nascent particle until the 125S stage. Most animal viruses also make 'empty' particles devoid of nucleic acid, indicating that the genome is not essential for virus particle formation.

Formation of enveloped virions is initiated by the formation of a nucleocapsid structure. This gains its lipid envelope through budding from an appropriate cellular membrane. The viral enve-

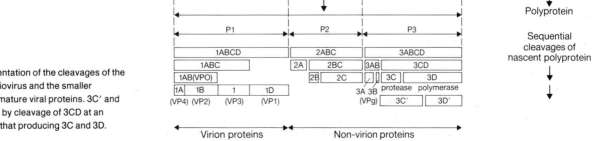

Fig. A34 Representation of the cleavages of the polyprotein of poliovirus and the smaller products to yield mature viral proteins. 3C′ and 3D′ are produced by cleavage of 3CD at an alternative site to that producing 3C and 3D. From [1].

lope proteins are first inserted into the appropriate membrane (which may be nuclear, Golgi, endoplasmic reticulum or plasma membrane, depending on the site of virus replication within the cell) and by a process which is not understood the nucleocapsid aligns with this to initiate the formation of a virion by budding. The mature particle is liberated by pinching off. Recognition between the nucleocapsid and envelope proteins is assumed to occur but this has been shown only for SINDBIS VIRUS (Togaviridae) [4]. Envelope proteins possess targeting sequences which direct them to specific membranes (*see* PROTEIN TARGETING); these sequences also permit the virus to distinguish between the apical and basolateral surfaces of polarized cells (*see* EPITHELIAL POLARITY), which in the respiratory or gut surface determines whether a virion is liberated 'outside' the body or within it. Most enveloped virions have some unknown means of excluding host membrane proteins. Retroviruses are the exception.

Viruses budding from the plasma membrane are automatically liberated from the cell, but the others still have that problem to overcome. Many escape without the disintegration of the cell. Some mature in vesicles which subsequently fuse with the plasma membrane (*see* EXOCYTOSIS). Others remain associated with the cell until its disintegration. There is little information on mechanisms of virus release.

Cellular pathology

Apart from the generalized inhibition of host protein synthesis referred to above, viruses are known to selectively inhibit the expression of host proteins (e.g. hormones, MHC MOLECULES) and host cell functions (e.g. activation of lymphocytes) often without causing any general cellular pathology. The realization that some viruses have the ability to bring a scalpel rather than a bludgen to bear on cellular function is a relatively new perception. It is also no coincidence that viruses can suppress the functioning of the immune system in a variety of ways.

A virus–cell interaction may be highly specific in the sense that the type of infection can vary according to a particular permutation. Five such interactions can be distinguished by the infection of cultured cells in the laboratory (lytic, persistent, latent, transforming and abortive). Although it is widely appreciated that viruses can kill cells, this is not an essential aspect of virus multiplication and it is perfectly possible for a virus to multiply without causing cell death or pathology. Mechanisms responsible for cell death are surprisingly poorly understood (*see* ANIMAL VIRUS DISEASE). For further information see [2,3].

N.J. DIMMOCK

See also: PAPOVAVIRUSES; *individual virus entries.*

1 Dimmock, N. J. & Primrose, S.B. (1993) *An Introduction to Modern Virology*, 4th edn (Blackwell Scientific Publications, Oxford).
2 Fields, B. N. & Knipe, D.M. (Eds) (1990) *Virology*, 2nd edn, Vols 1 & 2 (Raven, New York).
3 Francki, R.I.B. et al.(Eds) (1991) Classification and nomenclature of viruses. *Arch. Virol. Suppl. 2*, 5th edn (Springer-Verlag, Wien).
4 Branden, C. & Tooze, J. (1991) *Introduction to Protein Structure* (Garland, New York).
5 Lentz, T.L. (1990) The recognition event between virus and host cell receptor: a target for antiviral agents. *J. Gen. Virol.* **71**, 751–766.

Animal virus disease

ANIMAL VIRUSES are obligate intracellular parasites, and as with all parasites it is imperative that they do not kill or reduce the reproductive capacity of their host species and put their own continued existence at risk. Consequently most viruses do not cause serious disease; most give rise to 'subclinical' or 'silent' infections, or only minor disease. It is rare for viruses to kill their host; the mortality even from smallpox (variola) was only about 1% for infection with variola minor, a predominantly African strain, and ~25% with variola major, the Asian strain. Viruses that cause the death of their host are believed to represent a relatively new association in evolutionary terms which has not yet had time to evolve into (relatively) peaceful coexistence. This view is dramatically borne out by the current human immunodeficiency virus (HIV: Retroviridae) pandemic (*see* IMMUNODEFICIENCY VIRUSES) as the virus was almost unknown before 1970 except for rare infections in sub-Saharan Africa. Infection results in ACQUIRED IMMUNE DEFICIENCY SYNDROME (AIDS) which as far as we know is 100% lethal. On the other hand, the evolution of a new host–virus association from virulence to avirulence is well illustrated by the course of myxomatosis (caused by the myxoma virus: Poxviridae) amongst the rabbit population in Britain and Australia [1].

Some general principles of virus disease are covered here, with some examples of molecular mechanisms of virulence and pathogenesis. More information on specific molecular mechanisms of disease (where known) will be found in entries for individual viruses (*see* ADENOVIRUS; EPSTEIN–BARR VIRUS; HEPATITIS B VIRUS; IMMUNODEFICIENCY VIRUSES; INFLUENZA VIRUS; PAPOVAVIRUSES; RETROVIRUSES).

Classification of infections of cultured cells

Virus diseases are the result of a complex interaction between the infected tissue and the immune system [2]. It is helpful first to consider the variety of ways in which a virus can interact with cells in culture in the laboratory, without the complication of cellular differentiation, organization of cells into tissues, or immune responses.

A cell can only be infected if it possesses a receptor (virus receptor) recognized by a particular virus. Receptors are usually proteins but can be carbohydrates or rarely lipids. They are host cell-surface components often serving essential homeostatic functions [3] (*see* ANIMAL VIRUSES). A virus binds to its receptor by an attachment site on one of the viral surface proteins. This site is invariant (unlike antigenic sites) as any mutation is likely to be lethal, and is often hidden from the immune system. Entry of virus into the cell is achieved by ENDOCYTOSIS and/or, in the case of enveloped viruses, fusion between viral and cellular membranes (*see* ANIMAL VIRUSES). Although cells cultured in the laboratory are usually dedifferentiated and only distantly resemble their counterparts in the whole organism they allow five main virus–cell interactions to be distinguished (Table A10).

Table A10 Types of interactions between viruses and cultured cells

Lytic infection	The virus infects the cell, produces progeny and in so doing kills the host cell. The cycle takes ~6–12 hours. It is not known how the virus kills the cell or why, since the virus may usurp less than 1% of the cell's capacity for macromolecular synthesis
Persistent infection	This can have the same dynamics as lytic infection except that the cell is not killed; cells may produce virus for months while still multiplying. Persistent infections are host cell dependent as the same virus may be lytic in another cell type. A lytic infection may be modified by host factors such as ANTIBODY, INTERFERON or DEFECTIVE INTERFERING VIRUS, so that a balance is achieved between virus and cell and both survive
Latent infection	Members of the Herpesviridae in particular, such as herpes simplex virus, varicella-zoster virus, Epstein-Barr virus, and cytomegalovirus can cause lytic infections, but in certain cells, notably neurons, are normally latent and do not produce infectious progeny. Latency can be broken by external factors which disturb this balance and reactivate the virus. Latency is maintained by a balance between host and viral transcriptional factors. The switch from lytic to latent to lytic infection underlines the importance of the cellular environment
Transforming infection	Some members of the Papovaviridae, Adenoviridae, and Herpesviridae (DNA viruses) and of the Retroviridae (RNA viruses) can cause neoplastic TRANSFORMATION of infected cells—the cells become capable of unlimited cell division. This is the first stage in the complex production of a CANCER CELL and is accompanied by the INTEGRATION of the viral genome into that of the host
Abortive infection	Viruses do not multiply with equal efficiency in all the cells they can infect. There may be a reduction in the yield of virus particles (sometimes to zero) or an increase in the virus particle/infectivity ratio or both. These all reflect a deficiency of some cellular component necessary for multiplication, be it DNA, RNA, or protein. For example one cause of the production of noninfectious influenza virus particles is the absence of the appropriate cellular protease needed to cleave the haemagglutinin coat protein. A simple proteolytic cleavage is all that is needed to confer infectivity

How do viruses kill cells?

Although cells often die as a result of virus infection the example of some persistent infections shows that in some virus–cell interactions a virus can multiply without killing its host cell. What kills the cell is not clear [4]. The metabolic burden is estimated at around 1% and although many viruses can switch off host DNA, RNA or protein synthesis, the cell usually dies before these effects could have become lethal. The virus may possibly cause an ion imbalance which would rapidly have an effect on many cellular

reactions and it is interesting that ion channelling activities of certain proteins of HIV, influenza and polioviruses have recently been described, although not yet tied in with cell death. It seems that cell death is not essential to virus replication, nor is it always needed to allow progeny virus to escape from the cell. Many animal viruses are released by mechanisms such as budding which do not require dissolution of the cell's outer membrane. It is possible to see cell death as a host mechanism for destroying a site of virus production. Also there is now evidence that some viruses trigger APOPTOSIS—a cellular self-destruct mechanism which is part of normal development.

Classification of diseases

It is useful to impose somewhat arbitrary divisions upon what is a spectrum of disease processes. A particular virus may appear in several different categories, depending on the nature of the host–virus interaction. Any disease depends essentially on the interaction of the virus with the immune system. This is a two-way process and viruses have a variety of strategies — including concealment, camouflage, and suppression — which they use to evade immune responses.

Acute infections are the equivalent of lytic infection in cultured cells (above). After infection there is an incubation period during which the virus multiplies and which precedes clinical disease. Virus is cleared rapidly by immune responses and there is a solid long-lasting immunity against reinfection. Variola (smallpox) virus always, and measles virus nearly always, cause acute infections. Many other viruses (e.g. influenza: Orthomyxoviridae) can cause acute infections, but are often associated with inapparent or 'silent' infections.

Inapparent or silent infections are identical to acute infections except that no clinical disease accompanies virus multiplication. Poliovirus (Picornaviridae) is usually in this category and only exceptionally (< 1% of infections) causes disease.

The difference between acute and inapparent infections raises a question about the nature of 'disease': for example, the death of cells as a result of virus multiplication is of little consequence to the host when the cells are readily replaced, as they are in the gut or respiratory tract, but excessive loss or loss of irreplaceable cells, such as particular neurons of the brain, could cause lasting damage. Disease may result from the immune response to infection, which can increase tissue damage and release cytokines and other pharmacologically active molecules which can themselves be the cause of malaise.

Chronic and persistent infections are essentially acute or inapparent infections which are not terminated by the host. A chronic infection continues to produce virus or viral antigens in large quantities and differs only in degree from a persistent infection. They probably occur because the virus inhibits some aspect of the immune response (see below). Chronic infections may eventually lead to immune complex disease due to the presence of large amounts of viral antigen and reactive immunoglobulins whereas persistent infections are less prone to this complication [5].

Latent infections by formal definition do not produce infectious virus. They are always preceded by an acute or an inapparent infection. Acute infections by members of the Herpesviridae are followed by life-long latency (e.g. varicella-zoster virus which causes chickenpox). The latent conditions can revert to produce infectious virus and an acute infection; in the example cited this occurs about once in a lifetime and results in shingles, an entirely different disease.

Recently a common theme underlying the mechanism of latency has been proposed based on studies of herpes simplex virus, Epstein–Barr virus (EBV) and HIV [6]. In essence these viruses become latent in cells which lack host TRANSCRIPTION FACTORS needed for the expression of early virus gene products which in turn activate the acute infection. A variety of stimuli can turn on the expression of the host factors and activate an acute infection (Fig. A35).

The causes of reactivation are many and herpes simplex virus is commonly reactivated by high levels of UV irradiation found by the sea or at altitude, emotional disturbance or menstruation. There may also be an immune involvement; reactivated cytomegalovirus is a common and sometimes lethal complication of immunosuppression associated with organ transplantation. EBV, responsible for infectious mononucleosis, becomes latent in B lymphocytes. It is also associated with human cancer (see below).

Slowly progessive diseases usually have a relatively minor acute phase and then a long incubation period of several years before the manifestation of a second acute, often lethal, infection of a different kind. Measles virus (Paramyxoviridae) is responsible for the very rare disease subacute sclerosing panencephalitis (SSPE),

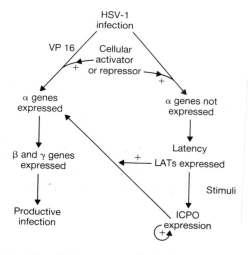

Fig. A35. Regulation of latency as exemplified by herpes simplex virus 1 (HSV-1) in neurons. VP16 is a virion protein which transactivates the early α genes and sets in motion the production of progeny virus and acute infection. It is thought that VP16 either needs an activating host cell factor or is repressed by one. Accordingly latency is established. In the latent state there is very little viral gene activity, and only latency associated transcripts (LATs) are expressed. Although the molecular stimuli of activation are unknown these probably result in the expression of the protein ICP0 and the resulting production of progeny virions. From [6].

a degenerative condition of the brain which occurs at the rate of 1–3 per 10^6 cases of acute measles; infection with measles virus at an unusually early age is thought to predispose to SSPE. There is on average a 5-year (range 2–20 years) incubation period. In contrast, AIDS is an almost invariable consequence of HIV infection. The reason for the long incubation period in SSPE is not known but isolates of measles virus from SSPE cases contain a considerable number of mutations. The enigmatic BORNA VIRUS causes subtle behavioural alterations in experimental infections in primates which affect their ability to breed.

Virus-induced tumours. All viruses which cause tumours have DNA as their primary genetic material in the cell (the RNA genome of the Retroviridae is converted on infection into DNA by the virion-associated reverse transcriptase). It is estimated that ~20% of all human cancers worldwide have a viral component [7]. The main DNA virus associated with human cancer is hepatitis B virus (HBV: Hepadnaviride) which is associated with hepatocellular carcinoma, particularly in the Far East where the incidence of virus infection is almost universal. Certain members of the human papillomavirus group (HPV: Papovaviridae) are associated with cervical, penile, and some anal carcinomas. Sequences of HPV16 in particular occur in 16% of cervical carcinomas. EBV (Herpesviridae) is associated with Burkitt's lymphoma and nasopharyngeal carcinoma. Of the Retroviridae, human T-cell leukaemia virus type 1 (HTLV-1) is associated with adult T-cell leukaemia/lymphoma syndrome and HTLV-2 probably with chronic T-cell malignancies; HIV is associated with non-Hodgkin's lymphoma and Kaposi's sarcoma, although it has not been demonstrated to be oncogenic *per se* (*see* IMMUNODEFICIENCY VIRUSES).

The genes of DNA viruses involved in carcinogenesis are usually also essential for virus replication; those of the RNA retroviruses are transduced cellular genes such as *src*, *myc* and *ras* which are involved in the control of cell division (*see* ONCOGENES; RETROVIRUSES). Viral genes may initiate tumours directly as a result of their expression, by acting as insertional mutagens and upsetting regulation of cellular division, by immunosuppression (such as the downregulation of class I MHC proteins by adenoviruses in rodents (see below) or of cell adhesion molecules by EBV) or by expressing products which can bind to and inactivate endogenous cellular TUMOUR SUPPRESSOR GENE products.

Cancer is a rare consequence (from 1 : 100 to 1 : 100 000) of infections. It is a multistage process and environmental and/or genetic cofactors play an essential part in its development. The incubation time between the initial infection and appearance of cancer is measured in years if not decades. An unusually early age of infection with HBV or HPV is thought to be one of the factors predisposing to later development of cancer. Because of the very high incidence of persistent HBV infection in the Far East much of the transmission of this virus is via the maternal route (*see* VERTICAL TRANSMISSION).

The discovery of human cancer viruses strongly suggests that immunization could prevent not only the primary infection but also the associated cancers. RECOMBINANT VACCINES comprising the surface antigen of HBV are commercially available and the World Health Organization has announced a scheme for worldwide immunization with a view to total eradication of the virus. The lymphoma caused by Marek's disease virus (Herpesviridae) in chickens can be prevented by administration of a live vaccine.

The molecular basis of disease: some examples

Poliovirus: molecular basis of neurovirulence

Poliovirus normally causes a subclinical infection of the gut but on rare occasions (<1%) it escapes, ascends the peripheral nerves and infects the cells of anterior horn cells of the dorsal root ganglia, particularly motor neurons which innervate muscles of usually the leg, or more rarely the chest, causing paralysis and eventual wasting of the muscles affected. Many infections cause only temporary paralysis. There are three SEROTYPES of virus; immunity to one does not protect against the other two. All cause identical infections although most disease is caused by type 1. Virulent strains of all three were attenuated empirically by serial growth in tissue culture at reduced temperature (31°C) by Albert Sabin in the 1950s to form the first vaccine against polio. The only caveat on Sabin's very effective and safe vaccine is reversion of the type 3 vaccine strain at the extremely low rate of one case per 10^8 doses of vaccine.

Sequencing of the original type 3 virulent strain Leon, the Sabin vaccine strain and virus isolated from vaccine-associated paralytic poliomyelitis has revealed the changes that have occurred in the single-stranded 7431-nucleotide RNA genome. The vaccine strain had undergone only 10 nucleotide changes during the attenuation process: two mutations in the 5' noncoding region, one in the 3' noncoding region, three leading to amino-acid changes and four silent mutations in coding regions (Fig. A36). By noting the changes that occurred when the vaccine strain re-

verted to the neurovirulent strain 119, it was possible to deduce which of the mutations was associated with virulence. There were in all seven nucleotide changes but only one, at position 472 in the 5' noncoding region, was involved in the acquisition of avirulence and subsequent reversion to virulence: this was a change from cytosine (C) to uridine (U) and back to cytosine (C). It occurs in all cases of vaccine reversion and significantly alters the computer-predicted SECONDARY STRUCTURE of the RNA, but despite having such precise information, the significance of the mutation in biological terms is not understood. However, it is noteworthy that only a single nucleotide stands between a virulent virus and a successful vaccine. The thinking now is to discover other nucleotides on which virulence depends and to mutate these so that the rate of reversion becomes insignificant [8,9].

Immunomodulation

In the simplest infections (acute or inapparent), the immune system orchestrates recovery and results in permanent protection from reinfection, but the situation is frequently more complicated. At one extreme it is the action of the immune system *per se*, rather than any cytopathic effects of the virus, which gives rise to cell and tissue damage and to signs and symptoms of disease. There is little precise molecular detail but this situation is exemplified by the measles rash which is due to the immune response to viral antigens in the skin. Malaise in influenza is believed to be due to the release of CYTOKINES (*see* LYMPHOKINES) since the systemic effects of influenza can be mimicked by administration of INTERFERON on its own.

More is known about immunosuppression by viruses [10], and it is possible that all virus infections modulate the immune

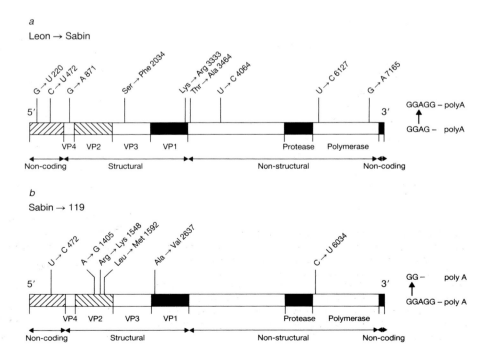

Fig. A36 Sequence changes occurring when (*a*) the original poliovirus type 3 wild-type neurovirulent Leon strain was attenuated to form the current Sabin vaccine and (*b*) when the Sabin vaccine strain reverts to the neurovirulent 119 strain. From [12].

response to some degree. One extreme is the destruction of CD4 T lymphocytes during HIV infection. These cells are involved in the activation of most B and T lymphocyte responses and are hence pivotal to the functioning of the immune system. In a more moderate way there are documented instances of nearly every component of the immune system suffering altered responsiveness, directly or indirectly, as the result of a virus infection. For the most part the molecular mechanisms are not known but effects of virus infection on the expression of MHC and other surface antigens involved in immune responses are described below.

Suppression of expression of MHC molecules

Class I and II MHC MOLECULES (*see* MAJOR HISTOCOMPATIBILITY COMPLEX) are borne on the outer surface of the plasma membrane and present foreign antigenic peptides to effector T cells (*see* ANTIGEN PROCESSING AND PRESENTATION; T CELL RECEPTOR). MHC class I proteins are constitutively expressed on nearly all cells but MHC class II only on cells of the immune system. Failure to express MHC molecules abrogates antigen presentation and both cell-mediated and humoral immune responses against protein antigens.

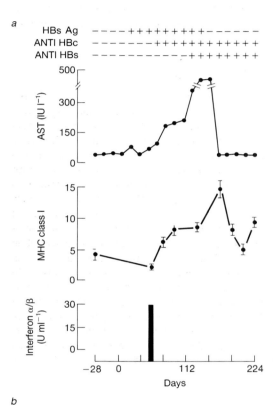

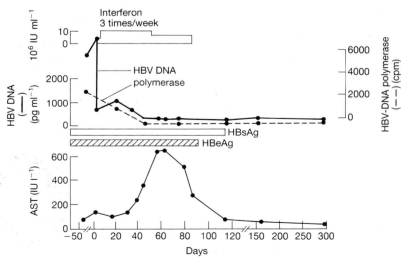

Fig. A37 *a*, Acute hepatitis B virus (HBV) infection in a chimpanzee inoculated on day 0. Markers of virus infection appear early, coinciding with the transient appearance of circulating interferon. This is followed by enhanced display of class I MHC proteins on hepatocytes. Liver damage (demonstrated by a rise in serum aspartate aminotransferase (AST) level) coincides with the appearance of antibodies. *b*, Successful clearance of a chronic human HBV infection by treatment with interferons α and β. HBs, virion surface antigen; HBc and HBe, internal virion antigens. From [11].

Expression of both classes of MHC molecule is responsive to a number of local stimuli, and class II proteins can be induced during virus infection on fibroblasts, epithelial and endothelial cells, and cells of the central nervous system, on which they are not normally expressed. Interferons α and β for example upregulate expression of class I MHC molecules, and interferon γ that of class I and II. This opens the way to immune intervention where the clinical situation results from an inability of T cells to clear infection due to insufficient expression of MHC molecules. Alternatively a T cell-mediated immune pathology may be reduced by downregulating the expression of MHC proteins.

Some viruses downregulate the expression of MHC proteins: adenoviruses downregulate expression of class I MHC proteins in a variety of ways. Type 2 adenovirus encodes a protein (E3gp19K) which is located in the luminal surface of the ENDOPLASMIC RETICULUM. In both human and mouse cells it binds to the nascent MHC class I heavy chain preventing its glycosylation and subsequent transport to the cell surface. Adenovirus-infected cells are therefore not recognized and not destroyed by CYTO-TOXIC T CELLS. Rats infected with type 2 adenovirus develop lung disease but this is much reduced on infection by mutant virus lacking functional E3gp19K protein.

Another serotype of adenovirus, type 12, inhibits the expression of MHC class I by a different mechanism which involves either a selective transcriptional block or failure to export mRNA from the nucleus. The viral protein E1A is implicated in this inhibition. Type 12 adenovirus is oncogenic in rodents and virus-induced tumours escape immune attack. However if such tumour cells are transfected with exogenous MHC class I genes whose expression is not influenced by E1A and reimplanted, the resulting tumours are rejected.

In humans, hepatitis B virus (HBV) causes an initial acute infection and in 10–90% of cases remains as a chronic infection. This depends on the age of the patient, with children infected neonatally being most susceptible. Acute liver disease caused by hepatitis B virus in primates correlates with increased expression of MHC class I proteins on hepatocytes as the result of the induction of interferons α and β during the immune response. Concomitant attack by cytotoxic T cells releases aspartate aminotransferase which serves as a marker of liver damage (Fig. A37a). Clearance of virus is incomplete in about 10% of infections, due it is thought to an acquired reduced responsiveness to interferons α and β and a resulting early fall in MHC class I expression; the viral infection then becomes chronic. Therapy with high doses of interferon α and β (Fig. A37b) upregulates MHC class I protein expression and once again this exposes hepatocytes to the action of cytotoxic T cells. There follows a transient exacerbation of disease due to the immune attack, with elevated transaminase levels followed by a fall in expression of viral antigens and viral DNA. In responsive cases the infection is clearly completely abolished whereas in others the virus may re-establish itself [11].

Inhibition of cellular 'luxury' functions

The classification of infections as lytic or nonlytic with the assumption that the latter viruses coexist peacefully with their host cell is an oversimplification, as viruses can alter the production of specialized cellular products (termed 'luxury' functions) while leaving the everyday 'housekeeping' functions of the cell unharmed. In one experimental system, infection of newborn mice with lymphocytic choriomeningitis virus (LCMV: Arenaviridae) was found to stunt their growth [13]. Fluorescent antibody staining located viral antigen in many tissues but, significantly, it was present in the anterior pituitary where GROWTH HORMONE is synthesized. The pituitary showed no sign of pathology when examined microscopically but a link between LCMV infection and growth retardation was established when virus was found only in those pituitary cells that make growth hormone. Growth hormone has major effects on growth and glucose metabolism, and infected mice die prematurely of severe hypoglycaemia at three weeks after infection. Transplantation of growth hormone-producing cells into infected mice ensured both their survival and normal development (Fig. A38). Reduction of growth hormone

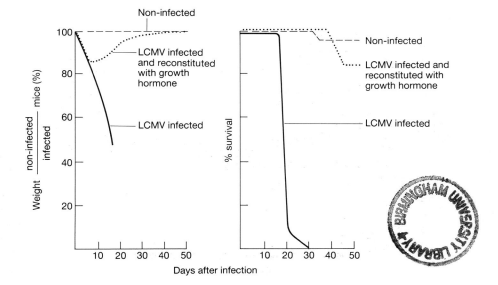

Fig. A38 Stunting of growth in newborn mice infected by the Armstrong 1371 strain of lymphocytic choriomeningitis virus (LCMV), and reversal of the trend by transplantation of cells secreting growth hormone. Without hormone replacement the mice die. From [13].

levels results from a gene-specific inhibition of initiation of transcription [5], presumably as a result of viral infection. LCMV can also switch off insulin production without damaging the β-cells of the pancreas.

Armed with this knowledge it seems profitable to re-examine human disease states characterized by hormone, lymphokine, or neurotransmitter deficiencies. Such cells would appear normal in all respects save for the loss of a 'luxury' function — luxury for the cell maybe, but not for the unfortunate patient.

N.J. DIMMOCK

1 Ross, J. & Sanders, M.F. (1984) The development of genetic resistance to myxomatosis in wild rabbits in Britain. *J. Hyg.* **92**, 255–261.
2 Mims, C.A. & White, D.O. (1984) *Viral Pathogenesis and Immunology* (Blackwell Scientific Publications, Oxford).
3 Lentz, T.L. (1990) The recognition event between virus and host cell receptor: a target for antiviral agents. *J. Gen. Virol.* **71**, 751–766.
4 Carrasco, L. et al. (1989) Modification of membrane permeability by animal viruses. *Pharmac. Ther.* **40**, 171–212.
5 Oldstone, M.B.A. (1991) Molecular anatomy of viral persistence. *J. Virol.* **65**, 6381–6386.
6 Garcia-Blanco, M.A. & Cullen, B.R. (1991) Molecular basis of latency in pathogenic human viruses. *Science* **254**, 815–820.
7 Dalgleish, A.G. (1991) Viruses and cancer. *Br. Med. Bull.* **47**, 21–46.
8 Evans, D.M.A. et al. (1985) Increased neurovirulence associated with a single nucleotide change in a non-coding region of the Sabin type 3 poliovaccine genome. *Nature* **314**, 548–550.
9 Stanway, G. et al. (1984) Comparison of the complete nucleotide sequences of the genomes of the neurovirulent poliovirus P3/Leon/37 and its attenuated Sabin vaccine derivative P3/Leon/12a,b. *Proc. Natl. Acad. Sci. USA* **81**, 1539–1545.
10 Maudsley, D.J. & Pound, J.D. (1991) Modulation of MHC antigen expression by viruses and oncogenes. *Immunol. Today* **12**, 429–431.
11 Thomas, H.C. (1990) Management of chronic hepatitis virus infection. In *Control of Virus Diseases* (Dimmock, N.J. et al., Eds) Society for General Microbiology Symposium 45, 243–259 (Cambridge University Press, Cambridge).
12 Dimmock, N. J. & Primrose, S.B. (1993) *An Introduction to Modern Virology*, 4th edn (Blackwell Scientific Publications, Oxford).
13 Oldstone, M.B.A. et al. (1984) Viral perturbation of endocrine function. *Nature* **307**, 278–280.

ankyrin A protein of the red cell membrane skeleton. *See*: MEMBRANE STRUCTURE.

Anlage (pl. Anlagen) *See*: PRIMORDIUM.

annealing (1) The reformation of a double-stranded nucleic acid, following separation of the two strands by denaturation. Annealing, also called renaturation or reassociation, occurs through complementary base pairing between the two separated strands and occurs most efficiently at low temperature and pH. That it can occur between only partially complementary DNA strands or between complementary RNA and DNA strands is a property that has been widely exploited in techniques for detecting a particular fragment or sequence of DNA or RNA. The number of copies of a DNA sequence in a genome can also be estimated from the kinetics of DNA annealing by COT ANALYSIS. *See*: HYBRIDIZATION; NUCLEIC ACID STRUCTURE.
(2) *See*: SIMULATED ANNEALING, a method used in MOLECULAR DYNAMICS.

annelid development The phylum Annelida comprises the poly-chaete worms (e.g. *Nereis*), the oligochaetes (e.g. earthworm), and the leeches. The polychaetes show a typical SPIRAL CLEAVAGE and a trochophore larva similar to that of molluscs (*see* MOLLUSC DEVELOPMENT). Oligochaetes and leeches show a cleavage pattern departing considerably from the archetypal spiral form. All however show INVARIANT CELL LINEAGES in the early stages and many important cell lineage studies have been performed on members of the group, starting with Whitman's study of the leech *Clepsine* in 1878, and continuing to the studies of Weisblat and others in recent years. Some classic microbeam ABLATION studies were performed by Penners on the oligochaete *Tubifex* in 1926.

For all annelids there are special regions of cytoplasm known as pole plasms which become localized after fertilization to the animal and vegetal regions of the egg. Both pole plasms enter the D lineage during cleavage, and much finds its way into important derivatives of the D blastomere: the somatoblast (2D) and the mesentoblast (4d). The somatoblast forms much of the ECTODERMAL part of the body and the mesentoblast much of the MESODERM. In oligochaetes, and particularly in leeches, these cells give rise to stem cell-like teloblasts whose progeny form the segmental repeating pattern in a clonally invariant, but complex manner. Although much emphasis has been placed on CYTOPLASMIC DETERMINANTS and the invariant early cleavage pattern, the more detailed the study of annelid development, the more decisions have been found to depend on INDUCTION. In this regard our view of the annelids has changed in a similar way to our view of molluscs and nematodes.

annexins A collection of proteins isolated from a variety of tissues which have many biochemical properties and have been given names indicative of their presumed function and the tissues in which they were found (Table A11). They include calcium- and phospholipid-binding proteins, and phospholipase A_2- and blood coagulation inhibitors. These apparently dissimilar proteins belong to a group that shares a conserved segment of some 80 amino acids repeated up to eight times in a single protein. At the time of writing there are 10 annexin groups, designated by roman numerals I–X. Some annexins have actin filament binding activity. *See also*: ACTIN-BINDING PROTEINS; CALCIUM-BINDING PROTEINS.

Crumpton, M.J. & Dedman, J.R. (1990) *Nature* **345**, 212.

annulate lamellae *See*: NUCLEAR ENVELOPE.

anomalous dispersion The variation of the magnitude of the ANOMALOUS SCATTERING effect sometimes measured in X-RAY CRYSTALLOGRAPHY with the incident X-ray wavelength. The advent of X-ray SYNCHROTRON RADIATION sources has enabled the incident wavelength to be chosen so as to maximize any anomalous scattering differences. Data collection at a few optimal wavelengths can provide information to solve the PHASE PROBLEM. Use of this MAD technique (multi-wavelength anomalous dispersion) avoids the need to find a HEAVY ATOM ISOMORPHOUS DERIVATIVE, but does require very accurate measurement of all the diffraction intensities.

Table A11 Annexins

Annexin	I	II	III	IV
Previous terminology	Lipocortin I p35 Calpactin II Chromobindin 9 GIF	Calpactin I Lipocortin II p36 Chromobindin 8 Protein I PAP-IV	Lipocortin III PAP-III 35-α Calcimedin	Endonexin I Protein II 32.5K calelectrin Lipocortin IV Chromobindin 4 PAP-II PP4-X 35-β Calcimedin

Annexin	V	VI	VII	VIII
Previous terminology	PAP-I IBC Lipocortin V 35K Calelectrin Endonexin II PP4 VAC-α 35-γ Calcimedin Calphobindin I Anchorin CII	p68, p70, 73K 67K Calelectrin Lipocortin VI Protein III Chromobindin 20 67K Calcimedin Calphobindin II	Synexin	VAC-β

anomalous scattering When electrons scatter X-rays of wavelength close to an absorption edge of the scatterer, the ATOMIC SCATTERING FACTOR (f) is modified by a component ($\Delta f'$), reducing the scattering in phase with f, and an additional component ($\Delta f''$) of scattering 90° ($\pi/2$) out of phase with the ordinary scattering. The magnitude of $\Delta f''$ is insignificant for atoms of low atomic number, but it can be detected for metals, halide ions, and, at the appropriate wavelength, for sulphur atoms. If there are anomalous scatterers amongst nonanomalous scatterers (e.g. a metalloprotein) then the usual equality between the STRUCTURE AMPLITUDES from a direct reflection (hkl) and its inverse ($-h -k -l$, which is denoted $\overline{hkl}$ and referred to as 'bar *h*, bar *k*, bar *l*'), known as Friedel's law, is violated. When the difference in diffracted intensity (I) from reflections hkl and $\overline{hkl}$ is measured, this anomalous scattering is utilized in the MIR (MULTIPLE ISOMORPHOUS REPLACEMENT) and MAD (multi-wavelength ANOMALOUS DISPERSION) phasing technique. *See*: X-RAY CRYSTALLOGRAPHY.

antagonist Any compound that inhibits a receptor.

antenna pigments Pigments whose function is to absorb light quanta and to pass the energy to photosynthetic reaction centres. Chlorophylls *b* and *c*, bacteriochlorophyll *c*, PHYCOBILINS, and CAROTENOIDS perform this light-harvesting function only. Chlorophyll *a* and bacteriochlorophylls *a* and *b* can be involved in the primary photochemical reaction, but also act as antenna pigments. *See also*: CHLOROPHYLL; CHLOROPHYLL *a/b* BINDING PROTEINS; LIGHT-HARVESTING COMPLEXES I AND II; PHOTOSYNTHESIS; PHYCOBILIPROTEINS.

Antennapedia complex (ANT-C) A cluster of genes in *Drosophila* which contains several HOMEOTIC GENES involved in the specification of segmental identity in the anterior thorax and posterior head regions (*labial*, *proboscipedia*, *Deformed*, *Sex combs reduced*, and *Antennapedia*), a SEGMENTATION GENE of the pair-rule class (FUSHI TARAZU), the gene for the MORPHOGEN responsible for the anterior–posterior axis (*bicoid*), and a gene involved in dorsoventral patterning (*zerknüllt*). *See*: DROSOPHILA DEVELOPMENT; HOMEOBOX GENES AND HOMEODOMAIN PROTEINS.

anterior group genes MATERNAL EFFECT genes in *Drosophila* required for the specification of the anterior–posterior axis of the egg. The key gene is *bicoid*. During OOGENESIS the NURSE CELLS synthesize *bicoid* RNA which then becomes localized at the anterior pole of the oocyte. After fertilization the RNA is translated and the protein forms a gradient with a high point anteriorly. The *bicoid* protein is a homeodomain-containing TRANSCRIPTION FACTOR and acts as a MORPHOGEN to specify the regional expression patterns of SEGMENTATION GENES of the gap class. Other anterior group genes — *exuperantia*, *swallow*, and *staufen* are required for the localization and stabilization of the *bicoid* RNA. *See*: DROSOPHILA DEVELOPMENT; GAP GENES; HOMEOBOX GENES AND HOMEODOMAIN PROTEINS.

anterior–posterior axis The long axis of an animal embryo, that is, cranial to caudal or head to tail. Also often written anteroposterior.

anterograde movement/transport The movement of material from the cell body to the more distal areas of a neuron.

antheridiogens A group of naturally occurring compounds (Fig. A39), produced by the prothallia of schizaeaceous ferns, which induce the formation of antheridia (male sexual organs). They also induce spore germination (which usually requires light) in the dark and inhibit the formation of archegonia (female sexual organs). Those characterized so far are all structurally related to the GIBBERELLINS, and include the methyl esters of simple gib-

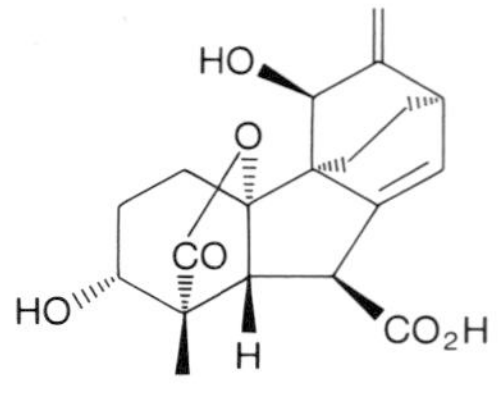

Fig. A39 Structure of antheridic acid, an antheridiogen produced by *Anemia* species.

berellins, such as GA$_9$ and GA$_{73}$, from *Lygodium japonicum*, and gibberellins in which the C and D rings have been rearranged, from *Anemia* species. They are among the most active plant hormones; GA$_{73}$ methyl ester induces antheridial formation in *L. japonicum* at 10^{-14} M.

Yamane, H. (1991) In *Gibberellins* (Takahashi, N. et al., Eds) 378–388 (Springer-Verlag, Berlin).

anthocyanins Water-soluble glycosides of hydroxylated 2-phenyl-benzopyrilium salts (anthocyanidins), pigments responsible for many of the pink, red, blue and black colours of flower petals, pollen, leaves and fruits.

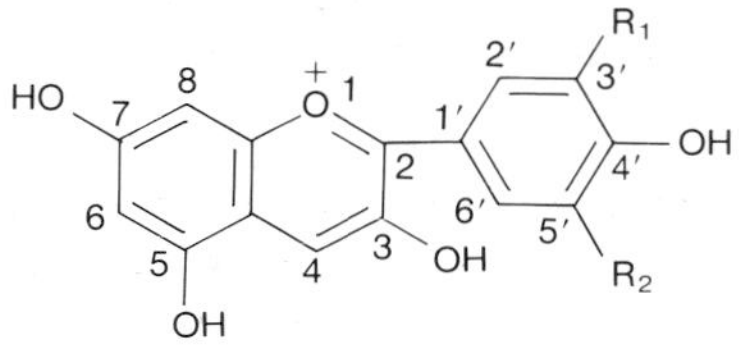

R$_1$, R$_2$ = H, OH, or OCH$_3$.

The 3-OH and, more rarely, 5-OH groups are glycosylated. The range of colours is due to different anthocyanidins (R$_1$, R$_2$ vary), different sap pH, complexes with metal ions and copigmentation with other flavonoids. They accumulate in VACUOLES of plant cells.

Antibodies

ANTIBODIES (immunoglobulins) are multifunctional GLYCOPROTEINS produced by the immune system of vertebrates and are essential for the prevention and resolution of infection by microorganisms. They are structurally highly variable molecules that carry out this function by recognizing and binding to particular molecular configurations on invading microorganisms and their products, each antibody being able to bind only one or a small number of related molecular configurations [1–3]. Any molecule or material bound specifically by an antibody is termed an antigen, and many different substances can act as antigens, not only those found on pathogenic microorganisms. The protective role of antibody is determined by the ability of specific antigen–

antibody complexes to activate one or more of the many effector mechanisms that contribute to the neutralization, destruction and elimination of the infecting microorganism.

Antibodies are secreted by the B CELLS of the immune system after terminal differentiation to PLASMA CELLS (*see* B CELL DEVELOPMENT). The molecular mechanisms that result in the generation of an antibody REPERTOIRE of almost infinite antigen recognition diversity are described in the articles on GENERATION OF DIVERSITY, IMMUNOGLOBULIN GENES and GENE REARRANGEMENT elsewhere in this volume. This article focuses on the basic structure and function of antibodies; further details of the structure of the immunoglobulin antigen-binding site at the molecular level will be found in IMMUNOGLOBULIN STRUCTURE.

The immunoglobulin fraction of blood consists naturally of a polyclonal mixture of many different types of antibodies specific for the different antigens the individual has been exposed to (i.e. they originate from many different clones of antibody-producing cells). MONOCLONAL ANTIBODIES are antibodies of a single specificity (i.e. they originate from a single clone of antibody-producing cells). Our current understanding of the structure and function of antibody molecules has resulted from study of specific antibody isolated from the sera of immunized experimental animals, or of polyclonal and monoclonal immunoglobulin of unknown antigen specificity. Long before the advent of recombinant DNA techniques, the isolation and purification of naturally occurring monoclonal immunoglobulins — the products of murine PLASMACYTOMAS or human MYELOMA — allowed the definition of the immunoglobulin isotypes (see below) and the generation of a library of amino-acid sequence data from which the structural basis for antibody specificity was determined.

The elucidation of antibody structure

The four-chain structure

The introduction of the technique of ion-exchange CHROMATOGRAPHY in the 1950s allowed rabbit IgG, the most abundant class of immunoglobulin in immunized animals, to be purified from serum in a single step. As specific antibody responses were also readily induced in rabbits, this laboratory animal was used in many of the early definitive investigations. Studies of the pattern of protein fragmentation following digestion of antibody with the enzyme papain, by Rodney Porter, and of the polypeptide chain components, by Gerry Edelman, led Porter to propose the four-chain structure for rabbit IgG in 1959 (see Fig. A40). This model proposed that each individual IgG molecule is composed of two light chains of M_r 25 000, of identical amino-acid sequence, and two heavy chains of M_r 50 000 also of identical amino-acid sequence. The light chains are covalently bound to the heavy chains through a single disulphide bridge and the heavy chains are similarly covalently linked to each other through one or more disulphide bridges (the number of disulphide bridges between heavy chains varies with isotype and species). Reduction of the disulphide bridges does not result in dissociation of the molecule, under physiological conditions, owing to the presence of multiple light–heavy and heavy–heavy interchain noncovalent bonds.

The 150K IgG molecule could be digested by papain to yield

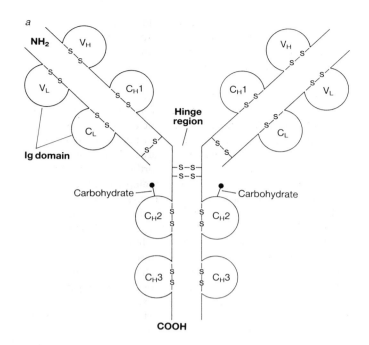

a

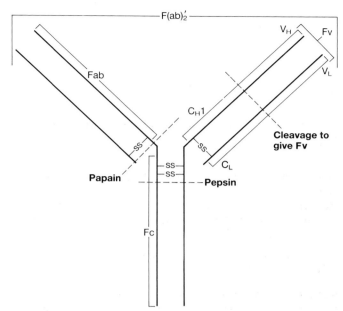

Fig. A41 Fragments of an antibody molecule obtained by enzymatic digestion. Digestion by papain produces two types of fragment: Fc, fragment crystallizable; Fab, fragment antigen binding (containing one antigen-binding site). Digestion with pepsin produces F(ab)′, a fragment containing two antigen-binding sites. Fv, the $V_L V_H$ heterodimer, can also be produced by cleavage or is now obtained by genetic engineering.

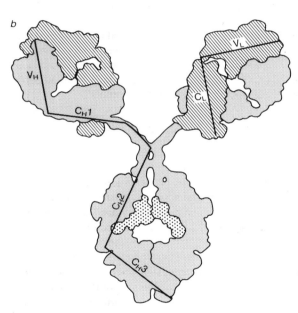

Fig. A40 *a*, Schematic representation of an immunoglobulin molecule (IgG). V_H and V_L are the variable domains of heavy and light chains respectively. C_L is the light chain constant region. C_H1, C_H2, and C_H3 are heavy chain constant region domains. S–S, disulphide bond. See text for further information. *b*, Diagrammatic representation of a human IgG molecule. The heavy chains are shaded in grey, the light chains hatched, and carbohydrate stippled (*see also* Plate 6*d*).

50K fragments (Fig. A41). Dialysis of the digestion products resulted in the formation of protein crystals that accounted for one-third of the original protein mass; this was named the Fc fragment as it was the 'fragment crystallizable'. The remaining fragments, accounting for two-thirds of the original protein mass,

were shown to bind antigen in an equivalent manner to the original molecule and these fragments were named the Fab fragments as they were antigen binding. Each antibody monomer has two identical antigen-binding sites and is thus divalent. The open and flexible polypeptide structure between the Fab and Fc regions is referred to as the hinge region. The equivalent regions of all antibody molecules are referred to by this common nomenclature although their Fc regions may never have been isolated in crystalline form. Rodney Porter and Gerry Edelman were awarded the Nobel Prize in 1972 for their pioneering studies that allowed the basic structure–function characteristics of antibody molecules to be understood.

Immunoglobulin (antibody) isotypes

In humans and the higher mammals five classes of immunoglobulin have been defined: IgG, IgM, IgA, IgD, and IgE and additionally, in humans, four subclasses of IgG and two subclasses of IgA. Since immunoglobulins of each class and subclass are present in all normal individuals they are referred to as isotypes (as distinct from the immunoglobulin ALLOTYPES and IDIOTYPES). The five different classes are determined by the type of heavy chain involved (termed γ, μ, α, δ, and ε respectively). Originally distinguished by immunological criteria (i.e. as distinct antigens), each isotype was later shown to be characterized by its amino-acid sequence and to be the product of a different gene segment (C genes, *see* IMMUNOGLOBULIN GENES).

The logic of immunoglobulin nomenclature may not be readily apparent and arises from the methods initially used to distinguish the different classes. Initially, two distinct forms of antibody were recognized — the high molecular weight macroglobulins and the lower molecular weight gammaglobulins — hence IgM and IgG (gamma refers to electrophoretic mobility). When a further class was recognized it was decided to systematize the nomenclature and to call this new class IgA and it was suggested that IgM and IgG might be redesignated IgB and IgC. Thus, discovery of two further classes gave us IgD and IgE. However, the suggestion to rename IgM and IgG was never adopted. Two types of human immunoglobulin light chain were also defined by their distinct antigenicity and named kappa (κ) and lambda (λ) to denote their discovery by Korngold and Lipari.

Constant and variable regions

In the 1960s it became appreciated that BENCE-JONES PROTEIN, excreted in the urine of patients with multiple MYELOMA, represents an overproduction of light chain by the malignant clone of cells. Consequently it was homogeneous and readily purified for structural analysis. Sequence analysis of Bence-Jones protein from different individuals revealed that amongst κ chains the amino-acid sequence of the N-terminal 110 residues was always different whereas the sequence of the C-terminal 110 residues was constant. These regions or domains were therefore termed the variable (V) region (in this case V_κ) and the constant (C) region (C_κ); λ chains are similarly composed of V and C regions. Comparison of the sequences of heavy chains of the same isotype showed that they also comprised a variable N-terminal region of 110 amino-acid residues whereas the remainder of the sequence is constant and defines the isotype (Fig. A40*a*). Minor sequence differences between the C regions of immunoglobulins of the same isotype from different individuals allow the definition of allotypes (allelic variants of each C gene), first defined serologically using ALLOANTISERA.

In 1969 Edelman's group published the first complete covalent structure of an antibody molecule (i.e. the amino-acid sequences of the light and heavy chains). Examination of the amino-acid sequence revealed repeating motifs of ~110 amino-acid residues. Each is termed an immunoglobulin homology region or immunoglobulin domain (Fig. A40) and is now known to be the product of an individual EXON (*see* IMMUNOGLOBULIN GENES). This finding also suggested that the prototype immunoglobulin molecule may have been a protein of 110 amino acid residues and that subsequent gene duplications and the evolution of mechanisms for joining (splicing) gene segments could account for the emergence of antibody molecules (*see* IMMUNOGLOBULIN SUPERFAMILY). The hinge region was seen to have a distinct characteristic sequence which included the inter-heavy chain disulphide bridges and to be rich in proline residues, thus accounting for the open structure. The Fab and Fc regions are compact globular structures each composed of four paired homology domains.

Each domain has a characteristic tertiary structure that is referred to as the immunoglobulin fold (Fig. A42). It is comprised of two surfaces of antiparallel β-pleated sheet linked through a disulphide bridge. Hydrophobic residues are packed into the internal space and hydrophilic residues are exposed on the surface of the molecule.

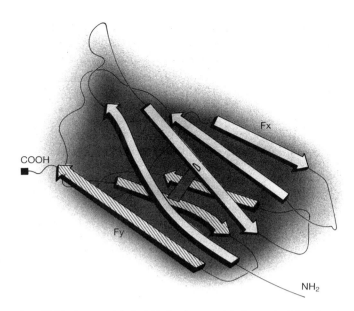

Fig. A42 The immunoglobulin fold of an immunoglobulin constant domain. The upper Fx face is formed by four strands of antiparallel β-pleated sheet; the lower Fy face by three. The hydrophobic core is stippled grey; the outer surface is hydrophilic.

Antigen recognition/binding

Comparisons of V region sequences showed that variability was not uniformly distributed but concentrated into three areas that were named the HYPERVARIABLE REGIONS. The crystal structure of an antigen-binding Bence-Jones protein demonstrated that the residues of the hypervariable regions (e.g. residues 25-35, 50-55, 95-100 in the light chain) are contact residues for antigen binding and hence they are now referred to as the COMPLEMENTARITY DETERMINING REGIONS (CDR) to indicate their functional role. It was appreciated that the more constant sequence of residues outside of the CDRs was required to maintain the essential immunoglobulin fold which results in the CDRs being brought into three-dimensional proximity with each other. These residues are referred to as the framework regions. The same pattern is observed for heavy chains, and in the intact antibody molecule the light and heavy chain CDRs form a continuous surface for contact with antigen (*see* IMMUNOGLOBULIN STRUCTURE).

Idiotypes

The unique structural features of the antigen-binding site of an antibody molecule, generated by the CDRs, may be antigenic within the same or heterologous species and result in the production of anti-idiotypic antibodies. The idiotype of an antibody molecule is the sum of all of the individual antigenic determinants (idiotopes) related to the binding site structure. The demonstration of anti-idiotypic responses within the same strain of mice and/or the production of auto anti-idiotypic responses led Niels Jerne to propose that immune regulation may be determined by a network of idiotype–anti-idiotype interactions [4].

Further analysis of framework sequences allowed the identification of immunoglobulins having a high degree of homology such that they could, conceivably, be encoded by a common germ-line gene requiring only POINT MUTATIONS to encode for the observed sequence. These related sequences were taken to define V gene subgroups (*see* IMMUNOGLOBULIN GENES) and antisera were also developed that allowed their serological detection.

Genetically engineered antibodies

Occasionally, it has been possible to derive a subfragment of the Fab fragment — the Fv fragment — composed of the light and heavy chain V regions only. It is a heterodimer formed by noncovalent bonds and having the same antigen-binding specificity and affinity as the parent molecule. Such molecules can now be readily constructed using genetic engineering techniques [5]. The finding that the V region of the heavy chain makes the major contribution to the antigen-binding affinity has led to the production of recombinant proteins comprising the variable region of the heavy chain only (Hv fragments) as antigen specificity reagents.

As evidenced above, GENETIC ENGINEERING techniques have been widely applied to antibody molecules. Our need to understand fully the profile of biological activities of each of the human antibody isotypes has been significantly advanced by studies using chimaeric immunoglobulins. They are constructed using murine genes encoding antibody specificity (V regions) and human C region genes encoding the human isotypes. A further development allows murine antibodies to be 'humanized' by 'grafting' the CDR of specific murine antibodies (*see* MONOCLONAL ANTIBODIES) into the desired human isotype background. This technology finds its most appropriate application when the antibodies are used for *in vivo* diagnostic or therapeutic purposes. Conservative PROTEIN ENGINEERING approaches are being applied to alter the specificity or affinity of antibody molecules and to modulate Fc effector functions. This will permit the rational design of 'customized' antibody molecules having a profile of activities considered to be optimal for a given application.

Antibody effector functions

Antibodies may be regarded as transducer molecules that relay a 'message' to an effector system. The message is the presence of antigen in the form of specific immune complexes (antigen–antibody complexes). Cellular receptors specific for the Fc region of each of the antibody classes (FC RECEPTORS) have been reported on a variety of cells. Those expressed on T CELLS are thought to be involved in immune regulation but they are structurally and functionally poorly defined. Also ill defined is the placental Fc receptor that facilitates specific transport of IgG from the mother to the foetus. By contrast Fc receptors expressed on leukocytes are well defined and characterized. There are three Fc receptors for human IgG, distinguished by the preferred notation huFcγRI, huFcγRII, and huFcγRIII (hu standing for human); receptors for IgE and IgA are similarly referred to as huFcεRI, huFcεRII, and huFcαR respectively. Interactions of antigen–antibody complexes with Fc receptors on macrophages may activate the release of

proteolytic enzymes, reactive oxygen species and phagocytosis (*see* ENDOCYTOSIS). Complexes of antigen with IgG, IgM or IgA may activate COMPLEMENT through one or both of two pathways; the classical pathway initiated by the activation of C1 and the alternative pathway initiated by the activation of C3. The ensuing cascade of reactions may result in the release of inflammatory mediators, phagocytosis via complement receptors, lysis of sensitized cells or bacteria by the production of holes in their membranes etc. The interaction site for the first component of complement, C1, has been mapped to the Fc region of IgG and IgM isotypes.

The activation of the IGE RECEPTOR on mast cells results in the release of chemical mediators such as histamine and is responsible for immediate hypersensitivity reactions and allergies.

R. JEFFERIS

See also: T CELL RECEPTORS; T CELL RECEPTOR GENES.

1 Burton, D.R. (1985) Immunoglobulin G: Functional sites. *Mol. Immunol.* **22,** 161–206.
2 Jefferis, R. (1991) Structure–function relationships in human immunoglobulins. *Netherlands J. Med.* **39,** 188–198.
3 Roitt, I. (1991) *Essential Immunology*, 7th edn (Blackwell Scientific Publications, Oxford).
4 Jerne, N.K. (1984) Idiotype networks and other preconceived ideas. *Immunol. Rev.* **79,** 5–24.
5 Winter, G. & Milstein, C. (1991) Man-made antibodies. *Nature* **349,** 293–299.

antibody combining site The antigen recognition and binding site on an ANTIBODY, the region of the molecule which binds a specific antigenic determinant (EPITOPE) owing to molecular complementarity. The combining site is formed by the noncovalent association of the N-terminal domains of a light and a heavy chain such that the hypervariable or COMPLEMENTARITY DETERMINING REGIONS are brought together in space to form a continuous protein surface. Each Fab region of an antibody molecule contains one combining site. *See*: IMMUNOGLOBULIN STRUCTURE.

antibody diversity *See*: GENERATION OF DIVERSITY.

antibody engineering The manipulation or mutation of the DNA that encodes antibodies to generate molecules not found in nature. A minimally engineered antibody may result from a single base change leading to a single amino acid replacement in the protein with a change in functional activity, for example, antigen specificity, affinity or an interaction site for triggering of effector functions. Humanizing of animal antibodies represents maximally engineered molecules in which the specificity of an animal antibody is 'transferred' to a human immunoglobulin by exchange of all the COMPLEMENTARITY DETERMINING REGIONS.

Winter, G. & Milstein, C. (1991) *Nature* **349,** 293–299.

anticoding strand The strand of a double-stranded DNA that is transcribed into RNA during TRANSCRIPTION.

anticodon The sequence of three nucleotides in a TRANSFER RNA molecule which is complementary to a specific CODON in mRNA.

Codon–anticodon recognition is by complementary BASE PAIRING but can involve some nonstandard base pairing between the base in the first position of the anticodon and the base in the third position of the anticodon (*see* WOBBLE). Some tRNAs contain modified bases in their anticodon which influence the specificity of codon–anticodon recognition. *See*: PROTEIN SYNTHESIS.

antidepressant drugs Drugs used in the treatment of depression (*see* AFFECTIVE DISORDERS). The first effective antidepressants were inhibitors of MONOAMINE OXIDASE (MAO). However the most widely used have been the 'tricyclic' antidepressants, named after the chemical structure of the original members of this group. These drugs block neuronal re-uptake of NORADRE-NALINE, and in some cases also 5-HYDROXYTRYPTAMINE (5HT), thereby diminishing the rate of removal of neurotransmitter from the synapse (*see* SYNAPTIC TRANSMISSION). However other actions, such as blockade of α_2-ADRENERGIC RECEPTORS, have been suggested to be important for some of these compounds. More recently introduced antidepressants include reversible inhibitors of a subtype of MAO, selective inhibitors of 5HT neuronal re-uptake and antagonists of various 5HT RECEPTORS. Tryptophan too has an antidepressant action, presumably by increasing the synthesis of 5HT.

antidiuretic hormone (ADH) VASOPRESSIN.

antigen Any substance or material that is specifically recognized by ANTIBODY or a T CELL RECEPTOR. A distinction may be made between an antigen and an IMMUNOGEN; the latter is a substance or material that induces a specific immune response whereas an antigen may be a structurally related substance or material that is recognized through a cross reaction. In practice the terms are often used synonymously. Secreted antibodies and their corresponding membrane-bound forms can recognize a wide variety of substances as antigens whereas T cell receptors can only recognize fragments of proteins complexed with MHC MOLECULES on cell surfaces. Antigens recognized by immunoglobulin receptors on B cells are subdivided into three categories: T cell-dependent antigens; type 1 T cell-independent antigens; and type 2 T cell-independent antigens (*see* B CELL DEVELOPMENT). *See also*: EPITOPE; HAPTEN.

antigen-binding site Site on antibody molecule which recognizes and complexes with specific antigen. Also known as the antibody combining site. *See*: ANTIBODIES; IMMUNOGLOBULIN STRUCTURE.

antigen presentation The formation of a noncovalent interaction between foreign antigen and MHC class I or class II molecules at the surface of a cell, which allows recognition of foreign antigen by MHC-restricted T CELLS and the generation of an antigen-specific immune response (*see* MAJOR HISTOCOMPATIBILITY COMPLEX; T CELL RECEPTOR). The term refers to the interaction of foreign antigen with class II MHC molecules on specialized ANTIGEN-PRESENTING CELLS such as macrophages, or describes the association of viral peptides (or other foreign antigens generated intracellularly) with class I MHC molecules on virus-infected cells. Formation of antigen–MHC complexes requires that large protein antigens be enzymatically processed within the antigen-presenting cell and re-expressed at the cell surface as smaller peptide fragments that then associate with class I or II MHC molecules. Peptide fragments bind to MHC molecules in a groove in the surface. Although each MHC molecule seems to be able to bind a great variety of different peptide fragments, allelic polymorphism in MHC class I and II protein sequences influences the affinity of a particular antigen fragment for a given MHC molecule, and thus also the ability of a particular combination of foreign antigen and MHC to elicit a T cell response. *See*: ANTIGEN PROCESSING AND PRESENTATION.

antigen-presenting cell (APC) Any of a number of different cell types that can process foreign protein antigens and express them as peptide fragments complexed with class II MHC molecules at the cell surface (*see* MAJOR HISTOCOMPATIBILITY COMPLEX) and are thus capable of activating antigen-specific T cells and generating an immune response (*see* T CELL ACTIVATION; T CELL RECEPTOR). Most 'professional' APCs are of the macrophage/monocyte lineage, but B cells, thymocytes and other specialized cells such as dendritic cells (dendrocytes) also express class II MHC molecules and act as antigen-presenting cells (*see* T CELL ACTIVATION; T CELL DEVELOPMENT). Whereas these cells express class II molecules constitutively, nearly all cells can be induced to express high levels of class II MHC molecules in culture by the addition of the LYMPHOKINE γ-interferon, suggesting that a locally high concentration of γ-interferon *in vivo* may permit otherwise class II-negative cells to function as APCs.

Antigen processing and presentation

THE T lymphocytes (T CELLS) of the immune system can only recognize ANTIGEN in the form of short peptides derived (processed) from the native protein antigen. In addition, they only recognize such peptides as antigenic when these are associated at a cell surface with self glycoproteins encoded by the MAJOR HISTOCOMPATIBILITY COMPLEX (MHC). These peptide–MHC complexes lie at the heart of immune, and most likely many autoimmune, responses. Much current research is therefore aimed at understanding the cell biological and biochemical basis of their generation and expression.

Class I and class II MHC molecules

The MHC encodes two main types of cell-surface glycoprotein: the MHC class I and class II molecules. These are highly POLYMORPHIC in mammals. Polymorphism in MHC antigens is medically inconvenient as it leads to difficulties in organ transplantation, but the polymorphism is presumed to have been selected to increase the peptide binding versatility of the population as a whole and thus its overall resistance to pathogens.

Class I MHC molecules are composed of a membrane-spanning subunit (the heavy chain) which is noncovalently associated with β_2-microglobulin, and is expressed on almost all cell types. The three-dimensional structure of a class I MHC molecule reveals a cleft in which most polymorphic residues are clustered and which binds peptide. Class II MHC consists of two membrane-spanning polypeptides which form a peptide-binding site very similar to that on class I (*see* Plate 6*d*). Normally, class II MHC expression is limited to certain types of so-called 'professional' ANTIGEN-PRESENTING CELL (APC) such as B lymphocytes (B CELLS), macrophages, dendritic cells, and in humans, activated T cells.

The affinity of T CELL RECEPTORS for antigen is thought to be relatively low, so productive interactions between T cells and MHC-expressing APCs also require one of the two co-receptor molecules that T cells may express (i.e. CD4 or CD8) as well as other adhesion and co-stimulatory molecules. T cells expressing CD4 interact with APCs expressing class II MHC whereas T cells expressing CD8 interact with class I.

These T cell subtypes execute different effector functions on ligation of their antigen receptors. CD8 cells are generally cytotoxic and can kill target cells through exocytosis of lytic granules, whereas CD4 cells have a helper/inducer function mediated at least in part by LYMPHOKINE secretion. It therefore makes sense to express class I on all cell types (as all cells can become virus infected) and to limit class II expression to those cell types with the capacity to proliferate and differentiate in response to T cell lymphokines.

Class I and class II MHC molecules present peptides from different cellular compartments

Early studies on antigen presentation revealed a distinct sensitivity of the process to metabolic inhibitors. Presentation on class I MHC molecules was usually sensitive to inhibitors of PROTEIN SYNTHESIS but was insensitive to drugs such as CHLOROQUINE which block proteolysis in LYSOSOMES. In contrast, presentation on class II MHC molecules was sensitive to chloroquine but rather insensitive to protein synthesis inhibitors [1,2]. This suggested that quite distinct mechanisms were responsible for the generation of peptides captured by class I compared with class II molecules.

It was found that class I MHC molecules were able to present peptides derived from proteins present in cytosolic, nuclear, and even mitochondrial compartments, but could not be loaded when the native protein was taken up into the endosome/lysosome vacuolar system [1]. In contrast, class II MHC molecules seemed only able to present peptides derived from proteins normally found within, or which could gain access to, the endosome system [3].

Presentation of cytosolic proteins on class II MHC molecules has subsequently been demonstrated, however (see Fig. A43), although it is not clear at what stage in class II MHC trafficking these peptides are loaded. Also, presentation of exogenous antigens on class I MHC molecules may be a specialized function of certain cell types and can be demonstrated by engineering the osmotic lysis of endosomes and thus delivery of antigen to the cytosol. MHC molecules therefore provide a constant update in peptide form of cellular protein content (on class I) and environmental protein composition (on class II) for screening by the T cell system.

Given the distinct effector functions of the T cells which recognize class I and class II MHC, this very striking discrimination among the cellular compartments that can contribute to the peptide repertoire of class I and class II makes sense: for example, if class I MHC molecules readily presented endocytosed viral proteins, uninfected cells as well as infected cells might easily become targets for killing by cytotoxic T cells. Similarly, presentation on class II MHC of peptides from circulating pathogens (which are taken up by ENDOCYTOSIS) is optimized by preventing loading of class II MHC with cytosolic peptides.

How class I and class II MHC molecules control which peptides become bound is now reasonably well understood and is outlined below. First, information on the physical characteristics of the peptides bound is briefly summarized.

The 'endpoint' of antigen processing

The peptides bound to class I and class II differ in length and in other respects (see below) which again suggests that different processing mechanisms are involved. Most naturally processed T cell EPITOPES presented on class I molecules are eight or nine amino-acid residues long, the exact length depending on the class I allele. Sequencing of bulk mixtures of peptides eluted from milligram amounts of purified class I molecules revealed a strong preference for certain residues at certain positions, and these have been termed 'anchor residues' [4]. For example, on the mouse K^b class I molecule, which binds nonamers, tyrosine was always found at position 2 and either isoleucine or leucine at position 9. Other *in vitro* studies show that increasing or decreasing the peptide length by even a single residue dramatically increases the dissociation rate of peptide from class I molecules and consequently increases the instability of the complex [5]. Co-crystals of mouse class I MHC molecules and peptide reveal the importance of the peptide N and C termini as well as the anchor residues in making interactions with the class I binding pocket [6]. Significantly, these crystal structures also underline the importance of interactions between the peptide backbone and side chains from the binding pocket; that is, features common to all peptides contribute significantly to stable binding. There is nonetheless a strong selection during the assembly and maturation of class I molecules for peptides of the right length which also carry appropriate anchor residues. Different alleles of class I select or 'restrict' binding to peptides with distinct anchor motifs. The extent to which peptide delivery for class I binding may also be controlled at the level of peptide production, transport into the ENDOPLASMIC RETICULUM (ER), loading within the ER, or at a post-binding trimming step is not known at the time of writing.

In contrast, class II MHC molecules bind longer peptides of variable length (13–25 residues). Unlike class I, a single peptide determinant can be expressed in different versions that share a common core sequence but display 'ragged' N and C termini [7]. This suggests that, unlike the case of class I MHC binding, peptides of different length but with a common core sequence are all stably bound to class II MHC molecules. Presumably, interactions with the peptide N and C termini do not contribute to stable binding as they do for class I-associated peptides. Heterogeneity

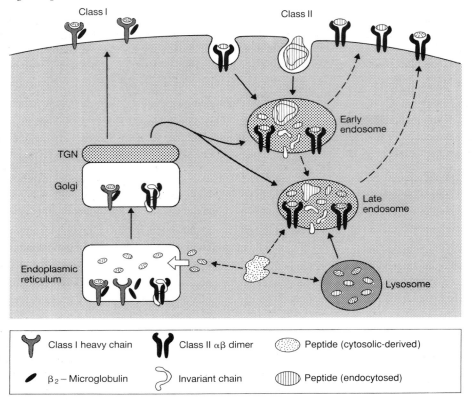

Fig. A43 Schematic representation of processing pathways in an antigen-presenting cell. Protein antigen produced in the cytosolic compartment of a cell (e.g. as a result of virus infection) is processed into peptides which enter the endoplasmic reticulum (ER) (and possibly lysosomes) (dashed arrows indicate pathways not yet established in detail). Class I MHC molecules fold together with β_2-microglobulin in the secretory pathway, an association which is greatly stabilized by binding by the specific peptides thought to be delivered into the ER by the TAP gene products (see text). The peptide–class I MHC complex is transported through the Golgi apparatus and from there to the cell surface and/or to endosomes. Antigen entering the cell by endocytosis is degraded into peptide fragments in endosomes, where it binds to class II MHC molecules. Class II molecules are $\alpha\beta$ dimers which assemble in the ER in a complex with the variant chain (Ii) which does not bind peptides. After transport through the Golgi apparatus the $\alpha\beta$Ii complex is targeted to the endosome/lysosome system where the invariant chain is removed by proteolytic cleavages. Capture of peptides by class II MHC molecules is favoured at the acidic pH of the endosomes. Stable complexes then appear on the cell surface. Neither the compartment in which peptides are loaded onto class II MHC molecules, nor the route that assembled complexes take to the cell surface is certain. Some cytosolic antigens can be presented on class II MHC although it is not clear at which point they feed into this pathway. Adapted from [22].

in peptide length is consistent with the possibility that the peptide-binding groove on class II molecules may be open at both ends, whereas that on class I is closed. Individual peptides bound to class II molecules [8] share some features analogous to the anchor residues identified on class I MHC molecules and these presumably contribute to stable binding.

Abundant peptides recovered from complexes with class II MHC molecules include those from extracellular proteins such as serum albumin, transferrin receptors, and peptides derived from class II MHC molecules themselves. Technical difficulties currently prevent analysis of the numerous peptides present in lower abundance. For class II MHC molecules this includes most T cell epitopes generated following antigen uptake and processing.

Cell biological basis of class I and class II MHC antigen presentation

Why do class I and class II MHC show a strong preference for peptides from endogenous and exogenous antigens respectively?

There is now a satisfactory general explanation although the detailed mechanisms have yet to be unravelled.

1 Class I MHC. A key point to emerge is that peptide binding to class I MHC molecules is obligatory for stable expression of the MHC molecule on the cell surface. Mutant cells in which antigenic peptides cannot be transported from the cytoplasm into the ER fail, in most cases, to express normal levels of surface class I molecules [9]. These cells have mutations in genes mapping within the MHC which are thought to encode 'transporters of antigenic peptides' (TAP genes). The dependence of class I transport on peptide binding can be explained by the finding that at 37°C class I heavy chain does not associate stably with β_2-microglobulin in the absence of an appropriate peptide. Consequently, the 'quality control' system (*see* BiP; MOLECULAR CHAPERONES) operating in the secretory pathway prevents such 'empty' molecules from reaching the cell surface (*see* PROTEIN SECRETION). Those that do are unstable and are quickly degraded [10].

The homology with other peptide transporters in the so-called

ABC (ATP-binding cassette) family (e.g. the CYSTIC FIBROSIS gene product, the multidrug resistance P GLYCOPROTEIN, and the STE6 gene product in yeast), together with evidence that their polymorphism in the rat gives rise to distinct sets of peptides bound to a single class I allele (see below), and the fact that expression of introduced *TAP* genes in the mutant cells restores surface expression of class I molecules [11], strongly suggests that the TAP proteins transport peptides for loading of class I molecules. Other genes, homologous to subunits of the multicatalytic cytosolic proteinase or 'proteasome' have also been found within the MHC (*see* PROTEIN DEGRADATION). The precise contribution of these proteins to production and delivery of peptides for loading of class I remains to be firmly established.

Class I molecules therefore arrive on the cell surface fully occupied and moreover are routed there directly from the *trans*-Golgi network [12]. Binding seems to be essentially irreversible, so that even if class I molecules subsequently enter the endocytic pathway they do not usually bind the peptides generated there. This protects cells from sensitization by passively acquired antigen and killing by cytotoxic T cells.

2 Class II MHC. As class II MHC molecules share their early biosynthetic route with class I molecules an explanation is required (1) for the lack of binding of peptides to class II at the stage when class I molecules bind peptides, and (2) for the ability of class II molecules efficiently to capture peptides in the endosome/lysosome system. Both these features of peptide binding seem to be explained by an association of class II molecules with a third polypeptide encoded outside the MHC locus and known as the INVARIANT CHAIN. The invariant chain associates with class II molecule α and β chains shortly after biosynthesis and chaperones the dimer through most of the secretory pathway but does not accumulate to significant levels on the cell surface. Experiments *in vitro* have shown that the invariant chain directly or indirectly prevents peptide binding to class II MHC molecules [13]. Moreover, it is responsible for targeting class II MHC molecules to their eventual site of peptide loading within the endocytic pathway. The targeting signal has been mapped to a short region of the cytosolic domain of the invariant chain [14]. The class II–invariant chain complex is therefore diverted out of the secretory pathway, probably at the *trans*-Golgi network, and delivered to the endosome system. Here, the invariant chain is removed by proteolysis thus generating for the first time a functional binding site for peptides on the αβ dimer. The acidic pH within the endosome/lysosome system may also optimize peptide binding. As with class I MHC, peptide binding seems to stabilize class II MHC molecule structure to such an extent that a proportion of mature cell-surface class II αβ dimers remain associated in normally denaturing detergents such as sodium dodecyl sulphate [15].

The essential features of class I versus class II traffic are summarized in Fig. A43 and explain how class I and class II MHC molecules bind peptides present within the secretory and endocytic pathways respectively. As in the case of class I MHC, most cell-surface class II MHC molecules are occupied by peptide and there is some evidence that failure to acquire peptide leads to degradation, in this case in lysosomes [15]. However, cells carrying deletions within the MHC have been described which fail to present native antigen to T cells but still express cell-surface class II MHC molecules [16]. The class II molecules on these cells are conformationally abnormal but the fact that they appear at all suggests that the coupling of cell surface expression to peptide capture and the attainment of a mature conformation is not as strict for class II as it is for class I molecules. These interesting mutants seem to have lost a crucial but as yet ill-defined function for normal class II maturation.

The overall picture from work on both the structure and the biosynthesis of MHC molecules shows that peptide should be considered an integral subunit of the assembled mature MHC molecule. The immune system has exploited features of the folding and maturation of MHC molecules and cellular trafficking pathways to ensure the display of appropriate peptides to either CD4 or CD8 T cells.

Factors modulating antigen processing and presentation

A central factor limiting antigen presentation on MHC molecules is their biosynthetic output. Antigen presentation on both class I and class II molecules depends on new MHC molecule biosynthesis [3,17]; a cell expressing MHC molecules but not actually making them is not able to present new antigens although it may continue to display the peptides from previously encountered antigens for many days. Unlike other peptide-binding proteins, such as ATP-regulated intracellular molecular chaperones (*see* MOLECULAR CHAPERONES), MHC molecules in living cells seem to have an essentially monogamous relationship with peptide [5,18]. The biological importance of this persistence of peptides on MHC molecules is illustrated by the behaviour of dendritic cells such as those in the epidermis (Langerhans cells). These cells seem to separate a class II MHC biosynthetic and antigen-processing phase from a later phase in which presentation of pre-existing peptide–MHC antigen complexes persists, but presentation of subsequently offered antigen is ineffective in spite of abundant cell-surface class II molecules. *In vivo* the later phase may correlate with migration of Langerhans cells from sites of antigen capture to sites in lymphoid tissue where antigen is presented to T cells [19].

Other factors modulating the range of peptides presented are poorly characterized at present. Nonetheless, with the basic features now established, modulating influences are of particular interest and importance to immune and autoimmune responses. In the context of class I MHC, a particularly interesting illustration of how the set of peptides presented on the same class I molecule may be modulated comes from work on the rat MHC complex. Here the MHC-linked putative peptide transporter genes show striking polymorphisms which correlate with distinct sets of peptides bound to the same class I molecule. These results provide the best evidence for interaction between the putative peptide transporters and the peptides actually bound, and reveal an unexpected modifying influence on the peptide repertoire expressed on class I molecules [20]. It remains to be seen, for both class I and class II, whether polymorphisms in proteins involved in the peptide generation and loading process are going to be of general importance.

The processing machinery which gives rise to peptides associated with class II molecules is also poorly characterized as yet and the precise compartmentation of the assembly of class II–peptide complexes is still being worked out. Studies using antigen-specific B lymphocytes reveal that high-affinity ANTIBODIES can affect the course of antigen processing and as a result, its outcome in terms of expressed T cell epitopes ([21] and unpublished work). Antibodies may therefore steer the T cell response in a particular direction and so may have an influence on, for example, the effectiveness of vaccines and the dominant representation of certain T cell epitopes in a response.

C. WATTS

1 Townsend A. and Bodmer H. (1989) Antigen recognition by class I restricted T cells. *Annu. Rev. Immunol.* **7**, 601–624.
2 Unanue E.R. (1984) Antigen presenting function of the macrophage. *Annu. Rev. Immunol.* **2**, 395–428.
3 Morrison, L.A. et al. (1986) Differences in antigen presentation to MHC class I and class II-restricted influenza virus-specific cytolytic T lymphocyte clones. *J. Exp. Med.* **163**, 903–921.
4 Rotzschke, O. & Falk, K. (1991) Naturally occurring peptide antigens derived from the class I MHC-restricted processing pathway. *Immunol. Today* **12**, 447–451.
5 Cerundolo, V. et al. (1991) The binding affinity and dissociation rates of peptides for class I MHC molecules. *Eur. J. Immunol.* **21**, 2069–2075.
6 Matsumura, M. et al. (1992) Emerging principles for the recognition of peptide antigens by MHC class I molecules. *Science* **257**, 927–932.
7 Hunt, D.F. et al. (1992) Peptides presented to the immune system by the murine class I MHC molecule I-A^d. *Science* **256**, 1817–1920.
8 Hammer, J. et al. (1992) Identification of a motif for HLA-DR1 binding peptides using M13 display libraries. *J. Exp. Med.* **176**, 1007–1013.
9 Townsend, A. et al. (1989) Association of class I MHC heavy and light chains induced by viral peptides. *Nature* **340**, 443–448.
10 Schumaker, T. et al. (1990) Direct binding of peptide empty MHC class I molecules on intact cells and in vitro. *Cell* **62**, 563–567.
11 Kelly, A. et al. (1992) Assembly and function of the two ABC transporter proteins encoded in the human major histocompatibility complex. *Nature* **355**, 641–644.
12 Neefjes, J.J. et al. (1990) The biosynthetic pathway of class II but not class I MHC molecules intersects the endocytic pathway. *Cell* **61**, 171–183.
13 Roche, P. & Cresswell, P. (1990) Invariant chain association with HLA-DR molecules inhibits immunogenic peptide binding. *Nature* **345**, 615–618.
14 Bakke, O. & Dobberstein B. (1990) MHC class II-associated invariant chain contains a sorting signal for endosomal compartments. *Cell* **63**, 707–716.
15 Germain, R.N. & Hendrix, L.R. (1991) MHC class II structure, occupancy and surface expression determined by post-endoplasmic reticulum antigen binding. *Nature* **353**, 134–139.
16 Mellins, E. et al. (1990) Defective processing and presentation of exogenous antigens in mutants with normal HLA class II genes. *Nature* **343**, 71–74.
17 Davidson, H.W. et al. (1991) Processed antigen binds to newly synthesized MHC Class II molecules in antigen-specific B lymphocytes. *Cell* **67**, 105–116.
18 Lanzavecchia, A. et al. (1992) Irreversible association of peptides with class II MHC molecules in living cells. *Nature* **357**, 249–252.
19 Steinman, R.M. (1991) The dendritic cell system and its role in immunogenicity. *Annu. Rev. Immunol.* **9**, 271.
20 Powis, S. et al. (1992) Effect of polymorphism of an MHC-linked transporter on the peptides assembled in a class I molecule. *Nature* **357**, 211–215.
21 Davidson, H.W. & Watts, C. (1989) Epitope-directed processing of specific antigen by B lymphocytes. *J. Cell Biol.* **109**, 85–92.
22 Long, E. (1992) Antigen processing for presentation to CD4$^+$ T cells. *New Biologist* **4**, 274–282.

antigen receptors *See:* ANTIBODIES; B CELL DEVELOPMENT; IGM RECEPTOR; IMMUNOGLOBULIN STRUCTURE; T CELL DEVELOPMENT; T CELL RECEPTORS.

antigen recognition The binding of an ANTIGEN by specific receptors on T CELLS and B CELLS of the immune system, which triggers their activation, proliferation, and terminal differentiation to effector cells. *See:* ANTIBODIES; ANTIGEN PROCESSING AND PRESENTATION; B CELL DEVELOPMENT; IMMUNOGLOBULIN GENES; IMMUNOGLOBULIN STRUCTURE; MAJOR HISTOCOMPATIBILITY COMPLEX; T CELL DEVELOPMENT; T CELL RECEPTORS; T CELL RECEPTOR GENES.

antigenic determinant Molecular configuration on the surface of an antigen that is recognized by an ANTIBODY. *See also:* EPITOPE.

Antigenic variation in African trypanosomes

AFRICAN trypanosomes are unicellular eukaryotic parasites of mammals transmitted by tsetse flies [1–4]. The trypanosomes *Trypanosoma brucei rhodesiense* and *T. b. gambiense* cause sleeping sickness in humans. Laboratory experiments are mostly done with *T. b. brucei* strains that grow well in rodents (after syringe transmission), but not in humans. In the mammalian host trypanosomes are entirely covered by a dense surface GLYCOPROTEIN coat and ANTIBODIES directed against this coat lyse the trypanosome. By drastically changing the composition of the coat — antigenic variation — a subfraction of the trypanosomes escapes immune lysis, allowing maintenance of a chronic infection and the continued presence of trypanosomes in the bloodstream. A single trypanosome and its progeny can produce more than 100 coats that have no ANTIGENIC DETERMINANTS in common. Coat switching is not induced by antibody and occurs spontaneously at a frequency of 10^{-6} per division in established laboratory strains, but at much higher rates in the field.

The surface coat

The coat of bloodstream trypanosomes [2,5] consists of a single species of protein, the variant specific surface glycoprotein (VSG). VSGs range in size from 400 to 500 amino-acid residues and are anchored in the plasma membrane by a complex glycosylphosphatidylinositol (GPI) tail (*see* GPI ANCHORS). The three-dimensional structure of two VSGs has been determined at 2.9 Å resolution and the complex GLYCAN part of the tail is completely known. A remarkable feature of the GPI anchor is that its fatty acids are exclusively myristic acid (a saturated C_{14} fatty acid). This makes the trypanosome vulnerable to myristic acid analogues [6].

When the trypanosome enters the tsetse fly, it shuts off VSG synthesis, sheds its VSG coat and replaces it with a coat of another glycoprotein, known as procyclin or procyclic acidic repetitive protein (PARP). After completing its developmental cycle in the fly salivary gland, the trypanosome regains a VSG coat.

Unusual features of gene expression

Each messenger RNA in trypanosomes starts with the same

capped sequence of 39 nucleotides, called the mini-exon or spliced-leader sequence [7]. This sequence is acquired by *trans*-splicing from a 140-nucleotide donor RNA (*see* RNA SPLICING). *Trans*-splicing allows the trypanosome to convert uncapped POLY-CISTRONIC precursor RNAs into a series of capped mRNAs. In fact, nearly all TRANSCRIPTION UNITS cover more than one gene and units with more than 10 genes exist [1–4]. It is probably *trans*-splicing that allows trypanosomes to use ribosomal RNA promoters for making transcripts that are efficiently converted into mRNA [8,9]. In contrast, in mammals pre-mRNA capping is tightly coupled to TRANSCRIPTION by RNA polymerase II and transcripts made by RNA polymerase I from rRNA promoters cannot be capped and used for mRNA synthesis. No internal INTRONS have yet been found in trypanosome genes and it is not known whether *cis*-splicing occurs at all.

As in other simple eukaryotes, chromosomes do not become condensed during MITOSIS in trypanosomes. However, all the linear DNA molecules contained in trypanosome chromosomes are small enough to be separated by pulsed-field gel ELEC-TROPHORESIS, allowing simple gene allocation to individual chromosomes by Southern hybridization (*see* HYBRIDIZATION) [10]. A special feature of *T. brucei* is the presence of some 100 minichromosomes of 50–150 kilobases (kb).

Switching coats

A trypanosome carries $\sim 10^3$ different genes for VSGs, but usually only one of these is expressed at any time [1–4]. The expressed gene is invariably located near the end of a chromosome. Switching can occur in several ways, as diagrammed in Fig. A44:

1 The gene in an active VSG gene expression site (ES) is replaced by a copy of a different VSG gene. This replacement probably occurs by a GENE CONVERSION process, in which short blocks of homology play a part as indicated in Fig. A44*a* and *b*. Most silent VSG donor genes are in long arrays in a chromosome-internal

position (Fig. A44*a*), but some 10^2 genes are near TELOMERES, mostly in the minichromosomes. The sequences flanking silent telomeric genes are very similar to those flanking the VSG gene in the expression site and these genes can usually enter the expression site more efficiently than the silent chromosome-internal genes (Fig. A44*b*).

2 Switching on another telomeric VSG gene expression site and switching off the previously active one (Fig. A44*c*). There are at least 6 and possibly as many as 20 different expression sites. How expression sites are switched on and off remains unknown.

The complex structure of VSG gene expression sites

Figure A45 shows a simplified VSG gene expression site. In addition to the VSG gene, each site contains at least eight other genes, the so-called expression-site associated genes (ESAGs), cotranscribed with the VSG gene [1,5–7]. The PROMOTER of the expression site has been defined by DNA TRANSFECTION studies and is active both in plasmids and after reinsertion into chromosomes. This indicates that shut-off of promoters is regulated negatively (i.e. by proteins preventing activity) and in *cis* (i.e. by flanking sequences, or chromosome topology). An unusual feature of this promoter is its recognition by a RNA polymerase that is highly resistant to α-amanitin, whereas transcription of other protein-coding genes in trypanosomes (with the exception of the procyclin gene) is sensitive to α-amanitin. Probably this polymerase transcribing VSG gene expression sites is RNA polymerase I (the polymerase that only transcribes rRNA genes in other eukaryotes), but this remains to be proven [8,9].

Timing of VSG gene expression

Efficient antigenic variation requires [4,11]:
1 A sufficiently large repertoire of VSG genes;
2 A mechanism to express these genes one by one;

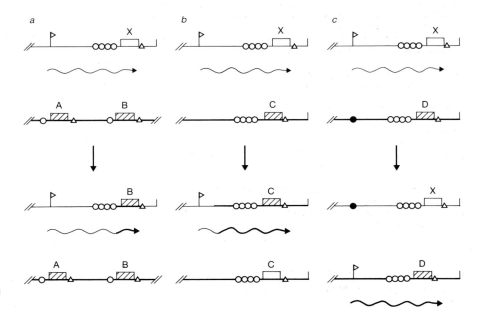

Fig. A44 The main pathways for activating a variant-specific surface glycoprotein gene in *T. brucei*. *a*, Duplicative transposition of a chromosome-internal gene; *b*, telomere conversion; *c*, activation *in situ* of an expression site. The boxes represent different VSG genes; A and B are chromosome-internal, C, D and X are telomeric genes. The flag indicates the active promoter of the telomeric VSG gene expression site, the filled circle an inactive promoter, the wavy line the pre-mRNA and the vertical bar the end of a chromosome. Open circles, Imperfect 70-bp repeats; triangle, short homology blocks in the 3′ end of the VSG genes and beyond.

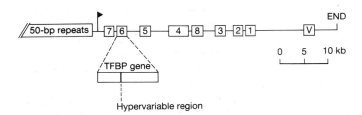

Fig. A45 Schematic representation of an active trypanosomal VSG gene expression site. The flag indicates the promoter, the numbers in the boxes indicate the ESAGs, V the VSG gene and END the chromosome terminus. ESAG-4 codes for an adenylate cyclase [16]; TFBP for the transferrin binding protein; the function of the other ESAGs and of the long block of imperfect 50-bp repeats is not yet known. After [17].

3 Mechanisms to express these genes in a certain order.

This order is illustrated in Fig. A46. After introduction of trypanosomes into a fresh host (in this case a rabbit), antigenic variants appear in a semi-defined order. Waves of parasitaemia are usually dominated by one antigenic type, but often contain several minor types ('heterotypes'). A always tends to come out first, whether the rabbit is infected with variant C, D or Z. This indicates that the imprecise order in which variants appear in a chronic infection is due to switching preference; the trypanosome switches preferentially to gene A, but once the host contains antibodies against VSG A, new variants A are killed and other variants can grow out. To a large extent this order is genetically programmed: initially VSG genes are expressed that reside in an active expression site that is easily switched on; then VSG genes are activated that are in telomeric locations with a high degree of homology to expression sites; then chromosome-internal and telomeric VSG genes are transposed that share good homology in their flanking sequences with expression sites; and finally VSG genes are activated that have poor homology with the expression site and that require the presence of a related gene in the expression site for entry. These late genes are often PSEUDOGENES, that is with incomplete reading frames, and only the usable segments of these genes end up in the expression site, resulting in mosaic expressed VSG genes.

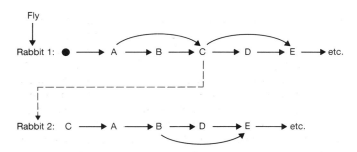

Fig. A46 A highly schematic representation of the orderly appearance of trypanosome variants in a chronic mammalian infection. The metacyclic coat (VSG gene) repertoire, indicated by the filled circle, is switched on in the salivary gland of the tsetse fly. Letters represent the major variants in subsequent waves of parasitaemia. The order is imprecise; for example A may go to C or D.

Probably other factors than this genetic programme contribute to the order in which consecutive VSGs appear in a chronic infection. During switching from A to B, there is complete mixing of VSGs and it has been postulated that some mixed coats are better than others, for example AB and AC are functional, but AG not. Direct experimental evidence for this postulate is not available.

The metacyclic repertoire

Once an infection is under way, it is in the best interests of the parasite to limit population heterogeneity, to use the VSG gene repertoire as economically as possible. Diversity is advantageous in the incoming inoculum, however. In Africa, most wild animals are already infected with trypanosomes and it is therefore difficult for a new strain to get a foothold. To orchestrate this diversity, the trypanosome uses a separate set of expression sites, the metacyclic expression sites, to switch on the production of VSGs in the salivary gland of the tsetse fly (see Fig. A46). As the 10^4 trypanosomes that enter the mammalian host by fly bite contain some 20 different variant antigen types (VATs) there must be 20 different metacyclic expression sites (M-ESs). Only two of these have been characterized to some extent [12]. The M-ES promoters have no homology with the promoters of the bloodstream repertoire expression sites and they seem to be located only a few kilobases upstream of the VSG gene. The M-ESs have only a few 70-base pair (bp) repeats upstream of the VSG genes, limiting their homology to the bloodstream repertoire expression sites. They also lack a full complement of ESAGs and where the ESAG transcripts found in the metacyclic trypanosomes come from has not been established. How the trypanosome manages to switch on only one of the 20 M-ESs in each trypanosome is also still a mystery. Switching from one M-ES to another is possible, as single cloned metacyclic VATs readily switch. The analysis of the metacyclic repertoire has been difficult because of this rapid switching rate and the small amounts of material available.

How does the trypanosome combine antigenic variation with food uptake?

In the mammalian host, trypanosomes have simplified their metabolism to the bare minimum required for ATP synthesis and growth. The substrates required are taken from the host. Obviously, efficient uptake of nutrients cannot be accomplished without receptors and transport systems exposed to the environment. How does the trypanosome combine antigenic variation with efficient food uptake?

This paradox appears to be resolved by two complementary devices [13]:

1 Substrates such as host LOW DENSITY LIPOPROTEIN and TRANSFERRIN are taken up by ENDOCYTOSIS in the flagellar pocket at the base of the flagellum. This pocket is accessible to macromolecules but of course not to host macrophages (that are similar in size to the entire trypanosome). Hence, the nutrient uptake machinery is protected from the cellular arm of the host immune system [14].

2 The receptors may undergo antigenic variation as well. Recent work indicates that one of the ESAGs, ESAG-6, is a transferrin-

binding protein (TFBP) and presumably the transferrin receptor [13]. TFBP genes located in different expression sites are homologous but differ in a hypervariable region (see Fig. A45). If this region were immunodominant, switching to another expression site would allow the trypanosome to escape from antibodies that have accumulated against the resident TFBP and that may interfere with transferrin uptake. If this is correct, vaccination against all variants of this and other receptors might afford protection against trypanosome infection.

It should be stressed that this interpretation of the available data is speculative. As it is now relatively easy to add genes to the trypanosome genome or to disrupt genes [15], a critical test of this speculation is within reach.

P. BORST

See also: CHROMOSOME STRUCTURE; EUKARYOTIC GENE EXPRESSION; EUKARYOTIC GENE STRUCTURE.

1 Pays, E. & Steinert, M. (1988) Control of antigen gene expression in African trypanosomes. *Annu. Rev. Genet.* **22**, 107.
2 Cross, G.A.M. (1990) Cellular and genetic aspects of antigenic variation in trypanosomes. *Annu. Rev. Immunol.* **8**, 83.
3 Van der Ploeg, L.H.T. (1991) Control of antigenic variation in African trypanosomes. *The New Biologist* **3**, 324.
4 Borst, P. (1991) Molecular genetics of antigenic variation. *Immunoparasitol. Today* **March**, A29–A33.
5 Down, J.A. et al. (1991) Crystallization and preliminary X-ray analysis of an intact soluble-form variant surface glycoprotein from the African trypanosome, *Trypanosoma brucei. J. Mol. Biol.* **218**, 679.
6 Doering, T.L. et al. (1991) An analog of myristic acid with selective toxicity for African trypanosomes. *Science* **252**, 1851.
7 Bangs, J.D. et al. (1992) Spectrometry of mRNA cap4 from trypanosomatids reveals two novel nucleosides. *J. Biol. Chem.* **267**, 9805–9815.
8 Zomerdijk, J.C.B.M. et al. (1991) Efficient production of functional mRNA mediated by RNA polymerase I in *Trypanosoma brucei. Nature* **353**, 772.
9 Rudenko, G. et al. (1991) RNA polymerase I can mediate expression of CAT and neo protein-coding genes in the parasitic protozoan *Trypanosoma brucei. EMBO J.* **10**, 3387.
10 Gottesdiener, K. et al. (1990) Chromosome organization of the protozoan *Trypanosoma brucei. Mol. Cell. Biol.* **10**, 6079.
11 Pays, E. (1989) Pseudogenes, chimaeric genes and the timing of antigen variation in African trypanosomes. *Trends Genet.* **5**, 389.
12 Graham, S.V. et al. (1990) Distinct, developmental stage-specific activation mechanisms of trypanosome VSG genes. *Parasitology* **101**, 361.
13 Borst, P. (1991) Transferrin receptor, antigenic variation and the prospect of a trypanosome vaccine. *Trends Genet.* **7**, 307.
14 Balber, A.E. (1990) The pellicle and the membrane of the flagellum flagellar adhesion zone, and flagellar pocket: functionally discrete surface domains of the bloodstream form of African trypanosomes. *Crit. Rev. Immunol.* **10**, 177.
15 Ten Asbroek, A.L.M.A. et al. (1990) Targeted insertion of the neomycin phosphotransferase gene into the tubulin cluster of *Trypanosoma brucei. Nature* **348**, 174.
16 Paindavoine, P. et al. (1992) A gene from the variant surface glycoprotein expression site encodes one of several transmembrane adenylate cyclases located on the flagellum of *Trypanosoma brucei. Mol. Cell. Biol.* **12**, 1218.
17 Revelard, P. et al. (1990) A gene from the VSG expression site of *Trypanosoma brucei* encodes a protein with both leucine-rich repeats and a putative zinc finger. *Nucleic Acid Res.* **18**, 7299.

antihaemophiliac globulin Blood fraction originally shown in 1937 to be lacking in classical HAEMOPHILIA A and now known to contain the blood coagulation factor (Factor VIII) which is deficient in this disorder. *See*: BLOOD COAGULATION AND ITS DISORDERS.

anti-idiotypic antibodies Antibodies having specificity for highly restricted or unique structural features of another antibody. The structural features recognized relate to the antigen-binding site and hence the COMPLEMENTARITY DETERMINING REGIONS. The epitopes (antigenic determinants) recognized are termed idiotopes and the sum of all the idiotopes of a given molecule is called the IDIOTYPE. XENOGENEIC anti-idiotypic antisera are readily produced and there is evidence for the production of auto anti-idiotypic antibody. These observations led Niels Jerne to propose that self recognition through a cascade of idiotype–anti-idiotype interactions may constitute an essential regulatory network within the immune system.

Kennedy, R.C. et al. (1986) *Sci. Am.* **255**, 40–48.

anti-Lepore haemoglobin The product of a FUSION GENE generated by unequal crossing over between the β- and δ-globin genes (*see* HAEMOGLOBIN AND ITS DISORDERS; GLOBIN GENES). The fusion gene has the N-terminal part of β and the C-terminal part of the δ gene. *See*: HAEMOGLOBIN LEPORE for the reciprocal event.

antimorphic allele (antimorph) A term in the classification of allele type devised by the geneticist H.J. Muller (*see* ALLELE for a full explanation of this classification). Antimorphic alleles possess activity which is opposite to that of the wild-type gene product.

anti-NP antibodies Antibodies specific for the 4-hydroxy-3-nitrophenyl acetyl group, which has been widely used as a hapten to provoke the production of a relatively homogeneous antibody response in the C57Bl/6 strain of mouse. As a result, antibodies raised in different individuals of this mouse strain express many apparently identical idiotopes (*see* ANTI-IDIOTYPIC ANTIBODIES). Interestingly, anti-NP antibodies bind the related antigen 4-hydroxy-5-iodo-3-nitrophenyl acetyl group with higher affinity than the NP group. This represents an example of a HETEROCLITIC ANTIBODY.

anti-oncogene TUMOUR SUPPRESSOR GENE. *See also*: ONCOGENES.

antiplasmin *See*: SERPINS.

antiport *See*: MEMBRANE TRANSPORT SYSTEMS.

antipsychotic drugs Drugs used in the treatment of SCHIZOPHRENIA and other psychotic disorders including mania. All the commonly used antipsychotics, also known as neuroleptics, are effective antagonists at the D_2 subtype of DOPAMINE RECEPTORS. This blockade of DOPAMINE function can cause PARKINSONISM, one of several unwanted side effects of treatment. The identification, by molecular cloning techniques, of other dopamine receptors with substantial structural and pharmacological similarities to the D_2 receptor has provided one approach to the search for selective antagonists that may be antipsychotic without exhibiting such substantial side effects *See*: SCHIZOPHRENIA.

Antirrhinum majus Snapdragon, one of the species in which the genetic basis of FLOWER DEVELOPMENT has been best studied. *See also*: TRANSPOSABLE GENETIC ELEMENTS: PLANTS.

antisense RNA RNA molecules generated by reversing the orientation of the transcribed region of a gene with respect to a suitable transcriptional PROMOTER. This results in generating a transcript of the ANTISENSE DNA strand. Such antisense RNA has the potential to form an RNA–RNA duplex with the natural 'sense' mRNA transcript of the gene, thereby preventing its translation. Antisense RNA provides a means of inactivating the expression of specific genes and can be applied to both simple and complex eukaryotes. It has been exploited in genetic engineering: copies of antisense RNA genes have been engineered into the genome of, for example, tomato plants to prevent the production of enzymes that hasten softening of the fruit (*see* PLANT GENETIC ENGINEERING).

Green, P.J. et al. (1986) *Annu. Rev. Biochem.* **55**, 569–597.

antisense strand The strand in a DNA duplex complementary to the sense strand, and therefore corresponding to the sequence of the mRNA. It is also called the coding strand.

antitermination A mechanism utilized by both bacteriophage and bacterial species to regulate gene expression (*see* BACTERIAL GENE EXPRESSION). It involves the by-passing of a naturally occurring transcriptional terminator by RNA polymerase as a consequence of the enzyme interacting with one or more ancillary proteins (antitermination factors). Such termination readthrough results in an mRNA transcript with an extended 3′ sequence containing one or more CISTRONS which can now be translated. The best characterized example of antitermination is in the activation of expression of the delayed early genes of the bacteriophage LAMBDA during establishment of the lytic cycle. The antitermination factor involved is the pN protein, the product of the *N* gene. pN binds to two different sequences (*nutL* and *nutR*) located 5′ to the terminator to be read through and becomes attached to RNA polymerase as it transcribes through the *nutL* and *nutR* regions towards the transcriptional terminators t_{L1} and t_{R1}.

Friedman, D.I. et al. (1987) *Annu. Rev. Genet.* **21**, 453–488.

antithrombin *See*: BLOOD COAGULATION AND ITS DISORDERS; SERPINS.

antithrombin III deficiency Heritable condition involving the lack of the SERPIN proteinase inhibitor — ANTITHROMBIN III — which inhibits the enzymatic action of blood COAGULATION FACTORS, in particular activated Factors XII, XI, IX, X, plasmin, and thrombin. Both the binding of antithrombin III to the target enzymes and its inhibitory activity are increased many fold by HEPARIN. The condition is inherited as an AUTOSOMAL DOMINANT trait, and leads to a high risk of thromboembolism from early adult life onwards, presumably from the unopposed action of thrombin. *See*: BLOOD COAGULATION AND ITS DISORDERS.

α₁-antitrypsin A SERPIN synthesized by the liver and present in human serum. More than 30 allelic forms of α₁-antitrypsin have been distinguished in humans by acid starch gel electrophoresis or isoelectric focusing (*see* ELECTROPHORESIS). The locus is designated *Pi* (protease inhibitor). The most frequent alleles are of the M subtype (*PiM*); two variants — *Z* and *S* — are associated with an appreciably reduced level of α₁-antitrypsin. No enzyme activity is detectable in homozygotes for the very rare NULL ALLELE *PiO*.

The heritable disorder α₁-antitrypsin deficiency is a condition of variable outcome, inherited in an autosomal CODOMINANT fashion. 70–80% of homozygotes for the Z allele develop obstructive emphysema. Environmental factors, especially smoking, influence the manifestation of disease, possibly by stimulating release of proteinases from phagocytes. Liver disease usually first detected in childhood may progress to cirrhosis. The frequency of ZZ homozygotes is one in 3400 in the United Kingdom. PRENATAL DIAGNOSIS may be made by sampling foetal blood. The Z mutation results in an abnormality of secretion of α₁-antitrypsin (*see* SERPINS).

AP endonucleases A class of enzymes involved in DNA REPAIR in both prokaryotes and eukaryotes. These enzymes recognize depurinated or depyrimidinated residues in DNA and hydrolyse the phosphodiester bond at either the 5′ or 3′ side of the modified residue depending on the type of AP endonuclease. Cleavage of the phosphodiester bond permits access to an EXONUCLEASE which removes the residue immediately adjacent to the damaged residue thereby allowing resynthesis of the excised sequence.

AP lyase *See*: DNA REPAIR.

AP site APURINIC or APYRIMIDINIC SITE. *See also*: DNA REPAIR.

AP1 Eukaryotic TRANSCRIPTION FACTOR. *See*: EUKARYOTIC GENE EXPRESSION; STEROID RECEPTOR FAMILY.

AP3 (amino-3-phosphonopropanoate), AP4 (2-amino-4-phosphonobutanoate) Antagonists at both NMDA and metabotropic EXCITATORY AMINO ACID RECEPTORS.

AP3

AP4

AP5 (2-amino-5-phosphonopentanoic acid) Selective antagonist at the NMDA EXCITATORY AMINO ACID RECEPTOR.

APC ANTIGEN-PRESENTING CELL.

apical ectodermal ridge (AER) ECTODERMAL structure that runs along the distal edge of the developing vertebrate limb bud. Removal of the ridge causes the loss of limb structures in the proximo-distal axis, and it is therefore necessary for correct PATTERN FORMATION in the developing limb. The ridge is thought to interact with the subjacent limb bud MESENCHYME — the progress zone.

apical membrane *See*: EPITHELIAL POLARITY.

apical meristem The dividing tissue at the tip of plant shoots and root, from which further growth occurs. *See*: PLANT HORMONES.

Aplysia A genus of marine molluscs, of the family Opisthobranchia, commonly known as the sea hares. The nervous system of *Aplysia californica*, the Californian sea hare, has been a favoured system for study for many years because of its large ganglionic nerve cells. *Aplysia* has been studied particularly in respect of neuronal circuits underlying simple reflexes such as the gill withdrawal reflex, and the long-term, activity-dependent synaptic modifications involved in habituation and sensitization of this reflex. *See also*: LONG-TERM POTENTIATION.

apocrine A form of secretion in which the apical surface of the secretory cell is shed, involving cytoplasmic loss. An example is in the secretion of fat by cells of the mammary gland, during which the fat droplet is surrounded by apical plasma membrane as it leaves the cell.

apoenzyme The protein part of an enzyme, without any nonprotein prosthetic groups that are normally required for full function.

apoinducer An ancillary protein required by RNA polymerase to initiate TRANSCRIPTION from some operon promoters (*see* BACTERIAL GENE EXPRESSION). The apoinducer binds to the DNA in a sequence-specific manner and is essential for transcription initiation.

apolipoprotein Protein part of a lipoprotein. *See*: PLASMA LIPOPROTEINS AND THEIR RECEPTORS.

apolipoprotein B (apoB) *See*: PLASMA LIPOPROTEINS AND THEIR RECEPTORS.

apoplast The nonliving part of the plant body usually considered to consist of the cell walls and intercellular spaces through which substances may be transported.

apoprotein Protein lacking any nonprotein prosthetic groups that are normally required for full function.

apoptosis Type of cell death which is thought to be under direct genetic control (*see* PROGRAMMED CELL DEATH), and which is distinct from the process of necrosis. During apoptosis, cells lose their CELL JUNCTIONS and microvilli, the cytoplasm condenses and nuclear CHROMATIN marginates into a number of discrete masses. While the nucleus fragments, the cytoplasm contracts and mitochondria and ribosomes become densely compacted. After dilation of the endoplasmic reticulum and its fusion with the plasma membrane, the cell breaks up into several membrane-bound vesicles — apoptotic bodies — which are usually phagocytosed by adjacent cells. Activation of particular genes (e.g. *ced* genes in *Caenorhabditis elegans*, TUMOUR SUPPRESSOR GENES in vertebrates) is thought to be necessary for apoptosis to occur. Apoptosis induced by numerous cytotoxic agents can be suppressed by expression of the gene *bcl-2* which produces a cytoplasmic protein Bcl-2. As fragmentation of chromatin into oligonucleotide fragments is characteristic of the final stages of apoptosis, DNA cleavage patterns can be used as an *in vitro* assay for its occurrence. CYTOTOXIC T CELLS induce apoptosis in their target cells by binding to a cell-surface protein Fas (Apo-1) but the signalling mechanism is not yet known.

Cory, S. (1994) *Nature* **367**, 317–318.

APP Amyloid precursor protein. *See*: ALZHEIMER'S DISEASE.

apurinic site (AP site) Position in DNA where a purine BASE has been lost by cleavage of the N-glycosylic bond; the polynucleotide backbone consisting of the phosphodiester bonds and deoxyribose sugar remains intact, but is subject to cleavage by an AP endonuclease, an AP lyase or chemical hydrolysis. *See*: DNA REPAIR.

APV (2-amino-5-phosphonovaleric acid) Older name for AP5.

apyrase Enzyme (EC 3.6.1.5) isolated from potato (*Solanum tuberosum*) with ATPase and ADPase activity. It is used to deplete ATP in studies aimed at assessing the energy requirements of specific steps in, for example, protein import into mitochondria or endoplasmic reticulum (*see* PROTEIN TRANSLOCATION.)

apyrimidinic site (AP site) Position in DNA where a pyrimidine BASE has been lost by cleavage of the N-glycosylic bond; the polynucleotide backbone consisting of the phosphodiester bonds and deoxyribose sugar remains intact, but is subject to cleavage by an AP endonuclease, an AP lyase or chemical hydrolysis. *See*: DNA REPAIR.

Arabidopsis thaliana Wall cress, a small dicotyledonous plant of the family Cruciferae which is a favoured model organism for plant genetic and developmental studies because of its short life cycle and small simple genome (haploid chromosome number 5), the smallest plant genome known (0.2 pg DNA per genome). Classical genetic studies began in *Arabidopsis* at the beginning of this century, and even in the 1940s it was being called the 'botanical *Drosophila*'. It has sprung into prominence in recent years because its small size and rapid reproduction, and small genome, make it eminently suitable for cultivating on a large scale and for molecular genetic studies. The haploid genome size is 7×10^7 bp with ~80% single-copy DNA. This is in marked contrast to many plants of interest whose genomes are polyploid and/or contain very large amounts of repetitive DNA. Because of the small genome size and paucity of dispersed repetitive sequences, the technique of CHROMOSOME WALKING and other gene isolation strategies, which are not practicable in other plants, can be applied in *Arabidopsis*. The *Arabidopsis* genome is being physically mapped in detail.

Meyerowitz, E.M. (1987) *Arabidopsis thaliana. Annu. Rev. Genet.* **21**, 93–111.

arabinogalactans Plant polysaccharides containing a linear or branched D-GALACTAN core, further substituted by L-arabinose either as single residues or as short chains. Their structures are often very complex. Typical cell-wall arabinogalactans of higher plants contain cores of $\beta 1,4$ linked D-galactopyranose irregularly substituted by L-arabinofuranosides or disaccharides of the form Ara(f)1,5Ara(f)1-. The branched $\beta 1,3 : \beta 1,6$ galactans of some woods can also be cores of arabinogalactans (such as arabinogalactan A of larch) carrying single L-arabinofuranosyl residues or disaccharides of the type Ara(p)$\beta 1,3$Ara(f)1-. In GUM ARABIC, saccharides of the forms DGal $\alpha 1,3$LAra(f)1,4DGlcA$\beta 1$-,DGal$\alpha 1$, 3LAra(f)1- and LAra(f)1- occur on such a galactan core. Little is known of their synthesis. The galactan cores must be synthesized first, probably in stages if they are branched and presumably act as primers for transfers from UDP-LAra, which is the only known donor of L-arabinose in plants, as well as other sugar nucleotides. Some arabinogalactans are biosynthetically related to PECTINS.

arabinose *See*: SUGARS.

arachidonic acid A polyunsaturated 20-carbon FATTY ACID with four double bonds (Fig. A47). It is an integral part of many membrane PHOSPHOLIPIDS in animal cells, from which it is cleaved by the action of PHOSPHOLIPASES. It can act as a second messenger and is further metabolized to give rise to a large family of local chemical mediators, the EICOSANOIDS.

$$CH_3(CH_2)_4(CH{=}CHCH_2)_4(CH_2)_2COOH$$

Fig. A47 Arachidonic acid.

archaebacteria One of three proposed primary kingdoms of organisms (the others are EUBACTERIA and EUKARYOTES); regarded as a class of PROKARYOTES in other classifications. Archaebacteria are microorganisms lacking a nucleus and other cellular organelles and distinguished from eubacteria by a distinctive cellular membrane that contains LIPIDS composed of branched (polyisoprenoid) carbon chains connected by an ether linkage, rather than an ester linkage, to glycerol and by the absence of muramic acid in their cell walls (*see* BACTERIAL ENVELOPE; SUGARS). They are often found in extreme environments (high temperatures and low pH, high salt, reducing atmosphere), which suggests that they are adapted to environments that might have been present during the early emergence of life. They are also distinguished from the other kingdoms on the basis of their rRNAs and tRNAs, on their susceptibility to various antibiotics, and on the characteristics of their metabolic pathways. There are three major groups of archaebacteria: the methanogens, the thermoacidophiles and the extreme halophiles.

archencephalic induction Regional NEURAL INDUCTION by the chorda mesoderm of the roof of the ARCHENTERON or by exogenously applied material, producing FOREBRAIN or archencephalic structures.

archenteron Future or primitive gut of the GASTRULA. It consists of a layer of cells lining a new cavity formed by the flattening and invagination of the vegetal part of the BLASTULA during GASTRULATION. During this process the previous cavity, the BLASTOCOEL, is progressively obliterated and the gastrula assumes a double-walled cup shape. The archenteron retains its contact with the exterior via the BLASTOPORE.

area opaca The outer yolky extraembryonic ring of the early avian embryo. Once extraembryonic blood vessels form, it is called the area vasculosa. *See*: AVIAN DEVELOPMENT.

area pellucida The inner embryonic region of the early avian embryo. It refers to a structure unique to avian embryos and should not be confused with the ZONA PELLUCIDA. *See*: AVIAN DEVELOPMENT.

Arenaviridae A family of enveloped RNA ANIMAL VIRUSES with spherical or pleomorphic particles 50–300 nm diameter, with club-shaped projections on the surface. The genome is composed of two molecules of (−)-sense single-stranded RNA (M_r 1.1–1.6×10^6, and $2.1–3.2 \times 10^6$) and the virus also contains three host RNAs (28S, 18S, 46S). Viruses of this family cause persistent infections in rodents (e.g. lymphocytic choriomeningitis (LCM)), and can be transmitted to humans (e.g. Lassa fever, Argentinian and Bolivian haemorrhagic fevers).

ARF proteins *See*: GTP-BINDING PROTEINS.

arginine (Arg, R) An AMINO ACID with a basic side chain.

arginosuccinate synthetase deficiency *See*: CITRULLINAEMIA.

arrestin A protein of M_r 45 000 found in photoreceptor cells over the phylogenetic range from insects to humans. It binds calcium and ATP but these are inessential for its binding to photoexcited RHODOPSIN in its phosphorylated form. The binding of arrestin terminates the activity of this photoreceptor in the signal transduction process. Similar proteins are thought to terminate other G protein-coupled receptors. *See also*: VISUAL TRANSDUCTION.

ARS The azophenylarsonate group, which is an immunogenic HAPTEN when attached to a protein carrier. It has been used to dissect ANTIBODY responses, particularly within the A/J strain of mice which produce antibodies of restricted heterogeneity that bear a common cross-reactive IDIOTYPE. The association between the ability of antibodies to bind arsonate and the presence of the major cross-reactive idiotype is due to the highly restricted use of genetic elements in the construction of the antibody molecules. The same V_κ, J_κ, V_H, D_H and J_H gene segments are always used and specific junctional amino acids such as arginine at position 96 ($V_\kappa J_\kappa$ junction) in the light chain and serine at position 100 ($V_H D_H$

junction) in the heavy chain are generally retained. Rare mutational events allow the effects of single amino-acid substitutions on antibody specificity and affinity to be evaluated.

ARS element Autonomously replicating sequence. First identified as a DNA sequence that confers the property of high frequency autonomous TRANSFORMATION of the yeast SACCHAROMYCES CEREVISIAE. Such sequences can be derived from the genome of *S. cerevisiae* (in which they may represent chromosomal ORIGINS OF REPLICATION) or from a wide range of eukaryotic organisms (from fungi to man). Structure–function analysis has implicated two important sequences for ARS function: domain A (5′ (A/T)TTTAT(A/G)TTT(A/T) 3′) and domain B (5′ CTTTTAGC(A/T)(A/T)(A/T) 3′) with domain B being located 50–100 bp 3′ to domain A. Domain A is absolutely required for ARS function whereas other flanking sequences may be important for function.

Williamson, D.H. (1985) *Yeast* **1**, 1–5.

artificial chromosomes *See*: CHROMOSOME STRUCTURE; YEAST ARTIFICIAL CHROMOSOMES.

aryl sulphatases Lysosomal aryl sulphatases are ACID HYDROLASES involved in removing sulphates from, for example GLYCOSAMINOGLYCANS. The absence of a specific aryl sulphatase can lead to lysosomal storage disease. Aryl sulphatase A (EC 3.1.6.8) deficiency is associated with late infantile metachromatic leukodystrophy and aryl sulphatase B (EC 3.1.6.12) deficiency with mucopolysaccharidosis type VI.

Ascaris A parasitic nematode worm in which CHROMOSOME DIMINUTION occurs.

ascidian development Ascidians are Urochordates, thought to be related to the vertebrates because they possess a free-living larva with neural tube, notochord, and muscle arranged in the typical vertebrate pattern. Although some work has been done on the regeneration of the compound ascidians, the chief significance of the group for development biology is the early embryonic development of simple ascidians. Like several other invertebrate groups these show an invariant cleavage pattern and the cell lineage of *Styela* was described to a high level of detail by Conklin in 1905. This has been refined in recent years by Japanese workers and is now known from egg to larva.

The egg contains a subcortical layer of cytoplasm that is rearranged after fertilization into a posterior position, and in *Styela*, where it contains yellow pigment granules, is known as the yellow crescent. This cytoplasm is mainly partitioned by the invariant cleavages into the muscle lineages and since the time of Conklin it has been supposed that it contains a muscle determinant. The evidence for this is as follows: isolated blastomeres containing the yellow cytoplasm will produce muscle on culture; removal of the blastomeres containing the yellow crescent prevents muscle formation in the rest of the embryo; transplantation of the yellow cytoplasm to blastomeres that do not normally contain it causes some to produce muscle. The molecular nature of the presumed muscle determinant remains unknown.

Although ascidian embryologists have tended to favour the view that the body plan of the embryo is specified entirely by localization of CYTOPLASMIC DETERMINANTS, more intensive study has shown that there are, as in all animal embryos, a number of inductive interactions (*see* INDUCTION). These include the formation of some muscle from cells lacking the yellow crescent, the formation of the neural tube, and the relative identity of two specialized sensory cells, the ocellus and the otolith.

ascites Peritoneal exudate which includes fluid and inflammatory cells. In experimental animals ascites may be produced in response to PLASMACYTOMA or HYBRIDOMA cells growing in the abdominal cavity. A single mouse may produce up to 10 ml of ascitic fluid, which can be tapped off with a syringe and which can contain 1–10 mg ml^{-1} of a MONOCLONAL ANTIBODY. Repeated injections of complete Freund's adjuvant may also be used to induce ascites. The fluid will contain approximately the same level of antibody as the animal's own serum. This may be used as a source of polyclonal immunoglobulin since the induction of ascites can be timed to coincide with elevated levels of specific antibodies following immunization. Clinically, ascites may be the result of tumours seeding into the peritoneum, obstruction of the portal circulation (e.g. cirrhosis) or hypoalbuminaemia.

Ascomycotina Fungi of the EUMYCOTA in which the sexual spores (ascospores) are housed in a structure called the ascus (pl. asci) from which they are forcefully ejected. The usual number of ascospores is eight (but four are produced in some yeasts). The ascus is sometimes housed in a fruiting body known as the ascocarp. This class includes the yeasts and the mould *Neurospora*, which are important model systems for genetics, as well as scores of important plant pathogens such as the *Erisyphe* mildews.

ascus (pl. asci) Sac-like cell that contains ascospores. Found in fungi of the class Ascomycotina. In the yeast SACCHAROMYCES CEREVISIAE an ascus is formed from a diploid cell subjected to nutrient starvation, with the four haploid ascospores formed within the ascus representing the products of a meiotic division. *See also*: NEUROSPORA CRASSA.

asexual reproduction Any form of reproduction that does not involve sex. It includes parthenogenesis and vegetative reproduction. Asexual reproduction in eukaryotes has been classified into two types: amixis and automixis. In amixis, MEIOSIS is absent and new individuals are produced by MITOSIS. The new individuals may be either unicellular buds (e.g. in yeasts and protozoa) or multicellular offshoots (e.g. most plants). In automixis, some aspect of a meiotic cycle occurs, but is often abortive and DIPLOIDY in the offspring is ensured by a number of different mechanisms. In some instances, one or the other of the meiotic divisions is suppressed (planaria); two haploid cleavage nuclei fuse to form a homozgous diploid embryo (coccids). In most Hymenoptera, males are produced asexually. They develop as haploid individuals from unfertilized eggs. The genetic consequence of asexual reproduction is that the genotype of the offspring is identical to the genotype of the parent, although some shuffling of loci between homologues, due to meiotic recombination in automixis, may occur. Asexual reproduction by cell division is the main mode of reproduction in prokaryotes, although a form of sex (involving the transfer of genes from one individual to another) does occur (*see* F FACTOR; PLASMIDS).

asialoglycoprotein GLYCOPROTEIN which has had its terminal SIALIC ACID residues removed by, for example, neuraminidase enzyme activity. Such desialylated proteins in serum can be bound by the hepatocyte cell surface asialoglycoprotein receptor for subsequent receptor-mediated ENDOCYTOSIS and degradation.

asialoglycoprotein receptor LECTIN-like receptor on hepatocytes that removes galactose-terminated glycoproteins from the blood.

Asn The AMINO ACID asparagine.

asolectin A misleading, and almost archaic, term for plant lecithin (phosphatidylcholine), which is occasionally used in the food industry. A common food additive. *See*: LIPIDS.

asparagine (Asn, N) A neutral AMINO ACID.

aspartic acid (Asp, D) A hydrophilic, acidic AMINO ACID.

Aspergillus nidulans A member of the class Ascomycotina, this filamentous fungus has a homothallic life cycle (Fig. A48). It grows as a colourless mycelium which is made up of multinucleate cells and which bears multicellular conidiophores. The uni-

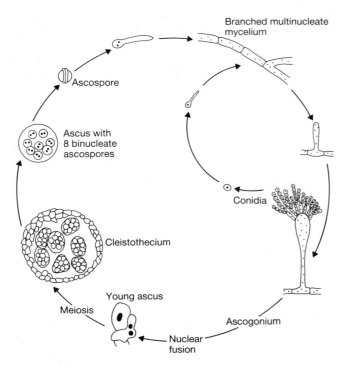

Fig. A48 The life cycle of *Aspergillus nidulans*.

nucleate conidia are usually dark green to dull grey-green in colour. *A. nidulans* was originally exploited as a model organism for classical genetic studies as parasexual crosses can be made between two different genetically marked strains. Both DIPLOIDS and HETEROKARYONS can be readily obtained in *A. nidulans*, facilitating the development of the mitochondrial genetics of this organism. More recently, genetic transformation with PLASMID DNA has been achieved, opening the way for the application of molecular genetic tools.

Pontecorvo, G. et al. (1953) *Adv. Genet.* **5**, 141–238.

association constant A quantitative measure of the AFFINITY of a receptor for its ligand (e.g. of the antigen-binding site of an antibody for its cognate antigen). It is determined by Scatchard analysis of equilibrium binding experiments and has the dimensions $l \, mol^{-1}$. *See*: SCATCHARD PLOT.

associativity In respect of LONG-TERM POTENTIATION the situation whereby weak tetanic stimulation of one afferent pathway, which does not induce LTP, can induce LTP when paired with strong afferent activation of another input to the same group of neurons. This means that the postsynaptic depolarization that is required in order to provide sufficient NMDA receptor activation to induce LTP can be conferred by another group of afferent inputs.

aster A radially symmetric array of MICROTUBULES that grows from a mitotic CENTROSOME. The MITOTIC SPINDLE in each metazoan cell will usually contain one aster at each spindle pole (*see* MITOSIS). Asters have a role in defining the place where an animal cell will form its cleavage furrow during CYTOKINESIS. They may also help to position the spindle relative to the rest of the cell during both symmetric and asymmetric cell divisions. The spindles of mitotic cells in higher plants lack asters and are often referred to as anastral. The spindles of some lower plants and protozoa form an array of microtubules that projects out from the spindle and probably plays the spindle-positioning role of the aster, but these structures lack the radial symmetry that gives a true aster its name. Most INTERPHASE animal cells contain an array of microtubules growing from their centrosomes, but these structures generally lack the radial symmetry of a mitotic aster. They also contain fewer microtubules (about one-fifth as many), and their microtubules are generally longer. The interphase aster-like arrays contribute to an animal cell's traffic of membrane-bounded vesicles (*see* ENDOCYTOSIS; PROTEIN SECRETION) and to the polarity of the cell structure and movement (*see* CELL MOTILITY).

astrocyte *See*: GLIAL CELLS.

Asx (B) Either of the AMINO ACIDS asparagine or aspartic acid.

asymmetric cell division Any situation where cell division generates two morphologically and/or developmentally nonequivalent daughter cells. It occurs during asexual reproduction (e.g. budding of yeast), during division of a STEM CELL to produce another stem cell plus a progenitor cell for a specific cell lineage, during meiotic divisions of OOGENESIS and during early development. Many eggs contain localized CYTOPLASMIC DETERMINANTS of cell character and their cellularization by a series of asymmetric cell divisions produces a direct way of specifying differences between cells in a spatially organized manner. For example, the first and second divisions of an amphibian egg are symmetric with respect to the animal–vegetal axis (i.e. parallel to it). The plane of the third cleavage is perpendicular to it, that is, an asymmetric division generating small cells from the animal pole region which will contribute to much of the embryo, and larger yolky cells from the vegetal pole which will form gut. In mammals there is no obvious localization of determinants in the egg but a contact-dependent polarity develops within individual blastomeres at the 8-cell stage. If the plane of division is orientated approximately perpendicular to the axis of polarity at the fourth cleavage, an asymmetric division occurs which will generate the progenitors of the two cell lineages found in the blastocyst. Asymmetric divisions also generate cell diversity in plants.

asymmetric unit   The basic repeating unit of a crystal, related to all other such units within the UNIT CELL by the operation of SYMMETRY elements, and to those in other unit cells by the lattice translations **a**, **b**, **c**. In the rare case of a crystal having no symmetry elements, the asymmetric unit is the same as the unit cell. *See*: MOLECULAR AVERAGING; NON-CRYSTALLOGRAPHIC SYMMETRY; STRUCTURE FACTOR; SYMMETRY; SYSTEMATIC ABSENCES.

ataxia telangiectasia (Louis–Bar syndrome)   Human disorder characterized by facial and ocular dilation of blood vessels (telangiectasias), progressive neuromuscular degeneration and cerebellar ataxia, sensitivity to ionizing radiation, immunodeficiency and a high incidence of lymphoreticular cancer. The disease is inherited in a RECESSIVE fashion and the gene for the most frequent form has been mapped to human chromosome 11q22-23. The overall frequency of the disorder is less than one in 30 000 births. DNA synthesis in cells from affected patients is inhibited less by ionizing radiation than is DNA synthesis in normal cells, but the precise nature of the biochemical defect is unknown. *See also*: DNA REPAIR.

atomic scattering factor (f)   The amplitude of X-rays scattered by an atom. At zero angle of DIFFRACTION the atomic scattering factor (f) approximately equals the atomic number of the atom but it falls off rapidly with increasing BRAGG ANGLE, or angle of diffraction. *See also*: ANOMALOUS SCATTERING; X-RAY CRYSTALLOGRAPHY.

atomic scattering length (b)   An atom, from the point of view of scattering neutrons, may be considered to be an object presenting an effectively impenetrable cross–section area, σ. Thus, if there is an incident flux of N neutrons per unit area per second, then $N\sigma$ neutrons are scattered in all directions by the atom where σ is the scattering cross-section for the atom in question. The scattering length b is related to the scattering cross-section σ by $4\pi b^2 = \sigma$. *See*: NEUTRON SCATTERING AND DIFFRACTION.

ATP   Adenosine 5′-triphosphate. *See*: NUCLEOSIDES AND NUCLEOTIDES.

ATP-binding cassette proteins (ABC proteins)   Also known as the ABC superfamily, a family of transport (and other) proteins all containing a structurally similar ATP-binding domain of ~200 amino acids containing two short ATP-binding motifs. They include the periplasmic binding proteins of bacteria, the P-GLYCOPROTEIN TRANSPORT SYSTEM associated with MULTIDRUG RESISTANCE in tumour cells, the product of the *STE6* gene in yeast which mediates export of the a-factor mating pheromone, the CYSTIC FIBROSIS transmembrane conductance regulator, and the TAP gene products that transport processed antigen peptides from cytosol to endoplasmic reticulum (*see* ANTIGEN PROCESSING AND PRESENTATION). *See also*: BACTERIAL PROTEIN EXPORT; MEMBRANE TRANSPORT SYSTEMS.

***atp* genes**   Genes found in chloroplasts, mitochondria and bacteria, and encoding some subunits of the ATP SYNTHASE complex. *See also*: CHLOROPLAST GENOMES; MITOCHONDRIAL GENOMES: PLANT.

ATP synthase   The enzyme of CHLOROPLAST, MITOCHONDRIAL, and bacterial membranes that couples the synthesis of ATP from ADP and inorganic phosphate to a flux of protons down the electrochemical gradient from one compartment to another (e.g. thylakoid lumen to stroma in chloroplasts and intermembrane space to matrix in mitochondria). The enzyme is composed of a membrane-embedded complex (CF_0 in the thylakoid membrane of chloroplasts; F_0 in the inner mitochondrial membrane), which forms a proton channel, to which is attached a large water-soluble complex (CF_1 or F_1-ATPase) exposed to the stroma or intermembrane space and which has the subunit composition α_3, β_3, γ, δ, ϵ. The three-dimensional structure of bovine heart mitochondrial F_1-ATPase has been determined to 6.5 Å resolution by X-RAY CRYSTALLOGRAPHY (Fig. A49). A PROTONMOTIVE FORCE set up by the action of the oxidative or photosynthetic ELECTRON TRANSPORT CHAIN drives protons through CF_0/F_0, causing a conformational change in CF_1/F_1. This conformational change then drives ATP synthesis. There is one active site for catalysis of the reaction $ADP + P_i = ATP$ in each β subunit. The CF_0/F_0 component contains three or four kinds of subunit, some present in multiple copies. In the mitochondrial enzyme this channel can be blocked by the classic inhibitor of the enzyme, OLIGOMYCIN. The chloroplast enzyme subunits γ, δ, and CF_0-II are encoded in the nucleus; the others are encoded in the chloroplast. Under appropriate conditions the enzyme can operate in reverse to hydrolyse ATP and set up a protonmotive force. *See also*: ATPASES; ELECTRON TRANSPORT CHAIN; PHOTOSYNTHESIS.

ATPase(s)   Adenosine triphosphatase, the enzymatic activity that hydrolyses ATP (adenosine 5′-triphosphate) to ADP (adenosine 5′-diphosphate) and inorganic phosphate. ATPase activity is a property of many different proteins. ATPase activity is generally coupled to an energy-requiring process such as the active transport of ions across membranes (*see* ION PUMPS; MEMBRANE TRANS-

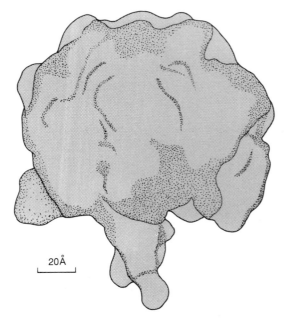

20Å

Fig. A49 The topography of bovine heart mitochondrial F$_1$-ATPase at 6.5 Å resolution. The F$_1$ domain is composed of 3α : 3β : 1γ : 1δ : 1ε and has an M_r of 371 000. The stem is presumed to be part of the stalk that connects F$_1$ with the membrane domain in the intact ATP synthase. (From Abrahams, J.P. et al. (1993) *EMBO J.* **12**, 1775–1780.)

PORT), the unwinding of DNA (*see* DNA GYRASE), muscle contraction (*see* MUSCLE; MYOSINS), and intracellular transport (*see* CELL MOTILITY; MICROTUBULE-BASED MOTORS), as well as to numerous energy-requiring metabolic reactions. *See also*: ATP SYNTHASE; CA^{2+}-ATPASE; H$^+$, K$^+$-ATPASE; NA$^+$,K$^+$-ATPASE; P-TYPE ATPASES; V-TYPE ATPASES; UNCOATING ATPASE.

Palme, K. (1992) *Int. Rev. Cytol.* **132**, 223–283.

atrazine A herbicide (2-chloro-4-ethylamino-6-isopropylamino-*s*-triazine) which acts by binding at or near the plastoquinone-binding site of the 32 kDa HERBICIDE BINDING PROTEIN (D1) of PHOTOSYSTEM II, thereby inhibiting photosynthetic electron transport. Atrazine-resistant mutants have arisen naturally in which there are identified changes of single amino-acid residues in the D1 polypeptide. Introduction of such mutations into crop plants is an objective of PLANT GENETIC ENGINEERING. *See also*: PHOTOSYNTHESIS.

***att* site** The bacteriophage LAMBDA attachment site, being a region between the *gal* and *bio* genes in the ESCHERICHIA COLI chromosome into which λ preferentially integrates. Integration is a result of recombination between the *att* site and an attachment site in the bacteriophage chromosome.

attachment plaque Focal contact, adhesion plaque. *See*: CELL JUNCTIONS; CELL MOTILITY.

attenuation A mechanism used to regulate several biosynthetic OPERONS in bacteria including the *Escherichia coli trp* operon (Fig. A50). It involves the regulation of TRANSCRIPTION through a transcription termination signal located, in the case of the *trp* operon, between the promoter and the first structural gene of the operon, and is closely coupled to translation. Transcription of this ATTENUATION region (*trpL*) generates a 141-nucleotide mRNA whose transcription is terminated before the first structural gene (*trpE*). Translation of this mRNA commences soon after transcription begins. The RNA encodes a short (14 amino acid) LEADER PEPTIDE which contains two adjacent tryptophan residues. When tryptophan levels are low in the cell, tryptophanyl-tRNA is limiting and the ribosome stalls at the two Trp codons. This allows a secondary structure to form in the RNA that allows RNA polymerase to by-pass the *trpL* terminator and continue to transcribe the *trp* structural genes. When tryptophan is not limiting, translation of *trpL* proceeds normally and an alternative secondary RNA structure allowing termination is formed. This mechanism allows RNA polymerase to 'sense' the cellular levels of tryptophan and gene expression is adjusted accordingly. Attenuation is also used to regulate the expression of the *his*, *leu*, *thr* and *ilv* operons of *Escherichia coli* and is found in other prokaryotes.

Yanofsky, C. & Kolter, R. (1982) Attenuation in amino acid biosynthesis operons. *Annu. Rev. Genet.* **16**, 113–134.

auditory system The anatomical apparatus pertaining to the sense of hearing. It includes the external ear, the tympanum, the ossicles, the fluid-filled cochlea, and areas of the central nervous system, such as the cochlear nuclei, lateral and medial superior olive, inferior colliculus, medial geniculate body, and the primary auditory cortex.

AUG The most commonly used translation INITIATION CODON that also acts as a codon for methionine within mRNA coding sequences. *See*: GENETIC CODE; PROTEIN SYNTHESIS.

autapses SYNAPSES formed between one part of a cell and another part of the same cell, typically axodendritic or axosomatic.

autoantibodies ANTIBODIES produced against the antigens of an individual's own body. The production of autoantibodies may be damaging in that they can set up a destructive immune response directed against the tissue bearing that antigen. *See*: AUTOIMMUNE DISEASES.

autocatalytic splicing (self-splicing) A mechanism of RNA SPLICING in which the excision of the intron is an autonomous property of the RNA itself. First described in *Tetrahymena* rRNA, it has also been found in mRNA precursors in mitochondria of yeast and other fungi and in some RNAs in chloroplasts. *See*: RIBOZYMES.

autocrine Applied to the secretion of signalling molecules such as GROWTH FACTORS and hormones that affect the secretory cell itself.

autogenous control A mechanism used to regulate gene expression in which the product of the gene (protein or RNA) regulates

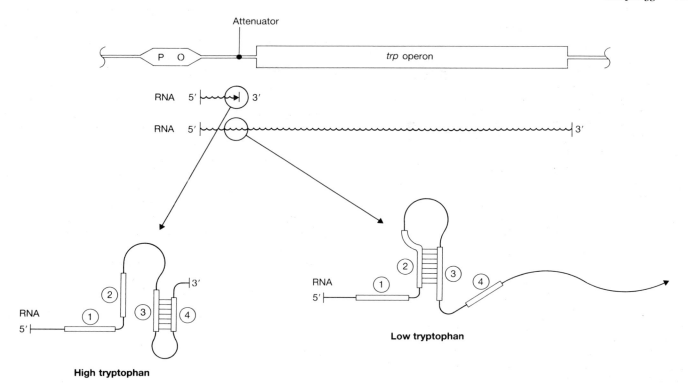

Fig. A50 Attenuation control of the *trp* operon of *Escherichia coli*. The structure of the *trp* operon indicating the location of the promoter (p) and operator (o) region, and the attenuator is shown at the top. The proposed secondary structures of the transcribed RNA in the region of the attenuator in conditions of low and high tryptophan are shown below.

its own synthesis. This form of regulation can be positive (i.e. activating gene expression) or negative (i.e. inhibiting gene expression) and is usually mediated at the post-transcriptional level. An example of this control in prokaryotes is provided by gene *32* of the bacteriophage T4 in which excess of the gene *32* protein (p32) binds to an AU-rich region in its own mRNA thereby preventing ribosomes from initiating translation. Autogenous control can also affect transcription; the cI repressor protein of bacteriophage LAMBDA regulates its own transcription. In eukaryotic cells the best characterized example of autogenous control is the regulation of TUBULIN synthesis. In this case, the cellular levels of tubulin mRNA are controlled by the free pool of tubulin monomers in the cell with excess free tubulin monomers inducing instability in the tubulin mRNA. A short sequence located immediately adjacent to the translation INITIATION CODON of the tubulin mRNA is needed for this regulatory mechanism, but the means whereby the mRNA becomes destabilized remain to be established.

autoimmune diseases Diseases in which the production of ANTIBODIES or T CELLS directed against a self-antigen is a major cause of the pathology of the disease. Those in which AUTOANTIBODIES are involved include BULLOUS PEMPHIGOID, Graves' disease (*see* THYROID HORMONE), some forms of DIABETES MELLITUS, and MYASTHENIA GRAVIS, SYSTEMIC LUPUS ERYTHEMATOSUS, pernicious anaemia, autoimmune haemolytic anaemia, glomerulonephritis, autoimmune thrombocytopenic purpura, and pemphigus vulgaris. Self-reactive T cells are thought to be involved in rheumatoid arthritis and insulin-dependent diabetes mellitus, and have also been found in other autoimmune diseases. Experimental allergic encephalomyelitis (EAE) can be induced in rodents by immunization with MYELIN basic protein and is mediated exclusively by autoreactive T cells.

autoimmunity An immune response directed against self-antigens. Autoimmunity results from the breakdown of the normal mechanisms of self-tolerance that prevent the production of functional self-reactive clones of B and T cells (*see* B CELL DEVELOPMENT; T CELL DEVELOPMENT). *See*: AUTOANTIBODIES; AUTOIMMUNE DISEASES.

autologous From the same individual; used to describe proteins, genes, cells, etc.

autonomic ganglia *See*: PERIPHERAL NERVOUS SYSTEM.

autonomic nervous system *See*: PERIPHERAL NERVOUS SYSTEM.

autonomously replicating sequence *See*: ARS ELEMENT.

autophagy Process by which cytoplasmic organelles are selectively destroyed. This is thought to involve the engulfing of a given organelle by membranes derived from the ENDOPLASMIC RETICU-

LUM to form an autophagosome which fuses with a LYSOSOME initiating digestion of the contents.

autophosphorylation The phosphorylation of a PROTEIN KINASE by itself, usually through an intramolecular reaction. In the case of a second-messenger-dependent protein kinase, the dephosphorylated autophosphorylation site may inhibit the protein kinase, the inhibition being relieved by binding of the SECOND MESSENGER. The protein kinase may then remain activated until the autophosphorylated site is dephosphorylated (as in the multifunctional calmodulin-dependent protein kinase for example). In the case of receptor-linked TYROSINE KINASES, autophosphorylation may create a site at which an interaction with an effector protein (containing an SH2 DOMAIN) occurs. *See:* GROWTH FACTOR RECEPTORS; PROTEIN PHOSPHORYLATION; SECOND MESSENGER PATHWAYS.

Schulman, H. & Lou, L.L. (1989) *Trends Biochem. Sci.* **14**, 62–66.

autoradiography Procedure widely exploited in cell and molecular biology to detect the location of radioactively labelled material on or within cells and tissues, or on gels. The emissions from the radioisotope make an image on photographic film or IMAGE PLATE laid over the material and thus the presence and positions of the radioactively labelled molecules are recorded. The radioactive isotope may be incorporated into cellular material in the form of a radioactively labelled metabolite or precursor to macromolecules such as proteins and nucleic acids. Radioactively labelled reagents (e.g. specific ANTIBODIES, or nucleic acid PROBES) may also be added to cells and tissues to detect and locate particular structures or nucleic acid sequences. The procedure is very sensitive, and is widely used in conjunction with HYBRIDIZATION with radioactive probes to identify a particular DNA on a gel or nitrocellulose filter and to locate the expression of specific mRNAs in tissue sections. The use of autoradiography with low energy emitters such as ^{3}H and ^{14}C requires the use of 'indirect' autoradiography (or FLUOROGRAPHY). *See also:* RECEPTOR AUTORADIOGRAPHY.

autoreceptors In the nervous system, presynaptic receptors which are acted on by the released NEUROTRANSMITTER to regulate the subsequent release of that transmitter system. *See:* LONG-TERM POTENTIATION.

autoregulation *See:* AUTOGENOUS CONTROL.

autosomal Describes a gene carried on an AUTOSOME, that is any chromosome other than a SEX CHROMOSOME.

autosomal dominant Single-gene trait that is encoded on an autosome (a chromosome other than a sex chromosome) and is expressed in dominant fashion in heterozygotes, that is, the trait is expressed if only one copy of the allele responsible is present. Examples of genetic diseases with autosomal dominant inheritance are rare, but include HUNTINGTON'S CHOREA and polycystic kidney disease. *See also:* MENDELIAN INHERITANCE.

autosomal recessive Single-gene trait that is encoded on an autosome (a chromosome other than a sex chromosome) and is expressed in a recessive fashion, that is, the trait is manifest only if two copies of the allele responsible are present. Many genetic diseases and inborn errors of metabolism are of this type. *See also:* MENDELIAN INHERITANCE.

autosome Any chromosome other than the SEX CHROMOSOMES. In the human there are 22 homologous pairs of autosomes and a pair of sex chromosomes.

autotomy (1) In higher vertebrates the self-mutilation that sometimes follows nerve damage.
(2) The casting off of a leg, tail, or joint, as seen in some lizards, and various crabs.

auxI/II *See:* IAAM/H GENE.

auxin A class of PLANT HORMONES, discovered between 1926 and 1928. The term was originally used as a generic name for substances which caused curvature (by cell elongation) when applied to the *Avena* (oat) coleoptile. K.V. Thiman defined them as 'organic substances which at low concentrations (<0.001 M) promote growth (cell enlargement) along the longitudinal axis, when applied to shoots of plants freed as far as practical from their own inherent growth-promoting substances, and inhibit the elongation of roots'. Auxins include naturally occurring and synthetic compounds, the most common example of the former being INDOLE-3-ACETIC ACID (IAA), which seems to be ubiquitous in higher plants, but also including 4-chloroindol-3-yl acetic acid (4-Cl IAA) and phenylacetic acid. Effects include promotion of shoot and, possibly, root extension at dilute concentrations and inhibition at stronger ones, inhibition of abscission, promotion of adventitious rooting, induction of seedless fruits (parthenocarpy), leaf epinastic curvature and tracheary differentiation, suppression of lateral bud outgrowth (apical dominance), and stimulation of ethylene biosynthesis. They cause rapid promotion of cell extension (in <15 min), which is currently explained in terms of increased cell wall extensibility arising from acidification due to accelerated proton extrusion that increases the electrical potential difference between the inside and outside of the cytoplasm (hyperpolarization). They also cause changes in the rate of expression of several genes (*see* SAURS), whose function is unclear.

Thiman, K.V. (1969) In *The Physiology of Plant Growth and Development* (Wilkins, M.B., Ed.) 1–45 (McGraw-Hill, London).
Estelle, M. (1992) *BioEssays* **14**, 439–444.

auxin-binding proteins (ABPs) Proteins that bind INDOLE-3-ACETIC ACID and other AUXINS specifically. The most extensively characterized is a membrane-associated GLYCOPROTEIN of M_r 20 000–22 000 isolated from maize and referred to as ABP-1. It is localized primarily in the ENDOPLASMIC RETICULUM and may exist *in vivo* as a dimer. Molecular cloning of ABP-1 from maize and *Arabidopsis* suggests that it is a member of a gene family and has revealed that the newly translated protein contains a 38-residue SIGNAL SEQUENCE and a C-terminal KDEL sequence that targets it to the lumen of the endoplasmic reticulum. Evidence based on the abolition of auxin-stimulated HYPERPOLARIZATION of the plasma membrane of PROTOPLASTS by antibodies against ABP-1 has indicated that some ABP may be present in this membrane and function as an auxin receptor. The plasma membrane ABPs may, however, be distinct proteins that share a common epitope with ABP-1. Other auxin-binding proteins have been identified by PHOTO-AFFINITY LABELLING. For example, proteins of M_r 23 000 and 40 000–42 000 located in the plasma membrane may be carrier proteins involved in polar auxin transport. A soluble ABP of M_r 65 000 has been identified using ANTI-IDIOTYPIC ANTIBODIES although its function is not yet known.

Jones, A.M. & Herman, E.M. (1993) *Plant Physiol.* 101, 595–606.

auxotroph A bacterial or fungal strain carrying a mutation that generates a nutritional requirement. In order to grow an auxotroph on a defined medium (i.e. a medium containing the minimum nutrients essential for the growth of the WILD-TYPE organism), the medium must be supplemented with the essential nutritional requirement. This simple growth requirement can be used as a means of distinguishing an auxotroph from a PROTOTROPH. Auxotrophs requiring particular amino acids and bases have been widely exploited in classical genetic studies of *Escherichia coli* and yeast, and auxotrophy may also be used as a selectable marker.

Avena sativa Oat.

AVG (aminoethoxyvinylglycine) L-2-Amino-4-(2-aminoethoxy)-*trans*-3-butenoic acid (Fig. A51) (M_r 232.4 as the dihydrochloride). It inhibits pyridoxal phosphate-mediated reactions, including the conversion of S-ADENOSYLMETHIONINE to ACC (1-aminocyclopropane-1-carboxylic acid), the precursor of the gaseous plant hormone ETHYLENE. It is frequently used at concentrations of ~0.5–10 mmol m^{-3}, to inhibit ethylene biosynthesis in attempts to establish ethylene involvement in physiological processes. Specificity of action is tested by attempting full reversal of AVG effects with ACC.

Yu, Y-B. & Yang, S.F. (1979) *Plant Physiol.* 64, 1074–1077.

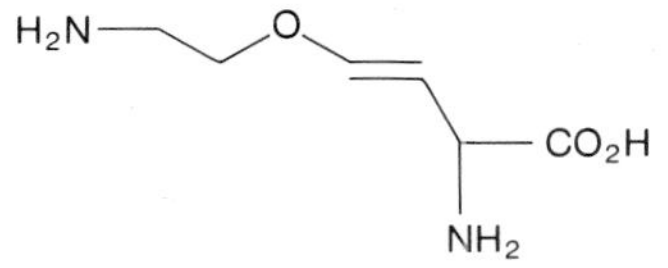

Fig. A51 Structure of AVG.

Avian development

THE avian embryo has a long and distinguished history as a subject of embryological study. The ancient Egyptians appear to have been the first to investigate its development in a systematic way; they incubated hens' eggs for varying periods of time and opened them to examine the state of development of the embryo. It was also a study of avian embryos that led William Harvey, in the 17th century, to the discovery of the circulation of the blood, and observations of the beating heart of the chick embryo that occupied René Descartes in part of his *Discours de la Méthode*. Notwithstanding this, the avian embryo is still one of the organisms of choice for modern developmental biology because it is easily obtained, large and relatively translucent, allows delicate microsurgical manipulations to be performed easily, and because its development is relatively well understood.

Birds are amniotes, like mammals, whose development they closely resemble (*see* MAMMALIAN DEVELOPMENT). The main differences are in the earliest stages; the avian embryo does not have a placenta and is a self-contained developing system.

To the modern developmental biologist, the avian embryo offers a very accessible system in which molecular studies can be combined with 'classical' embryology. Excellent staging systems are also available for the chick embryo. Transplantation, cell labelling, immunocytochemistry, and chemical treatments can now be combined with *in situ* HYBRIDIZATION and NORTHERN BLOTTING to examine changes in the patterns of gene expression resulting from experimental manipulations with well-studied effects. Moreover, with the advent of techniques for producing TRANSGENIC birds, which have recently become available, the developmental effects of targeted mutations can be studied in cellular as well as molecular detail.

The egg

Most of the material in the new laid egg (Fig. A52) [1,2] is nutrients, laid down by the mother to support the embryo. The embryo itself lies initially on the surface of the yolk, just under the vitelline membrane. The egg is designed to conserve water, to allow gaseous exchange and to prevent microorganisms from coming into contact with the embryo, as well as to provide nutrition and protection to the embryo.

It also provides a complex environment for the embryo to develop. For example, the pH of the yolk is slightly acid, while that of the albumen is quite alkaline. During early stages of development, the edges of the single-cell-thick embryo attach to the inner (yolk) face of the vitelline membrane, on which it expands, and it is thus poised in a pH gradient that may amount to as much as 3 pH units across a single cell.

Chick and quail embryos hatch 19–21 days after laying if incubated at 38°C; quail embryos develop slightly faster (19–20 days) than chicks (20–21 days).

Development *in utero*

After fertilization, which occurs internally in the mother, a hen's egg spends about 5 hours in the oviduct, and then moves to the uterus, where CLEAVAGE begins. This happens about 5.5 h after laying of the previous egg. The cleaving embryo remains in the uterus some 20 h before it is laid.

Cell division (cleavage) of the fertilized egg occurs in a planar way, with the daughter cells staying in the plane between the vitelline membrane and the yolk. Unlike mammalian embryos, which have holoblastic cleavage, cleavage in avian embryos is meroblastic. This means that the membranes separating new

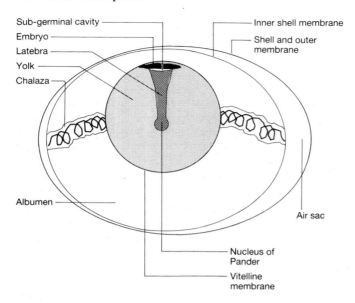

Fig. A52 Diagram of the principal components of a fertile hen's egg. The yolk is mainly PHOSPHOLIPIDS and is the main source of nutrients for the embryo. It is enveloped by a translucent, noncellular vitelline membrane and surrounded by albumen (egg white). The whole is contained within a calcareous shell, from which it is separated by two egg membranes. At the blunt end of the egg, between the two membranes, is an air sac. The embryo lies on the surface of the yolk, just under the vitelline membrane. The viscous albumen surrounding the vitelline membrane is connected with the inner membrane by two GLYCOPROTEIN threads called the chalazae, which allow the yolk to rotate so that the embryo always faces the top of the egg. In addition to nutritive components and GLYCOSAMINOGLYCANS to increase its viscosity, the albumen contains the enzyme LYSOZYME, a bacteriostatic agent. Egg albumen is therefore often used in embryo cultures to prevent the growth of microorganisms as well as to provide a source of nutrients.

cells arise as open cleavage planes, stretching into the yolk. As a result, the cells forming in the centre of the cleaving blastodisc are smaller than those around its periphery, and those around the edge contain more yolk. Because of this quirk, the MORULA and BLASTULA stages, as defined in lower vertebrates, are difficult to distinguish in avian embryos.

As the egg descends along the mother's reproductive system, it rotates while the albumen and shell components are secreted around it. During this rotation, the plane of the cleaving blastodisc remains at an angle of about 45° to the radius of the earth. This angle and the rotation are believed to be important in determining the polarity of the embryo. The edge of the blastodisc facing downwards will form the caudal (posterior) part of the embryo, under the influence of gravity.

It is possible to culture avian embryos out of the egg from late uterine stages until the formation of axial structures (head primordium, beating heart, somites, neural tube, notochord, etc.). Although it is still difficult to access the very earliest stages of development (cleavage) for experimental manipulation, several groups have recently developed techniques of 'total embryo culture' which allow access to the one-cell stage and incubation of the egg to the hatching-chick stage. Using this technique, it has been possible to inject a DNA construct into the single cell embryo and to study its expression at much later stages of development.

Formation of the embryonic axis and origin of the mesoderm

At the time of egg laying, the chick blastoderm consists of a disc of cells, some 2 mm in diameter, comprising an inner translucent area pellucida and an outer area opaca (Fig. A53). The latter region only contributes to extraembryonic structures. The first

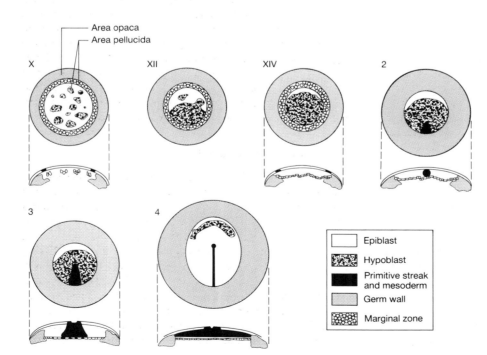

Fig. A53 Diagram illustrating the early stages of development of the chick embryo. Two staging systems are used [6]: Eyal-Giladi and Kochav's (in Roman numerals) for pre-primitive streak stages, and Hamburger and Hamilton's (in Arabic numerals) starting at stage 2 with the appearance of the primitive streak. Each diagram shows the embryo viewed from its dorsal (epiblast) side, with the posterior end of the blastodisc facing downwards. Some transverse sections through the blastodiscs are also shown.

cell layer to be present as such is the epiblast, which is continuous over both areae opaca and pellucida. It is an epithelium, one cell thick, which soon becomes pseudostratified and columnar, the apices of the cells facing the albumen. From this arises a second layer of cells, the hypoblast. At this time, the hypoblast is no more than several unconnected islands of 5–20 cells. However, by ~6 h of incubation, more cells are added to it and it becomes a loose but continuous epithelium. The source of these hypoblast cells is the deep (ENDODERMAL) portion of a crescent-shaped region, the marginal zone (Fig. A53), which separates areae pellucida and opaca at the future posterior (caudal) end of the embryo.

The embryo now consists of two layers: the epiblast proper, from which will arise all the embryonic tissues, and the hypoblast which will give rise only to extraembryonic tissues (mainly the yolk sac stalk) although it may also contain some PRIMORDIAL GERM CELLS. As the hypoblast continues to spread as a layer from posterior to anterior parts of the blastodisc, cells appear between the previous two germ layers. These are the first cells of the MESODERM [3]. As more of these accumulate, they coalesce in the first axial structure of the embryo, the **primitive streak**, which makes its appearance at the posterior margin of the area pellucida at about 10 h of incubation. Later, more mesodermal cells are recruited into the primitive streak by migration from the epiblast, as the streak elongates along the anteroposterior axis of the embryo.

All the mesoderm eventually migrates out of the streak to give rise to four mesodermal components: the lateral plates, intermediate mesoderm, paraxial mesoderm and the axial notochord. The lateral plates will give rise to the circulatory system, the lining of the coelom, the limb skeleton, and most of the remaining mesodermal organs. The intermediate mesoderm is a cord separating the lateral plate from the paraxial mesoderm, which gives rise to the transitory embryonic kidneys (pronephros and mesonephros). The paraxial mesoderm becomes subdivided into somites (see below), which give rise to the skeletal (voluntary) musculature, to the dermis, and to the skeleton of the trunk. The notochord forms a rod in the midline of the embryo and contributes to the vertebral bodies and intervertebral discs.

Before formation of the lateral plate, the mesoderm is packed densely at the primitive streak; as the lateral plate forms, it migrates massively, away from the axis of the streak. The left and right halves of the lateral plate later become separated from each other by the regression (shortening) of the primitive streak that occurs after the end of GASTRULATION. The notochord is laid down as a rod of mesoderm by the cranial tip of the primitive streak (Hensen's node), and elongates as the primitive streak regresses.

Some primitive streak cells insert into the hypoblast, displacing it towards the edges of the area pellucida (Fig. A52). These primitive streak derived, endodermal cells form the definitive endoderm, which will give rise to the lining of the gut. The elongation of the primitive streak and the expansion of the blastodisc, together with further recruitment of cells derived from the posterior marginal zone (this contribution now forming the junctional endoblast, which is also extraembryonic; Fig. A53), confine the original hypoblast to a crescent-shaped region underlying the anterior portion of the area pellucida; this region is known as the germinal crescent because it contains, transiently, the primordial germ cells, which will later migrate into the gonads.

Regulation and induction

Higher vertebrate embryos are capable of extensive REGULATION. If a pre-streak chick blastoderm is cut into several portions, most of the portions are able to produce a complete, albeit somewhat smaller, embryo. Furthermore, if pieces derived from several embryos are combined, a fairly normal embryo develops which receives contributions from all the donor fragments. These findings indicate that at this stage of development, groups of cells are not committed to a preordained programme of development. Development at this stage must, therefore, transcend the CELL-AUTONOMOUS interpretation of a genetic programme; cell interactions are important in determining the pattern of gene expression in individual cells.

There are, in theory, two possible interpretations for these phenomena: the first is that cells in the embryo are PLURIPOTENT, and become specified to their ultimate fates by cell interactions; when a fragment is separated from the embryo, cells in the fragment interact so that the correct proportions of different cell types are generated. The second interpretation of the phenomenon of regulation requires the embryo to consist of a fairly random mix of cells already specified to their ultimate fates, which later sort to their correct destinations. According to this second interpretation, isolation of a fragment of embryo leads to sorting of the cell types to newly defined sites, but the cells do not change fates. We do not yet know which of these two interpretations is correct, although the first is generally favoured.

The hypoblast layer appears to be involved in the specification of embryonic polarity. If it is rotated by 180° about its anteroposterior axis, the primitive streak now forms from the opposite (anterior) pole of the blastodisc. The nature of the interactions between hypoblast and presumptive primitive streak cells in the epiblast is not understood, although it has been suggested that the hypoblast induces the formation of mesoderm from the epiblast, as the vegetal pole cells induce mesoderm from the animal cap in amphibian embryos (*see* INDUCTION). However, it has not yet been possible to demonstrate mesoderm induction in isolated fragments of avian embryos, as has been done in amphibians.

Origin of the nervous system

The cells of Hensen's node have rather special properties. If a node is grafted into the lateral region of a host embryo, a second nervous system is formed (Fig. A54), just like a graft of the dorsal lip of the blastopore generates a second nervous system in amphibians (*see* AMPHIBIAN DEVELOPMENT; INDUCTION). This is neural induction. It can be demonstrated, using chick/quail chimaeras or other methods of cell marking, that the supernumerary nervous system is derived from cells of the host epiblast rather than of the grafted node. Some signal from Hensen's node can therefore change the fate of the overlying epiblast cells to make them become nervous system, rather than epidermis.

After neural induction, the epiblast that will become neuro-

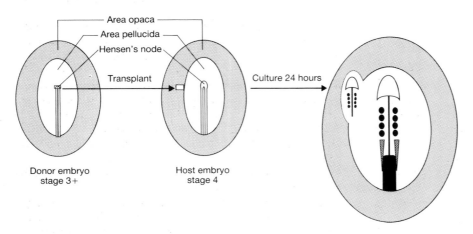

Fig. A54 When the tip of the primitive streak, known as Hensen's node, of a donor embryo is transplanted into another region of a host embryo, the transplant induces the formation of a secondary (ectopic) embryonic axis. The transplant itself differentiates into mesodermal cells (somites and notochord), while the nervous system comes from the epiblast of the host.

epithelium undergoes a series of dramatic changes. First, it becomes thickened. Then its edges elevate, forming the neural plate. This plate then rolls into a tube (Fig. A55). The whole process progresses roughly in an anterior-to-posterior direction so that the neural plate is still elevating in the tail region while it has already closed in cranial regions. The rims of the neural plate that meet during tube closure (the dorsal part of the neural tube) give rise to the neural crest, a special multipotent population of migratory cells that colonize many regions of the embryo and give rise to melanocytes, to the sensory component of the PERIPHERAL NERVOUS SYSTEM, to the autonomic ganglia (sympathetic and

parasympathetic), to glial (Schwann) cells that accompany peripheral nerves, to the adrenal medulla, and to skeletal elements of the cranial vault and face. The motor nerves do not arise from the neural crest; instead, axons emerge from the ventral portion of the neural tube, from where they extend outwards to find their peripheral targets.

This stage of development is amenable to experimentation. The neural plate and neural tube can be extirpated, rotated, and transplanted, as can the notochord underlying it. Individual cells can be marked and their descendants followed, or they can be injected with antibodies or nucleic acids. Chick/quail chimaeras

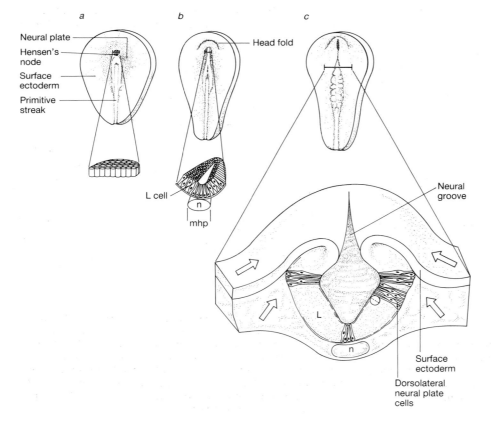

Fig. A55 Mechanism of formation of the neural tube in avian embryos. *a*, Stage 4; *b*, stage 6; *c*, between stages 6 and 8. Whole mounts of entire blastoderms and cross-sectional views are shown. The neural furrow begins to be generated (see *b*) as the neural plate becomes anchored to the notochord (n) and midline neural plate cells decrease in height and become wedge-shaped to form the median hinge point (mhp). Forces for these processes are thought to be intrinsic to the neural plate and to include microtubule-mediated cell elongation, cell division and cell rearrangement. Between stages 6 and 8 (*c*), bending of the neural plate is the predominant event. The dorsolateral neural plate becomes anchored to surface ectoderm, and dorsolateral neural plate cells increase in height and become wedge-shaped, resulting in neural plate furrowing, neural plate elevation around the mhp and neural fold convergence. Forces responsible are both intrinsic (cell wedging) and extrinsic (open arrows), the latter being provided by surrounding non-neuroepithelial tissues. Adapted from [5] with kind permission of Dr G.C. Schoenwolf.

can be constructed to study interactions between different groups of cells.

Segmentation

The segmental organization of vertebrate embryos is most obvious in the pattern of somites, from which derive the vertebrae and ribs of the axial skeleton, the dermis of the trunk, and all the voluntary musculature of the adult [4]. The metameric pattern of somites determines the segmental arrangement of other structures in the embryo, such as the peripheral nervous system.

In the chick embryo, some 55 pairs of somites form, each somite being, at first, an epithelial sphere that buds off the rostral (anterior) end of each of the paired segmental plates of paraxial mesoderm that appear about 1.5 days after the egg is laid. Each pair of somites takes about 1.5 h to form. Some 6–8 h after its initial appearance, each somite splits up into two further components (Fig. A56): the dermomyotome dorsally, which retains some epithelial characteristics and gives rise to the dermis of the trunk and to skeletal muscle, and the sclerotome ventromedially, which is a loose MESENCHYME that gives rise, along with the notochord, to the axial skeleton. Recent results suggest that the segmental pattern of somites is controlled at least in part by cell-autonomous properties depending on the cell division cycle.

Each sclerotome is subdivided into a rostral (anterior) and a caudal (posterior) half. Differences between the cells of the two halves determine the segmental organization of the peripheral nervous system: motor nerves and neural crest cells are restricted to the rostral half of each sclerotome and are unable to colonize the caudal half. There are numerous molecular differences between the two halves of the sclerotome, but it is not yet known which of these are directly responsible for the segmental pattern of components of the peripheral nervous system.

The subdivision of the sclerotome into rostral and caudal halves is maintained by mixing restrictions between the two kinds of cells. Rostral cells are able to mix with rostral cells, and caudal with caudal, but when rostral and caudal cells meet they generate a boundary that separates them. This is probably important in the morphogenesis of the vertebral column. Traditionally, it is thought that the rostral half of one sclerotome joins the caudal half of the preceding sclerotome to give rise to one vertebra, a model which accounts for the observation that the vertebrae are out of phase with the corresponding myotomes (axial muscle blocks). However, recent experimental work tracing the derivatives of individual somites has cast some doubt on this interpretation; it suggests that each somite gives rise to one vertebra, but that sclerotome cells may shift anteriorly by a half segment by migrating along the notochord.

Two portions of the CENTRAL NERVOUS SYSTEM are also organized in a segmental way (Fig. A57). The hindbrain (rhombencephalon) is subdivided into seven or eight segments (*see* VERTEBRATE NEURAL DEVELOPMENT) which display an alternating pattern of neuronal differentiation and other properties. These segments (rhombomeres), like the somites, are separated from each other by lineage restrictions that prevent cells from one rhombomere from crossing to the adjacent one. The diencephalon is similarly subdivided into four segments ('prosomeres' or 'diencephalic neuromeres'), again with an alternating pattern of neuronal differentiation and cell surface properties, and lineage restrictions maintaining the boundaries between adjacent prosomeres. However, the spinal cord does not appear to be truly segmented.

Limb development

The avian embryo has also been used extensively for studies on limb development. The limb buds begin to form at about 3 days

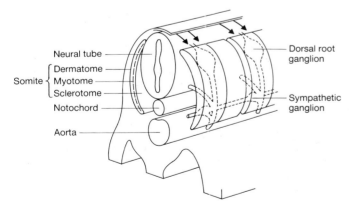

Fig. A56 Diagram illustrating the relationships between the developing nervous system in the trunk of avian embryos and the neighbouring mesodermal structures. All the components of the peripheral nervous system are confined to the anterior (rostral) halves of the somites. The thick arrows dorsally indicate the streams of migrating neural crest cells that will give rise to the dorsal root and sympathetic ganglia.

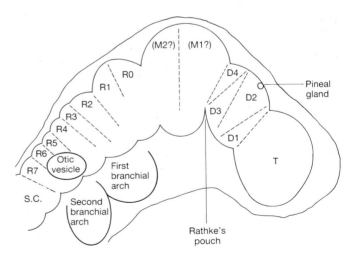

Fig. A57 Subdivisions of the central nervous system of the head of avian embryos. The diencephalon is subdivided into four neuromeres, D1–D4. The rhombencephalon (hindbrain) is subdivided into rhombomeres R1–R7, the first of which later subdivides into two further neuromeres R0 and R1. The midbrain may also be subdivided into two units (M1–M2) but the evidence for this is not yet strong. In the hindbrain, pairs of rhombomeres (R2–R3 and R4–R5) are aligned, respectively, with the first two branchial arches. The next pair of rhombomeres is aligned with the otic (ear) vesicle.

after laying, when the ectoderm opposite somites 15–21 (wing bud) and 27–33 (leg bud) begins to elevate from the surface of the body, enclosing a mass of loosely packed lateral plate mesoderm. In time, the cells of this mesoderm will condense to form the primordia of the limb skeleton. The muscles and dermis arise from somite-derived cells which migrate into the limb buds.

Grafting experiments (*see* PATTERN FORMATION) have shown that the posterior margin of the chick limb bud contains special cells, able to respecify the pattern of the whole limb. This region has been called the zone of polarizing activity (ZPA). If such a region is grafted into the anterior margin of a host limb bud, a mirror-image pattern of bone and muscle elements is formed. It is believed that the ZPA produces a diffusible MORPHOGEN (thought to be related to RETINOIC ACID), the local concentration of which at different positions in the limb determines the type of bone and muscle elements that will develop at those positions.

The distal part of the developing limb bud is also special (*see* PATTERN FORMATION). It consists of a thickened ridge of surface ectoderm (the apical ectodermal ridge (AER)) which covers a region of special mesoderm, called the progress zone. The AER is required for continued growth and development of the limb, while the progress zone appears to be involved in controlling the proximodistal sequence of skeletal elements.

Because the chick limb bud is very accessible to such transplantation experiments, it is an ideal system in which, in the near future, such operations might be combined with molecular techniques.

C.D. STERN

1 Burley, R.W. & Vadehra, D.V. (1989) *The Avian Egg: Chemistry and Biology* (John Wiley, New York).
2 Tullet, S.G. (1991) Avian incubation. In *Poultry Science Symposium No. 22* (Butterworth-Heinemann, London).
3 Stern, C.D. (1991) Mesoderm formation in the chick embryo, revisited. In *Gastrulation: Movements, Patterns and Molecules* (Keller, R.A. et al., Eds) 29–41 (Plenum, New York).
4 Keynes, R.J. & Stern, C.D. (1988) Mechanisms of vertebrate segmentation. *Development* **103**, 413–429.
5 Shoenwolf, G.C. & Smith, J.L. (1990) Mechanisms of neurulation: traditional viewpoint and recent advances. *Development* **109**, 243–270.
6 Stern, C.D. & Holland, P.W.H. (Eds) (1993) *Essential Developmental Biology: a Practical Approach.* (IRL Press, Oxford).

avian erythroblastosis virus (AEV) An acutely transforming RETROVIRUS.

avian leukosis virus (ALV) *See*: ONCOGENES; RETROVIRUSES.

avian sarcoma virus Rous' sarcoma virus. *See*: ONCOGENES; RETROVIRUSES.

avidin A tetrameric glycoprotein from egg white that binds to BIOTIN. *See*: BIOTINYLATION; IMMUNOELECTRON MICROSCOPY.

avidity The strength of the interaction of a multivalent ANTIBODY with a multideterminant ANTIGEN. The net avidity is a complex function of the AFFINITIES of the various determinants and of the valencies of both reactants involved. Thus, when multivalent polyclonal antibody reacts with polyvalent antigen the avidity is determined by the sum of all of the individual interactions taking place between individual antigen-binding sites and antigenic determinants. Taking an antibody with a given affinity for a certain antigen, the strength of binding will be higher for pentameric IgM than for divalent IgG antibodies.

avirulence genes *See*: PLANT PATHOLOGY.

avirulent Applied to strains or mutants of pathogens that cause no or only mild disease. *See*: PLANT PATHOLOGY.

avr A genetic locus in some plant pathogenic microorganisms that determines race/cultivar-specific expression of disease symptoms in conjunction with the functionally complementary resistance gene in the host. *See*: PLANT PATHOLOGY.

axial mesoderm Portion of the MESODERM underlying the midline of the embryo. In the strict sense, the axial mesoderm comprises only the HEAD PROCESS under the brain and the NOTOCHORD posterior to the ear vesicle. Some authors also include the PARAXIAL MESODERM as an axial component. In amphibian embryos the term is only rarely used. Instead, the term dorsal mesoderm is used to describe the axial components; this is because after folding of the embryonic body, the axial mesoderm comes to lie most dorsally in the embryo. The axial mesoderm arises from HENSEN'S NODE in higher vertebrates and from the dorsal lip of the blastopore in amphibians and is the tissue capable of eliciting NEURAL INDUCTION. *See also*: AMPHIBIAN DEVELOPMENT.

axis determination The major model systems for the analyis of the molecular basis of the determination of the anterior–posterior and dorsal–ventral axes in early embryogenesis have been *Xenopus laevis* (*see* AMPHIBIAN DEVELOPMENT) and *Drosophila* (*see* DROSOPHILA DEVELOPMENT).

axon Elongated cellular process that carries electrical impulses away from the cell body of a NEURON. The axon is often much divided, each branch making a synapse on the dendrites or other parts of another neuron. *See*: SYNAPTIC TRANSMISSION.

axon hillock *See*: NEURON.

axon terminal *See*: NEURON; SYNAPTIC TRANSMISSION.

axonal transport (axoplasmic transport) Transport of material along an axon from the cell body of a neuron towards the terminal or synapse (anterograde transport) or in the reverse direction (retrograde transport). Substances such as neurotransmitters carried within lipid vesicles are transported bidirectionally at fast speeds ($0.1–5\ \mu m\ s^{-1}$); CYTOSKELETAL components move *en bloc* only in the anterograde direction and more slowly ($0.001–0.01\ \mu m\ s^{-1}$). Fast transport is attributed to KINESIN or DYNEIN attaching to the vesicles and moving them in one or the other direction along microtubule tracks; slow axonal transport consists of a steady movement of cytoskeletal polymers from cell body to axon. Details of the mechanism are not understood. *See also*: MICROTUBULE-ASSOCIATED MOTORS.

Schnapp, B.J. & Reese, T.S. (1986) *Trends Neurosci.* **9**, 155–162.
Reinsch, S.S. et al. (1991) *J. Cell Biol.* **115**, 365–379.

axoneme Central core of eukaryotic cilium or flagellum. *See*: MICROTUBULES.

axoplasmic flow *See*: AXONAL TRANSPORT.

5-azacytidine Cytidine analogue (Fig. A58) which when incorporated into DNA cannot be methylated and leads to demethylated sites in DNA. Under certain circumstances it can induce changes in the state of cellular differentiation and activate genes on a silent X-chromosome, supporting a role for methylation in gene regulation. *See*: DNA METHYLATION; NUCLEOSIDES AND NUCLEOTIDES; X-CHROMOSOME INACTIVATION.

Azorhizobium Genus of RHIZOBIA which can carry out NITROGEN FIXATION in association with the roots of leguminous plants. *See also*: NODULATION.

AZT The deoxyribonucleoside analogue 3'-azido, 3'-deoxy-

Fig. A58 5-Azacytidine.

thymidine, also known as zidovudine. It inhibits the activity of the retroviral enzyme REVERSE TRANSCRIPTASE and is used in the chemotherapy of HIV virus infection (*see* IMMUNODEFICIENCY VIRUSES). It appears to improve the quality of life in patients in which the symptoms of AIDS have appeared, and delays death, but it is now thought not to delay the time after infection at which symptoms appear. *See also*: DIDEOXYRIBONUCLEOSIDES.

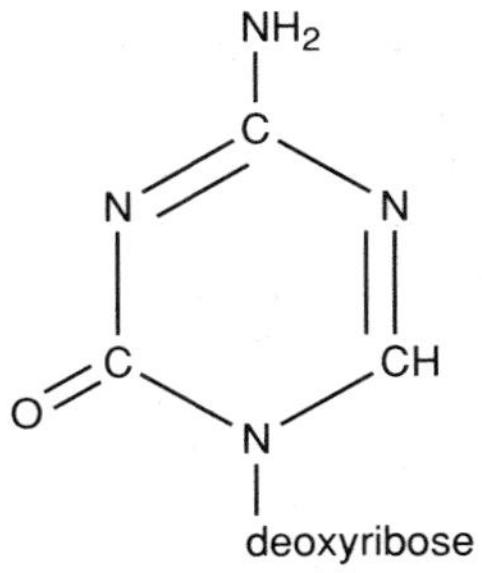

azurin *See*: MOLECULAR EVOLUTION: SEQUENCES AND STRUCTURES.

B

B Abbreviation for aspartate when found in an acid hydrolysate of a protein, when it may derive from either asparagine or aspartic acid residues.

B$_{max}$ *See*: SCATCHARD PLOT.

B antigen *See*: ABO BLOOD GROUP SYSTEM.

B cell (1) of pancreas, *see* ISLET CELLS.
(2) B lymphocyte. A subset of small LYMPHOCYTES found in SECONDARY LYMPHOID ORGANS and circulating in the blood, and characterized by the possession of antigen-specific cell surface IMMUNOGLOBULIN. When stimulated by the cognate antigen they proliferate and differentiate into antibody-producing plasma cells. *See*: ANTIBODIES; B CELL DEVELOPMENT.

B cell development and differentiation

B CELLS, a subset of LYMPHOCYTES, are the precursors of ANTIBODY-secreting cells. Each B cell carries on its surface IMMUNOGLOBULIN molecules of a single antigen-binding specificity which act as receptors for ANTIGEN. Interaction of antigen with cell-surface immunoglobulin contributes to the signals which induce B cells to proliferate and develop into antibody-secreting plasma cells. When this occurs, the immunoglobulin secreted as antibody has the same specificity as that found on the surface of the parent B cell. In addition, a proportion of B cells proliferating in response to antigen revert to small lymphocyte form as MEMORY B CELLS.

B cells differentiate from haematopoietic STEM CELLS in the bone marrow. The first phase of their development is independent of antigen and takes place in the bone marrow. Immunocompetent mature B cells leave the marrow for the secondary lymphoid organs, where they may encounter their cognate antigen and embark on the final phase of differentiation to antibody-secreting cells or memory cells.

Further information and a wide range of references on this subject can be found in [1].

The primary production of B cells

B cells are produced throughout life in the bone marrow from the stem cells that give rise to all the blood cells including red cells, phagocytes, and platelets as well as lymphocytes. Before birth, B cells are also produced in the liver and spleen. Primary B cell development is associated with the rearrangement of the genes encoding the antigen-combining variable (V) region of immunoglobulins from their germ-line configuration (Fig. B1) (*see* GENE REARRANGEMENT; IMMUNOGLOBULIN GENES). This process is responsible for generating much of the huge range of antibody specificities contained within the B cell REPERTOIRE (*see* GENERATION OF DIVERSITY). Further diversity in the V region is introduced by the HYPERMUTATION process (see below) which occurs in B cells activated by certain types of antigen.

During primary immunoglobulin gene rearrangement, B cell progenitors first rearrange the genes that encode the V regions of the immunoglobulin heavy (H) chains. If a productive rearrangement is achieved, IgM (*see* ANTIBODY) heavy chains are produced and the cells at this stage are termed pre-B cells. Some of the heavy chain is transported to the pre-B cell surface as a complex with two molecules, λ5 and Vpre-B, which form a surrogate

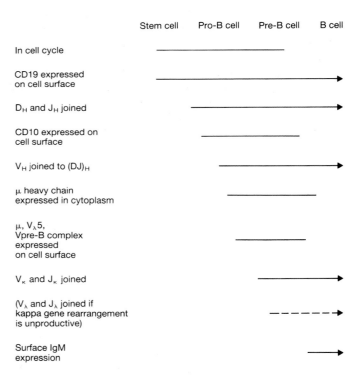

Fig. B1 The sequence of events during primary B cell production. V$_H$, D and J$_H$ are gene segments that together encode the immmunoglobulin heavy chain V region and V$_κ$ and J$_κ$ and V$_λ$ and J$_λ$ are gene segments together encoding the V region of immunoglobulin light chains κ and λ respectively (*see* ANTIBODIES; IMMUNOGLOBULIN GENES). CD10 and CD19 are B lineage-associated proteins. μ is the IgM heavy chain. The μ, λ5, Vpre-B complex is explained in the text.

immunoglobulin light chain. Provided this surface complex is produced the cells go on to rearrange their light chain genes, but if it is not they die. Once light chains are made, intact immunoglobulin is expressed at the cell surface marking the transition from pre-B cell to B cell. The common form of acute LEUKAEMIA, which occurs in childhood, is associated with genetic damage which impairs maturation in a clone of proliferating B cell precursors.

Productive rearrangement and expression of gene product at the cell surface leads to the cessation of gene rearrangement. In this way only one heavy chain allele and one light chain allele are expressed in B cells — the phenomenon of ALLELIC EXCLUSION — resulting in a B cell that expresses immunoglobulin of a single antigen specificity.

In adults, some 3–5% of the bone marrow mononuclear cells are involved in B cell production. Studies in rodents indicate that most newly produced virgin B cells (B cells that have not yet encountered foreign antigen) leave the bone marrow for the secondary lymphoid organs where they may be activated by antigen. If this does not happen most die within a day or two. About 10%, however, become mature peripheral B cells which have a life span of 4–6 weeks. This recruitment to the long-lived peripheral B cell pool does not seem to involve antigen-driven proliferation, although some degree of selection, based on the specificity of surface immunoglobulin of the recruited cells, does seem to be involved.

Most mature peripheral B cells are in a state of constant nonrandom migration between the secondary lymphoid tissues. These migrant cells are known as recirculating B cells. Most mature T cells also recirculate. Recirculating B and T cells pass from the blood into secondary lymphoid tissues through specialized small blood vessels. In the zones around these blood vessels T and B cells are in intimate contact with each other and ANTIGEN-PRESENTING CELLS involved in T and B cell activation. Newly produced virgin B cells also migrate into these areas but if they are not immediately activated by antigen, or recruited into the mature recirculating pool they die. In contrast the recirculating T and B cells, if not activated by antigen, move from this area into T cell-rich zones and B cell-rich zones respectively; the latter being termed follicles. They remain in these sites for some hours before leaving to re-enter the blood, usually via the lymphatic system. It seems that the majority of recirculating B cells are virgin cells in that they have not undergone antigen-driven proliferation. An important minority of the recirculating pool of B cells, however, are memory B cells.

B cell activation and terminal differentiation

An antibody response to a particular antigen involves: (1) a regulated expansion of the number of specific antigen-reactive B cells by proliferation and (2) the maturation of B cells to become antibody-secreting cells or memory cells (Fig. B2). These processes, in part, are initiated by the interaction of B cells with the antigen: but they also depend on a range of additional signals. Many of these additional signals are delivered by T CELLS (T cell help) which themselves have been activated by antigen (*see* T CELL DEVELOPMENT). Further signals are derived from accessory cells, which include certain MACROPHAGES and specialized antigen-presenting cells. Figure B2 illustrates the sequential signalling which drives antibody responses.

B cell activators include secreted messenger molecules known collectively as CYTOKINES, which normally act over short distances within the microenvironments in which B cells are activated. Cytokines involved in B cell activation include the interleukins (IL) IL-1, IL-2, IL-3, IL-4, IL-6, IL-7, and IL-10 (*see* LYMPHOKINES). The INTERFERONS, TUMOUR NECROSIS FACTORS (TNF), and shed soluble CD23 (*see* CD ANTIGENS) are other cytokines which can activate B cells (*see* LYMPHOKINES). Other signals are delivered by the direct interaction of molecules on B cells with those on HELPER T CELLS, macrophages or antigen-presenting cells.

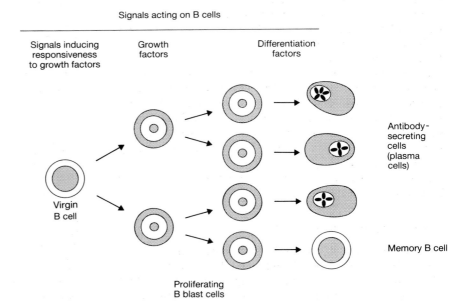

Fig. B2 Cellular events associated with an antibody response.

This complexity of signalling in part reflects the diversity of differentiated forms of B cell. In the case of antibody-secreting cells, for example, different combinations of signals control: (1) the class or subclass of antibody produced (*see* CLASS SWITCHING); (2) the site of antibody production, which is usually distant from the place where the parental B cell is activated; (3) the life-span of the antibody-secreting cell; and (4) the amount of antibody produced. The complexity of the system also reflects an ability to produce rapid but regulated antibody responses.

Inappropriate or defective antibody production can result in a number of diseases, such as: (1) allergic diseases (e.g. asthma and hay fever); (2) diseases associated with the deposition of antigen–antibody complexes in blood vessels, a process which can result in kidney failure; (3) overactivity of the thyroid as a result of stimulation by antibodies which specifically trigger the receptors in the thyroid for thyroid-stimulating hormone; (4) damage to tissues through the production of AUTOANTIBODIES; (5) bacterial infection through lack of protective antibody. Genetic mutation, translocation or deletion occurring in the B lymphocytes activated by antigen are responsible for myelomatosis, a neoplasm of bone marrow plasma cells, most of the chronic lymphocytic leukaemias and for the majority of non-Hodgkin's LYMPHOMAS.

There are three main microenvironments in which B cells are activated: (1) in association with macrophages; (2) in the zones where T and B lymphocytes enter lymphoid tissues from the blood or lymph; and (3) in the B cell follicles of lymphoid tissues. The nature of B cell activation in these sites depends on the type of antigen (see below) and whether the antigen has been encountered previously.

Categories of antigen

There are three different categories of antigen. These are distinguished by whether the core structure of the antigen is based on PROTEIN, POLYSACCHARIDE, or LIPOPOLYSACCHARIDE:

1 T cell-dependent (TD) antigens are based on proteins or polypeptides. These can be processed and presented to T cells in association with self MHC MOLECULES (*see* ANTIGEN PROCESSING AND PRESENTATION; MAJOR HISTOCOMPATIBILITY COMPLEX; T CELL RECEPTORS). Clinically important examples of TD antigens include viral antigens, tetanus and diphtheria toxoids, and antigens associated with pollens. In general, antibody responses to pure protein molecules as well as to small non-protein determinants (haptens) attached to proteins require accessory signals (help) from T cells. The signals required to induce B cells to respond to these antigens are delivered within the secondary lymphoid tissues — the lymph nodes, the spleen and the mucosa-associated lymphoid tissues, which include the tonsils, appendix and Peyer's patches.

2 Type 1 T cell-independent (TI-1) antigens are typified by bacterial lipopolysaccharides, particularly the ENDOTOXINS associated with Gram-negative bacilli. These antigens can induce antibody responses in the presence of macrophages and other accessory cells without T cell help. Macrophages have receptors for bacterial lipopolysaccharides and can be activated through these to produce cytokines which activate B cells. Many B cells have receptors for lipopolysaccharide. It is a characteristic of TI

antigens that they can evoke antibody responses as early as the second trimester of human foetal life.

3 Type 2 T cell-independent (TI-2) antigens are characteristically based on pure polysaccharide molecules. Clinically the most important are bacterial capsular polysaccharides. Encapsulated bacteria are particularly likely to cause pneumonia, septicaemia, and meningitis in children aged between 6 months and 5 years. It is probably significant that responsiveness to these antigens does not develop until well after birth and is not complete until about 5 years of age. The selective advantage of an immune system which delays the age at which responsiveness develops to this group of antigens remains obscure.

Antibody responses to TD antigens

Following the first exposure to a protein antigen the response is often slow to yield antibody and the amount of antibody produced is small — the primary immune response. On secondary challenge with the same antigen the response is more rapid and of greater magnitude — the secondary immune response. In the primary response, the main limiting factor is the availability of T cell help. If T help is made nonlimiting experimentally, the primary B cell response has a speed of onset and magnitude which is far more like a secondary than a primary response.

Somatic hypermutation

Characteristically, during TD responses the AFFINITY of the antibody produced increases markedly. This is brought about by a hypermutation mechanism which acts specifically on the genes encoding the variable regions of antibodies. Hypermutation does not seem to be activated during TI antibody responses and does not occur in the genes which encode the variable regions of T cell receptors. It is now clear that the hypermutation mechanism is switched on in activated B cells proliferating in GERMINAL CENTRES. These are structures that develop in the follicles of secondary lymphoid organs in the early phase of exposure to TD antigens and persist for about 3 weeks from the time of exposure to antigen. The B cells are induced to form germinal centres by interaction with antigen in the presence of specific T cell help. The B cells which have acquired mutations in their immunoglobulin V region genes seem to be selected on the basis of their capacity to be activated by antigen held on the surface of specialized antigen-retaining cells found in germinal centres.

Antibody responses to TI-1 antigens

These responses differ from TD responses in that they result in faster and greater antibody production on first exposure to antigen. This is attributable, at least in part, to the availability of accessory signals evoked by the interaction of lipopolysaccharide with both B cells and macrophages. Although protein associated with most TI-1 antigens may also evoke some T cell help, long-lived immunological memory for these antigens is poor and protection afforded by vaccines of TI-1 antigens is often only temporary.

Antibody responses to TI-2 antigens

Immunization with these antigens is generally ineffective in humans before 6 months of age and then only a few TI-2 antigens produce any response. Between 6 months and 5 years responsiveness gradually increases in terms of both the amount of antibody produced and the range of TI-2 antigens recognized. Because of the clinical importance of infections against encapsulated bacteria, vaccines are now being developed in which TI-2 antigens are linked to proteins like tetanus toxoid. These synthetic vaccines are able to evoke antibodies to certain bacterial capsular polysaccharides in children aged <6 months. Pure bacterial polysaccharides induce little or no B or T cell memory but these antigens may persist in the body and maintain B cell activation for many months.

Characteristically B cell activation outside follicles is confined to the first few days following exposure to antigen. All three classes of antigen seem to be able to activate B cells in these sites. By contrast the follicular response is mainly a property of TD antigens and can continue for months.

Marginal zone B cells

In addition to the recirculating B cells, a second subset of mature peripheral B cells is found in the marginal zones of the spleen and equivalent areas of other secondary lymphoid tissues. These marginal zone B cells also comprise a mixture of memory and virgin B cells. They do not recirculate but are located in sites where they are likely to encounter antigen. In the spleen, for example, they are perfused by blood sinusoids and in Peyer's patches they are located immediately beneath the gut epithelium. The responsiveness of these cells to activation signals differs from that of either recirculating B cells or newly produced virgin B cells. It seems that marginal zone B cells can respond to all three categories of antigen including TI-2 antigens. The other two types of B cell only seem to respond to TD and TI-1 antigens. Marginal zone B cells seem to be derived by maturation of recirculating B cells.

B1 B cells

In the foetus and during the first year of postnatal life the B cells show certain differences from the majority of those found later in life. These are now termed B1 B cells. Only a small proportion of B cells in adult life are B1 B cells. Many of the self-reactive antibodies which are produced in autoimmune diseases are, however, derived from B1 B cells. In addition, the proportion of leukaemias and lymphomas arising from B1 B cells in adults is greater than one would expect from the proportion of B cells belonging to this subset in adults. Any advantage associated with the delay in the development of those B cells which are numerically dominant after infancy remains a matter for speculation.

I. MACLENNAN

1 MacLennan I.C.M. et al. (1990) The evolution of B cell clones. *Curr. Topics Microbiol. Immunol.* 159, 65–78.

B cell differentiation factor *See*: IL-4.

B cell growth factor BCGF; IL-4.

B-DNA The most important form of the DNA double helix biologically (the other forms are called A-, C- and Z-DNA). In B-DNA the twisting antiparallel strands make a complete turn each of 10 bases, forming a helix with one narrow and one wide groove. This makes the bases partly accessible. B-DNA winds right handed, which means that looking along the axis of the double helix the strands turn clockwise. *See*: DNA; NUCLEIC ACID STRUCTURE.

B-factor *See*: TEMPERATURE FACTOR.

B lymphocytes *See*: ANTIBODIES; B CELL; B CELL DEVELOPMENT.

B subfibre Component of axoneme of eukaryotic cilia and flagella. *See*: MICROTUBULES.

Bacillus subtilis A Gram-positive, spore-forming, rod-shaped soil bacterium, widespread in nature. Endospore formation has been studied as an example of BACTERIAL DIFFERENTIATION. *B. subtilis* is nonpathogenic and can be consumed by mammals in large quantities without side effects; it also excretes large amounts of extracellular proteins. Both features make it attractive for use in biotechnological processes. Although it does not possess native PLASMIDS, a wide range of cloning VECTORS is now available, making it a suitable host system for the cloning and expression of foreign genes. *See*: GENETIC ENGINEERING.

***Bacillus thuringiensis* endotoxins** Insecticidal proteins produced by strains of the bacterium *Bacillus thuringiensis* in crystalline inclusion bodies. The proteins are degraded in the alkaline midgut of the insects to release toxic fragments. The insecticidal action of these proteins is highly specific: different strains of bacteria produce proteins active against different types of insect. *See*: δ-ENDOTOXIN.

bacitracin Polypeptide antibiotic (Fig. B3) produced by *Bacillus* species. It inhibits GLYCAN syntheses in Gram-positive bacteria by preventing the conversion of undecaprenyl pyrophosphate to undecaprenyl phosphate, so inhibiting the undecaprenol phosphate cycle (*see* UNDECAPRENOL AND DOLICHOL CYCLES). Undecaprenyl-phosphoryl sugars, for example, are rendered un-

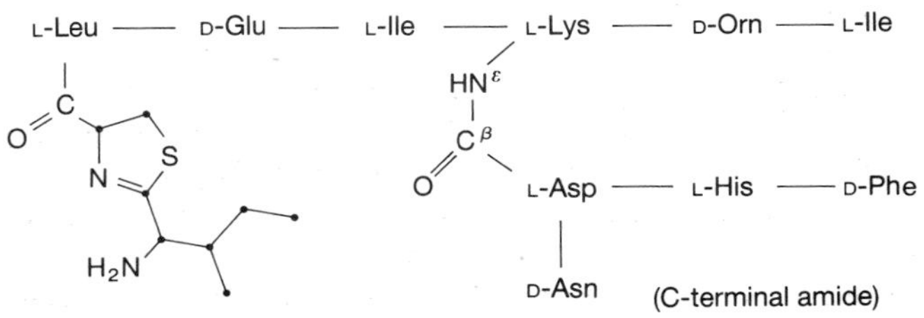

Fig. B3 Bacitracin A.

available for synthesis by being complexed with the antibiotic and held in membranes.

back-cross in animal and plant breeding, a cross of an F1 hybrid with one of the parental strains. *See:* MENDELIAN INHERITANCE.

back mutation A single base-pair mutation occurring in a mutant gene which restores the phenotype of the wild-type gene. A back mutation does not necessarily restore the wild-type DNA sequence.

bacterial cell *See:* BACTERIAL CHEMOTAXIS; BACTERIAL ENVELOPES; CYANOBACTERIA; PROKARYOTES.

Bacterial chemotaxis

MOTILE bacteria respond to chemical gradients by modulating their pattern of swimming so that progress is made toward a more favourable chemical environment. This behaviour, called **chemotaxis**, has been studied extensively in *Escherichia coli* and its near relative *Salmonella typhimurium* (see for example [1–3] and references therein). In these species the molecular components that mediate chemotaxis have been identified, and some mechanisms have been delineated. Where information is available about chemotaxis by other bacteria, there are many striking parallels with the system in *E. coli*, as well as some interesting variations [1,5,7]. The proteins that mediate intracellular signalling for chemotaxis are representatives of a large family of homologous bacterial sensory proteins, often referred to as two-component regulatory systems [1,5,6]. In these systems, environmental stimuli influence the activity of one component, a specific PROTEIN KINASE, which phosphorylates a second component, an appropriate effector protein [1,5,6]. In the chemotactic system the effector protein controls flagellar function; for most other two-component systems, the effector protein controls transcription of a specific gene or genes. These systems share many common features and thus a mechanistic understanding of the chemotactic system provides many insights into fundamental mechanisms by which prokaryotes sense and respond to their environment.

Chemotactic behaviour

A swimming bacterium traces a three-dimensional random walk. *E. coli* cells move in gentle curves, called runs, punctuated every second or two by brief episodes of uncoordination, called tumbles, that result in new, randomly chosen directions of swimming. The motor organelles are bacterial flagella (see Fig. B7 on p. 90), which are totally unlike eukaryotic flagella or cilia (for these *see* CELL MOTILITY; MICROTUBULES). In the bacterial flagellum, a long helical filament, about the diameter of a single microtubule, is turned like a propeller by a rotary motor. *E. coli* possesses around six flagella, randomly inserted over the cell surface, a pattern called peritrichous. The filament is attached to a complex struc-

ture of rings and rods, called the basal body, which spans the layers of the cell envelope (*see* BACTERIAL ENVELOPE) and includes the rotary apparatus [3,4].

Runs occur in wild-type bacteria when left-handed helical filaments rotate counterclockwise and thus exert a coordinated thrust. Tumbling occurs when peritrichous flagella rotate clockwise, causing uncoordinated pulling and deformation of the left-handed helices. A change in concentration of an active compound over time (a temporal gradient) results in an alteration of tumble frequency. Favourable changes suppress tumbles; unfavourable ones induce them. The bacterium senses spatial gradients by detecting the temporal gradient created as it swims. In a spatial gradient of attractant or repellent created by diffusion, the chemosensory system alters the pattern of swimming primarily by extending runs in favourable directions, thus creating a biased random walk that produces net progress.

The bacterial response to a temporal gradient is transient. Exposure of cells to a new chemical environment results in an altered probability of tumbles. However, after a time ranging from seconds to minutes, depending on the magnitude of the gradient, cells resume their initial behavioural pattern, even though the altered chemical environment persists. Thus, like most sensory cells, bacteria adapt. The biochemical mechanism of adaptation involves covalent modification of the receptor protein through which the stimulus passes. Specifically, several glutamyl residues on the cytoplasmic domain can be enzymatically methylated to create carboxyl methyl esters. An appropriate change in extent of methylation is mechanistically linked to adaptation.

Components of the chemotactic system

The chemotactic system in *E. coli* [1–3] includes four transmembrane receptor proteins, Tsr, Tar, Trg and Tap; six cytoplasmic proteins, CheR, CheB, CheA, CheY, CheZ and CheW, and three components of the flagellar switch, FliG, FliM and FliN.

Receptors are often referred to as transducers, because of their role in transducing ligand recognition into a sensory signal [7] and are sometimes known as methyl-accepting chemotaxis proteins or MCPs, because of their covalent modifications. Tsr and Tar mediate taxis towards serine and aspartate, respectively, as the result of sterospecific binding of those amino acids. Tar from *E. coli*, but not from *S. typhimurium*, also mediates taxis towards maltose by recognition of ligand-occupied, maltose-binding protein. Trg mediates galactose and ribose taxis by recognition of the two respective, ligand-occupied, sugar-binding proteins, and Tap performs the same function for dipeptides recognized by dipeptide-binding protein. Thus three of the four receptors in *E. coli* recognize polypeptide ligands. The binding proteins are water-soluble periplasmic proteins that also serve as components of transport systems for their respective ligands.

Each receptor also mediates response to at least one repellent. However, repellents seem to exert their effects by means other than sterospecific binding, for instance by altering intracellular pH and thus changing protonation of a crucial transducer residue [3].

CheR and CheB catalyse receptor methylation and demethylation, respectively, with *S*-adenosylmethionine as methyl donor.

CheA is a protein kinase that uses ATP to phosphorylate CheY and CheB, creating the active form of those proteins (*see* PROTEIN PHOSPHORYLATION). CheY is the effector protein that controls flagellar rotation; it seems that phospho-CheY interacts with the flagellar switch to induce clockwise rotation. The phosphoryl group on both CheY and CheB is short-lived. CheZ increases the rate of dephosphorylation of phospho-CheY.

The receptor proteins control flagellar rotation and methyl-esterase activity by influencing the level of phospho-CheA, a process in which CheW participates. The CheA kinase transfers phosphate from ATP to one of its own histidyl residues and subsequently to an aspartyl residue on CheY and presumably to an aspartyl residue on CheB. ATP may have an additional role in the chemotactic system [1].

Kinase–effector pairs with substantial sequence homology to CheA and the CheY/B pair are central components of environmental sensory systems in many bacteria [1,5,6], and where characterized, the phosphohistidine to phosphoaspartate chemistry is conserved. It is interesting that essentially the same chemistry was first described in the PHOSPHORYLATION CASCADE of the PHOSPHOTRANSFERASE TRANSPORT SYSTEMS that occur in a wide variety of bacteria. In *E. coli*, sugars taken up by such transport systems act as weak but distinct attractants. This effect is mediated by transfer of a phosphoryl group to the sugar being transported rather than by sterospecific recognition of attractant at a receptor site. Although the details are not entirely clear, it seems that the attractant action of a sugar transported by the phosphotransferase system reflects an interface between the phosphotransfer cascades of the chemotaxis and transport systems.

Excitation and adaptation

Events thought to occur at the receptor can be summarized by a model for a cycle of excitation and adaptation (Fig. B4). In the unstimulated state (**1**), the ligand-binding site is unoccupied and the methyl-accepting sites (four to six depending on the receptor) are continually methylated and demethylated, producing an average steady-state level of around one methyl group per protein. Addition of ligand creates a positive stimulus by interaction at a sterospecific site in the periplasmic domain of the receptor (**2**). Binding results in a shift in bias of flagellar rotation to counter-clockwise. This occurs in ~0.2 s. Transduction of information from binding site to flagellum can be considered in two steps: intramolecular — through the receptor from periplasmic to cytoplasmic domain — and intermolecular — through the cytoplasm from receptor to the flagellar switch by means of protein–protein interactions and altered rates of phosphotransfer. Up to this point, methylation is not involved. However, intramolecular transduction activates methyl-accepting sites for a net increase in methylation (symbolized in Fig. B4 by brackets around the methyl-accepting groups) and intermolecular transduction reduces methylesterase activity, thus setting the stage for adaptation. The system adapts to the continued presence of attractant (**3**). A new steady-state level of methylation is achieved, balancing the effect of ligand binding such that the cytoplasmic domain regains a null signalling state, functionally indistinguishable from the state

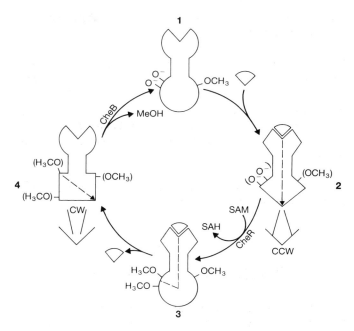

Fig. B4 Transducer states in the cycle of excitation and adaptation of the chemosensory receptors in *E. coli*. See text for explanation. SAM, *S*-adenosylmethionine; SAH, *S*-adenosylhomocysteine. CCW, counterclockwise signals; CW, clockwise signals. Adapted from [2].

before stimulation. Saturation of a ligand-binding site is balanced by a two- to threefold increase in the average number of methyl groups per receptor. Loss of ligand creates a negative stimulus (**4**), causing the periplasmic domain to assume its unoccupied conformation and thus eliminating the excitatory input from periplasmic to cytoplasmic domain. The change induced by methylation now has nothing to balance against and thus causes the cytoplasmic domain to assume an excitatory conformation that generates an intracellular signal for clockwise rotation. In this conformation, methyl esters are activated for net demethylation and intermolecular transduction increases methylesterase activity. The adapted state (**1**) is established in the unoccupied transducer by demethylation, restoring the cytoplasmic domain to the null signalling state.

Detection of a temporal gradient requires a comparison of present and past. Occupancy of the ligand-binding site is a measure of current concentration and covalent modification can reflect previous concentrations. The modification reactions are slow on the time scale of changes in occupancy. Thus, in an environment of changing concentration, the current extent of modification is the level that compensated for the extent of ligand occupancy a few moments previously. Thus balancing of ligand occupancy with modification is a comparison of current and past concentrations. In fact, cells seem to compute running averages of present and past concentrations. The result obtained for the past second is compared to the concentrations in the previous three seconds. As long as the values match, behaviour persists in the null state; when one value exceeds the other, an excitatory signal affects the flagellar switch and activities are set in motion to re-establish the balance.

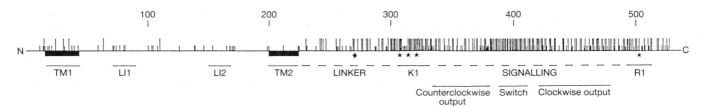

Fig. B5 Primary structure of chemotactic transducers from *Escherichia coli*. The horizontal line represents the aligned amino-acid sequences of Tar, Tsr, Tap and Trg as shown in [2]. Tall and short vertical lines indicate that the amino acid occupying that position is the same in all four or in three of the sequences, respectively. Notable features are: TM, transmembrane region; LI, sites of ligand interaction; K1 and R1, regions containing methyl-accepting sites (★); ♦, region sensitive to proteolysis. From [6].

Receptor structure and function

The chemoreceptor genes make up a homologous family, all coding for polypeptides of M_r ~60 000 [2,7]. Figure B5 shows an alignment of the four deduced amino acid sequences, indicating positions of residue identity. The polypeptides are disposed across the membrane by two transmembrane segments, creating a periplasmic domain of ~155 amino acids and a cytoplasmic domain of ~320. There is extensive conservation of amino acids in the cytoplasmic domains, which all interact with the same signalling proteins (CheA and CheW) and the same modification enzymes (CheR and CheB). Regions involved in excitation and adaptation are indicated in the figure. There is little residue identity among the aligned sequences of the transmembrane or periplasmic domains but residues involved in ligand interaction are in very similar positions in different receptors and it is likely that overall organization is conserved.

The native form of the receptors is a homodimer in which the two identical chains interact in all three domains, periplasmic, transmembrane and cytoplasmic. The three-dimensional structure of the periplasmic domain of Tar, the aspartate receptor, from *S. typhimurium* has been determined by X-ray crystallography [8]. The structure is a dimer of two four-helix bundles, forming a cylinder 70 Å long and 40 Å in diameter that may extend normal to the membrane in the intact receptor. Helix 1 from one subunit is in intimate contact with helix 1′ of the other, and aspartate is bound in the 1–1′ interface with additional interaction to residues of helix 4.

Structural differences between ligand-occupied and unoccupied receptor suggest that the conformational change that initiates intramolecular signalling is at least in part a scissors movement induced when ligand binding moves the membrane-distal ends of the two subunits closer together, causing potentially amplified movement in the transmembrane regions. Genetic evidence indicates that occupied binding proteins interact with receptors at regions near the identified aspartate-binding site, specifically the segments adjacent to the distal ends of helices 1 and 4 that connect helix 1 to 2 and helix 3 to 4. Thus binding protein interaction might also involve bridging the monomer–monomer interface and could induce a conformational change

similar to that caused by aspartate binding.

The conformational change induced in the periplasmic domain by ligand occupancy must be conveyed to the cytoplasmic domain. The mechanism of this transmission does not seem to involve creation or dissociation of dimers, as receptors persist as dimers under a variety of sensory states. However, it is possible that more subtle changes in subunit interactions are involved. In any case, the transmembrane segments, which appear to be organized in the dimer as a quasi four-helix bundle [9], seem to have more than a passive role in intramolecular signalling as single, conservative amino acid substitutions in those segments can induce constitutive signalling. Such effects might reflect a conformational change that conveys the sensory signal across the membrane by relative movements of the four transmembrane segments of the dimer.

Methyl-accepting taxis proteins have been detected, and in some cases characterized, in a wide variety of prokaryotes [6,10]. In many species, including the archaebacterium *Halobacterium halobium*, the methyl-accepting proteins are immunologically related, implying that these receptor proteins are related by a common ancestral protein that existed in an organism extant at a very early stage of evolution, possibly as much as four thousand million years ago.

Intermolecular signalling

The central component in the passage of excitatory signals from receptor to flagellar motor is the protein kinase CheA. This protein (M_r 73 000) has domains involved in autophosphorylation, phosphotransfer and receptor interaction. CheA transfers the γ-phosphate of ATP to its own His 48 and subsequently uses that phosphate to phosphorylate CheY on Asp 57. It is thought that phospho-CheY binds directly to the flagellar motor, interacting with the switch proteins to induce clockwise (tumble mode) rotation. In the absence of this interaction, the flagellum rotates exclusively counterclockwise, resulting in runs. In an unstimulated cell the balance between phosphorylation and dephosphorylation results in a steady-state level of phospho-CheY that produces an 80 : 20 ratio of runs to tumbles. Changes in ligand occupancy affect CheA activity which in turn causes altered levels of phospho-CheY. A decrease in occupancy at an attractant-binding site increases CheA kinase activity and the content of phospho-CheY, enhancing the probability of clockwise rotation and thus of tumbles induced by such rotation. This receptor action requires CheW and appears to involve a physical complex of the cytoplasmic domain of occupied receptor, CheA and CheW. An increase in occupancy at an attractant-binding site decreases CheA activity, phospho-CheY, and thus tumble-mode rotation, a process in which CheW may also be involved. These signalling changes are diagrammed in Fig. B6. As all receptors affect the same pool of CheA kinase, the system responds to multiple stimuli as if it were exposed to the algebraic sum of the environmental changes.

Excitatory changes in kinase activity do not persist indefinitely because two related effects mediate adaptation, restoring the stimulated receptor to its null signalling state by creating compensating alterations in methylation. One action affects stimulated receptors specifically and persistently, the other affects methylesterase activity globally and transiently. For example, positive stimuli reduce CheA kinase activity and thus the content

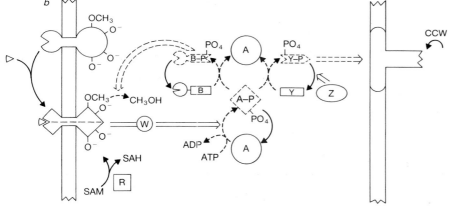

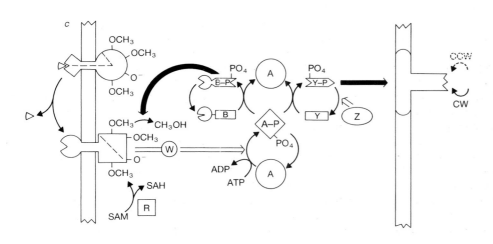

Fig. B6 Intermolecular signalling between chemosensory receptors and flagellum in bacterial chemotaxis. A schematic diagram of the protein–protein interactions and phosphotransfer reactions that link receptor to flagellum and that mediate sensory adaptation. *a*, The intermolecular circuitry. Phosphorylated CheY interacts with the flagellar machinery to produce clockwise (CW) rotation. Unphosphorylated CheY does not interact with the flagellar motor, leading to counterclockwise (CCW) rotation. In the unstimulated state a balance between phosphorylation and dephosphorylation of CheY results in a steady-state pattern of CCW rotation, leading to runs, and CW rotation, leading to tumbling. *b*, Binding of attractant to receptor leads to a decrease in CheY phosphorylation and a predominantly CCW rotation, thus decreasing tumbling. *c*, A decrease in occupancy of the receptor delivers a negative signal that results in increased CheY phosphorylation, and a predominantly CW rotation, increasing tumbling. The receptor protein is symbolized as in Fig. B4. Dotted arrows indicate decreased activities, thick arrows indicate increased activities. CheA, CheB, CheW, CheY and CheZ are indicated by the last letter of their names: phosphorylated forms are indicated by both a phosphate group on the shape representing the protein and a 'P' appended to the name.

of phospho-CheB, the active form of the methylesterase, without changing methyltransferase activity. This produces a net increase in methylation on all receptors, both stimulated and unstimulated. This general increase is transient; only the stimulated receptors maintain altered methylation past the period required for behavioural adaptation [1]. This occurs because the persistent effect of ligand occupancy stabilizes the change in methylation of stimulated receptor proteins.

The three-dimensional structure of CheY has been determined by X-ray crystallography [1,6]. The protein (M_r 14 000) is a compact, single domain, α-β structure with the phosphorylation site at one end of the molecule in a negatively-charged pocket made of aspartyl residues at the ends of two adjacent β strands. The three aspartates and a nearby lysine are conserved in the large family of related proteins from other two-component systems [1,5,6]. Interaction with the switch proteins occurs on a face of CheY separated from the phosphorylation site, so it is thought that the modification causes a conformational change that in turn exposes the site for flagellar interaction.

Flagellar structure and function

The flagellum is a complex structure (Fig. B7) that involves around 40 gene products in its structure, synthesis and assembly [3,4]. It is a true rotary motor, powered by an electrochemical ion gradient across the cytoplasmic membrane. In *E. coli* and *S. typhimurium* the energy source is the proton gradient — the PRO-

TONMOTIVE FORCE. In some other species a gradient of sodium ions is used. The flagellum includes a filament, hook, basal body and associated membrane proteins (Fig. B7). The filament is a helical polymer, several times the length of the cell, made up of a single type of protein — flagellin. It is a passive structure that serves as a propeller; rotation of the helix converts the torque of rotation into thrust. The filament is capped at the distal end by a specific protein, and its interface with the hook involves two other proteins. The helical hook structure serves as a universal joint, bending so that the several flagella of the cell can come together as a bundle behind the cell during a run. The basal body is the rotary apparatus. It contains a rod and a series of rings. The MS ring is embedded in the cytoplasmic membrane. The P ring is thought to pass through the peptidoglycan layer and the L ring is embedded in the outer membrane. The rod presumably transmits rotation from motor to hook and filament. The two outer rings are likely to be bushings for the rotating rod as it passes through the Gram-negative cell wall; these two rings do not occur in flagella in Gram-positive bacteria. Two cytoplasmic membrane proteins, MotA and MotB, couple the rotary motor to its source of energy and may cluster around the basal body in the membrane. It may be that protons pass through a pathway provided by MotA, generating rotational movement of a ring and the attached rod. The switch proteins are thought to cluster at the cytoplasmic side of the M ring, accessible for interaction with phospho-CheY.

G.L. HAZELBAUER

1 Bourret, R.B. et al. (1991) Signal transduction pathways involving protein phosphorylation in prokaryotes. *Annu. Rev. Biochem.* **60**, 401–441.
2 Hazelbauer, G.L. (1988) The bacterial chemosensory system. *Can. J. Microbiol.* **34**, 466–474.
3 Macnab, R.M. (1987) Flagella, and motility and chemotaxis. In Neidhart, F.C. et al. (Eds) Escherichia coli *and* Salmonella typhimurium: *Cellular and Molecular Biology*, 70–83, 732–759 (American Society for Microbiology, Washington, DC).
4 Macnab, R.M. (1992) Genetics and biosynthesis of bacterial flagella. *Annu. Rev. Genet.* **26**, 131–158.
5 Parkinson, J.S. & Kofoid, E.C. (1992) Communication modules in bacterial signalling proteins. *Annu. Rev. Genet.* **26**, 71–112.
6 Stock, J.B. et al. (1990) Signal transduction in bacteria. *Nature* **344**, 395–400.
7 Hazelbauer, G.L. et al. (1990) Transducers: transmembrane receptor proteins involved in bacterial chemotaxy. In Armitage, J.P. & Lackie, J.M. (Eds) *Biology of the Chemotactic Response*, 107–134 (Cambridge University Press).
8 Milburn, M.V. et al. (1991) Three-dimensional structures of the ligand-binding domain of the bacterial aspartate receptor with and without a ligand. *Science* **254**, 1342–1347.
9 Hazelbauer, G.L. (1992) Bacterial chemoreceptors. *Curr. Opinion Struct. Biol.* **2**, 505–510.
10 Morgan, D.G. et al. (1993) Proteins antigenically related to methyl-accepting chemotaxis proteins of *Escherichia coli* detected in a wide range of bacterial species. *J. Bacteriol.* **175**, 133–140.

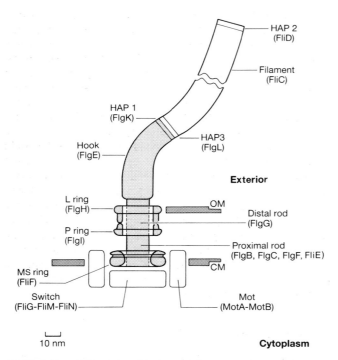

Fig.B7 Schematic illustration of the bacterial flagellum of *S. typhimurium*. Substructures and the proteins from which they are constructed are indicated. The locations shown for MotA, MotB and the three switch proteins are based on indirect evidence. CM, cytoplasmic membrane; OM, outer membrane; HAP, hook-associated protein. Adapted from [3].

bacterial differentiation Bacteria provide powerful model systems for the genetic analysis of differentiation and development. Two of the most informative have been polarity determination and flagellar production in *Caulobacter*, and sporulation in *Bacillus*.

In *Caulobacter* development, a predivisional cell divides to give rise to two distinct daughter cells — a swarmer cell and a stalked cell. These two cell types are a result of asymmetry generated before cell division in the predivisional cell. The swarmer cell

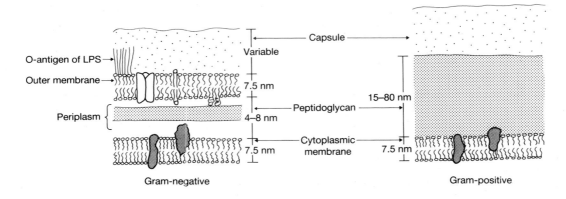

Fig. B8 Envelope structure of Gram-negative and Gram-positive bacteria.

possesses a flagellum and the flagellar gene products are specifically localized to the swarmer pole of the predivisional cell where the flagellum is assembled. This process involves the localization of both specific mRNAs and proteins. A hierarchy of transcriptional regulation ensures that the temporal order of the TRANSCRIPTION of the flagellar structural genes approximates the order of assembly of their products into the flagellum (*see also* BACTERIAL CHEMOTAXIS; BACTERIAL PATHOGENICITY).

In *Bacillus* sporulation, the developing cell is partitioned by an asymmetrically positioned septum into two distinct daughter cells, the mother cell and the forespore. Differential gene expression in the mother cell and forespore depends on the differential activity of specific SIGMA factors. These are regulatory proteins that bind to RNA polymerase and dictate specificity of PROMOTER recognition (*see* TRANSCRIPTION). A compartmentally restricted site-specific DNA rearrangement constructs a functional sigma K gene only in the mother cell. In addition, the expression of the sigma K gene is restricted to the mother cell by the compartment-specific expression of regulatory proteins.

Bacterial envelopes

THE cytoplasm of bacterial cells is surrounded by a bacterial envelope [1,2] that physically retains the cytoplasmic contents, functions as a site for interaction with the external environment, and performs numerous transport and biosynthetic processes essential for the viability of the cell. In 1884 Christian Gram developed a staining technique that distinguished bacteria as either Gram positive or Gram negative on the basis of their ability differentially to retain crystal violet/iodine precipitates when rinsed in alcohol, a difference that is due to the structure of their envelopes (Fig. B8). The envelopes of Gram-positive bacteria consist of a single membrane (the cytoplasmic membrane, similar to the plasma membrane of eukaryotic cells) surrounded by a thick cell wall comprised of PEPTIDOGLYCAN and TEICHOIC ACIDS which retains the crystal violet/iodine precipitates. Gram-negative bacteria have a much thinner cell wall composed only of peptidoglycan and from which the crystal violet/iodine stain easily

leaches away. This thinner cell wall is bounded by two membranes; an external outer membrane and an inner cytoplasmic membrane similar to those found in Gram-positive bacteria. The space or compartment between the two membranes of Gram-negative bacteria is the PERIPLASM, which contains not only the peptidoglycan but also soluble proteins and oligosaccharides. The envelopes of some bacteria have a variety of structures attached to or surrounding them, including capsular polysaccharides (see below), polymeric FIMBRIAE and PILI involved in bacterial adherence, and flagella required for chemotaxis (*see* BACTERIAL CHEMOTAXIS).

Capsular polysaccharides

Many bacterial species produce capsules consisting of either homo- or heteropolymeric polysaccharides (*see* GLYCANS) that surround the bacterial envelope. A variety of capsular types may be produced by different members of a single bacterial species, and this has led to a convenient system for serotyping bacteria on the basis of the interaction of specific ANTIBODIES with capsular antigens. In some species, such as ESCHERICHIA COLI, the capsule is referred to as the K-antigen. The strains of *E. coli* most commonly used for molecular genetic experiments are derivatives of *E. coli* K-12, which was unfortunately misnamed as this strain does not in fact produce any capsular polysaccharide. Capsules are often important for the virulence of pathogenic microorganisms (*see* BACTERIAL PATHOGENICITY), and the loss of ability to produce a capsule can render a strain avirulent. The large size and electrostatic charge of capsular polysaccharides inhibits engulfment of bacteria by phagocytes and thus contributes to the survival of the microorganism. It was studies on the transformation of nonencapsulated pneumococci to virulent capsulated forms that led Avery, Macleod, and McCarty to conclude that DNA was the hereditary material.

Outer membranes

The outer membranes of Gram-negative bacteria are unusual in that they have a very asymmetric bilayer structure, due to the presence of LIPOPOLYSACCHARIDE (LPS) in the outer leaflet of this membrane. Both the inner and outer leaflets contain three major phospholipids, including phosphatidylethanolamine (PE, which

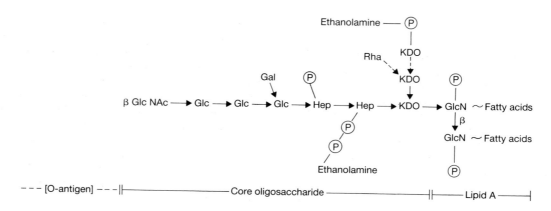

Fig. B9 Structure of LPS from *Escherichia coli* K-12. Dotted lines represent partial substitution. The O-antigen is absent from K-12 strains of *E. coli*. GlcNAc, *N*-acetyl-D-glucosamine; Glc, D-glucose; Gal, D-galactose; Hep, L-glycero-D-mannoheptose; P, phosphate; KDO, 2-keto-3-deoxyoctonic acid; Rha, L-rhamnose; GlcN, D-glucosamine. All hexose and heptose moieties in the core region are α-linked, unless indicated. Adapted from [4].

represents ~75% of the total phospholipid), phosphatidylglycerol (PG) and cardiolipin (CL). The LPS is composed of three covalently linked parts: a hydrophobic lipid A moiety that carries six or seven fatty acid chains that insert into the outer leaflet of phospholipids; a core of oligosaccharides that is linked to lipid A via 2-keto-3-deoxyoctonic acid (KDO); and an O-polysaccharide chain (or O-antigen) that consists of repeating units of different sugars (Fig. B9). Wild-type strains of *Salmonella*, *E. coli*, and other Gram-negative bacteria capable of synthesizing a complete lipopolysaccharide molecule are referred to as 'smooth' strains owing to the generally smooth appearance of their colonies on solid media. In contrast, strains that are defective in the production of the O-polysaccharide are called 'rough' strains as in many cases they have a rough colony morphology. The *E. coli* K-12 strains used for RECOMBINANT DNA studies lack an O-antigen and thus have a rough appearance on solid media. The loss of O-antigen by *Salmonella* sp. (i.e. *lpo* mutants) is associated with a loss of virulence. Loss of core oligosaccharides results in deep rough mutants which are exceptionally sensitive to antibiotics. Mutants in lipid A or KDO cannot be isolated (except as TEMPERATURE-SENSITIVE MUTANTS); thus this region of LPS must be considered essential for the structure and viability of the cell.

Members of the Enterobacteriaceae produce an enterobacterial common antigen, which in *E. coli* represents 0.2% of the dry weight of the cell. It is an acidic polysaccharide containing *N*-acetyl-D-glucosamine, *N*-acetyl-D-mannosaminuronic acid and 4-acetomido-4,6-dideoxy-D-galactose which is anchored to phospholipids in the outer membrane.

The protein composition of outer membranes is dominated by a few major outer membrane proteins (in some instances referred to as OMPs). One of the most abundant is the murein LIPOPROTEIN (Braun's lipoprotein, M_r 7200), associated with the inner leaflet of the membrane at a level of ~7×10^5 copies per *E. coli* cell. One-third of the population of lipoprotein is covalently linked to peptidoglycan via the ε-amino group of lysine present at the C terminus of the protein. A cysteine residue at the N terminus of lipoprotein is modified, with the sulphydryl group being substituted with a diglyceride and the amino group with a fatty acid. The protein appears to exist primarily in an α-helical form, possibly as trimers, and its main function is thought to be to maintain the structural integrity of the outer membrane–peptidoglycan complex.

Outer membrane porins are a group of very abundant proteins, representing 2% of total cell protein, which produce relatively nonspecific pores allowing small solutes to pass through. The porins of *E. coli* K-12 include OmpC and OmpF (M_r 37 083 and 38 306, respectively). The relative proportion of OmpC and OmpF in the outer membrane is determined by the osmolarity of the growth medium, with OmpC expression being derepressed by high osmolarity, and expression of OmpF being repressed under such conditions. The porin PhoE (M_r 36 782) is produced under conditions of phosphate starvation, whereas the LamB porin (M_r 47 393), which is involved in the uptake of maltose and maltodextrins, is induced by the presence of maltose in the growth medium. Tsx is a porin involved in nucleoside transport.

Porins are predominantly composed of β sheets; they have no α-helical content. The porins assemble into homotrimeric complexes that are unusual in being resistant to denaturation by sodium dodecyl sulphate (SDS). Soluble trimers can be obtained by SDS/NaCl extraction and can be dissociated into monomers by heat treatment. There are three diffusion channels per trimer.

BACTERIOPHAGES and COLICINS use porins as receptors. LamB for example, serves as the receptor for bacteriophage LAMBDA.

A number of other proteins in the outer membrane are involved in the uptake of vitamin B_{12} and SIDEROPHORES that chelate iron. These uptake systems, unlike the porins, are dependent on the function of the TonB protein to transduce energy to the transport process.

OmpA is another abundant outer membrane protein (M_r 35 159 in *E. coli* K-12). There are up to 10^5 copies of OmpA per cell and like the porins it is rich in β sheet. It can serve as a bacteriophage receptor and forms strong ionic interactions with the peptidoglycan layer. At present its function is unknown, although it may have a structural role as it clearly spans the outer membrane and interacts with the peptidoglycan layer. Mutants lacking OmpA serve as poor recipients for the conjugative transfer of DNA (*see* PLASMIDS).

Peptidoglycan

Peptidoglycan (murein sacculus or cell wall) is a polymer that contains about equal amounts of polysaccharide and peptide. The peptidoglycan of *E. coli* consists of alternating *N*-acetylglucosamine (GlcNAc or NAG) and *N*-acetylmuraminic acid (MurNAc or NAM) linked via β1,4-glycosidic bonds. A peptide, L-alanyl-D-isoglutamyl-L-*meso*-diaminopimelyl-D-alanine, is attached by an amide linkage to the carboxyl group of the D-lactyl moiety of MurNAc. A small percentage of the peptides lack a D-alanine and an even smaller percentage have an additional D-alanine. This type of peptidoglycan structural unit is common among all Gram-negative bacteria and some Gram-positive rods. Adjacent strands of peptidoglycan are covalently linked together via the side chains of the amino acids. Most of the cross-links are between the carboxyl group of D-alanine in one peptide and diaminopimelic acid (DAP) in an adjacent peptide. Only about half of the peptides are involved in intra-peptidoglycan cross-links. About 10% have a covalent link between the L-carboxyl group of DAP and the ε-amino group of the C-terminal lysine residue of lipoprotein. Such covalent interactions with lipoprotein as well as the ionic interactions with porins and OmpA provide over 400 000 contacts between the murein sacculus and the outer membrane. This explains why the peptidoglycan and the outer membrane remain at a fixed distance from one another in plasmolysed cells, whereas the cytoplasmic membrane shrinks and moves away.

A variety of enzymes participate in the assembly and elongation of peptidoglycan and are referred to as PENICILLIN-sensitive or penicillin-binding proteins (PBPs). Inhibition of these enzymes by penicillin results in defective cell wall synthesis and consequent cell lysis — the reason for the sensitivity of bacteria to penicillin.

Cytoplasmic membrane

The cytoplasmic membranes of bacteria have a 'conventional' bilayer structure (*see* MEMBRANE STRUCTURE) and are the site of dynamic activity. Unlike the outer membranes of Gram-negative organisms they act as osmotic barriers to the free diffusion of small solutes. Not surprisingly, many of the proteins in the cytoplasmic membranes of bacteria are involved in transport. In *E. coli* the cytoplasmic membrane contains 75% of the total cellular phospholipids, and 6–9% of the total cellular protein. Three different phospholipids contribute to the bilayer structure of the cytoplasmic membrane — phosphatidylethanolamine, phosphatidylglycerol, and cardiolipin. Approximately 200 different proteins, representing about 60% of the total protein found in the envelopes of Gram-negative bacteria, are present in the cytoplasmic membrane. This enormous array of proteins includes those involved in: (1) electron transport and oxidative phosphorylation (*see* ELECTRON TRANSPORT CHAIN); (2) solute transport via ion gradient, binding protein or phosphotransferase-dependent systems; (3) phospholipid, LPS and peptidoglycan biosynthesis; and (4) export of polypeptides across the cytoplasmic membrane (*see* BACTERIAL PROTEIN EXPORT). A PROTONMOTIVE FORCE across the cytoplasmic membrane drives the processes of oxidative ATP synthesis, transport of ions and certain solutes, and protein export.

Studies on the turnover of phospholipids in the cytoplasmic membrane have revealed that the hydrophilic phospho- head groups of phosphatidylglycerol are steadily lost whereas phosphatidylethanolamine is metabolically stable. This is the result of the phosphoglycerol head group being continuously transferred to a novel type of water-soluble oligosaccharide, known as membrane-derived oligosaccharides (MDOs) [2]. The phosphoglyerol moiety is covalently linked to 8–10 glucose units to give a molecule of M_r ~2400, with a considerable net negative charge. MDOs are located in the periplasm where they may account for up to 7% of the dry weight of a cell cultured in medium of low osmolarity. Indeed, the osmolarity of the periplasm may be regulated by MDOs.

The periplasm

The periplasm is sandwiched between the cytoplasmic and outer membranes of Gram-negative bacteria. Its volume and size have been estimated from the differential distribution of solutes of varying size, and from electron microscopic studies of fixed and frozen thin sections. The width is estimated to be between 4 and 7.5 nm under normal physiological conditions. The periplasmic volume and osmolarity respond to changes in the osmotic strength of the external medium. High concentrations of solutes such as 0.5 M sucrose can easily penetrate the outer membrane through porins, but cannot cross the cytoplasmic membrane. As a consequence the cytoplasmic compartment shrinks, eliminating the osmotic gradient across the cytoplasmic membrane, causing plasmolysis. In normal medium an osmotic gradient between the periplasm and the external environment exists which gives rise to an osmotic pressure equivalent to 3.5 atmospheres being exerted on the peptidoglycan cell wall.

The soluble contents of the periplasm, including periplasmic proteins, can be isolated by various procedures, including osmotic shock of plasmolysed cells or the formation of SPHEROPLASTS using lysozyme and ethylenediaminetetraacetic acid (EDTA) in hypertonic or isotonic conditions. These procedures release ~50 different proteins, which can be grouped according to their biological function.

The largest group is the periplasmic binding proteins, which deliver solutes to binding protein-dependent transporters that span the cytoplasmic membrane; some binding proteins are also involved in chemotactic responses to amino acids and sugars (*see* BACTERIAL CHEMOTAXIS). Another large group of periplasmic proteins includes scavenging enzymes responsible for hydrolysing a variety of substrates. The periplasm is also the site of a number of detoxifying enzymes such as β-lactamase, which is responsible for the hydrolysis of penicillin in penicillin-resistant bacterial strains.

The presence of cationic MDOs in the periplasm creates a significant Donnan equilibrium across the outer membrane which can be measured as an unequal distribution of Na^+ and Cl^- ions between the medium and periplasm. In a standard salts medium this represents a potential across the outer membrane of 30 mV. The physiological significance of the Donnan equilibrium remains uncertain.

Bayers junctions

Electron microscopic examination of thin sections of *E. coli* plasmolysed in 20% sucrose has revealed sites of adhesion (or Bayers junctions) between the cytoplasmic and outer membrane [3]. The existence and function of these adhesion sites remain the subject of some controversy since new techniques using frozen thin sections of *E. coli* have failed to detect Bayers junctions [2]. Nonetheless, it is generally accepted that Bayers junctions could provide a convenient route for export of envelope components, such as LPS and capsular polysaccharides. Moreover, such sites have been implicated in the export of proteins to the outer membrane and the infection of cells by bacteriophages.

T.R. HIRST

See also: BACTERIAL PROTEIN EXPORT.

1   Ingraham, J.L. et al. (1983) *Growth of the Bacterial Cell* (Sinauer, Sunderland, MA).
2   Neidhardt, F.C. et al. (Eds) (1987) Escherichia coli *and* Salmonella typhimurium, *Cellular and Molecular Biology* (American Society for Microbiology, Washington, DC).
3   Bayer, M.E. (1968) Areas of adhesion between wall and membrane in *Escherichia coli. J. Gen. Microbiol.* **53**, 395.
4   Nikaido, H. & Vaara, M. (1987) In Escherichia coli *and* Salmonella typhimurium, *Cellular and Molecular Biology*, 7–22 (American Society for Microbiology, Washington, DC).

Bacterial gene expression

GENE expression describes the complete process by which the information encoded in a GENE is converted into a functional structure — a PROTEIN or RNA molecule — which can then participate in cellular and physiological pathways. There are two main stages in the complex and energy-demanding chain of events involved in gene expression. The first is the TRANSCRIPTION of DNA into an RNA copy (a transcript). Transcripts of protein-coding genes — messenger RNAs (mRNAs) — are then used as templates in the second major step, translation, in which the information carried in the nucleotide sequence of the mRNA is decoded and converted into the amino-acid sequence of a protein (*see* GENETIC CODE; PROTEIN SYNTHESIS). Apart from mRNA, transcription produces RNA species that form part of the translational apparatus itself. RIBOSOMAL RNA molecules (5S rRNA, 16S rRNA and 23S rRNA) serve as the anchoring frameworks upon which ribosomal subunits are assembled. TRANSFER RNA (tRNA) species mediate at the interface between the RNA code and the polypeptide sequence during translation. Other small structural RNAs such as the RNA component of ribonuclease P (*see* RIBOZYMES) are also synthesized by transcription.

The direction of flow of genetic information during gene expression in bacteria is invariably observed to be:

$$\text{DNA} \xrightarrow{\text{transcription}} \text{RNA} \xrightarrow{\text{translation}} \text{protein}$$

However, REVERSE TRANSCRIPTASES, which catalyse the transfer of information from RNA to DNA, are also known to exist in bacteria, although their precise roles are not understood.

A large part of our knowledge of the component processes of prokaryotic gene expression derives from studies of the Gram-negative heterotrophic bacterium ESCHERICHIA COLI [1,2]. The circular 'chromosome' of this organism is approximately 4 million base pairs long and carries around 2500 genes (*see* BACTERIAL GENE ORGANIZATION).

In prokaryotes, unlike eukaryotes, the DNA is not separated from the cytoplasm by a nuclear membrane, and all stages of bacterial gene expression occur in the same cellular compartment, the cytoplasm. This means that translation of a bacterial mRNA molecule can begin before its synthesis has been completed. With a few exceptions [3] INTRONS are not found in prokaryotes, and therefore there is no need for an RNA SPLICING step before translation. Other eukaryotic mRNA processing steps such as cap formation (*see* RNA PROCESSING) also do not occur in prokaryotes. Transcription and translation in bacteria are therefore closely coupled, which has implications for the control of gene expression (see below).

The products of gene expression, on the other hand, may be directed to various locations, for example to either of the cellular membranes or to the periplasmic space, or may be secreted into the medium (*see* BACTERIAL PROTEIN EXPORT). PROTEIN TRANSLOCATION across the cytoplasmic membrane is most commonly directed by an extension of the polypeptide sequence at the 5′ end, called a signal peptide, which is cleaved off to yield the mature gene product at its final destination. However, alternative (and often more complex) transport mechanisms also exist. Both the transport and folding of proteins in the cell seem generally to be 'assisted' by so-called MOLECULAR CHAPERONES.

The levels of gene products in the cell are determined both by the rate of RNA and protein synthesis and by degradative processes (Fig. B10). The rates of each step can vary greatly from gene to gene. Thus control and regulation of the level of gene expression can be exerted at several different points in the overall scheme. The first control point is initiation of transcription — the point at which a gene is selected for expression, and whose efficiency largely determines the level of transcription. Also im-

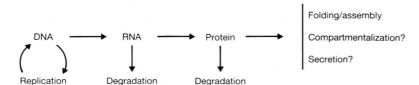

Fig. B10 Overall scheme of prokaryotic gene expression.

portant in the control of gene expression in bacteria are the rate of mRNA turnover, the control of initiation of translation, attenuation, and various post-translational processes.

Transcription

During transcription the DNA sequence is converted base for base into the equivalent RNA sequence. Transcription of all kinds of gene in *E. coli* is catalysed by a single type of DNA-dependent RNA POLYMERASE (*see* TRANSCRIPTION for details of the general mechanism).

Initiation of transcription

Transcription is initiated at positions defined precisely by sequence elements in the DNA known as PROMOTERS (Fig. B11). The promoter is the site at which the RNA polymerase binds in the correct position to start transcription. In most genes it is flanked by recognition signals for gene regulatory proteins — for example ACTIVATORS and REPRESSORS — which bind to specific sites in the DNA to increase or decrease the rate of initiation of transcription.

Analysis of the nucleotide sequences of more than 100 promoters in *E. coli* has identified two CONSENSUS SEQUENCES in short regions centred at positions respectively 35 and 10 base pairs upstream (with respect to the direction of transcription) of the point in the DNA sequence at which transcription begins. These sequences are involved in polymerase recognition and binding.

The topology of promoter DNA as well as the sequence motifs themselves is important in initiation. The coiling of the double helix brings the two sequence motifs into the correct positions for efficient recognition and binding of RNA polymerase. In some promoters, initiation of transcription is regulated by changes in the degree of SUPERCOILING of the DNA [4].

The RNA polymerase of *E. coli* is a multisubunit enzyme; the core enzyme ($\alpha_2\beta\beta'$) is capable of transcriptional elongation on its own, but requires the addition of a further subunit (σ) in order to bind specifically to promoter sites and thus initiate transcription in the correct place. The synthesis of alternative σ factors, for example in *E. coli*, *Bacillus subtilis* or *Streptomyces coelicolor*, allows the polymerase to be directed to different sets of promoters [5]. In *B. subtilis*, for example, several different σ factors are produced at different stages in sporulation, so that different sets of genes can be turned on at each stage.

In the absence of any ancillary regulatory proteins, the rate of initiation of transcription varies by up to a factor of at least 1000 over the whole range of promoters in *E. coli*. The efficiency of initiation is related to the sequence and topology of the promoter region. Gene expression occurring in the absence of any specific regulatory factors is known as constitutive. At many promoters, transcriptional initiation can be increased many-fold by the binding of specific regulatory proteins.

Operons

A feature of gene organization common in prokaryotes but rare in eukaryotes is the grouping of functionally related genes into

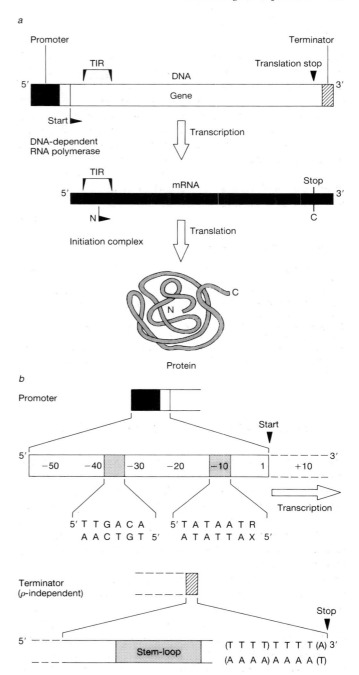

Fig. B11 Recognition signals in prokaryotic gene expression. *a*, Schematic depiction of the basic control signals in DNA and RNA in a hypothetical gene; *b*, detailed structure of promoter and terminator regions.

operons. In an operon, genes encoding, for example, the different enzymes of a metabolic pathway or the subunits of an enzyme complex, are clustered and are transcribed together into a POLYCISTRONIC transcript, under the control of a single promoter (*see also* BACTERIAL GENE ORGANIZATION). This transcript is then translated to give the individual proteins. Operons enable the rapid and efficient coordinate expression of a set of genes required to respond to a change in the external or internal environment.

In the classic model operon, the *lac* operon of *E. coli* (Fig. B12) [1,2], three proteins required to transport and metabolize lactose are synthesized only when lactose is present. A regulatory gene adjacent to the operon constitutively synthesizes a repressor protein which, in the absence of lactose, binds specifically to an OPERATOR region adjacent to the *lac* promoter and blocks transcription. Allolactose, a form of lactose, interacts with the repressor protein, changing its conformation so that it no longer binds to the operator (*see* ALLOSTERY) and allows transcription to be initiated at the *lac* promoter. Regulation of *lac* by its repressor is a negative regulatory circuit; positive regulatory circuits, involving activator proteins, also exist (see below). Many biosynthetic operons are controlled by FEEDBACK REGULATION involving the product of the metabolic pathway they encode (*see* TRYPTOPHAN OPERON).

A further control on expression of the *lac* operon is the binding of an activator protein — in this case a complex of CATABOLITE ACTIVATOR PROTEIN (CAP) and CYCLIC AMP — to the promoter which increases the level of transcription. Unlike the *lac* repressor, which is specific for the *lac* operator only, the CAP–cAMP complex can also bind to the promoters of other sugar-metabolizing operons. The CAP–cAMP complex is only formed when glucose, which is the easiest sugar for the bacterium to metabolize, is not available, and other sugars must be utilized. When glucose is available and cAMP levels are low, the CAP protein on its own cannot bind to these promoters and so the enzymes are not produced unnecessarily.

Termination of transcription

The termination of transcription occurs at specific sites, which are of two main types. One type of termination (factor-independent termination) occurs characteristically at sites defined by a series of U residues (in the mRNA) preceded by an INVERTED REPEAT that forms a STEM-LOOP structure at the 3′ end of the RNA transcript, which is thought to interfere with polymerase action, leading to release of the RNA. By contrast, factor-dependent termination, usually dependent on the interaction of protein factor rho (ρ) with the RNA polymerase, occurs at less

well-defined sites that are rich in cytosine and poor in guanosine.

Premature termination of transcription plays a part in regulating the expression of some operons. In some operons directing amino acid biosynthesis, such as the *trp* operon (*see* BACTERIAL GENE ORGANIZATION) expression is prevented in two ways when the amino acid product is plentiful: (1) a complex of repressor protein and amino acid binds to the operator and prevents transcription initiation; (2) any transcription that does occur is terminated prematurely at a factor-independent site in the LEADER region (ATTENUATION).

In the absence of amino acid, however, repression of the operon is lifted and efficient transcription is initiated; the stalling of ribosomes at Trp codons on the leader mRNA prevents formation of the terminator stem-loop, thus circumventing attenuation. This regulatory mechanism is restricted to prokaryotes because it requires the close coupling of transcription and translation.

Control of translation

The translation of a READING FRAME is initiated at an INITIATION CODON, which in *E. coli* is most commonly AUG (90%), less frequently GUG (8%) or UUG (1%), and in one case AUU. (*See* PROTEIN SYNTHESIS for details of the translation process.) The selection of a start codon is almost invariably dependent on the presence of a SHINE–DALGARNO REGION preceding the start codon by 4–12 bases. The Shine–Dalgarno region is complementary to part of the 3′ end of the 16S rRNA incorporated in the 30S ribosomal subunit and provides part of the RIBOSOME-BINDING SITE. The nature of the start codon, the extent of the homology between the Shine–Dalgarno region and the 16S rRNA, and the sequences of the regions flanking these two elements all influence the efficiency of initiation of translation, which can vary by up to a factor of 1000.

Additional specific recognition elements may be involved in determining the efficiency of translation [6]. This has given rise to the concept of a translational initiation region (TIR; Fig. B13), which extends beyond the limits of the classically defined ribosome-binding site [7]. Both the primary sequence of the TIR and the higher-order structures that it forms through intramo-

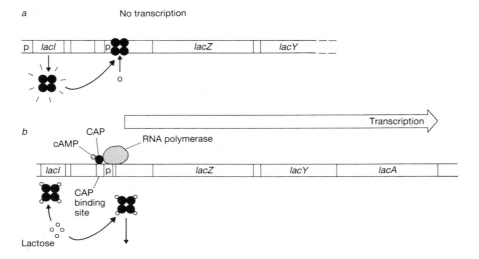

Fig. B12 Expression of the *lac* operon. The *lacZ*, *lacY* and *lacA* genes of the *lac* operon encode three proteins involved in the uptake and metabolism of the disaccharide lactose. In the absence of lactose (*a*) the operon is repressed by specific binding of a repressor protein (encoded by *lacI* and expressed constitutively) to the operator region (o). In the presence of lactose (*b*) repression is lifted (see text); if the transcriptional activator cAMP–CAP is also bound near the promoter (p) region (see text), the operon is expressed. The three genes are co-transcribed into a polycistronic mRNA, and translation begins before mRNA synthesis is completed. *lacZ* encodes β-galactosidase, *lacY* a permease, and *lacA* β-galactoside transacetylase.

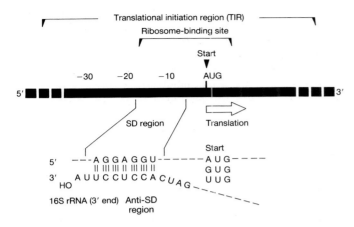

Fig. B13 The prokaryotic translational initiation region (TIR). This shows the relationship between the (functionally defined) TIR and the ribosome-binding site. The minimal definition of the ribosome-binding site comprises the Shine–Dalgarno (SD) region and the start codon plus the bases in between. The TIR will generally be larger than the ribosome-binding site. The SD region is complementary to a sequence (the anti-SD region, ASD) at the 3′ end of the 16S rRNA.

lecular base-pairing and stacking interactions influence the binding of 30S ribosomal subunits and the rate of translational initiation.

Once polypeptide elongation is underway, ribosomes continue translating the reading frame of the gene until they reach one of three NONSENSE CODONS: UGA, UAG, or UAA. The latter signal the site of translational termination. Two release factors, RFI (specific for UAG and UAA) and RF2 (specific for UGA and UAA), are involved.

Because of redundancy in the genetic code most amino acids can be encoded by more than one codon. Two forms of CODON BIAS can be distinguished: (1) a general bias in codon usage identifiable with a specific organism; (2) codon bias associated with different sets of genes within one organism, often apparently correlated with expression rates. However, codon bias is generally unlikely to be directly linked to the translational efficiency of specific genes [8].

In order to ensure the accurate translation of a gene's reading frame, it is essential that the triplet code is read sequentially from beginning to end. However, ribosomes can 'slip' into a different reading frame, thus recognizing codons out of phase with the original reading frame (frameshifting). There are examples of reading frames (e.g. *RF2* and *trpR*) where frameshifting is essential for the synthesis of functional proteins. In a more extreme form of this phenomenon, ribosomes 'jump' over codons that are not translated, to resume translation further down the mRNA (frame-jumping). A 50-nucleotide frame-jump is essential for the correct translation of the bacteriophage T4 gene *60*.

There is a wide range of regulatory circuits at the translational level. Most of these are negative regulatory systems involving proteins that bind to mRNA and prevent initiation of translation or complementary ANTISENSE RNA that hybridizes with mRNA, but others rely on activation by positive effectors [6].

mRNA stability

The degradation rates of mRNAs in *E. coli* are generally very high compared with eukaryotes. This allows the bacterial cell to shift rapidly the types and abundance of mRNA species, and thus of proteins, in response to changes in growth conditions. The overall degradation rates are also growth-rate dependent. Moreover, many other bacteria show generally slower degradation rates than *E. coli* which are adapted to their respective physiologies and growth rates. Neither the pathway nor the rate of mRNA degradation is identical for all genes. mRNA half-lives in *E. coli* vary from seconds to more than 20 minutes under laboratory conditions, so that there can be variation of at least a factor 50 in inactivation/degradation rates.

mRNA degradation involves the activities of both EXONUCLEASES and ENDONUCLEASES. There are at least two types of pathway. In the first, the mRNA is inactivated by endonucleolytic cleavage in the leader or coding regions. This is followed by further endonucleolytic cleavages and by 3′→5′ exonucleolytic degradation. No 5′→3′ exonucleases are known in *E. coli*, yet degradation has been observed to proceed effectively in the 5′→3′ direction in at least some cases. This apparent contradiction can be resolved by assuming that the initial endonucleolytic cleavage occurs in the TIR, and is followed by further cleavages occurring progressively and sequentially in the 5′→3′ direction along the gene. The resulting fragments are then degraded by exonucleases. In one model, the wave of endonucleolytic cleavage and exonucleolytic degradation follows the last ribosome that had initiated before the TIR was inactivated, but this model does not explain all the observed features of mRNA degradation. There may be different pathways subsequent to the initial cleavage of the mRNA. In the second type of pathway, exonucleolytic degradation from the 3′ end is responsible for inactivating the transcript [9,10].

Genes are more stable if endonucleolytic cleavage sites are not present and/or their 3′ ends are protected from exonucleases. However, only some of the endonucleolytic cleavage sites involved in mRNA degradation have been identified. In a number of messages, important steps in degradation are catalysed by known endonucleases, for example RNase III and RNaseE. Protection of the 3′ end of genes can be afforded by stem-loop structures in the mRNA [9]. Two exonucleases involved in mRNA degradation in *E. coli* have been identified so far: RNase II and polynucleotide phosphorylase.

Gene expression in the polycistronic environment

Regulation of bacterial gene expression in the post-transcriptional parts of the pathway takes on special significance in polycistronic operons. Both translational control and segmental differences in mRNA stability can result in differential gene expression from a polycistronic mRNA. Moreover, the grouping together of genes in an operon allows them to interact in ways that are impossible for MONOCISTRONIC messages. In particular, many genes are translationally coupled. This means that the translation of one gene is dependent on the translation of another (usually the upstream neighbouring gene). The coupling may take the form of reinitiation of ribosomes that have completed translation of the upstream gene. Alternatively, translation of this latter gene may be necessary to hold the adjacent downstream TIR open for the binding of fresh ribosomes that would otherwise be prevented from translating the downstream gene by the presence of higher-order structure in the mRNA (the facilitated binding mechanism).

J.E.G. McCARTHY

See also: EUKARYOTIC GENE EXPRESSION; EXPRESSION VECTORS.

1 Lewin, B. (1990) *Genes IV* (Cell Press, Cambridge, MA).
2 Neidhardt, F.C. et al. (Eds) (1987) Escherichia coli *and* Salmonella typhimurium: *Cellular and Molecular Biology* (American Society for Microbiology, Washington, DC).
3 Belfort, M. (1990) Phage T4 introns: self-splicing and mobility. *Annu. Rev. Genet.* **24**, 363–385.
4 Lilley, D.M. & Higgins, C.F. (1991) Local DNA topology and gene expression: the case of the *leu-500* promoter. *Mol. Microbiol.* **5**, 779–783.
5 Helmann, J.D. & Chamberlin, M.J. (1988) Structure and function of bacterial sigma factors. *Annu. Rev. Biochem.* **57**, 839–872.
6 McCarthy, J.E.G. et al. (1990) Post-transcriptional control in *E. coli*: the translation and degradation of mRNA. In *Post-transcriptional Control of Gene Expression* (McCarthy, J.E.G. & Tuite, M.F., Eds) 157–168 (Springer-Verlag, Berlin).
7 McCarthy, J.E.G. & Gualerzi, C. (1990) Translational control of prokaryotic gene expression. *Trends Genet.* **6**, 78–85.
8 Andersson, S.G.E. & Kurland, C. (1990) Codon preferences in free-living microorganisms. *Microbiol. Rev.* **54**, 198–210.
9 Belasco, J.G. & Higgins, C.F. (1988) Mechanisms of mRNA decay in bacteria. *Gene* **72**, 15–23.
10 Petersen, C. (1992) Control of functional mRNA stability in bacteria: multiple mechanisms of nucleolytic and non-nucleolytic inactivation. *Mol. Microbiol.* **6**, 277–282.

Bacterial gene organization

GENE organization, fine structure and expression have been studied in great detail for many years in bacteria and their phages using genetic techniques. In more recent times, RESTRICTION MAPPING, DNA CLONING, DNA SEQUENCING and other DNA manipulations have provided more comprehensive information and provided access to the genomes of a wider range of PROKARYOTIC microorganisms.

The genomes of many prokaryotes have been investigated: we know by far the most, however, about gene organization in the related heterotrophic bacteria ESCHERICHIA COLI and SALMONELLA TYPHIMURIUM. To a first approximation, general rules about gene organization deduced from *E. coli* appear to hold in other prokaryotes. However, gene manipulation techniques now give easy access to a diverse range of prokaryotic microorganisms and it can be expected that nature's diversity will provide exceptions to most generalizations.

Most bacteria contain circular genomes comprising several millions of base pairs of DNA: the DNA is replicated bidirectionally starting at a single REPLICATION ORIGIN and running to a TERMINATOR REGION [1]. Because replication starts before cell division, many bacteria, particularly those that are dividing rapidly, will contain more than one copy of the genome and DNA molecules that are partially replicated. Thus the GENE DOSAGE of genes near the replication origin will be greater than the dosage of those near the terminator region. In addition to a circular chromosome, many bacteria also carry genetic information on PLASMIDS: these are autonomously replicating segments of DNA which are of great importance in the passage of genetic information from one cell to another.

Genome organization

The *E. coli* genome contains around 4.7 million base pairs. A crucial feature of this and other prokaryotic genomes is that 98% or more of the DNA encodes 'useful' information and can be transcribed into RNA (*see* TRANSCRIPTION). Fine analysis shows that the genetic information in *E. coli* and other prokaryotes is packaged economically without gross repetition and apparent wastage, unlike the situation in the genomes of higher EUKARYOTES (*see* EUKARYOTIC GENE STRUCTURE; GENOME ORGANIZATION). The simple explanation for this is that it costs metabolic energy to replicate DNA; because prokaryotes have evolved to divide rapidly in hostile environments, genome size is limited and the appearance of useless DNA sequences has been selected against.

Notwithstanding this, some bacterial DNA may not serve a useful purpose and a small amount of repeated related sequences has been found, comprising around 1% of the genome. These repeated elements include a group of palindromic sequences (REP sequences [2]) which have no clear role.

Analysis of the number and abundance of different MESSENGER RNAs produced by a bacterial cell [3] shows that at least 98% of the *E. coli* genome is transcribed. Estimations using the average size of a protein give a coding capacity for the bacterial genome of some 2500 proteins. 1400 genes have so far been mapped and identified on the basis of phenotype [4]. Restriction maps of the entire *E. coli* genome have been deduced [5] and, at the time of writing, the nucleotide sequence of about a quarter of the *E. coli* chromosome has been determined [6] and the rest will follow by the end of the century. The overall picture [7,8] that has emerged from sequencing and mapping studies has been crucial in confirming points about gene organization that could only be guessed from previous genetic studies.

There appears to be little logic in the way the genome is organized. For example, genes encoding the ribosomal proteins are scattered around the chromosome. There is a definite lack of any clear correlation between the position of most genes and their function. There are no large regions assigned to genes of particular function and there are no large 'silent' regions (Fig. B14). Genes are encoded on both strands of the DNA and there is no rule dictating which genes are on which strand.

Operons

A particular feature of prokaryotic gene organization is the grouping of genes of related function together into an operon — a single TRANSCRIPTION UNIT under the control of a single PROMOTER. Genes encoding different enzymes of a particular metabolic pathway are often grouped in this way. Expression of the operon is regulated by regulatory proteins that bind to sites

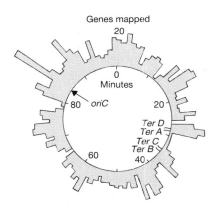

Fig. B14 Schematic representation of the circular *E. coli* chromosome showing the site of origin of bidirectional replication (*oriC*), the sites of termination of replication (Ter A–D) and the number of genes mapped at each segment (minute) of the chromosome, where the height of the bar at minute 1 represents 20 genes. Redrawn from [16].

within or flanking the promoter (*see* BACTERIAL GENE EXPRESSION). Typically, an operon consists of between 3 and 12 protein-coding sequences (cistrons) organized in tandem with a promoter upstream and a transcription terminator downstream of the protein-coding sequences (the structure of a 'typical' operon is shown in Fig. B15). The structural genes are transcribed into a POLYCISTRONIC mRNA which is then translated into the individual proteins.

This organization ensures that expression of all the genes in the operon is coordinated: in most cases, gene expression is tightly coupled to growth and development of the cell, and to the availability of nutrients, by regulatory proteins that interact at the promoter. In many cases, the gene encoding the regulatory protein that specifically shuts down or activates transcription from the operon is located next to its target promoter and is often transcribed in the opposite direction from the structural genes it controls; in many other cases the regulatory genes are unlinked.

Physical separation of a regulatory gene from its target may reflect an accident of evolution. Genes are typically transferred *en bloc* between bacteria either during conjugation (*see* PLASMIDS) or by TRANSDUCTION by phages such as LAMBDA, and separation of adjacent genes may occur during such a process. In some cases regulatory proteins have evolved to control expression at a large number of disparate promoters, creating a 'regulon'.

The genetic maps of related bacteria tend to be similar; the maps of *E. coli* and *S. typhimurium* can easily be aligned [9]: large differences may be accounted for by blocks of sequence that have been transferred into one of the organisms relatively recently. Severe constraints on evolutionary potential arise when a regulatory gene becomes physically separated from its target, for it cannot then be passed from one organism to another *en bloc* with its target genes.

In most bacteria, genes are close together. Genes within the same operon are commonly separated by less than 20 base pairs, although there are some exceptions where regulatory signals have been interposed between the genes of an operon. In many

cases the TERMINATION CODON of one gene overlaps the INITIATION CODON of the next. The coding sequences of different operons are typically separated by 50–200 base pairs, these sequences carrying transcription initiation and termination signals. There are very few regions with no function. However many operons contain OPEN READING FRAMES which are expressed but for which there is no known function. Common sense would dictate that such genes must have a function but in many cases there is no proof.

Prokaryotic coding sequences start at initiation codons and run in triplets to termination codons: the vast majority of genes contain no non-coding intervening sequences. However, INTRONS have been found in a small number of cases in BACTERIOPHAGE [10] and ARCHAEBACTERIAL genes, nullifying the dogma that introns are the sole property of eukaryotes.

Copy number

Most prokaryotic genes are present in a single copy per genome: however their effective concentration per cell may be higher as a result of gene dosage effects. The RIBOSOMAL RNA (rRNA) genes [11] are a notable exception to the general rule. *E. coli* carries seven loci encoding the three rRNAs — 5S, 16S, and 23S. Although these loci are unlinked they are organized such that transcription is in the same direction as replication, avoiding clashes between transcription and the replication fork. At each locus the three rRNAs are co-transcribed with a different tRNA, and the transcript is subsequently cleaved into its component functional RNAs. The requirement for multiple copies of the rRNA genes arises because rRNA is not amplified by translation.

Although other bacterial genes are rarely present in multiple copies, GENE DUPLICATION can arise and this is an important step in divergent evolution. Bacteria contain many families of related genes that must have evolved by such a process.

Genomic instability

It is clear that bacterial genomes are in a constant flux: bacteria readily exchange information by plasmids and by transduction, and additionally, bacterial populations can mutate rapidly as a consequence of very short generation times. This genetic flux is essential for the generation of diversity and the ability to respond rapidly to changing conditions. Genetic flux is also due to specific sequences in bacterial genomes that are programmed to create instability. One class of such sequences is the INSERTION SEQUENCES [12] which are often present in multiple copies in bacterial genomes (*see* TRANSPOSABLE GENETIC ELEMENTS). These are sequences of 700–1500 base pairs which encode one or more proteins allowing switching of sequence information from one site in the genome to another (transposition). A number of different insertion sequences have been fully characterized; typically a strain may carry between 5 and 10 copies of a particular sequence. Because many genes contain 'hotspots' for the insertion of such sequences their distribution is often not random. Transposons are genetic elements that consist of transposition functions plus extra sequences that may encode some selectable function such as drug resistance (*see* PLASMIDS). Insertion se-

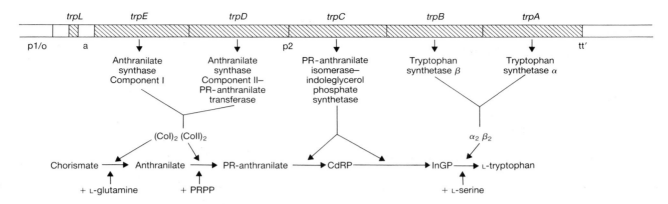

Fig. B15 Structural elements of a typical operon, the *E. coli trp* operon responsible for the biosynthesis of tryptophan. The location of the structural genes *trpE, D, C, B* and *A* are shown together with an indication of the reaction catalysed by the gene products. p1/o is the principal promoter and site of binding of a regulatory protein. p2 is a weak secondary promoter. A special feature of the *trp* operon is the existence of a small gene *trpL* that encodes a peptide that can cause premature arrest of transcription at the site a, the ATTENUATOR site (*see* BACTERIAL GENE EXPRESSION). PRPP, 5-phosphoribosylpyrophosphate; CdRP, carboxyphenylamino-1-deoxyribulose-5-phosphate; InGP, indole 3-glycerol phosphate. Redrawn from [17].

quences and the transposons constructed from them play a major part in the transfer of genetic information between different backgrounds. Transposable elements represent a programmed instability in the genome structure.

In some cases chromosomal rearrangements serve specific regulatory purposes [13]: for example, some promoters are located on genome segments that can be inverted, with inversion leading to the coupling of the promoter to particular genes and expression of those genes (*see* FLIP-FLOP PROMOTERS).

Bacterial 'chromosomes'

Although much is known about the sequence organization of bacterial genes, little is known about the physical organization of the bacterial genome within the cell. The circular DNA of a bacterium is typically around 1 mm long and must be compacted within a cell rarely more than 1 μm across. This is done by SUPERCOILING and packing the DNA with small DNA-binding proteins some of which are basic in nature like eukaryotic HISTONES and are known as HISTONE-LIKE PROTEINS (HLP) [14]. The total DNA is folded into around 50 domains but it is unlikely that these domains correspond to particular well-defined regions of sequence. In order to condense bacterial DNA into an organized compact structure considerable bending is required: it is likely that this is helped by specific DNA-bending sequences which are located in the repeated non-coding parts of the genome. The overall structure of the folded bacterial chromosome can alter according to the growth conditions and large changes occur in the stationary phase [15]. Presumably this is to allow the

shut-down of all but 'housekeeping' genes while the cell is not actively growing.

S.J.W. BUSBY

See also: BACTERIAL ENVELOPE; BACTERIAL PATHOGENICITY; BACTERIAL PROTEIN EXPORT.

1 Masters, M. (1989) The *Escherichia coli* chromosome and its replication. *Curr. Opinion Cell Biol.* **1**, 241–249.
2 Gilson, E. et al. (1984) A family of dispersed repetitive extragenic palindromic DNA sequences in *E. coli. EMBO J.* **3**, 1417–1421.
3 Hahn, W. et al. (1977) One strand equivalent of the *Escherichia coli* genome is transcribed: complexity and abundance classes of mRNA. *Science* **197**, 582–585.
4 Bachman, B. (1990) Linkage map of *E. coli*, Edition 8. *Microbiol. Rev.* **54**, 130–197.
5 Kohara, Y. et al. (1987) The physical map of the whole *E. coli* chromosome: application of a new strategy for rapid analysis and sorting of a large genomic library. *Cell* **50**, 495–508.
6 Kroger, M. (1990) Compilation of DNA sequences of *Escherichia coli* (update 1990). *Nucleic Acids Res.* **18**, 2549–2587.
7 Medigue, C. et al. (1990) Mapping of sequenced genes in the restriction map of the *E. coli* chromosome. *Mol. Microbiol.* **4**, 169–187.
8 Medigue, C. et al. (1990) *Escherichia coli* molecular genetic map: update 1. *Mol. Microbiol.* **4**, 1443–1454.
9 Riley, M. & Krawiec, S. (1987) Genome Organization. In Escherichia coli *and* Salmonella typhimurium: *Cellular and Molecular Biology* (Neidhardt F.C. et al., eds.) 967–981 (American Society for Microbiology, Washington DC).
10 Belfort, M. (1989) Bacterial introns: parasites within parasites? *Trends Genet.* **5**, 209–213.
11 Jinks-Robertson, S. & Nomura, M. (1987) Ribosomes and tRNA. In Escherichia coli *and* Salmonella typhimurium: *Cellular and Molecular Biology* (Neidhardt F.C. et al., eds.) 1358–1385 (American Society for Microbiology, Washington DC).
12 Deonier, R. (1987) Locations of native insertion elements. In Escherichia coli *and* Salmonella typhimurium: *Cellular and Molecular Biology* (Neidhardt F.C. et al., eds) 982–989 (American Society for Microbiology, Washington DC).
13 Craig, N. (1985) Site-specific inversion: enhancers, recombination proteins and mechanism. *Cell* **41** 649–650.
14 Drlica, K. & Rouviere-Yaniv, J. (1987) Histonelike proteins of bacteria. *Microbiol. Rev.* **51**, 301–319.
15 Higgins, C. et al. (1990) Environmental influences on DNA supercoiling: a novel mechanism for the regulation of gene expression. In *The Bacterial Chromosome* (Drlica, K. & Riley, M., eds.) 421–432 (American Society for Microbiology, Washington DC).
16 Masters, M. (1991) Prokaryotic chromosomes. *Curr. Biol.* **1**, 63–64.
17 Yanofsky, C. & Crawford, I. (1987) The tryptophan operon. In Escherichia coli *and* Salmonella typhimurium: *Cellular and Molecular Biology* (Neidhardt, F.C. et al., eds) 1453–1472 (American Society for Microbiology, Washington DC).

Bacterial pathogenicity: molecular aspects

BACTERIAL pathogenicity can be defined as the molecular mechanisms by which bacteria cause disease. Many bacteria can infect humans or animals, sustain themselves, and multiply on or in host tissues. Disease is an inadvertent but not inevitable consequence of such infection, depending as much on the nature of the host as that of the infecting bacterium. The pathogenicity of bacteria is complex and multifactorial, often involving a series of biochemical mechanisms acting in concert to produce disease [1,2]. The genes encoding the various virulence determinants involved may be located on the bacterial chromosome, but are frequently carried on PLASMIDS, for example the toxins produced by many pathogenic ESCHERICHIA COLI, or on BACTERIOPHAGES, for example DIPHTHERIA TOXIN in *Corynebacterium diphtheriae*.

Bacterial virulence factors can be divided broadly into those that assist colonization of the host (e.g. adherence to tissue surfaces and invasion of host cells) and those that assist survival in the hostile environment therein (e.g. resistance to host defences and the production of toxins). It is very rare for possession of only one of these factors to turn an otherwise harmless bacterium into a pathogen.

E. coli, although generally thought of as a commensal and a useful laboratory tool, includes strains capable of producing diseases of varying severity. These include intestinal diseases, notably diarrhoea, and extraintestinal conditions including urinary tract infections and meningitis in the newborn. Like many pathogens, *E. coli* strains produce adhesins — structures that mediate attachment to eukaryotic cells and which can be distinguished by their specificity for receptors on the target cell. Adhesins can represent the filamentous, hair-like structures known as FIMBRIAE or PILI, or they may be nonfilamentous components of the cell surface. Common F1A (type 1) fimbrial adhesins recognize the sugar α-mannose in GLYCOPROTEINS, whereas mannose-resistant (MR) adhesins bind to eukaryotic receptors other than mannose. A wide range of filamentous adhesins are produced by different *E. coli* strains with specificities for various receptors on human and animal tissues. Pathogenic strains may contain sets of genes encoding one or more types of fimbriae, sometimes in combination with nonfimbrial adhesins.

P pili of *E. coli*, which represent the main group of MR fimbriae associated with urinary tract infections, possess G-adhesins, which mediate the binding of uropathogenic *E. coli* to a digalactoside receptor on the epithelial surface. The biosynthesis and expression of P pili are encoded by the *pap* gene cluster present on the chromosome of P fimbriate *E. coli*. Each individual pilus is composed of a rigid helical rod made up of PapA protein subunits with a thin flexible fibrillar tip composed of PapE protein subunits and bearing at its extreme end PapG protein, which is the adhesin *per se* (Fig. B16) [3]. The tip structure also involves the minor PapF and PapK proteins, but the exact location of these components is not known. Each subunit comprising the complete fimbrial structure probably interacts with the MOLECULAR CHAPERONE PapD which targets assembly-competent complexes to the

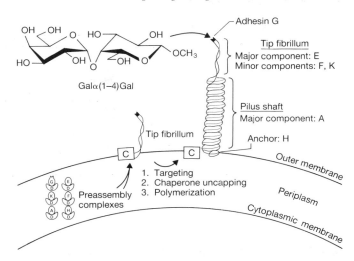

Fig. B16 Model of P pilus structure and biogenesis. The PapD chaperone protein interacts with each pilus protein (A, H, F, K etc.) to form assembly competent complexes. PapA subunits pack into a right-handed helical rod to form the shaft of the pilus. PapE subunits form open helical fibres called tip fibrillae. The PapG adhesin, located at the distal end of the tip fibrillum, mediates binding to the Galα (1–4) Gal receptor determinant. From [3].

PapC outer membrane assembly protein. Incorporation of a further protein called PapH stops the polymerization of PapA subunits and then anchors the mature pilus filament in the membrane. This complex structure can be viewed as a means of increasing the chances of successful contact between the adhesin molecule and its receptor on the host epithelial surface.

Bacterial capsules

Many bacteria are surrounded in a layer of capsular POLYSACCHARIDE which lies outside the cell envelope (*see* BACTERIAL ENVELOPES). The capsule protects the bacterium against desiccation, but also assists in resistance to phagocytosis (*see* ENDOCYTOSIS) and the lethal effects of serum. Bacteria of the same species may produce chemically distinct capsular polysaccharides, for example >70 types can be produced in *E. coli*. Blocks of genes encoding translocation of polysaccharide across the inner and outer membranes to the cell surface are conserved and flank a central variable group of genes responsible for biosynthesis of the specific capsule concerned.

The ability to produce particular chemical types of capsular polysaccharide may be associated with increased virulence and/or specific disease conditions. In both *E. coli* of capsular type K1, which causes newborn meningitis, and *Neisseria meningitidis* of serogroup B, which causes meningococccal meningitis and septicaemia, the capsular polysaccharide is composed of sialic acid (*N*-acetylneuraminic acid) which is also a surface component of human cells. Consequently the bacterial surface is not recognized as foreign by the host immune system due to the camouflaging effect of the polysaccharide layer. In *Haemophilus influenzae* the ability to produce the type b capsule is a critical but unstable virulence characteristic. The *cap* locus of *H. influenzae* which includes genes necessary for the synthesis and export of the polysaccharide contains two directly repeated duplicate 17-kb DNA segments separated by a unique bridge region of 1.2 kb. Spontaneous deletion of DNA spanning the junction between the cap repeats and including the bridge region disrupts a critical gene called *bexA* which is required for transport. Conversely, the number of tandem repeats of the *cap* genes may be increased by amplification within the chromosome. This results in increased capsule production and seems to be favoured in natural infections.

An alternative way of avoiding phagocytosis is exhibited by group A streptococci, which produce a surface protein known as M protein. This confers resistance to phagocytosis by preventing opsonization of the bacterium by COMPLEMENT. This is due in part to the ability of M protein to bind FIBRINOGEN and FIBRIN which sterically hinder access of complement to the bacterial surface.

Metabolic factors

Iron is an essential trace nutrient for bacteria and the ability of a pathogen to acquire iron from its host is an important determinant of virulence [4]. Extracellular iron in humans is bound primarily to the high-affinity glycoprotein iron carriers LACTOFERRIN and TRANSFERRIN. Pathogenic bacteria possess mechanisms for utilizing the iron associated with these proteins. In many bacteria, iron acquisition is mediated by SIDEROPHORES (e.g. enterochelin in *E. coli*), which chelate iron, and iron-regulated outer membrane proteins, some of which are involved in capturing iron from ferrisiderophore complexes. Siderophore-related genes are regulated by REPRESSOR molecules that utilize ferrous iron as COREPRESSOR. Some bacteria, such as *N. meningitidis*, do not produce siderophores but can acquire iron by direct contact between transferrin and iron-regulated outer membrane proteins that appear to function as specific receptors.

The acquisition of iron from haem or haemoglobin may be facilitated by the production of haemolysins or cytotoxins which lyse host cells and release intracellular iron complexes. The virulence of some strains of extraintestinal *E. coli* for example is enhanced by the production of haemolysin and the utilization of haem. In this and many other cases the synthesis of haemolysin is iron regulated, with derepression occurring under conditions of iron starvation.

Bacterial toxins

Enterotoxigenic *E. coli* elaborate at least two types of enterotoxin which induce secretory diarrhoea in humans and animals: a high molecular weight heat-labile toxin (LT), and a low molecular weight heat-stable toxin (ST). LTs belong to a family of cholera-like protein toxins (*see* CHOLERA TOXIN) which stimulate production of CYCLIC AMP by catalysing the ribosylation of the G_s protein (*see* GTP-BINDING PROTEINS) which regulates the activity of the adenylate cyclase complex in the PLASMA MEMBRANE (*see* ADP-RIBOSYLATION; SECOND MESSENGER PATHWAYS); the abnormally raised levels of cAMP result in active transport of Cl^- and inhibition of the uptake of Na^+ in cells of the intestinal epithelium, producing an osmotic gradient that causes the outflow of fluid into the lumen of the small intestine resulting in diarrhoea

and dehydration. These toxins are oligomers composed of a single A polypeptide noncovalently bound to an array of five B polypeptide molecules. Proteolytic cleavage of the A polypeptide and reduction of an intrachain disulphide bond activates fragment A1 which catalyses the ADP-ribosyltransferase reaction. The B polypeptides determine the binding of the toxins to GANGLIOSIDE receptors on the plasma membrane of target cells.

Methanol-soluble heat-stable enterotoxins (STA) are extracellular polypeptides of 18 or 19 amino acids which activate microvillus GUANYLATE CYCLASE. STAs have an unusual mechanism of secretion: a full-length precursor of 72 amino acids (preproSTA) contains a conventional N-terminal 19 amino acid SIGNAL PEPTIDE which is cleaved by signal peptidase allowing the resulting 53-amino acid protoxin (proSTA) to translocate to the periplasm from where it traverses the outer membrane by an as yet undefined mechanism and is proteolytically cleaved extracellularly to yield mature STA (*see* BACTERIAL PROTEIN EXPORT; PROTEIN TRANSLOCATION).

The expression of enterotoxin *per se* is not generally sufficient to convert an otherwise nonpathogenic *E. coli* strain into a pathogen. Normally enterotoxigenic strains additionally express adhesins and/or other aids to colonization of the host intestine.

Attaching-effacing lesions

Not all diarrhoeogenic *E. coli* rely on toxins to exert their effects. Enteropathogenic *E. coli* do not produce toxins but cause typical attaching-effacing lesions in the intestines. These lesions are characterized by close adherence to the enterocyte surface, the effacement of microvilli, and the disruption of the CYTOSKELETON of affected cells. Typically there is a dense accumulation of filamentous ACTIN beneath the attached bacterial cells. This phenotype is determined, at least in part, by the *eae* gene whose product is an outer membrane protein (M_r 94 000) which has similarities to invasion proteins in other bacteria.

Tissue invasion

Enteropathogenic *E. coli* generally stop short of actually invading tissue. However, many pathogens are able to enter host cells, in some cases using INTEGRINS as receptors. Most invasive pathogens exploit existing eukaryotic internalization mechanisms. For example, enteroinvasive *E. coli* and *Shigella* species enter animal epithelial cells by induced ENDOCYTOSIS requiring the expenditure of host energy and the active participation of host MICROFILAMENTS. Pathogens such as *Shigella flexneri* are able to induce 'nonprofessional' phagocytic cells to ingest the bacteria. In contrast to *Shigella*, which is confined to the epithelial layer, *Salmonella* is able to transcytose epithelial cells (*see* TRANSCYTOSIS) and penetrate to deeper tissue where it can enter endothelial cells.

Intracellular pathogens

Intracellular pathogens employ diverse strategies to survive when ingested by macrophages. For example, *Mycobacterium tuberculosis* inhibits phagosome fusion with LYSOSOMES, thereby preventing exposure to the toxic effects of their contents. In contrast,

Listeria monocytogenes and *Shigella flexneri* lyse the phagosomal membrane and escape into the cytoplasm.

Once free in the cytoplasm, *S. flexneri* polymerizes F actin (*see* ACTIN) on its surface because it expresses an actin nucleator protein specified by the *ipa* gene. These otherwise nonmotile bacteria are then able to move in the cytoplasm by a rearrangement of actin into bundles that form 'tails' to each bacterial cell and push them forwards. With the formation of protrusions into neighbouring cells and subsequent lysis of plasma membrane the bacteria are then able to pass from cell to cell. *Legionella* produces a PHOSPHATASE which blocks the formation of crucial SECOND MESSENGERS and also elaborates a cytotoxin. Together these disrupt oxidative killing mechanisms in neutrophils.

Proteases

Many bacteria produce proteolytic enzymes which inactivate the IMMUNOGLOBULINS of their hosts. Bacteria that colonize epithelial surfaces, such as *H. influenzae* and pathogenic *Neisseria*, secrete IgA proteases, presumably to combat secreted ANTIBODIES. Some other bacteria produce IgG-specific proteases.

Phase and antigenic variation

Particular virulence characters may be subject to phase variation where a PHENOTYPE is lost or gained reversibly. This tendency, which often involves genome rearrangement, may account for the genetic instability of certain virulence factors during laboratory subculture of bacterial pathogens. Phase variations are advantageous because they may, for example, allow otherwise adhesive bacteria to cease producing adhesin and to desorb from tissue to which they are attached. Subsequent re-expression of the adhesin then permits adherence elsewhere, assisting the progression of infection to new sites. Adhesiveness may be advantageous in colonization by binding to host epithelial cells but may confer disadvantages. For example, type 1 (mannose-sensitive) fimbriae of *E. coli* render the bacterium adhesive to phagocytes and hence more sensitive to phagocytosis. Thus a genetic switching system between two or more phases may have considerable advantages to pathogens.

Many pathogenic bacteria also exhibit qualitative changes in phenotype in the form of ANTIGENIC VARIATION of their dominant exposed surface components. This assists the pathogen to avoid antibodies elicited in the host by previous infections of the same type. In addition, the variability may alter functional properties of the component concerned, for example, tissue tropisms may be affected in the case of antigenic variation of adhesins. A good example of antigenic variation is provided by the related pathogenic bacterium *N. gonorrhoeae*, the causative agent of gonorrhoea, and *N. meningitidis*, the causative agent of meningococcal meningitis. Repeated infections by *N. gonorrhoeae* are possible owing largely to variation both within and between strains of two antigenically predominant components, hair-like proteinaceous filaments called pili (essentially the same as fimbriae in other bacteria) and so-called opacity (*opa*) or PII polypeptides embedded in the outer membrane of the bacterial envelope. Both pili and *opa* proteins are important adhesins in the pathogenic *Neisseria*

and failure to produce these proteins reduces the ability of the organism to initiate infection.

Both these critical virulence determinants are subject to phase and antigenic variation [5]. The molecular mechanisms driving variation are, however, fundamentally different in each case. Pilin, the major repeating subunit forming the neisserial pilus, is encoded by a single expression locus (*pilE*) which contains a PROMOTER. The pilin molecule may be divided into three approximately equal domains. The first 53 amino acids at the N terminus comprise a conserved (constant) region which is the same in pilins expressed by all strains. The central 53 amino acids constitute a semivariable domain where variation between strains is limited to amino-acid substitutions. The remaining C-terminal or hypervariable domain is dominant immunologically over the rest of the molecule and in different pilins contains numerous deletions and insertions of amino acids as well as substitutions.

Variation in the pilin expressed by any one strain is effected by RECOMBINATION between the *pilE* locus and one of a number of 'silent' pilin genes. The genome of *N. gonorrhoeae* contains a number of such truncated, nonexpressed *pilS* loci which are scattered at different sites and are frequently arranged in direct tandem repeats. Each copy of a silent pilin sequence is deleted with respect to the 5′ end of the *pilE* gene, lacks all or part of the constant and semivariable regions, and is without a promoter. There appear to be two routes by which new sequences may replace those at the expression locus: (1) internal or intragenomic recombination; and (2) TRANSFORMATION where DNA is taken up by the bacteria from the external environment from lysed fellow cells.

The process requires HOMOLOGOUS RECOMBINATION as cells that lack neisserial *RecA* gene function exhibit reduced pilus variation. In some instances, evidence suggests that a RECIPROCAL RECOMBINATION event is involved, but the majority of events are nonreciprocal. Usually, the donor *pilS* sequence replaces all or part of the 3′ end of the current *pilE* locus in a GENE CONVERSION event. A copy of the donor *pilS* sequence is retained at its original location and can participate in later recombination events.

The specific mechanisms involved in these genome rearrangements are still controversial. Some workers believe that recombination depends on the alignment of short conserved nucleotide sequences within the semivariable and hypervariable domains leading to the exchange of 'minicassettes' of sequence information between *pilS* and *pilE*. Additional recombination-promoting sequences have been implicated both 5′ and 3′ to the *pil* sequences. However, the lengths of homologous sequence that must participate in the DNA transactions involved are shorter than expected for homologous recombination. An alternative explanation for pilin variation has therefore been proposed. This is deletion repair, whereby the incoming *pilS* sequence acts as a template for repair of deleted regions of *pilE*.

Antigenic and phase variation in neisserial *opa* proteins occurs by a totally different mechanism. The opacity proteins are encoded in *N. gonorrhoeae* by a family of 11 complete genes which are essentially identical except for a short region encoding the variable surface-exposed portion of the polypeptide. These genes are switched on or off essentially at random by slipped strand mispairing of the pentameric sequence CTCTT which is present in tandem direct repeats in the coding sequence for the LEADER SEQUENCE needed for export of the *opa* protein to the bacterial outer membrane. The leader sequence is proteolytically cleaved to produce mature *opa* polypeptide. For a given number of repeats (*n*) the leader sequence coding region will be in frame with that for the mature *opa* polypeptide. However, if the number of repeats is $n \pm 1$ or $n \pm 2$, then the mature polypeptide following processing will be out of frame and nonfunctional. Hence any given bacterial cell can express either no *opa* proteins or up to as many as six different types simultaneously. This process appears to be regulated simply by mispairing of the repeats, probably during replication. Similar strand-slippage mechanisms are known to switch on and off expression of other virulence determinants, for example in the *pilC* gene of *N. gonorrhoeae*, whose product copurifies with pilin and may be a pilus component involved in receptor binding, and in the *lic* genes which determine the nature of the antigenically variable LIPOPOLYSACCHARIDE component of the outer membrane of *H. influenzae* (*see* BACTERIAL ENVELOPE).

Adaptation

During infections bacteria are obliged to colonize a series of microenvironments that differ with respect to temperature, osmolarity, nutrient availability, pH, and oxidative ability. Adaptation in response to the various stresses involved is a key component in bacterial pathogenicity and many virulence determinants are environmentally regulated. A number of virulence genes are known to be regulated by signal transduction involving a family of two-component sensory systems [6]. These include, for example, pilin production in *N. gonorrhoeae* where the product of the *pilB* gene, an outer membrane-spanning protein, acts as a sensor for environmental conditions together with the product of the *pilA* gene which behaves as a *trans*-acting response regulator of the *pilE* gene.

Changes in the environment also produce alterations in the superhelix density of the genome which can have consequential changes for gene expression as TRANSCRIPTION is favoured by the negatively supercoiled state. For example, the enterobacterial *osmZ* gene, which encodes the nucleoid DNA-binding protein H1/HNS and affects superhelicity, is a regulator of virulence in *S. flexneri* and is required for virulence in *Salmonella typhimurium* [7]. It appears that individual operons in bacteria, including those involved in pathogenicity, can be linked in global regulatory hierarchies which may be controlled by response to changes in superhelix density through the mediation of TOPOISOMERASES and DNA-packaging proteins.

Vaccine design

Molecular analysis of virulence determinants and the cloning of critical genes enables the rational design of safe bacterial vaccines. Cloned gene products may be used as the basis of SUBUNIT VACCINES consisting of one or more components of the pathogen that will elicit a protective immune response from the host without danger from other components of the bacteria. Genes encoding harmful components of vaccine strains can also be inactivated by recombinational replacement of the wild-type allele with a copy of the relevant cloned gene that has been inactivated by deletion and/or the insertion of a harmless marker gene.

Alternatively, genes encoding protective antigens can be expressed heterologously in a harmless or disabled host strain (e.g. in *aroA* mutants of *Salmonella typhimurium*, which are unable to invade intestinal tissue because they are deficient in the ability to make enterochelin, *p*-aminobenzoic acid, and aromatic amino acids).

J.R. SAUNDERS

See also: BACTERIAL GENE EXPRESSION; BOTULINUM TOXIN; PERTUSSIS TOXIN.

1 Finlay, B.B. & Falkow, S. (1989) Common themes in microbial pathogenicity. *Microbiol. Rev.* **53**, 210–230.
2 Neidhardt, F.C. et al. (Eds) (1987) Escherichia coli *and* Salmonella typhimurium: *Cellular and Molecular Biology* (American Society for Microbiology, Washington DC).
3 Kuehn, M.J. et al. (1992) P pili in uropathogenic *E. coli* are composite fibres with distinct fibrillar adhesive tips. *Nature* **356**, 252–255.
4 Crosa, J.H. (1989) Genetics and molecular biology of siderophore-mediated iron transport in bacteria. *Microbiol. Rev.* **53**, 517–530.
5 Meyer, T.F. et al. (1990) Variation and control of protein expression in *Neisseria*. *Annu. Rev. Microbiol.* **44**, 451–477.
6 Miller, J.F. et al. (1989) Coordinate regulation and sensory transduction in the control of bacterial virulence. *Science* **243**, 916–922.
7 Dorman, C. (1991) DNA supercoiling and environmental regulation of gene expression in pathogenic bacteria. *Infect. Immun.* **59**, 745–749.

Bacterial protein export

GRAM-NEGATIVE bacteria mediate their relations with the outside world through the numerous proteins they secrete or which are located in the periplasm and outer membranes of their envelopes (*see* BACTERIAL ENVELOPES). All these proteins, and the proteins secreted into the growth medium by Gram-positive bacteria, share the common trait of crossing a cytoplasmic membrane. All bacterial proteins begin their synthesis on ribosomes located in the cytoplasm, but those proteins destined to be exported across the cytoplasmic membrane are synthesized as longer precursors with N-terminal extensions called SIGNAL SEQUENCES or leader peptides [1]. The principles that govern the export of proteins in bacteria are similar to those involved in the secretion of proteins across the endoplasmic reticulum of eukaryotic cells (*see* PROTEIN SECRETION; PROTEIN TARGETING; PROTEIN TRANSLOCATION).

The signal sequences found on the precursors of *E. coli* envelope proteins have several features in common. They range in size from 15 to 26 amino acids, and although there is no strict homology, three regions can be identified. The first, comprising amino acids 2–10 at the extreme N terminus of the peptide, has a net positive charge that averages $+1.7$, due to the presence of Arg or Lys residues. This is followed by a region of nine or more predominantly hydrophobic amino acids that have been predicted to adopt an α-helical conformation in apolar environments such as the membrane. A Pro or Gly residue is often found at the end of the hydrophobic region followed by a sequence of 4–6 amino acids which are of a higher polarity. The amino acid directly adjacent to the site of cleavage of the precursor (often referred to as the − 1 position in signal sequences) as well as the next but one amino acid upstream of the cleavage site (i.e. the − 3 position) tend to be residues that have small side chains such as Ala, Gly, or Ser. This has been termed the − 1, − 3 rule and is useful in identifying suspected cleavage sites in exported proteins whose DNA has been cloned and for which the exact N terminus of the processed precursor is unknown.

Cleavage of most precursor proteins is carried out by LEADER PEPTIDASE [2], an enzyme located on the periplasmic face of the cytoplasmic membrane. A second peptidase, SIGNAL PEPTIDASE, removes signal sequences from lipoproteins (*see* BACTERIAL ENVELOPES).

Signal sequences appear to have at least two functions — to slow the rate of folding of precursor proteins as they emerge from the ribosome thus allowing sufficient time for the precursor to engage the export machinery, and to facilitate insertion of precursor proteins into the cytoplasmic membrane.

A number of *E. coli* genes have been identified which have a role in the export of proteins across the cytoplasmic membrane [3] (Fig. B17). These include *secA* (*prlD*), *secB*, *secD*, *secE* (*prlG*), *secF*, and *secY* (*prlA*), as well as the two genes that encode the signal peptidases, *lep* and *lspA*. Temperature-sensitive and cold-sensitive mutants of *secA* have been isolated that show a lethal secretion defect. SecA protein (M_r 102 000) is peripherally associated with cytoplasmic membrane or found in the cytoplasm. It has ATPase activity which is stimulated by the presence of membranes and precursor proteins and has been termed 'translocation ATPase' and is undoubtedly involved in coupling ATP hydrolysis to the events of protein export. SecY (M_r 49 000) and SecE (M_r 13 600) are integral cytoplasmic membrane proteins that are thought to constitute a proteinaceous complex that interacts with precursors as they are translocated across the membrane. SecD (M_r 67 000) and SecF (M_r 35 000) are integral membrane proteins with large periplasmic loops. They are thought to be involved in the late events of protein export.

SecB (M_r 16 600) and DsbA (M_r 21 000), which is a recently identified protein, seem to be involved in effecting the folding of precursor and mature proteins, respectively. SecB can interact with certain unfolded precursor proteins and maintain them in this state, thus allowing a more efficient interaction with the export apparatus. GroEL, another cytoplasmic protein which can also interact with unfolded proteins, has been shown to influence the export of certain proteins such as β-lactamase. These proteins thus exhibit properties characteristic of MOLECULAR CHAPERONES, in that they transiently interact with proteins and effect their folding (*see also* HEAT SHOCK). DsbA on the other hand is located in the periplasm and is required for efficient disulphide bond formation in exported proteins in *E. coli*.

The precise molecular mechanism by which the proteins constituting the export apparatus couple hydrolysis of ATP and utilize the protonmotive force to transfer precursor proteins efficiently across the cytoplasmic membrane remains a molecular enigma.

T.R. HIRST

1 Von Heijne, G. (1985) Signal sequences: the limits of variation. *J. Mol. Biol.* **184**, 99.
2 Dalby, R.E. (1991) Leader peptidase. *Mol. Microbiol.* **5**, 2855.
3 Schatz, P.J. & Beckwith, J. (1990) Genetic analysis of protein export in *Escherichia coli. Annu. Rev. Genet.* **24**, 215.

bacterial viruses *See*: BACTERIOPHAGES.

bacteriochlorophyll The major photosynthetic pigments of photosynthetic bacteria. *See*: BACTERIOPHAEOPHYTIN; CHLOROPHYLL; PHOTOSYNTHESIS.

bacteriochlorophyll-*a* protein The protein moiety of the bacteriochlorophyll–protein complex. *See*: Fig. P83 in PROTEIN STRUCTURE for an illustration of this protein.

bacteriophaeophytin Bacteriochlorophyll in which the central Mg^{2+} is replaced by $2H^+$. Two molecules occur in the photosynthetic reaction centre of PURPLE BACTERIA, one of which acts as an important intermediate in the primary charge separation. *See*: CHLOROPHYLL; PHOTOSYNTHESIS.

Bacteriophage(s)

BACTERIOPHAGES (bacterial viruses, phages) are, with about 4000 isolates studied by electron microscopy, the largest virus group. They occur throughout the bacterial world and in every conceivable bacterial habitat, even volcanic hot springs.

Phages destroy or modify bacteria. Destruction is caused by lysis of infected bacteria, and is detected by the resulting clear or veiled spots (PLAQUES) in a bacterial 'lawn' grown on solid media

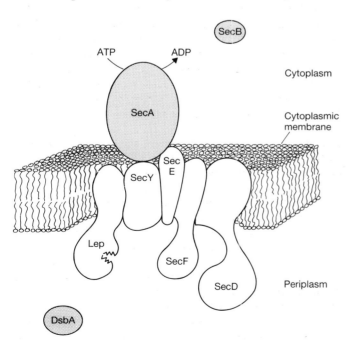

Fig. B17 Proteins involved in the export of envelope polypeptides across the cytoplasmic membrane of *E. coli*.

Table B1 Basic characteristics and frequency of phage families*

Shape	Nucleic acid		Lipids	Family	Representatives	No. of members‡
	Nature	Size†				
Tailed	DNA, ds, L	19–700	–	Myoviridae	T4	958
				Siphoviridae	1	2339
				Podoviridae	T7	556
Cubic	DNA, ss, C	4.4–6.1	–	Microviridae	φX174	40
	ds, C	8.4	+	Corticoviridae	PM2	2?
	ds, L	15.2	+	Tectiviridae	PRD1	15
	RNA, ss, L	3.6	–	Leviviridae	MS2	37
	ds, L	13.4	+, E	Cystoviridae	φ6	1
Filamentous	DNA, ss, C	4.4–7.4	–	Inoviridae	fd, MVL1	45
	ds, L	16	+, E	Lipothrixviridae	TTV1	4?
Pleomorphic	DNA, ds, C	11.8	+, E	Plasmaviridae	MVL2	5?
	ds, C	15.5	–, E	SSV1 group	SSV1	1

* Modified from [1].
† In kilobases (kb) for phages with ssDNA and kilobase pairs (kbp) for phages with dsDNA, respectively.
‡ See [2].
C, circular; ds, double stranded; E, envelope; L, linear; ss, single stranded; +, present; –, absent.

or by the inhibition of bacterial growth in liquid cultures. Alternatively, surviving bacteria can become phage infected without undergoing lysis and acquire new properties.

Many phages have practical applications, mostly in PHAGE TYPING of bacteria, as indicators of faecal pollution, and in DNA SEQUENCING and GENETIC ENGINEERING, where they are widely used as cloning and expression VECTORS (*see* DNA CLONING: LAMBDA). At one time, it was thought that they might prove useful in the therapy of human and animal bacterial diseases, but this has not proved to be the case and this use has practically been abandoned. By destroying valuable industrial bacteria, phages are a major nuisance in the dairy and fermentation industries.

Phages consist essentially of nucleic acid, a protein coat, and, in some cases, lipid. They belong to four structural groups: (1) isometric particles with cubic symmetry; (2) helical filaments or rods; (3) tailed phages with cubic CAPSIDS (the heads) and helical tails; and (4) pleomorphic particles without definite structure. Capsids with cubic symmetry are icosahedra or related bodies.

More than 95% of phages are tailed. Although a few tailed phages, such as phage λ (*see* LAMBDA), have been studied in exquisite detail, they are generally much less well known than cubic, filamentous and pleomorphic phages (which are also known as the CFP phages). For example, only five tailed phage DNAs have been fully sequenced.

Classification

Phages are classified into 10 families on the basis of gross morphology, nature of the nucleic acid, and presence of lipids or an envelope (Table B1, Fig. B18) [1,3,4].
1 Tailed phages include the Myoviridae, Siphoviridae, and Podoviridae families. All contain a single molecule of double-stranded (ds) DNA. Heads are icosahedra or elongated derivatives thereof. Many tailed phages have morphological features such as collars,

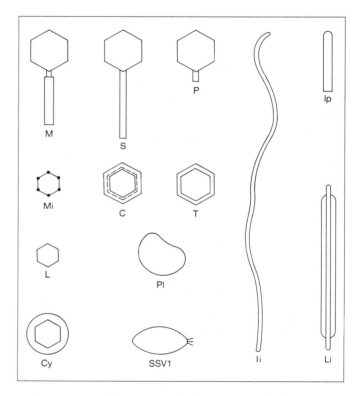

Fig. B18 Basic morphotypes of bacteriophages. C, Corticoviridae; Cy, Cystoviridae; Ii, Inoviridae, *Inovirus* genus; Ip, Inoviridae, *Plectrovirus* genus; L, Leviviridae; Li, Lipothrixviridae; M, Myoviridae; Mi, Microviridae; P, Podoviridae; Pl, Plasmaviridae; S, Siphoviridae; SSV1, SSV1 group; T, Tectiviridae. Modified from [1].

base plates or tail fibres. About 220 species have been recognized, but so far no genera.
2 Phages with cubic symmetry include five families with single-

stranded (ss) or ds DNA or RNA. Microviridae are very small and characterized by conspicuous capsomers (units of coat protein). Corticoviridae have a complex capsid with a LIPOPROTEIN layer located between two protein coats. Tectiviridae have two coats, namely an outer protein shell and an inner lipoprotein membrane. Leviviridae are small particles without any particular morphological features, resembling POLIOVIRUSES (*see* ANIMAL VIRUSES). Cystoviridae have a lipoprotein envelope and a dodecahedral capsid. They contain three molecules of dsRNA and an RNA POLYMERASE.

3 Filamentous and rod-shaped phages contain DNA. Inoviridae (ssDNA) are long filaments (genus *Inovirus*) or short rods (genus *Plectrovirus*). Lipothrixviridae (dsDNA) have a lipoprotein envelope and only one cultivable member, but similar particles have been observed electron microscopically [6,7].

4 Pleomorphic phages contain dsDNA. Plasmaviridae are rounded and consist of a lipoprotein envelope around a NUCLEO-PROTEIN granule without an apparent capsid. The SSV1 group has a single member, a lemon-shaped archaebacterial virus with bristles at one end [6,7].

Phage hosts

Tailed phages have been found in over 100 bacterial genera, including ARCHAEBACTERIA, CYANOBACTERIA, and budding, gliding, appendaged, or sheathed EUBACTERIA. CFP phages have a much more limited distribution. Leviviruses and filamentous inoviruses are limited to ENTEROBACTERIA and their close relatives, plectroviruses and plasmaviruses to MYCOPLASMAS, and lipothrixviruses and the SSV1 type to the archaebacterium, *Thermoproteus tenax*. In a general way, phages are host genus specific. Exceptions, infecting many genera, are common in enterobacterial phages and there are some plasmid-specific phages of Gram-negative bacteria.

Life cycles

Phages typically adsorb to specific RECEPTORS on the bacterial cell wall, but also to PILI, FLAGELLA, or the capsule (*see* BACTERIAL ENVELOPE). Most or all phages have fixation organelles such as fibres, spikes or adsorption proteins. Tailed phages adsorb by the tail. Leviviruses and filamentous inoviruses adsorb to the sides or ends of bacterial pili, respectively, and apparently reach the bacterial surface as pili are retracted.

Tailed phages and tectiviruses have devices for ejecting the DNA. The tails of tailed phages are permanent features often equipped with muralytic enzymes that digest the bacterial cell wall. In addition, the Myoviridae possess a contractile sheath which surrounds the tail tube and brings it into contact with the plasma membrane. In the Tectiviridae, the inner membrane transforms itself upon adsorption into a tail-like tube. In the *Inovirus* genus and cystoviruses, the capsid penetrates the cell wall, but not the plasma membrane. The process of infection itself — the entry of phage nucleic acid into the bacterial cell — is generally poorly understood.

A typical lytic cycle comprises a latent period of 20–40 minutes and ends in the release of 50–200 phages by lysis, or bursting, of the infected bacterium. Phages undergoing lytic cycles only are termed virulent. Latent periods and burst sizes vary widely according to the host and growth conditions. Some phages are defective and cannot multiply without the presence of a HELPER VIRUS which supplies some of its functions. At the late stages of infection, intracellular phages and even crystal-like inclusion bodies composed of masses of virus become visible in infected cells. Progeny phages are released by lysis of the host, slow extrusion (inoviruses, TTV1), or budding (plasmaviruses).

Phage morphogenesis is a complex, highly ordered process which may involve separate pathways for different phage constituents. In tailed and most cubic phages, the nucleic acid enters preformed capsids. In leviviruses, the capsid condenses around the RNA, and in inoviruses, capsid protein and phage DNA assemble at the cell periphery as the particle is extruded. Phage assembly is not error-proof and results, especially in tailed phages, in aberrant structures, for example head size variants, polyheads, polysheaths or multitailed particles. During assembly, phage capsids are able to package random fragments of bacterial DNA, which may then be transferred to new bacteria. This is the process of general TRANSDUCTION.

In the temperate cycle, phage DNA becomes a latent PRO-PHAGE. It integrates into bacterial DNA or persists as a PLASMID and is replicated at each bacterial division until it is liberated spontaneously or by INDUCTION, commonly by ultraviolet light or MITOMYCIN C. The phage GENOME is then expressed and a lytic cycle starts. A bacterium harbouring prophages is termed lysogenic. Lysogeny exists in all the bacterial kingdom. About half of tailed phages and members of the *Inovirus* genus, plasmaviruses, lipothrixviruses, and TTV1 are temperate. If integrated prophages are wrongly excised, progeny phages may carry fragments of bacterial DNA adjacent to their integration site and transfer it to new bacteria (specialized transduction). In lysogenic conversion, lysogenic bacteria acquire new properties that are specified by the phage genomes, for example new antigens or toxins.

In pseudolysogeny, also called carrier state, only part of the bacterial population is infected and produces phages. Steady-state infections are caused by noncytocidal viruses liberated by extrusion or budding.

Nucleic acid

All phages except one (cystovirus φ6) contain a single molecule of nucleic acid (13–~50% of the particle). All the principal types of nucleic acid occur in phages. Molecules are linear or circular (Table B1); those of corticoviruses, plasmaviruses, and TTV1 are SUPERCOILED. The RNA of leviviruses has a large number of secondary loop and HAIRPIN structures (*see* NUCLEIC ACID STRUCTURE).

Phage nucleic acid composition generally resembles that of the host. Phage genome sizes depend on capsid size and vary enormously. One tailed phage, a myovirus of *Bacillus megaterium*, contains 700 kilobase pairs (kbp) of DNA. This is the largest amount known in phages; genome sizes generally vary from 3 kb in leviviruses to >100 kbp, possibly more, in tailed phages. Tectiviruses, the largest cubic phages, have at least 33 genes. In tailed phages, genes with related functions often cluster together and genomes seem to consist of interchangeable modules.

Phage DNAs have particular features which are rare or do not occur in other viruses. Tailed phage DNAs may have circular permutations of sequence, terminal repeats, COHESIVE ENDS, terminal proteins, single-stranded gaps, unusual BASES (e.g. 5-hydroxymethylcytosine or 5-hydroxymethyluracil), or DNA-bound sugars. Terminal proteins and single-stranded gaps also occur in tectiviruses and plasmaviruses, respectively. OVER-LAPPING GENES and intergenic spaces are found in microviruses, inoviruses of the fd type, and leviviruses.

Replication

Double-stranded DNA is replicated uni- or bidirectionally in a semiconservative way, either via a ROLLING CIRCLE mechanism or by strand displacement (*see* DNA REPLICATION). In tailed phages, the infecting DNA remains linear, or is circularized and then replicates for a few rounds as a THETA RING (θ ring). The end-product is generally a giant DNA molecule or CONCATEMER, which is cut to size at fixed sites or to fit the phage head. Corticoviruses replicate as θ rings and rolling circles.

In microviruses and inoviruses, the viral ssDNA is transformed into dsDNA by synthesis of a complementary strand. This REPLI-CATIVE FORM (RF) is replicated as a rolling circle to generate progeny RF which, in turn, serves to synthesize novel viral DNA.

The (+) messenger-sense ssRNA of leviviruses codes for a REPLICASE and capsid protein and forms a replicating complex with ribosomal and other cellular proteins. A complementary (−) strand is then synthesized, which acts as a template for viral (+) RNA. In the dsRNA-containing cystoviruses, the RNA polymerase present in the virion transcribes three complementary strands from the three RNA molecules it contains, which serve as templates for viral RNA and as mRNA for viral proteins. Viral RNA is synthesized by strand displacement.

Transcriptional and translational controls

TRANSCRIPTION in tailed phages (and probably others) is largely regulated by the order of genes along the chromosome. Early genes, which initiate phage replication or shut off host metabolism, are transcribed first and genes encoding coat and other structural proteins last (late genes). Further regulatory factors include PROMOTERS, REPRESSOR proteins, transcription TERMINA-TORS and supercoiling of DNA (*see e.g.* BACTERIAL GENE EXPRESSION; LAMBDA; TRANSCRIPTION). In ssDNA, the distance of genes from a central transcriptional terminator sequence regulates the frequency of transcription, proximal genes being transcribed with the highest efficiency.

TRANSLATION is mainly controlled at the level of initiation or through ribosome binding. In leviviruses, the secondary structure of the RNA regulates its accessibility to ribosomes. Translation repressors have been described in tailed phages.

Proteins, lipids, and carbohydrates

The number of constitutive proteins (55–85% of the particle) depends on the complexity of the phage and varies between at least 42 in the tailed T-even phages and 2 in leviviruses, fd-type inoviruses and SSV1. Molecular weights (M_r) of individual proteins range between 5×10^3 and $>200 \times 10^3$ (generally about 50×10^3). Lipids constitute 23% in the enveloped cystoviruses and 10–15% in other lipid-containing CFP phages. Lipid contents of 10–15% have also been found in a few tailed phages of *Mycobacterium*. Phage lipids are quite diverse (up to seven or eight species per phage) and include phospholipids, neutral lipids, fatty acids and sterols (*see* LIPIDS). In some phages, the phospholipid composition depends on the phage and not the host. Glycolipids and GLYCOPROTEINS are minor components of a few tailed phages and of corticoviruses, tectiviruses, and plasmaviruses.

Origin and evolution of bacteriophages

Phages appear to be POLYPHYLETIC in origin. This is suggested by fundamental structural and biological differences between most phage families and also their confinement to certain hosts. The ubiquitous tailed phages probably originated first, apparently before the separation of eubacteria and archaebacteria and perhaps earlier than 3000 million years ago. Plasmid-specific inoviruses and leviviruses emerged much later as enterobacteria-like hosts evolved, possibly from cellular DNA or RNA that became infectious and provided with a coat.

Major avenues of phage evolution are point MUTATION, genome rearrangement (e.g. through duplication or inversion) and modular evolution through incorporation or exchange of genetic building blocks or 'modules'. Certain tailed phages seem to be true chimaeras, having incorporated elements from other phages, plasmids, or bacteria.

H.-W. ACKERMANN

1 Ackermann, H.-W. (1987) Bacteriophage taxonomy in 1987. *Microbiol. Sci.* **4**, 214–218.
2 Ackermann, H.-W. (1992) Frequency of morphological phage descriptions. *Arch. Virol.* **124**, 201–209.
3 Ackermann, H.-W. & DuBow, M.S. (1987) *Viruses of Prokaryotes*, Vols I & II (CRC Press, Boca Raton, FL).
4 Francki, R.I.B. et al. (Eds) (1991) *Classification and Nomenclature of Viruses.* *Arch. Virol.* (Suppl.2) (Springer, Vienna).
5 Klaus, S. et al. (1992) *Bakterienviren* (Gustav Fischer, Jena).
6 Maniloff, J. (1988) Mycoplasma viruses. *CRC Crit. Rev. Microbiol.* **15**, 339–389.
7 Zillig, W. et al. (1988) Viruses of archaebacteria. In *The Bacteriophages* (Calendar, R., Ed.), Vol. I, 517–558 (Plenum, New York).

bacteriophage lambda *See*: LAMBDA.

bacteriorhodopsin A membrane protein from the purple membrane of the halophilic bacterium *Halobacterium halobium*, which functions as a light-driven proton pump. Bacteriorhodopsin consists of a polypeptide of M_r 26 000 surrounding the chromophore all-*trans*-retinal. Bacteriorhodopsin is a seven transmembrane domain protein but has no sequence homology with vertebrate RHODOPSIN, despite its similar membrane topology, chromophore attachment, and photochemistry. Because of its regular crystalline arrangement in the purple membrane, a 6 Å three-dimensional structure of bacteriorhodopsin was determined in

1975 by electron microscopy and diffraction, providing the model topology for subsequent interpretation of the seven transmembrane domain families of receptor proteins (*see* G PROTEIN-COUPLED RECEPTORS). The three-dimensional structure of bacteriorhodopsin at 3.5 Å resolution has been further determined by electron cryomicroscopy and shows the helices roughly perpendicular to the membrane with the retinal set at about 20° to the membrane plane (Fig. B19). The retinal is covalently linked to the seventh transmembrane helix via a Schiff base to a lysine side chain. On absorption of a photon of light the all-*trans*-retinal isomerizes to 13-*cis*-retinal, resulting in a sequence of structural changes in the protein that triggers proton pumping. The mechanism of proton pumping has been studied in bacteriorhodopsin by site-directed mutation studies, and a photocycle involving proton transfer between the Schiff base in retinal and aspartic acid residues 85, 212, and 96 has been proposed.

Halobacterium also contains two closely related members of the bacteriorhodopsin family: halorhodopsin, which functions as a light-driven chloride pump, and sensory rhodopsin, which is a phototactic photoreceptor.

Henderson, R. & Unwin, P.N.T. (1975) *Nature* **257**, 28–32.
Henderson, R. et al. (1990) *J. Mol. Biol.* **213**, 899–929.

bacteroid *See*: NITROGEN FIXATION; NODULATION.

Baculoviridae Family of enveloped DNA viruses of arthropods, members of which are being developed as EXPRESSION VECTORS for producing recombinant proteins in insect cell cultures. The virion contains one or more rod-shaped nucleocapsids containing a molecule of circular supercoiled double-stranded DNA (M_r 54×10^6–154×10^6). The virus used as a vector is generally *Autographa californica* nuclear polyhedrosis virus (NPV). Expression of introduced genes is under the control of the strong promoter that normally regulates expression of the polyhedrin protein component of the large nuclear inclusions in which the viruses are embedded in the infected cell.

baculovirus expression system *See*: BACULOVIRIDAE; EXPRESSION VECTORS.

Bal31 nuclease NUCLEASE from the bacterium *Alteromonas espejiana* Bal31. It has three activities:
1 An EXONUCLEASE activity on both strands of double-stranded DNA.
2 An exonuclease activity on single-stranded DNA.
3 A random ENDONUCLEASE activity on single-stranded DNA.
The enzyme is mostly used in the manipulation of DNA *in vitro* to create DELETION MUTANTS by shortening DNA fragments of linearized VECTORS from both ends. The DNA fragments so generated can be religated, after FILLING IN the recessed ends using the KLENOW FRAGMENT of the enzyme DNA polymerase I. This is necessary as the double-stranded exonuclease activity of Bal31 is the sum of its 3′ exonuclease activity and its single-stranded endonuclease activity. The latter activity is slower than the former, leaving staggered ends.

balanced lethal LETHAL MUTATIONS can be maintained in a stock without the need for selection if the chromosome carrying one lethal mutation is 'balanced' by being HETEROZYGOUS with a chromosome carrying a nonallelic lethal mutation. Recombination would lead to the breakdown of this system and so these 'balancer' chromosomes contain INVERSIONS which suppress recombination.

balanced polymorphism A relatively permanent kind of genetic equilibrium in which ALLELES of a polymorphic locus (*see* GENETIC POLYMORPHISM) are present in the population at a steady state frequency. This occurs when the HETEROZYGOTES for the alleles under consideration have a higher adaptive value than either HOMOZYGOTE. Such a situation exists with the sickle cell haemoglobin gene in a malaria zone where the incidence of the advantageous heterozygous state balances the loss of homozygotes from the population (*see* HAEMOGLOBIN AND ITS DISORDERS).

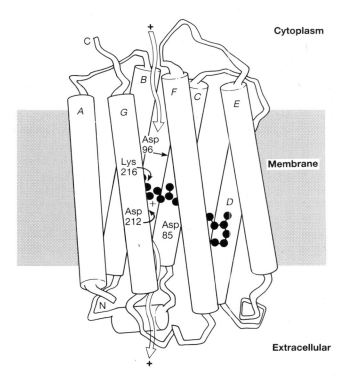

Fig. B19 Model of the bacteriorhodopsin proton pump. The retinal chromophore is in the middle of the molecule, within the membrane. A–G designate the seven transmembrane α-helices, of length 30–40 Å, connected by segments alternately at the cytoplasmic and extracellular surfaces of the purple membrane. Residues from helices A, B, C, F, and G form the proton channel and interact with the chromophore which is linked to Lys 216 on helix G. Ionized Asp residues 85 on C and 212 on G, located in the hydrophilic channel extending down from the retinal, are involved in transporting the proton from the Schiff-base nitrogen to the outside after photoisomerization. Protonated Asp 96 on C, located in the hydrophobic channel above the retinal, provides the proton that recharges the Schiff base. Reprotonation of Asp 96 from the cytoplasm completes the photocycle. From Caspar, D.L.D. (1990) *Nature* **345**, 666.

Balb/c cells A fibroblast CELL LINE derived from embryo of the Balb/c mouse strain.

Balbiani rings Greatly distended regions visible in the POLYTENE CHROMOSOMES found in the salivary glands of DROSOPHILA and other insects, and which are thought to be sites of genetic activity.

band (1) On chromosomes, *see* CHROMOSOME BANDING.
(2) Discrete section of a gel containing a purified DNA, RNA, or protein species separated out by size by ELECTROPHORESIS. Bands are visualized by staining or by the addition of radioactively (or otherwise) labelled PROBES. *See*: DNA SEQUENCING; DNA TYPING; NORTHERN BLOTTING; SOUTHERN BLOTTING.

band 4.1 protein *See*: ACTIN-BINDING PROTEINS.

band III (or 3) protein Single-chain transmembrane protein (M_r 106 000) in red blood cells which forms the anion channel through which bicarbonate ions formed inside the cell are exchanged for chloride ions from the extracellular medium, thus enabling the red cell to get rid of CO_2. The band III protein is linked on the cytoplasmic side to ankyrin, a component of the red cell membrane skeleton. It is also known as the anion exchanger. *See*: MEMBRANE STRUCTURE.

band retardation The differential migration in ELECTROPHORESIS of naked DNA and of the same sized DNA to which a protein is bound. A DNA–protein complex will migrate through a gel more slowly than naked DNA owing to its apparent higher molecular weight. Nonspecific binding between DNA and protein can be reduced by adding excess synthetic nucleic acid. The specificity of the interaction can be demonstrated by outcompeting bound labelled DNA with excess amounts of the same DNA unlabelled.

bandsharing coefficient An index of the similarity of two DNA fingerprint patterns (*see* DNA TYPING). The DNA fingerprint of an individual consists of several radioactive DNA bands of different intensities and containing DNA fragments of different molecular weights which are detected by autoradiography. The bandsharing coefficient, S_{xy}, of two DNA fingerprint patterns is the proportion of bands one individual shares with another, and is given by:

$$S_{xy} = (2n_{xy})/(n_x + n_y)$$

where n_x and n_y are the number of bands in the DNA fingerprints of individuals x and y, and n_{xy} is the number of bands which are apparently common to both patterns. The bandsharing coefficient is of interest in studies of POPULATION GENETICS and animal behaviour because in many populations it is a good measure of genetic relatedness. *See*: DNA TYPING: APPLICATIONS IN CONSERVATION AND POPULATION GENETICS.

Lynch, M. (1990) *Mol. Biol. Evol.* **7**, 478–484.

BAP The plant CYTOKININ BENZYLADENINE.

bare lymphocyte syndrome SEVERE COMBINED IMMUNODEFICIENCY resulting from failure of expression of MHC class II molecules on B CELLS, MACROPHAGES, and DENDRITIC CELLS and thus a failure in ANTIGEN PRESENTATION to helper T cells (*see* MAJOR HISTOCOMPATIBILITY COMPLEX). The underlying defect is an absence of or abnormality in a MHC-specific TRANSCRIPTION FACTOR.

Barr body A densely staining HETEROCHROMATIN inclusion body visible by light microscopy in the nuclei of somatic cells in female mammals. It was first described by Murray Barr in the cat in 1949. The number of Barr bodies is always one less than the number of X chromosomes in the GENOME, and it is the visible consequence of X-CHROMOSOME INACTIVATION.

Bart's hydrops syndrome *See*: THALASSAEMIA.

basal body *See*: MICROTUBULES.

basal ganglia *See*: PERIPHERAL NERVOUS SYSTEM.

basal lamina Thin sheet of specialized extracellular matrix that underlies all epithelia and also surrounds individual muscle cells, fat cells, and Schwann cells. It is composed mainly of type IV collagen, proteoglycans, and laminin. *See*: CELL–MATRIX INTERACTIONS; EXTRACELLULAR MATRIX MOLECULES.

base Derivative of the basic nitrogenous heterocyclic compounds pyrimidine and purine (Fig. B20). The pyrimidine bases found in nucleic acids are uracil (in RNA), thymine (in DNA), and cytosine; the purine bases are adenine and guanine (Fig. B21). Modified bases, such as 5-methylcytosine (*see* DNA METHYLATION), are also found in nucleic acids. TRANSFER RNA especially contains small amounts of unusual bases such as 5-methylcytosine, dihydrouracil, 6-methylaminopurine, 6,6-dimethylaminopurine, 1-methylguanine, 2-methylamino-6-hydroxypurine, and 2,2-dimethylamino-6-hydroxypurine and others.

Other purine bases that occur in nature are hypoxanthine, xanthine (Fig. B22), caffeine (1,3,7-trimethylxanthine), and theobromine (3,7-dimethylxanthine). Hypoxanthine is the base in the nucleoside inosine.

base analogue A purine or pyrimidine with a high degree of structural similarity to the bases adenine, cytosine, guanine, thymine or uracil. As a result it can become incorporated into DNA where it can become paired with the wrong nucleotide

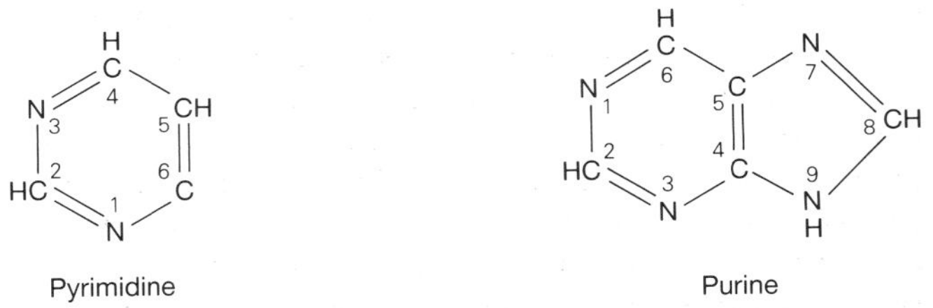

Pyrimidine   Purine

Fig. B20 The purine and pyrimidine ring systems from which the bases are derived.

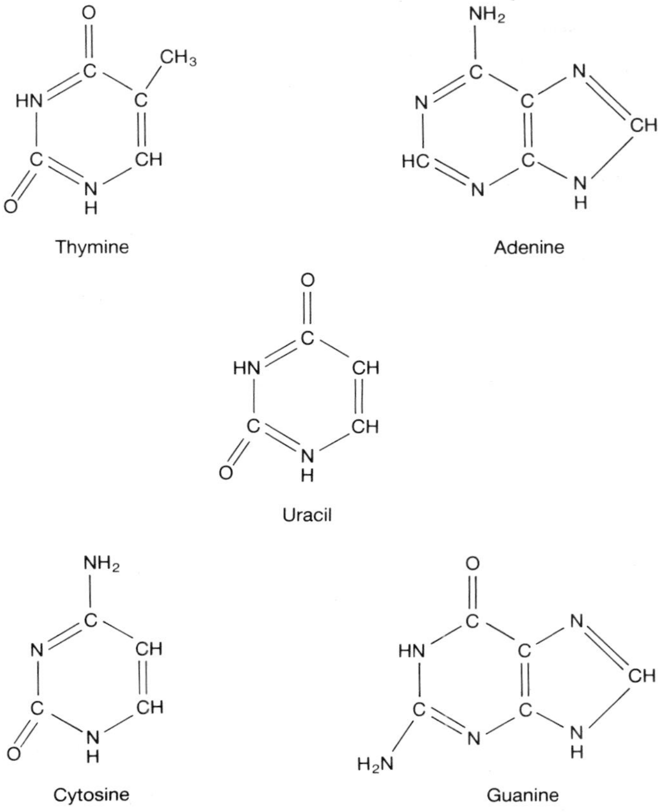

Thymine / Adenine

Hypoxanthine / Xanthine

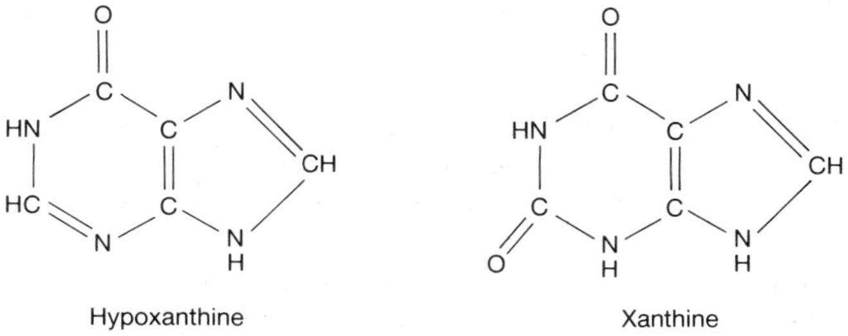

Fig. B22 Hypoxanthine and xanthine.

Uracil

Cytosine / Guanine

Fig. B21 The purine and pyrimidine bases in nucleic acids.

base pairing, other pairings are possible. The 28 possible base pairs for A, G, U(T), and C involving at least two (cyclic) hydrogen bonds are illustrated in Fig. A389. *See*: DNA; HYBRIDIZATION; NUCLEIC ACID STRUCTURE; RNA; TRANSFER RNA.

base stacking The stacking of successive planar BASE PAIRS one above another in the DNA double helix. It is in essence the result of hydrophobic interactions between nucleotides, and is a force perpendicular to the plane of the bases. The energy released by hydrogen bonding between base pairs and by stacking contributes to the thermodynamic stability of the α-helical structure of double-stranded DNA. *See*: DNA; NUCLEIC ACID STRUCTURE.

base substitution (base-pair substitution, nucleotide substitution) Alteration in DNA involving a change in a single base pair at a particular site. There are two categories: in a transition one

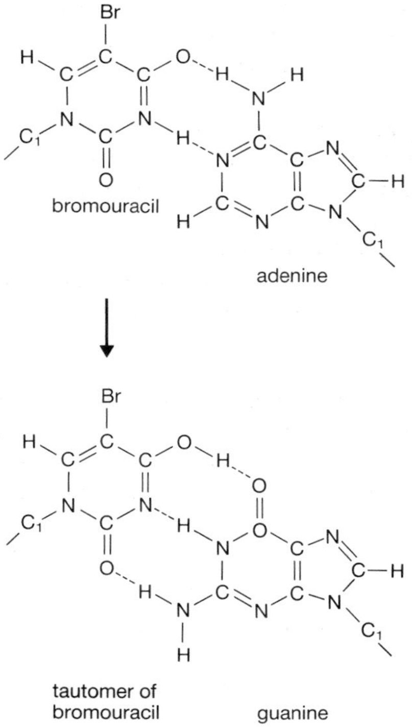

leading to a point MUTATION. For example, 5-BROMOURACIL is a thymine analogue which, in the rare enol state, allows for three hydrogen bonds rather than thymine's usual two. So the enol form of 5-bromouracil can, during DNA REPLICATION, direct the incorporation of guanine instead of adenine into the new complementary strand (Fig. B23). The adenine analogue 2-AMINOPURINE causes the change of an AT pair to a GC pair by pairing with thymine during one round of DNA replication and then directing incorporation of a C at the next round of replication.

base excision repair *See*: DNA REPAIR.

base pair, base pairing The common nucleic acid BASES adenine (A), guanine (G), thymine (T), cytosine (C), and uracil (U) pair selectively with each other in nucleic acids (*see* DNA; NUCLEIC ACID STRUCTURE; RNA). In Watson–Crick base pairing A pairs with T (or U) and G with C (Fig. B24, XIX and XX). Two hydrogen bonds can be formed between A and T, and A and U, while C and G are linked through three hydrogen bonds. The A–T (or T–A) and C–G (or G–C) base pairs form the 'rungs' of the DNA double helix, holding the two complementary strands together. Base pairing can also occur within single-stranded DNA or RNA molecules. The length of a double-stranded nucleic acid is generally given in base pairs (bp) or kilobase pairs (kbp). Apart from Watson–Crick

bromouracil / adenine / tautomer of bromouracil / guanine

Fig. B23 The substitution of guanine for adenine caused by the base analogue 5-bromouracil.

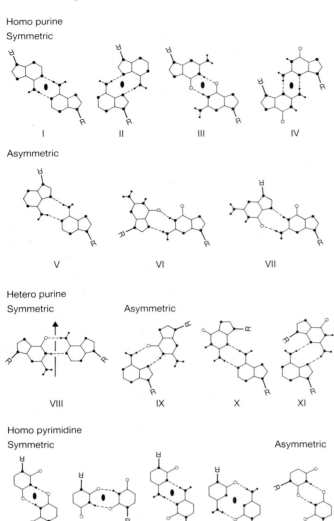

Homo purine
Symmetric

I   II   III   IV

Asymmetric

V   VI   VII

Hetero purine
Symmetric   Asymmetric

VIII   IX   X   XI

Homo pyrimidine
Symmetric   Asymmetric

XII   XIII   XIV   XV   XVI

Hetero pyrimidine
Symmetric only

XVII   XVIII

Purine–pyrimidine
Symmetric

Watson–Crick

Asymmetric

reversed
Watson–Crick

XXI   XXII

Hoogsteen   reversed
Hoogsteen

XXIII   XXIV

XXV   XXVI   XXVII   XXVIII

Fig. B24 The 28 possible base pairs for A, G, U(T), and C involving at least two (cyclic) hydrogen bonds. Hydrogen and nitrogen atoms are displayed as small and large filled circles, oxygen atoms as open circles, glycosyl bonds as thick lines with R indicating ribose C1′ atom. Base pairs are boxed according to composition and symmetry, consisting of only purine, only pyrimidine, or mixed purine–pyrimidine pairs and asymmetric or symmetric base pairs.

Symmetry elements ● and ↑ are twofold rotation axes (dyads) vertical to and within the plane of the paper. In the Watson–Crick base pairs XIX and XX and in base pairs VIII and XVIII, pseudosymmetry relating only glycosyl links but not individual base atoms is observed. From Saenger, W. (1984) *Principles of Nucleic Acid Structure* (Springer-Verlag, New York).

purine is substituted for another purine or a pyrimidine is substituted for another pyrimidine. In a transversion a purine is substituted for a pyrimidine or vice versa. *See*: MOLECULAR EVOLUTION; MOLECULAR PHYLOGENY; MUTATION.

base triples *See*: NUCLEIC ACID STRUCTURE.

basement membrane BASAL LAMINA. *See*: CELL–MATRIX INTERACTIONS; EXTRACELLULAR MATRIX MOLECULES.

basic fibroblast growth factor (bFGF) *See*: AMPHIBIAN DEVELOPMENT; GROWTH FACTORS; INDUCTION.

basic leucine zipper A protein motif involved in DNA binding. *See*: PROTEIN–NUCLEIC ACID INTERACTIONS.

basic zipper motif *See*: PROTEIN–NUCLEIC ACID INTERACTIONS.

Basidiomycotina Fungi of the EUMYCOTA in which the sexual spores (basidiospores) are developed from a structure called a basidium. The distinguishing characteristic of the basidium is that spores are not enclosed (as they are in the Ascomycotina) but are exposed, often terminal, cells of the basidium itself. This class includes the (usually saprophytic) mushrooms and toadstools, but also includes plant pathogenic species (e.g. the rust *Puccinia graminis*).

basolateral membrane *See*: EPITHELIAL POLARITY.

basophil White blood cell which contains large cytoplasmic granules staining with basic dyes. Basophils express high affinity FC RECEPTORS for IgE antibodies on their surface, cross-linking of which stimulates the release of granule contents (degranulation), and the synthesis of EICOSANOIDS such as prostaglandins and leukotrienes.

Bayers junctions Contact sites between the inner cell membrane and outer membrane of Gram-negative bacteria. *See*: BACTERIAL ENVELOPE.

BCGF B cell growth factor. LYMPHOKINE supporting B cell proliferation. At least two BCGF species have been characterized. The first, BCGFI, also referred to as B-cell stimulatory factor 1 (BSF-1) and now known as interleukin-4 (*see* IL-4), is able to induce DNA synthesis in highly purified resting B cells. The other, BCGFII, can induce immunoglobulin secretion from activated B cells and also stimulates murine B cells to proliferate.

bcl-2 A gene 25 kb in length, located at human chromosome 18q21. It is involved in the t(14;18)(q32;q21) translocation of FOLLICULAR LYMPHOMA. The breakpoint in malignant cells may fall in the 3′ half of exon III of *bcl-2* (the Major Breakpoint Cluster Region, MBCR), or ~13–20 kb 3′ of the MBCR, or in a minority of cases in a region 5′ of the MBCR. The function of *bcl-2* is related to APOPTOSIS in that it prevents programmed cell death.

bcr *See*: ABL; ONCOGENES.

BDNF BRAIN-DERIVED NEUROTROPHIC FACTOR.

Beckwith–Wiedemann syndrome A foetal overgrowth syndrome. *See*: CONTIGUOUS GENE SYNDROME; PARENTAL GENOMIC IMPRINTING.

belt desmosome Intermediate junction. *See*: CELL JUNCTIONS.

Bence-Jones protein Free immunoglobulin LIGHT CHAIN excreted in the urine by a majority of patients with MULTIPLE MYELOMA. The neoplastic plasma cells in this disease synthesize an excess of light chain relative to the HEAVY CHAIN. The excess light chain is secreted by the cells as monomers or dimers, of M_r 25 000 and 50 000 respectively. Proteins of this mass are readily filtered through the kidney into the urine.

The presence of Bence-Jones protein in the urine is a diagnostic feature of multiple myeloma. It is named after the physician who first described its occurrence as a protein that precipitated on heating urine to 60–70°C and redissolved on heating above 80°C. This behaviour was first noted by a Dr Watson who sent a sample to Henry Bence-Jones.

β-bend, β-turn (beta-bend, beta-turn) *See*: PROTEIN STRUCTURE.

bent DNA DNA characterized by the presence of several polyadenine (poly(A)) tracts, three to five nucleotides long, ~10.5 base pairs apart, which represent one turn of a B-form DNA helix. DNA with this general structure can bend at the 3′ end of the poly(A) tracts, probably facilitating the interaction between proteins binding to such regions. DNA fragments containing bent DNA characteristically move considerably more slowly in gel ELECTROPHORESIS (10% polyacrylamide) than would be predicted on the basis of their molecular weight. Bent DNA has been found in viral ORIGINS OF REPLICATION and in ARS ELEMENTS of SACCHAROMYCES CEREVISIAE. *See also*: DNA; NUCLEIC ACID STRUCTURE.

benzodiazepines Drugs used primarily in the treatment of anxiety, although they are also effective as anticonvulsants, sedatives and muscle relaxants. Benzodiazepines act at a subtype of receptors for GABA — the GABA$_A$ receptor. These receptors are associated with a chloride ION CHANNEL in the neuronal membrane which is activated by GABA binding. Benzodiazepines, by binding to a site on this receptor complex, enhance the effect of GABA at the receptor. *See*: GABA AND GLYCINE RECEPTORS.

benzyladenine (BAP) 6-Benzylaminopurine (Fig. B25). M_r 225.3. A CYTOKININ, which was first synthesized in 1955. It is used as a growth regulator in plant tissue culture to promote growth and induce shoot production. It has been reported to occur naturally as the free base or N^9-riboside in several higher plant species.

Skinner, C.G. & Shive, W. (1955) *J. Am. Chem. Soc.* **77**, 6692–6693.

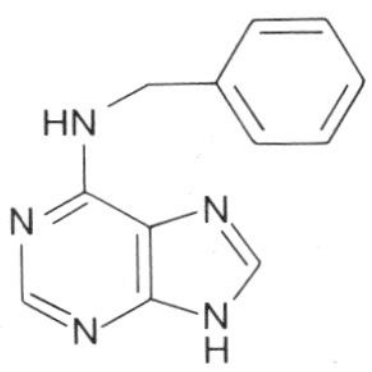

Fig. B25 Structure of benzyladenine.

6-benzylaminopurine *See*: BENZYLADENINE.

Bergman glia *See*: GLIAL CELLS.

best phase Defined in diffraction analysis as that phase which, when used to calculate an ELECTRON DENSITY MAP gives least error in the resultant map. Sometimes referred to as the centroid phase, being the centroid of the phase probability distribution, as distinct from the most probable phase. *See*: X-RAY CRYSTALLOGRAPHY.

beta Headwords with the prefix beta- or β- may also be indexed under the initial letter of the headword itself.

beta-barrel (β-barrel) *See*: PROTEIN STRUCTURE.

beta bend (β-bend) A hairpin turn between successive strands in a β-sheet. *See*: PROTEIN STRUCTURE.

beta-bulge (β-bulge) An irregularity within a β-sheet where one strand contains an extra residue which cannot be accommodated in the regular hydrogen bonding pattern. *See*: PROTEIN STRUCTURE.

beta chain (β chain) May refer to a subunit of any of numerous proteins in which subunits have been arbitrarily designated α, β, γ, δ, etc. Often refers to the β chain of HAEMOGLOBINS.

beta hairpin (β-hairpin, β-turn) A basic type of secondary structure in proteins in which two adjacent antiparallel β-strands are joined by a loop. *See*: PROTEIN STRUCTURE.

beta-sheet (β-sheet) An element of secondary structure found in many proteins formed of adjacent β-strands. *See*: PROTEIN STRUCTURE.

beta-strand (β-strand) One of the basic conformational structures in proteins, a short stretch of polypeptide chain in an almost fully extended conformation. *See*: PROTEIN STRUCTURE.

beta-turn (β-turn) A hairpin turn between successive strands in a β-SHEET. *See*: PROTEIN STRUCTURE.

Bessell function Mathematical function useful in the analysis of systems with an angular periodicity. *Cf.* FOURIER SERIES for systems with linear repetitions.

Beta vulgaris Sugar beet.

betacyanins Red pigments of the betalain group. Structure based on betalaminic acid, which is condensed with various imino and amino acids and amines. An example is betanin, the pigment of red beet (*Beta vulgaris*).

betalaminic acid

betanin

Other important betacyanins are amaranthin and muscaurin. They are found almost exclusively in the Centrospermae, and never occur with ANTHOCYANINS in the same species.

BGA BONE GLA PROTEINS.

BHK cells *Baby hamster kidney cells.* A fibroblast CELL LINE derived from Syrian hamster kidney.

bicoid A key gene in DROSOPHILA DEVELOPMENT defining the anterior domain of the anterior–posterior axis of the embryo. It is one of three localized maternal signals controlling the activation of gap genes in anterior, posterior, and terminal regions of the embryo. *bicoid* mRNA is localized to the anterior pole of the egg where its translation produces an anterior–posterior concentration gradient of *bicoid* protein. The *bicoid* protein contains a homeodomain and acts as a TRANSCRIPTION FACTOR for a number of gap genes. The specific genes activated depend on the concentration of *bicoid* protein. *bicoid* is probably the clearest example of a MORPHOGEN.

bidirectional replication Replication of DNA in both directions from an ORIGIN OF REPLICATION, in which two replication forks move away from the origin as replication proceeds. *See*: DNA REPLICATION.

Bijvoet pair A pair of DIFFRACTION intensities from a reflection *hkl* and its inverse $\overline{hkl}$. In X-RAY CRYSTALLOGRAPHY, measurement of the difference between these two intensities in either the presence or absence of ANOMALOUS SCATTERING provides information either to solve the PHASE PROBLEM or to estimate the accuracy of the intensity measurements respectively.

binary vector system Vector system composed of two complementary recombinant vectors used in PLANT GENETIC ENGINEERING. One vector carries a modified T-REGION derived from the Ti PLASMID and usually contains selection markers for plant transformation contained between two border regions. The T-region can be transferred into the plant cell genome with the aid of Ti plasmid-encoded virulence (*vir*) gene products acting in *trans* from the other vector, a HELPER PLASMID.

bindin An insoluble protein ($M_r \sim 30\,000$) which is released from the acrosome of sea urchin sperm at FERTILIZATION (*see* ACROSOME REACTION). Bindin is probably a LECTIN and binds in a specific manner to the GLYCOPROTEINS in the VITELLINE MEMBRANE of the egg.

biodegradation The use of respiring bacteria, resting bacterial cells or isolated enzymes to convert unwanted biological or industrial waste materials to more acceptable products. Examples are the dioxygenases which introduce pairs of hydroxyl groups into otherwise carcinogenic aromatic compounds and thus render them water soluble and less toxic; the lipases which may be used to hydrolyse the fatty waste which often solidifies and blocks the drains of fast food outlets, or to convert the very large quantity of waste saturated animal fat trimmed from carcasses at slaughter-houses into monoglycerides which have uses as food emulsifiers. Another use is to hydrolyse the very toxic cyanide in the waste water from gold leaching plants.

Kobayashi, M. et al. (1990) *J. Bact.* **172**, 4807–4815.

biological clocks Cellular and neural processes which oscillate endogenously and function as pacemakers to drive overt rhythms of physiology and behaviour. Biological clocks can oscillate with a wide range of periodicities, giving rise to many types of rhythm, including ultradian (minutes to hours), circumtidal (e.g. for shore-dwelling organisms), CIRCADIAN (daily), and circannual (yearly). The DROSOPHILA *period* 'clock' gene, identified originally by the short, long, or absent circadian rhythm mutations *per*^s, *per*^l, and *per*^o, encodes a 4.5-kb mRNA and a protein of 1226 amino acids. The *per* protein contains sequences homologous to known TRANSCRIPTION FACTORS, and is expressed in neural and non-neural tissues. *per* mRNA shows a circadian rhythm of abundance in adult *Drosophila* neurons which is several hours out of synchrony with cellular *per* protein levels, consistent with the idea that this gene's expression sets up a negative feedback loop in which *per* protein decreases the binding activity of positively acting transcription factors in both the 5′ flanking and intragenic region of the locus. Other target loci for the *per* product have not yet been identified. Moreover, several other loci affect the expression of circadian rhythms in *Drosophila* (*restless*, disconnected). The cellular basis of biological clocks probably reflects multiple hierarchies and interactions of genes.

biological containment The enfeebling by genetic means of phages, viruses and bacterial hosts used in GENETIC ENGINEERING and other RECOMBINANT DNA work, thus rendering them incapable of surviving outside the laboratory.

biotechnology Any technology which exploits the biochemical activities of living organisms or their products (e.g. isolated enzymes). Antibiotic production and brewing are among the long-established biotechnological industries but since the development of RECOMBINANT DNA TECHNOLOGY and the ability to transfer genes from one organism into another (heterologous gene transfer), the potential of biotechnology has expanded enormously. *See*: DNA CLONING; GENETIC ENGINEERING; MONOCLONAL ANTIBODIES; NANOTECHNOLOGY; PLANT GENETIC ENGINEERING; PROTEIN ENGINEERING; RECOMBINANT VACCINES; TRANSGENIC TECHNOLOGIES.

biotin Biotin (Fig. B26) is a vitamin which acts as a coenzyme in carboxylation reactions such as:

ATP + pyruvate + HCO_3^- = ADP + orthophosphate + oxaloacetate

catalysed by pyruvate carboxylase (EC 6.4.1.1). Biotin is linked through the carboxyl group of its aliphatic side chain to a lysine residue of the enzyme forming an amide bond. *See*: BIOTINYLATION.

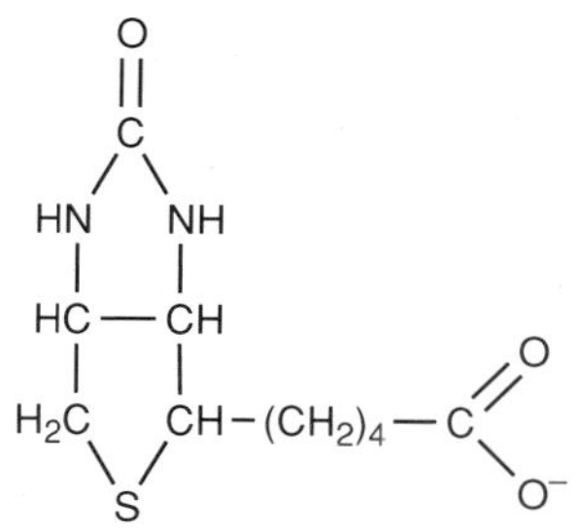

biotin–avidin/streptavidin labelling *See*: BIOTINYLATION.

biotinylation BIOTIN can be linked covalently to both proteins and nucleic acids and is used as a label in many techniques. An alternative method to radioactivity for labelling DNA *in vitro* uses biotinylated-dUTP as the label (Fig. B27). In this molecule biotin is bound through its aliphatic side chain to the N-3 of dUTP (*see* NUCLEOSIDES AND NUCLEOTIDES). In a NICK TRANSLATION reaction (or any other method for DNA labelling) it can become incorporated into DNA. DNA labelled in this way can be used in HYBRIDIZATION experiments. The detection of the biotinylated hybrids depends on the application of a detector complex which gives an enzymatic or chemoluminescent reaction. The detector complex includes the compound STREPTAVIDIN — a bacterial homologue of mammalian AVIDIN, which binds biotin tightly. The biotin–streptavidin (avidin) system is also used as an affinity reagent in IMMUNOELECTRON MICROSCOPY.

biotransformation The use of respiring live bacterial cells, resting bacterial cell suspensions ('bags of enzymes') or isolated enzymes to catalyse the transformation of one chemical into another. Traditional biotransformations are brewing (sugars into ethyl alcohol) and silage making (grass sugars into lactic acid). One of the newer biotransformations is the production of the monomer of acrylic plastics (acrylic acid) on a 50 000 tonnes per annum scale using the enzyme nitrilase from the bacterium *Rhodococcus rhodochorus* to catalyse the transformation:

$$CH_2{=}CH{-}C{\equiv}N + 2H_2O = CH_2{=}CH{-}COO^- + NH_4OH.$$

The nitrilase enzyme was mutated to be resistant to denaturation by acrylonitrile and to produce product at $650\,g\,l^{-1}$. The cost advantage of biocatalysis arises because only one product is formed regioselectively and enantioselectively, and this avoids much expensive purification to remove, for example, heavy metal catalysts from food grade chemicals or to isolate one enantiomer from a mixture. Most modern drugs now being introduced, for example, are enantiomerically pure.

biotrophy Growth of fungi that is entirely dependent on a living host. Biotrophy is associated with characteristic fungal structures

Fig. B26 Biotin.

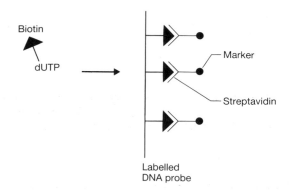

Fig. B27 Biotinylation and the biotin–streptavidin reaction.

which facilitate nutrient transfer from the host to the parasite or symbiont, such as HAUSTORIA. Biotrophic fungi do not generally survive in a defined culture medium but rather must be propagated on their hosts. Prominent among the biotrophs are the rusts (e.g. *Puccinia* spp.), mildews (e.g. *Erysiphe* spp.), and mycorrhizal fungi. *See:* PLANT PATHOLOGY.

biotype A grouping of strains of one species of bacterium showing similar biological properties, e.g. metabolic characters. Applied to *Agrobacterium* strains, denotes a system of classification based on metabolic properties. Biotype 1 strains are characterized by a positive 3-ketolactose reaction. Biotypes 2 and 3 do not give a 3-ketolactose reaction and are distinguished from each other by the ability of biotype 2 to catabolize erythritol, mucic acid, tartarate and malonate, and the inability of biotype 3 to utilize erythritol, melezitose, ethanol and mucic acid. Biotype 2 will also grow on DIM medium (which contains cellobiose and malachite green) whereas biotype 3 will not.

BiP A MOLECULAR CHAPERONE protein of M_r ~78 000 found in the ENDOPLASMIC RETICULUM. It is related to the Hsp70 family of HEAT SHOCK proteins and is thought to be involved in the completion of translocation of nascent secretory proteins into the ER from the cytosol (*see* PROTEIN TRANSLOCATION) and the stabilization of prefolded protein structure in the ER lumen. It also binds to abnormal, misfolded or foreign proteins whose exit from the ER is blocked. Mammalian BiP has also been known as the immunoglobulin heavy chain binding protein or the glucose-regulated protein grp78. In normal conditions, BiP is synthesized constitutively and comprises ~5% of luminal protein content.

bipolar disorders *See:* AFFECTIVE DISORDERS.

Birnaviridae Family of icosahedral nonenveloped RNA viruses, diameter 60 nm, whose genomes are composed of two molecules of double-stranded RNA (M_r 2.5×10^6 and 2.2×10^6). They infect fish and birds.

2,3-bisphosphoglycerate *See:* 2,3-DIPHOSPHOGLYCERATE.

bithorax complex (BX-C) The pioneering genetic studies of E.B. Lewis on the bithorax complex in *Drosophila* paved the way for our present understanding of the genetic regulation of development. The HOMEOTIC GENES of the bithorax complex are responsible for specifying the developmental programs of most thoracic and abdominal segments. In embryos lacking the bithorax complex the segments posterior to the second thoracic segment (T2) fail to make their normal characteristic structures but instead all appear similar to T2. The bithorax complex contains three genes — *Ultrabithorax*, *abdominal-A*, and *Abdominal-B* — which are required for the development of successively more posterior segments. These genes encode TRANSCRIPTION FACTORS of the homeodomain class and are expressd in overlapping domains along the antero-posterior axis of the developing embryo. Other genetic loci within the complex (e.g. *bithorax*, *postbithorax*) identify regulatory elements required for the correct spatial patterning of the expression of the protein-coding genes. *See also*: DROSOPHILA DEVELOPMENT; HOMEOTIC GENES.

Lewis, E.B. (1978) *Nature* **276**, 565–570.

bivalent In MITOSIS, chromosome which has been duplicated to form two sister chromatids held together at the centromere. In MEIOSIS, a pair of duplicated homologous chromosomes held together by chiasmata.

blast cell Immature cell of the haematopoietic system with morphological aberrations of cell structure in the cytoplasm and nucleus. Using electron microscopy, monoclonal antibodies, and investigation of gene rearrangement status, blast cells may be assigned to the myeloid or lymphoid lineage. *See* HAEMATOPOIESIS.

blast crisis Stage of clinical and haematological progression of CHRONIC MYELOID LEUKAEMIA characterized by refractoriness to therapy and haematological features of acute leukaemia of the myeloid, and more rarely, of the lymphoid type.

blastema (pl. blastemata) A region with a high rate of cell division, which acts as a source of cells. An example is the region at the distal end of a regenerating amphibian limb. From here arise the cells that will form the tissues of the regenerate. *See:* LIMB DEVELOPMENT.

blastocoel Cavity formed in the centre of the early embryo in mammals, some amphibians, and some marine invertebrates. The fertilized egg undergoes several cleavage divisions, forming a multicellular embryo. Ions and fluid begin to be pumped across the cell membranes into the intercellular spaces and gradually these become confluent, forming a single cavity — the blastocoel. The outermost cells of the embryo form an epithelium, with tight junctions that create a seal against the external medium. Mammalian embryos that reach this stage are called blastocysts and the blastocoel separates the inner cell mass, which gives rise to the embryo proper, from the trophoblast, which will form the extraembryonic tissues. In amphibians, the blastocoel forms in the centre of the animal hemisphere and embryos of this stage are called blastulas. *See also*: AMPHIBIAN DEVELOPMENT; EPITHELIAL POLARITY; MAMMALIAN DEVELOPMENT.

blastoderm A specific, early stage of development of some organisms. In *Drosophila* and other insects (*see* DROSOPHILA DEVELOPMENT), the blastoderm stage is characterized by a single layer of nuclei arranged on the surface of the embryo, before the movements of GASTRULATION begin. The stage is subdivided into two: the syncytial blastoderm stage, before cell membranes form between the individual superficial nuclei, and the cellular blastoderm stage after cells become defined by the appearance of membranes. In some vertebrates (see AVIAN DEVELOPMENT; ZEBRAFISH DEVELOPMENT), the blastoderm is the group of developmental stages during which the embryo is flat, before body folding begins, and multicellular. In avian embryos, for example, the blastoderm first comprises a single layer of cells, the EPI-

BLAST. Two further layers of cells are then formed: the HYPO-BLAST ventrally and the MESODERM between the two. This trilaminar embryo is still called the blastoderm, usually until the middle of the process of NEURULATION, when the embryo becomes more three-dimensional by the appearance of the HEAD FOLD.

blastomere Cells produced from the first cleavage divisions of the fertilized egg. These cells become smaller with each division. In the mouse, the blastomeres undergo a compaction process at the late 8-cell stage, after which the embryo is called a morula. Until this point, the blastomeres all have equal developmental potential and could each form a normally sized mouse. In the amphibian embryo, polarity is already defined at the one-cell stage and the development fate of subsequent blastomeres depends on the plane of cleavage. The first two cleavages are normally both vertical and perpendicular to each other, resulting in four cells of similar size. The third cleavage division is equatorial and separates the embryo into animal (upper) and vegetal (lower) halves. *See also*: AMPHIBIAN DEVELOPMENT; MAMMALIAN DEVELOPMENT.

blastopore A depression in the dorsal vegetal quadrant of the late BLASTULA in amphibian embryos. Its appearance marks the beginning of GASTRULATION. The blastopore elongates laterally and eventually becomes a complete circle surrounding a plug of yolk cells. Sheets of cells, which undergo temporary changes in shape, then move in around the lip of the blastopore, and into the interior of the embryo. This belt of invaginating cells is called the marginal zone and is the presumptive MESODERM. As invagination progresses a new space is created inside the embryo, the ARCH-ENTERON, which is the primitive gut. The cells of the animal pole, which do not invaginate, form the prospective ECTODERM and spread to release the cell sheets that have migrated inwards. Towards the end of gastrulation, this epithelium of the animal hemisphere extends to cover the whole exterior of the embryo and the blastopore shrinks to a very small circle.

blastula Stage in amphibian embryonic development analogous to the mammalian BLASTOCYST. As successive CLEAVAGE divisions give rise to a multicellular embryo, fluid begins to penetrate into the spaces between these cells and gradually the spaces become confluent, forming a single cavity — the blastocoel — in the centre of the animal hemisphere. Rapid and asymmetric cleavage continues for 12 to 15 more divisions, synchrony eventually being lost and the division rate slowing down towards the end of this stage. At this point, known as the mid-blastula transition, TRANSCRIPTION of the embryo's genome begins. *See also*: AMPHIBIAN DEVELOPMENT.

Blood coagulation and its disorders

WHEN the lining of a blood vessel is damaged a complex series of events takes place which is designed to prevent blood loss and,

ultimately, to restore the integrity of the vessel. Although short-lived vasoconstriction and physical factors such as the pressure of extruded blood on the vessel wall may play some part in haemostasis, the main players in the haemostatic mechanism are PLATELETS and the blood coagulation system.

Damage to the vascular endothelium exposes subendothelial structures which attract platelets and induce them to aggregate reversibly. The protein THROMBIN, formed during activation of the coagulation pathway, generates insoluble cross-linked fibrils of the protein FIBRIN and causes the platelets to aggregate irreversibly. The resulting platelet–fibrin clot is an effective barrier against loss of blood from the vascular system and also serves as a scaffold for subsequent repair of the lining of the blood vessel. Coincident with these events a variety of limiting reactions are brought into play to confine the haemostatic reaction to the site of the injury.

Blood coagulation

Blood coagulation is the result of the complex interaction of a number of protein clotting factors through a cascade (Fig. B28). Some of these factors normally exist as inactive enzyme precursors which can be transformed into a trypsin-like protease (a SERINE PROTEINASE) either by an alteration in their conformation or by the action of a converting enzyme that causes proteolytic cleavage at specific amino-acid residues. The inactive precursors are designated by roman numerals and the active state is designated by a following 'a' — Factor VIII and Factor VIIIa for example. Some of the factors of the coagulation pathway, notably VII, IX, and X, together with PROTHROMBIN, require vitamin K for their biosynthesis in the liver. The ultimate objective of the coagulation pathway is the activation of thrombin from its precursor prothrombin and hence the cleavage of FIBRINOGEN to produce an insoluble fibrin clot. The generation of thrombin is a consequence of two different reactions. The first involves the activation of Factor X, the second the conversion of prothrombin to thrombin. The transformation of Factor X to its activated form Xa occurs through two distinct pathways called the intrinsic and extrinsic coagulation cascades (see Fig. B28).

The intrinsic pathway

Activation of the intrinsic pathway is initiated by the exposure of subendothelial structures to blood by a contact mechanism that is not fully understood. This results in the activation of Factor XII to XIIa which contains an active serine centre. This enzyme is able to activate Factor XI and is also involved in activation of the fibrinolytic system (see below). Factor XIa interacts with Factor IX, converting it to the serine proteinase Factor IXa, which in turn is able to convert Factor X to its serine proteinase Factor Xa. The activation of Factor X must occur rapidly and be localized to the site of vessel wall injury. This is accomplished by the action of a cofactor called Factor VIII, in association with calcium. The formation of the Factor IXa–Factor VIIIa–Factor X–calcium complex is thought to require specific receptors on both endothelial cells and probably also on platelets. Factor VIII circulates in the blood complexed to another factor called von Willebrand factor.

The latter functions mainly by binding to specific platelet receptors and probably mediates the adhesion of platelets to subendothelial structures.

The extrinsic pathway

The activation of the extrinsic coagulation pathway is initiated by exposure of blood to a membrane-bound GLYCOPROTEIN called tissue factor. The latter is able to form a stable complex with Factor VII. This interaction, which produces a conformational change that converts Factor VII to VIIa, increases the affinity of Factor VII to Factor X.

Activation of the intrinsic or extrinsic pathways leads to the generation of sufficient amounts of Factor Xa to convert prothrombin to the two-chain serine protease thrombin. This reaction is accelerated by the action of Factor V in conjugation with calcium and an activated platelet surface. Its activated form binds tightly to a specific protein on the platelet membrane and this complex forms a receptor needed for recruitment of Factor Xa.

Fibrinogen and fibrin

Fibrinogen is a dimeric protein that consists of six polypeptide chains held together by disulphide bonds and folded into a trimodular structure. It is converted to fibrin by thrombin, which removes fibrinopeptides from the N-terminal ends of the Aα and Bβ chains. The resulting fibrin monomers polymerize into fibrils. Other fibrin–fibrin interactions lead to lateral association of protofibrils and branch-point formation. Stabilization of fibrin is facilitated by the action of Factor XIII. This is converted to its active

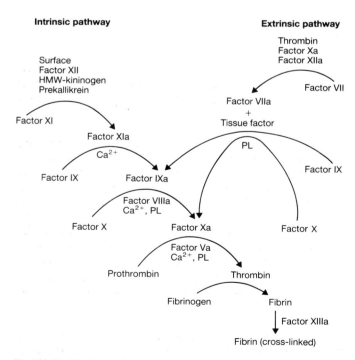

Fig. B28 The blood coagulation cascade. PL, Phospholipid; HMW-kininogen, high molecular weight kininogen.

form, XIIIa, by thrombin. It catalyses the formation of intermolecular γ-glutamyl-ε-lysine bonds between fibrin monomers.

Localization of clot formation

In order to localize clot formation and to prevent unwanted intravascular coagulation there is a variety of fail-safe mechanisms. One such involves protein C, protein S, and thrombomodulin. Protein C is a vitamin K-dependent protein that is converted to an activated form, a serine proteinase, by thrombin. This reaction is enhanced by an endothelial cell cofactor called thrombomodulin which forms a complex with thrombin, resulting in a potent activator of protein C and a marked decrease in the coagulant function of the enzyme. Activated protein C inactivates the coagulation pathway by inhibiting Factor Va and Factor VIIIa in the presence of phospholipid. The anticoagulant activity of activated protein C also requires the presence of protein S, another vitamin K-dependent coagulation factor, that enhances the activity of activated protein C. Protein S functions as a cofactor in the binding of activated protein C to phospholipid and increases the affinity of activated protein C to membrane surfaces. The intrinsic coagulation pathway is also inhibited by antithrombin III, a potent serine proteinase inhibitor that is normally present in plasma (*see* SERPINS). The antithrombin–proteinase reaction is accelerated by HEPARIN which binds to a specific site on antithrombin III. Heparin cofactor II is another serine proteinase inhibitor which inhibits thrombin in the presence of heparin.

The fibrinolytic system

Finally, there is a fibrinolytic system that is primarily responsible for the dissolution of fibrin deposited on vascular endothelium. This requires the action of PLASMINOGEN which can be enzymatically converted to PLASMIN — a potent two-chain serine proteinase. Just like the coagulation system the activation of plasmin can occur by both intrinsic and extrinsic pathways. Although these are not so well worked out as those of the coagulation pathway the extrinsic pathway is triggered by two major classes of serine proteases, a tissue-type activator (TISSUE PLASMINOGEN ACTIVATOR) and UROKINASE. The degradation of fibrin is a multistage process during which a variety of different molecular byproducts are produced.

In summary, haemostasis is a finely tuned mechanism involving blood platelets, vessel wall, and the intrinsic and extrinsic coagulation systems counterbalanced by an equally complex series of pathways designed to inhibit unwanted coagulation. This process should be looked on as a highly dynamic system; as well as controlling bleeding from major trauma it is constantly being activated owing to the minor trauma to which blood vessels are exposed.

The genes for most of the human and many animal coagulation factors have been cloned and sequenced and in many cases have been assigned to individual chromosomes (*see* DNA CLONING; DNA SEQUENCING; GENE MAPPING). This has been particularly useful in determining the molecular pathology of the coagulation disorders.

Coagulation disorders

Diseases of haemostasis and coagulation fall into several groups. First there are those which involve the vessel wall and platelet numbers or function. These conditions are associated with spontaneous bleeding due to lack of the first-line defence mechanism and are usually characterized by bruising or pinpoint haemorrhages under the skin called purpura. Second, there are the genetic deficiencies of the clotting factors that make up the cascade. By far the commonest are haemophilia A, haemophilia B or Christmas disease, and von Willebrand disease, due to deficiency of Factors VIII, IX and von Willebrand factor respectively.

Haemophilia A

Haemophilia A is a SEX-LINKED disorder characterized by a reduction or absence of Factor VIII in the blood. Many different MUTATIONS in the Factor VIII gene have been defined, including point mutations, insertions, deletions, and amino-acid substitutions that produce an abnormal protein product. The condition is characterized by bleeding after trauma, spontaneous bleeding into joints, and major bleeding after surgical procedures. It varies in severity depending on the level of Factor VIII and can be corrected by giving Factor VIII concentrate.

Other disorders

Christmas disease follows a similar clinical pattern and the molecular pathology is also similar to that of haemophilia A. Von Willebrand disease is characterized by a milder bleeding disorder with a tendency to spontaneous cutaneous bleeding. Genetic deficiencies of the other clotting factors are associated with bleeding disorders of variable severity. Factor XIII deficiency is characterized by bleeding and abnormal wound healing.

The molecular pathology of fibrinogen is complex. Many different abnormal fibrinogens have been described, mostly due to single amino-acid substitutions or deletions. They may be associated with abnormal bleeding or with an increased propensity to thrombosis depending on the specific site of the mutation.

Deficiencies of antithrombin III or proteins S and C have also been described. The molecular pathology of antithrombin III is very heterogeneous but most of the disorders that have been described to date are due to the production of abnormal protein due to point mutations. A defect in these proteins presents clinically as an increased tendency to form clots in veins and recurrent pulmonary emboli.

D.J. WEATHERALL

Stamatoyannopoulos, G. et al. (1987) *The Molecular Basis of Blood Diseases* (W.B. Saunders, Philadelphia).

blood groups Blood groups are antigens or polymorphic characters (systems) found in blood; conventionally the term refers particularly to those expressed on the red cell surface. Blood group antigens comprise a large and heterogeneous set of EPITOPES expressed on the surfaces of erythrocytes, against which ANTIBODIES occur naturally or can be raised. Cross-reactive (i.e. similar) epitopes may also occur on other cell types, within extracellular matrices and on secreted GLYCOPROTEINS. Where antibodies occur naturally, or are readily elicited in man, transfusion of unmatched blood may lead to serious consequences, such as haemagglutination and haemolysis. Most of the minor blood groups do not carry risk. Antigens analogous to the blood group antigens of man occur in many other mammalian species. Many, though not all, blood group antigens are GLYCANS, including the ABO BLOOD GROUP SYSTEM, the most important for blood transfusion, and its related LEWIS BLOOD GROUP. Other carbohydrate blood group antigens include the I,I SYSTEM, the P BLOOD GROUP SYSTEM, and the MNS BLOOD GROUP and the Tn antigen. The RH BLOOD GROUP SYSTEM is carried on protein epitopes. *See also*: DUFFY BLOOD GROUP; GENETIC POLYMORPHISM; GERBICH BLOOD GROUP; KELL BLOOD GROUP; XG^A BLOOD GROUP.

Bloom's syndrome Rare human disorder inherited in an AUTOSOMAL RECESSIVE fashion. Clinical features include characteristic erythematic facial lesions, photosensitivity, stunted growth and immunodeficiency. Affected individuals have an increased risk of developing cancer. Cells from patients have a 10-fold or greater increase in the frequency of spontaneous SISTER CHROMATID exchanges and other cytogenetic abnormalities. Bloom's syndrome cells exhibit delayed joining of large DNA replication intermediates, and the cells are moderately hypersensitive to several different types of DNA-damaging agents. An altered DNA LIGASE activity has been detected in extracts from Bloom's syndrome cells. *See also*: DNA REPAIR.

blotting, of nucleic acids *See*: NORTHERN BLOTTING; SOUTHERN BLOTTING.

blotting, of proteins *See*: WESTERN BLOTTING.

blue-green algae CYANOBACTERIA. Cyanobacteria were originally considered a special group of algae, marked by their different colour, resulting from the presence of PHYCOBILIPROTEINS.

blue light responses Many plants respond to light of wavelength 400–500 nm (blue light) by morphogenetic changes such as suppression of stem elongation and phototropic curvature. The specific initiation of transcription of certain genes is mediated by blue light (e.g. the *Cab* gene family in pea). Blue light photoreceptors (sometimes called cryptochrome) and the components of the signal transduction pathway have not yet been unequivocally identified, but there is evidence that the chromophore is flavin. It is not clear whether a separate receptor is needed for responses to near ultraviolet wavelengths. Blue-light activated G proteins have been identified as part of the signal transduction mechanism in several cases.

blunt ends The ends of a double-stranded DNA molecule cut straight across so that the DNA fragment is double-stranded over its entire length. Some RESTRICTION ENZYMES produce blunt ends (e.g. *Sma*I, *Eco*RI). Fragments of DNA with single-stranded extensions can be blunt-ended by treatment with, for instance, S1 NUCLEASE. Blunt ends are sometimes called flush ends.

BLV Bovine leukaemia virus. A slow transforming RETROVIRUS.

BM40 Osteonectin. *See*: CALCIUM-BINDING PROTEINS.

Bohr effect The alkaline Bohr effect is the decreased oxygen affinity of haemoglobin seen at alkaline pH (above pH 6 for human haemoglobin). It reflects a sensitivity of oxygen binding to changes in pH, a phenomenon of considerable importance because of its relationship to the transport of CO_2 in the blood. Carbon dioxide is transported largely as bicarbonate ion produced by reaction with water:

$$CO_2 + H_2O \rightarrow HCO_3^- + H^+$$

The resulting protons can combine with haemoglobin where they stabilize the deoxy state of the molecule thus favouring oxygen release in the tissues. For human haemoglobin the alkaline Bohr effect disappears at pH 6 and a reverse effect — the acid Bohr effect, of no apparent physiological significance, is seen at lower pH. *See also*: HAEMOGLOBINS.

Bolton–Hunter reagent *N*-Succinimidyl-3-(4-hydroxyphenyl)-propionate. It reacts under mild conditions with primary amines in, for example, proteins. The hydroxyphenyl group can be labelled with radioiodine (e.g. ^{125}I) which is then incorporated into the protein. Modified versions have introduced a sulphonate group into the succinimidyl ring making the reagent water soluble and membrane impermeable.

Bolton, A.E. & Hunter, W.M. (1973) *Biochem. J.* **133**, 529–539.

bone development Some bones (e.g. clavicle, skull) develop within connective tissue and are often known as membrane bones, but the majority involve deposition and mineralization within a cartilage model laid down in the embryo or foetus (this process is recapitulated during repair of fractures). Bone matrix (osteoid) is laid down by osteoblasts and mineralized with calcium phosphate (hydroxyapatite). Osteoblasts develop into osteocytes which remain within bone matrix and are thought to regulate bone turnover. Bone growth occurs by apposition, that is by deposition of bone matrix on to a free surface. Developing the shape of adult bone from the embryo cartilage model therefore requires remodelling, which involves resorption of bone matrix by osteoclasts (cells of the monocyte/macrophage lineage) and localized deposition by osteoblasts. This turnover process persists in adult bone. Bone deposition may involve an acidic protein of M_r 17 500, bone morphogenetic protein, which can stimulate connective tissue cells to differentiate into bone cells. This protein is related to the GROWTH FACTOR TGFβ.

bone Gla proteins Abundant vitamin K-dependent CALCIUM-BINDING PROTEINS of bone and dentine, not reported from other tissues. Two have been reported: osteocalcin (BGP) is secreted by osteoblasts and constitutes 15% of the soluble protein of demineralized bone matrix; it is a 50-residue polypeptide containing three γ-carboxyglutamic acid (Gla) residues; matrix Gla protein, comprising 3% of the soluble protein, is weakly homologous and is 79 residues long including five Gla residues. BGP is not required for mineralization of bone, but may regulate it as it inhibits crystallization.

bone marrow Haematopoietic tissue located inside bones and from which blood cells originate (*see* HAEMATOPOIESIS). It consists of a spongy stroma of reticuloendothelial cells filled with fat cells, haematopoietic STEM CELLS, and blood cells in various stages of development. Erythrocytes and nonlymphoid leukocytes develop to maturity in the bone marrow; B lymphocytes complete their differentiation in secondary lymphoid organs (*see* B CELL DEVELOPMENT) and T lymphocytes migrate to the thymus at an early stage in their development (*see* T CELL DEVELOPMENT). Bone marrow also contains mature T cells and plasma cells acquired through the blood stream. Stromal cells, T cells, and other leukocytes produce numerous cytokines (*see* LYMPHOKINES) which are involved in blood cell proliferation and differentiation. *See also*: CSF; GM-CSF; IL-1; IL-3; IL-7.

border repeat sequence Imperfect direct repeats of 25 bp which define the limits of the T-region in TI PLASMIDS and RI PLASMIDS. DNA between these two sequences is transferred to plant cells during the infection process to become the T-DNA of transformed cells.

Bordetella pertussis The causative bacterium of whooping cough, which was isolated and described in 1906 by Jules Bordet and Octave Gengou. It is a fragile, nonmotile, Gram-negative coccobacillus. The virulent form is described as the phase 1 organism, which possesses both a capsule and pili. The vaccine consists of either heat-killed phase 1 organisms or a crude extract prepared from them. Three doses are given at four-week intervals starting in the first year of life. *B. pertussis* has a marked ADJUVANT effect which is exploited to improve the immunogenicity of tetanus and diphtheria toxoids, by administering all three vaccines simultaneously. The component responsible for the lymphocytosis observed in patients with whooping cough is a four-subunit protein (M_r 73 000). Systemic changes, B-cell and macrophage functional alterations are presumed to result from endotoxins and peptidoglycan moieties. *See also*: PERTUSSIS TOXIN.

Borna disease virus Single-stranded positive-strand RNA virus (*see* ANIMAL VIRUSES) probably of the Coronaviridae, which produces slowly progressive disease in a number of experimental vertebrate model systems. In the tree shrew (*Tupaia glis*), a primitive primate, the virus causes a variety of changes in normal behaviour including hyperactivity and spatial and temporal disorientation. In breeding pairs the normal division between the active aggressive male and more passive female behaviour becomes blurred. The normal pattern of sociosexual activity which is a necessary preamble to successful mating is so disturbed that breeding no longer occurs although the animals remain in good physical health. Such changes are consistent with alterations to NEUROTRANSMITTERS. Rats inoculated at birth become persistently infected with large amounts of virus in the central nervous system but show no signs of severe neurological disease. However, they suffer learning deficiencies and subtle behavioural alterations. In addition there is now serological evidence (from serum antibody) that Borna disease virus infects man with the highest incidence in psychiatric patients.

VandeWoude, S. et al. (1990) *Science* **250**, 1278–1281.

Borrelia The spirochaete *Borrelia hermsii* is notable for undergoing antigenic variation in its PILUS antigens, that is, different antigenic variants of pilus proteins can be produced sequentially. It also alters a surface protein called the major variable protein (MVP). *Borrelia burgdorferi* is the agent of Lyme disease. *See*: BACTERIAL PATHOGENICITY.

***boss* gene** bride-of-sevenless. A gene involved in *Drosophila* EYE DEVELOPMENT.

botulinum toxin Protein neurotoxin produced by the bacterium *Clostridium botulinum*. Seven serotypes, A–G, are known. The mechanism of action of botulinum-B toxin has been elucidated. In the activated form it is a METALLOPROTEINASE which blocks NEUROTRANSMITTER release by cleaving the SNAP RECEPTOR protein synaptobrevin-2, an integral membrane protein of small synaptic vesicles. Botulinum-B toxin is produced as a single polypeptide chain which is cleaved to generate two chains linked by a single disulphide bond. The heavy chain binds specifically to neuronal cells and allows entry of the light chain which after reduction of the disulphide bond prevents neurotransmitter release by cleaving synaptobrevin-2.

Schiavo, G. et al. (1992) *Nature* **359**, 832–835.

boutons Presynaptic axon terminal swelling, containing vesicles, which discharge their contents of neurotransmitter molecules into the synaptic cleft following depolarization. *See*: NEURONAL CYTOSKELETON; SYNAPTIC TRANSMISSION.

bovine leukaemia virus (BLV) A slow-transforming RETROVIRUS occurring naturally in cattle, where it causes leukaemia.

bovine pancreatic trypsin inhibitor (BPTI) A nonserpin SERINE PROTEINASE inhibitor. This small protein of 58 amino acids has been widely used to study high resolution details of PROTEIN STRUCTURE and PROTEIN FOLDING pathways.

bovine papilloma virus (BPV) A PAPOVAVIRUS with a small circular double-stranded DNA genome (7.9 kb) which causes local skin tumours in cows. Its transmission is probably mechanical or via arthropod vectors. The virus will also replicate in a wide variety of mammalian cells, where it can maintain 50 to 200 copies per cell without lysis. This feature has been exploited in the development of SHUTTLE VECTORS based on BPV, allowing for replication of cloned genes in both *Escherichia coli* and mammalian cell lines.

bovine spongiform encephalopathy (BSE) *See*: TRANSMISSIBLE SPONGIFORM ENCEPHALOPATHIES.

BPTI BOVINE PANCREATIC TRYPSIN INHIBITOR.

BPV BOVINE PAPILLOMA VIRUS. *See also*: PAPOVAVIRUSES.

***Brachyury* locus** The *Brachury* (*T*) gene of the mouse is required for normal MESODERM formation and is involved in the organization of axial development. The gene has been cloned, and is expressed in nascent mesoderm of the PRIMITIVE STREAK and in presumptive and mature NOTOCHORD cells, but is downregulated in mesoderm cells that migrate away from the streak. Mutation gives rise to a graded series of defects along the body axis, the severity of which seems to correlate with GENE DOSAGE; the more posterior along the body axis, the higher the gene dose required for normal development. The principal defect in mutant embryos is disruption of the HEAD PROCESS and notochord precursor cells at the anterior end of the primitive streak. Mutant PHENOTYPES range from reduced tail length to disrupted trunk development and failure to form the allantois (sac-like outgrowth of alimentary canal in embryo), the latter resulting in embryonic death.

Bradyrhizobium Genus of RHIZOBIA which can carry out NITROGEN FIXATION in association with the roots of leguminous plants. *See also*: NODULATION.

Bragg angle (θ) Bragg's law states that, for radiation of wavelength λ incident on an object with a spacing *d*, DIFFRACTION is observed at an angle 2θ where $\lambda = 2d\sin\theta$.

brain areas *See*: CENTRAL NERVOUS SYSTEM.

brain-derived neurotrophic factor (BDNF) A member of the NERVE GROWTH FACTOR (NGF) family of trophic factors. It was originally isolated from brain tissue using its ability to increase the survival of sensory neurons as a purification assay. Subsequent cDNA cloning showed it to share ~60% amino acid identity with NGF. In the central nervous system BDNF has a trophic action on retinal, cholinergic, and dopaminergic neurons, and in the peripheral nervous system it acts on both motor and sensory neurons. It has a high-affinity receptor (trkB) distinct from but related to the NGF receptor (trkA), and shares a low-affinity receptor with NGF and other members of the family.

brain development *See*: AVIAN DEVELOPMENT; VERTEBRATE NEURAL DEVELOPMENT.

brain stem Region of brain at the base, which runs into the spinal cord. It consists of the MIDBRAIN, pons, and medulla oblongata.

brain vesicles Contiguous ballooninings or dilatations of the anterior part of the early vertebrate NEURAL TUBE, first clearly described by von Baer in the early 19th century. Three vesicles are initially recognizable: the future FOREBRAIN or prosencephalon, subdivided into telencephalon (including future cerebral cortex and basal ganglia) and diencephalon (future thalamus and hypothalamus), the MIDBRAIN or mesencephalon, and the HINDBRAIN, subdivided along the long axis into further segmental swellings (RHOMBOMERES) and into two broader regions, the metencephalon, or future pons/cerebellum, and myelencephalon or future medulla.

branch migration The mutual interchange of parallel strands between two double-stranded DNA helices which occurs during

RECOMBINATION and does not require DNA replication. After the initial crossing-over and the subsequent formation of cross-strand connections, branch migration continues the recombination event through a process of right-handed axial rotation. So during branch migration two double helices are linked, forming a cross-strand intermediate.

branched RNA An intermediate form of mRNA, produced during the process of RNA SPLICING. The free 5′ end of the INTRON folds round to interact with the branch site (a region ± 30 nucleotides upstream from the 3′ end of the intron). The 5′ end of the intron becomes covalently linked to an A in the branch site.

Brassica napus Oilseed rape.

brassinolide 2α,3α,22(R),23(R)-tetrahydroxy-24(S)-methyl-B-homo-7-oxa5α-cholestan-6-one (Fig. B29). M_r 482. The first member of the BRASSINOSTEROIDS to be isolated, from oilseed rape pollen, and characterized. It is one of the most biologically active brassinosteroids, displaying the characteristic brassin activity.

Grove, M.D. et al. (1979) *Nature* **281**, 216–217.

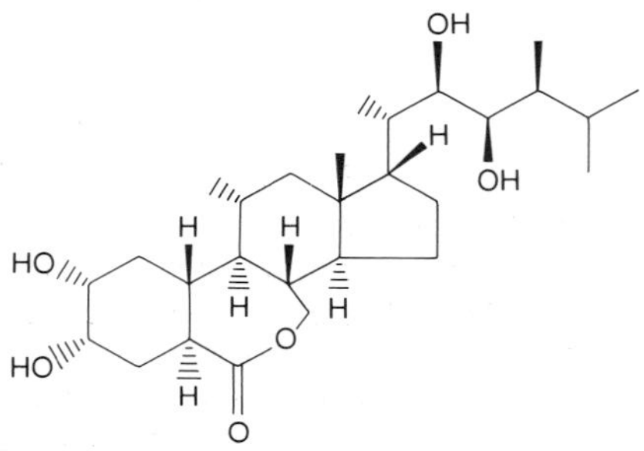

Fig. B29 The structure of brassinolide.

brassinosteroids A group of growth-promoting steroids that are probably widely distributed in higher plants. About 30 naturally occurring brassinosteroids have been fully characterized. Structurally they are derivatives of 5α-cholestane and differ in the extent of oxidation at C6, as well as in the nature of alkylation on C24 and in the number and orientation of hydroxyl groups in the A ring. The most biologically active compounds, and presumably the most biosynthetically advanced, contain 6-keto (e.g. castasterone) or 7-oxalactone (e.g. BRASSINOLIDE) structures. They exhibit brassin activity — the characteristic elongation, curvature, and splitting of the bean second internode and the bending of rice laminae — that is unique to this class of growth substances. In other systems they often mimic the effects of the other growth substances. Their function is still unclear, but they are thought to constitute a distinct class of plant hormones. *See also*: PLANT HORMONES.

Cutler, H.G. et al. (Eds) (1991) *Brassinosteroids: Chemistry, Bioactivity and Applications*. (ACS Symposium Series 474, American Chemical Society, Washington, DC).

Bravais lattice Any one of the 14 three-dimensional space-filling lattices which occur in crystals. *See*: SYMMETRY.

breathing The 'opening' of double-stranded regions in DNA to become single stranded, allowing, for example, interaction with proteins. Because of the dynamic condition of DNA, breathing occurs frequently. It is essential for the process of BRANCH MIGRATION.

brefeldin A A macrocyclic lactone fungal antibacterial agent which inhibits transport from the ENDOPLASMIC RETICULUM (ER) causing the Golgi stack to fuse with the ER. It works by inhibiting the exchange of GDP for GTP bound to the GTP-BINDING PROTEIN ARF so that COP proteins (COPs) cannot be recruited to Golgi membranes. *See*: COATED PITS AND VESICLES; GOLGI APPARATUS.

Donaldson, J.G. et al. (1992) *Nature* **360**, 350–352.
Helms, J.B. & Rothman, J.E. (1992) *Nature* **360**, 352–354.

bristle patterning Studies on the generation of bristle patterns in *Rhodnius* larvae and on metamorphosis in *Oncopeltus* showed that bristles are spaced in a pattern that maximizes the distance between them. When new bristles are added, they arise in the largest spaces between the pre-existing bristles. The basis of this seems to be a process of lateral inhibition, in which a developing bristle inhibits the differentiation of similar elements nearby. The molecular analysis of this mechanism in *Drosophila*, which is common to the specification of NEUROBLASTS in the embryo and the specification of bristle mother cells in the adult, shows that it involves the neurogenic genes NOTCH and *Delta* and the proneural genes of the ACHAETE-SCUTE COMPLEX. *See also*: DROSOPHILA DEVELOPMENT; INSECT NEURAL DEVELOPMENT.

bromophenol blue A dye with a blue colour at pH > 4.6. In a polyacrylamide or agarose gel it migrates in an electric current at the same rate as double-stranded DNA fragments of about 300 base pairs, and is therefore commonly used to monitor the migration of DNA in gel ELECTROPHORESIS.

5-bromouracil (5-BU) An analogue of the base thymine with bromine at the C-5 position in place of the methyl group found in thymine (Fig. B30). The normal keto form of 5-bromouracil base pairs with adenine but can spontaneously change at high frequency to the enol form which can base pair with guanine. It is therefore a BASE ANALOGUE mutagen causing AT → GC transition MUTATIONS during DNA replication.

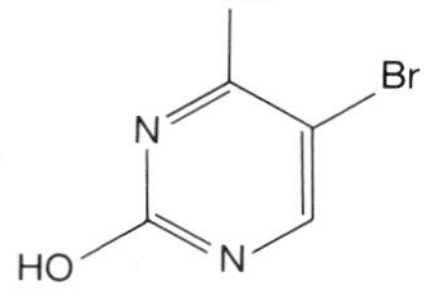

Fig. B30 5-Bromouracil.

bromovirus group Sigla from the type member, brome mosaic virus. MULTICOMPONENT VIRUSES with small isometric particles ($t = 3$) 26 nm in diameter, which contain four species of (+)-strand linear RNA (Fig. B31). RNA 1 and RNA 2 are encapsidated in separate particles and encode components of the viral polymerase. One molecule each of RNA 3 and RNA 4 are encapsi-

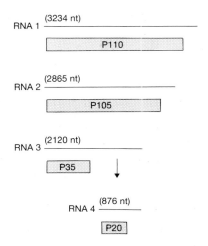

RNA 1 (3234 nt)
P110

RNA 2 (2865 nt)
P105

RNA 3 (2120 nt)
P35

RNA 4 (876 nt)
P20

Fig. B31 Genome organization of brome mosaic virus (BMV). The lines represent the RNA species and the boxes the proteins (P) with M_r given as $\times 10^{-3}$; nt, nucleotides.

dated in the same particle. RNA 3 is bicistronic, encoding at the 5′ end the putative cell-to-cell movement protein and, at the 3′ end, the viral coat protein. The coat protein is expressed from RNA 4. The infective genome comprises RNAs 1, 2 and 3; RNA 4 is subgenomic. *See also*: PLANT VIRUSES.

Francki, R.I.B. (1985) In *The Plant Viruses*, Vol. 1, 1–18 (Plenum, New York).

brown (b) A mouse COAT COLOUR GENE required for making black rather than brown eumelanin. The gene encodes a tyrosinase-related protein (TRP-1). Two dominant mutations at the *brown* locus produce hairs with a pigmented tip but unpigmented base resulting from premature melanocyte death.

BSE Bovine spongiform encephalopathy. *See*: TRANSMISSIBLE SPONGIFORM ENCEPHALOPATHIES.

β-bulge An irregularity within a β-sheet structure in a protein where one strand contains an extra residue which cannot be accommodated in the regular hydrogen bonding pattern. *See*: PROTEIN STRUCTURE.

bulge, bulge loops *See*: NUCLEIC ACID STRUCTURE.

bullous pemphigoid An autoimmune blistering skin disease involving separation of the basal layer of keratinocytes from the basement membrane. Autoantibodies recognize several hemidesmosome components and are probably responsible for the aetiology of the disease. *See*: BULLOUS PEMPHIGOID ANTIGENS; CELL JUNCTIONS.

bullous pemphigoid antigens (BPAG) ANTIGENS recognized by antisera from patients with the autoimmune blistering skin disease bullous pemphigoid, namely BPAGI (M_r 240 000), BPAGII (M_r 180 000) and BPAGIII (M_r 125 000). These antigens recognize hemidesmosomal plaque protein related to DESMOPLAKINS. BPAGII is a transmembrane protein with an extracellular

COLLAGEN-like domain. ANTIBODIES against extracellular EPITOPES of BPAGII are probably agents in the separation of cells from the basement membrane.

Giudice, G.J. et al. (1991) *J. Clin. Invest.* **87**, 734–738.
Tanaka, T. et al. (1991) *J. Biol. Chem.* **266**, 12555–12559.

bundle sheath Cylinder of PARENCHYMA or SCLERENCHYMA surrounding a vascular bundle especially in leaves.

α-bungarotoxin *See*: α-TOXINS.

Bunyaviridae A family of enveloped RNA viruses infecting vertebrates and arthropods, and which also have plant virus representatives. Four genera of animal viruses have been distinguished. The spherical or oval particles are 90–100 nm in diameter with a lipid envelope and glycoprotein surface spikes. The envelope encloses three ribonucleocapsids enclosing three (−)-sense single-stranded RNA species (L 3.5×10^6, M $1-2 \times 10^6$, S $0.4-0.8 \times 10^6$).

buoyant density centrifugation *See*: DENSITY GRADIENT CENTRIFUGATION.

Burkitt's lymphoma A higher grade malignant B-cell LYMPHOMA with a typical diffuse pattern of growth ('starry sky'). This tumour occurs in endemic (African) and nonendemic forms. Both types are associated with a chromosomal TRANSLOCATION involving a heavy or immunoglobulin light chain gene and the c-*myc* ONCOGENE, that is, t(8;14)(q24;q32) (in most cases), or t(2;8)(p11;q24), or t(8;22)(q24;q11). According to current pathological classification Burkitt's lymphoma is a small non-cleaved lymphoma. *See also*: EPSTEIN–BARR VIRUS.

bursa of Fabricius An organ in birds which develops as an outpouching of the dorsal hind gut and is the critical organ for the production of a normal B CELL repertoire. As the organ develops in embryonal life, it becomes organized into multiple internal folds of thickened epithelium, which become colonized with lymphoid progenitor cells. These cells then form into focal proliferations of lymphocytes predominately of the B cell lineage and virtually all these B cells rapidly express immunoglobulin heavy and light chains as well as the avian equivalent of the class II histocompatibility antigens (*see* MHC MOLECULES). Surgical bursectomy results in birds which are agammaglobulinaemic, have very low circulating levels of B cells and manifest immunological unresponsiveness to certain antigens.

burst size The number of virus particles produced by a single infected cell. The terminology comes from BACTERIOPHAGE infection of bacteria where it is calculated as the number of phage produced after LYSIS of the bacterial culture (determined by PLAQUE counting) divided by the number of infected cells. The latter is determined as the number of plaques formed by spreading the bacteria on a lawn of phage-sensitive cells after inactivating unadsorbed phage (by antibodies for example).

C

C (1) CYTOSINE.
(2) The sulphur-containing AMINO ACID cysteine.

C-banding Staining technique for constitutive heterochromatin which results in dense staining at centromeric regions on chromosomes. *See*: CHROMATIN; CHROMOSOME BANDING; CHROMOSOME STRUCTURE.

C cytoplasm Source of C-type cytoplasmic male sterility (cmsC) in maize. Distinguished from other types of male sterility by the nuclear genes that restore fertility. *See*: MITOCHONDRIAL GENOMES: PLANT.

C-DNA *See*: NUCLEIC ACID STRUCTURE.

C gene CONSTANT REGION GENE. *See*: IMMUNOGLOBULIN GENES; T CELL RECEPTOR GENES.

C-reactive protein A β-globulin present in the sera of healthy individuals in trace amounts but which is elevated in sera of patients suffering from inflammatory diseases such as rheumatic fever. It is a member of the PENTRAXIN family. It binds, in the presence of Ca^{2+}, to a polysaccharide component of pneumococcal cell walls, and activates the classical COMPLEMENT pathway by binding to complement component C1.

C region Constant region. Region of ANTIBODY or T CELL RECEPTOR of relatively constant sequence and which, in antibodies, is responsible for the effector functions of the molecule. In antibodies it varies with the class of the antibody molecule. *See also*: COMPLEMENT; IGA; IGE; IGG; IGM; IMMUNOGLOBULIN GENES; T CELL RECEPTOR GENES.

C terminus The carboxy-terminal end of a polypeptide chain, which bears the free carboxyl group. Protein biosynthesis proceeds from the N terminus to the C terminus. *See*: AMINO ACIDS; PROTEIN STRUCTURE; PROTEIN SYNTHESIS.

C-type animal lectins Calcium-dependent animal carbohydrate-binding proteins, such as MANNOSE-BINDING PROTEINS, endocytic receptors of hepatocytes and macrophages (*see* ASIALOGLYCOPROTEIN) and the SELECTIN cell adhesion molecules.

C-value A measure of GENOME size generally expressed in base pairs (bp) of DNA per HAPLOID genome, or in picograms (pg) of DNA per haploid cell. Each species has a characteristic C-value. Table C1 indicates the range of C-values found. *See also*: C-VALUE PARADOX; GENOME ORGANIZATION.

Table C1 C-values for representative species

Organism	C-value (bp DNA per haploid genome)
Mycoplasma pneumoniae (bacterium)	5.0×10^5
Escherichia coli (bacterium)	4.2×10^6
Saccharomyces cerevisiae (yeast)	2.3×10^7
Caenorhabditis elegans (nematode worm)	8.0×10^7
Drosophila melanogaster (fruit fly)	1.4×10^8
Mus musculus (mouse)	2.7×10^9
Xenopus laevis (toad)	3.1×10^9
Homo sapiens (human)	3.3×10^9
Pisum sativum (pea)	4.8×10^9
Lilium longiflorum (lily)	3.2×10^{10}
Protopterus aethiopicus (lungfish)	1.4×10^{11}

C-value paradox The failure to be able to correlate closely the total amount of DNA in a GENOME (the C-VALUE) with the genetic and morphological complexity of the organism in question. This paradox is evident both between species with apparently similar complexities, but very different C-values, and between species with similar C-values but very different complexities. The paradox also relates to the apparent absolute excess of DNA in the genomes of most eukaryotes compared with the predicted coding capacity of that genome; for example, the amount of DNA in the human genome could theoretically encode up to 300 000 genes, but recent estimates suggest the true value is nearer 30 000.

C1 COMPLEMENT component C1. C1 is a trimolecular complex of C1q, C1r, and C1s, which depends on Ca^{2+} ions for its integrity. C1q, which resembles a truncated PROCOLLAGEN, is composed of three different polypeptide chains and interacts with ANTIBODY. C1r becomes an active SERINE PROTEINASE which in turn cleaves and activates C1s. This is another serine proteinase and acts on C4 and C2. C1r and C1s are homologues of HAPTOGLOBIN, a serine 'non-proteinase'. The gene(s) for human C1q (a and b) are located on chromosome 1p , those for C1r and C1s together at 12p13.

C1-inhibitor A member of the SERPIN family of proteinase inhibitors, which inhibits complement component C1. It offers a 'bait' of exposed polypeptide, on whose cleavage the molecule undergoes a major conformational change, leading to an irreversible binding with the protease. C1-inhibitor is important in the homeostasis of the coagulation, fibrinolytic, and kinin systems (*see* BLOOD COAGULATION AND ITS DISORDERS). Heterozygous deficiency leads to marginal sufficiency of the protein and episodic attacks of the dominantly inherited disease hereditary angio-oedema.

C2 COMPLEMENT component C2. This is composed of a single polypeptide chain and is a serine proteinase PROENZYME. It is encoded by a slightly polymorphic gene encoded in the MAJOR HISTOCOMPATIBILITY COMPLEX (MHC) on human chromosome 6, adjacent to the genes for C4 and factor B. It has no structural relationship with the former, but is a close homologue of the latter. C2 forms a complex with C4 in the presence of Mg^{2+} and is then cleaved by active C1s to form the complex protease C42. This cleaves C3 and C5.

C3 COMPLEMENT component C3. This is the most abundant of the complement components, and has a central role in the whole complement system. Like C4, it is composed of two polypeptide chains encoded by a single significantly POLYMORPHIC gene which is located at p13 on human chromosome 19, and also contains a thioester bond allowing it to form covalent complexes with local molecules. The deposition of C3 on an invading organism has three consequences:

1 The organism is very much more readily phagocytosed by white blood cells.

2 Activation of the alternative pathway which leads to the formation of further C3-convertase enzyme and the amplification of the complement activation event.

3 The bridge is made to the formation of the membrane attack complex, which may damage the osmotic integrity of the organism's membrane. Probably the most important homeostatic mechanisms of the complement system act through the destruction of the active C3b fragment.

C3 plant Any of the majority of plants that fix CO$_2$ solely by way of the CALVIN CYCLE in which the product of ribulose-bisphosphate carboxylase is the three-carbon 3-phosphoglyceric acid. A characteristic feature of C3 plants is photorespiration, which results from an oxygenase side reaction catalysed by the carboxylase; consequently such plants have a relatively high compensation point (the light intensity at which photosynthesis and respiration just balance). *See*: PHOTOSYNTHESIS.

C4 COMPLEMENT component C4. C4 is composed of two polypeptide chains encoded by a single cistron, although most humans have two C4 ISOGENES located within the MAJOR HISTOCOMPATIBILITY COMPLEX (MHC) on chromosome 6. The C4 genes exhibit POLYMORPHISM both of gene number, with about 10% of HAPLOTYPES differing from the modal two-gene model, and of an extreme level of structural variation, which makes C4 the most polymorphic protein after the cell-bound MHC molecules themselves. C4 is activated by cleavage of the N-terminal portion of the α chain. It contains an internal thioester bond between different parts of the α chain which becomes available to form covalent bonds with either amino or hydroxyl groups on adjacent molecules (e.g. antigen or antibody) or, by default, with water. C4 interacts with C2 to form the next proteinase of the complement pathway.

C4 plant One of a substantial minority of plants (e.g. maize, many tropical grasses, some dicotyledons) that fix CO$_2$ initially into the four-carbon oxaloacetic acid in a reaction catalysed by phospho-enolpyruvate carboxylase located in mesophyll cells. Four-carbon products of the reaction are transported to the bundle-sheath cells (Kranz cells) which are characteristically large and contain many chloroplasts. Here a decarboxylation reaction (NAD- or NADP-linked malic enzyme, PEP carboxykinase) liberates CO$_2$ which is refixed by ribulose-bisphosphate carboxylase and the CALVIN CYCLE. This acts as a CO$_2$ concentrating mechanism and C4 plants do not show photorespiration. C4 photosynthesis is less efficient than C3, but gives higher rates at high temperatures. *See*: PHOTOSYNTHESIS.

C5 COMPLEMENT component C5. This has many similarities with C4 and C3 but lacks the thioester bond. It is encoded on human chromosome 9q. It is activated by the same enzymes that activate C3, but in addition C5 needs to be bound to C3b before it becomes susceptible to activation. The cleavage of C5 is the last enzymatic event in the activation of complement. Cleaved C5 forms a stable complex with C6, and this initiates the formation of the terminal complex.

C6 and C7 These COMPLEMENT components are composed of one polypeptide chain each, share many structural similarities, are encoded by adjacent but 3′ to 3′ genes on human chromosome 5 near the gene for C9, and combine in sequence with C5 to form a mobile but short-lived complex with affinity for membranes. They are a complex mosaic of structural features which are important in making the transition from a set of fluid-phase proteins to a membrane-penetrating complex.

C8 COMPLEMENT component C8. This is composed of three polypeptide chains (α, β, γ), which are encoded by two separate genes close to each other on human chromosome 1p and another on chromosome 9q. Each of the larger chains shares a considerable number of structural features with both C6 and C7 on the one hand, and with C9 on the other. Independent deficiencies of C8 α/γ and C8 β are found and are complementing *in vitro*, indicating that the molecule is made up of two loosely associated subunits.

C9 COMPLEMENT component C9. This readily polymerizes in a reaction initiated by the presence of membrane-bound C5–8, leading to the formation of a membrane-penetrating complex involving usually in excess of six C9 molecules. Once established in the cell membrane, the complex damages the osmotic integrity of the cell, and depending on the ability of the cell to maintain pumping and to isolate the damaged part of the membrane, leads to osmotic swelling and finally rupture. The polymerized C9-containing complex is very resistant to proteolysis. C9 is encoded by a gene on human chromosome 5p near the C6 and C7 genes.

Ca^{2+}-ATPases Membrane ion transport proteins of the P-type ATPase family mediating the active transport of CALCIUM, present in most eukaryotic cells. The Ca^{2+}-ATPase from the sarcoplasmic reticulum of muscle has been most intensively studied. It is a protein of M_r ~110 000 with sequence similarity to the α subunit of Na$^+$,K$^+$-ATPase. Hydrophobicity plots indicate seven transmembrane regions, with a large part of the polypeptide chain located on the cytosolic side of the membrane. Two Ca^{2+} are transported for each ATP hydrolysed. Calcium-dependent phosphorylation of an aspartate residue is presumed to lead to a conformational change in the enzyme in which two Ca^{2+} bound to the cytosolic face are transported across the membrane and released into the lumen of the sarcoplasmic reticulum. Subsequent dephosphorylation of the enzyme returns it to the original state. Vanadate inhibits Ca^{2+} transport by stabilizing the enzyme in the phosphorylated state. Ca^{2+}-ATPase reconstituted into calcium-loaded membrane vesicles *in vitro* can be induced to synthesize ATP from ADP and P_i in a reverse reaction. *See also*: MEMBRANE TRANSPORT SYSTEMS.

Ca²⁺/calmodulin-dependent protein kinase A Ca^{2+}/calmodulin-dependent, serine/threonine-specific PROTEIN KINASE of broad specificity which seems to phosphorylate numerous target proteins in response to a rise in cytosolic CALCIUM. Also known as the calmodulin-dependent multiprotein kinase or multifunctional calmodulin-dependent protein kinase. It exists as mixed dodecamers of three related subunits — α (M_r 50 000), β (M_r 58 000) and β′ (M_r 60 000) — with the α and β forms being different gene products and the β and β′ forms being produced by alternative RNA SPLICING. All three are maintained in an inactive state by binding to an inhibitory region. On binding of Ca^{2+}/calmodulin to an adjacent region, inhibition is relieved and the inhibitory sequence becomes an AUTOPHOSPHORYLATION site. This phosphorylation maintains the kinase in an active state even if Ca^{2+} is removed. This kinase may be involved in LONG-TERM POTENTIATION, as the specific inhibitor KN-62 (1-[NO-bis,1,5-isoquinolinesulphonyl]-N-methyl-L-tyrosyl-4-phenylpiperazine) prevents the induction of long-term potentiation.

Schulman, H. & Lou, L.L. (1989) *Trends Biochem. Sci.* **14**, 6266.

CAAT box A CONSENSUS SEQUENCE found in the PROMOTERS of many mammalian genes transcribed by RNA polymerase II. The core consensus sequence is:

5′ GGCCAATCT 3′

and is found UPSTREAM of the TATA BOX element. The CAAT box binds one or more TRANSCRIPTION FACTORS, some of which are expressed in all tissues (e.g. CTF/NF1 or CAAT box transcription factor), whereas others are only present in some tissues (e.g. C/EBP in liver). The CAAT box is an important *cis*-acting element for both constitutive and regulated transcription of many mammalian genes.

McKnight, S. & Tjian, R. (1986) *Cell* **46**, 795–805.

CaaX box Amino-acid motif involved in membrane localization of *ras* proteins and other proteins by lipid MEMBRANE ANCHORS (*see* GTP-BINDING PROTEINS). *See also*: INTERMEDIATE FILAMENTS; NUCLEAR ENVELOPE.

cab protein CHLOROPHYLL *a/b* BINDING PROTEIN.

CaBP CALCIUM-BINDING PROTEIN.

cachectin Tumour necrosis factor α. *See*: TNF α.

cadherins Family of CELL ADHESION MOLECULES which mediate strong cell–cell contacts through homophilic binding. N-cadherin, for example, is found on nerve, heart, and lens cells and expressed transiently in other cell types during development. It consists of 748 amino acids and in humans the gene has been mapped to chromosome 18q. *See*: CELL ADHESION MOLECULES; EPITHELIAL POLARITY; UVOMORULIN.

Caenorhabditis development

THE nematode *Caenorhabditis elegans* has been the subject of intensive investigations into the mechanisms of development [1, 2] because of its amenability to genetic analysis, the constancy of its development and anatomy, and its transparency, which allows all its nuclei to be observed by light microscopy. In addition, its genome has been physically mapped, greatly simplifying molecular genetic analyses. This soil-dwelling roundworm is readily propagated in the laboratory and exists in two forms (Fig. C1): self-fertilizing hermaphrodites that produce both male and female gametes, and males, which can mate with hermaphrodites to produce outcross progeny. Adult hermaphrodites contain 959 somatic nuclei, males contain 1031 somatic nuclei. The indeterminate number of germ nuclei is generally greater than 1000.

C. elegans development occurs optimally over three days in six stages: the embryo; four larval stages (L1 through L4); and the fertile adult. Larval and adult stages are separated by cuticular moults. Somatic development occurs during all stages except the adult. Under conditions of starvation or crowding an alternative developmental pathway can occur resulting in a specialized L3 larva, the dauer larva. Dauer larvae are long-lived forms arrested in development; appropriate environmental stimuli cause them to resume normal development. Hermaphrodites begin self-fertilization shortly after reaching adulthood and produce a brood of about 300 fertilized eggs.

Fertilization and embryogenesis

In hermaphrodites, sperm and oocytes derive from the same germ cell precursors although they differentiate at different times and locations in the gonad. Before fertilization, oocytes show no overt anterior-posterior (A/P) polarity. Sperm fertilize the oocyte at the prospective posterior pole of the embryo; it is not known whether sperm entry defines the posterior pole, or is a consequence of pre-existing A/P polarity in the oocyte. Upon fertilization, the oocyte nucleus, which is arrested at prophase of the first meiotic division, completes MEIOSIS and two polar bodies are extruded at the anterior.

Embryogenesis occurs over about 13 hours under optimal conditions. The first phase (about half of embryonic development) is characterized by rapid cell proliferation with little overt DIFFERENTIATION or MORPHOGENESIS. The embryonic genome becomes transcriptionally active sometime before the 28-cell stage.

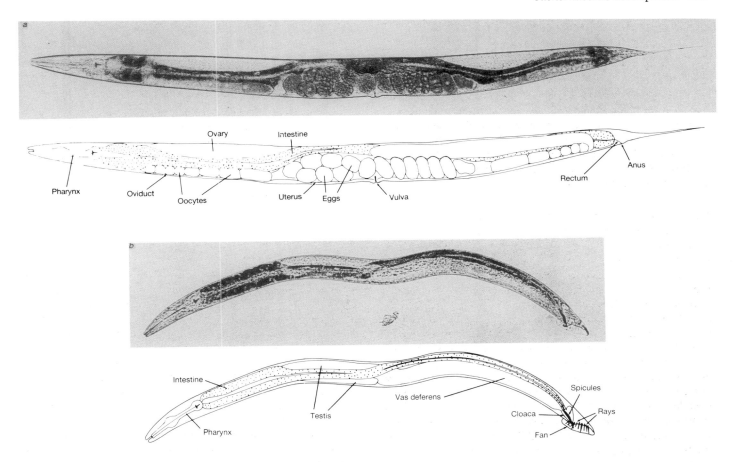

Fig. C1 Light micrographs of adult hermaphrodite (upper) and male (lower) *Caenorhabditis elegans* in left lateral views. Major organs and other anatomical structures are indicated on the drawings beneath each micrograph. The animals are approximately 1 mm long. From [4].

Cell division largely ceases in the second phase and extensive morphogenesis and differentiation ensue, terminating in a hatching L1 larva containing 558 nuclei (560 in males).

During the first cleavages, a set of six founder cells is produced by a series of asynchronous, asymmetric divisions. Each founder cell undergoes rounds of symmetric and synchronous cell division with its own characteristic periodicity, roughly in proportion to its size. Three founder cells produce only a single differentiated cell type: for example the 'E' founder cell produces the entire intestine (ENDODERM) and no other cell type. The other three engender a number of cell types that populate ECTODERM and MESODERM (Fig. C2).

The axes and left–right handedness are established during the first three rounds of cell division. The zygote divides asymmetrically along the A/P axis; the larger daughter (the AB founder cell) is at the anterior. During the next two rounds of AB division, the dorsal–ventral axis and handedness respectively are generated when mitotic spindles (*see* MITOSIS), initially perpendicular to the A/P axis, skew along this axis. This skewing *per se* determines the geometry of the animal as reversing the direction of skewing at

the appropriate AB division reverses the dorsal–ventral axis or handedness.

GASTRULATION begins at the 28-cell stage and continues through most of the proliferative phase. During gastrulation, individual cells sink from the surface into the interior of the embryo. Through the course of embryogenesis a few long-range cell migrations occur; however, most cells do not undergo extensive migrations, being born in the appropriate position.

Following the proliferative phase, the spheroid embryo undergoes morphogenesis. ACTIN filaments in the hypodermis, the external epithelium, provide the force to promote its circumferential contraction, causing the embryo to elongate fourfold into a cylindrical worm shape.

Post-embryonic development

From newly hatched larva to fertile adult, *C. elegans* grows approximately eightfold in length and develops further. Post-embryonic development is limited: only about 10% of the nuclei produced during embryogenesis divide post-embryonically, resulting primarily in increased numbers of neurons, muscles, sex-specific cells, and hypodermal and intestinal nuclei. There is no major remodelling of the body plan or organs, except those associated with sexual functions. At hatching, the sexes are nearly identical; the pronounced sexual dimorphism occurs post-embryonically with the production of sex-specific structures,

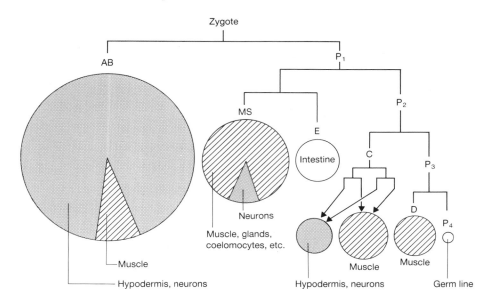

Fig. C2 Cell types produced by the embryonic cell lineages of *Caenorhabditis elegans*. The cell divisions which generate the embryonic founder cells are indicated. The area of each circle is proportional to the number of cells produced by the corresponding founder cell, and the area of each sector is proportional to the number of cells of the type indicated. Founder cells are denoted by capital letters. Cells classically regarded as mesodermal are indicated by striping, those usually considered as ectodermal by stippling. From [3].

including the formation of the gonad and mature gametes, the hermaphrodite vulva, and complex mating structures of the male tail. Cell and nuclear divisions are coordinated with growth: stereotyped cell lineages occur during each larval stage between moults. During post-embryonic development, some cells undergo cell or nuclear migration and others ENDOREPLICATE their DNA to produce polyploid nuclei.

Cell lineage

Knowledge of the cell LINEAGE of *C. elegans* has provided a means of examining developmental processes at the resolution of single cells. Changes in the lineage resulting from genetic or physical intervention can be used as a precise measure of changes in CELL FATES and histories.

The entire somatic cell lineage of *C. elegans* (Fig. C3) has been determined [3] and correlated with the cellular anatomy known from electron microscopic reconstructions. Lineages of developing animals were followed by observing nuclei with Nomarski MICROSCOPY; the lineage was assembled from observations of many individuals. With minor exceptions the lineage is invariant; that is, the pattern of cell divisions and fates of cells are nearly identical in all individuals. The entire post-embryonic somatic lineage of another nematode, *Panagrellus redivivus*, has been analysed and is also virtually invariant.

One of the striking features of the *C. elegans* lineage, particularly the embryonic lineage, is its complexity and the relative absence of regular patterns throughout it. In many cases, anatomical symmetry is generated by asymmetric lineages; for example much of the bilateral symmetry arising during embryogenesis is produced by non-bilaterally symmetric lineages. Similarly, the limited segmental reiteration of sets of hypodermal cells produced in the embryo does not arise by segmentally repeating lineages; rather, the repeated segments are generated in a piecemeal fashion. Often, cells of a specific type arise by quite different lineage patterns or from distantly related progenitors. Conversely,

very different cell types can arise from closely related lineages: for example, although the division between mesoderm and ectoderm generally occurs early in the lineage, in some instances differentiated muscles and neurons are SISTER CELLS.

Despite its complexity, there are a number of patterns in the wild-type lineage. The simplest pattern is the clonal lineage, in which a precursor produces a single cell type — the gut and germ-line lineages are conspicuous examples (Fig. C2). Very often a particular pattern of cell divisions, or sublineage, is repeated at various points in the lineage (e.g. a particular sublineage pattern is repeated 70 times in *P. redivivus*). A third pattern is the stem cell lineage in which one daughter of a precursor behaves like its mother and the other takes on a different fate. Certain complex lineages are converted to simpler stem cell-like lineages in an *unc-86* mutant (*unc*, for *uncoordinated*), indicating that complexity in the lineage may arise in part from modification of simpler underlying patterns. (*unc-86* is described further under CAE-NORHABDITIS NEURAL DEVELOPMENT.)

Comparisons of *C. elegans* and *P. redivivus* show how lineage patterns can be altered through evolution and provide clues to how genes may control lineages. The number of times that a particular cell division or sublineage is repeated and the A/P polarity of particular lineages are subject to phylogenetic variation, differing between the two species. In addition, terminal fates of cells produced by a lineage can differ between the two species, and in some cases a portion of the fate of one cell in one species is associated with its sister in the other.

The invariance of the lineage allows use of lineage analysis to assess aberrant development in mutants. A large number of mutants defective in post-embryonic lineages have been identified. In some, the difference between wild-type and mutant lineages are analogous to the phylogenetic differences described above. Some lineage genes encode apparent TRANSCRIPTION FACTORS, implying that cell division patterns are regulated in part by modulating the TRANSCRIPTION of other genes (*see* DROSO-PHILA DEVELOPMENT; EUKARYOTIC GENE EXPRESSION); other

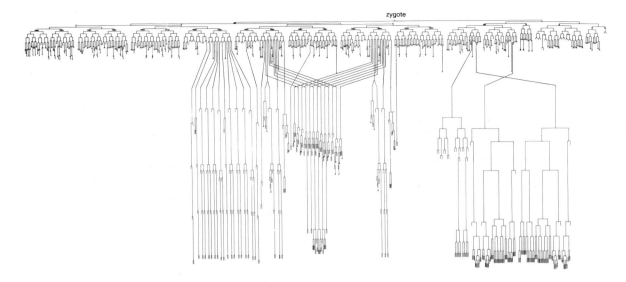

zygote

Fig. C3 The complete somatic cell lineage of the nematode *Caenorhabditis elegans*. The vertical axis indicates developmental time (covering ~50 hours from the first to the last cell division at 20°C), and the horizontal axis, the approximate direction of cell division (usually anterior to the left). The embryonic lineage is indicated at the top; the post-embryonic lineages involve a subset of the cells produced during embryogenesis and are indicated below the embryonic lineages. Adapted from [3–5].

lineage genes control lineage patterns through cell–cell interactions (see below).

Cell-autonomous development

Features of the lineage suggest that the fates of some cells are directed by CELL-AUTONOMOUS mechanisms that are unaffected by the environment of the cell. Identical sublineages sometimes arise from distantly related precursor cells. That such precursors reside in different environments suggests they may be intrinsically programmed to execute such sublineages. Cell autonomy has been investigated by eliminating cells with a laser microbeam (laser ablation) and determining whether the development of nearby cells is affected. The fates of most cells are unaltered by ablation of nearby cells in *C. elegans*, providing support for cell autonomy in much of its development.

Direct evidence for cell autonomy in the early embryo has been obtained by examining differentiation of physically isolated blastomeres. For example, isolated E cells undergo gut differentiation autonomously. The potential to differentiate gut can be transferred from the cytoplasm of a gut precursor to a blastomere that does not normally produce gut, consistent with the existence of a CYTOPLASMIC DETERMINANT for gut differentiation. When cleavage of early embryos is blocked, but nuclear division and differentiation allowed to progress, a number of differentiation markers are expressed in those cleavage-blocked cells whose descendants would give rise to the markers in normal embryos. Thus, the potential to express differentiated cell fates is asymmetrically

distributed to blastomeres during early cleavages, suggesting that founder cell fates are controlled by maternally encoded MOSAIC determinants that are segregated at determinative cell divisions (*see* MATERNAL CONTROL OF DEVELOPMENT; DETERMINATION).

The germ-line specific material, called P granules, behaves as would be expected for a cytoplasmic determinant. Immunodetection of these maternally encoded granules shows they are initially distributed uniformly throughout the zygote but are progressively segregated to the germ-line lineage before each determinative cleavage by an actin-dependent mechanism. P granules are probably analogous to POLAR GRANULES found in other organisms such as *Drosophila* and may represent a mosaic determinant for the germ-line fate.

Molecular components required for cytoplasmic segregation are encoded by the *par* (for *par*titioning) genes. Mutations in any of four *par* genes in the mother lead to death of the embryo (strict maternal-effect embryonic lethality) and prevent asymmetric germ-line segregation of P granules. Most of the *par* mutations perturb asymmetric cleavage of the zygote, resulting in two blastomeres of equal size that divide synchronously; thus, these genes are required to establish the A/P axis. Mutant embryos generally do not differentiate intestine, and adult progeny that have escaped lethality lack a germ line (grandchildless phenotype), suggesting that the mutants fail to partition determinants of intestinal and germ-line cell fates.

Cell-nonautonomous development

Although cell autonomy is prevalent, cell–cell communication controls a number of cell fates in *C. elegans*. Some of these interactions are INDUCTIONS, in which one type of cell controls the fate of a different cell type. In the embryo, the daughters of the AB founder cell are equipotential at birth; however, they have different fates. These differences are determined by their environment since their fates are reversed with a reversal in their positions. Production of pharynx muscles by one of these AB daughters requires the presence of non-AB-derived cells in the embryo. Thus, the differences in fate of the two AB daughters are the result of induction by non-AB-derived cells.

Post-embryonically, an inductive interaction leads to development of the hermaphrodite vulva in response to a signal from an underlying gonadal cell (see below). In addition, a post-embryonic inductive interaction occurs in germ cell determination: a cell at the distal end of the gonad promotes mitotic proliferation of germ cells in its vicinity. If this cell is eliminated by laser ablation, germ cells stop proliferating and become meiotic, differentiating into gametes. The product of the *glp-1* gene (*glp* for *germ line* proliferation), a putative cell-surface RECEPTOR (see below), is required in this induction and also in the interaction that induces pharynx development from one AB daughter in the embryo.

A second type of cell interaction occurs between cells of an EQUIVALENCE GROUP, sets of equipotential cells that can take on two or more alternative fates. The fate chosen is determined by lateral signalling between the members of the group. Laser ablation studies have revealed a number of equivalence groups in *C. elegans* and demonstrated the priority of the alternative fates. The simplest example is an equivalence group of two cells with two different fates: if either of the pair of cells is ablated, one fate, called 1°, will always be expressed by the single survivor (the other fate is called 2°). Thus, the two cells compete for the 1° fate and communication between two ensures that only one will express that fate. Lateral signalling between members of a number of different equivalence groups requires the *lin-12* gene product (*lin* for *lin*eage-abnormal), a putative cell-surface receptor.

Cell–cell communication is regulated to ensure that inappropriate interactions between cells do not occur. The *lin-12* gene product prevents an inappropriate inductive interaction in the gonad: in mutants lacking *lin-12* function, somatic cells can induce germ cells to undergo mitotic proliferation in the proximal region of the gonad. In wild-type embryos, this interaction is prevented and germ cells undergo meiosis. In another example, the *pal-1* gene, which encodes a homeodomain-containing protein (see HOMEOBOX GENES AND HOMEODOMAIN PROTEINS), is required within particular hypodermal cells in the posterior of the animal to block their response to a signal produced by neighbouring hypodermal cells.

Vulval development

The hermaphrodite vulva, a post-embryonic structure through which eggs are laid and copulation occurs, has proven a useful model for studying genetic and molecular control of the lineage. Its development is controlled by both cell-autonomous and nonautonomous mechanisms.

A set of hypodermal cells in early larvae, the vulval precursor cells (VPCs), constitute the vulval equivalence group. Genetic and cell ablation studies have led to a model of vulval development in which a diffusible inductive signal from a cell (called the anchor cell, or AC) in the underlying gonad causes VPCs near it to undergo vulval lineages; the most distant VPCs generate non-vulval hypodermal cells. The single VPC closest to the AC receives the highest level of this spatially graded AC signal, causing it to undergo the lineages of the central portion of the vulva. It then causes its immediate neighbours, by lateral signalling, to execute alternative vulval lineages. Thus, both inductive and lateral signalling cell interactions act in determining the fates of the VPCs.

Viable mutants lacking a vulva or which make ectopic vulva-like structures are readily identified. Lineage defects of mutants and genetic studies have made it possible to assign a number of genes required for vulval development to one of three stages in a genetic pathway for vulval development: generation of the VPCs, determination of the VPCs, and expression of the three VPC lineages once they have been determined.

Molecular and genetic analyses suggest a direct role for three of these vulva-affecting genes in the transduction pathway that directs vulval fates in response to the AC signal. The *lin-3* gene product appears to be a secreted GROWTH FACTOR related to epidermal growth factor (EGF) and is a likely candidate for the AC signal. *let-23* (*let*, for *let*hal), a gene required downstream of *lin-3* in the vulval signalling pathway, encodes a putative tyrosine kinase of the EGF receptor subfamily (*see* GROWTH FACTOR RECEPTORS; PROTEIN KINASES). A *ras* protein (*see* GTP-BINDING PROTEINS) is encoded by the *let-60* gene, which is required downstream of *let-23*. These studies have led to a model in which the *lin-3*-encoded signal molecule, secreted by AC, binds to and activates the *let-23*-encoded receptor expressed on the surface of the VPCs. The bound receptor activates the *let-60 ras* protein and activated *let-60 ras* protein then directs vulval development in the responding VPCs, presumably through a SECOND MESSENGER system.

Lateral signalling between the VPCs and other bipotential cells of equivalence groups requires the *lin-12* gene. *lin-12* encodes a putative TRANSMEMBRANE PROTEIN that contains EGF-like repeats in its presumptive extracellular domain. This gene acts cell autonomously and is therefore proposed to encode the receptor for lateral signalling. The *lin-12* gene is physically very close to *glp-1*, which is implicated in inductive cell–cell interactions (see above), and the products of these two genes are HOMOLOGOUS. These genes perform partially redundant functions, and it is probable that one arose by duplication of the other. Their products are also homologous to the *Notch* gene of *Drosophila melanogaster*, which is also implicated in lateral signalling (*see* DROSOPHILA DEVELOPMENT; INSECT NEURAL DEVELOPMENT). Once the VPC fates are determined, the lineages that they produce are probably controlled cell autonomously. Consistent with this is the finding that *lin-11*, which is required to express a particular vulval sublineage, encodes a homeodomain-containing protein (i.e. a probable transcription factor).

Heterochronic genes

The heterochronic genes regulate the timing of developmental events. Mutations in these genes lead either to retarded development, in which events that normally occur early are inappropriately repeated later, or precocious development, the converse transformation. Four heterochronic genes have been assigned to sequential stages in a genetic pathway that controls the switch from larval to adult behaviour of the hypodermis. The existence of single mutations that can lead to heterochronic modifications of the lineage implies that heterochrony can be a mechanism for rapid modification of development during evolution, and indeed, some of the differences in lineages between related wild-type nematode species are similar to those seen in *C. elegans* heterochronic mutants.

One of the heterochronic genes, *lin-14*, can alternately mutate to cause either a RECESSIVE precocious or a DOMINANT retarded phenotype. This gene encodes a nuclear protein which is expressed at the highest levels in the embryo and attenuated post-embryonically. Thus, it is likely that *lin-14* promotes early cell fates and its subsequent repression allows the switch from early to late cell fates.

Programmed cell death

During *C. elegans* development, 131 nuclei undergo a fate of programmed cell death. Shortly after dying, dead cells are engulfed by neighbouring cells; engulfment of the cell corpses is prevented in *ced-1, 2, 5, 6, 7, 8,* and *10* (*ced* for *cell death defective*) mutants.

Onset of the cell death programme is blocked in *ced-3* and *ced-4* mutants, in which cells normally destined to die survive. These survivors often differentiate into neurons, which can be functional. The *ces-1* and *ces-2* genes (*ces*, for *cell death specification*) affect execution of the *ced-3,4*-dependent cell death in a subset of cells that normally undergo programmed cell death. As all these mutants are viable, programmed cell death is a nonessential event in *C. elegans* development. The *ced-3* and *ced-4* genes act cell autonomously, implying that the cell death programme is intrinsic to the cells destined to die and not to their surviving neighbours. This programme is repressed by the *ced-9* gene: if *ced-9* function is removed from embryos, they die as a result of ectopic execution of the *ced-3,4*-dependent cell death programme.

Homeotic genes

A number of genes encoding homeodomain-containing proteins have been identified in *C. elegans*. At least four of these are clustered at a single locus in the genome, and two genes in this cluster, *mab-5* and *egl-5*, direct adjacent region-specific development, indicating that, as in *Drosophila* and vertebrates, *C. elegans* contains a cluster of homeobox genes whose relative position in the cluster corresponds to the region of the animal whose development they control (*see* DROSOPHILA DEVELOPMENT; HOMEOBOX GENES AND HOMEODOMAIN PROTEINS).

Genes controlling dauer larva development

Mutants that fail to form dauer larvae or that form them constitutively in the absence of the appropriate environmental cues are readily identified, and have been used to define a genetic pathway for dauer larva development. One such gene, *daf-1* (*daf* for *dauer formation*), encodes a putative cell-surface serine-threonine kinase (*see* PROTEIN KINASES), which may be a receptor for environmental stimuli that regulate dauer development.

J.H. ROTHMAN

1 Brenner, S. (1974) The genetics of *Caenorhabditis elegans. Genetics* **77**, 71–94.
2 Wood, W.B. (1988) The nematode *Caenorhabditis elegans* (Cold Spring Harbor Laboratory, Cold Spring Harbor, NY).
3 Sulston, J.E. et al. (1983) The embryonic lineage of the nematode *Caenorhabditis elegans. Dev. Biol.* **100**, 64–119.
4 Sulston, J.E. & Horvitz, H.R. (1977) Postembryonic cell lineages of the nematode *Caenorhabditis elegans. Dev. Biol.* **56**, 110–156.
5 Kimble, J. & Hirsch, D. (1979) The postembryonic cell lineages of the hermaphrodite and male gonads in *Caenorhabditis elegans. Dev. Biol.* **70**, 396–417.

Caenorhabditis neural development

THE nervous system of the nematode *Caenorhabditis elegans* is composed of 302 neurons in the adult hermaphrodite and 381 neurons in the adult male. In both sexes an almost independent circuit of 20 neurons is found in the muscular feeding organ (pharynx) which regulates its activity. *C. elegans* neurons are all individually identifiable and have simple morphologies, being frequently monopolar and with very few or no branches. The microanatomy of all neurons, as well as their synaptic connections with each other (their connectivity), have been established by reconstructing the entire nervous system of an adult hermaphrodite using electron micrographs of serial sections [1]. This reconstruction represents a unique reference tool to study the function of genes involved in the development and differentiation of the nervous system.

In spite of its apparent simplicity the nervous system of *C. elegans* controls a variety of behaviours: animals move forward and backward by propagating sinusoidal waves along their bodies; their heads perform complex exploratory movements; they respond by attraction or repulsion to sensory stimuli; and they possess two sex-specific behaviours, egg laying in the hermaphrodite and male mating. The latter complex behaviour requires a set of male-specific neurons. An important feature of *C. elegans* as a model system for the study of the nervous system is that these behaviours can essentially be dispensed with under laboratory conditions, in which the worms are maintained in ideal conditions on a lawn of bacteria and do not require movement to feed. Furthermore, *C. elegans*, being a self-fertilizing hermaphrodite, can reproduce without mating. Indeed, only two types of neurons have been shown to be necessary for survival: a pair of neurons called the canal associated neurons (CAN), which are thought to be involved in the function of the excretory cell, and the motor neuron M4 which is required for the proper functioning of the pharynx. This has made possible the isolation of a large number of viable mutations affecting behaviour [2]. Underlying developmental defects resulting in neuroanatomical abnormalities in these mutants can be identified and studied using electron microscopy and immunocytochemistry (*see* IMMUNOELECTRON MICROSCOPY).

A large number of *C. elegans* genes which have been shown genetically to be involved in the development of the nervous system are being analysed [3]. These include genes involved in neurogenesis, acquisition of neuronal identity, axonal transport, axonal pathfinding, axonal outgrowth, specific synapse formation, programmed cell death, neuronal degeneration, sex-specific development and sensory transduction. This article will mainly discuss those genes for which published molecular information is available.

Neurogenesis

Most neurons (222 in hermaphrodites and 224 in males) are born during embryogenesis and change very little, if at all, during later development. However, an important set of motor neurons found in the ventral nerve cord (Fig. C4) is generated at the end of the first larval stage (*see* CAENORHABDITIS DEVELOPMENT) and the incorporation of these cells in an already functional nervous system entails the reorganization of part of the juvenile connectivity. The latest phase of neuron production occurs only in the male during the third larval stage and adds the male-specific neurons required for mating.

The precursors of *C. elegans* neurons are not set aside from the ECTODERM as a group. Instead, they arise from precursor cells that also produce non-neuronal cells, including non-ectodermal cells (muscles) (*see* CAENORHABDITIS DEVELOPMENT), the segregation between neuronal and non-neuronal fates frequently occurring only at the terminal cell division. In addition, some postembryonic neurons arise from precursors which function as differentiated epidermal cells in earlier stages and participate in specialized functions such as cuticle formation.

In spite of their variable origins, *C. elegans* neurons represent a distinct cell type. They have long processes typical of neurons, their nuclei display a particular morphology, and they express tissue-specific antigens. To date no gene specifically involved in the production of all neurons has been identified and it is therefore not known if all *C. elegans* neurons use common mechanisms to acquire a neuronal fate.

Acquisition of identity

Several genes involved in the production of specific subsets of neurons have been identified. Among the best characterized is the gene *unc-86* (*unc*, from *uncoordinated*, indicating a locomotory defect). The expression and function of *unc-86* is specific to neurons and their precursor neuroblasts. In the affected neuroblast lineages, *unc-86* mutations alter the fate of one daughter of a division so that it generates cell types normally produced by its mother. This gene therefore appears to participate in the coupling of CELL LINEAGE to cell identity (*see* CAENORHABDITIS DEVELOPMENT). It encodes a putative TRANSCRIPTION FACTOR which shares homologies with two mammalian genes (*Pit-1* and *Oct-2*), thus defining the POU class of homeobox-containing genes (*see* HOMEOBOX GENES AND HOMEODOMAIN PROTEINS). Several other members of this class are also specifically expressed in neural tissue in mammals and insects (*see* VERTEBRATE NEURAL DEVELOPMENT).

In addition to its role in neuroblasts, the *unc-86* protein is also expressed in 57 differentiated neurons (one fifth of the 302 neurons in the adult hermaphrodite). Some of these neurons are descendants of neuroblasts expressing and requiring *unc-86* whereas others begin to express *unc-86* only after their genera-

tion. These neurons do not share any known common feature such as connectivity, NEUROTRANSMITTER phenotype, function or spatial location. The expression of *unc-86* in differentiated neurons does not therefore seem to have a specific consequence but rather to participate in a combinatorial manner in the specification of neuronal cell type.

In contrast to *unc-86*, the expression and function of another homeobox-containing gene, *mec-3*, is restricted to a group of cells which share common features. *mec-3* is expressed in eight sensory cells, the six microtubule cells (MCs,) which are involved in the response to light touch, and the FLP and PVD neurons. PVD mediates response to harsh mechanical stimuli, and FLP has an ultrastructure suggestive of a mechanoreceptor, but its function is unknown. The six MCs possess identical subcellular structures including a bundle of large MICROTUBULES, which have been shown by genetic analysis to be required for sensory transduction. *mec-3* mutations abolish the function of PVD and the MCs, which also lose their specific ultrastructural features. The cells affected in *mec-3* mutants are among the cells expressing *unc-86*, and, indeed, *mec-3* expression requires *unc-86* function. Although *mec-3* is expressed in a group of neurons sharing common characteristics, it is not sufficient to confer a stable set of characteristics on cells that express it. Maintenance of its expression requires the action of the *mec-17* gene, and the gene *mec-7*, which encodes a β-TUBULIN specific to the MCs, is expressed in early stages in the absence of *mec-3*.

Another homeobox gene, *unc-4*, is likely to be involved in the specification of motor neuron cell types in the ventral cord (Fig. C4). In *unc-4* mutants, a subset of one type of motor neuron (VA) receives synaptic connections appropriate for another type of motor neuron (VB). *unc-4* is expressed in the VAs, suggesting that the defect lies with the postsynaptic motor neurons rather than with the interneurons that synapse on them.

Many different genes affecting lineages which produce neurons as well as other cells have been studied and are being molecularly analysed (*see* CAENORHABDITIS DEVELOPMENT). The picture emerging from the analysis of genes exclusively involved in the specification of neurons is an apparent prevalence of homeobox genes. Numerous homeobox genes are also expressed in the developing nervous system of both insects and vertebrates. The requirement for these genes in nematodes demonstrates once more profound similarities in the developmental mechanisms of all animals with bilateral symmetry.

Axonal transport

The product of the gene *unc-104* has been implicated in the biogenesis of SYNAPSES. *C. elegans* neurons make synapses at multiple varicosities along their axons which contact, *en passant*, other axons that run in parallel. These varicosities are packed with transmitter-containing vesicles clustered at a zone of release. Several distinct morphological classes of vesicles can be distinguished with variations in size, shape and content. In *unc-104* mutants, there are no large varicosities, there is a reduction of the number of synapses, a shrinking of the active zones, and an extreme reduction in the number of synaptic vesicles in the remaining varicosities. At the same time, the neuronal cell bodies in these mutants contain large numbers of small vesicles which

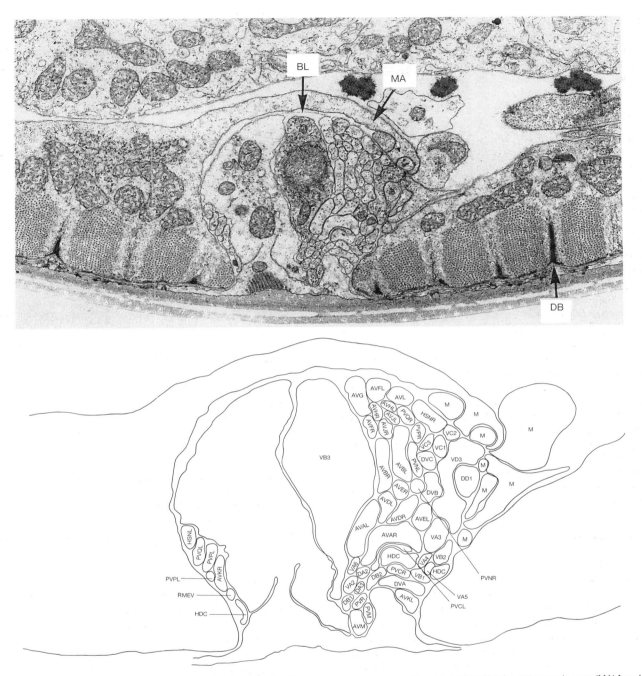

Fig. C4 Electron micrograph of transverse section through the ventral cord (upper panel) and identification of neuronal processes (lower panel). The ventral cord consists of a process bundle running alongside a longitudinal ridge of hypodermis; the whole structure is bounded by a thin basal lamina (BL). Axons of motor neurons arrange themselves next to the basal lamina on the right-hand side of the cord in a fixed arrangement. The usual sequence of motor neuron classes from dorsal to ventral is VCn, VDn, DDn, VAn, and VBn. A neuromuscular junction from a VD3 can be seen in this section; the motor neurons synapse through the basal lamina onto muscle arms (MA) from both left and right ventral muscle quadrants. The cell bodies of the motor neurons that innervate body muscles are arranged in a linear sequence in the ventral cord. This section contains the cell body of VB3. The ventral cord also contains the interneurons that synapse onto these motor neurons and other interneurons with little or no synaptic activity in the cord. M, muscle; HDC, hypodermis. From [1].

are not seen in nonmutant animals and which resemble small electron-lucent SYNAPTIC VESICLES. These observations indicate that in *unc-104* mutants, synaptic vesicles are not normally transported to their site of release.

Consistent with these observations is the finding that the *unc-104* gene encodes a protein with a KINESIN-like motor domain at its N-terminus but with no further homology to other known members of the kinesin superfamily (*see* MICROTUBULE-

BASED MOTORS). The *unc-104* protein may therefore represent a molecule specifically involved in the anterograde transport of synaptic vesicles. The other observed synaptic deficits in *unc-104* (reduction in size and number of varicosities as well as a reduction in the size of the active zones) may be secondary consequences of a decrease of the availability of vesicles. Such a developmental role in synaptogenesis and SYNAPTIC PLASTICITY for the transport and availability of synaptic vesicles is consistent with observations made in other systems.

Axonal pathfinding

The cell bodies of most neurons in *C. elegans* are assembled in ganglia in the head and tail. Only a few neurons are found at various positions along the body. The axonal processes form a nerve ring around the pharynx, as well as longitudinal process bundles joined by circumferentially running commissures. The nerve ring is the main region of neuropil together with the ventral cord which contains the cell bodies of a variety of motor neurons and interneurons involved in locomotion (see Fig. C4).

The problem of how pioneering axons as well as migrating cells find their correct position has been addressed by genetic studies and a number of mutations resulting in guidance defects have been identified. Three genes (*unc-5*, *unc-6* and *unc-40*) have been found to be required for guidance along the circumferential coordinate. *unc-6* is required for guidance in both the dorsal and ventral directions, *unc-5* is required for dorsal guidance and *unc-40* is required for ventral guidance. Mutations in these genes do not affect guidance along the longitudinal axis. One important conclusion arising from these studies is that guidance can be organized in coordinate-specific and direction-specific systems. The gene *unc-6* encodes a protein with partial homology to vertebrate LAMININ subunits (see EXTRACELLULAR MATRIX MOLECULES). The *unc-5* gene product is an integral membrane protein and possesses immunoglobulin- and thrombospondin-like domains. It may represent a receptor for the *unc-6* protein used by dorsally growing axons.

S. HEKIMI

See also: CELL–MATRIX INTERACTIONS; INTEGRINS.

1 White, J.G. et al. (1986) The structure of the nervous system of the nematode *Caenorhabditis elegans*. *Phil. Trans. R. Soc. Lond.* **B314**, 1–340.
2 Brenner, S. (1974) The genetics of *Caenorhabditis elegans*. *Genetics* **77**, 71–94.
3 Hedgecock, E.M. et al. (1987) Genetics of cell and axon migrations in *Caenorhabditis elegans*. *Development* **100**, 365–382.

caeruloplasmin A blue α_2-globulin glycoprotein which contains over 90% of the copper in human plasma. Its precise function is not known but serum levels are reduced in WILSON'S DISEASE.

caesium chloride A caesium salt (M_r 168.36) used to make the density gradient in equilibrium DENSITY GRADIENT CENTRIFUGATION procedures for preparing nucleic acids.

calbindin, calcineurin *See*: CALCIUM-BINDING PROTEINS. Calcineurin is a PROTEIN PHOSPHATASE.

calcitonin A peptide hormone of 32 amino acids, which is produced by the parafollicular (C) cells of the thyroid. Although its precise physiological role is unclear it has been shown to inhibit bone resorption and to decrease renal tubular reabsorption of calcium, sodium, phosphate, magnesium, potassium and some other ions. It also inhibits the secretion of a wide variety of gastrointestinal peptide hormones.

calcitonin gene-related peptide (CGRP) Potent vasodilatory peptide produced by alternative splicing (*see* RNA SPLICING) from the gene for calcitonin. It is released from axon terminals in some parts of the PERIPHERAL NERVOUS SYSTEM.

Calcium

CALCIUM is a divalent metal cation, halfway down Group IIa of the periodic table with an ionic radius of 0.99 Å. This should be contrasted with magnesium which is also a group IIa metal, but with an ionic radius of 0.67 Å so that it has a significantly higher surface charge density. These differences mean that biological systems can usually discriminate between these two cations. Calcium is typically found at concentrations of a few millimolar in extracellular physiological fluids, but the concentration of free calcium inside resting cells is kept at or below 100 nM. Although calcium has a number of important extracellular roles, for example in regulation of blood clotting, or maintaining the structural integrity of biological membranes, it is the intracellular roles which will be reviewed in this article, particularly the means of controlling its entry into cells. The importance of calcium in physiological solutions has been recognized for over a century following the pioneering work of Ringer and Locke. Early work focused on the role of calcium in MUSCLE contraction and secretion (*see* EXOCYTOSIS). More recently it has become obvious that calcium is involved in a whole range of biological processes, ranging from control of membrane permeability to metabolic regulation, cell shape and gene expression.

For calcium to function as an extracellular signal, a number of criteria must be fulfilled. The resting concentration of calcium inside cells must be carefully controlled. There must be a range of entry mechanisms so that this concentration can be increased on appropriate stimulation. There must be a suitable range of calcium-regulated effectors inside cells. Finally, there must be mechanisms for restoring calcium to its normal intracellular levels. These considerations will be discussed in turn.

Calcium entry mechanisms

Calcium can enter the general cytoplasm of a cell either from the extracellular fluid, or by release from intracellular stores. The mechanism used depends on the particular circumstances: the calcium required for NEUROTRANSMITTER release from nerve terminals comes mainly from outside the cell, but in skeletal muscle, twitch contractions use calcium from intracellular stores. Following agonist stimulation of many tissues there is frequently a biphasic rise in intracellular calcium; an initial transient liber-

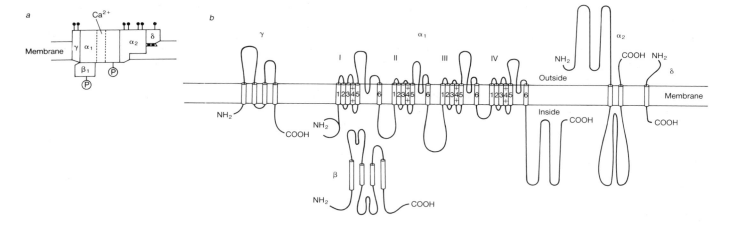

Fig. C5 Subunit structure of skeletal muscle calcium channels. *a*, Model of the subunit structure of the channel derived from biochemical data. P, Phosphorylation sites; ↑, sites of *N*-glycosylation. *b*, Models of the topology of individual subunits in the membrane. Cylinders represent predicted α-helical transmembrane segments as derived from HYDROPATHY ANALYSIS (for α_2, δ and γ) and from a combination of hydropathy analysis and analogy with current structures of Na$^+$ and K$^+$ channels (for α_1). The transmembrane arrangement of α_2 and δ is still tentative. β is a peripheral membrane protein. Adapted from [10].

ation from intracellular stores and then prolonged entry of extracellular calcium. Some of the major mechanisms of calcium release and entry are reviewed below.

Extracellular calcium entry

Voltage-dependent calcium entry. Voltage-dependent calcium entry involves the opening of voltage-sensitive calcium channels (*see* VOLTAGE-GATED ION CHANNELS). Voltage-dependent calcium channels are a heterogeneous group of molecules which may be distinguished both electrophysiologically and pharmacologically. Typically they can be divided into four classes: L, T, N, and P types [1].

L-type channels are activated by high voltage and are modulated by 1,4-dihydropyridines. They are found throughout excitable tissues and are also important in nonexcitable cells. They are the major pathway for calcium entry in cardiac and smooth muscle cells. They can be blocked by calcium antagonists such as verapamil, diltiazem, and some dihydropyridines. Interestingly, their openings are actually promoted by a subset of dihydropyridines — the calcium agonists. Although classed as voltage-dependent channels, their opening can be promoted by a range of conditions. In the heart, the well-known positive inotropic effect of CATECHOLAMINES is due to their actions in promoting the opening of this channel. This is mainly due to phosphorylation by the CYCLIC AMP-DEPENDENT PROTEIN KINASE, but direct modulation of the channels by the stimulatory GTP-BINDING PROTEIN, G$_s$, has also been reported. Calcium-dependent phosphorylation can also cause channel opening.

The structure of L-channels from skeletal muscle is known in

some detail (Fig. C5). Five subunits have been isolated and cloned: α_1, α_2, β, γ, and δ. The α_1 subunit is the largest, with a calculated M_r of 212 000. This is made up of what appear to be 24 transmembrane helices, arranged in four symmetric blocks, a structural arrangement very similar to that of the Na$^+$ channel (*see* VOLTAGE-GATED ION CHANNELS). The α_1 subunit contains the voltage sensor and the binding sites for all classes of calcium antagonists (and agonists). When expressed in the absence of other subunits it can function as a calcium channel. However, complete reconstitution of the calcium channel properties requires coexpression of the other subunits [1].

T-type calcium channels are activated by small depolarizations, in contrast to the L-type channel. They also show rapid inactivation. They are more sensitive to Ni^{2+} than L-type channels, but are rather insensitive to dihydropyridines. They have a widespread distribution, and their clearest physiological function is to support pacemaker activity in the heart.

N-type channels resemble L-type channels in being activated by high voltage, but are insensitive to dihydropyridines. They can be blocked by ω-conotoxin. They are largely restricted to neurons, where in some types at least they are responsible for the calcium entry required for neurotransmitter release.

P-type channels are found in cerebellar Purkinje cells and are distinguished pharmacologically by being insensitive to dihydropyridines and ω-conotoxin, but are blocked by funnel web spider venom. Much less is known of the structure of these channels, and they probably represent molecularly heterogeneous families [1].

Receptor-operated calcium channels. Although L-type voltage channels can be opened by hormones and neurotransmitters, this is an indirect action, relying either on phosphorylation or a G protein. Some channels have binding sites for ligands as an integral part of their structure, and these ligands can directly cause channel opening. One of the most interesting of these is the *N*-methyl-D-aspartate receptor (NMDA receptor, *see* EXCITATORY AMINO ACID RECEPTORS), which responds to excitatory amino acids such as glutamate, and allows calcium entry. This channel can only be opened when the cell has been depolarized — under resting conditions it is blocked by Mg^{2+}. The channel may be important in modulating neuronal activity under a variety of conditions, for example in LONG-TERM POTENTIATION.

An ATP-dependent calcium channel has been described in arterial smooth muscle which shows no voltage dependence, and thus is a pure receptor-gated channel.

Mechanically operated calcium channels. These channels are responsible for mechanotransduction in a number of situations. The channels are usually activated by stretch and are found on skeletal or smooth muscles.

Sustained entry of calcium in response to agonists. In many tissues, after agonists have caused discharge of intracellular calcium stores, they are able to cause a maintained calcium influx from outside the cell [2]. These ligands do not act on receptor-operated calcium channels, and as this influx occurs in nonexcitable cells it cannot occur via depolarization activating voltage-dependent calcium channels. Two possibly complementary mechanisms may be at work to cause this calcium entry. There is evidence that in some cells the mere fact that the intracellular calcium stores are empty is sufficient to trigger calcium entry to refill them. Initially it was thought that this calcium could pass directly to the stores without entering the general body of the cytoplasm, but this is controversial [1]. The nature of the signal sent back from the stores to the plasma membrane is unknown.

Second, agonists themselves may be able to communicate with plasma membrane calcium channels. This could be done through G proteins acting in the plane of the membrane, or indirectly by second messengers such as inositol phosphates (*see* SECOND MESSENGER PATHWAYS). A particular area of controversy is the role of inositol 1,3,4,5-tetrakisphosphate (InsP$_4$), a metabolite of inositol 1,4,5-trisphosphate (InsP$_3$). It has been suggested that InsP$_4$ is essential for the maintained phase of calcium elevation seen in many cells. Although there is some experimental evidence in favour, other workers have been able to produce maintained calcium elevation without the apparent presence of this compound. It is technically difficult to carry out the relevant experiments under realistic physiological conditions, and as a result, there is still no consensus on this issue [1,2].

Intracellular calcium release

Inositol phosphate-mediated calcium release. The chief means whereby hormones and neurotransmitters release calcium from internal stores is the generation of InsP$_3$ by the action of PHOSPHOLIPASE C (PLC) on PHOSPHATIDYLINOSITOL 4,5-BISPHOSPHATE (PtdIns$_2$, PIP$_2$). PLC is activated by a G-protein-mediated mechanism (*see* SECOND MESSENGER PATHWAYS). The InsP$_3$ pathway of calcium release (Fig. C6) is of significance in both excitable and nonexcitable cells. The InsP$_3$ produced rapidly diffuses into the cytoplasm, where it releases calcium from a nonmitochondrial store. The nature of this store is controversial. In some tissues it appears to be closely associated with the ENDOPLASMIC RETICULUM; however, in other tissues, especially neutrophils which contain little endoplasmic reticulum, it has been suggested that a separate organelle, the calcisome, represents the InsP$_3$-sensitive store. This appears to be located just below the plasma membrane. The calcium-release organelle must contain receptors for InsP$_3$, and should also have a calcium pump (CA^{2+}-ATPASE) to allow refilling, and an internally located CALCIUM-BINDING PROTEIN such as calreticulin to act as a calcium store. Unfortunately it is difficult to measure all three of these activities simultaneously. It may be that different cell types use progressively more specialized calcium-release organelles, depending on their function. Thus there may be a continuum from specialized endoplasmic reticulum to distinct calcisome. It is quite clear that in addition to the calcium store releasable by InsP$_3$, cells also contain an endoplasmic reticulum-like calcium store that is insensitive to InsP$_3$ [3,4].

A characteristic of InsP$_3$ calcium release is that it is 'quantal' or 'incremental'. It might be expected that a submaximal concentration of InsP$_3$ would eventually lead to a full release of calcium from the InsP$_3$-sensitive stores, albeit at a much slower rate than a supramaximal concentration. In fact, a submaximal concentration of InsP$_3$ will only release a certain fraction of the total labile calcium pool. This is not due to metabolism of InsP$_3$, and must either mean there is rapid negative feedback leading to inactivation of the release mechanisms, or that the calcium stores show a range of affinities towards InsP$_3$, and these release their calcium

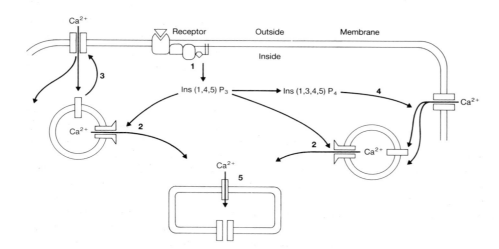

Fig. C6 Schematic representation of inositol phosphate-mediated calcium release. **1**, Agonist–receptor interaction causes hydrolysis of PIP$_2$ to produce inositol 1,4,5-trisphosphate (Ins(1,4,5)P$_3$). **2**, Ins(1,4,5)P$_3$ causes release of Ca^{2+} from internal stores (calcisome, endoplasmic reticulum). **3**, In some cells, the empty stores are able to send a signal (nature unknown) to plasma membrane Ca^{2+} channels, causing Ca^{2+} entry into the cytoplasm and Ca^{2+} stores. **4**, Inositol 1,3,4,5-tetrakisphosphate, a metabolite of Ins(1,4,5)P$_3$, may also send a signal to membrane Ca^{2+} channels, causing Ca^{2+} entry. **5**, Ca^{2+} is removed into vesicles from which it may be released by further Ins(1,4,5)P$_3$ or Ca^{2+}.

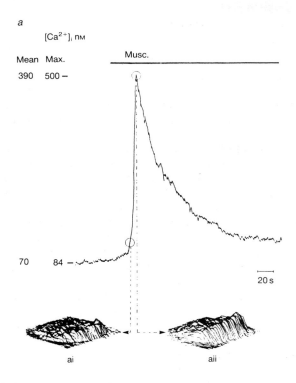

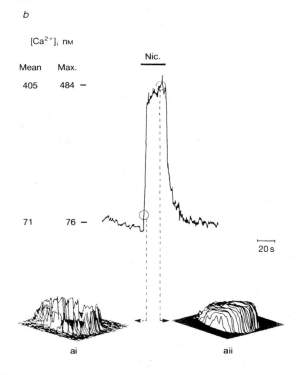

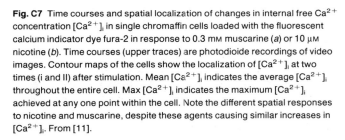

Fig. C7 Time courses and spatial localization of changes in internal free Ca^{2+} concentration $[Ca^{2+}]_i$ in single chromaffin cells loaded with the fluorescent calcium indicator dye fura-2 in response to 0.3 mM muscarine (*a*) or 10 μM nicotine (*b*). Time courses (upper traces) are photodioide recordings of video images. Contour maps of the cells show the localization of $[Ca^{2+}]_i$ at two times (i and II) after stimulation. Mean $[Ca^{2+}]_i$ indicates the average $[Ca^{2+}]_i$ throughout the entire cell. Max $[Ca^{2+}]_i$ indicates the maximum $[Ca^{2+}]_i$ achieved at any one point within the cell. Note the different spatial responses to nicotine and muscarine, despite these agents causing similar increases in $[Ca^{2+}]_i$. From [11].

in an all-or-none manner. A recent survey has concluded that the latter mechanism is the most likely [5]. Quantal release will have the effect of increasing the temporal resolution of any calcium signal; each burst of $InsP_3$ production will be faithfully mimicked by a short burst of calcium release.

The $InsP_3$ receptor itself is a tetramer of M_r 260 000 subunits which has been purified and cloned from cerebellar Purkinje cells. There are neuronal and non-neuronal forms of this protein, derived by alternative RNA SPLICING. It functions as an $InsP_3$-gated calcium release channel. Calcium itself has an inhibitory effect on release, and this requires an accessory protein, cal-medin.

Voltage-mediated calcium release. In skeletal muscle, calcium needed for fast twitch contraction is stored intracellularly in the sarcoplasmic reticulum. This is released by voltage changes transmitted over the surface plasma membrane, which forms specialized T tubules to bring it into close association with the sarcoplasmic reticulum. Calcium release from the sarcoplasmic reticulum can be modulated by the plant alkaloid ryanodine, and this compound has been used as a tool to purify the release

channel. The protein has a subunit M_r of 560 000 with significant homology to the $InsP_3$ receptor. Like the $InsP_3$ receptor, the native protein exists as a tetramer, and functions as a calcium channel. In the electron microscope a 'foot' can be seen extending out from the channel to the T tubules. This seems to interact with the skeletal muscle L-type calcium channels, which act as the voltage sensors, providing the link between membrane depolarization and calcium release. Myotubes from a certain strain of mice that lack normal voltage-dependent calcium release also lack the skeletal muscle L-type calcium channels. Normal release can be restored by TRANSFECTION with the cDNA for this protein.

The role of L-type calcium channels as calcium-transporting molecules in skeletal muscle is unclear, as is the role of the ryanodine receptor in nonskeletal muscle cells. This latter releases calcium in response to both caffeine and calcium itself. These properties characterize a subset of the $InsP_3$-insensitive stores found in a wide range of cells. Calcium-induced calcium release is a feed-forward mechanism of considerable physiological importance in the heart, and may also be important in smooth muscle and neurons [6].

Calcium waves and oscillations. Optical recording of calcium signals from single cells has demonstrated that these frequently show temporal and spatial heterogeneity. For example, when adrenal CHROMAFFIN CELLS are challenged with muscarine, the initial calcium rise is localized to just one pole of the cell. This slowly spreads out in a diffuse wave to the centre, before fading away. This signal is generated by $InsP_3$ production, and is in contrast to the spatial pattern produced by the alternative cholinergic agonist nicotine, which causes calcium entry through voltage-dependent calcium channels (Fig. C7). A typical pattern is for calcium release to be initiated at a single locus, and then spread slowly at between 5 and $100\,\mu\text{m s}^{-1}$. The wave must be propagated by the diffusion of a calcium-releasing substance which takes part in a feed-forward process. The diffusible message could be calcium itself acting to cause calcium-induced calcium release at InsP$_3$-insensitive stores, or it could be InsP$_3$, generated by a calcium-enhanced PLC.

Calcium waves can occur through populations of COUPLED CELLS, thereby synchronizing activity. It has been suggested that in the pancreas a calcium wave is responsible for the coordinated process of secretion followed by Cl$^-$ uptake. In addition to agonist-induced calcium waves, stationary calcium gradients have been reported in resting cells. In growing cultured neurons, a calcium gradient has been correlated with neurite outgrowth. However, static gradients are difficult to demonstrate in many situations [5].

A stage beyond calcium waves are oscillations of intracellular calcium. These have been observed in a wide variety of cell types, with periods varying from seconds to minutes. The frequency of oscillations normally increases with agonist concentration. The oscillations depend on intracellular calcium stores, at least in their early stages. Various models have been proposed to explain the oscillations. In some of these the PIP$_2$ breakdown oscillates, in others the oscillations arise at the level of calcium release. In certain models calcium has inhibitory effects (on either InsP$_3$-induced calcium release or PLC activation). In other models it feeds forwards to cause calcium-induced calcium release from InsP$_3$-insensitive stores or promotes PLC activity. In these latter models oscillations arise from the local exhaustion of releasable calcium leading to periodic cessation of release while stores are refilled. It has been suggested that oscillations allow transmission of activity in a digital form, represented by the frequency of oscillation. This would be a more robust signal than relying on the amplitude of a calcium wave. Calcium-responsive proteins such as PROTEIN KINASES would then act as integrators of this signal. However, the physiological function of many oscillations remains to be determined [1,7].

Calcium-sensitive effectors

A great many processes respond to calcium. Calcium activates a variety of protein kinases, for example phosphorylase kinase, through calmodulin (*see* CALCIUM-BINDING PROTEINS), PROTEIN KINASE C, as well as PROTEIN PHOSPHATASES, for example calcineurin. Calcium can also activate proteases, phospholipases and nucleases. These may be particularly important in programmed cell death — a process in which calcium has a key role. Through activation of kinases calcium can regulate activities such as gene expression, mitogenesis and metabolism. The calcium-induced polymerization of cytoskeletal elements can control cell shape and motility (*see* CELL MOTILITY). Calcium can activate a variety of ion channels controlling membrane permeability to potassium and chloride ions, thus regulating cell excitability. It plays a fundamental if still rather mysterious role in secretion (*see* EXOCYTOSIS). Frequently these actions are integrated to produce a coordinated biological response such as long-term potentiation in the hippocampus [8].

Removal of calcium

During an ACTION POTENTIAL, calcium can enter a cell at about 5 pmole per mg protein per ms. A cell can remove calcium in a variety of ways. In the cytoplasm there is a wide range of soluble calcium-binding proteins, some of which act as diffusible calcium buffers (e.g. parvalbumin in muscles). These can respond very rapidly to changes in calcium but are unable to cope with a sustained calcium load owing to their low capacity. Calcium can be sequestered into intracellular organelles: sarcoplasmic reticulum, InsP$_3$-sensitive and insensitive endoplasmic reticulum and calcisomes. These provide a spatially fixed buffer, and depending on the cell type can promote or hinder the propagation of the calcium signal in any given direction. The calcium is usually taken up into these by a calcium pump, at about 0.1 pmole per mg per ms. Although less rapid to respond than the cytosolic buffers, the intracellular organelles have much higher capacities. Mitochondria are able to accumulate calcium, but their uptake mechanisms are of low affinity (K_{Ca} 5 μM) and so are only likely to be important in pathological conditions.

Ultimately, all the calcium that enters a cell from outside must be removed again across the plasma membrane. As there is a considerable electrochemical gradient favouring calcium entry even at rest, calcium extrusion is a constant process. The cell has two proteins which can remove calcium: the plasma membrane calcium pump, and the sodium–calcium exchanger. The calcium pump of the plasma membrane is distinct from the intracellular pumps: it is regulated by calmodulin, extrudes one Ca^{2+} ion per molecule of ATP hydrolysed and has an M_{r} of 140 000. It has a high affinity for calcium (K_{Ca} 0.2 μM) but a low capacity. By contrast the sodium–calcium exchanger has a K_{Ca} of 0.5–1.0 μM but a much higher capacity. It will normally utilize the Na$^+$ gradient to move calcium out of cells; however at the height of the action potential the Na$^+$ gradient will be such that the exchanger will reverse and bring calcium into the cell. It is likely that in a resting cell, the passive leak of calcium is dealt with by the calcium pump; however following stimulated calcium entry, the sodium–calcium exchanger provides a more rapid way of removing excess calcium [9].

D. POYNER

1 Tsien, R.W. & Tsien, R.Y. (1990) Calcium channels, stores and oscillations. *Annu. Rev. Cell Biol.* **6**, 715–760.
2 Meldolesi, J. et al. (1991) Ca^{2+} influx following receptor activation. *Trends Pharmacol. Sci* **12**, 289–292.
3 Krause, K.-H. (1991) Ca^{2+}-storage organelles. *FEBS Lett.* **285**, 225–229.
4 Rossier, M.F. & Putney, J.W. (1991) The identity of the calcium-storing, inositol 1,4,5-trisphosphate-sensitive organelle in non-muscle cells: calcisome, endoplasmic reticulum, or both? *Trends Neurosci.* **14**, 310–314.
5 Cheek, T.R. (1991) Calcium regulation and homeostasis. *Curr. Opinion Cell Biol.* **3**, 199–205.
6 Ebashi, S. (1991) Excitation–contraction coupling and the mechanism of muscle contraction. *Annu. Rev. Physiol.* **53**, 1–16.
7 Berridge, M.G. (1990) Calcium oscillations. *J. Biol. Chem.* **265**, 9583–9586.
8 Kennedy, M.B. (1989) Regulation of neuronal activity by calcium. *Trends Neurosci.* **12**, 417–420. See also other articles in the same issue (No. 11).
9 Blaustein, M.P. (1988) Calcium transport and buffering in neurons. *Trends Neurosci.* **11**, 438–443.
10 Catterall, W.A. (1991) Functional subunit sructure of voltage-gated calcium channels. *Science* **253**, 1499–1500.

11 Cheek, T.R. et al. (1989) Spatial localization of the stimulus-induced rise in cytosolic Ca^{2+} in bovine adrenal chromaffin cells. Distinct nicotinic and muscarinic patterns. *FEBS Lett.* **247**, 429–434.

---

calcium-binding loop *See*: CALCIUM-BINDING PROTEINS.

Calcium-binding proteins

CALCIUM in the form of its ion Ca^{2+} is essential to the workings of many cellular phenomena (*see* CALCIUM) and many of its effects are mediated through calcium-binding proteins (CaBPs). There is a great variety of these proteins, some with a low affinity for calcium, binding it only at millimolar concentrations, and some of high affinity, binding it in the micromolar to nanomolar range. Outside cells, where Ca^{2+} levels are high (1–5 mM), Ca^{2+} is an essential cofactor for certain adhesion molecules (*see* section on cadherins in CELL ADHESION MOLECULES) and blood coagulation factors (*see* BLOOD COAGULATION AND ITS DISORDERS). Inside cells, in the cytosol, resting free Ca^{2+} levels are extremely low (~50 nM) and small increments of Ca^{2+} perform second-messenger functions (*see* CALCIUM; SECOND MESSENGER PATHWAYS).

There are many intracellular cytosolic CaBPs of moderate to high affinity, matching the physiological Ca^{2+} range; some are signalling proteins, some are calcium buffers, and many are of unknown function.

Most of the cytosolic CaBPs belong to the EF-hand superfamily, so called after the particular calcium-binding domain they contain. Several of the cytosolic CaBPs also associate with cell membranes, either directly or via the CYTOSKELETON. Proteins associated with SECRETORY VESICLES or CHROMAFFIN GRANULES include calpactin (annexin II), α-actinin, calmodulin, and the Ca^{2+}/calmodulin-dependent phosphorylation targets caldesmon and synapsin I (see below and Table C2). All of these may be involved in calcium-dependent EXOCYTOSIS.

Intracellular Ca^{2+} is of central importance in controlling cytoskeletal movement and muscle contraction (*see* CYTOSKELETON; MUSCLE), and many CaBPs are involved in these processes. They include calmodulin, the calpains, and many more of localized importance. Thus several proteins associated with ACTOMYOSIN or with the cytoskeleton have Ca^{2+}-dependent binding properties (including troponin C, spectrin and α-actinin) or are regulated by Ca^{2+}/calmodulin-dependent PROTEIN KINASES (including myosin light chains and caldesmon).

In intracellular organelles are CaBPs of both low and high affinity which contribute to the storage of calcium, notably in the ENDOPLASMIC RETICULUM and in the SARCOPLASMIC RETICULUM of MUSCLE cells (see Table C2). In cell membranes reside calcium pumps (ATPASES and NA^+/CA^{2+} EXCHANGERS).

This account concentrates on the EF-hand CaBPs. The other main groups of CaBPs are given in Table C2. Many CaBPs are abundant and have been identified independently by several laboratories, resulting in a confusing variety of names; the commonest synonyms are given in brackets. Dissociation constants for Ca^{2+} are given to an order of magnitude, but they may be strongly affected by the ionic environment, the presence of protein targets and the temperature. Further information on Ca^{2+}-effector mechanisms and the various groups of CaBPs can be found in [1–3].

The EF-hand superfamily

This is the largest group of CaBPs [4,5]. Almost all its members are cytosolic except for the INTEGRINS and OSTEONECTIN which have variant EF-hand domains, and possible candidates in the voltage-gated Na^+ and Ca^{2+} channels (*see* VOLTAGE-GATED ION CHANNELS). The members with known functions can apparently be divided into 'buffer' proteins which act merely by controlling the cytosolic calcium levels, and 'modulator' proteins which regulate other proteins in a Ca^{2+}-dependent manner. However, for many family members the functions are still unknown.

Members of the superfamily are defined by the presence of Ca^{2+}-binding domains with a helix-loop-helix structure, sometimes also called the calcium-binding loop. They can be recognized by the presence of the CONSENSUS SEQUENCE shown in Fig. C8. The conformation of this sequence is essentially the same in all the superfamily members whose structures have been determined — calmodulin, troponin C, parvalbumin, oncomodulin, and ICaBP-9 [6]. It is named the 'EF hand' from a simple mnemonic model of the structure as formed by the E and F helices of parvalbumin. The central, most conserved region is a loop containing the six calcium-coordinating residues (Fig. C9). Each contributes one oxygen atom from its carboxyl or hydroxyl side chain to coordinate the Ca^{2+} ion, except for the Glu at $-z$ which contributes two oxygens, and the side chain at $-x$ which is often short and bonds the Ca^{2+} via a water molecule. The hydrophobic residues on either side of this calcium-binding loop belong to the two α helices.

These domains almost always come in pairs (see Fig. C9), and the crystal structures show that the members of each pair are linked by β-sheet bonding between their calcium-binding loops, and by hydrophobic interactions between the conserved hydrophobic residues in their α helices. Longer range structure is specific to the individual proteins. The modulator proteins, cal-

Fig. C8 Consensus sequence of the EF-hand domain. O can be I, L, V or M (single-letter amino-acid notation) and a dot indicates a position with no strongly preferred residue. The six residues that coordinate the calcium ion are labelled x, y, z, −x, −y, and −z. This consensus is derived from 99 EF-hand domains which seem 'normal' in that the x and y positions have hydrophilic residues and no insertions or gaps are required to align them. They were taken from the list published in [4] together with the recently published sequences of chicken visinin, bovine recoverin, and human calretinin. Each residue listed is present in >25% of the sequences, and those underlined are present in >80% of the sequences. Many EF-hand proteins include one or more related domains which do not satisfy the criteria given above; some such domains can bind Ca^{2+}, but in many proteins it is not known whether they bind Ca^{2+} or not because binding has not been assigned to individual domains.

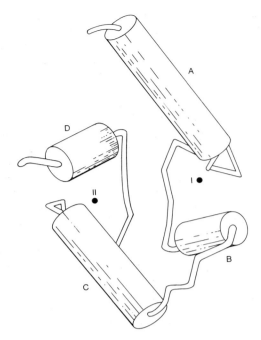

Fig. C9 A diagrammatic representation of a typical Ca²⁺-binding domain. There are two calcium-binding loops — helix A–loop I–helix B and helix C–loop II–helix D — which are related by a pseudo-twofold rotation axis between the loops. Ca²⁺ ions are represented by filled circles. From [6].

modulin and troponin C, undergo a conformational change on binding Ca²⁺, which apparently opens up a hydrophobic cup at the back of the paired calcium-binding domains; this is thought to be the site which interacts with the target proteins of calmodulin. In the crystal structures of parvalbumin and ICaBP-9, which are not known to interact with other proteins, the potential hydrophobic cup is blocked by extra N-terminal or linker sequences.

Gene structures

Unlike, for example, the IMMUNOGLOBULIN SUPERFAMILY in which the common immunoglobulin domain is almost always encoded by a single (or occasionally two) EXONS, the INTRON positions in the genes for EF-hand proteins do not correlate with the repeating pattern of domains. The introns do not map at homologous positions in different domains of a single protein, nor do they coincide in different protein families. Several intron positions are shared between the genes for calmodulin, troponin C, parvalbumin, and SPEC1, but a comparison of vertebrate and arthropod genes shows that there has also been intron insertion and intron deletion within this group. Calbindin-D28 and calretinin share a completely different pattern of introns, and calpain has a different pattern again. The S100 family members all have a shared gene arrangement with a single intron mapping between the two domains.

Members of the EF-hand superfamily

The EF-hand superfamily is continually expanding. The known members include the following proteins, many of which have several variants encoded by separate genes.

Calmodulin. This abundant protein is a ubiquitous mediator of the intracellular Ca²⁺ signal [7] and is found in all eukaryotes including yeast. Its sequence is invariant in mammals and birds and highly conserved in all animals and plants. With an M_r of 17 000 (17K), it has four EF hands, all of which bind Ca²⁺ with affinity 10^{-5} to 10^{-6} M, with some cooperativity and a weak dependence on the magnesium concentration (Fig. C10). On binding Ca²⁺, calmodulin becomes able to interact with a variety of target proteins. Their calmodulin-binding regions do not share a single sequence, but generally include a potential amphipathic α helix with clustered basic amino acids. The most important targets are a set of Ca²⁺/calmodulin-dependent PROTEIN KINASES, which are activated by Ca²⁺/calmodulin binding and have a wide range of substrates. Ca²⁺/calmodulin also directly affects the activity of other proteins, including some forms of CYCLIC NUCLEOTIDE PHOSPHODIESTERASE, ADENYLATE CYCLASE, and CA²⁺ ATPASE; thus it generates extensive 'cross-talk' to the other second-messenger pathways of the cell.

Recently, it has been shown that Ca²⁺/calmodulin binds its target by bending so that the two pairs of EF-hands form a symmetric clamp, with the α-helical target peptide running through the middle [10]. In the case of the Ca²⁺/calmodulin-dependent kinases, the binding serves to remove a PSEUDOSUBSTRATE inhibitory domain from the kinase active site.

There is also at least one protein that preferentially binds calmodulin in its calcium-free form; this is neuromodulin (B-50, GAP-43), a major membrane-bound protein of growing neurites (*see* NEURONAL CYTOSKELETON). Its calmodulin binding is regulated by its phosphorylation by PROTEIN KINASE C, so it could have a role in regulating calmodulin activity in the neuronal growth cone.

Troponin C. This is a similar protein to calmodulin but is restricted to muscle, where it mediates the Ca²⁺ activation of actomyosin through a Ca²⁺-dependent change in the structure of the actin–myosin–tropomyosin–troponin complex. Troponin C has two sites which competitively bind Ca²⁺ (10^{-7} M) and Mg²⁺ (10^{-3} M), and two sites which more weakly bind Ca²⁺ only (10^{-5} M).

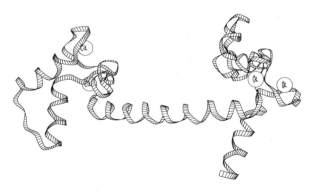

Fig. C10 Calmodulin.

Myosin light chains. MYOSIN includes two types of light chain, called 'essential' (LC-2, DTNB-extractable LC) and 'regulatory' (LC-1 and 3, alkali LC). There are several tissue-specific variants of each, and regulatory LC variants are diversified by alternative RNA SPLICING. All have four EF-hand domains but their function has changed during evolution. In invertebrate myosins, the regulatory LC binds several Ca^{2+} ions and this relieves its inhibitory binding to the myosin heavy chain. In vertebrates, however, the regulatory LC binds only one Ca^{2+} ion and the essential LC binds none; they do control myosin function but only through being substrates for Ca^{2+}/calmodulin-dependent phosphorylation, and in skeletal muscle this is not subject to regulation.

Spectrin and α-actinin. These two homologous cytoskeletal proteins have a pair of EF hands at the C terminus. α-Actinin binds actin via its N-terminal domain; this binding is abrogated in the presence of submicromolar Ca^{2+}, thus releasing actin filaments from bundles. DYSTROPHIN, although homologous to spectrin and α-actinin throughout its length, including the EF-hand domains, has lost most of the conserved Ca^{2+}-binding residues in those domains.

Parvalbumin. Found in fast-twitch muscle fibres, parvalbumin is present at particularly high concentrations in cold-blooded vertebrates whereas smaller amounts are present in mammals. There are several isoforms. Parvalbumin has three EF-hand domains. The affinity for Ca^{2+} is 10^{-7}–10^{-9} M, but Mg^{2+} also binds, competitively, under physiological conditions; the limiting factor in Ca^{2+} binding is deduced to be the off-rate for Mg^{2+} which is $\sim 1\,s^{-1}$. It is thought likely that parvalbumin terminates the muscle contraction by binding sarcoplasmic calcium, but it is unclear whether the time course of $\sim 1\,s$ is sufficient for this function. Parvalbumin is also present in some other tissues, notably brain, where it is found mainly in rapidly firing inhibitory neurons. As there is no evidence that it interacts with any target protein, it is thought that its function may be to soak up the Ca^{2+} that enters the cell during frequent firing. As in muscle, the parvalbumin would then buffer the cytosolic Ca^{2+} at a reduced level, while the Ca^{2+} load would be removed by the higher affinity but lower capacity Ca^{2+} pumps of the plasma membrane and endoplasmic (or sarcoplasmic) reticulum.

Oncomodulin. This close relative of parvalbumin was first identified in tumour cells but is naturally expressed in the very early embryo and trophoblast. It can act *in vitro* like calmodulin to activate cyclic nucleotide phosphodiesterase and calcineurin (see below). In rats, but not in humans, the oncomodulin gene has an unusual promoter and first exon that are provided by a retrotransposon long terminal repeat (*see* TRANSPOSABLE GENETIC ELEMENTS).

Recoverin, S-modulin, and visinin. Found in retinal photoreceptor cells (rods and cones), each of these proteins consists of two normal and two partial EF hands. Recoverin (in mammalian rods) is thought to mediate adaptation of the signal transduction mechanism, as does the homologous protein S-modulin in amphibian rods (*see* VISUAL TRANSDUCTION). In the resting (partially calcium-bound) form, S-modulin inhibits phosphorylation and desensitisation of rhodopsin [11]. Light leads to a delayed decrease in the Ca^{2+} concentration, detachment of S-modulin, and thus reduced sensitivity of the transduction system. Visinin (in cones) and two other proteins (in brain) are closely related.

Calbindin-D28 and calretinin. These closely related CaBPs are found in some neurons [12]. Each consists of six EF hands, but the sixth is aberrant and may not bind Ca^{2+}. It is not known whether they are modulator proteins or merely Ca^{2+} buffers, but the sequences are strongly conserved, more so than those of the Ca^{2+}-buffering proteins parvalbumin or ICaBP-9. The Ca^{2+} affinity is $10^{-6.5}$ M. Calbindin-D28, calretinin, and parvalbumin are found in diverse and largely nonoverlapping sets of neurons. Calbindin-D28 is also abundant in other tissues of some classes of vertebrates, notably the kidney and intestine of birds where it is the vitamin D-dependent 28K CaBP (cholecalcin). It is associated with Ca^{2+} absorption, and is thought to function by increasing the Ca^{2+} capacity of the cytosol so that diffusion of free Ca^{2+} across the epithelial cell is greatly supplemented by diffusion of the calcium-bound calbindin. However, it has also been reported to be associated with vesicular transport. In mammalian intestine, it is replaced by ICaBP-9 (see below).

Calpains. These Ca^{2+}-dependent NEUTRAL PROTEASES are present in many cell types [8]. The heavy chain (isoenzymes I and II) consists of a cysteine protease domain attached to a Ca^{2+}-binding domain which itself has four EF hands. The associated light chain is nonenzymatic but has a very similar Ca^{2+}-binding domain. Calpain is reversibly controlled by a cytosolic inhibitor protein called calpastatin. The native form of calpain is inactive except at high Ca^{2+} concentration, but it becomes translocated to the cell membrane and activated, either by means of an activator protein, or by autoproteolysis. The activated protein operates at Ca^{2+} concentrations down to 10^{-7} M.

There are many potential substrates for calpain, and its roles may be as wide-ranging as those of calmodulin. It is thought to act on many cytoskeletal and membrane proteins, with effects on local and whole-cell activities. It may also aid in degradation of surplus proteins in the cytosol, and is a possible agent of the cell damage that is caused by excessive Ca^{2+} influx.

Sorcin is a widely expressed CaBP very like the calpain light chain, subject to cyclic AMP-dependent phosphorylation, and of unknown function. It is greatly overproduced in MULTIDRUG-RESISTANT cells as its gene is co-amplified with the neighbouring gene that specifies the mechanism of drug resistance.

Diacylglycerol kinase. Another enzyme that includes EF hands (two) in the same polypeptide as the catalytic domain.

Calcineurin B. This is a Ca^{2+}-binding light chain bound to calcineurin A, which is the catalytic subunit of the Ca^{2+}/calmodulin-dependent PROTEIN PHOSPHATASE PP2B, which modulates other protein phosphatases and kinases in neurons. The role of the B chain is unclear.

The two-domain family

This group within the EF-hand superfamily has only been characterized from mammals. Nine members are known, which are abundant in different cell types [2,9]. They are 9–11K in size and have 34–60% homology with each other. Each consists of two EF-hand domains. The second domain is a typical EF hand but the first is variant, with the consensus sequence shown in Fig. C11. In the crystal structure of ICaBP-9, the residues labelled – z and – y coordinate the Ca^{2+} ion as in a normal EF hand, but the residues at the x, y, and z positions coordinate the Ca^{2+} ion only via their backbone carbonyl groups, and the ligand at the – x position is a water molecule.

Several members of this family are known to exist as homo- or heterodimers. Some are known to block phosphorylation of other proteins, and various functions for individual members have been proposed, but the overall function of this abundant protein family has not yet been identified. The members are as follows.

S100α and β.

Two very abundant proteins of GLIAL CELLS, also present in some other cell types. They exist as αα, ββ and αβ dimers. Each monomer binds two Ca^{2+} ions, with $K_d \sim 10^{-4.5}$ M in domain II and $10^{-3.5}$ M in the variant domain I. Binding Ca^{2+} induces the protein to change shape and become more hydrophobic, as with calmodulin. Ca^{2+}-dependent effects include promotion of MICROTUBULE dissociation, and inhibition of phosphorylation of several proteins, including microtubule-associated protein tau. S100 is also found outside cells, for example in the cerebrospinal fluid, and when applied in tissue culture medium, S100ββ promotes neuronal survival and neurite outgrowth. However, the biological significance of all these effects is not yet clear.

MRP-8 and 14 (cystic fibrosis antigen, calgranulins).

These are the most abundant soluble proteins in granulocytes, and are also expressed in macrophages during chronic inflammation. Like S100, they can form a dimer and are reported to inhibit protein phosphorylation. They appear in the serum in patients with CYSTIC FIBROSIS or rheumatoid arthritis. They can be isolated as a complex with 'macrophage migration inhibitory factor', and a fragment of one of them was previously described as a 'neutrophil immobilizing factor'. Again, their true biological roles are unknown.

p11 (42C).

This is a small subunit of annexin II (calpactin, see Table C2), present in many cell types. Whereas the annexin chain (p36) binds Ca^{2+}, p11 does not, having mutations in both the Ca^{2+}-binding loops. The p36 chain is always found in excess. Formation of the native complex, $p36_2.p11_2$, blocks phosphory-

```
E.AO..OO..FH.YS..EGDK..L.K.ELKELO..E
    x   y   z      -y   -z
```

Fig. C11 Consensus sequence of the first EF-hand domain of CaBPs of the two-domain family. O can be I, L, V, or M and a dot indicates a position with no strongly preferred residue. (Each residue shown is present in at least six of the nine known members.)

Table C2 Other main groups of CaBPs

Extracellular CaBPs

FIBRINOGEN; BONE Gla PROTEINS; OSTEONECTIN; THROMBOSPONDIN; INTEGRINS; cadherins (*see* CELL ADHESION MOLECULES); α-LACTALBUMIN. The extracellular CaBPs are a heterogeneous group of proteins

The SERINE PROTEINASES involved in blood coagulation are synthesized as proenzymes with noncatalytic domains of various types preceding the proteinase domain, and two of these domain types bind Ca^{2+}. The first is a small acidic domain that is the most N-terminal in the vitamin K-dependent factors VII, IX, X, prothrombin, protein S, and some other proteins. It includes several γ-carboxyglutamic acid (Gla) residues produced by vitamin K-dependent modification of the precursor protein, and this domain is responsible for the Ca^{2+}-dependence of these factors in blood coagulation. The second Ca^{2+}-binding domain is homologous to epidermal growth factor (EGF), and is found in factors VII, IX, X, XII and protein S. This domain has a different modified amino acid, β-hydroxyasparagine, and at least some copies bind Ca^{2+} with K_d ranging from 10^{-3} M to less than 10^{-7} M

Organellar CaBPs

Calsequestrin

This is thought to provide most of the calcium storage capacity in muscle fibres. It has an M_r 42K, high calcium capacity (40–50 Ca^{2+} per molecule) and low affinity (10^{-3} M). It is found in the terminal cisternae of the sarcoplasmic reticulum, at the enormous concentration of ~100 mg ml^{-1}, and thus could store up to ~100 mM calcium—far in excess of what could be maintained in the free ionic form. Being located in the terminal cisternae, this calcium is stored close to the sites where it must be released. The protein is extremely acidic (net charge – 75), but has no internal repeats or homologies with other proteins; it has one carbohydrate group. It has an elongated form with little secondary structure, and becomes more compact and helical on binding Ca^{2+}

Sarcalumenin

A 160K glycoprotein, this is a very acidic CaBP of the lumen of the sarcoplasmic reticulum

Calreticulin (calregulin)

This is the high-affinity CaBP of sarcoplasmic reticulum, and is also found in nonmuscle cells, apparently in the endoplasmic reticulum. It is a very acidic 55–60K glycoprotein, with no homology to other proteins. The C terminus has the sequence Lys-Asp-Glu-Leu (KDEL) which is characteristic of proteins retained within the endoplasmic reticulum (*see* PROTEIN TARGETING)

Calcium pumps, Ca-ATPases and Ca/Na exchangers

See: Ca^{2+}-ATPASES; CALCIUM; MEMBRANE TRANSPORT SYSTEMS

Intracellular CaBPs other than those of the EF-hand superfamily

PROTEIN KINASE C

ANNEXINS (lipocortins, calpactins, calelectrins)

This is a family of proteins which bind phospholipid and Ca^{2+}. At least seven members are known, sharing a pattern of repeated homologous domains; they have no homology to other families of proteins. At least one (annexin II, calpactin) seems to be involved in Ca^{2+}-dependent vesicle fusion in EXOCYTOSIS. *See also:* ACTIN-BINDING PROTEINS

Gelsolin, severin, and villin

Homologous CaBPs which, in the presence of Ca^{2+}, break up actin filaments. *See:* ACTIN-BINDING PROTEINS; CELL MOTILITY

The different families do not share any homology except where stated.

lation of the p36 chain, and this may perhaps modulate its Ca^{2+}-dependent role in EXOCYTOSIS.

p9Ka (42A, 18A2, Mts1) and calcyclin (2A9). Like S100β these are strongly induced by growth factors and are elevated in growing cells. S100β and calcyclin peak in G1 phase and p9Ka peaks in S phase (*see* CELL CYCLE). p9Ka is particularly abundant in metastatic tumour cells. On the other hand, calcyclin, p9Ka, and p11 are induced in PC12 cells by NERVE GROWTH FACTOR, as part of the induction of neural differentiation in these cells. p9Ka is also induced during the differentiation of mammary myoepithelial cells.

S100-L. This is the most recently discovered member of the 2-domain subfamily. It is abundant in lung and kidney.

ICaBP-9 (calbindin-D9). This is the intestinal vitamin D-dependent CaBP of mammals. It is more divergent than the other family members, and binds Ca^{2+} with higher affinity (10^{-6} M), and does not undergo a large conformational change on binding. It is thought to function in mammalian epithelial Ca^{2+} transport as calbindin-D28 does in birds (see above).

EF-hand proteins in other eukaryotes

All the proteins described so far are from vertebrates, although many of them are also represented in invertebrates. But some notable classes are so far known only from invertebrates and unicellular eukaryotes. They include the following, all of which have four EF-hand domains unless otherwise stated, although some domains are highly aberrant and may not bind Ca^{2+}.

Sarcoplasmic CaBPs are found in many invertebrates — molluscs, annelids, arthropods and protochordates — and are analogous to parvalbumins. Like them, they are abundant in 'fast' muscles, and have variable numbers of sites that usually bind both Ca^{2+} (10^{-7}–10^{-8} M) and Mg^{2+}, and they do not seem to interact with other proteins.

The **Ca^{2+} vector protein** from amphioxus (*Branchiostoma*) is a muscle protein in which domains III and IV resemble calmodulin, and bind two Ca^{2+} with K_d 10^{-7} M, but domains I and II are very aberrant. It binds to a sarcoplasmic protein of the immunoglobulin/titin family, whose function is unknown.

SPEC proteins (from sea urchin) form a diverse family found in embryonic ectoderm. **LPS1** (from another sea urchin) is a unique eight-domain variant.

Aequorin (from a jellyfish) is a luminescent protein that flashes on binding Ca^{2+} and thus can be used experimentally for measurement of intracellular Ca^{2+} changes.

Caltractin is associated with the basal body of the MITOTIC SPINDLE in *Chlamydomonas*. **CDC31** (from yeast) was identified genetically as being required for spindle pole body duplication, which is a calcium-regulated step in the cell division cycle. It is most closely related to caltractin and may be functionally similar.

EF-hand proteins in bacteria

A CaBP with four EF hands has been sequenced from *Strepto-myces erythraeus*. This typical EF-hand protein is the only one yet known from a bacterium.

In the galactose/glucose-binding protein of the periplasmic space of *Escherichia coli* (*see* BACTERIAL ENVELOPE), a single bound Ca^{2+} was discovered by crystallography. The binding site looks like a typical EF hand except that it is unpaired and lacks the second helix, suggesting that it could be a product of convergent evolution.

J.H. ROGERS

1 *Trends Neurosci.* **12**(11), 417–478 (1989). Special issue on calcium-effector mechanisms.
2 Heizmann, C.W. (Ed.) (1991) *Novel Calcium-binding Proteins* (Springer-Verlag, Berlin).
3 Pochet, R. et al. (Eds) (1990) *Calcium-binding Proteins in Normal and Transformed Cells (Adv. Exp. Biol. Med.)* **269** (Plenum, New York).
4 Moncrief, N.D. et al. (1990) Evolution of EF-hand calcium-modulated proteins, I: Relationships based on amino acid sequences. *J. Mol. Evol.* **30**, 522–562.
5 Persechini, A. et al. (1989) The EF-hand family of calcium-modulated proteins. *Trends Neurosci.* **12**, 462–467.
6 Strydnaka, N.C.J. & James, M.N.G. (1989) Crystal structures of the helix-loop-helix calcium-binding proteins. *Annu. Rev. Biochem.* **58**, 951–998.
7 Klee, C.B. & Vanaman, T.C. (1982) Calmodulin. *Adv. Protein Chem.* **35**, 213–321.
8 Melloni, E. & Pontremoli, S. (1989) The calpains. *Trends Neurosci.* **12**, 438–444.
9 Kligman, D. & Hilt, D.C. (1988) The S100 protein family. *Trends Biochem. Sci.* **13**, 437–443.
10 Ikura, M. et al. (1992) Solution structure of a calmodulin-target peptide complex by multidimensional NMR. *Science* **256**, 632–638.
11 Kawamura, S. (1993) Rhodopsin phosphorylation as a mechanism of cyclic GMP phosphodiesterase regulation by S-modulin. *Nature* **362**, 855–857.
12 Baimbridge, K.G. et al. (1992) Calcium-binding proteins in the nervous system. *Trends Neurosci.* **15**, 303–308.

---

calcium channels, calcium channel antagonists *See*: CALCIUM.

calcium phosphate coprecipitation A method for the efficient DNA TRANSFECTION of many types of mammalian cells. The DNA to be introduced into the mammalian cell is mixed with a phosphate solution and, upon the addition of calcium chloride, a fine precipitate of calcium phosphate and DNA is formed. The precipitate is added directly to a monolayer of mammalian cells which take up the precipitated DNA by a process analogous to ENDOCYTOSIS.

calcium polyamines *See*: PLANT WOUND RESPONSES.

calcium pumps *See*: CA^{2+}-ATPASES.

calcyclin *See*: CALCIUM-BINDING PROTEINS.

caldesmon *See*: ACTIN-BINDING PROTEINS.

calelectrins *See*: ANNEXINS.

calgranulins *See*: CALCIUM-BINDING PROTEINS.

Caliciviridae A family of small nonenveloped RNA viruses found in a wide range of mammals, with icosahedral particles of ~35 nm which show cup-shaped depressions in the surface when negatively stained for electron microscopy. The capsids are composed of 180 copies of a single subunit and enclose a single-stranded (+)-sense RNA (M_r ~2.8×10^6) with a VPg protein capping the 5′ end and a poly(A) tail at the other.

callose A β1,3 GLUCAN of higher plants, containing small amounts of glucuronic acid. It is abundant in the sieve tubes and sieve plates of phloem and is synthesized rapidly in response to wounding in many plant tissues. Traces are present in many normal plant cell walls as a component of the HEMICELLULOSE fraction. Callose synthetase is a membrane-bound enzyme which transfers glucose from UDP-Glc onto a primer that seems to be a GLYCOPROTEIN always present in higher plant cells. Cells also contain callose hydrolase, which rapidly degrades callose, but not CELLULOSE, releasing oligosaccharides that can activate callose synthetase without becoming incorporated into callose. The primer for cellulose synthesis may be related to that for callose synthesis and the two glucans are often closely associated: oligosaccharides derived from cellulose also activate callose synthetase. The mechanism of chain termination is not known. *See also*: PARAMYLON; PLANT WOUND RESPONSES.

Bowles, D. (1990) *Annu. Rev. Biochem.* **59**, 873–907.

callus Tissue composed of PARENCHYMA cells and developed as a result of meristematic activity near a wounded surface or in tissue culture. Formerly also used for a deposit of callose on sieve plates.

calmodulin An abundant CALCIUM-BINDING PROTEIN. *See*: CA^{2+}/CALMODULIN-DEPENDENT PROTEIN KINASE.

Calothrix Genus of CYANOBACTERIA within Section IV of the classification of Rippka et al. They are filamentous, capable of nitrogen fixation and show chromatic adaptation (the modulation of synthesis of different phycobiliproteins in response to changes in wavelength of incident light). *Calothrix* PCC 7101 is also known as *Tolypothrix tenuis*, and *Calothrix* PCC7601 is also known as *Fremyella diplosiphon*.

Rippka, R. et al.(1979) *J. Gen. Microbiol.* **111**, 1–61.

calpactins *See*: ANNEXINS. *See also*: ACTIN-BINDING PROTEINS.

calpains *See*: CALCIUM-BINDING PROTEINS.

calpastatin A cytosolic protein inhibitor of the CALCIUM-BINDING PROTEIN calpain.

calponin *See*: ACTIN-BINDING PROTEINS.

calregulin (calreticulin), calretinin, calsequestrin, caltractin *See*: CALCIUM-BINDING PROTEINS.

Calvin cycle Also known as the photosynthetic carbon reduction (PCR) cycle (Fig. C12). Ribulose 1,5-bisphosphate, utilized in the fixation of carbon dioxide by RIBULOSE-BISPHOSPHATE CARBOXYLASE/OXYGENASE (RUBISCO) to form 3-phosphoglycerate, is regenerated through a series of reactions, involving the light-dependent reduction of 3-phosphoglycerate to 3-phosphoglyceraldehyde and its subsequent transformation through a sequence of transferase and isomerase reactions involving C_6, C_4, C_7 and C_5 sugar phosphates. The fixed carbon is released from the cycle to form fatty acids, glycerol, amino acids and carbohydrate, the product depending on the species, stage of cell development and nutrition. The cycle is found in all photosynthetic and chemosynthetic organisms, with the exception of some photosynthetic bacteria.

CAM (1) CELL ADHESION MOLECULE.
(2) CRASSULACEAN ACID METABOLISM.

cambium The MERISTEM which forms the secondary XYLEM and PHLOEM of DICOTYLEDONS, GYMNOSPERMS and some lower vascular plants. It lies as a sheet of cells one or more cells thick between xylem and phloem and divides periclinally, contributing more cells to the xylem usually centripetally and to the phloem usually centrifugally. In woody plants it forms a complete cylinder in stems and roots. Also any similar meristem giving tissue by regular periclinal divisions, such as cork cambium. Formerly also the secondary thickening meristem of certain monocotyledons.

cAMP CYCLIC AMP. For headwords starting with 'cAMP' *see* 'cyclic AMP'.

CAMPATH™ A cell-surface antigen (CD52, *see* CD ANTIGENS) which is strongly expressed on virtually all human lymphocytes and monocytes but is absent from other blood cells. The antigen is a GPI-linked glycoprotein. HUMANIZED ANTIBODIES to this antigen are in therapeutic use.

L-canalin 2-Amino-4-aminooxybutyric acid, $H_2N.O.(CH_2)_2.CH(NH_2).COOH$. Produced by hydrolysis of L-CANAVANINE during germination of certain legume seeds.

L-canavanine 2-Amino-4-guanidooxybutyric acid, $HN : C(NH_2).O.(CH_2)_2.CH(NH_2).COOH$. Typically found in seeds of legumes, especially jack beans (*Canavalia ensiformis*), in which it accounts for up to 4% dry weight, and serves as a nitrogen store. Upon germination, it is broken down with release of ammonia. It is structurally analogous to the AMINO ACID arginine, and hence acts as a competitive inhibitor of enzymatic reactions utilizing arginine. Toxic to cattle.

cancer cells Cells that have become malignantly transformed so that they exhibit uncontrolled growth and invasive properties. Depending on the cell type involved they form a solid tumour (e.g. carcinoma, sarcoma) or cause various types of LEUKAEMIA. Apart from their uncontrolled proliferation and ability to metastasize (i.e. to invade other tissues), there are numerous other differences between cancer cells and the normal cells from which they are derived. They often appear relatively undifferentiated, and ex-

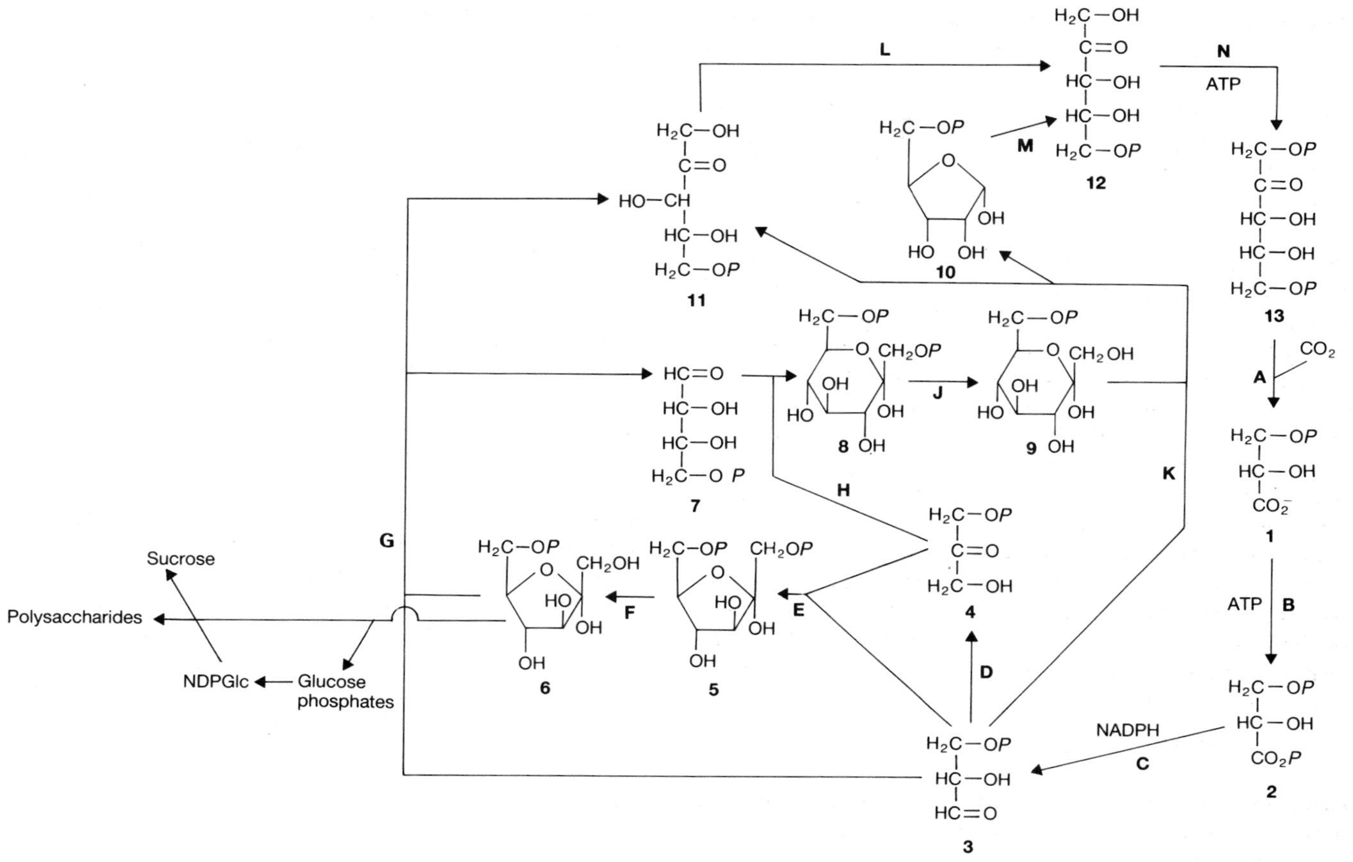

Fig. C12 The Calvin cycle. Key: *Intermediates*: **1** 3-phosphoglycerate; **2** 1,3-diphosphoglycerate; **3** glyceraldehyde 3-phosphate; **4** dihydroxyacetone phosphate; **5** fructose 1,6-bisphosphate; **6** fructose 6-phosphate; **7** erythrose 4-phosphate; **8** sedoheptulose 1,7-bisphosphate; **9** sedoheptulose 7-phosphate; **10** ribose 5-phosphate; **11** xylulose 5-phosphate; **12** ribulose 5-phosphate; **13** ribulose 1,5-bisphosphate. *Enzymes*: **A** ribulose-bisphosphate carboxylase (EC 4.1.1.39); **B** phosphoglycerate kinase (EC 2.7.2.3); **C** glyceraldehyde-3-phosphate dehydrogenase (NADP$^+$)(EC 1.2.1.9); **D** triosephosphate isomerase (EC 5.3.1.1); **E** fructose-bisphosphate aldolase (EC 4.1.2.13); **F** fructose-bisphosphatase (EC 3.1.3.11); **G** transketolase (EC 2.2.1.1); **H** fructose-bisphosphate aldolase (EC 4.1.2.13); **J** sedoheptulose-bisphosphatase (EC 3.1.3.37); **K** transketolase (EC 2.2.1.1); **L** ribulose-phosphate 3-epimerase (EC 5.1.3.1); **M** ribose-phosphateisomerase (EC 5.3.1.6); **N** phosphoribulokinase (EC 2.7.1.19).

press novel tumour antigens at the cell surface. The development of a fully malignant cancer cell is thought to be a multistep process, involving at some stage one or more mutations which alter, deregulate, or delete genes involved in the control of cell growth and differentiation (*see* CARCINOGENS AND CARCINOGENESIS; ONCOGENES; TUMOUR SUPPRESSOR GENES).

Cells of early embryos share several characteristics with cancer cells. Both types of cell possess a high growth rate, both appear relatively undifferentiated, in terms of the appearance under the microscope and in terms of the markers they express, and both are capable of invading other tissues. In addition, embryonic cells

forming tissues, and cancer cells forming tumours, are able to recruit their own blood supply by stimulating local ANGIOGENESIS. These observations have led to the proposal that cancer might be a kind of atavistic cell state, in which cells that ought to differentiate, revert to a permanent embryonic growth phase. Evidence that this view is at least partly justified comes from immunological and gene cloning studies. A number of cell-surface markers known as oncofoetal antigens or carcinoembryonic antigens have been found only on the surfaces of embryonic cells and cancer cells. Second, molecular biologists studying early embryos find that the normal versions of many oncogenes first discovered in mammals are also present in the genomes of organisms as disparate as yeast, DROSOPHILA, and frogs, and that they function during embryogenesis. The protein products of many of these proto-oncogenes have been identified as TRANSCRIPTION FACTORS, GROWTH FACTORS and GROWTH FACTOR RECEPTORS, or molecules involved in intracellular signalling. *See also*: CELL CYCLE; CHROMOSOME ABERRATIONS; TRANSFORMED CELL.

CAP *See*: CYCLIC AMP ACTIVATOR PROTEIN.

5′-cap *See*: RNA PROCESSING.

cap-binding proteins Proteins that bind to the cap structure at the 5′ end of mature eukaryotic mRNA (*see* RNA PROCESSING). *See*: PROTEIN SYNTHESIS.

capacitation Unlike other vertebrate sperm those of mammals are unable to fertilize eggs immediately on leaving the male body and require a period in the female genital tract — capacitation — to acquire this capacity. The details of this process are obscure but it seems to involve modifications to the sperm membrane that are a prerequisite for the ACROSOME REACTION.

capillary blotting *See*: SOUTHERN BLOTTING.

capillovirus group The name of this group of PLANT VIRUSES is derived from the Latin *capillus*, a hair; the type member is apple stem grooving virus. The particles are filamentous and flexuous, about 650×12 nm. They are composed of a single species of $(+)$-strand linear RNA (M_r 2.7×10^6) encapsidated in helically arranged subunits of a single coat protein species (M_r $\sim 27\ 000$). The genome organization has not yet been determined.

Milne, R.G. (Ed.) (1988) *The Plant Viruses*, Vol. 4 (Plenum, New York) 423 pp.

capping (1) Of RNA. *See*: RNA PROCESSING.
(2) Of membrane proteins, clustering of membrane proteins at one pole of a cell that can be induced in cells capable of locomotion by binding of a multivalent ligand such as antibody or lectins to the surface. The cap is always formed at the pole opposite to the CENTROSOME.

capping proteins *See*: ACTIN-BINDING PROTEINS.

capsaicin Chemical compound, N-(4·hydroxy-3-methoxybenzyl nor-*trans*-6-enamide), which is the pungent agent in hot peppers and is toxic for small diameter unmyelinated polymodal nociceptors. It is used in neurobiological research and as a counter-irritant in rubefacient ointments.

capsid The protein coat of a virus particle. It consists of virus-encoded protein subunits of one or several types, packed into a regularly symmetric structure (e.g. icosahedral or rod-like capsids). *See*: ANIMAL VIRUSES; BACTERIOPHAGES; PLANT VIRUSES; SELF-ASSEMBLY.

capsomer(e) Repeating structural unit of viral protein coat, consisting of one or more proteins.

capsule *See*: BACTERIAL ENVELOPE.

CapZ *See*: ACTIN-BINDING PROTEINS.

carbamoyl-phosphate synthetase deficiency AUTOSOMAL RECESSIVE condition due to a deficiency of the urea cycle enzyme carbamoyl-phosphate synthetase (ammonia producing) (EC 6.3.4.16) and characterized by severe hyperammonaemia (excess ammonia in the blood). It is commonly present at birth and the features, which include vomiting, irritability, seizures, and coma, are frequently precipitated by protein-rich feeds.

carbohydrates *See*: EXTRACELLULAR MATRIX MOLECULES; GLYCANS; PLANT CELL WALL MACROMOLECULES; SUGARS and individual entries.

carboxy-terminus The end of a polypeptide chain that bears the free carboxyl group. It is often abbreviated to C terminus (the abbreviation used in this book). Protein biosynthesis proceeds from the N terminus to the C terminus. *See*: AMINO ACIDS; PROTEIN STRUCTURE; PROTEIN SYNTHESIS.

carboxyl proteinase PROTEINASE, also known as acid proteinase, which has a pH optimum of below 5 owing to the participation of an acid residue in the catalytic process (EC 3.4.23). The group includes pepsin and cathepsin D.

γ-carboxylation The post-translational addition of a carboxyl group (—COOH) to the γ-amino group of glutamine residues in certain proteins. *See*: POST-TRANSLATIONAL MODIFICATION OF PROTEINS.

carboxypeptidase Proteolytic enzyme which hydrolyses the peptide bond linking the terminal amino acid at the free C-terminal end of a POLYPEPTIDE CHAIN, thus splitting off a C-terminal residue. EC 3.4.16, the serine carboxypeptidases, which contain a serine at the active site, and EC 3.4.17, the metallo-carboxypeptidases.

carcinogen Any agent — chemical, physical, or viral — that causes cancer. *See*: CHEMICAL CARCINOGENS AND CARCINOGENESIS; ONCOGENES; MUTATION; RETROVIRUSES; TUMOUR SUPPRESSOR GENES.

carcinoma A malignant tumour of epithelial origin.

cardiac muscle Specialized muscle tissue of the mammalian heart which differs in structure from skeletal and smooth muscle. It is composed of muscle fibres made up of cylindrical cells joined end to end.

cardiolipin *See*: MEMBRANE LIPIDS.

cardiovirus Genus of the PICORNAVIRIDAE containing encephalomyocarditis virus and Mengo virus. The RNA genome has a long poly(C) tract near the 5' end.

carlavirus group Sigla from carnation latent virus, the type member. The particles are rod shaped and slightly flexuous, $600 – 700$ nm long and ~ 13 nm in diameter. Each particle contains a single species of $(+)$-strand linear RNA (M_r 2.7×10^6) encapsidated in helically arranged subunits of a single species of coat protein (M_r $32\ 000$). The genome organization has not yet been determined. *See also*: PLANT VIRUSES.

Milne, R.G. (Ed.) (1988) *The Plant Viruses*, Vol. 4 (Plenum, New York).

carmovirus group Sigla from carnation mottle virus, the type member. Single component viruses with isometric particles ($t = 3$), 30 nm diameter. Each particle contains a single species of $(+)$-strand linear RNA which encodes five products (Fig. C13). The 86K (M_r $86\ 000$) and 98K proteins are potential read-through products from the 27K protein. These three proteins have amino-

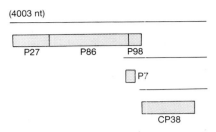

(4003 nt)

P27 P86 P98

P7

CP38

Fig. C13 Genome organization of carnation mottle virus (CarMV). The lines represent the RNA species and the boxes the proteins (P) with M_r given as $\times 10^{-3}$; nt, nucleotides.

acid motifs that suggest an involvement in viral replication. The function of the 7K product is unknown. The 3′ protein is the virion coat protein. The 7K and coat proteins are translated from subgenomic mRNAs. *See also*: PLANT VIRUSES.

Morris, T.J. & Carrington, J.C. (1988) In *The Plant Viruses*, Vol. 3, 73–112 (Plenum, New York).

carotenoid C40 polyene isoprenoid pigments which occur universally in plants, predominantly but not exclusively in PLASTIDS. The two main classes are carotenes, which are hydrocarbons, and xanthophylls which contain hydroxyl groups or epoxide rings and are therefore slightly polar. They usually occur in association with specific proteins. Their functions are varied, including light harvesting in PHOTOSYNTHESIS, protection against damaging effects of light and oxygen, and pigments of fruits and flowers.

carrier Individual HETEROZYGOUS for a RECESSIVE genetic trait. They are themselves unaffected but have the potential to pass the trait on to their children. The carrier may have a subclinical manifestation of the disease which enables them to be identified for the purposes of genetic counselling. For example, females heterozygous for one of the mutations responsible for DUCHENNE MUSCULAR DYSTROPHY have raised levels of a specific muscle enzyme, creatinine kinase.

carrier DNA The efficiency of uptake of specific DNA molecules in TRANSFORMATION or TRANSFECTION protocols can sometimes be increased by mixing the DNA with a nonspecific DNA sample, that is, a carrier DNA.

cartilage *See*: BONE DEVELOPMENT; CARTILAGE DEVELOPMENT; EXTRACELLULAR MATRIX MOLECULES.

cartilage development Cartilage is a tissue formed from mesenchymal cells that secrete large quantitites of collagenous extracellular matrix. Cartilage matrix is deformable, unlike bone, and consists of high concentrations of PROTEOGLYCANS and type II collagen. Cartilage cells (chondrocytes) are isolated from each other and occupy a small space or lacuna in the surrounding matrix. The cartilage is usually surrounded by a perichondrium, made up of specialized fibroblasts, in a matrix of type I and type II collagen and other proteoglycans. In long bone formation an initial cartilage matrix is replaced by bone matrix. *See also*: EXTRACELLULAR MATRIX MOLECULES.

cassette model *See*: YEAST MATING-TYPE LOCUS.

cassette mutagenesis A technique of SITE-DIRECTED MUTAGENESIS in which a stretch of DNA in the target is excised and replaced with a synthetic double-stranded oligodeoxyribonucleotide containing the required mutation.

CAT The bacterial enzyme CHLORAMPHENICOL ACETYLTRANSFERASE.

***cat* gene** The gene encoding CHLORAMPHENICOL ACETYLTRANSFERASE (CAT) and widely exploited as a REPORTER GENE in mammalian cells. The *cat* gene used for such studies was isolated from transposon Tn9.

Gorman, C.M. et al. (1982) *Mol. Cell. Biol.* **2**, 2044–2051.

catabolite activator protein CYCLIC AMP ACTIVATOR PROTEIN.

catabolite repression A phenomenon first described in *Escherichia coli* where, in the presence of glucose, there is a coordinate REPRESSION of the synthesis of enzymes involved in the catabolism of carbohydrates other than glucose (e.g. lactose, galactose, and arabinose). Repression is mediated at the level of TRANSCRIPTION. For the LAC OPERON the mechanism of catabolite repression involves a regulatory protein called CAP (CYCLIC AMP ACTIVATOR PROTEIN), with CYCLIC AMP acting as an EFFECTOR molecule. The CAP protein only binds to the promoters of the responsive operons when complexed with cAMP and repression is exerted by reduction of cellular levels of cAMP in the presence of glucose. This mechanism ensures that the cells utilize glucose as a carbon source in preference to other less efficiently metabolized carbohydrates such as galactose.

catalase Widely distributed oxidoreductase (EC 1.11.1.6) catalysing the breakdown of hydrogen peroxide (H_2O_2) to oxygen and water

$$H_2O_2 + H_2O_2 = O_2 + 2H_2O$$

and which can also use several organic substances such as ethanol and phenols as donors. It is a haem-containing protein of M_r 232 000 comprising four subunits and is part of a cellular mechanism whereby aerobic cells are protected against damage by oxygen free radicals. In mammalian cells it is found in PEROXISOMES.

catalysis Most catalysts for biochemical reactions are proteins (ENZYMES) but some nucleic acids (RIBOZYMES) also have catalytic activity. *See*: MECHANISMS OF ENZYME CATALYSIS.

catalytic RNA *See*: RIBOZYMES.

catecholamine neurotransmitters The monoamine neurotransmitters DOPAMINE, NORADRENALINE, and ADRENALINE, which share a 3,4-dihydroxyphenyl (catechol) structure (Fig. C14). They are synthesized in the neuron from the AMINO ACID tyrosine. The first step in this process is catalysed by tyrosine

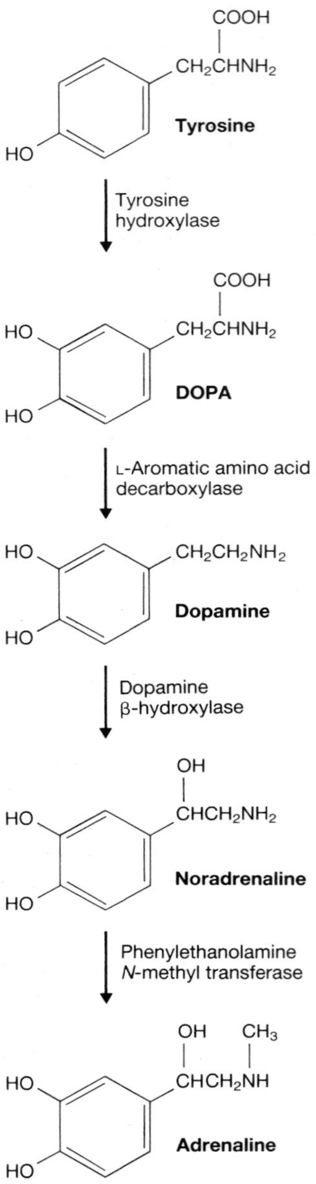

COOH
|
CH$_2$CHNH$_2$

Tyrosine

HO

↓ Tyrosine
hydroxylase

COOH
|
CH$_2$CHNH$_2$

HO

DOPA

HO

↓ L-Aromatic amino acid
decarboxylase

CH$_2$CH$_2$NH$_2$

HO

Dopamine

HO

↓ Dopamine
β-hydroxylase

OH
|
CHCH$_2$NH$_2$

HO

Noradrenaline

HO

↓ Phenylethanolamine
N-methyl transferase

OH   CH$_3$
| |
CHCH$_2$NH

HO

Adrenaline

HO

Fig. C14 Biosynthesis of catecholamine neurotransmitters.

hydroxylase which, being the rate-limiting enzyme in the pathway, is the site for inhibitory control of catecholamine synthesis. Decarboxylation of the resultant DOPA to form dopamine occurs rapidly, and this step is followed in noradrenergic neurons by hydroxylation to noradrenaline. Further *N*-methylation is responsible for the formation of adrenaline in the adrenal medulla and a few brain neurons. Inactivation of the catecholamine neurotransmitters occurs via active neuronal re-uptake or by metabolism via MONOAMINE OXIDASE and/or catechol-*O*-methyl transferase.

catenins Proteins associated with the cytoplasmic domain of cadherins (*see* CELL ADHESION MOLECULES) and mediating interaction with the actin cytoskeleton (MICROFILAMENTS). Three catenins (α, β, and γ) have been described so far (M_r ~102 000, 88 000, and 80 000 respectively). α-Catenin is related to vinculin (*See* ACTIN-BINDING PROTEINS) whereas β-catenin is related to PLAKOGLOBIN and to the product of the *Drosophila* SEGMENT POLARITY GENE *armadillo*. *See also*: CELL JUNCTIONS; DROSOPHILA DEVELOPMENT.

cauliflower mosaic virus (CaMV) *See*: CAULIMOVIRUS GROUP; PLANT VIRUSES; 35S PROMOTER.

caulimovirus group Sigla from cauliflower mosaic virus, the type member. This group has isometric particles (50 nm in diameter) which contain double-stranded DNA and which are found in cytoplasmic proteinaceous inclusion bodies. Each capsid contains a single species of DNA (8 kbp) (Fig. C15) which has single-stranded discontinuities at specific sites, one in one strand (the (–)-strand) and two or three in the other strand. Transcription is asymmetric giving two RNAs, one (1.9 kb, 19S) being the mRNA for the inclusion body protein and the other (8.2 kb, 35S) full length with a terminal redundancy of about 180 nucleotides. The genome encodes six identified products (Fig. C15): that of gene I is the cell-to-cell movement protein, of gene II the aphid transmission factor, of gene III, function unknown, of gene IV the coat protein precursor, of gene V the REVERSE TRANSCRIPTASE, and of gene VI the inclusion body protein; no product has been found for OPEN READING FRAME VII. Replication is biphasic with transcription taking place in the nucleus giving the 8.2-kb RNA which is the template for reverse transcription in the cytoplasm. *See also*: PLANT VIRUSES.

Covey, S.N. (1985) In *Molecular Plant Virology* (Davies, J.W., ed.) Vol. 2, 121 (CRC Press, Boca Raton, FL).

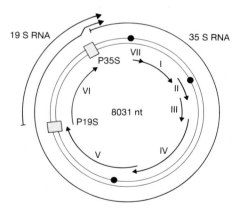

Fig. C15 Genome organization and transcription of cauliflower mosaic virus (CaMV). The two-lined circle represents the double-stranded DNA, the positions of the single-stranded discontinuities shown by solid circles and of promoters by boxes. The two transcripts are indicated outside the circles and the positions of the genes inside the circle.

***Caulobacter* differentiation** *See*: BACTERIAL DIFFERENTIATION.

caulonema The branched part of a moss PROTONEMA with few CHLOROPLASTS and oblique walls. It often has upright branches rich in chloroplasts and it bears the apical cells from which the gametophores (gamete-bearing branches) arise.

CAVD Congenital absence of vas deferens. *See*: CYSTIC FIBROSIS.

CCC CHLORMEQUAT CHLORIDE

CD CIRCULAR DICHROISM.

CD antigens

THE CD (cluster of differentiation or cluster designation) system of nomenclature has been used to standardize the naming and identification of human leukocyte DIFFERENTIATION ANTIGENS recognized by MONOCLONAL ANTIBODIES. This nomenclature is the result of five international workshops (Paris, 1980; Boston, 1984; Oxford, 1986; Vienna, 1989; Boston, 1993) and is recognized by the World Health Organization.

The most important aspect of the CD system is that the nomenclature is derived from the use of monoclonal antibodies recognizing differentiation antigens expressed on the surface of human leukocytes. The CD number (e.g. CD3) refers to a group of monoclonal antibodies (or cluster) with similar properties and has only a chronological basis. The antibodies within such a cluster will generally have the same pattern of reactivity with a panel of haematopoietic cells (*see* HAEMATOPOIESIS), show similar staining in tissue sections, and recognise an antigen of defined molecular mass (M_r) (using either immunoprecipitation and reducing SDS-PAGE (*see* ELECTROPHORESIS) or binding to a cell expressing the appropriate transfected gene). It is this molecule recognized by the cluster of monoclonal antibodies which is loosely termed the CD antigen.

The correct form of the nomenclature is:

CD (w) *N* (a) (R[A]) [Mol. Wt.] [Antigen name][Antibody name]

where CD is cluster of differentiation; (w) is used to indicate that a particular cluster is a 'workshop' designation requiring confirmation; *N* is the number of the cluster; (a) indicates the use of alternative genes (a, b, c, etc.) on different cell types, thereby splitting the cluster; (R) indicates a restricted 'subcluster' due to antigenic modifications during differentiation such as the alternative splicing of exons A, B, C, etc. to give the designation RA, RB, RC; [Mol. Wt] is the relative molecular mass (M_r) of the immunoprecipitated antigen measured by SDS-PAGE under reducing conditions; [Antigen name] is the common, or functional, name of the molecule recognized; [Antibody name] is the name (or clone number) of the individual monoclonal antibody being used to define the CD antigen in a particular application. The parentheses ( ) indicate that the item is not present in all CD names, and brackets [ ] show that the items are optional: neither is written as part of the actual nomenclature. The whole nomenclature is often abbreviated to CDN, which defines the antibody cluster. To avoid confusion it is recommended that the terms CDN antibody and CDN antigen (or molecule) should always be used explicitly.

Cluster analysis of monoclonal antibodies

In order to appreciate fully the part that the CD system has played in broadening our understanding of the structure and function of molecules both on leukocytes and the cells that they interact with, it is useful to summarize the basic principle of cluster analysis as applied in the Leukocyte Workshops. Cluster analysis allows the comparison and simplification of data from more than a thousand different monoclonal antibodies to define fewer than one hundred CD groups. This success is partly due to the wide range of cell types available, from normally differentiated blood and lymphoid cells to leukaemic- and growth factor-initiated monoclonal cell lines *in vitro*, which can be obtained as single cell suspensions. This allows simple indirect fluorescence labelling techniques to be applied, followed by semiquantitative analysis by flow cytometry (*see* CELL SORTING), which gives data usually expressed as the percentage of cells positive above a background threshold. The basic assumption made for the purpose of identifying clusters of monoclonal antibodies against the same differentiation antigen is that each antigen will have a unique pattern of expression, that is, it will be generally restricted to the cell types which actually require the function of that particular molecule. Although this may not be true for all differentiation antigens, it is likely to focus the analysis on those molecules that have a critical functional role at distinct stages in cell maturation and development. Once the monoclonal antibodies have all been tested with as many different cell types as possible, the data are entered into a computer which performs the necessary statistical calculations to compare all monoclonal antibody reactivity patterns with each other and computes some measure of the probability that any two or more antibodies have the same reactivity with all the cell types tested. This forms the basis of a cluster, which can then be confirmed if tissue section staining, immunoprecipitation, and any other available biochemical and functional data are consistent for the majority of clustering antibodies. In many cases the group of monoclonal antibodies that make up a cluster, as well as the identity of the recognized antigen, has been confirmed by identifying, cloning, and expressing the antigen cDNA in transfectants (*see* DNA CLONING). It should be noted, however, that antibodies to CD antigens have often been used to differentiate cells within specific haematopoietic lineages, so that the information on any particular cluster is frequently lineage biased. This is reflected in the complete listing of CD antigens in Table C3, which are grouped according to the perceived lineage reactivity of the antibodies.

CD antigens as complex antigenic molecules

The methods for arriving at CD antigen designations will clearly produce examples of all the different types of cell-surface molecules (see Table C3), but it should be noted that monoclonal antibodies recognize only a small part of most antigens, that is the EPITOPE. Although the CD system generally attempts to define surface differentiation molecules, the discrepancy between the antigenic epitope defined by the CD monoclonal antibodies, and the molecule (or molecules) expressing that epitope, has led to some complications. However, sorting out these complications

Table C3 CD antigens defined by monoclonal antibodies in the 3rd and 4th International Leukocyte Antigen Workshops

Cluster	M_r (reduced) (kDa)	Structure	Function	Main tissue distribution and comments
T-cell biased clusters:				
CD1	43–49	MHC I-like (+ β_2-microglobulin)	Antigen presentation	Cortical thymocytes, some epithelia. CD 1a,b,c,d genes differentially expressed
CD2	45–58	Two IgSF domains + TM	T cell activation signal: ligand is CD58	All T cells, thymocytes and NK cells. CD2R is an activation epitope
CD3γ	25–28	One IgSF domain + TM	T cell receptor complex	CD3 chains are associated with the
CD3δ	20	One IgSF domain + TM	T cell receptor complex	signal transduction functions of the
CD3ϵ	20	One IgSF domain + TM	T cell receptor complex	T cell receptor complex on T cells
CD4	55	Four IgSF domains + TM	Co-receptor for MHC II, HIV receptor	Thymocytes, MHC-II restricted T cells
CD5	67	Three ScR domains + TM	T cell activation signal: ligand is CD72	Thymocytes, T cells, B cell subset
CD6	100–130	Three ScR domains + TM	Not known	Thymocytes, T cells, BCLL, brain
CD7	40	One IgSF, hinge, TM	Not known	T cells at all stages of differentiation
CD8	α32–34, $\alpha\beta$68 unreduced	α, β S-S heterodimer, each with one IgSF, hinge, TM	Co-receptor for MHC I	Thymocytes, MHC-I restricted T cells
CD27	50–55	Homodimer, each with two NGF domains + TM	Not known	T cells, increases on activation
CD28	44	Homodimer, one IgSF + TM	T cell activation signal: ligand is B7 antigen	Most T cells, plasma cells
CDw60	Various	NeuAc2, 8NeuAc2, 3 Gal1, 4	Co-stimulatory signals	T cells, monocytes, platelets
B-cell biased clusters:				
CD10	100	Neutral endopeptidase	Hydrolysis of peptides e.g. fMet-Leu-Phe	CALLA antigen: early B cells, B cell blasts, also various epithelia
CD19	95	Two IgSF domains + TM	Signal transduction	All B cells and precursors but not plasma cells
CD20	33–37	Four TM domains	Possible Ca^{2+} channel	All B cells, not plasma cells
CD21	145	15/16 CCP domains + TM	C3d and EBV receptor	Mature B cells, FDC, some thymocytes, pharyngeal + cervical epithelia
CD22	130/140	Heterodimer of five and seven IgSF domains + TM	Adhesion molecule: ligands CD45R0, CD75	B cell subset (lost on activation), also hairy cell leukaemias
CD23	45	One CL domain, hinge, TM	IgE Fc receptor	Mature B cells, monocytes, macrophages, eosinophils, platelets, and dendritic cells
CD24	35–45	GPI sialoglycoprotein	B-cell signalling?	All B cells (not plasma cells), granulocytes, 2% of thymocytes, neuroblastomas
CD37	40–52	TM4SF	Not known	Mature B cells (not pre-B or plasma cells), also low on some T cells and myeloid cells
CD40	50	Four NGF domains + TM	Signal transduction?	Mature B cells, some epithelial cells
CD72	42	Homodimer CL, hinge, TM	Ligand for CD5	All B cells (not plasma cells), macrophages
CD73	69	GPI-linked	Ecto-5′-nucleotidase	70% B cells, subset of T cells, thymocytes
CD74	43/41/35/33	Four spliced forms, TM	MHC II invariant chain	B cells, monocytes, activated T cells
CDw75	53 + 87	Sialylated glycoproteins	Not known	B cells, epithelia, erythrocyte precursors
CD76	85,67	Sialylated glycoproteins	Not known	B cells (not follicles), melanocytes, epithelia
CD77	Various	Gal1, 4Gal1, Glc1 (Gb$_3$)	Not known	Germinal centre B cells, Burkitt's lymphoma
CDw78	Not known	Not known	Not known	B cells (up with activation), macrophages
NK-cell biased clusters:				
CD16	50–80	Two IgSF + TM or GPI	Aggregated Ig Fc receptor	NK cells and macrophages (TM form), neutrophils (GPI linked form)
CD56	175–185	Isoform of NCAM	NK/target cell adhesion	NK cells, some neural tissues
CD57	Various	Oligosaccharide determined	Not known	NK cells, some T cells, B cells, monocytes, neural cells
Granulocyte biased clusters:				
CD15	Various	3 Fuc-*N*-acetyllactosamine	Ligand for CD62	Granulocytes, monocytes, Reed–Sternberg cells
CDw17	Various	Lactoceramide	Granule function?	Neutrophils, monocytes, platelets
CD35	250	From 3 to 30 CCP domains	C3b + C4b phagocytosis	Neutrophils, monocytes, erythrocytes, etc.
CDw65		Ceramide dodecasaccharide	Neutrophil activation?	Granulocytes, some monocytes
CD66	170–200	Four IgSF domains + TM	Homotypic adhesion	Neutrophils, some myeloid progenitors, colonic epithelium
CD67	100–110	Three IgSF domains + GPI	Not known	Granulocytes

continued

Table C3 *Continued*

Cluster	M_r (reduced) (kDa)	Structure	Function	Main tissue distribution and comments
Monocyte/macrophage biased clusters:				
CD13	150–170	Aminopeptidase N	Regulatory peptide metabolism	All myelomonocytic lineage, renal epithelium, fibroblasts
CD14	53–64	Leucine rich, GPI linked	LPS receptor/signalling	Monocytes, macrophages, some granulocytes and B cells
CDw32	40	Two IgSF domains + TM	FcγRII Ig receptor	Monocytes, spliced B form on B cells, spliced A and C forms on neutrophils
CD33	67	Two IgSF domains + TM	Not known	Monocytes, G + M precursors, not stem cells
CD64	72	Three IgSF domains + TM	FcγRI mono-Ig receptor	Macrophages, monocytes, activated neutrophils
CD68	110	Glycosylated TM protein	Lysosomal membrane?	Macrophages, monocytes, neutrophils, basophils, large granular lymphocytes
Progenitor cell biased clusters:				
CD34	105–120	Unique TM protein	Not known	Immature haematopoietic cells, endothelia
Platelet biased clusters:				
CD9	22–27	TM4SF	Adhesion/signalling	Platelets, early B cells, activated T cells, eosinophils, basophils
CD31	130–140	Six IgSF domains + TM	Not known	Platelets, monocytes, granulocytes, endothelial cell junctions
CD36	88	Not known	Not known	Platelets, monocytes, umbilical endothelium
CD41	125/22	Integrin α chain: heterodimer with CD61	Platelet adhesion	Platelets, megakaryocytes
CD42a	17–22	Leucine rich + TM	Platelet adhesion:	Platelets and megakaryocytes
CD42b	22–25/135–145	Leucine rich + TM dimer	via von Willebrand factor	
CD51	125/24	Integrin α chain: heterodimer with CD61	Platelet adhesion: RGD ligands	Platelets, megakaryocytes, some B cells
CD61	105	Integrin β3 chain:	Adhesion molecule	Platelets, megakaryocytes, macrophages
CD62	140	CL, EGF, nine CCP + TM	Selectin adhesion	Activated platelets, megakaryocytes, also endothelial cells
CD63	53	TM4SF	Not known	Activated platelets, other leukocytes low
Activation antigens:				
CD25	55	α chain of heterodimer, two CCP domains + TM	IL-2 receptor (high affinity)	Activated T cells, B cells, and monocytes
CD26	110	Dipeptidyl peptidase IV	Collagen binding?	Increases with activation on lymphocytes, also epithelial and endothelial cells
CD30	105–120	Five NGF domains + TM	Not known	Mitogen activated T and B cells, plus large cell lymphomas
CD38	45	Small TM glycoprotein	Not known	Activated B and T cells, thymocytes, pre-B cells and plasma cells
CD39	78	Not known	Possible adhesion role	Activated NK, T and B cells
CD69	28,32	S-S linked homodimer	Triggers cytotoxicity?	Activated T + NK cells, thymocytes, platelets
CDw70	Not known	Not known	Not known	Activated B cells, some activated T + monocytes
CD71	90–95	S-S homodimer + TM	Transferrin receptor	All dividing cells, including activated leukocytes
Clusters not biased to any particular lineage:				
CDw12	90–120	Phosphoprotein	Not known	Myeloid cells, platelets
CD11a	180	Integrin α chain: dimer with CD18	Adhesion/signalling: ligands CD54, ICAM-2	Lymphocytes, granulocytes, monocytes, macrophages, increasing on memory T cells
CD11b	170	Integrin α chain: dimer with CD18	Adhesion/signalling: ligands iC3b, matrix, CD54	Myeloid cells, NK cells
CD11c	150	Integrin α chain: dimer with CD18	Adhesion/signalling: ligand fibronectin	Macrophages, hairy cell leukaemia, also other myeloid cells
CD18	95	Integrin β₂ chain	Adhesion/signalling	Most leukocytes
CD29	130	Integrin β₁ chain	Adhesion molecule	Most leukocytes, granulocytes weak, increased on memory T cells
CD49b	165	Integrin α chain: dimer with CD29	Adhesion: collagen/laminin	B cells, monocytes, and platelets

continued on p. 152

Table C3 *Continued*

Cluster	M_r (reduced) (kDa)	Structure	Function	Main tissue distribution and comments
CD49d	150	Integrin α chain: dimer with CD29	Adhesion: fibronectin	Thymocytes, subsets of lymphocytes
CD49f	120/25	Integrin α chain: dimer with CD29	Adhesion: laminin	Memory T cells, thymocytes, monocytes, platelets
CD43	95–135	Mucin-like + TM	Ligand: CD54	Activated B cells plus most leukocytes
CD44	80–95, 130	Cartilage link-like + TM	Hyaluronate adhesion	Many haematopoietic and other cells
CD45	180–240	Large TM glycoprotein	Regulation of signalling?	All leukocytes (not RBC)
CD45RA	200–240	Exon A isoforms of CD45	Not known	Subsets of T and B cells, other leukocytes predominantly naive T cells
CD45R0	180	A, B, C exon-lacking CD45	Not known	Subsets of T and B cells, other leukocytes predominantly memory T cells
CD46	51–58 + 59–68	Four CCP domains + TM	MCP: binds C3b or C4b, protection from complement	All leukocytes (not RBC), epithelia, sperm
CD47	47–52	Not known	Role in cation fluxes?	All haematopoietic cells (including RBC), epithelium, endothelium, sperm
CD48	40–47	Two IgSF + GPI link	Not known	All peripheral blood leukocytes (not RBC), thymocytes, 30% marrow, bronchial epithelia
CDw50	130	N-glycosyl monomer	Not known	Leukocytes, but not RBC or platelets
CDw52	21–28	Small GPI glycoprotein	Not known	All lymphocytes (T, B, NK) and monocytes
CD53	35–42	TM4SF	Possible transport role?	Leukocytes, including osteoclasts, but not RBC
CD54	85–110	Five IgSF + TM	Ligand for β2 integrins	Upon activated leukocytes and endothelia
CD55	60–70	Four CCP domains + GPI	DAF: protection from complement lysis	Most cells in contact with serum
CD58	55–70	Two IgSF + TM or GPI	Adhesion ligand for CD2	Germinal centre B cells, memory T cells, macrophages, endothelia amongst highest
CD59	19	Ly-6 like domain + GPI	Complement protection	Leukocytes (B cells low), broadly expressed on endothelia, epithelia, placenta, etc.

Antigens are classified by their main cellular reactivity, but are often also expressed on other haematopoietic or nonhaematopoietic cells.

BCLL, B-cell chronic lymphoblastic leukaemia; CALLA, common acute lymphoblastic leukaemia antigen; CCP, complement control protein; CL, C-type lectin; EBV, Epstein–Barr virus; EGF, epidermal growth factor; FDC, follicular dendritic cell; GPI, glycosylphosphatidylinositol anchor; HIV, human immunodeficiency virus; IgSF, immunoglobulin superfamily; MHC I, major histocompatibility complex class I; MHC II, major histocompatibility complex class II; NCAM, neural cell adhesion molecule; NGF, nerve growth factor; NK, natural killer; ScR, scavenger receptor; TCR, T cell receptor; TM, transmembrane region; TM4SF, four transmembrane region spanning domain superfamily.

For more detail, see: Barclay, A.N. (1993) *The Leucocyte Antigen FactsBook* (Academic Press, London).

Approximately 50 new clusters have been defined at the 1993 workshop, since this table was prepared.

has often been particularly revealing of how structural changes can be made to modify the functions of cell-surface molecules during differentiation. The following examples exemplify how the CD nomenclature has attempted to resolve the differences between epitope and antigen molecules on the cell surface.

The INTEGRINS are a family of molecules generally made up of a common β chain (e.g. CD18) which can be complexed to one of a number of different α chains (in this example, from three different gene products, CD11a, CD11b, and CD11c). Each unique αβ chain combination is expressed on different haematopoietic cells which gives rise to four patterns of expression which can be recognized by monoclonal antibodies to epitopes on the molecular complex. Antibodies which recognize epitopes on each α chain fall into the following clusters: CD11a, with a molecular complex equivalent to the lymphocyte function associated antigen LFA-1; CD11b which corresponds to the monocyte complement receptor 3 or MAC-1; and CD11c or p150/95 which has been used as a marker for tissue macrophages.

This highlights how many CD antigens also have other names (see Table C4), many of which were based on the name of the monoclonal antibody or the effect that the antibody had when added to some assay of function. Monoclonal antibodies that bind to epitopes on the β chain (CD18) will obviously tend to recognize all cells expressing any member of the β2 integrins, which will approximate the sum of CD11a, CD11b, and CD11c. There is also the potential for monoclonal antibodies to recognize conformational epitopes expressed only on the αβ chain complex, or epitopes shared between the related α chains, and these will not always fall into the recognized clusters.

Finally, antibodies can recognize carbohydrate determinants that modify the protein molecule, but which will often be shared with totally unrelated antigens. The branched pentasaccharide 3-fucosyl-*N*-acetyllactosamine is used by neutrophils to modify various surface molecules, including CD66 and the CD11/18 family. Monoclonal antibodies which recognize this structure cluster by their neutrophil specificity as CD15, but part of their binding includes this carbohydrate epitope on the CD11/18 integrin molecules.

S.P. COBBOLD

Barclay, A.N. (1993) *The Leucocyte Antigen FactsBook* (Academic Press, London).

Table C4 Index of alternative names for CD antigens

3-Fucosyl-*N*-acetyllactosamine	CD15	GPIX	CD42a	MIRL	CD59
3-FAL	CD15	GPIIIa	CD61	MLA1	CD63
AIM	CD69	GPIIb	CD41	MLR-3	CD69
Aminopeptidase N	CD13	GPIV	CD36	Mo-1	CD11b
B1	CD20	H19	CD59	MRC OX-44 (rat)	CD53
B220	CD45RA	Hermes	CD44	MRC OX-45 (rat)	CD48
B4	CD19	HNK-1	CD57	NCA-95	CD67
Ba	CDw78	HTA-1	CD1	NCA-160	CD66
BCM1	CD48	HuLy-m3	CD48	NCAM (isoform)	CD56
BGP-1	CD66	HuLy-m5	CD46	NKH-1	CD56
Ber-H2	CD30	Hutch-1	CD44	Neutral endopeptidase (NEP)	CD10
BLA	CD77	ICAM-1	CD54	OKM-1	CD11b
Blast-1	CD48	Ii or Iγ	CD74	OKM-5	CD36
Blast-2	CD23	In (Lu) related p80	CD44	P-18	CD59
BL-CAM	CD22	Interleukin-2 receptor	CD25	p24	CD9
Bp35	CD20	β1 integrin	CD49/29	p55	CD25
C3b/C4b receptor	CD35	β2 integrin	CD11/18	p85	CD44
C3d receptor	CD21	β3 integrin	CD41/51/61	p150	CD11c
CALLA	CD10	Invariant chain (MHC-II associated)	CD74	P-selectin	CD62
CAMPATH-1	CDw52	Ki-1	CD30	PADGEM	CD62
Ceramide dodecasaccharide 4c	CDw65	Ki-24	CDw70	PECAM-1	CD31
CGM6	CD67	L3T4 (mouse)	CD4	Pgp-1	CD44
CR1	CD35	Lacto-*N*-fucopentaeose III (LNF III)	CD15	Phagocytic glycoprotein 1	CD44
CR2	CD21	Lactosylceramide	CDw17	Platelet 53K activation antigen	CD63
CR3	CD11b	LCA	CD45	P^k blood group	CD77
DAF	CD55	Leu-1	CD5	Protectin	CD59
Dipeptidylpeptidase IV	CD26	Leu-19	CD56	PTLGP40	CD63
DPP IV	CD26	Leu-21	CDw78	P-selectin	CD62
E-rosette receptor	CD2	Leu-23	CD69	Sialophorin	CD43
EA-1	CD69	Leu7	CD57	SRBC receptor	CD2
EBV receptor	CD21	LeuCAMa	CD11a	T1	CD5
ECMRIII	CD44	LeuCAMb	CD11b	T4	CD4
Ecto-5′-nucleotidase	CD73	LeuCAMc	CD11c	T6	CD1
Enkephalinase	CD10	Leukocyte common antigen	CD45	T8	CD8
FcεRII	CD23	Leukocyte common antigen (misuse)	CD18	T9	CD71
FcγRI	CD64	Leukocyte common antigen of trophoblasts	CD46	T10	CD38
FcγRII	CD32	Leukosialin	CD43	T11	CD2
FcγRIII	CD16	Lewis x (Le^x)	CD15	T11.3	CD2R
Globotriaocylceramide (Gb$_3$)	CD77	LFA-1	CD11a	T12	CD6
GMP-140	CD62	LFA-2	CD2	T200	CD45
gp34/28	CD69	LFA-3	CD58	Tac antigen	CD25
gp40	CD7	LNFIII	CD15	TCR	CD3
gp41/35/33	CD74	LO-panB-a	CDw78	TLX	CD46
gp45-70	CD46	Ly-1 (mouse)	CD5	Tp44	CD28
gp50	CD40	Ly-24 (mouse)	CD44	Transferrin receptor	CD71
gp52-40	CD37	Ly-5 (mouse)	CD45	UM4D4	CDw60
gp55	CD14	Lyb-2 (mouse)	CD72	Very late activation antigens	CD49/29
gp67	CD33	Lyt2/3 (mouse)	CD8	VIM2	CDw65
gp80	CD39	MAC-1 (mouse)	CD11b	Vitronectin receptor α chain	CD51
gp100	CD10	MACIF	CD59	Vitronectin receptor β chain	CD61
gp115	CD43	Macrosialin (mouse)	CD68	VLA-β	CD29
gp120	CD26	MCP	CD46	VLA-α	CD49
gp150	CD13	ME491	CD63	W3/25 (rat)	CD4
GPIB	CD42b	Metalloendopeptidase	CD10		

CR, complement receptor; EBV, Epstein–Barr virus; SRBC, sheep red blood cell. See entries in body of encyclopedia for many of the other abbreviations.

CD3 complex Complex of invariant polypeptides associated with the CLONOTYPIC antigen-specific T CELL RECEPTOR (TCR) and which couples the receptor to the intracellular SECOND MESSENGER PATHWAYS. It consists of three polypeptide chains γ, δ, and ε, associated with two other polypeptides ζ and η which occur either as ζζ homodimers or ζη heterodimers in the TCR–CD3 signal-transducing complex. Expression of CD3γ, δ and ε is T-cell specific and their genes are closely linked (on chromosome 11q23 in humans) whereas expression of ζ and η, which occur in other signal-transducing complexes, is not T-cell specific.

CD4 antigen Cell-surface glycoprotein expressed on a subset of T CELLS. It has an accessory function in the recognition by T cells of peptide antigens bound in the cleft of class II MHC MOLECULES. It is a single-chain transmembrane glycoprotein (M_r ~55 000) with a single membrane-spanning region. The structure of the two most N-terminal domains of CD4 has been determined by X-RAY CRYSTALLOGRAPHY, confirming the presence of the immunoglobulin fold (*see* IMMUNOGLOBULIN SUPERFAMILY). The extracellular portion of this single-chain molecule is made up of four immunoglobulin domains, and the membrane-distal, N-terminal domains are thought to bind to the class II MHC molecule in close association with the T CELL RECEPTOR during engagement with the antigen peptide–MHC II complex. This is thought to bring the cytoplasmic domain, which interacts with the lymphocyte-specific TYROSINE KINASE p56^{lck}, into the T-cell receptor signalling complex to amplify the signal for activation. Monoclonal antibodies against CD4 inhibit the function and activation of CD4$^+$ T cells both *in vitro* and *in vivo*, although it is not known whether this is simply a blockade in the formation of the complete TCR/CD4/MHC antigen complex, or a form of negative signalling by cross-linking CD4 molecules away from associated TCR signalling components. The CD4 molecule also acts as the receptor for human immunodeficiency virus 1 (HIV-1) (the virus responsible for AIDS, *see* IMMUNODEFICIENCY VIRUSES).

CD4 is expressed on most thymocytes at the same time as CD8, but as T cells are positively selected in the thymus and are exported to the periphery, they express either CD4 or CD8, but not both (*see* T CELL DEVELOPMENT). In humans and rats, but not mice, CD4 is also expressed on monocytes, although its function on these cells is not known.

Parnes, J.R. (1989) *Adv. Immunol.* **44**, 265–311.
Wang, J. et al. (1990) *Nature* **348**, 411–418.

CD8 antigen Cell-surface glycoprotein expressed on a subset of T CELLS. It has an accessory function in the recognition by T cells of peptide antigens bound in the cleft of class I MHC MOLECULES. On peripheral human T cells CD8 may be present as homodimers (subunit CD8a, M_r 34 000) or as heterodimers or multimers (CD8aCD8β). The structure of the CD8a homodimer has been determined by X-RAY CRYSTALLOGRAPHY, although the majority of cell surface expression is thought to be in the heterodimeric CD8αβ form. However, the α chain homodimers alone are able to reconstitute function on CD8a cDNA transfection into variant T cell clones that have lost CD8 expression, so the role of the β chain remains obscure. The extracellular portions of each chain are made up of a single immunoglobulin domain (*see* IMMUNOGLOBULIN SUPERFAMILY) together with an extended hinge region rich in Pro, Ser, and Thr which has four O-linked glycosylated sites. The membrane-distal, N-terminal domains are thought to bind to the α3 domain of the class I MHC molecule in close association with the T CELL RECEPTOR during the engagement with the antigen peptide–MHC I complex. This is thought to bring the cytoplasmic domain, which interacts with the lymphocyte-specific TYROSINE KINASE p56^{lck}, into the T cell receptor signalling complex to amplify the signal for activation. Monoclonal antibodies against CD8 inhibit the function and activation of CD8$^+$ T cells both *in vitro* and *in vivo*, although antigen recognition by CYTOTOXIC T CELLS often appears to be less CD8 dependent than were their precursors.

CD8 is expressed on most thymocytes at the same time as CD4, but by the time that T cells are positively selected in the thymus and are exported to the periphery, they express either CD4 or CD8, but not both, in a ratio of ~2 : 1 (*see* T CELL DEVELOPMENT). CDα is also expressed to a variable extent on NATURAL KILLER cells.

Parnes, J.R. (1989) *Adv. Immunol.* **44**, 265–311.
Leahy, D.J. et al. (1992) *Cell* **68**, 1145–1162.

CD45 Cluster determinant 45, also designated LCA (leukocyte common antigen), T200, B220, and Ly-5. A family of transmembrane protein tyrosine phosphatases, expressed as major surface components of leukocytes and other nucleated haematopoietic cells, where they may comprise up to 10% of the cell membrane protein. Rat thymocyte LCA contains 118 amino-acid residues. Each member consists of a heavily glycosylated and cysteine-rich external segment, which is probably a ligand-binding domain, a single transmembrane region and an intracellular segment that contains two tyrosine phosphatase catalytic domains which show 40% and 33% amino acid identity to protein tyrosine phosphatase PTP1B (*see* PROTEIN PHOSPHATASES). A SPHINGOLIPID moiety is bound to the extracellular segment. At least eight ISOFORMS (M_r 180 000–220 000), result from alternative splicing of primary RNA transcripts in the region encoding the N terminus and consequent differences in glycosylation which might alter ligand-binding properties. The gene encoding members of the LCA family consists of ~30 EXONS, and LCA heterogeneity results from alternative RNA SPLICING between exons 3 and 7. The isoforms are differentially expressed on leukocyte subsets. CD45 has a role in T-lymphocyte activation and proliferation, where it is essential for the coupling of T CELL RECEPTOR stimulation to the production of the second messenger inositol trisphosphate and subsequent release of Ca^{2+} from intracellular stores (*see* SECOND MESSENGER PATHWAYS). In B lymphocytes, CD45 may be a component of the antigen receptor complex and therefore have a role in antigen-induced activation (*see* B CELL DEVELOPMENT).

Thomas, M.L. (1989) *Annu. Rev. Immunol.* **7**, 339–369.
Trowbridge, I.A. et al. (1991) *Biochim. Biophys. Acta* **1095**, 46–56.

CDC, cdc Cell division cycle genes in SACCHAROMYCES CEREVISIAE and SCHIZOSACCHAROMYCES POMBE respectively. *See*: CELL CYCLE.

CDC31 *See*: CALCIUM-BINDING PROTEINS.

cDNA Complementary DNA. A complementary DNA copy of an mRNA molecule synthesized *in vitro* by reverse transcription. The basic protocol, outlined in Fig. C16, is a three-step process and begins with total cellular RNA. In the first step a DNA strand complementary to the mRNA is enzymatically synthesized using an oligodeoxythymidine (oligo(dT)) PRIMER which anneals to the 3′ poly(A) tail found on most eukaryotic mRNA species (*see* RNA PROCESSING). RNA synthesis is catalysed by the enzyme REVERSE TRANSCRIPTASE. In the second step, the mRNA strands of the mRNA–DNA hybrid are destroyed either enzymatically or chemically. In the final step the full double-stranded DNA molecule is enzymatically generated using *Escherichia coli* DNA polymerase I and exploiting as a primer a HAIRPIN LOOP formed by reverse transcriptase during the synthesis of the first DNA strand. cDNA molecules generated in this way can then be cloned by standard protocols (*see* DNA CLONING). cDNAs represent double-stranded DNA copies of transcribed regions in cells and, if fully mature mRNA is used as the template, they will lack any introns and the promoter and terminator regions associated with the genomic copy of the transcribed region (*see* EUKARYOTIC GENE STRUCTURE). *See also*: DNA LIBRARIES.

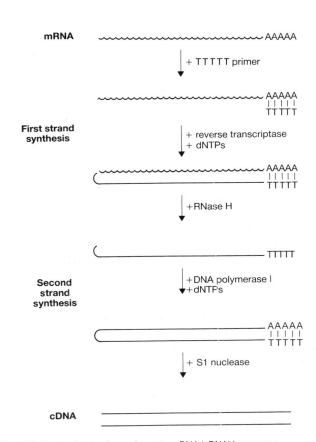

Fig. C16 The synthesis of complementary DNA (cDNA) by reverse transcription.

cDNA library *See*: DNA LIBRARIES.

CDP Cytidine 5′-diphosphate. *See*: NUCLEOSIDES AND NUCLEOTIDES.

CDR COMPLEMENTARITY DETERMINING REGION.

CDR grafting *See*: CHIMAERIC IMMUNOGLOBULIN GENES.

cell *See*: ANIMAL CELL; PLANT CELL; PROKARYOTES; UNIT CELL.

Cell adhesion molecules

CELL–CELL interactions are essential for the correct assembly of tissues during development. It is now clear that cells possess molecular recognition systems on their surfaces that allow them to interact in a specific way with similar and dissimilar cell types and to transduce specific signals across the membrane to the interior of the cell. Those cell surface proteins that are involved in the recognition and adhesion of one cell to another are known generally as cell adhesion molecules. Studies on the central nervous system (CNS) have been of crucial importance in defining and analysing cell adhesion molecules, and this review will concentrate on this tissue.

Two protein families in particular are responsible for specific patterns of cell adhesion in the nervous system. These are the immunoglobulin superfamily [1,2], and the cadherin superfamily [3,4]. The other main families of cell adhesion molecules — the INTEGRINS and the SELECTINS — are dealt with elsewhere in the encyclopedia.

Table C5 Immunoglobulin superfamily adhesion molecules found in tissues other than brain

CD2	Activated T cells	Two Ig domains, the N-terminal one lacking the S—S bond
LPA-3	Different cell types	Two Ig domains, the N-terminal one lacking the S—S bond. Two isoforms produced by RNA splicing, one with a GPI anchor, the other with a transmembrane region
ICAM-1	Inducible on a wide variety of cells by inflammatory mediators, e.g. LPS, interferon-γ, interleukin-1, and tumour necrosis factor	Five Ig domains. A receptor for the integrin LFA-1
ICAM-2	Constitutively produced on endothelial cells	Two Ig domains. A receptor for the integrin LFA-1
VCAM-1	Induced on endothelium by inflammatory mediators	Six or seven Ig domains. Receptor for the integrin VLA-4

Unlike NCAM, these members of the Ig superfamily have no fibronectin III repeats.

The immunoglobulin superfamily

A large number of CAMs in the nervous system and other tissues [5] are members of the IMMUNOGLOBULIN SUPERFAMILY (Table C5). Common to all is the presence of one or more IMMUNOGLO-BULIN DOMAINS in the extracellular region. These domains are thought to be the site of HOMOPHILIC and HETEROPHILIC binding (like-to-like and like-to-unlike binding, respectively). Each immunoglobulin domain is ~75–100 amino acids long and they can be divided into subgroups V, C1 and C2 on the basis of their sequence similarity to immunoglobulin variable or constant regions (*see* ANTIBODIES; IMMUNOGLOBULIN STRUCTURE). All immunoglobulin superfamily CAMs found in the brain also contain domains with sequence similarity to the protein fibronectin (the fibronectin type III repeat) (*see* EXTRACELLULAR MATRIX MOLECULES). Unlike the cadherins they operate by a calcium-independent mechanism.

The main structural features and family relationships of known immunoglobulin superfamily CAMs in the brain are shown in Fig. C17.

The neural cell adhesion molecule (NCAM)

The neural cell adhesion molecule (NCAM) (Fig. C17) is perhaps the best characterized CAM in the nervous system and operates via homophilic binding (sometimes called homotypic binding), that is, it binds to NCAM molecules on other cells. Like all immunoglobulin family CAMs found so far in the brain, it exists in several variant forms — isoforms — which vary in size, structure and method of attachment to the membrane. NCAM has no homologues in vertebrates, and the only other homologue found in the nervous system is the fasciclin II protein from *Drosophila*.

The largest NCAM isoforms of M_r 180 000 (180K) and 140K are transmembrane proteins, the smaller 120–125K forms are

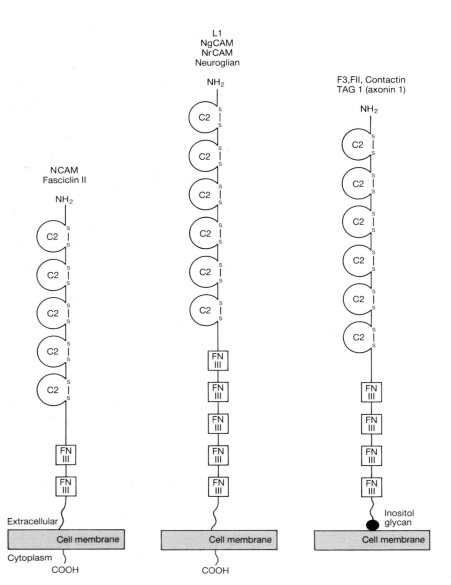

Fig. C17 Schematic diagram showing the subfamilies of the immunoglobulin (Ig) superfamily that are involved in recognition events in the nervous system. Three subfamilies have been characterized on the basis of the number of immunoglobulin domains and fibronectin repeats. No attempt has been made to represent the GPI form of NCAM which has the same basic structure in the extracellular domain as the transmembrane form shown.

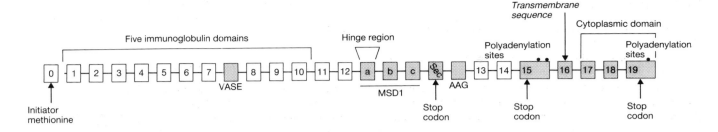

Fig. C18 Diagram showing the intron–exon structure of the NCAM gene. 26 exons have been identified and their relative positions are marked. In the diagram no attempt has been made to map the distances between exons accurately as these differ between species. The exons are numbered from the 5′ end of the gene (exon 0) to the 3′ end (exon 19). Alternative exon usage is indicated by shading. Exons 1–10 are responsible for the five immunoglobulin (Ig) domains; the hinge region is found in MSD1a; the GPI anchor in exon 15; transmembrane sequence in exon 16; and the cytoplasmic domain in exons 17–19. The MSD1 region are exons that were originally found as a unit in muscle and the VASE exon is an alternatively spliced exon in Ig domain 4.

attached to the cell membrane via a glycosylphosphatidylinositol anchor (*see* GPI ANCHORS) whereas the smallest isoforms of ~115K are soluble. There is also extensive microheterogeneity due to alternative splicing of the NCAM gene (*see* RNA SPLICING) and also differential GLYCOSYLATION.

Gene and protein structure. The NCAM gene is large and the EXONS comprising the coding region cover more than 50 kilobases (kb) of genomic DNA (Fig. C18). A total of 26 exons have now been identified and these are used in different combinations, by alternative splicing, to generate the different NCAM size classes. The five immunoglobulin domains on NCAM are encoded by two exons each. As the immunoglobulin molecule itself has only a single exon per domain this suggests that immunoglobulins may have evolved from an NCAM-like structure. It is not known why so many recognition molecules utilize the immunoglobulin domain but it may be a structure that is resistant to proteolysis. This could be a major advantage in events that occur at cell surfaces.

There is little structural variation between the five immunoglobulin domains of NCAM. This uniformity may be important as the immunoglobulin domains are believed to mediate homophilic binding and it is therefore likely that a common binding structure is required.

The domains responsible for NCAM homophilic interactions are not known but experiments using anti-immunoglobulin domain antibodies to block function suggest domain 3 is important.

Extensive alternative splicing occurs in the exons between the first and second fibronectin repeats. The five exons in this region are associated with at least eight individual splicing pathways. The region is called the MSD (muscle-specific domain) region as the first three exons were originally found in skeletal muscle NCAM.

The amino-acid sequence of the MSD region is unusual. The first exon encodes a string of proline residues whereas the other two are rich in serine and threonine. This has two consequences. The first is that the proline residues put a bend or kink in the

NCAM protein which could assist homophilic binding. Secondly, the serine and threonine residues are sites of O-linked carbohydrate attachment (*see* GLYCOPROTEINS AND GLYCOSYLATION). This has been analysed in most detail in muscle where the O-linked carbohydrate has been isolated and sequenced and has been found to be a sialylated form of the disaccharide galactose $\beta(1,3)$-N-acetylgalactosamine. The function of the O-linked carbohydrate may be to act as a spacer to increase the distance of NCAM immunoglobulin domains from the plasma membrane and put NCAM in a more favourable location to be involved in specific interactions. This may be important in cells such as skeletal muscle which have a particularly thick extracellular matrix.

Another exon in this region, SEC, was also first identified in muscle. When used, this exon has the effect of introducing a TERMINATION CODON into the READING FRAME, closing it prematurely and thus generating the short soluble NCAM isoform. The final exon in this region is unusual in being only three base pairs (AAG) long and when used changes an Arg to Gln-Gly. Its usage is tightly regulated.

An exon called VASE has been found between exons 7 and 8. When used, it inserts a sequence of 10 amino acids with a unique structure, which is found in an exposed region of the NCAM protein at the cell surface. The region is equivalent to a hypervariable region in an immunoglubulin molecule. The function of this region seems to be to shift NCAM from a protein that can function in cell adhesion and cell signalling to a form that functions in adhesive phenomena only. The high abundance of this form in the adult but not developing nervous system backs up this idea. It may also possibly alter the carbohydrate that is attached at a site near its insertion.

Membrane attachment. Downstream of the fibronectin repeats are the exons that dictate the different mechanisms of membrane attachment. NCAM is found either spanning the plasma membrane or attached to its outer face by a GPI anchor. The signals for the GPI anchor are contained in exon 15 which encodes a C-terminal region of 27 amino acids.

The use of exon 15 is tightly regulated and is likely to be associated with differing needs for transmembrane signalling. In cells such as skeletal muscle myotubes where there is a need for NCAM-mediated adhesion but not signalling the GPI anchor is used.

Transmembrane NCAM isoforms are generated by a different splicing pathway that uses exon 16, which encodes a transmembrane region, instead of exon 15. Combinations of exon 16 and other downstream exons generate the different transmembrane isoforms. Isoform 180K utilizes exons 16–19; the deletion of exon 18 generates the 140K transmembrane isoform. Insertion of exon 18 is highly regulated and is generally used by post-mitotic neurons only.

Glycosylation. Differences in carbohydrate side chains are also important in NCAM function. There are six putative sites of N-linked carbohydrate attachment on NCAM but which sites are used is not known. The carbohydrate is of the high-mannose core type but the terminal sugars are quite heterogeneous.

NCAM is an unusual glycoprotein in that under certain circumstances it contains long chains of the sugar polysialic acid (PSA) as a result of expression of the enzyme polysialyltransferase in the cell. These PSA chains can be up to 200 residues long and seem to be a key positive modulator of NCAM function. PSA expression is required for the growth of neurons during development. NCAM rich in PSA still binds in a homophilic manner but less strongly than NCAM low in PSA. NCAM from adult brain, where maximum binding strength is required, is low in PSA, whereas embryonic brain NCAM is high in PSA, corresponding to the time of maximum neuronal plasticity, where cell–cell connections are being broken and reformed and cell signalling is important.

L1 protein

The L1 protein (NILE in the rat and G4 in the chicken) is a CAM found on axon tracts in the nervous system and is involved in axon fasciculation (the bundling of axons into nerve fibres). It is a transmembrane protein of ~200K, composed of six immunoglobulin domains and five fibronectin type III repeats and with a large cytoplasmic domain (see Fig. C17). Ng-CAM from the chick is the product of a related gene. Other related proteins include the Nr-CAM from chick brain and neuroglian from *Drosophila*.

L1 appears to bind in both a homophilic and, under certain circumstances, a heterophilic manner. The molecular partner in the heterophilic interaction is believed to be the TAG 1/axonin 1 protein (see below). As NCAM and L1 colocalize at many sites it

has been suggested that they may physically interact but it is clear that each can operate in the absence of the other.

Other groups

A third structural subgroup of the immunoglobulin superfamily CAMs in the brain is involved in axonal growth and fasciculation. The first member to be identified is called variously F11, F3 or contactin (by different groups who isolated cDNA clones independently using EXPRESSION CLONING strategies). All three isolates have six immunoglobulin domains and four fibronectin repeats but whereas F11 and F3 are GPI-anchored, contactin seems to have a transmembrane domain. Its binding specificity has not yet been determined but it may mediate heterophilic binding.

The second glycoprotein in this subgroup is a transiently associated glycoprotein (TAG), whose structure has also been elucidated from cloned DNA. TAG 1 has been found on commissural neurons near their projection to the midline of the spinal cord and may be similar to a protein called axonin 1. TAG 1 has six immunoglobulin domains and four fibronectin repeats. It seems to be associated with the cell membrane by a GPI anchor and is also found in soluble form.

It seems likely that TAG 1 has both a homophilic and a heterophilic binding mechanism.

The cadherin superfamily

The cadherins [3,4] are adhesion molecules that operate via a calcium-dependent binding mechanism and are involved in the formation of strong cell–cell bonds. Strong bonding is of great importance in the generation of cellular specificity in cell sorting and in the formation and maintenance of CELL JUNCTIONS. Different cadherins are important in these events.

Three members of the cadherin family have so far been described in detail although there is evidence of a number of new cadherins. The three main family members are N-cadherin, which is expressed in brain, skeletal muscle, and cardiac muscle; E-cadherin (uvomorulin) in liver and other epithelial tissues (*see* EPITHELIAL POLARITY); and P-cadherin in placenta and mesothelium. All three are believed to be functionally important because

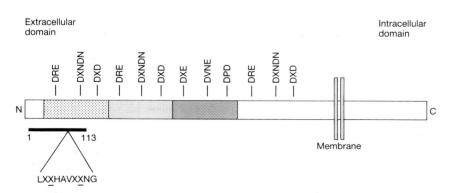

Fig. C19 Schematic diagram showing the structure of N-cadherin as a representative of the family of cadherins. Cadherins are generally >700 amino acids long with the bulk of the sequence in the extracellular region. There is a transmembrane region and a large cytoplasmic region that provides a linkage to the cytoskeleton. Repeating domains in the extracellular region are indicated by different shading and labelled with the type of repeat (in single-letter amino-acid notation). Within these domains there is evidence of duplication of specific sequence motifs although whether this arises by exon duplication is not known. The N terminus of the protein contains one of the sites of homophilic binding. This is modelled around the amino-acid sequence HAV at about amino acid 100. The other site of binding is not known. Further details can be found in [4].

of their very precise patterns of expression in individual tissues and because perturbation of their expression has drastic results.

Ectopic expression of N-cadherin in the nervous system via injection of N-cadherin messenger RNA, for example, leads to altered cell morphology and disorganization of the neural tube and other cell areas expressing N-cadherin.

The cadherins are in general transmembrane proteins of about 124K and seem to be the products of different genes. Ca^{2+} is required for their function in that it binds to the extracellular domain of the cadherin and stabilizes its structure. In its absence cadherins are rapidly degraded from the cell surface. Like NCAM the cadherins bind by a homophilic mechanism. The binding specificity is quite rigid and there is little evidence of the major cadherin species binding to each other or to other proteins via their extracellular domains.

The main features of cadherin structure are shown in Fig. C19. There is extensive sequence similarity in different species and between different cadherins, especially in the cytoplasmic domain; in chick and human N-cadherin there is only a single amino acid difference in this region. This high degree of similarity perhaps reflects a common function. Cadherins show no evidence of differential splicing or differential glycosylation.

A class of cytoplasmic proteins called the CATENINS is able to interact with the cadherins and may mediate their interaction with the CYTOSKELETON. One of these catenins was recently identified as a vinculin homologue, a 120K protein known to be associated with cell junctions, and which is believed to mediate interactions with the cytoskeleton.

Binding specificity. An N-terminal region of 113 amino acids is involved in mediating homophilic interactions and is also a site where blocking antibodies bind. A common cell adhesion recognition sequence modelled around the tripeptide His-Ala-Val (HAV) is important and will block cadherin-mediated interactions in a number of systems including neurite outgrowth. The binding specificity of cadherins is so precise that in gene transfer experiments it has been found that cells synthesizing different cadherins will sort out one from another. These studies emphasize the strong cell—cell bonds that these molecules can generate.

The second binding site on cadherin that specifies homophilic interactions is not yet known.

N-cadherin has an important role in a number of recognition events such as myoblast fusion and axonal growth. N-cadherin antibodies will block myoblast fusion although other CAMs such as NCAM are clearly also involved in this process. In terms of neurite outgrowth N-cadherin seems to be important in growth over a number of complex substrata. Gene transfer experiments have also shown that N-cadherin can promote neurite outgrowth [6]. In both N-cadherin- and NCAM-mediated neuronal growth intracellular SECOND MESSENGER PATHWAYS are involved consequential to the initial adhesive event. N-cadherin purified from chick brain is also active in supporting neurite outgrowth when coated on a culture dish. These model systems have been invaluable in defining some of the basic parameters associated with cadherin function and are likely to be of great value in further structure–function studies.

F.S. WALSH

See also: CELL–CELL INTERACTIONS; CELL–MATRIX INTERACTIONS; CELL JUNCTIONS.

1 Walsh, F.S. & Doherty, P. (1991) Structure and function of the gene for neural cell adhesion molecule. *Semin. Neurosci.* **3**, 271–284.
2 Rathjen, F.G. & Jessell, T.M. (1991) *Semin. Neurosci.* **3**, 297–307.
3 Ranscht, B. (1991) *Semin. Neurosci.* **3**, 285–296.
4 Takeichi, M. (1991) Cadherin cell adhesion receptors as morphogenetic regulators. *Science* **251**, 1451–1455.
5 Springer, T.A. (1990) Adhesion receptors of the immune system. *Nature* **346**, 425–434.
6 Doherty, P. & Walsh, F.S. (1991) *J. Cell Sci.* Suppl. **15**, 13–21.

cell-autonomous Applied to developmental genes or mutations whose immediate effects are restricted to the cell in which they are expressed.

cell body *See*: NEURON.

Cell–cell interactions in development

During the development and growth of animals and plants, cells become differentiated from each other through the activation of specific batteries of genes, and organize themselves in space and time to generate the morphological patterns that are the hallmark of every species. At any given moment the state of each cell, that is, the genes it is expressing, and the associated programme of protein modification, cell proliferation, differentiation, or death, depends on two variables. The first is an internal and to a certain extent autonomous pattern of gene expression; the second is the ability to receive and interpret signals from surrounding cells and, to a similar degree, to emit signals about its own state. The resulting exchange of information provides a constant feedback into the 'autonomous' programme and is commonly termed 'cell communication' or 'cell–cell interaction'.

Cell interactions have fundamental roles in many biological processes, from FERTILIZATION to the functioning of neural networks and the immune system, and are the driving forces that pattern cells during development and differentiation (*see* INDUCTION; PATTERN FORMATION). The importance of cell interactions in development was first clearly illustrated by the experiments of H. Driesch on early sea urchin embryos. He found that individual blastomeres of a sea urchin embryo separated at the two or four blastomere stage each produced a small complete embryo instead of the half or quarter embryos that might be expected if the development of individual blastomeres were completely autonomous. This experiment indicated that in the embryo, the development of each blastomere must be being influenced by the other cells. The ability of the cells to 'regulate' their behaviour when isolated and to produce whole animals instead of the parts they would have produced had they remained in their original place demonstrates a most important property of most developing systems. Interactions between cells allows the exchange of information about cell states and thus allows the coordination of these states during growth and differentiation that is essential to produce a multicellular organism. It is very likely that the molecular mechanisms that underlie cell interactions are the key elements of the homeostasis central to all biological systems during embryonic development.

The two essential components of a cell communication system are signals and RECEPTORS; in addition, there must be a mechanism to link the signalling event to appropriate effectors (*see* SECOND MESSENGER PATHWAYS). The signals vary from small molecules such as nucleotides (e.g. cAMP), to peptides (some neurotransmitters and hormones), to larger polypeptides and proteins (like growth factors) of varying size and composition. In addition, by-products of cellular metabolism such as steroids and retinoids are used as signals. Any of these molecules can be secreted by cells and thus may have long range effects by reaching distant cells. Alternatively, the signal can be anchored to the producing cells and act locally.

There is also a great molecular diversity amongst receptors. Many receptors are proteins anchored to the cell membrane which contain an external domain dedicated to the binding of the ligand/signal and an intracellular domain dedicated to relaying the signal elicited by ligand binding to signal transduction networks located inside the cell. A well studied group of such signalling receptors is the receptor protein tyrosine kinase family in which ligand binding activates a kinase activity contained in the intracellular domain which, in turn, modulates intracellular signal transduction pathways (*see* GROWTH FACTOR RECEPTORS). In other receptors the cytoplasmic domain does not have a signalling ability itself (in terms of an enzymatic capacity), but is coupled to other molecules with such potential, for example nonreceptor PROTEIN KINASES of the *src* gene family. The effectors of signalling events are, in general, proteins located in the nucleus or other subcellular compartments like the cytoskeleton or the secretory pathway. In order to reach these targets, the signal triggered by ligand binding to the receptor is relayed through signal transduction pathways which can be modulated in exquisite detail. In addition to membrane receptors, there are also nuclear receptors represented by DNA-binding proteins which become activated upon binding of the ligand; well-characterized examples of these are the STEROID RECEPTOR SUPERFAMILY. In this case the signalling molecule elicits a response by directly activating the TRANSCRIPTION of particular genes.

The basic molecular tool kit described above is used by cells to coordinate their behaviour by a variety of strategies that can be grouped into two main kinds:
1 Interactions between equivalent cells; and
2 Interactions between nonequivalent cells.
The latter can be further subdivided into interactions between neighbouring or distant cell populations.

Interactions between equivalent cells

In this situation, individual elements of an otherwise homogeneous cell population become different without the input of a signal from an external source (a cell or cell population). The process is usually initiated by the activation of a gene product (A), which might be (but need not be) a TRANSCRIPTION FACTOR, in all cells of the population. The activity of this product fluctuates

randomly and, above a certain threshold, may trigger the differentiation of a particular cell through the activation of specific batteries of genes. When the concentration of A in a cell reaches this threshold, it triggers a signal that results in the suppression of the amplification of A in the surrounding cells. This operation leads to the suppression of the development of this new phenotype in those cells in which the concentration of A had not reached the threshold for differentiation and thus ensures that only one cell of the group will acquire this fate.

In *Drosophila* neurogenesis cell interactions of this kind generate a stereotyped and evenly spaced pattern of neural precursors within an otherwise homogeneous sheet of ectodermal cells. In this particular case two proteins have been identified with important roles in the processing of information: Notch and Delta (*see* INSECT NEURAL DEVELOPMENT). These proteins contain multiple copies of a particular motif also present in the vertebrate epidermal growth factor (EGF repeats, *see* GROWTH FACTORS) which have been shown to be important in mediating their physical interaction in cultured cells. The data available at the moment indicate that concentration-dependent interactions between these two proteins, with Notch as a receptor for some inhibitory signal that depends on Delta, determine the fate of individual cells. The proteins that implement neural differentiation are a family of helix-loop-helix containing proteins (*see* PROTEIN-NUCLEIC ACID INTERACTIONS). Thus, in the general model discussed above, the proteins of the ACHAETE–SCUTE COMPLEX are one example of the fluctuating variable, Delta is the signal that will inhibit their transcription in cells other than the one that is becoming a neural precursor, and Notch is a receptor for this signal.

Theoretical models have been particularly successful at modelling the emergence of phenotypically different cells from a homogeneous population. The first was put forward by A. Turing and provided a tremendous stimulus for the understanding of the chemical basis of patterning in biological systems. More recently the Turing model has been extended to and integrated with modern phenomenology and problems through the work of A. Gierer and H. Meinhardt. Unfortunately although these models show that, given certain constraints that usually are built into the system, it is possible to generate spatial inhomogeneities using chemical reactions and thus tell us about general conditions for this to occur, they do not address the important issues of the molecular mechanisms involved.

Interactions between neighbouring nonequivalent cells

The initial conditions of this interaction establish a cell or a group of cells within an assembly which are different from the rest and will act as a source of signals to change the state of neighbouring cells. In a strict sense, this interaction refers to nearest neighbours, that is cells that are adjacent to each other, and represents an example of INDUCTION at the level of single cells. A cascade generated by the reiteration of this process can lead to a large amount of cell diversity because once a cell has become different it can itself become the source of differences for other cells. The lineage-dependent development of the nematode *Caenorhabditis elegans* (*see* CAENORHABDITIS DEVELOPMENT) and multiple inductive interactions during the development of the frog *Xenopus* (e.g. mesoderm or lens induction, *see* AMPHIBIAN DEVELOPMENT) provide examples of this kind of interaction.

The patterning of the ommatidial units of the compound eye of *Drosophila* (*see* EYE DEVELOPMENT) is an exquisite outcome of variations on the same theme, with a centrally located photoreceptor cell acting as a source of information for the development of seven other photoreceptor cells that cluster around it. The fate of each of these seven cells appears to depend on stereotyped nearest-neighbour interactions.

The restriction of the cell–cell interactions to nearest neighbours can occur either by restriction in the expression of receptors to cells that are adjacent to the signalling cells, or by restriction of the range of action of the signal, as occurs with many growth factors which have their range of action restricted by the physicochemical properties of the molecules themselves or by interactions with components of the extracellular matrix. The latter can alter the receptor binding ability of the signal; for example, the TGFβ receptor β-glycan — a membrane-bound PROTEOGLYCAN — contributes to defining the affinity of TGFβ for signalling receptors, by locally concentrating the activity of this growth factor.

Interactions between non-neighbouring nonequivalent cells

The basis for this type of interaction is also a cell or a population of cells that is different from others and which acts as a source of signals. In this case, however, the signalling element is not necessarily the nearest neighbour of the responding population but is located in a distant position in the animal or the plant. A classic example of this type of interaction is the action at a distance of hormones that are carried to their targets by the blood stream. An organ synthesizes a hormone and secretes it into the blood, where it may reach very distant organs which will capture it with specific receptors.

A less dramatic example of the same phenomenon is the action of growth factors within a field of cells in which a cell population strategically located within a sheet of cells secretes a polypeptide with properties that restrict its movement across the cellular field. In this manner it reaches a limited number of cells, perhaps with a decaying concentration as these cells lie further away from the source. The activity of the growth factor results in changes in the fate of the responding cells. It is likely that the assignation of cell fates during the early development of vertebrates relies on this type of mechanism. In *Xenopus*, for example, maternally localized growth factors diffuse locally to reach a subset of the embryonic cells and determine their developmental fate.

The range of this interaction is determined, in general, by the physicochemical properties of the molecules and those of the extracellular environment in which they move. This kind of interaction is the basis for concepts like POSITIONAL INFORMATION, in which the fate of particular cells within a field is determined by concentrations of a particular molecule.

At present, the molecular characterization of cell interactions is still at the stage of describing and analysing the elements that make up the signalling pathways that operate within cells. The next, more challenging, step will be the understanding of how the events generated by the functional networking of these elements lead to the coordinated regulation of gene activity and cellular behaviour that is characteristic of multicellular systems.

A. MARTINEZ ARIAS

See also: CELL–MATRIX INTERACTIONS IN DEVELOPMENT.

cell centre *See*: CENTROSOME.

cell coupling The metabolic or electrical coupling of adjacent animal cells that can occur through the passage of ions or small metabolites through gap junctions (*see* CELL JUNCTIONS).

The cell cycle

LIVING organisms are composed of cells, whose growth and division require a regular sequence of events and processes that comprise the cell cycle. Some of these processes are continuous, like the synthesis of most of the molecules in the cell including proteins and lipids. By contrast, the synthesis of DNA and the act of cell division are discontinuous processes; some authors consider them as 'events'. Studies of the cell cycle are largely concerned with the discontinuous processes: their nature, mechanism, control, and coordination. The interplay between continuous and discontinuous processes is of great importance; thus, speeding up the cell cycle without a corresponding increase in growth will produce smaller and smaller cells, while increasing the growth rate without speeding up the cycle would lead to a population of larger and larger cells. In most organisms, cells of a particular type have a typical size, which is maintained by homeostatic mechanisms that balance growth and division.

The two key discontinuous processes for cell survival are the replication of DNA and the segregation of chromosomes to the two daughters of cell division during MITOSIS. If either of these processes is performed inaccurately, the daughter cells will be different from each other, and will almost certainly be flawed. The flaws may be of two kinds. MUTATIONS in DNA molecules, which often occur during replication, may be fatal when they lead to loss of particular essential genes, but they usually affect one or at most a few genes. By contrast, the gain or loss of whole chromosomes by segregation errors during mitosis leads to much more widespread damage, because large numbers of genes are gained or lost at a single stroke. Diploid cells have a degree of tolerance for such damage but even so, there are limits (*see* CHROMOSOME ABERRATIONS).

Segregation of chromosomes occurs during mitosis, normally a relatively brief period in the cell cycle, which culminates in the highly visible act of cell division (CYTOKINESIS). The rest of the cell cycle comprises interphase, when nothing much appears to be happening as seen down a microscope. Growth usually occurs during interphase and temporarily stops during mitosis. In eukaryotic cells, chromosome replication occurs only during interphase, and replication and segregation are mutually exclusive processes. This is not true of all organisms, however: in rapidly growing bacteria, a relatively simple system allows DNA synthesis and chromosome segregation to proceed harmoniously at the same time.

Stages of the cell cycle

The usual description of the cell cycle concerns rapidly growing and dividing eukaryotic cells. It describes the cycle in terms of a series of phases — interphase and mitosis (M) — and the subdivision of interphase into the times when DNA synthesis is going on, known as S-phase (for synthesis phase) and the gaps that separate S-phase from mitosis. G1 is the gap after mitosis before DNA synthesis starts, and G2 is the gap after DNA synthesis is complete before mitosis and cell division. Further subdivisions are used to describe cells that temporarily 'leave' the cell cycle; for example, well-nourished fibroblasts in culture lacking GROWTH FACTORS pause after mitosis in a state that is often called G0. Such cells will only enter S-phase when they receive signals from outside, usually in the form of peptide growth factors.

Cells can also rest in G2 phase, and many do, in which case the arrival of a signal causes entry into mitosis or meiosis. A well-known example of the latter kind of cell are the OOCYTES and eggs of many animals, although resting in G2 is by no means confined to specialized germ cells. Many organisms that spend most of their life cycles as haploids, like the fission yeast *Schizosaccharomyces pombe* or the cellular slime mould *Dictyostelium discoideum* normally pass through S-phase immediately after mitosis without a significant lag, and spend the majority of the cell cycle between the end of S-phase and the next mitosis while the cells reach the critical size required for division. This makes sense in terms of preserving chromosome integrity, because haploid organisms only have a single copy of each chromosome. If that chromosome is damaged, there is an irretrievable loss of genetic information that may well be lethal. But after S-phase, there are two copies of each chromosome, and information lost from one by unlucky damage can be retrieved from the other copy, where it is unlikely that damage will have affected exactly the same stretch of DNA. The importance of this very simple consideration is revealed by studies of the sensitivity of haploid yeast to X-rays. Cells in G1 and S-phase are extremely sensitive to X-rays; a single broken chromosome is almost always a lethal event. By contrast, G2 cells are much more resistant, because RECOMBINATION between sister chromatids can reconstruct viable chromosomes from damaged pieces.

The 'Alternation' and 'Completion' problems

Two overriding problems need to be solved in order to understand the cell cycle and its control. The first is known as the 'Alternation Problem', the second as the 'Completion Problem'. The completion problem is concerned with the kind of issues raised in the preceding section. What mechanisms ensure that one process is complete before the next process starts? The answer in a word is 'Checkpoints'. Checkpoints are safe stopping points in the cell cycle where progress can be halted for the time it takes to check some condition. For example, cells do not normally enter mitosis

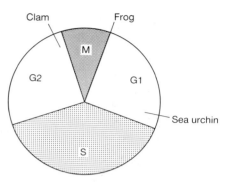

Fig. C20 Eggs awaiting fertilization are arrested in a variety of cell cycle stages.

with damaged or unreplicated chromosomes. But certain mutations, of which *rad9* in the budding yeast *Saccharomyces cerevisiae* is the best known, cause a failure to arrest in response to broken chromosomes. Cells with this mutation cannot delay the cell cycle and may thus enter mitosis with damaged chromosomes, with fatal consequences for the daughters of the ill-timed division. Provided the chromosomes are undamaged, however, the *rad9* yeast cells are perfectly viable.

Another good example of a checkpoint is found in eggs awaiting FERTILIZATION. Depending on the species, eggs are arrested in G1 (sea urchins), M (frogs) or G2 (clams) (Fig. C20). The arrival of the sperm causes a large, transient increase in intracellular CALCIUM that breaks the arrest and allows development to proceed. It is remarkable that different organisms can generate temporary arrest points at different places in the cell cycle while using the same signal to announce successful fertilization. In general, it seems to be possible to construct 'plug-in modules' that can detect almost anything that might be relevant to a particular cell at a particular time, and couple this to the cell cycle control machinery. The details of the construction and working of these modules are not yet well worked out. At a minimum they must contain a sensor and an effector, and in the cases where they have been studied, as in the case of the mating response of yeast, a complex signal transduction pathway often lies between. Probably all such modules use PROTEIN KINASES to effect regulatory PROTEIN PHOSPHORYLATION at some stage. It is not difficult to imagine, at least in principle, how such modules are constructed (Fig. C21).

Much less is known about the principles that underlie the solution to the alternation problem. The question is, what mechanisms normally ensure the strict alternation of S-phase and M-phase? There is no rule that cells must divide after replicating their chromosomes, or that cells must replicate their chromosomes after completing cell division, even though the vast majority of cells do, in fact, obey such rules. For example, many insect cells have enormous chromosomes (POLYTENE CHROMOSOMES) that are easily visible in interphase because chromosome replication continues while cell division is suppressed. Eggs and sperm, on the other hand, suppress the round of DNA synthesis that normally follows cell division when they go through the reduction cell division of meiosis in order to produce haploid gametes. In budding yeast at least, the beginnings of an idea are starting to

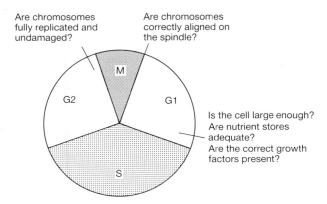

Fig. C21 Some well-known cell cycle checkpoints.

emerge that can explain this alternation and explain what prevents cells from returning to mitosis until they have undergone S-phase. In a way, the system relies on checkpoints that interconnect the cell division and the chromosome replication cycles so that neither is allowed to run free, and the circuitry is arranged to produce strict alternation (Fig. C22). We will return to this discussion after taking a closer look at the enzymes and molecules which control these cycles.

The early students of the cell cycle, from the late 19th century until well into the 1960s, focused their attention on the visible events of mitosis itself, and were especially concerned to understand the process in mechanical terms. In Mazia's famous simile, the chromosomes were regarded as the corpse at a funeral: the *raison d'être* for the proceedings, but playing no active part. Every so often, questions would be asked: what determines when mitosis occurs? Do cells contain some sort of 'master regulator'? The first evidence that such a thing existed came in about 1970, from two very different approaches. Johnson and Rao took cells that had been synchronized in different stages of the cycle and fused them together to produce HETEROKARYONS. The most striking results were obtained by fusing mitotic with nonmitotic cells. The newly introduced nonmitotic nuclei rapidly lost their nuclear envelopes and their chromosomes underwent various forms of condensation (if they happened to be in the course of replication, the condensed chromosomes looked pulverized).

A similar result was obtained when interphase nuclei were microinjected into metaphase-arrested frog eggs: each nucleus formed a mitotic spindle. Next, it was discovered that quite small amounts of cytoplasm from these eggs could induce G2-arrested frog oocytes to undergo meiotic maturation and pass into M-phase. The factor responsible for this effect was named MATURATION PROMOTING FACTOR (MPF). It turned out that MPF was present in all mitotic cells, from yeast to human. Almost 18 years were to pass, however, before biochemists purified and, together with geneticists, characterized MPF, which provided the first real key to understanding cell cycle transitions.

Transitions in the cell cycle are what provide puzzles and challenges: the transition from interphase to mitosis, and mitosis back to interphase; the transition from quiescence to growth and back to quiescence; the transition from being prepared for DNA synthesis to actually starting it. How can the molecular mechanisms underlying these transitions be studied?

Yeast cell cycle mutants

Budding yeast

The most important contribution to understanding the control of the cell cycle has come from genetic studies performed on simple yeasts. The first such studies were performed by Leland Hartwell and colleagues on the budding yeast, *S. cerevisiae*. In the course of isolating temperature-sensitive mutants for DNA, RNA, or protein synthesis, they realized that some of the strains were blocked at specific points in the cell cycle. If an asynchronous population of such a mutant was transferred from 23°C (permissive) to 36°C (nonpermissive), the cells stopped dividing and all showed the same morphology; all unbudded, for example, or all with two connected cells unable to complete cytokinesis.

One of these temperature-sensitive mutants proved central to understanding the transition from S-phase to mitosis. *cdc28* (cell *division cycle* mutant) fails to form a bud at the nonpermissive temperature, but shows no defect in macromolecular synthesis. Indeed, *cdc28* cells continue to grow at the nonpermissive temperature, and can complete an already started S-phase although they can never start a new one. Further analysis showed that a functional CDC28 gene was required for a cell cycle transition that Hartwell called 'Start'. (Start might just as easily have been

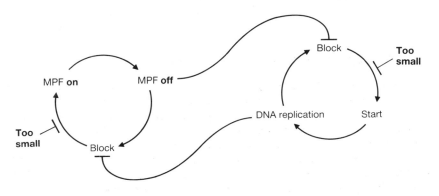

Fig. C22 Connections between the cell division cycle and the DNA replication cycle.

Cell division cycle   DNA replication cycle called 'Stop', which is in effect the name the equivalent transition has been given in cells of higher eukaryotes — the 'restriction point'.) Budding yeast halts the cell cycle at Start in response to a variety of different conditions: starvation for essential nutrients, being too small, or the proximity of a potential mating partner, signalled by a peptide pheromone.

Many other *cdc* mutants have been isolated, and classified according to their phenotypes, some of which are intriguing. For example, at the nonpermissive temperature *cdc4* mutants can pass Start but cannot initiate DNA synthesis, but they put out a new bud at regular intervals, as if a cell cycle clock were running independently of the events and processes it was meant to control.

Analysis of these mutants was originally intended to provide a set of pathways whose connections and intersections would provide a kind of metabolic roadmap of the cell cycle. It was inspired in part by the great success of bacterial and phage genetics in delineating the pathways of bacteriophage assembly, which had revealed how phage heads and tails were made separately and then assembled, and how certain key proteins formed a template for the assembly of others. One could imagine similar schemes for the cell cycle, with one process leading to another.

Fission yeast

Studies on the fission yeast *Schiz. pombe* provided important parallel insights into cell cycle control. This yeast grows as an elongating cylinder, which is periodically divided in two by the formation of a septum at the midpoint. Mitosis precedes septum formation. Paul Nurse and colleagues discovered cell cycle mutants, called wee mutants, that caused these cells to enter mitosis and divide at smaller-than-usual sizes. A key observation was the recognition of *wee* alleles of *cdc2*, a gene essential for cell cycle progression. The existence of these mutants argued that *cdc2* must be a particularly important regulator of mitosis, in that not only was it essential for initiation of the process, but it could also determine at what length of cell mitosis occurred. Genetic screens identified several other genes whose products appeared to interact with *cdc2*, including *wee1*, *cdc25*, and *cdc13*. (Any discussion of yeast cell cycle genes is bedevilled by the difference in conventions of nomenclature between *S. cerevisiae* and *Schiz. pombe*, *see* GENETIC NOMENCLATURE.)

The first hints of an underlying unity and order to cell cycle control came when *cdc2* and *CDC28* were cloned and sequenced, and turned out to be interchangeable between these two very distantly related yeasts. The amino-acid sequences of the predicted proteins showed strong homology with each other and to members of the PROTEIN KINASE family of enzymes. PROTEIN PHOSPHORYLATION mediated by protein kinases is one of the most common means of altering the activity of enzymes and other proteins in eukaryotic cells, and phosphorylation of a few key proteins could provide the means of orchestrating the profound alterations in cell architecture which occur during mitosis.

To add to the excitement came the discovery that human cells also contain a protein kinase, known as $p34^{cdc2}$, which is almost identical to Cdc2 and Cdc28, and which can rescue a *cdc2* mutant *Schiz. pombe* strain. The question then was: does this enzyme control mitosis in animal cells? The answer is yes, as evidenced by the finding that the protein kinase activity of $p34^{cdc2}$ is low during interphase, and high during mitosis, and that temperature-sensitive mutants of *cdc2* (the gene encoding $p34^{cdc2}$) in human cells cannot enter mitosis at high temperature. Several important proteins whose properties change during mitosis, like the nuclear lamins, for example (*see* NUCLEAR MATRIX), are substrates for $p34^{cdc2}$, and mutations in the key serine and threonine target residues abolish not only their phosphorylation, but also their altered properties.

A simplified picture of how mitosis is controlled can now be built up. A protein kinase (Cdc2, Cdc28, or $p34^{cdc2}$) is turned on and phosphorylates the appropriate proteins, whose properties alter such that the ensemble can then enact mitosis. This model may also serve for other cell cycle transitions.

Cyclins

Having identified $p34^{cdc2}$ and its homologues as crucial to the transition to mitosis, the next question was how their activity is regulated. $p34^{cdc2}$ is present at more or less constant levels throughout the cell cycle but is only active at mitosis. The proteins that regulate $p34^{cdc2}$ activity were first identified in sea urchin eggs. They are the cyclins, proteins whose levels oscillate during the cell cycle; in the case of mitotic cyclins, high during mitosis, low during interphase. The existence of such 'periodic proteins' makes it much easier to understand how cells can so completely alter their behaviour during mitosis, but it is surprising how long it took before this interesting class of protein was discovered.

Fertilized sea urchin eggs go through their first three cleavage divisions rapidly and synchronously, which made them an obvious choice of material in which to search for periodic proteins. When these embryos are labelled with radioactive amino acids, one of the most abundant of the newly synthesized proteins can be seen to disappear abruptly about 10 minutes before cell division. This is the result of the activity of a highly specific protease that is active for about 5 minutes in mitosis, starting a minute or two before the metaphase to anaphase transition. The protease's targets have been named cyclins as they undergo a sharp oscillation in concentration during the cell cycle.

Cyclins regulate the activity of $p34^{cdc2}$ and its yeast counterparts. They form 1 : 1 molecular complexes with protein kinase catalytic subunits of the family which includes Cdc28, Cdc2, and $p34^{cdc2}$, and which are known as the cyclin-dependent kinases or Cdc2 kinases. These catalytic subunits contain ~300 amino acids, and are thus similar in size to the catalytic domain of the well-known CYCLIC AMP-DEPENDENT PROTEIN KINASE. That kinase is active as a monomer but members of the Cdc2 family of protein kinases are devoid of activity without their cyclin partners. Purified MPF, for example, is now known to be a complex of $p34^{cdc2}$ and cyclin B.

Oocyte maturation: $p34^{cdc2}$, cyclin and the control of MPF activity

From biochemical work on MPF, it has been found that combination with cyclins is necessary, but not sufficient for $p34^{cdc2}$

activity. Another protein kinase, known as CAK (Cdc2 activating kinase), which is itself a member of the Cdc2 family, is required to phosphorylate threonine 161 of p34^{cdc2} before MPF can bind to its target substrates. In addition to this positive regulation, stringent negative regulation of MPF activity is provided by phosphorylation of threonine 14 and tyrosine 15 of p34^{cdc2}. The enzymes responsible for controlling phosphorylation at these sites were first identified genetically as the products of the *wee1* and *cdc25* genes. The *wee1* gene specifies a protein kinase that negatively regulates Cdc2, while the *cdc25* gene encodes a tyrosine phosphatase that can activate p34^{cdc2} that has been turned off by Wee1.

The meiotic maturation of frog oocytes provides one of the best understood examples of regulation of the activity of p34^{cdc2}. These cells contain close to 10^{12} molecules of p34^{cdc2}, of which roughly half are already bound to a B-type cyclin and are phosphorylated on Tyr 15 (by the inhibitory protein tyrosine kinase Wee1). In this state they possess no detectable protein kinase activity. But when the oocytes are exposed to the hormone progesterone, an activating tyrosine protein phosphatase (Cdc25) is activated, the inhibitory tyrosine kinase Wee1 is switched off, and the cyclin B : cdc2 complexes rapidly acquire protein kinase activity (MPF activity). This activation is self-promoting, for the Cdc25 phosphatase is activated, and the inhibitory Wee1 kinase is correspondingly inhibited by MPF activity (Fig. C23). This positive feedback control loop ensures that entry into mitosis proceeds rapidly and irreversibly once it has been started.

The state of mitosis is stable, and persists for as long as it takes the chromosomes to align on the metaphase plate, organized by the mitotic spindle. Once the chromosomes are all aligned, the special protease is activated and cyclin is degraded, which turns off MPF activity. Almost certainly, other targets for the protease exist, possibly among the structural elements that hold the sister chromatids together. The separation of the sister chromatids at anaphase may depend on dissolving the glue that holds them together. Other important elements in promoting the return to interphase are the protein phosphatases that reverse the phosphorylation catalysed by MPF. In both yeast and *Drosophila*, temperature-sensitive mutations in protein phosphatases can block cells in mitosis.

Other cell cycle transitions

How well does the mitotic paradigm apply to other cell cycle transitions? How does a knowledge of cyclin-dependent kinases help explain the alternation and completion problems? Studies of mating in *S. cerevisiae* have shed much light on this question. When two haploid yeasts of opposite mating type meet, they must synchronize their cell cycles at Start in order to mate. They do so by chemical signalling through peptide mating hormones called α and a. These mating factors bind to cell-surface receptors and induce the activation of a protein kinase (of the MAP KINASE family) which activates a protein called Far1 that binds to cyclin : Cdc28 complexes and inhibits their protein kinase activity. If the concentration of active Cdc28 is too high, Far1 is powerless to stop the cycle, and such cells are resistant to the growth-inhibitory effects of mating factor. This phenomenon forms the basis of an assay for cyclins active in the G1 phase in yeast, of which five are presently known: Cln1, Cln2, Cln3, Hcs26 and OrfD. These belong to three different families: Cln1 and Cln2 are similar, and Hcs26 has similarity to OrfD, whereas Cln3 is very distantly related to all the others. Selective deletion of the genes for these cyclins reveals a surprising degree of redundancy; yeast can survive with only one or two of the five G1 cyclins. In addition to the G1 cyclins, *S. cerevisiae* contains at least six different B-type cyclins which appear to function at different times in the cell cycle. Clb1, Clb2, Clb3 and Clb4 all appear to have roles in mitosis; like the G1 cyclins, they are redundant, and Clb2 alone suffices to keep yeast alive and happily dividing. Clb5 and Clb6 seem to be required for passage through S-phase, however, despite their strong family resemblance to the mitotic cyclins. Most of these cyclin mRNAs and proteins show periodic expression through the cell cycle; the exception is Cln3, which is present at a low but constant level throughout the cycle. Cln3 seems to be required for turning on synthesis of the main G1 cyclins, Cln1 and Cln2.

The yeast cell cycle

Although the details remain to be worked out, current thinking suggests the following description of the yeast cell cycle (Fig. C24).

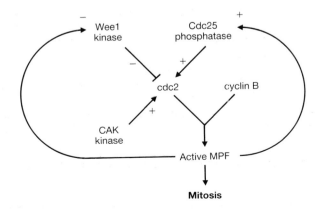

Fig. C23 The flip-flop activation switch that controls entry into mitosis.

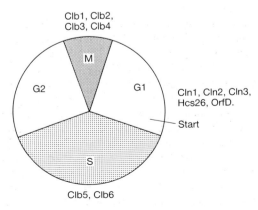

Fig. C24 In *S. cerevisiae* different cyclins combine with the same kinase subunit (Cdc28) to catalyse different cell cycle events.

At the end of mitosis, the mitotic cyclins are degraded by the special protease, which stays active until Start. This prevents the cell from immediately returning to mitosis. At Start, synthesis of Cln1 and Cln2 is activated, and these proteins combine with Cdc28 to form a protein kinase that initiates formation of the bud and prepares the cell for S-phase, including activation of the *CLB5* and *CLB6* genes. Clb5 and Clb6 replace Cln1 and Cln2 as partners of Cdc28, and in some way promote DNA synthesis. Towards the end of S-phase, transcription and translation of *CLB1* and *CLB2* starts up, which turns off transcription of the 'earlier' cyclin genes, and as these 'later' cyclins accumulate, they transform the Cdc28 kinase into its mitotic form, carrying the yeast into telophase, at which point the protease is activated and the cycle repeats itself.

The cell cycle of higher eukaryotes

The same general principles probably apply to the cell cycles of higher eukaryotes. MPF, which is composed of Cdc2 (p34^{cdc2}) and B-type cyclins (of which at least three are known in frogs), is active during mitosis, and is turned off by a special protease at the end of mitosis. While Cdc2 protein persists during the whole cell cycle in actively dividing cells, its activating partners only accumulate in G2 phase. The closely related protein kinase, Cdk2, is first found associated with cyclin E and kinase activity in late G1, and as S-phase starts up, active cyclin A : Cdk2 kinase is found in the nucleus. A steady stream of new cyclins and new Cdc2-like partners are being discovered, but it is not known if they are all cell cycle regulators. An idea of the complexity in higher eukaryotic cells is hinted by the list of cyclins: A1, A2, B1, B2, B3, C, D1, D2, D3, E, F, and G and an equally long list of Cdc2-like protein kinase subunits. Figure C25 is an attempt at tabulating the known relationships and functions of this protein kinase family.

The ability of a cell to perform different tasks at different times during the cycle thus depends in part on making new mRNAs and proteins under transcriptional control, in part by modifying the activity of existing enzymes and proteins by reversible protein phosphorylation, and in part by clearing away unwanted proteins by specific proteolysis. Although the main principles and some of the details are now clear, it would however be wrong to think that the problems of cell cycle control are anywhere near fully understood.

Further information can be found in [1–5] and the references cited therein.

T. HUNT

1 Hartwell, L.H. & Weinert, T.A. (1989) Checkpoints: controls that ensure the order of cell cycle events. *Science* **246**, 629–634.
2 Murray, A.W. & Hunt, T. (1993) *The Cell Cycle: An Introduction* (W.H. Freeman, New York).
3 Murray, A.W. & Kirschner, M.W. (1989) Dominoes and clocks: the union of two views of cell cycle regulation. *Science* **246**, 614–621.
4 Nasmyth, K. (1993) Control of the yeast cell cycle by the Cdc28 protein kinase. *Curr. Opinion Cell Biol.* **5**, 166–179.
5 Norbury, C. & Nurse, P. (1992) Animal cell cycles and their control. *Annu. Rev. Biochem.* **61**, 441–470.

cell death *See*: APOPTOSIS; PROGRAMMED CELL DEATH.

cell determination, cell differentiation *See*: DETERMINATION; DIFFERENTIATION.

cell division *See*: ASYMMETRIC CELL DIVISION; CELL CYCLE; CYTOKINESIS; MITOSIS; MEIOSIS.

cell fate *See*: FATE.

cell fractionation (subcellular fractionation) The isolation of particular subcellular organelles or fractions, after breaking open the cells, on the basis of distinct properties of the organelles — usually size and density as determined by centrifugation.

cell-free systems *See*: IN VITRO TRANSCRIPTION; IN VITRO TRANSLATION.

cell fusion The joining of two cells to give one cell containing two nuclei. Fusion of cells can occur naturally, as in the fusion of myoblasts to form the myotube during muscle development (*see* MYOGENESIS), or can be induced in culture by application of an electric field (electrofusion) or by chemical and biological agents such as polyethylene glycol and inactivated Sendai virus. Induced cell fusion has many uses in the formation of cell hybrids for various applications, such as gene mapping (*see* HUMAN GENE MAPPING; SOMATIC CELL HYBRIDS), the formation of HYBRIDOMAS for monoclonal antibody production, and the study of NUCLEO-CYTOPLASMIC INTERACTIONS during development. For example, in the ascidian *Ciona intestinalis*, in which each cell lineage displays a fixed fate, nuclei from late gastrula stages can direct the development of many highly differentiated tissues when fused with enucleated eggs. This experiment demonstrated that the early specification of cell fate in eggs with strictly defined lineage assignments is determined by CYTOPLASMIC DETERMINANTS acting on totipotent embryonal nuclei.

cell hybrids *See*: SOMATIC CELL HYBRIDS.

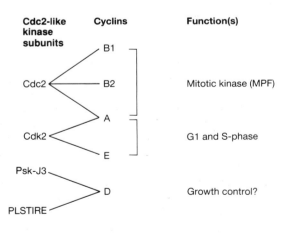

Cdc2-like kinase subunits	Cyclins	Function(s)
Cdc2	B1, B2, A	Mitotic kinase (MPF)
Cdk2	A, E	G1 and S-phase
Psk-J3, PLSTIRE	D	Growth control?
PSSALRE	25K subunit	Brain neurofilament kinase

Fig. C25 The relationships between cyclins and their partner kinase subunits in higher eukaryotes.

Cell junctions

CELL junctions are macromolecular plasma membrane specializations involved in contact between one cell and another, or between a cell and its environment, for example the extracellular matrix. These specializations are clearly visible by transmission electron microscopy and freeze fracture analysis. Such contacts occur in multicellular eukaryotic organisms. This definition will deal primarily with the junctions found in vertebrate cells, although some specific junctions are typical of invertebrates (*see e.g.* SEPTATE JUNCTIONS). Many cell junctions are adhesive in function, for example intermediate junctions and desmosomes [1], whereas others are occlusive, for example tight junctions [1], or communicating, for example gap junctions [2]. Junctions involved in synaptic transmission, such as the NEUROMUSCULAR JUNCTION, require the passage of signalling molecules (NEUROTRANSMITTERS) across a small but finite cleft (the synapse) to impinge on the recipient cell. These will be dealt with elsewhere (*see* SYNAPTIC TRANSMISSION).

In general, cell junctions are composed of three components which can be contributed by separate molecules, or by DOMAINS of the same molecule: (1) an extracellular component involved in recognition and binding of the target cell or molecule; (2) a transmembrane component; (3) a cytoplasmic component involved in coupling to the CYTOSKELETON or, as in synapses, an intracellular signalling pathway (*see e.g.* G PROTEIN-COUPLED RECEPTORS; SECOND MESSENGER PATHWAYS).

Adhesive cell junctions

Cell junctions can be involved in adhesion between cells or between a cell and its extracellular matrix. The former class, namely intermediate junctions and desmosomes, are symmetric, that is, each neighbouring cell contributes a mirror-image half of the junction to the final structure. In contrast, extracellular matrix junctions, that is focal adhesions and hemidesmosomes, are asymmetric; the attached cell contributes transmembrane and cytoplasmic cytoskeletal components whereas the extracellular matrix ligand can be secreted from the same cell, or from other cells in the tissue. There are also similarities between the two classes of adhesive junction, manifest in the use of some common components. Adhesive junctions not only function to maintain the integrity of organized tissues, but also have a role in the establishment of tissues during morphogenesis and organogenesis. Pathological changes in such junctions may also contribute to the aetiology of human diseases such as cancers and skin disorders.

Intermediate junction

Also called belt desmosome, zonula adh(a)erens or adh(a)erens junction, the intermediate junction is widely distributed in tissues, notably in the junctional complex of epithelia where its relationship to desmosomes and tight junctions is best seen (Fig. C26) [1]. It is found basolaterally relative to the tight junction in polarized epithelial cells and is characterized by close apposition

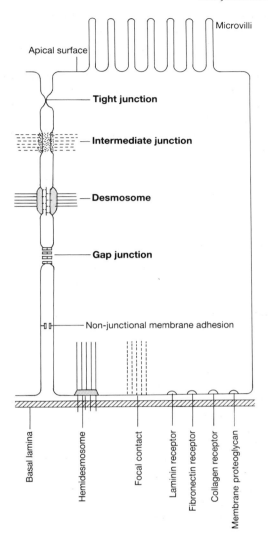

Fig. C26 Cell junctions in a stylized epithelial cell, showing the junctional complex (tight junction, intermediate junction, desmosome). From [5].

of neighbouring cell membranes (gap = 15–25 nm). The junction can extend over large areas and often has a belt-like appearance just below the tight junction, giving rise to one of its synonyms. Characteristically there is an intercellular region of fuzzy appearance (the adhesive material) and an electron-dense filamentous, mat-like region on the cytoplasmic side of each plasma membrane. Into this meshwork F-actin filaments (*see* MICROFILAMENTS) (~7 nm in diameter) apparently insert.

Some of the molecular components of the intermediate junction have been characterized. The adhesive element seems to be contributed by members of the cadherin family of calcium-dependent CELL ADHESION MOLECULES (CAMs) [3]. The assembly of the junctional complex depends on adequate levels of extracellular Ca^{2+}, which may act via its effect on cadherin-mediated adhesion. Cadherins are found not only in cell junctions, but also distributed on non-junctional membrane. Different cadherin family members are found in intermediate junctions, depending on the cell type. These transmembrane proteins all share similar extracellular calcium-binding adhesive domains exhibiting homotypic (homophilic, i.e. like to like) binding. Their cytoplasmic domains are highly conserved and apparently mediate binding to the actin filament system. A group of proteins called 'catenins' can associate with these cytoplasmic domains, and may be involved in junctional assembly, but it is not known whether they persist in the final structure. Fully formed intermediate junctions are associated on their cytoplasmic side with a rod-like protein 'tenuin' (M_r ~400 000, 40K), vinculin (see focal contacts below) and a ubiquitous junctional protein called plakoglobin (see desmosome below).

Desmosome

Also called spot desmosome or macula adh(a)erens, the desmosome is a plate-like adhesion restricted to epithelial cells and a few others [4] and is also basolaterally located. The apposed cell membranes have a gap of 20–35 nm with intercellular fuzzy material and often a dense midline. On the cytoplasmic side an electron-dense plaque of 15–20 nm thickness is found separated slightly from the plasma membrane by a less dense region. INTERMEDIATE FILAMENTS (10 nm), usually cytokeratins, loop into the cytoplasmic side of the plaque. The extreme insolubility and stability of the whole structure is believed to reflect its role, often likened to a pop-rivet, in stabilizing epithelial sheets against mechanical stress, for example in skin. However, recent data suggest that this junction can be dynamic.

Two types of transmembrane GLYCOPROTEIN are found in the desmosome and are related to the cadherin superfamily. The largest type (DESMOGLEINS, M_r ~130–150K) have a unique cytoplasmic extension which may contribute to the dense plaque [4]. The other type (DESMOCOLLINS, M_r ~100–120K) are more similar to the classical cadherins. Two large rod-like proteins (desmoplakins I and II (DPI, II); M_r 240K and 215K) consisting of coiled-coil homodimers are the major intracellular plaque components and PLAKOGLOBIN (M_r ~82K) is also present. The circumference of the plate-like plaque is delineated by a very large protein desmoyokin (M_r ~680K). Other cell-type specific components can also be present [4].

Focal contact

Also called focal adhesion or adhesion plaque, this is an extracellular matrix contact sharing some features in common with the intermediate junction. It is seen at sites of cell–matrix adhesion and has associated actin filaments. The transmembrane component is usually a member of the INTEGRIN family. The cytoplasmic tail of the integrin β subunit is believed to interact with a large cytoplasmic protein talin (M_r ~240K), unique to focal contacts. This in turn may bind vinculin and α-actinin and connect to the actin filaments.

Hemidesmosome

The hemidesmosome, like the morphologically similar desmosome, interacts on its cytoplasmic side with intermediate filaments. The two structures, however, do not appear to share any molecular components. The extracellular ligand for the hemidesmosome on the basement membrane of epithelia is not definitively known. Extracellular 'anchoring filaments' are seen to extend from the hemidesmosome into the basement membrane. Recently a member of the integrin family ($\alpha_6\beta_4$) has been localized to hemidesmosomes. The plaque contains a large protein bullous pemphigoid antigen I (M_r ~230K), so called because of the presence of autoantibodies to this protein in the sera of patients with the blistering skin disease of the same name (*see* BULLOUS PEMPHIGOID). This protein is related to DPI and II (see desmosome) and presumably has a similar role. Other proteins such as the transmembrane bullous pemphigoid antigen II (M_r ~180K) may also contribute.

Other cell junctions

Junctions whose function is not primarily thought to be adhesive include the tight junction and the gap junction.

Tight junction

The third component of the epithelial junctional complex is the tight junction, also called the zonula occludens or occluding junction, which provides an occluding barrier to the passage of proteins and electrolytes across the monolayer. Located at the junction of the apical and lateral plasma membranes it is characterized microscopically as a circumferential region of close membrane apposition, even fusion. In freeze-fracture studies it has a ribbon-like appearance, and is probably composed of protein and lipid in the hexagonal II phase (*see* MEMBRANE STRUCTURE). A transmembrane adhesive component is assumed but has not yet been identified. Two peripheral cytoplasmic components have been characterized, namely ZO-1 (M_r ~225K) and cingulin (M_r ~140K), but no association with the cytoskeleton has been observed.

Gap junction

Gap junctions occur widely and play a part in intercellular communication by allowing the passage of small molecules (M_r <1000) directly between the cytoplasm of interconnected cells [2]. This results in electrical and metabolic coupling. In thin-section electron micrographs gap junctions appear as closely apposed plasma membrane areas with a 2 nm gap. After freeze fracturing they display a regular array of intramembrane particles. These form hexagonal connexons with a central pore (Fig. C27). It is believed that small molecules can pass through the pore, the diameter of which can be regulated like the iris of a lens. Each partner cell contributes a mirror-image half of the junction.

Although the overall structure is similar in different cells, the proteins (connexins) that form the six subunits comprise a family of related molecules with tissue-specific patterns of expression and varying molecular weights (M_r 21–70K). They are integral membrane proteins which span the plasma membrane four times, having both their N and C termini on the cytoplasmic side. The variation in size and properties arises from differences in the

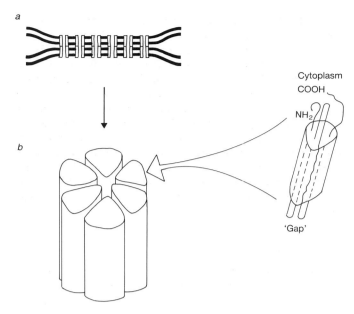

Fig. C27 Structure of a gap junction. *a*, Diagrammatic representation of two membranes forming a gap junction. Connexons form channels connecting the interiors of cells. *b*, Model of a connexon. In an intact junction two such connexons are orientated end on with a channel 1.6–2 nm in diameter extending down the centre. Inset, one of the connexon subunits (connexin) showing its proposed orientation in the membrane. Adapted from [6].

intracellular loop between transmembrane spans 2 and 3, and in the C-terminal tail. Connexins appear to be able to form heterologous as well as homologous pairings across the intercellular gap. Regulation of this property, as well as of the opening and closing of the pore, can be achieved by phosphorylation of cytoplasmic residues by CYCLIC AMP-DEPENDENT PROTEIN KINASE and the pp60^{src} TYROSINE KINASE, and PROTEIN KINASE C amongst others.

A.I. MAGEE

See also: ACTIN-BINDING PROTEINS; CELL MOTILITY; CELL–MATRIX INTERACTIONS; EPITHELIAL POLARITY; EXTRACELLULAR MATRIX MOLECULES.

1 Farquhar, M.G. & Palade, G.E. (1963) Junctional complexes in various epithelia. *J. Cell Biol.* **17**, 375–412.
2 Beyer, C. et al. (1990) Connexin family of gap junction proteins. *J. Membrane Biol.* **116**, 18–194.
3 Takeichi, M. (1990) Cadherins: a molecular family important in selective cell–cell adhesion. *Annu. Rev. Biochem.* **59**, 237–252.
4 Schwartz, M.A. et al. (1990) Desmosomes and hemidesmosomes: constitutive molecular components. *Annu. Rev. Cell Biol.* **6**, 461–491.
5 Garrod, D.R. (1986) Desmosomes, cell adhesion molecules and the adhesive properties of cells in tissues. *J. Cell Sci. Suppl.* **4**, 221–237.
6 Evans, W.H. (1988) Gap junctions: towards a molecular structure. *BioEssays* **8**, 3–16.

cell line A cell which will proliferate indefinitely in culture. Most cells isolated directly from nonembryonic tissues will only undergo a limited number of cell divisions in culture and thus cell lines cannot be established from them. Established cell lines are either derived from a tumour (e.g. HeLa cells, Friend leukaemia cells) or embryonic tissues (e.g. BHK cells), or have become immortalized as a result of having undergone a TRANSFORMATION which causes the cells to behave in some respects as tumour cells (e.g. CHO cells). Numerous cell lines representing many different cell types have been established. Collections of cell lines are kept, for example, at the European Collection of Animal Cell Cultures at Porton, UK, and the American Type Culture Collection (ATCC) at Rockville, Maryland, USA.

cell lineage The collection of descendants of a single cell during normal development. The term is more usually reserved for the experimental situation of following, or mapping, these descendants and their ultimate FATES by direct observation or by some marking method that allows the progeny of the parent cell to be identified. The expression cell lineage analysis does not formally differ from fate mapping, but in practice the first is used for mapping the progeny of single cells while the latter implies that the progeny of a group of cells is studied.

cell lineage analysis The progeny of an individual cell can be identified in various ways, such as simple observation (in the nematode *Caenorhabditis elegans*), by injecting a tracer molecule such as HORSERADISH PEROXIDASE (most systems), or by genetically marking cells by X-ray induced MITOTIC RECOMBINATION (in *Drosophila*). Such analysis has revealed that the nematode, in contrast to vertebrates and insects, has an almost invariant lineage. Lineage analysis can reveal when restrictions in cell fate are imposed in development. For example, before GASTRULATION in the zebrafish (*see* ZEBRAFISH DEVELOPMENT), individual cells give rise to progeny of various tissue types whereas after gastrulation the progeny of individual cells are restricted to a single tissue type. Analysis of cell lineage in *Drosophila* and vertebrates has revealed the existence of lineage restrictions or COMPARTMENTS. In EYE DEVELOPMENT, lineage analysis established that the ommatidia are not clonal units, thus initiating the investigation of the role of CELL–CELL INTERACTIONS in the development of this structure. *See also*: CAENORHABDITIS DEVELOPMENT; DROSOPHILA DEVELOPMENT.

cell locomotion *See*: CELL MOTILITY.

Cell–matrix interactions in development

FOR many years the insoluble material present between cells — the EXTRACELLULAR MATRIX (ECM) — was thought to have a simple structural role in maintaining the shape of the developing and adult animal. In recent years, however, it has become clear that the ECM comprises a complex mixture of PROTEOGLYCANS and GLYCOPROTEINS (*see* EXTRACELLULAR MATRIX MOLECULES) secreted by cells that also have important instructive influences on the behaviour not only of the secreting cells themselves but

Table C6 Extracellular matrix molecules

Glycoproteins
Collagen (types I–XIV)
Elastin
Fibrinogen
Fibronectin
J1160/180
L1
Laminin
Merosin
NCAM
Nidogen (also called ectactin)
Tenascin (also called cytoactin, hexabranchion,
    myotendinous antigen, J1200/220)
Thrombospondin
Vitronectin
von Willebrand factor

Proteoglycans
Chondroitin sulphate proteoglycans
Heparan sulphate proteoglycans
Keratan sulphate proteoglycans

also on other adjacent cell types. Widely differing aspects of cell behaviour ranging from migration to differentiation are influenced by the ECM. Here we illustrate its various roles with selective examples and highlight some of those areas in which significant unanswered questions remain.

Some of the best characterized ECM molecules are shown in Table C6 (*see also* EXTRACELLULAR MATRIX MOLECULES). Many other potential ECM components have been described recently and the list seems certain to expand in the next few years. Of the best characterized proteins, collagen and elastin were originally thought to have a largely structural role. However, some collagens have also been shown to stimulate cell migration, suggesting that the distinction between structural and adhesive ECM molecules is not clear-cut.

To alter cell behaviour, ECM molecules must interact directly with the cell, and these interactions are mediated via cell surface receptors which are often concentrated in specialized structures called focal adhesions or hemidesmosomes (*see* CELL JUNCTIONS). The best defined class of such receptors is the INTEGRINS which are heterodimeric glycoproteins composed of an α and a β chain. The integrins can be subdivided into four major and a number of minor groups, with each member of the group sharing the same β chain but being distinguished by distinct but homologous α chains. These different combinations of α and β chains confer specificity for different ECM components. By way of example, the specificities of the different αβ dimers in the β_1 group are shown in Table C7 (see [1] for a detailed review). Other receptors for ECM molecules have also been recognized, but are much less well defined. An interesting example is the enzyme galactosyltransferase which can mediate cell spreading on laminin substrates [2]. Cell surface homologues of ECM molecules themselves may also provide an important class of receptors. Proteoglycans have been shown to bind to collagen, fibronectin, tenascin and thrombospondin, and could therefore act as cell matrix receptors when attached to transmembrane core proteins.

Migration

There are many spectacular examples of precisely targeted cell migration over long distances during development. A particularly well studied example is neural crest cell migration in vertebrates, in which cells migrate from the neural tube to form much of the peripheral and autonomic nervous system and a range of structures including the melanocytes in the skin (*see* AVIAN DEVELOPMENT; VERTEBRATE NEURAL DEVELOPMENT).

ECM components are important in stimulating migration. This is well illustrated by the role of the ECM protein fibronectin in neural crest cell migration. Neural crest cells will migrate on fibronectin substrates in cell culture. ANTIBODIES or synthetic peptides that block the interaction of fibronectin with its cell-surface integrin receptor block migration of neural crest cells both in cell culture and *in vivo*, confirming the role of cell–matrix interactions in this migration (further information on fibronectin is provided in [3]).

The ECM glycoproteins laminin and thrombospondin can also stimulate cell migration. They have been implicated in the control of migration *in vivo* by virtue of their ability to stimulate migration in cell culture and their presence in the substrate of migratory cells. Laminin, for example, is present within the developing central nervous system (CNS) at the time of axon outgrowth (which, it could be argued, represents migration of the neuronal growth cone) and *in vitro*, laminin is a potent stimulator of neurite outgrowth [4].

Studies examining the effects of single ECM molecules in isolation may be misleading, as it is likely that more than one ECM component will be involved in any particular cell migration *in vivo*. Evidence for this has come from experiments using tissue slices to mimic the complex migratory substrate present *in vivo*. Granule cell neurons, which normally migrate into the developing cerebellum along radially oriented Bergmann glia (*see* GLIAL CELLS), will follow their normal migratory pathway when placed on tissue slices of cerebellum. Their migration can be inhibited by antibodies against the ECM glycoproteins tenascin and thrombospondin [5,6] as well as by antibodies against cell-surface CELL ADHESION MOLECULES such as NCAM. However, these antibodies block migration at different points along the pathway suggesting that migration is regulated by more than one ECM component and that each of these components acts at a different stage in migration.

As ECM molecules stimulate cell migration the question arises as to whether they are also responsible for the precise guidance of

Table C7 Ligand specificities for various members of the β_1 group of integrins

Receptor	Ligand
$\alpha_1\beta_1$	Collagen/laminin
$\alpha_2\beta_1$	Collagen/laminin
$\alpha_3\beta_1$	Collagen/fibronectin/laminin
$\alpha_4\beta_1$	Fibronectin*
$\alpha_5\beta_1$	Fibronectin*
$\alpha_6\beta_1$	Laminin
$\alpha_7\beta_1$	Laminin

* $\alpha_4\beta_1$ and $\alpha_5\beta_1$ bind to different regions of fibronectin.

migrating cells which is evident during development. The attractive hypothesis that an adhesive molecule such as fibronectin could form a simple track seems unlikely. Fibronectin is widespread during development, being found in areas adjacent to migratory pathways in regions not entered by migrating cells. Moreover, some migrating cell types synthesize fibronectin, which would enable them to move independently of such a track [7]. One alternative possibility is that the ECM provides physical cues, perhaps in the form of oriented fibrils, or that simple barriers exist made of closely opposed layers of cells. There is evidence of both these mechanisms operating in neural crest cell migration.

Some ECM molecules may, however, guide cell migration by an entirely different mechanism. Not all ECM glycoproteins associated with migration are adhesive. Tenascin, for example, is found in association with cell migration during both development and wound healing, but seems to be anti-adhesive for some cell types. When used as a substrate for CNS neurons in cell culture tenascin produces little adhesion and when added to fibronectin substrates it counteracts the adhesive effect of fibronectin on cell spreading [8]. An antiadhesive function has also been suggested for the proteoglycan chondroitin sulphate proteoglycan.

Anti-adhesive molecules could regulate migration in two ways. They might prevent migration by inhibiting adhesion, but, alternatively, they could stimulate migration by ensuring that cells do not adhere so tightly to their substrate that they are unable to move at all. A distinct localization of anti-adhesive molecules among a more general distribution of adhesive ECM molecules would then provide a precisely delineated pathway for cell migration. This attractive hypothesis will be tested as our understanding of this class of molecules grows.

Differentiation

Simple cell culture experiments show that the ECM can both stimulate and inhibit differentiation. Fibronectin, for example, inhibits the differentiation of myoblasts into myotubes (*see* MYO-GENESIS) but will stimulate the differentiation of neural crest cells into adrenergic neurons. Such experiments do not distinguish whether the role of the ECM is instructive or simply permissive, but they do prove that the ECM has an important effect on cell differentiation during development.

One case where the ECM is known to be involved in the control of differentiation *in vivo* is skin formation. The keratinocytes that form the superficial layer of the skin arise from a proliferative cell population adjacent to the basal lamina in the basal layer of the epidermis. Keratinocyte differentiation commences when one of these cells loses its attachment to the basal lamina, drops out of division and starts to synthesize keratin (*see* INTERMEDIATE FILA-MENTS). As the keratinocytes forming the outermost layer of skin are shed they are replaced by new keratinocytes developing from the proliferating cell population.

This differentiation is preceded by a loss in adhesiveness for fibronectin and other ECM components such as laminin and collagen types I and IV. This seems likely to be important in allowing the cells to become detached from the basal lamina. But loss of contact with the ECM molecules of the basal lamina may actually provide the stimulus for differentiation. Fibronectin can inhibit keratinocyte differentiation [9], suggesting that contact with this and other ECM molecules on or adjacent to the basal lamina may prevent differentiation *in vivo* until such time as adhesion is lost and the cells move to a superficial position in the epidermis.

As well as their role in development, ECM molecules may also be essential for the maintenance of the differentiated state. Differentiated mammary epithelial cells cease production of milk protein when placed in culture. If, however, they are cultured in the presence of basal lamina-like material they maintain milk protein production suggesting that it is the interaction with the basal lamina that maintains their differentiated state [10].

A spectacular example of the importance of the ECM in the maintenance of differentiation is provided by the process of repair in denervated muscles. Following denervation, the axon will regenerate, reinnervate the muscle at the original site of the previous NEUROMUSCULAR JUNCTION and form a new functional junction. The information for the precise targeting is contained in the basal lamina around the muscle fibre rather than in the fibre itself, as the same targeting is seen if the fibre is frozen and killed but the basal lamina left intact [11]. Guidance of the axon terminal to the site of the neuromuscular junction must therefore result from ECM-derived molecular clues present within the basal lamina. One promising candidate molecule is S-laminin, a variant of laminin concentrated within the ECM at the neuromuscular junction [12].

The ECM also plays a part in the subsequent formation of the new neuromuscular junction. Another protein associated with the basal lamina — agrin — is also found concentrated adjacent to the old neuromuscular junction. It plays a part in the aggregation of acetylcholine NICOTINIC RECEPTORS on the muscle surface into the newly formed junction [13].

Indirect effects

In addition to its direct effects, the ECM can act as a scaffold to bind non-matrix molecules that then act on the cell. This role of the ECM as a 'molecular pinboard' is being increasingly recognized as an important part of the interaction of GROWTH FACTORS with cells. Binding to the ECM could increase the effective local concentration of a growth factor, or could ensure that different growth factors can be presented together to the cell.

A good example of the indirect role of ECM components in the action of growth factors comes from work on the proliferation of haemopoietic precursor cells in bone marrow culture. Haemopoietic precursors are stimulated to proliferate by added growth factors but cell proliferation can be abolished by treatment with the enzyme heparinase, which degrades the heparin sulphate proteoglycans of the ECM [14]. This suggests that these ECM components are essential for the interaction of the growth factor with the cell.

Some clues as to how the ECM might interact with and potentiate growth factors are provided by work on fibroblast growth factors (FGFs) (reviewed in [15]). Heparin (which mimics heparan sulphate proteoglycan) enhances the activity of acidic FGF, possibly by altering its conformation, and such a mechanism is likely to operate in intact matrix. In addition the stability of basic FGF is enhanced by matrix binding, which would also be expected to lead to increased growth factor activity.

These examples of indirect effects are certainly only the tip of the iceberg and many other cases in which the ECM is used as a pinboard to deposit information during development will emerge. It is clear that the ECM can no longer be thought of as a static mixture of molecules but rather as a material that is constantly being altered by adjacent cells so as to provide appropriate instructive cues that act both immediately and later in development.

Conclusion

Substantial questions remain, however, as to the role of the ECM. First, different forms of many ECM molecules exist as a result of either alternative RNA SPLICING (fibronectin) or GENE DUPLICATION (laminin, thrombospondin). The function of these different forms remains poorly understood although, as described above, the laminin variant S-laminin seems to have a role in the formation of the neuromuscular junction. The differently spliced forms of fibronectin are expressed at different times and places during development suggesting that this molecular heterogeneity will turn out to be important. Second, the role of the different integrins during development remains to be established. We need to know how the different α chains are distributed, as the early studies using antibodies to the shared β chains are unable to distinguish integrins of widely differing specificities. Third, the mechanisms by which the ECM can alter gene expression (see section on differentiation above) are obviously a fundamental part of the effect of the ECM on cell behaviour but remain poorly understood.

A more general problem has also emerged from recent work. Our knowledge of the molecular biology of extracellular matrix molecules and their corresponding cell surface receptors has not been matched by an increased understanding of exactly how these molecules regulate cell behaviour *in vivo*. Many cell culture experiments have been performed on neoplastically TRANSFORMED CELL lines and it is difficult to extrapolate from these to embryonic cells that may change their phenotype very rapidly. Very few studies have attempted to address function by blocking specific ECM components *in vivo*, in part owing to the difficulty of ensuring that blocking by antibodies or by peptides is specific. The ability to knock out or overexpress specific ECM molecules or their receptors in TRANSGENIC animals offers a chance of overcoming this problem, and the application of these techniques should enable the rapid increase in our knowledge of the molecular biology of the ECM to be translated into an understanding of its complex role in the intact animal.

S. SCOTT-DREW

C.K. FFRENCH-CONSTANT

See also: GLYCANS.

1 Humphries, M.J. (1990) The molecular basis and specificity of integrin–ligand interactions. *J. Cell Sci.* **97**, 585–592.
2 Runyan, R.B. et al. (1988) Functionally distinct laminin receptors mediate cell adhesion and spreading: the requirement for surface galactosyltransferase in cell spreading. *J Cell Biol.* **107**, 1863–1871.
3 Hynes, R.O. (1990) *Fibronectins* (Springer-Verlag, New York).
4 Timpl, R (9189) Structure and biological activity of basement membrane proteins. *Eur. J. Biochem.* **180**, 487–502.
5 Chuong, C-M., et al. (1987) Sequential expression and differential function of multiple adhesion molecules during the formation of cerebellar cortical layers. *J. Cell Biol.* **104**, 331–342.
6 O'Shea, K.S. et al. (1990) Deposition and role of thrombospondin in the histogenesis of the cerebellar cortex. *J. Cell Biol.* **110**, 1275–1283.
7 ffrench-Constant, C. & Hynes, R.O. (1988) Patterns of fibronectin gene expression and splicing during cell migration in chicken embryos. *Development* **104**, 369–382.
8 Chiquet-Ehrismann, R. (1991) Anti-adhesive molecules of the extracellular matrix. *Curr. Opinion Cell Biol* **3**, 800–804.
9 Adams, J.C. & Watt, F.M. (1989) Fibronectin inhibits the terminal differentiation of human keratinocytes. *Nature* **340**, 307–309.
10 Li, M.L. et al. (1987) Influence of a reconstituted basement membrane and its components on casein gene expression and secretion in mouse mammary epithelial cells. *Proc. Natl. Acad. Sci. USA* **84**, 136–140.
11 Sanes, J.R. et al. (1978) Reinnervation of muscle fibre basal lamina after removal of myofibres. Differentiation of regenerating axons at original synaptic sites. *J. Cell Biol.* **78**, 176–198.
12 Hunter, D.D. et al. (1989) A laminin-like adhesive protein concentrated in the synaptic cleft of the neuromuscular junction. *Nature* **338**, 229–234.
13 Nitkin, R.M. et al. (1987) Identification of agrin, a synaptic organizing protein from *Torpedo* electric organ. *J. Cell Biol.* **105**, 2471–2478.
14 Roberts, R. et al. (1988) Heparan sulphate bound growth factors: a mechanism for stromal cell mediated haemopoiesis. *Nature*, **332**, 376–378.
15 Klagsbrun, M. (1990) The affinity of fibroblast growth factors (FGFs) for heparin; FGF-heparan sulfate interactions in cells and extracellular matrix. *Curr. Opinion Cell Biol.* **2**, 857–863.

cell-mediated immune response/immunity Immune response or immunity due to T CELLS. *Cf.* HUMORAL IMMUNE RESPONSE. *See also*: T CELL DEVELOPMENT.

cell migration While many cells remain essentially static for most of their life, other cells are capable of considerable movement with respect to their surrounding tissues. The process of cell migration is made possible only by the plasticity of the cell's cytoskeleton which is capable of sustained and rapid reorganization. Effective movement demands that the cell can be polarized such that the plasma membrane is quiescent everywhere except at the leading edge, where lamellipodia and microspikes project in a forward direction (*see* CELL MOTILITY). During migration, MICROTUBULES and actin MICROFILAMENTS act in concert, the microtubules functioning to recycle membrane vesicles towards the leading edge, whereas the actin-based motors help to extend lamellipodia. Migration also depends on adhesion to the extracellular substrate, and involves many different types of EXTRACELLULAR MATRIX MOLECULE. Cell migration is of great importance in development; examples are the vertebrate SOMITE cells that migrate into the developing limb buds, and the NEURAL CREST, a neuroectodermal population of cells which migrates from the dorsal part of the neural tube to form the ganglia of the peripheral nervous system and melanocytes. Specific cues in the extracellular matrix of developing tissues, along with the differential expression of adhesion molecules on the developing cells, are thought to guide migratory populations during development, and thus the twin roles of the extracellular matrix (substrate adhesion and guidance) overlap. *See also*: CELL–MATRIX INTERACTIONS.

Cell motility

MOTILITY is a broad term encompassing intracellular movements, cell shape changes, and the movement of portions of the cell, as well as the movement of cells from one place to another. But the mechanisms responsible for many of these movements are similar from a molecular viewpoint. Although the hardware is beginning to be well understood, the way in which motor systems are controlled continues to provide puzzles at a molecular level.

Types of motility

In addition to the 'passive' Brownian movement of molecules, there are many active motile processes within a eukaryotic cell: these include bulk cytoplasmic streaming, chromatid movement in MITOSIS, the transport of vesicles to or from the plasma membrane (*see e.g.* ENDOCYTOSIS; EPITHELIAL POLARITY; EXOCYTOSIS; PROTEIN TARGETING; PROTEIN SECRETION), as well as the active redistribution of organelles within the cell and of molecules within the plane of the plasma membrane. Cessation of active movement within the cytoplasm is usually a prelude to cell death.

Not only may appendages of eukaryotic cells, such as CILIA and MICROVILLI move, but the cell may change shape actively as in spreading, phagocytosis, cytokinesis (*see* MITOSIS) and MUSCLE contraction. In some cases protrusion of a portion of the cell may be a consequence of contraction elsewhere (blebbing), but in other cases there seems to be active local protrusion as, for example, in the production of filopodia or lamellipodia (see Fig. C29 below).

Locomotion, the movement of a whole cell from one location to another, is the most complex manifestation of motility and requires an intracellular motor system, the coupling of the motor to the environment, and the interpretation of 'traffic rules' by the cell. By moving, unicellular microorganisms can potentially improve their access to food or light. In multicellular organisms locomotion enables the complex reorganization of cell groups in morphogenesis, and important in wound healing, immune surveillance, and tumour invasion.

Motor systems

There are three well-studied and commonly occurring biological motor systems, the bacterial flagellum, actin-associated motors, and microtubule-associated motors. Other motor systems do exist but are rarely encountered. In all cases the basic problem is to transduce chemical energy into mechanical work and a variety of solutions have been devised. Some are direct, utilizing reversible conformational change in proteins with a hydrolysis step required in each cycle; others are very indirect, involving, for example, a local increase in colloid osmotic pressure which generates expansive hydrostatic forces within a delimited compartment (e.g. in the discharge of coelenterate nematocysts).

The mode of action of bacterial flagella and control of bacterial chemotaxis is considered in detail elswhere (*see* BACTERIAL CHEMOTAXIS) and this article will concentrate on the mechanisms of motility in eukaryotic cells.

Actin-associated motor systems

There seem to be two types of actin-associated motor mechanism, those that involve MYOSIN (an actin-activated ATPase) and those that involve ACTIN-BINDING PROTEINS. The actin–myosin motor in its most specialized and highly ordered form is responsible for the contraction of striated MUSCLE and depends on the sliding of thin actin filaments (F-actin, a polymer of globular G-actin, stabilized with tropomyosin) as a result of their interaction with thick bipolar myosin II filaments. The actin–myosin contraction cycle is depicted in Fig. C28.

A simpler form of myosin (myosin I) is found in nonmuscle cells as well as conventional myosin II. The head of myosin I behaves like that of myosin II, but the tail is attached to membranes that can therefore be dragged towards the barbed end of an actin filament (cytoplasmic microfilament). Whether the actin–myosin motor is highly ordered (as in striated muscle) or arranged as a loose two-dimensional mesh (e.g. the terminal web underlying microvilli in epithelial cells), the basic principle remains the same, although the calcium-based control mechanism may be myosin-associated in nonmuscle systems. Deletion of myosin II from cells of the cellular slime mould *Dictyostelium discoideum* does not inhibit locomotion, although it does affect the control of turning and the process of cytokinesis.

Other actin-associated motors depend on the formation of cross-linked microfilament meshworks or bundles and do not necessarily involve myosin. A similar meshwork of microfilaments underlying the plasma membrane (cortical meshwork) gives mechanical support to the cell surface. Whether movement is generated as a result of directed assembly of a gel or as a result of altered osmotic pressure when components of the gel are cleaved remains contentious. Deletion of some actin-binding proteins (α-actinin and severin) does not apparently inhibit movement, but there may be some redundancy. Bundles of parallel microfilaments may have a structural role (as in microvilli) or their rapid formation may cause protrusion (as in the acrosomal process of some sea urchins). In this latter case movement depends on assembly processes, in contrast to the sliding inter-

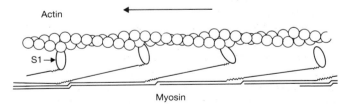

Fig. C28 Schematic diagram of 'sliding filament' model of muscle contraction involving a cycle of attachment and detachment of myosin from actin filaments. Conformational change in the S1 region of the myosin heavy-chain head while it is attached to the actin filament moves the myosin filament relative to the actin. Release of the actin requires ATP binding to the myosin and its hydrolysis. After ATP hydrolysis the myosin head returns to its starting conformation, rebinds to an actin monomer further along the filament and the cycle is then repeated. Because the two ends of the myosin filament pull in opposite directions they cause the antiparallel actin filaments of each half of the sarcomere to slide towards one another. The filaments themselves do not contract but slide relative to one another.

actions driven by reversible conformational change in the actin–myosin motor which causes contraction of membrane-attached antiparallel microfilament bundles (stress fibres) in tissue cells.

Actin-based motors in which the assembly and disassembly of microfilaments plays an important part — either directly as a mechanism or in order to relocate the motor as the cell proceeds — are generally sensitive to CYTOCHALASINS (drugs that inhibit F-actin polymerization), whereas systems in which the actin filaments are more or less permanent and stabilized by tropomyosin are insensitive. Thus cell locomotion and cytokinesis are cytochalasin-sensitive whereas contraction of striated muscle and stress fibres are not.

Microtubule-associated motors

Microtubule-associated motors are of two kinds: those in which the microtubule acts as a static 'rail' with which motor molecules analogous in function to myosin interact; and those in which assembly/disassembly may provide the motive force. In the first category are the two proteins DYNEIN and KINESIN. Dynein is an ATPase first described from the flagella and cilia of eukaryotic cells where it forms the side arms of the outer doublets of the axoneme (*see* MICROTUBULES). It moves towards the minus ' – ' or disassembly end of microtubules by repeated cycles of attachment to a microtubule, conformational change and release, energy being consumed in the cycle.

In the axoneme the dynein is attached to the A-tubule of one microtubule doublet and interacts with the B-tubule of the adjacent doublet; conformational change in the dynein brings about the sliding of doublets relative to one another and causes the cilium to bend (although a complex control mechanism must ensure that only half of the doublets are active at any one time). The cytoplasmic form of dynein (formerly known as microtubule-associated protein 1C or MAP 1C) brings about anterograde movement of vesicles and intracellular particles (*see* MICROTUBULE-BASED MOTORS).

A second motor molecule, KINESIN, moves attached particles towards the distal plus, ' + ', or assembly end of microtubules. The activities of dynein and kinesin are thought to be responsible for (bidirectional) AXONAL TRANSPORT, for the maintenance of cytoplasmic organization by continually directing the movement of membranous organelles through the cytoplasm, for many of the different movements of mitosis and, in the case of dynein, for ciliary movement.

Although the details are not yet clear it is possible that some of the movements of mitosis, particularly the poleward movement of chromatids at anaphase A, may depend on disassembly of microtubules at the kinetochore, although an active motor system in the kinetochore is required as well.

Other systems

The motors described above are the most widely distributed, but there are a number of other systems where the mechanism is different, or in some cases, not understood. Pure conformational change drives rapid contraction and expansion of the spasmoneme in the stalk of sessile protists such as *Vorticella*; the movement of the amoeboid sperm of nematodes does not seem to depend upon conventional motor molecules; and the gliding movement of bacteria and gregarine protozoans are brought about by unknown means.

Coupling the motor to the environment

Forward movement of any object necessitates an equal and opposite force acting upon the immediate environment. Thus the movements of organelles through the cytoplasm involve the interaction of motor molecules with static elements that must resist the forces acting upon them. This problem is particularly obvious when considering the locomotion of cells, which must find a means of coupling their internal motor systems to the outside world.

Two major modes of locomotion are recognized — swimming and crawling. In the former the cell moves forwards by causing a rearward acceleration of the fluid environment, either small amounts at high speed or large volumes at low speed. For microscopic objects like cells the major problem is viscous drag rather than inertial resistance (the Reynold's number is very small) and movement will cease almost instantaneously when effort stops; cells do not glide to a halt if they stop swimming. Bacteria swim by rotating their flagella (a mechanism analogous to that of a ship's propellor), eukaryotic cells by beating cilia or by generating waves in fields of cilia. In both cases there is a rearward flow of fluid and, if the cell is anchored, as in ciliated epithelia, for example, fluid will simply flow over the cell. Free-living unicells need to move to find the optimum environment, and the ability to move fluid over a sheet of cells is important in multicellular organisms from sponges to mammals.

Crawling motion requires that the cell form some sort of transient anchorage and by acting upon this fixed point the body of the cell may be moved forward (Fig. C29). The best understood crawling cells are tissue cells of multicellular organisms, in particular FIBROBLASTS and LEUKOCYTES, and the individual amoebae of cellular slime moulds. Anchorage may be mechanical (as for example the interaction of leukocytes with fibrillar elements of three-dimensional matrices such as connective tissue) or may be adhesive, as will almost always be the case for cells moving over a planar substratum.

For most eukaryotic tissue cells a more or less resistive substratum is essential for forward movement and swimming is impossible. Forward movement therefore usually requires that transient adhesions be formed with the substratum, adhesions that can support the contractile forces exerted by the motor machinery. Unless the mechanical properties of the substratum are sufficient to resist the forces exerted by the moving cell then the net result will be to drag material (or small particles) towards the cell rather than move the cell forward. New adhesions must continually be formed at distal sites and this requires protrusion from the front of the cell.

Formation of protrusions seems to depend upon an actin-based system and myosin II is apparently not involved; on the other hand, bringing the body of the cell towards the attachment site probably involves the sliding of antiparallel microfilaments mediated by myosin. Microfilaments are attached to the cytoplasmic

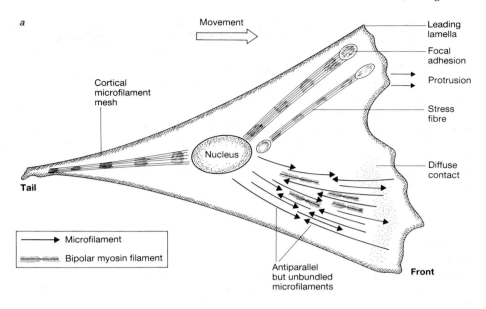

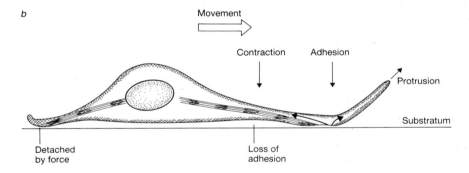

Fig. C29 Composite diagrams of a moving fibroblast. *a*, Viewed from top; *b*, viewed from the side.

side of the adhesion sites by a complex array of proteins including talin and vinculin (*see* ACTIN-BINDING PROTEINS); the antiparallel microfilaments have a perinuclear attachment site. In slow moving or stationary cells these attachment points may be discrete focal adhesions (*see* CELL JUNCTIONS) with large microfilament bundles attached; in fast moving cells like leukocytes the attachment points are diffuse and bundles do not form. Microfilament bundles are known to be contractile and have a periodicity reminiscent of sarcomeres, but whether contraction is important when bundles are not present is unclear.

For effective locomotion, an internal mechanism is needed to define an axis of polarity; random protrusions will not lead to efficient movement. Generally, the area of greatest protrusive activity will by default be the front of the moving cell. Some cells, notably epithelial cells, seem to lack the ability to restrict protrusions to a limited area, and in isolation will make protrusions in all directions leading to cell spreading rather than translocation. Other cells such as fibroblasts seem to be able to maintain their polarity and the commitment to a limited region for protrusion leads to a short-term tendency to persist in moving in one direction. How internal polarity is specified and maintained remains an interesting problem and although it can be perturbed

(e.g. by COLCHICINE treatment, which disrupts microtubules) the mechanism does not seem to depend simply upon microtubules.

Control of motile behaviour

As already implied, internal systems control motility — both the rate at which the motor operates and the direction in which force is generated. For bacteria the major determinant of movement is the direction of flagellar rotation and the frequency of reversal. In eukaryotic ciliates the distribution of the cilia over the cell surface and the direction of the effective stroke of each cilium will be the main factors, although with flagella (which are longer) it is the direction of wave propagation that determines whether the cell is pushed or pulled. For crawling cells a major factor is where protrusion occurs, determined by the intrinsic polarity of the cell, though the rate of protrusion is also regulated.

In the absence of any external cues (i.e. in an isotropic environment) movement is generally a random walk — in three dimensions for swimmers, in two dimensions for crawlers — with some short-term tendency to persistence. Factors that alter the rate of movement or the frequency of turning (kinetic factors) may act either upon the motor or upon its coupling to the environment. Thus a positively chemokinetic factor may increase the speed of movement (orthokinesis) or the frequency with which alterations in direction occur (klinokinesis) (Fig. C30). Reducing the adhesiveness of a surface may prevent spreading and inhibit movement; but making the surface very adhesive (or increasing the half-life of adhesions) may also inhibit movement by making detachment impossible. Such effects are non-directional and do not provide directional (vectorial) information.

Most environments are, however, anisotropic, and various factors influence the direction of movement. Gradients of many environmental properties may affect the cell in a vectorial fashion so that there is a tendency for the cell to move up or down the gradient. Perception of gradients requires a sensory mechanism within the cell, although the response is forced upon the cell by anisotropic stimulation of the motor, not by any cognitive effort. In haptotaxis (the response to a gradient of substratum adhesiveness) cells may become trapped and accumulate at the adhesive end of the gradient, but they may not have responded directionally and arguably this is not therefore a true tactic response. In the case of chemotaxis (response to a diffusible chemical) it seems likely that the region of the cell surface that binds the most factor will become the front because an increase in receptor occupancy

gives a positive feedback to the protrusive motor process. Moving up-gradient causes progressive increases in receptor occupancy and continued stimulation at the front; steady state levels lead to adaptation in the receptor system and, in the absence of positive reinforcement, movement reverts to a random pattern, if there is a basal level of motile activity, or ceases completely. Other types of gradient may elicit a response although the sensing mechanism is not always understood: phototaxis is a response to light, galvanotaxis to electric fields, magnetotaxis to magnetic fields (usually the Earth's). Tropic responses (not to be confused with trophic, growth, responses) are tactic responses in which only a part of the cell moves — for example the nerve growth cone in axonal extension.

In environments in which there are aligned fibrillar elements or regular topographic cues the direction of cell movement may be constrained to an axis (in 2-D) or a plane (in 3-D), a guidance phenomenon (sometimes called contact guidance) that is seen, for example, when fibroblasts crawl over surfaces with regular grooves or over aligned collagen fibres. In the latter case it is also possible to think of the surface as being anisotropically rigid with greater resistance to deformation along the axis of the fibres.

Another important determinant of the movement of metazoan cells is their interaction with other cells. In contact inhibition of locomotion cells cease moving following collision with another cell and will attempt to move in another direction. If surrounded by other cells movement ceases, and cells in culture tend to form monolayers rather than piling up. Cells that fail to obey this simple rule will be able to move into territory occupied by other cells — to invade. Contact inhibition of movement may be reciprocal (both cells respond to contact) or nonreciprocal (only one of the colliding cells reacts). Some tumorigenic cells show contact inhibition of locomotion when they contact one another (which will promote dispersion) but not when contacting normal cells (which will facilitate invasion). Because these behavioural responses are not based upon absolute rules but are probabilistic the phenomenon has been difficult to study and the mechanism remains uncertain.

Although the molecular basis for many motor systems is becoming much better understood, the control of motile activity remains unclear and many complex behavioural responses have not even been fully described. Much remains to be done before it will be possible to give a complete molecular explanation of how, for example, a particular group of cells migrates through a developing embryo to a specific location and then stops.

Further information on the topic of cell motility will be found in [1–6].

J.M. LACKIE

1 Bellairs, R. et al. (1982) *Cell Behaviour* (Cambridge University Press).
2 Bershadsky, A.D. & Vasiliev, J.M. *Cytoskeleton* (Plenum, New York).
3 Bray, D. (1992) Cell Movements (Garland, New York).
4 Heaysman, J.E.M. et al. (1987) Cell Behaviour: Shape, Adhesion and Motility. *J. Cell Sci. Suppl.* **8**.
5 Lackie, J.M. (1986) *Cell Movement and Cell Behaviour* (Allen & Unwin, London).
6 Preston, T.M. et al. (1990) *The Cytoskeleton and Cell Motility* (Blackie, Glasgow).

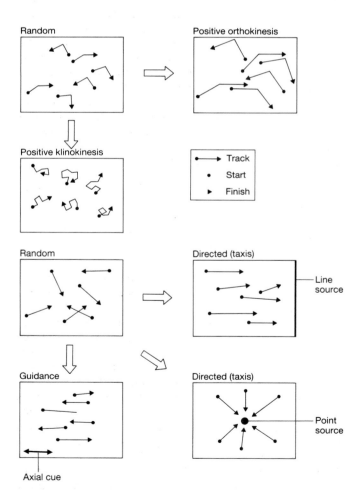

Fig. C30 Types of movement. Kinesis, taxis and guidance.

cell-nonautonomous In development, applied to genes and their products, or to a mutation, whose actions are not restricted to the cell in which they are expressed.

cell polarity *See*: EPITHELIAL POLARITY.

cell signalling Intercellular communication, the influencing of the behaviour of one cell by another, is carried out in two main ways. The first is by secreted molecules, such as hormones, growth factors, or neurotransmitters, which are secreted by one type of cell and act on distant or nearby tissues, depending on whether they are secreted into the bloodstream or directly into the intercellular medium. With the exception of the STEROID HORMONES and other lipid-soluble substances, these secreted signal molecules bind to cell-surface receptors on the target cell. The signal set up by binding of ligand to receptor is transduced across the cell membrane by the receptor to activate intracellular signalling pathways which lead to a specific response by the cell to the stimulus (*see* e.g. G PROTEIN-COUPLED RECEPTORS; GROWTH FACTORS; GROWTH FACTOR RECEPTORS; SECOND MESSENGER PATHWAYS; SYNAPTIC TRANSMISSION and individual named receptors).

The second main means of intercellular communication is by direct cell–cell contact by cell-surface adhesion receptors such as INTEGRINS. The nature of the signal transmitted by cell–cell contacts is not known in most cases, but is thought to involve the cytoskeleton (*see* CELL ADHESION MOLECULES; LECTINS; SELECTINS). *See also*: CELL–CELL INTERACTIONS.

cell sorting (1) In developmental biology, the *in vitro* phenomenon where heterogeneous populations of cells separate into aggregates of homogeneous cell type. Important conceptually in the formation of the differential adhesion theory of tissue morphogenesis.

(2) Technique for the preparative or analytic separation of cells of different types, usually on the basis of the presence or absence of particular membrane proteins to which specific antibodies can be raised. Exquisitely specific MONOCLONAL ANTIBODIES are usually used. There are several ways in which cells may be sorted. One method, also called cell panning, is to attach the appropriate antibody or other cell-specific reagent to a solid support such as a tissue culture dish. A heterogeneous cell suspension is then placed in the dish and cells bearing the membrane protein will adhere to the dish; those without it can be washed off. The second widely used method uses a fluorescence-activated cell sorter (FACS). In this procedure, the mixed population of cells is incubated with a fluorescently labelled antibody against a particular surface protein. The cell suspension is then drawn into a stream of fine droplets, each droplet containing no more than one cell. The FACS assays each droplet for the presence or absence of the fluorescent marker using a laser beam and detector, and confers a charge on any droplet containing a labelled cell. The stream passes between two charged plates, and the voltage difference between the plates draws the labelled cells into a separate stream. FACS analysis and other methods of cell sorting are widely used in immunology as a means of separating lymphocytes of different types and at different developmental stages, and in medicine to determine which cell type is affected in diseases such as leukaemia and immunodeficiency.

cell-surface receptors Receptors located in the plasma membrane of cells.

1 Signalling receptors. Receptor proteins for most hormones, neurotransmitters, growth factors, antigens, and other chemical and physical signals (e.g. light) are transmembrane proteins located in the plasma membrane of cells, with the receptor site exposed to the external environment. The signal generated by stimulation of the receptor by the binding of its specific ligand is transduced across the membrane by a conformational change in the receptor which activates the cellular response machinery in the cytoplasm. In all these cases entry of the ligand itself into the cell is not required (*cf.* STEROID RECEPTOR SUPERFAMILY). *See*: EXCITATORY AMINO ACID RECEPTORS; G PROTEIN-COUPLED RECEPTORS; GABA AND GLYCINE RECEPTORS; GROWTH FACTOR RECEPTORS; INSULIN AND ITS RECEPTOR; MUSCARINIC RECEPTORS; NICOTINIC RECEPTOR; OLFACTORY TRANSDUCTION; SECOND MESSENGER PATHWAYS; T CELL RECEPTOR; VISUAL TRANSDUCTION; VOLTAGE-GATED ION CHANNELS.

2 Adhesion receptors. Specific receptor proteins in the plasma membrane mediate adhesive interactions between cells, and between cells and extracellular matrix. In some cases such adhesive interactions may generate a signal which is transmitted via the receptor to produce a cellular response. *See*: CELL ADHESION MOLECULES; CELL–MATRIX INTERACTIONS; EXTRACELLULAR MATRIX MOLECULES; INTEGRINS; LECTINS; SELECTINS.

3 Transport receptors. Receptor proteins for large molecules such as cholesterol-carrying lipoproteins or the iron transport protein transferrin, which cannot pass the plasma membrane, are also located in the plasma membrane. Binding of ligand is followed by receptor-mediated ENDOCYTOSIS which internalizes the ligand. Receptor-mediated endocytosis of viruses bound to a variety of cell-surface receptors is also a means by which certain viruses gain entry to animal cells. *See*: ANIMAL VIRUSES; PLASMA LIPOPROTEINS AND THEIR RECEPTORS.

cell wall *See*: BACTERIAL ENVELOPE; PLANT CELL WALL MACROMOLECULES.

3T3 cells Mouse fibroblast CELL LINE derived from embryo; can be Swiss, Balb/c, or NIH type, etc.

cellular blastoderm A stage of embryogenesis in insects. In DROSOPHILA, embryos cleave in a syncytium: the nuclei are contained in one large cytoplasm ~400 µm long and ~160 µm in diameter. After nine divisions, when there are ~500 nuclei, they move out to the periphery of the embryo to form what is known as the syncytial blastoderm. After four more divisions, membranes grow inward from the surface of the blastoderm and enclose each nucleus. The embryo is now a cellular blastoderm. Transplantation experiments show that nuclei in the syncytial blastoderm are not yet DETERMINED and can give rise to any tissue type in the adult. Upon cellularization, however, the nuclei become determined to develop into particular cell types depending on their position in the embryo. The first cells to appear differentiated are the pole cells, a patch of cells lying at the extreme posterior end of the cellular blastoderm, and which give rise to the germ cells. This kind of development occurs in many insects, but the number of nuclear divisions between fertilization, syncytial, and cellular blastoderm stages may vary. In addition, in short germ band insects, the nuclei may not distribute themselves uniformly around the surface of the embryo at the syncytial blastoderm stage as they do in *Drosophila*, but may collect at a much higher concentration in a particular region which will give rise to the germ band.

cellular oncogene *See:* ONCOGENES.

cellular slime moulds Slime moulds of the Order Acrasiales, eukaryotic organisms of uncertain affinity. They exist as free-living amoebae which aggregate to form a pseudoplasmodium from which a cellular fruiting body develops. The cellular slime mould *Dictyostelium discoideum* is a favoured model organism for molecular biological studies of the development, cell biology, and genetics of a simple eukaryote. *See:* DICTYOSTELIUM DEVELOPMENT.

cellulase Enzyme (EC 3.2.1.4) catalysing the degradation of CELLULOSE by hydrolysing internal β1,4 glycosidic linkages.

cellulose Family of linear, unbranched polymers of β1,4 D-glucopyranosyl residues, probably containing a very small amount of other sugars, particularly mannose, within the chains. The relative molecular mass varies between M_r <10^5 and >10^7 depending on the source and degree of degradation during isolation. Cellulose is produced as an extracellular GLYCAN by some bacteria and is a major component of plant cell walls of all types. In higher plants it is found in both primary and secondary walls and is more abundant and of higher molecular weight in the latter. In fungi, it occurs in the cell walls of the Acrasiales, Oomycetes and Hyphochytridiomycetes, but not of higher fungi.

Cellulose is uncommon in animals, though it occurs in various invertebrates (e.g. tunicin in tunicates). It has been reported in traces in mammalian blood vessels. Cellulose is considered to be the most abundant biological macromolecule on earth. The residue left after extraction of lignin, pectins and hemicellulose from plant cell walls is α-cellulose, which contains mannose, xylose, arabinose, galactose and traces of other sugars. Some of this may represent residual XYLAN and ARABINOGALACTAN, but these sugars can only be removed under conditions which hydrolyse some glycosides. Cellulose I is such a material purified from cotton fibres: if dissolved in alkaline solutions of tetrammino-cupric salts and reprecipitated with alcohol it forms cellulose II.

By X-ray diffraction and infrared spectroscopy studies cellulose II shows a structure of alternate antiparallel chains, hydrogen-bonded to each other through the C4 and C6 hydroxyl groups of adjacent chains and each chain internally hydrogen-bonded between the C3 hydroxyl of one glucosyl residue and the ring oxygen of the residue preceding it (Fig. C31). The unit cell accommodates four 'forward' chains grouped about one 'backward' and a disaccharide unit in each. Parallel chains lie in the same plane (unlike CHITIN). Cellulose I has a similar antiparallel arrangement, but there is dispute as to how it arises on a larger scale: several possible forms of chain folding have been proposed.

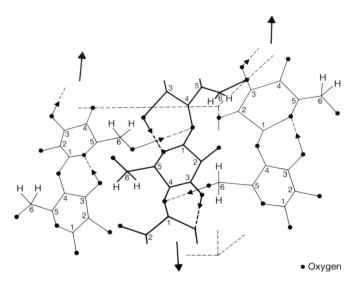

Fig. C31 Three adjacent chains in cellulose (arrows indicate direction of chains). Dotted lines indicate boundaries of the unit cell. From Cook, G.M.W. & Stoddart, R.W. *Surface Carbohydrates of the Eukaryotic Cell*, 173 (Academic Press, 1973).

In vivo, cellulose I is aggregated to form crystallites, between and around which are less regularly packed chains, the whole forming the cellulose fibre. Xylans are associated with the exterior of the fibre. It is thought, but not proven, that occasional mannosyl residues occur in chains of native cellulose which, by creating irregularities in the crystallites, prevent the propagation of dislocations under stress by restricting slippage between the chains, and so confer strength.

Cellulose synthesis has been analysed in detail in the alga *Prototheca zopfii* and the mechanism in higher plants appears very similar. Glucose is transferred from UDP-Glc onto dolichyl phosphate to form dolichyl pyrophosphoryl glucose, which then acts as an acceptor for two further glucosyl residues from UDP-Glc, giving dolichyl pyrophosphoryl cellotriose. This accepts further transfers of glucosyl residues, partly from UDP-Glc and partly from dolichyl phosphoryl glucose, to produce the β1,4 glucan primer chain. The chain is then transferred from dolichol to protein, to yield dolichyl pyrophosphate and a cellulose protein primer, which is resistant to β-elimination. COUMARIN inhibits this transfer. The cellulose synthetase complex then acts on the primer, at or near the cell surface, to transfer glucose from GDP-Glc to form the cellulose chains which appear eventually to be released from the protein. The mechanisms of folding, addition of sugars such as mannose and chain termination are unknown. GDP-Man stimulates the incorporation of GDP-Glc in some systems. The system is not stimulated by oligosaccharides, unlike callose synthetase.

Shafizadeh, F. & McGinnis, G.D. (1971) *Adv. Carb. Chem. Biochem.* **26**, 297–351.
Delmer, D.P. (1983) *Adv. Carb. Chem. Biochem.* **41**, 105–152.
Okuda, K. et al. (1993) *Plant Physiol.* **101**, 1131–1142.

celom *See:* COELOM.

CENP B Protein localized to mammalian centromeres. *See:* CHROMOSOME STRUCTURE.

centimorgan (cM) The unit of distance on a GENETIC MAP, equal to 1% recombination between two gene loci on the same chromosome.

central dogma A scheme first proposed by Francis Crick in 1958 which accounted for the apparent unidirectional flow of genetic information from DNA to protein via an RNA intermediate (Fig. C32). This 'Central Dogma of Molecular Biology' has had to be revised following the discovery of REVERSE TRANSCRIPTASES which partly reverse the flow of genetic information by synthesizing DNA from an RNA template.

Crick, F.H.C. (1970) *Nature* **227**, 561–563.

central nervous system (CNS) The central nervous system of vertebrates consists of the brain and the SPINAL CORD. Molecular neurobiological research on the central nervous system includes elucidation of the function and mechanisms of action of NEUROTRANSMITTERS and NEUROMODULATORS and their receptors (*see* LONG-TERM POTENTIATION; SYNAPTIC TRANSMISSION; and individual neurotransmitters and receptors) and their role in disease (*see* AFFECTIVE DISORDERS; SCHIZOPHRENIA); studies of the cell biology of neurons (*see* CELL ADHESION MOLECULES; NEURONAL CYTOSKELETON); studies of the molecular biology and molecular genetics of neurodegenerative diseases such as ALZHEIMER'S DISEASE (*see* NEURODEGENERATIVE DISORDERS); neuronal and nervous system development (*see* AVIAN DEVELOPMENT; VERTEBRATE NEURAL DEVELOPMENT); and work on the TROPHIC FACTORS produced by neurons and GLIAL CELLS and their role in neural development and possible applications for repairing damage caused by stroke or injury. *See also:* CEREBELLUM; CEREBRAL CORTEX; FOREBRAIN; MIDBRAIN; HINDBRAIN; RHOMBOMERES. For neurotransmitter and neuromodulator receptors *see* ADRENERGIC RECEPTORS; DOPAMINE RECEPTORS; EXCITATORY AMINO ACID RECEPTORS; GABA AND GLYCINE RECEPTORS; 5HT RECEPTORS; MUSCARINIC RECEPTORS; NICOTINIC RECEPTORS; OPIOID PEPTIDES AND THEIR RECEPTORS.

centric Applied to REFLECTIONS within a set of DIFFRACTION intensities which have a STRUCTURE FACTOR which may be real or imaginary (but not both) in consequence of the presence of one or more centres of SYMMETRY within the crystal. *See:* X-RAY CRYSTALLOGRAPHY.

centrifugation techniques *See:* DENSITY GRADIENT CENTRIFUGATION; SEDIMENTATION COEFFICIENT; SEDIMENTATION EQUILIBRIUM; SVEDBERG UNIT.

centriole Small cylindrical structure in eukaryotic cells ~0.2 μm across and 0.4 μm long. It is composed of nine bundles of three fused MICROTUBULES connected to each other and to a central core by other proteins. Centrioles often occur in pairs, one lying at right angles to the other. A pair of centrioles forms the central part of the CENTROSOME or cell centre in animal cells (*see* MITOSIS). The basal bodies of eukaryotic cilia and flagella are modified centrioles (*see* MICROTUBULES).

centroid phase *See:* BEST PHASE.

centromere The region on a CHROMOSOME responsible for its segregation during MITOSIS or MEIOSIS (*see* CHROMOSOME STRUCTURE). The centromere is usually at the site of the major chromosomal constriction. Centromeres can be defined and mapped genetically in organisms that make tetrads during meiosis. They accomplish two distinct functions, both of which are essential for accurate chromosome segregation:
1 They are replicated last during chromosome replication and hold SISTER CHROMATIDS together until ANAPHASE onset.
2 They form the functionally significant attachment between a CHROMATID and the MITOTIC SPINDLE.
The latter function is associated with a chromosomal specialization that can be identified under the microscope and is called the KINETOCHORE. Kinetochores bind spindle MICROTUBULES. The DNA at the centromere includes sequences that will bind the proteins that make up the kinetochore. In some fungi this is only a few hundred base pairs of DNA, but in most organisms the sequences appear to be longer and rather repetitive. The DNA at and near the centromere is often AT-rich and has a lower buoyant density than the bulk of DNA. As such it often bands in a centrifuge beside the majority and has been called SATELLITE DNA.

centrosome (mitotic centre, cell centre) A region of the cytoplasm, usually lying near the nucleus, that initiates the majority of a cell's MICROTUBULES. The centrosomes found in the cells of most metazoa usually contain a pair of CENTRIOLES surrounded by some structurally ill-defined material, the pericentriolar MATERIAL, that is the material active in microtubule initiation. The centrosome gets its name from the fact that it is commonly located near the cell's geometrical centre, but some cells, like those of columnar epithelia, contain a centrosome positioned near their apical pole. Centrosomes are important because they define the position of microtubule initiation and both the orientation and the number of the microtubules they initiate. The molecular mechanism by which the pericentriolar material acts to promote tubulin assembly has not been defined, but the process is controlled by the stage in the CELL CYCLE. Mitotic centrosomes initiate 5–10 times as many microtubules as INTER-

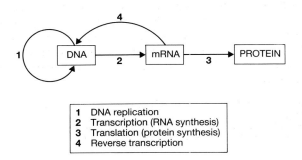

1 DNA replication
2 Transcription (RNA synthesis)
3 Translation (protein synthesis)
4 Reverse transcription

Fig. C32 The central dogma of molecular biology.

PHASE centrosomes and form the ASTERS of the mitotic cell. This transition is in part due to the action of a PROTEIN KINASE called MATURATION (or MITOSIS) PROMOTING FACTOR (*see* CELL CYCLE). Fungi contain structures that serve the role of centrosomes, but they are often built into the nuclear envelope. Because their first recognized functions pertained to mitosis, they are often called SPINDLE POLE BODIES. Higher plants and other cells that make ANASTRAL spindles lack centrosomes.

cephalosporins β-Lactam derivative antibiotics produced by strains of *Cephalosporium* and streptomycetes and active against bacteria. Cephalosporin C and its derivative cephamycin C are used as antibacterials and as the basis for synthesis of semisynthetic cephalosporins.

ceramide *See*: MEMBRANE LIPIDS.

Cercospora rosicola A deuteromycete fungal pathogen of roses (*Rosa* spp.) that, in liquid medium, secretes the plant hormone ABSCISIC ACID (ABA) in relatively large amounts (up to $60 \, \text{mg} \, l^{-1}$). The fungus has been used extensively to study the biosynthetic pathway for ABA, but hopes that this would help understand ABA synthesis in higher plants were discouraged by the finding that 1′-deoxy ABA, the precursor of ABA in the pathogen, is unlikely to be the precursor in leaves of *Xanthium* (cocklebur).

Creelman, R.A. & Zeevaart, J.A.D. (1982) *Plant Physiol.* **75**, 166–169.

cerebellar mutants More than 30 mouse mutations have been described that affect development of the CEREBELLUM in a variety of ways, and that cause functional defects of motor control. With the exception of *lurcher*, a DOMINANT mutation, they are AUTOSOMAL RECESSIVE. In each case the mutation affects CELL MIGRATION and/or DIFFERENTIATION rather than the earlier generation of cerebellar cells. Many have pleiotropic effects, for example the loss of dopaminergic cells in the *weaver* mutant. Despite the gross malpositioning of cells in several of the mutations, their inputs may still target accurately, for example on the misplaced Purkinje neurons in the *reeler* mutant and the granule cells in *weaver*. In the latter case, the primary defect may reside in an inability of the granule cells to recognize and migrate along the Bergmann GLIAL CELLS that normally guide them to their final positions within the cerebellar cortex. Cloning of these genes and their products should provide important insights into the mechanisms of VERTEBRATE NEURAL DEVELOPMENT.

cerebellum Part of the HINDBRAIN, a pair of finely convoluted hemispherical masses of tissue stituated behind the midbrain, concerned with regulation of muscle tone and posture and coordination of movement in relation to sensory signals received in other parts of the brain. The orderly neuroanatomy of the cerebellum, which contains only five different types of neuron, has been studied in considerable detail. Output from the cerebellar cortex is via Purkinje neurons, each Purkinje neuron receiving inputs from many thousands of neurons. *See also*: CEREBELLAR MUTANTS.

cerebral cortex The superficial layer of grey matter of the cerebral hemispheres, some 2 mm deep and consisting of several layers of nerve cell bodies and their complex interconnections. In humans and anthropoid apes the highly convoluted surface of the brain provides a much enlarged area of cortex. The cortex is the site of analysis and interpretation of sensory information, of the generation of all voluntary motor action, and of higher cognitive functions such as learning and memory and conscious perception. It receives input from cranial nerves and peripheral sensory receptors via various nuclei in other parts of the brain. Different functions, such as the analysis of visual, auditory, and other sensory information, and generation of motor signals, are localized to different areas of the cortex.

cesium chloride *See*: CAESIUM CHLORIDE.

CF (1) CYSTIC FIBROSIS.
(2) Coupling factor (chloroplast). A component of the chloroplast ATP SYNTHASE complex. *See also*: PHOTOSYNTHESIS.

CFI *See* CHALCONE FLAVANONE ISOMERASE.

CFTR Cystic fibrosis transmembrane conductance regulator. *See*: CYSTIC FIBROSIS.

CGD CHRONIC GRANULOMATOUS DISEASE.

cGMP CYCLIC GMP. *See also*: VISUAL TRANSDUCTION.

CGRP CALCITONIN GENE-RELATED PEPTIDE.

chain termination *See*: DNA SEQUENCING.

chalcone flavanone isomerase (CFI) Enzyme (EC 5.5.1.6; recommended name, chalcone isomerase) catalysing the interconversion of chalcones and flavanones, typically:

chalcone   flavanone

Metabolically, chalcones are made by the action of CHALCONE SYNTHASE, and the flavanones converted to a wide range of different FLAVONOIDS. This isomerization may proceed spontaneously at pH 8.0. CFI is induced in some plant cells by illumination and other environmental signals, resulting in FLAVONOID production.

chalcone synthase (CS) Enzyme (EC 2.3.1.74) catalysing the condensation of three molecules of malonyl-CoA with one molecule of 4-hydroxycinnamoyl-CoA or a derivative to form the chalcone at acid pH (Fig. C33). At pH 8.0 *in vitro*, the flavanone is the product, apparently because of the spontaneous isomerization

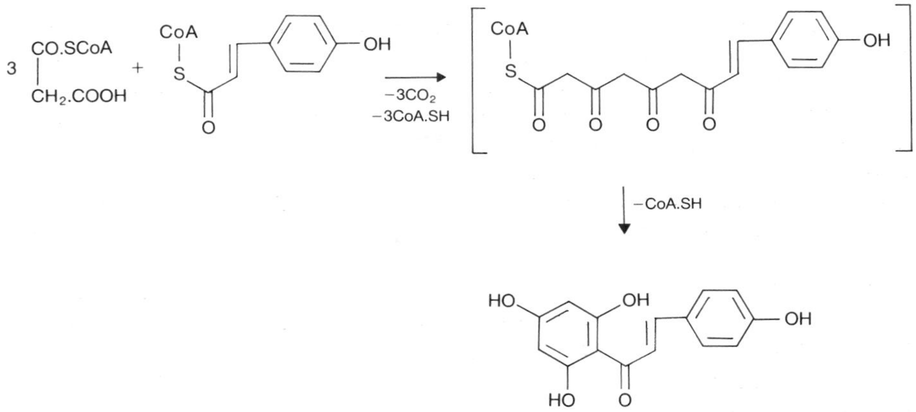

Fig. C33 The reaction catalysed by chalcone synthase.

of the chalcone. The reaction catalysed by chalcone synthase is important as the first unique step in FLAVONOID biosynthesis distinct from that of lignin, following the generation of 4-hydroxy-cinnamoyl-CoA. Induction of its synthesis in plants by light and pathogens has received special attention.

channel blockers *See*: ION CHANNELS.

chaperone proteins *See*: MOLECULAR CHAPERONES.

chaperonins A class of MOLECULAR CHAPERONE proteins homologous in structure to the bacterial chaperone protein GroEL (*see also* HEAT SHOCK) and which are found in prokaryotes and in eukaryotic organelles (mitochondria and chloroplasts) which are of endosymbiont origin. The GroEL protein and its homologues are now known as the chaperonin-60 proteins since they consist of oligomers (subunit M_r 60 000) of 14 subunits stacked in two rings of seven subunits. Bacterial chaperonin-10 is an oligomer composed of a ring of seven subunits of M_r 10 000 and is also a product of the *groE* operon. *See also*: BACTERIAL PROTEIN EXPORT.

Charcot–Marie–Tooth disease Also known by the descriptive term hereditary motor and sensory neuropathy (types I and II). This is a degenerative disease principally involving the peripheral nerves and inherited as an AUTOSOMAL DOMINANT trait. It usually presents during childhood or adolescence with difficulty in walking or because of foot deformity. Weakness and wasting occur in the lower limbs and later the hands may be involved. There is a variable degree of distal sensory loss.

Chargaff's rule Erwin Chargaff noted that in samples of DNA from various organisms the total CYTOSINE plus GUANOSINE content varied widely, but the amount of the two nucleotides was always equal. Hence, [G] = [C] and [A] = [T], and [purines] = [pyrimidines]. This rule is a consequence of complementary BASE PAIRING in DNA and was important in assisting the elucidation of the structure of DNA by Watson and Crick.

charon vectors A class of CLONING VECTORS based on bacteriophage LAMBDA which take their names from the old ferryman of Greek mythology who conveyed the spirits of the dead across the river Styx. They are primarily designed to ensure optimal BIOLOGICAL CONTAINMENT; for example, some of the charon vectors (e.g. charon16A) have AMBER MUTATIONS introduced into genes A and B of the lambda genome thereby ensuring that infective phage can only be assembled if an *Escherichia coli* host strain used encodes an AMBER SUPPRESSOR tRNA (e.g. *sup*E).

Blattner, F.R. et al. (1977) *Science* **196**, 161–169.

Chédiak–Higashi syndrome Rare AUTOSOMAL RECESSIVE disorder characterized by lack of resistance to pyogenic bacterial infections, lymphocyte infiltration into various organs, and partial albinism. It is thought to be due to a generalized cellular defect that leads to increased fusion of lysosomes, secretory granules, and other intracellular vesicles, resulting *inter alia* in defective neutrophil, natural killer cell, and phagocyte function. The *beige* mouse is an animal model.

Chemical carcinogens and carcinogenesis

A CAUSAL link between a human cancer and a chemical agent in the environment was first reported in 1775 by the surgeon Percival Pott, who deduced a relationship between the high incidence of scrotal cancer in chimney sweeps and their exposure to soot. Studies in the early decades of this century established that the active constituents of soot and coal tar are POLYCYCLIC AROMATIC HYDROCARBONS. In 1964, Brookes and Lawley showed [1] that when applied to mouse skin, polycyclic aromatic hydrocarbons become covalently bound to DNA, with the extent of binding correlating with carcinogenic potency. This experiment was a key element in support of the somatic mutation theory of cancer, and provided the impetus for many subsequent studies on the molecular basis of chemical carcinogenesis [2].

Among the compounds and agents that have been established as either producing chemical carcinogens or acting as carcinogens in their own right in humans [3] are cigarette smoke (lung cancer), benzene (leukaemia), 2-naphthylamine (bladder cancer), and benzidine (urinary cancer). Polycyclic aromatic hydrocarbons

(PAHs), which are probably the most extensively studied carcinogens in the laboratory, are typically produced by the combustion of fossil fuels and are found in, for example, coal tar and cigarette smoke. Thus it is widely believed that they may be major causative agents in human lung cancer. ^{32}P-postlabelling techniques (which detect very low levels of carcinogen–DNA adducts) have been used in a large number of studies to show that particular population groups, such as smokers, have elevated levels of PAH–DNA adducts [2,4], although it is not yet possible to use these data in a predictive manner for early diagnosis of malignant disease. Complete data on the relatively small number of chemicals that have been unequivocally identified as human carcinogens are available in the *Monographs on the Evaluation of the Carcinogenic Risk of Chemicals to Humans* published by the International Agency for Research on Cancer.

Action of chemical carcinogens

Most known chemical carcinogens react chemically with DNA, either directly or as derivatives produced by metabolic activation to chemically reactive electrophilic species, which can then form covalent adducts [2]. Reaction typically occurs with the phosphate groups as well as with the DNA bases themselves. The ring nitrogen atoms of purines are the most reactive. Interactions have been documented at N2, N3, O6, N7, and C8 of guanine; N1, N3, N6, and N7 of adenine; O2, N3, N4, and C5 of cytosine; and O2 and O4 of thymine (*see* NUCLEOSIDES AND NUCLEOTIDES). In general there are high correlations between extent of adduct formation and carcinogenicity. PAHs in particular produce a complex mixture of adducts. In many cases these adducts, as well as unreacted products of metabolism, have been separated by chromatographic techniques and characterized by spectroscopic and chemical methods. *In vitro* and *in vivo* studies with these metabolic products have enabled the 'ultimate carcinogens' to be identified for a number of carcinogens.

The PAH benzo[a]pyrene was first isolated in pure form from coal tar by Kennaway in 1933 [5], and has subsequently been the major focus of studies on chemical carcinogenesis. Benzo[a]pyrene itself is unreactive but is metabolized by microsomal CYTOCHROME P450 enzymes, mainly in the liver, to a number of oxygenated species. Among them are four isomeric 7,8-diol-9,10-epoxides, one of which, the (+)-*anti* isomer (7R,8S-dihydroxy-9S,10R-epoxy-7,8,9,10-tetrahydrobenzo[a]pyrene), is a significantly more potent carcinogen, in terms of initiation of tumours in mice, than any of the three others. It is also the most MUTAGENIC in mammalian cell lines, for example in Chinese hamster V79 cells, although not in typical bacterial mutagenicity test systems such as the TA100 strain of *Salmonella typhimurium* (*see* AMES TEST). In this system, the (+)-*syn* isomer is the most potent in terms of number of revertants produced.

Theoretical studies have shown that the enhanced reactivity of these diol epoxides compared with other positions of epoxidation on the benzo[a]pyrene ring system, can be rationalized by the 'bay region' theory which predicts an enhanced ease of carbonium ion formation. The diol epoxides react with a variety of sites on a DNA molecule. Analyses of the distribution of products from both benzo[a]pyrene-treated animals and cells have shown that the major covalent adduct is with the N2 position of guanosine (Fig. C34*a*). This DNA minor groove (*see* NUCLEIC ACID STRUCTURE) adduct is considered to be responsible for the carcinogenic activity of benzo[a]pyrene. Analogous behaviour, with carcinogenic specificity being confined to one or a few of the many products of metabolism (usually bay region diol epoxides), is shown by other PAHs such as benz[a]anthracene, chrysene, phenanthrene and 3-methylcholanthrene.

Aromatic amines such as *N*-acetoxy-2-acetylaminofluorene (AAF), 4-aminobiphenyl, 2-naphthylamine and benzidine (the last two of which were identified as causative agents in bladder cancer in workers in the dyestuffs and chemical industry by the pioneering epidemiological studies of R.M. Case in 1954), are *N*-oxidized by cytochrome P450 monooxygenases and/or flavin oxidases to powerfully electrophilic compounds that can interact with DNA. The active metabolite of AAF, the *N*-hydroxy derivative, can itself undergo a variety of further metabolic changes, depending on the target tissue. The major product of *N*-hydroxy-AAF attack on DNA is the C8-guanine adduct. This results in a major structural change in DNA as a result of the bulky C8 adduct producing a *syn* conformation for each affected guanosine, and has been shown to result in the formation of left-handed Z-DNA (*see* NUCLEIC ACID STRUCTURE) with certain DNA sequences [2]. 2-Naphthylamine forms adducts at the N2 and C8 positions of guanines and the N6 of adenines.

A number of naturally occurring carcinogens have been identified [3]. Many are of plant origin. The aflatoxins are fungal metabolites that are prevalent in, for example, mouldy peanuts. They are powerful initiators of liver cancer in animals and probably in humans. Major adduct formation is at the N7 position of guanine (Fig. C34*b*). This adduct is rapidly removed by

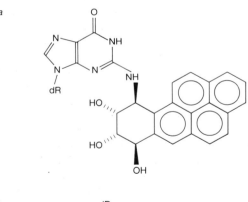

Fig. C34 The structure and stereochemistry of guanosine adducts with the active metabolites of (*a*) benzo[a]pyrene and (*b*) aflatoxin. dR, deoxyribose.

cellular enzymes to give a depurinated lesion in DNA (*see* DNA REPAIR).

Nitrosamines such as *N*-nitroso-*N*-methylurea are established animal carcinogens and may have a role in gastrointestinal cancers through their formation from nitrites in food preservatives. Metabolic activation of nitroso compounds produces alkyl carbonium ions, which can directly alkylate DNA, most commonly at the O6 position of guanines.

Repair of chemically damaged DNA (*see* DNA REPAIR) is normally the initial response of a cell to a chemical carcinogen. Mechanisms of repair are best understood in bacteria, especially *Escherichia coli*, where a large number of specific repair enzymes have been identified. Bulky PAH adducts can be removed by excision repair endonucleases. O6-alkyl damage can be efficiently repaired by specific methyltransferases. Certain genetic defects in humans can result in impaired repair mechanisms and thus an increased propensity to tumour formation. Individuals with XERODERMA PIGMENTOSUM have been shown to be defective in a DNA repair gene, ERCC-3, that has DNA HELICASE activity [6].

Mutagenicity and carcinogenesis

The majority of chemical carcinogens produce MUTATIONS in DNA. Benzo[a]pyrene, *N*-acetoxy-2-acetylaminofluorene and aflatoxin B1 cause GC to TA transversions in bacteria and mammalian cells, although benzo[a]pyrene in particular can produce base substitutions such as AT to TA. *N*-nitroso compounds cause GC to AT transitions. Alkylation of the O4 position of thymine, which probably also has an important role in tumour initiation induced by *N*-nitrosoureas, results in TA to CG transitions.

A high correlation between mutagenicity and carcinogenicity is the underlying assumption behind the widely-used AMES TEST [7], which speedily assays for mutants in a bacterial system, together with an added rat liver homogenate, which contains the microsomal cytochromes P450, to provide metabolic activation of carcinogens where needed. There are, however, a significant number of exceptions to the correlation, with the consequence that the simple Ames test alone cannot be relied on as a totally reliable indicator of a substance's safety for humans. Asbestos, for example, is an acknowledged human carcinogen yet it does not form DNA adducts or give a positive Ames test result.

Several chemical carcinogens have been found to produce mutations in the *ras* proto-oncogene [8,9] (*see* ONCOGENES), converting it into an active oncogene which leads to the initiation of neoplastic TRANSFORMATION in cell lines and to the production of experimental tumours in rodents. *N*-nitroso-*N*-methyl urea induces mammary, prostate and other carcinomas in rats, with the majority of these tumours showing a G to A transition at the second position in codon 12 (GGA to GAA) of the Ha-*ras* oncogene. Benzo[a]pyrene-induced skin tumours in mice have been found to contain A to T transversions in the second position of codon 61 (CAA to CTA) in the Ha-*ras* gene. Mutations at codons 12 and 61 in *ras* have been found in numerous human tumours and are well established as causing transformation *in vitro*, suggesting that *ras* activation can be a critical initiating step in the multistage process of tumorigenesis.

S. NEIDLE

1 Brookes, P. & Lawley, P.D. (1964) Evidence for the binding of PAH to the nucleic acids of mouse skin: relation between the carcinogenic power of hydrocarbons and their binding to DNA. *Nature* **202**, 781.
2 Singer, B. & Grunberger, D. (1983) *Molecular Biology of Mutagens and Carcinogens* (Plenum, New York).
3 Searle, C.E. (Ed.) (1984) *Chemical Carcinogens*, 2nd edn (American Chemical Society, Washington DC).
4 Phillips, D.H. (1989) Detection by ^{32}P-postlabelling of carcinogen–DNA adducts in animals and man. In *New Trends in Genetic Risk Assessment*, Jolles, G. & Cordier, A. (Eds) (Academic, New York).
5 Osborne, M & Crosby, N.T. (1987) *Benzopyrenes* (Cambridge University Press, Cambridge).
6 Brugge, J. et al. (Eds) (1991) *Origins of Human Cancer* (Cold Spring Harbor Laboratory Press, New York).
7 McCann, J. et al. (1975) Detection of carcinogens as mutagens in the *Salmonella*/microsome test: assay of 300 chemicals. *Proc. Natl. Acad. Sci. USA* **72**, 5135.
8 Vousden, K.H. et al. (1986) Mutations activating human c-Ha-*ras1* proto-oncogene (*HRAS1*) induced by chemical carcinogens and depurination. *Proc. Natl. Acad. Sci. USA* **83**, 1222.
9 Zarbl, H. et al. (1985) Direct mutagenesis of Ha-*ras*-1 oncogenes by *N*-nitroso-*N*-methylurea during initiation of mammary carcinogenesis in rats. *Nature* **315**, 382.

chemical synthesis The synthesis of biological polymers such as peptides or oligonucleotides by purely chemical means, without the use of enzymes to mimic the biological syntheses (*see* OLIGONUCLEOTIDE SYNTHESIZERS; PEPTIDE SYNTHESIS). Oligonucleotide PROBES, primers for the POLYMERASE CHAIN REACTION, synthetic genes for small peptides, and defined oligonucleotides to form unusual DNA structures are produced by chemical synthesis.

chemical transmission *See*: SYNAPTIC TRANSMISSION.

chemiosmotic theory Theory due to Peter Mitchell, originally formulated to explain the coupling of ATP synthesis to the transport of electrons along ELECTRON TRANSPORT CHAINS (*see also* PHOTOSYNTHESIS). In outline, protons (H^+) ejected across the inner mitochondrial membrane into the inter-membrane space (or from the thylakoid membrane into the thylakoid lumen in chloroplasts) as a result of the passage of electrons along the chain, set up an electrochemical potential gradient (the PROTONMOTIVE FORCE) across the membrane. Because these membranes are selectively permeable to ions, protons can only move back across the membrane down the electrochemical gradient through the transport system provided by ATP SYNTHASE, which uses the energy stored in the proton gradient to synthesize ATP from ADP and inorganic phosphate and to do work by energizing transport systems. In bacteria, chemiosmotic coupling via a proton gradient across the plasma membrane is used to power the inward transport of many nutrients (co-transport with H^+) and flagellar rotation. *See also*: CHLOROPLAST; MEMBRANE TRANSPORT SYSTEMS; MITOCHONDRION; OXIDATIVE PHOSPHORYLATION.

chemotaxis The ability of a motile cell to respond to chemical changes in its environment by directed movement. *See*: BACTERIAL CHEMOTAXIS; CELL MOTILITY.

chemotaxis receptors (1) In bacteria, *see* BACTERIAL CHEMOTAXIS.

(2) In inflammatory reactions, neutrophils, macrophages, and other leukocytes are attracted to sites of inflammation by a variety of chemoattractants. The actions of these molecules (e.g. complement component C5a) are mediated by specific cell-surface receptors.

chemotropism Directed growth of cells or axon growth cones towards a target in response to a diffusible material released from the target region. The term is used almost exclusively in relation to axon growth cones, being distinguished from chemotrophism, which refers to the survival-enhancing effects of growth factors on neurons, and CHEMOTAXIS, which refers to a directed movement by whole cells such as leukocytes or bacteria. Chemotropism is thought to be an important mechanism for axon guidance in the developing vertebrate nervous system, being demonstrable *in vitro* in the guidance of peripheral sensory axons, spinal cord commissural axons and corticopontine projections. The molecular nature of the factors responsible is unknown, and is now under active investigation.

Chi site A short DNA sequence which seems to stimulate unusually high levels of RECOMBINATION within its immediate environment. In *Escherichia coli* this form of recombination depends on the product of the *recA* gene. Such recombination hot spots were first identified in the bacteriophage LAMBDA and were called Chi after the greek letter χ, which looks like the cross-over structure formed during recombination. Chi sites also occur in the *E. coli* genome about once every 5–10 kb. All Chi sites contain the core Chi sequence:

5′-GCTGGTGG-3′
3′-CGACCACC-5′

Chi sites may identify targets for the enzyme EXONUCLEASE V whose subunits in *E. coli* are the products of the *recBCD* genes.

chiasma (*pl.* chiasmata) Cross-like configuration seen in MEIOSIS and the visible consequence of RECOMBINATION between non-sister chromatids. It represents a site at which two of the chromatids in a bivalent have been broken at corresponding points and the broken ends have been rejoined cross-wise to generate new chromatids.

chimaera (chimera) (*pl.* chimaerae, chimaeras, chimeras). A composite embryo containing cells or tissues with two or more different GENOTYPES (*cf.* MOSAIC). Chimaeras can be produced experimentally in order to follow genetically marked cells within the embryo. The genetic marker may consist of cells of a different species. The word derives from Greek mythology; the Chimaera was a monster whose parents were Typhon and Echidna, and which consisted of three parts: a lion's head, a she-goat's middle and a snake's tail. *See also:* ALLOPHENIC MICE.

chimaeric antibodies An ANTIBODY molecule that contains structural elements from two or more different antibody molecules, often from different animal species. The DNA encoding such antibodies is generated by the application of GENETIC ENGINEERING techniques to the DNA encoding each of the chains of the contributing antibodies (*see* CHIMAERIC IMMUNOGLOBULIN GENES). This technology is commonly exploited to 'clothe' the antigen-binding sites of specific murine MONOCLONAL ANTIBODIES (which are readily obtained from HYBRIDOMAS) in human material when they are to be used as *in vivo* therapeutic agents. *See:* HUMANIZED ANTIBODIES.

chimaeric immunoglobulin genes Novel rearranged IMMUNOGLOBULIN GENES constructed *in vitro* using recombinant DNA technology and comprising rodent variable (V) regions joined to human constant (C) regions. This allows for the production of chimaeric MONOCLONAL ANTIBODIES with a desired antigen specificity together with human effector functions (*see* HUMANIZED ANTIBODIES). Chimaeric monoclonal antibodies are more suitable than rodent monoclonal antibodies for human therapy as they elicit a reduced anti-antibody response. Chimaeric immunoglobulin genes have also been constructed in which the COMPLEMENTARITY DETERMINING REGIONS of a rodent V region coding sequence are grafted into the framework regions of a human V region coding sequence. Thus CDR grafting can confer the antigen specificity of an antibody raised in a rodent on an otherwise human monoclonal antibody.

chitins A large family of GLYCANS which are β1,4-linked, linear polymers of 2-deoxy-2-acetamido-D-glucose (*N*-acetyl glucosamine). They occur in the walls of higher fungi and in the exoskeletons of insects, arachnids and many other groups of invertebrates and as an extracellular polymer of some prokaryotes. It is not certain that all forms of chitin are truly related to each other. Chitins are extremely insoluble and resistant to hydrolysis.

Chitin is the second most abundant organic chemical (after CELLULOSE, which it resembles). In α-chitin, the unit cell has an antiparallel arrangement of chains, rather like α-cellulose, except that parallel chains do not lie in the same plane. α-chitin is exceedingly insoluble and may be cross-linked to polypeptide by polymeric phenols.

Chitin synthesis in eukaryotes does not closely resemble the mechanism of assembly of any other glycan. Its basic features appear common to all eukaryotes although there is interspecific variation. The prokaryotic process resembles the assembly of other bacterial homopolymers.

In eukaryotes, glucosyl diacylglycerol acts as the primer. *N*-acetyl glucosamine is transferred to this from UDP-GlcNAc and the lipid-linked disaccharide then acts as the acceptor for chitin synthetase, which also transfers the amino-sugar from UDP-GlcNAc. When the chain has been constructed, a glycosidase cleaves off glucosyl diacylglycerol and leaves homopolymeric α-chitin. Synthesis seems to depend on the continuous production of chitin synthetase. This is initially formed as a zymogen, from which the active synthetase is released proteolytically by 'activating factor'. A specific peptide inhibitor can prevent this release *in vivo*. CYCLOHEXIMIDE inhibits chitin synthesis by preventing the formation of the zymogen. Additionally, there is a specific protein inhibitor of the chitin synthetase reaction itself.

Many species synthesize chitin by way of 'chitosomes', which are membranous organelles often with a characteristic 'target-like' appearance under transmission electron microscopy and which give a positive Nadi reaction histochemically. Drugs inhibiting chitin synthesis show considerable interspecific variation *in vivo*, which seems to arise from variations in their pharmacokinetics rather than from basic differences in the synthetic mechanism. Among them are various polyoxins, diflubenzuron and several benzophenylureas.

chloramphenicol Antibacterial antibiotic (Fig. C35) obtained originally from *Streptomyces venezuelae*; commercial production is by chemical synthesis. It inhibits prokaryotic PROTEIN SYNTHESIS by blocking the peptidyl transferase reaction on the ribosome. In eukaryotic cells, chloramphenicol only inhibits mitochondrial and chloroplast protein synthesis.

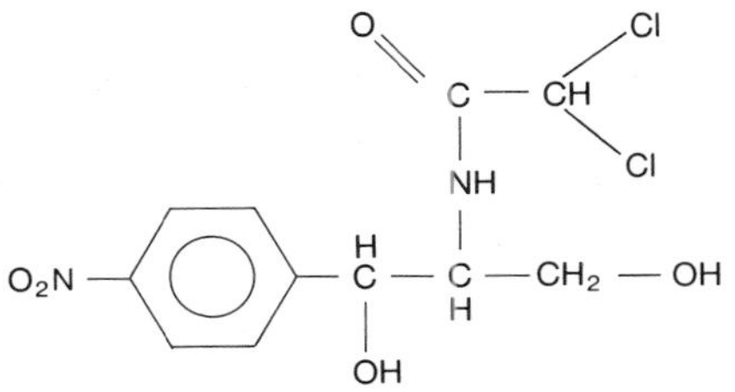

Fig. C35 The structure of chloramphenicol.

chloramphenicol acetyltransferase (CAT) A bacterial enzyme (EC 2.3.1.28), product of the *cat* gene which confers host cell resistance to the antibiotic CHLORAMPHENICOL. CAT inactivates chloramphenicol by acetylating the drug at one or both of its two hydroxyl groups. Because eukaryotic cells do not synthesize CAT, the *cat* gene has been widely exploited as a REPORTER GENE for *in vivo* analysis of eukaryotic PROMOTER sequences, particularly in mammalian cells. It is usually assayed by a method based on thin-layer CHROMATOGRAPHY (TLC) in which [^{14}C]chloramphenicol can be separated from the acetylated and inactive derivatives which are only synthesized in the presence of the CAT enzyme.

chloramphenicol amplification The numbers of copies per cell of the multicopy plasmids used for DNA CLONING (e.g. pBR322) can be increased (amplified) by growing the transformed cells in the presence of the antibiotic CHLORAMPHENICOL. The replication of the plasmid DNA is under RELAXED CONTROL and, although the host DNA REPLICATION machinery and protein synthesis is inhibited by chloramphenicol, plasmid DNA replication continues. A plasmid copy number in excess of 3000 per cell can be achieved by such amplification.

chloride channel *See*: CYSTIC FIBROSIS; GABA RECEPTOR; ION CHANNELS.

chlorina-f2 mutant A barley mutant of natural origin deficient in chlorophyll *b*. The deficiency is most likely to be in biosynthesis of chlorophyll *b*, but polypeptides of LHCII and LHCI are also deficient. The thylakoids lack grana stacks, but photochemical activities are normal. *See*: PHOTOSYNTHESIS.

chlormequat chloride (chlorocholine chloride, CCC) 2-Chloroethyltrimethylammonium chloride (Fig. C36); M_r 158. A PLANT GROWTH RETARDANT, first reported in 1960 and currently the most important in agronomic use. Its major application is to prevent lodging (stem collapse) in wheat, over 80% of which is treated with this retardant in western Europe. As a quarternary ammonium compound it is thought to inhibit the biosynthesis of GIBBERELLINS by acting on *ent*-kaurene synthetase. However, it inhibits this enzyme only at high concentration ($>10^{-4}$ M) and the gibberellin pathway may not be its sole site of action.

Tolbert, N.E. (1960) *J. Biol. Chem.* **235**, 475–479.

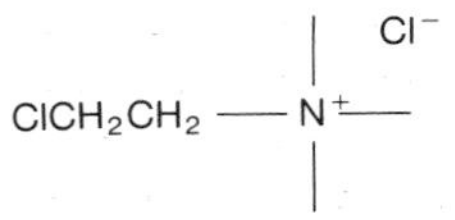

Fig. C36 Structure of chlormequat chloride.

chloronema The creeping part of a moss PROTONEMA whose cells are rich in chloroplasts and have transverse walls. Also applied to filamentous stages of liverworts and fern GAMETOPHYTES.

chlorophyll A group of macrocyclic Mg-tetrapyrrole pigments which have essential roles in PHOTOSYNTHESIS as major light harvesting pigments (*see* LHCI, LHCII) and as primary electron donors in the photochemical reactions (Fig. C37). Chlorophylls are closely related chemically and biosynthetically to haem, but are less polar because of esterification of side-chain carboxyls with phytol and methanol. In plants, reaction centre cores contain only chlorophyll *a*, whereas *a* and *b* occur together in accessory protein complexes. Most purple bacteria contain bacteriochlorophyll *a* which has a more saturated ring structure than chlorophyll *a*, although some (e.g. *Rhodobacter viridis*) contain bacteriochlorophyll *b*. Chlorosomes of green photosynthetic bacteria contain bacteriochlorophylls *c*, *d* or *e*. Chlorophylls and bacteriochlorophylls almost always occur *in vivo* as part of specific chlorophyll–protein complexes (*see* CP).

chlorophyll *a/b* binding (cab) proteins The accessory CHLOROPHYLL PROTEINS of plants and green algae. They contain non-covalently bound chlorophyll *a* and chlorophyll *b* in various amounts. The most abundant is LIGHT-HARVESTING COMPLEX II (LHCIIb), the main ANTENNA chlorophyll protein of PHOTOSYSTEM II (PSII). Minor cab proteins of PSII (CP29, CP24) are probable intermediates in energy transfer to the reaction centre. PHOTOSYSTEM I also contains an antenna cab protein complex (LHCI). All show considerable sequence similarity and are likely to contain three membrane-spanning helices. *See also*: PHOTOSYNTHESIS.

chlorophyll proteins The protein part of chlorophyll–protein complexes. *See*: CP.

chloroplast CHLOROPHYLL-containing organelle in plant and algal cells in which PHOTOSYNTHESIS takes place (Fig. C38). The chloroplast stroma is the site of the 'dark' reactions of photosyn-

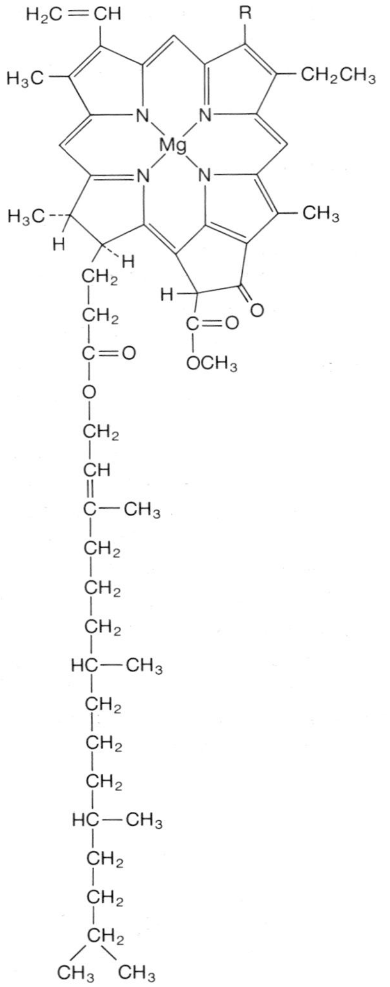

Fig. C37 Chlorophyll. In chlorophyll *a*, R = CH₃; in chlorophyll *b*, R = CHO.

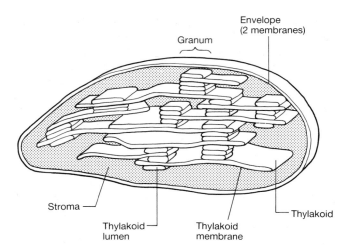

Fig. C38 Schematic cross section through a higher plant chloroplast. The light-harvesting complexes and photochemical reaction centres (photosystems I and II), the electron transport chain terminating in NADP reductase, and ATP synthase are embedded in the thylakoid membrane. Electron transport results in the formation of a proton gradient (protonmotive force) across the THYLAKOID membrane, the passage of protons down the gradient through ATP synthase and the consequent synthesis of ATP in the stroma. NADPH is also produced.

thesis (*see* CALVIN CYCLE), the synthesis of starch, the synthesis of all fatty acids and some amino acids, and the reduction of nitrite to ammonia. Chloroplasts contain DNA (*see* CHLOROPLAST GENOMES) and the machinery for PROTEIN SYNTHESIS. Subunits of various chloroplast enzymes are encoded in the chloroplast genome, the remaining chloroplast proteins are encoded in the nucleus and the proteins are imported into the chloroplast from the cytoplasm (*see* GRANA; PROTEIN TARGETING; PROTEIN TRANSLOCATION; THYLAKOID).

chloroplast DNA *See*: CHLOROPLAST GENOMES.

Chloroplast genomes

THE earliest evidence for the presence of a genetic system in CHLOROPLASTS came from the work of Baur and Correns early this century on the inheritance of leaf (and chloroplast) colour. They found that for some mutations the outcome of a cross was determined by the female parent, and this non-Mendelian transmission was subsequently termed MATERNAL INHERITANCE. These data were interpreted as being due to genetic determinants (although purely EPIGENETIC changes were postulated by some) in the chloroplasts, with the latter being transmitted through the egg cells and not the pollen. Although maternal inheritance is common for chloroplast-determined traits, it is not the only pattern. In some species, inheritance from either (but only one) parent is possible, and in others, such as conifers, strict PATERNAL INHERITANCE has been reported. It is not clear how far uniparental inheritance depends on simple physical exclusion of PLASTIDS from pollen or egg, rather than selective destruction.

It was several decades after the work of Baur and Correns that the presence of DNA and a protein synthetic system was demonstrated in chloroplasts. To find out which polypeptides are chloroplast-coded (or at least synthesized in chloroplasts), several experimental approaches were adopted. The main ones were: (1) classical genetic analysis to demonstrate NON-MENDELIAN INHERITANCE; (2) use of differential antibiotic sensitivity of chloroplast and cytoplasmic protein synthetic systems (*see* PROTEIN SYNTHESIS); (3) protein synthesis *in vitro* by isolated chloroplasts; (4) TRANSCRIPTION and translation *in vitro* of chloroplast DNA; (5) translation *in vitro* of chloroplast RNA; and (6) DNA SEQUENCING of chloroplast DNA. The chloroplast genomes of three plants have now been fully sequenced (the liverwort *Marchantia polymorpha*, tobacco, and rice), and there are large amounts of sequence data from many others [1].

Gene content

All polypeptides synthesized by chloroplasts are chloroplast encoded, and with relatively few exceptions (perhaps most notably among nonchlorophyte algae), all chloroplast DNAs contain the same set of about 120 genes. The genome is typically circular, and around 120–150 kilobase pairs (kbp) in length, with much less variation between species than in plant mitochondrial genomes.

The majority of plastid polypeptides are, however, encoded in nuclear DNA. They are synthesized outside the plastid, with an N-terminal extension (TRANSIT SEQUENCE) that directs import into the plastid and is subsequently cleaved (*see* PROTEIN TARGETING; PROTEIN TRANSLOCATION). These polypeptides will not be discussed further here.

All plastid types (chloroplasts, etioplasts, chromoplasts, amyloplasts, etc) appear to have the same genome. Genes have been identified for sufficient TRANSFER RNAS (*trn* genes) to recognize all CODONS, all the chloroplast ribosomal RNAs — 23S, 16S, 5S, and 4.5S — (*rrn* genes), and a number of polypeptides including: components of the large and small subunits of the RIBOSOME (*rpl* and *rps* genes respectively); PHOTOSYSTEM I (*psa* genes); PHOTOSYSTEM II (*psb* genes); the photosynthetic electron transfer chain (*pet* genes); the ATP synthase complex (*atp* genes); RNA polymerase (*pol* genes); a putative NAD(P)H dehydrogenase complex (*ndh* genes); and the large subunit of RIBULOSE BISPHOSPHATE CARBOXYLASE/OXYGENASE (*rbcL*). There are a few less well characterized OPEN READING FRAMES including some (*frx*) which may encode ferredoxin-related polypeptides, and some with no conclusive HOMOLOGY to other known sequences. Genes for the small subunit of ribulose bisphosphate carboxylase (*rbcS*) and the ELONGATION FACTOR EF-Tu (*tufA*) are among those encoded in plastids from some but not all species.

Figure C39 shows the current map of the chloroplast genome from rice (*Oryza sativa*).

Genome organization

With a few exceptions, including some legumes and conifers that have single copies of all sequences, the genome is composed of two regions of SINGLE-COPY SEQUENCES, typically of about 20 and 80 kbp, separated by inverted repeated sequences which are variable in size between species, but typically of about 20 kbp. It is the variation in extent of the INVERTED REPEATS that is responsible for most of the variation in plastid genome size between species. Both elements of the repeat are identical, suggesting the existence of copy-correction mechanisms. The rRNA genes are among those contained within the repeats.

The genome exists in a number of isomeric forms. RECOMBINATION between the arms of the inverted repeat inverts the relative orientation of the small and large single-copy regions. Both of these isomers are found in equal proportions in a plant, suggesting that the recombination is relatively frequent within its lifetime. Multimeric molecules, mainly dimers and trimers, also arise, probably by intermolecular recombination. They may have a head-to-head or head-to-tail configuration, depending on the involvement of the inverted repeats.

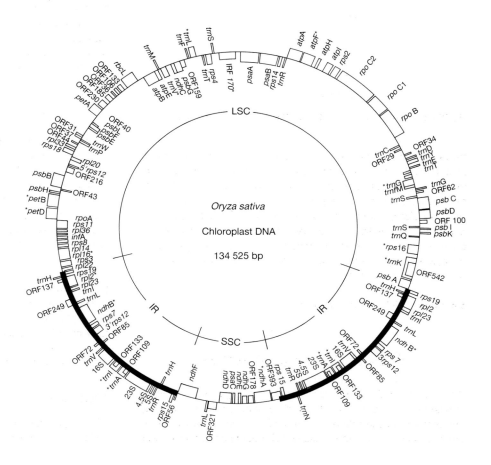

Fig. C39 Genetic circle map of the *Oryza sativa* chloroplast genome drawn to scale. Genes shown on the outside of the circle are encoded on the A strand and transcribed counterclockwise. Genes on the inside are encoded on the B strand and transcribed clockwise. Asterisks denote split genes. LSC, large single-copy region; IR, inverted repeat; SSC, small-single copy region. From [3].

These observations suggest that recombination does occur within chloroplasts of higher plants, although recombination between genomes of different chloroplasts (in cell fusion experiments, for example) seems to be rare, perhaps limited by the need for chloroplast fusion. Recombination between chloroplast genomes can be demonstrated in appropriate crosses with *Chlamydomonas*. Gene order is remarkably highly conserved between species, with differences generally being due to a small number of inversions. For example, a single inversion of roughly a quarter of the genome essentially interconverts the tobacco and liverwort gene orders. The cause of these inversions is not clear, although a number seem to be associated with tRNA genes. Species that have lost one arm of the inverted repeat appear to have a much higher level of other rearrangements, suggesting that the repeats help to stabilize the genome. High levels of rearrangement, including deletions, often occur during anther culture of some species.

Gene expression

The efficient expression of chloroplast genes in *Escherichia coli*, either *in vivo* or in extracts *in vitro*, suggested similarities in the DNA sequences directing and controlling gene expression (*see* BACTERIAL GENE EXPRESSION). Analysis of point and deletion mutation constructs by *in vitro* transcription in chloroplast extracts showed that sequences similar to the − 35 and − 10 regions of *E. coli* genes are indeed generally, although perhaps not universally, required for efficient expression.

The exact subunit composition of chloroplast RNA POLYMERASE is not clear, and it is possible that there are different forms of the enzyme with different specificities. The transcript patterns observed in NORTHERN BLOTS are usually very complicated, with many genes being transcribed into large POLYCISTRONIC transcripts that are subsequently processed into much shorter molecules. Short inverted repeated sequences observed at the ends of a number of genes were thought to represent transcription terminators, but it now seems likely that many represent RNA PROCESSING sites.

A small number of plastid genes contain INTRONS. Most of these are of types I and II and autocatalytic splicing has been demonstrated for a number of introns from *Chlamydomonas*. Given that transcription and translation must take place in the same physical compartment, and that a significant population of partially spliced transcripts exists, it will be very interesting to see how RNAs that have not been fully spliced are prevented from being translated.

Some plastid transcripts must be spliced in *trans*; separate RNA molecules must be joined together to generate the full coding sequence (TRANS-SPLICING). This has been shown for the *psaA* gene of *Chlamydomonas*, and the *rps12* gene of a number of species. In *Marchantia*, for example, this gene is encoded in three EXONS. The second and third are separated from each other by an intron of some 500 nucleotides, but from the first by some 66 kbp. Furthermore, the first exon is on the opposite DNA strand from the second and third exons. Synthesis of rps12 in the chloroplast must therefore require splicing together of two separate RNAs, and the existence of ligated molecules has been demonstrated by RNA sequencing. The boundaries of the first and second introns are similar to those of group II introns, so the splicing reactions may well be similar to those for more 'conventional' introns.

A limited amount of editing of chloroplast mRNA occurs (*see* RNA EDITING). Translation appears to follow the 'universal' GENETIC CODE, and shows many similarities to the process in *E. coli*, with genes possessing classical RIBOSOME-BINDING SITES. As mentioned earlier, there are similarities in antibiotic sensitivity and resistance between chloroplast and bacterial ribosomes. For example, cytoplasmic ribosomes are inhibited by cycloheximide but not chloramphenicol, while the reverse is true for ribosomes of plastids and *E. coli*.

Control of gene expression

The two best studied systems in this context are the response to changes in light levels (for example on greening of dark-grown plants) and, to a lesser extent, the conversion of chloroplasts to chromoplasts. Early work showed that the transcript levels of a number of genes increased in response to illumination. These were termed photogenes with *psbA* being a well-studied example. It was widely assumed that changes in transcript levels reflected changes in the rate of transcription initiation (perhaps mediated by changes in genome ploidy, SUPERCOILING, polymerase specificity, levels of regulator proteins, etc.), but more recently RUN-OFF ASSAYS have suggested that the modulation of RNA stability may be more important.

It seems likely that the potential STEM-LOOP structures in mRNAs are involved, and there is evidence for specific binding proteins associating with these stem-loops, which may regulate the rate of mRNA processing or degradation.

There is also evidence for the modulation of translation in response to light. For example, the alteration in the rate of synthesis of *psbA* protein in *Spirodela* chloroplasts in response to changing light levels cannot be satisfactorily explained by alterations in transcript levels. In *Chlamydomonas*, nuclear proteins have been implicated in the translation of specific chloroplast RNAs. A nuclear mutant has been described which abolishes the translation of the *psbD* mRNA, although the RNA itself is still present. This situation is similar to the *pet* mutations of yeast mitochondria (*see* MITOCHONDRIAL GENOMES: LOWER EUKARYOTES). The final level of control may be protein stability, with proteins surplus to requirements being degraded. In this regard it is interesting to note that most, and arguably all, plastid-encoded polypeptides are at some stage physically associated with nucleus-encoded polypeptides, and this might help in integrating nuclear and plastid gene expression. The extent and mechanism of integration of chloroplast and nuclear gene expression is at present one of the key problems in plastid molecular biology [2].

There is increasing evidence that protein turnover is an important process in plastids. For example, the psbA reaction centre polypeptide is very susceptible to damage during photosynthesis, and is removed proteolytically to be replaced by a functional polypeptide. In a number of other examples, when a subunit of a complex is lacking as a result of mutation, other subunits are synthesized but then degraded rather than assembled.

Evolutionary origins

It is now widely accepted that plastids are the outcome of an endosymbiotic relationship between a photosynthetic prokaryote and a nonphotosynthetic organism established in the distant evolutionary past. The ENDOSYMBIONT HYPOTHESIS was first postulated in the 19th century, on the basis of the suggestion that chloroplasts were reproduced by binary fission, rather than being formed *de novo*. A large body of biochemical and ultrastructural evidence has accumulated to support the theory, and debate now mainly concerns what sort of organisms were responsible for establishing the symbiosis.

Given the variation in ultrastructure as well as pigment composition, it seems likely that plastids are POLYPHYLETIC in origin, although the chlorophyte plastids (those of chlorophyte algae and green plants) as a group may well be MONOPHYLETIC. CYANOBACTERIA have been suggested as the original endosymbiont, but although they are indeed oxygenic photosynthetic prokaryotes, they differ in pigment composition from the chloroplasts of chlorophytes, lacking chlorophyll *b* and possessing PHYCOBILIPROTEINS. The PROCHLOROPHYTES, of which the best documented are *Prochloron didemni* and *Prochlorothrix hollandica*, have also been proposed as representing the chloroplast ancestors, on the grounds that they have chlorophyll *b* and lack phycobiliproteins. Attempts to ascertain their true evolutionary position using ultrastructural, biochemical, and molecular biological data have, however, led to inconsistent conclusions.

Since the establishment of endosymbiosis, there has been a loss of genes to the nucleus. Why some genes and not others should have been transferred is not clear. It has been suggested that the polypeptides whose genes are retained in the plastid are for some reason unable to cross the plastid envelope. This is unlikely to be the case, however. For example, it has been shown that a modified *psbA* gene with the protein-coding sequence fused to a sequence encoding a chloroplast envelope transit peptide, when introduced into and expressed in a plant cell nucleus produced a protein that could be imported into the chloroplast at levels sufficient to allow it to substitute for the endogenously coded protein.

Evidence for the more recent transposition of DNA between organelles is provided by the demonstration of promiscuous DNA, that is, sequences that are present in organelles other than the one in which they are normally expressed. Thus, sequences homologous to chloroplast DNA can be demonstrated in the mitochondrial and nuclear genomes of a large number of plants (*see* MITOCHONDRIAL GENOMES: PLANTS). What the mechanism of this transfer is, and whether or not it is similar to the mechanism for transfer of functional genes is not clear.

Chloroplast transformation

For a long time, one of the difficulties of working with the chloroplast genome was the lack of a suitable genetic TRANSFORMATION system to modify existing genes or introduce new ones. In spite of a report of *Agrobacterium*-mediated transformation of the chloroplast genome, this approach does not seem to be successful. More recently, *Chlamydomonas* chloroplasts were suc-

cessfully transformed by DNA carried on tungsten microprojectiles introduced by the 'biolistic' technique (*see* MICROBALLISTICS). A wild-type chloroplast *atpB* gene (encoding the β subunit of ATP synthase) was used as a selectable marker, with the recipient strain carrying a mutant *atpB* (making it unable to carry out photophosphorylation and requiring a source of fixed carbon in the medium). Transformation was selected by the ability to grow phototrophically, and was shown to result from the stable replacement of the mutant sequence by the transforming one. Transformation of plastids of higher plants is also possible. The availability of a reliable transformation system will be of great assistance in studying all the aspects of chloroplast biology described above.

C.J. HOWE

See also: ELECTRON TRANSPORT CHAIN; PHOTOSYNTHESIS.

1 Ohyama, K. et al. (1988) Structure and organisation of *Marchantia polymorpha* chloroplast genome. *J. Mol. Biol.* **203**, 281–372.
2 Gruissem, W. (1989) Chloroplast gene expression: How plants turn their plastids on. *Cell* **56**, 161–170.
3 Hiratsuka, J. et al. (1989) The complete sequence of the rice (*Oryza sativa*) chloroplast genome: intermolecular recombination between distinct tRNA genes accounts for a major plastid DNA inversion during the evolution of cereals. *Mol. Gen. Genet.* **217**, 185–194.

chloroquine A weak base (Fig. C40) which can accumulate in the interior of acidified organelles and raise their pH towards neutrality. The uncharged chloroquine molecules are membrane permeable, but when they enter an acid compartment they are converted to charged species that are membrane impermeable and thus raise intraorganelle pH. The functions of the GOLGI APPARATUS, ENDOSOMES and LYSOSOMES as well as membrane traffic pathways involving these organelles can be disrupted by chloroquine. The effects are reversible on removing (washing out) the drug. Chloroquine is an antimalarial agent.

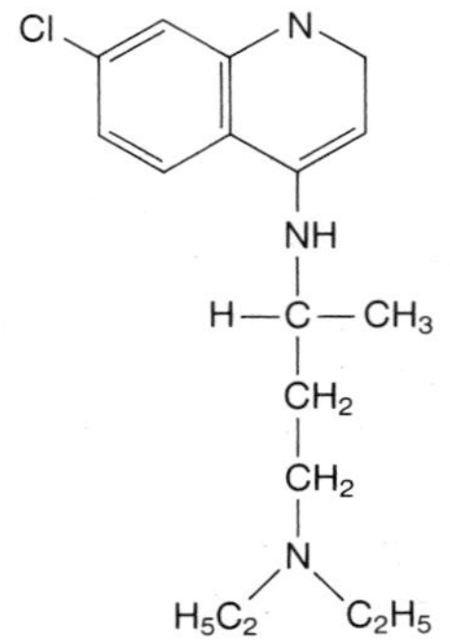

Fig. C40 Chloroquine.

chlorphonium chloride (Phosfon^{TM}) Tributyl-(2,4-dichlorobenzyl)-phosphonium chloride (Fig. C41); M_r 383.5. A PLANT GROWTH RETARDANT from the Mobil Chemical Co. It reduces the production of GIBBERELLINS by inhibiting activity A of *ent*-kaurene synthetase, with I_{50} ~10^{-6} M for the enzyme from

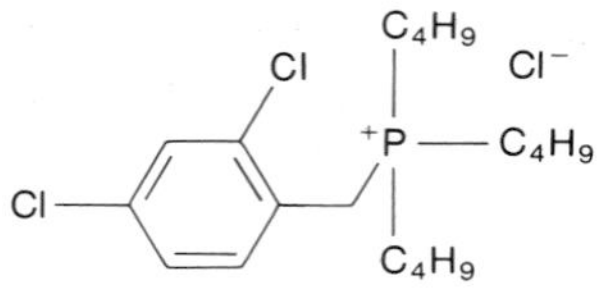

Fig. C41 Structure of chlorphonium chloride.

Marah macrocarpus endosperm. It has small-scale use in horticulture to control size of some ornamental plants.

Fall, R.G. & West, C.A. (1977) *Plant Physiol.* **59**, 22–29.

CHO cells Chinese *h*amster *o*vary cells. An epithelial CELL LINE.

cholera toxin An oligomeric protein of M_r 87 000 consisting of an A subunit and five B subunits, produced by the Gram-negative bacterium *Vibrio cholerae* and responsible for the massive loss of fluid from the body in cholera. The A subunit consists of an A_1 peptide (M_r 23 000) linked to an A_2 peptide by a disulphide bond. The B chains recognize and bind specifically to G_{M1} ganglioside on intestinal cell surfaces, and the A subunit is inserted into the cell. Proteolytic cleavage of the A_1 peptide and reduction of the disulphide bond activates the ADP-ribosyltransferase activity of the A_1 peptide. This catalyses the transfer of an ADP-ribose unit from NAD to Arg 201 of the α subunit of G_s protein (ADP-RIBOSYLATION), inactivating its GTPase activity and resulting in the overstimulation of ADENYLATE CYCLASE and the overproduction of cAMP (*see* GTP-BINDING PROTEINS; SECOND MESSENGER PATHWAYS). This in turn stimulates active transport of ions across the plasma membrane and leads to an outflow of Na$^+$ and water from the tissue into the gut. *See also*: BACTERIAL PATHOGENICITY.

cholesterol A sterol (Fig. C42), an essential neutral lipid constituent of most eukaryotic plasma membranes, and which is also required for steroid synthesis and as a precursor of vitamin D. Animals usually obtain cholesterol from the diet but there is the capacity from endogenous production to supply most of the daily requirement. Virtually all nucleated cells can synthesize cholesterol but the qualitatively important tissue is the liver. It is carried

Fig. C42 Cholesterol.

in the blood in the form of lipoproteins. *See also*: MEMBRANE STRUCTURE; PLASMA LIPOPROTEINS AND THEIR RECEPTORS.

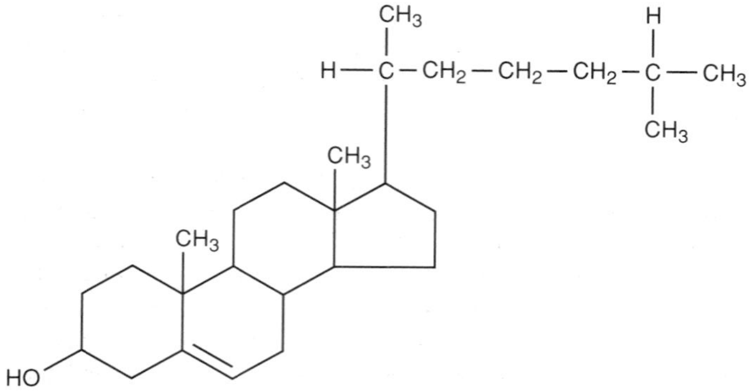

choline A quaternary amine, HOCH$_2$-CH$_2$N(CH$_3$)$_3^+$, which occurs in the PHOSPHOLIPID phosphatidylcholine and the neurotransmitter ACETYLCHOLINE. It is lipotropic, that is it can reduce liver fat content by increasing phospholipid turnover.

choline acetyltransferase Enzyme (EC 2.3.1.6) that catalyses the reaction between acetyl coenzyme A and choline to form ACETYLCHOLINE, by the transfer of an acetyl group from acetyl coenzyme A.

chondriome The MITOCHONDRIAL and CHLOROPLAST GENOMES of a cell. The term is also sometimes taken to refer to the mitochondrial genome alone, and sometimes to the mitochondria as a whole.

chondroblast, chondrocyte Cartilage cell (chondrocyte) and its precursor (chondroblast). Chondrocytes secrete large quantities of extracellular matrix, principally collagen type II and proteoglycans, which forms cartilage.

chondroitin sulphate A sulphated GLYCOSAMINOGLYCAN of the extracellular matrix (M_r 5000–50 000), present in cartilage, cornea, bone, skin, and arteries. *See also*: EXTRACELLULAR MATRIX MOLECULES; PROTEOGLYCANS.

chondronectin *See*: EXTRACELLULAR MATRIX MOLECULES.

chordamesoderm The rod of AXIAL MESODERM which characterizes the embryos of CHORDATES. It comprises the HEAD PROCESS in the head and the NOTOCHORD in the trunk behind the ear. This specialized mesoderm is thought to have appeared during evolution (in now extinct groups of protochordates such as the Solutes and the Mitrates) to confer rigidity on the animal. The term is only rarely used; more frequently the terms notochord and head process are used instead.

chordate The group of animals (phylum Chordata) whose embryos possess a NOTOCHORD. The group includes all vertebrates, and other forms such as the tunicates which do not possess an internal skeleton. All chordates other than the vertebrates are known as protochordates. There are many important differences of developmental significance between the two groups. For example, only the vertebrates possess an anterior brain, and while sense organs are shared by both groups, only vertebrates possess complex paired peripheral sense organs. The vertebrate pharyngeal skeleton is made of cartilage, in contrast to collagen and chitin in protochordates. Finally, water movement over the gills is driven by a muscular pharynx in vertebrates, whereas it occurs by ciliary movements in protochordates.

chorion The outermost of the EXTRAEMBRYONIC MEMBRANES. It functions as the major partition for the exchange of gas between the embryo and its environment. In mammals, the chorion becomes part of the placenta where, in addition to respiratory functions, it supplies nutrition and removes waste.

In insects, the chorion is a complex and physiologically important extracellular assembly composed of many different kinds of protein. Groups of proteins are deposited around the oocyte in overlapping succession and following an apparent developmental programme responsible for chorion morphogenesis. Accordingly, there are groups of early, middle, late middle, late, and very late proteins. The developmental regulation of the insect chorion genes has been intensively studied. *See also*: CHORION GENE AMPLIFICATION.

chorion gene amplification The CHORION gene clusters of DROSO-PHILA and some other insects are amplified during choriogenesis to provide the large quantities of proteins required during a brief period. The chorion proteins are synthesized by the ovarian follicle cells and before chorion gene expression the entire follicle cell genome undergoes additional rounds of DNA synthesis until it reaches 16 times the haploid DNA content. Following genome amplification, the chorion genes are selectively amplified a further 10-fold. *See also*: GENE AMPLIFICATION.

chorionic gonadotrop(h)in Glycoprotein hormone of M_r 30 000 produced by the developing TROPHOBLAST. It acts predominantly to maintain the corpus luteum which is essential for pregnancy survival until the placenta produces adequate oestrogen and progesterone at 7–8 weeks of pregnancy (in humans). In the male foetus it stimulates gonadal production of testosterone which is important for the male differentiation of the developing foetus. It is detectable from 8 days post-ovulation. In DOWN'S SYNDROME its serum concentration may be raised.

chorionic villus sampling Method of obtaining trophoblast tissue for cytogenetic and biochemical analysis of human foetal DNA. From 8–12 weeks gestation the chorion may be biopsied by either the transcervical or transabdominal route. The foetal loss rate is 1–2% in experienced hands. *See also*: PRENATAL DIAGNOSIS.

Chou–Fassman analysis One of several computational methods used for predicting the secondary structure of proteins. Chou and Fassman base their method on a statistical analysis of a database of known PROTEIN STRUCTURES, noting the frequency with which particular amino acid residues are found in α-helices, β-strands, and loop (or turn) conformation. The effect of neighbouring residues in the partamino-acid sequence is taken into consideration. A Chou–Fassman analysis of a protein sequence provides average probability values for each residue, these being the probability of its being α-helical, β-strand, and turn.

Chou, P.Y. & Fassman, G.D. (1974) *Biochemistry* **13**, 222–245.

Christmas disease *See*: HAEMOPHILIA B.

chromaffin cell Neural-crest derived CATECHOLAMINE-containing cell associated with the sympathetic nervous system. Chromaffin cells are found in the adrenal medulla, sympathetic ganglia, and paraganglia (*see* PERIPHERAL NERVOUS SYSTEM) and are used in culture, for example, to study the mechanisms of secretion of catecholamines (*see* EXOCYTOSIS). Fibroblast growth factor (*see* GROWTH FACTORS) can induce their differentiation into neurons.

chromatic adaptation The modulation of synthesis of different pigments in response to changes in the wavelength of incident light, as in certain CYANOBACTERIA.

chromatid One half of a replicated CHROMOSOME, which can be seen in late prophase of MITOSIS and in diplotene in the first division of MEIOSIS.

Chromatin

THE eukaryotic cell nucleus is typically ~1 μm in diameter. It contains a large amount of DNA (a total length of 1–2 metres) which must be efficiently packaged in such a way as to guarantee access to genetic information. Thus, each DNA molecule is packaged in the form of a CHROMOSOME. Chromosomes are composed of chromatin, a densely staining material initially recognized in two different forms: highly condensed HETERO-CHROMATIN and more diffuse EUCHROMATIN.

The composition and structure of fragments of chromatin (prepared by limited digestion of nuclei with micrococcal NU-CLEASE) have been well studied. Chromatin fragments contain DNA (a negatively charged polymer) complexed with stoichiometric amounts of highly positively charged proteins called HISTONES, and much smaller amounts of other DNA-binding proteins, collectively referred to as nonhistone proteins (*see e.g.* HMG PROTEINS). The histones organize DNA into a regular repeating structure, the basic unit of which is the nucleosome, and they are primarily responsible for packaging of DNA in chromatin.

The nucleosome

Decondensed chromatin viewed in the electron microscope resembles 'beads on a string'. Each bead is a chromatosome containing about two SUPERCOILS of DNA (~166 bp) wrapped around a core histone octamer and sealed by a single molecule of linker histone (H1) bound at the point where the DNA enters and exits. H1 also binds to the linker DNA (the string connecting one bead to the next). Nucleosome core particles are prepared by extensive digestion of chromatin with micrococcal nuclease. They contain 146 bp of DNA wrapped around the core histone octamer, and no H1. The crystal structure of the core particle (solved to ~7 Å resolution) shows that the DNA is in the B-form (*see* B-DNA). It is not smoothly wrapped around the histone octamer but has four kinks symmetrically located one and four DNA turns on each side of the pseudo-dyad axis of symmetry of the core particle [1].

The histone octamer is composed of two molecules each of the four core histones: H2A, H2B, H3 and H4. These are small (with relative molecular masses (M_r) of 11 000–15 000), very basic proteins which have been highly conserved during evolution. In the absence of DNA, the octamer decomposes into a tetramer (H3$_2$–H4$_2$) and two dimers (H2A–H2B). In the nucleosome core, the tetramer interacts with the central turn of DNA, and the dimers (on opposite sides of the tetramer) interact with the half turns above and below the tetramer. All four core histones have an N-terminal 'tail' domain (so-called because it has no secondary structure in the absence of DNA), a central globular domain, and a short C-terminal tail domain (H4 lacks a C-terminal tail). The globular domains are responsible for core histone–core histone interactions in the tetramer and the dimers, and for dimer–tetramer interactions in the octamer. They also bend DNA around the octamer; the core histone tail domains can be removed with trypsin without major disruption of the core particle. The tail domains contain about half of the positively charged residues and most of the POST-TRANSLATIONAL MODIFICATION sites of the core histones, but their functions are unclear.

Histone H1, the linker histone, is not part of the nucleosome core. It is very rich in lysine residues, less well conserved than the core histones, and is larger, having a M_r of ~22 000. A typical somatic H1 (~220 residues) contains 66–70 basic residues, nearly all of which are lysine (not arginine). H1 has a tripartite domain structure: a central globular domain of ~80 residues flanked by a short N-terminal tail domain and a long C-terminal tail domain. The globular domain recognizes the H1 binding site on the nucleosome core and seals the DNA turns; the N-terminal tail domain probably helps to position the globular domain; and the C-terminal tail domain is probably directed along the linker DNA. Most of the positively charged residues are in the C-terminal domain, and this domain binds most strongly to DNA. There are a number of tissue-specific variants of H1, such as H5, which have the same domain structure, but contain more arginine residues and have a higher positive charge density. They bind DNA much more strongly than somatic H1, and are found in cells containing inactive genomes, such as nucleated erythrocytes and sperm.

The length of DNA in the core particle is invariant (146 bp), but the average length of the linker DNA varies between species and tissues, giving rise to a characteristic repeat length (the average length of DNA in a nucleosome), observed on digestion of chromatin with micrococcal nuclease. Most tissues contain a typical 'somatic' H1 and have a repeat length of 180–200 bp. Unusually short repeat lengths (~160 bp, with little or no linker DNA) are associated with depletion of H1 (in neuronal chromatin) or possibly its absence (in yeast). Unusually long repeat lengths (>200 bp) are associated with variants of H1 such as H5 (in avian erythrocytes, repeat length ~ 212 bp) and sea urchin sperm H1 (~240 bp).

Linking number paradox

The core particle contains ~1.75 left-handed supercoils of DNA wrapped around the core histone octamer, but the measured change in linking number (*see* SUPERCOILING) induced in a closed circular DNA by a histone octamer is only – 1, not – 1.75. This discrepancy is the so-called linking number paradox. However, the observed change in linking number is the sum of changes in DNA twist as well as DNA writhe. Calculation of both twist and writhe depends on the geometry of the surface on which the DNA is wrapped, which in this case is the approximately cylindrical histone octamer. Therefore the writhe is the sum of the writhe of nucleosomal DNA (~ – 1.75) and the (cylindrical) surface writhe (~ + 0.2); and the twist is the sum of the twist of the nucleosomal DNA (the winding number) or number of base pairs per turn and the surface twist (~ – 0.2). Thus, the paradox may be resolved if the winding number of DNA in the nucleosome core particle is different from that of free DNA (~10.5 bp per turn), which is indeed the case. The measured winding number of most of the DNA in the core particle is ~10.0 bp per turn, but the three central turns apparently have ~10.6 bp per turn. This result was obtained using both the NUCLEASE DNase I and hydroxyl radical as probes of DNA periodicity. It has been argued that the three central turns also have 10.0 bp per turn, and that the apparent (observed) value of 10.6 bp per turn reflects steric inhibition of cleavage by DNase I at the expected sites, but this is unlikely to be true for hydroxyl radical. In conclusion, the change in twist provides a positive contribution to the linking number change, but not quite enough to resolve the paradox quantitatively.

Nucleosome positioning

Nucleosome cores can adopt a defined position with respect to underlying DNA sequence both *in vitro* and *in vivo*. This may involve a direct mechanism in which the core histone octamer prefers to bind to a particular sequence, often with base pair precision. Sequences near the centres of positioned nucleosomes are important in defining the position; they are often curved DNA (*see* DNA), although the rules for predicting positions have not been fully worked out. Also, analysis of CLONED DNA sequences obtained from nucleosome core particles revealed a tendency for minor grooves of DNA (*see* DNA) containing AT base pairs to face inwards and minor grooves containing GC base pairs to face outwards. Another mechanism directing positioning of the nucleosome core involves a fixed boundary (such as a sequence-specific protein at its binding site) which may influence the sites of deposition of histone octamers on either side during nucleosome assembly (statistical positioning). In this mechanism, nucleosome positions are independent of the underlying DNA sequence, depending only on the distance from the boundary. Examples of statistical positioning have been described. Nucleosome positioning may be particularly important in gene PROMOTER regions, where it could affect accessibility of the promoter to transcription factors and thus the ability of the gene to be expressed.

Modifications of nucleosome structure

In vivo, core histones can be acetylated, methylated, ADP-ribosylated, ubiquitinated, and phosphorylated (*see* POST-TRANSLATIONAL MODIFICATION OF PROTEINS). The sites of post-translational modification are almost exclusively located in the N-terminal tail domains. Little is known of their structural consequences. Modification often correlates with cellular processes such as DNA REPLICATION, TRANSCRIPTION, chromatin assembly, and DNA REPAIR. Acetylation has been most intensively studied.

Many of the lysine residues in the N-terminal tail domains can be acetylated, reducing their positive charge. Core histones in chromatin are mainly unacetylated or have just a few acetylated lysine residues. Hyperacetylation can be induced by treatment of cells with butyrate. The level of histone acetylation is determined

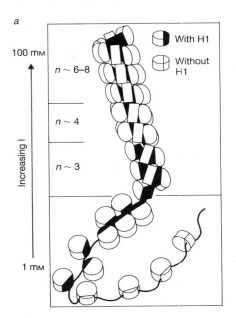

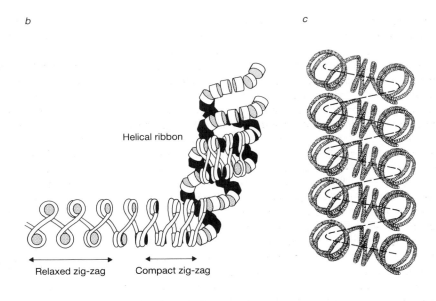

Fig. C43 Three models of the 30-nm chromatin fibre. In *a* and *b* the fully compacted structure is seen at the top of the figure and intermediate steps in the ionic strength-induced compaction are shown in the bottom part. In *c* only the DNA path is shown and chromatosomes on the further side of the solenoid are omitted for clarity. *a*, From [4]; *b*, from [5]; *c*, from [7].

by a dynamic equilibrium between the activities of histone acetyltransferase and deacetylase enzymes (the latter is inhibited by butyrate). The structure of hyperacetylated nucleosome core particles does not seem to be very different from that of unacetylated ones. The only clear difference is that hyperacetylated histone octamers constrain only about − 0.8 supercoil rather than the − 1 supercoil constrained in the unacetylated nucleosome core, but the structural basis of this is not clear. Core histones to be used for chromatin assembly in the early embryo are stored in acetylated form as complexes with NUCLEOPLASMIN (*see* MOLECULAR CHAPERONES) and N1/N2 proteins in *Xenopus* eggs.

There is some evidence for conformational changes in nucleosomes in response to modification or the binding of nonhistone proteins but this is controversial. The only proteins known to bind specifically to the nucleosome are the nonhistone HMG 14/17 proteins (*see* HMG PROTEINS), but they do not destabilize the nucleosome. Their functions are unknown, but they may be associated with transcriptionally active chromatin.

Chromatin folding

Most of the chromatin in the nucleus is in the form of a highly condensed filament about 30 nm in diameter (the 30-nm filament). Isolated chromatin fragments can be induced to form condensed structures similar in appearance to the filaments observed in nuclei by exposure to salt. Chromatin fragments become increasingly condensed as the salt concentration is in-

creased toward physiological. At concentrations of salt close to physiological (~0.1 M), chromatin fragments precipitate (a serious technical problem in the study of chromatin). Formation of these highly condensed filaments is dependent on stoichiometric amounts of H1 (i.e. one H1 molecule per nucleosome), and is influenced by the core histone tail domains. Most agree that the condensed 30-nm filament is the result of solenoidal (helical) folding of the beads-on-a-string nucleosomal (10 nm) filament, probably having six nucleosomes per turn. Others have suggested that the nucleosomal filament is a zigzag, which may then form a condensed ribbon and coil up to form a helical 30-nm filament with twelve nucleosomes per turn. It is agreed that the nucleosome cores (modelled as disks) are arranged with their flat faces close to parallel to the axis of the condensed 30-nm filament. The disposition of the linker DNA is unknown. The most popular models are depicted in Fig. C43 (see [2] for a review).

Chromatin loops

How chromatin filaments are organized in the nucleus is not understood, but much of the evidence supports the idea that INTERPHASE chromosomes are organized into loops of chromatin filament attached to the nuclear skeleton at their bases and projecting into the interior of the nucleus (*see* NUCLEAR ENVELOPE; NUCLEAR MATRIX). Each loop may contain a gene or related cluster of genes whose expression may in principle be regulated at the level of loop structure. Estimates of loop sizes vary widely, apparently depending on the method and the tissue source, but probably contain an average of 40 000–80 000 bp. On extraction of chromosomal proteins (including histones) from interphase chromatin, a residual interphase scaffold is observed in the microscope, which contains TOPOISOMERASE II and LAMINS A, B, C (*see* NUCLEAR MATRIX) and a few other (unidentified) proteins. Nuclease digestion of unprotected DNA in a preparation of scaffolds led to the identification of DNA sequences involved in attachment to the nuclear matrix (matrix attachment regions (MARs)). Whether these sequences really are matrix attachment regions or whether the scaffolds were disrupted during preparation (which involves quite harsh extraction methods) is controversial.

When the nuclear membrane breaks down during MITOSIS, chromatin is reorganized to form METAPHASE CHROMOSOMES in which a chromosomal metaphase scaffold, also containing topoisomerase II, is folded to form a quite regular helical coil, to which chromatin loops are attached at their bases. SISTER CHROMATIDS are usually of opposite helical handedness. Thus, the chromatin loop structure is apparently conserved, but the interphase and metaphase scaffolds are known to differ in protein content. Many nuclear proteins become highly phosphorylated during metaphase, including the lamins (which are released into the cytoplasm) and H1.

Chromatin structure and transcription

DNA must be packaged into chromatin such that genetic information is accessible. This is an obvious requirement for both transcription and DNA replication; transcription promoters and ENHANCERS, and ORIGINS OF REPLICATION must be accessible to DNA and RNA polymerases and their associated factors (*see* EUKARYOTIC GENE EXPRESSION). How this is achieved is not well understood. It is clear that assembly of such regulatory elements into chromatin *in vitro* is sufficient to repress their function completely, and there is some evidence that this is true *in vivo*. Thus, the organization of the genome into chromatin suggests a general mechanism for repression of genetic information, which must be circumvented for expression of particular sets of genes. This might occur as chromatin structure is transiently disrupted during DNA replication, possibly allowing specific factors (*see* TRANSCRIPTION FACTORS) to recognize their binding sites and bind to block their assembly into nucleosomes [3]. However, transcriptional activation of many inducible genes is accompanied by changes in chromatin structure which clearly do not require a round of DNA replication; this chromatin reorganization might involve interactions of specific factors with nucleosomes. DNA sequences (LOCUS CONTROL REGIONS) which direct the regulated expression of associated TRANSGENES independently of the site of integration in the chromosome have been identified; these might work through effects on chromatin structure and assembly [3].

Potentially transcriptionally active (e.g. inducible) genes and actively transcribing genes have a markedly different chromatin structure from that of untranscribed 'bulk' chromatin. The most obvious manifestation of this is the presence and appearance of sites in cell nuclei (both 5′ and 3′ of the TRANSCRIPTION UNIT) hypersensitive to DNase I. These sites usually include promoter and associated enhancer as well as other regulatory regions. For example, there are hormone-dependent DNase I hypersensitive sites in the chromatin of hormone-inducible genes. Hypersensitive sites are usually ~200 bp long, suggesting that their formation might involve the disruption or displacement of a single nucleosome, and the binding of general and activating transcription factors.

Since H1 is required for the formation of a highly condensed filament, one of the first steps in activating a gene might be loss of H1 with consequent local unfolding of the chromatin. Indeed, there probably is less H1 in actively transcribing chromatin than in inactive chromatin. Electron micrographs of the extremely active BALBIANI RING genes suggest that RNA polymerases in the act of transcription induce local unfolding or displacement of nucleosomes which rapidly reform after the polymerase has passed, and that the gene is condensed into the folded 30-nm filament only where transcribing polymerases are relatively far apart. These observations suggest a dynamic chromatin structure for transcriptionally active genes. In nuclei, chromatin containing an active or potentially active transcription unit is also more generally sensitive to DNase I than bulk or unexpressed chromatin (the phenomenon known as general DNase I sensitivity). The physical basis of this is unknown, but there are correlations with altered nucleosome structure, partial depletion of H1 and the binding of the nonhistone proteins HMGs 14 and 17. This sensitivity typically extends for many kilobases on both sides of the gene and so cannot be a consequence of transcription *per se*; it is thought to reflect an altered chromatin domain structure (see chromatin loops above).

Experiments *in vitro* show that a nucleosome core formed on the promoter of a gene is sufficient to prevent initiation of transcription. Whether this nucleosome-mediated repression is important *in vivo* has not yet been directly established, although incorporation of an origin of replication into a nucleosome core has been shown to repress replication effectively *in vivo*. In contrast, a nucleosome core can be transcribed through *in vitro* once transcription has been initiated, indicating that transcript elongation is not blocked by a nucleosome core. The mechanism by which bulky polymerases negotiate a gene assembled into nucleosomes is controversial; it is likely to involve nucleosome unfolding or displacement.

The relationship between chromatin structure and the regulation of gene expression is currently under intense study. For a thorough description and summary of the field of chromatin structure and function see [6].

D. CLARK

See also: PROTEIN–NUCLEIC ACID INTERACTIONS.

1 Richmond, T.J. et al. (1984) Structure of the nucleosome core particle at 7 Å resolution. *Nature* **311**, 532–537.
2 Felsenfeld, G. & McGhee, J.D. (1986) Structure of the 30-nm chromatin fiber. *Cell* **44**, 375–377.
3 Felsenfeld, G. (1992) Chromatin as an essential part of the transcriptional mechanism. *Nature* **355**, 219–224.
4 Thoma, F. et al. (1979) Involvement of histone H1 in the organization of the nucleosome and of the salt-dependent superstructures of chromatin. *J. Cell Biol.* **83**, 403–427.
5 Woodcock, C.L.F. et al. (1984) The higher-order structure of chromatin: evidence for a helical ribbon arrangement. *J. Cell Biol.* **99**, 42–52.
6 van Holde, K.E. (1989) *Chromatin* (Springer-Verlag, New York).
7 McGhee, J.D. (1983) Higher order structure of chromatin: orientation of nucleosomes within the 30 nm chromatin solenoid is independent of species and spacer length. *Cell* **33**, 831–841.

Chromatography

CHROMATOGRAPHY is a method of separating substances by allowing them to partition between two phases, one mobile and one stationary. Each substance to be separated moves as a zone which travels at a fraction of the speed of the moving phase. It separates from other substances because this fraction varies from substance to substance, depending on their partition coefficients between the two phases. As LeRosen pointed out [1] the fraction (R) of the substance in the moving phase at any time is equal to the fraction of the time that each molecule of the substance spends in the moving phase, and hence is also equal to the fraction of the speed of the moving phase with which the zone of substance moves.

The moving phase can be gas or liquid; the stationary phase can be a solid, or a liquid held immobile by a solid. The stationary phase may be packed in a column, as in the various types of column chromatography, or be spread in a sheet, as in paper chromatography and thin-layer chromatography. A great variety of systems is available.

Isocratic elution

In the simplest form of chromatography, which is the one with greatest resolving power, the moving phase has constant composition, so that the zone of substance being isolated moves at a fixed fraction of its speed. Hence the separation steadily improves all the time the chromatogram is running, as each zone becomes more and more separated from zones of substances that partition either more or less in favour of the moving phase.

Types of chromatography

The characteristics that give satisfactory separations depend rather little on the molecular interactions that determine the partition coefficients of the substances to be separated. Hence much of what can be said on the requirements for good separations applies irrespective of whether separations are based on relative volatility (gas–liquid chromatography), hydrophobicity ('normal' when the moving phase is more hydrophobic and 'reversed phase' where the stationary phase is more hydrophobic), or ionic interactions (ion-exchange chromatography). What the molecular forces are does not matter, provided that interactions are rapid. Thus hydrophobic interactions often play a large part in separations on ion exchangers. Slow interactions must be avoided; thus oxo acids chromatograph well in competition with acetate on strongly basic anion exchangers, which have substituted ammonio groups $-N^+R_3$, but badly on weakly basic ones, which have free or substituted amino groups, $-NH_2$ or $-NR_2$, presumably because hemiaminals form between carbonyl and amino groups and the formation equilibrates slowly.

Efficient running of chromatograms

Obviously, the good running of chromatograms depends crucially on the physical conditions. Equilibration between the phases must be rapid compared with the speed of the moving phase. This requires an appropriate physical form of the apparatus and of the stationary phase, represented, for example, by the particle size and distribution of the packing material of a column. Different column media require different packing regimes to achieve satisfactory columns.

Columns can be produced in factory conditions with better characteristics in these respects than can easily be packed in the laboratory. For this reason many high performance/pressure liquid chromatography (HPLC) and (for proteins) fast protein liquid chromatography (FPLC) systems are commercially available. These systems have many other advantages. They allow rapid flow optimized for the particular columns, and easy control of eluent composition. They also allow high sensitivity of detection and integration of peaks, partly by using narrow columns.

Another requirement for keeping zones sharp is more often overlooked. This is that the partition coefficient of the substance(s) being separated must not vary with load of the substance itself, since, if the substance runs at different speeds according to its concentration, its zone will spread. Then separation may be satisfactory up to a certain load, but above this the zones broaden (asymmetrically) and the system is described as overloaded. This may have little importance when separation is for analytical use, when the loads can be kept low, provided that sensitive techniques are available for detecting and measuring the substances being separated. But convenient preparative use requires appreciable loads to be handled.

Rational design can often increase the loads that can be handled. At the simplest level, the partition coefficient of a dissociating substance will depend on the ionic form of its groups, and hence on pH. The substance itself, however, has buffering power, and so will determine the pH of the zone. To avoid spreading on this account, the buffering of the medium must be high compared with any due to the substances being separated, so that the pH will not vary with sample load. Further, in ion-exchange chromatography, the ionic competition for the sites in the stationary phase must not alter with sample load. So the ionic concentration of the medium must be high compared with that due to the substances in the sample, since these themselves contribute to the ionic competition. Indeed, since few interactions are solely due to ionic attractions, the chemical nature of the competing ion may sometimes also need to be similar to that of the substances being separated.

Resolution

The substitution of group A for group B in a molecule is likely to make the same change in standard Gibbs energy of transfer of that molecule between two phases irrespective of the nature of the rest of the molecule. It therefore multiplies the partition coefficient by the same factor, whatever the substance it occurs in (of course this generalization breaks down if the rest of the molecule can hinder interactions of this group with the two solvents). Hence the resolving power of chromatography is appropriately judged by the factor by which the partition coefficient must change for the zones of two substances to be adequately resolved.

For column chromatography, it can be shown that resolution rises as the square root of the column length. Hence if two substances are just adequately resolved by a column, and a substance that moves halfway between them needs to be as well resolved from both, that is, the resolution of the column is to be doubled, a column four times as long is needed.

There is also a requirement for the partition coefficient to be suitable. Substances differing twofold in partition coefficient can be well separated if well retarded, but hardly separated at all if hardly retarded. For column chromatography, the resolution proves proportional to $(1 - R)^{1/2}$ (as above, R is the fraction of the speed of the moving phase with which the zone moves). Hence this fraction should not be more than about 0.3, if 84% of the maximal resolution is to be achieved (since $0.7^{1/2}$ is 0.84). Further retardation is inconvenient, since it involves larger running times and volumes of effluent, and there is little increase in resolution still to be gained. For paper and thin-layer chromatography, since the slower substances pass through less of the separating system, there is also a loss when R is low. Substances will then be best separated from others of similar structure if $R = 0.5$.

Gel permeation chromatography

This method, often known as gel filtration, is of immense value in protein chemistry by separating substances according to molecular size. The stationary phase, most often in a column, consists of particles in a gel. Large molecules, unable to enter the gel network, spend all their time in the moving phase, namely the interstitial water, so $R = 1$. Small molecules partition between the stationary phase, that is the water within the gel particles, and the moving phase, the interstitial water, and so exhibit a lower value of R.

The first use of this is simply transferring a substance with large molecules from one solvent into another. Thus the salt concentration of a protein or nucleic acid solution can be rapidly changed, without the time involved in DIALYSIS. Further, a protein can be passed through a zone of reagent, and the countercurrent effect means that reactions with highly unfavourable equilibrium constants can be carried out, for example the conversion of the pyridoxal phosphate form of an aminotransferase into its pyridoxamine phosphate form.

A much wider use, however, is the separation of chemically similar substances, usually proteins, on the basis of molecular size. The molecules are small enough to enter a fraction of the gel water, but because of the cross-linked gel matrix, they are unable to reach it all. The larger the molecules, the less of the gel water they can penetrate, and hence the nearer to unity the R values they exhibit.

The method can be used to obtain estimates of the M_r values of proteins, and to separate proteins by M_r value for either analytical or preparative purposes. Unlike gel electrophoresis in the presence of dodecyl sulphate (*see* ELECTROPHORESIS), the methods need not involve denaturing conditions, so biological activity can be measured after the separation, and the M_r value that is obtained may be of an oligomer rather than of dissociated subunits. Gels can be varied to cover a particular size range. The retardation varies, within limits, roughly linearly with $\log(M_r)$, and for a particular gel over a range of ~1.5 (i.e. 30-fold in M_r).

Gradient elution

Although the best separations of similar substances require isocratic elution, in which the substances have constant values of R throughout the separation, two circumstances often require a gradual (or even stepwise) change in eluent to be used, and a third may make it desirable. The first is a requirement, as in amino-acid analysis, for several substances to be separated on a single column. Since the most strongly retained will hardly move at all under the conditions for running the most easily eluted, the conditions have to be changed during the course of the run. A gradient in eluent composition is used therefore, so that it becomes more strongly eluting as the run proceeds.

The second circumstance, which applies to many proteins, is that no conditions can be found in which a finite partition coefficient can be observed between two phases. With ion exchangers, for example, although the partition coefficient of a monoanionic substance may vary linearly with the concentration of a competing monoanion in the eluent, that of a substance that exchanges with n univalent anions may vary with the nth power of the concentration of competing monoanion. With proteins there may be no simple relation with net charge, but many singly charged buffer ions may exchange for one protein molecule. Hence the partition coefficient changes sharply with concentration of competing ion. A minute change in this concentration effectively changes the protein from being completely adsorbed onto the surface of the exchanger to being negligibly adsorbed. Protein separations are still possible, since this desorption occurs for different proteins at different points in a gradient.

For such a separation there is no advantage in long, thin columns, and squat ones may be used with advantage. It is sometimes useful, however, to keep up the column length, because some isocratic separation may be occurring as the protein starts to desorb.

A third use of gradients is to avoid having to find the conditions in which a suitable value of R is achieved; the gradient passes through a range of conditions. This will lose resolution, since no separations of the substance from more strongly held ones occurs while it is stationary at the start of the gradient, and no separation from similar substances occurs while it moves with $R = 1$ at the end. But the excellence of modern chromatographic systems may be so high that perfectly adequate separations are obtained without the labour of finding precise conditions for isocratic elution.

Affinity chromatography

Chromatographic techniques can make use of the specific binding properties of some proteins, for example receptors for their agonists, antibodies for their cognate antigens, and enzymes for their substrates or effectors. An adsorbent can be constructed containing a group that is close enough to the natural ligand of the protein so that the adsorbent binds the protein specifically. A spacer arm may be attached so that the site on the protein can be reached, even if the protein's binding site is in a cavity in the molecule. Users should be aware that such adsorbents have limits to their specificity, since they will themselves be ion exchangers if the ligand analogue is charged. The adsorbed protein can be desorbed by increase of ionic strength or by competition for the site on the protein with a soluble ligand.

The same advantage can often be obtained without the labour of constructing a specific adsorbent, by desorbing a protein from an ion exchanger by addition of a specific ligand [2]. A preliminary experiment with the exchanger, chosen to have the same sign of charge as the ligand, finds conditions in which the protein is only just adsorbed. Addition of the ligand can then cause desorption.

Caution

Like any other method, chromatography repays thought. It has traps for the unwary. Thus substances added to the sample, for example chelating agents such as EDTA, may be removed by ion exchange, and may be present in too low a concentration for the exchanger to be equilibrated with them. An anion exchanger may steadily raise the pH of the effluent in a gradient of increasing ionic strength by anion–hydroxide exchange. These and similar processes may lead a gradient, designed to have steadily increasing eluting power, to have a temporary drop in it, so that a single protein may emerge in two peaks. Hence checks, such as rerunning and other methods of analysis are always desirable.

H.B.F. DIXON

See also: ELECTROPHORESIS.

1 LeRosen, A.L. (1945) The characterization of silicic acid–celite mixtures for chromatography. *J. Am. Chem. Soc.* **67**, 1683–1686.
2 Scopes, R.K. (1977) Purification of glycolytic enzymes by using affinity-elution chromatography. *Biochem. J.* **161**, 253–262; Multiple enzyme purifications from muscle extracts by using affinity-elution chromatography. 265–277.

---

chromatophores Small membrane vesicles obtained on disintegration of cells of photosynthetic bacteria. They contain all the photosynthetic pigments and are capable of photophosphorylation and light-induced reduction of NAD^+. They are 'inside out' compared with the intact cell, so that what was topologically the periplasmic space becomes the interior of the chromatophores. *See also*: PHOTOSYNTHESIS.

chromophore The nonprotein pigment prosthetic group of photoreceptors such as PHYTOCHROME, BACTERIORHODOPSIN, and RHODOPSIN which absorbs light.

chromoplast Nonphotosynthetic, carotenoid-containing PLASTID that is responsible for the yellow, orange, and red colours of many fruits and flowers, certain roots (e.g. carrot), and autumn leaves of deciduous trees. The carotenoid content is more varied and covers a wider range of components than that of chloroplasts. Chromoplasts are often, but not always, developed from chloroplasts and may have lost the plastid DNA.

chromosomal aberrations/abnormalities *See*: CHROMOSOME ABERRATIONS.

chromosome A single piece of cellular DNA, together with the proteins that help to define its structure and level of activity. Prokaryotes generally contain one circular chromosome, while most eukaryotes contain more than two, in some cases several hundred. In eukaryotic cells, chromosomes can be visualized in the light microscope as thread-like, deeply staining bodies located within the nucleus. At metaphase of MITOSIS their structure becomes highly compacted (condensed) and they appear as short rod-like bodies with a constriction (the centromere) at which they are attached to the mitotic spindle. In this state each chromosome has a characteristic shape and size and with suitable treatment (*see* CHROMOSOME BANDING) individual chromosomes can be identified (*see* KARYOTYPE). An acrocentric chromosome is one in which the centromere is located near the end of the chromosome, a metacentric chromosome is one in which the centromere is located near the middle of the chromosome. In a submetacentric chromosome the centromere is located just away from the centre of the chromosome while in a telocentric chromosome the centromere is near the telomere, that is the end of the chromosome.

Chromosomes contain both the GENES and the regulatory structures that help to define the level of gene expression. Chromosome structure is built up around the DNA double helix. In eukaryotic cells, the DNA is wound around octamers of HISTONES to make NUCLEOSOMES. The nucleosomes are in turn packed in a helix to make a solenoid of ~30 nm diameter. This CHROMATIN fibre probably unfolds briefly during periods of DNA REPLICATION and RNA synthesis. The organization of the 30-nm fibre varies during the CELL CYCLE as the DNA becomes dispersed for interphase and condensed for mitosis. The exact arrangement of this fibre in higher-order structures is still a matter of debate, but it is probably coiled into one higher level of helix and then bent to form loops that range in and out from a chromosome scaffold. Chromosome structure is regulated, at least in part, by its component proteins. Histones are subject to extensive POST-TRANSLATIONAL MODIFICATION, including both phosphorylation and acetylation. In addition there are many nonhistone chromosomal proteins that bind either to chromatin in general or to specific regions of individual chromosomes. The former may have a role in the regulation of chromosome condensation and the latter in the regulation of gene expression. *See also*: CHROMOSOME ABERRATIONS; CHROMOSOME STRUCTURE; NUCLEAR MATRIX.

Chromosome aberrations

CHROMOSOME aberrations or abnormalities are the gross changes in the genome that occur as a result of changes in the normal chromosome number of the species, or as a result of large-scale structural rearrangements in individual chromosomes. They involve whole chromosomes or large regions of a chromosome, and are often visible under the light microscope. The discovery of such chromosome aberrations in the early years of the century, and their correlation with changes in phenotype, were instrumental in the development of the chromosome theory of heredity, and they have been of great use in gene mapping (*see* HUMAN GENE MAPPING).

Chromosome structure

Chromosomes can best be studied at METAPHASE, when chromosome condensation is maximal. A metaphase chromosome consists of two sister chromatids that will separate in mitotic anaphase (*see* MITOSIS). Each chromatid contains one double helix of DNA running continuously from one end of the chromosome to the other. The DNA is complexed with histones to form CHROMATIN which is further complexed with nonhistone proteins. In the metaphase chromosome, the nucleoprotein thread is tightly coiled and compacted (*see* CHROMOSOME STRUCTURE) and thus becomes visible in the light microscope after suitable staining.

To visualize metaphase chromosomes, they must first be fixed on a slide, and then stained, either uniformly with a dye such as orcein, or for more resolution and detail, with banding techniques after gentle denaturing treatments (*see* CHROMOSOME BANDING). The two main banding techniques used are Giemsa banding (G banding) and reverse banding (R banding) which are grossly complementary and enable the detection of some 150–200 bands generally agreed upon and officially recognized by the ISCN (International System for Human Cytogenetic Nomenclature). A still higher resolution can be obtained by synchronizing cell division and blocking the chromosomes in a prometaphase stage, so that the chromosomes are more elongated and may show up to 600 and even 1000 bands.

A metaphase chromosome consists of two arms separated by a primary constriction which marks the location of the CENTROMERE or spindle attachment, essential for the normal movements of the chromosomes in relation to the spindle (*see* CHROMOSOME STRUCTURE; MITOSIS). The short arm is called p, the long arm q. Chromosomes with the centromere located in the middle are termed metacentric; with the centromere at one end of the chromosome, acrocentric or telocentric; and with a centromere located between these two extremes, submetacentric or subacrocentric. A chromosome fragment lacking a centromere is termed acentric.

An organism is characterized by a given number of chromosomes in its somatic cells — called the diploid number (2N). Germ cells of either sex contain a haploid number (N) of chromosomes. The diploid chromosome number of humans is 46, there being 22 pairs of autosomes and one pair of sex chromosomes. Females have two homologous X chromosomes, males have one X and one Y chromosome. Human chromosomes range in size from <1 μm to somewhat longer than 5 μm. A KARYOTYPE is a standard way of displaying the chromosomes of an individual in which they are lined up, starting with the largest, with the short arm pointing to the top. Table C8 gives some conventional karyotype symbols.

Terminology

Changes in the number of whole sets of chromosomes usually lead to POLYPLOIDY, in which the nucleus contains a multiple of the haploid chromosome number of the species. As a rule, haploidy itself is not a viable condition. A chromosome complement containing three haploid sets is termed triploid, with four sets, tetraploid. Changes can also involve individual chromosomes, and are then referred to as aneuploidy. The absence of

Table C8 Conventional karyotype symbols used in human genetics

A–G	Chromosome groups
1–22	Autosome designation
X, Y	Sex chromosome designations
p	Short arm of chromosome
q	Long arm of chromosome
ter	Terminal portion: pter refers to terminal portion of short arm; qter refers to terminal portion of long arm
+	Preceding a chromosome designation, indicates that the chromosome or arm is extra; following a designation, indicates that the chromosome or arm is larger than normal
−	Preceding a chromosome designation, indicates that the chromosome or arm is missing; following a designation, indicates that the chromosome or arm is smaller than normal
dup	Duplication
del	Deletion
inv	Inversion
t	Translocation

one chromosome from the diploid complement gives a monosomy (*see e.g.* TURNER SYNDROME). The presence of an extra one is trisomy (*see e.g.* DOWN'S SYNDROME), and of two additional copies of a chromosome, tetrasomy.

Structural changes occur as a result of chromosome breakage and rejoining. Sometimes all material will be conserved, but in other cases some may be lost. Chromosome breaks may occur spontaneously (i.e. for no known reason) or they may be induced by a so-called clastogenic agent such as ionizing radiation or DNA-damaging chemicals. Usually, broken ends rejoin and the break is repaired. However, a break may lead to a deletion of material (also known as a deficiency) or to complex structural rearrangements. Breaks may occur at all stages of the cell cycle, and during mitosis or meiosis.

Two breaks in the same chromosome may result in the formation of either a centric ring and an acentric fragment, or an acentric ring and an interstitial deletion (Fig. C44a). An interstitially deleted fragment may remain as an acentric fragment; very small fragments are called minutes. If the two breaks occur in the same arm, rejoining may be accompanied by an inversion of the intervening material (a paracentric inversion). If a break occurs in each arm, a pericentric inversion may result (Fig. C44b). Breaks occurring simultaneously in two different chromosomes may be accompanied by an exchange of material, leading to a translocation (Fig. C44c). DUPLICATIONS can occur in several ways, for example by unequal sister chromatid exchange (see below) or by simultaneous breakages in two homologous chromosomes (Fig. C44d), leading to nonreciprocal translocation. There are many more possible structural rearrangements, but in general the mechanism comes down to one or more breaks occurring during interphase and joining of the broken ends in various ways, with the results becoming visible in the next metaphase — which may be the last if the changes are such that disruption is inevitable in the next mitosis.

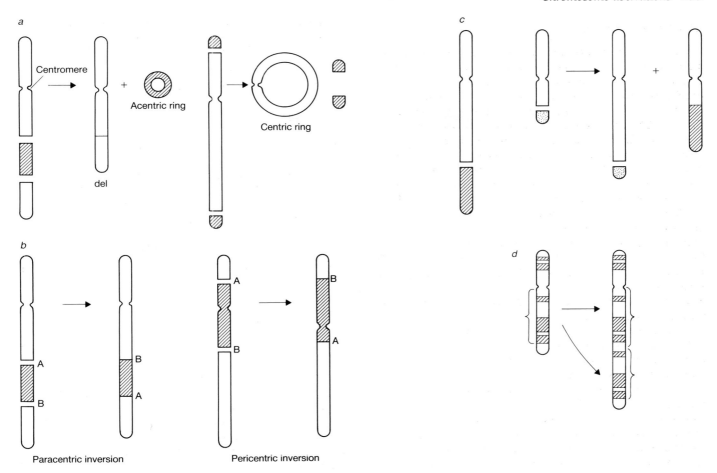

Fig. C44 Generation of some types of chromosome aberration. *a*, Interstitial deletion with ring formation; left-hand, formation of an acentric ring and an interstitial deletion; right-hand, formation of a centric ring and acentric fragments; *b*, paracentric and pericentric inversions; *c*, reciprocal translocation; *d*, nonreciprocal translocation resulting in a duplication. Once a duplication has occurred, further duplication may occur by UNEQUAL CROSSING-OVER.

Reciprocal translocations, in which material is exchanged between two chromosomes, have been observed for all human chromosome arms in nonmalignant as well as in malignant cells. If occurring constitutionally they may be balanced or unbalanced, that is, without or with loss of genetic material. In the latter case malformative stigmata and/or mental deficiency may be present. This will also be the case as a consequence of meiotic segregation with formation of unbalanced gametes in the germ cells of a translocation carrier. Hundreds of syndromes due to chromosome rearrangements, structural as well as numerical, have been described [1].

Breakpoints

Chromosome breakpoints do not occur randomly. Macroscopically they seem to occur in the dark regions (using quinacrine banding (Q-banding)), or the clear regions (with G-banding). Breaks are also distributed differently with particular chromosome-breaking agents, and the location of 'hot spots' depends on the breaking agent. Thus, in so-called secondary or treatment-induced myelodysplasias or leukaemias, breaks tend to cluster primarily in chromosomes 3q, 5q, 6p, 7q, 11q, 12p, 13q, and 17q; a distribution different from that seen in constitutionally occurring reciprocal translocations. An important question in this respect is whether the breaks and exchanges in reciprocal translocations are random, or whether specific combinations are preferred. It has become clear that the presence of certain sequences, on some chromosomes, some of them showing homology to each other, others highly repetitive, may facilitate the occurrence of nonrandomly occurring exchanges. In addition, chromosome breaks and rearrangements have been related to so-called fragile sites. Fragile sites appear as unstained or extended regions, which are a constant feature in an individual, but are expressed in only some of the cells. They are inherited in a Mendelian fashion. According to a recent listing [2], of 107 sites 83 were classified as 'common' and 24 as rare. Fragile sites have not been characterized molecularly, and their relation to specific breaks in malignant disease is controversial. They do not have any phenotypic expression, except maybe in the Xq27 FRAGILE X SYNDROME.

SISTER CHROMATID EXCHANGE (SCE) is a process which can only be visualized with a differential staining after two cell cycles in culture in which one of the chromatids is labelled, using BrdU or [³H]thymidine. As the latter also induces SCE, its spontaneous frequency is unknown, as is its biological significance, as the genetic make-up of the cell is not changed if the exchanges take place at identical points. Unequal SCE however may lead to DUPLICATION or deletion of genes. SCE can be increased slightly by X-rays and more dramatically by bifunctional alkylating agents such as mitomycin G, and is considered to be a sensitive test for mutagenesis. Its incidence is greatly increased in BLOOM'S SYNDROME.

Chromosome anomalies in lymphoid disorders

There are by now clear examples of nonrandomly occurring translocations in lymphoid disorders, caused by recombination between homologous sequences on the chromosomes involved. The involvement of a B-cell or T-cell receptor locus in the rearrangements characteristic of non-Hodgkin lymphomas and lymphocytic leukaemias indicates that these disorders arise at the pre-B and pre-T stages of lymphocyte differentiation, at which time the breaking and rejoining accompanying V-D-J recombination is occurring in immunoglobulin and T-cell receptor genes (*see* B CELL DEVELOPMENT; GENE REARRANGEMENT; IMMUNOGLOBULIN GENES; T CELL DEVELOPMENT; T CELL RECEPTOR GENES).

During this process a nonamer (GGTTTTTGT) and heptamer (CACTGTG) located 5′ to each of the J regions of the antigen receptor genes are used by specific recombinases for gene rearrangement. Forbidden recombinations between different chromosomes may occur using these sequences, as some proto-oncogenes have similar sequences, and this may lead to specific chromosomal translocations (*see* ONCOGENES). In fact, this mechanism has been found to be operative in any lymphoid malignancy, B or T, acute or chronic — lymphoid-specific differentiation genes or structurally related genes being involved in all these cases. For B cell proliferations, the chromosomes involved are 2p11 (light chain kappa locus), 14q32 (immunoglobulin heavy chain locus), and 22q11 (light chain lambda locus), and for T cell proliferations chromosome 7q35 (TCRβ), and 14q11 (TCRα and TCRδ). Remarkably, chromosome 7p which carries the TCRγ locus has not so far been shown to be involved. Otherwise, all lymphoid malignancies show a structural rearrangement of one or more of the chromosomes mentioned [3] (Table C9). Purely numerical changes in lymphoid malignancies are uncommon. The significance of trisomy 12 in some cases of CHRONIC LYMPHO-CYTIC LEUKAEMIA (CLL) and in some lymphoblast cell lines is unknown.

In nonmalignant lymphoid cells, chromosome changes have been found in ATAXIA TELANGIECTASIA, accompanied by re-arrangement of the TCRγ chain in clonal structural anomalies involving chromosomes 7 and 14. The translocation t(14;18) characteristic of follicular lymphoma has been found occasionally in presumably normal tonsil tissue. A spectacular and provoca-tive recent discovery is the presence in some tumour-infiltrating lymphocytes as well as in a high proportion of normal foetal and postnatal thymus of trisomy 7 accompanied by expression of the co-receptor molecule CD4, trisomy 10 with the CD8 phenotype, and a combination of those trisomies together with Y-chromo-

Table C9 Characteristic and nonrandom chromosome changes in lymphoma and some other lymphoproliferative disorders

Chromosome change	Disorder
t(2;3)(p12;q27)	NHL
t(2;5)(q23;q35)	Ki-1 NHL
t(2;8)(p12;q24)	BL
+3	NHL, T cell
t(3;14)(q27;q32)	NHL
t(3;22)(q27;q11)	Diffuse large B cell lymphoma
+5	NHL
6q −	CLL and HL
6p −	NHL, T cell lymphoma
+7	NHL
7p −	NHL (secondary change)
t(7;9)(q34;q34)	T-ALL
t(7;9)(q35;q34)	T-ALL
t(7;11)(q35;q13)	T-ALL
t(7;19)(q34;p13)	T-ALL
t(8;14)(q24;q32)	BL
t(8;22)(q24;q11)	BL
t(9;14)(p13;q32)	NHL
t(11;14)(q13;q32)	'Mantle zone' lymphoma
11q −	CLL and NHL
+12	CLL and diffuse small cell lymphoma
t(14;14)(p11;q32)	T-ALL
inv(14)(q11;q32)	T cell lymphoma
t(11;14)(p13;q11)	Childhood T-ALL
t(11;14)(p15;q11)	T-ALL
t(10;14)(q24;q11)	T-ALL
t(1;14)(p32;q11)	T-ALL
t(14;18)(q32;q21)	Follicular lymphoma
t(14;19)(q32.3;q13.1)	CLL
14q +	NHL
+18	NHL

ALL, acute lymphocytic leukaemia; NHL, non-Hodgkin lymphoma; BL, Burkitt's lymphoma; CLL, chronic lymphocytic leukaemia.
See Table C8 for terminology. The numbers in parentheses indicate the chromosomes and chromosome bands involved (*see* CHROMOSOME BANDING).

some loss in CD4 CD8 cells [4]. The significance of this discovery for tumour immunology, tolerance, or perhaps apoptosis and cell death remains to be clarified.

In myeloid proliferations no involvement of the heptamer, octamer or nonamer sequences has been found. However, in the translocation t(9;22) typical of chronic myeloid leukaemia (CML) (*see* PHILADELPHIA CHROMOSOME) a repetitive segment of 289 nucleotides may have a role in the *BCR* rearrangement with *ABL*, as not only is this repeat found in the 5′ region of the *BCR* gene but sequences very similar are also found in retroviral LONG TERMINAL REPEATS (LTR), HERV K10 human endogenous virus, and different ALU SEQUENCES. The latter may perhaps also play a major part in the mechanism of translocation in nonmalignant cells.

H. VAN DEN BERGHE

1 Therman, E. & Susman, M. (1993) *Human Chromosomes. Structure, Behavior and Effects*, 3rd edn (Springer-Verlag, Berlin).
2 Hecht, F. et al. (1990) A guide to fragile sites on human chromosomes. *Cancer Genet. Cytogenet.* **44**, 37–45.

3 Yunis, J.J. & Tanzer, J. (1993) Molecular mechanisms of hematologic malignancies. *Crit. Rev. Oncogenesis* **4**, 161–190.

4 Dal Cin, P. et al. (1992) Trisomy 7 characterizes tumour-infiltrating lymphocytes in kidney tumours and in the surrounding kidney tissue. *Proc. Natl. Acad. Sci. USA* **89**, 9744–9748.

chromosome banding (1) The pattern of transverse bands seen on eukaryotic METAPHASE CHROMOSOMES after treatment with certain fluorescent stains. The banding pattern of a particular chromosome is highly specific and constant and is used to distinguish similar sized chromosomes in the KARYOTYPE. Banding patterns are used in gene mapping, and to detect CHROMOSOME ABERRATIONS such as partial duplication. Typical stains used are GIEMSA (G-banding) and QUINACRINE (Q-banding). The bands on each chromosome are numbered according to a standard nomenclature (Fig. C45). The designation 3q13, for example, designates a position in band 3 of region 1 of the long arm (q) of chromosome 3. The short arm of the chromosome is conventionally designated p. *See*: HUMAN GENE MAPPING.

(2) The pattern of transverse light and dark bands seen on the POLYTENE CHROMOSOMES of, for example, the fruit fly *Drosophila melanogaster* in the light microscope. This type of banding is much finer scale than that in (1). *See*: DROSOPHILA.

chromosome diminution The fragmentation and elimination of germ line-specific CHROMATIN in early embryogenesis in the cells giving rise to somatic tissues. The phenomenon was discovered by T. Boveri in 1887 in the nematode *Parascaris* and provided the first proof for the early segregation and independent development of germ-line and somatic cells. It was later observed in other nematodes, and similar phenomena occur among crustaceans, insects, and ciliated protozoa. In *Ascaris lumbricoides* about 25% of the germ-line DNA is expelled from the presomatic cells. The expelled DNA consists largely of REPETITIVE DNA but also includes some specific single-copy sequences. Chromosome (or chromatin) diminution is a complex molecular process that includes site-specific chromosomal breakage, new telomere formation (*see* CHROMOSOME STRUCTURE), and DNA degradation. Both its cause and significance are unknown.

chromosome jumping A procedure for isolating clones which contain regions of the same chromosome which are not contiguous, as in the case of CHROMOSOME WALKING. The procedure outlined in Fig. C46 involves the generation of large DNA fragments (80–150 kb long) either by partial digestion with a restriction enzyme or by digestion with a low-frequency cutting

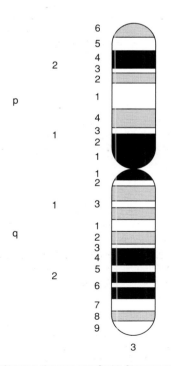

Fig. C45 Banding of human chromosome 3 with Giemsa staining.

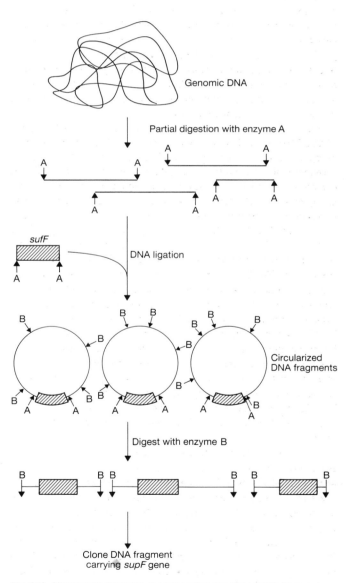

Fig. C46 Chromosome jumping; the generation of a cloned DNA fragment that contains two DNA sequences widely separated on a chromosome.

restriction enzyme, for example *Not*I. These DNA fragments are then circularized using DNA ligase in the presence of a DNA fragment containing the *Escherichia coli* tRNA AMBER SUPPRESSOR gene *supF*. Some of the circularized fragments will thus contain this *E. coli* sequence at the ligated junction. Digestion of these ligated products with an enzyme known not to cut within the *supF* gene (e.g. *EcoRI*) generates DNA fragments containing the *supF* gene flanked on either side by DNA sequences that were initially up to 150 kb apart on the chromosome. These *supF* fragments can then be cloned into bacteriophage vectors containing suppressible amber mutations in one or more essential genes (e.g. CHARON VECTORS). Using these cloned fragments as probes to screen a standard phage or cosmid DNA LIBRARY will identify clones that contain DNA fragments which are separated by up to 150 kb. This approach allows researchers to 'jump' down a chromosome rapidly towards a gene of interest. The CYSTIC FIBROSIS gene was cloned by a combination of chromosome jumping and chromosome walking from a known linked gene sequence.

Collins, F.S. & Weissman, S.M. (1984) *Proc. Natl. Acad. Sci. USA* **81**, 6182–6186.

chromosome map A genetic or physical map of a single chromosome. *See*: HUMAN GENE MAPPING.

chromosome painting *See*: HUMAN GENE MAPPING.

chromosome pair In diploid cells, two homologous chromosomes (one from each parent) that carry the same gene loci, are similar in size and shape, and pair at MEIOSIS.

chromosome polymorphism A POLYMORPHISM in a gross chromosomal feature.

chromosome rearrangement *See*: CHROMOSOME ABERRATIONS; GENE REARRANGEMENT.

chromosome sorting The separation of one type of chromosome from another. Chromosomes can be separated by pulsed-field gel electrophoresis or by fluorescence-activated sorting using chromosome-specific fluorescent markers. *See*: ELECTROPHORESIS; FACS; HUMAN GENE MAPPING.

Chromosome structure: molecular aspects

THE DNA of EUKARYOTES is organized as linear molecules into CHROMOSOMES. The chromosome is the level of organization at which processes such as RECOMBINATION and segregation of replicated copies of the DNA occur (*see* DNA REPLICATION; MITOSIS). The organization of DNA into the organelle which is a chromosome is most obvious at METAPHASE when the chromosome is condensed into a highly structured body. At this point in the CELL CYCLE important chromosomal elements are visible (telomeres and centromeres [1]) or can be made visible as trans-

verse bands on the chromosome by staining techniques. The functions of telomeres and centromeres are known in part and their structure is beginning to be elucidated. There is clear evidence for a level of chromosomal organization into functional regions at the ten to hundreds of kilobase level [2] but some levels of organization, such as the banding revealed on staining, remain of unknown functional significance. In sequencing the human genome (*see* DNA SEQUENCING; HUMAN GENOME PROJECT) a major landmark will be the complete sequence of a single chromosome. While this is some way off, the sequence of yeast (*Saccharomyces cerevisiae*) chromosome III has been determined by a European consortium [3], and substantial progress has been made with the chromosomes of the nematode *Caenorhabditis elegans*.

Chromosome function

The role of the chromosome is to deliver a copy of the replicated DNA to each daughter cell at cell division. Failures in either replication or delivery result in aneuploidy — deviation of the cell from a normal chromosome and hence DNA content. Aneuploidy has severe consequences; it is, for example, detected in about 70% of human spontaneous abortions. To avoid this fate a minimal chromosome must have three functional elements: origins of replication, telomeres and centromeres. All three of these are available as cloned DNA sequences in yeast and have been assembled *in vivo* and *in vitro* to form artificial chromosomes [4].

Origins of replication in yeast were first obtained as sequences which allowed PLASMIDS to replicate autonomously in the yeast cell (*see* ARS SEQUENCES). Some of these sequences have subsequently been shown to be replication origins in the chromosome, and proteins which interact with them are being isolated. A simple consensus sequence can be derived but additional (in some instances curved) DNA may be involved *in vivo*. Origins of replication in mammalian cells have been defined in some genes but their existence as particular DNA sequences or locations is not universally accepted.

Telomeres

The basic asymmetry of DNA replication means that the ends of the chromosome cannot be completely replicated by the cell's normal replication machinery. In the absence of a special mechanism a gradual but eventually lethal shortening of the chromosome would result. A common solution to this problem is the capping of the chromosomes with short repeated sequences which have a strand asymmetry in GC content such that one strand is G rich and one C rich. Protozoa, yeasts, plants, and vertebrates have all adopted this structure for their telomeres (chromosome ends) (Fig. C47).

In human, *Tetrahymena* and *Euplotes* a RIBONUCLEOPROTEIN (RNP) telomerase has been described [5] which can add terminal repeats onto existing G-rich sequences, including existing telomeres, and hence ensure their complete replication. Where the sequence of the RNA component of this RNP is known it contains the complement to the repeat that is added. In *Tetrahymena*, mutation of this RNA results in a change in the sequence of the added telomeric repeats. Such altered telomeres give rise to a

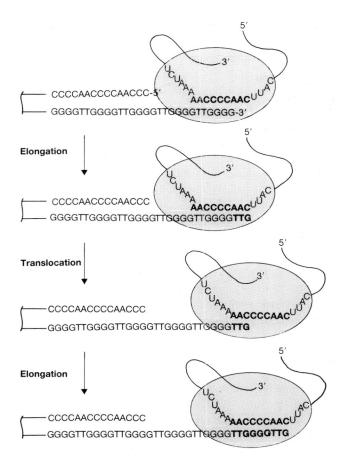

Fig. C47 A model for elongation of telomeres by telomerase.

senescent phenotype in the cell. A yeast mutation which results in continually shortening telomeres is also known to have a senescent phenotype.

Vertebrate telomeres have arrays of terminal repeats of the sequence TTAGGG [6]. The length of these arrays varies between species; mouse telomeric arrays are five to ten times longer than human. In human B cells and skin, the length of the telomere is known to decrease with increasing age of the individual. The telomeres of the germ cells are longer than those of somatic tissues. It has been postulated that the shortening of telomeres is a result of a lack of telomerase in somatic tissues and, coupled with the senescent phenotypes described in other species, this loss of telomeres and consequent loss and terminal degradation of chromosomes has been suggested as a possible cause of ageing and tumorigenesis. In the mouse (a short-lived species) a reduction in telomere length with increasing age has not, however, been detected.

The introduction of cloned telomeres or synthetic telomeric repeat arrays into mammalian cells results in the formation of new telomeres by a process which involves chromosome breakage. This implies that the terminal repeats are the only essential requirement for telomere function and that telomerase activity is present in cultured cells.

In addition to their role in replication of the ends of chromosomes, telomeres have other less well defined functions. They prevent the recombination events to which broken DNA and chromosome ends are prone. In meiosis they are confined to the nuclear envelope and in mammals are the sites from which pairing of chromosomes starts. It is likely that these functions are also provided by the terminal repeats and the proteins that interact with them. Although the DNA adjacent to the terminal repeats is not required for function of telomeres this region of the chromosome has some distinctive features. Immediately subterminal sequences are a complex set of repeated sequences which are polymorphic in location in both yeast and human genomes. The region of human chromosome 16 between the telomere and the α-globin locus is polymorphic in length by up to 150 kb. In humans, minisatellite repeats (*see* VARIABLE NUMBER TANDEM REPEATS) are concentrated in telomeric regions and the genetic maps of some chromosomes suggest that there may also be an increased rate of recombination in these regions. As some minisatellite loci are recombination hot-spots these two observations may be connected [7].

Centromere

The centromere is the genetic element responsible for chromosome segregation at cell division (*see* MITOSIS). At a structural level it is visible at metaphase as a constriction in mammalian chromosomes and is the site of attachment of spindle microtubules to the proteins that make up the kinetochore. By CHROMOSOME WALKING from flanking markers applying a genetic selection for segregation, *S. cerevisiae* centromeric DNA sequences have been cloned. When present on a plasmid they result in plasmid copy number limitation and an approximation to correct segregation.

Several centromeres have been cloned from different *S. cerevisiae* chromosomes and consensus features have emerged [8]. Three regions have been described: CDE I with a 7 bp consensus, followed by CDE II, a region consisting of 78–86 bp and an AT content of greater than 90%. The final region, CDE III, is 26 bp long and has bilateral symmetry. CDE I is essential for meiotic but not for mitotic segregation of plasmids or minichromosomes. CDE III is essential for centromere activity, and some point mutations (particularly at the centre of symmetry) can abolish its function. The shortest functional *S. cerevisiae* centromere is 125 bp long.

The next best characterized centromere is that of the fission yeast *Schizosaccharomyces pombe*. Here the centromeres contain repeated DNA sequences again arranged with some bilateral symmetry but the major difference is that of size. The smallest *S. pombe* centromere is about 60 kb in size. This difference between the amount of DNA involved in centromere function in the two organisms may be due to the attachment of only a single spindle fibre in the case of *S. cerevisiae* whereas multiple spindle fibres attach to the *S. pombe* centromeres. The centromeres of mammals are also the site of attachment of multiple spindle fibres. The nature of mammalian centromeric DNA is unknown; in many species the DNA around the centromere consists of satellite DNA, tandem repeats of varying length, which can constitute a large proportion of the genome.

Mammalian centromeres have been extensively examined using antibodies raised to nuclear proteins or occurring in patients with autoimmune diseases. A wide range of different proteins are localized to the centromere. One of these, CENP B, has been found to bind *in vitro* to a DNA sequence that is present in human alphoid DNA [9], a highly repeated sequence which co-localizes with centromeric antigens in stretched chromosomes. The same sequence is also present in mouse minor SATELLITE DNA and this observation has raised the possibility that binding of CENP B to these sequences is a necessary but not sufficient step for centromere formation *in vivo*. This is supported by the finding that nuclear injection of antibodies against CENP B disrupts chromosome segregation. Centromeres are complex assemblies of DNA and protein and their composition and the interactions between their components are at present unclear.

Artificial chromosomes

The three components described above are sufficient to confer chromosome function on a fragment of DNA. In the case of *S. cerevisiae*, where all the components are available they have been assembled *in vivo* and *in vitro* into artificial chromosomes. When additional DNA is added to take the overall size up to over 90 kb, rates of chromosome loss can approximate to those found for natural chromosomes. The availability of artificial chromosomes in this organism has made it possible to study not only the effect of the components but also the effect of their relative organization and position.

Chromosome structure

Both metaphase and interphase chromosomes show structural organization. The idea that the CHROMATIN is organized into discrete DOMAINS is widely accepted and is often envisaged as loops of 20–200 kb of DNA extending from a site of tethering. This site might be a scaffold or matrix attachment site (*see* NUCLEAR ENVELOPE; NUCLEAR MATRIX) and there is evidence that

matrix-attachment sites may correspond to the binding sites for TOPOISOMERASE II. The loops have been proposed to be a level of organization of GENE CLUSTERS such as the β-globin locus and its regulatory locus control region (*see* HAEMOGLOBIN AND ITS DISORDERS).

Perhaps the most clear cut compartmentalization is in the constitutive heterochromatin, regions of the genome which remain condensed during interphase. These are composed largely of repeated sequences and few if any genes. In *Drosophila*, chromosomal rearrangements which bring an active gene into close proximity to heterochromatin may result in silencing of the gene. The switching off of the silent mating type loci in yeast (*see* YEAST MATING-TYPE LOCUS) is also thought to involve this type of effect as these genes have their control regions intact.

Chromosome banding

At the metaphase stage of the cell cycle chromosomes can be banded by a variety of stains in combination with chemical or enzymatic treatments [10] (Fig. C48). These banding patterns have been used as a means of recognizing individual chromosomes and in the most favourable circumstances allow points with about a 10 megabase (Mb) separation to be distinguished. The association of disease phenotypes with abnormalities of chromosome banding formed the original basis of GENE MAPPING to the chromosome or chromosomal region. The molecular basis of a few of the banding methods is known to involve base composition. Many correlations have been made between banding and other aspects of genome function and organization. The *Kpn* family of repeated sequences is enriched in dark bands produced by GIEMSA BANDING (G banding) whereas the *Alu* repeat family is enriched in the opposite set of bands, R bands. There have been a number of suggestions that the distribution of genes between these two types of region is also nonrandom and that the pale regions of the chromosomes seen after G banding are most gene rich.

The molecular basis of chromosome structure and function is beginning to emerge, driven by systems such as yeast and

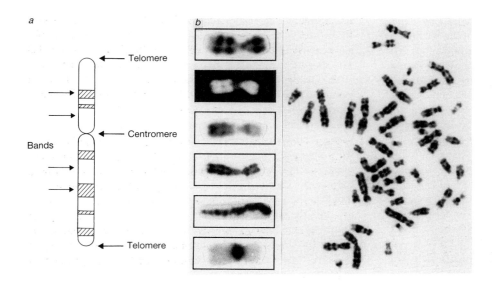

Fig. C48 *a*, An idealized mammalian metaphase chromosome after banding. *b*, right-hand, Giemsa-banded human metaphase spread; left-hand, human chromosome 9 after different banding treatments. Adapted from [10].

Drosophila where powerful genetics are available. Many of the ideas developed in these systems can be applied to the chromosomes of higher eukaryotes.

H. COOKE

See also: CHROMOSOME ABERRATIONS; DNA METHYLATION; EUKARYOTIC GENE EXPRESSION; FRAGILE X SYNDROME; TRANSCRIPTION.

1 Blackburn, E.H. & Szostak, J.W. (1984) The molecular structure of centromeres and telomeres. *A. Rev. Biochem.* **53**, 163–194.
2 Gasser, S.M. & Laemmli, U.K. (1987) A glimpse at chromosomal order. *Trends Genet.* **3**, 16–21.
3 Oliver, S. et al. (1992) The complete DNA sequence of yeast chromosome III. *Nature* **357**, 38–46.
4 Burke, D.T. et al. (1987) Cloning of large segments of exogenous DNA into yeast by means of artificial chromosome vectors. *Science* **236**, 806–812.
5 Greider, C.W. & Blackburn, E.H. (1985) The identification of a specific telomere terminal transferase activity in *Tetrahymena* extracts. *Cell* **43**, 405–413.
6 Moyzis, R.K. et al. (1988). A highly conserved repetitive DNA sequence (TTAGGG) present on the telomeres of human chromosomes. *Proc. Natl. Acad. Sci. USA* **85**, 6622–6626.
7 Jeffreys, A.J. et al. (1991) Minisatellite repeat coding as a digital approach to DNA typing. *Nature* **354**, 204–210.
8 Clarke, L. (1990) Centromeres of budding and fission yeasts. *Trends Genet.* **6**, 150–154.
9 Matsumoto H. et al. (1989) A human centromere antigen (CENP B) interacts with a short specific sequence in alphoid DNA. *J. Cell Biol.* **109**, 1963–1973.
10 Bickmore, W.A. & Sumner, A.T. (1989) Mammalian chromosome banding. *Trends Genet.* **5**, 144–178.

chromosome walking The systematic isolation of DNA clones which contain overlapping DNA fragments from the same chromosome (Fig. C49). One end of the chromosomal insert of a single recombinant phage or cosmid is used to probe an appropriate genomic DNA LIBRARY to identify additional recombinants which contain both that sequence and a region of the chromosome adjacent to it not carried by the original recombinant. This procedure is then repeated using the end of the newly identified chromosomal fragment as the probe, and so on until a collection of recombinant phage or cosmids have been identified which, between them, contain a contiguous portion of the target chromosome. These recombinants are sometimes referred to as CONTIGS. The DNA can then be mapped and sequenced to identify the gene of interest. Chromosome walking can be used to reach a gene whose position is still unknown, by starting from a nearby known sequence. *See also*: CHROMOSOME JUMPING.

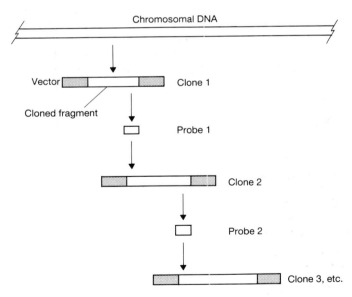

Fig. C49 A small region from the end of clone 1 (probe 1) is used to identify complementary DNA fragments in the genomic library (clone 2). One end of clone 2 is used to generate probe 2 which is then used to identify a further overlapping DNA fragment in the genomic library: clone 3. This is continued until a series of overlapping DNA fragments have been identified which cover the chromosomal region of interest.

chronic granulomatous disease (CGD) Very rare condition of impaired natural immunity in which around two-thirds of cases show X-LINKED RECESSIVE inheritance and the remainder AUTOSOMAL RECESSIVE inheritance. CGD is characterized by recurrent bacterial infection leading to the formation of granulomas. The defect is a failure by neutrophils and macrophages to generate bactericidal superoxide after phagocytosis of bacteria. The underlying defect in the X-linked form is a defect in neutrophil cytochrome *b*.

chronic lymphocytic leukaemia (CLL) A type of LEUKAEMIA with relatively good prognosis usually occurring in aged people. The hallmark of the disease is a higher white blood cell count, of which up to 90% may be a monoclonal population of mature lymphocytes.

chronic myeloid leukaemia (CML) A malignant disease characterized by over-representation of cells belonging to the three myeloid series (granulomonocytic, megakaryocytic, erythrocytic) in bone marrow and peripheral blood (*see* HAEMATOPOIESIS). Differentiation of myeloid precursors is maintained until the BLAST CRISIS of CML emerges.

***chv* loci** Chromosomal virulence loci. These are chromosomal loci in *Agrobacterium* which encode functions necessary for efficient transformation of plant cells. One (*chvE*) appears to affect the recognition of the phenolic inducer whereas two others (*chvA*, *chvB*) appear to affect the binding of the bacterium to the plant cell. *exoC* (*chvC*) and *chvD* appear to affect expression of plasmid-encoded virulence (*vir*) genes. *See also*: TI PLASMID.

chymotrypsin A SERINE PROTEINASE digestive enzyme from pancreatic juice (EC 3.4.21.1) consisting of three polypeptide chains held together by disulphide bonds, and which is produced from its inactive precursor, chymotrypsinogen (Fig. C50), by specific cleavage and excision of residues at two sites (*see* POST-TRANSLATIONAL MODIFICATION OF PROTEINS). Chymotrypsin cleaves preferentially at peptide bonds involving the carboxyl end of tyrosine, tryptophan, phenylanine, and leucine. Chymotrypsin is used to digest proteins into smaller peptides in preparation for PROTEIN SEQUENCING.

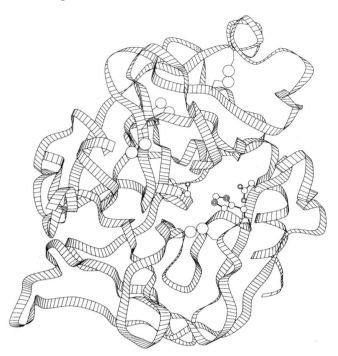

Fig. C50 α-Chymotrypsin A. Large circles represent disulphide bonds.

cI repressor A REPRESSOR protein encoded by the *cI* gene of the bacteriophage LAMBDA. It has an M_r of 27 000 and binds as a homodimer to two OPERATOR regions (O_L, O_R) within the lambda genome that control the transcription of lambda genes involved in LYTIC GROWTH. Synthesis of the cI repressor by an integrated lambda genome (the prophage) is necessary to ensure maintenance of the LYSOGENIC state and ensures that the prophage is transmitted during cell division. Phage with defective cI repressor cannot be maintained as lysogens.

ciliary neurotrophic factor (CNTF) A cytosolic (nonsecreted) protein of M_r 22 800 (200 amino acids) produced by astrocytes in the central nervous system and Schwann cells in the peripheral nervous system. It is thought to be released by GLIAL CELLS when damaged and to promote the repair of damaged or degenerating motor neurons. *In vitro*, CNTF has been shown to promote the survival of chick parasympathetic motor neurons and embryonic spinal cord neurons, and *in vivo* to prevent degeneration of motor neurons in the mouse mutant progressive motor neuropathy. The CNTF gene has been cloned.

Thoenen, H. (1991) *Trends Neurosci.* **14**, 165–170.

cilium (*pl.* cilia) Small hair-like motile proteinaceous appendages found on many eukaryotic cells. They are ~0.25 μm in diameter and are constructed from MICROTUBULES. *See*: CELL MOTILITY; MICROTUBULE-BASED MOTORS.

cingulin Protein associated with the tight junction. *See*: CELL JUNCTIONS.

Ciona *See*: ASCIDIAN DEVELOPMENT.

circadian rhythms Rhythmic cellular, physiological, or behavioural processes which continue with a periodicity close to 24 hours under constant environmental conditions (from the Latin, *circa diem*). By definition, such rhythms are endogenously generated, rather than driven by external environmental changes; thus they reflect the activity of underlying BIOLOGICAL CLOCKS. In many species, circadian rhythms are precisely synchronized to the environment by the ambient light–dark cycle, but nonphotic cues such as temperature cycles can also provide temporal information. Examples of circadian rhythmicity have been identified in most organisms including unicellular algae, fungi, higher plants, invertebrates, and vertebrates. A recent study in the CYANOBACTERIUM *Synechococcus*, in which the PROMOTER for the *psbA* gene which encodes the photosynthetic protein D1 was attached to a reporter gene producing bioluminescence, provides evidence for the existence of circadian rhythmicity in prokaryotes. Cellular circadian rhythms occur in the rate of both transcription and translation. In the mammalian suprachiasmatic nucleus of the hypothalamus, the location of the principal neural circadian pacemaker, circadian rhythms occur in the inducibility of IMMEDIATE EARLY GENES encoding TRANSCRIPTION FACTORS (e.g. c-*fos*, *egr-1*) by environmental light-activated pathways.

Circular dichroism

CIRCULAR dichroism (CD) is an optical characteristic of a molecule, which reflects asymmetric features of the molecular structure [1,2]. CD spectroscopy is very useful for characterizing structural features rapidly and with small amounts of material. If a typical cuvette is used, 1–2 ml of a protein or nucleic acid solution of around 0.05–0.5 mg ml^{-1} is sufficient for CD measurements. CD spectra allow characterization of the secondary structure of a protein, α-helix, β-sheet, β-turn etc. (*see* PROTEIN STRUCTURE), and the type of nucleic acid structure, A-form, B-form etc. (*see* A-DNA; B-DNA; NUCLEIC ACID STRUCTURE) [3]. CD spectra are also useful for monitoring such phenomena as PROTEIN FOLDING, protein–ligand binding, nucleic acid–ligand binding and PROTEIN–NUCLEIC ACID INTERACTIONS. Recent advances in producing native and mutant proteins (*see* GENETIC ENGINEERING; PROTEIN ENGINEERING), synthetic oligonucleotides and oligopeptides, as well as an increased understanding of the three-dimensional structures of DNA, RNA, proteins and their complexes, have made CD measurements much more useful and informative.

Principles

A molecule is optically active if it interacts differently with left- and right-circularly polarized light. Optical activity is observed for an inherently asymmetric CHROMOPHORE, such as a peptide bond. A symmetric chromophore has no intrinsic optical activity but it can obtain optical activity by its interaction with neighbouring groups as do purines and pyrimidines in nucleic acids. The sign and amplitude of the optical activity of a chromophore depend on the local environment. Optical activity can be detected either as the differential change in velocity of two beams through the sample, which is optical rotatory dispersion (ORD), or as the differential absorption of each beam, which is circular dichroism (CD). CD is more frequently used than ORD because of superior instrumentation and the higher resolution of CD curves.

CD spectra are characterized by ΔA (the differential absorption of the two beams) or θ_m (the molar ellipticity). ΔA is defined as

$$\Delta A = \Delta \varepsilon cL = A_L - A_R$$

where A_L and A_R are the absorbances of the left- and right-circularly polarized beams, respectively, $\Delta \varepsilon$ is the difference in the two extinction coefficients, c is the sample concentration in mol per litre, and L is the path length in centimetres. The two circularly polarized beams initially have equal amplitude, and the resultant beam is therefore a plane-polarized wave. After passing through optically active material, as the amplitudes of the two circularly polarized beams are now different, the resultant beam is elliptically polarized. The ellipticity in degrees, θ_{obs}, is measured as:

$$\theta_{obs} = \tan^{-1}(b/a)$$

where b/a is the ratio of two axes, minor to major, of the elliptically polarized beam. The molar ellipticity is obtained as

$$\theta_m = (\theta_{obs} \times 10)/cL \text{ (degree cm}^2 \text{ dmol}^{-1})$$

The molar ellipticity is related to $\Delta \varepsilon$ by the equation,

$$\theta_m = 3300 \Delta \varepsilon$$

As the CD amplitude of a protein largely depends on the number of peptide bonds, it is convenient to use the mean residue ellipticity, the ellipticity divided by the number of amino-acid residues in a protein, rather than the molar ellipticity itself.

CD of peptides and proteins

The predominant intrinsic optical activity of a protein arises from peptide bond absorption. CD spectra of proteins at wavelengths below 240 nm therefore reflect primarily the structural arrangements of peptide bonds, or the protein's secondary structure. CD spectra resulting from various types of secondary structures — α-helix, β-sheet, and so-called random coil — were first identified using polypeptides (Fig. C51). The CD spectrum of each type of secondary structure was also calculated from a set of CD spectra of whole proteins in which the percentages of secondary structure were known from corresponding crystal structures. Using these standard CD curves of secondary structure, the percentage of each secondary structure in a protein can be predicted from its CD spectrum. Recent studies have suggested that the characteristics of CD spectra in the far ultraviolet (UV) region (<200 nm) are important for secondary structure prediction.

However, this type of secondary structure prediction is reliable only in a semiquantitative way. There are some limitations. For example, the CD spectrum of the β-turn has not yet been established with certainty. In general, the term random coil is used for a state of residues not assigned to any other structure

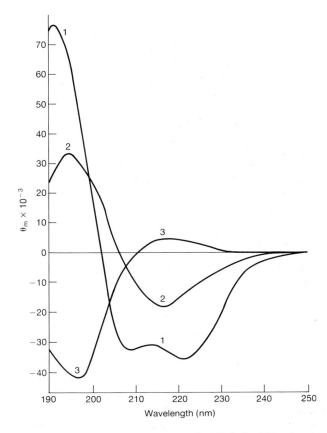

Fig. C51 CD spectra of poly-L-Lys in the α-helical (1), β-sheet (2) and random coil (3) conformations. From [4].

types. These residues are not, of course, random. In fact, the CD spectra for random coil structures show a negative CD band between 190 and 200 nm of considerably different amplitudes. Even the sign of a weaker band in the long-wavelength region is not always the same. Even within regions of well-defined structures, care must be taken in interpreting CD spectra. The amplitudes of CD bands of β-sheet structures depend upon the sequence, length, and the solvent. Similarly, short α-helices show CD spectra which are different from those of standard α-helices and the amplitudes depend on the number of amino-acid residues involved in the helix, although it is not clear whether these effects reflect the stability of an α-helix, or simply its size (Fig. C52).

Particular care must be taken to interpret CD spectra of proteins containing the residues Trp, Tyr, and Phe. The aromatic side chains of these amino-acid residues sometimes show induced CD at around 280 nm, and also at around 200–230 nm. Although the amplitude of these bands is typically small, the CD band of Trp at around 280 nm, and those of Tyr and Phe at around 200–230 nm can sometimes be significant, especially for short peptides. Disulphide bonds between two Cys residues also have a weak, and broad, intrinsic CD at 240–360 nm. Although the contributions of specific residues can complicate the secondary structure prediction, the CD band of Trp at around 280 nm often serves as a good spectroscopic measure for protein folding.

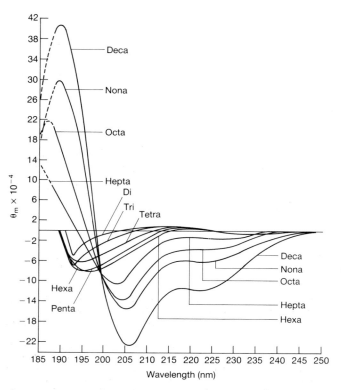

Fig. C52 CD spectra of *N*-carbobenzoxy-γ-ethyl-L-glutamate oligomer in trimethyl phosphate. From [5].

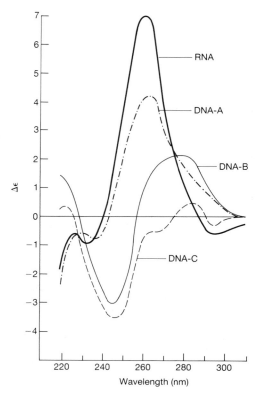

Fig. C53 CD spectra of A-, B-, and C-forms of DNA and of rice dwarf virus RNA. From [6].

CD of nucleic acids

The CD signals in nucleic acids arise predominantly from asymmetrically positioned purine and pyrimidine bases, as bonds in the sugars and the phosphodiester linkages do not absorb light in the wavelength range that is usually studied. CD spectra of nucleic acids therefore reflect the mode of base stacking. The interpretation of CD spectra of nucleic acids is more difficult than those of proteins, because, as explained above, CD spectra of proteins are primarily independent of amino-acid sequence, but those of nucleic acids depend on the sequence of bases stacking on each other. Differences in base stacking result from many factors, the sequence of nucleotides, the geometry (A-form, B-form etc.) of the strand, the type of nucleic acid (RNA or DNA) and the number of nucleotide strands (i.e. single, double or triple helices).

The classical B-form and A-form spectra have been obtained for a mixed sequence DNA in water and in ethanol respectively (Fig. C53). The classical B-form spectrum is characterized by a positive band (maximum at 270–275 nm, zero at 257–259 nm) and a negative band (minimum at 245–248 nm, zero at ~228 nm). The total positive area of the classical B-form spectrum is about the same as the total negative area. The classical A-form spectrum shows a much larger positive area (maximum near 260 nm, zero at around 240 nm) and a very small negative area (minimum at 210 nm). A recent study using synthetic oligonucleotides has shown that the negative band in the CD

spectra of AT-rich oligonucleotides is much sharper and larger than that in the classical B-form spectrum. This is probably due to real differences between two structures. However, it should be noted that separating the effects of sequence and those of structure is extremely difficult.

As described above, CD spectra of right-handed nucleic acids typically show negative bands in the shorter wavelength range and positive bands in the longer wavelength range; if a reversed spectrum is observed, a simple interpretation is that the structure is left-handed. This has been proved for artificial oligonucleotides, where each connection between a base and a sugar is fixed by two covalent bonds (Fig. C54). CD of some GC-rich and IC-rich nucleotides has been reported as reversed type. These DNAs are believed to fold into Z-form (*see* DNA; NUCLEIC ACID STRUCTURE).

The magnitudes of CD bands of single-stranded nucleic acids are greater than those of double- or triple-stranded ones (Fig. C55). That of RNA is usually larger than that of DNA.

A special type of CD spectrum (Ψ type) is observed when long DNA interacts with a variety of biological materials such as basic polypeptides, histone H1, and polyamines. The spectra are characterized by two very large peaks, one negative at 210 nm, and the other, positive (Ψ⁺) or negative (Ψ⁻), at around 270–290 nm. These spectra result from the aggregation of DNA, which causes light scattering. A similar spectrum is observed by adding some chemicals such as polyethylene glycol and trifluoroethanol, to DNA. A major difference between the classical A-form CD spec-

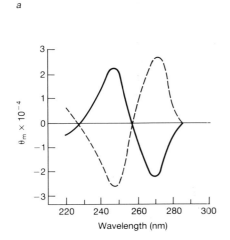

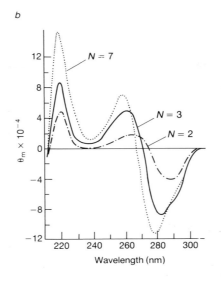

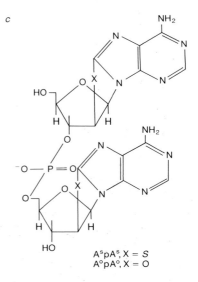

A^spA^s, X = S
A°pA°, X = O

Fig. C54 CD spectra of left-handed dinucleotides. Those of A°pA° (—), and ApA (----) (a) and A^s(pA^s)_n (b) are shown. ApA, L-adenyl-(3'-5')-L-adenosine. c, A°pA°, 8,2'-anhydro-8-oxy-9-β-D-arabinofuranosyladenine phosphoryl-(s'-5')-8,2'-anhydro-8-oxy-9-β-D-arabinofuranosyladenine; A^spA^s, 8,2'-anhydro-8-mercapto-9-β-D-arabinofuranosyladenine phosphoryl-(3'-5')-8,2'-anhydro-8-mercapto-9-β-D-arabinofuranosyladenine. From [7].

trum and the Ψ^+ spectrum is that the Ψ^+ spectrum has a large negative band at around 210 nm whereas the classical A-form spectrum does not.

Instrumentation and special types of CD

Unlike a normal spectrophotometer, a CD spectrometer has a linear polarizer and a quarter wave retarder to convert the linearly polarized light into circularly polarized light. The retarder is a stressed plate modulator which oscillates alternately to make left- and right-circularly polarized lights. The difference in absorption of the two circularly polarized lights, which is amplified by a lock-in amplifier, is very small compared with the absorption itself and this is the major problem of CD measurements.

Magnetic circular dichroism (MCD) can be measured by attaching a magnetic field to a sample compartment. Optical activity can be induced by the application of a magnetic field, which perturbs the energy levels of the system. The technique is useful in the study of METALLOPROTEINS. Fluorescence-detected circular dichroism (FDCD) is defined as the difference in fluorescence intensities excited by either of two differently circularly polarized lights. FDCD is useful for the studies of proteins containing Tyr, Trp, or Phe and transfer RNA containing the Y (wyosine) base. The FDCD spectrum, even of a complex system containing many chromophores, can be simple if only a few fluorophores are contained. Scattering artefacts are partly avoided by the use of fluorescence detection.

M. SUZUKI

1 Campbell, I.D. & Dwek, R.A. (1984) *Biological Spectroscopy* (Benjamin Cummings, San Francisco, CA).
2 Johnson, W.C. Jr (1988) Secondary structure of proteins through circular dichroism. *Annu. Rev. Biophys. Biophys. Chem.* **17**, 145–166.
3 Tinoco, I. Jr. & Bustamante, C. (1980) The optical activity of nucleic acids and their aggregates. *Annu. Rev. Biophys. Bioengng.* **9**, 107–141.
4 Greenfield, N. & Fasman, G.D. (1969) Computed circular dichroism spectra for the evaluation of protein conformation. *Biochemistry* **8**, 4108–4116.
5 Goodman, M. et al. (1969) Sensitive criteria for the critical size for helix formation in oligopeptides. *Proc. Natl. Acad. Sci. USA* **64**, 444–450.
6 Moore, D.S. & Wagner, T.E. (1974) Double-helical DNA and RNA circular dichroism. *Biopolymers* **13**, 977–986.
7 Ikehara, M. et al. (1972) Left-handed helical polynucleotides with D-sugar phosphodiester backbones. *Nature New Biol.* **240**, 16–17.
8 Brahms, J. (1965) Optical activity and the conformation of polynucleotide models of nucleic acid. *J. Mol. Biol.* **11**, 785–801.

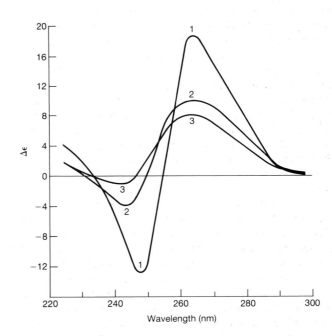

Fig. C55 CD spectra of poly(A) (1), the double-stranded poly(A) + poly(U) (2), and the triple-stranded poly(A) + poly(U) + poly(U) (3). From [8].

cis-acting A term used to describe a component of a DNA sequence which influences the activity of the same or an immediately adjacent DNA sequence. For example, the DNA sequences to which TRANSCRIPTION FACTORS bind within PROMOTERS are said to be *cis*-acting control elements as they affect the expression of the adjacent gene or a gene on the same chromosome. *Cis*-acting elements do not generally encode a polypeptide. *Cf.* TRANS-ACTING.

cisterna (*pl.* cisternae) Although applicable to any membrane-bound compartment the term is usually restricted to the flattened sacs and tubular extensions of the ENDOPLASMIC RETICULUM and Golgi stack (*see* GOLGI APPARATUS).

cistron The term originally coined by Seymour Benzer to describe the genetic unit identified by the *cis-trans* COMPLEMENTATION TEST. As the cistron is essentially equivalent to a GENE, the latter term is now used in preference.

citrate synthase Enzyme that catalyses the condensation of acetyl-CoA and oxaloacetate to form citrate and CoA in the tricarboxylic acid cycle. There are two types of citrate synthase, each specific for a different stereoisomer of citrate (EC 4.1.3.7 and EC 4.1.3.28). *See* PROTEIN STRUCTURE for illustration of this protein.

citrullinaemia A heritable condition associated with the deficiency of the urea cycle enzyme argininosuccinate synthetase (EC 6.3.4.5). Most patients die in the neonatal period of severe hyperammonaemia — excess ammonia in the blood. Some present later with psychomotor retardation, feeding difficulties, and episodic neurological disturbance. Asymptomatic cases are sometimes seen. The enzyme defect can be demonstrated in leukocytes, skin fibroblasts and liver. It is an AUTOSOMAL RECESSIVE trait mapping to human chromosome 9.

class switching The intrachromosomal recombinational process by which a rearranged immunoglobulin heavy chain variable region coding sequence can be progressively linked to different heavy chain constant region genes. It permits expression of ANTIBODY molecules with the same antigen specificity but different effector functions during an immune response. *See*: B CELL DEVELOPMENT; GENE REARRANGEMENT; IMMUNOGLOBULIN GENES.

classical pathway *See*: COMPLEMENT.

classification of protein structure *See*: PROTEIN STRUCTURE.

clathrin Structural protein forming the polyhedral lattice on clathrin-coated pits and vesicles (*see* COATED PITS AND VESICLES). Native clathrin consists of three copies of clathrin heavy chain (M_r 180 000) and three copies of clathrin light chain (M_r 30 000–40 000), forming a three-legged structure called a TRISKELION. The light chains are more heterogeneous than the heavy chains: there are two clathrin light chain genes in mammals, and both light chain messages are alternatively spliced in brain and other tissues (*see* RNA SPLICING). Clathrin is capable of *in vitro* assembly into empty cages resembling the coats on clathrin-coated vesicles.

Clavibacter A genus of Gram-positive bacteria which are typified by small irregular, nonsporing, rod-shaped cells. Members of this genus form short curved, rod-, or wedge-shaped cells. They are chemo-organotrophic, obligately aerobic, and nonmotile. All members of the genus cause disease in plants and many of the diseases are systemic, or become so once the pathogen reaches the plant vascular system. *Clavibacter* spp. survive poorly in soil and are usually only found in association with plants. They survive periods between crops by growing in or on seeds.

Davis, M.J. et al. (1984) *Int. J. Syst. Bacteriol.* **34**, 107–117.

cleavage The series of mitotic divisions extending from FERTILIZATION to development of a BLASTULA, which occur without intervening periods of growth. (The term may also be applied to the process of CYTOKINESIS during this period.) Cleavage serves to increase the nucleocytoplasmic ratio to approximately adult proportions and generates sufficient cells (blastomeres) for organization of the embryo into a complex multilayered structure. The pattern of cleavage may be either radial, bilateral, or SPIRAL and is also affected by the distribution of YOLK in the egg. Some eggs contain minimal yolk which is evenly distributed. These eggs (e.g. of mammals) cleave completely (holoblastic cleavage). Others have a concentration of yolk at one end (e.g. birds) and either cleave incompletely (meroblastic cleavage) or holoblastically but with cleavage being retarded at the vegetal pole (e.g. amphibians). In yet other eggs, yolk is concentrated away from both the centre and the periphery of the egg (e.g. insects). Nuclei divide in the nonyolky centre and then migrate to the periphery where they become separated by membranes (superficial cleavage).

cleavage nucleus Nucleus of ZYGOTE formed by the fusion of male and female pronuclei.

climacteric Period during fruit ripening, when respiration increases rapidly to a peak, before declining as senescence sets in. It is associated with the production of ethylene, disappearance of acid, conversion of starch to sucrose, appearance of ANTHOCYANINS and other pigments in the outer tissue layers, and softening of fleshy tissue owing principally to the production and action of polygalacturonase. Characteristic of apple, pear, peach, plum, tomato and banana; not found in citrus fruits, cherry, strawberry and pineapple. Similar changes may be observed in flower petal development, and in senescing leaves immediately before leaf fall.

clonal analysis A term used to refer both to CELL LINEAGE ANALYSIS and to the analysis of mutations restricted to particular CLONES of cells. In DROSOPHILA, for example, the technique of MITOTIC RECOMBINATION can generate homozygous mutant cells within a heterozygous embryo, allowing analysis of the cellular phenoype of recessive lethal mutations.

clonal restriction In the general sense, one or more CLONES (definition 1) of cells which are confined In practice, the term is usually used only to signify that the descendants of a particular group of progenitor cells are confined to well-defined spatial limits which they cannot cross. The expression is central to the definition of cell lineage COMPARTMENTS which are thought to be important in defining the boundaries of MORPHOGENETIC FIELDS. *See also*: CELL LINEAGE.

clonal selection The process that ensures the specificity of an immune response. According to Burnet's clonal selection theory, pre-existing clones of lymphocytes are each characterized by the specificity of their antigen receptor. One cell clone expresses only one type of antigen receptor of defined specificity (*see* B CELL DEVELOPMENT; T CELL DEVELOPMENT). On encounter with the cognate antigen, that clone is preferentially stimulated and expanded. More recent results have demonstrated that the antibody specificity of a B cell clone may change during development of the B cells because the variable regions of the antibody can be diversified by a HYPERMUTATION mechanism.

clone (1) In embryology, immunology, cell biology, etc., a group of cells derived from a single common progenitor cell.
(2) Two or more individuals of a given species with identical GENOTYPES. Thus, monozygotic twins are clonally related because they are both derived from the same fertilized egg.
(3) In molecular biology, an individual host/vector unit (e.g. a recombinant plasmid vector in a bacterial host) carrying a single insert DNA sequence (*see* CLONED DNA; DNA CLONING).

cloned DNA A fragment of DNA, either isolated directly from the genome of an organism (genomic DNA) or as a complementary DNA (cDNA), which has been covalently joined to a suitable CLONING VECTOR. The cloned DNA fragment does not necessarily contain an intact gene. *See*: DNA CLONING.

cloning strategy Before embarking on the cloning of a specific gene, a strategy must first be devised which will allow for its identification. Initially such strategies involve the construction of a suitable DNA LIBRARY and the identification of a suitable nucleic acid PROBE (or an ANTIBODY if one wishes to screen an expression library). Strategies based on the POLYMERASE CHAIN REACTION (PCR) are now favoured, in which a specific DNA sequence can be amplified *in vitro* without the need for cloning into a vector. But as with other strategies this requires some previous or deduced knowledge of the nucleotide and/or amino-acid sequence of the target gene and/or protein. Positional cloning strategies depend on knowing the area of the chromosome in which the gene resides.

cloning vector DNA molecule into which DNA fragments can be introduced *in vitro* using RESTRICTION ENZYMES and DNA LIGASES (*see* DNA CLONING). Such recombinant DNA molecules, usually based on either a bacterial PLASMID or on a BACTERIOPHAGE (e.g. LAMBDA or M13), can then be introduced into living cells by TRANSFORMATION or TRANSFECTION and be propagated. To ensure maintenance and replication of such vectors in the host cell they must also contain an origin of replication recognizable by the host cell, carry a SELECTABLE MARKER GENE (e.g. one coding for antibiotic resistance), and replicate in multiple copies per cell. Cloning vectors are used for gene isolation and manipulation.

Cloning vectors for general use are relatively small, as larger molecules are more readily degraded during preparation and manipulative procedures. Plasmid and phage vectors have quite low upper limits on the amount of inserted DNA that can be accommodated — 1500 bp for phage M13, ~5 kb for plasmid vectors, 25 kb for lambda vectors, and 40 kb for cosmids (see below). Yeast artifical chromosomes (YACs) which can carry very large inserts of DNA (up to hundreds of kb) have also been developed as cloning and EXPRESSION VECTORS. There are also cloning vehicles based on eukaryotic viruses such as SV40 and the cauliflower mosaic virus.

The nomenclature of plasmid cloning vectors follows standard rules. All names are of the form pXY123 (e.g. pBR322, pBR325, pEMBL8, pSC101, pUC8), where 'p' indicates that this vector is a plasmid, the letters identify the laboratory in which the plasmid was originally constructed (e.g. BR stands for Bolivar and Rodriguez, EMBL for the European Molecular Biology Laboratory, SC for Stanley Cohen — pSC101 was one of the plasmids used in early recombinant DNA experiments), and the number distinguishes the plasmid from others developed in the same laboratory. There are numerous plasmid cloning vectors in use, many of which are derived from pBR322.

Bacteriophage LAMBDA (λ) is the basis of a range of cloning vectors. It can accommodate pieces of DNA from 5–25 kb, which are too large for plasmid or M13 vectors. λ vectors are of two main types:
1 Insertion vectors (e.g. CHARON; λgt10), in which part of the λ genome has been deleted and new DNA is inserted at a separate site.
2 Replacement vectors (e.g. λWES.λB′, λEMBL4), in which two recognition sites for a restriction enzyme flank a dispensable 'stuffer fragment' which is replaced by the inserted DNA. Replacement vectors can usually carry larger pieces of DNA than insertion vectors (up to 23 kb) and selection is on the basis of genome size.

Vectors derived from the single-stranded bacteriophage M13 are widely used to obtain single-stranded cloned DNA for DNA SEQUENCING and other applications. Only small inserts of DNA can be accommodated in M13 vectors but hybrid plasmid–M13 vectors can accommodate inserts of up to 10 kb.

COSMIDS (e.g. pJB8) are plasmids into which a λ *cos* site has been inserted. On ligation with new DNA the linearized plasmids form concatemers in which the *cos* sites are separated by an appropriate length to be recognized by λ packaging enzymes. Recombinant plasmid DNA can then be packaged *in vitro* into phage heads for infection of *E. coli* host cells. *In vivo* the cosmid replicates as a circular plasmid. Using cosmid vectors of around 5–6 kb some 40 kb of DNA can be inserted.

Other vectors have been developed as dual purpose cloning and EXPRESSION VECTORS (e.g. SHUTTLE VECTORS). Cloning vectors have also been developed for use with other bacterial host cells, for example, *Streptomyces* (e.g. based on the phage phiC31), *Bacillus*, and *Pseudomonas*. See also: PHAGEMIDS.

Other important cloning vectors for use with host cells other than bacteria include the 2 μm PLASMID of the yeast SACCHAROMYCES CEREVISIAE and its episomal derivatives (e.g. YEp plasmids, *see also* SHUTTLE VECTORS). Vectors based on the 2 μm plasmid can also be used in the fission yeast SCHIZOSACCHARO-

MYCES POMBE. Yeast integrative plasmids (YIps) are bacterial plasmids containing a yeast gene which enables the plasmid DNA to be maintained by integration into the yeast chromosome by homologous recombination. Yeast replicative plasmids (YRps) carry a yeast chromosomal origin of replication and can multiply as extrachromosomal plasmids in yeast cells.

Yeast artificial chromosomes (YACs) (e.g. pYAC2) can be used to clone very large pieces of DNA in yeast cells. They are constructed from a bacterial plasmid vector (e.g. pBR322) into which are inserted an *S. cerevisiae* chromosomal origin of replication, selectable marker genes from yeast, including one which will be inactivated by insertion of the DNA to be cloned, DNA from the centromere region of a yeast chromosome, and TEL sequences on which telomeres will be constructed *in vivo* (*see* CHROMOSOME STRUCTURE; SACCHAROMYCES CEREVISIAE). YACs can accommodate pieces of DNA up to several hundred or more kb long, and can therefore be used to clone complete genes. Genomic libraries in YAC vectors also reduce the number of clones required for storing a complete genome. For the human genome, a cosmid library of more than 250 000 clones can be reduced to 60 000 clones of 150 kb each stored in YAC vectors.

Cloning vectors for mammalian cells have been based mainly on the PAPOVAVIRUSES SV40 and bovine papilloma virus. An SV40 vector — SVGT-5 — was used to first clone the mouse β-globin gene. Other animal viruses, notably VACCINIA VIRUS, are used as the basis of EXPRESSION VECTORS for use with mammalian cells.

For gene cloning and expression in higher plants, vectors based on the TI PLASMID, and on plant DNA viruses have been developed (*see also* PLANT GENETIC ENGINEERING; 35S PROMOTER). *See also*: DNA LIBRARIES; GENETIC ENGINEERING.

clonotypic Characteristic of a particular clone of cells. Used especially in immunology to describe the antigen-specific receptors of lymphocytes (*see* ANTIBODIES; B CELL DEVELOPMENT; IMMUNOGLOBULINS; T CELL DEVELOPMENT; T CELL RECEPTORS). These receptors are composed of the protein products of rearranged IMMUNOGLOBULIN GENES or T CELL RECEPTOR GENES. Each rearrangement is usually random in respect to the particular V, (D) and J elements used, and there is additional N-REGION DIVERSITY and also diversity introduced by somatic hypermutation (in immunoglobulins only). Expression of rearranged alleles follows the rule of ALLELIC EXCLUSION and so the surface immunoglobulin or T cell receptor molecules expressed are of a single type and unique to that lymphocyte precursor cell and the clone formed by its mitotic descendants.

closest tree Criterion for evolutionary relatedness. *See*: INVARIANTS.

closterovirus group Name derived from Greek *kloster*, thread, descriptive of the long very flexuous rod-shaped particles 600–2000 nm long and 12 nm in diameter. Each particle contains a single species of (+)-strand linear RNA (M_r 2.2×10^6–4.7×10^6) encapsidated in helically arranged protein subunits. There are two subgroups: 1 (type member beet yellows virus) have coat protein subunits of 23 000–25 000 and are aphid transmitted; 2

(type member apple chlorotic leafspot virus) have coat protein subunits of 27 000 and no known vectors. The genome organization has not yet been determined for any member of either subgroup. *See also*: PLANT VIRUSES.

Milne, R.G. (Ed.) (1988) *The Plant Viruses*, Vol. 4 (Plenum, New York).

Clostridium botulinum Gram-positive, spore-bearing, anaerobic rod-shaped bacterium, present in soil, and the causal agent of botulism, a fatal form of food poisoning caused by a protein exotoxin produced by the bacterium in contaminated food. The toxin causes progressive paralysis of muscles innervated by the parasympathetic system, by preventing the release of acetylcholine from stimulated nerves. The mechanism of action of BOTULINUM TOXIN has been elucidated.

clotting factors *See*: BLOOD COAGULATION AND ITS DISORDERS; COAGULATION FACTORS.

cM CENTIMORGAN.

CML CHRONIC MYELOID LEUKAEMIA.

CMP Cytidine 5′-monophosphate. *See*: NUCLEOSIDES AND NUCLEOTIDES.

cms (cmsC, cmsS, cmsT) Cytoplasmic male sterility. *See*: MITOCHONDRIAL GENOMES: PLANT.

CNQX (6-cyano-7-nitroquinoxaline-2,3-dione) Selective antagonist of non-NMDA EXCITATORY AMINO ACID RECEPTORS (Fig. C56). However, at higher concentrations it is also an antagonist at the glycine site on the NMDA excitatory amino acid receptor.

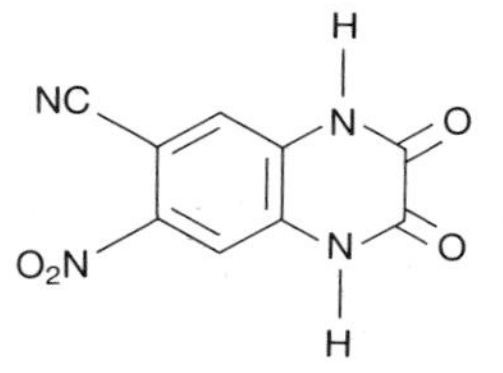

Fig. C56 Structure of CNQX.

CNTF CILIARY NEUROTROPHIC FACTOR.

coagulation factors (clotting factors, Factors I–XIII) A group of protease enzymes and cofactors that act in a cascade in blood clotting, each enzyme releasing the active factor from its inert precursor in the sequence. They are divided into two systems: the intrinsic system which is activated when blood comes into contact with nonendothelial surfaces; and the extrinsic system triggered by tissue injury and the consequent release of phospholipoproteins. The end result is the production of thrombin for the conversion of fibrinogen to fibrin to form a gel-like clot. *See*: BLOOD COAGULATION AND ITS DISORDERS.

coat colour genes In mice, more than 50 loci have been identified on the basis of their effects on coat colour when mutant. Of these, only a minority are directly involved in the synthesis of the melanic pigments. The key enzyme in the pathway of melanin synthesis, tyrosinase, is encoded at the *albino* (*c*) locus, whereas the *brown* (*b*) locus encodes a tyrosinase-related protein. Most coat colour mutants produce their effects on colour indirectly, either through interference with melanocyte development or morphology, or through effects on the interaction between melanocytes and cells in the hair bulb. Thus, the analysis of coat colour genes has provided insight into more general developmental processes. For example, the *Dominant spotting* (*W*) locus was initially identified on the basis of a pigmentation defect but, when homozygous, the mutation has severe pleiotropic effects on pigmentation, HAEMATOPOIESIS, and germ cell formation. These same processes are also affected by mutations at the *Steel* (*Sl*) locus, and recently *W* has been shown to encode the receptor tyrosine kinase c-Kit (*see* GROWTH FACTOR RECEPTORS) and *Sl* to encode its GROWTH FACTOR ligand. *See also*: AGOUTI.

Coated pits and vesicles

COATED pits and vesicles are found in all eukaryotic cells, from yeasts to mammals. Their function is to move membrane and the proteins it contains from one compartment of the cell to another. Thus, coated pits bud from a 'donor' compartment and pinch off into the cytoplasm as free coated vesicles. The coat is then lost from the vesicle before it goes on to fuse with the appropriate 'acceptor' compartment. Two major types of coated vesicle have so far been identified: clathrin-coated vesicles and non-clathrin-coated or COP-coated vesicles. They differ from each other in many respects, but recent findings have revealed some similarities in the proteins that make up their coats.

Clathrin-coated vesicles

A typical mammalian cell contains at least a thousand clathrin-coated pits and vesicles, with diameters of 100–300 nm (Fig. C57). They bud from two distinct membrane compartments: the PLASMA MEMBRANE and the *trans* Golgi network (TGN) (*see* GOLGI APPARATUS). The clathrin-coated vesicles that bud from the

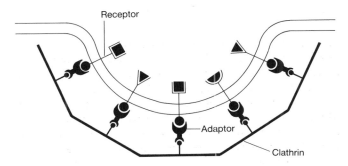

Fig. C57 Model of a clathrin-coated pit.

plasma membrane are responsible for the selective ENDOCYTOSIS of a number of transmembrane proteins, often receptors for extracellular ligands such as low density lipoproteins, transferrin, and epidermal growth factor (*see e.g.* GROWTH FACTOR RECEPTORS; LIPOPROTEINS AND LIPOPROTEIN RECEPTORS). Because such receptors may be up to a hundred times more concentrated in the coated pit than on the rest of the plasma membrane, uptake of the ligands is very efficient. Once the vesicle has been uncoated, it fuses with an early ENDOSOME where further sorting occurs.

The clathrin-coated vesicles that bud from the TGN are also involved in the concentration of ligands mediated by the MANNOSE 6-PHOSPHATE RECEPTOR, and the consequent segregation of lysosomal proteins. Newly synthesized lysosomal enzymes that have been phosphorylated on mannose residues earlier in the secretory pathway bind to mannose 6-phosphate receptors when they reach the TGN. The receptor–ligand complexes are packaged into clathrin-coated vesicles, which then lose their coats and fuse with prelysosomes. In this way lysosomal enzymes are diverted from the 'default' pathway to the plasma membrane taken by most other proteins (*see* LYSOSOMES).

Partial clathrin coats are also often observed on nascent secretory granules in cells with a regulated secretory pathway (*see* PROTEIN SECRETION), although the precise role of clathrin-coated vesicles in regulated secretion still needs to be established (*see also* EXOCYTOSIS).

Coat components

Thus, both plasma membrane-associated and TGN-associated clathrin-coated vesicles are able to recognize and concentrate a subset of transmembrane proteins in their respective compartments. This sorting process is thought to be achieved by the binding of coat proteins to the cytoplasmic tails of such membrane proteins (see Fig. C57). Genetic engineering experiments demonstrate that the signal for internalization of plasma membrane proteins often includes an aromatic amino acid, usually a tyrosine, in a tight turn configuration. The signal for incorporation of mannose 6-phosphate receptors into clathrin-coated vesicles budding from the TGN is not known but is thought to be different from the internalization signal. *In vitro* binding studies indicate that these signals work by interacting with components of the coat called adaptors (Fig. C57). As adaptors can bind to clathrin as well, they are believed to provide clathrin-coated vesicles with their characteristic selectivity.

Both clathrin and adaptors are protein complexes that cycle on and off the membrane. The soluble form of CLATHRIN consists of three copies of clathrin heavy chain (M_r 180 000) and three copies of a clathrin light chain (M_r 30–40 000), forming a three-legged structure called a TRISKELION. Under favourable conditions *in vitro*, clathrin triskelions will spontaneously assemble into empty cages with the same polyhedral lattice structure as that seen on coated vesicles. If adaptors are present, assembly conditions are less stringent and the adaptors become incorporated into the coat. Thus, the self-assembly properties of the coat proteins are thought to provide the mechanism for coated pit invagination, although *in vitro* studies suggest that pinching off as a coated vesicle and subsequent uncoating are both steps that require energy and additional cytoplasmic factors (*see* UNCOATING ATPASE).

Adaptors in solution are heterotetramers consisting of one copy of a β-type ADAPTIN (M_r 105 000), one copy of an α- or γ-adaptin (M_r 91–108 000), and one copy each of two smaller proteins of M_r ~50 000 and 20 000. Structurally, the adaptors appear as a brick-like 'head' flanked by two smaller 'ears', connected by protease-sensitive hinges. Two distinct types of adaptors have been identified, known as HA-I (or AP1) and HA-II (or AP2) because of their characteristic elution profiles on hydroxylapatite columns. The two adaptors are associated with the TGN and the plasma membrane respectively and are thought to account for the different specificities of the clathrin-coated vesicles that bud from these two compartments.

Most of the coat proteins associated with clathrin-coated vesicles have now been cloned and sequenced. So far, all organisms that have been studied have been found to have only a single clathrin heavy-chain gene. The genes encoding the other coat subunits appear to be more heterogeneous. None of the proteins shows any HOMOLOGY with any other proteins previously sequenced, but the three types of adaptins, α, β, and γ, show some homology with each other and are thought to be derived from a common ancestral gene. There are also structural similarities between the three types of adaptins: in all three cases, the N-terminal 60–70K is incorporated into the adaptor head and the C-terminal ~30K forms one of the two ears, connected by a stretch of 100–150 amino acids, rich in proline and glycine residues, shown to correspond to the hinge.

The cloning of the clathrin heavy chain gene from the yeast *Saccharomyces cerevisiae* has allowed the function of the protein to be investigated by targeted gene disruption. Surprisingly, some yeast strains that have lost their clathrin heavy-chain gene are still viable, although they grow very slowly and mislocalize an enzyme thought to reside in the TGN. Traffic through the constitutive secretory pathway is essentially unimpaired in such cells, however. Similarly, mammalian cells microinjected with antibodies against the clathrin heavy chain are still able to secrete proteins efficiently. Thus, traffic from the ER through the Golgi apparatus to the plasma membrane appears to be mediated by non-clathrin-coated vesicles.

COP-coated vesicles

A second population of coated vesicles is also associated with the Golgi apparatus, but their coats appear less highly structured than those on clathrin-coated vesicles. These non-clathrin-coated vesicles are associated with all of the Golgi compartments, including the various cisternae of the Golgi stack and probably the intermediate compartment or *cis* Golgi network as well. Unlike clathrin-coated vesicles, they do not appear to sort or concentrate the proteins that they transport; instead, the vesicles are thought to be involved in bulk flow of newly synthesized proteins through the constitutive secretory pathway (*see* PROTEIN SECRETION).

Isolation of the non-clathrin-coated vesicles has shown that their coats consist of four major proteins, called α-COP, β-COP, γ-COP, and δ-COP (M_r 160 000, 110 000, 98 000, and 61 000 respectively). β-COP is the best characterized of the four proteins:

it has been localized to non-clathrin-coated vesicles by IMMUNO-ELECTRON MICROSCOPY, and its cDNA has been cloned and sequenced. The deduced protein sequence of β-COP is significantly homologous to that of β-adaptin, although the homology is restricted to the N-terminal half of the protein and no hinge region is apparent. None of the other COPs have yet been sequenced, but their similarity in size to some of the other coat proteins on clathrin-coated vesicles suggests that there may be further homologies. A complex containing all four COPs, together with smaller proteins, has also been purified from cytosol, although the exact stoichiometry of the complex (or 'coatomer') has not yet been established.

The entire cycle of COP-coated vesicle formation, docking onto the appropriate acceptor membrane, uncoating and vesicle fusion has been reconstituted *in vitro* using isolated Golgi stacks. Reagents have been found that interfere with the normal cycle of vesicle coating and uncoating. In particular, the drug BREFELDIN A causes a rapid loss of β-COP from the Golgi membrane in living cells and also interferes with COP-coated vesicle formation *in vitro*. As this event precedes the drug-induced changes in the distribution of Golgi membrane proteins, it has been postulated that loss of the coat may lead to the later changes that occur. The coats are stabilized in intact cells by AlF_4^- and *in vitro* by the non-hydrolysable GTP analogue GTP-γ-S, indicating that GTP-BINDING PROTEINS are involved in the COP-coated vesicle cycle.

M.S. ROBINSON

1 Pearse, B.M.F. & Robinson, M.S. (1990) Clathrin, adaptors and sorting. *Annu. Rev. Cell Biol.* **6**, 151–171.
2 Serafini, T. et al. (1991) 'Coatomer': a cytosolic protein complex containing subunits of non-clathrin-coated Golgi transport vesicles. *Nature* **349**, 248–251.

***

cob gene Mitochondrial gene encoding the apocytochrome *b* component of the CYTOCHROME BC$_1$ COMPLEX.

Cockayne's syndrome Human disorder with a RECESSIVE pattern of inheritance, characterized by severe photosensitivity, distinct facial and skeletal abnormalities, and progressive neurological degeneration. Cells from Cockayne's syndrome patients are hypersensitive to ultraviolet light and to chemical carcinogens, resembling the disease XERODERMA PIGMENTOSUM in this respect. The two conditions are closely related genetically and patients can have symptoms of only one disorder, or of both simultaneously. Clinically, the syndrome is distinguished by a particular form of dwarfism and by the absence of the freckling, dermatosis, and cancerous lesions that are generally found in xeroderma pigmentosum. Cells from affected patients display alterations in nucleotide excision repair that range from subtle defects restricted to actively transcribed genes to severe overall repair defects. *See also*: DNA REPAIR.

coding joint Joint between the separate coding sequences of an assembled V-D-J region in IMMUNOGLOBULIN GENES and T CELL RECEPTOR GENES. It is formed during the recombination process (*see* GENE REARRANGEMENT; RECOMBINASE) and constitutes part of the final coding sequence of the rearranged gene. The recombination process results in a coding joint and a SIGNAL JOINT which does not form part of the rearranged coding sequence. The junction between assembled V, (D) and J segments is imprecise. Bases may be added or removed at the joint (*see* N-REGION DIVERSITY).

coding sequence Any DNA or RNA sequence that encodes genetic information, that is, the amino acid sequence of a protein or the nucleotide sequence of an RNA.

coding strand The strand of an RNA or DNA molecule that contains the genetic (sense) information. Within a single double-stranded molecule such as a chromosome the coding strand may differ for different genes. *See*: GENETIC CODE; TRANSCRIPTION; TRANSLATION.

codominance Situation where neither of two different alleles at a locus is dominant over the other, usually producing a phenotype different from, and sometimes intermediate between, the two homozygous parental types. *See also*: MENDELIAN INHERITANCE; MUTATION.

codon A sequence of three successive bases in nucleic acid (DNA and RNA) that specifies a particular amino acid or a translation termination signal (*see* GENETIC CODE). A codon formally refers to the sequence in RNA; the corresponding sequence in DNA is often known as a 'triplet'. The genetic code contains 64 codons of which 61 define one or other of the 20 amino acids known in proteins. The remaining three codons encode signals for the termination of translation. *See also*: PROTEIN SYNTHESIS.

codon bias, codon usage The propensity to use a particular codon to specify a particular amino acid. The GENETIC CODE is redundant, with 61 codons carrying the information for 20 different amino acids. Most amino acids are encoded by more than one codon; leucine, for example, has six different codons (the maximum number) — UUA, UUG, CUU, CUC, CUG, CUA. Analysis of DNA sequences from a wide variety of organisms has revealed that not all codons for a specific amino acid are used with equal frequency. Such bias in codon usage is both species specific and, within a given species, is more extreme for mRNAs that encode abundant proteins. Furthermore, the codon bias observed for a given species directly relates to the abundance of the corresponding isoacceptor tRNA species; the more frequently a codon is used, the more abundant the tRNA used to decode that codon. Codon bias may represent a means of ensuring optimal translational efficiency of mRNAs coding abundant proteins.

Ikemura, T. (1982) *J. Mol. Biol.* **158**, 573–597.

coelenterate development The coelenterates comprise the hydroids, which are usually colonial (although the well-known *Hydra* itself is a free-living polyp), together with the jellyfish and sea anemones. The latter have no developmental biology, save for some striking examples of metaplasia during regeneration. Following the celebrated work of Tremblay in the 18th century, the hydroids have been intensively studied from the point of view of

pattern regeneration and the results have had a far-reaching influence on developmental biology. Work on colonial hydroids by Child contributed to the concept of developmental GRADIENTS, the later work of Rose to the law of distal transformation and the still later work of Wolpert to the theory of POSITIONAL INFORMATION.

In *Hydra* there seem to be separate processes controlling regeneration of the head and the foot, but the rules for the two extremities are quite similar. If the head is grafted into the flank it can suppress regeneration from the cut surface. The characteristics of this interaction suggest that the head produces a diffusible inhibitor. If the normal head is removed then the inhibition is quickly lost and the tissue near the cut autonomously acquires the positional coding of a new head. A short segment from the midbody region may regenerate a head (or a foot) at both ends, to produce a 'bipolar form'. Various substances have been isolated from *Hydra* and other coelenterates that accelerate or retard regeneration from one or other extremity, and it is also known that substances interfering with signal transduction mechanisms can have severe effects on polarity and pattern.

The advantage of *Hydra* is that the microsurgery is quite easy, and that long-range inductive signalling appears to persist throughout the life of the animal. Colonial hydroids in particular seem to secrete MORPHOGENS in the medium so are favourable material for morphogen identification.

coelom The cavity between SOMATOPLEURE and SPLANCHNOPLEURE, the two layers of cells derived from the LATERAL PLATE MESODERM. It characterizes higher animals.

cofilin *See*: ACTIN-BINDING PROTEINS.

cohesive ends Sometimes referred to as 'sticky ends', these are the single-stranded extensions found in double-stranded DNA which has been digested with a RESTRICTION ENZYME which cuts in a staggered fashion. The single-stranded 5′ or 3′ extensions produced by a given restriction enzyme are complementary and thus can base pair via nominal hydrogen bonding (Fig. C58). Single-stranded extensions generated by different restriction enzymes are not usually complementary although there are some exceptions to this rule.

coiled coil Protein conformation found, for example, in collagen and laminin (*see* EXTRACELLULAR MATRIX MOLECULES) in which a stretch of α-helix is further helically coiled.

cointegrating plasmid A vector used in the cointegration method of gene transfer in plants. The vector is based on a standard *Escherichia coli* CLONING VECTOR (e.g. pBR322) which also carries a T-DNA region from the TI PLASMID of *Agrobacterium*. The 'foreign' gene to be introduced into the plant is inserted into this plasmid, which is then transferred via CONJUGATION or TRANSFORMATION into an *Agrobacterium* strain carrying an intact Ti plasmid. Recombination between the recombinant plasmid and the endogenous Ti plasmid occurs by way of the homologous T-DNA sequences, generating a new form of the Ti plasmid now carrying the required gene. Subsequent infection of the plant with

Bam HI

5′ —————— GGATCC —————— 3′
3′ —————— CCTAGG —————— 5′

↓ Digest

5′GATCC —————— 3′
  | | | | G —————— 5′
      G| | | |
3′ —————— CCTAG 5′

Hydrogen bonds

Cohesive end

↓ Reanneal

5′ —————— GGATCC —————— 3′
3′ —————— CCTAGG —————— 5′

Fig. C58 An example of cohesive ends generated by digesting a double-stranded DNA molecule with the restriction enzyme *Bam*HI.

the transformed *Agrobacterium* results in the transfer of the 'foreign' gene to the chromosomes of the plant.

Col plasmids A general class of bacterial PLASMIDS which encode bacteriocins such as COLICINS, for example the ColE1 plasmid which encodes the colicin E1. The Col plasmids can be divided into two groups: group I, which are nonconjugative plasmids with molecular weights of ~5–8 × 10^6; and group II, which are conjugative plasmids with molecular weights of 50–80 × 10^6.

colcemid A methylated derivative of COLCHICINE with similar antimitotic properties.

colchicine A plant alkaloid (Fig. C59) (derived from the autumn crocus *Colchicum autumnale* L.) which binds to TUBULIN mole-

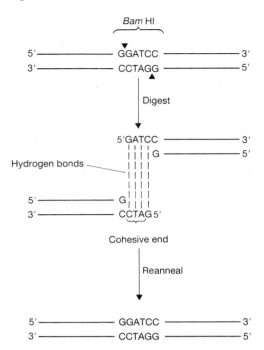

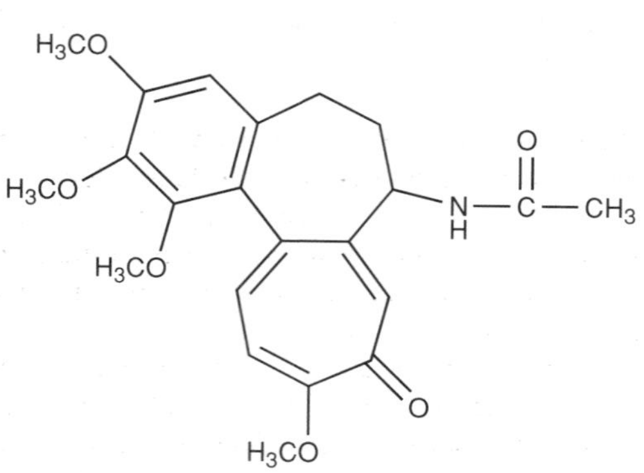

Fig. C59 Structure of colchicine.

cules, inhibiting their addition to MICROTUBULES and leading to microtubule depolymerization. In this manner colchicine causes the MITOTIC SPINDLE to disappear and blocks the cell in MITOSIS at metaphase.

cold shock proteins In the bacteria *Escherichia coli* and *Bacillus subtilis* an abrupt downshift in temperature leads to the production of specific cold shock proteins of which CS7.4 and CspB respectively are the major species. They share homology with the eukaryotic Y-BOX FACTORS. CS7.4 has been shown to be a transcriptional activator of two genes. The crystal structure of CspB has been determined.

ColE1 plasmid A Col plasmid that encodes COLICIN E1. The ColE1 plasmid, which is probably the best characterized member of the Col plasmid group, is 6646 bp long and has been used extensively in the construction of CLONING VECTORS for *Escherichia coli* (e.g. pBR322), by providing the necessary ORIGINS OF REPLICATION. It has a COPY NUMBER of ~15 per chromosome but this can be amplified by growing cells in the presence of chloramphenicol (*see* CHLORAMPHENICOL AMPLIFICATION). The replication of ColE1 seems to be entirely dependent on host-encoded functions. In addition to encoding colicin E1, the plasmid also encodes a protein which confers immunity to ColE1 and also a small protein which regulates its replication in *E. coli*.

coleoptile The hooded structure covering the shoot apex of embryos and seedlings of grasses and some other MONOCOTYLEDONS.

coleorhiza The sheath covering the primary root apex of grass embryos.

colicins Antibacterial proteins encoded by COL PLASMIDS. Cells which carry a related Col plasmid are usually insensitive to colicins because these plasmids encode not only a colicin (*see* COLE1 PLASMID), but also an immunity determinant. Colicin production is normally repressed, but a variety of agents, including ultraviolet light, can induce synthesis. Colicins mediate their bacteriostatic effects initially by binding to the cell wall, but different colicins may have quite different modes of action. For example, colicin E3 cleaves the large rRNA molecule (16S rRNA) in the large subunit of the *Escherichia coli* ribosome, whereas colicin E1 seems to uncouple energy-dependent processes by an interaction with the bacterial cell membrane.

collagen Fibril-forming protein which is a major component of some types of extracellular matrix (*see* EXTRACELLULAR MATRIX MOLECULES). It is secreted as a precursor, procollagen, which is processed by proteolytic cleavage to form mature collagen. Collagen contains the unusual AMINO ACIDS hydroxylysine and hydroxyproline. The mature collagen molecule is composed of three polypeptide strands wound around each other to form an elongated rod.

collagen disorders Diverse group of disorders variously caused by defects in the numerous genes encoding the different types of collagen chains, their transcriptional control, the post-translational modification of collagen chains and the cross-linking of the heterotrimeric collagen molecules into fibrils. *See*: EHLERS-DANLOS SYNDROME; EXTRACELLULAR MATRIX MACROMOLECULES; MARFAN SYNDROME; OSTEOGENESIS IMPERFECTA.

colloidal gold *See*: IMMUNOELECTRON MICROSCOPY.

colominic acid Extracellular polysaccharide of some strains of the bacterium *Escherichia coli*, consisting of linear chains of varying length, composed of α2,8-linked N-ACETYL NEURAMINIC ACID (i.e. a polysialic acid). It is synthesized by transglycosylation.

colony-stimulating factors *See*: G-CSF; GM-CSF; IL-5; M-CSF.

colour blindness The genetically determined inability to distinguish certain colours. *See*: CONE PIGMENTS; VISUAL TRANSDUCTION.

colour vision pigments *See*: CONE PIGMENTS.

combinatorial diversity The repertoire diversity in IMMUNOGLOBULINS and T CELL RECEPTORS that arises from the different combinations of variable region gene elements (V, D and J) during GENE REARRANGEMENT. The GERM-LINE DIVERSITY (the potential diversity due to the number of different gene segments in the germ-line genome) is enhanced by combinatorial diversity. For example, a given V gene may be rearranged to one of several different D segments and to one of several different J regions. By this means a given V gene segment may give rise to receptors of different specificity. *See*: GENERATION OF DIVERSITY.

combinatorial library A DNA LIBRARY in which both heavy and light ANTIBODY chains are expressed in *Escherichia coli* using bacteriophage VECTORS. Starting with mRNA from mouse spleen or human peripheral blood lymphocytes as template, cDNA is generated *in vitro* using REVERSE TRANSCRIPTASE. PCR is then used to amplify the Fd ($V_H + C_H1$) component of the heavy chains and the light chains which in turn are cloned into bacteriophage LAMBDA vectors to generate separate heavy chain and light chain libraries. By selective cleavage of the DNA from these two libraries followed by religation, a combinatorial library can be constructed in which both the heavy and light chains are co-expressed in a single *E. coli* cell.

Huse, W.D. et al. (1989) *Science* **246**, 1225–1281.

combining site An antibody combining site, the antigen-binding site of an antibody.

commitment Most authors use the terms commitment and DETERMINATION interchangeably, to mean an irreversible decision of a cell to DIFFERENTIATE in a particular direction. Some authors, however, use commitment to identify a determinative decision that is not irreversible.

comovirus group Sigla from cowpea mosaic virus, the type mem-

ber. MULTICOMPONENT VIRUSES which have small isometric particles, 28 nm in diameter ($t = 1$) which sediment as three components. The bottom component contains RNA 1, the middle component contains RNA 2 and the top component lacks nucleic acid. The RNAs have 5′ VPg (a viral protein attached to the 5′ end of genomic RNA) of M_r 4 000 and are 3′ POLYADENYLATED, and are each translated to give a POLYPROTEIN which is subsequently cleaved (Fig. C60). The final products for RNA 1 are: P32 protein (a protease), P58 (unknown function), VPg, P24 (also has protease function) and P87 (involved in the polymerase). Those for RNA 2 are: P48/58 (implicated in cell-to-cell spread) and CP37 and CP23 (capsid proteins). *See also*: PLANT VIRUSES.

Francki, R.I.B. et al. (1985) In *Atlas of Plant Viruses*, Vol. 2, 1 (CRC Press, Boca Raton, FL).

compaction *See*: MORULA.

compartment A cell lineage compartment, a term introduced in the mid-1970s by García-Bellido, Lawrence, and their colleagues. The formal definition of a compartment is: 'The collection of all surviving descendants of a group of founder cells which are also confined to a spatial domain within well-defined boundaries'. Thus, a compartment is a POLYCLONE, or a group of clones where cells never mix with cells of neighbouring compartments. The boundaries between compartments may or may not correspond to the boundaries between visible structures, but are thought to be defined by limits of expression of particular genes, such as that of ENGRAILED in *Drosophila*. The boundaries between adjacent compartments define the spatial limits of MORPHOGENETIC FIELDS. During development, embryos become subdivided into an increasing number of compartments. *See also*: CELL LINEAGE; CLONAL RESTRICTION; DROSOPHILA DEVELOPMENT.

García-Bellido, A. et al. (1979) *Sci. Am.* **241**, 102–111.

compatible host–pathogen interaction In plant pathology, an interaction between a host and a pathogen that results in disease. The term is typically used to describe a combination of host and pathogen for which a gene-for-gene resistance–avirulence interaction exists. Compatibility can be due either to the lack of dominant resistance (*R*) genes in the host or to the presence of recessive avirulence (*a*) genes in the pathogen. The term 'compatible' should not be used in describing the pathogen itself, but should be reserved for the interaction with the host. *See also*: PLANT PATHOLOGY.

competence (1) A physiological state that can either arise naturally in some bacterial species, or be artificially induced in others, which results in an increased ability by the bacterial cell to take up exogenous DNA. Although the underlying cellular changes are not well defined, acquisition of competence appears to be associated with changes in the bacterial cell wall that occurs at a specific stage of growth of a culture. A fraction of a bacterial culture that becomes naturally competent and the duration of the so-called competent state are dependent on the species and the growth conditions used. Naturally occurring competence can be found in *Pneumococcus*, *Bacillus*, and *Haemophilus* species, while

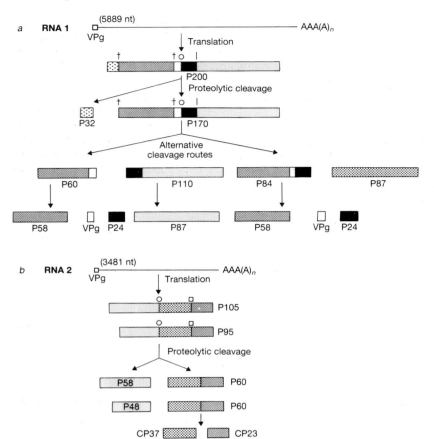

Fig. C60 Genome organization of cowpea mosaic virus (CPMV). The lines represent the RNA species and the boxes the proteins (P) showing the processing of the POLYPROTEINS; M_r given as $\times 10^{-3}$; nt, nucleotides; VPg, protein covalently linked to 5′ end of the nucleic acid.

for *Escherichia coli* a state analogous to the competent state can be artificially induced by incubating the cells in calcium chloride at 4°C. Once made competent, these cells can be stored at – 80°C in a cryoprotective agent such as glycerol or DMSO for many months. In this case the mechanism of DNA uptake is very different from that employed by naturally competent cells. Cells must be made competent before TRANSFORMATION with exogenous DNA (e.g. RECOMBINANT DNA).

(2) The ability of a cell or tissue to respond to a particular inducing signal (*see* INDUCTION). Competence is believed to depend largely on the expression of appropriate receptors specific for the inducing molecule(s) by the responding cells.

competitive In biochemistry and molecular biology, describes the relationship between two different molecules, generally the normal substrate and an inhibitor, where the actions of the inhibitor prevent the normal substrate binding to an active site. In many cases competitive inhibition is due to the two molecules being able to bind to the same binding site on an enzyme, receptor, etc., only one of the molecules being able to bind at any one time and thus excluding binding of the other. In enzyme reactions for example, competitive inhibitors are often those that compete with substrate for the substrate binding site on the enzyme. However, some competitive inhibitors of allosteric enzymes and other proteins may act by binding to sites other than the active site and by so doing alter the conformation of the active

site so that substrate binding is prevented. *See:* ALLOSTERIC EFFECTS.

Complement

COMPLEMENT is the collective term for a large number (about 20) of enzymes, proenzymes, and other proteins which form the principal effector mechanism of immunity in extracellular body fluids (especially blood plasma). Although antigen-nonspecific in itself, complement is the final effector mechanism of specific antibody-mediated immune responses (*see* ANTIBODIES) as well as being activated directly by some types of bacteria. In addition to the circulating complement components, there are a range of cell-membrane molecules which interact with the complement system, including COMPLEMENT RECEPTORS which have roles in phagocytic reactions and in the collection and transfer of immune complexes to phagocytes.

It is usual for the action of complement to be initiated by the binding of antibody (IgM or IgG) to antigen particles to form an immune complex, although many bacteria have surface properties which directly initiate complement activation. The consequences of complement activation are twofold: the more biologically important is the OPSONIZATION of particles so that they are more readily ingested by white blood cells. More obvious, much used in experimental systems, but less biologically important, is the formation of a molecular complex able to breach biological membranes and cause cells and bacteria to lyse. In addition, some of the activation fragments of the complement system have important pharmacological effects in mediating inflammation.

It is conventional to think of the complement system as three interlinking pathways, the classical, initiated via antibody, the alternative, a positive feedback pathway which may be independently initiated, and the terminal, involved in the formation of the lytic complex (*see also* C1–C9; FACTORS B, D, H, I, P; C1-INHIBITOR). Further reading on the complement system will be found in [1–3].

The nomenclature of the complement system is an undoubted but embedded barrier to its understanding. The components leading to lysis are prefixed with a 'C' and numbered in rough order of discovery, which is, unfortunately, not the order of action. The alternative pathway components are known as 'factors', as are some of the control proteins. There is a range of associated molecules with trivial or meaningful names or acronyms (Table C10). In general, individual polypeptide chains are given Greek letters and when complement components are cleaved the larger product is called the 'b' fragment, the smaller 'a' (but do not rely on it!).

The most interesting features of the complement system are the reaction pathways and their homeostasis, the range of specialist activities and physicochemical properties of the individual molecules, the clues to the evolution of this complex system from the structural elements which appear in the molecules and their genetical relationships, and its biological role in defence and disease.

Pathways

The classical pathway

This comprises the components C1, C4, C2, and C3 (*see* individual

Table C10 The components of the human complement system

Component	No. of polypeptide chains	No. of cistrons	Human chromosome	Polymorphism*	Deficiency*	Homologies
C1q	3	3	1p (b chain)	?	+	Procollagen
C1r	1	1	12p13.2	+	+	C1s, Haptoglobin. Serine proteinase (highly modified)
C1s	1	1	12p13.2	+		C1r, Haptoglobin. Serine proteinase (highly modified)
C4	3	1 but 2 genes	6 (MHC)	+ + +	+	C3, C5. α_2-Macroglobulin
C2	1	1	6 (MHC)	+	+	Factor B. Serine proteinase (highly modified)
C3	2	1	19p13	+ +	+	C4, C5. α_2-Macroglobulin
C5	2	1	9q	(+) (Melanesians)	+	C3, C4. α_2-Macroglobulin
C6	1	1	5p	+ +	+	C7, C8, C9 thrombospondin, LDL-r, EGF-p, SCR, factor I, perforin
C7	1	1	5p	+ (esp. in Orientals)	+	C6, C8, C9 thrombospondin, LDL-r, EGF-p, SCR, factor I, perforin
C8	3	3	1p (2 genes) and 9q	+	+ & +	C6, C7, C9 thrombospondin, LDL-r, EGF-p, perforin
C9	1	1	5p	−	+ + (Japanese)	C6, C7, C8 thrombospondin, LDL-r, EGF-p, perforin
Factor B	1	1	6 (MHC)	+ +	(+)	C2. Serine proteinase (highly modified)
Factor D	1	1	? (autosome)	(+) (Africans)	+	Serine proteinase. Same as adipsin?
Factor H	1	1	1q	+ (of length)	+	All SCR
Factor I	1	1	4	+ (Japanese)	+	Serine proteinase (highly modified)
Factor P	1	1	X	?	+	
C4b2			1q			SCR
DAF			1q		+	SCR
CR1			1q			SCR
CR2			1q			SCR
CR3 (α)			16			
C1-inhibitor	1	1	11	(+)	+ (dominant)	Serpin

* +, present; −, absent.

DAF, decay-accelerating factor; EGF-p, epidermal growth factor precursor; LDL-r, low density lipoprotein receptor; SCR, short consensus repeat.

entries) and is an enzyme cascade leading to the large-scale cleavage of C3 (Fig. C61*a*). C1 interacts with bound antibody via the C1q component, and this leads, through activation of C1r, to the generation of an active protease with fairly wide specificity — active C1s. C1s cleaves both C4 and C2, giving rise to the unstable complex protease C4b2, which, in turn, cleaves C3. A small pharmacologically active peptide is released from the N terminus of the α chain of C3, leaving the large active fragment C3b. This forms the pivotal point of the complement system, having the ability, due to an internal thioester bond, to form covalent complexes with local molecules (e.g. antigen). Bound C3b is recognized by cell receptors, especially on phagocytes, and this leads to increased phagocytosis (opsonization).

The alternative pathway

This is a positive feedback mechanism (Fig. C61*b*). The reaction mechanisms are analogous with those of the classical pathway, other than the initiating event. When C3b is generated, it can form a complex with factor B, and the latter is then cleaved (by factor D) to yield an active complex protease with specificity for C3. Thus more C3 is cleaved and the cycle amplifies itself until local depletion of reactants and homeostatic mechanisms bring it under control. The downregulating homeostatic mechanisms involve the dissociation of the C3bB complex and the subsequent cleavage of C3b to inactive fragments. Properdin (also known as factor P) binds and stabilizes the C3b–B complex on surfaces. Because C3 slowly spontaneously cleaves, it can, under favourable conditions, form a surface-bound stabilized complex with factor B, leading to an alternative pathway initiation event, independent of any other C3 convertase. This process is responsible for the direct activation of the alternative pathway by insoluble polysaccharides and endotoxin-bearing bacteria.

The terminal complement pathway

This comprises the components C5 to C9, which are named in reaction order. C5 is similar in many respects to C3 and C4 and is cleaved by either of the C3-cleaving enzymes C42 or C3bB, but additionally needs to be complexed with C3b. It then associates in turn with single molecules of C6, C7, and C8, which initiates the polymerization of C9 to form a membrane-inserting complex leading to lysis. The C567 complex has an affinity for cell (and other) surfaces. The formation of a transmembrane complex from soluble proteins is a notable and incompletely understood process.

Homeostasis

It will be apparent that the complexity and redundancy of forward reactions in the complement system are potentially hazardous to the animal in which they occur. There are a number of homeostatic mechanisms to control this.

The first is C1-INHIBITOR, a SERPIN protease inhibitor which acts stoichiometrically to bind active C1s and a number of other proteases belonging to the coagulation, fibrinolysis, and kinin systems (*see also* BLOOD COAGULATION AND ITS DISORDERS). It has no important influence on C1 bound to immune complexes, but it acts on any fluid phase C1s. There is no great excess of C1-inhibitor in plasma, and heterozygous deficiency leads to the dominantly inherited disease HEREDITARY ANGIO-OEDEMA.

Probably the most important homeostatic mechanisms are those that control the alternative pathway. These centre around the dissociation of the C3bB complex and the further cleavage of C3b to inactive fragments. Factor H competes for much the same site on C3b as factor B, and is an essential cofactor for factor I, a normally present and active SERINE PROTEINASE with apparent exquisite specificity for C3b (or C4b) complexed with factor H. There are a series of cleavages of the C3b β-chain, some also involving factor I and CR1 (a cellular complement receptor), while others can be performed by trypsin. The consequence is the removal of C3b from the system, the decay of factor Bb, but undiminished availability of both factor H and factor I. Hence, the size of the activation burst is limited by the availability of forward reactants, while the control components maintain their concentration. A somewhat similar system applies to the modulation of C4 activity, involving C4-binding protein, decay accelerating factor (DAF) and protein S, together with factor I.

The terminal complex is also subject to homeostasis. The plasma protein vitronectin (also called S protein) is a scavenger of terminal complexes which fail to reach their target, and prevents their involvement in bystander damage. Bystander or accidental damage to self cells is controlled principally by the cell-surface CD ANTIGEN CD59. This acts by interfering with the polymerization of the poly-C9. It is highly species specific, and probably largely explains why red cells are difficult to lyse with complement of the same species. It can be experimentally transferred between cells of different species, and confers protection on its host cells.

Evolution of the complement system

Many individual complement components belong to well-recognized families of proteins, some are a mosaic of structural

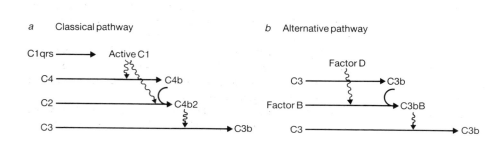

a   Classical pathway   *b*   Alternative pathway

Fig. C61 *a*, The classical pathway of complement fixation. *b*, The alternative pathway.

components common to other proteins, and many of the genes encoding them are found in clusters in the genome. The belief that an understanding of the genetics and the structure of the complement components would lead to a comprehensive understanding of the evolution of the system has been replaced, in the face of data, with a realization that what is now needed is a greater intelligence. Certain features of the evolution of the complement system are, however, evident:

1 Gene duplication (usually tandem) has had a major role.

2 Some structures are very commonly found: all the proteases belong to the serine protease family; the SHORT CONSENSUS REPEAT is common in all proteins which interact with C3 or its homologues C4 and C5; and there are common structures in the terminal pathway components.

3 The recruitment of features from completely unrelated systems is common.

4 Speculation about why a gene or cluster of genes are located where they are on 'philosophical' rather than 'evidential' grounds is foolish.

The data on structural features and locations are given in Table C10. We can reasonably surmise that C1r and C1s; the α, β, and γ chains of C1q; C2 and factor B; C3, C4 and C5; C6, C7, C8 (both α and β chains) and C9 all arose by duplication of genes which were part of the immediate predecessor(s) of the modern complement system, and in most cases the duplication was tandem. This implies that the predecessor was considerably simpler in a number of respects. Other proteins, particularly the serine proteases and their inhibitors, seem to have closer relations outside the complement system (allowing C1r/C1s and C2/factor B as combined cases) than within it.

It is now clear that the best explanation for the close linkage of C4, and C2/factor B within the MAJOR HISTOCOMPATIBILITY COMPLEX (MHC) III region is coincidence. Recent work has demonstrated that this region is packed with an assortment of genes, some of which are recognizable, others unknown, and most of them having no perceptible relationship with the immune system [4].

The complement system seems to be very rich in genetic diversity. Outstanding in humans is the colossal diversity of C4: the gene number varies between individuals and there are a large number of structural variants, making it the most polymorphic (expressed) system known after the MHC. Most of the other complement components have structural variants, some found only or predominantly in a particular racial group (e.g. C5, C7, factor D and factor I), others occurring in a BALANCED POLYMORPHISM in all populations (e.g. C6). These markers have been useful in genetics and forensic studies before the days of cloned genes and DNA TYPING, but no convincing claims for their evolution and maintenance have been put forward.

Deficiency states

One aspect of genetic diversity which is externally informative is the 'natural experiment' of deficiency. From inherited deficiencies and their associated diseases we can begin to confirm the biological function of the different parts of the complement system. What is clear is that the disease associations observed are not absolute: the complement and immune systems in their complexity provide a redundant mechanism for dealing with most infections and other insults. Deficiency of C3, either primary or secondary to defects in the regulation of the alternative pathway, has been associated with increased susceptibility to severe infection with bacteria (especially Gram-positive cocci), though it may also be virtually symptomless. Deficiency of the classical pathway components, including partial and secondary deficiency, seems to be associated with an increased susceptibility to immune complex diseases such as SYSTEMIC LUPUS ERYTHEMATOSIS (SLE), which is associated with C2, C4 and C5 deficiency. This may be because the complement system is involved in the solubilization of immune complexes and their transport to parts of the body where they are relatively unlikely to provoke an immune response (e.g. by breakdown in the liver). Deficiency of the alternative pathway has not been described (save for C3 deficiencies) and heterozygous deficiency of factor B. Deficiency of the terminal pathway is usually benign (C9 deficiency is relatively common in Japan), but has been associated with failure to kill bacteria of the genus *Neisseria*, leading to complications of gonorrhoea (caused by *N. gonorrhoeae*) and increased incidence of meningitis (caused by *N. meningitidis*).

It remains a surprise that such a complex system as complement should have evolved so apparently rapidly with the appearance of the vertebrates, and that it should have been maintained in such a recognizable form over the ensuing aeons, notwithstanding the often mild consequences of deficiency. Perhaps we shall, one day, understand better the mechanisms of evolution through the study of this 'luxury' system, with its great complexity and diversity.

M. HOBART

1 Reid, K.B.M. (1988) *Complement* (IRL Press, Oxford).
2 Morgan, B.P. (1990) *Complement — Clinical Aspects and Relevance to Disease* (Academic Press, London).
3 *Immunology Today* (1991) **12** (No. 9). A collection of brief reviews.
4 Kendall, E. et al. (1990) Human major histocompatibility complex contains a new cluster of genes between the HLA-D and complement C4 loci. *Nucleic Acids Res.* **18**, 7251–7257.

---

complement receptors Specific cell-surface receptors for COMPLEMENT components on a wide range of cells. At least nine types of receptor are thought to exist, of which CR1–4, which bind C3 already bound to a surface, are best characterized.

1 CR1 (C3b receptor, CD35) is a polymorphic transmembrane single-chain polypeptide (M_r 190 000–280 000) containing variable numbers of SHORT CONSENSUS REPEATS in the extracellular region. It is expressed on erythrocytes, where it is responsible for the clearance of circulating immune complexes, and other blood cells. It also stimulates the dissociation of classical and alternative pathway C3 convertases. Its deficiency is associated with the disease SYSTEMIC LUPUS ERYTHEMATOSUS.

2 CR2 (C3d receptor, CD21) is also a single-chain protein (M_r 145 000) with numerous short consensus repeats in the extracellular region. It is the receptor for the EPSTEIN–BARR VIRUS on B cells and nasopharyngeal epithelial cells.

3 CR3 (iC3b receptor, CD11bCD18) and CR4 (CD11cCD18) are INTEGRINS in which the β chain (CD18) is shared with LFA-1.

They are thought to be involved in the phagocytosis of particles coated with iC3b. A deficiency of CR3 leads to LEUKOCYTE ADHESION DEFICIENCY, which increases susceptibility to pyogenic infections.

Soluble complement components C3a, C4a, and C5a act at complement receptors on mast cells, smooth muscle cells, and endothelial cells, stimulating the release of histamine and other mediators of inflammation, muscle contraction, and an increase in vascular permeability.

complementarity determining regions (CDR) Sequences within the variable regions of immunoglobulin molecules that generate the antigen binding site (*see* ANTIBODY; IMMUNOGLOBULIN STRUCTURE). This binding site is complementary in shape and charge distribution to the EPITOPE recognized on the antigen. The CDRs are essentially synonymous with the HYPERVARIABLE REGIONS; however, CDR refers to a functional activity whereas the latter term refers only to a structural feature. There are three CDRs in each of the light and heavy chains which are separated from each other in the amino-acid sequence but are brought in close proximity to each other in the folded domain structures.

complementary DNA *See*: CDNA.

complementation The generation of the wild-type phenotype in an organism or cell containing two different mutations combined in a hybrid DIPLOID or a HETEROKARYON. Complementation shows that the mutations involved are nonallelic (i.e. do not occur in the same gene) and is the result of the provision of normal product from the corresponding wild-type allele on the other chromosome.

complementation group A collection of mutant ALLELES which, when examined for their ability to complement each other in pairwise combinations, fail to do so (*see* COMPLEMENTATION TEST), and must therefore be alleles of the same gene. This criterion was used to first define the basic genetic unit or cistron in bacteriophages and bacteria.

Benzer, S. (1959) *Proc. Natl. Acad. Sci. USA* **45**, 1607–1620.

complementation test Test used to determine whether two recessive mutations giving the same phenotype are in the same or different GENES. This is achieved by determining whether they complement each other and restore wild-type function if they lie in *trans* to each other in the DIPLOID state (Fig. C62). For example, when a pair of HOMOLOGOUS CHROMOSOMES carrying a mutation in gene A on one chromosome and a mutation in gene B on the other are combined in a diploid cell, one of the chromosomes will encode a functional B gene product, but a mutant (nonfunctional) A gene product, while the other chromosome encodes a functional A gene product, but a mutant (nonfunctional) B gene product. The resulting cell can therefore synthesize functional A and B gene products and complementation is said to have occurred. If both mutations had been in gene A, then such a cell would not be able to produce a functional A gene product. Complementation tests can be carried out in normally haploid

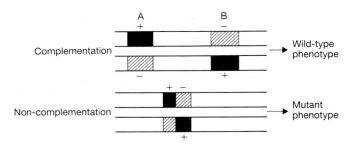

Fig. C62 Complementation. When mutant alleles at two different loci are in *trans*, complementation occurs as the cell has one functional allele of each type, but when the two mutations are in the same locus they fail to complement.

organisms such as bacteria by introducing the 'test' alleles on a plasmid or by TRANSDUCTION. In certain cases mutations at the same locus are able to complement each other (intragenic complementation). This is usually due to the functional gene product being composed of two or more subunits; the different mutations produce subunits with different defects which can nevertheless combine to produce a protein with some activity.

Fincham, J.R.S. (1966) *Genetic Complementation* (Benjamin, Menlo Park, CA).

complex I NADH-ubiquinone reductase. One of the three sites in the mitochondrial ELECTRON TRANSPORT CHAIN at which protons are transferred across the inner mitochondrial membrane.

complex II Succinate dehydrogenase, which catalyses the reduction of ubiquinase in the mitochondrial ELECTRON TRANSPORT CHAIN.

complex III The cytochrome bc_1 complex of the mitochondrial ELECTRON TRANSPORT CHAIN, one of the three sites at which protons are transferred across the inner mitochondrial membrane.

complex IV Cytochrome *c* oxidase, one of the three sites in the mitochondrial ELECTRON TRANSPORT CHAIN at which protons are transferred across the inner mitochondrial membrane.

complex number Mathematical entity which has both real and imaginary components: *cf*. real numbers and imaginary numbers (e.g. the square root of −1 usually denoted *i*). Complex quantities are used in many aspects of data analysis, for example DIFFRACTION METHODS.

complex oligosaccharide A type of N-linked oligosaccharide found on GLYCOPROTEINS. They are so named because of their relatively complex monosaccharide composition (*cf*. HIGH-MANNOSE OLIGOSACCHARIDE). Their common pentasaccharide core is elongated with two to five outer chains (antennae) that contain *N*-acetylglucosamine, *N*-acetylgalactosamine, fucose, galactose, neuraminic acid, or other less common monosaccharides, but not mannose. Complex oligosaccharides are the most highly processed and structurally diverse of the N-linked oligosaccharides. The largest complex oligosaccharides found on glycoproteins are the poly-*N*-acetyllactosaminoglycans, which have repeating *N*-acetyllactosamine (Galβ1,4GlcNAc) units.

complex structure factor This has both real and imaginary parts and is thus a COMPLEX NUMBER as distinct from a STRUCTURE FACTOR arising from a CENTRIC reflection with a set of diffraction data. See: X-RAY CRYSTALLOGRAPHY.

complexity The total length of unique sequence present within a nucleic acid sample as determined from RNA–DNA hybridization or DNA–DNA reannealing (*see* HYBRIDIZATION). *See also*: DNA COMPLEXITY.

composite transposon TRANSPOSABLE GENETIC ELEMENT composed of a structural gene (e.g. for antibiotic resistance) flanked by sequences such as insertion elements that mediate transposition.

compound heterozygote Organism that is heterozygous at more than one locus.

Con A Concanavalin A. LECTIN from the jack bean *Canavalia ensiformis*. See: PROTEIN STRUCTURE for illustration of structure.

concatemer Multiple copies of a GENOME covalently joined end to end in a tandem array. Many BACTERIOPHAGE genomes are in this form before packaging into empty viral heads. The concatemer is cleaved into genome-sized pieces during bacteriophage assembly (*see* COS SITE; LAMBDA).

concerted model The Monod–Wyman–Changeux model of cooperativity in proteins. *See*: ALLOSTERIC EFFECT; COOPERATIVITY.

concordance When a given phenotypic trait appears in both members of a pair of twins, they are said to be concordant for that trait.

conditional mutant A class of mutant carrying a mutation that affects an organism's PHENOTYPE only under defined restrictive conditions, and which has no phenotypic effects under permissive conditions. The most widely exploited examples of this type of mutant are the TEMPERATURE-SENSITIVE MUTANTS which express their mutant phenotypes at high (restrictive) temperatures, but not at a lower (permissive) temperature where an essentially WILD-TYPE phenotype is observed. Conditional mutants provide a means of analysing genes encoding products essential for cell growth and viability.

conditioned medium Tissue culture medium which has been used for the growth of cells. This medium probably contains small amounts of GROWTH FACTORS secreted by proliferating cells, and can support cell growth at a lower density than fresh medium.

conductance The inverse of electrical resistance or impedance. It is measured in siemens (S) and can be derived from Ohm's Law. Conductance in biological membranes is a measure of the ionic permeability of the membrane, and is mediated chiefly by ION CHANNELS. Membrane conductances fall into two broad categories: those gated by membrane voltage (*see*: VOLTAGE-GATED ION CHANNELS), and those activated directly by ligands binding to a cell-surface receptor (*see*: LIGAND-GATED ION CHANNELS).

Hille, B. (1992) *Ionic Channels of Excitable Membranes*, 2nd edn (Sinauer, Sunderland, MA).

cone pigments Opsin pigments found in the cones of the primate retina which are responsible for colour vision (and daylight vision generally). In humans there are three pigments, each comprising an opsin of slightly different sequence, encoded by a different gene, to which is covalently bound the chromophore 11-*cis*-retinal. Like RHODOPSIN, the corresponding photoreceptor pigment in retinal rod cells, the cone pigments are seven-span transmembrane proteins coupled to a G protein (*see* G PROTEIN-COUPLED RECEPTORS). The differences in spectral sensitivity of the three cone pigments (Fig. C63) both between each other and from rhodopsin are apparently determined entirely by the small differences in opsin sequence.

The cone pigments are encoded by gene clusters on the X chromosome. There are multiple copies of each gene and this can lead to unequal crossing-over and further duplication, deletion, or fusion of the various colour vision genes, resulting in defects in colour vision — 'colour blindness'. The existence of hybrid pigments of intermediate spectral sensitivity, produced by such a fusion gene, has been shown. The various forms of colour blindness can be explained by the deletion or fusion of particular genes. *See also*: VISUAL TRANSDUCTION.

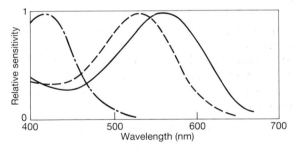

Fig. C63 The spectral sensitivities of the three normal colour pigments in the human eye.

confluent culture A cell culture in which adherent cells form a monolayer completely covering the surface of the substratum. The cell density at confluency is not necessarily the maximum density.

conformation of macromolecules See: DNA; GLYCANS; NUCLEIC ACID STRUCTURE; PROTEIN FOLDING; PROTEIN STRUCTURE; RNA.

congeneic (congenic) strains Strains of mice which have been bred to differ at only a single locus, and which are otherwise genetically identical, or nearly so. These strains are constructed by crossing two different strains of mice, one carrying the genetic background of interest, and the other carrying an allele, say A, of the locus which is to be transferred onto this background. Offspring of this cross are tested for the presence of allele A. Those carrying A are back-crossed to the background strain, and their progeny carrying A are again selected. After many (usually ~20) successive rounds of back-crossing and testing, a line of mice is obtained which is nearly identical to the background strain except at allele A. Congenic strains which differ only at the MAJOR HISTOCOMPATIBILITY COMPLEX (H-2) on chromosome 17 have been particularly useful for studies of the immune system.

congenital abnormality Any abnormality that is present in the newborn. It results from the interruption of normal foetal development by environmental or genetic factors.

congression (metakinesis) The movement of CHROMOSOMES to the METAPHASE PLATE that occurs during PROMETAPHASE. *See*: MITOSIS.

conjugation A form of bacterial 'mating' in which genetic material is transferred from one bacterial cell to another. Conjugation is mediated by plasmid-borne genetic determinants such as the F FACTOR and R FACTOR. In *Escherichia coli*, for example, cells carrying an F plasmid (F⁺ cells) bear long appendages — F pili — on their cell surface. When F⁺ cells come into contact with an *E. coli* cell lacking the F plasmid (an F⁻ cell) the F pili form a protoplasmic channel or conjugation tube between the two cells. The F⁺ cell will then donate a copy of the F plasmid to the F⁻ cell through the conjugation tube (Fig. C64). If the F factor has become integrated into the bacterial chromosome (Hfr strains), the chromosome itself will also be transferred. By stopping the transfer at different times and determining which genes have already been transferred, a genetic map of the bacterial chromosome can be constructed. R factor-mediated transfer of antibiotic-resistance genes on R plasmids within the Enterobacteriaceae has been important in the rapid spread of resistance to commonly used antibiotics.

connectin *See*: ACTIN-BINDING PROTEINS.

connective tissue Tissues constituting extracellular matrices and the cells within them, for example cartilage, bone, basement membrane. Cells include fibroblasts, macrophages, mast cells and other cells specialized for laying down matrices of various compositions, such as osteoblasts (in bone) and chondroblasts (in cartilage). The functions of the extracellular matrix are to support tissues and influence the development, polarity and behaviour of cells with which it is in contact. The diversity of its structure, which embraces bone, cartilage, teeth, cornea, tendons, and ligaments, is due to the versatility of its component molecules (*see* EXTRACELLULAR MATRIX MOLECULES). Any matrix will consist of one or more of the major fibrous proteins (collagen, elastin, and fibronectin, which respectively confer strength, confer elasticity, and mediate adhesion) within a resilient hydrated gel formed by GLYCOSAMINOGLYCANS and PROTEOGLYCANS (*see* GLYCANS). *See also*: CAENORHABDITIS NEURAL DEVELOPMENT; CELL–MATRIX INTERACTIONS; INSECT NEURAL DEVELOPMENT; VERTEBRATE NEURAL DEVELOPMENT.

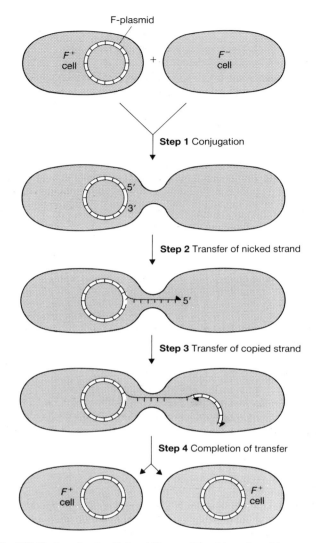

F-plasmid

F⁺ cell + F⁻ cell

Step 1 Conjugation

5′
3′

Step 2 Transfer of nicked strand

5′

Step 3 Transfer of copied strand

Step 4 Completion of transfer

F⁺ cell   F⁺ cell

Fig. C64 The transfer of an F plasmid from an F⁺ cell to an F⁻ cell by bacterial conjugation.

connexins Protein subunits of the connexons of gap junctions. *See*: CELL JUNCTIONS.

connexon Hexagonal arrangement of proteins (CONNEXINS) which constitutes the gap junction. *See*: CELL JUNCTIONS.

Connolly surface representation This shows a smoothed molecular surface as distinct from the actual van der Waals surface which has many cusps quite inaccessible to any atom. The smoothing process considers the path followed by a small sphere (having the size of a water molecule) rolling over the molecular surface. The result is a more pleasing representation of a molecule than its actual surface. The name derives from the author of a computer program used widely to generate these molecular pictures.

Connolly, M.L. (1983) *Science* **221**, 709–713.

consensus sequence When a large number of similar DNA sequences are compared, the consensus sequence represents the bases most often found at each position. For example, the consensus DNA sequence located approximately 10 bp UPSTREAM of the RNA start site in *Escherichia coli* transcriptional PROMOTERS is TATAAT although not all *E. coli* promoter sequences conform exactly to this sequence. Consensus sequences in DNA often define protein-binding sites, particularly when they are located within promoter regions.

conservative substitution The replacement of one amino acid in a polypeptide with another having similar properties. This kind of alteration is unlikely to change the structure or function of the resulting protein. *See also*: MOLECULAR EVOLUTION; MOLECULAR EVOLUTION: SEQUENCES AND STRUCTURES; MUTATION.

conservative transposition Type of transposition in which a copy of the transposon is left at the original site. *See*: TRANSPOSABLE GENETIC ELEMENTS.

conserve To preserve or retain a common feature. Conserved elements are sequences or structures that are retained among different organisms or molecules.

consistency *See*: MOLECULAR PHYLOGENY.

constant region (C region) Region of ANTIBODY or T CELL RECEPTOR of relatively constant sequence and which, in antibodies, is responsible for the effector functions of the molecule. *See also*: COMPLEMENT; IGA; IGE; IGG; IGM; IMMUNOGLOBULIN GENES; T CELL RECEPTOR GENES.

constant region genes (C genes) Gene segments encoding the invariant C-terminal domains of immunoglobulin and T cell receptor (TCR) molecules. The immunoglobulin light chain and TCR loci each contain a C gene of a single type although the J_λ–C_λ locus is present in multiple repeats and the TCR J_β–C_β region is duplicated in tandem. In contrast, the immunoglobulin heavy chain locus contains five different constant region genes (isotypes), μ, δ, γ, ϵ, and α. *See*: IMMUNOGLOBULIN GENES; T CELL RECEPTOR GENES.

constitutive Refers to an activity such as gene expression or enzyme activity that is present continually in a cell and does not appear to be subject to quantitative regulation, that is, constitutive genes are always 'turned on'. *Cf.* INDUCIBLE.

constitutive heterochromatin CHROMATIN that is permanently packaged in an inactive state (e.g. the inactivated X-chromosome in mammals, or regions around the centromere). *See also*: CHROMOSOME STRUCTURE.

constraint A fixed value of a quantity (e.g. bond length or angle) with which a structural model of a molecule is forced to agree. *Cf.* RESTRAINT, an ideal value to which a model is restrained to be compatible, and refined to approach more closely. *See*: MOLECULAR DYNAMICS; REFINEMENT OF MACROMOLECULAR STRUCTURES.

contact guidance The concept that directed cell or axon migration may result from the physical, as distinct from chemical, properties of the environment, for example by the presence of tissue barriers or channels. Paul Weiss (1934) showed that growing axons follow the orientation of the stress fibres of a plasma clot; similarly, muscle cells align with the stress folds created by fibroblasts cultured on a collagen substrate. Contact guidance has been suggested to be important in the guidance of axons along intercellular channels in the developing central nervous system, but its role *in vivo* remains unclear. *See*: CAENORHABDITIS NEURAL DEVELOPMENT; INSECT NEURAL DEVELOPMENT.

contact sites A and B *See*: DICTYOSTELIUM DEVELOPMENT.

contactin *See*: CELL ADHESION MOLECULES; IMMUNOGLOBULIN SUPERFAMILY.

contig A continuous sequence of DNA that has been assembled from overlapping cloned DNA fragments. The term is also applied to the section of chromosome bearing the contiguous genes involved in CONTIGUOUS GENE SYNDROMES.

Contiguous gene syndromes

THE name contiguous gene syndrome was coined by Schmickel [1] to describe an important set of clinical syndromes which do not fit any single pathogenic classification, but which by CYTOGENETIC analysis are associated with deletion of specific chromosomal segments, and by implication, a contiguous set of genes. They are thus distinct from conditions typically associated with chromosomal ANEUPLOIDY (*see* CHROMOSOME ABERRATIONS), such as DOWN'S SYNDROME or TURNER SYNDROME, or single-gene defects with pleiotropic effects, such as HURLER'S DISEASE or EHLERS–DANLOS SYNDROME. Contiguous gene syndromes show non-Mendelian inheritance, are typically sporadic, and familial inheritance is exceptional. Mental retardation is a common component of the constellation of clinical symptoms. Implicit in the term is the notion that by recognizing the condition as the consequence of chromosomal deletion, cytogenetic methods can be used to identify the map location and then molecular genetic methods applied to the isolation of each of the causative genes. Thus, this analytical approach to clinical genetics corresponds closely to the strategy of experimental DELETION MUTAGENESIS which has been vigorously applied to the genetics of the fruit fly *Drosophila* [2,3] and which in the mouse has been exploited for functional mapping of developmental genes, for example, those flanking the albino (c) locus [4].

Perhaps the simplest form of a contiguous gene syndrome in humans would be the subset of THALASSAEMIAS that are due not to point mutation or intragenic deletion, but to the partial or complete deletion of one or more genes within the β-globin gene complex on the short arm of human chromosome 11, or the α-globin complex on chromosome 16 [1]. Individuals have also been described with the HbH form of α-thalassaemia (*see* HERED-

ITARY PERSISTENCE OF FOETAL HAEMOGLOBIN) together with mild skeletal abnormalities and mental retardation and shown to be COMPOUND HETEROZYGOTES with subvisible deletion of the 16p13.3 region. Molecular analysis of these thalassaemias has only been possible because of the knowledge that the primary defect is in the globin gene cluster. In most cases, the molecular basis of the contiguous gene syndrome is completely unknown. Their selection for detailed cytogenetic and molecular genetic analysis is driven by the precept that POSITIONAL CLONING strategies (or 'reverse genetics') can reveal the biochemical basis of the pathology.

As the genetic aetiology is typically sporadic, independent affected individuals are all but certain to carry deletions with different end-points. Thus, a second hallmark of the contiguous gene syndromes (which distinguishes them from pleiotropic effects of single-gene disorders or aneuploidies) is that each affected individual displays a distinct but nonrandom spectrum of clinical symptoms. The observed spectra reflect the gene order within the deletion complex, centrally placed genes being deleted more consistently than proximal or distal genes. A caveat to this (which applies equally to single-gene disorders with pleiotropic effects, or aneuploidies) is that different genes within the contig, as the region involved is known, may vary with respect to their DOMINANCE, PENETRANCE, and age of onset, with genetic background and environmental factors also influencing overall expression.

It follows that a contiguous gene syndrome may present as a subspectrum of the pathology associated with chromosomal aneuploidy. Indeed, careful cytogenetic examination of atypical individuals may identify a subchromosomal deletion or rearrangement from which a critical region for major (or minor) components of the syndrome can be defined. This is the case for Down's syndrome, for which a well-defined constellation of clinical features typify the effect of trisomy 21. However, a few rare instances have been described associated with the salient features of Down's syndrome but lacking other characteristics. Cytogenetic and molecular genetic characterization of these anomalous patients suggests that the critical region involved is 21q22-qter [5].

The power of deletion mapping for gene isolation by positional cloning is now well recognized and emphasizes the potential value of tissue and cell lines from patients with contiguous gene syndromes for cytogenetic and molecular genetic analysis. Under favourable circumstances, the skilled cytogeneticist can detect changes in CHROMOSOME BANDING pattern resulting from interstitial deletions of the order of 5 million base pairs. Molecular markers known to map in the affected region may be available for immediate testing. If not, SOMATIC CELL HYBRID genetic techniques for enrichment cloning by chromosome-mediated gene transfer or radiation fusion can be used to generate new markers (*see* HUMAN GENE MAPPING). Alternatively, the corresponding region of a normal chromosome can be microdissected and cloned, or libraries made from FACS-sorted chromosomes screened exhaustively for deletion clones. An ordered map of the region can then be constructed by deletion mapping of independent patients (also with respect to other fortuitous chromosome breakpoints in any available somatic cell hybrids), by construction of a genetic map using polymorphic markers in family studies,

and by physical linkage of markers to large DNA restriction fragments resolved following pulsed-field gel ELECTROPHORESIS (PFGE) (*see* HUMAN GENE MAPPING). Candidate genes from the smallest region of overlap (SRO) can then be isolated by one or more of several strategies, including large DNA fragment cloning in YEAST ARTIFICIAL CHROMOSOME (YAC) vectors, CpG island cloning (*see* DNA METHYLATION), identification of conserved sequences by cross-hybridization to other species (a 'zoo blot'), exon trapping, or direct screening of cDNA libraries (*see* DNA LIBRARY) with genomic fragments (which may be whole YAC recombinants). Gene expression studies and DNA sequence analysis will help in assessing candidate genes. Ultimately, verification depends on the identification of independent, intragenic deletions, point mutations, and/or translocation breaks predicted to disrupt gene function, or functional COMPLEMENTATION or mutational rescue in appropriate culture systems or TRANSGENIC animals.

Syndromes

There is a temptation to label all clinical conditions associated with chromosomal rearrangements as contiguous gene syndromes, for example Langer–Giedion syndrome (del 8q24), Di George syndrome (del 22q11), Miller–Dieker syndrome (del 17p13), Beckwith–Weidemann syndrome (dup 11p15) and Prader–Willi syndrome (del 15q11) [1,6,7]. However, although the chromosomal aberrations characterizing different patients can undoubtedly be used to define the SRO in more precise molecular terms, until clinical evidence or experimentation proves otherwise, the formal possibility remains that the clinical pathology can be explained by the structural mutation and/or aberrant expression of a single gene. However, there are good examples of syndromes where an individual patient is affected by a set of normally independent clinical conditions, or where different patients have partially overlapping deletions and a corresponding partial overlap in pathology — the true hallmark of a contiguous gene syndrome.

Contiguous gene syndromes have been instrumental in providing clues to the location of several disease genes. A male with four normally independent X-linked disorders, CHRONIC GRANULOMATOUS DISEASE (CGD), DUCHENNE MUSCULAR DYSTROPHY (DMD), RETINITIS PIGMENTOSA and McLeod syndrome, was shown to have a barely visible cytogenetic deletion of Xp21 [8]. Because there was a complete deletion of Xp21 material from the single male X chromosome, it was possible to derive markers for the deletion region by a phenol-enhanced reassociation technique (PERT) for subtractive hybridization cloning and this led to the identification of the CGD and DMD genes [9].

Similarly, short stature, chondrodysplasia punctata, steroid sulphatase deficiency, and KALLMANN SYNDROME are X-linked disorders which have been found as isolated entities, or in various combinations associated with interstitial or terminal deletions of the short arm of the X chromosome [10]. By defining the patients in molecular terms, it was possible to deduce the trait order on the chromosome, and map the component genes with varying precision. The gene for Kallmann syndrome was pinpointed to a region of a few hundred kilobase pairs. The fact that Kallmann

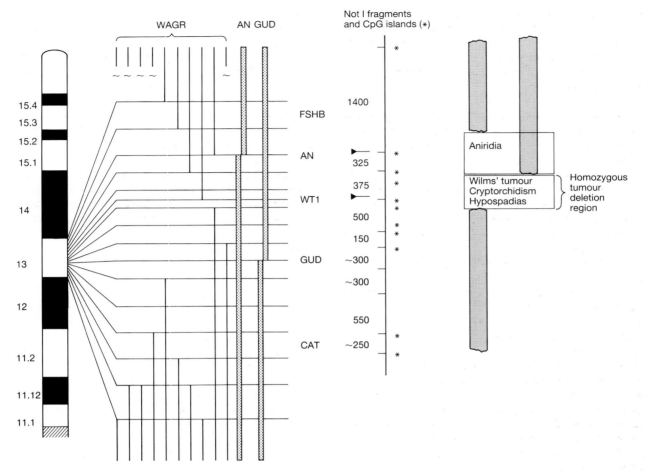

Fig. C65 A molecular and morbid anatomy of the WAGR region of human chromosome 11p13. Left, banding pattern of human chromosome 11 short arm; centre, expanded view of band 11p13 dissected by balanced translocation chromosomes (hatched) associated with genitourinary dysplasia (GUD) and aniridia (AN), and deletion chromosomes (solid lines, deletion intervals not to scale) from several independent WAGR patients; right, a long-range restriction map, with approximate distances shown in

kilobases and the positions of CpG islands marked with asterisks. The positions of the *WT1* and *AN* genes are shown. The smallest region of overlap for Wilms' tumour, cryptorchidism, and hypospadias as defined by the two WAGR deletion chromosomes and shown on the far right (broad hatching) coincides with the homozygous tumour deletion region by which the *WT1* gene was positionally cloned. Adapted from [13].

syndrome (which is associated with a lack of a sense of smell and defects in gonadotropin production) can also be inherited as a Mendelian trait strongly favours a unitary genetic explanation for the complex phenotype. Positional cloning, DNA sequencing, and expression analysis of a candidate gene for the X-linked form suggest that the gene encodes a protein related to the neural cell adhesion molecule NCAM (*see* CELL ADHESION MOLECULES) which controls neural cell migration, consistent with the notion that olfactory neurons and neurons synthesizing gonadotropin-releasing hormone share a common developmental pathway.

WAGR syndrome

WAGR syndrome, the association of WILMS' TUMOUR (a childhood nephroblastoma) with *A*niridia (failure to form an iris), a variety of *G*enitourinary abnormalities, and mental *R*etardation [11], is a well-studied and instructive example of a contiguous gene syn-

drome. Aniridia is usually inherited as an autosomal dominant trait showing complete penetrance and HAPLO-INSUFFICIENCY, but amongst sporadic cases, one-third also have Wilms' tumour. In contrast, Wilms' tumour shows low penetrance, is typically sporadic, but rare cases of familial Wilms' are inherited as an autosomal recessive. Two per cent of sporadic Wilms' tumour patients also have aniridia. Cytogenetic analysis of WAGR patients identified the short arm of chromosome 11, band 11p13, as the consistently deleted region [12]. WAGR patients often show intermediate levels of the enzyme catalase (CAT) and/or HEMIZY-GOUS dosage for the gene for follicle-stimulating hormone β subunit (*FSHB*). Analysis of independent deletion patients, with Wilms' tumour (*WT1*) but not aniridia (*AN*), and vice versa, established the following gene order:

centromere-*CAT-WT1-AN-FSHB*-telomere

The morbid map was given further precision by (1) isolation of many new molecular markers by exhaustive screening of FACS-sorted libraries, enrichment cloning from chromosome transfer and fragmentation hybrids selected for retention of markers on the short arm of chromosome 11; (2) construction of a long-range RESTRICTION MAP for the region by PFGE which identified the CpG islands likely to mark the 5′ ends of genes; (3) identification of patients with Potter facies, a form of genitourinary dysplasia (GUD), or with aniridia associated with distinct 11p13 chromosome translocations; and (4) identification of a Wilms' tumour patient with a microdeletion involving a single CpG island (Fig. C65). In this manner, candidate genes for *WT1* and *AN* have been isolated and analysed [13].

Correlation of genotype with phenotype suggests that constitutional mutation of the *WT1* gene alone (a 'zinc finger' gene encoding a putative TRANSCRIPTION FACTOR) can account for a wide variety of genitourinary abnormalities, including cryptorchidism (undescended testes) and hypospadias (urethral opening on underside of penis) and Denys–Drash syndrome, the rare association of Wilms' tumour with pseudohermaphroditism and progressive renal failure [13]. Recent comparative genetic studies show that the human homologue of the *Pax-6* gene (containing a PAIRED BOX and HOMEOBOX domain and likely to encode a transcription factor), which causes the small-eye phenotype in the mouse [14], maps close to *WT1* in band 11p13 in humans and is consistently deleted or rearranged in aniridia patients, and is therefore confirmed as an aniridia gene [15]. Thus, WAGR syndrome exhibits all of the characteristics of a contiguous gene syndrome and the molecular analysis of this condition illustrates the power of the concept for reaching a molecular understanding of complex clinical disorders, developmental defects and cancer.

D.J. PORTEOUS

See also: TUMOUR SUPPRESSOR GENES.

1 Schmickel, R.D. (1986) Contiguous gene syndromes: a component of recognizable syndromes. *J. Pediatr.* **109**, 231–241.
2 Lindsley, D.L. et al. (1972) Segmental aneuploidy and the genetic gross structure of the Drosophila genome. *Genetics* **71**, 157–184.
3 Cooley, L. et al. (1990) Constructing deletions with defined endpoints in Drosophila. *Proc. Natl. Acad. Sci. USA* **87**, 3170–3173.
4 Russell, L.B. et al. (1982) Analysis of the albino-locus region of the mouse: IV. characterization of 34 deficiencies. *Genetics* **100**, 427–453.
5 McCormick, M.K. et al. (1989) Molecular genetic approach to the characterization of the 'Down syndrome region' of chromosome 21. *Genomics* **5**, 325–331.
6 Emanuel, B.S. (1988) Molecular cytogenetics: towards dissection of the contiguous gene syndromes. *Am. J. Hum. Genet.* **43**, 575–578.
7 Ledbetter, D.H. & Cavenee, W.K. (1989) Molecular cytogenetics: interface of cytogenetics and monogenic disorders. In *The Metabolic Basis of Inherited Disease*, 6th edn (Scriver, C.R. et al., Eds) (McGraw-Hill, New York).
8 Francke, U. et al. (1985) Minor Xp21 chromosome deletion in a male associated with expression of Duchenne Muscular Dystrophy, chronic granulomatous disease, retinitis pigmentosa, and McLeod syndrome. *Am. J. Hum. Genet.* **37**, 250–267.
9 Orkin, S.H. (1986) Reverse genetics and human disease. *Cell* **47**, 845–850.
10 Ballabio, A. et al. (1989) Contiguous gene syndromes due to deletions in the distal short arm of the human X chromosome. *Proc. Natl. Acad. Sci. USA* **86**, 10001–10005.
11 Miller, R.W. et al. (1964) Association of Wilms' tumour with aniridia, hemihypertrophy and other congenital malformations. *N. Engl. J. Med.* **270**, 922–927.
12 Riccardi, V.M. et al. (1978) Chromosome imbalance in the aniridia–Wilms' tumour association; 11p interstitial deletion. *Pediatrics* **61**, 604–610.
13 van Heyningen, V. & Hastie, N.D. (1992) Wilms' tumour: reconciling genetics and biology. *Trends Genet.* **8**, 16–21.
14 Hill, R.E. et al. (1991) Mouse Small eye results from mutations in a paired-like homeobox-containing gene. *Nature* **354**, 522–525.
15 Jordan, T. et al. (1992) The human PAX6 gene is mutated in two patients with aniridia. *Nature Genet.* **1**, 328–332.

continuous variation Variation between individuals in a population in which differences are quantitative rather than qualitative and produce a continuous spectrum of variation over the whole population. Continuously variable traits (e.g. height in humans) in which some hereditary component is present, are presumed to be due to the actions of many different genes and also usually to be subject to considerable environmental influence. *See also*: MENDELIAN INHERITANCE; POLYGENIC INHERITANCE.

contour length The length of a DNA or RNA molecule, measured as for a fully extended (relaxed) molecule.

contractile ring A belt-like bundle of ACTIN FILAMENTS and MYOSIN which appears beneath the plasma membrane during division of animal cells. The constriction caused by this ring leads to the separation of the dividing cell into two daughter cells. *See*: CYTOKINESIS; MICROFILAMENTS.

contrast scattering-length density In a neutron SMALL-ANGLE SCATTERING experiment in solution, the difference between the MEAN NEUTRON SCATTERING-LENGTH DENSITY of a macromolecule ($\rho_{molmean}$) in that solvent and the solvent scattering-length density.

contrast variation A technique used in NEUTRON SCATTERING AND DIFFRACTION when a molecule is studied in a range of different solvents whose contrasting SCATTERING-LENGTH DENSITY varies relative to that of the molecule of interest. The most common example utilizes the contrast between the scattering densities of water, H_2O, and heavy water, D_2O. The scattering of proteins and nucleic acids is matched by that of a solvent containing 40% and 65% D_2O respectively. Thus, by studying the scattering of a protein–nucleic acid complex in solvents containing 0%, 40%, 65%, and 100% D_2O, the structure of the two components within the complex can be distinguished.

controlling elements *See*: TRANSPOSABLE GENETIC ELEMENTS: PLANTS.

convergence *See*: MOLECULAR PHYLOGENY.

convergent evolution Similarity between two structures, protein molecules or nucleic acid sequences due to independent evolution from different origins rather than the possession of a common ancestor.

Coomassie blue™ A blue-coloured dye used to detect proteins on polyacrylamide gel ELECTROPHORESIS (Fig. C66).

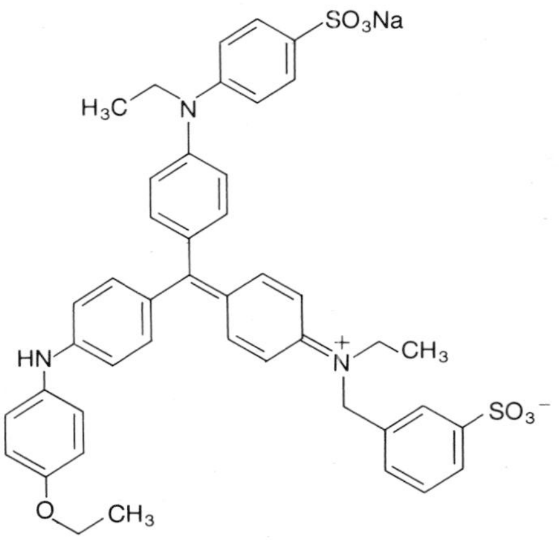

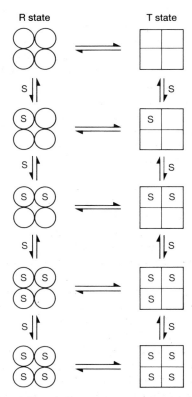

Fig. C66 The structure of Coomassie blue.

cooperativity (1) The phenomenon seen in many oligomeric proteins with multiple binding sites for a substrate or other ligand where the binding of one molecule of ligand influences the binding of subsequent molecules of the same ligand. Cooperativity is a particular case of an ALLOSTERIC EFFECT. Positive cooperativity describes cases where binding of ligand at one site increases the affinity of the other sites for the ligand, as in the oxygen binding of mammalian HAEMOGLOBIN; negative cooperativity describes cases where binding of ligand to one site decreases the affinity of the other sites for the ligand, as in the binding of tyrosine to one of the two tyrosine-binding sites in the enzyme tyrosyl-tRNA synthetase.

The most influential model for cooperative interactions is the Monod–Wyman–Changeux (MWC) concerted model which was first formulated to explain the cooperative oxygen binding of haemoglobin. It proposes that an oligomeric protein showing cooperative interactions between ligand-binding sites on different subunits exists as an equilibrium of two conformational states, the R-state, and the T-state. This nomenclature came from haemoglobin where the high affinity state is less constrained (relaxed) and the low affinity state is more constrained (tense). These descriptions are not appropriate for all allosteric systems but for consistency the R-state is the active, high affinity form and the T-state the inhibited form. Thus the T state is the predominant form when no ligand is bound and has a lower affinity for ligand than the R-state. The R- and T-states differ in the number and energies of the bonds between subunits (*see* HAEMOGLOBINS). In this model a partly ligated molecule is assumed to be in either the R- or the T-state, so that mixed states do not occur (Fig. C67). The model can be generalized to cover other allosteric proteins given the following assumptions:

1 The protein is an oligomer, made up of two or more identical 'protomers' and possesses at least one axis of symmetry. (The protomer is the smallest functional subunit of the protein, containing one specific binding site for each ligand. In homo-oligomeric proteins the protomer could be a single subunit, as in the lac repressor; in hetero-oligomeric proteins, it would generally represent one each of the different types of subunit.)

2 In each state, binding sites within the molecule for any particular ligand are all equivalent.

Fig. C67 The Monod–Wyman–Changeux concerted model of cooperativity.

3 The conformation of each protomer is constrained by its association with the other protomers. If the binding of a ligand to one protomer induces a conformational change in that protomer it will also cause an identical conformational change in all the other protomers.

The concerted model has been successful in interpreting structural and kinetic studies of many allosteric proteins. It cannot however accommodate the behaviour of proteins that show negative cooperativity.

The other influential model for cooperative interactions, the Koshland–Némethy–Filmer (KNF) sequential model, differs from the concerted model in that it assumes that progress from the T-(unliganded) to the R-state is sequential (Fig. C68), with the conformation of each subunit changing in turn as it binds ligand. This change may be transmitted to neighbouring 'empty' subunits via changes in tertiary structure at the subunit interfaces. In this model there will be as many different conformational states as there are subunits/binding sites. This model seems to fit the behaviour of some proteins better than the concerted model and also accommodates negative cooperativity. *See also*: HILL COEFFICIENT.

(2) In respect of LONG-TERM POTENTIATION the simultaneous activation of sufficient presynaptic afferent fibres to provide enough transmitter to depolarize the postsynaptic cell in order to allow sufficient NMDA receptor activation to induce LTP.

Perutz, M. (1990) *Mechanisms of Cooperativity and Allosteric Regulation in Proteins* (Cambridge University Press, Cambridge).
Evans, P.R. (1991) Structural aspects of allostery. *Curr. Opinion Struct. Biol.* **1**, 773–779.

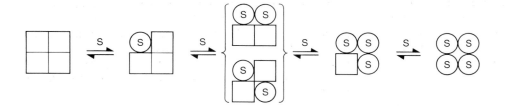

Fig. C68 The Koshland–Némethy–Filmer sequential model of cooperativity.

copia A TRANSPOSABLE GENETIC ELEMENT found in the genome of the fruit fly *Drosophila melanogaster* and which is the prototype of a family of several related transposable elements, the so-called copia-like elements. *Copia* itself is present in 20–60 copies per genome depending on the *D. melanogaster* strain, although *copia* and *copia*-like elements account for almost 1% of the total *D. melanogaster* genome. *Copia* is 500 bp long with identical direct terminal repeats of 276 bp and encodes an abundant mRNA. Like many other transposable elements, *copia* has a number of properties in common with RETROVIRUSES of vertebrates.

Mount, S.M. & Rubin, G.N. (1985) *Mol. Cell. Biol.* **5**, 1630–1638.

COPs COat Proteins found on Golgi transport vesicles (*see* COATED PITS AND VESICLES; GOLGI APPARATUS) and which are thought to be involved in the budding of non-clathrin-coated transport vesicles. There are four types of coat protein found on these vesicles: α (M_r 160K), β (110K), γ (98K) and δ (61K). β-COP shows homology to the clathrin-coated vesicle protein B-ADAPTIN.

Serafini, T. et al. (1991) *Nature* **349**, 215–220.

copy choice A model, originally proposed by John Belling in 1928, to explain the mechanism of CROSSING-OVER between homologous chromosomes in MEIOSIS in plants. He suggested that crossing-over occurs during chromosome duplication, with the crossing-over being restricted to the two supposedly new nonsister CHROMATIDS. For this model to be accepted one must assume that chromosome replication is conservative; we now know it is not (*see* DNA REPLICATION). Belling's model would predict that molecules of DNA in the process of being synthesized can switch from using the DNA of one homologous chromosome as template to using DNA of the other homologous chromosome as template, often at a different relative position (Fig. C69). The breakage and reunion model of crossing-over is currently the preferred model except in a few instances (*see* RECOMBINATION).

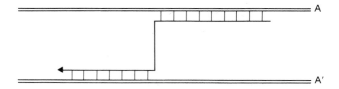

Fig. C69 The copy choice model of crossing-over. A and A′ are the template strands of the two respective homologous chromosomes.

copy number The average number of copies of a particular DNA sequence in a genome or of a DNA molecule (such as a PLASMID) in a single cell. Most often used in the context of plasmid copy number which indicates the average number of plasmid molecules per bacterial or yeast cell. *See also*: CHLORAMPHENICOL AMPLIFICATION.

co-receptor *See*: ANTIGEN RECOGNITION; CD4 ANTIGEN; CD8 ANTIGEN; T CELL RECEPTOR.

corepressor A small molecule (e.g. an amino acid), which facilitates the tight binding of a REPRESSOR to an OPERATOR sequence within a bacterial PROMOTER. For example, the *trp* repressor protein of *Escherichia coli* will only bind tightly to the *trp* operon operator if complexed with tryptophan, thereby ensuring that when there is an abundance of tryptophan in the cell, no synthesis of the tryptophan synthesizing enzymes encoded by the *trp* operon occurs. *See also*: BACTERIAL GENE EXPRESSION.

corona radiata An extracellular layer around the eggs of some mammalian species, lying between the ZONA PELLUCIDA and cumulus. In these cases the corona radiata refers to the follicle cells immediately adjacent to the zona pellucida, and the cumulus refers to cells lying further away from the egg and zona pellucida. Cells of the corona radiata nourish the growing oocyte through protoplasmic extensions which penetrate the zona pellucida. In humans during ovulation, the cumulus cells become loosened from one another by accumulation of fluid from the intercellular space. This enables them to separate easily, along with the enveloped ovum, from the follicle wall. Following ovulation, the cumulus cells become the corona radiata.

coronatine *See*: PLANT PATHOLOGY.

Coronaviridae A family of enveloped RNA viruses with pleomorphic particles 75–160 nm with the characteristic club-shaped projections on the surface. The genome consists of a single-stranded (+)-sense RNA (M_r 5-6 × 10⁶) with a poly(A) tail at the 3′ end. During replication a nested set of single-stranded RNAs is produced as well as the virion RNA. Coronaviruses are known to cause disease in humans, other mammals, and birds.

coronin *See*: ACTIN-BINDING PROTEINS.

cortical contractions The periodic waves of contraction that move over the surface of the egg at FERTILIZATION and before each CLEAVAGE. They are best studied in amphibia. Cortical contractions are also observed at the appropriate times in nucleated and anucleate fragments of activated eggs. The phenomenon has led to the concept of an autonomous cortical contractile programme that cooperates with nuclear components to generate the mitotic cycle.

cortical layer Layer of cytoplasm rich in MICROFILAMENTS immediately under the plasma membrane in eukaryotic cells. *See also*: CORTICAL REACTION.

cortical reaction Fusion of cortical granule membranes with the plasma membrane of the egg and release of cortical granule contents, which occurs on FERTILIZATION. The granules are membrane-bound vesicles which lie under the egg plasma membrane. Their contents include proteases and acidic glycoproteins. The fusion process is triggered by a sperm-induced rise (large and transient) in free CALCIUM in the cytoplasm. As released proteases and other enzymes alter egg coats and prevent entry of supernumerary sperm, this reaction is an important block to POLYSPERMY. In the sea urchin the reaction has two aspects: proteolytic enzymes from the granules destroy sperm receptors but also cause a delamination and hardening of an external egg coat (vitelline membrane) to form an impenetrable fertilization membrane. In mammalian eggs, the cortical granule reaction inactivates glycoproteins of the ZONA PELLUCIDA which are responsible for binding sperm and activating the ACROSOME REACTION.

corticotropin ADRENOCORTICOTROPIN.

Corticoviridae *See*: BACTERIOPHAGES.

cortisol 17α-Hydroxycorticosterone, the major glucocorticoid hormone synthesized by the adrenal cortex (*see* STEROID HORMONES). It has many and diverse effects on the metabolism of carbohydrates, proteins, and lipids. It is transported in the blood bound to a specific corticosteroid-binding protein. Cortisol and some other corticosteroids such as cortisone have immunosuppressive and anti-inflammatory activity and are used therapeutically. They are thought to act chiefly by blocking the release of CYTOKINES from activated mononuclear phagocytes and mast cells, and possibly also by the direct lysis of immature T cells. *See also*: STEROID HORMONE RECEPTORS.

COS cell A CELL LINE derived from the monkey CV-1 cell line which has DNA of the virus SV40 integrated into its genome as a result of TRANSFORMATION with a nonreplicative derivative of the virus. This cell line constitutively synthesizes the SV40-encoded T antigen and can be used to maintain plasmids carrying the SV40 origin of DNA replication at a high copy number. *See also*: PAPOVAVIRUSES.

Gluzman, Y. (1981) *Cell* **23**, 175–182.

cos site The DNA sequences at the end of the linear genome of bacteriophage LAMBDA that exist as complementary single-stranded tails or COHESIVE ENDS (Fig. C70). The circular form of the lambda genome is generated by base-pairing between the cohesive ends, the resulting double-stranded DNA region being called *cos*.

cosmid CLONING VECTORS that are hybrids between DNA sequences from a variety of PLASMIDS and the *cos* region of the LAMBDA genome (Fig. C71), and which thus combine some of the advantages of both types of vector. These vectors can be used to carry additional DNA sequences up to 45 kb long and are often used to construct genomic DNA LIBRARIES. For example, to construct a genomic library, a mixture of large DNA fragments (35–45 kb) obtained by digestion of genomic DNA by a suitable restriction enzyme can be ligated into cosmids following their digestion at a unique RESTRICTION SITE (e.g. *Bgl*II in Fig. C71). The cosmids are then introduced into *E. coli* by TRANSFECTION, where they replicate as autonomous plasmids. Recombinant cosmid DNAs recovered from the bacterial culture, by use of a

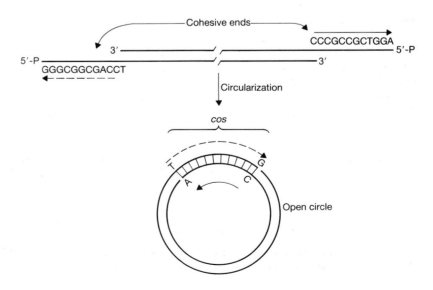

Fig. C70 Cos site. The cohesive ends of the linear lambda genome facilitate circularization of the genome to generate the cos site.

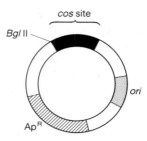

Fig. C71 The basic organization of a cosmid vector indicating the *cos* site, the selectable marker (Ap^r) and the origin of DNA replication *ori*.

selectable marker, can then each be packaged into a phage lambda *in vitro* as the required *cos* region is present (*see* LAMBDA). This enables a high recovery of recombinant DNAs, as each lambda virion can then be used to infect *E. coli*, producing a PLAQUE containing large amounts of the required DNA.

Collins, J. & Hohn, B. (1978) *Proc. Natl. Acad. Sci. USA* **75**, 4242–4246.

COSY *See*: NMR SPECTROSCOPY.

Cot analysis The term Cot (or C_0t) was invented by Britten and Kohne to quantify the kinetics of reassociation of two polynucleotide chains in solution (*see* HYBRIDIZATION). With a starting concentration (C_0), at any time (t) during isothermal reassociation, the fraction of single-stranded DNA molecules remaining is given by the formula:

$$C/C_0 = 1/(1 + kC_0t)$$

where k is the second-order reassociation constant. When half of the molecules have reannealed:

$$C_0t_{1/2} = 1/k$$

C_0t curves plot the fraction of DNA reassociated against log C_0t and reveal a direct relationship between $C_0t_{1/2}$ and DNA complexity (the total length of unique DNA sequence) (Fig. C72). For simple genomes (e.g. those of viruses and bacteria) the C_0t curve is of similar form, although as the genome size increases so the $C_0t_{1/2}$ value increases. The reassociation of eukaryotic DNA is anomalous because it contains three separate kinetic components. The fast component consists of millions of copies of small, highly repeated DNA sequences. The intermediate component renatures second and consists of MODERATELY REPETITIVE DNA 'sequence families'. The final slow component consists of nonrepetitive SINGLE-COPY DNA (*see* GENOME ORGANIZATION).

Britten, R.J. & Kohne, D.E. (1968) *Science*, **161**, 529–540.

co-transduction The TRANSDUCTION of two (or more) genetic markers in the same virus.

co-transfection *See*: CO-TRANSFORMATION.

co-transformation The simultaneous genetic TRANSFORMATION or TRANSFECTION of a cell with two different PLASMIDS each of which carries a different SELECTABLE MARKER gene. A DNA fragment carrying a nonselectable gene can also be introduced into cells by co-transformation in conjunction with a second DNA fragment or plasmid carrying a known selectable gene; the majority of cells acquiring the selectable gene also acquire the second nonselectable DNA fragment.

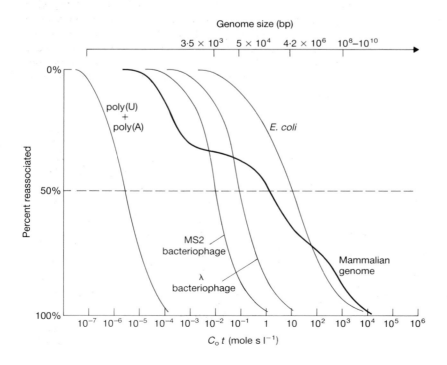

Fig. C72 C_0t curves demonstrate the relationship between $C_0t_{1/2}$ values and genomic complexity.

co-translational Events that occur to a polypeptide while it is still being elongated from the ribosome are known as co-translational. These include N-terminal acetylation, proteolytic removal of an N-terminal methionine residue, translocation across a membrane, cleavage of a SIGNAL SEQUENCE, covalent addition of core carbohydrates from a DOLICHOL intermediate (*see* GLYCOSYLATION), disulphide bond formation and folding (*see* PROTEIN FOLDING; PROTEIN STRUCTURE). If, during translation *in vitro*, in order to achieve translocation and removal of a signal sequence membrane vesicles have to be present during translation, then it is defined as a co-translational process. If translocation occurs when membrane vesicles are added after the precursor protein has been completed, and CYCLOHEXIMIDE added to prevent further elongation, then the process is post-translational. Co-translational translocation is shown in Fig. C73.

co-transmission The participation of two or more signals generated by different neurotransmitters in SYNAPTIC TRANSMISSION. Two different neurotransmitters released by the same nerve ending are known as co-transmitters.

co-transport *See*: MEMBRANE TRANSPORT SYSTEMS.

coumaromycin An antibiotic which inhibits DNA GYRASE (topoisomerase II) activity in bacterial cells.

countertransport *See*: MEMBRANE TRANSPORT SYSTEMS.

coupled cells Cells whose cytoplasms are connected by gap junctions so that the cells are metabolically coupled through the passage of small molecules and electrically coupled through the passage of ions. *See*: CELL JUNCTIONS.

coupling factor (CF$_1$, F$_1$) Protein complex required for the coupling of electron transport to ATP synthesis; the F$_1$ component of ATP SYNTHASE. *See also*: ELECTRON TRANSPORT CHAIN; PHOTOSYNTHESIS.

covalent modification Any chemical modification to a substance which involves formation or cleavage of a covalent bond. As applied to macromolecules such as proteins and nucleic acids, it indicates modifications such as the formation of S—S bonds and addition of phosphoryl groups to protein, or methylation of DNA, compared with modifications to macromolecular structure and interactions between macromolecules that involve noncovalent interactions such as HYDROGEN BONDING, the formation of SALT BRIDGES, and VAN DER WAALS FORCES. *See e.g.*: DNA METHYLATION; POST-TRANSLATIONAL MODIFICATION OF PROTEINS; PROTEIN FOLDING; PROTEIN–NUCLEIC ACID INTERACTIONS; PROTEIN STRUCTURE; NUCLEIC ACID STRUCTURE.

covalently closed circular DNA (cccDNA) Circular PLASMID DNA molecules which are tightly SUPERCOILED and thus have no broken phosphodiester bonds in either of the two polynucleotide chains. Plasmid DNA essentially exists in such a configuration in living cells. The existence of this conformation also facilitates the purification of plasmid DNA by caesium chloride DENSITY GRADIENT CENTRIFUGATION.

***cox* gene** Mitochondrial gene encoding a subunit of cytochrome *c* oxidase (*see* ELECTRON TRANSPORT CHAIN). The three largest subunits are encoded in the plant mitochondrial genome, by the genes *coxI*, *coxII* and *coxIII*. *See also*: MITOCHONDRIAL GENOMES: PLANT.

CP Chlorophyll–protein complex. CHLOROPHYLL almost always occurs *in vivo* as a noncovalent complex with specific proteins. These complexes can be separated by SDS (or LiDS) polyacrylamide gel ELECTROPHORESIS under mild conditions, when nearly all the chlorophyll remains associated with protein. Each type of photosystem contains its own characteristic chlorophyll protein(s) associated with the reaction centre core. These contain chlorophyll *a* in oxygenic organisms and usually bacteriochlorophyll *a* in photosynthetic bacteria (occasionally bacteriochlorophyll *b*). In addition, light-harvesting chlorophyll–protein complexes (*see* LHCI, LHCII) are associated with reaction centres in variable proportions. In chloroplasts but not CYANOBACTERIA these contain chlorophylls *a* and *b*. All types contain CAROTENOID. *See also*: PHOTOSYNTHESIS.

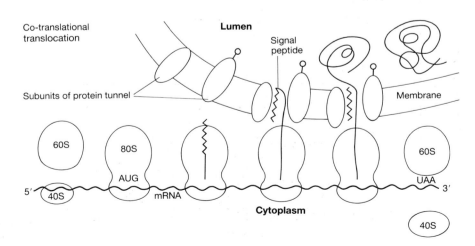

Fig. C73 Signal sequences may contact the translocation apparatus of a membrane while the nascent chain is still being elongated from the ribosome.

CP24 A minor chlorophyll *a/b* binding protein of the PHOTOSYS-TEM II antenna. *See also*: PHOTOSYNTHESIS.

CP29 A chlorophyll *a/b* binding protein more tightly associated with the reaction centre core of PHOTOSYSTEM II than is LHCII. *See also*: PHOTOSYNTHESIS.

cpDNA Chloroplast DNA. *See*: CHLOROPLAST GENOMES.

CPF phages *See*: BACTERIOPHAGES.

CpG island *See*: DNA METHYLATION.

CPI (P700–chlorophyll *a* protein) The reaction centre of PHOTO-SYSTEM I (PSI). The protein is composed of a heterodimer of very similar polypeptides of M_r ~80 000 encoded by the *psaA* and *psaB* genes. The complex contains one P700 per dimer, together with about 100 chlorophyll *a* molecules forming an ANTENNA, a small amount of β-carotene, two naphthoquinone molecules (one of which acts as acceptor A_1) and an iron–sulphur centre (centre X, or acceptor A_2). *See also*: PHOTOSYNTHESIS.

CR1–4 *See*: COMPLEMENT RECEPTORS.

crambin A SEED STORAGE PROTEIN. *See*: Fig. S6 for structure.

crassulacean acid metabolism (CAM) The process of fixation of carbon dioxide into malic acid at night by many stem and leaf succulents, followed by its release and reassimilation by photosynthesis into starch or other glucans during the day. Phosphoenolpyruvate (PEP) supplied by GLUCAN breakdown at night, is carboxylated by PEP carboxylase to oxaloacetate and then reduced to malate.

$$PEP + CO_2 \rightarrow oxaloacetate + P_i$$
$$Oxaloacetate + NADH + H^+ \rightarrow malate + NAD^+$$

PEP is recovered during the day from malic acid, temporarily stored in vacuoles, *either* through pyruvate *or* through oxaloacetate.

$$Malate + NAD(P)^+ \rightarrow pyruvate + CO_2 + NAD(P)H + H^+ \quad (1)$$
$$Pyruvate + ATP + P_i \rightarrow PEP + AMP + PP_i$$

$$Malate + NAD^+ \rightarrow oxaloacetate + NADH + H^+ \quad (2)$$
$$Oxaloacetate + ATP \rightarrow PEP + ADP + CO_2$$

PEP is converted to glucan by gluconeogenesis, and CO_2 assimilated by fixation and conversion through the reactions of the CALVIN CYCLE. The pathway allows carbon intake through open stomata at night, with closed stomata during the day. The consequent conservation of water allows the survival and growth of succulents in arid and semi-arid habitats.

CREB factor Cyclic AMP response element binding factor. A eukaryotic TRANSCRIPTION FACTOR responsible for the activation of certain genes by cAMP. *See*: EUKARYOTIC GENE EXPRESSION.

cretinism The syndrome of retardation of mental and physical development resulting from neonatal hypothyroidism. The clas-sical manifestation is large tongue, distended abdomen, umbilical hernia, and dry skin and hair. The majority of cases are associated with failure of development or abnormal location of the thyroid gland; a proportion (30%) are associated with a genetic defect in thyroid hormone synthesis.

Creutzfeld–Jakob disease *See*: TRANSMISSIBLE SPONGIFORM EN-CEPHALOPATHIES.

Crigler–Najjar syndrome A rare heritable cause of neonatal jaundice due to a deficiency of the enzyme glucuronyl transferase which leads to a reduced hepatic conjugation of bilirubin. There are two distinct groups. In type 1 there is complete absence of the enzyme resulting in severe unconjugated hyperbilirubinaemia in the neonatal period which leads to kernicterus and early death in most infants. Parental consanguinity is common suggesting an AUTOSOMAL RECESSIVE mode of inheritance. In type 2 there is partial deficiency of glucuronyl transferase activity perhaps associated with the homozygous state for a variant enzyme. Jaundice is less acute and less severe than in type 1.

crinophagy The fusion of SECRETORY GRANULES (from the regulated pathway of secretion) with LYSOSOMES or pre-lysosomal structures. It has been inferred from microscopic examination of several secretory tissues, though the process has not been reproduced in cell-free systems. It has been suggested that crinophagy occurs most often with immature or newly formed granules. One possibility is that they fuse with late ENDOSOMES as a result of the presence of some plasma membrane components in the endosome membrane. *See also*: EXOCYTOSIS.

Cro protein The product of the *cro* gene of bacteriophage LAMBDA. It is a small basic protein of 66 amino acids which forms a homodimer and acts as a regulatory protein by binding to a specific 17-bp OPERATOR sequence (oR3) in the lambda genome. Binding of Cro to the oR3 sequence prevents the synthesis of the CI REPRESSOR of lambda, thereby preventing the bacteriophage from entering the LYSOGENIC life cyle. Cro binding to its operator also inhibits expression of the early genes of lambda from both the P_L and P_R promoters.

Ptashne, M. (1987) *A Genetic Switch: Gene Control and Phage Lambda* (Blackwell Scientific Publications, Oxford).

CRP C-REACTIVE PROTEIN, an acute phase reactant. *See*: PENTRAXIN.

cross bridge Connection between thick (MYOSIN) and thin (ACTIN) filaments in MUSCLE. *See*: MUSCLE; MYOSIN.

crossing-over The physical exchange of parts of HOMOLOGOUS CHROMOSOMES during the process of MEIOSIS. This reciprocal exchange is responsible for genetic RECOMBINATION which results in a reassortment of ALLELES between the homologues. The frequency with which recombination occurs between two genes on a single chromosome can be used as a means of estimating the distance between them, as the further apart the two genes are, the greater the chance of a cross-over occurring between them. Crossing-over can be used to order genes on a chromosome, thereby generating a LINKAGE MAP or genetic map. Multiple cross-overs between an homologous chromosome pair can occur during a given meiotic event (seen as chiasmata) although the occurrence of a cross-over in one region suppresses the occurrence of a second crossing-over within that region, a phenomenon known as interference.

crown gall Tumorous disease of a wide range of higher, mainly dicotyledonous, plants, but gymnosperms can also be affected. Infection by *Agrobacterium tumefaciens*, normally at a wound near soil level (at the crown of the plant), results in neoplastic transformation of the plant cells and their proliferation to form a gall (crown gall). Gall cells have become permanently transformed as a result of the transfer of part of the TI PLASMID of *A. tumefaciens* into their genome.

cruciform structure Structure formed in a double-stranded nucleic acid by INVERTED REPEATS separated by a short sequence. *See also*: NUCLEIC ACID STRUCTURE.

cryoelectron microscopy (electron cryomicroscopy) Electron microscopy of specimens which have been fixed by hydration and rapid freezing and not by chemical fixatives, and which are examined at low temperature ($-160\,°C$) in the form of a thin vitrified water film without staining. In conjunction with computerized image analysis and reconstruction techniques, cryoelectron microscopy has proved particularly useful for determining the three-dimensional structure of large proteins and protein complexes and integral membrane proteins which can be obtained as crystalline arrays. *See also*: IMMUNOELECTRON MICROSCOPY.

Henderson, R. et al. (1990) *J. Mol. Biol.* **213**, 899–929.
Toyoshima, C. et al. (1993) *Nature* **362**, 469–471.

cryoenzymology The study of enzyme reactions at very low temperatures.

cryptic plasmid A plasmid which confers no new PHENOTYPE on the cell in which it resides. An example is the 2μM PLASMID of yeast.

cryptochrome Photoreceptor absorbing blue/ultraviolet light and ubiquitous in all living organisms. The chemical nature of the CHROMOPHORE is as yet uncertain, but may be either flavin, carotenoid, or haem. It is responsible for a wide range of photomorphogenetic effects, including PHOTOTROPISM, particularly in fungi and plants.

cryptovirus group Name derived from 'cryptic', which refers to the lack of symptoms induced by these viruses. There are two subgroups: A (type member white clover cryptic virus 1) have smooth isometric particles of 30 nm diameter, and B (type member white clover cryptic virus 2) have isometric particles of 38 nm diameter with prominent subunits. The particles contain double-stranded RNA, usually two species with M_r ranging from $1.6 \times$ $10^6 - 0.6 \times 10^6$. There are strong indications that the larger RNA encodes the viral polymerase which is found in virus particles and the smaller RNA encodes the coat protein (M_r 53 000–63 000). It has been suggested that these viruses do not have cell-to-cell movement proteins and spread through the plant is only at cell division. *See also*: PLANT VIRUSES.

Boccardo, G. et al. (1987) *Adv. Virus Res.* **32**, 171–214.

crystal structure The structure of a macromolecule as determined from X-RAY (or sometimes ELECTRON or NEUTRON) CRYSTALLOGRAPHY.

crystallin The major protein in the lens of the eye. In humans there are several types of crystallin (e.g. α, β, and γ), expressed at different stages of development, and encoded by a MULTIGENE FAMILY. For an illustration of the structure of γ-crystallin *see* PROTEIN STRUCTURE.

crystallization Growth of crystals suitable for use in X-RAY CRYSTALLOGRAPHY has been the subject of much recent study, being a necessary prerequisite for DIFFRACTION studies. Two critical steps are involved in crystallization: the nucleation of the initial seed and the growth of this seed. Many crystallization trials are sometimes required to find appropriate conditions of pH, concentration, and nature of precipitating agents (e.g. salts, polyethylene glycols of varying sizes) and optimum concentration of the macromolecule itself, which may need to be modified with ligands or specific cleavage(s) to obtain any crystals. Crystallization conditions often need to be screened on a very small scale; techniques in common use include dialysis in small buttons, or vapour diffusion using either a sitting drop or hanging drop arrangement (Fig. C74). These are often conveniently implemented in standard multi-well laboratory plates, to allow 24 conditions to be screened at once. When used with a 24-well plate, the volume of precipitant solution is about 1 ml with as little as 5 μl in the drop of protein plus precipitant.

Ducruix, A. & Giegé, R. (Eds) (1992) *Crystallization of nucleic acids and proteins: a practical approach* (IRL Press, OUP Oxford).

crystallography The application of the techniques of DIFFRACTION to crystals or crystalline arrays of molecules of biological interest. X-RAY CRYSTALLOGRAPHY has been the dominant technique, but for particular problems there are advantages in using neutron or electron crystallography (*see* ELECTRON CRYSTALLOGRAPHY; ELECTRON DIFFRACTION; ELECTRON MICROSCOPY; NEUTRON SCATTERING AND DIFFRACTION).

CS *See*: CHALCONE SYNTHASE.

CSF-1 Colony-stimulating factor-1. *See*: M-CSF.

CSIF IL-10.

ctDNA Chloroplast DNA. *See*: CHLOROPLAST GENOMES.

ctenophore development Ctenophores are more commonly known as comb jellies. They have a unique pattern of cleavage

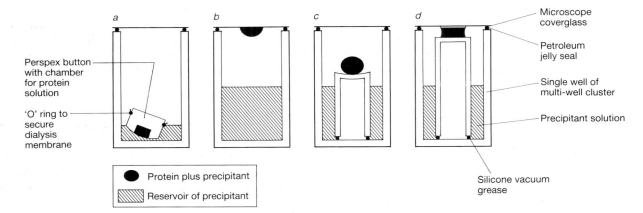

Fig. C74 Diagram to show types of crystallization arrangement for a protein. (a) Dialysis (b) Hanging drop (c) Sitting drop (d) Sandwich drop.

which is related to the ultimate biradial symmetry of the organism. An early example of a presumed CYTOPLASMIC DETERMINANT is the fluorescent ctenoplasm which is segregated into the micromeres of *Beroe*. A more recent study of localization was made by Freeman on *Mnemiopsis* and he concluded that determinants for the larval comb plates could be segregated into particular micromeres as late as the four-cell stage.

CTL Cytotoxic T lymphocyte. *See*: CYTOTOXIC T CELLS.

CTP Cytidine 5′-triphosphate. *See*: NUCLEOSIDES AND NUCLEOTIDES.

cucumopine An OPINE, the product of a condensation between 2-oxoglutarate and L-histidine, found in *Agrobacterium rhizogenes*-induced hairy root lines.

cucumovirus group Sigla from cucumber mosaic virus, the type member. MULTICOMPONENT VIRUSES with small isometric particles ($t = 3$) 29 nm in diameter. The particles contain four species of (+)-strand linear RNA (Fig. C75). RNA 1 and RNA 2 are encapsidated in separate particles and encode components of the viral polymerase. One molecule each of RNA 3 and RNA 4 (1049 nucleotides) are encapsidated in the same particle. RNA 3 is bicistronic, encoding at the 5′ end the putative cell-to-cell movement protein and at the 3′ end the viral coat protein. The coat protein is expressed from RNA 4. The infective genome comprises RNAs 1, 2 and 3; RNA 4 is subgenomic. *See also*: PLANT VIRUSES.

Francki, R.I.B. (1985) In *The Plant Viruses*, Vol. 1, 1 (Plenum, New York).

cucurbic acid *See*: PLANT WOUND RESPONSES.

CURL (receptosome, endosome) Acronym for Compartment of Uncoupling Receptor and Ligand, the term given to the endocytic organelles in which the low pH-induced dissociation of certain ligand–receptor complexes (e.g. ASIALOGLYCOPROTEINS and asia-

loglycoprotein receptors) occurs before recycling or sorting. More commonly known as ENDOSOME. *See*: ENDOCYTOSIS.

Geuze, H.J. et al. (1983) *Cell* **32**, 277–287.

Curtobacterium A genus of Gram-positive bacteria which are typified by small irregular, nonsporing, rod-shaped cells. Members of this genus are obligately aerobic chemo-organotrophs and are often associated with plants. The precise association between most *Curtobacterium* spp. and plants is not known and only one species, *C. flaccumfaciens*, is regarded as a plant pathogen. This organism causes vascular wilt and/or leaf spot of bean, red beet, tulips, and poinsettia.

Komagata, K. & Suzuki, K.-I. (1984) *Curtobacterium*. In *Bergey's Manual of Systematic Bacteriology* (Sneath, P.H.A. et al., Eds) 1313–1317 (Williams and Wilkins, Baltimore).

cutin *See*: PLANT CELL WALL MACROMOLECULES.

cyanelle A photosynthetic organelle within glaucophyte algae such as *Cyanophora paradoxa* and *Glaucocystis nostochinearum*. Cyanelles have many similarities to photosynthetic bacteria,

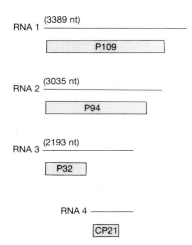

Fig. C75 Genome organization of cucumber mosaic virus (CMV). The lines represent the RNA species and the boxes the proteins (P) with M_r given as $\times 10^{-3}$; nt, nucleotides.

generally containing chlorophyll *a* and PHYCOBILIPROTEINS as photosynthetic pigments in a thylakoid membrane located peripherally. The organelle is bounded by a rudimentary peptidoglycan wall. It contains a genome slightly larger than that of plastids in higher plants, but with considerable similarities in content and organization. *See*: CHLOROPLAST GENOMES; CYANOBACTERIA.

Cyanobacteria

CYANOBACTERIA are microorganisms of enormous ecological importance in their own right which also serve in many respects as model systems for studying problems in the molecular biology of eukaryotic plants. This is because they are photosynthetic PROKARYOTES capable of oxygen-evolving (oxygenic) PHOTOSYNTHESIS resembling that of plants in many respects. They therefore contrast with other photosynthetic prokaryotes (except PROCHLOROPHYTES), whose photosynthesis is anoxygenic. (Some cyanobacterial species are, however, also capable of anoxygenic photosynthesis.) They were formerly known as 'blue-green algae', on account of the characteristic colour of many species caused by the presence of pigment–protein complexes called PHYCOBILIPROTEINS. These, together with CHLOROPHYLL *a*, are the main cyanobacterial photosynthetic pigments. Because of the possible risk of confusion with eukaryotic algae, the term 'cyanobacteria' is now preferred. The biology of cyanobacteria will be described briefly, followed by some selected areas of particular interest to the molecular biologist. For fuller accounts, consult [1,2].

General biology

The cyanobacteria were probably the first organisms on Earth to generate oxygen photosynthetically, and there is fossil evidence for their occurrence at least 3.5×10^9 years ago. They are represented in the fossil record by microfossils and stromatolites (structures arising from sedimentation and organic processes in microbial mats). They are largely aquatic and widespread, some living symbiotically with other organisms. They make a major contribution to oceanic productivity by their fixation of carbon dioxide and nitrogen.

The cells are bounded by an inner cell membrane (cytoplasmic membrane), a peptidoglycan cell wall, and an outer membrane, analogous to the arrangement in Gram-negative bacteria (*see* BACTERIAL ENVELOPES). Mucilaginous sheaths and fimbriae or pili are found in a number of species. Within the cell is an extensive membrane system, the thylakoid, which has a different pigment and protein composition from the cytoplasmic membrane, and is the site of the light reactions of photosynthesis. The polypeptides involved are very similar to those in CHLOROPLASTS. The main exceptions include the extrinsic polypeptides associated with oxygen evolution in PHOTOSYSTEM II (where the 23K (M_r 23 000) and 16K polypeptides of plants are lacking and an additional 9K polypeptide may be present) and the replacement of the LIGHT-HARVESTING COMPLEX containing chlorophylls *a* and *b* by the phycobilisome, a complex structure containing the phycobiliproteins phycoerythrin, phycocyanin and allophycocyanin (which have different spectral characteristics). Some cyanobacteria have the ability to modulate the levels of the different phycobiliproteins in response to light quality, a phenomenon called chromatic adaptation. In the species *Gloeobacter violaceus*, there is no thylakoid membrane and the cytoplasmic membrane has taken over the photosynthetic function.

As well as thylakoids, a number of other intracellular structures can be seen. These include:

1 Carboxysomes, which are polyhedral structures containing RIBULOSE BISPHOSPHATE and smaller amounts of various polypeptides. Ribulose bisphosphate also occurs in an apparently identical form in the cytoplasm.

2 Cyanophycin granules, which are nonribosomally synthesized polymers of arginine and aspartate, and probably function as storage material.

3 Gas vesicles, which are proteinaceous structures that can be filled with gas to provide buoyancy.

4 Polyphosphate granules, containing linear chains of phosphate groups functioning as a phosphate reserve.

Under some conditions, large accumulations of cyanobacteria ('blooms') appear on the surface of water bodies such as lakes and reservoirs. These are often toxic, either indirectly as a result of the decay of organic material as the bloom subsides, or directly as a result of toxin synthesis by the cyanobacteria. Cyanobacterial toxins include anatoxin a, produced by *Anabaena flos-aquae*, which may function by blocking transmission at mammalian neuromuscular junctions, and cyanoginosin (formerly called Fast Death Factor!), a cyclic peptide produced by *Microcystis aeruginosa*, with a wide range of toxic effects. Cyanobacterin, produced by *Scytonema hofmanni* (not usually a bloom former), is toxic only to other photosynthetic organisms and interferes with photosystem II function.

A number of cyanobacteria exhibit cellular DIFFERENTIATION. The most notable example is the formation of heterocysts, cells specialized for NITROGEN FIXATION, which will be discussed in more detail below. Others are akinetes (spore-like cells produced by a number of heterocystous strains), baeocytes (small, spore-like cells), and hormogonia (cell filaments involved in dispersal).

Nomenclature

The classification and nomenclature of cyanobacteria is not altogether straightforward, and a number of identical or closely related strains have acquired different names for largely historical reasons. The system proposed by Rippka and colleagues [3] is widely used, and assigns cyanobacteria to five sections. Briefly, Sections I and II contain the unicellular strains, and are distinguished from each other in the mode of division. Sections III, IV and V contain the filamentous strains. Those in Section III do not form heterocysts; those in Sections IV and V do, and are distinguished from each other by their mode of division. DNA : DNA hybridization data (*see* HYBRIDIZATION) have also been used quite widely in classification, but the results are not always easy to interpret.

Genetic manipulation and analysis

As would be expected, the development of techniques for the genetic manipulation of cyanobacteria has greatly assisted molecular genetic analysis. In the main, two approaches have been used — direct uptake (TRANSFORMATION) and conjugal transfer (*see* CONJUGATION) of DNA. Many strains exhibit natural competence (ability to be transformed), although *Anacystis nidulans* R2 and *Synechocystis* sp. PCC6803 are especially widely used. The conditions needed to maximize competence seem to vary quite widely, and may include treatments either in the light or in the dark. Many strains contain RESTRICTION ENZYMES which interfere with transformation. Treatment with ultraviolet light may enhance transformation frequencies, possibly by affecting the restriction systems. Transformation can be carried out with chromosomal or PLASMID DNA.

Homologous RECOMBINATION between incoming DNA lacking a replication origin and the recipient's chromosomal DNA leads to integration of incoming DNA into the chromosome, and this procedure has been exploited to disrupt resident gene function. But the use of this technique is often complicated by the presence of multiple copies of some genes in the chromosome. In addition, cyanobacteria are usually POLYPLOID, and it is often difficult to ensure that a particular gene has been disrupted on all the genomes present, especially if disruption causes a significant reduction in viability.

Cyanobacteria contain a wide range of plasmids, and many of these have formed the basis of *Escherichia coli*/cyanobacterial SHUTTLE VECTORS, with resistance to antibiotics such as ampicillin, streptomycin, kanamycin and chloramphenicol as selectable markers. A wide range of derivatives of these vectors has been constructed, including EXPRESSION VECTORS and REPORTER GENE constructs. Stimulation of DNA uptake by ELECTROPORATION has been demonstrated, and it seems likely that this technique will be applicable to a number of species previously considered refractory to transformation.

Conjugal transfer of DNA has also been used, particularly in filamentous species. This requires a conjugal plasmid (such as RP4) to encode the conjugation functions, a 'cargo' plasmid (to be transferred) and usually a helper plasmid. Transfer is carried out in a triparental mating between cells bearing (1) the conjugal plasmid, (2) cargo and helper, and (3) recipient cyanobacterium. Cyanobacteria are also susceptible to a number of BACTERIOPHAGES (cyanophages), but they have not been used to any great extent in genetic manipulation. The use of restriction enzymes that cut DNA very infrequently, combined with pulsed-field gel ELECTROPHORESIS, is being exploited in the production of chromosomal maps.

The power of genetic manipulation in cyanobacteria has been well illustrated in the dissection of the light reactions of photosynthesis, and photosystem II in particular. This has been possible because some cyanobacteria, and particularly the transformable *Synechocystis* sp. PCC6803, are capable of photoheterotrophic growth, that is, they are able to grow on a fixed carbon source supplied in the medium rather than on photosynthetically fixed carbon. (It may seem surprising that not all strains can grow heterotrophically, but in many cases it may simply be that an appropriate uptake system is lacking.) Light is still needed in photoheterotrophy, but photosystem II function is not, and there may still be some requirement for CYCLIC PHOTOPHOSPHORYLATION by PHOTOSYSTEM I. Genes for individual polypeptides from photosystem II can therefore be inactivated (by gene disruption techniques) and replaced by modified ones, without lethal effects. Partial inactivation of photosystem I also seems to be tolerated, although complete inactivation may be lethal or at least permit only very slow growth.

This kind of analysis of photosynthesis would be much more difficult, if not impossible, with plants because of (1) the difficulty of genetic manipulation of the chloroplast, where many of the relevant polypeptides are encoded, and (2) the problems incurred by inactivation of the light reactions of photosynthesis. It has yielded information on topics such as the assembly of photosystem II (by inactivating genes for specific polypeptides and examining which others are consequently missing from the complex) and the function of individual polypeptides and residues within them.

Gene expression

A large number of cyanobacterial genes have now been cloned (*see* DNA CLONING) and analysed at the molecular level, in many different species. They include those for rRNAs and tRNAs, and for components of ATP SYNTHASE, the 'Z-scheme' (photosystems I and II, etc.), the phycobilisome, enzymes of intermediary metabolism and of nitrogen and carbon fixation, the machinery of PROTEIN SYNTHESIS, gas vesicle proteins, transport proteins, and DNA recombination/repair proteins. Selection has mainly used the standard techniques for prokaryotes, such as direct selection for gene function (in cyanobacterial hosts or in heterologous hosts such as *E. coli*), hybridization with homologous or heterologous PROBES or oligonucleotides, and antibody screening of EXPRESSION LIBRARIES.

The successful expression of some genes in *E. coli* implies similarities in the processes of gene expression (*see* BACTERIAL GENE EXPRESSION; PROTEIN SYNTHESIS; TRANSCRIPTION), and classical −35 and −10 sequences upstream from the site of transcription initiation have been demonstrated in some cases. This is by no means generally true, however, and there may exist several different PROMOTER classes. Genes are not generally split (*see* INTRONS); an exception is the tRNA^{Leu} gene in which an intron has been found in several species so far (*see* RNA SPLICING). As with other prokaryotes, a number of transcripts are POLYCISTRONIC. Conventional ribosome-binding sites are associated with many genes, and translation follows the 'universal' GENETIC CODE.

Gene expression can be affected in response to a wide range of stimuli, including HEAT SHOCK, nutrient levels, light levels (where different members of a gene family, such as *psbA*, may show different responses) and light quality. It is generally assumed that amounts of transcripts are determined by the level of initiation of transcription, but there is at present rather little direct experimental evidence for this. If this is the case, it will also be important to ascertain the relative importance of modulation of factors such as RNA POLYMERASE specificity (by SIGMA FACTORS etc.) and promoter availability (by repressor proteins etc.). As with other bacteria, there is rather little information on TRANSLATIONAL CONTROL. Protein turnover is likely to prove important in determining the steady-state levels of some proteins such as the *psbA* reaction centre polypeptide of photosystem II (following damage during photosynthesis) and cyanophycins and phycobiliproteins, which are broken down under conditions of nitrogen starvation.

DNA rearrangements

Anabaena provides an unusual example of developmental rearrangement of DNA, during the formation of heterocysts under conditions of nitrogen starvation. Heterocysts are terminally differentiated cells specialized for nitrogen fixation, and are characterized by thick walls, loss of photosynthetic oxygen evolution and induction of NITROGENASE (from the *nifD*, *H* and *K* genes) and other enzymes. It is the extremely oxygen-labile nature of the nitrogenase that necessitates the development of specialized cells. In nonheterocystous nitrogen-fixing strains it seems that oxygenic photosynthesis and nitrogen fixation are separated on a temporal rather than a physical basis.

The developmental changes in heterocysts are associated with at least two genetic rearrangements. One involves the excision of an 11-kilobase-pair (kbp) sequence interrupting the *nifD* gene. This also has the consequence of bringing the *nifK* gene (otherwise at some distance from *nifH* and *nifD* which are adjacent to each other) under the control of the *nifH* promoter to form a *nifHDK* OPERON. This excision is brought about by recombination across an 11-base-pair (bp) repeated sequence mediated by the XisA protein encoded at one end of the excised region. The second rearrangement takes place in the vicinity of the *nifS* gene and involves deletion of a much larger segment, probably by a different recombination enzyme, as the base sequences at the endpoints of the two rearrangements are different. Increased levels of *nif* gene transcripts are associated with heterocyst formation, but the relationship of increased transcription to these rearrangements is not clear.

Protein sorting

Cyanobacteria have the problem that most prokaryotes do not face of sorting polypeptides (*see* BACTERIAL PROTEIN EXPORT; PROTEIN TARGETING) to one of two membrane systems (i.e. thylakoid or cytoplasmic membrane/periplasmic space/outer membrane). The LEADER SEQUENCES that direct polypeptides to the thylakoid (which closely resemble those targeting polypeptides to the thylakoid in chloroplasts) are similar to sequences that are believed to target polypeptides to the cytoplasmic membrane, yet the membranes have different polypeptide compositions. Furthermore, some polypeptides, such as the subunits of cytochrome oxidase, can be targeted to both membranes in proportions that can be modulated according to the physiological status of the cell. Elucidation of this paradox may well shed light on the corresponding problem of protein sorting within the plant cell chloroplast.

C.J. HOWE

1 Packer, L. & Glazer, A.N. (Eds) (1988) Cyanobacteria. *Meth. Enzymol.* **167** (Academic Press, London).
2 Fay, P. & van Baalen, C. (Eds) (1987) *The Cyanobacteria* (Elsevier, Amsterdam).
3 Rippka, R. et al. (1979) Generic assignments, strain histories and properties of pure cultures of cyanobacteria. *J. Gen. Microbiol.* **111**, 1–67.

cyanobacterin A compound excreted by the filamentous CYANOBACTERIUM *Scytonema hofmanni*. It is toxic to most cyanobacteria and some green algae and higher plants (if applied to the leaves). It is a diaryl substituted γ-ylidene, γ-butyrolactone, $C_{23}H_{23}ClO_6$, and appears to act by inhibiting PHOTOSYSTEM II. The biosynthetic pathway is poorly understood.

cyanocobalamin A form of vitamin B_{12} (cobalamin) present only in traces in nature but which is stable and used radioactively labelled for studies of B_{12} metabolism.

cyanogen bromide Reagent (CNBr) that specifically cleaves polypeptide chains on the carboxyl side of methionine residues. It is used to split long polypeptides into smaller peptides for sequencing by Edman degradation. *See:* PROTEIN SEQUENCING.

cyanoginosin A toxin from CYANOBACTERIA.

cybrid Transient anucleate 'cell' produced by fusion of two different cytoplasms. *See:* SOMATIC CELL HYBRIDS.

cyclic AMP (cAMP) Adenosine 3′,5′-cyclic monophosphate (Fig. C76), a nucleotide formed from ATP by the action of ADENYLATE CYCLASE. In eukaryotic cells (except, it seems, higher plants) it serves as a second messenger which is produced as a result of the activation of cell-surface receptors and which passes the stimulus to the cell interior through the activation of CYCLIC AMP-DEPENDENT PROTEIN KINASE. In bacteria, its level is regulated by nutrient supply and it controls gene expression through interaction with the CYCLIC AMP ACTIVATOR PROTEIN. In the cellular slime mould *Dictyostelium discoideum* cAMP also acts as a chemoattractant and signalling molecule in development. Cyclic AMP is converted to AMP by the enzyme cAMP phosphodiesterase. *See also:* BACTERIAL GENE EXPRESSION; CATABOLITE REPRESSION; DICTYOSTELIUM DEVELOPMENT; SECOND MESSENGER PATHWAYS.

cyclic AMP activator protein (catabolite activator protein, CAP, CRP) Also known as cyclic AMP receptor protein, a protein of two identical subunits of M_r 22 500 which positively regulates transcription from certain bacterial promoters (e.g. the *lac* or *gal* operons of *Escherichia coli*). It can only bind to the promoter when complexed with a single molecule of cAMP and when bound facilitates the binding of RNA polymerase at a site downstream of the CAP binding site. When cAMP levels are reduced, for example in the presence of glucose, CAP can no longer bind efficiently to the promoter. This in turn prevents RNA polymerase from binding, thereby turning off the expression of the operon. This mechanism is responsible for CATABOLITE REPRESSION by glucose in *E. coli*. *See also:* BACTERIAL GENE EXPRESSION.

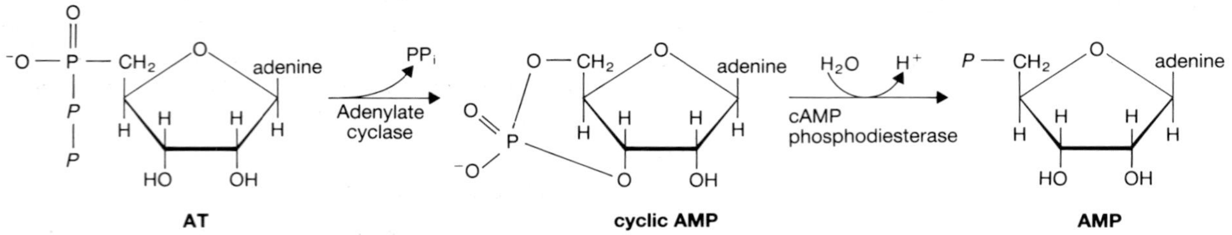

Fig. C76 Cyclic AMP, formation from ATP and conversion to AMP.

cyclic AMP-dependent protein kinases Serine–threonine PRO-TEIN KINASES that are activated by the second messenger CYCLIC AMP (*see* SECOND MESSENGER PATHWAYS). They are also known as the protein kinase A family. Many cAMP-dependent protein kinases contain a separate catalytic subunit and a regulatory subunit.

cyclic GMP Guanosine 3′,5′-cyclic monophosphate (Fig. C77), a nucleotide formed from GTP by the action of the enzyme GUANYLATE CYCLASE and which serves as a second messenger for some cell-surface receptors in certain eukaryotic cells (*see* SECOND MESSENGER PATHWAYS; VISUAL TRANSDUCTION).

3′,5′-cyclic GMP phosphodiesterase Hydrolytic enzyme (EC 3.1.4.35) that splits a phosphodiester bond in CYCLIC GMP to generate guanosine 5′-monophosphate (GMP).

cyclic nucleotide phosphodiesterase Ubiquitous enzyme that splits a phosphodiester bond in 3′, 5′ cyclic nucleotides (cAMP, cCMP, cGMP) to generate a nucleoside monophosphate. cAMP and cGMP phosphodiesterases are important in SECOND MESSEN-GER PATHWAYS involving these cyclic nucleotide second messengers as they rapidly degrade the second messenger, thus providing a sharp time-limited signal.

cyclic photophosphorylation *See*: PHOTOSYNTHESIS.

cyclin *See*: CELL CYCLE.

cyclin-dependent kinase *See*: CELL CYCLE.

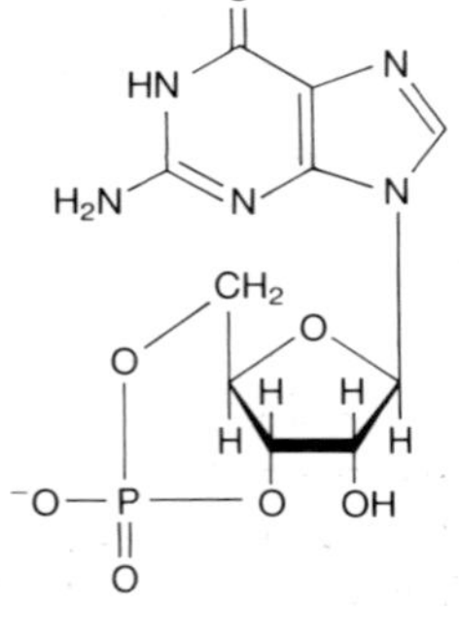

cycling and transport receptors *See*: CELL-SURFACE RECEPTORS; COATED PITS AND VESICLES; ENDOCYTOSIS; PLASMA LIPOPROTEINS AND THEIR RECEPTORS.

cycloheximide A glutarimide antibiotic (Fig. C78) from *Strepto-myces griseus* which is an effective inhibitor of cytoplasmic (but not mitochondrial) PROTEIN SYNTHESIS in eukaryotic cells. Cyclohex-imide binds to the 60S subunit of the ribosome and primarily blocks the translocation step of translation which is dependent on elongation factor EF-2.

cyclophilin A ubiquitous peptidyl-prolyl *cis-trans* isomerase (M_r 17 000) of 165 amino acids per monomer that catalyses the *cis-trans* isomerization of proline imidic peptide bonds in oligopep-tides and is thought to be involved in PROTEIN FOLDING. It is the receptor for the immunosuppressive drug CYCLOSPORIN A, bind-ing five drug molecules per pentameric cyclophilin but immuno-suppression is thought not to involve the enzymatic activity.

Pflügl, G. et al. (1993) *Nature* **361**, 91–94.

cyclosporin A A fungal undecapeptide peptide used as an immunosuppressive drug and particularly potent in the preven-tion of graft rejection. It causes its immunosupprevise effect by binding to members of the CYCLOPHILIN family of proteins. The drug–cyclophilin complex interferes with signal transduction in T cells, apparently acting downstream of receptor activation and second messenger formation and binding to and inhibiting the PROTEIN PHOSPHATASE calcineurin (*see* CALCIUM-BINDING PRO-TEINS). The manifest action of the drug is to downregulate production of the LYMPHOKINE interleukin-2 (IL-2) by T CELLS. The levels of several TRANSCRIPTION FACTORS, some of which are known to bind upstream of the IL-2 gene, are reduced after treatment of cells with cyclosporin A. Nuclear transcription factors AP-3 and NFκB are marginally downregulated, whereas the levels of AP-1 and the NFAT complex are markedly reduced.

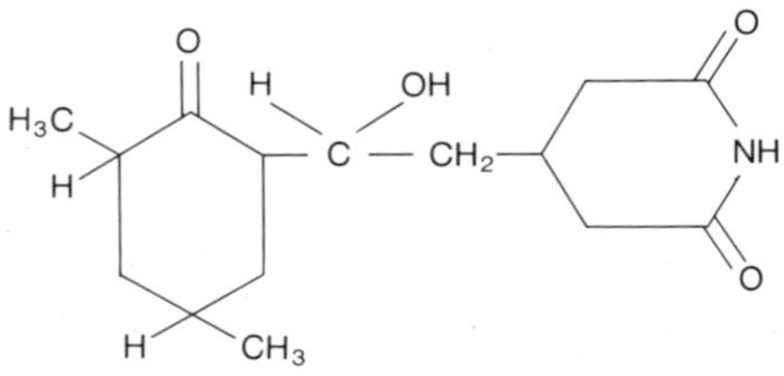

Fig. C77 Cyclic GMP.

Fig. C78 Cycloheximide.

cysteine, cystine (Cys, C) Sulphydryl-containing AMINO ACIDS. Two cysteine residues can be oxidized to form cystine, which forms a disulphide bond between two (parts of) polypeptide chains.

cysteine proteinases PROTEINASES with a cysteine residue at their active site.

Cystic fibrosis

CYSTIC fibrosis (CF, also occasionally called mucoviscidosis) is one of the most common AUTOSOMAL RECESSIVE diseases of Caucasians, with an incidence of around 1 in 2000 live births and a carrier rate of around 1 in 22. Clinically, the problems are a consequence of thick, sticky mucus secretions which are difficult to clear. Thus patients are prone to recurrent lung infections secondary to airway blockage, small bowel obstruction, pancreatic insufficiency and cirrhosis of the liver due to biliary tract obstruction. Males are almost always infertile. The diagnostic assay for CF is the sweat test: raised levels of sodium and chloride in stimulated sweat are characteristic of the condition, as the CF sweat gland duct is relatively impermeable to chloride anions.

The physiological defect

The most intensively studied aspect of the cellular physiology of CF has been the movement of chloride ions across epithelia [1]. A defective β-adrenergic-stimulated chloride conductance has been identified in CF tissues (cholinergic responses are intact). Initially, PATCH CLAMP experiments on lung epithelial cells suggested that an outwardly rectified apical membrane CHLORIDE CHANNEL (the ORCC) with a conductance of 26–50 pS was responsible for most of the chloride transport seen in CF epithelia. The channel has a very low probability of being in the open state at normal MEMBRANE POTENTIALS but the probability of opening increases dramatically with depolarizing (cell interior positive) potentials. In inside-out membrane patches, the channel from normal epithelia could be activated by PROTEIN KINASE A (PKA) and ATP, but that from CF sufferers could not. The presence of the channel in patches from CF tissue was confirmed by passing a depolarizing potential across the membrane; this procedure activated the channel. Activation of the channel by calcium also seemed to be intact in the CF preparations. The significance of these observations is unclear as it now seems that a lower conductance channel with a linear current–voltage relationship is primarily at fault in CF (see below). However, a defect in the regulation of an apical membrane chloride channel (of whatever biophysical characteristics) is considered to be one of the major consequences of the CF mutation.

Other abnormalities characterized at the physiological level include increased sodium reabsorption by the respiratory tract epithelium. This could result in increased water absorption and hyperconcentration of mucus. This defect is the target of a controlled trial of aerosolled amiloride (a sodium channel blocker) as a treatment for CF. Increased sulphation of mucus GLYCOPRO-TEINS has also been observed and may be a consequence of decreased acidification of ENDOSOMES secondary to an altered chloride channel. The cloning of the defective CF gene has allowed a more detailed analysis of these problems and a clearer picture of the biochemical and physiological defects has emerged.

Cloning the CF gene

The gene mutated in individuals with CF was isolated using a POSITIONAL CLONING strategy (i.e. without prior knowledge of the biochemical function of the gene product) [2]. CF was the first gene to be cloned by this method without any accompanying CHROMOSOME ABERRATION to help pinpoint the locus (*cf.* DUCHENNE MUSCULAR DYSTROPHY). Progress was assisted by the association between the main mutations and particular alleles of a closely linked polymorphic DNA MARKER on chromosome 7q31 (*see* HUMAN GENE MAPPING). The major challenges in CF research now are to understand the function of the normal gene product and how this is disturbed by the various mutations found within the gene, and to devise novel therapeutic strategies.

Gene structure and protein product

When it was first cloned, the gene at the *CF* locus was named the cystic fibrosis transmembrane conductance regulator (*CFTR*) because of its presumed role in chloride transport processes. The gene spans 260 kb of genomic DNA and is organized into 27 exons. The cDNA sequence predicts a protein of 1480 amino acids, M_r 170 000 and a pI of 8–9. Alternatively spliced products have been reported in some cell types, although there are no indications yet of a specific function for the alternative protein products.

The predicted protein has several features suggesting that it is a member of a family of membrane transport proteins. There are two domains comprising six membrane-spanning helices, two nucleotide-binding folds similar to those found in the ABC (ATP-binding cassette) family of proteins and two N-glycosylation sites on the largest of the extracellular loops. The nucleotide-binding folds (NBFs) contain Walker A and B consensus sequences for ATP binding. The protein appears to be internally duplicated around a further domain, the R domain, which contains many highly charged residues and potential PROTEIN PHOSPHORYLATION sites, suggesting a regulatory role for this region. Structures analogous to the R domain are not found in other members of the ABC family which includes the MULTIDRUG RESISTANCE (MDR) proteins (P glycoproteins), the product of the *STE6* gene of yeast, the bacterial periplasmic transport systems and the product of the White locus in *Drosophila*.

Mutations in CFTR

The first mutation described in CF patients is also the most common and is a deletion of 3 bp resulting in deletion of phenylalanine at amino acid position 508 (ΔF508). The frequency of this mutation varies widely between different countries, accounting for almost 90% of CF mutations in Denmark, but only ~25–30% in Macedonia and Turkey. It is not yet clear exactly how this mutation affects CFTR function, partly because the biochemistry of CFTR is poorly understood. The 3-bp deletion is in exon 10 which codes for part of the first NBF. ΔF508 lies between the Walker A and B motifs. It is possible that the mutation affects ATP binding, or that it gives rise to a conformational change which hinders interaction of the NBF with other parts of the molecule, for instance the R domain (see below). This latter hypothesis is supported by the fact that other single amino-acid deletions close to ΔF508 can also cause CF, and by physicochemical analyses which demonstrate that deletions such as these are likely to have a profound effect on the structure of the β-pleated sheet predicted for this region.

An international consortium has been established with the aim of collating data on the nature, frequency, population genetics and phenotypic effects of mutations within CFTR. Several interesting facts have emerged from this study now that more than 200 mutations have been described. There is a correlation between HOMOZYGOSITY for ΔF508 and the presence of pancreatic insufficiency; patients who are pancreatic sufficient usually have one 'mild' (non-ΔF508) allele. There appears to be little correlation between GENOTYPE and lung function. Homozygous null CF mutations (i.e. where mutations on both chromosomes lead to a truncated CFTR) can result in mild, moderate, or severe phenotypes. Mutations within the first NBF are much more common than within the second NBF. This may reflect an increased sensitivity of the CFTR molecule to structural changes in this region as suggested by studies involving TRANSFECTION of mutated DNA (see below) and POST-TRANSLATIONAL MODIFICATION studies. Despite extensive analysis only two MISSENSE MUTATIONS have been found within the R domain.

Mutations within CFTR have been shown to occur on both chromosomes 7 of male patients without CF but who suffer from congenital absence of the vas deferens (CBAVD). This birth defect is the basis of the male infertility seen in CF. Thus it would appear that CF and CBAVD are ALLELIC.

Transfection studies of normal and mutant CFTR

Creation of a full-length CFTR cDNA was retarded by the presence of an *Escherichia coli* 'poison sequence' (i.e. a sequence that cannot be cloned within *E. coli* as it disrupts cell function). This lies within exon 6 and has sequence similarity to prokaryotic PROMOTERS. However, using low copy number PLASMIDS, or *in vitro* MUTAGENESIS to alter the sequence slightly, full-length native CFTR cDNA could be stably propagated. Introduction of native CFTR into a CF pancreatic tumour cell line (by retroviral infection, *see* RETROVIRUS VECTORS) and into a CF epithelial cell line (by the vaccinia virus/T7 expression system) complemented the physiological defect seen in CF. These experiments proved (for those that were not convinced by the genetics) that CFTR was the gene defective in CF.

Although circulating lymphoblasts from CF individuals show no pathological phenotype they do demonstrate a defect at the biochemical level. CF lymphoblasts have a defective cAMP-activated chloride conductance which is limited to the G1 stage of the CELL CYCLE. ELECTROPORATION of native CFTR cDNA into CF lymphoblasts complements the defect. Interestingly, the chloride

channel produced in this case appears to be of the higher conductance (30 pS) type.

Transfection of CFTR cDNA into a variety of cell types has demonstrated that the protein undergoes at least two rounds of post-translational modification. One model proposes partial glycosylation in the endoplasmic reticulum, followed by further glycosylation in the Golgi apparatus (*see* GLYCOPROTEINS AND GLYCOSYLATION). The different forms of CFTR may be distinguished by their mobility on gel ELECTROPHORESIS. Most (but not all) of the naturally occurring CF mutations result in a defective glycosylation of CFTR. The immature forms of CFTR are recognized by a quality control mechanism and are rejected from the normal intracellular trafficking pathway (*see* PROTEIN DEGRADATION; PROTEIN TARGETING). The absence of mature CFTR at the cell surface would correlate with the lack of chloride transport activity seen in CF cells, so there may be more than one mechanism by which CFTR mutations cause the CF phenotype. Mutations within NBF1 are more susceptible to detection by the quality control mechanism than those in NBF2, but the meaning of this result is unclear.

Expression of CFTR

The POLYMERASE CHAIN REACTION (PCR) has been used to examine CFTR expression in different tissues, cell lines and at different stages of development. Airway epithelial cells from normal individuals and ΔF508 homozygotes express CFTR in similar amounts, probably one or two mRNA transcripts per cell. In the mid-trimester human foetus CFTR is expressed in lung, pancreas, gastrointestinal tract, and male genital ducts. CFTR transcription has been detected in several cell types of nonepithelial origin such as fibroblasts, U-937 histiocytic lymphoma cells, K-562 erythroleukaemia cells, fresh lymphocytes, neutrophils, monocytes, and alveolar macrophages. Absence of detectable transcripts on NORTHERN BLOTTING and nuclear RUN-OFF experiments shows that expression is at a low level in most of these cells. T84 colon carcinoma cells, for instance, transcribe 200–400 times more CFTR than fibroblasts.

In situ HYBRIDIZATION studies in tissue sections from the rat show a pattern of expression more in keeping with the distribution of the disease pathology. Transcripts are seen in the lungs (in the underlying lamina propria as well as the surface epithelium), pancreas, intestine, salivary glands, and testes.

Immunocytochemical studies with ANTIBODIES directed at synthetic peptides from CFTR have been used to examine the localization of the protein product. The protein was detected in various epithelia including the lung, sweat gland ducts, intestine, and pancreatic ducts. The level of expression in the lung was surprisingly low, especially when compared to that seen in the kidney tubules; no CFTR protein was evident in pancreatic acini.

Expression of CFTR as judged by the presence of mRNA and the protein product does seem quite specific for epithelia, but not all the organs involved have a major impact in the pathology of the condition. The low level of expression seen in nonepithelial cells may indicate that CFTR has a function in most cell types, but that this function is not essential or may be carried out by other systems. The CFTR promoter certainly has several characteristics suggestive of a HOUSEKEEPING GENE, for example it has no TATA box, a high G + C content with several possible Sp1 binding sites, and multiple transcription start sites.

CFTR function: channel or ATP-driven transporter?

Although it has been suggested that CFTR may be the chloride channel defective in CF, the CFTR protein structure predictions were interpreted as indicating that CFTR functions in a similar manner to the more than 40 known ABC transporters, that is CFTR transports some substrate or substrates in a vectorial fashion with a requirement for ATP. However, there are several lines of evidence suggesting that CFTR is a chloride channel:

1 Transfection of CFTR into a series of nonepithelial cell lines confers a cAMP-responsive chloride permeability on the recipient cells. This channel has a low conductance and a linear current–voltage relationship, properties which it is now thought are characteristic of the channel affected in CF.

2 *In vitro* mutagenesis has been used to replace positively charged amino acids within the transmembrane domains with negatively charged residues [3]. This results in a channel with altered anion selectivity, iodide being favoured over chloride. Lys 95 and Lys 335 seem particularly important as part of a high-affinity anion-binding site.

3 Deletion of the R domain results in a constitutively open channel which is no longer dependent on cAMP-stimulated protein kinase for activation [4].

In summary, CFTR looks like an ATP-dependent active transporter, but behaves as if it is a passive ion channel. One model of CFTR function proposes that it is a channel but that the R domain blocks the pore. Phosphorylation of the R domain removes the obstruction by way of a conformational change in the protein.

Interaction between the R domain, PKA, and the NBFs

What then of the function of the nucleotide-binding folds? The clustering of mutations (including ΔF508) within NBF1 certainly points to the importance of this region. One piece of evidence suggesting a regulatory role is provided by data demonstrating that deletion of most of the R domain suppresses the effect of a mutation in NBF2 (but not a mutation in NBF1). Stronger evidence for such a function comes from further patch-clamp studies of CFTR-transfected cells. As discussed above, ATP and PKA are required to activate the CFTR chloride channel in excised membrane patches from such cells. This PKA-induced activity is transient, but ATP added to the membrane patch can reactivate the channel even in the absence of PKA. This effect does not seem to be due to the presence of trace amounts of PKA (removed by sequential washes of the patch), as it is preserved in the presence of specific PKA inhibitors.

These data may be interpreted as showing that ATP binding (presumably to the NBFs) can maintain the open state of a chloride channel that has initially been activated by PKA-induced phosphorylation of the R domain. Use of ATP analogues suggests that ATP hydrolysis is required for this action, but that the effect is not dependent on a reversible phosphorylation (i.e. sequential action of a kinase and a PROTEIN PHOSPHATASE).

ATP regulation of chloride transport activity is seen in cells transfected with CFTR lacking the R domain, with CFTR bearing mutations in the Walker motif of NBF2, but not with CFTR with mutations in NBF1. ADP competitively inhibits the CFTR chloride channel by interacting with NBF2, but not NBF1, providing further evidence that the NBFs have functionally distinct roles. Together, these results suggest a model for CFTR function whereby PKA phosphorylation of the R domain is necessary for the chloride channel activity, and that this modification allows subsequent reactivation of the channel by ATP hydrolysis at NBF1 (Fig. C79). NBF2 could be involved in modulation of R domain activity. It is likely that certain phosphatases can be stimulated to deactivate the channel by dephosphorylation of the R domain.

It is important to bear in mind that CFTR may have functions besides that of a chloride channel. In the case of the MDR protein, transfection of MDR constructs can induce a cAMP regulated, 10 pS linear chloride current in the recipient cells. The MDR protein therefore appears to be bifunctional, at least in drug-resistant cell lines, with a passive anion channel and vectorial, energy-dependent drug transport activities.

Transfection of mutant (ΔF508) CFTR into fibroblasts and microinjection into a XENOPUS OOCYTE EXPRESSION SYSTEM seems to produce a CFTR which has residual activity [5]. Although no CFTR protein could be detected in fibroblasts transfected with the mutant gene (presumably because of abnormal processing), a chloride current 10–25% of wild-type values could be observed following stimulation with a cocktail of drugs which raises the intracellular cAMP level. The relevance of these findings to the *in vivo* abnormalities is unclear, but they do offer hope that in some tissues at least there may be sufficient mutant CFTR to act as a target for pharmacological agents which could boost residual channel activity. The low conductance CFTR channel is also capable of regulating the larger conductance ORCC as demonstrated in transfection studies. In cells lacking CFTR the ORCC cannot be activated by PKA, but this activity is restored on transfection of CFTR. The ORCC is therefore another potential target for pharmacological intervention, for example externally applied ATP or UTP can activate the ORCC. Despite the advances outlined above, much work will be needed to correlate the molecular and biochemical defects characterized so far with all the abnormalities seen at the physiological level.

Animal models

Since the identification of CFTR efforts have been made to create mouse models for CF via HOMOLOGOUS RECOMBINATION in embryonic stem cells. The mouse homologue of *CFTR* (*Cftr*) has been cloned and shown to have strong sequence similarity with its human counterpart (78% identical at the amino acid level). In particular, those amino acids at which naturally occurring CF mutations occur in man were found to be conserved (including ΔF508). Genomic clones containing various exons of *Cftr* were isolated and used to create constructs which 'knocked out' *Cftr* upon homologous recombination. The frequency of homologous recombination of *Cftr* sequences in embryonic stem cells was disappointingly low, but three laboratories have reported success-

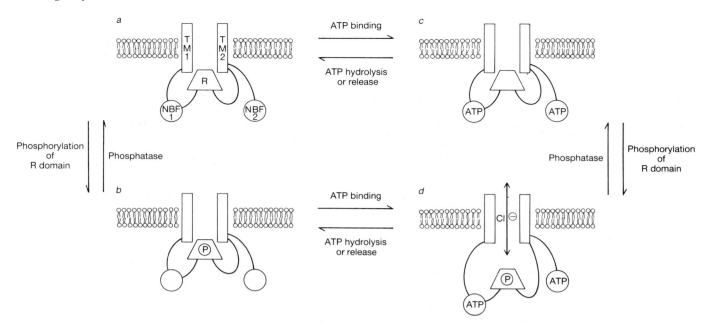

Fig. C79 A model for the regulation of CFTR. In its ground state (*a*) CFTR has an unphosphorylated R domain, and no ATP is bound to the nucleotide-binding folds (NBFs). β-Adrenergic stimulation or other second messenger systems activate protein kinase A and the R domain is phosphorylated (*b*). If ATP is bound to the NBFs before the action of phosphatases on the R domain, the chloride channel opens (*c*). Alternatively, ATP binding may occur before R domain phosphorylation (*d*). ATP hydrolysis at the NBFs deactivates the channel (*a* or *b*). TM, transmembrane region. The idea that the R domain may actually block the channel is taken from [4]. Adapted from [5].

ful generation of mice with homozygous mutations in the CF gene [6,7].

The mice with the *cf/cf* genotype have pathological features shortly after birth which closely resemble the syndrome as seen in humans, for example meconium ileus and electrophysiological changes. There are interesting and important differences in phenotype between mice produced in separate experiments. In one model [6] the mice died within 30 days of birth from the complications of intestinal obstruction, whereas in a second case [7] the mice survived and seemed relatively mildly affected. These differences may be due to factors such as genetic background or a low level of residual CFTR activity depending on the homologous recombination strategy used. The TRANSGENIC mice produced in these experiments will have several uses in the further elucidation of CFTR function and in the trial of new treatments for CF, including gene therapy. More refined experiments aimed at recreating the ΔF508 and other mutations are underway.

New therapeutic possibilities using CFTR

The complementation of the CF defect *in vitro* shows that it might be feasible to attempt to correct the basic defect of CF *in vivo* by introducing the normal gene into those cells affected by the disorder. As most of the morbidity and mortality of CF is second-ary to the pulmonary complications the first approaches to CF gene therapy have concentrated on targeting the lungs and airway epithelium. A major advantage is the accessibility of the affected tissue to agents introduced by aerosol inhalation.

Several vector systems are being assessed. Immunoliposomes are under development which will carry targeting, hydrophobic, and DNA-binding motifs. Retroviruses have been developed which can introduce reporter genes (e.g. α_1-antitrypsin gene) into murine respiratory tract epithelium and correct the biochemical defect of CF in freshly isolated CF respiratory tract epithelial cells. Expression seems stable over a period of months. Transfection of just 10% of CF respiratory tract epithelial cells *in vitro* is sufficient to provide electrophysiological correction of the defect, possibly because these cells are electrically coupled.

Another approach is to use viruses that are naturally tropic for the affected cell types, for example ADENOVIRUS and adeno-associated viruses. Human CFTR sequences have been transferred to rat lung epithelial cells *in vivo* using an adenovirus vector, and the human protein product was detectable 11–14 days after infection [8]. Transcripts from the donor gene were detectable for up to six weeks. A third system under investigation is the potential of receptor-mediated endocytosis using conjugates of DNA and various targeting moieties. However, the first success with gene therapy of an animal model has been achieved using the most simple delivery system, aerosolled liposomes [9].

It is not clear what the results of long-term treatment with such viruses would be. Particular concerns are the possible integration of transfected DNA into important host sequences (e.g. potential ONCOGENES), recombination of the donor viruses with naturally occurring viruses to produce more virulent strains, and the possible harmful effects of overexpression and ectopic expression of CFTR. Mice transgenic for CFTR under the control of the surfactant PROTEIN C promoter overexpress CFTR in several cell types of the lung without obvious adverse effects. Preliminary clinical trials of gene therapy for CF using both liposomes and adenovirus vectors are in progress.

P.J. SCAMBLER

1 Tsui, L-C. (1991) Probing the basic defect in cystic fibrosis. *Curr. Opinion Genet. Devel.* **1**, 4–10.
2 Rommens, J.M. et al. (1989) Identification of the cystic fibrosis gene: Chromosome walking and jumping. *Science* **245**, 1059–1065; Riordan, J.R. et al. Identification of the cystic fibrosis gene: Cloning and characterization of complementary DNA. *ibid.* 1066–1073; Kerem, B-S. et al. Identification of the cystic fibrosis gene: Genetic analysis. *ibid.*, 1073–1080.
3 Anderson, M.P. et al. (1991) Demonstration that CFTR is a chloride channel by alteration of its anion selectivity. *Science* **253**, 202–254.
4 Rich, D.P. et al. (1991) Effect of deleting the R-domain on CFTR-generated chloride channels. *Science* **253**, 205–207.
5 Wine, J.J. (1991) The mutant protein responds. *Nature* **354**, 503–504.
6 Snouwaert, J.N. et al. (1992) An animal model for cystic fibrosis made by gene targeting. *Science* **257**, 1083–1088.
7 Dorin, J.R. et al. (1992) Cystic fibrosis in the mouse by targeted insertional mutagenesis. *Nature* **359**, 211–215.
8 Rosenfeld, M.A. et al. (1992) *In vivo* transfer of the human cystic fibrosis transmembrane conductance regulator gene to the airway epithelium. *Cell* **68**, 1–20.
9 Hyde, S.C. et al. (1993) Correction of the ion transport defect in cystic fibrosis transgenic mice by gene therapy. *Nature* **362**, 250–255.

cystic fibrosis antigen *See*: CALCIUM-BINDING PROTEINS; CYSTIC FIBROSIS.

Cystoviridae *See*: BACTERIOPHAGES.

***cyt* gene** *See*: IPT GENE.

cytactin Tenascin. *See*: EXTRACELLULAR MATRIX MOLECULES.

cytidine The nucleoside containing the BASE cytosine. *See*: NUCLEOSIDES AND NUCLEOTIDES.

cytidine monophosphate, diphosphate, triphosphate (CMP, CDP, CTP) *See*: NUCLEOSIDES AND NUCLEOTIDES.

cytochalasin A family of fungal alkaloids (Fig. C80) which inhibit several kinds of cell movement, for example locomotion, phagocytosis, CYTOKINESIS, and ruffling, by binding to the growing end of actin filaments preventing the addition of further actin molecules. *See*: CELL MOTILITY; ENDOCYTOSIS; MICROFILAMENTS.

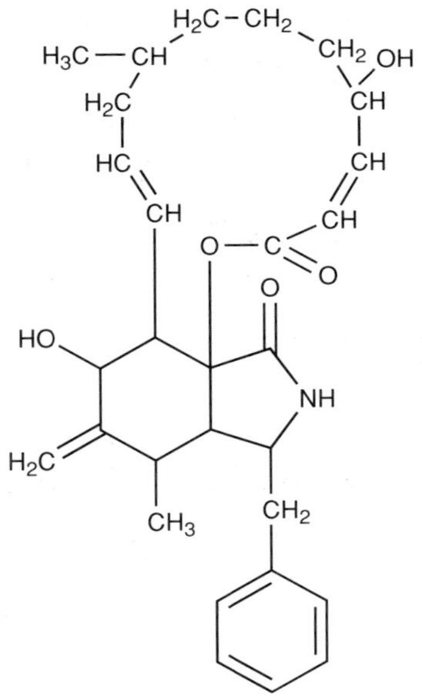

cytochrome(s) Haem proteins that catalyse electron transfer reactions, particularly in the respiratory and photosynthetic ELECTRON TRANSPORT CHAINS (*see also* PHOTOSYNTHESIS). Cytochromes are classified as *a*-type, *b*-type, or *c*-type according to the nature of the porphyrin in the HAEM prosthetic group. See individual entries.

cytochrome *a*, cytochrome *a*$_3$ Components of the cytochrome *c* oxidase of the mammalian mitochondrial ELECTRON TRANSPORT CHAIN.

cytochrome *b* Components of the QH$_2$-cytochrome *c* reductase complex of the mammalian mitochondrial ELECTRON TRANSPORT CHAIN. *b*-type cytochromes are also found in mammalian electron

Fig. C80 Cytochalasin B.

transport pathways in ENDOPLASMIC RETICULUM membranes (e.g. in CYTOCHROME P450 monooxygenases).

cytochrome *b*$_{245}$ One of the proteins of the ELECTRON TRANSPORT CHAIN involved in the oxidative burst that accompanies PHAGOCYTOSIS. The gene maps to the X-chromosome in humans. A defect at this locus causes CHRONIC GRANULOMATOUS DISEASE in which the phagocytes are unable to kill ingested microorganisms, which survive intracellularly. This leads to chronic local granulomatous inflammation with damage to the surrounding tissue. Despite the use of antibiotics, few children with this condition survive to adulthood. *See also*: CYTOCHROMES.

cytochrome *b*-559 Type of cytochrome *b* associated with the PHOTOSYSTEM II reaction centre. Oxygenic organisms contain two cytochrome *b*-559 components which are readily distinguished by their REDOX POTENTIALS. Cytochrome *b*-559$_{HP}$ (high-potential component, E_m = + 380 mV) consists of a heterodimer (α subunit, M_r 9000; β subunit, M_r 4000) containing one haem group. These subunits are encoded by the *psbE* and *psbF* genes of chloroplasts. They are closely associated with the D1 and D2 subunits of the photosystem II reaction centres (two haems per P680). The low-potential component (E_m = + 20 mV) occurs in stroma lamellae, possibly as a component of stroma photosystem II complexes, although the polypeptide concerned has not been positively identified. *See also*: CYTOCHROMES; PHOTOSYNTHESIS.

cytochrome *b*-563 The CYTOCHROME *b* of the CYTOCHROME *bf* complex in the thylakoid membranes of oxygenic organisms. It comprises an integral membrane polypeptide of M_r 24 000 with two haem PROSTHETIC GROUPS, which are distinguished by different REDOX POTENTIALS (− 50 and − 150 mV) and slightly different spectra, the high-potential component having a more asymmetric α-band. It is encoded by the *petB* gene. *See also*: CYTOCHROMES; PHOTOSYNTHESIS.

cytochrome b_6 The term originally applied by Hill and Davenport to the chloroplast CYTOCHROME *b* before it was realized that several components are present. Strictly speaking, therefore, it is both obsolete and ambiguous. Nevertheless, it is still sometimes used as a synonym for CYTOCHROME *b*-563.

cytochrome bc_1 complex This term is applied to the proton-pumping cytochrome *bc* complexes of both photosynthetic bacteria, in which it mediates the reoxidation of ubiquinol by the reaction centre, and mitochondria, in which it is the central part of the respiratory chain and is often known as complex III (*see* ELECTRON TRANSPORT CHAIN). The polypeptide composition of the complex in photosynthetic bacteria is much simpler than that of mitochondria, consisting of three main polypeptides, cytochrome c_1, cytochrome *b*, and the RIESKE IRON–SULPHUR PROTEIN. *See also*: CYTOCHROMES; ION PUMP; MEMBRANE TRANSPORT SYSTEMS; PHOTOSYNTHESIS.

cytochrome *bf* complex The protein complex of thylakoid membranes in oxygenic photosynthetic organisms that mediates electron transfer from PHOTOSYSTEM II to PHOTOSYSTEM I and pumps protons across the membrane (*see* PHOTOSYNTHESIS). It contains four main polypeptides — cytochrome *f* (M_r 34 000 (34K)), *petA* gene), the RIESKE IRON–SULPHUR PROTEIN (20K, nuclear encoded, *petC*), cytochrome *b*-563 (20K, *petB* gene) and a 17K polypeptide (subunit IV, *petD* gene) that contains no prosthetic group but may contribute to one of two quinone-binding sites and has sequence similarity to the C-terminal portion of mitochondrial cytochrome *b*. Antimycin and myxothiazol, inhibitors of the quinone-binding sites of the cytochrome bc_1 complex, have no effect on the cytochrome *bf* complex but similar effects are obtained with MOA-stilbene (methoxyacrylate-stilbene) and stigmatellin, respectively. *See also*: CYTOCHROMES.

cytochrome *c* Small haem-containing protein of M_r 12 400 (140 amino acids) which is a component of ELECTRON TRANSPORT CHAINS in mitochondria and bacteria (Fig. C81). It contains one atom of iron per molecule and, unlike haemoglobin, the haem is bonded to the protein via thioether links with cysteine residues. It is N-terminally acetylated. Cytochromes *c* from different species are highly conserved in sequence and structure.

cytochrome c_1 *See*: ELECTRON TRANSPORT CHAIN.

cytochrome *d* *See*: ELECTRON TRANSPORT CHAIN.

cytochrome *f* Cytochrome *f* (from the latin *folium*, leaf) is a *c*-type CYTOCHROME encoded by the *petA* gene, and is a component of the CYTOCHROME *bf* COMPLEX. It has a high redox potential ($E_m = +380$ mV), M_r 32 000, and in the *bf* complex receives electrons from the RIESKE IRON–SULPHUR PROTEIN and transfers them to plastocyanin (*see* PHOTOSYNTHESIS). Topological modelling based on sequence information and other evidence indicates that it possesses a single transmembrane helix, a large N-terminal haem-containing domain on the luminal side, and a short C-terminal domain on the stromal side.

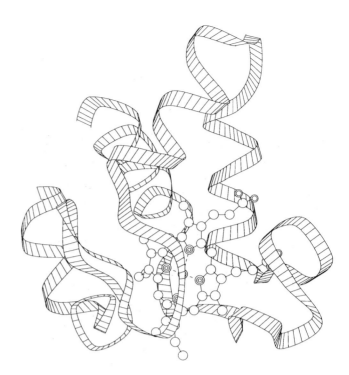

Fig. C81 Cytochrome c_{551}.

cytochrome *o* *See*: ELECTRON TRANSPORT CHAIN.

cytochrome P450 A collective term for an extensive family of haem-containing electron-transport molecules (*see* CYTOCHROMES) present in liver microsomes and involved in enzymatic oxidation of a wide range of substrates and their conversion to forms that are more easily excreted (Fig. C82). In some cases the metabolites produced may be carcinogenic (*see* CHEMICAL CARCINOGENS AND CARCINOGENESIS). Cytochromes P450 are also involved in the synthesis of compounds such as steroid hormones and prostaglandins. Many of the 200 or so genes that comprise this family are inducible by various exogenous agents. The inducibility shows genetic variability.

cytogenetics The science that links the study of the physical appearance of chromosomes with genetics. The study of KARYOTYPES. *See*: CHROMOSOME ABERRATIONS; HUMAN GENE MAPPING.

cytokeratins Members of the family of INTERMEDIATE FILAMENT proteins specific to epithelial cell types. Two types exist: acidic (type I) and basic (type II). One of each type is required to form a heterodimer which can assemble into higher order filamentous structures. At least 30 distinct cytokeratins exist which are differentially expressed. Defects in structure of cytokeratins are responsible for at least some inherited skin diseases, for example EPIDERMOLYSIS BULLOSA SIMPLEX.

Steinerts, P.M. & Roop, D.R. (1988) *Annu. Rev. Biochem.* **57**, 593–626.

cytokines Intercellular signals, usually protein or GLYCOPROTEIN, involved in the regulation of cellular proliferation and

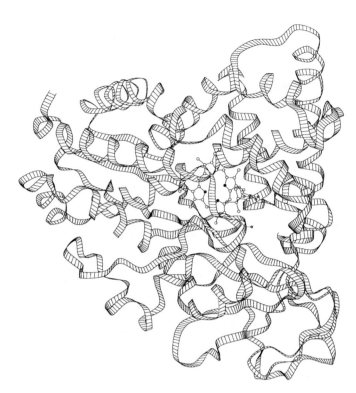

Fig. C82 Cytochrome P450.

function (*see also* GROWTH FACTORS). Those involved in hae-matopoiesis and lymphopoiesis have been most extensively studied, and the application of gene cloning techniques (*see* GENETIC ENGINEERING) has made available large quantities of highly purified novel materials. Many of the cytokine receptors are closely related, with the typical Trp-Ser-X-Trp-Ser motif in the extracellular domain, and several show TYROSINE KINASE activity in the cytoplasmic (C-terminal) domain (*see* GROWTH FACTOR RECEPTORS). Another family of cytokine receptors has been identified (e.g. for growth hormone), which have a single transmembrane domain and no intrinsic kinase activity. On stimulation they become associated with cytoplasmic tyrosine kinases. *See also*: INTERLEUKINS; LYMPHOKINES.

cytokinesis The portion of cell division that achieves a partitioning of the cytoplasm. It is distinguished from KARYOKINESIS, which is the division of the nucleus, usually achieved by MITOSIS. Cytokinesis usually begins during ANAPHASE and is completed during TELOPHASE, but in special circumstances mitosis will occur without cytokinesis and vice versa. Cytokinesis in most animal cells and protozoa is accomplished by cleavage, that is the formation of a furrow that constricts the diameter of the plasma membrane at the plane where the metaphase plate used to be and pinches the cell into two. In many cells the furrow forms in association with a contractile ring that is composed of actin MICROFILAMENTS, cytoplasmic MYOSIN, and associated proteins that gather at the cell equator to form a membrane-associated cinch. In higher plants, cytokinesis results from the action of the PHRAGMOPLAST, an array of MICROTUBULES that forms parallel to

the spindle axis in the region between the separating chromosomes. The phragmoplast moves many small vesicles to the cell's midplane where they fuse to form a large, flat vesicle that grows in diameter until it reaches the plasma membrane, where it fuses to form an extracellular space that transects the cell. New cell wall is then laid down to form the 'cell plate' that completes the process of division. Cytokinesis in some lower plants involves the 'phycoplast', another structure that uses microtubules to assemble membranous vesicles. In fungi and bacteria, cytokinesis involves the formation of a septum, achieved by the ingrowth of the plasma membrane and associated wall material. While this is mechanically analogous to the formation of a cleavage furrow, the mechanism appears to be different, and it is not yet well understood.

The site of cytokinesis is defined by aspects of the MITOTIC SPINDLE. In animal cells, a pair of ASTERS will induce cleavage at the cell membrane midway between them, regardless of the position of the asters. Thus an asymmetric position of the spindle is coupled to an asymmetric cleavage. In higher plants the phragmoplast forms at the plane defined by the 'pre-prophase band', a cortical ring of microtubules that anticipates the position of the metaphase plate. The mechanisms for this cross-talk between structures in the cell cortex and mitotic microtubules are not yet understood.

cytokinin One of the classes of plant hormones (not to be confused with mammalian CYTOKINES). The term was first proposed by Letham as a generic name for factors with similar activity to KINETIN, that is promoting cell division in plant tissue cultures in the presence of AUXIN. Although a number of compounds with diverse structures stimulate plant cell division and are said to have cytokinin-like activity, the term cytokinin is usually reserved for N^6-substituted adenine derivatives. As well as their activity in tissue cultures, cytokinins promote shoot initiation from callus, release lateral buds from dormancy, stimulate cell expansion in leaves, delay leaf senescence, and promote chloroplast development from etioplasts. The first cytokinin to be described was kinetin, which was isolated from autoclaved herring sperm DNA and is not a natural plant product. Natural cytokinins include ZEATIN, DIHYDROZEATIN, ISOPENTENYLADENINE, and BENZYLADENINE. Their N^9-ribosides and -ribotides, to which they are interconverted *in vivo*, are also physiologically active. Cytokinins, which are produced in root tips and young developing seeds, are derived from adenosine 5′-monophosphate (AMP) and isopentenyl-diphosphate by the action of AMP isopentenyl transferase. They are also components of tRNA. *See also*: PLANT HORMONES.

cytoplasm Cell contents within the plasma membrane (or cell membrane in bacteria) other than the nucleus. In eukaryotic cells it comprises the cytosol and the cytoplasmic organelles such as mitochondria, chloroplasts, etc.

cytoplasmic determinant A substance, organelle, or property which becomes allocated unequally between the two daughters of a cell division. This determinant is thought to be able to generate two daughter cells with different DEVELOPMENTAL POTENTIAL as a result of a single cell division. Although usually the term is used to refer to cytoplasmic substances or organelles, the determinant could be in the cell membrane or associated with the nucleus.

cytoplasmic inheritance Inheritance determined by the source of cytoplasm in a cross, usually mediated through mitochondrial or chloroplast DNA. It is therefore often determined by the female parent, and follows non-Mendelian patterns. *See*: CHLOROPLAST GENOMES; MITOCHONDRIAL GENOMES.

cytoplasmic male sterility (cms) Any of various types of MALE STERILITY in plants determined by the cytoplasm (usually by mitochondrial DNA). In maize, includes C-type (cmsC), S-type (cmsS), and T-type (cmsT). *See*: MITOCHONDRIAL GENOMES: PLANT.

cytoplasmic streaming Rapid streaming movement of cytoplasm seen especially in large plant and algal cells such as those of the charophyte alga *Nitella* in which the cytoplasm moves directionally and continuously around the cell. Movement of cytoplasm is thought to be due to the interaction of MYOSIN molecules with a stationary layer of actin filaments on the surface of a static cortical layer of cytoplasm. *See also*: CELL MOTILITY; MICROFILAMENTS.

cytoplast (1) A small anucleate fragment of a cell. *See*: SOMATIC CELL HYBRIDS.
(2) The living contents of a cell; cytoplasm and nuclei.

cytosine (C) A pyrimidine, one of the four nitrogenous BASES in DNA and RNA in which it pairs with guanine (*see* BASE PAIR). *See*: DNA; NUCLEIC ACID STRUCTURE; RNA. *See also*: NUCLEOSIDES AND NUCLEOTIDES.

cytoskeleton Dynamic filamentous networks of MICROFILAMENTS, MICROTUBULES, and INTERMEDIATE FILAMENTS, characteristic of eukaryotic cells, and involved in the control of cell shape, motility, intracellular organization, and cell division. *See*: ACTIN-BINDING PROTEINS; CELL JUNCTION; CELL MOTILITY; MEMBRANE STRUCTURE; MICROTUBULE-BASED MOTORS; MITOSIS; MUSCLE; MYOSINS; NEURONAL CYTOSKELETON; NUCLEAR ENVELOPE; NUCLEAR MATRIX.

cytosol In eukaryotic cells that part of the cytoplasm other than the cytoplasmic organelles and cytoskeleton. It is the site of much intermediary metabolism and biosyntheses, including PROTEIN SYNTHESIS on ribosomes.

cytotoxic T cells (T_C cells) Effector T lymphocytes, also widely known as cytotoxic T lymphocytes (CTL), which can destroy virus-infected cells and the cells of an allogeneic transplant. They recognize foreign antigen in the context of class I MHC MOLECULES (*see* MAJOR HISTOCOMPATIBILITY COMPLEX) and express the class I co-ligand CD8. CD4-positive MHC class II-restricted cytotoxic T cells have been reported, but they constitute only a minority of the population. Each cytotoxic T cell can kill several targets. It secretes the contents of secretory granules into the immediate vicinity of the target cell to which it is attached. The granules contain PERFORINS and various lysosomal enzymes (*see* LYSOSOMES). It takes in general less than 10 min for the cytotoxic T cell to deliver its lethal hit, and target cell death occurs by APOPTOSIS rather than by true lysis.

D

D (1) The charged AMINO ACID aspartic acid.
(2) Deuterium, the hydrogen isotope ^2_1H, mass 2.015.

D Originally identified by kinetic studies as an electron donor not involved in the oxidation of water, which donates an electron to P680 of the PHOTOSYSTEM II reaction centre, D has subsequently been identified with the substance that gives rise to EPR signal II_{slow} in its radical oxidized form, and as Tyr 160 of protein subunit D2 of the reaction centre. *See also*: ESR/ESP SPECTROSCOPY; PHOTOSYNTHESIS.

2,4-D (2,4-dichlorophenoxyacetic acid) A highly active, synthetic AUXIN (Fig. D1) whose effects on plants were first described in 1942. M_r 221. At high concentrations it is herbicidal, particularly against dicotyledons, and the free acid, its salts, or esters are used extensively in agriculture as selective systemic herbicides, mainly in cereal crops. Typical effects of herbicidal doses are inhibition of elongation growth, epinastic bending of leaves (perhaps as a result of increased ethylene biosynthesis), radial expansion, and, finally, disorganized cell proliferation.

Zimmerman, P.W. & Hitchcock, A.E. (1942) *Contrib. Boyce Thompson Inst.* **12**, 321–343.

Fig. D1 Structure of 2,4-D (2,4-dichlorophenoxyacetic acid).

D arm Part of a TRANSFER RNA molecule.

D-DNA *See*: NUCLEIC ACID STRUCTURE.

D–J joining The joining of a D segment to a J segment during rearrangement at the immunoglobulin heavy chain locus and at T cell receptor γ and δ loci. At the heavy chain locus, production of a functionally rearranged $V_H D_H J_H$ sequence occurs by initial D_H to J_H joining followed by recruitment of a germ-line V_H to the already juxtaposed $D_H J_H$. $D_H J_H$ joints are often found in T cells whereas $V_H D_H J_H$ joints are effectively specific to B cells. D to J joining also seems to precede DJ to V joining at the other loci. *See*: GENE REARRANGEMENT; IMMUNOGLOBULIN GENES; T CELL RECEPTOR GENES.

D-loop (1) A stable intermediate formed during the replication of double-stranded DNA genomes of organelles. For example, mito-

chondrial DNA REPLICATION (*see* MITOCHONDRIAL GENOMES) starts at a specific ORIGIN OF REPLICATION, but only one of the two DNA strands, the so-called H-strand, is copied. In mammalian mitochondria this new synthesis ceases after around 500 bases of the H-strand have been copied, but the newly synthesized strand is sufficient to displace the original parental strand as a displacement or D-loop (Fig. D2*a*). Subsequent DNA replication results in the unidirectional extension of the newly synthesized strand, thereby expanding the D-loop. Mitochondria from other organisms contain anywhere between one and six D-loops per genome. A similar D-loop structure has also been seen during chloroplast DNA replication (*see* CHLOROPLAST GENOMES).
(2) The DNA structure formed when one of the two strands of a double-stranded DNA molecule is displaced by an invading single-stranded region (Fig. D2*b*) such as is seen during HETERODUPLEX formation during RECOMBINATION.

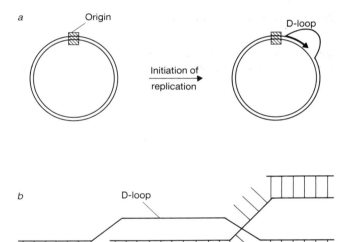

Fig. D2 *a*, Generation of a displacement (D)-loop during mitochondrial DNA replication; *b*, generation of a D-loop by strand displacement during recombination.

D segments (diversity segments) One of the three types of gene segment involved in assembly of the variable region coding sequence (V region) of immunoglobulin and T cell receptor polypeptide chains. D segments are present in the immunoglobulin heavy chain locus and in T cell receptor β and δ loci. In an immunoglobulin heavy chain, the D segment contributes between 1 and 17 amino acids of the third COMPLEMENTARITY DETERMINING REGION (CDR3) of the V region, located ~100 amino acids from the N terminus. D segments are unique in having heptamer–nonamer recombination signals on both sides of the coding sequence enabling joining to both an upstream V segment and a downstream J segment. *See*: GENE REARRANGEMENT; IMMUNOGLOBULIN GENES; T CELL RECEPTOR GENES.

D1, D2 subunits/polypeptides These two polypeptides (M_r 32 000 and 34 000 and encoded by the *psbA* and *psbB* genes respectively), together with CYTOCHROME B-559, form the reaction centre of PHOTOSYSTEM II. D1 is also known as the Q_B or herbicide-binding protein. They are responsible for binding the pigments and cofactors involved in the primary photochemical charge separation in PHOTOSYNTHESIS, the reduction of primary (Q_A) and secondary (Q_B) quinone acceptors, and possibly also the Mn atoms concerned with O_2 evolution. They show sequence similarity with each other, and functional and topological similarity with the L and M subunits of the reaction centre of purple bacteria.

D108 *See*: TRANSPOSABLE GENETIC ELEMENTS.

DAB 3,3'-Diaminobenzidine (or 3,3',4,4'-tetraaminobiphenyl) is a very sensitive substrate for horseradish peroxidase and is commonly used for immunohistological staining. An initially colourless solution of DAB turns an intense brown when oxidized by the enzyme in the presence of hydrogen peroxide. DAB staining is compatible with a wide range of common histological stains, since the reaction product is insoluble in both water and alcohol.

DAF DECAY ACCELERATING FACTOR.

DAG DYSTROPHIN-ASSOCIATED GLYCOPROTEINS.

daidzein 7,4'-Dihydroxyisoflavone.

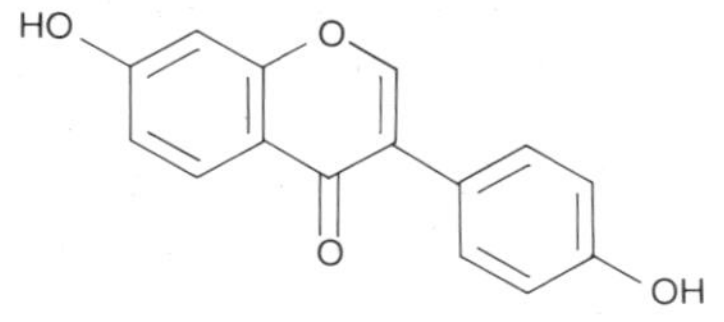

Found in soybean (*Glycine max*), species of *Cicer*, *Trifolium*, *Baptisia* and *Puerasia*, and in the Genistae.

dam methylase An enzyme from *Escherichia coli*, encoded by the *dam* (defective adenine methylation) gene, which modifies the adenine residue in the sequence 5' GATC 3' to N^6-methyladenine. Dam methylation of certain RESTRICTION SITES, for example those for *ClaI*, *BclI*, and *NruI*, prevents them from being recognized by the corresponding RESTRICTION ENZYME. However, the dam methylase is not part of a RESTRICTION AND MODIFICATION system, but rather is important during MISMATCH REPAIR of mismatched base pairs in newly replicated DNA. *See also*: DNA METHYLASE.

DAP L-2,6-Diaminopimelic acid (systematic name: 2,6-diaminoheptane-1,7-dioic acid), HOOC.CH(NH_2).(CH_2)$_3$.CH(NH_2).COOH. Important intermediate in the biosynthesis of L-lysine in bacteria,

cyanobacteria, green algae, higher plants and some fungi (not *Neurospora crassa* or yeasts). Converted to the *meso*-isomer, before decarboxylation to L-lysine.

DAPI 4',6-Diamidino-2-phenylindoledihydrochloride (Fig. D3). A fluorescent stain used to detect DNA in immunofluorescence microscopy. It binds preferentially to AT-rich DNA (e.g. mitochondrial DNA) and can be used in place of ETHIDIUM BROMIDE in DENSITY GRADIENT CENTRIFUGATION methods, particularly during the preparation of mitochondrial DNA.

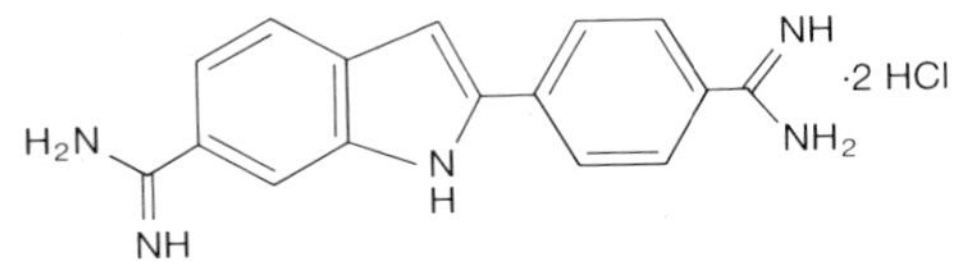

Fig. D3 Structure of DAPI.

DARPP Dopamine and cAMP-regulated *phospho*protein (M_r 23 000) is a potent and specific inhibitor of protein phosphatase PP1 (*see* PROTEIN PHOSPHATASES). Residues 9–50 of DARPP show 60% identity to the corresponding region of INHIBITOR 1 (another specific inhibitor of PP1), although after residue 50 there is little sequence similarity between the two proteins. Residues 8–38 constitute an active fragment and phosphorylation of threonine 34 by cyclic nucleotide-activated PROTEIN KINASES is a prerequisite for inhibition of PP1. DARPP is found at high concentrations in the basal ganglia, where its distribution correlates with dopamine-innervated neurons that possess D1 DOPAMINE RECEPTORS. Brown adipose tissue and kidney also contain high concentrations of DARPP. Dopamine inhibition of the Na^+, K^+-ATPase activity in renal tubule cells may be mediated through DARPP.

Darwinian evolution

THE theory of evolution published by Charles Darwin in 1859 [1] laid out a view of the natural world that caused an intellectual revolution in his day and whose implications we are still coming to terms with. Darwin's 'theory of evolution by means of natural selection' is in reality not one single theory but a complex set of interdependent subhypotheses [2,3]. The background for the development of the theory came largely from the five years (1831–1836) Darwin spent circumnavigating the globe as explorer, naturalist and companion to the captain of the *Beagle*, a Royal Navy survey ship whose primary task was to chart the coasts of South America.

During this time Darwin's scientific work was primarily in geology [4]: he wrote three books on the geology of the voyage whereas in biology he collected for others to interpret. On returning to London he was active in the Geological Society rather than, say, the Linnean Society, and his scientific articles were on the important geological topics of the day. However, the distinction between geology and biology was not as sharp as it is now.

For example, the second volume of Charles Lyell's three-volume *Principles of Geology* dealt with biological processes and how they affect geological processes such as rates of erosion. The volume included a long discussion on Lamarck's evolutionary theory together with Lyell's reasons for rejecting it. It also had extensive discussions of ecological principles including intra-and interspecific competition and the potential for increases in population numbers. It gave a mechanistic view of ecological processes (including extinction) which was largely derived from the Swiss plant ecologist De Candolle.

Darwin's geological background, and Lyell's approach in particular, was important for the development of Darwin's theory. Lyell was following the approach of an earlier Scottish geologist, John Hutton, who in the late 18th century had insisted on explaining past events strictly according to 'causes now in operation'. (This phrase is used in the subtitle of Lyell's *Principles*, namely, 'Being an enquiry how far the former changes of the earth's surface are referable to causes now in operation'.) It is now accepted that we have misled ourselves by applying the term 'uniformitarianism', with its emphasis on constant rates, to the ideas of Hutton and Lyell. If a name is required it is more accurate to use 'actualism', where the emphasis is on the 'actual' mechanisms being the same in the past as in the present. Lyell, and later Darwin, protested at any implication that rates would always be uniform — they saw no mechanism that would give constant rates. Lyell, in particular, emphasized that slow, continued small changes could sometimes lead to 'catastrophic' changes over a short period. Both Hutton and Lyell allowed a time scale of hundreds of millions of years and with such a time scale even slow processes would have major effects even if they only operated intermittently.

Darwin's work on the *Beagle* showed that for both South America and coral islands, major geological events could be understood as resulting from processes that are still operating in the present, and he was influential in helping to convince other geologists that Lyell's was a fruitful approach to geology.

The theory of evolution

Charles Lyell stopped at applying his mechanistic principles to the origin of species. To him, species were specially created, though once created were left to the operation of natural (scientific) laws and could, under those laws, become extinct. Darwin's early work extended Lyell's reasoning to the origin of species although the eventual implications went well beyond what either of them would have anticipated. His theory, combined with some important implications, may be summarized under the four headings of macroevolution (history), microevolution (mechanisms), whether microevolutionary mechanisms are *sufficient* to understand the past (actualism or Darwinism), and the implications of the mechanisms.

(1) Macroevolution: evolution as history

There is a continuity of living forms through geological time and this unifies all life in that all species share a common ancestor in the past. The relationship between species is best represented as a tree and this pattern arises from speciation and extinction. Darwin assumed that species can split into two or more lineages, thus increasing the number of species. However, unlike many earlier workers, he accepted that extinctions had occurred. Many earlier biologists assumed either that species created perfectly would not be allowed to become extinct, or that a mechanism would exist that always ensured species could adapt to a changing environment (Lamarck).

The processes of change, speciation and extinction continued over a time scale of hundreds of millions of years. This long time scale allowed relatively small forces to eventually have major effects. There was no end to the process, with change, extinction and speciation continuing all the time, albeit with varying rates at different times and places. In this area the long geological time scale was the essential contribution from Lyell's geology.

(2) Microevolution: mechanisms for evolution

The key to Darwin's theory was a testable mechanism that resulted in change over time. The two main features of this mechanism were the potential for increases in population numbers and the consequences of heritable diversity. It was well known before Darwin that species have the potential for an exponential increase in population numbers but also that the observed numbers for most species remain, over a number of years, roughly constant. Several authors had realized that because environmental resources are limited there must be both intra- and interspecific competition. Lyell, for example, quotes De Candolle's 'all plants are at war, one with another'.

But competition for survival would not result in any change if all individuals were identical. The second part of Darwin's mechanism considered variability within species. Again, variability within populations was well known and some of this variation was known to be inheritable. Before Darwin nearly all authors had assumed that selection could only eliminate variability which, by its very nature, was undesirable. Indeed, there was no real alternative to this view while it was assumed that species were perfectly created to fit their environment so that any variability was a 'deviation' from the unchanging and unchangeable 'ideal type'.

It is in this aspect of his theory that Darwin owes a particular debt to Lyell's insistence that all features of a mechanism to explain past events had to be 'causes now in operation'. Darwin fully accepted this line of reasoning and argued that all parts of his mechanism could be studied in the present — population numbers, inheritable variability and the higher diversity among sexually generated offspring. A difficult and crucial task was to show that variability could affect survival and here the experience of plant and animal breeders was important. As long as some of this inheritable variability affected the probability of survival and reproduction in the present environment (physical and biological) there would eventually be changes in the population.

The net result of all these processes is a change in frequency of variants between generations. The potential for excess population, together with limited resources, inheritable variability, and the effect of some part of this variability on survival and reproduction, are the processes resulting in natural selection. Darwin stressed that the resultant rate of change would be quite variable when examined on a geological time scale.

(3) Darwinism — or actualism

The processes of microevolution are *sufficient* to account for macroevolution, including the origin of humans with their mental processes and intellectual powers. Darwin was quite explicit that he considered mechanisms still operating to be both necessary and sufficient, 'If it could be demonstrated that any complex organ existed which could not possibly have been formed by numerous, successive, slight modifications, my theory would absolutely break down.' [1, p. 189].

He assumed that additional mechanisms for evolution would be found. The only limitation was that any new mechanism must be able to be studied in the present. Darwin considered sexual selection to be different from natural selection, though modern work (by concentrating on numbers of viable offspring, rather than survival of the original parents) usually considers sexual selection as an aspect of natural selection.

Whether currently operating mechanisms are sufficient to explain evolution has been the most controversial aspect of the theory. Many people accepted that the processes of natural selection would be necessary, but not sufficient — 'something else' would also be needed. Indeed, it was only with the development of the synthetic theory of evolution, neo-Darwinism, from the 1930s onwards, which incorporated Mendelian genetics into the theory, that the claim for sufficiency has been seriously recognized. In recent decades many new details of mechanisms have come from studying the evolution of DNA and protein sequences (*see* MOLECULAR EVOLUTION).

(4) Implications

The theory as outlined so far is largely an extension of Hutton's and Lyell's approaches to explaining events in the distant past. But the implications went far beyond. In several cases they contradict ideas that had been central to Western thought from at least medieval times, and in some cases from the ancient Greeks.

According to Darwin's theory, there was no need to postulate a plan, or ultimate cause, for organisms. There need be no 'purpose' for organisms, and especially no need for an animate world for humans 'to have dominion over'. The concept of ultimate cause had been influential in Western thought from at least the time of Aristotle. Darwin's theory did not exclude the possibility of purpose in the universe, only that it was unnecessary to postulate it.

There was a strong chance (stochastic) factor in evolution. It arises in many ways: in mutations (the origin of inherited variability), in survival and in selection (or the lack of it). There are deterministic factors also, but allowing a major role for nondeterministic factors in evolution was seriously criticized in Darwin's time, particularly by physicists imbued with Newtonian determinism.

The mechanisms explain the 'fit' between form, function and environment. Explaining the fit between organisms and their environment as proof of divine wisdom and design had been the main support for natural theology. This approach to nature (including the physical world) had been a major element in English thought since the late 17th century. Under Darwin's theory, in contrast, highly ordered structures can arise gradually from less ordered states and lead to new features: orchids can eventually arise from *Chlorella*, mammals and birds from unicells. Each intermediate form of the new structure has to be beneficial to its possessor (or at least not be harmful). A much discussed example is the origin of feathers where it is difficult to imagine a series of intermediates between scales and feathers that would assist flying. But if early feathers had another function (such as increasing insulation and thereby aiding temperature regulation) then intermediate states would benefit their possessors.

Because of the nature of inheritance the benefit (or liability) is conferred on the possessors of a particular combination of genes. Natural selection thus acts on individuals, and GENE FREQUENCIES change in populations. Such changes are not 'for the good of the species', for natural selection cannot be oriented to future goals. Evolutionary processes can lead to cooperation both within and between species and early biologists were well aware of this possibility. Coevolution between species occurs regularly, as in the 'arms race' between plants and the insects that feed on them. A corollary is that there is no unchanging 'essence' or 'ideal type' for any group or species, thus contradicting a powerful idea from Plato that was important in medieval thought.

Another important implication was that organisms were no longer expected to be perfectly 'designed' for their environment [5]. Many features will be less than perfect because of constraints from pre-existing forms, limited variability, the problem of conflicting requirements, and the requirement that all intermediate forms must be functional. Natural selection is expected to lead to local optima given the genetic diversity available, but not necessarily to the global optimum. Examples considered by Darwin were the explosive increase in numbers of some plants and animals introduced into new lands, suggesting that the previous inhabitants did not appear to be perfectly adapted to their own environment, eyes (including human) that do not give perfect vision and male nipples (which were unnecessary, and therefore imperfect to the Victorian mind). To Darwin, adaptation was relative rather than absolute.

Similarly there are no predetermined pathways of evolution. Earlier orthogenetic ideas envisaged complex organisms slowly unfolding over geological periods of time. In contrast, under Darwin's theory, if for example all vertebrates were removed from the Earth there is no expectation that the same body plans would develop again.

Taken overall, it was the implications of the theory that destroyed the concept of a purpose-built, human-centred universe and gave the theory such a powerful impact — an impact it still retains.

Application to macromolecules

The development of studies in MOLECULAR EVOLUTION and MOLECULAR PHYLOGENY has greatly strengthened the Darwinian view of evolution by revealing new mechanisms that help to explain past events. Today we have no trouble with genetic variability continually arising regardless of the needs of species or individuals, either short term or long term. Individual genes fit a model of 'descent with modification' and duplication of genes can lead to new gene functions. Stochastic procedures are now well accepted in science.

Our current view of evolution is much closer to that of Darwin than was common in the early stages of the synthesis of evolutionary theory and genetics (1930–1970). In the early literature of molecular evolution 'Darwinian evolution' was sometimes erroneously limited to cases where a mutation had been incorporated as a result of positive selection. We know many examples of positive selection at the molecular level but Darwin clearly stated that mutations could be neutral, 'neither beneficial nor injurious', and that such mutations could either remain a fluctuating element in the population (a polymorphism), or go to fixation. It would be more appropriate to call the neutral theory of evolution (*see* MOLECULAR EVOLUTION) 'ultra-Darwinian' rather than 'non-Darwinian'.

The Darwinian approach is currently being extended from living systems to search for simple chemical systems that could help explain the origin of life itself. In many ways the approach of Hutton, Lyell and Darwin in explaining past events by mechanisms that can be studied in the present has been so successful that we take much of it for granted. Increasingly, new mechanisms of evolution will be discovered from the study of macromolecules and in addition, molecular biology is now the prime source of information about the mechanisms of evolution.

D. PENNY

1 Darwin, C. (1859) *The Origin of Species by Means of Natural Selection* (John Murray, London).
2 Mayr, E. (1991) *One Long Argument: Charles Darwin and the Genesis of Evolutionary Thought* (Harvard University Press, Cambridge, MA).
3 Riddiford, A. & Penny, D. (1984) The scientific status of evolutionary theory. In *Evolutionary Theory: Paths into the Future* (Pollard, J.W., Ed.) 1–38 (Wiley, London).
4 Herbert, S. (1986) Darwin as a geologist. *Sci. Am.* May, 94–101.
5 Monod, J. (1977) Evolution as tinkering. *Science* **196**, 1161–1166.

Darwinism A descriptive term used to encapsulate Darwin's theory of evolution and its application. *See:* DARWINIAN EVOLUTION; SYNTHETIC THEORY OF EVOLUTION.

Databases and information handling in molecular biology

PROGRESS in basic and applied biological research depends increasingly on factual and bibliographic databases and on computing tools for organizing, analysing and modelling data. Current important databases include information on sequences of nucleic acids and proteins, structures of biological macromolecules, genome maps, genetic diseases, microbial strains, hybridomas, restriction enzymes, cloning vectors, industrial enzymes, toxicological data, and abstracts from biological journals. Table D1 presents a list of databases of particular interest to molecular biologists. These databases are produced by a variety of organizations or individual researchers, and vary considerably in size and timeliness and in the sophistication of their management,

structure and accessibility. Great challenges exist to improve existing databases, to create new ones, and to integrate these sources to facilitate flexible access to their diverse information.

Usage of databases

What are databases used for? A database user might view the database as a compendium of the scientific knowledge in a particular field of biology. Databases are therefore often used like encyclopedias for information retrieval, that is, to look up what information is known about a certain topic. A typical question is 'Find all sequences of mammalian mitochondrial genes containing protein-coding regions, published since 1990'. The entries fulfilling these requirements are then extracted from the database and can be further analysed.

Perhaps the most important application of sequence databases is for sequence similarity searches. A researcher can take a newly determined nucleotide or protein sequence and compare it to all sequences already stored in the databases. By applying appropriate search algorithms it is possible to find sequences with varying levels of similarity to the query sequence. Thus it is often possible to infer possible biological functions for unknown sequences or to get some indication of putative phylogenetic relationships between sequences (*see* DNA SEQUENCING; MOLECULAR EVOLUTION; MOLECULAR PHYLOGENY).

The databases themselves can also be used directly as objects for biological research. By thorough analysis of the information stored in the databases it is often possible to gain new insights into biological phenomena. As an example, comparison of all sequences from a gene or protein family can reveal sequence patterns which are characteristic for these families (*see e.g.* GENE FAMILY; GENE SUPERFAMILY; IMMUNOGLOBULIN SUPERFAMILY MOLECULAR EVOLUTION: SEQUENCES AND STRUCTURES). Similarly, particular functional sequences such as promoters, signal peptides or intron/exon boundaries can be identified and described. These investigations can have immediate effects on the experimental work. For instance, the analysis of the base composition of DNA sequences showed profound differences in CODON BIAS between sequences from different species and between strongly and weakly expressed genes and enabled significant improvements to be made to heterologous gene expression systems.

The usefulness of databases depends largely on the programs available for querying the database and the tools for manipulating and analysing the data. A large range of software products is available for all important hardware platforms and operating systems, ranging from comprehensive commercial packages to public-domain programs developed for special applications. Because of the complexity and breadth of this area the interested reader is referred to the literature for further information [2–4].

A common problem in the analysis of biological data is the current lack of standardization. Database formats are often incompatible, programs only run on certain machines, and exchanging information between different programs is difficult. Efforts are currently under way to encourage the use of standard data-exchange formats and to develop platform-independent software products, led by the American National Center for Biotechnology Information (NCBI).

Table D1 Some useful biological databases

Database	Supplier	Type of information contained
Sequence databases		
EMBL/GenBank/DDJB	EMBL, GenBank	DNA and RNA sequences
GenInfo	NCBI	
SWISS-PROT	EMBL	Protein and peptide sequences
PIR/NBRF	PIR	
VecBase	PIR	Cloning vector sequences
tRNA database	EMBL	Alignment of tRNA sequences
Sequence feature databases		
ECD	EMBL	*Escherichia coli* references, sequences and genetic map positions
EPD	EMBL	Eukaryotic promoters in EMBL/GenBank/DDBJ
TFD	NCBI	Target sites and sequence of eukaryotic transcription factors
PROSITE	EMBL	Protein sequence motifs
Structure databanks		
PDB	Brookhaven National Laboratory	Protein, DNA, and carbohydrate structure atomic coordinates
CARBBANK	Carbbank	Complex carbohydrate structures
Mapping databases		
GDB, OMIM	GDB	Human genetic and physical mapping data
CEPH	CEPH	
GBase	Jackson Laboratory	Mouse genetic and physical mapping data
Biological material collections		
ATCC databases	ATCC	Collections of cell lines, hybridomas, and microbial strains
MSDN	MSDN	Microbial and cell line information resources
Other		
REBASE	EMBL	Restriction enzyme data
ENZYME	EMBL	Enzyme EC numbers and reactions
LiMB	GenBank	Database of molecular biological databases

This table illustrates the range of databases available and is not comprehensive. For each database a supplier is indicated in the middle column. In many cases the databases can be obtained from other sources as well. See also the special database issues of *Nucleic Acid Research* [1].

Biological databases

Nucleotide sequence databases

The EMBL, GenBank and DDBJ database groups collaborate to collect nucleotide sequence data directly from the research community and from the scientific literature. EMBL and GenBank manage the increasing data volume within similar relational database management systems and produce and distribute separate databases of functional équivalence, though with different formats and some value-added features. The EMBL nucleotide sequence database in mid-1994 contained some 182 615 entries of sequence data comprising over 192 million base pairs. It is assumed that by the year 2000 the database will be at least a hundred times larger as a consequence of genome sequencing projects. Each entry is annotated with details such as sequence features (e.g. protein-coding regions, regulatory signals), unique identifiers (accession numbers), descriptive titles, taxonomy, references (e.g. journal citations), and cross-references to related databases (e.g. protein sequences, specialized annotation databases).

Protein sequence databases

Protein sequence databases are currently maintained separately from the nucleotide databases, although most data are derived from the nucleotide sequences of genes or messenger RNAs. They include annotation of features such as motifs and secondary structures. The PIR international collaboration consisting of groups at PIR (USA), MIPS (Germany) and JIPID (Japan) produce the PIR protein sequence database. The SWISS-PROT protein sequence database has a close relationship with the EMBL nucleotide databases and is produced by a collaboration between EMBL and the University of Geneva. SWISS-PROT is fully annotated and extensively cross-linked to other databases (e.g. EMBL nucleotide database, PIR protein database, Brookhaven structure databank, PROSITE pattern database).

Structural databases

Structural databases, which store the atomic coordinates of large and small biomolecules, are indispensable for protein structure analysis and drug design. The most important databases in this area are the Brookhaven crystallographic protein databank (PDB) and the Cambridge structural database of small molecules. By now, only a few hundred protein structures have been solved experimentally, so attempts have been made to derive protein structural information from protein sequences using the known structures of homologous proteins (HSSP database). A collection of structural data also exists for carbohydrates (CARBBANK).

Mapping databases

Owing to the rapid advance in genome research the interest in mapping databases has grown considerably. Physical maps are amongst the primary goals of most genome sequencing projects. Physical mapping facilitates the sorting of clone libraries and allows the mapping of new DNA sequences to their physical location on a chromosome (*see* HUMAN GENE MAPPING; HUMAN GENOME PROJECT). Most of these databases are currently not publicly available. Genetic linkage maps have been collected for many organisms such as humans, mouse, fruit fly (*Drosophila*) and microorganisms, and additional detail of gene function is often provided, as for example in the OMIM database of human genetic diseases. The most important mapping database is the genome database (GDB), maintained at the Johns Hopkins University, which provides information about the human genetic map, including data from OMIM and the mouse mapping database GBase.

Databases of biological function

Complementing the major databases mentioned above there are various more specialized data collections presenting detailed information on the biological roles and properties of genes or gene products. Scientists can obtain valuable information on, for instance, restriction enzymes (REBASE), promotor regions (EPD), transcription factors (TFD), or protein motifs (PROSITE). These databases are often closely linked to the major sequence databases, and there is now a network of pointers and cross-references between the general sequence databases and these specialized databases which allows the database users to locate related information in different databases quickly.

Other databases

These include: databases of two-dimensional protein gels which link genetic sequence information to gene products; information about microbial strains which can be found in catalogues offered via the Microbial Strain Data Network (MSDN) or from the American Type Culture Collection (ATCC) which also stores and distributes hundreds of cell lines, microbial strains, genetic probes and gene libraries; databases which contain information about enzymes and the reaction they catalyse; and databases of databases such as LiMB, the Listing of Molecular Biological Databases.

Of great importance to biologists are bibliographic databases like MEDLINE, CAS Online, BIOSIS, or EMBASE which try to cover all the articles published in the scientific literature. By providing on-line access to abstracts of these articles they constitute an indispensable information resource.

Access to databases

Several methods are used by database suppliers to provide access to their data. Some examples are given below, but as details tend to change over time database producers should be contacted directly for up-to-date information. In addition, databases are often reformatted and redistributed as part of commercial software packages.

Magnetic tape

Magnetic tape is the traditional medium for the regular distribution of large databases. A variety of tape formats exist to cover the range of end-user computer systems. Magnetic tape is normally used only as a distribution medium rather than as an on-line version of a database. Data would typically be copied off the tape to magnetic disk, then perhaps be reformatted for use with specific software packages. All the sequence and structure databases are available on magnetic tape from their producers. For example, EMBL distributes the EMBL nucleotide and SWISS-PROT protein sequence databases every 3 months on magnetic tape. GenBank and PIR distribute their databases similarly, with PIR also supplying query software for VAX/VMS systems.

CD-ROM

In recent years CD-ROM has become an attractive medium for distribution of large volumes of data due to its compactness, large capacity and standardization of format. CD-ROM readers are inexpensive and convenient peripherals compared to tape drives or hard magnetic disks. An international standard (ISO 9660) exists for CD-ROM format which enables the same disk to be used on a range of computer systems. It can be suitable simply as a distribution medium or further as a medium for direct working access. Although the speed of access to a CD-ROM is slower than to a magnetic disk, its large capacity encourages extensive indexing which can, for many applications, overcome the limitations of the inherent slow access. EMBL currently distributes a CD-ROM containing the EMBL nucleotide and SWISS-PROT protein sequence databases along with retrieval software for MS-DOS systems. The GenBank database is also now available on CD-ROM, and NCBI have produced a demonstration CD-ROM containing the GenInfo database.

File server

Many databases are available from a variety of file servers around the world. A file server is a computerized data archive that holds information available for anyone who can connect to this computer system via local or wide-area computer networks. These servers are typically operated by electronic mail requests to retrieve whole databases or specific database entries, or to perform certain analyses such as sequence homology searches. The international Internet computer network allows more direct interactive retrieval using a protocol called FTP which is becoming ever more popular. Currently the main file servers for access to molecular biology databases are maintained at EMBL, NCBI and GenBank. The EMBL file server, for example, contains the EMBL nucleotide and SWISS-PROT protein sequence databases, including the newest data created between releases, many other related databases, and a collection of free molecular biology software for UNIX, VMS, MS-DOS and Macintosh systems. Information about data access via file servers can be obtained from the database suppliers listed in Table D1 (addresses in Table D2).

On-line access

For most uses, it is desirable to work interactively with an on-line database system. Databases which are distributed on magnetic tape are typically installed on a user's local shared computer system. In many cases, the software to access the database is supplied separately. There are therefore hundreds of on-line versions of, for example, the nucleotide and protein sequence databases around the world. In many countries there are also database hosts available as commercial or publicly funded services. As the size of certain databases increases dramatically, and also as the need to access remote specialized hardware and software arises, the requirement for networked resources becomes important. One initiative being coordinated by EMBL is the development of a European molecular biology network (EMBnet) to link bioinformatics resources centres, a collaborative project involving some 14 European countries. EMBL maintains remote copies of the nucleotide database by distributing daily the newest sequence data via computer network to nationally mandated nodes. These nodes in turn undertake to render this data available to their user community, accessible either to analysis software as part of a general bioinformatics on-line service or by further redistribution. The GenBank On-line Service (GOS) is a more centralized example of this type of access.

P. STOEHR

R. FUCHS

1 *Nucleic Acid Res.* (1991) **19** (Supplement).
2 Doolittle, R.F. (Ed.) (1990) Molecular evolution: computer analysis of protein and nucleic acid sequences. *Meth. Enzymol.* **183** (Academic Press, San Diego).
3 von Heijne, G. (1987) *Sequence Analysis in Molecular Biology* (Academic Press, San Diego).
4 Bishop, M.J. & Rawlings, C.J. (Eds) (1987) *Nucleic Acid and Protein Sequence Analysis. A Practical Approach* (IRL Press, Oxford).

dATP Deoxyadenosine 5′-triphosphate, a nucleotide precursor for DNA synthesis. *See*: NUCLEOSIDES AND NUCLEOTIDES.

dauer larva A semidormant larval form of *Caenorhabditis* induced under conditions of starvation or overcrowding (*see* CAENORHABDITIS DEVELOPMENT). Entry into the state is triggered by a low molecular weight pheromone-like molecule which is secreted by the animals during growth. Several genes that function in the pathway of dauer larva formation have been identified.

Table D2 Supplier addresses

ATCC	American Type Culture Collection, 12301 Parklawn Drive, Rockville, MD 20852, USA (dmaglott@helix.nih.gov)
Brookhaven National Laboratory	Chemistry Department—Building 555, Upton, NY 11973, USA (pdb@bnlchm.bitnet)
Carbbank	Complex Carbohydrate Research Center, 220 Riverbend Road, Athens, GA 30602, USA (carbbank@uga.bitnet)
CEPH	Centre d'Etude du Polymorphisme Humain, 27 rue Juliette Dodu, Paris 75010, France (bc@frceph51.bitnet)
EMBL*	EMBL Data Library, Postfach 10.2209, 6900 Heidelberg, Germany (datalib@embl-heidelberg.de)
GDB	William H. Welch Medical Library, Johns Hopkins University, 1830 East Monument Street, Baltimore, MD 21205, USA (help@welch.jhu.edu)
GenBank	Mail Stop K710, Los Alamos National Laboratory, Los Alamos, NM 87545, USA (genbank@life.lanl.gov)
Jackson Laboratory	The Jackson Laboratory, Bar Harbor, ME 04609, USA
MSDN	Microbial Strain Data Network, University of Cambridge, 307 Huntingdon Road, Cambridge CB3 0JX, UK (CODATA/Dialcom: 42:cdt0001)
NCBI	National Center for Biotechnology Information, National Library of Medicine, Bethesda, MD 20894, USA (benson@ncbi.nlm.nih.gov)
PIR	National Biomedical Research Foundation, 3900 Reservoir Road, Washington, DC 20007, USA (pirmail@gunbrf.bitnet), or, in Europe MIPS, Max-Planck-Institut fuer Biochemie, 8033 Martinsried, Germany (mewes@vaxl.mips.mpg.dbp.de)

* Moving in September 1994 to the European Bioinformatics Institute, Cambridge, UK.

dcm methylase An enzyme from *Escherichia coli* encoded by the *dcm* gene that adds methyl groups to the C5 position of the internal cytosines in the sequences 5′ CCAGG 3′ and 5′ CCTGG 3′. Although not part of a host RESTRICTION AND MODIFICATION system dcm methylation of certain restriction enzyme sites, for example those for *Sfi*I, *Stu*I, prevents them from being recognized by the corresponding RESTRICTION ENZYME.

DCMU 3(3,4-Dichlorophenyl)-1,1-dimethylurea, the most potent of a series of inhibitors of PHOTOSYNTHESIS that act by competing with PLASTOQUINONE for the Q_B-binding site of PHOTOSYSTEM II. It is used as a nonselective herbicide (diuron). It is only a weak inhibitor of the reaction centre of purple bacteria. *See also:* PHOTOSYNTHESIS.

dCTP Deoxcytidine 5'-triphosphate, a nucleotide precursor for DNA synthesis. *See:* NUCLEOSIDES AND NUCLEOTIDES.

DEAD box, DEAH box Amino-acid motif (Asp-Glu-Ala-Asp or His) found together with other conserved sequences in proteins from a wide range of organisms. Those members of the family whose biochemical activities have been identified have RNA-dependent ATPase activity and/or are ATP-dependent RNA HELI-CASES. The eukaryotic INITIATION FACTOR eIF4A is the prototype member of the DEAD box family and some PRP PROTEINS (precursor RNA processing proteins) from the budding yeast *Saccharomyces cerevisiae* (e.g. PRP16, PRP22) are DEAH box proteins.

deafferentation The removal of afferent (incoming) nerve supply to a brain area, organ, limb, etc.

Debye–Waller factor Thermal motion of atoms in a crystal causes a decrease in the diffracted intensities by a factor

$$\exp[-B(\sin^2 \theta / \lambda^2]$$

where θ is the BRAGG ANGLE, λ is the wavelength and B the Debye–Waller temperature factor. This is related to the mean square total displacement U^2 by $B = 8\pi^2/3$ such that the root mean square total displacement of 0.5 Å and 1.0 Å correspond to B factors of about 6 and 27 respectively. *See:* X-RAY CRYSTALLOGRAPHY.

decapentaplegic (dpp) A *Drosophila* gene encoding a member of the transforming growth factor β superfamily (*see* GROWTH FACTORS) with many functions in development. It is required for the specification of cell fate in the dorso-ventral axis in the early embryo where the protein product forms a morphogenetic GRADIENT. Later in embryogenesis, *dpp* participates in an inductive interaction between the visceral mesoderm and endoderm in the morphogenesis of the larval gut. In imaginal disk development it is required for the correct elaboration of the proximo-distal axis. The genetics of *dpp* are complex and the gene has been functionally divided into three regions: shv, Hin, and disk. Although the gene produces multiple transcripts, all specify the same protein product from OPEN READING FRAMES located in the Hin region. The genetic complexity reflects large regulatory regions located both 5' and 3' to the transcripts. *See also:* DROSOPHILA DEVELOPMENT.

decay accelerating factor (DAF) Integral membrane protein (M_r 70 000) present on most blood cells and which acccelerates the dissociation of C3 convertases (*see* COMPLEMENT).

dedifferentiation A loss of differentiated cellular characteristics, which has been interpreted as a reversion to a DETERMINED progenitor cell state, that is the form the cell possessed prior to DIFFERENTIATION. This term can be misleading as it is often used to describe changes in cell behaviour in culture which may be difficult to distinguish from TRANSDIFFERENTIATION.

defective interfering virus (DI virus) Defective form of a viral genome with the ability to be replicated by factors provided in *trans* by the intact viral genome. The DI genomes may compete effectively with the intact genome for essential *trans*-acting factors and ameliorate the virus disease.

defective virus Virus in which functions required for replication have been lost or removed and which therefore cannot replicate without these functions being supplied by an accompanying HELPER VIRUS.

deficiency In cytogenetics, the loss of a chromosome or part of a chromosome. *See:* CHROMOSOME ABERRATIONS; DELETION.

defined medium Culture medium in which all of the chemical constituents are known.

dehydrogenases Oxidoreductase enzymes (of EC class 1) that catalyse oxidoreduction reactions involving the overall removal of hydrogen atoms (dehydrogenation) from the donor substrate in the general equation:

$$AH_2 + B \quad \rightarrow \quad A + BH_2$$
$$\text{donor} \qquad\qquad\qquad \text{acceptor}$$

See e.g. ALCOHOL DEHYDROGENASE; LACTATE DEHYDROGENASE; MECHANISMS OF ENZYME CATALYSIS.

delayed rectifier An ion channel whose opening is delayed in response to a stimulus and which acts to rectify the potential created by that stimulus, such as a DEPOLARIZING pulse. The classic delayed rectifier is the voltage-gated K^+ channel that opens in response to the change in potential induced during an ACTION POTENTIAL by increased Na^+ permeability. This channel remains open for as long as the membrane is depolarized. *See:* VOLTAGE-GATED ION CHANNELS.

deletion The loss of a segment of DNA. The extent of a deletion may vary from a single base to a section of a chromosome with the genes contained therein. Deletion mutants are organisms, cells, genes, etc. containing a deletion. *See:* CHROMOSOME ABERRATIONS; MUTATION.

deletion loop Single-stranded bubble formed in HETERODUPLEX DNA when one strand contains a deletion (Fig. D4). *See also:* HYBRIDIZATION.

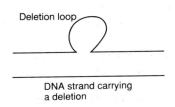

Deletion loop

DNA strand carrying
a deletion

Fig. D4 Formation of a deletion loop by the annealing of two partially complementary DNA strands, one of which has a deletion.

deletion mapping A GENE MAPPING technique that uses a panel of defined and overlapping DELETION mutants to map by COMPLEMENTATION the position of an unknown locus.

deletion mutagenesis Technique for producing a deletion *in vitro* in a cloned DNA. If the DNA contains two suitably placed RESTRICTION SITES, a precisely defined deletion may be made by cutting out the segment between them and ligating the cut ends. A deletion around a single unique restriction site may be made by cleaving the DNA at the site and degrading the cut ends further by nucleases such as BAL 31 and EXONUCLEASE III and S1 NUCLEASE. Rejoining of the two ends produces a set of DNAs with deletions of varying sizes around the restriction site. DNASE I may also be used to make an initial random cleavage and the cut ends degraded as above.

delta Headwords with the prefix delta- or δ- may also be indexed under the initial letter of the headword itself.

delta chain (δ chain) May refer to a subunit of any of numerous proteins in which subunits have been arbitrarily designated α, β, γ, δ, etc. Often refers to the δ chain of adult minor component haemoglobin (*see* HAEMOGLOBIN AND ITS DISORDERS).

delta deleting element (δ deleting element) Genetic element — ΨJ_α — at the 5′ end of the T cell receptor α locus and which is involved in deletion of the δ locus from within the α locus in the absence of α gene rearrangement (*see* T CELL RECEPTOR GENES). A site-specific RECOMBINATION occurs between δRec, which is located upstream of TCRδ, and ΨJ_α, which is a pseudo J_α element located downstream of C_δ and near the 5′ end of the J_α cluster. δRec encompasses at its 3′ end a consensus heptamer–nonamer signal sequence with a 23-base pair (bp) spacer, thus permitting, according to the 7-23 rule of VDJ recombination, rearrangement to ΨJ_α, which immediately 5′ carries a heptamer–nonamer signal with a 7-bp spacer. The developmental control mechanisms operating on this process are not yet known. The δRec–J_α recombination is formally equivalent to deletional V-J_α rearrangement, in which rearrangement of the TCRα genes is obligatorily associated with deletion of the intervening TCRδ locus.

delta virus *See:* HEPATITIS DELTA VIRUS.

dematin *See:* ACTIN-BINDING PROTEINS.

denaturation As applied to proteins and nucleic acids, the disordering of higher-level structure as a result of heating or other treatments that disrupt noncovalent interactions. Denaturation is accompanied by loss of function. *See:* NUCLEIC ACID STRUCTURE; PROTEIN FOLDING; PROTEIN STRUCTURE; RENATURATION.

dendrites Outgrowths from the neuronal cell body upon which incoming synaptic inputs generally terminate. *See:* DENDRITIC SPINE; NEURON; NEURONAL CYTOSKELETON; SYNAPTIC TRANSMISSION.

dendritic cells (dendrocyte) A particularly potent type of ANTIGEN-PRESENTING CELL. Very few dendritic cells, when compared with other types of antigen-presenting cell, are required to elicit a response from T CELLS. Dendritic cells are found primarily in the SECONDARY LYMPHOID ORGANS, although they can be found in the blood. The follicular dendritic cells of the lymph node do not express class II MHC MOLECULES (*see* MAJOR HISTOCOMPATIBILITY COMPLEX); they are therefore unable to present antigen to helper T cells. They bind IMMUNE COMPLEXES very effectively through COMPLEMENT RECEPTORS and are therefore effective in presenting antigens to B CELLS. *See also:* ANTIGEN PRESENTATION.

dendritic spine An outgrowth of a dendrite upon which excitatory synapses are thought to be located (*see* NEURON; NEURONAL CYTOSKELETON). The spine may therefore be the smallest unit which supports synaptic change. The spine proper is connected to the parent dendrite by a 'neck', whose properties may influence the degree of MEMBRANE POTENTIAL change which reaches the dendrites and hence the cell body (*see* LONG-TERM POTENTIATION; SYNAPTIC TRANSMISSION).

denervation The removal of the nerve supply to an area which can be achieved by cutting the axons projecting to that area (axotomy) or by destruction of the innervating neurons.

Denhardt's solution A solution used for blocking nylon or nitrocellulose membranes prior to the addition of a radioactive PROBE. It reduces nonspecific probe attachment to the membrane surface and is usually added as a 5× concentrate (0.1% (w/v) bovine serum albumin Fraction IV, 1% (w/v) Ficoll 400, 0.1% (w/v) polyvinylpyrrollidone in water). *See:* HYBRIDIZATION.

Denhardt, D.T. (1966) *Biochem. Biophys. Res. Commun.* **23**, 641–646.

density gradient centrifugation Centrifugation techniques which make use of the fact that molecules of a particular density centrifuged in a solution of varying density (the density gradient) will tend to collect in a band at that zone of the centrifuge cell at which their density and the density of the medium are exactly equal. Proteins of different SEDIMENTATION COEFFICIENTS can be separated by centrifugation in a sucrose density gradient. Sucrose density gradients are preformed before centrifugation by layering sucrose solutions of different concentrations and may either be continuous or form discrete steps in sucrose density. After centrifugation, the different fractions are collected through a hole in the bottom of the centrifugation tube. This method can also be used to check the purity of enzyme preparations by establishing that the protein and enzymatic activity coincide in the same fraction. Caesium chloride (CsCl) gradients can be used to separate proteins, RNA, and DNA in a mixture and to separate DNAs of different AT : GC content (and thus density). The mixture to be separated is added to a uniform solution of CsCl and subjected to high-speed centrifugation. A density gradient of CsCl is formed in the centrifuge tube, increasing towards the bottom. Supercoiled DNA (e.g. plasmid DNA) can be separated from linear or relaxed circular DNA by CsCl density gradient centrifugation in the presence of ethidium bromide.

deoxynojirimycin and deoxymannonijirimycin Inhibitors of glycosylation (*see* GLYCOPROTEINS AND GLYCOSYLATION) which interfere with glycoside hydrolysis during the processing of glycan side chains. Deoxynojirimycin inhibits glucosidase whereas deoxymannonijirimycin inhibits mannosidase.

deoxyribonuclease (DNase) A class of enzyme that enzymatically degrades DNA by hydrolysing the sugar–phosphate backbone between adjacent nucleotides. There are many different DNases in terms of both biological role and template specificity. DNase activity is involved in DNA REPAIR and DNA REPLICATION (in the excision of mismatched or chemically damaged bases), in RECOMBINATION and in DNA recycling by breakdown of DNA into its constituent nucleotides. Some DNases are able specifically to degrade single-stranded DNA (e.g. S1 NUCLEASE and MICROCOCCAL NUCLEASE), some are specific for double-stranded DNA (e.g. RESTRICTION ENZYMES), while others are able to degrade both single-stranded and double-stranded DNA (e.g. BAL31 NUCLEASE; DEOXYRIBONUCLEASE I). DNases can be further subdivided into two groups depending on their mode of action: ENDONUCLEASES, which cleave internal phosphodiester bonds within DNA, and EXONUCLEASES, which attack DNA from either the free 5′ end (5′ to 3′ exonuclease) or free 3′ end (3′ to 5′ exonuclease) and degrade the DNA strand one nucleotide at a time. Some enzymes can have both exo- and endonuclease activity (e.g. Bal31 nuclease).

deoxyribonuclease I (DNase I) An endonuclease originally isolated from bovine pancreas (EC 3.1.21.1) and which has a preference for double-stranded DNA, cleaving it endonucleolytically to 5′-phosphodinucleotide and 5′-phospho-oligonucleotide end products. It is widely used as a probe for so-called DNASE I HYPERSENSITIVE SITES in CHROMATIN, which are believed to indicate DNA available for TRANSCRIPTION.

deoxyribonucleic acid *See*: DNA.

deoxyribonucleoside, deoxyribonucleotide NUCLEOSIDE or NUCLEOTIDE in which the SUGAR is deoxyribose.

deoxyribose The SUGAR present in each nucleoside of DNA.

depactin *See*: ACTIN-BINDING PROTEINS.

depolarization A reduction in electrical potential difference across a membrane, that is, a change in MEMBRANE POTENTIAL towards zero, with the interior of the cell becoming less negative with respect to the exterior.

depression *See*: AFFECTIVE DISORDERS.

depurination Of nucleic acids, the removal of a purine base without breakage of the sugar–phosphate backbone. *See*: DNA REPAIR.

derepression The activation of TRANSCRIPTION of an individual gene or group of genes (e.g. an OPERON) following the dissociation of a REPRESSOR molecule from the OPERATOR region of a PROMOTER. Derepression is usually caused by the presence of an INDUCER which binds to the repressor. *See*: BACTERIAL GENE EXPRESSION; LAC OPERON.

dermatan sulphate A sulphated GLYCOSAMINOGLYCAN of the extracellular matrix (M_r 15 000–40 000) found chiefly in skin, blood vessels, and the heart. *See also*: EXTRACELLULAR MATRIX MOLECULES; PROTEOGLYCANS.

desensitization Entry of ligand-gated and voltage-gated ion channels into a refractory state, in which further stimulus has no effect, when exposed to high concentrations of the ligand. The process is not completely understood. In some channels desensitization seems to be mediated by PROTEIN PHOSPHORYLATION. *See also*: VOLTAGE-GATED ION CHANNELS.

desmin An INTERMEDIATE FILAMENT protein.

desmocollins Family of transmembrane CADHERIN-like proteins localized to desmosomes (*see* CELL JUNCTIONS). Three desmocollin genes (*DSC1*, *DSC2*, and *DSC3*) are currently known; *DSC1* and *DSC3* are located on human chromosomes 18 and 9 respectively. The genes are differentially expressed in desmosome-containing tissues, and during differentiation. Each gene can give rise to two protein products (a and b, M_r ~120 000 and ~100 000), differing only in their cytoplasmic tail sequences by virtue of an alternative RNA SPLICING mechanism.

Buxton, R.S. & Magee, A.I. (1992) *Semin. Cell Biol.* **3**, 157–167.
Buxton, R.S. et al. (1993) *J. Cell Biol.* **121**, 481–483.

desmogleins Family of transmembrane CADHERIN-like proteins localized to desmosomes (*see* CELL JUNCTIONS). They have a characteristic large cytoplasmic domain containing various numbers of repeats of a sequence of ~29 amino acids. Three desmogleins are currently known and they are differentially expressed in desmosome-containing tissues and during differentiation. They are desmoglein 1 (desmosomal glycoprotein I, DGI: M_r ~150 000), desmoglein 2 (human desmoglein C, HDGC, partial cDNA clone only), and desmoglein 3 (pemphigus vulgaris antigen, PVA: M_r ~130 000). Desmoglein 1 may be the target of AUTOANTIBODIES in the autoimmune blistering skin disease pemphigus foliaceous, whereas desmoglein 3 is the autoantigen in pemphigus vulgaris. Genes for all three known desmogleins are on human chromosome 18.

Buxton, R.S. & Magee, A.I. (1992) *Semin. Cell Biol.* **3**, 157–167.
Buxton, R.S. et al. *J. Cell Biol.* **121**, 481–483.

desmoplakins Desmoplakins I and II (DP; M_r ~240 000 and 210 000 respectively) are cytoplasmic plaque proteins of desmosomes produced from alternatively spliced mRNAs (*see* RNA SPLICING) derived from a single gene in humans (on chromosome 6). DPI contains an extended coiled coil rod domain not present in DPII. They may interact with INTERMEDIATE FILAMENTS. *See also*: CELL JUNCTIONS.

Jones, J. & Green, K. (1991) *Curr. Opinion Cell Biol.* **3**, 127–132.

desmosome *See*: CELL JUNCTIONS.

desmoyokin Protein present at eukaryotic CELL JUNCTIONS.

destrin The ACTIN-BINDING PROTEIN ADF.

desynapsis The premature separation of homologous chromosomes that have synapsed normally during the pachytene stage of MEIOSIS but have not formed chiasmata.

determination An irreversible decision made by an undifferentiated cell which restricts its DIFFERENTIATION into a smaller subset of cell types. A cell becomes determined when its DEVELOPMENTAL POTENTIAL no longer differs from its FATE. The determined state is passed on to its progeny. *See also*: COMMITMENT; TRANSDETERMINATION; TRANSDIFFERENTIATION.

deuterencephalic induction Regional neural INDUCTION by the chorda mesoderm of the ARCHENTERON roof or exogenously applied material, producing HINDBRAIN or deuterencephalic structures. *See*: AVIAN DEVELOPMENT; VERTEBRATE NEURAL DEVELOPMENT.

deuterium oxide Heavy water, D$_2$O. The replacement of some hydrogen atoms in proteins with deuterium (^{2}H) provides a means of distinguishing certain hydrogens (i.e. readily ionizable hydrogens and those in peptide bonds, which are termed 'exchangeable hydrogens') from others. It is also used in the contrast variation technique in NEUTRON SCATTERING AND DIFFRACTION.

Deuteromycotina Also known as the Fungi Imperfecti, the Deuteromycotina are species of EUMYCOTA which either lack a sexual stage or for which the sexual form is unknown, and thus have only been observed growing as a mycelium and propagating by asexual spores (conidia). They are classified according to the morphology of the conidia. Discovery of the sexual stage of such fungi may cause them to be reclassified in one of the other classes of Eumycota.

deuterostomes A major group of bilaterally symmetric animals (i.e. excluding the radially symmetric protozoans, sponges, and coelenterates) which together with the PROTOSTOMES represent two main evolutionary lines displaying characteristic basic plans of development. Deuterostomes include echinoids and chordates, and their development differs from that of protostomes in the following ways: (1) regulative (or indeterminate) form of development, in which the fates of cells are fixed relatively late in development and often depend on interactions between cells. The embryo as a whole retains the capacity to REGULATE and correct deficiencies previous to this time; (2) cleavage is often radial (i.e. the cleavage plane is parallel or perpendicular to the animal–vegetal axis); (3) the mouth forms from a region of the embryo away from the blastopore (or the equivalent region of invagination during GASTRULATION).

developmental mutants *See*: ANTERIOR GROUP GENES; ANTENNAPEDIA COMPLEX; BITHORAX COMPLEX; CAENORHABDITIS DEVELOPMENT; CAENORHABDITIS NEURAL DEVELOPMENT; DROSOPHILA DEVELOPMENT; FLOWER DEVELOPMENT; GAP GENES; HOMEOBOX GENES AND HOMEODOMAIN PROTEINS; INSECT NEURAL DEVELOPMENT; PLANT EMBRYONIC DEVELOPMENT; SEGMENTATION GENES; VERTEBRATE NEURAL DEVELOPMENT; ZEBRAFISH DEVELOPMENT and individual gene entries.

developmental pathway The FATE of a cell or a region of cells during normal or abnormal embryonic development. The term carries no significance regarding the state of cell COMMITMENT. Thus an uncommitted cell transplanted into a novel environment may adopt a novel developmental pathway.

developmental potential The range of all the possible PHENOTYPES that can be produced by the descendants of a cell or group of cells. The concept differs from FATE: the fate of a cell or group of cells is the range of phenotypes that actually are produced by those cells in normal development, while their developmental potential includes the phenotypes that the cells can give rise to in all possible circumstances. For example, consider a cell in the ventral portion of a newly formed SOMITE. The fate of this cell (in normal development) is to become part of the SCLEROTOME, and to contribute to the axial skeleton. However, if the cell is transplanted to the dorsal part of the somite, it will now contribute to dermis or muscle or both. The fate of this cell is to form sclerotome-derived axial skeleton, but its developmental potential also includes dermis and skeletal muscle. The difference between these two concepts is central to the definition of COMMITMENT and DETERMINATION: a cell becomes determined when its developmental potential no longer differs from its fate. *See also*: AVIAN DEVELOPMENT.

dexamethasone A synthetic GLUCOCORTICOID HORMONE which is able to bind to glucocorticoid receptors (*see* STEROID RECEPTOR FAMILY) and activate hormone-dependent PROMOTERS.

dextral cleavage During SPIRAL CLEAVAGE the orientation of the third cleavage plane determines whether the embryo is defined as cleaving dextrally (i.e. the cleavage planes are rotated in a clockwise direction when viewed from the animal pole). (In sinistral cleavage the cleavage planes would be rotated anticlockwise.) The orientation of cleavage planes is determined by the organization of the egg cytoplasm. In snails with coiled shells (e.g. *Lymnaea*) the direction of shell coiling and consequent organization of the viscera depends on the orientation of the third cleavage, all of which are controlled by a single gene in the female. Both dextral and sinistral forms are found within a population but the dextral form is dominant. This gene is an example of a MATERNAL EFFECT gene as it is only expressed during OOGENESIS and hence only manifest in the eggs of daughters.

dextrans Branched polymers of glucose found in yeast and bacteria in which glucose residues are joined by α1,6 linkages. Linkages at the branch point may be 1,2; 1,3; or 1,4 depending on strain and species. Dextrans have been studied extensively because of their potential as blood plasma substitutes in treatment of shock.

dGTP Deoxyguanosine 5′-triphosphate, a nucleotide precursor for DNA synthesis. *See:* NUCLEOSIDES AND NUCLEOTIDES.

DHFR DIHYDROFOLATE REDUCTASE.

DHP receptor DIHYDROPYRIDINE RECEPTOR.

DI virus DEFECTIVE INTERFERING VIRUS.

diabetes mellitus A heterogeneous disease due to a disorder of glucose metabolism leading to hyperglycaemia which in severe cases can lead to coma and death if not controlled. It probably arises due to an interaction of both enviromental and genetic factors. Formerly classified into either insulin-dependent (IDDM) or non-insulin dependent (NIDDM) kinds, it has more recently been divided into type I and type II. Type I is classically young onset and associated with diminished production of insulin. There is an association between type I diabetes and the expression of certain HLA antigens (*see* MAJOR HISTOCOMPATIBILITY COMPLEX) particularly B8, B15, DR3, and DR4. Type I diabetes mellitus is an autoimmune disease in which the insulin-producing β-cells of the pancreas are selectively attacked and destroyed. Type II is often late onset, there is appreciable insulin secretory capacity but it may be decreased or delayed and there is a varying degree of insulin resistance especially when obesity is a feature. There is a strong family aggregation. *See also:* INSULIN AND ITS RECEPTOR.

diacylglycerol Glycerol to which two FATTY ACID residues are esterified through carboxylic ester bonds: these are usually long-chain (C_{16}–C_{24}) fatty acyl residues in natural molecules, but are of various chain lengths in synthetic diacylglycerols. Diacylglycerols exist in three isomeric forms, but the key biological molecule is *sn*-1,2-diacylglycerol (1,2-DG) (Fig. D5). 1,2-DG is an intermediate in the biosynthesis of cell LIPIDS (including triacylglycerol, phosphatidylcholine, and phosphatidylethanolamine) and is also formed as a result of the hydrolysis of any glycerophospholipid by any phospholipase C. 1,2-DG formed by receptor-activated PHOSPHOINOSITIDASE C acts as a second messenger is the major natural activator of PROTEIN KINASE C, and 1,2-DG formed by other routes may also activate protein kinase C. *See:* SECOND MESSENGER PATHWAYS.

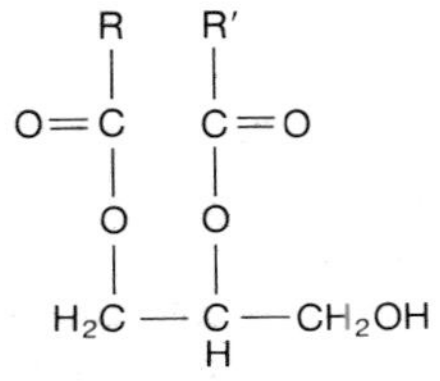

Fig. D5 Diacylglycerol.

diad Alternative spelling of DYAD, an axis of two-fold rotational SYMMETRY.

diageotropism The horizontal growth of stem structures, such as rhizomes and runners, in response to gravity.

diakinesis A stage of MEIOSIS.

diaminopimelic acid *See:* DAP.

dianthovirus group Named from *Dianthus*, the generic name of carnation, after the type member carnation ringspot virus. MULTICOMPONENT VIRUSES with isometric particles 31–34 nm in diameter. The particles contain two species of (+)-strand linear RNA (the genome organization of red clover necrotic mottle virus, the only member so far sequenced, is shown in Fig. D6). RNA 1 has three OPEN READING FRAMES. It is thought that P27 is read through into the next ORF to give P84, both of which have amino acid homologies suggestive of RNA polymerases. The 3′ gene is the coat protein. RNA 2 encodes the cell-to-cell movement function. *See also:* PLANT VIRUSES.

Francki, R.I.B et al. (1985) In *Atlas of Plant Viruses*, Vol. 2, 47 (CRC Press, Boca Raton, FL).

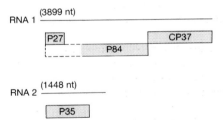

Fig. D6 Genome organization of red clover necrotic mottle virus (RCNMV). The lines represent the RNA species and the boxes the proteins (P) with M_r given as $\times 10^{-3}$; nt, nucleotides.

dicentric CHROMOSOME possessing two CENTROMERES, which can lead to chromosome breakage at MITOSIS.

dicotyledons, dicots A large group of flowering plants having an embryo with two cotyledons, parts of the flower usually in twos or fives or multiples, leaves with net veins, and vascular bundles in the stem in a ring surrounding the central pith. PLANT GENETIC ENGINEERING techniques are at present most advanced in dicot species, to which plants such as tobacco, petunia, and tomato belong. *Cf.* MONOCOTYLEDONS.

dictyosome The Golgi stack in plant cells. *See:* GOLGI APPARATUS.

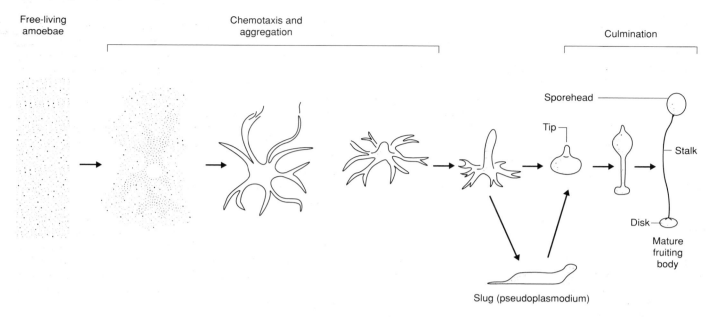

Free-living
amoebae

Chemotaxis and
aggregation

Culmination

Sporehead

Tip

Stalk

Disk

Mature
fruiting
body

Slug (pseudoplasmodium)

Fig. D7 Development and differentiation in *Dictyostelium discoideum*.

Dictyostelium development

THE cellular slime moulds — of which *Dictyostelium discoideum* is the most widely studied — are simple eukaryotes which have been used as model organisms for studies of eukaryotic cell signalling, gene expression, and development. Cells of *Dictyostelium* grow as free-living individual amoebae, normally feeding on bacteria. When their food source is exhausted, they undergo a multicellular development ending in formation of a fruiting body. This structure consists of a droplet of spore cells supported by a long tapering stalk embedded in a basal disk and can be up to a few millimetres in height. The spore cells are viable and ensure survival during periods of starvation. The stalk cells are dead but impart structure to the fruiting body, aiding spore dispersal. A fruiting body can contain between 10 and 10^5 cells, but cell proportions are maintained regardless of size. Development takes a minimum of 24 hours, during which there is little or no cell division. During this development, *Dictyostelium* exhibits many features of development seen in higher organisms, including cohesive cell contacts, cell signalling, chemotaxis, the secretion of extracellular matrices, and cellular differentiation.

Before the onset of development, the amoebae monitor cell density relative to food supply by measuring the concentration of secreted pre-starvation factor (PSF), a glycosylated, heat- and protease-labile molecule. High extracellular PSF concentration induces expression of gene products required for the aggregative phase of development. During aggregation, amoebae of *D. discoideum* are guided towards signal centres by pulses of CYCLIC AMP secreted initially by a small number of amoebae. In other species of cellular slime mould peptides or pterines serve as chemoattractants. Individual amoebae respond to the cAMP signal by move-

ment towards the signal source, changes in gene expression, and secretion of a cAMP pulse which relays the aggregation signal to cells further from the centre. As the cells approach the centre and come into contact with one another, cohesion through cell-surface molecules results in formation of multicellular streams (Fig. D7).

Aggregation results in a multicellular mound, which becomes more compact and then forms a distinct tip. The tip acts as a coordinating centre throughout the rest of development and is regenerated if excised. An extracellular matrix of GLYCOPROTEINS and CELLULOSE — the slime sheath — is secreted around the aggregate and increases its rigidity. Prespore and prestalk cells initially arise scattered throughout the aggregate. The prestalk cells sort to the tip and the base, leaving the prespore cells in the centre. An approximate 4 : 1 ratio of prespore to prestalk cells is maintained throughout the rest of development. Prestalk and prespore cells, although differentiated from each other, are not committed to their fates and TRANSDIFFERENTIATION may occur following cell loss to restore the appropriate prespore : prestalk ratio.

The tipped aggregate sometimes forms a freely motile pseudoplasmodium or slug. Slugs are long and thin, reaching up to 5 mm in length. They are photo- and thermotactic, and can move to sites more suitable for culmination — the development of the fruiting body. The duration of migration is controlled by ammonia. As the slug migrates it lays down a 'slime trail', a tube through which the main body of the slug passes.

In the slug, the tip acts in a similar manner to an embryonic ORGANIZER: when it is grafted to the side of another slug it induces a secondary axis. If the tip is removed it will regenerate from the remnant. It exerts tip dominance, preventing the formation of secondary tips.

Prespore cells are located in the slug posterior whereas the majority of prestalk cells are in the anterior. Prestalk cells are also found in a small region at the rear; these cells may be lost as the slug migrates. In addition, a small number of prestalk cells are present within the prespore zone; these are known as the anterior-like cells (ALC).

Prespore cells can be recognized by the presence of the prespore vesicles (PSV). These are specialized vesicles containing proteins destined for the spore coat. A number of prespore-specific genes have been characterized and these include the genes *psA*, *cotA*, *cotB*, and *cotC*. Cyclic AMP is required for both induction and maintenance of these markers.

Originally, prestalk (pst) cells were detected by preferential accumulation of the vital dye neutral red, and were shown to occupy the anterior fifth of the slug. The use of the *ecmA* and *emcB* genes as markers has shown the population to be more complex, comprising discrete populations of pstA, pstB, pstAB and pst0 cells.

PstA cells express *ecmA* and sort to the tip as it forms on the aggregate. They occupy the anterior tenth of the slug. PstB cells express *ecmB* alone and sort to the base where later they form part of the basal disk. PstAB cells express both *ecmA* and *ecmB* and are present in a small cone-shaped region within the slug tip. At culmination, the pstAB cells are the first cells to penetrate the prespore mass and make a substantial contribution to the basal disk. The pstA cells mature into pstAB. The pst0 cells, which also express *ecmA* but more weakly than 1pstA, occupy the second most anterior tenth of the slug and may interconvert with ALC. Both of these genes and other prestalk-specific markers are induced by differentiation-inducing factor (DIF) (see below).

At culmination, the prestalk cells undergo a series of morphogenetic movements likened to a 'reverse fountain'. They first move up the outside of the tip and then reverse direction to pass down the centre, penetrating through the prespore mass and embedding into the base. The prestalk cells then mature to form a rigid cellulose stalk tube. During maturation, they become vacuolized and deposit a cellulose cell wall.

The prespore cells are lifted up the newly synthesized stalk. As they rise, the prespore cells mature into spores. The ALC partition downwards into the base and upwards to the tip. They ultimately form part of the basal disk and a cradle to support the spore head.

The mature spores are ovoid and possess a thick cell wall containing cellulose and macromolecules released from the PSV. Spore maturation is triggered by activation of the CYCLIC AMP-DEPENDENT PROTEIN KINASE. Spores can resist a variety of environmental stresses. An inhibitor, discadenine, ensures that germination does not occur until spores have been dispersed.

Cell contact is not required for cell differentiation. Development of cells in low density monolayers or in shaking suspension has led to the identification of a number of diffusible molecules that are involved in induction of cell types and coordination of the developmental processes.

Mutants of *D. discoideum* development

Genetic estimates in *D. discoideum* suggest that as few as 300–400 genes control development, but are not essential for growth. Mutants have been isolated by conventional genetics, disruption of known genes (GENE KNOCKOUT) and random INSERTIONAL MUTAGENESIS.

The majority of mutants are unable to aggregate (agg⁻). Many of these lack the ability to make extracellular components and can be rescued by mixing with wild-type cells (e.g. the *synag* mutants which cannot produce cAMP signals but can respond to them). Some agg⁻ mutants, however, are CELL AUTONOMOUS (e.g. *frigid A*, which lacks a functional G protein; *see* GTP-BINDING PROTEINS) and cannot respond to cAMP signalling.

A second major phenotypic group of mutants arrest at the mound stage. These cannot develop further owing to a variety of deficiencies, for example in cell movement, gene expression, and production of DIF.

The phenotypes of culmination mutants are more diverse, ranging from those that have prolonged slug migration time (*slugger* mutants) to those that develop rapidly and exhibit precocious spore and stalk formation (*rde* and sporogenous mutants) and a mutant where all cells convert to stalk cells (*stkA*). In addition, there are many other mutants with more subtle phenotypes. These include mutants with abnormal aggregation, such as *streamer* mutants, and those that have an aberrant prespore : prestalk ratio.

Cyclic AMP in *D. discoideum* development

Cyclic AMP is an important extracellular signal throughout *Dictyostelium* development. Pulses of cAMP are required during aggregation, and higher levels build up during the multicellular stages. During these later stages, cAMP acts as a positive chemoattractant for pstA cells. A period of exposure to cAMP is necessary for the induction of all cell types, but prestalk cells then diverge and require DIF during later development. In the case of prespore cells, an extracellular cAMP signal is continuously required for gene induction and mRNA stability.

Differentiation inducing factor (DIF)

Differentiation Inducing Factor (DIF) is a central regulator of cell fate during *Dictyostelium* development. Historically, it was discovered as a substance released by developing cells and able to induce isolated amoebae to differentiate into stalk cells. It comprises a group of closely related chlorinated alkyl phenones among which DIF-1 is the predominant and the most active species (Fig. D8). In its ability freely to cross biological membranes and rapidly to induce gene expression it resembles the cholesterol-derived steroid hormones of higher organisms. Active at very low concentrations (0.1–1 nM), DIF-1 induces vegetative

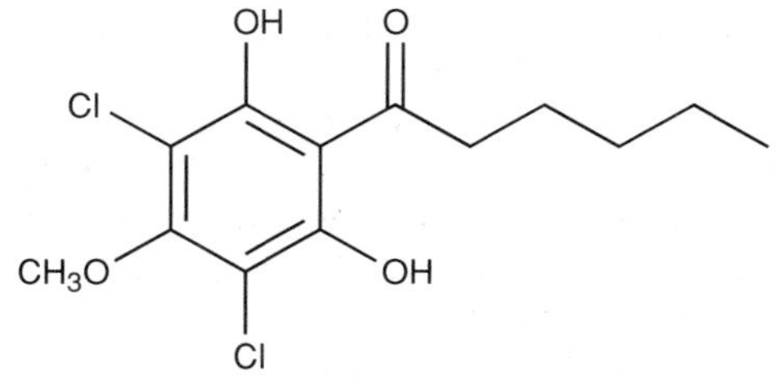

Fig. D8 Chemical structure of DIF-I.

amoebae to become prestalk and stalk cells, and inhibits prespore and spore differentiation.

DIF is actively synthesized and metabolized from the mound stage until completion of development. DIF-1 is metabolized almost exclusively in the front of the slug which is therefore a DIF-1 sink. This sink generates a DIF-1 gradient in the slug of unknown significance.

Ammonia in *D. discoideum* development

Large quantities of ammonia are produced during the development of *D. discoideum* as a consequence of protein degradation. Ammonia is known to inhibit the onset of culmination. A drop in extracellular ammonia concentration triggers the transition from slug migration to culmination. The exact mechanism by which ammonia exerts its effects is unknown, but high concentrations lower the level of intracellular cAMP, and may therefore control the activity of the cAMP-dependent protein kinase. High ammonia levels have also been shown to promote the conversion of prestalk cells to ALC.

Contact sites A and B

Cohesion molecules can be detected during *Dictyostelium* development. Contact site A (CsA) mediates a homophilic, EDTA-resistant adhesion and is induced during aggregation but persists up to the slug stage. It is not essential for development.

Contact site B mediates a EDTA-labile interaction between cells. It is induced earlier than A and is already present in axenically grown cells at high density. A group of small proteins (gp24) encoded by a MULTIGENE FAMILY is thought to be involved in this interaction.

A.J. HARWOOD
P. MORANDINI

1 Loomis, W.F. (1982) *The Development of* Dictyostelium discoideum (Academic Press, New York).
2 Raper, K.B. (1984) *The Dictyostelids* (Princeton University Press, Princeton).
3 Kimmel, A.R. & Firtel, R.A. (1991) cAMP signal transduction pathways regulating development of Dictyostelium discoideum. *Curr. Opinion Genet. Dev.* 1, 383–390.
4 Williams, J.G. (1991) Regulation of cellular differentiation during Dictyostelium morphogenesis. *Curr. Opinion Genet. Dev.* 1, 358–362.
5 Kay, R.R. (1992) Cell differentiation and patterning in Dictyostelium. *Curr. Opinion Cell Biol.* 4, 934–938.
6 Siu, C-H. (1990) Cell–cell adhesion molecules in Dictyostelium. *BioEssays* 12, 357–362.

Dictyostelium discoideum *See*: CELLULAR SLIME MOULDS; DICTYOSTELIUM DEVELOPMENT.

dideoxy chain termination Cessation of *in vitro* DNA synthesis as a consequence of incorporation of a DIDEOXYRIBONUCLEOSIDE triphosphate into the growing polyribonucleotide chain. Dideoxyribonucleoside triphosphates lack the 3'-OH group essential for the formation of a 5'–3' phosphodiester bond. This reaction is exploited in the dideoxy DNA SEQUENCING method first described by F. Sanger and his colleagues.

Sanger, F. et al. (1977) *Proc. Natl. Acad. Sci. USA* 74, 5463–5467.

dideoxyribonucleosides Analogues of deoxyribonucleosides and deoxyribonucleotides (*see* NUCLEOSIDES AND NUCLEOTIDES) in which the -OH group on the C3 of the ribose moiety is replaced by -H. Incorporation of a dideoxyribonucleotide into a growing DNA chain prevents further DNA synthesis and dideoxyribonucleotides are used as DNA chain terminators in the dideoxy chain termination method of DNA SEQUENCING. Dideoxyribonucleosides which inhibit the retroviral enzyme REVERSE TRANSCRIPTASE have been used as drugs against HIV (*see* IMMUNODEFICIENCY VIRUSES). They are ddC (zalcitabine, 2',3'-dideoxycytidine), and ddI (didanosine, 2',3'-dideoxyinosine).

DIF Differentiation inducing factor. *See*: DICTYOSTELIUM DEVELOPMENT.

difference Fourier synthesis A FOURIER SYNTHESIS in which the amplitudes of the set of coefficients are the differences between two sets of STRUCTURE FACTORS rather than the structure factors themselves. The resulting function, which would be an ELECTRON DENSITY MAP in the case of an X-RAY CRYSTALLOGRAPHY usage of this method, is less susceptible to error than a map produced by simple subtraction of the electron density maps corresponding to the two sets of structure factors.

The map from a difference Fourier synthesis reveals any discrepancies between the two electron density maps from which the sets of structure factors were calculated. The method is widely used for:

1 Improving a molecular model by minimizing the difference density in a map calculated from observed structure factors (based upon a model for the molecular structure).
2 Monitoring changes between two ISOMORPHOUS DERIVATIVES by analysing the difference density in a map calculated from structure factors of crystal 1 (which might be an unliganded protein) and crystal 2 (which might be liganded or modified in some other way).

The description here uses the terminology of crystallography, but the technique is useful wherever Fourier analysis is employed. *See also*: NEUTRON SCATTERING AND DIFFRACTION; NMR SPECTROSCOPY.

differentiation The expression or manifestation of the FATE of a cell. A differentiated cell expresses characteristic proteins and has a clearly defined morphology which identifies it as a member of a defined histological type. The process of differentiation is probably always irreversible in normal circumstances in most animal cells (but *see* DEDIFFERENTIATION and TRANSDIFFERENTIATION). In contrast to animal cells, many differentiated plant cells retain their ability to give rise to a complete new plant after suitable treatment (*see* PLANT GENETIC ENGINEERING; PLANT HORMONES).

Differentiation usually involves no loss of chromosomal material; rather it involves the selective expression of subsets of genes, which vary from cell type to cell type. Once achieved, the differentiated state is passed on to all progeny of the cell (in those cases where differentiated cells retain the ability to divide). The molecular mechanisms of differentiation are still incompletely understood but are being investigated in many systems (*see e.g.* B CELL DEVELOPMENT; CELL–CELL INTERACTIONS; CELL–MATRIX INTERACTIONS; DNA METHYLATION; EUKARYOTIC GENE EXPRESSION; EYE DEVELOPMENT; HAEMATOPOIESIS; LYMPHOKINES; MYOGENESIS; T CELL DEVELOPMENT).

differentiation antigens Cell-surface antigens that can be used to define different stages in cellular DIFFERENTIATION in a particular cell lineage.

differentiation factors Any protein or polypeptide that specifically induces cellular DIFFERENTIATION. *See e.g.*: GROWTH FACTORS; HAEMATOPOIESIS; LYMPHOKINES.

differentiation inducing factor (DIF) *See*: DICTYOSTELIUM DEVELOPMENT.

diffraction, diffraction methods When radiation strikes matter, it is scattered. If the scattering object comprises many similar elements arranged in a regular array then the scattered rays from these elements interfere with each other constructively in certain directions, giving enhanced scattering intensity or diffraction, and destructively in others, where the scattered rays cancel each other out. Diffraction is only detectable when the wavelength of the incident radiation is far less than the repeat distances within the array.

The directions of the diffracted rays are calculable from the spacings of the regular array of scattering elements and the incident wavelength; the intensities and phases of the diffracted rays are calculable from the detailed structure of an individual element. In biological structure analysis, the aim is to determine a detailed structure from the knowledge of its diffraction pattern or scattering, some parts of which may be unmeasurable.

The particular methods appropriate for diffraction analyses depend upon the nature of the sample: if it forms regular two- or three-dimensional arrays as single crystals the techniques of CRYSTALLOGRAPHY are available with diffraction in the directions permitted by BRAGG'S LAW, whereby a diffraction intensity can be envisaged as a REFLECTION; if it is ordered in only one direction, as a parallel array of fibres, the FIBRE DIFFRACTION patterns can still yield some three-dimensional information. Isolated molecules in solution exhibit only SMALL-ANGLE SCATTERING from which some structural information can be deduced.

A variety of radiation sources are used to study diffraction from biological systems, each with particular advantages. Since diffraction is only detectable when the wavelength of the incident radiation is far less than the repeat distances within the array, for any specific problem the wavelength must be roughly matched to the size of the regularly repeating unit. OPTICAL DIFFRACTION using wavelengths of about 5000 Å (or 0.5 µm) is useful in the analysis of photographic images (e.g. electron micrographs), being a very quick tool for locating the most regular and best preserved parts of a sample. X-RAY DIFFRACTION uses wavelengths of about 1.5 Å, which is of the same order as the covalent bonding distances within molecules such that the highest RESOLUTION potentially available, if the crystal is sufficiently well ordered,

gives details at an atomic level. Small organic molecules usually do diffract to the theoretical limit, but most biopolymers have a much lower diffraction limit as a result of crystal imperfections. X-rays are scattered by atoms in proportion to their number of electrons, such that hydrogen atoms are very weak scatterers and difficult to locate using X-rays while atoms of higher atomic number diffract up to two orders of magnitude more strongly at low diffraction angles but their X-ray scattering power falls off more quickly at higher angles. Conversely, neutron diffraction (*see* NEUTRON SCATTERING AND DIFFRACTION) depends upon the properties of an atomic nucleus and its spin. Tabulated values are far more uniform than for X-rays, with the exception of hydrogen which scatters out of phase from all other atoms including deuterium. Thus neutron diffraction is ideally suited for locating hydrogens and exchanges between hydrogen and deuterium — whether these are within the molecules of interest or in the surrounding solvent where CONTRAST VARIATION enables sample components of different scattering power (e.g. proteins and nucleic acids) to be distinguished. In ELECTRON MICROSCOPY the wavelength of electrons at the commonly used voltages is 0.05 Å or less, but the nominal resolution of an electron microscope is limited by instabilities and specimen properties to a few Å units. However, ELECTRON DIFFRACTION is useful for studying two-dimensional crystals, which may be only one, or a few, molecules thick in the third dimension. Three-dimensional structural data can be obtained from such thin crystalline arrays by analysis of a series of tilted specimens.

Diffraction of other types of irradiating beams is possible, but not generally used for biological structures. *See also*: IMAGE ANALYSIS; X-RAY CRYSTALLOGRAPHY.

DiGeorge syndrome A gross defect in human cell-mediated immunity associated with congenital absence of the thymus, and thus the non-development of T CELLS (*see* T CELL DEVELOPMENT). The parathyroid glands are also absent, leading to hypoparathyroidism. During embryonic life there is a failure of development of the third and fourth pharyngeal pouches from which the thymus and parathyroid glands arise.

digynic eggs/embryos Eggs with two sets of maternal chromosomes. Digynic eggs occur sporadically in nature by the suppression of one or the other meiotic divisions. They can also be produced experimentally by treatment of eggs with reagents such as CYTOCHALASIN, an actin-destabilizing drug which causes the suppression of polar body formation (*see* OOGENESIS), or by the technique of pronuclear transplantation. In the latter case, a fine glass needle attached to a micromanipulator is used to remove the female pronucleus from one fertilized egg and inject it into another egg. The male pronucleus is removed from the recipient egg so that the zygote contains two maternally derived genomes. *See also*: PARENTAL GENOMIC IMPRINTING.

dihydrofolate reductase (DHFR) NADPH-requiring enzyme (EC 1.5.1.3) which catalyses the synthesis of tetrahydrofolate, a metabolite essential for the synthesis of dTMP, glycine, and purines. The enzyme can be inhibited by METHOTREXATE or amethopterin, analogues of folic acid. Resistance to methotrexate in cultured cells can result from amplification of the *dhfr* gene (*see* GENE AMPLIFICATION). Inhibition of DHFR deprives cells of purines, blocks DNA synthesis and leads to a blockage in cell division.

dihydrophaseic acid (DPA) $(-)$-5-[S,8S-Dihydroxy-1R,5R-dimethyl-6-oxabicyclo(3,2,1)-octane]-3-methyl-2Z,4E-[r]-pentanoic acid (Fig. D9). M_r 282.3. A metabolite of the plant hormone ABSCISIC ACID, produced via PHASEIC ACID by reduction of the C-4′ ketone. Dihydrophaseic acid refers to the 4β-hydroxy compound; the 4α-hydroxy isomer, referred to as *epi*-dihydrophaseic acid, is also produced as a minor metabolite. Dihydrophaseic acid has no biological activity. It is metabolized mainly by formation of the 4′-O-glucosyl ether.

Milborrow, B.V. (1975) *Phytochemistry* 14, 1045–1053.

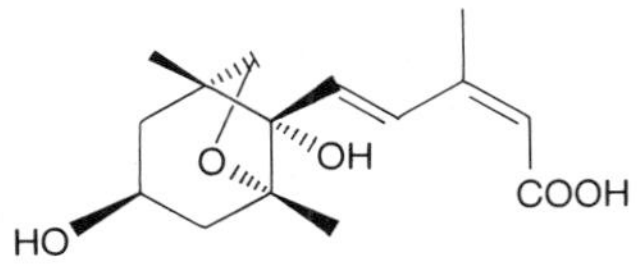

Fig. D9 Structure of dihydrophaseic acid.

dihydropyridine receptor (DHP receptor) Receptor for 1,4-dihydropyridines in the T-tubules of skeletal MUSCLE and in the membranes of other cell types (e.g. neurons and cardiac muscle cells) that functions both as a voltage sensor for excitation–contraction coupling in skeletal muscle and as an L-type calcium channel (*see* CALCIUM).

Tanabe, T. et al. (1988) *Nature* 336, 134–139.

dihydrozeatin ((diH)Z) (S)-6-[4-Hydroxy-3-methylbutylamino]-purine (Fig. D10). M_r 221.2. A naturally occurring CYTOKININ that was first isolated from immature lupin seeds, in which it is present in high concentrations (2 nmol per gram fresh weight). It occurs as the free base and N^9-riboside and -ribotide in plant tissues. It is formed from ZEATIN by the action of a soluble NADPH-requiring reductase that is specific for the free base. It is at least as biologically active as zeatin and, as it is not a substrate for cytokinin oxidase, is not metabolized as rapidly. It forms glucosyl conjugates in the same way as zeatin. *See also:* PLANT HORMONES.

Koshimizu, K. et al. (1967) *Tetrahedron Lett.* 1317–1320.

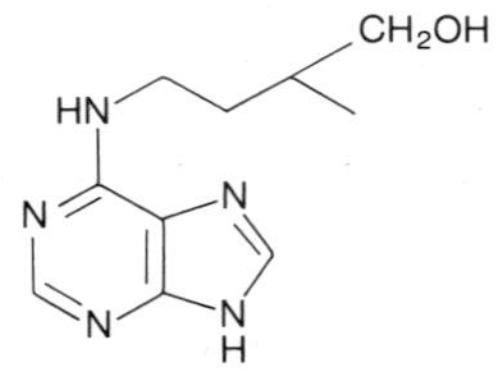

Fig. D10 Structure of dihydrozeatin.

2,4-dinitrophenol *See:* DNP.

dinucleotide repeat(s) Stretches of repeated dinucleotides, often GC, found throughout the genome and which are useful as markers for gene mapping. *See:* HUMAN GENE MAPPING.

dinucleotide repeat probes Synthetic DNA PROBES based on repeating dinucleotides (e.g. poly(CA), poly(GT)). Used in gene mapping (*see* HUMAN GENE MAPPING).

1,3-diphenylurea (DPU) N,N′-diphenylurea (Fig. D11). M_r 212.3. A compound with weak CYTOKININ activity first described as a factor in coconut milk that stimulated cell division in a tissue culture derived from carrot phloem. It is used as a growth regulator, for example in tissue culture. Related compounds in which one, but not both, of the phenyl residues is substituted are also active. Since diphenylurea and related active compounds, such as thidiazuron (N-phenyl-N′-1,2,3-thiadiazol-5-ylurea), inhibit cytokinin oxidase, an enzyme which catalyses cytokinin catabolism, their activity may be explained, at least in part, by an accumulation of endogenous purine cytokinins.

Shantz, E.M. & Stewart, F.C. (1955) *J. Am. Chem. Soc.* 77, 6351–6353.

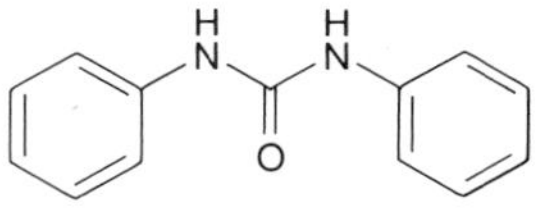

Fig. D11 Structure of 1,3-diphenylurea.

2,3-diphosphoglycerate, 2,3-bisphosphoglycerate (2,3DPG, 2,3BPG) The main intracellular regulator of oxygen uptake and release by HAEMOGLOBIN in humans. This molecule binds in the central cavity of the deoxy- (T-state) haemoglobin molecule, cross-linking the β chains, and stabilizing the molecule in that state. Thus 2,3DPG and oxygen binding are mutually exclusive and the overall effect is that 2,3DPG reduces the oxygen affinity of haemoglobin by stabilizing the deoxy form. The cellular concentration of 2,3DPG is modified by hypoxia, pH, and a number of other factors thus enabling oxygen transport to respond to a variety of physiological variables.

diphtheria toxin Phage-encoded protein toxin (M_r 61 000) produced by lysogenic strains of the bacterium *Corynebacterium diphtheriae*, the causal agent of diphtheria. It consists of a single polypeptide chain comprising two domains — A and B. The B domain is required to enable the A domain to cross the plasma membrane and enter the cell. Binding of the B domain to a specific cell-surface receptor stimulates RECEPTOR-MEDIATED ENDOCYTOSIS of the toxin, which is then cleaved into A and B fragments within endosomes and the A fragment transferred into the cytosol. The A domain inhibits eukaryotic PROTEIN SYNTHESIS by catalysing the ADP-RIBOSYLATION of a modified histidine residue (diphthamide) on the ELONGATION FACTOR EF-2 which blocks its function. The lethal effects of the toxin are due to its inhibition of protein synthesis.

diploid A cell or organism which contains homologous pairs of each chromosome. In complex multicellular organisms the SOMATIC CELLS are usually diploid. Sometimes abbreviated to *2n. Cf.* HAPLOID.

diplotene *See:* MEIOSIS.

diquat One of the low-potential ($E_m = -400$ mV) viologen dyes (1,1'-ethylene-2,2'-dipyridylium dibromide) that can act as an artificial electron acceptor for PHOTOSYSTEM I (closely related to PARAQUAT). One-electron reduction yields a violet-coloured, auto-oxidizable free radical. Reaction with oxygen yields the superoxide anion in the first instance; this may form the basis of its herbicidal action which depends on light and thus active turnover of the photosynthetic electron transport system. *See also:* PHOTOSYNTHESIS.

direct methods In X-RAY CRYSTALLOGRAPHY, the use of mathematical relationships between STRUCTURE FACTORS as a direct approach to solving the PHASE PROBLEM. These relationships all depend upon the assumption that ELECTRON DENSITY is always positive. Many also assume that within the molecular structure: (1) all atoms scatter equally (essentially true for carbon, nitrogen and oxygen; less good for metal atoms or phosphorus); and (2) all atoms are resolved (rarely true for macromolecules, but valid for small organic molecules). The relationships become statistically less powerful as the molecular weight increases. Thus in small molecule crystallography these methods are nearly always successful at solving structures, whereas in macromolecular crystallography their main use is in the HEAVY ATOM REPLACEMENT METHOD for determining the positions of a few (essentially equal) heavy atoms. The use of MOLECULAR REPLACEMENT and NON-CRYSTALLOGRAPHIC SYMMETRY is sometimes regarded as applications of direct methods applicable in special cases, whilst MAXIMUM ENTROPY is becoming available as a more sophisticated implementation of the traditional direct methods.

direct repeats Repeated nucleotide sequences that is oriented in the same direction along a DNA molecule, for example:

$$\overrightarrow{} \quad \overrightarrow{} \quad \overrightarrow{}$$
5′ ATCG (n)$_x$ ATCG (n)$_x$ ATCG 3′

In this case 5′ ATCG 3′ is the direct repeat. This arrangement is sometimes referred to as a 'head-to-tail' arrangement, particularly when the direct repeats form TANDEM REPEATS, that is with no nonrepeated nucleotides (n)$_x$ separating the repeated sequences. In eukaryotic genomes the length of repeated sequences ranges from two (dinucleotide) to several thousand nucleotides. *Cf.* INVERTED REPEATS. *See also:* MINISATELLITE DNA; RETROVIRUSES; TRANSPOSABLE GENETIC ELEMENTS; VARIABLE NUMBER TANDEM REPEATS.

displacement loop *See:* D-LOOP.

Dissociation (Ds) *See:* TRANSPOSABLE GENETIC ELEMENTS: PLANTS.

dissociation constant (K_d) A quantitative measure of the strength of binding of a ligand for its receptor. It can be determined by Scatchard analysis of equilibrium binding experiments and has the dimensions mol 1^{-1}. *See:* SCATCHARD PLOT.

distance geometry A method used to determine molecular structures by NMR SPECTROSCOPY.

distance methods *See:* MOLECULAR PHYLOGENY.

disulphide bond Covalent bond formed between the sulphur atoms of cysteine residues in proteins. *See:* AMINO ACIDS; PROTEIN FOLDING; PROTEIN STRUCTURE.

diuron A herbicide which inhibits PHOTOSYNTHESIS. *See:* DCMU.

divided gene A gene whose EXONS are transcribed into separate RNA molecules, which are then spliced. This contrasts with split genes, whose expression depends on intramolecular RNA SPLICING. Exons of divided genes may be separated by large distances, which can include other genes. *See:* CHLOROPLAST GENOMES; TRANS-SPLICING.

DM MYOTONIC DYSTROPHY.

DMSO Dimethylsulphoxide. A colourless hygroscopic organic liquid miscible in water and with an M_r of 78.1. Widely used as a cryoprotective agent for low temperature storage of bacteriophages and bacterial and mammalian cells. Also used in preparing competent *Escherichia coli* cells for TRANSFORMATION, although at high concentrations (>10% v/v) it is toxic to most types of cell.

DMSO shock Certain types of mammalian cell can be transfected at higher efficiencies if they are first exposed to 10% (v/v) DMSO (dimethylsulphoxide). Such a 'shock' can be applied to cells in either the calcium phosphate or DEAE-dextran TRANSFECTION protocols.

DNA (deoxyribonucleic acid)

DNA is the primary genetic molecule of life and carries within its sequence the hereditary information that determines the structure of PROTEINS in all eukaryotic and prokaryotic organisms. It contains the instructions by which cells grow, divide, and differentiate, and has provided a basis for the evolutionary process both within and between related species. DNA is the information-carrying material that comprises the GENES, or units of inheritance, which are arranged in linear arrays along the CHROMOSOMES of the cell.

DNA is a linear polymer made up of deoxyribonucleotide monomers (*see* NUCLEOSIDES AND NUCLEOTIDES). Each nucleotide consists of three components—a heterocyclic base, a sugar (deoxyribose), and a phosphate group [1]. DNA contains four types of bases, the purines adenine (A) and guanine (G), and the

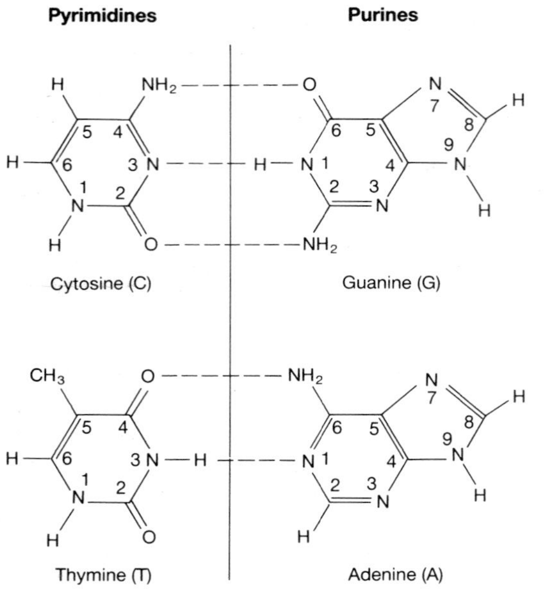

Pyrimidines   **Purines**

Cytosine (C)   Guanine (G)

Thymine (T)   Adenine (A)

Fig. D12 Bases in DNA showing the Watson–Crick hydrogen bonding between base pairs (---). Atomic numbering schemes for the purines and pyrimidines are shown.

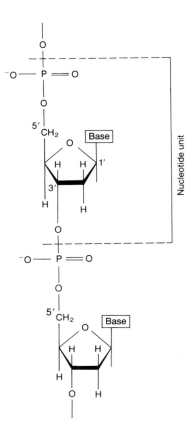

Fig. D13 Repeating nucleotide structure of a DNA chain showing the 5′–3′ phosphodiester linkage between individual nucleotides.

pyrimidines cytosine (C) and thymine (T) (see Fig. D12). The base is connected by a β-glycosyl linkage to C1 of the deoxyribose via the N1 of a pyrimidine or N9 of a purine base, to form a nucleoside. The 5-hydroxyl (OH) group of the sugar is phosphorylated to form a deoxyribonucleotide, which constitutes the primary building block of the polynucleotide, DNA.

The nucleotides are linked via the phosphate groups attached to the deoxyribose moieties to form the backbone of the polynucleotide. Specifically, the 3-OH of the sugar moiety of one deoxyribonucleotide is joined to the 5-OH of the adjacent sugar in the successive nucleotide by a phosphodiester bridge (Fig. D13). In DNA, nucleotides are linked together to form very long polynucleotide chains.

The DNA double helix

Historically, DNA was first identified as the hereditary material in cells by Oswald Avery (1944), who established that DNA was the transforming factor, capable of passing genetic information between cells and altering the pathogenicity of the *Pneumococcus* bacterium [2]. This observation was confirmed by Alfred Hershey and Martha Chase (1952), who differentially labelled T2 bacteriophage DNA with ^{32}P and its protein coat with ^{35}S and found that only the ^{32}P was transferred to progeny phage following infection of *Escherichia coli* [3]. It was at this time that the molecular structure of the first nucleoside was published, and the precise phosphate ester linkages that bound nucleotides together were determined [1]. Crucial additional information was then provided by Erwin Chargaff (1952), who found that although the amounts of DNA per cell and their sequence varied widely between organisms, the ratio of A : T or G : C bases was always unity (i.e. A = T and C = T; Chargaff's rules) [4]. The implication of

these observations did not become clear until the elucidation of the structure of the DNA molecule in 1953 [5,6]. James Watson and Francis Crick deduced the three-dimensional structure of DNA from X-RAY DIFFRACTION photographs of oriented polycrystalline DNA fibres, prepared by Rosalind Franklin and Maurice Wilkins. They deduced that DNA formed a right-handed double helix, containing two polynucleotide chains arranged in an antiparallel manner and coiled around a common axis. The sugar–phosphate groups formed the helical backbone, and the bases were disposed on the inside of the helix.

The bases in nucleic acids provide two important interactions, which confer stability on the DNA and are responsible for its overall structure. Firstly, there are hydrogen bonding interactions between the base pairs of the two polynucleotide chains in the plane of the bases and perpendicular to the helical axis — complementary base-pairing. Secondly, there is a base-stacking interaction between the bases on the same polynucleotide strand which is a vertical interaction between the bases in a direction parallel to the helical axis. Steric considerations mean that within the helix the purines can only bond to pyrimidines in order to maintain the regular shape of the helical backbone. Equally important is the selective hydrogen bonding of A to T and of C to G, using two and three hydrogen bonds respectively (Fig. D12). As a result of this specific BASE PAIRING, the two strands of DNA are complementary. The hydrogen bonds can be disrupted by heating, or by ionic changes in the bases at highly acid or alkaline conditions. Such reversible melting transitions are dependent upon the DNA sequence, and are useful for determining the complexity of the DNA (*see* HYBRIDIZATION) and are important for the denaturation/renaturation functions of the DNA during DNA REPLICATION and TRANSCRIPTION. Thus, whereas the sugar and the phosphate groups have a predominantly structural role in the DNA, the bases are responsible for enhancing the stability of the DNA helix and providing the specific DNA SEQUENCE encoding the genetic information for the production of proteins (*see* GENETIC CODE).

In 1962, Watson and Crick were awarded a Nobel Prize for their proposed structure of the DNA double helix, and it became generally accepted that DNA conformed to this regular helical structure, which was largely independent of the DNA sequence. Their model not only provided the basis for insight into the structure of DNA at the atomic level [5], but also suggested a model for the replication of this genetic material by virtue of the complementarity of its polynucleotide strands [6]. Matthew Meselson and Frank Stahl (1958) subsequently proved that the two strands of DNA separate during replication, and that each parent strand acts as a template for the production of progeny strands, to produce new DNA molecules containing one parent and one progeny strand — semiconservative replication [7] (*see* DNA REPLICATION).

Types of helices

With the advent of improved phosphotriester methods for the synthesis of oligonucleotides, both in solution and at an automated level, oligonucleotides of defined sequence could be synthesized in sufficient quantity for single crystal X-ray analysis (*see* X-RAY CRYSTALLOGRAPHY). Combined with improved spectroscopic techniques for the analysis of molecules in solution, it soon became clear that in contrast to the X-ray fibre data, from which a uniform rod-like DNA helix had been deduced, DNA was able to adopt a variety of different conformations depending on the DNA sequence, composition and environmental conditions (e.g. salt concentration, humidity) (*see also* NUCLEIC ACID STRUCTURE).

The crystal structures of defined oligonucleotides revealed systematic sequence-dependent structural modulations in the DNA helices, indicating that the DNA helix is highly polymorphic in nature. For example, there are the right-handed DNA helices which have been generally classified as A-type (A-DNA) or B-type (B-, B'-, C-, C'-, C''-, D-, E- and T-DNAs) [1], and the more recently discovered left-handed variety of DNA helices, known as Z-DNA [8]. Therefore, rather than being a static double-stranded molecule [5], DNA seems to be a rather dynamic structure comprising many different helical forms, which presumably may exist in equilibrium with each other. Both right-handed (A- and B-DNA) and left-handed (Z-DNA) helices are double-helical conformations with antiparallel chains held together by Watson–Crick hydrogen bonding between the bases. The helical parameters that determine the overall shape of the double helix for these different forms of DNA are summarized in Table D3 [9,10].

A-DNA is a more compact helix than the classical B-form of DNA, while Z-DNA is a much slimmer helix with a greater

Table D3 Comparison of A-, B-, and Z-DNA

Helix parameter	A-DNA	B-DNA	Z-DNA
Helix sense	Right-handed	Right-handed	Left-handed
Base pairs per turn	11	10.4	12
Rise per base pair (Å)	2.9	3.4	3.7
Helix diameter (Å)	25.5	23.7	18.4
Helical pitch (Å)	32	34	45
Sugar pucker	C3'-*endo*	C2'-*endo*	C2'/C3'-*endo*
Glycosidic bond	*anti*	*anti*	*anti/syn*

number of base pairs per helical turn (Fig. D14; *see also* Plate 1). Much of the structural differences between helices arise from the alteration in the puckering (folding) of the sugar ring. The sugar ring of the nucleotides is a nonplanar, five-membered furanose ring where changes in sugar pucker serve to minimize both torsional and steric interactions within the ring. Atoms which are displaced from the main sugar plane on the same side as the C5' atom are designated *endo* puckers, whereas those displaced on the opposite side of the ring to the C5' atom are designated *exo* puckers. In B-form DNA, the most usual sugar pucker is C2'-*endo*, where the C2' is displaced on the same side of the ring as C5'. In A-form DNA, which is typical of DNA helices at low ionic strength or low hydration, the sugar pucker is C3'-*endo*. In Z-DNA, there is an alternation of C2'-and C3'-*endo* puckering along the helical backbone. Another important parameter determining nucleotide geometry is the configuration of the glycosyl bond attaching the sugar and the base. Whereas all the nucleotides along the A- and B-DNA helices have the same conformation (*anti*), with the base and sugar located on different sides of the glycosidic bond, those along the Z-DNA helix alternate between the *anti* and *syn* conformations. The bases C and T adopt the *anti* conformation with the C2'-*endo* pucker, while the *syn* position adopted by G is associated with the C3'-*endo* pucker. In the latter, the pyrimidine ring lies directly above the sugar ring, and this conformation has been associated with the purines in Z-DNA. Thus, in contrast to the mononucleotide repeat unit in A-and B-DNA, the asymmetric unit in Z-DNA is a dinucleotide reflecting the alternation of the *anti–syn* conformation of the bases in the helical backbone as a result of the C2'-, C3'-*endo* puckering of the deoxyribose ring. These alternating nucleotide conformations are favoured by alternating purine–pyrimidine sequences, and result in a 'zig-zag' orientation of the sugar–phosphate residues in the DNA backbone which destroys the dyad symmetry between base pairs, hence the name 'Z-DNA' (Fig. D14).

Double-helical geometry

The preceding parameters define only the geometry of the monomeric nucleotide subunits of DNA however, and there are additional helical parameters which can be defined for the double helical DNA forms (see Table D3). These include (1) the rise, or the distance, between base pairs, which for B-DNA is 3.4 Å; (2) the number of base pairs (bp) per turn, which for B-DNA is 10.1–10.4 bp; and (3) the pitch of the helix, which is a combina-

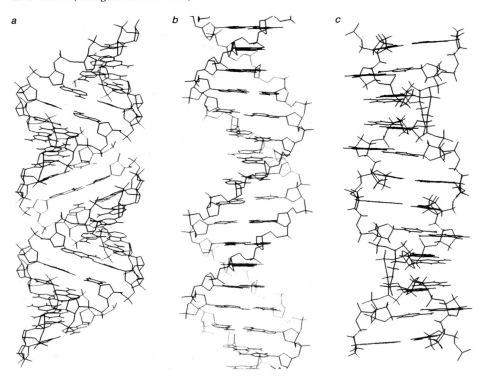

a   *b*   *c*

Fig. D14 Computer graphic representations of *a*, A-DNA; *b*, B-DNA; and *c*, Z-DNA. Courtesy of Paul Gane.

tion of parameters (1) and (2) for a 360° turn of the helix (i.e. the pitch is ~34 Å for B-DNA).

Also, owing to base–base interactions, the exterior of the DNA helix is discontinuous, displaying two types of groove — a major and a minor groove — which are important determinants of PROTEIN-NUCLEIC ACID INTERACTIONS. The major groove is defined as the side with C4 of the pyrimidine (pyr) and the N7 of the purine (pur) and the minor groove has O2 (pyr) or N3 (pur). In A-DNA, the C3'-*endo* puckering of the sugar residue causes the bases to tilt away from the normal helical axis, such that the minor groove becomes relatively broad and shallow and the major groove becomes narrower and deeper than in the regular B-DNA conformer. These relative changes in structure alter the hydration patterns of the DNA helix, specifically the interaction between water molecules and the phosphate backbone. Therefore, the A-DNA form is favoured by conditions of relatively low humidity compared to the B-DNA form, and the B- to A-DNA transition can be induced by dehydration. The irregularity of the Z-DNA backbone also alters the stacking pattern of the bases, and modifies the grooves in Z-DNA. In contrast to the right-handed B-DNA with its well-defined major and minor grooves, Z-DNA has one deep helical groove that is formally analogous to the minor groove of B-DNA. The concave major groove of B-DNA forms the convex outer surface of Z-DNA. The change in the relation of bases to the sugar–phosphate backbone in Z-DNA also results in considerable differences in the chemical reactivity of this conformer relative to B-DNA. For example, chemical modification of alternating CG polymers of the cytosines at position 5 (e.g. methylation, bromination) favours the Z-DNA conformation. Therefore, DNA can adopt several different conformations de-

pending on the relationship between environmental conditions such as counterions, chemical modification and hydration status and the DNA sequence.

Good examples of this are the homopurine–homopyrimidine DNA sequences such as poly(dA).poly(dT), which normally exhibit the classical B conformation [1]. If the environmental salt concentration is raised or the relative humidity decreased, the polynucleotide preferentially disproportionates into a triple helix (poly(dA).poly(dT)$_2$) and a single polynucleotide chain, poly(dA), in contrast to the normal B to A transition observed for other random sequence B-DNAs. Poly(dA).poly(dT) forms a regular A-DNA duplex in which the polynucleotides are oriented in an antiparallel fashion and held together by classical Watson–Crick hydrogen bonds. The extra polypyrimidine (poly(dT)) strand is accommodated within the deep major groove of this A-DNA duplex, and is hydrogen bonded to the polypurine strand by Hoogsteen base pairing, in which the two strands run in a parallel fashion (Fig. D15).

DNA structure *in vivo*

In vitro studies using oligonucleotides of defined sequence have revealed significant polymorphism in the structure of DNA, but to what extent does this reflect the structure of DNA *in vivo*? In the cell, not all of the DNA is chromosomal — mitochondria contain their own DNA and there may be additional extrachromosomal elements, such as autonomously replicating plasmids and viral DNA. Although most of the DNA in the genome is in the classical double-stranded B-form, viral DNA can exist in both a single-stranded infective form and a double-stranded replicative form.

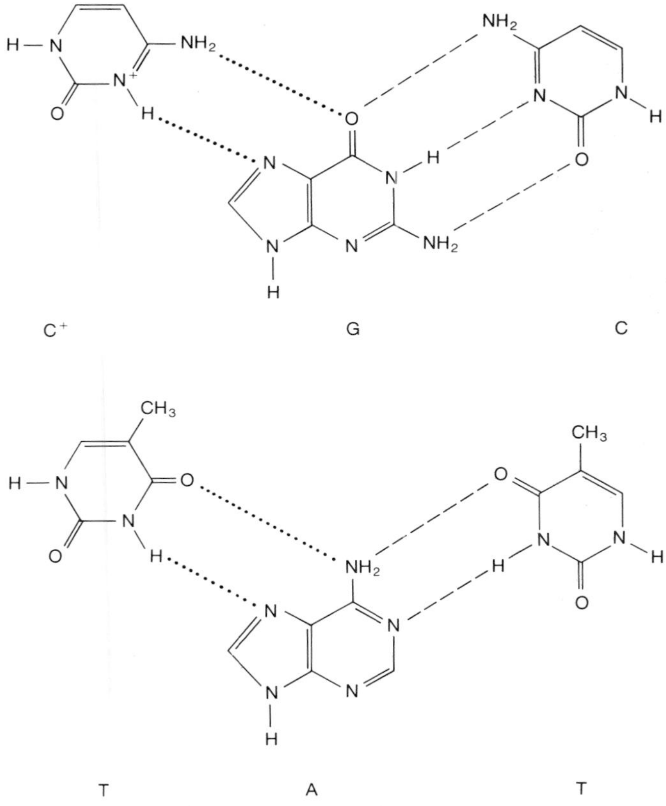

Fig. D15 Triple-stranded (triplex) DNA structures showing the regular Watson–Crick (---) and Hoogsteen (...) hydrogen bonding interaction between bases.

The total length of DNA contained within the genome of an organism varies depending on its genetic complexity, and may vary from micrometres (viruses) or millimetres (bacteria), to several centimetres (in higher eukaryotes). Therefore, the DNA must be organized and packaged into higher order forms within the cell, and DNA molecules *in vivo* may acquire a compact shape by existing in circular forms, where the two ends of a linear DNA are covalently bound to each other. These circular DNA molecules can be twisted into supercoiled molecules (*see* SUPERCOILING) by the action of DNA GYRASES, to adopt an even more condensed configuration than the 'relaxed' circular equivalent (*see* PLASMIDS). In prokaryotic cells, this supercoiling of the DNA is critical for the packaging of the DNA into the cells. The DNA is generally in an underwound state in these cells and exists in a negatively supercoiled form, where the superhelical density of the DNA varies between −0.03 and −0.09.

In prokaryotes, it is well established that the superhelical state of the DNA plays an important part in various biological processes including replication, transcription and recombination, by altering the accessibility of the DNA to proteins [10]. As the DNA supercoiling also determines the conformation of the DNA, alternative non-B DNA conformations have been implicated in gene regulation [9]. Of particular interest are the secondary structural forms of DNA which can be adopted as a function of both

supercoiling and DNA sequence (e.g. CRUCIFORMS in inverted repeat DNA, Z-DNA in alternating purine–pyrimidine sequences, and DNA triplexes in homopurine–homopyrimidine sequences), especially where such sequences occur in association with specific genes. Classically the problem has been relating the known supercoiling changes which occur during such processes with changes in the structure of the DNA *in vivo*. However, some of the methods available for mapping secondary DNA conformations *in vitro*, specifically the chemical, immunological and gel retardation assay techniques, have been successfully adapted for the mapping of DNA secondary structure *in vivo* [9]. The general concept that has emerged is that transcription itself is a major contributor to the supercoiling status of the DNA and as such regulates the dynamic equilibrium between B- and non-B-DNA conformations within the cell. Therefore, whereas nonphysiological levels of supercoiling may be required for the stabilization of secondary DNA structures *in vitro* (particularly cruciform and Z-DNA structures), protein–DNA interactions and/or changes in the topology of the DNA template mediated by such processes may facilitate the formation of non-B-DNA structures *in vivo*. Indeed, studies have confirmed that specific Z-DNA sequences may have either a negative or a positive role in the control of gene transcription depending on their local genomic environment, as a result of the twin supercoiling waves which are generated during transcription (positive supercoils ahead and negative supercoils behind the RNA polymerase complex) [11].

In eukaryotes, the relationship between supercoiling and transcription, and the topology of the template, are less well defined. In eukaryotes the DNA is packaged in the nucleus (0.5 μm in diameter) with the help of basic polypeptides called histones (*see* CHROMATIN), which neutralize the acidity of the DNA. Although the overall level of supercoiling in eukaryotes is therefore lower than prokaryotes, it is clear that similar fluctuations in the native level of supercoiling occur (albeit transiently) in association with gene transcription [12].

So although there is an abundance of information regarding the structure of DNA in solution, fibres and crystals, little is known about the structure of the DNA in its natural environment. The structure of DNA is clearly dynamic, and can be bent, kinked or unwound, or supercoiled to form secondary DNA structures including cruciforms, triplexes and Z-DNA [9]. As supercoiling alters the capacity of the DNA to unwind, such secondary structures may be formed transiently during various biological processes (e.g. DNA replication, transcription and recombination) and may have biological implications for the interaction of DNA with RNA or proteins. With the adaptation and development of new techniques for detecting DNA structures inside the cell, the role of DNA conformation in such processes may yet be elucidated and will provide a fruitful area for research in the future.

L.H. NAYLOR

See also: RNA.

1 Saenger, W. (1984) In *Principles of Nucleic Acid Structure* (Cantor, C.R., Ed.) (Springer-Verlag, Heidelberg/Berlin).
2 Avery, O.T. et al. (1944) Studies on the chemical nature of the substance inducing transformation of pneumococcal types. *J. Exp. Med.* **79**, 137–158.
3 Hershey, A.D. & Chase, M. (1952) Independent functions of viral protein

and nucleic acid in growth of bacteriophage. *J. Gen. Physiol.* **36**, 39–56.

4 Zamenhof, S. et al. (1952) On the desoxypentose nucleic acids from several microorganisms. *Biochim. Biophys. Acta* **9**, 402–405.

5 Watson, J.D. & Crick, F.H.C. (1953) Molecular structure of nucleic acids: a structure for deoxyribose nucleic acid. *Nature* **171**, 737–738.

6 Watson, J.D. & Crick, F.H.C. (1953) Genetical implications of the structure of deoxyribonucleic acid. *Nature* **171**, 964–967.

7 Meselson, M. & Stahl, F.W. (1958) The replication of DNA in *Escherichia coli. Proc. Natl. Acad. Sci. USA* **44**, 671–682.

8 Rich, A. et al. (1984) The chemistry and biology of left-handed Z-DNA. *Annu. Rev. Biochem.* **53**, 791–846.

9 Palecek, E. (1991) Local supercoil-stabilized DNA structures. *Crit. Rev. Biochem. Mol. Biol.* **26**, 151–226.

10 Stryer, L. (1988) *Biochemistry*, 3rd edn (Freeman, San Francisco).

11 Rahmouni, A.R. & Wells, R.D. (1989) Stabilization of Z-DNA *in vivo* by localized supercoiling. *Science* **246**, 358–363.

12 Giaever, G.N. & Wang, J.C. (1988) Supercoiling of intracellular DNA can occur in eukaryotic cells. *Cell* **55**, 849–856.

DNA abundance classes *See*: DNA COMPLEXITY; GENOME ORGANIZATION.

DNA-bending sequences *See*: NUCLEIC ACID STRUCTURE.

DNA-binding proteins Many proteins bind to DNA, including gene regulatory proteins, enzymes involved in DNA replication, recombination, repair, transcription, and degradation, and proteins involved in maintaining chromosome structure. They can be divided into two large groups:
1 Those that have some sequence-specific or secondary structure-specific requirement for DNA binding.
2 Those that bind DNA nonspecifically.
DNA binding is often mediated by distinct DNA-binding domains. For examples of sequence-specific DNA binding *see* HOMEOBOX GENES AND HOMEODOMAIN PROTEINS; PROTEIN–NUCLEIC ACID INTERACTIONS; RECOMBINATION; RESTRICTION ENZYMES; TRANSCRIPTION; TRANSCRIPTION FACTORS. For examples of nonspecific binding *see* CHROMATIN; DNA REPAIR; DNA REPLICATION; NUCLEASES; PROTEIN–NUCLEIC ACID INTERACTIONS; RECOMBINATION; TRANSCRIPTION.

DNA blotting The transfer of DNA, either intact or fragmented with RESTRICTION ENZYMES, from an agarose gel to a nylon or nitrocellulose membrane laid over the gel. The protocol, initially devised by E.M. Southern (*see* SOUTHERN BLOTTING), involved the use of paper towels to draw buffer through the agarose gel by capillary action. This has now been largely superseded by either ELECTROBLOTTING or VACUUM BLOTTING procedures which are more rapid.

DNA cloning The purification and isolation of individual fragments of DNA from a larger DNA or a mixture of DNA fragments by their incorporation into a carrier DNA (a VECTOR) to form a RECOMBINANT DNA that can be multiplied up in a bacterial host. The DNA to be cloned may be derived directly from a genome, or may be in the form of complementary DNA (cDNA) copies of the pool of mRNAs found in a particular tissue. A collection of cloned DNA fragments from a single source is known as a DNA library. GENE CLONING, the isolation of a DNA fragment containing a particular gene, can be achieved in a variety of ways. DNA libraries may be screened using various techniques (*see* DNA LIBRARIES; HUMAN GENE MAPPING; HYBRIDIZATION). The amplification of particular genes or DNA sequences from genomic DNA may now also be achieved directly through methods based on the POLYMERASE CHAIN REACTION (PCR). Expression cloning of genes, especially those for receptor proteins, may use unfractionated or fractionated mRNA from a particular tissue injected into a XENOPUS OOCYTE EXPRESSION SYSTEM or other suitable IN VITRO TRANSLATION system. ALLELIC RESCUE can be used to clone a mutant allele directly from the genome of the yeast *Saccharomyces cerevisiae*.

Most vectors for DNA cloning are based on either PLASMIDS or BACTERIOPHAGES which replicate independently of the host chromosome to produce high numbers of progeny molecules per host cell. Numerous cloning vectors, based mainly on the bacteriophages LAMBDA and M13 and on small plasmids from *E. coli* (*see* PBR322; COLE1), have been developed for different applications (*see also* CLONING VECTORS). YEAST ARTIFICIAL CHROMOSOMES (YACs) which can carry very large inserts of DNA (up to 1000 kb) have also been developed as cloning and EXPRESSION VECTORS. There are also cloning vehicles based on eukaryotic viruses such as SV40 and the cauliflower mosaic virus.

A cloning vector must possess certain basic requirements:
1 DNA sequences and/or genes (e.g. origin of replication, replication genes in phages) that allow it to be replicated in the host cell of choice; this is usually the bacterium *Escherichia coli*, but other bacteria and *S. cerevisiae* are also used as host cells for particular applications.
2 A site into which the DNA to be cloned can be inserted without interfering with replication and maintenance of the vector in the host cell.
3 One or more genes for some selectable phenotype (such as antibiotic resistance) which can be used to distinguish the resulting clones of host cells that have picked up a recombinant DNA.

Outline cloning procedure. A purified sample of the DNA to be cloned is first cleaved into smaller fragments either by RESTRICTION ENZYMES (endonucleases which cut DNA at specific sites) or by random SHEARING (for preliminary SHOTGUN CLONING). In one of the simplest cloning strategies, and the one used to produce the earliest recombinant DNAs, both the DNA to be cloned and the vector are cut with the same restriction enzyme, chosen so that it cleaves the vector only once and cuts in a staggered fashion leaving single-stranded COHESIVE ENDS. When treated samples of source DNA and vector DNA are mixed, the cohesive ends of the DNA fragments will transiently pair with at least a proportion of the complementary cohesive ends of the cleaved vector molecules. The gap between vector and insert DNA is joined by the enzyme DNA LIGASE. Although DNA ligase can also join blunt-ended DNAs the overall process is less efficient as the chance of two DNA ends coming together is less. The end result of ligation is a population of recombinant DNAs each containing a different insert (*see* Fig. G19 in GENETIC ENGINEERING for a general scheme). The resulting mixture is used to TRANSFORM bacterial cells at a ratio of vector molecules to cells that ensures that a host cell receives a maximum of one vector molecule. If a bacteriophage vector (e.g. lambda) is used, it is more efficient to package the recombinant DNA molecules into infectious phage particles in an *in vitro* packaging reaction and use the natural process of phage infection to introduce the recombinant DNA into host cells.

When a plasmid vector is used, the transformed bacteria are plated out on a solid agar medium at a concentration that ensures that each bacterial cell gives rise to a separate colony. Bacteria that have received a recombinant DNA are identified by the marker genes carried on the vector. At least two selectable marker genes are generally included on the vector, one of which is inactivated by the insertion of the foreign DNA. This provides a means of distinguishing between cells that have taken up a recombinant DNA and cells that have taken up a vector that has reformed without incorporating a DNA fragment. Marker genes commonly incorporated into plasmid vectors include resistance genes for the antibiotics ampicillin, tetracycline, and chloramphenicol (*see* CHLORAMPHENICOL ACETYLTRANSFERASE; β-LACTAMASE), and the *lacZ* gene which encodes β-GALACTOSIDASE (as in vector pUC8), which can be detected by the production of a coloured reaction product.

When a phage vector is used, bacteria containing recombinant phage are identified by the PLAQUES produced when the mixture of bacteria and phage is spread onto a solid medium and grown into a continuous lawn of bacteria. Each plaque is a clear area in the lawn produced as phages are released from an infected cell and infect and lyse neighbouring cells. At a suitable ratio of phage to bacterial cells in the initial mixture, each plaque will be the result of infection of a single bacterial cell with a single phage. Various methods are used to distinguish recombinant from nonrecombinant phages; for example: insertional inactivation of a *lacZ* gene carried by the vector so that recombinant plaques are clear whereas plaques produced by nonrecombinant phage are blue; insertional inactivation of the lambda *cI* gene (*see* LAMBDA) so that normal plaques appear turbid whereas recombinant plaques are clear; and selection on the basis of lambda genome size, where a lambda vector that has not received a DNA insert is too small to be packaged into the phage head.

As well as cloning by naturally cleaved cohesive ends, artificial cohesive ends (or synthetic sequences that can give rise to them on digestion with the appropriate restriction enzyme) can be added to DNA fragments (*see* ADAPTORS; HOMOPOLYMER TAILING; LINKERS; OLIGO(DG : DC) TAILING). Virtually any DNA fragment can therefore be cloned into a suitable matching vector.

Sambrook, J. et al. (1989) *Molecular Cloning* 2nd edn (Cold Spring Harbor Press, New York).

DNA complexity The total length of different DNA sequences present within a given DNA sample. The complexity of a DNA sample is usually defined by studying the rate of reassociation of single-stranded DNA chains obtained by denaturation of a double-stranded DNA (*see* HYBRIDIZATION). The parameter controlling the reassociation reaction is described as the C_0t value (*see* HYBRIDIZATION). The value required for 50% reassociation ($C_0t_{\frac{1}{2}}$) can be used to define, in base pairs, the complexity of the test DNA. Such analysis of complex eukaryotic genomes has revealed the presence of three kinetically distinct DNA components each with a characteristic complexity (*see* GENOME ORGANIZATION): a slow kinetic component that contains nonrepetitive DNA (SINGLE COPY DNA) which is unique to the genome

and has a complexity that corresponds closely to its actual length in base pairs; an intermediate kinetic component which consists of large repeated DNA sequences with a low repetition frequency (i.e. MODERATELY REPETITIVE DNA); and a fast kinetic component which consists of short highly repeated DNA sequences. The complexity and repetition frequency can be used to describe the properties of the different sequence components of a given organism's genome.

DNA Cot analysis *See*: COT ANALYSIS.

DNA-dependent DNA polymerase *See*: DNA POLYMERASES.

DNA-dependent RNA polymerase *See*: RNA POLYMERASES; TRANSCRIPTION.

DNA fingerprinting *See*: DNA TYPING.

DNA footprinting *See*: FOOTPRINTING.

DNA glycosylases Enzymes that catalyse the release of altered nucleic acid bases from DNA by cleaving the N-glycosylic bond linking the modified base residue to the deoxyribose sugar in the polynucleotide. *See*: DNA REPAIR; URACIL-DNA GLYCOSYLASE.

DNA gyrase Prokaryotic DNA gyrase (a type II DNA TOPOISOMERASE, EC 5.99.1.3) is a tetramer of M_r 400 000 (from *Escherichia coli*) comprising two different subunits A (M_r 97 000, encoded by the *gyrA* gene) and B (M_r 90 000, encoded by the *gyrB* gene) in the ratio $A_2 : B_2$. Upon binding to a covalently closed circular DNA (e.g. PLASMID DNA), the enzyme initiates a process of cutting and rejoining the DNA duplex and as a consequence negative supercoils are introduced (*see* SUPERCOILING). In the active protein–DNA complex, some 120 bp of DNA are wrapped around the protein core. During this process DNA gyrase undergoes a conformational change to an inactive form: ATP hydrolysis is required for restoration of the active conformation. The A subunit is responsible for DNA cutting and reunion; the B subunit contains the ATP hydrolysing activity. DNA gyrase introduces supercoiling processively and catalytically, introducing up to 100 supercoils per minute per molecule of DNA gyrase. *E. coli* gyrase can be inhibited by a variety of antibiotics (e.g. NOVOBIOCIN and NALIDIXIC ACID) and since these antibiotics are known to inhibit DNA REPLICATION, this implies a key role for DNA gyrase in DNA synthesis. DNA gyrase isolated from *Micrococcus luteus* can be used *in vitro* to introduce supercoils into plasmid DNA and to reversibly knot and catenate covalently closed circular DNA molecules. In the absence of ATP, gyrase will slowly relax negative supercoils, although *in vivo* this process is principally undertaken by topoisomerase I. Eukaryotic topoisomerase II is less well characterized and, unlike its prokaryotic equivalent, may require ancillary proteins to effect supercoiling.

DNA helicases Class of enzymes that unwind DNA to facilitate separation of the two strands of the duplex. Disruption of the hydrogen bonds is accomplished in a reaction that is coupled to the hydrolysis of a nucleoside 5′-triphosphate. A widely used assay for DNA helicase activity monitors the displacement of an oligonucleotide annealed onto a circular single-stranded DNA molecule. DNA helicases have roles in DNA REPLICATION, DNA REPAIR, and RECOMBINATION, and in the regulation of gene expression (*see* TRANSCRIPTION). In *Escherichia coli*, for example, the DnaB protein acts as a helicase to initiate unwinding from the origin during the initiation of DNA replication; the UvrD protein (DNA helicase II) acts to displace excised oligonucleotides during nucleotide excision repair and methyl-directed mismatch repair of DNA; and the RecBCD enzyme can unwind a duplex to produce a single-stranded region for homologous pairing in recombination.

DNA hybridization *See*: HYBRIDIZATION.

DNA libraries Large collections of cloned DNA fragments from a given organism, tissue, organ, or cell type.

Genomic libraries consist of a collection of cloned DNA fragments derived directly from chromosomal DNA. The usual way of creating a genomic DNA library, either of a single chromosome or of the whole genome, is to digest the starting material with RESTRICTION ENZYMES. The number of fragments produced will depend on which enzyme is used but, as an example, a restriction enzyme which cuts at a sequence occurring once in every 4000 nucleotides throughout the human genome will generate about 750 000 DNA fragments. In practice, conditions are adjusted so that a collection of overlapping fragments is obtained. The fragments are cloned using plasmid, cosmid, or phage VECTORS or as YEAST ARTIFICIAL CHROMOSOMES (YACs) (*see* DNA CLONING). Plasmids are used mainly for cloning small fragments of DNA (up to about 4000 nucleotides long). For larger DNA fragments (up to 40 000 nucleotides), COSMIDS and modified forms of phage LAMBDA are used while DNA fragments up to 1 million nucleotides long are carried by YACs.

Genomic libraries now exist for many species, and numerous human genomic libraries provide the source material for human gene mapping (*see* HUMAN GENE MAPPING; HUMAN GENOME PROJECT).

cDNA libraries are a collection of cloned DNAs representing cDNA (complementary DNA) synthesized from a complex mRNA population and inserted into a suitable cloning vector. To increase the efficiency of cloning, the cDNA must either be 'tailed' with a homopolynucleotide using TERMINAL DEOXYNUCLEOTIDYLTRANSFERASE (*see* HOMOPOLYMER TAILING) or have synthetic oligonucleotide LINKERS ligated to each end thereby introducing new RESTRICTION SITES which can be cleaved before insertion into the cloning vector. cDNA libraries have the advantage for many applications over genomic libaries in that they contain already spliced gene sequences transcribed in the cell type or tissue from which the original mRNA was extracted.

The techniques available for studying the contents of a DNA library divide into: (1) processes for screening the library to locate cloned fragments of particular interest; (2) methods of analysing cloned fragments to identify genes or other markers or to begin working out their DNA sequence; and (3) ways of assembling individual characterized fragments in their correct order to build longer stretches of DNA.

The most usual way of screening a genomic library is to use a radioactive or fluorescence-labelled PROBE — a short OLIGONUCLEOTIDE of sequence complementary to part of the DNA fragment whose position in the library is being sought. Under appropriate conditions the probe will bind only to the required cloned fragment (*see* HYBRIDIZATION). A sample of that clone can then be removed from the library and the structure of the fragment studied in detail.

cDNA expression libraries are also in general use. In these the cDNA is inserted into a suitable EXPRESSION VECTOR so that a cDNA of interest can be detected by its ability to direct synthesis of the gene product, usually as part of a FUSION PROTEIN, in *Escherichia coli*. *See also*: HYBRID-ARRESTED TRANSLATION; HYBRID-RELEASE TRANSLATION; IN VITRO TRANSLATION; XENOPUS OOCYTE EXPRESSION SYSTEM.

Winnacker, E-L. (1987) *From Genes to Clones: Introduction to Gene Technology* (VCH, Weinheim).

DNA ligases Enzymes (EC 6.5) that catalyse the formation of a PHOSPHODIESTER BOND between adjacent 5′-P and 3′-OH termini in a polynucleotide chain (Fig. D16). Their biological role is in DNA REPLICATION, in which they are required for covalently joining the discontinuous OKAZAKI FRAGMENTS generated on the lagging strand at the replication fork, and in DNA REPAIR. T4 DNA ligase (derived from T4-infected *Escherichia coli* cells) (EC 6.5.1.1) is a monomer of M_r 68 000. It is used *in vitro* to assemble RECOMBINANT DNAS from fragments of DNA with either blunt ends or compatible cohesive ends. *See*: DNA CLONING; GENETIC ENGINEERING.

Weiss, B. et al. (1968) *J. Biol. Chem.* **243**, 4543.
Lobban, P. & Kaiser, A.D. (1973) *J. Mol. Biol.* **79**, 453–471.

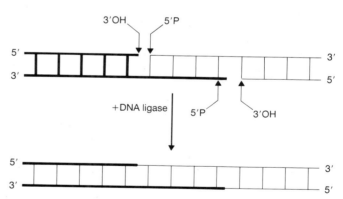

Fig. D16 Reaction catalysed by DNA ligase. An energy source such as ATP is also required.

DNA manipulation enzymes Enzymes which facilitate the *in vitro* modification of cloned DNA molecules. They include RESTRICTION ENZYMES for site-specific cleavage of the DNA, DNA LIGASES for covalently joining DNA fragments, and DNA METHYLASES which are used to methylate restriction enzyme recognition sites, thereby avoiding cleavage at these positions by the corresponding restriction enzyme. Although each class of enzyme occurs naturally, their biological roles do not necessarily dictate the uses made of them in DNA manipulation *in vitro*.

DNA methylation

LIKE many cellular macromolecules, DNA is subject to postsynthetic 'modification' by addition of small chemical moieties to the intact polymer. In a variety of organisms this involves enzymatic addition of methyl (—CH$_3$) groups to DNA, either at position C5 of cytosine (Fig. D17) or at position N6 of adenosine. The enzymes responsible for the addition are known as DNA methyltransferases or DNA methylases, and all use S-ADENOSYL METHIONINE as the methyl group donor.

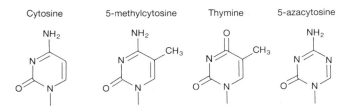

Cytosine   5-methylcytosine   Thymine   5-azacytosine

Fig. D17 The structure of cytosine and 5-methylcytosine (mC) compared to thymine and the drug 5-azacytosine (5-azacytidine). Note that thymine is the product of mC deamination. Azacytosine inhibits DNA methylation when incorporated into DNA, probably by covalently binding to the methyltransferase.

Prokaryotes

The biological significance of DNA methylation has preoccupied molecular geneticists for many years, and, in prokaryotes at least, a number of important questions have been answered. Best understood are the RESTRICTION AND MODIFICATION systems of bacteria [1], which protect against bacteriophage infection. Each system has two components: a methyltransferase which 'modifies' specific DNA sequences by addition of methyl groups, and a restriction endonuclease (RESTRICTION ENZYME) which can degrade DNA only when the specific sequences are methylation-free. The methyltransferase protects the bacterial genome from degradation by the restriction endonuclease. Invading viral DNA, however, is not protected and is therefore rapidly degraded.

Another role for DNA methylation in bacteria concerns DNA REPAIR. Mistakes in DNA REPLICATION, though rare, do occur, giving rise to unpaired bases in the double helix that are known as mismatches. The system that repairs the errors must be able to tell which base in a particular mismatched pair is incorrect. In other words, it must distinguish the newly synthesized DNA strand from the old ('parental') one. In some bacteria the distinction is achieved by delayed methylation of the new strand, which means that only the old strand carries methyl groups [2]. This asymmetry directs the mismatch repair process to the new strand.

Eukaryotes

Considerable efforts are being made to understand the significance of DNA methylation in animals, plants and fungi. In all these organisms it is established that the major (perhaps the only)

modified base is 5-methylcytosine (5-mC), although the sequence that is methylated shows some variation between kingdoms. Amounts of DNA methylation vary within each kingdom, being consistently low in the fungi tested so far. The highest levels occur in plants, where up to one-third of all cytosines in the genome can be methylated (e.g. in maize, *Zea mays*).

Most work has been done on animals, where the methylated sequence is CG (i.e. methylated cytosine is always followed by G on the 3′ side). CG is a self-complementary sequence, and methylatable cytosines therefore occur in pairs on opposite strands (Fig. D18). This is significant because the most efficient substrate for methyltransferases is not in fact a nonmethylated pair of CGs, but a pair of CGs, one of which is methylated. This kind of asymmetric structure arises each time methylated DNA is replicated, as the newly replicated strand is not yet methylated, whereas the parental strand contains methylated CG (mCG). Owing to its specificity for hemimethylated CG pairs, the methyltransferase only methylates CGs on the progeny strand that are paired with mCG on the parental strand. Thus the pattern of methylated and nonmethylated CGs is replicated at each cell division (Fig. D18).

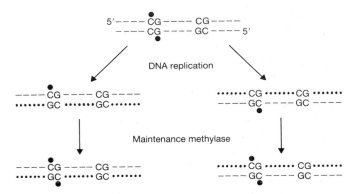

Fig. D18 Replication of the pattern of CG methylation by a maintenance methylase following DNA replication. The short segment of parental DNA (top) contains a symmetrically methylated CG pair and an unmethylated CG pair. Following DNA replication, a newly synthesized progeny strand (dotted line) is paired with each parental strand, but is not yet methylated. The maintenance methylase restores symmetry of the half-methylated sites, but is incapable of *de novo* methylation at the unmethylated CG pair. Thus a replica of the parental pattern of methylation is transmitted to each daughter cell.

By surveying the DNA of various animal species it has become clear that levels of mCG vary greatly. In some species (for example the fruit fly, *Drosophila*), the genome contains little if any CG in the methylated form, whereas in others (for example, mammals) most of the genomic CGs are methylated. The most heavily methylated genomes in the animal kingdom belong to the vertebrates. The difference between vertebrates (which account for only a few percent of animal species) and the rest is most striking when gene sequences are considered: nearly all vertebrate genes contain mCG, whereas nonvertebrate genes, as far as we know, contain none. These findings place firm limits on theories about the function of DNA methylation, as most animals appear to develop and live with little or no methylation near their genes. We must suppose either that DNA methylation has nothing to do with the control of gene expression during development, or that it has assumed a new control function in this respect with the evolution of the vertebrates. The fact that the phenomenon has developmental relevance in mammals is indicated by the finding that mice that are genetically deficient in the DNA methyltransferase are unable to complete development [3].

DNA methylation and transcription

There is experimental evidence that the methylation of CG has an effect on the expression of genes; specifically on TRANSCRIPTION. Several genes have been artificially methylated and then introduced into cells so that the effect could be tested. In nearly all cases the presence of mCGs led to repression of transcription. This observation fits with the finding that naturally methylated genes tend to lose their CG methylation when they are transcriptionally activated. It is also compatible with the observation that inactive genes that are methylated can be reactivated by treating with a drug, 5-AZACYTIDINE, which prevents the methylation of DNA (see Fig. D17). Taken together, the results encourage the belief that the removal of DNA methylation could be a controlling step in the activation of genes during development [4]. Like many theories, the idea that loss of methylation can control gene expression has its supporters and its doubters. Supporters can point to many examples correlating reduced methylation of a gene with its expression, and to the often dramatic effects of 5-azacytidine. Doubters, on the other hand, emphasize that much of the evidence implicating DNA methylation in gene control is circumstantial or suggestive, rather than conclusive. The critical experiments that could resolve the issue are yet to be done.

Although methylation loss is unproven as a gene activation mechanism, evidence for the opposite process, whereby active genes are suppressed through the addition of methylation, is well documented. For example, nonmethylated sequences that are associated with active genes on the mammalian X-chromosome become methylated during the process of X-CHROMOSOME INAC-TIVATION in female mammals. Artificial removal of the methylation with 5-azacytidine reactivates the genes, indicating that methylation does in fact contribute to repression. In a similar way, the replication of viral DNA can be suppressed by methylation of the viral or proviral genome. Again, removal of the methylation with azacytidine can reactivate the virus, causing infection. It should be emphasized that in neither of these cases (X inactivation or viral repression) is DNA methylation thought to be the primary inactivation mechanism. Rather, it seems that inactivation is initially achieved by other mechanisms, and is later stabilized by the methylation of CGs. Once methylated, the gene is repressed in an essentially irreversible manner, due to the heritability of CG methylation (see above).

How does DNA methylation contribute to transcriptional repression? Two sorts of mechanism have been considered: (1) mCG directly blocks access of TRANSCRIPTION FACTORS to their binding sites in the PROMOTER of a gene; (2) mCG interacts with nuclear components so as to alter CHROMATIN structure and thereby prevent transcription indirectly. There is evidence in favour of both mechanisms. Some transcription factors do have CG in their binding sites, and methylation of these sites does indeed prevent some (but not all) of these factors from binding. Strong evidence for the indirect mechanism has come with the discovery of a protein (MeCP1) that binds to DNA containing mCG. The protein binds regardless of the sequences surrounding mCG, but it requires the presence of at least 10 mCGs in the same molecule. By studying transcription of methylated genes in nuclear extracts and intact cells, it has been found that MeCP1 can repress transcription [5]. When MeCP1 is absent, due either to competition or to the use of cells that are MeCP-deficient, the methylated genes are transcribed.

CpG islands

Something about the functional role of DNA methylation can also be learned by examining regions of the mammalian genome that are constitutively free of methylation. Short regions of such nonmethylated DNA, known as CpG islands, account for about 2% of the mammalian genome [6]. Their location is interesting, as they are found at the 5′ ends of most mammalian genes. At present it seems that all mammalian 'housekeeping' genes, as well as a significant fraction of genes specifically expressed in particular tissues, have CpG islands at their promoters. Apart from their lack of methylation, CpG islands have a base composition that is more G + C rich than the bulk genome, and they also contain about 10 times more CpG. These characteristics sharply distinguish CpG-island DNA from the surrounding DNA, and have been used extensively to locate and clone genes in human DNA. It should be stressed that the islands are normally free of methylation whether the associated gene is transcribed or whether it is silent. The only known exception to this rule is the inactive X-chromosome in placental mammals. Here the CpG islands of inactivated genes become methylated in somatic cells (see above). In the extra-embryonic tissues and germ cells, however, X inactivation occurs without the benefit of CpG island methylation, and this is true even in somatic cells of marsupials. Interestingly, whenever inactivation occurs in the absence of CpG island methylation, it is leaky.

Mutability of mCG

Although most attention has been focused on the relevance of DNA methylation to transcriptional regulation, it is important not to overlook another dramatic effect of mCG, namely its mutability. Cytosine in DNA deaminates to give uracil at a significant rate and a repair system exists that repairs the damage. Methylcytosine also deaminates, but gives thymine (see Fig. D17) which, unlike uridine, exists elsewhere in the DNA, and may therefore be less easy to detect. Although there is an activity that repairs the resulting T:G mismatch in favour of C:G, it seems that this repair mechanism is less than 100% efficient. As a result, mCG is frequently replaced by TG. It has been calculated that mC is between 10 and 20 times more mutable than unmodified C. The consequences of this mutability are dramatically illustrated by analysis of the spectrum of point mutations that give rise to human genetic disease. It is estimated that about one-third of all such mutations are due to mCG to TG mutations [7]. It is a sobering thought that invertebrates, whose genes are methylation free, escape the burden of mutability that accompanies cytosine methylation.

A.P. BIRD

See also: EUKARYOTIC GENE EXPRESSION; MUTATION; NUCLEOSIDES AND NUCLEOTIDES.

1 Meselson, M. et al. (1972) Restriction and modification of DNA. *Annu. Rev. Biochem.* **41**, 447–466.
2 Glickman, B. et al. (1978) Induced mutagenesis in the dam mutants of *E. coli*: a role for 6-methyladenine residues in mutation avoidance. *Mol. Gen. Genet.* **163**, 307–312.
3 Li, E. et al. (1992) Targetted mutation of the DNA methyltransferase gene results in embryonic lethality. *Cell* **61**, 915–926.
4 Cedar, H. (1988) DNA methylation and gene activity. *Cell* **53**, 3–4.
5 Boyes, J. & Bird, A. (1991) DNA methylation inhibits transcription indirectly via a methyl-CpG binding protein. *Cell* **64**, 1123–1134.
6 Bird, A. (1987) CpG islands as gene markers in the vertebrate nucleus. *Trends Genet.* **3**, 342–347.
7 Cooper, D. & Youssoufian, H. (1988) The CpG dinucleotide and human genetic disease. *Hum. Genet.* **78**, 151–155.

DNA modification *See*: DNA METHYLATION; RESTRICTION AND MODIFICATION.

DNA nick translation *See*: NICK TRANSLATION.

DNA polymerases Enzymes that catalyse DNA synthesis by addition of deoxynucleotides to the 3′ end of a polynucleotide chain, using a complementary polynucleotide strand as a template (*see* DNA REPAIR; DNA REPLICATION; REVERSE TRANSCRIPTION). They comprise the DNA-dependent DNA polymerases (EC 2.7.7.7) and the RNA-dependent DNA polymerase, REVERSE TRANSCRIPTASE (EC 2.7.7.49). Most DNA polymerases cannot initiate DNA synthesis *de novo* and require the presence of a short DNA primer which is synthesized on the template DNA. DNA synthesis commences at the 3′ end of the primer.

In bacteria, as exemplified by *Escherichia coli*, three DNA polymerases — DNA polymerases I, II, and III — have been found. DNA polymerase I is used for excision repair and also fills in between OKAZAKI FRAGMENTS and removes the RNA primer during DNA replication. The DNA polymerase III holoenzyme synthesizes new DNA strands. DNA polymerase II is inducible by DNA damage and may be a DNA repair polymerase. DNA polymerase I is a single-chain protein containing three enzymatic functions — polymerase, 3′ → 5′ exonuclease (removal of mismatched bases), and 5′ → 3′ exonuclease (primer removal and excision of pyrimidine dimers). DNA polymerase III is a multisubunit protein (total M_r ~800 000) with processive polymerizing and 3′ → 5′ exonuclease activity (Table D4). A single DNA polymerase III enzyme can synthesize long stretches of DNA (several thousands of nucleotides) continuously at a rate of 1000 nucleotides per second. The maximum length synthesized at one time by DNA polymerase I is around 20 nucleotides, and synthesis proceeds at a rate of ~10 nucleotides per second.

In eukaryotic cells, three different enzymes — DNA polymerases α, δ, and ε — are implicated in nuclear DNA replication.

Table D4 The subunits of DNA polymerase III from *Escherichia coli*

Subunit		M_r	Activity
α	catalytically	140 000	Polymerizing
η	active	27 000	3′–5′ exonuclease
θ	core	10 000	
β		37 000	Required for processivity
τ		78 000	
γ		52 000	
δ		32 000	
δ′		32 000	

DNA polymerase α has a PRIMASE activity that can synthesize RNA primers. DNA polymerase γ is used for mitochondrial DNA replication. DNA polymerase β is used for base excision repair of DNA, whereas nucleotide excision repair uses DNA polymerases δ and/or ε (*see* DNA REPAIR). Polymerase α is a multisubunit assembly and polymerase β is a single-chain protein. *See also*: KLENOW FRAGMENT; SOS PATHWAY; TAQ POLYMERASE.

Kornberg, A. & Baker, T. A. (1991) *DNA replication* 2nd edn (Freeman, New York).

DNA polymorphism *See*: GENETIC POLYMORPHISM; RFLP.

DNA probe *See*: PROBE.

DNA profiling *See*: DNA TYPING.

DNA–protein interactions *See*: PROTEIN–NUCLEIC ACID INTERACTIONS.

DNA rearrangement *See*: ANTIGENIC VARIATION; BACTERIAL PATHOGENICITY; GENE REARRANGEMENT; PHASE VARIATION; YEAST MATING-TYPE LOCUS.

DNA repair

THE DNA in every cell is constantly damaged by environmental agents and by endogenous reactions. This damage interrupts the continuity of genetic information by inhibiting or preventing TRANSCRIPTION, DNA REPLICATION, and cell division. If the damage is unrepaired it can lead to death of the cell, or to the production of MUTATIONS by insertion of incorrect bases. As a result, organisms devote a considerable number of genes to enzymes and regulatory proteins that are involved in different pathways of DNA repair. Many relatively simple modifications to single damaged bases are dealt with by base excision repair, which sequentially removes the modified base and deoxyribose sugar. Lesions that cause greater disturbance to double helical DNA are corrected by nucleotide excision repair, which removes a segment of DNA surrounding the alteration. In both excision repair processes, the missing bases are resynthesized by a DNA POLYMERASE, using the complementary undamaged strand as a template. A few types of damage can be enzymatically repaired by direct reversal without removing bases from the DNA. Another

Table D5 DNA glycosylases used in base excision repair

Enzyme name	Modified bases recognized	Sources of base damage
Uracil-DNA glycosylase	Uracil	Deamination of cytosine; incorporation of dUTP in DNA
Hypoxanthine-DNA glycosylase	Hypoxanthine	Deamination of adenine
Hydroxymethyluracil-DNA glycosylase	Hydroxymethyluracil (enzyme found in mammalian cells, not in bacteria)	
Pyrimidine hydrate-DNA glycosylase	Thymine glycols; pyrimidine hydrates; ring-fragmented pyrimidines; urea	Ionizing and ultraviolet radiation; hydroxyl radicals and oxygen radical-generating agents
Formamidopyrimidine-DNA glycosylase	Purines with a fragmented imidazole ring; 8-hydroxyguanine	
3-Methyladenine-DNA glycosylase	Purines methylated at N3 or (at a slow rate) N7; pyrimidines methylated at the O^2 position	Intracellular S-adenosyl methionine; environmental alkylating agents such as methyl chloride
Pyrimidine dimer-DNA glycosylase	Cyclobutane pyrimidine dimers (enzyme encoded by the bacteriophage T4, and by the bacterium *Micrococcus luteus*)	Ultraviolet radiation

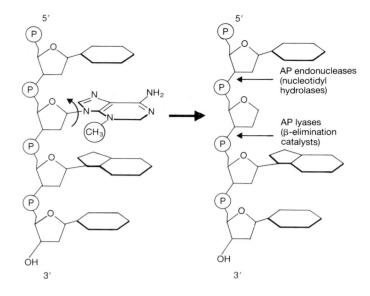

Fig. D19 Base excision repair. A segment of one strand of DNA is shown for simplicity. Left, a DNA glycosylase removes a modifed base (3-methyladenine in this case) by cleavage of the *N*-glycosyl bond. Right, the resulting AP site is cleaved in one of two ways. AP endonucleases cleave hydrolytically to yield a 5′-terminal residue of deoxyribose 5-phosphate and a 3′-OH terminal nucleotide residue. AP lyases cleave to yield a 3′ terminus which is an unsaturated aldehyde derived from the deoxyribose sugar, and a 5′-phosphate terminal nucleotide residue. Further processing removes the sugar residue so that a new nucleotide base can be inserted.

mechanism for tolerance of DNA damage is to exchange a damaged region for an undamaged one via homologous RECOMBINATION. Errors in DNA replication are a further source of DNA sequence changes. Several mismatch repair pathways exist that can correct potentially mutagenic mispairs in DNA.

Base excision repair

Some base alterations occur frequently enough to warrant repair by specialized proteins called DNA glycosylases [1]. These enzymes catalyse hydrolysis of the *N*-glycosylic bond linking a modified base residue to the deoxyribosephosphate backbone in DNA. Each DNA glycosylase can work only on a restricted range of structurally related substrates. For example, the base uracil (U) can arise in DNA by spontaneous deamination of the exocyclic amino group of cytosine (C). This is a constantly occurring hydrolytic process, with a rate that increases with temperature. The enzyme uracil-DNA glycosylase removes U residues from DNA, thus preventing a mutation that would have arisen in the next round of replication by pairing of the U with A. The known DNA glycosylases which function in repair of various different types of modified bases are summarized in Table D5.

After action of any of these DNA glycosylases, a baseless

deoxyribose sugar residue is left in the DNA, known as an AP (apurinic or apyrimidinic) site (Fig. D19). AP sites also arise in DNA by a continuous slow process of spontaneous hydrolysis of the glycosylic bond. It has been estimated that up to 10 000 purines are lost per human cell per day [2]. To repair these spontaneously arising AP sites, as well as those derived from base excision by DNA glycosylases, the phosphodiester backbone must be broken to allow an undamaged nucleotide to be incorporated. Cleavage of the backbone is accomplished by activities that fall into two classes: AP endonucleases (nucleotidyl hydrolases), and AP lyases (β-elimination catalysts) [3]. The first type of activity catalyses hydrolysis of the phosphodiester bond on the 5′ side of the deoxyribose sugar to leave a 3′-OH terminus. The remaining deoxyribosephosphate residue is then removed either enzymatically by a different enzyme, 5′-deoxyribosephosphodiesterase, or by β-elimination which occurs on the 3′ side of an AP site (Fig. D19). Some DNA glycosylases have an associated AP lyase activity, allowing the enzyme both to remove the altered base residue and to cleave the phosphodiester backbone. Incision at AP sites by β-elimination can also be promoted by basic proteins and polyamines. In any case, a one-nucleotide gap is introduced in one strand of the DNA, and a new base is inserted by DNA polymerase.

Nucleotide excision repair

Nucleotide excision repair is a versatile pathway that acts on a wide diversity of DNA lesions that distort the DNA helix or disrupt normal base pairing. The types of damage removed include pyrimidine dimer photoproducts induced by ultraviolet light (UV) (see Fig. D21 below), and DNA modifications produced by many chemical mutagens. For example, adducts produced by reactive derivatives of POLYCYCLIC AROMATIC HYDROCARBONS (found in cooked foods, burning tobacco and coal tar) are acted on by nucleotide excision repair (*see also* CARCINOGENS AND CARCINOGENESIS). Further substrates include DNA interstrand cross-links, and many other types of lesions.

A mechanism with the same basic features seems to operate in most organisms. Several proteins cooperate to incise DNA at the site of damage. This leaves a short gap which is filled in with undamaged nucleotides by DNA polymerase, and the gap is ligated (Fig. D20). Many details of the process in the bacterium *Escherichia coli* have been worked out [4,5]. Here, the UvrA, UvrB, and UvrC proteins assemble on DNA at a damaged site and form an incision nuclease. Two cuts are introduced on the damaged strand, usually one that is eight phosphodiester bonds away from the 5' side of the lesion and another that is four or five phosphodiester bonds away from the 3' side. The damaged oligonucleotide is displaced with the aid of DNA polymerase I and DNA HELICASE II (UvrD), and the resulting 12-nucleotide gap is filled in by DNA polymerase.

In some organisms there is an interaction of nucleotide excision repair with the transcriptional process. Damage is removed several times faster from the transcribed strand of active genes than from inactive genes, and the underlying mechanism is a subject of active investigation.

Inherited conditions associated with defective DNA repair

The consequences to a human being of a defect in nucleotide excision repair are vividly demonstrated by the disorder XERODERMA PIGMENTOSUM (XP) [6]. XP patients are hypersensitive to UV light and suffer from a high incidence of sunlight-induced skin and eye lesions, including melanoma and non-melanoma cancers. Severely affected individuals die as a consequence of multiple skin tumours. The sensitivity to sunlight and other mutagens is caused by inherited mutations in components of the nucleotide excision repair system. There are eight COMPLEMENTATION GROUPS of the disease, denoted XP-A through to XP-G, and XP-V. These groups represent different nucleotide excision repair genes, and so the process seems to be more complex in human cells than in bacteria. The XP-A gene codes for a ZINC-FINGER PROTEIN that binds damaged DNA. The XP-B and XP-D gene products are DNA helicases.

Another disease in which patients display sensitivity to sunlight is COCKAYNE'S SYNDROME [6,7], and this condition is associated with more subtle alterations in excision repair genes. FANCONI'S ANAEMIA is a further inherited condition that appears to be associated with defective repair [6]. Cells from people with this disease have a reduced ability to remove damage caused by agents that cross-link the two strands of DNA together, but the relationship of the defect in this disorder to the nucleotide excision repair process is uncertain. In addition to the genes associated with human diseases, alterations in excision repair genes have been recognized by studying cultured cell lines derived from other species. Studies of excision repair genes in rodent cell lines (*ERCC* genes) and of repair mutants in yeast (*RAD* genes) have been particularly productive in this regard [7].

Strand-break repair

A final step in both excision repair pathways is joining of the newly synthesized repaired region to the remainder of the DNA. Linking of 3' hydroxyl and 5' phosphate ends is enzymatically catalysed by a DNA LIGASE. Strand breaks in DNA can also arise when free radicals attack the phosphodiester backbone. Such radicals are produced by oxidative processes and by ionizing radiation. The resulting breaks usually require enzymatic action to produce ligatable termini.

In a few inherited conditions, aberrant joining of DNA is characteristic of the disorder. Cells from people with BLOOM'S SYNDROME have a high spontaneous and induced rate of SISTER CHROMATID EXCHANGE, and a slow joining of large DNA replication intermediates which may be the basis of other symptoms of the disease such as retarded growth and severe immunodeficiency (because of the involvement of DNA rearrangements in the differentiation of T and B lymphocytes, *see* B CELL DEVELOPMENT; T CELL DEVELOPMENT). Strains of mice known as *scid* (for severe combined immunodeficiency) are defective in repair of double-strand breaks in DNA, and this may similarly give rise to the immunodeficiency that is characteristic of these strains (*see* SCID; SCID MICE).

Repair of extensive DNA strand breakage takes time, and

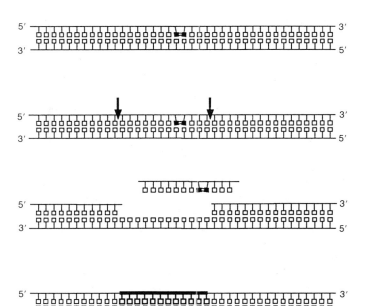

Fig. D20 Nucleotide excision repair. A damaged segment of DNA is recognized by a group of proteins which introduce incisions flanking the damage (arrowed), and a short oligonucleotide containing the damage is removed. The resulting gap is filled in by DNA polymerase, and the newly synthesized nucleotides are covalently ligated to the rest of the DNA.

proliferating eukaryotic cells have control mechanisms to stop DNA replication and cell division when strand breaks appear; growth is resumed after repair has taken place. One such distinct control point exists in the G2 phase of the CELL CYCLE. After radiation damage to DNA, cell-cycle progression is blocked to prevent entry into MITOSIS before repair is complete. In the yeast *Saccharomyces cerevisiae*, this response is controlled by a group of genes including *rad9*. Another level of control inhibits DNA synthesis after DNA damage. In the human disease ATAXIA TELANGIECTASIA, patients' cells are sensitive to killing by ionizing radiation and have a high incidence of chromosome aberrations. This is associated with a reduced ability of the cells to inhibit DNA synthesis after suffering damage.

Direct reversal of DNA damage

A few kinds of lesions in DNA can be corrected by direct enzymatic reversal, without the need to excise any nucleotides.

Methyltransferases

The oxygen atoms in DNA undergo occasional methylation by endogenous agents in the cell, or by environmental alkylating agents. O^6-methylguanine-DNA methyltransferases are enzymes found in many organisms that remove methyl groups from the susceptible positions (O^6 of guanine, O^4 of thymine, and oxygens in the phosphodiester backbone) [8]. Each enzyme transfers a methyl group to one of its own cysteine residues. The mechanism can also work, with lower efficiency, on ethylated oxygens. In some bacteria, one form of the enzyme becomes a transcriptional activator when methylated, and promotes synthesis of more DNA methyltransferase as well as a 3-methyladenine-DNA glycosylase and a few other protective gene products. Induction of proteins that defend against alkylation damage is known as the 'adaptive response' in bacteria. O^6-methylguanine DNA methyltransferase activity in human cells is sometimes suppressed at the transcriptional level, resulting in a methyltransferase-deficient state known as mex$^-$ or mer$^-$. Such cells are much more sensitive to alkylating agents than normal. This can have significant consequences in cancer therapy, as some widely used chemotherapeutic agents (such as the nitrosoureas) act by forming toxic alkylation damage in DNA.

Photolyases

Cyclobutane pyrimidine dimers are the most common type of lesion formed in DNA by UV light of wavelengths less than 320 nm. These photoproducts can efficiently block DNA replica-tion and transcription, and so organisms have developed several strategies to repair them. In the most direct mechanism, DNA photolyase (photoreactivating enzyme) binds to a cyclobutane dimer and catalyses monomerization back to two pyrimidines in a light-dependent reaction (Fig. D21) [9]. This is accomplished through the presence of CHROMOPHORE cofactors in the enzyme that absorb photons of 365–445 nm and transfer their energy to the cyclobutane ring of the dimer to facilitate bond rearrangement. DNA photolyases are widespread in bacteria and fungi. Each enzyme molecule has two chromophores. One is reduced flavin adenine dinucleotide ($FADH_2$), and the other is either a reduced pterin or a deazaflavin derivative, depending on the organism. These transfer light energy to the cyclobutane ring of the pyrimidine dimer in an electron cascade reminiscent of PHOTOSYNTHESIS. Photoreactivating activity has also been detected in fish, reptiles and marsupial mammals, although photolyases have not yet been purified from these organisms.

Mismatch repair

Because DNA polymerases are not completely accurate in copying a template, postreplicative repair systems exist that increase the overall fidelity of replication up to 10^4-fold by removing mispaired bases. Both prokaryotes and eukaryotes, including humans, have mismatch repair systems [10], but the process is best understood in bacteria. The core enzymes of the major mismatch repair system in *E. coli* are the products of the *mutH*, *mutL*, and *mutS* genes. A set of *hex* genes with similar functions exists in *Streptococcus*. Mismatch repair can only increase fidelity if the parental strand can be distinguished from the daughter strand, allowing repair to preserve the original information in the parental strand.

In *E. coli*, the strand discrimination signal is provided by adenine methylation in GATC sequences; newly replicated DNA is not immediately methylated on the daughter strand. MutH binds to DNA at GATC sequences and makes an incision on the unmethylated strand. MutS recognizes and binds to the mismatch, and the intervening region between the mismatch and the GATC sequence is excised. This leaves a single-stranded gap of several hundred to several thousand nucleotides which is filled in by DNA polymerase. MutL mediates communication between the distantly bound MutH and MutS products, perhaps bringing them together by looping out the intervening region. In *E. coli* strains with defective adenine methylation (*dam*), mismatch repair is no longer methyl-directed and becomes ineffectual in preventing mutation. In other organisms, mismatch correction does not appear to be methyl-directed. Instead, the strand discrimination signal is thought to be single-stranded nicks or gaps in newly

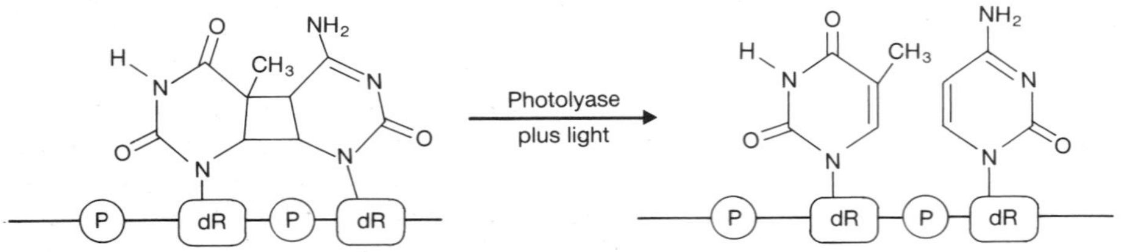

Fig. D21 Enzymatic photoreactivation. Cyclobutane pyrimidine dimers produced by UV light (a thymine–cytosine dimer is shown here) can be monomerized by the action of a photolyase which funnels light energy to the cyclobutane ring and catalyses bond rearrangement.

replicated DNA which have not yet been joined together by DNA ligase.

Some particular mismatches can also be repaired by alternative mechanisms. For example A–G mismatches in *E. coli* are repaired by the MutY DNA glycosylase which removes the A. Additionally, certain mismatches can arise in the absence of DNA replication. At CC(A/T)GG sequences in *E. coli*, the second cytosine is methylated at the 5 position. 5-Methylcytosine spontaneously deaminates at a low rate, transforming the base into a thymine. A mechanism exists which can recognize a T–G mispair in this specific sequence context, and the T is removed in a process known as very short mismatch repair. An analogous system exists in human cells to remove thymines which arise from deamination of 5-methylcytosine in methylated CG sequences.

A further general function of mismatch repair is to eliminate the mismatches in heteroduplex intermediates formed by recombination of two homologous pieces of DNA, and this can result in GENE CONVERSION.

Other mechanisms for tolerance of DNA damage

Cells have other strategies for tolerating DNA damage besides the different schemes for removal of lesions from DNA. A widespread and general mechanism is homologous recombination. A damaged segment of DNA can be replaced with a nondamaged segment derived from a homologous sequence present in the cell on a duplicate chromosome. This process is especially important in cases where the immediate complementary strand cannot be used as a template for repair. Such instances arise when a replication fork is blocked at a damaged site in DNA, or when the two DNA strands are cross-linked together by a chemical or radiation, or when breaks occur in both strands of the helix at the same position. In *E. coli* and many other bacteria, homologous pairing and strand exchange is mediated by the RecA protein [11].

In addition to its function in recombination, the RecA protein performs a regulatory function in some bacteria and indirectly controls the expression of genes that are involved in the cellular response to DNA damage. In *E. coli*, there are around 20 genes involved in this so-called SOS RESPONSE including *recA* itself, the excision repair genes *uvrA*, *uvrB*, and *uvrD*, the recombination genes *ruvA* and *ruvB*, and the *umuC* and *umuD* genes. The UmuC and UmuD proteins apparently aid in damage tolerance by allowing a DNA replication fork to proceed beyond damaged sites without requiring removal of the lesion [12]. This process is mutagenic or 'error-prone', as incorrect bases sometimes arise opposite the noncoding adducts.

R.D. WOOD

See also: DNA METHYLATION.

1 Sakumi, K. & Sekiguchi, M. (1990) Structures and functions of DNA glycosylases. *Mutat. Res.* **236**, 161–172.
2 Lindahl, T. (1993) Instability and decay of the primary structure of DNA. *Nature* **362**, 709–715.
3 Doetsch, P.W. & Cunningham, R.P. (1990) The enzymology of apurinic/apyrimidinic endonucleases. *Mutat. Res.* **236**, 173–202.
4 Grossman, L. & Yeung, A.T. (1990) The UvrABC endonuclease of *Escherichia coli*. *Photochem. Photobiol.* **51**, 749–755.
5 Selby, C.P. & Sancar, A. (1990) Structure and function of the (A)BC excinuclease of *Escherichia coli*. *Mutat. Res.* **236**, 203–211.
6 Cleaver, J.E. & Kraemer, K.H. (1989) *The Metabolic Basis of Inherited Disease*, 6th edn. In Scriver, C.R. et al., Eds 2949–2971 (McGraw-Hill, New York).
7 Weeda, G. & Hoeijmakers, J.H.J. (1993) Genetic analysis of nucleotide excision repair in mammalian cells. *Semin. Cancer Biol.* **4**, 105–117.
8 Lindahl, T. et al. (1988) Regulation and expression of the adaptive response to alkylating agents. *Annu. Rev. Biochem.* **57**, 133–157.
9 Sancar, G.B. (1990) DNA photolyases: physical properties, action mechanism, and roles in dark repair. *Mutat. Res.* **236**, 147–160.
10 Grilley, M. et al. (1990) Mechanisms of DNA-mismatch correction. *Mutat. Res.* **236**, 253–267.
11 West, S.C. (1992) Enzymes and molecular mechanisms of genetic recombination. *Annu. Rev. Biochem.* **61**, 603–640.
12 Echols, H. & Goodman, M.F. (1990) Mutation induced by DNA damage: a many protein affair. *Mutat. Res.* **236**, 301–311.

DNA replication

DUPLICATION of the genetic information encoded in DNA is the crucial step in the reproduction of living organisms and the growth of multicellular organisms. Before a cell can divide its DNA must be replicated completely and accurately. New copies of the two strands of DNA are synthesized by DNA POLYMERASES from deoxyribonucleoside triphosphates (*see* NUCLEOSIDES AND NUCLEOTIDES), releasing pyrophosphate at each step. The sequence of the new DNA strands is determined by BASE-PAIRING precursors to the unwound parental strands before polymerization (*see* DNA; NUCLEIC ACID STRUCTURE). Therefore the sequence of the parental molecule is copied precisely.

Replication is semiconservative so that each new DNA double helix consists of one parental template strand hydrogen bonded along its entire length by base-pairing to a newly synthesized strand. Replication initiates at specific replication origins (*see* ORIGIN OF REPLICATION), to generate two REPLICATION FORKS which are elongated bidirectionally by multienzyme replication complexes — the replisome. Prokaryotic circular DNAs are replicated from a single origin, forming a characteristic THETA STRUCTURE when replication intermediates are observed in the electron microscope. Eukaryotic DNAs have multiple origins of replication so that the large chromosomes can be replicated sufficiently rapidly. Unidirectional replication occurs in some phages and some prokaryotic circular DNAs replicate by a ROLLING CIRCLE mechanism (*see* LAMBDA; PLASMIDS). Fidelity of DNA replication is ensured by PROOFREADING and DNA REPAIR mechanisms. Regulation of DNA replication is a critical control point in cell proliferation.

Mechanisms

The two strands of a DNA double helix are antiparallel, that is, they run in opposite directions so that the terminal $5'PO_4$ of one strand is opposite the terminal $3'OH$ of the other. However, DNA polymerases can only add nucleotides to the $3'OH$ group of a polynucleotide chain. These constraints are overcome by the use of RNA PRIMERS and a semi-discontinuous mode of synthesis. With rare exceptions, DNA chains are initiated by synthesis of short RNA primers, which are initiated *de novo* by RNA polymerases known as primases using the DNA strand as a template. These provide a 3′ hydroxyl group from which DNA polymerases can synthesize new DNA. Since the strands are antiparallel and DNA polymerases can only add to the 3′ end, synthesis of both strands at a single replication fork requires two mechanisms. One strand, the leading strand, is synthesized continuously by the repeated addition of nucleotides to its 3′ end, whereas the other lagging strand is synthesized discontinuously in segments called OKAZAKI FRAGMENTS which are about 1000 nucleotides long in bacteria or 150 nucleotides in eukaryotes (Fig. D22). Gaps between the fragments are subsequently filled to form the second continuous DNA strand.

DNA polymerases

DNA polymerases polymerize deoxyribonucleoside triphosphates into DNA by a condensation reaction that forms a phosphodiester bond linking the 3-OH of the sugar component of one nucleotide and the 5-OH of the sugar of the next (*see* DNA; NUCLEIC ACID STRUCTURE) with the release of pyrophosphate. There are several types of DNA polymerase in any organism. In *Escherichia coli*, DNA polymerase III synthesizes both leading and lagging strands, acting as an asymmetric dimer in which the two catalytic subunits are accompanied by different combinations of accessory subunits on the two DNA strands. The gaps between Okazaki fragments are filled by DNA polymerase I.

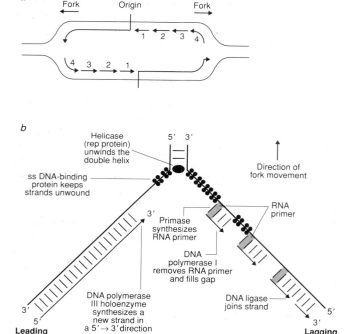

Fig. D22 *a*, As replication forks diverge from an origin of replication, the one new DNA strand is synthesized continuously (the leading strand), while the other is synthesized discontinuously in Okazaki fragments. Numbers indicate the order in which fragments are synthesized. *b*, Schematic diagram of events at the replication fork.

In eukaryotes, chromosomal DNA is replicated by three DNA polymerases α, δ, and ε. Polymerase α contains an integral primase. Polymerase α is required for lagging strand synthesis and possibly for the initial priming of leading strands too. Polymerase δ is required for leading strand synthesis. Its ability to synthesize very long stretches of DNA (called its processivity) is greatly increased by an accessory protein called PCNA (proliferating cell nuclear antigen). The role of polymerase ε is controversial at present, though it is generally agreed to be involved in eukaryotic DNA replication. Mitochondrial DNA is synthesized by polymerase γ. Most DNA polymerases, but probably not polymerase α, contain a proofreading EXONUCLEASE which excises misincorporated bases, increasing the fidelity of template copying.

Other replication proteins

In addition to DNA polymerases, many other proteins are involved in DNA replication, often in the form of multienzyme complexes (Table D6). For example, DNA HELICASES unwind the two strands of the double helix, hydrolysing ATP as they do so. The unwound strands are stabilized by single-strand binding proteins (SSB PROTEINS). DNA TOPOISOMERASES relax the torsional strain generated by unwinding the double helix. DNA topoisomerase I relaxes torsional strain without hydrolysing ATP; DNA topoisomerase II requires ATP hydrolysis and it is essential for the terminal stage of DNA replication to separate the progeny duplexes. Other proteins required for DNA replication include primases to synthesize RNA primers, DNA LIGASES to join adjacent Okazaki fragments after the gaps are filled, polymerase accessory proteins, such as PCNA in eukaryotes, and proteins involved in recognition of replication origins (see below).

Initiation of replication

Prokaryotes initiate DNA replication at unique sites, called origins of replication. In eukaryotes multiple origins are used, so that eukaryotic chromosomes are replicated by many replication forks simultaneously. In yeast the specific DNA sequences which function as replication origins have been identified and charac-

Table D6 Examples of replication proteins

Replication protein	Activity
DNA polymerases	Polymerize DNA strands
DNA helicases	Unwind the two stands of a DNA duplex
DNA topoisomerases	Relax torsional strain
Single-strand DNA-binding proteins	Stabilize unwound single strands
Origin-binding proteins	Bind and locally unwind origins of replication
Primases	Synthesize short RNA primers to initiate DNA synthesis
Ligases	Ligate fragments of DNA

terized, but replication origins in higher eukaryotes have been much more difficult to identify.

In prokaryotes and some animal viruses, and presumably elsewhere, replication origins are recognized by sequence-specific binding proteins, which can locally unwind a specific region of the double helix allowing replication to initiate. In *E. coli* this is the role of the protein encoded by the *dnaA* gene. DnaA protein binds to a short repeated DNA sequence at the origin. The DNA wraps around ~40 monomers of DnaA, causing an adjacent region to unwind and allowing binding of a DNA helicase encoded by the *dnaB* gene.

Termination and telomeres

When two converging replication forks meet, their nascent strands are joined. DNA topoisomerase II is required to unwind the two progeny DNA molecules from around each other in these final stages. The ends of the long linear chromosomes of eukaryotes, called telomeres, are replicated by a different mechanism. They consist of many copies of a short repeating sequence which are added by the enzyme telomerase. This enzyme contains an RNA template which it copies into DNA to complete the chromosome ends (*see* CHROMOSOME STRUCTURE).

Regulation

In prokaryotes the interval between consecutive initiations varies according to growth conditions. In *E. coli* three mechanisms appear to contribute to regulation of this time interval: availability of the active ATP-bound form of the DnaA protein, delayed methylation of the origin region of the nascent DNA strands, delaying re-initiation, and sequestration of the origin by binding to the bacterial membrane. In favourable growth conditions, a second round of DNA replication initiates before the first is complete.

Eukaryotic DNA replication is tightly coupled to the CELL CYCLE, so that DNA is replicated once and only once during a single S phase. Entry into S is regulated by a growth control network which involves GROWTH FACTORS, receptors, SECOND MESSENGERS, and specific TRANSCRIPTION FACTORS. Once replication has initiated, re-initiation within a single cell cycle is prevented.

R. LASKEY

See also: ADENOVIRUS; CHROMATIN; MITOSIS; NICK TRANSLATION; POLYMERASE CHAIN REACTION; ROLLING CIRCLE REPLICATION.

1 Darnell, J. et al. (1990) *Molecular Cell Biology* 2nd edn, Ch. 12 (Scientific American Books, New York).
2 Kornberg, A. & Baker, T.A. (1991) *DNA Replication* 2nd edn (Freeman, New York).

DNA restriction *See*: RESTRICTION AND MODIFICATION; RESTRICTION ENZYMES.

DNA satellite *See*: SATELLITE DNA.

DNA sequence The order of bases in a length of DNA. Also called base sequence or nucleotide sequence.

DNA sequence searching *See*: ALIGNMENT; DATABASES AND INFORMATION HANDLING.

DNA sequencing

THE sequence of nucleotides in DNA (or RNA) carries the genetic information encoding proteins and RNAs (*see* GENE; GENETIC CODE; PROTEIN SYNTHESIS; TRANSCRIPTION), and also determines to a large extent the secondary structures DNA and RNA may adopt (*see* DNA; NUCLEIC ACID STRUCTURE; RNA). The ability to determine the sequence of any piece of DNA rapidly and easily is therefore crucial to many problems in molecular biology. Two fundamental methods to determine the order of nucleotides in DNA were developed in the 1970s — Maxam–Gilbert chemical sequencing [1] and Sanger dideoxy chain termination sequencing [2]. Although contrasting in approach (the Maxam–Gilbert method is degradative, whereas the Sanger method depends on the synthesis of complementary-strand DNA) they both depend on the same principle: the generation of a nested set of single-stranded DNA molecules having one invariant end with the other ends differing by increments of single bases. A set of four reactions, specific for each base, generates a set of fragments which, when fractionated in parallel on the basis of length, allows the sequence to be deduced. RNA sequences (e.g. those of RNA virus genomes) can also be determined by these methods, after first converting the RNA sequence into complementary DNA (cDNA) using REVERSE TRANSCRIPTASE.

Maxam–Gilbert chemical sequencing

Chemical sequencing utilizes reactions that chemically modify bases in such a way that they can then be removed from the polynucleotide chain, which is therefore broken at that point. Two of the modifying reactions are specific for one base (G or C) while two others are specific for either the two PURINES or the two PYRIMIDINES. A complete set of fragments necessary for sequence determination is generated by limiting the reaction such that, on average, approximately only one site in each molecule is modified and cleaved. Although not so widely used as the Sanger dideoxy method for general sequencing, the Maxam–Gilbert protocol is very suitable for the study of modified bases and the interaction of proteins with DNA, and for sequencing OLIGONUCLEOTIDES.

Sanger dideoxy chain termination sequencing

While the chemical sequencing method has some specific uses, the dideoxy chain termination method has become the most widely used technique, particularly for large-scale sequencing projects. This technique depends on 2′,3′-dideoxyribonucleoside triphosphates (ddNTPs) (Fig. D23) being incorporated into an extending copy of a TEMPLATE DNA strand. Having been incorpo-

a

b

Fig. D23 *a*, 2′-deoxyribonucleoside-5′-triphosphate (dNTP); *b*, 2′,3′-dideoxyribonucleoside-5′-triphosphate (ddNTP).

rated, such a nucleotide cannot form the phosphodiester bond required to extend the chain further. Growth of the chain is thus terminated. In order to generate a complete set of such terminated chains, a small amount of a specific ddNTP is included in the reaction mixture in addition to the four deoxyribonucleoside triphosphates (dNTPs) required for normal DNA synthesis (*see* DNA REPLICATION). The frequency of chain termination, and thus the mean length of the terminated chains, depends upon the ratio of ddNTP to dNTP. A set of four reactions required for a sequence determination will thus each consist of a template DNA strand (the DNA whose sequence is required), a short complementary oligonucleotide PRIMER (to initiate chain extension at the desired point), the four dNTPs, and the correct ratio of one of the four ddNTPs. The chain is extended from the primer by a DNA POLYMERASE (Klenow, Taq, or T7) which lacks 5′ → 3′ EXONUCLEASE activity (as maintenance of a unique 5′ end is essential).

Fragment separation

The set of fragments generated by either method is fractionated by length by ELECTROPHORESIS through a denaturing polyacrylamide gel. To date, these have always been slab gels, 0.1–0.4 mm thick, usually of 6% acrylamide and containing 8 M urea as denaturant. Small-bore capillary gel electrophoresis has recently been proposed as a means to more rapid analysis.

Sequence reading

There are a variety of ways by which the products of the sequencing reactions can be detected and the sequence read. The starting material for chemical sequencing is radioactively labelled with ^{32}P either at the 5′ end by T4 POLYNUCLEOTIDE KINASE, or at the 3′ end by DNA polymerase I. Radioactive labelling of dideoxy-derived fragments is most often achieved by incorporating ^{32}P- or ^{35}S-labelled dNTPs in the extension reaction. Alternatively, the 5′ end of the primer can be labelled with ^{32}P by T4

polynucleotide kinase. To read the sequence, electrophoresis is stopped, usually when the shortest molecules have reached the bottom of the gel, and the gel is exposed to X ray-sensitive film or IMAGE PLATE (*see* AUTORADIOGRAPHY). The sequence (Fig. D24) can then be read either manually or by automated or semi-automated digitizing devices. Generally, sequences of 200–300 nucleotides can be deduced in this way.

Multiplex sequencing [3] takes advantage of indirect labelling procedures. The source DNA is subcloned into a number of libraries using a special set of vectors, each of which contains a unique oligonucleotide sequence. Template preparations and sequence reactions are carried out on mixtures of subclones, each mixture containing one clone from each library, and thus one member of each variety of vector. These reaction products are electrophoresed in unlabelled form, and the fractionated fragments are transferred to a nylon membrane. The sequence ladders are then revealed by sequential HYBRIDIZATIONS with oligonucleotide probes specific for the unique vector sequences. The labelling of the oligonucleotide may be radioactive or chemiluminescent. All the fragments for a particular sequence can thus be revealed from the complex mixture of fragments derived from many clones. Multiplex sequencing thus has the potential to reduce both the labour involved in sample preparation and reactions and the number of gel runs required for a large sequencing project.

Fluorescence detection

Radiolabelling is being increasingly replaced by the detection, in 'real' time, of fluorescence from laser-excited dye-labelled DNA fragments [4]. This approach is particularly applicable to dideoxy sequencing. The dyes can be incorporated either as 5′ end-labelled primer or as dye-labelled ddNTPs. In the currently most commonly used system, four dyes are used, one specific for each termination reaction and each emitting at a different wavelength. This has the advantage that the reaction products can be run in a single lane rather than four parallel lanes, allowing greater throughput. As the fluorescent molecules pass a detector near the bottom of the gel, data are output directly to a computer, allowing more rapid analysis and easier data retrieval. Devices based on this technology are currently being widely incorporated into schemes aimed towards the full automation of sequencing reactions and data handling.

Sequencing strategy

Ideally, single-stranded template DNA is used for dideoxy sequencing. This is most commonly obtained by the CLONING of fragments of up to 2 kilobase (kb) in VECTORS derived from the single-stranded DNA bacteriophage M13 [5]. Specific inserts may be generated by RESTRICTION ENZYME digestion of the source DNA. For large-scale sequencing projects, shotgun sequencing of randomly selected M13 subclones is an efficient starting point. The inserts for the M13 library (*see* DNA LIBRARIES) are usually generated by mechanical shearing of the DNA so that the distribution of the subclones is as random as possible, uninfluenced by the distribution of RESTRICTION SITES. Sequence can be produced

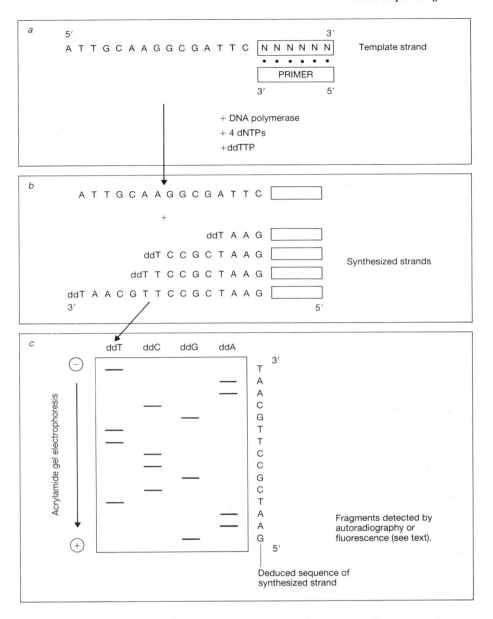

Fig. D24 Dideoxy chain termination sequencing. Schematic of: *a*, ddT reaction; *b*, reaction products; and *c*, subsequent fractionation by gel electrophoresis with ddC, ddG and ddA reactions.

from each randomly selected clone using the same universal primer (an oligonucleotide complementary to a region of the vector adjacent to the cloning site). The sequences obtained from the subclones can then be compared, overlapping regions identified, and the sequences assembled to form contigs (sets of overlapping clones). As a project proceeds, these contigs are linked by new data, but generally such a random approach is not pursued indefinitely and directed methods such as primer walking (in which specific oligonucleotides, deduced from sequence already determined, are used to initiate sequencing at particular positions) are used to finish the sequence. At the end of a project, all disagreements between readings should have been resolved, resulting in a definite assignment for each base in the final sequence.

Sequence analysis

Using computer programs the sequence is compared with sequences already determined and stored in the numerous sequence databases, and searched for features indicating the presence of genes, resemblances to known genes, control regions and other structural features of interest such as potential HAIRPIN LOOPS and repeats. Any predicted protein-coding sequences are compared with sequences in the protein sequence libraries and with libraries of known protein motifs. A match to either helps to validate the predictions and may give clues about the function of the protein. Because of the large amount of data generated, the majority of sequence interpretation is now performed using computer programs. However, predictions such as those of the positions of protein-coding genes from DNA sequences or of the secondary structure of proteins from amino-acid sequences are currently far from reliable (*see e.g.* PROTEIN ENGINEERING; PROTEIN FOLDING). Comparison of homologous DNA sequences from different species is now also widely used for determining phylogeny and tracing the evolution of gene families and super-families.

A.R. COULSON

R. STADEN

See also: ALIGNMENT; DATABASES AND INFORMATION HANDLING; HOMOLOGY; HUMAN GENOME PROJECT; MOLECULAR EVOLUTION: SEQUENCES AND STRUCTURES; MOLECULAR PHYLOGENY.

1 Maxam, A.M. & Gilbert, W. (1977) A new method of sequencing DNA. *Proc. Natl. Acad. Sci. USA* **74**, 560–564.
2 Sanger, F. Nicklen, S. & Coulson, A.R. (1977) DNA sequencing with chain-terminating inhibitors. *Proc. Natl. Acad. Sci. USA* **74**, 5463–5467.
3 Church, G.M. & Kieffer-Higgins, S. (1988) Multiplex DNA sequencing. *Science* **240**, 185–188.
4 Smith, L.M. et al. (1986) Fluorescence detection in automated DNA sequence analysis. *Nature* **321**, 674–679.
5 Messing, J. et al. (1977) Filamentous *coli* phage M13 as a cloning vehicle. *Proc. Natl. Acad. Sci. USA* **74**, 3642–3646.

DNA, single-stranded A single chain of deoxyribonucleotides which can be either linear, with free 5′ and 3′ ends, or circular. The genomes of certain BACTERIOPHAGES (e.g. M13), are circular single-stranded DNA molecules. Double-stranded DNA can be dissociated into two single strands of DNA by a variety of chemical and physical treatments (*see* HYBRIDIZATION).

DNA supercoiling *See:* SUPERCOILING.

DNA topology Double-stranded DNA is classically viewed as a linear right-handed double helix, but DNA is a very flexible molecule that can take up a number of forms, or topologies, both *in vivo* and *in vitro*. These include alternative structures to the classical B-DNA in which the number of base pairs per turn of the double helix is either increased (e.g. A-DNA) or decreased (e.g. C-DNA) from the B-form number of 10.4 base pairs per turn. Some duplex DNA molecules can also form left-handed helices (e.g. Z-DNA). Another topological variation which is only seen in circular DNA molecules is SUPERCOILING, which arises when a DNA duplex is twisted in space around its own axis and can form either negative supercoils, in which the DNA is twisted about its axis in the direction opposite to the clockwise turns of the right-handed helix, or positive supercoils which arise from twisting the DNA in the same direction as the clockwise turns of the helix. Discontinuities in the DNA structure (e.g. BENT DNA) may also occur. Localized changes in DNA topology can have a profound effect on both replication and transcription of the DNA. *See:* DNA; NUCLEIC ACID STRUCTURE.

DNA tumour viruses Members of the following families of viruses have been shown to have oncogenic potential or to be associated with particular cancers: Adenoviridae, Hepadnaviridae, Herpesviridae, Papovaviridae, Poxviridae. *See:* ADENOVIRUS; ANIMAL VIRUS DISEASE; EPSTEIN–BARR VIRUS; HEPATITIS B VIRUS; PAPOVAVIRUSES.

DNA typing

THE technique of DNA typing or DNA profiling, popularly known as 'DNA fingerprinting', uses analysis of the numerous hypervariable loci within the human genome to produce a genetic 'fingerprint' unique (or nearly so) to each individual.

A hypervariable locus, also known as MINISATELLITE DNA, consists of a block of tandem repeats of a short 'core' sequence. Hypervariable loci were first discovered in humans in 1980, when 15 alleles of different lengths were discovered at the locus designated D14S1 [1]. Subsequently, similar highly variable loci (HVRs) have been discovered at the insulin gene [2], the Harvey *ras* oncogene [3], the ζ globin pseudogene [4], the myoglobin gene [5], and the 3′ and 5′ hypervariable regions in the α-globin gene [6]. Apart from these, an additional 83 hypervariable repeat loci have been described in the human genome [7–9]. They all comprise tandem repeats of G-rich core sequences ranging in size from 11 to 60 bp.

The numbers of repeats at these loci vary between unrelated individuals and thus the number of repeats of each type present in each person constitutes a unique genetic 'fingerprint'. Analysis of such VARIABLE NUMBER TANDEM REPEATS (VNTRs) has until recently been the main approach to DNA typing; analysis of the slight variations in sequence in the tandem repeats within a single locus now also promises to be of great use.

The 'alleles' at each minisatellite locus are inherited in the same way as Mendelian genes (*see* MENDELIAN INHERITANCE) and so can be used both to identify a particular individual and to determine family relationships (*see* DNA TYPING: FORENSIC APPLICATIONS). Such DNA is present not only in the human genome, but also in that of other vertebrates, and DNA fingerprinting is also now widely used in ecological and evolutionary studies (*see* DNA TYPING: APPLICATIONS IN CONSERVATION AND POPULATION GENETICS).

The variability between individuals at the minisatellite loci is detected by restriction analysis of chromosomal DNA (see below). On digestion of human genomic DNA with RESTRICTION ENZYMES and probing with the core repeat sequence(s), fragments of different lengths are obtained from different individuals, the length of the restriction fragment being a function of the number of repeats present.

In 1985, Alec Jeffreys and colleagues published the first paper demonstrating the potential use of hypervariable loci as a set of genetic markers unique to each individual [10]. Using probes based on tandem repeats of the myoglobin locus, they were able to detect multiple hypervariable loci or 'fingerprints' in restriction digests of human genomic DNA when probes were hybridized under low stringency conditions (*see* HYBRIDIZATION). The probes (33.15 and 33.6, see Table D7), used together, detected up to 36 different bands (or alleles) of more than 3 kb in size in each individual [11]. Furthermore, the loci are randomly dispersed throughout the human genome and can be considered to be

Table D7 Comparison of selected minisatellites

Probe	Reference	Sequence
Core Jeffreys	[10]	GGAGGTGGGCAGGARG
Myoglobin	[5]	CTAAAGCTGGAGGTGGGCAGGAACGACCGARRT
33.15	[10]	AGAGGTGGGCAGGTGG
33.6	[10]	AGGGCTGGAGG
pλg3	[7]	AGAAAGGCGGGYGGTGTGGGCAGGGAGRGGCAGGAAT
λMS1	[8]	GTGGATAGG
YNH24	[9]	CAGCAGCAGTGGGAAGTACAGTGGGGTTGGTT
Insulin	[2]	ACAGGGGTGTGGGG
Harvey *ras* c1	[3]	GGGGGAGTGTGGCGTCCCCTGGAGAGAA
α-globin 3′HVR	[22]	AACAGCGACACGGGGGG
D14S1	[9]	GGYGGYGGYGGYGGYGGYGGY . . .
Core Nakamura	[9]	GGGNNGTGGGG

R, A or G; Y, C or T; N, any base.

independently inherited [12]. Jeffreys [11] quoted a band-sharing statistic of 0.25; hence the probability that two unrelated individuals share exactly the same pattern can be estimated as $0.25^{36} = 2 \times 10^{-22}$. The α-globin 3′ HVR locus also detects randomly distributed multiple loci which are not genetically linked to those discovered by Jeffreys and probably form a different family of minisatellites [13]. A number of hypervariable locus-specific probes have now been isolated. Table D7 shows the sequence of a selection of minisatellites.

The sequences of different hypervariable regions have been determined and compared [6,9,10,14] (Table D8). A certain amount of variation in the core typified by the core of Jeffreys and the core of Nakamura is found but the sequence GNNNNTGGG is almost invariant. The sequence GNGGGGNACAG (core of Jarman) has been proposed for the core sequence typified by the α-globin 3′ HVR.

The function of minisatellite DNA is unknown, although there is evidence that the core sequence may serve as a RECOMBINA-TION signal in human DNA and the recurring theme of G---GTGGG found in many different minisatellites may be important in this respect. The sequence is similar to the *chi* sequence in bacteriophage LAMBDA DNA (GCTGGTGG) which may function as a recombination hotspot in *recA*-mediated recombination in *Escherichia coli* [15,16].

Minisatellite DNA has a relatively high mutation rate involving unequal allelic length exchange (10^{-4} per kb of minisatellite [10] compared with an average recombination rate of 10^{-5} per kb of human DNA [17]). Mutations give rise to new minisatellites which have lost or gained kilobase stretches of tandem repeats, suggesting that minisatellites are recombination hotspots. Mutation rates have been shown to vary between different minisatellites [18];

hence for paternity analysis it is necessary to ensure that the mutation rate at the minisatellite(s) selected for analysis is not too great otherwise false exclusions could occur.

Although most minisatellite repeats are G-rich, hypervariable regions in the apolipoprotein B [19] and collagen type II [20] genes have a completely different AT-rich structure and probably comprise a different family.

Methodology of DNA typing

DNA extraction

DNA can be extracted for forensic and diagnostic purposes from whole blood, whole semen, vaginal fluid, hair roots, buccal cells, blood stains, and semen stains by overnight incubation in a detergent mixture, consisting of sodium dodecyl sulphate (SDS), proteinase K, and dithiothreitol (DTT). The detergent lyses cell nuclei, the proteinase K digests protein which would otherwise interfere with the recovery of DNA at a later stage, and the DTT reduces thiol or sulphur groups which are found in many proteins.

Digestion

After extraction, DNA is cut into fragments with a restriction enzyme. Usually, a restriction enzyme that recognizes a 4-bp sequence is used, for example *Hin*fI, *Alu*I, *Hae*III, or *Sau*IIIA. If the order of bases were random, these enzymes would, on average, recognize and cut at sequences $4^4 = 256$ bases apart. Because the minisatellite does not contain a restriction site for these enzymes, it remains intact; the enzyme will therefore cut DNA which flanks the actual minisatellite gene. Hence, many different restriction enzymes can be used to detect hypervariable regions provided that they do not cut within the minisatellite region itself.

Visualization using locus-specific probes

The DNA fragments produced by digestion are separated by agarose gel ELECTROPHORESIS and then Southern blotted onto a nylon membrane. Specific radioactive or nonisotopic probes rep-

Table D8 Comparison of core sequences

Probe	Reference	Sequence
Core Jarman	[6]	GNGGGGNACAG
α-globin 3′HVR	[22]	GGGGGG - AACAGACAC
Insulin HVR	[2]	TGTGGGG - ACAGGGG
ζ-globin HVR	[4]	GGAGGGG - ACAGTGG
Core Jeffreys	[10]	GGAGGTGGGCAGGARG

resenting core sequences are subsequently used to hybridize to the DNA bands transferred to the membrane (*see* HYBRIDIZATION). DNA fragments are visualized by autoradiography (for radioactively labelled probes) or other methods.

Multi-locus probes contain tandem repeats of a core sequence present at a number of different loci. The probe picks up all bands derived from these loci, producing the now familiar barcode pattern (Fig. D25). Because the repeats are present in different numbers in different individuals, restriction enzyme digestion will produce a different pattern of bands from each locus in each individual.

Single-locus probes, in contrast, contain repeats of a core sequence from a minisatellite locus of unique sequence and have become the favoured probes in forensic work (*see* DNA TYPING: FORENSIC APPLICATIONS).

Polymerase chain reaction (PCR)

The two main disadvantages of DNA restriction analysis of hypervariable DNA are that at least 60 ng is required for a successful analysis, and low molecular weight (degraded) DNA — such as that obtained sometimes from forensic samples — is usually not amenable to analysis. The solution to both these problems is the use of the POLYMERASE CHAIN REACTION (PCR) to amplify preselected regions of the genome (*see also* DNA TYPING: FORENSIC APPLICATIONS; DNA TYPING: APPLICATIONS IN CONSERVATION AND POPULATION GENETICS).

A novel application of PCR to DNA typing has recently been devised by Jeffreys [21]. Called minisatellite variant repeat mapping (MVR), this technique uses repeat unit sequence variation within alleles of the hypervariable minisatellite MS32 (locus D1S8). The repeated units contain internal variation at specific locations, where a base can be either A or T. Oligonucleotide primers can be used to detect the A/T variation within each repeat which is then converted into a digital code which can be recorded in a computerized database, giving each individual an identifier which is highly discriminatory.

P. GILL

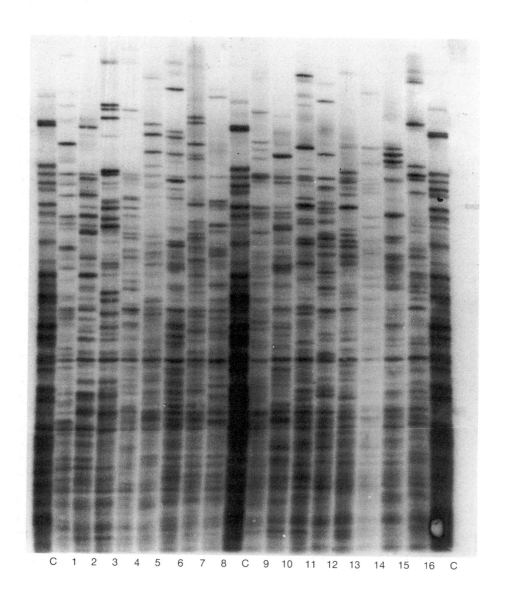

Fig. D25 Multi-locus DNA profiles from 16 randomly chosen individuals showing the great variation in patterns. The profile at either end and in the middle of the photograph is the same control sample.

See also: GLOBIN GENES; HAEMOGLOBIN AND ITS DISORDERS.

1 Wyman, A. & White, R. (1980) A highly polymorphic locus in human DNA. *Proc. Natl. Acad. Sci. USA*, **77**, 6754–6758.
2 Bell, G.I. et al. (1982) The highly polymorphic region near the human insulin gene is composed of simple tandemly repeated sequences. *Nature* **295**, 31–35.
3 Capon, D.J. et al. (1983) Complete nucleotide sequences of the T24 human bladder carcinoma oncogene and its normal homologue. *Nature* **302**, 33–37.
4 Proudfoot, N.J. et al. (1982) The structure of the human zeta-globin gene and a closely linked newly identified pseudogene. *Cell* **31**, 553–563.
5 Weller, P. et al. (1984) Organization of the human myoglobin gene. *EMBO J.* **3**, 439–446.
6 Jarman, A.P. et al. (1986) Molecular characterization of a hypervariable region downstream of the human α-globin gene cluster. *EMBO J.* **5**, 1857–1863.
7 Wong, Z. et al. (1986) Cloning of a selected fragment from a human DNA 'fingerprint': isolation of an extremely polymorphic minisatellite. *Nucleic Acids Res.* **14**, 4605–4616.
8 Wong, Z. et al. (1987) Characterization of a panel of highly variable minisatellites cloned from human DNA. *Annls Hum. Genet.* **51**, 269–288.
9 Nakamura, Y. et al. (1987) Variable number of tandem repeats (VNTR) markers for human gene mapping. *Science* **235**, 1616–1622.
10 Jeffreys, A.J. et al. (1985) Hypervariable 'minisatellite' regions in human DNA. *Nature* **314**, 67–73.
11 Jeffreys, A.J. (1987) Highly variable minisatellites and DNA fingerprints. *Biochem. Soc. Trans.* **15**, 309–317.
12 Jeffreys, A.J. et al. (1986) DNA 'fingerprinting' and segregation analysis of multiple markers in human pedigrees. *Am. J. Hum. Genet.* **39**, 11–24.
13 Fowler, S.J. et al. (1988) Human satellite-III DNA: an example of a 'macrosatellite' polymorphism. *Hum. Genet.* **312**, 142–146.
14 Jeffreys, A.J. et al. (1985) Individual-specific 'fingerprints' of human DNA. *Nature* **316**, 76–79.
15 Smith, G.R. et al. (1981) Structure of chi hotspots of generalized recombination. *Cell* **24**, 429–436.
16 Smith, G.R. (1983) Chi hotspots of generalized recombination. *Cell* **34**, 709–710.
17 Botstein, D. et al. (1980) Construction of a genetic linkage map in man using restriction fragment length polymorphisms. *Am. J. Hum. Genet.* **32**, 314–331.
18 Jeffreys, A.J. et al. (1988) Spontaneous mutation rates to new length alleles at tandem-repetitive hypervariable loci in human DNA. *Nature* **332**, 278–281.
19 Knott, T.J. et al. (1986) A hypervariable region 3' to the human apolipoprotein B gene. *Nucl. Acids Res.* **14**, 9215–9216.
20 Stoker, N.G. et al. (1985) A highly polymorphic region 3' to the human collagen gene. *Nucl. Acids Res.* **13**, 4613–4622.
21 Jeffreys, A.J. et al. (1991) Minisatellite repeat coding as a digital approach to DNA typing. *Nature* **354**, 204–209.
22 Higgs, D.R. et al. (1981) Highly variable regions flank the human alpha-globin genes. *Nucl. Acids Res.* **9**, 4213–4214.

DNA typing: applications in conservation and population genetics

APPLICATIONS of RECOMBINANT DNA TECHNOLOGY have become common in the study of POPULATION GENETICS and are of increasing importance in conservation studies. The technique of DNA profiling — the analysis of anonymous or noncoding repetitive, hypervariable DNA sequences, including multilocus DNA fingerprinting and single locus methods such as VARIABLE NUMBER OF TANDEM REPEAT (VNTR) analysis, RFLPs (restriction fragment length polymorphisms), and simple sequences — was originally developed for human medical genetics and forensic science (*see*

DNA TYPING; DNA TYPING: FORENSIC APPLICATIONS) and has been transferred with little difficulty to nonhuman eukaryote taxa (including many vertebrates, invertebrates, and plants).

Population genetics has been an integral component of evolutionary biology since the development of the SYNTHETIC THEORY OF EVOLUTION in the 1930s [1]. At that time, the application of Mendelian genetics to population-level phenomena, and the development of quantitative genetics to study polygenic systems, resolved the conceptual gap between evolutionary biologists studying empirical examples of natural selection and geneticists studying the inheritance of discrete differences in phenotypic markers or CYTOGENETICS.

The development of ALLOENZYME electrophoresis in the 1960s opened the way to the use of molecular genetic markers in nonhuman population genetics [2]. Sophisticated applications of alloenzyme methods, based on analysis of ALLELE frequencies at variable loci, were developed at every level of whole-organism study, from the analysis of parentage by genotype exclusion in birds [3] and mammals [4] as tests of empirical reproductive-success hypotheses, to molecular systematics studies [5]. Unfortunately, several factors limited the power of alloenzyme methods. Although inexpensive, alloenzyme electrophoresis requires preservation of tissue from field samples in liquid nitrogen, and only detects variation in net electric charge among alleles at coding loci. Thus, the great wealth of DNA sequence variation in noncoding regions is not accessible by alloenzyme electrophoresis.

Since the mid-1970s the ability to document differences in DNA sequence has increased greatly. One important development was RFLP analysis, combining the invention of SOUTHERN BLOTTING [6] and the use of RESTRICTION ENZYME digestion of DNA samples [7]. RFLP analysis makes use of differences among alleles as defined by the presence or absence of sites which are digested by restriction enzymes to generate length variation in DNA fragments [8].

The detection of variation in nuclear DNA sequence information as a source of genetic markers suitable for population genetics studies in nonhuman examples began with RFLP studies of the Lesser Snow Goose (*Anser caerulescens*) [9], and DNA fingerprinting in passerine birds [10, 11]. Since 1987 the species being studied by DNA methods, and the types of methods themselves, have been growing yearly.

The most frequent applications for hypervariable DNA profiling in wild species have been directed at resolving questions of individual reproductive success. Behavioural ecologists seek to explain observed behaviours among animals in terms of Darwinian fitness, or differential reproductive success. In many cases, documentation of the maternal identity is relatively simple. Field observations can demonstrate unambiguously which individual is the mother of a particular offspring if egg-laying or hatching can be observed. However, reproductive success by males cannot be determined in similar fashion. Observation of copulatory behaviour is not sufficient evidence of reproductive success in species with internal fertilization if there is any chance of more than one male copulating with a female during the fertile period, or if viable spermatozoa are stored by the female for any significant length of time. Both of these conditions are of importance in the mating systems of mammals, birds, and reptiles.

An example of the importance of biochemical determination of parentage can be found in studies of the Redwinged Blackbird (*Agelaius phoeniceus*) of North America. Observations of territories held by singing males led to the assumption that the males whose territories encompass more nests of females will be the fathers of all of the offspring within each territory. It was therefore assumed that variation in territory size (as number of offspring within the territory) would be a direct measure of variation in reproductive success among males [12]. Analysis of pedigrees within free-living Redwinged Blackbird nests in central Canada by use of multilocus DNA fingerprinting and RFLP analysis of sequence variation in the MAJOR HISTOCOMPATIBILITY COMPLEX (MHC) loci demonstrated significant extraterritorial fertilizations [13]. In that study no significant association between territory size and the number of offspring fertilized by the territory holder was found.

Applications of molecular methods to questions of population genetics have largely relied upon alloenzyme electrophoresis for the past 25 years. At present most applications of molecular methods to population genetics *per se* make use of alloenzyme methods or access DNA sequence variation in mitochondrial DNA (mtDNA). An extreme example of the use of mtDNA is as part of the definition of a new species of bird [14]. DNA profiling *per se*, the detection of highly variable DNA sequences, has not been applied frequently to population genetics, because the hypervariable nature of the sequences detected in DNA profiling means that allele frequency relationships among populations or species are obscured by the high 'noise' level of their mutation rates. In addition, multilocus DNA profiles are essentially PHENOTYPES rather than GENOTYPES, and so conventional analytical methods making use of allelic variation cannot be applied.

It is in the new field of conservation genetics that DNA profiling is having a major impact. Although unsuited to the analysis of allelic state, DNA profiling is a practical and powerful surrogate measure of genetic relatedness in situations where pedigrees are not available [15]. The most direct demonstration of this application is in the relationship between the mean BANDSHARING COEFFICIENT of multilocus DNA profiling (*see* DNA TYPING: FORENSIC APPLICATIONS) and other measures of genetic relatedness [16]. Measures of genetic relatedness are of critical importance in conservation programmes because many of the difficulties associated with the breeding performance of both captive and free-living populations of critically endangered species are traceable to high levels of inbreeding among the remaining individuals. This phenomenon has been applied to studies of genetic variation in free-living populations [17].

A direct demonstration of the relationship between genetic relatedness and breeding performance has been made in a highly endangered bird species, the Puerto Rican Parrot (*Amazona vittata*) [18]. Originally numbering many hundreds of thousands, by the early 1970s only 30 or so Puerto Rican Parrots were alive in the forests of Puerto Rico. At that point a captive breeding programme was begun, and numbers increased slowly for two decades. However, breeding performance was poor relative to the closely related Hispaniolan Parrot (*Amazona ventralis*) in the same aviary and held under the same conditions. Analysis of pedigrees revealed that all of the captive-bred individuals could be traced to four founder birds. Analysis with the DNA fingerprinting PROBES 33.15, 33.6, and *per* indicated that the genetic relatedness among

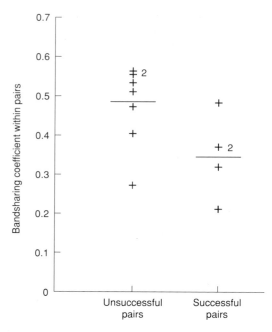

Fig. D26 DNA fingerprint bandsharing coefficients within successfully reproducing pairs and unsuccessful pairs of the Puerto Rican Parrot (*Amazona vittata*). Each cross represents a mated pair of birds, and numbers indicate more than one pair with the same bandsharing coefficient. The horizontal line indicates the mean bandsharing coefficient in each group. The higher bandsharing coefficients of unsuccessful pairs indicate a high genetic relatedness between mates which could not breed, evidence of the deleterious effects of inbreeding in this highly endangered species. Adapted from [18].

these founder birds, and among all known wild Puerto Rican Parrots closely resembles the relatedness among second-degree relatives among Hispaniolan Parrots, equivalent to an entire population of first cousins [18]. Furthermore, pairs of mated Puerto Rican Parrots which failed to reproduce displayed a significantly higher bandsharing coefficient than did successful breeding pairs, indicating severe inbreeding depression (Fig. D26).

D.A. GALBRAITH

1 Mayr, E. & Provine, W.B. (1981) *The Evolutionary Synthesis: Perspectives on the Unification of Biology* (Harvard University Press, Cambridge, MA).

2 Hubby, J.L. & Lewontin, R.C. (1966) A molecular approach to the study of genic heterozygosity in natural populations. I. The number of alleles at different loci in *Drosophila pseudoobscura. Genetics* **54**, 577–594.

3 Westneat, D.F. et al. (1987) The use of genetic markers to estimate the frequency of successful alternative reproductive tactics. *Behav. Ecol. Sociobiol.* **21**, 35–45.

4 Hayasaka, K. et al. (1986) Probability of paternity exclusion and the number of loci needed to determine the fathers in a troop of macaques. *Primates* **27**, 103–114.

5 Murphy, R.W. et al. (1990) Proteins I: isozyme electrophoresis. In *Molecular Systematics* (Hillis, D.M. & Moritz, C., Eds) 45–126 (Sinauer, Sunderland, MA).

6 Southern, E.M. (1975) Detection of specific sequences among DNA fragments separated by gel electrophoresis. *J. Mol. Biol.* **98**, 503–517.

7 Upholt, W.B. (1977) Estimation of DNA sequence divergence from comparison of restriction endonuclease digests. *Nucleic Acids Res.* **4**, 1257–1265.

8 Quinn, T.W. & White, B.N. (1987) Analysis of DNA sequence variation. In

Avian Genetics: A Population and Ecological Approach (Cooke, F. & Buckley, P.A., Eds) 163–198 (Academic Press, London).

9 Quinn, T. W. et al. (1987) DNA marker analysis detects multiple maternity and paternity in single broods of the lesser snow goose. *Nature* **326**, 392–394.

10 Wetton, J. H. et al. (1987) Demographic study of a wild house sparrow population by DNA fingerprinting. *Nature* **327**, 147–149.

11 Burke, T. & Bruford, M. W. (1987) DNA fingerprinting in birds. *Nature* **327**, 149–152.

12 Orians, G. H. (1980) Some adaptations of marsh-nesting blackbirds. *Monographs in Population Biology* **14** (Princeton University Press, Princeton, NJ).

13 Gibbs, H. L. et al. (1990) Realized reproductive success of polygynous red-winged blackbirds revealed by DNA markers. *Science* **250**, 1394–1397.

14 Smith, E.F.G. et al. (1991) A new species of shrike (Laniidae: Laniarius) from Somalia, verified by DNA sequence data from the only known individual. *Ibis* **133**, 227–235.

15 Amos, B. & Hoelzel, A.R. (1992) Application of molecular genetics techniques to the conservation of small populations. *Biol. Conserv.* **61**, 133–144.

16 Lynch, M. (1991) Analysis of population genetic structure by DNA fingerprinting. In *DNA Fingerprinting: Approaches and Applications* (Burke, T. et al., Eds) 113–126 (Berkhauser Verlag, Basel).

17 Gilbert, D. A. et al. (1990) Genetic fingerprinting reflects population differentiation in the California Channel Island fox. *Nature* **344**, 764–767.

18 Brock, M. K. & White, B. N. (1992) Application of DNA fingerprinting to the recovery program of the endangered Puerto Rican Parrot. *Proc. Natl. Acad. Sci. USA* **89**, 11121–11125.

DNA typing: forensic applications

THE introduction of DNA TYPING into forensic science has caused a revolution in the field — one of the spin-offs from the wider revolution in molecular biology due to the advent of the new techniques for DNA analysis (see techniques listed under RECOMBINANT DNA TECHNOLOGY). The basis of DNA typing — hypervariable DNA (minisatellite DNA) — was discovered in humans in 1980 but research into its use in forensic science and other applications did not begin in earnest until 1985. At that time, DNA analysis techniques were becoming much more accessible partly because the high quality molecular biology products needed for DNA analysis were beginning to be available commercially.

In 1985, Alec Jeffreys published his first papers on the analysis of minisatellite DNA in humans [1] (see DNA TYPING). The potential of this novel technique of 'DNA fingerprinting' for paternity testing was immediately apparent, and it was also soon used in immigration cases to prove disputed family relationships, but no one had so far investigated the potential of DNA analysis for forensic science. Clinical biochemists and geneticists dealt with large quantities of DNA extracted from fresh material such as blood and liver. Forensic scientists do not have the luxury of such pristine samples — they work almost exclusively on dry body fluid stains, which are rarely fresh and frequently contaminated with foreign material such as bacteria or dirt. In 1985, it was standard practice for the molecular biologist to isolate DNA from at least 1 ml of blood, but the forensic scientist may have much less than this to work from; hence techniques had to be scaled down.

The earliest experiments in establishing the methodology for forensic analysis [2,3] demonstrated conclusively that DNA could be retrieved and analysed not only from fresh body fluid stains but even from those several years old. DNA profiles can be obtained from blood and semen stains and hair roots, the most common forensic material, or any other body tissue available.

The principles behind DNA typing and the general procedure are outlined in the article on DNA TYPING above. This article discusses its particular application to forensic work. Figure D25 above (see DNA TYPING) illustrates the tremendous variation between different individuals in their minisatellite DNA; using a multi-locus probe, any band in the profile is found in roughly 1 in 4 (25%) of unrelated individuals — this means that the same two bands will be found in 1 in $4 \times 4 = 16$ individuals; the same three bands in 1 in $4 \times 4 \times 4 = 64$ individuals and so on. An average DNA profile contains 11 bands and any given combination will be found in 1 in 2.5 million individuals.

It is not true that DNA typing always gives a definitive answer in forensic work. Often with forensic samples, only a few bands may be observed because of degradation of the DNA; sometimes as few as four bands are reported, and the same four bands could occur in 1 in 200 unrelated individuals. The widely used term 'DNA fingerprinting' implies uniqueness, but this is not necessarily possible to establish in forensic work, and so forensic scientists prefer to use the term 'DNA profiling'.

Casework

DNA analysis was applied to a criminal case for the first time early in 1986, in the UK [4]. In 1983 a schoolgirl had been found raped and murdered on wasteground in a small village near Leicester. In 1986 a second girl was found in the same area. A man was arrested and charged with the murder of the second girl after making a confession. The police believed him also responsible for the first murder and approached Jeffreys to carry out DNA profiling. The suspect's DNA profile, however, did not match that of semen from both the scenes of crime — he could not have committed either offence. Tests by the forensic service confirmed these results and there then followed one of the most extraordinary man-hunts in British criminal history. Blood from all the males in the area (a total of 1500 individuals) was analysed, but without success. Towards the end of the exercise, a suspect was apprehended by police as a result of a conversation overheard by chance in a public house. The suspect's blood was tested and his profile matched that of the crime material.

Many thousands of cases have now used DNA profiling. Around 30% of suspects are exonerated by the technique, reflecting its use by the police as an investigative tool.

DNA analysis has been most commonly used in investigating rape. The forensic evidence here is the semen-contaminated vaginal swab taken after a rape. Vaginal swabs contain large amounts of DNA from the female which tends to obscure many of the bands from the sperm cells and so the two components need to be separated. Fortunately, sperm are surrounded by a coat of proteins rich in sulphur. This makes them resistant to chemicals normally used to solubilize DNA. This means that the vaginal cell DNA can normally be removed by the addition of a detergent which has no effect on the sperm. Once the female DNA has been removed, sperm DNA can be solubilized using chemicals that

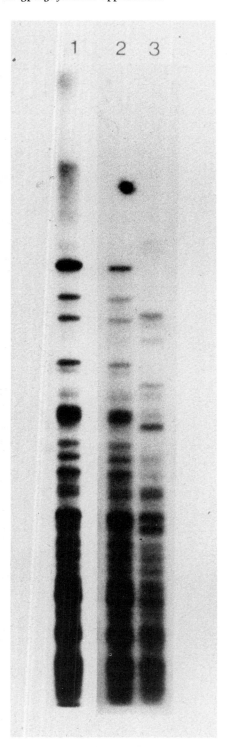

Fig. D27 The results of a differential extraction method for removing female DNA from a vaginal swab taken after rape. Multilocus probe analysis of 1, Vaginal swab from victim after removal of female DNA; 2, blood of male suspect; 3, blood of victim. Because preliminary experiments [3] showed that semen-contaminated vaginal swabs contained large amounts of DNA from the female, tending to obscure many of the bands from sperm, female cell nuclei are preferentially lysed by preliminary incubation in a detergent/ proteinase K mixture. Sperm nuclei are impervious to this treatment alone

break down the tough protein coat. The DNA profile obtained can be checked against the female's DNA to show that it comes from the sperm rather than the vaginal cells (Fig. D27).

Single-locus probes

A development of DNA profiling techniques much used in forensic work is the use of single-locus probes (SLPs). In contrast to multi-locus probes, these probes target sequences which contain tandem repeats of the core sequence of a minisatellite locus of unique sequence. Superficially, they give much simpler results than multi-locus probes; only one or two bands show up in the restriction analysis (Fig. D28). However, a large number of different SLPs are available, each giving a different pattern of bands and therefore enabling profiles to be built up. About three SLPs are needed to produce an analysis with the same statistical power as one MLP. Smaller quantities of DNA are needed using SLPs, and most important, the information obtained is easy to store digitally and retrieve from a database [5]. For these reasons, SLPs have superseded MLPs as the probes of choice in forensic work.

Statistical calculations tend to be more complex than for MLPs as each band in an SLP is associated with a different probability of occurrence. Figure D29 shows that some bands are very rare in the white Caucasian population; this must be taken into account when statistics are quoted. Furthermore, different ethnic groups have different frequency distributions, so this must also be taken into account. It has been necessary to generate databases on three different ethnic groups — white Caucasians, Asians and Afro-Caribbeans.

Databases

The reliability of DNA profiling and the validity of the probability estimates quoted in court have been intensively researched and discussed worldwide over the past few years [5,6]. One way of addressing this question is by reference to a large database of DNA profiles. The one created in the UK consists of a large number of random individuals and the entire database can be searched for chance matches. The largest search carried out by the forensic service in the UK so far has involved over half a million comparisons; using three different single-locus probes not one example of a chance match has been found to date. On average, one single-locus probe will produce a chance match in 1 in 100, two probes will produce a chance match in 1 in 10 000 and three probes a chance match in 1 in 1 million (conservative estimates).

The power of DNA profiling lies in its high discriminatory potential; hence its potential for producing databases of convicted individuals [4]. It is now routinely possible to search DNA profile

because they are ramified with cross-linked thiol-rich proteins which maintain the integrity of the sperm nucleus, and can therefore be separated from the female component by centrifugation. Female DNA can thus be selectively removed from the vaginal swab, leaving DNA bands which must have originated from the sperm cells, matching those of the male suspect.

databases in order to link together crimes in different areas. Checking against the DNA database enables scientists to confirm whether crimes have been committed by the same individual; this information can be passed back to the relevant police forces so that they can coordinate efforts, thus saving police time and money.

European standardization

With a technique which is evolving rapidly, it is difficult to ensure uniformity across international boundaries because there may be several different ways of carrying out any particular method. As DNA profiling became established, it was considered desirable to ensure that results from laboratories within Europe could be compared across boundaries; this was felt to be especially important for terrorist offences. As a result, the European DNA Profiling group (EDNAP) was set up in 1989. The aim of the group is to work towards standardization of methods to be carried out by collaborative experimentation [7]. Results of a latest series of experiments have clearly demonstrated that standardization is possible provided that laboratories follow similar protocols.

Polymerase chain reaction

A second revolution is currently taking place in forensic biology, as a result of the invention of the POLYMERASE CHAIN REACTION (PCR). This technique offers the possibility of analysing samples that are very small indeed. The method works by imitating in the test tube the natural process of DNA REPLICATION. PCR can take the DNA from just a single cell and replicate part of it to levels that allow easy detection. Loci that have been used for forensic studies with PCR include, as well as the hypervariable loci mentioned above, the variable DQ HLA locus (*see* MAJOR HISTO-COMPATIBILITY COMPLEX). There are obvious dangers in pushing such a technique to its extremes in forensic work. When samples are from an unknown source, care is needed to avoid contamination from the cells of laboratory worker, police officer or any other individual unconnected with the offence. Strict guidelines are needed to ensure that contaminating DNA cannot be amplified. For example, material from the scene of crime is always kept in a separate laboratory from the control blood samples from the suspect.

A bonus of PCR is its speed; it should be possible to reduce the time of analysis to less than 24 hours (conventional DNA profiling takes at least a week to obtain a result). The method is also cheaper and will ultimately be automated.

Samples can be analysed by PCR in several ways. A simple DOT BLOT method [8] has been used forensically to detect different alleles at one of the HLA loci (*see* POLYMERASE CHAIN REACTION)

Another adaptation of the PCR method is shown in Plate 2c. In the experimental example illustrated the products of PCR with DNA from two individuals in which three loci have been preferentially amplified, are then run on a gel to separate the DNA fragments. Superficially, the results are very similar to those obtained from SLPs, where only one pair of bands (one inherited from the mother and the other from the father) would normally be visualized per lane. The difference here is that the PCR method

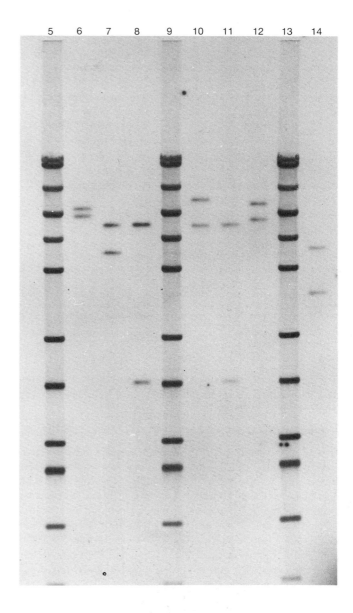

Fig. D28 Result of DNA typing using a single-locus probe. Lane 8 is a whole blood control and lane 11 a bloodstain from the same individual. Lanes 6, 7, 10, 12 and 14 are samples from other individuals. Lanes 5, 9 and 13 contain a ladder marker for sizing purposes.

enables us to attach different coloured marker dyes to each pair of bands. Each lane still corresponds to a single individual, but three loci can be identified by colour differences. Sizing is carried out by a computerized system.

The advantages are: cheapness, because only one reaction is needed instead of three; the results are easy to put onto a database; the method is fast and sensitive. The disadvantages are the cost of the hardware (~£80 000); also the results are not comparable with those obtained from the current methods, which means that new databases must be prepared. PCR

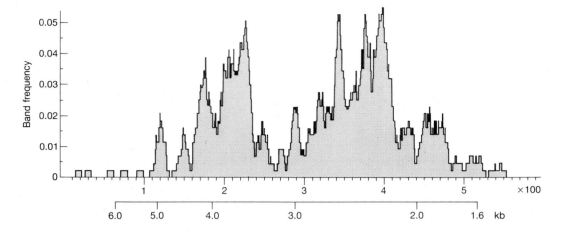

Fig. D29 Frequency histogram showing how common or rare particular bands are in a white Caucasian population. For example, bands which are 3.6 kilobases (kb) or 2.5 kb are more common than those greater than 5 kb in size. This type of information is used to determine how common DNA profiles are in populations. The locus used in this case was D2S44.

methods have been used on a surprising range of evidential material, including bone several hundreds of years old [9].

DNA sequencing: the ultimate test in forensic biology

Parts of mitochondrial DNA are highly variable between unrelated individuals, and the nucleotide sequence of these regions can be used to help identify a particular individual (Fig. D30). Mitochondrial DNA is particularly useful as it can be extracted successfully from very difficult evidential material such as tiny fragments of hair shaft and skin and bone from decomposed bodies [10]. In a recent case, the first of its kind in the UK, a sequence of 800 bases from the mitochondrial DNA of a decomposed body over 3 months old was analysed and was linked to a single living relative, a sister, who shared the same sequence.

P. GILL

1 Jeffreys, A.J. et al. (1985) Hypervariable 'minisatellite' regions in human DNA. *Nature* **314**, 67–73.
2 Gill, P. et al. (1985) Forensic applications of DNA 'fingerprints'. *Nature* **318**, 577–579.
3 Gill, P. et al. (1987) An evaluation of DNA fingerprinting for forensic purposes. *Electrophoresis* **8**, 38–44.
4 Gill, P. & Werrett, D.J. (1987) Exclusion of a man charged with murder by DNA fingerprinting. *For. Sci. Int.* **35**, 145–148.
5 Gill, P. et al. (1991) Databases, quality control and interpretation of DNA profiling in the Home Office Forensic Science Service. *Electrophoresis* **12**, 204–209.
6 Evett, I.W. & Gill, P. (1991) A discussion of the robustness of methods for assessing the evidential value of DNA single locus profiles in crime investigations. *Electrophoresis* **12**, 226–230.
7 Schneider, P.M. et al. (1991) Report of a European collaborative exercise comparing DNA typing results using a single locus VNTR probe. *For. Sci Int.* **48**, 1–15.
8 Saiki, R.K. et al. (1986) Analysis of amplified DNA with immobilized sequence-specific oligonucleotide probes. *Proc. Natl. Acad. Sci. USA* **86**, 6230–6234.
9 Paabo, S. et al. (1989) Ancient DNA and the polymerase chain reaction. *J. Biol. Chem.* **264**, 9709–9712.
10 Sullivan, K.M. et al. (1991) Automated amplification and sequencing of human mitochondrial DNA. *Electrophoresis* **12**, 17–21.

DNA viruses Viruses that contain double-stranded or single-stranded DNA as the genetic material (Table D9). *See*: ADENO-VIRUS; ANIMAL VIRUSES; BACTERIOPHAGES; EPSTEIN–BARR VIRUS; HEPATITIS VIRUSES; PAPOVAVIRUSES; PLANT VIRUSES.

DNase, DNAse *See*: DEOXYRIBONUCLEASE.

DNase footprinting *See*: FOOTPRINTING.

DNase I hypersensitive site When CHROMATIN is digested with very low concentrations of DEOXYRIBONUCLEASE I (DNase I) only those portions of the DNA molecule which are highly susceptible to the enzyme will be cleaved. Such hypersensitive sites are believed to coincide with DNA that is not tightly associated with NUCLEOSOMES and correspond to regions of around 200 base pairs. For a given gene the location of such DNase I hypersensitive sites is usually the PROMOTER region, but their location and/or frequency may vary from tissue to tissue. This variability reflects the transcriptional activity of the gene, as for the RNA polymerase to have ready access to its template (i.e. the promoter region of the gene), DNA must be essentially free of nucleosomes. The presence of such a hypersensitive site is not however sufficient to ensure transcription.

Elgin, S.C.R. (1981) *Cell* **27**, 413–415.

DNase protection assay *See*: FOOTPRINTING.

DNP 2,4-Dinitrophenol. Small lipophilic weak acid molecule, used to uncouple electron transport in mitochondria from the formation of a proton gradient and ATP synthesis. It acts as a proton IONOPHORE, carrying protons across the membrane and discharging the protonmotive force (*see* CHEMIOSMOTIC THEORY; ELECTRON TRANSPORT CHAIN; PHOTOSYNTHESIS).

dNTP Any or all of the four deoxyribonucleoside phosphates required for DNA synthesis. *See*: NUCLEOSIDES AND NUCLEOTIDES.

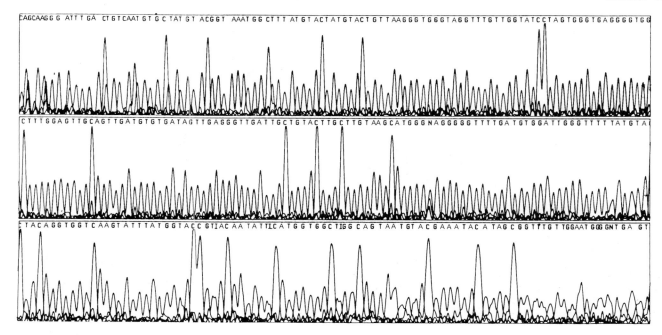

Fig. D30 An example of a computer-generated representation of a DNA sequence from human mitochondrial DNA. The string of bases can be compared with other sequences in the database in order to detect differences. The mitochondrial DNA was obtained from 3 cm of a single human hair shaft and the sequence is part of the noncoding region, bases 16 080–16 359.

docking protein (DP) Docking protein, or SRP (signal recognition particle) receptor, is a protein integrated into the rough ENDO-PLASMIC RETICULUM membrane, comprising two subunits, SRα (M_r 72 000) and SRβ (M_r 30 000), which recognizes and binds SRP when it is attached to arrested ribosomes synthesizing nascent chains containing SIGNAL SEQUENCES. SRα has a cytoplasmic domain of M_r 60 000, cleavable by proteases, and a M_r 12 000 domain which is involved in anchorage to SRβ and the membrane. SRα contains a consensus GTP-binding sequence, and GTP binding is required to displace SRP from the ribosome and release the nascent polypeptide for insertion into the mem-

brane. GTP hydrolysis may be required for release of SRP from its receptor. *See also*: PROTEIN TARGETING; PROTEIN TRANSLOCATION.

Meyer, D.I. et al. (1982) *Nature* **297**, 647–650.

dolichol The trivial name of eukaryotic polyisoprenoid alcohols, which differ from their bacterial counterparts (UNDECAPRENOLS) by having one of their isoprene units saturated. These lipids are found widely in eukaryotic cells and particularly in their mitochondria. About 5% of the total is involved in glycosylation reactions as dolichyl phosphoryl and pyrophosphoryl glycosides (*see* GLYCANS; GLYCOPROTEINS AND GLYCOSYLATION; GPI ANCHORS). Dolichylphosphorylmannose (Dol-P-Man) (Fig. D31) and dolichylphosphorylglucose (Dol-P-Glc) are used by glycosyltransferases in the lumen of the ENDOPLASMIC RETICULUM in the biosynthesis of the precursor of N-linked oligosaccharides, $Glc_3Man_9GlcNAc_2$-P-P-Dol. Dol-P-Man is also the source of the first mannose residue added to serine and threonine in yeast

Table D9 DNA virus families

	Bacteriophages	Plant viruses	Animal viruses
ds	Corticoviridae (e.g. PM2)	Caulimoviruses	Adenoviridae (adenovirus)
	Lipothrixviridae (e.g. TTV-1)		Baculoviridae (e.g. baculovirus)
	Myoviridae (T4 & the T-even phages)		Hepadnaviridae (e.g. hepatitis B)
	Plasmaviridae (e.g. MVL-2)		Herpesviridae (e.g. herpesviruses, Epstein–Barr virus)
	Podoviridae (T7 & the T-odd phages)		Iridoviridae
	Siphoviridae (e.g. lambda, P22)		Papovaviridae (papillomaviruses, SV40, polyomaviruses)
	SSV-1 group		Polydnaviridae
	Tectiviridae (PRD-1)		Poxviridae (vaccinia, smallpox)
ss	Microviridae (φX174, G4)	Geminiviruses	Parvoviridae
	Inoviridae (fd)		

ds, double-stranded; ss, single-stranded.

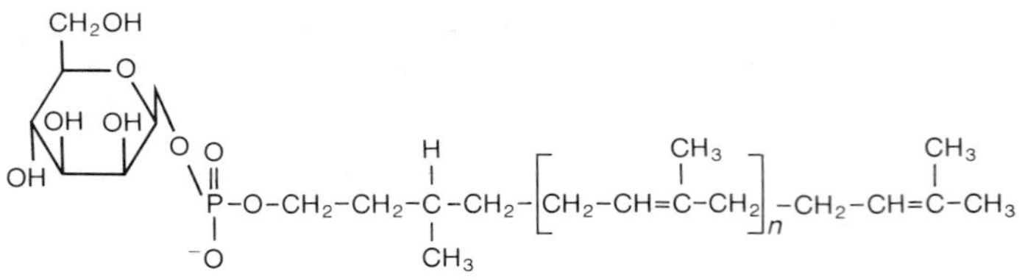

Fig. D31 Dolichylphosphorylmannose (Dol-P-Man). $n = 15$–19.

glycoproteins, and of the mannose residues in GPI anchors. Dolichyl phosphoryl glycosides (e.g. Dol-P-Glc, Dol-P-Man) are much more stable than the pyrophosphoryl glycosides (which include Dol-PP-Glc, Dol-PP-GlcNAc and glycosides of several larger saccharides) and are simple sugar donors to them. Pyrophosphoryl glycosides are mostly sugar acceptors, but some donate sugars to protein. The G-oligosaccharide is one such: it is a final product and is the major donor of glucose to protein during the synthesis of N-linked (type 2) glycans.

Related mechanisms are now known in a few prokaryotes. Dolichol is also found in some ARCHAEBACTERIA (i.e. some species of *Halobacterium* and *Methanothermus*).

domain A structurally and/or functionally discrete portion of a protein, nucleic acid, or membrane. In some genes domain boundaries correlate with EXONS. *See*: EPITHELIAL POLARITY; IMMUNOGLOBULIN DOMAIN; NUCLEIC ACID STRUCTURE; PROTEIN STRUCTURE.

dominant allele/mutation/trait ALLELE, MUTATION or trait in diploid organisms which is phenotypically expressed in the HETEROZYGOUS state in a manner indistinguishable from the HOMOZYGOTE. *See*: MENDELIAN INHERITANCE.

***dominant spotted* (*W*)** Mouse COAT COLOUR GENE encoding the receptor tyrosine kinase Kit (a potential ONCOGENE). *See also*: GROWTH FACTOR RECEPTORS.

domoic acid Neurotoxin originally isolated from infected shellfish (Fig. D32). Agonist with the highest potency for kainate-binding sites in the central nervous system. Domoic acid has been used as the immobilized ligand for affinity purification of non-NMDA EXCITATORY AMINO ACID RECEPTORS.

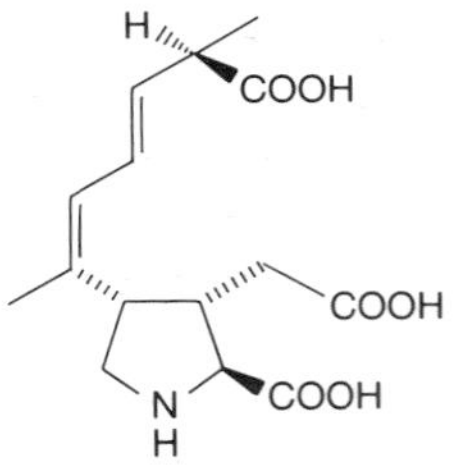

Fig. D32 Domoic acid.

donor splice site The junction between the 3′ (right) end of an

EXON and the 5′ (left) end of an INTRON in a precursor RNA. In transcripts from nuclear genes of eukaryotes the exon sequence at the junction can be variable, but the intron sequence invariably begins GU. This is the first junction to be cut during RNA SPLICING.

L-DOPA The biologically active form of dihydroxyphenylalanine. It is formed from tyrosine by tyrosine hydroxylase, and is a precursor of dopamine, in a reaction catalysed by the enzyme DOPA-decarboxylase (*see* CATECHOLAMINE NEUROTRANSMITTERS). L-DOPA is used in the treatment of Parkinson's disease (*see* NEURODEGENERATIVE DISORDERS) and manganese poisoning.

dopamine A CATECHOLAMINE NEUROTRANSMITTER (Fig. D33) particularly important in the control of movement. The great majority of brain dopamine is found in the striatum, and contained in neurons originating from a brain stem nucleus, the substantia nigra. The death of these cells, with a consequent loss of dopamine, is responsible for the symptoms of Parkinson's disease (*see* NEURODEGENERATIVE DISORDERS). Other dopaminergic neurons of the brain stem innervate the limbic system and cortex; abnormalities of these systems have been implicated in SCHIZOPHRENIA. Dopamine is also involved as a transmitter in renal, gastrointestinal and cardiovascular function. Dopamine acts at a number of different receptors (*see* DOPAMINE RECEPTORS).

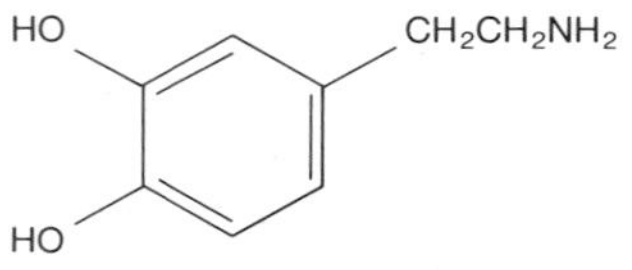

Fig. D33 Dopamine.

dopamine receptors Cell-surface receptors for the neurotransmitter and neuromodulator DOPAMINE. Classically differentiated into two subtypes, D_1 and D_2, on the basis of pharmacological differences. D_1 is a G PROTEIN-COUPLED RECEPTOR positively linked to adenylate cyclase, whereas D_2 receptors are either negatively linked to, or unassociated with, this SECOND MESSENGER PATHWAY. In the brain there are substantial interactions between the two receptor subtypes. Recently, molecular biological techniques have identified further dopamine receptor subtypes; D_3 and D_4 have many structural and pharmacological similarities to D_2, whereas D_5 receptors resemble D_1. Interest in dopamine receptors has been stimulated by the action of the ANTIPSYCHOTIC DRUGS, which have a common effect in blocking D_2 (and D_3 and D_4) receptors (*see* SCHIZOPHRENIA).

dorso-ventral axis (dorsal–ventral axis) One of the two primary embryonic axes which (together with the ANTERIOR–POSTERIOR AXIS) is established either within the egg or early during embryonic development. It is defined by the line joining the dorsal (back) and ventral (belly) regions of the animal. Most extensively studied in amphibians. *See also*: AMPHIBIAN DEVELOPMENT; DROSOPHILA DEVELOPMENT.

dorso-ventral pattern formation in *Drosophila* The DORSO-VENTRAL AXIS of the *Drosophila* embryo is established by a set of maternal gene products. A signal initiated in the somatic follicle cells that surround the oocyte leads, after fertilization, to the production ventrally of an extracellular ligand for the *Toll* gene product which is uniformly distributed in the plasma membrane of the egg. The ventral activation of *Toll* protein signals the nuclear accumulation of *dorsal* gene product in ventral nuclei and leads to the activation of zygotic genes *twist* and *snail* ventrally, and *zerknüllt* and DECAPENTAPLEGIC dorsally. How the initial extracellular signal is generated is not known, but the signal from the follicle cells appears to activate a SERINE PROTEINASE cascade involving the products of the genes *snake* and *easter* and leading to the ventral production of the ligand for the *Toll* protein. The extracellular portion of the *Toll* protein has leucine-rich repeats similar to those in the human platelet glycoprotein 1b involved in blood clotting. The cytoplasmic domain is similar to the intracellular portion of the receptor for interleukin-1. The activation of Toll creates a graded nuclear accumulation, with a high point ventrally, of the TRANSCRIPTION FACTOR dorsal which is homologous to the product of the proto-oncogene c-*rel*. This pathway is very similar to the regulation of the activity of the mammalian transcription factor NFκB in lymphocyte differentiation. *See also*: DROSOPHILA DEVELOPMENT; PATTERN FORMATION.

dosage compensation Mechanisms found in some species for maintaining a specific gene product at the same level in cells even when the number of copies of the gene varies between different cell types or between the sexes. For example, cells of male *Drosophila* are XY, but exhibit the same levels of most X-linked enzymes as female (XX) cells even though the female cell has twice the number of copies of the X-linked genes. This compensation is thought to be achieved by increased transcription of the X-linked genes in the male cell. In humans and other mammals, a similar problem is resolved in a different way, by the inactivation of one of the two copies of the X-chromosome in female cells (*see* X-CHROMOSOME INACTIVATION).

dosage effect The effect of the number of copies of a gene in a cell. *See also*: X-CHROMOSOME INACTIVATION.

dose–response curve *See*: SCHILD PLOT.

dot blot A rapid HYBRIDIZATION protocol for the detection and quantification of specific nucleic acid sequences (DNA or RNA). The aqueous nucleic acid sample is applied directly to a suitable hybridization membrane (nitrocellulose or nylon-based) as a dot of liquid and the sample filtered through the membrane by means of suction. This is usually achieved by use of a commercially available 'dot-blot' apparatus. Subsequent hybridization with the probe of choice is carried out essentially as for SOUTHERN BLOTTING (DNA–DNA) or NORTHERN BLOTTING (RNA–DNA). The technique can be adapted for quantitative rather than qualitative analysis; it can, for example, be used to determine the relative amounts of a particular DNA or RNA sequence in a mixture by utilizing control samples that contain known concentrations of the target sequence and employing densitometry to quantitate the relative hybridization signals from the dot blots. *See also*: DNA TYPING: FORENSIC APPLICATIONS; POLYMERASE CHAIN REACTION.

van Helden, P.D. & Olliver, C.L. (1981) In *Techniques in Molecular Biology*, Vol. 2 (Walker, J.M. & Gaastra, W., Eds) 178–186 (Croom Helm, London).

double helix *See*: DNA.

double minute chromosomes (DMC) Small extrachromosomal DNA elements that are found in the nuclei of METHOTREXATE-resistant (mtx) cell lines that show an unstable resistance phenotype. They are found in large numbers in such cell lines and each DMC carries between two and four copies of the DIHYDROFOLATE REDUCTASE (*dhfr*) gene. They are not true CHROMOSOMES because they lack a functional CENTROMERE and segregate erratically during MITOSIS. In the absence of methotrexate, DMCs are rapidly lost from the cell line which ultimately reverts to a methotrexate-sensitive (mtx^s) phenotype. The mechanism which generates DMCs is unknown, but is believed to involve multiple cycles of DNA REPLICATION concentrated in the region of the chromosome containing the *dhfr* gene followed by a RECOMBINATION-like event. *See*: GENE AMPLIFICATION.

Schimke, R.T. (1984) *Cell* **37**, 705–713.

double-sex (dsx) The *dsx* gene in DROSOPHILA has active but opposite regulatory functions in the two sexes; in females the *dsx* product represses the genes involved in the processes of terminal differentiation specific to males and in males it suppresses female development. In loss-of-function *dsx* mutations there is no repression of either kind of sexual differentiation, resulting in the formation of intersexual flies produced as a consequence of the simultaneous occurrence of male and female somatic sexual differentiation. The molecular basis for the sex-specific functions of *dsx* is differential processing of the primary transcript (*see* RNA SPLICING). In females, three non sex-specific exons are spliced to a fourth female-specific exon. In males, the first three exons are joined instead to two male-specific exons. The female splicing pathway is dependent on the genes *transformer* and *transformer-2*. The *dsx* protein products have no homology to any known protein class. *See also*: SEX DETERMINATION.

downregulation of receptors A decrease occuring *in vivo* or in culture in the number of molecules of a given receptor type at a particular location. This is often due to a decrease in TRANSCRIPTION of the gene(s) encoding the receptor protein, but it may also be due to a destabilization of the mRNA involved or of the assembled protein product, or to an increase in the removal of the receptor (e.g. by promotion of its internalization by ENDOCYTOSIS). Such a change is induced by various treatments for different receptors, especially, but not invariably, by long-term exposure (*in vivo* or *in vitro*) to an agonist of the receptor in question. The common phenomenon of receptor downregulation by agonist exposure is often initiated by a phosphorylation of specific sites on the cytoplasmic face of the receptor molecule.

Downregulation of receptors controlled by neural activity can also be important in the differentiation of the nervous system. Thus, in the case of the NICOTINIC RECEPTOR of skeletal muscle, for example, it has been shown that there is an activity-dependent inhibition of transcription of the receptor genes and a concurrent post-translational downregulation also.

Sibley, D.R. et al. (1987) *Cell* **48**, 913–926.
Laufer, R. & Changeux, J.P. (1989) *Mol. Neurobiol.* **3**, 1–54.

Down's syndrome (trisomy 21) Syndrome caused by TRISOMY of chromosome 21 and characterized by varying degrees of mental retardation and growth retardation and by a distinct facial appearance, which accounts for the syndrome's also being known in the past as mongolism. It is the commonest chromosomal abnormality in newborns. The trisomy 21 can be either complete, caused by a NONDISJUNCTION event at MEIOSIS, or incomplete, caused by the inheritance of a TRANSLOCATION involving part of chromosome 21. In 80% of cases of complete trisomies, the extra chromosome is maternal in origin. There is a significant association between the syndrome and maternal age: for a mother 20 years of age the risk is 1 in 2000 but by the age of 46 the risk rises to 1 in 25. However, around one-fifth of cases seem to be independent of maternal age. A less severe phenotype may be associated with MOSAICISM for trisomy 21.

Familial cases of Down's syndrome are caused by TRANSLOCATIONS involving the long arm of chromosome 21. One of the parents has a balanced translocation and is normal, but any children may inherit the translocated chromosome and be trisomic for part of chromosome 21. The commonest form of translocation consists of two long arms of chromosome 21 joined at the centromere (21q21q); any child who inherits the translocated chromosome 21 and a normal chromosome 21 will be trisomic for the long arm of chromosome 21 and will suffer from Down's syndrome.

downstream Applied to sequences on the CODING STRAND of DNA (or in mRNA) that are located 3′ to the region in question. For example, the translation INITIATION CODON of a gene is located downstream of the start site of transcription. *Cf.* UP-STREAM.

DPA DIHYDROPHASEIC ACID.

DPG 2,3-DIPHOSPHOGLYCERATE (2,3-bisphosphoglycerate).

dpp The *Drosophila* gene DECAPENTAPLEGIC.

DQ antigens In humans, class II MHC MOLECULES encoded by the HLA-DQ region of the MAJOR HISTOCOMPATIBILITY COMPLEX.

DR antigens In humans, class II MHC MOLECULES encoded by the HLA-DR region of the MAJOR HISTOCOMPATIBILITY COMPLEX. Various DR alleles (e.g. DR4 and DR3) are linked to an increased risk of developing certain immunological diseases (e.g. DR4: rheumatoid arthritis, pemphigus vulgaris, insulin-dependent diabetes mellitus; DR3: chronic active hepatitis, coeliac disease).

Drosophila

DROSOPHILA is a genus of Diptera (flies), family Drosophilidae. The family Drosophilidae includes over 3000 species. One of these, *Drosophila melanogaster* (Fig. D34), was chosen by T.H. Morgan in 1909 for studies on genetics. As an organism for laboratory study, *D. melanogaster* has the following advantages: a brief life cycle (10 days at 25°C); easy culture on an agar/yeast medium; a small number of chromosomes ($n = 4$); giant POLYTENE CHROMOSOMES (Fig. D35) in larval tissues; a small DNA content (0.18 pg DNA in the haploid genome, equivalent to 165×10^6 base pairs (bp)); high fecundity (a female may lay up to 3000 eggs in its life); and a large enthusiastic research community. There are now more than 40 000 research publications on aspects of the biology of *Drosophila*, making it one of the very best understood organisms. Much of our knowledge of the genetics of *Drosophila* is summarized in [1].

Most of the fundamental principles of genetics were established with *Drosophila* during the 'classical' period (around 1910–1950) (Table D10). Modern research with *Drosophila* covers the whole breadth of biology from ecology to neurobiology. Only a few highlights can be mentioned here.

Evolutionary biology

A remarkable feature of the *Drosophila* fauna is the existence of over 800 different species in Hawaii. The great majority are endemic to those islands and Carson has led a concerted attempt to use them for the study of evolutionary processes [2]. These studies have emphasized the role of founder events in speciation. The relationships between many species, as between continental *Drosophila*, can be established by analysis of the banding patterns of the giant polytene chromosomes. Many other studies of *Drosophila* evolution are now using molecular methods, in particular comparisons of nuclear and mitochondrial DNA sequences between species.

Development

Drosophila is being extensively studied for the analysis of the genetic control of development (*see also* DROSOPHILA DEVELOP-

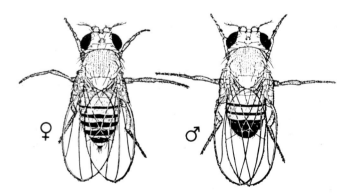

Fig. D34 *Drosophila melanogaster*. From [16].

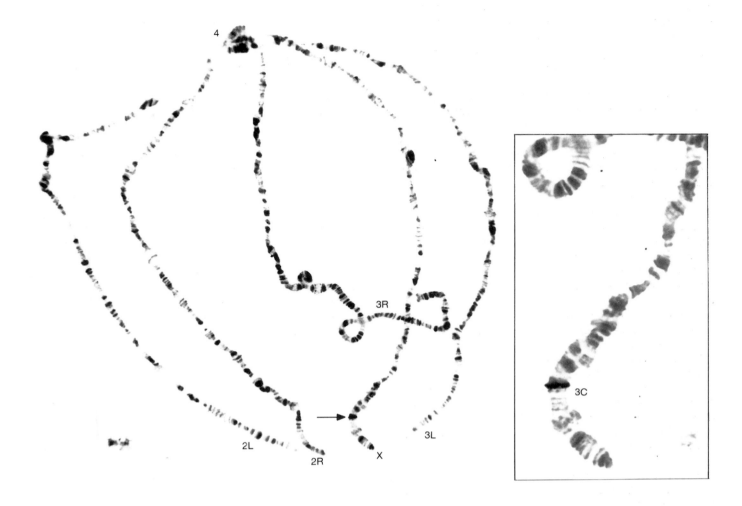

Fig. D35 Polytene chromosomes from the salivary gland of *Drosophila*. The arrow indicates the localization of the gene *near white* to the 3C region of the X-chromosome using a biotin-labelled probe. R and L, right and left arms. Photograph by M. Ashburner.

MENT). The relative contributions of maternal and zygotic gene products to embryogenesis (*see* MATERNAL CONTROL OF DEVELOP-MENT) have been determined by the study of mutations that disrupt specific developmental processes. Many genes must be active during oogenesis to establish the two main body axes of the embryo, the anterior–posterior and dorsal–ventral axes. Some of these genes are active in the germ-line cells and produce either RNA or protein products that are asymmetrically distributed in the embryo. Other genes are active in the somatic cells that surround the germ line and establish asymmetries in the oocyte. After fertilization, the activities of these maternally inherited gene products result in the specification of broad domains of the body. They do this by ensuring the region-specific activation of the so-called GAP GENES. The gap gene products in turn act to ensure the proper division of the body into its segmented components. Most of these genes, both maternal and zygotically acting, code for proteins that bind to DNA at specific sequences and which act on their target genes as transcriptional regulatory proteins (TRAN-SCRIPTION FACTORS).

Homeotic genes

One class of genes active in early development has attracted particular attention — the HOMEOTIC GENES. First identified because of their mutant phenotypes, the transformation of segment identities (e.g. the transformation of an antenna into a leg), these genes encode transcriptional regulators of a particular class, the homeodomain proteins (*see* HOMEOBOX GENES AND HOMEODOMAIN PROTEINS). The two major clusters of homeotic genes in *Drosophila* are the Antennapedia and Bithorax clusters. Each contains several genes that are expressed in different segments of the embryo. Remarkably, the anterior–posterior order in which these genes are expressed is the same as the order of the genes themselves in the chromosomes. Even more remarkably, this feature of segmentally expressed homeodomain genes is also found in vertebrates (*see* HOMEOBOX GENES AND HOMEODOMAIN PROTEINS; MAMMALIAN VERTEBRATE DEVELOPMENT).

Table D10 The contribution of *Drosophila* to classical genetics

Date	Contribution	Reference
1913	Chromosome non-disjunction	[4]
1913	Concept and practice of genetic mapping	[5]
1916	Chromosome theory of heredity	[6]
1917	Balanced lethal mutations	[7]
1925	Unequal crossing over	[8]
1926	Chromosomal polymorphisms in natural populations	[9]
1927	Mutagenic effect of ionizing radiation	[10]
1931	Genetic crossing over involves the physical interchange of chromosomal material	[11]
1932	Dosage compensation of sex-linked genes	[12]
1936	Mitotic recombination	[13]
1940	Genetic recombination can occur within genes	[14]
1946	Mutagenic effects of chemicals	[15]

Heat-shock genes

The discovery of homeodomain genes, and of their developmental role, is but one recent contribution of *Drosophila* research to biology. Another is the discovery of the HEAT SHOCK proteins. In 1962 F. Ritossa [3] found that a brief heat treatment of *Drosophila* larvae (e.g. at 37°C for 20 min) induced specific gene activities in the larval polytene chromosomes. These genes encode a family of proteins which function to ameliorate the consequences of stress. These proteins, and indeed the induction of these genes by stress, are now known to be universal, from bacteria to humans, and to have importance in the heat therapy of certain cancers.

Sex determination

Sex is determined in *Drosophila* by the ratio of the X-chromosomes to the autosomes. When this ratio is 1, a zygote develops as a female; when the ratio is 0.5, as a male. The genetic pathway from X : autosome ratio to sex differentiation is now quite well established: a key gene, called *Sex-lethal*, is functional when the ratio is 1 but not when it is 0.5. The product of this gene controls the expression of other genes that determine the sexual phenotype; interestingly, this control (and that of *Sex-lethal* itself) occurs at the level of RNA SPLICING (*see also* SEX DETERMINATION).

Behaviour

Drosophila exhibits complex behaviours, for example during mating. Several genes that are required for learning behaviour have been identified; the products of some of these are involved in cyclic nucleotide metabolism. Some behaviours of *Drosophila* are rhythmic. There are CIRCADIAN RHYTHMS in locomotor activity and in adult eclosion. One gene, called *period*, is essential for these rhythms, and flies mutant for this gene may be arrhythmic or display rhythms longer or shorter than the normal 24 hours. *per*

is also involved in determining the periodicity of the mating call of *Drosophila*.

Genetic manipulation

Many of the spectacular advances in modern *Drosophila* biology have resulted from technical breakthroughs. The most important of these has been a method for the genetic transformation of *Drosophila* with exogenous DNA. This method exploits a particular class of TRANSPOSABLE GENETIC ELEMENT natural to many populations of *D. melanogaster*, the *P* element. DNA carried on a *P* element VECTOR will, after injection into young embryos, stably incorporate into the fly's genome (*see* P ELEMENT-MEDIATED TRANSFORMATION).

M. ASHBURNER

See also: INSECT NEURAL DEVELOPMENT.

1 Ashburner, M. (1989) *Drosophila: A Laboratory Handbook* (Cold Spring Harbor Press, New York) 1331pp.
2 Carson, H. & Yoon, J.S. (1982) In *The Genetics and Biology of* Drosophila (Ashburner, M. et al., Eds) Vol. 3b, 298–344 (Academic Press, London).
3 Ritossa, F. (1962) *Experientia* **18**, 571–573.
4 Bridges, C.B. (1913) *J. Exp. Zool.* **15**, 587–606.
5 Sturtevant, A.H. (1913) *J. Exp. Zool.* **14**, 43–59.
6 Bridges, C.B. (1916) *Genetics* **1**, 1–52, 107–163.
7 Muller, H.J. (1917) *Genetics* **3**, 422–499.
8 Sturtevant, A.H. (1925) *Genetics* **10**, 117–147.
9 Sturtevant, A.H. (1926) *Biol. Zentrbl.* **46**, 697–702.
10 Muller, H.J. (1927) *Science* **66**, 84–87.
11 Stern, C. (1931) *Biol. Zentrbl.* **51**, 547–587.
12 Muller, H.J. (1932) *Proc. 6th Int. Congress Genetics* **1**, 213–255.
13 Stern, C. (1936) *Genetics* **21**, 625–730.
14 Oliver, C.P. (1940) *Proc. Natl. Acad. Sci. USA* **26**, 452–454.
15 Rapaport, J.A. (1946) *Dokl. Akad. Nauk USSR* **54**, 65–67.
16 Morgan, T.H. et al. (1925) *Bibliographica Genetica II* (Martinus Nijhoff, 's-Gravenhage).

Drosophila development

THE embryo of *Drosophila melanogaster* has been used extensively as a model to study the generation of pattern during development (*see also* PATTERN FORMATION). It is now possible to provide in outline a mechanistic model for the processes that translate maternal information laid down in the egg during oogenesis into the basic ground plan of the body of the embryo. This understanding is largely founded on the genetic analysis of developmental pathways and, in particular, on systematic screens for mutations affecting selected developmental processes, most notably segmentation.

The cloning of genes identified by developmental mutations (*see* DNA CLONING) has enabled the construction of specific nucleic acid PROBES and their use in *in situ* HYBRIDIZATION to determine the timing and site of gene expression during development. ANTIBODIES against the cloned gene products are also used to detect the site of synthesis and distribution of these proteins. A selection of genes involved in embryonic pattern formation is given in Table D11.

Table D11 Some genes involved in embryonic pattern formation in *Drosophila*

Gene name (symbol)	Maternal or zygotic	Classification by phenotype	Protein structural motifs/putative function	Comments
armadillo (*arm*)	m, z	Segment polarity	Cytoskeletal protein?	Sequence similarity to plakoglobin
bicoid (*bcd*)	m	Anterior pattern defects	Homeodomain/transcription factor	Anterior determinant encoding gradient morphogen
caudal (*cad*)	m, z	Loss/transformation of A10/11 (PS 15 or telson) + segmentation defects	Homeodomain/transcription factor	Maternal RNA is translated to yield protein gradient. Zygotic transcription in posterior region only
cubitus-interruptus (*ci*^D)	z	Segment polarity	Zinc fingers	Similar to oncogene GL1
dorsal (*dl*)	m	Dorsal group	Rel-related/transcription factor	Regulated by nuclear localization in dorso-ventral gradient
decapentaplegic (*dpp*)	z	Multiple defects in dorsal epidermis + appendages	TGF-β family/growth factor	Most similar to vertebrate bone morphogenesis protein BMP-2
Easter (*Ea*)	m	Dorsal group	Serine proteinase	Necessary for activation of ventral signal. Acts in perivitelline space
engrailed (*en*)	z	Pair rule/segment polarity	Homeodomain/transcription factor	Expression defines posterior compartment of each segment
even-skipped (*eve*)	z	Pair rule	Homeodomain/transcription factor	Responds directly to gap genes. Expression defines odd numbered parasegments
fused (*fu*)	m, z	Segment polarity	Serine-threonine protein kinase	
fushi-tarazu (*ftz*)	z	Pair rule	Homeodomain/transcription factor	Secondary pair rule gene. Expression defines even numbered parasegments
giant (*gt*)	z	Segment gap	Leucine zipper/transcription factor	
gooseberry (*gsb-d*, *gsp-p*)	z	Segment polarity	Homeodomain/transcription factor	Two closely linked genes with similar sequence and overlapping expression patterns
hairy (*h*)	z	Pair rule	Helix-loop-helix	Responds directly to gap genes. ?Lacks functional DNA binding domain. Analogue of ID protein
hunchback (*hb*)	m, z	Segment gap	Zinc finger/transcription factor	Specifies head and thorax. Regulated by bicoid protein
huckebein (*hkb*)	z	Terminal gap	Zinc finger/transcription factor	Regulated by terminal system
knirps (*kni*)	z	Segment gap	Cys-cys zinc finger/transcription factor	DNA binding domain similar to steroid hormone receptor family
Krüppel (*Kr*)	z	Segment gap	Zinc finger/transcription factor	
nanos (*nos*)	m	'Posterior group' gap gene	—	Posterior factor required for abdominal segmentation. RNA is localized to polar granules
Notch	m, z	Neurogenic gene	Transmembrane protein with EGF like repeats/adhesion molecule	Mediates cell interactions during many phases of patterning within epithelia, including neural/epidermal patterning
oskar (*osk*)	m	'Posterior group' gap gene	—	*oskar* RNA and protein are both localized to polar granules. Ectopic *oskar* expression induces formation of pole cells
paired (*prd*)	z	Pair rule	Homeodomain/transcription factor	Also carries second conserved DNA binding domain — 'paired box'
patched (*ptc*)	z	Segment polarity	Transmembrane molecule	

continued

Table D11 Continued

Gene name (symbol)	Maternal or zygotic	Classification by phenotype	Protein structural motifs/putative function	Comments
runt (*run*)	z	Pair rule	—	Nuclear protein. Pair-rule expression complementary to hairy
scute (*sc*)	z	'Pro-neural' gene	Helix-loop-helix/transcription factors	A cluster of five genes in the achaete–scute complex encode related proteins. Involved in patterning neural precursors and in sex determination
snail (*sna*)	z	Ventral pattern defects	Zinc finger/ transcription factor	Marker for presumptive mesoderm
snake (*snk*)	m	Dorsal group	Serine proteinase	Necessary for activation of ventral signal. Acts in perivitelline space
tailless (*tll*)	z	Segment gap	Cys-cys zinc finger/transcription factor	Steroid-receptor protein family. Regulated by terminal system
Toll (*Tl*)	m	Dorsal group	Transmembrane molecule	Receptor for ventral signal
torso (*tor*)	m	Anterior and posterior pattern defects	Transmembrane tyrosine kinase	Receptor for terminal signal
twist (*twi*)	z	Ventral pattern defects	Helix-loop-helix/transcription factor	Marker for presumptive mesoderm
vasa (*vas*)	m	'Posterior group' gap gene	'DEAD' family protein. ?RNA helicase	Protein is localized to polar granules
wingless (*wg*)	z	Segment polarity	'Wnt' family secreted molecule	Acts non-autonomously to maintain *en* expression
zen	z	Dorsal pattern defects	Homeodomain transcription factor	

Homeotic genes of antennapedia and bithorax complexes

abdominal-A (*abd-A*)	z	Segment transformation affecting A2–A8 (PS 7–14)	Homeodomain transcription factor	Antennapedia-like sequence
Abdominal-B (*Abd-B*)	z	Segment transformation affecting A5–A8/9 (PS 10–14)	Homeodomain transcription factor	Most similar to vertebrate *Hox* genes of group 9
Antennapedia (*Antp*)	z	Segment transformation affecting T1–T3 (PS 3–5)	Homeodomain transcription factor	Canonical antennapedia-like gene (most similar to vertebrate *Hox* genes of groups 6–8)
Deformed (*Dfd*)	z	Segment transformation affecting mandible/maxilla (PS 0/1)	Homeodomain transcription factor	Most similar to vertebrate *Hox* genes of group 4
labial (*lab*)	z	Head segment transformation and gut defects	Homeodomain transcription factor	Most similar to vertebrate *Hox* genes of group 1
proboscipedia (*pb*)	z	Segment transformation affecting maxilla labial palps (of adult only)	Homeodomain transcription factor	Most similar to vertebrate *Hox* genes of group 2
Sex combs reduced (*Scr*)	z	Segment transformation Labial/T1 (PS 2/3)	Homeodomain transcription factor	Most similar to vertebrate *Hox* genes of group 5
Ultrabithorax (*Ubx*)	z	Segment transformation T3–A7 (PS 5–13)	Homeodomain transcription factor	Antennapedia-like sequence

The *Drosophila* egg when laid is about 500 μm long and 170 μm in diameter. It is overtly polar, with the anterior–posterior (A/P) and dorsal–ventral (D/V) axes defined by its asymmetric shape and by the patterning of the egg shell. During normal development, the embryo always develops in the same orientation relative to the egg shell, implying that the embryonic axes are specified by cues provided during oogenesis. Embryonic development takes 22 hours (all timings are given as at 25°C) and terminates with the hatching of the first instar larva [1,2]. The larva feeds for 4 days, growing about 200-fold in mass and moulting twice before pupariation. In the pupa most of the larval tissues are broken down, to be replaced by tissues derived from clusters of previously undifferentiated cells that are carried within the larva. The development of these imaginal cells during the pupal period can be viewed as a delayed process of embryogenesis during which morphogenesis of the adult takes place.

Besides providing morphogenetic cues, oogenesis provides the *Drosophila* egg with an abundant supply of maternal RNAs and proteins (synthesized by the NURSE CELLS) as well as yolk. These allow development after fertilization to proceed very rapidly (Fig. D36). In the first 2 hours, 13 cycles of nuclear division occur without intervening cell division. Most of the resulting nuclei

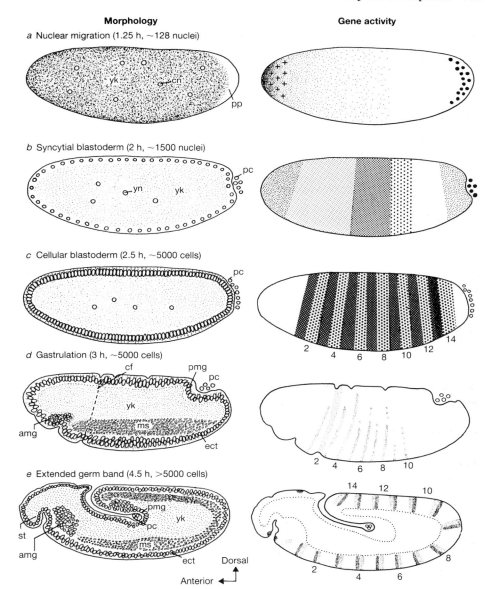

Morphology

a Nuclear migration (1.25 h, ~128 nuclei)

b Syncytial blastoderm (2 h, ~1500 nuclei)

c Cellular blastoderm (2.5 h, ~5000 cells)

d Gastrulation (3 h, ~5000 cells)

e Extended germ band (4.5 h, >5000 cells)

Gene activity

Fig. D36 Patterns of gene activity during early *Drosophila* development (modified from [14]). Diagrams on the left show the morphology of stages during early embryogenesis. Corresponding panels on the right show patterns of gene activity established at the corresponding stages. *a*, Localized maternal determinants: *bicoid* RNA (crosses); polar granules (dots). The *bicoid* protein gradient is shown by stippling; *b*, Gap gene expression: *hunchback*, *Krüppel*, *knirps* (shading, zones from anterior to posterior), *tailless* (stipple at both ends); *c*, Pair-rule stripes: *even-skipped* (dark) and *fushi-tarazu* (light). *d, e*, evolving pattern of segment polarity gene expression: *wingless* (dark) and *engrailed* (light). amg, anterior mid-gut; cf, cephalic furrow; cn, cleavage nucleus; ect, ectoderm; ms, mesoderm; pc, pole cells; pmg, posterior mid-gut; pp, pole plasm; yk, yolk; yn, yolk nucleus; st, stomodeum.

migrate to the surface of the egg to form a syncytial blastoderm; a few remain in the yolk to become vitellophages. The first nuclei to reach the posterior pole of the egg form cells precociously, enclosing around themselves a region of distinctive cytoplasm that contains POLAR GRANULES. These cells will form the GERM LINE, migrating during subsequent development through the mid-gut wall and into the gonads. With this exception, the egg remains syncytial for a further hour, until membranes pull down from the egg cortex and pinch off beneath the nuclei to complete cellularization [3]. It is during this transition from syncytial to cellular blastoderm that the outlines of the body plan are defined through the coordinated activity of an estimated 40–50 zygotic segmentation and patterning genes.

Tissue transplantation experiments show that, by the time of gastrulation (3 h), the 5000 cells of the blastoderm are SPECIFIED to form specific segments along the A/P axis, and within each

segment to form the major different tissues along the D/V axis (MESODERM, neurogenic region, epidermis and extra-embryonic membrane). GASTRULATION proceeds by the invagination of the mesoderm along a ventral furrow, and of the gut primordia near the anterior and posterior poles. Cell rearrangement causes the embryo to elongate, the posterior pole folding dorsally and anteriorly towards the head. The resulting extended germ band is not visibly segmented, but region-specific molecular markers detected by immunohistological techniques reveal a complex organization of distinct cell types.

Cell division resumes after gastrulation in a complex and stereotyped spatial pattern — one of the first visible signs of cell DIFFERENTIATION [4]. Segmentation becomes visible as a series of parasegmental grooves (see below) and tracheal pits within each metamere. In most tissues, cells divide only twice more before the larva hatches. In the second half of embryogenesis, cells divide only in the nervous system; morphogenetic movements lead to the final form of the larva, and tissue-specific cell differentiation occurs.

After hatching, the larva grows by a dramatic increase in cell size. Cells in most larval tissues do not divide. DNA replication occurs, leading to the formation of polyploid nuclei, and in some cases POLYTENE CHROMOSOMES, with up to 4000 copies of the genome in each cell. The imaginal cells remain diploid. Those in the head and thorax proliferate during the larval period, invaginating from the epidermis to form IMAGINAL DISKS [5]. Cells of the imaginal discs are not overtly differentiated, but by the end of the larval period they are already expressing molecular markers that presage the appearance of differentiated elements, such as bristles and veins, within the adult cuticle. Patterning within the eye disc is particularly apparent, forming the ordered cell clusters that give rise to ommatidia of the compound eye [6] (*see* EYE DEVELOPMENT).

The moult cycle of the larva is under hormonal control. Towards the end of the third larval instar, a pulse of ECDYSONE released in the absence of JUVENILE HORMONE triggers the events that lead to pupation. A puparium forms by tanning of the larval cuticle. About 5 hours later, the animal moults to form the pupa within the puparium. Within the pupal cuticle, the imaginal discs evert, fuse and undergo complex cell rearrangements to form the segments and appendages of the adult head, thorax and genitalia. The imaginal cells of the abdomen proliferate and spread across the pupal cuticle to form the adult epidermis. The nervous system is extensively restructured. Most larval polyploid cells die, to be replaced by the proliferation of nests of imaginal cells within the gut, salivary glands and other tissues [7].

Maternal specification of pattern

The study of maternal-effect mutations (*see* MATERNAL CONTROL OF DEVELOPMENT) has defined four independent maternal systems that define the POLARITY of the embryo. One of these systems defines patterning along the D/V axis. The other three define regions of the body pattern along the A/P axis [8,9]. These different systems use diverse molecular mechanisms: anterior and posterior body regions are specified by maternally encoded RNAs that are deposited within the egg, and act as localized CYTOPLASMIC DETERMINANTS. But the D/V polarity of the embryo, and the unique characteristics of its terminal regions, are specified by signals that are generated outside the egg and communicated to it by means of membrane RECEPTORS.

The anterior determinant is the product of the BICOID gene, which encodes a homeodomain TRANSCRIPTION FACTOR (*see* HOMEOBOX GENES AND HOMEODOMAIN PROTEINS). *bicoid* RNA is synthesized in the nurse cells, and localized within the anterior 10% of the egg during oogenesis. After fertilization, it is translated to generate a gradient of *bicoid* protein that extends throughout the anterior half of the egg. Different concentrations of this protein regulate the transcription of zygotic target genes at different distances from the anterior pole. Hence *bicoid* protein may be considered a MORPHOGEN.

The posterior determinants are associated with the polar granules and are necessary both for the formation of the germ line and for development of the abdomen. The polar granules comprise several localized maternal RNAs and proteins, including products of the *oskar*, *vasa*, and *nanos* genes. Of these, the *nanos* gene product is specifically required for abdomen formation. Unlike *bicoid*, *nanos* does not regulate transcription directly; instead its activity blocks the expression of a maternally encoded transcript in the posterior half of the egg. Its precise mode of action is unknown. The molecular basis for germ cell determination is also unknown.

The localized extracellular signals that specify D/V and terminal patterning are generated by the follicular epithelium — a somatically derived cell layer that surrounds the OOCYTE and secretes the egg shell [9].

For patterning of the terminal regions, zygotic target genes are activated via a signalling pathway that is dependent on the *torso* gene product, a putative transmembrane receptor TYROSINE KINASE that is uniformly distributed around the oocyte membrane. This receptor is activated only at the terminal regions by postfertilization release of its ligand stored locally in the egg membrane. D/V patterning is mediated by an analogous signalling pathway that is dependent on the *Toll* gene product. In this latter case an activity capable of inducing ventral structures in an apolar embryo has been shown to be released into the perivitelline fluid that lies between the oocyte and the egg membrane.

Patterning in the blastoderm

The activities of the maternal patterning systems direct the transcription of zygotic genes in specific regions of the blastoderm [10]. Because much of this patterning takes place in a syncytial environment, *Drosophila* can exploit an unusual patterning mechanism. Initial subdivision along the A/P axis is achieved by the localized expression of genes encoding transcription factors. Refinement of this spatial pattern appears to depend on the diffusion of these transcription factors from their sites of synthesis to generate intraembryonic gradients that constitute local POSITIONAL INFORMATION.

Many of the zygotic target genes have been identified by mutations that disrupt the pattern of structures formed by the larval cuticle. Those affecting segmentation were initially classified according to their phenotypic effects into gap, pair-rule and segment polarity classes [11]. Subsequent analysis has shown that these phenotypic classes correlate broadly with the position of each gene in a functional hierarchy. Segment gap genes are the immediate targets of the maternal patterning systems; they are expressed in relatively broad domains at defined positions along the axis of the egg (Fig. D36 and Plate 8*d* and *e*). The pair-rule genes are expressed slightly later, each in a series of stripes along the A/P axis spaced at double segment intervals (*see* Plate 8*a–c*, *e*). Shortly before gastrulation, the expression of the segment polarity genes defines a pattern with the periodicity of individual segments. Virtually all members of the gap and pair-rule gene classes encode transcription factors, including those with ZINC FINGER, HOMEOBOX, HELIX-LOOP-HELIX and LEUCINE ZIPPER motifs. The products of the segment polarity genes are more diverse, including transcription factors, membrane receptors and secreted molecules.

The origin of periodicity in the segment pattern is an intriguing problem. Models for the generation of repetitive patterns have invoked feedback mechanisms that intrinsically generate periodicity. The initial specification of segment pattern in *Drosophila* does not appear to use such a mechanism. Transcription of the gap genes subdivides the egg into a series of zones or annuli along the A/P axis, giving each position a unique combination of transcription factors [12]. Certain of the pair-rule genes (the 'primary' pair-rule genes) respond directly to these signals. For these genes, a different combination of the gap gene products activates each stripe of the pattern, using distinct elements within a complex PROMOTER for each stripe [13]. Such a mechanism is not intrinsically periodic. It has presumably been constrained by selection during evolution to generate regular stripes. Cross regulation between these primary genes and other ('secondary') pair-rule genes may constitute a feedback mechanism that refines and amplifies this initial pattern.

The patterns of expression of the gap and pair-rule genes are transient, decaying at or shortly after gastrulation (though many of the same genes are utilized again during later developmental processes). Their activity in the syncytial blastoderm may be viewed as elaborating a prepattern or scaffold upon which to build the definitive segment pattern. This prepattern provides information to define the pattern of expression of segment polarity genes. At least three distinct regions are specified across each segment — cells expressing the gene ENGRAILED, cells expressing *wingless* and cells expressing neither of these products. Further elaboration of pattern involves interactions between these distinct regions and probably occurs between cells, after gastrulation, rather than in the syncytium. The boundary between cells expressing *engrailed* and those expressing *wingless* becomes the morphologically visible PARASEGMENT boundary. By later stages of embryogenesis, commitment to express *engrailed* becomes essentially irreversible — and is associated with a clonal restriction into anterior and posterior COMPARTMENTS of each segment (*engrailed* off and *engrailed* on respectively). The segment boundary forms at the posterior margin of the stripe of cells that express *engrailed*.

Although segments have a common underlying pattern of segment polarity gene expression, each segment develops unique characteristics. The differences between segments are specified by the localized expression of HOMEOTIC GENES (Table D11), so called because mutations in these genes cause cells of one segment to develop so that they resemble another. The homeotic genes encode region-specific transcription factors. The correct segment-specific activation of these genes depends upon the same blastoderm prepattern that directs the repeating segment pattern, but in this case the unique combinations of segmentation gene products in each segment are interpreted to yield unique patterns of homeotic gene activation [14]. A process of molecular memory, as yet poorly understood, ensures that this specification of segment identity is stable throughout development.

M.E. AKAM

See also: DROSOPHILA; EUKARYOTIC GENE EXPRESSION; GROWTH FACTORS; GROWTH FACTOR RECEPTORS; INDUCTION; INSECT NEURAL DEVELOPMENT.

1 Campos-Ortega, J.A. & Hartenstein, V. (1985) *The Embryonic Development of Drosophila melanogaster* (Springer-Verlag, Berlin).

2 Bate, M. & Martinez-Arias, A. (Eds) (1993) *The Development of* Drosophila melanogaster. (Cold Spring Harbor, New York).

3 Foe, V.E. & Alberts, B.M. (1993) Studies of nuclear and cytoplasmic behaviour during the five mitotic cycles that precede gastrulation in *Drosophila* embryogenesis. *J. Cell. Sci.* **61**, 31–70.

4 Foe, V.E. (1989) Mitotic domains reveal early commitment of cells in *Drosophila* embryos. *Development* **107**, 1–23.

5 Ursprung, H. & Nothiger, R. (Eds) (1972) *The Biology of Imaginal Discs, Results and Problems in Cell Differentiation* (Springer-Verlag, Berlin).

6 Tomlinson, A. (1988) Cellular interactions in the developing *Drosophila* eye. *Development* **104**, 183–193.

7 Bodenstein, D. (1950) The post-embryonic development of *Drosophila*. In *Biology of* Drosophila (Wiley, New York).

8 Nüsslein-Volhard, C. et al. (1987) Determination of antero-posterior polarity in *Drosophila*. *Science* **238**, 1675–1681.

9 Stein, D.S. & Stevens, L.M. (1991) Establishment of dorsal–ventral and terminal pattern in the *Drosophila* embryo. *Curr. Opinion Genet. Dev.* **1**, 247–254.

10 Ingham, P.W. (1988) The molecular genetics of embryonic pattern formation in *Drosophila*. *Nature* **335**, 25–34.

11 Nüsslein-Volhard, C. & Weischaus, E. (1980) Mutations affecting segment number and polarity in Drosophila. *Nature* **287**, 795–801.

12 Lawrence, P.A. (1992) *The Making of a Fly: the Genetics of Animal Design* (Blackwell Scientific Publications, Oxford).

13 Small, S. & Levine, M. (1991) The initiation of pair-rule stripes in the *Drosophila* blastoderm. *Curr. Opinion Genet. Dev.* **1**, 247–254.

14 Akam, M. (1987) The molecular basis for metameric pattern in the *Drosophila* embryo. *Development* **101**, 1–22.

DRP DYSTROPHIN-RELATED PROTEIN.

drug-resistance factor/plasmid A term originally coined to describe 'factors' which conferred resistance to antibiotics on bacteria. Such factors were first detected in Japan in 1957 in strains of *Shigella*. Subsequent studies identified these factors as PLASMIDS which carry drug-resistance genes (*see* R FACTORS). Such plasmids can carry multiple resistance genes resulting in bacteria resistant to several different antibiotics.

drug-resistance genes Genes conferring resistance to antibiotics, antitumour drugs, etc. They variously encode enzymes that inactivate the drug in question, for example the bacterial genes *bla* (β-LACTAMASE) and *cat* (CHLORAMPHENICOL ACETYLTRANSFERASE), or may act in other ways (*see e.g.* MULTIDRUG RESISTANCE). *See also*: R FACTOR.

Ds (Dissociation) *See*: TRANSPOSABLE GENETIC ELEMENTS: PLANTS.

dsx The *Drosophila* gene DOUBLE-SEX.

Duchenne muscular dystrophy

DUCHENNE muscular dystrophy (DMD) is one of the most common and most devastating genetic diseases of childhood [1]. It is an X-LINKED DISEASE which affects around 1 in 3000 live male births per annum and is characterized by severe progressive muscle weakness. Affected boys are usually diagnosed at around five years old on the basis of their clinical presentation and muscle biopsy findings. The children are slow in learning to walk,

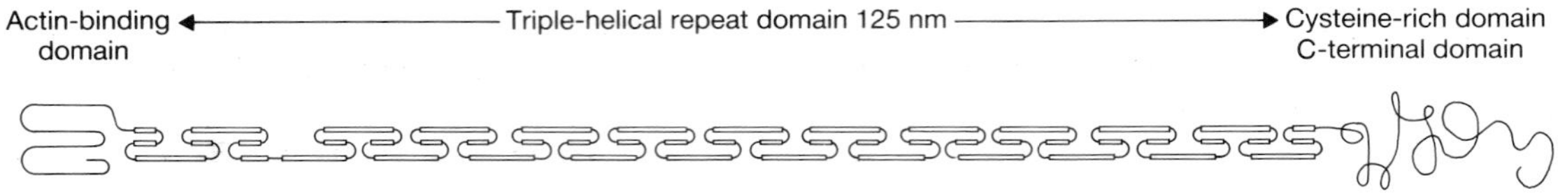

Actin-binding ◄──────── Triple-helical repeat domain 125 nm ────────► Cysteine-rich domain
domain   C-terminal domain

Fig. D37 A model of the putative structure of dystrophin. From [3].

employing a characteristic 'waddling gait', and have difficulty in climbing stairs and rising from a prone position. They often have enlarged calves caused by the replacement of diseased muscle with fat and connective tissue, and abnormally high levels in the bloodstream of muscle enzymes such as creatine kinase. Biopsies of affected muscles reveal characteristic dystrophic changes in the muscle fibres, including variation in fibre size, degeneration and cell death. Sufferers from DMD become wheelchair-bound by the age of 12 and die, usually from respiratory failure, in their early 20s. Around 30% of DMD boys have some degree of mental retardation. A milder disease, Becker muscular dystrophy (BMD), is now known to be allelic to DMD, being caused by mutations in the same gene. Boys affected with BMD have less severe muscle weakness, do not become wheelchair-bound until their teens or later, and can have normal life expectancy. BMD is much rarer than DMD, affecting ~1 in 30 000 births.

The DMD gene and its protein product dystrophin

The cloning and characterization of the gene responsible for DMD and BMD is one of the triumphs of modern molecular genetics. The pattern of inheritance of the disease indicated that the gene was located on the X-chromosome and the gene was further localized to chromosome band Xp21 using three complementary strategies. Linkage analysis using polymorphic DNA markers located on the X-chromosome indicated that both the DMD and BMD disease phenotypes were linked to markers in Xp21. The analysis of rare cases of girls with DMD revealed gross structural mutations of one of their X-chromosomes, which always involved Xp21. The identification of a small number of DMD patients with microscopically detectable deletions of part of their X-chromosome showed that these deletions always removed all or part of Xp21. Portions of the coding sequence of the DMD gene were cloned by comparing normal DNA with DNA from a male DMD patient with a large deletion involving Xp21, and from female patients with rearrangements of their X-chromosomes. The remainder of the gene was isolated from cDNA libraries (*see* DNA LIBRARIES) constructed from skeletal muscle tissue [2].

The DMD gene transcript is 14 kilobases (kb) in length and is expressed in a limited number of tissues, principally skeletal muscle, smooth muscle, and brain. The gene is composed of at least 76 exons spread over 2.3 megabases (2.3×10^6 base pairs) of DNA, making it the largest human gene characterized to date [3]. The DNA sequence of the human DMD gene is very similar to that of the equivalent genes in the mouse and in the chicken; the DMD gene is also similar to that of another human gene of unknown function localized on chromosome 6 [4].

The protein encoded by the DMD gene is named dystrophin. It is abundant in skeletal and smooth muscle and is detectable at lower levels in some other tissues. The protein has four predicted functional domains (Fig. D37): an N-terminal domain which is thought to bind to ACTIN, a central rod-shaped domain composed of 26 repeat units similar to those found in SPECTRIN and certain other structural proteins, a domain rich in the amino acid cysteine, and a C-terminal region which may bind to other proteins [5]. The function of dystrophin is as yet unknown; in the normal muscle fibre it is localized at the cell membrane, where it may have a role in maintaining structural integrity during contraction. Recently, dystrophin has been shown to be associated with a number of other proteins which form a GLYCOPROTEIN complex spanning the membrane; the role of dystrophin may be to anchor this complex to the cytoskeleton [6]. Dystrophic muscle fibres exhibit abnormal CALCIUM homeostasis, which could be caused by leakage across the damaged membrane, and atypical ion fluxes perhaps caused by the abnormal spatial position of the glycoprotein complex.

The molecular basis of DMD and BMD

DMD and BMD are caused by mutations in the DMD gene. In approximately 70% of cases the mutation is a DELETION which removes part of the coding sequence [7]. The difference between the severe DMD phenotype and the mild BMD phenotype is due to the type of deletion. In DMD the deletion disrupts the READING FRAME of the gene so that little or no dystrophin is produced; accordingly dystrophin in usually undetectable in muscle samples from DMD patients. In BMD, the mutation does not disrupt the reading frame but allows the production of a truncated but partly functional dystrophin molecule. The mutated dystrophin is detectable in muscle samples from BMD patients and is correctly localized in at least a fraction of muscle fibres [8]. The mildest BMD phenotypes are associated with deletions of the central repeat domain of the protein; mildly affected BMD patients have been reported who are missing more than half their DMD coding sequence. DUPLICATIONS and INVERSIONS of coding sequences, and POINT MUTATIONS which disrupt the reading frame have also been detected. In the mouse model of DMD, the *mdx* mouse, a point mutation in the mouse DMD gene introduces a TERMINATION CODON and thereby abolishes the production of dystrophin.

The cloning and characterization of the DMD gene has enabled the development of efficient strategies for PRENATAL DIAGNOSIS and CARRIER detection based on DNA analysis, and direct analysis of DNA or RNA using the POLYMERASE CHAIN REACTION (PCR) [9]. Around one-third of cases of DMD are not familial but arise as a result of new mutations. Much effort has therefore been directed towards the development of effective treatments for the disease, such as a method for replacement of the defective gene. Preliminary experiments have been carried out in the mouse, but a cure for DMD is still some way in the future.

B. BYTH

See also: MUSCLE.

1 Emery, A.E.H. (1987) Duchenne muscular dystrophy. In *Oxford Monographs on Medical Genetics*, No. 15 (Harper, P. & Bobrow, M.) (Oxford University Press, Oxford).
2 Love, D.R. & Davies, K.E. (1989) Duchenne muscular dystrophy: the gene and the protein. *Mol. Biol. Med.* **6**, 7–17.
3 Koenig, M. et al. (1987) Complete cloning of the DMD gene in normal and affected individuals. *Cell* **50**, 509–517.
4 Love, D.R. et al. (1989) An autosomal transcript in skeletal muscle with homology to dystrophin. *Nature* **339**, 55–58.
5 Hoffman, E.P. et al. (1987) Dystrophin: the protein product of the Duchenne muscular dystrophy locus. *Cell* **51**, 919–928.
6 Ervasti, J.M. & Cambell, K.P. (1991) Membrane organization of the dystrophin–glycoprotein complex. *Cell* **66**, 1–20.
7 Koenig, M. et al. (1989) The molecular basis of Duchenne versus Becker muscular dystrophy: correlation of severity with type of deletion. *Am. J. Hum. Genet.* **45**, 498–506.
8 Monaco, A.P. et al. (1988) An explanation for the phenotypic differences between patients bearing partial deletions of the Duchenne muscular dystrophy locus. *Genomics* **2**, 90–95.
9 Beggs, A.H. et al. (1990) Detection of 98% of DMD/BMD gene deletions by polymerase chain reaction. *Hum. Genet.* **86**, 45–48.

Duffy blood group The Duffy red cell antigens Fy^a and Fy^b are the products of alleles at the *FY* locus on the long arm of chromosome 1. The *FY* locus was the first marker locus to be localized to an autosome. The phenotype Fy(a – b –) is very rare in Caucasians but common in people of African origin. The Fy antigens are unimportant in blood transfusion but are thought to be associated with the invasion of red cells by the malaria parasite. Fy(a – b –) people are resistant to *Plasmodium vivax* malaria and Fy(a – b –) red cells are refractory to invasion *in vitro* by the simian malarial parasite *P. knowlesi*.

duplex formation The association or reassociation of two complementary single-stranded nucleic acid chains into a double-stranded molecule with complementary base pairing (*see* DNA; HYBRIDIZATION; NUCLEIC ACID STRUCTURE; RNA).

duplication A type of CHROMOSOME ABERRATION in which part of a chromosome is duplicated.

dwarfism, genetic The commonest form of heritable dwarfism is ACHONDROPLASIA, which is inherited as an AUTOSOMAL DOMINANT TRAIT with complete PENETRANCE and a frequency of 1 : 26 000. The condition is characterized by a reduction in size of limbs with a normal-sized head and trunk and distinctive facial features. Affected individuals have normal life span and IQ. Genetic dwarfism can also be caused by the rarer disease mucopolysaccharidosis type IV (Morquio syndrome), an AUTOSOMAL RECESSIVE heritable defect in carbohydrate metabolism (specifically, a deficiency of either galactose-6-sulphatase or β-galactosidase). The symptoms of the disease include a shortened trunk with normal-sized limbs, normal IQ and facies, and death in the third decade. A number of other rare recessive genetic diseases are also associated with dwarfism; examples include BLOOM'S SYNDROME and chondroectodermal dysplasia.

dyad symmetry, diad symmetry Rotational SYMMETRY about a two-fold axis.

dynamin A nucleotide-sensitive microtubule-binding protein of M_r 100 000; its GTPase activity is stimulated in the presence of MICROTUBULES. *In vitro*, it cross-links microtubules into bundles and, in the presence of GTP, generates intermicrotubule sliding. Dynamin is the product of the *shibire* gene in *Drosophila melanogaster*, which is needed for synaptic vesicle recycling, and is homologous with the *VPS1* locus in *Saccharomyces cerevisiae*, which seems to be involved in membrane protein sorting, and with the mammalian Mx proteins, which are involved in interferon-induced viral resistance.

Shpetner, H.S. & Vallee, R.B. (1992) *Nature* **355**, 733–735.

dynein Large motor molecule associated with MICROTUBULES. It was originally discovered as the protein that constitutes the cross-bridges between doublet tubules in flagellar and ciliary AXONEMES, but a more soluble form has been found in cytoplasm. Most kinds apparently move only towards the minus end of microtubules (retrograde movement in AXONAL TRANSPORT). Dynein differs most clearly from KINESIN-like proteins in having a very large (M_r >200 000) motor domain, which is susceptible to cleavage by a combination of ultraviolet irradiation and bound vanadate ions. Each 400K polypeptide, with its motor domain in the middle, forms a large globular structure with one or two visible projections. An N-terminal segment of the polypeptide probably acts as a stem or tail, which allows dynein to be bunched into bouquet-like molecules with two or three globular heads, depending on the form of dynein. Associated 70–90K chains, plus some light chains, are part of the stem complex. *See also*: MICROTUBULE-BASED MOTORS.

dynorphins *See*: OPIOID PEPTIDES AND THEIR RECEPTORS.

dysbetalipoproteinaemia *See*: PLASMA LIPOPROTEINS AND THEIR RECEPTORS.

dysgammaglobulinaemia A condition more commonly referred to as selective immunoglobulin deficiency in which there is a low level or absence of one or more, but not all, of the ANTIBODY classes. Isolated IgA deficiency is the most common form with an incidence of 1 : 300 among normal blood donors. A majority of IgA-deficient individuals are healthy, but recurrent sinusitis, otitis media and an increased incidence of asthma and other allergies occur. In humans, selective IgG subclass deficiency is also associated with recurrent respiratory tract and chest infections.

dystrophia myotonia (DM) *See*: MYOTONIC DYSTROPHY.

dystrophin A large (M_r 427 000) protein, abundant in smooth and skeletal MUSCLE and localized to the inner surface of the sarcolemma membrane in skeletal muscle, and whose absence or disruption results in Duchenne and Becker muscular dystrophy (*see* DUCHENNE MUSCULAR DYSTROPHY). It consists of four main domains — an N-terminal domain which binds actin, a central rod-like domain containing repeat units, and which is thought responsible for dystrophin self-association; a cysteine-rich domain, and a C-terminal domain. A complex of glycoproteins binds to dystrophin in part of the cysteine-rich region and adjacent C-terminal domain, and may link dystrophin with the extracellular matrix and the cytoskeleton.

dystrophin-associated glycoproteins (DAG) A complex of glyco-proteins associated with DYSTROPHIN in the sarcolemma of skeletal muscle cells. It includes cytoskeletal, transmembrane, and extracellular glycoproteins, and is much reduced in DUCHENNE MUSCULAR DYSTROPHY. The extracellular component (M_r 156 000) binds laminin and may thus provide a linkage with the extracellular matrix (*see* EXTRACELLULAR MATRIX MOLECULES).

Ibraghimova-Beskrovnaya, O. et al. (1992) *Nature* **355**, 696–702.

dystrophin-related protein (DRP, utrophin) Nonallelic homologue of DYSTROPHIN expressed in muscle and other cells. Analysis of the cDNA sequence indicates a protein of M_r 395 000 and 3433 amino acids long. The gene maps to human chromosome 6.

E

E The charged AMINO ACID glutamic acid.

E face of lipid bilayer In FREEZE-FRACTURED membranes, the face representing the hydrophobic interior of the outer half of the lipid bilayer.

E rosette The rosetting of XENOGENEIC erythrocytes around lymphocytes has been used to identify and separate B CELLS and T CELLS. Sheep red blood cells (SRBC) bind avidly to human T cells so that under appropriate conditions a single lymphocyte can be covered by perhaps 12 red cells. E rosettes have lowered buoyant density compared to normal lymphocytes and they can, therefore, be separated by centrifugation on a discontinuous gradient such as Ficoll. The SRBC receptor of human T cells is now known to be the CD2 antigen (*see* CD ANTIGENS).

E-site (exit site) The site at which the nascent polypeptide chain leaves the RIBOSOME. *See also*: PROTEIN SYNTHESIS.

EAA Excitatory amino acids. Usually acidic amino acids such as glutamate and aspartate which can activate specific receptors in the central nervous system and act as excitatory NEUROTRANSMITTERS. *See*: EXCITATORY AMINO ACID RECEPTORS; SYNAPTIC TRANSMISSION.

early genes *See*: EARLY REGION; LAMBDA; PAPOVAVIRUSES.

early light-induced protein (ELIP) A group of CHLOROPLAST proteins that are synthesized very early during greening of plants. Both proteins and RNA are degraded before chloroplast development is complete. They show considerable sequence similarity with the CHLOROPHYLL *a/b* PROTEINS and may be involved in the integration of pigments (CHLOROPHYLLS and CAROTENOIDS) into mature CHLOROPHYLL–PROTEIN COMPLEXES.

early region Region of a viral or bacteriophage genome which encodes 'early gene' products which are essential for establishment of infection (e.g inhibitors of bacterial restriction enzymes and cellular protein synthesis). Early proteins also control the expression of the virus-encoded LATE GENES. *See e.g.*: LAMBDA; PAPOVAVIRUSES.

EBNA Epstein–Barr virus nuclear antigen. *See*: EPSTEIN–BARR VIRUS.

EBV EPSTEIN–BARR VIRUS.

EC cells EMBRYONAL CARCINOMA CELLS.

ecdysone, ecdysterone The steroid moulting hormone ecdysterone is found in insects and crustaceans and is produced by activation of the prohormone, ecdysone. The molecular biology of its action has been studied in detail in *Drosophila* where pulses of ecdysterone (20-hydroxyecdysone) control growth and trigger larval moults, pupal development, and METAMORPHOSIS. The ecdysone response provides a model system for the temporal regulation of a programme of gene expression. Studies of ecdysone-induced salivary gland chromosome puffing indicate that binding of ecdysone by its receptor induces a set of early PUFFS, and represses intermoult and late puffs. The protein products of the early puffs are proposed to act as regulators that direct the sequential activation of the late puffs and also repress their own expression, thus switching off the response. The ecdysterone receptor (EcR) gene has been identified and encodes a member of the STEROID RECEPTOR SUPERFAMILY. Three early puff genes have been characterized and all encode DNA-binding proteins of various classes: the *Broad* complex encodes zinc-finger proteins, the *E74* puff products contain ETS DOMAINS and the *E75* proteins are members of the steroid receptor superfamily. The primary transcripts of these genes are unusually long (60–100 kb), and the time required for TRANSCRIPTION may contribute to the timing of gene expression in the regulatory hierarchy.

echinoderm development Sea urchin embryos have been studied by developmental biologists since the late 19th century, when the classical experiments of Driesch (1891) established that small but anatomically normal larvae could be produced from embryo fragments. Latterly they have also been used extensively in molecular studies.

The eggs average some 100 μm in diameter, and after fertilization the first two cleavages are vertical and the third is equatorial. In the fourth cleavage, the four cells at the vegetal pole cleave unequally (generating four micromeres and four macromeres) while the four at the animal pole cleave equally to produce eight mesomeres. Subsequently a BLASTOCOEL develops and STEREOCILIA form at the animal pole. Following invagination of the micromeres (to form the precursor of the skeleton) the remaining vegetal cells gastrulate, and the embryo elongates at all extremes to form the pluteus larva, with oral arms, mouth and anus, and skeletal rods. The adult animal finally develops after metamorphosis.

The FATE MAP, defining, for example, the fates of the micromeres, mesomeres, and macromeres, was established by Hörstadius in the 1930s, using vital staining of single blastomeres, and has been studied more recently with intracellular dyes. The animal has also been used in many studies of the process of REGULATION and DETERMINATION during embryonic development. All of the first four blastomeres, for example, can usually develop small plutei in isolation, but after this stage, and as regional specification takes place, separation along the animal–vegetal axis often produces defective larvae. The micromeres can also induce a secondary ARCHENTERON following implantation to novel sites in the vegetal hemisphere, and can induce a secondary axis if placed at the animal pole, and may therefore be the source of a graded signal during normal development. In a series of manipulations, Hörstadius showed that CELL–CELL INTERACTIONS are important for maintaining the normal distribution of cell fates along the animal–vegetal axis. Removal of vegetal cells causes animal cells to alter their normal fates and replace the missing cells. Molecular studies have shown that a high proportion of the mRNAs in the early embryo are maternally derived, being present in the oocyte (*see* MATERNAL CONTROL OF DEVELOPMENT). By GASTRULATION the number of different mRNA species has fallen from some 11 000 to around 8500, of which 10% are newly expressing ZYGOTIC GENES. Recent studies have also examined the regulation of expression of specific genes during early development, for example those coding for ectoderm-specific markers, using analysis of the expression of REPORTER GENE constructs following injection into eggs. It seems likely that the sea urchin will continue to provide an excellent model system for the study of development at the molecular level. It also continues to provide a favoured system for the study of gastrulation; recent experiments have shown, for example, that isolated micromeres in culture are able to perform their morphogenetic functions cell autonomously.

ectoderm One of the three GERM LAYERS of the early embryo. In early development of higher vertebrates it gives rise to cells of the other germ layers (embryonic ENDODERM and MESODERM). After GASTRULATION the ectoderm can give rise to the skin (epidermis) and to the nervous system, depending upon its position in the embryo and whether or not it has been subjected to the influence of NEURAL INDUCTION by AXIAL MESODERM.

Edman degradation *See*: PROTEIN SEQUENCING.

EDS EHLERS–DANLOS SYNDROME.

EDTA Ethylene diaminetetraacetate, a metal chelating agent commonly used as an enzyme inhibitor. It is also used to weaken bacterial cell walls preparatory to extraction of DNA. It chelates magnesium ions that are essential to maintenance of cell wall integrity.

Edward's syndrome *See*: TRISOMY 18.

EF ELONGATION FACTOR. *See also*: ACTIN-BINDING PROTEINS; GTP-BINDING PROTEINS; PROTEIN SYNTHESIS.

EF-hand A molecular fold common to a protein superfamily of CALCIUM-BINDING PROTEINS.

effector cells In the immune system the terminally differentiated B cell-derived PLASMA CELLS, which produce ANTIBODIES, and

the various functional classes of T CELLS. *See*: CYTOTOXIC T CELLS; HELPER T CELLS; SUPPRESSOR T CELLS; $\gamma\delta$ T CELLS.

EGF Epidermal growth factor. *See*: GROWTH FACTORS.

EGF receptor Epidermal growth factor receptor, the product of the cellular ONCOGENE *c-erbB1*. *See*: ACTIN-BINDING PROTEINS; GROWTH FACTOR RECEPTORS.

egg maturation *See*: OOGENESIS.

egg–sperm interactions *See*: FERTILIZATION; SPERM–EGG INTERACTIONS.

EGTA Ethylene glycol bis (β-aminoethyl ether)-*N,N,N′,N′*-tetraacetate, a chelating agent that binds Ca^{2+} specifically and with high affinity and which is used to reduce the extra-cellular concentration of free Ca^{2+}.

Ehlers–Danlos syndrome (EDS) A heterogeneous group of disorders (I to XI) sharing clinical features of abnormal velvety skin which heals poorly, hyperextensibility of the joints and lax ligaments. Different forms have additional features such as spontaneous or prolonged bleeding, rupture of large blood vessels, ocular abnormalities, and kyphoscoliosis. Where defined, the genetic defect is in the synthesis of various types of collagen. EDS type VII is caused by a defective gene for collagen type I; EDS type IV by errors in the α_1 chain gene of collagen type III. The distribution of features can be partly explained on the basis of the tissue distribution of different collagens. The frequency overall is one in 10 000 births and depending on the type may be inherited as a RECESSIVE, DOMINANT, or X-LINKED trait. *See*: EXTRACELLULAR MATRIX MOLECULES.

Ehrlich ascites *See*: ASCITES.

eicosanoids A large family of local chemical mediators, including the prostaglandins, thromboxanes, hydroperoxy fatty acids, leukotrienes, and lipoxins (Fig. E1), based on a 20-carbon fatty acid skeleton derived from ARACHIDONIC ACID and other 20-carbon polyunsaturated fatty acids. Depending on the tissue, eicosanoid synthesis is stimulated by local damage, hormonal stimuli, and cellular activation (e.g. of mast cells by IgE binding to FC RECEPTORS). Eicosanoids bind to specific cell-surface receptors to produce multifarious effects in a wide variety of tissues.

Prostaglandins can stimulate inflammation, inhibit or stimulate ADENYLATE CYCLASE, modulate SYNAPTIC TRANSMISSION, stimulate contraction of smooth muscle in some organs (hence their effects as abortifacients), act as vasodilators, and regulate membrane ion transport. There are nine main classes of prostaglandins (PGA–I), of which PGE_2 and $PGF_{2\alpha}$ are most commonly encountered. The subscript numeral denotes the number of C=C double bonds outside the cyclopentane ring. PGI_2 is also known as prostacyclin. The effect of aspirin (acetylsalicylate) as an anti-inflammatory agent is due to its inactivation of the enzyme prostaglandin synthase which catalyses the conversion of arachidonate to prostaglandin. Aspirin inhibits the cyclo-oxygenase

Fig. E1 *a*, Prostaglandin E$_2$; *b*, Thromboxane A$_2$; *c*, Leukotriene C$_4$; *d*, Lipoxin A.

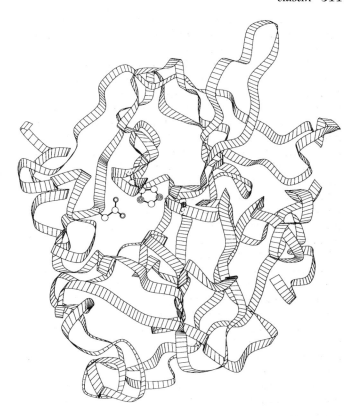

Fig. E2 Elastase.

activity of the enzyme, which forms prostaglandin G$_2$ from arachidonate, by acetylating the N-terminal amino group of the catalytic subunit.

Thromboxanes, produced mainly by platelets, are synthesized by further modification of prostaglandin H$_2$ by thromboxane synthetase. TXA$_2$ and its derivative TXB$_2$ are the only thromboxanes found in appreciable amounts. As in prostaglandins, the subscript denotes the number of double bonds outside the (six-membered) ring. Thromboxanes are involved in blood clotting through their ability to activate platelets.

Leukotrienes (LTA–E) are produced from arachidonic acid via the lipoxygenase pathway. Hydroperoxy fatty acids produced by lipoxygenase action on arachidonic acid are further metabolized to produce leukotrienes. LTC$_4$, LTD$_4$, and LTE$_4$ constitute the 'slow-reacting substance of anaphylaxis' and have been implicated as the main agents responsible for bronchoconstriction in allergic asthma.

Lipoxins are produced via the lipoxygenase pathway. They can cause contraction of smooth muscle, vasodilation, chemotaxis, hyperfiltration in the kidney, and inhibition of natural killer cell activity.

Eigen limit The upper limit on the size of a replicating RNA molecule as determined by the error rate of replication.

Elaeis guineensis Oil palm.

ELAM-1 Endothelial leukocyte adhesion molecule. *See*: SELECTINS.

elastase SERINE PROTEINASE (EC 3.4.21.11) (Fig. E2) which hydrolyses the EXTRACELLULAR MATRIX MOLECULE elastin (among other proteins). It is formed from an inactive precursor proelastase by specific cleavage and preferentially cuts a polypeptide chain at bonds involving the carbonyl groups of amino acids bearing uncharged nonaromatic side chains. It is specifically inhibited by the SERPIN α$_1$-antitrypsin. Elastase released from white blood cells is largely responsible for the damage to lung tissue seen in states of α$_1$-antitrypsin deficiency, which leads to emphysema.

elastase inhibitor α$_1$-Antitrypsin. *See*: SERPINS.

elastin *See*: EXTRACELLULAR MATRIX MOLECULES.

electric organ Modification of muscular or epithelial tissue found in certain fishes, such as the electric eel *Torpedo californica*, which discharges electrical energy. It is made up of plate-like structures (electroplax) the membranes of which are rich in receptors and ion channels involved in neurotransmission. *See*: NICOTINIC RECEPTORS.

electrical coupling, electrical synapse *See*: CELL JUNCTIONS; SYNAPTIC TRANSMISSION.

electroblotting The use of an electric field to transfer polypeptides (WESTERN BLOTTING), RNA (NORTHERN BLOTTING) or DNA (SOUTHERN BLOTTING) electrophoretically from a polyacrylamide or agarose gel onto an appropriate membrane. This method has the advantage of being faster and more efficient than CAPILLARY BLOTTING.

electron cryomicroscopy *See*: CRYOELECTRON MICROSCOPY.

electron crystallography The determination of the two-dimensional or three-dimensional structure of a macromolecule or macromolecular assembly by observation in the electron microscope and ELECTRON DIFFRACTION, followed by image analysis and reconstruction using methods derived from crystallography, principally FOURIER ANALYSIS. It can be applied to viruses, crystalline arrays of membrane proteins (*see* BACTERIORHODOPSIN), and large protein assemblies. The need for image analysis arises principally from the fact that in the transmission electron microscope image of, for example, a negatively stained virus particle, all features along the direction of view are superimposed, making details of structure difficult to interpret. Preparation and staining procedures and radiation damage also introduce artefacts. Image analysis corrects for the operating conditions of the microscope and separates the contributions to the image from different levels of the specimen.

Image analysis of negatively stained specimens, as in the early work on virus particles, can give resolutions down to 15 Å. For higher resolution, unstained specimens are required. Frozen hydrated specimens are now widely used (*see* CRYOELECTRON MICROSCOPY) and contrast enhanced by recording the image with the objective lens underfocused (phase contrast electron microscopy). To minimize specimen damage at high resolution a low dose of electrons is used.

Techniques of image analysis and reconstruction depend on enhancing the signal-to-noise ratio in the image, caused by individual variations between molecules, by averaging over a large number of copies of the molecule or particle, and/or by comparing large numbers of particles in different orientations.

Three-dimensional image reconstruction is based on the projection theorem of crystallography. A three-dimensional Fourier transform of an object can be built up section by section using two-dimensional transforms of images of the object from different views. Different orientations are obtained by tilting the microscope stage and/or by averaging across a regular array of particles in different but identifiable orientations. Electron diffraction patterns provide the amplitudes needed to compute the three-dimensional map in finer detail. The Fourier transform of the proposed three-dimensional map is compared with the electron diffraction patterns and the Fourier coefficients manipulated and corrected to approximate the observed amplitudes, and it is then transformed back into a reconstructed image. *See also*: ELECTRON MICROSCOPY.

Klug, A. (1983) 1982 Nobel Prize Lecture (The Nobel Foundation, Stockholm).
Henderson, R. & Unwin, P.N.T. (1975) *Nature* **257**, 28–32.
Unwin, P.N.T. (1993) *J. Mol. Biol.* **229**, 1101–1124.

electron density This is the atomic property responsible for the scattering of X-rays by molecules. Thus when X-rays strike a crystal, the resulting diffraction pattern is calculable from the known position and identity of all its constituent atoms, using standard tables for the spatial distribution of the electrons for each atomic species. When diffraction is recorded from a crystal of unknown structure, the primary result of an X-RAY CRYSTALLOGRAPHY analysis is an ELECTRON DENSITY MAP.

electron density maps An electron density map is the primary result of an X-ray crystallographic analysis, being calculated from the observed DIFFRACTION intensities and the phases, which may be derived in various ways — from related structures or from multiple sets of diffraction data obtained with appropriately different experimental arrangements. The map may be calculated directly or by a DIFFERENCE FOURIER SYNTHESIS. Its RESOLUTION, or level of fine detail, depends directly upon the extent of the diffraction data included in its calculation. Electron density maps are usually presented as three-dimensional contoured plots, often displayed on specialized molecular graphics hardware display systems, allowing real-time rotation and visualization of the density. Interpretation of an electron density map is necessary to reveal the crystal structure under investigation. Interpretation involves recognition of specific features (e.g. continuity along the course of a polypeptide backbone), assignment of these to particular structural elements, and assessment of the expected validity of these choices. This process is increasingly facilitated by the tools of computer software packages, but in general it is not yet completely automated. *See:* X-RAY CRYSTALLOGRAPHY.

electron diffraction Electron microscopes with appropriate settings of the microscope lenses and apertures can record DIFFRACTION patterns as well as images. Study of the resolution of an electron diffraction pattern gives a quick assessment of the extent of preservation of specimen detail and of the microscope performance. For a periodic sheet of molecules, the electron diffraction pattern is sampled in two dimensions. Combination of these diffraction amplitudes with their phases, which can be determined by image analysis of the corresponding micrograph, yields an image of the molecular structure in projection. Determination of a three-dimensional structure requires the further collection of a tilt series of corresponding images and diffraction patterns, sampling the continuous diffraction streaks — a much larger task, with far more stringent requirements on both specimen and instrument.

electron microscopy (EM) Electron microscopy allows the visualization of cell and macromolecular structure down to 1 nm or even less in ideal conditions, in comparison with the most powerful light microscopes, which have a point-to-point resolution of 0.1 μm. Two main types of electron microscopy are used by biologists: transmission electron microscopy (TEM) for viewing tissue sections and macromolecules; and scanning electron microscopy (SEM) for examining three-dimensional structure of surfaces. Electron microscopy has provided images of phage and virus particles, and of the ultrastructure of cells and their organelles, and has enabled the processes of transcription, replication, and recombination to be visualized. Of importance to molecular biology are the recently developed techniques of CRYOELECTRON MICROSCOPY (also known as electron cryomicroscopy) and IMMUNOELECTRON MICROSCOPY.

Electron microscopy gives greater RESOLUTION than light microscopy owing to the shorter wavelengths of electrons, which decrease with speed (0.4 Å at an accelerating voltage of 1000 V, 0.004 Å at 100 000 V). The actual limits on resolution are set by the limitations of electron magnetic lenses which have a much smaller numerical aperture than an optical lens, and problems with specimen preparation, contrast, and damage. More recently, the development of scanning tunnelling microscopy (STM) is beginning to enable the visualization of, for example, DNA structure on the atomic scale.

The first commercial transmission electron microscope was produced by Siemens in 1939. High-voltage electron microscopes (1 000 000 V), introduced in the late 1970s, enabled the viewing of thicker specimens and are of use in the three-dimensional imaging of tissues and macromolecules (*see* IMAGE ANALYSIS).

The transmission electron microscope follows the same general principles as a light microscope, with the electrons replacing photons. Electrons generated by a heated filament are accelerated and concentrated into a fine beam by passage through a small hole; they are then focused through a magnetic 'lens' onto the specimen. The image produced by electron scattering by the specimen is captured on a fluorescent screen or a photographic plate. Electron-dense areas in the specimen scatter electrons, which are lost from the beam, and thus show up as darker areas in the final image. The EM must operate in a high vacuum as electrons are scattered in air, so living specimens cannot be examined. Magnetic lenses also have great depth of focus, and this, together with the very weak penetrating power of electrons, means that only very thin sections of tissues can be examined. High energy electron beams can enable thicker tissues to be examined, but can cause excessive heating and specimen damage.

Tissues for TEM are prepared as semithin or ultrathin sections of a thickness of 50–100 nm. The tissue to be examined is first fixed (with glutaraldehyde or osmium tetroxide), dehydrated, and embedded in epoxide resin, and sections are then cut on an ultramicrotome. Sections are floated off onto a fine wire grid and stained with an electron-dense stain (e.g. uranyl acetate followed by lead) to enhance contrast. Different cellular components stain differently with a particular stain; lipids for example stain darkly with osmium tetroxide, which reveals cell membranes.

Shadowing techniques enhance the outline of macromolecules such as DNA, RNA, and proteins, and of virus particles. In directional shadowing the specimen is coated with a metal (gold, palladium, platinum, tantalum or tungsten) evaporated from a point source at an angle to the specimen so that the metal preferentially condenses on the high points or on one face of the sample surface. Electron microscopy of DNA–RNA hybrids (*see* R-LOOP MAPPING), for example, was used to confirm the discovery of introns in eukaryotic genes. Shadowing is also used for FREEZE-FRACTURE preparations and to visualize organelles. Rotary shadowing is used to increase the bulk of macromolecules such as DNA and RNA. Relaxed DNA is spread on a grid and attached to a platform which is turned rapidly while the metal is evaporated onto the surface of the specimen. The surfaces of cells and cellular structures or the interiors of membranes exposed by freeze-fracture can be examined by TEM after making a thin metal replica of the surface. The specimen is shadowed with metal, a reinforcing film of electron-transparent carbon added and the tissue dissolved away.

Negative staining is used in particular to examine virus morphology. The specimen is surrounded with a heavy metal stain (uranyl acetate, phosphotungstic acid, ammonium molybdate, or methylamine tungstate) which provides a dark background against which the particle stands out. The stain also penetrates open irregularities or crevices in the particle surface showing up features such as capsomer arrangement (*see* Fig. P36 in PLANT VIRUSES).

Scanning electron microscopy enables the surface of cells and cellular structures to be examined. It was introduced commercially in 1965. The surface of the specimen, coated with a heavy metal (usually gold or gold palladium), is systematically scanned by a narrow beam of electrons. The quantity of electrons scattered from the surface is monitored and used to control a second beam which builds up an image of the surface on a screen. Resolutions of ~ 10 nm are usual, giving an effective magnification of up to 20 000. Although resolution is not as high as for TEM, the great depth of focus of SEM provides striking and informative three-dimensional views of the surfaces of many sorts of biological structures, from individual cells to developing organs.

electron paramagnetic resonance *See*: ESR/EPR SPECTROSCOPY.

electron spin resonance *See*: ESR/EPR SPECTROSCOPY.

Electron transport chain

IN aerobic microorganisms and eukaryotic cells, ATP (adenosine triphosphate), the chief store of chemical energy, is mainly produced by OXIDATIVE PHOSPHORYLATION, in which ATP is synthesized from ADP with the concomitant oxidation of a reduced substrate derived from the breakdown of sugars and fatty acids.

The energy for ATP synthesis comes from the flow of electrons from the reduced substrate along an electron transport chain, a series of electron carriers embedded in a membrane (cell membrane in prokaryotes, the inner mitochondrial membrane in eukaryotic cells; *see* MITOCHONDRIA). The electron transport chain catalyses the transfer of electrons from reduced substrates

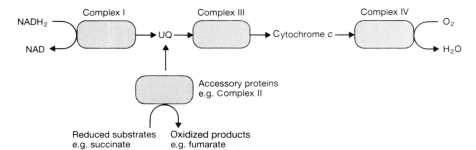

Fig. E3 Electron transport chain of mammalian mitochondria.

to molecular oxygen, resulting in oxidation of the substrate and reduction of oxygen to water. The energy available from this flow of electrons from negative to positive REDOX POTENTIAL is used to drive the transport of protons across the membrane so setting up a proton electrochemical potential difference across the membrane, often known as the PROTONMOTIVE FORCE (Δp). This in turn drives protons back through an ATP SYNTHASE embedded in the membrane, with formation of ATP as described by the CHEMIOSMOTIC THEORY.

Components of the electron transport chain

The electron transport chain of mammalian mitochondria consists of ubiquinone and four major proteins (complex I, complex III, cytochrome *c*, and complex IV), together with a number of accessory proteins (Fig. E3). All the proteins have PROSTHETIC GROUPS that are able to accept and donate electrons or hydrogen atoms; all except cytochrome *c* are hydrophobic proteins spanning the mitochondrial inner membrane.

The full chain oxidizes NADH, which is produced in the mitochondrial matrix by the dehydrogenases of the TRICARBOXYLIC ACID CYCLE or by the pathway for β-OXIDATION of fatty acids. The accessory proteins allow electrons from less strongly reducing sources to be fed into the chain at ubiquinone before passing along the rest of the chain to oxygen. There are many enzymes that catalyse reduction of ubiquinone; all contain a flavin adenine dinucleotide (FAD) prosthetic group. The most important examples are succinate dehydrogenase and acyl-CoA dehydrogenase, but electrons are also fed in by enzymes such as glycerol phosphate dehydrogenase and choline dehydrogenase. There are minor inputs of electrons to cytochrome *c*, via sulphite oxidase and cytochrome b_5 in the intermembrane space.

The electron transport chain in mitochondria from nonmammalian sources is very similar, but there are sometimes differences in the accessory proteins. For example, many types of plant mitochondria have an alternative oxidase that takes electrons from ubiquinone and passes them to oxygen without the involvement of complex III or complex IV. The electron transport chains of prokaryotes are similar in principle to those of eukaryotes, with quinones and protein complexes that are homologous or analogous to the eukaryotic ones. However, there is a much wider range of sources of electrons and of terminal electron acceptors. Depending on the growth conditions, *Escherichia coli* for example can contain lactate or formate dehydrogenases that feed electrons to ubiquinone or menaquinone, and cytochrome *o*, cytochrome *d*, nitrate reductase or fumarate reductase to pass the electrons to terminal acceptors. The photosynthetic electron transport chains of plants and microorganisms are closely related to the mitochondrial chains, and have many homologous components (*see* PHOTOSYNTHESIS). For example, the CYTOCHROME *bf* complex of chloroplasts has extensive structural and sequence homology with complex III. Other electron transport chains, such as the CYTOCHROME P450 involved in detoxification, are analogous to the mitochondrial electron transport chain, but do not pump protons.

Figure E4 shows the approximate sizes and shapes of the protein components of the electron transport chain. The structures of the large complexes have been determined at low resolution by image analysis of electron micrographs and by X-RAY CRYSTALLOGRAPHY of two-dimensional crystalline arrays of the proteins. Higher resolution structures are just beginning to be obtained.

Ubiquinone is the only nonprotein member of the chain. The quinone headgroup can be reversibly reduced to quinol, and the long hydrophobic lipid tail ensures that the molecule lies deep

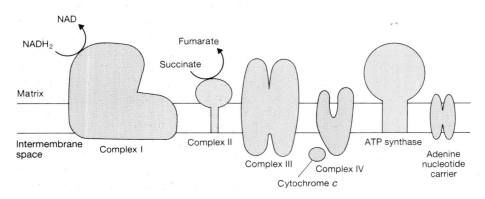

Fig. E4 Approximate shapes and sizes of the proteins involved in oxidative phosphorylation.

within the membrane. It collects reducing equivalents from complex I and from the accessory enzymes such as complex II, and diffuses in the plane of the membrane until it encounters complex III. There it passes on its reducing equivalents and is involved in the Q-cycle, the proton-pumping mechanism of complex III.

CYTOCHROME c is a small soluble protein (M_r 10 000–15 000) with a covalently attached HAEM group which can accept or donate electrons. It has been very extensively studied; the primary sequence from more than 100 species is known, and high-resolution crystal structures of both oxidized and reduced forms are available. In mitochondria its main role is to catalyse electron transport between the larger and more slowly diffusing membrane-bound complex III and complex IV. The electron transfer reaction has been investigated in detail using the reactions with cytochrome b_5 and with cytochrome c peroxidase.

Complex I (NADH-ubiquinone reductase) is the least well understood of the three major protein complexes of the electron transport chain. During the oxidoreduction reaction it catalyses the transport of four protons across the inner membrane per electron pair transferred (Fig. E5), but the mechanism by which it does this is unknown. The enzyme consists of about 25 subunits, of which seven are encoded in the MITOCHONDRIAL GENOME in mammals. It has a monomer M_r of perhaps 850 000. There is good evidence that it is a dimer when purified; it is less clear whether it is a dimer in its natural environment. The initial electron acceptor is the flavin mononucleotide (FMN) prosthetic group. There are 22 to 24 iron atoms contained in eight or nine iron–sulphur centres; three tetranuclear (4Fe–4S) and five or six binuclear (2Fe–2S). Altogether there are six distinct signals from these centres detectable by EPR spectroscopy (*see* ESR/EPR SPECTROSCOPY). The Fe–S centres are involved in transport of electrons from the flavin to the site at which ubiquinone is reduced, and may also be important in the proton translocation reaction. Rotenone, the classic inhibitor of complex I, acts by binding to the quinone-binding site, so preventing electron transport through the complex to ubiquinone.

Complex III, also known as **ubiquinol-cytochrome c reductase** or the **cytochrome bc_1 complex**, is better understood than complex I. The mammalian enzyme has five large subunits and six smaller ones, with a combined monomer M_r of ~275 000. It is thought to operate as a dimer. One of the large subunits, cytochrome b, is mitochondrially encoded. Complex III is structurally and functionally related to the cytochrome b_6f complex of chloroplasts (*see* PHOTOSYNTHESIS). It catalyses the electrogenic translocation of two protons across the mitochondrial inner membrane per electron pair; an additional two protons are released to the cytosol during the overall reaction of oxidation of the hydrogen carrier ubiquinol by the electron carrier cytochrome c. Proton translocation is widely accepted to be by a Q-cycle mechanism, in which ubiquinol is sequentially oxidized and reduced by cytochrome b. Ubiquinol is first oxidized to the semiquinone at a binding site on the outer, cytosolic side of the membrane (centre o), with the electron transferred to an unusual Fe–S centre (known as the Rieske centre) and then to cytochrome c_1 and cytochrome c. The semiquinone is oxidized by cytochrome b_{566}, which passes the electron across the membrane to b_{562}, a

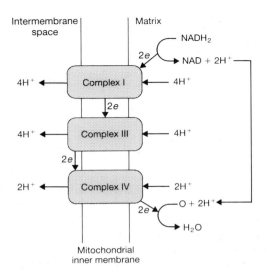

Fig. E5 Proton pumping by the mitochondrial electron transport chain.

second b-type haem on the same polypeptide. At a binding site on the inner surface of the membrane (centre i) the electron is used to reduce ubiquinone; after two such reductions the ubiquinol produced may rejoin the ubiquinone pool in the membrane and may be oxidized at centre o. This cyclic mechanism results in proton translocation as electrons pass from ubiquinol to cytochrome c. The classic inhibitor of complex III is antimycin, which binds to the ubiquinone-binding site in centre i and prevents electron transport through the complex. Myxothiazol is an even better inhibitor of the enzyme; it prevents ubiquinone binding at centre o.

Complex IV cytochrome c oxidase catalyses electron transport from cytochrome c on the cytosolic face of the membrane to molecular oxygen, effectively on the matrix side. This electrogenic movement of two charges across the membrane is accompanied by translocation of an additional two protons per electron pair. Together with the net uptake of two protons per electron pair on the inside when water is formed, this results in the disappearance of four protons from the matrix, the translocation of four charges across the membrane, and the appearance of two protons in the cytosol. The mechanism of the proton translocation remains unknown, although several models have been proposed. The enzyme has 12 or 13 subunits, and a monomer M_r of ~160 000. Once again, it is thought to operate as a dimer. The three largest subunits are mitochondrially encoded. Its low-resolution structure was elucidated before the other complexes, and is the best resolved. Electrons from cytochrome c are accepted by cytochrome a and a copper atom, Cu_A. They are passed across the membrane to the binuclear centre, consisting of cytochrome a_3 and Cu_B in very close association, where oxygen is reduced. Several of the intermediates formed during the catalytic cycle of oxygen reduction at the binuclear centre have been observed and identified, and there is agreement on the outlines of the catalytic mechanism of oxygen reduction. The enzyme also contains a third Cu, with a structural role. The classic inhibitors of complex

IV — cyanide, azide and carbon monoxide — bind to the haem of cytochrome a_3 and prevent binding of oxygen.

M.D. BRAND

1 Cramer, W.A. & Knaff, D.B. (1991) *Energy Transduction in Biological Membranes* (Springer-Verlag, New York).
2 *Curr. Topics Bioenerget.* **15** (1987).

Electrophoresis

ELECTROPHORESIS is the separation of substances achieved by applying an electric field to samples in solution. In its simplest forms it depends on the different velocities with which the substances move in the field.

Electrophoretic mobility in solution

Although electrophoresis is rarely performed in free solution (as opposed to within gels), consideration of the process in solution helps in understanding the mechanism of gel electrophoresis, so is considered first.

Factors determining mobility

The force on the molecule in the field is given by the field strength (i.e. the potential gradient) multiplied by the charge of the molecule. Since the particle migrates without accelerating, this force must be balanced by that due to viscous drag, which is proportional to the velocity of the particle. Hence the velocity is proportional both to the charge and to the electric field.

Substances are separated because their molecules differ both in charge and in their resistance to movement through the medium. The second of these is most simply expressed by the force required to give them unit velocity.

The charge that must be used for such a calculation is the charge of the particle that moves. Thus, with a small peptide, it will be the net charge of the molecule, derived from the ionized groups it contains. With a protein, however, we have to distinguish two types of ionized groups, using lysine residues as an example. Some may be on the surface of the molecule, and their ammonio groups, $-NH_3^+$, contribute to the net charge. Their counter ions, distributed in the solvent around them in the manner postulated by Debye and Hückel, are free to move in the opposite direction. But if such a residue, although just as external in the sense that it is freely available to solvent, is in a crevice in the molecular surface, its counter ions may be close enough so that they too are in the crevice and so are constrained to move with the particle. Hence this residue makes no contribution to the net force on the particle. A spectacular demonstration of this effect is the equality of the mobilities of tobacco mosaic virus and the rods formed by its protein. Removal of the highly charged RNA does not affect the mobility, as it is replaced by anions from the buffer, and these must move with the rods.

For molecules of similar shape the resistance to motion is proportional to their surface area, since it is over this surface that viscous drag occurs. This in turn is proportional to the two-thirds power of their volume. Thus Offord [1] found that a large number of small peptides showed mobilities proportional both to the net charge and to $M_r^{-2/3}$. For such small molecules an n-mer should therefore migrate faster than its monomer by a factor of $n^{1/3}$.

Maintenance of uniform velocity

If all the molecules of one substance are to have the same velocity, so that bands remain sharp, the electric field must be uniform. The electric current is the same across any section cut between the electrodes, so the field can only be uniform if the conductivity is uniform throughout. Since the substances being separated themselves contribute to conductivity, this condition cannot be met; it can be approached by keeping the conductivity of the medium high compared with any contribution from the substances in the sample. This has a major drawback, that the higher the conductivity for a given field, the higher the heat generated in the solution. This gives many difficulties: loss of uniformity as thermal gradients are set up, mixing of the solution by convection, and damage to heat-sensitive substances in the sample. It is for this reason that electrophoresis is predominantly an analytical method of separation; its preparative use is only convenient for very small quantities, since with large amounts of material high conductivities are required to prevent zones spreading, and hence large amounts of heat must be dispersed. Laminar flow at right angles to the electric field has sometimes been applied for preparative use.

Since electrical and thermal conductivities are correlated, it is hard to remove heat while maintaining the necessary electrical insulation. When potentials of some kilovolts are used, for example in paper electrophoresis of peptides, two methods predominate. One is to immerse the paper in a liquid hydrocarbon (white spirit), itself cooled in turn. Convection in this liquid removes the heat, although the paper is always surrounded by insulating substances. Only substances negligibly extracted by the hydrocarbon can be run. The other method is to press the paper between cooled metal plates, but separated from them by a thin sheet of plastic. This sheet sufficiently combines thermal conductivity with electrical insulation.

Prevention of convection

The classical approach to preventing convection was boundary electrophoresis. The sample was in a U-tube, and filled its bottom section up to the vertical parts. Sharp boundaries were made between the sample and the buffer solution above it. As the proteins moved at different velocities, each boundary broke up into a number of boundaries, equal to the number of components (Fig. E6). Each boundary was stabilized by gravity, since there was one extra component below it, and therefore a higher density. The process was conducted at 4°C, so that density changes with temperature were minimal; the sample and cell were ribbon shaped, so that cooling was efficient. Optical analysis of the boundaries, based on the changes of refractive index at each, allowed precise analysis of the composition of a mixture.

The drawbacks of this approach are obvious. Complex apparatus is required, and a considerable time is taken in setting up and

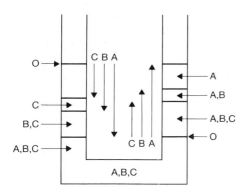

Fig. E6 Diagram of boundary electrophoresis of a mixture of three proteins. The proteins are called A, B, and C, in order of diminishing mobility. O is the origin, where the boundary is made before electrophoresis, with buffer above it and a solution of A, B, and C below. The arrows mark the distance migrated by each protein, and hence the distance moved by the boundary below which this protein is found.

running each sample. The sensitivity is low. Preparative use is impossible, since the zones of different substances do not separate, but only the boundaries that mark the limits of the zones.

When zones of different proteins are to be separated, mixing of the medium must be prevented. The liquid may be stabilized by wetting a solid medium (often paper or a thin layer), by containing a density gradient of an inert substance, for example a sugar, or by being set in a gel. Gel electrophoresis is probably the most valuable of the approaches, but it brings in new factors, which are dealt with below.

Gel electrophoresis

By running the electrophoresis in a gel, convection is abolished, and hence one cause of band broadening. The chains of substance that form the gel slow the migration of molecules, and do so progressively more as the molecular size increases. Hence molecular size is a predominant, and in some uses effectively the only, determinant of relative mobility.

Protein electrophoresis in dodecyl sulphate

This technique combines electrophoresis in a gel with the presence of a detergent (usually sodium dodecyl sulphate, SDS). The medium is a buffer solution, the gel usually polyacrylamide — hence the terms PAGE (polyacrylamide gel electrophoresis) and SDS-PAGE. Polyacrylamide is obtained by polymerizing acrylamide to give $-[CH_2-CH(-CO-NH_2)-]_n-$ chains. These are cross-linked by including some methylenebis(acrylamide), $(CH_2=CH-CO-NH)_2CH_2$, in the acrylamide solution, since on polymerization one molecule can get incorporated into two chains. The detergent dodecyl sulphate, $CH_3-[CH_2]_{11}-O-SO_3{}^-$, plasters the protein molecules, giving them an almost constant charge/mass ratio, overcoming their original differences in charge. It unfolds them to give a rod-like structure by mutual repulsion of charges. To ensure completion of this denaturation, the sample is heated with the detergent beforehand, in the presence of a thiol to destroy any DISULPHIDE BONDS. Under these conditions the relative mobility is a function of M_r, falling almost linearly as $\log(M_r)$ rises, and is little affected by other features of the molecules. Hence estimates of M_r can be obtained by comparison with known proteins. Sensitive methods of detection, often by one of many staining methods that can be quantified, allow this technique to be used to assess the relative amounts of different proteins and hence as a criterion of purity.

Protein electrophoresis without detergent

If it is necessary to correlate a labile, biological activity with a particular band in gel electrophoresis, the separation must be made in nondenaturing conditions. Mobility is then a function of both charge and size. It is found, however, that this method can be used to determine the size of the molecule, that is, with any subunits still associated. This is done by measuring the dependence of mobility on gel concentration; the larger the molecule the greater the fall in mobility as the gel concentration increases.

Electrophoresis of nucleic acids

Separations of nucleic acids by size form the basis of countless methods of molecular biology (*see e.g.* DNA TYPING; DNA SEQUENCING; RESTRICTION ANALYSIS; RESTRICTION MAPPING). Sequencing methods depend on the separation of a whole series of molecules, each differing from the next by addition of a single nucleotide. These procedures resemble the separation of proteins in the presence of dodecyl sulphate, in that each molecule has the same density of charge per unit length. For the mobility/length dependence to be typical all molecules must be linear, so cyclic forms must be broken and the formation of secondary structure inhibited by high urea concentrations. The cyclic form of a double-stranded DNA can also be separated into bands according to the degree of supercoiling (*see e.g.* PLASMIDS).

Fragments of nucleic acids up to some hundreds of nucleotide residues are conveniently separated in polyacrylamide gels. For much larger nucleic acid molecules the gels are usually of the algal polysaccharide agarose (consisting of alternating units of 3,6-anhydro-L-galactose, glycosylated on O-4, and of D-galactose, glycosylated on O-3, both pyranose). The agarose must be carefully purified from acidic polysaccharides that occur with it. It can form gels with enough rigidity even when the gel network is loose enough to be permeable to fairly large molecules.

For very large molecules of nucleic acid, such as complete chromosomes, the technique of pulsed-field gel electrophoresis (PFGE) has been highly successful. In conventional electrophoresis, all DNA molecules over a certain size move at the same rate, because it is only 'end-on' that they can pass through the gel. With the pulsed field, the molecules are forced to change their direction of migration by periodic changes in the direction of the applied field. Typically, it is applied for one pulse at 45° or more to the direction in which the nucleic acids are to travel, and the next pulse at the same angle to the other side of this direction (Fig. E7). The larger molecules take more time to adopt a conformation and orientation in which they can migrate in a new direction, and so move more slowly. There are many variants,

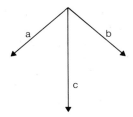

Fig. E7 Pulsed-field gel electrophoresis. Arrows a and b mark the directions of the field applied in each pulse, and hence the directions in which the nucleic acids move during the pulse. With pulses in each direction of equal duration, arrow c marks the direction of net movement of the molecules.

differing in pulse length, angle, etc. DNA molecules of up to 10^7 base pairs have been separated by this technique. Markers containing a series of DNAs, whose sizes are multiples of a single value, can be obtained by linear concatenation of well-characterized nucleic acids of relatively small size.

Sample addition and stacking

If sharp bands are to be obtained, not only must all identical molecules have the same velocity, but they must start at the same position. Hence the sample must be applied as a narrow zone. Originally slits in gels were made for sample addition, but it proved much more convenient to add samples to wells made at the top of a vertical gel, which is held between two glass plates. An inert substance, such as glycerol, is added to the mixture to raise its density above that of the buffer in the electrode vessel. Hence a sample of a few microlitres can be pipetted through this buffer into the well and it sinks onto the gel surface.

Although this is enough to get sharp samples for nucleic acids, say, further principles can be used in addition for proteins. The first is the use of discontinuous systems. For example, the gel may contain a buffer of Tris, the primary amine (HO-CH$_2$-)$_3$C-NH$_2$, and its hydrochloride. It is at a fairly high pH, >8.5, and as the pK of Tris is just over 8, only a small fraction of it is neutralized with HCl. The buffer in the electrode vessel, however, is of a lower pH, below 7, the Tris being largely neutralized by glycine. At this pH the ratio of the glycine zwitterion to glycinate anion is large, so the electrophoretic mobility of the glycinate is low, since the molecule spends only a small fraction of time carrying a net charge. When the field is applied, the chloride ions move into the gel. They cannot leave a gap between themselves and the glycinate, because their tendency to do so readjusts the potential gradient. Hence a sharp chloride–glycinate boundary moves into the gel. The complex of protein with dodecyl sulphate has intermediate mobility, and so concentrates at this boundary. Thus protein molecules left behind by local inhomogeneities catch up with this front and are concentrated. When the boundary enters the gel it comes to a region of higher pH, so the glycine–glycinate equilibrium readjusts, giving this ion greater mobility than the proteins. They therefore separate by the process described above, now travelling in a uniform Tris–glycine mixture of higher pH.

A further process can be added to improve the concentration of the sample into a narrow band. This is to have a stacking gel, a band of gel at the top with a low gel concentration. This gives a region in which the stacking can occur, without fear that the proteins will start to separate by size. It is at the lower pH of the electrode vessel, so that the glycine 'trailing ion' moves more slowly than the protein complexes, but it initially contains the chloride 'leading ion' as the anion. Hence the chloride–glycinate boundary passes through it, and the protein concentrates at this boundary. This boundary then reaches the higher pH and higher gel concentration of the 'running gel' for the separation to start.

Associated procedures

The sophistication of detection methods is a subject on its own, using techniques such as blots from the gel, immobilizing the nucleic acids on such blots, and probing them for desired sequences in many ways (*see e.g.* SOUTHERN BLOTTING). The increase in sensitivity and specificity of such methods has contributed as much to the usefulness of electrophoresis as the changes in the technique itself. Immunoelectrophoresis, for example, is often classed as a method on its own, allowing a gel to be used for precipitation of the materials it contains with ANTIBODIES after an electrophoretic separation.

Two-dimensional procedures allow a gel to be run first in one direction, and then under different conditions, at right angles, thus enabling the separation of large numbers of different proteins from a mixture. The method of two-dimensional gel electrophoresis using isoelectric focusing (see below) in one dimension followed by SDS-PAGE at right angles to it, devised by P.H. O'Farrell in 1977, has been used to estimate the total number of proteins synthesized by the bacterium *Escherichia coli* [2].

Related techniques

Isoelectric focusing

In this technique substances are separated in a pH gradient. They migrate, in either direction, until they come to the point on the gradient at which they have zero mobility — the isoelectric point.

The gradient is set up and maintained by applying an electric field to a solution of ampholytes (substances carrying both basic and acidic groups). The mixture contains substances with a range in the numbers of such groups, and their solution is placed between a cathode in strong base and an anode in strong acid. The ampholytes will thus be positively charged near the anode, and negatively charged near the cathode, and will be repelled by these electrodes and move away from them. They set up a pH gradient by migrating to the point where each is isoelectric.

An example of the power of the method is the separation of two forms of cattle growth hormone, of 191 and 190 residues, Ala-Phe-R and Phe-R, respectively, where the slight difference in pK of the terminal amino group was enough to change the isoelectric point by 0.33 [3].

Just as with electrophoresis, isoelectric focusing may be performed in gels or in density gradients. Similarly, laminar flow at right angles to the field can be used to obtain preparative separation; this is better with isoelectric focusing, because the substances approach a final position in the gradient, so that recycling improves the separation.

Isotachophoresis

As described above for stacking, conditions can be chosen so that a boundary, there between chloride and glycinate, is formed, and the substances in the sample, in that case a protein–detergent complex, concentrated at this boundary. Other systems can be set up where a series of such compounds like the glycine and chloride are chosen, so that many boundaries are present, all moving with the same velocity. The different substances in the sample concentrate at different boundaries and thus are separated.

Investigation of proteolytic digests

In 1977 Cleveland et al. [4] devised an ingenious method for characterizing proteins not only by molecular size, but also, in a second gel, by the sizes of specific peptides produced with specific peptidases. It elegantly combines several techniques. The proteins are first separated in a gel containing dodecyl sulphate. The blocks of gel containing the band or bands of protein under study, visualized by brief staining, are added to the wells of the second gel. This has the form described above, with a particularly deep stacking gel above the resolving gel. Buffer containing 20% glycerol is added round the gel fragments in the well, and further buffer, containing 10% glycerol and the peptidase to be used, is layered on top of this, below the buffer of the electrode vessel. Finally, the electric field is applied. The glycinate–chloride boundary enters the stacking gel, and both the protein, as it migrates out of the slices of the first gel, and the peptidase stack at this boundary, possibly with one slightly behind the other. The electric field is switched off before the boundary reaches the resolving gel. Since the peptidase and protein are so close, slight diffusion allows the proteolysis to occur. The field is reapplied after time for proteolysis, and the peptides produced separate when they enter the resolving gel.

H.B.F. DIXON

See also: CHROMATOGRAPHY.

1 Offord, R.E. (1966) Electrophoretic mobilities of peptides on paper and their use in the determination of amide groups. *Nature* **211**, 591–593.
2 Celis, J.E. & Bravo, R. (Eds) (1983) *Two-dimensional gel electrophoresis of proteins: methods and applications.* (Academic Press, New York).
3 Wallis, M. (1973) The pK_a values of the α-amino groups of peptides derived from the N-terminus of bovine growth hormone. *Biochim. Biophys. Acta* **310**, 388–397.
4 Cleveland, D.W. et al. (1977) Peptide mapping by limited proteolysis in sodium dodecyl sulphate and analysis by gel electrophoresis. *J. Biol. Chem.* **252**, 1102–1106.

electrophoretic mobility *See:* ELECTROPHORESIS.

electroplax *See:* ELECTRIC ORGAN.

electroporation A method of genetic TRANSFORMATION of animal cells and plant and fungal protoplasts, in which a mixture of the cells to be TRANSFECTED and the DNA which is to be introduced is subjected to a brief electrical pulse. This results in the formation of transient pores in the cell membranes which allow the DNA to pass through before they are repaired. Electroporation can be used to transform a wide range of cell types and can also be used to introduce macromolecules other than DNA (e.g. RNA, antibodies).

elicitor Compound that induces plant defence responses. In most cases elicitors have been characterized in terms of their ability to induce PHYTOALEXINS, although elicitors have also been isolated by their ability to induce other responses (the HYPERSENSITIVE RESPONSE, lignification). An elicitor may be derived from a pathogen (a biotic elicitor) or may be abiotic (e.g. some heavy metals such as Hg^{2+}). The term has been used to describe pure and crude preparations. Elicitors may be specific (inducing a response only on hosts carrying the appropriate *R* genes) or nonspecific (inducing a response regardless of the host's complement of *R* genes). Elicitor activity has been found in protein, lipid, and carbohydrate fractions of fungal and bacterial preparations. Elicitor molecules have also been isolated from plants themselves and termed 'endogenous elicitors'. *See also:* OLIGOSACCHARINS; PLANT PATHOLOGY.

ELIP EARLY LIGHT-INDUCED PROTEIN.

ELISA Enzyme-linked immunosorbent sandwich assay.

elliptocytosis The presence of oval or elliptical red cells in a peripheral blood film. The commonest cause is hereditary elliptocytosis which is inherited as an AUTOSOMAL DOMINANT TRAIT. The HOMOZYGOUS state causes a severe and sometimes fatal haemolytic anaemia in infancy. The HETEROZYGOUS state is much more variable in severity and the pattern seen tends to run in families. Elliptocytes may be observed occasionally in different forms of THALASSAEMIA and in iron-deficiency anaemia.

elongation factors Ancillary proteins required for polypeptide chain elongation during PROTEIN SYNTHESIS. Elongation factors undergo a cycle of association and dissociation from the RIBOSOME at the addition of each amino acid. In bacteria there are three elongation factors: EF-Tu (Fig. E8) which reacts with GTP and aminoacyl-tRNA (aa-tRNA) to form an aa-tRNA : EF-Tu : GTP complex before tRNA binding to the ribosome; EF-Ts which displaces GDP from EF-Tu by itself forming a complex with EF-Tu; and EF-G which mediates the movement of a peptidyl-tRNA from the A site to the P site in the ribosome. In eukaryotic cells there are two principal elongation factors: EF-1α (analogous to bacterial EF-Tu) and EF-2 (analogous to bacterial EF-G). In yeasts and other fungi a third elongation factor, EF-3, has been described whose functional role in protein synthesis remains to be defined. *See also:* GTP-BINDING PROTEINS.

embryo transfer Technique in which an early embryo, usually at the 2–8 cell stage, derived from *in vitro* fertilization is introduced into the oviduct or uterus of a recipient by the transcervical route.

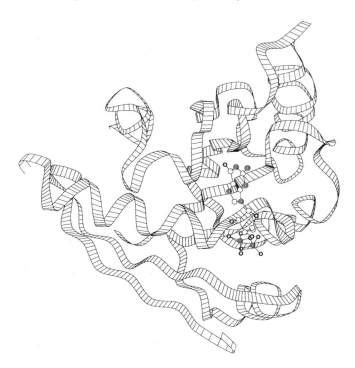

Fig. E8 Elongation factor EF-Tu.

embryonal carcinoma cells (EC cells) The STEM CELLS of TERA-TOCARCINOMAS, with molecular and developmental properties similar to the INNER CELL MASS. They can differentiate into a wide variety of adult tissue types, such as gut epithelium, muscle, cartilage, and bone. EC cells have been used experimentally to analyse cell DIFFERENTIATION and developmental gene expression. According to the EC cell line used, for example, treatment with retinoic acid (*see* RETINOIDS) may induce differentiation along the ENDODERM pathway, or may activate the expression of Hox genes *see* HOMEOBOX GENES AND HOMEODOMAIN PROTEINS in a concentration-dependent manner.

embryonic haemoglobin *See*: HAEMOGLOBIN AND ITS DISORDERS.

embryonic lethal Mutation that causes death at the embryo stage.

embryonic stem cells (ES cells) PLURIPOTENT cells of normal karyotype derived from the mouse INNER CELL MASS or BLASTO-CYST. When grown in culture on feeder layers of irradiated cells they proliferate, and when implanted into mice of identical genetic constitution they form tumours (TERATOCARCINOMAS) comprising a variety of differentiated tissue types. When injected into blastocysts, ES cells will contribute to the embryos, generating CHIMAERAS. When this contribution includes the GERM CELLS, intact mice can be bred from the original cells (*see* TRANSGENIC TECHNOLOGIES). Thus, the effects of targeted gene disruption (GENE KNOCKOUT) can be examined by first disrupting the gene in question in the ES cells, and then breeding intact mice from them. ES cells also offer the possibility of establishing permanent tissue

culture lines from embryos homozygous for genes with potentially important roles during normal development.

emphysema *See*: α₁-ANTITRYPSIN; SERPINS.

en The *Drosophila* gene ENGRAILED.

En Enhancer-inhibitor, a plant transposable element. *See*: TRANS-POSABLE GENETIC ELEMENTS: PLANTS.

3′ end The free end of a polynucleotide chain that bears a nucleotide with a 3′-OH group not linked to another nucleotide (*see* DNA; RNA). When used of genes, it conventionally denotes the end of the gene at which transcription terminates, which corresponds to the 3′ end of the RNA transcript (*see* TRANSCRIPTION). *Cf.* 5′ END.

5′ end The free end of a polynucleotide chain with a nucleotide bearing a 5′-OH or 5′-P group that is not linked to another nucleotide (*see* DNA; RNA). When used of genes, it conventionally denotes the end of the gene at which transcription begins, which corresponds to the 5′ end of the RNA transcript (*see* TRANSCRIP-TION). *Cf.* 3′ END.

end-labelling The addition of a radioactive or other label to one of the ends of a nucleic acid. DNA can be 5′ end-labelled using the enzyme T4 POLYNUCLEOTIDE KINASE which catalyses the transfer of the radioactively labelled γ phosphate of a deoxyribo-nucleotide donor (usually [γ-³²P]ATP) to the 5′ hydroxyl group of a polynucleotide, oligonucleotide, or nucleoside-3′-triphosphate. The end-labelled nucleic acid can be used as a radioactive PROBE for NORTHERN BLOTTING or SOUTHERN BLOTTING or for screening a DNA LIBRARY.

end-plate potential (EPP, e.p.p.) A rapid transient DEPOLARIZA-TION of a skeletal MUSCLE fibre membrane which is caused by ACETYLCHOLINE being released from motor neuron terminals at the MOTOR END-PLATE. The EPP can only be recorded in the vicinity of the motor end-plate because, unlike ACTION POTEN-TIALS, they do not propagate and their underlying current dissipates along the cell membrane. They are mimicked by acetylcholine applied only in the region of the end-plate: the acetylcholine acts on NICOTINIC RECEPTORS which gate cation channels that are predominantly permeable to sodium. The EPP has a rapid time course with a maximum duration of tens of milliseconds, and an amplitude that is graded according to the strength of the stimulus, and therefore the amount of transmitter released, of a few millivolts. If their amplitude is sufficient the threshold for generation of action potentials may be reached, and they will give rise to action potentials and subsequent mechanical contraction. *See also*: SYNAPTIC TRANSMISSION.

endocrine Applied to the secretion of hormones into the blood-stream from specialized glandular tissue (e.g. adrenaline and corticosteroids from the adrenals, insulin from pancreatic islet cells). Endocrine signalling typically can operate over quite large distances in the body.

endocytic pathway Collective term for organelles involved in ENDOCYTOSIS; includes COATED PITS AND VESICLES, ENDOSOMES, MULTIVESICULAR BODIES, pre-lysosomes, LYSOSOMES, secondary lysosomes, and phagosomes.

Endocytosis

ENDOCYTOSIS is the process by which eukaryotic cells internalize extracellular fluid, macromolecules, and particles into membrane-bounded vesicles. It involves invagination of the plasma membrane, followed by the release of free vesicles into the cytoplasm, and can occur by several biochemically distinct mechanisms [1]. It is believed to be essential for cell viability and is required for numerous functions including: the acquisition of essential nutrients [2] (*see e.g.* PLASMA LIPOPROTEINS AND THEIR RECEPTORS); the clearance of unnecessary, damaged (e.g. desialylated serum glycoproteins) or potentially harmful components (e.g. secreted lysosomal hydrolases or proteases) from the extracellular medium; the processing and presentation of MHC class II-dependent antigens [3] (*see* ANTIGEN PROCESSING AND PRESENTATION); the TRANSCYTOSIS of molecules across epithelia [4]; cellular responses to GROWTH FACTORS and CYTOKINES [5]; and the regulation of cell surface antigen expression. In addition, endocytosis is utilized by a range of pathogens to gain entry to the cell; examples include viruses (e.g. SEMLIKI FOREST VIRUS and INFLUENZA VIRUS), and bacteria (e.g. *Legionella pneumophila*) and protozoa (e.g. *Toxoplasma gondii* and *Leishmania donovani*). Certain plant and bacterial toxins (e.g. RICIN and DIPHTHERIA TOXIN) also enter the cell by endocytosis [1].

The different forms of endocytosis can be classified as follows.

Constitutive endocytosis

This occurs in virtually all nucleated eukaryotic cells and is responsible for:

1 Fluid phase endocytosis (pinocytosis) — the nonspecific uptake of extracellular fluid and solutes.

2 Receptor-mediated endocytosis — the highly efficient uptake of macromolecular and small particulate ligands that bind to specific cell surface receptors (e.g. low density lipoprotein (LDL), transferrin, epidermal growth factor (EGF), Semliki Forest virus) [1,2]. Receptor-mediated endocytosis can occur through the continual internalization of receptors, such as the LDL and transferrin receptors, that endocytose both in the presence and absence of ligand. Alternatively, receptors may be induced to internalize by binding ligand (e.g. EGF receptor, platelet-derived growth factor (PDGF) receptor).

In a number of cell types, fluid phase and receptor-mediated endocytosis are mediated primarily through clathrin-coated vesicles [2] (*see* COATED PITS AND VESICLES) (Fig. E9). Clathrin-coated vesicles form continually through the cell cycle (except during MITOSIS when membrane transport in general is shut down) independently of the binding of specific ligands. Endocytic coated vesicles have an internal diameter of about 100 nm (internal volume $\sim 5 \times 10^{-4}\ \mu m^3$, membrane area $\sim 0.03\ \mu m^2$) and, in cells such as baby hamster kidney cells, around 1000 form from the plasma membrane each minute. The fluid content of each vesicle is derived nonspecifically from the media surrounding the cells. By contrast, the membrane, which is derived from the plasma membrane, can be enriched in certain specific plasma membrane components (e.g. the LDL receptor and transferrin receptor) which are able to interact with components of the clathrin coat [6].

Coated vesicle formation is ATP dependent, and is not known to involve cytoskeletal components. Thus neither fluid phase endocytosis nor receptor-mediated endocytosis are significantly influenced by agents that disrupt the actin or microtubule-based CYTOSKELETONS.

Following internalization, the clathrin coat is removed and the vesicles fuse with early ENDOSOMES. Early endosomes are the first station on the endocytic pathway for sorting endocytosed membrane and ligands to different sites in the cell. Much of the membrane and fluid content is returned, or recycled, to the cell surface [2,4].

The receptors and ligands internalized by receptor-mediated endocytosis can undergo various fates [2,4]. Certain receptors,

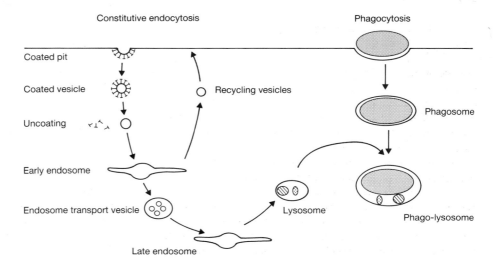

Fig. E9 The basic pathways of endo- and phagocytosis.

exemplified by the LDL receptor, dissociate from their ligand and recycle to the plasma membrane to mediate further rounds of endocytosis. LDL particles remain in the cell and are transported to the LYSOSOMES for degradation. Similarly, iron is removed from transferrin and is absorbed into the cell, while transferrin receptors and apotransferrin recycle to the cell surface. By contrast, EGF receptors and EGF do not recycle, but are directed to lysosomes and degraded. In epithelial cells, ligands and receptors involved in transcytosis are sorted into transcytotic vesicles [4] (*see* TRANSCYTOSIS).

The mechanisms involved in sorting specific receptors to these different pathways are unclear but involve, at least in part, the acidification of endosomes and the aggregation of individual membrane components. Thus the low pH in endosomes promotes the dissociation of LDL from the LDL receptor and the release of iron from diferric transferrin, whereas ligand-induced cross-linking of receptors can result in their being sorted to lysosomes.

The transfer of membrane components and ligands through the different compartments of the endocytic pathway is believed to occur through vesicular intermediates. However, the details of these processes remain to be established.

Fluid phase and receptor-mediated endocytosis are often treated as independent processes. But although the kinetics of internalization of fluid phase markers and receptor-mediated markers are very different, the vesicular mechanisms and the intracellular itinerary involved in their uptake are the same. Vesicles lacking a morphologically identifiable coat and distinct from clathrin-coated vesicles have been proposed to mediate fluid phase and receptor-mediated endocytosis in certain cells [7]. In contrast to clathrin-coated vesicles, these noncoated vesicles have not been characterized and their role in endocytosis remains to be clarified.

Phagocytosis

This is a ligand-induced process responsible for the uptake of particulate ligands in excess of 150–200 nm in diameter, for example bacteria and cell debris [1]. Phagocytosis is usually the property of specialized phagocytic cells such as macrophages, amoeboid protozoa, and neutrophils. However, many cells are capable of phagocytic activity when challenged with appropriately OPSONIZED ligands capable of binding to the cell surface.

The mechanism of phagocytosis involves the interaction of molecular components on the surface of the particle (e.g. an ANTIBODY to a bacterial coat protein) with receptors (e.g. FC-RECEPTORS) on the cell surface. The process is induced by the particle and is receptor dependent. However, the mechanism of internalization is distinct from the constitutive receptor-mediated endocytosis described above. The progressive interaction of the receptors with the opsonin (a process referred to as 'zippering' [8]), coupled to the polymerization of ACTIN FILAMENTS on the cytoplasmic side of the membrane adjacent to the bound particle, leads to the particle being drawn into the cytoplasm within a tightly apposed membrane invagination. The invagination eventually pinches off to form a free cytoplasmic vesicle termed a PHAGOSOME (Fig. E9). The process is ATP dependent and, in contrast to fluid phase and receptor-mediated endocytosis (see

above), is inhibited by reagents which inhibit microfilament polymerization (e.g. cytochalasin D). The phagosome may subsequently fuse with lysosomes to form a PHAGOLYSOSOME in which the internalized material is degraded.

Other types of endocytosis

Additional forms of endocytosis, such as macropinocytosis, can be observed under specific circumstances. For example, treatment of certain cells with growth factors or phorbol esters induces a transient increase in fluid phase endocytosis [9]. Similarly, macropinocytosis can be observed in cells microinjected with $p21^{ras}$ [10]. The increased uptake is due to endocytic events distinct from constitutive fluid phase endocytosis, occurs through the formation of large (>100 nm diameter) vesicles, without any morphologically visible cytoplasmic coat, and is sensitive to the inhibitor amiloride. Macropinocytosis is often associated with membrane ruffling, and frequently occurs at the leading edges of motile cells (*see* CELL MOTILITY) as membrane ruffles collapse back onto the cells and enclose a vesicle.

M. MARSH

See also: ABSORPTIVE ENDOCYTOSIS; EPITHELIAL POLARITY; EXO-CYTOSIS; MEMBRANE RECYCLING.

1 Steinman, R.M. et al. (1983) Endocytosis and the recycling of plasma membrane. *J. Cell Biol.* **96**, 1–27.
2 Goldstein, J.L. et al. (1985) Receptor mediated endocytosis: concepts emerging from the LDL receptor system. *Annu. Rev. Cell Biol.* **1**, 1–39.
3 Mellman, I. (1990) Endocytosis and antigen processing. *Semin. Immunol.* **2**, 229–237.
4 Rodman, J.S. et al. (1990) Endocytosis and transcytosis. *Curr. Opinion Cell Biol.* **2**, 664–672.
5 Backer, J.M. et al. (1990) Receptor-mediated internalization of insulin requires a 12 amino acid sequence in the juxtamembrane region of the insulin receptor β-subunit. *J. Biol. Chem.* **265**, 16450–16454.
6 Trowbridge, I.S. (1991) Endocytosis and signals for internalization. *Curr. Opinion Cell Biol.* **3**, 634–641.
7 van Deurs, B. et al. (1989) The ways of endocytosis. *Int. Rev. Cytol.* **117**, 131–177.
8 Griffin, F.M. Jr et al. (1975) Studies on the mechanism of phagocytosis. 1. Requirements for circumferential attachment of particle-bound ligands to specific receptors on the macrophage plasma membrane. *J. Exp. Med.* **142**, 1263–1282.
9 West, M.A. et al. (1989) Distinct endocytic pathways in epidermal growth factor-stimulated human carcinoma A431 cells. *J. Cell Biol.* **109**, 2731–2739.
10 Bar-Sagi, D. & Feramisco, J.R. (1986) Induction of membrane ruffling and fluid-phase pinocytosis in quiescent fibroblasts by ras proteins. *Science* **233**, 1061–1068.

---

endodeoxyribonuclease A NUCLEASE that acts preferentially on DNA, cleaving the chain internally.

endoderm One of the three GERM LAYERS of the early embryo. In vertebrates and other chordates, it gives rise to the digestive system and some of its associated organs. It should be distinguished from the HYPOBLAST of amniote embryos, which only contributes to EXTRAEMBRYONIC TISSUES.

endomembrane system The plasma membrane and the internal membranes of the eukaryotic cell (excluding those of mitochondria and chloroplasts) which are either physically continuous or are linked by transport vesicles that carry membrane from one to another. The system is generally held to comprise the ENDOPLASMIC RETICULUM, the GOLGI APPARATUS, LYSOSOMES, ENDOSOMES, and the plasma membrane. *See*: COATED PITS AND VESICLES; ENDOCYTOSIS; EXOCYTOSIS; GLYCOPROTEINS AND GLYCOSYLATION; PROTEIN SECRETION; PROTEIN TARGETING; PROTEIN TRANSLOCATION.

endoneurium The interstitial connective tissue in a peripheral nerve which separates individual nerve fibres. It is composed of longitudinally directed collagenous fibres, and an occasional fibroblast. Close contact exists between the collagen and the basement membrane of SCHWANN CELLS.

endonexin *See*: CALCIUM-BINDING PROTEINS.

endonuclease Enzyme that cleaves phosphodiester bonds within a nucleic acid chain (*see* NUCLEIC ACID STRUCTURE). *See*: ENDODEOXYRIBONUCLEASE; ENDORIBONUCLEASE; NUCLEASE; RESTRICTION ENZYME; RIBONUCLEASE; RIBOZYME.

endopeptidase A proteolytic enzyme which hydrolyses peptide bonds located within a POLYPEPTIDE CHAIN rather than the bonds linking the terminal amino-acid residues (*cf.* AMINOPEPTIDASE; CARBOXYPEPTIDASE). *See*: PROTEINASES.

endoplasmic reticulum (ER) In eukaryotic cells, an extensive internal membrane system forming an interconnected system of channels and vesicles (cisternae) and continuous with the NUCLEAR ENVELOPE. It is the site of lipid synthesis, the site of new membrane synthesis, and is involved in the synthesis and sorting of proteins that are destined for export from the CYTOSOL (*see* GOLGI APPARATUS; LIPID SYNTHESIS; LYSOSOMES; MEMBRANE SYNTHESIS; PROTEIN SECRETION; PROTEIN TARGETING; PROTEIN TRANSLOCATION). Most cells, especially those actively secreting protein, contain extensive regions of rough endoplasmic reticulum (RER). Under the electron microscope these regions appear to be 'studded' by ribosomes attached to the cytosolic face of the ER membrane. These ribosomes are engaged in the synthesis of membrane or secretory proteins. The initial stages of protein glycosylation take place within the ER (*see* GLYCANS; GLYCOPROTEINS AND GLYCOSYLATION). Cells specialized for lipid metabolism contain large areas of smooth endoplasmic reticulum (SER). SER lacks ribosomes on its cytosolic face and tends to be tubular rather than sheet-like. It may be continuous with the RER. This SER membrane is the site of lipid synthesis and metabolism (e.g. the production of steroid hormones from cholesterol in Leydig cells), and in hepatocytes, for example, contains the CYTOCHROME P450 enzyme family responsible for the detoxification of harmful substances.

Proteins and membrane entering the ER system and destined for the plasma membrane or secretion are mainly transported onward via the TRANSITIONAL ENDOPLASMIC RETICULUM to the GOLGI APPARATUS where the further processing of the oligosaccharide side chains of glycoproteins occurs. Transport is in the form of small membrane-bound vesicles which pinch off the ER.

Enzymes and membrane proteins are also directed to LYSOSOMES via the ER. Much evidence shows that the ER also functions as a 'quality control' point. Misfolded proteins and incompletely oligomerized multisubunit proteins are retained in the ER, in some cases by BIP, but also by other mechanisms (*see* MOLECULAR CHAPERONES). It also contains the enzyme PROTEIN DISULPHIDE ISOMERASE which is involved in disulphide bond rearrangement during protein folding. *See also*: ANTIGEN PROCESSING AND PRESENTATION; KDEL; RETENTION SIGNAL; SALVAGE COMPARTMENT; VESICLE-MEDIATED TRANSPORT.

Hurtley, S. & Helenius, A. (1989) *Annu. Rev. Cell Biol.* **5**, 277–307.
Helenius, A. et al. (1992) *Trends Cell Biol.* **2**, 227–231.

endoribonuclease Enzyme that cleaves internal phosphodiester bonds in an RNA polymer. It can be single- or double-strand specific. *See*: NUCLEASE; RIBONUCLEASE; RIBOZYME.

endorphins *See*: OPIOID PEPTIDES AND THEIR RECEPTORS.

endosomal transport vesicles Putative vesicles involved in transport between early and late ENDOSOMES. The mechanisms involved in transporting endocytosed ligands from endosomes to LYSOSOMES are not clearly established. One model predicts that transit involves the formation of endosomal transport vesicles. Morphologically, structures believed to be responsible for this transport are frequently observed to be MULTIVESICULAR BODIES. *See*: ENDOCYTOSIS.

Griffiths, G. & Gruenberg, J. (1991) *Trends Cell Biol.* **1**, 5–9.
Murphy, R.F. (1991) *Trends Cell Biol.* **1**, 77–82.

endosome (CURL, receptosome) Organelle of the ENDOCYTIC PATHWAY intermediate between the PLASMA MEMBRANE and LYSOSOMES. Endosomes are the first compartment to which the ligands, membrane components and fluid internalized by fluid phase and receptor-mediated endocytosis are delivered following ENDOCYTOSIS from the cell surface. Endosomes perform a sorting function, returning the bulk of the membrane to the cell surface while retaining certain specific components for TRANSCYTOSIS or transport to lysosomes.

Morphological and cell fractionation experiments indicate that endosomes comprise two functionally distinct populations. Early endosomes are located in the cell periphery and are the first organelles to be labelled with endocytic markers. Late endosomes (also sometimes called pre-lysosomes) are labelled with endocytic tracers destined for lysosomes and only after these tracers have entered early endosomes. The late endosomes have a more perinuclear distribution and a higher buoyant density. Both early and late endosomes have an acidic internal pH, which is important in dissociating and sorting internalized ligands and receptors, the late endosomes being more acidic than early endosomes. Both compartments are morphologically complex, comprising a system of interconnected membrane-bounded tubules (~50 nm in diameter), cisternae and vesicles. Late endosomes frequently contain internal membrane arrays and have a high density of the cation-independent MANNOSE 6-PHOSPHATE RECEPTOR. Membrane flow through the system is considerable with the result that the organelles are highly plastic and may form an extended reticulum with elements continually undergoing fusion and fission.

Endosomes are the site for penetration of a number of viruses, for example SEMLIKI FOREST VIRUS and INFLUENZA VIRUS, both of which use the low pH endosomal environment to trigger their membrane fusion activity.

Helenius, A. et al. (1983) *Trends Biochem. Sci.* **8**, 245–250.
Griffiths, G. et al. (1989) *J. Cell Biol.* **109**, 2703–2720.
Hopkins, C.R. et al. (1990) *Nature* **346**, 335–339.
Griffiths, G. & Gruenberg, J. (1991) *Trends Cell Biol.* **1**, 5–9.

endosymbiont hypothesis The DNA-containing organelles of eukaryotic cells — MITOCHONDRIA and CHLOROPLASTS — are believed to have evolved from prokaryotic ancestors which developed an endosymbiotic relationship with a host cell. Through time much of the organelle genome has been transferred to the nucleus. The evidence supporting the endosymbiont hypothesis comes largely from the striking similarities between some of the enzymes catalysing oxidative phosphorylation in mitochondria and the corresponding proteins from certain aerobic bacteria, the similarities in size and antibiotic sensitivity of mitochondrial and bacterial ribosomes and extensive similarities in the DNA sequences of genes from both sources. *See also*: MITOCHONDRIAL GENOMES.

Margulis, L. (1970) *Origin of Eukaryotic Cells* (Yale University Press, New Haven).

endothelium (*pl.* endothelia) A histological term originally used to describe an epithelium derived from the ENDODERM (*see* GERM LAYERS). Now more commonly used to refer to the innermost cell layer lining the blood vessels, which is of MESODERMAL origin.

endotoxin A toxin produced by and remaining within a bacterial cell, which is only released into the host when the bacterium dies or is broken down.

δ-endotoxins A family of insecticidal protein toxins produced by the bacterium *Bacillus thuringiensis* during sporulation. The bacterial genes encoding these toxins have been incorporated into plants in an attempt to confer pest resistance (*see* PLANT GENETIC ENGINEERING). The active toxins (M_r 60 000–70 000) are active against insect larvae of the orders Lepidoptera, Diptera, and Coleoptera, with particular toxins being specific for each group. δ-Endotoxins are synthesized as protoxins and crystallize as a parasporal inclusion of ~1 μm and are ingested in this form by the larva. The protoxin is cleaved by proteinases in the insect gut to release the active toxin. δ-Endotoxins bind to specific high-affinity receptors on the brush border of the larval gut epithelium creating pores in the cell membrane. This leads to lysis of the gut epithelium and death of the larva from starvation and septicaemia. All the δ-endotoxins have five conserved regions of amino-acid sequence; target specificity is determined by the divergent sequences. The crystal structure of δ-endotoxin specific for Coleoptera has been obtained. It consists of three distinct structural domains: domain I is a seven-helix bundle which is presumed to be responsible for pore formation in the insect membrane; domain II is a triangular arrangement of three β sheets containing

the receptor-binding site; and domain III consists of a sandwich of two β sheets.

Li, J. et al. (1991) Crystal structure of insecticidal δ-endotoxin from *Bacillus thuringiensis* at 2.5 Å resolution. *Nature* **353**, 815–821.

energy of activation In enzyme-catalysed reactions this is usually the Gibbs energy of activation ($\Delta G^{\ddagger}$) which is equal to the difference in free energy between the TRANSITION STATE and the ground state. *See*: MECHANISMS OF ENZYME CATALYSIS.

engrailed (*en*) A HOMEOBOX GENE of *Drosophila*, a member of the segment polarity class of SEGMENTATION GENES. It has a role in the specification of pattern both in the embryonic repeat units, the PARASEGMENTS, and also in the adult segments. In the embryo it is expressed in 14 stripes corresponding to the 14 parasegmental repeat units. This expression pattern is established by the PAIR-RULE GENES (e.g. FUSHI-TARAZU and *even-skipped*) and is then maintained by cell–cell interactions which involve the secreted product of the *wingless* gene. The *engrailed* product acts as a TRANSCRIPTION FACTOR and its target genes probably include the HOMEOTIC GENES. The *engrailed* gene played an important part in the development of the COMPARTMENT hypothesis as *engrailed* mutations cause the transformation of cells in the posterior compartment of the wing into anterior structures and also cause cells to disrespect the anterior–posterior clonal restriction boundary.

enhancer A *cis*-acting regulatory sequence involved in the transcriptional activation of eukaryotic genes (*see* EUKARYOTIC GENE EXPRESSION; TRANSCRIPTION). Activation of an enhancer results in an up to 1000-fold increase in the basal rate of transcription. From studies on the SV40 viral enhancer, it has been shown that an enhancer can operate when located either 5′ or 3′ to the transcriptional start site, can function in either orientation, and can operate even when placed at a distance >3 kb from the transcriptional start site. The nucleotide sequences of different eukaryotic enhancers show very little homology but a core CONSENSUS SEQUENCE has been identified: 5′ GTGAAG 3′. Enhancers are generally binding sites for transcriptional activating proteins (*see* TRANSCRIPTION FACTORS) and are tissue specific. Enhancer sequences are frequently incorporated into mammalian EXPRESSION VECTORS to optimize expression in a chosen cell line. *See also*: HEAVY CHAIN ENHANCER; LYMPHOID-SPECIFIC ENHANCERS.

Enhancer–inhibitor (En) *See*: TRANSPOSABLE GENETIC ELEMENTS: PLANTS.

enkephalins *See*: OPIOID PEPTIDES AND THEIR RECEPTORS.

entactin *See*: EXTRACELLULAR MATRIX MOLECULES.

enteric nervous system *See*: PERIPHERAL NERVOUS SYSTEM.

enterochelin A bacterial SIDEROPHORE. *See*: BACTERIAL PATHOGENICITY.

enterocyte Intestinal epithelial cell.

enterotoxins Toxins produced by enteric bacteria, such as toxigenic strains of *Escherichia coli*. *See*: BACTERIAL PATHOGENICITY.

Enterovirus Genus of the PICORNAVIRIDAE containing polioviruses 1–3 and Coxsackie A virus. *See*: ANIMAL VIRUS DISEASE; ANIMAL VIRUSES.

Entwicklungsmechanik This term is now loosely translated as 'experimental embryology'. Wilhelm Roux (1850–1924) chose the term quite deliberately, to emphasize the mechanism of development over the contemporary idea of 'developmental physiology', and to allow him to include in the study the twin aspects of form and function. Roux founded one of the earliest journals in this field, now known as *Wilhelm Roux Archiv für Entwicklungsbiologie*. The emphasis on mechanism has continued to the present day in developmental biology: it is currently accepted that development is best studied by perturbing embryonic development. This approach contrasts with the descriptive approach favoured for much of the 19th century, which sought understanding through the morphological description of parts. While the mechanical approach has been extremely successful, modern embryologists tend to be familiar with the development of far fewer species than their 19th century equivalents.

enucleate Lacking a nucleus.

env Viral gene encoding the protein of the envelopes of RETROVIRUSES. *See also*: IMMUNODEFICIENCY VIRUSES.

envelope *See*: BACTERIAL ENVELOPE; NUCLEAR ENVELOPE; VIRAL ENVELOPE.

enzyme Enzymes are the biological catalysts that specifically catalyse chemical reactions in living cells. Because of their central importance in brewing and wine making and the accessibility of secreted enzymes in cultures of microorganisms and animal digestive juices, enzymes were among the first biological macromolecules to be studied chemically. The term 'enzyme' (from the Greek 'εν ζύμη, in yeast) was proposed by W. Kühne in 1878 to denote 'something that occurs that exerts this or that activity, which is considered to belong to the class called fermentative. The name is not, however, intended to be limited to the invertin [invertase] of yeast, but it is intended to imply that more complex organisms, from which the enzymes pepsin, trypsin, etc. can be obtained, are not so fundamentally different from the unicellular organisms as some people would have us believe'.

A more modern definition — 'a protein with catalytic properties due to its power of specific activation' — has had to be amended in recent years as a result of the discovery of catalytic RNAs (*see* RIBOZYMES).

The existence of specific biological catalysts was first recognized with the discovery of 'diastase' in malt extract (amylase, which converts starch to sugar) by Payen and Persoz in 1833, but there was subsequent controversy over whether enzymatic activity was due to a chemical substance (Liebig) or whether it was inseparable from living cells (Pasteur). The controversy was resolved when E. Büchner obtained the yeast fermentation system in a cell-free extract in 1897. The first enzyme to be crystallized in relatively pure form was urease (by Sumner in 1926).

Unlike other chemical catalysts, enzymes are highly specific in the reactions they catalyse and the substrates utilized. In 1894 Emil Fischer proposed the influential 'lock and key' analogy for enzyme–substrate interaction which has more recently been extended to include the idea of 'induced fit' proposed by Koshland (*see* MECHANISMS OF ENZYME CATALYSIS).

In 1898 Duclaux proposed a codification of enzyme nomenclature using the suffix '-ase' to be added to a root indicating the nature of the substrate on which the enzyme acts. With the exception of some proteolytic enzymes which had already been given the suffix '-in' (papain, pepsin, trypsin, etc.) this general rule persists in present enzyme nomenclature, in which the common recommended name of an enzyme usually indicates the substrate and the nature of the reaction catalysed (e.g. lactate dehydrogenase).

The overwhelming majority of enzymes are proteins, but RNAs with enzymatic activity have also been discovered (*see* RIBOZYMES). A list of the main types of enzyme activity according to the EC classification is given in Table E1. Illustrations of the three-dimensional structures of some enzymes are to be found in MOLECULAR EVOLUTION: SEQUENCES AND STRUCTURES; PROTEIN ENGINEERING; PROTEIN KINASES; PROTEIN STRUCTURE and Plates 4 and 5. *See also*: MECHANISMS OF ENZYME CATALYSIS.

Dixon, M. & Webb, E.C. (1979) *Enzymes*, 3rd edn (Longmans, Harlow).
Enzyme Nomenclature 1984. Recommendations of the Nomenclature Committee of the International Union of Biochemistry (Academic Press, 1984).

enzyme catalysis *See*: MECHANISMS OF ENZYME CATALYSIS.

enzyme–substrate complex *See*: MECHANISMS OF ENZYME CATALYSIS.

eosinophil White blood cell which contains cytoplasmic granules staining with the acidic dye eosin. Eosinophils express FC RECEPTORS for IgE and IgA antibodies on their surface, and are the main effector of antibody-dependent cell-mediated cytotoxicity against multicellular parasites such as helminths which provoke IgE antibodies. The granules contain basic proteins, of which the two main ones are major basic protein, which is toxic to helminths, and major cationic protein. Eosinophil growth and differentiation are stimulated by interleukin 5 (IL-5). *See also*: HAEMATOPOIESIS.

eosinophil colony-stimulating factor/differentiation factor IL-5.

ependymal cells Cells forming the lining of the central canal of the spinal cord and the ventricular surfaces of the brain, and which derive from neuroepithelial cells after migration to their adult position. Epithelial cells of the choroid plexus are believed to be modified ependymal cells.

epiblast A term only used in amniote embryos (*see* AVIAN DEVELOPMENT; MAMMALIAN DEVELOPMENT). It is the amniote equivalent of the ECTODERM, the layer of cells that after gastrulation gives rise to the skin and nervous system.

Table E1 Categories of enzyme activity

EC 1 Oxidoreductases
Dehydrogenases
Reductases
Oxidases
Dioxygenases
Monooxygenases
Nitrogenase
Hydrogenase

EC 2 Transferases
Methyltransferases
Acyltransferases
Glycosyltransferases
Phosphorylases
Aminotransferases
Kinases
Nucleotidyltransferases
    (e.g. RNA polymerases, DNA polymerases)
Sulphotransferases

EC 3 Hydrolases
Esterases
Lipases
Phosphatases
Phosphodiesterases
Sulphatases
Nucleases
Glycosidases
Peptidases
    (e.g. carboxypeptidases, aminopeptidases)
Proteinases
    Thiol proteinases
    Acid proteinases
    Metalloproteinases
    Amidases
Deaminases

EC 4 Lyases
Decarboxylases
Aldolases
Dehydratases
Nucleotide cyclases

EC 5 Isomerases
Racemases
Epimerases
Mutases

EC 6 Ligases (synthetases)
Amino-acyl-tRNA synthetases
Acyl-CoA synthetases
Carboxylases

This list is not comprehensive.

epiboly The spreading of presumptive ECTODERM over the surface of the embryo which accompanies invagination of the presumptive ENDODERM and MESODERM during GASTRULATION. *See also*: AMPHIBIAN DEVELOPMENT.

epicotyl The shoot of plant embryos and seedlings above the level of the attachment of the cotyledons.

epidermal growth factor (EGF) *See*: GROWTH FACTORS.

epidermal growth factor receptor (EGF receptor) *See*: ACTIN-BINDING PROTEINS; GROWTH FACTOR RECEPTORS.

epidermolysis bullosa A group of blistering skin diseases involving separation or splitting of the epidermis. EB simplex can be caused by mutations in either of the basal keratinocyte-specific CYTOKERATINS K5 or K14. *See*: INTERMEDIATE FILAMENTS.

Coulombe, P.A. et al. (1991) *Cell* **66**, 1301–1311.
Bonifas, J.M. et al. (1991) *Science* **254**, 1202–1205.

epigenesis Historically, epigenesis was a doctrine (opposed to preformation) which maintained that the embryo develops *de novo* rather than from a preformed homunculus. C.H. Waddington greatly contributed to modern ideas concerning epigenesis by introducing the idea of an epigenetic landscape, a theoretical construct which modelled developmental pathways as rivers channelled by a system of valleys. The relative constancy of the developmental outcome was seen to result from a system of developmental checks and buffers rather than by a strictly determined process. More recently, the term epigenesis has been used to describe mechanisms that can control gene expression without themselves being under the immediate control of genes. One such example might be CYTOPLASMIC DETERMINANTS.

epiligrin *See*: EXTRACELLULAR MATRIX MOLECULES.

epimorphosis REGENERATION that requires cell division. Opposed to MORPHALLAXIS, which does not. The regeneration of the vertebrate limb is often cited as the archetypal example of epimorphic regeneration, while the regeneration of *Hydra* is cited as the example of morphallactic regeneration. However, the distinction does not necessarily correlate with the modes of regeneration found in higher and lower organisms.

epinectin *See*: EXTRACELLULAR MATRIX MOLECULES.

epinephrine An alternative name for ADRENALINE.

epineurium The dense collagenous external connective tissue sheath of all peripheral nerve trunks. The collagenous fibres are mainly longitudinally orientated, and are interspersed with elastic fibres and fibroblasts. *See also*: EXTRACELLULAR MATRIX MOLECULES.

episome A PLASMID or BACTERIOPHAGE which, although usually found as an autonomously replicating genetic element in the cytoplasm, can integrate into the bacterial chromosome, for example the F PLASMID of *Escherichia coli*.

epistasis The situation in which one gene (the epistatic gene) masks the expression of a gene at another LOCUS (the hypostatic gene). For example, in *Drosophila* the recessive gene *apterous* produces wingless homozygotes, so any other recessive gene present affecting wing morphology, such as the *curled* wing gene, will have its action masked.

Plate 1

Plate 1. Computer-generated images of different double-helical forms of DNA. *Upper row*, ray-traced van der Waals representations; *lower row*, ribbon and stick representations showing base pairing. Atoms are represented as: carbon, green; phosphorus, yellow; oxygen, red; nitrogen, blue; hydrogen, white. In the images of B-DNA and A-DNA only hydrogen atoms participating in hydrogen bonds are shown. Images were generated on a Silicon Graphics Indigo2 workstation from the molecular modelling package QUANTA (Molecular Simulations Inc, Burlington, Massachusetts). All images on this page courtesy of Paul Gane, University of Kent. *See also*: DNA; NUCLEIC ACIDS.

a. B-DNA. The classical Watson–Crick right-handed double helix with 10–10.4 bp per turn, and a rise (distance between base pairs) of 3.4Å, which is the most usual conformation of double-stranded DNA *in vivo*.

b. A-DNA. A right-handed helix with 10.9 bp per turn, and a rise of 2.6Å per bp, formed in conditions of low ionic strength or low hydration.

c. Z-DNA. A left-handed helix which can be formed by DNA with alternating purine–pyrimidine sequences, resulting in a 'zigzag' orientation of the sugar–phosphate residues in the DNA backbone. In Z-DNA the asymmetric unit is a dinucleotide (CG in the example illustrated), rather than a mononucleotide as in A-DNA or B-DNA. There are 12 base pairs per turn and the rise is 3.6Å. In alternating CG polymers, chemical modification of the cytosines at position 5, for example by methylation or bromination, favours the Z-DNA conformation.

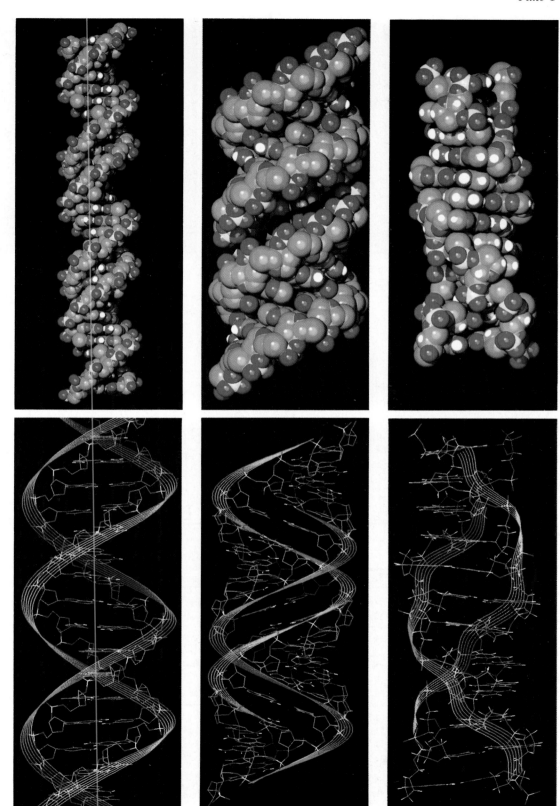

a   b   c

Plate 2

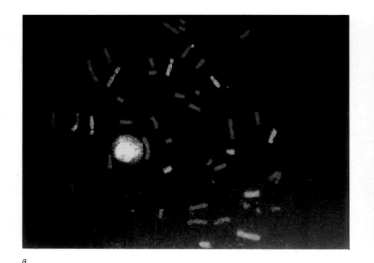

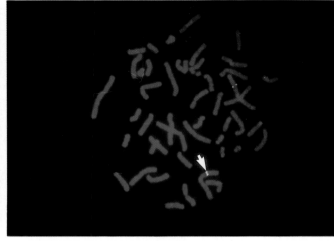

a

b

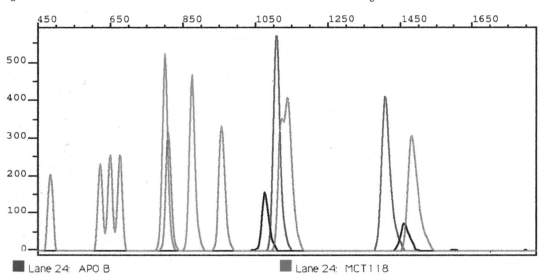

	Lane 24: APO B		Lane 24: MCT118
	Lane 24: YNZ22		Lane 24: INTERNAL STANDARD

Peak/Lane	Min.	Size	Peak Height	Peak Area	Scan #
1B, 24	221	678	590	13161	1109
2B, 24	282	896	416	12094	1410
1G, 24	160	426	542	9581	801
2G, 24	175	489	489	9085	875
1Y, 24	215	653	160	3663	1077
2Y, 24	292	931	76	2240	1463
1R, 24	97	137	206	3557	485
2R, 24	124	266	240	3569	622
3R, 24	130	291	264	3997	650
4R, 24	135	316	268	4013	677
5R, 24	162	434	324	5738	810
6R, 24	191	558	338	6912	958
7R, 24	224	689	356	6314	1123
8R, 24	228	702	417	9854	1140
9R, 24	296	945	311	10519	1483

c

Plate 2

a. Analysis of the human chromosome content of a somatic cell hybrid. DNA from the hybrid has been labelled with biotin and used to 'paint' the chromosomes of a normal human lymphocyte. The chromosomes which are illuminated green, in this case numbers 2,5,8,15,18,21 and X, must have been present in the hybrid. *See*: HUMAN GENE MAPPING. Photograph courtesy of S. Povey.

b. In situ hybridization of a cosmid clone containing the gene determining phosphoglucomutase onto normal human chromosomes, demonstrating the position of the gene *PGM1* on chromosome 1 (arrowhead). *See*: HUMAN GENE MAPPING. Photograph courtesy of S. Povey.

c. The use of the polymerase chain reaction (PCR) in DNA analysis for DNA profiling. A printout from the ABI GENE-SCANNER is shown. The DNA fragments obtained by selective PCR from genomic DNA are run in a single lane. They appear as a series of different coloured peaks where each colour corresponds to a different locus. The machine scans the positions of different coloured bands, automatically sizing them (size increases along the long axis of the graph). The sizes of the blue, green and black peaks are automatically measured relative to internal markers (red) and this information is recorded in the table below. Sizes of the peaks are given in base pairs. The peak height and peak area columns give information about the quantity of DNA present. *See*: DNA PROFILING; POLYMERASE CHAIN REACTION.

Plate 3

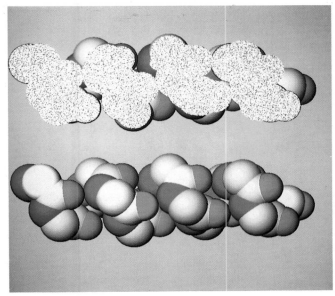

a

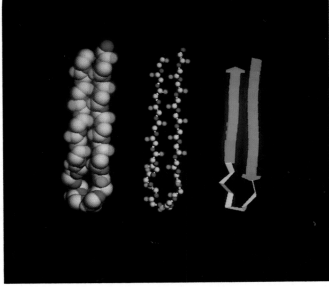

b

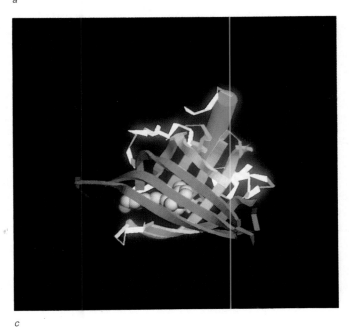

c

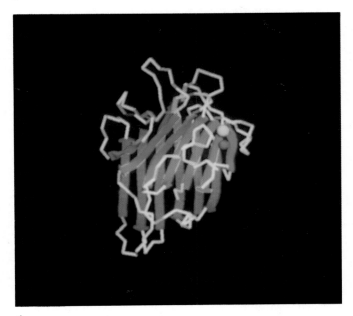

d

Plate 3

a. Space-filling model of the α-helix secondary structure motif in proteins. *Top*, view of the surface. *Bottom*, cheese-wire cut through centre of helix. *See also*: Fig. P77 in PROTEIN STRUCTURE.

b. Three depictions of the β-hairpin motif in proteins. β-Hairpins (β-turns) connect adjacent β-strands into antiparallel β-sheets, as can be seen in the protein structures below. *Left*, backbone atoms only; *centre*, backbone and side chains; *right*, schematic diagram with arrows representing strands.

c. Retinol-binding protein, an example of the 'up-and-down' β-barrel supersecondary structure, in which eight antiparallel β-strands connected by

β-hairpins form an open-ended flattened 'barrel', closed by hydrogen bonding between the first and last strands, and in this protein enclosing a retinol (vitamin A) molecule.

d. The plant lectin concanavalin A from jackbean, which contains a β-sheet 'jelly roll motif' in which eight antiparallel β-strands form a β-barrel with connections over and under the top of the barrel (*see also* Fig. P79 in PROTEIN STRUCTURE).

All images on this page courtesy of A.M. Lesk.
Cylinders depict α-helices, and broad arrows or ribbons β-strands.

Plate 4

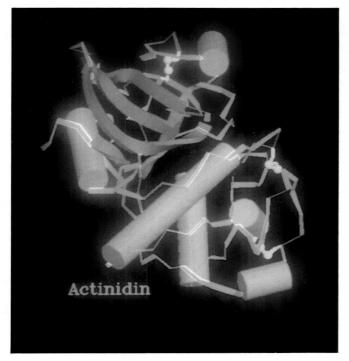

a

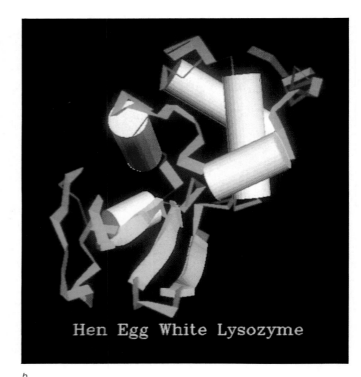

b

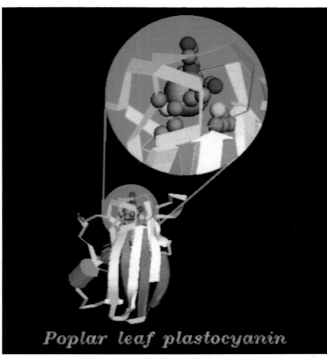

c

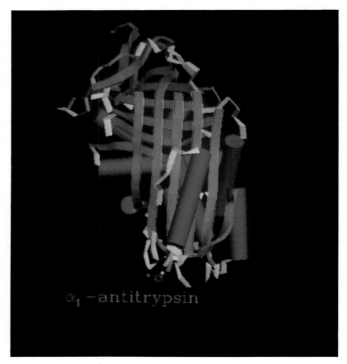

d

Plate 4

a. The thiol proteinase actinidin from kiwi fruit (Chinese gooseberry) (*see also* Fig. M33 in MOLECULAR EVOLUTION: SEQUENCES AND STRUCTURES).

b. The glycosidase lysozyme from hen egg white.

c. The copper-binding protein plastocyanin from poplar leaf. The copper-binding site is shown in detail in the circle. *See also*: Fig. M34 in MOLECULAR EVOLUTION: SEQUENCES AND STRUCTURES.

d. The serpin α_1-antitrypsin. *See also*: SERPINS.

All images on this page courtesy of A.M. Lesk.
Cylinders depict α-helices and broad arrows or ribbons β-strands.

Plate 5

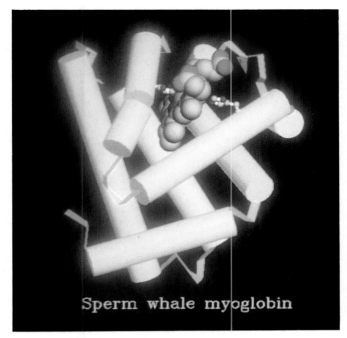

Sperm whale myoglobin

a

Lupin leghaemoglobin

b

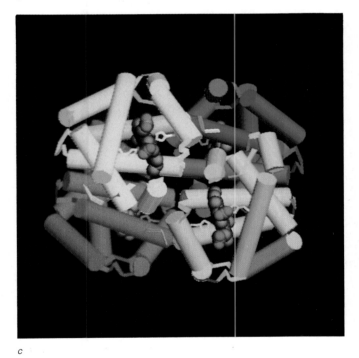

c

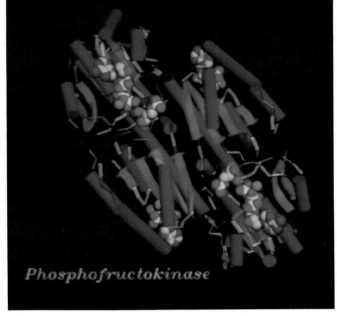

Phosphofructokinase

d

Plate 5

a. The oxygen-transport protein sperm whale myoglobin. The oxygen-binding haem is depicted in a space-filling representation. *See also*: HAEMOGLOBINS and Fig. M51 in MYOGLOBIN.

b. The root nodule protein leghaemoglobin from lupin. The oxygen-binding haem is depicted in a space-filling representation. *See also*: Fig. H6 in HAEMO-GLOBINS.

c. The oxygen-transporting protein haemoglobin A from human red cells. It is composed of four globin subunits, two α-globins and two β-globins. *See also*: HAEMOGLOBINS; HAEMOGLOBIN AND ITS DISORDERS.

d. Two of the four subunits of the allosteric enzyme phosphofructokinase from *Bacillus stearothermophilus*, in the R state, showing ATP and fructose 6-phosphate bound at the active sites (bottom right and top left) and the allo-steric effector ADP bound at the effector sites (bottom left and top right).

All images on this page courtesy of A.M. Lesk.
Cylinders depict α-helices and broad arrows or ribbons β-strands.

Plate 6

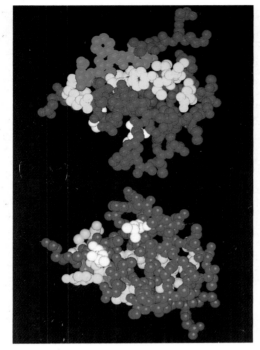

a

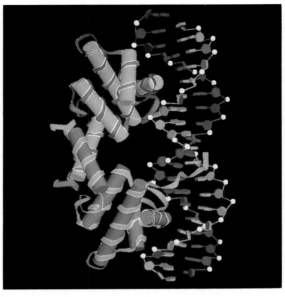

b

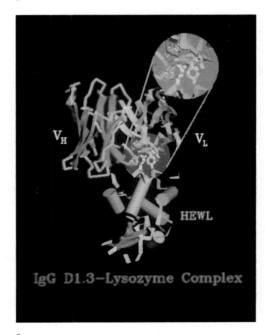

c

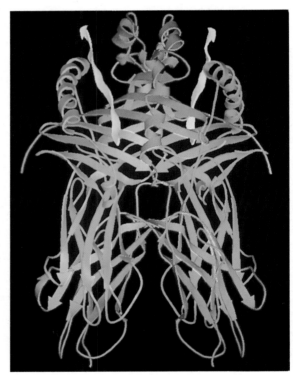

d

Plate 6

a. Top, three dimensional structure of the insulin monomer with the putative receptor-binding region in front, shown in green and yellow; residues essential for negative cooperativity are green. *Bottom*, back view showing in white residues LeuA13 and LeuB17, which may participate in a second binding site required for high affinity. *See also*: INSULIN AND ITS RECEPTOR.

b. Oestrogen steroid receptor binding to DNA.

c. Hen egg white lysozyme binding to the antigen-binding site of the immunoglobulin fragment (Fab) D1.3. *See also*: IMMUNOGLOBULIN STRUCTURE.

Images *b* and *c* courtesy of A.M. Lesk.

d. The three-dimensional structure of the human MHC class II histocompatibility antigen HLA-DR1. The molecule is represented here as a dimer of the αβ heterodimer (α-chain, blue; β-chain, orange), the form found in the crystals. The yellow arrows represent bound peptides. Photograph courtesy of D.C. Wiley, Howard Hughes Medical Institute, Cambridge, Massachusetts, from Brown, J.H. et al. (1993) *Nature* **364**, 33–39.

Plate 7

a

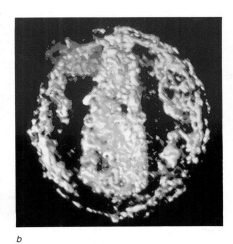

b

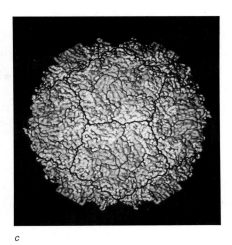

c

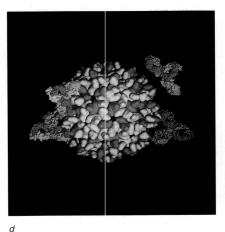

d

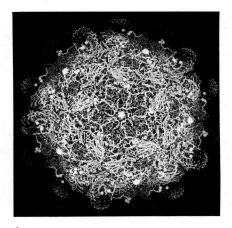

e

Plate 7

a. Capsid architecture of the tomato bushy stunt virus (*see* TOMBUSVIRUS GROUP). The icosahedral capsid is composed of 180 identical protein subunits (*see* Fig. T25 for subunit structure) arranged in three different symmetry-packing environments. Image courtesy of A.J. Olson, Scripps Clinic and Research Foundation, © 1984.

b. Human immunodeficiency virus (HIV) based on electron microscope data from U. Skoglund of the Karolinska Institute and S. Hoglund of Uppsala University, shows a cone-shaped core containing genetic material surrounded by a spherical envelope. Computer graphic image and photography by D.S. Goodsell & A.J. Olson, The Scripps Research Institute, © 1991.

c. Computer graphic model of the surface of poliovirus. This representation is based upon the results of X-ray diffraction studies at 2.9Å of the Mahoney strain of poliovirus solved in the laboratory of James M. Hogle, The Scripps Research Institute (*Science* **229**, 1358–1365 (1985)). The protein subunits are colour coded as follows: VP1, blue; VP2, yellow; VP3, pink. The fourth protein in the capsid, VP4, is not visible from outside the particle. The narrow black lines in the image delimit individual protomer units while the thicker black lines distinguish the pentameric assembly intermediates. The picture was produced using RAMS software developed by M.L. Connolly and modified by A.J. Olson to represent boundary information. Computer graphic image and photography by A.J. Olson, The Scripps Research Institute, © 1986. See *also*: PICORNAVIRUSES.

d. Computer graphic model of antibody interaction with viral capsid. The model uses X-ray crystallographic coordinates for the independently solved structures of an intact immunoglobulin and an intact spherical virus particle. Although the interaction depicted is purely hypothetical, it shows the size relationship between antibody and viral particle. Computer graphic image and photography by Arthur J. Olson, The Scripps Research Institute, © 1989.

e. Three-dimensional structure of foot-and-mouth disease virus (FMDV) at 2.9Å resolution. determined by X-ray diffraction. The image shows the α-carbon trace of the capsid polypeptides. The capsid is composed of 60 copies of each of four proteins VP1 (blue), VP2 (green), VP3 (red), and VP4 (yellow), of which VP1–VP3 are exposed on the surface. One of the most prominent features of the virus surface is a disordered protrusion (the FMDV loop) of the VP1 polypeptide (indicated by the blue dotted hemispheres), which is the most immunogenic portion of the virus capsid. The yellow balls at the centre of the image indicate the position of the disulphide ring around the 5-fold axis of symmetry, and the pale blue balls indicate the position of the C terminus of VP1 (residue 213) and the ends of the well-defined density for the FMDV loop (residues 136 and 157). Photograph courtesy of R. Acharya, E. Fry and D. Stuart, Laboratory of Molecular Biophysics, Oxford, from Acharya, R. et al. (1989) *Nature* **337**, 709–716. See *also*: PICORNAVIRUSES.

Plate 8

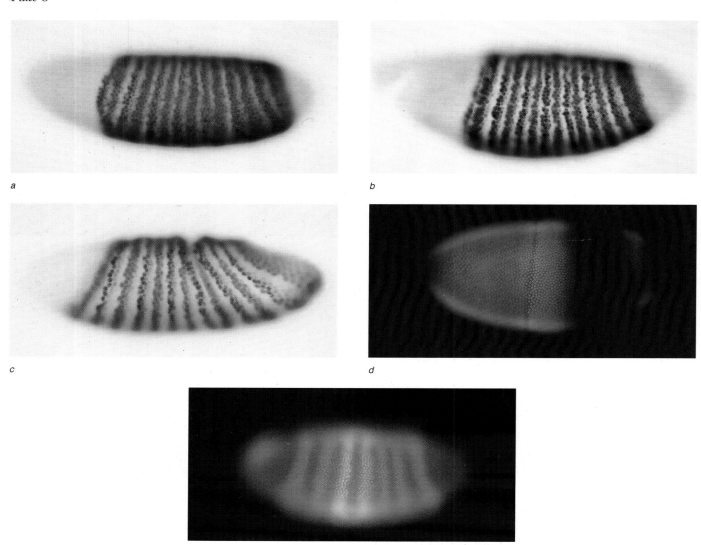

a

b

c

d

e

Plate 8

a, *b*, *c*. Expression of the pair rule genes *fiz* (brown) and *eve* (grey) in the *Drosophila* embryo through stages 5(2) to 8(1), showing expression in alternate parasegments. The stripes of *fiz* and *eve* expression narrow from the posterior margin and sharpen anteriorly as they intensify. *a*, Stage 5(2); *b*, stage 6; *c*, stage 8(1). *ftz* and *eve* proteins are visualized by antibody staining. From Lawrence, P.A. (1992) *The Making of a Fly* (Blackwell Scientific Publications, Oxford). Photographs courtesy of Peter Lawrence, MRC Laboratory of Molecular Biology, Cambridge. In these and the other photographs on this page the anterior end of the embryo is to the left.

d. Expression of gap genes in *Drosophila* embryogenesis. *Drosophila* embryo labelled with fluorescent antibodies against the *hunchback* gene product (red) and the *Krüppel* gene product (green). Yellow fluorescence indicates the zone of overlap of expression of the two genes. Photograph courtesy of K. Howard, Roche Institute of Molecular Biology.

e. Relationship of expression of the gap gene *Krüppel* and the pair rule gene *hairy* in *Drosophila* embryogenesis. *Drosophila* embryo labelled with fluorescent antibodies against the *Krüppel* gene product (red), and the *hairy* gene product (green). Yellow fluorescence indicates the zone of overlap of expression of the two genes. Photograph courtesy of K. Howard, Roche Institute of Molecular Biology, from Howard, K. & Struhl, G. (1990) *Development* **110**, 1223–1231.

See also: DROSOPHILA DEVELOPMENT.

Epithelial polarity

EPITHELIA are a diverse group of tissues that collectively represent one of the commonest cell types found in the animal kingdom. They form a boundary between different compartments within the body and at the interface between the organism and its environment. Epithelia occur on the body surface and line the cavities of all the major organ systems, and, during development, have a crucial role in morphogenetic processes and spatial patterning. In broad terms, the function of epithelial cells is to provide protection and regionalization to the body, and to regulate the molecular composition of, and exchange between, the compartments that they separate. This is achieved by cellular specializations allowing substances to be transported in particular directions across epithelia (vectorial transport). Transport processes in and across epithelia include directed absorption or secretion of organic molecules, and directed ion/solute and fluid transport. Epithelial cells must therefore acquire and maintain a polarized organization in respect of the membrane and cytoplasmic components involved in these processes (Fig. E10).

Simple epithelia composed of a monolayer of epithelial cells possess an apical surface facing the outside environment or the lumen of an internal cavity, a lateral surface at which adjacent cells are in contact, and a basal surface usually associated with a BASAL LAMINA (basement membrane) and access to MESENCHYME and the circulatory system. Some epithelia, such as the skin, are multilayered; here, only the outermost cells have an apical surface and innermost cells have contact with the basal lamina. Each surface constitutes a distinct domain with a particular structure and molecular composition. Lateral and basal membranes (basolateral) are often fairly similar in structure and composition and will be treated here as one domain.

Within the cytoplasm, the distribution of organelles (such as ENDOSOMES, LYSOSOMES, GOLGI APPARATUS, secretory vesicles and MITOCHONDRIA) and cytoskeletal elements (MICROFILAMENTS, MICROTUBULES) can also be polarized with respect to the apicobasal cell axis and vectorial transport route. Epithelial tissues can derive from all three GERM LAYERS during development and the mechanisms responsible for generating and maintaining this fundamental spatial asymmetry have been the subject of numerous investigations [1–4]. Most studies have been carried out on cultured epithelial cell lines such as the Madin–Darby canine kidney (MDCK) line, but both mature and developmental primary tissue cultures (e.g. mammalian hepatocytes, trophectoderm, and kidney) have also been used.

Epithelial surface specializations

Basolateral domain

The basolateral epithelial surface is specialized for cell–cell and cell–substratum (substrate) adhesion, signal reception and communication, and houses enzyme systems responsible for driving vectorial ion and fluid transport. Cell–cell adhesion along the lateral contact site is mediated by homophilic binding (i.e. like to like) of calcium-dependent CELL ADHESION MOLECULES —

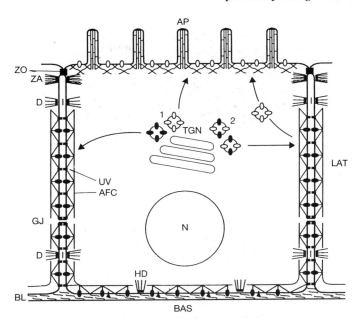

Fig. E10 Highly schematic diagram showing the principal features of surface polarity in a hypothetical epithelial cell. The apical membrane (AP) contains domain-specific proteins (open ovals) and is shown as microvillous, structured by actin bundles and terminal web filaments. The lateral membrane (LAT) contains uvomorulin-mediated adhesion sites (UV) and intercellular junctions (zonula occludens, ZO; zonula adhaerens, ZA; desmosomes, D; gap junction, GJ). The ankyrin–fodrin cytoskeleton (AFC) associates with basolateral membrane proteins (solid ovals) particularly Na^+,K^+-ATPase. The basal membrane (BAS) is in contact with a basal lamina (BL); both hemidesmosomes (HD) and receptors for extracellular matrix components are found at this domain. (1) Direct delivery of apical and basolateral membrane proteins to their respective domains in separate carrier vesicles, after sorting at the *trans* Golgi network (TGN) (e.g. MDCK cells); (2) default delivery of all membrane proteins to the basolateral membrane followed by sorting and transcytosis of apical membrane proteins in carrier vesicles to the apical domain (e.g. hepatocytes). N, nucleus.

cadherins — of which the principal member in mammalian cells is the integral membrane GLYCOPROTEIN, uvomorulin (E-cadherin) (M_r 120 000 (120K)). The extracellular N-terminal domain of uvomorulin contains specific amino-acid sequences necessary for calcium binding and homophilic adhesion.

Cell–cell contact is strengthened by an intercellular junctional complex (see CELL JUNCTIONS) at the apical extremity of the basolateral surface, comprising an outermost tight junction (*zonula occludens*), an intermediate junction (belt desmosome, *zonula adhaerens*), and innermost spot desmosomes (sing. *macula adhaerens*). Tight junctions are thought (1) to prevent the uncontrolled passage of molecules across the epithelium by 'seepage' between cells, and (2) to help maintain the segregation of apical and basolateral integral membrane proteins and outer leaflet lipids. In the more basal region of the lateral contact site gap junctions (see CELL JUNCTIONS) are also located providing hydrophilic channels (composed of hexagonally arranged connexon subunits) for the passage of low molecular weight solutes from cell to cell.

Cell adhesion to the basal substratum may be aided by specific receptors (e.g. INTEGRINS) for extracellular matrix components of the basal lamina (principally laminin, type IV collagen, proteoglycans; *see* EXTRACELLULAR MATRIX MOLECULES) and this can allow for indirect linkage with cytoplasmic MICROFILAMENTS. In addition, hemidesmosomes (molecularly distinct from desmosomes, *see* CELL JUNCTIONS) provide anchorage sites for cytokeratin filaments with the basal substratum.

In addition to these adhesive and junctional specializations, the basolateral domain of epithelia also contains receptors for growth factors and hormones, signal transduction systems (*see e.g.* G PROTEIN-COUPLED RECEPTORS; SECOND MESSENGER PATHWAYS), and is usually the site at which the transmembrane ION PUMP, Na^+,K^+-ATPase, is localized (*see* MEMBRANE TRANSPORT). This enzyme generates a polarized flow of Na^+ out of the cell into the basolateral compartment, contributing to an ion, and therefore electrical, gradient across epithelia. This electrochemical gradient is responsible for passive water flow across the epithelium and for vectorial transport of ions and solutes by way of apical membrane channels and cotransporters (*see* MEMBRANE TRANSPORT). It is dependent on the integrity of the tight junction for its maintenance. The basolateral surface also contains a distinct cytoskeleton (a membrane skeleton) immediately under the plasma membrane, consisting principally of a meshwork of fodrin (non-erythroid SPECTRIN) and ankyrin (*see* MEMBRANE STRUCTURE).

Apical domain

In contrast to the basolateral domain which has many molecular features in common with non-epithelial cells, the composition of the apical domain of epithelial cells is specific to this polarized cell type. In addition to the presence of ion channels and cotransporters required for vectorial transport, the apical surface is also usually rich in glycosphingolipids (*see* GLYCOLIPIDS) (localized in the EXOPLASMIC membrane leaflet) and is depleted in PHOSPHOLIPID. The sphingolipids contribute to membrane stabilization and, in intestinal epithelia, may also provide protection from lipid hydrolases present in the lumen.

The apical surface is also often folded into many microvilli, to a much greater extent than the basolateral surface, reflecting its greater capacity for uptake. A microvillous topography in turn necessitates a distinct membrane skeleton organization (ACTIN bundles linked to TERMINAL WEB) to provide structural support. In the intestinal brush border, myosin I, villin, tropomyosin, and a spectrin isoform (TW260/240) are important cytoskeletal proteins regulating microvillar and terminal web substructure.

Maintenance of polarization

Although epithelial cells, like other cells, undergo continual turnover of their membrane proteins and lipids, the polarized molecular organization of apical and basolateral domains must be maintained for their vectorial transport functions to be preserved. Studies on different epithelial systems have consistently shown that biogenetic pathways for newly synthesized membrane proteins and glycoproteins include a sorting compartment where apical and basolateral components are segregated. The subcellular location of this sorting compartment varies in different epithelia (Fig. E10).

Protein sorting

In MDCK cells, proteins destined for the apical domain are segregated from basolateral proteins at the *trans* Golgi network (TGN), a terminal compartment of the Golgi complex, following proteolytic processing and glycosylation (*see* GLYCOPROTEINS AND GLYCOSYLATION). Segregation here results in nascent proteins being delivered directly to their appropriate membrane face in carrier vesicles. Segregation and direct delivery of apical membrane lipids (glycosphingolipids) also occurs at the TGN in MDCK cells. In hepatocytes, on the other hand, all newly synthesized membrane proteins are delivered to the basolateral domain (which in liver lines the sinusoid lumen) where sorting takes place before TRANSCYTOSIS of apical membrane proteins in carrier vesicles to the apical domain (which lines the bile canaliculi). In intestinal cells, sorting of different apical domain proteins can take place either at the TGN (direct delivery) or at the basolateral membrane (indirect delivery).

These tissue-specific variations in the sorting site may reflect differences in the character and functional activity of the different epithelia. In hepatocytes, the extent of the basolateral membrane greatly exceeds that of the apical membrane and secretion only occurs in the basolateral direction. MDCK cells, however, have a larger apical domain, engage in both apical and basolateral secretion, and have evolved an intracellular targeting compartment for efficient maintenance of polarized domains.

Targeting signals

Correct sorting and delivery of membrane proteins during biogenesis presumably depends on the existence of targeting signals on proteins (*see* PROTEIN TARGETING; PROTEIN TRANSLOCATION), a means of signal identification by cellular receptors, and the capacity for transport to and 'docking' of carrier vesicles at the appropriate domain. Very little is known about these events at the molecular level in epithelia although some progress has been made in recent years. Targeting signals are apparently preserved on the proteins, as incorrectly localized proteins in MDCK cells can subsequently be rerouted to the correct domain by transcytosis. TRANSFECTION of cDNA encoding modified or chimaeric exogenous proteins into MDCK cells has been an important approach to identify the intramolecular targeting site on different proteins. These studies to date suggest that the signal specifying apical delivery of proteins is usually localized in the ectodomain. In contrast, many proteins that reside on the basolateral membrane surface possess an appropriate targeting signal on their cytoplasmic tail. This cytoplasmic site can include a tyrosine residue and be a signal similar to one promoting endocytosis of the protein from the cell surface. It is thought that targeting signals function by virtue of their effect on protein conformation rather than their primary sequence.

There is growing evidence that lipids and proteins destined for the apical domain are carried together by an integrated mechanism involving the same TGN-derived carrier vesicles. Glycosphingolipids in the exoplasmic leaflet of cellular membranes tend to self-associate by hydrogen bonding; this might lead to microdomains of glycosphingolipid enrichment within the TGN that then vesiculate to form apical membrane transport vesicles. One glycosphingolipid (glycosyl-phosphatidylinositol, GPI) has been shown to link covalently and exoplasmically with particular apical membrane proteins (*see* GPI ANCHORS). In transfection experiments, addition of GPI anchors to proteins that are normally basolateral or nonpolar in their distribution can indeed induce their apical delivery. Moreover, inhibition of GPI anchor synthesis can randomize the membrane delivery of proteins normally anchored and delivered apically. The GPI anchor therefore is at least one mechanism of signal recognition operating in polarized epithelial membrane biogenesis. GPI-anchored proteins may also act as receptors for the segregation and apical delivery of other non-anchored proteins required in this domain.

The cytoskeleton also seems to have a role in the transport of carrier vesicles to the apical domain. Drugs that depolymerize microtubules inhibit or reduce delivery to the apical membrane but apparently not to the basolateral membrane in MDCK cells, suggesting that a microtubule-based motor enzyme may mediate vesicular transport (*see* MICROTUBULE-BASED MOTORS). However, mechanisms by which carrier vesicles might engage in membrane fusion at particular domains are unknown. Apical and basolateral carrier vesicles have recently been isolated and their characterization may lead to an understanding of this process.

Establishment of polarity

The establishment of basolateral cell–cell adhesion is of fundamental importance in the development of epithelial polarity. For example, MDCK cells lose their polarity when they are isolated from each other but gradually regain it once cell–cell contact is established.

Uvomorulin-mediated adhesion

Uvomorulin-mediated adhesion is important in the maturation of the basolateral surface in at least two respects. First, uvomorulin binding between cells provides sufficient membrane apposition for the assembly of intercellular junctions. The adhesive function of uvomorulin depends on interaction of its cytoplasmic tail with cytoskeletal proteins such as the CATENINS which associate with actin. This complex is utilized during the maturation of the *zonula adhaerens* junction. Second, uvomorulin is fundamentally involved in membrane cytoskeleton organization and the segregation of Na^+,K^+-ATPase to the basolateral domain. The membrane skeleton protein ankyrin forms a complex with both fodrin (the predominant membrane skeleton component) and Na^+,K^+-ATPase in MDCK cells. The distribution and physical characteristics of these molecules are closely coordinated and can be modified by changing cell–cell contact patterns by removal or addition of extracellular calcium. In the absence of uvomorulin adhesion, fodrin and Na^+,K^+-ATPase are distributed diffusely, have short half-lives and are relatively soluble. When uvomorulin adhesion is initiated, these proteins gradually localize to the basolateral domain, turn over more slowly, and become insoluble

following nonionic detergent extraction, indicating that they have formed a stable macromolecular complex. Thus, cell–cell contact regulates the assembly and stabilization of the basolateral membrane cytoskeleton including, by complex formation, Na^+,K^+-ATPase localization. Uvomorulin appears to be directly involved in this assembly reaction by providing cytoplasmic binding sites for skeleton components. Moreover, uvomorulin expression in nonpolar fibroblastic cells that do not normally express it induces not only uvomorulin-mediated adhesion but also the polarization of fodrin and Na^+,K^+-ATPase to cell–cell contact sites. Polarization occurs in the absence of tight junction formation, suggesting that these junctions are not essential for initiating membrane protein segregation.

Uvomorulin adhesion is clearly a primary event in the generation of epithelial polarity, providing a molecular basis for the elaboration of the structural features of an epithelial sheet and for the polarized distribution of one important membrane component involved in vectorial transport. Uvomorulin adhesion will gradually lead to the formation of an associated membrane skeleton that stabilizes Na^+,K^+-ATPase and perhaps other basolateral membrane proteins. Only later again will the polarized transport pathways for newly synthesized membrane proteins (detailed above) become established.

How might the adhesive capacity of uvomorulin be regulated during development? During differentiation of mouse trophectoderm, uvomorulin adhesion between blastomeres occurs at the 8-cell stage coincident with cell polarization, but the glycoprotein is synthesized and is present on membranes from early cleavage (~24 hours earlier). Uvomorulin adhesion in this system is therefore regulated post-translationally, and appears to be triggered by a cell contact signal involving PROTEIN KINASE C activation. Interestingly, uvomorulin itself becomes phosphorylated for the first time in blastomeres coincident with the onset of adhesive capacity.

Role of cell–cell and cell–substratum contacts

Uvomorulin is not the sole arbiter of phenotypic change. For example, mouse 8-cell blastomeres can still generate a microvillous apical surface either in isolation or when uvomorulin adhesion is prevented by calcium depletion or blocking with specific antibody. In such cases, cell polarization is delayed and is oriented randomly, suggesting that in this system it is adhesion that specifies the epithelial axis. Single MDCK cells provided only with substratum contact can also rapidly polarize apical membrane marker proteins. Thus, aspects of apical membrane differentiation can take place independently of both uvomorulin adhesion and tight junction formation.

A role for substratum contact in epithelial polarity is best illustrated in embryonic kidney development. Tubule formation from the metanephrogenic mesenchyme coincides with expression of both uvomorulin and basal lamina macromolecules at cell surfaces. Inhibition of uvomorulin function by antibody treatment fails to block mesenchyme conversion to epithelium although maturation of the lateral membrane domain may be affected. However, if aggregating mesenchyme is prevented from interacting with laminin in the basal lamina by antibodies directed against the laminin A chain, epithelial polarization is prevented.

Distinct roles for cell–cell and cell–substratum adhesion have also been defined in MDCK cells grown as spherical aggregates that later form cysts. In these cysts the apical domain is formed on the outer face of the cyst. Here, cell–cell contact leads to membrane domain specificity (e.g. basolateral Na^+,K^+-ATPase at contact and central lumen sites) but full epithelial polarity (e.g. apicolateral tight junction) and central lumen expansion require substratum contact (provided by endogenous secretion of type IV collagen into the interior of the cyst). But, if the aggregates are grown in collagen gel, so that the substratum is now external, not internal, the axis of polarity reverses, with apical membrane and encircling tight junction now facing the central lumen.

These results help to clarify differences that occur in the orientation of developmental epithelia. Trophectoderm has apical membrane facing outwards towards the uterine environment, while the basolateral membrane faces the blastocyst lumen (blastocoel). Basal lamina components are secreted relatively late in trophectoderm differentiation, after uvomorulin adhesion has defined the epithelial axis. Conversely, in kidney development, the apical membrane faces inwards towards the tubule lumen. Here, the key substratum component, laminin A, is secreted early in differentiation on the outer surface of the mesenchyme aggregates, and thereby induces an opposite orientation in epithelial polarity. The combined activities of these different modes of adhesion therefore have a profound influence both on the spatial organization of individual epithelial cells and on their integration into three-dimensional structures during morphogenesis.

T.P. FLEMING

See also: CELL–MATRIX INTERACTIONS; MEMBRANE STRUCTURE.

1 Rodriguez-Boulan, E. & Nelson, W.J. (1989) Morphogenesis of the polarized epithelial cell phenotype. *Science* **245**, 718–725.
2 Wollner, D.A. & Nelson, W. J. (1992) Establishing and maintaining epithelial cell polarity: roles of protein sorting, delivery and retention. *J. Cell Sci.* **102**, 185–190.
3 Fleming, T.P. (Ed.) (1992) *Epithelial Organization and Development* (Chapman & Hall, London).
4 Rodriguez-Boulan E. & Powell S.K. (1992) Polarity of epithelial and neuronal cells. *Annu. Rev. Cell Biol.* **8**, 395–427.

epithelial–mesenchymal interactions Many inductive interactions taking place during vertebrate development (*see* INDUCTION) involve close interactions between two cell populations arranged, respectively, as an organized EPITHELIUM and as a loose MESENCHYME. Examples include induction in developing skin of epidermal structures such as feathers, scales, and claws by the underlying dermal MESODERM; in transplantation experiments in chick embryos, the region of origin of the mesoderm dictates which overlying epidermal structure will form. Further examples are seen during tooth, salivary gland, pancreas, and KIDNEY DEVELOPMENT. A variety of experiments, for example using filter barriers between the two cell populations, have shown that the induction signals can be transferred by direct cell contact (e.g. in metanephric kidney), by short-range diffusible molecules (e.g. in

pancreas) or via the local extracellular matrix (e.g. in vertebral cartilage). The molecular nature of many of these interactions remains, however, poorly understood.

epithelium (*pl.* epithelia) Sheet-like tissue composed of a single layer, or several layers, of tightly joined cells. Epithelia of different types cover all external and internal surfaces of the animal body, and serve protective and transport functions. In epithelia such as that lining the gut, the luminal face has different properties from the basal face, thus enabling selective directional transport of ions and other solutes across the epithelium. See: CELL JUNCTIONS; EPITHELIAL POLARITY; EPITHELIAL–MESENCHYME INTERACTIONS; INDUCTION.

epitope Antigenic determinant. The structure(s) on an ANTIGEN molecule that interact(s) with the combining site of an ANTIBODY or T CELL RECEPTOR as a result of molecular complementarity. Protein epitopes recognized by antibodies may be continuous or discontinuous depending on whether the amino-acid residues forming the epitope are in continuous peptide linkage or are in spatial proximity to each other as a consequence of the tertiary or quaternary structure of the molecule. These epitopes are expressed on the surface of the proteins. Protein epitopes recognized by T cell receptors are peptides generated by enzymatic degradation of the protein molecule and presented on the cell surface in association with class I or class II MHC MOLECULES (*see* ANTIGEN PROCESSING AND PRESENTATION).

EPS Extracellular polysaccharide. *See*: BACTERIAL ENVELOPE; GLYCOCALYX.

epsilon chain (ε chain) May refer to a subunit of any of numerous proteins in which subunits have been arbitrarily designated α, β, γ, δ, etc. Often refers to the ε chain of haemoglobins (*see* HAEMOGLOBIN AND ITS DISORDERS).

EPSP EXCITATORY POSTSYNAPTIC POTENTIAL.

EPSP synthase 3-Enolpyruvoylshikimate-5-phosphate synthase (EC 2.5.1.19; recommended name, 3-phosphoshikimate 1-carboxyvinyltransferase). Catalyses the condensation of phosphoenolpyruvate and shikimate 5-phosphate, with the release of orthophosphate.

It controls an important stage in the aromatic (shikimate) pathway in bacteria and plants, immediately preceding chorismate synthase, leading to phenylalanine, tyrosine and tryptophan. It is inhibited by the herbicide, GLYPHOSATE.

Epstein–Barr virus (EBV)

EPSTEIN–BARR virus (EBV) is a lymphocytic or γ-type human HERPESVIRUS which is carried as a persistent latent infection by the majority of the world's population. EBV was originally isolated by Epstein and Barr in 1964 from samples obtained from patients with BURKITT'S LYMPHOMA (BL) following the suggestion by Denis Burkitt that BL may be caused by an infectious agent. Primary EBV infection is usually by an oral route and is normally asymptomatic. If the initial infection is delayed to late childhood or early adolescence, an infectious mononucleosis may occur. EBV is associated with the development of two major malignancies, Burkitt's lymphoma and anaplastic nasopharyngeal carcinoma (NPC). The contribution of EBV to the aetiology of these conditions is unclear, but other cofactors are obviously required. With the development of more sensitive techniques, EBV has been found in a number of other conditions, including Hodgkin's lymphoma, salivary gland carcinoma, and several pathologies associated with immunocompromised individuals, including patients with AIDS.

Virus structure

In the electron microscope EBV shows a typical herpesvirus structure: a hexagonal nucleocapsid surrounded by a complex envelope. The viral particle is ~180 nm in diameter, and contains linear double-stranded viral DNA surrounding a protein core, which is enclosed in a capsid consisting of 162 hexagonal capsomeres. This, in turn, is surrounded by an envelope with external GLYCOPROTEIN spikes. The major envelope glycoprotein, gp350/220, is being used to develop SUBUNIT VACCINES against EBV.

Genome organization

The B95-8 strain of EBV, a prototype strain propagated in marmoset cells, was the first herpesvirus to be completely sequenced [1]. DNA from B95-8 contains ~172 kilobase pairs (kbp), with a composition of 59% G + C residues. Compared with other EBV strains, B95-8 has an 11.6-kbp deletion; the complete EBV genome is therefore ~184 kbp long. The overall arrangement of the DNA is grossly similar to that of other herpesviruses and contains multiple copies of TANDEM REPEATS of ~500 bp at each end of the linear genome and two unique regions, U_L and U_S, which are separated by a 3072-bp repeat sequence IR1 (internal repeat 1). There are no isomeric forms of EBV. Within the U_L segment are a number of smaller repeat sequences, several of which are contained within protein-coding sequences.

There is little cross-hybridization with other herpesvirus DNA, and limited overall sequence homology. There are regions of the EBV genome where gene arrangement is similar to other herpesviruses, notably the segment containing the EBV major DNA-binding protein, the glycoprotein gp110 gene, and the thymidine kinase and glycoprotein gp85 genes, implying that regions of EBV have arisen from an ancestral virus common to all herpesviruses.

Some regions of EBV are homologous to cellular sequences, notably the repeat sequence IR3 (Fig. E11) encoding the glycine/alanine copolymer region of the EBNA-1 protein (*EB Nuclear Antigen 1*), which has been described in cellular DNA, and the BCRF1 OPEN READING FRAME which has homology with the human interleukin 10 gene (*see* LYMPHOKINES). There is little strain variation in EBV, although two strains, EBV-A and -B, have been defined, primarily on the basis of serological differences in the EBNA proteins.

Life cycle

In vivo EBV is maintained as a latent infection (*see* ANIMAL VIRUS DISEASE) which is usually silent although virus shedding can be detected sporadically in saliva. The precise site of the latent infection *in vivo* is still unclear. EBV is detected in two main cell types, B CELLS and squamous epithelial cells. EBV infects mainly small resting B cells which have the highest density of the cell-surface receptor for C3d (CD21) to which EBV binds via the gp350 envelope glycoprotein. The virus is thought to bind to a similar molecule on epithelial cells.

Binding of EBV to CD21 on B cells results in activation of the cell from G0 into the G1 growth phase (*see* CELL CYCLE) and is accompanied by the synthesis of activation markers, such as CD23, a low-affinity receptor for the Fc fragment of IgE. Little is known of the early events following infection of epithelial cells. Following infection of B cells, a number of EBV genes are expressed (see below). At some point, the linear EBV genome

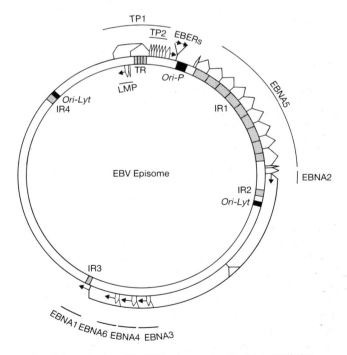

Fig. E11 Schematic view of the EBV episome, showing the transcription pattern of the latent genes in lymphoblastoid cell lines. The most 3′ exon of each RNA species is shown as an arrow to indicate the direction of transcription. Splicing is denoted by bent and curved lines joining the exons. The DNA repeat sequences (terminal (TR) and internal (IR)) are indicated, as are the latent and lytic origins of replication (Ori-P and Ori-Lyt respectively).

forms a covalently closed circular extrachromosomal episome. The single EBV genome is then subsequently amplified so that each cell contains many copies of the circularized genome.

The region of the DNA responsible for maintenance of the episome, the origin of plasmid replication, Ori-P, is found in the U_S region (Fig. E11). Ori-P consists of a series of 30-bp tandem repeat sequences, with a region of dyad symmetry ~1 kbp downstream. There is little evidence for consistent integration of EBV into cellular DNA, although cell lines have been identified which contain integrated viral DNA. Two regions of the EBV genome containing 97% nucleotide similarity have been identified adjacent to the IR2 and IR4 repeat regions. These are designated as duplicated sequence left (DSL) and duplicated sequence right (DSR), respectively. The origin of lytic DNA replication is contained within these regions.

Although little is known of EBV gene expression *in vivo*, EBV infection of B cells *in vitro* results in the establishment of continuously proliferating lymphoblastoid cell lines. EBV gene expression in these cell lines is usually referred to as the latent pattern of gene expression, that is, little or no virus is produced. In conditions such as BL and nasopharyngeal carcinoma, a more restricted pattern of gene expression is observed.

The latent pattern of gene expression is restricted to six nuclear proteins (EBNAs) and three membrane proteins: latent membrane protein (LMP) and terminal proteins 1 and 2 (TP1 and TP2). Two small, nonpolyadenylated RNA polymerase III transcripts, EBER 1 and EBER 2 (*EB Encoded RNAs*), are also synthesized. It has recently been shown that in NPC cells, a large amount of mRNA, not associated with any of the above products and synthesized in an ANTISENSE direction to known genes such as that for the viral DNA polymerase, is present, although no protein product has yet been associated with these transcripts.

Latent gene expression

Transcription of EBNA genes

All six EBNA proteins are transcribed from a large precursor TRANSCRIPTIONAL UNIT which covers 100 kbp of the EBV genome (Fig. E11). This is initiated from one of two PROMOTERS, one in the *Bam*HI C fragment (Cp) and the other in the *Bam*HI W fragment (Wp) of the EBV genome. During the initial stages of EBV infection transcription originates from Wp, and then switches to Cp. Functioning of the promoters seems to be mutually exclusive. ENHANCER regions have been described upstream of both promoters. The RNA SPLICING patterns of the EBNA species are complex [2] and are indicated in Fig. E11.

In NPC and BL cells, only EBNA-1 is expressed. This has been shown to originate from a promoter in the *Bam*HI F region, upstream of the EBNA-1 DNA-binding region III.

EBV latent proteins

EBNA-1. EBNA-1 is encoded by the BKRF1 open reading frame. EBV genes are designated according to the position and direction of the genes in the respective *Bam*HI fragments of the EBV genome: for example, BKRF1 is encoded in the *Bam*HI K fragment and is transcribed in a rightward direction, being the first gene in the *Bam*HI K fragment in this direction. BKRF1 contains the IR3 repeat region, which encodes a glycine–alanine copolymer accounting for about 30% of the protein. EBNA-1 is the only EBV protein that has been described in all types of EBV infection to date and is necessary for the maintenance of the EBV genome as an extrachromosomal episome. EBNA-1 interacts with the Ori-P region, binding to 20 copies of a 30 bp element and the 3′ dyad symmetry region. EBNA-1 binds at high affinity within each of the 30-bp elements and at four sites in and around the palindromic sequence. EBNA-1 also binds to sequences in the *Bam*HI Q fragment. The significance of this became clear recently when EBNA-1 was shown to be initiated from a promoter region (Fp), in BL and NPC cells in the *Bam*HI F fragment just upstream of the *Bam*H1 Q region, thus raising the possibility that EBNA-1 may be involved in the control of EBNA transcription in these cells.

EBNA-2. This antigen is encoded by the BYRF1 open reading frame as a protein of M_r ~80 000 which varies slightly among viral strains. EBNA-2 includes a polyproline region, a glycine–alanine repeat, and a highly charged acidic C terminus. It is thought to be necessary for B cell immortalization, although the mechanism by which it achieves this is unknown. EBNA-2 upregulates LMP and TP transcription. Antigenic differences between EBNA-2 variants have identified two distinct EBV strains (2A and 2B).

EBNAs-3c, -3b, and -3c. These are also called EBNA-3, -4, and -6, and are encoded by alternative splicing within the *Bam*HI E fragment, producing proteins of M_r >100 000. Although little is known of the functions of these proteins, there is some similarity in the first 200 residues, encoded by the BERF1, BERF2b, and BERF4 open reading frames; all also contain repeated domains at the C termini. EBNA-3c contains a number of motifs reminiscent of DNA-binding TRANSCRIPTION FACTORS, raising the possibility that it is involved in transcriptional regulation.

EBNA LP. Also known as EBNA-5, EBNA LP is encoded by two exons from the *Bam*HI W internal repeat, three exons from *Bam*HI Y and an exon from *Bam*HI Y. cDNA analysis has shown that EBNA-5 is encoded within bicistronic messages which may also contain other EBNA species. The protein is known as EBNA LP as it acts as a LEADER SEQUENCE of other EBNA species. It seems to be involved in immortalization of B lymphocytes but is not essential.

Latent membrane protein (LMP). This antigen is transcribed from the BNLF1 open reading frame and consists of three exons, a, b, and c. LMP is predicted to contain a short hydrophilic N terminus, six hydrophobic transmembrane domains, and a long hydrophilic C terminus. The predicted secondary structure shows similarity to ION CHANNELS, members of the rhodopsin family of receptors, and the *mas* ONCOGENE product. LMP has been shown to transform immortal, but nontumorigenic, rodent fibroblasts to a tumorigenic phenotype (*see* TRANSFORMATION).

Terminal proteins 1 and 2 (TP1 and TP2). Also known as LMP 2a and 2b, these proteins are the products of a highly spliced message in the left-hand region of EBV (on the conventional linear map of virion DNA), which is initiated from one of two promoters in the right-hand region of the genome. Circularization of the EBV genome is thus a prerequisite for TP synthesis. TP1 and TP2 initiate from different promoters.

EBERs

The most abundant RNA species in EBV-infected cells, the two EBER species are 166 and 172 bp long and are transcribed by RNA polymerase III, although the promoter regions for these species possess typical RNA polymerase II type motifs, including Sp1 and AP1 binding sites (*see* EUKARYOTIC GENE EXPRESSION; TRANSCRIPTION FACTORS). In common with other RNA pol III transcripts, they bind to the La protein and can functionally substitute for the similar small RNA species (VA RNAs) encoded by ADENOVIRUS. They are thought to be involved in regulating the cellular INTERFERON response to EBV. They are not necessary for cellular immortalization by EBV.

Viral replication

The lack of a cell system PERMISSIVE for EBV replication has hindered progress in the study of the viral replicative cycle. However, some cell lines can be induced to produce virus in a small proportion of cells. Use of the P3HR-1 strain of virus to superinfect cells has identified a region of DNA in the *Bam*HI Z region of the EBV genome which seems to disrupt latency. The gene involved is the product of the BZLF1 ORF. The steps responsible for inducing the lytic cycle are undefined. BZLF1 acts on two other regions of the EBV genome and upregulates the transcription of the BMLF1 and BRLF1 proteins. These products, together with BZLF1, then induce the transcription of other 'early' genes. EBV DNA replication is induced from the Ori-Lyt region which is found in the DSL and DSR regions of the EBV genome.

EBV and cancer

Following the original description of EBV in Burkitt's lymphoma samples, EBV has been implicated in the development of other malignancies, particularly anaplastic NPC and more recently, other lymphomas, including Hodgkin's lymphoma. Evidence linking EBV with BL was originally suggested by the isolation of the virus from BL cells kept in tissue culture. It was subsequently shown that BL biopsy material contained EBV DNA, although not infectious virus, and seroepidemiological studies show an elevated titre of antibodies to EBV antigens in BL patients. EBV also causes a lymphoma-like disease in New World primates. Although evidence implicates EBV in development of BL, the most striking feature of BL cells are the chromosomal TRANSLOCATIONS, resulting in the juxtaposition of immunoglobulin genes, usually on chromosome 14, more rarely from chromosomes 2 and 22, with the c-*myc* locus on chromosome 8 (*see* CHROMOSOME ABERRATIONS). The contribution of chromosome translocations and EBV infection to the development of BL is still unclear.

The association of EBV with NPC is less well characterized. Serological studies show an increased titre of ANTIBODIES to EBV antigens, particularly IgA antibodies to viral capsid antigen in patients before the onset of the tumour; EBV DNA is also found in all the tumour cells. In common with BL, there are geographical regions where the incidence of the tumour is dramatically increased. Thus EBV may be only one of the factors contributing to the development of these malignancies.

P.R. SMITH

1 Baer, R. et al. (1984) DNA sequence and expression of the B95-8 Epstein–Barr virus. *Nature* **310**, 207–211.
2 Speck, S.H. & Strominger, J.L. (1989) Transcription of Epstein–Barr virus in latently infected, growth transformed lymphocytes. *Adv. Viral Oncol.* **8**, 133–150.

equilibrium dialysis A technique which allows the ASSOCIATION CONSTANT for an ANTIBODY binding a monovalent HAPTEN to be measured. It employs a chamber divided into two compartments by a membrane which allows diffusion of the hapten but not the antibody. When a known amount of antibody is added to one chamber it binds hapten, thus effectively removing it from solution. Hence, hapten diffuses from the other chamber to restore the equilibrium. Sampling of the solutions and determination of the hapten concentrations allows equilibrium constants to be determined.

At equilibrium, the association constant K and the antibody valence (number of antigen combining sites), n, can be determined from the relationship:

$$r/c = Kn - Kr$$

where r is the ratio of moles of hapten bound per mole of antibody and c the concentration of unbound hapten. A plot of r/c against r at different hapten concentrations produces a straight line with a gradient of $-K$. *See*: SCATCHARD PLOT.

equilibrium potential In a biological system, the equilibrium potential (E_{ion}) for a given ion distributed across a semipermeable membrane is given by the Nernst equation:

$$E_{ion} = \frac{-RT \ln [ion]_i}{zF [ion]_o}$$

where R is the universal gas constant, T is the absolute temperature, z is the charge that the ion bears, F is the Faraday constant (i.e. the amount of electricity carried by a mole of electrons), and $[ion]_i$ and $[ion]_o$ are the concentrations of the ion on the different sides of the membrane, or inside and outside the cell, respectively. It can be seen that the electrical work ($-EzF$) done to prevent the ion movement balances the chemical work of the ion done by the ion flowing down its concentration gradient ($RT \ln [ion]_i / [ion]_o$).

The reversal potential of an ion is equivalent to its equilibrium potential. At a membrane potential above the ion's equilibrium potential the net ion flux will be in one direction, at the equilibrium potential there will be no net flux, but below the equilibrium potential the net ion flux will be in the opposite direction. Thus the ion flux reverses at the equilibrium potential: hence reversal potential. *See also*: SYNAPTIC TRANSMISSION.

equivalence group Group of cells with a common competence for further development. For example, in the nematode *Caenorhabditis elegans*, cell ABLATION experiments have shown that the ventral nerve cord cells P9.p, P10.p, and P11.p, although having different fates, are of equal developmental potential. Thus, if P11.p is destroyed, P10.p will adopt the fate of P11.p, and similarly P9.p will acquire the fate of P10.p. Another well studied example of an equivalence group occurs in the leech nerve cord. The RAS and CAS interneurons are situated on either side of the midbody nerve ganglion with homologues on the contralateral side. Both RAS and CAS interneurons become immunoreactive for a specific peptide at the same time, but this is lost from the contralateral homologue several days before development has ceased. However, unilateral ablation of the RAS and CAS cell lineages results in the permanent acquisition of peptide immunoreactivity in their homologues. *See also*: CAENORHABDITIS DEVELOPMENT; CAENORHABDITIS NEURAL DEVELOPMENT; NON-EQUIVALENCE.

ER ENDOPLASMIC RETICULUM.

erb The gene encoding the epidermal growth factor receptor (*erbB1*). *See*: GROWTH FACTOR RECEPTORS; ONCOGENES.

ERCC genes *See*: DNA REPAIR.

ERP EVENT-RELATED POTENTIAL.

Erwinia A bacterial genus belonging to the family Enterobacteraceae. Bacteria belonging to this genus are Gram-negative, chemo-organotrophic, motile, and facultatively anaerobic. They are often associated with plants as pathogens or saprophytes, or as constituents of the epiphytic flora. *See*: PLANT PATHOLOGY.

Lelliott, R.A. & Dickey, R.S. (1984) *Erwinia*. In *Bergey's Manual of Systematic Bacteriology* (Krieg, N.R. & Holt, J.G., Eds) 469–476 (Williams and Wilkins, Baltimore).

erythroblastosis fetalis A severe form of haemolytic disease of the newborn associated with the passage across the placenta of maternal IgG antibodies against the antigens of the Rh and, to a lesser extent, the ABO blood groups present on the foetal red cells. If the antibody is present in sufficient quantities, it leads to the destruction of the red cells and anaemia develops. The consequences are variable: at one extreme, hydrops fetalis, the baby is born prematurely, is pale and oedematous and usually perishes, at the other extreme anaemia develops during the second to eighth week of life. *See*: RH BLOOD GROUP SYSTEM.

erythrocyte Red blood cell. Small disk-shaped HAEMOGLOBIN-containing blood cell which transports oxygen. Mature adult mammalian erythrocytes have lost the nucleus and most of the other cellular organelles; haemoglobin synthesis takes place in the erythrocyte precursors. Birds have nucleated erythrocytes. Erythrocytes have a limited life-span (~120 days in humans) and the population is continually being renewed from stem cells in the bone marrow (*see* HAEMATOPOIESIS). Erythrocyte ghosts — empty erythrocyte plasma membranes — can be produced by lysing erythrocytes in a hypotonic salt solution, and have been one of the main sources of information on MEMBRANE STRUCTURE. *See also*: ABO BLOOD GROUP SYSTEM; HAEMOGLOBIN AND ITS DISORDERS.

erythroid lineage Lineage of developing blood cells that will give rise to red blood cells.

erythroid-colony stimulating factor ERYTHROPOIETIN.

erythromycin A macrolide antibiotic produced by *Streptomyces erythraeus* which inhibits PROTEIN SYNTHESIS in prokaryotes by binding to the 50S subunit and blocking the translocation reaction on the ribosome. It is also inhibit mitochondrial protein synthesis in eukaryotes such as yeast.

erythropoiesis The formation of erythrocytes (red blood cells).

erythropoietin Glycoprotein hormone (M_r ~50 000) that regulates the rate of formation of red blood cells. In the human adult 80% is produced by the kidneys. It acts on committed erythroid precursors to stimulate cell proliferation and increase the circulating red cell mass. It is produced in response to tissue hypoxia. *See*: HAEMATOPOIESIS.

ES cells EMBRYONIC STEM CELLS.

esc The EXTRA SEX COMBS gene in *Drosophila*.

Escherichia coli

THE study of *Escherichia coli* and its viruses has been central to the development of molecular biology. This species of bacterium was already well characterized and extensively studied by microbiologists and biochemists when, by good fortune, the strains chosen by the earliest workers in microbial genetics proved to have properties which made their experiments particularly rewarding. The interplay between the biochemical and genetical approaches which then developed has created the model by which other, more complex, organisms are now studied [1]. The development with *E. coli* of RECOMBINANT DNA TECHNOLOGY has provided a major new route to the study of the genes of higher organisms, one which is already creating a revolution in biological science. Most of the major ideas in molecular biology (with the principal exception of RNA SPLICING), were first developed in the course of work on *E. coli* and its BACTERIOPHAGES.

Initially, the two main reasons for the concentration on this one species were that it is especially easy to grow and that it was thought a safe, that is nonpathogenic, organism. *E. coli* grows rapidly in the simplest of media, requiring only glucose, ammonia, phosphate and a few cations, Na^+, K^+, Mg^{2+}, and Ca^{2+} and trace elements. Because it is so widespread — it is present in the lower small intestine and colon of all warm-blooded animals — it was thought safe to handle in bulk in the laboratory by nonmedical microbiologists, biochemists and students. Because of this, much was known about the species and its bacteriophages when the need for simple biological systems became apparent in the late 1930s.

Morphology and physiology

For the microbiologist, *E. coli* is a Gram-negative, rod-shaped bacterium ($1.2 \times 2.4\ \mu m$) with FLAGELLA, FIMBRIAE and, in some strains, PILI (Fig. E12). It does not form spores or show any form of differentiation during or after growth. Nor does it produce secondary metabolites in any quantity. As might be expected from its normal habitat, it is a heterotroph, that is, it grows by breaking down complex organic molecules, and a facultative anaerobe, that is, it can grow in the absence of oxygen by

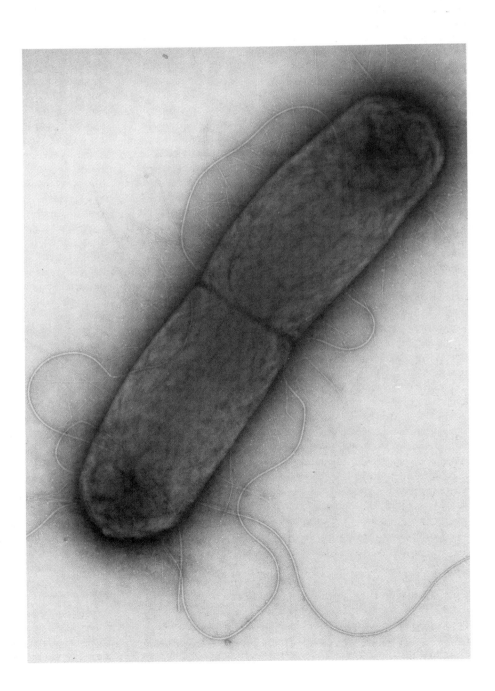

Fig. E12 An electron micrograph of *E. coli* embedded in an electron-dense material showing both flagella and pili.

fermentation. The genome contains $\sim 4.7 \times 10^7$ bp made up nearly equally of AT and GC base pairs.

Because of its ease of cultivation *E. coli* is readily isolated from natural sources. It can be obtained nearly pure in one step on simple media and was thus isolated and described early (1885). Often called *Bacterium coli* until the 1940s, the genus is now named after the German bacteriologist, Escherich. It is a member of the Enterobacteriacae, a family with many members of medical importance. The cause of bacillary dysentery, *Shigella dysenteriae*, is regarded by many bacteriologists as *E. coli* in all but name because of the relatedness of its DNA to that of *E. coli* by HYBRIDIZATION tests. The salmonellae, which include *Salmonella typhi*, the cause of typhoid, are thought to have diverged from the ancestral line which gave rise to the present species about a million years ago.

Although a minor component of the intestinal flora, its ease of isolation and identification has led water engineers and others to use the demonstration of the presence of *E. coli* as a diagnostic of faecal contamination.

Morphologically the cell has an outer membrane consisting of proteins, lipopolysaccharides and lipoproteins separated from the inner membrane by what is called the periplasmic space although it is largely occupied by proteins, many of them concerned with the uptake of metabolites. The PEPTIDOGLYCAN structure which provides the mechanical strength necessary to resist the pressure (~ 3 atmospheres) due to the osmotic strength of the cytoplasm is also found here (*see* BACTERIAL ENVELOPES). The outer membrane contains the antigens which provide the basis for the classification of individual isolates. A great number of variants are known, although most fall into a relative few major groups. These strains differ in other significant properties (e.g. enzymes, resistance to bacteriophages) to varying extents. The strain which has dominated microbial genetics — K12 — was isolated from a stool from a patient recovering from diphtheria in California in 1921. The use of more than one strain of *E. coli*, in particular, strains B and K12, and the rapid appreciation of the significance of strain-dependent phenomena several times proved of great value in the development of molecular biology. There is now a growing interest in detailed comparisons of *E. coli* strains as a route to its evolutionary history.

The outer membrane also contains several of the absorption sites for bacteriophages. These are often the structures associated with uptake of nutrients; for example, phage T1 absorbs to the receptor responsible for the uptake of Fe complexed with SIDERO-PHORES. There is no internal differentiation and no nuclear membrane, although the DNA appears to be organized into a looped structure; the nucleoid. *E. coli* can thus be regarded as similar to many other kinds of Gram-negative bacteria and their study can be based upon models developed for this species.

Safety

Long regarded as a harmless commensal in the normal healthy gut it is now clear that *E. coli* is not as 'safe' as first thought. It is an opportunistic pathogen and it can acquire PLASMIDS which confer upon their hosts the capacity to cause serious diarrhoea especially in young animals, because they produce toxins and colonization factors encoded by plasmid-borne genes (see BACTERIAL PATHOGENICITY). There are also plasmids which confer haemolytic properties. Nevertheless, the strains commonly used in the laboratory have been maintained in pure culture for upwards of 60 years and they appear to have lost the capacity to compete with established strains in animals (and man) when deliberately introduced. This point was of special importance with the development of recombinant DNA technology because *E. coli* was the obvious vehicle. It was thought wise to use only strains which have been further enfeebled by mutations in major pathways to reduce their competitiveness still further.

Genetics

Initially for those interested in what is now thought of as molecular biology, the attraction was the availability of well-characterized bacteriophages, notably phages T2 and T4 although later λ, P1, ΦX174, M13 and Mu came to play important roles in the development of the subject. Phages T2 and T4 were used by Delbruck, Luria, and Hershey in the 1940s and later by Benzer in the 1950s to demonstrate and study MUTATION and genetic RECOMBINATION in, at first sight, an especially simple system. Very rare events could be studied reproducibly because the number of individuals that could be examined at the same time ($\sim 10^9$) was so large, especially in comparison with the fruit fly, *Drosophila melanogaster*, and corn, *Zea mays*, the two species most studied by geneticists up to that time. The Hershey–Chase experiment [2] was influential as a simple and dramatic demonstration that all the genetic information of a virus was associated with its DNA and not its proteins. Brenner and Crick later deduced many of the features of the genetic code by experiments which relied upon the fact that *r*II mutants of T4 can grow in *E. coli* strain B but not in *E. coli* strain K12.

However, it was the simultaneous development of *E. coli* genetics by Lederberg and Tatum [3] which had the greater impact. The availability of stable auxotrophic mutants and the fortunate choice of *E. coli* K12 led to the demonstration that genetic information could pass from one bacterium to another by cell-to-cell contact (CONJUGATION) in an orderly manner which could be interpreted in terms of linkage groups as in eukaryotic organisms. Hitherto, *E. coli* (indeed all bacteria) had not been thought a suitable organism for genetics and there was, even, a debate as to whether bacteria when subject to selection, mutate or 'adapt'.

This discovery was rapidly developed. Hayes showed the transfer to be unidirectional and dependent on the presence in the donor strain of a factor, F (for fertility) (*see* F FACTOR). This factor was later shown to be a PLASMID, the first to be discovered of what has proved to be a large class of autonomous, small chromosomes widespread in nature and especially in Gram-negative organisms. It was fortunate that *E. coli* K12 is F^+, as less than one in twenty *E. coli* strains carry a sex factor. The development of the Hfr strains (mutants which give high yields of recombinants) as donors led to the demonstration that the *E. coli* chromosome is circular and not, as in eukaryotes, linear. All bacteria so far examined have circular chromosomes, as do all plasmids and most bacteriophages.

At the start, microbial genetics relied upon simple auxotrophic and antibiotic resistant mutants but the development of partial diploids and TEMPERATURE-SENSITIVE MUTANTS made it possible to study genetically, and thus often point the way for biochemists to study biochemically, the essential features of every aspect of the cell, its metabolism, growth and division. The major metabolic pathways and their control, DNA replication, repair, and recombination, cell wall synthesis, cell separation, protein export, mechanisms of action of antibiotics, motility and chemotaxis have all yielded, at least in part, to this approach. More than 1500 genes are now known, most have been mapped (Fig. E13), and the number for which the DNA sequence is known is increasing rapidly. It was over 400 by the end of 1992 and sequencing has become a route to the identification of new genes. The concentration on *E. coli* K12 became more intense as the accumulation of techniques, specially modified phages and plasmids and the detailed map of the chromosome made it progressively easier to tackle a complex problem in *E. coli* rather than any other bacterium. Only secondary metabolism and differentiation cannot be studied with *E. coli*. The final step was the development of DNA technology.

Although *E. coli* K12 became the strain of choice for most experiments from the 1950s it was fortunate that some experiments continued using other strains. A most significant observation involving a strain difference was that of Weiglé [4] who noticed that while phage λ grown in K12 could infect cells of strain C with high efficiency, the converse was not true. Only one in 10^4 phage grown in strain C could infect successfully cells of K12. This led to the study of host-controlled MODIFICATION and this in turn led to the discovery of RESTRICTION ENZYMES (restriction endonucleases) and their development as an essential tool of RECOMBINANT DNA TECHNOLOGY. *E. coli* is not normally capable of genetic TRANSFORMATION by naked DNA but cells starved in 0.1 M Ca^{2+} at 0°C and given a short heat shock (2 min at 40–42°C) take up plasmids efficiently. This technique led directly to the development of DNA CLONING.

E. coli are present in the gut within a few days of birth and throughout life. Thus even if only the population of *E. coli* in man is considered the number of individual cells present in the world at any one time, though difficult to estimate accurately, must be very large. The mutation rate, ~1 bp change per 10^{10} bp replicated (i.e. 1 bp per 200 replications), would suggest that there ought to be a great many variants but this is not found. Most of the isolates obtained in a short period of time can be grouped into relatively few major clones. This implies that the pressures which maintain the homogeneity of the species are highly selective. The same must be true of other bacteria but only for organisms associated with man are data accumulated on a world-wide scale. What these pressures are can only be guessed at but it is likely that in the near future ecological studies of *E. coli* will provide a major stimulus to the understanding of bacterial evolution.

K.A. STACEY

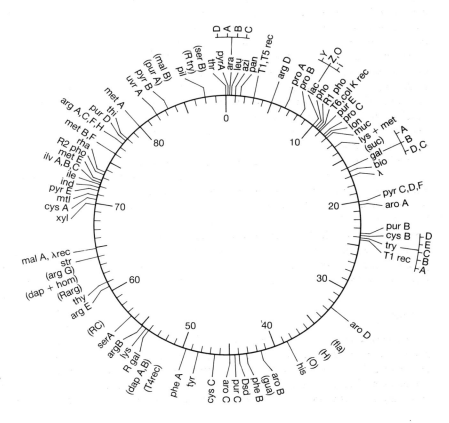

Fig. E13 The first map of the chromosome of *E. coli* to show the position of 100 genes (1964). The corresponding map in 1994 would show over 1500. Reprinted from [5].

1 Neidhardt, F.C. et al. (Eds) (1987) Escherichia coli *and* Salmonella typhimurium: *Cellular and Molecular Biology* (American Society for Microbiology, Washington, DC).
2 Hershey, A.D. & Chase, M. (1952) Independent functions of viral protein and nucleic acid in growth of bacteriophage. *J. Gen. Physiol.* **36**, 39.
3 Lederberg, E.M. & Tatum, E.L. (1946) Gene recombination in *E. coli. Nature* **158**, 558.
4 Bertani, G. & Weiglé J.J. (1953) Host-controlled variation in bacterial viruses. *J. Bact.* **65**, 113–121.
5 Taylor, A.L. & Thoman, M.S. (1964) The chromosome map of *Escherichia coli* K12. *Genetics* **50**, 659.

ESR/EPR spectroscopy

ELECTRON spin resonance (ESR) spectroscopy, also known as electron paramagnetic resonance (EPR) spectroscopy, is a magnetic resonance technique used for the study of paramagnetic materials, that is, those that contain unpaired electrons [1,2]. The paramagnetic species found in biological materials include free radicals and transition metal ions. ESR/EPR spectroscopy is analogous to nuclear magnetic resonance (NMR) spectroscopy (*see* NMR SPECTROSCOPY), in that it involves resonant absorption of electromagnetic radiation by the electron spins ($S = 1/2$), in which a splitting of energy levels is induced by an applied magnetic field. The principal differences between ESR/EPR and NMR arise from the relative scarcity of unpaired electrons, and from the much greater magnetic moment of the electron compared with those of protons and other nuclei. Thus ESR/EPR spectroscopy is more sensitive and selective.

The two terms, ESR and EPR spectroscopy, both refer to measurements made with the same instrument. The choice of term is at present a matter of personal preference. Some spectroscopists distinguish between ESR, which is applied to radicals, which have an identifiable electron spin (and generally have G-FACTORS close to the free-electron value of 2.00232), and EPR, which is applied to transition metal ions, where the distribution of electrons in the *d* orbitals leads to paramagnetism.

Radicals are produced during oxidation–reduction processes, such as reactions of reduced compounds with oxygen, and in the reactions of enzymes such as FLAVOPROTEINS. Radicals are also induced by visible, ultraviolet or ionizing radiation. Some of them, notably nitroxides, and some radicals in the solid state, are stable and can readily be detected by ESR/EPR. More short-lived radicals, such as superoxide or hydroxyl radicals, may be detected by adding a spin trap to the system [3]. Spin traps are diamagnetic compounds, usually nitroso compounds or nitrones, which react with the free radicals to produce more stable free radicals, usually nitroxides. The ESR/EPR spectrum of the radical adduct provides information about the chemical nature of the original radical [4].

The properties of the ESR/EPR spectrum may be exploited by introducing stable free radicals into biological systems [5]. They are described either as spin labels, which are introduced specifically into particular sites such as amino-acid residues of a protein to measure properties of the molecule such as rates of motion, or spin probes, which are attached noncovalently and serve to measure general properties of their environment. The most commonly used radicals for this purpose are nitroxides, for which

the theory of line-broadening is well developed. In the case of transition ions, ESR/EPR is able to detect oxidation states for which the electron spin, S, is an odd number: chromium(V), vanadium(IV), manganese(II), iron(III), cobalt(II), nickel(III), and molybdenum(V). Of these, only the first three are readily detectable at ambient temperatures in solution. Mn^{2+} will often substitute for magnesium in complexes with proteins and nucleotides, and is therefore a useful probe. ESR/EPR spectra of the other transition ions are broadened out at ambient temperatures, due to rapid electron-spin relaxation. These ions are observed at cryogenic temperatures, using liquid nitrogen or liquid helium. Metal clusters such as iron–sulphur clusters are also detectable, if the net spin is odd, as in reduced FERREDOXIN.

Principles of operation

In a typical ESR/EPR spectrometer (Fig. E14*a*), the sample is placed in a microwave cavity, and in a magnetic field of strength B_0. Resonant absorption of microwave radiation, frequency ν, by the electrons in the sample takes place, and is detected with a diode (Fig. E14*a*) [6]. The condition for resonant absorption is that the energy of the microwave quantum is equal to the Zeeman splitting between the electron energy levels:

$$\hbar\nu = g\mu_B B_0$$

where $\hbar$ and μ_B are Planck's constant and the Bohr magneton respectively. The factor g is treated as a spectroscopic variable which is a characteristic of the paramagnet. In contrast to most other forms of spectroscopy, the frequency of radiation is not varied, because the cavity resonates at a fixed frequency. Instead, the magnetic field is varied, in order to bring the different paramagnets into resonance (Fig. E14*b*). For a typical radical with $g = 2.0023$, and a microwave frequency of 9 GHz, the field required would be 321 millitesla (3210 gauss). It can be seen from this equation that the higher the g-factor, the lower the resonant magnetic field. The spectra are usually presented as the first derivative, produced by modulation of the magnetic field and phase-sensitive detection. The signal amplitude increases with higher microwave power (typically 1–200 mW), until the limit of microwave power saturation is reached. This depends on the electron spin-lattice relaxation time T_1; the signal is saturated more readily with longer relaxation time, that is, slower relaxation.

Sample preparation

Free radicals, including spin labels, and certain metal ions, notably Mn^{2+}, are detectable in solution at physiological temperatures. Because of microwave absorption by water, aqueous samples are examined in narrow flat cells or capillaries, with minimum sample volumes less than 100 µl. Minimal detectable concentrations are in the range 0.1–10 µM. For frozen samples, quartz tubes of ~3 mm internal diameter, containing a minimum volume of 120 µl, are used. Minimum detectable concentrations of transition ions vary from 0.1 to 100 µM, depending on the form of the spectrum. Spectra of ions such as Fe(III) are broad, and more difficult to detect. This is offset by the more favourable

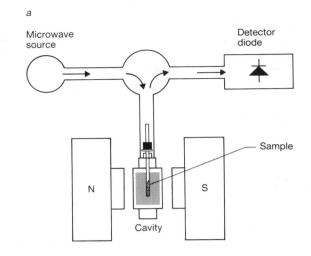

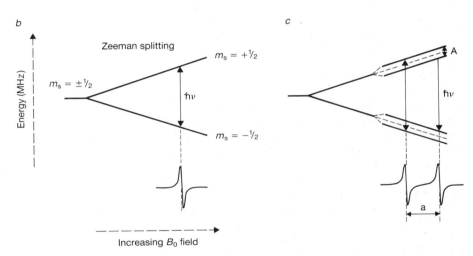

Fig. E14 *a*, Diagram of an ESR/EPR spectrometer. The microwave source is a klystron or Gunn diode. The paramagnetic sample is placed in a microwave cavity, which resonates at the microwave frequency, ν, and in a homogeneous magnetic field B_0. *b*, Resonance condition. Zeeman splitting of the electron energy levels, corresponding to spin quantum numbers $m_S = \pm 1/2$. *c*, Hyperfine splitting of the energy levels due to interaction with a nucleus, spin 1/2. Not to scale; the hyperfine splitting is typically three orders of magnitude smaller than the Zeeman splitting.

Boltzmann distribution, which determines the proportion of electrons which can contribute to the absorption, and varies inversely to the temperature (in Kelvin).

Information derived from ESR/EPR spectra

The spectra of different types of paramagnetic species have characteristic lineshapes. Examples are given in Figs E15 and E16. The spectra can be interpreted at a fundamental level, giving detailed information about molecular structure. Alternatively, ESR/EPR is frequently used as a 'fingerprint' technique to identify different paramagnetic species. ESR/EPR is quantitative, in that the integrated intensity of a signal is a direct measure of the number of unpaired electrons in the sample.

An important factor which affects the spectrum is the hyperfine interaction between the electron spin S, and nearby nuclear spins in the molecule. The hyperfine coupling, A (in MHz), leads to a splitting a (often expressed in mT) in the spectral lines (Fig. E14*c*). A second important interaction is that between electron spins. If there are several unpaired electrons in the same atom, as

in high-spin Fe(III), these are strongly coupled, by exchange interactions, to give a complex set of energy levels. Such interactions can give rise to spectra which are spread over a wide range of magnetic field (metmyoglobin in Fig. E16). Weaker interactions may be observed between electrons on adjacent paramagnetic centres, at distances up to 2 nm. These interactions are characterized by broadening of the spectrum, and enhancement of electron-spin relaxation rates.

Information obtained about radicals

ESR/EPR spectra of radicals in solution, as in Fig. E15, consist of numerous sharp lines, due to hyperfine splittings of the electron energy levels by protons and other nuclear spins.

1 The number and types of nuclei in the radical. Each nucleus, with spin I, causes a hyperfine splitting of the spectrum into $(2I + 1)$ lines, and the splittings are additive. Thus the benzoquinone radical is split into five lines by interaction with four protons (actually there are sixteen lines but some of these overlap as the protons are equivalent). The nitroxide spin-label spectrum is split

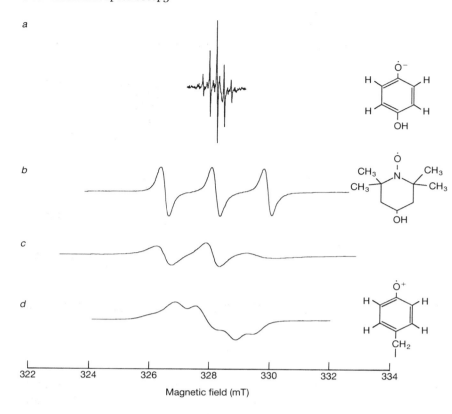

Fig. E15 Typical continuous-wave ESR/EPR spectra of free radicals. *a*, A photochemically generated benzoquinol radical in solution, with hyperfine splittings due to four equivalent protons; *b*, *c*, a spin-label nitroxide (TEMPOL) in ethanol and in glycerol, at 25°C. Note the splitting of the nitroxide radical into three lines by hyperfine interaction with the ^{14}N nucleus ($I = 1$), and the broadening of *c* due to slower motion in the more viscous solvent; *d*, tyrosyl cation radical in PHOTOSYSTEM II from spinach.

into three lines by the ^{14}N nucleus ($I = 1$). The hyperfine structure in the protein tyrosyl radical due to four protons (Fig. E15*d*) is less well resolved because of line broadening in the solid state.

2 The unpaired electron distribution on the radical. The magnitude of each hyperfine splitting depends on the nuclear moment, and the extent to which the electron density is delocalized onto the nucleus. This information is employed in molecular-orbital calculations of radical structure.

Information from spin labels and spin probes

1 Motion. The ESR/EPR spectra of paramagnets in solution, particularly radicals, are sensitive to rotational and translational motion, in the time regime 10^7–10^{10} s^{-1}. The spectra become broader as motion is restricted (Fig. E15*c*). By the use of special operating conditions, known as saturation-transfer ESR, it is possible to extend the time range of molecular motion to 10^4 s^{-1}, where comparisons can be made with NMR. Spin label studies have been notably successful in studies of the motion of lipids and proteins in membranes.

2 Polarity of the solvent. The spectra of nitroxides show greater line separations in a hydrophilic than in a hydrophobic environment.

3 Oxygen concentration. Oxygen is paramagnetic, and causes broadening of spectra of radicals. This effect is exploited in ESR oximetry, which can measure very low (micromolar) concentrations of oxygen. For this purpose, stable radicals with narrow-line ESR spectra are used.

Information from ESR/EPR spectra of transition ions

1 The concentration of a particular species. As not all states of a transition ion are detectable, ESR/EPR is not a suitable technique for measuring the *total* amount of a transition ion in a solution. However it can measure the amounts of specific paramagnetic species, in a complex mixture.

2 The types of metal ions may be determined by comparison of the *g*-factors with other complexes, or by hyperfine interactions with nuclear spins such as ^{55}Mn or ^{65}Cu. Identification may be assisted by substitution of the sample with stable isotopes having nuclear spins, such as ^{57}Fe ($I = 1/2$).

3 The oxidation state of the metal ions. Oxidation or reduction leads to change in the number of unpaired electrons. The paramagnetic state may change into a diamagnetic state, which has $S = 0$, and is undetectable by ESR/EPR, or a state with an even spin, such as high-spin Fe(II) ($S = 2$), which usually gives either very broad ESR/EPR spectra or is not detectable at all. Thus we can detect the Fe(III) in METHAEMOGLOBIN (see metmyoglobin in Fig. E16), but not the Fe(II) in HAEMOGLOBIN.

4 Spin states, such as high-spin Fe(III), ($S = 5/2$) and low-spin Fe(III), ($S = 1/2$).

5 Ligands to the metal ion can be detected by superhyperfine interactions with the nuclei of ligand atoms such as ^{14}N ($I = 1$). Here again, isotopic substitution may be applied.

6 Coordination geometry. The spectra are an indication of the symmetry properties of the coordination site, though not an infallible one in the distorted geometries found in proteins. The

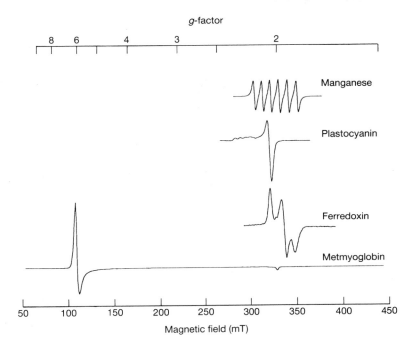

Fig. E16 ESR/EPR spectra of transition metals: Mn^{2+} in solution; a copper protein (PLASTOCYANIN); a non-haem iron cluster protein (reduced FERREDOXIN); a haem iron protein (METMYOGLOBIN). Note that the spectral range is greater than those for the free radicals (Fig. E15), and much greater than the parts per million range of NMR. The manganese spectrum is split into six lines by hyperfine interaction with the ^{55}Mn nucleus ($I = 5/2$), and the plastocyanin shows splitting into four lines in the low-field region due to hyperfine interaction with a mixture of the naturally occurring isotopes ^{63}Cu and ^{65}Cu nuclei ($I = 3/2$).

g-factors are often very sensitive to perturbations of the ligands around the metal ion.

A number of important biological electron-transfer agents owe their discovery to EPR spectroscopy. The iron–sulphur proteins of the mitochondrial ELECTRON TRANSFER CHAIN are an example. They have broad and inconspicuous optical absorption spectra but are predominantly observed in the low-temperature EPR spectra of mitochondria and even of whole heart tissue [7]. In chloroplasts, the primary processes of photosynthesis are light-driven reactions in which chlorophyll is oxidized to produce a radical, and the electron is transferred through a series of carriers including transition ion complexes and quinones. These reactions occur even at liquid helium temperature, and can be observed in the ESR/EPR spectrometer [8,9].

Types of ESR/EPR spectrometer

Various types of ESR/EPR spectrometer are in use [10]:

1 The conventional continuous-wave spectrometer (see Fig. E14), in which the sample is subjected to continuous microwave radiation. This type of spectrometer is used in the majority of biological applications. The method of operation is in some ways analogous to an optical spectrophotometer, except that the sample is in a magnetic field. Electromagnets are used rather than superconducting magnets, because of the requirement to sweep the field over a relatively wide range. Continuous-wave spectrometers normally work over a narrow range of microwave frequency; the commonest is X-band, 8.5–10 GHz.

2 Continuous-wave spectrometers operating at other frequencies, from 0.3 to 200 GHz. These permit the discrimination of different splittings of the electron energy levels, such as g-factors, nuclear hyperfine interactions and electron–electron interactions. Each step in frequency necessitates the replacement of most of the microwave circuit. Generally these are less sensitive than X-band spectrometers, and require higher concentrations of sample.

3 Pulsed, or time-domain, ESR/EPR spectrometers, in which the microwave power is applied to the sample as a series of intense pulses, of high power (up to 1 kW) and short duration (tens of nanoseconds), and a signal is detected in the form of a spin-echo. The pulsed method is analogous to pulsed NMR spectrometry. For instrumental reasons the enhancement in sensitivity offered by the pulsed method for NMR is not realized in ESR/EPR. The pulsed method is reserved for specialist applications. These include the detection of transient radicals produced by laser flashes, the measurement of electron-spin relaxation rates, and the measurement of very weak hyperfine interactions through electron spin-echo envelope modulation (ESEEM).

4 Electron-nuclear double resonance (ENDOR), in which the sample is irradiated simultaneously with saturating microwave radiation to excite the electron spins, and with radio-frequency radiation, to excite the nuclear spins. It may be considered as an NMR experiment with ESR/EPR detection. ENDOR is used to measure hyperfine couplings, and can provide detailed structural information about nuclear spins near to the paramagnetic centre. For spectra of radicals with several inequivalent nuclear hyperfine splittings, ENDOR has the effect of simplifying the spectrum. For spectra of solid-state samples it permits the observation of splittings which are concealed by the broad ESR/EPR lines. Pulsed ENDOR spectrometers also exist; they avoid some of the limitations of ENDOR, for example the requirement for microwave power saturation.

5 ESR/EPR imaging is an interesting potential application, which has not so far been widely used. It is being developed to examine the distribution of radicals or spin-labels in samples of material.

For imaging of biological materials, low frequencies, below 1.5 GHz, are preferred to avoid the microwave absorption by water.

R. CAMMACK

1 Kosman, D. J. (1984) Electron spin resonance. In *Physical Techniques in Biology and Medicine* (Rousseau, D. L., Ed.) 89–243 (Academic Press, Orlando, FL).
2 Sealy, R.C. et al. (1985) Electron spin resonance. In *Modern Physical Methods in Biochemistry, Part A* (Neuberger, A. & Van Deenen, L.L.M., Eds) 69–148 (Elsevier, Amsterdam).
3 Mason, R.P. & Mottley, C. (1987) In *Electron Spin Resonance* (Symons, M.C.R., Ed.) *R. Soc. Chem. Spec. Period. Rep.* **10B**, 185.
4 Buettner, G.R. (1987) Spin trapping: ESR parameters of spin adducts. *Free Radical Biol. Med.* **3**, 259–304.
5 Berliner, L.J. & Reuben, J. (Eds) (1989) *Biological Magnetic Resonance, Vol. 8: Spin Labels* (Plenum, New York).
6 Eaton, G. R. & Eaton, S. S. (1990) Electron paramagnetic resonance. In *Analytical Instrumentation Handbook* (Ewing, G.W., Ed.) 467–530 (Marcel Dekker, New York).
7 Beinert, H. (1978) EPR spectroscopy of components of the mitochondrial electron-transfer chain. *Meth. Enzymol.* **54**, 133–150.
8 Golbeck, J.H. Bryant, D.A. (1991) Photosystem I. *Curr. Topics Bioenerget.* **16**, 83–177.
9 Miller, A.F. & Brudvig, G.W. (1991) A guide to electron paramagnetic resonance spectroscopy of photosystem II membranes. *Biochim. Biophys. Acta* **1056**(1), 1–18.
10 Hoff, A.J. (1989) *Advanced EPR: Applications in Biology and Biochemistry* (Elsevier, Amsterdam).

estradiol Alternative spelling of OESTRADIOL.

ethephon Common name, approved by the American National Standards Institute, for 2-chloroethylphosphonic acid (Fig. E17) (M_r 144.5), a plant growth regulator that decomposes to generate the gaseous plant hormone ETHYLENE at pH ± 3 within treated tissues. Ethephon is used extensively in agriculture and horticulture to promote fruit ripening and abscission, to shorten shoot extension in arable crops, and to stimulate basal branching in roses and *Pelargonium*. Ethephon can be used as a substitute for ethylene gas; it is available commercially as a liquid concentrate (480 kg m^{-3}) and is active as an aqueous spray at 10–1000 g m^{-3}. In older literature, ethephon is referred to by its original trade name of 'Ethrel' or by the abbreviation CEPA.

Thomas T. H. (Ed.) (1982) *Plant Growth Regulator Potential and Practice* (British Crop Protection Council, Croydon).

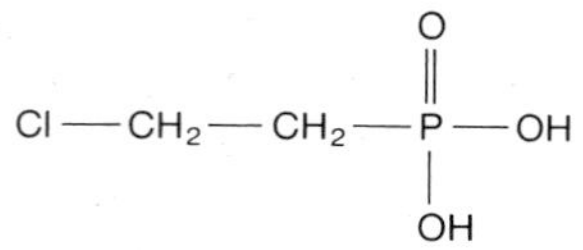

Fig. E17 Structure of ethephon.

ethidium bromide (Fig. E18) A compound which can intercalate into DNA causing the double helix to partially unwind; in so doing it affects the buoyant density of the DNA. This phenomenon is taken advantage of to increase separation of plasmid DNA from genomic DNA in caesium chloride–ethidium bromide BUOYANT DENSITY CENTRIFUGATION. Also used as a stain to visualize

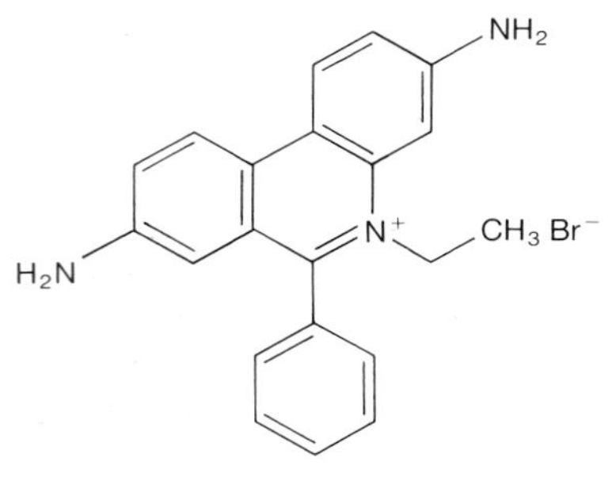

Fig. E18 Ethidium bromide.

DNA bands after agarose gel ELECTROPHORESIS. Once bound to the DNA fragments, ethidium bromide fluoresces under ultraviolet light. Ethidium bromide has also been shown to induce *petite* mutants in yeast.

Ethrel™ Original trade name for ETHEPHON (2-chloroethylphosphonic acid) when introduced in 1965 by Amchem Products Inc. as a plant growth regulator active by decomposing in treated tissue to form the gaseous plant hormone ETHYLENE.

ethyl maleimide (*N*-ethyl maleimide) Reagent used for studies of vesicle transport and PROTEIN TARGETING in cells as it inhibits a particular protein (the *N*-ethyl maleimide-sensitive fusion protein, the NSF protein) involved in the secretory pathway. *See also*: PROTEIN SECRETION; PROTEIN TRANSLOCATION.

N-ethyl maleimide-sensitive fusion protein *See*: NSF.

ethylene Ethene, M_r 28.05. The first plant growth substance to be discovered (1908), now recognized as a gaseous PLANT HORMONE with a varied spectrum of effects that are tissue and/or species specific. Activity increases approximately linearly with the $\log_{10}$ of the partial pressure in air from 0.0001 Pa up to 1.0 Pa. Larger amounts have little further effect. Growth responses commence within a few minutes of application. Physiological activities include the promotion of fruit ripening and abscission of leaves and fruit of dicotyledonous species, flower senescence, stem extension of aquatic plants, gas space (aerenchyma) development in roots, leaf epinastic curvatures, stem and shoot swelling (in association with stunting), femaleness in curcubits, fruit growth in certain species, apical hook closure in etiolated shoots, root hair formation, flowering in the Bromeliaceae, and DIAGEOTROPISM of etiolated shoots. Certain ethylene responses are known to involve increased gene expression. The biosynthetic pathway, from methionine, is well understood for most higher plants (*see* ACC; ACC OXIDASE; ACC SYNTHASE), but not for many ferns and other lower plants.

etioplast A form of PLASTID that develops in the leaves of dark-grown plants and develops further into a normal chloroplast upon illumination. It contains PROTOCHLOROPHYLL (an immediate biosynthetic precursor of CHLOROPHYLL) and some protein components of normal thylakoids organized into a paracrystalline array of lipid and protein known as the prolamellar body. Chlorophyll and most of the photosynthetic apparatus does not form until thylakoids develop during illumination of the leaf.

ETS domain A conserved DNA-binding domain of approximately 85 amino acids named after the first member of the *ets* gene family to be identified: the E26 ('E-twenty-six') acutely transforming RETROVIRUS of chicken. Ets proteins are regulators of TRANSCRIPTION, and some members, Ets1 and Ets2, act in cooperation with the AP-1 transcription factor; others interact with the SERUM RESPONSE FACTOR (SRF). Some Ets proteins possess transforming activity and *ets* family members, *fli-1* and *spi-1*, are activated by proviral integration in murine erythroleukaemias.

eubacteria One of three primary kingdoms of organisms (*see also* ARCHAEBACTERIA; EUKARYOTES) in some classifications; various classes of PROKARYOTES in others. They are widespread microorganisms that lack a nucleus and other cellular organelles. They are characterized by their susceptibility to various antibiotics, by their rRNAs and tRNAs, and by the presence of muramic acid within the cell walls (*see* BACTERIAL ENVELOPE; SUGARS). The eubacteria are generally considered to include the heterotrophic Gram-negative and Gram-positive bacteria (e.g. *Escherichia coli* and *Bacillus subtilis* respectively), the actinomycetes, the CYANOBACTERIA, and other photosynthetic bacteria.

euchromatin The bulk of CHROMATIN in the nucleus. During interphase it is not condensed and is therefore faintly staining. It becomes maximally condensed and darkly staining in METAPHASE.

eukaryotes Organisms whose cells have a complex compartmentalized internal structure in which different cellular functions are carried out in membrane-bounded organelles (*see* ANIMAL CELL; PLANT CELL). The DNA is associated with histones to form CHROMATIN and is organized into several or more CHROMOSOMES which are enclosed in a NUCLEUS. As a consequence of the presence of a nuclear membrane, gene TRANSCRIPTION and TRANSLATION in eukaryotes are separated in space and time, in contrast to prokaryotes. The eukaryotes comprise the unicellular fungi (e.g. yeasts), algae (e.g. *Chlamydomonas*), and protozoa (e.g. *Amoeba*) and all true multicellular organisms (algae, fungi, plants, and animals). Eukaryotes more complex than sponges and coelenterates are sometimes referred to as 'higher eukaryotes' while simple eukaryotic organisms (especially the yeasts) are called 'lower eukaryotes'.

eukaryotic cell *See*: ANIMAL CELL; EUKARYOTE; PLANT CELL.

Eukaryotic gene expression

THE vast array of different cells and tissues within multicellular plants and animals all synthesize different sets of proteins, some of which are unique to that particular cell. For example, immunoglobulins are produced only in B lymphocytes (B cells), insulin in pancreatic β cells, haemoglobin in red cell precursors and so on. Yet with very few exceptions, the set of genes is identical in all the different cell types and in the single-celled ZYGOTE from which they all arose during embryonic development. Clearly therefore, gene expression must be controlled in order to determine which genes will be active and produce proteins in each cell type (*see also* DIFFERENTIATION).

Levels of gene regulation

In principle, any of the stages by which the information in the DNA is converted into protein might be regulated so that specific proteins are produced only in certain tissues [1]. Thus regulation might take place by deciding which genes are transcribed into RNA (*see* TRANSCRIPTION) or by regulating which of these RNA products are correctly spliced to produce a functional messenger RNA (*see* RNA SPLICING). Gene regulation could also occur by deciding which of these fully spliced mRNAs are transported to the cytoplasm or by regulating which mRNAs are translated into protein (*see* PROTEIN SYNTHESIS).

Indeed, there is evidence that in particular cases gene regulation is regulated at each of these stages [2]. The production of many new proteins in the egg following fertilization is dependent on the translation into protein of fully spliced mRNAs pre-existing in the cytoplasm of the unfertilized egg but whose translation is blocked before fertilization occurs. This form of gene regulation is known as translational control.

By splicing the protein-coding regions (exons) of a single PRIMARY TRANSCRIPT in different combinations it is possible to produce two or more different mRNAs encoding different proteins in different tissues. This alternative splicing [3] is well illustrated in the case of the single gene that encodes both the calcium-modulating hormone calcitonin and a potent vasodilator — the calcitonin gene-related peptide (CGRP). This gene is transcribed into a primary RNA transcript in both thyroid gland and brain but a different combination of exons is spliced together in each tissue to produce calcitonin mRNA in the thyroid and CGRP mRNA in the brain.

But although a number of instances of regulation of gene expression after the first stage of transcription do exist, much evidence indicates that in most cases, gene regulation is achieved at the initial stage of transcription, by deciding which genes should be transcribed into a primary RNA transcript [2]. This type of control is referred to as transcriptional control. In these cases, once transcription has occurred, all the other stages in gene expression follow and the corresponding protein is produced. Even in the case of calcitonin/CGRP, alternative splicing is acting as a supplement to transcriptional control as the calcitonin/CGRP gene is only transcribed in the thyroid gland and the brain and not in other tissues. The regulation of gene transcription is therefore critical in the regulation of gene expression and the means by which this is achieved, in protein-coding genes especially, are the focus for the remainder of this entry.

Transcriptional control: regulatory DNA sequences

For a gene to be transcribed it is necessary for specific protein factors known as TRANSCRIPTION FACTORS [4,5] to bind to particular sites in the regulatory regions of the gene to induce its transcription by the enzyme RNA POLYMERASE.

In most eukaryotic protein-coding genes these binding sites are organized into four distinct elements (Fig. E19). Immediately adjacent to the start site of transcription is the PROMOTER, which in many genes contains the TATA box which is a binding site for the CONSTITUTIVELY expressed transcription factor TFIID (the TATA box binding factor, TBF). The binding of this factor to the promoter is critical to the assembly of a basal, stable transcriptional complex containing a variety of other transcription factors (TFIIA, B, E and F) as well as RNA polymerase II itself [6].

The level of transcription directed by the basal transcriptional complex bound at the promoter is greatly enhanced by the binding of other constitutively expressed factors to so-called upstream promoter elements (UPEs) which are usually located immediately 'upstream' (in terms of the direction of transcription) of the promoter itself [7]. Although the nature of the UPEs varies between different genes, very many genes contain either or both the CCAAT box which binds a variety of different transcription factors, and a GC-rich sequence which binds the constitutively expressed transcription factor Sp1.

The DNA sequences mentioned so far all bind constitutively expressed factors and therefore direct a similar level of transcription in all cell types. In addition, however, very many genes contain other regulatory DNA sequences, which are interdigitated with the UPEs and which bind transcription factors that only become active in specific cell types or in response to a particular signal. Thus the presence of these sequences can confer a specific expression pattern on a particular gene. For example, the presence of the specific binding site recognized by the HEAT SHOCK transcription factor will render a gene INDUCIBLE by raised temperature and the presence of the unrelated site which binds the glucocorticoid receptor (*see* STEROID RECEPTOR SUPERFAMILY) will render the gene inducible by treatment with glucocorticoid hormones.

Enhancers

All the DNA sequences described above are located close to the start site of transcription, with the promoter itself being immediately adjacent to the start site and the UPEs and the regulatory DNA sequences being interdigitated with one another upstream of the promoter. The transcription of eukaryotic genes can also be regulated however, by more distant elements known as ENHANC-ERS [8]. Enhancers contain binding sites for the same constitutively expressed or tissue-specific regulatory factors which bind immediately upstream of the promoter but often contain multiple copies of the binding site or binding sites for many different factors. This multiplicity of binding sites results in the binding of multiple transactivator proteins and therefore allows enhancers to function when located at great distances from the promoter, when placed upstream or downstream of it and when oriented in either direction relative to the promoter.

Although the enhancer cannot drive transcription itself, it can enhance the activity of the promoter by several orders of magnitude. Such enhancement may occur in all cell types if the enhancer contains binding sites for constitutively expressed transcription factors or may occur only in a specific tissue or in response to a specific signal if the enhancer contains binding sites for factors which are involved in gene regulation.

Activation of transcription

In many genes the binding of constitutively expressed transcription factors such as Sp1 and the CCAAT-box binding factors to UPEs will result in sufficient activation of the basal transcriptional complex for a high rate of transcription to occur. These genes will therefore be constitutively expressed in all tissues. In other genes however, further transcription factors must bind to specific regulatory DNA sequences upstream of the promoter or in the enhancer in order for high level transcription to take place. As the factors that bind to these regulatory DNA sequences are active only in specific cell types or following a specific signal, these genes will be expressed only in a limited range of cell types or following exposure to a particular stimulus.

It is clear therefore that processes which regulate the activity of specific transcription factors so that they stimulate the expression of particular genes only at the appropriate time or place are crucial to the control of eukaryotic gene expression. In some cases the transcription factor itself is synthesized in only one cell type and is absent from all other cell types (Fig. E20a). The IMMUNO-GLOBULIN GENES, for example, contain a binding site for the transcription factor Oct-2 (octamer binding transcription factor) upstream of the TATA box. Oct-2 is synthesized only in B cells and hence the immunoglobulin genes are only activated, leading eventually to antibody production, in B cells.

Similarly, genes expressed only in muscle cells such as creatine kinase contain binding sites for the MyoD transcription factor which is present only in muscle cells (*see* MYOGENESIS). This case

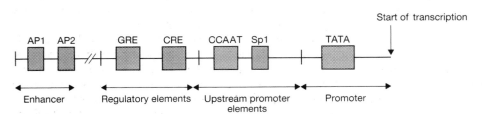

Fig. E19 Structure of a typical eukaryotic gene with a promoter containing a TATA box, upstream promoter elements (UPEs) such as the CCAAT and Sp1 boxes, regulatory elements inducing expression in specific cell types or in response to specific treatments such as cyclic AMP (CRE) or glucocorticoid hormones (GRE) and other elements within more distant enhancers. UPEs are often interdigitated with regulatory elements and the same regulatory elements are found both upstream of the promoter and in enhancers.

Start of transcription

| AP1 | AP2 | GRE | CRE | CCAAT | Sp1 | TATA |

Enhancer — Regulatory elements — Upstream promoter elements — Promoter

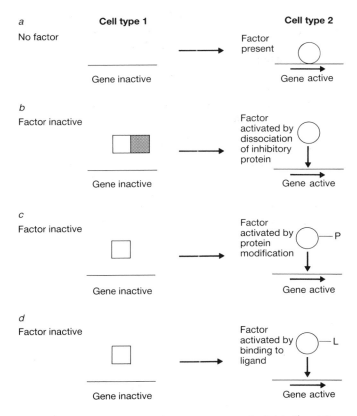

Fig. E20 Activation of transcription factors by new synthesis (*a*), dissociation of an inhibitory protein (*b*), protein modification (*c*) or ligand binding (*d*).

is even more dramatic however, as the artificial expression of MyoD in nonmuscle cells such as fibroblasts is sufficient to convert them into muscle cells, indicating that MyoD activates transcription of all the genes whose protein products are necessary to produce a differentiated muscle cell [9].

Unlike the case of MyoD, the expression of Oct-2 alone is not sufficient to produce differentiated B cells. This is because other transcription factors which are active only in B cells are also involved in producing the B cell-specific expression of genes such as those encoding the immunoglobulins. One such factor is NFκB which binds to a DNA sequence in the enhancer of the immunoglobulin κ light chain gene. Unlike Oct-2, however, NFκB is present in all cell types. But in most cells it is present in an inactive form in which it is complexed to an inhibitory protein, IκB, resulting in its being restricted to the cell cytoplasm [10]. In mature B cells however, NFκB is released from the inhibitory protein and moves to the nucleus where it can bind its DNA target sequence and activate transcription of the immunoglobulin κ light chain gene (Fig. E20*b*).

Hence, the action of transcription factors on gene expression can be controlled not only by regulating their synthesis (Fig. E20*a*) but also by regulating their activity (Fig. E20*b–d*). The combination of these two processes allows transcription factors to regulate the expression of numerous different genes in different cell types [4,5].

As shown in Fig. E20 several different mechanisms exist by which transcription factors can be converted from an inactive to an active form in response to a specific stimulus. In addition to the dissociation of an inhibitory protein (as in the case of NFκB (Fig. E20*b*)), transcription factors can also be activated by POST-TRANSLATIONAL MODIFICATIONS such as PROTEIN PHOSPHORYL-ATION (Fig. E20*c*). This is seen in the case of the CREB factor which mediates the induction of specific genes in response to CYCLIC AMP. This effect depends on the ability of cAMP to stimulate the activity of the enzyme PROTEIN KINASE A which phosphorylates CREB and converts it to an active form capable of stimulating the transcription of cAMP-dependent genes.

In addition to dissociation of protein–protein interactions and post-translational changes, transcription factors can also be activated by conformational changes induced by binding of a specific ligand (Fig. E20*d*). This is the case with the ACE1 factor in yeast which undergoes a conformational change following binding of copper. This conformational change converts ACE1 to an active form and results in the activation of specific ACE1-dependent genes in response to copper.

In particular cases more than one of these mechanisms of transcription factor activation are used together. Although NFκB is activated by dissociation from the inhibitory protein, IκB, this dissociation depends on the phosphorylation of IκB. Similarly, the activation of the glucocorticoid receptor by its steroid ligand is dependent on a ligand-induced conformational change; this change results in dissociation of the receptor from an inhibitory protein, known as Hsp90 (*see* HEAT SHOCK), allowing the receptor to move to the nucleus and activate gene expression.

In many cases, activation of a transcription factor occurs by conversion to a form which can bind to DNA and thereby activate transcription. Thus the conformational change induced in the ACE1 factor by copper binding converts it into an active DNA-binding form. Similarly, the dissociation of inhibitory proteins from NFκB or the glucocorticoid receptor allows these factors to move to the nucleus and bind to their specific recognition sequences in DNA. In contrast however, the CREB factor can bind to its specific sites in cAMP-responsive genes even before exposure to cAMP, and its phosphorylation does not enhance its ability to bind to DNA but stimulates the activity of its activation domain (*see* TRANSCRIPTION FACTORS).

Repression of transcription

Although most constitutively expressed or regulated transcription factors activate the transcription of specific genes, it is also possible for transcription to be specifically inhibited by the action of a transcription factor (Fig. E21) [11,12]. As with activating factors, the synthesis or activation of such an inhibitory factor in particular tissues or in response to a signal can specifically regulate the expression of particular genes.

In many cases the same factor can activate certain genes while repressing others. For example, the glucocorticoid receptor activates specific steroid-responsive genes by binding to a particular DNA-binding site, but can also repress other genes by binding to a distinct but related DNA-binding site. Following binding to this negative site, the receptor cannot activate transcription itself and

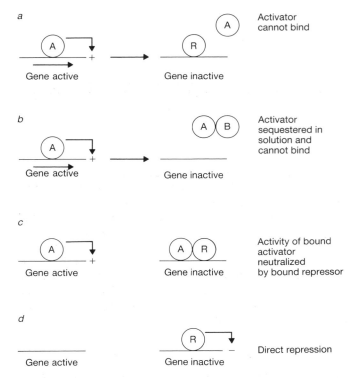

a

Gene active   Gene inactive   Activator cannot bind

b

Gene active   Gene inactive   Activator sequestered in solution and cannot bind

c

Gene active   Gene inactive   Activity of bound activator neutralized by bound repressor

d

Gene active   Gene inactive   Direct repression

Fig. E21 Potential mechanisms by which a transcription factor can inhibit transcription. *a*, Binding to DNA and preventing a positively acting factor from binding; *b*, interacting with the activator in solution and preventing it from binding; *c*, binding to DNA with the activator and preventing it from activating transcription; *d*, direct repression.

prevents another positively acting factor binding to this site, thereby inhibiting transcription (Fig. E21*a*). In a similar but distinct mechanism the glucocorticoid receptor also inhibits genes to which it cannot bind by complexing with the activating transcription factor AP1 in solution. This prevents AP1 from binding to DNA and thereby inhibits transcription (Fig. E21*b*).

These examples illustrate two mechanisms by which a factor can inhibit gene expression by preventing the binding to DNA of another activating factor. In addition however, cases are known in which inhibition is exerted by a factor which binds to another, DNA-bound, factor and prevents it activating transcription by masking its activation domain (Fig. E21*c*). In yeast, the GAL4 protein binds to DNA before exposure to galactose but can only activate transcription following the galactose-induced dissociation of the inhibitory GAL80 protein which exposes the GAL4 activation domain. Hence as with the activation of positively acting factors (see above), inhibitory factors can exert their effects by affecting the ability of a factor either to bind to DNA or to stimulate transcription following such binding.

In all the cases of inhibition discussed above, the inhibitory factor acts by neutralizing the activity of a positively acting factor. It is also possible that some factors inhibit eukaryotic gene expression directly (Fig. E21*d*), but in no case of transcriptional repression studied so far has the possibility of neutralization of a

positively acting factor been eliminated [11]. Nonetheless, even if their action is solely mediated by neutralizing positively acting factors, negatively acting factors have an important role in transcriptional regulation.

Chromatin

An additional complication of gene regulation in eukaryotes compared with prokaryotes is the fact that eukaryotic DNA is packaged into CHROMOSOMES in the form of a protein–DNA complex — CHROMATIN. In differentiated cells there is evidence that chromatin containing genes that are never expressed in that particular cell type is in a different physical state from that containing genes available for transcription, and most probably is inaccessible to transcription factors. How certain genes come to be 'packed away' permanently and become unavailable for transcription is not yet fully understood (*see also* DNA METHYLATION; LOCUS CONTROL REGION).

D.S. LATCHMAN

1 Nevins, J.R. (1983) The pathway of eukaryotic mRNA transcription. *Annu. Rev. Biochem.* **53**, 441–446.
2 Latchman, D.S. (1990) *Gene Regulation: A Eukaryotic Perspective* (Unwin Hyman, London).
3 Latchman, D.S. (1990) Cell-type specific transcription factors and the regulation of alternative RNA splicing. *New Biologist* **2**, 297–303.
4 Latchman, D.S. (1991) *Eukaryotic Transcription Factors* (Academic Press, London).
5 Johnson, P.F. & McKnight, S.L. (1989) Eukaryotic transcriptional regulatory proteins. *Annu. Rev. Biochem.* **58**, 799–839.
6 Saltzman, A.G. & Weinmann, R. (1989) Promoter specificity and modulation of RNA polymerase II transcription. *FASEB J.* **3**, 1723–1733.
7 Goodwin, G.H. et al. (1990) Sequence specific DNA binding involved in gene transcription. In *Chromosomes: Eukaryotic, Prokaryotic and Viral* (Adolph, K.W., Ed.) Vol. 1, 31–85 (CRC Press, Boca Raton, FL).
8 Hatzopoulos, A.K. et al. (1988) Enhancers and other cis-acting sequences. In *Transcription and Splicing* (Hames, B. & Glover, D.W., Eds) 43–96 (IRL Press, Oxford).
9 Olson, E.N. (1990) Myo D family: a paradigm for development? *Genes Devel.* **4**, 1454–1461.
10 Lenardo, M.J. & Baltimore, D. (1989) NF-kappa B: a pleiotropic mediator of inducible and tissue-specific gene control. *Cell* **58**, 227–229.
11 Levine, M. & Manley, J.L. (1989) Transcriptional repression of eukaryotic promoters. *Cell* **59**, 405–408.
12 Goodbourn, S. (1990) Negative regulation of transcriptional initiation in eukaryotes. *Biochim. Biophys. Acta* **1032**, 53–77.

Eukaryotic gene structure

EUKARYOTIC GENES are considered to comprise those found in the nuclear GENOME of EUKARYOTES together with those of the DNA viruses and RETROVIRUSES that infect eukaryotes (*see* ANIMAL VIRUSES; PLANT VIRUSES). The genes of mitochondria and chloroplasts are usually considered separately and are dealt with elsewhere in the encyclopedia (*see* CHLOROPLAST GENOMES; MITOCHONDRIAL GENOMES).

All genes comprise a coding region (encoding a polypeptide chain or a structural RNA such as transfer and ribosomal RNA) which is flanked by regulatory regions that control the initiation of TRANSCRIPTION (*see* EUKARYOTIC GENE EXPRESSION). Within

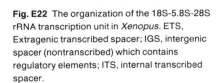

Fig. E22 The organization of the 18S-5.8S-28S rRNA transcription unit in *Xenopus*. ETS, Extragenic transcribed spacer; IGS, intergenic spacer (nontranscribed) which contains regulatory elements; ITS, internal transcribed spacer.

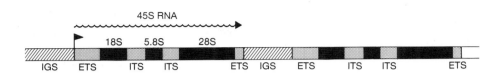

this basic framework, eukaryotic genes are very diverse in structure, reflecting their great diversity of function. There are a number of general differences between eukaryotic and prokaryotic genes (*see* BACTERIAL GENE ORGANIZATION). One notable difference is the presence of INTRONS (noncoding DNA sequences) which interrupt the coding sequence of many eukaryotic protein-coding genes and some tRNA genes, and which are removed from the primary RNA transcript to generate the functional mRNA (*see* RNA SPLICING).

There is considerable similarity between the eukaryotic and the prokaryotic genes encoding rRNAs and tRNAs; the main difference is that introns occur in some eukaryotic tRNA genes and there is a higher degree of repetition of both rRNA and tRNA genes in eukaryotes. Protein-coding genes in eukaryotes are extremely variable in size, intron number and PROMOTER organization.

In eukaryotes there are three main types of RNA POLYMERASE — I, III and II — each of which transcribes a particular set of genes, and eukaryotic genes may be conveniently classified into three groups depending on the RNA polymerase that transcribes them.

Genes transcribed by polymerase I: the rRNA genes

In virtually all eukaryotes (excluding the lower eukaryotes such as slime moulds and fungi) the 18S-5.8S-28S rRNA coding sequences (*see* RIBOSOMAL RNA) comprise a TRANSCRIPTION UNIT transcribed by RNA polymerase I (Fig. E22) whereas the 5S rRNA is separately encoded and is transcribed by RNA polymerase III. Both the 18S-5.8S-28S rRNA unit and the 5S rRNA genes are organized into gene clusters composed of tandem repeats of the transcription unit separated by a nontranscribed spacer [1,2].

Eukaryotes possess many more copies of the rRNA genes than do prokaryotes. The number of 5S rRNA clusters varies from one per haploid genome (in *Drosophila melanogaster*) to around one per chromosome (in *Xenopus borealis*). The total 5S rRNA gene copy number in eukaryotes varies from about 150 in fungi to more than 20 000 in *Xenopus* and the 18S-5.8S-28S rRNA repeat unit copy number is ~500 in *Xenopus*.

Some exotic arrangements of rRNA genes have been found. In amphibian oocytes, the rRNA gene repeat units replicate extrachromosomally. In the lower eukaryote *Tetrahymena*, the rRNA repeat units also replicate extrachromosomally, in this case inside a separate organelle called the macronucleus. Additionally, in some *Tetrahymena* strains, the 26S rRNA gene (the 28S rRNA

gene equivalent in this species) contains a class I intron which is removed by self-splicing (*see* RIBOZYMES). This was the first example of a self-splicing RNA to be discovered [3].

The transcriptional control regions of the 18S-5.8S-28S rRNA repeat unit are poorly conserved between eukaryotes; indeed they are the least well conserved eukaryotic promoter sequences known [4]. As a consequence, the transcriptional machinery of quite closely related eukaryotes (e.g. mouse and man) is incapable of transcribing each other's rRNA genes. 5S rRNA genes are transcribed by RNA polymerase III and their promoter will be discussed in the next section.

Genes transcribed by RNA polymerase III

These include the genes encoding almost all the small RNAs of the cell — tRNAs and small nuclear and small cytoplasmic RNAs — with the notable exception of all but one of the U RNAs (*see* RNA SPLICING; SNRNA). Additionally, some DNA virus genes (such as the VA RNA genes of ADENOVIRUS) fall into this category.

Some genes transcribed by RNA polymerase III, notably a subset of tRNA genes, contain an intron [1,2]. This is always small (between 14 and 60 nucleotides) and inserted one nucleotide 3' to the anticodon triplet (*see* TRANSFER RNA). The tRNA introns constitute a distinct class of intron which possesses no obvious consensus nucleotide sequences either within, or flanking, the intron. Furthermore, the splicing mechanism that removes such introns is distinct from the other splicing mechanisms (*see* RNA SPLICING). Most genes transcribed by RNA polymerase III have promoters that are embedded in the transcribed part of the gene (Fig. E23) [5], although there are some examples where either the promoters lie upstream of the transcriptional start site or promoter elements are found both upstream and inside the transcribed region. The internal promoters of different genes transcribed by RNA polymerase III possess conserved sequence elements that have been shown to be binding sites for TRANSCRIPTION FACTORS, notably TFIIIA [5].

Fig. E23 Organization of the 5S rRNA gene from *Xenopus* showing the internal promoter site.

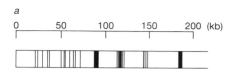

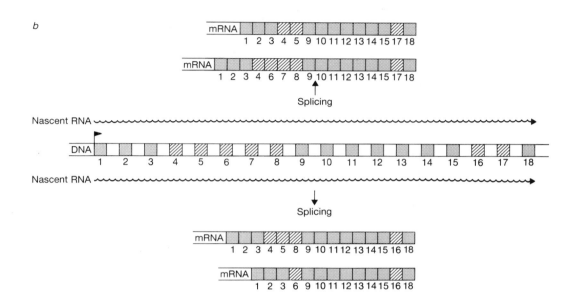

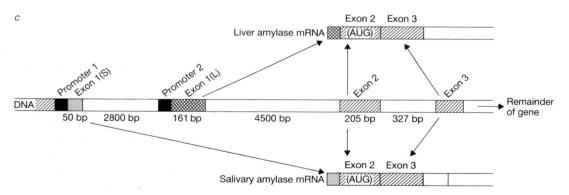

Fig. E26 Some examples of the range of eukaryotic gene structure. *a*, The human factor VIII gene, which contains 26 exons within a total length of 180 kb. Exons are black. *b*, Schematic diagram of the troponin T gene. Alternative RNA splicing can generate different mRNAs. Troponin T mRNA from fast skeletal muscle contains either exon 16 or exon 17 but not both. Within these two main classes of mRNA alternative forms are found containing different selections of exons 4–8. *c*, Mouse α-amylase gene. The use of alternative promoters and differential splicing can generate different levels of expression of a protein in different tissues. Promoter 1 is activated in salivary gland tissue only and produces a 30-fold higher level of transcription than from Promoter 2 which is active in both liver and salivary tissue. The resulting mRNAs have a different 5′ leader sequence but the same protein-coding sequence. In the salivary gland mRNA the second promoter region has been spliced out as part of a large intron. Not to scale. *d*, The pro-opiocorticotropin gene produces a polypeptide which may be cleaved in different ways to give several different functional peptides. Open arrows indicate sites of proteolytic cleavage. *e*, A gene within the intron of another gene. (▶) Sites of initiation of transcription.

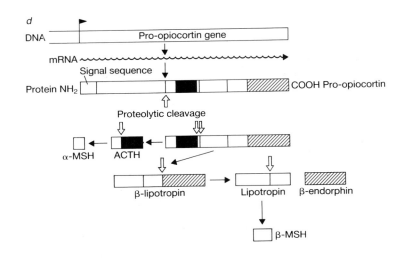

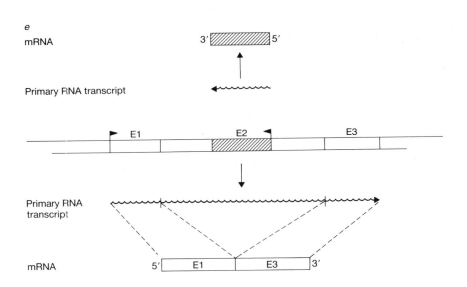

Fig. E26 *Continued*

Genes transcribed by RNA polymerase III are usually found in gene clusters, either scrambled or tandemly repeated [1,2]. An example of a tandemly repeated cluster (5S rRNA genes) has already been mentioned; many tRNA genes are found in scrambled gene clusters (Fig. E24).

Genes transcribed by RNA polymerase II

This class of eukaryotic genes exhibits the greatest degree of variety in structure, in keeping with the correspondingly huge diversity of function of their gene products. It includes all the protein-coding genes as well as some of the genes encoding U RNAs. Such genes vary in size from a few hundred bases (for many U RNAs) to more than two million bases (for the dystrophin

gene [6]; *see* DUCHENNE MUSCULAR DYSTROPHY). Most of the length of these huge genes is made up of introns.

A paradigm for eukaryotic protein-coding genes is the globin gene family [6], which illustrates many of the features of typical RNA polymerase II-transcribed genes. Each globin gene consists of a protein-coding sequence divided into several EXONS separated by introns (Fig. E25). The transcribed parts of vertebrate globin genes are all a few kilobase pairs long. The protein-coding sequence begins with an INITIATION CODON and this is preceded by an untranslated leader sequence.

The DNA sequences that regulate human β-globin gene expression are located both 5′ and 3′ to the site at which translation is initiated. At least five distinct regulatory elements have been identified in globin genes: a promoter element upstream (i.e. to the 5′ side) of the site of transcription initiation, a SILENCER element and two ENHANCER elements, one within the transcribed portion of the gene and one about 800 nucleotides downstream (i.e. to the 3′ side) of the gene. The fifth regulatory element — the

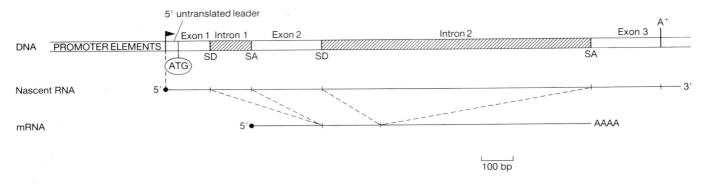

Fig. E25 A typical simple eukaryotic protein-coding gene: the human β-globin gene. The transcribed parts of the gene consist of three exons separated by two introns within a total length of 1.6 kilobases. Transcription commences at the beginning of exon 1 (▶) and the nascent RNA transcript extends through all exons and introns, terminating downstream of the POLYADENYLATION site (A +) (*see* RNA PROCESSING). In the mature cytoplasmic mRNA, the introns have been removed and the splice donor (SD) and splice acceptor (SA) sites fused . Cleavage and polyadenylation have also occurred on the RNA at the polyadenylation site and the mRNA is capped at the 5′ end.

LOCUS CONTROL REGION — lies far upstream of the human β-globin gene; it is in fact separated from it by three other globin genes, and its mode of action is not known. This last element was identified because it is required for the high-level expression of β-globin TRANSGENES in transgenic mice [7].

Not all genes transcribed by RNA polymerase II contain introns, although, in the higher eukaryotes, most do. Notable exceptions are many U RNAs and all HISTONE genes. In general, lower eukaryotes such as fungi possess fewer introns than higher eukaryotes such as animals and higher plants.

mRNA-encoding genes which are related to each other are frequently found in gene clusters. The β-globin gene in mammals, for example, is found close to the ε, δ, Gγ and Aγ globin genes (and to two defective β-globin PSEUDOGENES) [6,7]. This reflects their evolutionary origin by gene duplication (*see* MOLECULAR EVOLUTION). The coding regions of eukaryotic genes are sometimes evolutionary mosaics containing sequence elements, such as the IMMUNOGLOBULIN HOMOLOGY REGION and the FIBRONECTIN TYPE III DOMAIN, which are found in numerous other genes (*see*

IMMUNOGLOBULIN SUPERFAMILY; MOLECULAR EVOLUTION: SEQUENCES AND STRUCTURES: SUPERFAMILY).

Some representative examples of eukaryotic protein-coding gene structure are shown in Fig. E26. Huge genes may consist of several tens of exons (Fig. E26*a*). Even smaller genes, such as those for collagen (*see* EXTRACELLULAR MATRIX MOLECULES), may contain more than 50 exons, many of which are only nine base pairs long [2].

There are many examples known where a single gene can encode different mRNAs by the use of alternative splicing (*see* RNA SPLICING). The troponin T gene, for example, encodes two different protein isoforms (α and β) from a single gene by producing mRNAs which contain only one of two possible exons (Fig. E26*b*) [2]. IMMUNOGLOBULIN GENES of mammals also employ this strategy to produce secreted and membrane-bound forms of a particular immunoglobulin; in addition they carry the generation of diversity many steps further by providing many alternative protein-coding regions (in some cases hundreds) for the same part of the antibody molecule and the decision as to which is used in the final protein is made by DNA rearrangement in the developing B cell [1,2] (*see* B CELL DEVELOPMENT; GENE REARRANGEMENT; T CELL RECEPTOR GENES). There are also many examples of a gene possessing more than one possible transcriptional start site. If the start sites lie on different exons, as in the case of the amylase gene of rodents [2], (Fig. E26*c*), then alternative mRNAs are produced, containing different 5′ terminal exons.

Although truly polycistronic genes are only found in eukaryotic viruses, some eukaryotic genes, notably those for peptide hormones, encode more than one peptide species by directing the synthesis of a POLYPROTEIN, which is subsequently cleaved into various functional pieces (Fig. E26*d*).

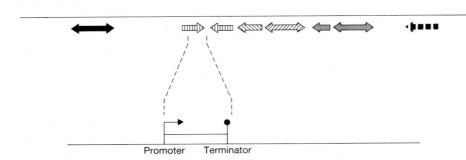

Fig. E24 Scrambled cluster of tRNA genes. Each arrow in the top part of the figure is a gene and the direction of the arrow refers to transcriptional orientation (5′ → 3′). Each type of tRNA gene (e.g. lysine, methionine, etc.) is represented by different shading of the arrows.

DNA viruses

The genes of eukaryotic DNA viruses obey the same general structural rules that appertain to the nuclear genes. Because viral genomes have rigid size constraints, alternative splicing leading to different protein-coding regions using a common 5′ untranslated LEADER SEQUENCE is correspondingly more common (*see* ADENO-VIRUS). In some cases, an entirely different gene is found in the intron separating the leader sequence from the protein-coding region (Fig. E26e). More examples of this arrangement are also being found in nuclear genes [2].

Eukaryotic viruses contain examples of POLYCISTRONIC genes. For example, many retroviruses (which although RNA viruses spend part of the life cycle as a DNA provirus integrated into the host cell genome) encode the structural components of the virus core particle, together with four enzymes required for the infectious cycle, on a single mRNA [8] (*see* RETROVIRUSES).

A. FLAVELL

See also: BACTERIAL GENE EXPRESSION; CHROMOSOME STRUCTURE; EUKARYOTIC GENE EXPRESSION; GENOME ORGANIZATION; HAEMO-GLOBIN AND ITS DISORDERS; PROTEIN SYNTHESIS.

1 Singer, M. & Berg, P. (1991) *Genes and Genomes*, 435–437, 459–562 (Blackwell Scientific Publications, Oxford).
2 Lewin, B. (1990) *Genes IV*, 482–517 (Oxford University Press, Oxford).
3 Kruger, K. et al. (1982) Self-splicing RNA: autoexcision and autocyclization of the ribosomal RNA intervening sequence of *Tetrahymena*. *Cell* **31**, 147–157.
4 Sommerville, J. (1984) RNA polymerase I promoters and transcription factors. *Nature* **310**, 189–190.
5 Pieler, T. et al. (1987) The 5S gene internal control region is composed of three distinct sequence elements, organized as two functional domains with variable spacing. *Cell* **48**, 91–100.
6 Davies, K.E. & Read, A.P. (1990) *Molecular Basis of Inherited Disease* (IRL Press, Oxford).
7 Grosveld, F. et al. (1987) Position-independent, high-level expression of the human β-globin gene in transgenic mice. *Cell* **51**, 975–985.
8 Weiss, R. et al. (1984) *RNA Tumor Viruses*, 2nd edn, 261–368 (Cold Spring Harbor Laboratory Press, Cold Spring Harbor, New York).

eukaryotic genome organization *See*: GENOME ORGANIZATION.

Eumycota One of the two divisions of the Kingdom Fungi (the other being the MYXOMYCOTA). The Eumycota are also known as 'true fungi', and include the following five groups: MASTIGO-MYCOTINA, DEUTEROMYCOTINA, ZYGOMYCOTINA, ASCOMYCOTINA, and BASIDIOMYCOTINA. All these fungi possess a cell wall and form mycelial strands, in contrast to the Myxomycota.

event-related potential (ERP) A potential resulting from the application of a stimulus. For example in electroencephalography, a weak electrical shock applied to a finger leads to an event-related potential in the brain.

evolution *See*: DARWINIAN EVOLUTION; MOLECULAR EVOLUTION; MOLECULAR EVOLUTION: SEQUENCES AND STRUCTURES; MOLECULAR PHYLOGENY.

evolutionary parsimony *See*: INVARIANTS.

EXAFS (extended X-ray absorption fine structure)

THE advent of intense X-ray beams from synchrotron radiation sources has allowed METALLOPROTEINS to be studied using X-ray absorption spectroscopy. The technique is capable of probing the immediate environment of metal atoms in proteins. A number of enzymes contain metal atoms and in many cases the metal site is essential for the catalytic function of the protein (*see* ELECTRON TRANSPORT CHAIN; MECHANISMS OF ENZYME CATALYSIS; NITROGEN FIXATION; PHOTOSYNTHESIS; PROTEIN STRUCTURE). EXAFS has been used to determine the local structure about the metal site in enzymes under a variety of conditions. Among other metalloproteins of interest that have been studied by this technique are the yeast TRANSCRIPTION FACTOR GAL4 and the nitrogenases of nitrogen-fixing bacteria.

When X-rays illuminate matter it is found that as the photon energy increases, the degree of absorption by the material decreases smoothly. In certain parts of the spectrum there are abrupt increases in absorption. This is termed the absorption edge and arises from the excitation of a core level electron which has received sufficient energy to be ejected from deep within the atoms stimulated. In studies on metalloproteins these electrons come from either the K or L shells which are closest to the nucleus of the metal atom. The amount of energy (measured in electronvolts) required to kick out these electrons is dependent on the nuclear charge and is thus characteristic for each element. Closer inspection of the absorption edge region reveals discontinuities just below the edge and on the edge which are due to transitions of electrons to valence levels which are involved in chemical bonding.

For condensed matter and molecular systems the profile on the high energy side of the absorption edge exhibits complex oscillatory behaviour extending over several hundred electronvolts, which is due to the presence of other atoms in the vicinity of the absorbing species. These modulations have been subdivided into two regions: (1) XANES (X-ray absorption near edge structure), which contains bound state transitions and reflects coordination geometry, symmetry of unoccupied atomic states and effective charge on the absorbing atom; and (2) EXAFS, which provides information on the number, type and distance of neighbouring atoms from the metal being probed.

Particular strengths of the technique which make it suitable for studies of biological systems are: (1) any state of matter may be used for measurements; (2) any element may be probed; (3) metal–ligand distances can be determined to high accuracy; and (4) the theory for describing the EXAFS phenomenon is well developed. Weaknesses include: (1) a lower limit of concentration of ~1 mM for 3d metals and 0.5 mM for 4d metals in proteins; (2) the EXAFS spectrum of an atom residing in several different sites is the average of all the sites; (3) COORDINATION NUMBERS are imprecise owing to the high correlation between the number of atoms and their respective DEBYE-WALLER type factor; (4) atoms of very similar atomic weight (e.g. oxygen, nitrogen and carbon) cannot be distinguished from one another; and (5) a ligand may be excluded from the theoretical fit versus experiment compari-

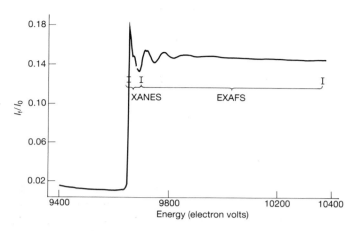

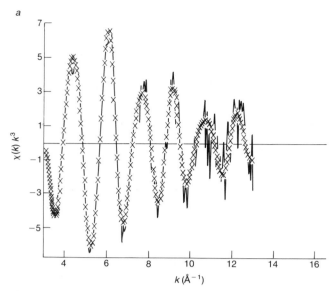

Fig. E27 A fluorescence excitation spectrum of yeast transcriptional activator GAL4 at the Zn K-edge. The XANES region is dominated by multiple scattering effects, whereas the EXAFS is a result mainly of single scattering. These data were recorded using an array of 12 germanium solid-state detectors and are the sum of eight scans.

son as a result of its very weak contribution to the EXAFS compared with the primary scatterer. But provided care is taken in the analysis, the technique is still an excellent means of obtaining structural information about the local environment of a metal atom in a biological system. Figures E27 and E28 provide an illustration of a typical EXAFS spectrum obtained from a metalloprotein and show the experimental data with its simulated theoretical fit.

Theory

To aid in the understanding of EXAFS, the ejected electron should be considered as a wave front spreading out like the ripples from a pond when a stone drops into it. When the expanding wave front encounters another atom this reflects back some of the wave (backscattering) which interferes with the outgoing wave. If this backscattering is in phase it increases the X-ray absorption, if out of phase it reduces it. Thus the intensity and frequency of the oscillations contain information on the radial environment of the absorbing atom.

Theoreticians have provided a complete theory for the description of the EXAFS phenomenon and have shown that with the exception of energies close to the absorption threshold a single-scattering formalism is sufficient to explain most of the experimental data. A simplified version of this equation is given below which neglects the curvature of the electron wave but is valid in the high energy region of the EXAFS spectrum. This may be written for the K-absorption edge as:

$$\chi(k) = -\sum_j \frac{N_j}{kR_j^2} \, |f_j(\pi)| \cdot \sin(2kR_j + 2\delta_1 + 2\Psi_j) \cdot$$

$$\exp(-2\sigma_j^2 k^2) \cdot \exp(-2R_j/\lambda)$$

This demonstrates that the oscillatory EXAFS function $\chi(k)$ is dependent on the type of scattering atom through the characteristic energy dependence of the backscattering amplitude $|f_j(\pi)|$,

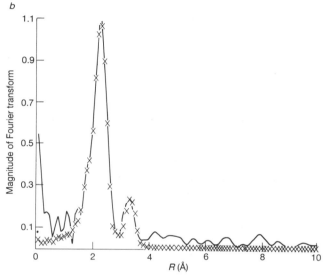

Fig. E28 Zinc K-edge EXAFS (*a*) and associated Fourier transform (*b*) of GAL4. The continuous line represents the experimental spectrum after background removal and normalization to the edge height. The pre-edge background is subtracted by fitting a linear function which is extrapolated over the whole energy range of the experimental spectrum. The post-edge background is removed by fitting a pair of polynomials which are matched at the knot. The Fourier transform of the EXAFS shows peaks which are due to atoms present at various scattering distances from the Zn atom. The experimental spectrum and its associated Fourier transform was simulated (shown as crosses) with one oxygen atom at 1.95 Å, three sulphur atoms at 2.3 Å and one sulphur atom at 3.34 Å from the Zn atom. The EXAFS data are presented as $\chi(k)k^3$ plotted against k where k is in Å^{-1}. k is the momentum of the photoelectron given by $k = [(E - E_-)2m/h^2]^{1/2}$ where m is its mass, E its energy and E_- the energy of the absorption edge.

number of scattering atoms N_j of type j, and the distance of these scattering atoms R_j from the primary absorber. $2\delta_1$ is the phase shift due to the potential of the emitting atom and ψ_j is the phase of the backscattering factor. The mean square variation in R_j is

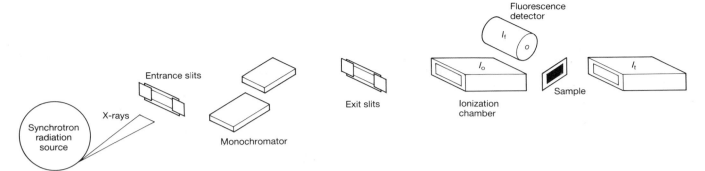

Fig. E29 An illustration of the general layout of an X-ray absorption spectrometer.

represented by the Debye–Waller type factor, σ_j^2, which assumes a harmonic distribution. λ is the elastic mean free path of the photoelectron and it is the damping term $(\exp(-2R_j/\lambda))$ which invariably limits the backscattering contribution to $\leqslant 6$ Å from the metal atom in a biological system. It is the low energy portion of the EXAFS spectrum which usually contains most of the information for biological systems because of the weak scattering power of the atoms of low atomic number (Z) (e.g carbon, nitrogen, and oxygen). In addition the more distant shells of atoms tend to contribute in this region. Thus the exact theory which does take into account the curvature of the electron wave, the spherical wave method, is now being used more frequently in the analysis of biological data. Its main drawback has been its mathematical complexity which has required a large amount of computational time to calculate a theoretical spectrum. Recently this has been greatly simplified without compromising its exact nature for polycrystalline and amorphous materials and it is now ideal to use for biological systems.

Experimental layout

Figure E29 illustrates the general layout of an X-ray absorption spectrometer. The source of X-rays is provided by a synchrotron which produces white radiation. This is collimated by entrance slits both vertically and horizontally. The X-ray beam is then intercepted by a two-crystal monochromator (typical crystal planes used are Si (111) and Si (220)) which provides discrete photon energies by monochromating the white radiation. By stepping through discrete angles an EXAFS spectrum can be recorded as a series of increasing energy points. Following the monochromator is a set of exit slits which are adjusted to allow only the monochromatic X-ray beam to pass. This part of the instrument is normally under vacuum or helium atmosphere to cut down air absorption. A beryllium window allows the X-ray beam to continue its course to the remaining instrumentation which is in air. The two most common configurations are the transmission and fluorescence geometries. The former uses two ion chambers with the sample placed in between. The first measures I_o, the incident intensity on the sample, and absorbs $\sim$20% of the monochromatic radiation, whereas the second ion chamber measures the amount of X-rays transmitted by the sample, I_t, and absorbs $\sim$80% of the X-ray beam.

For most biological applications, the fluorescence geometry is used because samples tend to be dilute, and this arrangement is more sensitive to the increased contrast between the background and the signal emitted from the element of interest. Contamination of the fluorescence signal by Compton and Rayleigh scattering means that the ideal position for the fluorescence detector is at 90° to the plane of incident radiation; this minimizes such effects, and the sample should be positioned at 45° to the radiation. The three most common detectors used are ion chambers, scintillation counters and solid-state detector arrays.

Application of EXAFS to some biological systems

A comprehensive review of the applications of EXAFS is beyond the scope of this entry (see [1] and [2] for reviews) but below are some brief illustrations of studies carried out on metalloproteins.

The iron K-edge EXAFS of the IRON-SULPHUR PROTEIN rubredoxin was the first biological system to be investigated and demonstrated that all the Fe–S bonds were equivalent at 2.26 Å. For the reduced protein the Fe–S distance was shown to increase by 0.5 Å over the oxidized form. No other technique at present can produce the same information with this degree of certainty.

An area of investigation where other spectroscopic techniques like optical absorption and ESR/EPR SPECTROSCOPY are ineffective is the study of zinc metalloproteins. This is because zinc has a filled 3d shell. Resolving this dilemma meant substituting the zinc atom by a more spectroscopically active metal such as cobalt or cadmium. These alternatives may not, however, always adopt the same environment as the zinc atom. This has been demonstrated by EXAFS studies on two proteins, PHOSPHOLIPASE C where cobalt has been substituted for zinc, and the transcriptional activator protein GAL4 in which the zinc was replaced by cadmium. Both proteins had a different coordination sphere for the substituted metal when compared with the native zinc atom, emphasizing the need to probe the native metal site directly whenever possible, which EXAFS can accomplish.

XANES and EXAFS have been used [3] to probe the copper and zinc sites in lyophilized and solution forms of the enzyme bovine SUPEROXIDE DISMUTASE. This protein exists as a dimer with each subunit containing one copper and one zinc atom separated by $\sim$6 Å (*see* Fig. S35 in SUPEROXIDE DISMUTASE). Investigations showed that the freeze-dried protein lost a coordinated water molecule from the copper site when compared to its solution form (characterized by four nitrogen atoms from imidazole ligands at 2.00 Å ± 0.02 Å with an oxygen ligated at 2.24 Å ± 0.05 Å from the copper site). The oxidized and reduced forms of the enzyme showed significant changes in both the XANES and EXAFS of the copper site. Reducing the enzyme produced a decrease in the number of ligands coordinated around the copper site to three histidines at $1.94 \text{ Å} \pm 0.02 \text{ Å}$, thus producing the first direct information on the reduced form of the protein. In all cases the zinc XANES and EXAFS spectra remained unperturbed showing the site remained unchanged.

Difficulties have been found in trying to analyse the radial contribution of the outer shell ligands from imidazole ring systems in metalloproteins. This was highlighted in the work on super-oxide dismutase where the outer shell carbon and nitrogen atoms were foreshortened by ~ 0.3–0.4 Å in the EXAFS analysis. It was clear that the single-scattering EXAFS formalism was inadequate in describing the contribution of these outer shell atoms to the experimental EXAFS spectrum. These anomalous results are due to the multiple scattering of the electron which occurs in imidazole rings and is not taken into account by the single-scattering theory. Such effects are dependent upon the geometry of the ligands in the metal site and are at a maximum when three or more atoms are arranged collinearly, for example in metal carbonyl systems. The magnitude of this phenomenon attenuates rapidly as the angle of scatter decreases from $180°$ and is negligible below $120°$. Other ligands that produce this effect in metalloproteins are porphyrin, tyrosine and inhibitors such as cyanide and azide. It is now standard procedure to include multiple scattering contributions in the analysis of EXAFS spectra.

The future progress of this technique is interwoven with advances in theoretical and experimental developments. At present very little use is made of XANES, which provides geometrical information and when combined with EXAFS will provide a three-dimensional picture of a metal site in a metalloprotein. Brighter X-ray sources and improvements in experimental apparatus and detectors will open up a whole new range of proteins to be studied and will increase the possibility of studying intermediates in a reaction by time-resolved and freeze-quench techniques.

G. DIAKUN

See also: CD SPECTROSCOPY; ESR/EPR SPECTROSCOPY; NEUTRON SCATTERING AND DIFFRACTION; NMR SPECTROSCOPY; X-RAY CRYSTALLOGRAPHY.

1 Scott, R.A. (1985) Measurement of metal–ligand distances by EXAFS. *Meth. Enzymol.* 117, 414–459.
2 Hasnain, S.S. (1987) Application of EXAFS and XANES to metalloproteins. *Life Chemistry Rpts* 4, 273–331.
3 Blackburn, N.J. et al. (1984) *Biochem. J.* 219, 985–990.

exchange factor *See*: GTP-BINDING PROTEINS.

excision repair *See*: DNA REPAIR.

excitatory amino acids (EAA) Amino acid NEUROTRANSMITTERS that cause excitation of the postsynaptic neuron (*see* SYNAPTIC TRANSMISSION). Glutamate and/or aspartate are believed to be the primary EAA candidates in the central nervous system. However, other amino acids still cannot be ruled out. *See also*: EXCITATORY AMINO ACID RECEPTORS; LONG-TERM POTENTIATION.

Excitatory amino acid receptors

EXCITATORY amino acid receptors (EAA receptors) are the major class of excitatory NEUROTRANSMITTER receptors in the vertebrate central nervous system. They are membrane-spanning proteins that mediate the stimulatory actions of glutamate and possibly other related endogenous acidic amino acids. EAA receptors are known to be crucial for fast excitatory neurotransmission and they have also been implicated in a variety of diseases including ALZHEIMER'S DISEASE and epilepsy. In addition, EAA receptors are integral to the processes of long-term potentiation (LTP) and long-term depression (LTD) (*see* LONG-TERM POTENTIATION), the likely synaptic mechanisms underlying learning and memory [1]. The recent cloning of the cDNAs encoding some of the subunits of these receptors and their deduced amino-acid sequences should lead to an increased understanding of the molecular and functional properties of this clinically important class of receptor proteins.

Types of EAA receptors

On the basis of their pharmacological and physiological profiles three main subtypes of EAA receptor have been recognized [2].
1 The metabotropic or *trans*-ACPD class of receptors are G PROTEIN-COUPLED RECEPTORS associated with the stimulation of inositol phospholipid metabolism or the modulation of adenylate cyclase (*see* SECOND MESSENGER PATHWAYS).
2 The ionotropic NMDA receptor, in which the agonist activates an intrinsic ion channel permeated by Na^+, K^+ and Ca^{2+} ions. This type has a high conductance ($\sim 50 \text{ pS}$) and is subject to a variety of distinct and specific ALLOSTERIC modulatory processes.
3 The non-NMDA receptors: (a) AMPA types of ionotropic EAA receptor are permeated by Na^+ and K^+ ions and in some circumstances (see below) the receptor complex also gates Ca^{2+} ions; (b) several cDNAs have been isolated that encode functional ionotropic receptor subunits with a high affinity for kainate but no appreciable AMPA binding [7].

Metabotropic receptor

Metabotropic receptors (mGluRs) are second messenger-coupled EAA receptors that do not have integral ion channels and which are activated by quisqualate and *trans*-1RS, 3RS-*cis*-1-amino-cyclopentyl-1,3-dicarboxylate (*trans*-ACPD) but not by AMPA [3,4]. Pharmacological evidence suggested the existence of several subtypes of mGluRs. In particular, the receptor sensitivity to the agonists ibotenate and quisqualate and the poorly selective antagonists amino-3-phosphonopropanoate (AP3) and 2-amino-4-phosphonobutanoate (AP4) (*see* AP3, AP4) differ in different brain regions in the rat. Following from the isolation of the first cDNA encoding a G protein-linked mGluR (mGluR1α), cross-hybridization and PCR techniques have revealed at least six subtypes of metabotropic receptors (mGluR1–mGluR6) [5]. The amino acid sequences of mGluR1–mGluR6 are highly conserved in the transmembrane domains and in the extracellular N-terminal region and the mRNAs for the different mGluRs show overlapping but distinctly different patterns of expression in rat brain. An alternative RNA splice variant of mGluR1 (mGluR1β) has also been identified. mGluR1β is smaller than mGluR1α due to truncation of the C-terminal domain by 292 amino acids. Overall, the primary sequence data and the postulated structure of these receptors appear similar to other previously characterized G protein-linked receptors in that they comprise a single polypeptide chain which contains seven hydrophobic, putative transmembrane, domains. The mGluRs are, however, considerably larger and have extremely long C- and N-terminal domains [5]. The different mGluR subtypes appear to couple to several different second messenger systems including phosphoinositide hydrolysis, activation of phospholipase D (*see* PHOSPHOLIPASE), up and down regulation of adenylate cyclase activity, and modulation of ion channel function [6].

NMDA receptors

NMDA receptors are so called from their stimulation by N-methyl-D-aspartic acid (NMDA), which distinguishes them from the AMPA and kainate types of ionotropic receptors. They also have different pharmacological profiles, ion-channel conductance characteristics and require glycine for receptor activation [7] (*see* GLYCINE MODULATION).

Owing to their crucial role in many neuronal processes and their implication in a wide variety of clinically important disease states, particularly their involvement in cell death following ischaemic damage, NMDA receptors represent a potentially important target for drug action. Although NMDA receptors are the most extensively characterized glutamate receptor subtype pharmacologically and electrophysiologically, for some time technical problems hindered their efficient solubilization and purification and the cloning of receptor cDNA.

Several NMDA receptor subunits have now been cloned [8–10]. The first (NMDAR1) was a protein of M_r 100 000, which was identified by expression cloning of a rat brain cDNA library in *Xenopus* oocytes (*see* XENOPUS OOCYTE EXPRESSION SYSTEM). The library was constructed from the rat brain poly(A)⁺ mRNA fraction that produced an apparently functional glutamate receptor activity when injected into *Xenopus* oocytes. This experimental approach depends for its success on the ability of a single type of subunit to form functional homo-oligomeric ligand-gated ion channels, even though (by analogy with other similar receptors) native receptors are likely to be hetero-oligomeric (*see* GABA AND GLYCINE RECEPTORS; NICOTINIC RECEPTOR). On its own, NMDAR1 produces functional receptors which are responsive to glutamate and glycine. A further family of putative NMDA receptor subunits —NMDAR2A–NMDAR2D -- has since been discovered by homology screening of mouse brain cDNA with various glutamate receptor probes. Although members of this subunit family do not appear to be able to form functional channels on

their own, combinations of NMDAR1 and these subunits give highly active NMDA receptor channels when expressed in oocytes. The actual subunit composition and stoichiometry of a native receptor has not yet been established however.

The deduced amino-acid sequences of the cloned NMDAR1 and NMDAR2A–D receptor subunits give polypeptides of between 938 and 1456 amino acids with four putative membrane-spanning regions and a long N-terminal extracellular domain (Figs E30 and E31). Comparisons with previously cloned non-NMDA receptor subunits are discussed below (see also Table E2 below).

Non-NMDA receptors

The other ionotropic EAA receptors are conventionally classified as non-NMDA receptors. Historically, within this class separate AMPA and kainate receptors have been recognized. AMPA and kainate receptors have been reported to differ in their rank orders of agonist potencies, some of their single-channel conductance levels and the neuroanatomical distributions of specific [³H]-kainate and [³H] AMPA binding sites. Competitive antagonists such as 6-cyano-7-nitroquinoxaline-2,3-dione (see CNQX) and 2,3-dihydroxy-6-nitro-7-sulphamoyl-benzo(F)quinoxaline (see NBQX) which are relatively specific for non-NMDA receptors can, in some cases, discriminate between the kainate and AMPA receptors. None the less, it is becoming increasingly widely accepted that AMPA and kainate can act at a single unitary non-NMDA receptor type [2].

AMPA Receptors: The GluR1–4 family

Several functional non-NMDA receptor subunits were cloned some time before those for NMDA receptors [9,11]. The first (GluR1) was obtained by expression cloning from a rat brain cDNA library (*see* GLUR1–GLUR4). The expressed homo-oligomeric receptor GluR1 is activated by kainate, domoate, glutamate and AMPA-type agonists but not by quisqualate. However, from their

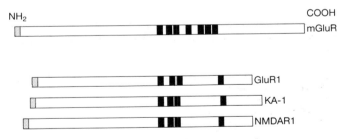

Fig. E30 Bar diagrams illustrating the sizes of, and number and relative distributions of, transmembrane regions (TM) in some of the EAA receptor subunits for which the cDNAs have been cloned. The sequences are aligned such that TM1 regions correspond. The black bars represent transmembrane regions designated by most of the original authors using hydrophobicity analysis. The hydrophobic region at the extreme N-terminal end (shaded grey) is thought to be the SIGNAL PEPTIDE. The metabotropic receptor (mGluR) has seven rather than four transmembrane regions as predicted for that class of receptor, but it does share the large N-terminal domain observed in sequences encoding functional ionotropic receptors.

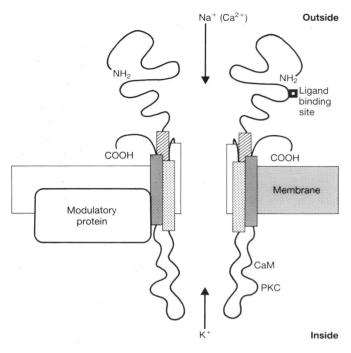

Fig. E31 Possible model for the generic subunit topology of ionotropic EAA receptors. The model is a composite derived from the molecular cloning and amino-acid sequence data and from assumed homology with other, better characterized, ligand-gated ion channels. The diagram is simplified to illustrate some of the salient features discussed in the text and two identical subunits are shown. There is a large N-terminal extracellular domain upstream of TM1 (hatched). In common with the nAChR, TM2 (unshaded) is proposed as the channel-lining segment of the subunit. There is a large intracellular loop between TM3 (stippled) and TM4 (grey) and the C terminus of the subunit is located on the external side of the neuronal membrane. Implications of this model are discussed in the text. It is not, as yet, possible to predict the stoichiometry of the receptor or to determine how many copies of individual subunits are present in ionotropic receptors *in vivo*. However, from biophysical considerations for the channel size needed to conduct mono- and divalent cations, it is assumed that at least four, and probably five, subunits of similar structure comprise the functional receptor. CaM, PKC, consensus phosphorylation sites for calmodulin-dependent protein kinase and protein kinase C respectively.

relative sensitivities to the various agonists GluR1–4 are now accepted as AMPA receptor subunits.

Homology screening has yielded the closely related clones GluR2, GluR3 and GluR4 (~70% conserved residues) from rat brain, and the mouse counterparts to GluR1 and GluR2 (α1 and α2) have been cloned from mouse central nervous system. GluR1–4 encode polypeptides of ~900 amino acids with apparent M_r ~100 000 and all respond to both AMPA and kainate. The deduced amino-acid sequence is only weakly homologous to known members of the ligand-gated ion channel superfamily such as the α2, α3, α4, β2, β3 or β4 subunits of the neuronal NICOTINIC RECEPTOR (nAChR); to the α or β subunits of the GABA$_A$ receptor; or to the 48K subunit of the glycine receptor [10] (*see* GABA AND GLYCINE RECEPTORS). Furthermore, the GluR1 protein, although possessing a proline and an aspartate residue characteristic of a ligand-gated ion channel extracellular cysteine

loop structure, contains only one cysteine residue located much closer to the proline than has been observed in the other receptors. In fact, the protein most closely related to the GluR group by sequence homology is the glutamine transporter binding protein from the bacterium *Escherichia coli*.

Alternative splicing. Two forms of each of the four clones GluR1–4 exist due to alternative splicing [12]. The proteins encoded by the mRNAs from these alternatively spliced forms (designated 'flip' and 'flop') show distinct expression patterns in the rat brain as determined by *in situ* hybridization. Both the alternatively spliced exons code for 38 amino acids which form a hydrophilic stretch. Although their binding properties are identical, the alternatively spliced forms show divergent functional characteristics with different rates of desensitization and steady-state current amplitude evoked by L-glutamate. No differences were observed in the kainate-evoked currents.

What is the likely role of multiple forms of non-NMDA receptor subunits alternatively spliced from the same gene product and displaying different desensitization kinetics and steady-state conductance levels? One suggested possibility is that insertion of receptors composed of these alternative forms into the post-synaptic membrane may affect the sensitivity of synaptic transmission. The switching between the flip and flop forms could therefore be of considerable importance in physiological processes such as LTP.

Calcium permeability. Expression studies with AMPA receptor subunits have shown that GluR1–GluR4 subunits assemble as hetero-oligomeric complexes which display different current–voltage (I–V) relations and Ca^{2+} permeability through the receptor channel. The presence of GluR2 both as a homomeric receptor and as a component of a heteromeric assembly results in near linear I–V plots and insignificant Ca^{2+} permeability. In the absence of GluR2 I–V plots are doubly rectifying and there is significant Ca^{2+} entry through the channel lumen. Thus the presence or absence of GluR2 dictates the characteristics of the AMPA receptor assembly. The molecular basis of the predominant role of GluR2 is the presence of an arginine residue in the putative second transmembrane domain (TM2) of the subunit. The TM2 regions of each of the subunits are thought to form the channel lining of the receptor. GluR1, 3, and 4 possess a glutamine residue at the corresponding position. The importance of this amino acid substitution has been demonstrated by point mutagenesis experiments. When the arginine residue in GluR2 is replaced by glutamine the presence of the mutated GluR2 does not prevent Ca^{2+} entry through the channel [9].

RNA editing. It is intriguing to note that the glutamine codon (CAG) is present in the putative TM2 region in the genomic DNA for all subunits even though all GluR2 mRNAs and a proportion of GluR5 (30%) and GluR6 (75%) mRNAs (see below) specify the arginine codon in their sequences. This has been explained by the existence of RNA EDITING. RNA editing is the enzymatic alteration of the RNA at some point after transcription from the genomic DNA. GluR2 appears to exist only in the edited form whereas GluR5 and GluR6 are detected in both the edited and unedited form at different stages of neuronal development suggesting some developmental switching mechanism may be involved.

Kainate receptors

High affinity kainate receptor subunits (GluR5–GluR7, KA-1, KA-2) constitute a different subfamily of glutamate receptors. GluR5 is only about 40% conserved with GluR1–GluR4 (see Table E2). When expressed in *Xenopus* oocytes only poor responses were detected to glutamate and none at all to AMPA or kainate. However, expression of GluR5 in mammalian 293 cells revealed very high affinity [³H] kainate binding sites and functional, but very fast, depolarizing responses were detected to domoate, kainate, glutamate, and AMPA. Two separate cDNAs have been isolated that encode slightly different forms of GluR5 protein but these do not correspond to the alternative splicing seen in GluR1–4. The subunit encoded by GluR51 contains a stretch of 15 amino-acid residues that are not present in the GluR52 form. Both forms give identical very weak glutamate-evoked depolarizations when expressed either separately or together in *Xenopus* oocytes. *In situ* hybridization of the GluR5 genes in the developing brain shows strong expression in areas of neuronal differentiation and synapse formation. In the adult brain the GluR5 genes have a distinct pattern of expression from GluR1–4 suggesting they do not constitute a subunit of hetero-oligomeric native receptors with those gene products. As GluR5 has few conserved structural motifs and no sequence identity with other ligand-gated ion channels it has been suggested that the GluR5 proteins represent a separate class of ligand-gated ion channel that may not be evolutionarily related to the general ligand-gated ion channel superfamily.

Other cDNAs from rat brain encoding receptor subunits that are recognized by kainate but not by AMPA have been reported [9] (see Table E2). One of these clones, KA-1, is expressed predominantly in the CA3 cells of the hippocampus, and its mRNA distribution closely matches the autoradiographic localization of high-affinity [³H] kainate binding. It has 30% sequence similarity to the GluR1–4 receptor subunits and *in vitro* shows a pharmacological profile similar to the mammalian high-affinity kainate-binding sites. Another clone, GluR6, has 80% sequence identity to GluR5 but less than 40% identity with GluR1–4. The expression pattern of the GluR6 gene corresponds well to the autoradiographic distribution of high-affinity [³H] kainate binding sites and, as for the KA-1 clone, high levels of expression were observed in the CA3 of the hippocampus.

Table E2 Homology (%) between clones encoding different EAA receptor subunits

	GluR1	GluR5	KA-1	NMDAR1
GluR2–GluR4	~70	~40	~35	~22
GluR6–GluR7	~38	~77	~42	~22
KA-2	35	44	70	22

Modified from [9].

Glutamate receptor stoichiometry

Polypeptides encoded by the GluR1–4 clones, GluR5, KA-1, GluR6 and NMDAR1 each form functional ligand-gated ion channels either in *Xenopus* oocytes or when transiently expressed in cell lines. However, the responses obtained from these homo-oligomeric receptors differ from native receptors or from combinations of different subunits expressed in *Xenopus* oocytes.

For the NMDA receptors, the greatest response to glutamate is found with a combination of NMDAR1 and NMDAR2 subunits. For the nonNMDA receptors, coexpression of GluR1, GluR2 and GluR3 gave the largest response to kainate — a 2.5-fold increase over the summed responses for each of the polypeptides alone. Other non-NMDA subunit combinations resulted in agonist profiles and current–voltage plots similar to those seen in native receptors. None the less, it remains to be determined if *in vivo* other structural subunits exist which do not contain both ligand-binding activity and channel-forming activity on the same chain.

From the currently available data it is not possible to determine the true stoichiometry of a native receptor. However, by analogy with other ligand-gated ion channels and from the *in vitro* evidence, it may be predicted that the glutamate receptors will consist of four or five subunits surrounding a central channel core, and are likely to be hetero-oligomers.

The brain distribution of the various subunits [13] also suggests that receptors with different combinations of subunits and different functional properties are present in different brain regions (*see e.g.* GABA AND GLYCINE RECEPTORS). For the NMDA subunits, NMDAR1 and NMDAR2A are widely distributed, whereas the other NMDAR2 subunits show different but overlapping expression.

Subunit topology

Possible secondary structures for NMDA and non-NMDA receptors have been deduced from primary amino-acid sequence data (Fig. E31). In accordance with the accepted model for the ligand-gated ion channel superfamily as a whole, the deduced protein sequences of the glutamate receptor clones show four hydrophobic putative transmembrane regions, and, in addition, they possess very large N-terminal extracellular domains which are characteristic of both ionotropic and metabotropic EAA receptors.

A number of other structural motifs are consistent with ligand-gated ion channel function. The region assigned as transmembrane region 2 (TM2) (the putative channel lining segment of the polypeptide) is flanked by negatively charged amino-acid residues that are believed to act as a charge selectivity filter for cationic channels. As both the glutamate and nicotinic acetylcholine receptors are cation selective, the lining of the ion channel pore must fulfil the necessary conditions to allow the passage of cations [14].

In NMDAR1 there is also a stretch of glutamic acid residues preceding TM2. This unusual motif may be associated with the voltage-sensitive Mg^{2+} blockade peculiar to NMDA receptors [15]. In non-NMDA receptors substitution of an arginine residue for a glutamine residue in the TM2 dictates the Ca^{2+} permeability of the channel [9]. NMDAR1, which preferentially gates Ca^{2+}, has an asparagine residue in the corresponding position. Ion permeation through the NMDAR1 channel may also be regulated by the serine and threonine residues present in the TM2 region, as these amino acids have been reported to be important for normal function of the nicotinic acetylcholine receptor channel. All of the GluR and NMDAR sequences contain consensus *N*-glycosylation sites in the extracellular domain and consensus phosphorylation sites for both CA^{2+}/CALMODULIN PROTEIN KINASE type II and PROTEIN KINASE C on the intracellular domains. This is intriguing as these enzymes are known to be involved in the induction and maintenance of LTP [16].

Native receptors

The molecular biological characterization of EAA receptor subunits is advancing at a remarkable rate but the determination of the subunit composition of native receptors is far from being resolved. Perhaps the most fruitful approach for determining the precise structure and subunit composition of EAA receptors as they occur in the central nervous system will come from the biochemical purification and functional reconstitution of native receptor proteins. Although this field has not advanced as rapidly as the molecular biology, the peptide sequences generated from the cloning work enable the synthesis of highly specific antipeptide antibodies which can be used for the immunopurification and subsequent biochemical analysis of distinct classes of native EAA receptor proteins.

J.M. HENLEY

1 Collingridge, G.L & Lester, R.A.J. (1989) Excitatory amino acid receptors in the vertebrate central nervous system. *Pharmacol. Rev.* **40**, 143–208.
2 Barnard, E.A. & Henley, J.M. (1990) The non-NMDA receptors: types, protein structure and molecular biology. *Trends Pharmacol. Sci.* **11**, 504–511.
3 Sugiyama, H. et al. (1987) A new type of glutamate receptor linked to phospholipid metabolism. *Nature* **325**, 531–533.
4 Masu, M. et al. (1991) Sequence and expression of a metabotropic glutamate receptor. *Nature* **349**, 760–765.
5 Nakanishi, S. (1992) Molecular diversity of glutamate receptors and implications for brain function. *Science* **258**, 597–603.
6 Schoepp, D.D. & Conn, P.J. (1993) Metabotropic glutamate receptors in brain function and pathology. *Trends Pharmacol. Sci.* **14**, 13–20.
7 Monaghan, D.T. et al. (1989) The excitatory amino acid receptors: their classes, pharmacology and distinct properties in the function of the central nervous system. *Annu. Rev. Pharmacol. Toxicol.* **29**, 365–402.
8 Moriyoshi, K. et al. (1991) Molecular cloning and characterization of the rat NMDA receptor. *Nature* **354**, 31–37.
9 Sommer, B. & Seeburg, P.H. (1992) Glutamate receptor channels: novel properties and new clones. *Trends Pharmacol. Sci.* **13**, 291–296.
10 Kutsuwada, T. et al. (1992) Molecular diversity of the NMDA receptor channel. *Nature* **358**, 36–41.
11 Barnes, J.M. & Henley, J.M. (1992) Molecular properties of excitatory amino acid receptors. *Prog. Neurobiol.* **39**, 113–133.
12 Sommer, B. et al. (1990) Flip and flop — a cell-specific functional switch in glutamate-operated channels of the CNS. *Science* **249**, 1580–1585.
13 Betz, H. (1990) Ligand-gated ion channels in the brain: the amino acid receptor superfamily. *Neuron* **5**, 383–392.
14 Unwin, N. (1989) The structure of ion channels in membranes of excitable cells. *Neuron* **3**, 665–676.
15 Nowak, L. et al. (1984) Magnesium gates glutamate-activated channels in mouse central neurones. *Nature* **307**, 462–465.
16 Bliss, T.V.P. & Collingridge, G.L. (1993) A synaptic model of memory: long-term potentiation in the hippocampus. *Nature* **361**, 31–39.

excitatory junction potential (EJP, e.j.p.) A rapid transient DEPOLARIZATION of a SMOOTH MUSCLE cell membrane due to the release of a NEUROTRANSMITTER from prejunctional autonomic nerve terminals. Although this is a form of SYNAPTIC TRANSMISSION the term 'junction' rather than 'synaptic' is preferred because of the relative lack of specialization of the nerve and muscle at the site of neurotransmission compared with a skeletal neuromuscular synapse or central nervous system synapse. The DEPOLARIZATION during the EJP renders the smooth muscle cell more excitable and more prone to contraction as voltage-dependent calcium channels (*see* VOLTAGE-GATED ION CHANNELS), responsible for ACTION POTENTIAL discharge, are more likely to open at the more positive MEMBRANE POTENTIAL. Time courses for EJPs tend to be slower than for excitatory synaptic potentials, and are in the range of 0.5 to 1.5 s; their amplitude is graded with strength of stimulus and amount of transmitter released, and is of the order of millivolts. In many tissues, such as the urinary bladder and vas deferens, EJPs are due to released ATP acting on postjunctional P_2 purinoceptors (*see* PURINERGIC SYSTEMS), resulting in an increased cation conductance. In many gastrointestinal smooth muscles the transmitter is ACETYLCHOLINE, which acts on postjunctional MUSCARINIC RECEPTORS, and in many sphincteric muscles it may be NORADRENALINE acting on α-ADRENERGIC RECEPTORS that is responsible for EJPs.

excitatory postsynaptic potential (EPSP, e.p.s.p.) The transient change in the MEMBRANE POTENTIAL (usually in a DEPOLARIZING direction) in a postsynaptic neuron caused by the binding of excitatory NEUROTRANSMITTERS released by the corresponding presynaptic neuron to postsynaptic receptors. The term is usually applied to SYNAPTIC TRANSMISSION in nerve cells, but may also be applied to other types of cells with well-defined synapses. The depolarization of the EPSP, during which the membrane potential becomes more positive, reflects an increased excitability of the cell as its membrane potential is brought closer to the threshold for generation of ACTION POTENTIALS. The amplitude of an EPSP is graded according to the intensity of the stimulus applied to the presynaptic nerve, and therefore the amount of transmitter that is released. If the depolarization is great enough and voltage-dependent sodium ion channels become operative (*see* VOLTAGE-GATED ION CHANNELS), action potentials will be generated. The amplitude of EPSPs is usually of the order of millivolts, with a latency of tens of milliseconds, and a duration of hundreds of milliseconds. Because of their rapid time course they are also known as fast EPSPs. In some cells EPSPs with a relatively long latency, 0.5 s or more, and prolonged time course can also appear: these are called slow EPSPs. In some cases the transmitter responsible for fast EPSPs is ACETYLCHOLINE, acting on postsynaptic NICOTINIC RECEPTORS; in some cases, but by no means all, slow EPSPs are due to acetylcholine acting on postsynaptic MUSCARINIC RECEPTORS.

exine The outer wall of mature pollen grains and some other spores, often of complex structure and resistant to decay.

exocrine Secretion from glands, via ducts, onto epithelial surfaces. Examples of exocrine glands are mammary glands, salivary glands, and sweat glands.

exocytic vesicle *See*: EXOCYTOSIS.

Exocytosis

EXOCYTOSIS is the terminal event of the secretory pathway in eukaryotic cells (*see also* PROTEIN SECRETION). It involves the fusion of the membrane of the secretory vesicle or granule with the cell plasma membrane. This takes place in a way that allows the selective exposure and release of the contents of secretory vesicles to the exterior without compromising cellular integrity, that is, without any loss of cytosol contents. Secretion is an important activity exhibited by almost all eukaryotic cells, allowing the release of substances to which cell membranes are impermeant, for example digestive enzymes, plasma proteins, hormones, neurotransmitters, and toxins. The phenomenon takes two forms:

1 Constitutive secretion, in which secretory products (such as serum albumin in liver cells or antibodies in plasma cells) are synthesized, processed and released in a continuous process. The rate is therefore rather constant and is limited mainly by the availability of mature secretory vesicles at the sites of release.
2 Regulated secretion, in which preformed secretory products are kept in readiness in secretory vesicles awaiting an appropriate signal.

In some cells both constitutive and regulated secretion occur side by side. While we understand that the mechanism of membrane fusion in each case is likely to be similar, most investigations of the terminal exocytotic step have been devoted to cells in which secretion is subject to regulation. The topic of regulated exocytosis has been comprehensively reviewed [1–4].

Molecular events leading to exocytosis

Historically, the investigation of exocytosis has involved the treatment of secretory cells or tissues with activating ligands (secretagogues) and measurement of the released materials. For example, it has long been recognized that sympathetic stimulation, or application of nicotinic cholinergic agonists to the adrenal medulla induces the release of CATECHOLAMINES. However, this approach does not generally allow a distinction to be made between so-called 'early events' of signal transduction and 'late events' specifically related to exocytotic fusion.

The early steps of this pathway, namely receptor activation and the generation of intracellular messengers such as Ca^{2+}, are typical of those that occur during the activation of many non-secretory cell types (*see* SECOND MESSENGER PATHWAYS). In particular, the level of free Ca^{2+} ions in the cytosol has long been considered to be of central importance in the regulation of exocytosis. In part this is because Ca^{2+} is both the most easily measured and the most easily manipulated of the possible intra-

cellular signals (*see* CALCIUM). In many cases, simply depriving the cell or tissue of extracellular Ca^{2+} is sufficient to prevent secretion. When the Ca^{2+}-carrying IONOPHORES became available in 1972 it became possible to bypass the initiating stages associated with receptor activation and to raise the cytosol concentration of Ca^{2+} directly. Most secretory cells treated in this way were found to respond in the predicted manner, releasing specific granule-contained materials and undergoing the characteristic morphological changes called degranulation. Later, with the introduction of Ca^{2+}-sensing fluorescent dyes such as quin2 in 1980 and subsequently fura2, it became simple to detect the rise in the cytosol level of Ca^{2+} that followed stimulation by secretagogues.

There has been much speculation concerning the mechanism by which Ca^{2+} might cause the specific membrane fusions that occur in the later stages of exocytosis. With the discovery of calmodulin, and latterly the Ca^{2+}-binding proteins of the AN-NEXIN series (*see* CALCIUM-BINDING PROTEINS), it is now understood that Ca^{2+} may act as an activating ligand for a protein- (enzyme-) mediated process rather than directly on the fusing surfaces. Recent developments, however, have questioned the universality of Ca^{2+} as a stimulus.

In the past 10 years, investigation and understanding in this area have been revolutionized by the development and application of the whole-cell PATCH CLAMP technique to measure changes in the electrical capacitance of the plasma membrane of single cells [5]. At $1\ \mu F\ cm^{-2}$, the capacitance per unit area is a remarkably constant parameter of biological membranes. A change in membrane capacitance thus registers a change in its area. Measurement of capacitance during secretion allows on-line monitoring of exocytosis as the plasma membrane area is incremented each time a granule fuses with it (Fig. E32). The patch-pipette in the whole-cell configuration also provides the means of controlling the composition of the cytosol and of introducing fluorescent indicator dyes for Ca^{2+} such as fura2, so that it is possible to measure the changes in cytosol Ca^{2+} while exocytosis is occurring.

Another recent development that also allows control of cytosol composition is the technique of selective permeabilization of the plasma membranes of cells in suspension [1,6]. Although lacking the precise resolution in space and time of the patch-clamp method, this technique, which can be applied to cells in suspension, is more flexible and is applicable to cells that are not amenable to analysis by the patch-clamp technique. Both approaches permit the experimental manipulation of the composition of the cytosol of the secreting cells and allow the investigation of the sequence of intracellular events that lead to secretion.

The application of these new techniques has shown that an increase in cytosol Ca^{2+} cannot be regarded as universally essential for exocytosis, and this has demanded a reassessment of the role of calcium. On the one hand, in adrenal CHROMAFFIN CELLS, one of the most widely investigated secretory systems, introduction of Ca^{2+} at physiologically relevant concentrations (i.e. around 10^{-6} M, buffered by the use of chelating anions such as EGTA) causes an increase in membrane capacitance; this occurs in a stepwise manner signifying the fusion of individual granules with the plasma membrane (see below). The introduction of Ca^{2+} into permeabilized chromaffin cells also elicits

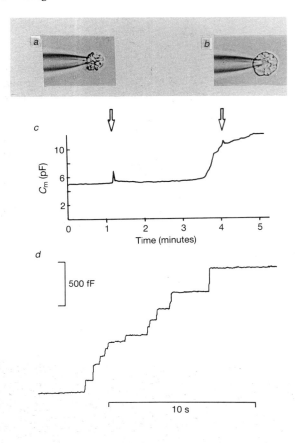

Fig. E32 Exocytosis from a horse eosinophil detected by whole-cell patch-clamp measurement of membrane capacitance. The time course of capacitance (C_m) was measured following the introduction of GTP-γ-S (a stable analogue of GTP that induces secretion in these cells) into the cell. *a, b*, Experimental set-up before (*a*) and during (*b*) release of GTP-γ-S. Photographs were taken at times indicated by the vertical arrows in *c*. *c*, Time course of capacitance following introduction of GTP-γ-S into the cell. The sharp increase in capacitance indicates the onset of exocytosis. (The spikes are artefacts due to shake induced by operation of the camera shutter.) *d*, An expanded capacitance trace during the most rapid stage of exocytosis. The steps indicate the fusion of individual granules. Courtesy of Susanne Scepek, Freie Universität Berlin.

CELLS, NEUTROPHILS, and EOSINOPHILS [1], exocytosis can be stimulated by introduction of GTP and it is now apparent that in these cells the role of Ca^{2+} is likely to be that of a modulator, not the direct effector of exocytosis. In patch-clamp and cell-permeabilization experiments, it has been found that synthetic, poorly hydrolysable analogues of GTP (e.g. GTP-γ-S) can induce exocytosis even when the level of Ca^{2+} is suppressed well below the normal resting level and ATP is absent. Moreover, introduction of GDP suppresses Ca^{2+}-induced secretion. These results indicate a role for a GTP-BINDING PROTEIN (called G_E) and suggest that in these cells, Ca^{2+} and guanine nucleotide act together, possibly in sequence, to induce exocytosis. Surprisingly, although ATP modulates secretion, a phosphorylation reaction does not comprise an essential step in the terminal events leading to exocytosis in these cells.

There is currently much discussion concerning the possible identity and function of the GTP-binding protein, G_E, that mediates exocytosis. Two main classes of guanine nucleotide-binding protein are known: the monomeric GTP-binding proteins related to the *ras* proto-oncogene product, and the larger heterotrimeric G proteins (*see* GTP-BINDING PROTEINS). Small GTP-binding proteins, identified as homologues of *ras*, are thought to ensure directionality and accuracy of destination in the traffic of intracellular vesicles in the early stages of the secretory pathway (from the ENDOPLASMIC RETICULUM through the GOLGI APPARATUS to the formation of mature secretory granules; *see* PROTEIN TARGETING; PROTEIN TRANSLOCATION). On the other hand, it has been argued that the cell physiology of guanine nucleotide-mediated (regulated) exocytosis is suggestive of a process controlled by an αβγ heterotrimeric, signal-transducing G protein [1].

It is now apparent that guanine nucleotides, and the associated GTP-binding proteins, also have a central role in the control of exocytosis in cells and tissues other than those of myeloid origin [1]. Good examples are the endocrine tissues involved in calcium homeostasis, which secrete hormones in response to low ambient levels of (extracellular) Ca^{2+} (e.g. parathyroid hormone (PTH) from the parathyroid and renin from the renal juxtaglomerular apparatus). Permeabilized cells derived from these tissues undergo exocytosis in response to introduction of guanine nucleotides and this is actually suppressed when Ca^{2+} is elevated into the micromolar range. Some inhibitory mechanisms are also expressed through GTP-binding proteins. For example, Ca^{2+}-induced exocytosis from permeabilized insulin-secreting cells (derived from pancreatic islet cell tumours), and ACTH-secreting cells (from a pituitary cell line) can be inhibited by somatostatin and α_2-adrenergic agonists (e.g. clonidine), in a manner dependent on the presence of a guanine nucleotide. Thus depending on the cell type, and the circumstances, both Ca^{2+} and GTP can mediate the stimulation and the inhibition of exocytosis. In some situations they seem to act in synergy, in others they act in opposition. A general scheme is given in Fig. E33.

Membrane fusion

Morphological aspects

The small size of most secretory granules (less than 0.1 μm in diameter) means that the morphological detail of the release

catecholamine secretion. For this cell anyway, Ca^{2+} seems to deliver a late signal for exocytotic fusion but it does not follow that Ca^{2+} is the unique stimulus for secretion from neuronal cells. Indeed, it has recently been shown that elevation of cytosol Na^+ can induce secretion of vasopressin (anti-diuretic hormone) from neurohypophyseal nerve endings under conditions in which Ca^{2+} is maintained at basal levels [7]. Although Ca^{2+} probably remains the major stimulating messenger for secretion from these cells, it is possible that under certain circumstances (prolonged high frequency stimulation), elevated Na^+ may be sufficient.

GTP-binding protein

A very different picture emerges when these same techniques are applied to certain granulocytic leukocytes. In permeabilized MAST

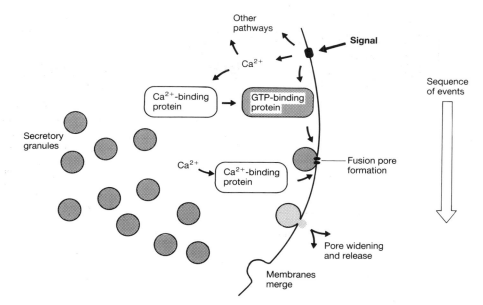

Fig. E33 Schematic illustration of various possible interactions involving Ca²⁺- and GTP-binding proteins in the stimulation of exocytosis. See text for further explanation.

process is beyond the resolving power of the light microscope. Indeed, the so-called 'degranulation' responses are only readily observed in those cells that possess large and abundant secretory vesicles and that release them abruptly. Moreover, each granule fusion is a discrete event, both spatially and temporally, and their transience, infrequency and molecular subtlety make observation by electron microscopy difficult. Chemical fixation gives rise to artefacts and to avoid these it is necessary to trap exocytotic events by rapid freezing [8]. A striking feature revealed by electron micrographs obtained in this way is that at no time does the surface of the secretory granule membrane form an extended area of close apposition with the cytosolic surface of the plasma membrane. Apart from the local point of contact, the two surfaces remain separated by a thick layer of cytoplasm. As membrane fusion is initiated and develops, there forms a structure called the fusion pore. Figure E34 shows a freeze-fracture image of a fusion pore. The surface of this structure has normal membrane morphology (a fracture surface with an abundancy of exposed intramembranous particles) and is continuous with the two fusing membranes. The focus of fusion pore initiation may be delimited by fibrillar structures which are sometimes seen to extend from the intervening layer of cytoplasm on to the granule membrane in both resting and stimulated cells.

Electrophysiological aspects

Application of the patch-clamp technique to measure the membrane capacitance of degranulating mast cells or eosinophils at high resolution reveals that exocytosis proceeds in steps (see Fig. E32d). The unitary events that give rise to this phenomenon consist of increments in membrane capacitance as a fusion pore forms between each granule and the plasma membrane. On occasion, an 'on-step' may be followed by an 'off-step' of similar amplitude due to the closure or annealing of a fusion pore shortly

Fig. E34 Freeze-fracture image of exocytosis in a stimulated rat mast cell. A small pore has formed between the plasma membrane and a granule (arrow). Bar = 0.2 μm. Courtesy of Douglas Chandler, Arizona State University.

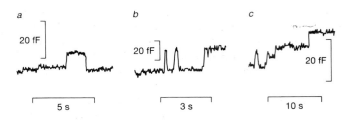

Fig. E35 Capacitance flicker indicating 'on' and 'off' steps in pore opening (see text). *a*, Rat mast cell; *b*, guinea-pig eosinophil; *c*, nerve terminal. Courtesy of Manfred Lindau, Freie Universität Berlin.

after it has opened. The pore may then open and close again a number of times without release of granule contents and this gives rise to capacitance flicker (Fig. E35). It indicates that in the early stages at least, membrane fusion is a reversible process. Once formed, the pores tend to enlarge and ultimately remain open. Although not observed in all secretory cells, capacitance flicker has also been recorded in nerve terminals derived from the posterior pituitary and in other membrane fusion processes such as fusion of red blood cells with fibroblasts mediated by a viral fusion protein.

The characteristics of fusion pores and their formation have been investigated in detail in the mast cells of *beige* mice, which have defects in secretion. Having only a few (typically less than 10) granules of relatively large size, these mutant cells give rise to particularly large capacitance steps. The molecular mechanism of pore formation, however, remains unknown, but the specificity, efficiency and characteristic non-leakiness of exocytosis make it very unlikely that it results from a spontaneous interaction between approaching membrane bilayers. A protein catalyst is required to make at least the initial contact and also perhaps to support an initial structure under which the necessary topological changes in the component membrane bilayers can take place.

B.D. GOMPERTS

P.E.R. TATHAM

See also: MEMBRANE STRUCTURE.

1 Lindau, M. & Gomperts, B.D. (1991) Techniques and concepts in exocytosis: focus on mast cells. *Biochim. Biophys. Acta* **1071**, 429–471.
2 Burgoyne, R.D. (1991) Control of exocytosis in adrenal chromaffin cells. *Biochim. Biophys. Acta.* **1071**, 174–202.
3 Almers, W. (1990) Exocytosis. *Annu. Rev. Physiol.* **52**, 607–624.
4 Gomperts, B.D. (1990) G$_E$: A GTP-binding protein mediating exocytosis. *Annu. Rev. Physiol.* **52**, 591–606.
5 Lindau, M. & Neher, E. (1988) Patch-clamp techniques for time-resolved capacitance measurements in single cells. *Eur. J. Physiol.* **411**, 137–146.
6 Gomperts, B.D. & Fernandez, J.M. (1985) Techniques for membrane permeabilization. *Trends Biochem. Sci.* **10**, 414–417.
7 Nordmann, J.J. & Stuenkel, E.L. (1991) Ca^{2+}-independent regulation of neurosecretion by intracellular Na$^+$. *FEBS Lett.* **292**, 37–41.
8 Chandler, D.E. & Heuser, J.E. (1980) Arrest of membrane fusion events in mast cells by quick-freezing. *J. Cell Biol.* **86**, 666–674.

exoenzyme Enzyme secreted by a cell, an extracellular enzyme.

exogastrulation Abnormal cell migration during GASTRULATION in which the outer vegetal surface of the gastrula fails to invaginate and folds outward instead. It is induced experimentally, especially in sea urchin and amphibian embryos, by treatment with hypertonic solutions. The archenteron is everted in these embryos and the endoderm encloses CHORDA MESODERM tissues, leaving the ectoderm as an empty sac. The chorda mesoderm sheet behaves as a self-differentiating unit (forming axial notochord associated with somites, pronephric (renal) rudiments, and other mesodermal structures) in contrast to the ectoderm which, in the absence of inductive influences from other tissues, remains as an undifferentiated sac. *See*: Fig. V4 in VERTEBRATE NEURAL DEVELOPMENT.

exon Block of nucleotide sequence within a eukaryotic GENE which is not removed from the initial RNA transcript by RNA SPLICING. Exons are separated from each other by INTRONS. As well as encoding sequences that will be part of the final gene product, exons can also encode leader sequences. In some genes the exons correspond to distinct modules (domains) of protein structure (*see* MOLECULAR EVOLUTION: SEQUENCES AND STRUCTURES; PROTEIN STRUCTURE). *See also*: EUKARYOTIC GENE EXPRESSION; EUKARYOTIC GENE STRUCTURE; RNA SPLICING; TRANSCRIPTION.

exon shuffling The incorporation of particular exons into different genes during molecular evolution, as evidenced by the presence of similar domains in many different proteins.

exonuclease A NUCLEASE that degrades nucleic acids by cleaving successive nucleotide residues (or short oligonucleotides) from the ends of the strands. 5′ and 3′ exonucleases are known.

λ exonuclease An EXONUCLEASE (EC 3.1.11.3) isolated from cells of *Escherichia coli* infected with bacteriophage LAMBDA, and possessing several catalytic properties which make it a useful reagent for the analysis and modification of DNA structure. λ Exonuclease processively degrades duplex DNA from the 5′ end ~100 times faster than for single-stranded DNA, with a preference for a 5′-P group over a 5′-OH group. The enzyme cannot cleave internal phosphodiester bonds in DNA even at a nick or gap. λ Exonuclease has many applications including the determination of DNA structure at cleaved termini, preparation of DNA substrates for TERMINAL DEOXYNUCLEOTIDYLTRANSFERASE, determination of the direction of transcription, and preparation of a DNA substrate before DNA SEQUENCING.

exonuclease III Exodeoxyribonuclease III (EC 3.1.11.2), an EXONUCLEASE from *Escherichia coli* which catalyses the release of mononucleotides from the 3′ end of double-stranded DNA. It also has a RIBONUCLEASE H-like activity, preferentially degrading RNA in an RNA : DNA hybrid. The enzyme will attack a nick in the nucleic acid backbone but fails to cleave either single-stranded DNA or double-stranded DNA with a recessed 5′ end. Exonuclease III treatment followed by S1 NUCLEASE treatment can be used to shorten double-stranded DNA in a controlled manner for the production of DELETIONS in regions of interest in a DNA fragment.

exonuclease V Exodeoxyribonuclease V (EC 3.1.11.5), an EXO-NUCLEASE from *Escherichia coli* with a preference for double-stranded DNA. It possesses DNA-dependent ATPase activity and is the product of the *recBCD* genes. It acts as an endonuclease on single-stranded DNA. *See also*: RECKLESS PHENOTYPE.

exonuclease VII Exodeoxyribonuclease VII (EC 3.1.11.6), an EXO-NUCLEASE from *Escherichia coli* which catalyses the hydrolysis of single-stranded DNA from both the 5′ and 3′ ends. It requires free 5′ and 3′ ends to initiate digestion and releases oligonucleotides rather than mononucleotides. The enzyme is widely used in transcript mapping experiments (in which the mRNA is HYBRIDIZED to its cognate genomic DNA) in conjunction with S1 NUCLEASE to determine the length of introns in a given gene.

exoplasmic On the extracellular face of a membrane.

exotoxin A bacterial toxin which is released from the bacterium into the host tissue, for example CHOLERA TOXIN, TETANUS TOXIN, DIPHTHERIA TOXIN, PERTUSSIS TOXIN and various bacterial HAEMOLYSINS.

expression cassette Genetic module comprising a gene and the regulatory regions necessary for its expression, which may be incorporated into a VECTOR.

expression cloning *See*: XENOPUS OOCYTE EXPRESSION SYSTEM.

expression library DNA LIBRARY in which cDNAs are cloned in vectors that allow their expression in a heterologous host system. *See also*: EXPRESSION VECTORS.

expression locus *See*: BACTERIAL PATHOGENICITY.

expression vector A specialized VECTOR which is constructed so that the gene it contains can be expressed in a (foreign) host cell. The coding sequence to be expressed is inserted in the correct relationship to host-specific PROMOTER and other transcriptional regulatory sequences and in the correct READING FRAME, so that the native protein is produced. The vector also contains sequences required for efficient translation (e.g. the SHINE–DALGARNO REGION for expression in bacterial cells). Expression vectors usually contain a transcript termination site 3′ to the inserted gene to ensure the mRNA produced is of the correct size.

Vectors for expressing or overexpressing a foreign gene in bacterial systems (*see* GENETIC ENGINEERING) are constructed from PLASMID or BACTERIOPHAGE vectors. Both reading frames and regulatory regions can be manipulated *in vitro* by SITE-DIRECTED MUTAGENESIS to maximize expression of the inserted gene. The most effective systems possess the recognition signals necessary to optimize all the major transcriptional and post-transcriptional stages in gene expression (*see* BACTERIAL GENE EXPRESSION), and are also regulatable. Some expression vectors, for example, possess both an inducible promoter and suitable RESTRICTION SITES (a MULTIPLE CLONING REGION) into which the required DNA can be cloned. Further DNA sequence encoding the SIGNAL PEPTIDE necessary for transport of a protein to the cell membranes or for secretion (*see* BACTERIAL ENVELOPE; BACTERIAL PROTEIN EXPORT) can be incorporated into the vector.

The use of expression vectors, especially of the relatively small circular expression plasmids (Fig. E36), greatly simplifies the process of overexpressing a chosen gene. Any DNA manipulations to enable overexpression can readily be made *in vitro* and the recombinant plasmid is introduced by TRANSFORMATION into a range of bacterial strains and amplified at will. Depending on the replication mechanism of the plasmid, it can be present at a high COPY NUMBER (30–50, or even more) in the cell. This,

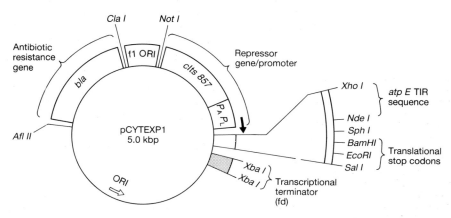

Fig. E36 A modular type of plasmid expression system used to achieve direct expression of cloned genes to high levels in the cytoplasm of *Escherichia coli*. The CYTEXP system is modular, in which each module bears an important component, and modules are readily exchangeable. In the example illustrated here, the vector carries promoters from bacteriophage lambda which are regulated by a plasmid-borne *clts857* repressor gene. The repressor itself is temperature labile, so that raising the temperature to 42°C inactivates it, thus switching on transcription. This runs through the translational LEADER SEQUENCES and MULTIPLE CLONING SITE and is terminated in the transcriptional terminator derived from bacteriophage fd. The translational leader sequence, derived from the *E. coli atp* operon, helps to encourage efficient translational initiation. The vector also carries two ORIGINS OF REPLICATION: one initiates replication of the double-stranded plasmid; the other (f1 ORI) allows generation, assisted by a helper phage, of single-stranded DNA that is useful as a template for *in vitro* mutagenesis. A further module bearing the *bla* gene (endcoding in β-LACTAMASE) is used to exert selective pressure on the bacterial cell to maintain the plasmid, which might otherwise be lost over generations of cell replication.

combined with the use of efficient expression recognition signals, usually ensures that a chosen gene can be expressed to yield large amounts of the encoded product, which can then be isolated from the bacterial culture. There are many applications of such procedures in both basic research and biotechnology (*see* GENETIC ENGINEERING).

There are disadvantages to *E. coli* as a host for expressing recombinant proteins of eukaryotic origin (*see* GENETIC ENGINEERING) and eukaryotic expression vectors have been developed for use in yeast (SACCHAROMYCES CEREVISIAE), filamentous fungi such as ASPERGILLUS NIDULANS, plant cells, insect cells, and mammalian cells.

Expression vectors for yeast are generally based on modified forms of the 2µM PLASMID, incorporating a strong regulatable promoter such as the *GAL* promoter. Plasmids that integrate into the genome are used for expression in *Aspergillus*. Vectors incorporating the T-DNA from the TI PLASMID of *Agrobacterium tumefaciens* are most generally used in PLANT GENETIC ENGINEERING; vectors incorporating the cauliflower mosiac virus (CaMV) promoter have also been developed for use in plants.

The insect cell/baculovirus vector system can provide a very high level of protein synthesis from the incorporated heterologous gene. Making use of the baculovirus polyhedrin gene promoter (which normally directs the synthesis of large amounts of a protein that forms large crystalline inclusion bodies in the late stages of infection), up to 50% of total cell protein can be the recombinant protein.

Vectors based on the mammalian PAPOVAVIRUSES SV40 and bovine papilloma virus (BPV) have been used to express genes in mammalian cells. SV40-based vectors include the origin of replication, the early genes involved in virus replication and part of the late region. High levels of protein production can be achieved with a combination of a modified SV40 vector and a COS cell line which contains a stably integrated portion of the SV40 genome that directs the production of the viral T antigen. This stimulates replication of the plasmid up to 100 000 copies or more per cell. Disadvantages to SV40 expression vectors are the relatively small DNA insert they can accommodate and their restriction to growth in monkey cell lines.

Expression vectors based on extensively modified VACCINIA VIRUS genomes exemplify the way a eukaryotic expression vector can be built up from modules from several different sources. The vaccinia virus genome is too large to manipulate directly *in vitro* and so such vectors start from an extensively modified plasmid containing a cloning site flanked by regulatory regions, which is then recombined into a vaccinia virus genome *in vivo*. One such vector is based on a plasmid, pTM1 (Fig. E37), that includes a gene for the RNA polymerase from the bacteriophage T7 and the corresponding strong constitutive promoter and transcriptional terminator from T7. The foreign gene to be expressed is placed in a multiple cloning site which includes a translation termination codon, under the control of the phage regulatory regions. Although vaccinia encodes its own RNA PROCESSING enzymes, capping of the transcripts from the foreign gene, and thus their stability and translatability, was found to be low. A module from the picornavirus encephalomyocarditis virus, comprising a long untranslated leader region that binds to the ribosome, was

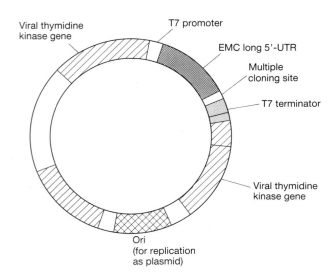

Fig. E37 Plasmid pTM1. Adapted from Moss, B. et al. (1990) *Nature* **348**, 91–92.

therefore incorporated immediately downstream of the T7 promoter to make the transcripts cap-independent. The modified plasmid also contains the vaccinia thymidine kinase gene sequence to provide a region of homology for recombination into the virus genome, and a single-stranded DNA origin of replication. Cells infected with vaccinia virus are then transfected with the plasmid and in a proportion of them a recombinant vaccinia vector will be produced that provides a high level of expression of the introduced gene as the virus replicates. This type of vector has been used successfully to express the CYSTIC FIBROSIS transmembrane conductance regulator protein in airway epithelial cells from cystic fibrosis patients with consequent correction of defective chloride channel regulation in these cells. *See also:* DNA CLONING; XENOPUS OOCYTE EXPRESSION SYSTEM.

Belev, T.N. et al. (1991) *Plasmid* **26**, 147–150.
McCarthy, J.E.G. (1991) In *Advances in Gene Technology II* (Greenaway, P.J., Ed.), Vol. 2, 145–175 (JAI Press, London).
Moss, B. et al (1990) *Nature* **348**, 91–92.
Rich, D.P. et al. (1990) *Nature* **347**, 358–363.

expressivity The range of phenotypes expressed by a particular genotype under given environmental conditions. *Cf.* PENETRANCE.

extensins Family of GLYCOPROTEINS of plant cell walls, which contain the arabinosyl-hydroxyproline linkage (*see* GLYCANS; PLANT CELL WALL MACROMOLECULES). Most are very insoluble and contain dityrosyl cross-links. The LECTIN of potato (*Solanum tuberosum*) is a soluble protein of this general class.

extension (e) Mouse COAT COLOUR GENE encoding the receptor for α-melanocyte-stimulating hormone (αMSH). Expression within the melanocyte is required for wild-type function. Recessive loss-of-function mutations lead to yellow coat colour (deregulated production of phaeomelanin); dominant alleles result in a completely black coat (wild-type is the AGOUTI colouring in which hairs are black with a yellow band towards the top).

extra sex combs (esc) The *esc* gene is involved in the regulation of the HOMEOTIC GENES in *Drosophila*. In embryos deficient in *esc* gene product, homeotic genes are released from their normal positional control and become expressed along the entire length of the body. This results in embryonic lethality with a phenotype of a repetition of posterior abdominal segment type along the body. This is due to the ubiquitous expression of the homeotic gene *Abd-B* which overrides the other homeotic genes to impose abdominal segmental development. In *esc* mutants, expression of the homeotic genes is initially normal but then degenerates at around 7 hours of development. The *esc* gene is thought to be involved together with the *Polycomb* group genes in the maintenance of homeotic genes in a repressed state.

Extracellular matrix molecules

EXTRACELLULAR matrix (ECM) is the noncellular material distributed throughout the body of multicellular organisms. Its remarkably diverse constituents include a wide variety of specialized GLYCOPROTEINS, PROTEOGLYCANS, complex CARBOHYDRATES, and other molecules. Its physical properties can vary from a clear gel to dense rigid bone; its organization can vary from locally homogeneous to elaborately structured.

The many types of ECM are comprised of differing sets of ECM molecules. The epithelia that cover the external and internal surfaces of the body rest on extracellular structures termed basement membranes or BASAL LAMINAE. These thin, sheet-like structures consist of an interconnected matrix of type IV collagen, laminin, heparan sulphate proteoglycan, and other molecules. Other types of ECM are organized quite differently. The loose connective tissue produced by fibroblasts is comprised of other ECM molecules such as fibrils of type I collagen and proteoglycans. Bone, cartilage, the sheets of collagenous material forming fascia, and the cords of collagen-forming ligaments and tendons, and even the clear vitreous humour of the eye all consist of different combinations of ECM molecules secreted by specific cells as structural materials.

Major functions of the ECM include: (1) providing structural support, tensile strength or cushioning; (2) providing substrates and pathways for cell adhesion and cell migration; and (3) regulating cellular DIFFERENTIATION and metabolic function in direct or indirect fashion, for example by modulating cell growth by binding of GROWTH FACTORS. The application of molecular biology to the analysis of the components of ECM has provided many new insights into both its structure and functions. In particular, the functions of ECM molecules can now be understood in terms of specific functional domains or binding and recognition sites. Certain sites involved in interactions with cells are bound by specific cell surface receptors; members of the INTEGRIN family of cell surface glycoproteins are thought to be important cellular receptors for many ECM proteins. Cell adhesion, cell migration, and the regulation of cellular differentiation are now beginning to be understood in terms of specific cell surface interaction between these diverse and important structural and regulatory ECM molecules and their specific receptors (*see* CELL–MATRIX INTERACTIONS).

Types and functions of ECM proteins

The many distinct ECM proteins identified to date are listed in Table E3. Further information about specific ECM molecules can be found in [1,2]. An important characteristic of some of these proteins is the ability to self-organize. Most notably, some types of collagens assemble into fibrillar structures, whereas others can form meshworks or sheet-like structures. Most ECM proteins bind to cells and/or to other ECM components.

ECM proteins such as fibronectin and laminin may help to provide pathways for cell migration. For example, studies of gastrulation in animal embryos reveal that fibronectin provides a substrate for the migration of cells during early development. Fibronectin and probably laminin are also important for the subsequent migration of embryonic NEURAL CREST cells throughout the body to form many of the structures of the face, the sympathetic and parasympathetic nervous systems, and skin pigment. Cells use specific receptors for adhesion and migration, which bind to key sites in these proteins. Inhibiting the receptor-mediated interaction of cells with such extracellular substrates halts migration.

Until recently, studies of ECM proteins and their genes generally concentrated on vertebrates. Many similar genes have now been identified in invertebrates, including collagen genes in *Drosophila*, *Caenorhabditis elegans*, and sea urchin, and a laminin homologue in *Drosophila*.

Repeating motifs within ECM protein genes are also present in other proteins. For example, the FIBRONECTIN TYPE III REPEAT has been found in many genes including certain collagens, CELL ADHESION MOLECULES (CAMs), the *C. elegans unc-22* gene, and even in the bacterial chitinase A1 gene.

Pathology

There are many human genetic disorders that affect the ECM, such as OSTEOGENESIS IMPERFECTA (a condition characterized by brittle bones), EHLERS–DANLOS SYNDROME (EDS, a heterogeneous syndrome characterized by loose joints, classified as types I to XI), and MARFAN SYNDROME (a condition characterized by skeletal, cardiovascular, and ocular abnormalities). To date, the following genes have been identified as responsible for these diseases: osteogenesis imperfecta is caused by mutations in either of the two collagen type I genes; EDS type VII by defective collagen type I; EDS type IV by errors in the α_1 chain gene of collagen type III; and Marfan syndrome by mutations in fibrillin. All are dominant mutations.

Collagens

Collagens are the most abundant animal proteins, constituting 25% of the total protein in mammals. They consist of three polypeptide chains, termed α chains, that associate with one

Table E3 A list of extracellular matrix proteins

Molecule	Molecular size	Notes
Collagens		Bind to cells, proteoglycans, fibronectin, laminin, vitronectin, entactin (nidogen)
At least 17 types of collagens are known		
Four major types are:		
Type I	$\alpha1(I)_2\alpha2(I)_1$, 95K each	Low hydroxylysine content. Constitutes 90% of body collagen
Type II	$\alpha1(II)_3$, 95K each	High hydroxyproline, high hydroxylysine content. Forms thinner fibrils than type I
Type III	$\alpha1(III)_3$, 95K each	High hydroxyproline, low hydroxylysine content. Forms fibrils
Type IV	$\alpha1(IV)_2\alpha2(IV)_1$, 170K each	Very high hydroxylysine content. Forms sheet-like structure in basal lamina
Elastin	Polymer of 85K*	Cross-linked between lysine residues
Fibronectin	Dimer of about 250K*	Binds to cells, collagens, proteoglycans, fibrinogen, amyloid P component
Laminin	Trimer of 400, 210, 200K	Cross-shaped heterotrimer. Binds to cells, proteoglycans, collagen type IV, entactin (nidogen), sulphatides
Proteoglycans	(Variety of types)	See text
Vitronectin	75K	Binds to cells, heparin, collagens
Thrombospondin	Trimer of 140K	Binds to cells, proteoglycans, collagens, laminin, fibrinogen, sulphatides, calcium
Tenascin (cytotactin)	Hexamer of 230K*	Contains epidermal growth factor (EGF)-like repeats and fibronectin type III repeats. Binds to cells, fibronectin
Entactin (nidogen)	150K	Binds to cells, laminin, collagen type IV
Osteonectin (SPARC)	33K	Binds calcium
Anchorin CII	34K	Binds to cells, collagen type II
Chondronectin	Trimer of 56K	Binds to cells, collagen type II
Link protein	39K	Binds to cartilage proteoglycan (aggrecan), hyaluronate
Osteocalcin	5.8K	Binds to hydroxylapatite
Bone sialoprotein	34K	Binds to cells, hydroxylapatite
Osteopontin	33K	Binds to hydroxylapatite
Epinectin	70K	Binds to epithelial cells, heparin
Hyaluronectin	59K	Binds to hyaluronate, neurons
Amyloid P component	25K (monomer)	Binds to fibronectin
Fibrillin	350K	Contains epidermal growth factor (EGF)-like repeat
Merosin	300K	Laminin A chain homologue. Covalently associated with laminin B1 and B2 chains
s-Laminin	190K	Homologous to laminin B chains
Undulin	Heterotrimer of 270, 190, 180K	Related to type XIV collagen
Epiligrin	Heterotrimer of 170, 145, 135K	
Kalinin	Heterotrimer of 200, 155, 140K	Chains are laminin homologues. Identical to nicein and possibly epiligrin

*Molecular weights vary owing to alternative splicing of precursor mRNA, which produces a family of closely related variants.

another to form a right-handed triple-helical structure. The helical domain of collagen has a characteristic tripeptide repeat sequence, Gly-X-Y, where the X and Y are frequently HYDROXY-PROLINE or proline. Collagens also contain HYDROXYLYSINE, which is required for cross-linking between and within collagen molecules and for lysine-linked glycosylation (*see* GLYCOPROTEINS AND GLYCOSYLATION). At least 17 genetically distinct types of collagen have been identified to date, each of which is composed of several functional DOMAINS as well as the triple-helical domain.

In collagen types I, II, III, V, and XI, the helical domain is an uninterrupted sequence of Gly-X-Y repeat units of about 1000 residues, which forms the long, linear shape. After secretion and triple helix formation, these procollagen molecules are cleaved at both ends, and then assemble into long, cable-like collagen fibrils. On the other hand, the helical domains of the nonfibrillar collagens (e.g. types IV, VI, VII, VIII, IX, X, XII, XIII, and XIV) have interruptions in the tripeptide repeat that provide mechanical flexibility, which probably contributes to their specific functional characteristics. One example is collagen type IV, which is found only in basement membranes, and which bends and binds to neighbouring type IV molecules to form meshworks and sheet-like structures.

Each type of collagen is usually encoded by several genes for the specific subunit chains, which generally combine into heterotrimeric molecules. The coding sequences of the collagen repeat consist of EXONS that range in size from ~30 to 150 base pairs (bp). In collagen types I and II, the length of each exon is a multiple of the 9-bp unit encoding a complete Gly-X-Y sequence; most exons are 54 bp long. This exon organization suggests that the coding sequences for helical domains are derived from duplications of a primordial exon. This rule does not apply to type IV collagen, which may have evolved differently.

Elastin

Elastin is found in many elastic tissues, such as in the lung and large blood vessels, in a fibrillar form. Elastin molecules are randomly coiled and are covalently cross-linked at lysine residues by peptidyl lysyl oxidase, and this network of elastin fibres gives tissues their elasticity. As judged from its cDNA sequence, the monomer elastin molecule termed tropoelastin is a hydrophobic nonglycosylated protein of about 800 amino acids with a high content of alanine, glycine, valine and proline. Unlike collagen, however, elastin has little hydroxyproline and no hydroxylysine. Structural analyses of the tropoelastin amino-acid sequence derived from cDNA and genomic DNA reveal that the tropoelastin molecule consists, for the most part, of alternating hydrophobic domains and lysine-rich domains; the latter participate in cross-links between molecules. These domains are encoded by relatively small exons (27–186 bp), of which there are a large number (34 in the human gene, and 36 in the bovine). There is one elastin gene, and sequence variations between elastin molecules arise from differential splicing of precursor messenger RNA (*see* RNA SPLICING).

Fibronectin

Fibronectin is a well-studied glycoprotein found in many types of ECM, in blood plasma, and on cell surfaces. The molecule consists of dimers or multimers of disulphide bond-linked polypeptides derived from a single gene. Three types of repetitive units (types I, II, and III) occupy most of the polypeptide. Differential splicing of the type III repeat sequences at two sites (ED-A or ED-B), or of three other sites within a third region (IIICS) gives rise to structural and functional diversity of this protein (Fig. E38*a*). Each fibronectin molecule contains a linear series of functional domains that bind to cells and to other key extracellular molecules including collagen, fibrin, and heparin/heparan sulphate. This organization can account for the ability of fibronectin to form cross-links between different ECM proteins or between cells and collagen or fibrin.

Fibronectin is at present the best studied of the ECM proteins in terms of its roles in cell adhesion and migration. Most interactions between fibronectin and cells are mediated by specific receptors that bind to certain key sequences in the molecule. A key recognition sequence in fibronectin is the amino-acid sequence Arg-Gly-Asp (RGD), although full binding specificity and affinity appear to require another polypeptide region as well. Short synthetic polypeptides containing an Arg-Gly-Asp-X sequence can competitively inhibit cell adhesion dependent on the RGD site. Cell surface integrins dependent on RGD for cell adhesion include $\alpha_5\beta_1$, $\alpha_3\beta_1$ and $\alpha_v\beta_3$. In certain cell types, such as cells derived from the embryonic neural crest, the adhesive activity of fibronectin can also be mediated by two other sites in the differentially spliced IIICS region containing the Leu-Asp-Val (LDV) or Arg-Glu-Asp-Val (REDV) sequence. The integrin $\alpha_4\beta_1$ is thought to be responsible for these interactions.

In addition to its striking adhesive activity, fibronectin can also be a substrate for the migration of many cell types, including embryonic gastrulating cells, neural crest cells and their derivatives, primordial germ cells, and a variety of tumour cells.

Fibronectin can also modulate the morphology, growth, and differentiation of certain cells. Many of its morphological effects — cell flattening, decreased cell surface microvilli and formation of intracellular actin MICROFILAMENT bundles — can be attributed to its adhesive activity. Fibronectin can, however, also stimulate certain cells to proliferate, as well as inhibiting differentiation. It can also promote the differentiation of certain neural crest cells to sympathetic nervous system cells. The mechanisms of these complex effects of fibronectin remain to be determined.

Laminin

Laminin is a large glycoprotein specific to basement membranes, where it is a major component. The best-studied form of laminin is a cross-shaped molecule as observed by rotary shadowing ELECTRON MICROSCOPY. This form is composed of three chains: an A chain (M_r 400 000) and two similar B chains (B1, 210 000 and B2, 200 000 in mouse), which are linked by disulphide bonds. The C-terminal end of the A chain forms a single large globular domain at the end of the long arm of the cross; the long arm is comprised of all three chains intertwined in a coiled coil structure. Each short arm contains two or three small globules (Fig. E38*b*). To date, at least four homologues of laminin polypeptide chains are known. Merosin is an A chain homologue found in brain and the peripheral nervous system. A homologue of the B1 chain termed s-laminin is concentrated at synaptic sites. Kalinin and K-laminin are recently described laminin homologues associated with basement membrane anchoring filaments. Distinct functions for these different types of laminin remain to be elucidated.

Laminin has a variety of biological activities, including promotion of cell adhesion, migration, neurite extension, and differentiation. The mechanisms by which cells interact with laminin seem to be complex, and the picture is still somewhat confused. A wide variety of adhesive recognition sequences for laminin, as well as distinct cell surface and pericellular receptors have been reported. An important region involved in interactions with cells is present in laminin near the junction of the long arm with the large globular domain. This 'neurite interaction' site can mediate adhesion of many cell types, as well as promote neurite extension. The cell surface integrin $\alpha_6\beta_1$ can interact with this region. On the other hand, there are some reports that the integrin $\alpha_3\beta_1$, as well as other non-integrin receptors or binding proteins including several different LECTINS, can also mediate cell adhesion to laminin. A second major region involved in cell interaction is

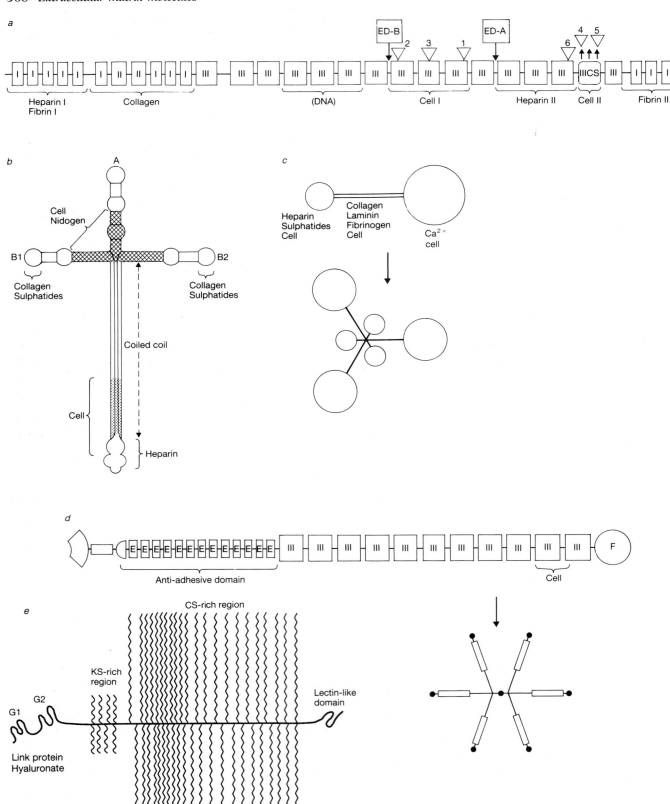

a

ED-B   ED-A

Heparin I
Fibrin I

Collagen

(DNA)

Cell I

Heparin II

Cell II

Fibrin II

b

A

Cell
Nidogen

B1

B2

Collagen
Sulphatides

Collagen
Sulphatides

Coiled coil

Cell

Heparin

c

Heparin
Sulphatides
Cell

Collagen
Laminin
Fibrinogen
Cell

Ca²⁺
cell

d

Anti-adhesive domain

Cell

F

e

CS-rich region

KS-rich
region

Lectin-like
domain

G1

G2

Link protein
Hyaluronate

located near the centre of the laminin cross; proteolytic fragments containing this region can mediate adhesion of a variety of cell types including tumour cells.

Proteoglycans

Proteoglycans consist of a core protein and one or more GLYCOS-AMINOGLYCAN side chains. Glycosaminoglycans are polymers of acidic disaccharides containing derivatives of the amino sugars glucosamine or galactosamine, which are often sulphated. Four main classes of glycosaminoglycan have been identified on the basis of the type of sugar residues and the linkages between these residues, and the number and location of sulphate groups: (1) heparan sulphate and heparin; (2) chondroitin sulphate and dermatan sulphate; (3) keratan sulphate; and (4) hyaluronic acid. All these glycosaminoglycans except hyaluronic acid are bound to proteins; hyaluronic acid has no sulphate and exists as a free polysaccharide. Chondroitin sulphate, dermatan sulphate, and heparan sulphate side chains are linked to the serine residues of core proteins via a specific trisaccharide sequence, (Ser)-xylose-galactose-galactose. The primary structures of a number of core proteins have been determined. Although they share some structural features, they display considerable diversity. For example, the cores of the heparan sulphate proteoglycans syndecan and perlecan are a transmembrane protein and a massively internally duplicated polypeptide, respectively.

Binding via their negatively charged side chains, proteoglycans interact with various ECM molecules (e.g. fibronectin and collagen), cell adhesion molecules, and growth factors. However, the core proteins of proteoglycans can also mediate interactions with other molecules. For example, the core protein of the major cartilage proteoglycan termed aggrecan has a hyaluronate-binding domain at the N terminus and a globular domain at the C-terminal end; the latter is a lectin with the capacity to bind to simple sugars (Fig. E38*e*). The N-terminal domain also interacts with cartilage link protein.

In cartilage, which is the richest source of proteoglycans, the side chains of proteoglycans provide high osmotic pressure and water retention, and thus contribute to the mechanical elasticity of cartilage. However, the functions of proteoglycans in other tissues are less clear. Proteoglycans can modulate cell adhesion to ECM either positively or negatively. Their side chains can bind to growth factors and may affect cell proliferation by providing a pool of sequestered growth factor that can be released by hydrolases. It has been suggested that some proteoglycans have roles in morphogenesis, such as in limb bud formation, as inhibition of proteoglycan biosynthesis by β-D-xyloside causes morphogenetic abnormalities and inhibition of cell differentiation.

Regulation and complexity of function

A rapidly expanding literature too broad to be reviewed here has been characterizing the regulation of expression of the genes for ECM proteins and their receptors. It is becoming clear that these molecules are individually regulated during embryonic development and that a host of growth factors and CYTOKINES such as TGF-β can modulate their expression and types of alternative splicing.

A second emerging generalization is that most ECM molecules undergo complex noncovalent interactions with other components of the matrix, and many can be bound by more than one cell surface receptor. Although the function of each molecule can be studied in isolation, most physiological processes will involve this sort of complex interaction. For example, cell adhesion by one molecule such as fibronectin can be modulated by proteoglycans and by anti-adhesive molecules such as tenascin. Complex regulatory pathways and molecular interactions are probably required to provide the level of sophistication needed for ECM molecules and their receptors to fulfil their roles in embryonic development, wound healing, and tissue organization.

S.-I. AOTA

K.M. YAMADA

See also: CELL ADHESION MOLECULES; CELL–MATRIX INTERACTIONS; CELL JUNCTIONS; GLYCANS.

1 Hay, E.D. (Ed.) (1991) *Cell Biology of Extracellular Matrix*, 2nd edn (Plenum, New York).
2 Sandell, L.J. & Boyd, C.D. (Eds) (1990) *Extracellular Matrix Genes* (Academic Press, New York).

---

extracellular polysaccharide *See:* BACTERIAL ENVELOPE; GLYCOCALYX.

extrachromosomal DNA *See:* CHLOROPLAST GENOMES; MITOCHONDRIAL GENOMES; PLASMIDS.

Fig. E38 (*opposite*) Schematic diagrams of some ECM proteins. *a*, Fibronectin. Three types of repetitive units (types I, II, and III) that occupy most of the molecule are shown by open boxes, and alternative splicing takes places at three sites (ED-A, ED-B, and IIICS). Important binding domains (e.g. for cells and for collagen) are also shown along the molecule. Small triangles show suggested cell-binding sites: 1, the RGD sequence; 2 and 3, two regions required for full cell-binding activity of the central cell-binding domain; 4, the LDV sequence; 5, the REDV sequence; 6, cell-binding sequences in the heparin-binding domain. *b*, Laminin. The A, B1, and B2 chains of laminin form a cross-shaped molecule with several globular domains in each arm. There are two major cell-binding regions as shown here by shading. *c*, Thrombospondin. This trimeric glycoprotein contains three dumbbell-shaped monomers which bind to various ECM components and cells. *d*, Tenascin. This hexagonal complex (hexabrachion) contains epidermal growth factor (EGF)-like repeats (E), a region of which is thought to have anti-adhesive activity, and the type III repeat (III) found in fibronectin. Six monomers form a huge star-shaped complex. *e*, Aggrecan. This chondroitin sulphate-containing molecule is one of the best-studied proteoglycans, and is the major cartilage proteoglycan. The N-terminal end of the core protein (~2200 amino acids) consists of two homologous globular domains (shown as G1 and G2, each 24 amino-acid residues in size); the G1 domain binds to hyaluronate and link protein. The C-terminal end has a globular domain homologous to lectins, but its function is presently unknown. The central region contains a keratan sulphate (KS)-rich region (~50 KS chains) and a chondroitin sulphate (CS)-rich region (~100 CS chains). The aggrecan molecule shown here is just one example of a proteoglycan. Proteoglycans are very heterogeneous as a class and differ in core protein structure, size, and the number and types of glycosaminoglycan side chains.

extraembryonic membranes Those tissues that develop from a fertilized mammalian egg, apart from the embryo proper, and which are essential for maintenance, nourishment, and protection of the developing foetus. The extraembryonic membranes consist of the placenta, parietal yolk sac (parietal endoderm and trophoblast), visceral yolk sac (visceral endoderm and mesoderm), CHORION, and AMNION (mesoderm and ectoderm).

In the mouse the amnion is generated from both ECTODERM and MESODERM at the posterior end of the PRIMITIVE STREAK. It appears first as a fold and then as a roof over the top of the cup-shaped primitive ectoderm and with further development of the embryo expands rapidly to form a thin membrane entirely surrounding it.

The parietal yolk sac is formed from TROPHOBLAST and parietal endoderm. Parietal endoderm is derived from primitive ENDODERM (the layer of cells appearing on the surface of the INNER CELL MASS, at 4½ days *post coitum* in the mouse). At implantation these primitive endoderm cells migrate as solitary cells onto the inner surface of mural TROPHECTODERM and differentiate into parietal endoderm which expresses vimentin, cytokeratins, and tissue plasminogen activator, and secretes BASEMENT MEMBRANE (including laminin, entactin, type IV procollagen, and heparan sulphate proteoglycan). This membrane (Reichert's membrane) forms one of the major barriers between maternal and foetal circulations during early development (<16 days gestation in the mouse).

The visceral yolk sac is formed from MESODERM and visceral endoderm. In the mouse, visceral endoderm is derived from those primitive endoderm cells which remain in contact with the EGG CYLINDER and become organized into a distinct squamous epithelial layer. Their function is to absorb products of the maternal circulation which have filtered through Reichert's membrane into the cavity of the parietal yolk sac and to synthesize substances needed by the foetus (e.g. α-foetoprotein, transferrin, high- and low-density lipoproteins, and α₁-antitrypsin).

The chorion is the outermost of the extraembryonic membranes and becomes part of the placenta where, in addition to respiratory functions, it supplies nutrition and removes waste. *See also*: MAMMALIAN DEVELOPMENT.

eye development (1) In *Drosophila*. The adult *Drosophila* eye consists of an array of about 800 ommatidia. Each ommatidium contains about 20 cells; eight photoreceptor neurons, R1-8, as well as four lens-secreting cone cells and eight other accessory cells. Each ommatidial unit is not formed by the clonal progeny of a single precursor cell but rather is built up by a series of CELL–CELL INTERACTIONS, recruiting members from a pool of equivalent cells. As cells join the cluster and DIFFERENTIATE they induce their immediate neighbours to adopt specific fates. The *Drosophila* eye is an important model system for the role of cell–cell interactions in developmental decisions. Many genes required for correct ommatidial development have been identified, and the specification of the last photoreceptor, R7, has been particularly well studied. The R7 cell is induced by a signal from the R8 cell. The R8 cell expresses the *bride-of-sevenless* (*boss*) gene product on its surface and this acts as a ligand for the *sevenless* (*sev*) product, a receptor tyrosine kinase (*see* GROWTH FACTOR RECEPTORS), on the responding cell. Activation of the *sevenless* kinase initiates a signalling pathway that involves a Ras-GTPase cycle using the products of the genes *Ras1*, *Son-of-sevenless* (*Sos*), and *GAP1* (*see* GTP-BINDING PROTEINS).

(2) In vertebrates. Eye development first becomes apparent with the evagination on each side of the optic vesicles from the side walls of the FOREBRAIN (diencephalon) vesicle. These become cup shaped, comprising two epithelial layers, and are connected to the diencephalon by the optic stalks. The outer layer becomes the pigmented retina and the inner layer generates the multilayered neural retina, comprising light-sensitive photoreceptors, glial cells, interneurons, and retinal ganglion cells. Axons of the latter cells project to the diencephalon along the optic stalk, so forming the optic nerve. Single, presumptive rat retinal cells can be labelled with replication-defective retroviral DNA (containing a β-GALACTOSIDASE gene for histochemical detection); the retroviral DNA incorporates into the genome of the retinal precursor cell, and its progeny, but does not infect neighbouring, clonally unrelated cells. This has shown that retinal cell phenotypes are determined late, at or after the terminal mitosis of the lineage.

Lens development results from an inductive interaction between the optic vesicles and the overlying ectoderm, the latter forming the lens epithelium, as originally shown by Hans Spemann (1901). Recent experiments in *Xenopus* indicate that the ectoderm becomes competent autonomously to form lens during late gastrula to mid-neurula stages, and that contact with the optic vesicle is necessary for the proper positioning and full differentiation of the lens. Lens differentiation involves further changes in cell shape and structure, as well as the synthesis of lens-specific proteins (crystallins).

ezrin A protein of M_r 81 000 found as a major component in the intestinal brush border microvilli. On stimulation of certain cell types with epidermal GROWTH FACTOR it becomes the major tyrosine phosphorylated protein and becomes localized to the microspikes and microvilli in these cells.

Bretscher, A. (1989) *J. Cell Biol.* **108**, 921–930.

F

F The aromatic AMINO ACID phenylalanine.

F$_{hkl}$ In CRYSTALLOGRAPHY the STRUCTURE FACTOR for the reflection with indices $h\ k\ l$. In the presence of ANOMALOUS DISPERSION, F^+_{hkl} and F^-_{hkl}, the members of a BIJVOET PAIR, are different.

F$^+$ The phenotype of *Escherichia coli* cells harbouring the F FACTOR.

F′, F-prime In an HFR STRAIN of the bacterium *Escherichia coli*, excision of the integrated F FACTOR from the bacterial chromosome occasionally occurs as a result of illegitimate recombination. As a consequence of this genetic exchange, the excised F factor incorporates segments of the bacterial genome both proximal and distal to its inserted location, and part of the F factor is retained in the bacterial genome. F factors modified in this way are denoted F-prime (F′) and may carry as much as 10% of the bacterial genome. If essential transfer and replication functions are retained by F′ plasmids, transfer into F$^-$ recipients is possible by the same mechanism as for native F factor.

F-actin The filamentous form of actin. *See:* MICROFILAMENTS.

F$_1$ ATPase *See:* ATP SYNTHASE; F-TYPE ATPASES.

F factor (F plasmid) The fertility (F) factor of the bacterium *Escherichia coli* is a single-copy PLASMID of ~94 kb. Through its possession of 19 transfer (*tra*) genes, F factor can transfer itself from donor F$^+$ cells into recipient F$^-$ cells lacking the plasmid, by the process of CONJUGATION. Through the replication of the F factor during transfer, donor cells retain their F$^+$ phenotype, and the F factor becomes distributed throughout the bacterial population. *See* F-PRIME; HFR.

F-type ATPases The ATP SYNTHASES of mitochondrial and chloroplast membranes, and of the bacterial cell membrane, consisting of a membrane-spanning subunit (F$_O$) and a catalytic subunit (F$_1$). *See* ELECTRON TRANSPORT CHAIN; PHOTOSYNTHESIS.

F1 generation The first hybrid generation produced by crossing two pure-bred lines. *See:* MENDELIAN INHERITANCE.

F2 generation The progeny of intercrosses between members of the F1 GENERATION. *See:* MENDELIAN INHERITANCE.

Fab fragment (fragment antigen binding) Protein fragment originally isolated and defined following digestion of rabbit IgG ANTIBODY by the enzyme papain. The enzyme cleaves the molecule within the HINGE REGION to release two univalent Fab fragments, each composed of an intact light chain disulphide-linked to the N-terminal half of the heavy chain. The light and heavy chains interact with each other through multiple non-covalent bonds to form a compact globular structure that binds antigen with an affinity equivalent to that of an antigen-binding site of the intact molecule. Similar fragments may be released from other immunoglobulin isotypes using a variety of enzymes.

F(ab′)$_2$ fragment A fragment originally isolated and defined following digestion of rabbit IgG ANTIBODY with the enzyme pepsin. The enzyme cleaves the molecule within the hinge region, to the C-terminal side of the inter-heavy chain disulphide bridges, to yield a divalent antigen-binding fragment. Mild reduction of this fragment yields the univalent Fab′ fragment. Similar fragments may be released from the other immunoglobulin isotypes using pepsin and other enzymes.

fabavirus group Named from Latin *faba*, bean, after the type member broad bean wilt virus. MULTICOMPONENT VIRUSES which resemble COMOVIRUSES in structure and composition but differ in cytopathology and transmission. There are three components of isometric particles 30 nm in diameter. The genome comprises two species of (+)-strand linear RNA, the bottom component containing RNA 1 (M_r 2.1×10^6) and the middle component RNA 2 (M_r 1.5×10^6); the top component lacks nucleic acid. The genome organization has not yet been determined. *See:* PLANT VIRUSES.

Fabry's disease (angiokeratoma) Rare X-LINKED genetic disease caused by a deficiency of the lysosomal enzyme α-galactosidase A. Affected individuals develop skin lesions and lipid changes in the nervous system. A common cause of death is renal failure. Around 70% of females heterozygous for the disease show mild symptoms; the disease therefore has intermediate X-linked inheritance. The α-galactosidase gene has been localized to the distal portion of the long arm of the X-chromosome (Xq22–24). *See:* LYSOSOMES.

facilitated diffusion *See:* MEMBRANE TRANSPORT SYSTEMS.

facilitation The increase in postsynaptic potential observed when one impulse is delivered shortly after another. Facilitation is caused by an increase in the amount of transmitter liberated by the presynaptic terminal. This effect can, if prolonged, lead to a decreased postsynaptic potential, caused by depletion of neurotransmitter in the presynaptic neuron. *See:* SYNAPTIC TRANSMISSION.

FACS Fluorescence-activated cell sorting/sorter. Technique for sorting cells of a particular type out of a mixture, by virtue of their specific cell-surface antigens. Fluorescently-labelled antibodies of the required specificity are used to label the appropriate cells with a fluorescent tag, which then allows them to be deflected from a main stream of cells (*see* CELL SORTING). FACS analysis has been much used to distinguish cells of the immune system (T cells and B cells) at the different stages of their development. The technique has also been modified to sort chromosomes (*see* HUMAN GENE MAPPING).

factor B A component of the alternative pathway of COMPLEMENT fixation. It is a homologue of C2 and is encoded by an adjacent gene in the MAJOR HISTOCOMPATIBILITY COMPLEX (MHC). It combines with activated C3 in the presence of Mg^{2+}, and is cleaved by FACTOR D to form an active C3-splitting protease — C3bBb. The forward reaction requires only the activation of C3 and the availability of factors B and D. The result is more C3 activation — the core of the alternative pathway feedback.

factor D Factor D is an always active SERINE PROTEINASE, awaiting the formation of its complex substrate, C3bB (*see* COMPLEMENT). It seems not to have any natural inhibitors. It seems to be identical with adipsin, a proteinase of fat tissue, and may have been 'recruited' to its complement function by some evolutionary accident.

factor H Factor H is a component of the homeostatic system for the alternative pathway of COMPLEMENT fixation. It competes with FACTOR B for C3b, and, if bound, makes the C3 available for breakdown by FACTOR I. It is a remarkable molecule structurally, composed only of disulphide-knotted SHORT CONSENSUS REPEAT motifs. This motif is common to most of the molecules which bind C3 or C4.

factor I Factor I is a specialized SERINE PROTEINASE responsible for initiating the breakdown of active C3b (and C4b) to inactive forms in the pathways of COMPLEMENT fixation. It is important in the homeostasis of the complement system, but requires that its substrates be combined with a cofactor molecule (FACTOR H or a cell-bound cofactor) before it can act.

factor P (properdin) Highly asymmetric protein which binds and stabilizes the surface-bound C3bBb complex of the alternative pathway of COMPLEMENT fixation. It is therefore an upregulator of the pathway and antagonizes the action of FACTOR H. It is not changed by its interaction with the complex, and if it dissociates, it is available for reuse. It is encoded on the X-chromosome.

Factor VIII A blood clotting factor that is absent or defective in classical HAEMOPHILIA A. *See also*: BLOOD COAGULATION AND ITS DISORDERS.

Factors V–XIII Coagulation factors. *See*: BLOOD COAGULATION AND ITS DISORDERS.

facultative heterochromatin CHROMATIN that is temporarily in a state that makes it unavailable for transcription. *See also*: CHROMOSOME STRUCTURE.

FAD Familial ALZHEIMER'S DISEASE. *See*: NEURODEGENERATIVE DISORDERS.

falciparum malaria The severest form of malaria, which is associated with *Plasmodium falciparum*. *See*: MALARIA.

familial adenomatous polyposis *See*: POLYPOSIS COLI.

familial defective abetalipoproteinaemia *See*: PLASMA LIPOPROTEINS AND THEIR RECEPTORS.

familial dysautonomia A hereditary condition characterized by defective lacrimation, discolouring of skin, disturbances of motor function, emotional instability, and a total lack of pain sensation.

familial hypercholesterolaemia Autosomal CODOMINANT inherited condition in which there is a genetic deficiency of cell-surface receptors involved in the removal of cholesterol-rich low density lipoprotein (LDL) from plasma (*see* PLASMA LIPOPROTEINS AND THEIR RECEPTORS). This leads to an increase in circulating LDL and total plasma cholesterol. The disease is caused by mutations in the LDL receptor gene located on chromosome 19 in humans. The HETEROZYGOUS form which affects between one in 300 and one in 500 individuals in the the UK and USA is characterized by the visible deposition of cholesterol in the skin and tendons and there is a substantial risk of ischaemic heart disease (50% die by 60 years). In the HOMOZYGOUS state the cutaneous manifestations and ischaemic heart disease may occur in childhood.

familial hyperchylomicronaemia *See*: PLASMA LIPOPROTEINS AND THEIR RECEPTORS.

Fanconi's anaemia A rare inherited disorder that causes a reduction in all circulating blood cells and is associated with multiple developmental abnormalities particularly of the skin and skeleton, hypersensitivity to agents that cross-link the two strands of DNA, and an increased incidence of myeloid leukaemia. It has an AUTOSOMAL RECESSIVE mode of inheritance. CHROMOSOMAL ABERRATIONS are common and consist mainly of chromatid breaks. A suggested cause is the failure of one of the DNA REPAIR systems. Four genetic complementation groups have been identified and a gene, *FACC*, involved in group C Fanconi's anaemia has been provisionally identified.

farnesylation *See*: MEMBRANE ANCHORS.

fasciclin *See*: CELL ADHESION MOLECULES; IMMUNOGLOBULIN SUPERFAMILY.

fasciculation The aggregation of individual nerve fibres to form bundles (nerves) which occurs during the development of the nervous system.

fascin *See*: ACTIN-BINDING PROTEINS.

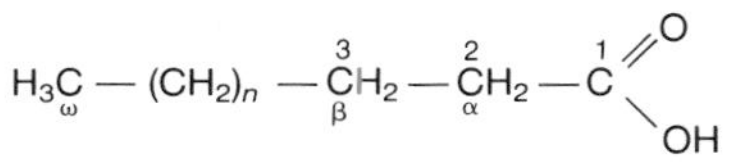

$$H_3\underset{\omega}{C} - (CH_2)_n - \underset{\beta}{\overset{3}{C}H_2} - \underset{\alpha}{\overset{2}{C}H_2} - \overset{1}{C}\overset{O}{\underset{OH}{\diagup}}$$

Fig. F1 General structure of a fatty acid.

fat cell ADIPOCYTE.

fate The range of tissues or cell types normally generated by a given cell or tissue in the embryo. Compare with the concept of DEVELOPMENTAL POTENTIAL, which is the range of tissues or cell types that a given cell is capable of generating.

fate map A spatial map of the FATES of different regions of an embryo at a particular stage of development. Usually applied to groups of cells. The term CELL LINEAGE is more often used to refer to the collection of fates generated by single cells.

fatty acids A class of organic compounds containing a long hydrocarbon chain and a terminal carboxylate group. They are the hydrophobic components of PHOSPHOLIPIDS and GLYCOLIPIDS, the main classes of membrane lipids (*see* MEMBRANE STRUCTURE). In addition, fatty acids are fuel molecules, being stored in cells as triacylglycerols (*see* LIPIDS). Fatty acid derivatives also act as signalling molecules (*see* EICOSANOIDS) and second messengers (*see* DIACYLGLYCEROL). The general structure of a fatty acid is illustrated in Fig. F1. The hydrocarbon chain may be saturated (no C=C bonds), monounsaturated (one C=C bond), or polyunsaturated (several C=C bonds). Fatty acids with C≡C bonds are also known. A C_{18} fatty acid, for example, may have one, two or several double bonds. $C_{18:0}$ denotes a C_{18} saturated fatty acid with no double bonds (stearate); $C_{18:1}$, a C_{18} fatty acid with one double bond (oleate) and so on (Table F1). Chain length and degree of saturation affect the melting point and thus the fluidity of fatty acids at biological temperatures. Unsaturated fatty acids are generally fluid at normal temperatures (e.g. those of plant oils like olive oil and sunflower oil) whereas saturated fatty acids of the same chain length have a higher melting point (e.g. the stearate of animal fat). Fatty acids with shorter chains have a lower melting point than those of similar degree of saturation but with longer chains.

fatty acylation *See:* N-ACYLATION.

favism A severe haemolytic reaction following exposure to the bean *Vicia faba* or even the pollen of the plant's blossom. Susceptibility occurs as a consequence of the X-linked condition GLUCOSE-6-PHOSPHATE DEHYDROGENASE DEFICIENCY. The condition is seen commonly in the Mediterranean region and the Middle East.

Fc (fragment crystallizable) A fragment originally isolated and defined following digestion of rabbit IgG ANTIBODY with the enzyme papain. (Fc is now more usefully thought of as 'fragment complement binding'.) On dialysis of the digestion products against distilled water, protein crystals were obtained. Fc is composed of the C-terminal half of the heavy chains including that part of the hinge region containing the inter-heavy chain disulphide bridges. Similar fragments may be released from the other immunoglobulin isotypes using a variety of enzymes and

Table F1 Some naturally occurring fatty acids in animals and plants

Number of carbons	Number of double bonds	Common name	Systematic name	Formula
12	0	Laurate	*n*-Dodecanoate	$CH_3(CH_2)_{10}COO^-$
14	0	Myristate	*n*-Tetradecanoate	$CH_3(CH_2)_{12}COO^-$
16	0	Palmitate	*n*-Hexadecanoate	$CH_3(CH_2)_{14}COO^-$
18	0	Stearate	*n*-Octadecanoate	$CH_3(CH_2)_{16}COO^-$
20	0	Arachidate	*n*-Eicosanoate	$CH_3(CH_2)_{18}COO^-$
22	0	Behenate	*n*-Docosanoate	$CH_3(CH_2)_{20}COO^-$
24	0	Lignocerate	*n*-Tetracosanoate	$CH_3(CH_2)_{22}COO^-$
16	1	Palmitoleate	*cis*-Δ^9-Hexadecenoate	$CH_3(CH_2)_5CH{=}CH(CH_2)_7COO^-$
18	1	Oleate	*cis*-Δ^9-Octadecenoate	$CH_3(CH_2)_7CH{=}CH(CH_2)_7COO^-$
18	one triple bond	Tarirate	6-Octadecinoate	$CH_3(CH_2)_{10}C{\equiv}C(CH_2)_4COOH$
18	2	Linoleate	*cis,cis*-Δ^9,Δ^{12}-Octadecadienoate	$CH_3(CH_2)_4(CH{=}CHCH_2)_2(CH_2)_6COO^-$
18	3	Linolenate	all *cis*-$\Delta^9,\Delta^{12},\Delta^{15}$-Octadecatrienoate	$CH_3CH_2(CH{=}CHCH_2)_3(CH_2)_6COO^-$
20	4	Arachidonate	all *cis*-$\Delta^5,\Delta^8,\Delta^{11},\Delta^{14}$-Eicosatetraenoate	$CH_3(CH_2)_4(CH{=}CHCH_2)_4(CH_2)_2COO^-$
18	1	Petroselinate (celery seed)	*cis*-Δ^6-Octadecenoate	$CH_3(CH_2)_{10}CH{=}CH(CH_2)_4COOH$
18	1	Ricinoleate (castor oil seed)	12-Hydroxy-*cis*-Δ^9-octadecenoate	$CH_3(CH_2)_5CHOHCH_2CH{=}CH(CH_2)_7COOH$
22	1	Erucate	*Cis*-Δ^{13}-Docosenoate	$CH_3(CH_2)_7CH{=}CH(CH_2)_{11}COOH$

In one system of nomenclature the position of a double bond is represented by Δ. *cis*-Δ^9- for example indicates a *cis* double bond between carbon atoms 9 and 10 counting from the carboxyl end. Most unsaturated fatty acids in biological systems contain *cis* double bonds. The systematic name of a fatty acid is derived from the name of its parent hydrocarbon and indicates the number of carbon atoms in the chain and the degree of saturation. In an alternative system of nomenclature, the carbons are numbered from the ω carbon (see Fig. F1) so that palmitoleate, for example, becomes ω-7-hexadecenoate.

are also referred to as Fc fragments although they have never been crystallized. The effector functions of antibody molecules are determined by receptor or ligand-binding sites within the Fc region.

Fc receptors Cell-surface molecules on a variety of cells which specifically recognize and bind the Fc regions of ANTIBODY molecules, thus mediating the antibody's effector functions. There is evidence for the existence of receptors for each of the immunoglobulin isotypes and the Fc receptors for human IgG, IgE, and IgA are well characterized. Fc receptors are involved in triggering phagocytosis of antigen–antibody complexes, inflammatory responses, and immune regulation. Individual cell types may have a characteristic profile of cellular Fc receptor expression which may vary according to the state of activation. *See*: COMPLEMENT; IGG, IGM, IGE; IGE RECEPTORS.

Unkeless, J.C. et al. (1988) *Annu. Rev. Immunol.* **6**, 251–281.
Ravetsch, J.V. & Kinet, H. (1991) *Annu. Rev. Immunol.* **9**, 457.

Fe protein Any protein containing iron, either as a HAEM group or in some other form.

Fe–S proteins *See*: IRON–SULPHUR PROTEINS.

feedback regulation The regulation of an enzyme reaction, or metabolic pathway, or gene expression by the final product of the reaction or pathway, a common mechanism of control in biological systems. Examples of feedback regulation are: the control of many biosynthetic operons by the final product of the metabolic pathway involved (e.g. the repression of the *Escherichia coli trp* operon by tryptophan) (*see* BACTERIAL GENE EXPRESSION), and the inhibition of enzymes by the end-products of the metabolic pathway in which they are involved (e.g. the inhibition of bacterial glutamine synthetase by AMP, histidine, tryptophan, glycine, and other end-products of glutamine metabolism.

feline leukaemia virus (FeLV) *See*: RETROVIRUSES.

FELIX *See*: PROTEIN ENGINEERING.

***fem* genes** Genes involved in determining sexual fate in the nematode *Caenorhabditis elegans*. The three feminization genes are all required for male somatic sexual differentiation. Mutations which induce loss of function in *fem-1*, *fem-2*, or *fem-3* loci result in individuals with XX (hermaphrodites) or XO (males) developing as females. The *fem* genes lie upstream of the TRANSFORMER gene *tra-1* and downstream of *tra-2* and *tra-3*. The key terminal regulator in the somatic sex determination pathway is *tra-1*, whereas in the germ line it is the three *fem* genes that appear most important. The *fem* gene products have not been identified yet and cloning of *fem-3* has not revealed any homology with other genes. *See also*: SEX DETERMINATION.

ferredoxin A small (M_r 11 000), nuclear encoded low-potential (E_m = – 400 mV) IRON–SULPHUR PROTEIN containing a single

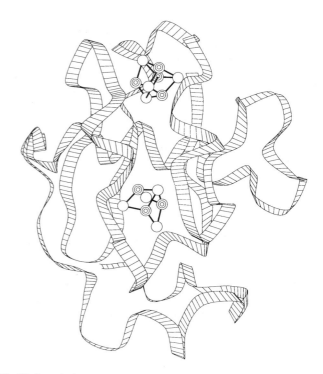

Fig. F2 Ferredoxin.

[2Fe–2S] centre and acting as the native electron acceptor of PHOTOSYSTEM I (Fig. F2). Its strongly acidic character is conserved throughout all oxygenic organisms. Its immediate reductant is the FeS-containing subunit C and in turn it acts as the reductant not only for $NADP^+$ (via the flavoprotein FERREDOXIN–NADP$^+$ REDUCTASE) but also for a range of other acceptors, including NO_2^-, SO_3^-, the glutamate synthase reaction, THIOREDOXIN (via FERREDOXIN–THIOREDOXIN REDUCTASE) and the cyclic electron transfer pathway (via FERREDOXIN–PLASTOQUINONE REDUCTASE). *See also*: PHOTOSYNTHESIS.

ferredoxin–NADP$^+$ reductase (FNR) FLAVOPROTEIN enzyme (M_r 36 000, prosthetic group, flavin adenine dinucleotide) with important regulatory properties, which catalyses the reduction of $NADP^+$ by FERREDOXIN in PHOTOSYSTEM I. In chloroplasts it occurs both in a bound state associated with photosystem I, and free in the stroma. It is nuclear encoded. The X-ray crystal structure shows that it is a two-domain protein. In CYANOBACTERIA the enzyme has an N-terminal extension giving it an M_r of 45 000. *See also*: PHOTOSYNTHESIS.

ferredoxin–plastoquinone reductase (FQR) Enzyme associated with PHOTOSYSTEM I where it catalyses the key electron transfer step associated with cyclic photophosphorylation (*see* PHOTOSYNTHESIS). It is the principal site of action of the inhibitor antimycin in chloroplasts. The activity is probably associated with a bound form of FERREDOXIN–NADP$^+$ REDUCTASE (although the free enzyme is inactive), but the binding protein has not been unequivocally identified. *See also*: PHOTOSYNTHESIS.

ferredoxin–thioredoxin reductase An enzyme in organisms with oxygenic PHOTOSYNTHESIS that catalyses the reduction of THIOREDOXIN by FERREDOXIN, thereby signalling a change from the dark to the light states. It is a heterodimer of $M_r \sim 30\,000$ which contains a catalytically active disulphide group and a [4Fe–4S] centre. One subunit is immunologically similar in all organisms, with M_r 13 000 and contains the active disulphide and probably the iron–sulphur centre. The other is variable, depending on species, with M_r between 7000 and 16 000. Sequencing of the gene from the CYANOBACTERIUM *Anacystis* suggests that this subunit binds FERREDOXIN. *See also:* PHOTOSYNTHESIS.

ferritin An iron storage protein found in tissues. The protein shell, apoferritin (M_r 460 000) comprises 24 subunits arranged with octahedral SYMMETRY. Each monomer is a four-helix bundle, a common domain in PROTEIN STRUCTURE. The protein surrounds a core of up to 4500 ferric ions. Because of the electron-dense nature of this iron core, ferritin can be used as a marker in electron microscopy (*see* IMMUNOELECTRON MICROSCOPY).

fertilization Process whereby the cytoplasm of two GAMETES of opposite sex becomes continuous. In animals reproducing by sperm and egg, this involves the fusion of plasma membrane surrounding the posterior of the sperm head with the membrane of the egg. In mammals, this process follows sperm binding to the ZONA PELLUCIDA, the sperm ACROSOME REACTION, and penetration of the zona pellucida by sperm. A projection on the egg surface (fertilization cone) often arises from the egg surface to meet and engulf the sperm head. In the fertilized egg, before nuclear fusion, the female and male nuclei are called pronuclei.

In the plant kingdom, fertilization is most complex in the flowering plants (angiosperms). In the majority of angiosperms, the mature female gametophyte — the embryo sac — comprises seven cells including the haploid egg cell, associated with two synergids, and a binucleate central cell. The embryo sac is surrounded by an integument to form the ovule. Fertilization occurs when a compatible pollen grain alights on the stigma of the flower and germinates to form a pollen tube. The haploid generative cell within the pollen tube divides to form two haploid sperm. The pollen tube grows down through the stigma and style towards the ovule and commonly enters it through the micropyle. It penetrates one of the two synergids and releases the two sperm nuclei and the pollen tube nucleus. One sperm nucleus enters the egg cell to fuse with the egg nucleus to form the diploid zygote from which the embryo develops. The other sperm nucleus unites with the two polar nuclei in the central cell of the embryo sac to form a triploid cell from which the endosperm of the seed develops. This double fertilization is restricted to angiosperms.

Various systems of SELF-INCOMPATIBILITY occur in flowering plants which prevent fertilization by self pollen. The molecular bases of these systems, which involve cell recognition between pollen and the stigma surface, are being elucidated.

fertilization membrane *See:* CORTICAL REACTION.

fes *See:* ONCOGENES.

fetal *See:* FOETAL.

fetoscopy *See:* FOETOSCOPY.

fibre diffraction This type of DIFFRACTION arises from elongated helical molecules (e.g. double-helical DNA, and fibrous proteins such as collagen and keratin) and complexes (e.g. muscle, microfilaments, and viruses such as tobacco mosaic virus (TMV)) which can be aligned with their fibre axes parallel, but are randomly oriented about that axis. The most famous structure in biology is the double helix of DNA which was determined in 1953 by comparison of the X-ray diffraction patterns observed from fibres of DNA with those calculated from trial models of the structure. A fibre diffraction pattern represents a cylindrical average of the transform of the sample molecule or complex. In contrast, globular molecules which crystallize yield a three-dimensional diffraction pattern from which the molecular structure can be determined by the techniques of X-RAY CRYSTALLOGRAPHY.

Fibre diffraction patterns are essentially two-dimensional; the extent to which three-dimensional structural details can be extracted from the data depends on the dimensions and symmetry of the fibre and the degree of orientation within the fibre. One particularly favourable pattern is that of TMV (Fig. F3). Here the fibre specimens are extremely well aligned such that the horizontal layer lines are widely separated, only broadening such that they start to overlap at the outer edge of the pattern where

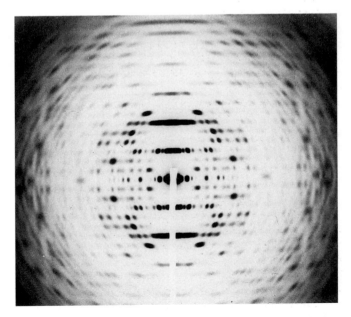

Fig. F3 Fibre diffraction photograph of a tobacco mosaic virus (TMV). The helical virus particle forms ordered gels in contrast to the single crystals formed by its protein disk. The repeat distance along the helix axis is 6.9 nm and every third layer line has strong intensity on the meridian. The top edge of the picture represents a resolution of approximately 0.42 nm. Courtesy of K.C. Holmes.

the higher RESOLUTION structural details are thus becoming obscured. TMV particles have an outer radius of about 80Å with 16⅓ subunits per helical turn, or 49 units in the axial repeat of three turns. In consequence of this three-layer repeat, strong diffraction intensity close to the meridian only occurs on every third layer line from the equator. On the intervening layer lines the position of strong intensity nearest to the meridian indicates the number of subunits per helical turn.

Along the layer lines the intensity is continuous, though modulated according to the detailed structure of the subunit; this contrasts with the case of diffraction from a crystal where the intensity is localized to the points of a Cartesian three-dimensional reciprocal lattice (*see* RECIPROCAL SPACE). Thus, crystalline diffraction is analysed by FOURIER SERIES, functions having a translational repeat over (x, y, z), whilst fibrous diffraction is analysed in terms of mathematical functions with a cylindrical repeat over the coordinates radius, azimuth, and axial height (r, θ, z); these Bessel functions, which are continuous along the discrete layer lines, are analogous to the discrete STRUCTURE FACTORS from crystals.

The PHASE PROBLEM of diffraction is even more severe in the case of these continuous intensity functions. It has been solved by the use of many HEAVY ATOM DERIVATIVES, as in the case of TMV, or by extensive model-building, as for the DNA double helix, muscle filaments, and other viruses.

Squire, J.M. & Vibert, P.J. (Eds) (1987) *Fibrous Protein Structure* (Academic Press, London).
Holmes, K.C. et al. (1975) *Nature* **254**, 192–196.

fibrillin *See*: EXTRACELLULAR MATRIX MOLECULES.

fibrin A blood protein, the end product of the coagulation cascade (*see* BLOOD COAGULATION AND ITS DISORDERS). Fibrin is produced from fibrinogen by the SERINE PROTEINASE thrombin (Fig. F4). Fibrinogen is a dimeric protein of M_r 340 000 in which three globular domains are connected by rod-like regions. It consists of three pairs of polypeptide chains, $(A\alpha)_2$, $(B\beta)_2$, and $(\gamma)_2$. Thrombin cleaves four peptide bonds in the central globular domain of fibrinogen to release an A peptide of 18 amino acids from each of the two α chains and a B peptide of 20 residues from each of the two β chains (fibrinopeptides). The resulting molecule is the fibrin monomer. Fibrin monomers self-assemble into an

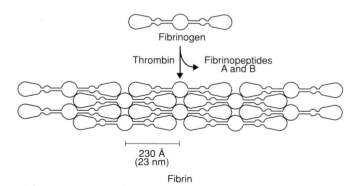

Fig. F4 The formation of fibrin from fibrinogen.

ordered fibrous protein array — fibrin — which is also stabilized by peptide bond cross-links between specific glutamine and lysine residues. In the electron micrograph fibrin has a repeating structure of period 230 Å. AFIBRINOGENAEMIA is a genetically determined defect resulting in the absence of fibrin in the blood.

fibrinogen Precursor to FIBRIN. It is cleaved by the proteinase THROMBIN to form fibrin in the last step of the blood coagulation cascade. It is also a CALCIUM-BINDING PROTEIN with three high-affinity calcium-binding sites ($K_d \sim 10^{-6}$ M) and many low-affinity sites which do not belong to any recognized family. *See also*: BLOOD COAGULATION AND ITS DISORDERS.

fibrinopeptides Short peptides (fibrinopeptide A and fibrinopeptide B, 18 and 20 residues respectively) released from fibrinogen after cleavage by THROMBIN to form a FIBRIN monomer. The fibrinopeptide sequence, having no biological function, shows one of the highest rates of change known for a protein-coding sequence over evolutionary time. *See also*: BLOOD COAGULATION AND ITS DISORDERS; MOLECULAR CLOCK.

fibroblast CONNECTIVE TISSUE cell, specialized for the secretion of collagenous extracellular matrix (*see* EXTRACELLULAR MATRIX MOLECULES). Fibroblasts have the capacity to differentiate into other members of the connective tissue cell family such as chondrocytes, osteocytes, smooth muscle, and adipocytes.

fibroblast growth factor (FGF) *See*: GROWTH FACTORS.

fibrocystic disease of the pancreas CYSTIC FIBROSIS.

fibroin The fibrous protein of silk.

fibronectin A glycoprotein found in the extracellular matrix, in blood plasma and in temporary association with cell surfaces, and which can mediate interactions between cells and extracellular matrix. Plasma fibronectin is a soluble dimer; extracellular matrix fibronectin is in the form of insoluble fibrils. Fibronectin can bind to other matrix components such as collagen and heparin and to specific cell surface receptors — INTEGRINS. *See*: CELL–MATRIX INTERACTIONS; EXTRACELLULAR MATRIX MOLECULES.

fibronectin receptors INTEGRINS present on the surfaces of many cells which act as receptors for the extracellular matrix protein fibronectin. *See*: CELL–MATRIX INTERACTIONS; EXTRACELLULAR MATRIX MOLECULES.

fibronectin type III domain/repeat Protein motif originally identified in the EXTRACELLULAR MATRIX MOLECULE fibronectin, and which is present in many other proteins. *See*: CELL ADHESION MOLECULES.

fibrous lamina Nuclear lamina. *See*: NUCLEAR ENVELOPE.

fibrous proteins Insoluble proteins that typically form fibrils such as fibrin (the protein of silk), some keratins (*see* INTERMEDIATE FILAMENTS), collagens I–III (*see* EXTRACELLULAR MATRIX MOL-

ECULES), and FIBRINOGEN (*see* BLOOD COAGULATION AND ITS DISORDERS).

field *See*: MORPHOGENETIC FIELD.

fijivirus group Named after Fiji disease virus of sugar cane. One of the two groups of plant viruses in the family REOVIRIDAE. The particles are isometric, 65–71 nm in diameter, with a knob 8–16 nm long 11 nm in diameter at each 5′ axis. Within the particle is a core, 54 nm in diameter which also has spikes, 8 nm long and 14–19 nm in diameter at each 5′ axis. The core contains ten species of double-stranded RNA (M_r ranging from 2.9×10^6 to 1.0×10^6). These viruses replicate in their plant hopper vectors as well as in plants. *See also*: PLANT VIRUSES.

Matthews, R.E.F. (1982) *Intervirology* 17, 85.

filaggrin An INTERMEDIATE FILAMENT protein.

filamin *See*: ACTIN-BINDING PROTEINS.

filensin An INTERMEDIATE FILAMENT protein.

fill-in reaction DNA synthesis carried out *in vitro* to convert the single-stranded tails of DNA fragments created by RESTRICTION ENZYMES that make staggered cuts, into double-stranded DNA. This reaction, which uses the protruding termini as templates, is achieved by incubation with KLENOW FRAGMENT of DNA polymerase I (for 3′ recessed termini) or T4 DNA POLYMERASE (for 5′ recessed termini). The BLUNT ENDS produced are of considerable utility during subcloning procedures as ligation to any other blunt-ended fragment is possible. As a consequence of the fill-in reaction the original RESTRICTION SITE is lost.

filopodium Finger-like outgrowth from the growth cone (the axon terminal) of a developing neuron.

Filoviridae A family of enveloped RNA viruses consisting of the viruses of Marburg fever and Ebola fever, highly virulent haemorrhagic fever viruses. The very long particles (on average 800–1000 nm, but sometimes much longer) contain a single-stranded (−)-sense RNA (M_r ~4.2×10^6).

filter hybridization HYBRIDIZATION technique in which single-stranded DNA or RNA bound to a suitable matrix (e.g. a nitrocellulose or nylon filter) is investigated for the presence of a particular sequence by immersion in a solution containing a single-stranded PROBE. The latter may be either DNA or RNA, and is labelled (often by a radioisotope) so that annealing between complementary sequences can be detected. Control over the specificity of annealing is achieved by varying the temperature and/or salt concentration of the hybridization reaction and subsequent filter-wash steps. *See also*: SOUTHERN BLOTTING.

fimbriae (*sing.* fimbria) Short straight proteinaceous appendages which protrude through the bacterial outer membrane. They are prominent in Gram-negative bacteria and have been implicated in adherence and bacterial aggregation. *See*: BACTERIAL ENVELOPE; BACTERIAL PATHOGENICITY.

Isaacson, R.E. (1985) In *Bacterial Adherence* (Savage, D.C. & Fletcher, M., Eds) 307–336 (Plenum, New York).

fimbrin *See*: ACTIN-BINDING PROTEINS.

fingerprinting *See*: DNA TYPING; PROTEIN FINGERPRINT.

FISH Fluorescent *in situ* hybridization. *See*: HYBRIDIZATION.

fish development *See*: ZEBRAFISH DEVELOPMENT.

FITC Fluorescein isothiocyanate. A fluorescent dye widely used to label proteins and nucleic acids. It is excited by light of wavelength 450–490 nm and emits in the range 520–560 nm.

fitness A measure of the advantage of a particular genotype over others in a population which is reflected by that genotype's higher frequency in the next generation.

flagellin Protein making up the bacterial flagellum. *See*: BACTERIAL CHEMOTAXIS.

flagellum (*pl.* flagella) Long whip-like motile projection from the surface of a bacterial or eukaryotic cell, which acts to propel the cell along. Bacterial and eukaryotic flagella have completely different structures and mechanisms of action.
(1) Bacterial. *See*: BACTERIAL CHEMOTAXIS.
(2) Eukaryotic. *See*: CELL MOTILITY; MICROTUBULES; MICROTUBULE-BASED MOTORS.

flanking sequences Sequences adjacent to the 5′ and 3′ termini of any given genetic element — in DNA or RNA.

flavanone synthase (FS) An alternative name for CHALCONE SYNTHASE (EC 2.3.1.74). When the reaction between malonyl-CoA and substituted 4-hydroxycinnamoyl-CoA, catalysed by chalcone synthase, is carried out *in vitro* at pH 8.0, the chalcone is isomerized to the flavanone, which then appears as the primary product.

Flaviviridae Family of enveloped RNA viruses infecting arthropods and vertebrates, with spherical particles 40–50 nm in diameter. The icosahedral nucleocapsid contains single-stranded (+)-sense RNA (M_r 4×10^6–4.6×10^6). Viral replication is in the cytoplasm. Members include dengue haemorrhagic fever, St Louis encephalitis, and yellow fever.

flavodoxin A low-potential FLAVOPROTEIN (M_r 17 000) that can substitute for FERREDOXIN in some CYANOBACTERIA and algae under growth conditions in which iron is not freely available.

flavohaemoglobins Proteins of M_r ~ 45 000 found in yeast and *Escherichia coli*, containing both flavin- and haem-binding domains. *See*: HAEMOGLOBINS.

flavonoid(s) Large group of compounds in plants, structurally based on the phenylchromane ring system.

The two phenyl rings associated with the pyran are referred to as rings A and B as indicated. The compounds are known as flavanones, flavones or anthocyanidins, depending upon the oxidation of the pyran ring. The group is sometimes extended to include chalcones, isoflavones, leucoanthocyanidins and stilbenes, which are structurally or biochemically related. Great diversity is achieved by hydroxylation of one, two or more of the carbon atoms of rings A and B, and especially C-3. These hydroxyl groups may be glycosylated with a wide range of sugars, or methylated. Unsubstituted carbon atoms may be glycosylated, acylated, alkylated or isoprenylated. Flavonoid glycosides are water-soluble, and usually occur in plant cell vacuoles. Some flavonoid aglycones can be secreted, for example, by bud scales.

flavonoid inducers Particular FLAVONOID molecules present in the root exudates of legumes which activate transcription of *nod* genes of rhizobia via the NodD regulatory protein. *See*: GENISTEIN: NODULATION.

flavoprotein Protein containing a flavin nucleotide (generally flavin adenine dinucleotide (FAD) or flavin mononucleotide (FMN)) as a PROSTHETIC GROUP. Flavin-containing proteins are often involved in electron transfer, for example all enzymes that catalyse the reduction of ubiquinone in the mitochondrial respiratory chain are flavoproteins, as is the initial electron acceptor in Complex I. *See*: ELECTRON TRANSPORT CHAIN.

flip and flop Alternatively spliced mRNAs from the GluR1–GluR4 genes encoding variants of AMPA receptor subunits (*see* EXCITATORY AMINO ACID RECEPTORS: RNA SPLICING). Flip variants produce larger postsynaptic responses than flop variants and therefore an increase in flip variants following LONG-TERM POTENTIATION is an attractive possible mechanism for its expression.

flip-flop The movement of a PHOSPHOLIPID molecule from one leaflet to the other of a bilayer membrane. It occurs chiefly as membrane is being synthesized in the ENDOPLASMIC RETICULUM by incorporation of newly synthesized lipids into the cytosolic face of the ER membrane. Here it is mediated by specific phospholipid translocators. It also occurs spontaneously at a low level in preformed membranes.

flip-flop promoters *See*: MU PHAGE: PHASE VARIATION.

floorplate An epithelial cell group that forms the ventral midline of the vertebrate NEURAL TUBE. It develops in response to local, as yet unidentified, inductive signals from the underlying notochord.

The floorplate influences the patterning of axon trajectories in the developing central nervous system. It releases a diffusible chemoattractant that orientates the growth of commissural axons towards the ventral midline of the spinal cord (CHEMOTROPISM). It also acts as an intermediate target for these axons, changing their trajectory on local contact. Lastly, the floorplate cells seem to provide a polarizing signal that determines the fate of other neuroepithelial cells (e.g. motor neurons) within the ventral neural tube. *See also*: NEURAL DEVELOPMENT: VERTEBRATE.

flow cytometry A technique used in CELL SORTING.

flower development In flowering plants (angiosperms), flowers develop from vegetative MERISTEMS which undergo a transition to a flower (or inflorescence) meristem in response to intrinsic developmental or environmental signals. Despite great differences in outward appearance, all flowers have a similar underlying organization, consisting of concentric whorls of different structures. In dicotyledonous plants there are usually four distinct whorls: from the outside in these are the sepals, the petals, the stamens, and the carpels. A number of homeotic floral mutations in which the structures of one whorl are absent or transformed into the structures of another have been identified, especially in the well-studied species ARABIDOPSIS THALIANA and *Antirrhinum majus*. Homeotic genes that specify the identity of meristems (e.g. *LEAFY* in *Arabidopsis*) and genes controlling the identity of floral organs (e.g. *APETALA2* and *AGAMOUS*) have been identified and cloned and their expression during development studied. Studies of double mutants and experiments with ectopically expressed transgenes suggest that floral organ identity is specified by a small set of genes, different combinations of which are expressed in the different organ primordia, with some members inhibiting the expression of others in certain whorls. The homeotic floral organ identity genes are thought to encode TRANSCRIPTION FACTORS and many of these genes contain a conserved DNA-binding domain — the MADS BOX.

Dennis, E. & Bowman, J.L. (1993) *Curr. Biol.* **3**, 90–93.

fluctuation analysis Also known as noise analysis. The technique is based on the fact that a part of the electrical noise in the MEMBRANE POTENTIAL or current is due to electrophysiological activity and increases when ION CHANNELS are activated (e.g. by the binding of a neurotransmitter agonist at its receptor). The membrane noise carries fundamental information about the underlying channels, for example the opening time and CONDUCTANCE of a single channel. These parameters can be revealed by determining the frequency spectrum (i.e. the Lorentzian) of the noise by computation based on Fourier transformation (see FOURIER SERIES). Fluctuation analysis complements data provided by PATCH CLAMP recording.

Cull-Candy, S.G. (1981) *Trends Neurosci.* **4**, 1–3.

fluctuation test Test carried out to demonstrate that bacterial variants in a culture have arisen by spontaneous genetic MUTATION rather than by an ADAPTIVE RESPONSE. Portions of several bacterial cultures grown under nonselective conditions are spread on plates containing a selective agent (e.g. an antibiotic). If the resistant bacteria that arise represent an adaptive response to physical contact with the selective agent then the same number of surviving bacterial colonies will be observed on each plate. A large fluctuation in the number of resistant colonies from plate to plate implies that spontaneous genetic mutations have arisen, their frequency depending on the stage of original culture growth at which the mutation occurred.

fluid mosaic model The description of the structure of biological membranes as a mosaic of membrane proteins which are free to move laterally in a lipid bilayer. *See*: MEMBRANE STRUCTURE.

fluid phase endocytosis (pinocytosis) Nonselective internalization of the fluid media surrounding cells. Fluid phase ENDOCYTOSIS is relatively inefficient compared to the receptor-mediated, or adsorptive, endocytosis of ligands which bind to the plasma membrane before internalization. Fluid phase endocytosis occurs constitutively through the continual formation of endocytic vesicles at the plasma membrane; in a number of cell types (e.g. baby hamster kidney cells) these vesicles are CLATHRIN coated, but noncoated vesicles may mediate uptake in some cell types (*see* COATED PITS AND VESICLES). Internalized fluid is delivered to endosomes from where a large proportion is recycled to the cell surface.

Steinman, R.M. et al. (1976) *J. Cell Biol.* **68**, 665–687.
Besterman, J.M. et al. (1981) *J. Cell Biol.* **99**, 716–727.

fluorescence-activated cell sorter/sorting *See*: CELL SORTING; FACS.

fluorescence energy transfer Technique which can be used for estimating the length of a macromolecular structure. A different fluorescent dye is attached to each end of the molecule and the transfer of energy between them on excitation is measured. Within a range 30–70 Å, the efficiency of transfer varies with the distance apart of the dyes. The technique has been used to study the HOLLIDAY STRUCTURE in DNA (*see* RECOMBINATION).

fluorography Technique in which ³H-labelled molecules can be detected in CHROMATOGRAPHY or ELECTROPHORESIS, by the introduction of a scintillator into the chromatogram or gel, which is then exposed to photographic film.

FMDV FOOT-AND-MOUTH DISEASE VIRUS.

fMet peptides Chemotactic peptides derived from breakdown of bacterial proteins at sites of infection and which attract white blood cells to the site of infection (*see also* LYMPHOKINES).

fms Gene encoding the receptor for the CYTOKINE M-CSF, which has tyrosine kinase activity. It is a potential ONCOGENE. *See*: GROWTH FACTOR RECEPTORS.

FNR FERREDOXIN NADP REDUCTASE.

focal contacts/plaques *See*: CELL JUNCTIONS.

focus formation In contrast to the property of limited density growth exhibited *in vitro* by cultured normal eukaryotic cells, neoplastically TRANSFORMED CELLS pile up in dense clusters, each of which is called a focus. From the number of foci produced, the potency of a particular transforming treatment (e.g. transfection with an ONCOGENE) may be assessed.

fodrin *See*: ACTIN-BINDING PROTEINS.

foetal haemoglobin *See*: HAEMOGLOBIN AND ITS DISORDERS; HAEMOGLOBIN F.

α-foetoprotein (AFP) A small protein (M_r 70 000) distantly related to serum albumin, produced in the mammalian foetus, but whose expression is switched off at birth. It is detectable in amniotic fluid and maternal blood. After correction for duration of pregnancy the level is helpful in PRENATAL DIAGNOSIS of various conditions. Levels are raised in open NEURAL TUBE defects (e.g. anencephaly and spina bifida) and defects of the abdominal wall, and reduced in trisomy 21 (DOWN'S SYNDROME).

The regulation of expression of the mouse α-foetoprotein gene is a model system for the study of gene regulation during mammalian embryonic development. *Cis*-acting regulatory elements activate TRANSCRIPTION of the gene in three cell types of the developing mouse embryo — the visceral endoderm of the yolk sac, the foetal liver, and the foetal gut — and include three upstream ENHANCERS. Transcription declines dramatically after birth, as a result of the activation of a negative control sequence lying between the most proximal enhancer and the gene PROMOTER (Fig. F5). The product of another gene, *raf*, also reduces α-foetoprotein postnatally, by altering transcript stability.

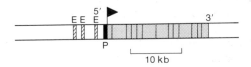

Fig. F5 Organization of the mouse α-foetoprotein gene. E, enhancer; P, promoter.

foetoscopy The insertion of a small endoscope through the abdominal wall and through the uterine wall into the amniotic cavity to observe the foetus directly in early pregnancy. It is used to facilitate the sampling of foetal blood via the umbilical cord vessels, for the biopsy of tissues, as an aid to a variety of therapeutic manoeuvres, and to confirm the presence of malformations in circumstances where ultrasonography has been inconclusive. It can be performed from 15 weeks gestation and on into the third trimester. Foetal loss is around 4%.

fold-back elements (FB elements) Long INVERTED REPEATS in DNA located either adjacent to one another or separated by up to several kilobases of intervening DNA. If the latter combination is denatured and reannealed the FB elements form a HAIRPIN.

Although the terminal 30 or so base pairs are well conserved, FB elements show considerable heterogeneity in their size and internally repetitious sequences. Studied most extensively in *Drosophila*, the involvement of FB elements in TRANSPOSITION and MUTAGENESIS is now apparent. *See also*: NUCLEIC ACID STRUCTURE; TRANSPOSABLE GENETIC ELEMENTS.

folic acid, folate As tetrahydrofolate, a cofactor required for the activity of some enzymes involved in one-carbon transfer reactions, in which it acts as a carrier for the carbon atom in the form of a hydroxymethyl, formyl, methyl, or formimino group.

follicle From the Latin *folliculus*, a little bag, the term describes the complex of cells and extracellular material that surrounds the OOCYTE in the ovary. During the menstrual cycle of mammals, the ovarian follicle undergoes a characteristic sequence of changes. In the primordial follicle the oocyte is surrounded by a single layer of epithelial cells, which are in turn surrounded by a loose mesenchyme. During maturation of the follicle, to the fully mature Graafian follicle, the epithelial cells proliferate as the oocyte enlarges, and a cavity forms amongst them. Most oocytes and follicles will die, while the rest proceed to ovulation.

follicle cells Somatic cells that form the follicular epithelium surrounding the oocyte in mammals, or the egg chamber in insects. In mammals, the follicular epithelial cells provide the oocyte with most of the metabolites required for growth, including protein and RNA, and precursors for phospholipid biosynthesis. These materials enter the oocyte through gap junctions between follicle cells and oocyte.

In insects, follicle cells also elaborate the chorion around each oocyte in a brief period during oogenesis.

In *Xenopus*, the fully grown (stage IV) oocyte is prepared for fertilization by the process of maturation. In response to the secretion of gonadotropins from the pituitary, the initiation signal for maturation, the follicle cells secrete progesterone, which triggers the conversion of the oocyte into an egg.

follicle-stimulating hormone (FSH) Glycoprotein hormone produced by the anterior pituitary, comprising an α-chain (92 amino acids) and a β-chain (115 amino acids). It stimulates growth of ovarian follicles and secretion of oestradiol in females and spermatogenesis in males.

follicular lymphoma A low grade malignant LYMPHOMA in which neoplastic cells in lymph nodes form aggregates resembling normal germinal centres. Around 80% of cases are associated with a t(14;18)(q32;q21) TRANSLOCATION involving the *bcl-2* gene and the immunoglobulin heavy chain locus. Cells are usually CD10 positive and CD5 negative. Variant translocations occur in rare cases.

foot-and-mouth disease virus (FMDV) A member of the Aphthovirus genus of the PICORNAVIRIDAE, the foot-and-mouth disease virus was the first virus to be recognized as a cause of animal disease, by Loeffler and Frosch in 1897. The icosahedral capsid is composed of 60 copies each of four proteins. VP1, VP2, and VP3

are partly exposed on the capsid surface and VP4 is internal. The three-dimensional structure of foot-and-mouth disease virus at 2.9Å has been determined (Plate 7e). Unlike polioviruses and rhinoviruses, the capsid surface has no deep depressions or 'canyons' in which the binding site for the virus receptor is located. In foot-and-mouth disease virus, the major antigenic determinant is the 'FMDV loop', a region comprising residues 140–160 approximately of VP1 which form a flexible loop on the capsid surface. A synthetic peptide based on the sequence of this loop can act as a powerful peptide vaccine, inducing high levels of neutralizing and protective ANTIBODIES. Part of this region also appears to form the receptor binding site, consisting of a conserved sequence Arg-Gly-Asp (RGD) surrounded by a region of more variable residues. The foot-and-mouth disease virus is more acid-labile than other picornaviruses.

Acharya, R. et al. (1989). *Nature* 337, 709–716.

footprinting Technique for determining the position and nature of DNA sequences bound by specific proteins. Following END-LABELLING of one of the DNA strands, this substrate is mixed *in vitro* with the DNA-binding protein(s) of interest. If the protein is bound sufficiently tightly to the DNA this prevents the region of DNA thus protected from being attacked by chemical or enzyme action. Most commonly used in this context is DEOXYRIBONUCLEASE I (DNase I) from bovine pancreas which is a nonspecific ENDONUCLEASE which, in excess, will randomly degrade both double-stranded and single-stranded DNA to mononucleotides. The basic protocol of DNase I footprinting is illustrated in Fig. F6. It requires that the DNA fragment to be studied be first END-LABELLED with a radioactive tag (e.g. ^{32}P), and then complexed either with a purified protein fraction or with a complex mixture of proteins (e.g. a nuclear extract). Limiting amounts of DNase I are then added such that, on average, only one phosphodiester bond per molecule is cleaved. In the absence of any bound protein this will give an uninterrupted ladder of labelled DNA fragments upon gel ELECTROPHORESIS and AUTORADIOGRAPHY. However, if a protein is bound to a specific sequence within the DNA fragment this will result in a gap or 'footprint' in the ladder. Analysis of the DNase I digestion products side by side, using an electrophoresis gel capable of separating DNA fragments differing in length by only a single nucleotide, allows the precise mapping of the region to which the protein has bound. Footprinting has been particularly useful in identifying *cis*-acting regulatory DNA sequences important for the control of TRANSCRIPTION (*see* BACTERIAL GENE EXPRESSION; EUKARYOTIC GENE EXPRESSION) and which bind *trans*-acting TRANSCRIPTION FACTORS.

forebrain The forebrain or prosencephalon arises from the most anterior part of the NEURAL TUBE (*see* VERTEBRATE NEURAL DEVELOPMENT), the latter being regionalized during its early development into forebrain, midbrain, hindbrain, and spinal cord. The forebrain is subdivided further into anterior telencephalon and posterior diencephalon. The telencephalon later forms the cerebral hemispheres and corpus striatum, while the diencephalon gives rise to the thalamic and hypothalamic regions, and to a portion of the pituitary gland.

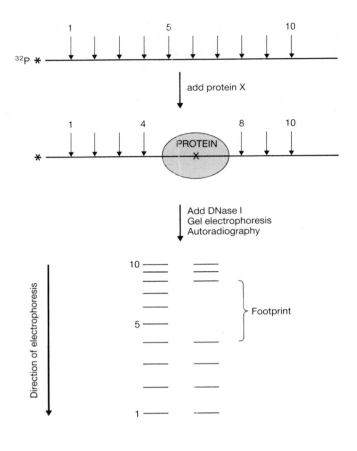

Fig. F6 DNase I footprinting. A single DNA molecule, end-labelled with ³²P, is shown which contains 10 potential DNase I cleavage sites (indicated by the arrows). Protein X binds at the position shown, thereby masking DNase I cleavage sites 5, 6, and 7. Upon electrophoresis of these two samples and subsequent autoradiography, the location of the bound protein is identified by the absence of DNA fragments cleaved at these three positions, the so-called 'footprint'.

fork head motif A DNA-binding motif found in certain TRAN-SCRIPTION FACTORS. It is named after the *Drosophila* fork head factor in which it was first identified.

formononetin 7-Hydroxy-4′-methoxyisoflavone. The 4′-methyl ether of DAIDZEIN.

Found in the chickpea (*Cicer arietinum*) and in most species in which daidzein is also found.

formyl peptides *See:* FMET PEPTIDES.

N-formylmethionine Derivative of the AMINO ACID methionine used to initiate a polypeptide chain in prokaryotes. *See:* PROTEIN SYNTHESIS.

Forssman antigen A classical example of a heterophile antigen — the same or very similar antigen expressed by a wide variety of organisms. The Forssman antigen is expressed on the surface of erythrocytes of many (horse, sheep, dog, cat, mouse, guinea-pig) but not all (rat, rabbit, cow, pig) mammalian species. It is also found in chickens, certain fish, plant tissues and bacteria, for example pneumococci and salmonellae. Antibody to the Forssman antigen is produced in patients with infectious mononucleosis (*see* EPSTEIN–BARR VIRUS) and its presence is diagnostic for the disease.

fos Proto-oncogene (c-*fos*) with a viral oncogene counterpart (v-*fos*) carried by a murine sarcoma virus. An 'immediate early gene' in mammalian cells, one of the first genes expressed in response to mitogenic and other stimuli. It encodes a nuclear TRANSCRIPTION FACTOR that forms homodimers, or heterodimers with Jun, which act at AP-1 sites. *See:* ONCOGENES.

fossil DNA DNA extracted from semi-fossilized or otherwise preserved ancient material. Using the POLYMERASE CHAIN REACTION minute quantities of DNA from such sources can be amplified and sequenced. 'Fossil DNA' has been extracted from semi-fossilized plant material, insects preserved in amber, frozen mammoth tissues, etc.

founder cells The first cells to arise within a lineage all of whose progeny contribute to a single territory. Thus, in the sea urchin blastula, there are four founder cells for the skeletal mesenchyme territory.

Fourier series The French mathematician Joseph Fourier showed that any periodic function f(*x*) can be represented by an infinite series of sine and cosine terms.

$$f(x) = a_0 + a_1\cos 2\pi (x) + a_2\cos 2\pi (2x) + a_3\cos 2\pi (3x) + \ldots$$
$$+ \, b_1\sin 2\pi (x) + b_2\sin 2\pi (2x) + b_3\sin 2\pi (3x) + \ldots$$

where *x* is a coordinate expressed as a fraction of the repeat period of the function f(*x*). This may be rewritten as

$$f(x) = \sum_{h=0}^{n} (a_h\cos 2\pi (hx) + b_h\sin 2\pi (hx))$$

or, using the nomenclature of exponentials and COMPLEX NUMBERS,

$$f(x) = \sum_{h=-n}^{+n} F_h \, e^{2\pi i (hx)}$$

and extending this explicitly to a three-dimensional periodic function f(*x,y,z*)

$$f(x,y,z) = \sum_{h=-n}^{+n} \sum_{k=-n}^{+n} \sum_{l=-n}^{+n} F_{hkl} \, e^{2\pi i (hx + ky + lz)}$$

Fourier series have many applications in data analysis, especially in DIFFRACTION and X-RAY CRYSTALLOGRAPHY, where the ELECTRON DENSITY of a molecule is a periodic function $f(x,y,z)$ within the crystal calculable from the STRUCTURE FACTORS $F(hkl)$ whose amplitudes — but not phases — are measurable from the diffraction pattern of the crystal.

Fourier synthesis A mathematical calculation to obtain the FOURIER TRANSFORM of a periodic function.

Fourier transform The set of coefficients in the FOURIER SERIES representation of a function.

FQR FERREDOXIN PLASTOQUINONE REDUCTASE.

fractionation of cells The separation out of cell organelles and other structures, usually on the basis of size and density by CENTRIFUGATION TECHNIQUES.

fragile chromosome site Site of abnormal chromosome structure, at which breakage can be induced by various treatments, and which appears as a nonstaining gap in METAPHASE SPREADS. *See:* FRAGILE X SYNDROME.

Fragile X syndrome

THE fragile X, or Martin Bell, syndrome is the most common single recognized form of inherited mental retardation. Fifty per cent of all X-linked mental retardation (XLMR) may be attributable to the fragile X syndrome. The disorder is found in all ethnic groupings with a frequency of 0.3–1 per 1000 in males and 0.2–0.6 per 1000 in females. The full clinical syndrome, which is found in approximately 60% of affected males, consists of moderate mental retardation with an IQ typically in the range 35 to 50, elongated facies with large everted ears, and macroorchidism. The syndrome is unusual in that it is associated with the appearance of a fragile site on the long arm of the X-chromosome at Xq27.3 [1,2]. This can be visualized cytogenetically in metaphase chromosomes prepared from lymphocytes of affected individuals which have been cultured under conditions of folate deficiency or thymidine stress (Fig. F7). The study of the segregation of polymorphic markers within fragile X families has confirmed that the mutation lies in the same region of the X-chromosome as that exhibiting cytogenetic fragility.

Unusually for an X-LINKED recessive disorder, up to 30% of carrier females show some degree of mental impairment, and the mutation may also be passed through males who do not show a fragile X site (normal transmitting males)[3,4]. Additionally, there is an imbalance of PENETRANCE of the phenotype in the different generations of kindreds in which the mutation is segregating. The likelihood of developing mental impairment depends on an individual's position in the pedigree. As the mutation progresses through the generations, the risk of mental impairment increases. These observations are not consistent with classical X linkage, and are collectively known as the Sherman paradox. Hypotheses

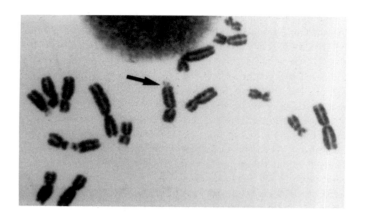

Fig. F7 Induction of the fragile X site in non-Giemsa banded metaphase chromosomes. The region of fragility is arrowed.

based on these observations have suggested that the mutation exists in two forms — a premutation and a full mutation form. Nonpenetrant individuals are said to carry a premutation chromosome, that is, a chromosome which has no abnormal phenotypic effect but which is capable of progressing to a fully penetrant mutation on passage through a female oogenesis.

The unusual segregation pattern and the lack of expression of the fragile site in nonpenetrant individuals have hampered both carrier and PRENATAL DETECTION by polymorphic marker segregation and cytogenetic analysis. Molecular cloning (*see* DNA CLONING) of the fragile X locus means that it is now possible to predict the phenotype of an individual with a high degree of accuracy by direct detection of mutations in the DNA [5–8]. Identification of these mutations has confirmed that the sites of chromosome fragility and gene mutation are indeed coincident.

Two mutations in the DNA at the fragile X site have been identified: abnormal hypermethylation (*see* DNA METHYLATION) of a CpG-rich sequence (a CpG island) and the AMPLIFICATION of DNA. The molecular basis of the amplification is the expansion of a CGG triplet into large arrays. In individuals expressing the full clinical phenotype, the DNA in this region becomes hypermethylated [9,10], leading to the shutdown of expression of the gene *FMR-1* (for *f*ragile X *m*ental *r*etardation), which is transcribed across this region [11]. It is the loss of this gene expression that is thought to account for the clinical phenotype.

Within the *FMR-1* CpG island of normal individuals is a variable number of repeats of the CGG triplet, ranging up to a maximum of 50 copies. These are faithfully replicated and segregated ensuring stable MENDELIAN INHERITANCE. On a fragile X-chromosome, however, the amplified CGG array can exist in two states which differ in the number of repeats. Individuals carrying a premutation chromosome and exhibiting a nonpenetrant phenotype have more than the 50 copies of the repeat found in the normal population but fewer than 150 copies. Affected individuals have more than 200 copies, the *FMR-1* CpG island is hypermethylated and the chromosome exhibits cytogenetic fragility.

A study of the genetic instability of this DNA amplification in fragile X families reveals its unusual nature. The number of repeats in the premutation range is somatically stable within an individual but is highly unstable on genetic transmission, mutating with an estimated frequency of 1 (that is in 100% of meioses) [8]. Ninety-five per cent of the changes result in an increase in the size of the array which means that the array progressively increases in size through the generations (Fig. F8), eventually passing the threshold level for the full mutation range. This progressive increase provides a molecular explanation for the Sherman paradox, as at each genetic passage the array increases in size, thus increasing the risk of a full mutation and mental impairment. The chance of developing the full mutation therefore depends upon the genotype and phenotype of the preceding generation, a phenomenon known as genetic anticipation. This form of DNA has been termed hereditary unstable DNA [12]. Although the mechanism of amplification is unknown, it most probably originates either as an artefact of unscheduled DNA replication or by unequal SISTER CHROMATID EXCHANGE recombination.

Hereditary unstable DNA has also been identified at several other human disease gene loci. In a manner similar to that in fragile X syndrome, the affected status is also associated with an increase in allele length due to amplification of a repeat array. In the case of KENNEDY'S DISEASE (X-linked spinal and bulbar muscular atrophy), a CAG trinucleotide in the first exon of the androgen receptor gene is increased in size [13]. In the case of MYOTONIC DYSTROPHY (MD), the trinucleotide CTG is amplified in affected individuals, the size of DNA amplification correlating with the expression of the full MD clinical phenotype [14]. Similarly, Huntington's disease is now known to be caused by the expansion of a CAG trinucleotide repeat within a gene, with the length of array correlating with the age of onset of the disease [15].

M. HIRST

1 Lubs, H.A. (1969) A marker X-chromosome. *Am. J. Hum. Genet.* **21**, 231–244.

2 Sutherland, G.R. (1977) Fragile sites on human chromosomes: demonstration of their dependence on the type of tissue culture medium. *Science* **197**, 256–266.

3 Sherman S.L. et al. (1984) The marker (X) syndrome: a cytogenetic and genetic analysis. *Ann. Hum. Genet.* **48**, 21–37.

4 Sherman, S.L. et al. (1985) Further segregation analysis of the fragile X syndrome with special reference to transmitting males. *Hum. Genet.* **69**, 289–299.

5 Kremer, E.J. et al. (1991) Mapping of DNA instability at the fragile X to a trinucleotide repeat sequence p(CCG)n. *Science* **252**, 1711–1718.

6 Nakahori, Y. et al. (1991) Molecular heterogeneity of the fragile X syndrome. *Nucleic Acids Res.* **19**, 4355–4359.

7 Oberle, I. et al. (1991) Instability of a 550bp DNA segment and abnormal methylation in fragile X syndrome. *Science* **252**, 1097–1102.

8 Fu, Y-H. et al. (1991) Variation of the CGG repeat at the fragile X site results in genetic instability: resolution of the Sherman Paradox. *Cell* **67**, 1–20.

9 Bell, M.V. et al. (1991) Physical mapping across the fragile X: hypermethylation and clinical expression of the fragile X syndrome. *Cell* **64**, 861–866.

10 Vincent, A. et al. (1991) Abnormal pattern detected in fragile X patients by pulsed field gel electrophoresis. *Nature* **329**, 624–626.

11 Verkerk, A.J.M.H. et al. (1991) Identification of a gene (FMR-1) containing a CGG repeat coincident with a breakpoint cluster region exhibiting length variation in fragile X syndrome. *Cell* **65**, 905–914.

12 Sutherland, G.R. et al. (1991) Hereditary unstable DNA: a new explanation for some old genetic questions? *Lancet* **338**, 289–292.

13 La Spada, A.R & Fischbeck, K.H. (1991) Variant androgen receptor gene in X-linked spinal and bulbar muscular atrophy. *Nature* **352**, 77–79.

14 Brook, J.D. et al. (1991) Molecular basis of myotonic dystrophy: expansion of a trinucleotide (CTG) repeat at the end of a transcript encoding a protein kinase family member. *Cell* **66**, 799–808.

15 MacDonald, M.E. et al. (1993) A novel gene containing a trinucleotide repeat that is expanded and unstable on Huntington's disease chromosomes. *Cell* **72**, 971–983.

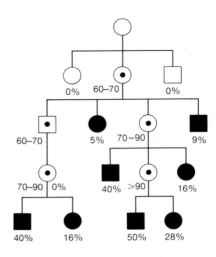

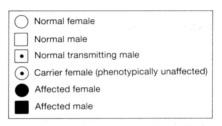

Fig. F8 Example of a fragile X pedigree showing genetic anticipation, passage of the fragile X mutation through phenotypically normal males and the progression of the trinucleotide array towards that corresponding to a fully penetrant mutation. Percentage risk of mental impairment of individuals is shown, as is the copy number of the CGG triplet. Based on data from [9].

fragmin *See*: ACTIN-BINDING PROTEINS.

frameshift MUTATION usually resulting from insertion or deletion of a single nucleotide (or two or four nucleotides), such that the correct READING FRAME of a protein-coding DNA sequence is destroyed. The altered reading frame usually results in the subsequent amino-acid coding sequence being changed or truncated and the production of an inactive mutant protein. Insertions or deletions may arise at a REPLICATION FORK in DNA if either the TEMPLATE DNA or the PRIMER exhibits unusual SECONDARY STRUCTURE; MUTAGENS that typically cause frameshifts are thought to act by stabilizing this structure. Suppression of frameshift mutations can be achieved either through intragenic suppression, when a further mutation restores the original reading frame, or by mutant TRANSFER RNAs which carry the appropriately altered number of bases in their anticodons. *See also*: GENETIC CODE; PROTEIN SYNTHESIS.

framework regions The relatively invariant peptide sequences within the variable (V) regions of immunoglobulin molecules (*see* ANTIBODY) that constitute ~75% of the V region. There are four framework regions interspersed with the hypervariable or COMPLEMENTARITY DETERMINING REGIONS (CDRs). They are enumerated as FR1, FR2, FR3 and FR4 starting at the N terminus of the light and heavy chains. Framework region determine the three-dimensional structure of the V region ensuring that the CDRs are brought into the correct steric relationship to each other. They also determine the noncovalent contacts between the light and heavy chain V regions and hence the quaternary structure of the Fab region and the generation of the antigen-binding site.

free energy change *See*: MECHANISMS OF ENZYME CATALYSIS.

freeze-etching A method of preparing cells for ELECTRON MICROSCOPY in which a frozen block of cells is first fractured (*see* FREEZE-FRACTURE). Before the exposed surfaces are platinum shadowed the specimen is freeze dried which exposes the cytoplasmic and exterior faces of the membrane and structures in the interior of the cell.

freeze-fracture A method of preparing membranes for ELECTRON MICROSCOPY which enables the interior of the membrane to be studied. A block of cells is frozen and then split; this often results in fracturing down the middle of the lipid bilayer of plasma membranes. A platinum replica of the exposed surface is prepared by shadowing and the organic material dissolved away. Freeze fracture enables visualization of transmembrane proteins and structures such as gap junctions which are left standing out from one of the interior faces of the membrane. The interior face of the exterior membrane leaflet is known as the E face and the interior face of the cytoplasmic leaflet is known as the P face.

French flag model A model used to illustrate the concept of POSITIONAL INFORMATION.

Friedrich ataxia The commonest form of heritable spinocerebellar degeneration with an incidence of one in 250 000. It is characterized by degeneration of large myelinated fibres in the spinal cord and spinocerebellar atrophy. The disease becomes manifest in childhood as a loss of balance and tendon reflexes. Affected individuals become increasingly ataxic (showing failure of muscle coordination) and are eventually chairbound. They do not usually survive beyond the third decade, with heart failure a common cause of death. The biochemical basis of the disease is unknown; it is inherited as an AUTOSOMAL RECESSIVE trait and the gene has been localized to the proximal long arm of chromosome 9.

Friend murine leukaemia virus A RETROVIRUS.

fructans Two unrelated groups of GLYCANS. Bacterial levans are large often branched, extracellular β2,6-linked polymers, closely related, biosynthetically, to the bacterial dextrans: both occur chiefly in Gram-positive cocci. Plant fructans are much smaller, usually linear, intracellular (intravacuolar) polymers. Two types are known: β2,1-linked inulins and β2,6-linked levans. Though widespread, they are common only in (e.g.) the Compositae (as inulins) and Graminae.

Bacterial levans are assembled by the reversible disproportionation of (e.g.) sucrose in the presence of levan sucrase to yield glucose and a triose, followed by further transfers to give free glucose and a terminally glucosylated fructan. Levan sucrase itself produces branches and a glycosyl enzyme intermediate is postulated.

Inulin synthesis is similar to that of levan. Sucrose is formed variously by the transfer of glucose from UDP-Glc onto fructose 6-phosphate to give sucrose phosphate, from which the phosphate is removed by sucrose phosphatase, or by direct transfer onto fructose. Sucrose : sucrose-1-fructosyl transferase then transfers fructose from one molecule of sucrose to another to form fructosyl β2,1 sucrose and glucose, by way of a fructosyl enzyme intermediate. Repeated transfers produce the polymer. The process ends at about 35 residues by an unknown mechanism: branching appears to need separate transferases. Plant levan synthesis is presumed to be similar.

fructose *See*: SUGARS.

fruit fly Flies of the genus *Drosophila*, usually referring to *Drosophila melanogaster*. *See*: DROSOPHILA; DROSOPHILA DEVELOPMENT.

fruit ripening The process of fruit maturation immediately preceding senescence. In some fruits it is accompanied by a respiratory CLIMACTERIC, with loss of acidity, the accumulation of sucrose, pigment changes in the outer layers and softening of the inner tissues, owing to cell wall breakdown especially under the action of polygalacturonase, cellulase and pectinesterase.

***frx* genes** A group of chloroplast genes coding for IRON–SULPHUR PROTEINS. The *frxA* gene codes for the 9000 M_r subunit C of PHOTOSYSTEM I, and is synonymous with *psaC*; *frxB* codes for a 20 000 M_r protein with two [4Fe–4S] centres that is part of the NADH dehydrogenase complex encoded by the chloroplast genome and expressed in *Chlamydomonas*, but probably only at a low level in higher-plant chloroplast; and the *frxC* gene product shows a high degree of homology to the iron-binding protein of nitrogen-fixing bacteria, such as the *nifH* protein of *Azotobacter vinelandii*. *See also*: PHOTOSYNTHESIS; NITROGEN FIXATION GENES.

FS FLAVANONE SYNTHASE.

FSH FOLLICLE-STIMULATING HORMONE.

ftz The *Drosophila* gene FUSHI TARAZU.

fucans A rare group of very acidic, usually sulphated polysaccharides found in marine eukaryotes. Many brown algae, especially of the Fucaceae, contain fucoidans which are α1,2 linear polymers of L-fucose, with α1,3 branches and sulphate ester at C4 of most fucosyl residues. Echinoderm eggs have jelly coats rich in sulphated fucan which is largely in the form of GLYCOPROTEIN. In some species the fucan is a heteropolymer with smaller amounts

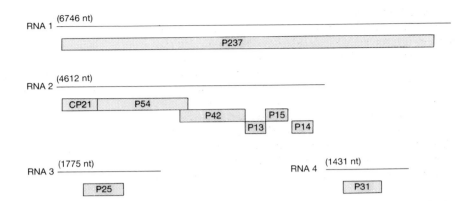

Fig. F9 Genome organization of beet necrotic yellow vein virus (BNYMV). The lines represent the RNA species and the boxes the proteins (P) with M_r given as $\times 10^{-3}$; nt, nucleotides.

of other sugars (galactose, mannose, hexosamine). Each fucosyl residue appears to be sulphated, usually at C4, and the glycosidic links appear to be at C2. Fucan phosphates have been reported in invertebrates.

Aspinall, G.O. (1970) *Polysaccharides* (Pergamon Press, Oxford).

fucose *See*: SUGARS.

fucosyltransferase *See*: GLYCOPROTEINS AND GLYCOSYLATION.

fungal mitochondrial genomes *See*: MITOCHONDRIAL GENOMES: LOWER EUKARYOTES.

6-furfurylaminopurine/⁶N-furfuryladenine *See*: KINETIN.

furovirus group Sigla from fungus-transmitted rod-shaped; type member is soil-borne wheat mosaic virus. MULTICOMPONENT VIRUSES with two sizes of rigid rod-shaped particles containing two species of ($+$)-strand linear RNA. The larger, 300 nm $\times$ 20 nm, contains an RNA of M_r 2.28×10^6, the smaller, 110–160 nm $\times$ 20 nm (varying between viruses), an RNA of 0.86×10^6–1.23×10^6. The genome organization has not yet been determined for any defined member but has for the possible member, beet necrotic yellow vein virus (Fig. F9). This virus has four nucleoprotein components, the smaller two of which are required not for viral replication but possibly for host determination and/or transmission by its fungal (Plasmodiophorales) vector. The protein encoded by RNA 1 has amino acid homologies with RNA polymerases. The 5′ gene of RNA 2 encodes the coat protein which has a weak termination codon; if read through this would give a product of 75 000 of unknown function. The P42, of unknown function, is probably translated from a subgenomic mRNA. The functions of the other three gene products from RNA 2 are unknown. *See also*: PLANT VIRUSES.

Brunt, A.A. (1986) In *The Plant Viruses* (van Regenmortel, M.H.V. & Fraenkel-Conrat, H., Eds) Vol. 2, 305 (Plenum, New York).

fushi tarazu (ftz) A member of the pair-rule class of SEGMENTATION GENES in *Drosophila*. Embryos homozygous for *ftz* mutations lack alternate segmental structures. The gene is expressed at the blastoderm stage in a simple pattern of seven stripes corresponding to the 14 PARASEGMENTAL repeat units. The *ftz* gene is located

in the ANTENNAPEDIA COMPLEX and contains a homeobox. The expression pattern is regulated by the primary pair-rule genes *hairy, even-skipped*, and *runt*. Regulatory elements for the generation of the stripe pattern — 'the zebra element' — and for autoregulation have been defined. The *ftz* gene product acts as a TRANSCRIPTION FACTOR to regulate the expression of segment polarity and HOMEOTIC GENES. *See also*: HOMEOBOX GENES AND HOMEODOMAIN PROTEINS.

fusicoccin A diterpene glucoside (Fig. F10) produced by the fungus *Fusicoccum amygdali*; M_r 680.8. It is toxic to plants, causing numerous physiological effects, including promotion of cell enlargement in growing tissues, extrusion of H^+ and Na^+, uptake of K^+ and Cl^-, and HYPERPOLARIZATION of the plasma membrane. Fusicoccin activates a H^+-ATPase in the plasma membrane resulting in an efflux of H^+. High-affinity receptor proteins for fusicoccin of M_r $\sim 30\,000$ have been isolated from plasma membranes of several species. Both the receptors and ATPases must be present in the membrane for fusicoccin to function and it is assumed that they are closely associated *in vivo*. The natural ligands for these receptors have not been identified. *See also*: PLANT WOUND RESPONSES.

Marré, E. (1979) *Annu. Rev. Plant Physiol.* **30**, 273–288.

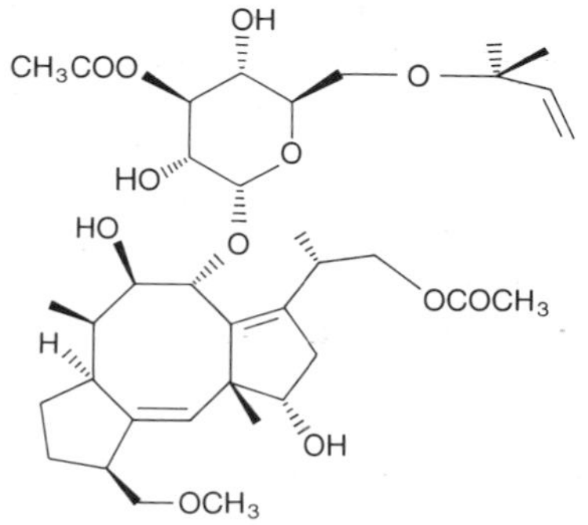

Fig. F10 Structure of fusicoccin.

fusidic acid Steroid obtained from the fungus *Fusidium coccineum* which inhibits both prokaryotic and eukaryotic PROTEIN SYNTHE-

SIS by preventing release from the ribosome of the ELONGATION FACTORS EF-G and EF-2, respectively. Ribosomes with either of these factors still bound are unable to accept aminoacyl-tRNAs at their A sites.

fusion, of cells *See*: CELL FUSION; SOMATIC CELL HYBRIDS.

fusion gene Gene which has been generated either by a spontaneous mutation event (e.g. UNEQUAL CROSSING-OVER which deletes material between two linked genes) or by manipulation of DNA *in vitro*, and which contains part of one gene and part of another. Fusion genes constructed from the coding region of a REPORTER GENE linked to the PROMOTER and other regulatory regions of another gene are widely used to studypromoter function and control. Fusion genes containing a heterologous protein-coding sequence joined to an inducible host cell promoter are used in GENETIC ENGINEERING to produce recombinant cells capable of synthesizing a novel gene product. *See also*: HAEMOGLOBIN FUSION GENES; RECOMBINATION.

fusion protein Protein product of a FUSION GENE.

(1) The product of a REPORTER GENE linked to a PROMOTER under study, when the normal protein product is itself difficult to assay directly. The 5′ UPSTREAM CONTROL ELEMENTS and a short N-terminal sequence of the gene under study are joined in the correct READING FRAME to a partially 5′-truncated sequence encoding an easily assayable product, such as β-GALACTOSIDASE or β-glucuronidase, which retains the activity and stability of the authentic enzyme.

(2) Protein produced from a fusion gene containing coding sequences from more than one gene. Such protein products may be of abnormal or altered function. *See e.g.* HAEMOGLOBIN FUSION GENES; ONCOGENES.

Fv fragment (variable fragment) A dimer composed of the non-covalently associated light and heavy chain V regions of a single 'arm' of an ANTIBODY molecule. It binds antigen with the same AFFINITY as an intact antibody molecule as it contains a single intact antigen combining site. It can be prepared, rarely, by pepsin digestion of monoclonal antibody, but can now be produced as a recombinant protein using GENETIC ENGINEERING techniques.

G

G (1) The purine BASE guanine.
(2) The small hydrophobic AMINO ACID glycine.

G-banding GIEMSA BANDING.

G-CSF Granulocyte colony-stimulating factor. Protein that stimulates predominantly the production and activity of mature neutrophilic polymorphonuclear leukocytes both *in vivo* and *in vitro*. The gene for human G-CSF is located on chromosome 17q11 and encodes a polypeptide of 174 amino acids, the mature protein having an M_r of 18 500. The murine equivalent is found on chromosome 11 and encodes a 19K protein of 178 amino acids. Both proteins may be glycosylated (*see* GLYCOPROTEINS) but retain activity after NEURAMINIDASE treatment. Recombinant material produced in the bacterium *Escherichia coli* is at least as effective as the native glycosylated molecule. The receptor for G-CSF does not possess TYROSINE KINASE activity, and is a member of the cytokine receptor superfamily (*see* GROWTH FACTOR RECEPTORS). The receptor has three fibronectin-like domains in the extracellular domain. G-CSF has been shown to be produced by macrophages, endothelial cells and fibroblasts after bacterial lipopolysaccharide stimulation; however, almost all murine tissues studied can produce active G-CSF *in vitro*. Clinically, recombinant human G-CSF is free from serious side-effects, at least in the short term, and has proven potential in patients at risk of infection as a result of depression in neutrophil numbers or function. *See*: LYMPHOKINES.

Nicola, N.A. (1990) In *Colony-stimulating Factors — Molecular and Cellular Biology* (Dexter, T.M. et al., Eds) Ch. 3, 77–109 (Marcel Dekker, New York).

_g_-factor In ESR/EPR SPECTROSCOPY the position of an absorption in a spectrum of a paramagnetic compound is a function of both microwave frequency and magnetic field. The *g*-factor is a parameter which defines the energy of the microwave quantum required to reverse the direction of the electron spin, in a particular magnetic field. For a free electron, *g* is very precisely known (g_e = 2.0023193044...). For the purposes of ESR/EPR spectroscopy the *g*-factor is considered to vary from this value to take account of effects such as spin-orbit coupling in molecules. In other words, *g* is treated as a variable whose value is a characteristic of the material being examined. Just as, in a spectrophotometer, a compound might be recognized from the wavelength of light which it absorbs, a compound in ESR/EPR is recognized by its characteristic *g*-factors.

G protein(s) Family of GTP-BINDING PROTEINS involved in signal transduction in eukaryotic cells. *See also*: G PROTEIN-COUPLED RECEPTORS.

G protein-coupled receptors

G PROTEIN-COUPLED receptors constitute probably the largest superfamily of cell-surface receptors. Each member of a given superfamily shares structural and/or sequence motifs and operates by a common transduction mechanism to mediate the transmission of extracellular signals into a biochemical or electrophysiological response within a cell.

G protein-coupled receptors are intrinsic membrane proteins, primarily located at the cell surface (*see* MEMBRANE STRUCTURE). It is thought that receptors of this type have seven membrane-spanning domains with the N terminus on the outside and the C terminus within, or facing, the cytoplasm. The binding of the signalling molecule to the outward facing part of the receptor structure results in the generation and transmission of a conformational change to the cytoplasmic face of the receptor. The receptor can then act as a catalyst and bind to an inactive heterotrimeric G protein (*see* GTP-BINDING PROTEINS), lowering the energy barrier for the binding/release of guanine nucleotides (GTP or GDP) at the G protein. Activation is associated with the formation of a G protein–GTP complex and subsequent association of one or more of its components with effector molecules (generally enzymes or certain ion channels) (*see* SECOND MESSENGER PATHWAYS).

The selectivity/specificity of the earliest steps of this type of catalytic signalling mechanism relies on three recognition processes: (1) the ability of the receptor to bind (generally) only one signalling molecule; (2) the ability of the conformationally changed receptor to bind selectively to one or more G-protein species; and (3) the ability of the activated G-protein species to bind to a specific effector molecule(s). It is obvious that coexpression of the specific receptor, G protein and effector molecule within a cell, as well as the availability of the signalling molecule for interaction with the receptor, are predeterminants for functional signalling.

In the case of G protein-coupled receptors, the signalling species may be endogenous molecules, (e.g. neurotransmitters or hormones), exogenous molecules (e.g. odorants; *see* OLFACTORY TRANSDUCTION), or light (in the case of VISUAL TRANSDUCTION). Many important drugs produce their therapeutic actions by binding to a specific G protein-coupled receptor and either producing a response or blocking the actions of the endogenous signalling molecule.

It should be noted that some other receptors (e.g. the insulin-like growth factor-II/mannose-6-phosphate receptor) can activate G proteins but are members of another receptor superfamily which has only a single transmembrane domain (*see* GROWTH FACTORS; GROWTH FACTOR RECEPTORS; LYSOSOMES; PROTEIN TARGETING).

Members of the G protein-coupled receptor superfamily

Estimates of the size of this family vary from about 400 to over 1000. These receptors would appear to constitute one of the largest gene families of the mammalian genome.

Over 240 members of this family had been cloned by February 1993. A partial list is shown in Table G1. Where possible the receptors are grouped under their generic name, which normally refers to the endogenous signalling species which activates the receptor. Sometimes the endogenous ligand is not known but the receptor has a known pharmacology; in other cases a candidate receptor has been cloned but its pharmacology is not known. In the latter instance such receptors are known as orphan receptors.

From Table G1 it can be seen that for each endogenous ligand many receptor subtypes can exist. In fact cloning studies have demonstrated the existence of more receptor subtypes than had been appreciated previously from pharmacological studies. The table does not include species variants or splice variants produced by alternative RNA processing. In the many instances of cloned G protein-coupled receptors, a very high degree of conservation of DNA and protein sequence of equivalent receptors has been found between mammalian species although greater differences are observed between mammalian sequences and those of birds and invertebrates. As might be expected, G protein-coupled receptors are found in many, if not all, organisms expressing heterotrimeric G proteins. These include yeast, where the mating factor receptors STE2 and STE3 regulate developmental processes including fusion and morphological changes, and the slime mould *Dictyostelium discoideum*, which possesses a receptor for CYCLIC AMP which acts as a chemoattractant (*see* DICTYOSTELIUM DEVELOPMENT).

DNA sequences corresponding to G protein-coupled receptors have also been discovered in the genomes of cytomegalovirus and herpesvirus saimiri. The ligands for, and functions of, such receptors are not known at present.

The largest subgrouping of G protein-coupled receptors is that of the odorant receptors [1]. This is a group of proteins whose mRNA is located predominantly in the olfactory epithelium (*see* OLFACTORY TRANSDUCTION). Estimates from SOUTHERN BLOTS of the numbers of these receptors range from 100 upwards. Although it is known that the nose can distinguish more than 10 000 different odours and that these odorant receptors operate by coupling to G proteins, the functional activation of only one of the cloned putative odorant receptors by an odorant molecule has been demonstrated. Messenger RNAs from an additional 48 members of this family have been found in germ cells of the testis. It has been suggested that these receptors could be involved in chemotaxis during fertilization. Recently, more than 60 receptors whose expression is restricted to the surface of lingual epithelia have also been cloned. These may be taste receptors.

Table G1 Cloned G protein-coupled receptors

ACTH	1
Adenosine	4
α-Adrenergic receptors	6
β-Adrenergic receptors	3
Angiotensin	2
Bombesin	3
Bradykinin	1
C5a	1
Cannabinoid	1
C-C Chemokine	1
CCK/gastrin	2
Dopamine	5
Endothelin	2
Formyl peptide	3
Gonadotropin-releasing hormone	1
Histamine	2
5-Hydroxytryptamine	9
Interleukin-8	3
LH/FSH/TSH	3
Mas	2
Melanocortin	1
Muscarinic	5
Neuropeptide Y	2
Neurotensin	1
Odorant (putative)	23*
Opioid	1
Opsins	4
PAF	1
Prolactin	1
Prostaglandin E	1
Somatostatin	3
Tachykinin	3
Taste (putative)	60*
Testis specific	48*
Thrombin	1
Thromboxine A$_2$	1
TRH	1
Tyramine/octopamine	1
Vasopressin	3
Glutamate (metabotropic)	6
Growth hormone-releasing hormone	1
Secretin	1
VIP	1
Parathyroid hormone	1
Calcitonin	1
cAMP	1
Yeast mating factor	2
Viral	4
Unknown	12

* Some partial sequences.
ACTH, adrenocorticotropin; CCK, cholecystokinin; FSH, follicle-stimulating hormone; LH, luteinizing hormone; PAF, platelet-activating factor; TRH, thyrotropin-releasing hormone; TSH, thyroid-stimulating hormone; VIP, vasoactive intestinal peptide.

Also illustrated in Table G1 are the wide variety of ligands that activate the different receptors. These range from high molecular weight hormones (e.g. luteinizing hormone (LH)/gonadotropin) and peptides, to polar or hydrophobic small molecules. A highly specialized case of receptor activation is that resulting from the light-induced isomerization of 11-*cis* retinal bound covalently to opsin.

Cloning strategies

The original approach used peptide sequences from purified receptor proteins to construct oligonucleotide PROBES with which to screen appropriate cDNA libraries (*see* DNA LIBRARIES). This strategy has been largely superseded by low-stringency homology cloning and POLYMERASE CHAIN REACTION (PCR) amplification of sequences from cDNA or genomic DNA. An alternative strategy is that of expression cloning from cDNA libraries using either *in vitro* RNA transcription and an electrophysiological assay of function in *Xenopus* oocytes (*see* XENOPUS OOCYTE EXPRESSION SYSTEM), or TRANSFECTION of cDNA libraries constructed in eukaryotic EXPRESSION VECTORS into COS cells and screening for binding activity with a radio-iodinated ligand [2–4].

Genomic organization

Many coding sequences of G protein-coupled receptors do not contain INTRONS but increasing numbers of receptors are now being cloned which contain up to 10 introns within the coding sequence. The location of exon–intron boundaries relative to the putative transmembrane domains is conserved within a subtype family (e.g. opsins, tachykinin receptors) but is highly variable between families. Similarly, the 5′-noncoding sequences of different G protein-coupled receptors contain variable numbers of introns. Different molecular forms of some G protein-coupled receptors (some functional, others nonfunctional) may be generated by alternative splicing (*see* RNA SPLICING).

A number of receptor genes lack a TATA box or a CAAT sequence upstream of the transcription initiation sites (*see* EUKARYOTIC GENE EXPRESSION), but contain putative cAMP response elements and binding sites for the TRANSCRIPTION FACTOR Sp1. Some of these features are characteristic of 'housekeeping' genes found in all cells, although such promoters in the D_1 and D_2 dopamine receptor genes have been shown to induce TRANSCRIPTION in a cell-specific manner. Studies of the control elements regulating G protein-coupled receptor gene expression are still at an early stage.

General features of receptor sequences

The predominant feature of the amino-acid sequences of G protein-coupled receptors is the presence of seven stretches of predominantly hydrophobic amino acids, each of sufficient length (typically 20–25 residues) to traverse the membrane as an α-helix. These hydrophobic sequences are generally considered to provide the core structure of these receptors by forming a seven-helix bundle which resides in the membrane and has a central pocket.

Considerable emphasis has been placed on the similarities between such a structure and the detailed model of BACTERIORHODOPSIN from *Halobacterium halobium* [5]. This is a light-driven proton pump and not a G protein-coupled receptor. However its structure, consisting of a 7-helical bundle, has been solved to a modest resolution. The coordinates of this structure have provided the basis for modelling studies of G protein-coupled receptors [6,7].

Evidence for such a structure for G protein-coupled receptors comes from peptide and EPITOPE mapping, chemical modification studies, and the application of a variety of physical biochemical techniques. These approaches have been applied primarily to mammalian RHODOPSIN and to the β-ADRENERGIC RECEPTOR but the conclusions have been considered to hold, at least in outline, for all G protein-coupled receptors. A low-resolution structure of the transmembrane helices of bovine rhodopsin has been obtained [8]. The projection structure is similar to but less elongated than bacteriorhodopsin and the helices are tilted differently from bacteriorhodopsin.

Interesting patterns emerge when the putative transmembrane sequences are arranged in a simple helical wheel model. This is illustrated for the sequences of the five muscarinic acetylcholine receptor subtypes [9] (Fig. G1) (MUSCARINIC RECEPTORS). The helical wheel projection shows the relative orientations of the amino acid side chains when looking down the axis of an α-helix. For each α-helix there is a hydrophobic face and a partially hydrophilic face. In the case of the muscarinic receptor, the amphipathic motif is pronounced for the third, sixth, and seventh helices. In addition, the hydrophobic face contains almost all the residues that are not conserved between different subtypes or between the same subtypes in different species. The helices can therefore be packed into a structure, analogous to that of bacteriorhodopsin, in which the nonconserved hydrophobic faces are orientated towards the lipid bilayer and the partly hydrophilic face points to a predominantly conserved central cavity. These patterns are found for all subtypes and species variants of the receptors in Table G1. At the most basic level, they are an intrinsic feature of the amino-acid sequence independent of the secondary structure. Generally, the putative transmembrane sequences are joined together by short sequences (loops), the exception being the third intracellular loop joining the fifth and sixth transmembrane sequences which is highly variable in length. The regions of this loop close to the transmembrane sequences as well the equivalent regions of the two short intracellular loops have been shown by mutagenesis and other studies to be important for G protein recognition and selectivity.

The C terminus of G protein-coupled receptors is of variable length and in most sequences contains a cysteine near the seventh transmembrane domain. In the case of rhodopsin and the β-adrenergic receptor this residue is palmitoylated, thereby providing an additional potential anchoring point of the C terminus to the membrane.

The C terminus and the third intracellular loop are often rich in serine and threonine residues. In the case of rhodopsin, β-adrenergic receptors, and some muscarinic receptors, some of these residues are targets for PROTEIN PHOSPHORYLATION by protein kinase A (CYCLIC AMP-DEPENDENT KINASE) and receptor kinases [10]. These phosphorylations which, in the case of the receptor kinases, are agonist dependent, result in desensitization and uncoupling of the receptor from the G protein.

Many G protein-coupled receptors contain two cysteine resi-

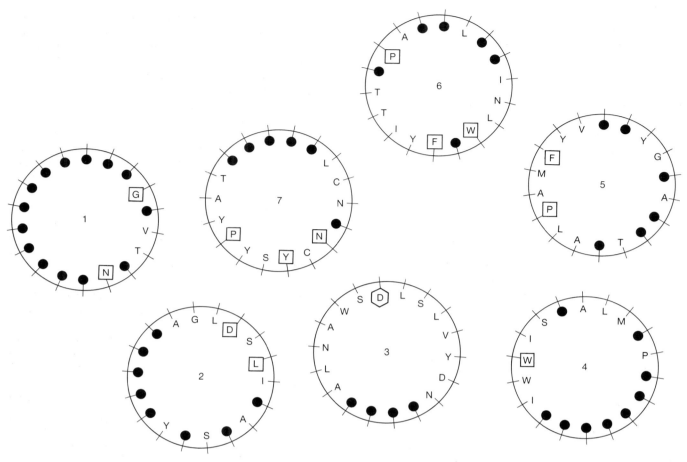

Fig. G1 A simple helical wheel model of the seven transmembrane domains of muscarinic receptors. Eighteen sequences of m1–m5 receptors from humans and other species, including the chick and *Drosophila* receptors are considered. Each helix is depicted as a helical wheel (angle 100°) with the short lines representing the orientation of the side chains. For simplicity of representation only the central 18 of the 20–25 residues thought to be present as α-helices are shown. The helices are viewed from the extracellular domain and are positioned approximately in the relative positions suggested by the projection map of rhodopsin [8]. Residues identical in all the sequences are shown by their single letter code and those found in most G protein-coupled receptors are enclosed in a square. In addition the aspartate residue in TM3, conserved in all monoaminergic receptors, is enclosed in a hexagon. Residues that are hydrophobic (including serine and threonine) and not conserved between muscarinic subtypes and species are depicted by filled circles. Both the conserved and hydrophilic residues tend to cluster at helix–helix boundaries and towards the interior of the structure, whereas the nonconserved and hydrophobic residues face outwards. Adapted from [9].

dues in conserved positions in the first and second extracellular loops. In the case of rhodopsin and muscarinic receptors these have been shown to form a disulphide bond. Mutation of either residue results in loss of binding and/or low levels of expression. The presence of a disulphide bond in these receptors may be important for the maintenance of a correctly folded structure. Otherwise the roles of the extracellular loops in receptor function have not been investigated in detail.

Most, but not all, G protein-coupled receptors are GLYCOPROTEINS. Consensus sites for N-linked glycosylation (NXS/T, where X is any amino acid) are concentrated at the N terminus. Mutation of these sites in β-adrenergic receptors and muscarinic receptors abolishes glycosylation but has essentially no effect on receptor expression or function. Only those receptors that have large N termini, such as the metabotropic glutamate receptors, the LH receptor, and the THROMBIN receptor, have the consensus

sequence for a SIGNAL PEPTIDE which may be cleaved during receptor processing and insertion into the membrane (*see* PROTEIN TRANSLOCATION).

The N terminus of thrombin receptors is important in receptor activation. The binding of thrombin, a proteinase, to its binding site on the receptor results in cleavage of part of the N terminus. The new N-terminal sequence generated on the receptor then acts as its own ligand and the receptor becomes self activated (*see* BLOOD COAGULATION AND ITS DISORDERS).

Subfamilies of G protein-coupled receptors

Even within the known sequences of mammalian G protein-coupled receptors it is difficult to identify a single amino acid that is conserved in position relative to the seven hydrophobic sequences. Of the receptors cloned to date there appear to be two

Table G2 Characteristic amino-acid sequence motifs found in most G protein-coupled receptors

Motif	Location
GN	First transmembrane domain
$\begin{Bmatrix} N \\ or \\ S \end{Bmatrix} L \begin{Bmatrix} A \\ or \\ S \end{Bmatrix} - \begin{Bmatrix} A \\ or \\ S \end{Bmatrix} D$	Second transmembrane domain
$\begin{Bmatrix} I \\ or \\ L \end{Bmatrix} \begin{Bmatrix} A \\ or \\ S \end{Bmatrix} - \begin{Bmatrix} D \\ or \\ E \end{Bmatrix} R \begin{Bmatrix} F \\ or \\ Y \end{Bmatrix}$	Intracellular end of third transmembrane domain
$W - - \begin{Bmatrix} A \\ or \\ S \end{Bmatrix}$	Fourth transmembrane domain
$\begin{Bmatrix} F \\ or \\ Y \end{Bmatrix} - - P$	Fifth transmembrane domain
F - - - W - P	Sixth transmembrane domain
N P - - Y	Seventh transmembrane domain
C	Two residues in the beginning of the third transmembrane domain and in the second extracellular loop

small subfamilies, the metabotropic glutamate receptor family [11] and the calcitonin/parathyroid hormone/secretin/growth hormone-releasing hormone receptor family, whose sequences bear essentially no sequence homology to each other and to the third large subfamily. The relative and absolute size of these subfamilies remains to be determined. Not surprisingly, because of evolutionary distance, the yeast mating factor receptor sequences have little homology with the sequences of mammalian or invertebrate G protein-coupled receptors.

Within each subfamily, homology tends to be concentrated within residues in the seven hydrophobic sequences or in residues close to these sequences, again pinpointing the importance of this core structure.

In the case of the largest subfamily, the receptor sequences contain most or all of the motifs shown in Table G2. Within this subfamily there are groupings of receptors that are more similar to each other than to other receptors. It is possible to construct dendrograms to visualize these groupings, which can be roughly linked to similarities in the chemical nature of the substances (e.g. proteins, peptides, monoamines) that are the naturally occurring ligands for the receptors.

The explanation for the existence of so many subtypes of a given receptor may lie in subtle differences in the structure of the receptor (e.g. the kinetics of ligand binding, the desensitization mechanism(s), or the ability to selectively interact with different G proteins). Alternatively, variations in the noncoding sequence may lead to differential regulation of the level of receptor expression in different tissues and different stages during development.

The ligand-binding site

All receptors that bind positively charged monoamines such as acetylcholine, histamine, and adrenaline have an aspartate residue towards the centre of the third transmembrane domain. The evidence from chemical labelling and mutagenesis studies is that this negatively charged residue is the primary site of binding of the positively charged ligand. In mammalian opsins a glutamate residue in a similar position acts as a counter ion to the positively charged Schiff's base formed from the binding of retinal to a lysine residue in the seventh transmembrane domain.

Hydrophilic residues in the seventh, sixth, and particularly the fifth transmembrane domains are thought to make hydrogen bonding interactions with appropriate groups on these monoamines. The picture is therefore of the neurotransmitter nestling in a hydrophilic pocket of the receptor, located in the extracellular half of the membrane.

In the case of the peptide receptors there is a conservation of sequence within the transmembrane domains analogous to that found for the monoamine receptors, but whether the binding site for these ligands is located within the transmembrane domains, the external facing loops, or both, is not known. Receptors for large hormones such as LH contain a large extracellular N terminus which is considered to be important in ligand binding. However, deletion of this region generates a receptor which can still be activated, albeit at lower affinity, by LH. It may well be that receptor activation by a specific conformational change on the inside of a G protein-coupled receptor is by interaction of ligands with the same binding pocket.

N.J.M. BIRDSALL

1 Buck, L. & Axel, R. (1991) A novel multigene family may encode odorant receptors: a molecular basis for odor recognition. *Cell* **65**, 175–187.
2 Dohlman, H.G. et al. (1991) Model systems for the study of seven-transmembrane segment receptors. *Annu. Rev. Biochem.* **60**, 653–688.
3 Evans, C.J. et al. (1992) Cloning of a delta opioid receptor by functional expression. *Science* **258**, 1952–1955.
4 Masu, Y. et al. (1987) cDNA cloning of bovine substance K receptor through oocyte expression system. *Nature* **329**, 836–838.
5 Henderson, R. et al. (1990) Model for the structure of bacteriorhodopsin based on high-resolution electron cryomicroscopy. *J. Mol. Biol.* **213**, 899–929.
6 Khorana, H.G. (1993) Two light transducing membrane proteins: bacteriorhodopsin and mammalian rhodopsin. *Proc. Natl. Acad. Sci. USA* **90**, 1166–1171.
7 Trumpp-Kallmeyer, S. et al. (1992) Modeling of G protein-coupled receptors: application to dopamine, adrenaline, serotonin, acetylcholine and mammalian opsin receptors. *J. Med. Chem.* **35**, 3448–3462.
8 Schertler, G.F.X. et al. (1993) Projection structure of rhodopsin. *Nature* **362**, 770–772.
9 Hulme, E.C. et al. (1990) Muscarinic receptor subtypes. *Annu. Rev. Pharmacol. Toxicol.* **30**, 633–673.
10 Strader, C.D. et al. (1989) Identification of two serine residues involved in agonist activation of the β-adrenergic receptor. *J. Biol. Chem.* **264**, 13572–13578.
11 Nakanishi, S. (1992) Molecular diversity of glutamate receptors and implications for brain function. *Science* **258**, 597–603.

G0 The resting or quiescent phase of the eukaryotic CELL CYCLE.

G1 The phase of the eukaryotic CELL CYCLE which occurs after MITOSIS (M phase) and before initiation of DNA synthesis (S phase). Successful completion of G1 — which tends to be the longest part of the cell cycle — is required for progression to S phase.

HOOC — CH₂ — CH₂ — CH₂ — NH₂

Fig. G2 γ-Aminobutyric acid (GABA).

G2 Phase of the eukaryotic CELL CYCLE immediately following DNA synthesis (S phase) and in which cells are TETRAPLOID. This phase lasts for several hours and is followed by MITOSIS (M phase).

G25K *See:* GTP-BINDING PROTEINS.

GA (1) GIBBERELLIN (*see also* PLANT HORMONES).
(2) GOLGI APPARATUS.

GABA (γ-aminobutyric acid) Inhibitory amino acid NEUROTRANS-MITTER in the central nervous system (Fig. G2). It acts via the GABA$_A$ and GABA$_B$ receptors to prevent neuronal depolarization and firing of an ACTION POTENTIAL. GABA$_A$ receptors are directly coupled to a Cl$^-$ channel and GABA$_B$ receptors to a K$^+$ channel via a G protein. *See:* GABA AND GLYCINE RECEPTORS; G PROTEIN-COUPLED RECEPTORS; SYNAPTIC TRANSMISSION.

GABA and glycine receptors

THE amino acids γ-aminobutyric acid (GABA) and glycine are the major inhibitory neurotransmitters in the vertebrate CENTRAL NERVOUS SYSTEM, mediating fast postsynaptic inhibition by their action at specific receptors (*see* SYNAPTIC TRANSMISSION). Glycine receptors and one class of GABA receptors (GABA$_A$) are structurally related ligand-gated ion channels belonging to the same superfamily as the NICOTINIC RECEPTOR for acetylcholine, and form the subject of this entry. A second class of structurally and pharmacologically distinguishable GABA receptors (GABA$_B$) are G PROTEIN-COUPLED RECEPTORS and are discussed elsewhere.

Binding of neurotransmitter to glycine and GABA$_A$ receptors is followed within milliseconds by the opening of a chloride ion channel in the postsynaptic membrane which results in general in the hyperpolarization of the postsynaptic neuron (*see* SYNAPTIC TRANSMISSION), and inhibition of transmission of an impulse. GABA$_A$ receptors predominate in the brain where they are widespread, whereas glycine receptors are found mainly in the spinal cord and brain stem, although they do occur in higher brain regions. GABA$_A$ receptors in particular have been studied intensively as they are the physiological sites of action of the benzodiazepine minor tranquillizers and of barbiturates, which both enhance the inhibitory action of GABA at these receptors.

GABA$_A$ receptors are characterized pharmacologically by their antagonism by the plant alkaloid bicuculline (Fig. G3), which competes with GABA, and by the convulsant picrotoxin, which acts noncompetitively with GABA to block chloride channel function. The action of glycine at the glycine receptor is blocked by the plant alkaloid strychnine, which acts as a convulsant. Benzodiazepines and barbiturates act as allosteric regulators of GABA$_A$ receptor activity, binding to sites other than the GABA-

binding site. GABA$_A$ receptors can also be regulated by certain neurosteroids (e.g. 5β-pregnan-3α-ol-20-one), and the anthelmintic agent avermectin (a glycosidic macrocyclic lactone from *Streptomyces avermitilis*), which seem to have distinct binding sites. Both these classes of compound potentiate GABA action.

Molecular structure

Glycine receptors have been shown to be pentameric glycoproteins composed of two types of subunit — α and β — each of which has been found in various ISOFORMS (Table G3). The GABA$_A$ receptor is also presumed to be a pentamer but its subunit composition is more complicated. Five different classes of receptor subunit — α, β, γ, δ and ρ — have been identified and each class contains several isoforms (Table G3). By analogy with the glycine receptors, pentameric GABA$_A$ receptors are believed to be assembled from combinations of these subunits, but the actual subunit composition and stoichiometry of a native GABA$_A$ receptor has not yet been established.

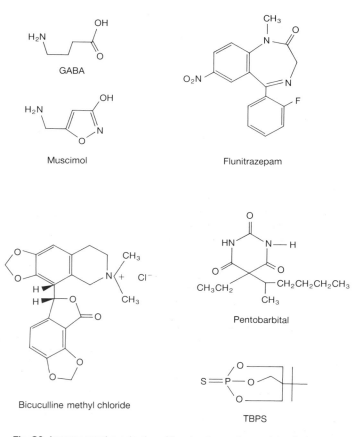

Fig. G3 A representative selection of the structures of some drugs that interact with the GABA$_A$ receptor. Muscimol and bicuculline are antagonists, competing with GABA. Flunitrazepam is a benzodiazepine and pentobarbital is a representative of the barbiturates. These regulate GABA action by binding to an allosteric modulatory site. TBPS, t-butylbicyclophosphoro thionate, is a noncompetitive antagonist of GABA$_A$ receptors which is thought to bind to the open state of the channel.

Table G3 Identified subunits of vertebrate GABA and glycine receptors and their localization in the central nervous system

Class	Isoforms	Distribution*
GABA$_A$ receptors		
α	1–6	Each isoform has a unique but overlapping distribution. α1, Throughout brain; α2, hippocampus and cortex; α3, olfactory bulb and cortex; α4, rare; α5, hippocampus; α6, cerebellar granule cells only
β	1–4 plus splicing variant b4′†	
γ	1–3, splicing variants γ2S and γ2L	
δ	—	Cerebellar granule cells
ρ	1, 2	Retina
Glycine receptors		
α	1–4 plus splicing variants α1′, α2′	α1, Spinal cord and brain stem; α2, α3, higher brain regions
β	—	Spinal cord and higher brain regions

* By *in situ* hybridization of mRNA with cDNA probes, giving the areas enriched in particular subtypes.
† In chick.

GABA$_A$ receptors

The GABA$_A$ receptor was first isolated by biochemical methods and shown to be a hetero-oligomeric glycoprotein of M_r ~250 000 [1,2]. Two types of subunit were identified — α (M_r 53 000) and β (M_r 58 000). Since then the picture has become more complicated. Cloning of receptor cDNAs and screening of cDNA libraries (*see* DNA CLONING) has identified multiple vertebrate GABA$_A$ receptor genes; the number stands at 16 at present [3,4]. Five classes of subunit have been distinguished on the basis of amino-acid sequence.

All the GABA$_A$ receptor polypeptides share structural features not only between themselves but also with the other members of the ligand-gated ion channel superfamily, including the glycine receptor subunits, the nicotinic acetylchoine receptor subunits, and the more recently identified serotonin$_3$ (5HT$_3$) receptor. They share domain organization but not amino-acid sequence homology with the glutamate receptor subunits (*see* EXCITATORY AMINO ACID RECEPTORS).

GABA$_A$ receptor polypeptides range in length from 424–521 amino acids with corresponding M_r values of 48 500 to 64 000 for the core (i.e. nonglycosylated) polypeptides. From the deduced amino-acid sequences they are predicted to have a large hydrophilic N-terminal extracellular domain, four hydrophobic membrane-spanning segments (M1–M4), and an intracytoplasmic loop between M3 and M4 (Fig. G4). At the level of primary sequence, isoforms of one subunit class share at least 75% amino-acid sequence identity whereas the identity between any two distinct subunit classes is around 35%. The most divergent regions between the polypeptides of one class are generally at the extreme N- and C-terminal domains and within the intracellular loops. Sequence similarity with the nicotinic acetylcholine receptor is found within the four transmembrane regions and within a region termed the Cys–Cys β loop in the N-terminal extracellular domain.

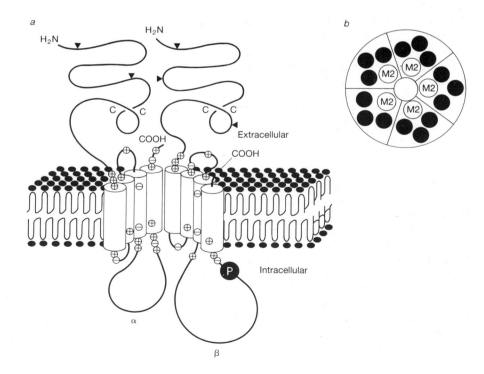

Fig. G4 Diagrammatic representation of the transmembrane organization of the mammalian GABA$_A$ and glycine receptors. *a*, Predicted transmembrane topology of the GABA$_A$ receptor α and β subunits seen from within the plane of the membrane. The same organization is predicted for all GABA$_A$ and glycine receptor subunits and their isoforms. All the polypeptides are predicted to have four membrane-spanning helices which are shown as cylinders; arrowheads represent the potential sites of N- glycosylation (*see* GLYCOPROTEINS), and P is the cAMP-dependent protein kinase site of the β subunit. The charged residues shown are only those located adjacent to the ends of the membrane domains and which are thought to be involved in chloride channel binding. *b*, View of the predicted model perpendicular to the plane of the membrane. It represents each of the five subunits of the receptors as a segment within the annular structure where the hole in the centre represents the chloride ion channel. The four membrane-spanning regions within each polypeptide are shown as filled circles with the M2 helix lining the wall of the channel. The subunit complements and their ordering around the rosette for either the GABA$_A$ or glycine receptors have not yet been determined. From [3].

Studies on recombinant receptors expressed in *Xenopus* oocytes (*see* XENOPUS OOCYTE EXPRESSION SYSTEM), show that all subunits contain (or can contribute to) a potential binding site for GABA, as each subunit type can form homo-oligomeric GABA-gated chloride ion channels when expressed on its own. GABA acts at the cell surface and cannot cross the plasma membrane, so the binding sites will be contained in the extracellular regions of the subunit polypeptides. The precise location and nature of the binding sites, or the number of binding sites per receptor is not known. From electrophysiological studies it is proposed that the binding of two molecules of GABA is required to open the chloride channel [5]. Binding sites for benzodiazepines are also contained in the extracellular domains.

Several GABA$_A$ receptor subunits contain consensus sequences for phosphorylation by various PROTEIN KINASES in their intracellular loop domains. The GABA$_A$ receptor is phosphorylated *in vitro* by PROTEIN KINASE A and PROTEIN KINASE C and phosphorylation is believed to be required to maintain function.

Oligomeric structure. Although GABA$_A$ receptors are believed to be pentameric structures, because of the large number of possible different subunits it has not yet proved possible to determine the subunit composition and stoichiometry of a native receptor. Indeed, pentameric permutations of the known subunits and their spliced variants could give rise to possibly hundreds of GABA$_A$ and, as below, glycine receptors, all perhaps with distinct functional properties.

In situ hybridization (*see* HYBRIDIZATION) shows that the messenger RNAs for the various subunits have specific distributions within the central nervous system. The α1 subunit is most abundant, occurring in many different cell types throughout the brain. Other subunits are more restricted in their distribution (Table G3). Overall the most abundant mRNAs are those for α1, β2 and γ2 subunits, and this correlates well with immunoaffinity purification of the different GABA$_A$ receptor subpopulations which shows that the most abundant receptor subunits are α1, β2/β3 and γ2. Thus this combination is a prime candidate for a native GABA$_A$ receptor but the subunit stoichiometry and the number of ligand-binding sites per receptor are not known.

Properties of cloned receptors. The expression *in vitro* in *Xenopus* oocytes of various combinations of subunits has not yet provided a receptor satisfying all the pharmacological criteria for a native GABA$_A$ receptor. Each type of GABA$_A$ receptor polypeptide when expressed alone can form a GABA$_A$-gated ion channel selective for chloride ions. In the cases where single-channel conductance of these homo-oligomers has been analysed, four conductance states have been found, as for native receptors, which indicates that this must be a property of the channel, rather than each conductance state reflecting the presence of one type of subunit. The expression of two different subunits generally produces enhanced channel activity.

Glycine receptors

Glycine receptors (Fig. G5) were also first isolated by biochemical methods [6]. The receptor thus identified was a pentameric glycoprotein structure of M_r 250 000. Two types of polypeptide subunits were identified — α and β — and a receptor composition of $α_3β_2$ was proposed. This is consistent with subsequent studies of the distribution of subunit mRNAs in the central nervous system (Table G3), although, as for GABA$_A$ receptors, the actual subunit composition of a native glycine receptor is still not unequivocally determined. cDNA cloning and expression and screening of cDNA libraries have identified multiple glycine receptor genes but so far only of the α and β classes (see Table G3) [7,8]. The binding site for glycine is contained in the α subunit.

Action of benzodiazepines

Some benzodiazepines potentiate the inhibitory action of GABA by increasing the frequency of chloride channel opening, and this is presumed to be the basis for the anxiolytic and tranquillizing effects of the classical benzodiazepine agonists such as Librium and Valium. In the absence of GABA, these benzodiazepines have no intrinsic effect on GABA$_A$ receptors but when GABA or GABA agonists are bound to the GABA-binding site they enhance benzodiazepine binding in a dose-dependent manner (the 'GABA

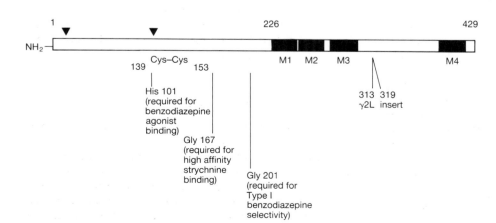

Fig. G5 Diagrammatic representation and interrelationship of important domains of GABA$_A$ and glycine receptor polypeptides. The numbered amino-acid sequence of the bovine GABA$_A$ receptor α1 subunit is shown, highlighting significant residues and regions within this and other GABA$_A$ and glycine receptor polypeptides. Arrowhead, Consensus sequences for N-glycosylation; M1–M4, transmembrane domains; Cys–Cys, conserved extracellular motif common to all members of the ligand-gated ion channel superfamily.

shift') by decreasing the dissociation constant for benzodiazepines [9]. Benzodiazepine pharmacology is complex, however, and other groups of benzodiazepines have antagonistic rather than potentiating effects. Yet others, although acting at the benzodiazepine regulatory site, are anxiogenic rather than anxiolytic. It is likely that some at least of these different effects are mediated through structurally distinct subclasses of GABA$_A$ receptor with different distributions in the brain. But, important though benzodiazepine regulation may be clinically, no endogenous ligand(s) for the benzodiazepine-binding sites has yet been identified and these receptor subclasses may have no physiological significance.

In vitro expression of various combinations of receptor subunits has provided some insight into the molecular basis for the actions of the various benzodiazepine compounds. Only receptors with a combination of γ2 or γ3 and an α and β combination show reproducible benzodiazepine regulatory activity; the type of α subunit then seems to determine the particular type of benzodiazepine receptor pharmacology. The inclusion of α1 or α2, for example, produces receptors responsive to the classical agonist benzodiazepines like Librium but with different patterns of responsiveness to the other types of benzodiazepines. α4 and α6, which do not bind the classical benzodiazepine agonists, can be made to do so by point mutations.

Action of barbiturates

Barbiturates such as (–)pentobarbital (see Fig. G3) also enhance the inhibitory action of GABA by binding to a distinct regulatory site on the receptor. They facilitate GABA neurotransmission by increasing the mean open time of the chloride ion channel. Barbiturates increase the binding of both GABA and benzodiazepines to their respective binding sites.

GABA$_A$ and glycine receptors in disease

Deficiencies in glycine receptors have been implicated in human diseases of motor function such as hyperkinaesia and spastic paraplegia by analogy with heritable disorders of motor function in the mouse mutant *spastic* and in Poll Hereford cattle in which a loss of glycine receptor function has been demonstrated. The mouse mutation, which maps to chromosome 3, selectively interferes with the accumulation of glycine receptor α1 subunits.

Malfunction or deficiency of GABA$_A$ receptors has been implicated in mechanisms of anxiety by virtue of the therapeutic effect of benzodiazepines. In animal models, pharmacological blockade of GABA neurotransmission can indeed elicit anxiety. Although other neurotransmitters seem to be implicated in the severe AFFECTIVE DISORDERS such as manic depressive illness, the gene for the GABA$_A$ receptor α3 subunit has been shown to map close to a region (on human chromosome X) associated with a heritable form of manic depression. Deletions of the β3 subunit gene are found in the rare neurological disorder ANGELMAN SYNDROME which is characterized by seizures and mental retardation.

F.A. STEPHENSON

1 Sigel, E. et al. (1983) A γ-aminobutyric acid/benzodiazepine receptor complex of bovine cerebral cortex: purification and partial characterization. *J. Biol. Chem.* **258**, 6965–6971.

2 Stephenson, F.A. (1993) In *New Comprehensive Biochemistry* **24** (Hucho, F., Ed.) 183–197 (Elsevier, Amsterdam).

3 Schofield, P.R. et al. (1987) Sequence and functional expression of the GABA$_A$ receptor shows a ligand-gated receptor superfamily. *Nature* **328**, 221–227.

4 Olsen, R.W. & Tobin, A.J. (1990) Molecular biology of GABA$_A$ receptors. *FASEB J.* **4**, 1469–1480.

5 Bormann, J. (1988) Electrophysiology of GABA$_A$ and GABA$_B$ receptor subtypes. *Trends Neurosci.* **11**, 112–116.

6 Pfeiffer, F. et al. (1982) Purification by affinity chromatography of the glycine receptor of rat spinal cord. *J. Biol. Chem.* **257**, 9389–9393.

7 Grenningloh, G. et al. (1987) The strychnine-binding subunit of the glycine receptor shows homology with nicotinic acetylcholine receptors. *Nature* **328**, 215–220.

8 Betz, H. (1991) Glycine receptors: heterogeneous and widespread in the mammalian brain. *Trends Neurosci.* **14**, 458–461.

9 Olsen, R.W. & Venter, J.C. (Eds) (1986) *Benzodiazepine/GABA receptors and Chloride Channels: Structural and Functional Properties* (Alan Liss, New York).

GAG GLYCOSAMINOGLYCAN, previously referred to as mucopolysaccharide.

gag–pol Designation of two (of the three) protein-coding regions typically found in RETROVIRUS genomes. The *gag* region encodes proteins which form the virion core and *pol* encodes a PROTEASE, a REVERSE TRANSCRIPTASE, and an INTEGRASE at its 5′, middle, and 3′ regions respectively. Following transcription of the retroviral genome into a POLYCISTRONIC mRNA, the host cell's protein synthetic apparatus produces a mixture of gag and gag–pol fusions. The latter gene product arises either by translational SUPPRESSION or FRAMESHIFTING past the *gag* TERMINATION CODON and is processed appropriately to give mature proteins.

GAL4 In the yeast SACCHAROMYCES CEREVISIAE the *GAL4* gene encodes a positive regulatory protein (Gal4) which binds to sequences (termed UAS$_{GAL}$) within the PROMOTERS of the galactose-utilization genes *GAL1*, *GAL7*, and *GAL10*, and in the presence of galactose, activates their transcription. The sequence-specific binding to DNA is mediated through the N-terminal portion of the protein and functions independently of the TRANSCRIPTIONAL ACTIVATION DOMAIN. *See*: EUKARYOTIC GENE EXPRESSION; PROTEIN–NUCLEIC ACID INTERACTIONS; TRANSCRIPTION FACTORS.

galactans GLYCANS which are largely or wholly polymers of galactose, although most are heteropolymers with other sugars as many are highly branched. There are several unrelated groups.

1 Linear galactans derived either from the repeating sequence

$$(-3_DGal(p)\beta 1,4_DGal(p)\alpha 1-)_n$$

or from

$$(-3_DGal(p)\beta 1,4_LGal(p)\alpha 1-)_n$$

are found in red algae (carrageenans, agarose). The former sequence is probably synthesized from UDP-Gal and the latter from UDP-Gal and GDP-Gal, each using two transferases alternately. Linear galactans and branched ARABINOGALACTANS containing linear galactan cores with the repeating sequence $(-4_DGal(p)\beta 1,4_DGal(p)\beta 1-)_n$ are common in terrestrial plants, though some may be artefacts of extraction (i.e. fragments of larger glycans). A single glycosyl transferase using UDP-Gal probably suffices for their assembly. *See also*: PECTINS.

2 Linear galactans based on (-3D/LGal(p)β1,3D/LGal(p)β1-)$_n$ occur in the egg jelly-coats of some echinoderms. They are probably synthesized from UDP-Gal and GDP-Gal depending on whether the polymer is based on D- or L-galactose, respectively.

3 Branched galactans based on linear chains of β1,3-linked D-galactopyranosyl residues, β1,6-linked branches and some short β1,6-linked side-chains are widespread in land plants, for example in wood of larch (*Larix* spp.) and in gum arabic of *Acacia* spp. Most occur as arabinogalactans and other sugars (e.g. GlcA) can be present. At least three enzymes must be involved in generating the main chains, the side chains and the links between them: there is no evidence for transglycosylation.

4 Highly branched galactans containing 1,3 and 1,6 linkages, in some cases purely of L-galactose, are known in gastropods (galactogen) and similar very branched galactans (but of D-galactose or a mixture of D and L-galactose) are present in mammalian lung (pneumogalactan). Such animal galactans are commonly sulphated and glucuronic acid residues may be present. Their synthesis is obscure. Galactan phosphates have been reported in invertebrates.

Aspinall, G.O. (1970) *Polysaccharides* (Pergamon Press, Oxford).
Stoddart, R.W. (1984) *The Biosynthesis of Polysaccharides* (Croom-Helm, London).

galactoglucomannan Family of GLYCANS from cell walls and gums of higher plants which contain cores consisting of either homopolymeric chains of β1,4-linked D-mannopyranosyl residues or irregular, nonrandom heterocopolymeric chains of β1,4-linked D-mannopyranosyl and D-glucopyranosyl residues [1]. The mannosyl and glucosyl residues are partially substituted at C6 by α-D-galactopyranosyl residues in many of these glycans, although glucomannans lacking galactose occur in some cell walls (e.g. *Iris* spp.) and root slimes (e.g. *Lilium henryi*). A 'pure' mannan, containing only traces of galactose, occurs in ivory nut. The physical properties of these glycans vary from extreme insolubility and hardness to great solubility and sliminess. It is not known how many glycosyl transferases are required for their synthesis or how priming occurs: the assembly of the core sequence is a particular biosynthetic problem. *See also*: CELLULOSE.

Dey, P.M. (1978) *Adv Carb. Chem. Biochem.* **35**, 341–376.
Stoddart, R.W. (1984) *The Biosynthesis of Polysaccharides* 236–239 (Croom-Helm, London).

galactosaemia An AUTOSOMAL RECESSIVE condition arising from a deficiency of the enzyme galactose-1-phosphate uridylyl transferase and which leads to galactose toxicity. It presents in the first few days following birth with failure to feed, diarrhoea, hypoglycaemia, acidosis, and dehydration. Liver dysfunction supervenes. Cataracts become obvious if the child survives more than a few weeks. Treatment consists of a diet excluding galactose as much as possible. Poor control may contribute to intellectual retardation. The incidence of the condition is one in 60 000.

galactose, galacturonic acid *See*: SUGARS.

β-galactosidase β-D-galactosidase. Enzyme (EC 3.2.1.23) which hydrolyses β-D-galactosides such as lactose into their component sugars by hydrolysis of terminal nonreducing β-D-galactose residues (*see* SUGARS). Lactose is hydrolysed into galactose and glucose. In the bacterium *Eschericha coli* the enzyme is found as a tetramer and is the translation product of the *LacZ* gene, which is itself part of the lactose operon (*see* LAC OPERON). The *LacZ* gene is transcribed as part of the polycistronic mRNA produced under the control of the *lac* promoter. The *E. coli LacZ* gene is often used as a REPORTER GENE in studies of promoter action. A translational in-frame fusion between the gene of interest and the *LacZ* gene puts *lacZ* expression under the control of the promoter under investigation. The activity of the promoter can then be assayed by measuring the β-galactosidase activity using, for example, ONPG (o-nitrophenyl-pyranogalactose) as a substrate.

galactosyltransferase A GLYCOSYLTRANSFERASE which transfers galactosyl groups from UDPgalactose to a glycan chain (*see* GLYCOPROTEINS AND GLYCOSYLATION). A cell-surface galactosyltransferase has been implicated in CELL–MATRIX INTERACTIONS.

GalNAc N-ACETYL-D-GALACTOSAMINE.

galvanotaxis Active directional migration of cells in response to an electric field. It is called positive if cells migrate towards the cathode and negative if cells migrate towards the anode. The term excludes ELECTROPHORESIS, which is the passive movement of cells in an electric field. The distinction is based on whether the cell's locomotory mechanism (cytoskeleton, attachment to substrate, etc., *see* CELL MOTILITY) is involved (galvanotaxis) or whether cells are moved in the field solely because of their net charge.

gamete In sexually reproducing species, a cell capable of fusing with a homologous cell produced by an individual of the opposite sex (FERTILIZATION) to form a ZYGOTE from which a new individual will develop. In higher animals, mature gametes (sperm in males and eggs in females) develop from PRIMORDIAL GERM CELLS which are set aside early in development and mature in specialized organs (e.g. testes in males, ovaries in females). The development and maturation of haploid gametes (SPERMATOGENESIS and OOGENESIS) from diploid germ cells involve synthesis and assembly of cytoplasmic components (especially in oogenesis) and MEIOSIS. Fusion of haploid gametes generates a diploid zygote.

gametophyte The generation of a plant that bears the gametes. In land plants it is derived from a spore, usually haploid. In PTERIDOPHYTES it is a separate free-living generation bearing male and/or female gametangia (organs in which gametes are produced). In seed plants the female gametophytes are much reduced and are retained on the plant (the SPOROPHYTE generation) while the male gametophytes are derived from the microspores dispersed as pollen to the neighbourhood of the female gametophytes. In algae, gametophytes occur as a separate generation from sporophytes in some species; in others the plant may bear both gametes and spores.

gamma Headwords with the prefix gamma- or γ- may be indexed under the initial letter of the headword itself.

gamma chain (γ chain) May refer to a subunit of any of numerous proteins in which subunits have been arbitrarily designated α, β, γ, δ, etc. Often refers to the embryonic and foetal γ chain of haemoglobins (*see* HAEMOGLOBIN AND ITS DISORDERS).

γδ receptors *See*: T CELL RECEPTORS.

gammaglobulin Originally, that fraction of serum which migrates with gamma mobility on ELECTROPHORESIS. In early studies it was shown that ANTIBODIES were contained in this serum fraction. The antibody purified from this fraction was thus termed gammaglobulin. It was essentially IgG and the terms became synonymous in common use. IMMUNOGLOBULIN used as passive immunotherapy is also referred to as gammaglobulin.

GA$_n$ Gibberellin A$_n$, n is currently 1–90. Trivial system of nomenclature proposed for the GIBBERELLINS. Numbers are allocated to new, fully characterized, naturally occurring compounds which possess the gibbane skeleton as they are discovered. Gibberellic acid is GA$_3$.

MacMillan, J. & Takahashi, N. (1968) *Nature* **217**, 170–171.

gangliosides Complex GLYCOLIPIDS most abundant in the plasma membranes of nerve cells but also found in other cell types in smaller amounts. More than 15 different gangliosides are known. They are glycosphingolipids containing one or more *N*-acetylneuraminic acid (NeuAc) residues (Fig. G6) which are negatively charged. In the widely used trivial nomenclature gangliosides are designated G$_{M1}$, G$_{M2}$, G$_{D1}$, G$_{T1}$ etc. where M, D, and T refer to the number of NeuAc residues (mono-, di-, tri-, etc.) and the subscript numeral is determined by subtracting the number of uncharged sugar residues in the head group (i.e. those excepting NeuAc) from 5. The lipid moiety (ceramide) is synthesized in the endoplasmic reticulum and the oligosaccharide head group is added in the Golgi apparatus. Ganglioside G$_{M1}$ in the membranes of intestinal epithelial cells acts as a receptor for CHOLERA TOXIN, facilitating its entry into the cell. A deficiency of a β-*N*-hexosaminidase, which results in a greatly raised level of ganglioside G$_{M2}$ in the brain, is the cause of TAY-SACHS DISEASE.

GAP GTPase activating protein. *See*: GTP-BINDING PROTEINS.

gap genes A class of SEGMENTATION GENES in *Drosophila* which are characterized by mutant phenotypes that show loss of groups of consecutive segments. These phenotypes and their expression domains show that the gap genes act to subdivide the embryo at the blastoderm stage into broad regions containing several segmental primordia. Their early activity is initiated by the MATERNAL EFFECT genes. The anterior and posterior groups of maternal genes establish a gradient of expression of the gap gene *hunchback* which, in turn, acts to position the expression of the gap genes KRÜPPEL, KNIRPS, and *giant*. The maternal terminal system specifies the expression of the gap genes *huckebein* and *tailless*. Subsequently, these patterns are refined by interactions among the gap genes. Molecular analysis reveals that the products of these genes are all likely to be transcriptional regulators and mostly have zinc-finger DNA-binding domains although the *giant* product is a member of the leucine-zipper family (*see* TRANSCRIPTION FACTORS). The gap genes act to specify the expression patterns of the PAIR-RULE and HOMEOTIC GENES. *See*: DROSOPHILA DEVELOPMENT.

gap junction *See*: CELL JUNCTIONS.

gap penalty The penalty in a scoring algorithm used in alignment of two sequences when one of these has a deletion or the other has an insertion. *See*: ALIGNMENT.

gastrula, gastrulation The stage of development during which the MESODERM begins to form and the embryo becomes three-layered. It is accompanied by massive cell movements that reorganize cells within the embryo, and by the appearance of a characteristic structure around which the mesoderm forms. In amphibians and lower organisms (including echinoderms and protochordates) this structure is the BLASTOPORE. In reptiles, birds, and mammals the structure is the PRIMITIVE STREAK. Gastrulation movements include INGRESSION, INVOLUTION, or INVAGINATION of cells derived from the ECTODERM to give rise to the mesoderm. *See*: AMPHIBIAN DEVELOPMENT; AVIAN DEVELOPMENT; MAMMALIAN DEVELOPMENT.

GAT A polymer of the synthetic tripeptide, Glu-Ala-Tyr. The antibody response to GAT in the BALB/c strain of mouse produces antibodies which bear a common idiotype, owing to the use of a single V$_H$ and V$_\kappa$ gene segment in the assembly of the heavy chain and kappa light chain genes. Antigen recognition in this system is dependent on variation in the D region. Thus, idiotype–anti-idiotype interactions in the GAT system are mediated through germ-line structures (presumed to be COMPLEMENTARITY DETERMINING REGIONS) of the V$_H$-V$_\kappa$ regions; whereas the D regions have a critical role in antigen recognition. This allows a germ-line idiotypic network to exist which can select V$_H$-V$_\kappa$ pairs, regardless of the underlying antigenic specificity.

Fourereau, M. & Schiff, C. (1988) *Immunol. Rev.* **105**, 69–83.

GATA-1 Erythroid-specific TRANSCRIPTION FACTOR. *See*: LOCUS CONTROL REGION.

Gaucher's disease An AUTOSOMAL RECESSIVE condition where the deficiency of the enzyme glucocerebroside β-glucosidase in

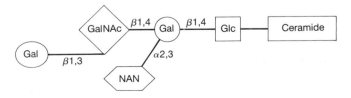

Fig. G6 Ganglioside G$_{M1}$.

macrophage LYSOSOME membranes leads to the abnormal accumulation of glucocerebroside in these cells. The reticuloendothelial system is most affected; liver and spleen become grossly enlarged, bone marrow extensively infiltrated and lungs, skin and nervous system are also affected. It is the most common of the sphingolipidoses, and there are three clinical subtypes differing in degree of neurological involvement. Chronic non-neuropathic Gaucher's disease occurs in the Ashkenazi Jewish population at a frequency of one in 2500 births.

GC box A common DNA sequence element found within the PROMOTERS of eukaryotic genes transcribed by RNA polymerase II. The element has the CONSENSUS SEQUENCE 5′ GGGCGG 3′. This sequence can occur in either orientation and in multiple copies: its presence correlates well with constitutive expression of the associated gene. *In vitro*, the mammalian TRANSCRIPTION FACTOR Sp1 has been shown to bind GC boxes. *See*: EUKARYOTIC GENE EXPRESSION; EUKARYOTIC GENE STRUCTURE.

GDP Guanosine 5′-diphosphate. *See*: GTP-BINDING PROTEINS; NUCLEOSIDES AND NUCLEOTIDES.

gel electrophoresis *See*: ELECTROPHORESIS.

gel filtration chromatography *See*: CHROMATOGRAPHY.

gel mobility shift assay, gel retardation assay *See*: BAND RETARDATION.

gelsolin *See*: ACTIN-BINDING PROTEINS.

geminivirus group Named from Latin *gemini*, twins, from the characteristic double particles. This is the only group of plant viruses to have single-stranded circular DNA as genomes; the DNA is contained in particles which are considered to be two isometric (*t* = 1) particles adhering at a 5′ axis. There are three subgroups. Members of subgroups A (type member maize streak virus, MSV) and C (type member beet curly top virus, BCTV) are transmitted by leafhoppers and have genomes of a single DNA species (2687–2749 and 2993 nucleotides respectively); those of subgroup A infect monocots and of subgroup C infect dicots. Members of subgroup B (type member African cassava mosaic virus, ACMV) are transmitted to dicots by whitefly and have two DNA species (2588–2779 and 2508–2724 nucleotides). Their genome organizations are shown in Fig. G7. MSV encodes four genes, two in the virion sense (V) and two in the complementary sense (C); V2 encodes the coat protein. The two genomic DNAs of ACMV have a region of sequence homology, the common region. ACMV DNA A encodes four genes, V1 being that for the coat protein. DNA B encodes two genes both of which are required for cell-to-cell movement. BCTV encodes four genes, V1 being for the coat protein. The functions of the other genes are unknown. Geminiviruses replicate in the nucleus via a complementary strand of DNA. *See also*: PLANT VIRUSES.

Harrison, B.D. (1985) *Annu. Rev. Phytopathol.* **23**, 55–82.

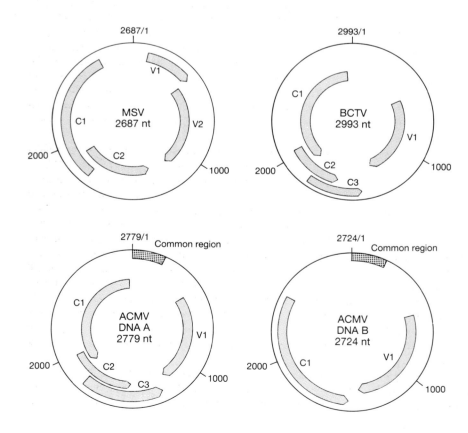

Fig. G7 *Genome organizations of geminiviruses. The circles represent the single-stranded DNA. The curved boxes are the* OPEN READING FRAMES *considered to give proteins, those in the virion sense labelled V and in the complementary sense C. ACMV, African cassava mosaic virus; BCTV, beet curly top virus; MSV, maize streak virus.*

Gene

THE gene is the basic unit of heredity and carries the genetic information for a given PROTEIN and/or RNA molecule. In biochemical terms a gene consists of a contiguous stretch of deoxyribonucleic acid (DNA) which in turn forms part of a much larger genetic unit, the CHROMOSOME. Even the simplest unicellular organisms (e.g. bacteria) require several thousand different genes and their respective gene products, while complex multicellular organisms such as ourselves may require as many as 50 000 different genes. An understanding of what genes are and how they store genetic information and then transmit that information when required by the cell has been researched for almost a century, but it is only in the last decade that we have come to appreciate the complexity of structures and mechanisms that contribute to what we now understand to be a gene.

The discovery of the gene

The existence of genes was first suggested by the work of Gregor Mendel [1] who demonstrated that the inheritance of single genetic traits could be explained in terms of particulate hereditary factors (see MENDELIAN INHERITANCE). These factors were faithfully passed down from one generation to the next and could exist in several different forms (ALLELES) within the population. Furthermore, Mendel demonstrated, in a series of elegant experiments in which he followed the inheritance of seven different traits in the garden pea, that some alleles of a given gene can be DOMINANT to other RECESSIVE alleles of the same gene. Most organisms, he concluded, carried two copies of each 'factor' in their cells except in gametes which only had one copy. This prediction led to the establishment of the Mendelian Laws of Inheritance.

The term 'gene' was first used in 1911 by Danish geneticist Wilhelm Johannsen as a convenient term for Mendel's particulate factors. A realization that genes were carried on chromosomes began to emerge at about the same time, principally from the studies of Thomas Hunt Morgan. In 1910 he provided the first definitive evidence for the so-called 'Chromosomal Theory of Inheritance', a theory first proposed in 1902 by Walter Sutton and Theodor Boveri. In a series of genetic experiments with the fruit fly *Drosophila melanogaster* Morgan convincingly demonstrated that the inheritance of the genes for eye colour, miniature wing and yellow body correlated with the inheritance of a specific chromosome, the X-chromosome. Because the X-chromosome carries the genes that determine sexual type in *D. melanogaster*, these genes were said to show sex-linked inheritance.

Over the next 20 years the chromosomal theory of inheritance became firmly established doctrine with studies on the inheritance of linked genes leading to a first description of the process of genetic RECOMBINATION between homologous chromosomes. This process results in the generation of new combinations of alleles on each chromosome during the formation of gametes and involves the reciprocal exchange of pieces of homologous chromosomes. Recombination ensures genetic variation between offspring of the same parents. (A direct physical demonstration of recombination between homologous chromosomes was provided in 1931 by Barbara McClintock in her cytogenetic studies with maize (*Zea mays*) chromosomes.) These crucial discoveries led Morgan and a colleague, A.H. Sturtevant, to propose that the frequency of recombination (or CROSSING-OVER) between two linked but physically separate genes on the same chromosome could be used as an estimate of their physical distance apart, and this in turn led to the development of GENE MAPPING techniques. The development of such techniques allowed for the first time the construction of LINKAGE MAPS which provide a linear representation of the order of genes on a specific chromosome. By the early 1930s it had become the accepted dogma that genes carried genetic information that determined a specific PHENOTYPE, and that such genes were carried as linear arrays on chromosomes which in turn were located in the nucleus. It was not until 1944 that the chemical nature of genes was confirmed.

The chemical nature of the gene

As soon as Mendel's ideas had been accepted, an earnest debate about the chemical nature of the genetic material began, with protein and DNA receiving equal favour. DNA had been first described by Friedrich Miescher in 1869 who identified it as a phosphorus-containing compound (which he called nuclein) that was present in nuclei. It was not until 1944 that direct experimental proof was obtained that DNA was the material of which genes were built. This proof came from Oswald Avery and his co-workers who demonstrated that DNA, but not protein or RNA, could be used to genetically transform an avirulent strain of *Streptococcus pneumoniae* to a virulent form [2]. The DNA used in such a genetic TRANSFORMATION experiment had to come from a virulent strain and the genetic transformation event that it mediated could be inherited through subsequent generations. Any doubts that still lingered about the chemical nature of the genetic material were finally laid to rest by the experiments of A.D. Hershey and Martha Chase who in 1952 showed by use of radioactive tracers that phage DNA and not protein was required for the BACTERIOPHAGE T2 to establish a lytic infection of *Escherichia coli* [3].

The three-dimensional structure of DNA and how that related to its role as the genetic material was established in 1953 by Francis Crick and James Watson in collaboration with Maurice Wilkins and Rosalind Franklin [4]. They found that DNA existed as a double helix consisting of two polynucleotide chains held together by hydrogen bonds (see DNA: NUCLEIC ACID STRUCTURE). The key components of the DNA molecule with respect to its role as the carrier of genetic information were the four BASES — adenine (A), thymine (T), guanine (G), and cytosine (C) — and it was the sequence of these bases along one of the two polynucleotide chains (the CODING STRAND) that carried the information in the form of a GENETIC CODE. Subsequently the three-base nature of the genetic code was elucidated and the amino acid assignment of each of the 64 CODONS inferred from IN VITRO TRANSLATION studies.

The two complementary strands of the double helix are held together by specific hydrogen bonding between A-T and G-C. The complementary nature of the two strands ensures that the information stored within a gene can be faithfully transmitted to subsequent generations by the process of DNA REPLICATION.

Not all genes are made of DNA; some animal and plant viruses and some bacteriophages have genetic systems based on RNA rather than DNA (*see* ANIMAL VIRUSES; BACTERIOPHAGES; PLANT VIRUSES). The ability of RNA to transmit genetic information was demonstrated in studies by A. Gierer and G. Schramm with the plant virus tobacco mosaic virus (TMV) in which they showed that the RNA component of the virus alone could induce the synthesis of new virus particles in infected leaves.

Genes encode RNA and protein

Genes are the 'blueprints' for all RNA and protein molecules found in the cell. Some genes encode RNA molecules as the final gene product (genes for RIBOSOMAL RNA, TRANSFER RNA, and other small RNAs) whereas others encode polypeptide chains, which are synthesized by way of the intermediate MESSENGER RNA (*see* GENETIC CODE; PROTEIN SYNTHESIS; RNA; TRANSCRIPTION). These blueprints can be modified by MUTATION which alters the genetic information encoded by the gene through alteration of the base sequence of the DNA. The consequences of a mutation on the activity of the gene product can vary; it may have no affect on its activity (i.e. be silent), or it may completely inactivate it. A further possibility is that the activity of the gene product may be modified in some way. Depending on the consequence of the mutation on the activity of the gene product such a gene mutation may result in an alteration in the phenotype of the organism.

Studies by Archibald Garrod at the beginning of this century first began to suggest that genes encode the information for specific polypeptides, but it was not until the work of George Beadle and E.L. Tatum in the 1940s that the 'One Gene–One Enzyme' hypothesis was experimentally established [5]. Using X-ray-induced mutants of the fungus NEUROSPORA CRASSA they identified a series of AUXOTROPHIC mutants with defects that could be corrected by the addition of a single metabolite, thereby allowing them to pinpoint a specific biochemical defect. Subsequent genetic analysis of these mutants revealed that in each case they were due to single gene defects. This finding led to the supposition that a single gene is responsible for the synthesis of a given enzyme and, if that gene is defective, so is the enzyme. Recessive alleles of a given gene thus represent mutations which completely inactivate the encoded enzyme since when present together with the normal WILD-TYPE allele in a HETEROZYGOUS diploid the enzyme deficiency is restored by the dominant wild-type allele. The hypothesis was later generalized and restated as 'one-gene–one-polypeptide chain'.

Proof that the 'one-gene–one polypeptide chain' hypothesis also applied to humans came from Vernon Ingram in 1957 when he showed that the single-gene trait SICKLE CELL ANAEMIA can be accounted for by a change in the amino acid composition of the β-globin subunit of the haemoglobin molecule (*see* HAEMOGLOBIN AND ITS DISORDERS).

Gene expression

The genetic information stored in a gene is downloaded in two sequential steps: transcription, whereby a linear portion of the gene is copied into a single-stranded RNA molecule, and translation (*see* PROTEIN SYNTHESIS). At the ribosomes the coded genetic message in mRNA is translated into an equivalent string of covalently linked amino acids — a polypeptide chain [6]. Once liberated from the ribosome, the polypeptide chain may be subject to POST-TRANSLATIONAL MODIFICATION (e.g. phosphorylation).

The flow of genetic information from DNA to RNA to protein is highly regulated in all cell types and also, in some situations, tightly coordinated. Such control of gene expression ensures that the cell does not expend unnecessary energy on synthesizing a gene product until that product is required by the cell. Thus the expression of a gene may be restricted to some point during the development of an organism, or it may only need to be expressed in certain cell types or in response to some external stimulus, or it may be required in all cell types at all times (i.e. it is constitutively expressed).

There also exist a family of gene-like sequences which for a variety of reasons never appear to be expressed, at least at the level of a functional gene product; such genes are referred to as PSEUDOGENES and by and large are highly mutated forms of a gene for which a fully functional counterpart also exists (e.g. the globin pseudogenes) [7].

Organization of the gene

The first attempts to probe the detailed structural organization of the gene were the work of Seymour Benzer in the late 1950s. Using an extension of gene mapping methods he developed a means of constructing intragenic maps utilizing different alleles of the *r*II gene of the bacteriophage T4 [8]. This provided new insights into how alleles are generated and provided evidence that the smallest unit of mutation and recombination was the base pair in DNA. The first attempt to compare such a 'fine-structure' genetic map to a physical one was made by Charles Yanofsky in the late 1960s using the tryptophan synthetase gene of *E. coli* [9]. He was able to demonstrate that the order of the mutation sites in the *trpA* gene was colinear with the order of the amino acid substitutions in the tryptophan synthetase polypeptides encoded by the mutants and could therefore approximately define the genetic boundaries of the *trpA* gene.

It was not until the mid 1970s, however, with the development of first DNA CLONING technology, and then DNA SEQUENCING methodologies that we were at last able to unravel the molecular organization of the gene to the level of the single base pair. Figure G8 highlights the basic structural features of a typical gene from both prokaryotic and eukaryotic cells. In each case the gene can be subdivided into two basic regions, the coding region and the noncoding region. The coding region contains the genetic code that is read by the translational machinery in the cytoplasm and is defined by the translation INITIATION CODON (usually AUG) and one or other of the three translation TERMINATION CODONS (UAA, UAG or UGA). The noncoding region represents those DNA sequences which are required for the expression of the genetic information but which are not translated into a polypeptide sequence. One exception to this later generalization is the INTRON sequences which are noncoding sequences found embed-

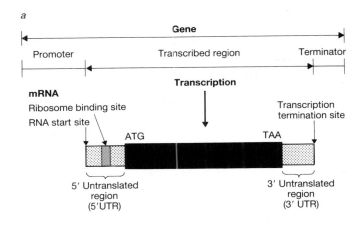

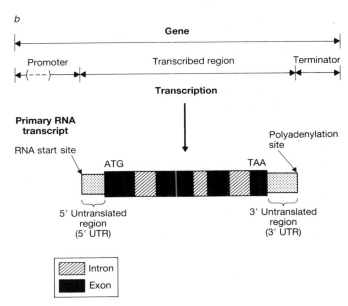

Fig. G8 Gene structure. *a*, Schematic diagram of the structure of the coding region and associated translational signals in a typical prokaryotic protein-coding gene. The coding region is continuous and is flanked by transcribed, but not translated regions, bearing a distinct ribosome-binding site and signals for the start and termination of translation. *b*, The corresponding region in a typical eukaryotic protein-coding gene. The coding region is interrupted by blocks of noncoding sequence (introns). These are subsequently spliced out of the primary transcript to give a functional mRNA.

ded in the coding sequences of the vast majority of eukaryotic (but not prokaryotic) genes (*see* EUKARYOTIC GENE STRUCTURE) [10]. These sequences appear to have no role in gene function or expression *per se* but rather are discarded during the processing of the primary RNA transcript (*see* RNA SPLICING). The PROMOTER region is required to promote the enzyme RNA POLYMERASE to bind to the gene at the correct position with respect to the region that needs to be transcribed, that is the TRANSCRIPTION UNIT. The prokaryotic promoter is relatively simple to define in physical terms whereas a typical eukaryotic promoter is generally much larger, with DNA sequences positioned many thousands of base pairs away still being able to have profound effects on the rate at which a gene may be transcribed (e.g. ENHANCER elements). The TERMINATOR region serves to terminate the migration of the RNA polymerase molecule once it has traversed the gene's transcriptional unit.

Genes in eukaryotes are MONOCISTRONIC in nature, that is they encode a single gene product, whereas bacterial genes can be part of a larger transcription unit, that is they can be POLYCISTRONIC in which case two or more gene products can be encoded by a single mRNA species (*see* LAC OPERON). In some BACTERIO-PHAGES, for example ΦX174, there are also examples of OVER-LAPPING GENES where two different gene products can be generated from the transcription and translation of the same region of DNA by utilization of different but overlapping READING FRAMES.

Not all genes ultimately encode a polypeptide chain as in some cases the RNA transcript can be the final gene product. This is the case for those genes which encode rRNAs, tRNAs and small nuclear RNAs. In eukaryotic cells these genes are transcribed by different RNA polymerases (RNA polymerase I and III) from that which transcribes protein-encoding genes (RNA polymerase II, *see* TRANSCRIPTION). Furthermore, the relative organization of the 'coding' and 'noncoding' regions differs; for example there are important elements of the transcriptional promoter of tRNA genes that lie within the 'coding' region.

Genes that encode polypeptides are generally present in one or two copies per HAPLOID genome, that is they are SINGLE-COPY GENES, whereas genes encoding tRNA, rRNA and histones are often present in multiple copies to ensure that sufficient gene product is available to the cell. Protein-encoding genes can also be subdivided into two classes based on the type of gene product; STRUCTURAL GENES, which encode the basic housekeeping proteins — metabolic enzymes, transport proteins, components of the cytoskeleton etc. — and REGULATORY GENES whose products regulate the expression of other genes, usually at the level of transcription.

Some genes can also be assigned to specific MULTIGENE FAMI-LIES whose members all encode polypeptides or RNAs with similar structures and/or biological functions (*see e.g.* HAEMOGLO-BINS; IMMUNOGLOBULIN SUPERFAMILY). Such multigene families may have evolved from a single ancestral gene by a process of GENE DUPLICATION and subsequent divergence by mutation.

The gene therefore represents the fundamental unit of inheritance in all living organisms and each individual has a unique genetic make-up resulting from the various permutations of alleles generated by mutation and recombination.

M.F. TUITE

See also: BACTERIAL GENE ORGANIZATION; BACTERIAL GENE EXPRESSION; EUKARYOTIC GENE EXPRESSION; DROSOPHILA.

1 Olby, R. (1985) *Origins of Mendelism*, 2nd edn (University of Chicago Press, London).
2 McCarty, M. (1985) *The Transforming Principle: Discovering that Genes are Made of DNA* (Norton, London).
3 Hershey, A.D. & Chase, M. (1952) Independent functions of viral proteins and nucleic acid in growth of bacteriophage. *J. Gen. Physiol.* **36**, 39–56.
4 Watson, J.D. & Crick, F.H.C. (1953) Genetic implications of the structure of deoxyribose nucleic acid. *Nature* **171**, 964.
5 Beadle, G.W. & Tatum, E.L. (1941) Genetic control of biochemical reactions in *Neurospora. Proc. Natl. Acad. Sci. USA* **27**, 499–506.

6 Watson, J.D. et al. (1987) *Molecular Biology of the Gene*, 4th edn (Benjamin Cummings, Menlo Park, CA).
7 Little, P.F.R. (1982) Globin pseudogenes. *Cell* **28**, 683–684.
8 Benzer, S. (1955) Fine structure of a genetic region in bacteriophage. *Proc. Natl. Acad. Sci. USA* **41**, 344–354.
9 Yanofsky, C. et al. (1967) The complete amino acid sequence of the tryptophan synthetase A protein (α-subunit) and its collinear relationship with the genetic map of the A gene. *Proc. Natl. Acad. Sci. USA* **57**, 296–298.
10 Chambon, P. (1981) Split genes. *Sci. Amer.* **244**, 48–59.

gene amplification The process whereby the COPY NUMBER of a particular gene is increased.

1 *In vitro*, successive rounds of the POLYMERASE CHAIN REACTION in the presence of appropriate primers will result in amplification of particular genes.

2 Insertion of a gene into a high copy number PLASMID will result in that gene becoming correspondingly abundant *in vivo*.

3 Certain environmental pressures can induce an adaptive response in eukaryotic cells in which appropriate protective genes are significantly amplified. These may be carried either chromosomally or extrachromosomally (*see* DOUBLE MINUTE CHROMOSOMES). For example, high doses of the drug METHOTREXATE result in amplification of the DIHYDROFOLATE REDUCTASE (DHFR) genes (Fig. G9), whose gene product is the target for methotrexate action.

4 Programmed gene amplification (e.g. of the rRNA genes in OOGENESIS in *Xenopus*) is sometimes utilized by eukaryotes as a means of producing higher levels of a gene product at particular stages in development. *See also*: CHORION GENE AMPLIFICATION; RRNA GENE AMPLIFICATION.

5 GENE DUPLICATION and amplification occurring over evolutionary time scales generates clusters of related genes in the genome (*see* MOLECULAR EVOLUTION).

gene cloning The process of isolating a particular GENE, usually from a DNA LIBRARY. The appropriate cloned DNA is detected and isolated in various ways including: HYBRIDIZATION with a labelled complementary nucleic acid PROBE; IN VITRO TRANSLATION and detection of the gene product by a labelled ANTIBODY; COMPLEMENTATION by the cloned DNA of a genetic defect in a host cell or organism; expression in a system such as the XENOPUS OOCYTE EXPRESSION SYSTEM. *See*: DNA CLONING; GENETIC ENGINEERING; HUMAN GENE MAPPING.

gene cluster In eukaryotes, related genes produced by GENE DUPLICATION and divergence are often found clustered together on a chromosome. Such clusters can range in complexity from two adjacent related genes to a tandem array of several hundred identical genes. Well-studied gene clusters in mammals include the GLOBIN GENES, the MAJOR HISTOCOMPATIBILITY COMPLEX (MHC), and the *Hox* genes (*see* HOMEOBOX GENES AND HOMEODOMAIN PROTEINS). The divergent protein products encoded by related members of a gene cluster may have acquired slightly different functions (as in the MHC) or be used during different stages of an organism's development (as in the globin genes) (*see* HAEMOGLOBIN AND ITS DISORDERS).

Evolution in bacteria has favoured the development of clusters of structurally unrelated genes (operons) encoding the enzymes of a particular metabolic pathway (*see*; BACTERIAL GENE EXPRESSION; BACTERIAL GENE ORGANIZATION). The genes in an operon are cotranscribed and thus expressed in a coordinated manner.

gene conversion (1) A DNA REPAIR process occurring occasionally at RECOMBINATION during the first division of MEIOSIS in heterozygotes, and which is detected by an unbalanced ratio of recombinant progeny (*see* MENDELIAN INHERITANCE). During recombination, stretches of HETERODUPLEX DNA are formed as a consequence of pairing and recombination between nonidentical ALLELES. Gene conversion results when the cell's DNA repair system removes one of the DNA strands from the mismatched region over a length of a few thousand base pairs and corrects it against the remaining strand (Fig. G10). As a consequence, unequal postmeiotic segregation of alleles is observed. Gene conversion is most conveniently studied in fungi, whose meiotic spore products can be easily isolated for genetic analysis (*see* TETRAD ANALYSIS).

(2) A gene conversion process can also occur during somatic GENE

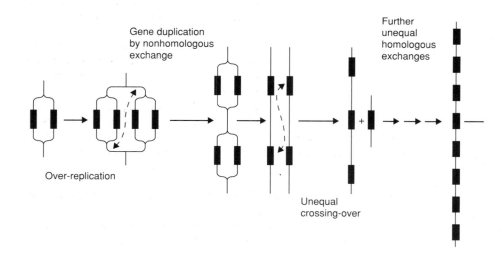

Gene duplication by nonhomologous exchange

Over-replication

Further unequal homologous exchanges

Unequal crossing-over

Fig. G9 Gene amplification as a result of unequal crossing-over after an initial gene duplication event. Possible mechanism for amplification of the *dhfr* genes.

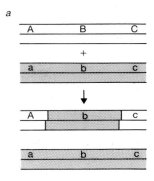

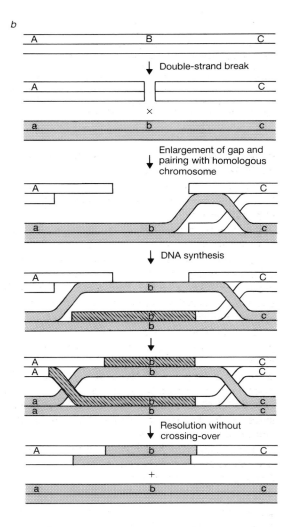

Fig. G10 Schematic (*a*) and one possible mechanism (*b*) of gene conversion during meiosis.

REARRANGEMENT. In some species, such as the chicken and rabbit, ANTIBODY diversity (*see* GENERATION OF DIVERSITY) is achieved by a sequential gene conversion process. In the chicken a number of PSEUDOGENES 5′ to the single functional V gene function as donors of short sequences which are transferred into the V gene by gene conversion during the development of B cells in the bursa.

gene disruption Disruption of the coding sequence of a gene and destruction of its function *in vivo* can be achieved by integrating a second gene into it. The latter — the insertion gene — usually has a selectable phenotype such as antibiotic resistance or the ability to complement a host AUXOTROPHIC requirement. For genes subcloned into appropriate plasmid or phage VECTORS disruption can be accomplished *in vitro*, and the disrupted gene then reintroduced into a host cell. Chromosomal genes, on the other hand, are selectively disrupted *in vivo* by ligating FLANKING SEQUENCES homologous to the target gene on to the insertion gene which then becomes integrated into the target by HOMOLO-

GOUS RECOMBINATION. The latter technique is used extensively in the genetic analysis of SACCHAROMYCES CEREVISIAE. Specific gene knockout has also been achieved in transgenic animals (*see* TRANSGENIC TECHNOLOGIES). *See also*: GENE TARGETING.

gene dosage The number of copies of a given gene present in the nucleus. For most genes in diploid organisms there are two copies, one on each homologous chromosome. Carriers of RECESSIVE MUTATIONS in which the recessive ALLELE is a loss-of-function mutation will have half the normal gene dosage. Genes located on the X-chromosome are present in two copies in mammalian females (XX) but only one in males (XY) and a specific mechanism of dosage compensation exists to overcome this imbalance (*see* X-CHROMOSOME INACTIVATION).

gene duplication The generation of an extra copy of a gene in the genome. Duplication events can theoretically occur as a result of several mechanisms: random breakage and reunion; UNEQUAL CROSSING-OVER at MEIOSIS between homologous repetitive sequences in the genome; retrotransposition (*see* RETROTRANSPO-SON); and over-replication (*see* GENE AMPLIFICATION). Once a gene has been duplicated, one of the copies can diverge over evolutionary time and assume a new but related function, and gene duplication is thought to be one of the main mechanisms by which functional diversity is generated in the genome. Examples in the human genome of closely related genes with similar functions that are presumed to have evolved by repeated duplications, are the genes for the colour vision pigments on the X-chromosome, the members of the α-globin and β-globin gene clusters on chromosome 16 and chromosome 11 respectively (in humans) (*see* HAEMOGLOBIN AND ITS DISORDERS), and the genes of the MAJOR HISTOCOMPATIBILITY COMPLEX (MHC) on chromosome 6. *See also*: MOLECULAR EVOLUTION; MOLECULAR EVOLUTION: SEQUENCES AND STRUCTURES.

gene expression The conversion of genetic information encoded in a gene into RNA and protein, by TRANSCRIPTION of a gene into RNA and (in the case of protein-coding genes) the subsequent translation of mRNA to produce a protein. *See*: BACTERIAL GENE EXPRESSION; EUKARYOTIC GENE EXPRESSION; GENE; PROTEIN SYNTHESIS; TRANSCRIPTION FACTORS.

gene family Genes of related structure and maybe function. *See*: MULTIGENE FAMILY; SUPERFAMILY.

gene-for-gene hypothesis *See*: PLANT PATHOLOGY.

gene frequency For any gene with multiple ALLELES, the probability that a given allele will be present at the gene locus in any individual within a particular defined population. For a gene involved in disease, the frequency of the mutant allele is expressed as a proportion of the frequency of the normal allele.

gene gun *See*: MICROBALLISTICS.

gene interaction The combined effect of two or more genes on a PHENOTYPE. There are three types of gene interaction:
1 One gene encodes a product that has a direct regulatory effect on another gene, as in the OPERONS of prokaryotes (*see* BACTERIAL GENE EXPRESSION).
2 A number of genes contribute towards a complex heritable phenotype such as height or IQ (such traits are termed multifactorial or polygenic).
3 Epistatic interactions, in which the actions of the genes are complementary but independent of each other.
The genes encoding the enzymes of a metabolic pathway can be epistatic to each other. If one gene ceases to function, the pathway becomes blocked and the products of the remaining genes (the enzymes) lose their substrates and may become downregulated as a result.

gene knockout The targeted disruption of a gene *in vivo* with complete loss of its function. It is usually achieved by producing a transgenic animal in which the gene in question has been rendered nonfunctional by an insertion targeted to the gene by HOMOLOGOUS RECOMBINATION. *See also*: GENE DISRUPTION.

gene library *See*: DNA LIBRARY.

gene locus The position of a GENE on a chromosome. Loci can be defined physically (e.g. by chromosome band or distance in base pairs from another marker locus) and genetically (distance in centimorgans (cM) from another marker locus). The locus is occupied by one of the ALLELES of that gene. *See also*: DNA TYPING; HUMAN GENE MAPPING; POLYMORPHISM; MAP DISTANCE.

gene mapping The construction of a map of the locations of genetic loci in relation to each other (a genetic map) or in absolute distances from each other (a physical map). Genetic maps can be obtained by measurement of the frequency of recombination between pairs of loci (*see*: LINKAGE MAP; MAP DISTANCE; MAP UNIT). Physical maps make use of a variety of techniques to determine the exact position of loci on a chromosome. *See*: HUMAN GENE MAPPING for examples.

gene phylogeny *See*: HOMOLOGY.

gene pool The total genetic information contained within a given population. The gene pool is not static: GENE FREQUENCIES within it can change under the influence of selective pressures, and as a result of interbreeding with other populations.

Gene rearrangement

FUNCTIONAL directed rearrangement of DNA occurs in somatic cells of eukaryotes in a small number of cases. Directed gene rearrangement is involved in the assembly of functional immunoglobulin and T cell receptor genes in vertebrates, which is covered here, the mating-type switch in yeast (*see* YEAST MATING-TYPE LOCUS), and ANTIGENIC VARIATION IN TRYPANOSOMES. Directed genomic rearrangements are also known in prokaryotes (*see* BACTERIAL PATHOGENICITY).

Gene rearrangements in the loci specifying the antigen receptors of B and T cells are of two main types: V-(D)-J recombination and immunoglobulin heavy chain class switching. In addition, other rearrangements at these loci include the deletion of the C_κ region, which is often found in B cells expressing λ light chains (*see* KAPPA DELETING ELEMENT; RS RECOMBINATION), and the deletion of the δ locus in T cells expressing αβ T cell receptors. Aberrant gene rearrangements involving the antigen receptor loci are often implicated in the generation of the chromosomal translocations which are found in many lymphoid malignancies.

V-(D)-J recombination

The genes encoding the IMMUNOGLOBULINS and the T CELL RECEPTORS — the antigen receptors in the immune system — are unique in that in both cases a functional gene must be assembled from a number of discrete noncontiguous gene segments (*see* IMMUNOGLOBULIN GENES; T CELL RECEPTOR GENES). This process of directed somatic gene rearrangement occurs early in the

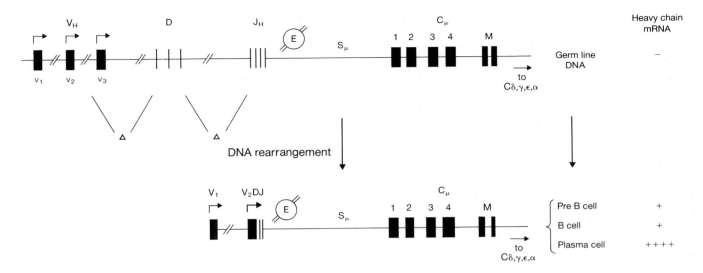

Fig. G11 A simplified view of the organization of the mouse immunoglobulin heavy chain locus and its rearrangement. The V_H segments are located ~100 kb upstream of a cluster of four J_H segments. The number of V_H segments varies greatly among different strains and is estimated to be 200–2000. About 12 D segments lie upstream of J_H with the bulk of them lying downstream of the cluster of V_H segments. The C region gene segments (only C_μ is illustrated here) begin ~7 kb downstream of the J_H segments. E, Heavy chain enhancer; S_μ, switch region. The arrows above the V gene segments indicate the direction of transcription.

differentiation of B cells and T cells (*see* B CELL DEVELOPMENT; T CELL DEVELOPMENT).

Immunoglobulins consist of a complex of two identical heavy chains and two identical light chains (of either the κ or λ type). Each polypeptide chain consists of an N-terminal variable (V) domain, which contributes to the antigen-binding site, and a C-terminal constant (C) domain, which determines antibody class and effector functions (*see* ANTIBODIES). The T cell receptor (TCR) is a heterodimer (either αβ or γδ) in which each chain contains immunoglobulin-related variable and constant regions (*see* IMMUNOGLOBULIN SUPERFAMILY; T CELL RECEPTORS).

For both immunoglobulins and TCRs, the sequence encoding the V regions is assembled from several gene segments. The immunoglobulin heavy chain V regions are encoded by variable (V), diversity (D) and joining (J) gene segments, light chain variable regions by corresponding V_L and J_L gene segments. An analogous set of V, D and J gene segments encodes the V regions of the TCR β and γ chains whereas the α and δ chains have only V and J segments. In order to create functional EXONS encoding the variable region, and therefore functional antibody or T cell receptor genes, the requisite gene segments must be brought together by a series of genomic rearrangements, referred to as V-(D)-J recombination.

V-(D)-J recombination is responsible for generating the highly diverse array of immunoglobulins and T cell receptors that is characteristic of the vertebrate immune system (*see* GENERATION OF DIVERSITY). The antigen recognition site of an antibody is formed by the interaction of three HYPERVARIABLE REGIONS, the so-called COMPLEMENTARITY DETERMINING REGIONS (CDRs). The variability of CDR1 and CDR2 arises mainly from the sequence heterogeneity of the multiple germ-line V genes. Further diversity in the antigen-binding site is generated in CDR3 by sequences that are formed by the juxtaposition of V, D and J segments and by nucleotide heterogeneity due to imprecise joining of those segments (see below). In this recombination process any J can usually pair with any D, which in turn can recombine with any V, thereby generating huge diversity.

Organization of antigen receptor genes

At each locus, a number of V, J (and D) segments are located close to exons encoding the C region (see Fig. G11, and IMMUNOGLOBULIN GENES; T CELL RECEPTOR GENES for further examples). The organization of the unrearranged loci is termed the germ-line configuration.

Sequence of rearrangement

In the immunoglobulin heavy chain locus (Fig. G11) and the TCRβ and TCRγ loci, the first rearrangement in the assembly of a VDJ exon is the joining of a D and a J segment. The resulting DJ sequence is then joined to one of the V genes. In the immunoglobulin light chain loci and the TCRα and TCRδ loci, a V region exon is assembled by the joining of a J segment to a V gene. Selection of gene segments for recombination is in general random, although in certain cases particular segments are used preferentially (*see* T CELL RECEPTORS). Rearrangement is preceded by low level transcription from the region to be rearranged and control of the rearrangement process is most probably by cell-type-specific ENHANCER, SILENCER, and PROMOTER elements found in the immunoglobulin and TCR loci.

In some rearranging loci (e.g. λ and TCRβ) the chromosomal order of the gene segments restricts the potential for recombination and certain VJ pairs, for example, are found preferentially.

The complete gene is expressed as a primary transcript including the rearranged V(D)J region and a C gene; the sequence intervening between V(D)J and C is then spliced out (*see* RNA SPLICING) to produce a messenger RNA.

Recombination signal sequences

Recombination takes place between specific sequences flanking each gene segment — recombination signal sequences (Fig. G12). The coding sequences of all germ-line V and D gene segments are immediately followed by a conserved palindromic heptamer (consensus 5′-CACAGTG-3′) and an AT-rich nonamer (consensus 5′-ACAAAAACC-3′) separated by a short, apparently nonconserved, spacer sequence. In the same way, all germ-line D and J gene segments are immediately preceded by a consensus nonamer and a consensus heptamer separated by a short spacer, with the heptamer adjoining the coding region. In any individual recombination sequence, the spacer has a length of either 12 base pairs (bp) or 23 ± 1 bp (the equivalent of one or two turns of the DNA helix, respectively). Thus, for example, the recombination signals flanking D_H segments have 12-bp spacers whereas V_H segments are followed by 23 ± 1-bp spacers and J_H segments preceded by 23 ± 1 bp spacers.

Functional rearrangements occur only between two gene segments each flanked by recombination signals containing in one case a 12-bp spacer and in the other a 23-bp spacer. This is known as the 12–23 rule; it results in an ordered and sequential assembly of V region gene segments and prevents aberrant joining that would lead to a nonfunctional rearrangement.

Imprecise joining

The joining mechanism of gene segments is imprecise. Many joins contain nucleotides at the junction (N sequences) that are not derived from known germ-line segments (*see* N-REGION DIVERSITY). Other rearrangements lead to the loss of nucleotides in the junctional region. This imprecision appears to generate significant diversity in the resulting V regions and thus in the antigen-binding capability of T cell receptor and immunoglobulin polypeptides (*see* GENERATION OF DIVERSITY). The expression of TERMINAL DEOXYNUCLEOTIDYLTRANSFERASE (TdT) activity in a cell correlates strongly with the ability to insert N-region sequences in the joins. This, and the fact that the nucleotide additions are usually GC rich, led to the conclusion that TdT is involved in the formation of the CODING JOINT. Other nucleotide additions are better explained by the usage of two instead of one D segments.

Productive and nonproductive rearrangements

Not all attempts at rearrangement produce a sequence that can be transcribed and translated to produce a functional polypeptide. Owing to the imprecise nature of the joining event and insertion of random base pairs (*see* N-REGION DIVERSITY), only one in three joins, on average, will have maintained the correct reading frame and constitutes a functional rearrangement. Thus in a given locus where a functional rearrangement is found on one allele, the second allele often displays a nonproductive or incomplete arrangement. For several of the immunoglobulin and T cell receptor loci, there is evidence that the locus may undergo multiple rounds of rearrangement by virtue of the large number of gene segments that remain available for use, until a productive rearrangement occurs.

The recombination mechanism

V-(D)-J recombination requires a variety of highly specific and regulated enzymatic activities (*see* RECOMBINASE). The DNA is cut at specific recognition sequences by an endonucleolytic activity. Then the resulting ends are modified by exonucleolytic, base addition and polymerase activities. Finally a ligase activity joins the free ends together.

Although the actual recombinase biochemical machinery seems identical in B and T cells, immunoglobulin genes are only rearranged in B cells, whereas TCR genes are only assembled in T cells. There are strong indications that, at a specific stage of lymphoid development, substrate-specific targeting factors and nonspecific recombination components work together as the V-(D)-J recombinase. This is consistent with the discovery that mice suffering from a defect causing SEVERE COMBINED IMMUNE DEFICIENCY show strongly impaired V-(D)-J rearrangement in B and T cells, although the *scid* mutation can be traced back to a defect in a ubiquitous DNA REPAIR mechanism. Flanking regulatory elements such as transcriptional ENHANCERS and possibly other as yet unidentified elements may contribute to the regulation of the rearrangement by modulating the accessibility of the gene segments to the recombinase.

Recombination-activating genes

The coexpression of the recombination-activating genes *RAG-1* and *RAG-2* is necessary and sufficient to generate V-(D)-J recombinase activity in fibroblasts transfected with artificial DNA rearrangement substrates. It is not yet clear whether *RAG-1* and

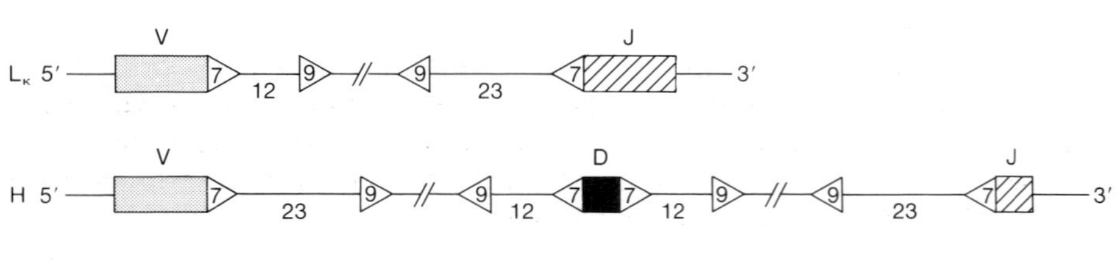

Fig. G12 Arrangement of recombination signal sequences. 7, 9, Conserved heptamer and nonamer sequences, respectively; 12, 23, nonconserved spacer sequences.

RAG-2 participate directly in the recombination process or whether they simply activate other components of the V-(D)-J recombinase.

Rearrangement models

Most of the rearrangement events can be described by the looping-out deletion model. In this case the intervening DNA including the two recombination signals is deleted from the chromosome (Fig. G13*a*). It is predicted from the looping-out deletion model that the deleted DNA can form circular molecules which contain the recombination signal sequences; these have indeed been detected. Other V-(D)-J recombinations are better explained by an inversion model (Fig. G13*b*). In contrast to looping out, inversion results in the retention of both the participating heptamer/nonamer signal sequences as a back-to-back joint as well as of the intervening DNA in the chromosome. By these criteria (as well as from studies of the rearrangement of artificial DNA substrates) joining by inversion has been shown to occur and is frequently detected in the immunoglobulin κ light chain and TCR loci. The looping out deletion model and the inversion model adequately explain the rearrangements of antigen receptor genes and are supported by most of the experimental data.

Class switching

In addition to V-(D)-J RECOMBINATION a second and distinct gene rearrangement event occurs in B cells, but not in T cells. Cells of the B-cell lineage that express an immunoglobulin heavy chain isotype other than μ or δ (*see* ANTIBODY; B CELL DEVELOPMENT) have undergone a class switch involving DNA recombination. This switch enables the cells to change the constant part of the immunoglobulin heavy chain, and therefore the immunological effector function mediated by the antibody, without changing the V region and the antigen specificity (Fig. G14).

The class switch is mediated by the switch (S) regions which are located at the 5′ end of all C_H genes except C_δ and involves the joining by recombination of two switch regions. The intervening DNA is deleted and a downstream constant region is brought into the vicinity of the variable region. Switch regions (switch sequences or S regions) are defined stretches of DNA 1–2 kb at the 5′ ends of each of the immunoglobulin heavy chain C region genes within which the recombinational events leading to class switching occur. The S regions consist of numerous tandem repeats of 'units' of 25–80 base pairs which are characteristic to each switch region. They include many occurrences of the pentanucleotides GAGCT and GGGGT, span long stretches of DNA (from 1 kb (S_ϵ) to 10 kb ($S_{\gamma 1}$)), and are homologous but not identical.

The μ C gene is invariably the first to be used (producing IgM in the PRIMARY IMMUNE RESPONSE), followed by switching to any of the γ, ε or α C genes (producing IgG, IgE or IgA respectively, in a secondary response). This class switching involves recombination and the deletion of intervening DNA (switch recombination). Switch recombination proceeds from the S_μ region to any of the S_γ, S_ϵ or S_α regions, resulting in the formation of an S_μ–$S_{\text{other isotype}}$ hybrid sequence. The site of recombination

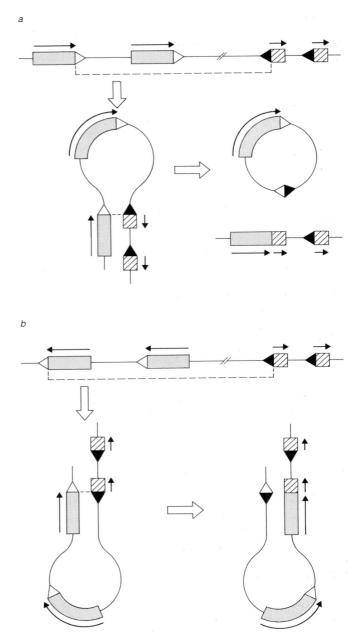

Fig. G13 *a*, Looping out deletion model of V-(D)-J recombination. *b*, Inversion model of V-(D)-J recombination. In this figure the triangles each represent a complete joining signal (i.e. nonamer, nonconserved spacer and heptamer, see Fig. G12). The deletion (*a*) or retention (*b*) of intervening sequences after a joint is dependent on the chromosomal orientation of the two gene segments (boxes) involved. Two fused triangles represent the reciprocal joint. Arrows indicate the transcriptional orientations of the gene segments.

within the S regions is not precise, in contrast to V-(D)-J recombination. The sequences TGGG and TGAG are often found adjacent to the recombination site. Regions upstream of the switch sequences have been implicated in binding factors which may regulate class switching.

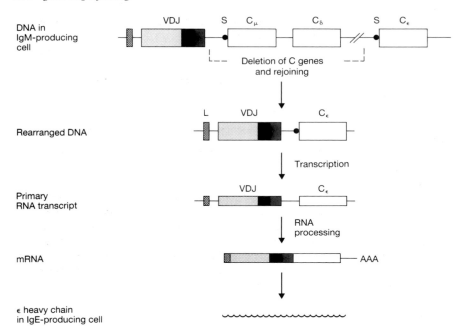

Fig. G14 Switch recombination in the immunoglobulin heavy chain locus. Recombination between switch regions upstream of each C_H gene results in deletion of the μ gene (and of genes downstream). The same VDJ region can then be expressed with a different C gene.

Class switch recombination is usually intrachromosomal. Because it occurs within intron sequences, switching does not affect the READING FRAME of the heavy chain gene. In contrast to the V-(D)-J rearrangement, the class switch event is not strictly site specific and locates to various sites within the S regions. The class switch is believed to be directed and can be induced by LYMPHO-KINES. Cytokines produced by T lymphocytes influence the choice of the constant region to which switch recombination occurs. For example, interleukin 4 is the major determinant of the generation of an IgE response.

The expression of the δ C gene (producing IgD), often occurring concomitantly with IgM expression, does not involve recombination and is achieved by differential RNA SPLICING. In human cell lines, however, μ to δ switching very occasionally occurs by switch recombination resulting in deletion of the $C_μ$ gene.

Further information on the topic of gene rearrangement in the immunoglobulin and T cell antigen receptor loci will be found in [1–5].

A. BETZ
M.S.NEUBERGER

1 Blackwell, T.K. & Alt, F.W. (1989) Mechanism and developmental program of immunoglobulin gene rearrangement in mammals. *Annu. Rev. Genet.* **2**, 605–636.
2 Calabi, F. & Neuberger, M.S. (1987) *Molecular Genetics of Immunoglobulin* (Elsevier, Amsterdam).
3 Davis, M.M. (1990) T cell receptor gene diversity and selection. *Annu. Rev. Biochem.* **59**, 475–496.
4 Esser, C. & Radbruch, A. (1990) Immunoglobulin class switching: molecular and cellular analysis. *Annu. Rev. Immunol.* **8**, 717–735.
5 Schatz, D.G. et al. (1992) V(D)J recombination: molecular biology and regulation. *Annu. Rev. Immunol.* **10**, 359–383.

gene superfamily *See*: SUPERFAMILY.

gene targeting, gene transplacement In some organisms (e.g. the yeast SACCHAROMYCES CEREVISIAE) a gene may be inserted into a particular chromosomal location by HOMOLOGOUS RECOMBINATION after TRANSFORMATION, provided that it is flanked within an INTEGRATION VECTOR by sequences homologous with the target region. The process, sometimes also called gene transplacement, can be used to substitute a particular chromosomal gene with a mutated form, or replace it with a new gene sequence. Almost certainly the chromosomal gene into which insertion occurs will be inactivated, so its function should not be essential for cell viability unless it can be substituted, for example by exogenous addition of a nutritional requirement. While this process is highly efficient for *S. cerevisiae*, it is a rare event in other fungi and in higher eukaryotes.

Gene therapy

GENE therapy is the term used to describe the correction of a disease by genetic manipulation. It is still at an early stage of development although a great deal of work has been done on the transfer of genes to experimental animals (*see* GENETIC ENGINEERING; TRANSGENIC TECHNOLOGIES).

There are several approaches to gene therapy. Somatic gene therapy involves the insertion of a gene into a cell other than a germ cell. For such a transfer to be effective throughout the life of the animal, the recipient cell must be renewable, and so experimental somatic gene therapy has involved mainly the haematopoietic STEM CELLS of the bone marrow, from which all blood cells derive, and, more recently, hepatocytes (liver cells), keratinocytes (skin cells), and endothelial cells. Genes inserted into cell populations of this type are, of course, not passed on to future generations. The other major form of gene therapy, germ cell therapy, which has not yet been carried out in humans, involves the injection of genes into fertilized eggs where they may be passed on to both somatic and germ cells during embryogenesis. The genetic change will then be transmitted to future generations.

In theory, there are two ways of attempting to cure a genetic disease. Replacement therapy entails inserting a gene into a cell in order to synthesize a gene product which is not being produced or is being synthesized in inadequate amounts. Hence this technique has its major application for RECESSIVE disorders, in which little or no normal gene product is usually present. The second approach is corrective therapy, that is, an attempt to correct the defective gene itself by lining up the defective and normal genes and inducing a recombinational event that replaces all or part of the defective gene with the normal DNA sequence (SITE-SPECIFIC RECOMBINATION). This would be of particular value for DOMINANT conditions in which the abnormal gene product synthesized by the defective allele was interfering with the product of the normal allele.

Mammalian cells will take up DNA directly either when it is in the form of a microprecipitate or when they are subjected to an electric current (electroporation) (*see* TRANSFECTION). These methods are inefficient, however, as in only a very small proportion of cells does the DNA enter the cell, reach the nucleus, integrate into the genome and become expressed successfully. A more efficient way of inserting a gene into the nuclear DNA is to use RETROVIRUS genomes as carriers. A variety of RETROVIRAL VECTORS have been constructed for this purpose in which the inserted DNA replaces some of the retroviral genes required for the production of viral structural proteins, but in which the viral sequences directing integration into the cell's DNA remain intact. Genes that are inserted via retroviral vectors are usually present as one copy per cell and integration into the genome is random. The major problems encountered with this approach have been efficiency of transfection, level of expression of inserted genes, and the lifespan of the transfected cells. However, many different human genes have been inserted into either cell lines or living animals using this approach and the conditions for applying this technique to treat human disease are gradually being worked out.

Because of continued concerns about the potential hazards of retroviral vectors a number of other approaches to gene transfer are being pursued. These include the complexing of appropriate DNA with proteins which will be bound to specific receptors and hence enter target cells, or the incorporation of DNA into LIPOSOMES which will fuse with cell membranes. Another approach is to construct minichromosomes.

So far there has been less success with gene correction therapy using site-specific recombination. This approach may, in the long term, be the best prospect for gene therapy because it provides a means of correcting a mutant gene without introducing a 'foreign' gene at random into the genome.

Currently, most work in this field is directed at short-term gene therapy to modify cells genetically. For example, lymphocytes from patients with ADENOSINE DEAMINASE DEFICIENCY (ADA deficiency) have been transfected with a normal ADA gene and reinjected into patients with some success. Lymphocytes which have a predilection for particular tumours — tumour-infiltrating lymphocytes (TIL) — are being modified to carry genes for LYM-PHOKINES with the objective of killing the tumours or modifying the immune response.

D.J. WEATHERALL

See also: CYSTIC FIBROSIS; TRANSGENIC TECHNOLOGIES.

Weatherall, D.J. (1991) *The New Genetics and Clinical Practice* 3rd edn (Oxford University Press, Oxford).

---

general acid–base catalysis One of the common MECHANISMS OF ENZYME CATALYSIS.

general amino-acid control The coordinated regulation of a large set of genes involved in amino acid biosynthesis. In the yeast SACCHAROMYCES CEREVISIAE such control is achieved through the GCN4-encoded TRANSCRIPTION FACTOR. During amino-acid starvation, Gcn4 levels increase, resulting in enhanced expression of target genes carrying the appropriate Gcn4 recognition sequence (5′ GTGACTC 3′).

general recombination *See*: LAMBDA; RECOMBINATION.

general transduction *See*: TRANSDUCTION.

Generation of diversity

THE immune system of vertebrates is one of the most resourceful defence mechanisms that organisms have ever evolved against foreign invaders. The uniqueness of the immune response lies in its specificity. In their immune systems, vertebrates possess a surveillance mechanism which can recognize and distinguish specific non-self molecular structures (ANTIGENS) on the surfaces of parasites, of cancer cells or of pathogenic microorganisms such as bacteria or viruses. Antigens are recognized by specific receptor molecules on the surface of the LYMPHOCYTES of the immune system. The number of different receptor specificities — the receptor REPERTOIRE — must be enormous to cope with all possible foreign antigens an organism may encounter. The problem is how this diversity is generated.

Antigen receptors

There are two types of antigen-specific receptors, which are expressed on different subsets of immunocompetent cells — IMMUNOGLOBULINS (Ig) on B lymphocytes (B CELLS) and T CELL RECEPTORS (TCR) on T lymphocytes (T CELLS). B cells also secrete their specific immunoglobulin as ANTIBODY molecules. Both types of immune receptor are members of the IMMUNOGLOBULIN SUPERFAMILY. The basic structure of all these receptors is a heterodimer in which receptor antigen specificity is contained in the N-terminal portions of the chains which are highly variable from molecule to molecule and are termed V regions, as compared to the C-terminal portions of the chains which have a more constant structure depending on the type of receptor, and are termed C regions. The T cell receptors are heterodimers com-

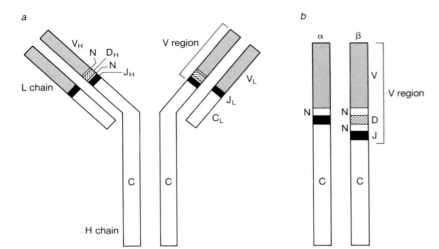

Fig. G15 Schematic of antibody (*a*) and T cell receptor (*b*) molecules.

posed of two chains, whereas the immunoglobulins are four-chain molecules composed of two identical heterodimers (Fig. G15).

A VARIABILITY PLOT comparing V regions from many different molecules shows that there are particular highly variable sub-regions, the so-called HYPERVARIABLE REGIONS. Crystallographic studies have demonstrated that these hypervariable regions form the antigen-binding pocket or surface. These parts of the receptors interact with antigen and are therefore referred to as COMPLEMENTARITY DETERMINING REGIONS (CDRs).

Generation of antibody diversity

The vertebrates have evolved various mechanisms which allow an individual to produce a large number of antibody molecules with different variable regions and hence of different antigen specificity.

An antibody molecule is composed of two identical heterodimers each composed of a heavy chain (H chain) and a light chain (L chain). Each polypeptide chain is encoded by a set of separate gene segments which are brought together to produce a complete gene (*see* GENE REARRANGEMENT; IMMUNOGLOBULIN GENES).

The V region of the H chain is made up from three gene elements — a variable region gene (V_H gene), a diversity segment (D segment) and a joining segment (J_H segment). The variable part of the L chain is encoded by two gene elements — a variable region gene (V_L gene) and a joining segment (J_L segment).

The mechanism of generation of diversity elucidated in mice and humans will be first discussed here, and some other mechanisms dealt with later. In both mouse and man antibody diversity is generated largely during the development of the pre-B cell into a mature immunocompetent cell (*see* B CELL DEVELOPMENT). V, D and J gene segments are rearranged by somatic recombination (*see* GENE REARRANGEMENT) to form a continous functional V-region gene.

Antibody diversity arises from the presence of multiple different copies of the V, D and J segments in the genome. A large part of

the diversity within the initial antibody repertoire present in virgin B cells in mice and humans (*see* B CELL DEVELOPMENT) results from the use of alternative combinations of these gene segments and is known as COMBINATORIAL DIVERSITY. In this way the diversity encoded in the germ line is amplified by the fact that a particular V_H region may be joined to one of several D and J_H segments, and a V_L to different J_L segments. The repertoire is further increased by the fact that a given H chain can associate with many different L chains.

A further source of diversity in the initial antibody repertoire is provided by segment joining at different nucleotide positions in the V, (D) and J gene segments. Thus V_H, D and J_H regions of different lengths can combine with each other. Different amino-acid sequences are generated, because the joining of the different gene elements does not take place at a precise nucleotide position. This JUNCTIONAL DIVERSITY is further enhanced by the fact that nucleotides can be inserted and deleted at the junctions during the process of gene rearrangement, a process known as N-REGION DIVERSITY. The diversity in the third complementarity determining region (CDRIII) of the H and L chains is due to junctional diversity. The structure of this part of the antibody molecule has a major impact on the antibody specificity. By increasing the diversity in the CDRIII regions of the H and L chain the size of the potential antibody repertoire is substantially increased. Thus a large antibody repertoire can be generated from a relatively small number of germ-line genes (~100 V_H, 20 D_H and 4 J_H and ~200 V_κ and 4 V_λ in the mouse). The diversity of the theoretical repertoire has been calculated as shown in Table G4.

The potential antibody repertoire of the mouse may be in the order of 10^{10} different molecules. Thus, the diversity generated by these mechanisms could produce a potential repertoire exceeding the number of lymphocytes. However, estimates for the actual repertoire are difficult to make. Genes may rearrange at different frequencies and not all rearrangements will result in the production of a functional antibody molecule. The combinatorial integration is not completely random. Furthermore, only a fraction of all developing immunocompetent cells are selected into the periphery (*see* B CELL DEVELOPMENT; T CELL DEVELOPMENT).

Table G4 Contribution of individual processes to potential repertoire diversity

κ chain	H chain
1 Combinatorial integration	
200 $V_κ$ × 4 $J_κ$ = 800	100 V_H × 20 D × 4 J_H = 8000
2 Junctional diversity	J_H length × 5
$V_κ$ to $J_κ$ × 2.5	V_H to D × 2.5
	D to J_H × 2.5
κ chain diversity = 2000	H chain diversity = 250 000

$$κ × H \text{ chain} = 5 × 10^8$$

$$\text{Factor due to N segment} > 10$$

$$\text{Grand total} \sim 10^{10} \text{ different receptors}$$

Generation of the secondary (memory) repertoire

Although most of the work in this area has been done in the mouse, it is believed that the situation is similar in man. Whereas the primary antibody repertoire is generated by germ-line-encoded sequences, in memory responses this repertoire is diversified by somatic mutations. During the course of a T cell-dependent antibody response (*see* IMMUNE RESPONSES), the AFFINITY of the antibody molecules for the antigen increases; this is referred to as 'the maturation of the immune reponse'. The change in antibody affinity is due to a hypermutation mechanism by which single-nucleotide substitutions are introduced into the variable regions of the antibody genes. In this way the antibody repertoire of antigen-activated B cells is further diversified. The few cells expressing variant immunoglobulin receptors of higher affinity for the stimulating antigen are then preferentially expanded and selected to differentiate into memory cells. Thus, the immune response matures through hypermutation and selection.

Somatic hypermutation occurs in cells expressing IgG or IgM antibodies. The hypermutation mechanism is active at the time that antigen-activated lymphocytes are proliferating in germinal centres in secondary lymphoid tissues (e.g. lymph nodes) (*see* B CELL DEVELOPMENT). It is a transient mechanism but can be reactivated following antigen stimulation of MEMORY B CELLS. The mechanism itself is not yet understood.

Diversity in the T cell receptor

There are two different types of T cell receptor, the αβ heterodimer and the γδ heterodimer (*see* T CELL RECEPTORS). The diversity of the T cell receptor is generated in essentially the same way as described above for the primary antibody repertoire. Like the V region of the antibody H chain, the variable regions of β and δ chains of T cell receptors are generated from three gene elements, a V, a D and a J segment. Like the V region of the antibody L chain, the variable region of α and γ chains are formed from two gene elements, a V and a J region (*see* T CELL RECEPTOR GENES). T cell receptor genes are rearranged during the development of mature functional T cells in the thymus (*see* GENE REARRANGEMENT; T CELL DEVELOPMENT).

Diversity is again the consequence of multiple copies of the various gene segments in the germ line and includes both combinatorial and junctional diversity. An interesting feature of the T cell receptor gene organization is the relative abundance of J segments compared to a relative scarcity of V segments. Thus, diversity in the T cell receptor is particularly enhanced in the third complementarity determining region (CDRIII).

Unlike an antibody molecule, the T cell receptor does not recognize free antigen, but only processed antigen which is presented on a cell surface in the form of peptides associated with an MHC MOLECULE (*see* ANTIGEN PRESENTATION; MAJOR HISTO-COMPATIBILITY COMPLEX). Hence the T cell receptor must interact both with constant determinants on the MHC molecule, and with variable peptides. This may be the reason why diversity is concentrated in the T cell receptor at one specific part of the variable region, the CDRIII. There is no hypermutation mechanism operating in T cells.

Different mechanisms of generating antibody diversity

Although the structure of the antibody molecule has been conserved in the evolution of vertebrates, changes in the organization of the immunoglobulin genes have occurred and different mechanisms of generating antibody diversity have evolved. In lower vertebrates, like the elasmobranch *Heterodontus fransci* (horned shark), the variable regions of the H chains are encoded by V, D and J region segments as in mouse or man. However, these gene segments are closely linked to constant region genes and are arrayed in multiple, individual clusters. In approximately 50% of these clusters gene segments are already joined (V_HD-J_H-C_H or even V_HDJ_H-C_H). The variable regions of the L chain are encoded by multiple V_L and J_L gene segments which again are closely linked to C_L genes. There are at least 40 V_L-J_L-C_L clusters. Thus antibody diversity is encoded by multiple copies of these gene clusters in the germ line of *Heterodontus* and is termed GERM-LINE DIVERSITY Fig. G16). V gene segments seem only to rearrange to their tandemly associated (D) and J gene segments, reducing the combinatorial possibilities. But germ-line-encoded diversity is enhanced by junctional diversity and extensive N-region insertion. Somatic hypermutation seems to play only a minor part in the generation of antibody diversity. A maturation of the antibody response (*see* B CELL DEVELOPMENT) has not been observed in this species.

A completely different mechanism of generating antibody diversity has evolved in the chicken. Only one functional V_L and J_L and one functional V_H, D and J_H gene segment are present. Diversity is generated by a GENE CONVERSION mechanism in which the unique rearranged V region is diversified by the introduction of sequences from a pool of V PSEUDOGENES which lie 5′ of the functional V gene. This occurs during development of B cells in the embryonic bursa.

Even among mammals there are several different strategies for generating diversity. Although rabbits have multiple functional germ-line V gene segments, only one is preferentially utilized. As in the chicken, antibody diversity is generated by a gene conversion mechanism. And whereas in mice the primary antibody repertoire is produced by rearrangement of germ-line-encoded

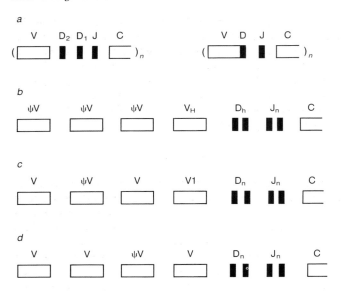

Fig. G16 Organization of the H chain locus in different species. *a*, Shark (*Heterodontus*), showing the clustered type configuration; *b*, chicken, with multiple pseudogenes (ψV) and only one functional V gene (V_H); *c*, rabbit, with multiple pseudo- and functional V_H genes and one gene (V1) used predominantly; *d*, mouse and human, with multiple pseudo- and functional V_H genes.

sequences and somatic hypermutation only occurs after antigen activation, in the sheep the primary repertoire is already diversified by somatic hypermutation.

C. BEREK

1 Tonegawa, S. (1983) Somatic generation of antibody diversity. *Nature* **302**, 572–581.
2 *Immunol. Rev.* (1987) Special issue on the 'Role of somatic mutation in the generation of lymphocyte diversity', **96**.
3 Davis, M.M. (1990) T cell receptor gene diversity and selection. *Annu. Rev. Biochem.* **59**, 475–496.
4 McCormack, W.T. et al. (1990) Avian B-cell development: generation of an immunoglobulin repertoire by gene conversion. *Annu. Rev. Immunol.* **9**, 219–241.
5 Reynaud, C.-A. et al. (1991) Somatic generation of diversity in a mammalian primary lymphoid organ, the sheep ileal Peyer's patches. *Cell* **64**, 995–1005.
6 Knight, K.L. (1992) Restricted V_H gene usage and generation of antibody diversity in the rabbit. *Annu. Rev. Immunol.* **10**, 593–616.

The genetic code

THE genetic code establishes the correspondence between the sequence of bases in NUCLEIC ACIDS and the sequence of amino acids in PROTEINS.

Flow of genetic information

The central dogma of molecular biology states that genetic information is contained in and perpetuated by the GENES which constitute particular sequences of nucleic acids [1]. DNA (deoxyribonucleic acid) is the primary repository of genetic information in all cells, and either RNA (ribonucleic acid) or DNA performs the same function in viruses. The information carried in nucleic acid is eventually expressed in the form of proteins which carry out all functions necessary to the replication and survival of cells. The conversion of genetic information into proteins is a two-step process:

1 TRANSCRIPTION of DNA into RNA. The RNA transcripts (messenger RNA (mRNA)) have the same sequence as the DNA CODING STRAND except that thymine in DNA is replaced by uracil in the RNA.

2 The translation step (*see* PROTEIN SYNTHESIS) converts the nucleotide sequence of mRNA into the sequence of amino acids constituting the polypeptide chain of a protein.

The triplet code

DNA is a linear polymer composed of four different building blocks (nucleotides) of which the variable parts are the bases adenine (A), thymine (T), guanine (G), and cytosine (C). Proteins are linear polymers composed of amino-acid building blocks, of which 20 different kinds are used in proteins (Table G5), and each kind of protein has a unique sequence. As the base sequence is the only variable element of DNA, the amino-acid sequence of a protein must be encoded by the sequence of bases in the corresponding DNA.

The four-letter alphabet of DNA is correlated with the 20-letter alphabet of amino acids by a three-letter nonoverlapping code. Each consecutive triplet of bases in DNA corresponds to an amino acid. The triplet nature of the code was first shown genetically by Francis Crick, Sidney Brenner and colleagues who demonstrated that each codon must consist of three (strictly speaking $3n$ with $n = 1, 2, 3...$) nucleotides [2].

Each triplet in mRNA is called a codon. The protein-coding sequence in mRNA starts at an initiation codon, runs consecutively and terminates in a termination or 'stop' codon (see Fig. G17). As the code is 'unpunctuated', correct translation of a sequence depends on starting at the right place so that the sequence is read in the correct reading frame.

Translation is a complex process whose fidelity relies on two

Table G5 The amino acids in proteins

Amino acids	Abbreviations	
Glycine	Gly	G
Alanine	Ala	A
Valine	Val	V
Leucine	Leu	L
Isoleucine	Ile	I
Serine	Ser	S
Threonine	Thr	T
Lysine	Lys	K
Arginine	Arg	R
Histidine	His	H
Aspartic acid	Asp	D
Asparagine	Asn	N
Glutamic acid	Glu	E
Glutamine	Gln	Q
Proline	Pro	P
Tryptophan	Trp	W
Phenylalanine	Phe	F
Tyrosine	Tyr	Y
Methionine	Met	M
Cysteine	Cys	C

classes of molecule. TRANSFER RNA (tRNA) is the adaptor molecule: within its nucleotide sequence is a three-nucleotide anticodon complementary to a particular codon, and at its 3′ end it carries the appropriate amino acid. The other class of molecules required to interpret the genetic code are the aminoacyl-tRNA synthetases, the enzymes that are responsible for attaching the correct amino acid to its corresponding tRNA. The specific recognition of tRNAs by aminoacyl-tRNA synthetases is often referred to as the second genetic code.

Deciphering the genetic code

The assignment of amino acids to their codons was originally made using the protein synthesis system from the bacterium *Escherichia coli*. In this *in vitro* system, Marshall Nirenberg showed that the polynucleotide poly(U) was translated into polyphenylalanine [3]. The translation of other synthetic polynucleotides of known sequence, as well as ribosome-binding assays *in vitro* soon led to the identification of 61 of the 64 codons. The entire genetic code has now been extensively confirmed by direct comparison of DNA SEQUENCES and the amino-acid sequences of the corresponding polypeptides.

The meaning of the code

Sixty-one out of the 64 codons of the genetic code correspond to an amino acid (Fig. G17). The codons AUG and sometimes GUG, which specify methionine (Met) and valine (Val) respectively, are also polypeptide chain initiation signals (*see* PROTEIN SYNTHESIS). The three remaining codons have no tRNA with a corresponding anticodon, and therefore act as stop signals or nonsense codons.

The code is not ambiguous — a given codon always designates a unique amino acid — but it is degenerate. Most amino acids,

with the exception of Met and tryptophan (Trp), are encoded by more than one triplet. Synonymous codons (i.e. codons defining the same amino acid) often differ only by the last base, which is very often degenerate. The second base relates amino acids with similar physicochemical properties (i.e. U indicates hydrophobic residues and A charged residues). This arrangement of the code table, which is not random, tends to minimize the effect of point mutations (*see* MUTATION).

Coding sequence organization

Coding sequences of prokaryotic genes are uninterrupted and the nucleotide sequences of genes and the amino-acid sequences of their encoded proteins are colinear, a feature established by genetic means by Yanofsky and colleagues in 1964 [4]. This cannot be generalized to other organisms. Only parts of eukaryotic genes code for proteins. Most eukaryotic genes are mosaics of long noncoding sequences (introns or intervening sequences) which interrupt the coding sequences (exons) (*see* EUKARYOTIC GENE STRUCTURE). The primary RNA transcript contains introns and exons; the introns are removed by RNA SPLICING during the RNA maturation process [5].

A particularly economical way of using DNA to encode information is demonstrated in some bacteriophages. These phage genomes contain OVERLAPPING GENES, in which one segment of DNA can be read productively in more than one reading frame. The DNA of bacteriophage φX174, for example, contains two genes completely 'buried' within larger genes of different reading frames.

The second genetic code

The existence of 'adaptor' molecules (tRNA) and their nature was first postulated by Crick in 1955 and only later were these small nucleic acids (between 75 and 90 nucleotides long) isolated [6]. Their secondary structure can be depicted as a 'clover leaf' (*see* TRANSFER RNA). The three-dimensional structure of yeast tRNA^{Phe} was the first to be unravelled by X-RAY CRYSTALLOGRAPHY in 1974. The ribose-phosphate backbone folds into a complex tertiary structure with a characteristic L-shape. An intricate three-dimensional network of hydrogen bonds between distant bases accounts for the fact that certain residues are conserved in the various sequences [7]. The three-dimensional structure of yeast tRNA^{Asp} confirmed the general character of the L-shape and all the key interactions [8].

The amino acid is attached to the ribose of the 3′ terminal adenosine of the molecule whereas the anticodon necessary for deciphering the genetic information in mRNA is located in an 'anticodon loop' at the other end of the molecule about 80 Å away. After being charged with an amino acid, tRNAs — aminoacyl-tRNAs — are carried to the ribosome where the codons in mRNA are deciphered (*see* PROTEIN SYNTHESIS).

Aminoacyl-tRNA synthetases

Codon recognition involves only the anticodon part of the tRNA and whatever amino acid is carried by that tRNA is automatically

Second position

First position	U	C	A		Third position
U	UUU Phe, UUC Phe	UCU Ser, UCC Ser	UAU Tyr, UAC Tyr	UGU Cys, UGC Cys	U, C
	UUA Leu, UUG Leu	UCA Ser, UCG Ser	UAA Stop, UAG Stop	UGA Stop, UGG Trp	A, G
C	CUU Leu, CUC Leu	CCU Pro, CCC Pro	CAU His, CAC His	CGU Arg, CGC Arg	U, C
	CUA Leu, CUG Leu	CCA Pro, CCG Pro	CAA Gln, CAG Gln	CGA Arg, CGG Arg	A, G
A	AUU Ile, AUC Ile	ACU Thr, ACC Thr	AAU Asn, AAC Asn	AGU Ser, AGC Ser	U, C
	AUA Ile, AUG Met	ACA Thr, ACG Thr	AAA Lys, AAG Lys	AGA Arg, AGG Arg	A, G
G	GUU Val, GUC Val	GCU Ala, GCC Ala	GAU Asp, GAC Asp	GGU Gly, GGC Gly	U, C
	GUA Val, GUG Val	GCA Ala, GCG Ala	GAA Glu, GAG Glu	GGA Gly, GGG Gly	A, G

Fig. G17 The universal genetic code

inserted into the growing polypeptide chain. The catalytic reaction by which aminoacyl-tRNA synthetases recognize and specifically charge their cognate tRNAs is therefore the key step in the translation process, and lies at the very basis of the genetic code. Nucleotides involved in the specific recognition by aminoacyl-tRNA synthetases are often referred to as the identity elements of a given tRNA molecule [9].

Aminoacylation of tRNAs is a two-step catalytic reaction involving three substrates: ATP and the amino acid form an intermediate adenylate (1) which subsequently reacts with the terminal adenosine of the CCA end of the tRNA to form an aminoacyl-tRNA (2):

$$\text{enzyme} + \text{amino acid} + \text{ATP}$$
$$\rightarrow \text{enzyme-(aminoacyl-AMP)} + \text{PP}_i \tag{1}$$

$$\text{tRNA} + \text{enzyme-(aminoacyl-AMP)}$$
$$\rightarrow \text{aminoacyl-tRNA} + \text{AMP} + \text{enzyme} \tag{2}$$

For a long time aminoacyl-tRNA synthetases were essentially characterized by their structural diversity: they exist in various oligomeric states and variable sizes and have very limited sequence homology with each other [10]. But owing to the progress in sequence analysis and crystallographic investigation the relation between structure and function in this family of proteins is now better understood.

The aminoacyl-tRNA synthetase family can be subdivided into two classes (Table G6) according to the structure of their active site, built around the ATP-binding domain [11]. The classification can be deduced from the primary sequence analysis of all known enzymes. Characteristic conserved sequence motifs constitute structural signatures of each subfamily. These motifs are associated with the ATP-binding domains which exhibit a different fold in each class.

The structural differences in the aminoacyl-tRNA synthetases are reflected in their function. Class I aminoacyl-tRNA synthetases attach the amino acid on the 2′ hydroxyl group of the terminal ribose whereas Class II enzymes attach the amino acid on the 3′ hydroxyl group of the terminal adenosine [12].

The crystal structures of five different enzymes, three from Class I (MetRS, TyrRS and GlnRS), and two from Class II (AspRS and SerRS) confirm the general classification. Two crystal structures of complexes formed between aminoacyl-tRNA synthetases and their cognate tRNAs, one for each group (glutaminyl and

Table G6 The two classes of aminoacyl-tRNA synthetases

Class I (2′ OH)*			Class II (3′ OH)*		
Glu Gln Arg			Gly Ala		
				Pro Ser Thr	
					Asp Asn
	Val Ile Leu Cys Met			His	Lys
		Tyr Trp	Phe		

*Primary site of attachment. The table is arranged to take into consideration the volume (increasing towards foot of table) and the chemical nature of the amino acids.

aspartyl respectively), provide a direct picture of this important intermediate in the translation process [13,14]. Two distinct modes of recognition of the tRNA molecules are revealed which can easily be generalized to the corresponding family of enzymes (Fig. G18).

Recognition of identity elements in the anticodon part of the tRNA involves distinct domains of the enzymes (N-terminal for aspartyl-tRNA synthetase, and C-terminal for glutaminyl-tRNA synthetase) specific for each system. In each case, the protein domains involved in the specific recognition of the tRNA molecule show no detectable homologies with other synthetases.

These data, together with the diversity of tRNA identity elements, point towards a complex recognition process, not compatible with a simple stereochemical correlation between the three-letter code and the synthetase or the amino acid, and do not support the idea of a second 'genetic code' within the usual accepted meaning of the term.

Universality of the code and evolution

For a long time the universality of the genetic code was accepted as dogma on the basis of the observation that eukaryotic genes can be accurately translated in *E. coli*. Genetic experiments largely support this general assumption but DNA sequencing studies have revealed a few cases of deviation from the usual code [15].

For example, UAG codons specify tryptophan in mammalian mitochondria and glutamic acid in ciliates instead of being normal 'stop' codons. A more complete examination shows that in non-plant mitochondria the following deviations from the universal code are observed: UGA for Trp, AUA for Met, AGR (R = A or G) for Ser and stop, AAA for Asn, CUN (N is any nucleotide) for Thr, and UAA for Tyr. A simplified code is used by these organelles, with 22 anticodons being sufficient for complete translation of mitochondrial mRNA.

In chloroplasts and in green plant mitochondria the universal code is used with a limited number of anticodons (31 for chloroplasts). These facts point towards the evolving character of the code, but do not answer the questions of direction and time scale involved. Codon usage in algae and ciliates (TAA and TAG are either termination codons or glutamine codons) is a good case for convergent evolution.

As for the origin of the genetic code, the discovery of the two classes of aminoacyl-tRNA synthetases quite distinct in their three-dimensional structures and with different functions sheds a new light on the problem. Are these two classes of enzyme indirect evidence of a pre-existing primordial RNA WORLD? This early RNA-based protein synthesis apparatus could have induced a primitive asymmetric recognition process which developed into two different classes of enzymes. It is clear that the recognition mechanism which emerged later is not a simple recognition code and today's systems exhibit sophisticated protein–RNA recognition processes with few key points of discrimination.

D. MORAS

A. POTERSZMAN

See also: PROTEIN–NUCLEIC ACID INTERACTIONS.

1 Crick, F.H.C. (1970) The central dogma of molecular biology. *Nature* **227**, 561–563.
2 Crick, F.H.C. et al. (1961) General nature of the genetic code for proteins. *Nature* **192**, 1227–1232.
3 Nirenberg, M.W. & Mattei, J.H. (1964) The dependence of cell-free protein synthesis in *E. coli* upon naturally occurring or synthetic polyribonucleotides. *Proc. Natl. Acad. Sci. USA* **47**, 1588–1602.
4 Yanofsky, C. et al. (1964) On the colinearity of gene structure and protein structure. *Proc. Natl. Acad. Sci. USA* **51**, 266–272.
5 Breathnach, R. & Chambon, P. (1981) Organization and expression of eukaryotic split genes coding for proteins. *Annu. Rev. Biochem.* **50**, 349–383.
6 Crick, F.H.C. (1957) *Biochem. Soc. Symp.*, 14–25.
7 Schimmel, R. et al. (Eds) (1979) *Transfer RNA: Structure, Properties and Recognition*, 83–160 (Cold Spring Harbor Laboratory, Cold Spring Harbor, New York).
8 Moras D. et al. (1980) Crystal structure of tRNA^{Asp}. *Nature* **288**, 669–674.
9 Normanly, J. & Abelson, J. (1989) tRNA identity. *Annu. Rev. Biochem.* **58**, 1029–1049.

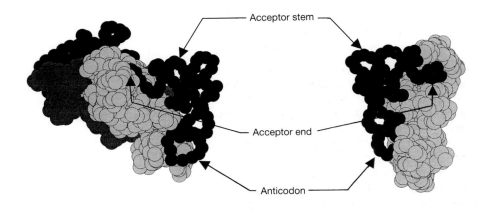

Fig. G18 Recognition of tRNA by aminoacyl-tRNA synthetases. Two distinct modes of recognition of tRNA have been observed in the crystal structures of complexes formed between aminoacyl-tRNA synthetases and their cognate tRNAs. A monomeric Class I enzyme, glutaminyl-tRNA synthetase is shown on the right. A dimeric Class II enzyme, aspartyl-tRNA synthetase is shown on the left. The glutaminyl enzyme approaches the acceptor stem of the tRNA from the minor groove side, forcing the CCA end to form a hairpin in order to bend towards the active site. In the aspartyl enzyme, the enzyme interacts with the acceptor stem of tRNA^{Asp} from the major groove side. The single-stranded CCA end dips naturally in the deep active site.

Acceptor stem

Acceptor end

Anticodon

Aspartyl-system
Dimeric
Class II

Glutaminyl-system
Monomeric
Class I

10 Schimmel, R. (1987) Aminoacyl-tRNA synthetases: general scheme of structure function relationships in the polypeptides and recognition of transfer RNAs. *Annu. Rev. Biochem.* **56**, 125–158.
11 Eriani G. et al. (1990) Partition of tRNA synthetases into two classes based on mutually exclusive sets of sequence motifs. *Nature* **347**, 203–206.
12 Moras, D. (1992) Structural and functional relationships between aminoacyl-tRNA synthetases. *Trends Biochem. Sci.* **17**, 159–164.
13 Rould, M. et al. (1989) Structure of E. coli glutaminyl-tRNA synthetase complexed with ATP at 2.8 Å resolution. *Science* **246**, 1135–1142.
14 Ruff, M. et al. (1991) Class II aminoacyl-tRNA: crystal structure of yeast aspartyl-tRNA synthetase complexed tRNA^Asp. *Science* **252**, 1682–1689.
15 Caron, F. (1990) Eukaryotic codes. *Experientia* **46**, 1106–1116.

genetic counselling In families where a genetic disorder has occurred, advice may be proffered with regard to the statistical risk of further affected offspring for that family and their relatives. If a particular disorder can be identified in a foetus then the couple may be offered the possibility of PRENATAL DIAGNOSIS followed, where appropriate, by a therapeutic abortion.

genetic disease Any heritable pathological condition which is caused by the presence of a mutant ALLELE — the disease gene. *See individual diseases.*

genetic distance A term used in two separate ways.
(1) A measure of overall evolutionary divergence between two populations or between different species. It may be calculated in various ways, including the number of nucleotide substitutions by which they differ at a particular locus, the degree of HYBRID-IZATION between DNA from the two samples, or by immunolog-ically detectable differences. *See*: MOLECULAR PHYLOGENY.
(2) Distance between two LOCI on the same chromosome segment as measured by the frequency of RECOMBINATION between them. *See*: MAP DISTANCE.

genetic drift The attrition of an allele in a small population in the absence of selection against that allele. It is associated with random sampling from a small population.

Genetic engineering

GENETIC engineering is an umbrella term for procedures that result in a directed and predetermined alteration in the GENOTYPE of an organism. It depends on a set of experimental techniques that enable individual GENES and DNA sequences to be isolated and purified by the procedure of DNA CLONING and to be further manipulated *in vitro*. RECOMBINANT DNA molecules generated by these manipulations (see below) carry novel combinations of DNA sequences and are designed so that they can be introduced into and become established in a new host cell. There they will direct the synthesis of a gene product not normally made by that cell, or will alter the expression pattern of a gene already present. In either case a genotypic change will result [1,2].

Techniques

The ability to manipulate and alter DNA *in vitro* in a predeter-mined way is central to genetic engineering. This ability depends on the activity of purified enzymes that cut and rejoin DNA molecules in very precise and predictable ways. The two most important classes of enzyme in genetic engineering are the bac-terial RESTRICTION ENZYMES (restriction endonucleases) and DNA LIGASE.

Gene isolation and cloning

Restriction endonucleases cut double-stranded DNA molecules at specific recognition sequences known as restriction sites (*see* Table R2 in RESTRICTION ENZYMES). In this way a predictable and reproducible set of DNA fragments (restriction fragments) is generated from a particular DNA molecule. DNA ligase can be used to join selected restriction fragments from different sources (Fig. G19). Further details of cloning procedures will be found under DNA CLONING.

Vectors

Restriction endonucleases and DNA ligase are used to construct recombinant DNA molecules, in which the required gene or other DNA fragment has been ligated into a vector DNA. The function of the vector is to allow the 'foreign' gene to be introduced into and become established within the desired host cell. Vectors used in the initial stages of genetic engineering to isolate, purify, and multiply up DNA fragments into usable amounts are called CLONING VECTORS. They are used to introduce DNA into a suitable fast multiplying host cell such as the bacterium *Escheri-chia coli*. In this way a rare sequence from a DNA mixture can be isolated and multiplied up into sufficient amounts for further manipulation. DNA fragments can also be rapidly multiplied up *in vitro* by the POLYMERASE CHAIN REACTION (PCR) without being incorporated into a vector. Vectors used to construct recombinant DNA molecules in which the introduced gene is to be expressed in the new host cell are called EXPRESSION VECTORS.

To act as a cloning vector a DNA molecule must be able to enter the host cell and, once inside, to be maintained in the cell, to replicate freely and to be passed on to daughter cells at cell division. Expression vectors must in addition contain suitable PROMOTER sequences and additional regulatory DNA sequences, which will allow the inserted gene to be correctly expressed in the host cell. When gene expression is required, the inserted gene is rarely in the original form but is replaced by a COMPLEMENTARY DNA (cDNA), which is a copy of the relevant messenger RNA and therefore contains an uninterrupted coding sequence for the required gene product, without INTRONS and other extraneous DNA sequences that might interfere, for example, with the expression of a eukaryotic gene in a bacterial host.

Naturally occurring DNA molecules that satisfy the basic re-quirements for a vector are PLASMIDS, and the genomes of BACTERIOPHAGES and eukaryotic viruses (*see* EXPRESSION VEC-TORS; VECTORS). Many cloning and expression vectors based on these DNAs have been developed by combining in the same vector molecule regulatory and other sequences from various sources.

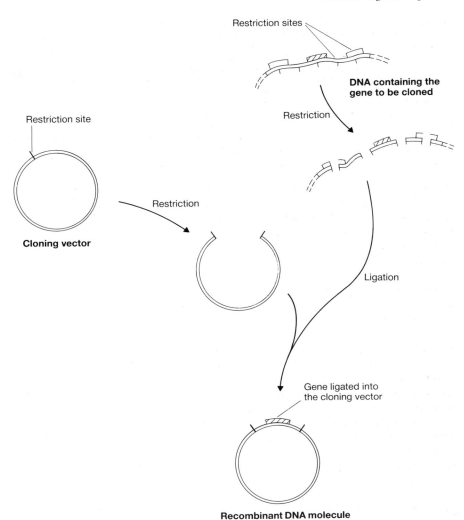

Fig. G19 Restriction and ligation in the construction of a recombinant DNA molecule.

Transformation and transfection

The method used to introduce the recombinant DNA molecule into the host cell depends on the type of organism involved. Bacteria will take up DNA molecules by the process of TRANS-FORMATION. A brief heat shock will induce the DNA to enter the cell more readily. Yeasts, filamentous fungi, and plant cells are often transformed after removal of their cell walls by degradative enzymes: the resulting PROTOPLASTS readily take up DNA and will re-form their cell walls spontaneously (*see* PLANT GENETIC ENGINEERING). Animal cells in culture are relatively easy to transform by introducing DNA by TRANSFECTION, which has been made more effective by the technique of ELECTROPORATION. Retroviruses are also being developed as highly efficient vectors to introduce genes into human cells destined for GENE THERAPY and to integrate the introduced genes stably into the genome (*see* RETROVIRAL VECTORS).

Alternatively, a physical method can be used, such as direct MICROINJECTION of the DNA into the cell nucleus, a technique often used to transform fertilized animal egg cells (*see* TRANSGENIC TECHNOLOGIES), or bombardment of the cell with gold or tungsten microprojectiles coated with DNA molecules.

Regeneration

If the genetic engineering project is successful the recombinant DNA will be replicated and passed to daughter cells when the initial genetically transformed cell divides, resulting in a pure culture of recombinant unicellular microorganisms, or, in the case of a plant, an engineered organism regenerated from a single transformed cell.

Techniques for transforming the fertilized egg cells or early embryos of mammals, and the raising of a genetically engineered embryo, are most highly developed for laboratory mice, and transgenic lines have been produced for developmental, immunological, genetic, and medical research (*see* TRANSGENIC TECHNOLOGIES). Genetic engineering of farm animals, fish, and birds is now being developed.

Applications

There are many practical applications of genetic engineering. A brief chronology of the development of genetic engineering and its applications is presented in Table G7. The ability to isolate genes and to read their DNA sequence (*see* DNA SEQUENCING) enables information to be gained about the structure of genes, the amino-acid sequence and possible structure and function of the proteins they encode, the processes involved in their expression, and the events that underlie the regulation of gene expression (*see* BACTERIAL GENE EXPRESSION: EUKARYOTIC GENE EXPRESSION). In addition, there are many actual and potential applications in industry, agriculture, and medicine.

Production of recombinant proteins

In industry, genetic engineering is leading to far-reaching advances in biotechnology by enabling animal proteins to be synthesized in microbial hosts. Many human proteins are important as proven or potential drugs, pharmaceuticals, and diagnostic reagents, but their preparation from animal tissues by traditional means is difficult, with low yields and problems with purity. However, it is now possible to isolate a gene for a human protein, to use recombinant DNA techniques to transfer the gene into a microbial host, and to obtain relatively large amounts of the pure protein from cell cultures. Proteins produced from recombinant DNA in a heterologous host are usually known as recombinant proteins (e.g. recombinant human insulin) to distinguish them from the naturally occurring protein.

At first this approach was attempted with the bacterium *E. coli* as the host but a number of problems arose. The steps involved in expression of a eukaryotic gene are more numerous and more complicated than the equivalent events in bacteria, so that even if an animal gene is placed under the control of an effective bacterial promoter, significant quantities of the protein product may not be synthesized. Intron splicing, for instance (*see* RNA SPLICING), will not occur in bacteria, and a bacterium will not process the initial translation product of an animal gene (*see* POST-TRANSLATIONAL MODIFICATION OF PROTEINS) in the correct way. So although a few animal proteins, notably human INSULIN and human GROWTH HORMONE [3], have been produced on a commercial scale in *E. coli*, other useful proteins have proved impossible to synthesize in this way.

The use of cultured eukaryotic cells as hosts is more promising, however. Eukaryotic microorganisms such as yeast and filamentous fungi have fewer problems in expressing animal genes and processing the translation products and have been used to synthesize mammalian proteins such as INTERFERONS, INTERLEUKINS, TISSUE PLASMINOGEN ACTIVATOR [4], and viral proteins such as the HEPATITIS B surface antigen. In addition, human Factor VIII, the blood-clotting protein that is defective in HAEMOPHILIA A, the commonest form of haemophilia, has been obtained from engineered hamster cells. An interesting innovation was the engineering of oilseed rape plants (*Brassica napus*) to produce human enkephalins (small neuropeptides with a number of potential clinical uses, *see* OPIOID PEPTIDES AND THEIR RECEPTORS) in their seeds [5], and simple animals such as the silkworm

Table G7 The development of genetic engineering

1960s–1970s	Isolation of restriction enzymes and their use to analyse DNA structure
1972–73	DNA cloning techniques involving recombinant DNA developed by Herbert Boyer, Stanley Cohen, Paul Berg, and others in the USA. A bacterial gene was the first gene to be cloned, in 1973
1974	First expression in bacteria of a gene from a different species
1975–77	Rapid DNA sequencing methods developed in Britain by F. Sanger and colleagues and in the USA by Allan Maxam and Walter Gilbert. The first DNA sequence of a complete genome (the 5375 bases of bacteriophage ΦX 174) was completed in 1977
1978	Bacteria produce human somatostatin from a chemically synthesized gene (*see* OLIGONUCLEOTIDE SYNTHESIS); later that year human insulin was also produced in bacteria from a chemically synthesized gene
1980	US Supreme Court rules that microorganisms can be patented
1982	Insulin (Eli Lilly's Humulin) is the first pharmaceutical made by genetically engineered bacteria to be approved for use in Britain and the USA
1981/2	First transgenic animals (mice) produced
1983	First transgenic plants expressing a gene of another plant species produced
1984	Stanford University is awarded a patent (the 'Cohen/Boyer patent') on the basic usage of recombinant DNA
1985	First transgenic farm animals produced (rabbits, pigs, and sheep)
1986	First controlled experimental releases of genetically engineered organisms into the environment
1989	US Patent Office announces that it will accept patent applications for genetically engineered plants and animals. First patented transgenic animal produced for pharmaceutical research — DuPont's 'oncomouse'
1990–92	First transgenic maize and wheat plants produced; genetic engineering of cereals becomes a reality
1992	Regulations for deliberate release of genetically engineered organisms established in the USA and EC
1992	First complete chromosome sequence (yeast chromosome III) published
1994	Genetically engineered tomato marketed in the USA

To date hundreds of enzymes, hormones, blood factors, cell proteins, viral proteins, etc., are available commercially for research purposes as reagents routinely manufactured from recombinant bacteria and eukaryotic cells. A smaller number of such products are available for human and veterinary use as drugs and vaccines. Many releases of genetically engineered plants, bacteria, animals, and viruses have been made worldwide since 1986, mostly in highly restricted field trials, but a few in larger-scale releases into the natural environment.

are also being explored as hosts for producing recombinant proteins [6].

Genetically engineered vaccines

Viral, bacterial, and parasite proteins for use in vaccines are also being mass-produced by the methods outlined above. In addition, live and killed vaccines in which the gene for the required antigen is incorporated into a viral carrier, such as VACCINIA VIRUS, or into a bacterial carrier, are also being developed (*see* RECOMBINANT VACCINES).

Genetically engineered plants

An important application of genetic engineering is in the development of novel varieties of crop plants for use in agriculture. For centuries crops have been improved by traditional breeding programmes but genetic engineering now offers a means of making directed alterations in the phenotypes of crop plants. One example has been the transfer from the bacterium *Bacillus thuringiensis* into maize plants of the gene coding for T-toxin, a bacterial protein that is a potent insecticide (*see* δ-ENDOTOXIN). The engineered maize plants synthesize T-toxin in their tissues and are unpalatable to caterpillars and other insect predators, resulting in an increased yield and less dependence on crop-spraying [7].

In similar projects resistances to herbicides or to viruses have been introduced into plants, and progress is being made towards engineering plants for tolerance to drought conditions and for the ability to grow on soils contaminated with seawater. Genes for these applications come from a number of sources, ranging from viruses and bacteria to wild relatives of crop plants.

Inactivation of resident genes as a result of the introduction of DNA encoding an ANTISENSE RNA which inhibits gene function is a novel development, and has already resulted in tomatoes that do not rot in storage so readily, owing to the inactivation of the gene for the fruit-softening enzyme, polygalacturonase [8] (*see* PLANT GENETIC ENGINEERING). On a lighter note, ornamental plants have been produced in which the pigment biosynthesis pathways have been engineered, resulting in novel petal colours and variegations [9].

T.A. BROWN

See also: DNA LIBRARIES; P ELEMENT-MEDIATED TRANSFORMATION (for genetic engineering of *Drosophila*); PROTEIN ENGINEERING.

1 Brown, T.A. (1990) *Gene Cloning*, 2nd edn (Chapman & Hall, London).
2 Watson, J.D. et al. (1992) *Recombinant DNA*, 2nd edn (Freeman, New York).
3 Goeddel, D. et al. (1979) Direct expression in *Escherichia coli* of a DNA sequence coding for human growth hormone. *Nature* **281**, 544–548.
4 Nyyssönen, E. et al. (1993) Efficient production of antibody fragments by the filamentous fungus *Trichoderma reesei*. *Biotechnology* **11**, 591–595.
5 Vanderkerchove, J. et al. (1989) Enkephalins produced in transgenic plants using modified 2S seed storage proteins. *Biotechnology* **7**, 929–932.
6 Reis, U. et al. (1992) Antibody production in silkworm cells and silkworm larvae infected with a dual recombinant *Bombyx mori* nuclear polyhedrosis virus. *Biotechnology* **10**, 910–912.
7 Koziel, M.G. et al. (1993) Field performance of elite transgenic maize plants expressing an insecticidal protein derived from *Bacillus thuringiensis*. *Biotechnology* **11**, 194–200.
8 Schuch, W. et al. (1989) Control and manipulation of gene expression during tomato fruit ripening. *Plant Mol. Biol.* **13**, 303–311.
9 Mol, J.N.M. et al. (1989) Genetic manipulation of floral pigmentation genes. *Plant Mol. Biol.* **13**, 287–294.

genetic fingerprinting *See*: DNA TYPING.

genetic fitness *See*: FITNESS.

genetic information The information encoded in the sequence of bases in DNA or RNA that directs and controls the synthesis of proteins and RNAs and determines the phenotype, survival, and reproduction of an organism or virus.

genetic load The average number of LETHAL ALLELES per individual within a given population.

genetic manipulation *See*: DNA CLONING; GENETIC ENGINEERING; PLANT GENETIC ENGINEERING; RESTRICTION ENZYMES; SITE-DIRECTED MUTAGENESIS; TRANSGENIC TECHNOLOGIES.

genetic map Map of the relative positions of LOCI on a chromosome estimated by the amount of RECOMBINATION between them. *See*: HUMAN GENE MAPPING; HUMAN GENOME PROJECT; LINKAGE MAP; MAP DISTANCE; RECOMBINATION FRACTION.

genetic marker Any gene or segment of DNA which can be identified and whose chromosomal location is known, and which can be used as a reference point for mapping other genes (*see e.g.* HUMAN GENE MAPPING). The term is also used for any gene of detectable phenotype that may be present or may be introduced into a cell in order, for example, to track its fate during embryonic development.

genetic nomenclature The conventions of genetic nomenclature vary between different groups of organisms. In most cases, however, it is generally recommended that the names of genes, alleles and their symbols, and genotypes, are written in italic, phenotypes in roman. It is also generally recommended that the names of mutations, as opposed to the specific designation of a mutant allele, should not be italicized. Many genes take their name from the mutation that first identified them. Although there are no particular rules governing the naming of mutations and genes, for historical reasons styles of naming in particular organisms are quite distinctive. Gene names in *Drosophila*, especially of developmental genes, tend to be descriptive and sometimes fanciful (e.g. *sevenless (sev)*, *bride of sevenless (boss)*; *son of sevenless (sos)*, *engrailed (en)*, *fushi tarazu (ftz)*), whereas in *Caenorhabditis*, genes which affect a particular general function are given the same name and distinguished by a numerical suffix (e.g. genes identified by cell lineage mutants *lin-3*, *lin-11*, *lin-12*, etc., genes identified by uncoordinated mutants *unc-86*, *unc-104* etc, and genes involved in programmed cell death *ced-1*, *ced-4*, etc.). Some general conventions of genetic nomenclature for organisms commonly encountered in the molecular biological literature are briefly outlined below.

1 *Escherichia coli* and other bacteria. In representing a genotype, each locus of a wild-type strain is designated by a symbol consisting of three lower case italic letters (e.g. *lac* for the loci at which mutations affect the cell's ability to utilize lactose). Different loci are distinguished by an italic capital letter suffixed to the locus designation (e.g. *lacA, lacY, lacZ, lacI*). The wild-type allele is designated by a plus sign (e.g. *lacZ*$^+$, or, if the context is clear, simply +). Phenotype symbols consist of three roman letters, the first of which is capitalized. Wild-type phenotypes are designated by a $^+$ (e.g. Ara$^+$), mutant phenotypes by a $^-$ (e.g. Ara$^-$), to denote absence of function, or other appropriate letter (e.g. Amp^r, for ampicillin resistance).

Demerec, M. et al. (1966) *Genetics* **54**, 61–76.

2 *Saccharomyces cerevisiae*. Gene loci are designated by a symbol consisting of three italic capital letters, which indicates the general function affected by mutations at the locus. Individual loci are distinguished from other loci involved in that function by a numerical suffix (e.g. *ADE2, GAL4, CDC2*). Recessive mutant alleles are written in lower case italic (e.g. *gal4, cdc2*); dominant mutant alleles are written in italic capitals. Unless the encoded product of a locus is a defined enzyme, it is designated by the locus name in lower case roman with an initial capital, followed by a lower case p (e.g. Gal4p, although the p is often dispensed with).

3 *Schizosaccharomyces pombe*. Loci and alleles are given three-letter italic lower case symbols indicating the general function affected, followed by a distinguishing number (e.g. *mat1*, the active mating-type locus; *cdc2*, a gene involved in regulating the cell cycle).

4 *Arabidopsis thaliana*. Gene loci are written in italic capitals (e.g. *APETALA2, AGAMOUS*) and given a symbol, usually a two- or three-letter abbreviation of the locus name (e.g. *AP2, AG*). Mutant alleles are written in lower case (e.g. *apetala2-1, apetala2-2 (ap2-1, ap2-2), agamous (ag)*).

5 Other plants. There are no general rules of gene nomenclature covering all plants, but conventions have been drawn up for crop plants by individual nomenclature committees (e.g. for maize, soybean, tomato, oats, wheat, barley, cotton). In general, loci and alleles are symbolized by a 1–3 letter italic abbreviation (e.g. the *Viviparous-1 (Vp1)* gene in maize); initial capital letters are often used for dominant mutant alleles (e.g. *Vp1*), all lower case for recessive; nonallelic genes with the same phenotypic effects are usually distinguished by a numerical suffix.

6 *Caenorhabditis elegans*. Loci and alleles are symbolized by a (generally) three-letter lower case italicized abbreviation, which is usually an abbreviation for the general function affected by mutants at the locus. Different loci affecting the same general activity are distinguished by numerical suffixes (e.g. *unc-4, unc-86, unc-104* are all genes which when mutant lead to uncoordinated movement).

7 *Drosophila melanogaster* and other insects. Apart from those encoding known enzymes, loci in *Drosophila* are generally given a name which describes the phenotype of the first mutant discovered in that gene, such as *antennapedia (antp), engrailed (en), even-skipped (eve), Notch (N)*. Loci and alleles and their symbols are italicized, phenotypes are roman. Recessive alleles and phenotypes are written with an initial lower case letter, dominant alleles and phenotypes with an initial capital. Genotypes are written using the allele symbols (with a + for wild-type alleles). Chromosomes are designated by roman numerals.

Lindsley, D.L. & Grell, E.H. (1968) *Carnegie Inst. Wash. Publ.*, **627**.

8 Mouse. Genotypes (symbols for loci and alleles) are italicized, phenotypes are roman. Symbols for dominant alleles or for the wild type are written in lower case with an initial capital (e.g. *Sry*, a male sex-determining gene); recessive alleles are all lower case (e.g. *mdx*). The locus symbol is the symbol of the first named mutant gene at that locus (e.g. *d* for the dilution locus, *Re* for the rex locus, *H-2* for the histocompatibility-2 locus). Names of genes and mutations are written in lower case roman whether dominant or recessive except where a proper name etc. requires a capital letter. In a genotype, wild-type alleles may be designated by a superscript + after the locus name, or simply by a + if the meaning is clear. Chromosomes are designated by arabic numerals.

Committee on Standardized Genetic Nomenclature for Mice (1963) *J. Hered.* **54**, 159–162.
Lyon, M.F. & Searle, A. (1989) *Genetic Variants and Strains of the Laboratory Mouse* (Oxford University Press).

9 Human. Genotypes (loci and alleles) are italicized, phenotypes and names of mutations are roman. Locus names are written in capitals (e.g. *RB-1*, the retinoblastoma gene; *SRY*, a male sex determining gene, *cf.* mouse nomenclature; *MYC*, the human proto-oncogene counterpart to the viral oncogene v-*myc*), recessive alleles (e.g. *myc*) in lower case. Chromosomes are designated by arabic numerals.

genetic polymorphism The occurrence in a population of two or more genetically determined variant forms of a particular characteristic at a frequency where the rarest could not be maintained by recurrent mutation alone. A genetic LOCUS is considered to be polymorphic if the variant form (ALLELE) is found in more than 1% of the population. Examples of genetic polymorphisms in humans are the loci of the MAJOR HISTOCOMPATIBILITY COMPLEX (HLA complex in humans) where numerous alleles of each gene exist, the ABO BLOOD GROUP SYSTEM, and many enzymes (*see* ISOENZYMES). Polymorphisms of DNA segments other than genes also exist and are of great value in gene mapping and DNA TYPING. DNA polymorphisms are due to the addition, deletion, or substitution of a single nucleotide at the site, or to variation in the number of tandem repeats of a DNA sequence. *See also*: BALANCED POLYMORPHISM; HUMAN GENE MAPPING; HYPERVARIABLE DNA; RFLP; VARIABLE NUMBER TANDEM REPEATS.

genistein 5,7,4'-trihydroxyisoflavone.

Found in the chickpea (*Cicer arietinum*), species of *Trifolium* and *Baptisia*, and in the Genistae.

genome The total DNA in a single cell (or unicellular organism, or virus). The total DNA content of diploid somatic cells is often termed the diploid genome in contrast to the haploid genome of the gametes. *See also*: GENOME ORGANIZATION; HUMAN GENE MAPPING; HUMAN GENOME PROJECT.

Genome organization

AN organism's characteristics are specified by its genetic information represented as a precise NUCLEIC ACID SEQUENCE, the sum total of this sequence information being its genome. Genome organization is, necessarily, both dictated by and related to, the functional requirements placed upon it by the organism's biological characteristics. In general, genome size (C-VALUE) is proportional to an organism's phenotypic complexity as dictated by the number of GENE PRODUCTS necessary for the replication and functional maintenance of new individuals, although departures from this principle are known (*see* C-VALUE PARADOX). In addition to 'coding' regions (transcribed into RNA), genomes must also possess sequences determining: (1) how coding regions are expressed and regulated; (2) the packaging, replication and correct division of the entire genome between newly produced cells (or unicellular individuals) so that each inherits the same parental genetic information; and (3) how changes to genetic information can be made, such as genetic information exchange (*see* RECOMBINATION). Further to these basic requisites, some genomes carry nucleotide sequences with, as yet, incompletely defined function and benefit, including repetitive DNA (see below) and TRANSPOSABLE GENETIC ELEMENTS. Further information on the topic will be found in [1–3].

Definition of the complete nucleic acid sequence of an organism's genome is expected to identify its every gene product, their relative arrangements and organization, and the sequences required for controlled gene expression and replication, thus leading to a much greater understanding of the organism's biology. Indeed, this is the hope behind the recent thrust to sequence the human genome (*see* HUMAN GENOME PROJECT) and the entire genomes of other genetically interesting and useful organisms such as the human gut bacterium (ESCHERICHIA COLI), baker's yeast (SACCHAROMYCES CEREVISIAE), the fruitfly DROSOPHILA and the laboratory mouse (*Mus musculus*).

Compositional and morphological genome organization

Double-stranded (ds) DNA is the genomic substance of most organisms and all cells. RNA viruses (*see* ANIMAL VIRUSES; BACTERIOPHAGES; PLANT VIRUSES; RETROVIRUSES) and VIROIDS, however, have ds or single-stranded (ss) RNA genomes. In all cases, each separate piece of genomic nucleic acid can be termed a CHROMOSOME. Prokaryotic genomes are usually single, closed circular dsDNA molecules, although extrachromosomal DNA in the form of circular PLASMIDS encoding beneficial but not essential gene products can be present. Eukaryotic genomes are usually encoded by several separate, linear pieces of dsDNA and the term nuclear genome signifies the genetic information sequestered as chromosomes, or sometimes also extrachromosomal DNA fragments, in the nucleus. The extranuclear genome of eukaryotic cells refers to the genomes of mitochondria and chloroplasts, which, like those of prokaryotes, are organized as single circular pieces of dsDNA (*see* CHLOROPLAST GENOME; MITOCHONDRIAL GENOMES).

The percentage guanine and cytosine nucleotide content (%G + C) is an established characteristic that distinguishes genomes from different organisms. In thermophilic bacteria, for example, the %G + C value can be very high because these nucleotides form a triple hydrogen-bonded BASE PAIR in DNA that stabilizes DNA to heat-induced denaturation. There are, of course, limits to how much of an organism's genome can be encoded using the G and C bases, as most proteins require CODONS for amino acids that are dictated using A and T bases and the %G + C content will influence CODON USAGE and CODON BIAS.

In eukaryotes, the nuclear membrane separates DNA REPLICATION and TRANSCRIPTION from messenger RNA translation (*see* PROTEIN SYNTHESIS) in the cytoplasm. Prokaryotes lack nuclei and must perform all three processes in the same (cytoplasmic) cellular compartment. Viruses and viroids require host cell machinery to express and replicate their genomes.

Genomes (chromosomes) can be packaged in a variety of ways. In viroids, the genome itself (chromosomal RNA) constitutes the organism. Addition of a protein coat (with or without lipid and carbohydrate) around the nucleic acid is usually sufficient for packaging viral genomes. In prokaryotes the genome is complexed with binding proteins so that the DNA is sufficiently packaged to 'fit' inside the cell as a NUCLEOID, although the genome is otherwise freely accessible in the cytoplasm. In eukaryotes, packaging of nuclear DNA into CHROMATIN and chromosomes is highly complex, requiring abundant special binding proteins (HISTONES and NONHISTONE PROTEINS) to fold up DNA into coils (NUCLEOSOMES), as well as scaffold proteins which hold the coiled DNA in a more tightly packed, chromosomal format and organize its distribution functionally within the confines of the nucleus (*see* NUCLEAR MATRIX). Packaging is not a static state, but depends on the functional activity of coding and regulatory regions during the CELL CYCLE. Lastly, eukaryotic chromosomes require special structures to cap their free DNA ends and to mediate their correct assortment and separation during cell division (*see* CHROMOSOME STRUCTURE).

Nucleic acid sequence organization of genomes

The simplest organisms (requiring the fewest gene products) have the smallest genomes and the least redundancy of sequence content (Table G8). Their gene organization is highly efficient and a high percentage of genomic sequence is transcribed. In most viruses for example, virtually every nucleotide encodes product or else regulates the expression, packaging or inheritance of coding nucleotides. Indeed, viral genes can often overlap one another somewhat to maximize the coding capacity of a nucleic

Table G8 Representative genome sizes

Group	Representative organism	Estimated genome size (kb)*	Approximate single-copy DNA content of genome (%)
Viruses	Bacteriophage lambda	48.5	
	Adenovirus	36	
Bacteria	*Escherichia coli*	4000	100
	Bacillus megaterium	30 000	
Fungi	*Saccharomyces cerevisiae* (baker's yeast)	16 500	90
Cellular slime mould	*Dictyostelium discoideum*	47 000	
Algae		37 500–190 000 000	
Protozoa		37 500–333 000 000	
	Tetrahymena pyriformis	190 000	90
Animals			
Nematoda		75 000–620 000	
	Caenorhabditis elegans	80 000	
Mollusca		375 000–5 100 000	
	Aplysia californica	1 700 000	55
Crustacea		660 000–21 250 000	
	Limulus polyphemus	2 650 000	70
Insecta		47 000–12 000 000	
	Drosophila melanogaster	165 000	60
	Musca domestica	840 000	90
Echinodermata	*Strongylocentrotus purpuratus*	845 000	75
Pisces		2 650 000–6 950 000	
Amphibia		950 000–78 500 000	
	Bufo bufo	6 600 000	20
	Xenopus laevis	2 900 000	75
Reptilia		1 600 000–5 100 000	
Aves	*Gallus domesticus*	1 125 000	80
Mammalia		2 800 000–5 200 000	
	Homo sapiens	2 800 000	64
	Mus musculus	3 300 000	70
Plants			
Gymnosperms		4 900 000–47 000 000	
Angiosperms		95 000–120 000 000	
	Arabidopsis thaliana	70 000	80
	Lycopersicon esculentum	700 000	
	Nicotiana tabacum	3 500 000	
	Zea mays	15 000 000	
	Lilium davidii	40 000 000	
	Fritillaria assyriaca	120 000 000	

* Per single chromosome set. Adapted from [4].

acid molecule the length of which is strictly prescribed (*see* OVERLAPPING GENES). Single control regions generally regulate expression of several coding sequences at once and this feature is retained in prokaryotic genomes, where sequences encoding related functional products may be organized as OPERONS producing POLYCISTRONIC mRNA under control of a single regulatory sequence (*see* BACTERIAL GENE EXPRESSION; BACTERIAL GENE ORGANIZATION).

Eukaryotic genes are MONOCISTRONIC, although a few can produce POLYPROTEINS, or possess alternatively transcribed regions. The distance between genes also tends to be increased in higher eukaryotic genomes compared with simpler organisms because of longer intragenic sequences. Further, many eukaryotic coding regions are interrupted by intervening sequences (INTRONS) which, though transcribed, generally play no direct part in coding for protein product (*see* RNA SPLICING) (Fig. G20).

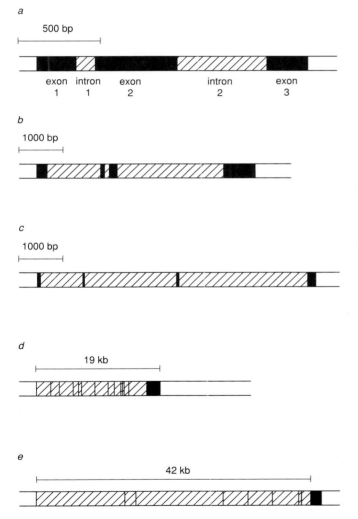

Fig. G20 Variation in length of some eukaryotic protein-coding genes. *a*, Human β-globin (1660 bp); *b*, human γ-interferon (4966 bp); *c*, human serum albumin (6682 bp); *d*, human tissue plasminogen activator (19 534 bp, exons vary from 43 to 757 bp long); *e*, human HGPRT (42 830 bp, exons vary from 18 to 640 bp long).

Introns explain, somewhat, the greater size of eukaryotic genomes compared with those of phenotypically simpler organisms and contribute to the much lower percentage of DNA sequence information that is actually expressed as mature RNA product (less than 1% in the human genome, which consists of $\sim 3 \times 10^9$ base pairs of DNA). Additionally, eukaryotic genomes can contain relatively high proportions of repetitive DNA, much of which has no coding function, although the extent of such repetition is highly variable among genomes, even those of phenotypically closely related organisms (*see* C-VALUE PARADOX).

Repetitive DNA

The presence of repetitive DNA in eukaryotic genomes can be demonstrated, and its extent measured, using DNA reassociation kinetics (*see* HYBRIDIZATION). Most STRUCTURAL GENES tend to be thus classified as belonging to the unique sequence, single-copy (nonrepetitive) category. A few, however, are encoded as moderately repetitive sequences (2–1000 copies). Examples include: (1) MULTIGENE FAMILIES, members of which are scattered throughout the genome (i.e. on different chromosomes); (2) GENE CLUSTERS, in which genes of closely related sequence and function are grouped close to one another on a single chromosome (Fig. G21); (3) tandem gene clusters, in which identical or closely related gene sequences are directly repeated in a 'head-to-tail' array at a single locus (e.g. the rRNA genes, and the histone genes in some animals, Fig. G22); and (4) PSEUDOGENES, representing extra copies of genes that are no longer viably expressed.

Moderately repetitive coding sequences usually produce RNAs or proteins required in large amounts (e.g. tRNAs, rRNAs, and histones). The bulk of the moderately repetitive sequence, however, is noncoding and tends to be organized as CONSTITUTIVE HETEROCHROMATIN at precise locations on chromosomes. It may have important, though incompletely defined, structural roles in genome organization and expression. Other, mainly noncoding, moderately repetitive sequences exist scattered throughout the genome, interspersed amongst genes and other DNA sequences (Fig. G23) (*see* ALU SEQUENCE; LINES; SINES). The role of these, too, remains to be completely defined, although some are likely to represent ORIGINS OF REPLICATION, scaffold attachment sites, and recombination 'hot-spots'. A distinct subfraction of the moderately repetitive, interspersed sequences represents TRANSPOSABLE GENETIC ELEMENTS, which can sometimes encode beneficial prod-

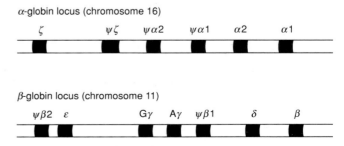

Fig. G21 The human α- and β-globin gene clusters. The symbol ψ denotes a pseudogene.

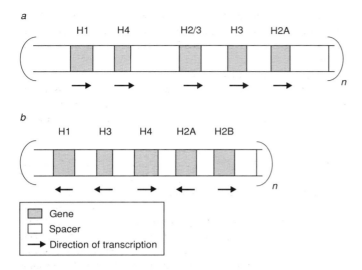

a

H1   H4   H2/3   H3   H2A

b

H1   H3   H4   H2A   H2B

☐ Gene
☐ Spacer
→ Direction of transcription

Fig. G22 Tandemly repeated genes. *a*, 'Early' histone genes (those expressed in very early embryos) from the sea urchin *Strongylocentrotus purpuratus*; *n* equals several hundred. *b*, Histone genes from *Drosophila melanogaster*; *n* = ~100. Each unit is repeated in tandem arrays of virtually identical copies. Similar histone repeat units occur in birds and mammals but are present in fewer copies and are dispersed throughout the genome.

ucts, replicate themselves and move from one chromosomal location to another. Mobile genetic elements are not limited to eukaryotic genomes.

In addition to unique sequence and moderately repetitive DNA sequences, many eukaryotic genomes, especially those of higher animals and plants, contain highly repetitive simple-sequence DNA. This can be sufficiently abundant and differ enough in %G + C composition from bulk genomic DNA, that it can be separated by DENSITY GRADIENT CENTRIFUGATION, as SATELLITE DNA. These sequences are often very simple (~6 bp long) and highly repeated (100 000–1 million times) in tandem. Such sequences contribute to forming structural elements of chromosomes such as telomeres, centromeres and pericentromeric heterochromatin (*see* CHROMOSOME STRUCTURE).

Variation in repetitive DNA among individuals is the basis of genetic fingerprinting (*see* DNA TYPING). Individuals of the same species can show wide diversity in the patterns (lengths and locations) of repetitive DNA clusters. This arises from nonhomologous recombination, the frequency of which increases when sequences are repeated many times in tandem at the same chromosomal location.

M.J. CARDEN

1 Lewin, B. (1980) *Gene Expression 2: Eucaryotic Chromosomes*, 2nd edn (John Wiley, New York/London).
2 Lewin, B. (1990) *Genes IV* (Cell Press, Cambridge, MA).
3 Singer, M. & Berg, P. (1991) *Genes and Genomes* (Blackwell Scientific Publications, Oxford).
4 Brown, T.A. (Ed.) (1991) Molecular Biology LabFax (Bios, Oxford).

genome rearrangement *See*: ANTIGENIC VARIATION; BACTERIAL PATHOGENICITY; GENE REARRANGEMENT; YEAST MATING-TYPE LOCUS.

genome sequencing *See*: HUMAN GENOME PROJECT.

genomic clone Cloned DNA isolated from a DNA LIBRARY prepared from chromosomal DNA (genomic DNA).

genomic DNA Chromosomal DNA, as opposed to complementary DNA (cDNA) copied from RNA transcripts.

genomic imprinting *See*: PARENTAL GENOMIC IMPRINTING.

genomic library One form of DNA LIBRARY.

genotoxic Able to damage DNA, a term used to distinguish CARCINOGENS that act directly on DNA from those that do not (nongenotoxic).

genotoxicology *See*: MUTATION.

genotype The actual genetic make-up of an organism (e.g. in terms of the particular ALLELES carried at a genetic LOCUS). Expression of the genotype gives rise to an organism's physical appearance and characteristics — the PHENOTYPE. *See*: MENDELIAN INHERITANCE.

geotropism Older term for GRAVITROPISM.

geranylgeranyl anchor *See*: GTP-BINDING PROTEINS; MEMBRANE ANCHORS.

Gerbich blood group Gerbich blood group antigens are located on glycophorin C (GPC) and glycophorin D (GPD), sialic-acid-rich red cell membrane GLYCOPROTEINS. These two glycoproteins are produced by one gene (*GYPC*) on chromosome 2, probably as a result of initiation of mRNA translation (*see* PROTEIN SYNTHESIS) at two different sites. GPC and GPD are attached to the cytoskeleton; their rare absence results in ELLIPTOCYTOSIS.

GERL (Golgi-endoplasmic reticulum lysosome) Structure found on the *trans* side of the Golgi stack and defined on the basis

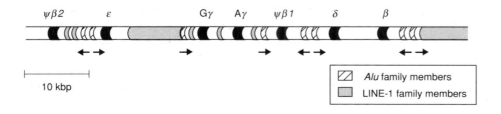

ψβ2   ε   Gγ   Aγ   ψβ1   δ   β

☐ *Alu* family members
☐ LINE-1 family members

10 kbp

Fig. G23 Interspersed repeated sequences in the human β-globin locus. Adapted from [3].

of acid phosphatase staining. It was argued to be a separate pathway for the biogenesis of LYSOSOMES directly from the ENDO-PLASMIC RETICULUM (ER) although the evidence for continuity with the ER was controversial. The Golgi part of GERL is now thought to be equivalent to the *trans* Golgi network. *See*: GOLGI APPARATUS.

Novikoff, A.B. (1976) *Proc. Natl. Acad. Sci. USA* **73**, 2781–2787.
Griffiths, G. & Simons, K. (1986) *Science* **234**, 438–443.

germ band In insect development, the embryonic primordium in post-blastoderm stages is referred to as the germ band to distinguish it from the extraembryonic ectoderm. The proportion of the egg length which contributes to the germ band varies widely among insects and insect development can usefully be subdivided into short, intermediate, and long germ band types.

germ cell development Germ cells, the embryonic progenitors of cells that will generate eggs and sperm during the life of an animal, appear to be set aside as a separate, specialized, cell population during GASTRULATION. In species where differential distribution of cytoplasm can take place owing to unequal cleavages (e.g. insects, amphibians), those cells that inherit a special type of cytoplasm (GERM PLASM) become determined as germ cells. In many species, ALKALINE PHOSPHATASE constitutes a useful marker expressed during early stages of their differentiation. In *Drosophila*, the *vasa* gene is expressed specifically in cells destined to become germ cells. In higher vertebrates, germ cells develop at the base of the allantois and later migrate to the genital ridges, either through the circulation or along the wall of the coelom (mesentery). It is generally believed that germ cells are attracted to the genital ridges by chemotaxis and/or selectively adhere there by specific cell adhesion. Germ cells appear to be multipotent, at least in *Xenopus*: if transplanted back into a BLASTULA stage embryo, they contribute to embryonic tissues that include NOTOCHORD and SOMITES. Epidermal growth factor (EGF), *Steel* factor (SF-1, the ligand of the mouse proto-oncogene product c-Kit) and the transforming growth factor TGF-β1 all seem to have a role in proliferation, migration, and differentiation of germ cells during early development. A mouse mutation affecting germ cell development (*germ cell deficient*) is available.

germ layers The three 'primary' layers of tissue in the early embryo: ECTODERM, MESODERM, and ENDODERM. In mammalian embryos, the term is usually used to describe only those tissues that contribute to the embryo proper and therefore excludes the TROPHECTODERM, which only contributes to placental structures. Early embryologists had an exaggerated view of the importance of the germ layers; this led to the introduction of terms like ENDO-THELIUM (originally used for epithelia derived from endoderm but *cf.* its present meaning) and mesothelium (epithelia derived from MESODERM).

germ line The continuity of GENETIC INFORMATION between one generation and the next in sexually reproducing organisms. In different contexts the term may refer to:
1 The cells that form the gametes of a sexually reproducing

organism, through which genetic information is transmitted from one generation to the next, as opposed to somatic cells. The germ line is usually segregated from the somatic cell lineage early in development.
2 The genetic information present in the ZYGOTE which is inherited by all somatic cells and (partially) by the sex cells of an individual: that is, the basic genetic constitution of an organism before any somatic GENE REARRANGEMENT or gene loss. *See also*: GENE THERAPY.

germ-line mosaicism The presence of a MUTATION (e.g. which results in a genetic disorder) in a proportion of the GERM CELLS of an individual, leading to the sporadic occurrence of the disorder among their offspring. If the mosaicism is due to a mitotic error having occurred during early gametogenesis, it will be restricted to the germ line in the parent. However, germ-line mosaicism may also originate because of a mutation in early embryogenesis, before the germ line segregates from other cell lines. In this case both germ cells and somatic tissues will be mosaic.

germ-line theory For many years a controversy raged as to whether the ANTIBODY repertoire is fully encoded in the germ line or whether a relatively small number of V genes (*see* IMMUNO-GLOBULIN GENES) can be diversified by a process of somatic mutation. We now know that various mechanisms have evolved to generate antibody diversity (*see* GENERATION OF DIVERSITY). The repertoire encoded in the germ line is enhanced by somatic mechanisms such as GENE REARRANGEMENT, HYPERMUTATION or GENE CONVERSION.

germ plasm (1) General term for cells or tissue from which a new organism can be generated. In plant conservation, may denote seeds or vegetative material; in general animal biology, denotes GERM LINE cells or their developmental precursors.
(2) In amphibian embryos, region of the fertilized egg or early embryo from which the germ line cells are derived. In anuran amphibia, for example, the germ plasm is marked by RNA-rich granules in the VEGETAL POLE of the fertilized egg, from which position it moves along the planes of subsequent CLEAVAGE to become localized at the floor of the blastocoel, thence contributing the PRIMORDIAL GERM CELLS. In urodeles, by contrast, the primordial germ cells are formed by INDUCTION within the mesoderm region, by interaction between the dorsal endoderm cells and the animal hemisphere cells. *See also*: AMPHIBIAN DEVELOPMENT.

germinal centres In a T cell-dependent immmune reponse, B cells activated by antigen start proliferating in the follicles of the peripheral lymphoid organs (e.g. lymph nodes). This leads to the formation of germinal centres which contain — in addition to B cells — HELPER T CELLS, follicular dendritic cells and tingible body macrophages. During proliferation of B cells in the follicles the B cell repertoire is diversified by a HYPERMUTATION mechanism. High affinity variants are selected to differentiate into MEMORY CELLS and perhaps also to PLASMA CELLS.

GFAP Glial fibrillary acidic protein. *See*: INTERMEDIATE FILA-MENTS.

GH GROWTH HORMONE.

***Gibberella fujikuroi* (Saw.) Wr.** Perfect or sexual stage of the fungus responsible for the *Bakanae* disease of rice and from which GIBBERELLINS were initially isolated. The imperfect or vegetative stage is *Fusarium moniliforme* (Sheldon). The fungus produces gibberellins as secondary metabolites and is grown in culture for the commercial production of gibberellic acid (GA$_3$), the major gibberellin metabolite. Certain minor products of the fungus, particularly GA$_4$ and GA$_7$, also have commercial application and their yield relative to that of GA$_3$ can be increased by judicious selection of fungal strain and culture conditions.

Jeffreys, E.G. (1970) *Adv. Appl. Biol.* **13**, 283–316.

gibberellin (GA) Member of a group of naturally occurring tetracyclic diterpene carboxylic acids possessing the *ent*-gibberellane (C$_{20}$) or *ent*-20-norgibberellane (C$_{19}$) skeletons. Currently about 90 structures have been identified in higher plants, fungi, and bacteria. Certain members act as endogenous growth substances in higher plants. Processes known to be influenced by gibberellin are stem elongation and leaf expansion, seed germination, fruit set and growth, sex expression in the flowers of some dioecious species, and flower induction in conifers. Gibberellins are present in growing plant tissues at 0.1 pmol–10 nmol g^{-1} fresh weight, with the highest concentrations in developing seeds. Gibberellin A$_1$ (*ent*-3α,10β,13-trihydroxy-20-norgibberella-16-ene-7,19-dioic acid-19,10-lactone) (Fig. G24) has been shown to be the primary endogenous gibberellin that regulates stem elongation in several species. The mechanism of gibberellin action has been most studied in the aleurone layer of germinating cereal grains where they induce the synthesis and secretion of hydrolytic enzymes into the endosperm.

Jones, R.L., & Jacobsen, J.V. (1991) *Int. Rev. Cytol.* **126**, 49–88.

Giemsa banding (G-banding) The pattern of dark and light bands seen on eukaryotic METAPHASE CHROMOSOMES subjected to treatment with Giemsa stain. The pattern is characteristic for each chromosome and can be used diagnostically for identification. Adjacent chromosome regions are thought to stain differently depending on the degree of compaction of the DNA. Genes replicated early in S phase (*see* CELL CYCLE) correlate with light staining bands. *See also*: CHROMOSOME BANDING; CHROMOSOME STRUCTURE; QUINACRINE BANDING; REVERSE BANDING.

gip A gene encoding the α$_{i2}$ subunit (*see* GTP-BINDING PROTEINS). It is a potential ONCOGENE and has been found mutated in tumours of adrenal cortex and endocrine tissues of the ovary.

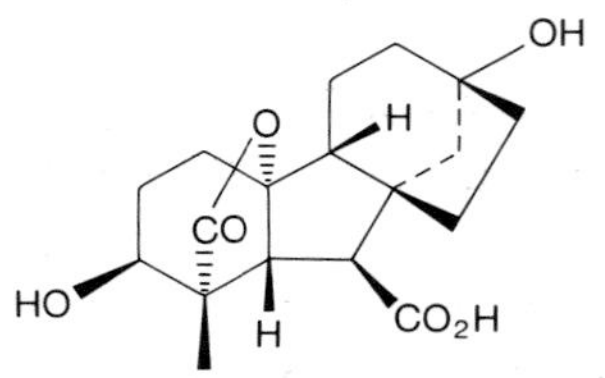

Fig. G24 The structure of gibberellin A$_1$.

Gla proteins *See*: BONE GLA PROTEINS.

Glanzmann's thrombasthenia Congenital bleeding disorder in which platelets are deficient in INTEGRIN α$_{IIb}$β$_3$.

GlcNAc N-ACETYL-D-GLUCOSAMINE.

glial cells The glial cells (or glia) are the supporting, non-neuronal cells of the central nervous system (CNS). 'They are as important as the neurons, which could not function without them. They wrap around and enclose the axons, dendrites, synapses and blood vessels. They probably control the composition of the interstitial medium' [1]. There are about ten times more glial cells than neurons in the CNS.

The term glia was coined by the German pathologist Rudolf Virchow in 1865, from the Greek word for glue, since he saw the glia as 'a binding substance... in which the nerve elements are embedded'. Golgi in 1885 was able to recognize several types of glia with his silver stain and showed their numerous radial processes. Cajal introduced in 1913 a specific stain for glial cells using gold/mercuric chloride, and recognized and named the astrocytes. By their morphological and staining characteristics, supplemented more recently by electron microscopy, the glia have been defined as several distinct cell types. The two broad classes of these are the neuroglia and the microglia. The latter are smaller and are either radial, with long processes, or are specialized subtypes such as the Bergmann glia of the cerebellum (which closely invest the dendrites of the Purkinje cells), the Müller glial cells of the retina, the ependymal cells (lining the cavities of the brain and spinal cord) and the tanycytes (differentiated, ciliated ependymal cells which line ventricles at certain specialized locations) [2]. One current view is that the radial microglia and the latter glial groups all have a common cell lineage [3]. Radial microglia often remain migratory and are also phagocytic, but must be distinguished from the resident macrophages which also exist in the central nervous system [4].

The neuroglia comprise the oligodendrocytes and the astrocytes. The former produce the MYELIN sheaths around central nervous system axons. They can be recognized by antibodies against galactocerebroside (GC, the main lipid in myelin) and against myelin components such as myelin basic protein and proteolipid protein. In their development, *in situ* hybridization to mRNAs for the myelin proteins gives positive signals. The oligodendrocytes initially extend processes to axons. They wrap them in sheaths of myelin layers and finally compact those sheaths. In mature oligodendrocytes very few ion channels or transmitter receptors have been detected except for GABA receptors, a rather low density of certain glutamate receptors and a prominent inwardly rectifying potassium channel [5].

The astrocytes were classically divided morphologically into the protoplasmic type in the grey matter of the brain and the fibrous type (with a large number of long fine radial processes) mostly in the white matter. Astrocytes contain characteristic filaments (*see* INTERMEDIATE FILAMENTS) and are identified using the antibody to the major filament protein, glial fibrillary acidic protein (GFAP). In cell cultures from developing rat optic nerve two types of astrocyte precursor have been recognized, the type-1A progenitor cell which produces the type-1 astrocytes common in the CNS, and the bipotential type-2A progenitor, which differentiates into the oligodendrocyte line and a type-2 astrocyte [6], and this has been extended to cultures from other brain regions. The type-2 astrocyte, however, has not been detected with certainty in postnatal brain *in vivo* and its significance remains uncertain. However, type-1 and type-2 astrocytes [8] and Bergmann glial cells [9] in culture express non-NMDA EXCITATORY AMINO ACID RECEPTORS, and these (but not NMDA receptors) have likewise been found in some brain astrocytes *in situ*. Astrocytes in cell culture express a wide range of ion channels and transmitter receptors. It is known that mature Bergmann glial cells (see above) *in situ* express only particular subtypes [10] of GABA$_A$ receptors (*see* GABA AND GLYCINE RECEPTORS) and some astrocytes are likely to do so too. Astrocytes are also well endowed with transmitter uptake systems and their functions may include serving as a sink for released transmitters and as a regulator of the ionic milieu of the adjacent neurons. Guidance by glia of migrating neurons is a primary function in the developing CNS. Furthermore, astrocytes in particular locations attach 'end-feet' to blood vessel endothelial cells or the pia and may control transmitter activity at these surfaces. Related to these structures, astrocytes (and perhaps microglia) associate with cerebral and/or endothelial cells to form the blood–brain barrier.

In the PERIPHERAL NERVOUS SYSTEM the equivalent of the oligodendrocyte of the CNS is the SCHWANN CELL (which has a completely different embryological origin). This cell wraps myelin layers around the axons of the myelinated nerves. Other types of peripheral glial cells occur at ganglia, sensory terminals and other specialized locations. *See also*: CILIARY NEUROTROPHIC FACTOR; VERTEBRATE NEURAL DEVELOPMENT.

1 Young, J.Z. (1991) The concept of neuroglia. *Ann. N.Y. Acad. Sci.* **663**, 1–18.
2 Risau, W. & Wolburg, H. (1990) Development of the blood–brain barrier. *Trends Neurosci.* **13**, 164–178.
3 Ransom, B.R. (1991) Vertebrate glial subtypes. *Ann. N.Y. Acad. Sci.* **663**, 19–26.
4 Graeber, M.B. & Streit, W.J. (1990) Perivascular glia defined. *Trends Neurosci.* **13**, 366.
5 Ketternman, H. Blankenfeld, G.V. & Trotter, J. (1991) Physiological properties of oligo dendrocytes during development. *Ann. N.Y. Acad. Sci.* **663**, 64–76.
6 Raff, M.C. (1989) Glial cell differentiation in the rat optic nerve. *Science* **243**, 1450–1455.
7 Wilkin, G.R., Marriott, D.R. & Cholewinski, A.J. (1990) Astrocyte heterogeneity. *Trends Neurosci.* **13**, 43–46.
8 Steplas, G.P., Djamgoz, M.B.A. & Wilkin, G.P. (1993) A patch clamp study of excitatory amino acid effects on cortical strocyte subtypes in culture. *Receptors Channel* **1**, 39–52.
9 Müller, T., Möller, T., Berger, T., Schnitzer, T. & Ketthmann, H. (1992) Calcium entry through kainate receptors and resulting potassium channel blockade in Bergmann glial cells. *Science* **256**, 1560–1563.
10 Wisden, W., McNaughton, L., Hunt, S.P., Darlison, M.G. & Barnard, E.A. (1989) Differential distribution of GABA$_A$ receptor subunit mRNAs in bovine cerebellum: localization of α2 in Bergmann glial cells. *Neurosci. Lett.* **106**, 7–12.

glial fibrillary acidic protein *See*: INTERMEDIATE FILAMENTS.

glial filaments *See*: GLIAL CELLS; INTERMEDIATE FILAMENTS.

glial growth factor (GGF) A basic protein of M_r 31 000, originally purified from bovine brain on the basis of its ability to stimulate proliferation of cultured rat Schwann cells. It also stimulates division of other cell types including astrocytes and fibroblasts. It may be involved further in the nerve dependence of amphibian LIMB REGENERATION, where it is detectable in the BLASTEMA.

glioblast (spongioblast) Embryonic cell type of the mantle layer of the neural tube which gives rise to astrocytes and oligodendrocytes of the neuroglia. *See*: GLIAL CELLS.

glioblastoma (gliocarcinoma, spongioblastoma multiforme) A highly malignant brain tumour composed of glial tissue. It contains poorly differentiated cells and areas of necrosis, vascular endothelial cells, haemorrhage, and invasive growth. Glioblastomas usually occur within the cerebral hemispheres of adults.

Gln The abbreviation for the AMINO ACID glutamine.

globin A monomeric protein or protein subunit of the large globin superfamily of oxygen-binding proteins all of which contain the characteristic globin fold; specifically, the various types of subunit of haemoglobin. *See*: HAEMOGLOBINS; HAEMOGLOBIN AND ITS DISORDERS.

α-globin *See*: HAEMOGLOBINS; HAEMOGLOBIN AND ITS DISORDERS.

globin genes Genes encoding the globin chains of HAEMOGLOBINS. In humans they comprise two related clusters (α and β) located on separate chromosomes (*see* HAEMOGLOBIN AND ITS DISORDERS). In humans the α cluster lies on chromosome 16 and the β cluster lies on chromosome 11. The β cluster extends over 50 kb and contains five functional genes and one PSEUDOGENE in the order ε, G$_γ$, A$_γ$, ψβ, δ, and β. The α cluster stretches over 28 kb and contains three active genes, three pseudogenes and the θ gene of unknown function in the order ζ, ψζ, ψα2, ψα1, α2, α1, θ. The evolution of the globin genes has been studied extensively (*see* MOLECULAR EVOLUTION; SEQUENCES AND STRUCTURES). The most primitive vertebrates — represented by the present-day cyclostomes (e.g. lamprey) — are presumed to have had a single globin gene. Jawed fishes have α and β clusters. In amphibia α and β genes are closely linked in tandem. In birds and mammals the α and β clusters are on separate chromosomes.

globin genes, regulation of expression and developmental regulation *See*: HAEMOGLOBIN AND ITS DISORDERS.

globin superfamily *See*: HAEMOGLOBINS.

globulin A general term for any globular protein insoluble or sparingly soluble in water but readily extractable in 5–10% NaCl.

glucagon Polypeptide hormone produced by the α cells of the pancreas. It antagonizes many of the actions of INSULIN. It acts predominantly on the liver, inhibiting glycolysis, promoting gluconeogenesis and thus raising blood glucose. To a lesser extent it has effects on lipid metabolism, reducing fatty acid synthesis and increasing fatty acid oxidation and ketogenesis. It acts at G PROTEIN-COUPLED RECEPTORS.

glucans Large and heterogeneous group of GLYCANS which are polymers exclusively or largely of glucose and which occur in both prokaryotes and eukaryotes. *See*: AMYLOIDS (PLANT); AMYLOPECTIN; AMYLOSE; CALLOSE; CELLULOSE; GALACTOGLUCOMANNANS; GLUCOMANNANS; GLYCOGEN; MIXED GLUCANS OF GRASSES; ONUPHIC ACID; PARAMYLON; PHYTOGLYCOGEN; SULPHATED GLUCANS.

glucomannan *See*: CELLULOSE; GALACTOGLUCOMANNAN.

glucosamine, glucose phosphates, glucose *See*: SUGARS.

glucose-6-phosphate dehydrogenase (G6PD) deficiency G6PD is a cytoplasmic enzyme found in all cells. It catalyses the first step in the hexose monophosphate pathway, and hence is involved in the generation of NADPH, which is required for a variety of biosynthetic pathways as well as for the stability of the enzyme CATALASE (*see also* PEROXISOMES) and, in particular, for maintaining levels of the reduced form of GLUTATHIONE (GSH). A deficiency of G6PD is of particular importance to the red blood cell as catalase and glutathione are essential for the breakdown of hydrogen peroxide and for the defence of cells against oxidative damage.

G6PD deficiency is a SEX-LINKED disorder and is the commonest enzyme deficiency known, with an estimated 400 million affected people worldwide. The highest prevalence rates are found in tropical Africa, the Middle East and Southeast Asia. These high GENE FREQUENCIES are thought to be the result of the defective alleles protecting against *Plasmodium falciparum* malaria.

G6PD deficiency is genetically heterogeneous. Over 300 different variants have been identified world-wide, and in addition, many cases of SILENT MUTATIONS in the gene have been discovered that are not associated with defective enzyme function. The gene for G6PD has been isolated and over 30 different point mutations have been identified as the basis for defective G6PD function.

Clinically, G6PD deficiency presents in a variety of ways. Many deficient individuals show no abnormality unless exposed to oxidant drugs such as CHLOROQUINE. Exposure to these agents results in oxidative damage to the red cell with haemoglobin precipitation, shortening of the cell lifetime and moderate to severe anaemia. Acute haemolytic anaemia can also be triggered in G6PD-deficient people by infection, or by the ingestion of fava beans (FAVISM). In some populations newborn G6PD-deficient babies develop severe jaundice. Finally, some G6PD variants are associated with chronic premature destruction of red cells and life-long anaemia.

Vulliamy, T. et al. (1992) *Trends Genet.* **8**, 138–143.

glucosylation The addition of a glucose residue. *See*: GLYCANS; GLYCOPROTEINS.

GluR Notation for cloned cDNAs that encode subunits of ionotropic EXCITATORY AMINO ACID RECEPTORS. Clones of non-NMDA receptor subunits have been named GluR1, GluR2, etc. or GluR-A, GluR-B, etc. by different workers. The terminology α, β, γ, δ, ε, and ζ is now also in use to denote subfamilies of ionotropic EAA receptor subunits. ε and ζ are NMDA receptor subunits. ζ1 is also called NMDAR1.

glutamate receptors *See*: EXCITATORY AMINO ACID RECEPTORS.

glutamate synthase (GOGAT) Glutamine : 2-oxoglutarate aminotransferase. Catalyses the transfer of -NH$_2$ from the amido group of glutamine to 2-oxoglutarate to yield two molecules of glutamate, with reducing power provided by NADPH (EC 1.4.1.12) or, in chloroplasts, reduced ferredoxin (EC 1.4.7.1).

$$H_2N.OC.(CH_2)_2.CH(NH_2).COOH + HOOC.(CH_2)_2.CO.COOH + 2H^+ + 2e \rightarrow 2HOOC.(CH_2)_2.CH(NH_2).COOH$$

The reaction has a major role in ammonia assimilation. Glutamine is first formed by glutamine synthase, and then, by the action of glutamate synthase, the -NH$_2$ is transferred into glutamate, and thence to form a wide range of amino acids by transamination or by direct glutamate transformation.

glutamic acid (Glu, E) An AMINO ACID (Fig. G25) found in proteins and which also, as glutamate, acts as a NEUROTRANSMITTER (*see* EXCITATORY AMINO ACID RECEPTORS).

$$HOOC - CH_2 - CH_2 - CH \begin{array}{c} NH_2 \\ \diagup \\ \diagdown \\ COOH \end{array}$$

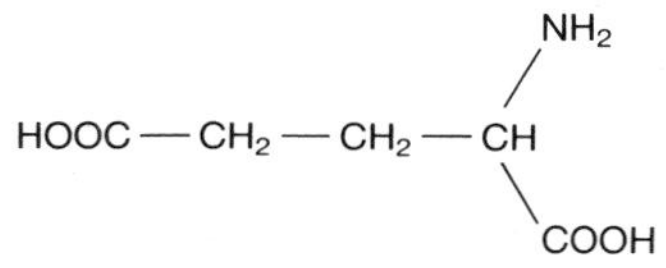

Fig. G25 Glutamic acid.

glutamine (Gln, Q) An uncharged AMINO ACID, the amide of glutamic acid.

glutathione Disulphide-bonded peptide which is reduced by NADPH and glutathione reductase in red blood cells to produce the tripeptide γ-glutamylcysteinylglycine (reduced glutathione). Reduced glutathione acts as a buffer to prevent oxidation of the cysteine residues in haemoglobin and other erythrocyte proteins. *See*: GLUCOSE-6-PHOSPHATE DEHYDROGENASE DEFICIENCY.

glutelins A general term for small alkali-soluble globular proteins, isolated from plant seeds. *See*: SEED STORAGE PROTEINS.

Glx (Z) Abbreviation indicating either of the AMINO ACIDS glutamic acid or glutamine.

Glycans

THIS general term includes (1) polymers composed wholly or largely of SUGARS and their simple derivatives (e.g. amino sugars and sugar alcohols) and (2) the carbohydrate parts of other types of molecule to which sugars are covalently linked, whether as monosaccharides, oligosaccharides or polysaccharides, to form glycoconjugates. The linkage of sugar is normally as a glycoside and the remainder of the molecule to which the sugar is linked is the aglycone.

Occurrence

In bacteria, cyanobacteria and related prokaryotes, glycan is found at the cell surface in association with both the membrane (LIPOPOLYSACCHARIDE; LIPOTEICHOIC ACID) and the cell wall (TEICHOIC ACID; PEPTIDOGLYCAN; TEICHURONIC ACID), and extracellularly (CELLULOSE; HYALURONIC ACID; ALGINIC ACID; COLOMINIC ACID; XYLAN; LEVAN; DEXTRAN). Extracellular glycan may be chemically related to wall components such as the O-ANTIGEN chains of lipopolysaccharide or may be entirely distinct (levan, dextran). A few bacteria, particularly ARCHAEBACTERIA, have been shown to contain true GLYCOPROTEINS, with features only distantly like those of eukaryotic glycoproteins; they appear to be membrane components. Viral envelopes often contain glycoproteins, but the glycan of these is wholly host-derived.

In eukaryotic cells glycan is particularly associated with membranes, including the plasma membrane and endomembranes, with extracellular matrices and cell walls, with cytoplasmic and vacuolar storage materials, and with the nuclear matrix (*see* EXTRACELLULAR MATRIX MOLECULES; GLYCOLIPIDS; GLYCOPROTEINS; MEMBRANE STRUCTURE; NUCLEAR MATRIX; PLANT CELL WALL MACROMOLECULES). Until recently glycan was thought to be absent from the cytoplasm, but some specific types of glycoconjugate have now been claimed in RIBOSOMES, TUBULIN and free within the cytoplasm.

At the eukaryotic cell surface, membrane glycolipid and glycoprotein are arranged so that their glycan is restricted to the outer face of the membrane, so as to form a GLYCOCALYX. Some glycoproteins (including the PROTEOGLYCAN type) can be attached by membrane anchors of glycosylated phosphoinositide (*see* GPI ANCHORS), so extending the glycocalyx far beyond the outer surface of the hydrophobic lipid bilayer of the membrane. In the endomembranes, most glycan is orientated in the topologically equivalent manner, though nascent glycan occurs at the opposite aspect of some membranes so that it has to cross the membrane during assembly. Glycans have been found in the membranes of NUCLEAR ENVELOPES, LYSOSOMES, and MITOCHONDRIA and are abundant in the ENDOPLASMIC RETICULUM and GOLGI APPARATUS, which are quantitatively the two major sites of their assembly. Autonomous synthesis occurs in nuclei and mitochondria.

Glycan less closely associated with membrane occurs as cytoplasmic storage GLUCANS (GLYCOGEN; STARCH), as intravacuolar storage glycans (LEVAN), as nuclear matrix glycan, as the glycoprotein of the mitochondrial intermembrane space, as cytoplasmic glycan and as soluble glycan destined for secretion.

Extracellular glycans of eukaryotes include a wide range of soluble glycoproteins as well as those glycoproteins (including proteoglycans), hyaluronic acid, and many polysaccharides which form the extracellular matrices of animal, plant, and fungal cells. Such matrices are of many different types, but their glycans are both structural and informational, in that they constitute a large part of the mass of most matrices and contribute to the organization and positional specification within the matrix and information exchange with cells (*see* CELL–MATRIX INTERACTIONS).

Structural analysis

Glycans may branch, their glycosidic bonds may be variously α or β anomers, their sugars may be pyranoses (p) or furanoses (f), and the positions of linkage may vary (*see* SUGARS); their structural analysis therefore presents problems not found with PROTEINS and NUCLEIC ACIDS. This is compounded by the extreme difficulty of purifying larger glycans and some glycoconjugates, the instability of some glycosides, and (until recently) the insensitivity of methods for separating and identifying sugars. Microheterogeneity is commonly found in glycans and can be of biosynthetic origin or artefactual.

The general strategy for sequencing large glycans is to purify them as far as possible and then to achieve fragmentation by partial acid hydrolysis, by use of endoglycosidases (*see* GLYCOSIDASES) and by β-elimination in alkali. The great variation in acid stability of glycosides is often helpful.

The fragments released are then separated and sequenced individually, after further partial degradation, if necessary. Sequencing and determination of anomeric configuration is usually by means of successive degradations with exoglycosidases, with monitoring of changes in molecular size, charge, and interaction with LECTINS or (rarely) MONOCLONAL ANTIBODIES. Additional information on position of linkage and sugar conformation can be obtained by METHYLATION ANALYSIS and SMITH DEGRADATION.

Recent technical developments have made detailed microanalysis possible. Important newer methods include: (1) mass spectrometry (especially with fast-atom bombardment); (2) the use of ion-exchange high performance liquid CHROMATOGRAPHY (HPLC) systems at high pH on pellicular resins, coupled with pulsed amperometric and conductivity detectors; (3) the use of lectins and lectin-HPLC methods; (4) the development of a range of endoglycosidases; and (5) the use of gel ELECTROPHORESIS to analyse GLYCOSAMINOGLYCAN sequences in a manner analogous to DNA sequencing. The problem of glycan configuration is becoming accessible to NMR methods (*see* NMR), including natural abundance ^{13}C-NMR [2]. Glycans absorb ultraviolet only at short wavelengths, but the monitoring of columns at 205 nm is now sometimes used.

Structural types

Glycans can be classified into a number of different general types, (Table G9), on the basis of chemical structure and biosynthetic mechanism. No generally standardized nomenclature for all of them exists. In many glycans there is (at least at initial synthesis)

Table G9 Classification of glycans

1 Phosphorylated glycans

Glycans in which phosphate esters form a regularly repeating part of the backbone, either occurring between all sugar residues or between every other pair of sugars. There are three types of sequence:

(a) -A-P-A-P-A, where A is a sugar alcohol; found in some teichoic acids (of bacteria, etc.) and in onuphic acid (from a gastropod).

(b) -A-P-B-P-A-P-B-, where A and B are dissimilar (usually A is a sugar alcohol and B a sugar); found in some teichoic acids.

(c) -AB-P-AB-P-, again found in some teichoic acids. In some cases additional sugars are added as side chains after the polymer has formed. (Note that glycans containing occasional phosphate esters in their chains, in irregular arrangement are unrelated to these; see 7 below).

2 Simple linear glycans

(a) *Simple linear glycans of the form -A-A-A-A-, in which all residues and linkages are alike.* Forms may arise in which local modification occurs after polymer assembly to produce 'blocks' of different structure (e.g. alginic acid).

(i) Synthesis is by the addition at the nonreducing terminal (e.g. AMYLOSE; eukaryotic cellulose; CHITIN).

(ii) Synthesis is in a retrograde direction (e.g. bacterial cellulose).

(b) *Simple linear glycans with a dissimilar reducing terminal* of the form B-A-A-A-A- (e.g. levan, bacterial dextran, inulin). Such glycans are typically found as extracellular bacterial polysaccharides and occur in vacuoles of some plants: they arise by transglycosylation (see below).

3 Glycans based on or derived from alternating disaccharides of the form -A-B-A-B-A-B-

(a) *Synthesized by addition at the nonreducing terminal.* Either the sugars, their anomeric configuration or position of their linkage may alternate, or any combination of these. Examples include glucosaminoglycans (HEPARAN SULPHATES, HEPARIN) and galactosaminoglycans (CHONDROITIN SULPHATES) of animals, some rhamnogalacturonans of plants (*see* PECTINS) and teichuronic acids of bacteria. Though glycosaminoglycans are synthesized as regular, alternating disaccharides, they are normally modified after chain assembly. In the galactosaminoglycans, additional sulphate residues may be inserted irregularly and in segments of chondroitin 4-sulphate, residues of D-glucuronic acid may be 5-epimerized *in situ* to L-iduronic acid, to give domains of DERMATAN SULPHATE. In the glucosaminoglycans, deacetylation, sulphation at various positions, and 5-epimerization can generate very irregular polymers without extensive domain structure and with an information content.

(b) *As above but synthesized in a retrograde direction* (i.e. by addition at the reducing terminal), for example HYALURONIC ACID.

(c) *As above but having peptide side chains on some sugars at the time of assembly of the glycan*; these side chains later cross-link. Peptidoglycans form this group.

4 Branched homogeneous glycans

(a) *Homogeneous branched glycans with 'backbones' of the general form*

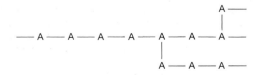

such that there are more unbranched than branched sugar residues and main 'backbone' sequences (-A-A-A-A-) can be identified. Theoretically the branches can be arranged in several ways, that is laminated

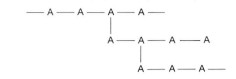

random

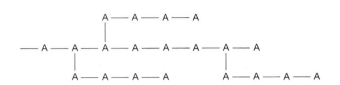

or random arborizing, in which the branches themselves branch further. This last is the closest approximation to the structures of glycogen, PHYTOGLYCOGEN and AMYLOPECTIN. The only variation from homogeneity is in the existence of two positions of linkage, to give branches. Typically, such branches are generated by TRANSGLYCOSYLATION, using a 'BRANCHING ENZYME'.

(b) *Highly branched homogeneous glycans* are a special case of the type above, where the degree of branching is so great that 'backbone' sequences cannot be defined, that is of the form

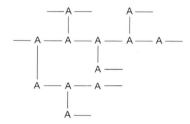

5 Heterogeneously branched glycans

(a) *Heterogeneously branched regular glycans* occur where the branches differ in sequence from the repeating backbone. The difference may be in the sugars themselves, their anomeric configuration, their position of linkage, their configuration, their conformation, or any combination of these. Commonly, such branches are short and of non-repeating structure, that is

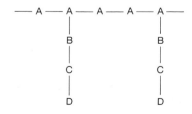

A single glycan may have one or several different types of side-chain and their spacing may or may not be regular. Examples include some TYPE 1 PECTINIC ACIDS and XYLANS.

(b) *Heterogeneously branched irregular glycans* are similar to the above, but also show irregularity in the backbone sequence, that is, they are of the form

continued

Table G9 *Continued*

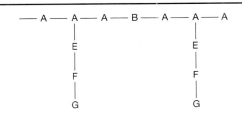

This is a more common type of plant polysaccharide than the regular form and most xylans and GALACTOGLUCOMANNANS are examples.

6 Heterogeneously superbranched glycans

(a) *Heterogeneously superbranched regular glycans* resemble the type of 5(a) (above), but the branches are themselves branched and may have either regular or irregular cores. If regular, they may be like the backbone, or unlike it, that is various types have forms such as

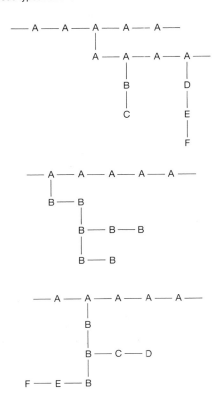

Examples come mostly from plant cell walls and gums and include some GALACTANS and some ARABINOGALACTANS.

(b) *Heterogeneously superbranched irregular glycans* correspond to the above, but have irregular backbones. They include the most complex types of glycan such as TYPE II PECTINIC ACIDS.

7 Homogeneously branched glycans with phosphate bridges

These are a distinctive group of polymers of fungal walls, particularly found in some yeast MANNANS. The basic structure is that of a branching mannan with $\alpha1,2$, $\alpha1,3$ and $\alpha1,6$ links, with occasional phosphate bridges from C6 or one mannosyl residue to C1 of the next. In some species other sugars may occur as outer chains.

8 'Modular' glycans

These form a distinct group found chiefly in Gram-negative bacteria. They are distinguished by having a repeating unit of up to about six sugars (which may be linear or branched) and by showing retrograde synthesis (i.e. they grow at the reducing terminal). Two subsets can be recognized.

(a) *'Modular' polysaccharides* which are pure glycans of forms such as

$$(\ —\ A\ —\ B\ —\ \underset{\underset{E}{|}}{C}\ —\ D\ —\ )_n$$

Each 'module' is synthesized in a 'forward' direction on undecaprenyl pyrophosphate and the modules are then spliced together 'backwards' to yield an overall retrograde assembly. Branches arise where the splicing is at a subterminal sugar.

(b) *O-antigen chains of lipopolysaccharides*, in which the 'modular polysaccharides' are added to non-repeating glycan cores which are, in turn, covalently linked to lipid A. These are major components of the outer membranes of many Gram-negative bacteria (*see* BACTERIAL ENVELOPE).

9 Irregular glycans

These contain little or no repeated structure and constitute a large group of saccharides that are typically found as glycoconjugates of lipid or protein. Most tend to be smaller than those glycans with a large content of repeated structure. They may be linear or branched.

(a) GLYCOLIPIDS.

(b) GLYCOPROTEINS can contain more than one type of glycan per molecule and so cannot be easily classified by saccharide content. Individual glycoproteins are classified primarily upon the basis of their linkage to protein and secondarily upon features of the glycan distal to this link. A brief listing of the different types is given here. More information will be found in the entry for GLYCOPROTEINS AND GLYCOSYLATION.

(i) *N-linked glycoprotein glycans* (of animals, plants and fungi) contain an N-aspartamido linkage between *N*-acetyl glucosamine (i.e. 1-deoxy, 2-acetamido D-glucopyranose) and the amino acid asparagine (Fig. G26) [10]. There is also recent evidence for N-linked glycoprotein glycan in several species of eubacteria and archaebacteria, especially the latter [11]. N-linked glycans are alkali-stable.

(ii) *O-linked glycoprotein glycans* are of several different unrelated types: (1) *mucin type glycans* of animals; (2) *'classical' glycosaminoglycan (or GAG) type glycans* of animal proteoglycans (*see* GLYCOSAMINOGLYCANS); (3) *O-mannosyl linked glycans*; (4) *O-galactosyl linked glycans*; (5) *'classical' collagen-type glycans* of mammalian collagen and collagen-related proteins (e.g. C1q); (6) *O-2 deoxy, 2-acetamido glucosyl proteins* which have been identified in cytoplasm, within nuclei and at the nuclear pores [12] and may be involved in the transport of mRNA; (7) *arabinosyl-hydroxyprolyl links* (Fig. G27): these occur with galactosyl serine, galactosyl hydroxyproline, and arabinosyl serine in plant cell wall glycoproteins (extensins); (8) *glycogen primer protein*, also called glycogenin (*see* GLYCOGEN) which contains the structure Glc-O-Tyr on a polypeptide of M_r ~38 000 which is also a subunit of glycogen synthase; (9) *O-fucosyl glycans* in which fucose or Glcβ1,3Fuc are linked to threonine have been found in urine, but the native glycoprotein is unidentified.

(iii) *S-linked glycoprotein glycan* has been found in a fragment from human urine: a glucosyl trisaccharide was linked to cysteine. Its source is unknown.

(iv) *Others*. Linkages of dermatan sulphate to L-lysine and, through xylose to L-alanine have recently been detected. Their chemistry is still imperfectly known.

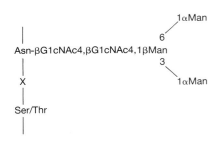

Fig. G26 The *N*-aspartamido linkage in N-linked glycoproteins (see Table G9, **9**).

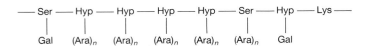

Fig. G27 Arabinosyl-hydroxyprolyl links found in plant cell wall glycoproteins (see Table G9, **9**).

a large, relatively homogeneous or repetitive 'polysaccharide' domain which is attached to protein or lipid by a smaller, distinct, domain called a link sequence, the formation of which is by a different mechanism and may occur at a different cellular site from that of the polysaccharide. In the classification given in Table G9, link sequences are considered under **9**.

Three-dimensional structures

Glycans with a high content of repeated sequence and little branching tend to form regular quasi-crystalline structures, with intrachain and interchain hydrogen bonding. Antiparallel arrangements of chains occur in cellulose and chitin, while helical structures are widely found (e.g. CALLOSE and XYLANS). Generally 1,4 or 1,6 linked glycans have helices of larger pitch than 1,3-linked types.

Branched glycans can have similar, orderly 'cores' from which side-chains project (as xylans) or can be formed into extended, flexible or random-coiled arrangements (as in plant amyloids). Self association can occur (as in type II pectinic acids and some glucosaminoglycans) and extensive hydration is common.

Typical pyranoside rings are about 6.4 Å in length, so that a 'typical' sugar is much larger than a 'typical' amino acid, especially if it is sulphated or has a bulky side chain (as in acetamidosugars and sialic acids). Hence a protein that 'recognizes' a glycan with high specificity must have an extended binding site, and upper limits exist on the size of glycan that a given size and shape of protein could accommodate (see LECTINS).

Recently, evidence has been obtained for specific glycan–glycan interaction between dissimilar glycoprotein glycans, suggesting that the recognition of glycan signal structures might be mediated with high accuracy by glycan.

Biosynthesis

Glycans can form from several different types of glycosyl transfer. Direct, nonenzymatic glycosylation (glycation) from free sugar is very rare, but occurs pathologically in DIABETES MELLITUS and galactosaemia. More usually, a sugar is activated by phosphorylation at the position at which the future glycoside will form (e.g. C1 in aldoses). Sugar phosphates are rarely direct donors for glycan formation *in vivo*; they are usually converted to a SUGAR NUCLEOTIDE of the nucleotide diphosphate–sugar (NDP-S) type;

sialic acids and the bacterial sugar 2-keto-3-deoxyoctonic acid (KDO) are converted to sugar nucleotides of the nucleotide monophosphate–sugar (NMP-S) type by reaction of the free sugar with the nucleoside triphosphate. Interconversion of sugar skeletons occurs at the sugar phosphate and sugar nucleotide levels. Sugar nucleotides may act directly as glycosyl donors, with or without a glycosyl enzyme intermediate, or give rise to glycosides of isoprenyl lipid phosphates and pyrophosphates.

In prokaryotes the isoprenoids involved in glycan synthesis are generally undecaprenyl phosphates and pyrophosphates; in eukaryotes they are the corresponding dolichyl compounds and retinyl phosphate (*see* DOLICHOL; RETINOL; UNDECAPRENYL AND DOLICHOL CYCLES; UNDECAPRENOL). Dolichyl phosphate and pyrophosphate have recently been shown to participate in glycan assembly in the membrane glycoprotein of *Halobacterium halobium*, an archaebacterium, and dolichol has also been found in *Methanothermus*. Undecaprenyl and dolichyl pyrophosphoryl sugars act as acceptors for other glycosyl transfers, variously from sugar nucleotides or undecaprenyl and dolichyl phosphoryl sugars. They also act as donors of 'prefabricated' glycan units or chains of such units to other glycan units (in prokaryotes) or protein (in eukaryotes) (Fig. G28) (*see also* GLYCOPROTEINS AND GLYCOSYLATION).

Retinyl phosphate serves as a carrier for shuttling sugars from sugar nucleotides across the membranes of the Golgi apparatus to sites of glycan assembly. Once assembled, some glycans can be rearranged by branching enzymes and other forms of transglycosylation. Most direct glycosyl transfers reverse the anomeric configuration of the sugar: preservation of configuration often implies a double transfer [3].

In bacterial glycan assembly at the cell membrane the isoprenoids serve as shuttles to transfer sugar from sugar nucleotides across the membrane and as 'handles' to locate the growing glycan at the membrane (see UNDECAPRENOL AND DOLICHOL CYCLES). Glycans synthesized in a wholly extracellular site, such as levans and dextrans, are formed by transglycosylation from disaccharides.

In eukaryotes, dolichol-dependent glycan assembly is particularly associated with the rough endoplasmic reticulum, with the isoprene acting as both a shuttle and a means of location in the membrane [4,5]. Autonomous synthesis in the nucleus and mitochondria is probably also dolichol-dependent.

Glycosyl transfers in the Golgi apparatus are dolichol-independent, though retinyl phosphate acts as a carrier in some

Fig. G28 (*opposite*) The dolichol phosphate cycle and the synthesis of the side chains of glycoproteins. G, Glucose; Gn, *N*-acetylglucosamine; M, mannose.

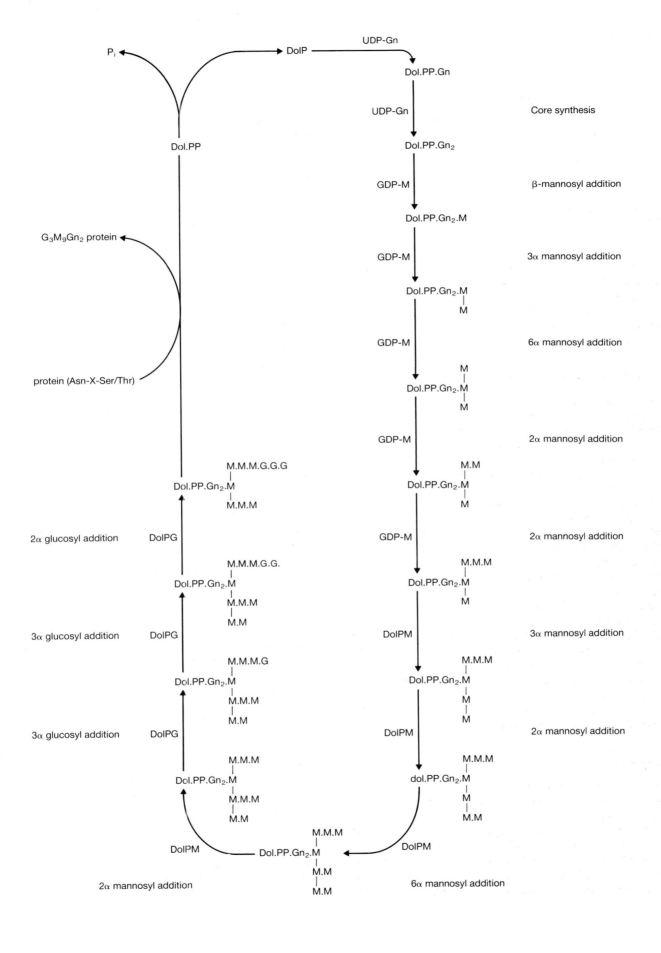

of them. They involve glycosyl transfers from sugar nucleotides by means of membrane-bound enzymes which are highly specific for the sugar transferred, the nucleoside base, the acceptor glycan sequence, the position of the glycoside formed, its anomeric configuration, and the conformation of the sugar attached.

From this comes the one-linkage–one-enzyme hypothesis, which postulates that each glycosidic linkage in a glycan arises from the action of an enzyme specific for the formation of that linkage and no other, and that the specificity also defines both the donor sugar nucleotide and the acceptor sequence (*see* GLYCOPROTEINS AND GLYCOSYLATION).

Very few exceptions are known to this rule (but see ABO BLOOD GROUPS). Hence glycan sequences are primarily defined by the successive action of glycosyl transferases and not by any form of simple encoding (e.g. as in polynucleotide or polypeptide chains). Certain structures are forbidden and most glycoprotein and glycolipid glycans appear to have singular pathways to them. Factors other than the presence or absence of glycosyl transferases, glycosyl donors, and acceptor glycans are also involved, as lectin-resistant mutants show, implying that the larger-scale arrangement of the reactants in membranes is also of central importance. Chain-termination may occur because of a transfer which precludes any other additions, or as a result of the removal of (e.g.) glycan from the transferase complex [3]. In many cases the exact mechanism is unknown. It should be noted that transfers commonly regarded as chain-terminating in mammals (e.g. α-L-fucosyl addition) are not necessarily so in lower animals.

The specificity for the glycan acceptor sequence, though high, is inevitably limited by the size of the individual sugars comprising it relative to proteins, so that specificity is seldom for more than about a tetrasaccharide. Hence, a single transferase may act at more than one point in assembly of a single glycan without breaching the one-linkage–one-enzyme hypothesis. Where a glycosyl transferase produces its own acceptor sequence it gives rise to a simple repeating glycan. Where two transferases act alternately, each producing the acceptor for the other, a repeating disaccharide results. In such cases mechanisms of chain termination must exist, though it is seldom known what they are.

Glycan assembly in the Golgi apparatus may be primed within the organelle, or elsewhere. Assembly of complex N-linked glycans involves the assembly of the core via dolichol-dependent transfers in the rough endoplasmic reticulum, translocation of the glycoprotein to the Golgi apparatus, removal of sugars by specific glycosidases (at and near the forming face of the Golgi apparatus) and construction of the outer chains as the protein moves further into the Golgi cisternae.

Glycosyl transfers at eukaryotic cell surfaces are relatively uncommon. Late stages in cellulose assembly occur there and the peculiar mechanisms of hyaluronate formation are surface-associated. Lysosomal glycans are primarily formed in the rough endoplasmic reticulum, but receive their phosphate residues in the Golgi apparatus, where they are segregated for lysosomal insertion (*see* LYSOSOMES; PROTEIN TARGETING). A failure of this mechanism leads to I-CELL DISEASE in humans.

Nothing is known of the origin of most cytoplasmic glycan, other than GLYCOGEN.

Transglycosylation within and between glycans occurs as a mechanism independent of sugar nucleotide or isoprenoid-related processes. It produces no net increase in the number of glycosidic linkages, but rearranges molecules or enlarges some at the expense of others. It is the major method of synthesis of bacterial levans, dextrans and of intravacuolar glycans of plants (e.g. levan and inulin).

Initiation of glycan synthesis

This can occur by several mechanisms. Disproportionation between two molecules of isoprenyl (or dolichyl) pyrophosphoryl glycan, two molecules of disaccharide or two (dissimilar) molecules of sugar nucleotide initiates the formation of O-antigenic polysaccharide, some types of bacterial glycoprotein glycan formation, other bacterial polysaccharide (e.g. levan) and hyaluronic acid respectively. The enzymes which initiate the chains also extend them: that is, no *specific* initiating mechanism exists. The initiation of inulin synthesis is rather similar. Specific 'core' synthesis, through production of a distinct priming or linking glycan sequence by specific glycosyl transfer, occurs with lipopolysaccharides and is probably general with eukaryotic glycans, though it has not been proved for many polysaccharides (*see* CELLULOSE; CHITIN; GLYCOGEN; GLYCOSPHINGOLIPIDS; GLYCO-GLYCEROLIPIDS).

It is not clear whether specific primers always exist for teichoic acids and peptidoglycan, as any bacterium will have at least a small amount of the glycan present under almost all circumstances. Lipoteichoic acid may serve as an initiator for teichoic acid assembly, but peptidoglycan chains are normally initiated by a failure of transglycosylation after successful transpeptidation, leaving an oligosaccharide unit unattached and available as a primer.

Processing reactions

These modify the N-linked glycan initially produced by the (classical) dolichol phosphate cycle of eukaryotic cells, to produce the various categories of N-linked glycoprotein glycan found in them. The enzymes involved are membrane-bound or membrane-associated and are mostly located in the Golgi apparatus (*see* GLYCOPROTEINS AND GLYCOSYLATION).

In fungi, the synthesis of cell-wall MANNAN requires a variant of the processing reactions [6], in which $Asn-(GlcNAc)_2(Man)_9(Glc)_3$ is converted to $Asn-(GlcNAc)_2(Man)_9$, as in other systems, but then receives three transfers from GDP-Man to form $Asn-(GlcNAc)_2(Man)_{12}$ which acts as a core. Further transfers from GDP-Man then build up the outer mannan chains. In some species occasional phosphoryl links occur in these and the outermost parts of the structure may bear sequences of other sugars.

In *Saccharomyces cerevisiae* a series of mutants are known with defects in the synthetic and processing pathways (Table G10).

Sulphation of high mannose structures can also occur, but its mechanism has been little studied.

Glycan synthesis can be inhibited by a wide range of mechanisms (Table G11).

Table G10 Yeast mutants defective in the synthesis and processing of glycans

Mutant	Reaction affected
alg1	$Dol.PP.Gn_2 \rightarrow Dol.PP.Gn_2.M$
alg2	$?Dol.PP.Gn_2.M_2 \rightarrow Dol.PP.Gn_2.M_3$
alg3	$?Dol.PP.Gn_2.M_5 \rightarrow Dol.PP.Gn_2.M_6$
alg4	Not known; global defect leading to accumulation of $Dol.PP.Gn_2.M_{5-8}$, (i.e. not of a single transferase)
alg5	$Dol.P + UDP\text{-}G \rightarrow Dol.P.G + UDP$
alg6	$Dol.PP.Gn_2.M_9 \rightarrow Dol.PP.Gn_2.M_9.G$
alg7	$Dol.P. + UDP\text{-}Gn \rightarrow Dol.PP.Gn + UMP$
alg8	$Dol.PP.Gn_2.G \rightarrow Dol.PP.Gn_2.M_9.G_2$
mnn1	$\alpha1,3$ mannosyl addition to core and outer chains
mnn2	$\alpha1,2$ mannosyl addition to $\alpha1,6$ mannan of outer chains
mnn4	Addition of phosphoryl mannose linkages
mnn5	$\alpha1,2$ mannosyl addition to $\alpha1,2$ mannosyl residues of outer chains
mnn6	As *mnn4*
mnn9	$\alpha1,6$ mannosyl addition to initial outer chain backbone (core is normal)
gls1	$\alpha1,2$ glucosidase deficit
fus mutants	Fusion mutants, some probably involving O-glycosyl mannan
sec mutants	Secretion mutants, some probably involving glycan (*sec53* is an allele of *alg4*)

Functions

In prokaryotes, the primary function of peptidoglycan is to form an inextensible macromolecular lattice, investing the cell, which prevents osmotic swelling and rupture of the surface membrane. Teichoic and teichuronic acids act as ion exchangers and regulate the microenvironment of the membrane. Lipopolysaccharides are structurally part of the cell surface and carry antigenic glycans which may also be environmental regulators, recognition structures and means of adhesion. In pathogens, they may be toxic to the host and modify host responses. Bacterial polysaccharides also influence the microenvironment, may form protective capsules and may interact with host defences so as to protect pathogens. Glycans of viral envelopes and of some bacteria can influence the uptake of the pathogen into cells to be parasitized.

In eukaryotes, glycans have very diverse functions and many have no known role. Energy storage glycans are common and are most usually cytoplasmic α-glucans (glycogen) or are associated with plastids (starch). Fructans (levan, inulin) are uncommon and intravacuolar. Other storage glycans are rare.

Structural glycans occur on two scales. Small oligosaccharides can stabilize protein folding and polypeptide chain association (e.g. in immunoglobulin hinge regions), or, by mutual repulsion, force the assumption of an extended, flexible configuration (e.g. antifreeze glycoprotein of arctic fish) [7]. Such glycans can confer

Table G11 Inhibition of glycan synthesis

Inhibition of primer synthesis. A common consequence of the inhibition of protein synthesis (e.g. by cycloheximide) leading to a slowly developing blockade of glycoprotein glycan production

Pseudo-priming. Can occur in cell-free and intactcell systems. Phenyl phosphate acts as an analogue of dolichyl phosphate and para-nitrophenyl xyloside acts as a false acceptor for glycosaminoglycan synthesis. Glycan typically forms, at least to some extent, but it is not produced as a normal glycoconjugate

Abortive synthesis. Analogous to lethal synthesis. Sugar analogues are incorporated into glycan and block its further synthesis by forming inactive acceptor analogues, so blocking further transfers. 2-Deoxyglucose, several fluoro-analogues of glucose and mannose, and glucosamine act this way on glycoprotein *N*-glycan synthesis, by interfering with the dolichol phosphate cycle. Abnormal dolichyl pyrophosphoryl oligosaccharides accumulate

Sequestration of intermediates. Occurs with bacitracin and a few related inhibitors. Undecaprenyl-phosphoryl sugars, etc. are rendered unavailable for synthesis by being complexed with the antibiotic and held in place in membranes

Inhibition of glycosidases. Inhibits synthesis of glycan by blocking processing reactions. It is an important mechanism for preventing synthesis of complex-type *N*-linked glycoprotein glycans. Commonly used inhibitors include 2-deoxynojirimycin and its relatives — swainsonine, castanospermine and australine. Their sites of action differ slightly: all lead to accumulation of metabolically 'earlier' glycans and loss of those which are 'later'

Genetic defects in acceptor proteins. Lead to a loss of sequons at sites of glycan attachment, with consequent production of aglyco-forms of protein (which may be degraded abnormally rapidly)

Premature termination. Occurs in the synthesis of repeating polymers: (a) if membrane systems are behaving abnormally, leading to early scission and release of the polymer (as appears possible with hyaluronic acid); or (b) if a termination signal is wrongly added too early. Altered membrane flow can lead to: (a) a failure of processing or packaging of glycan in the Golgi apparatus (e.g. the effect of monensin); or (b) a failure to bring glycosyltransferase(s), donors and acceptors together (e.g. in some lectin-resistant mutant cell lines)

Rare forms of inhibition include:

(a) Blockade of transferase synthesis, which is limiting in the action of cycloheximide on chitin synthesis;
(b) Inhibition of glycosyl transfer. Specific inhibitors for most transferases are not known, but important exceptions include the tunicamycins, which prevent synthesis of Dol-PP-GlcNAc and coumarin, which prevents transfer of oligoglucan from dolichyl pyrophosphate to protein, in cellulose synthesis. Such inhibition may be rapid or slow to appear, depending upon product pool sizes. Hence, indirect effects can be misleading;
(c) Inhibition of sugar nucleotide synthesis, etc., which is rarely limiting

resistance to proteolysis. Larger oligosaccharides are structural components of matrices in themselves and can have specific roles. In cell walls of higher plants [8] cell–cell adhesion requires the integrity of type I pectinic acids, while the cohesion of each cell wall as a whole, but not adhesion between them, requires cellulose. Cell wall extension involves rearrangement of xylans around cellulose fibres, modification of type II pectinic acids, new synthesis of several types of polysaccharide and EXTENSIN and changes in the sequestration of calcium ions by glycan.

In cartilaginous matrices of animals [9], highly hydrated hyaluronic acid interacts with a binding site on each molecule of the core protein of proteoglycan to give a noncovalent link reinforced by a link glycoprotein, so that the whole structure forms a vast hydrated lattice in which glycosaminoglycans have a space-filling function, as well as carrying informational domains with which other molecules can interact. A collagenous capsule and internal collagen restrains the indefinite swelling of the structure. Other variants of matrix exist, essentially based upon the same components (e.g. the vitreous humour of the eye). Animal matrices are chemically very diverse, but in all of them glycan is structurally important and can influence processes such as mineralization by its capacity for ion exchange and coordination.

Glycans are known to act as signals in some specific instances and they are thought to do so much more widely. Exposure of β-galactosyl residues on effete, circulating glycoproteins and blood cells, by loss of terminal sialyl residues, leads to their removal in the liver. α-Mannosyl residues on particulate surfaces will stimulate phagocytosis by macrophages, without opsonization. 6-Phosphoryl mannosyl terminals on glycoprotein glycan act as intracellular signals for targeting to lysosomes and extracellular signals for uptake by fibroblasts. The microheterogeneity of the glycans of transferrin and other glycoproteins influences the survival time of the different glycoforms in the circulation and their targeting to specific cell types. Glycoforms of membrane proteins also occur and can be tissue-specific (e.g. Thy-1 glycoprotein).

R.W. STODDART

1 Biermann, C.J. (1988) Hydrolysis and other cleavages of glycosidic linkages in polysaccharides. *Adv. Carb. Chem. Biochem.* **46**, 251–271.
2 Sweeley, C.C. & Nunez, H.A. (1985) Structural analysis of glycoconjugates by mass spectrometry and nuclear magnetic resonance spectrometry. *Annu. Rev. Biochem.* **54**, 765–801.
3 Stoddart, R.W. (1984) *The Biosynthesis of Polysaccharides* (Croom Helm, London/Sydney).
4 Kornfeld, R. & Kornfeld, S. (1985) Assembly of asparagine-linked oligosaccharides. *Annu. Rev. Biochem.* **54**, 631–664.
5 Hirschberg, C.B. & Snider, M.D. (1987) Topography of glycosylation in the rough endoplasmic reticulum and Golgi apparatus. *Annu. Rev. Biochem.* **56**, 63–87.
6 Kukuruzinska, M.A. et al. (1987) Protein glycosylation in yeast. *Annu. Rev. Biochem.* **56**, 915–944.
7 Rademacher, T.W. et al. (1988) Glycobiology. *Annu. Rev. Biochem.* **57**, 785–838.
8 McNeill, M et al. (1984) Structure and function of the primary cell walls of plants. *Annu. Rev. Biochem.* **53**, 625–663.
9 Kjellen, L. & Lindahl, U. (1991) Proteoglycans: structure and interactions. *Annu. Rev. Biochem.* **60**, 443–475.
10 Montreuil, J. (1980) Primary structure of glycoprotein glycans. Basis for the molecular biology of glycoproteins. *Adv. Carb. Chem. Biochem.* **37**, 158–223.
11 Lechner, J. & Weland, F. (1989) Structure and biosynthesis of prokaryotic glycoproteins. *Annu. Rev. Biochem.* **58**, 173–194.
12 Hart, G.W. et al. (1989) Glycosylation in the nucleus and cytoplasm. *Annu. Rev. Biochem.* **56**, 915–944.

glyceollins Isoflavonoid PHYTOALEXINS produced by some leguminous plants as part of the defence response. The elicitation of glyceollins has been particularly well defined at the molecular level in French bean (*Phaseolus vulgaris*), where it is associated with *de novo* induction of the phenylpropanoid pathway.

glycine (Gly, G) *See*: AMINO ACIDS; GABA AND GLYCINE RECEPTORS.

glycine modulation The requirement for the amino acid glycine for NMDA EXCITATORY AMINO ACID RECEPTOR activation. Glycine binds to a distinct site on the NMDA receptor and its binding is essential for receptor activation by NMDA or glutamate, that is, glycine is a cotransmitter at this receptor.

glycobiology That part of cellular and molecular biology referring to the structure, synthesis and biological role of GLYCANS. That science that deals with abnormality in these and the role of glycans in clinical disorders would be glycopathology.

glycocalyx Extracellular layer composed of carbohydrate. In bacteria it is usually a polysaccharide layer secreted by the cell and is also referred to as the slime layer or capsule (*see* BACTERIAL ENVELOPE). In eukaryotic cells, the glycocalyx refers to the layer formed by the oligosaccharide side chains of membrane GLYCOLIPIDS and GLYCOPROTEINS.

glycoconjugate *See*: GLYCANS.

glycoforms Variant forms of GLYCOPROTEINS in which heterogeneity arises, biosynthetically, by structural variation in the GLYCAN component. All glycoforms of a protein have identical peptide sequences, but the glycan varies both by the presence and absence of individual glycans and/or by their particular, individual structures. A similar, but distinct, term is needed for the equivalent variability in whole cells: 'glycotype' would be an appropriate usage.

glycogen Family of storage GLUCANS of animal tissues, which contain linear chains of α1,4 glucan, branching by way of α1,6 linkages. The polymers are continuously remodelled and have indefinite structures, with the average inter-branch distance varying with source.

Synthesis occurs in three stages. A membrane-associated glycogen initiator synthetase complex adds glucosyl residues from UDP-Glc to a GLYCOPROTEIN acceptor, which is a subunit of glycogen synthase (M_r 34 000) to form a primer containing about nine glucosyl residues at each of several primer sites. These are linked by Glc-O-Tyr sequences. Glycogen synthase, which is membrane associated, then adds further glucose, from UDP-Glc, to form larger chains of α1,4 glucan. This process is regulated by the interconversion of glycogen synthase between its Glc-6-P independent (active) and Glc-6-P dependent (relatively inactive) forms by the actions of synthase kinase and a specific phosphatase (*see* PROTEIN KINASES; PROTEIN PHOSPHATASES; PROTEIN PHOSPHORYLATION). The former is itself regulated by cAMP levels. Finally, a soluble branching enzyme severs short α1,4 terminal sequences and transfers them to positions a few residues before the (new) terminal positions, to form α1,6 branches by internal transglycosylation. Chain extension then proceeds again. A similar mechanism generates STARCH.

Physiologically the α1,4 glucan chains of glycogen are degraded by phosphorylases to produce glucose 1-phosphate and the branch points removed by debranching enzymes which release glucose. Mammalian liver and muscle phosphorylases are different enzymes, although they are similarly regulated by phosphorylase kinases (which are in turn regulated by cAMP levels, either directly or via kinase kinases). Regulation is in the opposite direction from that of glycogen synthetase. Free glycogen can also be degraded to glucose by lysosomal α-glucosidase.

Defects in the synthesis or degradation of glycogen or in the handling of sugar phosphates lead to glycogenoses (Table G12). The structure of glycogen is altered towards very long chains in branching enzyme defects and towards limit dextrin in debranching enzyme defects. Only an abnormality of lysosomal α-glucosidase leads to a true glycogen storage disease. Synthesis of glycogen chains by reversal of the phosphorylase reaction does not occur physiologically, but can occur *in vitro*. *See also*: AMYLOPECTIN; AMYLOSE.

Table G12 Glycogenoses

Type	Defective enzyme	Product abnormality
I	Glucose-6-phosphatase	Excessive quantity of normal glucan
II	Lysosomal acid maltase	Glucan in LYSOSOMES
III	Debranching enzyme	Limit dextrin
IV	Branching enzyme	Long outer chains/few branches
V	Muscle phosphorylase	Very large molecular size
VI	Liver phosphorylase	Very large molecular size
VII	Phosphofructokinase	Excessive quantity of normal glucan
VIII	Liver phosphorylase kinase	Very large molecular size

glycogen phosphorylase A transferase (EC 2.4.1.1) that catalyses the reversible transfer of a glucose residue from the non-reducing end of the glycogen chain to orthophosphate to form glucose 1-phosphate. The activity of mammalian glycogen phosphorylase is controlled by PROTEIN PHOSPHORYLATION by PHOSPHORYLASE KINASE (*see also* PROTEIN KINASES). Glycogen phosphorylase, which is active as a homodimer or homotetramer, is an ALLOSTERIC enzyme.

glycogen synthase A transferase (EC 2.4.1.11) involved in GLYCOGEN synthesis, transferring a glucose residue from UDP-glucose to the glycogen chain with the formation of a α1,4 linkage.

glycogenin The glucosylated primer for GLYCOGEN synthesis.

glycogenoses Disorders of GLYCOGEN metabolism. *See*: MCARDLE DISEASE; POMPE DISEASE.

Glycolipids

GLYCOLIPIDS comprise a LIPID moiety to which is attached a GLYCAN. They can be subdivided into several unrelated groups. Dolichyl phosphoryl and pyrophosphoryl glycans, undecaprenyl phosphoryl and pyrophosphoryl glycans, and retinyl phosphoryl glycans are all intermediates in the biosynthesis of other types of glycan (*see* DOLICHOL; RETINOL; UNDECAPRENOL; UNDECAPRENOL AND DOLICHOL CYCLES), while glycosphingolipids, glycoglycerolipids, steroidal (and related isoprenoid) glycosides, and glycosylated phosphoinositides are metabolic end-products.

Glycosphingolipids

These comprise a large group of glycolipids in which glycan is attached glycosidically to *N*-acylated sphingosine (*N*-acyl 4-sphingonine), or one of its derivatives, by the primary hydroxyl group of the long-chain base. When linked through amides to fatty acyl chains the aglycone is termed a ceramide. The structure of the lipid has no known influence upon the saccharide sequence of the attached glycan. Glycosphingolipids are the most abundant glycolipids of mammalian tissues and are widespread components of surface membranes and many endomembranes (but not the nuclear membrane) where they occur in the exterior leaflet with the polar oligosaccharide head groups on the extracellular (or luminal) face of the membrane. They vary widely in hydrophobicity and behaviour on partition into organic solvents, depending upon the size of the glycan. Mammalian glycosphingolipids fall into two groups, differing in the first sugar linked to the aglycone.

Galactosphingolipids

This is the smaller group and is based on Galβ1-ceramide. They seldom carry more than five sugar residues and are often sulphated at C3 of galactosyl residues to give sulphatides (i.e. sulphated glycolipids). Small sialoglycolipids are known in the galactosphingolipids.

Glucosphingolipids

These are based on Glcβ1-ceramide and are a large and diverse group. They seldom carry sulphate ester groups. There are three main biosynthetically distinct subsets:
1 Those having linear, nonrepeating, nonpolar, nonfucosylated glycans.
2 Those nonpolar forms having linear and sometimes repeating backbones, with fucosyl branches.
3 Those bearing sialylated glycans or their precursors. The sialylated glycosphingolipids are termed GANGLIOSIDES.

The ABO and Lewis blood-group antigens are expressed only on certain of the fucoglycolipids (*see* ABO BLOOD GROUP SYSTEM; LEWIS BLOOD GROUP).

The nomenclature of glycosphingolipids is complex and several older, trivial names persist in common usage. The principle of the systematic nomenclature is to assign a prefix to the name of the

Table G13 System for encoding sequences of glucosphingolipids

Prefix	Sequence	Symbol
lacto	Gaβ 3Gnβ 3Gaβ 3Gnβ 3Gaβ 4G-	Lc
lactoiso	Gaβ 4Gnβ 3Gaβ 3Gnβ 3Gaβ 4G-	Lci
lactoneo	Gaβ 4Gnβ 3Gaβ 4Gnβ 3Gaβ 4G-	Len
lactoisoneo	Gaβ 3Gnβ 3Gaβ 4Gnβ 3Gaβ 4G-	Lcni
globo	Ganα 3Gaα 4Gaβ 4G-	Gb
globoiso	Ganα 3Gaα 3Gaβ 4G-	Gbi
muco	Gaβ 3Gaβ 4Gaβ 4G-	Mc
ganglio	Gaβ 3Ganβ 4Gaβ 4G-	Gg

lipid on the basis of its first four sugars and a suffix to describe the number of sugars in it.

Thus

$$Gal\beta1,3GlcNAc\beta1,3Gal\beta1,4Glc\beta1\text{-Cer}$$

is lactotetraglycosyl ceramide (LcOseCer), and

$$GalNAc\beta1,3Gal\alpha1,3Gal\beta1,4Glc\beta1\text{-Cer}$$

is globoisotetraglycosyl ceramide (GbiOseCer) (Table G13). When extended beyond four sugars, the prefixes grow longer and the variants more numerous. Branched glycans are indicated by a Roman numeral to denote the sugar at which the branch occurs, numbering from the reducing end, and a superscript Arabic number to specify the position of the branch. The anomeric configuration of the branch glycoside is indicated if known. Thus $V^3\alpha Gal$, placed before the rest of the code for the lipid would indicate a branch galactosyl residue at C3 of the fifth sugar, in the α-anomeric configuration.

Lower animals can contain various unusual lipids, including classes of glycosphingolipids which do not fit to the standard system of nomenclature. Some of these contain sugars not found in mammalian glycosphingolipids (e.g. mannose, L-arabinose) or linked in ways which are considered impossible in mammalian glycoconjugates (e.g. internal L-fucosyl residues, internal sialyl residues linked to distal sugars other than sialic acid). Plant glycolipids also diverge from the structures found in mammals.

Glycoglycerolipids

This is a small family of lipids in which glycan is glycosidically linked to the C3 hydroxyl of DIACYLGLYCEROL, usually by way of a galactosyl residue. They occur widely in plant tissues and are found in a few animal tissues, notably brain. Monogalactosyl diacylglycerol has been implicated in myelination. They tend to be more water soluble than the corresponding glycosphingolipids.

Steroidal and other related isoprenoid glycosides

This family of plant glycosides is generally not considered as glycolipids, though they are amphipathic. Typically, sugars are linked glycosidically to hydroxyl groups of STEROIDS, at C3 of the A-ring, or in the equivalent position to other nonsteroidal polycyclic isoprenoids, in which the D-ring is six-membered. Mono- to triglycosyl steroids usually show cardiotonic activity although the

sugar moiety is not involved in this activity, while pentaglycosyl steroids, etc. behave as saponins. Examples include OUABAIN (β-L-rhamnosyl strophanthidin G), digoxin (Digβ1,4Digβ1, 4Digβ1-digoxigenin, where Dig is digitoxose) and digitonin (Glcβ1,4Galβ1,2(Xy1β1,3)Glcβ1,4Galβ1-digitogenin).

Glycosylated phosphoinositides (GPIs)

A family of recently identified, membrane-associated GLYCOLIPIDS of protozoa, fungi, mammals, and probably most other classes of eukaryotes, which are characterized by the possession of glycan linked glycosidically to *meso*-inositol [1]. There are four major classes recognized so far, which share the conserved structure: $Man\alpha1,4GlcNH_2\alpha1,6$-*meso*inositol-1-$PO_4$-lipid. All glycosylated phosphoinositides are thought to have the capacity to lock together by way of the core glycan to form 'plates' which cover and protect the lipid bilayer of the underlying membrane.

1 GPI ANCHORS of cell surfaces are linked to the C-terminal residue of the polypeptide of a glycoprotein or proteoglycan by way of the conserved, common core structure:

ethanolamine phosphate-6Manα1,2Manα1,6Man-α1,4GlcNH$_2$α1,6-*meso*inositol-1-PO$_4$-lipid.

Proteins to be linked to this structure have a hydrophobic domain of 17 amino acids at their C terminus when first synthesized. This contains a signal for GPI addition, which is cleaved from the protein as the GPI is attached at the (new) C terminus. Proteins that can be attached in this way appear to be of a limited number of families, which are not interrelated. Failure of the mechanism leads to protein accumulation in the Golgi apparatus.

2 Lipophosphoglycans (LPG) occur in *Leishmania* spp. and probably other protozoa. They have the general core structure

Glcα1,PO$_4$
|
6
|
X-PO$_4$-6Galα1,6Galα1,3Gal(f)1,3Manα1,3Manα-1,4GlcNH$_2$α1,6*meso*inositol-1-PO$_4$-lipid

The outer glycan (X) consists of a repeat sequence and a terminating cap sequence, together of the form

Galβ1,4Manα1,2Manα1-(PO$_4$-6Galβ1,4Manα1-)$_n$-
|
3
|
R

cap   repeat

where *n* is between 10 and 50 and R varies with species (in *L. major* it can be H, Ara(p)1,2(Galβ1,3)Galβ1-, or (Galβ1,3)O-3Galβ1-). In *L. mexicana* it is H or βGlc1-, and in *L. donovani* it is always H) and the nonreducing terminal galactosyl residue may be absent. The cap sequence is a ligand for the serum mannose-binding protein which can opsonize the parasite at certain stages of its life-cycle and facilitate its entry into cells.

3 Lipopeptidophosphoglycans (LPPG) occur in *Trypanosoma* spp.
4 Glycoinositol phospholipids (GIPLs) are of two types: type I correspond to the membrane anchors, but lack polypeptide; type

II correspond to the LPGs, but lack the extensive repeat sequences. They seem to be peculiar to protozoa and may confer resistance to host attack on the cell membrane of the parasite by (e.g.) the terminal membrane-attack complex of the COMPLEMENT system.

Biosynthesis of GPIs occurs by the sequential addition of sugars to *meso*-inositol, with dolichyl phosphoryl mannose being the donor for the core. The glucosamine residue is added as the acetamido-sugar and is then de-N-acetylated. The initiation of synthesis is thought to occur in the ENDOPLASMIC RETICULUM, with any later sugar additions beyond the core occurring in the Golgi apparatus.

R.W. STODDART

See also: MEMBRANE STRUCTURE.

1 Cross, G.A.M. (1990) *Annu. Rev. Cell Biol.* **6**, 1–39.
2 Mason, P. & Caras, I.W. (1992) *J. Cell Biol.* **119**, 763–772.

---

glycophorin Transmembrane protein of red blood cells. *See*: GERBICH BLOOD GROUP; MEMBRANE STRUCTURE; MNS BLOOD GROUP.

Glycoproteins and glycosylation

GLYCOPROTEINS are proteins in which carbohydrate is covalently linked to the polypeptide through the side chains of the amino-acid residues. Proteins to which carbohydrate is attached in this way are said to be glycosylated and the attachment of carbohydrate is termed glycosylation. The carbohydrate chains (GLYCANS) of glycoproteins are classified as either N- or O-glycans, depending on the amino-acid residues to which they are linked. N-Glycans contain the amino sugar *N*-acetylglucosamine (GlcNAc) (*see* SUGARS) linked to the side-chain nitrogen of asparagine. O-Glycans are linked to the hydroxyl groups of amino-acid side chains, and when they contain *N*-acetylgalactosamine (GalNAc) linked to serine or threonine are also called mucin-type glycans. However, mucin-type glycans are not restricted to the mucins, which are heavily O-glycosylated proteins that contain up to 50% serine and threonine and are found in epithelial secretions. Both N- and O-glycans can occur in the same glycoprotein, for example in bovine fetuin and human immunoglobulin A.

Glycoproteins are usually classified separately from PROTEOGLYCANS (*see* EXTRACELLULAR MATRIX MOLECULES; GLYCANS). The carbohydrate chains of proteoglycans, the GLYCOSAMINOGLYCANS (GAGs), have several structural features that have not been found together in a single glycoprotein glycan. GAGs contain a hundred or more monosaccharide residues per chain, that is, they are polysaccharides. They are made up of repeating disaccharide units that always contain either glucosamine or galactosamine and, with the exception of keratan sulphates, a uronic acid. Finally, all protein-linked GAGs contain sulphate. The carbohydrate chains of glycoproteins typically have fewer than 20 monosaccharide residues, that is, they are oligosaccharides, and are not composed of repeating units. The closest overlap between GAGs and glycoprotein glycans is seen with corneal keratan sulphate and the poly-*N*-acetyllactosaminoglycans — both are N-glycans and both contain disaccharide repeat units, but poly-*N*-acetyllactosaminoglycans usually are not sulphated nor are they as large as the keratan sulphates. Proteoglycans and free GAGs are components of the extracellular matrix.

Glycoproteins occur throughout the cell and its surroundings. They can be extracellular as are SECRETORY PROTEINS, can occur at the cell surface as components of the PLASMA MEMBRANE, of cell walls (*see* BACTERIAL ENVELOPE; PLANT CELL WALL MACROMOLECULES) and of the GLYCOCALYX, and can also be intracellular as components of the cytosol and organelles. Glycoproteins are made by both prokaryotes (bacteria) and eukaryotes (fungi, algae, plants and animals). Most mammalian secreted proteins, with the notable exception of serum albumin, are glycoproteins. Secreted glycoproteins include hydrolytic enzymes such as pancreatic ribonuclease and fungal α-amylase, hormones like follicle-stimulating hormone and chorionic gonadotropin, CYTOKINES like interferon and erythropoietin, transport proteins like transferrin and caeruloplasmin, and serum components such as COMPLEMENT proteins and blood clotting factors (*see* BLOOD COAGULATION AND ITS DISORDERS). Plasma membrane glycoproteins are functionally diverse and include CD ANTIGENS, IMMUNOGLOBULINS and RECEPTORS. The carbohydrate chains of glycoproteins (and glycolipids) are a predominant feature of the cell surface and occupy large volumes in relation to their molecular weight.

The glycoproteins of viral envelopes (*see* ANIMAL VIRUSES) are encoded by the viral genome, but synthesized in the infected host cell and embedded in a host-derived envelope membrane (*see* SEMLIKI FOREST VIRUS). The resident proteins of organelles such as the endoplasmic reticulum and lysosomes, are often glycoproteins. Cytosolic (cytoplasmic) and nuclear glycoproteins were discovered relatively recently and arise by an unorthodox form of eukaryotic glycosylation.

Cytoplasmic and nuclear glycoproteins are unorthodox because their existence is inconsistent with the mechanism by which most eukaryotic glycoproteins are synthesized. Glycoproteins usually acquire their carbohydrate chains in the lumen of the rough endoplasmic reticulum (RER) (*see* ENDOPLASMIC RETICULUM) and the GOLGI APPARATUS, which means that their glycans are excluded from the cytosol by the membranes that surround organelles. The carbohydrate chains are then processed as glycoproteins pass through the RER and Golgi apparatus to their final destinations. Some glycoproteins, on the other hand, acquire their carbohydrate, a single O-linked GlcNAc residue (O-GlcNAc), in the cytoplasm. These glycoproteins include some TRANSCRIPTION FACTORS and proteins of the nuclear pore (*see* NUCLEAR ENVELOPE), nonhistone chromosomal proteins, and proteins associated with the CYTOSKELETON. The significance of cytoplasmic glycosylation is unknown.

Bacteria do not have organelles such as the ER and seem to add carbohydrate to proteins at the cell surface. Most bacterial glycoproteins are components of the cell envelope (S-layer) or extracellular enzyme complexes. Their N-glycans differ from those of eukaryotes in having GalNAc, glucose (Glc), and perhaps rhamnose linked to asparagine. Bacterial O-glycans are linked to serine and threonine, but are not of the mucin-type because galactose (Gal) rather than GalNAc is the first residue (*see* GLYCANS).

Structure and biosynthesis

Hundreds of different glycan structures have been isolated from glycoproteins. This vast array of structures results from different combinations of comparatively few structural elements. All but a handful of N-glycans share a common pentasaccharide core sequence that contains mannose (Man) and GlcNAc (Fig. G29). Variations in core structure arise by the addition of extra monosaccharides (xylose (Xyl), fucose (Fuc), or GlcNAc) to the conserved sequence. Further N-glycan diversity is generated by extension of the cores with outer chains, or antennae, of varying lengths. Finally, the greatest diversity results from addition to the outer chains of different termination sequences and non-carbohydrate substituents such as phosphoryl, sulphuryl, acetyl, and methyl groups (Fig. G30). O-Glycans of the mucin type have in common only the peptide-bound GalNAc, but there are at least 12 different cores based on different combinations of substitution of the GalNAc at carbons 3 and 6 (Fig. G29). Many of the termination sequences of N-glycans, including N-ACETYL-NEURAMINIC ACID (NeuAc or sialic acid), the blood-group antigens (*see* ABO BLOOD GROUP SYSTEM), and sulphate, also occur in O-glycans (and glycolipids).

The N-glycans of eukaryotic glycoproteins share a common core because they are derived from a common precursor oligosaccharide. The precursor (Fig. G30) is synthesized by the stepwise addition of monosaccharides to a lipid carrier called DOLICHOL. The topology of the reactions has been difficult to determine, but

Types of N-glycans

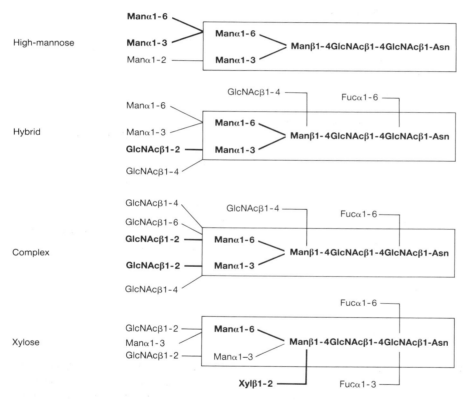

Types of O-glycans

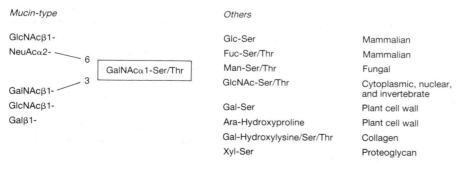

Fig. G29 The different types of carbohydrate chains of eukaryotic glycoproteins. The four types of N-glycan all share a common $Man_3GlcNAc_2$ pentasaccharide core linked to asparagine (large boxes). However, a few of the xylose-type oligosaccharides, which are found in plants and snails, lack the Manα1–3 residue. Certain monosaccharide residues may or may not be present (thin lines). High-mannose oligosaccharides with a core Fucα1–6 residue have been reported to occur in lysosomal enzymes, and some high-mannose oligosaccharides from protozoa have fewer than five mannose residues. Complex oligosaccharides have from two to five outer chains to give bi-, tri-, tetra- and penta-antennary structures. All mucin-type O-glycans share a common GalNAc core residue linked to serine or threonine (small box). Examples of larger structures, in which the outer chains are elongated, are shown in Fig. G30. α and β indicate anomeric configurations, and hyphens following numerals represent glycosidic linkages. Abbreviations: Ara, L-arabinose; Asn, asparagine; Fuc, L-fucose; Gal, D-galactose; GalNAc, *N*-acetyl-D-galactosamine; GlcNAc, *N*-acetyl-D-glucosamine; Man, D-mannose; NeuAc, 5-*N*-acetylneuraminic acid; Ser, serine; Thr, threonine; Xyl, L-xylose; D- and L-denote absolute configurations.

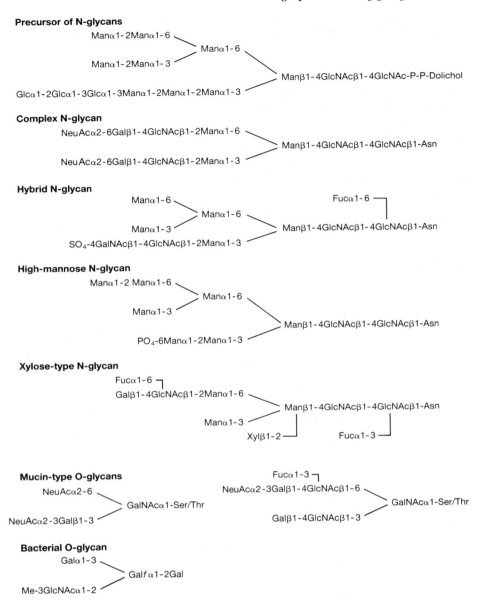

Precursor of N-glycans

Manα1-2Manα1-6 ⟩ Manα1-6

Manα1-2Manα1-3 ⟩ Manα1-6

Glcα1-2Glcα1-3Glcα1-3Manα1-2Manα1-2Manα1-3 ⟩ Manβ1-4GlcNAcβ1-4GlcNAc-P-P-Dolichol

Complex N-glycan

NeuAcα2-6Galβ1-4GlcNAcβ1-2Manα1-6 ⟩ Manβ1-4GlcNAcβ1-4GlcNAcβ1-Asn

NeuAcα2-6Galβ1-4GlcNAcβ1-2Manα1-3

Hybrid N-glycan

Manα1-6 ⟩ Manα1-6   Fucα1-6 ¬

Manα1-3   Manβ1-4GlcNAcβ1-4GlcNAcβ1-Asn

SO₄-4GalNAcβ1-4GlcNAcβ1-2Manα1-3

High-mannose N-glycan

Manα1-2 Manα1-6 ⟩ Manα1-6

Manα1-3

PO₄-6Manα1-2Manα1-3 ⟩ Manβ1-4GlcNAcβ1-4GlcNAcβ1-Asn

Xylose-type N-glycan

Fucα1-6 ¬

Galβ1-4GlcNAcβ1-2Manα1-6

Manα1-3 ⟩ Manβ1-4GlcNAcβ1-4GlcNAcβ1-Asn

Xylβ1-2 ⟩   Fucα1-3 ⟩

Mucin-type O-glycans

NeuAcα2-6 ⟩ GalNAcα1-Ser/Thr

NeuAcα2-3Galβ1-3

Fucα1-3 ¬

NeuAcα2-3Galβ1-4GlcNAcβ1-6 ⟩ GalNAcα1-Ser/Thr

Galβ1-4GlcNAcβ1-3

Bacterial O-glycan

Galα1-3 ⟩ Galƒα1-2Gal

Me-3GlcNAcα1-2

Fig. G30 Examples of N- and O-glycans of glycoproteins. The complex N-glycan is a biantennary structure found on many glycoproteins. GalNAc-4-sulphate and Man-6-phosphate are protein-specific sequences found on pituitary hormones and lysosomal enzymes, respectively. A third protein-specific sequence is the polysialic acid chain NeuAcα2-8(NeuAcα2–8)₂₀₋₂₀₀NeuAcα2-3Gal, which is only found on the complex N-glycans of the neural cell adhesion molecule NCAM. The xylose-type N-glycan is from the plant glycoprotein miraculin. The bacterial O-glycan is from an extracellular cellulase complex. The larger mucin-type glycan bears the sialyl Lewis x sequence NeuAcα2–3Galβ1–4(Fucα1–3)GlcNAc, which is also found on glycolipids and some N-glycans (poly-*N*-acetyllactosaminoglycans). The Lewis x sequence Galβ1–4(Fucα1–3)GlcNAc is also the stage-specific embryonic antigen-1 (SSEA-1). Abbreviations and symbols as in Fig. G29, except *f*, furanose (5-membered ring); Me, methyl; –P–P–, pyrophosphate.

the two GlcNAc and first five mannose residues seem to be added to dolichol on the cytosolic side of the rough endoplasmic reticulum membrane, while the remaining four mannose and the three glucose residues are added in the lumen of the RER. The mechanism by which the intermediate Man₅GlcNAc₂ oligosaccharide is presumably translocated across the RER membrane and into the lumen for further elongation is unknown.

The complete Glc₃Man₉GlcNAc₂ precursor of N-glycans is transferred to an asparagine residue in a still-growing polypeptide chain in the lumen of the RER (Fig. G31). This is the only instance in glycoprotein biosynthesis in which the transfer of more than one monosaccharide at a time is known to occur. Asparagine residues must occur within the tripeptide sequences Asn-X-Ser or Asn-X-Thr, in which X represents an amino acid other than proline, in order to accept the precursor oligosaccharide. After its

transfer to a protein, the precursor is enzymatically trimmed by the action of GLYCOSIDASES. The three glucose residues and up to four of the mannose residues are removed to form HIGH-MANNOSE OLIGOSACCHARIDES. The smallest of these, with five mannose residues, can be converted into a hybrid oligosaccharide by the enzyme GlcNAc transferase I, and the further removal of mannose residues produces other hybrid structures. The smallest hybrid oligosaccharide can be converted into the first COMPLEX OLIGOSACCHARIDE by the action of GlcNAc transferase II, after which more outer chains can be initiated by the action of GlcNAc transferases IV–VI. Elongation and termination of the outer chains of hybrid and complex oligosaccharides by various galactosyl-, *N*-acetylglucosaminyl-, fucosyl-, sialyl- and sulphotransferases occur in the *trans*-Golgi compartment (*see* GOLGI APPARATUS).

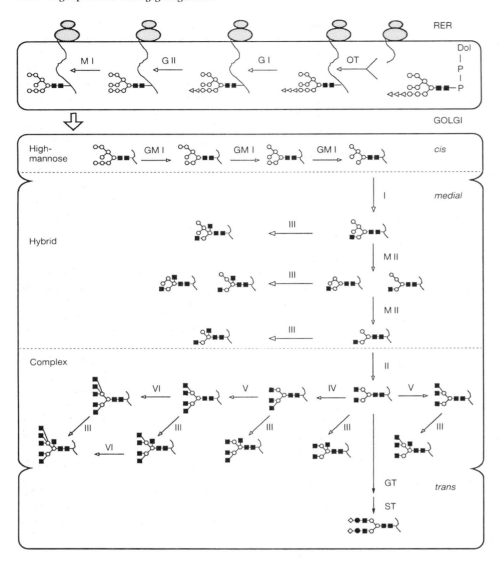

Fig. G31 Schematic pathway of eukaryotic N-glycan biosynthesis. The precursor is transferred co-translationally to elongating polypeptides in the rough endoplasmic reticulum (RER) from dolichol pyrophosphate (Dol–P–P) in a reaction catalysed by the enzyme oligosaccharyltransferase (OT). Subsequent processing reactions are catalysed in the RER by α-glucosidase I (GI), α-glucosidase II (GII) and α-mannosidase I (MI); in the *cis* Golgi by Golgi α-mannosidase I (GMI); in the *medial* Golgi by α-mannosidase II (MII) and *N*-acetylglucosaminyltransferases I–VI (I–VI); and in the *trans* Golgi by a variety of glycosyltransferases including β1,4-galactosyltransferase (GT) and α2,6-sialyltransferase (ST). The elongation and termination of only one (complex biantennary) oligosaccharide in the *trans* Golgi is shown, but all of the terminal *N*-acetylglucosamine residues of hybrid and complex oligosaccharides (except the 'bisecting' residue added by *N*-acetylglucosaminyltransferase III) can be elongated. Note that *N*-acetylglucosaminyltransferase IV is present in avian, but not mammalian, cells. Details of glycan structure are given in Figs G29 and G30. Adapted from [2] with additional information from [4].

■ *N*-acetyl-D-glucosamine
○ D-mannose
◁ D-glucose
● D-galactose
◇ Sialic (neuraminic) acid

The xylose-containing oligosaccharides found on some plant and molluscan glycoproteins constitute a fourth type of N-glycan. Many of these xylose-type structures could not have arisen by the conventional biosynthetic pathway described above, particularly those that lack the Manα1–3 residue or have only the Manα1–6 residue elongated. Such structures must arise by an alternative pathway, because the Manα1–3 residue must be elongated by GlcNAc transferase I (I in Fig. G31) before any subsequent processing reactions can occur. As plants transfer the same Glc₃Man₉GlcNAc₂ precursor to proteins, different processing enzymes are probably involved. (It should be mentioned that enzymes other than those in the conventional scheme have been detected in various cells and tissues, but their roles are not known.) Some protozoa also produce N-glycans whose structures are inconsistent with the conventional pathway, including high-mannose oligosaccharides that are not the specific isomers produced by Golgi α-mannosidase I and some that contain glucose. In these organisms, the lipid-linked precursors are different; they do not contain glucose and can have fewer than nine mannose residues.

The biosynthesis of mucin-type O-glycans, like that of N-glycans, occurs as glycoproteins pass through the RER and Golgi apparatus. However, O-glycans are not assembled as an oligosaccharide precursor that is added to the protein and then trimmed by glycosidase, and the transfer of O-glycans is not CO-TRANSLATIONAL. Transfer of the first GalNAc to serine or threonine occurs near the time of protein movement from the RER to the Golgi apparatus, and all of the monosaccharides in O-glycans are transferred one by one to the elongating glycan on the glycoprotein. As with the biosynthesis of N-glycans, the elongation and termination of O-glycans occur in the Golgi apparatus, and many of the same factors control N- and O-glycan biosynthesis.

Spatial segregation of the enzymes involved in glycoprotein biosynthesis forms a kind of assembly line through which the maturing glycans must pass and helps to ensure that the processing reactions occur in their proper sequence. For example, the premature action of elongating and terminating enzymes is prevented by their localization in the *trans*-Golgi compartment. Spatial segregation is particularly important in the case of β1,4-galactosyltransferase (GT in Fig. G31), because the addition of galactose residues by this elongation enzyme inhibits most of the processing reactions that occur in the *medial*-Golgi compartment. A second factor that influences glycoprotein biosynthesis is the relative levels (activities) of processing enzymes, especially those that compete for common substrates. For example, high levels of GlcNAc transferase III, which adds the so-called 'bisecting' GlcNAc residue, could lead to a higher proportion of hybrid glycans being formed, since most of the other processing reactions of the *medial*-Golgi compartment are inhibited by the presence of this residue.

These two examples illustrate the importance of a third factor controlling glycoprotein biosynthesis, the substrate specificities of the processing enzymes. In many cases, the action of a glycosidase or glycosyltransferase either inhibits or is required for the action of another enzyme, and this has the effect of limiting the possible reaction sequences.

Only a subset, from one to dozens, of glycan structures is usually associated with a particular glycoprotein, even though most cells have the capacity to produce a much wider array of structures. Even different glycosylation sites within the same glycoprotein can carry different subsets of structures. Many Asn-X-Ser or Asn-X-Thr sites do not carry an N-glycan at all. The variation in glycan structures associated with the same glycosylation site in a population of glycoproteins with the same amino-acid sequence is termed microheterogeneity. Glycoproteins with the same amino acid sequence but different glycans are known as glycoforms. Some carbohydrate structures appear to be polypeptide-specific, that is they are associated with only one glycoprotein or family of glycoproteins (see Fig. G30). In addition, different proteins expressed in the same cell line, e.g. CHO cells, can acquire a different subset of N-glycans. (In a converse situation, the same glycoprotein (ovalbumin) acquired hybrid glycans whether produced naturally in chickens or in a mouse cell line.) These observations indicate that the amino-acid sequence of a glycoprotein influences the processing of its carbohydrate chains.

The mechanisms by which amino-acid sequence influences glycosylation are poorly understood. Glycosyltransferases involved in the synthesis of polypeptide-specific glycans presumably must be able to recognize specific features of both the protein and carbohydrate portions of their substrates. Steric hindrance, rather than specific recognition, can sometimes account for the absence of carbohydrate at a potential glycosylation site, or the occurrence of different subsets of glycans at different sites within the same protein, especially when these subsets represent different degrees of processing (e.g. high-mannose versus complex). However, site-specific glycosylation may involve more subtle structural differences, such as the presence or absence of a bisecting GlcNAc on glycans that are otherwise highly processed and presumably equally accessible to glycosyltransferases. A possible explanation for subtle site-specific differences is that the conformation of an oligosaccharide is influenced by neighbouring polypeptide, and that glycosyltransferases favour one conformation of an oligosaccharide substrate over another. Consistent with this hypothesis, at least one glycosyltransferase (GlcNAc transferase V) does preferentially recognize one of two conformations of a model substrate.

Glycosylation is influenced further by the type of cell or tissue in which a glycoprotein is produced and by the developmental state during which it is produced. An example of cell type-specific glycosylation is that of human interferon-β_1, which acquires the terminal sequence NeuAcα2–3Galβ1- when produced in certain hamster cells but the sequences NeuAcα2–6Galβ1- and Galα1–3Galβ1- when produced in certain mouse cells. Examples of developmental variation in glycosylation are the stage-specific embryonic antigens (SSEA) of mice and the polysialylation of embryonic, but not adult, NCAM (*see* CELL ADHESION MOLECULES). Gross (not necessarily protein-specific) changes in glycosylation also occur as a result of malignancy. Both the size and extent of branching of N- and O-glycans can change as a result of oncogenic transformation, in which healthy cells become cancerous. Carbohydrate structures that are present at embryonic stages of development often reappear on the glycoproteins (and glycolipids) of tumour cell surfaces. In some cases, changes in the levels of certain glycosyltransferases have been correlated with the specific changes in glycan structure observed in a particular disease state.

Inhibitors of glycosylation and glycoprotein processing

Inhibitors are widely used to study the biosynthesis of glycoproteins and the roles of their N-glycans. Tunicamycin prevents N-glycosylation altogether by inhibiting formation of the lipid-linked precursor. Castanospermine and deoxynojirimycin do not prevent glycosylation, but inhibit the α-glucosidases that remove glucose residues from the precursor after it is transferred to protein. Deoxymannojirimycin interferes with N-glycan processing by inhibiting Golgi α-mannosidase I, leading to an accumulation of $Man_9GlcNAc_2$ structures on glycoproteins. Swainsonine inhibits α-mannosidase II and thus prevents the conversion of hybrid to complex oligosaccharides.

Functional aspects

The physical properties of many, but not all, glycoproteins are altered by glycosylation. An increased susceptibility to protease degradation and/or heat denaturation upon deglycosylation is often observed. In the absence of N- or O-glycosylation, some glycoproteins do not attain their proper tertiary structure (con formation) and are degraded either intracellularly or at the cell surface. The role of N-glycosylation in PROTEIN FOLDING stems from the fact that transfer of the $Glc_3Man_9GlcNAc_2$ precursor oligosaccharide to a polypeptide occurs before PROTEIN TRANSLOCATION and folding are complete. A high degree of O-glycosylation, as in mucins and the mucin-type domains of some plasma membrane glycoproteins, can force a polypeptide to adopt an extended, rod-like conformation. In some cases, the contribution by carbohydrate to the physical properties of a glycoprotein is profound. The high charge density, hydrophilicity, and viscosity of mucins are due to their carbohydrate chains and allow them to act as lubricants and protective agents. Likewise, the antifreeze glycoproteins of arctic fish depend on carbohydrate for their unique properties.

The effects of glycosylation on the biological activity of glycoproteins (e.g. enzymatic activity and ligand binding) are of increasing interest largely due to the potential therapeutic uses of glycoproteins that are produced by recombinant DNA technology in heterologous cell types. Glycoproteins that lack some or all of their carbohydrate can be obtained by enzymatically removing the glycans from purified glycoproteins, by isolating proteins produced in the presence of glycosylation inhibitors, by using mutant cell lines or bacteria to express proteins, or by elimination of glycosylation sites by site-directed mutagenesis. In some, but certainly not all, cases that have been examined, the absence of glycans or the presence of particular glycans have been found to influence biological activity. Although it seems reasonable to suppose that the presence of bulky and often charged glycans could influence the biological activity of a glycoprotein by altering tertiary structure and/or sterically hindering ligand binding, direct experimental evidence in support of this concept is still scarce.

Besides influencing the proteins to which they are attached, glycoprotein glycans also mediate specific intermolecular interactions. They can act as protein targeting signals. Glycoproteins bearing terminal Gal, GlcNAc or GalNAc-4-SO_4 residues are cleared from mammalian blood by hepatocytes and macrophages, respectively, by binding to specific receptor proteins (e.g. the hepatocyte ASIALOGLYCOPROTEIN receptor). The protein-specific mannose-6-phosphate sequence directs proteins to LYSOSOMES via binding to MANNOSE-6-PHOSPHATE RECEPTORS in the Golgi apparatus.

Cell–cell interactions can also be mediated through the recognition of carbohydrate sequences by receptor proteins. Sperm–egg interaction in mice occurs through binding of a sperm receptor to O-glycans on the egg surface. Binding of circulating leukocytes to vascular endothelial cells at sites of inflammation, and compaction of mouse embryonic cells at the 8–16 cell stage, also depend on specific glycan sequences, the sialyl Lewis x sequence (*see* LEWIS BLOOD GROUP) and SSEA-1, respectively (see Fig. G30). Most of the ligands recognized by human pathogens (bacteria, viruses and mycoplasmas) in their attachment to host cells are carbohydrates. The classical example is the binding of INFLUENZA VIRUS to N-acetylneuraminic acid. It should be noted that the relative importance of glycolipid versus glycoprotein

glycans is not known for all cell–cell interactions in which carbohydrate is involved, as the same sequences often occur in both classes of glycoconjugate.

For further reading see [1–3].

J. THOMAS

See also: LECTINS.

1 Ginsburg, V. & Robbins, P.W. (Eds) (1984) *Biology of Carbohydrates*, Vol. 2 (Wiley, New York).
2 Kornfeld, R. & S. Kornfeld (1985) Assembly of asparagine-linked oligosaccharides. *Annu. Rev. Biochem.* 54, 631–664.
3 Paulson, J.C. (1989) Glycoproteins: what are the sugar chains for? *Trends Biochem. Sci.* 14, 272–276.
4 Schachter, H. (1991) The 'yellow brick road' to branched complex N-glycans. *Glycobiology* 1, 453–461.

---

glycosaminoglycan (GAG) A term used with slightly variable meaning, but more precise than 'mucopolysaccharide' which it has replaced. It often includes HYALURONIC ACID and KERATAN SULPHATE, and sometimes includes CHITIN and various polymers of lower animals, but all of these are better considered as different classes of GLYCAN. True glycosaminoglycans are polysaccharides found in animals and have the distinctive (or type 1) link sequence to the core protein (with which they occur as PROTEOGLYCAN):

$$Ser/Thr\text{-}\beta1Xyl4,1\beta Gal3,1\beta Gal3,1GlcA4\text{-}$$

They contain alternating residues of uronic acid and hexosamine and, usually, contain sulphate.

Two general categories are known, which diverge biosynthetically at the fifth glycosyl transfer in their synthesis, that is, in the nature of the hexosamine added to the link sequence above. The two groups have different core proteins, but it is not known how these influence the fifth transfer or whether they do so directly. In galactosaminoglycans, the second galactosyl residue carries sulphate at C4 or C6: in the glucosaminoglycans this is not seen. In both groups some xylosyl residues are phosphorylated at C2. Some glucosaminoglycan appears to be attached to a N-linked sequence of a type susceptible to endoglycosidase F.

1 Galactosaminoglycans are based on the repeating disaccharide $(\text{-}4DGlcA(p)\beta1,3GalNAc(p)\beta1\text{-})_n$, which is CHONDROITIN. The N-acetyl galactosamine residue has a sulphate ester group at C6 in chondroitin 6-sulphate (Fig. G32) and at C4 in chondroitin 4-sulphate.

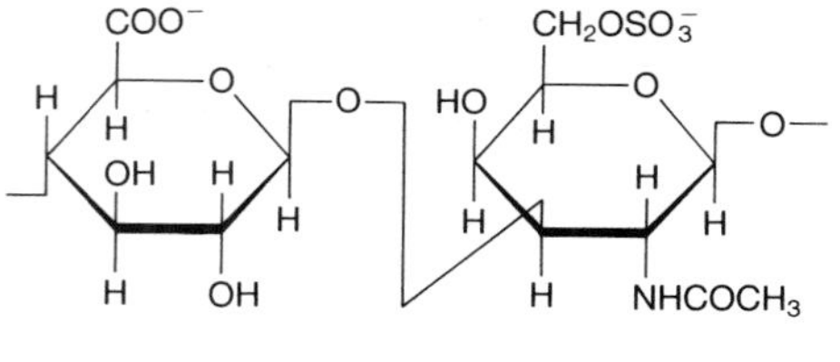

Fig. G32 Repeating unit of chondroitin 6-sulphate.

Various forms showing some residues sulphated in both positions, or carrying sulphate elsewhere, are known (e.g. in cartilaginous fish and in all mast cells). Chondroitin 4-sulphate can undergo 5-epimerization at glucuronic acid residues after the polymer is formed, to produce domains containing L-iduronic acid; such domains are of DERMATAN SULPHATE (Fig. G33).

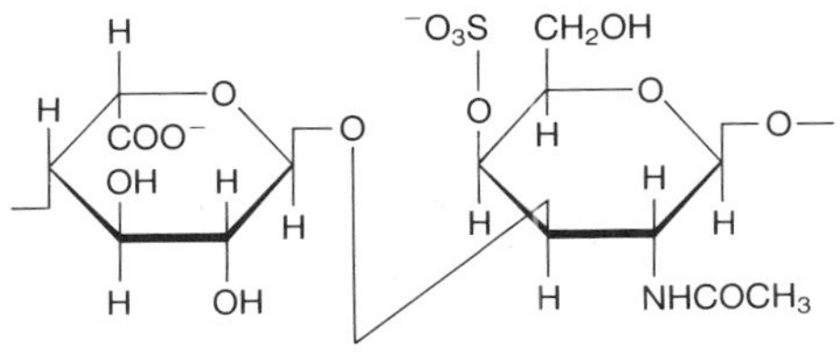

Fig. G33 Repeating unit of dermatan sulphate.

The mechanism resembles that found in ALGINIC ACID synthesis. **2 Glucosaminoglycans** or **heparinoids** are based on the repeating disaccharide (-4DGlcA(p)β1,4GlcNAcα1-)$_n$, but are subject to extensive modification after assembly. This occurs in a series of stages and each can occur to a variable extent. The modifications are: (a) deacetylation of N-acetyl glucosamine residues; (b) N-sulphation of unsubstituted glucosamine residues; (c) epimerization of D-glucuronic acid residues adjacent to N-sulphated glucosamine to form L-iduronic acid; (d) addition of sulphate ester groups at C2 of iduronic acid residues; and (e) 6-sulphation of the glucosamine residues and insertion of any other sulphate residues present (i.e. at C3 of glucosamine and C2 and C3 of glucuronic acid).

In HEPARIN, of connective tissue mast cells, most residues become modified, while the HEPARAN SULPHATES, which are widespread at animal cell surfaces and in matrices, are less altered and have highly irregular final sequences which are neither regular nor random, but have some degree of domain structure. In general, heparan sulphates have shorter chains than heparin; no pure homotypic heparan sulphate has yet been isolated, even from tissue cultures, suggesting that the synthetic mechanism is either very sensitive to the cellular environment or is imprecise. The modification of the chains begins while their assembly is still in progress and a terminal sulphation may be the limiting step in chain growth.

Heparinoids may carry information that determines the rate or form of matrix assembly, or influences cell–matrix or cell–cell association. Some contain oligosaccharides that interact with components of the blood clotting, fibrinolytic and complement cascades, and they are known to be important in the presentation of cytokines (especially those involved in fibrosis) to their target cells. The coding sequences for the attachment of antithrombin III and thrombin are known to be adjacent. *See also*: CELL–CELL INTERACTIONS; CELL–MATRIX INTERACTIONS; EXTRACELLULAR MATRIX MACROMOLECULES.

glycosidase An enzyme that catalyses the cleavage of a glycosidic linkage. Glycosidic linkages are formed by the reaction of sugars (the simplest of which are the MONOSACCHARIDES) with alcohols (to form O-glycosides), amines (to form N-glycosides), or other molecules. Glycosidases can be exoglycosidases, which

cleave the glycosidic linkage of unsubstituted (or terminal) sugar residues, or endoglycosidases, which cleave the glycosidic linkage of substituted (or internal) residues. Glycosidases are specific with respect to the sugar and the anomeric configuration (α or β) of the glycosidic linkage that is cleaved, and some are specific with respect to the molecule to which the sugar is linked (the aglycone). Glycosidases (α-glucosidases and α-mannosidases) are involved in processing the N-linked oligosaccharides of GLYCOPROTEINS. Purified glycosidases are used to release carbohydrate chains from glycoproteins and for the structural characterization of carbohydrate chains. The two enzymes most widely used for the removal of carbohydrate chains from glycoproteins are endo-β-N-acetylglucosaminidase H (EC 3.2.1.96) (Endo H), which releases high-mannose and hybrid oligosaccharides, and peptide-N⁴-(N-acetyl-β-glucosaminyl)asparagine amidase F (peptide : N-glycosidase F, PNGase F, N-glycanase) (EC 3.5.1.52), which, though strictly speaking is not a glycosidase, releases most N-linked oligosaccharides.

glycoside *See*: GLYCANS.

glycosphingolipid *See*: GLYCOLIPIDS.

glycosylation *See*: GLYCOPROTEINS AND GLYCOSYLATION.

glycosylphosphatidylinositol anchor *See*: GPI ANCHOR.

glycosyltransferases Enzymes (EC 2.4.1) transferring glycosyl groups from one compound to another. They are involved in GLYCAN synthesis and degradation and in the glycosylation of proteins and lipids. *See e.g.*: GLYCANS; GLYCOLIPIDS; GLYCOPROTEINS AND GLYCOSYLATION; GOLGI APPARATUS.

glyoxysome Plant cell organelle, found especially in germinating seeds, in which fatty acids are broken down and converted into acetyl-CoA for participation in the glyoxylate cycle. *See also*: PEROXISOME.

glyphosate N-(phosphonomethyl)glycine.

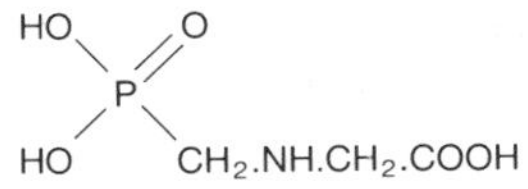

Organophosphorous herbicide for foliar application to many annual and perennial weeds. Readily absorbed and translocated in plants; inactivated by soil. Acts by inhibiting EPSP SYNTHASE in the shikimate pathway, and hence prevents the biosynthesis of aromatic amino acids (phenylalanine, tyrosine, and tryptophan) and secondary products derived from them, including lignin and flavonoids.

glypiation *See*: GPI ANCHORS.

GM-CSF Granulocyte-macrophage colony-stimulating factor.

Protein of 127 amino acids in humans (murine, 124 amino acids) with an M_r of 14 000. It is encoded by a gene on chromosome 5q21 (murine chromosome 11), and is normally glycosylated (to an M_r of ~27 000), although nonglycosylated recombinant GM-CSF shows increased biological activity and higher affinity for the GM-CSF receptor. Many cell types including macrophages, fibroblasts and T lymphocytes are capable of producing GM-CSF under the appropriate conditions. The GM-CSF receptor(s) has yet to be fully understood; proteins of different K_d and M_r have been reported, but it would seem that at least two species, one of high affinity (K_d >20 pM), and a second, more numerous receptor of lower affinity (K_d 0.7–1.2 pM) exist. GM-CSF activities include stimulation of the *in vitro* growth of progenitors of various haematopoietic lineages (macrophage, neutrophil, erythroid, eosinophil, megakaryocyte) in a dose-related manner. *In vivo*, GM-CSF induces increases in leukocyte counts (including lymphocytes), and has found clinical use in patients with suppressed myelopoiesis where infection may be a significant risk. *See*: HAEMATOPOIESIS; LYMPHOKINES.

Gough, N.M. & Nicola, N.A. (1990) In *Colony-stimulating Factors — Molecular and Cellular Biology* (Dexter, T.M. et al., Eds) Ch. 4, 111–153 (Marcel Dekker, New York).

goblet cell Specialized mucus-secreting cell found in intestinal epithelium.

GOGAT Glutamine : 2-oxoglutarate aminotransferase. *See*: GLUTAMATE SYNTHASE.

Golgi apparatus

THE Golgi apparatus (GA, Golgi complex) of the eukaryotic cell receives newly synthesized proteins and lipids from the ENDOPLASMIC RETICULUM (ER) and delivers them to their correct destination in the cell. These molecules pass, in sequence, through an ordered array of Golgi compartments each capable of making a specific, yet different, set of covalent modifications to the lipids and proteins that pass through. Macromolecules entering the Golgi apparatus share a common pathway up to the *trans* Golgi network (TGN) where they are separated from each other in preparation for final delivery. The Golgi apparatus in plants is also the site of cell-wall polysaccharide synthesis, with the exception of cellulose, which is synthesized at the cell wall by enzyme complexes exported via the Golgi apparatus (*see* PLANT CELL WALL MACROMOLECULES).

References [1–8] and references therein provide an overview of work on the various aspects of the Golgi apparatus from its discovery until the present day.

Morphology

First described in 1898 by the Italian anatomist Camillo Golgi (1843–1926) using a silver impregnation technique, the Golgi apparatus is found in all eukaryotic organisms and, in most animal cells, comprises a single, compact reticulum near to one side of the nucleus. In plants and fungi there are multiple copies of the Golgi dispersed throughout the cytoplasm, a situation so far found only during telophase in animal cells, after the organelle has divided and before it has fused to form the single, interphase reticulum.

In cross section, at the level of the electron microscope (Fig. G34), the central feature of the Golgi apparatus is a stack of closely apposed and flattened cisternae, 0.5–1 μm across, and comprising, typically, 3–6 cisternae, although up to 40 cisternae have been reported in some instances. In plants the stack is termed the dictyosome. Between one side of the stack, the *cis* side, and the transitional element region of the ER, there lies a tubulovesicular compartment, the *cis* Golgi network (also known as the intermediate compartment). On the opposite side of the stack, the *trans* side, lies the TGN, another tubulovesicular structure which is often closely apposed to the *trans*-most cisterna of the stack.

All Golgi compartments are embedded in a matrix termed the 'zone of exclusion' because other cytoplasmic structures down to the size of ribosomes appear to be excluded. The composition and function of this matrix are unknown.

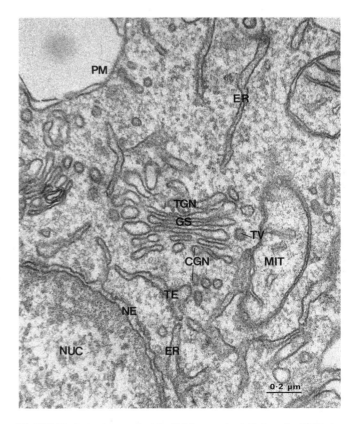

Fig. G34 Electron micrograph of the Golgi apparatus in HeLa cells. CGN, *cis* Golgi network; ER, endoplasmic reticulum; GS, Golgi stack; MIT, mitochondrion; NE, nuclear envelope; NUC, nucleus; PM, plasma membrane; TE, transitional elements; TGN, *trans* Golgi network; TV, transport vesicle. Photo courtesy of John Lucocq.

Compartmentation

Most but not all of the proteins and lipids that pass through the Golgi apparatus are covalently modified by some of the many processing enzymes which line the walls of the Golgi compartments. Modifications include acylation, phosphorylation, sulphation, and glycosylation. Glycosylation is the best characterized of these processes, especially in the case where oligosaccharides are linked to proteins through asparagine residues (N-glycosylation).

Mannose-rich oligosaccharide side chains are first linked to proteins in the ER, and after addition and partial processing in that compartment, are further trimmed of mannose residues in the Golgi followed by addition of the sugars that make up the complex oligosaccharides found on many mature proteins (glycoproteins). The events of protein glycosylation are described in more detail in the entry on GLYCOPROTEINS AND GLYCOSYLATION.

The processes that occur in the Golgi are strictly sequential and occur in different Golgi compartments: mannose trimming, the earlier event, is restricted to *cis* and *medial* Golgi cisternae; construction of complex oligosaccharides begins in the *medial* cisternae and is completed in the *trans* cisterna and TGN. Acylation (normally palmitoylation) is an early Golgi event whereas sulphation is a late event, occurring in the TGN.

The assembly of GLYCOLIPIDS is similarly ordered. Gangliosides, for example, are assembled from ceramide which is synthesized in the ER and then passed to the *cis* cisterna where conversion to glucosylceramide and then lactosylceramide occurs. Later steps occur in the *trans* cisterna and TGN.

The Golgi apparatus can therefore be thought of as an ordered series of compartments each carrying out a specific yet distinct set of operations on the proteins and lipids that pass through them. This means that each compartment must contain the appropriate set of enzymes needed to carry out the modifications characteristic of that compartment. It is, however, worth noting that, with the exception of the lysosomal modifications described below, none of the characterized post-translational modifications carried out by the Golgi is required for either sorting or transport (*see* PROTEIN TARGETING).

The function of stacking is also unknown as vesicular transport from cisterna to cisterna can occur between separated stacks. Gradients of both cholesterol and pH exist across the stack in a *cis* to *trans* direction but whether stacking is required for their maintenance is unknown.

Transport pathways

The default pathway

Transport of newly synthesized proteins from the ER to the plasma membrane occurs by default — once correctly folded in the ER no further address tags are needed to travel from compartment to compartment and none has been found on either constitutively secreted or plasma membrane proteins. The default pathway to plasma membrane or secretion is further described in the entry PROTEIN SECRETION.

Proteins not destined for the plasma membrane need tags to divert them from the default pathway or retention tags to prevent them moving along the pathway, past the point where they normally function. If they do move past this point, some proteins are equipped with retrieval tags which are used to return them to the correct compartment. This is illustrated schematically in Fig. G35 and discussed below.

The retrieval pathway

A number of ER proteins appear to function only if they are free within the lumen so they do, on occasion, escape. They are salvaged from the *cis* Golgi network and the Golgi stack because they bear a retrieval tag, a C-terminal tetrapeptide (-KDEL in mammals (single-letter amino-acid notation)), recognized by a receptor which returns them to the ER.

Some ER membrane proteins also have a retrieval tag in the cytoplasmic tail (-K-X-(X)-K-X-X in mammals (single-letter amino-acid notation where X = any amino acid)). The receptor that salvages these proteins from the *cis* Golgi network and Golgi stack has still to be identified.

Transport to lysosomes

The *cis* Golgi network is involved in constructing the address tag which, in mammalian cells, enables lysosomal enzymes to reach their destination (*see* LYSOSOMES). The tag, mannose 6-phosphate, is constructed from some of the terminal mannose residues on protein-bound N-linked oligosaccharides by two enzymes which recognize protein features unique to lysosomal enzymes. The tagged enzyme then binds to one of two mannose-6-phosphate receptors, located initially in the *cis* cisterna, and the complex moves, probably by bulk flow, to the TGN where it is diverted from the default pathway (see below) and sent, in several stages, to lysosomes.

Lysosomal membrane proteins have an address tag in the cytoplasmic tail. In yeast, the information for transport to vacuoles, the plant and fungal equivalent of lysosomes, is contained within a short sequence at the N terminus.

Transport to secretory granules

Secretory proteins destined for granules (*see* PROTEIN SECRETION) move, by default, to the TGN and although many are covalently modified during transport, none of these modifications seems to be required for subsequent diversion to secretory granules. Condensation, in which the secretory proteins clump together, begins in the TGN in a specific process which probably requires the lower pH of this organelle and Ca^{2+}. Specificity is indicated by the separation of granule proteins from constitutively secreted proteins and even, in some cells, from other condensing granule proteins destined for different secretory granules. The outside of the condensing granule is thought to bind to as-yet-unidentified receptors on the inner wall of the TGN, thereby enveloping it with a membrane which buds to form an immature secretory granule. Further condensation is accompanied by proteolytic processing of the secretory proteins and retrieval of excess membrane as immature granules fuse together, until the mature granule is formed.

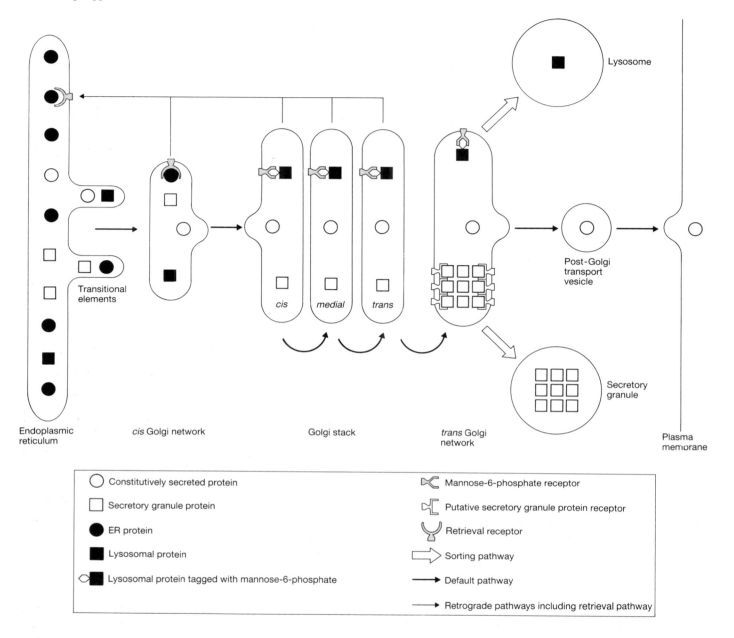

Fig. G35 Sorting in the Golgi apparatus.

Transport to the plasma membrane in polarized cells

Many cells have a plasma membrane divided into two or more domains which differ both compositionally and functionally. The means by which proteins reach these different domains is still unclear although transport to one domain is usually the default pathway (*see* EPITHELIAL POLARITY). Sorting to different domains occurs in the TGN and the address tags for some of the membrane proteins are found in the cytoplasmic tails.

Transport to the Golgi apparatus

Golgi membrane proteins move by default from the ER but stop when they reach the correct Golgi compartment. They must therefore contain an appropriate retention tag. The tag is contained within the membrane-spanning sequence but the way in which it operates is unknown.

Transport mechanism

Both membrane proteins and soluble luminal proteins move from compartment to compartment in transport vesicles which bud from one compartment and fuse with the next compartment on

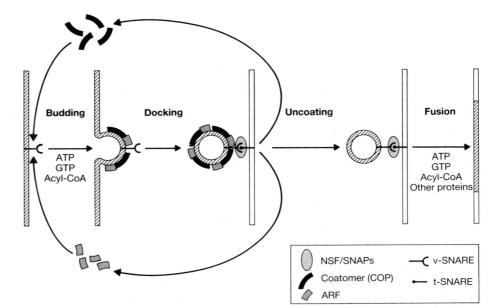

Fig. G36 COP-mediated transport. Model for vesicular transport between Golgi compartments. A v-SNARE is incorporated into the budding vesicle as ARF-GTP recruits COP proteins onto the membrane. The v-SNARE docks with the t-SNARE on the target membrane forming a complex with NSF/SNAPs. Hydrolysis of ARF-GTP triggers uncoating and fusion delivers the vesicle contents to the target compartment.

Legend in figure:
Budding · Docking · Uncoating · Fusion
ATP GTP Acyl-CoA
ATP GTP Acyl-CoA Other proteins
NSF/SNAPs · Coatomer (COP) · ARF · v-SNARE · t-SNARE

the pathway. At least two types of vesicle-mediated transport have been identified, which differ in their selectivity of uptake. COP-mediated uptake is non-selective whereas clathrin-mediated uptake is selective.

COP-mediated transport

Intra-Golgi transport by COP-coated vesicles (*see also* COATED PITS AND VESICLES) is the best characterized although a similar process is thought to occur from the ER to the *cis* Golgi and possibly from the TGN to the plasma membrane. Figure G36 provides a schematic illustration of this process as worked out using cell-free assays. ARF-GTP (ADP ribosylation factor with bound GTP) binds to Golgi membranes and recruits COP proteins which cause vesicles to bud from the cisternal rims, consuming ATP and acyl-CoA in the process. The COP-coated vesicles contain v-SNAREs (vesicle SNAP RECEPTORS) which dock with t-SNAREs on the target membrane, the interaction being mediated by NSF (N-ethylmaleimide-sensitive fusion protein), SNAPs (soluble NSF attachment proteins) and a small GTP-BINDING PROTEIN of the rab1/ypt class. Hydrolysis of GTP bound to ARF leads to uncoating and fusion follows, consuming more ATP and acyl-CoA.

Clathrin-mediated transport

Clathrin is involved in the sorting of lysosomal enzymes in the TGN. It binds, via adaptors, to the cytoplasmic tails of mannose-6-phosphate receptors, and polymerizes to form a coated patch (*see* COATED PITS AND VESICLES). Other proteins are sterically excluded from the patch thereby purifying the lysosomal enzymes away from other proteins. Clathrin is also involved in recovering membrane from immature granules as they condense to form secretory granules.

G. WARREN

1 Golgi, C. (1898) Sur la structure des cellules nerveuses. *Arch. Ital. Biol.* **XXX**, 60–71.
2 Palade, G. (1975) Intracellular aspects of the process of protein synthesis. *Science* **189**, 347–358.
3 Griffiths, G. & Simons, K. (1986) The *trans* Golgi network: sorting at the exit site of the Golgi complex. *Science* **234**, 438–443.
4 Pfeffer, S.R. & Rothman, J.E. (1987) Biosynthetic protein transport and sorting by the endoplasmic reticulum and Golgi. *Annu. Rev. Biochem.* **56**, 829–852.
5 Burgess, T.L. & Kelly, R.B. (1987) Constitutive and regulated secretion of proteins. *Annu. Rev. Cell Biol.* **3**, 243–293.
6 Pelham, H.R. (1989) Control of protein exit from the endoplasmic reticulum. *Annu. Rev. Cell Biol.* **5**, 1–23.
7 Kornfeld, S. & Mellman, I. (1989) The biogenesis of lysosomes. *Annu. Rev. Cell Biol.* **5**, 483–525.
8 Rothman, J.E. & Orci, L. (1992) Molecular dissection of the secretory pathway. *Nature* **355**, 409–416.

Golgi-endoplasmic reticulum lysosome *See*: GERL.

Golgi staining A silver impregnation method of staining of brain tissue, devised by the neuroanatomist Camillo Golgi in the 19th century. It is still used today because it only stains a small percentage of neurons (5–10%), and allows the visualization of the whole cell, including dendrites and axon.

gonadotropins (gonadotrophins) Any hormone having a stimulatory effect on the gonads. In humans they include the anterior pituitary protein hormones FOLLICLE-STIMULATING HORMONE (FSH) and LUTEINIZING HORMONE (LH), both of which are active in both sexes but with differing effects, and CHORIONIC GONADOTROPIN, produced by the placenta.

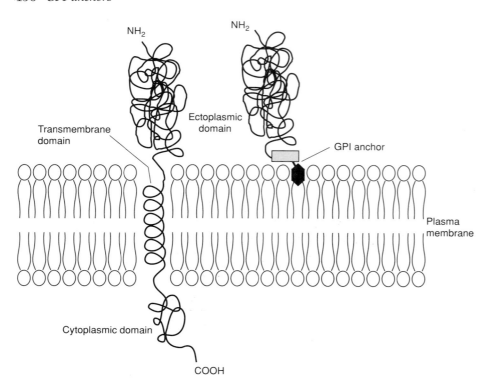

NH₂

NH₂

Transmembrane
domain

Ectoplasmic
domain

GPI anchor

Plasma
membrane

Cytoplasmic domain

COOH

Fig. G37 Comparison between transmembrane and GPI-anchored plasma membrane proteins. A class 1 transmembrane protein (left) has an N-terminal ectoplasmic domain, a transmembrane domain consisting of an α-helix of about 20 hydrophobic amino acids, and a C-terminal cytoplasmic domain. A GPI-anchored protein (right) has an N-terminal ectoplasmic domain and is embedded in the outer leaflet of the lipid bilayer solely via the lipid portion of the GPI anchor.

GPI anchors

GLYCOSYL-PHOSPHATIDYLINOSITOL (GPI) membrane anchors are complex glycophospholipids found covalently attached to a wide variety of externally disposed PLASMA MEMBRANE proteins in eukaryotes [1–3]. Some luminally disposed SECRETORY GRANULE proteins are also GPI anchored, but there are no examples of cytoplasmically oriented GPI-anchored proteins. The primary function of GPI anchors is to afford a stable association of protein with the membrane lipid bilayer. They should be thought of as an alternative anchoring mechanism to the use of a hydrophobic peptide transmembrane domain by class 1 membrane proteins (Fig. G37). All GPI anchors have a conserved basic core structure:

ethanolamine-PO₄-6Manα1,2Manα1,6Manα1,4GlcNH₂-
α1,6*myo*-inositol-1-PO₄-lipid

with species- and tissue-specific variations in substituents and lipid structure (Fig. G38). The *myo*-inositol-1-PO₄-lipid component is loosely described as a 'phosphatidylinositol' (PI). The lipid moiety is quite variable; most commonly it is *sn*-1-alkyl-2-acyl-glycerols, followed by *sn*-1,2-diacylglycerols, *sn*-1-acyl-2-*lyso*glycerols, and ceramides (Fig. G38). In many instances the inositol ring contains an additional lipid modification in the form of an ester-linked palmitic acid (C₁₆:₀) (*see* FATTY ACIDS for nomenclature). This inositol-palmitoylation renders the anchor resistant to the action of bacterial phosphatidylinositol-specific PHOSPHOLIPASE C (PI-PLC), an enzyme widely used to release GPI-anchored proteins from the surface of cells. It should be noted that the PI-PLC enzymes that cleave phosphatidylinositol

4,5-bisphosphate (PIP₂), as part of the inositol 1,4,5-trisphosphate (IP₃) second messenger system (*see* SECOND MESSENGER PATHWAYS), and those that cleave GPI structures are mutually exclusive. The GPI is attached to the protein through an amide bond between the C-terminal amino acid α-carboxyl group and the amino group of the ethanolamine phosphate bridge. The presence of extra ethanolamine phosphate substituents does not occur in protozoa or in yeast GPIs, but is a ubiquitous modification in the GPIs of higher eukaryotes. In some anchors there can be two or three of these extra units.

Identification of GPI anchors

The presence of a GPI membrane anchor on a protein can be inferred indirectly from the analysis of cDNA sequences (see below), or directly by structural analysis for the components ethanolamine and *myo*-inositol using amino-acid analysis and gas chromatography–mass spectrometry respectively. As an alternative to chemical analysis, biosynthetic radiolabelling of proteins with [³H] ethanolamine and [³H] *myo*-inositol may be used for cells in tissue culture. Another approach is the use of *Bacillus thuringiensis* PI-PLC. This enzyme is not cytotoxic and releases GPI-

Fig. G38 (*opposite*) The chemical structure of GPI anchors. All GPI anchors contain a conserved core structure (shaded box) with species- and tissue-specific variations in the lipid structure and core substituents, as indicated. AChE, acetylcholinesterase; DAF, decay accelerating factor; PARP, procyclic acidic repetitive protein; Prp, prion protein; PsA, pre-spore antigen; PSP, promastigote surface protease; VSG, variant surface glycoprotein.

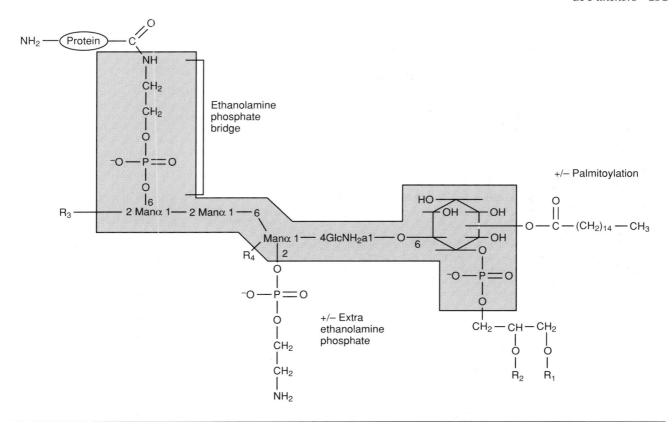

Example	R_1	R_2	R_3	R_4	Extra ethanolamine phosphate	Palmitoylation
Trypanosoma brucei VSG	$C_{14:0}$ fatty acid (exclusively)	$C_{14:0}$ fatty acid (exclusively)	—	± α-Gal1 — 2 α-Gal1 — 6 $\searrow$ α-Gal1 — 3 ± α-Gal1 — 2 $\nearrow$	—	—
T. brucei PARP	$C_{18:0}$ fatty acid (exclusively)	—	[NANAs, Galq, GlcNAcq]		—	+
Trypanosoma cruzi 1G7	$C_{16:0}$ alkyl	$C_{18:0}$ fatty acid	α-Man	—	—	—
Leishmania major PSP	$C_{24:0}$ alkyl	$C_{16:0}$ fatty acid	—	—	—	—
Yeast proteins	Ceramide		α-Man $_{1-2}$	—	—	—
Dictyostelium discoideum PsA	Ceramide		± α-Man	—	+	—
Torpedo AChE	$C_{16:0}$ fatty acid	$C_{16:0}$ fatty acid	α-Glc	β-GalNAc1 — 4	+	—
Human erythrocyte AChE	$C_{18:0}$ alkyl	$C_{22:4}$ fatty acid	—	—	+	+
Human erythrocyte DAF	Alkyl	$C_{22:4}$ fatty acid	?	?	+	+
Rat brain Thy-1	? alkyl	$C_{18:0}$ fatty acid	± α-Man	β-GalNAc1 — 4	+	—
Hamster Prp	?	?	± α-Man [NANA-Gal-GalNAc]		+	—

anchored proteins from the surface of living cells. The release of a particular protein can be monitored either by following the appearance of solubilized protein in the medium, or by following its loss from the cell surface. The latter criterion is often conveniently measured by fluorescence flow cytometry (*see* CELL SORTING) using a fluorescently labelled ANTIBODY to the protein of interest. The PI-PLC approach can also be used on cell lysates and purified proteins. For example, the release of the lipid moiety by PI-PLC changes the phase-partition behaviour of GPI-anchored proteins. In a Triton X-114 (TX-114) phase-separation experiment GPI-anchored proteins partition into the detergent phase before PI-PLC treatment, and into the aqueous phase after PI-PLC treatment. In addition, PI-PLC-cleaved GPI-anchored proteins express an epitope called 'the cross-reacting determinant' (CRD) which can be recognized by anti-CRD antibodies on WESTERN BLOTS. Although positive results with PI-PLC are powerful criteria for GPI anchor identification, negative results are ambiguous as many GPI anchors are inherently resistant to PI-PLC because of inositol-palmitoylation.

Biosynthesis and transfer to protein

The GPI anchors are synthesized as precursors in the ENDOPLASMIC RETICULUM. This process involves the sequential transfer of monosaccharides to a phosphatidylinositol (PI) acceptor and the transfer of ethanolamine phosphate from phosphatidylethanolamine to the terminal mannose residue [4] (Fig. G39). In higher eukaryotes the extra ethanolamine phosphates are probably added to the GPI intermediates during the synthesis of the GPI precursor. The other carbohydrate substituents (Fig. G38) are probably added after the GPI precursor has been attached to protein, most likely in the GOLGI APPARATUS. In mammalian cells, the precursor species, and its intermediates, are generally inositol-palmitoylated. Thus the PI-PLC-sensitive GPI anchors commonly expressed in these cells have undergone an extra processing step, involving the removal of the palmitate group from the inositol ring.

For a protein to receive a GPI anchor it must contain two pieces of information. First, it must have a SIGNAL SEQUENCE at its N terminus to direct its translocation across the endoplasmic reticulum membrane (*see* PROTEIN TARGETING; PROTEIN TRANSLOCATION). Second, it must have a GPI addition signal peptide (GPIsp) at its C terminus. This GPIsp sequence of 17–30 amino acids is highly variable and there is no CONSENSUS SEQUENCE. However, there is a general pattern, illustrated in Fig. G40. A row of 12–20 hydrophobic residues at the extreme C terminus of a predicted protein sequence is highly indicative of a GPIsp. This GPIsp sequence is exchanged for the preassembled GPI precursor in what is believed to be a transamidation reaction, where the ethanolamine amino group displaces the peptide (amide) bond at the GPIsp cleavage site. Functional GPIsp sequences can be spliced onto recombinant gene constructs for expression in eukaryotic cells.

In some circumstances a single gene can encode both GPI-anchored and transmembrane forms of the same protein (e.g. NCAM and LFA-3; *see* CELL ADHESION MOLECULES). This arises by differential RNA SPLICING, such that some messenger RNAs en-

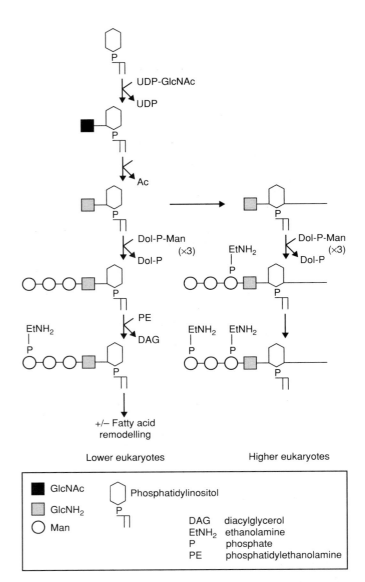

Fig. G39 The biosynthesis of GPI precursors. The scheme on the left relates to GPI precursor biosynthesis in *Trypanosoma brucei*. The precursor species in this case undergoes a process called fatty acid remodelling which is unique to this species. This involves the removal of each fatty acid and its replacement with myristic acid ($C_{14:0}$) donated from myristoyl-CoA. The scheme on the right is the consensus model for GPI precursor biosynthesis in mammalian cells. In this case the $GlcNH_2$-PI intermediate is modified by palmitoylation of the inositol ring before the addition of mannose and ethanolamine phosphate. The donor of the ethanolamine phosphate groups is unknown. In both cases the biosynthesis of the GPI precursor and its transfer to newly synthesized protein is believed to occur in the endoplasmic reticulum. Further modifications to the core structures are believed to occur after the transfer of the GPI anchor to the protein. DAG, diacylglycerol; Dol, dolichol; $EtNH_2$, ethanolamine; P, phosphate; PE, phosphatidylethanolamine.

code a transmembrane domain followed by a cytoplasmic domain at the C terminus, whereas other mRNA species encode a GPIsp sequence at the C terminus. In other cases, transmembrane and GPI-anchored versions of the same protein are encoded by different genes (e.g. for the FcγRIII receptor, *see* FC RECEPTORS).

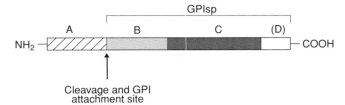

Fig. G40 Features of the GPI addition signalling peptide (GPIsp). GPIsp cleavage and GPI anchor attachment occurs at a small amino-acid residue. The amino acid which accepts the GPI anchor can be Ser, Gly, Ala, Asp, Asn, or Cys. The region (A) to the N-terminal side of the GPI anchor addition site contains no information necessary to the GPI addition/GPIsp cleavage transamidation reaction. Following the cleavage/attachment site is a moderately polar region of ~7–12 amino acids (B) with no other significant features except that the first amino acid after the GPIsp cleavage site cannot be Pro and the second amino acid after the cleavage site is a small polar residue, such as Gly, Ala, Ser. Following the moderately polar domain is a hydrophobic domain of 10–20 random hydrophobic amino acids (C). In most GPIsp sequences the hydrophobic domain terminates the sequence. However, polar domains beyond the hydrophobic domain can be tolerated. These polar domains (D) are usually different from conventional cytoplasmic domains in that they lack two or more positively charged amino acids within 12 residues of the hydrophobic domain. In the rare cases where more than one positive residue is found at the C terminus of the GPIsp, the hydrophobic domain is disrupted in the middle by a negatively charged amino acid.

These alternatively anchored forms of the same protein have the same extracellular functions, for example cell–cell adhesion (NCAM and LFA-3) and immune complex binding (FcγRIII), but can differ in their ability to mediate intracellular events (see below).

Functions and properties

The basic function of a GPI anchor is to afford a stable association of protein with the membrane, but possession of a GPI anchor has many consequences.

1 In mammalian cells GPI-anchored proteins are found in specialized membrane microdomains which are insoluble in the neutral detergent Triton X-100 (TX-100). These specialized microdomains appear to be rich in sphingolipids, glycosphingolipids and probably cholesterol [5].

2 The presence of a GPI anchor generally causes the targeting of proteins to the apical membrane of polarized cells (*see* EPITHELIAL POLARITY).

3 GPI-anchored proteins are generally excluded from the clathrin-mediated endocytic pathway (*see* COATED PITS AND VESICLES; ENDOCYTOSIS). Their levels of surface expression are generally very high and their turnover rates are generally low. However, in some cases, for example the GPI-anchored folate binding protein, a novel clathrin-independent form of endocytosis (called potocytosis) is known to occur.

4 Although GPI-anchored proteins have no cytoplasmic domain they can effect transmembrane signalling. In T CELLS, for example, cross-linking of any GPI-anchored protein into patches causes a rise in intracellular CALCIUM, and MITOGENESIS in the presence of PHORBOL ESTERS. The mechanism by which the transmembrane signal is mediated is unknown.

5 The mobility of GPI-anchored proteins in the plane of the membrane is dictated by their interactions with other membrane components and can range from extremely fast to essentially immobile.

6 There is evidence that some GPI-anchored proteins are released from membranes by the action of GPI-specific phospholipases. However the generality and the mechanism of control of this phenomenon is unknown. The normal sera of all mammals are known to contain GPI-specific phospholipase D which is capable of cleaving between the inositol ring and the phosphate of the PI moiety of all GPI anchors in detergent solution (*see* PHOSPHOLIPASES). However, the action of this enzyme on the surface of living cells has not been demonstrated.

7 The expression of alternatively anchored forms of the same protein (GPI versus transmembrane) through mRNA splicing or differential gene expression may be used to modulate the function of some proteins. For example, ligation of the transmembrane form of the FcγRIII receptor (CD16) with immune complexes directly triggers cytotoxic reactions and phagocytosis in monocytes, whereas ligation of the GPI-anchored form of the same receptor in neutrophils only potentiates these cells to respond to signals via other transmembrane receptors.

8 Defects in GPI anchor biosynthesis are associated with the human disease paroxysmal nocturnal haemoglobinuria (PNH). Patients with PNH lack GPI-anchored proteins on a range of cells (erythrocytes, platelets, granulocytes, monocytes, natural killer cells, and some B cells and T cells) owing to the proliferation of mutant haematopoietic stem cells. The lack of complement-regulating proteins, such as the GPI-anchored CD59 and DECAY ACCELERATING FACTOR (DAF), on the erythrocytes of PNH patients is thought to be responsible for the autohaemolysis characteristic of this disease.

9 The use of GPI anchors is more common in lower eukaryotes such as protozoa and yeast than in higher eukaryotes such as mammals. The parasitic kinetoplastid protozoa (e.g. *Trypanosoma brucei*, *T. cruzi*, and *Leishmania* spp.) are known to express extremely high levels of GPI anchors and/or GPI-related molecules which are not linked to protein. One of these GPI-related molecules is the lipophosphoglycan (LPG) of *Leishmania* parasites which contains a large repeating phosphosaccharide chain linked to a GPI-like glycolipid core. The LPG is known to confer virulence on *Leishmania* in both the mammalian host and the insect (sand-fly) vector. The presence of novel parasite-specific GPI-related molecules may reflect evolutionary elaboration of the conserved GPI anchor biosynthetic pathway.

M. FERGUSON

See also: CD ANTIGENS; GLYCANS; GLYCOLIPIDS; MEMBRANE STRUCTURE.

1 Cross, G.A.M. (1990) Glycolipid anchoring of plasma membrane proteins. *Annu. Rev. Cell Biol.* **6**, 1–39.

2 Ferguson, M.A.J. (1991) Lipid anchors on membrane proteins. *Curr. Opinion Struct. Biol.* **1**, 522–529.

3 Ferguson, M.A.J. & Williams, A.F. (1988) Cell surface anchoring of proteins via glycosyl-phosphatidylinositol structures. *Annu. Rev. Biochem.* **57**, 285–320.

4 Englund P.T. (1993) The structure and biosynthesis of glycosyl-phosphatidylinositol protein anchors. *Annu. Rev. Biochem.* **62**, 121–138.

5 Brown, D.A. et al. (1992) Interactions between GPI-anchored proteins and membrane lipids. *Trends Cell Biol.* **2**, 338–343.

gradient(s) In developing tissues, if a source of MORPHOGEN is positioned at one end of a sheet of cells, a concentration gradient of the morphogen — the morphogenetic gradient — will be established across the sheet. Cells may then differentiate according to the local concentration of morphogen they sense. *See*: GRADIENT THEORIES OF DEVELOPMENT; PATTERN FORMATION.

gradient theories of development Theories of development first proposed in the 1940s by C.H. Waddington, and in the 1950s by A. Turing. Both approached the problems from the perspective of field theories in which the parts of the field had to be simultaneously patterned. Most recently, gradient theories are of especial relevance to four areas of developmental biology: (1) cockroach limb regeneration; (2) control of gene expression in the early *Drosophila* embryo; (3) patterning of the vertebrate limb; and (4) *Hydra* development. Most gradient/diffusion theories, especially those in *Drosophila*, invoke a twofold system involving an activator and an inhibitor which diffuse at different rates. In each system, MORPHOGENS diffuse throughout the entire MORPHOGENETIC FIELD. Cells respond to the concentrations of activator and inhibitor, thus deriving POSITIONAL INFORMATION. *See also*: DROSOPHILA DEVELOPMENT; HYDRA DEVELOPMENT; LIMB DEVELOPMENT; PATTERN FORMATION; POLAR COORDINATE MODEL; REGENERATION.

graft-versus-host (GVH) reaction Destructive immunological reaction mounted by cells of a tissue graft against the host tissue. The GVH reaction is a CELL-MEDIATED IMMUNE RESPONSE mounted against the host's MHC antigens by immunocompetent T cells within the graft (*see* MAJOR HISTOCOMPATIBILITY COMPLEX). Severe GVHs are a particular problem in BONE MARROW transplantation in which mature T cells will be introduced with the new bone marrow.

grafting In experimental developmental biology a time-honoured technique for the analysis of developmental mechanisms in the early embryo. Organs or groups of cells, individual cells, or even portions of cells (e.g. cytoplasm or nucleus), can be grafted surgically into new or altered locations in early embryos, yielding many important results. Outstanding examples include the recognition of embryonic INDUCTION by Spemann and Mangold in 1924 who induced a secondary embryonic axis by grafting the dorsal lip of the blastopore to a position overlying presumptive ventral epidermis in amphibian embryos; and the analysis of restriction of cell potency by NUCLEAR TRANSPLANTATION into amphibian oocytes.

gramicidin A Peptide antibiotic of 15 amino acids isolated from *Bacillus brevis* (Fig. G41). A channel-forming IONOPHORE which forms transient dimers in membranes allowing the passage of cations.

grana Stacks of THYLAKOID disks which occur in CHLOROPLASTS of many, but not all, photosynthetic cells of higher plants; bundle sheath chloroplasts of some C4 plants contain only unappressed thylakoids. Grana stacks contain a variable number of disks (which is particularly large in shade plants) which are linked by stroma lamellae. PHOTOSYSTEM II units are concentrated in the appressed membranes of grana stacks, whereas PHOTOSYSTEM I units and ATP SYNTHASE complexes are concentrated in the stroma lamellae and the unappressed parts of the grana membranes. The CYTOCHROME *bf* COMPLEX occurs in both appressed and unappressed membranes. *See also*: PHOTOSYNTHESIS.

grandchildless *See*: MATERNAL EFFECT.

granulocyte Class of white blood cells of the myeloid lineage, which typically contain abundant cytoplasmic granules, and which include EOSINOPHILS, BASOPHILS, and NEUTROPHILS — distinguished histologically on the staining properties of their granules. They act as effector cells in inflammatory reactions and natural immunity, and are also stimulated by CYTOKINES produced by activated antigen-specific T CELLS. *See*: HAEMATOPOIESIS; LYMPHOKINES; G-CSF; GM-CSF.

granulocyte colony-stimulating factor (G-CSF) *See*: G-CSF.

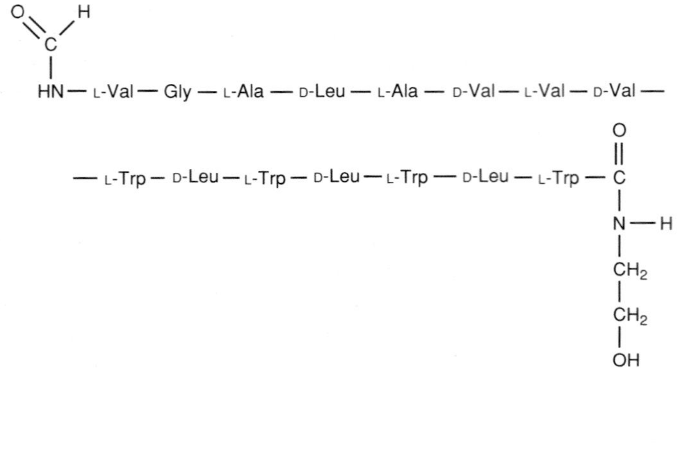

Fig. G41 Gramicidin A.

granulocyte-macrophage colony-stimulating factor (GM-CSF)
See: GM-CSF.

gravitropism Directional growth of plant organs in response to gravity. Roots, growing vertically downwards, are positively gravitropic, stems vertically upwards (in the absence of light), negatively gravitropic. Gravitropism has replaced the older and less explicit term, geotropism. *See also*: DIAGEOTROPISM.

Greek key motif A common secondary structure motif in proteins in which adjacent antiparallel β-strands in a β-sheet are arranged in a pattern that topologically resembles the key motif commonly used for decoration in ancient Greece (Fig. G42). *See also*: PROTEIN STRUCTURE.

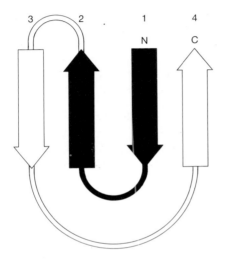

Fig. G42 The Greek key motif.

grey crescent In amphibians, shortly after FERTILIZATION, the egg cortex rotates about 30° relative to the internal cytoplasm, towards the site of sperm entry. In certain species, such as *Rana*, a region of the egg previously covered by the dark cortical cytoplasm is now visible, and appears grey due to its content of pigment granules. This region lies near the equator of the egg, opposite the site of sperm entry, and marks the point where gastrulation begins. *See also*: AMPHIBIAN DEVELOPMENT.

gRNA Guide RNA. *See*: RNA EDITING.

group I, group II introns *See*: RIBOZYMES; RNA SPLICING.

growth cone The tip of a growing or regenerating neuronal process, most commmonly an axon. This structure is highly dynamic and extends filopodia and lamellipodia which apparently investigate the surrounding environment seeking signals which govern the direction of axonal growth. The growth cone contains many actin MICROFILAMENTS and is capable of independent motility if severed from the axon. *See*: CELL MOTILITY; NEURONAL CYTOSKELETON; NEURON.

Growth factor(s)

GROWTH factors can be broadly defined as multifunctional, locally acting, intercellular signalling POLYPEPTIDES which control both the ontogeny and maintenance of tissue form and function.

A multicellular organism can be considered as a community of different types of cells whose individual proliferation, differentiation and physiological function must, in some way, be coordinated for the overall function of both individual tissues and the whole organism. This is achieved by specific intercellular signals which control cell multiplication, differentiation and behaviour.

These intercellular signals can be of two types: long-range (or endocrine) signals which are released into the circulation or other body fluids, and which modify the behaviour of physically remote cell types; and short-range (or paracrine) signals which act locally within tissues. Classical hormones, such as INSULIN or GLUCAGON, are examples of long-range signalling molecules. Growth factors are, by contrast, predominantly short-range, locally acting, intercellular signalling molecules.

Growth factors share a number of common biological properties apart from their predominantly local mode of action. They often exert their biological actions at very low (typically 10^{-9}–10^{-11} M) concentrations. This is because their action is mediated by their association with specific, high affinity receptors expressed by the target cell type (*see* GROWTH FACTOR RECEPTORS). The function of the growth factor receptor is not only to interact specifically with the ligand on the outside of the cell but also to generate an intracellular signal on the inside of the cell [1]. It is this generation of growth factor receptor-mediated intracellular signals, and their 'interpretation' by the responding cell that leads to the modification of target cell behaviour. As a result, the biological actions of growth factors are not (as their name might suggest) confined to the regulation of cell multiplication but can extend into a wide variety of aspects of cell function including DIFFERENTIATION, migration and gene expression.

In addition, growth factors exhibit cell-type specificity in their action. In other words, the same growth factor can have very different biological effects depending on the type of cell with which it interacts. As target cell response depends, in part, on the expression of specific receptors it is not surprising that growth factors exhibit wide differences in 'host range': some growth factors (particularly those involved in the regulation of HAEMATOPOIESIS) can be highly restricted in the range of cells on which they act, whereas others can regulate the behaviour of many different cell types. In this case it is not surprising that some growth factors at least (in contrast to endocrine hormones) are found to be widespread in their expression and distribution in the embryo and adult (see below).

Growth factor families

There are currently about 80 or more known genes in both vertebrates and invertebrates, encoding proteins which can be considered, on the basis of their biological function, to be growth factors. It is also very likely that more growth factor genes remain to be discovered. Growth factors therefore represent a large set of

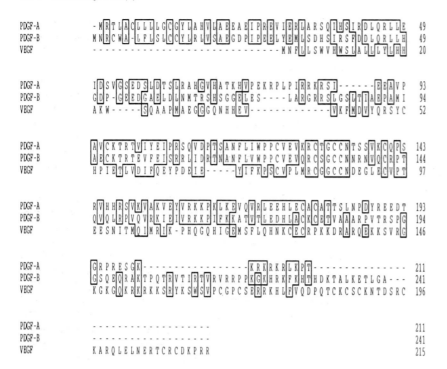

PDGF-A	- M R T L A C L L L L G C G Y L A H V L A E E A E I P R E V I E R L A R S Q I H S I R D L Q R L L E	49
PDGF-B	M N R C W A - L F L S L C C Y L R L V S A E G D P I P E E L Y E M L S D H S I R S F D D L Q R L L H	49
VEGF	- M N F L L S W V H W S L A L L L Y L H H	20

PDGF-A	I D S V G S E D S L D T S L R A H G V H A T K H V P E K R P L P I R R K R S I - - - - - E E A V P	93
PDGF-B	G D P - G E E D G A E L D L N M T R S H S G G E L E S - - - - L A R G R R S L G S L T I A E P A M I	94
VEGF	A K W - - - - - S Q A A P M A E G G G Q N H H E V - - - - - - - - - - - V K F M D V Y Q R S Y C	52

PDGF-A	A V C K T R T V I Y E I P R S Q V D P T S A N F L I W P P C V E V K R C T G C C N T S S V K C Q P S	143
PDGF-B	A E C K T R T E V F E I S R R L I D R T N A N F L V W P P C V E V Q R C S G C C N N R N V Q C R P T	144
VEGF	H P I E T L V D I F Q E Y P D E I E - - - - - Y I F K P S C V P L M R C G G C C N D E G L E C V P T	97

PDGF-A	R V H H R S V K V A K V E Y V R K K P K L K E V Q V R L E E H L E C A C A T T S L N P D Y R E E D T	193
PDGF-B	Q V Q L R P V Q V R K I E I V R K K P I F K K A T V T L E D H L A C K C E T V A A A R P V T R S P G	194
VEGF	E E S N I T M Q I M R I K - P H Q G Q H I G E M S F L Q H N K C E C R P K K D R A R Q E K K S V R G	146

PDGF-A	G R P R E S G K - - - - - - - - - - - - - - - - - K R K R K R L K P T - - - - - - - - - - - - - - -	211
PDGF-B	G S Q E Q R A K T P Q T R V T I R T V R V R R P P K G K H R K F K H T H D K T A L K E T L G A - - -	241
VEGF	K G K G Q K R K R K K S R Y K S W S V P C G P C S E R R K H L F V Q D P Q T C K C S C K N T D S R C	196

PDGF-A	- - - - - - - - - - - - - - - - - -	211
PDGF-B	- - - - - - - - - - - - - - - - - -	241
VEGF	K A R Q L E L N E R T C R C D K P R R	215

Fig. G43 The PDGF family.

polypeptides. These proteins, as a whole, do not share any common structural features but many can be grouped, on the basis of amino-acid sequence and tertiary structure, into MULTI-GENE FAMILIES. It is likely that, as new growth factors are discovered, the size and extent of growth factor multigene families will increase. Rather than attempting to describe every known growth factor, this review will focus attention on the most prominent growth factor families.

Platelet-derived growth factor

Platelet derived growth factor (PDGF) was discovered as the major component of serum required for the multiplication of a variety of cell types in culture [2]. PDGF was first purified from PLATELETS and is a heterodimeric molecule composed of two disulphide-linked polypeptide chains (each of M_r 12 500 (12.5K)) — PDGF-A and PDGF-B. Subsequent purification of PDGF-like activities from a number of different sources showed that PDGF exists in three forms: an AB heterodimer and AA and BB homodimers (Fig. G43). PDGF-A and PDGF-B are widespread in their expression, being particularly prominent in the early embryo, placenta and central nervous system. PDGF-A and PDGF-B exhibit distinct patterns of expression *in vivo* suggesting that each form of PDGF may have distinct biological functions.

Although PDGF was first identified by its ability to induce the multiplication of fibroblasts and smooth muscle cells, it has many other biological activities including chemotactic effects and regulation of differentiation of glial cells in the optic nerve. More recently, the PDGF family has been extended by the discovery of vascular endothelial cell growth factor (VEGF) which was identified as a MITOGEN for cultured endothelial cells. VEGF proves to be related to PDGF in sequence and has the characteristic

two-chain structure, although it exhibits different biological functions and target cell specificity *in vitro*.

Epidermal growth factor family

Epidermal growth factor (EGF) was one of the first growth factors to be discovered, as a result of its striking effects on the maturation of various epithelia in the newborn mouse [3]. Injection of male mouse submaxillary gland extracts into newborn mice was found to lead to accelerated eyelid opening and tooth eruption. The male submaxillary gland proved to be a rich source of EGF and this greatly facilitated its purification. EGF is a polypeptide of M_r 6000 which is cleaved from a much larger transmembrane precursor protein in the submaxillary gland [4]. EGF is mitogenic for a wide variety of cell types, particularly those of epithelial origin, as well as having nonmitogenic effects such as inhibition of gastric acid release in the stomach. EGF itself is restricted in its expression to the submaxillary gland and the gut but it emerges that EGF is but one example of a large family of related molecules with EGF-like function.

Transforming growth factor alpha (TGFα) was isolated from the conditioned media of virus-transformed cell lines as an EGF-like bioactivity [5]. TGFα is also cleaved from a larger transmembrane precursor to yield the mature protein (M_r 6K) which is related in both primary sequence and three-dimensional structure (Fig. G44) to EGF. It is of interest that the transmembrane precursor of TGFα is also biologically active; the 'tethering' of TGFα precursor in the membrane may therefore represent a mechanism for confining TGFα action to cells in immediate physical contact. Unlike EGF, however, TGFα is relatively widespread in its expression, being found in a wide variety of foetal and adult tissues. TGFα seems to bind to the same receptor as EGF and is, to all

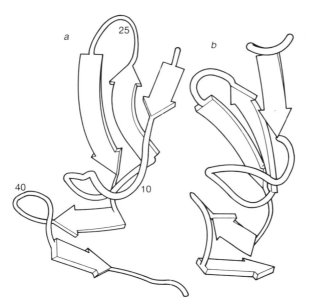

Fig. G44 Structures of *a*, epidermal growth factor (EGF) and *b*, transforming growth factor alpha (TGFα). Numbers indicate amino-acid positions.

appearances, identical to EGF in biological function.

Further EGF-like polypeptides have been found encoded by a variety of poxviruses (*see* ANIMAL VIRUSES) including VACCINIA VIRUS and myxomavirus. The existence of virally encoded EGF-like growth factors raises the possibility that poxviruses may mimic the biological actions of host-derived EGFs in the process of viral infection and reproduction.

More recently, a number of additional cellular proteins have been isolated which resemble EGF/TGFα in sequence. These include amphiregulin (AMPH) and heparin-binding epidermal growth factor (HB-EGF). These molecules both have a long N-terminal extension which appears to confer affinity for heparin on the protein, suggesting that both AMPH and HB-EGF may be bound to heparin sulphate-containing GLYCOSAMINOGLYCANS within the extracellular matrix (ECM) *in vivo* (*see* EXTRACELLULAR MATRIX MOLECULES). The localization of the growth factor in the ECM by glycosaminoglycans seems to be another common mechanism for confining the biological action of growth factors to a strictly local mode of action (see below). The full biological actions of AMPH and HB-EGF remain to be explored but they may differ significantly from the EGF/TGFα prototypes.

Fibroblast growth factor family

Fibroblast growth factor (FGF) was discovered, as its name suggests, from its ability to induce the proliferation of fibroblast cells *in vitro*. A key feature of the purification of FGF was the exploitation of the fact that FGF (like AMPH and HB-EGF) binds avidly to immobilized heparin. Initial purification of FGF bioactivity from brain revealed two closely related molecules, with very similar biochemical and biological properties, termed (on the basis of their pI and affinity for heparin) basic FGF (bFGF) and acidic FGF (aFGF). aFGF and bFGF are powerful mitogens for a wide variety of cell types, including fibroblasts, neurectodermal cells and, most

notably, capillary and large vessel endothelial cells. Expression studies revealed that bFGF exhibits extremely widespread expression in both adult and foetal tissues as well being found in a large number of both normal and tumorigenic cell lines (including large vessel endothelial cells). The expression of aFGF is, by contrast, restricted principally to cells of the central and peripheral nervous system. An additional curious feature of both aFGF and bFGF is that they lack a characteristic secretory SIGNAL SEQUENCE at their N termini (*see* PROTEIN SECRETION) and neither are, in fact, usually secreted from cells. This would tend to indicate that they are not normally made available to responsive cells unless released by mechanisms such as cell damage.

aFGF and bFGF are the prototype members of a larger family of interrelated growth factors [6]. The first of these, int-2, was identified as a gene located close to a frequent insertion site of RETROVIRUSES in virus-induced mammary tumours. Cloning and sequencing of the *int-2* gene revealed a molecule with significant sequence similarity to the aFGF and bFGF prototype genes (Fig. G45). Further studies of genes capable of inducing the proliferation of fibroblasts led to the discovery of K-FGF/Hst (FGF-4) and FGF-5. Molecular cloning of genes related to FGF-4 led to the discovery of FGF-6, a molecule sharing 80% identity in amino-acid sequence with FGF-4. Finally, an additional FGF-like growth factor KGF (FGF-7) was discovered as a major growth factor for keratinocyte-derived cell lines in culture. An important feature of these other FGF family members is that they all contain functional secretory signal sequences, and are therefore readily exported from cells. This suggests that at least one rationale for the existence of multiple related family members may be, in this case, to provide a means of controlling the dissemination and export of these molecules *in vivo*, by means of selective expression of individual family members.

Aside from their mitogenic properties, FGF family members have striking effects in early development. In the amphibian, *Xenopus laevis*, the INDUCTION of mesodermal cell types such as muscle and notochord has been experimentally demonstrated to occur as a result of one or more 'inducing signals' which are generated by cells in the lower (vegetal) part of the embryo and act upon cells in the equatorial region of the embryo to cause the differentiation of mesodermal cell types (*see* AMPHIBIAN DEVELOPMENT; INDUCTION; PATTERN FORMATION). Members of the FGF family of growth factors, in particular aFGF, bFGF, and FGF-4, can be shown to reproduce at least of part of the effects of the endogenous vegetal hemisphere inducing signal on isolated explants from the upper (animal) pole of the embryo. These findings strongly suggest that FGF-like growth factors may be 'natural' signals involved in the control of cell differentiation and pattern formation in early development.

Insulin-like growth factors

With the insulin-like growth factors (IGFs) the distinction between growth factors and endocrine hormones becomes blurred. The existence of IGFs was first predicted in the form of the 'somatomedin hypothesis', which argued that the effects of pituitary GROWTH HORMONE on skeletal growth were mediated by means of an intermediate class of bioactive peptides, the somatomedins.

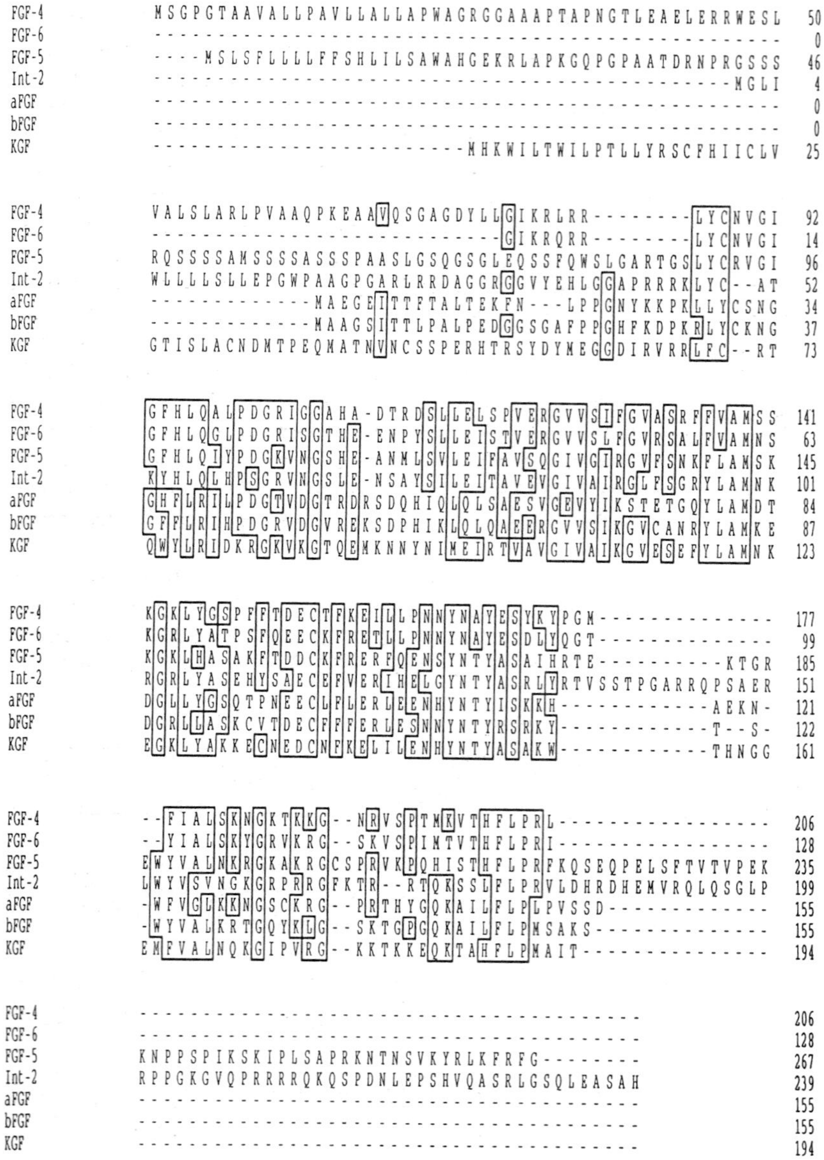

```
FGF-4   MSGPGTAAVALLPAVLLALLAPWAGRGGAAAPTAPNGTLEAELERRWESL   50
FGF-6   --------------------------------------------------    0
FGF-5   ----MSLSFLLLLFFSHLILSAWAHGEKRLAPKGQPGPAATDRNPRGSSS   46
Int-2   ---------------------------------------------MGLI     4
aFGF    --------------------------------------------------    0
bFGF    --------------------------------------------------    0
KGF     -------------------------MHKWILTWILPTLLYRSCFHIICLV   25

FGF-4   VALSLARLPVAAQPKEAAVQSGAGDYLLGIKRLRR--------LYCNVGI   92
FGF-6   GIKRQRR--------LYCNVGI   14
FGF-5   RQSSSSAMSSSSASSSPAASLGSQGSGLEQSSFQWSLGARTGSLYCRVGI   96
Int-2   WLLLLSLLEPGWPAAGPGARLRRDAGGRGGVYEHLGGAPRRRKLYC--AT   52
aFGF    MAEGEITTFTALTEKFN---LPPGNYKKPKLLYCSNG   34
bFGF    MAAGSITTLPALPEDGGSGAFPPGHFKDPKRLYCKNG   37
KGF     GTISLACNDMTPEQMATNVNCSSPERHTRSYDYMEGGDIRVRRLFC--RT   73

FGF-4   GFHLQALPDGRIGGAHA-DTRDSLLELSPVERGVVSIFGVASRFFVAMSS  141
FGF-6   GFHLQLPDGRISGTHE-ENPYSLLEISTVERGVVSLFGVRSALFVAMNS   63
FGF-5   GFHLQIYPDGKVNGSHE-ANMLSVLEIFAVSQGIVGIRGVFSNKFLAMSK  145
Int-2   KIYHLQIHPSGRVNGSLE-NSAYSILEITAVEVGIVAIRGLFSGRYLAMNK 101
aFGF    GIHFLRILPDGTVDGTRDRSDQHIQLQLSAESVGEVYIKSTETGQYLAMDT  84
bFGF    GFFLRIHPDGRVDGVREKSDPHIKLQLQAEERGVVSIKGVCANRYLAMKE   87
KGF     QWYLRIDKRGKVKGTQEMKNNYNIMEIRTVAVGIVAIKGVESEFYLAMNK  123

FGF-4   KGKLYGSPFFTDECTFKEILLPNNYNAYESYKYPGM--------------  177
FGF-6   KGRLYATPSFQEECKFRETLLPNNYNAYESDLYQGT--------------   99
FGF-5   KGKLHASAKFTDDCKFRERFQENSYNTYASAIHRTE---------KTGR   185
Int-2   RGRLYASEHYSAECEFVERIHELGYNTYASRLVRTVSSTPGARRQPSAER  151
aFGF    DGLLYGSQTPNEECLFLERLEENHYNTYISKKH-----------AEKN   121
bFGF    DGRLLASKCVTDECFFFERLESNYNTYRSRKY------------T--S-  122
KGF     ECKLYAKKECNEDCNFKELILENHYNTYASAKW-----------THNGG  161

FGF-4   --FIALSKNGKTKKG--NRVSPTMKVTHFLPRL----------------  206
FGF-6   --YIALSKYGRVKRG--SKVSPIMTVTHFLPRI----------------  128
FGF-5   EWYVALNKRGKAKRGCSPRVKPQHISTHFLPRFKQSEQPELSFTVTVPEK  235
Int-2   LWYVSVNGKGRPRRGFKTR--RTQKSSLFLPRVLDHRDHEMVRQLQSGLP  199
aFGF    -WFVGLKKNGSCKRG--PRTHYGQKAILFLPLPVSSD------------  155
bFGF    -WYVALKRTGQYKLG--SKTGPGQKAILFLPMSAKS------------   155
KGF     EMFVALNQKGIPVRG--KKTKKEQKTAHFLPMAIT--------------  194

FGF-4   -------------------------------------------------  206
FGF-6   -------------------------------------------------  128
FGF-5   KNPPSPIKSKIPLSAPRKNTNSVKYRLKFRFG-----------------  267
Int-2   RPPGKGVQPRRRRQKQSPDNLEPSHVQASRLGSQLEASAH           239
aFGF    -------------------------------------------------  155
bFGF    -------------------------------------------------  155
KGF     -------------------------------------------------  194
```

Fig. G45 The human homologues of the FGF family (see text).

Both IGF-I and IGF-II are made as secreted PROHORMONES (M_r 9K and 14K respectively) and require proteolytic cleavage to achieve their 6K form. Although similar in their biological function, IGF-I and IGF-II exhibit significant differences in their pattern of expression *in vivo*. In particular, IGF-I is expressed in juvenile life and is almost exclusively synthesized in the liver under the control of growth hormone (as predicted by the original hypothesis). IGF-II, by contrast, is expressed predominantly in the embryonic and foetal stages of mammalian development in a wide variety of different tissues. This suggests that the IGFs may have both paracrine and endocrine functions in controlling the growth of many tissue types *in vivo* [7].

Unlike many of the molecules considered so far, both IGF-I and IGF-II are present in the circulation and can be readily detected in plasma. As might be predicted from their patterns of synthesis, circulating IGF-I levels rise during juvenile life and then decline after puberty, whereas circulating IGF-II levels are highest in the foetal circulation and decline after birth. An important feature of the circulating IGFs is that they are not found free in plasma but are associated with a specific set of binding proteins. The function of the IGF-binding proteins is obscure but their primary function may be to block the bioavailability of circulating IGFs, thereby providing an additional mechanism for local control of growth factor action.

The biological importance of IGF-II for growth of the whole organism has been dramatically demonstrated in the living animal. Mice harbouring IGF-II genes which have been genetically inactivated show significantly reduced foetal and neonatal growth rates and body mass compared to their normal wild-type counterparts (*see also* PARENTAL GENOMIC IMPRINTING).

Transforming growth factor beta (TGFβ) family

The members of the transforming growth factor beta (TGFβ)

family exemplify, in many respects, the characteristic features of growth factors defined above. They have multiple effects on cell function and are widespread in expression [8].

In the course of the purification of TGFα from the culture supernatant of tumorigenic cells it was noted that, in the early stages of purification, fractions which contained most of the EGF-like bioactivity could also induce normal fibroblast cells to proliferate in semisolid medium without requiring attachment to a solid substrate. This polypeptide-induced 'anchorage independent growth' results from the combined action of TGFα and a second, structurally unrelated molecule, TGFβ. TGFβ is a homodimeric, disulphide-bonded protein of M_r 25K made up from two 12.5K polypeptide chains. Three closely related TGFβ genes exist (TGFβs 1, 2 and 3; Fig G46), creating three distinct homodimeric proteins and the hypothetical possibility of creating additional species by heterodimeric combinations of individual monomeric species. TGFβs 1–3 are widespread in their expression, TGFβ1 being almost ubiquitous, and in many tissues the different TGFβ family members are coexpressed.

A very important feature of the TGFβs is that they are secreted from cells in latent form, as a biologically inactive complex formed from one molecule of TGFβ and an additional 'latency-associated protein'. The consequence of latency is that TGFβ is biologically inactive until the latent complex is broken down. The biochemical mechanisms involved in the release from latency are uncertain but may involve either specific PROTEASES or GLYCOSIDASES. The phenomenon of latency again illustrates how growth factor action can be constrained to a local sphere of action.

A significant feature of TGFβ action is that its biological effects are often most clearly manifest in the presence of other growth factors. This is clearly demonstrated in the phenomenon of anchorage-independent growth of fibroblasts described above. In other contexts, however, TGFβ can block the mitogenic action of other growth factors. An important target of TGFβ action is the extracellular matrix. TGFβ alone, or in combination with other growth factors, has striking effects on the promotion of ECM deposition in a wide variety of cell types. This is brought about both by enhanced expression of ECM structural components, such as FIBRONECTIN and INTEGRINS, and by inhibition of enzymes involved in ECM turnover. Again the effects of TGFβ on ECM deposition are most clearly seen in the presence of other growth factors.

Thus it can be seen that the action of TGFβs is strictly dependent upon their context, in terms of both the physiological state (and phenotype) of the responding cell and the presence of other growth factors. The biological function of the TGFβs can

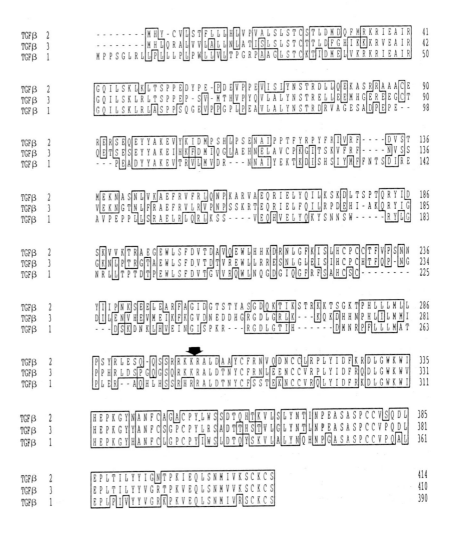

Fig. G46 Sequence alignment of human TGFβ 1, 2, and 3. The arrow marks the junction between proTGFβ and the mature peptide.

accordingly be complex in the whole organism and highly dependent upon the exact context in which it acts.

The TGFβ superfamily

TGFβs 1–3 are members of an extended SUPERFAMILY of structurally related molecules, many of which have important biological activities. The activin/inhibin family is composed of homo- and heterodimers of two protein chains derived from three genes α, βA and βB. Activins are comprised of β chain dimers and inhibins are comprised of α/β chain heterodimers. The activins and inhibins were first identified as reproductive hormones, but their most prominent function is an involvement in the process of mesoderm induction in early amphibian development described above. Activins act on isolated animal pole cells to induce mesodermal differentiation including the production of cell types (such as notochord) which are not generated in the presence of FGFs. Other TGFβ superfamily polypeptides include the family of MORPHOGENETIC PROTEINS BMPs 2–7 which induce bone deposition *in vivo* and mullerian inhibitory substance (MIS) which is a major mediator of secondary sexual differentiation in mammals. Additional TGFβ-like genes have been isolated from other species such as flies which may well also prove to have significant functions as intercellular signalling molecules in development.

J.K. HEATH

1 Heldin, C-H. & Westermark, B (1984) Growth factors: mechanism of action and relation to oncogenes. *Cell* **37**, 9–20.
2 Ross, R. et al. (1986) The biology of the platelet-derived growth factor. *Cell* **46**, 155–169.
3 Cohen, S. (1962) Isolation of a mouse submaxillary gland protein accelerating incisor eruption and eyelid opening in the new born animal. *J. Biol. Chem.* **237**, 1555–1661.
4 Carpenter, G. & Cohen, S. (1979) Transforming growth factor. *Annu. Rev. Biochem.* **48**, 193–216.
5 Derynck, R. (1988) Transforming growth factor alpha. *Cell* **54**, 593–594.
6 Burgess, W. & Maciag, T. (1989) The heparin-binding (fibroblast) growth factor family. *Annu. Rev. Biochem.* **58**, 575–606.
7 Froesch, E. et al. (1985) Actions of insulin-like growth factors. *Annu. Rev. Physiol.* **47**, 443–467.
8 Massagué, J. (1990) The transforming growth factor beta family. *Annu. Rev. Cell Biol.* **6**, 597–641.

Growth factor receptors

GROWTH factor receptors are cell-surface proteins whose purpose is to receive information from outside cells and to convey it across the cell membrane (signal transduction). The information resides in the concentration of a peptide or protein called a GROWTH FACTOR which binds to the receptor and converts the receptor to an active form. The receptor then interacts with proteins on the inner surface of the membrane and alters their properties or subcellular location resulting in changes in the cell's behaviour. These changes are various and occur at different times after receptor activation. Some alterations such as ion fluxes occur in fractions of seconds, others such as changes in the cell's shape take several minutes. The most fundamental consequence of receptor activation is to stimulate a programme of events which results in cell division (*see* CELL CYCLE; MITOSIS). This process may take 18 hours or longer to be completed.

Three classes of growth factor receptors have been described [1,2]. One consists of proteins formed of a core structure of seven transmembrane α-helical sequences. The next family is formed of large GLYCOPROTEINS which generally possess a single transmembrane sequence and TYROSINE KINASE activity [1]. A third receptor family consists of molecules which span the cell membrane once, but do not have kinase activity themselves [2]. These three superfamilies are characterized both by their structure and by their mechanism of signal transduction.

Seven-transmembrane-domain receptors

Despite their overall similarities, much diversity of structure and function exists within the members of the receptor superfamilies. It is possible, however, to define groups of receptors that form subfamilies. The seven-transmembrane-domain receptors can be divided into two major subfamilies. The first includes the adrenaline (epinephrine) receptors (α_1, α_2 and β_1, β_2 ADRENERGIC RECEPTORS), serotonin (5-hydroxytryptamine) receptors, the substance K receptor, the angiotensin receptor (the *mas* ONCOGENE), the muscarinic acetylcholine receptors, the bradykinin receptor, the endothelin receptor, and the bombesin receptor, all of which consist of seven transmembrane sequences with little extracellular structure (*see* G PROTEIN-COUPLED RECEPTORS). The second subfamily is defined by the luteinizing hormone receptor, and may also include the follicle-stimulating hormone receptor and the thyroid-stimulating hormone receptor. The former protein consists of a large extracellular domain of 333 amino acids followed by the characteristic seven transmembrane sequences.

Tyrosine kinase receptors

The receptors with tyrosine kinase activity, often known as receptor tyrosine kinases (RTKs), have been grouped into families on the basis of three criteria — the sequence homology of their kinase domains, their general structural layout, and the structural similarity of their ligands (Fig. G47). The type I family

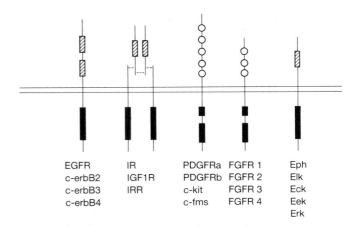

Fig. G47 Tyrosine kinase growth factor receptor families. EGFR, epidermal growth factor receptor; IR, insulin receptor; IGF1R, insulin-like growth factor I receptor; IRR, insulin receptor related receptor; PDGFR, platelet-derived growth factor receptor; FGFR, fibroblast growth factor receptor. Solid shading, tyrosine kinase sequences; hatched shading, regions rich in cysteine residues; circles, immunoglobulin-like domains.

consists of the c-erbB1 protein (the epidermal growth factor receptor), c-erbB2, c-erbB3, and c-erbB4 (also known as HER-1, 2, 3 and 4 respectively). c-erbB1/EGFR binds epidermal growth factor (EGF), transforming growth factor-α (TGFα), amphiregulin and heparin-binding EGF; c-erbB2 may bind the newly described ligand heregulin. No ligands are yet known to interact with the c-erbB3 protein. These receptors consist of a single glycosylated subunit whose extracellular domain contains two regions rich in cysteine residues. The intracellular tyrosine kinase domain is not interrupted by any nonhomologous sequences.

The type II receptors consist of insulin receptor, insulin-like growth factor I receptor, and the insulin receptor related receptor. The former two interact in a complex way with insulin, IGF-I, and IGF-II. These receptors are heterotetramers of two entirely extracellular glycosylated α subunits and two β subunits which span the cell membrane and contain the tyrosine kinase domain. The subunits are held together by disulphide bonds (*see* INSULIN AND ITS RECEPTOR).

The type III receptors consist of platelet-derived growth factor (PDGF) receptors A and B, c-Kit and c-Fms (also known as colony-stimulating factor (CSF) receptor). The PDGF receptor ligands are PDGF AA, BB, and AB; the ligand for c-Kit is the product of the gene *steel*, and for c-Fms CSF-1.

Some authors also include the fibroblast growth factor (FGF) receptors within the type III receptor family. This group of molecules consists of four receptors FGFR1–4 and seven ligands, FGF1–7. Their interactions are not yet well defined. The type III receptors are made up of glycosylated extracellular domains composed of variable numbers of immunoglobulin-like loops (*see* IMMUNOGLOBULIN SUPERFAMILY) and an intracellular domain in which the tyrosine kinase sequence is interrupted by stretches of unrelated amino-acid sequence.

The type IV receptors are formed of the Eph, Elk, Eek, Eck and Erk (putative) receptors whose ligands are not yet known. Members of this family of molecules consist of a single transmembrane glycoprotein subunit whose extracellular domain contains one cysteine-rich region and an intracellular domain containing a contiguous kinase sequence.

Several other tyrosine kinase receptors have been described which have not been included in the above families. These include *ret*, *trk* and *met* proteins, the last being the receptor for hepatocyte growth factor.

An additional level of complexity, particularly found in the FGF receptor family, is that alternative RNA SPLICING of receptor messenger RNA can create up to 12 receptor proteins from a single gene. The significance of these variant forms is currently being explored.

Non-tyrosine kinase receptors

Several of the third superfamily of growth factor receptors are expressed in haematopoietic cells. Those whose structures have been defined include the interleukin 2 (IL-2) receptor, the receptors for IL-3, IL-4, IL-6, and IL-7, the granulocyte–macrophage colony-stimulating factor receptor (*see* LYMPHOKINES), the erythropoietin receptor, growth hormone receptor, prolactin receptor, and the granulocyte colony-stimulating factor receptor. These have similar sized extracellular domains which contain several

conserved cysteine residues and variable length intracellular domains. In some cases, such as the IL-2 and IL-6 receptors, they consist of more than one subunit. These associated chains do not, however, seem to be structurally related.

Receptor mechanisms

The mechanisms by which each superfamily of growth factor receptors convey growth stimulatory signals across the cell membrane have been partially determined. The seven-transmembrane receptors interact with G proteins and stimulate adenylate cyclase, phospholipase C and probably phospholipase A2 (*see* G PROTEIN-COUPLED RECEPTORS; GTP-BINDING PROTEINS; SECOND MESSENGER PATHWAYS). Each receptor, however, seems to interact with a different selection of second messenger-generating systems. Most receptors have been shown to undergo both homologous (ligand-induced) and heterologous (via other systems) desensitization. This may be due both to reduction in receptor affinity for ligand (probably as a consequence of phosphorylation of intracellular serine or threonine residues, *see* PROTEIN PHOSPHORYLATION) and to a chronic reduction in receptor numbers.

The tyrosine kinase receptors are, with the exception of the type II subfamily, activated by ligand-induced dimerization. Dimerization stimulates tyrosine kinase activity leading to transphosphorylation of some of the receptor's intracellular tyrosine residues. These phosphorylated sequences now act as binding sites for intracellular proteins which contain SH2 DOMAINS. SH2 stands for *src*-homology region 2, a sequence of amino acids first identified in the c-*src* kinase. Proteins known to interact with these receptors include phospholipase Cγ, GAP (GTPase activating protein, *see* GTP-BINDING PROTEINS), phosphatidylinositol 3-kinase (*see* SECOND MESSENGER PATHWAYS), and the *crk* protein. All these become phosphorylated on tyrosine, and in the case of phospholipase C it is known that this increases its enzyme activity. Again, different receptors appear to interact with subsets of these proteins and no doubt more substrates remain to be identified. Ligand binding often leads to receptor DOWNREGULATION. Evidence also exists for some degree of receptor heterodimerization such as between c-erbB1/EGF receptor and c-erbB2.

The mechanism of action of the third superfamily of receptors is very poorly understood. Binding of IL-2 to its receptor activates the kinase of the *lck* protein (*see* LYMPHOKINES) and it may be that other receptors of this family interact with other members of the *src* family of tyrosine kinases. The role of the associated proteins seems to be to induce high affinity ligand binding. It is not known if they convey intracellular signals.

Growth factor receptors as oncogenes

Growth factor receptors are involved not only in promoting the growth of normal cells but also in the aberrant growth of many types of human cancer cells [3]. In particular, the c-erbB1/EGF receptor and c-erbB2 protein have been found to be mutated or overexpressed in human carcinomas. The c-erbB1/EGF receptor is mutated to an active form by deletion of extracellular sequences in about 30% of glioblastoma multiforme brain tumours. In addition, the gene is amplified, leading to overexpression of the receptor protein. Overexpression occurs in other tumours of the breast, stomach, lung, bladder, head and neck, and in gynaeco-logical cancer and oesophageal cancer, sometimes as a result of amplification and sometimes owing to increased mRNA tran-scription. Mutation seems to be rare in these tumour types [4]. The c-*erb*B2 gene is overexpressed as a result of amplification or transcriptional deregulation in about one-fifth to one-third of breast, stomach, lung, ovary, and bladder tumours [5]. Evidence from several studies suggests that receptor overexpression is an indicator of poor clinical behaviour in cancer patients, being associated with short relapse-free interval and overall survival. Such altered receptors represent targets for new forms of chemo-therapy.

W.J. GULLICK

See also: PROTEIN KINASES.

1 Ullrich, A. & Schlessinger, J. (1990) Signal transduction by receptors with tyrosine kinase activity. *Cell* **61**, 203–212.
2 Aaronson, S.A. (1991) Growth factors and cancer. *Science* **254**, 1146–1153.
3 Cross, M. & Dexter, T.M. (1991) Growth factors in development, transfor-mation and tumorigenesis. *Cell* **64**, 271–280.
4 Gullick, W.J. (1991) Prevalence of aberrant expression of the epidermal growth factor receptor in human cancers. *Br. Med. Bull.* **47**, 87–98.
5 Lofts, F.J. & Gullick, W.J. (1992) c-*erb*B-2 amplification and overexpression in human tumours. In *Genes, Oncogenes and Hormones: Advances in Cellular and Molecular Biology of Breast Cancer* (Dickson, R.B. & Lippman, M.E., Eds) Vol. III, 161–179 (Kluwer, Norwell).

growth hormone (GH, somatotropin) Mammalian protein hor-mone (191 amino acids) produced by the anterior pituitary. It is released from the pituitary in response to stimulation by the hypothalamic peptide growth hormone-releasing factor. Release is inhibited by somatostatin. Growth hormone acts indirectly to promote skeletal growth by stimulating the production of somatomedin-1 (insulin-like growth factor 1, *see* GROWTH FAC-TORS) from the liver. Growth hormone is also thought to promote the DIFFERENTIATION of fat cells and chondrocytes. Human GH contains two receptor binding sites, one of high affinity and one of low affinity. In the active signal transduction complex GH binds two molecules of the receptor in a sequential manner. Mutant GH lacking a low affinity binding site cannot stimulate signal trans-duction.

growth hormone deficiency In children this manifests itself as shortness of stature whereas in adults few clinical features result other than fine wrinkling of the skin and a tendency to hypogly-caemia. It may be congenital or acquired. Occasionally a genetic cause exists which is usually autosomal recessive although X-linked and dominant types have been described.

GRP 78 *See*: BIP.

gsp A potential ONCOGENE encoding the α_s subunit of the GTP-BINDING PROTEIN G_s. It is found mutated in human pituitary tumours.

GT A polymer of glutamic acid and tyrosine which induces ANTIBODIES with the same IDIOTYPES in certain strains of mice. It has been demonstrated that clones producing antibodies binding

to GT and to self antigens with similar affinity constants can be activated by foreign or ANTI-IDIOTYPE ANTIBODIES carrying the internal image of the antigen. The interaction of such foreign antigens with B cell clones producing AUTOANTIBODIES may play a part in breaking self-tolerance, leading to the occurrence of autoimmune phenomena.

GT–AG rule The striking sequence conservation around INTRON-EXON boundaries, found in eukaryotic genes from a variety of sources. Introns almost invariably begin and end with the dinu-cleotides GT and AG respectively. The aberrant pre-mRNA splic-ing observed when the GT–AG dinucleotides are absent underlines their importance. *See also*: EUKARYOTIC GENE STRUC-TURE; RNA SPLICING.

GTP Guanosine 5′-triphosphate. *See*: NUCLEOSIDES AND NUCLEOTIDES.

GTP-binding proteins

GTP-BINDING proteins are regulatory proteins found in all cells. They are versatile molecular switches, involved in the control of a wide range of biological processes — protein synthesis, signal transduction pathways, growth and differentiation. They all act through a common molecular mechanism based on their ability to bind the guanine nucleotides GTP and GDP selectively and with high affinity (K_d 10^{-7}–10^{-11} M). When GDP is bound these proteins exist in an inactive conformation. Exchange of GDP for GTP induces a conformational change to produce an active state which can then stimulate downstream processes such as the activation of an enzyme or the promotion of particular molecular complexes. An intrinsic guanosinetriphosphatase (GTPASE) activity converts the bound GTP to GDP to reinstate the inactive GDP-bound form. This cycle of guanine nucleotide exchange and GTP hydrolysis (Fig. G48) constitutes the molecular switch.

All GTP-binding proteins contain conserved amino-acid se-quence motifs. The sequences GXXXXGK(S/T) and DXXG are involved in binding phosphate groups whereas (N/T)(K/Q)XDinteracts with the nucleotide's purine ring [1] (abbreviations are in single-letter amino-acid notation; X is any amino acid).

GTP-binding proteins and protein synthesis

Many GTP-binding proteins are involved in the initiation, elon-gation, translocation and termination steps associated with prokaryotic and eukaryotic PROTEIN SYNTHESIS but most are poorly characterized. The best understood is the elongation factor EF-Tu from the bacterium *Escherichia coli*. In its active GTP-conformation this protein recognizes both aminoacylated transfer RNAs (aminoacyl-tRNAs) and the ribosome and promotes the formation of a complex of EFTu.GTP, aminoacyl-tRNA and the ribosome. Binding to the ribosome stimulates the otherwise very slow GTPase activity of EF-Tu to produce EF-Tu.GDP, which dissociates from the ribosome. To complete the cycle, a constitu-tively active exchange factor, EF-Ts, reconverts EF-Tu.GDP back

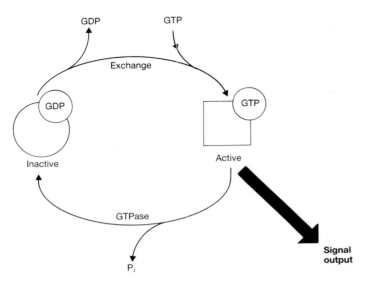

Fig. G48 A molecular switch. The cycle of guanine nucleotide exchange and GTP hydrolysis in GTP-binding proteins.

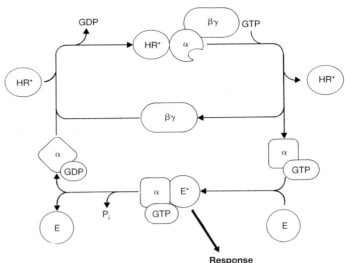

Fig. G49 The GTPase cycle in G proteins. HR*, Hormone–receptor complex (which catalyses guanine nucleotide exchange); E and E*, inactive and active effector proteins (targets of G protein regulation) respectively. Redrawn from [5].

to EF-Tu.GTP. As with all GTP-binding proteins, the energy released by GTP hydrolysis is not utilized in the downstream signalling process itself, but is used to induce the conformational switch in the regulatory protein.

The GTP-binding protein EF-Tu acts as a carrier to ensure that only aminoacyl-tRNA binds to the ribosome (binding of tRNA without amino acid would inhibit protein synthesis). However, once bound, the carrier must immediately fall off (leaving aminoacyl-tRNA behind) to ensure a rapid throughput for protein synthesis, and this is achieved by the ribosome-stimulated GTP hydrolysis.

A GTP-binding protein, EF-2, is also involved in the elongation steps of mammalian protein synthesis. The action of DIPHTHERIA TOXIN in blocking protein synthesis is mediated by monoADP-ribosylation of EF-2 (see ADP-RIBOSYLATION), which locks the EF-2 into its GTP-conformation. This prevents the formation of EF-2.GDP and its release from the ribosome and thus blocks protein synthesis.

G proteins and signal transduction

The interaction of HORMONES, NEUROTRANSMITTERS, GROWTH FACTORS and other signalling molecules with receptors on the external surface of cells triggers a cascade of intracellular events that eventually leads to a biological response. Many hormones (e.g. ADRENALINE) bind to receptors to stimulate the activity of ADENYLATE CYCLASE, generating increased levels of the intracellular SECOND MESSENGER, cyclic AMP. The transmembrane receptors do not, however, interact directly with adenylate cyclase; the activation event is mediated through a GTP-binding protein [2]. The particular family of GTP-binding proteins involved in coupling receptors to intracellular signalling pathways are called G proteins (see also G PROTEIN-COUPLED RECEPTORS). They are heterotrimeric and consist of an α subunit (of M_r about 40 000

(40K)) that binds GDP/GTP, and two intimately associated subunits β and γ. Activation of transmembrane receptors stimulates guanine nucleotide exchange on G proteins and (α.GDP)(βγ) is converted to (α.GTP)(βγ). α.GTP then dissociates from the βγ regulatory subunits and interacts with a downstream protein to produce a signal (Fig. G49). The life time of this interaction is determined by the intrinsic GTPase activity of the α subunit. The βγ subunits are common between different G proteins, and it is mainly the α subunits that confer specificity for downstream signalling although the βγ subunits are the effectors in a few cases.

Receptor activation of adenylate cyclase is mediated through the α subunit of one particular G protein, G_s. The GTPase activity of $α_s$ is such that several thousand molecules of cAMP are produced through activation of a single receptor. Other receptors (e.g. for opiates) lead to the inhibition of adenylate cyclase and this is mediated through the α subunit of a distinct G protein, G_i. Over 20 different α subunits of G proteins have been identified by purification or by cDNA cloning (see DNA CLONING). Two α subunits (the transducins $α_{tr}$ and $α_{tc}$) are found exclusively in the retina and mediate the activation of a cyclic GMP phosphodiesterase by the photoreceptor protein rhodopsin (see VISUAL TRANSDUCTION). Activation of rhodopsin by one photon can catalyse guanine nucleotide exchange on several hundred $α_t$ subunits. Each of these can activate a molecule of phosphodiesterase to hydrolyse several thousand molecules of cGMP, producing a massive amplification of the input light signal.

G proteins are known to regulate hydrolysis of plasma membrane PHOSPHOLIPIDS such as phosphatidylinositides (PI) and phosphatidylcholine (PC) (see SECOND MESSENGER PATHWAYS). G_q regulates PI hydrolysis through phospholipase Cβ (PLCβ, see PHOSPHOLIPASES) but G proteins controlling PC turnover have not yet been identified. Other G proteins are involved in the regulation of ION CHANNELS such as those for K$^+$ and Ca^{2+}. At least four β genes and three γ genes have been identified, adding yet more diversity and specificity to the G protein family.

G proteins have also been identified in the yeast *Saccharomyces cerevisiae*. They regulate intracellular signals induced by mating factor receptors. In this case, however, it has been shown genetically that the guanine nucleotide binding α subunit acts as a negative regulator of the βγ subunits. On binding GTP, α.GTP dissociates from the heterotrimer and it is βγ that activates downstream signals.

Alterations in G protein activity are important in human disease. CHOLERA TOXIN ADP-ribosylates the α subunit of G$_s$ at Arg 201 and blocks its intrinsic GTPase activity, leading to high constitutive activation of adenylate cyclase. G$_i$ is the site of action of PERTUSSIS TOXIN, which ADP-ribosylates the α$_i$ subunit and prevents its activation. Again this leads (indirectly) to high adenylate cyclase activity.

G proteins are also involved in cancer. Amino acid substitutions have been found in α$_s$ in human pituitary tumours. The mutations, found at either codon 201 or 227, block its GTPase activity and lead to constitutive activation of adenylate cyclase and high levels of cAMP. Pituitary cells are one of the few cell types in which cAMP is a proliferative signal; hence in these cells the mutated α$_s$ gene acts as an ONCOGENE (referred to as *gsp*). Mutations in the α$_{i2}$ gene (the *gip* oncogene) have been found in tumours of the adrenal cortex and endocrine tissues of the ovary, but it is not known which signals this particular α subunit regulates. There is much anticipation that analysis of the α$_q$ subunits regulating phospholipase C activity will reveal more mutations relevant to human cancer.

Small GTP-binding proteins

The prototype for this family is the *ras* proteins (Ras) [3] (*see also* ONCOGENES). The three mammalian Ras proteins, Ha-Ras, Ki-Ras and N-Ras, are expressed in most if not all cell types. They are monomeric proteins of M_r 21 000 with ~90% amino acid identity to each other. Highly conserved Ras proteins have been found in all eukaryotes and the genes *RAS1* and *RAS2* in *S. cerevisiae* and *ras1* in *Schizosaccharomyces pombe* have been characterized genetically in some detail. The real interest in these proteins has stemmed from the observation that amino-acid substitutions are found in Ras in many human cancers [4]. Alterations at codons 12, 13 or 61 have been detected in each of the proteins; for example 50% of colon CARCINOMAS and 95% of pancreatic ADENOCARCINOMAS have a mutant Ki-Ras, whereas 40% of acute MYELOID LEUKAEMIAS have an activated N-Ras.

Ras proteins are localized to the plasma membrane through a complex series of post-translational modifications signalled by a so-called CaaX box (C, Cys; a, aliphatic amino acid ; X, any amino acid) at their C terminus. This motif has also been found at the C terminus of many other proteins such as yeast mating factors, nuclear lamins (*see* NUCLEAR MATRIX) and G protein α and γ subunits. In the case of Ras, a C-15 farnesyl lipid is attached to the cysteine residue, the 'aaX' is proteolytically removed and the new C terminus is methylated. These modifications are not, however, sufficient to localize Ras to the plasma membrane and a

second signal is required. In the case of Ha- and N-Ras, this involves palmitoylation, the addition of a second lipid moiety palmitic acid to another cysteine two or five residues from the new C terminus. Ki-Ras does not have a second cysteine in this region; instead a polybasic domain of six lysine residues constitutes the additional signal for plasma membrane localization. Recent experiments have suggested that the CaaX box is yet more complicated. The α subunit of G$_i$, for example, terminates in CGLF yet this protein is not post-translationally modified. G protein γ subunits, on the other hand, end in CAIL and have a C-20 geranylgeranyl lipid addition instead of farnesyl.

A great deal of effort has been put into looking at the biochemical signals controlled by Ras. In *S. cerevisiae* the RAS proteins regulate the activity of adenylate cyclase in response to nutrients, although the details of this activation are not fully characterized. However adenylate cyclase is not regulated by Ras in mammalian cells or in *Schiz. pombe*. Mammalian Ras proteins are involved in cellular growth control; microinjection of neutralizing anti-Ras ANTIBODIES can block the growth of fibroblast cell lines in serum, and microinjection of mutant Ras proteins can stimulate cell growth even in the absence of growth factors. It is now clear that in mammalian cells, at least one action of Ras is to bind to and activate the protein kinase Raf which leads to the activation of MAP KINASE.

GTPase activating protein

Proteins that regulate the activity of Ras proteins have been identified. A protein of M_r 120 000 (120K) found in the cytoplasm of all mammalian cells can stimulate the GTPase activity of Ras several thousandfold. This GTPase *activating* protein, rasGAP, downregulates the active GTP form of Ras. A second downregulator of rasGTP has been identified; the product of the mammalian neurofibromatosis (*NF1*) gene. This protein, neurofibromin, has an M_r of ~300 000 (300K), and a 40K domain has been shown to stimulate the GTPase activity of Ras. In addition to downregulating Ras proteins, there is some evidence that GAP and neurofibromin may also contribute to downstream signalling from Ras proteins, although they have no known enzymatic activity other than stimulating Ras GTP hydrolysis. The N-terminal region of rasGAP has domains of homology with the Src protein (SH2 DOMAINS), which are thought to be the sites of interaction between proteins phosphorylated on tyrosine (*see* GROWTH FACTOR RECEPTORS; PROTEIN PHOSPHORYLATION; TYROSINE KINASES). GAP has been shown to bind to activated growth factor receptors (e.g. platelet-derived growth factor receptor) and is itself phosphorylated on tyrosine in response to growth factor stimulation. At least two proteins have been reported which can catalyse nucleotide exchange on Ras, though they are not well characterized. These proteins are candidate upstream regulators of Ras although in resting T CELLS at least, these ras exchange factors appear to be constitutively active and activation of Ras (conversion to the GTP form) occurs by inhibition of GAP activity.

Around 30 *ras*-related proteins have been identified by cDNA cloning and by protein purification [3]. They fall into three subgroups:

Close relatives (R-Ras, Rap1A, Rap1B, Rap2A, Rap2B, Ral, TC21)

These seven proteins are very similar to Ras with around 50% amino-acid sequence identity. At least three of these proteins, Rap1A, Rap1B and R-Ras, interact with rasGAP, though stimulation of GTPase activity occurs only with R-Ras. A GAP protein specific for Rap1 has also been identified. The function of none of these Ras-related proteins is known, though Rap1A has been shown to revert the phenotype of a Ras-transformed cell when overexpressed. This suggests that it too might have a role in growth control. However, Rap protein is found at high levels in non-proliferating cells such as platelets and neutrophils and is likely to have other signalling roles.

Rho family (Rho (A,B,C), Rac (1,2), G25K, TC10)

These seven proteins each have around 30% identity to Ras but 50% identity to each other. All members of the close Ras relatives and the rho family have a C-terminal CaaX box but, unlike Ras, the exact cellular location of these proteins has not yet been determined. In addition, Rac and G25K have been shown to contain the C-20 geranylgeranyl lipid. GAP proteins specific for rho (ABC), Rac (1,2) and G25K have been reported. The three rho proteins (A, B and C) regulate ACTIN polymerization and CYTOSKELETAL organization. Rho proteins are substrates for ADP-ribosylation on Asn 41 by the C3-exoenzyme of the bacterium *Clostridium botulinum*. Ribosylation inactivates the rho protein by blocking its interaction with its target and when the C3 enzyme is introduced into cells, cellular actin depolymerizes.

Rab family

Around 15 mammalian Rab proteins are known, each with around 30% amino-acid identity to Ras. Many Rab-like proteins have also been found in yeast and genetic analysis of one of these, SEC4, has suggested a role for these proteins in vesicle trafficking. SEC4 is localized on secretory vesicles and appears to target these vesicles, after leaving the GOLGI APPARATUS, to the plasma membrane (*see also* PROTEIN TARGETING). It has been speculated that the many mammalian Rab proteins are also involved in targeting vesicles between the many different compartments within a cell. In agreement with this, the Rab proteins are almost all ubiquitously expressed and are localized to specific intracellular compartments: for example Rab6 is in the medial/trans Golgi, Rab1 is in the ENDOPLASMIC RETICULUM, Rab4 and 5 are in early ENDOSOMES and Rab7 is in late endosomes. Rab3 proteins are expressed only in neural tissues and have been localized on SYNAPTIC VESICLES. The Rab proteins do not have C-terminal CaaX boxes; instead they have one of two motifs, -CC or -CXC. In the case of Rab3, it has been shown that a C-20 geranylgeranyl lipid is attached to a cysteine residue in the protein. Clearly, however, other signals are needed to target these proteins to their particular intracellular compartments.

Other GTP-binding proteins

Several other GTP-binding proteins are thought to be involved in protein transport and sorting. The ARF proteins were first recognized as being cellular cofactors for cholera toxin ribosylation of G_s. They are small GTP-binding proteins but unlike Ras-related proteins they do not have a C-terminal CaaX box. These proteins have been localized to the Golgi and are required for the correct assembly of transport vesicles. A GTP-binding protein, **VSP1**, involved in yeast vacuole protein sorting has at least two related mammalian proteins, Mx1 and Mx2, both identified originally as interferon-induced genes. TUBULIN, the major protein constituent of MICROTUBULES, is a GTP-binding protein. Tubulin cannot polymerize in the GDP form but exchange of GDP for GTP promotes self-assembly into microtubules. It is quite likely that many more biological processes await discovery that utilize the GDP/GTP conformational switch as a regulator.

A. HALL

See also: ENDOCYTOSIS; EXOCYTOSIS; OLFACTORY TRANSDUCTION; PROTEIN SECRETION.

1 Bourne, H.R. et al. (1991) The GTPase superfamily: conserved structure and molecular mechanism. *Nature* **349**, 117–126.
2 Gilman, A.G. (1987) G proteins: transducers of receptor-generated signals. *Annu. Rev. Biochem.* **56**, 615–649.
3 Hall, A. (1990) The cellular functions of small GTP-binding proteins. *Science* **249**, 635–640.
4 Barbacid, M. (1987) Ras genes. *Annu. Rev. Biochem.* **56**, 779–827.
5 Bourne, H. et al. (1990) The GTPase superfamily: a conserved switch for diverse cell functions. *Nature* **348**, 125–132.

GTPase Guanosine triphosphatase. Enzyme activity that hydrolyses guanosine 5'-triphosphate (GTP) to produce GDP and orthophosphate. *See*: GTP-BINDING PROTEINS; MICROTUBULE-BASED MOTORS.

GTPase activating protein (GAP) Protein that activates the GTPase activity of GTP-BINDING PROTEINS.

guanine (G) A purine which is one of the four nitrogenous BASES in DNA and RNA and which forms part of various NUCLEOSIDES AND NUCLEOTIDES. *See also*: BASE PAIRING; CYCLIC GMP; GTP-BINDING PROTEINS; NUCLEIC ACID STRUCTURE.

guanine nucleotides *See*: GTP-BINDING PROTEINS; NUCLEOSIDES AND NUCLEOTIDES.

guanine nucleotide binding proteins *See*: GTP-BINDING PROTEINS.

guanosine The nucleoside derived from the nitrogenous BASE guanine. *See*: NUCLEOSIDES AND NUCLEOTIDES.

guanosine monophosphate, diphosphate, triphosphate (GMP, GDP, GTP) *See*: GTP-BINDING PROTEINS; NUCLEOSIDES AND NUCLEOTIDES. *See also*: CYCLIC GMP.

guanosine triphosphatase *See*: GTPASE; GTP-BINDING PROTEINS.

guanylate cyclase Enzyme that catalyses the production of 3′,5′ cyclic guanosine monophosphate (CYCLIC GMP, cGMP) from GTP. Unlike ADENYLATE CYCLASE it is a soluble enzyme and thus not associated with the membrane. *See also*: SECOND MESSENGER PATHWAYS; VISUAL TRANSDUCTION.

guard cell One of the specialized pair of cells bordering a stomatal pore on the epidermis of leaves and some stems. Changes in the turgidity of the guard cells open and close the stoma thus varying access of gases into or out of the tissue.

GUG The CODON 5′ GUG 3′ normally specifies the AMINO ACID valine but occasionally in bacteria N-FORMYLMETHIONINE. A TRANSFER RNA species (tRNA^{fmet}) can bind to this triplet and initiate PROTEIN SYNTHESIS. This interaction requires WOBBLE to occur at the first base of the GUG codon and is thought to be facilitated sterically by the structure of tRNA^{fmet}.

guide RNA (gRNA) Small (50–80 nucleotides long) RNA molecules that hybridize to mRNA transcripts and function either to insert or to delete nucleotides (generally uridines) from the mRNA in the process of RNA EDITING in the mitochondria of kinetoplastid protozoans. Guide RNAs contain regions complementary to the edited sequence of the mRNA (if G : U base-pairing is allowed), and a 3′ tail of ~15 nonencoded uridines which are added post-transcriptionally. Nucleotides are thought to be inserted into or deleted from the unedited portion of the mRNA by a two-step transesterification reaction in which the mRNA strand is cleaved, a uridine (or uridines) is transferred from or to the 3′ end of the gRNA, and the mRNA strand is religated. The 5′ portion of the gRNA is encoded within the catenated maxi- and minicircles found within the mitochondria of trypanosomatids.

Guinier plot A particular form of graph relating the intensity of SMALL-ANGLE SCATTERING in a given direction to the deviation of the scattered beam from the incident radiation. The slope of a Guinier plot is proportional to the RADIUS OF GYRATION of the scattering object, while the intercept is related to the product of its volume and its contrast with the solvent. *See*: NEUTRON SCATTERING AND DIFFRACTION.

gums Heterogeneous group of plant polysaccharides all of which can be hydrated to yield viscous, sticky solutions, in which the polysaccharide chains assume highly flexible, extended configurations. Many are variants of normal cell wall GLYCANS, especially PECTINIC ACIDS and ARABINOGALACTANS, but a few resemble no other known glycans. *See also*: PLANT CELL WALL MACROMOLECULES; SLIMES.

gymnosperms Seed-bearing woody plants in which the seeds are not enclosed in an ovary and are carried on the surface of cone scales or their equivalent. They include the conifers, and a few other smaller groups (the cycads, *Ginkgo*, and the gnetophytes). PLANT GENETIC ENGINEERING is currently much less advanced for gymnosperms than for angiosperms.

gynaecium *See* GYNOECIUM.

gynandromorph mapping A FATE MAPPING method in *Drosophila* invented by A.H. Sturtevant. It employs an unstable ring-X-chromosome which is often lost in early cleavage divisions leading to an embryo which develops as a mosaic of XX (female) and XO (male) tissue. The orientation of the mosaic boundary is random. The frequency of separation of adult structures by a mosaic boundary is used to compute the distance between these adult primordia in the blastoderm and thus to create a fate map.

gynoecium The parts of a flower (the carpels) in which the female GAMETOPHYTES are formed. Also a group of female gametangia in mosses.

gypsy TRANSPOSABLE GENETIC ELEMENT of *Drosophila*.

gyrase *See*: DNA GYRASE.

gyration *See*: RADIUS OF GYRATION.

H

H The AMINO ACID histidine.

H^+-ATPases ATP-dependent proton pumps found in a variety of cellular membranes, including those of endocytic and exocytic organelles in eukaryotic cells, where they pump protons (H^+) into the organelle to maintain an acidic pH. *See*: V-TYPE ATPASE.

H chain *See*: HEAVY CHAIN.

H^+,K^+-ATPase A P-TYPE ATPASE responsible for acid secretion in the stomach by exchange of H^+ ions for K^+ ions.

H-*ras*, Ha-*ras* ONCOGENE of the *ras* family (*see* GTP-BINDING PROTEINS) found in the Harvey rat sarcoma virus (*see* RETROVIRUSES).

H substance (H antigen) The precursor oligosaccharide of the ABO BLOOD GROUP SYSTEM antigens. It is unmodified on group O cells; the addition of different sugar residues gives rise to A or B antigens. The *H* gene encodes a glycosyltransferase and maps to human chromosome 19.

H-subunit The 'heavy' polypeptide subunit of the photosynthetic reaction centre of purple bacteria (M_r 28 000) (*see* PHOTOSYNTHESIS). It has a single membrane-spanning helix and a hydrophilic domain lying on the cytoplasmic side of the L/M dimer, but binds no pigments; its function is not fully understood.

H-Y antigen Male-specific protein encoded by a gene on the mammalian chromosome, which appears very early in embryonic development.

H-2 The MAJOR HISTOCOMPATIBILITY COMPLEX (MHC) of the mouse, a region spanning ~1.5 centimorgans of DNA on murine chromosome 17 and containing a complex of closely linked polyallelic loci encoding three different classes of proteins involved in the generation of humoral and cellular immune responses. The relatively high frequency of genetic RECOMBINATION within the H-2 complex has produced many intra-H-2-recombinant mouse strains, facilitating fine mapping and functional analysis of the H-2 loci.

habituation A decrease in a behavioural reflex response to a repeated non-noxious stimulus, which was first described by Pavlov. In more general terms it is the gradual adaptation to a stimulus or to the environment. At the synaptic level, habituation is believed to involve the prolonged inactivation of Ca^{2+} channels in the presynaptic terminal.

haem (heme) Iron–porphyrin reddish pigment (Fig. H1) which forms the prosthetic group in a number of different proteins (haemproteins). It is the oxygen-binding moiety of myoglobin, haemoglobin, leghaemoglobin, etc. (*see* GLOBIN SUPERFAMILY; HAEMOGLOBIN AND ITS DISORDERS), and is also found in CYTOCHROMES where it is involved in electron transfer (*see* ELECTRON TRANSPORT CHAIN; PHOTOSYNTHESIS), and in some enzymes (e.g. catalase and peroxidase). It consists of one ferrous iron atom at the centre of protoporphyrin IX.

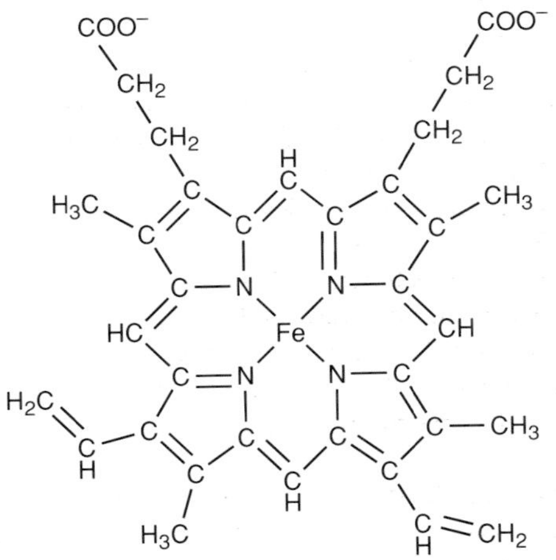

Fig. H1 Haem (Fe-protoporphyrin IX).

haemagglutination The aggregation and sedimentation of erythrocytes by specific ANTIBODIES (or other agents) directed against antigens on the erythrocyte surface. Haemagglutination has been used as an assay for the presence of antibodies. In the direct technique the antigens detected are expressed naturally on the erythrocyte surface (e.g. the ABO antigens, *see* ABO BLOOD GROUP SYSTEM). Indirect (conditioned or passive) haemagglutination requires soluble antigens specific for the antibodies to be assayed to be adsorbed or covalently linked to the erythrocyte surface. Haemagglutination can also be produced by some LECTINS and viruses (*see* INFLUENZA VIRUS). Any agent that causes haemagglutination is termed a haemagglutinin.

haemagglutinin Any agent that can cause HAEMAGGLUTINATION. The haemagglutinin (HA) of the INFLUENZA VIRUS is a trimeric protein that forms hollow cylindrical spikes projecting from the viral envelope. Each identical monomer is produced from a haemagglutinin precursor of 550 amino acids which is cleaved

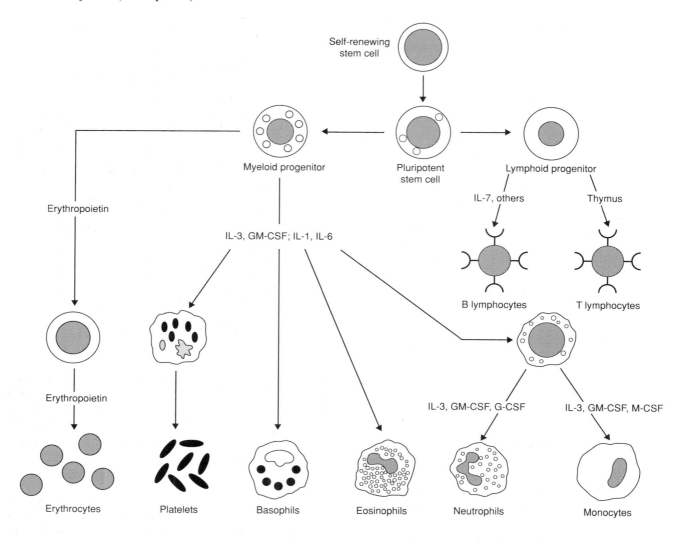

Fig. H2 Outline of the pathways of haematopoiesis. CFU, colony-forming unit; G-CSF, granulocyte colony stimulating factor; GM-CSF, granulocyte–macrophage colony stimulating factor; IL, interleukin; M-CSF, macrophage colony stimulating factor.

after insertion into the endoplasmic reticulum membrane to form HA_1 and HA_2 which are linked by an S—S bond. Each monomer has an elongated stem and a globular 'head'. Influenza haemagglutinin recognizes and binds to cell-surface receptors containing a terminal sialic acid (neuraminic acid) and mediates the fusion of viral membrane with host cell membrane in ENDOSOMES and the penetration of viral nucleocapsids into the cytoplasm.

haematopoiesis (haemopoiesis) The process of blood cell formation and DIFFERENTIATION (Fig. H2). All blood cells (erythrocytes, lymphocytes and other leukocytes), as well as osteoclasts, dendritic cells, Langerhans cells and Kupffer cells are derived from a common STEM CELL in the bone marrow. The assignment of cells to different lineages and their subsequent differentiation is controlled by numerous protein factors (haematopoietic factors)

and the specialized microenvironment of the bone marrow stroma within which haematopoiesis occurs.

During embryonic development the site of haematopoiesis changes. Initially blood formation occurs in the primitive yolk sac. By the fifth to sixth gestational week (in humans) it becomes established in the liver. In mid-trimester some blood formation occurs in the spleen as well as the liver. Haematopoiesis starts in the bone marrow by the fourth to sixth month and this becomes the major organ of blood production from the sixth month onward. Shortly after birth all vestiges of haematopoiesis are lost from the spleen and liver. By puberty most haematopoiesis occurs in the marrow of the flat bones — the sternum, vertebrae, iliac bones and ribs.

For the haematopoietic factors involved in blood cell differentiation *see* ERYTHROPOIETIN; G-CSF; GM-CSF; GROWTH FACTORS; LYMPHOKINES; M-CSF and the interleukins IL1–11. *See also*: B CELL DEVELOPMENT; T CELL DEVELOPMENT.

haematopoietic growth factors CYTOKINES involved in the DIFFERENTIATION and maturation of the progenitors of the blood cells (see HAEMATOPOIESIS). *See*: ERYTHROPOIETIN; G-CSF; GM-CSF;

GROWTH FACTORS; LYMPHOKINES; M-CSF; STEM CELL FACTORS and individual interleukins IL1–11.

haemerythrin A haem-containing protein with a single domain, belonging to the four-helix bundle family of structures. *See*: PROTEIN STRUCTURE for an illustration of this protein.

haemochromatosis A rare disorder of iron metabolism in humans in which excessive absorption of iron leads to abnormal tissue deposition and organ damage. The symptoms include enlargement and cirrhosis of the liver, skin pigmentation, diabetes and cardiac failure. The locus responsible for haemochromatosis is on the short arm of chromosome 6 close to the MAJOR HISTOCOMPATIBILITY COMPLEX (MHC), and the disease is inherited as an AUTOSOMAL RECESSIVE trait. It is an example of a sex-influenced disease: clinically affected females are ~10 times more common than affected males, probably because of differences in iron metabolism due to menstruation.

haemocyanin Copper-containing oxygen-carrying protein found in the blood plasma of some groups of invertebrates including molluscs and crustaceans. Haemocyanin from the snail *Helix* is a giant molecule of M_r 9 million, composed of ~180 subunits of M_r 50 000 each containing two copper atoms and binding one molecule of oxygen. Other haemocyanins form smaller polymeric molecules.

Haemoglobin(s)

HAEMOGLOBINS are HAEM-containing oxygen-binding proteins found in an enormous variety of organisms including bacteria, yeast, plants, invertebrates and vertebrates. Although the structures of relatively few have been determined, and despite the modest sequence similarity between haemoglobins of different phyla, all of these proteins appear to be built from one or more units which share a common tertiary structure. This fold, exemplified by MYOGLOBIN, appears to have arisen early in evolution, and different organisms have utilized it in different ways. Monomeric, dimeric, oligomeric, and polymeric haemoglobins are known, most of the variety in higher structure being found among invertebrates. This article presents a brief overview of the haemoglobin superfamily; disorders of human haemoglobin and human globin gene organization and expression are covered elsewhere in this volume (*see* HAEMOGLOBIN AND ITS DISORDERS). For reviews of vertebrate and invertebrate haemoglobins the reader should consult [1] and [2] respectively.

Sperm whale myoglobin was the first protein for which a complete secondary and tertiary structure was determined (by X-RAY CRYSTALLOGRAPHY) by John Kendrew and colleagues in late 1950s and early 1960s. In 1968 Max Perutz determined the structure of haemoglobin.

The myoglobin fold consists of seven or eight α-helices wrapped around the haem which lies in a hydrophobic cleft (Fig. H3) [3,4]. The helices are labelled A to H from the N terminus, the corners between them AB to GH, and the N- and C-terminal coil regions

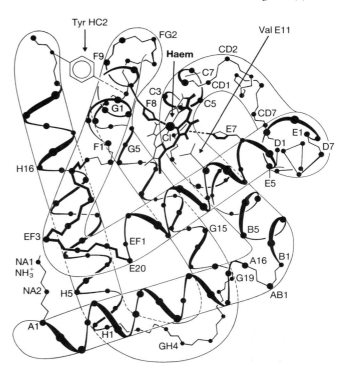

Fig. H3 The tertiary structure of the β-chain of human haemoglobin, typical of haemoglobins and myoglobins of all other species. The figure also shows the proximal and distal histidines, marked F8 and E7, the distal valine E11, and the tyrosine HC2 which ties down the C terminus by its hydrogen bond to the main chain carbonyl of valine FG5. The exact numbers of residues in the different segments are the same in all mammals, but vary in other vertebrates and especially invertebrates. From [11].

are called NA and HC respectively. Individual residues are numbered, beginning at the N-terminal end of the helix or corner of which they form a part. Thus residue F4 is the fourth residue from the N-terminal end of the F helix. The only invariant residues are histidine F8, which forms a covalent bond to the haem iron atom, and phenylalanine CD1, which packs against the haem. The two sides of the haem are referred to as proximal (the His F8 side) and distal, the side to which ligands such as oxygen can bind. From an alignment of 226 globin amino-acid sequences the restraints imposed by the myoglobin fold on the protein sequence have been analysed, and a 'template' constructed which allows globin sequences to be distinguished from other proteins [5]. (*See also*: Table M5 in MOLECULAR EVOLUTION: SEQUENCES AND STRUCTURES.)

Vertebrate haemoglobins

Human haemoglobins, and the haemoglobins of all vertebrates except some primitive fish such as lampreys, consist of four myoglobin-like subunits (termed globins), two α-type and two β-type, which associate noncovalently to form tetramers (*see* Plate 5). In vertebrate haemoglobins, two distal residues, valine E11 and histidine E7, are highly conserved. These residues lie in VAN DER WAALS contact with bound ligands, and in the case of a

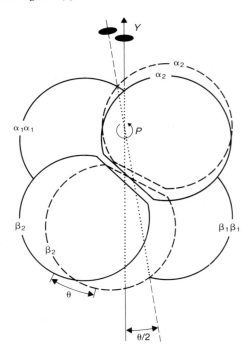

Fig. H4 Changes of quaternary structure of mammalian haemoglobin on transition from deoxy or T (solid lines) to oxy or R (dashed lines). The molecule is viewed along the rotation axis P, which is perpendicular to the plane of the paper and to the molecular dyad of deoxyhaemoglobin (Y) and intersects it at the point shown. The dimer $\alpha_2\beta_2$ turns by ψ = 13° about the P axis and shifts about 1Å along the axis towards the dimer $\alpha_1\beta_1$. Redrawn from [11].

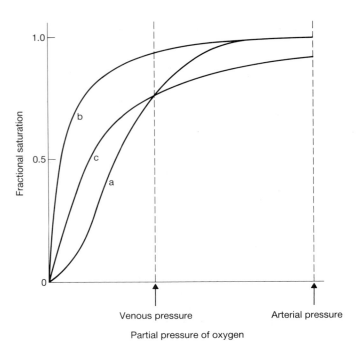

Fig. H5 Typical oxygen equilibrium curves for haemoglobin (a) and myoglobin (b). Curve c indicates a myoglobin-type oxygen equilibrium curve with the same percentage saturation as haemoglobin at venous partial oxygen pressure. From [12].

globin or myoglobin the histidine forms a HYDROGEN BOND with bound oxygen. This bond is much weaker or absent in β-globin. Human haemoglobin is an ALLOSTERIC protein, capable of forming two different arrangements of the four subunits (quaternary structures) [4]. In the deoxy state with no ligand bound to the haem groups the protein favours the T (tense) state; when oxygen binds to haemoglobin the protein switches to the R (relaxed) state, which has higher oxygen affinity. These changes occur because ligand binding pulls the iron atom towards the haem plane and flattens the haem. This triggers changes at the contacts between the subunits causing one αβ dimer to rotate about 15° relative to the other (Fig. H4). Hydrogen bonds and SALT BRIDGES found only in the T-state oppose the switch to the R-state and so lower the oxygen affinity. Since the R-state becomes predominant as more oxygen binds, the oxygen affinity increases with oxygen concentration. Oxygen binding is therefore cooperative, and the oxygen equilibrium curve shows a sigmoidal shape (Fig. H5).

The oxygen affinity is also modified by other ligands which bind to haemoglobin. The human red cell contains roughly 1 mM DPG (2,3-diphosphoglycerate) which binds only to T-state haemoglobin (one molecule of DPG per tetramer). The DPG binding site is inside a water-filled cavity in the protein which is narrower in the R-state than in the T-state. The binding site in the cavity of T-state haemoglobin is therefore lost on switching to the R-state. Since DPG binds only to T-state haemoglobin it moves the allosteric equilibrium towards the low oxygen affinity form of the protein. Other 'allosteric effectors' such as Cl^-, H^+ and CO_2 also stabilize the T-state. The increase in proton binding on moving from the R- to the T-state is called the alkaline Bohr effect. For human haemoglobin the effect disappears at about pH 6, below which the effect is reversed. This acid Bohr effect has no known physiological role.

Not all mammalian haemoglobins use DPG to lower oxygen affinity; a number (mainly in ruminants and cats) have evolved which have an intrinsically low oxygen affinity unaffected by DPG. This is achieved by two mutations at the N terminus of the β chains. DPG binding to human haemoglobin involves two salt bridges to valine NA1β and histidine NA2β. The first of these residues is usually missing in low affinity haemoglobins, and the second is large and hydrophobic such as leucine or phenylalanine. These changes severely weaken DPG binding. It has been suggested [6] that the hydrophobic side chains probably serve to stabilize the T-state by holding the ends of the A helices in place. Other animals use different effector molecules in place of DPG. Birds for example use inositol pentaphosphate (IPP). The principal changes in avian haemoglobin (relative to human haemoglobin) appear to be histidine H21β → arginine, alanine H13β → arginine and asparagine H17β → histidine [1]. These increase the positive charge within the central cavity of the protein, and balance the negative charge of IPP. Fish haemoglobins use ATP and GTP instead of DPG, a change apparently brought about by the mutation of histidine NA2β → glutamine. The precise conformation of IPP and ATP or GTP in the bound state is uncertain as no X-ray structure of a haemoglobin bound to these molecules has been solved.

Crocodile haemoglobin, unlike other animal haemoglobins, shows a lower oxygen affinity in the red cell than when purified and stripped of ions, yet it does not bind organic phosphates. Instead, the T-state is strongly stabilized by bicarbonate ions. Two bicarbonate ions bind to each haemoglobin tetramer, lowering the oxygen affinity as much as inositol hexaphosphate (IHP) does in human haemoglobin. This unique allosteric property is probably related to the ability of crocodiles to remain submerged for long periods. Unlike diving mammals such as sperm whales, crocodiles do not have unusually high concentrations of myoglobin in the muscle, and so rely on extracting as much oxygen as possible from the blood. A binding site for the two bicarbonate ions has been proposed [7], which suggests that the DPG site of human haemoglobin has been modified by three or four mutations to create two bicarbonate binding sites close to and related by the molecular dyad axis. Experimental tests of this hypothesis by engineering the proposed sites into human haemoglobin by PROTEIN ENGINEERING have failed to transplant the bicarbonate effect. An alternative bicarbonate binding site has been proposed at the β C helices. These form a sliding contact with the FG corners, and all crocodilian haemoglobins uniquely show two more positive charges than human haemoglobin in this highly conserved region of the molecule.

Animals living at high altitudes have evolved haemoglobins with high oxygen affinity to extract oxygen from the thinner atmosphere. Some birds have adapted to extreme altitudes, a Ruppell's griffon (a type of vulture) once being hit by an airliner at more than 10 000 metres over the Ivory Coast. The ability of vultures to fly at such altitudes has not yet been explained, but a simple mechanism of altitude adaptation has been found for a species of goose. The greylag goose lives in lowland plains, and has blood with unremarkable oxygen affinity. The closely related bar-headed goose migrates over the Himalayas at an altitude of about 9000 metres, and has blood of unusually high oxygen affinity. The increase in oxygen affinity was predicted [1] to be due to a single mutation relative to greylag goose haemoglobin, proline H2α → alanine. This proline residue forms a contact with D6β (which is leucine in bar-headed goose but methionine in human haemoglobin) at the α1β1 subunit interface. All mutations which disrupt such contacts destabilize the T-state relative to the R-state. This prediction has been verified [8] by the engineering of this mutation into human haemoglobin, causing a large increase in oxygen affinity. Interestingly, the high-flying Andean goose has evolved a high affinity haemoglobin by mutating residue D6β to serine, which breaks the same contact and leads to the same increase in oxygen affinity.

Many bony fish haemoglobins show an exceptionally strong alkaline Bohr effect, oxygen binding becoming noncooperative and the oxygen affinity dropping sharply as the pH is lowered to about 6 or less. This is known as the Root effect, and is used to supply oxygen to the swim-bladder and the eye. Both of these organs secrete lactic acid into the blood which then discharges oxygen. Despite much study the molecular basis of this effect is not understood. The X-ray crystallographic structure of a Root effect haemoglobin from the Antarctic fish *Pagothenia bernachii*

was recently solved in the carbonmonoxy form, but as the molecule is in the R-state the structure gives no insight into the mechanism of the Root effect.

Unlike other vertebrate haemoglobins, lamprey haemoglobin is a homodimer rather than a heterotetramer. The dimer tends to split into monomeric globin subunits on oxygenation, increasing the oxygen affinity, so that oxygen binding is cooperative. Lampreys and their haemoglobin seem to have changed little over the last 500 million years, during which the α- and β-globins of other vertebrates diverged. The structure of lamprey haemoglobin has been solved crystallographically in the cyanide-bound form.

Invertebrate haemoglobins

The haemoglobins of invertebrates show far more variety in structure than those of vertebrates [2]. Not all invertebrate blood haemoglobins are encapsulated in red cells, some being large extracellular proteins with M_r of up to 8 million. The intracellular haemoglobins are much smaller, generally no bigger than human haemoglobin. X-ray structures have been determined for the dimeric haemoglobin of the clam *Scapharca inaequivalvis* and the tetrameric haemoglobin of the worm *Urechis caupo*. *Scapharca* haemoglobin is a highly cooperative dimer with a totally different mechanism to vertebrate haemoglobins. There is no large conformational change on ligand binding, but the haems communicate with each other via a tyrosine residue which forms different contacts with the neighbouring subunit depending on the ligation state of the haem. The intersubunit contacts are made by the E and F helices (instead of the G and H helices as in human haemoglobin), an arrangement which brings the haem groups much closer to each other than in human haemoglobin. The homotetrameric haemoglobin of *U. caupo* has few intersubunit contacts and binds oxygen noncooperatively. The structure is distinct from both vertebrate and clam haemoglobins, the B and D helices facing inwards. No structure is available for the giant extracellular haemoglobins, but the major haem-containing subunits of haemoglobin from the earthworm *Lumbricus terrestris* have been sequenced. Each molecule of this haemoglobin has a molecular weight of about 3 800 000 (3800K) and contains roughly 200 myoglobin-like subunits of several types, held together by intra- and intersubunit disulphide bridges. Other proteins of 33–38K, which may be composed of two haem-free myoglobin-like domains, are needed for assembly of the subunits.

The brine shrimp *Artemia* produces three polymeric haemoglobins which are homo- and heterodimers of two different chain types called α and β. The α and β subunits each have an M_r of ~130K and consist of nine tandemly repeated myoglobin-like domains separated by short linker sequences. Despite weak sequence similarity each domain appears to be a functional oxygen-binding unit with the myoglobin fold. The genes for the α and β subunits seem to have arisen from several duplication and fusion events more than 200 million years ago.

Leghaemoglobin

LEGHAEMOGLOBIN, which is found in the root nodules of legumes, has an overall structure very similar to that of myoglobin (Fig.

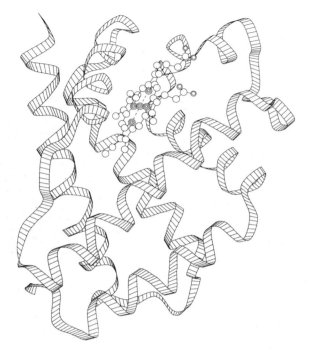

Fig. H6 Lupin leghaemoglobin.

H6 and Plate 5*b*). Amino-acid sequences and gene structure suggest that leghaemoglobin and animal globins diverged from a common ancestor more than 10^9 years ago [2]. Leghaemoglobin has a very high oxygen affinity and serves to protect the oxygen-sensitive nitrogenase of the root-nodule nitrogen-fixing bacteria (*see* NITROGEN FIXATION). However, haemoglobin has also been found in non-nitrogen fixing plants, leading to the suggestion that it may be present in all plants.

Bacterial haemoglobin

The bacterium *Vitreoscilla* produces a similar protein (sharing 24% sequence identity with lupin leghaemoglobin), whose intracellular concentration rises if the cells are grown under low oxygen tension. Expression of the protein in *Escherichia coli* has been shown to allow the host bacteria to grow faster to higher density. The precise function of the protein in bacteria is unknown, but it may serve to deliver oxygen to the respiratory ELECTRON TRANSPORT CHAIN. Yeast and *E. coli* both produce closely related flavohaemoglobins, proteins of M_r ~45 000 which contain both flavin- and haem-binding domains [9].

It is impossible to consider the enormous quantity of sequence data on oxygen-binding proteins without speculating on their evolution and ancestry [3]. To judge from their distribution, the myoglobin-like proteins have certainly been around for a long time. It is interesting that the principal difference between vertebrate α- and β-globins, the lack of a D helix in α-globin, appears to be of no functional significance. Deleting the β subunit D helix, and subsequently inserting it into the α subunit, leaves the oxygen equilibrium curve of human haemoglobin unaffected [10].

There are numerous variants of human haemoglobin in the population, many of which do not lead to any symptoms or illness (*see* HAEMOGLOBIN AND ITS DISORDERS). Clearly the protein is extremely robust, and this tolerance of mutation without loss of function has permitted nature to construct quite distinct oxygen-binding proteins from a common building unit to fill a variety of needs in different organisms.

J. TAME

See also: COOPERATIVITY; PROTEIN STRUCTURE.

1 Perutz, M.F. (1983) Species adaptation in a protein molecule. *Mol. Biol. Evol.* **1**, 1–28.
2 Riggs, A.F. (1991) Aspects of the origin and evolution of non-vertebrate hemoglobins. *Am. Zool.* **31**, 535–545.
3 Dickerson, R.E. & Geis, I. (1983) *Hemoglobin: Structure, Function, Evolution and Pathology* (Benjamin/Cummings, Menlo Park, CA).
4 Perutz, M.F. et al (1987) Stereochemistry of cooperative mechanisms in hemoglobin. *Acc. Chem. Res.* **20**, 309.
5 Bashford, D. et al. (1987) Determinants of a protein fold. Unique features of the globin amino acid sequences. *J. Mol. Biol.* **196**, 199–216.
6 Perutz, M.F. & Imai, K. (1980) Regulation of oxygen affinity of mammalian haemoglobin. *J. Mol. Biol.* **136**, 183–191.
7 Perutz, M.F. et al. (1981) Allosteric regulation of crocodile haemoglobin. *Nature* **291**, 682–684.
8 Jessen, T.-H. et al. (1991) Adaptation of bird haemoglobins to high altitudes: Demonstration of molecular mechanism by engineering. *Proc. Natl. Acad. Sci. USA* **88**, 6519–6522.
9 Zhu, H. & Riggs, A.F. (1992) Yeast flavohemoglobin is an ancient protein related to globins and a reductase family. *Proc. Natl. Acad. Sci. USA* **89**, 5015–5019.
10 Komiyama, N.H. et al. (1991) Was the loss of the D helix in α globin a neutral mutation. *Nature* **352**, 349–351.
11 Perutz, M.F. (1990) *Mechanisms of Cooperativity and Allosteric Regulation in Proteins* (Cambridge University Press).
12 Fermi, G. & Perutz, M.F. (1981) *Atlas of Molecular Structures in Biology: Haemoglobin & Myoglobin* (Clarendon Press, Oxford).

Haemoglobin and its disorders

HAEMOGLOBIN is the oxygen-carrying protein of the red blood cell in vertebrates, and is responsible for the transport of oxygen from lungs to tissues. Vertebrate haemoglobins are members of the large and widespread globin superfamily of proteins (*see* HAEMOGLOBINS). All vertebrate haemoglobins are tetramers, composed of two pairs of different GLOBIN chains, each chain associated with one HAEM molecule (Fig. H7; see Fig. H3 in HAEMOGLOBINS for the detailed structure of the β-globin subunit). Oxygen binds reversibly to haemoglobin by binding to the iron atom of the haem, which is in the ferrous (Fe^{2+}) state. Each haem can bind one oxygen molecule.

In all mammals, structurally different haemoglobins are produced during different periods of development, an adaptive process designed to meet differing oxygen transport requirements. In humans, sheep and goats there are embryonic, foetal and adult haemoglobins; human foetal haemoglobin, for example, has a higher oxygen affinity than adult haemoglobin, an adaptive response to the oxygen requirements of the foetus. In other species there is a direct transition from the embryonic to the adult form of haemoglobin.

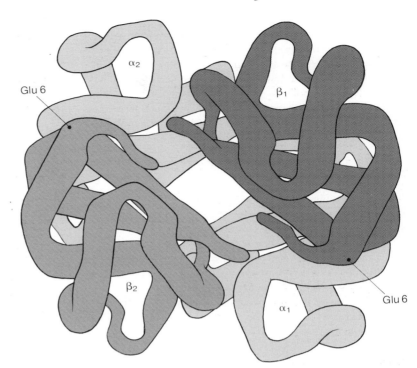

Fig. H7 Schematic veiw of human haemoglobin A viewed down the central axis. It consists of two identical α chains (each 141 amino acids long) and two identical β chains (146 amino acids long). Each globin subunit contains a haem prosthetic group located in a crevice or 'haem pocket' formed by the folded polypeptide chain (not shown; *see* Fig. H3 in HAEMOGLOBINS for a detailed structure of the β-subunit). Each α chain is in contact with two β chains; these interactions are defined as $\alpha_1\beta_1$ and $\alpha_1\beta_2$. Glu 6 is the amino acid that is mutated to valine in sickle cell haemoglobin. Parts of the α chains are omitted for simplicity. From [3].

Heterogeneity of human haemoglobin

Human adult haemoglobin is a heterogeneous mixture of proteins consisting of a major component, haemoglobin A (HbA), and a minor component, haemoglobin A2, constituting about 2.5% of the total. Except for some of the embryonic haemoglobins (see below), all the normal human haemoglobins contain one pair of α-globin chains. In HbA they are combined with two β-globin chains ($\alpha_2\beta_2$), in HbA$_2$ with δ chains ($\alpha_2\delta_2$), and in foetal haemoglobin (HbF) with γ chains ($\alpha_2\gamma_2$). HbA can also be glycosylated post-translationally at the terminal valine of the β chain to produce a minor component HbA$_{1c}$.

Human haemoglobin shows further heterogeneity, particularly in foetal life. HbF is a mixture of two molecular forms which differ by only one amino-acid residue — either glycine or alanine at position 136 in their γ chains. They are designated $\alpha_2\gamma_2^{136\,\mathrm{Gly}}$ and $\alpha_2\gamma_2^{136\,\mathrm{Ala}}$, and are called $^{\mathrm{G}}\gamma$ chains and $^{\mathrm{A}}\gamma$ chains respectively. At birth the ratio of molecules containing $^{\mathrm{G}}\gamma$ chains to those containing $^{\mathrm{A}}\gamma$ chains is about 3 : 1; this ratio varies widely in the trace amounts of HbF present in normal adults. The foetal to adult haemoglobin switch, which reflects a switch from predominantly γ- to predominantly β-chain production, starts before birth and is complete by the end of the first year of life. In some cases a point mutation in or deletion of the β-globin gene results in the benign condition of HEREDITARY PERSISTENCE OF FOETAL HAEMOGLOBIN in which γ rather than β chains are produced in adult life.

Before the eighth week of intrauterine life there are three embryonic haemoglobins, haemoglobins Gower 1 ($\zeta_2\varepsilon_2$), Gower 2 ($\alpha_2\varepsilon_2$), and Portland ($\zeta_2\gamma_2$). The ζ and ε chains are the embryonic counterparts of the adult α, and the β, γ and δ chains respectively.

During foetal development there is an orderly switch from ζ to α chain production, and from ε to γ to β chain production. δ chain production commences after birth.

Structure–function relationships

Haemoglobin is an allosteric protein, undergoing alterations in configuration that are essential to its function of taking up and releasing oxygen (*see* ALLOSTERY). Its oxygen-binding properties are reflected in a sigmoid oxygen dissociation curve; it binds oxygen tightly in the lungs, where there is a high partial pressure of oxygen, and releases it rapidly when it encounters a low partial pressure of oxygen in the tissues. Furthermore, the oxygen affinity of haemoglobin can be modified according to physiological needs, the curve shifting to the left or right in response to pH (*see* BOHR EFFECT), temperature, and carbon dioxide levels. Some of these adaptive changes are the result of binding small molecules, notably 2,3-DIPHOSPHOGLYCERATE.

The haemoglobin molecule exists in two quaternary states (*see* ALLOSTERY; PROTEIN STRUCTURE) — oxy or relaxed (R) and deoxy or tense (T). There are no stable intermediate forms. There is a free energy change of ~42–50 kJ mol^{-1} of tetramer on the transition from the oxy to deoxy states.

Haemoglobin binds oxygen cooperatively; the binding of one oxygen molecule facilitates the uptake of oxygen by the three remaining haem molecules. This phenomenon depends on re-arrangements in the protein which occur on oxygen uptake. The α and β subunits change their position relative to each other, with most of the movement occurring at the $\alpha_1\beta_2$ contact (*see* Fig. H4 in HAEMOGLOBINS), and in the process the β-chain haems move closer together by about 7 Å. These conformational changes are caused by alterations in the conformation of the haem molecule consequent on the presence or absence of bound oxygen, which is transmitted to the protein structure and causes a change in the tertiary structure of the subunit. This in turn causes the abrupt transition of the quaternary structure from the T to the R configuration as SALT BRIDGES between the subunits break apart and the constraints on the $\alpha_1\beta_2$ interface are relaxed. The shift from T to R structure opens up clefts in the unliganded β-chain haems, greatly increasing their affinity for oxygen.

The human globin genes

Each type of globin chain is the product of a specific gene. In humans, the α and ζ genes form a cluster on chromosome 16, whereas the γ, β and δ gene cluster is on chromosome 11. The different human haemoglobins together with the arrangement of these gene clusters is shown in Fig. H8. The two gene clusters are thought to be derived from the duplication and divergence of an ancestral globin gene (*see* MOLECULAR EVOLUTION). The α-like genes resemble each other more closely than they do the β-like genes and vice versa, suggesting that duplication of genes within the clusters has occurred more recently than the divergence between α and β.

The α-globin gene cluster

The α gene cluster usually contains one functional ζ gene and two functional α genes — α2 and α1. It also contains four PSEUDOGENES, ψζ, ψα1, ψα2, and θ. These are gene loci with homology to the α or ζ genes but which contain mutations that render them functionless. They are thought to be evolutionary remnants of once-active genes. The θ gene has only been discovered recently and is remarkably conserved among different species. Although it seems to be expressed in early foetal life its function is unknown.

Each α gene is located in a region of homology, about 4 kilobases (kb) long, interrupted by two small nonhomologous regions. The homologous regions have resulted from gene duplication and the nonhomologous segments may have arisen subsequently by insertion of DNA into the noncoding regions round one of the two genes. All the globin genes have two INTRONS and three EXONS. The exons and the first intron of the two α-globin genes are identical in sequence but the second intron of α1 is nine bases longer and differs elsewhere by three bases from that in the α2 gene. Despite their close homology, the sequences of the two α-globin genes diverge in the 3′ untranslated regions 13 bases beyond the TAA stop signal. These differences make it possible to distinguish their messenger RNAs and assess the relative output of the two α genes; the production of α2 mRNA exceeds that of α1 by a factor of 1.5–3.

The ζ and ψζ genes are also highly homologous. The introns are much larger than those of the α-globin genes, and, in contrast to the latter, the first intron is larger than the second. In each ζ gene, intron 1 contains several copies of a repeated 14-base pair (bp) sequence which is similar to those located between ζ and ψζ and also near the human insulin gene. Compared with ζ there are three base changes in the coding sequence of the first exon of ψζ, one of which gives rise to a premature stop codon, thus turning it into an inactive pseudogene.

The regions separating and surrounding the α and α-like structural genes have been analysed in detail. This gene cluster is highly polymorphic. There are five so-called HYPERVARIABLE REGIONS in the cluster, one downstream from the α1 gene, one between ζ and ψζ, one in the first intron of both ζ and ψζ, and one 5′ to the cluster. These regions have been sequenced and found to consist of tandem repeats of nucleotide sequences which vary in number between individuals. Taken together with numerous single-base RFLPs (restriction fragment length polymorphisms) in the region, the genetic variability of the α gene cluster reaches a HETEROZYGOSITY of ~0.95. In the majority of persons therefore, it is possible to distinguish each parental α-globin gene cluster by its pattern of RESTRICTION SITES.

The β-globin gene cluster

The arrangement of the β-globin gene cluster on the short arm of chromosome 11 is ε, ^{G}γ, ^{A}γ, ψβ, δ, β. Each of the individual genes and their flanking regions have been sequenced. Like the α1 and α2 gene pair, the ^{G}γ and ^{A}γ genes share a similar sequence. In fact the ^{G}γ and ^{A}γ genes on the same chromosome are identical in the region 5′ to the centre of the second intron, yet show some divergence on the 3′ side. At the boundary between the conserved and divergent regions there is a sequence which may be a 'hot spot' for the initiation of RECOMBINATION events that have

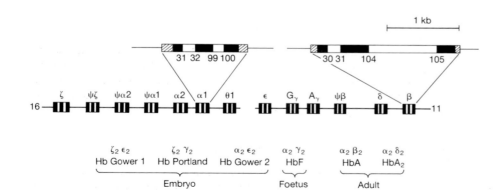

Fig. H8 The human haemoglobins and the globin gene clusters on chromosomes 11 and 16. The extended figures of the α1 and β genes show the introns in dark shading, the exons in open boxes, and the flanking regions in lined shading.

led to unidirectional GENE CONVERSION during evolution, keeping the 5′ sequence of the two genes identical.

Like the α-globin genes the β gene cluster contains a series of single-base RFLPs, although in this case no hypervariable regions have been identified. The distribution of RFLPs in the β-globin gene cluster falls into two domains. On the 5′ side of the β gene, spanning about 32 kb from the ε gene to the 3′ end of the ψβ gene, three common patterns of RFLPs (HAPLOTYPES) are found. In the region encompassing about 18 kb to the 3′ side of the β-globin gene there are also three different patterns common in different populations. Between these regions there is a sequence of about 11 kb in which there is randomization of the 5′ and 3′ domains and hence where a relatively higher frequency of recombination may occur.

Recent studies indicate that the β-globin gene haplotypes are similar in most populations, although there is much more variation between individuals of African origin. This suggests that these haplotype arrangements were laid down very early during human evolution and is consistent with data obtained from mitochondrial DNA POLYMORPHISMS which point to the early emergence of a relatively small human population from Africa, with subsequent divergence into other racial groups.

Structure of noncoding regions of the globin genes

The regions flanking the coding regions of the globin genes contain a number of conserved sequences that are involved in their expression. To the 5′ side of the coding region is the TATA BOX, which serves to locate accurately the site of transcription initiation, usually about 30 bases downstream, and which also seems to influence the rate of transcription (*see* EUKARYOTIC GENE EXPRESSION). In addition, there are two so-called upstream promoter elements; 70 or 80 bp upstream from the site of initiation is a second conserved sequence, the CCAAT BOX, and further 5′, ~80–100 bp from the initiation site, is a GC-rich region with a sequence that can be either inverted or duplicated. These upstream promoter sequences are also required for optimal transcription; mutations in this region of the β-globin gene lead to defective expression. The globin genes also have conserved sequences in their 3′ flanking regions, notably AATAAA which is the POLYADENYLATION signal site.

Expression and regulation of the globin genes

Red blood cell precursors synthesize the various types of globin subunit in a strictly regulated manner; for each haemoglobin the two types of subunits of which it is composed must be synthesized synchronously and a regulatory system has also evolved which ensures that the different haemoglobins are produced at appropriate times during foetal and adult life. α and β chain synthesis is almost synchronous although a slight excess of α chains is produced; these are degraded by proteolytic enzymes in the red cell precursor.

Very little is known about the regulation of globin gene transcription. It is clear that the methylation state of the genes plays an important part in their ability to be expressed (*see* DNA METHYLATION); in human and other animal tissues the globin genes are extensively methylated in non-erythroid organs and are relatively undermethylated in haematopoietic tissues. There is a change in the methylation pattern and CHROMATIN configuration round the globin genes at different stages of human development.

Various *trans*-acting protein TRANSCRIPTION FACTORS influence the expression of the globin genes. Apart from the general transcription factors required for gene expression, tissue-specific and developmental stage-specific factors must be involved. Binding sites for the haematopoietic-specific DNA-binding proteins GATA-1 and NF-E2 have been found; no developmental stage-specific factors have been identified. Potential binding sites for many ubiquitous DNA-binding transcription factors have been identified in both the α- and β-globin gene clusters. In addition to general promoter sites common to many eukaryotic genes, several ENHANCER sequences have been identified, which may be involved in tissue-specific and developmental stage-specific regulation. Another important control element, the LOCUS CONTROL REGION (LCR), has been identified many kilobases upstream from both the α and β gene clusters, and seems to regulate transcriptional activity throughout the clusters.

Developmental changes in human globin gene expression

The synthesis of β-globin commences early during foetal life, at around 8–10 weeks gestation. Subsequently it continues at a low level, forming ~10% of total non-α chain production, up to about 36 weeks gestation, after which it is considerably augmented. At this time γ-globin chain synthesis begins to decline, so that at birth approximately equal amounts of γ- and β-globin chains are being produced. Over the first months of life there is a gradual decline in γ-chain synthesis and by the end of the first year it amounts to less than 1% of the total non-α-globin output. In adults the small amount of HbF is confined to an erythrocyte population called 'F cells'.

It is not known how this series of developmental switches is regulated. They are not organ-specific, but are synchronized throughout the developing haematopoietic tissues. Although environmental factors may be involved, there is also some form of 'time-clock' built into the haematopoietic stem cell. At the chromosomal level it seems likely that regulation occurs in a complex manner involving the interaction of developmental stage-specific *trans*-activating factors with specific sequences in the γδβ gene cluster.

Molecular pathology

The genetic disorders of haemoglobin are the commonest single gene disorders in the population world-wide. They are caused by mutations that either alter the structure of a globin chain or result in a drastic reduction in the rate of production of one or more types of globin chain.

Structural haemoglobinopathies

Over 400 structural variants of human haemoglobins have been described. Most of these structural variants result from a single base change in one or other of the globin genes that leads to the production of an abnormal haemoglobin with a single amino-acid substitution. There are rarer structural variants with shortened or elongated subunits (e.g. HAEMOGLOBIN CONSTANT SPRING). Variant haemoglobins are associated with disease only if they alter the function, solubility or stability of the haemoglobin molecule. The commonest and clinically most important are HAEMOGLOBIN S (sickle cell haemoglobin), HAEMOGLOBIN C, and haemoglobin E, which occur at polymorphic frequencies in many populations.

Sickle cell haemoglobin results from the substitution of glutamic acid by valine in the sixth position of the β-globin chain, as a result of a mutation of GAG to GTG in the β gene. The change from the polar glutamic acid to the nonpolar amino acid valine in this position, which is on the surface of the molecule, causes an alteration in the surface properties of the β subunits resulting in aggregation of haemoglobin molecules in the deoxygenated state (*see* HAEMOGLOBIN S). They form insoluble fibres inside the red cell which deform the cell into the sickle shape characteristic of this condition. The sickling deformity in turn leads to shortening of the red cell survival time and aggregation of sickled erythrocytes in small blood vessels with subsequent death of tissue due to a reduced oxygen supply.

A severe anaemia — sickle cell anaemia — occurs in people homozygous for the sickle cell gene. Sickle cell trait is the usually symptomless condition seen in people heterozygous for the sickle cell gene. Despite the severity of sickle cell anaemia, the gene is common in parts of Africa and in populations of black African descent. This is because, in the heterozygous state, it affords some protection against the most severe form of malaria.

Some haemoglobin variants, such as haemoglobin Zürich, cause molecular instability; the haemoglobin molecule precipitates in the red blood cell causing its premature damage in the circulation. Yet others interfere with oxygen transport (*see* HAEMOGLOBIN M).

Thalassaemias

The genetic disorders of haemoglobin that are characterized by a reduced rate of production of the α- or β-globin chains are called thalassaemias. There are many forms, the commonest being α- and β-thalassaemia which are characterized by a reduced rate of production of the α- or β-globin chains respectively. This leads to unbalanced globin chain synthesis with the precipitation of the chain that is produced in excess, damage to the developing red blood cell, and hence a varying degree of anaemia. Over 100 different mutations have been described in the globin genes of patients with thalassaemia. They include partial or complete deletions, point mutations that produce premature termination codons within exons, mutations that cause a shift in the reading frame of the genetic code and hence premature termination of chain synthesis, a wide variety of different mutations that involve the critical splice junctions and hence cause abnormal splicing of mRNA, and point mutations that lie in or near to the promoter boxes and cause defective transcription of the globin genes (*see also* HAEMOGLOBIN BART'S; HAEMOGLOBIN CONSTANT SPRING; HAEMOGLOBIN H; HAEMOGLOBIN LEPORE; HEREDITARY PERSISTENCE OF FOETAL HAEMOGLOBIN; THALASSAEMIA).

The thalassaemias are most common among people living in malarial and previously malarial areas around the Mediterranean, the Middle East and India, and like the sickle cell gene, the prevalence of genes for thalassaemia is thought to be due to the protection they afford in the heterozygous state against malaria.

D.J. WEATHERALL

See also: HAEMATOPOIESIS.

1 Bunn, H.F. & Forget, B.G. (1986) *Haemoglobin: Molecular, Genetic and Clinical Aspects* (W.B. Saunders, Philadelphia).
2 Weatherall, D.J. et al. (1992) The hemoglobinopathies. In *The Metabolic Basis of Inherited Disease*, 7th edn (Scriver, C.R. et al., Eds) (McGraw Hill, New York).
3 Branden, C. & Tooze, J. (1991) *Introduction to Protein Structure* (Garland, New York/London).

haemoglobin A (HbA) Predominant normal adult human haemoglobin, composed of two α-globin chains and two β-globin chains. *See*: HAEMOGLOBIN AND ITS DISORDERS.

haemoglobin Bart's A tetramer of γ-globin chains formed during foetal haematopoiesis in the condition α-THALASSAEMIA where there is a total or relative absence of α-globin chains. γ_4 is physiologically useless.

haemoglobin C A variant form of human HAEMOGLOBIN found commonly in West Africa. In the β chain lysine is substituted for glutamic acid at position 6. Haemoglobin C has relatively low solubility and leads to red cell rigidity and premature destruction in the microcirculation. HOMOZYGOTES have a mild haemolytic anaemia.

haemoglobin Constant Spring Human haemoglobin containing a α-globin variant with 31 additional amino-acid residues at the C-terminal end. This results from a point mutation in the termination codon of the α_2 globin locus. The translation of the elongated mRNA is inefficient and hence haemoglobin Constant Spring is produced in very low quantities and is associated with the phenotype of α-thalassaemia.

haemoglobin F (HbF, foetal haemoglobin) The main HAEMOGLOBIN in foetal life, composed of α-globin chains combined with γ-globin chains as $\alpha_2\gamma_2$. HbF is a mixture of two different molecular forms which differ by one amino acid in the γ chains at position 136. The two γ chains are referred to as G_γ (containing glycine) and A_γ (containing alanine) respectively. *See also* HAEMOGLOBIN; HEREDITARY PERSISTENCE OF FOETAL HAEMOGLOBIN.

haemoglobin fusion genes Genes resulting from UNEQUAL CROSSING-OVER between linked related genes, and which contain part of one gene and part of another. They produce for example HAEMOGLOBIN KENYA and HAEMOGLOBIN LEPORE.

haemoglobin H (HbH) A tetramer of normal β-globin chains, β_4. It is produced when there is a marked reduction of α-globin synthesis as occurs in α-THALASSAEMIA. It is relatively unstable and precipitates in the red cells in the circulation forming large inclusion bodies.

haemoglobin homotetramers Haemoglobin formed in the absence or relative absence of α-globin chains. Excess γ chains associate to form γ₄ (HAEMOGLOBIN BART'S), excess β chains to form β₄ (HAEMOGLOBIN H).

haemoglobin Kenya HAEMOGLOBIN product of the FUSION GENE that results from UNEQUAL CROSSING-OVER between A_γ and β. It contains the N-terminal sequence of the A_γ gene and the C-terminal sequence of the β gene. The intervening genes are deleted (including the δ gene). The overall phenotype is that of HEREDITARY PERSISTENCE OF FOETAL HAEMOGLOBIN, the modified foetal haemoglobin entirely compensating for the lack of δ- and β-globin chains.

haemoglobin Lepore HAEMOGLOBIN product of the FUSION GENE that results from UNEQUAL CROSSING-OVER between β- and δ-globin genes. The intervening material is deleted. The fusion gene encodes a single β-like chain that consists of the N-terminal sequence of δ joined to the C-terminal sequence of β. Several types of haemoglobin Lepore are known, the difference between them lying in the point of transition from δ to β sequences. The phenotype is classified as δ–β thalassaemia and in the homozygous state there is a mild degree of anaemia.

haemoglobin M (HbM) An abnormal human HAEMOGLOBIN associated with inherited methaemoglobinaemia, a rare condition seen only in heterozygotes and characterized by cyanosis which is present from early life. The variant haemoglobin has the iron atoms fixed in the Fe^{3+} state and may arise from structural alteration in either the α- or β-globin chain caused by amino-acid substitutions near the haem pocket.

haemoglobin S (HbS, sickle-cell haemoglobin) A variant human HAEMOGLOBIN occurring very frequently in black African populations and sporadically throughout the Mediterranean region and the Middle East. It differs from haemoglobin A by the substitution of valine for glutamic acid at position 6 in the β-globin chain. This leads to surface changes, which result in haemoglobin S tending to form long stable polymers in the deoxygenated state. If in sufficiently high concentration this leads to deformation and the characteristic sickling of the red cells in conditions of low oxygen tension. *See also*: HAEMOGLOBIN AND ITS DISORDERS; SICKLE CELL ANAEMIA.

haemoglobinopathies *See*: HAEMOGLOBIN AND ITS DISORDERS and entries for individual haemoglobins.

haemolysin A bacterial protein EXOTOXIN which lyses red blood cells. Produced especially by pathogenic streptococci and staphylococci.

haemolytic anaemia, hereditary If the lifespan of red cells is shortened and this is not adequately compensated by increased production of red cells by the bone marrow then haemolytic anaemia occurs. There are diverse hereditary forms of this condition reflecting the different mechanisms by which red cell survival is shortened. Genetic abnormalities in the structure of the red cell membrane occur in hereditary spherocytosis, elliptocytosis, stomatocytosis, pyropoikilocytosis and acanthocytosis. Alterations in the structure or synthesis of haemoglobin, as in SICKLE CELL ANAEMIA and THALASSAEMIA respectively, also cause haemolytic anaemia as do inborn errors of metabolism that interrupt energy pathways, as in GLUCOSE-6-PHOSPHATE DEHYDROGENASE DEFICIENCY and pyruvate kinase deficiency.

haemophilia A (classical haemophilia) A bleeding disorder with an X-LINKED recessive mode of inheritance, caused by defects in the gene for the coagulation factor Factor VIII, which is on the distal long arm of the human X-chromosome. There is a deficiency of a small part of the Factor VIII molecule (Factor VIII : C) which is functionally active in the coagulation cascade. The severity of the bleeding tendency closely relates to the plasma concentration of Factor VIII coagulant activity and the degree of severity tends to run true in families. The condition is treatable by transfusion of normal Factor VIII. There is considerable variation in the incidence of the disease worldwide (1 : 4000 in the UK). *See*: BLOOD COAGULATION AND ITS DISORDERS.

haemophilia B (Christmas disease) A heritable bleeding disorder due to deficiency of the coagulation factor Factor IX as the result of defects in the gene for Factor IX. It is an X-LINKED recessive condition and the locus has been mapped to the long arm of the human X-chromosome. It is six to seven times less frequent than haemophilia A in the general population. *See also*: BLOOD COAGULATION AND ITS DISORDERS.

Haemophilus Genus of small Gram-negative aerobic heterotrophic bacteria with the form of slender rods. *H. influenzae* is a secondary pathogen in acute respiratory infections and is also found in chronic bronchitis. *See also*: BACTERIAL PATHOGENICITY.

haemopoiesis *See*: HAEMATOPOIESIS.

haemoprotein Protein containing a HAEM group. *See*: CYTOCHROMES; HAEMOGLOBINS.

hairpin A double-stranded nucleic acid structure formed by BASE-PAIRING between two regions of the same strand of a nucleic acid molecule. The two complementary regions are arranged inversely and can be either immediately adjacent or separated by a noncomplementary DNA sequence (Fig. H9). In the latter case the hairpin structure has an additional loop and is said to form a hairpin-loop structure. *See*: DNA; NUCLEIC ACID STRUCTURE; RNA.

hairpin turn (β-turn, β-hairpin) *See*: PROTEIN STRUCTURE.

hairy root disease Tumorous plant disease caused by *Agrobacterium rhizogenes* and characterized by proliferation of roots. The tumour phenotype is due to transfer of DNA from the RI PLASMID in the bacterium into the plant cell genome.

a

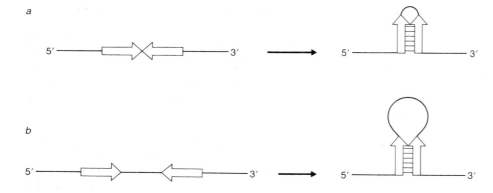

b

Fig. H9 The formation of a hairpin (*a*) and a hairpin-loop (*b*) structure in nucleic acid.

half-life (1) For macromolecules *in vivo*, the time in which a given amount of the substance in question is reduced to 50% of its initial concentration by normal turnover (e.g. the normal degradative and metabolic processes of the cell). *See e.g.* PROTEIN DEGRADATION AND TURNOVER.

(2) For radioisotopes, the average time required for one-half of the original atoms of a radioactive material to undergo radioactive decay. It can also refer to the time required in a living system or in an ecosystem for half of a radioactive tracer or contaminant to disintegrate by radioactive decay or to be eliminated by natural processes.

Halobacterium halobium Halophilic purple bacterium, which contains the light-driven membrane proton pump BACTERIORHODOPSIN and the related proteins HALORHODOPSIN and SENSORY RHODOPSIN.

halorhodopsin Purple membrane protein from the halophilic bacterium *Halobacterium halobium* which functions as a light-driven chloride pump. It is closely related to BACTERIORHODOPSIN, with 30% homology in the amino-acid sequences and similar membrane topology. The replacement of two of the key aspartic acid residues in bacteriorhodopsin by alanine and threonine in halorhodopsin is responsible for the change in the ion specificity of the pump.

Oesterhelt, D. & Tittor, J. (1989) *Trends Biochem. Sci.* **14**, 57–61.

hammerhead A class of self-cleaving RNAs characterized by the presence of 13 conserved nucleotides and by the ability to be folded into a characteristic RNA secondary structure. The structure consists of three base-paired regions that radiate from a single-stranded region. The secondary structure so drawn is reminiscent of the head of a hammer. *See:* RIBOZYMES.

hanging drop A technique used in the CRYSTALLIZATION of macromolecules.

haplo-insufficiency Situation where one normal copy of a gene alone (in a diploid cell) is not sufficient to support normal function.

haploid Cells that contain only a single copy of each particular CHROMOSOME. In higher eukaryotes, which are largely DIPLOID, this situation is restricted to the GAMETES, while in bacteria the haploid state predominates. The abbreviation *n* is used to identify the haploid state and to differentiate it from the diploid state (*2n*).

haplotype A means of denoting the collective genotype of a number of closely linked loci, such as those of the MAJOR HISTOCOMPATIBILITY COMPLEX (MHC), which are usually inherited as a unit. Each haplotype defines the sequence of alleles along one of the homologous chromosomes. Most commonly it refers to the constellation of alleles present at the H-2 or HLA regions (*see* MAJOR HISTOCOMPATIBILITY COMPLEX) in a given individual, or a designation for the particular allele at a given H-2 or HLA locus. Murine MHC haplotypes are usually designated by a single-letter superscript; for example, the Balb/c mouse has an H-2^d haplotype, and is H-$2K^d$, H-$2L^d$ at these two class I MHC loci. Although haplotypes were originally defined by differential immune responsiveness of mouse strains to synthetic polypeptide IMMUNOGENS, most haplotype designations are now made serologically or by molecular characterization of individual MHC alleles and their protein products.

hapten From the Greek word *haptein* meaning 'to fasten' and referring to the ability of a simple chemical molecule to bind antibody. A hapten is not able to induce an immune response when administered alone but may do so when attached to a carrier molecule (e.g. albumin or Ficoll). A substance may be a complete antigen in some species but function only as a hapten in others. *See:* ANTI-NP; ARS; GAT; GT; PC.

haptoglobin Protein circulating in blood which binds free haemoglobin as a means of removing it from the serum. The gene encoding haptoglobin in humans is located on the distal long arm of chromosome 16. It is a member of the SERINE PROTEINASE superfamily and has some similarities to plasminogen, but does not have proteolytic activity as it cannot be cleaved to an active form by tissue plasminogen activator because of a mutation at the cleavage site.

haptotaxis Active directional migration of cells up or down a gradient of adhesion. *See also:* CELL-MATRIX INTERACTIONS.

Harvey murine sarcoma virus Murine RETROVIRUS which is the source of the H-*ras* ONCOGENE.

HAT medium, HAT selective system A selective medium introduced in 1967 by Littlefield, containing hypoxanthine, aminopterin (amethopterin, methotrexate), and thymidine. Aminopterin is a folate antagonist, which binds strongly to the enzyme folic acid reductase, effectively preventing *de novo* synthesis of guanosine-5'-phosphate (GMP). To grow in this environment a cell must be able to synthesize nucleotides for DNA via the salvage pathway. This requires the enzyme hypoxanthine : guanine phosphoribosyl transferase (HGPRT or HPRT) which catalyses reactions of the bases hypoxanthine and guanine with 5-phosphoribosyl-1-pyrophosphate to form the nucleotides inosine-5'-phosphate (IMP) and GMP. HAT medium is used to select SOMATIC CELL HYBRIDS in gene mapping and other studies and HYBRIDOMA cells produced by the fusion of normal antibody-producing lymphocytes and an immortal PLASMACYTOMA cell line which lacks HGPRT. When cultured in HAT the plasmacytoma cells die because they are unable to synthesize DNA, and unfused B cells also die because they do not possess the immortal property of the plasmacytoma or hybridoma cells. Only the hybridoma cells, possessing both HGPRT and immortality, survive.

haustorium (*pl.* haustoria) A specialized fungal hyphal structure which is believed to function in the absorption of nutrients from the living host plant. A haustorium is formed after a hypha penetrates the host cell wall, and causes the invagination of the host plasma membrane. Haustoria are generally associated with BIOTROPHIC pathogenic fungi, such as the rusts.

Hb Haemoglobin. *See*: HAEMOGLOBINS; HAEMOGLOBIN AND ITS DISORDERS, and individual haemoglobin variants.

HB101 A strain of *Escherichia coli* derived from the original *E. coli* B strain by Herbert Boyer. HB101 has the genotype *supE44 hsdS20 recA13 ara-14 proA2 lacY1 galK2 rpsL20 xyl-5 ptl-1* and has been widely used as a host strain for TRANSFORMATION and GENETIC ENGINEERING. It is a RECA-deficient host.

HBV HEPATITIS B VIRUS.

HCD HEAVY CHAIN DISEASE.

hcf mutant High chlorophyll fluorescence mutant. A maize mutant which was selected by giving a high level of fluorescence from photosystem II (a technique first applied to *Chlamydomonas*). The high fluorescence indicates inefficient use of light energy and can be caused by deficiencies in almost all aspects of the photosynthetic apparatus, except those influencing the oxidizing side of photosystem II.

HDL High-density lipoprotein. *See*: PLASMA LIPOPROTEINS AND THEIR RECEPTORS.

HDV HEPATITIS DELTA VIRUS.

head fold In amniote embryos, a ventrally directed movement of all three GERM LAYERS in the head region, which defines the entrance to the foregut and the outline of the head. Formation of the head fold also defines the septum transversum, the precursor of the diaphragm, which subdivides the trunk into thorax and abdomen and which also carries the primordia of the liver and heart. It should not be confused with HEAD PROCESS which is the cephalic part of the NOTOCHORD.

head process The portion of the NOTOCHORD in the head of amniote embryos, lying anterior to the otic (ear) vesicle. It is derived from cells in HENSEN'S NODE, and extends anteriorly as a rod. In amniote embryos it underlies the developing brain between the middle of the hindbrain (boundary between RHOMBOMERES 4 and 5) and the diencephalic part of the forebrain. It should not be confused with HEAD FOLD.

Heat shock

CELLS have different stress responses. Each is triggered by one or more kinds of environmental insult or stress, the main consequence of its induction being increased stress tolerance levels. The heat shock response is one of the best-characterized of these stress responses. In 1962 Ritossa reported that exposure to high but nonlethal temperatures (heat shock) and to certain chemicals causes the appearance of new PUFFS on the salivary gland POLYTENE CHROMOSOMES of the fruit fly *Drosophila busckii*. These puffs, clearly visible in the light microscope, represent high transcriptional activity on heat shock-inducible genes, an RNA synthesis that generates the messenger RNAs for the heat shock proteins (Hsps or HSPs). More recently a wide range of organisms, from bacteria to the higher vertebrates, has been shown to display similar dramatic changes in gene expression with heat shock [1].

The heat shock response

The response to heat shock entails both strong INDUCTION of genes for heat shock proteins and REPRESSION of most genes that were being expressed previous to this induction. There are marked changes both to patterns of gene transcription and to the way in which mRNAs are selected for translation by the protein-synthesizing machinery of the cell.

Changes in protein synthesis

Protein synthesis is altered so that unique structural features of the heat-induced mRNAs are recognized, thus ensuring that these mRNAs are preferentially translated. Those mRNAs existing prior to the shock are no longer translated, yet they are not always degraded and can be stored in inactive form to be translated yet again after the response has been switched off [1].

Induction of gene transcription

The transcriptional changes with heat shock are due to the

presence of a heat shock element (HSE) in heat shock gene PROMOTERS. This element is a DNA sequence needed for specific induction of transcription in response to heat shock. In the heat-inducible promoters of the bacterium *Escherichia coli* the CONSENSUS SEQUENCE for the heat shock element is CTGCCACCC at nucleotide positions − 44 to − 36 relative to the transcription initiation site. In the heat-inducible promoters of eukaryotes it is contiguous repeats of the 5-bp sequence NGAAN arranged in altenating orientation (N is any nucleotide), positioned upstream of the TATA BOX element [2].

Mechanism of heat shock gene induction

The *E. coli* heat shock element is recognized by RNA polymerase only when this polymerase has a special 32K (M_r 32 000) sigma (σ) regulatory subunit (σ^{32}; the product of the *rpoH* gene) bound in place of its usual 70K σ subunit (σ^{70}; the *rpoD* product) (*see* TRANSCRIPTION). *rpoD* is itself stimulated by a heat shock, the resulting elevation of σ^{70} levels contributing to the switch-off of the heat shock response by enabling σ^{70} to compete efficiently against σ^{32} for the σ-binding site on RNA polymerase. In eukaryotic cells the heat shock element is the binding site for a transcription activating factor (*see* TRANSCRIPTION FACTORS). This factor, one of the best-studied transcription activators of eukaryotes, needs the operation of the heat shock trigger before it can efficiently promote RNA initiation [2].

Heat-inducible promoters often have other activator elements in addition to those directing their induction by heat shock. These other elements direct the induction of specific heat shock proteins during normal developmental processes (e.g. moulting of *Drosophila*; sporulation of yeast) and viral infections (e.g. λ phage multiplication in *E. coli*; adenovirus infection of human cells) [1].

Heat shock proteins

The small number of proteins induced by heat shock are remarkable for their very high degree of structural conservation during evolution, a conservation that must reflect perpetuation of functions necessary for cell survival [3]. All eukaryotes induce heat shock proteins with sizes of 80–90K (Hsp90), 68–74K (Hsp70), and 58–60K (Hsp60), in addition to a variety of 'small Hsps' (Table H1).

The Hsp90, Hsp70 and Hsp60 induced by heat shock are extremely similar in structure to other proteins that are CONSTITUTIVELY expressed and which have essential roles in unstressed cells [3, 4]. Hsp90 is essential for the viability of yeast cells. Even though its levels increase with heat shock it is found in association with many regulatory proteins (e.g. STEROID RECEPTORS and various PROTEIN KINASES) in mammalian cells under normal conditions of growth (Table H1).

Hsp90, Hsp70 and Hsp60 all have ATPase activity. Constitutively synthesized Hsp70-related (or 'Hsc70') proteins normally account for about 1% of total protein in the absence of heat shock. They transiently associate with and dissociate from specific target proteins, in so doing catalysing polypeptide chain unfolding and refolding reactions which are essential for correct formation of many protein tertiary structures (*see* PROTEIN STRUCTURE); also the assembly of multisubunit proteins and protein assemblies [4].

Table H1 The major eukaryotic heat shock proteins

Heat shock protein	Occurrence	Role	Homologous *E. coli* protein (amino-acid homology to equivalent human Hsp in brackets)
Hsp110	Not universally present; concentrated in nucleolus of mammalian cells after heat shock	Unknown	—
Hsp90	Abundant in cytoplasm, a small fraction translocates to nucleus after heat shock. Vertebrates have a form of Hsp90 in the lumen of the endoplasmic reticulum	Complexes with either the inactive or active forms of several regulatory proteins; also cytoskeletal proteins. Possibly a form of 'chaperone'	*htpG* protein (42%)
'Hsp70 family' proteins	Diverse species (nine members in yeast, only three of which are induced by heat shock). Each form is located in a specific cellular compartment, much of the heat-induced Hsp70 being transported to the nucleolus	'Chaperonins', especially of nascent polypeptides. Probably in all cellular compartments. Those involved in import of proteins into organelles may have an unfolding and disassembly role	*dnaK* protein (50%)
Hsp60	Localized inside cytoplasmic organelles (mitochondria and chloroplasts)	CHAPERONINS catalysing correct folding and assembly of organelle proteins	*groEL* protein (54%)
'Small Hsps' (15–30K)	In both cytoplasm and organelles	Unknown. No apparent enzymatic activity, but homologous to crystallins. Can aggregate as large granular arrays	—
Ubiquitin and ubiquitin ligases	Cytoplasm	Intracellular turnover of proteins (*see* PROTEIN DEGRADATION)	—

These proteins also catalyse polypeptide chain unfolding in preparation for translocation across membranes (*see* PROTEIN TRANSLOCATION); uncoating of CLATHRIN-coated vesicles and possibly (see Fig. H10) the repair of misfolded or denatured proteins. Because of these properties, proteins of the structurally unrelated Hsp90, Hsp70 and Hsp60 families are frequently termed MOLECULAR CHAPERONES [4].

In eukaryotes certain genes of the ubiquitination system (including some encoding UBIQUITIN and ubiquitin ligases) are activated by heat shock (Table H1). This probably reflects the demand for increased activity of the ubiquitination pathway for intracellular protein turnover in heat-shocked cells (*see* PROTEIN DEGRADATION). It is known that stress-enhanced proteolysis involves sequestration of proteins destined for breakdown as high molecular weight multiubiquitinated complexes. These complexes then serve as substrates for the multicatalytic–multifunctional protease particle (proteasome) present in all eukaryotic cells [5]. Although bacteria do not have the ubiquitin pathway and proteasome, one of the heat shock genes of *E. coli* (*lon*) encodes a protease. In addition, an *E. coli* mutant that cannot induce the heat shock response (*rpoH⁻*) is defective in degrading normally unstable proteins.

The heat shock trigger

For experimental purposes the heat shock response is usually induced by a temperature upshift. The optimal temperature for induction is species dependent, but usually a degree or two above the maximum that permits growth. Also, the response is generally transient, heat shock protein synthesis becoming repressed in most systems just a few minutes after its induction by either upshift to moderate temperatures or an upshift followed by a return to normal temperatures [1].

There are other inducers of heat shock protein synthesis besides thermal stress [1]. These include several potentially cytotoxic chemicals (e.g. ethanol, heavy metal ions such as Cd^{2+},

sodium arsenite and amino-acid analogues) and physiological states which may cause generation of highly reactive free radicals (e.g. refeeding after glucose starvation and recovery from anoxia). These inducers probably share with high temperatures the ability to cause intracellular accumulation of aberrant or partially denatured protein. It is therefore thought that damaged, denatured or otherwise improperly folded protein is the trigger for induction of the heat shock response (see Fig. H10). The response serves to increase the ability of cells to cope with increased intracellular levels of such protein. With synthesis of Hsp70 they have additional capacity for sequestering denatured protein as complexes with Hsp70, while the enhanced synthesis of ubiquitin and ubiquitin ligases allows more degradation of aberrant protein by the ubiquitination pathway. Hsp70 may even be able to use its ATPase activity to effect the repair of misfolded protein.

Biological effects

A mild, nonlethal heat shock renders most cells much more resistant to short exposures to still higher, normally lethal temperatures. This induced resistance to heat killing (thermotolerance) is the most marked and rapid cell physiological change associated with induction of the heat shock response. Thermotolerance at normally lethal temperatures, the maximum temperature for growth, and even the minimum temperature for growth (cold tolerance) can all be affected by mutational inactivation of specific yeast genes for heat-induced or Hsp70-related proteins [1, 3]. However these same properties are unaffected by inactivation of certain other heat shock genes in yeast. In general, cells in rapid growth are much less thermotolerant than nondividing cultures. The lowered thermotolerance of tumours as compared to surrounding nontumorigenic tissues and the heat enhancement of the cytotoxic actions of widely used chemotherapeutic agents form the basis of the use of heat stress (hyperthermia) in cancer therapy. Stress proteins are also prominent among the dominant ANTIGENS recognized in the course of the immune

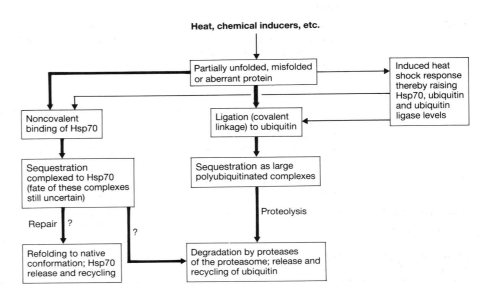

Fig. H10 Induction of the heat shock response probably assists the eukaryotic cell to renature or degrade partially damaged proteins.

response of humans to bacterial and parasitic infections. In addition there is evidence that they have a role in autoimmune disease [6].

P.W. PIPER

1 Craig, E.A. (1986) The heat shock response. *CRC Crit. Rev. Biochem.* **18**, 239–280.
2 Sorger, P.K. (1991) Heat shock factor and the heat shock response. *Cell* **65**, 363–366.
3 Lindquist, S. & Craig, E.A. (1988) The heat shock proteins. *Annu. Rev. Genet.* **55**, 1151–1191.
4 Rothman, J.E. (1989) Polypeptide chain binding proteins: catalysis of protein folding and related processes in cells. *Cell* **59**, 591–601.
5 Heinemeyer, W. et al. (1991) Proteinase yscE, the proteasome/multi-catalytic-multifunctional proteinase: mutants uncover its function in stress induced proteolysis and uncover its necessity for cell survival. *EMBO J.* **10**, 555–562.
6 Young, R.A. & Elliot, T.J. (1989) Stress proteins, infection and immune surveillance. *Cell* **59**, 5–8.

heat shock element (HSE) *See*: HEAT SHOCK.

heavy atom isomorphous derivative A method of solving the PHASE PROBLEM in X-RAY CRYSTALLOGRAPHY whereby a crystal is modified by the addition of a few heavy atoms, which scatter X-rays strongly in comparison with organic molecules. The modified crystal may be obtained by soaking a crystal in a suitable heavy atom solution or by co-crystallizing in the presence of a heavy atom. Provided that the modified crystal is isomorphous with the original crystal, and that the positions of the heavy atom(s) can be determined by DIRECT METHODS or Patterson search methods, then the additional diffraction data from the heavy atom derivative crystal can provide useful phasing information for the STRUCTURE FACTORS from the unmodified or native crystal.

heavy chain (H chain) (1) The larger of the two types of polypeptide chain of which immunoglobulin molecules are composed (*see* ANTIBODIES; IMMUNOGLOBULIN STRUCTURE). Two identical heavy chains and two identical light chains combine to form a four-chain structure. The heavy chains of the molecules contribute, together with light chains, to antigen specificity and are responsible for effector functions of antibodies such as antibody-dependent cell-mediated cytotoxicity and complement fixation. The five classes of antibodies (IgM, IgD, IgG, IgE, and IgA) differ in their heavy chains (μ, δ, γ, ε, and α respectively) and in their effector functions. Heavy chains are produced in either a membrane-bound or a secretory form. These forms differ only in their C terminus, the membrane-bound form including a transmembrane and cytoplasmic portion lacking in the secretory form. (2) MYOSIN heavy chain.

heavy chain disease (HCD) NEOPLASMS of B CELLS characterized by the expression of surface IMMUNOGLOBULIN heavy chains without association with light chains. The expressed heavy chain is truncated in size, generally lacking most or all of the variable region and CH_1 domain. In the majority of cases studied at the molecular level, the shortened heavy chain arises as a result of DNA MUTATIONS which introduce or remove RNA splice sites,

resulting in abnormal RNA SPLICING of the heavy chain RNA. Truncated heavy chains also occur as a result of DNA deletions and aberrant GENE REARRANGEMENT events within the expressed heavy chain gene. Most γ and α HCD cells express no immuno-globulin light chain, whereas μ-type HCD cells usually secrete free light chain (BENCE-JONES PROTEIN).

heavy chain enhancer ENHANCER element in the J_H-C intron of immunoglobulin heavy chain genes (Fig. H11, *see also* IMMUNO-GLOBULIN GENES). Enhancer activity is lymphoid specific and has been located in mice to a 700-base pair RESTRICTION FRAGMENT. Activity and lymphoid specificity of the heavy chain enhancer is largely regulated by the availability and binding of sequence-specific DNA-binding proteins. One of these sequence elements, the octamer motif (ATTTGCAT), is present both in the heavy chain enhancer and in the V_H gene promoter. Promoter–enhancer interaction follows assembly of V_H, D, and J_H DNA segments to create a functional immunoglobulin heavy chain gene (*see* GENE REARRANGEMENT). In some B cell MYELOMAS, the level of heavy chain expression is unaffected if the enhancer is deleted following the onset of transcription initiation. This may be because the enhancer is required to initiate high level transcription in early B cells but not to maintain it in later stages.

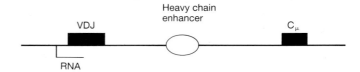

Fig. H11 Heavy-chain enhancer.

Hebbian D.O. Hebb proposed that conjoint pre- and postsynaptic activity are required to cause changes in synaptic strength. The induction of LONG-TERM POTENTIATION (LTP) is therefore 'Hebbian' because both pre- and postsynaptic sides of the synapse need to be active. LTP in the dentate gyrus is also Hebbian and like LTP in CA1 is NMDA receptor dependent. LTP in CA3 is neither NMDA receptor dependent nor Hebbian (*see* Fig. L8 in LONG-TERM POTENTIATION).

Hebb, D.O. (1949) *The Organization of Behaviour* (John Wiley, New York).

HeLa cell An undifferentiated human CELL LINE that was one of the first to be grown continuously under laboratory conditions. It was originally obtained in 1952 from a CARCINOMA of the uterine cervix of a human patient Henrietta *Lacks*. Unlike most human cells it has approximately four copies of each chromosome, that is it is a TETRAPLOID. This cell line has been widely used in the study of gene structure and function, in particular the study of TRANSCRIPTION FACTORS.

helicases Enzymes that unwind nucleic acid duplexes in an ATP-dependent and processive manner. See: DNA HELICASES; RNA HELICASES.

alpha-helix (α-helix) *See*: PROTEIN STRUCTURE.

helix-loop-helix DNA-binding secondary structure feature found in some DNA-binding proteins, such as MyoD1 and myogenin and adenovirus E12 and E47. *See*: PROTEIN–NUCLEIC ACID INTERACTIONS; TRANSCRIPTION FACTORS.

helix-turn-helix DNA-binding secondary structure feature found in some DNA-binding proteins. *See*: HOMEOBOX GENES AND HOMEODOMAIN PROTEINS; PROTEIN–NUCLEIC ACID INTERACTIONS; TRANSCRIPTION FACTORS.

3_{10}-helix A right-handed polypeptide helix which is wound more tightly than the usual α-helix. It occurs in natural proteins at the ends of α-helical runs and in some peptides that span biological membranes. The name 3_{10}-helix indicates three amino-acid residues per helical turn with 10 atoms in the ring formed by closing the hydrogen bond (*cf.* α-helix which is the 3.6_{13}-helix, *see* PROTEIN STRUCTURE).

helper plasmid In PLANT GENETIC ENGINEERING, a Ti (or Ri) plasmid from which the T-region has been removed and which can provide the *vir* gene products in *trans*. *See*: TI PLASMID.

helper T cells (T_H cells) T lymphocytes that cooperate with other lymphocytes to expedite an immune response. Helper T cells use their αβ T CELL RECEPTOR to recognize antigen when it is associated with a class II MHC MOLECULE (*see* MAJOR HISTOCOMPATIBILITY COMPLEX (MHC)); they therefore express the co-ligand CD4 (*see* CD ANTIGENS). They produce LYMPHOKINES, which can give positive signals to other effector cells; thus they can aid the generation of a CYTOTOXIC T CELL response and cooperate with B CELLS to aid ANTIBODY production.

helper virus Viruses that can allow DEFECTIVE VIRUSES present in the same host cell to replicate and generate infective viral particles. These functions are provided in *trans* during a mixed infection. In some cases helper viruses can allow viruses that lack almost all the genome to replicate; for example, some transforming RETROVIRUSES contain no intact viral genes, but are able to replicate in the presence of a helper virus.

heme Alternative spelling for HAEM.

hemibiotrophy Growth of fungi that is partly dependent on a living host. Hemibiotrophs initially establish a biotrophic interaction upon colonizing a host plant, and derive their nutrients from the host without killing it. At a later stage, once the host plant has died, the hemibiotroph may grow as a necrotroph, living off dead host plant tissue. A classic hemibiotroph is the oocmycete pathogen of potato, *Phytophthora infestans*, the causal agent of late blight.

hemicellulose A mixture of GLYCANS and glycoconjugates of plant cell walls extracted by alkali from the residue left after extraction of PECTINS by cold and hot water and chelating agents (i.e. it is the noncellulosic part of the residue). Its composition varies with the source and the exact method of extraction. The major components are GALACTANS, some ARABINOGALACTANS, XYLANS, XYLOGLUCANS, GALACTOGLUCOMANNANS and related glycans. *See*: PLANT CELL WALL MACROMOLECULES.

hemidesmosome *See*: CELL JUNCTIONS.

hemizygote A DIPLOID cell or individual in which one of the two copies of a given gene has been lost either by gene deletion or by chromosome loss. Human males (and the heterogametic sexes of other species) are also said to be hemizygous for genes located on the X-chromosome but not present on the Y (and vice versa).

Hensen's node A bulbous accumulation of cells at the anterior (rostral) tip of the PRIMITIVE STREAK, appearing during GASTRULATION in avian and mammalian embryos. It contains cells whose progeny contribute to the HEAD PROCESS, NOTOCHORD, EMBRYONIC ENDODERM, and the medial portion of the SOMITES. Its superficial cells also contribute to the FLOORPLATE of the NEURAL TUBE. Hensen's node comprises cells of all three embryonic GERM LAYERS. It is considered to be the amniote equivalent of the amphibian dorsal lip of the BLASTOPORE because if it is transplanted to another region of a host embryo it will induce a supernumerary embryonic axis or supernumerary neural tube (*see* AVIAN DEVELOPMENT; INDUCTION). It is named after Viktor Hensen, who first described this structure in 1876. *See also*: AVIAN DEVELOPMENT; MAMMALIAN DEVELOPMENT.

Hepadnaviridae A family of enveloped DNA animal viruses with an unusual genomic structure and a mode of replication that involves an RNA intermediate. Hepadnaviruses have been isolated from humans (HEPATITIS B), woodchucks, ground squirrels, and ducks. Although the viral genome is circular, there is a single gap in both DNA strands so each strand is linear. One strand is full length and the gap is bridged by an incomplete shorter strand. Protein is covalently bound to the 5′ end of the full-length strand. After uncoating, a full-length (+)-strand RNA is transcribed from the viral DNA. This 'pregenome' is then packaged into a core-like particle with a virus-encoded REVERSE TRANSCRIPTASE and a presumed primer protein. Within the core, a (−)-strand DNA is synthesized by reverse transcription of the RNA and the pregenome degraded. The incomplete (+)-DNA strand is synthesized, and mature virions are surrounded by a membrane. *See also*: ANIMAL VIRUS DISEASE.

heparan sulphate *See*: EXTRACELLULAR MATRIX MOLECULES; GLYCOSAMINOGLYCANS.

heparin A GLYCOSAMINOGLYCAN of M_r 6000–25 000 produced by lung, liver, skin, and mast cells, and sometimes linked to protein to form a PROTEOGLYCAN. The disaccharide repeat of D-glucuronic acid or L-iduronic acid and *N*-acetyl-D-glucosamine, has two or three sulphates per repeat unit. Heparin acts as an anticoagulant by stimulating the formation of a complex between ANTITHROMBIN III and thrombin (and other serine proteinases in the clotting cascade) thus inactivating it. Heparin also binds to FIBRONECTIN. *See*: BLOOD COAGULATION AND ITS DISORDERS; EXTRACELLULAR MATRIX MOLECULES.

hepatitis B virus (HBV) A member of the HEPADNAVIRIDAE, the hepatitis B virus particle (the Dane particle) is spherical, about 42 nm in diameter, and consists of an enveloped icosahedral nucleocapsid (the core particle) containing a circular DNA (M_r 1.6×10^6) which is partly double-stranded and partly single-stranded. The envelope protein (hepatitis B surface antigen, HBSAg) is also found as small particles in the plasma of infected patients and is the antigen used in vaccines against hepatitis B. The surface of the virion carries particles composed of at least seven subunits. Chronic infection with hepatitis B is associated with the development of hepatocellular carcinoma in later life. Hepatitis B DNA is able to integrate into host cell DNA. The integration of a proviral genome into the host cell genome is thought to be responsible for its oncogenic potential, possibly as a consequence of the activation of transcription of an adjacent host gene by the virus. HBV encodes a transcriptional activator (HBx) which has been shown to be involved in the development of liver cancer in mice. *See:* ANIMAL VIRUSES; ANIMAL VIRUS DISEASE.

hepatitis delta (δ) virus (HDV) A defective RNA virus forming particles of ~36 nm diameter and which requires the presence of hepatitis B or other HEPADNAVIRIDAE to replicate. When present with HEPATITIS B VIRUS HDV appears to lead to a greater incidence of severe chronic active hepatitis and liver cirrhosis. The genome is a circular single-stranded RNA of 1678 nucleotides. *See:* ANIMAL VIRUSES.

hepatitis viruses The various viruses causing the symptoms of hepatitis belong to different families. Hepatitis A (infectious hepatitis) is a member of the PICORNAVIRIDAE, hepatitis B (serum hepatitis) belongs to the HEPADNAVIRIDAE, and non-A, non-B hepatitis is thought to be caused by members of the CALICIVIRIDAE (water-borne hepatitis), or TOGAVIRIDAE (post-transfusion hepatitis).

hepatocyte Liver epithelial cell.

hepatocyte growth factor (scatter factor, HGF, SF) A secreted polypeptide consisting, in its active form, of two chains, A and B. The A chain contains four KRINGLE domains with homology to the kringle domains of plasminogen and plasminogen activators, and is disulphide-linked to the shorter B chain. It is secreted as an inactive, uncleaved precursor of M_r ~90 000. Urokinase-like proteolytic enzymes then cleave it into the active two-chain form, of M_r ~60 000. Its receptor is the transmembrane TYROSINE KINASE proto-oncogene product c-Met, which requires the first two kringles for activation. HGF/SF is expressed in the liver of patients with fulminant hepatic failure and is secreted by various fibroblastic cell lines (e.g. MRC-5, human lung fibroblasts). When the protein is added to certain epithelial cells, it causes them to scatter (hence its alternative name) and to adopt fibroblast-like morphology. It also has angiogenic activity. It causes tube-like morphogenesis of kidney epithelial cells and has activity as a neural inducer in the chick embryo.

heptamer The palindromic heptamer recognition signal for the RECOMBINASE which mediates GENE REARRANGEMENT of IMMUNO-GLOBULIN GENES and T CELL RECEPTOR GENES lies 12 bp (one helix turn) or 23 bp (two helix turns) from the NONAMER on either side of a V-D, V-J or D-J gene segment. The heptamer is always adjacent to the coding region. Fusion of the heptamer sequences during gene rearrangement makes the SIGNAL JOINT. The heptamer and nonamer sequences are important in determining the specificity of the joining process.

herbicide-binding (or receptor) protein Also called 32 kDa herbicide-binding protein. Protein of M_r 32 000, a constituent of the PHOTOSYSTEM II (PSII) reaction centre. It is the product of the *psbA* gene. The protein contains a binding site for a molecule of PLASTOQUINONE (Q_B) which can thereby undergo photoreduction in two successive one-electron transfers. Herbicides acting on PSII bind at or close to this site. Analogous to the L subunit of the bacterial reaction centre. *See also:* PHOTOSYNTHESIS.

32 kDa herbicide-binding protein (D1) *See:* HERBICIDE-BINDING PROTEIN.

hereditary angio-oedema Genetic disease caused by a deficiency of C1-INHIBITOR.

hereditary disease (genetic disease) A heritable pathological condition which is caused by the presence of a mutant ALLELE (the disease gene). *See individual diseases.*

hereditary persistence of foetal haemoglobin (HPFH) Benign condition characterized by the continued production of foetal γ-globin into adult life; these γ-globin chains combine with α-globin chains to form foetal haemoglobin (HbF, $\alpha_2\gamma_2$) (*see* GLOBIN GENES; HAEMOGLOBIN AND ITS DISORDERS). The γ-globin genes are normally switched off late in gestation and in the adult HbF accounts for only 1% of total haemoglobin. HETEROZYGOTES for the trait express HbF as 20–30% of total haemoglobin; in HOMOZYGOTES it constitutes 100%. If HPFH is present with a β-thalassaemia determinant (*see* THALASSAEMIA), the production of γ-globin chains can compensate for the absent β-globin chains. Most cases of HPFH can be attributed to mutations at the β-globin locus. Deletional forms of HPFH are associated with deletions within the gene cluster which remove various parts of the genes and control elements but leave one or both of the γ-globin genes intact. In nondeletional cases of HPFH the β- and δ-globin genes are intact and the overproduction of γ-globin is caused by mutations within the gene PROMOTERS. In other cases, the trait is not linked to the β-globin locus, suggesting the existence of a (presumably) regulatory locus elsewhere in the genome. The identification of this HPFH determinant may potentially lead to the development of useful therapies for β-thalassaemia.

hereditary spherocytosis A genetic abnormality of the red cell membrane which results in an abnormal spherical shape and leads to chronic haemolysis. Both autosomal dominant and recessive types are known; associated defects of the red cell membrane proteins ANKYRIN and SPECTRIN have been described. *See also* GLUCOSE-6-PHOSPHATE DEHYDROGENASE DEFICIENCY; HAEMOLYTIC ANAEMIA.

heritability Statistical measure of the degree to which a phenotypic trait (including a disease) is genetically determined. Many traits have both genetic and environmental components.

Herpesviridae A family of animal DNA viruses (*see* ANIMAL VIRUSES). They include the type 1 and 2 herpesviruses, the varicella-zoster virus, which causes chickenpox and shingles, the EPSTEIN–BARR VIRUS and cytomegalovirus. Type 1 herpesvirus causes cold sores, type 2 is associated with genital lesions, and types 1 and 2 have been implicated in human cancers. The mature virion is 100–180 μm in diameter and enclosed in a double membrane. The viral capsid is ICOSAHEDRAL and composed of 162 capsomers. Genome size varies from 120–200 kb, encoding around 50 genes. Genome organization varies between different members of the family. In the herpesviruses, the genome consists of two parts separated by inverted repeats. Recombination between the repeats at the ends of the genome and the internal inverted repeats gives rise to different forms of the genome in which one or both of the parts have become inverted. Herpesviruses multiply inside the nucleus of the host cell. Three classes of genes (α, β, and γ) have been identified which are expressed very early, early, and late (after DNA replication) during infection.

hetero-oligomeric Applied to proteins composed of different types of subunit. Applied for example to members of the LIGAND-GATED ION CHANNEL SUPERFAMILY. *See*: EXCITATORY AMINO ACID RECEPTORS; GABA AND GLYCINE RECEPTORS; NICOTINIC RECEPTOR. *Cf.* HOMO-OLIGOMERIC.

heterochromatin *See*: CHROMATIN.

heterochronic mutations Mutations that alter the timing of developmental events. The nematode (*Caenorhabditis elegans*) is particularly useful for studies of mutation which cause heterochrony because the lineages of all the cells in the adult are known and are more or less invariant among normal individuals. Heterochronic mutations in *C. elegans* cause accelerated or retarded development of particular cell lineages or whole larvae. For example, the mutation *lin-14* causes cells to differentiate early, skipping the divisions that normally occur in the first larval stage. The mutation *lin-4* delays differentiation by causing lineages to reiterate the divisions which give rise to the first larval stage. It has been proposed that these kinds of mutations may also be important in an evolutionary perspective for establishing differences in behaviour and anatomy between closely related species. *See also*: CAENORHABDITIS DEVELOPMENT.

heterochrony A term used in evolutionary genetics to describe the change in onset of a particular developmental process so that the growth rate of a particular feature is changed in a descendant relative to an ancestor.

heteroclitic antibody An ANTIBODY induced to one antigenic determinant that binds with higher affinity to a different, but usually structurally related, determinant.

heterocyst Specialized nitrogen-fixing cells produced by CYANO-BACTERIA of Sections IV and V of the classification of Rippka et al. Some other genera can fix nitrogen without forming heterocysts. Heterocysts are characterized by the synthesis of a thick envelope, the induction of nitrogenase synthesis and the loss of photosynthetic oxygen evolution. Differentiation is associated with a genetic rearrangement, which activates nitrogenase synthesis.

Rippka, R. et al. (1979) *J. Gen. Microbiol.* **111**, 1–61.

heterodimer Protein made up of two different polypeptide chains, associated either covalently or noncovalently.

Heterodontus fransci Horned shark. *See*: GENERATION OF DIVERSITY.

heteroduplex A hybrid duplex nucleic acid (usually DNA) formed by the annealing of two partially or totally complementary polynucleotide chains derived from different parental molecules. Heteroduplex DNA is formed naturally at the point of RECOMBINATION between homologous chromosomes.

heteroimmune *See*: LAMBDA.

heterokaryon A single cell, syncytium, or multinucleate fungal mycelium which contains HAPLOID nuclei of two or more different GENOTYPES within the same common cytoplasm. Heterokaryons are formed when the haploid hyphal cells of certain filamentous fungi fuse to produce a single cell with an $n + n$ PLOIDY as opposed to the diploid $2n$ generated by KARYOGAMY. Heterokaryons can break down to produce haploid cells. The heterokaryon test can be used to confirm the CYTOPLASMIC INHERITANCE of a genetic trait (e.g. genetic traits determined by mitochondrial genes).

heterologous Derived from a different source, applied for example to: ANTIBODIES raised against proteins of another species; a GENE introduced into a cell of a different species or type from that of its original source.

heterologous gene expression systems A system (*in vivo* or *in vitro*) enabling the expression of a gene in a species or cell type different from that from which it was originally isolated. *See e.g.*: EXPRESSION VECTORS; GENETIC ENGINEERING; IN VITRO TRANSLATION; PLANT GENETIC ENGINEERING; TRANSGENIC TECHNOLOGY; XENOPUS OOCYTE EXPRESSION SYSTEM.

heterophilic binding Binding of one molecule to another different molecule. Applied to the binding of adhesion molecules. *Cf.* HOMOPHILIC binding.

heterosis The phenomenon of greater FITNESS and survival of hybrid organisms, particularly those formed by crossing highly inbred lines. Heterosis is usually a reflection of an increased HETEROZYGOSITY, and the increased vigour is thought to be due to the masking of RECESSIVE ALLELES deleterious in the HOMOZYGOUS state.

heterotypic Like to unlike. Applied to binding of adhesion molecules.

heterozygosity A measure of the amount of GENETIC VARIATION at a given LOCUS within a population. The measure is the frequency of HETEROZYGOTES at the locus.

heterozygote, heterozygous A cell or organism with a DIPLOID genotype in which the two ALLELES of a given LOCUS are different. *Cf.* HOMOZYGOTE. *See*: MENDELIAN INHERITANCE.

heterozygote advantage The advantage of the heterozygote over either of the homozygotes, seen for example in the case of SICKLE CELL TRAIT.

Hex *See*: DNA REPAIR.

hexad Sixfold axis of SYMMETRY.

hexosaminidase β-*N*-Acetylhexosaminidase (EC 3.2.1.52). An enzyme that hydrolyses terminal nonreducing *N*-acetyl-D-hexosamine residues in *N*-acetyl-β-D-hexosaminides.

Hfr Strains of *Escherichia coli* first identified by L. Cavalli-Sforza which show a *H*igh *f*requency of *r*ecombination. These strains have a copy of the F PLASMID integrated at random within their genome. During bacterial CONJUGATION with an F⁻ strain the entire chromosome of the Hfr strain is mobilized by the integrated F plasmid and transferred to the F⁻ cell. The mechanism of transfer is believed to involve ROLLING CIRCLE REPLICATION of the F plasmid-containing chromosome. Hfr strains have proven useful in the construction of the GENETIC MAP of *E. coli*.

HGF HEPATOCYTE GROWTH FACTOR.

HGPRT (hypoxanthine : guanine phosphoribosyltransferase) Enzyme (EC 2.4.2.8) important in the salvage of free PURINES derived from nucleotides for reuse in nucleotide and nucleic acid biosynthesis. It converts the free purine to the corresponding purine nucleoside 5′-phosphate. 90% of free purines are recycled by such a pathway in humans. The gene for this enzyme maps to the long arm of the human X-chromosome. *See*: LESCH–NYHAN SYNDROME for the genetic disorder associated with the absence of this enzyme.

high-density lipoprotein (HDL) *See*: PLASMA LIPOPROTEINS AND THEIR RECEPTORS.

high-mannose oligosaccharide Type of N-linked oligosaccharide found on GLYCOPROTEINS. The biosynthetic precursor of hybrid and COMPLEX OLIGOSACCHARIDES, produced by trimming of the common $Glc_3Man_9GlcNAc_2$ precursor of N-linked oligosaccharides catalysed by α-glucosidase and α-mannosidase. High-mannose oligosaccharides typically range in size from $Man_5GlcNAc_2$ to $Man_9GlcNAc_2$. However, some lower eukaryotes transfer smaller precursors to their glycoproteins, and have mature high-mannose oligosaccharides with unusual structures,

some of which contain glucose. Yeasts make large high-mannose structures known as MANNANS. *See also*: GLYCANS.

high-mobility group proteins *See*: HMG PROTEINS.

high performance/pressure liquid chromagraphy (HPLC) *See*: CHROMATOGRAPHY.

highly repetitive DNA A distinct component of the genome of many higher eukaryotic cells which consists of large numbers of short DNA sequences extensively repeated as a TANDEM ARRAY. The repeated elements are identical or near identical in DNA sequence within a given array, and so are also known as simple-sequence DNA. This component can be identified by examining the kinetics of DNA reassociation as such simple-sequence DNA reassociates very rapidly (*see* COT ANALYSIS; HYBRIDIZATION). In addition, if the base composition of the repeat element is different from the overall base composition of the genome in which it is found, this fraction will form a discrete band upon BUOYANT DENSITY GRADIENT CENTRIFUGATION in caesium chloride. This additional fraction usually represents no more than 5% of the genome and is referred to as satellite DNA. The function of highly repetitive DNA remains obscure but *in situ* hybridization studies suggest that it is mainly found as HETEROCHROMATIN located around the CENTROMERES of mitotic chromosomes. *See*: CHROMOSOME STRUCTURE; GENOME ORGANIZATION; MINISATELLITE; VARIABLE NUMBER TANDEM REPEATS.

Hill coefficient The exponent n_H in the Hill or logistic equation used for example in fitting data for the binding of a ligand or for a measured response thereto:

$$[B] = \frac{B_{max}[L]^{n_H}}{[L]^{n_H} + K_d^{n_H}}$$

where [B] is the concentration of ligand bound to the receptor when [L] is the concentration of free ligand. (For other terms *see* SCATCHARD PLOT.) Binding data obtained in the competition of an unlabelled ligand for receptor sites occupied by a RADIOLIGAND can also be analysed using the Hill equation.

The plot of $\log[B]/(B_{max}-[B])$ against $\log[L]$ will have, by rearrangement of the Hill equation, a positive slope of n_H: this is the Hill plot. In fact, it will depart from that slope at high and low extremes of [L]. In practice, a direct fit is now generally made by iterative computer fitting of the binding data to the equation, to give a calculated value of n_H. If there is only one set of uniform binding sites present, with no COOPERATIVITY, then n_H is 1. Values greater than 1 denote positive cooperativity between multiple binding sites on one receptor complex.

When a response is being measured instead of B (e.g. a current flow *I* due to receptor/channel activation by the binding of an agonist A), then the Hill plot is of $\log I/(I_{max} - I)$ against $\log[A]$, and n_H tends towards (but does not attain) the number of binding sites per molecule that must be occupied together for the channel to open. The latter number is given, in fact, by the slope of the simpler log–log plot, that is the plot of the log *I* against log [A] at low values (only) of [A]. These plots have been used for native

NMDA receptors which fitted a model with two independent binding sites for NMDA, and where n_H was 1.4 and the log–log plot slope was 1.93. Likewise, n_H approaches two molecules per receptor for agonist activation of the NICOTINIC RECEPTORS and also of the GABA AND GLYCINE RECEPTORS in the native state; however, when only the binding of the agonist is measured, n_H is 1 in each case.

A value of n_H significantly below 1 may denote either negative cooperativity within a single population of binding sites or that two or more classes of binding site are present, differing in their K_d values.

Patneau, D.K. & Meyer, M.L. (1990) *J. Neurosci.* **10**, 2385–2399.

Hill plot, Hill slope *See:* HILL COEFFICIENT.

hindbrain Posterior region of the brain, also called the rhombencephalon. It develops caudal to the forebrain and midbrain. In the 19th century attempts were made to find homologies between it and the segmentation of the head MESODERM. Although the hindbrain is clearly segemented (into RHOMBOMERES), neither its floorplate nor the neural tube immediately caudal to it are. Recent studies on the seven complex rhombomeres of the higher vertebrate hindbrain have focused on their relationship to development of the cranial nerves, segment-specific patterns of HOX GENE expression (*see* HOMEOBOX GENES AND HOMEODOMAIN PROTEINS; VERTEBRATE NEURAL DEVELOPMENT), and the possibility that each hindbrain segment border represents a lineage restriction boundary. The cranial neural crest derived from the hindbrain is thought to be responsible for patterning of the branchial arches, and as such, of much of the mammalian head. The region is therefore of considerable importance in terms of both developmental biology and the evolution of the vertebrate head. *See also:* MAMMALIAN DEVELOPMENT.

hinge region Region of the immunoglobulin heavy chain located between the CH_1 and CH_2 domains (*see* ANTIBODY; IMMUNOGLOBULIN STRUCTURE). It is open in structure, rendering the antibody molecule susceptible to enzymatic cleavage within this region. The hinge region is thought to confer segmental flexibility on the antibody molecule, allowing the antigen-binding region to move relative to the Fc region. This flexibility is important in allowing the two antigen-binding sites to engage antigen and for the Fc region to express its effector activities through interaction with ligands and receptors. The length of the hinge region has been proposed to be directly correlated with hinge region flexibility. Hinge regions are present in immunoglobulin γ, δ, and α chains but not in immunoglobulin μ or ε chains. Most hinge regions are encoded by a single short EXON between the CH_1 and CH_2 exons. The hinge region of the human δ chain is encoded by two exons and that of the human $γ_3$ chain by four exons. In human and mouse α chains the hinge region is encoded by the start of the CH_2 exon.

hippocampus Phylogenetically ancient brain structure which is involved in the formation and consolidation of memories, and in the generation of emotions. *See:* LONG-TERM POTENTIATION.

histamine A basic amine (2-(4-imidazolyl)-ethylamine) produced by decarboxylation of the AMINO ACID histidine. The highest levels are found in the lungs, skin, and gastrointestinal tract. At the cellular level it is found mainly in mast cells and basophils, but also occurs in the brain in histaminergic neurons. Histamine has a wide range of effects, which include capillary dilatation, contraction of smooth muscle, increased gastric acid secretion, and acceleration of heart rate. Histamine released from mast cells on stimulation by specific antibody of IgE bound to Fc receptors is responsible for many of the symptoms of immediate hypersensitivity reactions (allergic reactions).

histidinaemia An inborn error of amino-acid metabolism inherited as an AUTOSOMAL RECESSIVE trait. The defect lies in the enzyme histidine ammonialyase and the condition is characterized by elevated levels of histidine and histidine metabolites in body fluids such as urine. The clinical picture is variable and the original description of mental handicap and speech defect has not been confirmed in prospectively studied cases.

histidine (His, H) One of the polar AMINO ACIDS.

histidine protein kinase A group of PROTEIN KINASES found in bacteria which phosphorylate proteins at histidine residues. They are involved in various two-component signalling systems in bacteria. *See:* BACTERIAL CHEMOTAXIS; NITROGEN FIXATION; REGULATION OF GENE EXPRESSION; TI PLASMID.

histochemistry The study of tissue structure and arrangement using specific stains.

histocompatibility In tissue transplantation, the genetic identity between donor and recipient tissues which is required for acceptance of donor grafts. Genes controlling graft rejection are encoded within the MAJOR HISTOCOMPATIBILITY COMPLEX (MHC). Although MHC genes were first discovered through their influence on tissue graft survival, this is a secondary function of genes whose primary function is the general regulation of immune responsiveness. A number of non-MHC-encoded minor histocompatibility loci exist in the mouse whose influence appears to be limited to graft rejection.

histone(s) A set of small basic proteins which form the core structure of NUCLEOSOMES (*see* CHROMATIN), by binding directly to negatively charged chromosomal DNA. They are present in all eukaryotic cells. Histones are classified on the basis of the relative proportions of the basic amino acids lysine and arginine (Table H2). This group of proteins represents some of the most highly conserved proteins known. Two molecules each of the histones H2A, H2B, H3, and H4 form an octamer which constitutes the core of the nucleosome. In addition to their structural role, the core histones may also play an important part in regulation of gene expression. A single molecule of histone H1 is located external to the core histone octamer and can be removed without altering the structure of the nucleosome. Histone H1 appears to be absent from yeast and can be found in a variety of forms within a given species. In avian red blood cells H1 is replaced by another

Table H2 Histone classification

Histone type	% Basic amino acid		Approx. M_r
	Lys	Arg	
H1	29	1	23 000
H2A	11	9	14 000
H2B	16	6	13 800
H3	10	13	15 300
H4	11	14	11 300

type of histone designated H5. The biological functions of H1 and H5 are unknown. The histone genes are among the few known eukaryotic genes to lack introns and in most species are clustered into multiple repeating units (*see* EUKARYOTIC GENE STRUCTURE) *See also*: PROTEIN–NUCLEIC ACID INTERACTIONS.

Kornberg, R.D. & Klug, A. (1981) *Sci. Am.* **244**, 52–79.

histone-like proteins (HLP) Small basic DNA-binding proteins which are involved in packaging bacterial chromosomal DNA. *See e.g.* HU protein in PROTEIN–NUCLEIC ACID INTERACTIONS.

HIV Human immunodeficiency virus (*see* IMMUNODEFICIENCY VIRUSES).

HIV receptor The human immunodeficiency virus (HIV) (*see* IMMUNODEFICIENCY VIRUSES) selectively infects T CELLS that express the surface molecule CD4. The virus binds directly to CD4 as ANTIBODIES to CD4 can block binding and subsequent infection. There are some cell lines in which the CD4 molecule cannot be detected that are susceptible to infection. Because anti-CD4 antibodies can still block this infection, it seems likely that the CD4 molecule is present albeit at a very low level. Many human cell lines become susceptible to HIV infection when CD4 is transfected into them (*see* TRANSFECTION). However, the virus does not penetrate mouse cell lines after such transfection even though it becomes bound to the cell surface. This indicates that CD4 is not the only factor that determines susceptibility to infection.

HLA *Human Leukocyte Antigen*, the designation for the human MAJOR HISTOCOMPATIBLITY COMPLEX (MHC), encompassing ~2.5 centimorgans of DNA on the short arm of chromosome 6. The HLA complex includes loci encoding human class I and II MHC molecules and various proteins encoded by region III, and is the human counterpart of the mouse H-2 complex. The combination of molecular alleles at the clustered HLA loci defines the tissue type of an individual.

HLP *See*: HISTONE-LIKE PROTEINS.

HMG proteins High mobility group proteins. A class of proteins distinct from HISTONES which are found in CHROMATIN and represent a subclass of the NON-HISTONE PROTEINS. The biological role of the HMG proteins is unknown, but is presumed to include gene regulation and maintenance of chromosome structure.

HnRNA *Heterogeneous nuclear RNA*, a term used to described the primary RNA transcripts generated by RNA polymerase II which are found in the nucleus (*see* TRANSCRIPTION). It is, as the name implies, a heterogeneous collection of RNA molecules which are on average some four to five times larger than mature mRNAs and in some cases more unstable. HnRNA largely represents unprocessed mRNAs still containing INTRON sequences, but only about 25% of these molecules are actually processed to give mature mRNA which is transported out of the nucleus. The remaining 75% of the HnRNA is completely broken down within the nucleus. Some HnRNAs are also found associated with a number of different proteins forming stable ribonucleoprotein (HnRNP) complexes. *See*: RNA; RNA SPLICING.

Hogness box The conserved DNA sequence element TATA which is found in virtually all PROMOTERS transcribed by RNA polymerase II in eukaryotes and in the majority of functional bacterial promoters.

Holliday structure A DNA intermediate formed during RECOMBINATION whose existence was first proposed by Robin Holliday in the mid-1960s. The structure is formed by the reciprocal invasion of a broken single-strand of DNA from one chromosome homologue to the DNA helix of the other homologous chromosomal partner.

Holliday, R. (1964) *Genet. Res.* **5**, 282–304.

holoblastic cleavage Complete CLEAVAGE of the type seen in eggs whose YOLK is minimal and/or evenly distributed (e.g. sea urchins and mammals).

holoenzyme A complete functional enzyme, containing all the structural domains, cofactors, and subunits required for full function.

homeobox, homeodomain *See*: HOMEOBOX GENES AND HOMEODOMAIN PROTEINS.

Homeobox genes and homeodomain proteins

THE homeobox is a semiconserved DNA sequence of 180 nucleotides found within the coding region of many eukaryotic genes, which constitute a MULTIGENE FAMILY of homeobox genes or homeogenes. The sequence encodes a protein motif of 60 amino acids — the homeodomain, which forms part of a homeodomain protein or homeoprotein. The terms derive from the original identification of the homeobox as a region of sequence similarity between three genes of the fruitfly *Drosophila melanogaster* — *Antennapedia*, *Ultrabithorax* and *fushi-tarazu* — the first two of which are HOMEOTIC GENES controlling position-specific differentiation of body regions. Evolutionary conservation of the homeodomain sequence is not absolute, and definition is usually based on 12 highly conserved residues, a distinctive predicted secondary structure, plus overall sequence similarity to other homeodomains [1–4]. The sequence variation creates difficulties in defining the limits of the gene family: currently included are several hundred genes from animals, fungi and plants, some of which share less than 25% identity over the homeodomain. However, classes or subfamilies of homeobox/homeodomain can be recognized which share much higher sequence similarity: for example *Antennapedia* (*Antp*)-type (also referred to as Class I or A-type), *engrailed* (*en*)-type (also called Class II), *paired* (*prd*)-type (the paired domain), *muscle segment homeobox* (*msh*)-like, and POU homeoboxes/homeodomains [1,3]. In the POU class, the homeodomain and an adjacent POU-specific domain together comprise a semiconserved region of around 150 amino acids — the POU domain (named after its original discovery in the mammalian *Pit-1*, *Oct-1* and *Oct-2* genes and the nematode *unc-86* gene) [3].

Biochemical, immunocytochemical, and genetic analyses suggest that homeodomain proteins are TRANSCRIPTION FACTORS. This function involves site-specific DNA sequence recognition and binding, mediated by the homeodomain (or POU domain). Secondary structure prediction, nuclear magnetic resonance (NMR) and X-ray crystallography studies reveal that the homeodomain contains three α-helices, the second and third of which form a helix-turn-helix motif resembling that of prokaryotic REPRESSOR proteins (*see* PROTEIN–NUCLEIC ACID INTERACTIONS). However, the mode of interaction with DNA differs from the latter, with only residues of homeodomain helix 3 making specific contacts with bases in the major groove of the DNA, and residues N-terminal to helix 1 making contacts in the minor groove [5] (Fig. H12). Consistent with this model, several regions of the homeodomain contribute to the specificity and/or affinity of DNA binding [6]. Interestingly, a few genes have been described which encode proteins with two different DNA-binding domains: for example, a homeodomain plus a zinc-finger motif, or a homeodomain plus a 'paired' domain.

The mechanisms by which homeodomain proteins regulate transcription, following DNA binding, are poorly understood. Protein–protein interactions have been implicated and, at least for the POU class, have been shown to modulate both transcriptional effects and DNA-binding properties [3].

For most homeodomain proteins, the target genes they are presumed to recognize and regulate have not been identified. Some homeobox genes regulate their own transcription and that of other regulatory genes; candidate target genes for a *Drosophila* homeotic gene, *Ultrabithorax*, have also been recently identified by an elegant biochemical strategy [7]. Target gene identities are clearer for some of the mammalian POU homeodomain proteins identified by biochemical purification of known transcription factors. For example, the Pit-1 protein is a transcriptional activator regulating expression of the prolactin and growth hormone genes, whereas Oct-2 activates the transcription of immunoglobulin genes [3].

Genetic studies in flies and nematodes suggest that homeobox genes are involved in the control of a vast diversity of cellular and developmental processes, including spatial patterning, cell fate determination and cell differentiation [1] (*see* CAENORHABDITIS DEVELOPMENT; DROSOPHILA DEVELOPMENT). They include the *Drosophila* homeotic SELECTOR GENES of the Antennapedia Complex and Bithorax Complex which contain *Antennapedia*-type homeoboxes, the maternally transcribed anterior determinant *bicoid*, as well as several genes controlling the formation and maintenance of segments (*even skipped*, *paired*, *fushi tarazu*, *engrailed*, *gooseberry*).

Many *Drosophila* homeobox genes have been evolutionarily conserved, such that structural homologues can be identified in vertebrates. These include the mammalian *En-1* and *En-2* genes related to *engrailed*, and the genes of the four *Hox* clusters related to the homeotic genes.

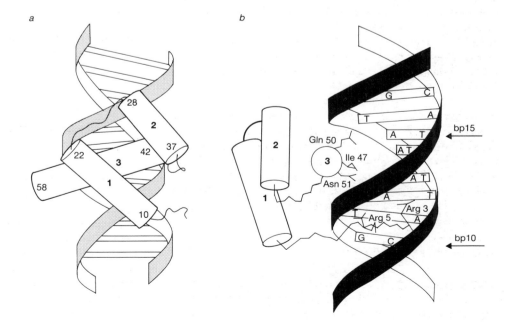

Fig. H12 Two views of the three-dimensional structure of a homeodomain–DNA complex determined by X-ray crystallography. Cylinders show the positions of α helices. Small numbers are amino-acid positions. From [5].

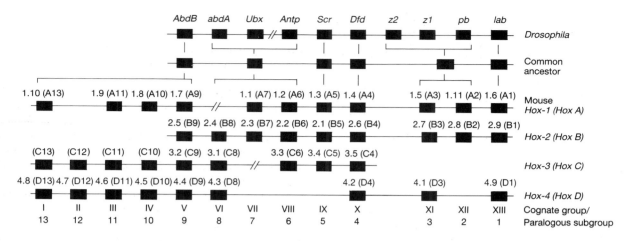

Fig. H13 Genomic organization of mammalian *Hox* cluster genes and *Drosophila* homeotic genes. Closely related *Hox* genes are aligned vertically to form 13 cognate groups; the possible evolutionary relationships to the *Drosophila* homeotic genes of the ANT-C and BX-C are also shown. The proposed revised names are given in parentheses; for example, *Hox-2.1* is *HoxB5*. From [9].

The mammalian *Hox* genes

The *Hox-1*, *-2*, *-3* and *-4* (= *HoxA*, *B*, *C* and *D*) clusters each contain at least nine tightly linked homeobox genes with the same transcriptional orientation, giving a total of at least 38 mammalian *Hox* genes [2,8,9]. Sequence comparisons allow the genes to be classified into 13 'cognate groups' or 'paralogous subgroups', each containing at most one gene from each cluster, presumably reflecting evolution by serial duplication of an ancestral *Hox* gene cluster (Fig. H13). The organization of each *Hox* cluster is also similar to that of the *Drosophila* homeotic gene complexes, implying common ancestry; however, a major difference is the presence of several additional homeobox genes, related to *Drosophila* *Abd-B*, at the 5′ end of the *Hox-1*, *-3* and *-4* (= *HoxA*, *C* and *D*) clusters.

Original nomenclature of mammalian *Hox* genes was complex: each gene was ascribed two numbers, for example: *Hox-2.1*. The first number denotes the gene cluster; the second reflects order of discovery and does not necessarily reflect chromosomal order or the relationship to genes from other clusters. A revised nomenclature has been proposed whereby the clusters are referred to as *HoxA*, *B*, *C* and *D*, and each gene is ascribed a subsequent number denoting the paralogous subgroup (1 to 13 going from 3′ to 5′).

Many mammalian homeobox genes were isolated by virtue of sequence conservation and not by genetic analyses; hence gene expression patterns detected by *in situ* nucleic acid HYBRIDIZATION have often been used to infer putative functions. Best characterized are the *Hox* cluster genes, which are each expressed within precisely delineated, overlapping, anteroposterior domains of the embryonic mesoderm, the neural crest derivatives and the developing central nervous system [8,9]. Adjacent *Hox* genes within a cluster respect different anterior boundaries to their expression; the spatial order of these boundaries generally being colinear with chromosomal gene order (more 3′ genes having more anterior limits to expression). For example, within the developing mouse central nervous system (*see* VERTEBRATE NEURAL DEVELOPMENT), *HoxB2* expression has an anterior boundary between rhombomeres 2 and 3, *HoxB3* between rhombomeres 4 and 5, and *HoxB4* between rhombomeres 6 and 7. The precise correlation between *Hox* gene expression boundaries and rhombomere boundaries suggests that the developmental fate of hindbrain segments is controlled by precise differential gene activity. A relationship has also been reported between *HoxB* gene order and responsiveness to RETINOIC ACID *in vitro*, and between *HoxD* gene order and the level and timing of expression during vertebrate limb and genital development [8,9]. Interestingly, *Drosophila* homeotic genes also respect colinearity between chromosomal position and region of activity, suggesting that this feature has been conserved since the evolutionary divergence of the lineages leading to vertebrates and insects [8,9].

The expression patterns, and the similarities to *Drosophila* homeotic genes in expression and organization, have led to the hypothesis that *Hox* genes are expressed in response to positional and temporal signals within vertebrate embryos, and their products regulate position-specific morphogenesis, spatial patterning and differentiation. Experimental evidence in favour of this hypothesis has come from injection of ANTIBODIES to homeodomain proteins into amphibian embryos and production of transgenic mice aberrantly expressing *Hox* genes [7]. Homologous recombination in embryonic stem (ES) cells has been used to engineer mice with mutations in specific homeobox genes: their phenotypes have confirmed important developmental roles for *En* and *Hox* genes, and suggested there may be functional overlap between related homeobox genes [2,8,9].

P.W.H. HOLLAND

See also: PROTEIN STRUCTURE; PROTEIN–NUCLEIC ACID INTERACTIONS.

1 Scott, M.P. et al. (1989) The structure and function of the homeodomain. *Biochim. Biophys. Acta* **989**, 25–48.
2 Wright, C.V.E. (1991) Vertebrate homeobox genes. *Curr. Opinion Cell Biol.* **3**, 976–982.
3 Rosenfeld, M.G. (1991) POU-domain transcription factors: pou-er-ful developmental regulators. *Genes Dev.* **5**, 897–907.

4 Vollbrecht, E. et al. (1991) The developmental gene *Knotted-1* is a member of a maize homeobox gene family. *Nature* **350**, 241–243.

5 Kissinger, C.R. et al. (1990) Crystal structure of an engrailed homeodomain–DNA complex at 2.8Å resolution: a framework for understanding homeodomain–DNA interactions. *Cell* **63**, 579–590.

6 Damante, G. & DiLauro, R. (1991) Several regions of Antennapedia and thyroid transcription factor I homeodomains contribute to DNA binding specificity. *Proc. Natl. Acad. Sci. USA* **88**, 5388–5392.

7 Gould, A.P. et al. (1990) Targets of homeotic gene control in *Drosophila*. *Nature* **348**, 308–312.

8 McGinnis, W. & Krumlauf, R. (1992) Homeobox genes and axial patterning. *Cell* **68**, 283–302.

9 Holland, P.W.H. (1992) Homeobox genes and vertebrate evolution. *BioEssays* **14**, 267–273.

homeogenetic induction INDUCTION of a cell or tissue by a previously induced cell or tissue. It is usually used only to describe induction of the same cell type as the inducing tissue. A commonly cited example is the induction of NEURAL PLATE from recently induced neural plate: if a piece of amphibian or avian ECTODERM that has been exposed to the influence of inducing mesoderm is placed adjacent to a piece of ectoderm that has not, the induced property (ability to form a neural plate) is transferred to the host ectoderm.

homeotic genes Genes in which mutation results in the transformation of body parts into structures normally found elsewhere. For example, in the *Drosophila* mutant *Antennapedia*, the antenna is transformed into a leg. In *Drosophila*, many of the characterized homeotic genes lie in two clusters, the ANTENNAPEDIA COMPLEX and the BITHORAX COMPLEX. These genes are responsible for the specification of segmental identities in the developing fly. The genes are expressed in overlapping domains along the antero-posterior axis and their products are homeodomain proteins which act as transcriptional regulators. Mutations in homologous genes (Hox genes) in the mouse have recently been shown to cause homeotic transformations. Homeotic mutations are also known in plants. *See*: DROSOPHILA DEVELOPMENT; FLOWER DEVELOPMENT; HOMEOBOX GENES AND HOMEODOMAIN PROTEINS.

homing receptor Leukocyte homing receptor. Cell-surface protein on lymphocytes and neutrophils, that mediates adhesion to endothelial cells. *See*: SELECTINS.

homo-oligomeric Composed of identical subunits. Applied for example to neurotransmitter receptor complexes produced from the expression in *Xenopus* oocytes of the cDNA for a single subunit. *See*: EXCITATORY AMINO ACID RECEPTORS; GABA AND GLYCINE RECEPTORS; LIGAND-GATED ION CHANNEL SUPERFAMILY; NICOTINIC RECEPTOR. *Cf*. HETERO-OLIGOMERIC.

homocystinuria An AUTOSOMAL RECESSIVE inborn error of amino-acid metabolism. There is a deficiency of cystathionine β-synthase and homocysteine accumulates proximal to the metabolic block and is oxidized to homocystine. The clinical picture is of mental handicap in 50%, seizures in 10%, downward dislocation of the lens, multiple venous and arterial thromboembolic lesions, and a Marfan-like syndrome (*see* MARFAN SYNDROME).

homoimmune *See*: LAMBDA.

homokaryon A multinucleate fungal mycelium or other SYNCYTIUM containing nuclei of only one GENOTYPE.

homologous, homology Two sequences are homologous if they share a common ancestor at some point in the past. Homology is usually detected by estimating the probability of finding, by chance, two sequences with a given level of sequence identity (*see* ALIGNMENT). Because sequences continue to diverge with time not all homology can be found this way. With more distant relationships homology can often be inferred at the protein level by similarity of three-dimensional structure or similarity of the active site and mechanism of action.

Homology arises in two ways: when a lineage splits into two species, and when a gene is duplicated within a genome. In the first case the genes are said to be orthologous and in the second paralogous. This can be illustrated by comparing α-globin and β-globin from humans and horses (Fig. H14). All vertebrate GLOBINS are homologous in that their genes are derived from a single globin gene in early vertebrates. This gene became duplicated (*see* GENE DUPLICATION), one copy leading to α-globin and one to β-globin. These two molecules thus share a common ancestral sequence.

α-Globins from human and horse are orthologous; they have a common ancestor and a tree from these sequences would estimate the time of divergence of humans and horses. In contrast, α-and β-globins are paralogous; they also have a common ancestor but a tree which included, say, human α-globin and horse β-globin would estimate the time of the gene duplication, not the time of separation of the human and horse lineages. Trees reconstructed from orthologous genes are said to be species phylogenies (*see* MOLECULAR PHYLOGENY). Trees reconstructed from paralogous genes are gene phylogenies and indicate the history of a gene family (*see* SUPERFAMILY). Examples occur where only parts of two sequences are homologous or when two or more parts of the same molecule may be homologous (*see* MOLECULAR EVOLUTION; MOLECULAR EVOLUTION: SEQUENCES AND STRUCTURES).

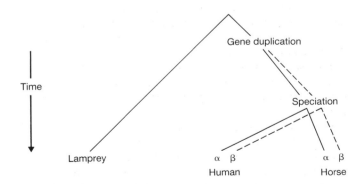

Fig. H14 A tree illustrating two forms of homology — orthology and paralogy. Any pair of α-globin sequences or any pair of β-globin sequences are orthologous because they result from a speciation event and so reflect a species phylogeny. An α- and a β-globin sequence are paralogous, reflecting a gene duplication, and their comparison thus shows a gene phylogeny.

homologous chromosomes The maternally and paternally derived chromosomes present in a diploid cell that bear equivalent genetic information, are similar in morphology, and pair during MEIOSIS.

homologous recombination Genetic RECOMBINATION that occurs between DNAs with long stretches of HOMOLOGY (e.g. between homologous chromosomes at MEIOSIS) and which is mediated by enzymes that show no particular sequence specificity (*cf.* site-specific recombination). It is also known as general recombination. Homologous recombination can be used to target introduced DNA to particular regions of the chromosome, thus disrupting selected genes. *See*: GENE KNOCKOUT.

homophilic binding (homotypic binding) Like-to-like binding. The concept has arisen from work on the NCAM CELL ADHESION MOLECULE, which binds to other NCAM molecules on other cells. This was originally shown by studies of liposomes containing the purified protein. Since then a number of proteins have been shown to exhibit this unusual type of interaction. In mechanistic terms it is likely that there are at least two sites involved in the binding, each of which could bind to its complementary site on the other molecule. Binding is therefore analogous to the lock-and-key mechanism (or heterophilic ligand–receptor binding) except that the two complementary sites are carried on the same molecule. The details are not well worked out in most systems, but in NCAM the IMMUNOGLOBULIN DOMAINS are clearly important.

homopolymer tailing A general method for joining DNA molecules *in vitro*. Complementary homopolymer sequences (e.g. oligo(dA), oligo(dT)) are added onto the respective 3′ termini of each DNA molecule to be joined. Such homopolymeric extensions are catalysed by TERMINAL DEOXYNUCLEOTIDYLTRANSFERASE.

Deng, G. & Wu, R. (1981) *Nucl. Acids Res.* **9**, 4173–4188.

homotypic Like to like. Applied to binding of cell-surface adhesion molecules.

homozygote A DIPLOID cell or organism in which the two ALLELES at a given locus are identical. *Cf.* HETEROZYGOTE. *See*: MENDELIAN INHERITANCE.

homozygous *See*: HOMOZYGOTE.

Hoogsteen base pairing A nonstandard form of BASE PAIRING.

hordeivirus group From Latin *hordeum*, barley, after the type member barley stripe mosaic virus. MULTICOMPONENT VIRUSES with rigid rod-shaped particles 100–150 nm long and 20 nm in diameter. The infective genome comprises three species of (+)-strand linear RNA (Fig. H15). The product of RNA 1 (α) has homologies to RNA polymerases. The 5′ gene of RNA 2 (β) encodes the coat protein; the functions of the other three genes are unknown. RNA 3 (γ) varies in size according to virus strain (e.g. type strain, 3164 nucleotides; ND18 strain, 2791 nucleo-

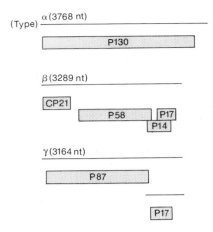

Fig. H15 Genome organization of barley stripe mosaic virus (BSMV). The lines represent the RNA species and the boxes the proteins (P) with M_r given as $\times 10^{-3}$; nt, nucleotides.

tides) and is bicistronic, the 5′ protein (87K or 74K depending on strain) having homologies to RNA replicases and the 3′ protein (unknown function) being translated from a subgenomic RNA 4 which is sometimes encapsidated. *See also*: PLANT VIRUSES.

Atabekov, J.G. & Dolja, V.V. (1986) In *The Plant Viruses* (van Regenmortel, M.H.V. & Fraenkel-Conrat, H., Eds) Vol. 2, 397 (Plenum, New York).

Hordeum vulgare Barley.

horizontal transmission Transmission of viruses from one individual to another (*cf.* VERTICAL TRANSMISSION). Most horizontal transmissions occur by the respiratory route or the faecal–oral route, as viruses cannot penetrate the skin unaided and most easily enter the body through the naked cell surfaces available for gaseous exchange (lungs) or absorption of food (gut). Both respiratory viruses and some that cause systemic infections (measles: Paramyxoviridae; smallpox: Poxviridae) are spread by the respiratory route. Sexual transmission is important for a small number of viruses (HIV: Retroviridae; herpes simplex viruses types 1 and 2: Herpesviridae) when either free virus or infected cells are transmitted. Viruses are only transmitted through the skin when there is some 'mechanical' means of penetration. Mosquitoes transmit yellow fever virus (Flaviviridae) to primates; fleas and mosquitoes transmit myxoma virus (Poxviridae) to rabbits in the UK and Australia respectively. Virus-contaminated hypodermic needles, used medically or by intravenous drug abusers, or needles used for acupuncture or tattooing can also spread virus. *See also*: ANIMAL VIRUSES; ANIMAL VIRUS DISEASE.

hormone(s) Substances produced and released by one tissue which have an effect on another tissue. In humans they include the pituitary hormones (e.g. ADRENOCORTICORTICOTROPIN, FOLLICLE-STIMULATING HORMONE, GROWTH HORMONE, LUTEINIZING HORMONE, PROLACTIN, THYROID-STIMULATING HORMONE), and hormones produced by endocrine glands (e.g. THYROXINE, ADRENALINE, the STEROID HORMONES, INSULIN, and GLUCAGON).

hormone conjugates A term used for compounds in which plant hormones are metabolically coupled to low molecular weight compounds, most commonly sugars, but also inositols and amino acids. Glucosyl conjugates in the form of ethers, esters, or in the case of the cytokinins, *N*-glucosides, are common for most plant hormone classes. Although conjugate formation is usually thought of as an irreversible deactivating process, some conjugates may be cleaved to the free hormone and their formation is considered to be involved in regulating hormone concentration. In this case they may function as storage forms or be important for long-distance hormone transport (*see* PLANT HORMONES).

Sembdner, G. et al. (1980) In *Hormonal Regulation of Development I. Molecular Aspects of Plant Hormones. Encyclopedia of Plant Physiology. New Series Volume 9.* (MacMillan, J., Ed.) (Springer, Berlin).

hormone receptors *See*: ADRENERGIC RECEPTORS; G PROTEIN-COUPLED RECEPTORS; INSULIN AND ITS RECEPTOR; STEROID RECEPTOR SUPERFAMILY.

horseradish peroxidase (HRP) Haemoprotein enzyme (EC 1.11.1.7) catalysing oxidation reactions involving the reduction of peroxide to water. The brown reaction product resulting from the HRP-catalysed oxidation of the substrate DAB (3,3′-diamino-benzidine) is used extensively as a marker stain in light and electron microscopy (i.e. with the peroxidase conjugated to a specific antibody, *see* IMMUNOELECTRON MICROSCOPY, or to streptavidin–biotin). It is also used as a fluid-phase marker for the study of ENDOCYTOSIS in cells, to trace the fate of embryonic cells, and as a marker for tracing neuronal connections. In the latter case, HRP applied to a nerve ending is taken up by ENDOCYTOSIS and transported up the axon (retrograde transport) to the distant cell body and dendrites of the NEURON.

hot spot A site in DNA or chromosome that is subject to a high frequency of RECOMBINATION or MUTATION.

housekeeping genes GENES which encode polypeptides or RNAs whose function is required by all cell types in a multicellular organism, for example cytoskeletal proteins such as actin and tubulin, RNA polymerases and ribosomal proteins. These genes are usually constitutively expressed and may number several thousands within a single cell type in higher eukaryotes.

Hox genes/Hox loci *See*: HOMEOBOX GENES AND HOMEODOMAIN PROTEINS; LIMB DEVELOPMENT; VERTEBRATE NEURAL DEVELOPMENT.

HPLC High pressure liquid chromatography. *See*: CHROMATOGRAPHY.

HPRT Hypoxanthine: guanine phosphoribosyltransferase. *See*: HGPRT.

HPV Human papilloma viruses. *See*: PAPOVAVIRUSES.

HR HYPERSENSITIVE RESPONSE. *See also*: PLANT PATHOLOGY.

HRGP Hydroxyproline-rich glycoprotein. *See*: PLANT CELL WALL MACROMOLECULES.

hrp Genetic locus involved in virulence determination in plant pathogenic pseudomonads. Bacteria containing a defective *hrp* locus are unable to incite disease in susceptible hosts or the hypersensitive response in non-host plants. *See*: PLANT PATHOLOGY.

Hsc70 Heat shock cognate protein 70, a constitutively expressed member of the Hsp70 family of proteins. *See*: HEAT SHOCK.

HSE HEAT SHOCK element.

HSP, Hsp Abbreviations for HEAT SHOCK proteins.

Hsp60, Hsp70, Hsp90 *See*: HEAT SHOCK; MOLECULAR CHAPERONES.

HSTF Heat shock transcription factor. *See*: HEAT SHOCK.

5HT 5-HYDROXYTRYPTAMINE.

5HT receptors Receptors for 5-hydroxytryptamine (5HT, serotonin) have been subdivided on the basis of pharmacological differences into $5HT_1$, $5HT_2$ and $5HT_3$; the $5HT_1$ group, originally defined on the basis of high affinity agonist binding is now further divided into several subtypes, labelled a to d. A $5HT_4$ receptor, linked positively to the adenylate cyclase SECOND MESSENGER PATHWAY, has also been identified, as have subtypes of the $5HT_3$ receptor. $5HT_1c$ and $5HT_2$ are linked to phosphoinositide metabolism, other $5HT_1$ sites are negatively linked to adenylate cyclase. The ion-channel linked $5HT_3$ receptor is the site of action for the anti-emetic drug ondansetron; this and antagonists at some of the other 5HT receptors are under investigation as potential anxiolytic or ANTIDEPRESSANT DRUGS. *See*: G PROTEIN-COUPLED RECEPTORS.

HTLV I, HTLV II Human T cell leukaemia viruses I and II. *See*: RETROVIRUSES.

HTLV III (LAV) Human T-cell lymphotropic virus type III, the human IMMUNODEFICIENCY VIRUS now known as HIV-1.

HU Protein complexed with DNA in prokaryotes. *See*: PROTEIN-NUCLEIC ACID INTERACTIONS.

human development The details of development of mammals vary between species, particularly as regards the EXTRAEMBRYONIC MEMBRANES, but the essential features of tissue and organ differentiation are the same (*see* MAMMALIAN DEVELOPMENT). The amniotic cavity of the human embryo develops by cavitation within the inner cell mass shortly after the blastocyst has begun implantation, and expands to obliterate almost completely the extraembryonic cavity. Unlike the mouse, the human embryonic axis arises within a flat plate of primitive ectoderm, and not within an egg cylinder. The molecular basis for unique features of human development, such as the development of particular brain regions, is not yet understood.

Human gene mapping

HUMAN gene mapping is the process by which the position of each LOCUS on the chromosome is determined. Classically, the loci making up the genetic map indicated the positions of GENES, which may be roughly defined as recognizable functional units. However, in modern terminology the term locus includes not only genes but also any defined DNA segments of unknown (or no) function which have been assigned to definite chromosomal positions, and human genetic maps contain many loci defined only in this way.

Within each chromosome the genes are linearly arranged in one long string of DNA which is supercoiled to fit into the length of a chromosome. The positions of loci may thus be defined in several ways. One of these is to assign the locus to a CHROMOSOME BAND recognized by standard cytogenetic techniques. The resolution of such banding varies but provides between 400 and 1000 bands in the human GENOME depending on the techniques used. A second definition of position would be in distance in base pairs from the end of the chromosome. Because of difficulties in defining the exact end of the chromosome this is not practicable so far in most cases. There are, however, several techniques which, within a short region of perhaps several million bases (megabases, Mb), allow an estimation of the order of the genes and their physical distances apart, usually expressed in kilobases (1000 bases, kb).

A quite different way of expressing map position is that of the genetic (or meiotic) map, where the distances between loci are determined by the amount of RECOMBINATION between them that occurs at MEIOSIS. The unit of recombination most frequently used is the centimorgan (cM), which is defined as that distance in which there is a 1% chance of observing a crossover on a single chromosome strand (chromatid). Direct observation of chiasmata (regarded as equivalent to crossovers) visible in human testicular biopsies has led to an estimate of the total genetic length of the AUTOSOMES as 2650 cM. Although direct observation of chiasmata in females is technically more difficult, it has been found from family studies that recombination is generally higher in female meiosis. The sex-averaged autosomal map length has been estimated as 3300 cM. The total amount of DNA in a haploid genome is estimated as 3×10^9 base pairs (bp). Hence a very rough approximation is that 1 centimorgan is equivalent to 1 million base pairs of physical distance but the relation of genetic distance to physical distance varies in different parts of the genome.

As well as the chromosomal DNA, an egg contains many copies in the cytoplasm of a small circular piece of DNA, about 16 000 bp in length, found in the mitochondria. This tiny 'chromosome', the mtDNA, contains 37 genes which have already been completely mapped, and which are inherited entirely from the mother. Some rare maternally inherited diseases, mostly causing muscle problems, are the result of mutations in mtDNA

(see MITOCHONDRIOPATHIES). The complete nucleotide sequence of human mtDNA is known. Its investigation is not usually considered to be part of human gene mapping and it is covered elsewhere in this volume (see MITOCHONDRIAL GENOMES).

Approaches to human gene mapping

The charting of the position of all the genes on the 22 autosomes and the two sex chromosomes can be approached either by physical methods or by family studies [1–3]. Within these two approaches different techniques are available depending on the scale of map required. Different degrees of resolution of map are obtained by the different methods, each of which will be considered in some detail. Although family studies for gene mapping originated long before adequate physical methods were available, today's family studies have been revolutionized by information obtained from physical methods, so these will be considered first.

Physical mapping methods

Somatic cell hybrids

The first problem in mapping is to assign the gene in question to its appropriate chromosome. The most widely used physical method has been the use of panels of human–rodent SOMATIC CELL HYBRIDS. Human–rodent hybrids useful for human gene mapping each contain all the rodent chromosomes together with a subset of human chromosomes, or in some cases single human chromosomes or parts of chromosomes. The first genes assigned to human chromosomes in this way were those which are expressed in cultured cells and whose gene product can be distinguished in human and rodent. By testing a panel of hybrids for the expression of a human gene product, such as an ISO-ENZYME, the presence of this human gene can be correlated with the presence of a particular human chromosome. The human chromosomal content of the hybrids was initially assessed with considerable difficulty by direct KARYOTYPE analysis. By 1978, when at least one gene had been found on each autosome, it was also possible to characterize the hybrids by the presence or absence of known genetic markers.

The presence of a particular genetic marker does not necessarily mean that a whole human chromosome is present, and it was essential to develop accurate methods for assessing the exact human chromosome content of hybrids. The most recent method for doing this is known as CHROMOSOME PAINTING, which in this context means that DNA extracted from a hybrid cell can be labelled with biotin (see BIOTINYLATION) and hybridized to a normal human METAPHASE SPREAD (see HYBRIDIZATION). The conditions can be adjusted so that only those human chromosomes or parts of chromosomes present in the hybrid are illuminated (see Plate 2a). Careful characterization of a panel of hybrids allows the establishment of a permanent resource from which an indefinite amount of material can be obtained.

Since the advent of cloning and DNA technology it is of course not necessary that the gene to be mapped is expressed in the hybrids, only that the gene can be recognized and distinguished from its rodent counterpart. Initially this was usually done by

RESTRICTION ENZYME digestion of DNA from the hybrids followed by Southern blotting and detection of the human gene using a radioactively labelled PROBE (*see* HYBRIDIZATION). However, the most economic way to use cell hybrids for human gene mapping is to identify PRIMERS which can be used in the POLYMERASE CHAIN REACTION (PCR) to amplify specifically the human and not the rodent gene. The primers are used to amplify DNA from a panel of somatic cell hybrids and the presence of the human-specific band correlated with the presence of a particular chromosome. Only a very small amount of DNA (perhaps 100 ng) is required from each hybrid. The limitation of this method is that some sequence must be known from the gene to be mapped. In general, it is best to employ sequence from INTRONS or from the 3′ untranslated region of a gene as these are much less likely than coding regions to be identical in human and rodent.

There are several ways in which somatic cell hybrids can be used to obtain information about regional localization of genes on a chromosome. One is to construct hybrids with known well-characterized parts of human chromosomes, for example using as a human 'parent' a healthy person in whom the chromosomes have been rearranged so that part of one chromosome is translocated into another (*see* SOMATIC CELL HYBRIDS). Another approach is the statistical one of taking a hybrid which contains only one human chromosome, shattering this chromosome with X-rays, and then constructing a large number of new hybrids which contain only very small pieces of the original chromosome. Any pair of genes which are close together are likely to be retained or lost together in these irradiation fragment hybrids; those further apart will be retained or lost independently. This method allows the construction of a linear map of genes similar to that obtained from family studies.

Chromosome sorting

Somatic cell hybrids used for human gene mapping are really just a way of separating the human chromosomes from each other. Recently it has become possible to do this more directly using a fluorescence-activated cell sorter (FACS) (*see* CELL SORTING). First developed to sort different types of cells tagged specifically with fluorescent dyes, this technique has been adapted for sorting chromosomes. A suspension of chromosomes is labelled simultaneously with two fluorescent dyes — a Hoechst dye which is specific for AT-rich chromosomal regions, and a chromomycin dye which binds most effectively to GC-rich regions. By exposing each chromosome in turn to excitation from two different laser beams, one appropriate for each dye, and then passing the chromosome through an electric field whose strength depends on the fluorescence emitted by the chromosome, it is possible to separate most chromosomes uniquely from each other, and collect different chromosomes on different regions of filter paper or into different tubes. The limitations of this technique are not only the complexity of the equipment and the inability to resolve certain groups of chromosomes (for example 10, 11, 12) but also the relatively small number of chromosomes that can be sorted in a day. The development of PCR, which with suitable primers allows a gene to be amplified from only a few target molecules, has greatly increased the potential for flow sorting as a tool in

human gene mapping. The main advantage of this method will probably be in mapping genes in relation to breakpoints in TRANSLOCATIONS by sorting the translocation chromosomes, as sorting is much quicker than making new somatic cell hybrids.

Direct *in situ* hybridization

If the gene which it is hoped to map has been isolated in a COSMID or phage clone or even as a cDNA clone (*see* DNA CLONING) it is possible to visualize its chromosomal position directly by *in situ* hybridization of the cloned DNA to a metaphase spread of chromosomes (*see* HYBRIDIZATION). Originally, the probe DNA was radioactively labelled, for example with tritium (^{3}H), and detected by AUTORADIOGRAPHY. More recently, nonradioactive methods known as FISH (fluorescent *in situ* hybridization) have been developed. These involve the incorporation of a substance such as biotinylated UTP into the DNA in a NICK-TRANSLATION reaction as the probe is being synthesized. The probe DNA is then denatured and allowed to hybridize under appropriate conditions with the denatured chromosomal DNA. The site of hybridization is detected by a fluorescently labelled avidin (FITC-avidin) which binds to the biotin. It may be necessary to increase the signal by one or more rounds of amplification in order to visualize what is effectively one molecule on a chromatid. Plate *2b* shows a cosmid clone encoding the major human gene for phosphoglucomutase (an enzyme in the glycolytic pathway) labelled by FITC and hybridizing to the band 1p22 on the short arm of chromosome 1. Note that all four chromatids are labelled. Using different types of fluorescent label it is now possible to look at the position of several different probes at the same time and order their position with respect to the centromere and to each other. By hybridizing to cells in metaphase, genes ~2–5 Mb apart can be distinguished by this method; if interphase cells are used the chromosomes are much longer and so resolution is finer (claimed to be about 50 kb) but analysis is more difficult.

Family studies

Physical mapping methods are only applicable to those genes which have either been cloned or whose effects can be recognized at the cellular level. For many diseases this is of course not the case and family studies are essential. In fact the ultimate goal of many people who map disease genes is POSITIONAL CLONING, the isolation and characterization of a gene causing inherited disease from its map position rather than through the more traditional approach from the biochemical abnormalities found in the disease.

Linkage analysis

The most important contribution of family studies to human gene mapping has been the study of genetic linkage, that is the tendency of two genes which are close together on the same chromosome to segregate nonrandomly in offspring. The more closely linked two genes are, the smaller is the probability that they will be separated at meiosis. In contrast, if two genes are far apart on the same chromosome, or are on different chromosomes, there is a 50% chance that they will be passed on together and the two genes are said to be unlinked. A family is only useful for a linkage study if a parent in that family is HETEROZYGOUS for both genes being studied.

An example of an informative family is shown in Fig. H16. In this case the investigation is to test whether the gene determining an autosomal dominant disease is linked to another gene (a genetic marker) which is known to exist as two ALLELES (alternative forms) A or B. The father (II.1) has the AUTOSOMAL DOMINANT disorder (indicated by shading) and he also carries the normal allele (not shaded). In this case the information available from the grandparents makes it clear that he has inherited both the disease and the A allele from his father (I.1). Individual II.1 has passed the disease allele on to three of his children and the healthy allele to two of them. Two of the children with the disease have also received the A allele and the two healthy children have received the B allele. Hence in these four children there has been no recombination between the disease gene and the genetic marker. However the fifth child has the disease and the B allele and is clearly a recombinant. In this family then, there is a 20% recombination between the disease and the genetic marker.

One family like this might well occur by chance and in order to be sure that segregation is really nonrandom it is usually necessary to study many families. The best families from a genetic point of view are those with many children and with grandparents available for testing. The latter fact is important because it allows one to deduce the 'phase', that is the way the genes must have been arranged in the parents. Because so many human families are very small, especially if some of their members suffer from severe inherited disease, mathematicians have spent much time working out how to extract the maximum information from a collection of even quite small families. This has led to the concept of a LOD SCORE which refers to the chance of obtaining given results if loci are linked at a certain RECOMBINATION FRACTION

(known as theta or θ) compared with the chance of obtaining these results if the loci are not linked. The term 'lod' means 'log of the odds' and is a logarithm to the base 10. The maximum lod score for the example given above is 0.418 at a theta of 0.2, and means that the result obtained is about 2.6 times more likely if the disease and the marker are genetically linked with 20% recombination than if they are truly unlinked. Although it is of course possible to work lod scores out from first principles or by looking particular combinations of recombinants and nonrecombinants up in tables, there are now several computer programs (for example LIPED and LINKAGE) widely available which accept primary family data and which are almost universally used to generate lod scores. Because each family is independent, the probabilities for each family can be multiplied together, that is the lod scores can be added. A lod score of +3 or more at any recombination fraction implies that the results are 1000 times more likely if the loci are linked at this θ than if they are not linked and is generally taken as significant. This test is in fact not as stringent as it might seem as the prior chance that two autosomal loci are linked is only 1 in 50. In practice about 97% of lod scores which reach 3.0 are subsequently shown to be correct.

The analysis of family data does not stop at pairwise lod scores and there are many computer programs which purport to turn new family data into multipoint genetic maps. For these, and for methods of analysis of other intractable problems such as complex genetic traits which are not determined by a single major gene, the specialist literature should be consulted.

The X-chromosome is by far the best mapped chromosome and this is because an inherited disease which is caused by a defective gene on the X-chromosome is easily recognized by its pattern of inheritance. The disease affects mostly boys, is always inherited through the mother and never passed from father to son. Because of this, many X-linked genes had been recognized before the first direct assignment of a gene to an autosome was made in 1968. Almost at the same time, the DUFFY BLOOD GROUP was assigned to chromosome 1 by its segregation with a visible chromosome abnormality, and the plasma protein HAPTOGLOBIN was assigned to chromosome 16 by a similar strategy. For many years few genetic linkages were found, in spite of considerable effort. However, in the past few years this has changed dramatically. Three factors have been particularly important in this: the wealth of excellent genetic markers now available; the ability to map many genes physically, directly to their respective chromosomes; and the availability of the large reference collection of CEPH families (see below).

New genetic markers

The minimum requirement for the start of a linkage study is that a parent is doubly heterozygous for each gene under consideration, that is that the forms inherited from father and mother are different. In the search for a disease gene the families have probably been collected because they are informative for the disease, but the family is only useful if it is also informative for the genetic marker to be tested. If there is no previous information about the position of the disease gene, about 200 highly informative genetic markers scattered evenly through the genome are

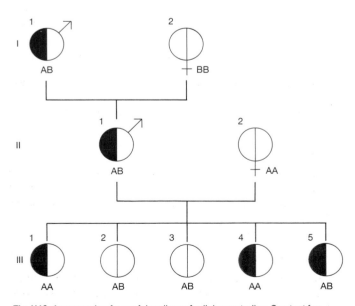

Fig. H16 An example of a useful pedigree for linkage studies. See text for details.

needed to make the search realistic. For many years the only markers available were BLOOD GROUPS, and enzyme and protein polymorphisms (*see* GENETIC POLYMORPHISM). Many of these were unmapped themselves and were in any case only informative in about 10% of families.

The first breakthrough was in 1980 with the discovery of RFLPs (restriction fragment length polymorphisms). Single base changes which alter the susceptibility of a particular stretch of DNA to cutting by a specific restriction enzyme are recognized by enzymatic digestion of DNA followed by Southern blotting. Drawbacks of this technique are that there are usually only two different alleles (the enzyme cuts or it does not) and the procedure takes several days and uses quite a lot of DNA. Two other types of variation have subsequently been found to be very useful in linkage studies. One is the detection of many chromosomal regions where different individuals have different numbers of tandem repeats of a defined short sequence — VARIABLE NUMBER TANDEM REPEATS (VNTRs). The number can vary from one to several hundred, so that almost everybody tested has a different pattern and all are heterozygous. Many of the most informative VNTRs are identified by locus-specific probes which have been derived from the core probes used to generate the complex 'DNA fingerprints' much used in forensic and legal work (*see* DNA TYPING). (The DNA fingerprint itself is the product of very many loci and is not useful for linkage studies.) Although these VNTRs, first recognized in 1985, are useful for linkage they do require Southern blots and there is some evidence that they are not evenly distributed along the chromosome.

The type of variation which seems at present to be closest to the ideal genetic marker was first described in 1990 and is the dinucleotide repeat. It appears that blocks of repeats such as $(CA)_{18}$ are found throughout the genome, perhaps in 10 000 different positions. The number of CA dinucleotides at any one position varies, for example one person may have $(CA)_{17}$ and $(CA)_{15}$ at the same position on this chromosome as his spouse has $(CA)_{13}$ and $(CA)_{16}$. These differences in length can be demonstrated by choosing single-copy primers from either side of the block, amplifying DNA from the people concerned in a PCR, and separating out the fragments by electrophoresis on a high resolution gel to examine their size. Figure H17 shows the inheritance of a CA block close to the arginino-succinate synthetase gene on chromosome 9q34. In this case the parents have four different alleles so that all four copies of chromosome 9q34 segregating in the children are effectively marked. Because such repeats are found quite close to any region of interest it should be possible to make virtually all families totally informative for every part of every chromosome. CA repeats have recently been used to construct a linkage map of the human genome [4].

CEPH families

The small size of human families is a major problem in genetic analysis. For families carrying disease genes there is nothing one can do about it, but is is possible to construct reference genetic maps of markers using normal families and these can be a tremendous help in planning the search for a disease gene. The Centre d'Etude Polymorphisme Humaine (CEPH) in France has

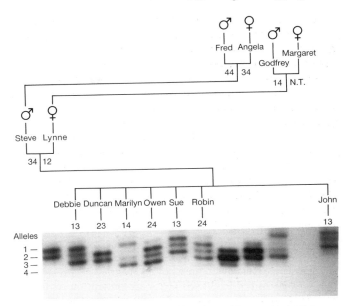

Fig. H17 A highly informative dinucleotide polymorphism used to track a particular region of chromosome 9q34 through a family. In this example Debbie, Duncan, Sue, and John have inherited the arginino-succinate synthetase (ASS) gene from their paternal grandmother, Angela, whereas the corresponding gene in Marilyn, Owen, and Robin came from their grandfather Fred. Even though Margaret was not tested it is possible to deduce that one of her ASS genes was passed through Lynne to Duncan, Owen, and Robin. Bands for Steve are not shown.

provided, to many investigators worldwide, DNA from permanent cell lines grown from 40 normal large families, having an average of eight children and all grandparents alive. As these families have now been tested for more than a thousand markers they constitute an extremely valuable resource for the construction of reference maps. They also have intrinsic interest for human gene mapping, for example in answering questions such as differences in male and female recombination fractions.

Other methods of gene mapping

Although the main approaches to human gene mapping have been outlined above, a variety of other techniques have been used to improve the human gene map or are potentially valuable.

Dosage

Occasionally, individuals with chromosome abnormalities can provide direct gene mapping information. A correlation can be made between the amount of a gene product and extra or missing chromosomal material.

Comparative mapping

Although mouse chromosomes appear very different from human ones there are many highly conserved regions and over short distances there seems to be some conservation of distance as well as gene order. Genetic mapping in mice is of course easier because of the ability to set up large-scale crosses which may involve scoring hundreds of offspring. Information obtained can often give a clue to the position or to the likely homology of a human disease gene. Examples of genes whose identification has been helped by comparative mapping include the skeletal deformity (Greig's syndrome on chromosome 7) and a syndrome involving deafness and white forelock (Waardenberg's on chromosome 2).

Fine detail mapping

In the search for a particular gene it may be necessary to make a very fine map of a small area of a chromosome, for example 1 or 2 Mb. This requires a range of different techniques which are not described in detail here. They include pulsed field gel electrophoresis (*see* ELECTROPHORESIS) by which large fragments of DNA (up to several megabases in size) are separated after the genomic DNA has been treated with 'rare-cutting' restriction enzymes. It is then possible to test for which genes are present on each fragment. Another approach is to isolate all the DNA from a particular region either in cosmids (fragment length ~40 kb) or YEAST ARTIFICIAL CHROMOSOMES (YACs) which may include fragments greater than a megabase in length. The idea is then to order the overlapping clones into a physical map or CONTIG so that all the genes in that region are included. This approach has recently been successful both for chromosome 21 and for the Y chromosome. In both cases a complete map of the region was assembled using sequence tagged sites (STS). These are short sequences of DNA which can be amplified by PCR. One hundred and ninety-eight STS from chromosome 21 were used to screen 70 000 YACs. The final map contains 810 YACs and spans a region of about 50 megabases on the long arm of chromosome 21 [5]. A YAC contig such as this provides a valuable resource for identifying genes known to lie in a particular chromosomal region. It could also be utilized as a starting point for sequencing the whole genome.

Fine detail meiotic mapping is a still more difficult problem. Even the CEPH families do not provide enough meioses to order genes which are separated by about 1 cM. It has been shown that it may be possible to amplify two genes simultaneously from single sperm. If this could be done reliably it would be possible to look at several thousand meioses in the same individual.

The present state of the human gene map

In July 1994 there were nearly 4700 genes and over 50 000 DNA segments whose physical position on a chromosome was known at least approximately. For many chromosomes sophisticated multipoint linkage programs have been used to construct consensus genetic maps in which genetic markers, and in some cases disease loci, are ordered with some estimate of genetic length in centimorgans. Figure H18 shows a summary of the state of knowledge of one particular chromosome, chromosome 9. It is possible that about 5% of human genes have now been identified but this is really guesswork. What is clear is that enormous progress has been made and the speed of advance in knowledge is increasing all the time. It is increasingly difficult to include all

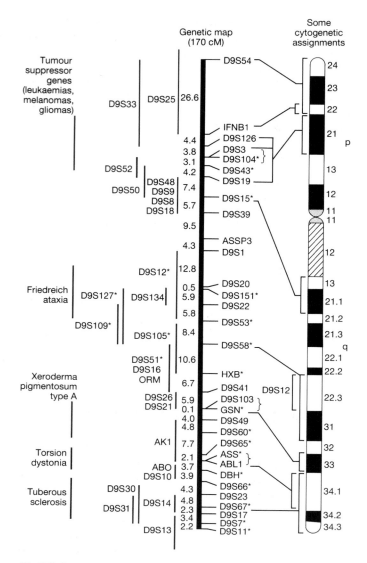

Fig. H18 Genes and genetic markers on chromosome 9 (as of October 1992), which has been estimated to contain 145 Mb of DNA and to be about 170 cM in sex-averaged genetic length. Markers prefixed D9 are DNA fragments of unknown (or no) function. The other abbreviations, e.g. ABO, ASS, IFNB1 refer to gene loci (in this case the loci for the ABO blood group antigens, the enzyme argino-succinate synthetase, and interferon β_1 respectively). Those loci marked * are excellent genetic markers showing a high degree of normal variation in the population. On the far left are marked the positions of the loci for some serious inherited diseases and also the position of some suspected TUMOUR SUPPRESSOR GENES which have not yet been precisely identified. Adapted from [7].

the known data relevant to human gene mapping in a single publication. The *Proceedings* of the 11th International Workshop on Human Gene Mapping, held in London in 1991 [6], occupied more than 2000 pages split into two large volumes. Most people in the field refer to GDB, a genome database hosted at the Johns Hopkins University (*see* DATABASES) which has nodes allowing access by researchers in many countries. Recent developments have also enabled human data to be displayed graphically on software (ACeDB) originally designed for the nematode worm *Caenorhabditis elegans*, whose small genome is much nearer being completely mapped than is that of humans.

It is now possible to envisage a time when the human gene map will be complete and each chromosome will be completely sequenced (*see* HUMAN GENOME PROJECT). As with the already mapped mitochondrial DNA, interest will then change to establishing the significance of differences between individuals.

S. POVEY

See also: CHROMOSOME ABERRATIONS; CONTIGUOUS GENE SYNDROME; CYSTIC FIBROSIS; DUCHENNE MUSCULAR DYSTROPHY; FRAGILE X SYNDROME; SEX DETERMINATION.

1 Weatherall, D.J. (1991) *The New Genetics and Clinical Practice* (Oxford University Press).
2 Davies, K.E. (Ed.) (1988) *Genome Analysis* (IRL Press, Oxford).
3 Davies, K.E. (Ed.) (1986) *Human Genetic Diseases: A Practical Approach* (IRL Press, Oxford).
4 Weissenbach, J. et al. (1992) A second generation linkage map of the human genome. *Nature* **359**, 794–801.
5 Chumakov, I. et al. (1992) Continuum of overlapping clones spanning the entire human chromosome 21q. *Nature* **359**, 380–387.
6 Human Gene Mapping 11. (1991) *Cytogenet. Cell Genet.* **58**, 1–2200.
7 NIH/CEPH Collaborative Mapping Group (1992) A comprehensive genetic linkage map of the human genome. *Science* **258**, 67–86.

The Human Genome Project

OVER the past 20 years the development of powerful new techniques for isolating, copying, manipulating, and analysing fragments of DNA has transformed medical and biological research (*see* DNA CLONING; DNA SEQUENCING; GENETIC ENGINEERING; POLYMERASE CHAIN REACTION; RECOMBINANT DNA TECHNOLOGY; RESTRICTION ENZYME). The causes of hereditary single-gene disorders are now much more easily studied by first mapping and identifying the responsible GENES (*see* HUMAN GENE MAPPING) rather than having to identify the faulty proteins they produce in the diseased state (which are often still unknown). Identification and analysis of the normal and defective forms of the gene can then give some clue to the biochemical defect. The sequence of nucleotides in any piece of DNA can now be read, opening the way for a greater understanding of how genes work. Many biological problems can now be investigated at the level of the one molecule ultimately responsible for controlling the chemistry of life.

Human cells are thought to contain between 50 000 and 100 000 genes. The genes themselves — the functional informational DNA — comprise probably no more than 5% of the total chromosomal DNA in humans and yet account for individuality, control development from conception to grave, and sustain life.

As DNA technology developed, it became possible to envisage a complete description of the human GENOME, including the locations of all the genes, thus creating an invaluable reference for medical and biological science — a 'handbook of life' which would increase enormously the potential for diagnosing and treating human diseases and for understanding biological pro-

cesses. The creation of such a handbook is the goal of the Human Genome Project.

The primary aim of the Project is to locate all the genes within the 24 chromosomes of the HAPLOID human genome and then to sequence the DNA at these locations. The secondary, and very much longer term, aim is to sequence all the chromosomal DNA, including the 95% currently described as 'nonfunctional'. The magnitude of these tasks is difficult to comprehend. Current estimates are 10 years to locate and sequence the functional genes and 15 years or more to sequence all the chromosomal DNA. Success in the latter enterprise will depend on the development of existing and new technologies and on adequate funding being made available to make that development possible.

Gene mapping and identification

The process of identifying a gene and establishing its position on a particular chromosome is demanding (*see* HUMAN GENE MAPPING)[1]. Human genes are being mapped at an exponential rate but the number mapped so far is still under 4000, less than 5% of the estimated total (although many genes fall into related MULTIGENE FAMILIES so those already mapped may represent a much higher percentage of the functional classes into which genes may be divided).

The genetic 'handbook' that will eventually be produced will contain two sections. One will consist of genetic and physical maps of the human genome showing the locations of all the genes; the other will contain the nucleotide sequence along the DNA of each chromosome.

Genetic mapping

Genetic mapping defines the location of a GENE LOCUS relative to another whose position is already known, on the basis of the frequency of RECOMBINATION that is observed between the two loci (*see* GENETIC DISTANCE; HUMAN GENE MAPPING). In the past, genetic linkage maps in humans could only be obtained by studying the inheritance of easily identifiable traits in large families over several generations, and suitable traits and families are rare. This type of investigation can now be supplemented by analysis of the chromosomal DNA of family members.

An important development in genetic mapping has been the exploitation of the POLYMORPHISM present in human DNA, not only within genes, but also in the apparently nonfunctional sequences. Such polymorphism can be detected in various ways, for example as the absence (or introduction) of a restriction site (a site at which a particular RESTRICTION ENZYME cuts), which leads to variability in the lengths of particular restriction fragments in digests from different individuals (*see* RFLP). Other useful sources of polymorphism are the VARIABLE NUMBER TANDEM REPEAT (VNTR) sequences distributed throughout the genome. Any segment of chromosomal DNA which varies in sequence between individuals can therefore be used as a reference (marker) segment in genetic mapping. There are major efforts within the Human Genome Project to identify polymorphic segments of DNA at frequent, evenly spaced intervals along each chromosome. The intention is to create a series of 'milestones' along the chromosomes which can be used as distance markers and as reference points for identifying more precisely the relative positions of genes and the approximate distance between them.

Mapping technology has revolutionized the study of single-gene hereditary disease by using linkage analysis with polymorphic markers to locate and isolate the gene responsible and then to deduce the protein sequence and possible structure from the nucleotide sequence of the gene. This reverses the earlier approach of identifying the gene from knowledge of the biochemical defect.

The basic functional defect in several genetic diseases has been defined in this way. The location and isolation of the genes for CYSTIC FIBROSIS and DUCHENNE MUSCULAR DYSTROPHY, for example, has led to identification of their protein products. This has provided important clues to the biochemical defects in these diseases and this knowledge could lead to new approaches to therapy, including the possibility of treatment at the genetic level (*see* GENE THERAPY). Isolating the gene and its mutant alleles also enables tests to detect carriers and affected foetuses (*see* PRENATAL DIAGNOSIS).

Physical mapping

Physical mapping, which builds on the framework of information provided by genetic mapping, relies on a variety of techniques which analyse DNA directly. Restriction fragments from human genomic DNA LIBRARIES are analysed for the presence of restriction sites, known genes and sequences that overlap with other fragments, and are then pieced together gradually in the correct order and assigned to a particular location on a chromosome. In this way a part of the genome from a defined chromosomal region can be referred to and studied in the laboratory for its informational content.

Deducing the sequence of a long length of DNA from its smaller fragments depends on being able to identify fragments with sequences in common. Computer databases containing information about the nucleotide sequences of the fragments are searched to find two fragments that contain an identical sequence somewhere along their length. Lining these fragments up alongside each other, with the identical sequences adjacent, gives the correct sequence for the longer length of DNA (the CONTIG). By repeatedly applying this procedure — known as contig mapping — fragments can be positioned correctly relative to each other over a long length of DNA.

Sequencing DNA

The Human Genome Project aims first to map the human genome and then to sequence it. The mapping process provides a description of the genome to which the detail of nucleotide sequence can subsequently be added. But the order of these tasks also reflects the current state of technology and financial resources. The two techniques most commonly used today for sequencing DNA (*see* DNA SEQUENCING) were developed over 15 years ago and, despite automation, are still neither fast enough nor cheap enough to meet the needs of the Project.

Using the most advanced automated machine, one person can now sequence about 2000 nucleotides per day. As the human genome contains about 3 billion nucleotides it has been estimated that even if 500 such machines could run for 250 days per year (allowing time for confirming results, maintenance, etc.), it would still take 12 years to sequence the genome at a total cost of between US$3 and 6 billion. A substantial increase in speed and reduction in cost is needed to make large-scale sequencing feasible and new technologies are being explored.

Whose genome?

The map of the human genome produced from the Human Genome Project will be a composite from many sources, mostly long-established laboratory cell lines. The sequence eventually obtained will be generic, containing information derived from many individuals. This is appropriate, as the Project is itself a collective rather than an individual initiative. The laboratory work is being done in an increasing number of countries, mainly through nationally funded programmes of research. At present, well-defined directed programmes of work are being carried out in the United States, the United Kingdom, Italy, Russia, the Netherlands, Japan and France, and by the Commission of the European Community. Other countries, including Israel and Canada, have recently initiated human genome programmes. However, even the most successful national programme cannot hope to meet the objectives of the Human Genome Project on its own.

The Human Genome Organisation (HUGO)

The importance of coordinating genome activities internationally, and of establishing a scientific body to do this, was first acknowledged by those working on genome analysis in April 1988 during a meeting at Cold Spring Harbor in the USA. It was recognized that, although many lines of communication already existed within the international community of human genome mappers and with government agencies and funding bodies, there was no existing organization that was sufficiently international, independent and broadly based to integrate genome activities into a truly global effort. With the support of the genome community, the Human Genome Organisation (HUGO) [2] was formally established as the global coordinating organization in 1989, with a remit to create the channels through which information, initiatives and ideas could flow. There are now HUGO members in more than 30 countries.

HUGO's enabling role is being fulfilled through a growing portfolio of activities designed to foster collaboration between genome scientists in order to: maximize the return from their efforts and avoid unnecessary competition or duplication; assess their needs and coordinate the development of services and facilities to meet those needs; facilitate the exchange of data and biomaterials; encourage the development and spread of relevant technologies; match the needs of genome scientists with the interests of funding agencies; channel genome information to appropriate destinations; and provide a focus for international consideration of the ethical, social, legal and commercial implications of the Human Genome Project.

Global coordination

The first international move towards coordinating the results and activities of genome researchers dates back to 1973 when, in response to the increasing amount of data about the locations of genes that was being generated by the newly developed techniques of SOMATIC CELL GENETICS, a series of international Human Gene Mapping Workshops was initiated. These aimed to summarize and validate the available data on the positions of genes along the human chromosomes. From 1973 onwards, successful mapping workshops were held every 2 years or so, but soon came the recombinant DNA revolution. There was an explosive increase in the number of markers (as well as genes) that could be mapped on the genome as well as a rapid merging of the interests of classical and cell geneticists with those of molecular biologists.

Following the seventh HGM Workshop in 1983 it became clear that a computerized approach to assembling, accessing and reporting on genome data was essential. Direct computer support was provided for the ninth Workshop in Paris in 1987 and a fully computerized, interactive and multiuser database was developed for the first time at Yale University for the tenth Workshop (HGM 10) in 1989.

For the next major stage in human genome mapping, represented by HGM 11, a more sophisticated system was needed that was based on internationally accepted, relational database standards. The aim was to have a widely accessible, on-line database that could be developed continuously and would include, at any given time, validated, up-to-date data on human genes and their map locations. This would provide the essentials needed to apply POSITIONAL CLONING techniques to human disease genes.

HGM 11 took place in London in August 1991 and probably brought together for the first time all the different groups of people involved in the Human Genome Project. The Workshop also provided confirmation that the new Genome Database (GDB) software, which had been developed at Johns Hopkins University in Baltimore, had the potential for handling the types and amounts of genome data that were being generated.

By the end of HGM 11 it had become clear that there should be a move away from a massive entry of data at the biennial Workshops in favour of continuous entry of data into an internationally accepted database, of which GDB would form the core. It was also recognized that the scale of the work required a subdivision into manageable packages — hence the development of Single Chromosome Workshops. HUGO will develop a coordinated programme of Single Chromosome Workshops, with an associated series of annual Chromosome Coordinating Meetings and biennial Human Genome Mapping Workshops over the next few years.

Single Chromosome Workshops are already widely regarded as a key feature of the Human Genome Project. Their aim is to bring together each year those actively working on a particular chromosome. The annual Chromosome Coordinating Meetings will bring together representatives of all the different Single Chromosome Workshops as well as those, such as database and nomenclature experts, whose interests span the whole genome. The intention is to provide a forum for the discussion of problems and policies, and to produce a detailed annual report, chromosome by chromosome, on the progress of the genome map.

Human Genome Mapping Workshops will be held biennially, beginning in 1993. Like the earlier series of HGM Workshops, these will be meetings for assessing and advancing the progress of the genome map overall (and therefore the progress of the Human Genome Project).

As the need arises, HUGO is establishing advisory groups to review present knowledge (for example, in the areas of informatics, intellectual property rights, ethics, or model organisms such as mouse) or to propose international programmes of research in defined areas. One such group is currently considering the question of human diversity and is drawing up proposals for establishing DNA/cell banks to hold materials for future studies from populations at risk of disappearing.

Information handling

As the Human Genome Project develops, the problems of handling the amounts of information that will be generated will increase. Faster computer processors will need to be developed as well as new disk systems that can store a higher density of information. Systems will also need to be developed for managing the data generated by the Project. Current work on mapping and sequencing smaller genomes such as that of *Caenorhabditis elegans* and *Escherichia coli* will provide useful developments.

Other aspects of handling the data from the Project must be considered. Who owns the data? Who should have access to it? How should that access be controlled? At what price? Should the knowledge arising from the Project be exploited commercially? Patent applications covering short sequences of human DNA of no known function were first filed in the United States towards the end of 1991 and have already drawn attention to some of these issues. There are not yet any clear-cut answers but the questions have evoked an overwhelming response from the scientists involved in the Project — whatever else is decided, the knowledge arising from the Project is too significant to belong to any one person, group or country.

Society and the Human Genome Project

The fascination of genes for biological scientists is easy to understand. But the importance of understanding the human genome goes far beyond the interests of research scientists. It impinges on every member of society. Knowing the location and identity of all the human genes increases enormously the possibility of being able to predict the occurrence of thousands of known (and yet to be identified) diseases with a strong genetic component. It opens the door to the earlier detection of many diseases, which should in turn increase the options for prevention and treatment and allow each individual a more positive role in managing their own health. It should lead to the development of therapies that treat the cause, rather than the symptoms, of illness.

How this information is handled within society — whether it is understood, welcomed and used for the common good, or misunderstood, feared and abused — is a matter that concerns everyone. Society must consider who should have access to the information emerging from the Human Genome Project. An individual's right to privacy must be protected while ensuring that information is freely available to those who need it. The decision making within society must be informed, and HUGO is taking steps to initiate discussions of the social, moral, legal and ethical issues arising from the Human Genome Project. The Human Genome Project is already providing unprecedented insights into human disease and development. The question is how best to use this rapidly increasing knowledge, and realize its fullest potential.

E.M. EVANS

1 McKusick, V.A. (1991) Current trends in mapping human genes. *FASEB J.* **5**, 12–20.
2 McKusick, V.A. (1989) The Human Genome Organisation: history, purposes and membership. *Genomics* **5**, 385–387.

human immunodeficiency virus (HIV) *See*: IMMUNODEFICIENCY VIRUSES.

human T cell leukaemia viruses *See*: RETROVIRUSES. For HTLV III *see*: IMMUNODEFICIENCY VIRUSES.

humanized antibodies ANTIBODIES of animal origin that have been modified, using GENETIC ENGINEERING techniques, to replace constant region and/or variable region framework sequences with human sequences, while retaining the original antigen specificity. They are commonly derived from rodent antibodies with specificity for human antigens and are to be used for *in vivo* therapeutic applications. This strategy reduces the host response to foreign antibody and allows selection of the human effector functions that are activated. Initially only the Fc regions of human antibodies were substituted but nowadays all but the COMPLEMENTARITY DETERMINING REGIONS of the rodent antibody can be replaced by human sequences.

Winter, G. & Milstein, C. (1991) *Nature* **349**, 293–299.

humoral immune response/immunity Immune response or immunity due to the production of circulating ANTIBODIES.

hunchback (*hb*) A member of the gap class of SEGMENTATION GENES in *Drosophila*: *hunchback* mutants exhibit loss of gnathal and thoracic segments and fusion of posterior abdominal segments. There is both maternal and zygotic transcription, initiated at separate promoters, with both transcripts encoding the same protein. The maternal transcripts are initially uniformly distributed in the egg but, soon after fertilization, are preferentially translated anteriorly as a result of translational repression by the posterior group gene *nanos*. The zygotic expression is activated by the anterior group gene *bicoid* and gives rise to a gradient of *hunchback* protein which patterns the expression of other gap genes. There is also a posterior band of expression. The *hunchback* protein contains two clusters of zinc-finger DNA-binding motifs (*see* PROTEIN–NUCLEIC ACID INTERACTIONS).

Hunter syndrome (mucopolysaccharidosis type II) X-LINKED sulphatase deficiency disorder with an incidence of around one in 100 000 births leading to a lysosomal storage disease. The deficiency is specifically of iduronate 2-sulphatase (EC 3.1.6.13), the enzyme catalysing the first step in the degradation of heparan sulphate.

Huntington's chorea (Huntington's disease) AUTOSOMAL DOMINANT inherited, late-onset, progressive disease, usually of middle life, characterized by abnormal movements (chorea) and dementia, and leading to death usually within 20 years of the onset of symptoms. The incidence is one in 20 000 births. New mutations are virtually unknown. The disease locus maps to chromosome 4 16.3 (short arm). The gene (*IT15*) was isolated in 1993 and encodes a putative protein of M_r 348 000 unrelated to any known protein. At the 5′ end of normal copies of the 210 000-kb gene are between 11 and 34 copies of a TRINUCLEOTIDE REPEAT CAG. The trinucleotide repeat region is amplified in disease genes, in which from 42 to 100 copies of the repeat are present. *See also*: NEURODEGENERATIVE DISORDERS.

The Huntington's Disease Collaborative Research Group (1993) *Cell* **72**, 971–983.

Hurler's disease (mucopolysaccharidosis type I) Lysosomal storage disease with an incidence of around one in 100 000 births resulting from an AUTOSOMAL RECESSIVE deficiency of α-L-iduronidase (EC 3.2.1.76), the enzyme catalysing the second step in the degradation of heparan sulphate (i.e. the hydrolysis of α-L-iduronosidic linkages in desulphated heparan).

Hy antigen A MINOR HISTOCOMPATIBILITY ANTIGEN. As such it can be the focus of a transplantation reaction. However, in contrast to differences at the MAJOR HISTOCOMPATIBILITY COMPLEX (MHC), minor differences are characterized by delayed rejection kinetics. Hy is encoded by the Y-CHROMOSOME and is therefore a male-specific antigen. The nature of the gene product is unclear, but it is likely to be located intracellularly, as an Hy-incompatible graft does not elicit an antibody response.

hyaline membrane A membrane formed around the egg immediately after FERTILIZATION. It is secreted by the cortical granules which lie under the plasma membrane of the unfertilized egg and release their contents on fertilization (*see* CORTICAL REACTION). The hyaline membrane consists mainly of hyaline protein, which is a large (estimated M_r 3 million) filamentous protein which probably has several ISOFORMS. Its function is thought to be to hold the BLASTOMERES together in the BLASTULA, a hollow ball of cells, all of which are bound to the hyaline membrane at their apical surfaces. At the commencement of GASTRULATION, presumptive mesenchymal cells detach from the hyaline layer and migrate inwards. This is the result, in part, of a loss of affinity of these cells for hyaline protein.

hyaluronectin *See*: EXTRACELLULAR MATRIX MOLECULES.

hyaluronic acid Family of unbranched, unsubstituted linear GLYCANS containing the repeating unit

(-3GlcNAc(p)β1,4GlcA(p)β1-)$_n$. The only variation is in chain length, which is usually $>10^4$ residues. Hyaluronic acid is a product of some bacterial and animal cells. In the latter, it is a major component of many connective tissue matrices, where it interacts with the core protein and 'link protein' of PROTEOGLY-CAN to form large proteoglycan aggregates. In mammals, hyaluronic acid is synthesized at the cell surface by successive transfers from UDP-GlcNAc + UDP-GlcA, without specific priming, as a 'giant SUGAR NUCLEOTIDE' from which it is probably released by 5′-nucleotidase. Two transferases act alternately. Synthesis is by addition at the reducing terminal, unlike other animal glycans (but like many bacterial glycans). Hence, it should not be classified as a GLYCOSAMINOGLYCAN, as formerly. *See also*: EXTRACELLULAR MATRIX MACROMOLECULES.

Prehm, P. (1983) *Biochem. J.* **211**, 191–198.

hyaluronidase Hyaluronidase (EC 3.2.1.35) splits (β1,4)-*N*-acetylglucosaminide links in HYALURONIC ACID and other glycoconjugates. The enzyme is found in the LYSOSOMES of many mammalian tissues but also extracellularly in body fluids. The lysosomal hyaluronidase has a sharp pH optimum in the range pH 3.5–4.1, distinguishing it from testicular hyaluronidase (deriving from the acrosomes of spermatozoa) which has a broad pH optimum extending up to pH 6.0.

hybrid antibodies Artificially constructed ANTIBODY molecules having two, or more, antigen-binding sites of differing specificities. Hybrid antibodies may be produced following reduction and reoxidation of a mixture of antibodies of different specificities, or of their F(ab′)$_2$ fragments. Only a proportion of the oxidized products will be hybrid antibodies but they can be purified by appropriate immunoaffinity techniques. Hybrid antibodies may also be produced following the fusion of two HYBRIDOMA cell lines that produce antibodies of differing specificity. Random association of light and heavy chains results in the formation of several antibody populations. Hybrid antibodies of the required dual specificity can be isolated by immunoaffinity techniques.

hybrid-arrested translation A technique used to identify and isolate the protein product of a cloned gene. The cloned double-stranded DNA is denatured and annealed to a mixture of mRNAs known to contain mRNA encoded by the required gene. mRNA : DNA hybrids are formed, which renders the relevant mRNA unavailable for translation. Translation of the annealed mixture in an IN VITRO TRANSLATION system thus results in the absence of the translation product of the test gene. Comparison with the translation products of an untreated sample can then identify the missing protein.

hybrid dysgenesis *See*: P ELEMENT.

hybrid hybridomas Hybridomas produced by the fusion of two hybridomas that express unique properties dependent on the combined characteristics of the parent hybridomas. They may be produced following T–T cell or B–B cell fusions.

hybrid nucleic acid A double-stranded NUCLEIC ACID in which the two strands are from different sources.

hybrid promoter A functional transcriptional PROMOTER containing DNA sequence elements from two (or more) well characterized promoters. Such hybrid promoters are constructed *in vitro* and are used to either probe the role of specific *cis*-acting DNA sequences in the control of TRANSCRIPTION, or to optimize the efficiency and/or regulation of the original promoter. An example of the former would be the various mammalian hybrid promoters which contain ENHANCER-like elements from the SV40 promoter. A well studied example of the latter is the *tac* promoter from *Escherichia coli* which contains the – 35 region of the *trp* promoter and the – 10 region of the *lac* promoter, and is widely used to express HETEROLOGOUS genes in *E. coli*.

hybrid-release translation A technique used to identify and isolate the protein product of a cloned gene. It is the converse of HYBRID-ARRESTED TRANSLATION but has the same overall objective. The cloned DNA is denatured and attached to a matrix of, for example, NITROCELLULOSE. A complex mixture of mRNAs is then added and hybrid mRNA : DNA allowed to form. Unannealed mRNAs are then removed by washing and the annealed mRNA released from the hybrid by denaturation. This mRNA is then translated in an IN VITRO TRANSLATION system to identify its protein product.

Hybridization

NUCLEIC ACID hybridization is a powerful and widely used technique which exploits the ability of complementary sequences in single-stranded DNAs or RNAs to pair with each other to form a double helix. Hybridization can take place between two complementary DNA sequences, between a single-stranded DNA and a complementary RNA, or between two RNA sequences. The two polynucleotide strands are held together in an antiparallel configuration by hydrogen bonding between the bases G and C and between A and T (or U) (*see* BASE PAIR). The structure is stabilized by base-stacking interactions (*see* DNA; NUCLEIC ACID STRUCTURE; RNA). The reformation of a double-stranded DNA (or RNA) from its two original single strands is also known as reannealing or renaturation.

Hybridization can occur between two separate strands (intermolecular hybridization), or between inverted repeat sequences within a single strand of nucleic acid (intramolecular hybridization; *see* HAIRPIN; PALINDROME). The helix-to-coil transition that occurs as the double helix is denatured can be monitored by measuring the absorbance at 260 nm, as the absorption of light at this wavelength by single-stranded nucleic acid is greater than by double-stranded nucleic acid (*see* HYPERCHROMIC EFFECT).

Nucleic acid hybridization has been exploited in a wide range of experimental procedures to address many questions about the synthesis, structure and function of DNA and RNA.

Stability

The thermodynamic stability of hybridized sequences (hybrids) is expressed in terms of the temperature at which the strands separate (the melting, or transition temperature, T_m). The T_m (expressed in °C) is dependent on several factors which are divisible on the one hand into those relating to the nucleotide sequences involved in the hybridization, and on the other into the experimental conditions under which the hybridization takes place. The effect of the (G + C) content of DNA on the T_m of a hybrid in a solution containing 0.2 M Na$^+$ is given by the following equation [1]:

$$T_m = 69.3 + 0.41 \text{ (percentage (G + C) content)}$$

The effect of ionic strength (I) upon the T_m is given by the following equation [2]:

$$T_{m2} - T_{m1} = 18.5 \log_{10} (I_2/I_1)$$

Hybridization of similar but not identical sequences

Nucleic acid hybridization may be used to measure the degree of similarity (degree of HOMOLOGY) between two nucleic acid sequences as hybridization can occur between sequences which contain some complementarity, but which are not identical; that is, hybrids can tolerate mismatched base pairs (mismatches) in the double helix. DNA : DNA hybridization of this sort has been used to measure the broad similarity between different genomes for taxonomic and evolutionary studies (*see* MOLECULAR PHYLOGENY).

The degree of sequence homology between the two strands in the hybrid influences the stability of the hybrid. The T_m of duplex (double-stranded) DNA decreases by 1°C for every 1–1.5% of mismatches [3] but this relationship is dependent on the distribution of mismatches thoughout the sequences. The above equations assume that the length of the DNA is not less than about 150 base pairs, and therefore, an empirical guide has been developed to estimate the stability of hybrids formed by short DNA sequences (for example synthetic OLIGONUCLEOTIDES) of between 11 and 23 bases long in 1 M Na$^+$ [4]:

$$T_m \text{ (in °C)} = 2(\text{number of A + T residues}) \\ + 4(\text{number of G + C residues})$$

The stability of hybrids also depends on whether DNA or RNA is involved in the hybridization:

$$T_m \text{ of RNA : RNA} > \text{RNA : DNA} > \text{DNA : DNA}$$

pH has relatively little effect on T_m in the range pH 5–9, but the molarity (M) of Na$^+$ influences hybrid stability significantly, and

hybridization conditions can be manipulated by the addition of formamide which destabilizes hybrids [5]:

$$T_m = 81.5 + 16.6 \log_{10} [M] + 0.41 \text{ (\% content of G + C)} \\ - 500/\text{length} - 0.62 \text{ (\% of formamide)}$$

Hybridization conditions which allow only the hybridization of identical or very similar sequences are known as stringent conditions; these include high temperature, low ionic strength and high formamide concentrations. Those that allow the hybridization of less similar sequences are known as relaxed conditions; they include lower temperatures, ionic strengths and low or zero formamide concentrations.

Solution hybridization

The rate at which hybridization occurs depends on the concentration of the nucleic acid sequences capable of forming hybrids. (An exception to this is intramolecular hybridization by INVERTED REPEAT sequences, the renaturation of which is independent of concentration.) This feature has been exploited in the analysis of DNA and RNA sequence complexity (the number of different sequences present and the relative concentration of these sequences) by solution hybridization where all molecules are freely diffusible.

By separating genomic DNA into single strands and then following the kinetics of its renaturation, several broad classes of sequence have been found in the typical eukaryotic genome: (1) highly repetitive DNA (satellite DNA); (2) moderately repetitive DNA; and (3) unique DNA sequences (*see* GENOME ORGANIZATION). The parameter used to measure the kinetics of renaturation is the $C_0 t$ value:

$C_0 t$ = initial concentration (C_0) in moles multiplied by the time (t) in seconds it takes for the DNA to reanneal (*see* COT ANALYSIS) [6].

Similarly, the analysis of renaturation kinetics for specific RNA populations with synthetic complementary DNA (cDNA) made from the RNA ($R_0 t$ values) has allowed the sequence complexity of messenger RNA populations to be determined (i.e. the proportion of the population that is made up of mRNAs of high or low abundance, and the number of sequences present in each abundance class). Solution hybridization, followed by digestion with S1 nuclease or ribonuclease (RNase) A, which under the appropriate conditions cleave only single-stranded DNA or RNA respectively, has been applied to the mapping of TRANSCRIPTION UNITS within cloned DNA sequences (*see* S1 MAPPING). Transcription units can also be mapped at lower resolution by the analysis of RNA : DNA heteroduplexes (double-stranded nucleic acids containing strands of different origin) by electron microscopy (*see* R-LOOP MAPPING).

Filter hybridization

In filter hybridizations, one of the nucleic acid components of the hybridization is immobilized on a membrane filter (originally nitrocellulose, but more recently, charged nylon membranes have also been used).

Southern hybridization (Southern blotting)

This technique, named after its inventor, Ed Southern, enables very low concentrations of specific sequences to be detected within complex DNA populations and is of wide application. Hybridization of a labelled probe directly to DNA fragments embedded in a gel is insensitive. So, after separation of RESTRICTION FRAGMENTS of DNA by ELECTROPHORESIS on agarose gels, they are transferred and fixed to a membrane filter (blotting), and then hybridized with a radioactively labelled DNA or RNA PROBE containing the required sequence [7] (Fig. H19). Among its many applications are the study of RFLPs (restriction fragment linked polymorphisms) in human disease and its use in DNA TYPING.

Northern blotting

This technique is similar to Southern blotting, and was named by analogy, and is used to analyse RNA sequences. RNA is electrophoresed under denaturing conditions on agarose gels, transferred and fixed to the membrane filter, and then hybridized with a radiolabelled DNA or RNA probe [8]. This procedure has been used to analyse specific messenger RNAs within complex RNA populations, and hence to study the differential expression of specific genes in response to particular physiological or environmental stimuli.

Dot blotting

Accurate and rapid measurement of the abundance of specific RNA or DNA sequences is often performed by dot blotting. Here, samples are dotted directly onto the filters without previous separation by electrophoresis. Filters are then treated as for Southern or northern hybridizations.

In situ hybridization

This is used to detect and locate specific DNA or RNA sequences in tissues or on chromosomes. A radioactively or fluorescently labelled DNA or RNA probe of the required sequence is applied to fixed tissue or chromosome preparations, where it hybridizes with any complementary sequences present. Unhybridized probe is then washed off. In the case of radioactively labelled probes, images of probe distribution are then superimposed upon the microscopic images of the chromosomes or tissue sections to determine the distribution of hybridized radioactive probe, and hence the complementary sequences, within the samples. Fluorescently labelled chromosomes, etc. are viewed directly.

In situ hybridization of chromosomes, especially fluorescent *in situ* hybridization (FISH), is widely used to map particular sequences in the genome (*see* HUMAN GENE MAPPING), or to study CHROMOSOME ABERRATIONS. *In situ* hybridization of tissue sections has been used extensively to study tissue-specific gene expression especially in the study of the developmental regulation of genes involved in pattern formation during *Drosophila* embryogenesis (*see* DROSOPHILA DEVELOPMENT).

Hybridization probes

Radioactively labelled DNA probes are frequently synthesized by

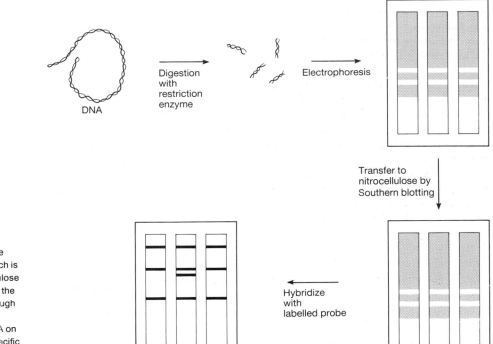

Fig. H19 Southern blotting. DNA fragments are first separated by electrophoresis on a gel which is then placed in contact with a sheet of nitrocellulose (or other suitable membrane) the same size as the gel. An appropriate solution is then drawn through the gel and nitrocellulose, eluting the DNA and trapping it on the nitrocellulose sheet. The DNA on the nitrocellulose sheet is then probed with specific probes. A similar procedure is followed for RNA in northern blotting.

nick-translating, random priming, or end-labelling specific DNA fragments or plasmids (*see* END-LABELLING; NICK TRANSLATION; POLYNUCLEOTIDE KINASE; RANDOM PRIMING). Radioactively labelled RNA probes are often made by *in vitro* transcription of the relevant DNA sequence. The sensitivity of the hybridization techniques described above depends upon the specific activity of the radioactive probe.

A.J.P. BROWN

1 Marmur, J. & Doty, P. (1962) Determination of the base composition of deoxyribonucleic acid from its thermal denaturation temperature. *J. Mol. Biol.* **5**, 109–118.
2 Dove, W.F. & Davidson, N. (1962) Cation effects on the denaturation of DNA. *J. Mol. Biol.* **5**, 444–467.
3 Bonnet, T.I. et al. (1973) Reduction in the rate of DNA reassociation by sequence divergence. *J. Mol. Biol.* **81**, 123–135.
4 Schulze, D.H. et al. (1983) Identification of the cloned gene for the murine transplantation antigen H2Kb by hybridization with synthetic origonucleotides. *Mol. Cell. Biol.* **3**, 750–755.
5 McConaughy, B.L. et al. (1969) Nucleic acid reassociation in formamide. *Biochemistry* **8**, 3289–3295.
6 Adams, R.L.P. et al. (Eds) (1986) *The Biochemistry of the Nucleic Acids*, 10th revised edn (Chapman & Hall, London).
7 Southern, E. (1975) Detection of specific sequences among DNA fragments separated by gel electrophoresis. *J. Mol. Biol.* **98**, 503–517.
8 Thomas, P.S. (1980) Hybridization of denatured RNA and small DNA fragments transferred to nitrocellulose. *Proc. Natl. Acad. Sci. USA* **77**, 5201–5205.

hybridoma Clone of cells produced by the fusion of a somatic cell with a cell line that has the capacity to survive and proliferate *in vitro*. Hybridomas derived from the fusion of an antibody-producing B cell from a specifically immunized animal with a PLASMACYTOMA cell line (Fig. H20) were originally developed for the production of MONOCLONAL ANTIBODY of defined specificity. Hybridomas may be stored frozen in liquid nitrogen and recovered when further product is required. T cell hybridomas are generated in a similar manner.

Milstein, K. (1980) *Sci. Am.* **243**, 66–74.

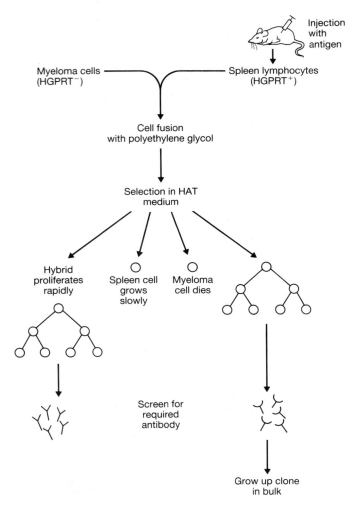

Fig. H20 Production of hybridomas.

***Hydra* development** This small freshwater coelenterate (class Hydrozoa) is used as a simple system in which to study regeneration, and the role of morphogenetic GRADIENTS in development. *Hydra* development involves the twin process of head and foot formation at opposite ends of a tubular body column. As the animal grows, new animals bud off from the side of the adult, in a region called the budding zone. Hydroids have considerable powers of regeneration, and new animals can be generated from fragments as small as 1/50th the size of a normal adult. The head and foot are thought to be specified in development by two pairs of morphogenetic gradients; one gradient of each pair has an activating role, while the other is inhibitory. The 'head gradient' (activating) has its maximal value at the head end of the animal, the 'foot gradient' is maximal at the foot. A piece of donor head tissue will give rise to a new bud when transplanted to the foot of a host animal because it possesses a high amount of 'head activator', whereas transplantation to another head region produces no head induction. The two gradient pairs are used to explain the results of such grafting experiments. To increase the frequency with which a middle piece of body tissue induces a new

head, it is first necessary to remove the head region of the host; head induction occurs in this case because the head inhibitor is no longer so abundant. It is thought to be the existence of the two inhibitory gradients that determines the location of the mid-body budding zone. *Hydra* development can be seen as a paradigm of PATTERN FORMATION through morphogenetic gradients.

hydrazinolysis The cleavage of DNA with hydrazine usually as part of the Maxam and Gilbert DNA SEQUENCING method. In the absence of added salt, hydrazine will cleave at both pyrimidine residues (C, T) whereas in the present of salt, cleavage will only occur at C residues.

hydrogen bond Weak noncovalent bond in which a hydrogen atom covalently bonded to a very electronegative atom interacts with another atom. In biological macromolecules the shared hydrogen is covalently linked to an oxygen, a nitrogen or, occasionally, a sulphur atom (the donor), and the other partner (the acceptor) is also oxygen or nitrogen or, occasionally, sulphur. The hydrogen is attracted to the acceptor by a partial negative charge on the latter and the bond is highly asymmetric. Hydrogen bonding is involved in BASE PAIRING in nucleic acids, the stabilization of protein structure and the binding of substrate to enzymes. Hydrogen bond lengths vary between 2.70 Å and 3.10 Å (Table H3) with bond energies ranging from ~12.5 to 29 J mol^{-1}. The strongest hydrogen bonds are those in which the donor, the hydrogen, and the acceptor lie on a straight line. *See*: BASE PAIR; NUCLEIC ACID STRUCTURE; POLAR SIDE CHAINS; PROTEIN FOLDING; PROTEIN STRUCTURE.

Table H3 Hydrogen bond lengths

Bond	Length (Å)
O—H···O	2.70
O—H···O$^-$	2.63
O—H···N	2.88
N—H···O	3.04
O$^+$—H···O	2.93
N—H···N	3.10

hydrogen ion pump *See*: PROTON PUMP.

hydrolases One of the main classes of ENZYMES (EC 3), which catalyse the hydrolysis of a wide variety of bonds. They include the esterases, phosphatases, phosphodiesterases, sulphatases, nucleases, glycosidases, peptidases, proteinases, amidases, and deiminases, and numerous other enzymes. *See also*: ACID HYDROLASE.

hydropathy plot/profile An analysis of the amino-acid sequence of a protein which is used to predict the position of highly nonpolar transmembrane regions. Successive 'windows' of 20 contiguous residues (the average number of residues making up a transmembrane α-helix) are taken (moving one residue at a time) and the free energy of the transfer of the sequence from the interior of the membrane to water is calculated. A peak of ⩾84 J mol^{-1} based on a 20-residue sequence in the hydropathy plot represents a possible transmembrane region.

hydrops fetalis Condition occurring in the presence of severe HAEMOLYTIC ANAEMIA in the newborn. The infant, if not stillborn, is usually born prematurely and characteristically both infant and placenta are grossly oedematous. The child is very pale and usually dies within the first hour. The commonest causes are Rh blood group incompatibility (*see* RH BLOOD GROUP SYSTEM) and homozygous α$_o$ THALASSAEMIA.

α-hydroxyacetosyringone Hydroxylated derivative of ACETOSYRINGONE with similar inductive capabilities.

hydroxylysine Unusual amino acid (Fig. H21) found in collagen (*see* EXTRACELLULAR MATRIX MOLECULES) and some plant cell wall proteins. It is formed by post-translational hydroxylation of lysine side chains in the ENDOPLASMIC RETICULUM.

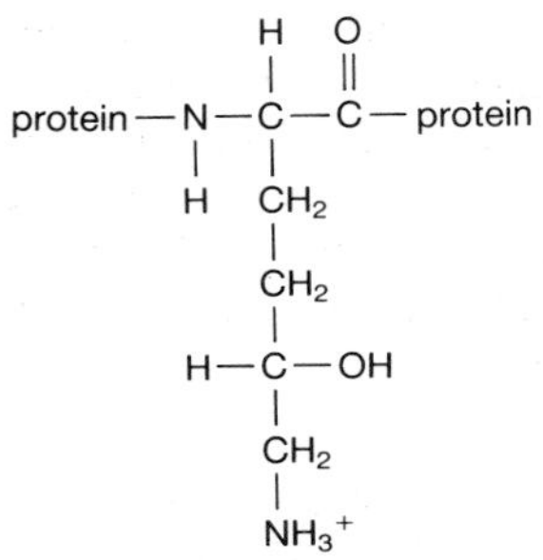

Fig. H21 Hydroxylysine.

hydroxyproline Unusual amino acid (Fig. H22) found in collagen (*see* EXTRACELLULAR MATRIX MOLECULES) and some plant cell wall proteins (*see* EXTENSINS). It is formed by post-translational hydroxylation of proline side chains in the ENDOPLASMIC RETICULUM.

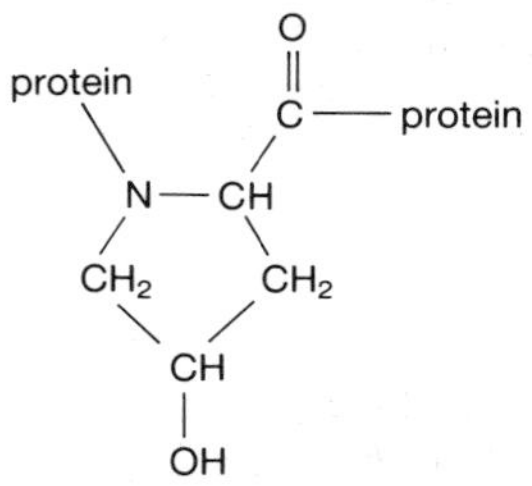

Fig. H22 Hydroxyproline.

5-hydroxytryptamine (5HT, serotonin) A major monoamine neurotransmitter (Fig. H23), synthesized from the AMINO ACID tryptophan by hydroxylation and decarboxylation processes analogous to those involved in CATECHOLAMINE NEUROTRANSMITTER synthesis. Synthesized by cells of neural crest origin (part of the amine precursor uptake and decarboxylation (APUD) system) and in the brain, and also by nonneural cells, 5HT synthesis is normally limited by the concentration of tryptophan; its metabolic removal is via MONOAMINE OXIDASE. In the brain and spinal cord 5HT is contained in neurons originating primarily in the raphe nuclei of the brain stem. These systems are involved in arousal, sleep and consciousness, aggression and control of mood, and thus are implicated in disturbances of these behaviours such as anxiety and depression (*see* AFFECTIVE DISORDERS). 5HT is also a major neurotransmitter in the enteric nervous system, controlling motility and secretion in the gastrointestinal tract (*see* PERIPHERAL NERVOUS SYSTEM). It is also one of the principal substances

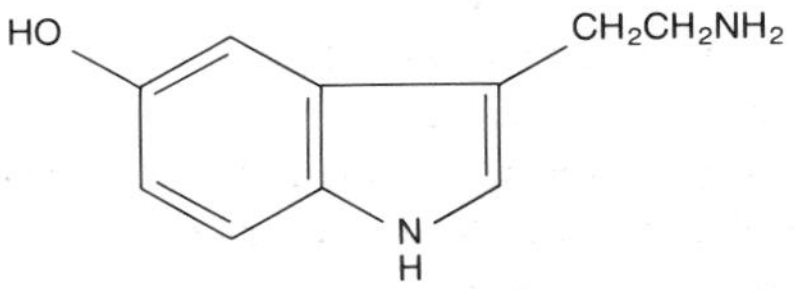

Fig. H23 5-hydroxytryptamine.

released from carcinoid tumours, which most commonly arise in the foregut. *See also*: 5HT RECEPTORS.

hyperammonaemia All INBORN ERRORS OF METABOLISM affecting the urea cycle lead to some degree of hyperammonaemia, excessive amounts of ammonia in the blood, but this is most severe in CARBAMOYL-PHOSPHATE SYNTHETASE DEFICIENCY and ORNITHINE CARBAMOYLTRANSFERASE DEFICIENCY. The former is inherited as an AUTOSOMAL RECESSIVE condition whereas the latter is an X-LINKED DOMINANT condition. They commonly present at birth and the features, which include vomiting, irritability, seizures and coma, are frequently precipitated by protein-rich feeds.

hypercholesterolaemia Conditions with elevated plasma cholesterol. *See*: FAMILIAL HYPERCHOLESTEROLAEMIA; HYPERLIPIDAEMIA; PLASMA LIPOPROTEINS AND THEIR RECEPTORS.

hyperchromic effect The change in the absorbance of a DNA solution upon denaturation, usually by increasing temperature. The effect arises as a consequence of interactions between the electron systems of the bases stacked in the double helix. Denaturation of DNA and the resulting strand separation reduces these interactions and increases the measured absorbance towards the higher value of free bases. The hyperchromic effect can be used to define the base composition of a DNA sample.

hypergammaglobulinaemia Elevated levels of serum IMMUNOGLOBULIN. Hypergammaglobulinaemia may be evident due to the presence of monoclonal immunoglobulin (e.g. PARAPROTEIN in multiple MYELOMA) or polyclonal immunoglobulin (e.g. in patients with rheumatoid arthritis or acquired immunodeficiency syndrome) (*see* ANTIBODIES). It may also result from continuous stimulation of antibody production following repeated exposure to antigen, for example after hyperimmunization of experimental animals.

hyperlipidaemia (hyperlipaemia) Condition associated with raised blood levels of CHOLESTEROL or TRIGLYCERIDE (*see* LIPIDS). It may be primary, due to a number of inherited conditions, or secondary to dietary, hormonal, or drug factors. Hyperlipidaemia is often associated with an increased risk of cardiovascular disease. *See*: PLASMA LIPOPROTEINS AND THEIR RECEPTORS.

hypermorphic allele (hypermorph) A term in the classification of allele type devised by the geneticist H.J. Muller in the 1930s (*see* ALLELE for a full explanation of this classification). Hypermorphic alleles possess greater than wild type activity. They are relatively difficult to detect because most genes are fully recessive and a homozygous hypermorph would be indistinguishable from wild type.

hypermutation A somatic mutational mechanism which can be activated in B CELLS and which introduces nucleotide substitutions specifically into the variable regions of the IMMUNOGLOBULIN GENES and adjacent regions. A mutation rate of 10^{-3} bp per generation has been calculated. The mechanism is not yet understood. Hypermutation seems to be dependent on cell prolifera-

tion. The activation of the mechanism is independent of the immunoglobulin CLASS SWITCH (*see* B CELL DEVELOPMENT).

hyperpolarization An increase in the electrical potential difference across a membrane, the inside of the cell becoming more negative with respect to the exterior.

hyperprolinaemia Abnormally high levels of proline in the bloodstream. There are two types of heritable hyperprolinaemia, both rare and inherited as AUTOSOMAL RECESSIVE traits: type I is caused by a deficiency of the enzyme proline oxidase and is associated with renal disease; type II involves a deficiency of the enzyme 1-pyrroline-5-carboxylate dehydrogenase, and is associated with seizures and mental retardation.

hypersensitive response (hypersensitive reaction, HR) The rapid death of host cells in response to challenge with fungal, bacterial, or viral pathogens, a common defence mechanism in plants. Such suicide necrosis creates an unfavourable environment which localizes the pathogen, thereby conferring resistance against infection. Although the HR can be considered a defence mechanism in its own right, it is generally associated with the induction of secondary defence responses (e.g. PHYTOALEXIN accumulation, lignification) in neighbouring living cells. Although the mechanism of host death during the hypersensitive response is unknown, loss of membrane function has been identified as an early event. *See also*: PLANT PATHOLOGY.

hypersensitivity (1) *See*: HYPERSENSITIVE RESPONSE; PLANT PATHOLOGY.
(2) Heightened and inappropriate immune response to an antigen, as in an allergy or ANAPHYLAXIS.

hypervariable regions A nucleotide sequence or segment of polypeptide chain which exhibits extremely high sequence diversity when homologous sequences from individual proteins or homologous sequences from different individuals are compared. In immunology the term was originally introduced to describe the variability observed within the sequences of immunoglobulin light chains (*see* ANTIBODIES; IMMUNOGLOBULIN GENES) and later of heavy chains (*see* COMPLEMENTARITY DETERMINING REGIONS).

hyphae (*sing.* hypha) Filaments of protoplasm enclosed by a wall which comprise the fungal or actinomycete vegetative phase — the mycelium — and some algal vegetative bodies. Hyphal structure differs considerably between prokaryotes (actinomycetes) and eukaryotes (fungi and algae) and between the different eukaryotic groups. In the higher fungi (Ascomycotina and Basidiomycotina) the hyphae are septate — divided into compartments by transverse walls (septa).

hypobetalipoproteinaemia *See*: PLASMA LIPOPROTEINS AND THEIR RECEPTORS.

hypoblast In amniote embryos (reptiles, birds, mammals), a transitory layer of cells which contributes only to EXTRAEMBRYONIC STRUCTURES. The term is not equivalent to ENDODERM which is usually used for the embryonic layer that gives rise to the digestive system of the adult. In avian embryos, the hypoblast gives rise to the yolk sac stalk. *See also*: AVIAN DEVELOPMENT; MAMMALIAN DEVELOPMENT.

hypocotyl The part of the axis of plant embryo or seedling below the insertion of the cotyledons and above the primary root. At germination it elongates in some species pushing the cotyledons above ground. In other species it stays below ground and, in a few, it forms tubers.

hypogammaglobulinaemia (agammaglobulinaemia) Lowered serum immunoglobulin levels, a clinical IMMUNODEFICIENCY state characterized by repeated pyogenic bacterial infections. The term agammaglobulinaemia is commonly used in cases where the serum GAMMAGLOBULIN level is very low ($<1 \text{ g l}^{-1}$). Hypogammaglobulinaemia is reserved for cases where the concentration of serum IgG is $>1 \text{ g l}^{-1}$, but below normal levels. The cause may be primary, for example the failure of B CELL DEVELOPMENT in Bruton's X-LINKED AGAMMAGLOBULINAEMIA or secondary, as a consequence of another condition, for example antibody loss into the gut in Crohn's disease.

hypomorphic allele (hypomorph) A term in the classification of allele type devised by the geneticist H.J. Muller in the 1930s (*see* ALLELE for a full explanation of this classification). It generally implies that the allele has reduced activity compared with wild type.

hypophosphatasia Rare heritable deficiency of tissue alkaline phosphatase. The symptoms include skeletal abnormalities similar to those caused by rickets, defective teeth, and the presence of phosphoethanolamine in urine. In its severe forms, manifest in infancy, the disease is inherited as an AUTOSOMAL RECESSIVE trait. The disease is probably linked to the alkaline phosphatase gene locus at the distal end of the short arm of chromosome 1.

hypothalamus Brain structure, a part of the diencephalon, forming the floor and part of the lateral wall of the third ventricle. The hypothalamic nuclei serve to activate, control and integrate peripheral autonomic mechanisms, endocrine activities and many somatic functions. Small peptides (e.g. LHRH, GRH) secreted by hypothalamic neurons act on anterior pituitary cells to stimulate the release of pituitary hormones.

hypoxanthine *See*: BASES.

hypoxanthine : guanine phosphoribosyl transferase *See*: HGPRT.

I

I (1) The hydrophobic AMINO ACID isoleucine.
(2) The ribonucleoside INOSINE.

I band *See*: MUSCLE.

I-cell disease Very rare human storage disease of the metachromatic leukodystrophy type, in which a generalized accumulation of many types of GLYCAN and LIPID occurs within LYSOSOMES. This is accompanied by very high levels of lysosomal acid hydrolases in serum. The underlying lesion is a deficiency in the PHOSPHOTRANSFERASE necessary for the addition of 6-phosphoryl residues to the high-mannose glycan side chains of the acid hydrolases which are required to target them to lysosomes. The disease is inherited as an AUTOSOMAL RECESSIVE trait. Fibroblasts are mainly affected and contain inclusion bodies (hence I-cells). Although other tissues share the enzyme defect, there is less decrement of lysosomal enzyme content. This suggests the presence of an alternative pathway of lysosomal enzyme targeting independent of MANNOSE 6-PHOSPHATE RECEPTORS.

I-J In the mouse, a chromosomal subregion within the H-2 I region of the MAJOR HISTOCOMPATIBILITY COMPLEX originally identified serologically and thought to encode polymorphic determinants expressed selectively on SUPPRESSOR T CELLS, perhaps as part of the antigen receptor, and on soluble SUPPRESSOR FACTORS secreted by these cells. Failure to identify a T_s-specific I-J gene locus by extensive molecular analysis of the H-2 I region as well as more recent serological analyses has suggested that I-J suppressor molecules are not encoded by a novel gene mapping in the I region, but rather that they represent determinants present on T cell receptors that recognize self-MHC class II molecules and/or receptors for such molecules.

Ia antigens Immune-*a*ssociated antigens, also called class II MHC molecules (*see* MAJOR HISTOCOMPATIBILITY COMPLEX). They are polymorphic cell-surface glycoprotein heterodimers, expressed predominantly on macrophages and other ANTIGEN-PRESENTING CELLS, and function as the restricting elements in MHC-restricted immune responses (*see* T CELL RECEPTOR).

IAA INDOLE-3-ACETIC ACID. *See*: AUXINS; PLANT HORMONES.

iaaM/H genes (auxI/II, tmsI/II, shiI/II) Genes for the conversion of tryptophan to indoleacetamide and to indoleacetic acid (an AUXIN), which are carried in the T-region of the TI PLASMID of many *Agrobacterium* strains. When transferred to plant cells they are responsible for the accumulation of auxin which helps maintain the tumorous phenotype.

IAP A TRANSPOSABLE GENETIC ELEMENT of mice.

IBA INDOLE-3-BUTYRIC ACID.

ICaBP-9 Calbindin-D9. *See*: CALCIUM-BINDING PROTEINS.

ICAM-1, ICAM-2 Intercellular adhesion molecules. Cell adhesion molecules of the IMMUNOGLOBULIN SUPERFAMILY. ICAM-1 is a receptor for the INTEGRIN LFA-1. *See*: CELL ADHESION MOLECULES.

ice nucleation factor A factor which causes supercooled water to freeze by ordering of water molecules into a lattice resembling ice. Frost damage on many plants is caused by the presence of certain bacteria, particularly *Pseudomonas syringae* pathovars, which are able to trigger ice formation. Three classes of ice nucleation structure have been identified in *P. syringae*. Common to each class is an ice nucleation protein of M_r 120 000 which is the product of the *inaZ* gene. This protein is membrane bound and has unique N and C termini and a central octapeptide repeat which is thought to be the template for ice crystal formation. Post-translational attachment of phosphatidylinositol and mannose and possibly glucosamine result in the most active ice nucleating structure. Attachment of mannose and glucosamine to the ice nucleation protein yields two other classes with lower ice nucleating activity. It is thought that the sugar moieties are involved in aggregating the lipoglycoprotein compound into large aggregates which maximizes activity.

Lindow, S. E. (1983) *Annu. Rev. Phytopathol.* **21**, 363–384.
Green, R.L. & Warren, G.J. (1985) *Nature* **317**, 645–648.
Turner, M.A. et al. (1991) *J. Bacteriol.* **173**, 6528–6536.

icosahedral symmetry A particular arrangement of dyad, triad and pentad rotational SYMMETRY axes, whereby 60 equivalent positions are generated on a spherical surface. This gives the greatest possible number of strictly equivalent positions upon any closed surface but, if the positions are required to be only QUASI-EQUIVALENT, then multiples of 60 units can be accommodated. *See*: ANIMAL VIRUSES; PLANT VIRUSES.

ICV injection Any injection that directly delivers its load into the ventricular spaces of the cerebral hemispheres.

IDDM Insulin-dependent DIABETES MELLITUS.

idiotype A set of antigenic determinants — idiotopes — that characterize an individual ANTIBODY molecule. Some idiotopes are unique to a given antibody molecule and are termed private idiotopes; they reflect the unique combination of the light and heavy chain variable regions. Other idiotopes may reflect structural homologies common to a small subpopulation of antibody molecules and are termed cross-reactive idiotopes. *See also:* ANTI-IDIOTYPIC ANTIBODIES.

iduronic acid *See:* GLYCOSAMINOGLYCANS.

IF (1) INTERMEDIATE FILAMENTS.
(2) Initiation factors. *See:* PROTEIN SYNTHESIS.

IFN INTERFERONS.

IgA Class A immunoglobulin (*see* ANTIBODY; IMMUNOGLOBULIN GENES), composed of two identical light chains and two identical α heavy chains (total M_r ~160 000). It comprises ~10–15% of total human serum immunoglobulin (1.5–3 mg ml^{-1}) and is the predominant antibody in secretions (saliva, tears, milk, nasal mucus, and gastrointestinal and respiratory secretions). It provides immune defence against organisms that invade mucosal surfaces. Subclasses IgA1 and IgA2 are determined by heavy chain subclasses α1 and α2 respectively. 80–90% of serum IgA is of the IgA1 subclass, whereas most IgA in secretions is subclass IgA2. It may exist as monomer, dimer and higher multimeric forms. Also, the predominant form in secretions is associated with secretory component (*see* SECRETORY IGA).

IgD Class D immunoglobulin (*see* ANTIBODY; IMMUNOGLOBULIN GENES), composed of two identical light chains and two identical δ heavy chains (total M_r ~175 000). IgD is a relatively minor component of serum, comprising ~0.3% of total serum immunoglobulin in humans (0.04 mg ml^{-1}). Most IgD is expressed on B CELL surfaces. A single mature B cell may express both IgM and IgD on its surface, both isotypes sharing a common variable region. They result from alternative splicing (*see* RNA SPLICING) of a single RNA that contains both μ and δ constant region sequences. The biological role of IgD remains unclear.

IgE Class E immunoglobulin (*see* ANTIBODY; IMMUNOGLOBULIN GENES), composed of two identical light chains and two identical ε heavy chains (total M_r ~190 000). Least abundant immunoglobulin class in serum, comprising ~0.003% of human serum immunoglobulin (0.1–0.5 mg ml^{-1}). IgE is medically important as the major mediator of immediate type HYPERSENSITIVITY, which is due to the binding of IgE to high affinity Fcε receptors (FcεRI) on mast cells and basophils. Binding of specific antigen to IgE on FcεRI receptors causes IgE cross-linking and receptor clustering. Cell degranulation results, with pharmacologically active substances including HISTAMINE and 5-HYDROXY-TRYPTAMINE (serotonin) being released. These substances induce the clinical manifestations of ANAPHYLAXIS. Allergic individuals may have higher than normal levels of serum IgE and an increased number of high affinity IgE receptors on their basophils and mast cells.

IgE receptor The FC RECEPTOR for IGE. There are two types of IgE receptor: high affinity receptor FcεRI on mast cells and basophils, and low affinity receptor FcεRII (human CD23 antigen) on B CELLS, monocytes, platelets, and eosinophils. IgE bound to FcεRI receptors is the major mediator of immediate type hypersensitivity (*see* IGE). The FcεRI receptor is composed of four polypeptide chains (αβγ$_2$; total M_r ~76 000). The two γ chains are disulphide linked. Fc′εRII (CD23) is a single, glycosylated polypeptide chain of 321 amino acids. It comprises a 274-residue C-terminal extracellular domain, a 24-residue transmembrane domain and a 23-residue N-terminal cytoplasmic domain (total M_r ~45 000). It is present on all mature μ$^+$δ$^+$ B cells but is not expressed on immature B cells in bone marrow or on B cells following CLASS SWITCHING. The role of human CD23 antigen, if any, remains unclear although its developmentally limited expression suggests it is a B cell-specific DIFFERENTIATION ANTIGEN.

IgG Class G immunoglobulin (*see* ANTIBODY; IMMUNOGLOBULIN GENES), composed of two identical light chains and two identical γ heavy chains (total M_r ~150 000). Comprises ~75–85% of total human serum immunoglobulin in adults (9–15 mg ml^{-1}) and is the predominant immunoglobulin of SECONDARY IMMUNE RESPONSES. There are four IgG subclasses in humans (IgG1–4) and four in mice (IgG1, IgG2a, IgG2b, IgG3). The subclasses display different biological properties and are represented in varying amounts in serum. For example, in mice, IgG1 is normally the predominant subclass expressed against protein antigens, whereas IgG2a is dominant during viral infections. All four subclasses bind to specific FC RECEPTORS on a variety of cell types, including macrophages, mast cells, polymorphonuclear cells, and lymphocytes. Fc receptor binding may induce phagocytosis (*see* ENDOCYTOSIS), release of chemical mediators, antibody-dependent cellular cytotoxicity (ADCC), and the regulation of lymphocyte function. IgG is unique in its ability to cross the placenta. Apart from human subclass IgG4, all IgG subclasses can activate COMPLEMENT fixation.

IgH Immunoglobulin heavy chain. *See:* ANTIBODIES; IMMUNOGLOBULIN GENES.

IgM Class M immunoglobulin (*see* ANTIBODY; IMMUNOGLOBULIN GENES), composed of two identical light chains and two identical μ heavy chains. Membrane-expressed IgM and secretory IgM differ in the amino-acid sequences at their μ chain C termini due to alternative splicing (*see* RNA SPLICING) of the μ chain RNA. In mammals, secretory IgM is a pentamer of μ$_2$L$_2$ monomers linked by disulphide bridges between the heavy chains and complexed with a J chain molecule. The total M_r of the pentameric complex is ~950 000, comprising 5–10% of total serum immunoglobulin in adults (~1 mg ml^{-1}). IgM is the first immunoglobulin isotype to be expressed following GENE REARRANGEMENT and is the predominant immunoglobulin class present in a PRIMARY IMMUNE RESPONSE. Following interaction with antigen, membrane-bound IgM on mature B cells is involved in signalling the activation, proliferation, and further differentiation of the B cell. This may include increased production of secretory IgM and CLASS SWITCHING to other heavy chain isotypes (*see* B CELL DEVELOPMENT).

IGS Internal guide sequence. *See:* RIBOZYMES; RNA SPLICING.

IgS, IgSF IMMUNOGLOBULIN SUPERFAMILY. A group of proteins that contain segments of sequence showing significant sequence similarities with immunoglobulin domains.

IHF Integration host factor. *See:* LAMBDA.

I,i blood group system BLOOD GROUP antigens defined by internal GLYCAN sequences. The i antigen has the sequence -Galβ1,3GlcNAcβ1,4Galβ1,3GlcNAcβ1,4Galβ1-, while I antigen is at the branchpoint of the structure,

$$-Gal\beta1,4GlcNAc\beta1,3(-Gal\beta1,4GlcNAc\beta1,6)Gal\overset{*}{-}\beta1,4GlcNAc\beta1,3Galp1-$$

in which the asterisked galactosyl residue bears branches at C3 and C6. I erythrocytes are agglutinated by a cryoglobulin produced in some forms of haemolytic anaemia and are much the more common type in adults. Infants generally express i much more strongly their erythrocytes than I and the developmental switch is unusually late, at about eleven years of age. Secreted glycoproteins show I-antigenicity from foetal life.

IL-1 Human interleukin-1. A (nonglycosylated) protein with diverse roles in immunity and inflammation. It is produced from a human gene at chromosome 2q14. The original M_r of 15 000–17 000 has been complicated by the demonstration of biological activity in molecules of higher and lower molecular weights. IL-1α has ~154 amino acid residues, and IL-1β ~153; both forms bind to the same receptor, which is a member of the IMMUNO-GLOBULIN SUPERFAMILY. Cellular sources of IL-1α and β include macrophages, endothelial cells, lymphocytes, fibroblasts, epithelial cells, astrocytes, keratinocytes and osteoblasts, and responding cell types include haematopoietic progenitor cells, neutrophils, hepatocytes and muscle cells in addition to many of the producer cell types. Documented effects include the stimulation of the immune and haematopoietic systems considered to be of importance in inflammation and also in host defence against infection. *See:* LYMPHOKINES.

Dinarello, C.A. (1991) *Blood* **77**, 1627–1652.

IL-2 Interleukin-2, previously known as T cell growth factor. A protein of 133 amino acids, typically O-glycosylated (*see* GLYCO-PROTEINS) at the threonine residue at position 3, although non-glycosylated material retains biological activity. The protein has a calculated M_r of 15.5K, the natural product 19–22K, and is encoded by a single gene located on human chromosome 4. IL-2 is typically produced by activated T lymphocytes and promotes proliferation of T cells and other immune cells such as natural killer cells (NK cells), B lymphocytes and macrophages. Therapeutic indicators include immunotherapy in certain cancers. *See:* LYMPHOKINES.

Muukerjee, B.K. & Pauly, J.L. (1990) *J. Clin. Lab. Anal.* **4**, 138–149.

IL-2 receptor (CD25, TAC) The IL-2 receptor exists in at least two forms. One, accounting for ~10% of all IL-2 receptors, is a high affinity (K_d 10^{-11} M) receptor and mediates T cell responses to IL-2; the other is of lower affinity (K_d 10^{-8} M). IL-2 receptors

are formed from two proteins (M_r55 000 and 75 000) the former of which represents the lower affinity receptor, is recognized as CD25 (TAC) (*see* CD ANTIGENS), has been recently cloned and is produced from a gene on human chromosome 10. The combination of both 55 and 75K proteins represents the high affinity receptor. It has been suggested that binding of IL-2 to the 75K protein may initiate induction of the gene transcribing the 55K protein and hence the production of the high affinity receptor. The 75K protein produces the signal for cell growth, whereas IL-2 binding is facilitated by the 55K protein. *See also:* LYMPHOKINES.

Leonard, W.J. et al. (1990) *Phil. Trans. R. Soc. Lond.* **B327**, 187–192.

IL-3 (multi-CSF, BPA, PSF, MCGF) Interleukin-3. A LYMPHOKINE produced by activated T cells which supports the growth and development of myeloid and erythroid cells and the maintenance of cell lines of primitive haematopoietic cells and mast cells. Murine IL-3 is a GLYCOPROTEIN of 166 amino acids (protein M_r 15 500, four glycosylation sites, M_r of natural glycoprotein 28 000) which is produced from a gene on chromosome 11, whereas human IL-3 (152 amino acids, M_r 15 000, two glycosylation sites) is produced from a gene located on chromosome 5q. Human and murine IL-3 share relatively little homology and there is little cross-species reactivity. The IL-3 receptor has been found on numerous murine and human cell lines, both transformed and normal, and also on murine neutrophils and both human and mouse monocytes. It is a member of a cytokine receptor superfamily (*see* GROWTH FACTOR RECEPTORS).

Morris, C.F. et al. (1990) In *Colony-stimulating Factors — Molecular and Cellular Biology* (Dexter, T.M. et al., Eds) Ch. 6, 177–214 (Marcel Dekker, New York).

IL-4 Interleukin-4. A LYMPHOKINE produced by activated HELPER T CELLS. It is a growth factor for B and T cells and mast cells and exerts other effects on haematopoietic cells and their precursors. Murine IL-4 is a GLYCOPROTEIN of M_r 13 200 (nonglycosylated protein), 120 amino acids, six cysteine residues, with three glycosylation sites and a glycosylated M_r of 20K. Nonglycosylated material is still active. Human IL-4 is a glycoprotein of 129 amino acids (seven cysteines, two glycosylation sites) which shares 70% homology with the murine DNA sequence and 50% homology at the amino-acid level, yet demonstrates no cross-species reactivity. Human IL-4 is produced from a gene with four exons and three introns around chromosome location 5q31, close to the IL-5 gene. The IL-4 receptor is a member of the cytokine receptor super-family (*see* GROWTH FACTOR RECEPTORS), and has been shown to exist on both transformed and nontransformed cell lines of human and mouse origin. Normal human fibroblasts have very high numbers of receptors, neutrophils very few, and monocytes and T cells intermediate numbers.

Arai, N. (1989) *J. Immunol.* **142**, 274–282.

IL-5 (eosinophil colony-stimulating factor/differentiation factor) Interleukin-5. Glycoprotein produced by HELPER T CELLS from a gene at human chromosome 5q31 within 90–240 kb of the IL-4 gene, and 500 kb from the IL-3 and GM-CSF genes. The natural material is N-glycosylated (at two sites in human and three sites in mouse) to yield an M_r of 42–66K, although this is a dimer of

20K subunits in humans, 22K in the mouse. Activity is retained in nonglycosylated material. The human (115 amino acids) and mouse (125 amino acids) molecules share 67% homology at the amino-acid level and show limited cross-species reactivity (mouse has 50% activity on human targets, human has 10% activity on mouse cells). Murine IL-5 stimulates the proliferation and development of eosinophils and B cells. Human material influences eosinophil growth, though it has been recently reported also to support basophil differentiation. The receptor has been cloned, and is a 45K member of the cytokine receptor superfamily (*see* GROWTH FACTOR RECEPTORS).

Yokota, T. et al. (1987) *Proc. Natl. Acad. Sci. USA* **84**, 7388–7392.

IL-6 (IFNβ$_2$, BSFII, HSF) Interleukin-6. LYMPHOKINE produced by activated T cells and fibroblasts. It is known to stimulate the growth of B cells, HYBRIDOMAS, and megakaryocyte precursors and has some limited antiviral activity. The gene for human IL-6 is located on chromosome 7q and encodes a protein of 284 amino acids (M_r22 000–29 000) which has two glycosylation sites and four cysteine residues perhaps involved in disulphide bonding of the mature protein. In common with some other cytokine receptors the IL-6 receptor requires association with a second protein (gp130) to exhibit high affinity for its ligand and subsequent signal transduction. The receptor contains motifs in common with the IMMUNOGLOBULIN SUPERFAMILY and cytokine receptor superfamily (*see* GROWTH FACTOR RECEPTORS). Raised levels of IL-6 have been noted in rheumatoid arthritis, multiple MYELOMA and SYSTEMIC LUPUS ERYTHEMATOSUS, indicating a role in pathogenesis.

Kishimoto, T. (1989) *Blood* **74**, 1–11.

IL-7 Interleukin-7. Glycoprotein of M_r ~20 000 derived from bone marrow or thymus stromal cells. It comprises 177 amino acids (25-amino acid signal sequence, three potential glycosylation sites, protein M_r ~17 400). The gene is on human chromosome 8 and IL-7 is produced from six exons spread over 33 kb. It stimulates B cell growth and is also involved in the production of T cells, lymphokine-activated killer cells (LAK cells) and platelets. A high level of expression of IL-7 messenger RNA in the normal thymus has been noted, although IL-7 was originally identified, purified, and cloned on the basis of its ability to support the growth of pre-B cells *in vitro*. IL-7 may be involved in the growth of B and T cell acute lymphoblastic LEUKAEMIAS. The receptor for IL-7 is thought to have an M_r of ~75–79K although 159–162K has also been shown. At least two different affinities have also been noted. The receptor is a member of the cytokine receptor superfamily (*see* GROWTH FACTOR RECEPTORS). *See*: LYMPHOKINES.

Widmer, M.B. et al. (1990) *Int. J. Cell Cloning* **8** (Suppl. 1), 168–170.

IL-8 (NAF, NAP-1, MONAP, MDNCF, LYNAP) Interleukin-8. A LYMPHOKINE produced by activated T cells, activated monocytes, fibroblasts and endothelial cells. It acts as a potent neutrophil activator and is a chemoattractant for T cells and neutrophils. IL-8 is produced from a gene of four exons and three introns on human chromosome 4q12–21. The mature protein (of 72 amino acids) has four cysteines and no N-glycosylation sites. It is highly homologous to platelet basic protein, platelet factor 4, melanoma-stimulating factor and other members of the macrophage inflammatory protein (MIP) group. It has been cloned.

Matsushima, K. et al. (1988) *J. Exp. Med.* **167**, 1893.

IL-9 (P40) Interleukin-9. A glycoprotein LYMPHOKINE (protein M_r 14 000; glycoprotein M_r 32 000–39 000) produced by activated T cells. The human IL-9 gene is located at 5q31-32 and consists of five exons spread over 4 kb of DNA. The protein has 140 amino acids. IL-9 promotes the growth of some T cell clones and adult thymus and spleen subpopulations, but is inactive on CYTOTOXIC T CELLS. Human and mouse IL-9 are 56% homologous, both have 10 cysteine residues and multiple N-glycosylation sites (human four, mouse three). The spectrum of activity of IL-9 may also include stimulation of IL-3-dependent mast cells and human megakaryoblastic leukaemia cells.

Yang, Y.C. et al. (1990) *Blood* **74**, 1880–1884.

IL-10 (CSIF) Interleukin-10. A LYMPHOKINE produced by type 2 HELPER T CELLS and B cells, which suppresses CYTOKINE production by type 1 helper T cells. It is a polypeptide of M_r 17 000 which is naturally N-glycosylated and bears a degree of similarity to the protein BCRF1, encoded by the EPSTEIN–BARR VIRUS.

MacNeil, I.A. et al. (1990) *J. Immunol.* **145**, 4167–4173.

IL-11 Interleukin-11. Protein (M_r 23 000, ~180 amino acids) derived from bone marrow stromal cells which enhances IL-3-stimulated megakaryocyte growth, and shows some overlap with the biological activities of IL-6 and IL-7. It is distinct from both these molecules and unlike most CYTOKINE proteins has no cysteine residues. *See also*: LYMPHOKINES.

Paul, S.R. et al. (1990) *Proc. Natl. Acad. Sci. USA* **87**, 7512–7516.

ilarvirus group Sigla from isometric labile ringspot; type member tobacco ringspot virus. MULTICOMPONENT VIRUSES with quasi-isometric particles 26–35 nm in diameter. The four (+)-strand linear RNA species are encapsidated in separate components. The bottom component contains RNA 1 (1.1×10^6–1.3×10^6) which, together with RNA 2 (1.18×10^6–0.89×10^6) encapsidated in the middle component, are thought to be monocistronic mRNAs encoding subunits of the RNA-dependent RNA polymerase. RNA 3 (0.91×10^6–0.7×10^6) is bicistronic, encoding at the 5′ end the putative cell-to-cell movement protein and at the 3′ end the viral coat protein; the coat protein is expressed from the subgenomic RNA 4 (0.3×10^6). As with ALFALFA MOSAIC VIRUS, RNAs 1, 2 and 3, together with coat protein molecules or RNA 4 are required for infection. *See also*: PLANT VIRUSES.

Francki, R.I.B. (1985) In *The Plant Viruses*, Vol. 1, 1–18 (Plenum, New York).

Ile (I) The AMINO ACID isoleucine.

image analysis, image reconstruction Methods for extracting the maximum amount of information from an electron micrograph of a macromolecule or macromolecular assembly such as a virus particle in order to aid construction of a two- or three-dimensional structure (*see* ELECTRON CRYSTALLOGRAPHY; ELECTRON DIFFRACTION). Image analysis has been applied to electron micrographic images of virus particles, structures such as MICROTUBULES and MUSCLE filaments, core particles of nucleosomes (*see* CHROMATIN), crystalline arrays of membrane proteins (*see* BACTERIORHODOPSIN; NICOTINIC RECEPTORS), and large multisubunit proteins such as CHAPERONINS.

image plate, imaging plate An integrating area detector using a storage phosphor. X-ray or autoradiograph images recorded on an image plate are read out in digitized form by a laser scanning device and erased by bright light such that the image plate can be used repeatedly, unlike photographic film which can be used only once, needs chemical processing and has a significant level of background. Image plates have a wide dynamic range, low intrinsic background, high sensitivity, and efficient X-ray detection over the wavelengths commonly used for clinical or laboratory purposes. The use of image plates is increasingly superseding that of traditional film for many forms of radiography and in X-RAY CRYSTALLOGRAPHY.

Amemiya, Y. et al. (1987) *Science* **237**, 164–168.

imaginal disks Insect larval structures which give rise to adult tissues when stimulated by the hormone 20-hydroxyecdysone (ecdysterone) during METAMORPHOSIS (*see* ECDYSONE RESPONSE). Larvae of DROSOPHILA have 10 pairs of imaginal disks which give rise to the eyes/antennae, the legs, the halteres, the wings, and various head structures, and a single genital disk (Fig. I1). Imaginal disks develop from patches of cells on the epidermis of the first instar larva. These cells proliferate extensively during larval development, and by the third instar have developed into a folded invaginated epithelial sac. At metamorphosis, the epithelium evaginates, unfolds, and extends to form the adult structure.

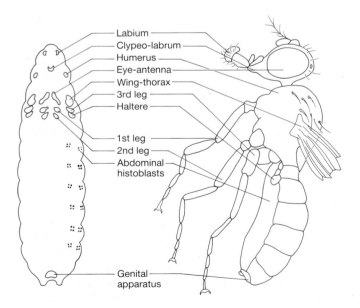

Fig. I1 Larval imaginal disks in *Drosophila* and the adult structures to which they give rise.

The adult abdominal epidermis arises from clusters of cells in the abdomen of the larva called ABDOMINAL HISTOBLASTS.

immediate early genes (IEGs) A class of genes that appear to be activated first in response to a wide range of external stimuli, for example mitogens, hormones or neurotransmitters, which activate SECOND MESSENGER PATHWAYS. Expression of these genes may regulate the subsequent expression of other genes, termed late response genes.

immediately subterminal sequences *See*: CHROMOSOME STRUCTURE.

immortal cell line *See*: CELL LINE.

immune complex Complex of ANTIGEN and ANTIBODY.

immune response *See*: ANTIBODIES; ANTIGEN PROCESSING AND PRESENTATION; B CELL DEVELOPMENT; IMMUNOGLOBULIN GENES; LYMPHOKINES; T CELL DEVELOPMENT; T CELL RECEPTOR; T CELL RECEPTOR GENES.

immune response genes (Ir genes) The class II genes of the MAJOR HISTOCOMPATIBILITY COMPLEX (MHC). These are polyallelic loci that control the ability to mount selective immune responses against protein antigens. In the mouse, Ir genes are encoded within the I-region of the H-2 complex on chromosome 17; human Ir genes are encoded within the HLA complex on chromosome 6. Ir genes encode the MHC class II proteins (Ia antigens).

immunity *See*: ANTIBODIES; CELL-MEDIATED IMMUNITY. Specific resistance to the onset of disease after infection by bacteria, viruses, or other pathogens.

immunoadsorption, immunoaffinity purification A process used for the specific purification of ANTIGENS or ANTIBODIES by binding them to complementary antibodies or antigens attached to an inert insoluble matrix. The bound antigen or antibody is subsequently released from the adsorbent by exposure to mildly denaturing solvents (e.g. low pH buffers or high salt concentrations).

immunoblotting *See*: WESTERN BLOTTING.

immunocytomas Malignant proliferation of lymphoplasmacytic cells usually associated with production of a monoclonal IGM, Waldenstrom's macroglobulin.

immunodeficiency A clinical state in which there is a deficiency of immune responsiveness, and which may arise for many different reasons. Broadly these states can be categorized into primary or inherited immunodeficiency, and secondary or acquired immunodeficiency. As a result of the multiple cell lineages of the immune system, there is a broad spectrum of primary and secondary immunodeficiency states which may involve either specific or nonspecific limbs of the immune response. Defects in phagocytic cells (e.g. CHRONIC GRANULOMATOUS DISEASE of childhood) and deficiencies of various COMPLEMENT components are examples of the latter category. Defects of specific immunity involving the development and maintenance of the T cell-mediated response or the humoral response are examples of the former (e.g. X-LINKED AGAMMAGLOBULINAEMIA and ATAXIA TELANGIECTASIA). Acquired immunodeficiency may result from some medical treatments (e.g. radiation therapy and the use of immunosuppressant drugs in organ transplant recipients) or as a result of infection (e.g. AIDS, *see* IMMUNODEFICIENCY VIRUSES). A considerable amount is known about the categories of immunodeficiency in humans and other vertebrates. *See also*: ADENOSINE DEAMINASE DEFICIENCY; AGAMMAGLOBULINAEMIA; DYSGAMMAGLOBULINAEMIA; HYPOGAMMAGLOBULINAEMIA; SEVERE COMBINED IMMUNODEFICIENCY.

Immunodeficiency viruses

THE human immunodeficiency viruses (HIV) and the related simian immunodeficiency viruses (SIV) are members of the lentivirus group of the RETROVIRUSES. As such, they share the basic retroviral life cycle: an RNA genome is replicated via reverse transcription into a DNA provirus which becomes integrated into the host cell genome (*see* RETROVIRUSES). The structural viral particle components, products of the *gag*, *pol* and *env* genes, are essentially similar to those of other retroviruses. The organization of the principal structural proteins of an infectious virion of HIV is shown in Fig. I2 and the genes in Fig. I3.

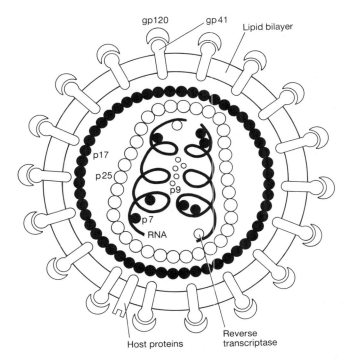

Fig. I2 Structure of an infectious virus particle of the human immunodeficiency virus (HIV). gp120 and gp41 are envelope proteins; p17, p25, p9 and p7 are structural viral proteins.

Clinical picture

The lentiviruses are pathogenic retroviruses that induce degenerative diseases in their animal hosts. Characteristically, they cause a mild primary syndrome, followed by a prolonged and unremitting secondary disease. In humans, infection with HIV is initially associated with a mononucleosis-like disease with localized or general lymphadenopathy and frequently with diarrhoea. In the longer term (perhaps tens of years), HIV-infected patients develop the acquired immunodeficiency syndrome (AIDS), characterized by generalized immune suppression, multiple opportunistic infections and neurological disease. HIV is now the widely recognized primary causative agent of AIDS, although it is the multiple co-infecting viral and bacterial pathogens which are responsible for the clinical syndrome of diseases seen in AIDS patients.

Characteristically, HIV infection leads to a progressive depletion of CD4 T CELLS (T lymphocytes carrying the cell-surface glycoprotein CD4), for which the virus is tropic both *in vivo* and *in vitro*. The major cellular receptor for the virus is the CD4 molecule. CD4 cells are gradually lost, probably through direct cytopathic activity of the virus itself as well as through immune responses directed against the virus-infected cells. CD4 T cells are crucial regulatory cells of the human immune system and their loss may be the main reason for the increased susceptibility to opportunistic infections and rare cancers, such as Kaposi's sarcoma, seen in AIDS patients.

AIDS is also associated with neurological dysfunctions, leading to AIDS-related dementias, some of which may be directly attributable to infection by HIV of brain cells carrying CD4 or to infection of other cell types via a second receptor that has not yet been molecularly characterized.

Molecular biology of HIV

Lentiviruses have the most complex genomes of all the retroviruses. In addition to the standard *gag*, *pol* and *env* genes of all replication-competent retroviruses, the HIV genome contains genes that regulate proviral gene expression in both a temporal and a quantitative way (see Fig. I3). This genomic intricacy is responsible for the ability of the virus to exist as a latent infection in many cells and for the sudden emergence of viral progeny in particular circumstances (usually immune activation of the host). The interplay between these viral gene products and cellular factors is at the heart of the mechanisms of viral replication and, therefore, pathogenesis.

The HIV genome

HIV type 1 (HIV-1, formerly called LAV and HTLV-III) contains more than six genes not found in other groups of retroviruses. The *tat* gene encodes a protein of 86 amino acids which is a positive regulator of viral gene expression by accelerating synthesis of the proviral RNA transcripts encoding viral structural proteins. Viruses mutant in *tat* cannot replicate by themselves. The TAT protein acts in the nucleus, on a short region of nucleic acid — the transactivating responsive element or TAR — both to

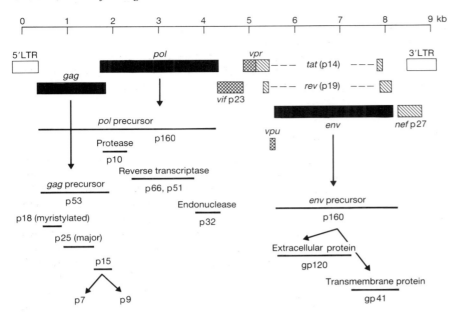

Fig. I3 The genome of HIV-1. Genes common to all retroviruses are shaded in black; those specific to HIV are hatched. The encoded proteins and their post-translational processing are shown below the genes.

increase the levels of messenger RNAs (mRNA) containing the TAR and to increase the stability of such mRNA molecules. It seems, therefore, that TAT can act both transcriptionally (on the DNA copy of TAR) and post-transcriptionally (on the RNA copy of TAR). Because TAR is located at the 5′ end of the proviral DNA long terminal repeat (LTR) (*see* RETROVIRUSES), around the site of transcription initiation, all viral transcripts contain it. In the absence of TAT protein, the TAR sequence (which forms a complex secondary structure) will inhibit protein synthesis from mRNA molecules containing it; with added TAT, this inhibition is overcome and viral proteins are readily synthesized. TAT may bind directly to RNA TAR sequences, but the mechanism by which TAT acts on the DNA form of TAR is unclear. TAT has been proposed to act by overcoming transcription ATTENUATION and possibly by preventing the rapid degradation of viral RNA.

The second positive regulator of viral protein expression is the *rev* gene product (*see* Fig. I3). The REV protein acts as a positive post-transcriptional regulator of specifically structural protein expression. HIV proviral gene expression involves variously spliced mRNAs. Full length genomic messages encode all the viral proteins and a singly spliced message encodes the 3′ half of the genome including *env* (similarly to the singly spliced *env* message of other retroviruses). Doubly spliced messages exclusively encode the regulatory, but not the structural, proteins.

REV acts by stimulating the nuclear export of the singly spliced and unspliced RNAs. HIV structural genes contain sequences that direct the retention of the RNA transcripts within the nuclear compartments in which splicing occurs freely. In the absence of REV these RNAs become fully spliced into messages encoding only the regulatory proteins. However, in the presence of REV, a specific sequence within the *env* gene — the *cis*-acting antirepression sequence (CRS) — binds REV and this prevents the routing of RNA to the splicing compartment. Instead, the RNA is exported to the cytoplasm in the largely unspliced form, where it can direct synthesis of the viral structural proteins.

Expression of REV therefore reduces the levels of fully spliced mRNAs which include those specifying REV itself and TAT. The subsequent reduction in levels of TAT will lead to a downregulation of overall viral gene expression. Therefore, using just two of the regulatory genes of HIV, a fine balance can be established between the relative levels of viral mRNA species.

The largest of the HIV regulatory genes is the *nef* gene (see Fig. I3). The NEF protein (M_r 27 000) is translated from a fully spliced mRNA. Its role *in vivo* is not as clearly defined as that of TAT or REV. Viruses lacking functional NEF are reported to replicate more rapidly in culture than NEF$^+$ viruses (hence the name nef, 'negative factor'). NEF may promote the switch into viral latency by shutting down viral expression until stronger stimulatory factors intervene. NEF is located at the inner surface of the plasma membrane in the infected cell, tethered there by a myristic acid modification at its N terminus. This location, and its reported GTP binding and GTPase activity, suggest it may be a viral analogue of signal-transducing cellular proteins (*see* GTP-BINDING PROTEINS). In simian models of AIDS (see below), *nef* function seems to be crucial for progression from SIV infection to AIDS, suggesting that the gene is crucial for the pathogenicity of the lentiviruses. The apparent paradox of its repressive effects on multiplication *in vitro* and its requirement for pathogenicity *in vivo* can be reconciled by assuming that its putative repressive effect on viral gene expression and thus on the production of viral antigens may help virus-infected cells to remain hidden from the immune system.

Two further genes — *vif* and *vpu* — are required for virion mrphogenesis and maturation. The cytoplasmic virion infectivity protein (VIF) seems important for efficient cell-to-cell spread of HIV in culture, and possibly in and between individuals. The mechanism of action is unknown. VIF seems to act together with the VPU protein to ensure production of large amounts of highly infectious virus, when the molecular switch controlled by TAT and REV and their interaction with cellular factors is pushed in favour of viral replication. VPU facilitates the budding and assembly of maturing virions. VPU, which is unique to HIV-1, appears to increase the efficiency of viral morphogenesis by degrading CD4 when it is still in the endoplasmic reticulum. This prevents association of CD4 with the envelope protein gp160 at the cell surface and liberates gp160 to be included in viral particles.

Finally, HIV contains a second transactivator gene, *vpr*. The VPR protein can accelerate the rate of production of viral proteins and can transactivate several heterologous genes *in vitro*. However, its function *in vivo* is unknown.

From genes to disease

The self-limiting balance of TAT and REV expression, their interaction with cellular factors and the interactions of the other regulatory proteins of HIV play a major part in controlling the viral latency/expression switch and the subsequent development of disease. Any activation of viral transcription will increase levels of TAT and REV by transcription initiated at the LTR. The action of REV will soon lead to decreased expression of TAT and thus to eventual shutdown of proviral expression. In this way short bursts of viral production can be achieved to disseminate virus in the body. At the same time, however, the very limited expression of the provirus may prevent the infected cell from being 'seen' by the immune system and cleared from the body. Either cells harbouring virus can then continue to produce virus chronically at low levels or the virus can survive in a totally latent state within infected lymphocytes, macrophages and other cells without being detectable by the immune system. In particular, emerging data suggest that DENDRITIC CELLS can act as a major reservoir for HIV infection, fuelling chronic infection of T cells.

There is much evidence that immune activation of infected T cells leads to the production of cellular factors that act on the viral LTR to stimulate viral gene expression. Therefore, chronic exposure to other foreign antigens — bacteria or viruses that activate the normal immune response — can throw the latency/expression switch in favour of viral replication. If the body is subject to high levels of immune challenge, virus production can increase dramatically. More CD4 cells become infected, more virus is produced and more CD4 cells are lost, partly as a result of the cytopathic effects of the virus. Further cells are destroyed as a result of the immune response directed against cells which have emerged from latency and are producing high levels of viral antigens or against cells which have bound shed viral proteins (mainly the envelope proteins and VIF). Several different mechanisms have been proposed for how HIV infection may lead to loss of function of CD4 cells other than by direct cell killing by the virus. These include suggestions that certain viral particles may mimic host class II MHC molecules so that antiviral immune responses lead to an autoimmune blockade of 'self' MHC molecules, inhibiting normal antigen presentation and helper T cell function. Alternatively, gp120 may deliver a negative signal to CD4 cells by binding CD4, leading to T cell ANERGY, making them refractory to further stimulation or even inducing APOPTOSIS. Finally, so-called SUPERANTIGENS may exist in HIV infection which are either endogenous or exogenously encoded by HIV itself. These superantigens may either be potent activators of T

cells, making them more susceptible to infection, or they may induce anergy or deletion of a subset of CD4 T cells which have a specific β chain of the T cell receptor. Whatever the mechanism, the increasing loss of CD4 cells leads to immune deficiency, and progression to full blown AIDS is assured as the body's capacity to clear infection becomes reduced.

Variability of HIV

Many different strains of HIV can coexist even within one individual and there is the potential for enormous variation in certain genes. Variability between isolates is greatest in the *env* region which probably forms the main site of the antigenic variation that allows the virus to evade the emerging immune response. The variability derives from the reverse transcription process. Reverse transcriptase is highly error-prone and has notoriously low levels of fidelity when copying RNA into DNA. Mistakes are frequently incorporated into the proviral copy that forms the template for the next generation of virus. If the mutations destroy the function of a critical protein, such as TAT, the new provirus will be unable to replicate and the change will not become fixed in the virus population. However, if the mistake can be tolerated then the mutation will contribute to the constant evolution of the virus population. Frequent changes in certain hypervariable regions of the envelope protein are undoubtedly beneficial to the virus because these are often the binding sites for neutralizing ANTI-BODIES. Continual alteration of these attachment sites generates, within an infected individual, a population of 'quasi-species' of viruses, all slightly different. The host immune response, which may be able to keep pace with such variation early in infection, will probably become overwhelmed as it becomes weakened late in the progression to disease.

Variability is a major contribution to the pathogenesis of HIV and provides a huge obstacle to developing a vaccine. Any vaccine based on envelope-derived peptide antigens, for example, will have to focus on regions of the envelope proteins that are prevented from varying by functional constraints, such as the need to bind the CD4 receptor molecule at specific invariant residues.

Variability also poses a considerable barrier to the development of effective and long lasting antiviral drugs. The first licensed anti-HIV drug, the thymidine analogue 3'-azido-3'-deoxythymidine (AZT), acts as a chain terminator in the synthesis of viral DNA from RNA during reverse transcription. Viral reverse transcriptase incorporates AZT significantly more often into growing DNA strands than does the cellular DNA polymerase, making it an effective virus-specific drug. However, drug-resistant virus strains are emerging from AIDS patients who have been on AZT treatment for several months. It seems likely that such strains may develop *in vivo* as a result of the error-prone reverse transcription of the gene for reverse transcriptase itself, leading to mutant reverse transcriptases. Mutations that led to rejection of AZT by the reverse transcriptase, for example, would be selected and rapidly become fixed in the virus population within a patient undergoing AZT therapy. These considerations point towards the use of combination drug therapies in the future using drugs related to AZT such as ddC or ddI (*see* DIDEOXY-

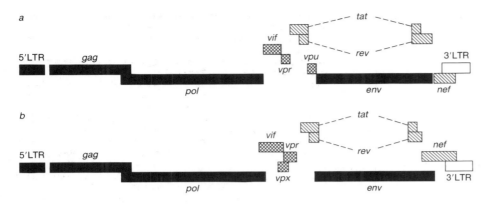

Fig. I4 Comparison of the genomes of HIV-1 (top) and HIV-2 (bottom). HIV-2 contains one gene, *vpx*, which is not present in HIV-1, but lacks the HIV-1 *vpu* gene.

NUCLEOSIDES) so that resistant strains can only emerge if two or more mutations, in different genes, are fixed simultaneously into a progeny virus genome. However, viruses are now being recovered from patients treated with combinations of AZT and ddI or ddC and there is some evidence that such combination therapy, while reducing the individual toxicities, is not markedly more effective in preventing emergence of AZT-resistant virus.

Origins

HIV-1 was the first human immunodeficiency virus to be characterized, in 1983. In 1985 the sequence of a lentivirus clearly different from HIV-1, but more closely related to it than to any other known lentivirus was reported. The new virus was named HIV-2 and differs from HIV-1 in its genomic organization (Fig. I4).

Epidemiological studies have now shown that each HIV virus is associated with one of the two foci of the current AIDS pandemic in Africa. HIV-1 infection predominates in central African cases of AIDS, whereas the outbreak centred on West Africa is characterized by HIV-2. There is preliminary evidence that disease may take longer to emerge after HIV-2 infection compared with HIV-1. This has yet to be confirmed, however, and the genetic difference which determines this decrease (if any) in pathogenicity remains to be determined.

Shortly after the discovery of HIV-2, the sequence of a lentivirus isolated from a captive macaque monkey was reported to be similar to HIV-2. This virus — simian immunodeficiency virus (SIV) — is now thought to have originated in a sooty mangabey monkey and entered the macaque by cross infection. The sooty mangabey is indigenous to West Africa, the location of the HIV-2 epidemic. Other simian immunodeficiency viruses have since been discovered and the current opinion is that the progenitor viruses of HIV may have evolved from SIVs of African monkeys.

In their natural monkey hosts the various SIVs (Table I1) do not cause AIDS-like disease, but can do so when transferred into other monkey species. It seems probable therefore that the natural hosts have had a very long period in which to adapt to the viruses. Disease is only produced when virus is transferred to a nonadapted host. If this argument can be extended to HIV infection of humans, it would suggest that HIV has been transferred across the species barrier from monkey to human relatively recently. AIDS in humans would then be a consequence of the

very short time that humans have had to adapt to the viral infection.

AIDS may therefore have originated in two separate horizontal infection events which could have occurred up to 50 years ago. In one, an SIV harboured in an African green monkey in central Africa was passed to a human; in a second independent event another simian to human transmission occurred in West Africa from, perhaps, a sooty mangabey. Two foci of novel retroviral infections in a nonadapted host were produced, leading to disease. Subsequent evolution of the transmitted viruses has produced the viruses we now know as HIV-1 and HIV-2 respectively, both of which are still evolving rapidly within their new human hosts. According to this scenario HIV-1 and HIV-2 probably do not have a common ancestor in humans.

Although this currently seems the most likely explanation of the observed molecular relationships between SIVs and HIVs, there are other explanations, including the possibility that the ancestral virus originally came from a non-primate.

Co-factors in AIDS

Recently there have been reports of a very few patients in which there is no evidence of HIV infection but who have AIDS-like diseases. The possibility that there is another organism/virus that causes AIDS cannot therefore be ruled out. Co-factors for disease progression, such as mycoplasma infection, have been frequently suggested and some workers have proposed that HIV is not the causative agent of AIDS. However, although it is clear that several

Table I1 Examples of different isolates of SIV and their natural hosts

SIV isolate	Host species
SIV_{mac}	*Macaca mulatta* (rhesus monkey)
SIV_{stm}	*M. arctoides* (stump-tailed macque)
SIV_{fas}	*M. fascicularis* (cynomolgus macaque)
SIV_{smm}	*Cerecebus atys* (sooty mangabey)
SIV_{agm}	*Cercopithecus aethiops* (African green monkey)
SIV_{mnd}	*Papio sphinx* (mandrill)

different factors influence the speed and severity of disease development, the evidence for the central role for HIV in AIDS is currently overwhelming.

Vaccine development

HIV has several attributes which make it extremely difficult to develop an effective vaccine.

1 The extreme genetic variability which leads to immune evasion.

2 The retroviral mode of replication whereby it conceals its genetic material in the host cell genome and can lie latent or active at very low levels for long periods.

3 The target cells, which are generally CD4 T cells, are the very cells needed to fight infection.

4 Its mode of transmission. It is principally transmitted through sexual intercourse, by injection of infected cells, or by transmission from an infected mother to the child before, during or after birth. This means that it is virtually impossible to knock out the virus before it has entered cells and become hidden.

All these factors make many of the approaches to vaccine development that have proved successful against other pathogens seem less applicable to HIV.

Vaccines based on attenuated live virus or killed virus, and SUBUNIT VACCINES based on intact viral proteins are under study. The ability to induce AIDS in monkeys by infection with non-adapted SIVs is being exploited as an animal model to test potential vaccines. Although protection from homologous virus challenge following immunization with killed SIV has been shown, no protective immunity to HIV has yet been demonstrated in humans and the mechanisms of immunological protection remain unknown.

Neither vaccination nor any of the therapies currently or likely to be available within the next few years will provide a cure for HIV infection. Until a safe and reliable vaccine or therapy is available, avoidance of infection remains the only solution.

R. VILE

See also: ANIMAL VIRUSES; RNA SPLICING; T CELL DEVELOPMENT; TRANSCRIPTION.

1 Sheppard, H.W. & Ascher, M.S. (1992) The natural history and pathogenesis of HIV infection. *Annu. Rev. Microbiol.* **46**, 533–564.

2 Haseltine, W.A. (1991) Molecular biology of the human immunodeficiency virus type 1. *FASEB J.* **5**, 2349–2360.

3 Cullen, B.R. (1991) Regulation of HIV-1 gene expression. *FASEB J.* **5**, 2361–2368.

4 Vaisnav, Y.N. & Wong-Staal, F. (1991) The biochemistry of AIDS. *Annu. Rev. Biochem.* **60**, 577–630.

5 Spiegel, H. et al. (1992) Follicular dendritic cells are a major reservoir for human immunodeficiency virus type 1 in lymphoid tissues facilitating infection of CD4$^+$ T-helper cells. *Am. J. Pathol.* **140**, 15–22.

6 Pantaleo, G. et al. (1993) The immunopathogenesis of human immunodeficiency virus infection. *New Engl. J. Med.* **328**, 327–335.

7 Habeshaw, J. et al. (1992) Does the HIV envelope induce a chronic graft-versus-host-like disease? *Immunol. Today* **13**, 207–210.

8 Amadori, A. et al. (1992) CD4 epitope masking by gp1210/antigp120 antibody complexes: a potential mechanism for CD4$^+$ cell function down-regulation in AIDS patients. *J. Immunol.* **148**, 2709–2716.

9 Janeway, C. (1991) Mls: makes a little sense. *Nature* **349**, 459–461.

10 Ameisan, J.C. (1992) Programmed cell death and AIDS: from hypothesis to experiment. *Immunol. Today* **13**, 388–391.

11 Ramsay, A.J. (1992) Diversity and variation in human immunodeficiency virus: implications for immune control. *Immunol. Cell Biol.* **70**, 215–221.

12 Biberfeld, G. & Emini, E.A. (1991) Progress with HIV vaccines. *AIDS 1991*, **5**, S129–S133.

13 Wainberg, W.A. et al. (1993) Clinical correlates and molecular basis of HIV drug resistance. *J. AIDS* **6**, S36–S46.

14 Smith, D.K. et al. (1993) Unexplained opportunistic infections and CD4$^+$ T lymphocytopenia without HIV infection. An investigation of cases in the United States. *N. Engl. J. Med.* **328**, 373–379.

15 Duesberg, P.H. (1992) AIDS acquired by drug consumption and other noncontagious risk factors. *Pharmacol. Ther.* **55**, 201–277.

16 Chanock, R.M. et al. (Eds) (1991) *Vaccines '91: Modern Approaches to New Vaccines Including Prevention of AIDS.* (Cold Spring Harbor Laboratory, New York).

17 Arnon, R. & Van-Regenmortel, M.H. (1992) Structural basis of antigenic specificity and design of new vaccines. *FASEB J.* **6**, 3265–3274.

18 McCune, J.M. (1991) The infective process *in vivo. Cell* **64**, 351–363.

19 Cullen, B.R. (Ed.) (1993) *Human Retroviruses* (Oxford University Press, New York).

Immunoelectron microscopy

IMMUNOELECTRON microscopy is a method of revealing the distribution of molecular components of biological specimens at the ultrastructural level. The principle is to incubate whole specimens, or sections, with reagents such as ANTIBODIES with specific affinity for a particular component. These reagents have narrow specificities for particular sets of molecules and their position on the specimen is revealed using electron-dense markers such as colloidal gold particles. The markers are linked either directly to the primary affinity reagent (single-step labelling) or to a second affinity reagent that binds specifically to the primary one (two-step labelling). After contrasting procedures which display the ultrastructure in the electron microscope the distribution of the electron-dense marker, and therefore that of the molecular component, can be studied. A typical two-step labelling sequence for locating a protein antigen is illustrated in Fig. I5.

Specimen preparation

Preparation starts with fixation, performed usually but not always with aldehyde cross-linking agents such as glutaraldehyde. Fixation ensures stability of structures and components during processing. Fixed specimens are not freely permeable to proteins and so access of affinity reagents to molecular components of the specimen is ensured either by permeabilization or by sectioning of the specimens. After such procedures the specimens or ultra thin sections thereof are labelled in aqueous solution. All modern preparation methods aim to limit denaturation of the specimen by limiting the use of conventional treatments such as fixation in osmium tetroxide or exposure of the specimen to organic solvents at room temperature.

Sectioning methods

Thawed frozen sections [1] are prepared from frozen aldehyde-fixed specimens in special microtomes cooled to around −100°C. The formation of damaging ice crystals during freezing is prevented by cryoprotectants such as sucrose. The sections are thawed by picking them up on a droplet of liquid cryoprotectant

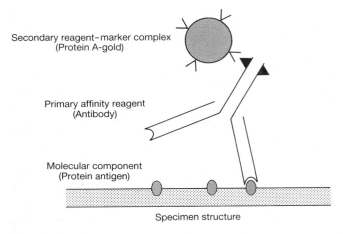

Secondary reagent–marker complex
(Protein A–gold)

Primary affinity reagent
(Antibody)

Molecular component
(Protein antigen)

Specimen structure

Fig. I5 A typical labelling sequence in immunoelectron microscopy.

solution. The method is rapid but serial sections are difficult to prepare and tissue blocks must be stored in liquid nitrogen.

In progressive lowering of temperature (PLT) technique [2] unfrozen aldehyde-fixed specimens are dehydrated in organic solvents and embedded in resin at low temperatures. Formation of ice crystals is prevented by keeping the temperature above the freezing point of the solvent–water mixtures present in the specimen. The method uses Lowicryl or London (LR) resins. These are specially designed methacrylate-based resins that have low viscosity and can be polymerized using ultraviolet light at temperatures around – 30°C or lower. Although the technique may take days, conventional ultramicrotomes can be used, serial sections are prepared easily and both resin blocks and sections can be stored at room temperature.

An alternative to chemical fixing is rapid freezing without cryoprotection. Even with the fastest freezing rates detectable damage to the ultrastructure caused by ice crystals is only prevented in surface layers about 10–15 µm thick. Since the specimens are unstable on thawing they must be embedded in resin before sectioning. This is done at low temperature (e.g. at – 85°C) by first exchanging organic solvents for ice in the still unfrozen specimen by a process called freeze substitution [3]. This is followed by resin infiltration and polymerization, either at low temperature using Lowicryls or at room temperature using epoxy resins. Chemical fixatives and/or heavy metal ions can be included in the solvent. The structural preservation and labelling densities can be superior to the PLT technique.

Nonsectioning methods

Solid specimens can be made permeable to reagents without sectioning by freezing and thawing or exposure to detergents. Such 'pre-embedding methods' often use the enzyme HORSERAD-ISH PEROXIDASE (HRP) as a marker. The enzyme, covalently linked to primary or secondary affinity reagents, produces electron-dense reaction products whose location is identified after conventional processing procedures using osmium tetroxide and epoxy resin embedding. The technique is prone to false negatives produced by variable penetration of labelling reagents and also to

imprecise localization caused by diffusion of electron-dense reaction products and/or translocation of antigens produced by permeabilization agents.

Permeabilization can be avoided if components lie on the specimen surface. They may be labelled with affinity reagents before or after aldehyde fixation followed by conventional processing for transmission or scanning electron microscopy (*see* ELECTRON MICROSCOPY).

Labelling

Affinity reagents

Antibodies are the main class of primary reagents. Others include LECTINS, used for oligosaccharide localization, and PROBES used for *in situ* NUCLEIC ACID HYBRIDIZATION such as cDNA. Primary reagents can be linked to electron-dense markers directly but are more commonly localized with secondary reagents of broad specificity. Examples of secondary reagents for antibody localization include PROTEIN A, PROTEIN G or second antibodies, all of which can be complexed with gold particles.

One special class of reagent includes AVIDIN and STREPTAVIDIN, both of which bind to biotin (*see* BIOTINYLATION). The latter can be covalently linked to proteins which can then be localized using avidin or streptavidin–marker complexes. Uses of this technique include localization of biotinylated primary reagents such as antibodies, lectins or cDNA, and localization of proteins which have been applied to living cells before processing, such as endocytosed ligands.

Markers

Gold particles are the most useful markers in immunoelectron microscopy because they are very electron dense, are easily prepared by reduction of gold salts, form stable complexes spontaneously with most protein affinity reagents and can be counted easily (Fig. I6). Particles in the size range 3–15 nm are most popular and are best prepared using tannic acid/citrate reduction [4]. The small variation in particle diameters in these preparations means that particles of different sizes (e.g. 3, 5 or 8 nm in diameter) can be used to label different components on the same specimen.

Protein–gold complexes are prepared by mixing suspensions of gold particles (a colloid) with a solution of the protein affinity reagent, under appropriate conditions [5]. The complexes are purified easily by density gradient ultracentrifugation and can be stored frozen for many years.

Another electron-dense marker is FERRITIN. Currently it is used less often than gold particles because it is less electron dense and the conjugates are more complicated to prepare than gold–reagent complexes.

Labelling procedure

Ultrathin sections are usually attached to special electron microscope support grids which are then floated on buffer droplets containing the reagents. Typically, protein antigen localization

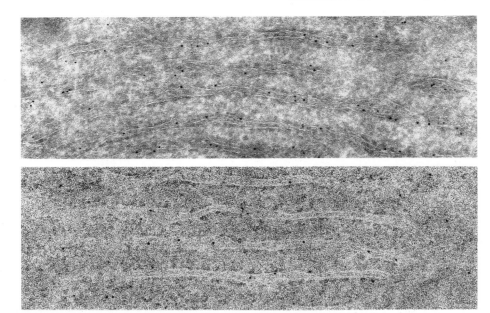

Fig. 16 Localization of the enzyme glucosidase II in thin sections of hepatocytes using polyclonal antibodies followed by protein A complexed to 6-nm gold particles. The gold particles are located predominantly on the luminal aspect of the rough ENDOPLASMIC RETICULUM. Thawed frozen section (top) and PLT technique (bottom).

using an antibody is as follows. The first steps include (1) prevention of nonspecific binding of affinity reagents by incubation on solutions containing foetal calf serum or gelatine from cold-water fish, and (2) blocking of residual free aldehyde groups in the specimen using solutions of glycine or ammonium chloride. Subsequently the sections are incubated on suitably diluted antibody and a suitable secondary reagent–marker complex such as protein A–gold. Each step is followed by washes in buffer. Finally, the specimens are stained, dried and observed in the electron microscope.

In resin sections, contrast is enhanced by applying heavy metal 'stains' such as uranium and lead salts. Thawed frozen sections show disruption of structures after air drying [6], however this problem has been overcome by including the heavy metal stains in solutions of water-soluble embedding media. After drying down onto the grid, these solutions form thin films that help to support the structures.

Quantitation

The local concentration of gold labelling can be found by relating particle counts to the size of a structure or its profile which can itself be estimated using morphometric and stereological techniques. While the local concentration of gold particles can be used in a semiquantitative manner to identify structures in which antigen is concentrated it has proved difficult to use gold labelling to estimate the absolute amount of antigen present. The reason is that many variables in immunocytochemical method combine to determine how many gold particles will finallly becomes associated with each antigen in the section. One proposed solution is to use calibration gels containing known concentration of purified, soluble proteins [7]. In this method the concentration of antigen in a structure can be found by comparing the labelling densities over the gel and the specimen. Another method estimates the number of antigens associated with each gold particle directly by combining immunoelectron microscopy and biochemistry under standard, precisely controlled conditions [8].

J. LUCOCQ

1 Tokuyasu, K.T. (1973) A technique for ultracryomicrotomy of cell suspensions and tissues. *J. Cell Biol.* **57**, 551–565.
2 Carlemalm, E. et al. (1982) Resin development for electron microscopy and an analysis of embedding at low temperature. *J. Microsc.* **126**, 123–143.
3 Humbel, B.M. & Schwarz, H. (1989) Freeze-substitution for immunocytochemistry. In *Immuno-gold Labelling in Cell Biology* (Verkleij, A.J. & Leunissen, J.L.M., Eds) Ch. 7, 115 (CRC Press, Boca Raton).
4 Slot, J.W. & Geuze, H.J. (1985) A new method for preparing gold probes for multiple-labelling cytochemistry. *Eur. J. Cell Biol.* **38**, 87–93.
5 Slot, J.W. & Geuze, H.J. (1985) The sizing of protein A colloidal gold probes for immunoelectron microscopy. *J. Cell Biol.* **90**, 553–556.
6 Griffiths, G. et al. (1984) On the preparation of cryosections for immunocytochemistry. *J. Ultrastruct. Res.* **89**, 65–78.
7 Slot, J.W. et al. (1989) Quantitative aspects of immuno-gold labeling in embedded and nonembedded sections. *Am. J. Anat.* **185**, 271–281.
8 Griffiths, G. & Hoppeler, H. (1986) Quantification in immunocytochemistry: correlation of immunogold labelling to the absolute number of membrane antigens. *J. Histochem. Cytochem.* **34**, 1389–1398.

immunoelectrophoresis Technique of gel ELECTROPHORESIS in which proteins are identified by precipitation with specific antisera after electrophoretic separation of the material to be analysed.

immunofluorescence microscopy Methods for light microscopy which make use of specific ANTIBODIES tagged with fluorescent dyes to visualize the distribution of particular proteins or other antigens in cells and thus the structures of which they form part. The locations of several different proteins in the same cell can be visualized using antibodies tagged with differently coloured dyes.

immunofluorescent techniques Any technique that depends on the binding of an antigen by a specific antibody labelled with a fluorescent dye. This general technique has many applications including cell sorting and chromosome sorting (*see* FACS), the microscopic study of cell structure and gene expression by IMMU-NOFLUORESCENCE MICROSCOPY, and the staging of cells at different phases of development and differentiation (by their differential expression of cell-surface antigens).

immunogen A substance capable of provoking an immune response. The term is often used as a synonym for ANTIGEN, but its emphasis is on the ability to elicit a response, rather than to bind to antibody. Hence, all immunogens are antigens, but not all antigens are necessarily immunogens.

immunogenicity The ability of a substance to stimulate an immune response under suitable conditions. Immunogenicity is dependent on the properties both of the putative IMMUNOGEN and of the recipient. The properties which confer immunogenicity to a substance are not known precisely; however it must be recognized as foreign. Generally, increasing molecular weight and complexity improve the immunogenic properties of a substance. The mode of presentation of an antigen to a responsive cell or animal also determines whether it is immunogenic. The same ANTIGEN may be immunogenic to varying degrees in different species; for example pneumococcal capsular polysaccharides are immunogenic in man but not in rabbits.

immunoglobulin(s) Structurally highly variable GLYCOPROTEINS that specifically recognize and bind ANTIGEN and are involved in the specific immune response against pathogenic microorganisms. They are produced exclusively by the B CELLS of the immune system (*see* B CELL DEVELOPMENT) and occur in both membrane-bound and secreted (ANTIBODY) form. The immunoglobulin monomer is composed of four polypeptide chains, two identical light chains and two identical heavy chains (*see* ANTIBODY). The N-terminal 110 amino acid residues of both the light and heavy chains are variable between immunoglobulin molecules of differing antigen-binding specificity and determine that specificity. There are two types of light chain, κ and λ, which are defined by the differences in the amino-acid sequence of the C-terminal 110 residues.

In humans and higher mammals there are five main structural and functional classes (isotypes) of immunoglobulin — IgM, IgD, IgG, IgA and IgE (*see individual entries*) — which are defined by differences in the C-terminal region (constant region) of the heavy chain. Within the IgG class four subclasses are defined that have highly homologous heavy chains that may differ in important functional activities. Similarly, there are two subclasses of IgA. The heavy chains of the isotypes are designated μ, δ, γ3, γ1, α1, γ2, γ4, α2, and ε; the order given is that of the heavy chain constant region genes on human chromosome 14.

Each individual can produce antibodies of almost any antigen specificity required. This extreme variability is achieved by a unique mechanism of immunoglobulin gene assembly which occurs in B cells early in their development (*see* GENE REARRANGE-MENT; GENERATION OF DIVERSITY; IMMUNOGLOBULIN GENES). *See also*: IMMUNOGLOBULIN STRUCTURE.

immunoglobulin binding protein, immunoglobulin heavy chain binding protein *See*: BiP.

immunoglobulin classes/subclasses Denoted IgA, IgD, EgE, IgG, and IgM (*see also individual entries*). *See*: ANTIBODIES.

immunoglobulin domain The folded globular structure of an IMMUNOGLOBULIN HOMOLOGY REGION. It is composed of two 'sheets' of protein structure referred to as antiparallel β-pleated sheet (*see* PROTEIN STRUCTURE). It is typical of proteins of the IMMUNOGLOBULIN SUPERFAMILY and may also be referred to as the immunoglobulin fold. *See*: ANTIBODIES; IMMUNOGLOBULIN STRUCTURE.

immunoglobulin fold (Ig-fold) A common folding pattern found in both immunoglobulin V and C domains consisting of two β-sheets built up from antiparallel β-strands and termed the Ig-fold. A conserved disulphide bond links the two sheets. V domains are distinguished by two extra β-strands in one sheet. The presence of the Ig-fold was predicted in other members of the IMMUNOGLO-BULIN SUPERFAMILY on the basis of sequence similarities and has now has been confirmed by structural studies of MHC class I molecules, and CD2, CD4, and CD8 antigens. *See also*: ANTIBOD-IES; CD ANTIGENS; IMMUNOGLOBULIN STRUCTURE.

Immunoglobulin genes

THE IMMUNOGLOBULINS function as cell-surface antigen receptors on B CELLS and as secreted ANTIBODIES. An immunoglobulin molecule is composed of two identical heavy (H) chains and two identical light (L) chains. Light chains may be either the kappa (κ) type or lambda (λ) type (*see* ANTIBODIES). Each chain of the immunoglobulin molecule consists of two functionally distinct regions: an N-terminal variable (V) region, and a C-terminal constant (C) region. The antigen-binding site is formed by the pairing of heavy and light chain V regions, whereas the C region domains mediate characteristic biological activities (effector functions) associated with immunoglobulin molecules (*see* ANTIBOD-IES; IGA; IGD; IGE; IGG; IGM; IMMUNOGLOBULIN STRUCTURE).

In mammals, there are three separate sets of immunoglobulin genes — the heavy chain (H) locus, the light (L) chain κ locus and the L chain λ locus, situated on different chromosomes (Table I2). In their germ-line state immunoglobulin genes are nonfunctional, consisting of arrays of multiple gene segments. At each locus, the coding sequence of the V region is assembled during B CELL DEVELOPMENT from this large pool of gene segments by a site-specific recombination mechanism, referred to as V-(D)-J recombination (*see* GENE REARRANGEMENT).

Table I2 Chromosomal locations of immunoglobulin loci

	Murine	Human
Heavy chain (H)	12	14 q32.3
Kappa (κ)	6	2 p12
Lambda (λ)	16	22 q11

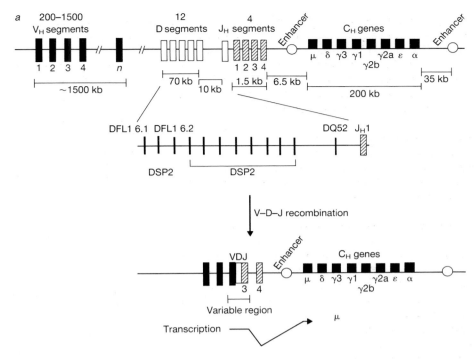

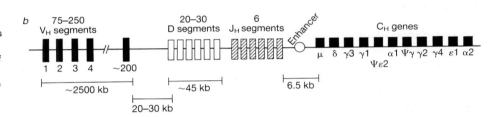

Fig. 17 Immunoglobulin heavy chain loci. *a*, Murine heavy chain locus showing gene organization, V-(D)-J recombination, and expression of μ heavy chain. After a V region exon has been assembled, a primary transcript is made that includes the VDJ sequence and the μ gene. Sequences intervening between the end of the assembled J region and the beginning of the μ gene (the J-C intron) are spliced out to produce a functional heavy chain mRNA. *b*, Organization of the human heavy chain locus.

Three types of gene segment are joined to form a heavy chain V-region coding sequence: variable (V_H), diversity (D), and joining (J_H) segments (see Fig. 17). Functional light chain V-region EXONS are assembled from two segments, a V gene (either V_κ or V_λ) and a J segment (either J_κ or J_λ) (see Fig. 18).

The V_H segment encodes a LEADER PEPTIDE of 16–30 amino acids plus 95–100 amino acids of the V region, including two HYPERVARIABLE REGIONS (CDR1 and CDR2, *see* COMPLEMENTARITY DETERMINING REGIONS). A 130-bp INTRON separates the leader peptide and variable region EXONS. The 10–17 bp D segment plus the sequences at the V_H-D and D-J_H boundaries encode the 3–7 amino acids of the third hypervariable region (CDR3); the remaining 12–21 amino acids of the V region are encoded by the J_H segment.

κ and λ light chains are each ~214 amino acids long. The rearranged $V_\lambda J_\lambda$ exon encodes a leader peptide plus the 98–120 amino acid V region.

Heavy chain C genes

In mammals, there are five major classes of heavy chain C region as defined by their amino-acid sequence: μ_H, δ_H, γ_H, ε_H and α_H.

Murine and human γ_H chains each include four subclasses. Each class, or subclass, is encoded by a separate C gene (C_H gene) — these lie downstream of the J_H gene segments (Fig. 17). Each C_H gene is composed of 3–4 exons, encoding separate structural and functional domains, plus 1–2 additional exons ~2 kilobases (kb) downstream, which when included in the messenger RNA by differential RNA SPLICING, specify membrane localization (*see* ANTIBODY; B CELL DEVELOPMENT).

Rearranged V regions are always first expressed in combination with the C_μ gene, to yield μ heavy chain (secreted form μ_s, or membrane-bound form μ_m). Association of the μ heavy chain with a functional light chain yields antigen-specific IgM molecules (*see* ANTIBODY; B CELL DEVELOPMENT).

V region gene segments

Human H chain locus

An estimated 75–250 contiguous V_H gene segments lie within a 2000–3000-kb region of human chromosome 14 [1]. On the basis of amino-acid sequence homologies, these are divided into three V_H subgroups and on the basis of DNA sequence similarities, into

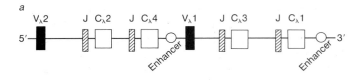

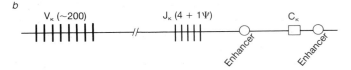

Fig. I8 Organization of murine κ (*a*), and λ (*b*) light chain loci.

six gene families. Members within a family display 80% or more DNA sequence similarity. Around 20–30 D segments are clustered within a 40–50-kb region between the V_H and J_H segments (Fig. I7). The distance between the most downstream V_H segment and the nearest D segment is approximately 33 kb, and from the same V_H segment to the first of the J_H segments about 100 kb. Six functional J_H segments are interspersed with three J_H PSEUDOGENES.

A second cluster of D segments (the Dminor locus) has been reported, which is possibly located immediately upstream of, or near to the 5′ end of, the entire V_H locus [2]. However, there is no evidence that sequences from the Dminor cluster are ever used during V region formation. A number of D segments were also reported to be interspersed with V_H gene segments [3] although it is unclear whether these D segments are actually located within the Dminor locus.

Murine H chain locus

The total number of murine V_H gene segments is probably between 200 and 1500 (Table I3). These have been divided into three protein subgroups (V_H subgroups); they are further divided into nine gene families on the basis of DNA sequence homologies. Unlike human V_H segments, members of different murine V_H gene families are in general clustered. On the basis of coding or flanking sequence homologies, the twelve murine D segments are divided into three gene families: DSP2 (9 members), DFL16 (2 members), and DQ52 (1 member). The DQ52 segment is approxi-

Table I3 Number of variable region gene segments

	Murine	Human
V_H	200–1500	75–250
D	12	20–30
J_H	4	6
V_κ	~200	~80
J_κ	5	5
V_λ	2	40–70
JC_λ	4	6

mately 700 bp 5′ of the first J_H segment, and the other 11 segments within a 70-kb cluster from 10 to 80 kb further upstream (Fig. I7). There are four functional J_H segments.

Light chain loci

In humans, ~80 V_κ gene segments lie within a 2000–3000-kb region on chromosome 2; members of the four V_κ gene families (V_κ I–IV) are intermingled throughout the entire V_κ locus. Around half of the sequenced V_κ gene segments are nonfunctional pseudogenes. Also, at least 10 more V_κ pseudogenes exist on other chromosomes. Five functional J_κ segments are located 20–30 kb downstream of the nearest V_κ segment; J_κ5 is separated from the single C_κ exon by a 2.5-kb intron. The murine V_κ locus (Fig. I8) is less well characterized, although there are probably at least 200 V_κ gene segments. A cluster of four functional J_κ segments plus one J_κ pseudogene ($\psi 1J_\kappa 3$) lie upstream of a single C_κ exon.

The λ light chain locus is very small. Most laboratory strains of mice have only two V_λ gene segments and four C_λ genes; each C_λ gene has an associated J_λ segment (Fig. I8). The $JC_\lambda 4$ unit is nonfunctional. In humans, greater diversity of λ V region sequences than in mice probably reflects a larger number of V_λ gene segments (probably between 40 and 70 segments), although this has not been fully characterized. Six contiguous C_λ genes, each with an associated 5′ J_λ segment, are located downstream of the V_λ segments. Additionally, three unlinked C_λ genes have been identified by nucleic acid HYBRIDIZATION analyses.

Gene rearrangement

The assembly of V region gene segments at H and L chain loci is ordered and developmentally regulated (*see also* B CELL DEVELOPMENT). Site specificity of the V-(D)-J recombination process is mediated by short signal sequences which lie immediately adjacent to each of the unrearranged coding segments. These are composed of a palindromic heptamer (consensus 5′-CACAGTG-3′) and an AT-rich nonamer (consensus 5′-ACAAAAACC-3′), separated by a nonconserved spacer of 12 bp (12 signal) or 23 bp (23 signal) (*see* GENE REARRANGEMENT; RECOMBINASE). The first recombination involves D to J_H joining at both heavy chain alleles. At this stage (the pro-B cell), it is observed that most D-J_H joins represent a single READING FRAME. A V_H segment is then joined to the DJ_H sequence on one chromosome. If the joining creates a functional VDJ coding sequence, a negative feedback mechanism prevents V_H to DJ_H recombination at the other allele. This negative regulation, which ensures the expression of only one heavy chain specificity per cell, is termed ALLELIC EXCLUSION. Infrequently, V_H to DJ_H junctions are observed at both H alleles but, in these cases, one of the two is nonfunctional. The regulation of allelic exclusion at heavy chain loci is unclear, but may involve expression of the product of the functionally rearranged gene (μ chain) in a complex with two accessory proteins — Vpre-B and λ5 [4]. These proteins structurally resemble immunoglobulin light chains. Cells with rearranged heavy chain genes and unrearranged light chain genes mark the pre-B cell stage.

V to J joining at the κ locus follows formation of a functional heavy chain gene. A similar allelic exclusion mechanism operates to limit formation and expression of only a single light chain allele per cell. Finally, if both κ gene rearrangements are nonfunctional, λ light chain gene VJ joining takes place; the expressed κ or λ light chains then associate with the existing μ heavy chains to form IgM immunoglobulin molecules. Expression of IgM on the cell surface defines the B cell stage.

An estimated 10^{12}–10^{14} different antigen-binding specificities can potentially be generated within a single individual, owing to sequence diversity within the V regions. This diversity is generated by a combination of COMBINATORIAL DIVERSITY and JUNCTIONAL DIVERSITY (*see* GENERATION OF DIVERSITY).

Immunoglobulin gene expression

Immunoglobulin gene expression is restricted to B lymphocytes and is developmentally regulated. It is mainly controlled at the transcriptional level, although the large rise in heavy and light chain mRNA levels which occurs during B cell to PLASMA CELL maturation also involves increased mRNA stability. Initiation of high level TRANSCRIPTION follows completion of V region exon assembly, and is controlled by a PROMOTER and one or more ENHANCERS (*see* EUKARYOTIC GENE EXPRESSION). Enhancer elements lie within the heavy chain J_H-C_μ intron (the heavy chain enhancer), κ gene J_κ-C_κ intron, and downstream of C_κ (Figs I7 and I8). B-cell-specific enhancer elements are also observed 25 kb downstream of the C_α heavy chain gene in rats and mice. The murine λ locus contains enhancer elements 35 kb downstream of $C_\lambda 1$ and 15.5 kb 3′ of $C_\lambda 4$ (Fig. I8). Promoter sequences are located upstream of all V gene segments, and are therefore retained during V_H to DJ_H or V_L to J_L joining.

The promoters and enhancers contain multiple DNA sequence elements which bind nuclear proteins. Binding of proteins to specific sites positively or negatively regulates the level and cell-type specificity of immunoglobulin gene transcription (*see* EUKARYOTIC GENE EXPRESSION; TRANSCRIPTION FACTORS). Some of these sites are indicated in Fig. I9. The octamer motif (consensus sequence 5′-ATTTGCAT-3′) is present in the heavy chain enhancer, the J_κ-C_κ intron enhancer, and in the promoter of all immunoglobulin genes. Octamer sequences within

heavy chain promoter regions are in reverse orientation (5′-ATGCAAAT-3′), relative to other immunoglobulin gene octamer sites. Optimum, tissue-specific transcription of rearranged heavy chain genes is largely dependent on the J_H-C_μ intron enhancer plus the octamer site within the V_H promoter [5].

Two classes of octamer binding activity, initially characterized by electrophoretic mobility of DNA–protein complexes, have been demonstrated in B cells. The first, referred to as Oct-1, is found in all cell types tested; the other, Oct-2, is expressed specifically in lymphocytes, and is generally presumed to be a key factor in controlling activation of high level, B cell-specific immunoglobulin gene transcription [6].

Several other sites which regulate heavy chain enhancer activity have been characterized (Fig. I9a): four 'E-motifs' (μE1, 2, 3, 4), with the consensus sequence 5′-CAGGTGGC-3′, were originally identified as sites that bind nuclear proteins in whole lymphocytes [7]. Functionally, all these sites seem to contribute to enhancer activity. Inhibition of heavy chain enhancer activity in nonlymphoid cells is partly mediated by a fifth E-motif, site μE5.

The C1, C2, and C3 sites (Fig. I9a) show homology to the SV40 enhancer core motifs (consensus 5′-GTGAAAG-3′), although they are not essential for heavy chain enhancer activity — mutation of these sites only slightly reduces heavy chain enhancer activity in lymphocytes. Site μB influences both the activation and lymphoid specificity of the enhancer, possibly through cooperative interaction with site μE3 [8].

A number of different DNA sequence elements which regulate heavy chain gene transcription have been identified in the promoter region (Fig. I9b). These include a conserved heptamer sequence (consensus 5′-CTCATGA-3′) located immediately upstream of the octamer, and a pyrimidine-rich region. Oct-1 and Oct-2 proteins can also bind to the heptamer sequence. The κ gene J-C intron enhancer contains three E-motifs: sites κE1, κE2, and κE3 (Fig. I9c). Site κE3 seems to bind the identical protein (NFμE3) that binds to site μE3 in the heavy chain enhancer. J_κ-C_κ intron enhancer activity is largely dependent on the binding of protein NFκB to its recognition site, κB (Fig. I9c).

Class switching

Following interaction with antigen, a lymphocyte may express an immunoglobulin with identical antigen specificity to the original IgM (the first type of immunoglobulin made) but with a different C region sequence, changing the class of immunoglobulin molecule produced to IgD (δ chain), IgG (γ), IgE (ε), or IgA (α). With the exception of the switch to IgD expression, this heavy chain class switch is caused by a second, B cell specific, somatic recombination event — switch recombination (Fig. I10). This occurs within or near to switch regions located upstream of each C_H gene, and results in the deletion of the C_μ gene and one or more downstream C_H genes. The first C_H gene beyond the 3′ end of the deletion is therefore brought adjacent to the rearranged V region and is expressed with it. In IgM to IgG2a class switching, for example, recombination between switch regions upstream of C_μ and $C\gamma_{2a}$, deletes the C_μ-C_δ-$C_{\gamma 3}$-$C_{\gamma 1}$ region and places the $C_{\gamma 2a}$ exons immediately downstream of the V region exon: 5′-VDJ-C_γ2b-C_γ2a-C_ε-C_α-3′.

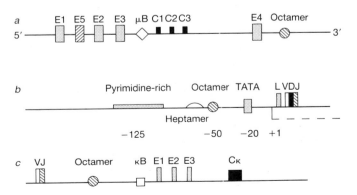

Fig. I9 Regulatory sites in the immunoglobulin genes. *a*, Heavy chain enhancer; *b*, heavy chain promoter; *c*, J-C_κ intron enhancer.

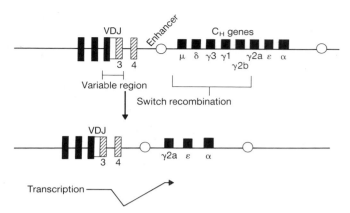

Fig. I10 Switch recombination. Recombination between switch regions upstream of each C_H gene results in deletion of the μ gene (and of genes downstream). The same VDJ region can then be expressed with a different C gene.

The IgD isotype comprises light chains plus δ_H chains; δ_H chain expression results from alternative RNA splicing, rather than switch recombination.

Immunoglobulin gene disorders

In several diseases of the immune system, aberrant rearrangements of the immunoglobulin loci are observed, including deletions, inversions, and chromosomal translocations (*see* CHROMOSOME ABERRATIONS). BURKITT'S LYMPHOMA, a human B cell neoplasm, is characterized by reciprocal translocations between chromosome 8 and the immunoglobulin locus on chromosome 14 (heavy chain), 2 (κ light chain), or 22 (λ light chain). The translocation breakpoint on chromosome 8 is within, or adjacent to, the cellular ONCOGENE c-*myc*, and results in juxtaposition of c-*myc* with immunoglobulin gene sequences. Although the level of c-*myc* expression is elevated in Burkitt's lymphoma cells, which probably contributes to the malignant phenotype, a causal relationship between the translocation and c-*myc* activation has not been established.

Translocation involving the heavy chain locus and a region of chromosome 11 containing the *bcl-1* oncogene is observed in chronic lymphocytic LEUKAEMIAS (CLL). In follicular LYMPHOMAS, translocation joins the *bcl-2* oncogene on chromosome 18 to the heavy chain locus (chromosome 14). Translocation breakpoints often occur near to sequences that resemble V-(D)-J recombination signals (*see* GENE REARRANGEMENT; RECOMBINASE) or heavy chain switch regions, suggesting that, in some cases, translocation may result from erroneous V-(D)-J joining or class switching. B cell neoplasms caused by incorrect RNA processing of immunoglobulin gene transcripts have also been described (*see* HEAVY CHAIN DISEASE).

J.F. MCBLANE

See also: IMMUNOGLOBULIN STRUCTURE; T CELL RECEPTORS; T CELL RECEPTOR GENES.

1 Blackwell, T. K. & Alt, F.W. (1988) Immunoglobulin genes. In *Molecular Immunology*, 1–60 (IRL Press, Oxford).
2 Buluwella, L. et al. (1988) The use of chromosomal translocation to study human immunoglobulin gene organization: mapping D_H segments within 35kb of the C_μ gene and identification of a new D_H locus. *EMBO J.* **7** 2003–2010.
3 Matsuda, F. et al. (1988) Dispersed localization of D segments in the human immunoglobulin heavy-chain locus. *EMBO J.* **7**, 1047–1051.
4 Karasuyama, H. et al. (1990) *J. Exp. Med.* **172**, 969–972.
5 Jenuwein, T. & Grosschedl, R. (1991) Complex pattern of immunoglobulin μ gene expression in normal and transgenic mice: nonoverlapping regulatory sequences govern distinct tissue specificities. *Genes Devel.* **5**, 932–943.
6 Staudt, L. M. & Lenardo, M.J. (1991) *Annu. Rev. Immunol.* **9**, 373–398.
7 Ephrussi, A. et al. (1985) B lineage-specific interactions of an immunoglobulin enhancer with cellular factors *in vivo. Science* **227**, 134–140.
8 Nelsen, B. et al. (1990) *Mol. Cell. Biol.* **10**, 3145–3154.

immunoglobulin homology region (immunoglobulin domain) Homologous sequences of ~110 amino-acid residues that are repeated throughout the light and heavy chains of immunoglobulin molecules (*see* ANTIBODIES; IMMUNOGLOBULIN STRUCTURE; IMMUNOGLOBULIN SUPERFAMILY). Each homology region is encoded in a distinct EXON and its folded structure in the immunoglobulin molecule constitutes a domain. It is suggested the immunoglobulins originated as recognition molecules of 110 amino acid residues and gene duplication events in evolution have resulted in the more complex structures observed in the present day.

immunoglobulin-like (Ig-like) domain Many proteins contain segments of sequence with similarities to immunoglobulin V or C domains and these are predicted to have a folding pattern similar to that in immunoglobulins — the immunoglobulin fold. These domains are called Ig-like to distinguish them from the domains of immunoglobulins themselves. See: IMMUNOGLOBULIN SUPERFAMILY.

immunoglobulin pseudogenes Regions of DNA with close sequence similarity to functional immunoglobulin genes (*see* PSEUDOGENES). In mammals, pseudogenes have been identified in mouse and man at both heavy and light chain loci (*see* IMMUNOGLOBULIN GENES). There are three human C_H pseudogenes: ψγ, ψε1 and ψε2. The ψγ and ψε1 sequences lie within the main C_H gene cluster on chromosome 14 whereas ψε2 is a PROCESSED PSEUDOGENE translocated to chromosome 9. No well-conserved mouse C_H pseudogenes have been identified. In chickens, there are 25 V_λ pseudogenes upstream of a unique functional V_λ gene and a larger number of V_H pseudogenes within a 60-kb region immediately upstream of the single, functional V_H gene. Following assembly of functional heavy chain and λ light chain V region exons by recombination, these pools of pseudogenes are used to introduce sequence diversity into the downstream recombined V gene by GENE CONVERSION (*see* GENERATION OF DIVERSITY). The mammalian V pseudogenes have not been clearly shown to contribute to somatic diversity by gene conversion mechanisms.

Immunoglobulin structure

IMMUNOGLOBULINS are a family of highly variable GLYCOPROTEINS which bind specifically, and in some cases very tightly, to molecules foreign to the organism (ANTIGENS). Immunoglobulins are produced only by the B CELLS (B lymphocytes) of the immune system and are synthesized in both membrane-bound and secreted form. As cell-surface receptors for antigen they mediate the antigen-specific activation of B cells by antigen and the subsequent production of specific ANTIBODIES — the secreted form of immunoglobulin (*see* B CELL DEVELOPMENT). Immunoglobulins are members of a large superfamily of proteins, which have a variety of roles in the organization and control of the immune response (*see* IMMUNOGLOBULIN SUPERFAMILY; T CELL RECEPTORS).

The basic structure, function, and classification of immunoglobulins and their biological role as antibodies are covered in the article on ANTIBODIES elsewhere in this volume. This article focuses on the detailed structure of these molecules, in particular the antigen-binding site and antigen–antibody interactions [1–3].

Common features of immunoglobulins were first recognized physicochemically and serologically (*see* ANTIBODIES), then described in terms of their amino-acid sequences, and only much later revealed in atomic detail by X-RAY CRYSTALLOGRAPHY. Limited proteolytic digestion and treatments that cleave disulphide bonds showed that antibodies are composed of multiple polypeptide chains. The chains are distinguished by size as light (L) chains (M_r ~23 000) and heavy (H) chains (M_r ~60 000). A typical immunoglobulin (IgG) contains two identical L chains and two identical H chains.

Systematic internal homologies in the amino-acid sequences of H and L chains suggested that immunoglobulins are composed of multiple related copies of a basic folding unit of ~100 amino acids. Each of these units forms an individual, quasi-independent, three-dimensional structure — a domain — with a common folding pattern (the immunoglobulin fold). Light chains contain two such domains, and heavy chains contain four or five domains.

An apparently limitless number of different antibody specificities can be generated by any healthy individual by generating immunoglobulins with variability in the amino-acid sequences in appropriate regions to create antigen-binding sites of diverse structure. The N-terminal domains of the light and heavy chains are called the variable (V) regions because they exhibit far greater sequence variation than the other domains, called the constant (C) regions (Fig. I11). Within the variable domains there are regions of still greater variability — called the hypervariable regions or COMPLEMENTARITY DETERMINING REGIONS (CDRs) — which make up the antigen-binding site. At the atomic level, the hypervariable regions correspond to six loops of polypeptide chain joining strands of a double β-sheet structure.

Immunoglobulins fall into five main classes — IgG, IgA, IgM, IgD and IgE (*see individual entries*) — which differ in the amino-acid sequence and number of domains in the constant regions of the H chain. Among light chains, there are two different ISOTYPES κ and λ (*see* ANTIBODIES) which differ in amino-acid sequence and in the structure of certain regions in the antigen-binding site.

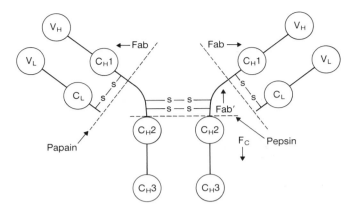

Fig. I11 Schematic diagram of the structure of IgG. The molecule contains four polypeptide chains: two identical light chains each containing one V and one C domain, denoted V_L and C_L, and two identical heavy chains each containing one variable and three constant domains, denoted V_H, C_H1, C_H2 and C_H3. The C_H1 and C_H2 domains are linked by a short hinge region. The angle between the V_L–V_H domain pair and the C_L–C_H1 domain pair is called the elbow angle. Immunoglobulins also contain carbohydrate moieties not shown in this figure. S-S, Interchain disulphide bridges; Fab, Fab′, Fc, fragments produced by proteolytic cleavage.

The chains are held together by disulphide bridges, and in addition, corresponding domains — V_L–V_H and C_L–C_H1 and C_H3–C_H3 (but not C_H2–C_H2) — form interfaces packed together by extensive van der Waals contacts. It is characteristic of proteins that interior interfaces are formed by the packing of complementary surfaces. This complementarity of fit fixes the relative spatial disposition of the pieces that interact. A tendency to conserve the residues involved in these interfaces explains why many different L and H chains can pair freely to form immunoglobulin molecules. In the case of the V_L–V_H interaction, the conservation of the relative geometry has the additional important consequence that, to a reasonable first approximation, the double β-sheet frameworks of V_L and V_H domains form a scaffolding of nearly constant structure on which the antigen-binding site is erected.

All heavy chains can be produced in secreted and membrane-bound form as a result of developmentally regulated alternative RNA SPLICING (*see* B CELL DEVELOPMENT; IMMUNOGLOBULIN GENES). The membrane-bound forms have a transmembrane region of ~26 uncharged hydrophobic amino-acid residues and a variable length cytoplasmic tail.

X-ray crystal structure analyses of immunoglobulins have been carried out on intact antibodies, BENCE-JONES PROTEINS (consisting of light-chain dimers, *see* ANTIBODIES), Fc fragments and — most commonly — Fab or Fab′ fragments (see Fig. I11). In some cases, the structures of molecules in different states of ligation are known: fragments of the antibody D1.3 have been solved in the unligated state, in complex with its antigen (lysozyme) and in complex with the ANTI-IDIOTYPIC ANTIBODY E225. The structure of Fab B1312 has been solved in an unligated state and in a complex with its antigen, a peptide from myohaemerythrin.

Figure I12 shows the structure of a typical V domain and a

a

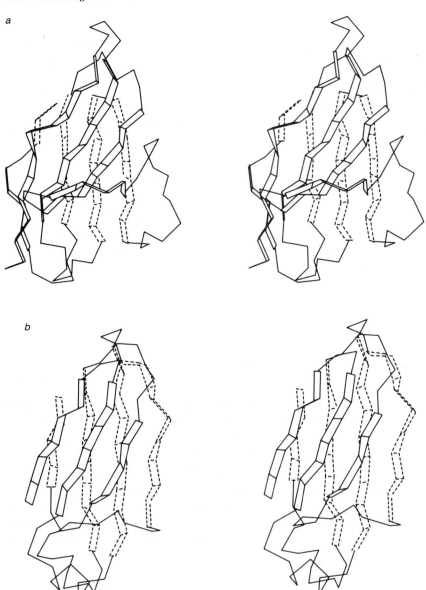

b

Fig. I12 Stereopairs showing folding patterns of (*a*) variable and (*b*) constant domains. The domains shown here are the V_L and C_H1 domains of Fab IgG KOL. Strands of β-sheet are shown as polygonal ribbons, and the loops joining them as lines joining the Cα atoms of successive residues. Each domain contains two β sheets; one is shown in solid lines and the other in broken lines.

typical C domain. Each has the form of a double β-sheet 'sand-wich'. (A β-sheet is one of the two basic types of secondary structure in proteins, *see* PROTEIN STRUCTURE.) In immunoglobu-lin domains the two sheets are oriented with the strands approx-imately parallel, and the two sheets are pinned together by a disulphide bridge between conserved cysteine residues.

Figure I13 shows the structure of the Fab fragment: $V_L + C_L$ and $V_H + C_H1$ of IgG KOL. The antigen-binding site is at the top of the figure.

The antigen-binding site

Six loops — three from the V_L domain and three from the V_H domain — form the antigen-binding site. The three loops from the V_L domain are called L1, L2 and L3 (or CDR1, CDR2, and CDR3)

in order of their appearance in the amino-acid sequence (*see* Fig. I14). V_H domains contain three corresponding CDRs, loops H1, H2 and H3.

The L2, H2, L3 and H3 regions are 'hairpins'; they link successive antiparallel strands of a single sheet; that is, they connect strands of sheet that are hydrogen bonded to each other. In contrast, L1 and H1 bridge a strand from one of the two sheets to a strand in the other.

Figure I15 shows the antigen-binding loops of the myeloma protein McPC603 and its ligand phosphorylcholine (PC). There is a rough symmetry in the arrangement of the loops: L1 is opposite H1, L2 is opposite H2, and L3 opposite H3. The central position of H3 in the site should be noted. Figure I16 shows the same structure, with main-chain atoms drawn in the ball-and-stick representation as in the previous figure, but with the side-chain

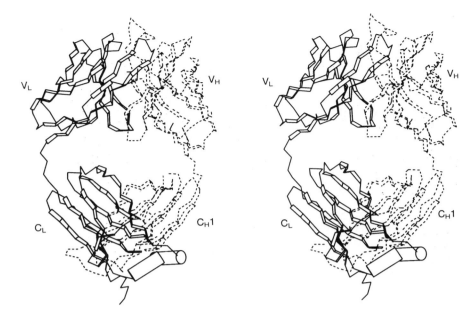

Fig. 113 Stereopair illustrating the Fab fragment of IgG KOL, showing the spatial relationships between V_L, C_L, V_H, and C_H1 domains. Solid lines, V_L and C_L domains; broken lines, V_H and C_H1 domains.

atoms shown as van der Waals spheres. This picture is designed to give the viewer the impression that the surface of an antibody, as seen by an antigen, is created mostly but not entirely by side-chain atoms. Some main-chain atoms are 'visible' to the antigen, and, correspondingly, it is characteristic of antibody–antigen interaction that a majority but not all of the contacts involve side-chain atoms of the antibody.

Different loops have their own characteristic sets of conformations, which can depend on their lengths, sequences and interactions with their surroundings.

For short hairpins in proteins, the conformation is usually determined primarily by the amino-acid sequence of the loop itself. For a polypeptide chain to reverse direction in a 'tight' turn, it usually must take advantage of a special residue such as Gly or

Pro which allows a nonstandard main-chain conformation. Therefore, the conformations of many short hairpins depend on their length and on the position of such a special residue. This principle applies to many of the shorter CDR hairpins — L2, L3 and in some cases H3; H2 however seems to be exceptional in that a framework residue (71 of the heavy chain) strongly influences its conformation.

Medium-sized loops tend to fall into two classes. The conformations of those that have ends close together in space are usually determined by hydrogen-bonding interactions of polar atoms that point into the interior of the loop. (The typical $V_\kappa L3$ loop, discussed below, illustrates this case.) The conformations of more extended loops, such as H1, depend on the packing of bulky side-chains against or into the core of the protein. In H1, for

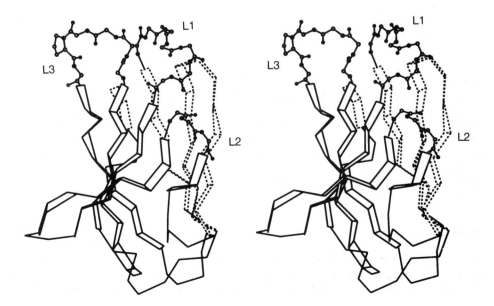

Fig. 114 Stereopair of the V_L domain V_κREI, showing the antigen-binding loops L1, L2 and L3.

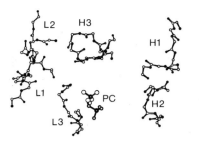
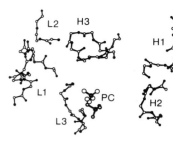

Fig. I15 The spatial distribution of antigen-binding loops or complementarity determining regions (CDRs) in the antigen-binding site of myeloma protein McPC603. The orientation of the stereopair is chosen to give a view down onto the antigen-binding site. Phosphorylcholine (PC) is the ligand.

example, residue 29 of the heavy chain is usually a Phe, which is packed into the β-sheet framework.

Relation between sequence and structure in antigen-binding loops

The study of antigen-binding sites of antibodies lies at the intersection of two topics of interest. One is the biology of the immune response: what is the relation between the amino-acid sequences of immunoglobulins, as generated by the genetic combinatorial and somatic mutational mechanisms (*see* B CELL DEVELOPMENT; GENE REARRANGEMENT; IMMUNOGLOBULIN GENES) and the three-dimensional conformation of the resulting antigen combining sites? The second is the question of how the recognition and binding of an antigen by antibodies relates to our general understanding of protein–ligand interactions, and the extent to which we can give a quantitative explanation of affinity and specificity at the level of interatomic interactions.

Analysis of the antigen-binding loops in known structures has shown that in most cases their main-chain conformations are determined by a few particular residues and that, among corresponding loops of the same length, only these residues need to be conserved to maintain the conformation of the loop. The conserved residues may be those that can adopt special main-chain conformations or that form special hydrogen-bonding or packing interactions. Other residues in the sequences of the loops are thus left free to vary, to modulate the surface topography and charge distribution of the antigen-binding site.

The ability to isolate the determinants of loop conformation in a few particular residues in the sequence makes it possible to analyse the distribution of loop conformations in the many known immunoglobulin sequences. At least five out of the six hypervariable regions of antibodies — L1, L2, L3, H1 and H2 — have repertoires of only a few main-chain conformations or

'canonical structures'. (For H3, which shows the largest variations in length and conformation, the situation is still unclear.) Most sequence variations only modify the surface by altering the side chains on the same canonical main-chain structure. Sequence changes at a few specific sets of positions switch the main chain to a different canonical conformation.

As an example Fig. I17 shows the L3 loop from V_κMcPC603. There is a *cis*-proline at position 95. Hydrogen bonds between the side chain of the residue at position 90, just N-terminal to the loop, and main-chain atoms of residues in the loop, stabilize the conformation. The side chain is an Asn in McPC603 (this residue can also be a Gln or a His in other V_κ chains). The combination of one of these polar side chains at position 90 and the proline at position 95 constitutes the 'signature' of this conformation in this loop, from which it can be recognized in the amino-acid sequence of an immunoglobulin for which the structure has not been experimentally determined.

Figure I18 shows a set of canonical structures for the three antigen-binding loops of V_κ light chains, and for the H1 and H2 of V_H domains. Although this is not a complete catalogue, the loops shown here account for ~90% of the hypervariable regions in V_κ domains and ~70% of the H1 and H2 regions of V_H domains.

An antigen–antibody complex: HyHEL-5 and hen egg white lysozyme

The construction of the antigen-binding site and its interaction with antigen is illustrated in a series of pictures of the complex between Fab HyHEL-5 and hen egg white lysozyme (*see also* Plate 6c). The orientation chosen looks down onto the antigen-binding site — that is, an 'antigen's-eye view'.

The spatial disposition of the antigen-binding loops in this view is shown in Fig. I19. In this figure the loops are represented simply by line segments joining successive Cα atoms.

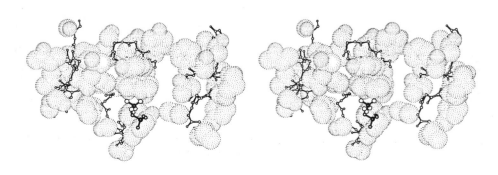

Fig. I16 Stereopair of the antigen-binding loops of McPC603. Main-chain atoms and phosphorylcholine are shown in ball-and-stick representations as in the previous figure; side-chain atoms are shown as van der Waals spheres.

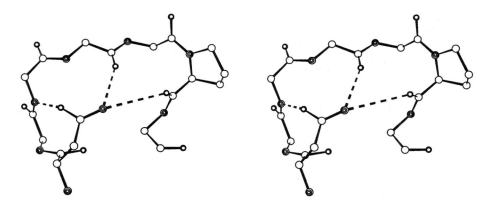

Fig. I17 Stereopair of the L3 region of McPC603, showing the hydrogen bonding to the Asn side chain of the residue adjacent to the loop, and the *cis*-proline.

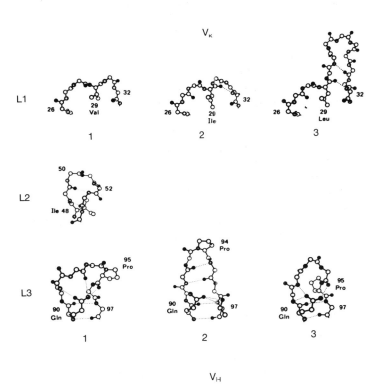

V_κ

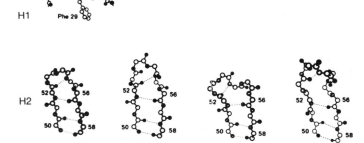

Fig. I18 Canonical structures for the antigen-binding loops L1, L2 and L3 of V_κ domains and H1 and H2 from V_H domains.

Figures I20–I23 show the same structure in increasing detail and complexity. Figure I20 shows all nonhydrogen atoms of the antigen-binding loops. The backbone appears in thicker lines with the side chains in thinner lines. Figure I21 shows the same skeletal representation of the antigen-binding loop as Fig. I16 but also includes a chain tracing of the antigen–lysozyme. Figure I22 shows the interaction of HyHEL-5 with lysozyme in more detail. Residues in the lysozyme that make contact with the antibody are shown in bold ball-and-stick representation. Residues in the antibodies that make contact with the lysozyme are shown in skeletal representation. Broken lines indicate hydrogen bonds between antigen and antibody.

We next show the nature of the packing of lysozyme in the complex, to illustrate the complementarity of the surfaces. In Fig. I23 the lysozyme side chains that make contact with the antibody are shown in bold ball-and-stick representation. Other

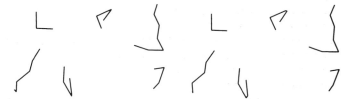

Fig. I19 Chain-tracing of the antigen-binding loops of HyHEL-5. Lines joining successive Cα atoms indicate the overall shape and disposition of the loops. This standard 'antigen's-eye view' will be retained in Figs I20–I23 (all stereopairs).

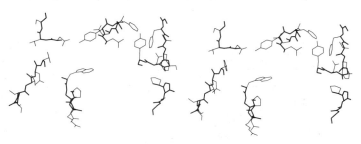

Fig. I20 Skeletal models of the antigen-binding loops of HyHEL-5. Thick lines represent the backbone and thin lines the side chains.

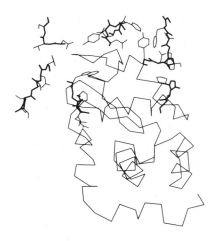

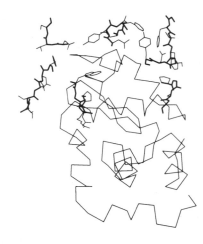

Fig. 121 The HyHEL-5–lysozyme complex: skeletal representation of the antigen-binding loops and chain-tracing of hen egg white lysozyme.

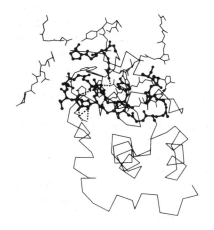

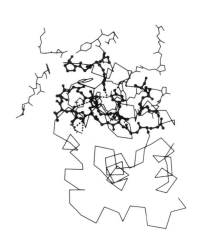

Fig. 122 Interactions between HyHEL-5 and hen egg white lysozyme. Residues in the lysozyme that make contact with the antibody are shown in bold ball-and-stick representation. Residues in the antibody that make contact with the lysozyme are shown in skeletal representation. Broken lines indicate hydrogen bonds between antigen and antibody.

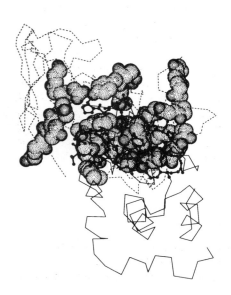

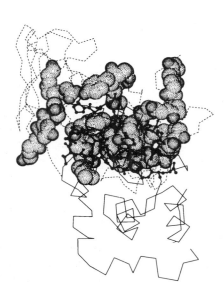

Fig. 123 Packing of lysozyme against the surface of the antibody combining site in the HyHEL-5 complex. Side chains of the antigen making contact with the antibody are shown in bold ball-and-stick representation. Other portions of the lysozyme are shown as a chain trace with thin solid lines. Residues of the antibody in contact with lysozyme are shown in space-filling representation of the van der Waals surfaces of the atoms. Other residues of the antibody are shown as a chain trace with broken lines.

portions of the lysozyme are shown as a chain trace (segments joining Cα atoms of successive residues) with thin solid lines. Residues of the antibody in contact with the lysozyme are shown in space-filling representation of the van der Waals surface of the atoms. Other residues from the antibody are shown as a chain trace with broken lines. This figure gives some idea of how the antibody is nestled into the antigen-binding site, and the nature of the complementarity of the surfaces.

A.M. LESK

A. TRAMONTANO

1 Alberts, B. et al. (1989) *Molecular Biology of the Cell*, Ch. 18 (Garland, New York).
2 Marquart, M. & Deisenhofer, J. (1982) The three-dimensional structure of antibodies. *Immunology Today* **3**, 160–166.
3 Chothia, C. et al. (1989) The conformations of immunoglobulin hypervariable regions. *Nature* **342**, 877–883.

Immunoglobulin superfamily

THE immunoglobulin superfamily (IgSF) is a large group of proteins, currently including about 100 different polypeptides, that contain segments of sequence of ∼100 amino acids which have significant similarities to those of IMMUNOGLOBULIN DOMAINS. These sequences are believed to have arisen by gene duplication and divergence from a primordial immunoglobulin-like domain and they are predicted to have a similar structure to that of immunoglobulin domains, namely the immunoglobulin fold (*see* ANTIBODIES).

Both immunoglobulin variable (V) and constant (C) domains have a common folding pattern termed the immunoglobulin fold (Fig. 124). The fold consists of a sandwich of two β sheets consisting of antiparallel β strands (*see* PROTEIN STRUCTURE). The two sheets are stabilized by a disulphide bond between a pair of conserved cysteine residues. V domains are distinguished from C domains by extra amino acids in the middle of the sequence (Fig. 124b) that makes up the β strands C' and C'' on one of the sheets (Fig. 124a). The members of the immunoglobulin superfamily have been identified on the basis of sequence similarities with immunoglobulin V or C domains. The level of similarity is usually of the order of 20–30% identity. Most of the similarities are in the regions that form the β strands with fewer in the loops between the strands. Note that when the term V domain is used to describe the possible folding pattern in immunoglobulin superfamily molecules, it refers only to the proposed arrangement of β strands and does not imply variability in sequence, which has only been found in immunoglobulin and T CELL RECEPTOR V domains. A detailed discussion of the sequence similarities in the immunoglobulin superfamily and its members is given in [1,2].

Criteria for inclusion of a sequence in the immunoglobulin superfamily

Sequences are usually identified as being immunoglobulin related when a search of the protein database with a novel protein sequence shows some reasonable scores with members of the immunoglobulin superfamily. The sequence is then analysed by identifying possible domains of about 100 amino acids. In most immunoglobulin superfamily domains there is a conserved disulphide bond that joins the two β sheets and the pair of cysteine residues involved can be identified by the characteristic patterns of residues around them. Each domain extends about 20 residues either side of these cysteine residues and this feature provides a good method of comparing immunoglobulin-related domains. Thus the putative domain extending 20 residues either side of the predicted positions of the conserved disulphide is compared with a large number of other immunoglobulin and immunoglobulin-related domains with a program such as ALIGN [2,3]. Overall the level of identities is typically about 20% and the scores are often not high enough to make a relationship obvious. However if moderate scores of 3–4 standard deviations are obtained against many members of the immunoglobulin superfamily, a good case for inclusion in the immunoglobulin superfamily can be made as this finding points to conserved features of the immunoglobulin superfamily being recognized in the new molecule.

The difference in domain length means that V domains score poorly with C domains; thus new sequences tend to score better with one group rather than with all sequences. Analysis of the sequences showed that many contained patches of sequence that were typical of V domains, for example the sequence Asp-Xaa(Gly or Ala)-Xaa-Tyr-Xaa-Cys in β strand F (Fig. I24b) but their length was similar to C domains. Thus two categories of C domains have been distinguished: a C1 set that contains the typical C domains of immunoglobulin, T cell receptors and the immunoglobulin-related domains of MHC antigens (*see* MAJOR HISTOCOMPATIBILITY COMPLEX); and a second C2 set that has many of the characteristic sequence patterns of V domains but which are not of sufficient length to form a V-domain fold.

Thus it is likely that a domain is immunoglobulin related if it gives scores of 3 or more standard deviations in the ALIGN analysis with the majority of sequences in one of the groups in the immunoglobulin superfamily, that is C1 set, C2 set, or V set. A further test is that the sequence should contain residues compatible with the formation of the β strands. Thus the cores of the β strands are usually made up of alternating hydrophobic residues and it should be possible to predict the approximate positions of the β strands by comparison with other sequences.

Although the conserved disulphide is a hallmark of immunoglobulin C and V domains, it is not essential for the immunoglobulin fold and an active antibody has been characterized where this disulphide is lacking [4]. There are several sequences that lack the conserved disulphide but show statistically significant similarities with other immunoglobulin superfamily domains; in all cases so far these domains are found in structures that already have one or more immunoglobulin-related domains with the conserved disulphide, strengthening the argument for their inclusion as immunoglobulin related.

Genomic structure of members of the immunoglobulin superfamily

The structure of immunoglobulin genes shows that each domain is encoded by one exon except for the V domains where the sequence for the final β strand is formed by gene rearrangement

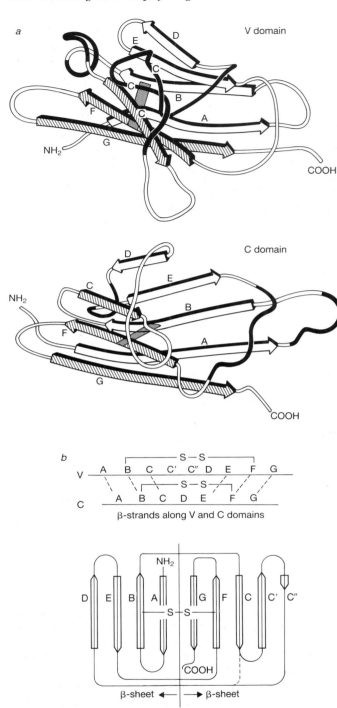

Fig. I24 *a*, The immunoglobulin fold in immunoglobulin V and C domains. *b*, Schematic representation of folding pattern for V and C domains. The upper diagram shows the notation of the β strands along V and C domains (note the extra C′ and C″ strands in the V domain); the lower diagram the folding pattern in the β-sheets.

from a small J region segment (and in heavy chains also a smaller D region). This is important in increasing the diversity of immunoglobulin and also T cell receptor V domains, and has not been observed in any other immunoglobulin-related structures (*see* GENE REARRANGEMENT; GENERATION OF DIVERSITY; IMMUNOGLOBULIN GENES; T CELL RECEPTOR GENES) [5]. The boundaries of the exons coincide with the ends of the domains defined from the structure obtained by X-RAY CRYSTALLOGRAPHY.

The one exon per domain relationship has been found for the majority of immunoglobulin-related domains although there are some exceptions where there is an intron approximately in the centre of the domain. In one case two domains are encoded by one exon. In all cases so far the introns are of Phase 1 type, which splits the codon between the first and second base. Phase 1 introns are also common to genes for other superfamily domains, such as FIBRONECTIN TYPE III DOMAINS. This allows proteins to be built up in evolution by duplication of exons for immunoglobulin-related domains and to include exons from more than one superfamily type. The presence of several immunoglobulin and fibronectin type III domains is found particularly in some membrane glycoproteins from the brain (*see* CELL ADHESION MOLECULES), and has also been found more recently in some cytoplasmic proteins.

Structure of members of the immunoglobulin superfamily

Figure I25 illustrates the domain organization of some members of the superfamily. The structures of immunoglobulin V and C domains, β$_2$-microglobulin, MHC class I antigen and the first two domains of the lymphocyte surface marker CD4 have been determined by X-ray crystallography. β$_2$-Microglobulin and the immunoglobulin-related domain of MHC class I antigen show strong sequence similarity to immunoglobulin C domains and a good case for their inclusion in the immunoglobulin superfamily could be made even before determination of their structure. However the case for CD4 was more controversial. The first domain looked most like an immunoglobulin V domain and this was established by the structure [6,7]. However the second domain was less convincing and its inclusion as immunoglobulin related was borderline. In order to get the best alignment a disulphide was predicted within the β sheet rather than between the sheets and given that the length of the sequence was less than that of known immunoglobulin C domains, it would have to have a truncated domain. The X-ray crystallography structure of the first two domains of CD4 established this to be the case. This makes a strong argument that those sequences that satisfy the criteria discussed above will have an immunoglobulin-like fold.

In one case a structure very similar to an immunoglobulin fold has been determined by X-ray crystallography. The PapD protein is a MOLECULAR CHAPERONE in the bacterium *Escherichia coli*. It has two immunoglobulin-like domains but shows no significant sequence similarity to immunoglobulins.

Functions of the immunoglobulin superfamily

Most of the members of this large superfamily are present at the

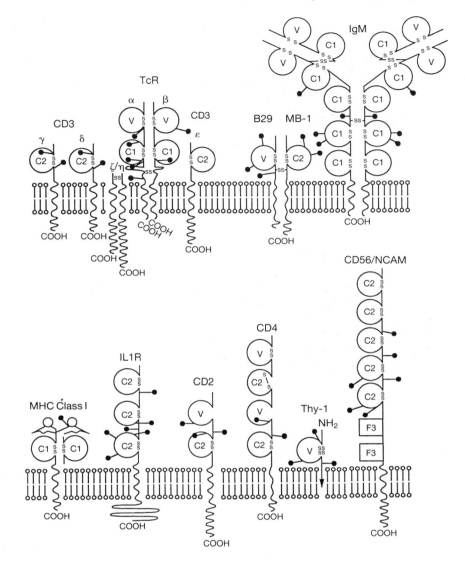

Fig. I25 Some members of the immunoglobulin superfamily. NCAM also contains fibronectin type III superfamily domains (F3). CD3γ, δ, ζ, η, ε, Cell-surface glycoproteins of the CD3 complex; TcR, T cell receptor; B29 and MB-1, glycoproteins of the antigen recognition complex on B cells; IgM, membrane-bound immunoglobulin M; MHC class I, class I MHC molecule, comprising a membrane-linked polypeptide chain encoded in the MHC region complexed with β$_2$-microglobulin; IL1R, interleukin 1 receptor; CD2, CD4, T-cell surface antigens; Thy-1, brain/lymphoid antigen; CD56/NCAM, neural cell adhesion molecule. V, C1, C2, Immunoglobulin superfamily domains of the V set, C1 set and C2 set respectively. F3, fibronectin type III superfamily domains. Small black circles indicate glycosylation sites. See [3,8] for more examples of this and other superfamilies present at cell surfaces.

cell surface (Fig. I25). In addition to the obvious roles in antigen recognition — immunoglobulins as antigen receptors on B cells and in body fluids, the T cell antigen receptor, and the role of MHC antigens in the presentation of antigenic peptides to T lymphocytes — immunoglobulin superfamily members are known to mediate many cell–cell interactions in other tissues. Examples include the Fc receptors that interact with immunoglobulin C regions; HOMOPHILIC interactions between adhesion molecules such as the neural cell adhesion molecule (NCAM) (*see* CELL ADHESION MOLECULES) and the carcinoembryonic antigen (CEA); HETEROPHILIC interactions between pairs of immunoglobulin-related molecules such as the CD2 and LFA-3 (CD58) molecules on lymphocytes and other cells (*see also* CD ANTIGENS); the receptors for cytokines such as interleukins 1 and 6 (IL-1 and IL-6) (*see* LYMPHOKINES); and receptors for GROWTH FACTORS such as platelet-derived growth factor (PDGF) and colony-stimulating factor 1 (CSF-1) (*see* GROWTH FACTOR RECEPTORS). Several

immunoglobulin superfamily proteins have accessory roles in the activation of T cells, for instance the CD8 and CD4 antigens, which bind to MHC class I and II antigens respectively, and the CD3 chains that form part of the antigen recognition complex on T cells (*see* T CELL RECEPTOR).

There are also other superfamily molecules of unknown function including many in the nervous system such as the Thy-1 antigen, myelin-associated glycoprotein, and contactin. Other superfamily members are known to act as receptors for viruses, for example CD4 for human immunodeficiency virus (HIV), ICAM-1 for rhinovirus, and a protein defined on its ability to bind the poliovirus. Until recently all immunoglobulin superfamily members had been found at the cell surface or secreted into the extracellular space but now they have been found in the cytoplasm, for example the large proteins TITIN and twitchin (see below) which are involved in muscle contraction.

Evolution of the immunoglobulin superfamily

The sequence similarities between immunoglobulin V and C domains have led to the proposal that these domains arose by gene duplication and divergence from a single primordial immunoglobulin domain. At first it was thought that this family of molecules was uniquely involved in the recognition of foreign antigens. However the finding of immunoglobulin superfamily domains in molecules that have clearly no role in antigen recognition such as the brain molecules showed that this superfamily has a much wider range of functions. Thus the antigen recognition role represented the adaptation of a widely used domain type to a highly sophisticated genetic process involving recombination to obtain the range of antigen recognition specificities that is the hallmark of the antigen receptors on B and T lymphocytes. The repertoire of specificities in immunoglobulin is further enhanced by somatic mutation.

Immunoglobulin superfamily members have been identified in invertebrates, thus predating the vertebrate immune system. These molecules include a number of proteins in nervous tissue such as fasciclin II and amalgam in *Drosophila* and twitchin in *Caenorhabditis elegans*. This latter protein is of considerable interest as it is a very large protein made up of large numbers of immunoglobulin superfamily and fibronectin type III domains; it is present in the cytoplasm and thought to be involved in muscle contraction and relaxation. Another enormous cytoplasmic protein — titin — also contains immunoglobulin-like domains. This finding raises the question as to whether the immunoglobulin superfamily arose from a primordial domain in the cytoplasm and was then adapted to the cell surface or vice versa. One feature of immunoglobulin superfamily domains is their relative stability to protease digestion and this may be the major reason why the immunoglobulin superfamily domain was utilized so widely for interactions at the cell surface. Thus it seems more likely that the immunoglobulin superfamily domain arose as a single domain at the cell surface to mediate cell interactions and then lost its SIGNAL SEQUENCE and gave rise to the cytoplasmic proteins.

A.N. BARCLAY

1 Williams, A.F. (1987) A year in the life of the immunoglobulin superfamily. *Immunol. Today* **8**, 298–303.
2 Williams, A.F. & Barclay, A.N. (1988) The immunoglobulin superfamily — domains for cell surface recognition. *Annu. Rev. Immunol.* **6**, 381–403.
3 Dayhoff, O. et al. (1983) Establishing homology in protein sequences. *Meth. Enzymol.* **91**, 524–545.
4 Rudikoff, S. & Pumphrey, J.G. (1986) Functional antibody lacking a variable-region disulphide bond. *Proc. Natl. Acad. Sci. USA* **83**, 7875–7878.
5 Honjo, T. et al. (Eds) (1989) *Immunoglobulin Genes* (Academic Press, New York).
6 Wang, J. et al. (1990) Atomic structure of a fragment of human CD4 containing two immunoglobulin-like domains. *Nature* **348**, 411–418.
7 Rhu, S-E. et al. (1990) Crystal structure of an HIV-binding recombinant fragment of human CD4. *Nature* **348**, 419–426.
8 Barclay, A.N. et al. (1993) *The Leucocyte Antigen FactsBook* (Academic Press, London).

immunoperoxidase techniques Any technique that depends on the identification of an antigen with a specific antibody conjugated to HORSERADISH PEROXIDASE.

immunotoxins Antitumour ANTIBODIES coupled to toxic compounds, radioisotopes, or drugs, which are intended to direct the cytotoxic agent specifically to tumour cells. Such conjugates are being used in immunotherapy trials in cancer patients and in experimental animals. Toxins such as RICIN or DIPHTHERIA TOXIN are highly potent inhibitors of PROTEIN SYNTHESIS and in theory, could be given in extremely low doses if they were bound to tumour-specific antibodies to form immunotoxins.

imprinting Except in the special case of maternal CYTOPLASMIC INHERITANCE, most genes behave in the same way with respect to DOMINANCE in an individual offspring, or PENETRANCE among the offspring of a cross, irrespective of which parent transmits the gene. In some cases, however, gene expression is influenced by parental origin. This PARENTAL GENOMIC IMPRINTING is thought to be a form of reversible EPIGENETIC suppression of gene expression. It is suggested to explain the failure of PARTHENOGENESIS in all mammals, the selective inactivation of certain genetic loci from the male parent in female mammals and in some insect species, and the parental origin of effects seen in the inheritance of some genetic diseases in humans (e.g. Prader–Willi and Beckwith–Wiedemann syndromes). Homologue behaviour may differ with respect to RNA synthesis, DNA methylation, amount of heterochromatin, or the ability to cause developmental abnormalities.

in organello Within an organelle.

in situ **hybridization** (1) Technique used widely in cell and developmental biology to determine the distribution of a particular mRNA transcript in an organism, a particular tissue, or within a cell, by HYBRIDIZATION of the tissue with a labelled probe of the required specificity. It is routinely carried out on whole embryos (in DROSOPHILA or *Caenorhabditis*), on thin sections (1–10 μm) of whole avian, amphibian or mammalian embryos, or on embryonic or adult tissues or cultured cells.
(2) Similar technique used to localize gene sequences to particular chromosomes or regions of chromosomes. *See*: HUMAN GENE MAPPING.

in vitro **fertilization (IVF)** Technology for fertilizing mammalian eggs in culture outside the body, after which the fertilized egg is replaced in the mother, or a surrogate mother, to complete its development; it has developed rapidly over the past decade as a treatment for human infertility. In humans, after hormonal stimulation of ovulation oocytes are collected from the ovary by laparoscopy under general anaesthetic. An oocyte is then incubated with sperms in culture medium. The fertilized egg is allowed to divide and when the embryo reaches the 2–8-cell stage it is introduced into the mother's uterus or oviduct. The IVF pregnancy rate is around 18% (of embryo transfer procedures).

IVF is also used to propagate valuable stocks of farm animals such as sheep and cows more rapidly, by increasing the number of progeny available from a valuable mother by rearing embryos obtained by IVF in a surrogate mother. Culture conditions enabling IVF vary significantly between species.

in vitro **mutagenesis** *See:* SITE-DIRECTED MUTAGENESIS.

in vitro **transcription systems** Cell-free systems used to transcribe either endogenous or cloned DNA sequences into complementary RNA sequences using RNA polymerases (*see* TRANSCRIPTION). There are four types of *in vitro* transcription system:

1 Systems which use isolated nuclei to transcribe endogenous DNA templates.

2 Systems which use soluble whole-cell or nuclear extracts to transcribe exogenous DNA templates.

3 Systems which use purified RNA polymerases (e.g. those from BACTERIOPHAGES T7 or SP6).

4 Systems which are directly coupled to IN VITRO TRANSLATION. These coupled transcription–translation systems are only applicable to bacterial extracts, particularly those of *Escherichia coli*. *In vitro* transcription is used to study the mechanism of transcription and its control, and to generate mRNA templates for IN VITRO TRANSLATION studies or for use as probes for SOUTHERN BLOTTING or NORTHERN BLOTTING (RIBOPROBES).

In vitro translation

IN VITRO translation is the process by which proteins may be synthesized in the test tube, reproducing to a certain extent the processes of PROTEIN SYNTHESIS that occur in living cells. Such *in vitro* translation is carried out in a cell-free lysate which is essentially a crude cytosolic fraction prepared from disrupted cells with the cell debris (cell walls, membranes etc.) removed by centrifugation. The resulting cell-free lysate contains all of the macromolecular components required for protein synthesis: ribosomes, mRNA, translation factors, tRNAs and aminoacyl-tRNA synthetases, together with the amino acids. When the lysate is supplemented with a suitable chemical means of generating the necessary energy requirements for protein synthesis (i.e. ATP and/or GTP), together with the correct ionic environment, translation of the endogenous mRNA already associated with ribosomes will occur.

In vitro translation systems have been prepared from both prokaryotes and eukaryotes and used to probe the basic mechanism of protein synthesis and its control. By using such systems various parameters of the translation system can be manipulated in a way that is not possible with intact cells. For instance, ionic conditions in the lysate can be altered by changing K^+ and Mg^{2+} concentrations. The ability to carry out such manipulations also means that a wide range of exogenously supplied mRNA templates can be translated *in vitro*; for example, if the Mg^{2+} concentration is increased to a level significantly higher than is observed in the living cell (i.e. from ∼2 mM to 12 mM) then ribosomes can translate a synthetic polynucleotide template such as polyuridylic acid (poly(U)) even in the absence of an encoded translation INITIATION CODON or in the absence of translation INITIATION FACTORS. This ability to translate defined synthetic templates is an important property of *in vitro* translation systems and contributed to the elucidation of the GENETIC CODE. *In vitro* translation systems can also be used to translate a wide range of

naturally occurring mRNAs or mRNAs generated by *in vitro* transcription from a plasmid containing the BACTERIOPHAGE SP6 or T7 transcriptional promoters (*see* EXPRESSION VECTORS). Such *in vitro* synthesized mRNAs can be added to a cell-free lysate and their ability to be translated assessed both qualitatively and quantitatively.

The first *in vitro* translation systems to be developed were from the bacterium ESCHERICHIA COLI [1], but currently the most widely used cell-free lysate for *in vitro* translation studies is the mRNA-dependent rabbit reticulocyte lysate system [2]. This lysate is prepared from red blood cells collected from rabbits that have been made anaemic and then allowed to recover. The reticulocyte is not a typical cell as it lacks a nucleus but this property aids *in vitro* translation as no endogenous transcription can occur in the lysate. The lysate is treated with MICROCOCCAL NUCLEASE, a Ca^{2+}-activated ribonuclease, to remove any endogenous mRNA. Before the addition of an exogenous mRNA, the micrococcal nuclease is inactivated by the addition of EDTA to chelate the Ca^{2+}. This ensures that the added mRNA is then translated against a very low background of endogenous translation (Fig. I26).

A defined set of ionic conditions is required for optimal translation of mRNAs in a cell-free translation system; for the reticulocyte lysate these are 2 mM Mg^{2+} and 100 mM K^+. ATP and GTP are usually added at 1 mM and 0.2 mM respectively to allow aminoacyl-tRNA synthetases to aminoacylate the endogenous tRNAs, and to ensure optimal initiation and elongation factor activity. Creatine phosphate and creatine phosphokinase are added to allow for the regeneration of the essential nucleotide triphosphates from their hydrolysed diphosphate forms. Cell-free lysates are also usually supplemented with the naturally occurring amino acids together with one or more radiolabelled amino acids (e.g. [^{35}S]methionine or [^{3}H]leucine). As the radiolabelled amino acid becomes incorporated into the polypeptides synthesized during the cell-free translation, it allows for both measurement of the efficiency of translation *in vitro* by scintillation counting and qualitative analysis of the proteins synthesized by SDS-polyacrylamide gel ELECTROPHORESIS and AUTORADIOGRAPHY.

The reticulocyte lysate has a relatively high *in vitro* translational activity at around one amino acid polymerized per second, a rate comparable to that in the intact reticulocyte. The lysate can be used to translate a wide range of heterologous mRNAs both accurately and efficiently, but does have certain practical disadvantages; it is difficult to maintain high translational activity in the lysate during long-term cryopreservation and it is very sensitive to a lack of HAEMIN. This sensitivity to the absence of haemin is due to the activation of an inhibitor of the initiation of translation, a PROTEIN KINASE called the haem-controlled repressor (HCR). The sensitivity to haem deficiency is possibly unique to reticulocytes.

Gel filtration of the cell-free lysate enables greater control of the conditions of translation as essential low molecular weight components (e.g. amino acids, nucleotide triphosphates, salts, etc.) are removed during this step and these can be added back to the lysate to control precisely the ionic conditions during translation.

The reticulocyte lysate can also be used as a high output *in vitro*

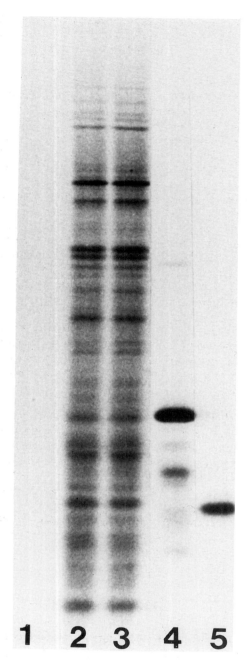

Fig. 126 Translation of a natural mRNA in a rabbit reticulocyte cell-free translation system. Synthesis was carried out in the presence of [35S]methionine and the labelled translation product detected after SDS-polyacrylamide gel electrophoresis by autoradiography. Lane 1, no exogenous mRNA added; lane 2, natural mRNA added.

translation system in which the translation reaction is carried out in a reaction chamber containing a relatively small volume of lysate (~250 µl). This can be used to translate a defined mRNA template (e.g. globin mRNA) for up to 100 hours continuously and can yield up to 2 mg of the target polypeptide [3]. This 'continuous' *in vitro* translation system relies on an ultrafiltration

membrane to keep the translation apparatus in the reaction chamber while spent reactants and proteins synthesized in the lysate are eluted by a continuous stream of a buffer containing fresh nucleotide triphosphates and creatine phosphate. This *in vitro* expression system has the potential to generate milligram quantities of any protein from its mRNA template in an easily purified form. It would be particularly useful for synthesizing proteins normally toxic to a growing cell or those that are very labile or easily degraded.

Numerous other cell-free translation systems have been developed from prokaryotic and eukaryotic sources. An efficient cell-free translation system from wheat germ can be prepared by grinding fresh wheat germ with sterilized sand and centrifuging the resulting slurry [4]. The lysate can then be either gel filtered or dialysed to remove low molecular weight components. This system has a low endogenous background level of translation and is easy and cheap to prepare. However, it is prone to premature termination of translation.

Systems from Ehrlich ascites tumour cells and HeLa cells have also been used, but the system offering the greatest potential is the SACCHAROMYCES CEREVISIAE cell-free translation system [5]. The power of the *S. cerevisiae* system lies in the range of mutants with differing genetic backgrounds that are available and the ability to generate lysates from strains bearing deletions in genes encoding key translation factors. The factors can then be added back to the system to assay their function in translation.

D. COLTHURST

1 Lengyel, P. & Soll, D. (1969) Mechanism of protein synthesis. *Bact. Rev.* **33**, 264–301.
2 Pelham, H.R.B. & Jackson, R.J. (1976) An efficient mRNA-dependent translation system from reticulocyte lysates. *Eur. J. Biochem.* **67**, 247–256.
3 Ryabora, L.A. et al. (1989) Preparative synthesis of globin in a continuous cell-free translation system from rabbit reticulocytes. *Nucleic Acids Res.* **17**, 4412.
4 Roberts, B.E. & Paterson, B.M. (1973) Efficient translation of tobacco mosaic virus RNA and rabbit globin mRNA in a cell-free system from commercial wheat germ. *Proc. Natl. Acad. Sci. USA* **70**, 2330–2334.
5 Gasior, E. et al. (1979) The preparation and characterisation of a cell-free system from *Saccharomyces cerevisiae* that translates natural mRNA. *J. Biol. Chem.* **254**, 3965–3969.

inborn error of metabolism Term invented by the English physician Sir Archibald Garrod in the early 20th century. It describes genetically determined biochemical defects which cause abnormalities of metabolism, such as a block in a metabolic pathway caused by the absence of a specific enzyme. These metabolic defects in turn cause defined heritable disorders (e.g. ALKAPTO-NURIA; HISTIDINAEMIA; HOMOCYSTINURIA; PHENYLKETONURIA).

incompatible host–pathogen interaction In plant pathology, an interaction between host and pathogen that does not result in disease. The term is typically used to describe a combination of host and pathogen for which a gene-for-gene resistance–avirulence interaction exists. Incompatibility is due to dominant resistance genes in the host. The term 'incompatible' should not be used in describing the pathogen itself, but only its interaction with the host. *See also*: PLANT PATHOLOGY.

incomplete (partial) dominance Incomplete, or partial, dominance occurs when individuals HETEROZYGOUS for a particular gene exhibit a PHENOTYPE intermediate between those of the two HOMOZYGOUS states. It is caused by partial expression of the gene in the heterozygous state and the term is usually used in preference to CODOMINANCE in cases where one of the alleles at the locus is recessive (i.e. loss of function).

indel For two sequences being compared an insertion in one sequence or deletion in the other; a neutral term useful when these two cases cannot be distinguished. *See*: ALIGNMENT.

indole-3-acetic acid (IAA) The principal naturally occurring plant hormone of the AUXIN class and the first nongaseous plant hormone to be chemically characterized (in 1934–35) (Fig. I27). M_r 175.19. It was originally isolated from human urine and the fungus *Rhizopus suinus* as rich sources, and its activity on higher plants compared with synthetic IAA. The presence of IAA in higher plants was confirmed by gas chromatography–mass spectroscopy in 1971. It is usually present at concentrations of $1–1000$ pmol g^{-1} fresh weight. It is physiologically active in bathing solutions ($>10^{-7}$ mol m^{-3}) and gives a wide range of effects (*see* AUXIN). IAA is transported from cell to cell preferentially from apical to basal ends of stems or roots. This polar transport is mediated by unidirectional protein carriers and contributes to maintaining the correct pattern of organ development, and the appropriate orientation of roots and shoots in relation to gravity and the direction of light.

Jacobs, W.P. (1979) *Plant Hormones and Plant Development*. (Cambridge University Press, Cambridge).

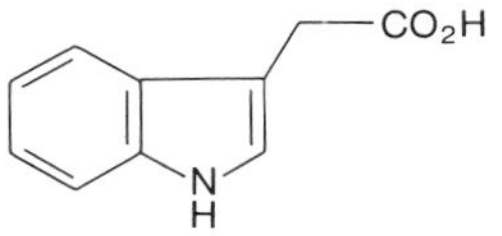

Fig. I27 Structure of indole-3-acetic acid.

indole-3-butyric acid (IBA) A synthetic growth-regulating substance of the AUXIN class (Fig. I28) rarely reported to occur naturally in plants. IBA possesses most of the physiological properties of the naturally occurring auxin INDOLE-3-ACETIC ACID (IAA) and may sometimes be more active than IAA, for example in stimulating the rooting of cuttings. The activity of IBA may be due to its breakdown to IAA through the action of ubiquitous β-oxidation enzymes

Thimann, K.V. (1977) *Hormone Action in the Whole Life of Plants* (University of Massachusetts Press, Cambridge, MA).

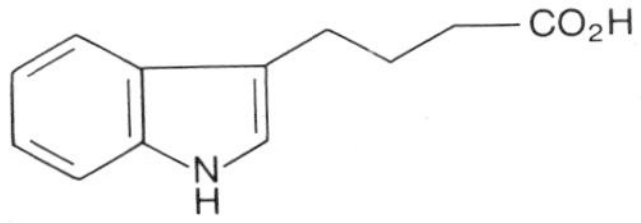

Fig. I28 Structure of indole-3-butyric acid.

induced fit The change in conformation of the active site of an enzyme to fit its substrate which occurs only after binding of the substrate. *See*: MECHANISMS OF ENZYME CATALYSIS.

inducer (1) In development; *see* long entry INDUCTION.
(2) In gene expression, a small molecule which will activate the TRANSCRIPTION of a gene whose product can metabolize the inducer (*see* BACTERIAL GENE EXPRESSION). The classic example of an inducer is allolactose (a metabolite of lactose) which induces expression of the bacterial LAC OPERON and thus the synthesis of the enzyme β-GALACTOSIDASE which is required for the metabolism of lactose. Inducers generally exert their effects by binding to a REPRESSOR, a gene regulatory protein which prevents transcription by binding to DNA in the absence of the inducer. IPTG (isopropyl-β-D-thiogalactopyranoside) is another inducer of the *lac* operon, although in this case the inducer is said to be a gratuitous inducer as it is not metabolized by β-galactosidase.
(3) Any agent that causes the activation of a PROPHAGE or PROVIRUS and its excision from the chromosome and entry into the lytic cycle (*see e.g.* LAMBDA; RETROVIRUSES).

inducible Applied to genes that are activated only in response to a specific stimulus (*cf.* CONSTITUTIVE). In its original sense as used in bacterial and fungal genetics, it applied to enzymes whose synthesis was activated in response to specific small molecule INDUCERS.

induction (1) In development; *see* long entry INDUCTION.
(2) In bacteria and other microorganisms, the synthesis of one or more enzymes in response to a specific INDUCER molecule. This phenomenon is largely restricted to free-living unicellular microorganisms such as bacteria or yeast. Induction is usually achieved by the activation of TRANSCRIPTION of the gene(s) encoding the enzyme(s) (*see* BACTERIAL GENE EXPRESSION).
(3) Of BACTERIOPHAGES and other viruses, the activation of a PROPHAGE or PROVIRUS thereby inducing entry into the lytic cycle (*see e.g.* LAMBDA; RETROVIRUSES).

Induction

THERE are two ways in which diversification of cell type can occur during embryonic development [1]. The first is by means of CYTOPLASMIC DETERMINANTS — molecules arranged asymmetrically within the egg which direct the development of cells which inherit them during cleavage. Cytoplasmic determinants seem to have an important role in, for example, the development of ascidian embryos. In other animals, such as vertebrates, cell diversity arises chiefly through induction. This has been defined as 'an interaction between one (inducing) tissue and another (responding) tissue, as a result of which the responding tissue undergoes a change in its direction of differentiation' [2]. Inductive interactions are frequently followed by cell movements which result in different tissues being apposed. This allows new inductive interactions and thus further cell diversification.

In some embryos, such as those of *Drosophila melanogaster* or

Caenorhabditis elegans, inductive interactions can be identified by developmental genetics (*see* CAENORHABDITIS DEVELOPMENT; DROSOPHILA DEVELOPMENT; INSECT NEURAL DEVELOPMENT). In others, where genetic analysis is not possible, inductive interactions are analysed by the methods of experimental embryology. This first requires an accurate FATE MAP, so that the normal DIFFERENTIATION of a group of cells can be followed. It is then necessary to abolish the presumed interaction between inducing and responding tissues, perhaps by separating them and following their development in isolation. If the fate of the responding tissue under these conditions is different from its normal fate one might infer that it has been deprived of an inductive signal. It is also possible to juxtapose inducing tissue with cells from elsewhere in the embryo, to discover whether other cells are responsive to induction and how they respond.

As well as accurate fate maps, this type of analysis requires culture media that closely resemble the normal embryonic environment so that cells are provided with all essential nutrients and other factors. In practice, these conditions are rarely met, and interpretation of such experiments requires caution.

Instructive and permissive inductive interactions

In spite of these difficulties, distinctions have been drawn between several types of inductive interaction. One such distinction is between permissive and instructive interactions. In the former, the inducing signal does not determine the nature of the response; rather, the responding cells are already 'predetermined' in some way to form a particular cell type, and the signal is only required to 'permit' the cells to complete development along their chosen pathway. In this case it is assumed that the nature of the signal need not be specific; a variety of nonspecific stimuli may suffice. Permissive inductions will not be covered in any depth here.

Instructive interactions can themselves be divided into two types. In the first, and simpler, the inducing tissue instructs some of the responding cells to become a third cell type (Fig. I29*a*). Although this is conceptually different from a permissive interaction, the two are difficult to distinguish experimentally because *a priori* there is no way of knowing whether the cells that form the third cell type were predetermined to do so. If, however, the responding tissue can differentiate in different ways in response to different signals, then the interaction must be instructive. The second, and more complicated, interaction results in the formation of several different cell types and, in addition, may transmit information about cells' positions within the embryo. This represents an example of the transmission of POSITIONAL INFORMATION [3]. One way of demonstrating that this type of inductive interaction occurs is to show that the responding tissue differentiates in different ways in response to different intensities of a signal (Fig. I29*b*).

Competence of the responding tissue

In discussing induction there is a tendency to concentrate on the nature of the signal produced by the inducing tissue. Equally important, however, is the competence of the responding tissue:

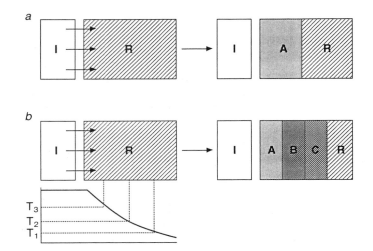

Fig. I29 The distinction between induction and the transmission of positional information. *a*, An example of induction, where the signal derived from an inducing tissue (I) causes nearby cells within the responding tissue (R) to form cell type A. Those cells out of range of the signal remain as cell type R. *b*, An example of positional information. The signal from the inducing region forms a concentration gradient within the responding tissue. Cells in the responding tissue interpret the local concentration of signal according to three thresholds (T_1, T_2, and T_3). Cells exposed to a concentration higher than T_3 form cell type A, those experiencing a concentration between T_3 and T_2 form type B, and those experiencing a concentration between T_2 and T_1 form type C. Where the local concentration of signal is lower than T_1, cells remain as type R.

its ability to recognize the inducing signal and to react appropriately to it. The nature of the response depends on the developmental history of the responding cells and their genotype. As an example of the former one can consider the animal pole cells of the amphibian embryo (*see* AMPHIBIAN DEVELOPMENT). During BLASTULA stages these cells are competent to respond to signals from the vegetal hemisphere of the embryo by forming MESODERM. By the early GASTRULA stage, however, the cells can no longer form mesoderm but become competent to respond to signals from gastrulating dorsal mesoderm by forming neural tissue. These interactions are discussed in some detail below.

The influence of the genotype of the responding tissue is exemplified by the induction of the mouthparts of amphibian embryos by head MESECTODERM and ENDODERM. This experiment was conceived by Spemann and performed 11 years later by Schott (see [4]), who transplanted ventral ECTODERM of the frog gastrula to the mouth region of a salamander gastrula and vice versa. The salamander larvae that resulted from this experiment developed frog-like horny jaws and suckers, while the frog larvae formed salamander-like balancers and teeth. Thus the inducing signal merely says 'form mouthparts'; the type of mouthparts formed depends on the genotype of the responding tissue.

We can now describe induction in the languages of cell and molecular biology. This is illustrated below with two inductive interactions that occur during early amphibian development. The two interactions are involved in the induction of the mesoderm and the nervous system.

Mesoderm induction

Much of the recent work on inductive interactions in amphibia has used the South African clawed frog *Xenopus laevis*. *Xenopus* develops very rapidly, and reaches the blastula stage about 4 hours after fertilization. It is around this time that the first inductive interaction is believed to occur; a signal from blastomeres at the vegetal pole of the embryo, which themselves eventually form endodermal tissue, appears to cause cells around the equator of the embryo to become mesoderm rather than ectoderm. This interaction can be illustrated by juxtaposing vegetal pole cells with animal pole tissue (Fig. 130). The animal pole cells are normally too far from vegetal cells to be induced to become mesoderm and they form ectoderm instead. When juxtaposed with vegetal cells, however, the animal pole blastomeres form notochord, muscle, blood and a variety of other mesodermal cell types. This interaction is known as mesoderm induction.

There is quite good evidence that mesoderm induction occurs in normal development as well as in this artificial situation. Firstly, if equatorial blastomeres are dissected from the embryo at an early stage they do not form mesoderm, whereas if isolated at blastula stages they do. Secondly, if the blastomeres of the embryo are dispersed during cleavage stages and only reaggregated at the early gastrula stage, when the animal pole cells are no longer competent to respond to the signal (see above), muscle, the major mesodermal cell type, does not develop. Both these experiments imply that cell interactions are required for the formation of mesoderm.

Is mesoderm induction permissive or instructive? Two pieces of evidence suggest that it is instructive. Firstly, different regions of the vegetal hemisphere induce different cell types from responding animal pole tissue. Ventral and lateral vegetal cells induce ventral mesodermal cell types such as blood and mesothelium. Dorso-vegetal cells, by contrast, induce dorsal mesodermal structures such as notochord and muscle. Furthermore, they make responding tissue behave as Spemann's 'organizer'; the cells acquire the ability to induce a secondary embryo when they are grafted to the ventral side of a host (Fig. 131). The formation of such a secondary embryo results from two inductive interactions that occur after mesoderm induction. The first is dorsalization, in which a signal from the organizer acts on ventral mesoderm and causes it to become more dorsal in nature. This can be regarded as a 'fine-tuning' of the dorsoventral differences set up by the vegetal hemisphere. The second interaction is neural induction, which is discussed below.

The second piece of evidence suggesting that mesoderm induction is instructive is that different concentrations of the polypeptide ACTIVIN, a potent 'mesoderm-inducing factor', will induce different cell types from responding animal pole cells.

Mesoderm-inducing factors

Experiments in which animal and vegetal pole regions of the *Xenopus* embryo are separated by filters which allow passage of molecules but not of cell processes show that mesoderm induction is mediated by soluble factors. Recently, great progress has been made in identifying the mesoderm-inducing factors which are believed to be secreted from vegetal pole cells. The candidates fall into two classes: one consists of basic FIBROBLAST GROWTH FACTOR (bFGF) and other members of the FGF family; and the other class comprises members of the TRANSFORMING GROWTH FACTOR β (TGFβ) family, of which the activins are the most potent (*see* GROWTH FACTORS; GROWTH FACTOR RECEPTORS). Both bFGF and, for example, activin A, can induce isolated animal pole ectoderm to activate mesoderm-specific genes and eventually to differentiate as mesodermal cell types.

In the same way that different regions of the vegetal hemisphere induce different types of tissue, members of the two families of inducing factor activate different mesoderm-specific genes and cause the formation of different types of tissue. bFGF tends to activate posterior- and ventral-specific genes and tissue types whereas activin induces anterior- and dorsal-specific structures. Activin also causes responding cells to behave as Spemann's organizer. This is further evidence that mesoderm induction is an instructive interaction, but the result is also interesting because it suggests that the inducing signal produced by dorso-vegetal blastomeres resembles activin and that the signal secreted by ventro-vegetal blastomeres resembles FGF.

Activin induces a wider range of cell types than FGF and this makes it possible to ask whether different concentrations of a mesoderm-inducing factor induce different cell types. When different concentrations of a *Xenopus* activin called XTC-MIF are applied to animal pole tissue, there is a tendency for dorsal and

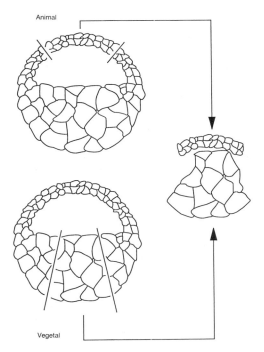

Fig. 130 Demonstration of mesoderm induction. Cells from the animal pole region of a *Xenopus* blastula-stage embryo are juxtaposed with tissue from the vegetal hemisphere. In normal development and when cultured alone, the animal pole tissue forms ectodermal derivatives. In combination with cells from the vegetal hemisphere it forms mesodermal cell types.

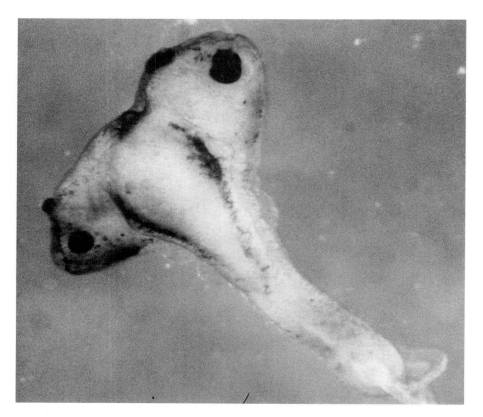

Fig. I31 A duplicated embryo resulting from an organizer graft.

anterior cell types to be induced in response to high concentrations and ventral cell types in response to low concentrations. This again implies that mesoderm induction is an instructive process, but it also suggests that activin can act as a MORPHOGEN, and impart positional information. This conclusion is reinforced by experiments in which animal pole cells are dispersed before being exposed to activin; this ensures that all cells are exposed to the same concentration of activin for the same time, and under these conditions the transitions from one cell type to another are very sharp. This result raises the possibility that patterning of cell types in the mesoderm is due to gradients of inducing factors [5] (*see* PATTERN FORMATION), although it should be borne in mind that the situation may be very complicated, with interactions occurring between different inducing and inhibiting molecules. Furthermore, recent evidence suggests that there is some 'predisposition' in the response of the animal cap to activin such that the dorsal region preferentially forms anterior structures and the ventral region posterior.

In addition to their dramatic effects on isolated animal pole regions, evidence is accumulating that activin and FGF-like molecules may be natural mesoderm-inducing factors. bFGF mRNA and protein are present in the *Xenopus* embryo at the appropriate stages and in sufficient amount to cause induction. Furthermore, if the function of the FGF receptor in the early embryo is abolished, the embryos that develop lack posterior structures, consistent with the proposed role for FGF in mesoderm induction.

Activin-like protein has also been detected in the *Xenopus* egg and early embryo. This is likely to be derived from maternal tissue, because transcription of activin genes in *Xenopus* does not begin until after the blastula stage. The role of activin in mesoderm formation has recently been confirmed in experiments in which the function of the activin receptor is abolished; in this case, the resulting embryos lack mesoderm completely.

Competence to respond to mesoderm-inducing signals declines at the beginning of gastrulation, and this coincides with a decrease in expression of cell-surface receptors for FGF. By this time activation of several mesoderm-specific genes has occurred, and some of these may be responsible for the dramatic cell movements associated with gastrulation. These movements convert the spherical and radially symmetric blastula into a three-layered structure with dorsoventral and anteroposterior axes. Gastrulation also marks the onset of neural induction, with the dorsal mesoderm moving beneath the responsive dorsal ectoderm.

Neural induction

Neural induction was discovered in 1921 by Spemann, who exchanged prospective epidermis and prospective neural plate between newt embryos at the early gastrula stage. He found that each graft behaved according to its new, rather than its original, position. However, if the same operation was performed at the late gastrula stage, the opposite result was obtained, with the grafted tissue differentiating according to its origin. This indicated that neural tissue is induced during gastrulation, and two further experiments demonstrated that the source of the inductive signal is dorsal mesoderm. The first experiment was the organizer graft, mentioned above, in which dorsal mesodermal tissue was grafted to the ventral side of a host embryo. This induces the formation of a secondary embryo, which includes neural tissue derived from host ventral ectoderm. The second experiment was done by Holtfreter, who maintained axolotl blastulae in a hypertonic salt solution and found that the presumptive mesoderm and endoderm rolled away from the ectoderm rather than involuting beneath it. This type of movement is called 'exogastrulation' and as a result, in the absence of involuting mesoderm, the entire ectoderm developed as epidermis, with no neural tissue.

Although the organizer graft shows clearly that ventral ectoderm can respond to neural inducing signals, it is also clear that dorsal ectoderm — that which normally forms the neurectoderm — responds much more efficiently. The origin of this predisposition is not clear. It may reflect the predisposition of the dorsal region of the animal cap to form anterior structures in response to activin, as mentioned above. Alternatively, it may represent a response to an activin-like mesoderm induction signal. The intensity of this signal would be too weak to induce Spemann's organizer, or muscle or notochord, but not so weak as to be ineffective and simply permit formation of ventral epidermis.

Regional neural induction

Neural induction also involves transmission of regional information, suggesting that it is an instructive interaction. This was first shown by the results of Spemann, who found that an organizer dissected from an early gastrula would induce head, trunk and tail structures, with appropriate nervous tissue, whereas one derived from a late gastrula, containing the last mesoderm to involute, would induce only trunk and tail. One suggestion from this result is that the anteroposterior character of the nervous system is specified by that of the underlying mesoderm. Unfortunately, our knowledge of the mechanism by which the anteroposterior pattern of the mesoderm is specified is extremely limited.

One view has it that cells acquire their anteroposterior specification as a result of mesoderm induction; that is, before gastrulation begins. The alternative idea is that anteroposterior position is established during gastrulation, with cells that have migrated furthest, or for the longest time, becoming most anterior in character. There is little evidence to support either scheme, but what there is favours the latter.

Neural inducing factors

Little progress has been made in identifying neural inducing factors. One reason may be that neural induction involves at least three steps: one to 'predispose' dorsal ectoderm to neural induction, one to induce neural tissue, and another to regionalize it. Thus application of a single factor, such as activin, may not be sufficient to cause neural induction.

Some progress has been made, however, in identifying the SECOND MESSENGER PATHWAYS used by the neural inducers. The results suggest that two second messenger systems are involved, one that acts through CYCLIC AMP (cAMP) and one that acts through PROTEIN KINASE C (PKC). PKC activation alone, unlike

cAMP elevation alone, is sufficient to induce neural structures. However, elevation of cAMP greatly enhances the response to activation of PKC. Although it seems probable that these pathways are involved in neural induction, it remains to be seen what activates them *in vivo*.

Finally, it is possible that some progress has been made in the identification of the 'transformation factor'. Application of RETINOIC ACID to *Xenopus* embryos causes the loss of anterior neural structures apparently without reduction in the amount of neural tissue. Retinoic acid is present in the embryo at the appropriate concentration (about 1.5×10^{-7} M) and it is possible that it is produced in the posterior region of the embryo and diffuses anteriorly, creating a gradient along the anteroposterior axis. A similar role for retinoic acid has been proposed for the development of the anteroposterior axis of the chick limb (*see* PATTERN FORMATION).

Later inductions

Mesoderm induction and neural induction establish the basic body plan of the vertebrate embryo. Later interactions are involved in establishing and patterning individual organs such as the lens, the limb and the heart. Here it is interesting and encouraging that the molecules used in early development, such as members of the TGFβ family, and retinoic acid, are also involved in later interactions. The involvement of retinoic acid in limb development is mentioned above, and TGFβ has been shown to be involved in several aspects of heart development. In the next few years great advances in our understanding of many inductive interactions are to be expected.

J.C. SMITH

1 Slack, J.M.W. (1991) *From Egg to Embryo. Regional Specification in Early Development*, 2nd edn (Cambridge University Press, Cambridge).
2 Gurdon, J.B. (1987) Embryonic induction — molecular prospects. *Development* **99**, 285–306.
3 Wolpert, L. (1969) Positional information and the spatial pattern of cellular differentiation. *J. Theor. Biol.* **25**, 1–47.
4 Hamburger, V. (1988) *The Heritage of Experimental Embryology: Hans Spemann and the Organizer.* (Oxford University Press, Oxford).
5 Green, J.B.A. & Smith, J.C. (1991) Growth factors as morphogens: do gradients and thresholds establish body plans? *Trends Genet.* **7**, 245–250.

induction ratio The degree to which an INDUCER increases the synthesis of an enzyme over and above the levels observed in the absence of the inducer.

influenza virus Virus of the family Orthomyxoviridae, which causes influenza and pneumonia. Three genera (A, B, C) and numerous serotypes have been identified in humans, with different strains in other warm-blooded vertebrates. The virus particle is roughly spherical (80–120 nm in diameter) or filamentous, with the helical ribonucleoprotein capsids enclosed by a membrane composed of M protein, which is in turn covered by a lipid envelope into which are inserted virus-encoded haemagglutinin (HA) and neuraminidase (NA) molecules. The genome is composed of eight molecules of linear (–)-sense single-stranded RNA (M_r 0.2×10^6–1.0×10^6).

The virus enters cells by binding via the trimeric haemagglutinin (M_r 250 000) to sialic acid-containing receptor molecules on the cell surface. The haemagglutinin molecule carries four separate epitopes (antigenic determinants) on its surface, some or all of which differ between strains. The sialic acid binding site, however, located in a depression in the molecule's surface, remains invariant. The bound virus is taken into the cell in endosomes, and the viral nucleic acid–transcriptase complex transferred from the endosome into the cytoplasm, and thence to the nucleus where the RNA is replicated and transcribed. Fusion of the virus and cell membranes is mediated by the haemagglutinin in the acidic conditions of the endosome.

In influenza A viruses the membrane protein M_2 is thought to function as an ion channel both when part of the virus in endosomes, when it allows acidification of the virus core, and later when incorporated into the membrane of Golgi vesicles during the transport of viral membrane proteins to the cell surface for envelope formation, where it raises the pH of the normally acidic trans-Golgi compartment. The channel function of M_2 is blocked by the drug amantadine (rimantadine), which is the basis for the action of this drug against influenza A viruses.

Influenza neuraminidase (sialidase) is a homotetramer (subunit M_r 60 000) which is thought to aid the release of newly synthesized virus particles from infected cells and also to assist the passage of the virus through the mucus lining the respiratory surfaces. Although influenza neuraminidases from different strains are antigenically distinguishable the active site is found to be invariant in all strains of A and B viruses characterized so far.

A unique feature of influenza viral mRNA synthesis is the phenomenon of 'cap stealing' — the transfer of a 5′ cap from newly synthesized host mRNA to the viral messenger sense RNA.

Influenza virus is highly variable antigenically and exhibits both antigenic drift (small changes that result in different serotypes emerging in the course of an epidemic) and antigenic shift in which radically different strains emerge to cause the next epidemic. Antigenic drift is thought to be caused by mutation and recombination between genomes within an infection, whereas the anti-genetic shifts are thought to be due to recombination of the human virus with strains from other species (e.g. birds).

Krug, R.M. (Ed.) (1989) *The Influenza Viruses* (Plenum, New York.)

ingression The penetration of superficial cells into the interior of the embryo. Usually the term implies that the cells penetrate individually rather than as a sheet (*cf.* INVAGINATION or INVOLUTION). Characterized by the presence of bottle cells, this process accompanies the early stages of GASTRULATION.

inherited disease/disorder *See*: GENETIC DISEASE.

inhibin *See*: GROWTH FACTORS.

inhibitor 1 A specific thermostable protein inhibitor (M_r 19 000) of protein phosphatase PP1 (*see* PROTEIN PHOSPHATASES) found in most mammalian tissues and in the liver of some species (e.g. rabbit) but not others (e.g. rat). Residues 9–54 are identical in rat

and rabbit and constitute the active fragment. Phosphorylation of threonine 35 by PROTEIN KINASE A is a prerequisite for inhibition, and the activity of inhibitor 1 is therefore regulated by hormones that raise the concentrations of CYCLIC AMP. Dephosphorylation can be mediated *in vitro* by protein phosphatases PP2A and PP2B.

inhibitor 2 A specific thermostable protein inhibitor (M_r 23 000) of protein phosphatase PP1 (*see* PROTEIN PHOSPHATASES) found in all mammalian tissues investigated, *Drosophila*, and starfish oocytes. Inhibitor 2 forms a complex with and inactivates PP1. The inactive complex can be reactivated by ATP, Mg^{2+} and glycogen synthase kinase 3, which phosphorylates threonine 72 on inhibitor 2. Prior phosphorylation of inhibitor 2 by casein kinase II enhances glycogen kinase 3 phosphorylation. Oscillation of the level of inhibitor 2 during the CELL CYCLE has been reported.

inhibitory junction potential (IJP, i.j.p.) A rapid transient HYPERPOLARIZATION of a SMOOTH MUSCLE cell membrane due to the release of a NEUROTRANSMITTER from prejunctional autonomic nerve terminals. Although this is a form of SYNAPTIC TRANSMISSION the term 'junction' rather than 'synaptic' is preferred because of the relative lack of specialization of the nerve and muscle at the site of neurotransmission compared with a skeletal neuromuscular synapse or central nervous system synapse. The hyperpolarization during the IJP renders the smooth muscle cell less excitable and more prone to relaxation as voltage-dependent calcium channels, responsible for ACTION POTENTIAL discharge, are less likely to open at the more negative membrane potential. Time courses for IJPs are very variable, according to the tissue in which they occur, but are in the range of 0.5 to 4.5 s; their amplitude is graded with strength of stimulus and amount of transmitter released, and can be up to ~30 mV. In most cases the hyperpolarization results from the opening of potassium ion channels due to the action of the neurotransmitter, which in most cases is undefined but may be ATP or the free radical of NITRIC OXIDE.

inhibitory postsynaptic potential (IPSP, i.p.s.p.) Transient HYPERPOLARIZATION of a postsynaptic cell membrane, brought about by the release of a NEUROTRANSMITTER from the corresponding presynaptic nerve terminal (*see* SYNAPTIC TRANSMISSION). In the central nervous system it is generally due to the action of the inhibitory amino acid neurotransmitters, usually GABA and GLYCINE. The hyperpolarization of the IPSP, during which the membrane potential becomes more negative, reflects a decreased excitability of the cell as its membrane potential is moved away from the threshold for generation of ACTION POTENTIALS. The amplitude of an IPSP is graded according to the intensity of the stimulus applied to the presynaptic nerve, and therefore the amount of transmitter that is released. Especially in the spinal cord, the amplitude of IPSPs can be up to ~30 mV, with a latency of tens of milliseconds, and a duration of hundreds of milliseconds; because of the rapid time course they may be called fast IPSPs. These are probably due to the release of glycine as a neurotransmitter from the presynaptic nerve terminal, activating chloride ion channels in the postsynaptic membrane (*see* GABA AND GLYCINE RECEPTORS). Some cells, particularly peripheral autonomic ganglia, have IPSPs with a relatively long latency, perhaps 0.5 s or more, and a prolonged time course: these are called slow IPSPs. In many cases the transmitter responsible for these slow IPSPs is ACETYLCHOLINE, acting on postsynaptic MUSCARINIC RECEPTORS coupled to potassium ION CHANNELS.

initial segment *See:* NEURON.

initiation codon The three-base CODON used to signal the beginning of a protein-coding sequence in a mRNA. In the majority of cases the initiation codon is AUG although in bacteria GUG and UUG can also be used. A specific INITIATOR TRNA (tRNA$_i^{Met}$) recognizes the initiation codon and inserts the amino acid methionine (N-formylmethionine in bacteria) at that position. *See:* BACTERIAL GENE EXPRESSION; BACTERIAL GENE ORGANIZATION; EUKARYOTIC GENE STRUCTURE; GENE; GENETIC CODE; PROTEIN SYNTHESIS.

initiation complex *See:* PROTEIN SYNTHESIS.

initiation factors (IF, eIF) A group of accessory proteins required for the initiation of PROTEIN SYNTHESIS. In prokaryotes (where they are designated IF) there are three initiation factors, each of which interacts with the 30S ribosomal subunit (*see* RIBOSOME) to ensure binding of the INITIATOR TRNA to the ribosome and of the ribosome to mRNA. In eukaryotes (where they are designated eIF) there are upwards of 10 different initiation factors many of which are multisubunit proteins. A trimeric factor, eIF-2, has a crucial role in the initiation of protein synthesis by bringing the initiator tRNA to the AUG codon. The activity of several initiation factors, including eIF-2, is controlled by phosphorylation (*see* PROTEIN PHOSPHORYLATION). Several initiation factors are also GTP-BINDING PROTEINS.

initiator tRNA The tRNA^{Met} species involved in the initiation of PROTEIN SYNTHESIS. In both prokaryotes and eukaryotes this tRNA is unable to participate in elongation of the polypeptide chain during translation even though it decodes the AUG (methionine) codon (and in bacteria, GUG and UUG). In bacteria, the initiator tRNA (designated tRNA$_f^{Met}$) carries a methionine residue that has been formylated on its amino group to give N-formylmethionyl-tRNA. In eukaryotes the initiator tRNA (designated tRNA$_i^{Met}$) does not carry formylated methionine although it is a distinct species from the tRNA involved in insertion of methionine during polypeptide chain elongation.

inner cell mass (ICM) The cells of the mammalian BLASTOCYST which give rise to the embryo proper. The mouse blastocyst consists of a spherical hollow vesicle with walls a single cell thick (the TROPHECTODERM), with a small localized thickening of cells inside — the inner cell mass. The inner cell mass is derived from the inner cells of the MORULA, and gives rise to all the embryonic tissues and the mesoderm of the placenta, as well as a layer of cells called the primitive endoderm, which forms part of the yolk sac (*see* EXTRAEMBRYONIC MEMBRANES). *See also:* MAMMALIAN DEVELOPMENT.

inosine Purine ribonucleoside containing the base hypoxanthine (Fig. I32) which is found in the anticodons of some TRANSFER RNAS. It occurs usually in the first position of the anticodon, where it can pair with adenine, cytosine, or uracil in the third position of the CODON (*see* WOBBLE). Inosine 5′-monophosphate is the biosynthetic precursor of AMP and GMP. *See:* BASES; NUCLEOSIDES AND NUCLEOTIDES.

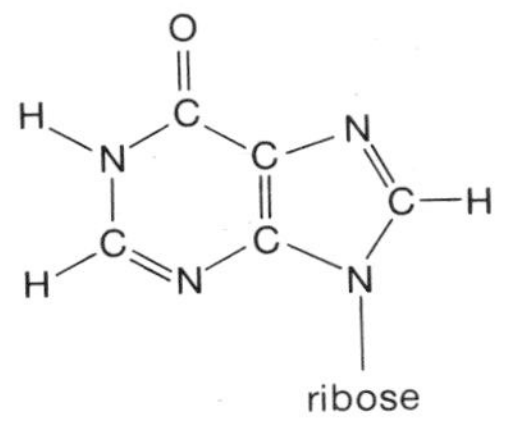

Fig. I32 The nucleoside inosine.

inositol The generic name for any isomer of hexahydroxycyclohexane: several inositols occur in nature. However, the term is most often used to refer specifically to *myo*-inositol, the isomer that in the preferred chair configuration (*see* SUGARS) has one axial hydroxyl (at position 2) and 5 equatorial hydroxyls (Fig. I33). *Myo*-inositol is the inositol found in natural PHOSPHOINOSITIDES and INOSITOL PHOSPHATES. *See also:* GPI ANCHORS.

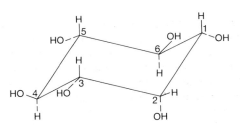

Fig. I33 *Myo*-inositol.

inositol lipid 3-kinase (phosphatidylinositol-3 kinase) A receptor-controlled enzyme capable of phosphorylating the 3-OH group of the inositol ring of phosphatidylinositol (PtdIns), phosphatidylinositol 4-phosphate (PtdIns4P), or PHOSPHATIDYLINOSITOL 4,5-BISPHOSPHATE (PtdIns(4,5)P$_2$). It is activated by many receptor tyrosine kinases in response to growth factors (*see* GROWTH FACTOR RECEPTORS), and also by some G PROTEIN-COUPLED RECEPTORS. *See also:* SECOND MESSENGER PATHWAYS.

inositol phosphates Derivatives of *myo*-inositol bearing various numbers of monoesterified and/or cyclic phosphate substituents. Best known are inositol 1,4,5-trisphosphate (Ins(1,4,5)P$_3$, IP$_3$) (Fig. I34), the Ca^{2+}-mobilizing second messenger formed by PHOSPHOINOSITIDASE C-catalysed hydrolysis of PHOSPHATIDYLINOSITOL 4,5-BISPHOSPHATE, and its metabolite inositol 1,3,4,5-

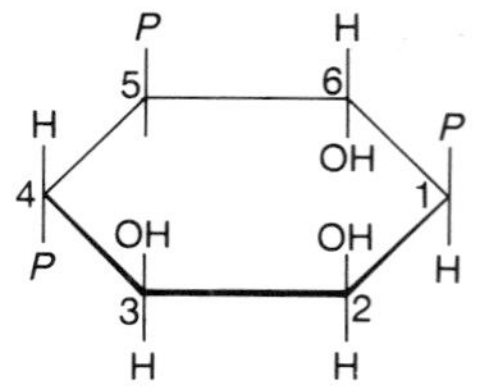

Fig. I34 Inositol 1,4,5-trisphosphate.

tetrakisphosphate (Ins(1,3,4,5)P$_4$) (which may also help to regulate cellular Ca^{2+} homeostasis). Cells contain several metabolites of Ins(1,4,5)P$_3$ and Ins(1,3,4,5)P$_4$, but also substantial quantities of various other *myo*-inositol polyphosphates (especially Ins(3,4,5,6)P$_4$, Ins(1,3,4,5,6)P$_5$ and InsP$_6$) which are of largely unknown function. In birds, inositol pentaphosphate is used in red cells to lower the oxygen affinity of haemoglobin (*see* HAEMOGLOBINS); InsP$_6$ is found mainly in plant seeds, where it is known as phytic acid, and as the magnesium calcium salt — phytin — is considered to be a phosphate storage product. *See*: SECOND MESSENGER PATHWAYS.

inositol phospholipids *See*: PHOSPHOINOSITIDES.

Inoviridae *See*: BACTERIOPHAGES.

Ins(1,4,5)P$_3$ Inositol 1,4,5-trisphosphate. *See*: INOSITOL PHOSPHATES; SECOND MESSENGER PATHWAYS.

insect development Development differs considerably beween the two main insect groups. The Holometabola undergo complete metamorphosis; the feeding larval stage and the adult are quite different and are separated by a pupal stage in which the body of the larva is almost completely broken down and reorganized. The Hemimetabola, in contrast, hatch from the eggs as larvae which are similar in morphology to the adults and are called nymphs. Embryonic development also differs and can be classified as long, intermediate, or short GERM BAND. In long germ band insects the entire body plan is established early in development whereas in short germ band insects development begins with only the rudiment of the head and then the segments of the thorax and abdomen are produced gradually from a posterior growth zone. Holometabolans tend to have long or intermediate germ band embryos whereas hemimetabolans are intermediate or short germ band. The molecular biology of development has been most intensively studied in *Drosophila* which is holometabolous with a long germ band embryo. *See*: DROSOPHILA DEVELOPMENT; INSECT NEURAL DEVELOPMENT.

Insect neural development

PERHAPS the most beguiling questions in development concern the formation of the most complex tissue in metazoans — the central nervous system (CNS). We would like to be able to describe, in molecular terms, how early embryonic cells give rise

to the precursors of the CNS, and how the extraordinary cellular diversity of the brain is achieved with such unerring accuracy. Invertebrates represent a simpler system with which to approach this enormous complexity, and *Drosophila* in particular allows a genetic dissection of the developmental decisions nerve cells are required to make. The great advantage in using mutations to look at neural development is that we can ask the fly what is important for its nervous system, rather than us telling it what we think is important. In addition, modern molecular biology has made the transition from detecting a gene by its phenotypic effect to isolating the gene product relatively easy. These advantages begin to permit the realistic expectation of a complete molecular description of developmental processes in a complex animal. The extrapolation of this information to vertebrate development is not only at the theoretical level — in very many cases, DNA sequence homologies suggest that the same or very similar molecules perform similar functions in vertebrates.

The problems involved in building the insect nervous system have been described at the cellular level in sizeable insects such as grasshoppers whose larger cells are more easily observed and manipulated by dye injection or laser ablation. These experiments have led to models for neuronal determination and recognition which are testable by mutation and molecular biology in *Drosophila*. From parallel studies on the formation of the peripheral nervous system, it seems that similar mechanisms are at play, often involving the same players.

The temporal sequence of neural development can be envisaged as a series of choices nerve cells and their precursors must make in order to achieve the cellular diversity and complex architecture of the CNS. These choices take place within the areas of neurogenesis, specification of neuronal identity, pathfinding, target recognition and synaptic differentiation (Fig. I35). The mechanisms underlying the first three of these choices are beginning to be revealed, particularly in *Drosophila*, but mechanisms of target recognition and synapse formation remain a mystery.

Neurogenesis

The insect CNS derives from a group of precursor cells, called NEUROBLASTS, which segregate individually from a sheet of ECTODERMAL cells. The remaining ectodermal cells give rise to a portion of the epidermis. When grasshopper neuroblasts are ABLATED using a laser microbeam, they can be replaced by adjacent ectodermal cells (REGULATION). This suggests that cells sort out their respective fates by intercellular communication — neuroblasts prevent neighbouring cells from adopting the neural fate and in this way, allow the excluded cells to adopt a secondary, epidermal fate.

What molecules mediate this communication process? In *Drosophila*, two classes of mutations exist which disrupt the fate of cells during neurogenesis [1]. One class, the neurogenic mutants, cause a failure of cells to choose between the neural and epidermal fates; instead all cells adopt the default, neural, fate resulting in embryos with a massive disorganized CNS and no cuticle. This phenotype is reminiscent of a defect in the intercellular communication required for grasshopper neurogenesis. Indeed, the cloning and study of two of the *Drosophila* neurogenic genes, NOTCH

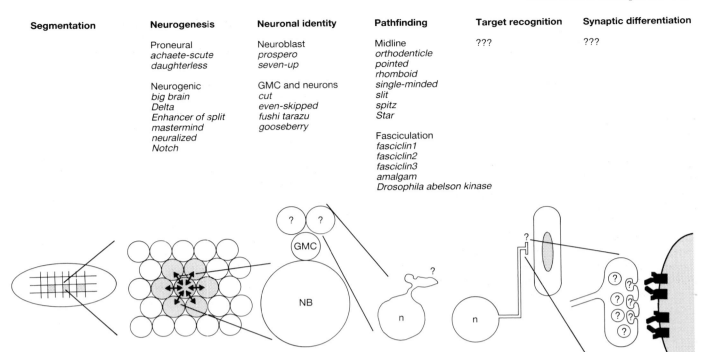

Segmentation	Neurogenesis	Neuronal identity	Pathfinding	Target recognition	Synaptic differentiation
	Proneural *achaete-scute* *daughterless*	Neuroblast *prospero* *seven-up*	Midline *orthodenticle* *pointed* *rhomboid* *single-minded* *slit* *spitz* *Star*	???	???
	Neurogenic *big brain* *Delta* *Enhancer of split* *mastermind* *neuralized* *Notch*	GMC and neurons *cut* *even-skipped* *fushi tarazu* *gooseberry*	Fasciculation *fasciclin1* *fasciclin2* *fasciclin3* *amalgam* *Drosophila abelson kinase*		

Fig. 135 The problems faced by developing *Drosophila* neurons and the genes implicated in their solution. The initial segmentation of the ectoderm is achieved at gastrulation by the cascade of early patterning genes. Neurogenesis is the process by which neural precursors (neuroblasts, NB) are set aside from remaining ectodermal cells which give rise to the epidermis. This process is achieved by cell–cell interactions directed and/or mediated by the proneural and neurogenic genes. Neuronal identity determines the thousands of different cellular fates open to neurons, and is directed at several levels: in the neuroblast (NB), the ganglion mother cell (GMC) or the neuron (n). The stereospecific route that axons follow (pathfinding) is achieved by selective neuronal recognition of both non-neuronal cells, for example in the midline, and neuronal cells by selective fasciculation. Genes which govern the terminal stages of neural development — target recognition and synaptic differentiation — have not been described.

and *Delta*, show that both are transmembrane proteins that specifically interact with each other.

What is the cellular machinery that *Notch* and *Delta* might be instructing in order to turn on specific fates? The second class of mutations affecting the fate of cells during neurogenesis are those of the proneural genes, defined principally as members of the ACHAETE-SCUTE complex (AS-C). Mutants of the AS-C have a qualitatively opposite phenotype to neurogenic mutants, so that instead of a larger than normal CNS, they have one much smaller than normal, suggesting that the normal genes are necessary to define the neural fate. The AS-C gene products are all members of the helix-loop-helix class of transcriptional regulators and DNA-binding proteins (*see* PROTEIN–NUCLEIC ACID INTERACTIONS; TRANSCRIPTION FACTORS). Their expression precedes the appearance of neuroblasts in the ectoderm, hence the term 'proneural'. This pattern of expression has led to the idea that the AS-C dictates neural potential to a group of cells (termed a 'proneural cluster') whose subsequent refinement to include just a single neuroblast depends on the cell–cell interactions mediated by the neurogenic genes. The epidermal counterpart to the AS-C may be the neurogenic gene complex, *Enhancer of split (E(spl))*. The gene products of *E(spl)* include some more helix-loop-helix proteins which are expressed in the epidermal precursors but not the neuroblasts. Thus, it seems likely that *Drosophila* neurogenesis, like grasshopper neurogenesis, depends on cell–cell interactions which are mediated by the *Notch* and *Delta* gene products whereas different cell fates are dictated by the *achaete-scute* and *Enhancer of split* gene products.

Neuronal identity

The bewildering complexity of cell states within the CNS is derived initially from a stereotypic array of neuroblasts, whose pattern is conserved between the grasshopper and the fruit fly. Laser ablation experiments on grasshopper ganglion mother cells indicate that the fate of the two daughter neurons they will produce is determined by the position of the neuroblast they were born from. However, the exact choice of fate between sibling neurons is initially enforced by their interaction. These experiments indicate at least a two-tiered level of control of neuronal identity — first the specification of a neuroblast lineage by the position the neuroblast occupies, and second the cell–cell interaction between sibling neurons to distinguish between their two alternative individual fates.

In *Drosophila*, mutants affecting both entire neuronal lineages and specific neurons have been described [2]. Thus, *prospero* mutants change the identity of some identified neurons whereas the gene itself is expressed in a subset of neuroblasts and ganglion mother cells. This gene is therefore a candidate for specifying neuroblast lineage fates.

What of those genes directing individual neuronal fates? A large number of genes involved in earlier aspects of ectodermal patterning are apparently redeployed in the nervous system, in ganglion mother cells or neurons. There are some examples, notably FUSHI TARAZU and *even-skipped*, which clearly are not just gratuitously expressed in the CNS but are helping to determine the identity of the neurons that express them.

Pathfinding

Differentiation of nerve cells leads to the elaboration of their remarkable structure — the axon. How do axons navigate across extracellular terrain towards their individual targets? Anatomical reconstructions (aided by the use of antibodies or of *lacZ* REPORTER GENES as lineage tracers) of both grasshopper and fruit fly nervous systems demonstrate a remarkable conservation in position and projection of identified neurons. These analyses highlight a two-tiered solution to axonal navigation — first the pioneering of nerve tracts by early axons along non-neuronal substrates, and second, selective addition of later axons to existing routes. These solutions involve specific recognition of non-neuronal cues in the initial pathfinding and neuron–neuron interaction in the process of fasciculation — the building up of nerve tracts and cords from many different axons.

A detailed cellular analysis of part of the *Drosophila* CNS shows that certain GLIAL CELLS may aid the very first pioneering axons to form the two commissural nerves in each segmental neuromere. Antibody staining of growth cones combined with *lacZ* lineage tracers show that the growth cones track along midline glial cells. Axons which pioneer the longitudinal nerve tracts may similarly be guided by six longitudinal glia which form a roof over developing NEUROPIL.

What sort of molecules guide these first axons? A series of mutations affecting the development of *Drosophila* midline-derived cells have been described — in the cases of *single-minded* and *slit*, these mutations lead to a complete absence of commissural nerves. The *slit* gene product is of particular interest as it is a cell-surface or extracellular matrix protein and seems to be picked up by axonal surfaces from midline glia [3]. As in vertebrate neurite extension, the extracellular matrix of insects may have a permissive role in pathfinding (*see* CELL–MATRIX INTERACTIONS; EXTRACELLULAR MATRIX MOLECULES). The expression of the extracellular matrix protein laminin, for example, predicts the formation of major axon pathways in both the central and peripheral nervous systems.

Axons which arrive later in development choose their pathways by selectively adhering to pre-existing routes, a practice known as fasciculation. In *Drosophila*, molecules which mediate fasciculation — fasciclins — are a class of CELL ADHESION MOLECULES, including some members of the immunoglobulin gene superfamily homologous to the mammalian NCAM. Mutations

that abolish some of these fasciclins do not affect the nervous system, but double mutants for both *fasI* and the *Drosophila abelson tyrosine kinase* have dramatic phenotypes of greatly diminished or absent nerves. This phenotype seems to arise from the misrouting of some axons. Thus, at least in this case, there is evidence of a combinatorial basis for selective neuronal recognition — a mechanism which can help neurons to encode diverse cell surface functions with a limited number of cell-surface proteins.

D.A. HARTLEY

See also: CAENORHABDITIS; CAENORHABDITIS NEURAL DEVELOPMENT; DROSOPHILA DEVELOPMENT; NEURON; NEURONAL CYTOSKELETON; VERTEBRATE NEURAL DEVELOPMENT.

1 Campos-Ortega, J.A. & Jan, Y.-N. (1990) Genetic and molecular bases of neurogenesis in *Drosophila melanogaster*. *Annu. Rev. Neurosci.* **14**, 399–420.
2 Doe, C.Q. (1991) The generation of neuronal diversity in the *Drosophila* central nervous system. In *Determinants of Neuronal Identity* (Shankland, M. & Macagno, E., Eds) (Academic Press, New York).
3 Klämbt, C. et al. (1991) The midline of the *Drosophila* central nervous system: a model for the genetic analysis of cell fate, cell migration and growth cone guidance. *Cell* **64**, 801–815.
4 Hortsch, M. & Goodman, C.S. (1991) Cell and substrate adhesion molecules in *Drosophila*. *Annu. Rev. Cell Biol.* **7**, 505–507.

insertin *See*: ACTIN-BINDING PROTEINS.

insertion Any MUTATION caused by the insertion of a nucleotide or stretch of nucleotides into a gene. Many naturally occurring insertion mutations are the result of the transposition of TRANSPOSABLE GENETIC ELEMENTS.

insertion loop *See*: DELETION LOOP.

insertion mutagenesis The induction of MUTATION by the insertion of another piece of DNA into the target GENE. This can be achieved by the random integration of retroviral proviruses in mammalian cells (*see* RETROVIRUSES), or by insertion sequences (IS elements) or transposons in bacteria and other organisms (*see* TRANSPOSABLE GENETIC ELEMENTS). Targeted insertion mutations intended to inactivate specific genes can be achieved, especially in yeast and mammalian embryonic stem cells, by HOMOLOGOUS RECOMBINATION between the target gene and the incoming DNA (*see* GENE KNOCKOUT; TRANSGENIC TECHNOLOGIES). Mutants generated by insertional mutagenesis are usually NULL MUTANTS showing a complete loss of function of the gene.

insertion sequences *See*: TRANSPOSABLE GENETIC ELEMENTS.

insertional inactivation The inactivation of a GENE by the insertion of another piece of DNA into its coding sequence (*see* INSERTION MUTAGENESIS). It is widely used to identify PLASMIDS carrying pieces of 'foreign' DNA (RECOMBINANT DNA) inserted into a cloning site located within an antibiotic-resistance gene thereby eliminating the encoded antibiotic-resistance phenotype. *See*: DNA CLONING; VECTORS.

instructive signal An inductive signal (*see* INDUCTION) that causes the responding tissue to develop along a pathway different from that followed in its absence. A classic example comes from early AMPHIBIAN DEVELOPMENT. Animal cap cells cultured in isolation adopt an ectodermal fate, but they adopt a mesodermal fate when cultured in combination with vegetal cells. The degree of instruction is limited by the range of options of the responding tissue (its competence), in the latter case comprising ectoderm and mesoderm only. Instructive signals are distinguished from permissive signals, which are necessary for full differentiation of a tissue but do not influence the developmental fate of the cells.

Insulin and its receptor

INSULIN is a peptide hormone secreted by the β cells of the islets of Langerhans of the pancreas in response to an elevation in blood glucose or other secretagogues. It plays a crucial role in glucose homeostasis, by regulating the uptake and metabolism of glucose by peripheral tissues and the production and storage of glucose by the liver. Insulin also regulates the metabolism of lipids and proteins, the synthesis of nucleic acids and the expression of certain genes. In some cells, and perhaps in foetal life, insulin also has a less well-defined role as a GROWTH FACTOR.

Alteration in the production and/or action of insulin is the major pathogenic factor in the diverse forms of DIABETES MELLITUS, a syndrome that features a pathological elevation in blood sugar and complications thereof.

Structure and biosynthesis of the insulin molecule

The mature insulin molecule is made of two polypeptide chains: an A chain, usually containing 21 amino acids, and a B chain of 30 amino acids. The chains are linked by two disulphide bridges, between Cys A7 and Cys B7, and between Cys A20 and Cys B19. An intrachain disulphide bridge between Cys A6 and Cys A11 creates a loop in the A chain.

Insulin is a member of a larger family of structurally and evolutionarily related peptides, which comprises the insulin-like growth factors I and II (*see* GROWTH FACTORS), relaxin, and the invertebrate bombyxins and molluscan insulin-like peptides [1].

The single-copy insulin gene is located in humans on chromosome 11. Some species (rat, mouse, *Xenopus laevis*, some fishes) have two active insulin genes. All known genes contain two introns except rat and mouse insulin I, which have only one [1].

Insulin is synthesized as a single chain precursor, preproinsulin, of 110 amino acids. The 24-residue hydrophobic N-terminal SIGNAL PEPTIDE is rapidly cleaved after transfer of the nascent polypeptide through the membrane of the rough ENDOPLASMIC RETICULUM (*see* PROTEIN TRANSLOCATION). There, the proinsulin molecule folds and the disulphide bonds are formed. The primary sequence of proinsulin is shown in Fig. 136. Proinsulin then migrates through the Golgi complex and ends up packaged in crystalline form in clathrin-coated secretory granules (*see* COATED PITS AND VESICLES; PROTEIN SECRETION). The 31 amino acid C-peptide which links A1 to B30 is then cleaved by two subtilisin-related endoproteases — convertase PC3, which cleaves preferentially at a basic dipeptide at the B/C junction, and convertase PC2, which cleaves selectively at a second basic dipeptide at the C/A junction [2].

The insulin crystal, present in the secretory granules, is made of hexamers, a symmetric assembly of three dimers arranged around two atoms of zinc. Insulin was the first protein to have its structure determined by X-RAY CRYSTALLOGRAPHY, and the three-dimensional structure [3] of the molecule is now known with a resolution of 1.5 Å (*see* Plate 6a). After its secretion by exocytosis from the β cell, insulin at the low concentration present in blood rapidly dissolves into monomers, which are the biologically active form of the molecule.

Rare mutations in the insulin or proinsulin molecule have been found in families of patients with diabetes mellitus or glucose intolerance, resulting from single nucleotide mutations [4]: Phe B25 to Leu (Insulin Chicago), Phe B24 to Ser (Insulin Los Angeles), Val A3 to Leu (Insulin Wakayama), Arg 65 to His (Proinsulin Tokyo/Boston/Denver), and His B10 to Asp (Proinsulin Providence). The latter two result in a deficient conversion of proinsulin to insulin (familial hyperproinsulinaemia).

Insulin actions at the cellular level

The principal target tissues for insulin action are muscle, liver, and fat. In sensitive cells (e.g. cardiac and skeletal muscle, and adipose tissue) insulin stimulates glucose transport along its gradient from outside to inside the cell by triggering the translocation of a certain type of glucose transporter (GLUT 4, so called because it is the fourth member of a transporter family [5] to have been cloned) from intracytoplasmic vesicles into the plasma membrane. In liver, brain, and other tissues where glucose transport is not sensitive to insulin, different transporters are present.

Insulin also regulates the activity of key regulatory enzymes, often by stimulating covalent modification such as reversible phosphorylation or dephosphorylation (*see* PROTEIN PHOSPHORYLATION). In addition, the level of some enzymes is regulated by insulin's action on gene expression.

All these intracellular actions are initiated by the binding of circulating insulin to a specific cell-surface receptor (*see* GROWTH FACTOR RECEPTORS). The binding of the hormone to the extracellular domain of the receptor induces a conformational change in both molecules, resulting in the generation of intracellular signals. The intracellular portion of the receptor is endowed with protein TYROSINE KINASE enzymatic activity [6]. Unlike most cellular PROTEIN KINASES, which phosphorylate serine or threonine residues, these kinases phosphorylate tyrosine residues. Receptors which are tyrosine kinases represent a large family of related molecules which includes the receptors for several growth factors (e.g. epidermal growth factor, insulin-like growth factors, platelet-derived growth factor).

The tyrosine phosphorylation of intracellular substrates by the insulin receptor is thought to initiate a cascade of phosphorylations which results in the regulation of key metabolic enzymes. The first phosphorylating event after insulin binding is in fact an autophosphorylation of the intracellular portion of the receptor

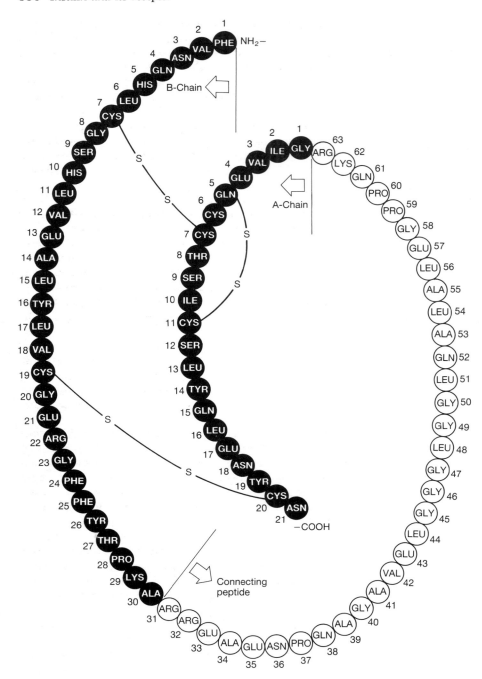

Fig. I36 Primary sequence of pig proinsulin. From [18].

itself, which markedly enhances its ability to phosphorylate other substrates. Few substrates have so far been identified with certainty. One is the so-called IRS-1 (insulin receptor substrate 1), which seems to be a docking protein featuring nine sequence motifs known to bind signal-transducing molecules containing SH2 or SH3 (Src homology) domains, such as PHOSPHATIDYL-INOSITOL 3-KINASE [7] (*see also* SECOND MESSENGER PATHWAYS). The binding of such molecules to IRS-1 would presumably make them available as substrates for the receptor tyrosine kinase.

Although a large body of evidence (including alterations to the kinase by SITE-DIRECTED MUTAGENESIS) supports the concept that the tyrosine kinase activity of the insulin receptor is an essential first step for many of the intracellular actions of the hormone, it does not preclude other mechanisms such as direct association of effector molecules with the receptor. It is not clear yet if, and where, the downstream signalling pathways for the metabolic and mitogenic effects of insulin diverge.

The tyrosine kinase is subject to negative regulation by serine/threonine phosphorylation of the receptor, as well as by PROTEIN PHOSPHATASES that undo its tyrosine phosphorylation.

Following insulin binding, the hormone–receptor complex is internalized through clathrin-coated pits, then vesicles, by a process known as ligand-mediated ENDOCYTOSIS [8] (*see also* COATED PITS AND VESICLES). This process is important in the regulation of receptor concentration at the cell surface. A long term increase in ambient insulin concentration results in a decrease in available receptors (down-regulation).

Structure and biosynthesis of the insulin receptor

The mature insulin receptor ($M_r \sim 350\,000$) is made up of two identical extracellular α subunits ($M_r \sim 135\,000$), and two identical transmembrane β subunits ($M_r \sim 95\,000$) (Fig. 137).

The α subunits (made of either 719 or 731 amino acids depending on the insertion of a 12-residue segment by alternative RNA SPLICING) contain the insulin-binding domain. Only one insulin molecule binds with high affinity to the whole receptor; when more insulin binds, the dissociation rate of the first bound molecule becomes faster, and the binding affinity decreases, a phenomenon called negative cooperativity [9]. It appears that the binding domain comprises at least two sequences, one N terminal, and one closer to the C-terminal end of the receptor's α chain [10]. The α subunit is heavily N-glycosylated (*see* GLYCANS; GLYCO- PROTEINS AND GLYCOSYLATION). It contains a cysteine-rich domain (26 Cys) between residues 155 and 312. The two α subunits are linked together and to the extracellular portion of the β subunit by disulphide bridges (possibly not more than one).

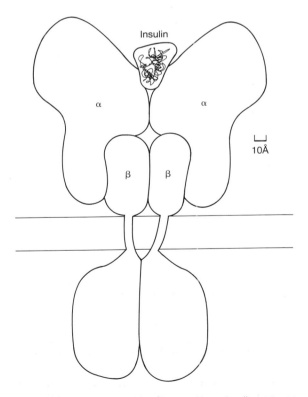

Fig. 137 Schematic view of the insulin receptor with one insulin monomer bound, drawn approximately to scale. Modified from [3].

The region of the insulin molecule that binds to the receptor has been mapped, as have the residues involved in the negative cooperativity [9] (*see* Plate 6a). This region includes the residues B24–B26, also involved in insulin dimerization. Several pieces of evidence suggest that the B20–B30 C-terminal portion of the B chain unfolds on receptor binding, exposing the underlying sur- face of the A chain (Ile A2, Val A3) for receptor contact [11]. Recent data suggest the existence of a second binding site on the hexamerization surface of insulin [12], opening up the possibility that one insulin molecule could bind through these two binding sites to two distinct receptor domains, one on each α subunit, thereby creating the high-affinity binding [13].

The β subunits (620 amino acids) comprise an extracellular portion, a transmembrane domain, and an intracellular portion that contains the tyrosine kinase. The ATP-binding site has been mapped to the consensus sequence Gly 991-X-Gly 993-X-X- Gly 996-(X)$_{21}$-Lys 1018 (the numbering corresponds to the vari- ant with the 12-residue insert present). The tyrosines that are autophosphorylated are distributed in three groups: Tyr 953, 960, and 972; Tyr 1146, 1150, and 1151; and Tyr 1316 and 1322. The phosphorylation of the first group (thought to be involved in some substrate binding) is not universally accepted. The phosphorylation of the second group is important for the metabolic effects of insulin, but maybe not for its mitogenic action. The phosphorylation of the third, C-terminal group, is suggested to inhibit the tyrosine kinase. The phosphorylation occurs through a *trans* mechanism: one β subunit phosphorylates the other.

The receptor is synthesized as a single-chain preproreceptor precursor. The N-terminal signal peptide is then removed by proteolytic cleavage. Two identical chains then assemble to form the receptor as a dimer, and a second cleavage occurs at a basic tetrapeptide between the α and β subunits, resulting in their separation. The mature $\alpha_2\beta_2$ structure is commonly refered to as a heterotetramer.

The insulin receptor shares a strong structural and sequence homology with the type I insulin-like growth factor receptor (which binds IGF-I and IGF-II with high affinity, and insulin weakly), and hybrid receptors made of one αβ moiety of each are formed in cells where both are expressed. The function of these hybrids is unknown.

The single gene encoding the insulin receptor, located on the short arm of human chromosome 19, spans more than 130 kilobase pairs and has a complex intron–exon organization (22 exons, 21 introns) [14]. The α and β subunits are each encoded by 11 exons. Exon 1 encodes the signal peptide. Exon 2 encodes a portion of the binding domain. Exon 3 encodes the cysteine-rich region. Exon 11 is alternatively spliced and encodes the 12 amino acids inserted in one of the two receptor variants. This insertion decreases the insulin binding affinity two- to threefold, but the functional significance of the alternative splicing is otherwise unknown. The region encoded by exon 12 contains the tetrabasic prorecptor processing site. Exon 15 encodes the transmembrane domain, and exons 17 to 21 the tyrosine kinase. The receptor gene promoter lacks a TATA box, a characteristic of HOUSEKEEP- ING GENES.

Nearly 50 insulin receptor gene mutations have been discovered in patients with various disorders associated with extreme insulin resistance [15]. These mutations may result in impaired synthesis and transport of the receptor to the cell membrane, impaired insulin binding, impaired transmembrane signalling, impaired endocytosis and recycling, or a combination of these. The analysis of these mutations has helped in the understanding of the functional role of certain receptor domains.

However, both insulin gene mutations and insulin receptor gene mutations represent only an infinitesimal fraction of the cases of diabetes mellitus.

Interestingly, in the patients with either insulin gene mutations or insulin receptor gene mutations, the severity of the glucose intolerance or diabetes mellitus was quite variable, even among bearers of the same mutation, suggesting that more than one single-gene defect is necessary to offset glucose homeostasis in non-insulin-dependent diabetes mellitus. A combination of several defective gene products ('diabetogenes') may be required in order to reach a threshold above which glucose homeostasis collapses [16,17].

P. DE MEYTS

1 Shuldiner, A.R. et al. (1991) Insulin. In *Insulin-like Growth Factors: Molecular and Cellular Aspects* (LeRoith, D., Ed.) 181–219 (CRC Press, Boca Raton, FL).
2 Steiner, D.F. et al. (1992) The new enzymology of precursor processing endoproteases. *J. Biol. Chem.* **267**, 23435–23438.
3 Baker, E.N. et al. (1988) The structure of 2Zn pig insulin crystals at 1.5 Å resolution. *Phil. Trans. R. Soc. Lond.* **B319**, 369–456.
4 Steiner, D.F. et al. (1990) Lessons learned from molecular biology of insulin-gene mutations. *Diabetes Care* **13**, 600–609.
5 Pessin, J.E. & Bell, G.I. (1992) Mammalian facilitative glucose transporter family: structure and molecular regulation. *Annu. Rev. Physiol.* **54**, 911–930.
6 Rosen, O.M. (1987) After insulin binds. *Science* **237**, 1452–1458.
7 Sun, J.S. et al. (1991) Structure of the insulin receptor substrate IRS-1 defines a unique transduction protein. *Nature* **352**, 73–77.
8 Carpentier, J.L. (1989) The cell biology of the insulin receptor. *Diabetologia* **32**, 627–635.
9 De Meyts, P. et al. (1978) Mapping of the residues of the receptor-binding region of insulin responsible for the negative cooperativity. *Nature* **273**, 504–509.
10 Andersen, A.S. et al. (1992) Identification of determinants that confer ligand specificity on the insulin receptor. *J.Biol.Chem.* **267**, 13681–13686.
11 Xin Hua, Q. et al. (1991) Receptor binding redefined by a structural switch in a mutant human insulin. *Nature* **354**, 238–241.
12 De Meyts, P. et al. (1990) Receptor negative cooperativity and insulin dimerization. *Diabetologia* **33**, Suppl. 1, 227.
13 Schäffer, L. (1993) The high-affinity binding site of the insulin receptor involves both α-subunits. *Exp. Clin. Endocrinol. Leipzig* **101**, Suppl. 2, 7–9.
14 Seino, S. et al. (1989) Structure of the human insulin receptor gene and characterization of its promoter. *Proc. Natl. Acad. Sci. USA* **86**, 114–118.
15 Taylor, S.I. et al. (1992) Mutations in the insulin receptor gene. *Endocr. Rev.* **13**, 566–595.
16 Granner, D.K. & O'Brien, R.M. (1992) Molecular physiology and genetics of NIDDM: importance of metabolic staging. *Diabetes Care* **15**, 369–395.
17 De Meyts, P. (1993) The diabetogenes concept of NIDDM. In *New Concepts in the Pathogenesis of Non Insulin-dependent Diabetes Mellitus* (Ostenson, C.G. et al., Eds) *Adv. Exp. Med. Biol.* **334**, 89–100 (Plenum, New York).
18 Chance, R. (1971) *Diabetes* **21**, Suppl. 2, 461–467.

insulin-like growth factors (IGF) *See*: GROWTH FACTORS; PARENTAL GENOMIC IMPRINTING.

int-1, int-2 *See*: GROWTH FACTORS; ONCOGENES.

integral membrane protein Membrane protein embedded within the membrane (e.g. a transmembrane protein), as opposed to being attached to one or other face. Also called an intrinsic membrane protein. *See*: MEMBRANE STRUCTURE.

integrase An enzyme encoded by a temperate BACTERIOPHAGE which mediates the INTEGRATION of the bacteriophage genome into the host bacterial chromosome. The best known example of an integrase is that of the bacteriophage LAMBDA; in this case the integrase binds as a dimer to both the phage genome and the host chromosome at specific attachment sites (*att* sites) to generate a staggered cut required for effecting DNA strand transfer (*see* RECOMBINATION). The event mediated by the lambda integrase is an example of site-specific recombination. *See also*: RETROVIRUSES.

integration The covalent insertion of one piece of DNA into another in a process analogous to genetic RECOMBINATION, and which results in a continuous piece of DNA. The integration of BACTERIOPHAGE and viral genomes, and various TRANSPOSABLE GENETIC ELEMENTS (e.g. bacterial insertion sequences) requires an enzyme (integrase or transposase) to mediate the process. *See*: LAMBDA; PAPOVAVIRUSES; RETROVIRUSES.

integration host factor (IHF) Protein from the bacterium *Escherichia coli* required for the integration of phage LAMBDA into the bacterial chromosome. *See also*: PROTEIN–NUCLEIC ACID INTERACTIONS; RECOMBINATION.

integration site The site at which a DNA sequence has become inserted into a chromosome or other DNA molecule. It is usually used in the context of PROPHAGE or PROVIRUS integration and can be either a random or a specific site. An example of the latter is the bacteriophage LAMBDA attachment site in the *Escherichia coli* genome; in contrast, RETROVIRUSES integrate into vertebrate genomes at random positions. The integration site can usually be identified by a short duplicated region of the host DNA sequence at either end of the integrated DNA sequence.

integration vector A PLASMID-based VECTOR for inserting genes and other DNA sequences into a chromosome. To ensure integration the vector does not contain an ORIGIN OF REPLICATION and thus can only be maintained during subsequent cell division of the host cell by integration into a stable genetic element, namely the chromosome. In the yeast SACCHAROMYCES CEREVISIAE a class of integration vector (designated YIp for *Y*east *I*ntegrating *p*lasmid) has been developed for integrating modified genes into the *S. cerevisiae* genome.

Integrins

INTEGRINS are cell-surface protein complexes forming one of the largest classes of cell-surface molecules mediating adhesion of cells to each other and to their surroundings. They appear to be present throughout the animal kingdom, in both vertebrates and invertebrates.

Cells need to adhere to each other and to molecules in their environment in many developmental and physiological processes. Obvious examples are the creation of tissues and organs and the maintenance of their integrity (*see* CELL JUNCTIONS), the formation of tightly adhesive structures like the myotendinous junction (muscle insertion), and the precise migration and targeting of cells and cell processes such as axons during development (*see* CAENORHABDITIS NEURAL DEVELOPMENT; CELL–MATRIX INTERACTIONS; INSECT NEURAL DEVELOPMENT; VERTEBRATE NEURAL DEVELOPMENT).

Many adhesive properties of cells are attributable to integrins. They act in effect as transmembrane connectors, linking the CYTOSKELETON of the cell to molecules of the surrounding environment. Many integrins recognize extracellular ligands, usually extracellular matrix or serum proteins (*see* EXTRACELLULAR MATRIX MOLECULES) and so act as substrate adhesion molecules (Table I4). Some mediate cell–cell interactions; these can involve either ligand-mediated or direct heterotypic cell–cell recognition (Fig. I38). (Members of two other major classes of adhesion molecule — NCAM and other adhesion molecules of the IMMUNOGLOBULIN SUPERFAMILY and cadherins — can mediate homophilic (like to like) cell–cell adhesion molecules and are discussed elsewhere in this volume, *see* CELL ADHESION MOLECULES.)

Integrins probably act also to transduce signals across the cell membrane, leading to changes in gene expression, cell behaviour and DIFFERENTIATION.

Structure

Integrins make up a large superfamily of transmembrane heterodimers, each composed of an α and a β subunit. New family members are continually being added. The 20 or more integrins identified to date fall into several subfamilies, each defined by the β subunit they contain. Each class of β chain can associate with any of a subset of α chains, the specific αβ combination determining the specificity of ligand recognition (Table I4). However, a growing number of promiscuous αβ associations are now subverting this rigid classification. The eight or so vertebrate β chains characterized so far show HOMOLOGY at the amino-acid level of some 37–55% one to another: the large extracellular portion of each (>90% of the molecule) has a well conserved pattern of cysteine residues and four repeats of a cysteine-rich domain; the short cytoplasmic domain (40–50 amino acids) often, though not invariably, contains a phosphorylatable tyrosine residue (Fig. I39). One β chain, β_4, although showing good homology to other β subunits in its extracellular domain, differs dramatically in its large cytoplasmic tail, which is about 1000 amino acids in length and includes four FIBRONECTIN TYPE III DOMAINS. Invertebrate β chains identified so far show good structural and sequence homology to their vertebrate counterparts (up to 45% homology to mammalian β_1).

The different vertebrate α subunits show less homology at the amino acid level (in the main 25–45%) and invertebrate α chains show as much homology to some vertebrate α chains as the vertebrate chains do to each other. However, all the α chains have very strongly conserved structural homology. The major features

Table I4 The integrin family

β chain	α chain	Ligands
β_1	α_1	CO I, CO IV, CO VI, LN(P1, E1)
	α_2	CO I(CB3), CO IV, LN, FN
	α_3	FN(RGD), CO I, LN(E3), epiligrin
	α_4	FN(CS1, CS5), VCAM-1 ($\alpha_4\beta_7$ = Peyer's patch addressin)
$\beta_{7(=\beta P)}$	α_5	FN(RGD)
β_4	α_6	LN(E8) ($\alpha_6\beta_4$ = ? ligand)
	α_7	LN(E8)
	α_8	FN
β_2	α_L	ICAM-1, ICAM-2
	α_M	C3bi(RGD?), ICAM-1, FB, Factor X, LPS, β glucan, *Leishmania* gp63
	α_X	C3bi?, FB, LPS, β glucan
β_3	α_{IIb}	FN(RGD), FB, vWF, VN, TSP, fibrin
	α_V	VN, vWF, FB, TSP?, SP, LN(P1), FN(RGD) [$\alpha_V\beta_1$ = FN, VN(RGD)]
β_5 (? = β_1)		FN(RGD), VN
β_6		FN, VN?, RGD
β_N (? = β_1)		FN, CO
β_8		?
(D. melanogaster)		
β_{PS}	α_{PS1}	?
	α_{PS2}	?

At least three other β and one other α chains have been defined at a biochemical level or cloned. A role for RGD has been shown for α_3, α_5, α_M, α_{IIb}, $\alpha_V(\beta_1, \beta_3, \beta_5, \beta_6)$. Alternate RNA splicing has been shown for α_3, α_4, α_6, α_{PS2} (exon 8), β_1, β_3, β_4, β_5, and β_{PS}. Integrins have been demonstrated in a number of mammalian (primate, rodent), avian (chicken), *Xenopus*, and invertebrate (*D. melanogaster* and *C. elegans*) species.

The cellular distribution of the majority of the integrin receptors is broad and is not detailed here. β_2 (LFA) integrins are restricted to leukocytes and $\alpha_{IIb}\beta_3$ is only found in cells of the megakaryocyte/platelet lineage.

The list of ligands for the integrins is abbreviated. CO, collagen; FB, fibrinogen; FN, fibronectin; LN, laminin; SP, bone sialoproteins; TSP, thrombospondin; VN, vitronectin; vWF, von Willebrand factor.

are a largely invariant pattern of cysteine residues, the presence of three or four repeats (one is sometimes degenerate) of a divalent cation binding motif, a fairly well conserved transmembrane domain and, again, short cytoplasmic domains which show little similarity in sequence (Fig. I39). The cation-binding domain shows homology to the E-F hand loop sequences typified by calmodulin (*see* CALCIUM-BINDING PROTEINS). A short stretch of five amino acids immediately following the transmembrane domain is almost absolutely conserved in the available α chain sequences (GFFKR in all α chains except *Drosophila* αPS2, which

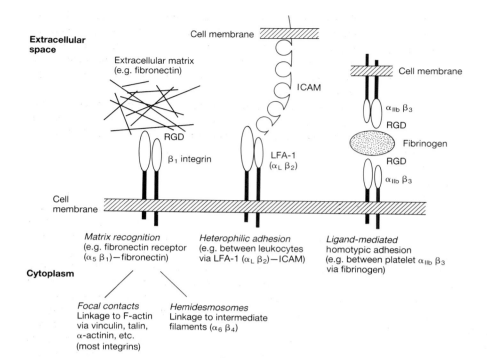

Fig. I38 The integrin family of cell adhesion molecules and types of recognition they mediate. RGD, Arg-Gly-Asp; LFA-1, see text.

has GFFNR; single-letter amino-acid notation). This may provide an alternative linkage to cytoplasmic proteins.

In some α chains, proteolytic cleavage of the extracellular domain creates a dimer of heavy and light chains joined by disulphide bonds (Fig. I39). Some α subunits, particularly those of the β_2 (LFA) family, also contain a domain (the I domain) of 180–200 amino acids 'inserted' at the N-terminal side of the first metal-binding domain. The I domain shows homology to repeats in a number of extracellular proteins including collagen and von Willebrand factor; its function is unknown, but may also confer specific ligand binding properties.

There is evidence for extensive association of the N-terminal regions of the extracellular domains of the α and β chains.

Binding specificity

Both integrin chains seem to cooperate in binding to ligands or to receptors on other cells and it seems likely that the specificity of ligand recognition stems from an association of particular α and β chains. Some α chains (α_v, α_4, α_6) can associate with two or more

different β chains, yielding receptors with different ligand recognition properties and tissue distributions.

The correlation between chain composition and specificity is, however, still problematical, possibly due to artefacts of the *in vitro* systems in which interactions are analysed. There is evidence, for example, that specificity *in situ* may involve certain GANGLIOSIDE lipids (which may be missing under *in vitro* conditions), and the role of specific divalent cations in modulating specificity is yet to be fully evaluated.

Differential exon splicing (alternative splicing) (*see* RNA SPLICING) leading to differences in the extracellular regions has been found in several integrin chains (e.g. α_6, β_1, β_3, β_4, β_5) and might also lead to modified affinity or specificity for ligand. The possible significance of such splicing during development is discussed below.

Recognition sequences in ligands

The sequence Arg-Gly-Asp (RGD) is found in the integrin-binding domains of a number of ligands (Table I4); it is presumed that

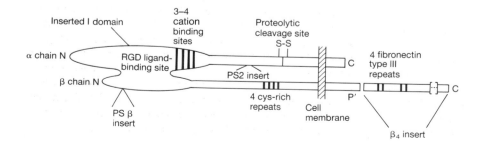

Fig. I39 Schematic diagram of integrin structure. The I domain is found in α_1, α_2, α_L, α_M and α_X chains. The proteolytic cleavage site is found in α_3, α_5, α_6, α_{IIb}, α_V and $\alpha PS2$.

sequences flanking this tripeptide determine the exact binding specificity. Several alternative integrin recognition sequences have been identified — in fibronectin (LDV in the CS1 segment of the alternatively spliced IIICS domain), in FIBRINOGEN (H12 sequence at the N terminus of the γ chain), and in laminin (the E8 region) for example.

Connections to cytoskeleton

On migrating cells *in vitro* integrins are usually fairly dispersed, but in the integrin-mediated, tight adhesion of cells to extracellular matrix substrates like fibronectin or vitronectin, the integrins are clustered into focal contacts (focal adhesions) (*see* CELL JUNCTIONS) which serve also as the points where actin cytoskeleton stress fibres terminate. There is direct evidence now to suggest that, *in vitro* at least, integrins do connect to the F-actin cytoskeleton. Many different molecules are localized to focal contacts, but the cytoplasmic tail of the integrin β chain may link directly to α-ACTININ (also an actin-binding protein) and to talin (which binds vinculin, yet another actin-binding protein). At least one integrin, $\alpha_6\beta_4$, which is localized to the basal surfaces of stratified epithelia, is probably linked to hemidesmosomes and thence to INTERMEDIATE FILAMENTS, providing an alternative transmembrane connection for integrins.

Functions

In most cases integrins seem to act as very specific glues — specific in terms of the position (and the duration) of their functioning. This specificity may be due to transcriptional control of integrin synthesis, or via the activation of already synthesized integrins, as in T CELL interactions or in platelet activation (see below).

Developmental roles

Direct evidence of a role for integrins in development comes mainly from *Drosophila* where members of the β_1-like PS1 and PS2 integrin family play a part in the maintenance of muscle attachments during embryogenesis and in later development. The PS1 and PS2 integrins are always expressed in a complementary way, on the tendon cells and on muscle, respectively, which make up the attachments, suggesting that the two may form a heterotypic ligand-mediated adhesion (Fig. I39). In null mutants in the βPS gene, where no β chain is produced, the αPS chains, although produced normally, fail to be transported to the cell surface; as a consequence of all this, muscle attachments, which initially form, collapse immediately on use leading to paralysis and death of the embryo. There is also evidence from *in vitro* studies for a role for a β_1 integrin in muscle insertions in chick.

In *Drosophila*, and presumably equally in vertebrates, integrin function is not confined to the myotendinous junction. The PS integrins are also involved in the adhesion of apposed epithelia (again the two integrins show a complementary distribution, suggesting that they cooperate in these processes), in the migration of cell sheets and in eye development; in these last two cases, integrin–matrix, rather than cell–cell adhesions are indicated.

Developmentally regulated alternative splicing of the αPS2 transcript, which occurs in the region of one of the postulated ligand-binding domains, may relate to the variable functions of the PS2 integrin during development. Recent results suggest that integrins also function during the development of another invertebrate, the nematode worm, *Caenorhabditis*.

What seems clear from these studies is that integrin expression depends little on cell type or lineage but, rather, on positional specificity — the integrin is expressed only where and when it is required.

T cell interactions

There is evidence that intracellular changes can effect a transition of an integrin from a form which has low affinity for its extracellular ligand (ground state) to a high affinity (functionally activated) form. A well studied example is LFA-1 (the integrin $\alpha_L\beta_2$) in T lymphocytes (T cells). This integrin is usually inactive in T cells, but is rapidly and transiently activated following cross-linking of the T CELL RECEPTOR to antigen presented by B cells or monocytes. LFA-1 then recognizes its ligand or 'counter receptor' (ICAM, *see* CELL ADHESION MOLECULES) on the associated cell, thus helping to stabilize the cellular interaction. In this case, and perhaps in others, integrin activation appears to stem from the action of second messengers.

Physiological roles

Two human genetic disorders have pointed up roles for integrins in platelet function and neutrophil migration. A critical role for the integrin $\alpha_{IIb}\beta_3$, the receptor for fibrinogen, in platelet physiology is illustrated by the congenital bleeding disorder, Glanzmann's thrombasthenia, in which the patient's platelets are deficient in $\alpha_{IIb}\beta_3$, or express an abnormally functioning protein. Affected individuals show mucocutaneous bleeding due to abnormal platelet function despite a normal platelet count.

The disease may be caused by a variety of mutations in either the α_{IIb} or β_3 gene leading to suppressed integrin α or β chain synthesis and hence reduced surface expression on platelets. One form of Glanzmann's thrombasthenia is caused by a point mutation in the ligand-binding site of β_3 and leads apparently to a failure of the mutant integrin to undergo the conformational change, induced by ligand binding, necessary for it to become active as a receptor.

Another human genetic disease — leukocyte adhesion deficiency (LAD) — shows the involvement of β_2 integrins in the migration of neutrophils to sites of inflammation, a key stage in host defence against bacterial infection. Neutrophils (and other leukocytes) from patients with LAD lack β_2 integrins as a consequence of mutations in the β_2 gene which result in reduced or absent synthesis of β_2. The normally associated α chains (α_L, α_M and α_X) then fail to be transported to the cell surface, as seen with mutations in the *Drosophila* βPS gene. The clinical syndrome is characterized by recurrent bacterial sepsis due to abnormal leukocyte function, caused by lack of integrin-mediated adhesion to the β_2 ligand ICAM on endothelia (among other tissues), thus preventing leukocytes migrating to sites of infection.

Pathological functions

Integrin-mediated cell adhesion seems to be involved in some disease states, and the integrin–ligand interaction may thus provide a target for therapeutic intervention.

Blood vessel wall injury leads to platelet adhesion to damaged vascular endothelium, platelet activation and aggregation, initiation of the clotting enzyme cascade, and eventual thrombus formation with the attendant risk of thrombosis. Platelet aggregation and adhesion mediated by $\alpha_{IIb}\beta_3$ and other platelet integrins in part depends on the presence of RGD in the ligand. MONOCLONAL ANTIBODIES that block $\alpha_{IIb}\beta_3$ activity are currently on clinical trial in patients with coronary artery thrombosis. Short RGD-containing peptide analogues and naturally occurring snake venom proteins, which contain conformationally restrained RGD sequences, are also being tested as blocking agents with the aim of developing compounds that are highly active, stable *in vivo* and exhibit selectivity for $\alpha_{IIb}\beta_3$ over other RGD-dependent integrins. Some of these may eventually find a place in therapy.

Integrins are also involved in the disruption of normal cell behaviour that accompanies neoplastic transformation and META-STASIS in cancer. One of the main consequences of neoplasia (uncontrolled cell proliferation) is the unregulated spread of the proliferating progeny of the clone of TRANSFORMED CELLS. Both local spread (migration across endothelium into the blood-vascular system) and distant metastasis depend on disruption of the normal extracellular matrix of the tissue and modification of the integrin repertoire of the tumour cells. Matrix deposition is usually reduced (and probably qualitatively different) and matrix-destroying proteolytic enzymes are induced in tumours. In some tumours, for example CARCINOMA of the breast or lung, integrins of the β_1 family are usually decreased in amount; correlation with experimental assessment of metastatic potential has been found in some systems (e.g. increased levels of $\alpha_4\beta_1$ and $\alpha_v\beta_3$ resulting in increased tumorigenicity in MELANOMA).

Anti-integrin antibodies and RGD peptides have been found to affect tumour cell behaviour *in vivo*. Simultaneous injection of antibody or peptide with tumour cells can block local tumour growth and distant metastasis in experimental animals. The treatment of cancer and the modification of its spread by altering integrin behaviour is an exciting possibility for the future.

Other areas of current interest involve attempts to modify leukocyte migration in inflammatory disease by blocking β_2 (LFA) integrins with monoclonal antibodies, the development of RGD peptide analogues to block pathological bone resorption and the use of synthetic ligands for integrins to enhance wound healing by stimulating keratinocyte migration and proliferation.

Signal transduction

It seems likely that a major role of integrins as signal-transducing receptors is to mediate the translation of cell–extracellular matrix interactions into organizational signals for the cell, leading to changes in cytoskeletal attachments, gene expression or state of differentiation.

The differentiation of neurons and consequent process outgrowth which occurs on laminin, but not fibronectin, substrate requires the function of an integrin laminin receptor (although such data do not rigorously demonstrate that the integrin is directly involved in the signal transduction rather than simply providing necessary adhesion).

Signal transduction via integrins is also evident in the effect of RGD peptides on cell–matrix interaction *in vitro*. RGD peptides binding to integrins both inhibit cell adhesion and lead to the rapid detachment of the cytoskeleton from the integrin inside the cell and the eventual dissociation of the focal adhesion plaques.

Transmembrane signalling involving integrins also seems to be able to change the expression of genes which are not directly related to integrin function. Antibody-induced interference with ligand–integrin recognition (using an anti-integrin antibody which does not perturb cell or cytoskeletal morphology) causes the fairly rapid activation of collagenase and stromelysin genes in fibroblasts; these two enzymes are pericellular METALLOPROTEIN-ASES which may partake in the extracellular matrix remodelling which is thought to be a necessary accompaniment of many differentiative and morphogenetic changes, probably including detachment of integrins from the matrix.

Future prospects

Our understanding of the regulation of these multicomponent interactions is still at a very early stage. Perhaps the main problem at present is that most data on integrin function and possible roles come from *in vitro* analysis. It is often assumed that *in vitro* integrin functions reflect accurately the situation *in vivo*. We believe that the inherent complexity of the system and our relatively primitive knowledge of integrin function makes this unlikely; indeed much of the data may bear little relationship to true physiological roles.

The challenge for the future is to progress from artificial and simplistic *in vitro* systems to examine cell receptor–matrix interactions *in vivo*.

M. WILCOX†
M.A. HORTON

† This article is dedicated to the memory of Michael Wilcox who died on February 21, 1992.

1 Ruoslahti, E. & Pierschbacher, M. (1987) New perspectives in cell adhesion: RGD and integrins. *Science* **238**, 491–497.
2 Hynes, R.O. (1987) Integrins: a family of cell surface receptors. *Cell* **48**, 549–554.
3 Hemler, M.E. (1990) VLA proteins in the integrin family: structures, functions and their role in leukocytes. *Annu. Rev. Immunol.* **8**, 365–400.
4 Juliano, R.L. & Haskill, S. (1993) Signal transduction from the extracellular matrix. *J. Cell Biol.* **120**, 577–585.

intercalary regeneration The 'filling-in' of missing parts of the pattern by REGENERATION after normally separate parts of the body are grafted together (e.g. in amphibian and insect limbs). The form taken by the regenerated tissue is compatible with the notion that cells at each level along the proximo-distal or circumferential axis have a distinct character ('positional value') and that these values are arranged in an incremental series along each axis. Any discontinuity in positional value induced by experimental manipulation provokes local growth and these proliferating cells adopt positional values which eliminate the discontinuity in the sequence and differentiate accordingly. *See also*: PATTERN FORMATION; POLAR COORDINATE MODEL.

intercalation The insertion of a molecule between the stacked bases of the DNA double helix (*see* CHEMICAL CARCINOGENS AND CARCINOGENESIS; DNA; NUCLEIC ACID STRUCTURE). This can result in partial unwinding of the DNA helix leading to errors during DNA REPLICATION and the generation of + 1 and − 1 FRAMESHIFT mutations. The best studied intercalating agents are the ACRIDINE DYES (e.g. proflavine), ETHIDIUM BROMIDE, and propidium di-iodide. Some intercalating agents (e.g. benzpyrene), chemically modify the DNA prior to intercalation.

interferons (IFN) Antiviral and antiproliferative CYTOKINES. IFNα, produced in virus-infected B lymphocytes is a non-glycosylated polypeptide of 166 amino acids produced from ~20 different genes on human chromosome 9. IFNβ is produced by virus-induced fibroblast/endothelial-type cells. It is encoded by a single gene on human chromosome 9, has 166 amino acids and is normally a glycosylated dimer. IFNβ$_2$ is a quite distinct protein encoded by a gene on chromosome 7 and is now known as interleukin-6 (*see* IL-6). IFNα and β are type I interferons and share a common receptor of M_r 100 000–130 000. IFNγ is a type II interferon, produced by T cells from a single gene on human chromosome 12. Normally glycosylated, it is a tetramer of a 146-amino acid subunit, and like α and β demonstrates antiviral and antiproliferative activities. IFNγ has its own receptor of predicted M_r 54 000, which is distinct from the IFNα/β receptor. INFγ has antiviral effects, inhibitory effects on cell proliferation, and immunoregulatory effects and is often known as 'immune interferon'.

The antiviral effects of interferon are the result of a 'priming' effect of interferon treatment on cells. Interferon synthesis is itself induced by virus infection (double-stranded RNA molecules are particularly effective) or by treatment with double-stranded poly-ribonucleotides (e.g. poly(I):poly(C)). By its action at specific cell-surface receptors interferon then induces the expression of high levels of two enzymes, a 2′,5′-oligoadenylate (2,5-A) synthetase and a protein kinase. Subsequent infection of these 'primed' cells by virus then triggers separate biochemical pathways involving the two enzymes, which lead to viral (and cellular) RNA degradation and an inhibition of PROTEIN SYNTHESIS. Interferon treatment increases the level of 2,5-A synthetase in the cell several thousand-fold by inducing specific gene expression. Viral dsRNA activates the pre-existing synthetase which catalyses the production of 2,5-A$_n$ (n = ~2–15 residues) from ATP. 2,5-A dependent RNase L or RNase F then cleaves viral and cellular RNAs (e.g. mRNAs and rRNAs). The interferon-induced protein kinase also requires the presence of dsRNA for activity. The activated kinase phosphorylates the α subunit of the eukaryotic protein synthesis INITIATION FACTOR eIF-2, preventing the recycling of eIF-2 bound to the initiation complex and thus inhibiting protein synthesis. Together, these two actions of interferon block protein synthesis and cell growth, and thus virus multiplication. *See*: LYMPHOKINES.

Balkwill, F.R. (1989) *Lancet* **I** (No. 8646), 1060–1063.

interleukins Group of structurally unrelated proteins and GLYCO-PROTEINS which act as intercellular signals mediating reactions between immunoreactive cells. The term has become more loosely applied recently as many interleukins are now known to have effects on diverse cell types such as lymphocytes, endothelial cells, epithelial cells, fibroblasts, and haematopoietic cells. The recent development of cell lines which both produce and react to interleukins has facilitated the identification of many different molecules and the application of gene cloning technology in this area has been particularly fruitful. Many interleukins are available in highly purified recombinant forms which has in turn allowed the definition of their cellular targets and mode of action *in vitro*. *See*: IL-1; IL-2; IL-3; IL-4; IL-5; IL-6; IL-7; IL-8; IL-9; IL-10; IL-11; LYMPHOKINES.

Intermediate filaments

INTERMEDIATE filaments or 10 nm filaments were so named because they are intermediate in diameter between that of MICROTUBULES (25 nm) and MICROFILAMENTS (5–7 nm). On their initial description in the 1960s and 1970s as 10 nm filaments, they were thought to be derived from already identified cytoskeletal elements such as microtubules, microfilaments and myofilaments [1]. The subsequent identification and characterization of this MULTIGENE FAMILY has revealed a widespread and diverse group of at least 40 proteins with a relative molecular mass (M_r) range of 40 000–125 000 (40–125K).

Intermediate filament (IF) proteins form a completely separate cytoskeletal network from those of microtubules and microfilaments which serves a number of unique functions. IFs are prominent components in the cytoplasm of cells, providing physical links between specific domains on the plasma membrane as well as between the plasma and nuclear membranes. Also, the recent discovery that the nuclear lamina comprises proteins — the lamins — which are also members of the IF family, raises the interesting possibility of an integrated network between the nuclear and cytoplasmic compartments of the cell (*see also* NUCLEAR MATRIX; NUCLEAR ENVELOPE).

This large group of proteins has been classified into six different types for vertebrates on the basis of sequence comparisons, each protein type also exhibiting a characteristic cell-specific expression profile (Table I5). The recent discovery of a completely new type of vertebrate IF protein — nestin — despite a previous decade of intensive research, probably indicates that the present classification is not yet definitive. The discovery of two eye-lens-specific IF proteins, filensin and phakosin, which do not conform to the current IF types, gives credence to this view.

Diversity of IF proteins

Each cell lineage is characterized by a particular evolution and final selection of IF protein expression. This clear relationship between IF proteins and their tissue-specific expression has been widely exploited as a diagnostic tool for the identification of cell origins in pathologies [2]. An extensive range of immunological

Table I5 The intermediate filament protein family

Type	IF proteins	Tissue specificity
Vertebrate IFs		
Type I	Keratins — acidic, neutral	Epithelia
Type II	Keratins — basic	Epithelia
Type III	Vimentin	Mesenchyme and most cultured cells
	Desmin	Myogenic cells
	GFAP	Astrocytes, some glia
	Peripherin	Neurons, PNS and CNS
Type IV	Neurofilament proteins	
	NF-H, NF-M, NF-L	Neurons
	α-Internexin	Neurons
Type V	Nuclear lamins	Nuclear envelopes
	A-type, lamins A and B	
	B-type, lamins B_1 and B_2	
Type VI	Nestin	Neuroepithelial stem cells
Invertebrate IFs		
nnIF	*Ascaris* proteins A and B	Giant body muscle cells
nnIF	*Helix* proteins A and B	Oesophagus
nnIF	*Aplysia* protein A	

reagents specific to the different IF protein types as well as to individual proteins, is commercially available and these are widely used and accepted as invaluable tools for research and diagnostic purposes.

Epithelia express type I and II IF proteins which are collectively called keratins and comprise some 30 different proteins. The pattern of keratin expression follows the histological classification of the different epithelia. For example, keratins 1 and 10 are found exclusively in the epidermis, a stratified epithelium, whereas keratins 8 and 18 are found typically in simple epithelia such as that composed of hepatocytes in the liver. Even within the same epithelium, the keratin profile can change during cell differentiation as in the epidermis where the basal cells express keratins 5 and 14 but the supra-basal layers, derived from the basal layer, express keratins 1 and 10.

Differentiation-related changes in IF protein expression are seen also in other cells and often involve the co-expression of two different IF types. During neuronal differentiation, for example, combinations of type III and IV IF proteins follow cell-specific patterns of expression. The neurofilament type IV proteins are expressed in all neurons but peripherin, a type III protein, is expressed in a specific subset of neurons. These are those neurons derived from the neural crest and those derived from the neural tube that project axons to the periphery as well as a subset of neurons contained within the central nervous system. This illustrates the ability to define subsets of differentiated cells on the basis of IF protein expression.

The switching of IF protein type during development and cellular differentiation is also frequently observed and appears to coincide with major developmental events. For instance, keratin filaments seem to be the only cytoplasmic IFs expressed in the early embryo and in the preimplantation embryo [3] but the primary mesenchymal cells, which are derived from the embry-

onic ectoderm at late day 8 in mouse embryogenesis, express vimentin, a type III protein. Similarly, during neurulation the type VI protein, nestin, is expressed in neuroectodermal cells but upon initiation of terminal differentiation, the appropiate type III and IV IF proteins are expressed [4]. These observations suggest functional distinctions for the different IF protein types.

IF function

Since the beginning of the 1980s, IFs have been allotted a largely structural role [5]. The challenge has been to design experiments which reveal the specific function of the different IF types in the tissue and in the individual cell. So far, most success has come from elucidating the genetic basis for several human skin diseases which have highlighted the necessity of an intact IF network to tissue function.

In vertebrates, keratins are the most stable IFs and they are expressed in those sites in the body exposed to the greatest physical and environmental stress. As a mark of the extra stabilization required, keratin IFs are fasciated and physically link desmosomes, structures that are important to the maintenance of epithelial tissue integrity (*see* CELL JUNCTIONS). Very recently, the importance of these proteins to tissue function has been graphically illustrated by studies on the skin conditions epidermolysis bullosa simplex and epidermolytic hyperkeratosis [8]. In the epidermis of people with these conditions, cellular IF organization is severely disrupted by a single point mutation in the highly conserved sequence motifs at either end of the central α-helical domain. The physical properties of the epidermis can be severely compromised by such mutations, in some cases fatally so.

In axons, neurofilaments help determine the physical dimensions of the axon as well as positively contributing to axonal function. Alteration either to the number, or to the extent of neurofilament protein phosphorylation or to the neurofilament subunit composition results in significant axonal changes in animal and bird experimental models [7]. Indeed, the overexpression of either NF-L [8] or NF-H [9] in transgenic mice results in progressive neuropathies which very closely resemble human motor neuron diseases. In the case of elevated NF-H it provides an animal model to study AMYOTROPHIC LATERAL SCLEROSIS [9].

Glial fibrillary acid protein (GFAP), the astrocyte-specific IF protein, appears to be directly correlated to the extension of cell processes in response to co-culture with neurons [7]. In these examples of IF function, the IFs are essential to cells in the context of tissues. This does not preclude important roles for IFs in intracellular processes because, for example, the nuclear lamins appear to contribute functionally to nuclear division, gene expression, and DNA replication [10], which are central cell biological events.

The nuclear lamins, which underlie the inner nuclear membrane, are important to the structural integrity and organization of the nucleus. As a consequence of depleting lamins from *in vitro* nuclear reconstitution assays, DNA replication is no longer accomplished and the nuclei become extremely fragile. The role of the lamina in differentiated somatic cells seems to be different in that immunodepletion of the lamins prevents nuclear assembly.

These differences may be a function of the lamin isotypes expressed in the two different types of nuclei. Both A- and B-type lamins are found in the nuclei of differentiated cells whereas the B-type lamins are the only form found in the nuclei of undifferentiated cell types. A-type lamins associate more strongly with chromatin than B-type lamins, indicative of a functional role in the nuclear organization of chromatin in differentiated cells [10].

IF structure and protein sequences

Keratins were the first group of IF proteins to be studied and were recognized as fibrous and α-helical in nature from X-ray studies in the 1950s. The α-helical nature of the proteins means that coiled coil interactions are important to subunit formation but in the case of the higher order interactions ionic and hydrophobic interactions are also equally important. A hierarchy of filament substructure has been described, each filament comprising proto-filaments which associate to form protofibrils. There are believed to be three or four protofibrils per filament diameter, so that although filaments can be many micrometres long, the filament substructures need not necessarily be continuous. A lumen is seen in some IFs. The precise details of the packing geometries for the filament subspecies and protein subunits have not yet been elucidated although many models have been proposed [5,11]. Protein domains important to filament integrity and interfilament interactions have been identified and are discussed below.

Keratin filaments are obligate heteropolymers of type I and II proteins which are present in an equimolar ratio in the filament. Other IF proteins do not share this characteristic, forming copolymers only when two or more are present in the same cell.

Keratins comprise a large proportion of hair, nail and wool as well as other 'hard' keratin structures such as horn and quill. These proteins are, however, quite different from feather keratins which are predominantly β-sheet, and are not considered as IF proteins.

In the nucleus, the lamin filaments can form a highly ordered filament network displaying a square lattice array in *Xenopus* oocyte nuclei. In somatic cells, this order may not be so well preserved such that spatial differences in the lamina network may exist [10].

IF proteins are identified on the basis of protein sequence homologies (Fig. I40) and the protein domain that exhibits the greatest degree of homology across the whole IF protein family is in the central α-helical domain [5]. Within this domain there are two sequences with a very high degree of conservation, one at the beginning of helix 1A and the other at the end of helix 2B. These sequences seem to delineate the central α-helical domain. The type V proteins differ from all the other IF protein types in that they possess an extra 42 residues at the end of helix 1B. This feature has also been found in many of the invertebrate IF proteins sequenced to date. The discovery of conserved intron–exon boundaries in these invertebrate IF genes has led to the suggestion that these lamin-like proteins indicate the ancestral origins of the vertebrate IF proteins [10].

The N and C termini, which flank the central α-helical domain, differ substantially between the various IF protein types and it is in these regions that the diversity in molecular weight as well as protein function exists. These domains have varying effects upon filament assembly and filament architectures [6,11].

IF assembly

IF assembly generally requires a neutral pH buffer, 100 mM salt, 2 mM Mg^{2+}, and reducing agent. Nucleotides are not required. A critical concentration in the order of $40\,\mu g\,ml^{-1}$ has been calculated for the *in vitro* assembly of several IF proteins. Filament assembly can be monitored by fluorescence spectroscopy using appropriately labelled protein, sedimentation assay, change in optical density and viscometry. IF proteins are soluble in low ionic strength buffers (<10 mM), at alkaline pH (>8) but in order to solubilize an existing filament network urea, guanidinium chloride or sodium dodecyl sulphate (SDS) must be used. IFs are stable to high salt and nonionic detergent extraction and this is a frequently used protocol to selectively purify the IF-rich fraction from cells and tissues.

The initial stages in the assembly process have been characterized by *in vitro* assembly studies [5,11]. Two protein chains associate in a parallel, in-register fashion to produce a protein molecule in a coiled-coil conformation. Association of two molecules to form a four-chain unit or tetramer is the next stage of assembly and it is believed that at this point the molecules are antiparallel and staggered with respect to one another. The protein molecule is ~45–50 nm long, and this is attributable to the central α-helical domain (Fig. I40). The coiled-coil is stabilized by hydrophobic residues, the α-helical domain being arranged in a heptad repeating sequence. The N- and C-terminal domains are

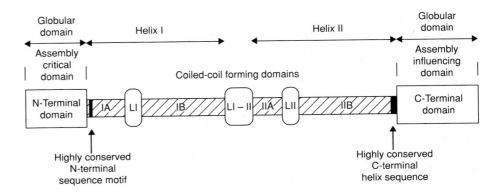

Fig. I40 Intermediate filament protein domains. See text for explanation.

not α-helical but are required for filament assembly, although their relative importance differs. The N terminus seems to be absolutely required whereas the C terminus can modulate assembly and also contribute to filament–filament interactions. Assembly *in vitro* and *in vivo* can be modulated by phosphorylation at sites in the N-terminal domain [11]. In the case of the nuclear lamina, phosphorylation by kinases including the cdc2 kinase (*see* CELL CYCLE; PROTEIN KINASES) and then subsequent dephosphorylation (PHOSPHATASES) control the disassembly/assembly process that occurs during the cell cycle and this is thought to involve an additional phosphorylation site adjacent to the conserved C-terminal sequence in helix 2B [10].

The neurofilament proteins NF-M and NF-H both possess long C-terminal domains that are extensively phosphorylated, containing multiple repeats of the sequence KSP (single-letter amino acid notation). It is believed that the phosphorylation of these domains and the resulting unusually high charge density cause the extensions to electrostatically repel each other and create a zone of exclusion around the neurofilament [9]. Perturbation of this packing can result in motor neuron diseases.

The two highly conserved α-helical sequences in the central domain (Fig. I40) are also important to subsequent interfilament interactions as point mutations in these regions cause filaments to aggregate laterally [6]. Deletion of these sequences produces DOMINANT mutations that induce filament disassembly, even of existing networks. Networks can also be disassembled or radically rearranged using MONOCLONAL ANTIBODIES, or peptides derived from assembly-important domains, or by induced hyperphosphorylation.

IFs are dynamic structures and a soluble pool for vimentin IFs has been described. The four-chain unit, or tetramer, seems to be the stable form found in equilibrium with the polymer pool. More recently, BIOTINYLATED proteins have been microinjected into cells to monitor the dynamic nature of the IF network [11]. These labelled proteins appear to integrate along the filament length and previous *in vitro* assembly data had suggested that lateral association/dissociation of IF subunits as well as end-to-end mechanisms could operate. Analysis of FRAP (fluorescence recovery after photobleaching) data supports an apolar arrangement of subunits in IFs and wall exchange of subunits.

In the case of the lamins, processing of the newly translated protein at the C terminus is required. This domain contains a NUCLEAR LOCATION SEQUENCE (*see* PROTEIN TARGETING) for initial transport into the nucleus, but the protein must also be isoprenylated and carboxymethylated before incorporation into the nuclear lamina. There is a conserved recognition sequence at the very C terminus of vertebrate lamins (except for lamin C, a splice variant of lamin A), called the CaaM motif that directs this post-translational modification [10]. In the case of lamin A, this signal is removed during normal processing. The CaaM signal is a general mechanism for the lipid modification of proteins.

GLYCOSYLATION of lamin A and of keratins 8, 13 and 18 have been reported but the function of this post-translational modification is unknown.

IF–membrane interactions and IF-associated proteins

Interactions with either the plasma and/or nuclear membranes

are essential to IF function whether considering either the cytoplasmic or the nuclear IFs.

In epithelia, the plasma membrane attachment site for keratin IFs is a clearly defined structure, the desmosome or hemidesmosome (*see* CELL JUNCTIONS). Here, it is the desmocollins rather than the desmogleins of the desmosome that direct the attachment of the keratin filaments to the plasma membrane [12]. Both desmogleins and desmocollins are glycosylated integral plasma membrane proteins. In other cases the attachment site is not defined structurally although the identity of some of the plasma or nuclear membrane associated proteins is known. Two such candidates are ankyrin and spectrin which interact with vimentin and desmin via the N-terminal non-α-helical domain. These proteins have a widespread distribution and could therefore represent a frequent mechanism for IF binding to plasma membranes. At the nuclear membrane, it has been proposed that lamin B could act as a type III IF-protein binding site although the *in vivo* significance of this observation has yet to be understood [10].

A number of integral nuclear proteins on the inner nuclear membrane bind to the lamins termed LAPs (lamina associated polypeptides) which are thought to be receptor proteins for nuclear lamins [13]. The relative importance of the various mechanisms to attach the lamins to the nuclear membrane has yet to be decided but there is clear evidence that the lamins themselves are important in directing the nuclear membrane location of some of these other proteins [10].

Plectin is another well characterized IFAP (IF-associated protein) and has been identified in a wide range of tissues [14]. The functions of this 300K protein are varied, from forming its own network and cross-linking IFs, to linking IFs to other cytoskeletal networks. Other IFAPs that are found in specialized cell types include filaggrin (37K), loricin, astrocytic 48K protein, synemin (230K) and paranemin (280K) to name but a few [14].

IF gene structure and gene expression

The isolation, characterization and subsequent cloning of IF proteins from invertebrates and their comparison with vertebrate strongly suggests a pathway of common ancestry. Most of the intron–exon boundaries are conserved, even in the C-terminal non-α-helical domain. Those exons that specify the nuclear location signal and the CaaM signals are lost from the invertebrate gene sequence and suggest that these proteins arose from a lamin-like ancestor [10].

The type III vertebrate IF genes show more homology in terms of intron–exon positions to the invertebrate and lamin IF genes. Differences in the intron–exon positions for the type I, II, IV and VI protein genes suggest a model involving both the gain and loss of introns. The type IV and VI protein gene structures are the most diverse, the NF-H gene having only one intron which is conserved with the other IF genes. Their place in the IF gene ancestry must await the elucidation of invertebrate neurofilament gene structure. The present model also does not accommodate the completely different gene structure for the *Drosophila* lamin.

Mammalian IF genes seem to be dispersed throughout the genome in conserved chromosomal locations. For instance the type I genes are found largely on chromosome 17, and type II genes on chromosome 12 in humans. Exceptions exist, for instance keratin 18, which is located to chromosome 12. This represents the only naturally occurring pair of keratins to be found on the same chromosome. Vimentin, desmin, GFAP, NF-L and NF-H are found on chromosomes 10, 2, 17, 8 and 22 respectively in humans [15].

The regulation of IF gene expression is thought to be largely at the transcriptional level. Typical regulatory sequences such as *cis*-acting DNA sequences upstream from the transcription start site, as well as the usual TATA box, have been identified (*see* EUKARYOTIC GENE EXPRESSION; TRANSCRIPTION). Enhancer elements are not restricted to upstream regions. For example, in the mouse gene for keratin 8, one element is located 1 kb downstream of the coding sequences and in the keratin 18 gene an enhancer has been found in the first intron. No single element is responsible for controlling the expression of the various IF genes in their correct tissue or cell type as demonstrated by studies on vimentin, desmin, GFAP, NF-L, and several keratins. Positive and negative regulatory elements have been identified in GFAP, desmin, and vimentin genes. In the case of vimentin, specific protein binding is associated with silencer and anti-silencer elements. This complicated mechanism for the control of regulation for vimentin is perhaps necessary to accommodate the spectrum of cell types in which vimentin can be expressed, but it does not necessarily follow that the control of expression of the other IF genes will be any simpler [16].

R. QUINLAN

1 Ishikawa, H. et al. (1968) Mitosis and intermediate-sized filaments in developing skeletal muscle. *J. Cell Biol.* **38**, 538–555.
2 Osborn, M. et al. (1985) Differentiation of the major human tumour groups using conventional and monoclonal antibodies specific for individual intermediate filament proteins. *Ann. N. Y. Acad. Sci.* **455**, 649–668.
3 Jackson, B.W. et al. (1980) Formation of cytoskeletal elements during mouse embryogenesis. Intermediate filaments of the cytokeratin type and desmosomes in the preimplantation embryos. *Differentiation* **17**, 161–179.
4 Lendahl. U. et al. (1990) CNS stem cells express a new class of intermediate filament protein. *Cell* **60**, 585–595.
5 Steinert, P.M. & Roop, D.R. (1988) Molecular and cellular biology of intermediate filaments. *Annu. Rev. Biochem.* **57**, 593–625.
6 Coulombe, P.A. (1993) Cellular and molecular biology of keratins. *Curr. Opinion Cell Biol.* **5**, 17–29.
7 Liem, R.K.H. (1993) Molecular biology of neuronal intermediate filaments. *Curr. Opinion Cell Biol.* **5**, 12–16.
8 Xu, Z. et al. (1993) Increased expression of neurofilament subunit NF-L produces morphological alterations that resemble the pathology of human motor neuron disease. *Cell* **73**, 23–33.
9 Cote, F. et al. (1993) Progressive neuropathy in transgenic mice expressing the human neurofilament heavy gene: a mouse model of amyotrophic lateral sclerosis. *Cell* **73** 35–46.
10 Dessey, G.N. (1992) Nuclear envelope structure. *Curr. Opinion Cell Biol.* **4**, 430–435.
11 Stewart, M. (1993) Intermediate filament structure and assembly. *Curr. Opinion Cell Biol.* **5**, 3–11.
12 Troyanovsky, S.M. et al. (1993) Contributions of cytoplasmic domains of desmosomal cadherins to desmosome assembly and intermediate filament anchorage. *Cell* **72**, 561–574.
13 Foisner, R. & Gerace, L. (1993) Integral membrane proteins of the nuclear envelope interact with lamins and chromosmes, and binding is modulated by mitotic phosphorylation. *Cell* **73**, 1267–1279.
14 Foisner, R. & Wiche, G. (1991) Intermediate filament-associated proteins. *Curr. Opinion Cell Biol.* **3**, 75–81.
15 Pendleton, J.W.S. et al. (1991) The peripherin gene maps to mouse chromosome 15. *Genomics* **9**, 369–372.
16 Zehner, Z. (1991) Regulation of intermediate filament gene expression. *Curr. Opinion Cell Biol.* **3**, 67–74.

intermediate junction *See*: CELL JUNCTIONS.

intermediate mesoderm The portion of the mesoderm of the trunk of vertebrate embryos lying between the PARAXIAL MESODERM and the LATERAL PLATE MESODERM. The intermediate mesoderm contains cells that give rise to the two types of embryonic kidney, pronephros and mesonephros and their ducts. In higher vertebrates, the pronephros and mesonephros are both transitory, but their ducts (Wolffian and Müllerian) contribute to the adult urogenital system. There is no intermediate mesoderm in the head of higher vertebrates.

intermembrane space *See*: MITOCHONDRIA.

internal guide sequence (IGS) *See*: RIBOZYMES.

interneuron Small NEURON in the central nervous system that connects two neurons.

internode *See*: NODE OF RANVIER.

interphase Phase in the eukaryotic CELL CYCLE in which DNA synthesis occurs.

interspersed repeated sequences *See*: GENOME ORGANIZATION; LINE; SINE.

intervening sequence (IVS) An alternative name for INTRON, that is, a DNA sequence within a gene which is transcribed during the synthesis of the primary RNA transcript and then removed from the RNA by RNA SPLICING to produce a mature mRNA, rRNA, or tRNA. *See also*: RIBOZYME.

intracellular recording The measurement of the electrical activity of a cell by the placement of the recording electrode within the cell, by piercing the membrane. The main drawback of this technique is that it considerably disrupts the cell.

intracellular transport *See*: ANTIGEN PROCESSING AND PRESENTATION; ENDOCYTOSIS; EXOCYTOSIS; MICROTUBULES; MICROTUBULE-BASED MOTORS; NUCLEAR CYTOSKELETON; PROTEIN TARGETING; PROTEIN TRANSLOCATION; SECRETION; VESICLE-MEDIATED TRANSPORT.

intrachromosomal recombination *See*: SISTER CHROMATID EXCHANGE.

intrachromosomal translocation The transposition of a chromosomal segment from one location to another on the same chromosome. In order for this to happen three breaks must occur within that chromosome.

intragenic complementation The occasional case where two allelic mutations (i.e. occurring at the same locus) complement each other in hybrids (*see* COMPLEMENTATION). This can occur if the locus normally encodes a subunit of a homomeric protein and if each mutation causes a defect in the protein which can be compensated for by the presence of the other subunit in the multisubunit protein.

intragenic recombination A RECOMBINATION event in which the point of exchange between recombinant chromosomes lies within a gene. This may also give rise to a second local recombinational exchange in which case the first episode is said to confer negative interference.

intragenic suppression Phenotypic correction (suppression) of a FRAMESHIFT mutation when a compensating mutation within the same gene restores the original reading frame.

intramembrane particles Particles visible on the internal surface of a membrane fractured along the middle of the lipid bilayer in FREEZE-FRACTURE ELECTRON MICROSCOPY and which represent large integral membrane proteins (*see* MEMBRANE STRUCTURE).

intron A region within a gene which is transcribed but which is subsequently excised from the primary RNA transcript by RNA SPLICING to generate the mature RNA. Introns (or intervening sequences) were first discovered in 1977 in the genome of ADENOVIRUS (a eukaryotic DNA virus) and in the β-globin gene of mouse and rat. Largely (but not exclusively) confined to eukaryotic genes, introns are present in genes encoding proteins, tRNAs, and rRNAs. Few prokaryotes have intron-containing genes, while the simple eukaryote SACCHAROMYCES CEREVISIAE has relatively few intron-containing genes compared with higher eukaryotes. The number and organization of introns varies from gene to gene with some genes having in excess of 50 introns separating the coding (EXON) sequences (e.g. the chicken pro-α_2 collagen gene). The sizes of introns also show remarkable variation, from 31 nucleotides in a gene of the SV40 virus to an intron of more than 210 000 nucleotides in the human dystrophin gene.

Introns in genes for nuclear-encoded proteins have conserved 5' and 3' SPLICE SITES (5'GU ------- AG3') and are removed from the primary transcript by SPLICEOSOMES (*see* RNA SPLICING). A number of introns are 'self-splicing', that is they are capable of catalysing their own removal in the absence of any protein (*see* RIBOZYMES). Self-splicing introns fall into two groups: group I introns (e.g. the intron in the *Tetrahymena* rRNA gene) require only a guanosine cofactor for autocatalytic splicing; group II self-splicing introns do not require a cofactor but do require the hydroxyl group of an adenine at the 3' end of the intron at the 5' splice site.

The evolutionary history and biological role of introns has been the subject of much debate, and their true significance remains to be defined. In many genes introns appear to demarcate functional domains of the encoded protein indicating that introns may play a role in the evolution of proteins (*see* EXON SHUFFLING: MOLECULAR EVOLUTION: SEQUENCES AND STRUCTURES). *See also*: EUKARYOTIC GENE STRUCTURE; MITOCHONDRIAL GENOMES: LOWER EUKARYOTE.

Berget, S.M. et al. (1977) *Proc. Natl. Acad. Sci. USA* **74**, 3171–3174.
Chow, L.T. et al. (1977) *Cell* **12**, 1–8.
Jeffreys, A.J. & Flavell, R.A. (1977) *Cell 12*, 1097–1108.

intron transposition *See*: MITOCHONDRIAL GENOMES: LOWER EUKARYOTES.

inulin Fructose homopolymer from various plant species (e.g. Jerusalem artichoke). Composed of fructose residues in furanose form, linked linearly by β2,1 glycosidic linkage. *See*: FRUCTANS.

invagination The penetration of a layer of cells into the interior of the embryo. The term implies that part of the surface of a more or less spherical embryo is pushed into the interior of the sphere, as if poking a finger into a rubber ball. It should be distinguished from INGRESSION and INVOLUTION. Invagination characterizes the process of GASTRULATION in lower animals, such as echinoderms.

invariants Invariants are increasingly being developed to infer evolutionary trees from sequences. An invariant is a general concept for any function which will take the same value (i.e. be invariant) for a specific scientific model, irrespective of the particular values of some parameters in the model. An example from MOLECULAR PHYLOGENY uses an evolutionary tree model. In the present context the model (*see* MAXIMUM LIKELIHOOD) is the tree, the specified mechanism of evolution, and the varying parameters are the unknown edge weights (lengths) of the tree. Invariants are then calculated from sequences to indicate which edges are in the correct evolutionary tree, irrespective of the rate of evolution on the tree. Linear invariants are a subclass that are linear combinations of parameters in the data. Thus it is valid to apply the usual statistical tests of significance.

Evolutionary parsimony is an invariant method for reconstructing evolutionary trees from nucleotide sequences. It is applied to groups of four taxa at a time and gives three values that correspond to the support for each of the three trees that link four taxa. It is expected that two of these values, those corresponding to the incorrect trees, should be zero (invariant). The third value is expected to be positive and significantly greater than zero. Evolutionary parsimony is susceptible to error when nucleotide frequencies vary between sequences and an extension, 'compositional statistics', is then available though still only for four sequences at a time. A Hadamard transform, a discrete FOURIER TRANSFORM, can be used for up to 20 taxa but does not yet work well for four-character states such as nucleic acid sequences. The 'closest tree' criterion is often used with the Hadamard transform to identify the tree which is closest to the transformed sequences in a multidimensional space.

Penny, D. et al. (1991) *Trends Ecol. Evol.* **7**(3), 73–79.

invasin *See*: BACTERIAL PATHOGENICITY.

inversion A CHROMOSOME ABERRATION where a segment of chromosome has been rotated through 180° relative to its adjacent regions. If an inversion includes the centromere it is termed pericentric; if it excludes the centromere it is termed paracentric.

invertase β-D-Fructofuranosidase (EC 3.2.1.26). Catalyses the hydrolysis of terminal nonreducing β-D-fructofuranosides, most notably sucrose. Found in yeast, fungi and higher plants. Yeast and *Neurospora* invertases are glycoproteins. The invertase of higher plants is often found predominantly in cell walls, but also in vacuoles. Invertase may also show limited fructotransferase activity.

inverted repeat Two identical copies of a sequence of bases of DNA on the same DNA molecule that are repeated in reverse orientation but on opposite strands (Fig. I41). If the two repeated sequences are immediately adjacent they are said to form a PALINDROME. If the repeated sequences are separated by three or more bases they have the potential to form CRUCIFORM structures via complementary base pairing.

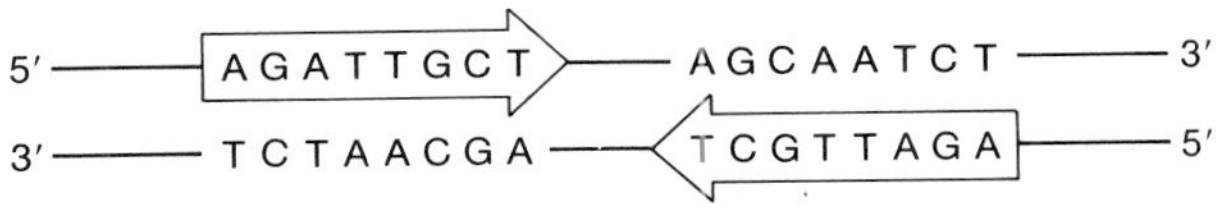

Fig. I41 An inverted repeat sequence in a double-stranded DNA molecule.

involution The penetration of a layer of cells into deeper regions of the embryo by rolling against the inner surface of another cell layer (EPIBOLY; *cf.* INVAGINATION). It characterizes the process of GASTRULATION in amphibian and other embryos. Should not be confused with INGRESSION or INVAGINATION.

ion channel A protein-lined pore in a membrane which provides an aqueous environment through which ions can traverse the hydrophobic lipid bilayer. Ion channels are specific for anions or cations and many are specific for one type or a small set of ions. Some ion channels are permanently open (e.g. the K^+ 'leak' channel involved in generating the resting MEMBRANE POTENTIAL), allowing passive transport of ions down their electrochemical gradients. Most, however, are 'gated' and open only in response to a stimulus. The two main types of gated ion channels are:
1 The ligand-gated channels, which open in response to binding of a ligand such as a neurotransmitter (*see* CALCIUM; EXCITATORY AMINO ACID RECEPTORS; GABA AND GLYCINE RECEPTORS; NICOTINIC RECEPTOR).
2 The VOLTAGE-GATED ION CHANNELS, whose opening is controlled by changes in the membrane potential.

ion-exchange chromatography *See:* CHROMATOGRAPHY.

ion pumps Transport proteins that actively transport ions across the membrane against a concentration gradient. *See e.g.:* CA^{2+}-ATPASE; H^+,K^+-ATPASES; MEMBRANE TRANSPORT SYSTEMS; NA^+,K^+-ATPASE; P-TYPE ATPASES; V-TYPE ATPASES; VACUOLAR H^+-ATPASE.

ionophore Small hydrophobic molecules that dissolve in membranes and increase their permeability to ions. There are two main types: (1) those that complex with the ion to form a mobile ion carrier (e.g. valinomycin); and (2) those that form aqueous channels in the membrane (e.g. GRAMICIDIN A).

ionotropic receptor A LIGAND-GATED ION CHANNEL (as opposed to a second messenger-coupled receptor for the same NEUROTRANSMITTER). On activation these receptors open integral ion channels allowing the passage of ions that can, depending on the receptor type, depolarize or hyperpolarize the cell membrane. Examples include:
1 GABA AND GLYCINE RECEPTORS which possess chloride channels. Activation of these receptors hyperpolarizes the cell membrane and they are therefore inhibitory.
2 NICOTINIC RECEPTORS and non-NMDA EXCITATORY AMINO ACID RECEPTORS which gate the cations Na^+ and K^+. These receptors depolarize the cell membrane and they are therefore excitatory.
3 The NMDA EXCITATORY AMINO ACID RECEPTOR which gates Ca^{2+} as well as Na^+ and K^+.
Cf. METABOTROPIC.

iontophoresis The transference of ions by an electromotive force which is commonly used as a means to deliver chemicals into cells.

IP₃ Inositol 1,4,5-trisphosphate. *See:* INOSITOL PHOSPHATES; SECOND MESSENGER PATHWAYS.

IPP INHIBITORY POSTSYNAPTIC POTENTIAL.

***ipt* gene (also known as *cyt, tmr, roi*)** Gene for the synthesis of the plant cytokinin hormone isopentenyladenine. The gene is present in the T-region of the TI PLASMID of many *Agrobacterium* strains and is transferred to plants during the infection process. Accumulation of the cytokinin in transformed cells helps maintain the tumorous phenotype. When the *ipt* gene is mutated, resulting tumours have high auxin and low cytokinin, favouring the production of roots: hence the rooty phenotype of mutants. *See also:* PLANT HORMONES.

IPTG Isopropyl-β-D-thiogalactopyranoside, a non-metabolizable INDUCER of the *Escherichia coli* LAC OPERON (Fig. I42). It achieves this by binding directly to the *lacI* REPRESSOR but without an initial requirement for functional β-galactosidase. IPTG is widely used as an inducer in connection with CLONING VECTORS and EXPRESSION VECTORS that utilize the *lac* promoter.

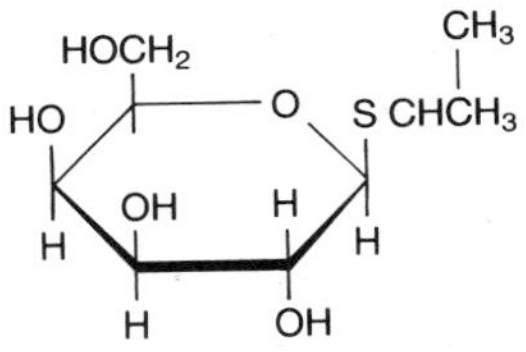

Fig. I42 Isopropyl-β-D-thiogalactopyranoside.

Ir genes IMMUNE RESPONSE GENES.

Iridoviridae Family of icosahedral nonenveloped DNA viruses infecting vertebrates (amphibia and fish) and invertebrates, and which produce iridescent virus inclusions in infected cells. Four genera are distinguished — Iridovirus (the small iridescent insect viruses), Chloriridovirus, Ranavirus, and Lymphocystivirus. The genome is a linear double-stranded DNA (M_r $100-200 \times 10^6$) and

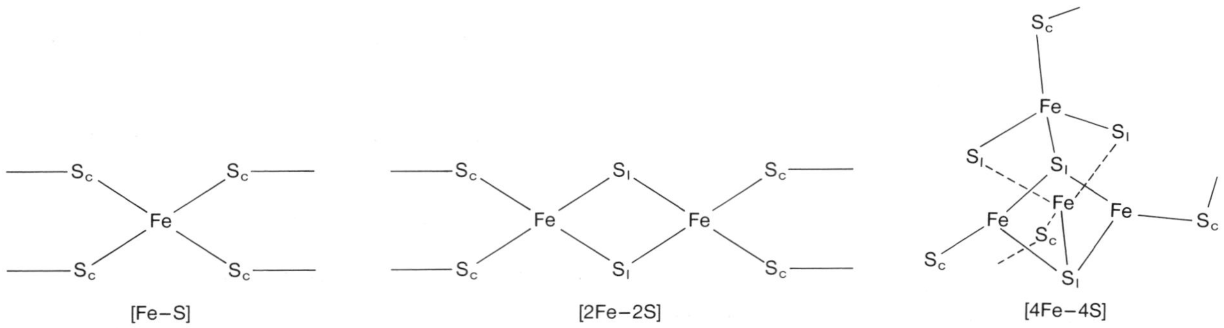

[Fe–S] [2Fe–2S] [4Fe–4S]

Fig. I43 Iron–sulphur clusters. S_C, Sulphur atom of a cysteine residue; S_I, inorganic sulphide.

in some strains is terminally redundant and circularly permutated. The insect iridoviruses are used as viral insecticides.

iron–sulphur proteins (Fe–S proteins, non-haem iron proteins)
Non-haem proteins containing a cluster (or clusters) of iron atoms bonded to sulphur atoms (Fig. I43). [Fe–4S], [2Fe–2S], [3Fe–4S], and [4Fe–4S] clusters are known as well as more complex arrangements (as in the molybdenum-containing reaction centres of nitrogenase, *see* NITROGEN FIXATION). Fe–S proteins include aconitase ([4Fe–4S]), succinate dehydrogenase ([2Fe–2S], [4Fe–4S], [3Fe–4S]), and NADH dehydrogenase ([2Fe–2S], [4Fe–4S]) (*see* ELECTRON TRANSPORT CHAIN). Fe–S clusters as prosthetic groups are involved in a wide range of reduction reactions, the iron atoms cycling between Fe^{2+} and Fe^{3+} states.

IS1, IS2, etc. Insertion elements. *See*: TRANSPOSABLE GENETIC ELEMENTS.

islet cells Endocrine cells which form islands in the pancreas known as the islets of Langerhans, and which are distinct from the surrounding digestive enzyme-secreting exocrine cells. There are two types of islet cells; α cells which secrete GLUCAGON and β, or B, cells which secrete INSULIN.

isoacceptor tRNAs TRANSFER RNAs with different codons but which accept the same amino acid.

isoalleles ALLELES which are identical in their phenotypic effects but can be distinguished at protein or DNA level.

isochromosome A type of CHROMOSOME ABERRATION in which one of the arms of a particular chromosome is duplicated because the centromere divides transversely and not longitudinally during cell division. The two arms of an isochromosome are therefore of equal length and contain the same genes.

isoelectric focusing *See*: ELECTROPHORESIS.

isoelectric point (pI) For any protein the pH value at which the protein will not move in an electric field. *See*: ELECTROPHORESIS.

isoenzymes (isozymes) Variants of an enzyme occurring within a single species, often showing tissue-specific distribution, or differential distribution in different cellular compartments. Isoenzymes have the same general catalytic activity, but differ in physical properties (e.g. stability, optimum pH, isoelectric point), and may be separated by, for example, gel ELECTROPHORESIS or isoelectric focusing. Isoenzymes may represent:
1 Genetically distinct proteins encoded at different loci.
2 Heteropolymeric enzymes of two or more different polypeptide chains (allelic or nonallelic) in different combinations.
3 Multiple alleles at a single genetic locus (alloenzymes).
In addition, isoenzymes can result from alternative splicing or the use of alternative start sites of transcription in a gene, and differential POST-TRANSLATIONAL MODIFICATION. *See*: ALCOHOL DEHYDROGENASE; LACTATE DEHYDROGENASE.

isoflavonoids Plant phenolics structurally derived from 3-phenylchromane, of which the most important are the isoflavones

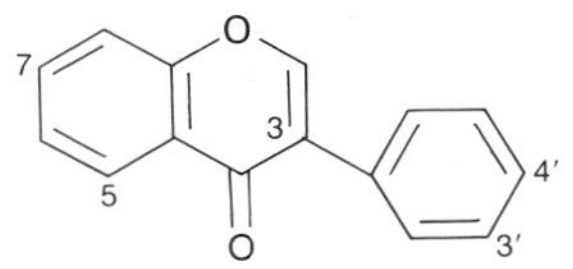

Within the group, structural variation is due to different hydroxylation patterns; carbon atoms 3′, 4′, and 7 are most commonly hydroxylated. The hydroxyl groups may be glycosylated or methylated. They are biosynthesized by B-ring migration following chalcone formation. Isoflavonoids also include the more complex rotenoids (e.g. rotenone), pterocarpans and coumestans.

isoforms Different forms of a protein that may be produced from different genes, or from the same gene by alternative RNA SPLICING.

isogenic Synonym for HOMOZYGOUS.

isolectins Different forms of heteromeric LECTINS, with different combinations of subunits.

isoleucine (Ile, I) A neutral aliphatic AMINO ACID.

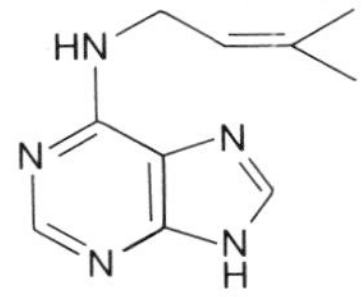

Fig. 144 Structure of isopentenyladenine.

Fig. 145 Two pairs of isoschizomers and their recognition sites.

*Cfo*I	G C G C		*Sma*I	C C C G G G	
*Hha*I	G C G C		*Xma*I	C C C G G G	

isomerases One of the main classes of ENZYMES, (EC 5) which catalyse the racemization or epimerization of a centre of chirality, for example, alanine racemase which catalyses the interconversion of D- and L-alanine.

isomorphous derivative A modified molecular species, for which the derivative crystals are isomorphous with the native crystals; i.e. both the unit cells of the two crystals and their contents are essentially the same. HEAVY ATOM ISOMORPHOUS DERIVATIVES are used in X-RAY CRYSTALLOGRAPHY. Often a derivative is isomorphous only at low resolution, becoming non-isomorphous at higher resolution, where more precise equivalence is necessary, due to slight differences in unit cell dimensions or in relative rotations of one or more domains of the molecule.

isopentenyladenine (2iP) 6-[3-Methylbut-2-enylamino]purine or γ,γ-dimethylallylaminopurine (Fig. 144); M_r 203.2. A CYTOKININ first isolated from tRNA^{Ser} from yeast, but subsequently shown to occur in higher plants. The N^9-riboside and -ribotide of 2iP also occur naturally and have cytokinin activity. In the first step of cytokinin biosynthesis the ribotide is formed from adenosine-5′-monophosphate (AMP) and isopentenyldiphosphate by the action of AMP-isopentenyl transferase, after which it is reversibly hydrolysed to the riboside and free base. Isopentenyladenine is hydroxylated to ZEATIN. AMP-isopentenyl transferase is encoded by the *tmr* and *tzs* genes of *Agrobacterium tumefaciens*; the former gene is part of the T-DNA transferred to plants infected with this bacterium and its expression is responsible for the elevated cytokinin levels in crown galls (*see* TI PLASMID).

Helgeson, J.P. & Leonard, N.J. (1966) *Proc. Natl. Acad. Sci. USA* **56**, 60–63.

isopropyl-β-D-thiogalactopyranoside IPTG.

isoschizomers RESTRICTION ENZYMES isolated from different microorganisms which have the same recognition sequence (Fig. 145). Isoschizomers do not necessarily cut within the recognition sequence at the same position (e.g. *Sma*I and *Xma*I).

isotype(s) Nonallelic variant forms of a biological molecule that are all found in each individual of all members of a species (*cf.* ALLOTYPE, IDIOTYPE). In human ANTIBODIES it refers to the immunoglobulin classes and subclasses that constitute the nine isotypes, each encoded by a distinct gene sequence.

isotype switching CLASS SWITCHING. *See*: GENE REARRANGEMENT; IMMUNOGLOBULIN GENES. Also occasionally used to describe the developmental stage in pre-B cells (*see* B CELL DEVELOPMENT) in which a cell that has been undergoing rearrangement at the immunoglobulin κ light chain locus now proceeds to rearrangements at the λ light chain locus.

isotypic exclusion The expression by B CELLS of light chains of either the κ or λ type but not both. *See*: ANTIBODIES; B CELL DEVELOPMENT; GENE REARRANGEMENT; IMMUNOGLOBULIN GENES.

isozyme *See*: ISOENZYME.

IVS INTERVENING SEQUENCE.

J

J chain Small protein chain holding the monomers together in class M immunoglobulin. *See*: IGM.

J segments (joining segments) DNA segments which encode the distal 12–21 residues of B and T cell antigen receptor variable (V) regions (*see* ANTIBODIES; IMMUNOGLOBULIN GENES; T CELL RECEPTORS; T CELL RECEPTOR GENES). In immunoglobulin heavy chains, the DJ boundary encodes the V region CDR3 domain (*see* COMPLEMENTARITY DETERMINING REGIONS) and the remainder of the joining (J) segment encodes part of the HINGE REGION. In mouse germ-line DNA there are four heavy chain J (J_H) segments, five functional J_κ segments and four $J_\lambda C_\lambda$ units. In humans, a cluster of six functional J_H segments is interspersed with three J_H sequence PSEUDOGENES. There are five functional J_κ segments and three tandemly arranged functional $J_\lambda C_\lambda$ units. The multiplicity of J segments contributes to the diversity of immunoglobulins (*see* GENERATION OF DIVERSITY).

jasmonic acid (JA) 3-Oxo-2-(2Z-pentenyl)cyclopentane-1-acetic acid (Fig. J1). M_r 306.3. It is ubiquitous in higher plants and especially concentrated in young fruit ($3.10\ \mu g\ g^{-1}$ fresh weight in *Vicia faba*). JA and its more active methyl ester (METHYL JASMONATE) are metabolites of linolenic acid (*see* FATTY ACIDS). The predominant natural form is the *cis* (1*R*,2*S*) isomer, which is more active than the co-occurring *trans* (1*R*,2*R*) compound. The two forms are easily interconverted. Jasmonic acid and its methyl ester are effective at $0.05\ mol\ m^{-3}$ as inhibitors of chlorophyll formation, cereal organ extension, and soybean callus growth, and as promoters of leaf senescence, stomatal closure, tuber formation, and of the production of numerous leaf proteins, including PROTEINASE INHIBITORS important in plant defence against insects. Related compounds include tuberonic acid, the glucosyl ether of which is a tuber-inducing factor in potatoes, and cucurbic acid, a growth retardant isolated from seeds of *Cucurbita*

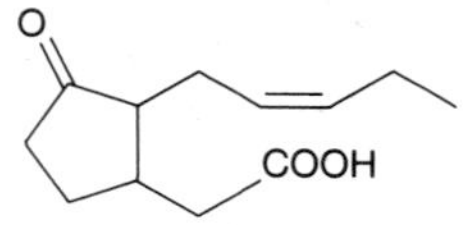

Fig. J1 Structure of jasmonic acid.

pepo. *See*: PLANT HORMONES; PLANT WOUND RESPONSES.

Koda, Y. (1992) *Int. Rev. Cytol.* **135**, 155–199.

JC virus *See*: PAPOVAVIRUSES.

jelly roll motif Supersecondary structure eight-stranded barrel motif in proteins. It is found in the plant LECTIN concanavalin A (*see* Plate 3*d*), in coat proteins from many icosahedral viruses, influenza haemagglutinin, and other proteins.

joining segments J SEGMENTS.

jumping genes A colloquial name sometimes given to TRANSPOSABLE GENETIC ELEMENTS.

jun Cellular ONCOGENE whose product Jun contains a leucine zipper motif, and in a complex with the product of the *fos* gene (Fos) forms the activating TRANSCRIPTION FACTOR AP-1 (*see* EUKARYOTIC GENE EXPRESSION).

junction *See*: CELL JUNCTIONS; NUCLEIC ACID STRUCTURE.

junctional complex The specific arrangement of intercellular junctions at the apicolateral region of epithelial cells. The tight junction is most apical, with the intermediate junction and desmosomes respectively more basally located. *See*: CELL JUNCTIONS; EPITHELIAL POLARITY.

junctional diversity Diversity of antigen specificity in the antigen receptor repertoire which is due to imprecision at the point of junction between V, D, and J gene segments in rearranged immunoglobulin and T cell receptor genes, as well as to the nontemplated insertion (*see* N-REGION DIVERSITY) and deletion of bases at the points of junction during the rearrangement process. *Cf.* COMBINATORIAL DIVERSITY. *See*: GENERATION OF DIVERSITY.

junk DNA Term occasionally used to denote the repetitive sequences of no apparent function in the genomes of higher eukaryotes. *See*: GENOME ORGANIZATION.

K

K The single-letter abbreviation for the AMINO ACID lysine.

K$^+$ *See*: POTASSIUM.

K_a, pK_a In physical chemistry K_a is the dissociation constant of an acid or base and is defined for a base as $K_a = [B][H^+]/[BH^+]$ where B is the base and BH$^+$ its conjugate acid, and for an acid as $K_a = [A^-][H^+]/[HA]$, where HA is the acid and A$^-$ its conjugate base. The pH of half neutralization of an acid or base is pK_a which is equal to $-\log K$.

K_a is also sometimes used in molecular biology as the symbol for the association constant of a receptor for its ligand (e.g. in the measurement of the AFFINITY of an antibody for its cognate antigen). *See*: SCATCHARD PLOT.

K_d The equilibrium binding (dissociation) constant for a ligand and its receptor. *See*: SCATCHARD PLOT.

K_m Michaelis constant. *See*: MECHANISMS OF ENZYME CATALYSIS.

kainate Selective agonist of non-NMDA EXCITATORY AMINO ACID RECEPTORS.

Fig. K1 Kainic acid.

kallikrein A SERINE PROTEINASE found in plasma (EC 3.4.21.34) and tissues (EC 3.4.21.35). Plasma kallikrein cleaves peptide bonds between Lys-Arg and Arg-Ser in kininogen to produce bradykinin, and also activates coagulation Factors XII, VII, and plasminogen (*see* BLOOD COAGULATION AND ITS DISORDERS) by specific cleavage. It is formed from prekallikrein by activated Factor XII. Tissue kallikrein cleaves Met-Lys and Arg-Ser bonds in kininogen to produce lysyl-bradykinin.

Kallmann syndrome A syndrome defined by the association of hypogonadotropic hypogonadism with anosmia (no sense of smell) and with a variety of central nervous system disorders and renal aplasia. It may be inherited as an autosomal dominant, recessive, or X-linked form. It also occurs sporadically (*see* CONTIGUOUS GENE SYNDROMES).

kanamycin Aminoglycoside antibiotic (Fig. K2) produced by *Streptomyces kanamyceticus* which inhibits bacterial PROTEIN SYNTHESIS by binding to the 30S ribosomal subunit. It is widely used as a drug for selection in the transformation of bacterial and plant cells (*see* GENETIC ENGINEERING; PLANT GENETIC ENGINEERING) when used in combination with the bacterial neomycin phosphotransferase gene from the TRANSPOSABLE GENETIC ELEMENT Tn5 which confers resistance to kanamycin.

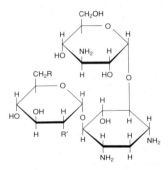

Fig. K2 Kanamycin A. R = NH$_2$, R′ = OH. Kanamycin B. R = R′ = NH$_2$.

kappa deleting element (KDE) A cryptic recombination signal sequence (*see* GENE REARRANGEMENT; IMMUNOGLOBULIN GENES) found ~24 kilobases downstream of the human C_κ gene in the immunoglobulin gene cluster. It can rearrange with sequences upstream of the C gene, located either in the J_κ-C_κ intron or adjacent to the J_κ or V_κ segments. This recombination, seen in many B cells that produce λ light chains, occurs by a looping-out mechanism and leads to complete deletion of the C_κ gene. The analogous process at the mouse κ locus is referred to as RS RECOMBINATION. *See also*: DELTA DELETING ELEMENT.

kappa (κ) light chain One of the two mutually exclusive types of LIGHT CHAINS found in ANTIBODIES, the other being called lambda (λ). *See also*: IMMUNOGLOBULIN GENES.

KAR2 gene The gene from the yeast SACCHAROMYCES CEREVISIAE encoding the immunoglobulin binding protein (BiP). Mutations in this gene also cause a defect in KARYOGAMY.

Kartagener's syndrome A hereditary condition causing infertility in males due to abnormal flagella on sperm which render the sperm immotile. The flagellar defect has been identified as a complete absence of DYNEIN arms in the axonemes of all cilia and flagella in the body (*see* MICROTUBULES; MICROTUBULE-

BASED MOTORS). Affected patients also suffer from chronic sinusitis and bronchiectasis, and from situs inversus, a left–right inversion in the location of organs such as heart and liver. Women with the condition are fertile.

karyogamy The fusion of the two gametic nuclei to form the zygotic nucleus.

karyokinesis Nuclear division. *See:* MITOSIS.

karyoplast A nucleus isolated from a eukaryotic cell. The nucleus is surrounded by a thin layer of cytoplasm and plasma membrane.

karyotype A display of the METAPHASE CHROMOSOMES of a somatic cell of an individual to show their number, shape, and size. The karyotype is normally shown as a photomicrograph of the chromosomes arranged in a standard way (e.g. in humans, chromosomes 1–22 and the sex chromosomes). The process of preparing such a photomicrograph is known as karyotyping. Individual chromosomes are identified by CHROMOSOME BANDING and in a formal karyotype, photographs of chromosome pairs are aligned to provide a visual representation of the organism's chromosomal constitution. Karyotypes have many uses in CYTOGENETICS — for example, to distinguish different species and to detect visible CHROMOSOME ABERRATIONS. *See also:* HUMAN GENE MAPPING; SOMATIC CELL HYBRIDS.

ISCN (1985) *An International System for Human Cytogenetic Nomenclature* (Harnden, D.G. & Klinger, H.P., Eds) *Cytogenet. Cell Genet.* (Karger, Basel).
Birth Defects (1985) Original Article Series, Vol. 21, No. 1 (March of Dimes Birth Defects Foundation, New York).

kb The abbreviation for kilobase, 1000 bases, a unit in which the length of a single-stranded RNA or DNA is often measured. The abbreviaton kbp (kilobase pair) is sometimes used in relation to double-stranded nucleic acids.

kdal, kD, kDa KILODALTON, a widely used unit of molecular weight. *See:* RELATIVE MOLECULAR MASS.

KDEL Soluble proteins of the ENDOPLASMIC RETICULUM (ER) such as protein disulphide isomerase, BiP and endoplasmin, contain the C-terminal sequence of amino acids KDEL (Lys-Asp-Glu-Leu), which acts as a RETENTION SIGNAL. Mutations of this sequence cause proteins to leave the ER; attachment of the sequence to secretory proteins results in their retention in the ER. Minor variations of the sequence have been found, but it must be at the C terminus to function, hence the true signal is 'KDEL-Stop'. In the budding yeast SACCHAROMYCES CEREVISIAE the signal is HDEL; in the fission yeast SCHIZOSACCHAROMYCES POMBE it is ADEL. Proteins containing these retention signals are not retained by binding to the membrane of the ER. Instead they leave it by vesicle-mediated transport along with secretory proteins, but are selectively retrieved from a subsequent compartment of the secretory pathway and returned to the ER. The receptor responsible for this retrieval may be the *ERD2* gene product.

Munro, S. & Pelham, H. (1987) *Cell* **48**, 899–907.

Kearns–Sayre syndrome *See:* MITOCHONDRIOPATHIES.

Kell blood group The original Kell antigen, K or KEL1, has a frequency of about 9% in Caucasians. Anti-K may cause severe haemolytic transfusion reactions or haemolytic disease of the newborn. Kell is a complex BLOOD GROUP system comprising at least 21 antigens located on a 93 000 M_r GLYCOPROTEIN. The *KEL* complex locus is located on chromosome 7 and the gene has been cloned. Kell antigen expression is partially dependent on the action of an X-borne gene (*XK*) which produces Kx antigen. Absence of Kx is associated with weak expression of Kell antigens, acanthocytosis, and various muscular and neurological defects (McLeod syndrome), and sometimes with CHRONIC GRANULOMATOUS DISEASE resulting from an Xp21.1 deletion.

Kennedy's disease X-linked spinal and bulbar muscular atrophy (SBMA). It is an adult onset motor neuron degenerative disease characterized by progressive muscle weakness and atrophy. Some affected males also show gynaecomastia and reduced fertility. It is associated with and probably caused by amplification of a CAG repeat in the first exon of the androgen receptor gene located on the proximal long arm of the X-chromosome at Xq11-12. *See:* NEURODEGENERATIVE DISORDERS.

keratan sulphate A sulphated polymer based on the poly *N*-acetyl lactosamine repeating unit of keratan, (-3DGal(p)β1, 4DGlcNAc(p)β1-)$_n$ (Fig. K3). Most of the hexosamine residues have sulphate ester at C6. Variable amounts of fucose and sialic acid are present and the polymer acts as a blood group antigen. Corneal keratan sulphate (type I keratan sulphate) is linked to protein via an *N*-glycan of the glycoprotein type, while skeletal keratan sulphate (type II keratan sulphate) is linked via a mucin-type O-glycoside. Hence keratan sulphates are specialized 'conventional' glycoproteins, rather than true GLYCOSAMINOGLYCANS. Their biosynthesis is, therefore, initiated either by the transfer of the glycan produced by the dolichol phosphate cycle to protein or by the addition of *N*-acetylgalactosamine to serine or threonine. Chain extension implies the alternate action of two glycosyltransferases and chain termination is thought to occur by a sialyl or possibly a fucosyl transfer. Keratan sulphate chains normally occur on the same core proteins that carry glycosaminoglycan chains. *See also:* EXTRACELLULAR MATRIX MACROMOLECULES; GLYCANS.

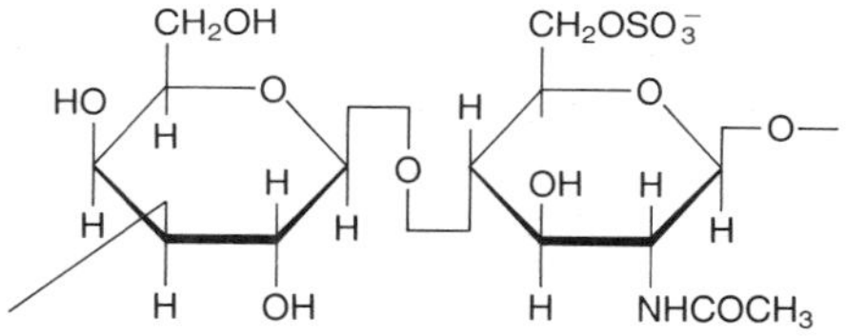

Fig. K3 The repeating unit of keratan sulphate.

keratin, keratin filaments INTERMEDIATE FILAMENT proteins found in epidermal cells and structures derived from them (hair, wool, nails, horn).

KEX genes Genes of the yeast SACCHAROMYCES CEREVISIAE, mutations in which cause a defect in the secretion of the KILLER TOXIN.

keyhole limpet haemocyanin (KLH) This oxygen-transporting protein, and HAEMOCYANINS of other molluscs, are excellent IMMUNOGENS in most vertebrates. Because of this, KLH has been used extensively as a carrier protein in experiments designed to investigate the immune response to HAPTENS. KLH tends to self-associate, so an M_r of 3×10^6 is assigned for experimental purposes. Removal of aggregates tends to impair or reduce its immunogenicity.

Ki-*ras* Oncogene of the *ras* family, isolated from the Kirsten rat sarcoma virus. *See*: GTP-BINDING PROTEINS; ONCOGENES; RAS; RETROVIRUSES.

Kidd blood group BLOOD GROUP defined by the red blood cell antigens (JK^a and JK^b) encoded by the *JK* gene which maps to the short arm of human chromosome 2. There are no naturally occurring antibodies to these antigens.

kidney development Kidney development is one system for the study of INDUCTION. The kidney develops from the INTERMEDIATE MESODERM (between somites and lateral plate) in three stages, comprising pronephros, mesonephros, and metanephros (the definitive kidney). The mesonephros is the permanent excretory organ of the cartilaginous fishes and amphibians, but is superseded by the metanephros in the amniotes. In birds, the pronephros forms between the lateral plate and the somite mesoderm, forming paired buds which grow dorso-laterally towards the ECTODERM. The distal portion of each fuses with the tubules of the next segment, and this series of connecting cells forms the pronephric duct, which eventually degenerates after glomeruli have formed. The Wolffian duct, which develops caudal to the pronephros, is necessary for proper mesonephric development. The anterior mesonephric tubules differentiate simultaneously (unlike those of the pronephros) and open into the COELOM, whereas posterior tubules do not. The metanephros, the third avian kidney to be formed, begins activity before that of the mesonephros ceases, and continues to function throughout the life of the animal. It develops from the Wolffian ducts after the latter open into the cloaca. Beginning as the ureteric bud (or metanephric diverticulum) it induces differentiation of secretory tubules and renal corpuscles from the surrounding nephrogenous tissue. The ureteric bud develops into the ureter, while mesodermal cells surrounding the tubules form the stroma and capsule of the kidney. The molecular basis for the induction of the definitive tubular excretory units is poorly understood.

killer toxin A polypeptide secreted by strains of the yeast SACCHAROMYCES CEREVISIAE which carry the M double-stranded RNA virus. The toxin is toxic to yeast strains which lack the M double-stranded RNA since this RNA also encodes an immunity determinant. The toxin consists of two subunits (α and β) linked by a pair of disulphide bonds. Many other yeast species secrete a similar toxin.

kilobase A length of 1000 bases of polynucleotide chain. abbreviated to kb, the abbreviation often also used instead of kbp (1000 base pairs) as a measure of a double-stranded nucleic acid molecule.

2.2 kilobase gene Chloroplast gene for the β subunit of ATP SYNTHASE (*atpB*). It was initially recognized as a gene adjacent to that for the large subunit of RIBULOSE BISPHOSPHATE CARBOXYLASE/OXYGENASE, transcribed in the opposite direction into a 2.2 kilobase transcript, and its product was identified later.

kilodalton (kdal, kD, kDa) A widely used unit for the molecular weight of macromolecules. It equals 1000 daltons. *See*: RELATIVE MOLECULAR MASS.

kinases Enzymes that catalyse the transfer of a phosphate group from ATP or other nucleoside triphosphate to a substrate, for example hexokinase (EC 2.7.1.1) which catalyses the reaction ATP + D-hexose = ADP + D-hexose 6-phosphate. *See also*: PROTEIN KINASES.

kindling The production of long-term electrophysiological and morphological changes in local neuronal tissue following repeated electrical stimulation. For example, months after stimulation an intense seizure can be elicited by relatively few electrical stimuli. The mechanisms underlying kindling may be similar to LONG-TERM POTENTIATION, in which a brief period of intense activity results in a persistent change in synaptic strength.

kinesin Motor protein associated with MICROTUBULES and probably present in all eukaryotic cells. The form discovered originally is a soluble rod-shaped molecule composed of two polypeptide chains, which travels towards the plus ends of microtubules (anterograde movement in AXONAL TRANSPORT). It has ATPase/GTPase activity which resides in N-terminal segments (M_r 35 000–45 000) of the two chains and these segments form a pair of globular 'heads' at one end of the rod. The molecule is similar to two-headed MYOSIN, apart from being smaller and having a globular domain at the tail-end of the rod. Various kinesin-like molecules possess homologous motor domains attached to different 'tails'. In some cases, the motor domains are at the C-terminal end of the polypeptide and one such protein, at least, moves along microtubules in the opposite direction from the original form. *See also*: MICROTUBULE-BASED MOTORS.

kinetic complexity When DNA reassociates from single strands in solution after melting the kinetics of this reassociation reflect the relative amounts of unique and repetitive DNA sequence — the complexity (*see* COT ANALYSIS; DNA COMPLEXITY; HYBRIDIZATION).

kinetin 6-Furfurylaminopurine or N^6-furfuryladenine (Fig. K4). M_r 215.2. The first CYTOKININ to be characterized, although it is not a natural plant product. It was isolated as a factor from autoclaved herring sperm DNA that stimulated cell division in a tissue culture derived from tobacco stem internodes. Its structure was confirmed by synthesis. Kinetin and its N^9-riboside are used

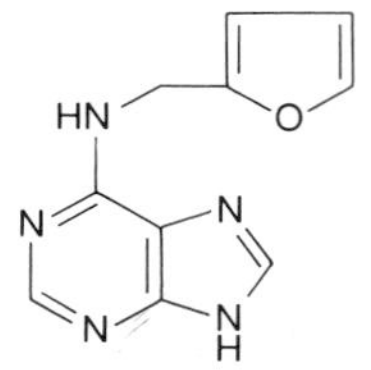

Fig. K4 Structure of kinetin.

in tissue culture to stimulate growth by cell division and to initiate shoot formation.

Miller, C.O. et al. (1955) *J. Am. Chem. Soc.* **77**, 1329.

kinetochore The specialization on each mitotic CHROMATID that attaches it to the MITOTIC SPINDLE (*see* MITOSIS). Each METAPHASE CHROMOSOME has two kinetochores at its CENTROMERE, or primary constriction (*see* CHROMOSOME STRUCTURE). They are composed of proteins bound specifically to the centromeric DNA. Kinetochores contribute to the accurate segregation of chromosomes during mitosis. Kinetochore structure is visible in the electron microscope. Many cells contain kinetochores that are built from two or more layers of proteinaceous plaques, ~0.5 μm in diameter and 0.02 μm thick. Kinetochores seem to be made by loops of DNA that bind proteins and assemble microtubule-binding domains. Some kinetochore proteins remain associated with centromeric DNA throughout the CELL CYCLE. One such is a histone-like protein of M_r ~18 000. Others have M_r of 80 000 and 140 000. Beginning in PROMETAPHASE, kinetochores display the ability to move over the surface of spindle MICROTUBULES. This motor activity of kinetochores is probably due in part to cytoplasmic DYNEIN which is concentrated there during mitosis, but dynein is not bound to interphase kinetochores. Kinetochores therefore assemble slowly during the cell cycle, beginning during S-phase and continuing until mitosis, when they become fully functional to bind and move along microtubules. *See also*: MICROTUBULE-BASED MOTORS.

kinetochore fibre *See*: MICROTUBULES; MITOSIS.

kinetoplast DNA An alternative name for the mitochondrial DNA of certain protozoa, for example *Trypanosoma*, which contain a single large mitochondrion. *See*: MITOCHONDRIAL GENOMES: LOWER EUKARYOTE.

kinins (1) BRADYKININ.
(2) Cytokinins. *See*: PLANT HORMONES.

Kirsten rat sarcoma virus RETROVIRUS which is the source of the Ki-*ras* ONCOGENE.

kit A receptor tyrosine kinase (*see* GROWTH FACTOR RECEPTORS) whose ligand is the product of the murine *steel* gene (*see* COAT COLOUR GENES). A potential ONCOGENE.

Klebsiella Genus of nonmotile bacteria of the family Enterobacteriaceae, present in soil and water. They include the opportu-

nistic pathogens *K. pneumoniae* and *K. aerogenes*. *K. pneumoniae* is also of interest as a free-living nitrogen fixing bacterium (*see* NITROGEN FIXATION). VECTORS for use in *Klebsiella* have been developed from BACTERIOPHAGE P4, which can replicate as a phage or a PLASMID.

Klenow fragment A polypeptide fragment of M_r 76 000 derived by proteolytic cleavage of *Escherichia coli* DNA polymerase I. This fragment contains both the 5′ → 3′ polymerase and the 3′ → 5′ EXONUCLEASE activities, but lacks the intrinsic 5′ → 3′ exonuclease activity of DNA polymerase I. Its uses include: the filling in of 3′ recessed termini generated by digestion of DNA with RESTRICTION ENZYMES; the generation of truncated DNA fragments in the Sanger dideoxy chain termination DNA SEQUENCING method; and the conversion of single-stranded cDNA to double-stranded DNA for cloning.

Jacobsen H., Klenow H. & Ovagaard-Hansen K. (1974) *Eur. J. Biotechnol.* **45**, 623.

KLH KEYHOLE LIMPET HAEMOCYANIN.

Klinefelter syndrome One of the commonest chromosome abnormalities, occurring at a frequency of one in 600 live births. The karyotype is XXY. The phenotype is male with defective sexual maturation, small firm testes, gynaecomastia, and, in a proportion of cases, minor mental retardation.

knirps (kni) A member of the gap class of SEGMENTATION GENES in *Drosophila*; the mutant phenotype is the loss of several abdominal segments. The gene is expressed in a broad band in the abdominal region at the blastoderm stage and the position of this expression is determined by the gap genes *hunchback* and *tailless*. The *knirps* product contains a zinc-finger DNA-binding motif characteristic of the STEROID HORMONE RECEPTOR SUPERFAMILY.

knob (1) In cytogenetics, a conspicuously darkly staining CHROMOMERE that may allow identification of a particular chromosome. (2) In parasitology, cells parasitized with some strains of the malaria parasite *Plasmodium falciparum*, which produce knobs that attach to endothelial cells and block cerebral vessels giving rise to serious cerebral malaria.

Koshland model The Koshland–Némethy–Filmer model of co-operativity in proteins. *See*: ALLOSTERIC EFFECT; COOPERATIVITY.

Kozak consensus sequence *See*: PROTEIN SYNTHESIS.

Kozak scanning model *See*: PROTEIN SYNTHESIS.

Kpn family *See*: CHROMOSOME STRUCTURE.

Kranz anatomy The organization of photosynthetic cells in leaves of C4 PLANTS into two distinct structural and functional types: the mesophyll cells containing granal chloroplasts which fix CO_2 by phosphoenolpyruvate carboxylase into a C4 acid; and large bundle sheath cells (Kranz cells) which import the C4 acid, decarboxylate it, and refix the liberated CO_2 by the CALVIN CYCLE. The latter contain chloroplasts that are often agranal. *See also*: PHOTOSYNTHESIS.

kringle domain A common protein domain of about 85 amino-acid residues with a characteristic pattern of three internal disulphide bridges, found in blood clotting and fibrinolytic proteins (Fig. K5).

Krüppel (Kr) A gap class SEGMENTATION GENE in *Drosophila* whose mutant phenotype shows defects in thoracic and anterior abdominal segments. Krüppel transcripts are expressed in a broad band in the central region of the blastoderm embryo. The limits of this band are specified by the anterior determinant *bicoid* and by the gap genes *hunchback* and *giant*. The *Krüppel* product is a TRANSCRIPTION FACTOR with zinc-finger DNA-binding motifs and acts, in combination with other gap genes, to regulate the patterns of expression of PAIR-RULE and HOMEOTIC genes.

Kupffer cell A phagocytic cell of the liver, responsible for the ingestion and removal of material including old red blood cells.

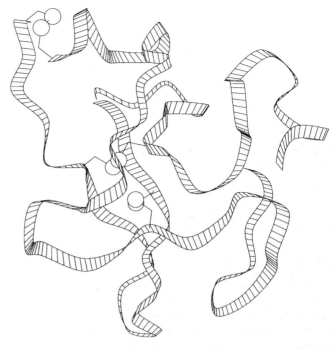

Fig. K5 A kringle domain. The paired circles represent disulphide bonds.

L

L The single-letter abbreviation for the AMINO ACID leucine.

L1 A mammalian LINE. L1 is present at ~20 000–50 000 copies per genome and is typically around 6500 bp in length.

L-19 *See:* RIBOZYMES.

L cell An established CELL LINE derived from a mouse SARCOMA.

L, M, H subunits/polypeptides The three subunits of the reaction centre of purple photosynthetic bacteria such as *Rhodobacter sphaeroides* (*see* PHOTOSYNTHESIS) named according to the order of appearance on SDS-polyacrylamide gel electrophoretograms. The true M_r values are L, 30 500; M, 34 000; H, 28 000. The reaction centre from *Rhodobacter viridis* contains in addition a tightly bound four-haem CYTOCHROME subunit.

L subunit The 'light' subunit of the photosynthetic reaction centre of purple bacteria (M_r 30 500) (*see* PHOTOSYNTHESIS) which shows sequence and structural homology with the M subunit; the LM heterodimer binds the pigments and cofactors involved in the primary photochemical charge separation. It is topologically and functionally similar to subunit D1 of PHOTOSYSTEM II.

lac operon *lac*, in *Escherichia coli*, is the classic example of an inducible bacterial OPERON. It consists of a single PROMOTER controlling the TRANSCRIPTION of three downstream structural genes: *lacZ* (β-galactosidase), *lacY* (permease), and *lacA* (acetylase). *lacY* and *lacZ* enable the uptake and utilization of lactose from the medium (see Fig. A156 in BACTERIAL GENE EXPRESSION). The three genes are transcribed into a single POLYCISTRONIC mRNA. Gene expression requires INDUCTION or DEREPRESSION of the *lac* promoter. The *lac* operon is negatively regulated: the operon is CONSTITUTIVELY transcribed unless repressed by the *lac* REPRES-SOR protein, LacI. LacI is encoded by a gene (*lacI*) which lies just upstream of the *lac* promoter, and is transcribed independently from its own promoter. The gene product, LacI, forms a repressor tetramer that binds specifically to the OPERATOR region of the *lac* promoter, thereby coordinately repressing the transcription of the *lacZ*, *lacY*, and *lacA* genes. A small molecule inducer (naturally allolactose, but the nonmetabolizable analogue isopropyl-thiogalactoside (IPTG) is generally used experimentally) binds to the repressor, producing an ALLOSTERIC change in its conformation which releases it from the operator, and allows transcription to be re-established. Full expression of *lac* also requires the action of transcriptional activating proteins (*see* BACTERIAL GENE EXPRES-SION). The *lac* operon has also been widely exploited in the development of CLONING VECTORS that allow for identification of recombinants by insertional inactivation of the *lacZ* gene.

Miller, J. & Reznikoff, W. (Eds) (1980) *The Operon* (Cold Spring Harbor Laboratory, New York).

lac repressor A homotetrameric protein (M_r 152 000) encoded by the *lacI* gene (*see* LAC OPERON) that binds to the operator region of the *lac* promoter to prevent transcription. The *lac* repressor has been intensively studied as an example of ALLOSTERY and as a DNA-binding protein (*see* PROTEIN–NUCLEIC ACID INTERACTIONS).

α-lactalbumin Milk protein of 123 amino acids (Fig. L1) which is the B protein of the two components (A and B) of the lactose synthetase complex. α-Lactalbumin has a high-affinity (k_d 10^{-9} M) calcium-binding site. Its structure superficially resembles the EF hand (*see* CALCIUM-BINDING PROTEINS) but has a striking similarity with the structure of lysozyme, with which it shares a common evolutionary precursor.

Acharya, K.R. et al. (1989) *J. Mol. Biol.* **208**, 99–127.

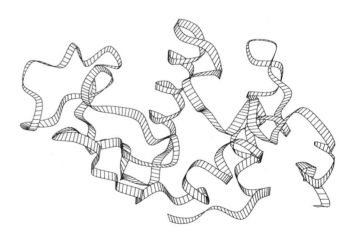

Fig. L1 α-Lactalbumin.

β-lactam antibiotic An ANTIBIOTIC characterized by the presence of a β-lactam ring (Fig. L2), which is in essence a cyclic amide with three carbons and one nitrogen. The β-lactam ring carries a variable side chain. The nature of this side chain determines the susceptibility of the antibiotic to inactivation by β-LACTAMASES. A chemically modified side chain may render a β-lactam antibiotic resistant to β-lactamases. β-Lactam antibiotics include the PENI-CILLINS.

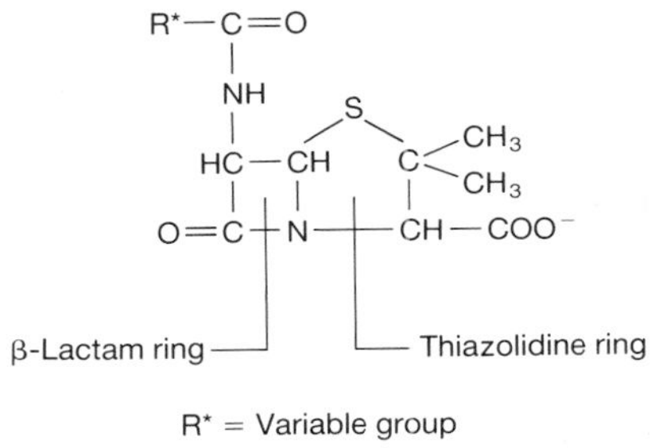

R* = C=O
|
NH
| S
| /
HC — CH   C — CH₃
| | \
O=C — N — CH₃
| |
        CH — COO⁻

β-Lactam ring — Thiazolidine ring

R* = Variable group

Fig. L2 β-Lactam structure.

β-lactamase A class of enzymes (EC 3.5.2.6) which inactivate β-LACTAM ANTIBIOTICS through cleavage of the β-lactam ring. A wide variety of bacteria produce β-lactamases. The *bla* gene encoding a β-lactamase is often used as a SELECTABLE MARKER in plasmid VECTORS, as in the general cloning plasmid for *Escherichia coli* — pBR322.

lactase deficiency A deficiency in the production of the enzyme lactase, which is relatively rare in Europeans but common in some racial groups, and which results in lactose intolerance, the inability to digest milk, which produces symptoms of nausea, vomiting and abdominal pain. Lactase production is controlled by a gene with three alleles: L, l_1 and l_2. Individuals carrying LL, Ll_1 or Ll_2 produce lactase both as adults and as children. l_1l_2 individuals produce lactase as children but not as adults, but can digest milk that has been soured or turned into cheese. Those with l_2l_2 do not produce lactase even in infancy.

lactate dehydrogenase (LDH) Enzyme (EC 1.1.1.27) involved in basic cellular metabolism, catalysing the reversible reaction

$$\text{L-lactate} + NAD^+ \rightleftharpoons \text{pyruvate} + NADH + H^+.$$

LDH provides a good example of an ISOENZYME, having five different forms performing the same function. The active enzyme (M_r 135 000) is composed of four subunits (subunit M_r ~27 000) of two types, A and B, encoded by different genes, and the five isoenzymes are formed by different combinations of subunits — AAAA, AAAB, AABB, ABBB, and BBBB. The serum level of the various isoenzymes is a useful diagnostic tool in myocardial infarction. Duck LDH is a CRYSTALLIN (*see* MOLECULAR EVOLUTION: SEQUENCES AND STRUCTURES). *See also:* NAD-DEPENDENT DEHYDROGENASES; PROTEIN ENGINEERING.

β-lactoglobulin Protein found in milk, usually existing as a dimer with monomer M_r 18 000 (around 162 amino acids). It is particularly acid-stable, resisting denaturation at pH 2. It has a marked structural homology to RETINOL-BINDING PROTEIN.

Papiz, M.Z. et al. (1986) *Nature* **324**, 383–385.

lactoperoxidase Enzyme (EC 1.11.1.7) isolated from cow's milk which is used to catalyse the incorporation of radioiodine (^{125}I) into tyrosine groups on proteins and cell surfaces.

lactose intolerance *See:* LACTASE DEFICIENCY.

LAD LEUKOCYTE ADHESION DEFICIENCY.

LAF Lymphocyte- or leukocyte-activating factor. An early term for interleukin 1α (IL-1α). *See:* IL-1; LYMPHOKINES.

lag phase Period in the growth of a microbial or cell culture in nonlimiting media which precedes exponential doubling. When microorganisms are added to media in which environmental conditions are significantly different from their previous state, they take time to adjust to these new conditions before entering a period of exponential growth (LOG PHASE). This adjustment period, the lag phase, represents the time taken for the molecular and biochemical processes of the cell to monitor and adapt to the new growth environment. A LAG PHASE can also be brought about by a change in temperature or substrate in a culture, but is not used to describe changes that lead to a limiting or total cessation of growth (i.e. stationary phase).

LAM-1 Leukocyte adhesion molecule. *See:* SELECTINS.

Lambda

LAMBDA (λ) is a bacterial virus (BACTERIOPHAGE or phage) isolated by E.M. Lederberg in 1951 from *Escherichia coli* K-12, the bacterium with which J. Lederberg and E.L. Tatum began their work in bacterial genetics (see [1] for a general review). The phage has a linear DNA genome packaged within an icosahedral head, with a long, non-contractile tail ending in a single tail fibre (Fig. L3). This fibre adsorbs to receptor sites in the outer membrane of the bacterium. The physiological role of the receptor sites, encoded by the bacterial *lam*B gene, is the uptake of maltose. In addition, an inner membrane protein, encoded by *pts*M, is essential for the injection of λ DNA into the host.

The genome of λ is a double-stranded DNA molecule of 48 502 base pairs (bp) with 5′ extensions of 12 bases. These single-stranded projections of complementary sequence, the COHESIVE ENDS, enable the λ genome to circularize when it enters the

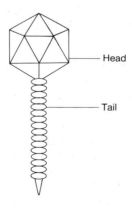

Fig. L3 Bacteriophage λ.

bacterial cell. The resulting covalent sequence that includes the 12 bp of the cohesive ends is called *cos*.

Lysogeny versus lysis

Lambda, being a TEMPERATE PHAGE, may follow either the temperate (LYSOGENIC) or productive (LYTIC) pathway (*see* Fig. L4). In either case, TRANSCRIPTION of the phage genes is initiated from two 'early' PROMOTERS, p_L and p_R, to provide functions essential for DNA REPLICATION, genetic RECOMBINATION, establishment of lysogeny and expression of 'late' genes.

In the temperate response most phage functions become repressed, either directly or indirectly, by λ REPRESSOR (the product of the *cI* gene, *see* Fig. P63 in PROTEIN–NUCLEIC ACID INTERACTIONS), which binds to the flanking OPERATOR regions (o_L and o_R) associated with P_L and P_R (*see* BACTERIAL GENE EXPRESSION for the basic principles of regulation of bacterial and phage gene expression). If repression occurs in time to prevent expression of the LATE GENES, lysis is avoided and lysogeny may ensue. Stable lysogeny normally requires INTEGRATION of the λ genome into the host chromosome so that the integrated phage genome (PROPHAGE) is replicated as part of the *E. coli* chromosome.

Lytic infection, on the other hand, requires a switch from transcription of the early genes to transcription of the late genes. Moderation of the early promoters is achieved by the binding of Cro protein, rather than repressor, to o_L and o_R. Late transcription is dependent on the product of the *Q* gene (see Fig. L4) and provides the proteins necessary to make the heads and tails of the phage particle (the virion). The transition from early to late transcription is associated with a switch in the predominant mode of DNA replication from THETA REPLICATION to ROLLING CIRCLE

REPLICATION. The λ-encoded enzyme terminase cuts the resulting CONCATEMERIC (multimeric) genomes within *cos* sequences to generate linear λ genomes with cohesive ends; cutting is an integral part of the process of packaging the genomes into phage heads. Lysis of the cell releases about 100 infective bacteriophage.

Establishment of lysogeny

Bacteriophage are assayed by their ability to lyse cells and as a consequence produce PLAQUES in a lawn of bacteria; if all infected cells are lysed the plaques will be clear. The temperate nature of λ is manifest by the survival and growth of those cells in which repression is achieved, and the consequent TURBID PLAQUE morphology. In each cell infected with λ the choice between lysogeny and lysis is determined by the eventual occupation of o_L and o_R by either repressor or Cro, the 'anti-repressor' (Fig. L5). Insight into the control of lysogeny began with the isolation of mutants that produce clear plaques. These mutants identified *cI*, the repressor gene itself, and two genes (*cII* and *cIII*) whose products (CII and CIII) aid the establishment of lysogeny. CII, protected from host protease by CIII, activates transcription of *cI* from the promoter for repressor establishment (p_E or p_{RE}). Repressor subsequently controls its own synthesis (autogenous control); in binding to o_R it not only represses p_R but it activates transcription of its own gene from p_M (or p_{RM}) (*see* Fig. L4), and thereby maintains the level of repressor. Lysogeny is favoured by high MULTIPLICITY OF INFECTION and poor growth conditions.

Phage immunity

As prophages are kept dormant by repressor bound to o_L and o_R,

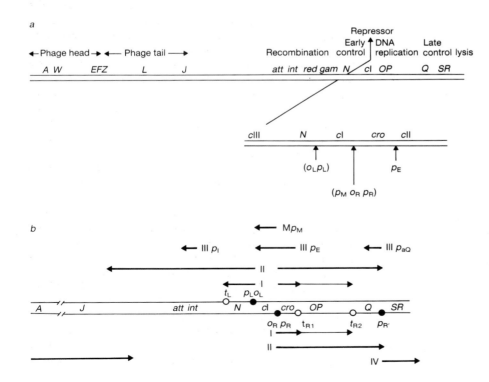

Fig. L4 The λ genome and its main transcripts. *a*, Genes that encode related functions are clustered. *b*, Transcription of the circular genome proceeds initially from the promoters, p_L and p_R. Early transcripts (I) terminate at t_L and t_{R1} respectively (some escape termination at t_{R1} and reach t_{R2}). Early transcription provides the *N* gene product, an antiterminator, which alters RNA polymerase so that transcription proceeds through t_L, t_{R1} and t_{R2}. N-dependent transcription (II) is essential for both the lysogenic and lytic pathways. In the lysogenic pathway CII activates three promoters (III); transcription from p_I provides integrase, transcription from p_E (often called p_{RE}, for *R*epressor *E*stablishment) a burst of repressor synthesis, and transcription from p_E and from p_{aQ} delays expression of the late genes by convergent transcription or ANTISENSE RNA. Repressor bound to o_L and o_R blocks all transcription other than that of the *cI* gene; repressor bound to o_R activates transcription from p_M (sometimes called p_{RM} for *R*epressor *M*aintenance) (M). Alternatively, in the lytic pathway Q permits transcription of the late genes by antitermination of a constitutive transcript from $p_{R'}$ (IV), and Cro occupies the operators. G, Major promoters; o, major termination signals.

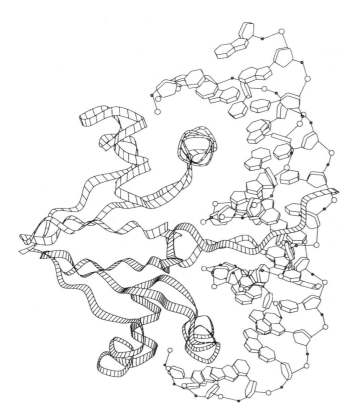

Fig. L5 Interaction of the cro protein with lambda DNA.

lysogenic cells include free repressor molecules that will bind to, and repress, incoming λ genomes. Lysogenic cells are therefore immune to superinfection by λ and the region including the *cI* and *cro* genes and their cognate operators is termed the immunity region. Immunity to superinfection indicates that the repressor of the resident prophage is able to bind to the operators of the superinfecting phage. Two different phages are homoimmune if their repressors recognize the same operator system or heteroimmune if they have a repressor system with a different operator sequence specifity. A phage λ with mutant operators to which repressor is no longer able to bind shows CONSTITUTIVE transcription from p_L and p_R. It can, as a consequence, propagate in a homoimmune lysogen and is said to be virulent.

Phage induction

The normally quiescent prophage is induced by agents that elicit the SOS PATHWAY (e.g. ultraviolet light). Under these conditions the product of the host *recA* gene stimulates cleavage of the λ repressor into its two functional domains: an N-terminal domain that dictates specificity of DNA binding and a C-terminal domain responsible for dimerization of repressor molecules and the co-operative interactions between repressor dimers which facilitate the occupancy of adjacent binding sites in each operator (*see* PROTEIN–NUCLEIC ACID INTERACTIONS). The N-terminal domain by itself can bind DNA but cannot bind another repressor molecule to form a dimer. In the absence of repressor dimers and

the cooperative interactions between them, the concentration of repressor needed to occupy o_L and o_R is greatly increased; hence p_L and p_R are derepressed and the lytic pathway results. Cro is the first gene product to be made and it binds to the operators, initially blocking transcription of *cI* from p_M, hence maintaining the control switch set for the lytic mode.

λ as a transducing phage

The nature of the λ prophage was explained by the Campbell model in which a reciprocal crossover between two circular genomes (the phage genome and the bacterial genome) generates one larger circular genome [1]. λ integrates into the bacterial chromosome at a specific site, the *att* site (*see also* RECOMBINATION). The site-specific integration of the λ genome is catalysed by a phage-encoded enzyme, integrase, but also requires a host protein. Excision of the prophage from the bacterial genome needs an additional phage product, thereby providing the means of directional control of integration and excision. Aberrant excision of circles of DNA including part of the phage genome together with neighbouring bacterial DNA can generate transducing phages: phages able to transfer bacterial DNA from one strain to another (a process called TRANSDUCTION). As the bacterial DNA is limited to sequences that flank the integrated prophage, λ is a specialized transducing phage. Specialized transducing phages offered the first means of molecular cloning of bacterial DNA, before recombinant DNA could be made *in vitro*.

λ as a vector for DNA cloning

The λ genome was readily adapted to serve as a vector for cloning fragments of DNA generated *in vitro* [2] (*see* DNA CLONING; GENETIC ENGINEERING; VECTORS). λ vectors take advantage of the fact that about 40% of the λ genome encodes no function essential for a lytic infection, and phage λ is tolerant of deviations from normal genome size; from 38 to 52 kilobases (kb) is acceptable. The simpler λ vectors (insertion vectors) are deleted for some nonessential DNA (up to ~9 kb) and they retain single targets (cloning sites) for one or more restriction enzymes. A λ genome deleted for all nonessential genes is too small to be packaged and full advantage of the cloning capacity of a λ vector requires a derivative genome with a dispensable 'stuffer' fragment flanked by restriction targets. The stuffer fragment of a λ replacement vector is excised and must be replaced with a donor DNA fragment if the genome is to exceed 38 kb and be large enough to be packaged.

One advantage of λ as a cloning vector is that the *in vitro* process of packaging recombinant λ genomes in phage proheads and the addition of tails, thereby permitting adsorption to *E. coli*, provides excellent recovery of recombinant phage genomes (the efficiency can reach 10% as compared with 0.1% for TRANSFECTION). Packaging extracts include the phage-encoded terminase which can generate linear monomeric genomes from concatemeric DNA; cutting and packaging require that *cos* sequences are 38–52 kb apart. *In vitro* packaging thus offers an efficient route for the recovery of libraries of recombinants. Each recombinant genome packaged *in vitro* can produce a plaque, that contains a clone of recombinant phage. Libraries of plaques can be screened in all the usual ways (*see* GENETIC ENGINEERING), but hybridization with a PROBE sequence is very effective as plaques contain much free DNA in addition to ~10^6 phage particles.

λ as a model system

The study of λ and its interactions with *E. coli* has uncovered many findings of general relevance to molecular biology. The identification, isolation and characterization of the first restriction and modification enzymes (*see* DNA METHYLATION; RESTRICTION AND MODIFICATION; RESTRICTION ENZYMES) followed the detection of the phenomenon of host controlled restriction as a barrier that reduces the efficiency of transfer of phage from one bacterial strain to another. Integration host factor (IHF), a protein that binds to and bends DNA, was recognized by its essential role in site-specific integration, and the CHAPERONINS, encoded by the bacterial genes *groEL* and *S*, by their role in phage morphogenesis. The capacity to recover recombinant λ genomes by packaging them into phage capsids *in vitro* also resulted from studies of phage morphogenesis. The use of λ to study recombination identified special nucleotide sequences (CHI SEQUENCES) in phage and bacterial DNA which are essential if DNA is to be a good substrate for the major general RECOMBINATION system of the host (the RECBCD PATHWAY).

The analysis of gene regulation in λ has revealed different levels and aspects of gene control, for example, the role of TRANSCRIPTIONAL ACTIVATOR proteins (CI and CII), and the importance of ANTITERMINATION (*see* Fig. L4) and of specific downstream sequences in RETROREGULATION at a post-transcriptional level.

Study of the repressors of λ and other LAMBDOID PHAGES has pioneered the analysis of TRANSCRIPTION FACTORS on the basis of their subdivision into distinct functional and structural DOMAINS. λ repressor can be separated into two domains, one involved in DNA recognition and binding and the other in cooperative interactions with other λ repressor molecules. It is now known that the DNA-binding domain also interacts with RNA polymerase to activate transcription. Subdivision into functional domains and cooperative interaction between regulatory proteins have since been found for other transcription factors in both bacteria and eukaryotes. Detailed analysis of the interaction of λ repressor with DNA has also revealed how these proteins can recognize specific nucleotide sequences by interaction with amino acids within the helix-turn-helix motif typical of many DNA-binding proteins (*see* PROTEIN–NUCLEIC ACID INTERACTIONS; TRANSCRIPTION FACTORS).

Key features of a genetic switch, as exemplified by the switch that determines whether lysogeny or the lytic pathway will ensue, are now understood in terms of the different affinities with which the different regulatory proteins, repressor and Cro, bind to the same tripartite operators [3].

N.E. MURRAY

See also: COSMIDS; HEAT SHOCK; MOLECULAR CHAPERONES; PHASMIDS.

1 Hendrix, R. et al. (Eds) (1983) *Lambda II* (Cold Spring Harbor Laboratory, New York).

2 Murray, N.E. (1991) Special uses of λ phage for molecular cloning. *Methods Enzymol.* **204**, 280 (Academic Press, New York).

3 Ptashne, M. (1987) *A Genetic Switch* (Cell Press, Cambridge MA/Blackwell Scientific Publications, Oxford).

lambda Headwords with the prefix λ may also be indexed under the initial letter of the headword itself.

lambda (λ) light chain One of the two mutually exclusive types of LIGHT CHAINS found in ANTIBODIES, the other being called kappa (κ). *See also*: IMMUNOGLOBULIN GENES.

lambdoid phages BACTERIOPHAGES related to and including bacteriophage LAMBDA. As originally defined, they share three properties: they recombine when intercrossed, their DNA molecules have identical pairs of COHESIVE ENDS, and their PROPHAGES are inducible by ultraviolet irradiation. The lambdoid family is now extended to all phages that are able to exchange genetic information with λ by homologous recombination; this includes the *Salmonella* phage P22.

lamellipodium *See*: CELL MOTILITY.

lamin An INTERMEDIATE FILAMENT PROTEIN. *See also*: NUCLEAR ENVELOPE.

lamina *See*: BASAL LAMINA; NUCLEAR ENVELOPE.

laminin A protein of the extracellular matrix (*see* EXTRACELLULAR MATRIX MOLECULES), a component of basal laminae, which binds to type IV collagen, heparan sulphate, and cell-surface laminin receptors. *See also*: CAENORHABDITIS NEURAL DEVELOPMENT; CELL–MATRIX INTERACTIONS; INSECT NEURAL DEVELOPMENT.

lampbrush chromosome Type of CHROMOSOME visible at MEIOSIS in some species, and most easily detected and studied in amphibian oocytes. The chromosomes are in the stage at which two pairs of SISTER CHROMATIDS are held together at CHIASMATA. Lateral loops of chromosomal material extrude from the structure at certain positions giving the chromatid the appearance of an old-fashioned 'lampbrush'. These structures are one of only a few states in which transcription can be visualized directly, as the chromosomes are in a highly extended form permitting the observation of individual TRANSCRIPTION UNITS.

Langer–Giedion syndrome *See*: CONTIGUOUS GENE SYNDROMES.

Langerhans cells ANTIGEN-PRESENTING CELLS found in the skin and parts of the intestinal tract.

LAR LEUKOCYTE COMMON ANTIGEN RELATED PROTEIN.

large subunit binding protein (LSU-BP) A CHLOROPLAST protein that is thought to act as a MOLECULAR CHAPERONE in the assembly of RIBULOSE BISPHOSPHATE CARBOXYLASE/OXYGENASE (RUBISCO) from its subunits. It is closely related to the GroEL protein of the bacterium *Escherichia coli* (46% sequence similarity), is nuclear encoded, and is synthesized as a precursor form in the cytoplasm. *See also*: HEAT SHOCK; PROTEIN TRANSLOCATION.

large T antigen *See*: PAPOVAVIRUSES.

lariat structure A loop-like structure formed during the RNA SPLICING of mRNAs. Following cleavage of the 5′ junction of the intron, the lariat is formed by the 5′ end of the intron joining via an unusual 5′–2′ PHOSPHODIESTER bond to a site near the 3′ intron–exon junction. The 3′ junction is then cleaved and the exons are covalently linked. The excised intron lariat is released by the reaction but is rapidly turned over.

late genes/late region Genes in BACTERIOPHAGES and other viruses which encode functions required after genome replication, such as the production of viral coat proteins and cell lysis. *See also*: PAPOVAVIRUSES.

late-replicating X-chromosome *See*: X-CHROMOSOME INACTIVATION.

lateral inhibition Developmental mechanism for spacing structures at regular intervals which depends on the structure (e.g. a bristle) producing an inhibitor which prevents formation of a similar structure within a certain area around it. *See*: BRISTLE PATTERN; INSECT NEURAL DEVELOPMENT; PATTERN FORMATION.

lateral plate mesoderm The portion of the MESODERM of the trunk of vertebrate embryos lying lateral to the INTERMEDIATE MESODERM. The lateral plate mesoderm starts to form during GASTRULATION and later subdivides into two plates: one dorsal, called the SOMATOPLEURE and one ventral, called the SPLANCHNOPLEURE. The gap between them is called the COELOM, a cavity which characterizes higher animals. The lateral plate mesoderm gives rise to the circulatory system including the heart, and contributes to many other organs and tissues.

lateral signalling Cell–cell signalling between members of an EQUIVALENCE GROUP. *See*: CAENORHABDITIS DEVELOPMENT.

lattice A crystal can be considered as a three-dimensional lattice of points, each in an identical environment to all others. The spacings between lattice points are the repeat distances (a, b, c) of the regular crystalline stacking in three dimensions; these are the edges of the primitive unit cell. Translational SYMMETRY in a crystal may give more than one lattice point per conventional unit cell; in such cases the crystal has a centred lattice, generating a centred unit cell, as can be recognized from the pattern of SYSTEMATIC ABSENCES within the diffraction pattern. The coordinate system of a conventional unit cell reflects its full symmetry; there are 14 distinct Bravais lattices in three dimensions. All centred lattices can generate primitive unit cells by suitable choice of coordinate axes (which may then be no longer parallel to any of the symmetry elements). *See*: DIFFRACTION; MOLECULAR AVERAGING; NON-CRYSTALLOGRAPHIC SYMMETRY; POINT GROUP; RECIPROCAL SPACE; SPACE GROUP; SYMMETRY; X-RAY CRYSTALLOGRAPHY.

Laue diffraction An experimental arrangement to study DIFFRACTION with a wide wavelength spectrum, so enabling the simultaneous measurement of many more REFLECTIONS. *See*: X-RAY CRYSTALLOGRAPHY.

LAV (HTLV III) Lymphoadenopathy-associated virus, the human IMMUNODEFICIENCY VIRUS now known as HIV-1.

Lb LEGHAEMOGLOBIN. *See also*: HAEMOGLOBINS.

LB Left border. DNA sequences around the left-most limit of a T-region in the TI PLASMID. It contains a copy of the border repeat sequence and possibly associated sequences which may inhibit T-strand initiation.

LCA Leukocyte common antigen, a protein found on the surface of white blood cells. *See*: CD45; PROTEIN PHOSPHATASES.

LCAM Uvomorulin, a CELL ADHESION MOLECULE of the cadherin family, which is involved in cell–cell binding in epithelial and other tissues, and in embryonic development. *See*: EPITHELIAL POLARITY.

LCAT deficiency *See*: PLASMA LIPOPROTEINS AND THEIR RECEPTORS.

LCR LOCUS CONTROL REGION.

LD50 The median lethal dose, the dose of virus, toxic chemical, etc., at which 50% of test animals die.

LDH LACTATE DEHYDROGENASE.

LDL Low-density lipoprotein. *See*: PLASMA LIPOPROTEINS AND THEIR RECEPTORS.

LDP LONG DAY PLANT.

leader peptidase Enzyme that cleaves leader peptides from the protein precursors of exported bacterial proteins. *See*: BACTERIAL PROTEIN EXPORT.

leader peptide Sequence of ~16–30 amino acids at the N terminus of newly synthesized secretory or membrane proteins which mediates transport across the ENDOPLASMIC RETICULUM membrane or across the bacterial cell membrane, and is subsequently cleaved. Also known as signal peptide. *See*: BACTERIAL PROTEIN EXPORT; PROTEIN TARGETING; PROTEIN TRANSLOCATION; SIGNAL SEQUENCE.

leader sequence Nucleotide sequence at the 5′ end of sequences encoding protein or structural RNA which is transcribed, and in some cases translated, but which does not encode part of the final functional protein or RNA. *See*: PROTEIN TARGETING; PROTEIN TRANSLOCATION; SIGNAL SEQUENCE.

leaky mutation A MUTATION within a gene that fails to abolish the activity of the gene product completely. For example, a leaky AUXOTROPHIC MUTATION would result in the mutant cell still being able to grow in the absence of the essential nutrient but not to the same extent as the WILD-TYPE cell. Leaky mutations are usually MISSENSE mutations.

least squares method A statistical approach for estimating the 'true value' of one or more parameters from a set of observed values. The principle of least squares states that the most probable value of any observed quantity is such that the sum of the squares of the deviations of the observations from this value is the least. A simple application of this principle shows that the most probable value of a quantity measured many times is the arithmetic mean of these measured values. An extension of it shows that, where observations have different errors, the weighted mean is appropriate: where each measurement is weighted by the reciprocal of the square of its standard error.

The least squares approach is widely used to fit a model to a set of observed data, or graphically to fit a straight line, or a curve, to a set of points. When the data variables (x, y) have a linear relationship, for example:

$$y_i = ax_i + b$$

then the least squares values of *a* and *b* can be precisely determined, given a choice as to whether the observations have errors in their values of y_i or x_i or both. When the relationship between the variables is nonlinear, the parameters can only be determined by successive approximation starting from a set of trial values. This can become an enormous computational problem if there are many observations and parameters to be fitted. For example, in the refinement of a crystallographic protein structure there may be 10^5 observations and several thousand parameters describing the atomic model to be fitted to these observations. *See:* X-RAY CRYSTALLOGRAPHY.

Leber's hereditary optic neuropathy *See:* MITOCHONDRIOPATHIES.

LECAMs *See:* SELECTINS.

lecithin Phosphatidylcholine. *See:* MEMBRANE STRUCTURE; PHOSPHOLIPIDS.

Lectins

LECTINS are proteins, normally GLYCOPROTEINS, that bind SUGARS and agglutinate red blood cells. They should be distinguished, however, from the numerous immunoproteins (ANTIBODIES) and enzymes that may also bind sugars. (However, some lectins are now said to have enzyme activity as glycosidases.)

Agglutination occurs because there are at least two binding sites for the sugar on each lectin molecule and this results in the formation of cross-bridges between cells. Lectins may, however, bind to cells without agglutination. Although binding to red blood cells has traditionally been one of the ways of distinguishing

lectins, a few do not in fact agglutinate red blood cells and some agglutinate bacterial and other cells rather than red blood cells [1]. It is the sugar-binding property that is the predominant feature of lectins and is responsible for their biological actions and their use in biological experimental techniques. Sugar binding, in conjunction with the related agglutination action, serves to identify them in tissue extracts and facilitates their subsequent isolation.

Statements to define lectins are fairly straightforward for general purposes, but the term can be extended to accommodate a whole class of sugar-binding proteins. A definition must make the term useful by grouping together materials with similar properties and distinguishing them from other substances. Some groups will overlap, with some substances falling into more than one group; some substances will be more conveniently grouped together by an extension of the definition so that they are all included and thereby increase the usefulness of the term.

Thus, although the original definition of lectin specified the agglutination of red blood cells, the term now incorporates those proteins that agglutinate other cells and also some proteins that are not known at present to agglutinate any cells at all, but which do bind sugars and have stretches of amino-acid sequence in their polypeptide subunits which are similar to those of more characteristic lectins.

The definition proposed by Goldstein et al. [2] and adopted by the Nomenclature Committee of the International Union of Biochemistry states that 'a lectin is a sugar-binding protein of non-immune origin that agglutinates cells or precipitates glycoconjugates'. This definition provides positive and easily testable properties for identifying possible lectins. Individual lectins are usually named after the organism, in most cases a plant, from which they were obtained (Table L1).

Source

Lectins were originally found in seeds of leguminous plants, where they occur mainly in the protein bodies of the cotyledons of the embryo at the later stages in seed maturation. Similar sugar-binding proteins have now been found in the seeds of a large number of plants and also in plant tissues other than seeds. Wheat-germ agglutinin, for example, is localized at the external surface of the embryo, in the embryonic roots and coleorhiza, the root cap of the radical, the scutellum and other tissues of the embryo and developing seedling; it is not found in the endosperm. The amounts and distribution of lectins vary with the growth and development of the seed and the plant. In vegetative tissue some lectins are localized in the cell wall. Lectins are also found in animal and bacterial cells (*see* ANIMAL LECTINS; SELECTINS).

Structure and properties

Lectins usually consist of two or four identical polypeptide subunits. When differences between the subunits are found, however, they can be quite marked. There is usually one sugar-binding site per subunit and these binding sites are normally for the same sugar, are all of the same type and do not interact with each other.

Lectins composed of different subunits can be found in different

Table L1 Lectins from various organisms

Organism	Specificity of sugar bound	No. of subunits
Anguilla anguilla (eel)	α-L-Fuc	2
Arachis hypogaea (peanut)	β-D-Gal(1-3)-D-GalNAc	4
Canavalia ensiformis (jack bean) Concanavalin A	α-D-Man, α-D-Glc	4
Datura stramonium (thorn apple)	(D-GlcNac)$_n$	2
Glycine max (soybean)	D-GalNAc	4
Helix pomatia (edible snail)	D-GalNAc	6
Lens culinaris (lentil)	α-D-Man	2
Limulus polyphemus (horseshoe crab)	NeuNAc	–
Pisum sativum (pea)	α-D-Man	4
Pseudomonas aeruginosa	D-Gal	–
Ricinus communis (castor bean)	β-D-Gal	4
Triticum vulgaris (wheat) Wheat germ agglutinin	(D-GlcNAc)$_n$, NeuNAc	2
Vicia faba (broad bean)	D-Man, D-Glc	4
Vigna radiata (mung bean)	α-D-Gal	4

Fuc, fucose; Gal, galactose; GalNAc, *N*-acetylgalactosamine; Glc, glucose; GlcNAc, *N*-acetylglucosamine; Man, mannose; NeuNAc, *N*-acetylneuraminic acid.

forms — isolectins — arising from various combinations of the monomers in the complete dimer or tetramer. Subunits may differ in their amino-acid sequence and, if the lectin is a glycoprotein, as most of them are, it may also differ in the nature and linkages of the sugars in the attached oligosaccharide side chains.

Differences in primary structure (amino-acid sequence) of the subunits can arise by transcription of separate genes on the same or different chromosomes, or by POST-TRANSLATIONAL MODIFICATION. The latter is the means of producing certain lectins in which there is extensive HOMOLOGY between the two subunits (e.g. a light α chain and a heavy β chain). These are produced from an initial precursor polypeptide of the type NH$_2$–signal sequence–β chain–α chain–COOH which is cleaved into the two subunits after translation. This proteolysis can take place in protein bodies where some of the seed lectins are stored. When homologies occur in the amino-acid sequence of the subunits it is likely that the separate genes have been formed by GENE DUPLICATION. In certain families of plants rich in lectins, such as the legumes, extensive homologies occur in the primary sequences of the different lectins, as for example between pea and lentil lectins and concanavalin A from the jack-bean.

Those lectins that are glycoproteins usually have an attached carbohydrate side chain of the type acetyl-*N*-glucosamine attached to a mannose-rich oligosaccharide core which carries peripheral sugars such as galactose and fucose. The carbohydrate side chain is linked to the protein by an *N*-glycosidic bond via asparagine. These sugars are added to the protein within the lumen of the ENDOPLASMIC RETICULUM and GOLGI APPARATUS.

The core oligomer is donated via a dolichol derivative to the protein during its translation and passage into the endomembrane system (*see* GLYCOPROTEINS). Modifications and additions to this core are subsequently completed in the Golgi apparatus [3]. Other lectins such as those of potato have oligosaccharides of arabinofuranose glycosidically linked by β-bonds to the hydroxyl group of hydroxyproline and also galactose units linked by *O*-glycosidic bonds to serine. The structure of these lectins is very similar to that of the hydroxyproline-rich proteins found as structural components of plant cell walls. The presence or absence of the sugar oligosaccharide side chains seems to have little effect on the lectin activity.

Nearly all lectins contain metal ions, usually Mg^{2+} and/or Mn^{2+}, which are bound to the protein at specific sites. The agglutination and sugar-binding properties of the lectin are in some cases dependent on the presence of these metal ions.

Sugar binding

Binding of lectins to sugars represents one of many interactions between sugars and proteins. Sugars are bound to proteins for the function of carbohydrate-metabolizing enzymes, SUGAR TRANSPORTERS, CHEMOTAXIS receptors, HORMONES and INTERFERONS. An important factor in the association is the conformation of the oligosaccharide that is bound. Proteins bind the sugar molecules in a definite groove within their three-dimensional conformation (Fig. L6). Generally, lectins bind to oligosaccharides rather than to single sugars. The carbohydrates are held in a ribbon-like configuration within the binding groove by van der Waals forces and hydrogen bonding, involving the —OH groups of the sugars as both donors and acceptors. Precise and stereospecific interactions are formed and maintained by the orientation of hydrogen-

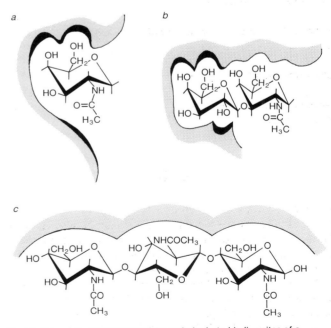

Fig. L6 Schematic presentation of the carbohydrate-binding sites of *a*, soybean; *b*, peanut; and *c*, wheat-germ agglutinins. After [1].

bonding residues which are in turn fixed by complex hydrogen bond networks with other residues within the binding sites. Extensive networks of hydrogen bonds are thereby formed that extend outward from the carbohydrate ligand to at least three shells of amino-acid residues in the protein. During binding, conformational changes may occur which allow the carbohydrate to be oriented for binding to progress and which result in specific interaction between protein and carbohydrate [4].

Binding sites for a particular carbohydrate can occur on different proteins — on several enzymes and lectins for example. Very often, all these proteins will have a common homologous amino-acid sequence, suggesting that this sequence represents the site for the binding property of these proteins.

Additional binding sites specific for non-sugar ligands and responsible for precise protein–protein interaction are known to be present, sometimes as hydrophobic domains, on a number of lectins. A great many lectins have nonpolar regions near the sugar-binding sites and these can participate in the binding of sugar derivatives with hydrophobic substituents such as aglycones. Since these sites have been conserved during the evolution of the lectins they may have some functional significance. It might be possible, for instance, for some control of biological activity to be exerted at these domains. Both gibberellic and abscisic acids affect seed germination (*see* PLANT HORMONES) and they also influence the production of some seed lectins. The hydrophobic regions may also give rise to wrong interpretations as to the specificity and existence of lectin-like activity if hydrophobic glycosides are used as probes of activity.

Biological functions

The biological functions of lectins are by no means certain but all suggestions for their biological activity are based mainly on their capacity to recognize and target particular oligosaccharides. This makes them very obvious candidates for attachment and recognition at cellular membranes as the sugar portions of glycoproteins and glycolipids are exposed at the outer surface of the protein–lipid bilayers that constitute these membranes (*see* MEMBRANE STRUCTURE).

Various phenomena of cell–cell recognition are attributed to lectin–sugar binding. These include bacterial and fungal attachment to specific cells during early stages of infection or initiation of plant–symbiont relationships (*see* NODULATION; PLANT PATHOLOGY). Paradoxically, the suggestion has also been made that binding of plant lectins to the cell walls of parasitic microorganisms could provide protection against bacteria and fungi.

Mitogenic stimulation of some cells is triggered after the lectin binds to the cell surface and this is very apparent when phytohaemagglutinin, from *Phaseolus vulgaris*, is attached to quiescent, nondividing lymphocytes. However, no examples of this phenomenon are known for the plant cells in which the lectin is formed. The toxicity of some lectins such as ricin is due to the different activities of the two different polypetides of their subunits. One is a lectin and binds to the cell surface, bringing about the entry of the other, toxic, polypeptide into the cell.

Incomplete glycoproteins are removed from the bloodstream of mammals by 'acceptors' (lectin-like properties) on the external surface of liver cells, which recognize the atypical terminal sugars of the incomplete carbohydrate side chains and bring about their removal by pinocytosis after attachment (*see* ASIALOGLYCOPROTEIN; C-TYPE ANIMAL LECTINS). A family of proteins with lectin activity — the SELECTINS — are also involved in cell adhesion between leukocytes and endothelium in inflammatory reactions.

Applications

In spite of the questions surrounding the biological function of lectins their properties make them extremely important components of many techniques in cell biology and biochemistry. They are used extensively, for example, in the procedures for glycoprotein isolation, blood typing, glycoprotein identification at membrane surfaces, the identification and separation of cells and cytochemical localization of glycoproteins in cells, and also as MITOGENS.

D.H. NORTHCOTE

See also: PROTEIN STRUCTURE.

1 Lis, H. & Sharon, N. (1986) Lectins as molecules and tools. *Annu. Rev. Biochem.* **55**, 35–67.
2 Goldstein, I.J. et al. (1980) What should be called a lectin? *Nature* **285**, 66.
3 Kornfield, R. & Kornfield, S. (1985) Assembly of asparagine-linked oligosaccharides. *Annu. Rev. Biochem.* **54**, 631–664.
4 Quiocho, F.A. (1986) Carbohydrate-binding proteins: tertiary structures and protein–sugar interactions. *Annu. Rev. Biochem.* **55**, 287–315.

leech development *See*: ANNELID DEVELOPMENT.

leech nervous system The nervous system of the leech (Phylum Annelida, Class Hirudinea) contains almost identical segmental ganglia, each containing around 350 cells. Individual sensory and motor neurons have been identified, the synaptic terminations are known, and their fields of innervation have been accurately mapped. Because of this detailed knowledge, and for many other technical reasons, the leech nervous system is frequently used as an experimental model. *See also*: ANNELID DEVELOPMENT.

leghaemoglobin A plant-encoded member of the globin superfamily (*see* HAEMOGLOBINS) found in the nitrogen-fixing root nodules of leguminous plants. It is very similar in overall structure to MYOGLOBIN (Fig. M51) but has only a 15% similarity in amino-acid sequence. Leghaemoglobin has a very high oxygen affinity and protects the oxygen-sensitive nitrogenase of the nitrogen-fixing bacteria (*see* NITROGEN FIXATION). Similar proteins have also been found in nonlegumes. Leghaemoglobin genes in both legumes and nonlegumes differ from the vertebrate globin genes in having a third intron positioned between the two introns characteristic of genes of the globin superfamily.

Legionella Only genus in the bacterial family Legionellaceae. *Legionella* spp. are small motile Gram-negative rods, widely distributed in most environments. *Legionella pneumophila* is the causal agent of Legionnaire's disease, an acute purulent pneumonia. *See*: BACTERIAL PATHOGENICITY.

Leigh syndrome *See*: MITOCHONDRIOPATHIES.

Leishmania Genus of parasitic protozoa. *See*: GLYCOLIPIDS.

lens development *See*: EYE DEVELOPMENT.

lentiviruses *See*: RETROVIRUSES.

Lepore gene *See*: HAEMOGLOBIN LEPORE.

leptotene A stage in MEIOSIS.

Lesch–Nyhan syndrome A rare X-LINKED RECESSIVE condition characterized by neurological symptoms — mental retardation, spastic cerebral palsy and self-mutilation — and by excessive accumulation of uric acid in the blood resulting in urinary tract stones, arthritis, and gout. The disorder is due to mutations in the gene encoding the enzyme HGPRT (hypoxanthine : guanine phosphoribosyltransferase) located on the long arm of the X-chromosome. The uric acid linked symptoms are explicable by the breakdown of metabolites that accumulate owing to the absence of HGPRT; the neurological symptoms are of unknown origin. Affected individuals rarely live beyond early childhood.

lethal giant larva (l(2)gl) locus In *Drosophila*, mutations of the recessive tumour gene *l(2)gl* cause failure of larvae to metamorphose and overgrowth of some imaginal primordia such as the brain and imaginal disks. The product of the gene (a protein of M_r 130 000) is expressed on the cell surface with a wide tissue distribution. It shows homology to the vertebrate cadherin family of CELL ADHESION MOLECULES but lacks a transmembrane domain.

lethal mutation A gene that has undergone a lethal mutation is typically incapable of producing an active form of an indispensable protein. This is incompatible with survival in a haploid organism. In a diploid organism lethal mutations are characterized as those that kill the organism directly, usually early in development (embryonic lethal) or prevent it from reproducing. Lethal mutations may be DOMINANT or RECESSIVE.

leucine (Leu, L) An AMINO ACID with an uncharged hydrophobic side chain.

leucine zipper A protein sequence motif which is involved in protein–protein interactions and protein–DNA binding. The region involved in protein dimerization consists of leucine residues at intervals of seven residues, and is flanked by a basic region involved in DNA binding. *See*: PROTEIN–NUCLEIC ACID INTERACTIONS; TRANSCRIPTION FACTORS.

leucinopine An OPINE made by the conjugation of leucine and 2-oxoglutarate. Unlike most ketoglutarate-based opines which exist in the D,L- form, leucinopine exists in the L,L- form. A minor opine, apparently of poor nutritional value to *Agrobacterium*.

leucoplast A form of PLASTID that lacks pigments and occurs in nonphotosynthetic cells of higher plants. Called an amyloplast when it contains starch grains.

leukaemia A malignant disease due to an abnormal proliferation of white blood cells which can be detected in the peripheral blood and/or bone marrow. The disorder may be rapidly fatal in absence of treatment (acute leukaemia) or slowly progressing (chronic leukaemia). The myeloid or lymphoid lineages may be involved (so-called myeloid leukaemia or lymphoid leukaemia). In some acute cases, however, malignant cells may have mixed characteristics from the two main lineages. *See also*: CHROMOSOME ABERRATIONS; ONCOGENES.

leukocyte White blood cell. The term covers all cells of the myeloid and lymphoid lineages. *See*: HAEMATOPOIESIS.

leukocyte-activating factor (LAF) Interleukin 1α. *See*: IL-1; LYMPHOKINES.

leukocyte adhesion deficiency (LAD) Very rare AUTOSOMAL RECESSIVE disorder characterized by recurrent bacterial and fungal infection and impaired wound healing. The underlying defect is a deficiency in expression of the β-INTEGRIN CD18 which causes impairment in adhesive and phagocytic functions of leukocytes.

leukocyte common antigen (LCA) *See*: CD45; PROTEIN PHOSPHATASES.

leukocyte common antigen related protein (LAR) A transmembrane glycoprotein of 1881 amino acids, a tyrosine phosphatase found in a wide variety of cells. It has an extracellular segment with three IMMUNOGLOBULIN-LIKE DOMAINS and nine FIBRONECTIN TYPE III REPEATS, a transmembrane segment and an intracellular segment with two tyrosine phosphatase catalytic domains (*see* PROTEIN PHOSPHATASES). As its extracellular section is structurally similar to the neural adhesion molecules NCAM and fasciculin II (*see* CELL ADHESION MOLECULES), it has been suggested that LAR may have a role in the contact inhibition of cell growth.

leukopenia Reduction in the number of circulating white blood cells (leukocytes), reflecting predominantly a reduction in numbers of NEUTROPHILS and LYMPHOCYTES.

leukosis (leukocytosis) An increase in circulating white blood cells. This may reflect an expansion in the numbers of any of the five types of leukocyte — NEUTROPHIL, LYMPHOCYTE, EOSINOPHIL, BASOPHIL, or MONOCYTE — but usually refers to increased numbers of neutrophils or lymphocytes.

leukotrienes *See*: EICOSANOIDS.

levan Fructose homopolymer (e.g. inulin), derived from various plant species. *See*: FRUCTANS.

Leviviridae *See*: BACTERIOPHAGES.

Lewis blood group Lewis antigens are carbohydrate antigens not intrinsic to red cells, and are carried on secreted GLYCOLIPIDS secondarily adsorbed onto red cells and other tissues. Red cell phenotypes Le(a + b −), Le(a − b +), and Le(a − b −) depend on the interaction of genes at the *LE* locus and *SE(FUT2)* locus; these loci are at different locations on chromosome 19. The *LE* gene product, an α-4-L-fucosyltransferase, acts on Type 1 precursor oligosaccharide molecules in the ABO biosynthetic pathway (*see* ABO BLOOD GROUP SYSTEM), to produce Le^a, Le^b, ALe^b, or BLe^b antigens. Lewis antibodies are more important in tissue transplantation than in blood transfusion. The type 2 isomers of Le^a and Le^b antigens are encoded by an α-3-L-fucosyl transferase, the product of the *X* gene which acts on a Type 2 structure to produce x and y antigens. x antigen (CD15) and sialylated x antigen appear to be ligands for molecules mediating leukocyte adhesion to endothelium and platelets (*see* SELECTINS).

Leydig cell Testosterone-producing CONNECTIVE TISSUE cells lying between the tubules of the testis. These cells have extensive smooth ENDOPLASMIC RETICULUM in the membranes of which are located the enzymes required for the synthesis of CHOLESTEROL and its steroid-hormone derivative, testosterone.

LFA-1 *See:* INTEGRINS.

LFA-3 (CD58) A cell-surface transmembrane protein (M_r 55 000–70 000) expressed on a wide variety of cell types. It has been identified as a ligand for the CD2 cell-surface glycoprotein of T cells and thymocytes. *See:* CD ANTIGENS.

lgp Lysosomal glycoprotein, a major component of lysosomal membranes. *See:* LYSOSOMES.

LH *See:* LUTEINIZING HORMONE.

LH1/LH2 Light-harvesting bacteriochlorophyll proteins of purple photosynthetic bacteria. LH1 is closely associated with the reaction centre and contains bacteriochlorophyll molecules absorbing at 875 nm, whereas LH2 occurs in variable stoichiometry with the reaction centre and contains bacteriochlorophyll molecules absorbing at 800 and 850 nm. They consist of heterodimers of small α and β subunits of M_r 5000–7000. *See also:* CHLOROPHYLL–PROTEIN COMPLEXES; LIGHT-HARVESTING COMPLEXES; PHOTOSYNTHESIS.

LHC LIGHT-HARVESTING COMPLEX; **LHCI** LIGHT-HARVESTING COMPLEX I; **LHCII** LIGHT-HARVESTING COMPLEX II.

LHCII kinase A thylakoid-bound enzyme that catalyses the phosphorylation of a fraction of the LIGHT-HARVESTING COMPLEX II (LHCII) in plant chloroplasts. The phosphorylated complexes move out into the stroma lamellae where they probably associate functionally with PHOTOSYSTEM I units. The enzyme is activated in the light by reduction of the PLASTOQUINONE pool (and probably a component of the CYTOCHROME BF COMPLEX). This phenomenon is the basis of the transition between 'State I' and 'State II', which is involved in the regulation of the distribution of light energy between the two photosystems. *See also:* PHOTOSYNTHESIS.

LHCP LIGHT-HARVESTING CHLOROPHYLL PROTEIN; an older name for LIGHT-HARVESTING COMPLEX II.

Li-Fraumeni syndrome *See:* ONCOGENES.

LIAC LIGHT-INDUCED ABSORPTION CHANGE.

library *See:* DNA LIBRARY.

ligand-gated ion channel(s) ION CHANNELS whose opening or closing is stimulated by binding of a ligand (*cf.* MECHANICALLY GATED ION CHANNELS; VOLTAGE-GATED ION CHANNELS). They include neurotransmitter receptors (*see* EXCITATORY AMINO ACID RECEPTORS; GABA AND GLYCINE RECEPTORS; NICOTINIC RECEPTOR) which are responsive to extracellular ligands, and channels that open in response to intracellular ligands, such as ions (*see* CALCIUM) or nucleotides (*see* VISUAL TRANSDUCTION).

ligand-gated ion channel superfamily A group of cloned receptor subunits that share common structural and functional motifs. Members include subunits of NICOTINIC RECEPTORS, GABA$_A$ AND GLYCINE RECEPTORS and some EXCITATORY AMINO ACID RECEPTORS. Where evidence is available, receptor complexes within this group are hetero-oligomers of five membrane-spanning subunits — in the case of the nAChR, two α, and one each of a β, δ, and γ (or ε) subunit. The complete receptor complex forms an ion channel which is opened or closed by the action of specific ligands (e.g. the neurotransmitters glutamate or acetylcholine).

Each subunit of the ligand-gated ion channel superfamily is thought to have four membrane-spanning stretches. At least one transmembrane region must form the lumen of the ion channel and the second transmembrane region is believed to fulfil this function, and is indeed the region of highest conservation in most known ligand-gated ion channels. The excitatory amino-acid receptors, however, are less conserved than other members of the family.

ligand-induced endocytosis Type of receptor-mediated ENDOCYTOSIS in which ligand binding to cell surface receptors induces receptor internalization. Unoccupied receptors are not internalized.

ligase(s) One of the main classes of enzymes (EC class 6) in the EC classification. They catalyse the joining together of two molecules coupled with the cleavage of a pyrophosphate bond of ATP or a similar triphosphate. They include the DNA LIGASES, RNA LIGASES, and AMINOACYL-tRNA SYNTHETASES.

ligation The process whereby a 5′–3′ PHOSPHODIESTER BOND is formed between two DNA fragments; usually mediated by a DNA LIGASE.

ligation chain reaction A technique using the POLYMERASE CHAIN REACTION.

light chains (1) The smaller of the two types of polypeptide chain present in immunoglobulin molecules, with an M_r of 25 000. There are two antigenically and genetically distinct forms of human light chain (L chain) — named kappa (κ) and lambda (λ) to denote their discovery by Korngold and Lipari. The Greek letters are used in accordance with the nomenclature of immunoglobulin chains (*see* ANTIBODIES). An immunoglobulin monomer contains two identical light chains, and light chain type is independent of the immunoglobulin class (determined by the heavy chain). Light chains have an N-terminal variable region (V_L), which contributes to the antibody combining site, and a C-terminal constant region (C_L). Light chains are covalently bound to the heavy chains by a single disulphide bond between the C_L and C_{H1} domains. There is ~40% sequence homology between the C_L regions in human κ and λ chains, suggesting a common evolutionary origin.
(2) MYOSIN light chain.
(3) CLATHRIN light chain.

light-harvesting chlorophyll protein (LHCP) The various kinds of chlorophyll–protein complexes involved in light harvesting (the absorption of light energy) in PHOTOSYNTHESIS. Each photosynthetic reaction centre usually contains 50–300 CHLOROPHYLL molecules whose function is to absorb light quanta and to pass the excitation energy to the special chlorophyll pair in the reaction centre where the primary charge separation takes place (*see* ANTENNA PIGMENTS). *See*: LIGHT-HARVESTING COMPLEXES I and II.

light-harvesting complex (LHC) Any pigment–protein complex that is relatively loosely associated with a photosynthetic reaction centre and which may occur in a variable ratio to it, and whose function is to absorb light energy and pass it on rapidly to pigments closer to the reaction centre. *See*: LIGHT-HARVESTING COMPLEXES I and II; PHOTOSYNTHESIS.

light-harvesting complex I (LHCI) The CHLOROPHYLL *a/b* protein complex that acts as the light-harvesting antenna for PHOTOSYSTEM I in green plants. It is encoded by a family of nuclear genes related to those of other CHLOROPHYLL *A/B* BINDING PROTEINS. It contains four polypeptides of M_r 24 000, 21 000, 17 000 and 10 000, and has a chlorophyll *a* : *b* ratio of 3 : 4. *See also*: PHOTOSYNTHESIS.

light-harvesting complex II (LHCII) The CHLOROPHYLL *a/b* protein that acts as the light-harvesting ANTENNA for *photosystem II* in green plants. It is encoded by a family of nuclear genes related to those of other CHLOROPHYLL *A/B* BINDING PROTEINS. It contains two major polypeptides and a minor one (all in the range M_r 25 000–28 000). Each polypeptide contains 5–7 chlorophyll *a*, 4–6 chlorophyll *b*, and 2–3 xanthophyll molecules (chlorophyll *a* : *b* ratio = 1.2 : 1). The amount is variable, but it is usually the major chlorophyll protein and can represent up to 50% of the thylakoid membrane protein. *See also*: PHOTOSYNTHESIS.

light-induced absorption change (LIAC) Phenomenon in which the absorbance of a preparation at a specific wavelength is altered (increased or decreased) by exposure of that preparation to radiation of a similar or different wavelength. Usually, the term LIAC is used for observations of cell-free preparations containing membrane components, in which changes in the absorption properties of CYTOCHROMES or FLAVOPROTEINS may be induced by light. Such LIACs commonly indicate electron transport events (*see e.g.* ELECTRON TRANSPORT CHAIN; PHOTOSYNTHESIS).

lignans Formed structurally from the linkage of two C6–C3 units, to give the general structure (I), which may further cyclize.

(I)   podophyllotoxin

They occur widely in the plant kingdom in heartwood, leaves and other organs, and resinous extracts. Lignans are of pharmacological interest because of their cytotoxic and tumour-inhibiting activity. Podophyllotoxin, isolated from *Podophyllum peltatum*, has cytotoxic action like that of COLCHICINE.

lignin *See*: LIGNANS; PLANT CELL WALL MACROMOLECULES.

likelihood A concept central to the field of statistical inference and hypothesis testing. A standard situation in a scientific enquiry is that several hypotheses can be entertained regarding a phenomenon, each of which leads to predicting a different probability distribution for certain measurable quantities. Once a measurement of these quantities has been performed, the likelihood of each hypothesis is the probability it had assigned to the values actually observed. Likelihood is a natural figure of merit by which to rank hypotheses in the light of observed data. It can be combined with a measure of the *a priori* probability of each hypothesis by means of Bayes's theorem. Likelihood is used in the MAXIMUM ENTROPY METHOD of direct phase determination to test alternative phase assumptions against each other and to detect significant phase indications. *See also*: MAXIMUM LIKELIHOOD.

Bricogne, G. (1993) *Acta Cryst.* **D49**, 37–60.
Lindley, D.V. (1965) *Introduction to Probability and Statistics from a Bayesian Viewpoint*. 2 Vols. *Part 1: Probability, Part 2: Inference* (Cambridge University Press, Cambridge).

LIM domain Zinc-binding protein domain first identified in transcription factors Lin-II (*Caenorhabditis elegans*), Isl-1 (rat) and Mec-3 (*C. elegans*).

limb development (1) Invertebrate. Limb development in invertebrates has been studied in particular using the regeneration of the cockroach limb as a model system. Studies in this system indicate a proximo-distal system of positional values. Cutting out a portion of the tibia and joining the cut ends puts cells from different levels in apposition, growth is stimulated and the missing parts are replaced. The new cells acquire positional values smoothly interpolated between those of the two sets of cells brought into confrontation. This intercalary regeneration is readily modelled by a gradient of positional information. The molecular basis of this gradient is unknown but genes involved in limb formation, such as *Distal-less*, are being characterized in *Drosophila*.

(2) Vertebrate. The developing vertebrate limb has provided a popular model system for the study of PATTERN FORMATION. The limb originates from the LATERAL PLATE MESODERM as a bud of MESENCHYME cells with an ectodermal covering. Proximo-distal growth of the bud is controlled by the APICAL ECTODERMAL RIDGE (AER), a prominent region of the ectodermal epithelium at the distal tip of the bud, the AER in turn controlling the growth and maintenance of the subjacent population of mesoderm cells, the PROGRESS ZONE. Tissue patterning along the anterior–posterior axis (thumb-to-little finger axis) is orchestrated by the ZONE OF POLARIZING ACTIVITY (ZPA), a subpopulation of mesoderm cells positioned at the posterior margin of the progress zone. Polarizing activity is demonstrated by the effects of grafting ZPA tissue to new positions in the limb bud; in chick embryos, for example, adding a second ZPA to the anterior margin of the bud, opposite the host ZPA, leads to a mirror image duplication of the limb pattern. It has been proposed that the ZPA is the source of a MORPHOGEN, perhaps retinoic acid; RETINOIDS are detectable in the limb bud, and application of retinoic acid to an anterior position can mimic the effect of a ZPA graft.

Recent evidence suggests that expression of some of the vertebrate Hox genes (*see* HOMEOBOX GENES AND HOMEODOMAIN PROTEINS) in defined areas of the developing limb field could be important for REGIONAL SPECIFICATION of the tissue pattern. Genes of the Hox-1 and Hox-4 complexes in the mouse are activated sequentially, such that more 5′ genes are expressed both after and in successively more postero-distal positions than 3′ genes, resulting in a collection of nested expression domains. This colinearity of gene activation is similar to that described for the HOMEOTIC GENES of *Drosophila*. Mirror image pattern duplication, produced as above, is preceded by parallel duplication of the Hox-4 expression domains, supporting the possibility that the Hox genes encode positional information in the limb. The expression of a particular combination of Hox genes, leading to the activation of further (unidentified) downstream genes, is suggested to mark each cell with a particular phenotype as it leaves the progress zone.

limb regeneration *See*: REGENERATION.

limbic system Loosely defined system in the central nervous system consisting of primitive cortical tissue encircling the upper brain stem and which also includes the amygdala and hippocampus. It is concerned with the generation of emotions.

lin **mutants** *See*: LINEAGE MUTANTS.

lincomycin Glycoside antibiotic from *Streptomyces* sp. which in-hibits bacterial PROTEIN SYNTHESIS, apparently by inhibition of the ribosomal PEPTIDYL TRANSFERASE.

lineage *See*: CELL LINEAGE.

lineage (*lin*) mutants The complete knowledge of both the cellular anatomy and the cell lineage of the nematode *Caenorhabditis elegans* has allowed the definition of developmental defects at the level of resolution of single cells. Mutants altered in cell lineage have been identified in a variety of screens; for example mutants with vulval defects can be identified as viable mutants defective in egg laying, and then examined for cell-lineage defects. Other cell-lineage mutants have been discovered by direct observation of lineage defects in living embryos or from the collection of mutants with abnormal cell numbers. Lineage mutants have identified some key developmental control genes including HOMEOTIC GENES such as *lin-12*, HETEROCHRONIC GENES such as *lin-14*, and genes involved in programmed cell death such as *ced-3*. See: CAENORHABDITIS DEVELOPMENT.

LINE(s) Acronym for *Long INterspersed DNA sequence Elements* which are a class of interspersed repetitive RETROTRANSPOSON-like sequences found in mammalian genomes. They seem to be derived from RNA polymerase II-generated transcripts (Fig. L7) and differ from SINES (short interspersed sequences) in that the latter seem to be derived from transcripts of RNA polymerase III. One ubiquitous family — the LINE-1 family — contains sequences up to 7 kb long which are also often found in truncated and rearranged form. They are present in thousands of copies per genome. Both LINEs and SINEs are members of a class of retrotransposons that appear not to encode proteins that have transposition functions, although they contain features that suggest they originated from RNA sequences. The mechanism whereby these RNA sequences have become incorporated into a DNA copy within the genome remains unclear but may involve REVERSE TRANSCRIPTION.

Hutchinson, C.A. et al. (1989) In *Mobile DNA* (D.E. Berg & M.M. Howe, Eds) 593–618 (American Society of Microbiology, Washington DC).

Fig. L7 Organization of a typical mammalian LINE element.

linkage LOCI that tend to be inherited together more often than would be expected by chance (*see* MENDELIAN INHERITANCE) are said to show linkage. Genetic linkage is a reflection of the physical location of the loci on the same chromosome segment. Unlinked loci have a 50% chance of being separated by a RECOMBINATION event and assort independently at meiosis. Loci which are close together are less likely to be separated by recombination and are therefore more likely to be inherited together. The distance between linked loci is measured in terms of the frequency of recombination events occurring between them and is expressed in centimorgans (cM) (*see* MAP DISTANCE). Testing for linkage between loci is a powerful tool in GENETIC MAPPING as it allows the positioning of a new locus relative to the established gene map (*see* HUMAN GENE MAPPING). *See also*: LOD SCORE.

linkage disequilibrium The situation when a particular combination of ALLELES at two closely linked loci appears more frequently than would be expected by chance (*see* MENDELIAN GENETICS). Linkage disequilibrium is often observed between a disease gene (a mutant allele) and a flanking allele (which can then be used as a marker for the disease), and may reflect the common ancestral origin of the mutation. The disequilibrium can persist for several generations, depending on the GENETIC DISTANCE between the two loci. Loci which are very close together will tend to remain in disequilibrium for a greater number of generations than those which are further apart.

linkage group A group of LOCI which can be shown to segregate together with predictable frequency. This genetic linkage is a direct result of their physical location on the same stretch of DNA.

linkage map Genetic map that gives the order of gene LOCI on a chromosome and the relative distances separating them based on the frequency of RECOMBINATION between them. *See*: HUMAN GENE MAPPING; LINKAGE.

linker DNA (1) DNA connecting one NUCLEOSOME to the next. *See*: CHROMATIN.
(2) OLIGODEOXYNUCLEOTIDE linkers. Short synthetic DNA sequences that are used to 'link' two fragments of DNA during ligation in the construction of RECOMBINANT DNAS. This is a common method for inserting new RESTRICTION SITES into a cloned DNA sequence.

linker scanning *See*: SCANNING MUTAGENESIS.

linking number (*L*) In closed circular DNA, the linking number L_k is an integer which describes how one polynucleotide strand winds about the other in double-stranded DNA. It is invariant under all topological forms; positive for right-handed winding and negative for left-handed. In relaxed, unstrained circular DNA the linking number is equal to the number of double-helical turns, or twists (T_w) and the DNA shape is that of a flat circle with $L_k = T_w$. In supercoiled DNA the twist changes (with the number of base pairs per turn) and the supercoiling is described by the WRITHING NUMBER (W_r) where $L_k = T_w + W_r$. *See also*: SUPERCOILING.

Brock Fuller, F. (1971) *Proc. Natl. Acad. Sci. USA* **68**, 815–819.

linoleic acid, linolenic acid *See*: FATTY ACIDS.

lipase(s) A group of hydrolytic enzymes acting on ester bonds in lipids. Commonly refers to:
1 Pancreatic lipase (EC 3.1.1.3) which catalyses the hydrolysis of triacylglycerol in the lumen of the gut to monoacylglycerol and fatty acid.

2 Lipoprotein lipase (3.1.1.34) which catalyses the hydrolysis to glycerol and fatty acids of triacylglycerol present in chylomicrons and low-density lipoproteins and is present on endothelial cells of the capillaries of target tissues, and in blood and milk.
3 The hormone-sensitive triacylglycerol lipase of adipose tissue, which removes a fatty acid residue from position 1 or 3 of the glycerol backbone, the first step in the pathway leading to complete hydrolysis of triacylglycerols.
See also: PHOSPHOLIPASES.

lipid bilayer, lipid membrane *See*: MEMBRANE STRUCTURE.

lipidoses Lysosomal storage diseases where ACID HYDROLASES responsible for lipid breakdown are deficient. In a subset of lipidoses, the sphingolipidoses, enzymes responsible for sphingo-lipid catabolism, are absent. *See*: GAUCHER'S DISEASE.

lipids A heterogeneous class of organic compounds (Table L2) which are extractable from biological material by nonpolar solvents such as ether, chloroform, benzene, etc. but not by aqueous solvents. The word 'lipid' is derived from the Greek *lipos*, fat. It connotes solubility in organic rather than aqueous solvents, and is used to refer to molecules that contain long acyl chains, and by extension, to co-occurring molecules like steroids with similar physicochemical properties. Lipids have numerous functions: membrane constituents (*see* MEMBRANE LIPIDS; MEMBRANE STRUCTURE); storage compounds (e.g. TRIACYLGLYCEROLS in plant seeds and animal adipase tissue); fuel molecules (*see* β-OXIDATION); hormones (*see* STEROID HORMONES; STEROID HORMONE RECEPTOR SUPERFAMILY); and protective coverings (e.g. waxes, *see* PLANT CELL WALL MACROMOLECULES). Some lipids (e.g. the sterol CHOLESTEROL) may be found complexed with protein to form LIPOPROTEINS (*see e.g.* PLASMA LIPOPROTEINS AND THEIR RECEPTORS). *See also*: DOLICHOL; RETINOL; UNDECAPRENOL.

lipocortins (calpactins) Local polypeptide chemical mediators produced by leukocytes which inhibit the cleavage of membrane phospholipid by phospholipase to produce arachidonic acid, and

Table L2 The main types of lipid in biological material

FATTY ACIDS

Fatty-acid containing
Neutral fats (TRIACYLGLYCEROLS)
PHOSPHOLIPIDS
   GLYCEROPHOSPHOLIPIDS
   SPHINGOPHOSPHOLIPIDS
GLYCOLIPIDS
   Derivatives of sphingosine e.g. cerebrosides, gangliosides
   Derivatives of glycerol

Others
Aliphatic alcohols and waxes
TERPENES
STEROIDS and STEROLS

Phospholipids and glycolipids are the main lipid constituents of biological membranes.

thus inhibit EICOSANOID synthesis. *See*: ACTIN-BINDING PROTEINS; ANNEXINS; CALCIUM-BINDING PROTEINS.

lipodepsinonapeptide A LIPODEPSIPEPTIDE molecule which contains nine amino acids in the peptide chain. *See*: PLANT PATHOLOGY.

lipodepsipeptide A family of molecules containing an N-terminal lipid moiety, a short peptide chain, and a lactone linkage at the C terminus. The lactone linkage renders the molecule partially cyclic. Tolaasin, a lipodepsipeptide toxin produced by *Pseudomonas tolaasii*, a pathogen of the cultivated mushroom (*Agaricus bisporus*), has a M_r of 1985 and consists of a peptide of 18 amino acid residues with a β-octanoic acid group at the N terminus. A sequence of seven D-amino acids gives part of the molecule a novel left-handed α-helical configuration. Lipodepsipeptides are produced by a range of microorganisms, especially the phytopathogenic pseudomonads, and many are potent membrane active toxins. *See*: PLANT PATHOLOGY.

Nutkins, J. et al. (1991) *J. Am. Chem. Soc.* **113**, 2621–2627.

lipofection The introduction of a gene into a cell via LIPOSOMES into which the DNA has been packaged.

lipofuscin granules Autofluorescent lysosomal RESIDUAL BODIES also known as age pigments, containing various fat-soluble pigments. Accumulation of lipofuscin is generally thought to reflect the ageing process and is found in many neurons of older people. The main orange-fluorescing pigment in granules isolated from retinal pigmented epithelium is an amphoteric quaternary amine *N*-retinylidene-*N*-retinylethanolamine.

lipo-oligosaccharides A family of molecules that can induce root hair curling and nodulation on legumes at very low concentration. The first to be found was a tetraglucosamine which is decorated with acetyl groups, a sulphate, and a lipid side chain. *See*: NODULATION.

lipopeptidophosphoglycans, lipophosphoglycans *See*: GLYCO-LIPIDS.

lipopolysaccharide (LPS) A diverse family of specialized GLYCO-LIPIDS of Gram-negative microorganisms, in which saccharide forms repeating O-ANTIGEN chains, attached through a non-repeating core glycan to lipid A, which is integrated into the outer membrane (*see* BACTERIAL ENVELOPES). They are synthesized at the cell surface. Many act as surface antigens and, if shed, behave as bacterial toxins; they are of great clinical importance. *Salmonella* lipopolysaccharide illustrates all the features of such molecules; in other species they can be much simpler. Its lipid A consists of the disaccharide -4GlcNH$_2$β1,6GlcNH$_2$α1- with fatty acyl chains substituting the otherwise unoccupied hydroxyl groups and phosphate ester groups attached to the α1- and -4 positions indicated.

The α1-phosphate is linked to the sugar 3-oxo-3-deoxymanno-octulosonic acid (KDO, formerly ketodeoxyoctulo sonic acid). This,

the first sugar of the core sequence, is linked in turn to two further molecules of KDO to give a branched trisaccharide; one arm of this is linked through phosphate to ethanolamine and the other is linked to three molecules of heptose in the sequence Hep1,7Hepα1,3 Hepα1,5(KDO). The heptose linked to KDO is connected to ethanolamine by a pyrophosphoryl residue at C4. The entirety of the core sequence to the third heptose forms the inner core, which can be cross-linked to other molecules of lipopolysaccharide via its ethanolamine residues. The outer core, in *Salmonella typhimurium*, consists of the saccharide GlcNAcα1,2Glcα1,2Galα1,3(Galα1,6)Glc1- which is linked to C3 of the second heptose of the inner core. The O-antigenic chain is glycosidically linked to the outer core at C4 of the subterminal α-glucosyl residue.

Synthesis of the entire core sequence occurs by glycosyl transfers from SUGAR NUCLEOTIDES, with addition at the nonreducing terminals and catalysis by highly specific glycosyl transferases. Mutants are known (rough mutants) which have a variety of defects in core synthesis and so lack O-antigenic outer chains. They are more common in outer core synthesis, reflecting the importance and relatively invariant structure of the chains closest to the membrane. Defects in lipid A synthesis are lethal, leading to difficulty in its analysis.

The O-antigenic outer chains are synthesized separately as modular glycans, with each module being assembled in the same direction as the core (i.e. conventionally), but on undecaprenyl pyrophosphate. The modules are then spliced together in a retrograde sense before being transferred to the core sequence. In species other than *Salmonella*, there may be no clear separation of an inner and outer core and the whole sequence can be smaller. The O-antigenic chains can also be simpler, but there is very little variation in lipid A and the core sugars immediately adjacent to it.

lipoprotein Complex of lipid and protein. *See*: BACTERIAL ENVE-LOPES; PLASMA LIPOPROTEINS AND THEIR RECEPTORS.

liposome Small vesicle bounded by a bilayer lipid membrane made artificially from phospholipids by sonication or by removal of detergent from phospholipid–detergent complexes. DNA, proteins, and other materials can be enclosed within the liposome and can be introduced into animal cells by liposome fusion with the plasma membrane (lipofection).

lipoteichoic acid A form of lipid-bound TEICHOIC ACID found in bacterial cell membranes rather than in cell walls.

Lipothrixviridae *See*: BACTERIOPHAGES.

β-lipotropin Peptide produced in the pituitary by the processing of pro-opiomelanocortin, and which is further processed to β-endorphin. *See*: OPIOID PEPTIDES AND THEIR RECEPTORS.

lipoxins *See*: EICOSANOIDS.

5-, 12-, and 15-lipoxygenases Enzymes involved in EICOSANOID synthesis.

13-lipoxygenase *See:* PLANT WOUND RESPONSES.

Listeria Genus of bacteria of the family Corynebacteriaceae. *Listeria monocytogenes* is a rare cause of meningo-encephalitis in adults and of granulomatosis infantiseptica in the newborn. It occurs as Gram-positive nonsporing rods, sometimes slightly curved, often in pairs end to end at an acute angle. *See:* BACTERIAL PATHOGENICITY.

lithium *See:* AFFECTIVE DISORDERS.

liver cell *See:* HEPATOCYTE.

LLG Log likelihood gain. *See:* MAXIMUM LIKELIHOOD.

local circuit neuron Neurons with short axons which generally remain within an area close to the cell body. Also known as a Golgi type II neuron. *Cf.* PROJECTION NEURON.

locomotion *See:* CELL MOTILITY.

locus (*pl.* loci) The position of a GENE on a CHROMOSOME. The term may also be used for the chromosomal location of any characterized DNA sequence. *See also:* ALLELE; HUMAN GENE MAPPING; MENDELIAN INHERITANCE.

locus control regions (LCR) Regulatory regions of DNA associated with the α- and β-globin gene clusters and which are required for their expression and which regulate transcription throughout the cluster. LCRs are formally defined as regions able to confer position-independent expression on a linked globin gene in transgenic mice. In the β-globin cluster the LCR is spread over ~10 kb between 5 and 18 kb upstream of the ε-globin gene. It includes four DNASE I HYPERSENSITIVE SITES each representing a putative regulatory region of ~300 bp. The α-globin LCR is a region of ~300 bp lying 40 kb upstream of the embryonic ζ globin gene.

LCRs are thought to interact with the regulatory elements of the individual globin genes to activate, enhance, and developmentally regulate their expression. The regulatory regions of the LCR contain binding sites for the erythroid-specific transcription factors GATA-1 (GATA site) and NF-E2 (an 'AP-1 site'), and a GACC/GGTG site which can be bound by ubiquitous factors such as Sp1.

Crossley, M. & Orkin, S.H. (1993) *Curr. Opinion Genet. Dev.* **3**, 232–237.

lod score Statistical measure of the degree of LINKAGE between two loci. It is calculated as the logarithm of the odds for linkage against the odds of no linkage between two loci. (The term 'lod' is derived from 'log of odds'.) A lod score of 3.0 therefore represents odds of 1000 : 1 in favour of linkage and is taken as the criterion for accepting the hypothesis of linkage. Similarly a lod score of – 2.0 is taken as the value at which linkage can be formally rejected (odds of 100 : 1 against linkage). The lod score varies with the RECOMBINATION FRACTION between the two loci: a linkage calculation provides both a measure of the degree of certainty with which two loci can be said to be linked and a measure of the most likely GENETIC DISTANCE between them.

log phase A step following LAG PHASE in the growth cycle of microorganisms or mammalian cells in a nonlimiting growth medium. It refers to the logarithmic or exponential growth of the cell population with time, and is usually followed by a nongrowing STATIONARY PHASE.

logistic equation *See:* HILL COEFFICIENT.

long day plant (LDP) A plant that is induced to flower by exposure to a regime in which the nights (i.e. periods of darkness) are shorter than a certain critical value.

long terminal repeat (LTR) A DNA sequence directly repeated at both ends of an integrated RETROVIRUS genome (the provirus), which is not found in the retroviral RNA genome. LTRs are generated through a replication process occurring during integration and consist of three distinct regions: U3, R, and U5. LTRs also contain an active RNA polymerase II PROMOTER which allows transcription of the integrated provirus by host cell RNA polymerase II to generate new copies of the retroviral RNA genome.

Long-term potentiation

LONG-TERM POTENTIATION (LTP) is an enduring enhancement in the efficiency of SYNAPTIC TRANSMISSION and is a property of many excitatory synapses in the nervous system. Interest in LTP lies in the widely held belief that the modifications underlying this form of SYNAPTIC PLASTICITY may be involved in learning and memory and in the development of synaptic connections in the vertebrate central nervous system [1]. Although LTP is exhibited by synapses in many regions of the brain, and also at certain synapses in the PERIPHERAL NERVOUS SYSTEM, the process is most pronounced in higher centres involved in cognitive function, in particular the cerebral cortex and hippocampus [2]. Thus, the considerable interest in LTP, particularly as studied using slices of hippocampus maintained *in vitro*, is driven by the hope that it will provide increased understanding of the molecular basis of learning and memory in vertebrates, including humans.

Properties

LTP (Fig. L8) is expressed as an increase in the size of the EXCITATORY POSTSYNAPTIC POTENTIALS (EPSPs) which are evoked by low-frequency (typically <0.1 Hz) stimulation of single afferent fibres or, more commonly, groups of fibres. LTP can last many hours *in vitro*, and days or weeks *in vivo*. The process is normally induced by delivering one or more brief periods of high-frequency stimulation (typically 100 Hz) to the afferent pathway of interest. Such TETANIC STIMULATION usually comprises 20–100 stimuli; however, as few as three shocks can be sufficient to induce LTP, provided that an optimal pattern of stimulation (PRIMED BURST PARADIGM) is used. This pattern closely approximates the natural THETA RHYTHM of hippocampal neurons *in situ*.

LTP in many pathways of the brain displays the properties

a

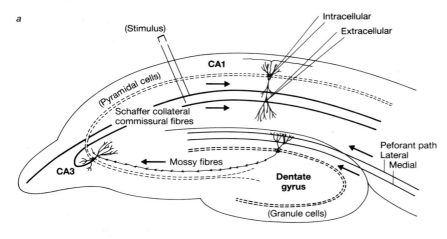

b

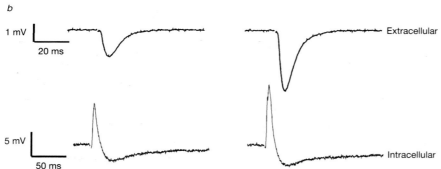

Fig. L8 Long-term potentiation (LTP) in the hippocampal slice. A schematic representation of a hippocampal slice illustrating the primary synaptic connections. Low-frequency electrical stimuli of the Schaffer collateral commissural fibres evokes synaptic responses which can be recorded extracellularly or intracellularly in the CA1 pyramidal cell region. Following a brief period of high-frequency stimulation (100 Hz, 1 s) a long-lasting enhancement of synaptic transmission can be observed following a return to low-frequency stimulation. Adapted from [3].

of SPECIFICITY, COOPERATIVITY, and ASSOCIATIVITY. These properties, together with the HEBBIAN nature of its induction, have strengthened the belief that LTP and associative learning in vertebrates may share the same basic molecular mechanisms.

Induction

For a fuller review of the mechanisms regulating the induction of LTP see [3]. The pathways in the brain that exhibit LTP are believed to use L-GLUTAMATE as an excitatory NEUROTRANSMITTER. The three EXCITATORY AMINO ACID RECEPTOR subtypes — the AMPA receptor and the NMDA receptor (which are ligand-gated ion channels) and the metabotropic receptor (a G PROTEIN-COUPLED RECEPTOR) — each have a distinct role in glutamate-mediated synaptic transmission. (The molecular biology of these receptors is described elsewhere in the encyclopedia, *see* EXCITATORY AMINO ACID RECEPTORS.)

Activation of the AMPA receptor mediates the normal, fast EPSP which is responsible primarily for low-frequency excitatory synaptic transmission, both before and after the induction of LTP (Fig. L9). NMDA receptors on the other hand do not contribute appreciably to low-frequency synaptic transmission because the ion channels which they gate are subject to a VOLTAGE-DEPENDENT BLOCKADE by Mg^{2+} present in the synaptic cleft. Under normal conditions coincident inhibitory synaptic mechanisms cause membrane hyperpolarization, which results in an intensification of the Mg^{2+} block of the NMDA receptor-operated

channels. During tetanic stimulation however, the membrane becomes depolarized for sufficient time to allow the Mg^{2+} block to be reduced and this enables the NMDA receptor system to contribute significantly to the synaptic response. Under these conditions NMDA receptor activation mediates a slow EPSP which, like AMPA receptor-mediated EPSPs, is generated mainly by the influx of Na^+ ions. However, NMDA receptor-operated channels are also permeable to Ca^{2+}, and Ca^{2+} entry is believed to be the essential trigger for the induction of LTP at most synapses (Fig. L9*b*). At such synapses, NMDA receptor antagonists, such as AP5 as well as intracellular Ca^{2+} chelators, such as EGTA, prevent LTP induction. In a few pathways, such as the hippocampal mossy fibre pathway, which connects dentate gyrus granule cells to CA3 pyramidal neurones, the induction of LTP does not involve NMDA receptors. This form of LTP is a quite distinct form of plasticity in that it is neither Hebbian nor associative in nature.

The sustained depolarization during tetanic stimulation is provided, in part, by summation of AMPA receptor-mediated EPSPs. An additional factor allowing depolarization is that the degree of inhibition alters during high frequency transmission. Inhibitory synaptic transmission is provided by γ-aminobutyric acid (GABA) which acts on postsynaptic GABA$_A$ and GABA$_B$ receptors (*see* GABA AND GLYCINE RECEPTORS) to generate INHIBITORY POST-SYNAPTIC POTENTIALS (IPSPs) which normally limit the synaptic activation of NMDA receptors. During high-frequency synaptic transmission the IPSPs are depressed. This is caused by GABA

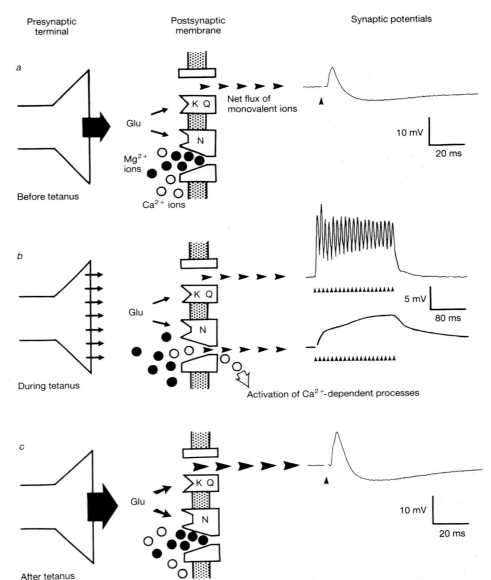

Presynaptic terminal   Postsynaptic membrane   Synaptic potentials

Fig. L9 NMDA receptor activation is required for the induction of LTP. *a*, Under standard conditions, during low-frequency stimulation, the excitatory postsynaptic potential (EPSP) is mediated by the influx of sodium ions through the AMPA receptor-operated channel (KQ). The NMDA receptor-operated channel (N) is blocked by magnesium ions. *b*, During high-frequency stimulation, the postsynaptic neuron is maintained in a depolarized state. This alleviates the magnesium block of NMDA receptor-operated channels. The influx of calcium ions through the NMDA receptor-operated channels is considered an essential trigger for the induction of LTP. *c*, Following the induction of LTP the enhanced synaptic response, like the normal EPSP, is mediated by AMPA receptor activation. From [3].

acting on GABA_B AUTORECEPTORS on the terminals of the inhibitory interneurons to limit the amount of GABA released by subsequent stimuli. This effect on GABA release takes at least 20 ms to begin and lasts for a few seconds; therefore, it is effective during high- but not low-frequency transmission. The auto-inhibition of GABA release is necessary for the induction of LTP induced by primed burst stimulation [4].

In general, it is considered that NMDA receptor activation induces LTP and AMPA receptor activation expresses the modification. This is an oversimplification, however, as NMDA receptors contribute to synaptic transmission, particularly during high-frequency transmission, and the NMDA receptor-mediated component of synaptic transmission can also be modified long term [5]. This has the important consequence that LTP can result in an enhancement in the ability of synapses to support plasticity (by enhancement of NMDA receptor-mediated synaptic trans-

mission), as well as resulting in an increase in synaptic efficiency *per se*.

Activation of metabotropic glutamate receptors (ACPD receptors) triggers various intracellular second messenger cascades (*see* SECOND MESSENGER PATHWAYS). These receptors may also be involved in the induction of LTP as ACPD receptor activation is known to enhance LTP [6], which may be due to the enhancement of NMDA receptor-mediated responses by ACPD receptor activation. In addition, the activation of ACPD receptors alone can result in LTP by bypassing the requirement for the activation of the NMDA receptor system [1]. It has recently been shown, using a selective ACPD receptor antagonist, that the synaptic activation of ACPD receptors is required for the induction of both NMDA receptor-dependent and NMDA receptor-independent forms of LTP in the hippocampus [7].

Signal transduction in LTP

There is considerable uncertainty as to the mechanisms triggered following the transient activation of NMDA receptors. PROTEIN KINASES, and certain PROTEASES and PHOSPHOLIPASES, have all been implicated in the generation of LTP. Some of the best evidence to date points to the involvement of PROTEIN KINASE C (PKC) [2]. This enzyme may be activated by the rise in intracellular Ca^{2+} concentration, initiated by Ca^{2+} entry through NMDA channels, and by the activation of ACPD receptors, some subtypes of which couple to phosphoinositide-specific phospholipase C (PLC). With respect to LTP, it is presently unclear whether PKC is activated transiently (with the resultant phosphorylation providing a persistent signal, *see* PROTEIN PHOSPHORYLATION) or whether the kinase is activated constitutively. Neither the ISO-FORMS of PKC nor the substrates involved in these processes are known. Inhibitor studies indicate that CA^{2+}/CALMODULIN-DEPENDENT PROTEIN KINASE II (CaMKII) and protein TYROSINE KINASES are also involved in LTP. Recently, it has been shown that mutant mice with a deletion of the gene encoding CaMKIIα [8] show impaired LTP.

Although the details remain elusive, most workers in the field agree that phosphorylation of one or more substrates is involved in LTP. However, phosphorylation alone is unlikely to provide for changes that last for the entire duration of LTP. Ultimately, therefore, alterations in protein synthesis and/or gene expression are likely to be involved, but there is at present only very preliminary information. There are indications that the IMMEDI-ATE EARLY GENES *zif/268* and c-*fos* are expressed after both the activation of NMDA receptors and the induction of LTP [9,10]. However, the transcriptional inhibitor actinomycin D does not impair LTP, measured for at least 3 hours following its induction. Thus, these immediate early genes must be coding for factors which are involved (if at all) in later stages of the expression of LTP. It has been reported that translational inhibitors, such as anisomycin [11] and cycloheximide [12], shorten the duration of LTP to a few hours, provided that they are present at the time of the tetanus. Together, these data suggest that protein synthesis, controlled by messenger RNAs existing at the time of tetanus, may be involved in LTP from about 3 hours after induction onwards.

Expression of LTP

There is considerable controversy as to how the enhancement of synaptic transmission is caused. One possibility is that there is an increase in the amount of synaptically released L-glutamate accessing the postsynaptic receptors. This could be due, for example, to an increase in the amount of L-glutamate released per nerve impulse or to a decrease in the efficiency of L-glutamate reuptake. LTP might also be maintained by changes in the number or conductance properties of postsynaptic receptors. A third possibility is that there are structural changes in the dimensions of dendrites or DENDRITIC SPINES (*see also* NEURONAL CYTOSKELETON) which may enable greater postsynaptic depolarization for a given amount of presynaptically released L-glutamate. It is probable that a combination of some or all of the above factors are involved in the expression of LTP.

The cloning of subunits of the glutamate receptor families (*see* EXCITATORY AMINO ACID RECEPTORS) provides suggestions for possible postsynaptic modifications. In the AMPA receptor family there are four known subunits (GluR1–4 or A–D) which can exist in alternatively spliced variants termed 'flip' and 'flop' [13]. In theory, LTP could be associated with a change in the subunit composition of the AMPA receptor. In particular, as 'flip' variants generate larger glutamate-induced currents than 'flop' variants, an alteration in the flip/flop ratio is an attractive idea. Another possibility is a regulation of the RNA EDITING which affects the Ca^{2+} permeability and conductance of the GluR2 subunit. At the present time, however, there is no reason to believe that Ca^{2+}-permeable AMPA receptors are involved in LTP.

An alteration in the subunit composition of NMDA receptors could, in theory, underlie the modification of this component of synaptic transmission. In this respect it is noteworthy that recently cloned subunits [14] have many different properties including distribution, conductance and susceptibility to blockade by Mg^{2+}.

Future prospects

The mechanisms of induction of NMDA receptor-dependent LTP are now well established. However, the biochemistry of the signal transduction mechanisms is still far from being understood and will remain difficult to clarify until more selective and potent antagonists of the intracellular enzymatic reactions which may be involved have been developed. Another area of great controversy concerns the mechanisms of the expression of LTP. Some of the recent developments in molecular biology outlined above may aid investigation of these problems.

Z.I. BASHIR
G.L. COLLINGRIDGE

1 Bliss, T.V.P. & Collingridge, G.L. (1993) A synaptic model of memory: long-term potentiation in the hippocampus. *Nature* **361**, 31–39.
2 Bashir, Z.I. & Collingridge, G.L. (1992) Synaptic plasticity: long-term potentiation in the hippocampus. *Curr. Opinion Neurobiol.* **2**, 328–335.
3 Collingridge, G.L. (1992) The mechanism of induction of NMDA receptor-dependent long-term potentiation in the hippocampus. *Exp. Physiol.* **77**, 771–797.
4 Davies, C.H. et al. (1991) GABA$_B$ autoreceptors regulate the induction of LTP. *Nature* **349**, 609–611.
5 Bashir, Z.I. et al. (1991) Long-term potentiation of NMDA receptor-mediated synaptic transmission in the hippocampus. *Nature* **349**, 156–158.
6 McGuinness, N. et al. (1991) Trans-ACPD enhances long-term potentiation in the hippocampus. *Eur. J. Pharmacol.* **197**, 231–232.
7 Bashir, Z.I. et al. (1993) Induction of LTP in the hippocampus needs synaptic activation of glutamate metabotropic receptors. *Nature* **363**, 347–350.
8 Silva, A.J. et al. (1992) Deficient hippocampal long-term potentiation in α-calcium-calmodulin kinase II mutant mice. *Science* **257**, 201–206.
9 Cole, A.J. et al. (1989) Rapid increase of an immediate early gene messenger RNA in hippocampal neurons by synaptic NMDA receptor activation. *Nature* **340**, 474–476.
10 Jeffery, K.J. et al. (1990) Induction of Fos-like immunoreactivity and the maintenance of long-term potentiation in the dentate gyrus of unanesthetized rats. *Mol. Brain Res.* **8**, 267–274.
11 Krug, M. et al. (1984) Anisomycin blocks the late phase of long-term potentiation in the dentate gyrus of freely moving rats. *Brain Res. Bull.* **13**, 39–42.

12 Stanton, P.K. & Sarvey, J.M. (1984) Blockade of long-term potentiation in rat hippocampal CA1 region by inhibitors of protein synthesis. *J. Neurosci.* **4**, 3080–3088.

13 Sommer, B. et al. (1990) Flip and Flop: a cell-specific functional switch in glutamate-operated channels of the CNS. *Science* **249**, 1580–1585.

14 Kutsuwada, T. et al. (1992) Molecular diversity of the NMDA receptor channel. *Nature* **358**, 36–41.

15 Bliss, T.V.P. & Collingridge, G.L. (1987) NMDA receptors — their role in long-term potentiation. *Trends Neurosci.* **10**, 288–293.

loricin An INTERMEDIATE FILAMENT protein.

Lou Gehrig's disease AMYOTROPHIC LATERAL SCLEROSIS.

Louis–Bar syndrome ATAXIA TELANGIECTASIA.

low melting point agarose *See*: AGAROSE.

low-density lipoprotein (LDL) *See*: PLASMA LIPOPROTEINS AND THEIR RECEPTORS.

Lp(a) Lipoprotein (a), a low-density plasma lipoprotein comprising apolipoprotein (a) (apo(a)) linked by a disulphide bond to apolipoprotein B. Apo(a) is a member of the SERINE PROTEINASE superfamily with a resemblance to plasminogen but no enzymatic activity, found only in primates and the hedgehog. *See*: PLASMA LIPOPROTEINS AND THEIR RECEPTORS.

LPS LIPOPOLYSACCHARIDE.

LSU The large, catalytic, subunit (M_r 56 000) of RIBULOSE BISPHOSPHATE CARBOXYLASE/OXYGENASE (RUBISCO). *See also*: PHOTOSYNTHESIS.

LSU-BP LARGE SUBUNIT BINDING PROTEIN.

LT Large T antigen. *See*: PAPOVAVIRUSES.

LTD Long-term depression. Prolonged decrease in synaptic efficacy following a rapid train of electrical stimulation. Mediated, at least in part, by changes in numbers and/or sensitivity of EXCITATORY AMINO ACID RECEPTORS. *See*: LONG-TERM POTENTIATION.

LTP LONG-TERM POTENTIATION. A long-lasting enhancement of SYNAPTIC TRANSMISSION which is reflected by an increase in the size of the excitatory postsynaptic potential following appropriate conditioning of the afferent pathway. LTP is believed to utilize mechanisms which are involved in learning and memory in higher animals.

LTR LONG TERMINAL REPEAT in integrated retroviral genomes.

lucifer yellow Aldehyde-fixable fluorescent tracer used following microinjection to analyse cell–cell communication via junctions (*see* CELL JUNCTIONS) and following uptake by pinocytosis to study the ENDOCYTIC PATHWAY (*see also* ENDOCYTOSIS).

luciferase Any of various enzymes from bacteria, fireflies, coelenterates, molluscs, and deep-sea fish which catalyse a light-emitting reaction. The enzymes from animals (EC 1.13.12.5–8) are monooxygenases which catalyse a reaction that oxidizes a species-specific LUCIFERIN to oxyluciferin which emits light in a stoichiometric manner. The firefly enzyme requires ATP and is used as a highly sensitive quantitative assay for ATP and as a chemiluminescent detection reagent in nonradioactive nucleic acid and protein labelling kits. Genes encoding luciferase (the *lux* genes) have also found use as exquisitely sensitive REPORTER GENES for studies on PROMOTER regulation, etc. Bacterial luciferase (EC 1.14.14.3) catalyses a reaction between a long-chain aliphatic aldehyde, reduced flavin mononucleotide, and oxygen, which produces an activated $FMN.H_2O$ complex that breaks down with the emission of light.

luciferins Heterogeneous group of compounds found in fireflies, molluscs, coelenterates, and some luminous deep-sea fishes that emit light when oxidized with molecular oxygen in a reaction catalysed by LUCIFERASES. Firefly luciferin is (*S*)-4,5-dihydro-2-(6-hydroxy-2-benzothiazolyl)-4-thiazolecarboxylic acid; *Cypridina* (fish) luciferin is [3-[3,7-dihydro-6-(1*H*-indol-3-yl)-2-[(*S*)-1-methyl-6-propyl]-3-oxoimidazo-[1,2-*a*]pyrazin-8-yl]propyl]-guanidine.

lumen The cavity within membrane-bounded tubular or sac-like organelles such as the ENDOPLASMIC RETICULUM, GOLGI APPARATUS, and LYSOSOMES.

lunularic acid A natural PLANT GROWTH RETARDANT originally isolated from the liverwort *Lunularia cruciata* (Fig. L10). M_r 258.2. It is thought to be ubiquitous in liverworts, in which it inhibits the growth of thalli. Its production is enhanced in long days and may be under the control of PHYTOCHROME. It is accompanied by prelunularic acid, its presumptive biosynthetic precursor from which it may be formed nonenzymatically. Its specificity as a growth inhibitor has been questioned and its function is still uncertain.

Valio, I.F.M. et al. (1969) *Nature* **223**, 1176– 1178.

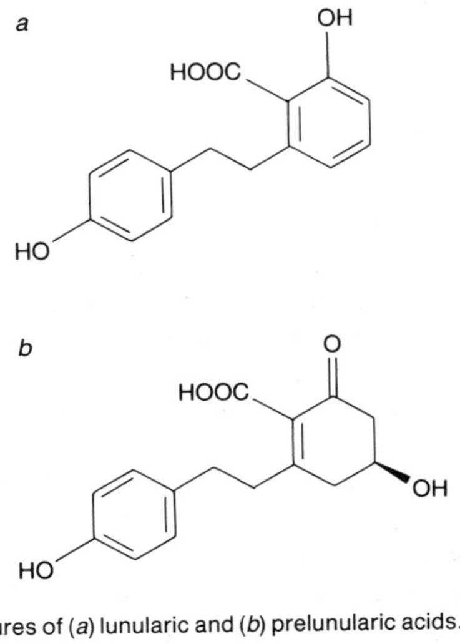

Fig. L10 Structures of (*a*) lunularic and (*b*) prelunularic acids.

luteinizing hormone (LH) Glycopeptide (M_r 28 000) secreted by basophilic staining cells in the anterior pituitary. It stimulates steroid synthesis in all cells of the ovary and stimulates the increased synthesis and secretion of testosterone from the Leydig cells of the testis at puberty. It works in concert with FOLLICLE-STIMULATING HORMONE to induce ovulation. Mutations in the G PROTEIN-COUPLED RECEPTOR for LH are responsible for familial precocious puberty in boys.

luteovirus group From Latin *luteus*, yellow, leaf yellowing being a common symptom; type member barley yellow dwarf virus. Single component viruses with isometric particles 25–30 nm diameter. The genome comprises (+)-strand linear RNA (Fig. L11). The 5′ gene is thought to frameshift to the P60 gene to give two products (P39 and P99) which have amino acid homologies to RNA polymerases. The coat protein gene, translated from a subgenomic RNA, is thought to read through to give the P72, the function of which is not known. The 3′ P7, also translated from a subgenomic RNA, has unknown function. *See also*: PLANT VIRUSES.

Casper, R. (1988) In *The Plant Viruses*, Vol. 3 (Koenig, R., Ed.) 35–258 (Plenum, New York).

Lutheran blood group BLOOD GROUP determined by red cell antigens (Lu^a, L^b) specified by the *Lu* gene on human chromosome 19.

luxury gene Gene encoding a specialized function and which is expressed at a high level in a particular cell type. Examples of luxury genes include the α- and β-globin genes expressed in erythrocyte precursors and the IMMUNOGLOBULIN GENES expressed in B cells.

lyases Enzymes that catalyse the removal from or addition of a group to a double bond, or other cleavages involving electronic rearrangement. Lyases include decarboxylases, such as oxaloacetate decarboxylase which catalyses the reaction oxaloacetate = pyruvate + CO_2, hydratases and dehydratases, which eliminate water, and ADENYLATE CYCLASE.

lycopene A deeply coloured CAROTENOID (ψ,ψ-carotene) that occurs in the CHROMOPLASTS of ripe fruit, especially tomatoes, but does not occur in functional CHLOROPLASTS.

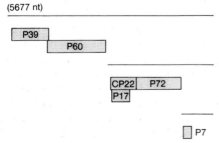

Fig. L11 Genome organization of barley yellow dwarf virus (BYDV). The lines represent the RNA species and the boxes the proteins (P) with molecular weights given as $\times 10^{-3}$; nt, nucleotides.

Lycopersicon esculentum Tomato.

lymphoblastoid cell line B CELLS immortalized by a transforming agent, generally EPSTEIN–BARR VIRUS, and thus capable of growing in continuous cultures.

lymphocyte Small mononuclear white blood cell present in large numbers in lymphoid tissues and circulating in the blood, and which mediates the antigen-specific immune response. There are two main types, B CELLS and T CELLS, which have different and complementary roles in immune responses. *See*: ANTIBODIES; B CELL DEVELOPMENT; HAEMATOPOIESIS; IMMUNOGLOBULIN STRUCTURE; IMMUNOGLOBULIN GENES; T CELL DEVELOPMENT; T CELL RECEPTORS.

lymphocyte-activating factor Interleukin 1α. *See*: IL-1; LYMPHOKINES.

lymphocyte growth factors *See*: LYMPHOKINES.

lymphoid lineage Lymphocytes, which comprise B CELLS and T CELLS. *See*: B CELL DEVELOPMENT; HAEMATOPOIESIS; T CELL DEVELOPMENT.

lymphoid-specific enhancers ENHANCERS have been identified in the murine and human IMMUNOGLOBULIN GENES — in the J-C$_\kappa$ intron, 3′ of the C$_\lambda$ gene, in the heavy chain gene J$_H$-C$_H$ intron (see HEAVY CHAIN ENHANCER), and 25 kb 3′ of the C$_\alpha$ heavy chain gene in rat and mouse. Immunoglobulin gene enhancers display both cell-type specificity and developmental cell-stage specificity in their activity. These specificities are thought to be largely regulated by the availability and binding of sequence-specific DNA-binding proteins (*see* NF-ΚB).

lymphoid transformation/tumours Ill-controlled clonal proliferation of lymphoid cells related to a genetic change accompanied by ONCOGENE deregulation. Similar mechanisms apply to both human (e.g. BURKITT'S LYMPHOMA) and mouse (e.g. PLASMACYTOMA) systems. Rearrangement of immunoglobulin genes in B cells and of T cell receptor genes in T cells plays a major part in oncogene deregulation.

Lymphokines

THE growth, differentiation, development and function of a variety of cell types can be regulated *in vitro* by CYTOKINES that exert their effects by binding to specific receptors on the cell membrane of the target cells. The lymphokines are a group of soluble polypeptide cytokines produced by LYMPHOID CELLS which have a major role in mobilizing and activating haematopoietic cells in response to a variety of infective agents and inflammatory stimuli [1–4]. Lymphokines also have a role, directly or indirectly, in the regeneration and repair of haematopoietic and nonhaematopoietic tissues.

Most lymphokines were originally detected and characterized through their effects on haematopoietic cells *in vitro*. The effects observed on target cells can include: (1) promotion of cell survival by suppression of APOPTOSIS; (2) stimulation of MITOGENESIS and cell proliferation; (3) inhibition of proliferation; (4) induction of DIFFERENTIATION and acquisition of a mature cell phenotype (*see e.g.* B CELL DEVELOPMENT; T CELL DEVELOPMENT); and (5) modulation of the function of mature cells. While lymphokines can be produced from normal helper T lymphocytes (HELPER T CELLS), either constitutively or following mitogenic stimulation, some can also be produced and secreted by cells of the MONOCYTE/ MACROPHAGE lineage, endothelial cells and fibroblasts. Cell lines derived from malignant tissues (lymphoid or myeloid LEUKAEMIAS as well as solid tumours) are a particularly rich source of many lymphokines and this has made their biochemical characterization and molecular cloning much easier. The recent availability of recombinant material has allowed detailed structure–function analysis of some of these molecules and their receptors and has also encouraged studies on their potential therapeutic uses.

Colony-stimulating factors

Because lymphokines were originally characterized using *in vitro* assay systems that defined some biological effect, their nomenclature is not without problems (Table L3). The colony-stimulating factors (CSFs) for example, are named by their ability to stimulate the *in vitro* proliferation and development of haematopoietic progenitor cells and allow them to form colonies when immobilized in a soft gel matrix. Thus, granulocyte/ macrophage CSF (a 'true' lymphokine produced by activated T cells and also by other cell types), macrophage CSF and granulocyte CSF (mainly produced by monocytes and various tissue cells and not considered to be classical lymphokines) all stimulate the development of colonies containing neutrophils (G) and/or macrophages (M) from their appropriate bi-or unipotent progenitor cells (*see* GM-CSF; M-CSF; G-CSF). But, in addition to their growth-stimulating effect on progenitor cells derived from bone marrow, GM-CSF, M-CSF and G-CSF can also synergize with each other, or with other lymphokines, to stimulate proliferation and development of more primitive multipotent haematopoietic stem cells.

Furthermore, each of the CSFs has powerful effects on mature post-mitotic cells of the blood: GM-CSF and M-CSF enhance the phagocytic ability of macrophages and increase their ability to kill various protozoa and tumour cells, while GM-CSF and G-CSF increase the production of superoxide anions in neutrophils in response to the peptide formylMet-Leu-Phe and enhance their phagocytic ability and cellular cytotoxicity against antibody-coated tumour cells (antibody-dependent cell cytotoxicity, ADCC). Eosinophils also respond to GM-CSF in a similar fashion. As GM-CSF, M-CSF and G-CSF also stimulate the migration of mature phagocytic cells, the CSFs probably play an important part in mobilizing and activating macrophages and neutrophils at the site of infection. In this respect it is worth noting that the production of CSFs by various tissue cell types is directly or indirectly stimulated by infective organisms or their products.

Table L3 Nomenclature

Preferred name	Other names
Granulocyte-macrophage colony stimulating factor (GM-CSF)	Macrophage-granulocyte inducer (MGI) Colony stimulating activity (CSA)
Macrophage colony stimulating factor (M-CSF)	Colony stimulating factor-1 (CSF-1)
Granulocyte colony stimulating factor (G-CSF)	Pluripoietin Differentiation factor
Interleukin-3 (IL-3)	Multi-colony stimulating factor (multi-CSF) Haemopoietic cell growth factor (HCGF) Mast cell growth factor (MCGF) Persisting cell-stimulating factor (PSF)
Interleukin-1 (α and β) (IL-1)	Lymphocyte activating factor (LAF) Haemopoietin-1 Endogenous pyrogen
Interleukin-2 (IL-2) Interleukin-4 (IL-4)	T cell growth factor (TCGF) B cell growth factor (BCGF-1) B cell stimulating factor-1 (BSF-1)
Interleukin-5 (IL-5)	B cell growth factor-II (BCGF-II) Eosinophil differentiation factor (EDF)
Interleukin-6 (IL-6)	B cell stimulating factor-II (BSF-II) Plasmacytoma growth factor Interferon β_2 (IFNβ_2) Hepatocyte stimulating factor (HSF)
Interleukin-7 (IL-7)	Lymphopoietin-1 (LP-1)
Interferon α-I and II (IFNα)	Leukocyte interferon Type I interferon
Interferon β (IFNβ)	Fibroblast interferon Type I interferon
Interferon γ (IFNγ)	Immune interferon Type II interferon Macrophage activating factor (MAF)
Tumour necrosis factor β (TNFβ)	Lymphotoxin

A fourth CSF (multi-CSF), which was also originally detected by its ability to stimulate the *in vitro* clonogenic proliferation of multipotent and lineage-restricted myeloid progenitor cells and the growth of mast cells, is now more commonly known as interleukin-3 (*see* INTERLEUKINS; IL-3). Unlike the other CSFs, the production of IL-3 is restricted to normal T cells and to certain leukaemic cell lines. Initially, it was suggested that IL-3 also acted on precursor T cells but this has not been substantiated and its role in lymphopoiesis remains an enigma. Indeed, its role in myelopoiesis and maintaining steady-state haematopoiesis in the bone marrow is also controversial (as indeed it is for the other CSFs) and in common with GM-CSF, G-CSF, and M-CSF, the primary function of IL-3 may well lie in modulating the function of mature phagocytic cells. The CSFs and IL-3, because of their growth-stimulatory activity on myeloid progenitor cells, are often referred to as the haematopoietic cell growth factors.

Interleukins and their effects

Other cytokines, which were originally characterized by the ability to promote growth and/or differentiation of T or B lymphocytes or their precursors, can also influence myeloid cells.

Interleukin-1

Interleukin-1 (*see* IL-1) has also been called haemopoietin-1 by virtue of its ability to synergize with the CSFs and potentiate the growth and development of primitive myeloid progenitor cells. IL-1 however, has a broad spectrum of activities: it is produced by monocytes, endothelial cells, fibroblasts and other cell types, in response to bacterial products, in two forms (IL-1α and IL-1β) which bind to the same cell-surface receptors. In addition to its direct effect on haematopoietic progenitor cells, IL-1 also stimulates the production of CSFs and other cytokines such as platelet-derived growth factor (*see* GROWTH FACTORS) from a variety of cell types and of interleukin-2 (*see* IL-2) from T cells. Furthermore, IL-1 has been implicated as a cofactor in T-cell proliferation driven by a variety of antigens. Cells capable of producing and responding to IL-1 are ubiquitously distributed and production is stimulated by a wide range of substances (bacteria and their products, immune complexes, tumour-promoting agents (*see* PHORBOL ESTERS), CSFs and other cytokines, and even by itself) and it is clear that IL-1 has a variety of roles in immune function and tissue repair and that its effects are not restricted to the haematopoietic system. The most obvious consequence of systemic production, or administration, of IL-1 is the induction of fever.

Interleukin-2

Perhaps the most widely studied activity of IL-1 is in the immune system, particularly in the proliferation of T cells (*see* T CELL DEVELOPMENT). While the role played by IL-1 in antigen-mediated proliferation of T cells is still unclear, it is likely to be related to the ability of IL-1 to induce the production of IL-2 by T lymphocytes and the associated expression of high-affinity IL-2 RECEPTORS on the T cells. The binding of IL-2 to its receptor leads to stimulation of DNA synthesis. Clearly this can occur in either an autocrine fashion (the cells producing IL-2 respond to their own growth factor) or a paracrine fashion (the IL-2 acts on adjacent cells).

Following activation of T cells, a series of intracellular events occurs commonly seen during stimulation of mitogenesis in other cell systems, including translocation and activation of PROTEIN KINASE C, mobilization of CALCIUM, phosphorylation of intracellular proteins and expression of certain cellular PROTO-ONCOGENES (e.g. c-*fos*). The production of IL-2 and the response of the cells to IL-2 leads to intracellular alkalinization via activation of the Na^+/H^+ EXCHANGER — an event that is also seen on stimulation of myeloid progenitors with the CSFs and which seems to be essential for DNA synthesis to take place subsequently — and expression of the proto-oncogene c-*myb*.

The initial effects of IL-2 are thought to occur via intracellular binding to a relatively low affinity receptor (IL-2Rβ). In a way not understood, this then leads to expression of the IL-2Rα (*Tac*) gene and ultimately to the expression of the high affinity heterodimeric (IL-2Rαβ) receptors on the cell surface. In this respect, it is worth noting that the IL-2 receptor is becoming a paradigm for receptors for other lymphokines and CSFs, where the biological significance of 'high' and 'low' affinity receptors is rapidly becoming established.

Other interleukins

Once activated by antigens and driven by IL-2, the proliferating helper T cells then begin to produce a range of other lymphokines. These include interleukins-3, -4, -5, -6, -7, -8, -9 and -10 (*see* IL-4, IL-5, IL-6, IL-7, IL-8, IL-9, IL-10), interferon-γ (*see* INTERFERONS, GM-CSF and tumour necrosis factor (*see* TNF). Some of these have direct effects on T lymphocytes or their precursors whereas others influence the growth and/or function of B lymphocytes and nonlymphoid cells.

IL-4, for example, stimulates the synthesis of IgE in B cells (*see* CLASS SWITCHING), exerts a proliferative effect on certain mature T cells and on CD4⁻ CD8⁻ thymocytes, influences the response of NATURAL KILLER CELLS (NK cells) and lymphokine-activated killer cells (LAK cells) to IL-2, potentiates the tumoricidal activity of macrophages and their expression of MHC MOLECULES, synergizes with IL-3 to promote the growth of connective tissue mast cells, influences the response of myeloid progenitor cells to CSFs, and inhibits growth of some connective tissue cells. Such multiple targets and pleiotropic activity are not uncommon for cytokines. With IL-1 and IL-4 in particular, where production is likely to occur locally in response to tissue damage, inflammatory stimuli, infective agents and so on, a coordinated immune and tissue response is necessary and the actions of lymphokines should be seen in this context.

A corollary of this is that persisting high circulating levels of lymphokines, because of their pleiotropic effects, may have disastrous pathological consequences. This has indeed been found to be the case: chronic inflammatory reactions, multiple organ failure, respiratory distress syndromes, arthritis and so on have been linked to persistent excessive production of cytokines such as IL-1 and lymphokines such as TNF and interferon. The severe life-threatening side effects seen on administration of IL-2 to patients can also be attributed not only to the direct effects of IL-2 on T cells and perhaps other target cells such as the anterior pituitary, but also to the variety of other lymphokines produced by the proliferating T cells. Moreover, aberrant constitutive production of some lymphokines has also been linked to oncogenic processes via autocrine growth stimulation.

Like IL-4, IL-5 and IL-6 also have effects on lymphoid cells and myeloid cells. Murine IL-5, for example, is a B-cell DIFFERENTIATION FACTOR that enhances IgA synthesis in the presence of B-cell activating agents and also promotes proliferation of activated B cells. It also facilitates the generation of cytotoxic T cells in the presence of IL-2 and promotes the survival, migration and functional activities (e.g. ADCC, phagocytosis, superoxide production) of mature eosinophil granulocytes. Again, the ability of IL-5 to stimulate several distinct cell types is compatible with the idea that the production of IL-5, by activated T cells, is the stimulus for a coordinated response to infection of the host by parasites and helminths (the 'targets' of eosinophils). The downside is that production of IL-5 is probably responsible, at least in part, for the eosinophilia and resulting pathology seen in allergies. It is important to note that human IL-5, although retaining the effect on eosinophils seen with murine IL-5, does not seem to possess the B cell activities of its murine counterpart.

Classically, the proliferation and differentiation of B cells *in vitro* required the presence of T cells and it was the search for growth factors (lymphokines) released from T cells that led to the characterization of IL-6 as a 'T-cell replacing factor'. Before this, however, the molecule had been independently discovered as a cytokine released from fibroblasts treated with poly(I:C) and named interferon-β_2 (IFNβ_2). Subsequent cloning demonstrated, however, that IFNβ_2 had little sequence homology with the other interferons and also lacked their spectrum of activities. The term IL-6 was subsequently adopted for this molecule and, although a lymphokine, IL-6 can clearly be produced by several cell types (monocytes, fibroblasts, endothelial cells) in response to bacterial products, IL-1 and other agents.

The biological effects attributed to IL-6 now include: (1) maturation of precursor B cells to antibody-producing cells (*see* B CELL DEVELOPMENT); (2) growth stimulation of myeloma and plasmacytoma cells; (3) an accessory factor for T-cell activation and proliferation of thymocytes; (4) differentiation of cytotoxic T cells in the presence of IL-2; (5) enhancement of the lytic function of NK cells; (6) interaction with CSFs to promote the growth and differentiation of primitive myeloid progenitor cells; (7) induction of differentiation in some myeloid leukaemic cells lines; and (8) induction of acute phase proteins from hepatic tissue.

The 'benefits' of the multitude of responses that can be elicited by IL-6 are perhaps difficult to understand at first glance: like many other cytokines, however, in normal situations the IL-6 is probably produced locally where it then exerts its effects. The mechanism of signal transduction by IL-6 (like IL-4) is unclear and does not seem to involve 'classical' second messenger systems such as inositol phospholipid turnover or rapid Ca^{2+} fluxes. What is interesting however is that the receptor for IL-6 belongs to the IMMUNOGLOBULIN SUPERFAMILY which also includes the receptors for PDGF, M-CSF, IL-1, G-CSF, GM-CSF and a number of well-characterized cell membrane proteins such as class I MHC molecules, Thy-1, CD1, CD2, CD4, CD8 and LFA-3 (*see* CD ANTIGENS). Like the IL-1, G-CSF, and GM-CSF receptors, the IL-6 receptor does not possess intrinsic TYROSINE KINASE activity (*see* GROWTH FACTOR RECEPTORS). However, binding of IL-6 triggers an association between the IL-6 receptor and a membrane glycoprotein (gp130). This association is involved in the signalling processes and has led to the suggestion that receptors with no intrinsic tyrosine kinase activity are functionally linked to a protein tyrosine kinase. Some recent data on the receptors for GM-CSF and for IL-3 give credence to this hypothesis.

Whereas IL-6 is concerned with terminal maturation of B cells, another cytokine, IL-7, mediates the survival, proliferation, and differentiation of their precursor cells — the pro-B cells (*see* B CELL DEVELOPMENT). The discovery of IL-7 stemmed from the observation that the maintenance and differentiation of pro-B cells *in vitro* required the presence of bone marrow stromal cells. One of the active factors produced by the stromal cells was subsequently molecularly cloned and designated IL-7. It is possible that the complete range of biological activities of IL-7 remains to be determined, but evidence to date suggests a role of IL-7 in supporting the growth of IgM-negative pre-B cells and their differentiation to surface IgM-positive cells, and proliferation of thymocytes and mature T cells in response to LECTINS — an effect probably mediated (like that of IL-1 and IL-6) via stimulation of IL-2 and IL-2 receptors. It is perhaps significant that memory T cells respond better to IL-7 than do unprimed T cells and that although the IL-2 receptor is induced, antibodies to the receptor do not block proliferation.

So far, this entry has emphasized the role of lymphokines and other cytokines in recruiting proliferation and differentiation of myeloid and lymphoid progenitor cells and enhancing the functional activity of the mature cells. The CSFs however also act as chemotactic stimuli and their production locally can mobilize neutrophils and macrophages to the sites of infection. A recently described factor — interleukin-8 (*see* IL-8) — is also a potent chemotactic and activating factor for T cells in addition to its previously known chemotactic activity for neutrophils. As IL-8 is produced by monocytes/macrophages in response to bacterial products, it is possible to imagine a scenario where, following tissue injury by a variety of mechanisms, a cytokine cascade is set in motion that results in recruitment of immune effector cells to the site of injury and the release of a variety of cytokines and lymphokines that then play a part in host defence and tissue repair.

The primary stimulus (Fig. L12*a*) evokes a tissue response characterized by the release of cytokines: the range of cytokines produced will be in part determined by the nature of the stimulus, but will include a variety of peptides that are chemoattractants. These establish concentration gradients and the appropriate target cells then undergo migration towards the site of tissue damage. At the same time, various cytokines are produced that activate cells involved in cellular and humoral immunity and that facilitate the proliferation of cells involved in normal healing and repair. Following the presentation of antigen by macrophages to infiltrating T cells and in the presence of cofactors, a lymphokine cascade is established whereby a variety of cells involved in immune function are activated, undergo proliferation and develop into effector cells (Fig. L12*b*). On most occasions, the cellular reaction is self-limiting and is localized to a specific region. In this respect, the production of TGFβ by T lymphocytes as a 'late' event in immune stimulation may have an important role as it reduces the expression of the IL-2 receptor, inhibits immunoglobulin synthesis and inhibits the synthesis and release of IL-1. Interleukin-10 (*see* IL-10) also has modulatory effects on lymphokine production. However, following severe infections, trauma or in certain pathological conditions, systemic effects can be seen — fever elicited by IL-1, acute phase response stimulated by IL-6, and stimulation of HAEMATOPOIESIS in the bone marrow mediated by CSFs and the interleukins.

T.M. DEXTER

See also: GROWTH FACTORS; GROWTH FACTOR RECEPTORS.

1 Webb, D.R. & Goeddel, D.W. (Eds) (1982) *Lymphokines, Vol. 13: Molecular Cloning and Analysis of Lymphokines* (Academic Press, New York).
2 Meager, A. (1990) *Cytokines* (Open University Press, Milton Keynes).

a

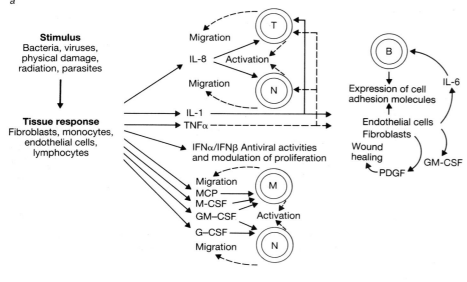

b

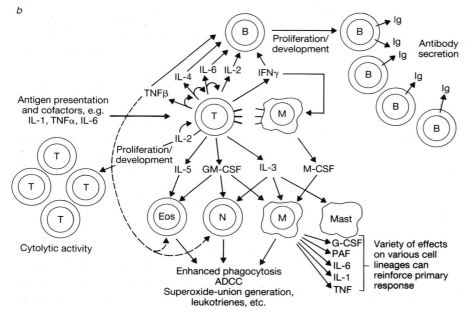

Fig. L12 *a* Roles of lymphokines and other cytokines in tissue damage and repair. B, B cell; M, monocyte/macrophage; MCP, macrophage chemotactic protein; N, neutrophil; T, T cell. See Table L3 for other abbreviations. *b*, Role of lymphokines in the antigen-mediated activation of T cells and the progress of the immune response. Eos, Eosinophil; Mast, mast cell; PAF, platelet activating factor.

3 Waterfield, M.D. (Ed.) (1989) Growth factors. *Br. Med. Bull.* **45** (Churchill Livingstone, London).
4 Whetton, A.D. & Dexter, T.M. (1989) Myeloid haemopoietic growth factors. *Biochim. Biophys. Acta* **989**, 111–132.

lymphoma Malignant disorder preferentially involving secondary lymphoid organs such as spleen and lymph nodes. The term non-Hodgkin lymphoma is used to define an heterogeneous group of such malignancies other than Hodgkin's disease. B-cell and T-cell lymphomas can be distinguished by cell-surface antigens. Classifications based on pathological and clinical criteria refer to three main groups: high-, intermediate-, and low-grade. *See also*: BURKITT'S LYMPHOMA; FOLLICULAR LYMPHOMA.

lymphopoiesis Generation of lymphocytes from the haematopoi-etic stem cell of the bone marrow. *See*: B CELL DEVELOPMENT; HAEMATOPOIESIS; T CELL DEVELOPMENT.

lymphotoxin (LT) Also called tumour necrosis factor-β (TNFβ), a glycoprotein CYTOKINE of M_r 21 000–24 000 produced exclusively by activated T CELLS and encoded by a gene within the MAJOR HISTOCOMPATIBILITY COMPLEX. It is a secreted cytokine, activating neutrophils and endothelial cells. It has 30% homology with TNFα and can bind at the same receptors. *See*: TUMOUR NECROSIS FACTOR.

lyophilization Also known as freeze drying, this technique involves the evaporation of an aqueous sample to complete dryness in a vacuum. It is generally used to preserve material for long-term storage, or for the removal of unwanted solute. The material is kept frozen throughout the process.

lysine (Lys, K) An AMINO ACID with a basic side chain.

lysogen, lysogenic, lysogeny A bacterium containing a latent PROPHAGE integrated into its genome as a consequence of infection by a temperate BACTERIOPHAGE is termed a lysogen. The prophage can be later released from the genome by INDUCTION which enables it to enter the LYTIC CYCLE of infection (*see* LAMBDA).

lysophosphatides Phospholipids from which one fatty acid has been removed.

lysopine A member of the OCTOPINE family of OPINES in which alanine is the amino acid conjugated to pyruvate.

lysosomal enzymes Principally the several dozen ACID HYDRO-LASES found in LYSOSOMES which contribute to the degradation of a wide range of biological macromolecules.

lysosomal storage diseases *See*: ARYL SULPHATASE; GAUCHER'S DISEASE; HUNTER SYNDROME; HURLER'S DISEASE; I-CELL DISEASE; LIPIDOSES; MUCOPOLYSACCHARIDOSES.

Lysosomes

LYSOSOMES are intracellular organelles with an acid interior (~pH 5) bounded by a single membrane and containing many hydrolytic enzymes which are optimally active at an acid pH [1]. They are present in almost all animal cells and in plant cells the VACUOLE is the equivalent organelle. Lysosomes were initially identified and characterized during the development of subcellular fractionation techniques by de Duve and co-workers in the 1950s. Fractions enriched in ACID HYDROLASES were obtained and the enzyme activity found to be latent, that is, substrates were only accessible to the hydrolases after membrane disruption.

Lysosomes are heterogeneous in morphology, but characteristically they appear, in the electron microscope, to be organelles of ~0.5 μm diameter, often with electron-dense cores. As they are difficult to define by simple morphological features of size and shape, additional criteria are required. The histochemical reaction products of ACID PHOSPHATASE (EC 3.1.3.2) have often been used to locate lysosomes, but the best technique is IMMUNOELEC-TRON MICROSCOPY to show the presence of specific acid hydrolases and the absence of mannose 6-phosphate receptors (see below). There are often a few hundred lysosomes in a single cell, accounting for 0.5–5% of the cell volume. The majority are found in the juxtanuclear region close to the GOLGI APPARATUS, MICROTUBULE-ORGANIZING CENTRE and late ENDOSOMES. The relationship of lysosomes to some other subcellular compartments is shown in Fig. L13.

The function of lysosomes is the intracellular digestion of all main classes of biological macromolecules. Material for degradation in lysosomes can be delivered via ENDOCYTOSIS or AUTO-

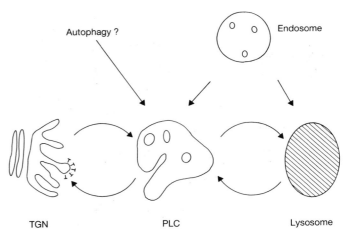

Fig. L13 The relationship of lysosomes to other subcellular compartments. Arrows show possible routes of membrane traffic. TGN, *trans* Golgi network; PLC, pre-lysosomal compartment.

PHAGY. Although other intracellular organelles, including endosomes, contain degradative enzymes, it is useful to consider lysosomes as the terminal compartment for the degradation of endocytosed or autophagocytosed material (*see also* PROTEIN DE-GRADATION).

Lysosomal enzymes and membrane proteins are synthesized on membrane-bound polysomes and delivered to lysosomes after passage through the ENDOPLASMIC RETICULUM (ER) and the Golgi complex. As this is the same route used for PROTEIN SECRETION and the delivery of plasma membrane proteins to the cell surface, sorting is necessary to target lysosomal components to their correct destination. There is convincing evidence that much of this sorting occurs in the *trans* Golgi network (TGN) (*see* GOLGI APPARATUS; PROTEIN SECRETION; PROTEIN TARGETING).

Delivery of intralysosomal enzymes

The synthesis and delivery to lysosomes of many intralysosomal enzymes is now well understood. In each case the nascent protein is co-translationally and vectorially transported across the ER membrane (*see* PROTEIN TRANSLOCATION), the N-terminal signal peptide is cleaved and N-glycosylation occurs in the ER lumen (*see* GLYCOPROTEINS AND GLYCOSYLATION). They then move by vesicular transport to the Golgi complex where their covalently linked oligosaccharides are enzymatically modified by glycosyltransferases resulting in the acquisition of terminal mannose 6-phosphate residues. This allows binding to one of two distinct MANNOSE 6-PHOSPHATE RECEPTORS which target the enzymes to lysosomes [2].

The best-studied of these receptors is the cation-independent receptor (M_r variously reported as ~215 000–300 000) which binds mannose-phosphorylated ligands at the slightly acid pH occurring in the lumen of the Golgi complex and releases them at pH <5.5. The primary amino-acid sequence of this receptor as deduced from molecular cloning suggests that it is a type I MEMBRANE PROTEIN with 15 repeats of a sequence of ~150 residues in the luminal domain and a C-terminal cytoplasmic tail of 163 amino acids (Fig. L14*a*). The receptor sorts mannose-phosphorylated lysosomal enzymes from the TGN into COATED VESICLES whose coats contain clathrin and γ-adaptin.

Immunoelectron microscopic studies have shown that the cation-independent mannose 6-phosphate receptor is not present in lysosomes themselves but is found concentrated in a smooth membrane compartment termed the pre-lysosomal compartment (PLC) which is also a late endosomal compartment (i.e. extracellular ligands taken up by receptor-mediated endocytosis appear in this compartment from ~15 min after addition to cells). The PLC contains intraorganelle membranes that may be vesicles or invaginations from its boundary membrane. The structure of the PLC suggests that it may be equivalent to or part of the population of multivesicular bodies (MVBs) observed in morphological studies of late endosomes.

Lysosomal enzymes are found at a lower concentration in the PLC than in lysosomes and it is suggested that the PLC is an intermediate compartment between the TGN and lysosomes which receives newly synthesized lysosomal enzymes associated with mannose 6-phosphate receptors from the TGN-derived coated vesicles. Membrane recycling occurs both between lysosomes and PLC and between the PLC and the TGN. This latter recycling has been reconstituted *in vitro* and requires GTP hydrolysis and an *N*-ethylmaleimide (NEM)-sensitive cytosolic factor (NSF). It also differs from vesicular transport within the Golgi complex as it is insensitive to addition of the weak base PRIMAQUINE. At present even less is known about the mechanism of delivery of lysosomal enzymes from the PLC to the lysosome. In some cells the PLC may be an extensive tubular network with occasional or intermittent continuity with lysosomes.

In addition to the high molecular weight cation-independent mannose 6-phosphate receptor, there is a second receptor with similar binding specificity but in which binding of phosphorylated oligosaccharides is cation dependent. This receptor is therefore called the cation-dependent mannose 6-phosphate receptor. Like the cation-independent receptor it is a glycosylated type I membrane protein of similar intracellular distribution, being enriched in the PLC. It has an M_r of ~46 000, exists in the membrane as a dimer, and consists of a luminal domain of 159 amino acids, a single transmembrane region and a C-terminal cytoplasmic tail of 67 amino acids (Fig. L14*b*). The transmembrane regions and cytoplasmic tails of the two receptors show no obvious HOMOLOGY, but the entire luminal domain of the cation-dependent receptor shares homologies with the repeating domains of the cation-independent receptor. Both receptors bind 1 mole of mannose 6-phosphate per mole of monomeric subunit.

Although the route of lysosomal enzymes from the TGN to lysosomes appears to be the major delivery pathway it is not the only possible route. Mannose 6-phosphate receptors are found throughout the endocytic pathway and at the cell surface the cation-independent receptor is identical to the receptor for insulin-growth factor II (IGF-II) suggesting a role in growth factor signalling when present at the cell surface (*see* GROWTH FACTOR RECEPTORS; PARENTAL GENOMIC IMPRINTING). Around 5–20% (depending on cell type) of mannose phosphorylated lysosomal enzymes are secreted from the cell before delivery to lysosomes. These enzymes may be recaptured by the surface mannose 6-phosphate receptors and delivered to lysosomes via the en-

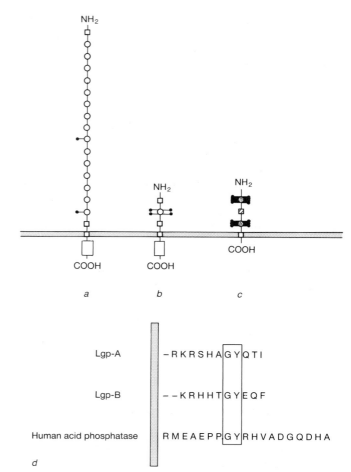

Fig. L14 Membrane proteins of the pre-lysosomal compartment and lysosomes. *a*, Cation-independent mannose 6-phosphate receptor showing 15 luminal repeats, a transmembrane domain and a C-terminal cytoplasmic tail; *b*, cation-dependent mannose 6-phosphate receptor; *c*, lgp-A showing repeat luminal domains seperated by a proline-rich region and a transmembrane domain with a short C-terminal cytoplasmic tail; *d*, the sequence of the C-terminal cytoplasmic tails of rat lgp-A, rat lgp-B and human acid phosphatase aligned relative to the membrane (shaded) to show the conserved GY motif. In *a* and *b*, known glycosylation sites are indicated by filled circles, although other putative sites are available. In *c*, putative glycosylation sites are shown by filled circles.

docytic pathway. The various intracellular and cell-surface pools of mannose 6-phosphate receptors seem to be in equilibrium and intracellular trafficking is independent of ligand binding.

Molecular mutagenesis and transfection experiments have suggested that motifs in the cytoplasmic tail of the cation-independent receptor are important in targeting — a tyrosine-containing motif for internalization from the cell surface and a tyrosine-independent region within the C-terminal 40 amino acids for sorting from the TGN to the PLC.

Other targeting signals

There is some evidence that signals other than those encoded in mannose 6-phosphate receptors may be important in targeting soluble enzymes to lysosomes. The cells from patients with I-CELL DISEASE and PSEUDO-HURLER POLYDYSTROPHY are deficient in the PHOSPHOTRANSFERASE necessary for synthesis of the mannose 6-phosphate moiety. The cells can make normal amounts of lysosomal enzymes but these are not correctly targeted to lysosomes. Patients with I-cell disease have high levels of lysosomal enzymes in their blood and their cells contain large inclusions (hence I-cell disease) containing undigested substrates. The analysis of this disease provided an early clue to the role of mannose 6-phosphate as a recognition marker for lysosomal enzyme targeting.

But despite the lack of the phosphotransferase, many cells (especially hepatocytes) in I-cell patients contain near normal levels of lysosomal enzymes, indicating the presence of a targeting pathway to lysosomes that is independent of the mannose 6-phosphate receptors. All soluble lysosomal enzymes studied so far are synthesized as preproenzymes with an N-terminal signal sequence cleaved in the ER. Conversion of proenzyme to enzyme by proteolysis occurs later in the membrane traffic pathway and it is possible that the pro-piece contains sorting information. This may be analogous to the N-terminal sequence of 30 amino acids in carboxypeptidase Y which directs this protein into the lysosome-like vacuole compartment of yeast, and which when fused to a secretory protein is sufficient to redirect the chimaeric protein to the vacuole.

Delivery of lysosomal membrane proteins

Lysosomal membranes contain an ATP-dependent PROTON PUMP responsible for maintaining the low intraorganellar pH and various specific transport proteins for amino acids, sugars and other metabolites and nutrients. They also contain, as major components, a family of highly glycosylated, acidic membrane proteins of unknown function. Several members of this family, known as lysosomal glycoproteins (lgps) have been cloned and their primary sequence deduced, showing them to be type I membrane proteins [3]. Two distinct but highly homologous proteins have been described in various species: lgp-A and lgp-B (though other names and abbreviations have been used). Individual lysosomes contain both types of lgp and they each consist of large glycosylated luminal domains, single membrane-spanning segments and short C-terminal cytoplasmic tails of 10–11 amino acids (Fig. L14c). Overall, there is ~70% similarity between the primary sequences of known lgps although the cytoplasmic domains show much higher conservation.

A further type I lysosomal membrane protein, the precursor of soluble acid phosphatase (the mature form is obtained by proteolytic cleavage), has also been cloned and the primary sequence deduced. This protein has a cytoplasmic tail of 18 amino acids which may be aligned with the cytoplasmic tails of lgps to reveal a conserved GY motif (single-letter amino-acid notation) in a similar position relative to the membrane-spanning segment (Fig. L14d). It is now clear that the lgp cytoplasmic tail contains sufficient information for lgp targeting to lysosomes. There is no glycosylation requirement for targeting in this pathway nor any interaction with the mannose 6-phosphate receptor. Compelling evidence from molecular 'cut-and-paste' experiments has shown

that attaching the lgp-A cytoplasmic tail to a cell-surface receptor glycoprotein is sufficient to target the chimaeric protein to lysosomes. Moreover SITE-DIRECTED MUTAGENESIS has shown that the Y residue in the GY motif is required for efficient targeting and its position relative to the membrane-spanning segment and C terminus must be conserved.

In addition to being concentrated in lysosomes the lgps are also found in the PLC, endosomes and, in some cells, on the cell surface. Their intracellular distribution differs with cell type, but in several cells trafficking between the various pools has been observed. Two routes of lgp delivery to lysosomes are possible, one from the TGN to the lysosomes via the PLC and the other from the TGN to the cell surface and thence to lysosomes via the endocytic pathway. The time course of delivery in PULSE-CHASE EXPERIMENTS favours the former as the major pathway, but it is clear that when lgps are expressed on the cell surface they can be internalized via clathrin-coated pits and delivered to lysosomes by the endocytic route.

The availability of antibodies to different lgps has allowed experiments using cell fusion systems to demonstrate the exchange and intermixing of lgps on different lysosomes confirming data using lysosomal content marker and underlining the dynamic nature of lysosomes.

Targeting machinery

Although there has been great progress in identifying intrinsic molecular signals required for targeting soluble proteins and lgps to lysosomes very much less is known about the cytosolic machinery that is responsible for achieving this targeting. This problem relates to the controversial issue of how endocytosed proteins are delivered to lysosomes. At one extreme it has been proposed that early peripheral endosomes can mature all the way to lysosomes, presumably collecting specific lysosomal proteins en route. At the other extreme it is proposed that all transfers on the endocytic pathway are achieved by carrier vesicles budding from one compartment and fusing with the next.

Carrier vesicles with appropriate properties have been identified in various cells and tissues and a cell-free system has been developed from rat liver to show the interaction of putative carrier vesicles with lysosomes. This has demonstrated the need for cytosolic proteins, NSFs, ATP and GTP for maximal interaction. It is likely that cell-free systems will prove as important in developing a knowledge of delivery to lysosomes as they have been in understanding events early in the endocytic pathway and the mechanism of trafficking through the Golgi complex.

Other routes

Even less is known about other delivery routes to lysosomes. In the case of particles that are phagocytosed, intracellular fusion of the PHAGOSOME occurs with one or more lysosomes to form a PHAGOLYSOSOME. Phagosomes also fuse with endosomes. It has so far only been possible to reconstitute phagosome fusion with endosomes *in vitro* and not phagosome fusion with lysosomes.

The development of an autophagic vacuole by smooth membrane surrounding a portion of cytoplasm seems to be an early event in autophagy. However the route of delivery from this organelle to the lysosome is not well described though it may involve interaction with late endosomal compartments or direct fusion with lysosomes.

At least one pathway by which cytosolic proteins can enter lysosomes directly has been described and subjected to some molecular analysis. This pathway occurs in cultured cells that are responding to withdrawal of serum growth factors. There is selective degradation of cytosolic proteins containing peptide motifs related to KFER (*see* PROTEIN DEGRADATION).

There is also evidence that some lysosomal membrane proteins enter lysosomes from the plasma membrane.

Lysosomal storage diseases

The important part that one lysosomal storage disease (I-cell disease) played in defining the normal pathway of targeting lysosomal enzymes has already been mentioned. Various other lysosomal diseases exist. Most are storage diseases in which the absence of a specific lysosomal enzyme leads to accumulation of a particular substrate in the lysosomes (*see* GAUCHER'S DISEASE; HUNTER SYNDROME; HURLER'S DISEASE; LIPIDOSES). Recessive mutations are responsible and the diseases are rare, although they can be more frequent in particular ethnic groups. Other abnormalities include disorders in transport across the lysosome membrane and disorders which may be explained, for example, by the inability of normal lysosomes to cope with abnormal substrates. Together these diseases emphasize the importance of lysosomes in the normal life of cells and organisms.

Lysosomal fate

Whereas we now know a great deal about the biogenesis of lysosomes by studying defined protein components, considerable mystery surrounds their fate. On the basis of microscopic evidence it has been claimed that lysosomes can become RESIDUAL BODIES. These are dense bodies containing acid hydrolases, materials resistant to degradation and degradation products which are slow to escape. It is not known whether such bodies always fuse with other lysosomes and endocytic vesicles or whether there is another route of demise. In liver the discharge of lysosomal contents into bile has been reported and it is possible that excretion into the extracellular environment is a more general route for discharge of undigestible lysosomal contents.

J.P. LUZIO

1 Holtzmann, E. (1989) *Lysosomes* (Plenum, New York).
2 Kornfeld, S. (1987) Trafficking of lysosomal enzymes. *FASEB J.* 1, 462–468.
3 Kornfeld, S. & Mellman, I. (1989) The biogenesis of lysosomes. *Annu. Rev. Cell Biol.* 5, 483–525.

lysozyme Small (M_r 14 600) hydrolytic enzyme (EC 3.2.1.17) of ubiquitous distribution, found especially in tears and other animal secretions, and in hen egg-white, which hydrolyses 1,4-β linkages between *N*-acetylmuramic acid and *N*-acetyl-D-glucosamine in peptidoglycans of bacterial cell walls (*see* BACTERIAL ENVELOPES) and between *N*-acetyl-D-glucosamine residues in fragments of CHITINS. *See:* Fig. P83*b* in PROTEIN STRUCTURE for illustration.

lytic cycle/pathway Pathway of virus replication that results in multiplication of viral particles and lysis of the infected cell. *See:* ANIMAL VIRUSES; BACTERIOPHAGES; LAMBDA.

lyticase *See:* ZYMOLYASE.

M

M The single-letter abbreviation for the AMINO ACID methionine.

M_r RELATIVE MOLECULAR MASS.

M13 A BACTERIOPHAGE of *Escherichia coli* which infects only F$^+$ strains as infection occurs through the F sex PILUS. The viral genome consists of single-stranded DNA, although a double-stranded genome is formed at an intermediate stage during viral replication (*see* Fig. R35 in ROLLING CIRCLE REPLICATION). During the infective cycle, virus particles are secreted into the culture medium by a membrane-budding process. The culture supernatant represents a fairly pure source of single-stranded viral genome, a property exploited during the preparation of single-stranded DNA templates for dideoxy chain termination DNA SEQUENCING. The single-stranded/double-stranded nature of M13 is also exploited in the method of OLIGONUCLEOTIDE-MEDIATED MUTAGENESIS.

M-CSF (CSF-1) Macrophage colony-stimulating factor. In humans it exists in two forms; both are homodimers which do not retain activity if dissociated. The 'long form' (M_r 70 000–90 000) exists as a GLYCOPROTEIN complex of two 35–45K subunits of ~223 amino acids each, whereas the 'short form' (40–50K natural glycoprotein) is a dimer of 20–25K subunits each of ~145 amino acids. M-CSF is normally produced by various cell types, including stromal cells and macrophages, and is present in urine. The human M-CSF gene is located on chromosome 1p13-21 in numerous exons spread over 22 kilobases of DNA and produces two species of mRNA, the translated polypeptides of which share a common C-terminal sequence and are then glycosylated and associate as homodimers.

M-CSF stimulates the growth of monocyte precursors and can promote the continued growth of mature macrophages. Purified recombinant M-CSF retains its biological activity *in vivo*. The M-CSF receptor (M_r 165K) is encoded by the c-*fms* proto-oncogene (*see* ONCOGENES) which is found on human chromosome 5, and is a member of the IMMUNOGLOBULIN SUPERFAMILY. Rearrangements of this region are associated with refractory anaemia, myelodysplastic syndromes, and therapy-associated acute myeloid LEUKAEMIA.

Kawasaki, E.S. & Ladner, M.B. (1990) In *Colony-stimulating Factors — Molecular and Cellular Biology* (Dexter, T.M. et al., Eds) Ch. 5, 155–176 (Marcel Dekker, New York).

M phase Stage of eukaryotic CELL CYCLE in which MITOSIS occurs.

M subunit The 'medium' subunit of the photosynthetic reaction centre of purple bacteria (M_r 34 000) which has sequence and structural homology with the L subunit; the LM heterodimer binds the pigments and cofactors involved in the primary photochemical charge separation. It is topologically and functionally similar to subunit D2 of plant cell PHOTOSYSTEM II. *See also*: PHOTOSYNTHESIS.

machlovirus group Sigla from maize chlorotic dwarf virus, the type member. Single component viruses with isometric particles 30 nm in diameter. Each comprises a molecule of (+)-strand linear RNA (M_r 3.2×10^6) with the capsid made up of two species of protein (M_r 18 000 and 30 000). There is no information yet on genome organization. *See also*: PLANT VIRUSES.

Gingery, R.E. (1988) In *The Plant Viruses*, Vol. 3 (Koenig, R., Ed.) 259–272 (Plenum, New York).

macroautophagy *See*: PROTEIN DEGRADATION.

macromere *See*: SPIRAL CLEAVAGE.

macromolecule General term for a large complex organic molecule such as a protein, nucleic acid, or large polysaccharide.

macrophage Mononuclear phagocyte, a phagocytic cell of the myeloid lineage found in all organs and connective tissue. A 'professional' ANTIGEN-PRESENTING CELL, it phagocytoses and kills bacteria, viruses, and foreign particles, and processes them for presentation to T CELLS (*see* ANTIGEN PROCESSING AND PRESENTATION). Macrophages are also involved in clearing antigen–antibody complexes from the tissues (*see* ENDOCYTOSIS). Cytokines produced by activated helper T cells stimulate phagocytosis and other macrophage functions. *See also*: HAEMATOPOIESIS; MONOCYTE.

macrophage colony-stimulating factor *See*: M-CSF.

macropinocytosis *See*: ENDOCYTOSIS.

macula adh(a)erens Desmosome. *See*: CELL JUNCTIONS.

MAD Multi-wavelength anomalous dispersion. A technique useful in X-RAY CRYSTALLOGRAPHY whereby measurements of ANOMALOUS DISPERSION, usually made at a SYNCHROTRON, provide information to solve the PHASE PROBLEM.

MADS box A conserved presumed DNA-binding sequence present in the products of some plant HOMEOTIC GENES involved in specifying floral identity (*see* FLOWER DEVELOPMENT) and in the human SRF and yeast McM1 TRANSCRIPTION FACTORS.

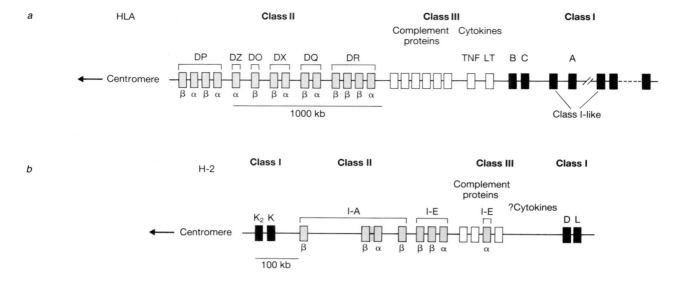

Fig. M1 Schematic maps of the MHC regions of (*a*) human and (*b*) mouse. The number of α and β genes in each human class II locus varies with different alleles. Class I and class II genes are each composed of multiple exons. The α₁, α₂, and α₃ domains are each encoded in a single exon.

magic spot The name given to a compound synthesized when *Escherichia coli* cultures are starved of amino acids. Under such conditions the cells respond by shutting down production of rRNA, ribosomal proteins and purine nucleotides, the so-called STRINGENT RESPONSE. *rel* (or relaxed) mutant strains of *E. coli* do not respond in this manner to amino acid starvation, and do not synthesize magic spot. Magic spot is composed of two guanosine polyphosphates, ppGpp and pppGpp. These are thought to act in a complex pathway regulating transcription of rRNA.

magnesium Mg^{2+}. Cation essential for the activity of many enzymes. ATP is usually complexed with magnesium within the cell to form MgATP.

magnesium blockade Voltage-sensitive block of the NMDA receptor channel by the divalent cation Mg^{2+}. At normal cell resting potentials the channel is blocked by Mg^{2+} but this blockade is relieved when the membrane is DEPOLARIZED by activation of non-NMDA receptors. *See*: EXCITATORY AMINO ACID RECEPTORS.

The major histocompatibility complex (MHC)

THE major histocompatibility complex (MHC) is a cluster of closely linked genetic loci encoding three different classes (class I, class II, class III, see below) of polypeptide products involved in the generation and regulation of immune responses [1]. The MHC is divided into a series of regions or subregions (Fig. M1) and each region contains multiple loci. An MHC is present in all vertebrates, and the mouse MHC (*see* H-2) on chromosome 17 and the human MHC (*see* HLA) on chromosome 6 are the best characterized. MHC protein products — often simply termed MHC MOLECULES — were initially detected serologically as antigens controlling tissue transplant rejection between genetically nonidentical individuals and were thus often called MHC antigens. Subsequently, genes mapping to this chromosomal region were shown to encode proteins that control T cell killing of virus-infected cells, immune surveillance and the ability to mount an ANTIBODY response against certain foreign antigens.

Within each region, the gene products encoded by each locus are related structurally and functionally. Class I and II genes are multiallelic and highly POLYMORPHIC and are now known to encode cell-surface proteins involved in the presentation of protein antigens to T cells during the generation of an immune response (*see* ANTIGEN PROCESSING AND PRESENTATION). Class III genes encode various serum proteins, including some components of the COMPLEMENT system. The genes for the putative peptide transport proteins also map to this region.

Class I MHC

Genetic organization

Class I MHC genes encode the ubiquitously expressed classical TRANSPLANTATION ANTIGENS, as well as the murine Qa and TL cell-surface antigens. Transplantation antigens are encoded by loci in the mouse H-2K, D, and L subregions and human HLA-A, -B, and -C subregions (see Fig. M1). Class I loci are polymorphic and as a result of heterozygosity, more than one of each type of class I molecule is often expressed by one individual (*see* HAPLOTYPE). Nucleotide sequence comparison of different murine class I alleles suggests that the extensive sequence polymorphism may have been generated by multiple GENE CONVERSION and GENE DUPLICATION and deletion events, presumably via homologous but unequal crossing-over (*see* MOLECULAR EVOLUTION; RECOMBI-

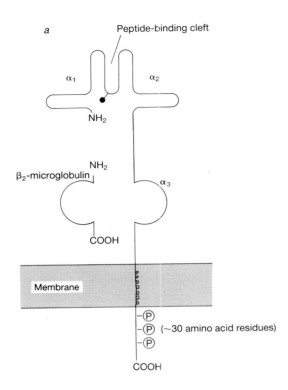

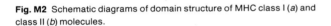

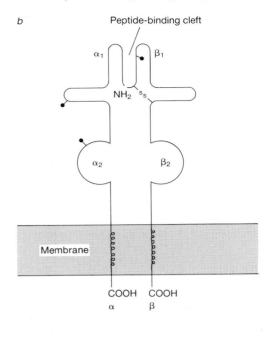

Fig. M2 Schematic diagrams of domain structure of MHC class I (*a*) and class II (*b*) molecules.

NATION). Alleles at each class I locus exhibit predictable patterns of amino acid variability, with the N-terminal regions being more variable than the C-terminal regions.

Class I MHC protein structure

Class I MHC molecules are integral membrane GLYCOPROTEINS composed of two polypeptide chains: an MHC-encoded transmembrane polypeptide of M_r 45 000 (45K) noncovalently associated with a nonpolymorphic extracellular polypeptide, β_2-MICROGLOBULIN (β_2-m) of M_r 12 000, encoded on chromosome 2 in the mouse and chromosome 15 in the human. Expression of class I proteins and β_2-m are coordinately regulated, probably at the transcriptional level. The MHC gene encoding the 'heavy' polypeptide chain is composed of multiple exons, each corresponding to a separate polypeptide domain. The class I polypeptide is thus divided into three external structural domains (α1–α3, each ~90 residues), a hydrophobic transmembrane region (40 residues) and a short cytoplasmic tail (30 residues); the complete cell-surface class I antigen is composed of four extracellular protein domains, three contributed by the class I MHC-encoded polypeptide, and the fourth domain from β_2-m (Fig. M2). Protein sequence comparisons between class I molecules and immunoglobulin heavy and light chain constant domains and the X-ray crystallographic structure determination of an HLA class I molecule [2] show that the α3 domain of class I MHC molecules and the single domain comprising β_2-m exhibit immunoglobulin-like secondary and tertiary folding and are thus considered

members of the IMMUNOGLOBULIN SUPERFAMILY. Interaction of the class I polypeptide with β_2-m occurs through this external domain, and the T cell co-receptor molecule CD8 also binds to the α3 domain. An immune control function has not been ascribed to β_2-m but the molecule appears at least to provide structural support for the folded class I molecule. The other two class I external protein domains determine the binding of endogenously derived protein antigens and binding to the T CELL RECEPTOR.

The X-ray crystal structure of class I MHC molecules shows a deep peptide-binding groove along the surface of the molecule, formed by the folding of the α1 and α2 protein domains (Fig. M3), and most class I polymorphic residues (i.e. those that differ from one allelic form to another) map either within this groove or along its outside edge. This clustering of polymorphic residues at the antigen-binding site increases its potential for structural variation and thus its potential for binding a wide spectrum of peptide fragments with differing relative affinities.

The peptide-binding site on any individual MHC molecule can apparently bind a wide range of peptide fragments. Class I molecules bind peptides around nine amino acids long. The peptide lies in an extended conformation along the groove, making contact with certain residues only.

Interaction between the class I molecule and T cell receptor is thought to occur through the polymorphic regions at the outer edge of the antigen-binding groove.

Function of class I MHC molecules

The normal biological function of class I MHC molecules is to present, on the cell surface, antigenic peptide fragments of intracellularly generated foreign protein antigens in a form that T

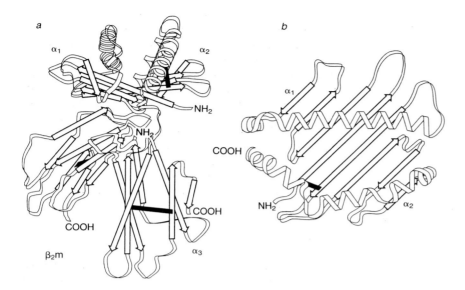

Fig. M3 Crystal structure of a class I MHC molecule, HLA-A2. *a*, Side view; *b*, view looking down on the peptide-binding pocket. $\beta_2 m$, β_2-microglobulin. From [2].

cells can recognize (*see* ANTIGEN PROCESSING AND PRESENTATION; MHC RESTRICTION; T CELL RECEPTOR). The class I molecule forms a complex with a specific viral antigen, processed and degraded intracellularly to a short peptide fragment and it is this complex that is recognized as 'altered self' MHC and bound by a T cell receptor on a CYTOTOXIC T CELL as the first step in triggering lysis of a virus-infected cell.

Some tumour-associated protein antigens also bind to class I molecules on the surfaces of neoplastically transformed target cells, and recognition of these complexes by cytotoxic lymphocytes and the ensuing lysis of the transformed cell is a mechanism of tumour surveillance. Many cytotoxic T cells can also lyse uninfected target cells bearing a foreign (allo) class I MHC allele (*see* ALLO-MHC). The non-self class I allele is recognized by the cytotoxic T cell receptor in much the same manner as the self MHC–antigen complex. This recognition of foreign class I alleles, also known as alloreactivity, is one reason for the rejection of transplants between unmatched MHC individuals (*see* HISTOCOMPATIBILITY; TRANSPLANTATION ANTIGEN).

Qa and TL antigens

The mouse Qa and TL lymphoid DIFFERENTIATION ANTIGENS are also encoded by multiple loci but, unlike other class I products, are expressed only on some cell types: Qa antigens are expressed only on cells of haematopoietic lineages, and TL antigens are expressed only on thymocytes and some leukaemic cells. The Qa and TL antigens show structural homology to transplantation antigens: a 'heavy' MHC-encoded transmembrane polypeptide of M_r 45 000 is joined noncovalently to β_2-m. The MHC-encoded subunit exhibits a domain structure similar to other class I products.

Class II MHC

Class II MHC molecules have a more restricted tissue distribution than class I and are involved in presenting antigens mainly to

HELPER T CELLS, a crucial step in the generation of an immune response. The class II MHC genes have therefore been termed IMMUNE RESPONSE GENES (Ir genes), as before their mechanism of action was discovered they were recognized as critical in controlling the body's ability to make an immune response to certain synthetic peptide antigens.

There are two biochemically defined class II molecules in the mouse — I-A and I-E — encoded by polymorphic loci mapping to the I region of the murine MHC (Fig. M1). These molecules are also known as Ia antigens. In the human, there are at least three types of class II MHC molecules: HLA-DR, DQ, and DP (Fig. M1). DZ, DO, and DX may be pseudogenes or encode products not yet discovered. There are numerous class I-like genes beyond HLA-A outside the MHC region. Sequence homologies show that murine I-E and I-A gene complexes are homologous to the human DR and DQ complexes respectively.

Structure of class II MHC molecules

Class II MHC molecules are heterodimeric glycoproteins. The complete cell-surface class II molecule is composed of two noncovalently associated integral membrane polypeptide chains, a 'heavy' α subunit (M_r 34 000) and a 'light' β subunit (M_r 29 000). Unlike class I MHC molecules, both polypeptide subunits are encoded within the MHC.

Class II molecules have an overall domain structure similar to that of class I MHC molecules. For class II molecules, however, both subunits are structurally similar, with two external domains (each ~90 residues), a transmembrane region (20–25 residues), and a cytoplasmic tail (15–20 residues). Thus, like the folded class I molecule, the assembled cell-surface class II molecule comprises four extracellular protein domains ($\alpha 1$, $\alpha 2$, $\beta 1$, $\beta 2$) (Fig. M2). $\alpha 2$ and $\beta 2$ domains show sequence homology to immunoglobulin constant regions, and are therefore probably folded into immunoglobulin-like domains (*see* IMMUNOGLOBULIN SUPERFAMILY).

The α1 and β1 domains of class II polypeptides adopt a similar tertiary structure to the class I molecule, so that folding of these domains creates a 'pocket' to bind antigenic peptide fragments and present them to appropriate T cell receptors [3] (*see* Plate 6*d*).

Class II molecules show extreme polymorphism: two alleles may encode polypeptides that differ by up to 20 amino acids. This polymorphism is localized largely to 'hypervariable' regions within the N-terminal α1 and β1 domains, mapping either within or adjacent to the putative antigen-binding site of the folded molecule. It is this variability that provides the structural basis for class II-mediated regulation of immune responsiveness: antigenic peptides are bound differentially by the products of different class II alleles, and elements of the combined tertiary structure of this class II–antigen complex are recognized by specific T cell receptors.

A third non-MHC-encoded polypeptide subunit, the INVARIANT CHAIN (I$_i$), is intracellularly associated with the class II α and β chains. This protein (M_r 31 000) is not polymorphic and has not been detected in association with cell-surface MHC II molecules. Its function seems to be to direct intracellular transport and assembly of α and β chains to the cell surface (*see* ANTIGEN PROCESSING AND PRESENTATION).

Class II gene organization

The structural genes encoding both subunits, α and β, of the murine I-A complex, as well as the β subunit of the I-E complex, map to the I-A subregion of the murine H-2 complex (Fig. M1), while a locus within the H-2 I-E subregion encodes the Eα subunit of the I-E complex. The human HLA-DR and DP subregions contain multiple expressed DR and DP β loci and the DP subregion also contains multiple DP α loci (Fig. M1). This multiplicity of class II heavy and light chain loci within an antigen family increases the combinatorial diversity of class II MHC molecules in the human.

Function of class II MHC molecules

Class II MHC molecules are expressed predominantly on the surfaces of B lymphocytes and antigen-presenting macrophages (*see* ANTIGEN-PRESENTING CELL) and are also found on a substantial proportion of activated T cells. Secreted, soluble factors involved in immune regulation have also been shown to carry class II or other I region-encoded determinants (*see* I-J).

Class II molecules regulate the activation of antigen-specific, MHC-restricted helper T cells which are required for the activation of cytotoxic T lymphocytes and of antibody-producing B cells. Stimulation of helper T cells by foreign protein antigens requires that the antigen be first taken up by an antigen-presenting cell and enter an intracellular acidic compartment in which it is processed into peptide fragments. These are then presented on the cell surface in a complex with a class II molecule.

Presumably owing to the structural complexity of the antigen-binding site on the class II MHC molecule and its high degree of polymorphism, class II MHC molecules can discriminate subtle differences in the structures of antigens such as insulin and lysozyme; whether an individual is an immunological 'responder' or 'nonresponder' to a particular foreign antigen is determined by its class II genotype, which in turn dictates the fine tertiary structure of its class II proteins.

Like class I responses, MHC class II-directed immune responses are restricted by the requirement for T cell recognition of self-MHC molecules on antigen-presenting cells. T cell clones able to recognize self MHC molecules are selected during ontogeny in the thymus (*see* T CELL DEVELOPMENT).

T cell receptors for antigen can also recognize non-self class II MHC molecules, either bound to self peptide so that the T cell receptor simultaneously recognizes both self antigen and MHC molecule, or through the polymorphic residues on allelic class II molecules alone.

Class III MHC

Class III genes encode several serum protease components of the complement cascade. The MHC contains loci encoding three complement components — C2, C4, and factor B.

Structurally and functionally, these MHC products and the genes encoding them are distinct from the genes and products of the MHC class I and II loci, suggesting that perhaps class III genes should not be viewed as an integral part of the MHC, but may instead be comparable to the several MHC-linked genes encoding enzymes, such as neuraminidase, uninvolved in immune regulation.

The polyallelic loci encoding C2, C4, and factor B lie between the HLA-D and HLA-B loci in the human and within the H-2S region of the murine MHC. Pulsed-field gel ELECTROPHORESIS has shown that the class III region spans ~1100 kilobases (kb) in both mouse and human and class III genes are separated from flanking class I loci on the telomeric side by at least 400 kb and from class II loci on the centromeric side by about 400 kb.

C2 and factor B are encoded by single closely linked loci in both human and mouse. The C2 gene encodes the proenzyme C2, an early-acting serine protease active in the classical complement pathway; the gene for proenzyme factor B, a serine protease in the alternative pathway, maps within 1 kb of the C2 gene. C4 is synthesized as a propolypeptide, which is post-translationally processed into its final active multichain form.

The three-chain C4 molecule is an early-acting enzymatic component of the classical complement pathway. In the human, C4 is encoded by two genes, C4A and C4B, which map within 50 kb centromeric of the C2 and factor B gene cluster, and within 10 kb of one another. C4A and C4B encode two isotypic forms of C4 protein that differ by <1% of nucleotide sequence. In the mouse, there are two C4-like genes, C4 and Slp, that map within 80 kb of one another and within 50 kb of C2 and factor B. The two isotypic protein forms share ~94% homology. However, Slp is not activated in the complement cascade as is C4, and has no other known function.

The peptide transporters thought to be involved in the transport of antigenic peptides from the cytoplasm into the endoplasmic reticulum also appear to be encoded within the MHC III (*see* ANTIGEN PROCESSING AND PRESENTATION).

L. MENGLE-GAW

1 Abbas, A.K. et al. (1991) *Cellular and Molecular Immunology*, Ch. 5 (W.B. Saunders, Philadelphia), and references therein.
2 Bjorkman, P.J. et al. (1987) Structure of the human class I histocompatibility antigen HLA-A2. *Nature* **329**, 506–512.
3 Brown, J.H. et al. (1988) A hypothetical model for the foreign antigen-binding site of class II histocompatibility molecules. *Nature* **332**, 845–850.

malaria Disease caused by four species of the protozoan parasite *Plasmodium*: *P. falciparum*, *P. vivax*, *P. ovale*, and *P. malariae*. *P. falciparum* causes the greatest morbidity and mortality. The vectors for malaria are anopheline mosquitoes which transmit the parasite in its sporozoite form to humans during a blood meal. The parasite then proliferates in a complex life cycle, first in the liver, forming merozoites, and then in red blood cells (forming merozoites and gametocytes). Fever, anaemia, and reduced tissue blood flow are consequences of parasitaemia, with cerebral complications in some cases. The infection predominantly occurs in tropical countries of Africa, Asia and Latin America. Research into possible vaccines against malaria has increased with the advent of recombinant DNA techniques and the ability to clone parasite antigens. People who are HETEROZYGOUS for the sickle-cell mutant of haemoglobin show increased resistance to malaria (*see* HAEMOGLOBIN AND ITS DISORDERS; HAEMOGLOBIN S). The malaria parasite exists for part of its life cycle in the red blood cell, where it reduces the pH of the cell slightly, but sufficiently to increase the tendency of the cell to sickle in patients carrying a mutation for sickle cell haemoglobin. The mutant haemoglobin forms fibres which deform the cell, leading to sickling. The cell membrane becomes more permeable to potassium ions on sickling, resulting in a leakage of potassium out of the cell leaving a lowered intracellular concentration of potassium which kills the parasite. This resistance to malaria in individuals heterozygous for the sickle cell mutation has led to the increased frequency of the mutation in populations exposed to malaria, even though in the homozygous state the mutation is generally lethal if the condition is untreated.

male fertile Lines in which the male plant is fertile. Generally used in contrast to MALE STERILE.

male precocious puberty The rare AUTOSOMAL DOMINANT genetic defect familial male precocious puberty is a gonadotropin-independent disorder restricted to males, in which signs of puberty develop as early as the age of 4. It is caused by a mutation in the G PROTEIN-COUPLED RECEPTOR for LUTEINIZING HORMONE which in males is expressed in the Leydig cells of the testis. The mutant receptor is constitutively active, resulting in Leydig cell hyperfunction and hyperplasia.

male sterile Lines in which the male plant is sterile, but the female plant is fertile. Fertility can often be restored by other mutations. Some of the best studied occurrences of male sterility are determined by the mitochondrial genome. *See*: MITOCHONDRIAL GENOMES: PLANT.

malignant Term applied to tumour cells that have the capacity for uncontrolled growth and the ability to metastasize (to invade underlying tissues and move to new sites in the body).

malonyl ACC (MACC) 1-(Malonylamino)cyclopropane-1-carboxylic acid is the product of cytosolic malonylation by D-amino acid malonyltransferase of ACC, the precursor of the gaseous plant hormone ETHYLENE. It is sequestered in the vacuole after ATP-dependent, carrier-mediated transport across the tonoplast. Vacuolar MACC is not metabolized further although acylhydrolases may effect this reversal in the cytoplasm.

Yang, S.F. et al. (1985) In *Ethylene and Plant Development* (Roberts, J.A. & Tucker, G.A., Eds) 9–21 (Butterworths, London).

malting The process of preparing barley or other grain for brewing or spirit distillation by steeping, germinating under carefully controlled conditions of moisture and temperature, usually on a malting-floor, and finally kiln-drying. During the treatment, α-amylase and maltase activities increase markedly, to make glucose available from starch in the subsequent fermentation.

Mammalian development

THOSE features special to mammalian development [1,2] are determined largely by three aspects of the reproductive process in eutherian mammals.

1 Internal fertilization. Spermatozoa deposited in the female reproductive tract must pass through the uterus to the site of fertilization. During this passage, they are matured in uterine secretions. However, the embryo, after fertilization, must return by the same route to implant in the uterus and this process has quite different secretory requirements. Several days are required to convert the female tract from a sperm-supporting to an embryo-supporting state. The early period of development is in consequence relatively extended such that by 3–4 days the mouse embryo, for example, has only about 120 cells (Fig. M4, see p. 608).

2 Young are born at a relatively mature post-embryonic stage. The advanced development, growth and maturity of the mammal at birth requires extended maternal nutrition via the placental membrane system. The oocyte is small (50–100 μm) and nonyolky and the early stages of development (to about day 7 in the mouse and day 14 in the human) are dominated by the development of these extraembryonic support tissues (which comprise some 99% of the total conceptus). The formation of a discrete group of cells with an exclusively embryonic developmental potential does not occur until this period of extraembryo formation is over. In consequence, the development of embryonic axes is late (day 7 in mouse and 14 in human; Fig. M4).

3 Obligate sexual reproduction. Activation of mammalian oocytes by means other than spermatozoa can induce development to post-implantation stages in mouse and pre-implantation stages in human but further development has not been observed. Reversion to an asexual reproductive strategy as has occurred in other vertebrates does not occur in mammals. Pronuclear transplantation studies have revealed that some GENE LOCI on the chromosome sets derived from the oocyte differ in their accessibility for TRANSCRIPTION during embryogenesis from the same loci derived from the sperm. These parental imprinting effects are a consequence of the distinctive experiences of spermatogenesis and oogenesis and mean that one set of paternal and one set of maternal chromosomes are essential for complete development to occur (*see also* PARENTAL GENOMIC IMPRINTING).

Mammalian development is difficult to study because it is internal, it is relatively slow, generation times are long, few interesting mutations are available for pre-embryonic and embryonic stages, and, at early stages, material available for conventional biochemical analysis is limiting. Some progress has been made for a few species using *in vitro* fertilization and culture to study early development. Limited culture of mouse and rat embryos from the post-implantation stage through basic embryogenesis has been achieved. Microbiochemical and molecular biological techniques (e.g. two-dimensional gel analysis of proteins (*see* ELECTROPHORESIS), metabolic assays, the amplification of selected transcripts by REVERSE TRANSCRIPTION and POLYMERASE CHAIN REACTION) are being applied to early embryonic stages and the detection of specific gene expression by *in situ* HYBRIDIZATION is being achieved. The identification in mouse and human of important developmental genes HOMOLOGOUS to those identified in *Drosophila* by mutation studies has opened up genetic approaches to mammalian development (*see e.g.* HOMEOBOX GENES AND HOMEODOMAIN PROTEINS). This approach is being enhanced by the ability to target mutations to genes of interest in embryonal stem cells *in vitro* and then to incorporate the mutated cells into the germ lines of embryos (*see* TRANSGENIC TECHNOLOGIES). Advantage has been taken of the relatively slow pace of early mammalian development to apply techniques of cell biology to the study of CELL–CELL INTERACTIONS and CELL–MATRIX INTERACTIONS and the CELL CYCLE in early development. The successful study of development, largely in the mouse, has been translated to the clinic to provide therapy for infertility, study of early human developmental abnormalities, and the potential for preimplantation diagnosis of genetic disease (*see* PRENATAL DIAGNOSIS), and GENE THERAPY.

Progress in the basic scientific study of mammalian development and the use of new technologies is likely to increase the pace of transfer from laboratory to clinic. A final justification for the study of mammalian development is the intrinsic fascination of studying a process through which each of you reading this entry passed.

M.H. JOHNSON

See also: AMPHIBIAN DEVELOPMENT; AVIAN DEVELOPMENT; CAENORHABDITIS DEVELOPMENT; DROSOPHILA DEVELOPMENT; INDUCTION; PATTERN FORMATION; VERTEBRATE NEURAL DEVELOPMENT; ZEBRAFISH DEVELOPMENT.

1 Kaufman, M.H. (1992) *The Atlas of Mouse Development* (Academic Press, New York).
2 Johnson, M.H. & Everitt, B.J. (1995) *Essential Reproduction*, 4th edn (Blackwell Scientific Publications, Oxford).

manic-depressive illness *See*: AFFECTIVE DISORDERS.

mannan Polymer of mannose residues. *See*: CELLULOSE; GALACTOGLUCOMANNAN; GLYCANS; GLYCOPROTEIN.

mannopine (MOP) An OPINE, the product of conjugating mannose and glutamine. The gene determining its synthesis is found in the T_R-regions of octopine TI PLASMIDS and of RI PLASMIDS. Mannopinic acid (MOA) is the product of conjugating mannose and glutamic acid.

mannose *See*: SUGARS.

mannose 6-phosphate receptor One of two related type I membrane GLYCOPROTEINS, which mediate the sorting of lysosomal enzymes that have been phosphorylated on mannose residues (*see* LYSOSOMES). The larger of the two receptors, the cation-independent mannose 6-phosphate receptor (M_r 215 000), has a noncytoplasmic domain consisting of 15 repeating units and has also been identified as the receptor for insulin growth factor II (IGFII) (*see* GROWTH FACTOR RECEPTORS). The smaller receptor, the cation-dependent mannose 6-phosphate receptor (M_r 46 000), has only one such unit. There are two pathways for receptor-mediated lysosomal enzyme delivery to the prelysosome, both involving clathrin-coated vesicles: directly from the *trans* Golgi network, or via the plasma membrane (*see* COATED PITS AND VESICLES). The direct pathway is thought to be the major one.

mannose-binding proteins In mammals, serum proteins that bind to oligomannose cell-surface oligosaccharides of bacteria and other pathogenic microorganisms, inducing COMPLEMENT-mediated lysis or opsonization. They are members of the ANIMAL LECTIN family.

mannosidases Enzymes (α-D- and β-D-mannosidases, EC 3.2.1.24 and 25) that catalyse the hydrolysis of the glycosidic linkage between a mannose residue and another molecule. α-Mannosidases of the ENDOPLASMIC RETICULUM and GOLGI APPARATUS are involved in processing the N-linked oligosaccharides of GLYCOPROTEINS.

MAP Microtubule-associated protein. *See*: MICROTUBULES; NEURONAL CYTOSKELETON; MICROTUBULE-BASED MOTORS. *See also*: MAP KINASE; MAP KINASE KINASE.

map distance The distance between two loci on the genetic map is measured in terms of the likelihood of a RECOMBINATION event occurring between them. Map distances are measured in centimorgans (cM), with 1 cM approximating to a recombination frequency of 1%. The correlation between genetic distances (in cM) and physical distances (in base pairs) is not exact, as rates of recombination are not constant throughout the genome.

MAP kinase A serine/threonine-specific PROTEIN KINASE which is activated by treatment of cells with agents promoting proliferation and/or differentiation (e.g. insulin, epidermal growth factor, nerve growth factor, antigen (for T cells)). The protein kinase is normally assayed using microtubule-associated protein-2 or MYELIN basic protein as substrate, and was originally termed microtubule-associated protein-2 kinase (MAP2 kinase). However, with the realization that it is activated by numerous extracellular messengers, many of which are MITOGENS, the

608

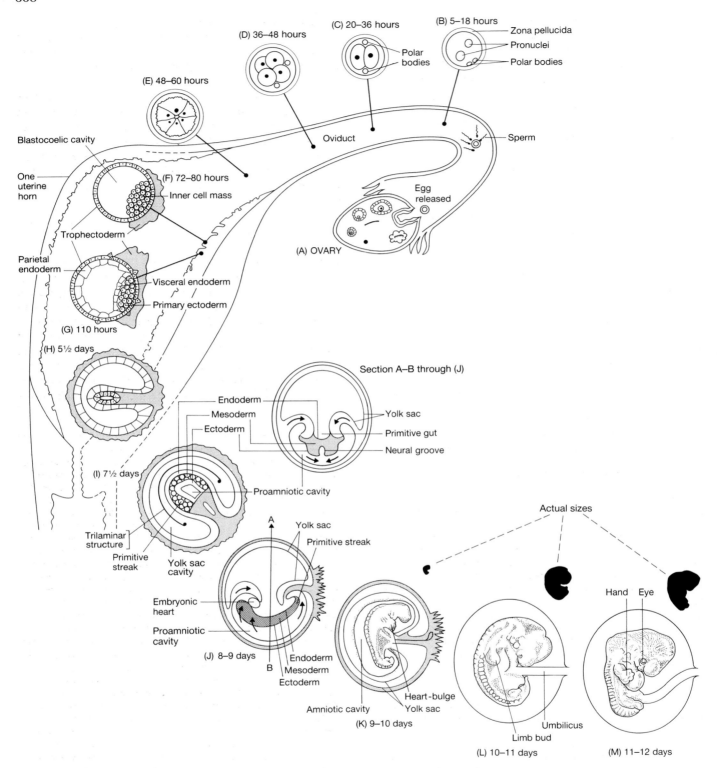

Fig. M4 The development of the mouse (see p. 606). Mature oocytes are released from the ovarian follicle (A) to enter the oviduct (Fallopian tube). They are covered by a translucent acellular membrane called the zona pellucida which is retained for 3–4 days until just before implantation. At ovulation, oocytes are arrested in the second meiotic metaphase division (see MEIOSIS). Only when fertilized by a spermatozoon is meiosis resumed and the second polar body extruded; 5–9 h later (B), two pronuclei (one from each gamete) form and the first round of DNA synthesis commences. This process lasts about 9 h as the two pronuclei migrate centrally and 18–20 h after fertilization the pronuclear membranes break down, a mitotic spindle forms (see MITOSIS) and the two sets of parental chromosomes come together for the first time on the metaphase plate.

name has been changed to mitogen- or messenger-activated protein kinase (MAP kinase). It exists as at least two ISOFORMS of M_r 43 000 and 42 000, which are also known as ERK1/ERK2 (extracellular signal-regulated kinase 1/2). MAP kinase is only active after phosphorylation on both tyrosine and threonine residues, which seem to be phosphorylated by a single downstream kinase (MAP kinase kinase). MAP kinase is likely to be a central component in the response to growth and/or differentiation stimuli, and its unusual requirement for phosphorylation on both tyrosine and threonine residues may be to prevent accidental activation by other protein kinases.

Boulton, T.G. et al. (1991) *Cell* **65**, 663–675.

MAP kinase kinase A PROTEIN KINASE which activates the protein kinase MAP KINASE by phosphorylating it on both tyrosine and threonine residues. It is itself activated by phosphorylation by the Raf protein. *See also*: PROTEIN PHOSPHORYLATION; RAF; RAS.

map unit The unit of distance used in genetic mapping is the centimorgan (cM) (*see* MAP DISTANCE). One centimorgan approximates to a RECOMBINATION frequency of 1%.

mapping *See*: HUMAN GENE MAPPING; HUMAN GENOME PROJECT; RESTRICTION MAPPING.

mapping function Mapping functions are used to interconvert physical distances (measured in base pairs) and GENETIC DISTANCES (measured in centimorgans). There are three different functions (Kosambi's, Haldane's, and Ott's) which are applied in different contexts and give slightly different values.

marafivirus group Sigla from maize rayado fino virus, the type member. They are single nucleoprotein component viruses with isometric particles 31 nm in diameter; virus preparations also have particles which contain no nucleic acid. The capsids are made up of two polypeptide species (M_r 29 000 and 22 000) and contain one species of (+)-strand linear RNA (M_r 2.0×10^6–2.4×10^6). There is no information yet on genome organization. *See also*: PLANT VIRUSES.

Gomez, R. & Leon, P. (1988) In *The Plant Viruses*, Vol. 3 (Koenig, R., Ed.) 212–233 (Plenum, New York).

MARCKS *Myristoylated Alanine Rich C Kinase Substrate*, a protein first identified as a major substrate for PROTEIN KINASE C. Activation of protein kinase C by any of a number of agonists in several cell types leads to the phosphorylation of MARCKS with subsequent redistribution from the cytoplasmic face of the plasma membrane to the cytosol. More recently it has been discovered that MARCKS is a calcium/calmodulin-sensitive actin filament cross-linking protein (*see* ACTIN-BINDING PROTEINS) also controlled by phosphorylation. The affinity of the protein for actin is approximately halved on phosphorylation. Weak bundling activ-

---

Fig. M4 *Continued* Up until the 2-cell stage (C), no significant transcription occurs, but two bursts of messenger RNA synthesis occur in G1 (0–1 h post division) and G2 (8–16 h post division), and maternal mRNA is largely destroyed by the late 2-to early 4-cell stage. Cleavage divisions (D) thereafter occur at about 10–12 h intervals, and this period is often called the **morula** stage. At the 8-cell stage the process of compaction occurs (E), in which cells flatten on each other to maximize cell contact and each cell undergoes a major shape change from being a symmetrical sphere to a columnar epithelial cell. Over the next two cell cycles (F), the inner (basolateral) portions of these cells contribute largely to a cluster of cells called the **inner cell mass** (ICM), whilst the outer (apical) portions contribute to the **trophectoderm**. Fluid accumulates between the trophectodermal cells, its escape limited by zonular tight junctions (*see* CELL JUNCTIONS) formed apico-laterally, so that the ICM is displaced to one side of a blastocoelic cavity. This stage is called the **blastocyst** (F). The cells of the ICM form two populations. One is called **primary endoderm** (or **hypoblast**) and fronts the blastocoelic cavity and some of its cells start to migrate over the inner surface of the trophectoderm to form parietal endoderm, those remaining over the ICM being called visceral endoderm (G). The remaining ICM cells, sandwiched between the visceral endoderm and the trophectoderm, are primary **ectoderm** (or **epiblast**). The trophectoderm over the ICM proliferates to form a special trophectodermal tissue, the ectoplacental cone. The proliferating ectoplacental cone spreads both outwards and inwards, so that by 5.5 days (H) the primary ectoderm is pushed to the centre of the blastocoelic cavity. The cells of the ectoderm become columnar and form a cyst-like structure within which a small space appears and enlarges to generate the (pro-)amniotic cavity (I). By 7–8 days, in a restricted region of the primary ectoderm, cells start to proliferate (the **primitive streak**), some of these cells spilling out of the ectodermal epithelial layer to migrate between the ectoderm and the visceral endoderm layers, so forming the **mesoderm**. Other proliferating ectodermal cells pass through the developing mesoderm to enter the endodermal layer, displacing the visceral endoderm cells to replace them completely with the so-called definitive endoderm. The result is a trilaminar structure of definitive endoderm, mesoderm and ectoderm, all derived from the primary ectoderm of the ICM (I). This trilaminar structure grows in extent as the cells in the primitive streak region continue to proliferate. Some of the mesoderm cells migrate beyond the confines of the trilaminar structure between the trophectoderm and the parietal endoderm. The trilaminar structure is the true starting material from which the embryo proper and then the foetus will develop. All the tissues not included within this structure, which make up the bulk of the conceptus, are called extra-embryonic (i.e. 'outside' the embryo), and form the support and anchoring tissues that link both functionally and structurally the developing embryo/foetus to the mother's uterus. Across these placental and amniotic tissues must pass nutrients and waste products of metabolism. The position of the primitive streak marks the posterior end of the embryo and so with its appearance clear axes (anteroposterior, dorsoventral, left–right) are established. The trilaminar structure undergoes a rather complex growth process in which it, in essence, turns inside out (J) so that the outward-facing endoderm comes to face inwards and develops into the lining of the gut and its derivatives such as trachae, bronchi, pancreas etc. Conversely, the inward-facing ectoderm is turned to face outwards, and gives rise to the epidermis. Along the midline of the anteroposterior axis of the growing trilaminar structure, the ectoderm invaginates into the underlying mesoderm (section AB of J) to form a groove which pinches off to give a tube — the **neural tube**. From this tube, the neural tissues form, including brain and spinal cord. The mesoderm forms such tissues as muscle, kidney, connective tissue and blood. By 9–10 days (K), the embryo has a clear head developing and an obvious heart starting to pump blood around the embryo. Growth and development is rapid, with limb formation and the appearance of the nostrils, external ears and eyes (L,M). Throughout, the embryo is connected to the placental tissues by the umbilicus carrying blood vessels. In the mouse, the young are born at 19–20 days, the last half of pregnancy being concerned with growth in size of the rudimentary organ systems, and their functional maturation. Adapted from [1].

ity is seen at high concentrations of MARCKS and this bundling activity is also seen with a synthetic lysine-rich peptide corresponding to residues 155–173 of the protein. MARCKS is found at the focal contacts of cultured cells. *See also*: CELL JUNCTIONS; CELL MOTILITY.

Hartwig et al. (1992) *Nature* **356**, 618–622.

Marfan syndrome A DOMINANT heritable disorder affecting skeleton, cardiovascular system, and eyes. The extremities are long and thin, the fingers and toes spidery (arachinodactyly), and height is often increased. Scoliosis is common. The palate is high arched. Aortic incompetence and dissecting aneurysm may occur. Upward and sideways dislocation of the lens is characteristic. Joint capsules, ligaments, tendons, and fascia may be weak leading to dislocation and herniae. Mutations in the gene for the extracellular matrix protein fibrillin have been identified. Prevalence is one in 20 000. *See also*: EXTRACELLULAR MATRIX MOLECULES.

marginal zone (1) A crescent-shaped region which separates areae pellucida and opaca at the future posterior end of the early avian embryo.
(2) Region of blastula giving rise to mesoderm in amphibian embryos.
See: AVIAN DEVELOPMENT; AMPHIBIAN DEVELOPMENT.

marker exchange mutagenesis A MUTATION technique whereby cloned genetic determinants of a particular trait are mutated *in vitro* and then reintroduced into the original strain to replace the wild-type ALLELE. Mutagenesis is usually accomplished by insertion of a genetic marker, such as an antibiotic-resistance gene, into the cloned DNA, which is then 'exchanged' by a recombinational event for the wild-type allele. Recombination is facilitated by cloning the mutated DNA on a PLASMID which will not replicate when transferred back into the wild-type strain (a suicide plasmid), or by destabilization of the vector plasmid by the introduction of an incompatible plasmid. Marker exchange mutagenesis provides a means of creating precisely defined mutants and, importantly, can be used to demonstrate that the cloned gene is responsible for a presumed trait (*see e.g.* PLANT PATHOLOGY).

marker gene Gene or DNA segment of known chromosomal location whose presence can easily be determined either by its gene product or by other means (e.g. HYBRIDIZATION) and which can be used as a point of reference when mapping new loci (*see* HUMAN GENE MAPPING).

Martin Bell syndrome FRAGILE X SYNDROME.

***mas* gene** Mannopine synthase gene. Found in the T_R-regions of octopine Ti PLASMIDS and of Ri PLASMIDS.

Mason–Pfizer monkey virus *See*: RETROVIRUSES.

mast cell The tissue counterpart of the circulating BASOPHIL and involved in inflammatory and hypersensitivity reactions. Mast cells bear high affinity receptors for IgE and other immunoglobulins. Stimulation through binding of ligand to these receptors leads to the release of numerous chemical mediators, including histamine, heparin and prostaglandins. *See*: EICOSANOIDS; FC RECEPTORS; IgE; IgE RECEPTORS.

master chromosome The mitochondrial DNA molecule containing all the sequence information of the plant mitochondrial genome (except for plasmid-like sequences of variable occurrence). In most species, recombination across directly repeated sequences in the master chromosome fragments it into smaller circular molecules, which persist in conjunction with the master chromosome. *See*: MITOCHONDRIAL GENOME: PLANT.

Mastigomycotina Fungi of the division EUMYCOTA which produce zoospores. Three classes are identified according to the type of flagellum carried by the zoospore. The Chytridiomycetes have one posterior whiplash-type flagellum; the Hyphochytridiomycetes have one anterior tinsel-type flagellum; the Oomycetes have two flagella, a posterior whiplash-type and an anterior tinsel-type.

maternal control of development Immediately after FERTILIZATION, the cells in the embryos of all organisms divide (*see* CLEAVAGE) and become differentiated from each other more rapidly than at any other time. RNA synthesis is often negligible or absent during this very early phase of development, at least in part because the cycles of DNA replication and cell division proceed with very short interphases, the time in the CELL CYCLE when TRANSCRIPTION occurs. However, the egg is one of the largest cells in the organism, and during OOGENESIS it is invested with all the materials, in the form of RNA and protein, needed for early growth and development. These maternally derived stores consist not only of 'housekeeping' factors, such as rRNA, histone and tubulin protein, and mRNA and other molecules required for protein synthesis, DNA replication, and cell division, but also of morphogenetic factors, which have more specific roles in early development. These morphogenetic factors (preformed mRNAs and protein) are thought to function only in early development to aid in establishing the body plan of the embryo.

In organisms such as the nematode *Caenorhabditis elegans* and the fruit fly *Drosophila melanogaster*, which are amenable to genetic study, it is possible to identify mutations in genes encoding maternal morphogenetic factors. A simple test for such mutations is to cross females homozygous for a mutant gene with normal males. If the female is able to mate and lay normal eggs, but the wild-type gene product supplied by the father is not sufficient to ensure normal development of the offspring, the mutation is called a maternal effect mutation (*see* MATERNAL EFFECT).

The degree to which morphogenetic factors derived from the mother control development differs widely from species to species. In mice, most of the RNA deposited during oogenesis is degraded by the 2-cell stage, and from then on, development is thought to be under zygotic control. In *Drosophila*, on the other hand, maternal factors have a number of crucial roles in PATTERN FORMATION. The anterior–posterior axis is established by a large number of maternal gene products which ultimately serve to localize the activities of three proteins: the *bicoid* gene product in the anterior regions of the embryo, the *nanos* gene product in the posterior, and the *torso* gene product at either end. Similarly, the dorsal–ventral axis is determined by a group of ~20 maternal effect genes, whose ultimate function is to localize the *dorsal* protein to the nuclei of cells in the ventral regions of the embryo.

In many species, GERM CELLS are also specified by maternal factors. The cytoplasm of many eggs, including those of nematodes, insects, and amphibians, but not mammals, contains a substance called the GERM PLASM. The germ plasm consists of protein and RNA, often in the form of granules which are localized in a particular region of the egg. Any cells which have inherited the germ plasm at the end of cleavage will give rise to germ cells. *See also*: AMPHIBIAN DEVELOPMENT; DROSOPHILA DEVELOPMENT.

maternal determinants *See*: CYTOPLASMIC DETERMINANTS; MATERNAL CONTROL OF DEVELOPMENT; MATERNAL EFFECT; OOGENESIS.

maternal effect A maternal effect gene is one whose function is required exclusively in the maternal genome for the development of the embryo. Thus, the PHENOTYPE of maternal effect mutations is dictated exclusively by the GENOTYPE of the mother and hence assumed to be due to composition or cytoarchitecture of the egg arising during OOGENESIS. Many examples are known in *Drosophila*: *grandchildless* (sterile offspring); *bicaudal* (defects in anterior–posterior axis, lethal); *dicephalic* (doubled anterior segments, lethal); *dorsal* (lacking ventral structures, lethal). In some spirally cleaving molluscs, direction of cleavage and consequent direction of coiling of the shell and viscera (dextral or sinistral) are controlled by a maternal effect gene (*see* DEXTRAL CLEAVAGE). *See also*: DROSOPHILA DEVELOPMENT; MATERNAL CONTROL OF DEVELOPMENT.

maternal inheritance Inheritance of a trait through the female parent, usually determined by chloroplast or mitochondrial DNA. *See*: CHLOROPLAST GENOMES; MITOCHONDRIAL GENOMES.

mating factors *See*: YEAST MATING FACTORS.

mating type locus LOCUS in the yeasts SACCHAROMYCES CEREVISIAE and SCHIZOSACCHAROMYCES POMBE that determines mating type. Cells of some yeast strains can switch from one mating type (**a**) to the other (**α**). *See*: YEAST MATING TYPE LOCUS.

mating type switch *See*: YEAST MATING TYPE LOCUS.

matrix (1) The innermost aqueous compartment of MITOCHONDRIA, surrounded by the inner membrane.
(2) Extracellular matrix. *See*: EXTRACELLULAR MATRIX MOLECULES.
(3) A mathematical concept for an array of elements, manipulated by the rules of matrix algebra which allow a concise notation for handling such arrays which may have very many elements in a multidimensional space. In least squares refinement algorithms (*see* LEAST SQUARES METHOD) a normal matrix is established and then inverted, this latter step being an enormous computational problem for arrays with thousands of elements in each direction.

matrix attachment regions Sites in the NUCLEAR MATRIX to which DNA is attached.

matrix Gla protein *See*: BONE GLA PROTEINS.

Matthews number (V_m) Crystalline proteins of known molecular weight were surveyed by B.W. Matthews who found that solvent occupied between 27% and 65% of the crystal volume, the usual fraction being close to 43%. The Matthews number represents the UNIT CELL volume per molecular weight (V_m) being within the range from 3.53 $Å^3$ per dalton to 1.68 $Å^3$ per dalton with a usual value of 2.4 $Å^3$ per dalton. Knowledge of the unit cell dimensions and density of a protein crystal enables calculation of the weight of one ASYMMETRIC UNIT. If the molecular weight of the protein, or its subunits, is known, then the number of protein subunits per asymmetric unit of the crystal can be deduced. This must be a whole number, and not a fraction, such that there is often only one possible number that is consistent with the normal values of V_m. For oligomeric proteins it is often found that the whole oligomer occupies more than an asymmetric unit: for example, a tetramer may crystallize with 1/2 or 1/4 of the whole molecule asymmetric unit and a hexamer with 1/6 or 1/3 or 1/2. This simplifies crystal structure determination, often to that for a single subunit, not the oligomer.

Average values of V_m have been used to estimate the molecular weight of an unknown protein (a more accurate determination requires careful measurements of both the solvent content and the precise density of the crystal) and to determine the subunit structure of an oligomer.

Matthews, B.W. (1968) *J. Mol. Biol.* **33**, 491.

maturation *See*: OOGENESIS; SPERM MATURATION.

maturation promoting factor (MPF, mitosis (or meiosis) promoting factor) A PROTEIN KINASE initially defined by its ability to induce oocytes to complete maturation and enter metaphase of meiosis I (*see* MEIOSIS) and which is now known to have a central role in the CELL CYCLE. It contains a catalytic subunit of M_r ~34 000, which was first identified as the product of cell division cycle gene 2 (*cdc2*) of the fission yeast, *Schizosaccharomyces pombe*, and hence is called p34^{cdc2}. The catalytic subunit is activated in part by forming a complex with the protein CYCLIN, which is synthesized during INTERPHASE and degraded at the onset of ANAPHASE. There are several different cyclins, and the substrates recognized by the kinase may depend in part on the cyclin in the complex. The activity of the kinase is also regulated by the binding of additional subunits and by the state of phosphorylation of the catalytic subunit, a modification that is subject to both positive and negative regulation. MPF can phosphorylate HISTONES *in vitro*, and *in vivo* it probably phosphorylates many proteins whose activities must be modified at the onset and termination of mitosis. *See also*: MITOSIS; PROTEIN PHOSPHORYLATION.

Mauthner cell A large neuron located bilaterally in the medulla of teleost fish and amphibians. It is notable for its large size, and unusual shape, and the fact that it combines all known mechanisms of SYNAPTIC TRANSMISSION. Its function relates to the ear and the lateral line system.

Max A DNA-binding TRANSCRIPTION FACTOR that forms a hetero-oligomer with the oncoprotein Myc (*see* ONCOGENES), enabling the complex to bind to DNA.

Maxam–Gilbert method A method of rapid sequence analysis of DNA based on controlled chemical degradation of DNA. *See*: DNA SEQUENCING.

maxicircles *See*: MITOCHONDRIAL GENOMES: LOWER EUKARYOTES.

maximum entropy method A computational procedure well suited to the recovery of signals or the reconstruction of two-dimensional or three-dimensional objects from incomplete and noisy data. The criterion of maximum entropy, or minimum added information, is due to Jaynes. It ensures that the final model fits the data only to within their experimental accuracy, while remaining 'maximally non-committal' with respect to the missing data. The maximum entropy method can give superb results when applied to reconstruction problems for which the setting of information theory is natural rather than contrived; and for which the statistical properties of the noise are well described. It can however create artefacts if either of these requirements is not met. It is used in DIRECT METHODS of phase determination in X-RAY CRYSTALLOGRAPHY through its connection with the saddle-point method.

Bricogne, G. (1984) *IEEE Trans. Acta Cryst* **A40**, 410–455.
Bricogne, G. (1991) In *Maximum Entropy in Action* (Buck, B. & Macaulay, V.A., Eds) 187–216 (Oxford University Press, Oxford).
Jaynes, E.T. (1978) *IEEE Trans.* **SSC-4**, 227–241.

maximum likelihood A standard statistical procedure to estimate the values of parameters influencing the distribution of measurable quantities from observed values of these quantities. The LIKELIHOOD of a set of parameter values is the probability they assign to the actual observations, and the maximum likelihood method of parameter observation consists in choosing the values which maximize it. If the probability law governing the distribution of the observables is a multivariate Gaussian with known covariance MATRIX, the maximum method simplifies to the LEAST SQUARES METHOD [1].

In molecular biology this approach is used:
(1) As an optimality criterion for selecting an evolutionary tree for DNA sequences [2]. In molecular evolutionary studies, the criterion is to find the evolutionary model that has the highest likelihood of generating the observed sequences. A scientific model generally consists of three parts: the physical description, mechanisms, and initial conditions. In an evolutionary model the physical description is usually a tree. The most common mechanism for changes to a sequence is that mutations are 'independent and identically distributed' (i.i.d). This means that changes to the sequence are independent of those on any other lineage or at any other position in the sequence (independent) as well as being equally likely to occur at any position (identically distributed). The initial conditions are the weights (edge lengths or probabilities of change on edges of the tree) on each edge of the tree. From a specific model (tree, mechanism and edge lengths) the likelihood of generating the observed data is calculated. It is usual to assume one mechanism and to alter the tree and edge lengths to find the model which has the highest likelihood, hence the name maximum likelihood. The approach has the advantage of converging to the correct tree as longer sequences are used, provided the assumptions about the mechanism are valid. The primary disadvantage is that the excessive amount of computing time means that it is usual to examine only a small proportion of all possible trees. (2) As a statistical approach to solving the PHASE PROBLEM in X-RAY CRYSTALLOGRAPHY.

1 Lindley, D.V. (1965) *Introduction to Probability and Statistics from a Bayesian Viewpoint*. 2 Vols. Part 1: Probability, Part 2: Inference (Cambridge University Press, Cambridge).
2 Goldman, N. (1991) *Syst. Zool.* **39**, 345–361.

***mbp* gene** A pair of related genes (*mbpX* and *mbpY*) that occur in the CHLOROPLAST GENOME of the liverwort, *Marchantia polymorpha*, but not that of tobacco (*Nicotiana tabacum*). Their function is uncertain, but they show sequence similarity to the histidine transport genes of the bacterium *Salmonella typhimurium* and the maltose transport system of *Escherichia coli*. The derived amino-acid sequence of the *mbpX* gene shows a nucleotide-binding site near the N terminus, suggesting that it may be involved in nucleotide-driven transport.

McArdle disease GLYCOGEN storage disease arising from a deficiency of muscle glycogen PHOSPHORYLASE. Patients present in adult life with muscle cramps after exertion and these may be accompanied by the presence of MYOGLOBIN in the urine. Proximal muscle wasting and weakness develop with age. The phosphorylase is encoded by a gene on chromosome 11. The disease frequency is one in 500 000.

McLeod syndrome *See*: KELL BLOOD GROUP.

MCPs Methyl accepting chemotaxis proteins. *See*: BACTERIAL CHEMOTAXIS.

MDCK cells Madin–Darby canine kidney cells. *See*: EPITHELIAL POLARITY.

mdx A mouse mutant for the dystrophin gene, and which provides an animal model of human DUCHENNE MUSCULAR DYSTROPHY.

ME method MAXIMUM ENTROPY METHOD.

mean neutron scattering-length density Sum of the scattering lengths for all atoms divided by their total volume. *See also*: NEUTRON SCATTERING AND DIFFRACTION.

mechanically gated ion channels ION CHANNELS whose opening is regulated by mechanical forces, as in the cation channels at the tips of stereocilia on the sound-detecting hair cells of the cochlea, which open when the bundles of stereocilia are tilted in response to vibration.

Mechanisms of enzyme catalysis

THE overwhelming majority of chemical reactions in living cells are subject to rate enhancement by protein ENZYMES, one exception being the recently discovered RIBOZYMES. A knowledge of the means by which precisely folded polypeptide chains accelerate energetically favourable chemical reactions has come from both chemical and biological evidence. The term 'mechanism' encompasses conclusions arising from consideration of protein structures, kinetic studies, the use of reversible and irreversible inhibitors, spectroscopic experiments, and genetic analysis. Further information on the topic can be found in [1–4].

Few enzymes can be said to be completely understood as chemical catalysts at the atomic level. None the less, a useful understanding of enzyme catalysis has been achieved for relatively simple transformations, such as those involving bond breakage by water (hydrolysis, mediated by the vast and diverse group of HYDROLASES), oxidation/reduction reactions (mediated by e.g. DEHYDROGENASES), the addition or removal of water (hydration/dehydration, mediated by e.g. HYDRATASES), and intramolecular proton migrations (mediated by e.g. ISOMERASES).

In any discussion of enzymes it is axiomatic that the protein, in common with all chemical catalysts, cannot alter solution equilibria of reactants (substrates) and products. Reactions for which the sign of Gibbs free energy change (ΔG) for the chemical transformation is positive (for defined reaction conditions such as temperature, pH, and concentrations of reactants and products) are unfavourable and therefore their rates are unaffected by the presence of a catalyst. Catalysis by enzymes is thus concerned not with the *possibility* of a reaction taking place (determined by *thermodynamic* considerations) but with the *frequency* of its occurrence in a convenient unit of time — a matter for *kinetic* analysis.

Considerations of the general problem of enzyme mechanisms have commonly started with separate discussions of the two most striking features of enzyme action, specificity and rate acceleration. Explanations for the former, the most striking special property of enzymes, which distinguishes them from the majority of nonbiological catalysts, have been more accessible with the advent of many high resolution crystallographic studies of enzyme–substrate complexes. Such structures reveal the number, type, and geometries of specific pairwise interactions which underlie the affinity of an enzyme for its substrate(s). It has proved convenient to separate them into ELECTROSTATIC INTERACTIONS, HYDROGEN BONDS (charged and uncharged), nonpolar interactions involving VAN DER WAALS FORCES, and considerations of HYDROPHOBICITY in instances where nonpolar regions of an enzyme active site offer energetically favoured contacts (compared with solvent) for a ligand with similar properties.

Rate acceleration in enzyme reactions is inseparable from substrate binding (and specificity) in that one of the most obvious determinants of effective catalysis is the approximation of reactants which accompanies their binding. In a two-substrate reaction of the form $A + B \rightarrow C + D$ the sequestration from solvent and the precise alignment of the reactive moieties of A and B is achieved by the structural complementarity of the active site surface. Three other mechanisms which are usually invoked to explain rate enhancement include general acid/base catalysis, the formation of covalent enzyme–substrate intermediates, and a phenomenon variously described as distortion, strain, or 'induced fit'. In most well-studied enzyme reactions at least two of these three mechanisms can be seen to have roles in rate acceleration — in addition to that provided by the approximation of reactants.

General acid/base catalysis makes use of the chemical properties of amino-acid side-chain substituents with pK_a values poised near the pH of biological environments and thereby able to accept or donate a proton. Examples include the imidazole of histidine, the ε-amino group of lysine, and the carboxyl groups of glutamate or aspartate (*see* AMINO ACIDS). In addition, many reactions proceed via the formation of one or more intermediates which are covalent adducts of enzyme and substrate. Such examples of so-called covalent catalysis call attention to the need not only to ensure the proximity and stereochemical orientation of reactants but also to fractionate the overall chemical transformation into energetically discrete steps, wherein, for example, electronic or charge stabilization is most readily accomplished via the formation of a covalent intermediate. A further mechanistic explanation of rate acceleration by enzymes is that of strain or distortion of bound substrate *en route* to the transition state wherein it is postulated that the complementarity between protein and reactants is optimal. The knowledge that enzymes are usually much larger than their substrates and that they may undergo changes in conformation on the addition of substrate(s) favours the view that something akin to distortion or 'induced fit' must play a part in bringing an enzyme-bound substrate from its relatively unreactive ground state to one which is energetically more favourable — the transition state (see below).

Implicit in all discussions of enzyme mechanism is an acceptance of the fact that concepts of catalysis are deeply rooted in reversible thermodynamics (energetics). The schematic diagram in Fig. M5 resembles most textbook examples which aim to point out that formation of the enzyme–substrate (ES) complex in the transition state (ES‡) has, in effect, lowered the activation energy for the process under study and, having done so, increased the probability that it will occur — thereby increasing its rate. The power of transition state theory in explaining enzyme catalysis rests in its ability to relate rate effects to the difference (ΔG) between the Gibbs energy of enzyme and reactants in the transition state as compared with that of the ground state.

With real enzyme systems, even those involving relatively simple transformations, the situation is both more complex and more interesting. Figure M6 shows the energy diagram for the reaction catalysed by TRIOSE PHOSPHATE ISOMERASE, an enzyme of glycolysis (*see* Fig. P81 in PROTEIN STRUCTURE for structure), wherein the transformation of dihydroxyacetone phosphate (DHAP) to R-glyceraldehyde 3-phosphate (GAP) proceeds via an enediol intermediate and involves only a deprotonation of DHAP

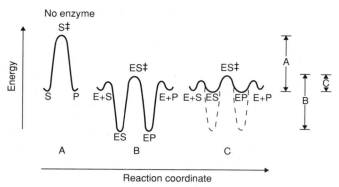

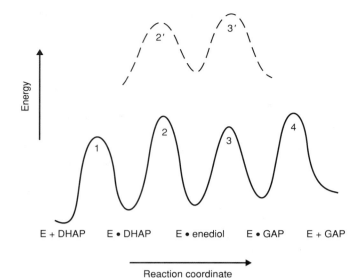

Fig. M5 Effect of enzyme on transition state (‡) energy of ES complex with stabilization (B) and destabilization (C) of substrate and product complexes. Stabilization of ES and EP as well as ES‡ leads (B) to no net reduction in energy of activation (S‡) in (A) whereas marked reduction (C) accompanies destabilization of ES and EP.

Fig. M6 Energy diagram for the conversion of dihydroxyacetone phosphate (DHAP) to glyceraldehyde 3-phosphate (GAP) via an enediol intermediate. The dashed line represents the transition state barriers to and from the enediol intermediate in the absence of enzyme.

and reprotonation of the enediol (Fig. M7). Proton transfers to or from C-1 and C-2 in the forward and reverse reactions are mediated by the precisely positioned carboxylate of glutamate-165, the 'general base' of the mechanism. It has been suggested that triose phosphate isomerase is a fully evolved enzyme in that it has optimized the relationship between overall rate (k_{cat} or 'turnover number') and affinity for substrate, the latter given by $1/K_m$ where K_m is the Michaelis constant for the productive binding of substrate by enzyme. That triose phosphate isomerase may have attained evolutionary 'perfection' in substrate capture and rate enhancement is suggested by the ratio of k_{cat}/K_m for DHAP in the forward reaction. The term k_{cat}/K_m has been referred to as the specificity constant, which, for relatively simple transformations such as DHAP → GAP can be incorporated into the rate expression $v = k_{cat}/K_m.[DHAP].[E]$. The importance of such a formulation is that it relates the initial velocity (v) of an enzyme reaction to the concentrations of free substrate [DHAP] and free enzyme [E] and a term (k_{cat}/K_m) which is in effect an apparent second order rate constant (in $s^{-1} M^{-1}$) for the encounter of the two reactive components of the system, DHAP and E in this case. As the value of k_{cat}/K_m approaches the range of 10^8 to $10^9 s^{-1} M^{-1}$, the limit of so-called diffusion-controlled chemical processes in solution, any evolutionary pressures acting to increase rate (k_{cat}) on the one hand or affinity ($1/K_m$) on the other must be accompanied by corresponding changes in the other parameter. In short, the likelihood of productive encounters,

yielding product(s), cannot increase further.

The case of triose phosphate isomerase illustrates the principles of enzyme catalysis in a well-studied system which possesses most of the features noted above. It does not, however, employ covalent catalysis. The latter is most lucidly illustrated by the SERINE PROTEINASE family of proteolytic enzymes wherein the covalent intermediate involves an ester ($R_1 COOR_2$) between the carbonyl of the —C—N— peptide bond cleaved by the enzyme and the hydroxyl of a conserved serine residue (R_2 above) at the active site. Subsequent attack of the ester intermediate by water leads to release of the peptide fragment $R_1 COOH$ and regeneration of HO—R_2 (serine).

Essential to all formulations of enzyme mechanism in the broadest sense is the notion of physical complementarity of enzyme with the chemical transition state, the transitory structural entity arising from substrate and present at the energy barrier for the reaction in Fig. M5. Attainment of the reduction in peak energy of enzyme-bound transition state (S‡) results from the precise stereochemical (and stereoelectronic) approximation of enzyme and the transition state structure. It can be appreciated

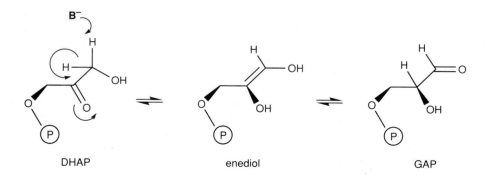

Fig. M7 The conversion of dihydroxyacetone phosphate (DHAP) to glyceraldehyde 3-phosphate (GAP) via an enediol intermediate. The general base (B⁻) is the carboxylate of Glu 165.

intuitively that complementarity of enzyme and *substrate* (rather than transition state) would lead to high affinity sequestration of substrate instead of catalysis. For those enzymes which have been studied in sufficient detail it is clear that the results are those predicted by theory. Rate enhancement via a decrease in activation energy for the chemical step can only be achieved when stabilization of the transition state is appreciably greater than for substrate. Indeed, as shown in Fig. M5, the most important consideration for effective catalysis is the *relative* destabilization of bound substrate in the ES complex (and bound product in EP complex). In each case the binding energy (or stabilization energy) of S (or P) in their respective complexes arises in the main from a decrease in entropy (ΔS), the thermodynamic description of disorder or randomness of S (or P) in going from free solution to the bound state.

Although chemical theory can now satisfactorily account for enzyme catalysis, it should be noted that two types of experiment have further focused attention on the structural complementarity of the enzyme active site with that of the transition state. The first concerns the extraordinarily tight binding of enzyme inhibitors which are analogues of predicted transition state structures. The second and more recent example is the deliberate isolation of ANTIBODIES (immunoglobulins) raised against compounds which are transition state 'mimics' of energetically favourable chemical reactions. Such antibodies (ABZYMES) often possess catalytic activity whereas antibodies raised against potential substrate(s) do not. The latter complexes most closely resemble the nonproductive ES complexes of Fig. M5.

W.V. SHAW

See also: MOLECULAR EVOLUTION: SEQUENCES AND STRUCTURES; PROTEIN STRUCTURE.

1 Fersht, A.R. (1985) *Enzyme Structure and Mechanism*, 2nd edn (Freeman, London).
2 Knowles, J.R. (1991) Enzyme catalysis: not different, just better. *Nature* **300**, 121–124.
3 Fersht, A.R. & Gani, D. (Eds) (1991) *Enzymic Catalysis* (The Royal Society, London) 78pp.
4 Evans, P.R. & Benkovic, S.J. (Eds) (1992) Catalysis and regulation. *Curr. Opinion Struct. Biol.* **2**, 713–722.

mechanoreception The detection of movements in the environment of a sensory cell, for example, the detection of vibration by the hair cells of the inner ear and the detection of muscle tension by stretch receptors in muscle.

medulla (1) Of brain, the medulla oblongata, the lowermost region of the brainstem, running into the spinal cord (*see* CENTRAL NERVOUS SYSTEM).
(2) Of adrenal gland, thymus, and other organs, the innermost part (*cf.* cortex, the outer layer).

megakaryocyte White blood cell of the myeloid lineage, with a large multilobed nucleus, from which anucleate PLATELETS are produced. *See*: BLOOD COAGULATION AND ITS DISORDERS; HAEMATOPOIESIS.

meiosis Type of nuclear division that occurs in eukaryotes during the formation of HAPLOID cells (e.g. eggs or sperm) from a DIPLOID cell, and in which the chromosome number is exactly halved (*cf.* MITOSIS in which the chromosome number remains unchanged). Genetic RECOMBINATION, the exchange of material between homologous chromosomes, occurs during meiosis in most species. The cellular machinery of meiosis, for example the microtubule spindle on which the chromosomes become aligned and segregate to the poles, is essentially similar to that in MITOSIS.

Meiosis is divided into two distinct phases — division I, in which the reduction in chromosome number takes place and which is therefore also known as the 'reduction division', and division II, which is analogous to a mitotic division in somatic cells. Meiosis typically produces four haploid progeny from the original diploid cell.

DNA synthesis and chromosome replication occur before division I commences and so the cell enters meiosis with its chromosomes replicated, although this fact does not become apparent in the light microscope until later. Division I comprises four stages — prophase I, metaphase I, anaphase I, and telophase I. Prophase I is conventionally subdivided further into leptotene (in which the chromosomes become visible as long thin threads), zygotene, pachytene, diplotene, and diakinesis. In zygotene, homologous chromosomes pair up lengthwise (synapsis); in pachytene they become more compact and chiasmata (*see* RECOMBINATION) are seen. In diplotene, the individual chromatids become apparent and each synapsed pair of chromosomes is seen to be composed of four chromatids (a tetrad). The homologous chromosomes separate but are still held together by chiasmata, and each pair of chromatids is still joined at the centromere. During diakinesis the chiasmata move to the ends of the chromosomes (the resolution of recombination) and the chromosomes become more compact, the nuclear envelope breaks down and the synapsed chromosomes start to become aligned on the meiotic spindle, which has been forming outside the nuclear envelope. In metaphase I the chromosomes have become randomly oriented on the METAPHASE PLATE. Homologous chromosomes then separate and move to opposite poles (anaphase I). A nuclear membrane is reformed around each set of chromosomes and the cell divides (telophase I). A short interphase now generally supervenes, but unlike the interphase of a somatic cell, there is no further DNA replication. The random orientation of homologous pairs on the metaphase plate and subsequent segregation is the cause of the classic Mendelian law of the independent assortment of alleles (*see* MENDELIAN INHERITANCE.)

During meiotic division II, the already replicated chromosomes go through the stages corresponding to a mitosis, in which they separate into individual chromosomes which are segregated to opposite poles to form two haploid nuclei from each product of the first stage of meiosis.

Developing germ cells often become arrested for long periods of time in one or other of the stages in the first meiotic division (*see* OOGENESIS). The protein factor (MATURATION PROMOTING FACTOR) that is instrumental in inducing frog oocytes arrested in metaphase I to complete meiosis has turned out to be central to the general control of cell cycles, mitotic as well as meiotic (*see* CELL CYCLE).

meiotic map Genetic map. *See*: LINKAGE MAP.

melanocyte A melanin-pigmented LYMPHOCYTE, found in mammalian epidermis.

melanocyte-stimulating hormone (melanocortin, MSH) Pituitary peptide hormone produced in two forms (α and β). α-MSH stimulates synthesis of the black pigment eumelanin in melanocytes. The melanocyte-stimulating hormones are synthesized by cleavage of the POLYPROTEIN pro-opiomelanocortin. *See*: COAT COLOUR GENES; OPIOID PEPTIDES AND THEIR RECEPTORS.

melanoma Dark mole on the skin arising from the pigmented melanocytes in the basal layer of the epidermis. Malignant melanoma can arise from such benign papillomas and is often a highly invasive metastatic cancer.

melting temperature *See*: TRANSITION TEMPERATURE.

membrane anchors Some proteins are associated with the cytoplasmic face of the PLASMA MEMBRANE by a lipid moiety added post-translationally to the N terminus of the protein or to an amino acid side chain. Myristoyl, geranyl, palmitoyl, and farnesyl anchors have been found (*see e.g.* ADRENERGIC RECEPTORS; GTP-BINDING PROTEINS; PAPOVAVIRUSES). Some proteins are tethered to the external face of the plasma membrane by GPI ANCHORS added post-translationally within the ENDOPLASMIC RETICULUM.

membrane(s) The cytoplasm of all cells is bounded by a lipid bilayer membrane — the plasma membrane — in which are embedded numerous different types of proteins (*see* MEMBRANE STRUCTURE). In eukaryotic cells, similar membranes also bound the various cellular organelles (*see* ANIMAL CELL; PLANT CELL). Because of the intrinsic impermeability of the lipid bilayer to most ions and solutes, and to large macromolecules, the passage of most ions and solutes across a membrane, and the internal environment of the cell and of its organelles is controlled by the various membrane proteins, which include ion channels, ion pumps, and nutrient transporters (*see* MEMBRANE TRANSPORT SYSTEMS). *See also*: BACTERIAL ENVELOPE; COATED PITS AND VESICLES; ENDOCYTOSIS; EXOCYTOSIS; MEMBRANE LIPIDS; MEMBRANE POTENTIAL; MEMBRANE RECYCLING; MEMBRANE STRUCTURE; MEMBRANE SYNTHESIS; MEMBRANE TOPOLOGY; MEMBRANE TRANSPORT SYSTEMS; MITOCHONDRIAL BIOGENESIS; NUCLEAR ENVELOPE; PROTEIN TRANSLOCATION; PROTEIN SECRETION.

membrane domain *See*: EPITHELIAL POLARITY.

membrane fusion Two lipid bilayers brought into contact with each other have the ability to fuse into a single lipid bilayer. This property forms the basis of many cellular phenomena such as EXOCYTOSIS, intracellular VESICLE-MEDIATED TRANSPORT, and the uncoating of enveloped viruses in ENDOSOMES and transfer of the nucleocapsid into the cytoplasm (*see e.g.* INFLUENZA VIRUS; SEMLIKI FOREST VIRUS). *See also*: COATED PITS AND VESICLES; ENDOCYTOSIS; PROTEIN SECRETION; SYNAPTIC TRANSMISSION.

membrane lipids The classes of LIPIDS which constitute the lipid bilayer of biological membranes (*see* MEMBRANE STRUCTURE). The main groups found in most membranes are PHOSPHOLIPIDS, GLYCOLIPIDS, and STEROLS, the relative abundance of any particular lipid varying between different cell types and tissues, and between different types of organisms. Membrane lipids are amphipathic with a hydrophobic tail and a polar hydrophilic head and thus spontaneously form a lipid bilayer in an aqueous environment, with the hydrophobic tails towards the centre of the membrane. In the glycerol-based phospholipids (*see* Table P1), the hydrophobic unit is formed by the hydrocarbon tails of two fatty acids esterified to glycerol; in sphingomyelin and the sphingosine-based glycolipids it is formed by the hydrocarbon chain of sphingosine and the fatty acid chain attached to the amino group of sphingosine. The hydrophilic headgroup in the glycerol-based phospholipids may be choline, ethanolamine, serine, inositol, or an additional glycerol. In sphingomyelin it is phosphorylcholine; in glycolipids it is a sugar or glycan.

In animal cells, lipids of the endomembrane system are mainly synthesized at the ENDOPLASMIC RETICULUM (ER). The glycerol-based membrane PHOSPHOLIPIDS phosphatidylcholine, phosphatidylserine, phosphatidylethanolamine, and phosphatidylinositol are synthesized on the cytoplasmic face of the ER (Fig. M8) and enter the cytoplasmic leaflet of the ER membrane. After synthesis, phosphatidylcholine is preferentially transferred to the luminal leaflet of the ER membrane by the action of phospholipid translocator proteins. SPHINGOLIPIDS are synthesized in the ER membrane as ceramide (*see* Fig. S16*b*) which is then exported to the GOLGI APPARATUS. There it is either glycosylated to form

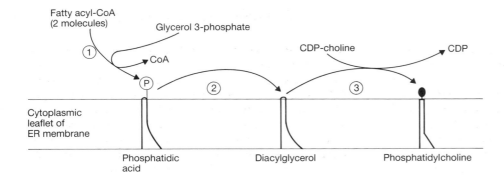

Fig. M8 The main pathway of phosphatidylcholine synthesis in mammals. 1, Acyl transferase; 2, phosphatase; 3, choline transferase. Phosphatidylethanolamine may be synthesized by a similar scheme involving CDP-ethanolamine. Phosphatidylinositol and phosphatidylserine are synthesized by the addition of inositol and serine respectively to CDP-diacylglycerol formed by the reaction of CTP and phosphatidate.

glycosphingolipids (*see* GLYCOLIPIDS), or receives a phosphorylcholine headgroup derived from phosphatidylcholine to form sphingomyelin. Mitochondrial membrane lipids are thought to be imported from the ER via specific phospholipid exchange proteins as phosphatidylcholine and phosphatidylserine, which may be converted within the mitochondrial membrane into diphosphatidylglycerol (also known as cardiolipin), and phosphatidylethanolamine.

The chloroplasts and proplastids of plant cells are the site of fatty acid synthesis and, depending on species, lipids of chloroplast membranes may be synthesized entirely *in situ* or may be partly imported from the ER. Chloroplast membrane lipids synthesized in the organelle have a characteristic fatty acid composition (the fatty acid esterified to position 2 of glycerol is almost exclusively C_{16} and that esterified to position 1 is either C_{16} or C_{18}) which resembles that of prokaryotic membrane lipids rather than eukaryotic membrane lipids (position 2 primarily C_{18}; position 1 C_{18} or C_{16}).

Plant cell membranes also contain lipids based on phytosphingosine. *See also*: CHOLESTEROL; GANGLIOSIDES.

membrane permeability *See*: MEMBRANE TRANSPORT SYSTEMS.

membrane potential The voltage difference across a membrane, usually but not exclusively taken to be the PLASMA MEMBRANE of a cell. Probably every type of cell has such a potential. As the extracellular space is usually at earth potential (i.e. 0 mV), the membrane potential refers to the value on the inside of the plasma membrane. A membrane potential will result when two conditions are met simultaneously: (1) there must be a concentration gradient of an ion across the membrane (the electrochemical gradient); (2) the membrane must be at least partially permeable to that ion. For a membrane permeable to a single ion, the value of the membrane potential can be predicted from the Nernst equation. Where multiple ions are involved, the Goldman–Hodgkin–Katz theory must be used. The values of the 'resting' membrane potentials found in cells vary from a few to several tens of millivolts, with the interior of the cell negative with respect to the exterior. The membrane potential has a crucial role in cell function by affecting voltage-dependent CONDUCTANCE and transmembrane ion transport. *See also*: MEMBRANE TRANSPORT SYSTEMS; NA⁺,K⁺-ATPASE; SYNAPTIC TRANSMISSION; VOLTAGE-DEPENDENT ION CHANNELS.

Shepherd, G.M. (1988) *Neurobiology*, 2nd edn (Oxford University Press, Oxford).

membrane protein Protein associated with a biological membrane. Intrinsic or integral membrane proteins are embedded in the lipid bilayer of the membrane, and can only be extracted by detergent solubilization; extrinsic or peripheral membrane proteins are more loosely associated with one or other surface of the membrane by electrostatic and hydrogen bond interactions. Transmembrane proteins, which span the lipid bilayer, are anchored in the membrane by one or more hydrophobic TRANSMEMBRANE REGIONS. Proteins may also be tethered to either face of the membrane by a lipid MEMBRANE ANCHOR. In eukaryotic cells proteins are incorporated into most cellular membranes (except those of mitochondria and chloroplasts) at the site of new membrane synthesis in the membranes of the ENDOPLASMIC RETICULUM. *See*: BACTERIAL ENVELOPES; GLYCOPROTEINS AND GLYCOSYLATION; MEMBRANE STRUCTURE; MEMBRANE SYNTHESIS; MEMBRANE TRANSPORT SYSTEMS; PROTEIN TRANSLOCATION.

membrane recycling Replacement or reutilization of membrane components. In both the endocytic and exocytic pathways (*see* ENDOCYTOSIS; EXOCYTOSIS) membrane is continually removed from organelles to form TRANSPORT VESICLES that carry membrane components to the next stage(s) of the pathway. To maintain the size and biochemical composition of individual compartments much of this membrane must be recycled. For example, in cells undergoing constitutive endocytosis a membrane area equivalent to the entire plasma membrane is endocytosed every 0.5–2 hours, depending on the cell type, delivered to ENDOSOMES and recycled back to the cell surface. In the order of 100 (in lymphocytes) to 1000 (in baby hamster kidney cells) coated vesicles endocytose from the cell surface each minute (*see* COATED PITS AND VESICLES), and an equivalent number of vesicles are required to recycle the membrane.

Recycling provides the means to reutilize specific membrane components. The low density lipoprotein (LDL) receptor (*see* LDL RECEPTOR) and TRANSFERRIN RECEPTOR for example mediate multiple rounds of ligand binding and endocytosis, the cation-independent MANNOSE 6-PHOSPHATE RECEPTOR delivers newly synthesized lysosomal hydrolases to the endocytic pathway by cycling between the *trans* Golgi network (*see* GOLGI APPARATUS) and late endosomes, secretory granule membrane is recycled after exocytosis, and ENDOPLASMIC RETICULUM luminal proteins containing a C-terminal amino acid sequence (Lys-Asp-Glu-Leu; KDEL) are recycled to the ER by a cycling KDEL receptor (*see* PROTEIN TARGETING).

Steinman, R.M. et al. (1976) *J. Cell Biol.* **68**, 665–687.
Steinman, R.M. et al. (1983) *J. Cell Biol.* **96**, 1–27.
Pelham, H.R.B. (1989) *Annu. Rev. Cell Biol.* **5**, 1–23.
Besterman, J.M. et al. (1981) *J. Cell Biol.* **99**, 716–727.
Thilo, L. & Vogel, G. (1980) *Proc. Natl. Acad. Sci. USA* **77**, 1015–1019.

membrane skeleton Cytoskeletal framework underlying the plasma membrane. In the red cell it forms a regular structure (*see* Fig. M10 in MEMBRANE STRUCTURE).

Membrane structure

BIOLOGICAL membranes have evolved to provide compartments within which the complex chemistry of life can proceed undisturbed by changes in the outside world. In eukaryotic cells, cellular compartments surrounded by a PLASMA MEMBRANE are further subdivided by internal membranes which define organelles such as nuclei and mitochondria. The membranes are flexible, and are capable of growth and subdivision, and of mediating specific communications within and between cells without loss of integrity [1,2]. In spite of the complexity of these requirements the basic structural principles involved are straight-forward.

The key to reconciling the dynamic nature of the membrane with maintenance of its structural integrity lies in the physical properties of the PHOSPHOLIPIDS which constitute 15–50% of almost all membranes [3]. The geometry and polarity of the phospholipid molecules are such that in aqueous media they spontaneously form extended bilayers with a hydrophobic core. This bilayer is permeable to nonpolar molecules but not to ions or most polar molecules. Its mechanical stability, flexibility, electrical resistance, and other physical characteristics account for many of the general properties of biological membranes. More specific functions of membranes, such as selective transport of metabolites and signalling triggered by a variety of agents, are mediated by proteins [4], which are retained in the membrane by interaction of hydrophobic amino acids with the fatty acid chains of the phospholipids. Membrane proteins exist in enormous variety, and resemble cytoplasmic proteins in being folded into compact globular structures. This minimizes the surface of interaction with the lipid, so that although protein may account for 30–80% of the weight of the membrane it immobilizes only a small proportion of lipid and does not affect the basic physical properties of the lipid bilayer.

Membrane phospholipids

Although hundreds of different phospholipids are known, they all share physical characteristics which enable them to form stable membranes. These are seen at their most typical in lecithin (Fig. M9*a*). A polar headgroup of glycerol esterified in position 1 by phosphorylcholine is balanced by twin hydrophobic tails of two long-chain (C_{14}–C_{22}) FATTY ACID groups in positions 2 and 3. The polar headgroups and the hydrophobic tails have similar cross sections, so that when the molecules are packed side by side, they produce a flat sheet with a polar face and a hydrophobic face. In an aqueous medium the hydrophobic surfaces interact to give a stable bilayered structure, which forms spontaneously when water is added to solid lecithin. Mechanical agitation, especially by ultrasonic vibration (sonication), breaks up these sheets to yield closed vesicles (LIPOSOMES) which have proved valuable models for studying many aspects of membrane function. The fatty acid chains vary in length and in unsaturation, containing from zero to six double bonds. A pure phospholipid containing a single species of saturated fatty acid forms rather rigid quasi-crystalline bilayers (a gel phase) which melt at a characteristic temperature to a liquid crystalline phase. The more heterogeneous the fatty acid chains, and, more especially, the higher the unsaturation, the lower and less well defined is the transition temperature; for dipalmitoyl lecithin it is 37°C, for dioleyl lecithin it is – 22°C. The fatty acid composition of natural membranes varies widely but there is always sufficient unsaturation to give a fluid lipid phase; organisms which live at low temperatures have a higher proportion of unsaturated fatty acid.

In other types of phospholipid the choline base of lecithin is absent or is replaced by ethanolamine, serine, glycerol, or inositol (*see* LIPIDS). In GLYCOLIPIDS choline phosphate is replaced by a variety of oligosaccharides. SPHINGOLIPIDS contain the long-chain amino alcohol sphingosine in place of glycerol. Most membranes contain complex mixtures of these types (Fig. M9*b*). The exact composition depends on the source, but quite wide variations are tolerated and the physical properties of the membrane remain much the same. Very little is known about specific functions of

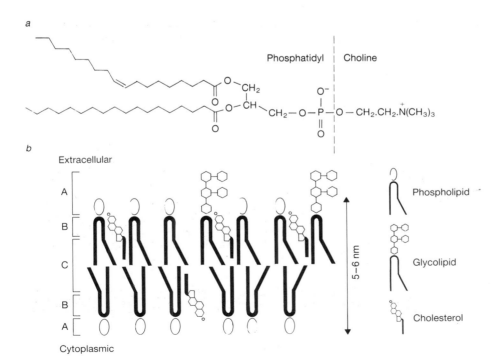

Fig. M9 Membrane lipids interact to form a stable bilayer. *a*, Lecithin (phosphatidylcholine). In other phospholipids the choline base is replaced by ethanolamine, serine, or inositol. *b*, A typical bilayer. A, Polar head group region; B, hydrophobic surface region, condensed by cholesterol; C, fluid hydrophobic interior. Adapted from [1].

individual lipid classes, with the major exception of the phosphatidylinositol phosphates, from which inositol trisphosphate and diacyl glycerol are liberated following activation of the enzyme PHOSPHOLIPASE C by a number of G PROTEIN-COUPLED RECEPTORS. These products act as second messengers in cell activation systems mediated by PROTEIN KINASE C and CALCIUM (*see* SECOND MESSENGER PATHWAYS).

CHOLESTEROL is a further important lipid component, present in animal cell membranes but absent from prokaryotes. It is particularly abundant in the plasma membrane where it may reach one mole per mole phospholipid. It contributes to a more condensed structure and decreases the nonspecific permeation of hydrophobic molecules into cells. It is much less abundant in intracellular organelles (5–10%) and is replaced by ergosterol in plants.

The physical state of most of the phospholipids in biological membranes is close to that in model bilayer or liposome systems as judged by techniques such as differential scanning calorimetry, NMR SPECTROSCOPY, ESR/EPR SPECTROSCOPY, and partition of solutes into the bilayer. The interior of the bilayer behaves as a medium with a viscosity similar to that of castor oil and solvent properties similar to those of octanol. The lipid molecules are free to move within the bilayer (at a speed of $1–2$ μm s^{-1}) but rarely leave it and rarely reorient from one surface to the other spontaneously. The solubility of dipalmitoyl lecithin in water is very low (10^{-10} M) and the rate of flipping across the membrane is 10^{-4}–10^{-6} s^{-1} ($t_{1/2}$ about 1 day). The fluidity and the solvent powers are both decreased by cholesterol. In contrast, the protein molecules which may account for up to 80% by weight of the membrane have much less effect on lipid mobility because of their compact structure. This was clearly demonstrated in electron micrographs of freeze-fractured preparations of biological membranes, which in contrast to the smooth fracture faces of pure phospholipids, showed many large (50–150 Å) particles randomly distributed in the membrane, which could be identified as membrane-associated protein.

Membrane proteins

Membrane proteins can be divided experimentally into extrinsic proteins, removable by extremes of pH or ionic strength, or by urea and other disrupters of protein interactions, and intrinsic or integral membrane proteins, which are released only when the lipid bilayer is dissolved by detergents (Fig. M10). Both types show great diversity of structure and function. Many intrinsic membrane proteins are glycoproteins with their extramembrane domains glycosylated by branched oligosaccharides of varied structure (*see* GLYCOPROTEINS AND GLYCOSYLATION). These together with the glycolipids form a loose network of carbohydrate — the glycocalyx — which interacts with proteins and PROTEOGLYCANS of the extracellular matrix, stabilizing cellular interactions (*see* CELL–MATRIX INTERACTIONS; EXTRACELLULAR MATRIX MOLECULES).

The extrinsic proteins are exemplified by constituents of the CYTOSKELETON consisting of cytoplasmic proteins such as SPECTRIN and ACTIN which contribute to a network, just inside the plasma membrane, connected to structural elements within the cell. Spectrin and its relatives are also bound to intrinsic proteins of the membrane either directly or via other proteins such as ankyrin (*see* MEMBRANE SKELETON). The cytoskeleton interacts with motor proteins such as MYOSIN, KINESIN, and DYNEIN and this allows control of the shape and dynamics of the cell membranes (*see* CELL MOTILITY; MICROTUBULE-ASSOCIATED MOTORS).

Most of the known intrinsic membrane proteins are anchored by transmembrane α-helices, each containing about 20 amino acids of predominantly hydrophobic character. The simplest have only a single helix, often linking an extracellular catalytic or recognition domain to the membrane and sometimes connecting it to another functional domain within the cell (Fig. M11*a*) (*see* GROWTH FACTOR RECEPTORS; IMMUNOGLOBULIN SUPERFAMILY; T CELL RECEPTORS).

Most intrinsic membrane proteins have more complex transmembrane domains containing between 2 and 20 α-helices,

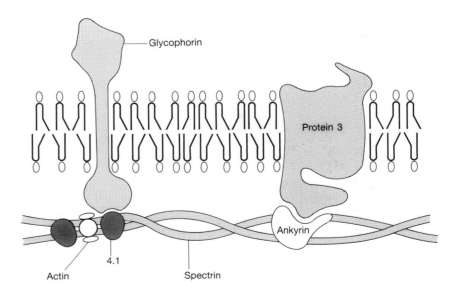

Fig. M10 Interactions between intrinsic membrane proteins and the cytoskeleton of the erythrocyte. Glycophorin is a small protein carrying some of the carbohydrates which determine blood group specificity (*see* GERBICH BLOOD GROUP; MNS BLOOD GROUP). Protein 3 is an anion exchanging protein with 12 transmembrane segments. Spectrin and its relatives form extended networks which interact with the cytoskeletal elements including actin, microfilaments (actin filaments), and microtubules. Adapted from [1].

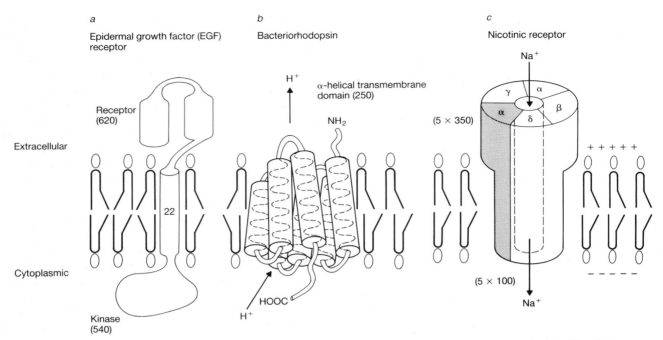

a Epidermal growth factor (EGF) receptor

b Bacteriorhodopsin

c Nicotinic receptor

Fig. M11 Structures of intrinsic membrane proteins. The number of amino-acid residues which form each part of the structure are shown in brackets. *a*, The structure of epidermal growth factor receptor is typical of a number of growth factor receptors, including that for insulin. A disulphide-rich extracellular domain is linked by a single helix to a protein tyrosine kinase domain which can phosphorylate the receptor itself and intracellular substrates. Receptors of the immunoglobulin superfamily, particularly important in intercellular recognition, are also built on this pattern, but lack the kinase domain. *b*, Bacteriorhodopsin has a particularly informative structure. It not only provides the basis for a detailed understanding of light-activated proton transport but is a prototype structure for a family of hundreds of seven-helix proteins, including visual rhodopsin and a variety of receptor proteins which recognize adrenaline, dopamine, 5-hydroxytryptamine (serotonin), neuropeptides, and olfactory stimulants. The effects are transmitted to the cytoplasm by GTP-binding proteins. *c*, The nicotinic acetylcholine receptor is responsible for transmission of a nerve impulse across many synapses, especially at skeletal muscle neuromuscular junctions. Acetylcholine released from a nerve ending binds to the α-subunit of the receptor and opens the channel which allows Na^+ to flow into the cell and depolarize the membrane. Each subunit has four transmembrane regions as judged from hydrophobicity plots. Receptors for glycine, glutamate, and GABA have a similar structure. Voltage-activated Na^+, K^+, and Ca^{2+} channels are related in overall architecture but are composed of a single polypeptide chain folded into discrete transmembrane proteins, each with a six-helix structure. Adapted from [1].

joined by short or extended cytoplasmic and extracellular loops. A large group of proteins with seven transmembrane helices are the G PROTEIN-COUPLED RECEPTORS. In cell-surface receptor proteins, activity of the cytoplasmic functional domain is controlled by occupancy of the recognition domain (the receptor) by a hormone, growth factor, neurotransmitter, or other signal molecule. The mechanisms of signal transmission across the membrane are only just beginning to be understood. Some receptors, such as those for some growth factors (*see* GROWTH FACTOR RECEPTORS) have a cytoplasmic protein kinase domain which can phosphorylate themselves and other proteins; others — the G protein-coupled receptors — utilize separate GTP-BINDING PROTEINS, which can control the formation of cyclic nucleotides, inositol phosphates or other intracellular messengers.

Many other complex transmembrane proteins are transport proteins in which several helices surround a channel through which ions or metabolites can pass (*see* ION CHANNELS; MEMBRANE TRANSPORT) (Fig. M11*c*). Passage down a concentration gradient may be controlled by ligands (e.g. acetylcholine, *see* NICOTINIC RECEPTOR), or by MEMBRANE POTENTIAL (voltage-gated Na^+, K^+, or Ca^{2+} channels, *see* CALCIUM; VOLTAGE-GATED ION CHANNELS). Transport against a gradient can be energized by ATP,

by oxidative metabolism, or by counterflow of ions (*see* MEMBRANE TRANSPORT).

Structure of membrane proteins

Progress in determining the three-dimensional structure of membrane proteins has been very slow because the simultaneous presence of extended polar and hydrophobic surfaces in the same molecule makes it difficult to find solvents for purification and crystallization. After 20 years of work with varied detergent systems no general strategy has been discovered, and only two or three structures, mostly of proteins involved in PHOTOSYNTHESIS, have been determined by X-RAY CRYSTALLOGRAPHY. A single high-resolution structure, that of BACTERIORHODOPSIN, has also been solved by electron crystallography of two-dimensional arrays (see Fig. M11*b*).

Knowledge of the structure of other membrane proteins comes mainly from analysis of thousands of amino-acid sequences in the light of known protein structures. Transmembrane helices can usually be identified because long runs of hydrophobic residues are rare in soluble proteins and so a preliminary definition of the membrane domain can be obtained. The extramembrane domain can sometimes be isolated and crystallized as a separate unit (e.g. in membrane-bound members of the IMMUNOGLOBULIN SUPER-FAMILY).

A small group of integral membrane proteins is known which lack transmembrane α-helices. These are the porins of the outer envelope of Gram-negative bacteria (*see* BACTERIAL ENVELOPE; MEMBRANE TRANSPORT SYSTEMS). The helices are replaced by β-strands whch form a hydrogen bonded sheet around an aqueous pore. This limitation of transmembrane protein structure to packed hydrophobic helices or closed β-sheets follows from the necessity of satisfying all the potential peptide hydrogen bonds internally. No hydrogen bonds can be formed to the hydrophobic membrane core and, since an unbonded peptide hydrogen is energetically expensive, structures which contain one are very rare and connections between elements of secondary structure are confined to the aqueous phase or the polar head group region.

Electrical properties of membranes

A less tangible aspect of membrane structure, of particular importance for cell function, is the electrical structure defined by ion distribution. Because of the insulating properties of the lipid bilayer, membranes can support a potential of a few hundred millivolts, provided that they do not contain proteins of high ion permeability. The plasma membrane outer envelope of a living cell fulfils this condition and the interior of the cell is maintained at a negative potential relative to the exterior by cation pumps (*see* MEMBRANE POTENTIAL; MEMBRANE TRANSPORT SYSTEMS). This potential can modulated by ion channel proteins specific for Na$^+$, K$^+$, or Ca^{2+}, the operation of which may itself be potential dependent (voltage gated). This system gives rise to an excitable membrane which provides the basis for nerve conduction and the synchronization of muscle contraction (*see* ACTION POTENTIAL; MUSCLE; SYNAPTIC TRANSMISSION).

N.M. GREEN

See also: EXCITATORY AMINO ACID RECEPTORS; GABA AND GLYCINE RECEPTORS; OLFACTORY TRANSDUCTION; PROTEIN TRANSLOCA-TION; PROTEIN STRUCTURE; VISUAL TRANSDUCTION.

1 Evans, W.H. & Graham, J.M. (1989) *Membrane Structure and Function* (IRL Press, Oxford).
2 Gennis, R.B. (1989) *Biomembranes* (Springer-Verlag, Berlin).
3 Marsh, D. (1992) Lipids in membrane structure. *Curr. Topics Struct. Biol. 2*, 497–502.
4 Kuhlbrandt, W. (1992) Membrane proteins. *Curr. Topics Struct. Biol. 2*, 503–510.

membrane synthesis The synthesis of new membrane in cells occurs by the insertion of new material into pre-existing membranes. In eukaryotic cells, the site of synthesis of both the protein and lipid components of most membrane (*see* MEMBRANE STRUC-TURE) is the ENDOPLASMIC RETICULUM (ER). The ER membrane contains the enzymes required for the synthesis of PHOSPHO-LIPIDS, SPHINGOLIPIDS, and CHOLESTEROL. Lipid synthesis takes place at the cytosolic face of the membrane. For phospholipids, the phospholipid precursor phosphatidic acid is synthesized by the fatty acylation of glycerol phosphate. The hydrophobic fatty

acyl chains remain embedded in the outer leaflet of the membrane bilayer, and the various headgroups (e.g. choline, serine, inositol) are added at the cytosolic face. Newly synthesized membrane lipids are added into the cytosolic leaflet of the ER membrane and equilibrate across the membrane by a 'flip-flop' mechanism which is thought to be mediated by the action of phospholipid translocators. Further modification of sphingolipids to form GLYCOLIPIDS and SPHINGOMYELIN occurs on the luminal side of the GOLGI APPARATUS, and thus these lipids are always found on the noncytosolic face of membranes (e.g. the extracellular face of the plasma membrane, where the glycosylated headgroups, along with the oligosaccharide side chains of GLYCO-PROTEINS form the glycocalyx).

Most membrane proteins appear to be delivered to the ER membrane attached to ribosomes before synthesis is complete, and are then incorporated into the membrane (*see* PROTEIN TRANSLOCATION). Once the orientation of a membrane protein across the membrane is established it does not change.

Membrane synthesized in the ER is carried to the other membranes of the ENDOMEMBRANE SYSTEM by transport vesicles (*see* GOLGI APPARATUS; LYSOSOMES; PROTEIN TARGETING). Synthesis of the membranes of mitochondria and chloroplasts, which are not part of the endomembrane system, involves some synthesis of protein within the organelle and also the import of proteins synthesized in the cytosol. Lipids are believed to be imported from the ER and other cell membranes by phospholipid exchange proteins, which pick up a lipid molecule from one membrane and discharge it into another.

The various membranes within a eukaryotic cell have distinct protein and lipid compositions, as do different membrane domains in polarized cells (*see* EPITHELIAL POLARITY). Specific targeting signals that direct proteins to particular organelles are known (*see* PROTEIN TARGETING; PROTEIN TRANSLOCATION).

membrane topology The two faces of a cellular bilayer membrane are not identical, as a result of, for example, different lipid composition and the particular orientation of glycoproteins and glycolipids in the membrane. The difference in the two faces is largely established at the site of membrane synthesis (in the ENDOPLASMIC RETICULUM in eukaryotic cells) and does not change as membrane is transported through the cell. In particular, once inserted into a membrane, the orientation of transmembrane proteins remains the same. *See*: MEMBRANE STRUCTURE; MEM-BRANE SYNTHESIS; PROTEIN SECRETION; PROTEIN TRANSLOCATION.

membrane traffic *See*: ENDOCYTOSIS; EXOCYTOSIS; MEMBRANE RE-CYCLING; SECRETION.

Membrane transport systems

AN essential property of living cells is their ability to control their solute composition by the selective transport of molecules and ions across the cell membrane. The transport of selected species is facilitated by a network of membrane-spanning proteins, collectively termed transporters. Generally, these systems can serve simply to select the solutes entering the cell, or can be involved in the active accumulation of certain solutes, or in the efflux of toxic and other solutes. Many different transport systems have evolved to carry out these functions [1,2].

Structure and structural analysis

With the advent of cDNA cloning techniques (*see* DNA CLONING) it has been possible to sequence the genes for many transport proteins and to predict their topology within the membrane. Structurally, they are thought to fall into two classes:

1 Those composed of β-strands arranged in a β-BARREL (e.g. porins).

2 Those composed of a cylindrical arrangement of α-helices (e.g. passive transporters, active transporters, ion channels, and receptor proteins).

The methods for predicting membrane-spanning β-strands are based on looking for alternating hydrophobic and hydrophilic residues, but this is difficult as the membrane-spanning region is only 7–9 residues long. However, the known three-dimensional structures of a number of porins have shown that this is not a strict requirement. More success has been obtained in predicting the turns of β-barrels by CHOU–FASSMAN ANALYSIS. The procedure for predicting membrane-spanning α-helices is based on identifying a stretch of 21 hydrophobic amino acids, as this represents the minimum number of amino acids capable of spanning the membrane as an α-helix (see HYDROPATHY PROFILE). In a number of cases the putative structures have been confirmed by biophysical techniques such as infra-red and CIRCULAR DICHROISM spectroscopy to approximate the relative amounts of α-helix and β-structure, vectorial proteolysis, and sequence-specific antibody binding (see [3] for a review), and by fusing genes to the transporters at predicted points in the putative topology and the testing for intra- and extracellular activity (e.g. through β-GALACTOSIDASE or alkaline phosphatase activity) [4]. Bacterial alkaline phosphatase (PhoA) has the important property that it is enzymatically active only after export across the bacterial inner (cytoplasmic) membrane into the periplasmic space (*see* BACTERIAL ENVELOPES). In studies of bacterial transporter proteins using fusions to PhoA, fusions to periplasmic segments of cytoplasmic membrane proteins show high PhoA activity, while fusions to cytoplasmic segments show low activity.

More recently, an EPR technique (*see* ESR/EPR SPECTROSCOPY) has been used to identify which residues are involved in forming α-helices [5]. The residues down a putative helix are systematically replaced with cysteine residues, which are labelled with an EPR probe. The exposure of the probe to the membrane, protein, or aqueous medium is tested with different relaxation agents (e.g. oxygen, chromium, or oxalate). In the case of one particular helix of BACTERIORHODOPSIN (for which the three-dimensional structure is known) the O_2 accessibility of the residues was found to alternate, between exposure to the protein and membrane, with a period of 3.6, symptomatic of an α-helix.

In contrast to the preponderance of sequence data available for transporters there is little detailed three-dimensional information. Some β-barrel proteins have been crystallized and their three-dimensional structures determined (see below), but our under-standing of the structure of those transporters composed of α-helices is largely based on the known structures of bacteriorhodopsin and the bacterial photosynthetic REACTION CENTRE (*see also* PHOTOSYNTHESIS). However, neither of these 'transporters' serves such a complex function as the translocation of carbon substrates across the membrane. The bacterial photosynthetic reaction centre is composed of three protein subunits (denoted L, M, and H). The L and M subunits, each consisting of five helices, are arranged into a cylindrical structure of elliptical cross-section. The cylinder has an outer ring of six helices, three contributed by each subunit, and an inner ring composed of the remaining helices. The H subunit consists of a single helix and a large globular domain, exposed to the cytoplasm, which is thought to have a role in stabilizing the L–M complex.

General mechanisms

Most transporters are believed to operate by similar mechanisms, in which the transporter alternates between two principal conformations [1,3]. For both passive and active transporters, the substrate translocation event involves a conformational change in the transporter such that the binding site is alternately presented at the two faces of the membrane. This conformational reorientation of the transporter is invariably the rate-limiting step in the transport cycle. The situation is somewhat different for ion channels, because the transporter cycles between open and closed conformations. While in the open conformation, ions are allowed to pass through the channel, as opposed to being bound and translocated to the opposite membrane face, as for passive and active transporters. Porins are mechanistically similar to ion channels, but reside largely in the open conformation. Some porins have been shown to be voltage-dependent (see below).

Porins

The outer membrane of Gram-negative bacteria (see BACTERIAL ENVELOPES) acts as a protective barrier to a variety of harmful reagents such as antibiotics and toxins. Small nutrient molecules and ions can permeate this barrier via membrane-spanning proteins, called porins. These porins provide a large open water-filled pore or channel through the membrane that is sufficiently small to exclude larger toxins. The concentration gradient of each transported species is used to drive their translocation through the porin.

A number of porins, such as the nonspecific OmpF and OmpC (outer membrane proteins F and C) and the anion selective PhoE (phosphoporin) porins from *Escherichia coli*, have been studied in detail. A number of ligand-specific porins have also been identified; such as LamB, the transporter for maltose and maltodextrins, and Tsx, the nucleotide transporter, in *E. coli*. There seem to be many specific porins in the outer membrane of *Pseudomonas aeruginosa*; including D1 (OprB, the glucose transporter), D2 (OprD, the basic amino acid transporter), and P (OprP, the phosphate transporter).

The most significant advance in our understanding of porin function has been the recent determination of the three-dimensional structures of several nonspecific porins [6,7]. In contrast to most other membrane proteins, which have a largely α-helical structure, each subunit of these trimeric porins is composed of a 16-stranded antiparallel β-barrel containing a pore. The strands are tilted (35–50°) from the barrel axis. The β-strands are joined at the intracellular face of the membrane by short β-turns, to create a smooth surface, whereas at the extracellular surface they are connected by long loops (L1 to L8), to give a rough surface. A particularly long loop (L3) folds inside the barrel and contributes to a constriction of the pore halfway through it. The pore size decreases, from 15×22 Å to 7×11 Å at the constriction. The amino acid residues of L3 are critical in determining the ion selectivity of the porins.

The entrance to the pore, provided by the rough end of the β-barrel, probably acts as an initial screen of the size and charge of translocated solutes. The constricton zone then provides a more rigorous discrimination of translocated solutes, on the basis of their size and charge.

Passive transporters (uniporters)

Certain small molecules and ions, such as sugars, nucleosides, amino acids and anions, are taken up by cells in a process that does not require energy. The translocation of these hydrophilic species across an essentially hydrophobic membrane barrier is facilitated by specific transporters, proteins which span the membrane. These transporters simply catalyse the movement of these molecules and ions down their concentration gradient, which would otherwise be negligibly slow. This type of transport is commonly termed facilitated diffusion.

The glucose transporters found in most, if not all, mammalian cells are a classic example of a passive transport system. To date, seven isoforms have been identified (GLUT1 to 7), which differ in their tissue distribution. These transporters are glycoproteins of around 500 amino acids (M_r 55 000), which are predicted to be arranged into three major domains:
1 Twelve α-helices, which span the membrane.
2 A large, highly charged, cytoplasmic domain, bridging helices 6 and 7.
3 A smaller external domain, bridging helices 1 and 2, which bears a carbohydrate moiety.
A number of biophysical studies support this putative topological model. The structure is common to a number of other transporters (e.g. bacterial sugar symporters). For many of these transporters there is sequence homology between the N- and C-terminal halves of the protein, suggesting that the 12 α-helix structure has arisen by the duplication of a gene encoding a 6-helix structure.

Another well characterized example is the anion-exchanger, which catalyses the one-for-one electroneutral exchange of anions, principally chloride for bicarbonate, across the plasma membrane. Three isoforms of this 911 amino acid (M_r 90 000–100 000) (erythrocyte transporter) glycoprotein have been identified, which differ in their tissue specificity. This transporter can be divided into two distinct structural domains: a highly negatively charged N-terminal domain of about 400 residues, facing the cytoplasm, and a membrane-spanning domain of about 500 residues. The cytoplasmic domain is a highly extended structure possessing high-affinity binding sites for several glycolytic enzymes, haemoglobin, and the membrane cytoskeletal protein ankyrin (*see* MEMBRANE STRUCTURE). The membrane domain is thought to consist of 10–14 α-helices. The cytoplasmic domain of the transport protein can be cleaved, by trypsin, to leave the membrane domain which can still catalyse anion exchange. Recently, the membrane fragment has been crystallized and a low resolution two-dimensional structure has been determined by electron diffraction [8]. This exciting achievement holds promise for the future determination of the three-dimensional structure of a transport protein.

Most examples of this type of transporter are from mammalian cells, but a few examples are known from other organisms. The glycerol transporter (GlpF) of *E. coli* is the only known example in bacteria. The sequence of this 281 amino-acid protein suggests that it contains six α-helices.

Cotransporters

Cotransporters couple the movement of an ion, down its electrochemical gradient, to that of another molecule against its concentration gradient. The ions involved are principally Na^+, in mammals, and H^+ and K^+ in bacteria. The electrochemical gradients of these ions are created by the ubiquitous Na^+, K^+-ATPase (in mammalian cells) and H^+-ATPase. Cotransport in which the ion and the transported species move in the same direction is termed symport, while transport systems in which the ion and transported species move in opposite directions are termed antiporters.

A number of H^+-driven cotransporters have been cloned and sequenced from a variety of organisms, and shown to have homology with the passive mammalian glucose transporters described above [9,10]. This superfamily of transporters can be subdivided into five families:
1 Sugar uniporters and symporters, including the mammalian passive glucose transporters, the bacterial symporters for arabinose (AraE), galactose (GalP), and xylose (XylE), and the yeast glucose (Hxt2 and Snf3), and galactose (Gal2) transporters.
2 Bacterial oligosaccharide symporters, for lactose (LacY), sucrose (ScsB), and raffinose (RafB).
3 Drug-resistance antiporters of bacteria and yeast, for tetracycline (TetA-G, Tcr1-2), methylenomycin A (Mmr), quinalones (NorA), antiseptics (QacA), and multidrugs (EmrA and Bmr).
4 Bacterial phosphate ester antiporters, for phosphoglycerate (PgtP), glycerol 3-phosphate (GlpT) and hexose phosphate (UhpT).
5 Bacterial citrate (CitA and H), α-ketoglutarate (KgtP) symporters, and the proline/betaine (ProP) symporter involved in osmoregulation.
All these transporters are predicted to have 12 membrane-spanning α-helices, but only the members of family **1** are predicted to have a large central cytoplasmic domain.

Two major families of Na^+-driven cotransporters have been identified [11]:
1 Those homologous to the mammalian Na^+/glucose cotransporter (SGLT1) found in kidney and intestine, including the mammalian kidney nucleoside (SNST1) and *myo*inositol (SMIT1) transporters and the *E. coli* proline (PutP) and pantothenate (PanF) transporters.

2 Those homologous to the mammalian Na$^+$/GABA (γ-aminobutyric acid) cotransporter (GAT1) from brain, including the glycine (GLYT1), dopamine (DAT1), serotonin (SERT1), noradrenaline (NET1) and proline (PROT1) transporters from brain and the betaine (BGT1) transporter from kidney.

Although the Na$^+$-driven cotransporters are also predicted to have 12 α-helices, they are not homologous to the passive glucose transporters or the homologous proton-driven transporters. In the SGLT1 family, the loops connecting the extracellular ends of helices are generally larger than those connecting the intracellular ends, and larger than the corresponding extracellular loops of the GLUT proteins. Moreover, there is a particularly large hydrophilic loop, of about 90 residues, between helices 11 and 12, and there is no C-terminal hydrophilic loop. The GAT1 family are characterized by large loops at the N-terminal and between α-helices 3 and 4. In contrast to the SGLT1 family, they have a moderately large C-terminal loop.

Two physiologically important Na$^+$-driven antiporters are the Na$^+$, Ca^{2+}-exchanger of cardiac muscle, which helps to regulate the cytosolic Ca^{2+} level and, therefore, muscular contraction of the heart; and the Na$^+$, H$^+$-exchanger, involved in cytoplasmic pH regulation.

Ion pumps

Many cells maintain concentration gradients of certain ions, such as Na$^+$, K$^+$, Ca^{2+}, and H$^+$, across both their plasma and their intracellular membranes. These ion gradients are maintained by ion pumps, which couple the hydrolysis of ATP to the movement of ions against their electrochemical gradient. Two extensively studied ion pumps are the mammalian Na$^+$, K$^+$-ATPase and the Ca^{2+}-ATPase [12].

The Na$^+$, K$^+$-ATPase (or sodium pump) is responsible for maintaining the high internal K$^+$ and low Na$^+$ concentration characteristic of most eukaryotic cells. This is generally accomplished by the pump exchanging three Na$^+$ ions from the cytosol for two K$^+$ ions from the extracellular medium. The pump consists of two noncovalently linked polypeptides: a 1016 amino-acid α-subunit (M_r 100 000) and a 302 amino-acid glycosylated β-subunit (M_r 55 000). Most of the functions of the pump have been localized to the α-subunit. It contains the binding sites for Na$^+$, K$^+$, and ATP, and also for OUABAIN, a potent inhibitor of the pump. Ion translocation is coupled to phosphorylation of this subunit. The exact function of the β-subunit remains unknown.

The α-subunit is thought to be arranged in two major domains: a membrane-spanning domain consisting of eight α-helices, and a large cytoplasmic domain. The cytoplasmic domain is composed of a large loop connecting helices 4 and 5, and two smaller loops connecting helices 2 and 3, and helices 7 and 8. An aspartate residue of this domain is phosphorylated during ion translocation. The β-subunit consists of a single membrane-spanning α-helix, linking a short cytoplasmic N-terminal loop with a larger extracellular loop containing several glycosylation sites.

The pump is thought to operate by the following sequence of events [13]. Cytoplasmic Na$^+$ binds to a high affinity site on the pump catalysing its phosphorylation. Two distinct events then follow. There is a minimal conformational change in the protein which separates the ions from the cytoplasmic medium. A more substantial rearrangement of the protein then spontaneously occurs, as the pump undergoes the ion-translocating E1 to E2 conformational transition. The affinity for Na$^+$ is decreased and that for K$^+$ increased. The Na$^+$ ions are replaced by K$^+$ ions from the extracellular medium, inducing dephosphorylation of the pump and a reversal of the ion-translocating conformational change (E2 to E1).

The Ca^{2+} pump of the SARCOPLASMIC RETICULUM membrane is another well characterized example of an ion pump [14]. It serves the function of maintaining a low cytosolic Ca^{2+} concentration in muscle cells by pumping Ca^{2+} into the sarcoplasmic reticulum, allowing muscle relaxation. The pump consists of a single 997 amino-acid polypeptide (M_r 110 000). On the basis of the amino-acid sequence, the pump has been proposed to be arranged into three major domains: a stalk, consisting of five α-helices, which connects a large cytoplasmic domain with a membrane domain consisting of 10 α-helices. The cytoplasmic domain is the site for ATP binding and phosphorylation. The stalk, which contains a preponderance of charged amino-acid residues, is probably involved in funnelling the Ca^{2+} ions to the binding site formed by the transmembrane helices.

Recently, the low resolution (14 Å) structure of the Ca^{2+} pump has been determined by three-dimensional CRYOELECTRON MICROSCOPY [15]. The cytoplasmic part has a complex structure consisting of several domains, and resembles the head (ATP-binding and phosphorylation domain) and neck (stalk domain) of a bird. The α-helices of the transmembrane part of the pump are also arranged into several distinct domains: a linear domain, an inclined domain, and a curved domain.

Ion channels

Ion channels have a vital role in intracellular communication, in that they allow the free movement of selected ions down their electrochemical gradient when they open in response to either the binding of a ligand, or a change in MEMBRANE POTENTIAL [16]. This role is typified by the sequence of events leading up to the contraction of a vertebrate skeletal muscle. The initial stimulation of a sensory neuron triggers a wave of membrane DEPOLARIZATION down the motor neuron, mediated by voltage-dependent sodium channels (*see* VOLTAGE-GATED ION CHANNELS). These channels open in response to a localized change in the membrane potential maintained by the Na$^+$, K$^+$ pump. The change in potential deriving from the flux of Na$^+$ ions through a particular channel causes neighbouring channels to open, thus propagating a wave of potential change (depolarization). Arrival of the depolarization wave at the neuromuscular synapse (MOTOR END-PLATE) activates voltage-dependent Ca^{2+} channels within the neuron. The resulting dissipation of the Ca^{2+} gradient, usually maintained by the Ca^{2+} pump, triggers release of the neurotransmitter acetylcholine into the synaptic cleft. Free acetylcholine diffuses across the cleft and binds to specific receptors on the muscle cell membrane. These receptors then act as ligand-dependent cation channels, permeable to both Na$^+$ and K$^+$ ions. The flux of these ions causes depolarization of the muscle cell membrane, activating voltage-dependent Ca^{2+} channels, largely located within the sarcoplasmic reticulum. The resulting increase in cytosolic Ca^{2+} triggers muscular contraction as the Ca^{2+} ions binds to the troponin component of the muscle filament (*see* MUSCLE).

The two types of ion channel can be distinguished by their differing structure: ligand-dependent ion channels are characteristically made up of several polypeptides, whereas voltage-dependent ion channels principally consist of a single polypeptide. However, the single polypeptide of voltage-dependent channels usually contains several strongly homologous repeat sequences, so that the two types of ion channel have essentially parallel structures.

For example, the nicotinic acetylcholine receptor on skeletal muscle (*see* NICOTINIC RECEPTORS) is a pentameric complex of five similar glycoprotein subunits. Each subunit consists of around 500 amino acids (M_r 55 000–66 000) and they are distinguished from one another by a series of insertions and deletions. Each subunit is predicted to have four transmembrane regions, an extracellular globular N-terminal domain, and a globular cytoplasmic domain connecting helices 3 and 4. The principal difference between the subunits is in the size of the globular domains.

On the other hand, the voltage-dependent Na^+ channel is essentially composed of one large glycoprotein (1820 amino acids, M_r 260 000), although several smaller glycoproteins (M_r 30 000–40 000) can be associated with the large glycoprotein at the extracellular surface of the membrane. Analysis of the sequence of the large subunit has shown that there are four homologous multiple membrane-spanning domains, interconnected by three large globular cytoplasmic domains. Three of the transmembrane domains have been predicted to consist of six fully, and two partially membrane-spanning helices. A small globular extracellular loop connects helix 5 with the first partial transmembrane helix (helices 6 and 7). The fourth domain is differentiated from the others in that the partial transmembrane helices are thought to be replaced by a β-strand, because this segment contains several prolines, which would make a helical conformation unlikely.

The three-dimensional structure of the acetylcholine receptor has been determined at low resolution by electron diffraction [17]. The receptor is clearly partitioned into three domains. There is a long cylindrical domain, 25 Å wide with an internal diameter of about 10–13 Å, which extends about 60 Å into the synaptic cleft. The receptor subunits form the walls of this structure. The diameter of the receptor rapidly decreases as it traverses the membrane, until it reaches a value of less than 10 Å towards the centre of the lipid bilayer. This structure probably represents the central ion channel, which has been shown to have a minimal diameter of about 7 Å by studying the conductance of ions of known size. In fact, it has been shown that a tight packing of the five α-helices with a 20° tilt would produce a structure of the correct dimensions.

Periplasmic binding-protein dependent transport systems

A number of solutes, including sugars, amino acids and ions, are transported across the cytoplasmic membrane of Gram-negative bacteria (*see* BACTERIAL ENVELOPES) via high affinity transport systems in which the solute first interacts with a globular protein in the periplasm. A number of binding proteins have been crystallized and their three-dimensional structure determined [18,19]. These soluble proteins are essentially divided into two lobes that engulf the substrate upon its binding. This protein then interacts with a membrane-associated transporter complex, generally composed of two hydrophobic proteins and two hydrophilic proteins [20]. The hydrophobic proteins are predicted to be composed of six membrane-spanning α-helices, which must be involved in translocating the substrate across the membrane. The hydrophilic proteins are known to bind ATP and are presumed to hydrolyse the ATP in driving the translocation of the substrate. The mechanism(s) of the interaction of the binding protein with the membrane components, and the subsequent translocation of the substrate across the membrane, are unknown.

This type of bacterial transport system belongs to a superfamily of homologous transporters, which share the common feature of having an ATP-binding cassette and are consequently called ABC systems [21]. This superfamily includes the human MDR (multi-drug resistance) or P-glycoprotein, which is involved in the extrusion of anticancer drugs from the cytoplasm of cancer cells; the human CYSTIC FIBROSIS transmembrane regulator protein (CFTR); the α mating factor export protein of the yeast *Saccharomyces cerevisiae*; and the Hly complex of *E. coli*, which is involved in the secretion of haemolysin. The transport systems differ in whether or not the various membrane components are fused together.

Phospho-transferase systems

The phospho-transferase systems (PTS) found in bacteria catalyse the translocation of sugars and sugar derivatives across the cytoplasmic membrane. The movement of these sugars is coupled to their phosphorylation by phosphoenolpyruvate (PEP). The transported sugar is phosphorylated during transport in order to maintain the concentration gradient of the sugar across the membrane. Hence, the PT systems catalyse the active uptake of sugars [22].

The translocation of sugars involves at least two soluble proteins (Enzyme I and HPr), which are common to phospho-transferase systems for different sugars, and one or more membrane-associated proteins (Enzyme II), which are sugar-specific. The reaction begins with the phosphorylation of Enzyme I by PEP, which then transfers its phosphate to HPr. HPr is a small heat-stable protein, for which the three-dimensional structure has been determined [23]. HPr interacts with and phosphorylates the membrane-associated Enzyme II protein(s). The Enzyme II for a particular sugar usually consists of three domains, two hydrophilic domains (IIA and IIB) that become phosphorylated successively, and a hydrophobic domain (IIC) which probably spans the membrane. For some PT systems the three domains have dissociated and exist as independent proteins. The sequences of several IIC domains suggest that they contain 6–8 α-helices. The PhoA fusion technique has been used to determine the topology of the *E. coli* mannitol Enzyme II. This was shown to have six α-helices. This Enzyme II is known to exist as a tight dimer, indicating that 12 α-helices are required for functionality. The IIC domain can catalyse the facilitated diffusion of sugars across the membrane but the IIA and IIB domains are required for phosphorylation of the sugar.

A. WALMSLEY

See also: MEMBRANE STRUCTURE; PROTEIN STRUCTURE.

1 Walmsley, A.R. (1991) Cell Membrane Transport. In *The Encyclopedia of Human Biology*, Vol. II (Dulbecco, R., Ed.) (Academic Press, New York).
2 Nikaido, H. & Saier, M.H. (1992) Transport proteins in bacteria: common themes in their design. *Science* **258**, 936–942.
3 Walmsley, A.R. (1988) The dynamics of the glucose transporter. *Trends Biochem. Sci.* **13**, 226–231.
4 Manoil, C. & Beckwith, J. (1986) A genetic approach to analyzing topology. *Science* **233**, 1403–1408.
5 Altenbach, C. et al. (1990) Transmembrane protein structure spin labeling of bacteriorhodopsin mutants. *Science* **248**, 1088–1092; Millhauser, G.L. (1992) Selective placement of electron spin resonance spin labels: new structural methods for peptides and proteins. *Trends Biochem. Sci.* **17**, 448–452.
6 Weiss, M.S. et al. (1990) The three-dimensional structure of porin from *Rhodobacter capsulatus* at 3 Ångstrom resolution. *FEBS Lett.* **267**, 268–272.
7 Cowan, S.W. et al. (1992) Crystal structures explain functional properties of two *E. coli* porins. *Nature* **358**, 727–733.
8 Wang, D.N. et al. (1993) Two-dimensional structure of the membrane domain of human band 3, the anion transport protein of erythrocyte membrane. *EMBO J.* **12**, 2233–2239.
9 Griffith, J.K. et al. (1992) Membrane transport proteins: implications of sequence comparisons. *Curr. Opinion Cell Biol.* **4**, 684–695.
10 Marger, M.D. & Saier, M.H. (1993) A major superfamily of transmembrane facilitators that catalyse uniport, symport, and antiport. *Trends Biochem. Sci.* **18**, 13–20.
11 Wright, E.M. et al. (1992) Sodium cotransport proteins. *Curr. Opinion Cell Biol.* **4**, 696–702.
12 Carafoli, E. (1992) The Ca^{2+} pump of the plasma membrane. *J. Biol. Chem.* **267**, 2115–2118.
13 Gadsby, D.C. et al. (1993) Extracellular access to the Na, K pump: pathway similar to ion channel. *Science* **260**, 100–103.
14 Inesi, G. et al. (1992) Long-range intramolecular linked functions in the calcium-transport ATPase. *Adv. Enzymol.* **65**, 185–215.
15 Toyoshima, C. et al. (1993) Three-dimensional cryo-electron microscopy of the calcium ion pump in the sarcoplasmic reticulum membrane. *Nature* **362**, 469–471.
16 Miller, C. (1992) Ion channel structure and function. *Science* **258**, 240–241.
17 Unwin, N. (1993) Nicotinic acetylcholine receptor at 9 Ångstrom resolution. *J. Mol. Biol.* **229**, 1101–1124.
18 Adams, M.D. & Oxender, D.L. (1989) Bacterial periplasmic binding protein tertiary structures. *J. Biol. Chem.* **264**, 15739–15742.
19 Quiocho, F.A. (1990) Atomic structures of periplasmic binding proteins and the high-affinity active transport systems in bacteria. *Phil. Trans. R. Soc. Lond.* **B 326**, 341–351.
20 Ames, G.F-L. (1988) Structure and mechanism of bacterial periplasmic transport systems. *J. Bioenerg. Biomemb.* **20**, 1–18.
21 Hyde, S.C. et al. (1990) Structural model of ATP-binding proteins associated with cystic fibrosis, multidrug resistance and bacterial transport. *Nature* **346**, 362–365.
22 Meadow, N.D. et al. (1990) The bacterial phosphoenol-pyruvate: glucose phosphotransferase system. *Annu. Rev. Biochem.* **59**, 497.
23 Jia, Z. et al. (1993) Active centre torsion angle strain revealed in 1.6 Å resolution structure of histidine-containing phosphocarrier protein. *Nature* **361**, 94–97.

membrane vesicle Any small membrane-bounded vesicle present in eukaryotic cells. *See*: EXOCYTOSIS; GOLGI APPARATUS; LYSOSOME; PROTEIN SECRETION; PROTEIN TARGETING.

memory cell Long-lived T CELL or B CELL specific for the immunizing antigen which is produced after exposure to antigen and which is immediately activated on a subsequent exposure to the antigen to give a rapid and generally stronger SECONDARY IMMUNE RESPONSE. The formation of memory B cells, but not of memory T cells, also involves somatic HYPERMUTATION in the V regions of the IMMUNOGLOBULIN GENES so that antibodies of higher affinity for the antigen are produced in the subsequent immune response (affinity maturation). CLASS SWITCHING also occurs in the formation of memory B cells, so that the circulating antibodies produced in a secondary immune response are predominantly of the IgG class.

Mendelian inheritance

MENDELIAN inheritance is the name given to the mode of inheritance of traits determined by gene variants of discrete effect carried in the nuclear GENOME of sexually reproducing EUKARYOTIC organisms (*cf.* CONTINUOUS VARIATION; NON-MENDELIAN INHERITANCE).

Gregor Mendel (1822–1884) was the first to analyse inherited traits in terms of unit differences determined by discrete genetic factors. In experiments beginning in the 1850s he made controlled crosses between different true-breeding varieties of garden pea that showed a number of clear-cut heritable differences with respect to plant height, seed colour, flower colour and seed shape. From the results, he attributed each difference to a genetic 'factor' that could exist in either of two forms. In modern terminology, the factors are called GENES and the alternative forms of a gene are referred to as ALLELES.

For example, a seedling resulting from Mendel's cross between tall and short pea varieties inherits one allele, symbolized as *T*, from the tall parent, and the alternative allele, *t*, from the short parent. The hybrid seedling will therefore have the genetic constitution — GENOTYPE — *Tt*. (The parents would have had the genotypes *TT*, tall, and *tt*, short, respectively.)

Dominant and recessive

Mendel found that the first generation progeny from the tall × short cross, the F1 generation, were all tall, that is, they all had the tall PHENOTYPE (the appearance or outward manifestation of a trait, as opposed to the actual genetic constitution, the genotype, of the organism). The observation was rationalized by attributing to *T* an overriding dominant effect, *t* being termed recessive (i.e. not contributing to the phenotype in this combination of alleles). The dominant/recessive relationship was found by Mendel in respect of each of the unit differences he studied (e.g. round vs wrinkled seeds, purple vs white flowers), but it is not universal (*see e.g.* CODOMINANCE; DOSAGE EFFECT; INCOMPLETE DOMINANCE).

Mendel's First Law of Segregation

Recessive alleles in F1 hybrids are not lost; they retain their

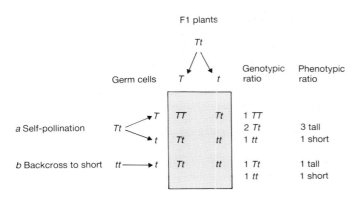

Fig. M12 Explanation of phenotypic ratios in progeny of F1 pea plants from Tall (*T T*) × Short (*tt*) cross.

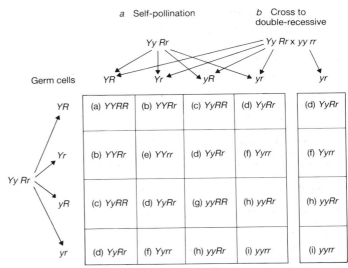

Fig. M13 Ratios given by simultaneous segregation of two gene differences in pea: Y/y (yellow/green seed) and R/r (round/wrinkled seed). Each box will constitute 1/16 of the total under *a* or 1/4 of the total under *b*. The nine different genotypes, distinguished by letters a–i, correspond to phenotypes as follows:

a, b, c, d, yellow round: 9/16 in *a*; 1/4 in *b*
e, f, yellow wrinkled: 3/16 in *a*; 1/4 in *b*
g, h, green round: 3/16 in *a*; 1/4 in *b*
i, green wrinkled: 1/16 in *a*; 1/4 in *b*

identity and can reassert their effects in later generations. When Mendel raised a second (F2) generation from the tall F1 hybrids about 25% of the plants were short, a frequency explained by Mendel's First Law of Segregation. This states that allelic factors present together in an F1 hybrid are segregated from one another in a 1 : 1 ratio in the formation of the male and female germ cells. Thus, a *Tt* F1 plant (HETEROZYGOUS for the *T* gene, as opposed to the HOMOZYGOUS *TT* and *tt* parents) produces statistically equal numbers of *T* and *t* ova in its ovules and *T* and *t* pollen grains in its anthers. Mendel's concept of segregation stood in sharp contrast to the ideas previously prevailing, according to which the distinctive contributions of the parents were irretrievably blended in the offspring.

The well-known Mendelian ratios with respect to traits determined by single genes (single-gene traits or single-factor differences) follow from the basic 1 : 1 segregation in the F1 germ cells. Thus, the 3 : 1 ratio of tall to short plants found in the F2 progeny is simply explained on the basis of the random union of ova and pollen grains (Fig. M12*a*). The theory predicts that one-third of the F2 plants showing the dominant (i.e. tall) phenotype will be of genotype *TT* and give nothing but tall progeny, whereas two-thirds will be *Tt* and repeat the 3 : 1 ratio in an F3 generation if crossed between themselves. This is indeed found to be the case. The most direct verification of the 1 : 1 F1 germ cell ratio comes from the back-cross of the heterozygous F1 to the recessive homozygous parent (i.e. *Tt* × *tt*); this gives a 1 : 1 ratio of dominant to recessive phenotypes among the progeny (Fig. M12*b*).

Mendel's Second Law of Independent Assortment

Mendel's Second Law refers to situations where an individual is heterozygous in respect of two or more allelic differences. It states that in such a case, the segregation of one pair of alleles is independent of that of any other pair. An example from Mendel's experiments concerns the inheritance of colour and shape of the pea seed.

The differences between green and yellow and between round and wrinkled in the ripe seed are each inherited as if due to a single allelic difference, with yellow (*Y*) dominant to green (*y*) and round (*R*) to wrinkled (*r*). The F1 progeny of a cross between

pure-breeding yellow round and green wrinkled strains are of genotype *Yy Rr*; the germ cells that they form are *Y R*, *Y r*, *y R* and *y r* in statistically equal numbers. This explains the 1 : 1 : 1 : 1 ratio of round yellow, wrinkled yellow, round green and wrinkled green seeds seen in the progeny of the back-cross to the double-recessive parent (Fig. M13*a*) and the 9 : 3 : 3 : 1 ratio obtained by self-pollination of the F1 (Fig. M13*b*).

Genes and chromosomes

Mendel's Laws apply as much to animals as to plants, and can be understood as reflecting the chromosomal location of the genes and the alternation of HAPLOID and DIPLOID states which is the universal hallmark of sexual reproduction. Animals and higher plants carry 2*n* microscopically visible CHROMOSOMES (comprising *n* pairs) in the nuclei of their body cells but, in germ cell formation, the chromosome number is halved through the process of MEIOSIS, which consists of two divisions of the cell nucleus accompanied by only one division of the chromosomes. In meiosis each kind of chromosome is necessarily segregated from its partner into a different germ cell, and the pairs of alleles carried on a pair of chromosomes are thus segregated too. Each chromosome pair segregates independently of any other chromosome pair, which accounts for Mendel's Second Law so long as different pairs of alleles are associated with different chromosome pairs.

Linkage

In fact, a mass of evidence accumulated since Mendel shows that

Triple heterozygote *A/a B/b C/c* × triple homozygous recessive *a/a b/b c/c*

Germ cells formed by meiosis in the triple heterozygote:

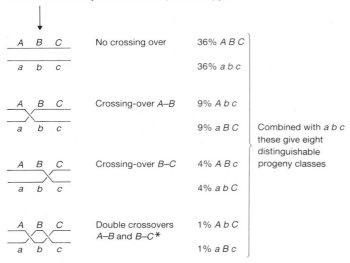

A B C	No crossing over	36% *A B C*	
a b c		36% *a b c*	
A B C	Crossing-over *A–B*	9% *A b c*	
a b c		9% *a B C*	Combined with *a b c* these give eight distinguishable progeny classes
A B C	Crossing-over *B–C*	4% *A B c*	
a b c		4% *a b C*	
A B C	Double crossovers *A–B* and *B–C**	1% *A b C*	
a b c		1% *a B c*	

Fig. M14 A hypothetical example of linked inheritance: three allelic differences *A/a*, *B/b*, *C/c*, linked on the same chromosome in order *A–B–C* with 20% crossing-over between *A* and *B* and 10% crossing-over between *B* and *C*. The triple heterozygote *A/a B/b C/c* is crossed to the triple homozygous recessive *a/a b/b c/c*. For simplicity, the crossover diagrams are drawn as if the exchanges were between whole chromosomes. In fact, crossing-over occurs after the chromosomes have divided into chromatids, and only one chromatid from each chromosome is involved in each crossover (*see* RECOMBINATION).

* The infrequency of the products *AbC* and *aBc* identifies them as double crossover products and defines the order *A–B–C*. The percentages are calculated here on the basis of no interference between crossovers. In fact, doubles are usually somewhat less frequent than predicted on this basis.

different pairs of alleles are not always reassorted independently but frequently show linkage, that is, a greater or lesser tendency for the parental combinations of alleles to outnumber the reassorted combinations (recombinants). Linkage is easily understood in terms of the location of genes on the same chromosome, so that they will tend to be inherited together. The fact that linked genes show any RECOMBINATION at all is explained by the microscopically observable exchange of corresponding segments (CROSSING-OVER) between chromosomes at the first division of meiosis. The likelihood of linked genes being recombined depends on the frequency of crossing-over between their chromosome loci, which in turn depends on the distance between them (Fig. M14).

Gene conversion

The ratios from which Mendel's Laws were deduced were statistical in character and subject to random sampling error. However, in certain fungi, the products of meiosis can be recovered as tetrads of spores, each of which can be cultured and its genetic composition determined. Here, segregation of alleles generally gives the precise 2 : 2 ratio in each tetrad. Tetrad analysis, while it permits this exact verification of Mendelian segregation as the general rule, has shown that the rule has occasional exceptions,

referred to as GENE CONVERSION. In this phenomenon, which occurs at frequencies of up to a few per cent in some fungi, genetic information is transferred between different alleles at the first division of meiosis giving 3 : 1 or 1 : 3 ratios in some individual tetrads, but only rarely affecting the 1 : 1 ratio overall. There is good reason to believe that gene conversion occurs in higher plants and animals also, but at very low frequencies (1 in 10 000 or less).

J.R.S. FINCHAM

See also: POLYGENIC INHERITANCE; RECOMBINATION; SEX LINKAGE.

Fincham, J.R.S. (1983) *Genetics* (John Wright, Oxford) 643pp.

Menkes disease An X-LINKED genetic disorder resulting in deteriorating brain disease and white, twisted, brittle hair. The underlying defect is an abnormality in copper (Cu^{2+}) transport due to an abnormality in a gene encoding a P-TYPE ATPASE copper transport protein.

Nature Genet. (1993) **3**, 7–13, 14–19, 20–25.

mercaptoethanol A thiol reagent ($HO-CH_2-CH_2-SH$) generally used for the reduction of disulphide bonds in proteins.

mercaptopurine Purine analogue capable of blocking DNA synthesis. The inactive precursor is converted to ribonucleotide by the enzyme hypoxanthine : guanine phosphoribosyl transferase (HGPRT). This ribonucleotide analogue then blocks DNA synthesis by competitively inhibiting the conversion of inosinic acid to adenylosuccinic acid and xanthylic acid which in turn are precursors of adenine and guanine deoxyribonucleotides. It is used as a cytotoxic agent.

MERFF syndrome *See*: MITOCHONDRIOPATHIES.

meristem Plant tissue in which the cells proliferate by MITOSIS. In vascular plants the main meristems are those at the apices of roots and shoots responsible for all growth in length and much growth in thickness. The CAMBIUM is also an important meristem in conifers and woody dicotyledons.

meroblastic cleavage An incomplete form of CLEAVAGE restricted to eggs in which large amounts of YOLK are restricted to one end (e.g. eggs of reptiles, fish, and birds). Cleavage furrows are unable to penetrate the yolk and cleavage is restricted to an area of nonyolky cytoplasm, the blastodisk.

merodiploid Usually applied to bacteria, this term defines a partially diploid cell. Such a cell is generated when a fragment of bacterial chromosome is introduced into a host bacterium, usually by CONJUGATION or TRANSDUCTION thus generating a fraction of the host chromosome with an homologous counterpart. Recombination can occur between the whole chromosome and the introduced chromosome fraction.

merogone Experimentally produced egg fragments. If a sea urchin egg is split surgically along the meridian both merogones contain cytoplasm from the vegetal and animal poles and, if fertilized, develop into small but normal blastulae. Conversely, if the merogones are obtained by splitting the egg into vegetal and animal halves only the animal merogone gives rise to a near-normal blastula. These classic experiments of Hörstadius (1928) demonstrate MOSAICISM along the animal–vegetal axis of the egg in this species.

meroistic oocyte OOCYTE in certain insect species with NURSE CELLS. In DROSOPHILA, the oocyte is interconnected to 15 nurse cells by ring canals. A meroistic oocyte does not enter an active transcriptional phase as transcription is restricted to the nurse cells from which RNA is transported into the oocyte. *Cf.* PANOISTIC OOCYTE.

merosin Homologue of the A chain of the laminin, found in the nervous system. *See*: EXTRACELLULAR MATRIX MOLECULES.

mesaxon A pair of parallel membranes marking the line of edge-to-edge contact of SCHWANN CELLS encircling an axon.

mesenchyme (pl. mesenchymata) A rather vague term used by histologists to describe embryonic tissues consisting of stellate or fibroblastic cells loosely arranged in extracellular matrix. It does not have to be of MESODERMAL origin, nor is it necessarily a single tissue type.

mesoderm One of the three GERM LAYERS of the early embryo. In vertebrates and other CHORDATES, it gives rise to most of the musculoskeletal system, to the circulatory system, and to most of the internal organs of the adult. It arises during the process of GASTRULATION when it forms a layer situated between the other two germ layers, ECTODERM and ENDODERM. It is usually subdivided into tissues with different developmental FATES depending on their proximity to the embryonic axis: AXIAL MESODERM; HEAD PROCESS AND NOTOCHORD; PARAXIAL MESODERM; INTERMEDIATE MESODERM; and LATERAL PLATE MESODERM. *See*: AMPHIBIAN DEVELOPMENT.

mesoderm induction *See*: INDUCTION.

mesophyll The main tissue of the leaf, composed of cells usually rich in CHLOROPLASTS, bounded by the upper and lower epidermises. It can often be subdivided into one or more layers of closely packed, elongated palisade cells, lying under the upper epidermis, and the spongy mesophyll below, which contains less regular cells and large air spaces.

mesosome An invagination of the bacterial plasma membrane.

mesothorax Segment in larva of *Drosophila* corresponding to adult thoracic segment T2.

messenger RNA (mRNA) RNA copy of a protein-coding GENE which is translated at the ribosome to produce a polypeptide chain. *See*: BACTERIAL GENE EXPRESSION; BACTERIAL GENE ORGANIZATION; EUKARYOTIC GENE EXPRESSION; EUKARYOTIC GENE ORGANIZATION; PROTEIN SYNTHESIS; RNA; RNA PROCESSING; RNA SPLICING; TRANSCRIPTION.

Met (M) The AMINO ACID methionine.

met The gene encoding the HEPATOCYTE GROWTH FACTOR receptor, and which is a potential ONCOGENE. *See*: GROWTH FACTOR RECEPTORS.

metabolic coupling Coupling of cells via gap junctions which allows the passage of small molecules such as ATP and metabolites from one cell to another. *See*: CELL JUNCTIONS.

metabotropic A term applied to certain of the receptors for NEUROTRANSMITTERS to distinguish them from receptors for the same ligands which contain an internal ion channel, that is from the ionotropic receptors. A metabotropic receptor is one that is coupled to separate cellular effector proteins, usually through a G protein (*see* G PROTEIN-COUPLED RECEPTORS) and therefore uses one of the SECOND MESSENGER PATHWAYS (e.g. via cAMP, inositol trisphosphate, arachidonic acid, etc.). The term is most commonly used to distinguish the metabotropic EXCITATORY AMINO ACID RECEPTORS (e.g. the TRANS-ACPD RECEPTORS) from the ionotropic glutamate receptors. It was introduced in this sense by John Eccles, and the first demonstration that glutamate can stimulate a second messenger pathway via a separate set of such receptors was made by Sladeczek et al. in 1985. However, the term is equally applicable to distinguishing other such pairs, for example the GABA receptors (*see* GABA AND GLYCINE RECEPTORS), where the $GABA_B$ receptors are metabotropic and the $GABA_A$ receptors are ionotropic.

Schoepp, D.D. & Conn, P.J. (1993) *Trends Pharmacol. Sci.* **14**, 13–20.
Sladeczek, F. et al. (1985) *Nature* **317**, 717–719.

metacentric Chromosome with the centromere about halfway along it.

metalloproteinases PROTEINASES (EC 3.4.24) with a metal (usually zinc) PROSTHETIC GROUP. The group includes proteinases from snake venom, numerous microbial proteinases, and vertebrate and bacterial collagenases.

metalloproteins Proteins containing metal atoms. The metal atoms may variously be functional prosthetic groups, as in HAEMOGLOBIN and many enzymes; bound for transport, as in METALLOTHIONEIN and TRANSFERRIN; or form part of a structural feature in the protein, as in the zinc-finger TRANSCRIPTION FACTORS (*see also* PROTEIN–NUCLEIC ACID INTERACTIONS). Metals found in proteins include iron (Fe) (in HAEM and nonhaem proteins, *see* IRON–SULPHUR PROTEINS); zinc (Zn); copper (Cu), as in the oxygen transport protein HAEMOCYANIN; manganese (Mn); molybdenum (Mo) (in some bacterial nitrogenases); and vanadium (V) (in some bacterial nitrogenases) (*see* NITROGEN FIXATION). In addition, numerous enzymes and other proteins require the divalent metal cations Ca^{2+} or Mg^{2+} for activity (*see e.g.* CALCIUM-BINDING PROTEINS).

metallothionein Collective name for a group of cysteine-rich, low molecular weight (M_r 6600) mammalian proteins that bind heavy metals (especially zinc ions) and have a role in their homeostasis and detoxification. Each molecule is thought capable of binding up to six Zn^{2+} ions. The probable function of the protein is in complexing unwanted metals or as a redox buffer. The upstream regulatory sequences of the metallothionein gene PROMOTER contain metal-responsive regulatory elements which can be used to control the expression of heterologous genes introduced into animal cells.

Stuart, G.W. et al. (1974) *Nature* **317**, 828–831.

metamerism Term defined by E. Ray Lankester in 1904, with respect to the bodies of arthropods, which are characterized by division into a series of 'rings, segments, somites', which can be shown to be repetitions of each other. The same phenomenon is true for the vertebrate body, where metamerism is seen in the repetitions of the SOMITES, myotomes, and dermotomes in the trunk, and in the neuromeres (RHOMBOMERES) of the segmented hindbrain. Recent interest in metamerism stems from the discovery of a number of genes expressed in the early development of the embryos of both invertebrates and vertebrates either in metamerically repeated patterns or which have their boundaries of expression at specific morphological (segment) boundaries. *See also*: DROSOPHILA DEVELOPMENT; HOMEOBOX GENES AND HOMEO-DOMAIN PROTEINS; SEGMENTATION; VERTEBRATE NEURAL DEVELOPMENT.

metamorphosis In many species, embryonic development leads to larval stages which are very different in form from the adult organism. Metamorphosis is the process of transition from larva to adult.

Amphibian metamorphosis involves considerable restructuring which is most dramatic in the anurans with the development of limbs and degeneration of the tail. In addition to alterations in form, metamorphosis causes a wide variety of coordinated changes including the developmental maturation of liver enzymes, haemoglobin, and eye pigments, as well as the remodelling of the nervous, digestive, and reproductive systems. All these processes are under hormonal control. During larval stages prolactin acts as a larval growth hormone and inhibits metamorphosis. As the hypothalamus–pituitary–thyroid axis matures, the secretion of the thyroid hormones, thyroxine and triiodothyronine, rises and prolactin levels decline and metamorphosis ensues.

In insect metamorphosis, in response to the hormone ECDYS-TERONE, larval tissues are largely destroyed and are replaced by an entirely separate population of imaginal cells (*see* DROSOPHILA DEVELOPMENT; ECDYSONE RESPONSE; IMAGINAL DISKS).

metaphase The stage of MITOSIS when the fully condensed CHROMOSOMES are arranged at the equator of the MITOTIC SPIN-DLE. The beginning of metaphase is difficult to define, because the chromosomes are brought slowly and irregularly to the midplane of the spindle. The end of metaphase occurs with the onset of ANAPHASE when the SISTER CHROMATIDS separate and start to move toward the spindle poles. Metaphase stages also occur in MEIOSIS.

metaphase chromosomes Chromosomes at the METAPHASE stage of MITOSIS, when they are very compact and are visible with suitable staining in the light microscope, the preparation often being known as a metaphase spread. At this stage individual chromosomes can be distinguished by their size, shape, and banding patterns. *See also*: CHROMOSOME BANDING; KARYOTYPE.

metaphase plate The ordered cluster of CHROMOSOMES that forms during METAPHASE of MITOSIS and MEIOSIS. It forms at the plane that lies midway between the poles of the MITOTIC SPINDLE and is perpendicular to the pole-to-pole axis.

metaphase spread *See*: METAPHASE CHROMOSOMES.

metastasis The ability of tumour cells to detach from the initial tumour mass, seed themselves in a separate site, and establish a new colony.

metathorax Segment in larva of *Drosophila* corresponding to adult thoracic segment T3.

methionine (Met, M) A sulphur-containing AMINO ACID.

methionine tRNA A TRANSFER RNA species which is charged with the amino acid methionine. There are two types of tRNA^{Met} found in cells. The first form is responsible for translating methionine codons other than the initiator codon and is often designated the elongator tRNA^{Met}. The second form specifically binds to the AUG initiator codon at the beginning of the coding sequence and is often designated the initiator tRNA$_i^{Met}$. It is involved in the formation of the initiation complex on the RIBOSOME — the first step in the association of ribosomal subunits before translation of mRNA. In prokaryotes the initiator tRNA^{Met} is charged with methionine which is then formylated to *N*-formylmethionine, to form *N*-formylmethionyl-tRNA$_i$ (fmet-tRNA$_i^{Met}$). Eukaryotic tRNA$_i^{Met}$ is charged with unmodified methionine. *See*: PROTEIN SYNTHESIS.

methotrexate (MTX) 4-Amino,10-methyl folic acid, also known as methylaminopterin. An inhibitor of the enzyme DIHYDROFO-LATE REDUCTASE (DHFR), an enzyme essential for mammalian cell growth owing to its role in the synthesis of dTMP, a precursor nucleotide of DNA. Prolonged treatment of cells with methotrex-ate results in amplification of the *DHFR* gene to as many as 100 copies (*see* DOUBLE MINUTE CHROMOSOMES; GENE AMPLIFICATION). A gene linked to the *DHFR* gene can also be co-amplified as a result of treatment with methotrexate.

methyl accepting chemotaxis proteins (MCPs) *See*: BACTERIAL CHEMOTAXIS.

methyl jasmonate (JA-Me) An odoriferous component of the essential oils of *Jasminum* (jasmine) and *Rosmarinus* (rosemary) that is also present in most plants and organs (Fig. M15). It possesses physiological activity when applied to plants at 0.05 mol m^{-3} similar to that of JASMONIC ACID (JA) but is some-what more active than JA. Since it is released by plants in

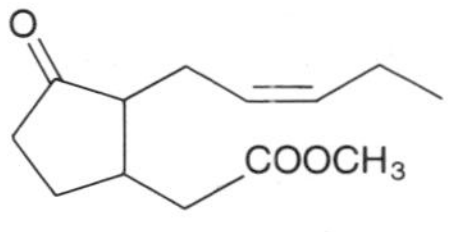

Fig. M15 Structure of methyl jasmonate.

response to wounding or pathogen attack and induces the synthesis of PROTEINASE INHIBITORS, it may act as an interplant signal. *See also*: PLANT HORMONES; PLANT WOUND RESPONSE.

Farmer, E.E. & Ryan, C.A. (1990) *Proc. Natl. Acad. Sci. USA* **87**, 7713–7716.

methyl mercuric hydroxide A compound used in molecular biology for the denaturation and disruption of the secondary structure of nucleic acids. In particular, it is used to denature RNA before and during gel ELECTROPHORESIS in NORTHERN BLOTTING although it has now been largely superseded by formaldehyde. However, despite the disadvantage of its toxicity, methyl mercuric hydroxide is unrivalled as a nucleic acid denaturant.

methyl mercuric nitrate Methyl mercuric salts (commonly nitrate) are used as a HEAVY ATOM ISOMORPHOUS DERIVATIVE in X-RAY CRYSTALLOGRAPHY of proteins. The small size of the nitrate renders accessible cysteine residues which are often inaccessible to larger mercurial reagents.

methylases *See*: METHYLTRANSFERASES.

methylation Addition of a methyl group ($-CH_3$) to a side chain of an amino acid in a protein (*see* POST-TRANSLATIONAL MODIFICATION) or to a nucleotide in DNA (*see* DNA METHYLATION) or RNA (*see* TRANSFER RNA).

methylation analysis Widely used technique for determining positions of branching in GLYCANS. Methyl ethers of unsubstituted hydroxyl groups are produced by the treatment of glycan with dimethyl sulphate and alkali or methyl iodide and silver oxide. Free reducing terminals, if present, are first protected by conversion to methyl glycosides using methanolic hydrogen chloride. Subsequent hydrolysis yields partially methylated mono- or oligosaccharides, in which free hydroxyl groups represent positions at which glycosidic linkages formerly existed.

methylation protection The chemical modification of DNA by addition of a methyl group to a nucleotide at or near the recognition site of a RESTRICTION ENZYME, thus preventing cleavage of the DNA at that site by the relevant restriction enzyme. *See also*: DNA METHYLATION.

5-methylcytosine BASE formed by methylation of cytosine, found in TRANSFER RNA, and, as a result of DNA METHYLATION, in the genomic DNA of many species.

N-methyl-D-aspartate *See*: NMDA.

methyltransferases (methylases, transmethylases) Enzymes

(EC 2.1.1) that transfer methyl groups from one compound (the donor) to another (the acceptor). A common donor of methyl groups in biochemical reactions is the compound S-ADENO-SYLMETHIONINE. *See*: BACTERIAL CHEMOTAXIS; DNA METHYLATION; DNA REPAIR; POST-TRANSCRIPTIONAL MODIFICATION OF PROTEINS; RESTRICTION AND MODIFICATION SYSTEMS.

mevinolin Also known as monocolin K (Fig. M16). Cytotoxic metabolite of the fungi *Aspergillus terreus* and *Monasius ruber*. It is a competitive inhibitor of HMG-CoA reductase, used widely to inhibit isoprenoid biosynthesis in plants and animals. It lowers CHOLESTEROL levels in man and thus has important pharmacological potential.

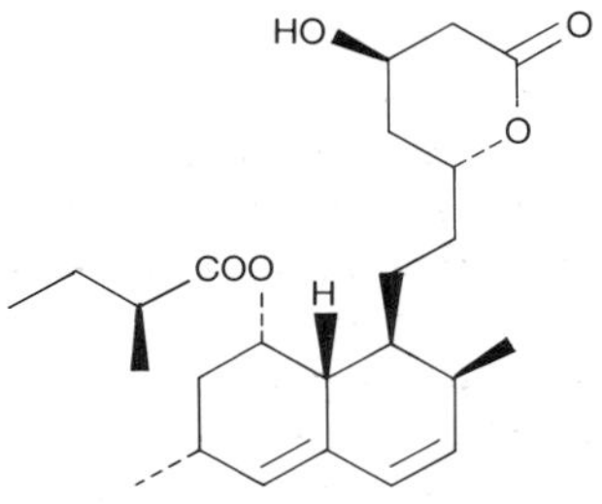

Fig. M16 Mevinolin

MF MALE FERTILE.

MHC, MHC class I, MHC class II *See*: MAJOR HISTOCOMPATIBILITY COMPLEX; MHC MOLECULES.

MHC class III Nonpolymorphic proteins encoded within the MAJOR HISTOCOMPATIBILITY COMPLEX (MHC) and which are primarily serum proteinases functioning as components in the classical and alternative pathways of COMPLEMENT activation. C2, C4, and factor B map to the H-2S region in the mouse, and to the HLA class III region between the HLA-D and HLA-B loci in the human. Class III gene products are structurally distinct from the MHC class I and II products and, unlike these proteins, serve no role as restricting elements in the regulation of immune responsiveness, suggesting that these loci did not evolve by divergence from other MHC loci.

MHC molecules Cell-surface glycoproteins encoded (or partly encoded) by loci of the MAJOR HISTOCOMPATIBILITY COMPLEX (MHC), and which are essential to the initiation of IMMUNE RESPONSES against most protein antigens. Their function is to complex with fragments of foreign proteins, thus presenting foreign antigens at the cell surface in a form that is recognizable by T CELLS (*see* T CELL ACTIVATION; T CELL DEVELOPMENT; T CELL RECEPTOR). There are two main classes of MHC molecule, and the various members of each class are present in each individual, usually in two different allelic forms as a result of the exceptionally large number of alleles at each MHC locus.
1 Class I MHC molecules are polymorphic cell-surface molecules that can be divided into two distinct categories. The classical

transplantation antigens (H-2K, D, and L in the mouse, and HLA-A, -B, and -C in humans) are present on all nucleated cells and when complexed with antigen function as targets for T cell-mediated immunosurveillance of virus-infected or cancerous cells. Recognition of non-self allelic forms of these antigens on transplanted tissue by T cells also initiates graft rejection. The second category of class I antigens includes the less polymorphic haematopoietic differentiation antigens, TL and Qa, which are expressed only on a subset of bone marrow-derived cells and whose function is unknown. Molecules of both class I categories are glycoproteins composed of one MHC-encoded integral membrane glycoprotein with a single membrane-spanning region and M_r 45 000, associated at the cell surface with a nonpolymorphic polypeptide, β_2-MICROGLOBULIN (M_r 12 000), encoded outside the MHC.

2 Class II MHC molecules are highly polymorphic membrane glycoproteins expressed on the surfaces of immunocompetent cells, such as B cells, activated T cells in humans and monocytes/macrophages, as well as some specialized cells such as endothelial cells. Two heterodimeric class II complexes have been identified in the mouse, I-A and I-E, both encoded within the H-2 I region on chromosome 17. At least three families of expressed class II complexes exist in humans: HLA-DR, HLA-DQ, and HLA-DP. Genes encoding the α and β chains of these complexes map centromeric of the human class I and III genes on chromosome 6. *See also*: ANTIGEN PROCESSING AND PRESENTATION.

MHC restriction In the most general sense, the restriction of antigen recognition by a T CELL RECEPTOR to a complex of specific antigen with a particular MHC MOLECULE (*see also* ANTIGEN PROCESSING AND PRESENTATION). In the normal course of events, T cells from a particular individual will only recognize antigen complexed with that individual's self MHC molecules and the term MHC restriction is often used in this sense. The requirement for self MHC antigens arises from the circumstances of T CELL DEVELOPMENT in the thymus. *See also*: MAJOR HISTOCOMPATIBILITY COMPLEX.

MIC Minimum inhibitory concentration. The minimum concentration of virus, antibiotic, etc. that inhibits growth of cells or microorganisms in culture.

micelle Spherical structure formed by some lipids (e.g. salts of fatty acids) in an aqueous medium, in which the hydrophobic tails of the molecules are oriented towards the interior, with the polar headgroups on the outside.

microautophagy *See*: PROTEIN DEGRADATION.

microballistics Method of delivering DNA into (plant) cells by projection of DNA-coated particles into cells or tissue. The particles may be projected ballistically using nail gun cartridges or compressed gas (helium), or electrostatically.

microbody PEROXISOME.

microcalorimetry The measurement of the enthalpy of molecular processes by measuring heat production. Applications of the technique include the study of redox reactions of the respiratory pathways, protein unfolding, oxidative yields from energy storage molecules, and enzymatic catalysis. Microcalorimetric measurements are now expressed as kilojoules (kJ) mol^{-1} or, traditionally, as kilocalories (kcal) mol^{-1} (where 1 cal = 4.184 J).

micrococcal nuclease An endonuclease (EC 3.1.31.1) produced by the bacterium *Staphylococcus aureus*, which cleaves single-stranded nucleic acids. This enzyme is active only in the presence of calcium ions and its activity can be terminated by the addition of a calcium chelator. It hydrolyses the 5′-phosphodiester bonds of RNA (and to a lesser extent DNA). One of its primary uses is in the treatment of reticulocyte lysates for IN VITRO TRANSLATION to destroy endogenous mRNAs, thus generating a lysate completely dependent on added mRNAs.

microcystin-LR A cyclic heptapeptide, which specifically inhibits PROTEIN PHOSPHATASES PP1 and PP2A at nanomolar concentrations. It is a potent hepatotoxin and is produced by certain species of CYANOBACTERIA such as *Microcystis*, found in freshwater lakes and reservoirs.

Microcystis aeruginosa A CYANOBACTERIUM.

Microfilaments

MICROFILAMENTS are linear unbranched assemblages of the protein ACTIN (monomer M_r 43 000) of diameter 5–7 nm. They are one of the components of the CYTOSKELETON present in all eukaryotic cells [1,2]. In muscle cells they are also known as thin filaments, and are found in association with tropomyosin and troponins (*see* MUSCLE). In situations where the type of filament in question is unambiguous (*cf.* INTERMEDIATE FILAMENTS) they are known simply as 'filaments'.

The structure of actin and the microfilament

Actin is the most abundant protein in typical eukaryotic cells, accounting for as much as 15% of total protein. It is a highly conserved protein: the amino-acid sequence of actin from *Acanthamoeba*, a small soil amoeba, is 95% identical to vertebrate ISOFORMS of actin. The arduous task of solving the structure of G-actin (globular actin, the monomer) by X-RAY CRYSTALLOGRAPHY has recently been completed [3]. These studies have revealed that the actin monomer is approximately pear shaped, and when viewed conventionally with the more pointed end uppermost, both the N and the C termini are seen in the bottom right hand corner (Fig. M17). Actin is composed of four domains with a large cleft almost bisecting the molecule. This cleft forms a binding site for both a divalent cation (most probably magnesium in cells) and nucleotides (ATP or ADP).

When viewed in the electron microscope microfilaments appear as double-stranded helices. However, diffraction and crystallographic reconstructions indicate that the microfilament may

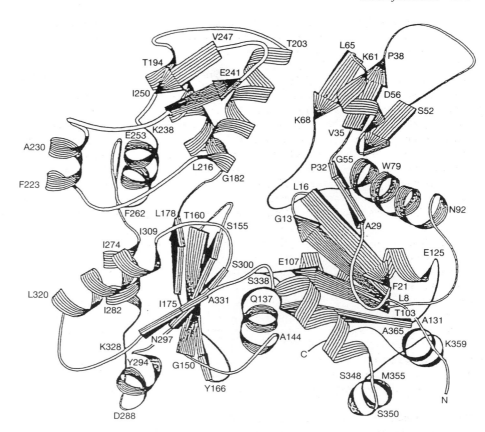

Fig. M17 A 'ribbon' tracing representation of the actin monomer. The molecule can be divided into two main domains, the large domain on the left and the small domain on the right. Between these two domains, toward the bottom of the cleft, is the metal and nucleotide-binding site. Both the N and C termini are near the bottom of the small domain. From [3].

be more accurately described as a single-stranded helix. Each monomer is rotated 166° with respect to neighbouring subunits which means that every 36 nm, or every 13 subunits, subunits eclipse each other at what appears to be a crossover (Fig. M18). There is some degree of disorder within the filament and these measurements are known to vary especially when filaments are bound by ACTIN-BINDING PROTEINS. From knowledge of the G-actin structure, a much more detailed picture of the structure of the microfilament has been deduced [4].

Because the actin subunit has polarity, the microfilament is also polar. Traditionally, the ends of the microfilament have been referred to as 'pointed' and 'barbed'. This nomenclature arises from the resemblance of microfilaments 'decorated' with fragments of myosin II (*see* MYOSIN) to arrowheads in the electron microscope, and in reconstructions. Happily, this nomenclature coincides with the pointed appearance of the actin monomer.

In physiological conditions, actin molecules (G-actin) spontaneously self-associate to form microfilaments (F-actin). This polymerized form of actin is in constant equilibrium with G-actin. Monomers are able to add to the ends of filaments, form nuclei with another two monomers to create new filaments, and to leave filaments. The polarity of actin subunits within microfilaments means that the two ends are topologically different: the narrow face of actin subunits is exposed at the pointed end while the more bulbous face containing the N and C termini is exposed at the barbed end.

The overall rates of monomer assembly at the two ends are

different, although monomers can assemble and disassemble from both ends. The barbed end is preferred for monomer addition over the pointed end, where net disassembly is more favoured. The possibility has been raised that a net flux of monomers may occur along the length of the filament — adding to the barbed end, travelling the length of the filament, and leaving from the pointed end. This has been called treadmilling [5]. There is evidence that treadmilling may indeed occur close to the leading edge of motile lamellipodia of fibroblasts and neuronal growth cones (*see* CELL MOTILITY).

Interestingly, a microfilament in the process of treadmilling may actually do work, as ATP is hydrolysed in the process. If, for example, an organelle were attached by an actin-binding protein to a particular subunit within a filament, it is possible that the organelle might be forced to move down the length of a filament from the barbed to the pointed end, a direction contrary to that generated from interaction with myosin.

The organization of microfilaments within cells

Actin in skeletal muscle cells is organized almost entirely as microfilaments, decorated with an assemblage of ACTIN-BINDING PROTEINS such as tropomyosin, the troponins and, at the Z disk, α-actinin. Between these microfilaments, or thin filaments, are thick filaments composed of myosin II. The interaction of these two filament types produces the contractile force (*see* MUSCLE).

In nonmuscle cells, the situation with respect to actin is much

Fig. M18 Model of an actin filament showing helical symmetry. This was constructed using coordinates derived from a best fit of the monomer structure to fibre diffraction data from orientated filaments [3]. The helical symmetry can be considered in two ways: (1) as a tight left-handedly wound, single helix in which light-dark monomers alternate around the axis; (2) as two, more slowly winding helices (dark, light strands) that wind in a right-handed sense. An important feature of the model is that the bonds between molecules in the same strand are more extensive than those across the axis, between different strands. Model courtesy of Paul J. McLaughlin, MRC Laboratory of Molecular Biology, Cambridge, U.K.

more dynamic: actin is found in a large variety of structures as well as existing as monomers. Smooth muscle cells are intermediate between these two extremes. Given the ionic conditions within cells one would expect to find that very nearly all actin would be in filamentous form. This however is not the case:

around half the actin is present in the G-form. It is thought that this is due to the presence of a variety of small actin-binding proteins such as profilin and actin depolymerizing factor (ADF), and peptides such as thymosin β4. These are thought to bind individual monomers and thus keep a large pool of sequestered monomers in existence. It is suspected that because of this large monomeric pool, actin-containing structures within cells (in culture at least) are constantly changing. It remains to be seen whether actin is as dynamic in cells within tissues, but because of the energy involved in rapidly turning over the actin cytoskeleton, one might expect a more quiescent situation.

Microfilaments within typical cultured vertebrate cells exist as fine meshworks of individual microfilaments cross-linked into gels by a variety of actin-binding proteins. Such fine isotropic gels are especially common in the rapidly changing lamellae of moving cells. In some cell types such as neurons, these areas of fine gel exist between bundles of actin filaments (actin cables) which are associated with microspikes, pointed extensions of the lamellae (*see* CELL MOTILITY). Towards the nucleus, microfilaments tend to be formed into bundles. Some of these bundles are very large and may contain myosin, giving the structure a sarcomeric appearance. These bundled arrangements are known as stress fibres and have been shown to be contractile. Stress fibres are typically associated with focal contacts, areas in molecular contact with the culture vessel substrate on the media side, and protein-rich areas on the cytoplasmic side of the membrane (*see* ACTIN-BINDING PROTEINS; CELL JUNCTIONS; CELL MOTILITY).

The function of microfilaments

In muscle cells the function of the microfilament is almost certainly to develop contractile force in concert with myosin II. In the nonmuscle cell the same function is fulfilled at the CONTRAC-TILE RING in dividing cells and in the stress fibre, which develop contractile force on substrates that may be important in wound healing. In addition to these obvious functions, microfilaments are involved in a number of other cellular processes such as the control of EXOCYTOSIS, the intracellular transport of some bacteria (*see* BACTERIAL PATHOGENICITY) and localization of RNA, many of which are subtle but none the less important. Microfilaments also perform a variety of specific task in particular cell types, such as maintaining the structure of stereocilia in cochlear cells and of microvilli in intestinal epithelial cells.

The study of microfilaments and their involvement in cellular processes has been greatly facilitated by two fungal toxins, CYTOCHALASIN and PHALLOIDIN. The effect of these drugs on actin is in many ways opposite: cytochalasin binds to G-actin and prevents polymerization whereas phalloidin binds only to F-actin and stabilizes the filament against disassembly. Phalloidin labelled with fluorescent markers has proved a specific and convenient tool to visualize F-actin structures within cells. Cellular functions inhibited by either of these toxins are assumed to be microfilament dependent. However it has been suspected that because cytochalasin is also known to interfere with sugar transport, not all effects represent perturbation of the actin cytoskeleton. Recently, an elegant study has demonstrated the specificity of cytochalasin. Cells expressing a novel mutant actin which is unable to bind cytochalasin are unaffected by concentrations of the drug found to be disruptive in wild type cells [6].

S.K. MACIVER

See also: INTERMEDIATE FILAMENTS; MICROTUBULES.

1 Preston, T.M. et al. (1990) *The Cytoskeleton and Cell Motility* (Blackie, Glasgow).
2 Bray, D. (1992) *Cell Movements* (Garland, New York).
3 Kabsch, W. et al. (1990) Atomic structure of the actin: DNase I complex. *Nature* **347**, 37–49.
4 Bremer, A. & Aebi, U. (1992) The structure of the F-actin filament and the actin molecule. *Curr. Opinion Cell Biol.* **4**, 20–26.
5 Wegner, A. (1976) Head to tail polymerization of actin. *J. Mol. Biol.* **108**, 139–150.
6 Toyama, S. & Toyama, S. (1988) Functional alterations in β'-actin from a KB cell mutant resistant to cytochalasin B. *J. Cell Biol.* **107**, 1499–1504.

microglia (Hortega cells) Small cells of nonectodermal origin that form one of the three major families of GLIAL CELLS in the central nervous system. They are believed to be mesodermal in origin and may have several sources, such as circulating monocytes and perivascular pericytes. Following infection and damage to brain tissue, they become actively phagocytic.

β₂-microglobulin (β₂-m) Polypeptide of M_r 12 000 noncovalently associated with class I MHC MOLECULES (*see* MAJOR HISTOCOMPATIBILITY COMPLEX (MHC)), particularly through interactions with the α_3 domain of the MHC molecule. It is also found free in serum. It is encoded outside the MHC. The molecule is a member of the IMMUNOGLOBULIN SUPERFAMILY; it has the prototype single-domain structure, stabilized by an intrachain disulphide bond and is therefore a candidate for the primordial gene from which the rest of the family is descended.

microinjection A technique for introducing nucleic acids (for heterologous gene expression) into large cells such as the oocytes of *Xenopus* via a thin glass needle and micromanipulator (*see* XENOPUS OOCYTE EXPRESSION SYSTEM). A similar procedure is used to introduce cloned genes into the nucleus of an egg cell *in vitro* as a first step to generating a TRANSGENIC organism (*see* GENETIC ENGINEERING; TRANSGENIC TECHNOLOGIES).

micromere *See*: SPIRAL CLEAVAGE.

microscopy *See*: ELECTRON MICROSCOPY; IMMUNOELECTRON MICROSCOPY.

microsome Homogenized membrane vesicles which sediment by centrifugation less rapidly than intact mitochondria but more rapidly than ribosomes. Operationally they form pellets after 1 hour at 100 000*g* and contain membranes derived mostly from the ENDOPLASMIC RETICULUM and GOLGI APPARATUS.

Microtubule(s)

MICROTUBULES represent one of the classes of filamentous structures which make up the CYTOSKELETON found within the cytoplasm of eukaryotic cells (*see also* INTERMEDIATE FILAMENTS; MICROFILAMENTS). They are cylinders with a diameter of 25 nm composed of repeating protein subunits of TUBULIN. Microtubules have many functions in cells, including directed transport of small cytoplasmic vesicles (*see* MICROTUBULE-BASED MOTORS), chromosome movement at meiosis and MITOSIS, provision of cell shape, and movement of flagella and cilia. In some microtubular structures the single cylinder may be replaced by doublet microtubules (e.g. in cilia and flagella) and triplets (in basal bodies and centrioles).

Structure

The wall of the microtubule is composed of heterodimers of α- and β-tubulin. The heterodimer subunits are around 8 nm in length and are arranged in longitudinal rows called protofilaments (Fig. M19). End-on views of most microtubules show the wall to be composed of 13 protofilaments, although some microtubules in nematodes have been found with 11, 12, or 15 protofilaments.

The protofilaments of a microtubule wall are juxtaposed such that adjacent tubulin subunits are staggered, forming a three-start left-handed helix. The protofilaments may not be entirely straight along the length of the microtubule, rather they may be

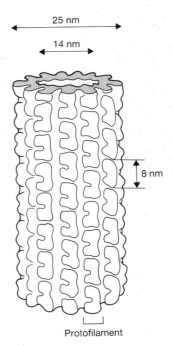

Fig. M19 Diagram showing the main features of a microtubule. Dumbbell-shaped units represent 6S tubulin heterodimers. They are arranged head to tail to form longitudinal protofilaments. These are arranged in staggered fashion side by side to form the wall of the microtubule.

capable of a long supertwist along the microtubule axis. Tubulin subunits align head to tail along a protofilament in a repeating fashion and this imposes an intrinsic polarity on the microtubule. Essentially, one end has α-tubulin subunits exposed whereas the other has β-tubulin subunits exposed. Microtubule polarity can be determined experimentally by addition of exogenous tubulin molecules to the side wall of existing microtubules. When viewed in cross-section these form hooks whose curvature gives an indication of the polarity of the microtubule. Tubulin subunits add at a faster rate to one end of a microtubule than to the other. This fast growing end of a microtubule is termed the plus (+) end and the slow growing end, the minus (–) end. This polarity is important in understanding microtubule assembly dynamics in the cell (see below).

Assembly

Cell extracts, particularly of tubulin-rich tissues such as brain, incubated at 37°C under appropriate buffer conditions can assemble the tubulin subunits into microtubules. These microtubules can be depolymerized in the cold at 4°C. Cycles of temperature-dependent assembly, disassembly, and centrifugation can lead to the purification of microtubule protein containing mainly tubulin but also some MAPs (microtubule-associated proteins). This *in vitro* assembly of tubulin occurs in buffers containing Mg^{2+} and GTP and can be measured by quantitative assays such as light scattering. Such assays can form a valuable way of studying both the assembly of tubulin and the inhibitory effects of various drugs. The tubulin molecule binds GTP at two sites. GTP or GDP can be bound to the β-tubulin site, the so-called E site or exchangeable site. This site is so called because the bound nucleotide can exchange with nucleotide in the medium. GTP bound at this E site is hydrolysed to GDP during the polymerization of tubulin. Another nucleotide binding site exists on the α-tubulin subunit and GTP bound here is nonexchangeable, hence the site is termed the N site of the dimer. GTP bound at this site may have a structural role as it is not hydrolysed during the assembly process. The divalent cation Mg^{2+} also binds to the tubulin dimer and potentiates assembly whereas Ca^{2+} is a potent inhibitor of microtubule assembly. GTP hydrolysis is not an absolute prerequisite for assembly. Nonhydrolysable GTP analogues can also support microtubule polymerization.

Analysis of *in vitro* assembly kinetics by various criteria that measure polymer mass show that the process reaches a stable plateau. However, more direct measurements show that individual microtubules in the population can be shrinking while others are growing. This concept of dynamic instability of individual microtubules has been visualized microscopically. In free microtubules in solution, studies have shown that there are cycles of polymerization and depolymerization at both ends of the microtubule. However, both the addition and loss of tubulin dimer subunits is faster at the + end and the catastrophic depolymerization of the + end occurs less frequently. In the cell, the dynamic instability of microtubules is influenced by the location of organizing centres or anchoring organelles at the ends of microtubules (see below and MITOSIS). One unproven explanation of this phenomenon suggests that hydrolysis of tubulin-bound GTP may

Table M1 Chick and mouse β-tubulin isotypes

Isotype	Pattern of expression	Overall amino acid identity
Class I	Constitutive; many tissues	
Class II	Major neuronal; many tissues	
Class III	Minor neuronal; neuron specific	
Class IVa	Major neural; brain specific	92–97%
Class IVb	Major testes; many tissues	
Class V	Minor constitutive; absent from neurons	
Class VI	Major erythrocyte/platelets	78%

lag behind assembly such that a rapidly polymerizing microtubule may be stabilized by a cap of GTP-tubulin subunits. If GTP-tubulin was released more slowly than GDP-tubulin dimers then such GTP capped ends would grow whereas those with exposed GDP-tubulin dimers would shrink. It has not been possible to obtain definitive proof of such a GTP cap; however the concept of the dynamic nature of individual microtubules is important in explaining many microtubule functions in the cell.

Tubulins, isotypes, and genes

Tubulin subunits have an M_r of ~50 000. The primary sequences of α- and β-tubulins are around 36–42% identical, indicating that they may well have evolved from a single ancestral gene. Both α- and β-tubulin can exist in a variety of isotypes (Table M1). These can be produced by the expression of individual genes within a MULTIGENE FAMILY, and/or by the action of a variety of POST-TRANSLATIONAL MODIFICATIONS.

Some eukaryotic microbes have small tubulin multigene families (yeast has one β- and two α-tubulin genes) whereas vertebrates have more. For instance in chickens there are six genes encoding different β-tubulin isotypes. The expression of individual members of the multigene family can vary with cell type and developmental stage of the organism, for instance a particular gene encoding a β-tubulin isotype in *Drosophila* is only expressed in the male testis. Mutations in this gene have been found to render males of the organism sterile.

Equivalent isotypes expressed in particular tissues of different species are extremely highly conserved suggesting that expression of isotypes may be important in providing some microtubule functions (binding of particular MAPs etc.). Many experiments show that tubulin isotypes do in fact mix within an individual microtubule. However, only a few experiments suggest that there are functional differences within the tubulin isotypes. A particular *Drosophila* β-tubulin isotype appears not to be able to perform all of the functions of another, indicating some level of specialization of function.

Recently a third member of the tubulin superfamily has been described. This is γ-tubulin. It is present in much smaller amounts in cells than either the α- or β-tubulin proteins and does not appear to be a component of the microtubule itself; rather it appears to be localized at microtubule organizing centres. γ-Tubulin was identified through molecular genetic analysis. It is around 35% homologous to both α- and β-tubulin. It is therefore possible that γ-tubulin acts as an important component of the cell's nucleating centre for the assembly of microtubules.

Tubulin isotypes can also be generated by post-translational modifications. A unique modification is the reversible detyrosination and tyrosination of α-tubulin. Many α-tubulins are synthesized with a C-terminal tyrosine encoded by their gene. This tyrosine can be removed by a tubulin carboxypeptidase and can be added in an ATP-requiring reaction by the enzyme tubulin tyrosine ligase. A second modification of α-tubulin is the reversable acetylation of a lysine residue at position 40. In some cells, acetylated and/or detyrosinated α-tubulin is preferentially present in stable microtubules. At present this is an association and there is no evidence for a causal link between a modification and a particular microtubule location or function. Monoclonal antibodies have been found which can detect both detyrosinated tubulin and acetylated tubulin respectively. These have proven extremely useful in immunofluorescence studies as they can detect the subsets of microtubules which contain these particular post-translationally modified tubulins. β-Tubulin can be post-translationally modified by phosphorylation, and finally, both α-tubulin and β-tubulin can be modified by glutamylation of residues in the C terminus. Up to six glutamic acid residues can be added to the γ-carboxyl group of glutamates at the C terminus of each polypeptide. Again, the functional consequences of these modifications of tubulin are unknown.

Antimicrotubule drugs

The tubulin molecule binds a large number of drugs whose action is often described as being antimitotic. The classic case is the drug COLCHICINE which binds with a 1 : 1 stoichiometry to the tubulin heterodimer. Another group of drugs, including the anticancer vinca alkaloids, vinblastine and vincristine, interact at another region on tubulin distinct from the colchicine-binding site. Although colchicine is the classic antimicrotubule agent used with mammalian cells it is often not effective against the microtubules of lower organisms. This appears to be a direct result of an evolutionary difference in the tubulins since colchicine does not inhibit the *in vitro* assembly of these lower organism tubulins.

Another class of microtubule inhibitors is comprised of the benzimidazole carbamate group of compounds (e.g. nocodazole and benomyl). These compete with colchicine for the same binding site. Some of these drugs can be effective on microtubules of many different types of organisms. Others, such as Benomyl, are extremely effective against fungal microtubules but not against those of mammalian cells. Benomyl is therefore a useful commercial fungicide.

Antimicrotubule agents such as colchicine inhibit assembly by binding to the tubulin dimer. Inhibition appears to be a substoichiometric phenomenon. It appears that only a few dimer molecules with bound colchicine need to be added to the end of a microtubule to lead to inhibition of assembly.

The vinca alkaloids can not only disassemble microtubules and inhibit assembly but can also form polymorphic forms of tubulin such as paracrystals and helical arrangements. The drug taxol, isolated from the Pacific yew tree (*Taxus*), inhibits microtubule function in an unique way. It hyperstabilizes microtubules and promotes tubulin assembly.

MAPs and microtubule motors

In vitro polymerization of microtubules from cell extracts has identified a group of proteins that copurify with tubulin. These were called MAPs. They bind to the outside wall of microtubules and have been visualized in the electron microscope as wispy projections. The number of identified MAPs is growing rapidly and they appear to be very diverse, differing both between species and between different cell types. The brain MAPs are the best characterized and have the general property of promoting tubulin polymerization and effecting the stability of microtubules (*see* NEURONAL CYTOSKELETON).

The catalogue of brain MAPs includes the tau proteins, MAP1, MAP2, MAP1C and DYNAMIN. The tau proteins have an M_r range of 36–45K by sequence analysis and are long rod-like molecules. MAP2 is also a large extended molecule whose sequence reveals multiple copies of a tubulin-binding motif. MAP1C is an ATPase and appears to be a cytoplasmic DYNEIN, and dynamin can bundle microtubules yet appears to have a GTPase activity. *In vivo* these proteins often show restricted localization; tau protein, for example, is concentrated in the axons.

A classic group of MAPs are represented by the axonemal dyneins. These proteins form the outer and inner arms of the outer doublet microtubules of eukaryotic cilia and flagella (see below). They are very large, multisubunit ATPase proteins with a molecular form resembling a stem or stalk-like structure with multiple heads (*see* MICROTUBULE-BASED MOTORS).

Some MAP proteins, such as the dyneins and KINESIN, have the unique property of being able to function as microtubule motors. This property can be revealed by *in vitro* assays in which kinesin, flagellar or cytoplasmic dyneins are bound to the surface of glass microscope slides and can cause microtubules to slide over the surface in the presence of ATP. Kinesin can also bind to polystyrene beads and move them along microtubules. This motility has been described as having a directionality, kinesin moving particles towards the + end of a microtubule, dynein directing the opposite end motility. However, it now appears that this may be too simplistic a view, as many additional members of these families of microtubule motor proteins have been identified. Molecular genetic studies reveal that these motor proteins have functions in diverse cellular events such as mitosis, premeiotic events, organelle transport etc.

Microtubules in cells

Although free microtubules can be polymerized in the test-tube from tubulin dimer subunits, this is not the mechanism that is used under most conditions in cells. Rather, microtubules are polymerized from, and nucleated by, particular structures inside cells. These structurally diverse organelles are termed MICROTUBULE ORGANIZING CENTRES (MTOCs). One example is the CENTROSOME found in the cytoplasm of mammalian cells; a network of microtubules radiates from the centrosome, which is located close to the nucleus. The centrosome comprises a pair of CENTRI-

OLES surrounded by an amorphous cloud of pericentriolar material. Evidence suggests that this amorphous material is the nucleating centre for the microtubules. The centrosome components duplicate and separate before cell division to form the two poles of the MITOTIC SPINDLE. This mammalian cell centrosome is just one example of the range of structures that have been termed MTOCs. Others are flagella basal bodies, spindle plaques, and spindle pole bodies in fungi. In almost all cases studied microtubules nucleated by an MTOC have their + ends distal from the MTOC and their − ends located at the MTOC. There is evidence for a fixed number of nucleation sites per MTOC and the above influence of the MTOC on polarity of microtubules in cells has profound implications for microtubule function (*see* MICRO-TUBULE-BASED MOTORS; MITOSIS).

Microinjection of fluorescently labelled tubulin into these cells shows that tubulin subunits are added to microtubules at the distal (+) end. However, mammalian cell cytoplasmic microtubules are highly dynamic and are constantly growing and shrinking, such that the lifetime of an individual microtubule may be less than 20 minutes. Other microtubules in cells are extremely stable (the microtubules of the cilium axoneme and cortical microtubules in protozoa for example). The dynamic nature of microtubules is important in that it allows the cell to redefine its cytoskeleton. Such modulations of polymerization and depolymerization are important in flagellum growth and retraction, remodelling the cytoskeleton in moving cells, mitosis, cytokinesis, nuclear orientation, and centrosome positioning.

Cilia and flagella

Although the cilia and flagella of eukaryotic cells have different mechanics of movement they possess a remarkable conservation of structure. This typical pattern of microtubules has become known as the 9 + 2 arrangement (Fig. M20). Cilia and flagella possess a membrane — a specialized region of the plasma membrane — which surrounds a core of precisely arranged and cross-linked microtubules, the axoneme. In the axoneme is an outer ring of nine doublet microtubules surrounding two inner singlet microtubules, the central pair. Doublet microtubules represent a specialized, elaborated type of microtubule whereby one complete, 13-protofilament microtubule (the A tubule or sub-fibre) shares a common wall with another incomplete, 10-protofilament, microtubule (the B tubule or subfibre). The packing of the tubulin dimers differs in the A tubule and the B tubule. The A tubule protofilaments have helically symmetrical, axially staggered arrangements of tubulin dimers, whereas the B tubule protofilaments have an unstaggered arrangement of dimers. The junctions between the A and B tubule are complex, with some evidence for an eleventh thinner protofilament at the inner junction. There is also evidence for insoluble filaments in the walls of the A and B tubules composed of the protein tektin. The A tubule bears an outer and inner dynein arm. The dynein arms are capable of extending to the B tubule of the adjacent axonemal doublet and generating the force for ciliary bend formation and movement. All of the dynein arms in a flagellum point in the same clockwise direction when the axoneme is viewed from base to tip. Adjacent axonemal doublets are also linked by an elastic filamentous connection involving the protein nexin. The two central pair singlet microtubules are surrounded by an elaborate central sheath and small projections or radial spokes serve to make connections between each of the A tubules and the outer face of this central sheath. This organization of the

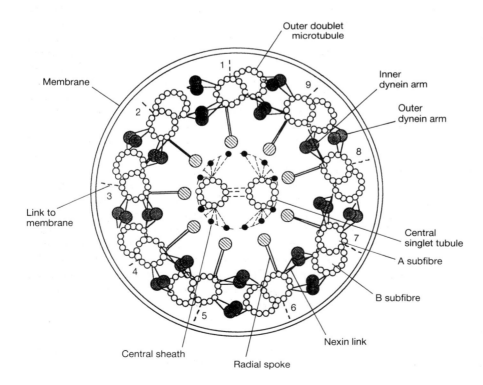

Fig. M20 *a*, Cross-sectional view of an idealized flagellum showing the main structures of the axoneme. Viewed from the tip of the flagellum. The dynein arms and radial spokes occur at intervals along the longitudinal axis. *b*, Structure of a doublet microtubule showing the conventional numbering of protofilaments in the A and B subfibres.

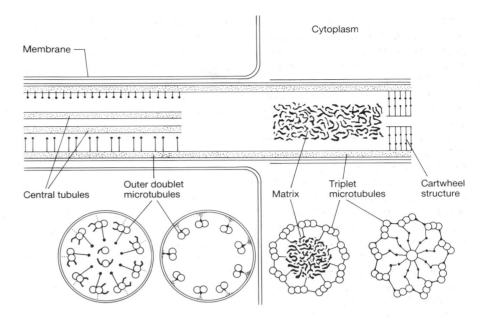

Fig. M21 Diagram of the structure of a basal body and axoneme of a eukaryotic flagellum.

axonemal microtubules has been essentially conserved in evolution from free-living protozoa to mammalian cells.

Axonemal microtubules have polarity in that their + ends are at the distal tip of the cilium or flagellum. At the base of each cilium or flagellum is a basal body composed of nine triplet microtubules (Fig. M21). Triplet microtubules comprise one complete 13-protofilament A tubule sharing a common wall with an incomplete B tubule, which in turn shares a common wall with an incomplete C tubule. The A and B tubules of the nine outer doublet microtubules of the axoneme are continuous with the A and B tubules of the basal body. The C tubules of the basal body triplets, however, end in the transition zone between the basal body and the axoneme, located at the point where the flagellum or cilium protrudes from the cell. The proximal (–) ends of the central pair, singlet microtubules of the axoneme originate in this transition zone. The nine triplet microtubules of the basal body are interconnected by complex sets of fibres and a cartwheel structure of radial spokes. The basal body therefore acts as the template (MTOC) for the elaboration of the axonemal microtubules.

Centrioles have essentially the same ultrastructural organization as basal bodies. They are usually found in pairs in the cytoplasm, oriented orthogonal to each other. Centrioles are associated with amorphous pericentriolar material that often acts as the main nucleating material (MTOC) for polymerization of the cytoplasmic microtubules. However, in many cells the centrioles can convert to basal bodies when differentiation to a flagellated cell type is required. Sometimes one of the centrioles present in the centrosome of an animal cell produces an axonemal-like extension of microtubules. This primary cilium is only present during interphase and is immotile.

K. GULL

1 Amos, L.A. & Amos, W.B. (1991) *Molecules of the Cytoskeleton* (Macmillan, London).
2 Bray, D. (1992) *Cell Movements* (Garland, New York).

microtubule-associated protein (MAP) *See*: MICROTUBULES; NEURONAL CYTOSKELETON; MICROTUBULE-BASED MOTORS.

Microtubule-based motors

THE types of motility involving MICROTUBULES include those where microtubules slide relative to one another, as in the axonemes of CILIA and FLAGELLA, and the movement of other objects along microtubule tracks as, for example, in the fast transport of vesicles along nerve axons (AXONAL TRANSPORT). Proteins that interact directly with microtubules appear to serve as molecular motors in all these cases [1,2]. In contrast to the motor proteins associated with actin MICROFILAMENTS (i.e. the MYOSIN group) all of which seem to travel in the same direction along an actin filament (towards the 'plus' end, the one more active during filament growth or shrinkage), microtubule-associated motors provide movement in both directions (Fig. M22).

There are at least two major families of microtubule-associated motor molecules, known as dyneins and kinesins. Although kinesins appear to move primarily towards the plus ends of microtubules and dyneins towards the minus ends, there is evidence of exceptional behaviour in both families.

For many years, only the dyneins of cilia and flagella were known but, in recent years, cytoplasmic extracts from many sources have been shown to contain soluble forms of both dynein and kinesin, molecules which bind to microtubules, pellet with them under centrifugation, and can then be released into solution when ATP is added. Other forms of kinesin, which may not necessarily be extractable as soluble components of cytoplasm, have been identified by isolation and sequencing of their genes (*see* DNA CLONING; DNA SEQUENCING). A GTPase named DYNAMIN [3] may be a third type of microtubule-based motor but further studies are needed to establish its status.

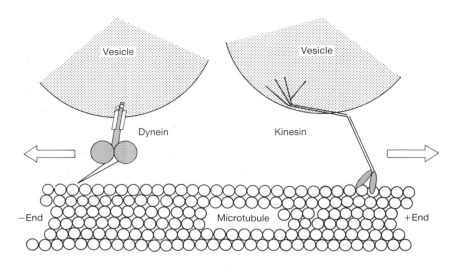

Fig. M22 Schematic diagram of the interaction of dynein and kinesin with microtubules to provide movement of vesicles along microtubule tracks.

An important technological advance that was crucial for the discovery and assay of the new motor proteins was the development of video-enhanced light microscopy. With the greatly improved contrast provided, it is possible to see individual microtubules lying on a glass coverslip. If the surface has been coated with a layer of motor molecules, the microtubules may be observed sliding. Alternatively, motor molecules can be attached to small organelles or to synthetic objects such as plastic beads or gold particles and these will move along microtubules pre-fixed to the glass surface. If the two ends of a microtubule can be distinguished by seeding its polymerization with a recognizable structure such as a CENTROSOME (spindle pole) (*see* MITOSIS) or a piece of axoneme, it is possible to discover whether a given motor molecule moves towards the plus or the minus end. The use of the assembly-promoting drug TAXOL, from the Pacific yew, which keeps microtubules stable during all these operations, has also been an important factor in this work.

ATPase activity

The known molecular motors can be distinguished from one another by their specificities for ATP and other nucleotides; even different polypeptides of dynein react differently. Compared with myosins, dyneins have a high specificity for ATP and hydrolyse other natural nucleotides slowly. Kinesin, on the other hand, has a broad substrate specificity and hydrolyses both ATP and GTP equally well. Dyneins are rather sensitive to the ATPase inhibitors vanadate (see below) and EHNA (*erythro*-9-[3-(2-hydroxynonyl)] adenine), kinesins much less so. Agents that react with sulphydryl groups such as NEM (*N*-ethylmaleimide), invariably inactivate dyneins but usually do not affect kinesin.

If axonemal dynein molecules are isolated under mild conditions, their intrinsic ATPase is low compared with the activity of beating axonemes but there is a marked increase (referred to as 'latent' ATPase) after harsher treatments. This uncoupling of ATPase activity from motility may represent a conformational change in the dynein heavy chain or the loss of a regulatory polypeptide chain. In contrast, no treatment has yet been found to increase the ATPase activity of cytoplasmic dynein, apart from contact with microtubules.

Kinesin tetramers have a very low intrinsic ATPase activity, but in the presence of microtubules have an activity comparable to dynein or myosin. Loss of the β chains also seems to increase the level. The activity is not affected by low levels of vanadate (though it is inhibited by levels above 100 μM, as is myosin) nor can the polypeptide be cleaved by UV irradiation. ATPase activity can be uncoupled from motility by, for example, the binding of MONOCLONAL ANTIBODIES to a site on the heads, or photobleaching of fluorescently labelled molecules.

Kinesin was initially purified from squid axoplasm by making use of the fact that it binds tightly to microtubules in the presence of AMPPNP, a nonhydrolysable analogue of ATP. This is in contrast to dynein and myosin, which are released by AMPPNP. However, the properties of kinesins purified from other sources vary. AMPPNP produces rigor in squid or mammalian kinesins but not when added to extracts from sea urchin eggs or the slime mould *Dictyostelium*; *Dictyostelium* kinesin is detached by AMP-PNP. Presumably the effect depends on whether the analogue, when binding to a particular enzyme, more closely mimics intact ATP (causing detachment) or the combined products of hydrolysis, ADP + P_i (causing strong binding).

Molecular structure

Kinesin

As purified from a variety of sources including mammalian brain, squid axoplasm and sea urchin egg cytoplasm, kinesin consists of 'heavy' (α) chains of M_r 100 000–140 000 (100–140K) and smaller (β) chains of 65–75K. The molecular mass of the molecule is consistent with its being a heterotetramer of two α and two β chains. Electron microscope (EM) images show rod-shaped molecules, with some resemblance to two-headed myosin, though the rod domain is shorter (60–70 nm, rather than 150–160 nm) and the two heads at one end are smaller than those of myosin. Also, there is a significant structure at the tail end of the kinesin rod. The tail-end structure ranges in appearance from a small knob to a fairly large structure, which may look like a single globular domain or a fan with a filamentous substructure.

The first (cDNA) sequence for a kinesin α chain came from the fruit fly *Drosophila melanogaster*, identified by means of an antibody to sea urchin kinesin used to assay for the gene product. A 45K N-terminal stretch of sequence and a shorter sequence at the C terminus are predicted to form globular domains. The N-terminal domain, which includes a sequence characteristic of ATPases, has the capacity for ATP-dependent interaction with microtubules and is responsible for the motile activity. From its sequence, the central segment is predicted to form an α-helical coiled coil, with a characteristic near-longitudinal helical band of hydrophobic residues thought to be responsible for holding two polypeptides together as a helical rod. A nonhelical interruption in the centre may account for some EM images showing bent rods.

Antibody labelling suggests the β chains, when present, lie at the opposite end of the kinesin rod from the force-producing heads, giving rise to the larger versions of the tail-end structure. They may, in some unknown way, inhibit the ATPase activity of the α chains but their position on the molecule suggests they could also control interactions with particles that are transported along microtubules. As all crossbridges between vesicles and microtubules appear shorter than the length of a kinesin rod, it has been proposed that half of the rod (up to the kink) may also be involved in binding to vesicles.

Several genes, sequenced after being identified as the sites of mutations causing mitotic defects in various organisms, show homology with the *Drosophila* kinesin α chain; each predicted gene product contains a segment of 350 amino acids homologous to the N-terminal force-producing 'head' domain of standard kinesin but they have unrelated 'tail' sequences. The genes include: *ncd* and *nod* from *Drosophila*, where mutations cause defects in meiotic or mitotic chromosome segregation; *KAR3* from the yeast *Saccharomyces cerevisiae*, where mutations interrupt the process by which the haploid nuclei of two conjugating cells move together along microtubules and fuse (karyogamy); and *bimC* from the mould *Aspergillus nidulans*, in which a mutant cannot separate its spindle pole bodies during mitosis.

As most, if not all, of these mutations produce defects in mechanisms involving relative sliding between different sets of microtubules, the kinesin-like proteins involved may have TUBULIN-binding sites at both ends. The *bimC* sequence suggests a monomer molecule with a non-α-helical C-terminal tail. The *ncd* and *KAR3* proteins, which may also occur as monomers, both have a reverse orientation, with the 'head' domain at the C-terminal end of the polypeptide. The behaviour of *ncd* protein expressed *in vitro* indicates that reversing the structure can produce a motor which works in the reverse direction.

Both the *KAR3* and *bimC* genes hybridize to numerous DNA fragments from their respective genomes; these observations, together with the occurrence of at least six different kinesin-like proteins in *Drosophila*, show that individual genomes contain many members of the kinesin MULTIGENE FAMILY. As with the myosin superfamily, members of the kinesin superfamily share a common motor domain, but tails with varying properties allow them to carry out different specific roles in cells.

Dyneins

An extract of dynein from flagellar or ciliary axonemes consists of six or seven distinct large polypeptides ('heavy' chains), with molecular masses in excess of 400K, and various smaller chains including 'intermediate chains' of 70–90K and 'light chains' of around 20K. The inner and outer rows of arms in an axoneme consist of different sets of polypeptides. A flagellar heavy chain, the outer arm β chain, and a cytoplasmic one from *Dictyostelium* have been sequenced and are significantly homologous.

The effects of vanadate on all known dyneins seem sufficiently characteristic to provide a diagnostic assay for dynein and its involvement in cellular processes. Not only is the ATPase activity particularly sensitive to low concentrations of vanadate ions, which probably act as a phosphate (P_i) analogue, but, surprisingly, in the presence of vanadate and MgATP, each heavy chain is cleaved by ultraviolet (UV) irradiation into two large pieces and the ATPase activity is destroyed.

Proteolytic enzymes produce a different cleavage pattern. Most heavy chains break down into a large (200–300K) central fragment (containing the UV-vanadate cleavage site) and two smaller terminal fragments. The central fragment sequence includes four predicted sites of ATPase activity. The small N-terminal domain, which is most easily removed *in vitro*, also includes a possible ATPase site, in flagellar β dynein at least.

Isolated molecules contain one, two or three of the heavy chains. In the EM, a single large polypeptide appears as a single globular 'head', usually with some sort of tail; the roughly spherical head is much larger (9–12 nm in diameter) than a myosin head (a 15 nm × 4.5 nm rod). Larger molecules can resemble small bunches of flowers; those consisting of two or three heavy chains have two or three globular heads. EM images of specimens shadowed with metal after rapid freezing show most detail. Each head has a very fine stalk, 13–16 nm long, projecting from it which may represent the N-terminal domain. The bunched stems of a bouquet are 20–40 nm long and may represent the C-terminal domains; they carry small globular subunits, thought to be the intermediate and light chains.

In an axoneme (Fig. M23), the stems and associated subunits (the 'A-end' of the molecule) condense down on to the surface of the A-tubule to which they are fixed. The heads of each bouquet associate closely to form one large head and their stalks combine to form a single projection (the 'B-end' of the molecule) that reaches out to contact the B-tubule of the neighbouring doublet. When activated, the dynein arms move over the surface of the neighbouring B-tubule towards the basal (minus) end, pushing this doublet tubule in the plus-end direction. Details of the interaction are unclear. The protruding stalks, which interact with the second microtubule but are not released by ATP, may slide passively or, alternatively, may inhibit unwanted sliding. During ATP-dependent sliding, the globular head itself may interact directly with the microtubule surface.

Cytoplasmic dynein, which was first identified for certain in mammalian brain extracts, is a minor component of standard microtubule preparations, originally called MAP1C but unrelated to any of the major high molecular weight microtubule-associated proteins (MAPs) (*see* NEURONAL CYTOSKELETON). But in the

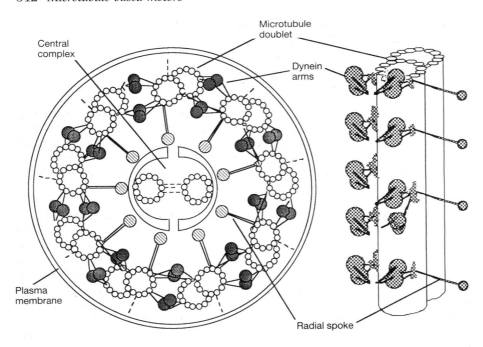

Central complex

Microtubule doublet

Dynein arms

Plasma membrane

Radial spoke

Fig. M23 Diagram of axoneme structure.

absence of ATP, the amount of MAP1C that pellets with microtubules is comparable with that of the structural MAPs. Similar proteins have been purified from nematodes, slime moulds and protists, as well as from various vertebrate tissues and cell types.

Cytoplasmic dynein is structurally indistinguishable from two-headed flagellar dynein. Polypeptides of around 75K copurify with the 400K heavy chains and are assumed to be associated with the stem of the bouquet. The stem domain binds to microtubules, but much more weakly than flagellar dynein binds to A-tubules. The stem must also bind, indirectly (see below), to membranous vesicles. Stalks, apparently identical to those protruding from flagellar dynein heads, also bind weakly to microtubules in a non-ATP-dependent manner; thus both ends of brain dynein can bind to microtubules and crosslink them into bundles. As in the case of axonemal dynein, it is presumed that an interaction of the globular head with microtubules produces the sliding force.

Function

Cilia and flagella

Early work by Summers and Gibbons [4] first directly demonstrated the dynein-induced sliding of microtubules; after a light trypsin treatment, the doublet tubules of isolated axonemes could be observed, by dark-field light microscopy, in the process of sliding over one another. Like the normal beating movement of undigested axonemes, the motility depended on the presence of MgATP. It is postulated that, normally, a programme of restricted sliding movements between adjacent doublet tubules, each producing a bend, leads to axonemal beating.

The inner and outer dynein arms of axonemes (Fig. M23) have

no polypeptides in common; there is even evidence for differences between neighbouring inner arms. If demembranated axonemes are dialysed in low ionic strength medium, all the arms are extracted and the axonemes become immotile. Washing in 0.5 M KCl extracts only outer arms; the axonemes are still motile but show half the beat frequency. Cilia of mutant *Chlamydomonas* cells lacking outer arms show a similar drop in beat frequency, compared with wild-type. In contrast, a mutant possessing outer arms but no inner arms is paralysed, though isolated axonemes produce interdoublet sliding. It seems that inner arms normally govern *in vivo* motility and are essential for converting sliding into periodic bending, whle the faster outer arms determine the overall speed.

Axonal transport

A major suggested role for the soluble forms of kinesin and dynein is in bidirectional particle transport (Fig. M22). The transport rates of 1–5 μm s^{-1} observed in cells are reasonably consistent with the sliding speeds produced *in vitro* by purified kinesin and dynein. In axons, essentially all microtubules are arranged with their plus ends away from the cell body. Retrograde but not anterograde transport is inhibited by vanadate and EHNA at concentrations comparable to those effective *in vitro*, supporting the proposal that kinesin transports organelles away from, and dynein back towards, the cell body.

Surprisingly, however, the addition of AMPPNP to extruded axoplasm not only stops movement in both directions but also leaves all particles firmly attached. Similarly, a monoclonal antibody to kinesin has been reported to inhibit bidirectional organelle transport. These and other observations suggest there may be some sort of cooperative interaction between the two motors.

Neither purified motor protein will rebind to stripped vesicles, so accessory proteins are presumably involved. A number of possible candidates have been identified by various research groups but have not been well characterized. There may be a large range of different accessory proteins that are used for different specific purposes.

Other cytoplasmic transport

Anterograde and retrograde motors may act quite independently under some circumstances. There is evidence that kinesin and microtubules are responsible for the distribution of the ENDO-PLASMIC RETICULUM (ER) in cells; *in vitro*, membrane extracts have been observed spreading out along microtubules attached to a glass surface, producing a branching network similar to the arrangement *in vivo*. Assembled microtubules also seem to be required to keep the elements of the GOLGI APPARATUS concentrated near the cell centre; presumably a retrograde motor of some kind is associated with this system.

An unusual example of a dynein-like protein that powers anterograde movement has been isolated from the cytoplasm of the colonial amoeba *Reticulomyxa*. The purified protein produces bidirectional movement *in vitro* and movement in both directions is equally affected by a variety of ATP analogues and ATPase inhibitors. It is not yet known whether the anterograde and retrograde motors are different polypeptides or whether they are identical apart from being phosphorylated or not.

Mitosis

Mitosis requires a variety of microtubule-associated movements. At different times chromosomes move bidirectionally on kinetochore microtubules and both activities have been exhibited *in vitro* by chromosomes isolated under different conditions. There is a flux of tubulin along the kinetochore microtubules themselves, suggesting they are pulled in towards the spindle pole, where subunits are lost from their minus ends whilst growth occurs at their plus ends. And, during anaphase, relative sliding between two sets of microtubules emanating from opposite poles pushes them further apart. Antibodies have located both a dynein-like and a kinesin-like antigen on kinetochores and kinesin appears concentrated in the vicinity of the mitotic poles. Mitotic and meiotic defects resulting from mutations in various kinesin-like proteins are also known (see above). Intensive studies of spindles in the next few years are likely to unravel the details of these fascinating intracellular machines.

L.A. AMOS

1 Porter, M.E. & Johnson, K.A. (1989) Dynein structure and function. *Annu. Rev. Cell Biol.* **5**, 119–151.
2 Vallee, R.B. & Shpetner, H.S. (1990) Motor proteins of cytoplasmic microtubules. *Annu. Rev. Biochem.* **59**, 909–932.
3 Vale, R.D. (1992) Microtubule motors—many new models off the assembly line. *Trends Biochem. Sci.* **17**, 300–304.
4 Summers, K.E. & Gibbons, I.R. (1971) ATP-induced sliding of tubules in trypsin-treated flagella of sea urchin sperm. *Proc. Natl. Acad. Sci. USA* **68**, 3092–3096.

microtubule-organizing centre (MTOC) Structure in cells of ani-mals and higher plants from which the MICROTUBULES of the cytoskeleton appear to radiate, also known as the CELL CENTRE. *See*: CENTRIOLE.

Microviridae *See*: BACTERIOPHAGES.

mid-blastula transition In amphibians, the point in early development (after the first 12 or so cell divisions, the exact number depending on the species) when cell division slows down and different regions of the embryo proliferate at different rates and begin to differentiate from one another. TRANSCRIPTION from the zygote genome begins. *See also*: AMPHIBIAN DEVELOPMENT.

midbody (spindle remnant, Flemming body, Flemming Koerper) A structure formed during CYTOKINESIS in animal cells by the action of the cleavage furrow on the MICROTUBULES of the MITOTIC SPINDLE that remain between the separating ANAPHASE chromosomes. It is composed of two families of microtubules, one that grew from each spindle pole, which overlap for a short distance near the midplane of the cell (*see* Fig. M30 in MITOSIS). At this zone of overlap, the space between the microtubules is filled with an amorphous, osmiophilic matrix that binds the microtubules together and makes them stable to many of the treatments that normally depolymerize microtubules, for example, subunit dilution, temperatures of ~0°C, and COLCHICINE. During its formation, the midbody is part of the spindle, but its stability in many cells leads it to persist after the rest of the spindle is gone. In these cases it forms the structural basis for an isthmus that connects the two daughter cells after cell division appears to be complete. Small molecules can pass through this connection, even when the microtubules are present, and in some cells, like the spermatids that are the products of the mitotic and meiotic division of a sperm mother cell in *Drosophila*, the microtubules dissolve but the connections persist, maintaining a continuity of the cytoplasm between the apparently distinct cells. Cells connected by a persistent midbody are not permanently linked, because a special separation process begins sometime during the subsequent interphase, extruding the midbody from both of the cells it joined. The structure is ultimately sloughed off into the extracellular space.

Several midbody antigens have been identified, and many of them are parts of the nucleus or the chromosomes prior to midbody formation. Their incorporation into the midbody may be a way of getting rid of these proteins at mitosis, so the daughter cells can begin their CELL CYCLES without them and synthesize them anew as needed.

midbrain The midbrain (mesencephalon) develops from the region of the NEURAL TUBE between the forebrain and hindbrain. Besides containing a variety of long axon tracts that interconnect the various centres of the forebrain with the hindbrain and spinal cord, it also contains collections of neuronal cell bodies. The substantia nigra, for example, is a prominent motor nucleus in the human midbrain, and its degeneration results in the symptoms and signs of Parkinson's disease (*see* NEURODEGENERATIVE DISEASES). Other nuclei include centres for visual and auditory processing, as well as motor nuclei innervating the eye muscles.

middle repetitive DNA *See*: MODERATELY REPETITIVE DNA.

Miller–Dieker syndrome *See*: CONTIGUOUS GENE SYNDROMES.

Miller indices (*hkl*) Three-dimensional coordinates which identify a Bragg reflection in DIFFRACTION. In reciprocal space they behave as normal Cartesian coordinates but in real space they describe intercepts made by planes or, originally, the planar faces observed in crystal morphology. *See also*: X-RAY CRYSTALLOGRAPHY.

mineralocorticoid(s) STEROID HORMONES such as aldosterone produced by the adrenal cortex and involved in the regulation of electrolyte (e.g. sodium) and water balance in the body. They act at intracellular receptors (*see* STEROID RECEPTOR SUPERFAMILY)

mineralocorticoid receptor (MR) A member of the STEROID RECEPTOR SUPERFAMILY specific for MINERALOCORTICOIDS.

miniature end-plate potential (MEPP, m.e.p.p.) Small depolarization, of ~0.5 mV, which is recorded in skeletal muscle fibres in the vicinity of a MOTOR END-PLATE, and which has a time course similar to that of an END-PLATE POTENTIAL. It is a spontaneous event, caused by the release of ACETYLCHOLINE from the presynaptic motor nerve terminal. The amplitude of MEPPs seems to vary in discrete multiples of a single unitary amplitude, representing the quantal release of acetylcholine. *See also*: SYNAPTIC TRANSMISSION.

minichromosome Viral DNA chromosome which is packaged into chromatin using host histones. This type of viral chromosome is found in the PAPOVAVIRUSES.

minicircles *See*: MITOCHONDRIAL GENOMES: LOWER EUKARYOTES.

minigene A cloned gene which has been subject to internal deletion *in vitro* (*see* SITE-DIRECTED MUTAGENESIS). Cloned eukaryotic genes containing numerous INTRONS are difficult to propagate and manipulate using current molecular biological techniques because of their large size. However, investigation of possible regulatory sequences within INTRONS requires that a genomic, rather than cDNA, clone is used. The solution to this problem is to delete some internal exons, reducing the size of the DNA and creating a 'minigene'. Possible regulatory sequences are preserved, with the advantage that the minigene transcript is distinguishable by size from any endogenous homologous transcripts.

Scott, R.W. et al. (1984) *Nature* **310**, 562–567.

miniprep A common name for any method for rapidly purifying small quantities of PLASMID DNA from a bacterial culture. Most methods yield relatively pure plasmid DNA but heavily contaminated with RNA.

Birnboim, H.C. & Dolly, J. (1979) *Nucleic Acids Res.* **7**, 1513.

minisatellite DNA composed of tandemly repeated very short sequences. *See*: CHROMOSOME STRUCTURE; DNA TYPING; VARIABLE NUMBER TANDEM REPEAT.

minisatellite variant repeat mapping *See*: DNA TYPING.

minor histocompatibility antigens Antigens that provoke late rejection in tissue transplantation and are not encoded within the MAJOR HISTOCOMPATIBILITY COMPLEX.

MIR MULTIPLE ISOMORPHOUS REPLACEMENT.

mismatch repair Although the enzyme complex responsible for DNA REPLICATION has mechanisms of self-correction, a low level of errors remain in newly replicated DNA. This error rate, ~1 bp per 10^8 bp replicated or ~50 errors per *Escherichia coli* chromosome, is reduced by postreplication correction to ~1 bp per 10^{10}. The study of MUTATOR strains, that is strains with high rates of spontaneous mutations, has led to the identification of these corrective mechanisms. The postreplication repair of mismatched base pairs has much in common with the repair of DNA damaged by ultraviolet light or DNA-reactive chemicals (*see* CHEMICAL CARCINOGENS AND CARCINOGENESIS; DNA REPAIR). The strand carrying the 'wrong base' is cut and degraded to create a gap which is filled in by DNA polymerase I and the final link made by DNA LIGASE (*see* EXCISION REPAIR). Mismatch repair also affects the outcome of genetic RECOMBINATION and mutant strains with a defect in this form of repair show increased levels of recombination.

The major DNA mismatch repair mechanism in *E. coli* requires the combined actions of the MutH, MutL, MutS, and UvrD (helicase II) proteins. It recognizes G-T mismatches with >90% efficiency as well as most other mismatches with lower but still high efficiencies (except C-C mispairs). Since the 'wrong' nucleotide is in the daughter strand, any mechanisms of postreplication repair must distinguish the new from the old (parental) strand. This is achieved in *E. coli* through the actions of the DAM METHYLASE, an enzyme that methylates the A in 5′-GATC-3′ sequences. The action of this enzyme is slow, so that for 2–5 minutes after replication the daughter strand is undermethylated. Recognition of a mismatch by the MutL and MutS proteins is thought to activate an ENDONUCLEASE, MutH protein, which cleaves and degrades the unmethylated strand (removing as many as 1000 nucleotides) with the aid of helicase II. The combined actions of these proteins together with others, including DNA polymerase II holoenzyme and DNA ligase, have been shown *in vitro* to eliminate mismatches and restore the correct sequence *in vitro*.

However, even in *E. coli* it is clear that strand discrimination must also be achieved in some other way because the mutation rates of *dam* mutants which lack the methylase are lower than those of mutants which lack one of the MutH, L, or S proteins. How this is achieved is not clear. Only the Enterobacteriaceae have Dam methylases and in most other bacteria the level of methylation is low or zero, but their mutation rates are no higher than those of *E. coli*. It has been suggested that in these bacteria the daughter strand is picked out by the presence of gaps or breaks in newly replicated DNA.

A second mechanism of mismatch repair, specific for G-A mismatches, is the function of the MutY protein. This mechanism is unidirectional: the strand with the A is always the one corrected. MutY protein is a DNA-adenine glycosylase which creates an apurinic site which is then processed as in other methods of DNA repair. The mismatch is presumably recognized by the unusual orientation of the guanine (*syn* as opposed to *anti*, *see* DNA; NUCLEIC ACID STRUCTURE). This enzyme must complement the MutHLS system in recognizing what must be the most common mistake of the replication process, mispairs with G. MutM protein also prevents G-A mismatches by elimination of damaged guanine residues, again by glycolytic removal of the base and the creation of an apurinic site. It is also involved in the removal of other damaged bases.

In addition, a third form of mismatch repair is found in strains of *E. coli* that possess a cytosine methylase which methylates the internal C in 5′-CC^A/rGG-3′ sequences. Deamination of cytosine is normally corrected by DNA-uracil glycosylase but in this case deamination yields thymine. The mismatch is recognized by the Vrs protein which initiates repair, but because it involves the removal of fewer than five nucleotides it is called very short patch repair (VSP repair). The *vrs* gene is in the same operon as the methylase. VSP repair was recognized by anomalies in recombination caused by the creation of G-T mismatches in HETERODUPLEX DNA in the context of a sequence of which one strand was methylated.

Mismatch repair is responsible, at least in part, for the very low level of recombination between DNA of different species. A mutation in one of the *mutH*, *L*, or *S* genes allows a thousand-fold increase in the yield of recombinants from crosses between *Salmonella typhimurium* and *E. coli*.

Wildenberg, J. & Meselson, M. (1975) *Proc. Natl. Acad. Sci. USA* **72**, 2202–2206.

missense mutation Point MUTATION in which a codon is changed into one encoding another amino acid. *See*: GENETIC CODE.

Mitochondrial biogenesis

MITOCHONDRIA differ from most organelles in that they depend upon two genetic systems for their growth and propagation. The MITOCHONDRIAL GENOME is relatively small but is essential for the biogenesis of enzymes required for oxidative phosphorylation (*see* ELECTRON TRANSPORT CHAIN). A few polypeptides of the mitochondrial ATP synthase complex and of respiratory chain enzymes such as cytochrome *c* oxidase are encoded on mitochondrial DNA and are synthesized on ribosomes in the mitochondrial matrix. In animals and in yeasts, only about 12–15 polypeptides are synthesized within the mitochondrion and the vast majority of mitochondrial polypeptides are encoded on nuclear chromosomes and synthesized in the cytosol from where they must be transported into or across the mitochondrial membranes. The structure and function of the mitochondrial genome are discussed elsewhere in this volume and the present article is largely concerned with our current understanding of mitochondrial protein transport.

The central features of mitochondrial biogenesis in general and protein transport in particular appear to be well conserved throughout evolution. Biochemical studies with animal mitochondria have given results very similar to those obtained with mitochondria from yeasts and filamentous fungi, and proteins from one source are generally efficiently transported into mitochondria from another. Because of this similarity in transport pathways the results obtained with lower eukaryotes are relevant to mitochondrial biogenesis in animals and plants. Baker's yeast, *Saccharomyces cerevisiae*, has long been used for the investigation of mitochondrial biogenesis and much of our understanding of mitochondrial protein targeting is derived from biochemical analysis using yeast as a convenient source of material. More significantly, though, yeast is able to grow independently of oxidative phosphorylation — by fermentation. It has therefore been possible to isolate mutants with specific mitochondrial defects, for example in the biogenesis of enzymes of oxidative phosphorylation.

The well-developed systems of genetic analysis and genetic engineering in *S. cerevisiae* have allowed several genes encoding mitochondrial polypeptides to be identified, cloned and sequenced, thus making available cloned target sequences for testing in experimental *in vitro* systems. Several essential genes whose products catalyse different steps in the targeting pathway have also been identified.

Targeting sequences

Most cytoplasmically synthesized mitochondrial proteins contain an N-terminal presequence (transit sequence) which is required for targeting and is removed proteolytically within the organelle. Presequences lack acidic amino acid residues and are rich in basic residues that are apparently arranged such that the presequence has been predicted to form an amphiphilic α helix. Mitochondrial targeting sequences must clearly be recognized by components of the cytosol and the target membrane as being distinct from secretory signal sequences and other intracellular targeting signals (*see* LYSOSOMES; PROTEIN SECRETION; PROTEIN TARGETING; PROTEIN TRANSLOCATION).

Mitochondrial proteins are synthesized with information targeting them not only to the mitochondrion but to a specific subcompartment within the organelle. Mitochondrial proteins have to be targeted either to the inner or outer membrane or to one of the two aqueous compartments delineated by the mitochondrial membranes — the matrix and the intermembrane space (IMS) (Fig. M24). Elegant studies using FUSION PROTEINS containing targeting information from the N termini of several precursor polypeptides have shown that, no matter which mitochondrial subcompartment the parent protein is normally targeted to, the extreme N terminus contains information for targeting to the mitochondrial surface, and that a default pathway involving a matrix-targeting sequence found in essentially all imported mitochondrial proteins so far carries the proteins across both membranes into the matrix [1]. Polypeptides that are transported to either of the mitochondrial membranes or to the intermembrane space contain additional intramitochondrial sorting signals downstream (i.e. to the C-terminal side) of the matrix-targeting sequence. Thus, as with the general secretory pathway, mitochondrial protein transport uses a hierarchical system of target-

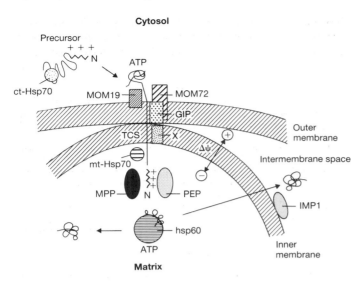

Fig. M24 Model of the general mitochondrial protein import pathway. The import pathways of a precursor carrying a positively charged targeting sequence from the cytosol via contact sites into the matrix space of a mitochondrion are shown. ct-Hsp70, 70K cytosolic heat-shock protein; $\Delta\psi$, membrane potential across the inner membrane; Hsp60, 60K heat-shock protein; IMP1, inner membrane protease 1; MOM19, 19K mitochondrial outer membrane protein in *Neurospora crassa* with receptor fucntion; MPP, mitochondrial processing peptidase; mt-Hsp70, mitochondrial Hsp70; PEP, processing enhancing protein; TCS, translocation contact sites; X, unidentified component(s) of contact sites in the inner membrane.

ing signals to direct proteins to one of four subcompartments.

Outer membrane localization is specified by a STOP-TRANSFER SEQUENCE immediately adjacent to the matrix-targeting sequence, whereas targeting to the inner membrane generally depends on hydrophobic membrane insertion sequences.

Soluble components of the yeast mitochondrial IMS, such as flavocytochrome b_2, cytochrome c peroxidase and an inner membrane protein, cytochrome c_1, which has its bulk exposed to the IMS, undergo a rather more complex targeting pathway involving proteolytic removal of the matrix-targeting signal within the mitochondrial matrix and re-export to the IMS. Each of these three polypeptides contains secondary sorting information between the site of removal of matrix-targeting sequence and a second cleavage site. These sorting signals either operate as stop-transfer sequences, allowing the extreme N terminus but not the bulk of the protein to cross the inner membrane [2] or, according to an alternative model, direct export of the processed intermediate form out of the matrix to the IMS where a second proteolytic cleavage releases the mature polypeptide [3]. This latter model has been suggested to have evolved from the protein export system of the prokaryotic ancestors of mitochondria (*see* BACTERIAL PROTEIN EXPORT).

Components of the targeting pathway

Several components of the mitochondrial protein transport pathway have been identified in both *Saccharomyces cerevisiae* and *Neurospora crassa* [4,5]. A polypeptide (MOM19) of M_r 19 000 (19K) in the outer membrane of *Neurospora* mitochondria acts as a receptor for many imported mitochondrial proteins, including the β subunit of ATP SYNTHASE, CYTOCHROME c_1 and porin. A second receptor of 72K (MOM72) is involved in the recognition and transport of other proteins, including the adenine nucleotide translocator. Similar receptors are found in yeast mitochondria. These two receptors are thought to form a complex with two further polypeptides, MOM38 and MOM22, which is involved in the recognition, membrane insertion and translocation of precursor proteins. MOM38 apparently corresponds to at least part of the functionally defined general import protein (GIP), a protein that acts subsequent to receptor-mediated binding and is common to MOM19- and MOM72-dependent pathways [3]. An integral outer membrane polypeptide of 42K (ISP42) has also been implicated in mitochondrial protein transport.

Models of transport mechanisms

A possible mechanism for transport of proteins across the mitochondrial membranes has been suggested [6] whereby precursor polypeptides synthesized in the cytosol are recognized by receptors on the mitochondrial surface. In this scheme translocation is initiated by insertion of the N-terminal targeting peptide into the outer membrane and in response to the transmembrane electrical potential difference ($\Delta\psi$) generated by the respiratory chain the N terminus of the targeting sequence crosses the inner membrane, resulting in unfolding of the membrane-spanning domain. By binding Hsp70 proteins (*see* HEAT SHOCK) or other MOLECULAR CHAPERONES in the cytoplasm, at least some precursor proteins are maintained in a translocation-competent state, that is they are prevented from misfolding. As polypeptides appear in the mitochondrial matrix they become associated, during translocation, with mitochondrial Hsp70 and this interaction may provide the driving force for transport subsequent to the translocation initiation event mediated by the targeting sequence. Proteins are then passed from Hsp70 to Hsp60 in the mitochondrial matrix, which assists in correct folding. ATP hydrolysis is associated with release of polypeptides from each of these heat-shock proteins. This model may have relevance for the transport of proteins across other eukaryotic membranes and heat-shock proteins are certainly known to be involved in other transport pathways.

Translocation contact sites

The mitochondrial outer and inner membranes can be observed in electron micrographs to come into close contact at several distinct sites. Under special conditions cytoplasmic ribosomes accumulate on the outer membrane at these sites of contact and it was initially suggested in 1975 that proteins might enter mitochondria through these sites [7]. Several lines of evidence now substantiate this view. The polyribosomes attached to the mitochondrial surface are involved in the synthesis of polypeptides destined for mitochondria, with attachment almost certainly mediated by the nascent polypeptide chain. The involvement of contact sites in protein transport has been shown more directly by experiments involving intermediates trapped in the targeting pathway. Proteins that have been allowed to interact with the mitochondrial surface and begin their transport have been trapped by low temperature or by preventing the unfolding of the bulk of the protein through interaction with antibodies and other techniques. Under these circumstances the trapped intermediates span both membranes, with the N terminus available in the matrix where it can be proteolytically processed and the C terminus trapped externally. Trapped intermediates have been shown by IMMUNOELECTRON MICROSCOPY to be located at contact sites and submitochondrial fractionation on sucrose density gradients has isolated a distinct fraction that is intermediate in density between those of the outer and inner membranes. These contact sites are enriched in some of the components of the targeting pathway, including MAS70 in yeast and MOM72 in *N. crassa*.

It has recently been suggested [8] that translocation contact sites are not fixed structures but are formed transiently by the interaction of targeting components diffusing laterally in inner and outer membranes. This view is supported by the finding that some polypeptides involved in transport are not concentrated in contact sites. Proteins can also be efficiently transported directly across the inner membrane, indicating that they can be independently recognized and translocated in a manner that clearly does not depend on intact contact sites.

When mitochondria are depleted of ATP but maintain $\Delta\psi$, precursor proteins begin to be transported, with the N terminus being inserted across the inner membrane. Further transport is blocked however, as this requires the ATP-dependent mHsp70 in the mitochondrial matrix to 'pull' the protein across the inner membrane. These trapped precursor proteins have been transported completely across the outer membrane and their bulk released into the IMS. This may have been achieved by the two segments of the transport channel diffusing away from the contact sites.

This model may help to explain the localization of proteins to the IMS. If these proteins are trapped during transport across the inner membrane because of the presence of a stop-transfer sequence (see above) then diffusion away from a contact site would release the 'mature' part of the protein into the IMS where it would then be released by proteolytic cleavage.

Other targeting pathways

Most of the proteins synthesized within the mitochondrion are targeted to the inner membrane but the mechanism of their insertion is poorly understood. Nevertheless, subunit II of yeast cytochrome *c* oxidase has been shown to be synthesized with an N-terminal presequence that is removed by a protease on the outer face of the inner membrane that also catalyses the second cleavage in the maturation of flavocytochrome b_2. Thus targeting of endogenously and exogenously synthesized proteins share at least one component and may share more.

Exceptions to the general pathway for protein transport into mitochondria have been found, the most notable being cytochrome *c*, a soluble component of the IMS. This protein does not contain a typical N-terminal targeting sequence, does not undergo proteolytic cleavage and does not require $\Delta\psi$ for transport.

Instead, apocytochrome *c* synthesized in the cytosol is transported in a receptor-mediated process that depends upon the covalent attachment of protohaem IX by cytochrome *c* haem lyase.

G.A. REID

1 Hurt, E.C. & van Loon, A.P.G.M. (1986) How proteins find mitochondria and intramitochondrial compartments. *Trends Biochem. Sci.* 11, 204.
2 Reid, G.A. et al. (1982) Import of proteins into mitochondria. Import and maturation of the mitochondrial intermembrane space enzymes. *J. Biol. Chem.* 257, 13068.
3 Hartl, F.-U. & Neupert, W. (1990) Protein sorting to mitochondria: evolutionary conservations of folding and assembly. *Science* 247, 930.
4 Baker, K.P. & Schatz, G. (1991) Mitochondrial proteins essential for viability mediate protein import into yeast mitochondria. *Nature* 349, 205.
5 Pfanner, N. et al. (1991) *Trends Biochem. Sci.* 16, 63.
6 Neupert, W. et al. (1990) How do polypeptides cross the mitochondrial membranes. *Cell* 63, 447.
7 Kellems, R.E. et al. (1975) Cytoplasmic type 80S ribosomes associated with yeast mitochondria. *J. Cell Biol.* 65, 1.
8 Glick, B. et al. (1991) Protein import into mitochondria: two systems acting in tandem. *Trends Cell Biol.* 1, 99.
9 Wienhues, U. & Neupert, W. (1992) Protein translocation across mitochondrial membranes. *BioEssays* 14, 17.

mitochondrial DNA (mtDNA) *See:* MITOCHONDRIAL GENOMES.

Mitochondrial genomes: animal

ALL except the most primitive eukaryotes contain MITOCHONDRIA, the intracellular organelles responsible for the terminal stages in the degradation of fats and sugars, including the reactions of the TRICARBOXYLIC ACID CYCLE, the RESPIRATORY CHAIN and OXIDATIVE PHOSPHORYLATION. Mitochondria invariably contain their own multicopy DNA genome, which encodes a small subset of their functions. Typically these comprise 6–15 polypeptide subunits of the respiratory chain and ATP SYNTHASE complexes, some or all of the RNA components of the mitochondrial translation apparatus (including at least two ribosomal RNAs (*see* PROTEIN SYNTHESIS), and in some instances, a few ribosomal proteins. Mitochondrial DNA (mtDNA) is replicated and transcribed by a separate enzymatic machinery from that of the eukaryotic cell nucleus, and is rightfully regarded as a semi-autonomous genome within the cell (*see also* MITOCHONDRIAL GENOMES: LOWER EUKARYOTE; MITOCHONDRIAL GENOMES: PLANT).

Metazoan mtDNAs are typically circular, 15–20 kilobases (kb) in size and organized in a highly economical fashion [1], with little or no noncoding information between genes, and no INTRONS (Fig. M25). Some pairs of genes also overlap by short distances. Animal mitochondrial messenger RNAs contain almost no untranslated sequence, and in some cases TERMINATION CODONS are created post-transcriptionally, by POLYADENYLATION at a 3′ terminal U or UA. Usually there is only a single extended noncoding region in the genome, in which are located the origin of leading-strand replication and other control information. This region may be as short as 120 bp, as in some sea urchins, or as long as many kilobases, as in some flies and amphibia.

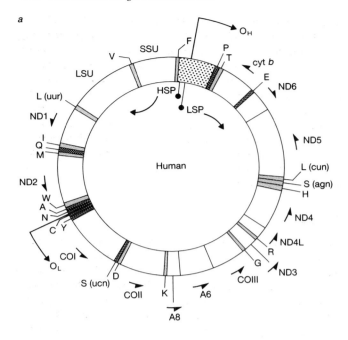

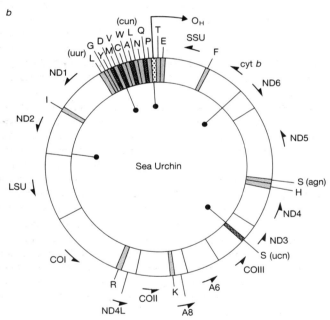

Fig. M25 Physical maps of the human (*a*) and sea urchin (*b*) mitochondrial genomes, drawn approximately to scale. Protein-coding and rRNA genes are shown as open boxes, and denoted as follows: SSU, LSU, small and large subunit rRNAs; COI, COII, COIII, subunits 1–3 of cytochrome *c* oxidase; A6, A8, subunits 6 and 8 of ATP synthase; cyt *b*, apocytochrome *b*; and ND1-6, ND4L, subunits 1–6 and 4L of NADH dehydrogenase. tRNA genes are shown as hatched boxes (diagonal hatching for those encoded on the major coding (H) strand, cross-hatching for those encoded on the L-strand), and denoted by the one-letter amino-acid notation to indicate decoding specificities, with codon recognition groups indicated for the leucine and serine tRNA genes; half-arrows denote transcriptional sense for the rRNA and protein-coding genes. The noncoding region, in which is located the origin of leading-strand replication (O_H), is indicated by dotted shading.

DNA replication

Metazoan mtDNAs replicate by an unusual mechanism, similar in some respects to that of many prokaryotic extrachromosomal DNAs (*see e.g.* PLASMIDS). The best understood example is mammalian mtDNA, but mtDNA replication is probably similar in all metazoans. The model proposed for mammalian mtDNA replication, some details of which remain to be rigorously proven, is as follows [2]. An extended RNA primer, synthesized by mitochondrial RNA POLYMERASE (i.e. without a requirement for a specific PRIMASE), is processed by a specific ENDONUCLEASE, designated MRP, creating the 3′ end for extension by the mitochondrial (γ) DNA polymerase at the origin of heavy-strand (H-strand) synthesis (O_H). The MRP endonuclease is a RIBONUCLEOPROTEIN (RNP), the RNA component of which is nuclear encoded, and RNA–RNA base pairing with the substrate is required for its activity. Primer processing occurs close to a set of conserved sequence elements (CSB1, 2 and 3), which are thought to confer MRP site recognition, and to favour primer–template base-pairing. The processed primer is extended unidirectionally, by strand displacement, over the noncoding region. In molecules which are not committed to replication, the extension of the nascent strand is aborted close to the boundary between noncoding and coding DNA, forming a D-loop (triplex) structure: nonreplicating mtDNA seems to be predominantly in this D-loop form. The nascent strand (D-strand) is continuously resynthesized and degraded, and may have a role in transcriptional regulation. Short conserved termination-associated sequences (TAS) demarcate the 3′ end(s) of the D-strands in mammalian mtDNA. Productive replication involves either readthrough of these termination signals, or reinitiation on a preformed D-strand (the mechanism remains unclear in either case), and proceeds unidirectionally by strand displacement.

The initiation of DNA synthesis on the lagging strand (the light or L-strand) is by an entirely distinct mechanism, involving a specific primase, also apparently an RNP, which recognizes a conserved stem-loop structure in the displaced strand. This origin of light-strand synthesis (O_L) is located in a short transfer RNA (tRNA) gene cluster, approximately two-thirds of the way around the genome. Because of this unusual mode of replication, a large part of the molecule remains single stranded for a considerable portion of the replication cycle, possibly contributing to the high MUTATION rate of mtDNA, as this single-stranded state precludes repair.

Inheritance of mtDNA

In metazoans, mtDNA is effectively inherited through the maternal line: this seems to be due to the enormous disparity in mitochondrial genomic copy number in the male and female

Fig. M25 *Continued* The origin of lagging-strand synthesis (O_L), is shown for human mtDNA. In sea urchin mtDNA, there seem to be multiple lagging-strand origins, with one prominent origin located close to the A6/COIII gene boundary. The direction of DNA strand synthesis from the various origins is indicated. The major promoters for human mtDNA (HSP and LSP) are shown as filled circles, with the direction of transcription from them indicated by arrows. The same symbol is used in sea urchin mtDNA to denote the putative, bidirectional promoters.

gametes, rather than to any active mechanism for the exclusion of paternal mtDNA. In mammalian oogenesis, mtDNA is amplified >1000-fold in the absence of nuclear DNA replication, and by an even greater factor in the large yolky eggs of species such as amphibia, teleosts or arthropods. Although mammalian mtDNA is propagated as a multicopy genome, studies of lineages of heteroplasmic individuals (i.e. those harbouring more than one kind of mtDNA, distinguishable by DNA sequence polymorphisms), have revealed a strong founder effect in each generation, with the oocyte mtDNA population regenerated from a small number of founder genomes. In organisms such as *Drosophila*, with a much higher egg mtDNA copy number, the number of founder genomes is probably much larger. There is no evidence for a generalized RECOMBINATION system in animal mtDNA, and somatic cell fusion experiments typically result in the recovery of a single parental mitochondrial genotype, by mitotic segregation.

Transcription

TRANSCRIPTION of animal mtDNA is POLYCISTRONIC. In the simplest case, that of vertebrate mtDNAs [2], two PROMOTERS, one for each strand, are located close together, adjacent to the D-loop, and provide for the synthesis of two full genome-length precursor transcripts, which are rapidly processed to create the full complement of rRNAs, tRNAs and mRNAs encoded by the genome. Differential gene expression is essentially achieved by post-transcriptional mechanisms, although the relative rates of synthesis of the two rRNAs are maintained at a higher level than that of the other transcripts encoded on the H-strand, by virtue of a transcriptional ATTENUATION mechanism, acting immediately downstream of the rRNA genes.

The mitochondrial RNA polymerase is composed of a single catalytic subunit (M_r ~140 000) and is related to the BACTERIO-PHAGE T3/T7 family of RNA polymerases (on the basis of the sequence of the yeast homologue). This polymerase is assisted by a TRANSCRIPTION FACTOR, denoted mtTF1, which binds to multiple sites upstream of the promoters, enhancing the level and specificity of transcriptional initiation. mtTF1 is a member of the HMG-1 family of DNA-binding proteins. A further protein factor binds to the attenuator sequence downstream of the large subunit rRNA gene of mammalian mtDNA, and has been shown to effect transcriptional termination *in vitro*. Other potential protein-binding sites have been identified in the noncoding region of mammalian mtDNA which are similar to binding sites found in the upstream regulatory regions of some nuclear genes coding for mitochondrial proteins, and which bind a transcriptional activator (NRF-1). It has been suggested that this may represent a system for coordinating nuclear and mitochondrial gene expression, although a role for NRF-1 in the mitochondrion has not yet been assigned. Other sequences related to mtDNA are found commonly in nuclear DNA, although these seem to consist mostly of transposed PSEUDOGENES.

In invertebrate mtDNAs the mechanism of transcription is less well documented. Echinoderm mtDNAs, for example, seem to contain multiple, bidirectional promoters (see Fig. M25); however, the observed overlaps between various pairs of genes, implying mutually exclusive processing pathways, suggest that,

here also, regulation is predominantly post-transcriptional.

In most animal mtDNAs, tRNA sequences act as processing sites in the polycistronic primary transcripts (the so-called punctuation model), allowing for further economy, in that the enzymes involved in tRNA maturation (RIBONUCLEASE P and a 3′ processing endonuclease) are also involved in the biosynthesis of mRNA and rRNA. Some mRNA termini are not adjacent to tRNA sequences, and the mechanics of processing at such sites is unknown. Animal mitochondrial mRNAs are polyadenylated, although the conventional poly(A) addition signals of nuclear DNA are absent: rRNAs are also apparently 3′ (oligo-) adenylated, at least in some taxa.

Translation

Metazoan mitochondrial translation is unusual in several respects: the GENETIC CODE shows several differences from the 'universal' code. UGA, a stop codon in the standard code, specifies tryptophan in animal mitochondria. Other altered assignments seem to be taxon-specific: AGA, for example, specifies arginine in the universal code, serine in invertebrate mtDNAs and a termination codon in mammalian mtDNAs. A 'minimal' set of 22 tRNAs is used, with each amino acid (apart from serine and leucine) represented only once: hence in many cases, all four synonymous codons representing a given amino acid are decoded by a single tRNA, with enhanced WOBBLE properties. A single methionyl acceptor tRNA is used for both initiation and elongation, although the amino acid is formylated to distinguish the initiator tRNA (tRNA^{fMet}).

Mitochondrial tRNA structures are also highly unorthodox, with many otherwise conserved residues absent (especially those that function as polymerase III recognition elements in nuclear DNA), frequent nonstandard base pairings in stems, shortened extra stems, and even some structures which do not conform to the canonical clover-leaf, notably the serine tRNA (anticodon AGY, Y is a pyrimidine, or AGN, N is any nucleotide), which lacks a D arm (*see* TRANSFER RNA) in all metazoans. Non-clover-leaf tRNAs are, in fact, the rule in some taxa, such as nematodes.

Translational initiation occurs usually at the first AUN, but the absence or extreme paucity of 5′ untranslated information, plus the results of protein-binding studies, have led to the suggestion that ribosomes and initiation factors recognize and bind to structural features found within the coding sequence, at the 5′ ends of mitochondrial mRNAs.

Coding capacity

The coding capacity of animal mtDNAs is very highly conserved phylogenetically, with the same 13 protein-coding genes, two rRNAs and 22 tRNAs present in all taxa, although the gene for ATP synthase subunit 8 has not been identified in nematode mtDNA. The remaining proteins encoded are apocytochrome *b*, seven subunits of NADH dehydrogenase (complex I of the respiratory chain), cytochrome *c* oxidase subunits I, II and III and ATP synthase subunit 6. All are key integral membrane components of the proton-translocating complexes of the inner mitochondrial membrane (*see* RESPIRATORY CHAIN) [3].

Despite the rapid rate of sequence evolution at nonselected sites, many of these protein-coding sequences, notably those for apocytochrome *b* and cytochrome *c* oxidase subunit I, are highly conserved at the amino-acid level. These features of mtDNA evolution, coupled with its high copy number, ease of isolation and simplified pattern of inheritance, have made mtDNA an ideal marker of genetic distance and diversity, with applications in fields as diverse as population biology, ecology, anthropology, taxonomy (at all levels), archaeology and forensic medicine [4–6] (*see* DNA TYPING: FORENSIC APPLICATIONS).

The order of rRNA and protein-coding genes in mtDNA has also been highly conserved in metazoan evolution, with, on average, fewer than one rearrangement per 100 Myr in any given lineage, since the divergence of the metazoan phyla. Thus, the echinoid and mammalian genomes (see Fig. M25) differ only by the transpositions of two genes (LSU rRNA and NADH dehydrogenase subunit 4L). The positions of tRNA genes are more variable, however, with a number of differences between orders of insects, for example. In echinoderms, the majority of tRNA genes are clustered around the replication origin (tRNA genes are also clustered in some fungal mitochondrial genomes).

Evolutionary origin

The generally accepted view of the evolutionary origin of mitochondria, the ENDOSYMBIONT HYPOTHESIS, proposes that the organelle began as a free-living organism, akin to modern-day aerobic EUBACTERIA, which invaded or was engulfed by a primitive eukaryotic anaerobe, to mutual advantage [7,8]. The progressive surrender of most of the genetic information of the endosymbiont to the 'host' nucleus is presumed to have left behind the mitochondrial genome as a small remnant of its former genetic autonomy. Support for this view has come from analysis of the rRNA gene sequences of mitochondrial, nuclear and prokaryotic DNAs, which has demonstrated that the mitochondrial sequences are most closely related to those of a particular group of eubacteria (the *alpha* subdivision of the purple nonsulphur bacteria).

A major unsolved puzzle concerns the reason for the survival of mtDNA as a separate genome over a vast period of evolutionary time. The most popular view is that some feature of the inner membrane enzyme complexes requires the synthesis of at least some of their subunits at the site of their incorporation into the membrane, via an ancient conserved assembly pathway (*see* MITOCHONDRIAL BIOGENESIS). Two alternative views are that it is an evolutionary accident, resulting from GENETIC DRIFT between the two systems leading to their eventual functional incompatibility, or that it has resulted from an active, selfish mechanism for the maintenance of an independent genome, similar to those used by many prokaryotic extrachromosomal DNAs. None of these views is entirely supported from available data: the mystery of the mitochondrial genome remains.

H.T. JACOBS

See also: MITOCHONDRIOPATHIES.

1 Attardi, G. (1985) Animal mitochondrial DNA: an extreme example of genetic economy. *Int. Rev. Cytol.* **93**, 93–145.

2 Clayton, D.A. (1991) Replication and transcription of vertebrate mitochondrial DNA. *Annu. Rev. Cell Biol.* **7**, 453–478.

3 Chomyn, A. & Attardi, G. (1987) Mitochondrial gene products. *Curr. Topics Bioenerget.* **15**, 295–329.

4 Avise, J.C. (1991) Ten unorthodox perspectives on evolution prompted by comparative population genetic findings on mitochondrial DNA. *Annu. Rev. Genet.* **25**, 45–69.

5 Harrison, R.G. (1989) Animal mitochondrial DNA as a genetic marker in population and evolutionary biology. *Trends Ecol. Evol.* **4**, 6–11.

6 Moritz, C. et al. (1987) Evolution of animal mitochondrial DNA: relevance to population biology and systematics. *Annu. Rev. Syst. Ecol.* **18**, 269–292.

7 Cavalier-Smith, T. (1987) The simultaneous symbiotic origin of mitochondria, chloroplasts and microbodies. *Ann. NY Acad. Sci.* **503**, 55–71.

8 Gray, M.W. (1989) Origin and evolution of mitochondrial DNA. *Annu. Rev. Cell Biol.* **5**, 25–50.

Mitochondrial genomes: lower eukaryote

THE enormous genetic diversity of the lower eukaryotic kingdoms makes any generalizations concerning their mitochondrial genomes (or any other feature) somewhat problematic (see e.g. [1]). However, it is convenient to consider these kingdoms together, if only because of this diversity, as the mitochondrial DNAs (mtDNAs) of animals and of plants each have their own narrowly defined sets of properties (*see* MITOCHONDRIAL GENOMES: ANIMAL: MITOCHONDRIAL GENOMES: PLANT). The lower eukaryote whose mitochondrial genome has been the most extensively studied is the yeast SACCHAROMYCES CEREVISIAE. This has been a popular organism in which to investigate the genetics of respiratory function, because of its ability to grow, in defined media, in the complete absence of aerobic respiration. For this reason, studies in yeast are the cornerstone of cytoplasmic genetics, and it was the first organism in which the existence of a mitochondrial genome was demonstrated. In addition, the manipulability of its nuclear genome has permitted the isolation and characterization of a large number of nuclear genes whose products have direct or indirect roles in respiration, notably through their effects on the maintenance and expression of mtDNA [2].

Yeast respiratory mutants

Yeast respiratory mutants form small colonies in glucose-rich media (hence their original name *petite*), and are unable to grow on nonfermentable substrates such as glycerol. These mutants fall into four classes: *mit*$^-$ mutants, with point MUTATIONS or small defined deletions in mtDNA, usually inactivating a single gene; *rho*$^-$ mutants, in which most of the mitochondrial genome is deleted, leaving behind a short, 'selfishly' replicating, tandemly repeated genome, often coding for no identifiable proteins (*see* SELFISH DNA); *rho*^{0} mutants, in which the mitochondrial genome is entirely absent; and *pet* mutants, in which nuclear genes required for respiration are inactivated or modified [3].

Yeast mitochondrial genome structure and coding capacity

The mitochondrial genome of wild-type (*rho*$^+$) yeast is about 80 kilobases (kb) in size (Fig. M26a) [4]. It is one of the largest fungal

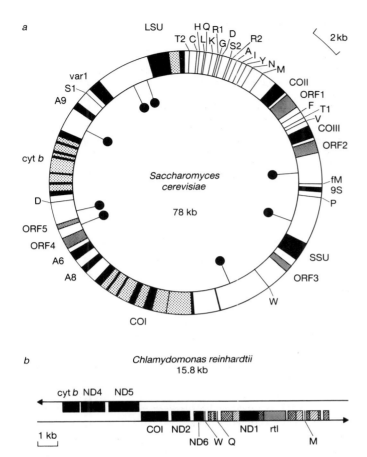

a

LSU

var1
S1
A9

cyt *b*

D

ORF5
ORF4
A6
A8

COI

Saccharomyces cerevisiae

78 kb

2 kb

COII
ORF1
F T1
V
COIII
ORF2

fM
9S
P

SSU
ORF3

W

b

Chlamydomonas reinhardtii
15.8 kb

cyt *b* ND4   ND5

COI ND2   ND1 rt1
    ND6 W Q   M

1 kb

Fig. M26 Physical maps of the mitochondrial genomes of (*a*) a yeast, *Saccharomyces cerevisiae* and (*b*) a unicellular green alga, *Chlamydomonas reinhardtii*, drawn approximately to the same scale. Identified protein-coding and rRNA exons are shown as black boxes, denoted as follows: SSU, LSU, small and large subunit rRNAs; COI, COII, COIII, subunits 1–3 of cytochrome *c* oxidase; A6, A8, A9, subunits 6, 8, and 9 of ATP synthase; cyt *b*, apocytochrome *b*; ND1, ND2, ND4, ND5, ND6, subunits 1–6 and 4L of NADH dehydrogenase; with introns shown as light grey shaded boxes. Dark grey boxes denote unidentified open reading frames (ORFs), including the reverse transcriptase-like gene of *Chlamydomonas* (rt1). rRNA genes in pieces are denoted by diagonally hatched boxes (small subunit) or cross-hatched boxes (large subunit). For *Chlamydomonas*, the genes on the upper bar are transcribed right to left, those on the lower bar left to right. All yeast genes except tRNA-thr1 (T1) are transcribed in the clockwise direction. tRNA genes are shown as black lines, and denoted by the single-letter amino acid notation to indicate decoding specificities, with isoacceptors denoted as S1, S2 etc. The 'GC clusters' (*ori* sequences) of yeast mtDNA are flagged ( ). The arrowheads denote the terminal repeats of *Chlamydomonas* mtDNA. Adapted from [4,12].

ribosomal protein, *var* 1; and a somewhat variable set of functions involved in RNA SPLICING and mtDNA metabolism, which are believed to derive from selfish DNA elements.

Mitochondrial introns

The enhanced (and variable) mitochondrial genome size of yeast compared with that of animals is due in part to the presence of multiple INTRONS in some genes, notably *COB* and *COX1*, and in part to the presence of long stretches of highly AT-rich intergenic sequence, punctuated with short clusters of GC-rich DNA. These GC clusters seem to contain functional or (in some cases) defective replication origins. The *var* 1 gene seems to have been recently recruited from intergenic DNA of this type. The number and positions of introns are strain dependent, and some of them contain OPEN READING FRAMES (ORFs). These commonly encode proteins that function in splicing — the so-called maturases — which are required in *trans* for the removal of the introns that encode them and sometimes for that of other introns. Other intronic ORFs seem to code for DNA ENDONUCLEASES or for proteins resembling REVERSE TRANSCRIPTASE, which are involved in intron transposition (see below). Some intron-encoded proteins have dual roles in splicing and intron transposition.

Fungal mitochondrial introns in general fall into two recognized classes distinguishable by secondary structure: group I and group II (*see* RNA SPLICING). These are *cis*-spliced by distinct mechanisms, which involve specific, conserved structural domains. Group II intron splicing involves lariat formation, which suggests that group II introns may resemble the ancestor of nuclear pre-messenger RNA (pre-mRNA) introns. Both group I and group II introns are also found elsewhere; for example group II introns are common in chloroplast DNA (*see* CHLOROPLAST GENOME) and group I introns are found (rarely) in nuclear DNA, (such as in the ciliate *Tetrahymena*), and also in prokaryotic phage genomes. Although introns of both classes may be induced to self-splice *in vitro* under defined conditions, mitochondrial introns seem to require nuclear-encoded functions, as well as maturases, for efficient splicing *in vivo*. Nuclear-encoded factors which facilitate mitochondrial splicing frequently seem to be RNA-binding proteins, often with other (proven or putative) roles in mitochondrial protein synthesis, such as aminoacyl-tRNA synthetases, or translation factors [6]. This has been deduced from the phenotypes of mutants in such genes, in specially constructed yeast strains containing no mitochondrial introns.

Intron transposition

Some mitochondrial introns in yeast are capable of transposition [7]. The best characterized intron-encoded endonuclease involved in intron transposition in yeast mitochondria is that encoded by the *omega* intron of the LSU rRNA gene. Because of its great target-site specificity (the endonuclease recognizes an 18 base pair (bp) sequence), the intron always 'homes' to a single target in a substrate (intronless) mtDNA molecule. The mechanism seems to involve a double-strand cut, followed by exonucleolytic degradation and strand invasion, leading to co-conversion of flanking markers, and is similar to the action of the HO endonuclease in mating type switching (*see* YEAST MATING-TYPE LOCUS).

mitochondrial genomes, which range in size down to about the size of animal mtDNA (~20 kb). Although larger than animal mtDNAs, yeast mtDNA actually encodes fewer genes [5]: a large and small subunit (LSU, SSU) RIBOSOMAL RNA (rRNA); three subunits of ATP SYNTHASE (6, 8 and 9); apocytochrome *b* (*COB*); three subunits of cytochrome *c* oxidase (*COX1*, *COX2*, *COX3*); 24 TRANSFER RNAS (tRNAs) required for intramitochondrial protein synthesis; the RNA component of mitochondrial RIBONUCLEASE P (the enzyme that catalyses the 5′ maturation of tRNAs); a single

Other fungi

Other fungal mitochondrial genomes have a broadly similar structure and gene content to that of yeast [5], although the amount of noncoding DNA is nowhere as high as in *S. cerevisiae*. The mtDNAs of filamentous fungi encode several subunits of NADH dehydrogenase (complex I of the respiratory chain), also specified in plant and animal mtDNAs: this enzyme is totally lacking in yeast mitochondria, and is substituted by a nuclear-encoded, non-proton pumping NADH dehydrogenase. The tRNA genes of fungal mtDNAs are predominantly grouped (in one or two clusters), and the rRNA genes are nonadjacent, and fall into separate TRANSCRIPTION UNITS. The gene for ATP synthase subunit 9 is nuclear encoded in filamentous fungi, but an apparently silent copy remains in mtDNA, with unknown function.

Transcription

Transcription of fungal mtDNA is polycistronic, but there are many PROMOTERS in the genome, of varying strengths, unlike the situation in animal mtDNAs [8]. The mitochondrial RNA POLYMERASE [9] is a completely distinct enzyme from its nuclear counterparts, and is composed of a single catalytic subunit of M_r ~140 000 (the product of the *RPO41* gene), structurally related to the RNA polymerases of the T-odd BACTERIOPHAGES. This functions in combination with a specificity factor (M_r 43 000), encoded by the *MTF1* gene, which is required for correct promoter recognition and which resembles prokaryotic σ factors (*see* TRANSCRIPTION; BACTERIAL GENE EXPRESSION). The yeast homologue of the mammalian mitochondrial TRANSCRIPTION FACTOR (mtTF1), a member of the HMG-1 family of DNA binding proteins encoded by the *ABF2* gene, is also required for efficient transcription of mtDNA. The mitochondrial RNA polymerase recognizes a short AT-rich sequence overlapping the transcriptional start site, which is well conserved within species, but less so between long-diverged taxa. Experiments in which the transcription of nuclear DNA sequences has been compared in isonuclear *mit⁻*, *rho⁻* and *rho⁰* strains have indicated that nuclear transcription is sensitive to the mitochondrial genotype. Sequence elements in nuclear DNA which are similar to the mitochondrial promoter sequence seem to be involved, and this may reflect a system for coordinating nuclear and mitochondrial gene expression, although the mechanism remains mysterious.

Recombination

Yeast mtDNA is highly prone to RECOMBINATION, and intron-encoded proteins and some free-standing ORFs in the genome which are related to them may be involved in this. Markers as close as 1 kb apart can behave in crosses as if genetically unlinked. Inheritance in yeast mtDNA is biparental, with mitotic segregation following mating and recombination resulting in the purification of the progeny mitochondrial genotype. In many other fungi, essentially uniparental inheritance of mtDNA is observed, though its mechanism is unknown. Segregation distortion, resulting from mechanisms similar to the action of the *omega* endonuclease, or from properties conferred by extragenomic PLASMIDS may be involved.

Programmed senescence

Many fungal species seem to undergo programmed senescence, and this phenomenon is invariably accompanied by catastrophic mitochondrial genomic rearrangements. The senescence phenotype is generally inherited as a cytoplasmic trait, and seems to be associated either with mitochondrial plasmids or with the presence of (presumed) selfish, intronic elements in the mitochondrial genome. In *Neurospora*, the senescence plasmids *kalilo* and *maranhar* each encode a mitochondrial-type RNA and DNA polymerase, and senescence is accompanied by the destructive insertion of these plasmids into mtDNA. In *Podospora* a *COX1* intron is amplified as an extragenomic plasmid during senescence, and is thought to cause genomic rearrangements either by aberrant reintegration events, or by virtue of the reverse transcriptase-like protein it encodes. Other fungal mitochondrial plasmids, such as *Varkud* or *Mauriceville* in *Neurospora*, also encode reverse transcriptases, but are essentially benign.

DNA replication

The replication mechanism of fungal mtDNA is poorly understood. Although the GENETIC MAP of *S. cerevisiae* mtDNA is circular, no circular DNA molecules have been detected by any physical method, even those involving the gentlest disruption. This has been construed as evidence in favour of a ROLLING CIRCLE REPLICATION mechanism, although it is possible that the genome is maintained as a set of bidirectionally replicating, overlapping, linear molecules, constantly rearranged by intermolecular recombination.

The GC clusters most commonly found in the genomes of *rho⁻* petites have been identified by various assays as active replication origins, with initiation on the two strands occurring only several nucleotide pairs apart. The two origins in each of these *ori* (or *rep*) sites are distinguishable, however, because only one is associated with an immediately upstream promoter recognizable by the mitochondrial RNA polymerase. Initiation on the lagging strand is presumably by a distinct mechanism, possibly involving an as yet unidentified primase. An *ori* sequence is not, however, absolutely essential for a functional *rho⁻* genome to replicate: some very degenerate genomes seem to consist entirely of short, tandemized blocks of AT sequence. In addition, it remains unclear whether just one or many *ori* sequences are functional in each individual *rho⁺* genome.

Molecules with putative roles in mtDNA metabolism in yeast, such as a TOPOISOMERASE, the mtDNA polymerase, DNA HELICASE and single-strand DNA-binding protein (SSB PROTEIN), have now been identified genetically and/or biochemically, and in general are distinct from the proteins performing these roles in the nucleus, and, as predicted by the ENDOSYMBIOTIC HYPOTHESIS (*see also* MITOCHONDRIAL GENOMES: ANIMAL), are closer in sequence to their prokaryotic counterparts.

RNA processing and translation

RNA processing in fungal mitochondria is unusual in respects other than splicing, in that a conserved dodecanucleotide 3′ processing signal (totally unrelated to those used for POLYADENY-

LATION in higher organisms) is involved in mRNA maturation, and some nuclear-encoded enzymes seem to have specific roles in the 5′ processing and/or stabilization of particular mRNAs. Positively acting, mRNA-specific factors are also a feature of mitochondrial translation in yeast, with commonly a number of such proteins required for the efficient initiation of translation on a single message. At least three nuclear-encoded gene products are required, for example, for *COX2* mRNA translation, as deduced from experiments in which suppressors of mutations in such genes have been generated by rearrangements, juxtaposing the affected coding sequence with the 5′ leader of another mitochondrial gene. In general, such factors seem to interact directly with 5′ untranslated sequence information to form a translation-competent complex. It has been proposed that these proteins may have roles in the co-translational assembly of the respiratory chain complexes, perhaps reflecting a conserved ancient pathway for membrane biogenesis (*see also* MITOCHONDRIAL BIOGENESIS).

Genetic code

Mitochondria use variant GENETIC CODES, although deviations from the 'universal' code are minor, and taxon specific, and some groups of organisms seem to use the standard code. In *S. cerevisiae* for example, CUN (N is any nucleotide) specifies threonine, not leucine. Like animal mtDNAs, yeast uses a reduced set of tRNAs with enhanced WOBBLE properties, such that fourfold degenerate codons are translated by a single isoaccepting tRNA species. In some groups of Protozoa, and in the green alga *Chlamydomonas*, the set of tRNA genes in mtDNA seems to be drastically reduced (Fig. M26*b*): in trypanosomes none at all has been identified, and it seems that the missing tRNAs are nuclear-encoded and imported into the organelle. The mechanism of RNA import is unknown.

rRNA biosynthesis

Algal and protozoan mitochondria are also notable for two other bizarre features of RNA synthesis. In *Chlamydomonas*, and probably in many protozoans, the mitochondrial rRNAs are not transcribed as contiguous molecules, but instead synthesized in short blocks, in some cases with other sequences (e.g. coding for tRNAs) insterspersed between them. Some of the rRNA modules are also arranged in a different 5′–3′ order from that seen in the rRNA genes of other taxa. These rRNA genes in pieces do, however, seem to give rise to functional ribosomes, and are believed to do so by the formation of intermolecular base pairings, which allow the modular rRNAs to adopt secondary structures remarkably similar to those of conventional rRNAs.

Protozoan mtDNAs

Protozoan mtDNAs exhibit particularly striking examples of RNA EDITING [10], the post-transcriptional alteration of transcribed sequences to generate mature, functional RNA molecules. In trypanosomes, this phenomenon can be extreme. The mRNA for cytochrome *c* oxidase subunit III in *Trypanosoma brucei* is modified such that over 50% of the final coding sequence is unrepresented in the primary transcript. This appears to be achieved by a mechanism involving GUIDE RNAs, short transcripts imperfectly base-paired with the substrate RNAs, and which are believed to be used by the editing machinery (editosome) as a template for correcting the transcribed sequences.

The functional mRNAs are encoded, unedited, in the so-called maxicircle genome, which coexists in the kinetoplast, the highly modified single mitochondrion, in an interlinked network with many copies of shorter molecules (the minicircles), which lack bona fide coding sequences, but which encode many of the guide RNAs. Both maxicircles and minicircles replicate by a ROLLING CIRCLE mechanism [11], and the whole network somehow unravels topologically at cell division, to release two separate daughter networks of interlinked circles.

Editing involves the addition and deletion of uridine residues, proceeds essentially in a 3′ to 5′ direction, and is responsible for the elimination of FRAMESHIFTS, the creation of TERMINATION and INITIATION CODONS (and perhaps RIBOSOME-BINDING SITES), and the reconstruction of a coding sequence from what would otherwise be nonsense sequences. The patterns of RNA editing differ greatly between even closely related taxa, and at different developmental stages, and its function remains mysterious. Other groups of protistans, such as slime moulds, use an entirely different form of RNA editing, involving the insertion of individual C residues at multiple sites.

Protozoan mtDNAs are much more variable in gene content and organization compared with those of higher eukaryotes, which is not surprising given the very ancient diversification of the organisms grouped under this name. Both circular and linear mitochondrial genomes are found. A core of respiratory chain genes (apocytochrome *b* and cytochrome *c* oxidase subunit I), together with at least two rRNA genes, seem to be common to all mtDNAs. A greater number and variety of ribosomal protein genes appear to have been preserved in lower eukaryotic mtDNAs than in animals, plants, and fungi.

A further unusual feature of protozoan mtDNAs is the presence of blocks of sequence related to the chloroplast DNAs of higher plants (*see* CHLOROPLAST GENOME). In *Paramecium* such sequences appear even to include photosystem genes. In the malarial parasite *Plasmodium* and its relatives, a second organellar genome is present, with plastid-like rRNA genes arranged in inverted repeat configuration, and open reading frames resembling those of some plastid genes. It is not known where in the cell this genome is located, although it seems likely that it resides in a different compartment from the linear mitochondrial genome, encoding respiratory chain subunits and rRNA genes in pieces. It seems possible that some unicellular eukaryotes may be complex genetic mosaics, resulting from repeated symbiotic associations which became irreversible (*see* ENDOSYMBIONT HYPOTHESIS), although some of these events may be ancient.

H.T. JACOBS

1 Gray, M.W. (1989) Origin and evolution of mitochondrial DNA. *Annu. Rev. Cell Biol.* **5**, 25–50.
2 Grivell, L.A. (1989) Nucleo-mitochondrial interactions in yeast mitochondrial biogenesis. *Eur. J. Biochem.* **182**, 477–493.
3 Tzagoloff, A. & Dieckmann, C.L. (1990) *PET* genes of *Saccharomyces cerevisiae*. *Microbiol. Rev.* **54**, 211–225.
4 De Zamoroczy, M. & Bernardi, G. (1986) The primary structure of the mitochondrial genome of *Saccharomyces cerevisiae* — a review. *Gene* **47**, 155–177.

5 Chomyn, A. & Attardi, G. (1987) Mitochondrial gene products. *Curr. Topics Bioenerget.* **93**, 93–145.

6 Lambowitz, A.M. & Perlman, P.S. (1990) Involvement of aminoacyl-tRNA synthetases and other proteins in group I and group II intron splicing. *Trends Biochem. Sci.* **15**, 440–444.

7 Perlman, P.S. & Butow, R.A. (1989) Mobile introns and intron-encoded proteins. *Science* **246**, 1106–1109.

8 Costanzo, M.C. & Fox, T.D. (1990) Control of mitochondrial gene expression in *Saccharomyces cerevisiae*. *Annu. Rev. Genet.* **24**, 91–114.

9 Schinkel, A.H. & Tabak, H.F. (1989) Mitochondrial RNA polymerase: dual role in transcription and replication. *Trends Genet.* **5**, 149–154.

10 Stuart, K. (1991) RNA editing in trypanosomatid mitochondria. *Annu. Rev. Microbiol.* **45**, 327–344.

11 Ryan, K.A. et al. (1988) Replication of kinetoplast DNA in trypanosomes. *Annu. Rev. Microbiol.* **42**, 339–358.

12 Michaelis, G. et al. (1990) *Mol. Gen. Genet.* **223**, 211–216.

Mitochondrial genomes: plant

EARLY studies on the plant mitochondrial genome were hampered by its size, which is larger than that of chloroplasts or of mitochondria of other organisms (*see* CHLOROPLAST GENOME; MITOCHONDRIAL GENOMES: ANIMALS; MITOCHONDRIAL GENOMES: LOWER EUKARYOTES) and ranges from about 200 kilobase pairs (kbp) in some species of *Brassica* to some 2500 kbp in muskmelon. There is considerable variation between related species, with the watermelon having a genome of only 330 kbp — one-seventh the size of that of muskmelon. Furthermore, electron microscopic analysis of mitochondrial DNA (mtDNA) preparations indicates the presence of large numbers of circular and linear molecules of varying sizes, while agarose gel ELECTROPHORESIS of RESTRICTION ENZYME digests shows complex profiles, with widely varying amounts and relative proportions of different restriction fragments.

Genome structure

The situation became much clearer with the more detailed analyses carried out on *Brassica campestris* and on maize [1]. The analysis of *B. campestris* showed that the mitochondrial genome had a tripartite structure composed of a 'master' circle of 218 kbp, and two subgenomic circles of 135 kbp and 83 kbp. The master circle contains two direct repeats of 2 kbp, and homologous recombination across these generates the subgenomic circles. In maize, the situation is more complicated (Fig. M27) with six sets of repeats in the master chromosome of 1, 2, 3, 10, 12 and 14 kbp. All but the 14-kbp repeats were found to be in the direct configuration, and all but the 10-kbp repeat were shown to be recombinationally active, generating a large population of molecules of varying sizes which correlated well with the CONTOUR LENGTHS of circular molecules measured in the electron microscope. Not all plant mitochondria have a multipartite genome however — the genome of *Brassica hirsuta* is present as a single circle.

There is also evidence that a low level of abnormally rearranged genomes, termed sublimons, may exist. It is suggested that under certain conditions a sublimon may become selectively (perhaps stochastically) amplified and come to represent the predominant genomic form. This might account for the changes in profile of mitochondrial DNA molecules observed during a range of tissue culture regimes.

It is assumed that a mechanism must exist normally to ensure that all mitochondria receive a full set of genetic information at division, otherwise a proportion of permanently nonfunctional mitochondria might be generated. One solution might be for the master chromosome to be the only replicating molecule, but this has not been established. RECOMBINATION between DNA molecules from different mitochondria is also possible, and is a frequent occurrence in SOMATIC CELL HYBRIDS.

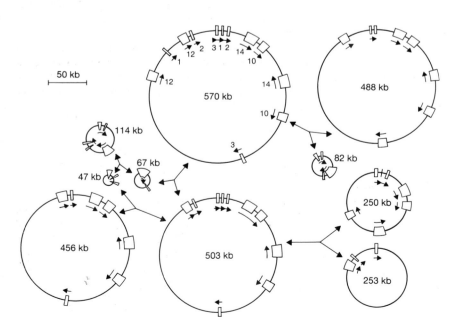

Fig. M27 Circular forms of the mitochodrial genome of maize. The master circle of 570 kilobases (kb) contains the entire sequence complexity of the mitochondrial genome. Intramolecular recombinations between pairs of homologous direct repeats result in the breakdown of the master circle into a population of subgenomic circles. Subgenomic circles containing the same direct repeats may then undergo intermolecular recombination with each other to form larger circles. Many more circular permutations, resulting from intra- and intermolecular recombination events, are feasible than are illustrated. Adapted from [1].

Mitochondrial plasmids

As well as the molecules derived by recombination from the master chromosome, mitochondria of a large number of plants contain PLASMIDS, which may be linear or circular double-stranded DNA or single- or double-stranded RNA. A number of these are associated with cytoplasmic male sterility (cms) (see below). Linear plasmids S1 and S2 are abundant in the cmsS line of maize. They are 6.4 and 5.4 kbp long respectively, and have terminal inverted repeats of 0.2 kbp, with proteins attached to the 5′ ends. They contain a number of expressed OPEN READING FRAMES. Related plasmids are also found in the mitochondria of male-fertile South American maize plants of the RU (Racimo de Uva) group. Recombination between the terminal repeats of the S plasmids and homologous sequences in the chromosome results in the linearization of the latter. Another linear plasmid in maize mitochondria, with some sequence homology to S2, may contain at least one essential transfer RNA gene. Some of the RNA plasmids have been shown to replicate in a DNA-independent fashion, probably using an RNA-DEPENDENT RNA POLYMERASE.

Coding content

Ascertaining which RNA and protein species are mitochondrially synthesized or coded has proved much more difficult than with chloroplasts, mainly owing to the larger size of the mitochondrial genome and the failure of mitochondrial DNA and RNA to be transcribed and translated efficiently *in vitro*. The major approaches used have included HYBRIDIZATION to mitochondrial DNA (either of mitochondrial RNA or heterologous DNA), DNA SEQUENCING, and the analysis of proteins synthesized by isolated mitochondria. Figure M28 shows a genetic map of normal and cmsT (see below) mitochondrial genomes from maize.

rRNA and tRNA genes

The RIBOSOMAL RNAs, 26S, 18S and 5S, are mitochondrially encoded, and the 26S and 18S species are analogous to the rRNAs of other mitochondria. A number of TRANSFER RNAs are mitochondrially encoded, although it seems that in some organisms (e.g. *Chlamydomonas*) a few are encoded in the nucleus and imported post-transcriptionally. Import of RNA into mitochondria has been demonstrated in other organisms, and the mechanisms

may be similar. In spite of the large genome complexity, only a small proportion of mitochondrial polypeptides are mitochondrially encoded. The rest are encoded in the nucleus, and synthesized on cytoplasmic ribosomes with N-terminal extensions that direct import of the polypeptides into the mitochondrion and are subsequently cleaved (*see* MITOCHONDRIAL BIOGENESIS; PROTEIN TARGETING).

Other genes

Two-dimensional electrophoretic analysis suggests that isolated mitochondria synthesize 30–50 polypeptides. These may, however, include incomplete or modified forms of several proteins and so far less than half this number of genes has actually been identified. The genes identified mainly encode components of the inner membrane, including CYTOCHROME B of the cytochrome bc_1 complex (from the *cob* gene), subunits I, II and III of the CYTOCHROME OXIDASE complex (*cox* genes), and subunits α, 6, 8 and 9 of the ATP SYNTHASE complex (*atp* genes). The α subunit is encoded in the nucleus in other organisms; the coding site of subunit 9 is variable. At least one subunit, possibly more, of the NADH dehydrogenase complex (*nad* genes) and an uncertain number of RIBOSOMAL PROTEINS are also mitochondrially coded.

Promiscuous DNA

As well as *bona fide* genes, plant mitochondria also contain sequences derived from other organelles. The phenomenon, termed promiscuous DNA, is not unique to plant mitochondria, but is perhaps best characterized there. In the first example from plants, it was shown that mitochondria from a number of maize lines contained a 12-kbp sequence derived from the chloroplast and including the chloroplast 16S rRNA gene and tRNA^{Ile} and tRNA^{Ala} genes. The chloroplast still retained copies of the sequences. A further promiscuous sequence was identified containing the *rbcL* gene. The similarity between mitochondrial and chloroplast sequences suggests that the transpositions were relatively recent.

It is not clear to what extent promiscuous sequences are transcribed. It is possible that some may now be exploited by the mitochondrion to produce functional tRNAs. Some promiscuous sequences are found in regions of mitochondrial DNA associated with cytoplasmic male sterility.

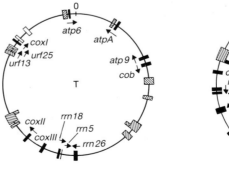

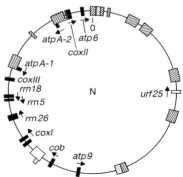

Fig. M28 Genetic maps of the highly rearranged mitochondrial master chromosomes of cmsT (T) and normal (N) maize. Positions and orientations of genes (black boxes), and locations of repeated sequences (hatched, stippled and open boxes) are shown. Adapted from [4].

Gene expression

Gene expression in plant mitochondria is much less well understood than gene expression in chloroplasts [2]. There is disagreement on what proportion of the mitochondrial genome is transcribed. It seems likely that at least 61 kbp of the 218-kbp genome of *B. campestris* is actively transcribed, generating 24 major transcripts, although data from this and other organisms suggest that a larger proportion of the genome may be represented in low abundance transcripts that are rapidly turned over. Use of individual gene PROBES in NORTHERN BLOTS often shows the presence of several transcripts for any one gene, and RNA capping experiments suggest that this is due to the presence of multiple initiation sites. These sites are often close to sequences resembling a 9-base sequence believed to be important in transcription initiation in yeast mitochondria. Most genes appear to be transcribed individually, although the 18S and 5S rRNA genes are within a few hundred base pairs of each other, and are cotranscribed. Stem–loop structures may mediate termination.

RNA PROCESSING is required for the expression of some genes. For example, the 18S–5S rRNA transcript must be cleaved and trimmed to generate the mature species. Some genes, such as *coxII* of maize, are split by INTRONS apparently belonging to group II. Little is known of the details of the splicing process (*see* RNA SPLICING), and there is no indication of the TRANS-SPLICING seen in some chloroplast genes. Comparison of cDNA and genomic DNA sequences shows that RNA EDITING occurs, altering a number of Cs to Us in the transcripts of the wheat *nad3*, *cob*, *coxII* and *coxIII* genes [3]. The process may be similar to that seen in the kinetoplasts (mitochondria) of trypanosomes. POLYADENYLATION of RNA does not take place.

Translation (PROTEIN SYNTHESIS) is rather poorly characterized. As in chloroplasts and *Escherichia coli*, and in contrast to the cytoplasm, it is insensitive to cycloheximide and sensitive to chloramphenicol. However, it is also resistant to erythromycin and streptomycin. UGA is believed to function as a TERMINATION CODON, in contrast to other mitochondrial systems. It has been proposed that CGG encodes tryptophan rather than arginine, but the sequence data might also be explained by editing of a CGG codon to UGG before translation. Sequences analogous to bacterial ribosome-binding sites, complementary to the 3′ end of 18S rRNA, have been identified in some transcripts upstream from the AUG initiation codon. It is not clear how and to what extent gene expression is modulated.

Cytoplasmic male sterility

A number of mutations have been reported in plant mitochondria. The NCS (nonchromosomal stripe) mutations produce a range of growth defects but are relatively poorly understood and will not be discussed further here. Much more is known about the cytoplasmic male sterility mutations which, although differing from each other in detail, have many basic similarities. The most fully studied examples are in maize, sorghum and petunia, although they occur in a very wide variety of plants. Transmitted maternally (cytoplasmically), they lead to a failure to generate viable pollen. They are therefore of considerable agricultural importance, as they can be used to block self-fertilization and hence force cross-fertilization and the production of hybrid seed. The usual alternative would be hand-emasculation, which is at best time consuming and for some crops impossible.

A particular cms line, cmsT, was widely used in maize breeding in the USA. By 1970, some 85% of hybrid maize had the cmsT cytoplasm, and this probably greatly contributed to a disastrous epidemic of southern corn leaf blight (caused by a fungus variously known as *Helminthosporium maydis*, *Drechslera maydis*, *Bipolaris maydis* and *Cochliobolus heterostrophus*) which resulted in the loss of crops worth over a billion dollars. The race T of this fungus produces a toxin, T toxin, to which cmsT mitochondria are especially sensitive, becoming uncoupled and leaky.

In addition to cmsT, two other cms lines have been well studied in maize, cmsS and cmsC. They can be distinguished by the fact that different nuclear genes, *Rf* genes, can restore fertility to the different lines. Thus the genes *Rf1* and *Rf2* will together restore fertility to cmsT lines but not to cmsS or cmsC. Likewise *Rf3* will restore fertility to cmsS, and *Rf4* to cmsC. Also, mitochondria isolated from the different cms lines have slightly different polypeptide synthesis profiles, and this has given important clues to understanding the condition.

Mitochondria of cmsT plants produce a novel 13K (M_r 13 000) polypeptide, the overall rate of synthesis of which is greatly reduced in the presence of the nuclear *Rf1* and *Rf2* genes. A region of the mitochondrial DNA of cmsT plants was found to contain two unidentified open reading frames for polypeptides of 25K and 13K (urf25 and urf13 in Fig. M28). This region of the genome has arisen as a result of at least seven aberrant recombination events involving regions from or near the genes for ATP synthase subunit 6, the 26S rRNA and a promiscuous chloroplast gene for a tRNA^{Arg}. Transcripts for the 25K open reading frame were also found in male-fertile lines, but the 13K open reading frame appeared to be unique to cmsT, and its transcription profile was altered in the *Rf1Rf2* background. Antibodies to synthetic peptides encoded by the 13K open reading frame showed that this protein was the same as the novel one synthesized by isolated mitochondria. This suggests that the 13K polypeptide is responsible for the male sterility, although exactly how is not clear. There is some evidence that it is associated with the ATP synthase complex, and it is possible that some stage of pollen formation places heavy bioenergetic demands which cannot be met in the presence of the 13K polypeptide.

The polypeptide also appears to confer susceptibility to the T toxin. When cultured cells were selected *in vitro* for resistance to the T toxin, the lines generated had deletions or other rearrangements in the 13K open reading frame. In addition they were male fertile. Furthermore, expression of the 13K polypeptide in other organisms, such as *E. coli*, makes them sensitive to the T toxin, which brings about similar symptoms to those observed in cmsT mitochondria. In summary, it therefore seems that T-type cms is caused by the presence of a novel polypeptide causing pollen failure and sensitivity to T toxin, whose gene arose from aberrant recombination events.

Other examples of cms in maize are less well understood. As outlined earlier, mitochondria of cmsS plants contain linear episomes (plasmids) that recombine with the genome and linearize it.

Although the episomes encode high molecular weight polypeptides that are synthesized by cmsS mitochondria, they are probably not directly responsible for the trait, since they are still found in fertile Rf lines and some fertile revertants. Mitochondria of cmsC plants also synthesize a novel polypeptide, of 17.5K, which may replace a 15.5K polypeptide. The synthesis profile is unchanged by the presence of the fertility-restoring *Rf4* gene, though.

In sorghum, cms is again associated with alterations in mitochondrial polypeptide synthesis, including a change in the size of the coxI polypeptide resulting from aberrant recombinations. Although the change in *coxI* may not be the direct cause of the cms, it is likely at least to be indicative of the genetic rearrangements that are responsible. Expression of a MOSAIC GENE has also been correlated with cms in petunia. Sublimons may be involved in the generation of cms. Male-fertile maize lines have been shown to contain low levels of genome arrangements similar to those that are abundant in male-sterile lines. One can therefore speculate that novel molecules are continually being generated by a range of recombination events, some of which may involve episomal elements. Some of these molecules might encode polypeptides with potentially profound phenotypic effects such as male sterility. Amplification could then give rise to a male-sterile plant, and the condition could revert by a further amplification of molecules from which the chimaeric genes had been deleted. Nuclear genes could affect the expression of cms at any of a number of levels from transcription through to protein stability. The plant mitochondrion is an organelle with an extraordinary degree of plasticity in genome organization, and understanding the mechanism and control of genetic rearrangement will be at least as important as understanding the mechanism and control of gene expression.

C.J. HOWE

See also: MITOCHONDRION.

1 Lonsdale, D.M. et al. (1984) The physical map and organization of the mitochondrial genome from the fertile cytoplasm of maize. *Nucleic Acids Res.* **12**, 9249–9261.
2 Levings, C.S. & Brown, G.G. (1989) Molecular biology of plant mitochondria. *Cell* **56**, 171–179.
3 Gualberto, J.M. et al. (1989) RNA editing in wheat mitochondria results in the conservation of protein sequences. *Nature* **341**, 660–662.
4 Palmer, J. (1990) Contrasting modes and tempos in evolution in land plant organelles. *Trends Genet.* **6**, 115–120.

mitochondrial mutation *See*: MITOCHONDRIOPATHIES.

mitochondrion (*pl.* mitochondria) Semi-autonomous organelle in EUKARYOTIC CELLS in which ATP is generated by OXIDATIVE PHOSPHORYLATION. A liver cell, for example, contains about 1000 mitochondria, occupying 10–15% of the cell volume. Mitochondria contain DNA which encodes some mitochondrial proteins including those involved in DNA replication and transcription and some components of the protein synthetic apparatus. It also encodes rRNAs and tRNAs, a few of which depart from the universal genetic code in their anticodon/amino acid specificities. Other mitochondrial proteins are encoded by the nuclear genome and imported into the organelle (*see* MITOCHON-DRIAL BIOGENESIS). Mitochondria are bounded by a double membrane. The outer membrane contains a nonspecific PORIN that forms aqueous channels through the membrane and allows passive diffusion of solutes of M_r up to ~10 000. The inner membrane is folded into numerous cristae, is impermeable to ions and small molecules and contains the enzymes of the ELECTRON TRANSPORT CHAIN and the ATP SYNTHASE along with numerous transport proteins for small molecules, such as pyruvate.

The aqueous compartment between the outer and inner membranes of the mitochondrion is known as the intermembrane space, whose ionic and small molecule composition is very similar to that of the cytosol. It contains several soluble enzymes associated with the respiratory chain and a few others, including ADENYLATE KINASE. Cytochrome *c* is the best characterized protein of the intermembrane space. Cytochrome *c* peroxidase and flavocytochrome b_2 are soluble components of the intermembrane space of yeast mitochondria but are not found in higher eukaryotes.

Inside the inner membrane is the mitochondrial matrix. The mitochondrial DNA is located here along with all the components required for DNA replication, transcription and translation on mitochondrial ribosomes. The matrix also contains the enzymes of the tricarboxylic acid (TCA) cycle, those involved in β-oxidation of fatty acids and some of those involved in the biosynthesis of amino acids, nucleotides, and other molecules. Several proteins required for the transport, proteolytic processing, and assembly of functional mitochondrial proteins have also been identified in the matrix. *See also*: CHEMIOSMOTIC THEORY; ENDOSYMBIONT HYPOTHESIS; MITOCHONDRIAL BIOGENESIS; MITOCHONDRIAL GENOMES.

Mitochondriopathies

THE human mitochondrial genome and its products (*see* MITOCHONDRIAL GENOMES: ANIMAL) have long been suspected of involvement in a number of multisystem degenerative diseases, principally affecting the musculature and nervous system. These suspicions were originally based on: (1) the observation of ultrastructural and biochemical abnormalities of mitochondria in association with several such pathologies; (2) the predominance of maternal inheritance (although some of these mitochondrial pathologies occur sporadically); and (3) the fact that the most commonly affected tissues are those known or believed to be most dependent on respiratory metabolism.

Analysis of mitochondrial DNA (mtDNA) in affected individuals and their families (by for example, RESTRICTION ANALYSIS and DNA SEQUENCING) has provided convincing evidence that MUTATIONS in mtDNA are involved in these diseases, although not necessarily in a simple manner. In most cases a mutant genome is present alongside an ostensibly wild-type genome in the affected individuals (i.e. they are heteroplasmic), and the representation of the mutant genome is sometimes, but not always, correlated with the severity of the condition and with the tissue specificity of symptoms. The genetic basis of the mitochondriopathies is reviewed in [1–4].

Sporadic mitochondriopathies

Large heteroplasmic deletions of mtDNA were first reported in muscle biopsy samples from patients suffering from a relatively common, sporadically occurring form of mitochondrial myopathy. In its severe and usually fatal form this disease is described as Kearns–Sayre syndrome (KSS), comprising muscle weakness, ataxia, pigmentary retinopathy, heart conduction block, ophthalmoplegia (paralysis of the eye muscles), deafness, and progressive dementia. The less severe variant of this disease (CPEO) usually comprises no more than chronic, progressive, external ophthalmoplegia. Partial duplications of mtDNA have also been observed in some severely affected patients and in these cases the rearranged molecules are detectable in blood.

Deletions have subsequently been reported in blood cell mtDNA of patients with Pearson's syndrome (a disease affecting pancreatic, spleen and blood functions), in heart muscle of patients with cardiomyopathy, and, more controversially, in the brains of patients with Parkinson's disease (*see* NEURODEGENERATIVE DISORDERS).

Genetic defects

In general, the genes covered by these deletions or duplications do not correlate in any simple fashion with the nature or severity of measurable biochemical abnormalities, and it remains an open question how these genetic lesions result in the development of a pathological state. The deleted molecules seem to be functionally dominant, in the sense that diseased muscle fibres contain both types of mtDNA. Cells with only the deleted mtDNA would presumably be nonviable, and hence selected against during development.

Deletions and duplications commonly involve apparent recombinational events, occurring between various pairs of short, directly repeated sequences in the genome, typically of 10–15 base pairs (bp). Although many such rearrangements have been found, a particular ('common') deletion of 5 kilobases (kb) is found in a significant minority (~35%) of KSS/CPEO patients. The mechanism of RECOMBINATION is unknown, but could involve replication slippage, strand invasion, or protein-mediated strand exchange. Almost all such deletions preserve the separate origins of replication for each of the two strands (*see* MITOCHONDRIAL GENOMES: ANIMAL) leading to the suggestion that deleted molecules might be maintained by a replication advantage, even if they confer deleterious effects on the cell or organelle harbouring them. Pathogenic rearrangements of mtDNA might also be generated, as in plants and fungi, by extragenomic mitochondrial PLASMIDS (*see* MITOCHONDRIAL GENOMES: PLANT; MITOCHONDRIAL GENOMES: LOWER EUKARYOTE) although none has yet been found in metazoans.

Heritable mitochondriopathies

Heritable rearrangements of mtDNA are much rarer, but a maternally inherited large deletion (removing the light strand origin, which is presumably replaced functionally by a cryptic origin elsewhere on the DNA strand) has been reported in a pedigree suffering from a syndrome comprising diabetes and deafness, but without muscle pathology. Point mutations of pathological significance have also been identified by painstaking sequence analysis of mtDNA from affected families and from many normal controls or, in some cases, by the chance occurrence of a restriction site polymorphism (*see* RFLLP) as a result of the mutation. Such mutations are also commonly heteroplasmic and have been found to map to a variety of sites in the mitochondrial genome, including conserved amino-acid residues of protein-coding genes, several tRNA genes, and also in potential regulatory regions, such as the attenuator sequence located downstream of the rRNA genes (*see* MITOCHONDRIAL GENOMES: ANIMAL).

Each identified point mutation tends to be associated with a distinct, tissue-specific pathology. Thus, any one of a number of nonconservative substitutions in the genes for NADH dehydrogenase subunits 1 or 4 (ND1 or ND4) is found in association with Leber's hereditary optic neuropathy (LHON), a disease which is characterized by maternally inherited, bilateral, adult-onset blindness, resulting from optic nerve degeneration. A point mutation in the $tRNA^{Lys}$ gene correlates with the MERRF syndrome, a severe form of mitochondrial myopathy accompanied by myoclonic epilepsy.

The mechanism whereby heteroplasmy, involving a mutant mitochondrial genome, may be maintained between generations is unclear. One possibility is that there is very little selection for respiratory competence during oogenesis and early development. Another is that the apparently wild-type genomes which coexist with mutant mtDNAs also bear one or more point mutations, requiring that the two genomes complement one another to support efficient respiration. The maintenance of heteroplasmy would be required because of the absence of a generalized recombination system in mammalian mitochondria.

Nuclear gene involvement in mitochondriopathies

Nuclear genes are also clearly involved in a number of these diseases. For example, familial cases of a KSS-like disease are associated with multiple deletions of mtDNA, but the predisposition to this condition is transmitted in a manner consistent with AUTOSOMAL DOMINANT inheritance. Nuclear mutations in fungi, which promote mitochondrial genomic instability, may provide an instructive model. In addition, LHON exhibits incomplete PENETRANCE, with a significant sex bias suggestive of an X-LINKED component, although other explanations are possible. Pedigree studies to test this idea have given inconsistent results, and the question remains open.

Defects in nuclear genes seem to be the predominant cause of some rare mitochondriopathies. Two likely examples are Leigh syndrome and Reye syndrome, each with a devastating (but quite distinct) encephalopathy, the latter also associated with hepatic failure. Reye syndrome manifests most commonly as an acute post-viral syndrome in children, and has been linked to the therapeutic use of aspirin. However, a very similar pathology has been reported in adults, associated with a generalized defect in intramitochondrial protein metabolism. The characteristic symptoms of Leigh syndrome (subacute necrotizing encephalomyelopathy) may also manifest in association with a defect in mitochondrial protein import and assembly (*see* MITOCHONDRIAL BIOGENESIS), or alternatively as a result of a mitochondrial mutation. Leigh symptomology is found, for example, in severely affected members of a maternal pedigree carrying a (heteroplasmic) nonconservative substitution in the mitochondrial gene for subunit 6 of ATP synthase, a mutation which is minimally associated with a pathology of neurological ataxia and RETINITIS PIGMENTOSA (NARP).

A deficiency of a foetal/neonatal ISOFORM of a subunit of cytochrome *c* oxidase has been postulated as a cause of benign infantile mitochondrial myopathy, a transient pathology from which the affected infants usually make a complete recovery.

Mutant mitochondrial genomes have been successfully passaged from patient-derived cell lines to recipient cells of a different nuclear background, achieved by CYTOPLAST fusion to a *rho*^{0} cell line cured of mtDNA by prolonged growth in ethidium bromide. These experiments have demonstrated that abnormalities in the patterns of mitochondrial protein synthesis and/or in respiratory chain functions are determined by the mitochondrial genotype, at least for the point mutations associated with MERRF and NARP.

Muscle pathology

Conditions in which the respiratory chain is affected are commonly associated with a characteristic muscle pathology, denoted as 'ragged red fibres', and marked by accumulations of abnormal mitochondria often with paracrystalline inclusions, in the diseased fibres. This symptomology is seen with MERRF (hence its name), KSS/CPEO, and with another syndrome denoted MELAS (mitochondrial encephalomyopathy with lactic acidosis and stroke-like episodes), but is absent in LHON and NARP. It has been suggested that this muscle pathology may correlate with alterations to mtDNA which interfere in some way with the mitochondrial translational apparatus. These would include the tRNA point mutations seen in MERRF (in tRNA^{Lys}) and MELAS (in tRNA^{Leu}), as well as the gross rearrangements found in KSS/CPEO, which would be predicted to lead to over- or underrepresentation of a variable number of tRNAs. The large deletion associated with diabetes and deafness is not associated with a muscle pathology, however.

Tissue specificity

The striking and variable tissue specificity of the mitochondriopathies is unexplained. The mutations associated with LHON, for example, have at most only a small effect on the NADH dehydrogenase activity of blood platelet mitochondria, although it is unknown whether the level of this enzyme is affected more drastically in the optic nerve and retina. One popular view is that mtDNA-encoded respiratory chain subunits may interact with nuclear-encoded polypeptides represented as different tissue-specific isoforms. A number of subunits of cytochrome *c* oxidase, and also the ATP/ADP translocator protein, are known to be encoded by MULTIGENE FAMILIES variably expressed in tissue-specific patterns. If this were to apply to the critical

nuclear-encoded subunits with which ND1 and ND4 must interact, then the LHON pathology could result from the inability of the mutant polypeptides to function properly in concert with those nuclear isoform(s) of their partner protein(s) most highly expressed in the optic nerve.

Alternative hypotheses regarding the pathogenic role of a number of mitochondrial mutations have been proposed, and one of the goals of current research in this area is to test these. For instance, the tRNA^{Leu} mutation most commonly found in MELAS patients also affects the binding of the attenuator protein at the 16S rRNA/tRNA^{Leu} gene boundary (*see* MITOCHONDRIAL GENOMES: ANIMAL). The mutation reduces the efficiency of transcriptional termination *in vitro*, although it is unclear as yet if this has any significance in pathogenesis. In addition, the pathology of KSS/CPEO might be explained as the result of the synthesis of FUSION PROTEINS, encoded across the abnormal junctions created by deletions or duplications. The synthesis of such a protein has been demonstrated in cultures of cells harbouring mitochondria derived from a KSS patient, and there is a precedent for this as a pathogenic mechanism in cytoplasmic male sterility in plants (*see* MITOCHONDRIAL GENOMES: PLANT). It may be hypothesized that the abnormal proliferation of mitochondria which characterizes ragged red fibres may result from a feedback mechanism, responding to the disruption of the respiratory membrane by such fusion peptides. A similar argument may be made for missense proteins, which might be created if tRNA mutations cause increased translational misreading.

Ageing

The high mutation rate of mtDNA and the metabolic importance of its products has led some researchers in the field to suggest that it is an important mutational target in the ageing process. Mitochondria are the main site of free radical generation in cells, and it has long been supposed that the most significant DNA damage is caused by such agents. A pathogenic involvement of mtDNA is therefore now plausible in many degenerative (especially neurodegenerative) conditions, at least as one component of a potentially heterogeneous disease process.

H.T. JACOBS

1 Harding, A.E. (1991) Neurological disease and mitochondrial genes. *Trends Neurosci.* **14**, 132–138.
2 Vanbrunt, J. (1991) Mitochondrial defects and disease. *Bio/Technology* **9**, 329.
3 Wallace, D.C. (1989) Mitochondrial DNA mutations and neuromuscular disease. *Trends Genet.* **5**, 9–13.
4 Shoffner, J.M. & Wallace, D.C. (1990) Oxidative phosphorylation diseases: disorders of two genomes. *Adv. Hum. Genet.* **19**, 267–330.

mitogen Any agent that induces MITOSIS and (usually) cell division. *See*: CELL CYCLE; LECTINS.

mitomycin C Antibiotic produced by *Streptomyces caespitosus*. It inhibits nuclear division, DNA and protein synthesis in mammalian cells, and is used clinically as an anti-tumour agent.

Mitosis

MITOSIS is the process of CHROMOSOME segregation that occurs during the division of EUKARYOTIC cells. It is part of the CELL CYCLE and alternates with INTERPHASE, during which the chromosomes and other essential cell parts are duplicated. Its name derives from the Greek word for thread, because the chromosomes are threadlike during its early stages. Mitosis comprises five stages, called PROPHASE, during which the chromosomes condense, PROMETAPHASE, during which they become organized, METAPHASE, during which they are arranged at the cell's midplane in a cluster called the METAPHASE PLATE, ANAPHASE, during which each duplicate chromosome separates into two genetically equivalent pieces that move apart, and TELOPHASE, during which the chromosomes decondense into the interphase condition.

During the interphase that precedes mitosis, both the DNA and the CENTROSOMES duplicate. The cell doubles in mass and volume, so all synthetic events essential for cell division are completed. Mitosis and the companion process of CYTOKINESIS then accomplish the mechanical division of this double cell into two viable parts.

Prophase

Chromosome condensation

Chromosome condensation during prophase is an essential part of mitosis because the piece of DNA that comprises each chromosome is long compared with the diameter of the nucleus. During interphase it becomes entangled with itself and others like it, so separating the duplicated chromosomes into two identical sets without some previous sorting would be impossible. The process of chromosome condensation is not well understood, but it involves the tightening of gyres and loops of the chromosome (*see* CHROMOSOME STRUCTURE) to form an object whose length is less than the diameter of the nucleus and whose width is usually ~1/10 its length. During condensation, the NUCLEAR ENVELOPE is always present to separate the nucleus from the cytoplasm. PROTEIN KINASES act to hyperphosphorylate the HISTONES in the CHROMATIN, while other enzymes de-acetylate them; the rate of RNA synthesis drops to essentially zero. Meanwhile, the nucleolus usually disperses into evenly distributed particles of ribonucleoprotein. In most higher eukaryotes the NUCLEAR LAMINA disperses as the LAMINS become phosphorylated.

Cytoplasmic changes

During prophase the cytoplasm also undergoes a significant rearrangement. The interphase MICROTUBULES and MICROFILAMENTS that make up the CYTOSKELETON depolymerize and/or break up into short pieces. The GOLGI APPARATUS disperses into many small vesicles that become distributed throughout the cytoplasm. Animal cells tend to round up during this time, presumably because their infrastructure is weakened with the dissolution of the cytoskeleton, but INTERMEDIATE FILAMENTS

generally do not break down; they often form a cage around the nucleus. The already duplicated centrosomes generally separate during prophase so as to lie on opposite sides of the nucleus, and at about this time they change so as to initiate 5–10 times as many microtubules as during interphase. The two resulting structures are radial arrays of short and labile microtubules called the mitotic ASTERS. In cells of higher plants, the prophase microtubules often form a sheath around the nucleus, defining the line that will become the axis of the MITOTIC SPINDLE.

Prometaphase

Prophase ends and prometaphase begins with the onset of interactions between the chromosomes and the microtubules of the forming spindle. In the cells of all higher plants and animals, this process follows the breakdown of the nuclear envelope, which allows the chromosomes from the nucleus to interact with the microtubules from the cytoplasm. In many lower eukaryotes, the nuclear envelope stays intact throughout mitosis (*see e.g.* SACCHAROMYCES CEREVISIAE), and the microtubules that assemble within the nucleus are said to form a closed spindle. Many fungi contain one or two structures called SPINDLE POLE BODIES that serve as centrosomes, but these are built into a large pore within the nuclear envelope. The mechanism by which TUBULIN and other proteins essential for mitotis get into the mitotic nucleus of a fungus has not yet been studied.

Early in prometaphase, chromosomes begin to interact with the microtubules initiated by the centrosome. The two KINETOCHORES, located at the CENTROMERE of each chromosome (*see* CHROMOSOME STRUCTURE), are the principal sites of this interaction. Kinetochores bind to the surfaces of microtubules and move over them at 10–50 μm min^{-1} toward their centrosome-associated ends (the 'minus' or slow-growing ends of the microtubules). These movements often promote a clustering of the chromosomes around the centrosomes, but as every chromosome has two kinetochores and each cell has two centrosomes, each at the centre of its own array of microtubules, the chromosomes are soon being pulled in two directions at once, putting them under tension. This tension is important, because it has been shown to be essential for the formation of a stable association between a chromosome and the spindle. So long as both kinetochores of a chromosome are pulled toward the same pole, the chromosome's association with the spindle is labile and the chromosome is likely to reorient, giving it another chance to attach correctly. This property assures that each chromosome becomes oriented in such a way that when its CHROMATIDS separate at the onset of anaphase, one copy of the genetic information it contains will go to each spindle pole.

After a chromosome has become oriented so that SISTER CHROMATIDS are associated with opposite centrosomes, it moves slowly toward and away from the pole it lies near. Over time, these erratic movements add up to a CONGRESSION to the spindle equator, so the chromosomes gradually accumulate at the metaphase plate. Once they are there, the cell is said to be in metaphase.

The molecular mechanisms of prometaphase are beginning to be understood. The initial, microtubule-dependent motions of kinetochores toward centrosomes are probably due to microtubule-dependent motor enzymes like DYNEIN and KINESIN, which are bound to kinetochores, centrosomes and the microtubules that connect them. Dynein is an ATP-dependent motor enzyme that moves over a microtubule surface towards its minus end; kinesin and its relatives are analogous enzymes that generally move in the opposite direction (*see* MICROTUBULE-ASSOCIATED MOTORS). Their locations in the mitotic spindle implicate them in the motions seen. Isolated chromosomes will move over microtubules in an ATP-dependent fashion in either the minus or the plus direction, depending on the state of phosphorylation of kinetochore components. The ways in which these motor activities are controlled and orchestrated to bring about the complex motions of prometaphase are now under study.

The metaphase spindle

The metaphase spindle is constructed from microtubules that emanate from the cell's two centrosomes, each of which resides at one of the spindle poles. Some of these microtubules end on the kinetochores of the chromosomes, forming a bundle called a chromosomal (or kinetochore) SPINDLE FIBRE. Other microtubules end free in the body of the spindle. Some of the latter are long enough to interdigitate with their counterparts that emanate from the opposite centrosome (Fig. M29). Spindles that form in cells possessing centrosomes generally have microtubules that project out into the cytoplasm, forming the metaphase asters. Spindles like those of higher plants, which lack centrosomes, are said to be ANASTRAL.

Anaphase

Anaphase begins with the separation of each centromere into two, allowing the sister chromatids to part and move to the spindle pole with which each is associated. The control of anaphase onset is not yet understood, but it is important for cell viability. If a cell enters anaphase before all its chromosomes are attached to the spindle, chromosome segregation is likely to be imprecise, and one of the daughter cells will probably die. Indeed, cells in which one chromosome is not yet attached to the spindle delay anaphase for a while, suggesting that there is a control on the quality of prometaphase that regulates mitotic progression.

Excessive delay in metaphase, however, is also deleterious, because time is wasted that could otherwise be spent in the activities of interphase, either growing to divide again or carrying out a specialized function, such as that of a neuron.

Anaphase is usually thought of in two parts: the motion of the chromosomes to the poles, called anaphase A and the elongation of the distance between the spindle poles, called anaphase B. During anaphase A, the microtubules associated with each kinetochore shorten. Shortening results from loss of tubulin subunits; about 3/4 of this loss occurs at or near the kinetochores and the remainder occurs near the poles.

Chromosomes move at 0.5–3 μm min^{-1}, depending on cell type and temperature. The force that must be applied to a chromosome to stop it is $\sim 5 \times 10^{-5}$ dynes. The rate of chromosome movement seems to be regulated in part by the rate at which the kinetochore-associated microtubules disassemble. Anaphase B is usually about one-half the speed of anaphase A. The extent of anaphase B is highly variable, ranging from almost zero in some cells to ~ 10 times the length of the metaphase spindle in others. In general, a long anaphase B is characteristic of large cells with small spindles.

The mechanisms of anaphase are partially understood. Both microtubule disassembly and microtubule-associated motor activities are likely to contribute to the forces that pull chromosomes to the poles. During anaphase A, many of the spindle's microtubules shorten. The spindle elongation of anaphase B, however, is accompanied by a polymerization of the microtubules that were long enough at metaphase to interdigitate with similar microtubules from the opposite centrosome. These microtubules add tubulin at their centrosome-distal ends and simultaneously slide over one another as the poles move apart. In diatoms the force for this sliding is generated where the interdigitating microtubules overlap. The same may be true of other cells, but when asters are present, they may contribute to the forces that promote anaphase B.

Telophase

Late in anaphase, the chromosomes contract and the nuclear envelope begins to reform around them. The nuclear lamina is re-established, and once the envelope is complete, chromosome decondensation begins and the cell is said to be in telophase. RNA

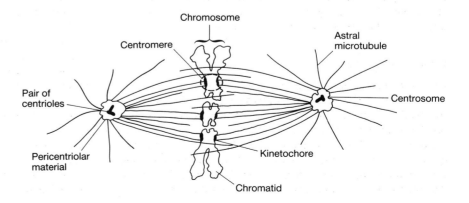

Fig. M29 The metaphase spindle of an animal cell.

synthesis begins as soon as decondensation is underway, and in the rapidly dividing cells of early embryos, DNA replication can do the same. Thus the distinction between telophase and the onset of interphase is hard to define.

Cytokinesis

In animal cells the cleavage furrow characteristic of cytokinesis usually begins to form during anaphase. The furrow pinches in the cell's equator, bundling the spindle microtubules that run past the separating chromosomes into a shaft called the MIDBODY (Fig. M30). The midbody persists for different times in different cells, but it can last though interphase until the next mitosis. It probably serves to anchor the cleavage furrow at its greatest extent of contraction, and the isthmus it defines can be retained as a channel of communication between the daughter cells. In cells of higher plants, the PHRAGMOPLAST forms late in anaphase to assemble a new cell membrane and wall, effecting cytokinesis.

During telophase, an animal cell will begin to flatten back down and re-establish its interphase structure. Each centrosome begins to initiate fewer, longer microtubules, the Golgi apparatus reforms in the vicinity of the centrosome, and the cell recommences pinocytosis and its various forms of actin-dependent motility.

J.R. MCINTOSH

See also: MEIOSIS.

1 Kirschner, M. & Mitchison, T. (1986) Beyond self-assembly: from microtubules to morphogenesis. *Cell* **45**, 329–342.
2 McIntosh, J.R. & Hering. G.E. (1991) Spindle fiber action and chromosome movement. *Annu. Rev. Cell Biol.* **17**, 403–426.
3 Mitchison, T.J. (1988) Microtubule dynamics and kinetochore function in mitosis. *Annu. Rev. Cell Biol.* **4**, 527–550.
4 Nicklas, R.B. (1988) The forces that move chromosomes in mitosis. *Annu. Rev. Biophys. Chem.* **17**, 431–449.

mitotic centre *See*: CENTROSOME.

mitotic chromosomes Generally refers to chromosomes at the METAPHASE stage of MITOSIS when they are fully condensed and, with suitable staining, are visible in the light microscope so that individual chromosomes can be identified. *See*: CHROMOSOME BANDING; KARYOTYPE.

mitotic domains In DROSOPHILA embryogenesis, the first 13 nuclear divisions are virtually synchronous and take place in a single multinucleate cell — the syncytial blastoderm. Thereafter the cycles slow down and cell membranes form around the nuclei. At this stage groups of cells in different regions of the embryo continue to divide in synchrony with each other, and such a group is called a mitotic domain. Mitotic domains were first identified using antibodies to detect MITOTIC SPINDLES in dividing cells. The cells that contained spindles at any given time, and were therefore in METAPHASE, were found in clusters, usually in bilaterally symmetric positions in the embryo. Cells within a mitotic domain often bear important features in common, such as spindle orientation, cell shape, or the larval structure to which they will give rise. *See also*: DROSOPHILA DEVELOPMENT.

mitotic index A parameter used to describe the fraction of a population of cells that is in the process of cell division at any given time. The mitotic index is the number of cells in MITOSIS divided by the total number of cells in the population.

mitotic recombination A technique, also called somatic recombination, used in the study of DROSOPHILA genetics and development. Clones of cells HOMOZYGOUS for a RECESSIVE mutation with a visible phenotype are produced on a HETEROZYGOUS wild-type background. This technique is useful for studying mutations which are lethal when homozygous in the whole organism but not in small clones of cells. It is also useful for marking clones in order to determine how they grow and become distributed during development. To produce marked cells, a population of flies heterozygous for a recessive mutation is treated with a dose of radiation sufficient to induce occasional exchanges between homologous chromosomes. These exchanges usually occur during the G2 stage of the CELL CYCLE when there are four CHROMATIDS; the ensuing MITOSIS will have one of two possible outcomes: it will produce either two heterozygous daughter cells identical to the mother cell, or one cell with no copies of the recessive mutation and one with two copies. The latter cell will express the recessive phenotype, and go on to give rise to a clone of mutant cells as development proceeds.

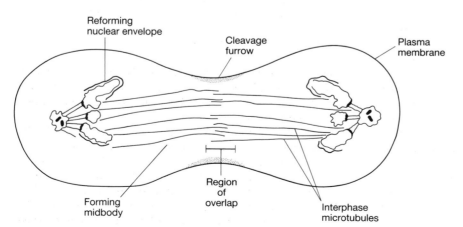

Fig. M30 Late anaphase and the onset of cytokinesis.

Reforming nuclear envelope

Cleavage furrow

Plasma membrane

Forming midbody

Region of overlap

Interphase microtubules

mitotic spindle The bipolar array of MICROTUBULES and associated proteins that interacts with the duplicated CHROMOSOMES during MITOSIS to organize them and segregate them into two equal sets.

mixed glucans of grasses An obscure group of mixed β1,3/β1,4 GLUCANS with structural features resembling CELLULOSE and CALLOSE; they are of interest in relation to the effects of oligosaccharides derived from β1,3 and β1,4 glucans on callose synthetase and the relationship between the synthesis of callose and that of cellulose.

MK-801 (+)-5-Methyl-10,11-dihydro-5H-dibenzo [a,d] cyclohepten-5,10-iminemaleate. Noncompetitive antagonist (channel blocker) at the NMDA receptor (Fig. M31). *See:* EXCITATORY AMINO ACID RECEPTORS.

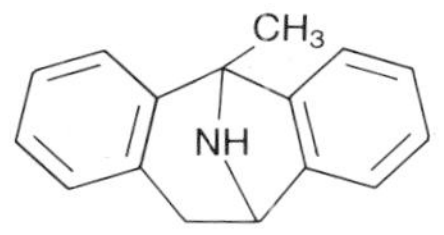

Fig. M31 MK-801.

ML MAXIMUM LIKELIHOOD.

MLO MYCOPLASMA-LIKE ORGANISM.

Mls system Endogenous SUPERANTIGENS present in certain mouse strains.

MLV Murine leukaemia virus. *See:* RETROVIRUSES.

MMTV Mouse mammary tumour virus. *See:* ONCOGENES; RETROVIRUSES.

M-MuLV Moloney murine leukaemia virus. *See:* RETROVIRAL VECTORS; RETROVIRUSES; TRANSGENIC TECHNOLOGIES.

Mn A transposable genetic element found in maize. *See:* TRANSPOSABLE GENETIC ELEMENTS: PLANTS.

MNS blood group The MN and SsU blood group antigens are carried on the sialic acid-rich red cell membrane GLYCOPROTEINS glycophorin A (GPA) and glycophorin B (GPB) respectively. They are encoded by closely linked genes on human chromosome 4. Red cells of rare phenotypes lacking these glycoproteins are refractory to invasion by the malarial parasite *Plasmodium falciparum*.

The precursor of the M and N blood-group antigens is the T antigen (or Thomsen–Friedenreich antigen) Galβ1,3GalNAc(p)α1-Ser-Thr. This derives from the Tn antigen which has the structure GalNAc(p)α1-Ser/Thr and is the precursor of all mucin-type glycans. In Tn syndrome there is a failure to add galactose to the Tn antigen and hence to produce sialyl derivatives of T antigen. There is also no 6-sialylation of the Tn antigen, so severe hyposialylation of erythrocyte surfaces occurs.

mobile DNA, mobile genetic elements *See:* TRANSPOSABLE GENETIC ELEMENTS.

mobile ion carrier *See:* IONOPHORE.

mobilization The transfer of a bacterial PLASMID to a new host, usually by CONJUGATION. The product of a *mob* (mobilization) gene is one of a number of factors required for such conjugal transfer. Plasmid vectors in common laboratory use do not possess all elements required for mobilization. PBR322, for instance, requires the presence of a ColK plasmid for conjugal transfer.

moderately repetitive DNA Class of eukaryotic genomic DNA differentiated from HIGHLY REPETITIVE DNA by its physical REASSOCIATION KINETICS after melting of the DNA duplex (*see* HYBRIDIZATION). In contrast to highly repetitive DNA the sequence repeats are dispersed throughout the genome (*see* INTERSPERSED REPEATED DNA). The sequences present vary widely from species to species, numbers of repeats range from hundreds to 10^6 (in the case of the ALU SEQUENCE in the human genome) and the length of the sequences from a few hundred to a few thousand base pairs. The function of much moderately repetitive DNA has yet to be defined. *See also:* GENOME ORGANIZATION; LINE; SINE.

modification and restriction *See:* RESTRICTION AND MODIFICATION.

modulatory component A high molecular weight protein associated with non-NMDA EXCITATORY AMINO ACID RECEPTORS and which ALLOSTERICALLY regulates the affinity of AMPA binding. When coupled to the receptor the modulatory component downregulates the affinity of AMPA binding; when the modulatory component is removed (either by radiation inactivation or by biochemical means) the affinity of AMPA binding is increased. No ligands are yet known for this protein.

modulatory site Distinct site of action on (as opposed to associated with) an IONOTROPIC receptor complex where a modulator binds and allosterically regulates the activity of the receptor. Examples include the polyamine site and the glycine site on the NMDA receptor where the binding of spermidine potentiates the channel activity. *See:* EXCITATORY AMINO ACID RECEPTORS.

MOI MULTIPLICITY OF INFECTION.

molecular averaging An approach towards improving the phases used in X-RAY CRYSTALLOGRAPHY in the calculation of initial ELECTRON DENSITY MAPS, in order to facilitate the interpretation of an unknown molecular structure.

If a crystal contains two or more identical molecules in the ASYMMETRIC UNIT, it may be possible from a preliminary map — or from electron micrographs — to identify similar molecular features and so to define both the surface envelope of each molecule and its relative orientation within the unit cell. (For oligomeric structures, the molecular symmetry may already be known; for example, many tetramers have their subunits related by three orthogonal DYAD axes.) The electron density representing the two or more molecules should be identical for all atoms (excepting only those involved in intermolecular contacts within the crystal lattice) such that any differences between the respective densities can be attributed to errors in either the measured STRUCTURE AMPLITUDES or the present set of trial phases. Construction of a modified electron density map containing averaged values between the molecules inside each of their envelopes and the original values elsewhere enables a new set of phases to be calculated by means of a FOURIER TRANSFORM. Combination of these new phases and the previous trial phases, with suitable weighting of the two phase sets, can be expected to yield an electron density map which is more readily interpretable.

The success of the method depends on many factors, including the number of molecules being averaged, the validity of their molecular envelopes, and the accuracy of the relative molecular orientations. The procedure of averaging and phase combination is usually used iteratively, often at successively higher RESOLUTION as the coordinate transformation relating the molecules becomes more precisely defined.

Molecular averaging has been used most powerfully for viruses and their coat proteins: 17-fold averaging was used for the coat protein disk of TOBACCO MOSAIC VIRUS and fivefold or higher averaging has been used for many icosahedral viruses.

Bricogne, G. (1976) *Acta Cryst.* **A32**, 832–847.

Molecular biology

THE term ''molecular biology'' seems first to have been used by Warren Weaver in his 1938 report to the Rockefeller Foundation:

> Among the studies to which the Foundation is giving support is a series in a relatively new field, which may be called molecular biology . . .

Astbury used the term a year later, in 1939, and it became increasingly common as time went on. It was apparently first used in the name of an institution in 1956, for the Medical Research Council Laboratory of Molecular Biology at Cambridge (originally called the Unit for the Study of the Molecular Structure of Biological Systems); in that of a journal in 1959, the *Journal of Molecular Biology*; in that of an international organization in 1963, the European Molecular Biology Organization; and in that of an encyclopedia in the present volume.

There have been many attempts to formulate an adequate definition of molecular biology, many of them of course reflecting the intellectual preoccupation of their authors. Thus Erwin Chargaff wrote of it as:

> the practise of biochemistry without a licence . . .

a definition ironically intended, but which did indicate that its devotees in the early days were generally trained in other fields. Indeed few of the early molecular biologists were brought up as

biologists; most were chemists or physicists (James Watson was an exception).

Francis Crick, who was educated as a physicist, emphasized the interdisciplinary nature of the subject, an extremely important element especially in the early stages:

> I . . . was forced to call myself a molecular biologist because when enquiring clergymen asked me what I did, I got tired of explaining that I was a mixture of crystallographer, biophysicist, biochemist and geneticist . . . .

Perhaps the most satisfactory brief definition is that of Jacques Monod:

> What is new in molecular biology is the recognition that the essential properties of living beings could be interpreted in terms of the structures of their macromolecules.

At the beginning there were two schools of molecular biology. The first comprised those primarily interested in the *three-dimensional* structure, or *conformation*, of biologically important macromolecules, especially the proteins, and whose technique of choice was X-RAY CRYSTALLOGRAPHY. Early practitioners were Astbury, Bernal, and Pauling. Perhaps the first experiment in molecular biology was Astbury's demonstration in 1929 that the X-ray pattern, and therefore the structure, of a human hair changed when it was stretched. Then, a little later, came those interested in biological *information* and its replication. These made up what may be called the *one-dimensional* school; early practitioners were Delbrück and Luria. The two schools, which at first had rather little to do with one another, began to have active connections in the early 1950s when it became clear that DNA was the genetic material, and when one of those trained in the informational school — Watson — came to the Cambridge laboratory which was a centre of the conformational school, and worked with Crick on the structure of DNA. Since that time the two schools have become increasingly intertwined with one another and with other fields, and indeed today the boundaries between biochemistry, genetics, molecular biology, and biophysics have become less and less well defined.

J.C. KENDREW

Molecular chaperones

MOLECULAR chaperones are a diverse group of proteins that mediate the correct assembly of other polypeptides, but are not themselves part of the functional assembled structure [1]. Molecular chaperones are involved in many macromolecular assembly processes, using assembly here in its broadest sense to include folding and refolding of proteins during their synthesis and transport, as well as the association of polypeptides with each other and other macromolecules to form oligomeric complexes. They have been implicated in the folding of newly synthesized polypeptide chains, in the partial unfolding or subunit disassociation that may occur when a protein carries out its function, in the transport of proteins across membranes (*see* PROTEIN TRANSLOCATION), and in the repair of proteins partially denatured by exposure to environmental stresses such as high temperature (*see* HEAT SHOCK).

The term molecular chaperone was used first [2] as a succinct way of describing the function of one protein — nucleoplasmin. This protein is found in the nucleus where it is believed to mediate the assembly of nucleosomes (*see* CHROMATIN). Nucleosomes consist of histone proteins bound to DNA by electrostatic bonds. If purified DNA and purified histones are mixed together at physiological ionic strength there is a spectacular failure of self-assembly — an instant precipitate forms instead of nucleosomes. If the histones are first mixed with nucleoplasmin, however, complexes form in which the strong positive charge of the histone molecules is reduced by combination with the acidic nucleoplasmin. Addition of DNA to the mixture results in the formation of nucleosomes and the release of nucleoplasmin. Nucleoplasmin does not bind to DNA or to nucleosomes, nor is it a component of nucleosomes. The steric information for nucleosome assembly resides in the histones and not in nucleoplasmin, as nucleosomes can assemble *in vitro* without it under certain unphysiological conditions. The function of nucleoplasmin, therefore is to prevent incorrect interactions between histones and DNA by transiently reducing the positive charges on the histone proteins. The term molecular chaperone was proposed for nucleoplasmin by analogy with the function of human chaperones, which is essentially to prevent incorrect interactions between potentially complementary surfaces.

Nucleoplasmin is the archetypal molecular chaperone, and it was the realization that this term could usefully be extended to a much larger range of unrelated proteins that function in a variety of cellular processes that spawned the more general chaperone concept [4]. This concept is forcing a re-examination of the principle of protein SELF-ASSEMBLY as the number of proteins that can be regarded as chaperones is steadily increasing (Table M2).

This principle of self-assembly holds that all the information needed for assembly of a functional protein resides in the aminoacyl sequence of its component polypeptides. On the other hand, the chaperone concept postulates that in many cases interactions within and between polypeptides need to be controlled by some outside agency (the chaperone protein) in order to reduce the probability that incorrect interactions will produce a nonfunctional structure [3]. The chaperone concept does not, however, violate the principle of self-assembly, as no known chaperone actually conveys steric information for either polypeptide folding or subunit association — they do not determine the basic folding pathways but rather determine which one will be taken.

Molecular chaperones act by binding noncovalently to specific structural features in their target that are accessible only during assembly, and so inhibit unproductive assembly pathways that would otherwise act as kinetic dead ends and produce an incorrect, non-functional structure. Binding is reversed under as yet undefined circumstances which favour the formation of the correct structure; in some cases, but not all, dissociation of the chaperone from its target requires ATP hydrolysis. Members of some families of chaperone, such as the bacterial CHAPERONINS and the 70K heat shock proteins (Hsp70s) (*see* HEAT SHOCK) can each bind to a wide range of unrelated polypeptides and hence seem to recognize some aspect(s) of secondary or higher order structure in the target polypeptides.

Chaperone function is required because many cellular processes involving protein assembly carry an inherent risk of malfunction as a result of the large number, variety and flexibility of the noncovalent interactions that hold proteins in their functional conformations. A number of essential cellular processes involve the transient exposure of interactive protein surfaces to the intracellular environment with the consequent risk of errors due to incorrect interactions. (The term interactive surfaces refers to any regions of intra- or intermolecular contact between parts of an asembled protein-containing structure that are important in maintaining that structure.) Such processes include:

1 PROTEIN SYNTHESIS. Polypeptide chains are made by cells in a vectorial fashion so that the N-terminal region appears first, the middle regions later and the C-terminal region last. PROTEIN FOLDING is more rapid than protein synthesis, so if the normal fate of the N-terminal region in the final polypeptide is to combine, for example, with the C-terminal region, it might interact with something else in the cell before the C-terminal region is made. Some chaperones are known to bind to many polypeptides during synthesis of the latter.

2 Protein transport. Proteins that enter and function within organelles such as MITOCHONDRIA, CHLOROPLASTS, ENDOPLASMIC RETICULUM (ER) and bacterial periplasm, traverse the membrane in an unfolded or partially folded conformation (*see* BACTERIAL PROTEIN EXPORT; MITOCHONDRIAL BIOGENESIS; PROTEIN TRANSLOCATION). Such proteins are made on free or ER-bound ribosomes in the cytosol and have to be prevented by chaperones from folding into a conformation that is incompatible with translocation. During translocation, regions of the protein appearing in the lumen of the organelle before transport is complete are also bound by intraorganellar chaperones, and there is some evidence that this binding may provide the driving force for transport.

3 Protein function. There are cases where the normal functioning of oligomeric complexes involves changes in the interactions between the component polypeptides so that regions previously involved in intermolecular contacts are transiently exposed. Cases of this sort include DNA REPLICATION, the recycling of CLATHRIN cages (*see* COATED PITS AND VESICLES), and the assembly of MICROTUBULES.

4 Organelle biogenesis. Many of the proteins found inside organelles such as mitochondria and chloroplasts consist of some polypeptides made inside the organelle itself and other polypeptides imported into the organelle after synthesis in the cytosol. Modulation of their binding properties by chaperones may be required before they are all present together in the same compartment in order to prevent incorrect interactions.

5 Stress responses. Environmental stresses such as high temperature cause the denaturation of proteins and the formation of protein aggregates (*see* HEAT SHOCK). To protect against such stresses and to repair the denatured proteins, cells accumulate proteins called heat shock or stress proteins (Hsps), many of which

Table M2 Proteins regarded as molecular chaperones and their functions

Family	Members/Location	Functions
Nucleoplasmins	Nucleoplasmin Protein XLNO-38 Protein Ch-NO38 Nucleoplasmin S	Assembly of nucleosomes and possibly ribosomes, and ribonucleoprotein particles in eukaryotes
Chaperonins	Chaperonin 60 Chaperonin 10 TCP1 TF55	Folding of newly synthesized and transported proteins in bacteria, plastids, mitochondria, and eukaryotic cytosol
Heat shock proteins 70	Several homologues in the nucleus, cytosol, plastid, mitochondrion, endoplasmic reticulum, and bacteria	Assembly and disassembly of newly synthesized, transported, and oligomeric proteins
Heat shock proteins 90	Several homologues in the nucleus, cytosol, endoplasmic reticulum, and bacteria	Masking of binding sites Folding of polypeptides?
DnaJ protein	Homologues in bacteria and eukaryotic cytosol	Interaction with Hsp 70 and GrpE
GrpE protein	Homologues in bacteria and eukaryotic cytosol	Interaction with Hsp 70
SecB protein	Bacterial cytoplasm	Protein transport in bacteria
Signal recognition particle	Bacterial cytoplasm and eukaryotic cytosol	Prevention of incorrect folding of transported precursor proteins
Prosequences	Several proteases	Correct folding
Ubiquitinated proteins	Precursor ribosomal proteins	Ribosome assembly in eukaryotes
PapD protein	Piliated bacteria	Assembly of bacterial pili
PrtM and PrsA	Bacilli and lactococci	Folding of secreted bacterial proteins
Lim protein	Pseudomonads	Folding of bacterial lipase
Rb protein	Nucleus	Binding of transcription factors
Small heat shock proteins	Many members in eukaryotes	Folding of polypeptides? Prevention of aggregation of stress-denatured proteins?
ExbB protein	Bacteria	Assembly of TonB protein
Prions	Infectious protein particles	Rogue molecular chaperones?

act as molecular chaperones. Many proteins originally identified as stress proteins have turned out to be molecular chaperones which are abundant even in the absence of stress (e.g. the chaperonins and Hsp70 and Hsp90, see Table M2). Thus the stress response can be viewed as an amplification of the basic chaperone function. Not all molecular chaperones are stress proteins, however (e.g. nucleoplasmin and the signal recognition particle).

Molecular chaperones are expected to occur in all parts of the cell where protein assembly occurs, and their precise nature, scope and mechanism of action are subjects of intense current research [7,8]. There is already evidence for the existence of co-translational chaperones, for example the PROSEQUENCES of the bacterial protease subtilisin and the alpha-lytic protease, which assist in the folding of the rest of the polypeptide before they are removed. Thus chaperone function need not necessarily be carried out by a separate molecule. Recent work also suggests that members of different classes of chaperone cooperate in mediating such processes as the transport of polypeptides into mitochondria and assembly into their final functional conformations (*see* MITOCHONDRIAL BIOGENESIS) [6].

R.J. ELLIS

1 Ellis, R.J. & Hemmingsen, S.M. (1989) Molecular chaperones: proteins essential for the biogenesis of some macromolecular structures. *Trends Biochem. Sci.* **14** 339–342.
2 Laskey, R.A. et al. (1978) Nucleosomes are assembled by an acidic protein which binds histones and transfers them to DNA. *Nature* **275**, 419–420.
3 Ellis, R.J. (1990) The molecular chaperone concept. *Semin. Cell Biol.* **1**, 1–9.
4 Ellis, R.J. (1990) Molecular chaperones: the plant connection. *Science* **250**, 954–959.
5 Ellis, R.J. & van der Vies, S.M. (1991) Molecular chaperones. *Annu. Rev. Biochem.* **60**, 321–347.
6 Kang, P-J. et al. (1990) Requirement for hsp70 in the mitochondrial matrix for translocation and folding of precursor proteins. *Nature* **348**, 137–143.
7 Gething, M-J. & Sambrook, J. (1992) Protein folding in the cell. *Nature* **355**, 33–45.
8 Ellis, R.J. et al. (Eds) (1993) Molecular chaperones. *Phil. Trans. Roy. Soc. Lond. B* **339**, 255–373.

molecular clock One of the earliest, but unexpected, results from the comparison of amino-acid sequences of proteins was reported in the mid-1960s by Emil Zuckerkandl and Linus Pauling. This was that the rate of evolution (amino-acid substitution rate) was similar for a given gene on different lines of descent. Why should CYTOCHROME C change at about the same rate in mammals, birds, fungi and plants? As shown in Table M3, the rates for different genes could vary by orders of magnitude but this observation only heightened the mystery.

There are thus two phenomena: a given gene shows a more or less constant rate on different lineages; but genes have different rates within the same lineage. A summary of Kimura's argument is as follows.

Let k^+ be the proportion of mutations in a gene that increase fitness of the organism (i.e. show positive selection), k^0 the proportion that are effectively neutral, and k^- the proportion that have a negative effect on fitness, and $k^+ + k^0 + k^- = 1$. Kimura's neutral theory (*see* MOLECULAR EVOLUTION) suggests $k^0 > k^+$ and for the calculation it is assumed $k^0 \gg k^+$. The value for k^- is immaterial as these mutations will normally be eliminated.

The number of neutral mutations per generation is $\mu k^0 N$, where μ is the mutation rate per generation, and N the size of a haploid population. A simple probabilistic argument shows that the probability of a neutral mutant being fixed in the population is $1/N$. The number of neutral mutations fixed into the population is their number, divided by the probability of being fixed, that is, $\mu k^0 N/N$. This gives the rate of molecular evolution as μk^0, the total mutation rate times the proportion of mutations that are neutral. As long as k^0 and μ are reasonably constant then an approximate molecular clock will result — with the proviso that the 'clock' is stochastic, not deterministic. Any change in function of a gene, including loss of function such as in a PSEUDOGENE, is expected to change the proportion of mutations that are neutral (k^0) and accelerate/decelerate its rate of evolution relative to other genes in an organism. A change in mutation rate is expected to change the rate of all genes.

Reasons for the variation in rates between genes involve the number of functional and structural constraints on the gene product. If there are few constraints on a gene, or on a portion of a gene, then k^0 is high (and thus k^- low) and it will be a rapidly evolving gene. If k^0 is low (and k^- high) it will be a slowly evolving

Table M3 The rate of evolution of different genes

k_{aa}	Rate ($\times 10^9$ year^{-1})
Fibrinopeptides	8.3
Pancreatic RNase	2.1
Lysozyme	2.0
α-Globin	1.2
Myoglobin	0.89
Insulin	0.44
Cytochrome *c*	0.3
Histone H4	0.01

The rate k_{aa} is the probability per year of a single amino acid being substituted.

gene. A consequence is that the functionally most important part of the molecule is the most highly conserved. This conclusion is now used routinely to identify important parts of a new sequence of unknown function when the sequence is available from several organisms.

Kimura, M. (1983) *The Neutral Theory of Molecular Evolution*, 74 (Cambridge University Press).

molecular drive *See*: MOLECULAR EVOLUTION.

molecular dynamics Calculations of the many configurations describing the trajectories of all atoms in a macromolecule were first used to study molecular thermal motion and the structural dynamics of proteins and its relevance to their function. Both computational and theoretical approaches have been used to simulate these structures, usually based on empirical energy functions from which the forces between atoms can be calculated — including both bonded (covalent) and non-bonded (e.g. van der Waals, electrostatic) interactions.

More recently these methods have been increasingly used not to simulate the molecule but as computational tools in the REFINEMENT OF MACROMOLECULAR STRUCTURES. The significant advance was the incorporation of an additional penalty term into the energy function such that, for example, alteration of the distance between two atoms from its target value — or non-planarity of a peptide bond — is energetically unfavourable. Thus molecules need not be rigidly constrained to the target values exactly, but can be restrained to fit them as closely as possible when the ideal values conflict with experimental observations.

Structures determined by NMR methods (*see* NMR SPECTROSCOPY) use experimental distance restraints between identified atoms, which fit directly into restrained molecular dynamics algorithms. Crystallographic structures use observed STRUCTURE FACTORS, with which the calculated structure factors should agree optimally; the incorporation into molecular dynamics algorithms of an additional penalty function (the X-ray term) representing the difference between the observed and calculated structure factors (in RECIPROCAL SPACE) — or equivalently the observed and calculated ELECTRON DENSITY MAPS (in real space) — enabled crystallographic models to be refined simultaneously against both energy and observational (X-ray) restraints.

The relative weighting of the X-ray and energy terms is a parameter needing careful optimization as a refinement proceeds. Other critical parameters include: the effective temperature at which the calculations are performed (*see* SIMULATED ANNEALING) since higher temperatures search a larger area of conformational space but need smaller time intervals in the calculations to avoid excessive geometrical distortions and instabilities; the force field employed and especially the charges and dielectric constant used to mimic the bulk solvent atoms omitted from the dynamics calculations; and varying the RESOLUTION range of the X-ray data which influences the radius of convergence from a starting model. The combination of molecular dynamics with simulated annealing is an extremely powerful analytical tool, now used very widely for effective macromolecular refinements.

Van Gunsteren, W.F. (1983) *Proc. Natl. Acad. Sci. USA* **80**, 4315–4319.

Molecular evolution

THE discoveries of molecular biology can only be understood in an evolutionary context and, in addition, molecular biology has made major contributions to the study of evolution. Amino acid and nucleotide sequences provide the primary data for the study of evolution at the molecular level. They provide large amounts of information and allow quantitative testing of many evolutionary ideas [1].

Molecular studies cover all aspects of DARWINIAN EVOLUTION. The major contributions have included: quantitative studies of the relationships between all forms of life (*see* MOLECULAR PHYLOGENY); showing the nature and amount of genetic variation in populations; demonstrating mechanisms for increasing both the amount of genetic information and molecular complexity; supporting a strong chance (stochastic) element in evolution; helping identify important functional parts of sequences; demonstrating the opportunistic nature of evolution with both a lack of perfection and lack of long-term goals; understanding human origins and evolution; and finally in allowing realistic, and in principle testable, theories for the origin of life. Indeed the molecular perspective is an important reason that many controversies about evolution have decreased markedly within scientific circles. The following sections highlight the main aspects of molecular evolution. For a more detailed treatment of the evolution of protein sequence and structure *see* MOLECULAR EVOLUTION: SEQUENCES AND STRUCTURES.

The neutral theory of evolution

The determination of the amino-acid sequences of proteins, which began with that of insulin in 1955, rapidly provided important information for population genetics and evolutionary biology. Two discoveries in the early 1960s had particular impact. First, there was the finding of several forms of an enzyme in a population (ISOENZYMES) as measured by variation in the electrophoretic mobility of proteins in tissue extracts — indicating a high degree of genetic diversity (GENETIC POLYMORPHISM). Second, there was the discovery from the comparison of amino-acid sequences that the rate of evolution (as judged by amino-acid substitutions) for a particular protein is fairly constant even on different lines of descent (the MOLECULAR CLOCK).

Both observations were unexpected. It had been assumed that, aside from the relatively short periods when one ALLELE was replacing another, there would be a single 'most fit' allele for any gene. Under this assumption selection would be necessary to maintain heterozygosity (polymorphism) in a population. The classic case of such a BALANCED POLYMORPHISM is SICKLE CELL ANAEMIA where the heterozygote has a selective advantage over both homozygotes in malarial conditions and the sickle-cell allele is maintained in the population despite its lethality when homozygous. Calculations showed there should be a limit to the number of such polymorphisms because there would be a selective cost to the population — most individuals would be well below optimum fitness. If there were many such polymorphisms in a population then no individuals would have maximum fitness — hence the concept of GENETIC LOAD. The high level of genetic diversity at the molecular level, though unexpected, is certainly consistent with the Darwinian assertion that new variation is constantly being generated, irrespective of any 'needs' of the organism.

Equally surprising was the similar rate of evolution for a given gene on quite different lines of descent. For morphological evolution it is accepted that some species (e.g. tuatara, king crabs, *Peripatus*, and the maidenhair tree *Ginkgo*) remain virtually unchanged morphologically after hundreds of millions of years whereas other species have changed markedly. It was assumed, incorrectly we now know, that evolution at the molecular level would vary similarly with evolution of macromolecules being slowest in the lineages that changed little morphologically and perhaps also that some proteins would evolve faster in some lineages than in others.

Motoo Kimura demonstrated that both observations are explained if most of the variability within a species, and most of the amino-acid changes between species, are neutral with respect to selection [2]. That is, any two genetic variants (alleles) are equally suitable. This is definitely not to say the variant alleles have no function or that they have no effect on the protein. All that was assumed is that they were equally suitable, they were neutral with respect to each other. An illustration could be two amino acids in the active site of an enzyme which have to be a certain distance apart but it does not matter which amino acids fill the space between. The spacer sequence would be essential to the activity of the protein, but neutral with respect to which amino acids fill the gap. A summary of Kimura's calculation leading to the molecular clock is given under the entry MOLECULAR CLOCK.

Similar arguments show that neutral mutations can explain the high degree of genetic diversity in populations. A neutral allele may persist in the population for many generations until lost by random processes. Conversely, a neutral allele that is eventually fixed into the population takes more generations to become fixed if the population is large. The theory predicts the observed high levels of heterozygosity (genetic variability) in populations and allows estimates of its variance and effects of population size (larger populations are expected to be more variable). Thus the neutral theory accounts for the two unexpected observations — the high genetic diversity and the molecular clock. For Darwinian theory, neutralism supports the significance of stochastic processes in evolution and that there is no necessary purpose for genetic variability — both points were controversial initially.

The neutral theory is, from a scientific view, a very powerful theory and many testable predictions can be made from it. Some of the most important are listed in Table M4.

Although exceptions can always be found, the predictions are all generally supported. The neutral theory has been outstandingly successful in understanding evolution. All that is specified is that neutral mutations outnumber advantageous mutations ($k^0 > k^+$, *see* MOLECULAR CLOCK) and that a mutation being neutral is quite different from its being functionless. The concepts of neutralism are now applied to morphological evolution, in particular human variability, and could also be applied to the evolution of human languages.

Table M4 Some predictions of the neutral mutation theory

1. Genes with a stable function and in organisms with a similar mutation rate will show a molecular clock; the expected variance in the rates can also be estimated
2. Conversely, a change in, or loss of, function for a gene will lead to a change in the rate of evolution
3. A high level of genetic polymorphism will exist in populations, it will increase with population size, and its variance can be estimated
4. The total rate of nucleotide change is much greater than predicted by genetic load arguments
5. Evolution will be faster in proteins that are highly variable in existing populations
6. Evolution will be faster in the parts of genes with fewer functional constraints; conversely the rate will be slowest in the functionally important part of molecules
7. Genes no longer being expressed (PSEUDOGENES) must eventually lose recognizable homology to the original sequence (loss of disused functions)
8. In coding sequences the third position in triplets should evolve faster than the first two positions, and this rate should approach that of pseudogenes
9. The ratio of changes in the third position of the triplet, to the first and second positions, should be higher in proteins which are evolving more slowly
10. The overall proportion of amino acids in proteins should be similar to the frequency of triplets coding for them
11. Sequences, or portions of sequences, evolving faster than the neutral rate for the species must be the subject of positive selection
12. The rate of evolution will be increased (or decreased) by increases (or decreases) in the mutation rate (this is particularly obvious when studying RNA viruses)

Genome structure

Only a small portion of the genome in multicellular eukaryotes codes for functional molecules such as proteins or tRNA and rRNA. While some of the remainder presumably has a regulatory role, much of it appears to be without a discernible function that is beneficial to the genome or organism as a whole. Much of it, particularly in organisms with a longer life cycle, is taken up by repeated sequences, both clustered and dispersed (*see* GENOME ORGANIZATION). More detailed analyses suggest that there exist within lineages (species) mechanisms for both sequence homogenization and, on an evolutionary timescale, for rapid expansion or contraction of these repeated sequence families. Because little or no specific function can be assigned to these sequences it is suggested they are parasitic or 'selfish' DNA.

Mitochondria and chloroplasts both have small amounts of their own DNA (*see* CHLOROPLAST GENOMES; MITOCHONDRIAL GENOMES). The sequences of this DNA show clear prokaryotic affinities, confirming the endosymbiotic origin of these organelles.

Origin of new information

A major problem for classical evolutionary theory was how genetic information increased with time. This was called by Darwin the 'oyster to an Alderman' problem and more recently by Karl Popper the 'amoeba to an Einstein' problem. Again, studies of molecular evolution have provided a mechanism.

Some of the earliest proteins to be sequenced were the α-and β-globin subunits of HAEMOGLOBIN and the related protein MYO-GLOBIN. These polypeptide chains are similar both in sequence and in three-dimensional structure (*see* HAEMOGLOBINS; MOLECU-LAR EVOLUTION: SEQUENCES AND STRUCTURES). The simplest hypothesis for their evolution is that a duplication of a gene for a globin molecule occurred, followed by divergence of the copies to give a myoglobin and a single-chain haemoglobin. Repeating the cycle of gene duplication and divergence with the simple haemoglobin gene gave the α and β chains of present-day tetrameric vertebrate haemoglobins (*see* HOMOLOGY). It has been pointed out that this strategy follows Darwin's model of 'descent with modification'. This mechanism allows the increase in genetic information over time, with duplicated copies of a gene eventually evolving to a very different function in some cases. The existence of numerous MULTIGENE FAMILIES, some of which have been analysed in great detail at the protein and DNA level to unravel the mechanisms of their evolution, supports the idea of gene duplication and divergence as a major mechanism of evolution.

Mutations that would be lethal if only a single copy of the gene were present will usually be neutral if there is a second unaffected copy of the gene and may therefore become fixed into the population by the random process described by the neutral theory. On rare occasions a duplicated gene may incur mutations that are useful and then positive selection may lead to two (paralogous) copies of the original gene, each with a slightly different function.

One prediction of the model is that we should find duplicated copies of genes that have failed to evolve to a new function; these should be in the majority if the duplications are occurring without 'purpose' (*see* DARWINIAN EVOLUTION). Indeed, sequences homologous to existing genes but which, for a variety of reasons, cannot produce a functional gene product are found. Many of these PSEUDOGENES apparently result from a gene duplication but others (which lack INTRONS) may be the result of a reverse transposition event in which an mRNA has been copied into DNA by the enzyme REVERSE TRANSCRIPTASE and the DNA has become integrated into the genome.

Duplication of whole genes is not the only mechanism for increasing genetic information. Duplication of a part of a gene and recombination between genes can lead to novel and/or improved products. The first mechanism is illustrated by the fibrous protein collagen where a motif of three amino acids (the collagen repeat) has apparently been duplicated many times to give the characteristic long helical portion of a collagen polypeptide chain. The basic collagen gene structure has become duplicated and modified to give rise to a large family of these ubiquitous animal proteins (*see* EXTRACELLULAR MATRIX MOLE-CULES). A similar mechanism has been suggested for the origin of the electron carrier FERREDOXIN, a globular protein.

With the discovery that many eukaryotic genes are composed of exons and introns (*see* EUKARYOTIC GENE STRUCTURE), together with the observation that exons often correspond to protein structural and functional domains, came the suggestion that new proteins and functions could arise by duplication and/or re-arrangement of exons. The best studied examples include the protein factors involved in the cascade leading to cleavage of fibrinogen and blood clotting (*see* BLOOD COAGULATION AND ITS DISORDERS), and also the serine proteinase inhibitors (SERPINS). In both examples the proteins are composed of strings of domains encoded by exons already recognized in other proteins.

Finally, numerous microevolutionary studies of microbes in the laboratory demonstrate that genes, by simple stepwise mutational processes, may evolve into altered, or even new, functions. These studies show that a few point mutations can lead to modified enzymatic activity, many mutations are disadvantageous, and that others (as predicted by the neutral theory) have no discernible effect. When taken together with the potential for gene duplication, the molecular evolutionary basis for expansion and diversification of the genome is clearly laid out.

Many families and superfamilies of proteins are now known. In addition to the globins they include the IMMUNOGLOBULIN SUPERFAMILY, CALCIUM-BINDING PROTEINS, ACTINS, several families of proteases (e.g. SERINE PROTEINASES), G-proteins) (*see* GTP-BINDING PROTEINS) and various receptor superfamilies (*see e.g.* G PROTEIN-COUPLED RECEPTORS). New functions have arisen by duplication and divergence to give insulin-like protein hormones, the eye lens crystallins, and snake venoms. Individual proteins in these latter two groups have arisen independently from different precursors. Many newly discovered coding sequences are found to be related to known gene families although clearly many interesting ones remain to be discovered and their functions understood.

Metabolism

The comparative study of metabolic pathways has provided a variety of useful insights into relationships between organisms. A complete pathway is an integrated feature of an organism and consequently the possession of the same pathway by two organisms is an indication of relatedness. However the lack of a pathway is a less powerful diagnostic of unrelatedness as it may be that one species has simply lost that feature. For microorganisms, in the absence of much morphological data, microbiologists have relied heavily on comparative metabolic data.

Comparative data of this kind is used to deduce the metabolism of the earliest living organisms and has allowed the conclusion that they would have been able to synthesize tetrapyrroles and terpenes but possibly not fatty acids. The universal presence of ribonucleotides in cofactors for electron and group transfer is consistent with their early utilization in metabolism.

Going back to even earlier forms of life, the origin of metabolism remains contentious, with competing scenarios based on heterotrophic, phototrophic and chemotrophic sources of energy for the origin of life. One version of the heterotrophic model envisages a 'hot dilute soup', particularly of amino acids formed by known mechanisms under the nonoxidizing conditions of the early earth. Metabolic pathways could develop stepwise as the oceans became depleted of intermediates. More recent scenarios focus on energy-yielding cycles based on iron- and sulphur-containing organic molecules bound on surfaces.

Consideration of the nature of early organisms focuses on the problem of why some reactions proceed by a particular mechanism when other mechanisms would seem from a 'design' perspective to be better. This observation has led to the suggestion that some features of contemporary cell chemistry must be relics of earlier states, rather than being the optimal chemical mechanism for the reaction. Our current biochemical knowledge is that proteins are the most effective biological catalysts but there is continuing discovery of ribonucleotides and oligoribonucleotides playing vital roles in metabolism. This has led to the suggestion of an early RNA-based form of life.

Origin of life: an RNA world?

Nucleic acids were once considered to be involved mainly in information storage with proteins carrying out all catalytic functions. There were early suggestions that RNA would have been particularly important in the early stages of the origin of life but the dramatic discovery of the catalytic activity of some RNA molecules (*see* RIBOZYMES) gave a major impetus to the idea that the first organic living system was composed of RNA molecules rather than proteins — the RNA world. The 2′ hydroxyl group of ribose, which is absent from DNA, is very reactive and important for catalytic activity.

The existence of an early RNA world is consistent with the present-day roles for RNA and its precursor ribonucleotides in: information storage (mRNA); translation (tRNA); structure (rRNA); primers for DNA synthesis (new fragments of DNA require an RNA primer to initiate DNA synthesis); precursors of DNA synthesis (as ribonucleotides for deoxyribonucleotides); catalytic activity (ribozymes); ribonucleotide enzyme cofactors (FAD, NAD, NADP); and the biosynthesis of some functionally important amino acids (such as histidine) [3].

The existence of a possible RNA world increases interest in earlier theoretical work on properties of RNA systems [4]. Calculations show that the error rate of replication places an upper limit on the size of RNA molecules (the Eigen limit). If the molecules exceed a threshold number of errors per replication then randomization would occur. Below the threshold a stable 'quasispecies' will be maintained with a population of sequences varying around the consensus or 'master' sequence. Other biochemical 'problems' are lessened if an RNA system was the first form of life. The average size of exons (30–40 amino acids) is similar to the Eigen limit for RNA replication in the absence of catalysis. The exon structure of eukaryotic genes is accounted for (it is difficult to see advantages in developing them later). The problem arising from mirror image D- and L- forms of amino acids is removed in that proteins would only be integrated into early metabolic systems after RNA catalytic activity could distinguish the mirror images.

Large gaps remain in our knowledge of the many steps from an inorganic world to an organic living system. Nevertheless the information from molecular biology changes the problem from one that appeared insoluble to one where the expectation is that it will be solved experimentally.

Human evolution

At the other end of the scale, molecular evolutionary studies have thrown light on human evolution. They have contributed to knowledge of the origin and relationships of the human species with other primates and have also been used to analyse the more recent migrations of *Homo sapiens* across the world. From the early 1960s Morris Goodman and then Allan Wilson reported the high degree of biochemical similarity between humans (*Homo*), chimpanzees (*Pan*) and gorillas (*Gorilla*). This early work compared the strength of immunological reactions between proteins and antibodies but soon sequences of protein, and later of nucleic acids, reinforced the conclusion. In general there is 98.5–100% identity between gene sequences in these three species. This difference is similar to that found between species within a single genus of mammals, such as within *Equus* (horse, donkey, zebras) or *Mus* (mice). The current hypothesis is that humans and chimpanzee last shared a common ancestor 4–7 million years ago and that the gorilla separated from this lineage somewhat earlier.

The other aspect of human evolution that is amenable to analysis is the expansion of *Homo sapiens*. Mitochondrial DNA is only about 1 part in 10^5 of the human genome but is important for comparing populations because it has a high mutation rate, and therefore high variability. It is also transmitted from generation to generation through the maternal line without genetic recombination. The results have led to the conclusion that, although the earlier *Homo* species, *H. erectus*, spread out of Africa a million years ago, *H. sapiens* arose from within African populations and only spread to the rest of the world in the past 150 000 years or so. This is the 'Out of Africa' hypothesis. The data is still incomplete but additional sequences strengthen the idea [5]. The calculations assume the variant DNA sequences are 'neutral' and it is considered just chance which populations arrived at those parts of Africa that allowed them to migrate to other continents.

Studies of molecular evolution are particularly important for elucidating those aspects of evolutionary theory that are difficult to establish convincingly from other types of evidence. These include the common ancestry of all life; the high level of genetic diversity in populations; the strong stochastic (chance) element in evolution; a mechanism for the increase in genetic information and therefore of complexity; the lack of perfect design in organisms; the lack of any long-range plan or purpose in biology and no predetermined pathways of development; and finally that humans are very much a part of the biological world.

It is now apparent that DNA and protein sequences carry a large amount of information about their history, information about common ancestry, about past events (such as an RNA world), about duplications and about development of new functions by members of a gene family. Decoding this information continues to reveal insights into the evolution of life from its very beginning. With molecular data, evolutionary studies have become a 'normal' quantitative science.

D. PENNY

E. TERZAGHI

See also: GENETIC CODE.

1 Terzaghi, E. et al. (1984) *Molecular Evolution: An Annotated Reader* (Jones and Bartlett, Boston).

2 Kimura, M. (1983) *The Neutral Theory of Molecular Evolution* (Cambridge University Press, Cambridge).

3 Joyce, G.F. (1989) RNA evolution and the origins of life. *Nature* **338**, 213–224.

4 Eigen. M. et al. (1981) The origin of genetic information. *Sci. Am.* **244**(4), 78–94.

5 Vigilant, L. et al. (1991) African populations and the evolution of human mitochondrial DNA. *Science* **253**, 1503–1507.

Molecular evolution: sequences and structures

WITHIN the broad biochemical unity of life on Earth there is at the molecular level a great diversity. This has been brought into sharp focus by studies of the relationships among nucleotide sequences of GENES, amino-acid sequences of proteins and the three-dimensional structures and functions of proteins (*see* PROTEIN STRUCTURE). DNA SEQUENCING techniques now provide a direct window on the genome, and X-RAY CRYSTALLOGRAPHY a direct view of the atomic structures and functional mechanisms of proteins. The availability of very large amounts of quantitative information in each of these categories now enables detailed studies of the mechanism of evolutionary change on the molecular level [1–4].

Studies of molecular evolution include some questions that relate molecular data to the results of traditional macroscopic evolutionary studies: in what ways do molecular data confirm, supplement or modify our understanding of traditional phylogenetic relationships derived from comparative anatomy, embryology and the fossil record? (*See* MOLECULAR EVOLUTION; MOLECULAR PHYLOGENY.) Other questions are meaningful only in terms of molecular data: how do nucleotide sequences of DNA and amino-acid sequences of proteins vary within and between individuals, populations and species? What kinds of selective constraints can be observed to apply to DNA and protein sequences and structures?

We now recognize that the primary events in the generation of biological diversity are the substitution, insertion or deletion of nucleotides in DNA and the larger-scale transposition of pieces of genetic material (*see* CHROMOSOME ABERRATIONS; MUTATION; TRANSPOSABLE GENETIC ELEMENTS), and that selection reacts to protein function as determined by protein structure. Thus if a 'wild-type' gene produces a functional protein product, a mutant gene may produce either an alternative protein of equivalent function (a neutral mutation), a protein that carries out the same function but with an altered rate or specificity profile, a protein with an altered function, or a protein that does not function — or even fold — at all. Changes in the genome that alter molecules involved in control of genetic activity rather than changing the sequence of the mature gene product — regulatory mutations — are also very important in evolution.

Examination of HOMOLOGOUS genes and proteins in different species shows that evolutionary variation and divergence occur very generally at the molecular level, and that to a certain extent they have taken place in parallel with the divergence of macroscopic features of species. Proteins from related species have similar but not identical amino-acid sequences. These sequences determine similar but not identical protein structures. Even 30 years ago, comparison of the structures of haemoglobin and myoglobin — the first protein structures determined by X-ray crystallography — showed that although the amino-acid sequences had diverged, the basic qualitative 'fold' of the globin polypeptide chain was retained, although there were quantitative changes in structural details (*see* HAEMOGLOBINS).

Direct access to the genome: nucleotide sequences

The determination of DNA sequences has revealed many new and unsuspected features of the contents and organization of the genome.

Most eukaryotic genes are interrupted by stretches of nontranslated DNA, called intervening sequences or introns, which are deleted from the messenger RNA; the regions expressed are called exons (*see* EUKARYOTIC GENE STRUCTURE; RNA SPLICING). Introns seem to be fairly stable components of genes. Perhaps the best-studied family of genes is that of the globins. All functional vertebrate GLOBIN GENES — including those of humans, rabbit, mouse, chicken, and the frog *Xenopus laevis* — contain two introns at homologous positions. The intron–exon pattern of the globin genes has therefore been constant since before the divergence of myoglobin and haemoglobin, about 600–800 million years ago. However, the gene for LEGHAEMOGLOBIN in soybean has three introns, two of which correspond in position to those of the vertebrate globin genes; and the genes for insect globins have none.

In some cases, exons represent structural components of proteins that can be recombined in different contexts, as a mechanism of development of new proteins; this is called exon shuffling.

Evolutionary changes in protein sequence

Figure M32 shows sets of aligned globin sequences (*see* ALIGNMENT). Such tables are the basic tool of molecular evolutionary studies. The patterns of conservation and variation at the individual positions provide clues to the nature of the selective constraints on the molecule, more directly even than the structure itself does. Figure M32*a* shows the sequence of five mammalian globins: human and horse α and β chains, and sperm whale myoglobin. Figure M32*b* contains globins from much more diverse species.

The five mammalian globin sequences shown in Fig. M32*a* each contain around 140–150 residues. There are 25 positions at which the residue is conserved in all five sequences, including the two histidine residues that interact with the haem iron. Other positions show changes only among residues with very similar physicochemical properties: for example, position 3 contains only Ser or Thr; position 119 only Val, Ile, or Leu. But other positions contain residues varying widely in side-chain size and polarity; for example position 32 contains Glu, Gly, or Ile.

The many similarities leave no doubt that the sequences are related. Moreover, on the basis of the patterns of residue conservation and change at different positions, there is an obvious hierarchical classification. Position 32, for example, contains Glu in the human and horse α chains, Gly in the human and horse β chains, and Ile in myoglobin. The reader can easily identify other such positions. These show that the α chains are more similar to

each other than to the β chains or to myoglobin, and the β chains are more similar to each other than to the α chains or to myoglobin, but the α and β chains are more similar to each other than either class is to myoglobin. Given another mammalian globin sequence, it would be easy to identify it as a haemoglobin α chain, a β chain or a myoglobin.

Another interesting feature of the patterns of conservation is the numerous pairs of conserved residues separated by three, four or seven residues in the sequence. This indicates that they are in α-helices (but note that this inference is available only through the comparison of related sequences).

The sequences in Fig. M32*b* include the same five mammalian globins and additional homologues from an insect, plant and bacterium. These are much more diverse — the bacterial globin is the most distant from the others — and indeed only three positions are conserved in all eight sequences. The distal histidine (position 66) is one of several residues conserved in all the sequences shown except bacterial globin; however, there are numerous other globins known in which this residue has changed.

It is interesting to examine the residue variation at several positions in Fig. M32*b*. Position 3, which was limited to Ser and Thr in the mammalian globins, is still limited to Ser and Thr in all but the bacterial globin. Position 119, which varied only conservatively (Val, Ile or Leu) in the five mammalian sequences, contains no additional residues in the nonmammalian sequences shown. Position 32, which contained Glu, Gly, and Ile (already showing minimal constraint) is occupied in the nonmammalian sequences shown by Tyr, Ile, or Lys.

It is possible to construct 'evolutionary trees' from tabulations of related sequences (*see* MOLECULAR PHYLOGENY). Phylogenies derived from different families of proteins from the same range of species are generally mutually consistent, and also consistent in branching order, with phylogenetic trees based on classical methods. It is, of course, essential to choose functionally equivalent proteins; an attempt to derive mammalian phylogenetic relationships from globin sequences would have to be based on haemoglobin α chains taken by themselves, β chains by themselves or myoglobins by themselves (or even better on all three sets separately) but could not be carried out by mixing α chains from some species with β chains from others.

A subject that has received much attention in studies of molecular evolutionary trees is that of the MOLECULAR CLOCK — the idea that amino-acid substitution proceeds at a constant rate within individual protein families. If this were true, not only could we infer from substitution patterns such as those shown in Fig. M32 the topology of the evolutionary tree, but we could translate the extent of the divergence of the sequence into actual times of divergence from the latest common ancestor. Not only could such a calibration of evolutionary clocks be used to date events in biological history, it would also imply that rates of amino-acid substitution seem to be independent of rates of morphological change and of variations in selective pressure, and therefore appear to be nonadaptive. Indeed, recent studies of protein evolution based on analysis of three-dimensional structures show that adaptive changes in closely related species are brought about by only a few amino-acid substitutions, which amount to so small

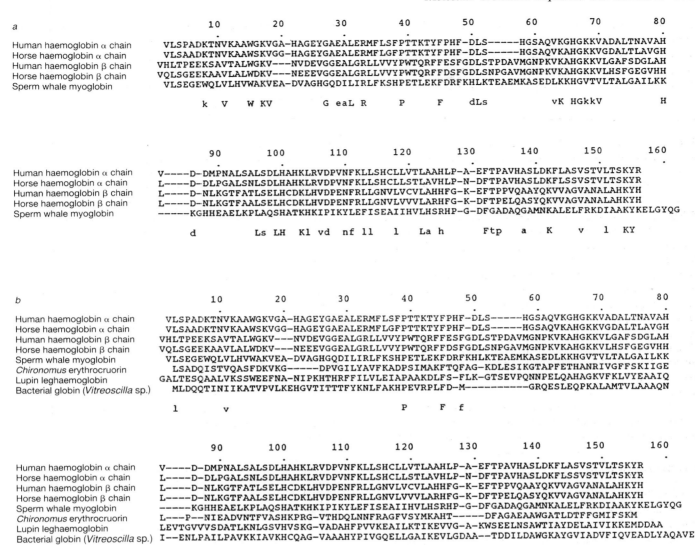

Fig. M32 Comparison of homologous amino-acid sequences of different members of the globin superfamily of proteins. *a*, Conserved residues are indicated in the line below the tabulation. Upper-case letters indicate residues that are conserved in all sequences, lower-case letters residues conserved in the haemoglobins but not the sperm whale myoglobin. The numbers refer to positions in the alignment, not to sequential residue indices in any particular molecule. Single-letter amino acid notation. *b*, In the line below the tabulation, upper-case letters indicate residues that are conserved in all eight sequences, and lower-case letters indicate residues that are conserved in all but the bacterial globin.

a fraction of the total number of substitutions between species that they do not significantly affect the statistics of amino-acid sequence divergence (see below). A conspicuous success of molecular clocks was the correct dating of the time of the human–African ape divergence, in apparent disagreement with palaeontological evidence; it was later accepted that the date derived from sequence analysis was correct. Nevertheless, considerable evidence has accumulated against the constancy of molecular clock rates across an entire protein family over longer periods of geological time.

Variation in selective constraints in protein molecules

What is undoubtedly true is that evolution runs at different rates in different protein families, and even in different regions in a single protein structure. Residues in proteins that are subject to strict functional constraints or that have crucial structural roles can accommodate mutations less easily than other residues. Different overall rates of evolution thus depend on the fraction of residues to which the structure and function are sensitive. For example, FIBRINOPEPTIDES, regions excised and discarded during the maturation of FIBRINOGEN and therefore subject to minimal functional constraint, have the fastest rate of evolution of any known family of proteins. This point is also seen in Table M5 which shows aligned sequences of mammalian proinsulins. The C peptide sequences, which are excised and do not appear in the mature functional hormone, show much higher variability than

Table M5 Alignment of the amino-acid sequences of the proinsulins of humans, pig, cow, guinea-pig, and rat

B-chain sequences

Human	FVNQHLCGSHLVEAL Y LVCGERGFFYTPKT
Pig	FVNQHLCGSHLVEAL Y LVCGERGFFYTPKA
Cow	FVNQHLCGSHLVEAL Y LVCGERGFFYTPKA
Guinea-pig	FVSRHLCGSNLVETL Y SVCQDDGFFY IPKD
Rat	FVKQHLCGPHLVEAL Y LVCGERGFFYTPKS

C-peptide sequences

Human	RREAEDLQVGQVELGGGPGAGSLQPLALEGSLQKR
Pig	RREAENPQAGAVELGGG - - LGGLQALALEGPPQKR
Cow	RREVEGPQVGALELAGGPGAGGL - - - - - EGPPQKR
Guinea-pig	RRELEDPQVEQTELGMGLGAGGLQPL - - QGALQXX
Rat	RREVEDPQVPQLELGGGPEAGDLQTLALEVARQKR

A-chain sequences

Human	G IVEQCCTS ICSLYQL ENYCN
Pig	G IVEQCCTS ICSLYQL ENYCN
Cow	G IVEQCCASVCSLYQL ENYCN
Guinea-pig	G IVDQCCTGTCTRHQL QSYCN
Rat	G IVDQCCTS ICSLYQL ENYCN

The proinsulin chain, a single polypeptide comprising the B, C, and A sequences, folds within the lumen of the endoplasmic reticulum, and disulphide bonds are formed between its A and B chains. The double basic residues at the ends (RR and KR) signal cleavage of the (connecting) C peptide, which has no hormonal activity. The alignment shows the close similarity of the A and B chains which make up the mature and functional hormone, and the somewhat lesser similarity in the C peptides.

the A and B chains, which make up the active molecule.

These cases are fairly straightforward. However, it is not always possible to say, *a priori*, what the effect of a mutation will be. The mutation in human haemoglobin responsible for SICKLE-CELL ANAEMIA (*see* HAEMOGLOBIN AND ITS DISORDERS) is a change from Lys to Val on the molecular surface, the clinical consequences of which could not be inferred from knowledge of the protein sequence and structure.

The existence in the genome of multiple copies of genes, capable of independent evolutionary variation, also affects rates of change. Multiple copies may be in the form of different alleles in diploid organisms or in the form of MULTIGENE FAMILIES. Gene families are often closely linked multiplets of similar DNA sequences which are differently expressed in different tissues or at different stages of development (*see* HAEMOGLOBIN AND ITS DISORDERS).

Evolution of protein structures

Included in the approximately 1000 protein structures now known are several members of families in which the molecules maintain the same basic folding pattern over ranges of sequence homology from near identity down to less than 20%. In both closely and distantly related proteins, the general response to mutation is conformational change. The maintenance of function in widely divergent sequences requires the integration of the response to mutations over all or at least a large portion of the molecule.

The structure of a protein depends on interactions among residues buried in its interior which form elements of secondary structure — helices and sheets (*see* PROTEIN STRUCTURE). Hydrophobic residues of helices and sheets form complementary surfaces that can pack efficiently against each other. In the globins, for example, there are five major helix–helix interfaces buried in each monomer.

The folding of all or at least most of the chain is necessary even if only a small minority of residues is actually responsible for function, because it is essential to bring the crucial residues into the proper relative spatial disposition. Thus function is typically achieved by the creation of a binding site of suitable size, shape and charge distribution to interact with ligands, and in some cases by creating a pathway for conformational changes that the protein undergoes during its active cycle. Examples of such conformational changes are the allosteric transitions of haemoglobin, phosphofructokinase, phosphorylase, and aspartate transcarbamylase (*see* ALLOSTERIC EFFECT; COOPERATIVITY), and the closure of interdomain clefts upon changes in state of ligation in hexokinase, citrate synthase and certain other enzymes.

Natural variations in families of homologous proteins that retain a common function reveal how the structures accommodate changes in amino-acid sequence. The ability of protein structures to accommodate mutations in nonfunctional residues permits a large amount of apparently nonadaptive change to occur. Residues active in function, such as the proximal histidine of the globins or the catalytic serine, histidine and aspartate of the SERINE PROTEINASES, are resistant to mutation because changing them would interfere, explicitly and directly, with function. Most buried residues are in the well-packed interfaces between helices and sheets. For structure to be maintained the buried residues must remain hydrophobic, but can change in size. Mutations that change the volume of buried residues generally do not change the conformations of individual helices or sheets, but produce distortions of their spatial assembly. These tend to take the form of rigid-body shifts and rotations, which may be as large as 7 Å, but more typically are 3–5 Å. Surface residues not involved in function are usually free to mutate. Loops on the surface can often accommodate changes by local refolding.

The nature of the forces that stabilize protein structures sets general limitations on these conformational changes; other constraints derived from function vary from case to case. In some protein families large movements are coupled to conserve the structure of the active site (e.g. in the globins); in others, active sites of alternative structure are found (e.g. in the CYTOCHROMES *c*). In proteins that for functional reasons cannot tolerate conformational change — such as those with multiple binding sites that must maintain a relative spatial disposition — amino-acid sequences are highly conserved.

Families of related proteins tend to retain similar folding patterns. If one examines sets of related proteins (see Figs M33 and M34) it is clear that although the general folding pattern is preserved, there are distortions which increase as the amino-acid sequences progressively diverge. These distortions are not uniformly distributed throughout the structure. Instead, in any family of proteins there is a core of structure that retains the same qualitative fold, and other parts of the structure that change conformation radically. To explain the idea of the common core of the two structures, consider the letters B and R. Considered as structures they have a common core, which corresponds to the letter P. Outside the common core they differ: at the bottom right B has a loop and R has a diagonal stroke.

Figure M33 illustrates the enzymes actinidin and papain, two quite closely related structures. These molecules have 49% residue identity in the common core, which consists of almost the entire structure except for small loop regions on the surface. The structural deviation is very small: the Cα atoms of the residues of the common cores of the two proteins can be superposed to within an average deviation of 0.77 Å.

Figure M34 shows the two distantly related copper-binding proteins, plastocyanin and azurin. The common core here is limited to <50% of the structure and the long loop at the left has entirely refolded. Nevertheless, the selective constraint on function has preserved the geometry of the copper-binding site.

Systematic studies of the structural differences between pairs of related proteins have defined a quantitative relationship between the divergence of the amino-acid sequence of the core of a family of structures and the divergence of the three-dimensional conformation. As the sequence diverges, there are progressively greater distortions in the main-chain conformation, and the fraction of

the residues in the core decreases. Until the fraction of identical residues in the sequence drops below about 40–50%, these effects are relatively modest: almost all the structure remains in the core, and the deformation of the main-chain atoms is on average no more than 1 Å. Actinidin and papain illustrate this regime (Fig. M33). With increasing sequence divergence, some regions refold entirely, reducing the size of the core, and the distortions of the residues remaining within the core increase in magnitude. Plastocyanin and azurin illustrate this effect (Fig. M34).

Figure M35 shows a comparison of pairs of homologous proteins from related families, including globins, cytochromes *c*, immunoglobulin domains, serine proteases, lysozymes, sulphydryl proteases, dihydrofolate reductases, and plastocyanin and azurin. Each point corresponds to a pair of proteins. After determining the core of the structure, and the number of identical residues in the aligned sequence of the core, the root-mean-square deviation of the main-chain atoms of the core was calculated. Figure M36 shows that change in the fraction of residues in the core as a function of sequence divergence. Distantly related proteins can vary; in some cases the fraction of residues in the core remains high, in others it can drop to less than 50% of the structure.

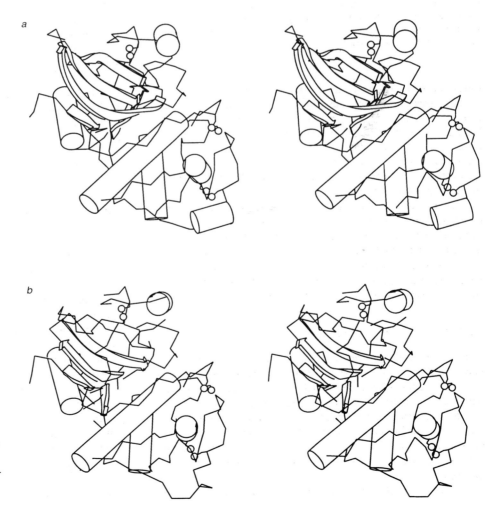

a

b

Fig. M33 Stereopair images of two closely related proteins. *a*, Actinidin (crystal structure by E.N. Baker); *b*, papain (crystal structure by I.G. Kamphuis, J. Drenth and E.N. Baker). The amino-acid sequences of these molecules have ~50% identical residues. (*See also* Plate 4*a*).

a

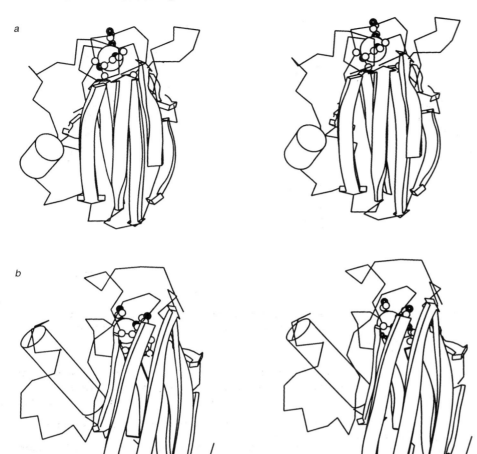

b

Fig. M34 Stereopair images of two distantly related proteins. *a*, Poplar leaf plastocyanin (crystal structure by J.M. Guss and H.C. Freeman); *b Alcaligenes denitrificans* azurin (crystal structure by G.E. Norris, B.F. Anderson and E.N. Baker). The large circle near the top of the structure marks the position of the copper. In this case the double β-sheet portion of these molecules retains the same fold, but the long loop at the left changes its conformation completely. Note that the helices in the loop are nonhomologous.

Evolution of proteins with altered function

There is some information about the way in which protein structures alter existing functions or develop new ones, although it tends to be more anecdotal than systematic. It is clear that many point mutations do have functional significance. A classic case is the specificity of the serine proteases. These proteins contain a surface cleft complementary in shape and charge distribution to the side chains next to which they cleave the polypeptide chain. At the bottom of the specificity pocket in trypsin there is an aspartic acid that is complementary in charge to the lysine and arginine side chains; trypsin cleaves peptide bonds adjacent to these amino acids. In contrast, the pocket of chymotrypsin is lined with hydrophobic residues, corresponding to the cleavage specificity for peptide bonds adjacent to large hydrophobic side chains such as phenylalanine. Other mutations exert more subtle effects in changing the shape of the pocket.

Haemoglobin provides another example of evolution by means of point mutations. The allosteric properties of haemoglobin, especially the regulatory responses to ligands other than oxygen, are brought about by substitution of a few amino-acid residues in key positions in the 'basic' globin unit represented by myoglobin. Most amino-acid substitutions in vertebrate haemoglobin evolution appear to be functionally neutral. For example, human adult and foetal haemoglobins differ by the replacement of Ser (position 143 in adult β-globin) for His (in foetal haemoglobin). As a result, the lower affinity of foetal haemoglobin for the regulatory ligand 2,3-DIPHOSPHOGLYCERATE (DPG) results in a higher effective oxygen affinity than adult haemoglobin and promotes the transfer of oxygen across the placenta to the foetus. (The intrinsic oxygen affinity of foetal haemoglobin in the absence of DPG is in fact lower than that of adult haemoglobin.)

There are numerous examples of recruitment of old structures to perform new functions. A folding pattern found in a large number of enzymes is the closed eight-stranded β barrel. This structure contains eight β-α units with the strands closed into a roughly cylindrical sheet and the helices packed on the outside (*see* PROTEIN STRUCTURE). First seen in triose phosphate isomerase, this fold, or very similar variants, has now appeared in more than 20 enzymes. In many cases the sequence similarity is

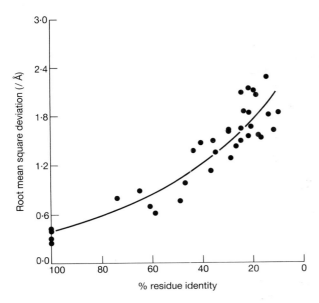

Fig. M35 The relationship between the divergence of the amino-acid sequence of the core of proteins and the divergence of the main-chain conformation. The points corresponding to 100% residue identity are proteins for which the structure was determined in two or more crystal environments, and the deviations show that crystal packing forces can modify slightly the conformation of the proteins.

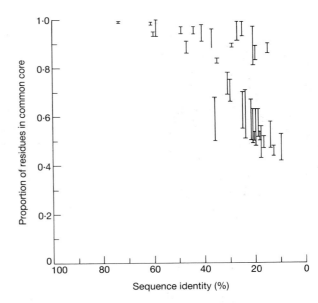

Fig. M36 The relationship between the divergence of the amino-acid sequence of the core of proteins and the relative size of the core.

so low that it is impossible to say whether the proteins are genuinely related by evolution, or are CONVERGENT. However, certain pairs of enzymes sharing this fold clearly show the evolution of new functions. Mandelate racemase and muconate lactonizing enzyme have ~30% residue identity in their sequences but catalyse different reactions. In spinach glycolate oxidase and yeast flavocytochrome *c* also, a few mutations in the active site have sufficed to change the function.

The recruitment of proteins as lens CRYSTALLINS shows another model of evolution: a novel function is acquired preceding divergence. In the duck, an active lactate dehydrogenase and an α-enolase serve as crystallins. In other cases crystallins are closely related to other proteins but gene duplication and some divergence has already occurred.

Gene duplication and divergence also provide mechanisms for generating regulatory control over function, by the development of an asymmetric oligomeric protein. In haemoglobin the tetrameric allosteric structure makes the transfer of oxygen from lungs to tissue more efficient, and creates binding sites between the subunits that can respond to regulatory ligands such as hydrogen or chloride ions and diphosphoglycerate. In contrast, bovine glutamate dehydrogenase, an enzyme containing distinct catalytic and regulatory subunits, has an internal homology that suggests its descent from an earlier molecule containing a catalytic activity only.

Although sequences have revealed a great deal about the course and mechanism of evolution at the molecular level, a basic challenge remains: understanding the relation between the static picture of the genome as revealed in the DNA sequence of an individual cell and the complex set of events involved in the development of the organism for which it is a blueprint.

A.M. LESK

1 Li, W.-H. & Graur, D. (1991) *Fundamentals of Molecular Evolution* (Sinauer, Sunderland, MA).
2 Terzaghi, E.A. et al. (1984) *Molecular Evolution, an Annotated Reader* (Jones and Bartlett, Portola Valley, CA).
3 McEntyre, R.J. (1985) *Molecular Evolutionary Genetics* (Plenum, New York).
4 Evolution of catalytic function (1984) *Cold Spring Harbor Symp. Quant. Biol.* LII.

molecular motors *See*: BACTERIAL CHEMOTAXIS; CELL MOTILITY; DYNEIN; KINESIN; MICROFILAMENTS; MICROTUBULE-BASED MOTORS; MUSCLE; MYOSINS.

Molecular phylogeny

MOLECULAR phylogeny infers evolutionary mechanisms and relationships of organisms, sometimes extinct, by comparing sequences of their nucleic acids and proteins. It does with the GENOME what the traditional comparative biological approach has done with anatomical, morphological, biochemical, and developmental features. Molecular phylogeny is based on the fact that DNA of organisms on different lines of descent accumulates changes resulting from random MUTATIONS. As a general rule, the more divergent two sequences are from each other, the more distant in time it is since they shared a common ancestor.

This general principle can be used to reconstruct evolutionary trees (Fig. M37). The simplest methods of measuring divergence just count the number of differences between pairs of sequences and reconstruct trees from this estimate of GENETIC DISTANCES.

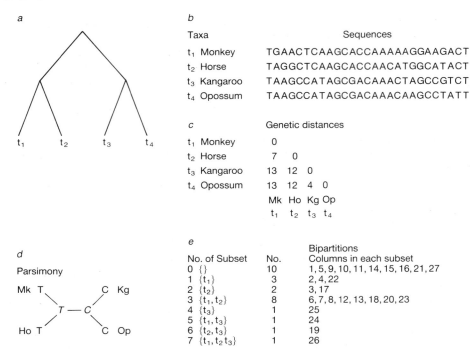

Fig. M37 Basic terminology for tree building. *a*, The terminology used is from graph theory, a field of mathematics. Graphs (including trees) consist of points and edges. Edges may have weights, and points may have names or data associated with them. Common synonyms for points are vertices and nodes. For edges synonyms are lines, internodes or links. A tree is a connected acyclic graph (all points are connected but there are no cycles in the graph). Phylogenetic trees have their pendant points (leaves) labelled, in this case, with names of the taxa from which sequences were derived. In this example the tree is rooted so there is a direction (time) on the edges. Most tree building methods do not give the root directly. *b*, A small example of sequence data with 27 nucleotide positions for four taxa. *c*, Genetic distances derived from the sequences by counting the number of differences between pairs of taxa. Many methods correct these observed distances to allow for multiple changes (including reversals to the original state). For *n* taxa there are $n(n-1)/2$ distance values; with 10 taxa there will thus be 45 values. *d*, The parsimony optimality criterion illustrated with position 6 in *b*. For a particular tree the nucleotides are specified at the terminal points and the nucleotides selected (shown in italics) at the internal nodes which minimize the number of mutations on the tree. Parsimony selects the tree that minimizes the number of mutations summed over all nucleotide positions. *e*, Bipartitions from *b* by counting the number of times subsets of taxa have a different code to the last taxon (t_4). For *n* taxa there are 2^{n-1} bipartitions even with characters having only two states (say purines and pyrimidines); with 10 taxa there will thus be 512 values. The bipartitions are used for a Hadamard (discrete Fourier) analysis method of tree reconstruction which uses the values under 'No.' to calculate the rates of evolution on all possible edges of a tree. Additional information on terminology is given in [5].

But early results from molecular studies found that variations in the methods used could lead to different results and that some of these alternatives would result in different biological conclusions.

Reconstructing evolutionary trees accurately is an exciting challenge for molecular phylogeny. Problems are both computational and theoretical, starting with the selection of sequences and their optimal ALIGNMENT. The computational difficulty results from the large number of possible trees; the theoretical difficulty is in testing whether the tree found is indeed correct. It is not possible to go back in time to check the tree and so ingenuity is required to estimate reliability. Tree reconstruction and evaluation is an interdisciplinary field where computer scientists, statisticians and mathematicians cooperate with biologists to devise more powerful methods.

Advantages of molecular data

The success of molecular biology means that, for the following reasons, sequence data are rapidly supplanting morphology as the primary data for reconstructing the course of evolution.

Wide scope or domain of a character. Any character used in a comparative study has a scope or domain over which it provides useful information. The scope of morphological characters is always limited — they are often helpful within the biological groups where they occur, but outside these groups they do not help. Flowers are useful for classifying flowering plants but botanists have spent centuries arguing about how the parts of the flower could be related to morphological structures in earlier groups of plants. The morphological characters did not solve the problem of the evolutionary relationship of flowering plants to these earlier groups. In contrast, some gene sequences of comparable function and demonstrable sequence and structural similarity can be found in all living organisms.

Large number of characters. Sequences of macromolecules potentially offer large numbers of characters. We are a long way from knowing the 3 billion nucleotides of the human genome (*see* HUMAN GENOME PROJECT) but recent evolutionary studies have used sequences of more than 10 000 base pairs. Even positions that are constant give information on the mechanisms and rates of evolution. Using longer sequences results in more stable trees; they are less likely to change as more data are collected.

Supplying useful characters. Whenever a new sequence is obtained it provides some useful phylogenetic information. With morphological data it is possible for a time-consuming study to fail to find useful characters.

Independence from the fossil record. Some groups are sufficiently well represented in the fossil record to allow reasonably complete estimates of phylogeny. But for most groups, either the fossil record is incomplete and/or there are so few morphological features, particularly for unicellular organisms, that evaluation of detailed relationships is not possible. All living organisms can, however, be studied from their sequences. In addition, DNA sequences can be obtained from the bones and skin of many species (including quaggas, Tasmanian 'tigers', moas, and mammoths) that have become extinct in the past 10 000 years (subfossil remains). In one case DNA has been sequenced from fossils of *Magnolia* leaves 17 million years old. The fossil record, combined with radioactive dating methods, is, however, essential in placing times of divergence on the results of molecular phylogeny.

Easier data handling. With sequences it is easier to develop objective, quantitative and testable methods for data handling and tree building. Using sequences does not remove the problem of ensuring that the two sequences are homologous and are aligned correctly (*see* ALIGNMENT; HOMOLOGY). Ideally, the optimum tree and alternative alignments should be determined simultaneously.

Range of evolutionary rates. Faster evolving sequences are required for taxa which have diverged relatively recently, slower evolving sequences for taxa that are only distantly related. Within any genome there are a wide variety of rates of evolution. Vertebrate mitochondrial DNA evolves fast, as do noncoding regions including PSEUDOGENES. Other sequences, such as conserved regions of ribosomal DNA (rDNA), evolve very slowly and are useful for discerning relationships between taxa that have diverged more than a billion years ago.

Mechanisms of evolution. Most methods for building trees can, for some data sets, give the wrong tree. The more that is known about the genetic mechanisms of nucleotide change (including the random replacement of neutral characters) the easier it is to design tree building methods that avoid these consistent errors.

Cost of obtaining data. In practice this may end up as being the most important. The POLYMERASE CHAIN REACTION (PCR) and the rapid automation of sequencing techniques means that the cost of sequence data keeps diminishing. The labour (time) costs for other sources of data are higher.

Methods for reconstructing evolutionary trees

The large number of approaches that have been used for reconstructing trees can be grouped on the basis of four properties. The first is whether they use (a) genetic distances, or (b) sequences directly.

With genetic distances the amount of change between each pair of species is estimated. Early results involved measurements of the strength of immunological cross-reactions between proteins; now DNA–DNA HYBRIDIZATION measurements are more common. A limitation is that such data may not be symmetric — using an antigen from species A to compare against the protein from species B may not give the same result as using the antigen from species B and comparing it to A. Genetic distances are now commonly estimated directly from sequence data by counting the number of differences between sequences (as illustrated in Fig. M37c). These values will be symmetric but there is a loss of information in going from sequences to distances (Fig. M37).

Distance methods have the apparent advantage of being faster and simpler than those using sequences directly. But this apparent simplicity is largely a result of the loss of information in converting sequences to distances and this loss gets worse with more taxa. The number of distance values increases in proportion to n^2 (where n is the number of sequences) whereas the number of classes in sequence data increases in proportion to $n!$ As such it is not possible to recover sequence information from genetic distances. Distance methods are not the only ones that omit information, parsimony (see later) omits positions where there has been little change though these sites are important for understanding the rates and mechanisms of evolution. Information from insertions and deletions (indels) is often ignored. Transversions may be less common than transitions but this information may be neglected, and so on.

The second property for classifying tree building methods is whether they (c) use an optimality criterion (or objective function) for choosing the best tree, or (d) follow an algorithm without using an optimality criterion (algorithmic methods). The most commonly used optimality criteria are parsimony (for sequences or distances), sum of squares of differences between a tree and observed distances, maximum likelihood, closest tree, and statistical tests based on invariants. The parsimony criterion, illustrated in Fig. M37c, finds the tree requiring the fewest mutations to fit the observed sequences. For maximum likelihood the criterion is, given a mechanism of nucleotide evolution: find the tree and rates of evolution that maximize the probability of getting the observed sequences. In contrast, algorithmic methods follow a set of rules (e.g. join the two species with the smallest genetic distance) rather than use an optimality criterion. They are usually very fast but cannot decide whether other trees are almost as good.

The third property divides tree-building methods into those which (e) evaluate the optimality criterion over all taxa, or (f) evaluate it on subsets of taxa (usually four taxa or quartets) and then look for a tree compatible with all subsets. Most early methods selected the optimal tree based on all sequences in the data set but some recent approaches use quartets of taxa, decide the relationship for each quartet, then reconstruct a tree from these.

The fourth property for tree building divides methods into (g) exact or (h) heuristic. Exact methods consider all trees and so guarantee (given sufficient time) to find the best tree(s) for the optimality criterion being used. This does *not* guarantee a tree is correct, just that it is the best tree for the data and optimality criterion used. A branch and bound algorithm (see later) greatly reduces the time required to find an exact solution but the reduction in time varies with the data. Heuristic methods are approximations, they run quickly and give an answer which, based on previous experience, should be 'close' to the optimum solution.

Why is it that heuristic methods are used when exact methods are available? Most tree-building methods work by a two-step optimization procedure. The first optimization is on a single tree and evaluates the optimality criterion for this tree. The second is to evaluate it for every tree. The real problem is the number of trees. Even ignoring the position of the original ancestor (root), there are $(2n - 5)!!$ trees where n is the number of taxa (sequences) being studied. The double factorial notation (!!) is equal to $1 \times 3 \times 5 \times ... \times (2n - 5)$. An approximation is that $(2n - 5)!!$ $1.41421 \times ((2n - 4)/e)^{n - 2}$. For $n = 20$ there are thus 2×10^{20} trees. A computer program that evaluated 10^6 unrooted trees per second would still require several million years for a complete search of all trees!

The simplest 'exact' method is a total search over all trees, but this is not practical for more than about 10 taxa with 2 027 025 trees. More effective are branch and bound methods. These programs combine a 'branch' procedure that potentially searches all trees and a 'bound' (or limit) which is the best value for any tree known at a particular stage of the search. The methods calculate the optimality criterion (say parsimony) on subtrees (trees with either some taxa and/or some nucleotide positions omitted). If the subtree exceeds the bound then all trees which contain this subtree can be eliminated from the search. This concentrates the search on a small subset of trees. But as the number of taxa increases, except for perfect data, even this approach becomes too slow and is currently limited to about 20 sequences. Branch and bound methods have mainly been used with parsimony but can be used with other optimality criteria.

Heuristic methods cannot guarantee to find the global optimum tree but, based on experience, are expected to find a good tree. How good will remain unknown for a particular case. Many approaches are used for heuristics. Cluster analysis methods, because they are algorithmic methods (see d) can only be used heuristically. Exact methods, such as branch and bound, can be used as heuristics by taking only the best choice at any stage and not examining the less likely options.

'Hill-climbing', or adaptive walks, are a common strategy in heuristics. A small change is made to the existing tree, possibly by 'branch swapping' in the tree. This new tree is accepted if it is better than the existing tree. Branch swapping continues until no better tree is found. One version accepts any better tree and resumes with the new tree. Another version, 'steepest ascent', only accepts the best solution from a round of branch swapping and then resumes the search with this tree. Both approaches are susceptible to being caught in local optima. To broaden the search

the program can restart many times with different trees in an attempt to find the global optimum.

Another group of methods can escape from local optima and accept an apparently worse tree in an attempt to eventually find a better tree. In general, heuristics that allow worse interim solutions are slower than simple hill-climbing methods but do allow the program to explore a wider range of options and do find better trees than a single run with only hill climbing.

Evaluating different methods

Is one of the numerous tree-building methods likely to be the best for most data sets? Or are they effectively equivalent? This is the theoretical/conceptual problem mentioned earlier. The two general approaches to the problem are the pragmatic, which estimates the reliability of trees from a particular study, and the theoretical which studies general properties of tree-building methods.

1 Pragmatic approaches to reliability. The accuracy of results from a particular study cannot be determined directly. The simplest approach, congruence, uses different sets of data from the same taxa and tests whether the same tree is generated. With molecular data different genes can be used, or a sequence split into first, second and third positions in codons, or first and second halves of the molecule. This occasionally works well when the same tree is found in each case. More commonly, different (but very similar) trees are found and then tree comparison methods are used to identify the portions of the trees that are most stable [1].

Random resampling methods are more powerful than simple congruence. Subsets of data are formed by randomly selecting nucleotide positions (columns) from the data. Columns can be sampled with replacement (bootstrapping), or without replacement (jack-knifing). Many subsets are formed, trees calculated from each, and the trees compared. The most powerful versions take subsets of longer and longer sequences and compare the results. As subsets of longer sequences are used we expect that the tree will eventually be stable: this is convergence. Even if convergence has not occurred it is possible to estimate the number of trees close to optimal that are expected to include the correct tree. Resampling methods require extensive computing time, though less than the time required to collect, collate and align the sequences.

2 General properties of a method. Progress has been made on understanding the basis of tree-building methods. An additional criterion, consistency, is now required. Will a particular method always give the correct tree if the sequences are sufficiently long? One of the earliest contributions of theoretical studies was demonstrating, with a model of just four taxa and some combinations of rates of evolution, that parsimony and some distance methods would converge to the wrong tree (Fig. M38). These methods are said to be inconsistent for the model. With five or more taxa, parsimony is inconsistent even with equal rates of evolution, and even with equal edge lengths for larger numbers of taxa. The performance of parsimony can be improved by including taxa that join into what otherwise would be long edges on the tree. A method is said to be inconsistent if there are any cases under a

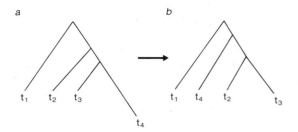

Fig. M38 Convergence to the wrong tree. *a*, The tree used to generate sequence data; it has taxa 3 and 4 joining first. If, however, either the rate of evolution has increased in the lineage leading to taxon 4, or the GC content has increased on both lineages going to taxa 1 and 4, then most methods for reconstructing trees can give the tree shown in *b* with taxa 2 and 3 joining first, or alternatively 1 and 4 joining first. There is a paradox here (Felsenstein's paradox) that with short sequences, because of sampling error, it may be possible to get the correct tree but with longer sequences the wrong tree will always be selected. There is thus convergence to the wrong tree with longer sequences and such methods are inconsistent for the model of evolution used.

given model where it will converge to the wrong tree. It is a separate, but important, question as to whether such cases are common.

The analyses start with a model of evolution and calculate the expected data. Tree-building methods can then be tested to see if they reconstruct the original tree. A model has the three parts: a tree (or graph); a mechanism for change to the sequences; and weights (lengths, or probabilities of change) along edges of the tree. The usual mechanism is that changes are 'independent and identically distributed' (i.i.d.). This means that changes to the sequence anywhere on the tree, and at any site, are independent, and also that all positions (nucleotide or amino acid) have the same chance of changing (identically distributed). Some mechanisms also assume the same rate of evolution along each edge of the tree — the MOLECULAR CLOCK (*see also* MOLECULAR EVOLUTION).

If an optimality criterion is consistent then exact methods using it will converge to the correct tree, provided the assumptions of the model are valid. If the model is too simple then a 'consistent' method could converge to the wrong tree! It will seldom be true, except possibly for PSEUDOGENES, that changes will be identically distributed over all nucleotide positions. The rate of nucleotide interconversion can vary on different lineages, and the frequency of nucleotides (GC content) can change. The important implications of these variables for tree reconstruction are largely unexplored.

The maximum likelihood and closest tree criteria are consistent under the appropriate model. The closest tree algorithm uses a form of discrete Fourier transforms (Hadamard) to estimate the corrected number of changes on each edge of the tree. The method selects the tree which minimizes the distance from the observed values to a tree. Because it works most naturally with two-state characters, its main application thus far has been in studying general properties of tree-building methods but a version for four-state characters (such as nucleotides) finds the optimum tree for up to 20 taxa.

A tree is an integral part of the model. Most methods output a

tree, even if a tree is inappropriate. Events such as lateral transfer of genes, hybrid origin of taxa, 'gene corrections' and recombination between viruses, could all lead to non-tree models. Another problem arises from polymorphism in populations — because the time of separation of two ALLELES (versions of the same sequence) may be older than the time of splitting of species lineages and this can lead to different genes implying different trees. A scientific method must allow the possibility of the data rejecting a tree model.

A limitation of the usual maximum likelihood approach is that it selects either (1) the best tree for a given mechanism of evolution, or (2) the best mechanism to fit a given tree. In practice we are seldom confident of either the tree or the mechanism of change. To illustrate this point we have made a simple model where both simple maximum likelihood and closest tree criteria will converge to the wrong tree because two separate lineages have independently acquired the same increase in GC content (Fig. M38). Problems like these can be detected by comparing properties of the observed sequences with those predicted by the tree that fits the data best.

Two recent approaches, evolutionary parsimony and closest tree (based on Hadamard transforms), at least partially overcome this problem. Evolutionary parsimony may lead to the conclusion in a particular case that *none* of the three binary trees spanning four taxa can be accepted.

Studies with molecular sequences have markedly increased our understanding of the 'Tree of Life'. The twin problems remain of the large number of possible trees and the difficulties of evaluating the reliability of trees. More powerful programs continue to be developed but care is still needed to avoid overstating conclusions. It is not possible yet to conclude if there is a single best method for all data sets. There is a trade-off between accuracy and speed. A much slower exact method may give a marginally better tree than a fast heuristic method.

Several well-tested computer packages are available that implement many of the methods discussed here. The three main ones for trees are Phylip (Phylogeny Inference, J. Felsenstein [2]), PAUP (Phylogenetic Analysis Using Parsimony, D. Swofford [3]) and Hennig86 (J.S. Farris). Of these the first two are particularly designed for sequences. Information on these packages is given in [4]. These make it easier to construct trees from sequences but the newest methods are still difficult to find on computer packages.

D. PENNY

1 Penny, D. et al. (1982) Testing the theory of evolution by comparing phylogenetic trees constructed from five different protein sequences. *Nature* **297**, 197–200.
2 Felsenstein, J. (1988) Phylogenies from molecular sequences: inference and reliability. *Annu. Rev. Genet.* **22**, 521–565.
3 Swofford, D. & Olsen, G. (1990) In *Molecular Systematics* (Hillis, D.M. & Moritz, C., Eds) 411–501 (Sinauer, Sunderland, MA).
4 Sanderson, M.J. (1990) Flexible phylogeny reconstruction: a review of phylogenetic inference packages using parsimony. *Syst. Zool.* **39**, 414–420.
5 Penny, D. et al. (1992) Progress with methods for evolutionary trees. *Trends Ecol. Evol.* **7**, 73–79.

molecular replacement An approach to solving the PHASE PROBLEM in X-RAY CRYSTALLOGRAPHY which relies on knowing the atomic structure of a molecule closely related to that whose structure is being newly determined. If the known structure can first be orientated within the new unit cell (by comparing all possible relative orientations of the two PATTERSON FUNCTIONS — by means of a rotation function) and can then be positioned within the new unit cell (by a search procedure or translation function) a set of trial phases can be calculated for the known molecule in the new cell. Such phases are often sufficient to solve all or part of the new structure.

Rossman, M.G. & Arnold, E. (1993) *International Tables for X-ray Crystallography* Vol. B, Reciprocal space (Shmueli, E., Ed.) 230–263 (Kluwer, Dordrecht).
Rossmann, M.G. & Blow, D.M. (1963) *Acta Cryst.* **16**, 39–45.

molecular weight The mass of a molecule expressed in atomic mass units. *See also*: RELATIVE MOLECULAR MASS.

mollusc development Molluscs, other than cephalopods, all show the characteristic SPIRAL CLEAVAGE also found in the polychaetes and some minor phyla. This similarity of embryonic development is one of the strongest pieces of evidence for the existence of the PROTOSTOME superphylum. Most embryological work has been done using a variety of marine and freshwater gastropods, and also the scaphopod *Dentalium*.

Molluscs (and to a lesser extent, polychaetes) were the subjects of celebrated blastomere isolation experiments by Wilson in 1904, which defined the characteristics of MOSAIC DEVELOPMENT. In contrast to sea urchins, which were known to be able to form a whole larva from a single blastomere of the two-cell stage (*see* ECHINODERM DEVELOPMENT), molluscan blastomeres tended to self-differentiate into partial larvae. The former were called 'regulative' and the latter 'mosaic'. Although still useful as descriptions of the behaviour of part-embryos, the significance of this distinction for mechanism is not now thought to be very great, as all animal embryos are now known to show some degree of localization of CYTOPLASMIC DETERMINANTS, and all have some regional specification by INDUCTION.

The spiral cleavage of gastropods is almost always right handed (dexiotropic) when viewed from the animal pole. But there are occasional local races showing left handed (laevotropic) cleavage and this early difference of symmetry determines the symmetry (right- or left-handed respectively) of the adult snail shell. The inheritance of handedness was studied by Boycott in the 1920s and was one of the first and clearest examples of a MATERNAL EFFECT mutation, as it is the genotype of the mother that determines the sense of cleavage of the egg.

In many gastropod species there is a cytoplasmic protrusion that appears during the early cleavages and is finally incorporated into the D lineage. This is called the polar lobe. Sometimes, as in *Ilyanassa* or *Dentalium*, the lobe at the two-cell stage is as large as the blastomeres, so forming the characteristic 'trefoil' configuration. Since Crampton in 1896, embryologists have been studying the effect of lobe removal, which causes a symmetrization of the embryo and the loss of many structures, some of which are normally formed by the D lineage and others are not. It is generally thought that the lobe cytoplasm contains a cytoplasmic determinant that confers an organizer function on the D lineage.

Despite a number of histochemical and electron microscope studies, the molecular nature of this determinant is quite unknown.

Moloney murine leukaemia virus (MoMLV) A slow-transforming RETROVIRUS. It is widely used as a tool in mammalian genetic studies for insertional inactivation of endogenous genes (INSERTIONAL MUTAGENESIS). However, because the mouse genome contains >50 copies of sequences showing strong homology to the MoMLV genome, detection of the native prophage using HYBRIDIZATION is not possible. A derivative of the virus is therefore used containing a cloned bacterial suppressor gene, enabling hybridization detection of the prophage.

molten globule *See*: PROTEIN FOLDING.

monensin A carboxylic IONOPHORE which exchanges K^+, Na^+, and H^+ in a one to one manner and causes dilation of the GOLGI APPARATUS and blocks transport of many newly synthesized proteins often at the *medial* to *trans* cisternal step.

Tartakoff, A.M. (1983) *Cell* **32**, 1026–1028.
Griffiths, G. et al. (1983) *J. Cell Biol.* **96**, 835–850.

monoamine neurotransmitters *See*: ADRENALINE; CATECHOLAMINE NEUROTRANSMITTERS; DOPAMINE; 5-HYDROXYTRYPTAMINE; NORADRENALINE.

monoamine oxidase (MAO) Monoamine O_2 : oxidoreductase (EC 1.4.3.4) is the enzyme responsible for the oxidative metabolism, and hence deactivation, of the monoamine neurotransmitters 5-HYDROXYTRYPTAMINE (5HT, serotonin), NORADRENALINE, ADRENALINE, and DOPAMINE. MAO also has a role in providing protection from other exogenous (i.e. dietary) amines that might otherwise have adverse effects on, for example, cardiovascular or neuronal function. It exists in two main forms, types A and B; MAO-A (specific inhibitor clorgyline) is responsible for the oxidative removal of 5HT and noradrenaline, whereas in humans MAO-B (specific inhibitor selegiline) is the enzyme by which dopamine is mainly metabolized. Both forms of MAO are encoded on the X chromosome. MAO-B has also been implicated in the neurotoxic effects of certain chemicals that provide a model for the neurodegenerative mechanism of Parkinson's disease (*see* NEURODEGENERATIVE DISEASES); MAO inhibition by selegiline treatment has been reported to slow disease progression.

monocentric Chromosome with one centromere.

monocistronic Applied to an RNA transcript derived from a single gene and thus encoding the information for only one polypeptide chain or RNA. *Cf.* POLYCISTRONIC.

monoclonal antibody Antibody derived from a single clone of B lymphocytes (B CELLS) and which is therefore homogeneous in structure and antigen specificity. These antibodies are useful as highly specific markers and detector molecules in studies of cellular ultrastructure and in immunoassays. They were initially obtained as PARAPROTEINS present in the sera of patients with certain lymphoproliferative diseases. Antibodies of a single, preselected specificity can now be produced from HYBRIDOMAS. The exquisite specificity and high titre of monoclonal antibodies has led to many applications. They are widely used in biological research as markers for specific antigens (e.g. gene products, cellular structures, cell-surface antigens, etc.) (*see e.g.* IMMUNO-ELECTRON MICROSCOPY). They are also used in diagnostic assays and in other biomedical applications, and are being developed for therapeutic use. HUMANIZED ANTIBODIES, monoclonal antibodies containing rodent COMPLEMENTARITY DETERMINING REGIONS on a human antibody framework, are being produced by PROTEIN ENGINEERING, as are monoclonal antibodies with enzymatic activity (ABZYMES).

Winter, G. & Milstein, C. (1991) *Nature* **349**, 293–299.

monocotyledons, monocots A large group of flowering plants having an embryo with one cotyledon, parts of the flower usually in threes, leaves with parallel veins, and vascular bundles scattered throughout the stem. Monocot plants of commercial interest for which PLANT GENETIC ENGINEERING techniques are being developed include the cereals (e.g. barley, rice, wheat), maize, and ornamental bulbs.

monocyte Mononuclear phagocyte, incompletely differentiated phagocytic white blood cell which circulates in the blood. Monocytes settle in tissues and mature into tissue MACROPHAGES. Circulating monocytes are involved in clearing antigen–antibody complexes from the circulation. *See*: HAEMATOPOIESIS.

Monod–Wyman–Changeux model *See*: COOPERATIVITY.

monogenic disease/disorder A genetic disease that follows one of the classical patterns of MENDELIAN INHERITANCE, and consistently shows genetic linkage to a single LOCUS. Under such circumstances it may be anticipated that MUTATIONS in a single GENE can account for the major features of the disease's natural history.

monooxygenases Enzymes that catalyse oxidoreduction reactions in which one atom of oxygen is incorporated into one of the hydrogen donors, for example the nonspecific microsomal monooxygenase CYTOCHROME P450 in which one atom of oxygen is introduced into a hydrogen donor (e.g. an aryl hydrocarbon) with oxidation of a flavoprotein in the general reaction:

$$RH + reduced\ flavoprotein + O_2 = ROH + oxidized\ flavoprotein + H_2O$$

monophyletic Arising from a single common ancestor.

monosaccharide *See*: SUGARS.

monosomy The absence of a single chromosome or part of a chromosome, for example, TURNER SYNDROME is 45,X (i.e. 44 autosomes and 1 X-chromosome in females), a monosomy for the X-chromosome.

monospermy Fertilization by only one spermotozoon.

monozygotic Arising from the same zygote, and therefore genetically identical. Monozygotic twins develop from a single fertilized ovum through fission occurring shortly after fertilization. The incidence of monozygotic twin births is 0.3–0.5% and varies little between races. Dizygotic twins, which arise from two separately fertilized ova, outnumber monozygotic twins twofold in Caucasians.

morphactins Synthetic plant growth-regulating derivatives of fluorene-9-carboxylic acid which inhibit growth, interfere with directional growth responses to gravity and light, promote fruit ripening and have a range of other effects, none of which have proved commercially viable in practical agriculture. The morphactin 2-chloro-9-hydroxyfluorene-9-carboxylic acid methyl ester (Merck) is a useful inhibitor of polar AUXIN transport (*see also* INDOLE-3-ACETIC ACID) in physiological studies.

Schneider, G. et al. (1965) *Nature* **208**, 1013.

morphallaxis REGENERATION that does not require cell division; opposed to EPIMORPHOSIS, which does. The regeneration of the vertebrate limb is often cited as the archetypal example of epimorphic regeneration, while the regeneration of *Hydra* is cited as the example of morphallactic regeneration. However, the distinction does not necessarily correlate with the modes of regeneration found in higher and lower organisms.

morphogen Any substance thought to impart or alter POSITIONAL INFORMATION in a developmental morphogenetic GRADIENT. RETINOIDS have been implicated as the morphogen in the developing vertebrate limb bud on account of the fact that application of retinoic acid to wing tissue buds can cause discrete mirror duplications of the forming digits, similar to those generated when the ZONE OF POLARIZING ACTIVITY (ZPA) is transplanted. The product of the *bicoid* gene in *Drosophila* is considered to be a clear example of a morphogen. *See*: DROSOPHILA DEVELOPMENT; LIMB DEVELOPMENT; PATTERN FORMATION.

morphogenesis Literally, the word is a Greek equivalent of PATTERN FORMATION. However, in embryology the two terms are usually used differently. Morphogenesis is the formation of embryonic pattern that includes considerable cell rearrangements (an example is GASTRULATION), whereas pattern formation is embryonic form arising from cells changing their fates *in situ*, without much cell rearrangement (an example is the formation of the bones of the vertebrate limb, which condense from the loose mesoderm cells within the limb bud).

morphogenetic field A portion of an embryo contained within well-defined boundaries, which can develop independently without instructive influences from the rest of the embryo. An important property of morphogenetic fields is that they are capable of REGULATION, that is, any portion of the field can regenerate the whole field. As development proceeds, fields subdivide, becoming smaller and more numerous. C.H. Waddington first defined the term in the 1920s. *See also*: COMPARTMENT.

morphogenetic gradient *See*: GRADIENTS; PATTERN FORMATION.

morula Stage in mammalian embryonic development when the embryo consists of a ball of between 8 and 50 cells, depending on the species. The morula stage is thought to be the last stage at which all cells in the embryo are identical to each other in respect of developmental potential. During the morula stage compaction occurs (at the 8-cell stage in mice). Indvidual blastomeres of the morula flatten and increase their surface of contact with one another. This is accompanied by many changes in the blastomere cell membranes, including the redistribution of gap junctions (*see* CELL JUNCTIONS) between blastomeres, and the redistribution of the cell adhesion molecule UVOMORULIN (also called LCAM or E-cadherin) to the junctions between blastomeres (*see* CELL ADHESION MOLECULES; EPITHELIAL POLARITY). Blastomeres also become polarized internally. Microvilli form at the sites of blastomere contact, and this is thought to be brought about by a redistribution of actin MICROFILAMENTS, and perhaps other internal organelles. Compaction is thought to be the first stage at which cells in the mammalian embryo become developmentally restricted. Experiments with AGGREGATION CHIMAERAS have shown that before compaction, all the cells of the 8-cell embryo can give rise to either embryonic or EXTRAEMBRYONIC tisssues, whereas after compaction, cells on the surface of the embryo are committed to extraembryonic tissues, while those located internally are more likely to give rise to the embryo proper. The outer cells will give rise to the TROPHECTODERM, the precursor of the embryonic portion of the placenta, while the inner cells give rise to the INNER CELL MASS, the progenitor of the embryo proper and the yolk sac (*see* EXTRAEMBRYONIC MEMBRANES).

mos Potential ONCOGENE encoding a serine/threonine PROTEIN KINASE (pp39^{mos}), also known as cytostatic factor, which is responsible for meiotic arrest in vertebrate eggs by preventing the degradation of cyclin (*see* CELL CYCLE). Cytostatic factor is synthesized during OOCYTE maturation and accumulates as a hyperphosphorylated form with protein kinase activity. On fertilization it is destroyed by a Ca^{2+}-dependent calpain proteinase.

mosaic An embryo containing cells of two different PHENOTYPES (*cf.* CHIMAERA). The term is used in two different contexts.
(1) A naturally occurring condition in which gene expression varies among cells of the same histological cell type. An example is the mammalian female body, in which one of the two X-chromosome is inactivated and the genes on it are not expressed. If the inactivation affects a different X-chromosome of the pair in different cells, some cells will express genes on one of the chromosomes of the pair, other cells will express genes situated on the other member of the pair. The effect of mosaicism can sometimes be visible externally, as in individuals with differently coloured eyes. *See*: X-CHROMOSOME INACTIVATION.
(2) An embryo incapable of REGULATION. The term 'mosaic' is used because the lack of ability to regulate implies that different cells are already DETERMINED to different FATES.

mosaic analysis A method of analysing the cell autonomy of, and sites of requirement of, specific genes. In *Drosophila*, homo-zogous mutant patches of tissue in a background of wild-type cells can be generated either in GYNANDROMORPHS or by MITOTIC RECOMBINATION. If a homozygous mutant patch is concurrently marked with a known CELL-AUTONOMOUS mutation, then the mutant phenotype can be seen to be confined to mutant cells (autonomous) or more or less widely distributed (nonautonomous). Genes required for the generation or transmission of diffusible signals are likely to be nonautonomous whereas those encoding TRANSCRIPTION FACTORS and surface RECEPTORS should be autonomous. Mosaic analysis can also reveal where genes are required; for example mosaics in the developing eye have been valuable in allocating the requirement for specific genes to particular cells in developing ommatidia (*see* EYE DEVELOPMENT).

mosaic embryos *See*: MOSAIC (2).

mosaic viruses PLANT VIRUSES that cause mottling on the leaves of infected plants.

Mössbauer spectroscopy *See*: NEUTRON SCATTERING AND DIFFRACTION.

most probable phase The phase which has the highest probability, as distinct from the BEST PHASE which is more normally used in solving the PHASE PROBLEM of X-RAY CRYSTALLOGRAPHY.

motoneuron MOTOR NEURON.

motor end-plate (neuromuscular junction) The specialized region of the skeletal MUSCLE fibre membrane that is the site of neurotransmission from somatic motor neurons (*see* SYNAPTIC TRANSMISSION). At its terminal the motor neuron axon is devoid of myelinating SCHWANN CELLS and branches many times before forming synapses with the skeletal muscle. The nerve terminal contains thousands of synaptic vesicles, around 50 nm in diameter, which contain ACETYLCHOLINE and also ATP. The synaptic cleft, across which the acetylcholine diffuses, is ~50 nm wide. Opposite the nerve terminal the skeletal muscle membrane is thrown up into folds that provide the structure with a large surface area, ~2000 μm². NICOTINIC RECEPTORS for acetylcholine are clustered at the end-plate (there are few, if any, on the skeletal muscle fibre membrane away from the end-plate), with a very high density of ~10 000 per μm². The plasmalemma of the end-plate is also associated with the enzyme ACETYLCHOLINESTERASE which degrades released acetylcholine and thus limits its action.

motor neuron (motoneuron) A neuron involved in the carrying out of motor function, in that it conveys efferent motor impulses directly or indirectly to muscle groups. There are several types, including lower motor neurons, whose terminals end in muscle (*see* PERIPHERAL NERVOUS SYSTEM), and the upper motor neuron which includes all the descending fibre systems that can influence the lower motor neuron especially the corticospinal pathway from the motor cortex. NEURODEGENERATIVE DISEASES involving motor neurons include AMYOTROPHIC LATERAL SCLEROSIS and KENNEDY'S DISEASE.

Fig. M39 Schematic diagram of the genome of phage Mu. The Mu genome is a linear double-stranded DNA of ~39 kbp. The region G, encoding tail-fibre protein, can undergo an inversion that alters host specificity. Not to scale.

motor neuron disease *See*: AMYOTROPHIC LATERAL SCLEROSIS; KENNEDY'S DISEASE.

mouse mammary tumour virus (MMTV) A slow-transforming RETROVIRUS.

moving window analysis Method of analysis used to identify hydrophobic regions of a protein chain. *See*: HYDROPATHY PLOT.

MPF MATURATION PROMOTING FACTOR. *See also*: CELL CYCLE.

MPMV Mason–Pfizer monkey virus. *See*: RETROVIRUSES.

MPTP 1-Methyl-4-phenyl-1,2,3,6-tetrahydropyridine, a drug which can induce long-term disorders of motor function with symptoms resembling those of PARKINSON'S DISEASE (*see also* NEURODEGENERATIVE DISORDERS). In humans and non-human primates administration of MPTP leads to destruction of the dopamine-containing cells of the brain's substantia nigra and the appearance of a Parkinson-like syndrome. Parkinsonism induced by MPTP has been alleviated by transplantation of foetal mesencephalic tissue, which is rich in developing dopaminergic neurons which are lost in Parkinson's disease.

MR (1) MOLECULAR REPLACEMENT.
(2) Mineralocorticoid receptor. *See*: STEROID RECEPTOR SUPERFAMILY.

mRNA Messenger RNA. *See*: BACTERIAL GENE EXPRESSION; BACTERIAL GENE ORGANIZATION; EUKARYOTIC GENE EXPRESSION; EUKARYOTIC GENE ORGANIZATION; PROTEIN SYNTHESIS; RNA; RNA PROCESSING; RNA SPLICING; TRANSCRIPTION.

MRP-8, MRP-14 Cystic fibrosis antigen. *See*: CYSTIC FIBROSIS.

MSH MELANOCYTE-STIMULATING HORMONE.

MT Middle T antigen. *See*: PAPOVAVIRUSES.

mtDNA Mitochondrial DNA. *See*: MITOCHONDRIAL GENOMES.

MTOC MICROTUBULE-ORGANIZING CENTRE. *See*: CENTRIOLE; MICROTUBULES.

Mu-MI A transposable genetic element found in maize. *See*: TRANSPOSABLE GENETIC ELEMENTS: PLANTS.

Mu phage A transducing double-stranded DNA BACTERIOPHAGE of *Escherichia coli*, the DNA of which is able to integrate into the host bacterial chromosome at almost any point. Any bacterial gene at the site of integration is usually inactivated, hence the name of the phage (Mutator phage). This property of almost random integration is used in *E. coli* gene mapping studies. On induction the integrated copy of Mu remains in place and copies replicated *in situ* undergo transposition into other sites on the chromosome. When the host chromosome is destroyed, Mu genomes (Fig. M39), linked to fragments of *E. coli* DNA at each end, are packaged into phage particles. *See*: FLIP-FLOP INVERSIONS; TRANSPOSABLE GENETIC ELEMENTS.

mucin Vague and largely outdated term (derived from the Latin *mucidus*, pertaining to nasal slime) for GLYCANS and, particularly, GLYCOPROTEINS of secretions, sometimes wrongly applied to plant and fungal slimes and bacterial capsules. Mucin-type glycans are saccharides linked to protein by the sequence Ser-O-α1-GalNAc-; they are not restricted to secretory glycoprotein, nor does secretory glycoprotein only contain linkages of this type so the term is misleading. Histologists still use the term (wrongly) and distinguish acid mucins (which bind basic dyes), neutral mucins (which do not), sulphomucins (containing sulphate) and sialomucins (containing sialic acid), even though the glycans concerned may not be of the 'mucin' type.

mucopolysaccharide An old term encompassing such diverse GLYCANS as HYALURONIC ACID, GLYCOSAMINOGLYCANS, PROTEOGLYCAN, KERATAN SULPHATE, CHITIN and several other less well-defined polymers. Although now falling into disuse in biochemistry and molecular biology it survives in some medical terminology.

mucopolysaccharidoses Lysosomal storage diseases arising from deficiencies of lysosomal enzymes involved in the degradation of GLYCOSAMINOGLYCANS (mucopolysaccharides) which result in their intracellular accumulation. *See*: DWARFISM; HUNTER SYNDROME; HURLER'S DISEASE.

muE motifs (μE1–4) Cluster of homologous DNA sequences in the immunoglobulin heavy chain ENHANCER region (consensus sequence 5′-CAGGTGGC-3′) (*see* IMMUNOGLOBULIN GENES). Sites E1–4 were originally identified as sequences which bound B cell-specific proteins in whole cells. However, experiments performed using nuclear protein extracts show that proteins from nonlymphoid cells can also bind to these sites. These include proteins which bind to all four μE motifs and others which show site preference. For example, protein NFμE1 binds exclusively to site μE1 and protein NFμE3 to site μE3. Functionally, the μE1–4 sites are involved in the regulation of lymphoid cell-specific enhancer activity. This is thought to be mediated, in part, by the μE-motif binding proteins. In TRANSFECTION experiments, deletion or mutation of any single μE site has little effect on transcription. However, simultaneous mutation of multiple μE sites (for example μE1 plus μE3 and μE4) can virtually abolish heavy chain enhancer activity. Sequences which show close similarity to the μE motif also occur in the immunoglobulin κ gene enhancer (sites κE1–4).

Muller cells Specialized large radial GLIAL CELLS, found in the retina.

Muller's ratchet The hypothesis put forward by H.J. Muller in 1964 that in an asexual population the load of deleterious mutations increases in a ratchet-like manner as mutation-free individuals become rare and are lost from the population. Muller's ratchet has been verified experimentally in bacteriophage populations and in ciliate protozoa. It has been put forward as one explanation of the evolution of sexual reproduction, which can increase the fitness of the population by regenerating mutation-free individuals by RECOMBINATION.

Chao, L. (1990) *Nature* **348**, 454–455.

multicatalytic-multifunctional protease particle The PROTEASOME. *See also*: PROTEIN DEGRADATION.

multicomponent viruses Viruses in which the genome is segmented into more than one RNA molecule each of which is packaged into a separate particle. *See*: ALFALFA MOSAIC VIRUS GROUP; BROMOVIRUSES; COMOVIRUSES; HORDEIVIRUSES; PLANT VIRUSES; TOBRAVIRUSES.

multi-CSF A haematopoietic colony-stimulating factor now known as interleukin-3. *See*: IL-3.

multidrug resistance Simultaneous multiple cross-resistance to a variety of chemically distinct drugs which can be shown by cell lines and tumour cells following exposure to a single cytotoxic agent. It is mediated by the P-GLYCOPROTEIN transport system. The P-glycoprotein is found to be expressed at unusually high levels in certain tumours at time of relapse. Related transport systems have been implicated in CHLOROQUINE resistance in malaria, and the gene defect in CYSTIC FIBROSIS.

multigene family A family of genes of related structure and usually related function, which have evolved by tandem GENE DUPLICATION, and sometimes have also become translocated or transposed to different chromosomes. Familiar examples include the MHC class I genes (*see* MAJOR HISTOCOMPATIBILITY COMPLEX), the GLOBIN GENES, the red and green cone pigment genes involved in red-green colour blindness (*see* VISUAL TRANSDUCTION), and the COLLAGEN gene family. *See also*: MOLECULAR EVOLUTION: SEQUENCES AND STRUCTURES; PROTEIN STRUCTURE; SUPERFAMILY.

multi-ion channel An ion channel which allows the passage of more than one ion at a time (e.g. the CYSTIC FIBROSIS transmembrane regulator).

multilocus probe *See*: DNA TYPING.

multiple isomorphous replacement (MIR) A method of solving the crystallographic PHASE PROBLEM by using crystals of two or more HEAVY ATOM ISOMORPHOUS DERIVATIVES which are isomorphous to the crystal under study. This was the first, and for many years the only, method for solving completely unknown macromolecular structures. It relies upon the modification of a native structure by a small number of heavy atoms (so-called because for

X-ray diffraction, the scattering power depends on the atomic number) at positions which can be determined, to give a derivative that is isomorphous with the native crystal, that is, molecules in both derivative and native crystals have similar orientations and positions within similar unit cells. If all the similarities are very close then this derivative will provide information at higher RESOLUTION, whereas weaker similarities will give only low resolution phases or no useful information at all.

The method is most easily understood graphically although all the necessary calculations are now made numerically. Measuring the DIFFRACTION pattern of both protein and heavy atom derivative crystals respectively gives the sets of STRUCTURE AMPLITUDES, which for any given reflection *hkl* are the radii of the phase circles for protein and derivative (F_P and F_{PH}).

From a knowledge of the position(s) of the heavy atom(s) (often determined from the PATTERSON FUNCTION of the derivative) the heavy atom STRUCTURE FACTOR is calculable, giving both the

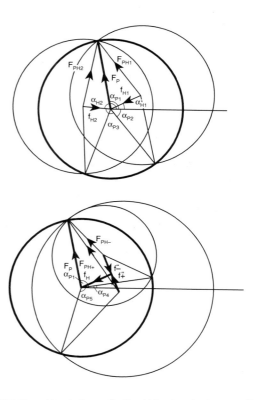

Fig. M40 The phase triangle for a reflection *hkl* having structure amplitudes from the protein and the first derivative (i.e. protein plus heavy atom) denoted F_P and F_{PH1} respectively, where the heavy atom structure factor $\mathbf{f}_{H1}$ has known amplitude and phase angle. From the first derivative, the phase angle for the protein structure is seen to be ambiguously determined as either α_{P1} or α_{P2}. This uncertainty is resolved by use of either (top) a second heavy atom derivative wherein the heavy atom structure factor $\mathbf{f}_{H2}$ and derivative amplitude F_{PH2} indicate a protein phase angle of either α_{P1} or α_{P3}; or (bottom) ANOMALOUS SCATTERING from the heavy atom derivative, where reflections *hkl* and $\bar{h}\,\bar{k}\,\bar{l}$ have measurably different structure amplitudes from the derivative, denoted F_{PH+} and F_{PH-} respectively. These arise because the anomalous component f″ is 90° out of phase with the normal scattering factor from the heavy atom. This is located either at coordinates (x, y, z) indicating protein phase angles α_{P1} or α_{P4}, or at coordinates (−x, −y, −z) indicating α_{P1} or α_{P5}.

amplitude F_H and the phase α_H of the heavy atom contribution to diffraction from the derivative. For each reflection *hkl* the derivative structure factor is simply the vector sum of those from protein and heavy atom, namely: $\mathbf{F_{PH}} = \mathbf{F_P} + \mathbf{F_H}$. From Fig. M40 it can be seen that for a single derivative there are only two possibilities for the protein phase α_{P1} or α_{P2}. This ambiguity is resolved by information from a second derivative (indicating α_{P1} or α_{P3}) or from ANOMALOUS SCATTERING of the single derivative (indicating α_{P1} or α_{P4} and α_{P1} or α_{P5}).

Complications arise when the measured intensities contain errors or when the model for heavy atom scattering is inaccurate. Use of additional derivatives, and/or progressive refinements of the heavy atom model, together with sophisticated weighting schemes, enable the BEST PHASE set to be determined for subsequent calculation of an ELECTRON DENSITY MAP. *See*: X-RAY CRYSTALLOGRAPHY.

Stout, G.H. & Jensen, L.H. (1989) *X-ray Structure Determination: a Practical Guide* (Wiley, New York).
Vijayan, M. & Ramaseshan, S. (1993) In *International Tables for X-ray Crystallography* Vol. B Reciprocal Space (Shmueli, U., Ed.) 264–279 (Kluwer Academic Publishers, Dordrecht).

multiplicity of infection (MOI) A term used in virology to define the ratio of infectious particles to target cells.

multiubiquitination The addition of multiple UBIQUITIN molecules to a protein. *See*: POST-TRANSLATIONAL MODIFICATION; PROTEIN DEGRADATION AND TURNOVER.

multivesicular body (MVB) Membrane-bounded cytoplasmic organelles which enclose one or more smaller membrane-bound vesicles. MVBs can be formed by AUTOPHAGY, but some seem to be components of the ENDOCYTIC PATHWAY and may function either in endosomal sorting or in the transport of membrane and ligands from early to late ENDOSOMES. In this latter capacity MVBs may also be termed ENDOSOMAL TRANSPORT VESICLES. *See*: ENDO-CYTOSIS.

Felder, S. et al. (1990) *Cell* **61**, 623–634.
Gruenberg, J. et al. (1989) *J. Cell Biol.* **108**, 1301–1316.

mung bean nuclease An ENDONUCLEASE (EC 3.1.30.1) isolated from mung bean sprouts and widely used in molecular biology. The enzyme preferentially attacks single-stranded DNA and RNA, producing 5′-monophosphate and 3′-hydroxy ends, whereas nucleic acid duplexes are relatively resistant to degradation. Uses include mapping gene transcripts in NUCLEASE PROTECTION ASSAYS, in transcript prevalence assays and in DNase I FOOTPRINTING assays.

muramic acid A carbohydrate component of the PEPTIDOGLYCAN of BACTERIAL ENVELOPES. *See also*: SUGARS.

murine leukaemia virus (MLV) *See*: ONCOGENES; RETROVIRUSES.

muscarine Toxic alkaloid found in some species of basidiomycete fungi (e.g. *Amanita muscaria, Inocybe* spp.). It acts at MUSCARINIC RECEPTORS in the peripheral nervous system.

muscarinic receptors One of the two classes of receptors which utilize ACETYLCHOLINE as an endogenous ligand, the other class being the NICOTINIC RECEPTORS. Muscarinic receptors are widespread throughout the central and peripheral nervous systems and mediate, for example, parasympathetic control of certain heart, smooth muscle and exocrine gland functions as well as being important in, for example, storage and retrieval of memory.

Muscarinic receptors are G PROTEIN-COUPLED RECEPTORS and can couple via G proteins to a number of closely or directly linked effector mechanisms, including inhibition of ADENYLATE CYCLASE, stimulation of INOSITOL 1,4,5-TRISPHOSPHATE production via a specific PHOSPHOLIPASE C, opening of a certain class of POTASSIUM CHANNELS, and activation of PHOSPHOLIPASES A_2 and D (*see* SECOND MESSENGER PATHWAYS).

Muscarinic receptors can be distinguished from nicotinic receptors by their pharmacology. For example, muscarinic receptors are activated by the AGONISTS muscarine and oxotremorine and are blocked by the ANTAGONIST atropine.

Five muscarinic receptor subtypes (m1–m5) have been cloned and sequenced. They exhibit a qualitatively similar pharmacology and conservation of sequence in the putative seven transmem-

Table M6 Muscarinic receptor nomenclature

Pharmacological characterization					
Subtype	M_1	M_2	M_3		
Other names used previously	$M_1\alpha$	$M_2\alpha$, cardiac M_2, C	M_2, $M_2\beta$, B, glandular M_2	M_2	—
Selective antagonists	Pirenzepine, (+)-telenzepine	AF-DX 116, himbacine, methoctramine, gallamine*	*p*-Fluorohexahydro-siladifenidol, hexahydrosila-difenidol,	—	—
				—	—
	$M_1 > M_3 \geqslant M_2$	$M_2 > M_1 \geqslant M_3$	$M_3 \geqslant M_1 > M_2$		
Molecular characterization					
Sequences	m1	m2	m3	m4	m5
Other names used previously	mAChRI, M1	mAChRII, M2	mAChRIII, M4	mAChRIV, M3	
Numbers of amino acids	460	466	589/590	478/479	531/532

* Not competitive.

brane domains that contain the acetylcholine-binding site. However, they couple to different preferred effector mechanisms (m1, m3, m5 to phospholipase C; m2, m4 to adenylate cyclase and to K^+ channels). In contrast to the five known molecular subtypes, only three pharmacological subtypes — M_1, M_2, and M_3 — have been defined on the basis of the action of selective antagonists (Table M6). The available evidence strongly suggests that the m1 sequence corresponds to the M_1 receptor, m2 to the M_2 receptor and m3 to the M_3 receptor. There is a candidate M_4 receptor, corresponding to the m4 gene, and found in the striatum and rabbit lung. A characteristic pharmacology for this receptor is emerging. The physiological role of the m5 species, apparently present in very low abundance in the central nervous system, is not known. *See also*: PERIPHERAL NERVOUS SYSTEM; SYNAPTIC TRANSMISSION.

Levine, R.R. & Birdsall, N.J.M. (Eds) (1989) *Trends. Pharmacol. Sci.* **10**, Suppl.
Hulme, E.C. et al. (1990) *Annu. Rev. Pharmacol. Toxicol.* **30**, 633–673.

Muscle

A FUNDAMENTAL property shared by all animals is one of movement, and any form of movement — walking, running, flying, or swimming — is made possible by the combined action of a group of muscles. The structure and function of various types of muscle provide a fascinating insight into the conversion of chemical energy, derived from food, into mechanical work.

Despite the large number of muscles involved in movement, it is possible to divide them into three main categories: skeletal, smooth, and cardiac muscle. Skeletal muscle, as the name implies, is attached to the skeleton through tendons and is responsible for movement and is under the voluntary control of the nervous system; it is also known as striated muscle due to the characteristic banding pattern produced by regularly spaced regions of higher refractive index along the fibre when viewed in a light microscope. The banding pattern is absent in smooth muscle, owing to a less ordered arrangement of the muscle proteins. Smooth muscle is, in general, not under voluntary control and is found in blood cell walls, the stomach and the intestines, etc. Cardiac muscle, although similar in appearance to skeletal muscle, is also not under voluntary control. Despite superficial differences between different muscle types, the fundamental mechanism of force generation is very similar.

The regularity in the structure of skeletal muscle makes it more amenable to structural techniques like ELECTRON MICROSCOPY and X-ray FIBRE DIFFRACTION compared with either smooth, or to a lesser extent, cardiac muscle, and it is the best understood. This section is therefore mainly devoted to striated muscle with only brief mentions of other types.

Structure

Striated muscle shows a very ordered structure which is evident when viewed on a macroscopic or indeed a molecular scale. Figure M41*a* shows the arrangement of single fibres in a whole muscle. The single muscle fibre, which is also the smallest unit of muscle capable of sustaining contraction, has diameters ranging

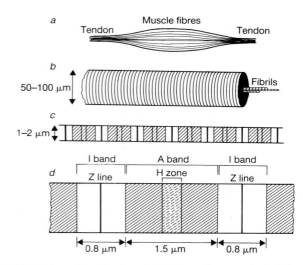

Fig. M41 Diagram of the structure of striated muscle as seen in the light microscope. The whole muscle (*a*) is made up of fibres, which contain cross-striated myofibrils (*b*). *c*, Isolated myofibrils showing the regular banding pattern. *d*, The banding pattern of a myofibril at resting length. From [6].

from 10 to 200 µm, depending on the species, but is typically 100 µm for the commonly used frog sartorius muscle; the length of fibre in this muscle is several centimetres. A single fibre contains a number of myofibrils (diameter 1–2 µm) (Fig. M41*c*). Myofibrils have a banded appearance when viewed in a light or electron microscope owing to periodic variations of density along the fibre axis; each repeating unit is known as a sarcomere (Fig. M41*d*). There are two clearly defined regions within the sarcomere: a region of higher density which appears darker is called the A-band, and a region of lower density, the I-band. Centred on the A-band is a lighter region known as the H-band with a relatively sharp dark band in the centre, called the M-band. The sarcomere is bounded by dark lines known as the Z-lines. The sarcomere length in resting frog muscle is typically 2.3 µm, the A-band ~1.5 µm and each of the (half) I-bands ~0.37 µm.

The higher resolution of the electron microscope provides much finer detail. Additionally, any regular features can be measured with good precision from the X-ray DIFFRACTION pattern. Electron microscopy of thin sections of striated muscle shows that the I-band consists of thin filaments, 5–7 nm in diameter, which extend into the A-band, interdigitating with the thick filaments of the A-band, which are 10–12 nm in diameter (Fig. M42). Most of the surface of the thick filament has regularly arranged projections, occasionally seen making contact with the thin filament and named cross-bridges. Electron microscopy of thin cross-sections of striated muscle shows lateral order in the filaments, that is at right angles to the filament axis. The images show that, although the thin filaments are poorly ordered in the I-band when there is no overlap with the thick filaments, much improved order is maintained in the overlap region where the thick filaments are arranged on a hexagonal lattice with the thin filaments occupying the trigonal positions. Thick and thin filaments from striated muscle of many other species are also arranged on a hexagonal lattice but the thin filaments have a different arrangement, with varying thin filament to thick fila-

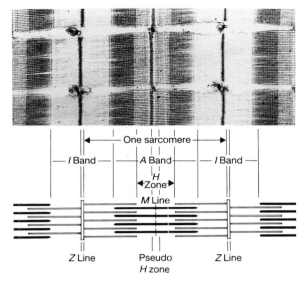

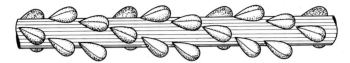

Fig. M43 Arrangement of myosin molecules in the thick filament of vertebrate muscle. From [2].

Fig. M42 The arrangement of thick and thin filaments in striated muscle and its relation to the banding pattern.

ment ratio, presumably to meet the different needs of different muscles.

Muscle proteins

The main constituents of the thick and thin filaments are MYOSIN and ACTIN respectively (proteins found in all muscle and in most non-muscle cells as well) along with smaller quantities of a number of accessory proteins. The complex of actin and myosin found in muscle is often called actomyosin.

Myosin and the thick filament

Myosin is an extremely asymmetric hexameric molecule (M_r 500 000) consisting of two globular heads, known as the myosin subunit 1 (S1) attached to one end of a long rod-like tail. It also contains two light chains (M_r 18 000–21 000) on each S1 which are known to play an important part in the regulation of vertebrate smooth muscle and scallop muscle (another well-researched system). The S1 is the most important part of the myosin molecule regarding force generation as it contains binding sites for both ATP, which fuels contraction, and actin and is believed to be the site for the myosin 'motor' — the site of force generation. As seen in the election microscope myosin-S1 has a shape approximating to an Indian club or a pear truncated at the wide end (*see* MYOSIN). It is 15–20 nm along the long axis and ~4–7 nm at its widest cross-section. The rod part of the myosin molecule is ~150 nm long and 2 nm in diameter and plays an important part in the assembly of myosin filaments; it seems as if assembly is a prerequisite for force generation as individual

myosin molecules are not involved in this process. Myosin filaments in striated muscle are relatively stable structures which are used repeatedly in force generation. In contrast, there are indications that in smooth muscle, myosin filaments are only assembled when required and subsequently disassembled. Although myosin forms the largest part of the thick filaments, there are a number of accessory proteins also present in the thick filament.

Thick filaments from a number of species have some common design features. For example, thick filaments from vertebrate muscle have several hundred myosin molecules arranged so that the myosin heads describe a helix of pitch close to 42.9 nm with a corresponding axial repeat at a third of the helical repeat, that is close to 14.3 nm with three pairs of myosin heads at each axial level (Fig. M43).

Actin and the thin filament

The thin filament consists mainly of G-actin — a globular protein of M_r 42 000 — polymerized into a double helical array known as F-actin. There are thirteen actin molecules for every six turns of the helix, which has a pitch of ~36 nm. In addition, the thin filament contains two proteins with important roles in the regulation of striated muscle contraction — troponin and tropomyosin (*see* ACTIN-BINDING PROTEINS). Tropomyosin is a rod-like molecule, 40 nm long and 2 nm in diameter, consisting of two α-helical chains and lying in the grooves of the actin helix as shown in Fig. M44, and bound to the troponin complex. The complex of troponin (Tn), spaced at 38.5 nm intervals along the thin filament, consists of three subunits: TnC (M_r 18 000), which can bind Ca^{2+} ions, TnI (M_r 22 000), which can inhibit actomyosin interaction, and TnT (M_r 38 000), which binds to tropomyosin.

Titin filaments

The existence of a third class of myofilament has been established recently. These filaments consist of the protein titin and differ from both the thick and thin filaments in being fairly elastic. Titin (M_r 3×10^6) is a long molecule spanning over half the sarcomere — from the M-band to the Z-line (*see also* ACTIN-BINDING PROTEINS). There are suggestions that the part of titin which overlaps the thick filament is used as a template for controlling the assembly of myosin filaments, whereas the part extending between the end of the thick filament and the Z-line is elastic.

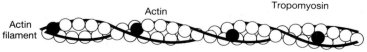

Fig. M44 The thin filament of striated muscle.

Contraction

Sliding filament model

A description of muscle as a static structure, however comprehensive, cannot convey the details of the mechanisms central to the dynamic aspects of force generation. The key observations, made in the 1950s, were that the A-band remained at constant length both in contracting muscle despite the fact that the sarcomere had shortened, and in stretched muscle, where the sarcomere was lengthened, although the I-band decreased in these experiments. The variable length I-band suggested that there was relative sliding between the thick and thin filaments, where both filaments were of constant length, but one end of the thin filaments was tethered to the Z-lines. Higher resolution electron micrographs of thin sections of striated muscle showed the two sets of filaments more clearly than had been possible with optical microscopes. It was suggested that the myosin heads had a central role in force generation by attaching to actin and producing a relative movement between the thick and thin filaments. This came to be known as the swinging cross-bridge model. It was proposed that the cross-bridges acted as independent force generators, force being directed towards the centre of the sarcomere from the two halves. Accurate measurements made subsequently on the amount of tension (or force) generated by the muscle as a function of sarcomere length, using muscle which may be stretched and shortened reversibly, have substantially corroborated this model.

The cross-bridge cycle

When myosin interacts with actin, ATP is hydrolysed at a high rate, the energy provided by the hydrolysis being used for filament sliding. A simplified scheme describing the primary events in the cross-bridge cycle is shown in Fig. M45. Several refinements have been proposed to the simple cycle described here but, for clarity, these are not dealt with here. In the absence of ATP myosin-S1 forms rigid and inextensible links with actin,

known as a rigor complex; addition of ATP, which binds to myosin-S1, releases the head from actin (step 1). ATP is hydrolysed very rapidly, leaving the products ADP and P_i still attached to the head, which is now in a 'high energy' state (step 2). The head binds to actin (step 3) followed by release of the hydrolysis products P_i and ADP and resulting in a conformational change in S1 which drives the actin filament by a distance of between 4 and 10 nm. The head stays locked in the rigor conformation with actin until a new molecule of ATP binds and releases it. The S1 is now ready to start a new cycle of attachment and force generation at a new actin site which has become accessible subsequent to thin filament movement.

The precise mechanism whereby the independent force generators or myosin cross-bridges generate force has been the subject of intensive investigations involving diverse techniques. To name but a few, biochemistry is used to study the kinetics of actomyosin interactions, and three-dimensional reconstruction of the S1–actin complex with electron microscopy and time-resolved X-ray diffraction is used to take snapshots of intermediate states in contracting muscle. Despite all this effort, the precise mechanisms by which force is generated in the interaction between actin and myosin have not been elucidated; the main reasons are that one is looking at and attempting to measure a complex phenomenon, involving relatively small structures (the myosin-S1 which contains the myosin motor is ~20 nm by 5 nm) in which one has to measure structural changes on a very rapid (millisecond) time scale. The movement of myosin cross-bridges on a millisecond scale can be visualized, somewhat indirectly, by recording the diffraction pattern, which is related to the structure, every millisecond for the period of interest. Such methods, which rely on a high intensity X-ray source generating SYNCHROTRON RADIATION and electronic detection methods, provide very good information on the 'average' behaviour of cross-bridges during contraction. It has been shown, for example, that a cross-bridge assumes during contraction a position much closer to the thin filament than its position very close to the thick filament in resting muscle. This general shift suggests that links between the thick and thin filaments are formed during contraction and subsequently bro-

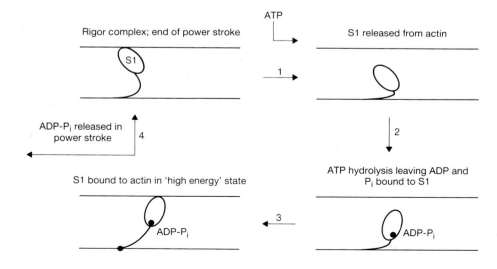

Fig. M45 *a*, The sliding filament model of contraction. *b*, The myosin cross-bridge cycle. See text for details.

ken. There is very good correlation between the formation of cross-bridges and the generation of force. Force is generated more slowly than cross-bridge formation, implying an intermediate attached state (called the weak binding state) in which very little force is generated which converts to a high force generating state, also known as a strong binding state. There is a great deal still to learn about the way muscle works and a combination of techniques are probably going to be needed to obtain better understanding.

Regulation of contraction

All muscle must perform its activity in a controlled fashion. It must be regulated and be able to switch on the force-generating mechanism when required and to switch it off when not needed. The regulation mechanism is based on different processes in different types of muscle, although they all seem to be mediated through calcium ions. There are two main types of regulation: thin filament based and thick filament based, although in some types of muscle both processes may act together.

Thin filament based

An elaborate membrane system runs through striated muscle, known as the SARCOPLASMIC RETICULUM (SR), which contains a store of CALCIUM ions. With the arrival of nervous impulses (*see* MUSCLE END-PLATE; SYNAPTIC TRANSMISSION) the sarcoplasmic membrane becomes depolarized resulting in the release of Ca^{2+} into the cytosol and leading to contraction. At the end of the stimulation, the Ca^{2+} ion pump (CA^{2+}-ATPASE) in the SR membrane removes Ca^{2+} from the cytosol and stores it back in the SR. It is thought that the free Ca^{2+} binds to TnC, resulting in a conformational change in the molecule. The change in TnC conformation causes an azimuthal shift in the tropomyosin molecule, which normally occupies a position blocking the interaction between actin and myosin-S1. For obvious reasons this is referred to as the steric blocking model. Recent evidence from time-resolved X-ray diffraction indicates that the tropomyosin shift in response to Ca^{2+} release is the first observable structural change during the early phases of contraction and seems to be a prerequisite for actomyosin interaction to proceed. When nerve impulses cease and Ca^{2+} is sequestered back into the storage compartments, tropomyosin moves back to its original position, and the actomyosin interaction is again inhibited. There is some disagreement about the detailed interpretations of the X-ray results and, according to an alternative hypothesis, the initial attachment of myosin heads distorts the local geometry of the actin filament and allows further interaction to proceed.

Thick filament based

Myosin-linked regulation may be direct or indirect; both processes are mediated by Ca^{2+} but are quite different from thin filament-based regulation. In the indirect process, which occurs in smooth muscle when Ca^{2+} levels increase above a critical level, an enzyme known as MYOSIN LIGHT CHAIN KINASE is activated. This promotes phosphorylation of the myosin regulatory light chain. It is thought that unphosphorylated light chain prevents myosin heads from making the conformational change which results in the force-generating step, possibly by inhibiting the relative movement of the head domains. In the direct form of myosin regulation, employed by scallop muscle, Ca^{2+} acts by binding directly to myosin which is then able to carry out interactions with actin.

A.R. FARUQI

See also: CELL MOTILITY; MICROFILAMENTS; MYOGENESIS.

1 Huxley, A.F. (1974) Muscular contraction. *J. Physiol., Lond.* **243**, 1–43.
2 Squire, J. (1981) *The Structural Basis of Muscular Contraction.* (Plenum, New York).
3 Bagshaw, C.R. (1982) *Muscle Contraction* (Chapman & Hall, London).
4 Huxley, H.E. & Faruqi, A.R. (1983) Time-resolved X-ray diffraction studies on vertebrate striated muscle. *Annu. Rev. Biophys. Bioeng.* **12**, 381–417.
5 Amos, L.A. & Amos, W.B. (1991) *Molecules of the Cytoskeleton* (Macmillan, London).
6 Huxley, H.E. (1983) Molecular basis of contraction in cross-striated muscles and relevance to motile mechanisms in other cells. In *Muscle and Nonmuscle Motility* (Academic Press, New York).

muscle satellite cell STEM CELL from which new skeletal muscle cells are produced. *See*: MYOGENESIS.

muscular dystrophy *See*: DUCHENNE MUSCULAR DYSTROPHY; MYOTONIC DYSTROPHY.

***mut* genes** *See*: DNA REPAIR.

mutable site Any site in a heritable nucleic acid that can be mutated. *See also*: MUTATION.

mutagen Any chemical or physical agent capable of causing a change in DNA such that the information in the genetic code is altered and a MUTATION produced. Physical mutagens include ionizing radiation and ultraviolet light. Mutagenic agents also include RETROVIRUSES and TRANSPOSABLE GENETIC ELEMENTS which insert into the genome. *See also*: CHEMICAL CARCINOGENS AND CARCINOGENESIS; DNA REPAIR.

mutagenesis The introduction of MUTATIONS *in vivo* or *in vitro*. Mutations can be induced in living organisms or in cultured cells by a variety of MUTAGENS, including ionizing radiation, ultraviolet radiation, or chemical mutagens, or by infection with certain viruses which integrate into the host genome (*see e.g.* RETROVIRUSES) or by the introduction of DNA previously mutagenized *in vitro*. *See also*: INSERTIONAL MUTAGENESIS; OLIGONUCLEOTIDE-MEDIATED MUTAGENESIS; P ELEMENT-MEDIATED TRANSFORMATION; SITE-DIRECTED MUTAGENESIS; TRANSGENIC ANIMALS; TRANSPOSABLE GENETIC ELEMENTS; TRANSPOSON TAGGING.

mutant Any organism, cell, virus, gene, etc. in which a MUTATION has occurred.

Mutation

MUTATION is the process whereby changes occur in the quantity or structure of the genetic material of an organism. Mutations are permanent alterations in the genetic material which may lead to changes in PHENOTYPE. An organism, gene, DNA sequence, etc. in which a mutation has occurred is called a mutant.

Mutation can involve modifications of the nucleotide sequence of a single gene, blocks of genes or whole chromosomes. Changes in single genes may be the consequence of point mutations, which involve the removal, addition or substitution of a single nucleotide base within a DNA sequence, or they may be the consequence of changes involving the insertion or deletion of large numbers of nucleotides (Fig. M46). Modifications of whole chromosomes include both changes in number or structural changes involving CHROMOSOME ABNORMALITIES. Numerical chromosome mutations may involve multiples of the complete KARYOTYPE, termed polyploidy, or they may involve deviations from the normal number of chromosomes, termed aneuploidy.

Point mutations

Transition mutations involve the substitution of one purine in the DNA by another purine or one pyrimidine by another pyrimidine, that is A by G and vice versa, or T by C and vice versa. Transversions involve the replacement of a purine by a pyrimidine and vice versa.

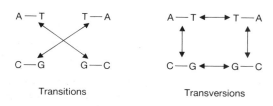

Such base substitutions (also termed base-pair substitutions or nucleotide substitutions) may affect the correct functioning of the product of the modified gene; the extent of the effect can, however, range from the undetectable to the severe. As the GENETIC CODE is degenerate, so that most of the amino acids inserted into a growing polypeptide chain (*see* PROTEIN SYNTHESIS) are coded for by more than one triplet, a base substitution may simply convert one codon for a particular amino acid to another codon for the same amino acid. Even when the substitution results in the insertion of a different amino acid into a polypeptide, a missense mutation, the amino acid may be an acceptable substitute and thus not lead to any significant change in the activity of the polypeptide.

However, some missense amino-acid changes can have drastic effects upon the folding of polypeptide chains or upon the configuration of the active site of an enzyme. Other base substitutions may have drastic effects because they convert a triplet coding for an amino acid into one of the three termination signals which lead to the premature termination of polypeptide synthesis. Such

a Original sequence

```
. . ATG GTG CTC AGC ATA GCT TAT AGC . . .
. . Met Val Leu Ser Ile Ala Tyr Ser . . .
```

b Point mutation (missense)

```
. . ATG GTG GTC AGC ATA GCT TAT AGC . . .
. . Met Val Phe Ser Ile Ala Tyr Ser . . .
```

c Insertion leading to frameshift with premature termination

```
. . ATG GTG CTC AGC ATA GCT TAT TAG C . . .
. . Met Val Leu Ser Ile Ala Tyr STOP
```

d Insertion leading to frameshift with altered amino-acid sequence and premature termination

```
. . ATG GTG C[GA TAT CTC TGT GT]T   CAG CAT AGC TTA TAG C . . .
. . Met Val Arg Tyr Leu Cys Val     Gln His Ser Leu STOP
```

e Deletion leading to frameshift

```
. . ATG GT[G CTC AGC ATA G]CT TAT AGC . . .
. . ATG GTC TTA TAG C . . .
. . Met Asp Leu STOP
```

f Silent mutation no change

```
. . ATG GTG CTA AGC ATA GCT TAT AGC . . .
. . Met Val Leu  Ser Ile Ala Tyr Ser . . .
```

Fig. M46 Types of mutation.

nonsense changes are usually accompanied by the loss of function of the gene product.

Frameshift mutations

The addition or deletion of base-pairs from the coding sequence of a gene may lead to the production of frameshift changes in the messenger RNA transcribed from that sequence and a severe effect is nearly always produced in the resultant polypeptide. Additions or deletions of one (or two or four) bases cause a change in the READING FRAME of the mRNA which can lead to premature termination as the result of generation of a stop codon in the new reading frame, or to the insertion of a number of incorrect amino acids in the growing polypeptide with a high probability of producing a defective gene product (*see also* FRAMESHIFT MUTATIONS.

Alleles

Different forms of a gene which may be produced by mutation are termed ALLELES. In diploid organisms such as *Homo sapiens* the chromosomes of each homologous pair may carry different alleles at a particular GENE LOCUS and are then termed heterozygous; when they carry identical alleles they are termed homozygous. Heterozygosity leads to complications in gene expression. Sometimes two different alleles express themselves independently of each other; this condition is called CODOMINANCE, and is seen, for example in the *A* and *B* pair of alleles of the human ABO BLOOD GROUP SYSTEM. Often, one of the two alleles dominates development and function to the full or partial exclusion of the effects of the other allele. In those cases the alleles are respectively DOMINANT (or partially dominant) and RECESSIVE (or partially recessive). When a dominant allele is present in a diploid organism its phenotype will be observed irrespective of the status of the other allele. However, the phenotype of a recessive allele will be observed only when two copies are present in the same cell (the homozygous state), or when it is present as a single copy with no homologous allele (the hemizygous state), as for alleles carried on the X chromosome in male mammalian cells (*see* MENDELIAN INHERITANCE).

Mutagenesis

Mutations can arise spontaneously as a result of events such as errors in the fidelity of DNA REPLICATION or the movement of TRANSPOSABLE GENETIC ELEMENTS within genomes. They are also induced following exposure to chemical or physical MUTAGENS. Such mutation-inducing agents include ionizing radiations, ultraviolet light and a diverse range of chemicals such as the alkylating agents, and polycyclic aromatic hydrocarbons, all of which are capable of interacting either directly or indirectly (generally following some metabolic biotransformations) with nucleic acids [1]. The DNA lesions induced by such environmental agents may lead to modifications of the base sequence when the affected DNA is replicated or repaired and thus to a mutation (Fig. M47).

The realization that some chemicals and radiations may mutate genes has led to the development of the discipline of genotoxicology, which involves the identification and study of environmental agents capable of producing mutational changes in exposed organisms. Such studies involve the use of short-term *in vitro* and *in vivo* test systems, of which the best known is the AMES TEST which makes use of specialized strains of the bacterium *Salmonella typhimurium* [2].

Inherited mutations

When spontaneous or induced mutations are present in the germ cells of sexually reproducing organisms they may be transmitted to offspring. Such inherited mutations contribute to the background MUTATIONAL LOAD where they contribute to disease and disability. Inherited disease may be produced by both chromosome abnormalities and by mutations in single genes, although the influence of these two categories upon the observed MUTATION RATE will be uneven.

Chromosome abnormalities involving numerical and structural chromosome alterations are most frequently dominant and their effects are seen in the first generation following their production. Point mutations of dominant effect will also be observed in the first generation, whereas recessive mutations will not be observed until later generations when they become homozygous.

Most chromosomal abnormalities (around 4 per 1000 live births) are numerical, of which the most common in humans is Down's syndrome, which involves the production of an individual with three copies of chromosome 21 (TRISOMY 21). Human structural chromosome abnormalities are present at a frequency of about 2 per 1000 live births of which between 15 and 22% are new mutations, while the remainder are inherited, more often from the mother. The prevalence of chromosome abnormalities in spontaneous abortions is substantially greater than that observed in live births, for example, at 12 weeks up to 50% of the spontaneous abortuses are chromosomally abnormal. The substantial differences between the frequency of chromosome mutations in abortuses and live births indicates the powerful selection during

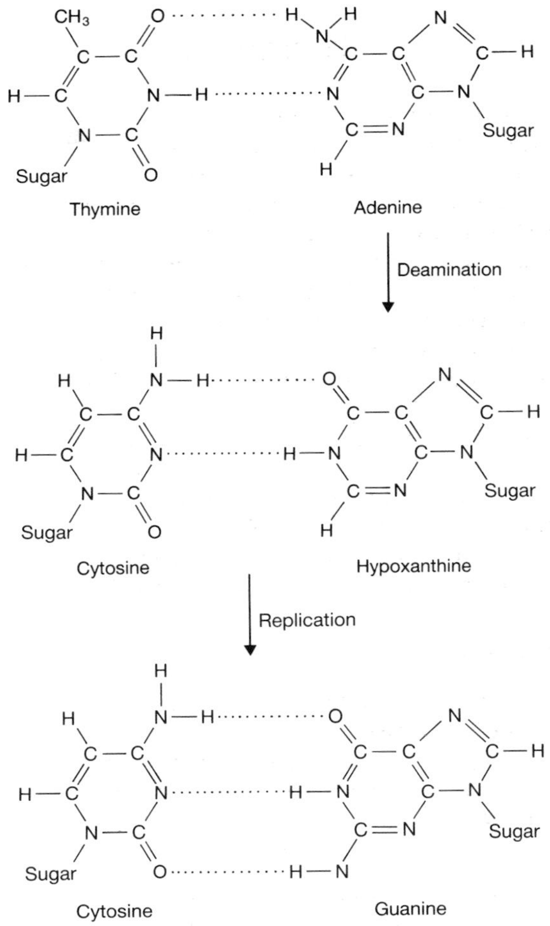

Fig. M47 An example of the action of a chemical mutagen that reacts directly with DNA. The mutagen nitrous acid reacts with bases that contain amino groups, causing their oxidative deamination. Adenine is converted to hypoxanthine, cytosine to uracil and guanine to xanthine. In subsequent rounds of replication, hypoxanthine pairs with cytosine rather than thymine, and uracil with adenine rather than guanine, causing base sequence changes in the new DNA.

pregnancy for offspring which are chromosomally 'normal'.

Mutations in single genes are found at frequencies of approximately 7 to 9 per 1000 live births, of which one-third are recessive mutations in the autosomes (non-sex chromosomes), one-third are dominant mutations in the autosomes and one-third are sex-linked mutations associated with the X CHROMOSOME. The range of human conditions involving mutant single genes is extensive and includes conditions such as ADENOSINE DEAMINASE DEFICIENCY, CYSTIC FIBROSIS, DUCHENNE MUSCULAR DYSTROPHY, GLUCOSE 6-PHOSPHATE DEHYDROGENASE DEFICIENCY, the HAEMOGLOBINOPATHIES, HAEMOPHILIA, HUNTINGTON'S CHOREA, MYOTONIC DYSTROPHY, and the THALASSAEMIAS.

Somatic mutations

Mutations that occur in somatic cells are not transmitted to the sexually produced offspring. However, such somatic mutations may be transferred to descendant daughter cells and mutations in some specific genes have been implicated in cancer [3].

It is now clear that mutations may lead to the induction of cancer when they occur in one or more of a battery of normal genes referred to as the PROTO-ONCOGENES. These proto-oncogenes may be modified by a variety of mutational changes to produce the cancer-causing ONCOGENES of which more than 40 have been identified [4]. Proto-oncogenes play an essential part in the control of cell growth and differentiation and disruption of their normal activity by mutational events may lead to the aberrant growth characteristics observed in cancer cells.

The existence of proto-oncogenes which can be converted or activated to oncogenes by mutational events indicates a role for exposure to environmental agents (radiation and chemicals) in cancer induction. However, there are other types of genes, the TUMOUR SUPPRESSOR GENES or anti-oncogenes, which contribute to cancer development only when they are inactivated or deleted. The normal forms of such genes appear to suppress the development of tumours [5,6] whereas when only mutant forms are present tumours may develop.

Somatic mutation is also involved in the generation of antibody diversity in the immune system (*see* GENERATION OF DIVERSITY; HYPERMUTATION).

J.M. PARRY

See also: CHEMICAL CARCINOGENS AND CARCINOGENESIS; DNA REPAIR; MOLECULAR EVOLUTION. Also mutations referred to under the organism involved, *see e.g.* CAENORHABDITIS DEVELOPMENT; DROSOPHILA; DROSOPHILA DEVELOPMENT; ESCHERICHIA COLI; SACCHAROMYCES CEREVISIAE; named mutations and genes; named human genetic diseases.

1 Auerbeck, C. & Robson, J.M. (1947) *Science* **105**, 243–247.
2 Ames, B.N. et al. (1975) *Mutat. Res.* **31**, 347–364.
3 Boveri, T. (1914) *Zur Frage der Enstehung maligner Tumoren*, 1–64 (Gustav Fischer, Jena).
4 Bishop, J.M. (1987) The molecular genetics of cancer. *Science* **235**, 305–311.
5 Friend, S.H. et al. (1986) A human DNA segment with properties of the gene that predisposes to retinoblastoma and osteosarcoma. *Nature* **323**, 643–646.
6 Marshall, C.J. (1991) Tumour suppressor genes. *Cell* **64**, 313–326.

mutation distance The smallest number of MUTATION events needed to generate one DNA sequence from another. *See also*: MOLECULAR PHYLOGENY.

mutation event The origin of a MUTATION, as opposed to its phenotypic manifestations, which may occur much later.

mutation frequency The frequency of occurrence of a given MUTATION in a population. When applied to recurrent mutations, it is usually defined as the proportion of allele A_1 that mutates to allele A_2 between one generation and the next.

mutation pressure The effect of a recurrent MUTATION on GENE FREQUENCY in a population.

mutation rate The number of MUTATION events in a particular unit of time, for example the number of mutations per cell per generation, or the rate of mutation per locus per gamete. In the study of heritable diseases, the mutation rate is expressed as mutations per locus per gamete, which is effectively mutations per locus per generation. It can be measured directly in AUTOSOMAL DOMINANT diseases (as the number of cases born to unaffected parents) and in diseases where carrier detection is possible (e.g. DUCHENNE MUSCULAR DYSTROPHY and CYSTIC FIBROSIS). For AUTOSOMAL RECESSIVE diseases where carrier status cannot be ascertained, the mutation rate is measured indirectly and involves estimates of the effect of carrier status on fitness.

mutational load The deleterious effects sustained by a particular population following the accumulation of deleterious genes which result from recurrent MUTATION.

mutator gene Any gene which when mutated enhances the spontaneous MUTATION rate of other genes in the same organism. The spectrum of mutational events induced can be mutator specific; for example, in *Escherichia coli* the *mutT* mutation specifically induces AT-CG transversions.

Coulondre, C. et al. (1978) *Nature* **274**, 775–780.

MVB MULTIVESICULAR BODY.

MVR-PCR Minisatellite variant repeat-polymerase chain reaction. *See*: DNA TYPING; POLYMERASE CHAIN REACTION.

Mx proteins *See*: GTP-BINDING PROTEINS.

myasthenia gravis Neuromuscular AUTOIMMUNE DISEASE with symptoms of muscle fatigue in which ANTIBODIES against the NICOTINIC RECEPTOR for acetylcholine are produced. Binding of autoantibodies to the receptor on muscle fibres results in degradation of receptor protein and a decrease in the efficiency of SYNAPTIC TRANSMISSION at the neuromuscular junction.

myb Potential ONCOGENE which has been found rearranged in human colon and bone marrow tumours. It encodes a nuclear protein involved in the control of cell proliferation. The viral counterpart (v-*myb*) is carried by the avian myeloblastosis virus.

myc Cellular counterpart to the ONCOGENE v-*myc* carried by an acutely-transforming strain of the RETROVIRUS avian leukosis virus. It is involved in the chromosome TRANSLOCATION t(8;13) (q24;q32) found in Burkitt's lymphoma where it is translocated into the immunoglobulin heavy chain gene (*see* IMMUNOGLOBULIN GENES). It encodes a TRANSCRIPTION FACTOR, forming a DNA-binding hetero-oligomer with the transcription factor Max.

mycelium The vegetative growth phase of fungi and actinomycetes, formed of an extensive system of branched HYPHAE.

Mycobacterium Genus of slender rod-like bacteria, occurring mostly in pairs, with a tendency to form branched cells. They are aerobic, nonmotile, noncapsulated, and nonsporing, and their cell walls contain waxy lipids which make them difficult to stain, and acid-fast when stained. *Mycobacterium tuberculosis*, the causal agent of tuberculosis, and *M. leprae*, the agent of leprosy, grow intracellularly.

mycoplasma-like organism (MLO) Very small plant pathogenic, motile prokaryotes lacking cell walls. They were discovered in 1967 infecting phloem tissue of plants with 'yellows' disease, which had previously been thought to be of viral aetiology owing to the ability of the infectious agent to pass a bacterial filter. MLOs cause gross distortions of flower morphology and colour. Unlike the similar SPIROPLASMAS they are not helical. They have not been cultured *in vitro*.

mycorrhiza A symbiotic or weakly pathogenic association of a fungus with the roots of a plant. Two main types are recognized: ecto- and endomycorrhiza. In ectomycorrhizae the fungus exists on the surface of the roots forming the so-called Hartig net. Endomycorrhizal fungi invade the roots where the mycelium is confined to a well-defined layer. An additional type is the ectoendomycorrhiza in which there is a Hartig net, but in which the fungus also penetrates the root.

myelin A fatty electrically insulating sheath of varying thickness around a nerve cell axon produced by SCHWANN CELLS in the periphery, and oligodendrocytes in the central nervous system (*see* GLIAL CELLS). The sheath is composed of alternate concentric layers of protein (tangentially oriented) and lipid (radially oriented). The myelin sheath is believed to be formed by a double-layered infolding of the glial cell membrane. *See also*: NODES OF RANVIER.

myeloid lineage Lineage of developing blood cells that will give rise to all white cells of the blood except lymphocytes. *See*: HAEMATOPOIESIS.

myeloma Malignant proliferation of PLASMA CELLS, mainly in the bone marrow. In more than 90% of cases a monoclonal immunoglobulin is secreted (*see* ANTIBODIES). Associated manifestations of disease are osteolysis, humoral immunodeficiency, bone marrow and renal failure, and secondary AMYLOIDOSIS.

myelopoiesis Generation of blood cells other than lymphocytes and red blood cells from the common haematopoietic stem cell of the bone marrow. *See*: HAEMATOPOIESIS.

Myf A TRANSCRIPTION FACTOR involved in MYOGENESIS.

***myo*-inositol** *See*: INOSITOL.

myoblast Immature muscle cell. *See*: MYOGENESIS.

MyoD A member of the helix-loop-helix family of TRANSCRIPTION FACTORS which is thought to regulate a large number of genes involved in MUSCLE differentiation (*see* MYOGENESIS). When *myoD* expression is induced in certain fibroblast cell lines, by TRANSFECTION or by treatment with 5-AZACYTIDINE (*see also*: DNA METHYLATION), some of the lines acquire characteristics of myoblasts, such as the expression of muscle-specific isoforms of actin and myosin. In some cases *myoD* expression alone confers on the cell line the ability to fuse and form elongated, multinucleate myotubes under appropriate culture conditions.

myofibril *See*: MUSCLE.

Myogenesis

MYOGENESIS, the development of MUSCLE, is one of the best understood examples of cellular DIFFERENTIATION [1]. The aim of the study of myogenesis is to find out how cells proceed from their first specification to become muscle to their differentiation as contractile muscle fibres. Of the many kinds of muscles in the animal kingdom, most work on myogenesis has been concerned with the differentiation of skeletal muscle in vertebrates, and it will form the core of this entry. The concentration on vertebrate skeletal muscle is due in part to its direct relevance to understanding human muscular dystrophies (*see* DUCHENNE MUSCULAR DYSTROPHY), but also to its being an unusually convenient cell type to study for several reasons: it comprises a large part of the body; it is extremely specialized; and, most importantly, it has proved possible to grow precursors of muscle cells in culture.

Development and differentiation of vertebrate skeletal muscle

Skeletal muscle develops from the myotomes of the SOMITES, which form in response to MESODERM INDUCTION (*see also* INDUCTION). The myotomes themselves become the embryonic axial musculature and some muscles of the adult trunk; other skeletal muscles, for example of the limbs, are derived from cells that migrate out from the somites. The earliest representatives of the myogenic lineage that have been isolated are the myoblasts, or muscle STEM CELLS: they proliferate when muscles are dissociated and the cells cultured. Myoblasts, which are typically not overtly differentiated, withdraw from the CELL CYCLE in G1, and fuse to form multinucleate muscle fibres, or myotubes. Differentiating myoblasts are called myocytes.

The differentiation of myoblasts into myotubes is a spectacular biochemical and structural transformation (Fig. M48) involving massive and coordinate accumulation of the sarcomeric proteins, such as ACTIN, MYOSIN, tropomyosin, troponin and titin (*see* ACTIN-BINDING PROTEINS; MUSCLE). Synthesis of other proteins also begins, such as the NICOTINIC RECEPTORS for acetylcholine that are necessary for nerve–muscle interactions, and muscle-specific ISOENZYMES of, for example, creatine kinase, a critical enzyme of muscle energy metabolism. Sometimes the term 'myogenesis' is used to refer just to the myoblast to myotube transition.

In addition to primary cultures of dissociated myoblasts, much use has been made of the several clonal CELL LINES that conveniently and reproducibly (though incompletely and not always faithfully) mimic the behaviour of primary myoblast cultures. A

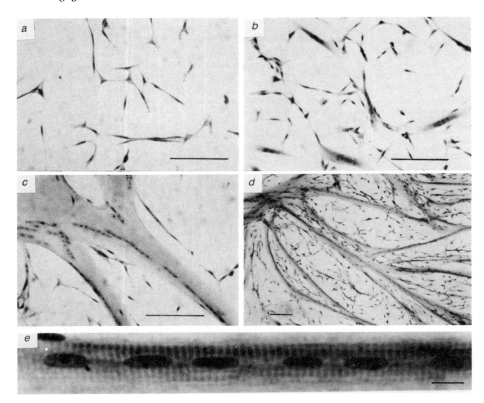

Fig. M48 Photomicrographs of quail muscle cells in culture. All cultures were inoculated with the same number of cells, and fixed, stained, and photographed on representative days, *a*, Day 2 culture consisting of characteristically bipolar mononucleated myoblasts. *b*, Day 3 culture in which fusion is first observed as the formation of short, 'stubby' multinucleated myotubes. *c*, Day 5 culture demonstrating the progressive increase in the number of nuclei within multinucleated myotubes. *d*, Lower magnification photograph of a day 5 culture showing the extensiveness of fusion. *e*, Higher magnification photograph of a multinucleated myotube illustrating the characteristic cross striations. *a–d*, bar = 0.1 mm; *e*, bar = 0.01 mm. From [5].

hypothetical earlier stage of myogenesis, the transition from a pluripotential mesodermal stem cell to a myoblast, has been studied using the C3H10T1/2 cell line. 10T1/2 cells can be converted by treatment with DNA METHYLATION inhibitors such as 5-azacytidine into myoblasts, adipoblasts, and chondroblasts.

There are different kinds of skeletal muscle fibres, defined by their physiology (e.g. fast or slow twitch) and also by the presence of different sets of the muscle-specific protein ISOFORMS that are encoded by small gene families. There is evidence that fibre type is specified both by intrinsic differences between myoblasts and by activity in response to innervation.

Myogenic factors

What determines a myoblast to make muscle, and how is its differentiation orchestrated? Experiments in which muscle and non-muscle cells were fused showed that there are proteins that can act dominantly to activate muscle-specific gene expression. That these myogenic factors are encoded by one or a small number of genetic loci was suggested by the high frequency of conversion of 10T1/2 cells into myoblasts by TRANSFECTION with unmethylated genomic DNA.

The identification of myogenic factors was a landmark in the study of cell differentiation [2]. The first of them, called MyoD for 'myogenic determination gene', was discovered by differential screening of cDNA libraries (*see* DNA CLONING; DNA LIBRARIES) for sequences expressed in myoblasts, but not in non-muscle cells. MyoD was defined as a myogenic factor by its ability in transfection experiments to convert fibroblasts into myoblasts (Fig. M49).

Three other mammalian proteins can do the same: myogenin, Myf5 and MRF4. These four factors form a family of related DNA-binding proteins (*see* PROTEIN–NUCLEIC ACID INTERACTIONS; TRANSCRIPTION FACTORS) that contain the helix-loop-helix dimerization domain and can interact with sequences needed for TRANSCRIPTION from muscle-specific PROMOTERS (see below).

Two other factors can convert non-muscle cells into muscle: the uncharacterized genomic locus *myd* can convert 10T1/2 cells; and the ONCOGENE v-*ski*, whose cellular counterpart is not expressed in a muscle-specific fashion, can produce muscle cells from a population of quail embryo fibroblasts.

Different roles for the different members of the MyoD family are suggested by their different periods of expression during differentiation in tissue culture and in embryogenesis. MyoD and Myf5 are expressed in proliferating myoblasts, and at lower levels in myotubes; myogenin synthesis is activated transiently following the stimulus to differentiate; MRF4 is present in primary musc' cell cultures, but absent from most established lines.

Fig. M49 Elongated, mononucleated myoblast produced by forced expression of MyoD in a dermal fibroblast. Striated myofibrils are seen by staining with an antibody against titin. From [6].

Although all of these factors are skeletal muscle-specific, they differ considerably in their periods of expression in embryogenesis. Some of the MyoD family are expressed at the right time and in the right place to be active in initiating muscle differentiation in the myotomes, others to play a part in maintaining or modulating it, or possibly in later stages of COMMITMENT. Evidence for a role in the earliest stages of muscle differentiation has been obtained in *Xenopus* embryos, in which the number of transcripts encoding MyoD and Myf5 homologues (XMyoD and XMyf5) begins to increase from a very low maternal level in early GASTRULAE (Fig. M50), about two hours before muscle-specific actin transcripts first accumulate. The concentration of XMyoD transcripts falls after somites form, and that of XMyf5 transcripts even sooner.

Transcripts that encode a quail homologue of MyoD — qmf1 — are detectable in the myotomes about as early as the contractile proteins; in mice, myogenin messenger RNA has first been detected at 8.5 days *post coitum*, also around the time when contractile protein mRNAs first accumulate in the myotomes, but MyoD is not expressed until two days later. MyoD and myogenin are thus expressed in the reverse order in mouse myotomes as compared to myoblasts in culture. MRF4 transcripts, which do not appear until much later, are the major species in adult rodent muscle.

A key question about the status of the MyoD family is whether or not the expression of these factors is sufficient to initiate all of the events of myogenesis. MyoD has been described as a 'master regulatory gene' on the basis of its ability to convert some fibroblasts and other cell types into apparently normal myoblasts. However, in transfection experiments with other kinds of cell, such as MELANOMA and NEUROBLASTOMA cell lines, some muscle-specific gene expression could be activated by MyoD, but not full

myogenesis. Some cell lines appear absolutely refractory to MyoD.

It is of particular interest to examine the effects of expressing MyoD family members in early embryonic cells like those that are normally induced to make muscle. Microinjection of synthetic mRNA encoding either XMyoD or XMyf5 into early *Xenopus* embryos causes ectopic activation of normally muscle-specific genes in animal cap cells (*see* ANIMAL POLE), indicating that these factors, together with components already present in animal caps, are sufficient at least to activate muscle-specific gene expression in normal embryonic cells. More recently, GENE KNOCKOUT experiments in mice have shown that, whilst MyoD-null or Myf5-null mice have near normal skeletal muscle, double mutants in these genes apparently have none, and that the myogenin gene is required for normal myogenesis [3].

Regulation of muscle-specific gene expression

Although implicated in effecting DETERMINATION, the most clearly documented activity of the MyoD family is as transcription factors for muscle-specific gene expression during myoblast differentiation. The regulation of the genes that encode the large number of muscle-specific isoforms of structural proteins and enzymes has been studied extensively, largely in experiments in which the expression of the genes and mutants derived from them is assayed by transfection into cell lines.

Normally these genes are expressed strongly in myocytes or myotubes, but not in myoblasts or non-muscle cells, and several promoter and ENHANCER regions capable of conferring this myocyte-specific expression on HETEROLOGOUS promoters have been identified. The importance of some of these sequences has also been tested more rigorously in TRANSGENIC MICE or frogs. No universal characteristic of these promoters and enhancers is yet clear. They contain multiple sequence motifs, variously shared with other muscle-specific promoters and enhancers, and with non-muscle promoters. In several cases these motifs have been shown to interact with muscle-specific binding activities, some of which have been identified as proteins related to MyoD, but others of which apparently are not. The following summary of the state of our knowledge of the control of two of the most thoroughly understood muscle-specific genes provides examples.

The gene that encodes the mouse muscle isoenzyme of creatine kinase (MCK) has extensive and complex control elements. It contains two muscle-specific enhancers, one about 1 kilobase upstream of the transcription start site and the other in the first INTRON, either of which is sufficient to drive myocyte-specific expression of a chloramphenicol acetyltransferase (CAT) transcription unit, acting as a REPORTER GENE. A myocyte-specific binding activity, MEF1, which is antigenically related to MyoD, interacts with both enhancers, but its sites of interaction with the 5′ enhancer, though necessary, are not sufficient for enhancer function. A second myocyte-specific factor, MEF2, and a myoblast-specific factor bind a quite different sequence in the 5′ enhancer, which is required for its full activity. Understanding the elements involved in regulating the expression of the MCK gene is further complicated by the observation that in transgenic mice 700 base pairs (bp) of promoter (lacking both 5′ and first intron enhancers) are sufficient to direct highly muscle-specific CAT

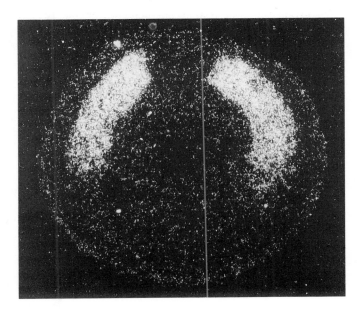

Fig. M50 *In situ* HYBRIDIZATION showing expression of the myogenic factor XMyf5 in the somitic mesoderm of a *Xenopus* gastrula. The gene is expressed before overt differentiation of the myotomes. From [7].

expression, albeit at a much lower level than in the presence of the 5′ enhancer.

A 36-bp sequence centred 100 bp upstream of the start of transcription of the chick acetylcholine receptor (AChR) α-subunit gene is an enhancer that stimulates expression of both the α subunit and the heterologous SV40 promoter (*see* PAPOVAVIRUSES) specifically in differentiating muscle cells. It contains two sites that can bind MyoD, and in which small mutations abolish activity. Although this picture currently appears unusually simple, the existence of other elements has not been ruled out, and it is not known if this short sequence is sufficient also to mediate the dramatic activation of this gene that occurs in response to denervation.

MyoD family members are implicated in the regulation of these two genes and several others. However, the mechanisms that activate genes specifically in skeletal muscle seem more complex and varied than could be explained in terms simply of the properties of the MyoD family, though it is, of course, possible that some of this complexity arises through indirect actions of MyoD family members. Be that as it may, since the MyoD family are the only muscle-specific transcription factors to have been cloned, detailed biochemical studies have necessarily been confined to this group. The sequences with which MyoD interacts in the MCK and AChR α-subunit gene are representatives of the E2 box motif that was originally described in immunoglobulin enhancers (*see* IMMUNOGLOBULIN GENES). On its own, MyoD binds with low affinity, but normally does so tightly as part of a hetero-oligomer, probably with the ubiquitous helix-loop-helix proteins encoded by the E2A gene.

The ability of helix-loop-helix proteins to form hetero-oligomers greatly increases the potential for regulation of their activity. The helix-loop-helix domain is responsible for oligomerization; DNA binding, which requires oligomerization, is mediated by a basic region immediately N-terminal to the helix-loop-helix domain. A helix-loop-helix protein (Id) that lacks a basic domain can inhibit the DNA binding of MyoD by forming nonfunctional complexes with it, or more probably its partner, for which it has a higher affinity. Id is expressed at higher levels in myoblasts than in myotubes, so the presence of Id, or another negative regulator, could explain why myoblasts (which contain MyoD) do not differentiate until appropriate signals relieve this repression.

Distinct biochemical activities of the different members of the MyoD family have not been described, except that MRF4 binds very poorly to E2 boxes. It is difficult to assign specific effects *in vivo*, because members of the MyoD family can activate their own and each other's expression. Knockout experiments in mice have, however, helped to distinguish the contributions of the different MyoD family members [3].

Other muscles, other animals

No myogenic factors have yet been identified even for cardiac muscle, during the development of which many of the same muscle-specific genes are expressed as in skeletal muscle. There is currently much more evidence of conservation across evolutionary time than between different kinds of muscle in the same species. In *Drosophila*, a gene (*nautilus*) that encodes a protein with a basic helix-loop-helix domain very similar to those of the vertebrate factors is expressed in subsets of somatic muscle precursor cells; in *Caenorhabditis*, a member of the family (CeMyoD) marks the body wall musculature, but is not expressed in the pharyngeal muscles. Several *Drosophila* homeobox-containing genes have also been identified, the patterns of expression of which implicate them in muscle development. *S59* is expressed from an early stage in a small subset of somatic muscle cells, suggesting that it could be involved in specifying which particular muscles they make. *H2.0* marks the early visceral mesoderm, and *msh-2* is transiently expressed in all of the segmented mesoderm, before it is restricted to cells that will form the visceral musculature and heart.

Myogenesis in ascidians (protochordates) occurs largely through the influence of a historically distinguished and unusually well documented, but as yet unidentified, CYTOPLASMIC DETERMINANT [4]. Muscle develops in the cells that, naturally or experimentally, inherit a particular region of yellow cytoplasm from the egg, although there are also secondary muscle lineages that are specified by other, possibly inductive, mechanisms.

Questions

The evidence has been outlined here that myogenesis proceeds to a significant extent through the influence of dominant myogenic factors, in particular, though not necessarily exclusively, those of the MyoD family. One major task of current work in the field is to deepen our understanding of the ways in which this influence is mediated, such as by interaction with other proteins. How is the type and specific identity of a muscle established? How does, for example, electrical activity modulate the expression of muscle proteins? How can the present tendency to reduce the problem of muscle cell differentiation to one of gene activation be enriched to explain the generation of complex structures that it entails?

But perhaps the key remaining question is to ask how cells first become determined as a consequence of mesoderm induction to make muscle, rather than, for example, fat or cartilage. This question can be framed in terms of the establishment of the synthesis of MyoD family members. Do the first myoblasts arise directly in response to mesoderm induction, or via a pluripotential mesodermal cell type, as suggested by the 10T1/2 model? To gain experimental access to the earliest stage of myogenesis will require the refinement of techniques for culturing and manipulating early mesodermal cells.

N.D. HOPWOOD

See also: AMPHIBIAN DEVELOPMENT; ASCIDIAN DEVELOPMENT; CAENORHABDITIS DEVELOPMENT; DROSOPHILA DEVELOPMENT; EUKARYOTIC GENE EXPRESSION; EUKARYOTIC GENE STRUCTURE.

1 Kedes, L.H. & Stockdale, F.E. (Eds) (1989) *Cellular and Molecular Biology of Muscle Development*. UCLA Symp. Mol. Cell. Biol. **93** (Liss, New York).
2 Weintraub, H. et al. (1991) The myoD gene family: nodal point during specification of the muscle cell lineage. *Science* **251**, 761–766.
3 Olson, E.N. & Klein, W.H. (1994) bHLH factors in muscle development: dead lines and commitments, what to leave in and what to leave out. *Genes Dev.* **8**, 1–8.
4 Whittaker, J.R. (1987) Cell lineages and determinants of cell fate in development. *Am. Zool.* **27**, 607–622.

5 Konigsberg, I.R. & Buckley, P.A. (1974) Regulation of the cell cycle and myogenesis by cell–medium interaction. In *Concepts of Development* (Lash, J. & Whittaker, J.R., Eds) 179–193 (Sinauer, Stamford, CT).

6 Choi, J. et al. (1990) MyoD converts primary dermal fibroblasts, chondroblasts, smooth muscle, and retinal pigmented epithelial cells into striated mononucleated myoblasts and multinucleated myotubes. *Proc. Natl. Acad. Sci. USA* **87**, 7988–7992.

7 Hopwood, N.D. et al. (1991) Xenopus Myf-5 marks early muscle cells and can activate muscle genes ectopically in early embryos. *Development* **111**, 551–560.

myogenin A TRANSCRIPTION FACTOR of the basic-helix-loop-helix class which is required for effective muscle development. It appears to act in the differentiation of myoblasts rather than in their generation. *See also*: MYOGENESIS.

myoglobin The monomeric haem-containing, oxygen-carrying protein of vertebrate muscle, which facilitates diffusion of oxygen in muscle tissue (Fig. M51). The protein is a member of the globin superfamily (*see* HAEMOGLOBINS). It was one of the first proteins for which a detailed three-dimensional structure was obtained by X-RAY CRYSTALLOGRAPHY. The gene encoding human myoglobin comprises three exons and is located on chromosome 22. Multiple dispersed copies of the middle exon exist in the human genome.

Kendrew, J.C. (1961) *Sci. Amer.* **205**, 91–111.

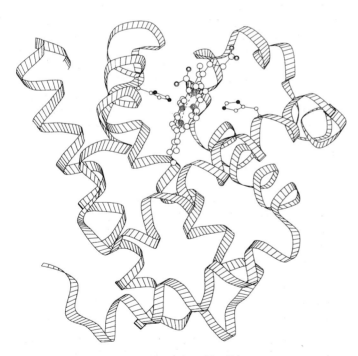

Fig. M51 Sperm whale myoglobin (*see also* Plate 5*a*).

Myosin(s)

MYOSINS are molecular motors which generate force and movement by interacting with actin filaments (*see* MICROFILAMENTS). They occur in a number of different forms in almost all eukaryotic cells, where they drive such diverse processes as muscular contraction, CYTOKINESIS and the transport of cellular organelles (*see* CELL MOTILITY; MITOSIS; MUSCLE).

All myosins contain at least one myosin head, or motor domain, which can generate force in the piconewton range. For example, the conventional myosin IIs (e.g. from skeletal muscle myofilaments) contain two heads and a long tail whereas myosin Is have a single head and a short tail (Fig. M52). The maximum velocity of motion varies from $\sim0.1\ \mu m\ s^{-1}$ (for vertebrate epithelial brush border myosin I) to $\sim70\ \mu m\ s^{-1}$ (for myosin from the alga *Nitella*). Each myosin head (motor domain) contains binding sites for both ATP and actin and utilizes the energy derived from ATP hydrolysis to produce force and movement when bound to actin.

While the head is common to all myosins, it is the tail domains associated with it which determine its role in the cell. For example, the tails of the two-headed myosin IIs enable them to assemble into filaments which are involved in muscular contraction, cytokinesis and the maintenance of cell polarity and tension, while the tails of single-headed myosin Is bind to membranes and they are involved in vesicle transport and membrane movement (see Fig. M53). Thus, myosin Is and IIs are present in the same cell where they generate a variety of movements and also have a structural role, maintaining tension for long periods of time.

We shall first consider the myosin IIs, which have been most extensively studied [1], before discussing some of the novel myosins which have been discovered more recently.

Myosin II structure and function

Although most of our information on the structure and function of myosin IIs has been gained by studies on muscle, all the myosin IIs so far studied from muscle and nonmuscle cells are basically similar, with two heads and a long tail (Fig. M52) [1]. They have a hexameric structure composed of two heavy chains (chain M_r $\sim200\ 000$ (200K)) which fold to form two globular heads joined onto an α-helical coiled-coil tail and two pairs of light chains (chain M_r 16–25K); the regulatory light chains and the essential light chains (Fig. M52). Myosin can be cleaved by the proteinase papain to yield the heads (also known as subfragment-1 or S1) which contain the actin- and ATP-binding sites, and the tail, which is involved in filament assembly. One pair of each light chain is noncovalently associated with each head near the junction with the tail. In vertebrate nonmuscle, smooth muscle, and molluscan myosins, the light chains regulate the force-generating interaction of the myosin head with ATP and actin. In the nonmuscle and smooth muscle myosins this regulation is controlled by phosphorylation of the regulatory light chains by a Ca^{2+}-activated MYOSIN LIGHT CHAIN KINASE (*see also* PROTEIN KINASES), whereas in molluscan myosins Ca^{2+} acts directly by binding to the myosin head. In vertebrate striated muscle myosins

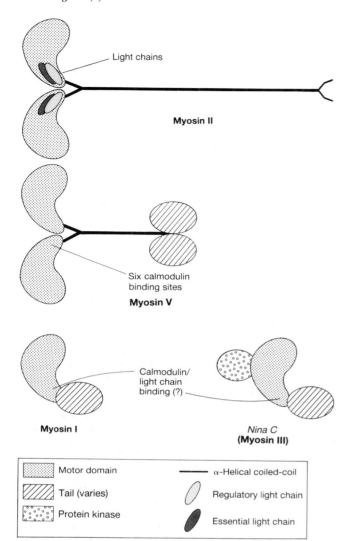

Fig. M52 Diagram showing the four of the main classes of myosins. (Myosins I and II may be further divided into subclasses [5].) All the myosins have a common motor domain (head), containing the ATP- and actin-binding sites, joined onto a variable C-terminal tail region. In the myosin IIs and myosin Vs there are two heavy chains which fold to form the two motor domains whereas the myosin Is (and probably the *NinaC* proteins) have just one heavy chain and therefore one motor domain. The heavy chains in the myosin II tail form an α-helical coiled coil which is involved in filament assembly. This tail region is flexible and can bend into a folded conformation which blocks assembly (see Fig. M55). In the tails of myosin Vs there are small regions of coiled-coil and C-terminal domains of unknown function. In the myosin Is, the tails contain a membrane-binding site and may also contain an actin-binding site and additional motifs. The *NinaC* proteins are unique in having a protein kinase chain domain at the N terminus. The myosin IIs have a pair of light chains bound to each motor domain whereas the other classes of myosin may have variable numbers of light chains or calmodulins associated with them.

the light chains have no apparent regulatory role; in these muscles Ca^{2+} regulates actin–myosin interaction by acting on the troponin–tropomyosin complex which is associated with the actin filament (*see* MUSCLE).

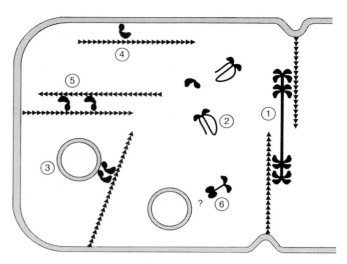

Fig. M53 Cartoon showing possible roles of myosins in a typical eukaryotic cell. The different classes of myosins are represented by the same forms as in Fig. M52. Actin filaments are represented by rows of triangles which indicate their polarity (pointed end (minus) and barbed end (plus)). The barbed ends of the actin filaments are inserted into the membranes (shown in grey) and the myosins move towards these ends (plus-end directed motors). The myosins are believed to be involved in the following cellular events.

1 Myosin IIs assemble into filaments and are involved in muscular contraction, cytokinesis and similar processes where they interact with membrane-attached actin filaments to generate contraction and/or tension. They may also have a structural role in maintaining the shape and polarity of the cell.

2 A pool of nonphosphorylated inactive 10S myosin II monomers may occur in the cytoplasm in equilibrium with filaments. Inactive myosin I molecules also probably occur free in the cytoplasm.

3 All the tail domains of myosin Is seem to contain a membrane-binding site and may be involved in the transport of membranous organelles such as vesicles along actin filament tracks around the cell.

4 Myosin Is may associate with the plasma membrane and bind actin filaments, as for example in the microvillus of epithelium brush border cells where they cross-link the central actin filament bundle to the membrane.

5 Some myosin Is contain an additional, ATP-independent actin-binding site in the tail and may therefore be able to cross-link actin filaments and generate a sliding force between them.

6 The *MYO2* protein appears to be involved in directed vesicle transport along actin filaments in yeast. Its mammalian and avian homologues may have similar roles. However, it is not clear whether they directly associate with vesicles (through a lipid-binding site or a specific receptor) or whether they have a less direct role in this transport process.

The myosin head (motor domain)

Electron micrographs of the myosin head show a large distal globular region and a narrower neck region. The light chains bind to the neck region, extending from the globular region to the neck-tail junction. The amino acid sequence GESGAGKT (single letter notation), conserved amongst all the myosins, is believed to be part of the ATP-binding site. The demonstration, using an *in vitro* motility assay, that the head alone is able to develop force and move actin filaments at velocities comparable to native molecules has proved that all the structural features necessary for force generation and motion are contained within this motor domain. The recently determined crystal structure of myosin-S1 [2] should aid elucidation of the mechanism underlying motor function.

Studies on the basic repeating unit of striated muscle myofibrils, the SARCOMERE, have contributed most to our knowledge of how ATP hydrolysis by the myosin head produces force and motion upon interaction with actin. The highly organized, almost crystalline, sarcomere structure of interdigitating myosin and actin filaments enables maximal force production with a very short response time (millisecond time scale). During muscular contraction, the myosin heads (cross-bridges) act asynchronously, undergoing a nucleotide-dependent cyclic association and disassociation with the actin filaments which leads to the unidirectional translocation of the actin relative to the myosin, as described by the sliding filament hypothesis. For more details see MUSCLE.

The simplified kinetic cycle (cross-bridge cycle) shown in Fig. M54 describes the interactions of the myosin head with ATP and actin [1,3]. In muscle and cytoskeletal assemblies such as the CONTRACTILE RING in the cleavage furrow, the actin concentration relative to the myosin is high and so the pathway through the cycle proceeds mainly via the bound actomyosin (AM) states, that is by steps $1 \to 2 \to 3$. The energy required for the generation of force and movement is provided by ATP binding to the head and is stored as conformational changes in the AM.ATP and AM.ADP.P_i states (steps 1 and 2). It is released as the power stroke driving the movement of the actin filaments relative to the myosin when P_i dissociates from the AM.ADP.P_i complex (step 3). A similar basic reaction mechanism is believed to operate within the motor domains of all the myosin ISOFORMS. It remains unclear, however, how the myosin head produces force and motion. Models have been proposed where the attached myosin head produces force by tilting about its site of attachment to actin, or by actively changing its shape (bending), or by undergoing some other more subtle structural change.

Myosin II tail

The α-helical coiled-coil tails of myosin IIs assemble into filaments by a two-step process; first there is a slow rate-limiting nucleation step followed by a more rapid elongation step. Although vertebrate striated muscle and smooth muscle/nonmuscle myosins assemble into filaments with clearly distinct structures, they all have a prominent 14.3 nm periodicity indicating that the tails are packed in these filaments staggered relative to each other by 14.3 nm. Striated muscle myosin filaments are bipolar, about 1.5 μm long, with a central bare zone flanked by ends in which the heads point in opposing directions. In these filaments, the heads project outwards from the filament backbone in a regular helical array to allow maximal cross-bridge formation with actin filaments (*see* MUSCLE). Smooth muscle and nonmuscle myosins assemble into flexible side polar or small bipolar filaments which can be readily assembled and disassembled in response to cellular requirements, for example for events such as cytokinesis and cell motility.

In vitro it has been shown that phosphorylation of the regulatory light chains of smooth muscle and nonmuscle myosins

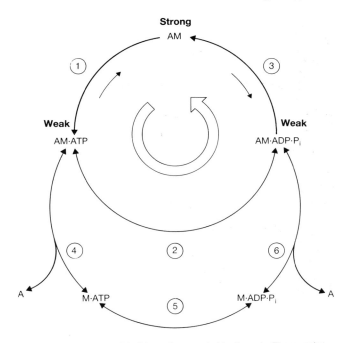

Fig. M54 A simplified model of the actin–myosin kinetic cycle. The myosin head (M) binds strongly to actin (A) in the absence of ATP ('rigor' state K_d ~10^{-10} M) whereas with ATP or products bound at the active site it has a weak affinity for actin (K_d ~10^{-5} M). Thus the myosin–nucleotide intermediates (M.ATP/M.ADP.P_i) are in rapid equilibrium with actin dissociating and rebinding at least 50 times per second (steps 4 and 6). Depending on the actin concentration there are two main pathways through the cycle. At low actin concentrations (*in vitro* conditions) the cycle will be mainly $1 \to 4 \to 5 \to 6 \to 3$ with ATP leading to the dissociation of the weakly bound AM.ATP and AM.ADP.P_i states. As the rate-limiting step (step 3, the release of products) is accelerated about 200-fold when the myosin head is bound to actin, rebinding to form AM.ADP.P_i occurs before ADP and P_i release. When the actin concentration is high (as in muscle) the equilibrium between the weakly bound A-M intermediates will be shifted towards the bound states where AM.ATP and AM.ADP.P_i predominate, and the cycle will proceed mainly via steps $1 \to 2 \to 3$. The energy derived from the release of products, mainly P_i release (step 3), operates the power stroke in the myosin head which drives the relative movement of the actin filaments. Adapted from [3].

promotes filament assembly as well as switching on myosin–ATP–actin interaction (*see* PROTEIN PHOSPHORYLATION) [4]. Light chain phosphorylation regulates filament assembly by controlling an equilibrium between assembly-blocked monomers with a folded tail (sedimentation coefficient 10S), assembly-competent monomers with an extended tail (sedimentation coefficient 6S) and filaments (Fig. M55). Although this equilibrium has not yet been demonstrated to occur *in vivo*, it is believed that in a resting cell a pool of inert folded myosin monomers exists in the cytoplasm. Upon activation, phosphorylation of the regulatory light chains by the specific CA²⁺/CALMODULIN-ACTIVATED KINASE would unfold these myosin monomers which could then assemble into filaments primed to interact with actin filaments to generate the required motile activity (see Fig. M55).

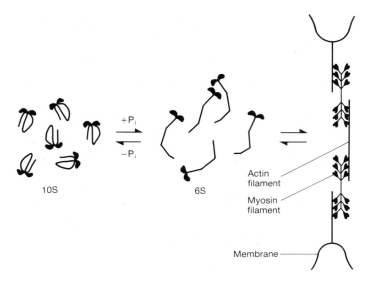

10S   6S

Actin
filament

Myosin
filament

Membrane ——

Fig. M55 Model proposed for the regulation of myosin filament assembly and contractility in nonmuscle cells. In a resting cell it is proposed that myosin II is present in an inert folded monomer conformation (sedimentation coefficient 10S). *In vitro* data have shown that the tails of these folded 10S monomers are bent in two places to generate a compact molecule which cannot assemble into filaments or interact with actin and has ADP.P$_i$ trapped in its active sites. The folded monomer conformation appears to be stabilized by the tail binding to the neck region of the head, near where the light chains are located. In this inert, compact form the myosin should be able to diffuse freely around the cell. When the cell is activated, the increase in Ca^{2+} concentration will activate the myosin light chain kinase and lead to phosphorylation of the regulatory light chains on the myosin. Phosphorylation destabilizes the folded conformation and the myosin tails unfold to form monomers with extended tails (sedimentation coefficient 6S) which rapidly assemble into filaments. Preliminary data suggest that the kinase may be located near or on the actin filaments in the cell. Thus the 'activated' myosin filaments can interact with actin filaments with the correct polarity and geometry to drive contractile events such as cytokinesis.

Role of myosin II in nonmuscle cells

Although the role of myosin II in muscular contraction is well documented, its functions in nonmuscle cells are less clearly understood. The most revealing study of the multiple functions of nonmuscle myosin II has been the molecular genetic manipulation of the single-celled haploid eukaryote *Dictyostelium discoideum* [5]. Using this approach, the single myosin II gene of *Dictyostelium* was disrupted by HOMOLOGOUS RECOMBINATION to generate a NULL MUTANT completely lacking this myosin. *Dictyostelium* null mutant cells in suspension fail to divide and become large and multinucleate. These observations provided the first direct evidence that myosin II is required for CYTOKINESIS (cell division) but not for karyokinesis (nuclear division). Myosin II-minus *Dictyostelium* mutants are able to extend cell surface projections and migrate on a surface, although ruffling occurs all over the membrane rather than at the leading edge and migration is far more haphazard than in wild-type cells. Myosin II therefore, is not required for cell motility but is necessary for the development of polarity in motile cells. These studies also demonstrated that myosin II is involved in the CAPPING of cell surface receptors but not in intracellular particle movement (Fig. M53). The

myosin Is present in these cells are believed to be responsible for cell motility and vesicle transport. It remains to be established whether the deduced roles of myosin in *Dictyostelium* apply to other cells.

Novel myosins

In recent years, a great diversity of myosin-like proteins has been found in many eukaryotic organisms. Their common feature is a conserved motor domain (myosin head), which may be associated with a variety of other domains. These other domains seem to determine the functional roles of these novel myosins in the cell. Phylogenetic analyses of the sequences of motor domains, and comparisons of tail sequence, suggests that there are at least nine separate classes of myosin. These are designated by Roman numerals [6]. In many cases, these myosins have been identified by molecular genetic techniques, which means that little is known about their biochemical properties. Some of the best characterized types are discussed below.

Myosin I

The best known class of novel myosins are the myosin Is which were first identified in the soil amoeba *Acanthamoeba* [6]. These are single-headed myosins, that is, they have only one heavy chain. Myosin I heavy chains all possess a myosin head-like sequence, or motor domain, at their N terminus, and a variable C-terminal tail domain (Fig. M52). Myosin I tails contain a basic region which binds to PHOSPHOLIPIDS and enables them to associate with membranes. They may also contain an ATP-independent actin-binding site, which may enable the molecule to drive sliding movements between actin filaments (Fig. M53). Although the role(s) of myosin Is in cells are not entirely clear, they are believed to be involved in motile processes involving membranes, such as vesicle transport, phagocytosis (*see* ENDOCYTOSIS) and cell motility.

A number of different mechanisms have been identified for regulating myosin I interactions with actin. Protozoan myosin Is are regulated by phosphorylation of their heavy chains by a specific kinase. They also have light chains, but the function of these has not yet been determined. All the vertebrate myosin Is which have been characterized to date contain a number of binding sites for calmodulin (*see* CALCIUM; CALCIUM-BINDING PROTEINS), and their interaction with actin seems to be regulated by Ca^{2+}/calmodulin binding to these sites.

NinaC (Myosin III)

The *NinaC* proteins were identified genetically in the eye of the fruitfly DROSOPHILA. Their unique feature is a protein kinase domain at their N terminus (Fig. A423). This is followed by a myosin-like head domain which is the most divergent known, and has not yet been shown to be a functional motor. They are present as two isoforms (chain M_r 174 and 132K) derived from a single gene by alternative RNA SPLICING, which endows each isoform with a distinct tail sequence. The functions of the tails are unknown, but both contain a basic region like that of the myosin Is.

Myosin V

This class of myosin-like proteins is represented by four recently discovered molecules; the mouse *Dilute* gene product, p190 protein from chicken brain and the yeast SACCHAROMYCES CEREVISIAE *MYO2* and *MYO4* proteins. The chicken brain protein has been biochemically characterized and is therefore known to have a chain M_r of 190K with an actin-activated Mg^{2+}-ATPase. It forms dimers and binds calmodulin, but appears not to assemble into filaments (Fig. M52). The *MYO2* and *Dilute* proteins were both identified by genetics. Yeast *MYO2* appears to be involved in the transport of vesicles to the site of bud formation. The phenotype of the mouse *Dilute* mutation varies; one allele simply causes the mice to have a pale coat colour (hence the name) whereas many others cause severe neurological defects leading to premature death. This suggests that the *Dilute* protein may play a part in vesicle transport in melanocytes and neurons.

The sequences of these molecules reveal a common plan (Fig. M52). They all have N-terminal motor domains, followed by six repeats of a putative calmodulin-binding motif. In the tail they contain a region of heptad repeats which probably allows them to dimerize by forming an α-helical coiled coil, as in myosin II. At the C terminus is a globular domain of unknown function.

Rapid progress is being made in identifying new myosins, by genetic, molecular genetic and biochemical approaches. As well as new members of the above classes, new classes are being discovered continually. The properties of these new molecules should clarify the roles of all motor proteins in the cell.

Prospects

Genetic manipulation of *Dictyostelium* has shed new light on the function of myosin in nonmuscle cells. Furthermore, the expression of functional myosins in *Dictyostelium* and other organisms in quantities that may be purified biochemically will enable structure–function relationships to be resolved. The development of *in vitro* assays to measure movement and force production, coupled with the expression of engineered motor domains, should enable the dissection of the myosin head in terms of motor function. Electron microscopic studies on muscle rapidly frozen in different active states, coupled with improved time-resolved X-RAY DIFFRACTION measurements on activated single muscle fibres, should allow the structural events underlying energy transduction and force production in the head (which occur on the time scale of a fraction of a millisecond) to be analysed. Integral to these studies of myosin motor function is the structure of the head at atomic resolution, which is now complete.

A.E. KNIGHT

T. ROWE

J. KENDRICK-JONES

1 Bagshaw, C.R. (1993) *Muscle Contraction* 2nd edn (Chapman and Hall, London).
2 Rayment, I. et al. (1993) Three-dimensional structure of myosin subfragment-1. *Science* **261**, 50–58.
3 Pollard T.D. (1987) The myosin crossbridge problem. *Cell* **48**, 909–910.
4 Citi, S. & Kendrick-Jones, J. (1987) Regulation of nonmuscle myosin structure and function. *BioEssays* **7**, 155–159.
5 Spudich J.A. (1991) In pursuit of myosin function. *Cell Regul.* **1**, 1–11.
6 Cheney, R.E. et al (1993) Phylogenetic analysis of the myosin superfamily. *Cell Motil. Cytoskeleton* **24**, 215–223.

myosin light chain kinase A Ca^{2+}/calmodulin-dependent, serine/threonine-specific PROTEIN KINASE which phosphorylates the P light chain of MYOSIN. In smooth MUSCLE, this triggers contraction in response to Ca^{2+}-elevating agents such as ACETYLCHOLINE (acting at the MUSCARINIC RECEPTOR). In skeletal muscle, this does not trigger contraction but seems to regulate the force of contraction. The tissue-specific ISOFORMS of myosin light chain kinase in smooth and skeletal muscle are specific for the P light chain from the same tissue. In the absence of Ca^{2+}/calmodulin, both are inhibited by a PSEUDOSUBSTRATE SITE, which is C-terminal to the catalytic domain. Binding of Ca^{2+}/calmodulin at a site overlapping this sequence relieves the inhibition.

Pearson, R.B. et al. (1991) *Eur. J. Biochem.* **200**, 723–730.

myotonic dystrophy (DM) An AUTOSOMAL DOMINANT muscular dystrophy with an estimated incidence varying from one in 50 000 to one in 7500, characterized by progressive muscular wasting, myotonia, cataracts, frontal balding and sometimes mental impairment. Onset is usually in adult life, but offspring of affected females who inherit the gene may be severely affected (congenital myotonic dystrophy). Age of onset and severity of symptoms are highly variable, and the disease can become progressively more severe in succeeding generations. A locus for the disease on human chromosome 19 has been identified. At this locus the trinucleotide CTG is amplified in affected individuals, the size of the DNA amplification correlating with the expression of the full DM clinical phenotype. *See also*: FRAGILE X SYNDROME; HUNTINGTON'S DISEASE; KENNEDY'S DISEASE for other examples of genetic disease involving unstable DNA.

Brook, J.D. et al. (1991) *Cell* **66**, 799–808.

Myoviridae *See*: BACTERIOPHAGES.

myristylation The post-translational addition of the long-chain FATTY ACID myristic acid to a protein, by which the protein is anchored to the cytoplasmic face of the plasma membrane. *See*: MEMBRANE ANCHORS; PAPOVAVIRUSES.

Myxomycota One of the two divisions of the Kingdom Fungi (the other being the EUMYCOTA) and which comprises the ACELLULAR SLIME MOULDS and similar organisms. They lack a cell wall and at some stage of development form a plasmodium, a multinucleate amoebic mass of protoplasm enclosed by the plasmalemma. It is uncertain whether the Myxomycota is closely related to the Eumycota, or even whether the various subgroups within it are closely related to each other. Although most of the Myxomycota feed on decaying material, there are some plant pathogenic classes which form plasmodia within host cells (e.g. the agent of club root disease of brassicas, *Plasmodiophora brassicae*).

N

N The single-letter abbreviation for the AMINO ACID asparagine.

N The abbreviation for the *Drosophila* gene NOTCH.

N-end rule A proposed correlation between the metabolic stability of a protein and the identity of its N-terminal amino acid. The degradation signal comprises a destabilizing N-terminal residue such as arginine and a specific internal lysine residue. *See also*: POST-TRANSLATIONAL MODIFICATION OF PROTEINS; PROTEIN DEGRADATION AND TURNOVER.

Bachmair, A. et al. (1986) *Science* **234**, 179–186.

N-linked oligosaccharide Carbohydrate chains attached to GLYCOPROTEINS through the side-chain (amide) nitrogen of asparagine (also called asparagine- or Asn-linked). In eukaryotic organisms they are added co-translationally to proteins in the ENDOPLASMIC RETICULUM as a precursor oligosaccharide, which is then processed as the glycoprotein moves through the GOLGI APPARATUS. All but a few N-linked oligosaccharides thus contain a common $Man_3GlcNAc_2$ core sequence, with *N*-acetylglucosamine (GlcNAc) linked directly to asparagine (Fig. N1), but there are many different sequences with which their outer chains can be elongated and terminated. There are four types of N-linked oligosaccharides: high-mannose, hybrid, complex, and the xylose-type.

N-*ras* A cellular *ras* gene originally isolated by virtue of its homology with the viral *ras* ONCOGENES. *See*: GTP-BINDING PROTEINS.

N-region diversity Diversity in the final amino-acid sequence of the immunoglobulin heavy chain (*see* ANTIBODIES) and T CELL RECEPTOR polypeptide chains which is introduced by the addition and deletion of nucleotides at the junctions between gene segments during GENE REARRANGEMENT and is not encoded by the genome (*see* GENERATION OF DIVERSITY). In the T cell receptor chains, N-diversity is introduced at the VJ, VD, DD, and DJ junctions. It results from the deletion of nucleotides at the extremities of the coding V, D and J region by the action of an EXONUCLEASE and the addition, at random, of nucleotides by the enzyme TERMINAL DEOXYNUCLEOTIDYLTRANSFERASE (TdT). The N regions can be delimited precisely once the sequences of the germ-line V genes, and D and J segments are known. For the T cell receptor loci, the absence of HYPERMUTATION (*cf.* IMMUNOGLOBULIN GENES) allows the use of oligonucleotides, corresponding to the V and J regions, as primers for the DNA amplification *in vitro* by the POLYMERASE CHAIN REACTION (PCR), and the sequencing of the N regions of the rearranged genes. This method can be applied to the identification of the N region of malignant clones of lymphocytes. Oligonucleotide PROBES specific for the N regions can then be used to detect residual malignant cells during the treatment of leukaemia or in bone marrow samples before an autograft.

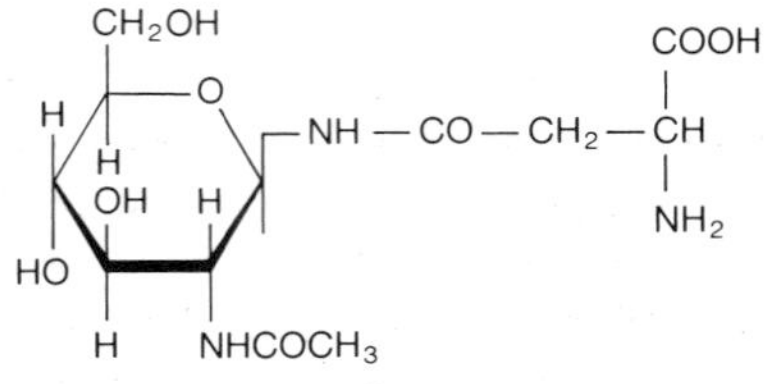

N terminus The amino-terminal end of a polypeptide chain, which usually bears a free α-amino group. Protein biosynthesis proceeds from the N terminus to the C terminus. *See*: AMINO ACIDS; PROTEIN STRUCTURE; PROTEIN SYNTHESIS.

N1-DNA, N2-DNA Larger and smaller respectively of two plasmid-like linear DNAs found in the mitochondria of a male-sterile line (IS1112C) of *Sorghum bicolor*. *See*: MITOCHONDRIAL GENOMES: PLANT.

NA NORADRENALINE.

Na^+–Ca^{2+} exchanger A sodium-driven antiport system in many eukaryotic cells which regulates the cytosolic level of free CALCIUM. One Ca^{2+} ion is exported for each three Na^+ ions that enter. *See also*: MEMBRANE TRANSPORT SYSTEMS.

Na^+–Ca^{2+}, K^+ exchanger Sodium- and potassium-driven antiport system (*see* MEMBRANE TRANSPORT SYSTEMS) in retinal photoreceptor cells which transports CALCIUM out of the outer rod segment during light adaptation. *See*: VISUAL TRANSDUCTION.

Na^+–H^+ exchanger A sodium-driven antiport system in mammalian and other eukaryotic cells, which is involved in cytoplasmic pH regulation. *See also*: MEMBRANE TRANSPORT SYSTEMS.

Na^+,K^+-ATPase Also known as the sodium pump, a P-TYPE ATPASE ubiquitous in the plasma membrane of most eukaryotic cells and which is involved in generating and maintaining the MEMBRANE POTENTIAL. It acts as an electrogenic ION PUMP, transporting three Na^+ ions out of the cytosol in exchange for two K^+ ions from the extracellular medium and producing an elec-

Fig. N1 N-linkage of *N*-acetylglucosamine to asparagine.

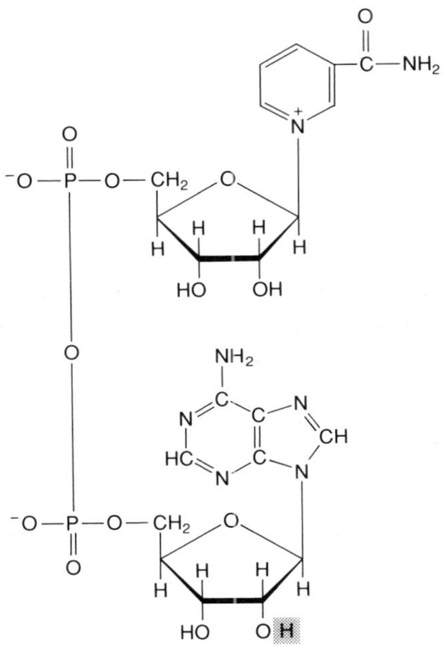

Fig. N2 Structure of the oxidized form of nicotinamide adenine dinucleotide (NAD). In nicotinamide dinucleotide phosphate ($NADP^+$) the group attached to the shaded oxygen atom is PO_3^{2-}.

trochemical gradient of Na^+ across the plasma membrane. *See:* MEMBRANE TRANSPORT SYSTEMS.

nAChR The nicotinic acetylcholine receptor. *See:* NICOTINIC RECEPTOR.

NAD, NADH, NADP, NADPH Nicotinamide adenine dinucleotide, a soluble dinucleotide that can be reversibly reduced by addition of two reducing equivalents to the nicotinamide ring in the form of a hydride H^- ion (Fig. N2). The oxidized form is abbreviated to NAD or NAD^+, the reduced form to, respectively, $NADH_2$ or NADH. The superscript $^+$ denotes a positive charge on the nicotinamide ring nitrogen; the NAD molecule as a whole is negatively charged. NAD is used as an acceptor of reducing equivalents in catabolism, particularly glycolysis, the tricarboxylic acid cycle and β-oxidation of fatty acids, and NADH is reoxidized by complex I of the ELECTRON TRANSPORT CHAIN during oxidative metabolism or by LACTATE DEHYDROGENASE or other dehydrogenases during anaerobic metabolism. The many proteins with NAD-binding domains of known structure all share a common fold — the ROSSMANN FOLD.

The closely related $NADP^+$/NADPH is phosphorylated on the 2′ position of the adenosine ribose ring; although it has similar redox properties it is kept very reduced in cells and is used as a source of reducing equivalents for biosynthetic reactions.

NAD^+- or $NADP^+$-dependent dehydrogenases A large class of oxidoreductase enzymes (EC 1.1.1) acting on alcohols and hemiacetals with NAD^+ or $NADP^+$ as the hydrogen acceptor. May be classified as types A and B on grounds of reaction mechanism, or

as long-chain dehydrogenases such as mammalian ALCOHOL DEHYDROGENASE and LACTATE DEHYDROGENASE, which are metallo-enzymes, and the non-metalloprotein NAD^+-dependent short-chain dehydrogenases, such as an insect alcohol dehydrogenase and some mammalian and bacterial steroid dehydrogenases. All contain a characteristic fold, the ROSSMANN FOLD, which comprises the nucleotide cofactor binding site.

Some enzymes are homotetramers, some homodimers, and each subunit (of around 350 amino acids) is divided into an NAD-binding domain and a catalytic domain. There is little amino acid homology between many of the NAD-dependent dehydrogenases (e.g. liver alcohol dehydrogenase, lactate dehydrogenase and glyceraldehyde 3-phosphate dehydrogenase) but their NAD-binding domains have a similar three-dimensional structure. The catalytic domains are quite different in each subfamily of enzymes.

The NAD-binding domain is composed of a six stranded twisted β-sheet flanked by α-helices. The two halves of the structure are symmetric, each half forming a mononucleotide-binding site.

NAD kinase Enzyme catalysing the phosphorylation of NAD^+ to NADP, by a transfer of a phosphoryl group from ATP to the 2′-hydroxyl group of the ribose moiety of NAD^+.

NADH-cytochrome b_5 reductase Enzyme that forms a complex with cytochrome b_5 and a desaturase on the cytoplasmic face of the endoplasmic reticulum. This complex is involved in the introduction of double bonds into long-chain fatty acyl-coenzyme A_5 during the formation of saturated fatty acids. Electrons are transferred from NADH to the FAD prosthetic group of NADH-cytochrome b_5 reductase, resulting in the reduction of the haem iron atom of cytochrome b_5 and the conversion of the non-haem iron of the desaturase into Fe^{2+}. This then interacts with molecular oxygen and the saturated fatty acyl-CoA, with the formation of a double bond.

NADH dehydrogenase complex Complex I of the mitochondrial ELECTRON TRANSPORT CHAIN, which reoxidizes NADH during oxidative metabolism. It is a large transmembrane protein complex of around 25 subunits, bearing a binding site for NADH on the side facing the mitochondrial matrix. Two electrons from NADH are transferred to a flavin mononucleotide (FMN) prosthetic group in the complex, reducing it, and then to a series of iron–sulphur clusters. From there the pair of electrons is passed to ubiquinone (coenzyme Q), reducing it to QH_2.

NADH-Q reductase NADH DEHYDROGENASE COMPLEX.

NADH shuttle The indirect transfer of electrons from cytosolic NADH into mitochondria with the consequent reoxidation of the NADH formed during glycolysis in the cytosol to NAD^+. In the cytosol, oxidation of NADH to NAD^+ is linked to the reduction of dihydroxyacetone (DHAP) to glycerol 3-phosphate (G3P) which diffuses across the outer mitochondrial membrane, with the transfer of an electron pair to the FAD prosthetic group of the enzyme. The electrons are then transferred to ubiquinone (Q) and enter the ELECTRON TRANSPORT CHAIN as QH_2.

NAF Older name for interleukin-8. *See*: IL-8.

nalidixic acid A member of a family of antibiotics which inhibit the activity of bacterial DNA GYRASE and thus inhibit DNA replication and bacterial growth. More specifically it inhibits the ATP-independent DNA breakage and reunion activity of the enzyme, located in the larger of the two subunits of DNA gyrase.

naloxone An alkaloid that antagonizes the actions of morphine and other opiates, and the endogenous opioids.

NANA N-ACETYLNEURAMINIC ACID.

nanotechnology The construction of devices from active components of nanometric (i.e. 10^{-9} m) dimensions. Modern electronics has become more powerful as the size of devices has been reduced from centimetres (the vacuum triode valve) to millimetres (the transistor) and then to micrometres (10^{-6} m) in present day very large scale integrated circuits. Each miniaturization decreases cost, increases speed and information per unit volume, and lowers heat dissipation per unit of information stored. The next step is to devices with nanometre dimensions. The physical route to these is the atomic force field microscope to move individual atoms into a pattern on the surface (a single row of metal atoms is the finest conceivable wire); the biological route is to exploit the property of existing subunits of proteins to self-assemble into a regular pattern with nanometre dimensions. The proteins required are copied from nature, but with extra stabilizing elements and organic conductors built in either by chemical synthesis (e.g. α-dimethylglycine) or by PROTEIN ENGINEERING. The ease of new protein design and construction by protein engineering techniques is facilitating rapid progress.

1-naphthylacetic acid (NAA) 1-Naphthalene acetic acid (Fig. N3). M_r 186.2. A synthetic AUXIN, the biological activity of which was first reported in 1939. It has practical application as a plant growth regulator in horticulture, being used for rooting cuttings, as a thinning agent on young apple and pear fruit, and to prevent pre-harvest drop of mature fruit. It is also used experimentally as a more stable alternative to INDOLE-3-ACETIC ACID, the principal natural auxin. For example, it binds with high affinity to the AUXIN-BINDING PROTEIN and induces HYPERPOLARIZATION of the plasma membrane in tobacco mesophyll protoplasts. *See also*: PLANT HORMONES.

Gardner, F.E. et al. (1939) *Science* **90**, 208–209.

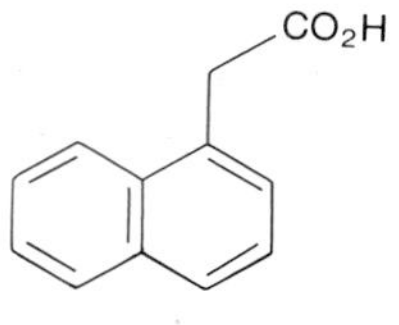

Fig. N3 Structure of 1-naphthylacetic acid.

naphthylphthalamic acid (NPA) N-1-naphthylphthalamic acid (Fig. N4) (M_r 291.3), known also as Naptalam, is a plant growth regulator with herbicidal, fruit thinning, and anti-insect-feeding activity. Its mode of action involves, in part, inhibition of cell-to-cell directional (polar) transport of the endogenous hormone INDOLE-3-ACETIC ACID and other AUXINS that exhibit such transport by binding to a protein subunit that may be a membrane-bound carrier. One consequence of this is an interference in orientation of extension growth with respect to gravity and light. The carrier complex to which NPA binds may be preferentially located at the basal end of auxin-transporting cells.

Rubery, P.H. (1987) In *Plant Hormones and their Roles in Plant Growth and Development* 341–362 (Martinus Nijhoff, Dordrecht).

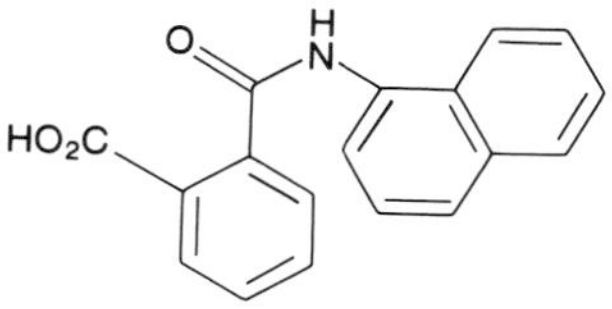

Fig. N4 Structure of *N*-1-naphthylphthalamic acid.

naringenin 5,7,4'-Trihydroxyflavanone (Fig. N5). It is found in peel of *Citrus* fruits, and in species of *Acacia*, *Centaurea*, *Crataegus* and *Helichrysum*, and is the first FLAVONOID formed in flavonoid biosynthesis, by isomerization of 4,2',4',6'-tetrahydroxy-chalcone. Among the glycosides, the 7-neohesperoside and 7-rutinoside are known for their exceptionally bitter taste.

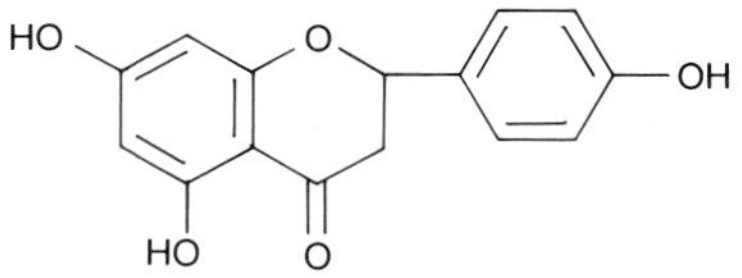

Fig. N5 Naringenin.

nasopharyngeal carcinoma Adult tumour of the nasopharynx, rare except among Chinese people. Predisposition is thought to have both genetic and enviromental components and the EPSTEIN–BARR VIRUS (EBV) may also be involved in its aetiology, EBV having been found in almost all biopsies.

natural killer cells (NK cells) Large granular cells thought to have a role in innate immunity and immune surveillance. NK cells can kill a wide range of target tumour cells regardless of whether they carry the same MHC MOLECULES, but neither the nature of the target molecule nor the receptor on the NK cell is known. NK cells have unrearranged T CELL RECEPTOR GENES. The best surface marker for NK cells is the CD16 antigen (*see* CD ANTIGENS), but some CD3$^+$ lymphocytes (T CELLS) also express CD16. In addition, ~60% of NK cells express the CD2 antigen.

natural selection The differential selection of genetic characteristics (ultimately through differences in reproductive success) mediated by the variation in FITNESS between organisms, which is in part genetically determined. A term coined by Charles Darwin, and central to his evolutionary theories. *See*: DARWINIAN EVOLUTION.

NBD NORBORNADIENE.

NBQX **(6-nitro-7-sulphamobenzo[f]quinoxaline-2,3-dione)** Selective non-NMDA receptor antagonist (Fig. N6). *See*: EXCITATORY AMINO ACID RECEPTORS.

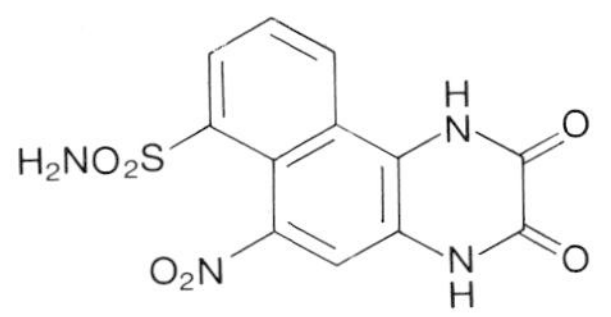

Fig. N6 NBQX.

NCAM Neural CELL ADHESION MOLECULE.

NCS NONCHROMOSOMAL STRIPE MUTATION.

***ndh* genes** Genes for subunits of the mitochondrial NADH DEHYDROGENASE COMPLEX. CHLOROPLAST GENOMES also contain a number of homologous OPEN READING FRAMES, which are believed to encode subunits of a NAD(P)H dehydrogenase (NAD(P)H-plastoquinone oxidoreductase), and are also designated *ndh* genes. *See*: ELECTRON TRANSPORT CHAIN; MITOCHONDRIAL GENOMES.

NE Norepinephrine. An alternative name for the neurotransmitter NORADRENALINE.

nebulin *See*: ACTIN-BINDING PROTEINS; MUSCLE.

necrovirus group Named from the common symptom, necrosis; type member tobacco necrosis virus. They are single component viruses with isometric particles ($t = 3$), 28 nm in diameter. They comprise coat protein subunits of M_r 22.6 000 and each contain a single species of (+)-strand linear RNA (M_r 1.3×10^6–1.6×10^6). There is no information yet on genome organization. *See also*: PLANT VIRUSES.

Matthews, R.E.F. (1982) *Intervirology* **17**, 146.

negative control Type of regulation of gene expression (generally used in connection with INDUCIBLE bacterial operons) in which a gene (or operon) is normally kept in a repressed state in which

transcription is prevented, by binding of a specific repressor. Repression is lifted when an inducer binds to the repressor and prevents its binding to the regulatory region of the operon. The LAC OPERON is the classical example of negative control.

negative staining *See*: ELECTRON MICROSCOPY.

negative strand RNA viruses Single-stranded RNA viruses in which the genomic RNA is of the opposite sense to the mRNA. *See*: ANIMAL VIRUSES; PLANT VIRUSES.

Neisseria Genus of Gram-negative bacteria which includes the human pathogens *Neisseria gonorrhoeae* and *N. meningitidis*. They are nonmotile, aerobic rods and cocci which grow intracellularly. *See also*: BACTERIAL PATHOGENICITY.

nematode development *See*: CAENORHABDITIS DEVELOPMENT; CAENORHABDITIS NEURAL DEVELOPMENT.

neodarwinism The orthodox modern synthesis of evolutionary theory, in which the principles of natural selection and evolution originally proposed by Charles Darwin are interpreted in the light of subsequent discoveries concerning the mechanism of inheritance. Also known as the Synthetic Theory of Evolution. *See*: DARWINIAN EVOLUTION; MENDELIAN INHERITANCE.

neomorphic allele (neomorph) A term in the classification of allele type devised by the geneticist H.J. Muller in the 1930s (*see* ALLELE for a full explanation of this classification). Neomorphic alleles possess activity which is qualitatively different from wild type. Such alleles are almost always dominant to both wild type and a complete deficiency allele.

neomycin An aminoglycoside antibiotic produced by *Streptomyces fradiae* widely used as a selective agent in molecular biology. It inhibits the initiation of prokaryotic PROTEIN SYNTHESIS by binding to the 30S ribosomal subunit. It is used to select for PLASMIDS carrying the NEOMYCIN PHOSPHOTRANSFERASE gene. Phosphorylation of neomycin by this enzyme renders it incapable of binding ribosomal components.

neomycin phosphotransferase (NPT) Kanamycin kinase (EC 2.7.1.95). Enzyme that catalyses the phosphorylation and inactivation of the antibiotics NEOMYCIN, KANAMYCIN, and related compounds. It confers resistance to these antibiotics and the gene is thus widely used as a SELECTABLE MARKER in DNA cloning and transfection.

neoplasm (tumour) Localized population of cells which has become independent of normal cellular growth controls and proliferates. Noninvasive neoplasms of limited growth (e.g. warts) are termed benign; malignant neoplasms (cancers) proliferate in an ill-controlled manner, become invasive, and often spread to other tissues.

neoplastic transformation The conversion of a normal cell into a tumour cell.

nepovirus group Sigla from nematode (-transmitted) polyhedral viruses; type member tobacco ringspot virus. They are multicomponent viruses with isometric particles 28 nm in diameter which sediment as three components. The capsids are composed of a single protein species, M_r 55 000–60 000. The bottom component contains RNA 1, the middle component RNA 2 and the top component lacks nucleic acid. The RNAs have 5′ VPg (a viral protein attached to the 5′ end of genomic RNA) (M_r 36 000) and are 3′ polyadenylated, and are each translated to give a POLYPROTEIN which is subsequently cleaved (Fig. N7). By analogy with cowpea mosaic virus, P63 is probably a protease, P72 has no known function, P2 is the VPg, P23 is probably a protease and P92 is probably part of the polymerase. For RNA 2, P93 is probably involved in cell-to-cell spread. *See also*: PLANT VIRUSES.

nerve cell *See*: NEURON.

nerve growth factor (NGF) A highly conserved basic protein of 118 amino acids comprising a small homodimer with three disulphide bridges. Nerve growth factor was originally described in the pioneering studies of Levi-Montalcini and Hamburger in the early 1950s. The protein was purified and sequenced using the male mouse salivary gland as a source. cDNA cloning has shown that NGF is a member of a larger family of TROPHIC FACTORS (including brain-derived neurotrophic factor and neurotrophin-3) acting upon distinct, overlapping populations of responsive neurons. At least four mRNA transcripts arise from differential splicing of the gene. NGF stimulates axon outgrowth and neuronal survival, and responsive neurons include sensory neurons, post-ganglionic sympathetic neurons, and cholinergic neurons of the basal forebrain. NGF mRNA and protein are present in the target tissues of responsive neurons, probably in limiting amounts so that neurons compete for the protein for survival during normal development.

High and low affinity forms of the NGF receptor have been cloned and sequenced, the former being critical for the trophic effects of NGF. It is identical to the proto-oncogene *trkA*, a transmembrane tyrosine PROTEIN KINASE, so PROTEIN PHOSPHORYLATION is presumed to be an important mechanism in NGF action.

nerve sprouting, nerve regeneration Refers to either the compensatory sprouting of one intact nerve into the territory of another or the attempted regeneration of a damaged nerve. This is usually only successful in the PERIPHERAL NERVOUS SYSTEM.

nerve terminal The axon terminal of a NEURON. *See also*: SYNAPTIC TRANSMISSION.

nestin An INTERMEDIATE FILAMENT protein.

NeuAc N-ACETYLNEURAMINIC ACID.

neural cell adhesion molecule (NCAM) *See*: CELL ADHESION MOLECULES.

neural crest In vertebrate embryos, a group of cells arising from the dorsal margins of the NEURAL PLATE as they approach to form the NEURAL TUBE. From this position, the cells of the neural crest migrate away from the neural tube in a ventral and lateral direction and give rise to many different types of cells. In the trunk, the pigment cells (melanocytes), the dorsal root ganglia (sensory nervous system), the autonomic nervous system, and the adrenal medulla all arise from neural crest cells. In the head, neural crest derivatives include some skeletal elements of the cranium. It is believed that neural crest cells are PLURIPOTENT, each cell being able to generate several derivatives. However, differences between cranial and trunk neural crest are probably determined very early in development (*see* DETERMINATION). *See also*: AMPHIBIAN DEVELOPMENT; AVIAN DEVELOPMENT; MAMMALIAN DEVELOPMENT; PERIPHERAL NERVOUS SYSTEM; VERTEBRATE NEURAL DEVELOPMENT.

neural development *See*: AVIAN DEVELOPMENT; CAENORHABDITIS NEURAL DEVELOPMENT; INSECT NEURAL DEVELOPMENT; VERTEBRATE NEURAL DEVELOPMENT.

neural folds During NEURULATION, the NEURAL PLATE, a thickened portion of the midline ectoderm, gradually curves or bends into a 'U' or 'V' shape. The arms of the 'U' or 'V' are continuous with the lateral ECTODERM. The region of contact between the 'U' or 'V' with the lateral ectoderm is known as the NEURAL FOLD. It is from this region that the NEURAL CREST arises. *See also*: VERTEBRATE NEURAL DEVELOPMENT.

neural induction Also known as 'primary induction' (although not the first INDUCTION to take place in the embryo), this phenomenon was first investigated by the classical experiments of Spemann and Mangold in amphibian embryos (1924). Early involuting MESODERM, forming during the process of GASTRULATION, was transplanted to lie adjacent to flank ECTODERM (i.e.

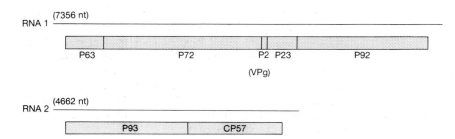

Fig. N7 Genome organization of tomato black ring virus (TBRV). The lines represent the RNA species and the boxes the proteins (P) with the suggested processing positions shown; M_r given as $\times 10^{-3}$. nt, nucleotides.

ectoderm that forms only epidermis during normal development), inducing it to form a secondary nervous system as a result. AXIAL MESODERM is therefore thought to induce the overlying (dorsal, midline) ectoderm to adopt a neural FATE during normal development. There is also evidence that an inducing signal passes within the plane of the ectoderm itself. The molecular nature of the inducing signals, and their relation to identified developmental signal molecules such as the ACTIVINS, is an area of active enquiry. *See also*: AMPHIBIAN DEVELOPMENT; VERTEBRATE NEURAL DEVELOPMENT.

neural plate As one of the consequences of NEURAL INDUCTION in vertebrate embryos, the ECTODERM in the midline of the embryo overlying the NOTOCHORD thickens, the cells becoming columnar and elongated. This thickened plate of cells is called the NEURAL PLATE. Its lateral edges then elevate to form the NEURAL FOLDS and eventually meet in the midline, where they seal, converting the plate into a NEURAL TUBE. *See also*: AVIAN DEVELOPMENT; VERTEBRATE NEURAL DEVELOPMENT.

neural tube A tube derived from the ECTODERM in vertebrate embryos, which constitutes the PRIMORDIUM of the brain and spinal cord. Initially it is a hollow cylinder of epithelial cells, with walls one cell thick, which arises by folding of the NEURAL PLATE. The dorsal margins of the closing tube produce the NEURAL CREST cells which migrate around the embryo. In the head, the neural tube develops characteristic bulging regions corresponding to the various subdivisions of the brain. *See also*: VERTEBRATE NEURAL DEVELOPMENT.

neuraminic acid (sialic acid) A nine-carbon, acidic, amino sugar (5-amino-3,5-dideoxy-D-*glycero*-D-*galacto*-nonulosonic acid), which is a common component of GLYCOLIPIDS and GLYCOPROTEINS. The amino group is always substituted with either an *N*-glycolyl or *N*-acetyl group, yielding *N*-glycolylneuraminic or *N*-acetylneuraminic acid, respectively. The hydroxyl groups are often substituted by acetyl, lactoyl, methyl, sulphate, and phosphate residues. 4-*O*-Acetylation makes neuraminic acid more resistant to the action of NEURAMINIDASE. Neuraminic acid is produced biosynthetically by the condensation of *N*-acetyl-mannosamine and pyruvate to give *N*-acetylneuraminic acid, which is then converted to *N*-glycolylneuraminic acid. Neuraminic acids are transferred to glycolipids and glycoproteins by membrane-bound sialyltransferases using the cytidine-monophosphate (CMP)-linked species as donor.

neuraminidase A glycosidase enzyme (EC 3.2.1.18) which catalyses the hydrolysis of the glycosidic linkage between NEURAMINIC ACID and another molecule. Neuraminidases catalyse the release of neuraminic acid from the carbohydrate chains of GLYCOPROTEINS. *See also*: INFLUENZA VIRUS.

neurectoderm Ectoderm destined to become neural tissue.

neurite A broad term used to describe both axons and dendrites produced by a nerve cell. Most commonly used to cover all cellular processes produced by neurons in culture, the term is also convenient for application to other cells whose processes are of an ambiguous nature.

neuroblast A neuronal precursor cell, that is a cell destined to differentiate into a neuron.

neuroblast differentiation In neuronal development, the DIFFERENTIATION (as opposed to the SPECIFICATION) of a NEURON. It involves the development of the appropriate NEUROTRANSMITTER system, acquisition of a mature NEURONAL CYTOSKELETON and arborization pattern (after migration of the cell body and axon, and dendrite growth to the requisite regions of the body), and formation of the correct synaptic connections with target cells. Many genes are involved, and in the nematode *Caenorhabditis elegans* mutations in some genes have been identified that affect a single aspect of the differentiation of an identified neuron, such as its transmitter phenotype, while others affect several traits simultaneously. Details of the genetic control of neuroblast differentiation have yet to be clarified for vertebrate systems. In at least one case (a subgroup of rat sympathetic neurons) soluble target-derived factors are thought to determine the mature transmitter phenotype. *See also*: CAENORHABDITIS NEURAL DEVELOPMENT; INSECT NEURAL DEVELOPMENT; PERIPHERAL NERVOUS SYSTEM.

neuroblastoma Malignant tumour of sympathoblast derived from the neural crest, occurring mainly in young children.

Neurodegenerative disorders

NEURODEGENERATIVE disorders comprise those diseases characterized by specific patterns of neural degeneration which are usually of late onset and progressive and which either precede or accompany clinical symptoms. They may have a genetic, environmental, or mixed aetiology.

Alzheimer's disease

This was first described neuropathologically by Alois Alzheimer in 1907 from material taken from patients dying with dementia. The histopathology is characterized by the presence of extracellular plaques and largely intracellular tangles within the cerebral cortex, hippocampus, and the diffuse subcortical projection system. Plaques are made up of a rim of dystrophic neurites surrounding a core of AMYLOID β-protein formed from abnormally processed amyloid precursor protein (APP), a membrane-spanning protein found in all nerve cells but of unknown function. Tangles are formed largely from abnormally phosphorylated tau protein, a MICROTUBULE-ASSOCIATED PROTEIN (*see* NEURONAL CYTOSKELETON). A significant number of AUTOSOMAL DOMINANTLY inherited Alzheimer's disease pedigrees have been described, but with genetic heterogeneity. Mutations in the APP gene have been found in a small number of families but excluded in others. Duplication of the APP gene is found in trisomy 21 (DOWN'S SYNDROME) and leads to 'Alzheimer-like' pathological change in the cerebral cortex [1].

Amyotrophic lateral sclerosis (ALS)

Also called motor neuron disease or Lou Gehrig's disease, this is a late onset, inexorably progressive, and ultimately fatal degeneration of upper and lower MOTOR NEURONS. Ten per cent of cases are inherited as an autosomal dominant trait with late onset. In some but not all ALS pedigrees, the disease is linked to a mutation in the gene (*SOD1*) for the Cu/Zn metalloenzyme SUPEROXIDE DISMUTASE, which is located on chromosome 21 [2].

Huntington's chorea (Huntington's disease)

This disease is characterized by progressive, late but variable onset degeneration of the projection neurons of the basal ganglia of the forebrain. It shows autosomal dominant inheritance with complete PENETRANCE of a gene located at the end of the short arm of chromosome 4. The mutation has recently been shown to be due to the insertion of a CAG triplet repeat into a gene of unknown function [3].

Kennedy's disease

Also called spinal and bulbar muscular atrophy, this is a rare late onset form of motor degeneration that may be accompanied by mental retardation and show X-LINKED inheritance. The mutation is an expanded CAG repeat in the androgen receptor gene [3].

Parkinson's disease

A highly specific degeneration of DOPAMINE-containing pigmented cells of the SUBSTANTIA NIGRA of the MIDBRAIN, which leads to symptoms of muscular rigidity, weakness, and tremor at rest (parkinsonism). The disease has no apparent genetic basis but can be simulated by intravenous injection of the drug MPTP which causes a similar loss of dopamine-containing cells and results in a comparable range of clinical symptoms. It is one of the few neurodegenerative diseases that can be managed pharmacologically, in this case by the administration of L-DOPA, a precursor of dopamine (*see* CATECHOLAMINE NEUROTRANSMITTERS). Attempts at treating Parkinson's disease by the transplantation into the brain of foetal dopamine-producing neurons have produced amelioration of symptoms in a few cases [4].

S.D. DAVIES

S.P. HUNT

1 Rossor, M. (1993) Molecular pathology of Alzheimer's disease. *J. Neurol. Neurosurg. Psych.* **56**, 583–586.
2 Rosen, D.R. (1993) Mutations in Cu/Zn superoxide dismutase gene are associated with familial amyotrophic lateral sclerosis. *Nature* **362**, 59–62.
3 Ross, C.A. et al. (1993) Genes with triplet repeats: candidate mediators of neuropsychiatric disorders. *Trends Neurosci.* **16**, 254–260.
4 Gage, F.H. (1993) Parkinson's disease: fetal implants put to the test. *Nature* **361**, 405–406.

neurofibromatosis (von Recklinghausen's disease) An AUTOSOMAL DOMINANT disease with complete PENETRANCE and a frequency of one in 3500, characterized by patches of 'cafe-au-lait' skin pigmentation, cutaneous tumours (neurofibromas), and small nodules in the iris (Lisch nodules). Brain or spinal cord tumours develop in about 10% of individuals, and learning difficulties are present in 30%. The gene responsible (*NF1*) has been localized to the proximal long arm of chromosome 17. As many as half the cases of neurofibromatosis are thought to be due to new mutations. *See also*: TUMOUR SUPPRESSOR GENES.

neurofibromin The product of the *NF-1* TUMOUR SUPPRESSOR GENE, which when mutated gives rise to the disease NEUROFIBROMATOSIS type 1. Neurofibromin is a GTPase-activating protein also known as NF1-GAP (*see* GTP-BINDING PROTEINS).

neurofilament INTERMEDIATE FILAMENT of NEURON. *See*: NEURONAL CYTOSKELETON.

neurogenesis The generation and development of NEURONS. *See*: ACHAETE-SCUTE COMPLEX; AVIAN DEVELOPMENT; CAENORHABDITIS NEURAL DEVELOPMENT; INSECT NEURAL DEVELOPMENT; NEUROBLAST DIFFERENTIATION; NEURON OUTGROWTH; NEURONAL CYTOSKELETON; NOTCH; VERTEBRATE NEURAL DEVELOPMENT.

neuroglia *See*: GLIAL CELLS.

neurohypophysis The posterior lobe of the pituitary which contains secretory nerve endings from the hypothalamus.

neuroleptics *See*: ANTIPSYCHOTIC DRUGS.

neuromedins A family of neuropeptides found in various neural tissues including the spinal cord, brain, and gut. There are many subtypes (e.g. neuromedin-B, neuromedin-C, neuromedin-U). Neuromedin-B shows sequence homology with the neuroactive peptide bombesin, and may have a role in smooth muscle contractility, and the regulation of thyroid-stimulating hormone release.

neuromodulation A general term that covers the regulation of the activity of neuronal cells other than by SYNAPTIC TRANSMISSION mediated by classical NEUROTRANSMITTERS. Neuromodulatory changes are usually considered to be those that affect a neuron's future behaviour in response to stimulation. Many compounds are known to act as neuromodulators, including dopamine and many neuropeptides. An example of neuromodulation by a neuropeptide is the effect of endorphins on the conduction and processing of pain signals (*see* OPIOID PEPTIDES AND THEIR RECEPTORS).

neuromodulin Membrane protein found in growing neurites, which binds calmodulin in its Ca^{2+}-free form (*see* CALCIUM-BINDING PROTEINS).

neuromuscular junction *See*: MOTOR ENDPLATE; SYNAPTIC TRANSMISSION.

neuron Nerve cell. Electrically excitable cell specialized for the conveyance of electrical impulses. MOTOR NEURONS carry motor

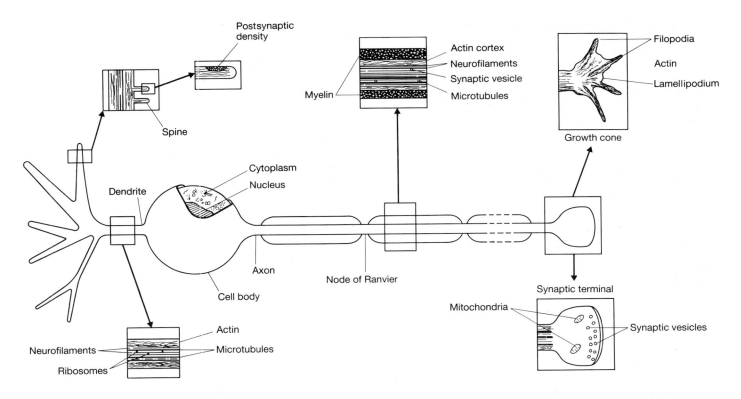

Fig. N8 A schematic diagram of a 'typical' neuron.

signals from the motor cortex within the central nervous system and from the central nervous system to muscles. SENSORY NEU-RONS carry signals from a wide range of sensory receptors in the periphery to the central nervous system. INTERNEURONS convey signals between neurons in the central nervous system.

Neuronal shape and architecture is highly diverse. The illustration (Fig. N8) shows some of the basic features common to most neurons. Most nerve cells have a cell body of between 8 and 30 µm in diameter which contains the nucleus and much of the synthetic machinery required to maintain this highly active cell. Extending from the cell body or soma are a variable number of cellular processes. The axon, which may or may not be encased in an insulating sheath of MYELIN, is generally the largest process, often unbranched, and transmits nerve impulses to target tissues. Axon terminals form SYNAPSES with the dendrites (and also the cell body and axon) of other neurons and muscle. Inside the axon are a variety of cytoskeletal elements — actin microflaments, neurofilaments and microtubules (*see* NEURONAL CYTOSKELETON). Associated with the microtubules will be a variety of organelles undergoing translocation, in particular synaptic vesicles being moved to the synaptic terminal at the end of the axon. The synaptic terminal is relatively free of cytoskeletal elements, with the exception of actin, and is packed with synaptic vesicles. In growing nerves, the end of the axon, the growth cone, is a highly motile area with long finger-like projections or filopodia, with web-like regions of cytoplasm — lamellipodia — stretched between.

Dendrites are usually shorter, more highly branched processes than axons and are specialized for receiving signals. Within dendrites, the cytoskeleton is sparser than that of the axon and there are no synaptic vesicles. Instead, RIBOSOMES may often be seen associated with microtubules and as the dendritic cytoplasm more closely resembles that of the cell body, the neuron is often divided into a somato-dendritic and an axonal compartment. At the ends of the fine dendrite branches, many small protuberances are seen. These dendritic spines contain actin filaments associated with a specialized region of membrane called the postsynaptic density where many neurotransmitter receptors are concentrated.

With the exception of some sensory receptors (e.g. olfactory neurons) differentiated neurons do not divide. However, damaged peripheral neurons can repair themselves and regrow, and spinal cord neurons have also been shown to have the potential to regrow. *See also*: ACTION POTENTIAL; CAENORHABDITIS NEURAL DEVELOPMENT; INSECT NEURAL DEVELOPMENT; NEUROBLAST; NEU-RODEGENERATIVE DISORDERS; OLFACTORY TRANSDUCTION; PE-RIPHERAL NERVOUS SYSTEM; SENSORY NEURON; SYNAPTIC TRANSMISSION; VERTEBRATE NEURAL DEVELOPMENT; VISUAL TRANSDUCTION; VOLTAGE-GATED ION CHANNELS and individual neurotransmitter receptors.

neuron outgrowth The formation of the elaborate circuitry of the mature nervous system involves the guidance of growing axons and dendrites to their target cells by multiple environmental cues. Three main mechanisms have been identified that influence the direction taken by neuronal GROWTH CONES:

1 Selective adhesion to neighbouring cells via CELL ADHESION

MOLECULES (e.g. of the cadherin family or immunoglobulin super-family) or to extracellular matrix via substrate adhesion molecules (e.g. laminin, fibronectin, tenascin, *see* CELL–MATRIX INTERACTIONS).

2 Repulsion from certain body regions via cell repulsion molecules.

3 Concentration gradients of soluble molecules released by pathway or target cells.

In DROSOPHILA, there is good evidence that neighbouring cells or axon tracts bear adhesion molecules, fasciclins, that mediate axon guidance. Removal of a fasciclin in the developing fly nervous system by genetic ABLATION disrupts normal axon pathfinding. Identifying the mechanisms by which growth cones transduce these various external signals to allow changes in direction of growth is an important area for future research. *See*: CAENORHABDITIS NEURAL DEVELOPMENT; INSECT NEURAL DEVELOPMENT; VERTEBRATE NEURAL DEVELOPMENT.

The neuronal cytoskeleton

THE neuronal CYTOSKELETON, the dynamic intracellular framework of protein tubules and fibres, is in principle much the same as that of any other cell in the body, only more so. No other cytoskeleton displays such intricacy and variation in structure as that of neurons and much of the information available on the cytoskeleton in general relates to studies on nervous tissue.

Most neuronal cell types contain the three main cytoskeletal components — MICROTUBULES, MICROFILAMENTS, and INTERMEDIATE FILAMENTS (neurofilaments). Associated with these elements is a bewildering array of accessory proteins involved in controlling cytoskeletal assembly and interactions. Some primary components and accessory proteins are specific to neurons, and some cytoskeletal subunits are restricted to different areas of individual neurons. Most important, the composition of the cytoskeleton of any one neuron alters markedly as it grows, responds to its environment during development and becomes mature.

Here we will follow a neuron from birth to the establishment of mature connections (synapses) and try and relate changes in the cytoskeleton to the varying demands made upon the cell. Neurons are long-lived cells, built for permanence as well as plasticity. A neuron is 'born' when it ceases to take part in the mitotic cell cycle. The new cell (the post-mitotic neuron) must then perform a number of complex tasks before it becomes fully functional. It must migrate from the site of its birth, extend different types of processes or NEURITES, and navigate these through a maze of other developing cells and neurites to form connections with its correct partner(s). The cytoskeleton is of critical importance for both CELL MOTILITY and shape change and this is nowhere more true than for neurons.

As the neuron grows, extends neuronal processes and changes shape its cytoskeleton becomes increasingly complicated. There are many causes of this complexity. Microtubules and intermediate filaments in particular are made up of a variety of subunit proteins which are each present in a number of different forms in the nervous system. Which subunits are used varies both with different neuronal populations and stage of development.

Microtubules

Microtubules are hollow tubes with an outer diameter of 24 nm and are comprised of α- and β-TUBULIN dimers (*see* MICROTUBULES for illustration). Both α- and β-tubulin are present in neuronal tissue in a variety of forms. These arise from the expression of several different tubulin genes (five β-tubulins and four α-tubulins are found in rodent neuronal tissue), the production of more than one mRNA from some of these, and POST-TRANSLATIONAL MODIFICATION of the resulting gene products. All the α-tubulins are substrates for enzymes that reversibly add and remove a tyrosine to the C terminus. In addition, both α- and β-tubulin can be acetylated, polyglutamylated and phosphorylated. In all, about 21 different forms of tubulin (α + β) are present in adult nervous tissue.

The number of different α- and β-tubulins expressed increases as post-mitotic neurons develop, with modified forms predominating as the neurons mature, and there are also changes in the amounts of the different types. The trend seems to be towards producing a more stable population of microtubules as the neuron matures. In some cases microtubules become enriched in a particular tubulin variant. Mature neurons, for example, contain a stable class of microtubules enriched in detyrosinated, acetylated α-tubulin, whereas in younger cells the microtubules contain much unmodified α-tubulin still possessing the terminal tyrosine.

Why the brain expresses so many types of tubulin is not self-evident. It is reasonable to assume that the pattern of tubulin subunits produced by a neuron results in assembled microtubules with some generalized properties common to all neurons — that they should be fast growing during early periods of neurite growth or that they should transport membrane-bound vesicles — but that the general function of the tubules should be fulfilled in a neuron type-specific way. Twenty-one types of tubulin could not allow for the subtle variations required for such complexity and the microtubule-associated proteins (MAPs) and other cytoskeletal proteins must now be considered.

Microtubule-associated proteins

MAPs are readily separated from tissue homogenates and co-purify with microtubules. Three classes of MAPs are generally identified — the high molecular weight MAPs (M_r 200 000–300 000), the tau proteins (M_r 50 000–60 000) and microtubule motor proteins (*see* MICROTUBULE-BASED MOTORS). High molecular weight MAPs and tau proteins are microtubule assembly regulators (predominantly promoters) and cross-linking proteins. Motor proteins such as KINESIN and DYNEIN are used by cells to transport membrane-bound organelles around the cell cytoplasm, to produce contraction and to produce CELL MOTILITY through their involvement in specialized structures such as cilia and flagella.

Neurons express a number of types of both high molecular weight MAPs and tau proteins and expression varies with devel-

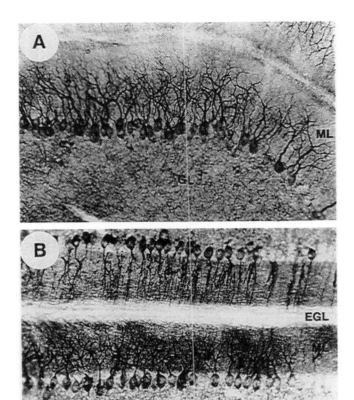

Fig. N9 Localization of the microtubule-associated protein MAP1A to the dendrites of Purkinje cells in the rat cerebellum. In these sections through brains of 3-week-old rats MAP1A is stained to show up as a black reaction product. The elaborate finely branched tree-like structures heavily labelled in *a* and *b* are the cell bodies and dendrites of Purkinje neurons. In the molecular layer (ML) of *a*, relatively little background stain is noticeable; in *b* there is substantially more. The background label represents staining of granule neuron axons which is not seen in normal rats by the time they reach this age, but is seen in animals with stunted brain development (e.g. hypothyroid animals).

opmental stage. For example, one form of high molecular weight MAP — MAP1A — increases in the brain during development whereas another — MAP2c — declines. Some MAPs are restricted to particular regions within the neuron (Fig. N9); most notably, MAP2a and MAP2b are found predominantly in dendrites.

Neurofilaments

Neurofilaments are the form of intermediate filament found within neurons of the central and peripheral nervous system. They are comprised of a variety of subunits and are associated with accessory proteins. They form 10-nm diameter fibres resembling ships' cables in fine structure and run longitudinally along axons from the cell body to the synaptic region (or growth cone in the developing neuron). The dendrites and the axonal tips of mature or growing neurons have relatively few neurofilaments.

The best understood neurofilament subunits are three related proteins termed the low molecular weight, medium molecular weight and high molecular weight subunits (LNF, MNF, and HNF). LNF, MNF, and HNF proteins are encoded by three separate genes which yield, in the mouse, polypeptides of M_r 62 000, 96 000 and 116 000 respectively. Extensive and variable phosphorylation causes large increases in the observed molecular weights of the purified proteins. All three proteins contain a conserved central rod domain and these are believed to co-assemble and form the major structural component of the filament (*see* INTERMEDIATE FILAMENTS). MNF and especially HNF proteins also have a C-terminal region which may form cross-links between individual filaments and to other cytoskeletal elements. Many mature neurons of the central nervous system contain neurofilaments assembled from all three subunits. In the peripheral nervous system and in some central neurons an additional subunit closely related to LNF, MNF and HNF, called peripherin, is also found.

Not all neurons contain neurofilaments. In particular, small neurons such as the numerous cerebellar granule cells do not. This may reflect an important role of neurofilaments in maintaining the integrity of wider diameter, long axons. Indeed, the tough, durable fibres produced by the neurofilament proteins are the closest to a stable, rigid, skeletal structure that the dynamic cytoskeleton achieves.

Neurofilament proteins are altered by post-translational modification. Neurofilaments are most prone to phosphorylation, which occurs primarily in axons rather than in the cell body or dendrites. Neurofilament subunits can be very heavily phosphorylated and this may have a role in stabilizing assembled subunits or in determining interfilament spacing. The spacing between neurofilaments has been proposed to be a determinant of axonal diameter which is itself of great importance to the speed with which an axon conducts an impulse.

Our imaginary growing neuron will eventually contain neurofilaments if its axon is to be of a reasonable diameter but its first intermediate filaments may well be comprised of VIMENTIN. Vimentin filaments are characteristic of cells such as fibroblasts and endothelial cells in the mature animal but are expressed in many types of neuronal and non-neuronal cells during development. An examination of intermediate filament gene expression in neurons would then demonstrate that there is a shift from vimentin expression to that of LNF and MNF followed some time later by HNF. Thus, during development and the production of axons and dendrites, neuronal intermediate filaments as well as microtubules show complex changes in their subunit composition and properties.

Intermediate filament associated proteins (IFAPs) are comparatively poorly understood, but they, like MAPs, also show marked changes in expression and distribution as the cell processes change from immature to mature form. For example, IFAPa-400 is expressed by very young neurons while they contain vimentin, but not by older, neurofilament protein-containing neurons. IFAPs in general are thought to cross-link neurofilaments to each other and also to microtubules and to the other principal cytoskeletal component — the microfilament.

Microfilaments

Microfilaments are 7-nm diameter polymers of ACTIN. They are found throughout the neuron but are arranged in specific regions of the cell to perform defined functions. Neuronal microfilaments are assembled from β and α isotypes of actin (two out of the six actin genes expressed in higher vertebrates). During development the ratio of β : α alters from 1 : 1 to 2 : 1. Actin itself may be modified by post-translational modifications such as methylation, phosphorylation and acetylation.

In common with many other cell types, neurons possess a cortex or girdle of microfilaments located under the PLASMA MEMBRANE. The cortex is attached both to the plasma membrane and to other cytoskeletal components, frequently through accessory proteins. In some regions of the neuronal membrane the association of microfilaments with other cellular components becomes extremely intricate. This is most notable at the pre- and postsynaptic regions of the axon and dendrites.

Cell movement

The microfilament network is believed to be important in creating tension in the cortex, tension which is involved in both the maintenance of cell shape and the morphological alterations which occur in cell migration and neurite production. In the newly forming neuronal process, the cortical microfilament meshwork of the tip of the neurite, the growth cone, is critically involved in extension and guidance. Thick rods of bundled microfilaments jut out from the growth cone to form the core of a structure termed a filopodium (Fig. N10). These extend from the growth cone to investigate the surrounding environment. Where suitable marking points are found, the filopodia will attach and active tension created by microfilament and MYOSIN interaction pulls the growing process in the direction of attachment (*see also* CELL MOTILITY). Localized tension at the growth cone, transmitted by the cortical actin mesh of the shaft of the process, can guide not only the elongating neurite but also the cell body itself if the neuron is one of those that migrates during axon formation. Specific growth-related proteins such as GAP43 are believed to be associated with the microfilament network and may alter the assembly or organization of the actin network. An assembled system of microfilaments is required for neuronal processes to navigate correctly to their target cells. Disruption of polymerized actin by toxins such as CYTOCHALASIN disturbs neurite navigation but not overall growth.

Synapses

Once the new dendrites and axons have grown and met their relevant partners, the domains on the sub-membranous microfilament cortex now change to allow for the consolidation of the new synapse. Specialized regions, the pre- and postsynaptic assemblies, form in the new axons and dendrites and microfilaments are essential to their structure. The axonal side of a synapse, the presynaptic terminal, is an elaborate swelling at the axon tip, which contains many neurotransmitter-containing synaptic vesicles and which possesses a complex, thick meshwork of cytoskeletal and associated proteins, the presynaptic web, abutting a corresponding region on the dendritic side, the postsynaptic density (*see* Figs S36, S37, S38 in SYNAPTIC TRANSMISSION).

Electron microscope examination of the presynaptic web shows it to be comprised predominantly of microfilaments together with a range of thinner filaments which bind and cross-link the actin microfilaments and synaptic vesicles. The cross-linking proteins, which are believed to include SYNAPSIN I and FODRIN, are critical in governing the assembly of the pre-terminal web during the release of neurotransmitters. Synapsin I is believed to cross-link synaptic vesicles to microtubules and microfilaments as well as to each other. Activity in the presynaptic web before the release of neurotransmitter causes the dissociation of synapsin I from the synaptic vesicles through Ca^{2+} and phosphorylation-dependent processes. The net result of this phenomenon is to increase the mobility of synaptic vesicles, presumably to improve their availability for fusion with the plasma membrane and subsequent transmitter release (*see* EXOCYTOSIS). Before fusion and release can occur, the synaptic vesicles must penetrate the network of actin filaments. Work on a close cellular relative of the neuron, the adrenal CHROMAFFIN CELL, strongly suggests that the dissolution of actin meshworks caused by Ca^{2+} influx is essential if neurotransmitter-containing vesicles are to reach the plasma membrane.

Once rearrangements in the cytoskeleton of the presynaptic terminal have permitted the release of neurotransmitter and it has diffused across the synaptic cleft, it reaches receptors on the postsynaptic membrane of a dendrite. The receptors are not randomly scattered around the dendrite but are highly concentrated in postsynaptic densities which are themselves most commonly found on tiny knob-like swellings or dendritic spines which protrude from dendritic branches. Postsynaptic densities form anchorage sites for neurotransmitter receptors and are extreme examples of another role of microfilaments, which is the positioning of membrane proteins, in particular, transmembrane proteins such as receptors and ion channels. The mechanics of how receptors are bound into the cell membrane is not absolutely clear but involves their attachment to proteins which are often members of the family of ACTIN-BINDING PROTEINS. A candidate for a major role in the anchoring of membrane proteins is fodrin. Fodrin is one of a family of related proteins found throughout the body which are believed to bind either directly or via other proteins to receptors and ion channels and to attach them to microfilaments of the submembranous cortical mesh. In the case of receptors in the postsynaptic density, the actin microfilaments pack the dendritic spine. In the case of the Na^+ channels of the axon, actin filaments lie beneath the membrane in a similar manner to elsewhere along the process. However, the Na^+ channels are restricted to the nodes of Ranvier by their association with a protein, ankyrin, which indirectly attaches to microfilaments (*see* MEMBRANE SKELETON). In both these situations, the microfilaments cross-link to the other principal cytoskeletal components — microtubules and neurofilaments.

The developing neuron has now passed from being a newly post-mitotic cell, no more than a sphere of cytoplasm and its nucleus, to its mature, highly complex and diverse form. Many proteins have been transiently expressed as the cytoskeleton

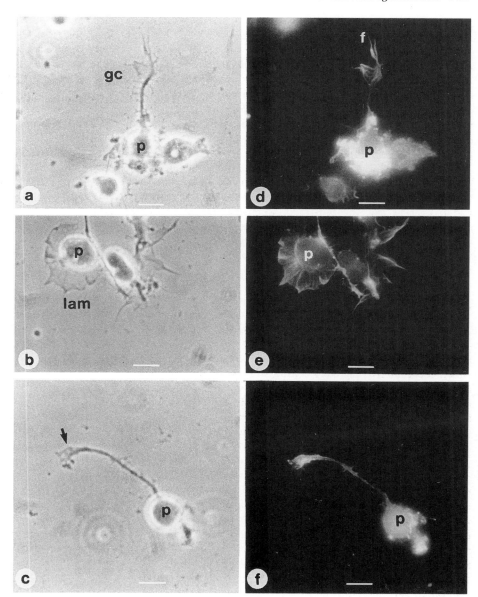

Fig. N10 Cerebellar granule neurons in culture stained to show actin. For each pair of micrographs (*a, d; b, e;* and *c, f*) the left-hand phase contrast micrograph shows the general cell structure, the right-hand fluorescence micrograph the location of actin (bright white rods). The rod-like structures are bundles of actin microfilaments; individual actin fibres are too small to resolve by this method. p, Cell perikaryon; gc, growth cone; f, filopodia; lam, lamellipodia. Scale bar, 5 μm.

changes from an immature plastic form to the more fixed type required by the adult neuron. But microfilaments, microtubules and neurofilaments are all undergoing constant assembly, disassembly and modification to maintain the structural integrity of the cell and its normal function. For example, the mature synapse is the most plastic region of the neuron in adult animals. Both the presynaptic terminal and the dendritic spine apparatus are thought to be reshaped as a result of the experiences involved in learning and memory. Proteins involved in the growth mechanisms of young neurites, such as GAP43, also have roles in models of learning such as LONG-TERM POTENTIATION in the hippocampus. The mature cytoskeleton is still responsive to the environment. Activation of transmitter receptors, for example, alters the association of accessory proteins with microtubules. In a more extreme case, in the regenerating nerve recovering from damage, the developmental programme of cytoskeletal protein expression is substantially re-expressed.

The neuronal cytoskeleton is a constantly changing yet highly organized structure whose roles are of fundamental significance to the most basic functions of any neuron — the reception, integration and transmission of messages throughout the body.

M. CAMBRAY-DEAKIN

See also: SYNAPTIC TRANSMISSION.

1 Burgoyne, R.D. (Ed.) (1991) *The Neuronal Cytoskeleton* (Wiley-Liss, New York).

neuronal filling techniques Techniques used to introduce marker dyes and stains, and other chemicals, into neurons. They fall into two main groups:

1 Those that deliver chemicals such as lucifer yellow into individual cells through an electrode.

2 Those that utilize natural uptake mechanisms on axons and dendrites to get molecules such as fluorogold and HORSERADISH PEROXIDASE conjugates into the cell, which are then retrogradely transported through the neuron.

neuropeptide Any of a very large number of bioactive peptides secreted by neurons in both central and peripheral nervous systems. They have a wide range of activities, acting as NEURO-TRANSMITTERS (e.g. the LHRH-like decapeptide acting in frog sympathetic ganglia), NEUROMODULATORS (e.g. opioid peptides), and RELEASING FACTORS for pituitary hormones (e.g. corticotropin-releasing factor). *See*: OPIOID PEPTIDES AND THEIR RECEPTORS; PERIPHERAL NERVOUS SYSTEM.

neuropil Areas of neuronal dendrites, axons (*see* NEURONS), and GLIAL CELLS found in the central nervous system. The neuropil is distinct from those areas with high densities of neuronal cell bodies.

neurosecretion The secretory activity and secretory products of nerve cells. *See*: EXOCYTOSIS; NEUROMODULATORS; NEUROSECRE-TORY VESICLES; NEUROTRANSMITTERS; SYNAPTIC TRANSMISSION.

neurosecretory vesicles Small membranous sac-like structures, found in neurons, which contain products to be released from the nerve cell. NEUROTRANSMITTER-filled neurosecretory vesicles are found, for example, in the presynaptic terminal at a chemical synapse (*see* SYNAPTIC TRANSMISSION). Upon stimulation, vesicles fuse with the presynaptic membrane and release neurotransmitter into the synaptic cleft. Vesicles are thought to be the structural unit of quantal neurotransmitter release. *See also*: EXOCYTOSIS.

Neurospora crassa Red bread mould. A filamentous fungus of the ASCOMYCOTINA notable particularly as the organism studied by Tatum and Beadle in the 1940s, studies which led to their formulation of the ONE GENE–ONE ENZYME HYPOTHESIS. The mould is heterothallic, growing as a HAPLOID vegetative mycelium, and thus potentially RECESSIVE mutations are immediately expressed. Furthermore, the two mating types can easily be crossed on solid agar medium, to give a diploid zygote from which eight haploid ascospores are produced by MEIOSIS. The arrangement of the eight ascospores within the ascus reflects the order in which nuclear divisions have taken place. It is thus possible to distinguish whether reassortment of any two alleles of a given gene occurred at the first or second division of the meiosis (tetrad analysis). This property makes possible the calculation of the distance of a given gene from the centromere of a chromosome, simply from the numbers of ascospores of a given genotype. *See also*: MENDELIAN INHERITANCE.

neurotoxin Any substance that exerts its toxic effects on the body by its action on nerve cells or is destructive to nerve cells. *See e.g.*

BOTULINUM TOXIN; CAPSAICIN; TETANUS TOXIN; TETRODOTOXIN; α-TOXINS.

neurotransmitter Signalling molecule that mediates SYNAPTIC TRANSMISSION in the nervous system. The 'classical' neurotransmitters mediating fast synaptic transmission between neurons and between neurons and muscle include the CATECHOLAMINE NEUROTRANSMITTERS (adrenaline, noradrenaline, and dopamine), EXCITATORY AMINO ACIDS, 5-HYDROXYTRYPTAMINE and GABA. Other compounds that are known to mediate intercellular communication between neurons include nucleotides and nucleosides (e.g. ATP, adenosine), free radicals of NITRIC OXIDE, and numerous small peptides (neuropeptides) (*see e.g.* OPIOID PEPTIDES; SUBSTANCE P). *See also*: PERIPHERAL NERVOUS SYSTEM.

neurotransmitter receptors Receptors for neurotransmitters comprise both LIGAND-GATED ION CHANNELS and G PROTEIN-COUPLED RECEPTORS (Table N1). *See individual entries.*

neurotransmitter transporters/reuptake systems Specific transport proteins for certain NEUROTRANSMITTERS (e.g. GABA) on GLIAL CELLS and the presynaptic terminal at a synapse which bind and internalize neurotransmitter, thus limiting the lifetime of the signal. *See*: XENOPUS OOCYTE EXPRESSION SYSTEM.

neurotrophin-3 Protein trophic factor from neural tissue and skeletal muscle which promotes the survival of some sensory neurons and motor neurons. It is related to NERVE GROWTH FACTOR and BRAIN-DERIVED NEUROTROPHIC FACTOR and binds to TrkC as a receptor.

neurulation The developmental process in vertebrate embryos which follows GASTRULATION, which begins with NEURAL INDUC-

Table N1 Neurotransmitter receptors

Ligand-gated ion channels	G protein-coupled receptors
	ADRENERGIC RECEPTORS (adrenaline, noradrenaline)
	DOPAMINE RECEPTORS
GABA$_B$ receptors (*see* GABA AND GLYCINE RECEPTORS)	GABA$_B$ receptors
Glycine receptor	
Glutamate receptors (NMDA) (*see* EXCITATORY AMINO ACID RECEPTORS)	Metabotropic (glutamate) receptors
5HT$_3$ RECEPTOR	5HT RECEPTORS
NICOTINIC RECEPTORS (acetylcholine)	MUSCARINIC RECEPTORS (acetylcholine)
P2X purinoceptors (*see* PURINERGIC SYSTEMS)	P1 & P2Y purinoceptors
	Neuropeptides (*see e.g.* OPIOID PEPTIDES AND THEIR RECEPTORS)

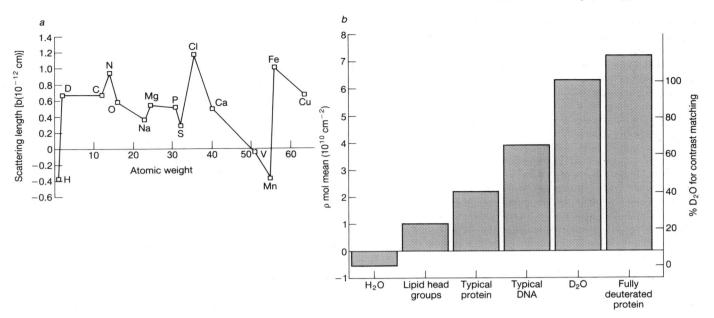

Fig. N11 *a*, Neutron atomic scattering lengths for atoms common in biological macromolecules. *b*, Mean neutron scattering-length densities for different biological macromolecules. The scattering-length densities change with the D_2O/H_2O mixture of the solvent (or mother liquor) due to D–H exchange; therefore in the figure the levels are given to correspond to solvents containing matching D_2O/H_2O scattering-length densities. Note the intense labelling which can be achieved by deuterating proteins at nonexchangeable sites.

TION and continues with the formation of the NEURAL PLATE, terminating with its closure into the NEURAL TUBE. *See also*: VERTEBRATE NEURAL DEVELOPMENT.

neutral evolution, neutralism *See*: MOLECULAR EVOLUTION.

neutral mutation A genetic change which results in a phenotype which has no effect on the organism's fitness.

neutral proteases, neutral proteinases PROTEINASES that function at neutral pH (*cf.* ACID PROTEINASE). *See*: METALLOPROTEIN-ASES.

neutron crystallography *See*: NEUTRON SCATTERING AND DIFFRACTION.

Neutron scattering and diffraction

IN crystallography and scattering studies, molecules are visualized by their component atoms, each atom having its own ATOMIC SCATTERING LENGTH. Atomic scattering lengths for X-rays increase progressively as the atomic number of the element in-creases and therefore light elements such as hydrogen are not easily detectable in ELECTRON DENSITY MAPS by X-RAY CRYSTAL-LOGRAPHY and their positions have to be implied from stereochemistry using the known locations of other atoms.

The atomic number dependence of scattering lengths for neutrons are completely different from those for X-rays and, as shown in Fig. N11*a*, even low atomic number elements such as H and D (hydrogen and deuterium respectively) scatter strongly; indeed, in the case of H the scattering length is significant and negative and scattering is in phase (as opposed to the usual 180° out of phase) with reference to the incident beam of neutrons.

Molecules for diffraction studies may be visualized as contours of equal SCATTERING-LENGTH DENSITIES, where the scattering-length density at a position in a molecule is calculated by counting the scattering lengths in a very small volume at the position in question and dividing by the small volume. MEAN SCATTERING-LENGTH DENSITIES for regions of molecules, complete molecules, and for solvents (or mother liquors) can be calculated for coherent scattering by dividing the total scattering length of the region by the total volume of the region. Figure N11*b* shows neutron mean scattering-length densities for common biomolecules, H_2O and D_2O (water and heavy water respectively). Several of the advantages of neutron diffraction lie in the exploitation of this distribution of scattering lengths and corresponding distribution of scattering-length densities.

Many features of the neutron scattering process are of value in molecular biology. Replacement of nonexchangeable H by D in a region of the molecule provides a completely isomorphous heavy-atom replacement. Hydrogen is almost impossible to detect in macromolecules by X-ray methods so this facility is of considerable value as structured water and protons, near the active site of an enzyme for example, are important in enzyme action.

Further, contrast variation may be used to study molecules in solution (or crystals in their mother liquors) containing different H_2O/D_2O mixtures. This varies the contrast of the solvent relative to the scattering-length densities of different regions of molecules. At low resolution a region can be contrast matched to the solvent so that unmatched regions can be studied separately. The absolute atomic scattering lengths on average are an order of magnitude less for neutrons than for X-rays, giving less beam damage and greater penetration but lower scattering signals in these experiments.

Neutrons are particles which can exchange energy and momentum with a scattering specimen and so an investigation of the inelastic neutron scattering inelastic scattering intensities) gives information about the dynamics of processes occurring in the sample over an important range of characteristic times. Neutrons can be scattered coherently and incoherently; in the former case there can be constructive and destructive interference between waves scattered by different atoms to give scattering in different directions which depend on the molecular structure (for example Bragg diffraction from crystals). It is the nuclear potential distribution in an atom which scatters the neutrons, a process which is localized to atomic nuclei, as opposed to the scattering of X-rays from the electron density distribution. Therefore neutron diffraction and scattering affords a potentially higher resolution in determining atomic positions. Lastly, critical reflection of neutrons from surfaces, for example from membranes and other lamellar structures of biological interest, provides information on the scattering-length density profiles through the thickness of the layers.

The following sections review the neutron facilities and types of experiment which can contribute to our understanding of biomolecular structure, dynamics, and function.

Sources of neutrons

The advent of high flux neutron reactor sources, in the early 1970s in particular, led to great advances in the use of neutrons to study biomolecules. Reactor sources produce neutrons by nuclear fission in the reactor core and they move out radially from the core, with wavelengths in the range given by a Maxwell distribution of velocities corresponding to the reactor temperature. Neutrons grouped in different angular directions from the core are passed through moderators and suffer one or two collisions with atoms from the moderator to be thermalized to the moderator temperature. So, for example there may be a hot source where the wavelengths are shifted to smaller values, a thermal source, and a cold source with wavelengths shifted to higher values.

Several neutron beams at different directions from each moderator are used in different types of experimental arrangement. For example, for small-angle scattering studies, a beam from the cold source is guided to a building several tens of metres away from the reactor by accurately positioned reflectors or guide tubes, and the neutron wavelength is further defined by a velocity (and hence wavelength) selector. The beam then passes through defining slits before being scattered by molecules in solution (in a flat quartz tube for example). A neutron detector, positioned after a chosen scatter path length, detects the angular distribution of scattered neutrons.

Recently, new types of source, spallation neutron sources (SNS), have been developed for the study of condensed matter. The SNS at the Rutherford Appleton Laboratory, UK for example comprises:

1 A source of H^- ions which are produced in 50 pulses per second of about $10\,\mu s$ duration;

2 A linear accelerator taking the particles to 70 MeV of energy;

3 A thin foil through which the pulses of particles pass, which strips electrons from the particles so that they become protons;

4 A proton synchrotron to take them to 800 MeV;

5 A 'kicker' magnet, which deflects the particles from their circular path so that they pass down a straight flight tube; and

6 A target composed of uranium (or titanium) where the 800-MeV proton pulses impact. Spallation neutrons are produced as a result of collisions with the target and, in the case of uranium, some fission neutrons are also produced.

Moderators and experimental arrangements for different types of study are arrayed around the target in a broadly similar way to the arrangements around a reactor source, but the precise design of equipment takes into account the different characteristics of the source, particularly its pulsed nature.

The continuous reactor source with its particular distribution of neutron fluxes as a function of wavelength has complementary advantages (and some overlapping facilities) compared with the spallation source. For example, some experiments using cold neutrons are particularly effective on reactor sources, and these sources are extremely stable and produce very high neutron fluxes. The relatively high flux of spallation neutrons at small wavelengths and the pulsed nature of the neutrons give different advantages, for example the wavelength of each neutron detected can be calculated from its flight time from the target for elastic scattering, since the timing may be made from the time of generation of the neutron pulse at the target.

Contrast variation in neutron diffraction solution studies: Stuhrmann analysis

The procedures are illustrated by the example in Figs N13 and N14 below (see also [1,2]). Theoretically it should be possible to use mathematical inversion calculations (FOURIER SYNTHESIS) directly to determine the positions of atoms in the unit cell of a crystal if all the X-ray (or neutron) diffraction spots from the crystal are observed. This is not possible in practice because the X-ray (or neutron) sources are not laser sources and the phases of the diffraction spots cannot be directly measured. However the *intensities* of the spots can be measured and a Fourier synthesis used directly to determine the PATTERSON FUNCTION for the unit cell.

In studying the scattering of X-rays or neutrons from solutions of biomolecules there is ideally no constructive or destructive interference of waves scattered from the independent molecules. However, interference does occur owing to scattering from atoms *within* each molecule, which therefore gives its own resultant contribution to the scattered intensity. Here again, it is the intensity of scattering that is important and the scattering intensity as a function of scattering angle can be subject to a sin-Fourier synthesis procedure, this time to calculate a spherically integrated Patterson function. This function is called the PAIR-DISTRIBUTION FUNCTION for the molecules in solution and the

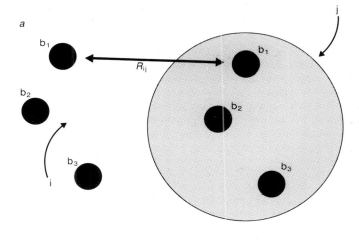

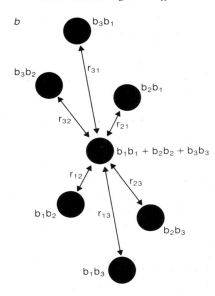

Fig. N12 The Patterson function $P(R)$ for a molecule containing only three atoms, of scattering lengths b_1, b_2 and b_3 respectively. $P(R)$ is generated by translating one representation of the atoms over a second representation, as in *a*, retaining the orientation but moving in three dimensions. Whenever the translation becomes equal to the distance vector between two atoms (say atom i and atom j), R_{ij}, then atom i of representation 1 overlaps atom j of representation 2. A point of weighting, $b_i.b_j$, is then plotted at a position R_{ij} relative to the centre of the Patterson plot. In this way, the Patterson function of *b* is generated as a function of interatomic distance vectors. The pair-distribution function, $D(R)$, is obtained by summing the Patterson values for all directions having the same interatomic distance (i.e. over a sphere of radius R in Patterson space). Fig. N14*b* shows the pair-distribution functions for nucleosome core particles. The peak of the curve occurs at the most probable interatomic distance, while the largest R value gives the maximum interatomic distance.

spherical integration arises because the molecules in solution take up random orientations as opposed to the definite orientations of unit cells in crystals.

The relationship of the Patterson function and the pair-distribution function to the positions of atoms in the molecule are explained in Figs N12 and N14*b*.

The small-angle scattering contrast-variation procedures are as follows:

1 In the neutron small-angle scattering experiment described in the preceding section, neutrons are scattered from a solution of biomolecules and the angular distribution of scattered neutrons is detected by an area detector, so that the data to be analysed are a plot of scattered intensity as a function of the scattering parameter Q, which is related to the angle of deviation (2θ) of a scattered beam from the incident beam ($Q = (4\pi\sin\theta)/\lambda$, where λ is the wavelength of the incident neutrons). These raw data are collected over as wide a range of Q as possible and the scattered intensities are scaled to the scattering from a standard and reduced to scattering per macromolecule (in what follows this term also encompasses macromolecular assemblies) in absolute units, $S(Q)$, using the known concentration of the solution and the molecular weight of the macromolecules.

2 The scattered intensities depend on the excess scattering-length density in the macromolecule, which is the difference between the scattering-length density distribution in the macromolecule itself and the mean scattering-length density of the solvent, ϱ_{sol}. Thus the contrast of the macromolecules is changed by changing the D_2O/H_2O ratio in the solvent and the data acquisition described in (1) is repeated for a range of solvent mixtures of H_2O and D_2O, each of which has its own ϱ_{sol}.

3 Each of the data sets of $S(Q,\varrho_{sol})$ is corrected for incoherent neutron scattering, particularly that coming from the solvent, which gives a nearly constant level of scattering with Q. Guinier plots of $\ln(S(Q,\varrho_{sol}))$ versus Q^2 (see Fig. N13*a*) are plotted for the small Q regions of the spectra, and if the solutions behave ideally (usually this means dilute solutions correctly buffered and inter-particle charge effects minimized) these plots are linear.

The slope of each Guinier plot is equal to one-third of the square of the RADIUS OF GYRATION of the macromolecule $(R_g(\varrho_{sol})^2/3)$ for the solvent concerned. The significance of the R_g values is explained in the example in Fig. N13 and in [1,2].

The square root of the $S(O,\varrho_{sol})$ values, (*see* SCATTERING VECTOR) derived from the intercept of each Guinier plot on the $\ln(S(Q,\varrho_{sol}))$ axis, is equal to the product of the volume of the macromolecule and the contrast. The contrast is the difference between each ϱ_{sol} value and the ϱ_{sol} value which matches the mean scattering-length density of the macromolecule, $\varrho_{molmean}$.

4 A graph of the square root of $S(O,\varrho_{sol})$ is plotted against ϱ_{sol} (see Fig. A463*b*) and, in keeping with the above discussion, this is a straight line of slope proportional to the effective volume of the macromolecule and an intercept on the ϱ_{sol} axis giving $\varrho_{molmean}$. Once the latter has been determined, the contrast $\varrho_{molmean} - \varrho_{sol}$ ($\bar{\varrho}$) can be assigned to each spectrum and to each corresponding Guinier plot.

5 A graph of $R_g(\bar{\varrho})$ squared versus the reciprocal of $\bar{\varrho}$ (the Stuhrmann plot, Fig. A463*c*) is then plotted and Stuhrmann's theory shows that this is concave looking from below. The slope at the origin gives information on whether or not higher or lower

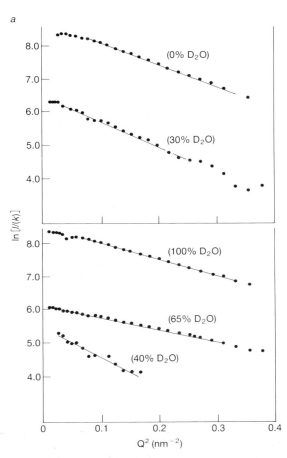

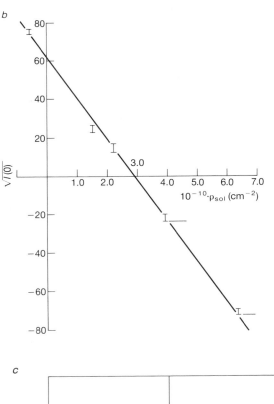

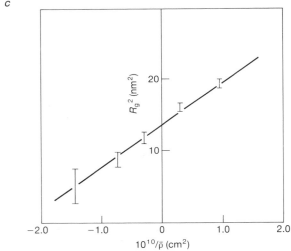

Fig. N13 *a*, Guinier plots for nucleosome core particles at the indicated D_2O percentages in the solvent (from [1]). In the figure, *I* is proportional to $S(Q)$. The slope of a line is $R_g^2/3$, where R_g is the radius of gyration at the appropriate solvent D_2O/H_2O ratio. Radius of gyration squared depends on the sum of each of the scattering lengths multiplied by the square of their distance from the centre of the macromolecule, all divided by the total scattering length for all atoms in the macromolecule. R_g therefore is large if the atoms are far out from the centre of the macromolecule and small if the

atoms are close to the centre. The large R_g of the DNA, when the protein is contrast matched at 40% D_2O, and the small R_g of the protein, when the DNA is contrast matched at 63% D_2O, is seen by the slopes of the lines. This proved that the DNA is wound around the protein core in the nucleosome core particle. *b*, Variation of the square root of the intensity at zero angle, $I(0)$, against solvent scattering-length density, ϱ_{sol}. $I(0)$ values were obtained by extrapolating the plots in *a* to zero angle. *c*, R_g^2 as a function of the reciprocal of contrast (the Stuhrmann plot) for nucleosome core particles from *a* and *b*.

scattering regions are more on the outside of the macromolecule (see later). The curvature at the origin depends on asymmetry in the regions of different scattering-length density in the macromolecule.

6 Each $S(Q,\bar{\varrho})$ spectrum can be processed as described above to produce corresponding pair-distribution functions, $D(R,\bar{\varrho})$ versus

R (see Fig. N12). This information is analysed as described in the example given in Figs N12 and N13*b*, but a most important rationalization of all the data at different contrasts makes the data interpretation easier and allows a proper refinement procedure. This is the fundamental scatter function analysis of Stuhrmann, which shows that at each value of Q the scatter intensities, $S(Q,\bar{\varrho})$,

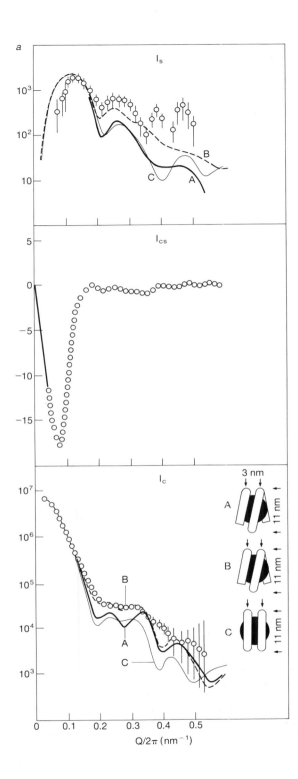

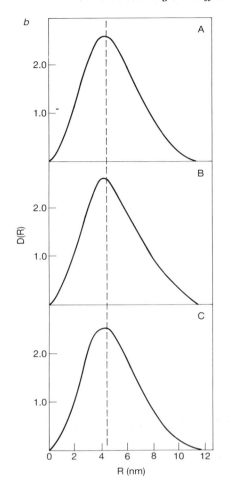

Fig. N14 *a*, The fundamental scatter curves for nucleosome core particles obtained as described in the text (from [2]). Models with the DNA coiling in different ways round the histone protein core were used to predict the scatter curves and the fits to the data are shown together with explanatory dimensions of the models concerned. *b*, Pair-distribution function for nucleosome core particles prepared in different ways in D$_2$O buffers (from [1]): A, control core particles; B, core particles in 20 mM NaCl buffers; C, reconstituted core particles. In all cases the largest interatomic distance was 11.5 nm and the most probable distance was 4.3 nm.

have a parabolic dependence on contrast $\bar{\varrho}$. So a parabolic regression carried out on the experimental data yields three parameters only which depend on Q (see Fig. N14*a*): an intensity contribution from the shape of the macromolecule proportional to $\bar{\varrho}$ squared; a contribution due to the internal structure; and a contribution from a cross-product of shape and internal struc-

ture, proportional to $\bar{\varrho}$. Each intensity function has its own corresponding pair-distribution function.

Small-angle neutron scattering from the NUCLEOSOME core particle serves to illustrate the contrast variation technique. The results of these studies are summarized and explained in Figs N13 and N14 respectively and [1,2].

Contrast variation neutron crystallography

Low-resolution neutron scattering from macromolecular assemblies in solution using the D_2O/H_2O contrast variation method (see above) allows regions of the assembly to be contrast matched, enabling the remaining region(s) to be studied separately. The method is applicable to resolutions where one can consider the mean scattering-length density in the regions and not the internal structure of scattering-length variations. The same method can be used for large single crystals of macromolecular assemblies where the D_2O/H_2O ratio of the mother liquor is varied. Clearly, the data from such studies contain the full information about the oriented three-dimensional structure of the non-contrast-matched components of the system.

This method has now been applied to several important biological systems and its use is expanding as more facilities become available. Two-component systems, such as the protein and DNA in nucleosome core particles or in viruses, may be studied in this way. One of the earliest valuable applications was the neutron low-resolution contrast variation study of the nucleosome core particle [3], where the DNA wrapped around the octamer histone protein core was characterized (at 37% D_2O to match protein) and the shape of the octamer derived (at 65% D_2O to match DNA).

Label triangulation studies of molecular assemblies

Many bacteria can be grown in D_2O-deuterated media to produce fully deuterated multimolecular structures such as RIBOSOMES and proteins expressed from recombinant DNA. Some eukaryotic cells can also be partially deuterated. In the case of some structures, the ribosome being the prime example, deuterated subunits and protonated subunits from different sources can be reconstituted into hybrid structures. The mean scattering-length density of a protein in the reconstituted assembly can be chosen by controlling its level of deuteration.

In label triangulation experiments with ribosomes two proteins are selected for a given experiment (two out of 34 for the 50S subunit and two out of 21 for the 30S). Thus the particle has three regions of scattering-length density: region 1 for protein 1, region 2 for protein 2 and region 3 for the rest of the particle. To calculate the Patterson function for a particle, a process called convolution of the scattering-length density is carried out (see Fig. N12) and this has the symbol $*$ which commutes rather like a multiplication symbol $\times$. The Patterson function therefore is of the form $1*1 + 2*2 + 3*3 + 1*2 + 1*3 + 2*3$ and each of the starred terms has a pair-distribution function giving rise to a contribution to the solution scattering intensities $S11 + S22 + S33 + S12 + S13 + S23$. Each of the six terms in the above Patterson function is weighted by the product of the two regions either side of the $*$ and this is controlled by the level of deuteration.

In theory four small-angle scatter experiments are carried out: (1) With all regions protonated to produce $Sa(Q)$; (2) with the two proteins deuterated and the rest protonated, $Sb(Q)$; (3) with only protein 1 deuterated, $Sc(Q)$; and (4) with only protein 2 deuterated, $Sd(Q)$. By careful consideration of the Patterson function contributions above it is easy to see that if we subtract $Sa(Q) + Sb(Q)$ from $Sc(Q) + Sd(Q)$, all terms cancel out except the $S12$ term.

If the $S12$ term, the result of one label triangulation experiment, is analysed it is seen to be due to the term $1*2$, that is the cross-correlation of protein 1 with protein 2 and no other parts of the particle contribute. This term contains information which enables the distance between the centres of gravity of the two proteins to be determined and it contains information about the shape of the two proteins.

The positions of the centres of scattering-length density of the proteins in the 30S ribosomal subunit have been determined [4] using triangulation with all the distances between the proteins determined as above.

Enhanced contrast variation using polarized protonated and deuterated targets

Stuhrmann and colleagues, at the research reactor at Geesthacht, in Germany, are developing methods of producing enormous enhancement in contrast for use in the contrast-variation methods described previously [5]. In this technique, solutions of macromolecules are frozen to temperatures of ~ 1 K and protonated labels in the deuterated matrix (or deuterated labels in the protonated matrix) are polarized for polarized neutron scattering experiments. Expansion of facilities using this technique should be available in the mid to long term.

High-resolution neutron crystallography

Although it is possible, in theory, to carry out a *de novo* analysis of a protein structure using neutrons, all neutron studies undertaken to date have followed earlier crystallographic work in which the protein structure has been solved and refined at high resolution using X-ray data [6]. Compared with X-ray analysis, relatively few high-resolution neutron diffraction studies of large biologically interesting molecules have been reported. These include studies on bovine pancreatic trypsin inhibitor (BPTI), crambin, haemoglobin, insulin, lysozyme, myoglobin, ribonuclease A, and trypsin. The main reason for the restricted use of neutron diffraction is the relatively low neutron flux currently obtainable, compared with X-ray sources. Thus, for high-resolution neutron studies, large crystals (5 mm^3 minimum) and long counting times (5 minutes per reflection) are required. However, as protein crystals do not suffer radiation damage from neutrons (cf. X-rays), often one large crystal is sufficient for measuring a high-resolution data set.

Neutron diffraction yields valuable additional information about macromolecular structures not readily determined by X-ray methods because of the weak scattering of X-rays from the hydrogen atom compared with carbon, nitrogen, and oxygen. Neutron diffraction provides: (1) more precise location of hydrogen atoms (hydrogen-bonding patterns within proteins); (2) a powerful tool for studying hydrogen exchange within crystals (dynamic behaviour of proteins); (3) a comprehensive description of the bound solvent molecules (protein–water interactions); and (4) information on water–water networks within the crystal lattice (bulk solvent structure).

Understanding the mechanism of action of an enzyme (e.g. the serine proteases) may depend on the identification of a key moiety at the active site which functions as the chemical base during the catalytic process. Neutron diffraction experiments can provide direct evidence of the involvement of particular hydrogen atoms in the catalytic mechanism, for example His 57 has been identified as the catalytic base in trypsin by neutron studies.

A neutron study on insulin to 2.2Å resolution has provided useful new information on the hydrogen positions, H/D exchange within the amide groups, solvent positions, and disorder in some side chains.

Neutron diffraction from fibres

If well oriented, long-chain biomolecules such as collagen, myofibrils, chromatin fibres, and DNA give diffraction patterns with spots that can be Miller-indexed as in single crystal diffraction. But although there is usually good orientation along the fibre axis, there are various degrees of crystallinity in the packing of the molecules in crystalline fibrils and various arrangements of fibrils in the total fibre.

Multicomponent systems such as chromatin can be hydrated in solvents of different D_2O/H_2O ratio so that different components of the system can be contrast matched. In this way neutron fibre diffraction from the protein in the chromatin (DNA contrast matched) and from the DNA (protein contrast matched) can be obtained and the diffraction studied as a function of the level of hydration. Early studies were able to assign a ~10 nm maximum in the diffraction to the repeat along the fibre axis direction of a coil of nucleosomes and to assign other maxima to diffraction from the core particles themselves. Indeed, a 3.5 nm maximum was present at all hydrations in D_2O even in the dry state and was due to the hydrophobic core of the core histone octamer in the nucleosomes.

The conformation of DNA depends not only on the sequence of bases, but more strikingly on the water content, which can be controlled by immersing fibres of DNA in atmospheres of different relative humidity. A, B, C, D, and Z conformations of DNA have been identified, the Z form being a left-handed helix (*see* DNA: NUCLEIC ACID STRUCTURE). Transitions between many of these conformations may be observed simply by changing the relative humidity and the transitions also depend on the positive counterions present.

As more and more base sequence-dependent functions of DNA in the cell are understood, it is becoming clear that the versatility of DNA in its bendability, phasing onto nucleosomes, and other conformational aspects is of vital importance (*see* CHROMATIN; PROTEIN–NUCLEIC ACID INTERACTIONS). It is valuable therefore to study the location of the water and counterions in the different DNA fibre conformations. Isomorphous replacement methods with fibre diffraction on synchrotron radiation sources are being used to study the location of several counterions in different DNA conformations, but high-angle neutron fibre diffraction has had to be used to locate the water molecules in the different conformations [8]. The experiments take advantage of the difference in the atomic scattering lengths of hydrogen and deuterium by measuring the neutron fibre diffraction first by hydration with H_2O and then by hydration with D_2O after allowing some hours for D–H exchange. Difference Fourier methods using this sort of data have enabled the location of water in some of the structures, notably

the D and A structures. In the A form, the identification of water linking the phosphate groups along each DNA strand, water in the major groove, and water in the minor groove, has been made.

Specular reflection of neutrons from surfaces and interfaces

Slow neutrons can be reflected from surfaces. The reflection spectrum gives information on the neutron refractive index profile perpendicular to the surface or interface, which is related in a simple way to the scattering-length density profile. Experiments of this kind can be done with X-ray reflection, but the different atomic scattering amplitudes for neutrons, with the advantages of big changes with isotopic substitution including deuterium for hydrogen, provide unique opportunities for surface structural analysis. The potential for this technique for surface chemistry has recently been realized after the construction of the critical surface reflection spectrometer, CRISP, at the Rutherford Appleton Laboratory's spallation neutron source.

Several groups have carried out measurements of layered structures of biological interest, for example phospholipid layers and phospholipid layers with bound polylysine [9]. The potential of this technique for the study of biomembranes and of detergents used in biochemistry is considerable.

Macromolecular dynamics by inelastic neutron scattering

The neutron experiments described so far probe the structures of biological macromolecules in a manner complementary to X-ray experiments, exploiting the different atomic coherent elastic scattering lengths provided by neutrons. However, static molecular structure determination is only one aspect of understanding function and one must also study molecular dynamics in order to probe active processes such as PROTEIN FOLDING, enzyme activity, transcription, and the dynamics of protein and DNA hydration.

The dynamics of biomolecules can be studied by inelastic neutron scattering, where the scattering from molecules, $S(Q,\omega)$, is studied as a function of frequency, ω, as well as Q. Neutrons exchange energy and momentum with the scattering molecules so the momentum transfer to the neutrons after scattering, indicated by Q, is important and, since the neutron has a large mass, the neutron velocities are only of the order of $1000 \, \text{ms}^{-1}$ and so are measurable. Various ingenious methods have been developed to analyse the energy of the scattered neutrons, both on reactor sources and on pulsed neutron sources where there are advantages in being able to measure velocities of individual detected neutrons from the time of flight from the instant of neutron pulse production at the target.

The potential of inelastic neutron scattering in the study of molecular dynamics is very great and significant developments have been achieved already. Facilities are becoming available for this kind of work, which is both difficult experimentally and has been hampered by the requirement for huge quantities of material (typically hundreds of milligrams of a protein) and limited use of scarce neutron facilities. Inelastic neutron scattering can study dynamic processes on a time scale of $\sim 10^{-7}$–10^{-13} s which covers a range difficult to achieve by other methods and so provides unique information in a range of characteristic times amenable to computer simulations.

Recently, myoglobin has been studied as a powder hydrated at 0.33 g D_2O g^{-1} protein in the low frequency range 1–150 cm^{-1} [10]. Between 100 and 180 K, there was a well-resolved peak in the incoherent inelastic scattering and no broadening of the elastic scattering, suggesting harmonic behaviour in the dynamics over this temperature range. Above 180 K, quasi-elastic scattering occurs and gradually causes the peak to become less distinct. The onset of the quasi-elastic scattering above 180 K indicates a dynamic transition of the system which correlates with observations using MÖSSBAUER SPECTROSCOPY and other inelastic scattering studies. Similar results on hydrated lysozyme powders suggest that the low frequency dynamics of globular proteins have common features.

J.P. BALDWIN
C.D. REYNOLDS
S.J. LAMBERT

1 Sibbet, G.J. et al. (1983) Neutron scattering studies of accurately reconstituted nucleosome core particles and the effect of ionic strength on core-particle structure. *Eur. J. Biochem.* **133**, 393–398.
2 Suau, P. et al. (1977) A low resolution model for the chromatin core particle by neutron scattering. *Nucl. Acids Res.* **4** (11), 3769–3786.
3 Bentley, G.A. et al. (1981) Neutron diffraction studies on crystals of nucleosome cores using contrast variation. *J. Mol. Biol.* **145**, 771–784.
4 Capel, M.S. et al. (1988) Positions of S2, S13, S16, S17, S19 and S21 in the 30S ribosomal subunit of *Escherichia coli*. *J. Mol. Biol.* **200**, 65–87.
5 Stuhrmann, H.B. et al. (1986) Dynamic nuclear polarization of biological matter. *Eur. Biophys. J.* **14** , 1–6.
6 Wlodawer, A. (1982) Neutron diffraction of crystalline proteins. *Prog. Biophys. Mol. Biol.* **40**, 115–159.
7 Schoenborn, B.P. (Ed.) (1984) *Neutrons in Biology* (Plenum, New York).
8 Forsyth, V.T. et al. (1989) Neutron fibre diffraction study of DNA hydration. *Int. J. Biol. Macromol.* **11**, 236–240.
9 Penfold J. & Thomas R.K. (1990) The application of the specular reflection of neutrons to the study of surfaces and interfaces. *J. Phys.: Condens. Matter* **2**, 1369–1412.
10 Cusack, S. & Doster, W. (1990) Temperature dependence of the low-frequency dynamics of myoglobin: measurement of the vibrational-frequency distribution by inelastic neutron scattering. *Biophys. J.* **58**, 243–251.

neutrophil White blood cell, also called a polymorphonuclear leukocyte, which contains a multilobed nucleus and granular cytoplasm. It is involved in inflammatory reactions, migrating from the bloodstream into tissues in response to chemotactic factors produced from bacterial degradation products and by activated T cells. Neutrophils express Fc RECEPTORS for antibodies and receptors for COMPLEMENT proteins and phagocytose opsonized bacteria and viral particles. *See also*: HAEMATOPOIESIS; LYMPHOKINES; SELECTINS.

nexin Protein found in eukaryotic cilia where it forms elastic links between adjacent microtubule doublets. *See*: MICROTUBULES. *See also*: PROTEASE NEXIN.

nexus Gap junction. *See*: CELL JUNCTIONS.

NF-AT Nuclear factor induced in T CELLS stimulated through the T CELL RECEPTOR–CD3 complex and which is required for induction of interleukin-2 gene expression (*see* IL-2; LYMPHOKINES). It is thought to be a TRANSCRIPTION FACTOR.

NF-E2 Erythroid specific TRANSCRIPTION FACTOR (*see* LOCUS CONTROL REGION). The erythroid-specific component of NF-E2 has been cloned and is a basic leucine zipper protein (*see* PROTEIN–NUCLEIC ACID INTERACTIONS). *In vivo* it is thought to form a dimer with some ubiquitous protein.

NF-κB Protein specific to B CELLS which binds to a specific DNA sequence (5′-GGGGACTTTCC-3′) within the immunoglobulin light chain κ locus ENHANCER region in mice and humans (*see* IMMUNOGLOBULIN GENES). The NF-κB binding site is essential for optimal κ enhancer activity. NF-κB DNA-binding activity is seen only in mature B cells and not in the earlier pre-B cell stage (*see* B CELL DEVELOPMENT). This cell-stage specificity correlates positively with the pattern of κ gene expression. Both the DNA-binding activity of NF-κB and κ gene expression can be induced in pre-B cells by MITOGENS such as bacterial lipopolysaccharide and phorbol esters. This does not require *de novo* protein synthesis, suggesting that induction activates a nonfunctional precursor form of NF-κB. PHORBOL ESTER induction of NF-κB activity has also been found in some non-B cells. Activity in these cells is undetectable without induction. It is possible that synthesis of the inactive precursor occurs in a wide variety of cell types, whereas production of active NF-κB is restricted to later stages of the B cell lineage.

NF-L, NF-H, NF-M INTERMEDIATE FILAMENT proteins from neurons. See: NEURONAL CYTOSKELETON.

NF1 A TUMOUR SUPPRESSOR GENE implicated in the development of some familial forms of NEUROFIBROMATOSIS. It encodes a protein resembling the GAP proteins that activate the GTPase activity of the products of the *ras* proto-oncogenes. See also: GTP-BINDING PROTEINS; ONCOGENES.

NGF NERVE GROWTH FACTOR.

nick translation A method of radioactively labelling double-stranded DNA (Fig. N15). Nicks on one strand of the DNA are introduced by the endonuclease DEOXYRIBONUCLEASE I (DNase I). The enzyme DNA POLYMERASE I is then used to add nucleotide residues to the 3′ end of the nick. DNA polymerase I also has a 5′–3′ EXONUCLEASE activity, and this eliminates nucleotides from the 5′ end of the nick. Polymerase I thus progresses along one strand of the duplex, incorporating radioactively labelled nucleotides as it does so.

Nicotiana tabacum Tobacco, a widely used host plant in experimental PLANT GENETIC ENGINEERING. *See also*: PLANT VIRUSES; TOBACCO MOSAIC VIRUS.

nicotinamide adenine dinucleotide *See*: NAD.

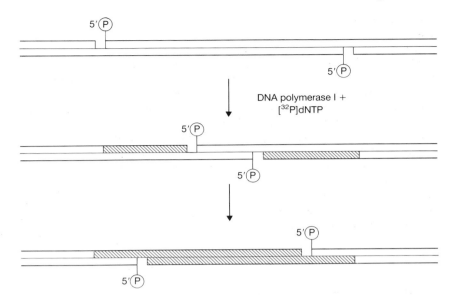

Fig. N15 Nick translation method of preparing highly radioactive DNA fragments. Starting with a nicked DNA duplex, the combined action of the 5′-to-3′ exonuclease and polymerase activity of DNA polymerase I from *Escherichia coli* in the presence of α-^{32}P-labelled deoxynucleoside triphosphates results in the movement of the nick down the strand and the synthesis of ^{32}P-labelled DNA in the wake of the nick.

Nicotinic receptors

NICOTINIC acetylcholine receptors (nAChRs) are found predominantly at vertebrate NEUROMUSCULAR JUNCTIONS where they mediate the response elicited by the neurotransmitter ACETYLCHOLINE (ACh). The interaction of acetylcholine with the nAChR results in the rapid opening of a cation-selective integral ION CHANNEL which leads ultimately to the contraction of skeletal muscle. The nAChRs are members of the LIGAND-GATED ION CHANNEL SUPERFAMILY. Indeed, their structures are the best characterized of this protein family and they have served as prototypes in the study of other members, which include the GABA$_A$, the glycine, and the serotonin (5HT$_3$) receptors (*see*: GABA AND GLYCINE RECEPTORS; SEROTONIN RECEPTORS).

Acetylcholine also interacts with a second class of structurally and pharmacologically distinct acetylcholine receptors, the MUSCARINIC RECEPTORS, which are G PROTEIN-COUPLED RECEPTORS and are discussed elsewhere in the encyclopedia.

Binding of neurotransmitter to the nAChR is followed within milliseconds by the opening of a cation channel which permits entry of Na$^+$ and K$^+$ into the recipient cells, leading to DEPOLARIZATION. Prolonged exposure to acetylcholine leads to a waning of this response and is termed desensitization. nAChRs found at the vertebrate neuromuscular junction and in the electric organs of certain fishes are single receptor subtypes and are called peripheral nAChRs. nAChRs are also found in the CENTRAL NERVOUS SYSTEM and in autonomic ganglia (*see* PERIPHERAL NERVOUS SYSTEM). Activation of these receptors also results in the increase in cation permeability, yielding depolarization and excitation. These nAChRs are termed neuronal receptors, and in contrast to the receptors at neuromuscular junctions, they are heterogeneous in structure. nAChRs have also been described in invertebrate nervous systems where they also mediate depolarizing responses.

Peripheral nAChRs are characterized pharmacologically by their activation by acetylcholine and the classic agonist, nicotine. They are antagonized by *d*-tubocurare and the basic polypeptide α-neurotoxins isolated from the venoms of certain snakes. Examples are α-BUNGAROTOXIN from *Bungarus multicinctus* and α-neurotoxin from the venom of the cobra, *Naja naja siamensis*. Additionally, certain agents including histrionicotoxin, chlorpromazine and phencyclidine are noncompetitive antagonists of the nAChR and act at the level of the ion channel to block cation translocation. Neuronal nAChRs differ from the peripheral receptor in that most are insensitive to the action of α-neurotoxins and are often referred to as α-bungarotoxin-insensitive receptors. However, a minority of neuronal nAChRs are antagonized by α-bungarotoxin and additionally, receptors in the autonomic ganglion are antagonized by the neuronal neurotoxin, κ-neurotoxin, a minor component of the *B. multicinctus* venom. nAChR pharmacology is fully reviewed in [1].

Molecular structure

The nAChR protein was the first neurotransmitter receptor to be isolated. It was discovered that neurotransmission between adjacent cells in the electric organ of the electric fishes (e.g. the electric ray *Torpedo californica* and the electric eel, *Electrophorus electricus*) is cholinergic with the identical pharmacology to that found at the vertebrate neuromuscular junction (i.e. nicotinic cholinergic). The electric organ is therefore a highly enriched source of homogeneous receptor, which is found at concentrations of 1000 pmol [^{125}I]α-bungarotoxin-binding sites per gram tissue, and constitutes 50% of the total membrane protein. This is in contrast to concentrations in innervated muscle of 0.5–1 pmol α-bungarotoxin-binding sites per gram tissue. Also, the availability of the α-neurotoxins provided a means of isolating the nAChR. First, α-bungarotoxin, a basic polypeptide of M_r 9000, could be labelled with [^{125}I] with retention of its biological activity. [^{125}I] α-bungarotoxin binds to nAChRs pseudoirreversibly thus providing a means of assaying for receptor activity. The α-neurotoxin from cobra venom has a lower affinity for the nAChR (K_d 10^{-10} M); this activity is, however, more appropriate for use in affinity column CHROMATOGRAPHY. Thus the receptor was initially isolated to homogeneity from electric organ of the eel and the ray. Purification from vertebrate muscle first used denervated muscle as the starting tissue. Denervation increases the nAChR concentration 24–30-fold as skeletal muscle compensates for the lack of acetylcholine input by the synthesis of increased numbers of nAChRs which are located over the entire muscle membrane surface instead of only at the crests of the muscle membrane in juxtaposition to the neurone.

From all the three tissues mentioned above, a single acidic membrane glycoprotein of M_r 250 000 and isoelectric point pI 5, was isolated. The receptor was shown to be hetero-oligomeric and composed of five polypeptide chains assembled as a pentameric complex $\alpha_2\beta\gamma\delta$. In SDS-polyacrylamide gel ELECTROPHORESIS, these subunits have M_r 40 000 (α), 50 000 (β), 55 000 (γ), and 60 000 (δ). In calf muscle, a developmental switch occurs such that the embryonic form is $\alpha_2\beta\varepsilon\delta$. Also, the nAChR of the electric ray can exist as a dimer via a disulphide bridge between the δ subunits of adjacent nAChRs.

The high-affinity acetylcholine-binding site was localized to the α subunit using the covalent affinity probes [^{3}H]-bromoacetylcholine, and 4-(N-maleimideo)-^{3}H benzyl trimethylammonium iodide (MBTA), and there are therefore two sites per receptor oligomer. More detailed studies have shown that it is the two adjacent cysteine residues, the vicinal cysteines 192 and 193 of the α subunit, which are labelled by these reagents. Reconstitution of the purified receptor into liposomes or planar lipid bilayers yields acetylcholine-gated ^{22}Na$^+$ cation fluxes which are blocked by α-bungarotoxin and which desensitize.

Neuronal nAChRs have been purified by immunoaffinity chromatography using antibodies raised against the fish nAChR. The oligomeric structures of these receptors are less well characterized although a recent study showed that the cloned chick α4/non-α1 subunit complement neuronal nAChR is pentameric. In SDS-polyacrylamide gel electrophoresis, two subunits have been distinguished on the basis of their difference in M_r. These are named the α subunit (M_r 48 000) and the β subunit (M_r 75 000). But note that it is the higher molecular weight subunit that is specifically labelled by MBTA, and therefore that this nomenclature for neuronal nAChRs is now inconsistent with peripheral nAChR subunits (see below for description of neuronal nAChR genes).

Molecular biology of nAChR polypeptides

Peripheral nAChR genes were first cloned from the electric fish mainly as a result of the pioneering work of Shosaku, Numa and colleagues in Japan [2]. The resulting cDNAs were later used to identify vertebrate peripheral nAChRs, and mammalian and invertebrate neuronal nAChRs. The deduced primary structures of all the nAChR subunits share structural features not only between themselves but with other members of the ligand-gated ion channel superfamily. The N-terminal sequence of each polypeptide is preceded by a short, mainly hydrophobic sequence of 24 amino acids. This is characteristic of the signal peptide

Table N2 A summary of the properties of nAChR polypeptides

Subunit	Classification	Characteristic features	
α1	Peripheral		Coexpression of α1β1γδ required for channel activity
α2			
α3		High-affinity acetylcholine binding subunit	Coexpression of α and β required for α-bgt-insensitive channel activity
α4	Neuronal		
α5			
α6			
α7			Forms α-bgt-sensitive homo-oligomers
ARD	Invertebrate Neuronal		
β1	Peripheral		
β2			
β3		All termed structural subunits	
β4	Neuronal		
β5			
γ	Peripheral		
δ	Peripheral		

α-bgt, α-Bungarotoxin.

sequence which determines membrane insertion. The exact molecular weights of the *Torpedo californica* nAChR subunits are 50 116 (347 residues, α), 53 681 (469 residues, β), 56 279 (489 residues, γ), and 57 565 (501 residues, δ). There is a high degree of homology in primary structure between the subunits with an average of 40% amino-acid sequence identity between all four chains. There is a closer homology between the α and β polypeptides, and between γ and δ, suggesting that the subunits originated from a single ancestral gene with a first branching for the α and β chains and a second branching for the γ and δ chains. Hydrophobicity analysis of the respective primary structures predicts that each subunit is polytopic, with four transmembrane α-helical segments, M1, M2, M3, and M4, each of at least 20 hydrophobic amino acids. The N terminus of the protein is predicted to be extracellular. It is at least 200 amino acids in length and is hydrophilic, with consensus sequences for N-glycosylation. All subunits contain within this region the Cys-Cys loop domain characteristic of all the ligand-gated ion channels with the conserved motif, Cys-X-X-X-X-X-X-hydrophobic-Pro-hydrophobic-Asp-X-X-X-Cys. The α subunit is characterized by two adjacent cysteine residues, 192 and 193, which contribute to the high-affinity acetylcholine-binding site. The ion channel is thought to be formed by the transmembrane regions, with each of the subunit M2 domains forming the inner lining of the channel. The M2 amino-acid sequence contains amino acids with hydrophilic side chains requisite for ion transport and indeed, site-directed mutagenesis of these residues showed decreases in single channel currents. Furthermore, the noncompetitive blocker of the nAChR, chlorpromazine, specifically labels residues Ser 262 of the δ subunit M2 and both Ser 254 and Leu 257 of the β subunit M2. Cation selectivity of the channel is determined by the negatively charged groups, particularly between the M2 and M3 domains. Indeed, recently an α7 nAChR homo-oligomer was converted from an acetylcholine-gated cation to an acetylcholine- gated anion channel by a single amino-acid point mutation [3].

The peripheral nAChR α and β subunits are given the nomenclature α1 and β1. The mammalian neuronal nAChR genes identified to date are α2–α7 and β1–β5 and in the invertebrates, the homologues of these are termed α-like and non α-like nAChR subunits. Each of these gene products shares the same structural features detailed above, again characteristic of the ligand-gated ion channel superfamily. The α subunits are designated as such because they contain the vicinal cysteines found in the peripheral nAChR α subunit but lacking in the other polypeptides. The overall amino-acid sequence similarity between fish and other vertebrate species for the same subunit type is high for mammals (i.e. of the order of 55–80%, the latter value for the α subunits) but falls to 50% when vertebrate and invertebrate sequences are compared. Figure N16 is a diagrammatic representation of the structure of the nAChR. For a more detailed discussion of receptor structure see [4–6].

For the peripheral nAChR, the co-expression of all four sub-units is required for acetylcholine-gated channel activity. For the neuronal nAChRs, in general, the co-expression of an α and a β subunit is required for cation flux. The α7 subunit in chick (alternative nomenclature, α-BgtBPα1 and α-BgtBPα2) however forms robust homo-oligomers and importantly, it is this class of neuronal nAChR that forms acetylcholine-gated channels which are antagonized by α-bungarotoxin (i.e. the α-bungaro-toxin-sensitive neuronal nAChR). The subunit complements of natural neuronal nAChRs are not yet known, but the expression studies suggest that they consist of a pentameric combination of an α and a β subunit. In addition, immunoprecipitation studies have shown that for at least some subpopulations of receptors, different types of α subunit are found within the same receptor. The salient properties of the nAChR genes are summarized in Table N2.

Three-dimensional structure

The X-ray crystallographic structure of the nAChR has not yet been determined, largely owing to the difficulties associated with the crystallization of integral membrane proteins. However, the availability of high concentrations of homogeneous nAChRs from

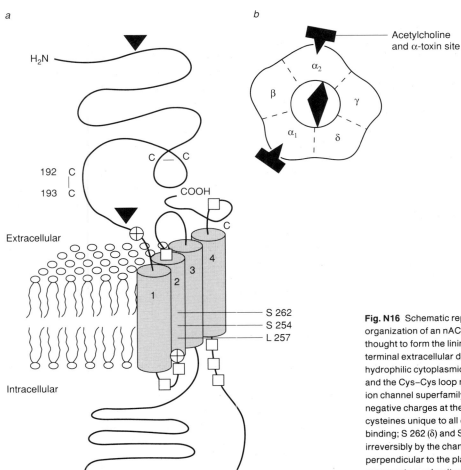

Fig. N16 Schematic representation of nAChR. *a*, The transmembrane organization of an nAChR α subunit. The M2 regions of each subunit are thought to form the lining of the channel. Pertinent features are the large N-terminal extracellular domain, the four transmembrane regions 1–4, the hydrophilic cytoplasmic loop between transmembrane regions M3 and M4 and the Cys–Cys loop motif characteristic of all members of the ligand-gated ion channel superfamily. ▼, Sites of N-glycosylation; □ distribution of the negative charges at the mouth of the channel. C–C 192, 193 are the adjacent cysteines unique to all α subunits and thought to be involved in agonist binding; S 262 (δ) and S 254 L 257 (both β) are the residues in M2 labelled irreversibly by the channel blocker chlorpromazine. *b*, The receptor as viewed perpendicular to the plane of the membrane. Each of the segments represents a subunit and the ordering of the polypeptides around the rosette was determined by methods reviewed in [7].

the electric fish has enabled both negative stain electron microscopy studies and a structural determination to a resolution of 17 Å using electron image analysis of polymerized receptor molecules rapidly frozen and observed at low temperatures (i.e. nAChR cryo-images — summarized in [7]). Thus, the nAChR, as viewed perpendicular to the plane of the membrane, appears as an 80 Å diameter rosette with a central pit and five peaks of electron density symmetrically arranged around the central pore. Each of these peaks is thought to be a subunit and the ordering around the rosette is αβαγδ in a clockwise direction. Viewed within the plane of the membrane, the nAChR appears as a cylinder of length 110 Å which spans the bilayer but protrudes mostly on the extracellular surface and thus into the synaptic cleft. This view of the nAChR also shows that the molecule is virtually at right angles to the membrane. It is suggested that when the channel opens in the presence of acetylcholine, the M2 transmembrane helices tilt such that the hydrophobic amino acids within this region move from the central pore (i.e. closed state) into the space at the interface between the subunits thus creating the open, conducting state. A model of the nAChR based on these types of study is shown in Fig. N17.

The nAChR and myasthenia gravis

The nAChR has a role in the neuromuscular AUTOIMMUNE disease myasthenia gravis. Patients with this disorder show

muscle fatigue which is exacerbated by exercise; the most prevalent symptom is the drooping of the eyelids, which can be relieved by administration of inhibitors of the enzyme ACETYL-CHOLINESTERASE. Following the first isolation to homogeneity of the nAChR, antibodies against the protein were raised in rabbits. It was found that these animals showed a flaccid paralysis similar to that found in myasthenic patients. A screen for the presence of anti-nAChR antibodies circulating in the serum of these patients yielded positive results and indeed, the progression of the disease in an individual can now be monitored by determination of the antibody titre. The pathogenesis of the disease is explained by anti-nAChR antibodies binding to the peripheral receptor, resulting in an increased degradation of receptor protein and a concomitant decrease in the efficiency of neuromuscular transmission. The cause of the specific antibody production is not known but the mapping of the antigenic determinants of the nAChR has given important insights into the receptor structure. Interestingly, the human nAChR α subunit exists as two isoforms, one of which has an insertion of 25 amino acids in the N-terminal extracellular domain. The functional properties of this isoform have not yet been delineated. A more detailed discussion of the role of the nAChR in the pathogenesis of myasthenia gravis can be found in [8].

F.A. STEPHENSON

See also: XENOPUS OOCYTE EXPRESSION SYSTEM.

1 Changeux, J-P. (1981) *Harvey Lect.* **75**, 85–254.
2 Noda, M. et al. (1982) Primary structure of α-subunit precursor of *Torpedo californica* acetylcholine receptor deduced from cDNA sequence. *Nature* **299**, 793–797.
3 Galzi, J-L. et al. (1992) Mutations in the channel domain of a neuronal nicotinic receptor convert ion selectivity from cationic to anionic. *Nature* **359**, 500–505.
4 Stroud, R.M. et al. (1990) Nicotinic acetylcholine receptor superfamily of ligand-gated ion channels. *Biochemistry* **29**, 11009–11023.
5 Changeux, J-P. et al. (1987) The nicotinic acetylcholine receptor: molecular architecture of a ligand-gated ion channel. *Trends Pharmacol Sci.* **8**, 459–465.
6 Deneris, E.S. et al. (1991) Pharmacological and functional diversity of neuronal nicotinic acetylcholine receptors. *Trends Pharmacol Sci.* **12**, 34–40.
7 Unwin, N. (1989) The structure of ion channels in the membranes of excitable cells. *Neuron* **3**, 665–676.
8 Newsom-Davis, J. et al. (1993) Autoimmune disorders of the neuromuscular junction. In *Clinical Aspects of Immunology* (Lachmann, P.J. et al., Eds) 2091–2113 (Blackwell Scientific Publications, Oxford).

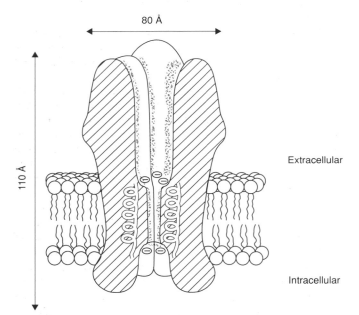

Fig. N17 Schematic representation to scale of the nAChR based on the three-dimensional structure determined by helical reconstruction from cryo-images of *Torpedo californica* post- synaptic membranes. The figure shows a section of the nAChR through the plane of the membrane. Depicted are four of the five subunits. A part of the α-helical segment M2 from each subunit lines the narrow pore spanning the hydrophobic part of the membrane bilayer. Rings of negatively charged groups near or at the end of M2 are probably located adjacent to the regions where the pore widens and the channel is formed. Note that the majority of the protein is extracellular. From [7].

NIDDM Non-insulin dependent DIABETES MELLITUS.

nidogen Entactin. *See*: EXTRACELLULAR MATRIX MOLECULES.

Niemann–Pick disease Rare heritable disease characterized by the accumulation of SPHINGOMYELIN and other lipids in various cells of the body. The symptoms include anaemia, enlarged liver and spleen, retarded physical and mental growth, and varying degrees of neurological disturbance. The disease is caused by a deficiency of the enzyme sphingomyelinase which converts sphingomyelin to phosphocholine and ceramide. There are five forms of the disease which have different progressions and tissue involvements. All forms are inherited as AUTOSOMAL RECESSIVE traits.

nif genes Bacterial genes encoding nitrogenase and other proteins required for nitrogen fixation. *See*: NITROGEN FIXATION: REGULATION OF GENE EXPRESSION.

NIH-3T3 cell *See*: 3T3 CELL.

ninaC proteins Proteins isolated from *Drosophila* which contain an N-terminal PROTEIN KINASE domain and a domain resembling a myosin head. *See*: MYOSIN.

NIP 4-Hydroxy-5-iodo-3-nitrophenyl acetyl group. *See*: ANTI-NP ANTIBODIES.

Nissl substance Rough ENDOPLASMIC RETICULUM and associated free ribosomes.

Nitella A genus of charophyte algae (stoneworts) with very large elongated internodal cells, much used in studies of cytoplasmic streaming and CELL MOTILITY. *See*: MYOSIN.

nitric oxide (NO) Gaseous signalling molecule and NEUROTRANSMITTER. Its biological roles were first discovered in 1987 when it was identified as the 'endothelium-derived relaxing factor' — the substance released by vascular endothelium in response to mechanical force and neurohumoral mediators such as acetylcholine, bradykinin and histamine, and which mediates relaxation of the underlying smooth muscle of the arterial wall with consequent vasodilation. In this system NO stimulates the muscle guanylate cyclase, leading to increased levels of cGMP and muscle relaxation. NO is the neurotransmitter involved in intestinal relaxation in peristalsis and penile erection. It is also thought to be involved in various effects of the neurotransmitter glutamate acting at NMDA receptors in the brain. NO can be synthesized from L-arginine in neurons and other cells by the enzyme nitric oxide synthase, and free radicals of NO act as a neurotransmitter in nonadrenergic noncholinergic (NANC) nerves in the PERIPHERAL NERVOUS SYSTEM. NO can exist in several oxidation–reduction states: the reduced form $NO^{\bullet}$, which is properly called nitric oxide; NO^{+}, the nitrosonium ion; and NO, nitrogen monoxide. Nitric oxide has been reported to mediate neuronal destruction in stroke, and this may be due to the production of the $NO^{\bullet}$ form which can react with superoxide anion to form neurotoxic peroxynitrite.

nitrocellulose A material used in molecular biology in sheet form principally for binding nucleic acids and proteins in blotting procedures. *See*: NORTHERN, SOUTHERN and WESTERN BLOTTING.

Nitrogen fixation

THE ability to utilize atmospheric nitrogen gas (dinitrogen, N_2) as a source of nitrogen for growth and biosynthesis is restricted to a relatively few species of prokaryotes, which are called diazotrophs. Diazotrophy is found in almost all genera of EUBACTERIA, in some ARCHAEBACTERIA, and is widespread among the CYANOBACTERIA. Biological nitrogen fixation, the reduction of N_2 to NH_3 by living organisms, contributes significantly to the replenishment of fixed nitrogen in the biosphere and makes up for nitrogen loss to the atmosphere as N_2 and N_2O, a consequence of the activities of denitrifying bacteria. The enzyme system responsible for biological fixation is nitrogenase, a complex enzyme whose structure is highly conserved between the different diazotrophic species [1].

As most ecosystems are low in available organic carbon and energy sources, free-living heterotrophic aerobic diazotrophs make only a small contribution to nitrogen input to agricultural systems, but do contribute to the maintenance of fertility in undisturbed ecosystems. In contrast, free-living cyanobacteria using PHOTOSYNTHESIS as an energy source do not face the same constraints, and can contribute significantly to soil fertility.

Some diazotrophs form symbiotic associations with plants, and under these conditions make a positive contribution to the nitrogen nutritional status of the plant, as most of the nitrogen fixed is not assimilated by the bacterial partner but is exported to the benefit of the plant. Some filamentous heterocystous cyanobacteria, the actinomycete *Frankia*, and most important from an agricultural point of view, members of the family Rhizobiaceae, form such associations. The rhizobia form nitrogen-fixing nodules on the roots of leguminous plants and the symbiotic association between different rhizobia and their leguminous hosts is highly specific. The molecular biology of the recognition process and nodulation is described elsewhere (*see* NODULATION). This article will describe the nitrogen fixation reaction itself and the enzymes involved.

All diazotrophs have a nitrogenase that contains molybdenum (Mo). The component proteins of Mo-nitrogenase from a number of diazotrophs of quite different physiology show only minor differences from each other. This nitrogenase consists of a complex of two extremely O_2-sensitive redox proteins, both of which are required for nitrogenase activity. But this is not the only nitrogenase; in 1986 it was established unequivocally that some diazotrophs have additional nitrogenase systems that are based not on molybdenum, but on vanadium (V) and iron (Fe) [1,2]. The structural genes for these latter nitrogenases have been cloned from *Azotobacter chroococcum* and *A. vinelandii* and have been sequenced (*see* DNA CLONING; DNA SEQUENCING). They show considerable homology with each other and Mo-nitrogenase.

All known nitrogenase systems have similar basic structures and requirements for activity — relatively large amounts of MgATP, a low potential electron donor ($E_m \sim -450$ mV) and the absence of O_2 (dioxygen), even in aerobic microorganisms. Each system is made up of one distinct FE PROTEIN complexed with one of three possible proteins — a MoFe protein, a VFe protein, or an analogous proteins (FeFe protein) that contains Fe but only low levels of Mo or V (see Table N3). When sufficient Mo is provided only Mo-nitrogenase is synthesized; under Mo-deficient conditions in a medium otherwise sufficient in V and Fe V-nitrogenase is synthesized; when both Mo and V are deficient the third nitrogenase is present (see NITROGEN FIXATION: REGULATION OF GENE EXPRESSION). The routine inclusion of Mo in growth medium delayed the discovery of Mo-independent nitrogenases for many years.

Table N3 Properties of MoFe, VFe, and FeFe component proteins of nitrogenases

Protein	Subunit M_r	Native M_r	Metal and sulphide content (g atom mol^{-1})			
			Fe	Mo	V	S^{2-}
MoFe protein $\alpha_2\beta_2$	α 55 288 β 59 459	229 494	32	2	0.05	30
VFe protein $\alpha_2\beta_2\delta_2$	α 53 877 β 52 775 δ 13 372	240 048	21	0.06	2	19
FeFe protein $\alpha_2\beta_2\delta_2$	α 58 326 β 51 153 δ 15 369	249 996	24	0.01	0.08	18

Mo-nitrogenase

The MoFe proteins are tetramers of native $M_r \sim 220\,000$, containing about 32 Fe and 30 acid-labile sulphur atoms organized in two types of redox centre. Each Mo atom is part of a polynuclear cluster containing Fe, S^{2-} and homocitrate (*R*-2-hydroxy-1,2,4-butanetricarboxylic acid). X-ray absorption spectroscopy (EXAFS) has provided most detail of the chemical environment of Mo in this cluster, and indicates a distorted octahedral environment with three S atoms, three Fe atoms, and three O, N or C atoms in the first coordination sphere. There is convincing evidence that these FeMoco centres are the site of N$_2$ binding and reduction, and are quite distinct from the Mo-cofactor of other molybdoenzymes (*see* METALLOPROTEINS).

SITE-DIRECTED MUTAGENESIS studies are consistent with the α subunit providing the binding domain for FeMoco. The FeMoco centres can be extruded from the FeMo protein under denaturing conditions without significant changes in the environment of the Mo occurring. The extracted cofactor retains the ability to activate the MoFe protein, which accumulates in mutants unable to synthesize FeMoco. The other type of redox centre found in MoFe proteins are FE-S CENTRES with unique spectroscopic properties P clusters). Preliminary X-ray structural analysis based on anomalous scattering data, which allows clusters containing Mo to be distinguished from those that do not, indicates that the FeMoco centres are 70 Å apart and are each 17 Å away from a cluster of similar size (presumably two linked 4Fe–4S centres or a single larger cluster). The P clusters have a suggested role in nitrogenase function as centres in which electrons are stored transiently during enzyme turnover. The recent 2.7 Å crystal structure confirms the earlier work and shows the P clusters to be two bridged 4Fe–4S cubes [3]. It also reveals FeMoco to have an extended structure formed from two cluster fragments with the composition [4Fe : 3S]–[3Fe : 3S : Mo] bridged by three nonprotein ligands. Homocitrate is ligated to the Mo atom at one end of the cluster.

The Fe proteins are dimers of native $M_r \sim 62\,000$ and have a single 4Fe–4S redox centre (E_m – 400 mV). Preliminary X-ray structure analysis shows the Fe–S cluster bridging the subunits, with two cysteine residues from each subunit providing ligands to the cluster, a model proposed earlier on the basis of chemical modification studies. Putative nucleotide-binding sites have been identified in the intersubunit cleft. On binding MgADP or MgATP the protein undergoes a conformational change and the midpoint potential of the Fe–S centre becomes more negative by approximately 100 mV. The Fe protein functions as a one-electron donor to the MoFe protein during nitrogenase turnover.

Under optimum conditions nitrogenase catalyses the reaction:

$$N_2 + 8H^+ + 8e^- + 16\,\text{MgATP} \rightarrow 2NH_3 + H_2 + 16\text{MgADP} + 16P_i \tag{1}$$

As discussed below, the apparent inefficiency of the system in the formation of H$_2$ at the expense of MgATP hydrolysis, is thought to be of mechanistic significance in terms of the catalytic reduction of N$_2$. Surprisingly, in view of the high energy requirement of nitrogenase, activity is not subject to product inhibition by either NH$_3$ or glutamine, the assimilation product formed from oxoglutarate by glutamine synthetase:

$$NH_3 + \text{2-oxoglutarate} + ATP \rightarrow \text{glutamine} + ADP + P_i \tag{2}$$

In addition to reducing N$_2$, nitrogenase will reduce a number of other substrates which contain triple bonds, and will also reduce protons. In the absence of any added reducible substrate, for example under an atmosphere of argon, protons are reduced to give H$_2$. This reaction, the simplest reduction catalysed by nitrogenase, has been used extensively in mechanistic studies. The reduction of acetylene (C$_2$H$_2$) to ethylene (C$_2$H$_4$) is also of particular importance, since this reaction is catalysed by nitrogenase *in vivo* and ethylene can be detected at extremely low concentrations. The acetylene reduction assay has been used extensively in genetic and ecological studies of nitrogen fixation.

The path of electron transfer to nitrogenase *in vivo* is best understood in *Klebsiella pneumoniae*. The oxidative decarboxylation of pyruvate is catalysed by the *nifJ* product and the electrons are accepted by the *nifF* product — a FLAVODOXIN. The semiquinone–hydroquinone redox couple of the flavin mononucleotide (FMN) redox centre (E_m – 422 mV) functions as the MgATP-dependent electron donor to the Fe protein of nitrogenase (Fig. N18).

A comprehensive model of nitrogenase function has been developed from pre-steady state kinetic analysis of the Mo-nitrogenase of *K. pneumoniae*. In this model the FeMoco centres of the MoFe protein are each reduced independently by a series of one-electron transfers from the Fe protein, with concomitant hydrolysis of two MgATP molecules for each electron transferred. After each electron transfer the two proteins dissociate, and the oxidized Fe protein with MgADP bound is re-reduced before MgADP dissociates and is replaced with MgATP. Three electrons are transferred to the MoFe protein before N$_2$ can bind to the reduced FeMoco centre, displace H$_2$, and so account for the stoichiometry of reaction (1). By analogy with chemical nitrogen fixation catalysed by metals, where the majority of dinitrogen transition metal complex binding is 'end-on' terminal, the reduction of the enzyme-bound dinitrogen is proposed to involve progressive protonation of the terminal (β) N atom, as multiple bond character builds between the α N atom and a metal atom (putatively Mo) at the substrate-binding site. During the reduction of N$_2$ free hydrazine (N$_2$H$_4$) is not detected, but an enzyme-bound

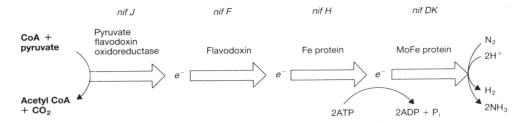

Fig. N18 Electron transport to nitrogenase in *Klebsiella pneumoniae*.

intermediate can be made to yield hydrazine under denaturing conditions. This contrasts with V-nitrogenase (see below), where hydrazine has been shown to be a minor product of N_2 reduction.

Mo-independent nitrogenases

The Fe proteins associated with the Mo-independent nitrogenases of *Azotobacter* have been purified. They are dimers with physico-chemical properties very similar to those of the Fe proteins of Mo-nitrogenase, and are able to form functional nitrogenases with MoFe protein.

VFe proteins from *A. chroococcum* and *A. vinelandii* have been purified and show similarities with MoFe proteins, but have an additional δ subunit type (see Table N3). They contain a vanadium and iron cofactor (FeVco) analogous to FeMoco and Fe–S clusters with spectroscopic properties similar to P clusters. EXAFS shows the V atom to be in a very similar chemical environment to that of Mo in FeMoco. The similarity between these cofactors is sufficient to allow FeVaco to activate MoFe protein isolated from mutants unable to synthesize FeMoco. Complementation is not complete however, as although protons and C_2H_2 are reduced, N_2 is not, suggesting that very specific interactions between the cofactor and polypeptide are required for N_2 reduction, which are missing in this hybrid protein.

Studies on the mechanism of substrate reduction by V-nitrogenase have shown that the pre-steady state rate of electron transfer between the two proteins is similar to that of Mo-nitrogenase, and is too fast to be rate limiting in enzyme turnover. Characteristics of the reduction of N_2 by V-nitrogenase are the production of a small amount (1%) of N_2H_4 as a product, and the effective reduction of N_2 at lower temperatures than with Mo-nitrogenase. In addition, when C_2H_2 is reduced a small amount of C_2H_6 is formed in addition to C_2H_4. Such differences presumably reflect minor changes in the geometry at the active site, as the reduction of C_2H_4 to yield some C_2H_6 is observed when MoFe protein is altered by single amino-acid substitutions at specific sites, or when a chemically modified FeMoco is present.

The FeFe proteins of the third nitrogenase of *A. vinelandii* and *Rhodopseudomonas capsulatus* have been purified. These proteins have the δ subunit, and contain Fe but only low levels of Mo or V (Table N3). Although direct evidence is lacking, the conservation of amino-acid residues implicated in binding the redox centres of Mo-nitrogenase, and the requirement for genes involved in FeMoco biosynthesis (see below), strongly suggest that analogous

redox centres will be shown to be present in this system.

In *A. vinelandii*, *A. chroococcum* and *R. capsulatus* there is an additional order of complexity in the genes involved in nitrogen fixation, as these microorganisms have both Mo-nitrogenase and the related Mo-independent nitrogenase systems. Genes that have a role in Mo-nitrogenase function are designated as *nif* genes, those with a role in V-dependent nitrogenase as *vnf* genes, and those involved with the third Fe-dependent nitrogenase as *anf* genes. Mutational analysis has revealed that some genes (*nifM*, *nifV* and *nifB*) are essential for the functioning of all three nitrogenases. Other genes (e.g. *vnf*E and *vnf*N) are required for the function of the V and third systems. As *nifV* and *nifB* have roles in FeMoco biosynthesis, their involvement in Mo-independent systems can be rationalized as an involvement in the biosynthesis of homologous cofactor centres in the other systems. It seems probable, given their sequence homology with *nifEN* (genes involved in FeMoco biosynthesis), that the *vnfEN* genes are essential for the biosynthesis of cofactors analogous to FeMoco, which are associated only with the V and third nitrogenase systems.

R.R. EADY

1 Dilworth, M.J. & Glenn, A.R. (Eds) (1991) *The Biology and Biochemistry of Nitrogen Fixation* (Elsevier, Amsterdam).
2 Eady, R.R. (1991) The Mo, V, and Fe based nitrogenase systems of Azotobacter. *Inorg. Chem.* **36**, 77–100.
3 Palacios, R. et al. (Eds) (1993) *New Horizons in Nitrogen Fixation* (Kluwer Academic, Dordrecht).

Nitrogen fixation: regulation of gene expression

THE genetic control of NITROGEN FIXATION is mediated by a complex system of genes required for the synthesis, activity and regulation of nitrogenase. Genetic analysis of the system was initiated in the free-living facultative anaerobic bacterium *Klebsiella pneumoniae*, which is relatively amenable to study by the methods of classical bacterial genetics and by the newer methods of molecular genetic analysis. Analysis has revealed an intricate system in which a cluster of 20 nitrogen fixation (*nif*) genes are arranged in eight coordinately regulated operons. The specific functions of nine of these genes are known and the roles of five others may be inferred from the phenotypes caused by mutating them (*see* NITROGEN FIXATION). The functions of the remaining six genes are presently unknown. The majority of these 20 *nif* genes are common to most (probably all) nitrogen-fixing organisms (diazotrophs), the exceptions being genes for the electron-transfer proteins NifF and NifJ and the regulatory protein NifL, which are not conserved.

Nitrogen fixation is regulated primarily at the level of *nif* gene TRANSCRIPTION and despite the taxonomic diversity found among nitrogen-fixing organisms, there is a considerable degree of uniformity in the mechanisms underlying this regulation in many, though not all, diazotrophs [1].

Nitrogen fixation is a very energy-demanding process and consequently diazotrophs tightly regulate both the synthesis of the nitrogenase enzyme and its subsequent activity in response to three major environmental effectors. Nitrogenase is markedly oxygen sensitive and hence environmental oxygen tension is one major regulatory factor. Likewise the availability of fixed nitrogen is a significant regulatory effector in free-living diazotrophs but is of less importance for symbiotic organisms, such as the rhizobia, which are adapted to exporting fixed nitrogen to their host. Finally, as nitrogenases are METALLOPROTEINS, the availability of the requisite metals can also have marked regulatory effects, particularly in those organisms that can synthesize 'alternative' nitrogenases with different cofactors (e.g. *Azotobacter*; *see* NITROGEN FIXATION).

Transcriptional activation of *nif* gene expression

nif gene expression is positively controlled in a wide range of diazotrophs and this control requires the products of two genes, *rpoN* and *nifA*. DNA SEQUENCING of many *nif* gene PROMOTERS has shown that in most cases these promoters have a highly conserved sequence (consensus TGGCAC[N]$_5$TTGC) between positions -26 and -12 with respect to the transcription start point. This sequence is quite unlike the consensus TTGACA[N$_{17}$]TATACA found in the -35, -10 region of most prokaryotic promoters (*see* BACTERIAL GENE EXPRESSION). It identifies a class of prokaryotic genes which are dependent for their transcription on a minor form of RNA polymerase containing the sigma factor σ^{54} (encoded by *rpoN*) (*see* TRANSCRIPTION). All σ^{54}-dependent promoters are positively controlled and require a specific class of activator protein which binds to an upstream activator sequence ~100 bp upstream of the $-26/-12$ region. Transcription activation seems to require DNA loop formation to bring the activator into contact with the RNA polymerase bound in the $-26/-12$ region.

The *nif*-specific activator protein is encoded by *nifA* which has been identified in a wide range of diazotrophs including free-living organisms (e.g *K. pneumoniae*, *Azotobacter vinelandii* and *Rhodobacter capsulatus*) and the three major groups of rhizobia (*Rhizobium*, *Bradyrhizobium* and *Azorhizobium*). Similar proteins (VnfA and AnfA) activate expression of genes for the vanadium (vnf) and third (anf) nitrogenases of *Azotobacter*.

The NifA amino-acid sequences suggest three structural domains in the protein. The function of the N-terminal domain is unclear but it may facilitate interaction with other proteins which sense various environmental factors. The central domain is characteristic of all σ^{54}-dependent activators and probably allows interaction with RNA polymerase or σ^{54} itself. This domain contains a consensus nucleotide-binding site, consistent with a role for ATP hydrolysis in transcriptional activation, which has been proposed to be a property of all σ^{54}-dependent activators [2].

The C-terminal domain contains a helix-turn-helix motif characteristic of DNA-binding proteins (*see* PROTEIN–NUCLEIC ACID INTERACTIONS) and this domain is required for specific binding to the upstream activator sequences.

nif gene regulation by NifA and σ^{54} is not universal and is not found, for example, in the CYANOBACTERIA or in ARCHAEBACTERIA. In these cases the mechanisms of *nif* gene regulation are not yet understood.

Oxygen control

Oxygen control can occur at the level of both NifA activity and *nifA* expression. The NifA protein family can be divided into two groups on the basis of the presence or absence of a Cys-X-X-X-X-Cys motif which can occur in a spacer sequence between the central and C-terminal domains. The presence of this motif, notably in all rhizobial NifA proteins and in NifA from *Rhodobacter*, identifies a subclass of NifA proteins whose activity is oxygen sensitive. It has been proposed that this motif may bind a metal ion which in some way mediates the oxygen sensitivity of the protein. In contrast, *K. pneumoniae* NifA activity is subjected to oxygen control by interaction with the product of a cotranscribed gene, *nifL*. NifL apparently senses the intracellular oxygen tension and regulates NifA activity accordingly, but the mechanism of this control is unknown.

In *Rhizobium*, oxygen control of *nifA* expression, and hence regulation of all other *nif* genes, is mediated by the action of two gene products, FixL and FixJ. These proteins are members of a large family of prokaryotic two-component regulatory systems in which a HISTIDINE PROTEIN KINASE (e.g. FixL) functions as a sensor protein and acts to phosphorylate a transcriptional regulator (e.g. FixJ) in response to certain stimuli [3] (*see* BACTERIAL CHEMOTAXIS for another example of such a system). FixL is a membrane-bound HAEMOPROTEIN which senses the environmental oxygen tension and activates FixJ by phosphorylation when the oxygen tension is low (Fig. N19). FixJ in turn activates *nifA* expression thereby ensuring that *nif* gene expression only occurs microaerobically, for example in the conditions which occur during symbiosis. Expression of *K. pneumoniae nifA* is likewise

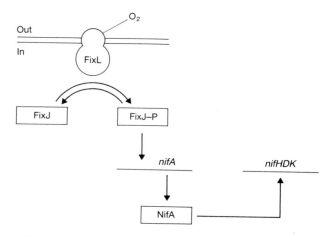

Fig. N19 Oxygen control of *nif* gene expression in *Rhizobium* by FixLJ.

subject to oxygen control, although in this case the effect is mediated by the level of DNA SUPERCOILING at the nifLA promoter. In many prokaryotes anaerobiosis induces an increase in negative DNA supercoiling and some promoters, including that of the nifLA operon in *K. pneumoniae*, are dependent on such supercoiling for transcription.

Nitrogen control

The requirements for nitrogen control of *nif* expression are largely determined by the physiological and ecological factors which influence nitrogen fixation in particular diazotrophs. In free-living diazotrophs, nitrogen status is a major regulatory factor and these organisms only fix atmospheric nitrogen when other sources of fixed nitrogen are limited.

In enteric bacteria a general nitrogen regulation (*ntr*) system controls the expression of many genes concerned with nitrogen metabolism including the *K. pneumoniae nif* genes [1,4]. The *ntr* system comprises four gene products; a uridylyltransferase (UTase) encoded by *glnD*, a small tetrameric effector protein (P_{II}) encoded by *glnB*, and a pair of regulatory proteins NtrB and NtrC which are also members of the two-component regulatory family [3].

UTase is the primary sensor of the cellular nitrogen status and responds to the intracellular ratio of the α-ketoglutarate to glutamine pools. When cells are nitrogen-limited UTase mediates the uridylylation of P_{II} by transfer of a uridylyl group onto a tyrosine residue on each of the four P_{II} subunits (Fig. N20). The uridylylated form of P_{II} (P_{II}-UMP) promotes the phosphorylation of the transcriptional activator NtrC by the histidine protein kinase NtrB. Phosphorylation of NtrC occurs in a two-step process involving autophosphorylation of a histidine residue in NtrB followed by transfer of the phosphate to an aspartate residue in the N terminus of NtrC. Phosphorylation of NtrC stimulates its DNA binding properties and also renders it able to function as a transcriptional activator.

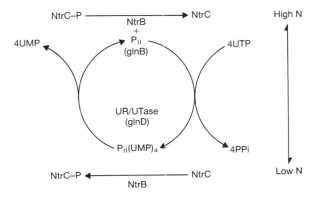

Fig. N20 Regulation of NtrC phosphorylation by the nitrogen control (*ntr*) system in enteric bacteria.

Under conditions of nitrogen excess this cascade of events is reversed. UTase now acts as a uridylyl-removing enzyme, con-

verting P_{II}-UMP to its deuridylylated form. P_{II} no longer stimulates the kinase activity of NtrB, and NtrB now promotes the dephosphorylation of NtrC (Fig. N20). As a result, the activator and DNA-binding properties of NtrC are diminished and expression from NtrC-dependent promoters is switched off.

In some free-living diazotrophs (e.g *K. pneumoniae* and *R. capsulatus*) NtrC regulates expression of *nifA* and consequently allows nitrogen control of nitrogenase synthesis. In *K. pneumoniae* the NifL protein can also respond to changes in fixed nitrogen status and provides a second level of nitrogen control by regulating NifA activity (Fig. N21).

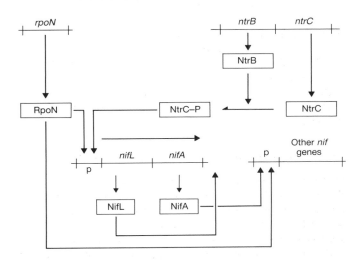

Fig. N21 The regulatory control circuits for *nif* genes in *Klebsiella pneumoniae*.

Metal regulation

The availability of certain metals is a major regulatory factor in *A. vinelandii* which can synthesize three different nitrogenase enzymes — a conventional molybdenum-containing enzyme, a vanadium enzyme and an iron enzyme (*see* NITROGEN FIXATION). Molybdenum represses synthesis of the vanadium and iron enzymes at the level of transcription of the *vnf* and *anf* genes and likewise vanadium represses *anf* gene expression, but the mechanisms of this control are not yet fully understood.

M.J. MERRICK

See also: NODULATION.

1 Merrick, M.J. (1992) Regulation of nitrogen fixation genes in free-living and symbiotic bacteria. In *Biological Nitrogen Fixation* (Stacey, G. et al., Eds) 835–876 (Chapman and Hall, New York).

2 Kustu, S. et al. (1989) Expression of σ^{54} (*ntrA*)-dependent genes is probably united by a common mechanism. *Microbiol Rev.* **53**, 367–376.

3 Stock, J.B. et al. (1989) Protein phosphorylation and regulation of adaptive responses in bacteria. *Microbiol Rev.* **53**, 450–490.

4 Reitzer, L.J. & Magasanik, B. (1987) Ammonia assimilation and the biosynthesis of glutamine, glutamate, aspartate, asparagine, L-alanine, and D-alanine. In *Escherichia coli and Salmonella typhimurium: Cellular and Molecular Biology*, Vol. 1 (Neidhard, F.C., Ed.) 302–320 (American Society for Microbiology, Washington, DC).

nitrogenase Bacterial enzyme that catalyses the reduction of molecular nitrogen to ammonia. *See:* NITROGEN FIXATION.

NK cells NATURAL KILLER CELLS.

NMDA (N-methyl-D-aspartate) Synthetic AMINO ACID which is the prototypic agonist for one subtype of glutamate receptor (Fig. N22) (*see* EXCITATORY AMINO ACID RECEPTORS). The receptor subtype activated by NMDA is therefore referred to as the NMDA receptor. Activation of the NMDA receptor system results in the induction of certain forms of synaptic plasticity (*see* LONG-TERM POTENTIATION) and may also be involved in epileptiform activity and neurodegenerative states.

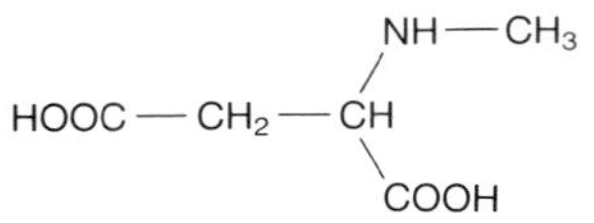

$$HOOC - CH_2 - CH \begin{array}{c} NH - CH_3 \\ | \\ \\ COOH \end{array}$$

Fig. N22 *N*-methyl-D-aspartate (NMDA).

NMDA receptor Class of ionotropic glutamate receptors (*see* EXCITATORY AMINO ACID RECEPTORS) that are specifically activated by NMDA with glycine as an essential cotransmitter.

NMR spectroscopy

NUCLEAR magnetic resonance (NMR) spectroscopy is one of a very limited number of techniques available to determine the structure of biological molecules at atomic resolution. Its usefulness extends far beyond structure determination — for example, to the probing of PROTEIN FOLDING and protein dynamics — but we shall concentrate here on the elucidation of the structure of globular proteins. It should be noted that the techniques discussed below are also applicable to the study of other biopolymers, including oligonucleotides and oligosaccharides (*see e.g.* NUCLEIC ACID STRUCTURE)[1]. PROTEIN STRUCTURE determination by NMR does not have as generic a range of applicability as X-RAY DIFFRACTION methods owing to its limitation to molecules in a relatively low molecular weight range, but in many ways the two techniques are complementary [2]. NMR experiments are normally performed on the protein in solution, and thus have provided overwhelming evidence that in general protein structures in the crystalline state are unchanged from in the solution state. The applicability of the experiments is not restricted by the limitations of crystallization; for example, NMR methods are able to tolerate considerable regions of flexibility in the protein that often preclude regularity in crystal lattices. NMR methods are particularly well suited to small proteins, for which CRYSTALLIZATION has been notoriously difficult.

Single phase radiofrequency electromagnetic radiation may be used to interchange atomic nuclei in a static magnetic field between magnetic spin-states. NMR spectroscopy utilizes the coherence of the absorbed energy to measure the resonance frequency of the nuclei involved. Different atomic isotopes absorb in quite distinct regions of the electromagnetic spectrum and furthermore, the exact frequency of absorbance of the same isotope depends upon the electronic environment of the individual nuclei. The spread of frequencies of one isotope is normally described relative to a reference frequency and is quoted as a fraction of that reference frequency. This is known as the chemical shift. For example, for the ^{1}H nuclei (commonly termed protons) of proteins in an 11.7 Tesla field-strength magnet, the spread of absorbance frequencies is ~5 kHz around a reference frequency of 500 MHz. Hence protons tend to absorb in the range 0 to 10 parts per million (p.p.m.) of the reference frequency. The dimensionless nature of the p.p.m. scale means that its values are independent of the field strength of the magnet.

Not all isotopes have the appropriate nuclear properties to be suitable for the type of NMR studies discussed here. Fortunately for molecular biologists, isotopes of most of the common atom types are represented in the list of suitable nuclei where the numbers in parentheses represent the natural abundance of these isotopes. ^{1}H (99.98%), ^{13}C (1.11%), ^{15}N (0.37%), ^{19}F (100%), ^{31}P (100%). Clearly, the utilization of carbon and nitrogen nuclei will benefit greatly from artificial isotopic enrichment of these rare isotopes, and strategies based on isotopic enrichment have considerably broadened the applicability of NMR methods. The most notable absentee from the above list is an isotope of oxygen. This, for example, limits the ability of NMR to describe hydrogen bonds directly in as much detail as would be desirable. However, there are indirect methods of obtaining the same information. For numerous reasons, some of which will be discussed below, the ^{1}H nucleus normally forms the basis of the NMR characterization of proteins; the other nuclei are used in a supporting role. In summary, the ^{1}H NMR spectrum of a protein reflects the absorbances of all its hydrogens spread according to their resonance frequencies. Each hydrogen in the molecule normally contributes a resonance of peak area one unit to the spectrum.

The upper molecular weight limit for NMR structure determination is the result of two factors. First, the more residues in the protein the more resonances are packed into the same chemical shift range and hence the greater the resonance overlap. Second, the width of the NMR resonances depends on the tumbling motion properties of the protein; smaller proteins have narrower resonances. This latter feature produces the following problems: (1) the broader the resonances the less the ability to resolve resonances of similar frequencies; (2) the broader the resonance the lower the peak height and hence the lower the signal-to-noise ratio of the spectrum (defined and practically utilized as peak height and not peak area); (3) the ability to transfer NMR information (termed coherence) through the chemical bonding network of the molecule is dependent on the lifetime of the information after absorbance by the nuclei. It is the reduction in the lifetime of the information that gives rise to broader lines as the rotational frequencies of the protein reduce.

The relatively low sensitivity of NMR compared with other spectroscopies that utilize higher frequency electromagnetic radiation requires a minimum sample concentration for structure determination of the order of 1 mM (15 mg ml^{-1} for a protein of M_r 15 000). However, the use of NMR to probe structural *changes*, for example, analogously to other forms of spectroscopy, is commonly carried out with samples an order of magnitude more

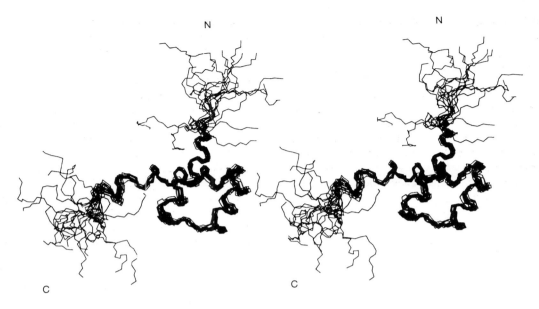

Fig. N23 A stereoview of 19 structures for residues 0–67 of the polypeptide backbone of the *Antennapedia* homeodomain from *Drosophila*. As no medium-range or long-range constraints could be detected for the terminal residues, a unique backbone fold is defined only for residues 7–59. From [4].

dilute. The typical sample size of 0.5 ml dictates the use of relatively large quantities of protein although the technique itself is nondestructive and the sample recoverable. The solubility of the protein may be a limiting factor. The width of NMR resonances are sensitive to restrictions in molecular rotations (see above) and are thus sensitive to oligomerization and nonspecific aggregation phenomena. The purity of the sample should be greater than 95%, although high molecular weight impurities are rarely observable owing to their relative immobility. More of a problem are small molecule impurities, for example residual buffer components such as tris, glycine and EDTA, that have very sharp lines that often give rise to spectral artefacts.

Experiments can be performed over a wide range of temperatures. Normally as high a temperature as possible is used to provide the sharpest resonances. Variation in temperature is a very useful source of resolving overlapping signals as there is some differential behaviour of the chemical shift of individual resonances with temperature. In terms of pH, experiments are often performed in the range 4–5. This reflects the exchange rate of amide N—H protons with the solvent being minimized around pH 4. At higher pH values more technically demanding experiments are often required to observe solvent exposed amide N—H proton resonances, although it is feasible to study most proteins at neutral pH. Again, variation of pH over a small range is an important means of resolving spectral overlap.

Structure elucidation

The elucidation of structures depends on the determination of numerous structural constraints. In order to identify which individual nuclei are involved in the constraints it is necessary to know the resonance frequency of the nuclei. The process of identification is known as resonance assignment; this is a crucial first step to any NMR investigation. Both resonance assignment and the elucidation of constraints depend on the transfer of information between nuclei. There are two means by which information is transferred: the nuclear Overhauser effect (the NOE) which is the result of transfer through space; and J-coupling (or scalar coupling) which permits transfer through the chemical bonding system. The NOE enables the identification of interproton distances of less than 5 Å as the rate of build-up of the NOE is proportional to the inverse sixth-power of the interatomic distance. In practice J-couplings are normally only resolvable between protons two or three bonds apart. The structural constraints of J-couplings resides in the relationship between the values of three-bond J-couplings and the corresponding torsional angle mapped out by those three bonds. There is a third type of constraint sometimes used in structure calculations relating to internal hydrogen bonding. The exchange with the solvent of amide protons involved in internal hydrogen bonds is retarded relative to those exposed to the solvent. This information cannot be used directly as the hydrogen-bond acceptor cannot be identified, but often after the first round of structure calculation all but one hydrogen-bond acceptor can be eliminated. Thus, such constraints may be introduced at the refinement stage (see below) to assist the structure calculation algorithms to converge.

As NMR structures depend on the number and precision of the constraints that define the relative position of individual atoms and torsional angles, they are unlike X-ray crystallographic studies in that there is no uniform resolution over the structure. Typically, NMR structures are represented by a family of coordinates that are equally valid so long as all the NMR-derived constraints are satisfied. The structure calculation algorithms (see below for a more detailed discussion) are repeated many times to ensure as far as possible that all space is sampled. although perfect sampling is probably not achievable. Figure N23 shows a

a 2D NOESY

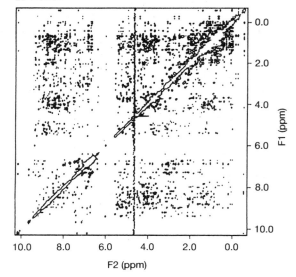

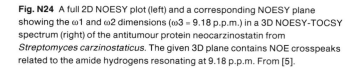

b 3D NOESY

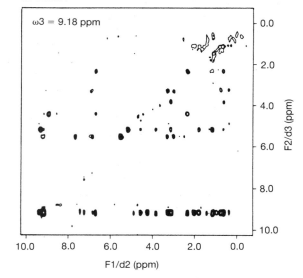

Fig. N24 A full 2D NOESY plot (left) and a corresponding NOESY plane showing the ω1 and ω2 dimensions (ω3 = 9.18 p.p.m.) in a 3D NOESY-TOCSY spectrum (right) of the antitumour protein neocarzinostatin from *Streptomyces carzinostaticus*. The given 3D plane contains NOE crosspeaks related to the amide hydrogens resonating at 9.18 p.p.m. From [5].

superimposition of 19 calculated protein structures. The close overlay of the central region of the structures (where a large number of constraints define the structure) contrasts with the two mobile termini (where there are no constraints to relate these regions to the core of the protein).

Resonance assignment

The 1980s saw many important advances in the techniques of NMR spectroscopy, culminating in the ability to determine the three-dimensional structure of globular proteins. Along with the development of magnet technology for more uniform and higher field-strength magnets, and the advances in computing for faster data processing and structure calculation, there was a major development in methods of sequence-specific resonance assignment. This allowed for the first time the identification of resonance frequencies for individual protons without recourse to a three-dimensional structure determined by other means.

Homonuclear experiments

The majority of NMR studies are carried out on unlabelled protein using only resonances of the hydrogen nuclei. This is the simplest case in terms of sample preparation — the protein or nucleic acid can be isolated or produced by recombinant techniques in the most efficient manner available. Indeed, it was for homonuclear studies that the first and still widely used sequential resonance assignment strategy was developed. The revolution

in resonance assignment paralleled the development of two-dimensional (2D) NMR experiments and the corresponding improvement in the ability to resolve resonances. In 2D experiments, the resonance frequencies of the nuclei are measured twice: either side of an event that transfers information between nuclei via J-coupling or NOEs. The primary experiments used are:
1 COSY. The transfer via J-coupling in a single step between nuclei two or three bonds apart.
2 TOCSY. The transfer via J-coupling between a network of nuclei within two or three bonds of another member of the network.
3 NOESY. The transfer via NOEs between nuclei less than 5 Å apart in space.

There are numerous variations on the basic experiments to suit specific circumstances, and hence the spectroscopist is armed with a battery of techniques. Theoretically, it is possible to improve resolution further through the recording of the resonance frequencies more than twice; for example, in a 3D experiment three frequency measurements are made separated by two information-transfer processes. Figure N24 shows a typical example of a contour plot of a 2D spectrum. In this case the off-diagonal 'crosspeaks' result from NOEs between hydrogen nuclei. Figure N24 also contains a corresponding 3D spectrum that highlights some of the resolving capability of a third dimension. In practice, the sensitivity of the information transferred twice along the desired pathway is relatively low, which limits this aspect of 3D spectra to more concentrated samples. However, 3D homonuclear experiments do in general provide numerous peaks with which to confirm conclusions from 2D experiments.

In sequential resonance assignment the individual amino-acid residues are identified according to the topology of their hydrogen network (termed spin-systems) as defined by the J-coupling. For example, in aspartic acid the amide proton is J-coupled to the α-proton which in turn is J-coupled to both β-protons. In glutamic acid residues (but not of course aspartic acid residues) this

J-coupling network is extended between the β-protons and the γ-protons. These network topologies are mapped out for each residue using COSY and TOCSY type experiments. However, not all residues have unique topologies. For example, it is impossible to distinguish aspartic acid from phenylalanine, or glutamic acid from methionine, if only two- or three-bond J-couplings may be observed. Hence residues are categorized into topology groups. These spin-systems are then linked in a sequence-specific manner using NOEs from NOESY spectra as there is a minimum of four bonds between the hydrogens of one residue and the next.

Adjacent amino-acid residues inevitably have hydrogens close in space to each other owing to the restrictions of the covalent geometry. Which hydrogens are involved depends on the peptide backbone dihedral angles and hence the local secondary struc-ture. Figure N25 shows an example of how adjacent amide hydrogens in helical conformations may be connected using sequential NOE crosspeaks in a 2D-NOESY spectrum. Indeed, the sequential NOEs, supported by NOEs across the turn of α-helix or across β-sheets provide a description of the secondary structure of the protein almost immediately on completion of the sequential assignment. However, the connection of the spin-systems of sequential residues using NOEs is complicated by the necessity to distinguish sequential NOEs from all other NOEs.

Heteronuclear experiments

Owing to the practicalities of sensitivity, virtually all assignment strategies including non-hydrogen nuclei (termed heteronuclei) require the use of isotopically enriched proteins. Thus the protein

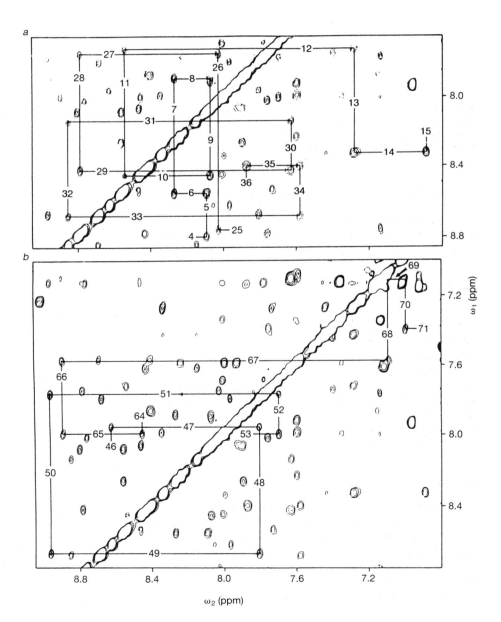

Fig. N25 Sequential resonance assignments in the amide hydrogen region of a NOESY spectrum of calbindin, showing polypeptide segments in helical conformations. Lines are drawn between crosspeaks arising from sequential residues and labelled with sequence-specific assignments. The experimental data are shown for four polypeptide segments; the sequences Glu 4–Ala 15 and Lys 25–Phe 31 are given in *a* and the segments Leu 46–Leu 53 and Glu 64–Lys 71 in *b*. From [6].

Table N4 Types of NMR experiment

Experiment	Transfer of coherence
Experiments on unlabelled proteins	
COSY	Single ^{1}H–^{1}H J-couplings
TOCSY (HOHAHA)	Multiple ^{1}H–^{1}H J-couplings
NOESY	Via NOEs
Experiments with single-labelled proteins (all may be extended with NOESY, TOCSY, etc.)	
HMQC	^{1}H–X J-couplings via multiple quantum states
HSQC	^{1}H–X J-couplings via single quantum states
HCCH	Between ^{1}H via attached ^{13}C nuclei
Experiments with double-labelled protein backbone resonances	
HNCA	Records H̲N, N̲H and Cα frequencies
HNCO	Records H̲N, N̲H and C̲O frequencies
H(CA)NH	Records HCα, N̲H and H̲N frequencies
HCACO	Records HCα, Cα and C̲O frequencies
HCA(CO)N	Records HCα, Cα and N̲H frequencies
HCANNH	Records HCα, Cα, N̲H and H̲N frequencies
HCA(CO)NNH	Records H̲Cα, Cα, N̲H and H̲N frequencies via the C̲O resonance
CBCANH	Records Cβ, Cα, and H̲N frequencies
CBCA(CO)NH	Records Cβ, Cα, and H̲N frequencies via the C̲O resonance

The simpler experiments may, of course, be performed on single- and double-labelled proteins.

must be recombinant (or at least produced from an organism for which it is possible to control the food source). The major advantages of heteronuclear experiments are that the resolution afforded by the heteronucleus is independent of that afforded by the hydrogens. Heteronuclear experiments for protein resonance assignment fall into two classes [3].

First, with ^{15}N-labelled proteins, the resonance assignment strategy mirrors that described above but with the advantage of the extra resolution of the ^{15}N resonances. The basic experiments that transfer information between the ^{15}N nuclei and their directly bonded amide proton, either the HMQC or HSQC experiments (Table N4) are combined with homonuclear experiments such as the TOCSY and NOESY experiments and two or three frequency measurements are made (e.g. 2D or 3D TOCSY-HMQC experiments). At least one of the frequency measurements is necessarily that of the ^{15}N nuclei resonances. The main techniques used currently are the NOESY-HMQC, the TOCSY-HMQC and the HMQC-NOESY-HMQC experiments. Similarly to homonuclear resonance assignment, the assignment procedure (as illustrated in Fig. N26) depends on the transfer of information via relatively small three-bond J-couplings. This limits the molecular weight range for the applicability of these techniques. Again there is the necessity to distinguish sequential NOEs from all other NOEs.

The second heteronuclear assignment method requires the labelling of all nitrogens as ^{15}N and all carbons as ^{13}C. In proteins labelled in this manner, resolution is improved through the measurement of three independent resonance frequencies (^{1}H, ^{15}N and ^{13}C). Further, it is now possible to link together the

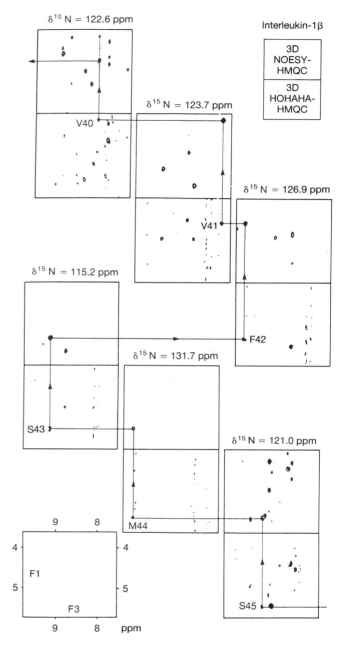

Fig. N26 Pairs of matching 3D NOESY-HMQC (upper) and TOCSY-HMQC (lower) planes showing the sequential connectivities for the sequence Gln 19 to Ser 45 of interleukin-1β. Sequential assignments are made by 'hopping' from one ^{15}N plane of TOCSY/NOESY pairs to a pair at the ^{15}N frequency of the neighbouring amino-acid residue using sequential NOEs (in an analogous manner to the analysis of 2D spectra). From [7].

resonances of individual residues and sequential residues via the relatively large one-bond C—C and C—N J-couplings. The combination of these two improvements allows these methods to be applied successfully to larger proteins. The major experiments currently favoured — HNCA, HNCO, H(CA)NH, HCACO, HCA-(CO)N (see Table N4) — use predominantly one-bond J-couplings

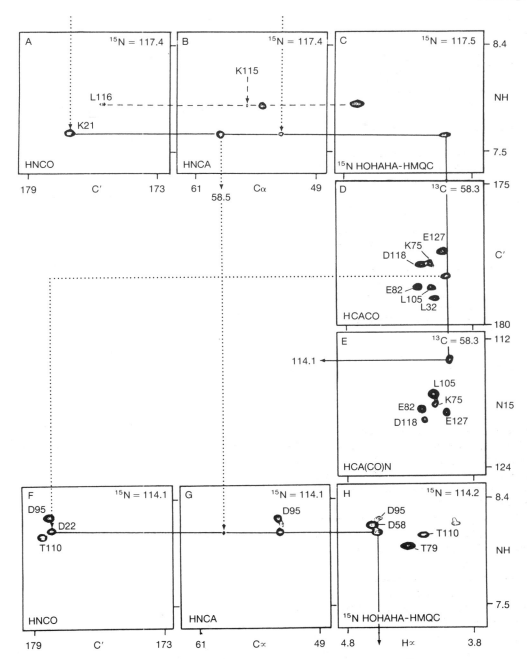

Fig. N27 Selected regions of slices from five separate triple resonance 3D NMR experiments illustrating the J-correlation between Lys 21 and Asp 22 of calmodulin uniformly labelled with ^{15}N and ^{13}C. The connectivity patterns for these two residues are traced by the solid and dotted lines, whereas the broken lines correspond to parts of the connectivity patterns observed for other residues. Slices A–C are taken at the ^{15}N chemical shift of Lys 21, and slices F to H are taken at the ^{15}N frequency of Asp 22. Slices D and E are taken at the Cα chemical shift of Lys 21, as observed in slice B, and are used to connect together the two sequential residues. Note that slice G shows a weak two-bond connectivity between the amide nitrogen of Asp 22 and the Cα carbon of Lys 21, thus providing further confirmation of the connectivity between the two residues. From [8].

to transfer information between peptide backbone nuclei. The degenerate nature of some of the information content of these spectra allows the independent confirmation of resonance assignments. An example of backbone resonance assignment using this methodology is illustrated in Fig. N27. A wealth of experiments are being devised to assist the assignment of double-labelled proteins, including the 4D HCANNH and HCA (CO)NNH experiments to connect residues in a single spectrum and the 3D CBCANH and CBCA(CO)NH experiments where the β-carbon is used to resolve *HC*α and HCα degeneracy. The sidechain resonances may be assigned using HCCH-COSY and HCCH-TOCSY experiments. An important consequence of double-labelling is the considerable increase in resolution available with which to identify NOEs, using for example 3D or 4D HCCH-NOESY experiments. The resulting increase in structural constraints is of significant benefit to the quality of calculated structures.

Structure determination

The unique ability of NMR among spectroscopic techniques to determine structures at atomic resolution results from a combination of the relationship of spectral parameters — the NOE and J-coupling — to internuclear distances and torsional angles, and the numerous simultaneous measurements of these parameters that results from the widespread distribution of the probes (hydrogen nuclei) throughout the molecule. Neither NOEs nor J-couplings produce a precise measurement of distance or angle; rather their interpretation leads to limits for constraints that may be applied during the generation of the molecular structure. Correspondingly, the number of constraints with which the position of one atom may be defined relative to its neighbours dictates the resolution that may be determined for the structure in that region. For example, regardless of mobility (see below) surface residues are normally less well defined than interior residues in proteins as their number of neighbouring residues is reduced.

By far the largest number of constraints result from the observation of NOEs. Hence it is important for the NOESY spectra (or their heteronuclear edited counterparts) to be as resolved as possible. NOE measurements are converted to distance ranges through the relationship of the initial build-up rate with the inverse sixth power of the internuclear distance. This relationship makes two assumptions: (1) that any two nuclei involved in a specific NOE may be considered in isolation from all others; (2) that the molecule is conformationally rigid. Deviation from the former assumption is addressed by the refinement procedure of back-calculation (see below). Deviation from the latter assumption is considerably more difficult to address. The breakdown of the latter assumption manifests itself most obviously when a single hydrogen shares NOEs with two or more other hydrogens to which it cannot possibly be close simultaneously. The population of more than a single conformation inevitably leads to a short estimate for the internuclear distance. In the case where interconversion between conformations is faster than the overall correlation time of molecular reorientation, apparent shortening of the estimate of internuclear distance may be opposed by a reduction in the NOE build-up rate resulting from the change in the effective correlation time associated with the internuclear distance that gives rise to the NOE.

A further problem arises when using NOEs as constraints. Not all NOEs will be resolved, which potentially leads to a biased constraint set. Related to this is the common situation that a number of resonances corresponding to individual hydrogens in the same group (e.g. the three hydrogens in a methyl group) share the same frequency. In these cases the distance estimate must be made to a pseudo-atom at the geometric average of the positions of the nuclei involved, and an appropriate correction factor applied to the upper bound of the constraint. Hence, it is important to define uniquely as many resonances as possible; the

stereospecific assignment of the prochiral methyl groups of valine and leucine residues, and of β-methylene groups leads to a considerable improvement in calculated structures.

The use of J-coupling derived constraints is limited to the observation of extreme values (large or small). These values may be used to restrict the torsional angles to one or more small ranges. The population of single conformations in specific ranges cannot be distinguished from averaging over a large range of torsional angles if mid-range J-coupling values are observed. Torsional constraints are commonly applied to both ϕ and χ_1 angles, and considerably improve the success of structure calculations.

There are two main ways of converting distance and torsional constraints into atomic coordinate sets — distance geometry algorithms and restrained molecular dynamics simulations.

Distance geometry calculations fall into two categories: metric matrix and variable target function algorithms. Only the former is true distance geometry, working in distance space. The latter works in dihedral angle space. Metric matrix algorithms firstly use triangle and tetrangle inequalities to produce a complete set of distance bounds. A metric matrix is generated (termed 'embedding') from initial distances chosen from within the distance bounds. The atomic coordinates are obtained from the eigenvalues and eigenvectors of the metric matrix. Variable target function algorithms select a series of dihedral angles to minimize the target function. Both distance geometry algorithms give rise to equivalent structures but may vary in their success rate of convergence.

Restrained molecular dynamics procedures work in Cartesian space and include the NMR-derived constraints as pseudo-energy potentials in the force field. Normally, the force field is stripped down to only its bonding potentials for computational expediency (termed restrained simulated annealing). Many of the force field potentials may be removed as it is the ability of the algorithm to sample space that is the principal quality required of the dynamics algorithm. Distance geometry algorithms commonly result in coordinate sets with poor local geometry; distance geometry structures are typically subject to a simulated annealing refinement. The structure calculation algorithms, whichever method is chosen, are repeated many times using different random starting selections to ensure a significant sampling of space.

The quality of NMR structures may be judged by three criteria. First, the structures are tested for residual constraint violations which would indicate that the coordinate set did not fully satisfy the input data. Secondly, root-mean-squared distance (r.m.s.d.) variations may be calculated between members of the family of generated structures. These two measurement criteria principally test the performance of the algorithms and are not ideal measures of the quality of the structures. Indeed, they are often interrelated — the more conservative the choice of constraint limits the fewer the residual violations but the worse the r.m.s.d. values and vice versa. It should also be noted that choosing too tight a series of constraint limits may fortuitously lead to a very precise series of structures that are inaccurate. The third measurement of quality of structure involves calculating the residual error between the primary experimental data (normally a series of NOESY spectra) and spectra calculated from the generated structures. This procedure, known as back-calculation, takes into account all the protons in the molecule simultaneously (avoiding the isolated spin pair approximation), but is still subject to errors arising from mobility effects. However, this method may be used to create a residual error target function against which the structures are refined iteratively.

J. MARTIN

J. WALTHO

1 Wüthrich, K. (1986) *NMR of Proteins and Nucleic Acids* (Wiley, New York).

2 Wagner, G. et al. (1992) NMR structure determination in solution: a critique and comparison with X-ray crystallography. *Annu. Rev. Biophys. Biomol. Struct.* **21**, 167–198.

3 Clore, G.M., & Gronenborn, A.M., (1991) Applications of three-and four-dimensional heteronuclear NMR spectroscopy to protein structure determination. *Prog. NMR Spectrosc.* **23**, 43–92.

4 Billeter, M et al. (1990) Determination of the three-dimensional structure of the *Antennapedia* homeodomain from *Drosophila* in solution by ^{1}H nuclear magnetic resonance spectroscopy. *J. Mol. Biol.* **214**, 183–197.

5 Gao, X. & Burkhart, W. (1991) Two- and three-dimensional proton NMR studies of apo-neocarzinostatin. *Biochemistry* (1991) **30**, 7730–7739.

6 Kördel, J. et al. (1989) ^{1}H NMR sequential resonance assignments, secondary structure, and global fold in solution of the major (*trans*-Pro34) form of bovine calbindin D_{9k}. *Biochemistry* **28**, 7065–7074.

7 Marion, D. et al. (1989) Overcoming the overlap problem in the assignment of ^{1}H NMR spectra of larger proteins by use of three-dimensional heteronuclear ^{1}H-^{15}N Hartmann–Hahn–multiple quantum coherence and nuclear Overhauser–multiple quantum coherence spectroscopy: application to interleukin 1β. *Biochemistry* **28**, 6150–6156.

8 Ikura, M. et al. (1990) A novel approach for sequential assignment of ^{1}H, ^{13}C, and ^{15}N spectra of larger proteins: heteronuclear triple-resonance three-dimensional NMR spectroscopy. Application to calmodulin. *Biochemistry* **29**, 4659–4667.

NO NITRIC OXIDE.

nociception A term coined by Sherrington in 1906, which describes the identification of potentially injurious physical stimuli. It should not be confused with the definition of pain offered by Merskey in 1979 as 'an unpleasant sensory and emotional experience associated with actual or potential tissue damage, or described in terms of such damage'.

***nod* genes** Genes in rhizobia involved in NODULATION.

Nodaviridae A family of insect RNA viruses. *See*: ANIMAL VIRUSES.

nodes of Ranvier Periodic constrictions of the MYELIN sheath surrounding large axons (*see* NEURON). They demarcate the junctions between adjacent SCHWANN CELLS, and are the sites of electrical impulse generation.

Nodulation

THE nitrogen-fixing symbiosis between leguminous plants and the bacteria known collectively as RHIZOBIA represents a finely tuned system of biochemical, physiological, and morphological changes in both partners. Because of its importance in agriculture (many legume crops such as beans, peas, clover, and peanuts require no nitrogen fertilizer) and its attraction as an example of coupled DIFFERENTIATION, it has received much attention, most of the recent significant work having been on the molecular biology and genetics of the system.

In most rhizobia–legume interactions, the first observed effect is the deformation (curling or branching) of root hairs by the bacteria which enter the hair and induce a plant-specified organelle, the infection thread. Through this tube the bacteria enter the plant, and ahead of it, plant cells divide within a novel MERISTEM. As nodules develop, the infection thread branches, growing through and between incipient nodule cells. Bacteria are released from infection thread tips into the plant cell cytoplasm and it is these bacteroid forms that fix N_2 to form NH_3 (*see* NITROGEN FIXATION) which is assimilated by the plant, and used to synthesize nitrogenous compounds (Fig. N28).

This plethora of events is reflected in, and caused by, the differential expression of several genes in both plant and bacterial partners. The bacterial *nif* genes encoding nitrogenase and ancillary proteins required for nitrogen fixation are activated in bacteroids (*see* NITROGEN FIXATION; NITROGEN FIXATION GENES) and the plant makes a series of proteins termed nodulins, which are expressed specifically in nodules.

In this article, only the genetic analysis of the early steps in the infection process will be considered. A more complete description of nodulation can be found in [1,2]. One fascinating aspect of the interaction will be highlighted, namely its specificity — particular legume groups are nodulated only by a limited range of rhizobial strains, species or genera [3]. Thanks to molecular biology, the genetic, functional, and chemical bases of the early steps of infection and of host range specificity have been established, at least for the bacterial partner.

In some rhizobia at least, *nod* genes that specify early stages of infection and determine the host range of a given strain are located on large PLASMIDS. The transfer of such a plasmid from a strain that nodulates (for example) peas to one that nodulates clover results in bacteria fully capable of nodulating the former host [4]. Such 'sym' (for symbiosis) plasmids also sometimes contain the *nif* genes.

Analysis of the molecular genetics of *Rhizobium* has been facilitated by the advent of wide host range cloning VECTORS (*see* DNA CLONING). These allow the construction of recombinant plasmids containing cloned *Rhizobium* DNA. Cloning is usually done in *Escherichia coli* for convenience and such recombinant plasmids can then be transferred from *E. coli* to *Rhizobium* so that the cloned DNA can be expressed and its function established. Other important tools are transposons (*see* TRANSPOSABLE GENETIC ELEMENTS), which are used as MUTAGENS. These elements, usually marked by an antibiotic resistance gene, when inserted into a gene cause a MUTATION. Thus, to map the target gene, it is necessary only to map the transposon either genetically (using the antibiotic resistance as the marker) or physically, by *in situ* HYBRIDIZATION (*see* TRANSPOSON TAGGING). Transposon mutagenesis is particularly valuable in the analysis of 'silent' genes (such as *nod* and *nif*), in which mutations have no discernible phenotype in the free-living bacteria but only become apparent when the bacteria are used to inoculate plants [5].

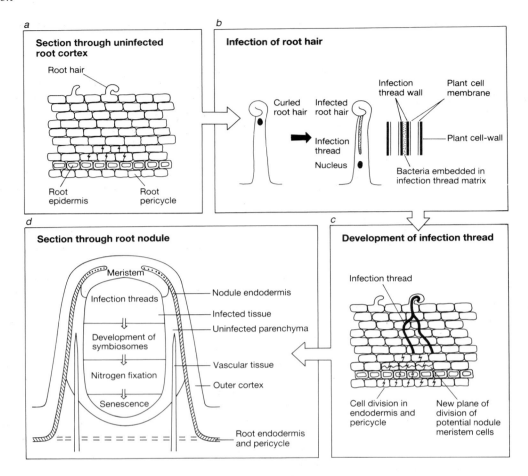

Fig. N28 Stages in nodulation. Rhizobia attach to legume root hair cells. Flavonoids secreted by legume roots induce *nod* gene expression in the rhizobia, resulting in production of a lipo-oligosaccharide signal molecule that induces root hair curling and division of root cortical cells to form a nodule meristem. Root hair curling entraps the rhizobia. An infection thread develops following redirection of root hair cell wall growth. The bacteria grow along this intracellular channel which elongates towards the developing nodule meristem. Bacteria infect cells of the nodule meristem and a nodule develops.

Sym plasmids

In *R. leguminosarum* biovar (bv.) *viciae* (which nodulates peas and vetches) the *nod* genes seem to be clustered within a region of DNA of ~15 kb as the transfer of cloned DNA no more than 15 kb long from 'pea' rhizobia to 'clover' or 'bean' rhizobia is sufficient to allow the recipients to nodulate peas [6]. Similar clustering of *nod* genes has also been demonstrated in the symbionts of alfalfa and of clover (*R. meliloti* and *R. leguminosarum* bv. *trifolii* respectively). Figure N29 illustrates the organization of *nod* genes in a strain of *R. leguminosarum* bv. *viciae*. Genes similar in terms of sequence and products have been identified in other rhizobia (*see* [5] but there are also strain- or species-specific genes. For example *nodO* (Fig. N29) appears only in *R. leguminosarum* bv. *viciae* whereas *R. meliloti*, which nodulates alfalfa, has three genes, *nodHP* and *Q* (see below), which are essential for nodulation of alfalfa but are absent from *R. leguminosarum*.

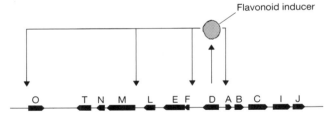

Fig. N29 *Nod* gene map from *Rhizobium leguminosarum* bv. *viciae* showing induction of the *nodABCIJ*, *nodFEL*, *nodMNT*, and *nodO* operons. Flavonoids secreted from legume roots induce *nod* gene expression after interacting with the regulatory NodD protein (see text).

Role of *nod* genes in nodulation and host range specificity

In species examined to date, mutations in *nodA*, *B* or *C* completely abolish nodulation and all aspects of the infection process. Strikingly, the Nod⁻ phenotype of such mutants can be corrected by introducing the corresponding genes from other rhizobial species. Thus *nodABC* are functionally equivalent in rhizobia with different host ranges.

In *R. leguminosarum* bv. *viciae*, *nodD* mutations also block nodulation but in other species (e.g. *R. meliloti*) *nodD* mutations have only a slight effect. This is because in the former, there is only one copy of *nodD* but in the latter, there are at least three *nodD*s.

Mutations in other *nod* genes lead either to a delay (*nodI, J, F, E,* and *L*) or else have no apparent effect on nodulation (*nodM, N, T,* and *O*). The reason for this is that different *nod* genes may compensate for each other or else there may be chromosomal 'housekeeping' genes with the same function as 'dedicated' *nod* genes (see below).

In contrast to *nodABC* mutations, those in *nodFEL* cannot be corrected by corresponding genes from other rhizobial species, indicating that they have a role in host range specificity. This has been confirmed directly; transfer of *nodFEL* from a 'pea' strain to a 'clover' strain allows the latter to nodulate peas [7]. Strikingly, mutations in these genes in the 'clover' strain, while delaying nodulation on clover itself, actually allowed the mutants to nodulate non-hosts such as pea or vetch [8]. Thus *nodFE* enhance the nodulation of the compatible legume but cause some form of 'rejection' by heterologous hosts. Recently, the molecular basis for this has been revealed, at least in part (see below).

nodFE are very similar in rhizobia with different host ranges. Their role in determining host range must therefore rely on subtle differences in the specificity of action of their products. A different situation obtains with the *nodHP* and *Q* genes of *R. meliloti*. These have no counterparts in other rhizobia but are essential for nodulation of alfalfa. In this case, they represent 'extra' genes that are specifically needed for the nodulation of this host. In work reminiscent of that with the *nodFE* genes, it was found that *R. meliloti* wild-type strains evoke no root hair curling on (for example) vetches. However *nodH* mutants, though having lost the ability to induce curling on the 'proper' host, alfalfa, now elicit that response on vetch. Again, recent work has pinned down the reason for this curious observation in chemical terms (see below).

Association with the cell envelope

As might be expected, in such an intimate interaction involving cell-to-cell contact, several *nod* gene products have been shown to be located at or near the rhizobial cell surface — associated with the membranes and/or the periplasmic space between them (*see* BACTERIAL CELL ENVELOPE).

The products of *nodE, nodD, nodI, nodT,* and *nodJ* are associated with the inner membrane. NodC is unusual in that it traverses the periplasmic space, having one part in the inner membrane and one in the outer. Finally, the product of *nodO* is excreted from the cells altogether.

Flavonoid activation of *nod* gene expression

The transcriptional control of the *nod* genes has been analysed with the aid of FUSION GENES in which a REPORTER GENE with an easily assayable product, such as the β-GALACTOSIDASE gene (*lacZ*), is fused to the various *nod* genes. By analysing these *nod–lacZ* fusions the following understanding of *nod* gene regulation has been obtained (see [2]).

Under normal growth conditions, only *nodD* is expressed at significant levels. When legume root exudate is added, the ex-

pression of all the other *nod* gene transcripts is significantly enhanced but expression of *nodD* remains unchanged. The active components in root exudates are particular FLAVONOIDS (flavones, isoflavones, or flavanones) which act at very low (100 nM) concentrations (Fig. N30). Different rhizobial strains and species display different sensitivities to individual flavonoid molecules.

nodD is the controlling regulatory gene. It is required for flavonoid-dependent activation of the other *nod* genes and in some species is also autoregulatory, repressing its own transcription in the presence or absence of flavonoid inducers. It binds to conserved regulatory sequences (called *nod* boxes) preceding those operons that are activated (Fig. N29). This binding occurs in both the presence and absence of *nod* gene inducers. Although not formally proven, it seems that flavonoid binds to *nodD*, causing an allosteric shift to 'open up' the *nod* box regulatory sequences, allowing initiation of transcription.

In some strains of rhizobia there is one copy of *nodD* (e.g. *R. leguminosarum* bv. *viciae*) whereas in others (e.g. *R. meliloti*) there are multiple copies of the regulatory gene. In the latter cases the different NodD proteins display different sensitivities to the spectra of different flavonoids in the root exudates of different potential host plants.

There is no evidence that NodD acts in concert with another regulatory protein which senses the environment (the two-component system). The behaviour of NodD is similar to that of members of the *E. coli* lysR family of regulatory proteins, whose sequence it slightly resembles. The 'lysR family' comprises a group of regulatory genes which, like *nodD* have the ability to repress their own transcription, and in the presence of appropriate small molecular weight molecules can induce the transcription of other genes (e.g. diaminopimelic acid causes *lysR*-dependent activation of other genes involved in lysine biosynthesis).

The rhizobial *nod* regulatory system is a good adaptation to conditions that the bacteria face in the field. The bacteria sense the presence of a potential host (most nonlegume roots do not exude flavonoids) and so only commit themselves to the 'expense' of making Nod proteins if there is a chance of initiating an infection.

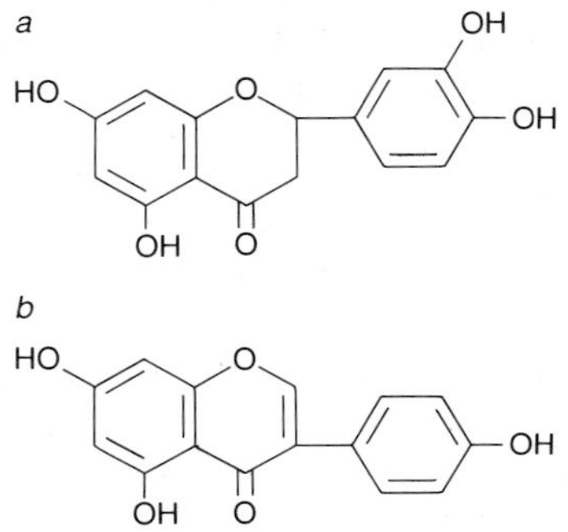

Fig. N30 The chemical structure of the plant-derived flavanone eriodictyol (*a*) and the isoflavone genistein (*b*) which induce *nod* gene expression in *R. leguminosarum* bv. *viciae* and *Bradyrhizobium japonicum* respectively.

Nodule-inducing factors

If *nod* gene inducers are added to cultures of rhizobia *in vitro*, the resulting cell-free supernatant can induce root hair curling on the corresponding host legume. This observation paved the way for the chemical identification of the active component(s). This was achieved in 1990 when a sulphated lipo-oligosaccharide (Fig. N31) made by *R. meliloti*, was shown to induce root hair deformation on alfalfa seedlings. This molecule is active at astonishingly low concentrations; at 10^{-11} M it induces root hair deformation and, indeed at 10^{-9} M it elicits nodule-like structures on alfalfa roots. Thus, it is active at concentrations several orders of magnitude lower than 'classical' PLANT HORMONES such as gibberellins, cytokinins, auxins, and ethylene [9,10].

Thanks to a combination of different disciplines there is now becoming available a genetic and functional dissection of the action of this novel phytohormone. Many of the answers have come from predictions of *nod* gene function from similarities in the sequences of their products and these of known biochemical function present in databases (*see* DATABASES). There have been very satisfying 'matches' between such proposed functions, the chemistry of the lipo-oligosaccharide and the observed plant responses.

NodM looked remarkably similar to the glucosamine synthase of *E. coli* and has in fact been shown to have that enzymatic function. The sugar in the oligosaccharide portion of the lipo-oligosaccharide is indeed glucosamine. *nodM* mutants still nodulate because the bacteria can also use its own 'housekeeping' glucosamine synthase to make this sugar.

In *R. leguminosarum* bv. *viciae*, the *nodFEL* operon encodes proteins with striking sequence similarity to, respectively, acyl carrier protein, fatty acid synthetase, and acetylase. It was gratifying, then, to find that *nodF* and *E* mutants are defective in the synthesis of the fatty acid side chain of the active lipo-oligosaccharide molecule and that *nodL* mutants fail to acetylate the sugars. As mutations in these genes only delay nodulation and do not seem to be affected in root hair curling it is apparent that these two decorations of the tetraglucosamine are not absolutely required for the infection process.

nodFE are important in the determination of host range specificity (see above). It seems likely that the NodF and E proteins of different rhizobia, although related in their sequence, have slightly different enzymatic specificities, affecting the precise structure of the lipid, and, so, of the particular host range.

nodI and *nodJ* make proteins similar to a series of polypeptides in other bacteria which are involved in the import or export of various small molecules, such as phosphate, oligosaccharides, and amino acids. It seems likely that NodI and NodJ , which are located at the bacterial cell surface, assist in the export of the lipo-oligosaccharide.

As described earlier, *R. meliloti*, but not other rhizobia, contain three genes — *nodH*, *P*, and *Q* — and mutations in these abolish nodulation and root hair curling ability on the normal host, alfalfa. Paradoxically, the wild-type *R. meliloti* could not induce root hair curling on heterologous hosts but *nodH*, *P* or *Q* mutants could. This curious observation now has a neat chemical explanation. Sequence analysis of these genes indicates similarities to genes in *E. coli* which are involved in sulphur metabolism. Consistent with this, then, was the finding that mutations in each of these genes either reduced (*nodP* and *Q*) or abolished (*nodH*) sulphation of the oligosaccharide; it was found that the desulphated form of the signal molecule had lost its ability to induce root hair curling on the homologous host, alfalfa, but had gained the property of curling root hairs of non-hosts such as vetches.

This dual approach of chemistry and genetics has dissected the exquisitely precise mechanism by which subtly different molecules elicit specific morphogenetic changes in particular species of legumes. There are few, if any, other cases in higher eukaryotes where the ability of a particular molecule to induce a specific organ in such a genus-specific way is understood in such chemical and genetic detail. As such, the chemical signalling between *Rhizobium* and legumes is indeed a paradigm for the studies of intimate interactions between prokaryotes and eukaryotes (*see also* PLANT PATHOLOGY).

Lectins and host range determination

So far, this article has concentrated on the bacterial genes that are involved in the host range specificity of nodulation. But what about the plant?

Less is known about this partner of the symbiosis; although several nodulins have been identified and characterized, their precise roles are mostly unknown. However, recent reports show that in one case at least, a LECTIN is of key importance in determining host range specificity.

Lectins are a family of sugar-binding proteins made by higher organisms; conceptually though not mechanistically they can be thought of as antibody-like in their behaviour. Using transgenic technology (*see* PLANT GENETIC ENGINEERING), a gene from pea which encodes the lectin characteristic of that legume has been introduced into clover. The transgenic clover plants could then be nodulated by *R. leguminosarum* bv. *viciae* which normally nodulates peas but not clover [11]. Whether the lectin binds to a specific surface molecule on a particular species of *Rhizobium* remains to be seen.

Future prospects

As in other areas of biology, the use of molecular techniques has revolutionized understanding of the *Rhizobium*–legume symbiosis. Having established the genetic basis (at least on the bacterial side) for two aspects of chemical signalling between the two partners

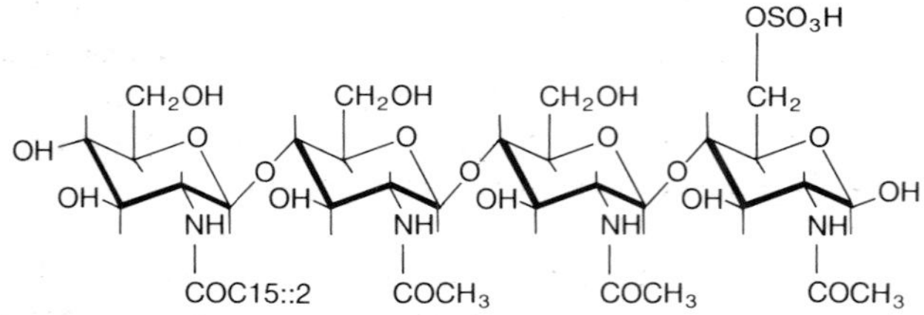

Fig. N31 Substituted lipo-oligosaccharide signal molecule produced by *R. meliloti* after *nod* gene activation.

(i.e. the flavonoid *nod* gene inducers and the lipo-oligosaccharide morphogen) a major challenge for the future is to determine how these factors actually work. In particular, the mechanism by which such a complex and precise programme of cell division and differentiation in the plant can be elicited with such exquisite specificity by a relatively simple molecule will lead us into exciting new areas of receptor studies, signal transduction, and organogenesis in plants.

A.W.B. JOHNSTON

1 Long, S.R. (1989) *Rhizobium*–legume symbiosis; life together in the underground. *Cell* **56**, 203.
2 Kondorosi, A. et al. (1991) The role of nodulation genes in bacterium-plant interaction. In *Genetic Engineering*, Vol. 13 (Setlow, J.K., Ed.) 115–136 (Plenum, New York).
3 Young, J.P.W. & Johnston, A.W.B. (1989) The evolution of specificity in the legume–*Rhizobium* symbiosis. *Trends Ecol. Evol.* **4**, 341–349.
4 Johnston, A.W.B. et al. (1978) High frequency transfer of nodulation ability between strains and species of *Rhizobium*. *Nature* **276**, 634.
5 Ruvkun, G.B. & Ausubel F.M. (1987) A general method for site-directed mutagenesis in prokaryotes. *Nature* **289**, 85.
6 Downie, J.A. et al. (1985) Cloned nodulation genes of *Rhizobium leguminosarum* determine host-range specificity. *Mol. Gen. Genet.* **190**, 359.
7 Surin B.P. & Downie, J.A. (1988) Characterization of the *Rhizobium leguminosarum* genes *nodLMN* involved in efficient host specific nodulation. *Mol. Microbiol.* **2**, 173.
8 Djordjevic, M.A. et al. (1985) Tn5 mutagenesis of *Rhizobium trifolii* results in mutants with altered host-range ability. *Mol. Gen. Genet.* **200**, 463.
9 Lerouge, P. et al. (1990) Symbiotic host specificity of *Rhizobium meliloti* is determined by a sulphated and acetylated glucosamine oligosaccharide signal. *Nature* **344**, 787.
10 Truchet, G. et al. (1991) Sulphated lipo-oligosaccharide signals of *Rhizobium meliloti* elicit root organogenesis in alfalfa. *Nature* **351**, 670.
11 Diaz, C.L. et al. (1989) Root lectin as a determinant of host-plant specificity in the *Rhizobium*–legume symbiosis. *Nature* **338**, 579.

nodulins Plant-encoded proteins expressed in root tissues in leguminous plants in response to invasion by rhizobia and which are involved in NODULATION.

NOE Nuclear Overhauser effect. *See*: NMR SPECTROSCOPY.

NOESY *See*: NMR SPECTROSCOPY.

noise analysis *See*: FLUCTUATION ANALYSIS.

nonallelic Produced by or occurring in different gene LOCI; applied to different forms of a protein, to mutations with similar phenotype, etc.

nonamer A conserved sequence of 9 bp — consensus 5′-GGTTTTTGT-3′ — which together with a conserved HEPTAMER sequence makes up the RECOMBINATION SIGNAL SEQUENCE flanking individual gene segments in immunoglobulin and T cell receptor loci (*see* GENE REARRANGEMENT; IMMUNOGLOBULIN GENES; T CELL RECEPTOR GENES). The exact sequence of the nonamer can vary but certain bases are generally conserved. These joining signals flank the V, D, and J segments at antigen receptor loci and mediate the activity of the RECOMBINASE.

non-autonomy *See*: MOSAIC ANALYSIS.

nonchromosomal stripe mutation (NCS) Maternally inherited mutation in maize, showing leaf striping and poor growth. Mutants also show rearrangements of mitochondrial DNA, and alterations in protein synthesis. The generation of NCS mutations appears to depend on the nuclear background.

non-crystallographic symmetry Additional symmetry present within some molecular crystals which is local symmetry within one asymmetric unit and not consistent with the long-range symmetry of the crystal lattice. The commonest example is probably that of fivefold symmetry, which can never be compatible with lattice symmetry but which is found in all icosahedral viruses, many of which crystallize with five (or a multiple of five) subunits in each asymmetric unit of the crystal. The existence of non-crystallographic symmetry provides constraints which are useful in overcoming the PHASE PROBLEM of X-RAY CRYSTALLOGRAPHY. *See also*: MOLECULAR AVERAGING.

nondisjunction The failure of a pair of homologous chromosomes to separate during MEIOSIS I or of two chromatids of a chromosome to become separated at meiosis II or MITOSIS. The daughter cell inherits either both homologous chromosomes or neither.

non-equivalence The cells of a particular tissue type expressing very similar differentiation markers are nevertheless termed non-equivalent if they have distinct developmental histories. For example, cells forming an invariant homologous structure in different body segments may have arisen from precursors which differed in their expression of the homeotic pattern-forming genes.

non-insulin dependent diabetes mellitus (NIDDM) Also called type II diabetes mellitus, it is a disorder of glucose homeostasis characterized by hyperglycaemia, peripheral insulin resistance, impaired hepatic glucose metabolism, and diminished glucose-dependent secretion of insulin from pancreatic β-cells. This latter defect may lie in the glucose signalling pathway in β-cells involving metabolically regulated POTASSIUM CHANNELS, which are the targets of sulphonylurea drugs commonly used in the treatment of NIDDM.

non-Mendelian inheritance Pattern of inheritance other than that generally displayed by nuclear genes in a diploid cell, for example the inheritance of mitochondrial or chloroplast genes (cytoplasmic or maternal inheritance). *See also*: MENDELIAN INHERITANCE.

non-NMDA receptor Ionotropic EXCITATORY AMINO ACID RECEPTORS for glutamate that are not activated by NMDA. Multiple types of non-NMDA receptor exist. Historically they have been classified as kainate or AMPA types but subtypes which are activated by both AMPA and kainate have also been identified.

nonpermissive cell In virology, cells in which a particular virus cannot multiply, for example, monkey cells are permissive for the virus SV40 (*see* PAPOVAVIRUSES) and are lysed, whereas mouse and rat cells (i.e. nonpermissive cells) are not.

nonpolar direction The direction in a crystal which is equivalent to one or more other directions under the action of crystal SYMMETRY.

nonpolar side chains Side chains of AMINO ACIDS which do not form hydrogen bonds, also often termed hydrophobic side chains. The amino acids alanine, valine, leucine, isoleucine, phenylalanine, proline, and methionine have nonpolar side chains (Table A5, Fig. A22 in AMINO ACIDS). They are predominantly found in the interior of protein molecules.

nonproductive rearrangement Rearrangement of immunoglobulin or T cell receptor genes that does not give rise to a functional gene product. *See*: B CELL DEVELOPMENT; GENE REARRANGEMENT; T CELL DEVELOPMENT.

nonreciprocal recombination RECOMBINATION in which the exchange of material between the two DNA molecules is unequal.

nonreciprocal translocation *See*: CHROMOSOME ABERRATIONS.

nonrepetitive DNA SINGLE-COPY DNA.

nonsense codon (termination codon) Any CODON specifying the termination of translation (*see* PROTEIN SYNTHESIS). In most organisms this will be one of the codons UAA, UAG, or UGA. Exceptions to this rule include lower ciliates and mitochondria where one or more of these nonsense codons can be recognized as a sense codon (e.g. in mitochondria UGA is recognized as a codon for tryptophan). *See*: GENETIC CODE.

nonsense mutation Any MUTATION that converts a codon specifying an amino acid into one coding for termination of translation (NONSENSE CODON). *See*: GENETIC CODE.

nonsense suppressor Any mutation that results in the translation (read-through) of a premature termination codon (NONSENSE CODON), and thus the production of a functional protein. A common form of nonsense suppressor is one that alters the anticodon of a TRANSFER RNA molecule in such a way that it can base pair with a nonsense codon, thus causing the insertion of an amino acid into the growing polypeptide chain at that point. *See also*: GENETIC CODE.

nonspherocytic haemolytic anaemia A broad term which covers any congenital haemolytic anaemia whose cause is not easily identifiable (excluding, for example, HEREDITARY SPHEROCYTOSIS). Identifiable causes include abnormalities of red cell enzymes (the commonest being GLUCOSE-6-PHOSPHATE DEHYDROGENASE DEFICIENCY) and some haemoglobinopathies (*see* HAEMOGLOBIN AND ITS DISORDERS).

nopaline An OPINE, the condensation product of arginine and 2-oxoglutarate, characteristic of *Agrobacterium tumefaciens* strains such as C58 and T37 (nopaline strains). Also the name given to a family of opines all of which are based on 2-oxoglutarate condensed with an amino acid (e.g. in nopalinic acid, the amino acid is ornithine).

nopaline strains Strains of *Agrobacterium* which induce the formation of, and which can also utilize, opines of the NOPALINE type.

nopaline synthase (NS) Enzyme encoded by a gene in the T-region of nopaline-type TI PLASMIDS which, in transformed plant cells, catalyses the condensation of arginine and 2-oxoglutarate to NOPALINE.

noradrenaline (NA) (norepinephrine, NE) A CATECHOLAMINE NEUROTRANSMITTER and hormone important in the central and peripheral nervous systems (Fig. N32). In the brain it is particularly concentrated in the hypothalamus, where it is involved in a variety of hormonal control mechanisms. Noradrenaline-containing neurones also innervate other brain regions, mainly from cell bodies in the brain stem. These systems contribute to the control of attention and mood (*see* AFFECTIVE DISORDERS). In the periphery it is the major post-ganglionic transmitter of the sympathetic nervous system (*see* PERIPHERAL NERVOUS SYSTEM). It shares with ADRENALINE a hormonal role following release from the adrenal gland.

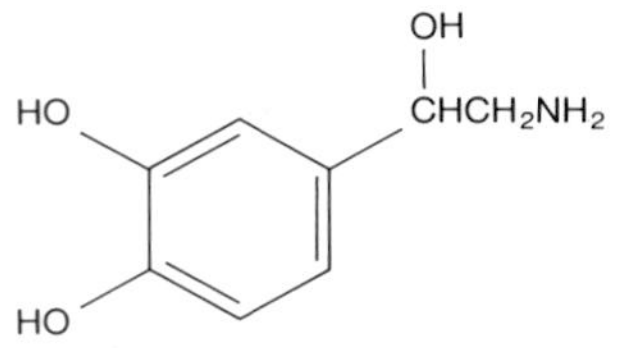

Fig. N32 Noradrenaline.

norbornadiene (NBD) 2,5-Norbornadiene (Fig. N33) (M_r 92.14) is a cyclic olefin, volatile at room temperatures, and used in plant physiological studies to inhibit competitively the action of the gaseous plant hormone ETHYLENE in growth and senescence processes. For example, ~200 Pa NBD will overcome the promoting action on citrus leaf abscission of 0.2 Pa ethylene.

Sisler, E.C. et al. (1985) *Physiol. Plant.* **63**, 114–120.

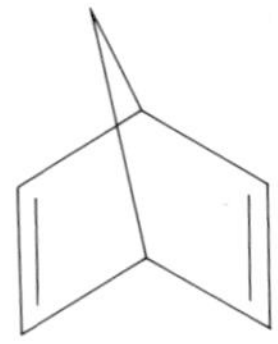

Fig. N33 Norbornadiene.

norepinephrine An alternative name for the neurotransmitter and hormone NORADRENALINE.

normal matrix *See*: LEAST SQUARES METHOD; MATRIX.

northern blotting A method of transferring denatured RNA onto a nitrocellulose or nylon filter for subsequent use in a HYBRIDIZATION assay. RNA is electrophoresed in a denaturing agarose gel (*see* ELECTROPHORESIS) before being transferred onto a membrane either by capillary action or under the action of an electrical field. A radioactively labelled DNA or RNA PROBE is hybridized to the filter-bound RNA to detect specific sequences. The technique was named by analogy with SOUTHERN BLOTTING.

Thomas, P.S. (1980) *Proc. Natl. Acad. Sci. USA* **77**, 5201–5205.

Notch (N) Gene in *Drosophila* that appears to have a key role in the decision between the alternative epidermal and neural cell fates in embryonic development. Mutations at the *Notch* locus display a 'neurogenic' phenotype with hyperplasia of the nervous system and concomitant loss of epidermal structures. The gene product is a large transmembrane protein with an external domain of 1700 amino acids containing 36 tandemly repeated sequence elements with homology to a sequence motif in epidermal growth factor (*see* GROWTH FACTORS). In neurogenesis, one cell of the proneural cluster will follow a neural fate while the other cells will become epidermal. MOSAIC ANALYSIS indicates that the *Notch* gene is required in the presumptive epidermal cells in order for them to receive a 'lateral inhibitory' signal from the presumptive neuroblast, and thus the *Notch* product may act as the receptor for this signal. *Notch* mutants are PLEIOTROPIC, and *Notch* may be involved in many cell fate decisions. *See*: INSECT NEURAL DEVELOPMENT.

notochord Predating the vertebral column in evolution, the notochord was ancestrally the mechanical strengthening rod of the protochordates (*see* CHORDATES). In amphioxus (*Branchiostoma*), which has no vertebral column, the notochord stems from the most caudal part of the animal to its most rostral tip. This is in contrast to higher vertebrates, where the most rostral extent of the notochord is the infundibulum. The notochord usually disappears during development in the head region of higher vertebrates. The notochord is derived from cells involuting in the anterior region of the primitive streak during GASTRULATION (Hensen's node in birds). In addition to its structural role, the notochord plays an important part in early development, for example in the induction of the floorplate of the central nervous system and the vertebral column. *See also*: AVIAN NEURAL DEVELOPMENT; VERTEBRATE NEURAL DEVELOPMENT.

novobiocin Dibasic acid antibiotic produced by various *Streptomyces* species (Fig. N34), active against Gram-positive bacteria. Acts by inhibiting the enzyme DNA GYRASE by blocking the binding of ATP to the enzyme.

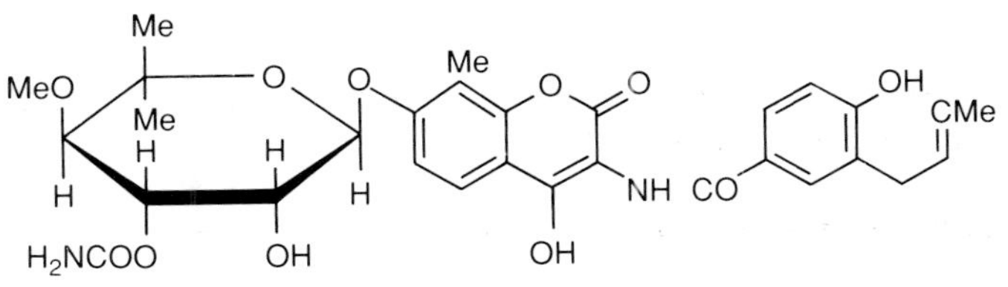

Fig. N34 Novobiocin.

Novozym™ A commercial preparation of cell wall lytic enzymes. Two types are available, Novozym 234 derived from *Trichoderma harzianum*, and Novozym 239 from *Aspergillus* species. The former enzyme has been used in the digestion of the yeast cell wall (especially that of *Schizosaccharomyces pombe*) before procedures such as TRANSFORMATION of spheroplasts (PROTOPLASTS).

NPA NAPHTHYLPHTHALAMIC ACID.

NPT NEOMYCIN PHOSPHOTRANSFERASE.

NS NOPALINE SYNTHASE.

NSF N-Ethyl maleimide-sensitive fusion protein. A homotetramer of subunit M_r 76 000 which is found both in cytoplasm and on membranes. It is involved in a number of membrane fusion processes including fusion of Golgi transport vesicles with Golgi cisternae, fusion of ENDOPLASMIC RETICULUM vesicles with Golgi cisternae, and fusion of endocytic vesicles. NSF is homologous to Sec18p, a protein identified from one of the secretory (*sec*) mutants in the yeast *Saccharomyces cerevisiae*. *See also*: ENDOCYTOSIS; GOLGI APPARATUS; PROTEIN SECRETION.

Wilson, D.W. et al. (1991) *Trends Biochem. Sci.* **16**, 334–337.

NT-3 NEUROTROPHIN-3.

NTP Any or all of the four ribonucleoside triphosphate precursors for RNA synthesis — ATP, GTP, CTP, and UTP.

Nuclear envelope

THE nucleus is the defining feature of eukaryotic cells. It contains the DNA in the form of chromosomes and the major cellular activities of TRANSCRIPTION and DNA REPLICATION occur within the nucleus, separated from the cytoplasm which is the site of PROTEIN SYNTHESIS. The nucleus is delimited from the cytoplasm by the nuclear envelope. The nuclear envelope and its organization is reviewed in [1,2] and references therein.

The nuclear envelope is a double membrane system. There are two lipid bilayers, the outer and inner nuclear membranes, separated by a space known as the perinuclear cisterna. The outer nuclear membrane is continuous with the membrane of the rough ENDOPLASMIC RETICULUM and its cytoplasmic face is studded with RIBOSOMES. The perinuclear cisterna is continuous with the lumen of the endoplasmic reticulum and a number of secreted proteins can be detected within the perinuclear cisterna presumably en route to the GOLGI APPARATUS. The overall thickness of the double membrane system is ~50–70 nm, and each membrane is 5–10 nm thick. The gross biochemical composition of the nuclear membrane system is similar to that of the endoplasmic reticulum but there may be less CHOLESTEROL and more esterified cholesterol in the nuclear envelope. There may also be differences between the outer and inner nuclear membranes in terms of lipid and protein composition. In addition there is evidence that the nuclear envelope contains RNA species that are distinct from nuclear and cytoplasmic RNAs. The evidence for these differences is not unequivocal owing to the difficulty of obtaining pure membrane preparations.

The two major proteinaceous components of the nuclear envelope are the nuclear pore complexes and the nuclear lamina.

Nuclear pores

Nuclear pores are among the most complex macromolecular assemblies in cells. They are large structures about 120–140 nm in diameter with an estimated relative molecular mass (M_r) of 1.1×10^8–1.2×10^8 and probably contain several hundred different polypeptides. They span the nuclear membrane and seem to be the major, and possibly only, channel for the transport of molecules between nucleus and cytoplasm. Proteins are imported into the nucleus from the cytoplasm to carry out nuclear functions and RNA is exported from the nucleus. A number of proteins and small nuclear ribonucleoprotein particles (snRNPs) shuttle between the nucleus and cytoplasm. These transport processes require energy and depend on the presence of specific signals in the transported molecules that are recognized in the transport process (*see* PROTEIN TARGETING; PROTEIN TRANSLOCATION). There is evidence that a single pore complex can simultaneously import a protein molecule and export an RNA molecule. Small molecules (proteins of M_r <65 000, ions and nucleotides) seem to be able to diffuse freely through the pore complex. The channel through which this free diffusion occurs is about 9 nm in diameter. The number of pore complexes per nucleus varies from a few hundred or thousands in somatic cells of higher eukaryotes and yeast to 5×10^7 in the nuclear envelope of the giant nuclei of amphibian oocytes, giving pore densities of 30–80 per μm^2.

Different features of nuclear pore structure are revealed by different techniques. Ultrastructural studies of the isolated amphibian oocyte nuclear envelope have led to a model of pore complex structure in which each pore is made up of two rings. One ring lies in the plane of the outer nuclear membrane whereas the other lies in the plane of the inner nuclear membrane. Each ring is made up of eight subunits. The rings are connected to eight spokes. The spokes point towards the centre of the pore complex and lie between the two rings. A number of pore complexes have a centrally located granule or plug at the level of the spokes whereas others do not. It has been proposed that the central plug or granule is a transporter assembly made up of two iris-diaphragm-like structures each of eight subunits. Transport of molecules through the pore complex is thought to be brought about by the alternate opening and closing of the iris diaphragms.

The other view of nuclear pore structure has come from the images seen in sectioned material rather than isolated nuclear envelopes. A section through the nuclear envelope at right angles to the plane of the membrane reveals fibrillar structures that are attached to both the cytoplasmic and nucleoplasmic surfaces of the nuclear pore. The diameters of these fibres are ~3–7 nm and on the nucleoplasmic face of the pore complex the fibres form a cylindrical unit or cage which extends for about 50 nm into the nucleus. The fibres attached to the cytoplasmic face of the pore

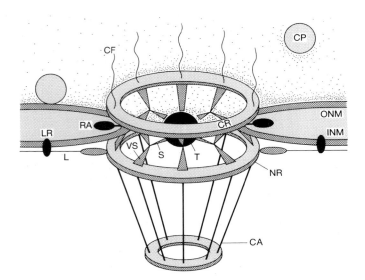

Fig. N35 A schematic overview of the nuclear pore complex and associated fibrous assemblies which incorporates features from a number of different models. CA, Cage assembly; CF, cytoplasmic fibrils; CP, cytoplasmic particle; CR, cytoplasmic ring; INM, inner nuclear membrane; L, lamina; LR, lamina receptor; NR, nuclear ring; ONM, outer nuclear membrane; RA, radial arm; S, spoke; T, transporter/plug; VS, vertical support. Courtesy of Chris Akey.

complex extend for a shorter distance into the cytoplasm but this may be due to difficulties in tracing their paths rather than real differences in length. These fibres have been implicated in the binding of proteins that are to be transported into the nucleus. Figure N35 gives a schematic view of the pore complex incorporating information from these two views.

A family of 8–10 GLYCOPROTEINS called nucleoporins have been identified as peripheral components of the nuclear pore complex and a large body of evidence suggests that they are involved in mediating nucleocytoplasmic exchange through the pore. Nucleoporins bind to LECTINS such as wheat germ agglutinin and carry *O*-linked *N*-acetylglucosamine; the enzyme that carries out the attachment of *N*-acetyl glucosamine is present in the cytosol. Comparison of the amino-acid sequences of a number of nucleoporins identifies a pentapeptide repeat (Gly-Phe-Ser-Phe-Gly GFSFG) which seems to be unique to nuclear pore proteins.

In some cells nuclear pores are also found in annulate lamellae which are stacks of pore-studded membranes that closely resemble nuclear envelopes and may represent a storage form of nuclear envelope and pore components.

Nuclear lamina

The nuclear lamina forms a proteinaceous layer of variable thickness that coats the inner (nucleoplasmic) surface of the nuclear envelope. The lamina provides a framework for organizing nuclear envelope structure and nuclear pore attachment and also provides anchorage sites at the nuclear periphery for interphase CHROMATIN. In the *Xenopus* oocyte nuclear envelope, the lamina appears as a dense mesh in which there are two sets of approximately orthogonally arranged filaments; the size of the mesh corresponds very closely to the length of the helical rod domain of the proteins (lamins) involved (see below). In contrast a three-dimensional analysis of the nuclear lamina in *Drosophila* and HeLa cells reveals a highly discontinuous, apparently fibrillar network with large voids in the nuclear periphery containing little or no lamin. Also, only a small fraction of the chromatin in the nuclear periphery in these cells is sufficiently close to a lamin fibre to be in direct contact.

The major proteins of the nuclear lamina are the lamins which form a family of related polypeptides. The complexity of this family may be great as lamin precursor proteins and minor lamin species have been reported. In somatic cells the major lamins are termed lamins A, B, and C with M_r values ~70 000, 68 000 and 60 000 respectively. In *Xenopus* and mouse the oocyte and post-fertilization nuclei have a single embryonic species of lamin. Later in development, when the nuclei stop dividing synchronously and the cells begin transcription and differentiation the adult forms of the lamin proteins are expressed. Undifferentiated cells such as embryonal carcinoma cells and HL60 promyeloblasts contain only the B lamin in their nuclear envelopes. When these cells are induced to differentiate *in vitro* the cells express new proteins indicative of differentiation but they also begin to synthesize A and C lamins. In *Drosophila* and perhaps other invertebrates only a single nuclear lamin has been found.

The amino-acid sequences of these proteins reveal that they are intermediate filament-like proteins (*see* INTERMEDIATE FILAMENTS) with a central helical rod domain of 360 amino acids displaying a heptad repeat pattern of amino acids. This region drives the formation of coiled coil helical dimers. These dimer–dimer interactions form the basis of the 10-nm thick filaments observed on the inside of the surface of the nuclear envelope. These filaments can be formed *in vitro* from purified intact protein or rod domain fragments. Under these conditions the A and C lamins predominantly occur as rods, 52 ± 2 nm long with two globular heads. The helical rod domain of the lamins is longer than the corresponding domain of the intermediate filament proteins (46 nm) because of the presence of an additional 42 amino acids in the second α-helical subdomain encoded by an additional exon.

Lamins undergo a number of post-translational modifications. Lamin A is proteolytically processed by the removal of 18 amino acids from the C-terminal end after assembly into the nuclear envelope. All lamin proteins identified to date (with the exception of lamin C) have a sequence motif at the C terminus commonly refered to as the CaaX box (C, cysteine; a, aliphatic amino acid; X, any amino acid). Such motifs are conserved features at the C termini of several other proteins, most notably members of the *ras* family of proto-oncogenes, small GTP-BINDING PROTEINS and the yeast mating pheromone. This tetrapeptide is the target for isoprenylation and carboxymethylation, modifications that are important for the correct targeting of B-type lamins to the nuclear membrane.

In higher eukaryotes the nuclear envelope breaks down during mitosis (open mitosis) but in yeast (*Saccharomyces* spp.) the nuclear membrane remains intact throughout the cell cycle (closed mitosis). At nuclear envelope breakdown the nuclear lamina is transiently disassembled most probably through hyperphosphorylation of lamin proteins by the PROTEIN KINASE that is a key regulator of the CELL CYCLE. This disassembly of the lamina is necessary but not sufficient for complete nuclear envelope breakdown.

The occurrence of lamins and intermediate filament proteins in different cells and the structures of the corresponding genes has led to the suggestion that the lamins predate, and are thus the evolutionary precursors of the intermediate filaments.

C. DINGWALL

1 Niggs, E.A. (1989) The nuclear envelope. *Curr. Opinion Cell Biol.* **1**, 435–440.
2 Gerace, L. & Burke, B. (1988) Functional organization of the nuclear envelope. *Annu. Rev. Cell Biol.* **4**, 335–374.

nuclear import signals (nuclear localization signals) Proteins whose destination is the cell nucleus are synthesized in the cytoplasm, but have short regions in their amino-acid sequence which act as signals for import into the nucleus (*see e.g.* PAPOVAVIRUSES). These regions contain basic residues (lysine or arginine) but the precise sequence of the import signal is not conserved between different nuclear proteins. In some cases two separate regions of the protein contribute towards efficient nuclear import. The process of import has two stages: binding of the protein to the NUCLEAR PORE, which is energy independent, and requires cytoplasmic factors, and energy-dependent transport through the nuclear pore. Some of the ribonucleoproteins involved in RNA SPLICING, the snRNPs, contain a different type of nuclear import signal. These particles are assembled in the cytoplasm, where the RNA component is trimethylated at its 5′ end. This 'cap' structure and a region of the protein together constitute a nuclear import signal for the snRNP.

nuclear lamina *See:* NUCLEAR ENVELOPE; NUCLEAR MATRIX.

nuclear magnetic resonance *See:* NMR SPECTROSCOPY.

Nuclear matrix

THE term nuclear matrix defines a family of structures isolated by sequential extraction of nuclei with non-ionic detergents, NUCLEASES and high concentrations of salt. The product, an apparently highly structured residual framework, probably represents a snapshot of the dynamic networks found *in vivo*.

The nuclear matrix is believed to play a central part in CHROMATIN organization, providing the axis around which complex arrays of looped domains of chromatin are formed. These domains fulfil crucial roles in DNA management but can also have far-reaching consequences for the control of gene activity, providing boundaries that define domains of gene expression. In addition, the nuclear matrix has been shown to exert a strong organizational influence on nuclear activities including DNA REPLICATION, TRANSCRIPTION, RNA PROCESSING and transport (*see also* NUCLEAR ENVELOPE).

During interphase (*see* CELL CYCLE), eukaryotic nuclei display few characteristic features. Condensed, inactive, HETEROCHRO-MATIN usually lies opposed to the nuclear periphery whereas decondensed, EUCHROMATIN occupies the nuclear interior. The only organelles — NUCLEOLI — provide the best structural detail with defined regions of RIBOSOMAL RNA (rRNA) transcription, processing, and RIBOSOME biogenesis. Yet despite this apparent lack of positional information individual CHROMOSOMES appear to occupy preferred sites and maintain spatially distinct 'territories'. These facts, together with the observed patterns of transcription by RNA polymerase II, localized sites of RNA SPLICING, defined routes for RNA transport and highly organized centres of DNA replication suggest that nuclei are highly ordered structures.

If so, how might nuclear organization be achieved?

The first hint of a structural component influencing DNA organization appeared in the 1950s when Mirsky and Ris showed that chromosomes retained their shape after extraction of the major chromosomal proteins — histones. The existence of a pro-teinaceous chromosome core was proposed. Later, Georgiev and Chentsov used electron microscopy to describe the interphase counterpart of the chromosome core, in nuclei digested with deoxyribonuclease (DNase) and extracted with 2 M NaCl. Some 15 years later, Berezney and Coffey coined the term 'nuclear matrix' to describe the nuclear components that remained when nuclei were extracted with a nonionic detergent, nuclease and hypertonic salt in turn [1–3].

Over the years, the original procedure has been modified extensively. Detergent-washed nuclei are usually prepared (commonly with Triton X-100 or NP 40), though in some laboratories cells are used directly. The extent of nuclease digestion and strength of the hypertonic extraction vary considerably (see Fig. N36). DNase digestion is used most commonly but may be omitted; the integrity of RNA is usually preserved. NaCl can be replaced by other salts such as $(NH_4)_2SO_4$.

Composition and morphology

The original preparation, from rat liver nuclei, used Triton X-100, endogenous nucleases and 2 M NaCl extraction and, like similar preparations from most cells or tissues (Fig. N36) contained the following major components:

1 Residual elements of the nuclear envelope, principally remnant pore complexes associated with the nuclear lamina.
2 Residual nucleoli.
3 A 'fibrillogranular' network of the nuclear interior.

Having defined the components, it is important to stress that the appearance of the nuclear matrix, and in particular its internal components, depend on the type of extraction used. Many pro-tocols comply with the general scheme for nuclear matrix prep-aration; apparently minor changes can have profound effects on the composition of the product, to the extent that each nuclear matrix preparation should be deemed the product of a clearly defined extraction protocol.

Depending on cell type and the isolation procedure used, typical preparations retain roughly 2% of the DNA, 30–60% of the RNA and 10–15% of the protein originally present in nuclei. Simple gel electrophoresis suggests some 40–50 major polypep-tides, with the LAMINS predominating (usually about 25% of the protein present). Two-dimensional gel electrophoresis shows a more complex picture, with as many as 500 polypeptides. Of these, some are clearly present in all cells of a chosen organism, as well as in different organisms, whereas others are restricted to specialized cell types.

The nucleic acid component of nuclear matrix preparations is also influenced by details of the extraction protocol used. RNA is believed to represent an important structural component in most nuclear matrix preparations; ribonuclease (RNase) digestion causes dramatic changes in nuclear matrix morphology. In prep-arations that are not treated with RNase, almost all heteroge-neous nuclear RNA (hnRNA) remains. In some preparations this is also true for small nuclear RNA (snRNA) molecules — however, certain snRNAs are not specifically associated with the nuclear matrix if other preparations are used. Because ribonucleoprotein particles containing RNA polymerase III transcripts, such as pre-tRNA (*see* EUKARYOTIC GENE STRUCTURE; TRANSFER RNA), are not matrix-associated it seems likely that any quantitative inter-actions reflect structures existing *in vivo*.

DNA content varies according to the nuclease digestion used. It is usual to use a level of DNase digestion such that 1–5% of the original DNA remains. But by using lower levels of digestion (perhaps with different enzymes) preparations containing any DNA concentration of choice may be obtained. Analysing DNA that remains associated with the nuclear matrix has indicated that specific, or at least nonrandom, arrangements exist (see later). For example, specific matrix attachment regions (MARs) are thought to define chromatin domain borders. In other cases important functional associations are implied — replication ori-gins, replication forks and transcribed sequences all show prefer-ential affinities for the nuclear matrix.

Although still controversial, it seems probable that some form

---

Fig. N36 (*opposite*) Morphology of the nuclear matrix isolated from HeLa cells. Nuclei isolated from HeLa cells (*a*) under hypotonic conditions (10 mM NaCl, 1.5 mM MgCl$_2$) were washed with Na-deoxycholate (0.5%) and Tween 40 (1%) and chromatin removed by treatment with DNase (500 μg ml^{-1} for 15 min at 20°C) and 0.4 M (NH$_4$)$_2$SO$_4$. A characteristic nuclear matrix is shown (*b*), after staining and visualization by transmission electron microscopy. Far greater detail can be seen using resinless electron microscopy techniques (*c–e*); stained and shadowed samples (about 100 nm thick) are shown. The nuclear matrix shown in *c* was prepared from cells directly, using buffers containing Triton (0.5%), DNase (100 μg ml^{-1} for 30 min at 20°C) and then 0.25 M (NH$_4$)$_2$SO$_4$. The complex structures seen in *c*

can be extracted with 2 M NaCl to reveal underlying core filaments (*e*). A similar intermediate filament-like network can be visualized in cells extracted under physiological conditions and treated with *Hae*III if cells are first encapsulated in agarose microbeads (*d*). These filaments clearly support the remnant chromatin structures; in this example 10% of the chromatin remains (electrophoresis was used to remove material solubilized by the restriction endonuclease digestion). An idealized view of nuclear structure is shown in *f*. The bars are 1 μm in *a–d* and 0.2 μm in *e*. Cy, cytoplasm; N, nucleolus; L, nuclear lamina; IF, cytoplasmic intermediate filaments; M, remnant mass that results from collapse of nuclear matrix component during extraction with 2 M NaCl. *b*, From [8]; *c*, *e*, *f*, from [9]; *d*, from [10].

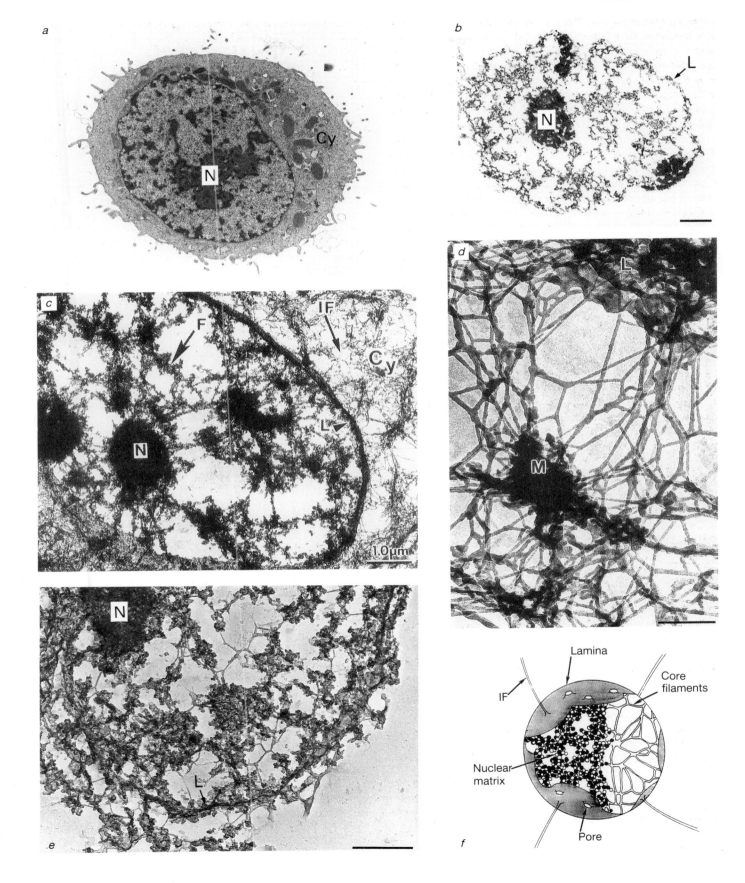

of skeletal network, analogous to the cytoplasmic network, the CYTOSKELETON, will form the structural basis for nuclear organization. This framework, together with an associated ribonucleoprotein component and residual elements of the nuclear membrane/lamina (*see* NUCLEAR ENVELOPE) and nucleoli should be considered the minimal matrix structure. In addition, polypeptides from complexes involved in activities such as replication, transcription and RNA processing remain associated with the minimal matrix in most nuclear matrix preparations. These represent a variable component that will be influenced by the growth characteristics of a cell.

Of the minimal matrix components the nuclear lamina is best defined (*see also* NUCLEAR ENVELOPE). This structure forms from a network of anastomosing 8–10-nm filaments made up of lamin polypeptides (usually three types) which possess the characteristic INTERMEDIATE FILAMENT (IF) repeat motif. This network underlies the nuclear membrane *in vivo* but seems to be discontinuous, with large voids; only a fraction of chromatin at the nuclear periphery actually lies in contact with the lamina. The core filaments of the internal matrix — the nucleoskeleton — may also be related to the IF family of proteins (see Fig. N36), but whereas VIMENTIN and CYTOKERATINS are found commonly in nuclear matrix preparations a skeletal network like the cytoskeleton has not been unambiguously defined.

The nuclear matrix and nuclear organization

Each diploid human cell contains some 6×10^9 base pairs (bp) of DNA, measuring roughly 2 m when extended, folded so as to occupy a nucleus of 5–10 μm across (nuclear size is quite variable depending on the biochemical activity of individual cell types). As cells divide, the 46 chromosomes of this diploid set must duplicate and identical cohorts separate before cell division; this must occur precisely to avoid cell death. Structural elements of the nuclear matrix and its presumed mitotic counterpart, the chromosome core, direct the higher order arrangement that satisfies this organizational complexity, by providing an axis around which arrays of chromatin loop assemble. In HeLa cells, these structures have an average size of roughly 75 kilobases (kb) with a broad distribution ranging from 5–250 kb. They maintain the same spatial sequence organization as the state of chromatin condensation changes through the cell cycle, implying that the primary interactions persist throughout.

Loops clearly have a crucial role in DNA management but can also define chromatin 'domains' of functional importance. Chromatin must decondense, allowing RNA polymerase access, if a gene is to be expressed. Relative condensation states are clearly stable, persisting from one cell generation to the next, and dominant; foreign genes introduced into inert chromatin are generally switched off also. To override the influence of host signals responsible for this POSITION EFFECT it is necessary to introduce sequences which themselves define an active chromatin state. One attractive way of achieving this is to construct chromatin domains (loops) that behave as independent operational units. Stable interactions between attachment sites in the chromatin and structural elements of the nuclear matrix define domain borders which isolate internal sequences from local influences. To date, the best candidate attachment sites are the matrix attachment region from the chicken lysozyme locus and a class of specialized chromatin structure (SCS) sequences from the *Drosophila* 87A7 HEAT SHOCK locus. Constructs containing these sequences express internal genes at the natural efficiency in almost all cases tested. In the case of the former, an attachment region binding protein (ARBP) (M_r 95 000) and its recognition sequence have been characterized [4].

Another, perhaps related, class of sequences has been shown to influence gene expression in the α- and β-globin loci. Like the sequences described above, these LOCUS CONTROL REGIONS have the ability to direct position-independent gene expression, in some way overriding the influence of host heterochromatin. The globin locus sequences have not, however, been shown to interact preferentially with the nuclear matrix; their mode of action remains unclear.

At present, most attachment sites reported in the literature contain AT-rich motifs related to the DNA topoisomerase II binding site. In almost all cases these have been defined by structural criteria; their functional importance remains unconfirmed. Unfortunately, studies that have attempted to relate these sequences to functional domains *in vivo* have proved inconclusive. It seems probable that this class of structural attachment, defined *in vitro*, is an experimental artefact that reflects some dynamic interaction inside the cell.

Role of the nuclear matrix in nuclear function

The nuclear matrix has been reported to have a pivotal role in most aspects of nuclear function [5]. Perhaps the best evidence, and certainly the most widely accepted functional involvement, comes from the analysis of DNA synthesis. Early experiments showed that when some 95% of bulk chromatin was removed from the nuclear matrix of cells pulse-labelled for 2 min with [^{3}H]thymidine, most of the [^{3}H]DNA remained. Extended pulse times and pulse-chase experiments suggested that the matrix provided fixed centres of DNA synthesis. Subsequently, this notion was reinforced by the identification of discrete hotspots of DNA synthesis each containing from roughly 25 (somatic cell) to as many as 500 (*in vitro* reconstituted nuclei using *Xenopus laevis* egg extracts) clustered replication complexes. As these structures persist throughout S-phase (*see* CELL CYCLE) the DNA polymerase complex must be immobile; the nuclear matrix provides a structural framework upon which active replication centres (factories), containing the necessary polymerases and auxiliary factors, are assembled.

Gene expression and transcription can also be influenced by nuclear matrix-dependent aspects of nuclear organization. For example, both active genes and the nascent products, pulse labelled *in vivo*, are tightly associated with the nuclear matrix. Moreover, if less stringent hypertonic conditions are used (say 0.5 M NaCl) the active RNA polymerase II remains in the nuclear matrix fraction. As with replication, these observations support the idea that genes are transcribed by enzyme complexes immobilized through their association with the nuclear matrix. The arrangement of DNA domains with respect to transcribed sequences reinforces this idea. For example, in a 163-kb region of the *Drosophila* X-chromosome, five anchorage sites and four loops have been defined, organized in a specific fashion with respect to

DNA sequence. Here, anchorage points demarcate transcriptionally active genes, implying that domain formation relates to function; perhaps protein complexes — transcription activating factors assembled at enhancers and promoters — operate to locate genes close to the nuclear matrix where transcription complexes are assembled (*see* EUKARYOTIC GENE EXPRESSION: TRANSCRIPTION FACTORS). In a limited number of cases genes have been shown to associate with the nuclear matrix before the onset of transcription.

Nuclear organization also seems to exert an influence post-transcriptionally. There is now good evidence that individual active genes are located at preferred (but not precise) sites *in vivo* and that RNA moves away from the gene and site of transcription according to some predefined programme. Specific RNA 'tracks' have been characterized, which seem to define the transit routes of nascent RNA from the gene to the cytoplasm. Recent experiments show that this spatial phenomenon is not chromatin dependent, but is related directly to some aspect of nuclear organization, presumably the nuclear matrix. As localization of the RNA processing machinery is also influenced by interactions with the nuclear matrix, hnRNA could remain associated with the nuclear matrix from its inception until being passed to the cytoplasm.

Nuclear matrix-like preparations

It is worth remembering that each nuclear matrix preparation is the product of a particular type of extraction and so may be subject to experimental influences. Consequently, whether the results described above are a true reflection of cellular organization or some by-product of human enterprise (i.e. an artefact) should not be overlooked. During a typical nuclear matrix preparation protein rearrangements can occur, histone cross-links can be formed, and at physiological temperatures a nonphysiological heat shock-like reaction that causes protein coagulation may be observed.

It is therefore worth considering other approaches that have been used to examine nuclear organization. These nuclear matrix-like preparations all probably contain the critical elements of nuclear organization.

Nucleoids, isolated from living eukaryotic cells by lysis in a hypertonic (2 M NaCl) solution containing Triton X-100, have the significant advantage of avoiding rearrangements associated with the preparation of nuclei. As the preparation includes no obligatory nuclease digestion, nucleoids allowed the first demonstration of intact DNA domains — loops. Sequences within these loops were later shown to be positioned nonrandomly; elements common to nucleoids and nuclear matrix preparations were assumed to provide points of DNA attachment [6].

Nuclear scaffolds, the product of a detergent (LIS) extraction procedure, eliminate high salt-dependent artefacts [7]. In these preparations, scaffold attachment regions (SARs), containing AT-rich motifs related to the DNA topoisomerase II consensus binding site, define DNA loops ranging from 5–100 kb. DNA topoisomerase II remains the best candidate 'loop fastener' in these and the related mitotic counterparts — chromosome scaffolds — but is prone to aggregation problems so may not be physiologically relevant.

Encapsulated nuclei, prepared under physiological conditions, allow the analysis of nuclear organization while eliminating known sources of artefact [5]. By encapsulating cells in microbeads before permeabilization the extracted cells are protected from aggregation and mechanical damage; even after washing in strong detergents residual encapsulated structures remain perfectly manipulable. This system has allowed a detailed analysis of structure–function relationships in higher eukaryotic cells, reinforcing many of the results from conventional nuclear matrix preparations. Moreover, morphological analyses of encapsulated nuclei retaining only 5% of chromatin display an internal filamentous network reminiscent of the intermediate filament network of the cytoplasm (see Fig. N36). At present this is the best candidate for the structural foundations of nuclear organization — the nucleoskeleton.

The nuclear matrix as a coordinating structure

In the 15 years since Berezney and Coffey first coined the term 'nuclear matrix' our perceptions of nuclear organization have continued to develop in complexity and sophistication. The possible organizational extremes are self-evident. In the first, nuclei could be considered a random or chaotic assortment of genes each competing for the attention of soluble components, factors, polymerases and so on, needed for their purposes. Here, a gene's only points of reference would be the nuclear border, nucleoli and adjacent genes of the same chromosomal DNA strand. Alternatively, nuclei might be so highly structured that specific coordinates would define a gene's location, and transcription, processing and transport would follow predetermined pathways to the cytoplasm. Not surprisingly, the reality lies somewhere between these extremes.

However, the fact remains that many aspects of nuclear function can only be explained by invoking some form of coordinating structure upon which nuclear organization is based. The nuclear matrix fulfils this role, providing sites of attachments that specify structural as well as functional chromatin domains and, at the same time, influencing DNA replication, RNA transcription, processing and transport in ways which are likely to be crucial in achieving the observed levels of efficiency and control.

It is important, however, not to underestimate the complexity of this matrix, and it should be remembered that the term is operationally defined and probably includes a number of interrelated structures. It is also important to remember that different types of nuclear matrix preparations can have a profound effect on its presentation. This probably reflects its dynamic nature *in vivo*, but has also allowed the criticism that major facets of nuclear structure, most notably the internal networks, are preparative artefacts with no counterpart *in situ*. As this now seems unlikely, the notion that skeletal networks — the nucleoskeleton, cytoskeleton and extracellular matrix — allow communication not only within a cell, between the nucleus and cytoplasm, but also between adjacent cells, perhaps even whole tissues, begins to gain appeal (*see also* CELL–MATRIX INTERACTIONS).

D.A. JACKSON

1 Verheijen, R. et al. (1988) The nuclear matrix: structure and composition. *J. Cell Sci,* **90,** 11–36.
2 Getzenberg, R.H. et al. (1991) Nuclear structure and the three-

dimensional organization of DNA. *J. Cell. Biochem.* **47**, 289–299.

3 Berezney, R. (1991) The nuclear matrix: a heuristic model for investigating genomic organization and function in the cell nucleus. *J. Cell. Biochem.* **47**, 1009–1123.

4 Bonifer, C. et al. (1991) Dynamic chromatin: the regulatory domain organization of eukaryotic gene loci. *J. Cell. Biochem.* **47**, 99–108.

5 Jackson, D.A. (1991) Structure–function relationships in eukaryotic nuclei. *BioEssays* **13**, 1–10.

6 Jackson, D.A. et al. (1984) Replication and transcription depend on attachment of DNA to the nuclear cage. *J. Cell Sci. Suppl.* **1**, 59–79.

7 Gasser, S.M. & Laemmli, U.K. (1987) A glimpse at chromosomal order. *Trends Genet.* **3**, 16–22.

8 van Eekelen, C.A.G. & van Venrooij, W.J. (1981) hnRNA and its attachment to a nuclear protein matrix. *J. Cell Biol.* **88**, 554–563.

9 He, D. et al. (1990) Core filaments of the nuclear matrix. *J. Cell Biol.* **110**, 569–580.

10 Jackson, D. & Cook, P.R. (1988) Visualization of a filamentous nucleoskeleton with a 23-nm axial repeat. *EMBO J.* **7**, 3667–3678.

nuclear membrane, nuclear pore *See*: NUCLEAR ENVELOPE.

nuclear Overhauser effect (NOE) *See*: NMR SPECTROSCOPY.

nuclear scaffold *See*: NUCLEAR MATRIX.

nuclear skeleton *See*: CHROMATIN; NUCLEAR MATRIX.

nuclear transplantation The transfer of a nucleus from one cell to another. If the recipient cell is an enucleated egg the technique can be used to assess the developmental potential of donor nuclei from embryonic or differentiated cells. Extensive experiments in amphibians have shown that nuclei from either embryonic or adult cells transplanted into enucleated eggs can give rise to essentially normal tadpoles. Before the advent of gene cloning, this was the strongest evidence available that genetic material is not lost upon differentiation. The technique also led to the discovery in mice of the non-equivalence of the maternal and paternal genomes (*see* PARENTAL GENOMIC IMPRINTING). Commercially, nuclear transplantation is used in the propagation of valuable stocks of cows and sheep.

nucleases Enzymes catalysing the hydrolytic cleavage of a polynucleotide chain, by cleaving the phosphodiester linkage between nucleotide residues (*see* DNA; RNA). Nucleases can be classified as either EXONUCLEASES or ENDONUCLEASES, and may be specific for single-stranded or double-stranded nucleic acids or both. Exonucleases cleave nucleotides from the end of a chain whereas the endonucleases cleave within the chain. Some nucleases (RIBONUCLEASES, RNases) act only or preferentially on RNA, others (DEOXYRIBONUCLEASES, DNases) attack DNA, and yet others attack both. Most nucleases are sequence nonspecific, some are specific for a particular type of base, and one group, the RESTRICTION ENZYMES, are sequence-specific DNA-cutting enzymes. *See also*: BAL 31; DEOXYRIBONUCLEASE I; EXONUCLEASE III; EXONUCLEASE V; EXONUCLEASE VII; λ EXONUCLEASE; MICROCOCCAL NUCLEASE; MUNG BEAN NUCLEASE; PANCREATIC RIBONUCLEASE; RIBONUCLEASE H; RIBONUCLEASE P; RIBOZYMES; RNA RESTRICTION ENZYMES; S1 NUCLEASE.

nuclease protection assay Technique for detecting binding of a protein to a nucleic acid, which utilizes the fact that the sequences to which the protein binds are protected from digestion by NUCLEASES. *See*: FOOTPRINTING.

nucleic acid *See*: DNA; NUCLEIC ACID STRUCTURE; RNA; TRANSFER RNA.

nucleic acid hybridization *See*: HYBRIDIZATION.

Nucleic acid structure

THE nucleic acids DNA and RNA are involved in both encoding and decoding the hereditary information in a cell. DNA and RNA are polymers of deoxyribo- or ribonucleotides respectively (*see* NUCLEOSIDES AND NUCLEOTIDES). The BASE of the nucleotide is glycosidically linked to a phosphorylated sugar (in cyclic furanoside form: ribose in RNA and deoxyribose in DNA) (Fig. N37). The O5′ of one sugar is connected through a phosphodiester linkage to the O3′ position of the next sugar, and so on, to form a single strand of DNA or RNA. The single strand composed of the sugar–phosphate backbone and sequence of bases is the primary structural unit from which the different nucleic acid secondary and tertiary structures are formed [1]. Electrostatic interactions dominate the forces involved in structuring the highly negatively charged and relatively flexible sugar–phosphate backbone, whereas a variety of forces control associations among the nucleic acid bases. A balance of these interactions serves to direct the formation of secondary and tertiary structures and acts to stabilize a 'folded' relative to a 'denatured' state in solution or crystalline phases of nucleic acids.

The most common bases guanine (G), adenine (A) and cystosine (C) (in both DNA and RNA), and thymine (T) and uracil (U) (in DNA and RNA respectively) are essentially functionalized heterocycles (aromatic ring systems). As such, they can form both base stacks in the vertical direction as well as hydrogen bonds to one another in the horizontal plane or other directions. The stabilization of particular nucleic acid structures arises from hydrophobic interactions (base stacking, π–π interactions) and hydrogen bonding patterns (*see* BASE PAIRS), and varies because of the chemical nature (charge distribution) of the different bases so that a large number of structures are possible. Covalent modifications of the bases, for example alkylation of amino and ring nitrogens as well as saturation of double bonds and substitutions in both DNA and RNA bases, further increase the structural diversity of nucleic acids. As a result, numerous hydrogen bonding schemes have been observed, and many structures have been predicted and shown to exist *in vitro*.

DNA exists primarily as a double helix *in vivo*, whereas RNA, although it also forms double-helical segments, frequently folds to form more complicated tertiary structures. A number of these structures are described here; there is, however, evidence for only some of them *in vivo* (*see* RNA; TRANSFER RNA). In addition, nucleic acids are often associated with proteins, and in many cases the structure is likely to be mediated by proteins *in vivo* (*see* CHROMATIN; PROTEIN–NUCLEIC ACID INTERACTIONS).

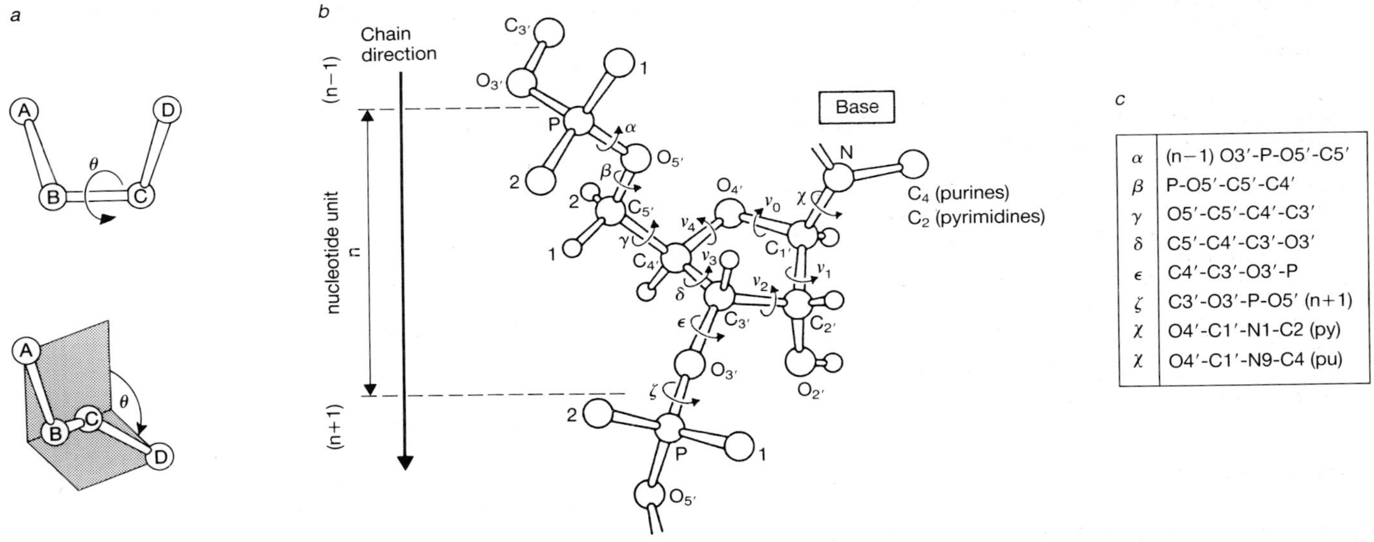

Fig. N37 Definitions of the nucleotide unit and torsion angles. *a*, The angle θ denotes the torsion angle associated with bond B–C. The magnitude of θ is zero when A–B and C–D are eclipsed and is positive as C–D moves in a clockwise direction relative to A–B. *b*, Diagram showing the atomic numbering of a nucleotide unit and the schematic definition of nucleotide torsion angles. The chain direction is 5′ to 3′ which refers to the connectivity of the ribose within the sugar–phosphate backbone. The sugar represented here is the ribose, which has the 2′-OH of RNA. In DNA, the deoxyribose lacks the 2′-OH. The C1′ of the sugar is attached to the N9 position of a purine (pu) base and to the N1 position of a pyrimidine (py) base. *c*, Atoms (equivalent to atoms A-B-C-D of *a*) which are involved in the definitions of the corresponding sugar-phosphate backbone torsion angles. Compiled from information in [1].

Nucleic acid torsion angles and helix parameters

The details of DNA and RNA structure are best represented using a universal set of torsion angles and helix parameters that unambiguously define structural relationships within the nucleic acid [2]. The torsion angles used to describe the structure of a nucleotide unit are shown in Fig. N37. Two of these have special significance and are useful in distinguishing nucleic acid structures; the torsion angle delta (δ) is related to the sugar pucker (see Table N5), and the angle chi (χ) describes the disposition of the sugar relative to the attached base about the glycosidic bond. The torsion angles are useful for characterizing any nucleic acid structure, whereas the helix parameters listed in Table N5 and illustrated in Fig. N38 are useful primarily for representing nucleic acid helices and relationships between base pairs (bp) and base steps.

Duplex DNA and RNA

The double helix is composed of two antiparallel complementary strands of nucleic acid wound into a helix with the sugar–phosphate backbone on the outside and the hydrogen-bonded Watson–Crick base pairs (C–G, A–T, and A–U) stacked on top of one another on the inside. Several canonical structures of nucleic acid double helices, such as the A-, B-, C-, D-, E-, S-, and Z-forms, have been identified from early fibre diffraction experiments. They fall into three main families, B-, A-, and Z-form, and oligonucleotide crystal structures of representatives of each family have been solved [1,3] and are depicted in Fig. N39. The B- and A-type helices are both right-handed, but A-DNA is shorter and fatter than B-DNA (*see* Plate 1*a* and *b*); in addition, the helical twists differ, and the base pairs are more inclined and are displaced (by ~4 Å) from the central helix axis in the A-form

Table N5 Helix parameters for nucleic acids

Parameters used to describe the disposition of individual base pairs relative to the helix axis (in °)
   Inclination (η)
   Tip (θ)
   Opening (σ)
   Propeller twist (ω)
   Buckle (κ)

Parameters used to describe the relationship of two successive base pairs (in °)
   Helical twist (Ω)
   Roll (ρ)
   Tilt (τ)

Translational parameters to describe the structure of individual base pairs (in Å)
   Displacement (d_y, d_x)
   Stagger (S_z)
   Stretch (S_y)
   Shear (S_x)

Translational parameters for successive base pairs (in Å)
   Rise (D_z)
   Slide (D_y)
   Shift (D_x)

A convention for these parameters was accepted in 1989, but previously there was ambiguity in the signs and even in the definitions of various terms. Not everyone has adopted the 1989 convention, which will add to the confusion [2].

Translation   **Rotation**

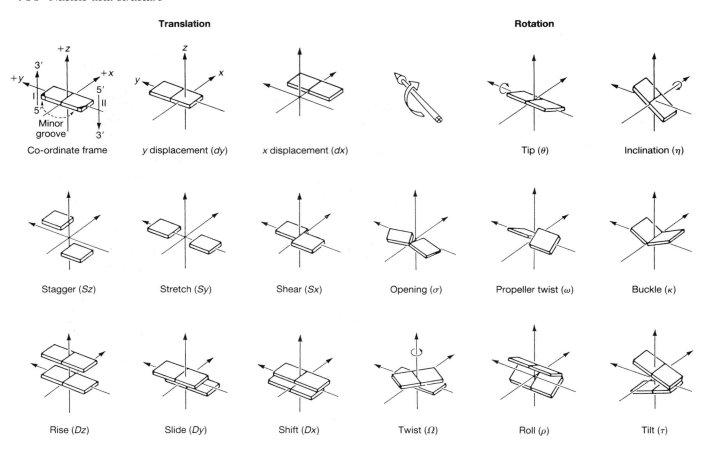

Fig. N38 Nomenclature and definitions of nucleic acid structure translational and rotational parameters (see also Table N5). These parameters describe the positions of individual bases within base pairs, and base pairs relative to the helix axis, and relative to each other in base pair steps. The positive direction is defined by the right-hand rule. Adapted from [2].

relative to the B-form (∼0 Å). Therefore, the major (M) and minor (m) grooves are different in shape (width and depth), electrostatic potential, and disposition of atoms used in molecular recognition (hydrogen bond donors and acceptors) [1,3]. In RNA, the A-form helix is favoured because the ribose 2'-OH (which the deoxyribose lacks) interferes with the formation of the B-type helix.

A comparison of the three canonical forms of double-helical DNA and A-RNA is shown in Table N6 [1,3]. One notable difference between the A and B helices is the sugar pucker, which affects interphosphate distances and backbone conformation. Furanose rings are not planar, but are puckered so that ring substituents can avoid steric clashes, and electronegative substituents can be in axial positions. The magnitude and type of sugar pucker can be calculated and expressed using the endocyclic torsion angles, υ_0–υ_4, pseudorotation parameters, and terms such as C2'-*endo*, or the exocyclic torsion angle δ. Two ranges of sugar puckers are preferred for nucleotides; they are C2'-*endo* and C3'-*endo*, with a δ of about 145° and 80° respectively. The most stable arrangement of a sugar with hydroxyls at both the 2' and 3' positions of the ribose (RNA) appears to be the C3'-*endo*

configuration of A-type helices, rather than C2'-*endo* (or other) as in B-DNA [1]. Therefore, DNA can form a variety of duplex structures, whereas the A-form and variants of it dominate the double-helical structures of RNA and DNA–RNA hybrids, although there are examples of extreme variants such as Z-RNA [3,4]. Hydration is important in stabilizing duplex nucleic acid structures [1–3]. The degree of hydration, which depends on the alcohol or salt concentration as well as the base composition, determines the structures a particular molecule adopts. Ordered water molecules appear as a single 'spine of hydration' in the narrow minor groove of A/T-rich B-DNA, whereas other patterns of water molecules appear in DNA of different groove widths and composition [3].

Z-forms of nucleic acids are extreme conformational variants of the double helix which are stable under dehydrating conditions (high salt concentrations) and alternating pyrimidine- (C,T,U)/purine- (G,A,I) sequences. The helix is left-handed (Fig. N39 and Plate 1*c*) and rather than having two well-defined grooves, Z-DNA has a narrow and deep minor groove, and virtually no major groove because the bases are displaced from the helix axis so that they create a surface in place of the major groove. The alternating character of the sequence is reflected in the helix parameters such as helical twist and the glycosidic torsion angles χ and δ which alternate as *syn* (pyrimidines) or *anti* (purines), and produce the zig-zag effect after which the structure is named [1,3,4].

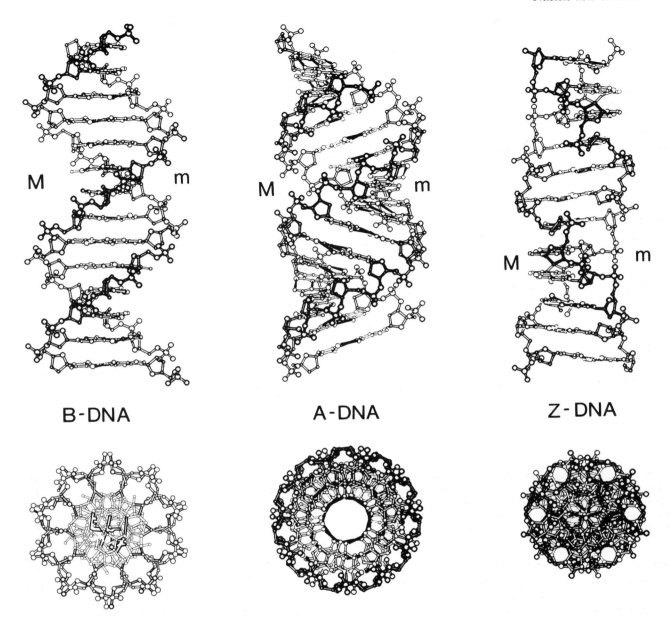

Fig. N39 DNA double-helical structures. *a*, B-DNA; *b*, A-DNA; *c*, Z-DNA. The upper structures are viewed perpendicular to the helix axis, the lower structures along the helix axis. M, Major groove; m, minor groove of the double helix. Adapted from [1]. *See also*: Plate 1.

Polymorphism and conformational flexibility of nucleic acids

Oligonucleotide crystal structures and studies of nucleic acids in solution have shown that DNA and RNA duplexes are not uniform molecules: there are local, global and dynamic variations on the canonical structures, sometimes even within the same molecule [3]. Such sequence-dependent structural variations occur because adjacent base pairs maximize base stacking. An example of structural polymorphism that occurs *in vivo* involves a structural variant of B-DNA formed by homopolymer adenine sequences [5]. The structure formed by tracts (≥4) of adenines or thymidines has an unusually high propeller twist and bifurcated hydrogen bonds in the major groove between adjacent base pairs which stabilize a straight helix with a narrow minor groove [3]. Mixed sequence DNA containing segments of these 'A-tracts' in phase with the DNA helical screw (every 10–11 base pairs per turn) gives a molecule that is intrinsically bent and which has biologically interesting properties [3,5] (*see* BENT DNA).

Not all nucleic acid molecules fit neatly into a defined structural family. In fact, most probably have conformations that lie somewhere between the canonical B-, C-, D-, and A-form structures [1,3,5]. One such example is a purine-rich portion of the TFIIIA-binding site in the *Xenopus* 5S RNA gene, which seems to lie between the B- and A-DNA conformations [3,5]. Nucleic acids can also adopt different structures depending on conditions such

Table N6 A comparison of helical parameters for B-, A-, and Z-DNA and A-RNA.

Parameter	B-DNA	A-DNA	Z-DNA (pG, pC)	A-RNA
Base pairs per turn	10	10.9	12	11
Rise per bp D_z (Å)	3.4	2.6	3.7	2.8
Helical twist (°)	36	33	− 50, − 10	32.7
Displacement D_x (Å)	0.6	− 4.4	3.2	− 4.4
Inclination (°)	− 2	22	− 7	16.7
Propeller twist (°)	− 13	− 6	∼0	− 13.8
Sugar pucker				
δ (°)	142	79	76, 147	84
Descriptor	C2′-endo	C3′-endo	C3′-, C2′-endo	C3′-endo
Glycosidic angle				
χ (°)	− 139	− 159	89, − 159	− 166
Descriptor	anti	anti	syn, anti	anti
Groove width				
Major (Å)	11.4	2.4	8.8	4.1
Minor (Å)	6.0	11.0	2.0	11.3
Groove depth				
Major (Å)	8.5	13.5	3.7	∼13
Minor (Å)	7.5	2.8	13.8	∼3

These values were obtained primarily from fibre diffraction experiments from several different sources and are based on van der Waals radii [1–5].
χ, Glycosidic torsion angle, δ, sugar pucker.

as those arising from crystal packing and protein or ligand binding.

Certain DNA sequences, such as TpA and CpA, have intrinsically more torsional flexibility than others and can adopt several conformations [5]. Many proteins, for example CAP (the catabolite gene activator protein) and FIS (factor for inversion stimulation) from the bacterium *Escherichia coli*, and DNase I, may recognize this intrinsic flexibility or rigidity and induce further local helix deformations (kinks) as well as global deformations (bends) of the DNA in recognition of their preferred binding sites [5]. Combinations of sequences that have distinct structural preferences with those that are deformable provide many possible nucleic acid structures for protein recognition [5].

Alternative base pairing schemes

In the duplex structures mentioned so far, Watson–Crick base pairing directs interstrand interactions. However, a wide variety of alternative hydrogen bonding patterns is possible and under the appropriate conditions provides the basis for many other nucleic acid structures [1]. Assuming that the formation of a base pair requires at least two hydrogen bonds, there are at least 28 possible base pairs including Watson–Crick, Hoogsteen and others (*see* BASE PAIRS). The modified bases and unusual tautomeric forms of bases increase the diversity of interactions possible.

Structures based on alternative base–base interactions have been observed *in vitro*. Non-Watson–Crick base pairs have been studied crystallographically to understand the effect of base–base mismatches, such as G–A, on DNA structure [3]. In ribosomal RNA for example, non-Watson–Crick base pairs are used frequently, and seem to be conserved and tolerated in regions of

duplex RNA. Crystal structures of mismatched base pairs such as U–C in an RNA duplex show that water can aid in interbase hydrogen bonding and increase the stability of the base pair [4,6]. The effect of these mismatches on the structure of the double helix is noticeable and may serve to direct protein binding either for purposes of repair (in the case of DNA) or specific recognition (in the case of the rRNA). An alternative base pairing arrangement has also been invoked to explain how nucleic acids can form duplexes with parallel strands [4,7] as well as multiple-stranded structures [8,9].

Triple-helical structures

Triple helices form in nucleic acids where a third strand makes non-Watson–Crick hydrogen bonds with an A-form duplex in the major groove. Figure N40*a* illustrates the directionality of the strands, and Fig. N40*a* and *d* show examples of base triple schemes that commonly occur. The C–G–C$^+$ base triple (*d*) is sensitive to pH, because the cytosine must be protonated, whereas the U(T)–A–U(T) base triple (*c*) is not. Under conditions of superhelical tension and acidic pH, the homopurine–homopyrimidine tract in a DNA duplex can dissociate to form a triple-helical structure known as H-DNA [8], where a segment of the pyrimidine strand forms Hoogsteen base pairs in the major groove of the remaining DNA duplex, leaving part of the purine strand unpaired and accessible to nucleases. In RNA there are many possible arrangements of base triples that have been proposed or observed in catalytic INTRONS and tRNA, and a few of these are illustrated in Fig. N40*c–f*. There is now evidence for many of the possible base triples from biochemical and NMR experiments [8].

Quadruple-helical structures

At neutral pH the major tautomeric form of guanine in DNA and RNA can form cyclic base tetrads such as those illustrated in Fig. N40*b*. Guanine (and possibly guanine/adenine) tetrads stack on top of one another to form helical structures which are stabilized by metals such as potassium and sodium, presumably because the metal can coordinate to the O6 of the guanine bases in the central cavity of the stacked base tetrads. Telomeric DNA sequences have conserved single-stranded regions consisting of repeats of tracts (∼3–8) of guanines separated by short tracts of A or T, but the structures that these sequences adopt *in vivo* are not known (*see* CHROMOSOME STRUCTURE). DNA fragments containing these sequences have been shown both to tetramerize and dimerize *in vitro* [9]. A general role for quadruple helices in the dimerization or protein-mediated dimerization of RNA and DNA is possible as these unusual four-stranded structures are formed by both four single strands as well as two strands which are folded back (hairpin-like) which interact to form a quadruple helix [9].

FIBRE DIFFRACTION studies and model building suggest that guanine homopolymers form quadruple helical structures of four parallel strands each with all of the guanines in the *anti* conformation (glycosidic angle χ ∼ − 160°) (Fig. N40*g*) [9]. The right-handed helix has a helical repeat of 11.5 base steps per turn, a

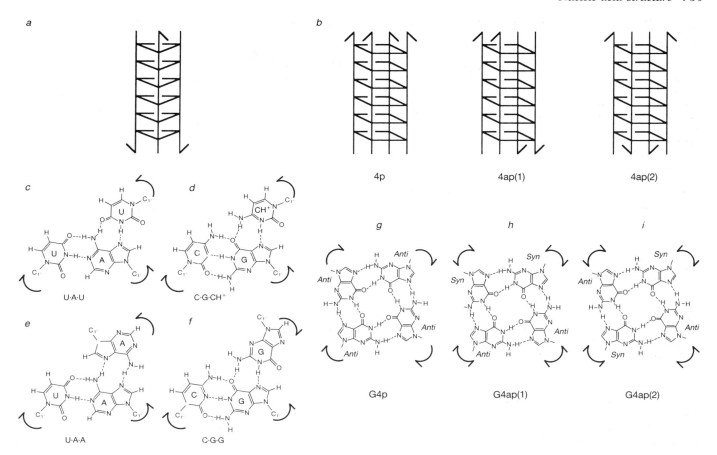

Fig. N40 Schematic diagrams of unusual triple and quadruple stranded nucleic acid structures. *a*, Strand arrangement in the triple helix. Examples of base triples are shown in *c–f* [12]: *c* and *d* occur in homopolyribonucleotides where the third pyrimidine strand is parallel to the purine strand and makes a Hoogsteen base pair with the purine in the major groove of the Watson–Crick (W–C) duplex; in the examples from tRNA (*e*, *f*) the second purine strand is parallel to the purine strand of the W–C duplex (*e*), and the second purine strand is antiparallel to the W–C duplex purine strand (*f*). The direction of the arrows refers to the direction of the corresponding strand. *b*, Strand arrangements are shown for three quadruple helical structures: one parallel stranded (*g*) having four parallel strands, and the two antiparallel stranded structures (*h*) and (*i*) having different arrangements of the parallel strands, either adjacent to (*h*) or opposite (*i*) each other. The hydrogen bonding scheme of the guanine base tetrad is shown in *g–i* [1,9,10]. The glycosidic torsional angle χ can be either *syn* or *anti*, and the arrangement of the *syn* and *anti* conformation is different in the various structures. All are *anti* in the 4p (*g*) model, whereas two are *anti* and two are *syn*, although arranged either adjacent (*h*) [11] or opposite each other (*i*) [10] in the two antiparallel structures. The double pointed arrows refer to the ability of alternate base tetrads in the antiparallel structures to flip over so that χ along each strand alternates from *syn* to *anti*.

rise of 3.4 Å, and a twist of 31° per base step. However, a crystal structure [10] and NMR study [11] of oligonucleotides containing $(G_4T_4G_4)_2$ show two different antiparallel arrangements of strands in the quadruple helix (Fig. N40*h*, *i*). The common feature of the antiparallel guanine tetrads is that the conformation of the guanine alternates between *syn* and *anti* along each strand [10,11]. Among other differences, the overall symmetry of the two antiparallel quadruple helices differs as a consequence of the adjacent (Fig. N40*h*) versus opposite (*i*) disposition of the *syn* and *anti* guanines, strand polarity, and position of T-spacers; whereas the structure *i* has two wider and two narrower grooves, the arrangement in *h* has three different groove shapes. However, these structures represent only a subset of possible quadruple helical structures that could arise from telomeric DNA (or RNA) of different sequences and lengths of T-spacer regions.

RNA folding

RNA rarely occurs in the cell in the duplex form. The behaviour of RNA is in some ways more reminiscent of proteins than of DNA. RNA has a sugar 2′-OH, and the bases are often more extensively modified than in DNA which also adds to the variety of interactions it can undergo [1]. These differences have dramatic consequences in the chemistry and in the different structures DNA and RNA prefer to adopt. For example, RNA is much more labile to chemical degradation than DNA and is particularly sensitive to hydrolysis under basic conditions [1]. RNA performs chemical reactions including those carried out in RNA SPLICING and in other catalytic RNA structures (*see* RIBOZYMES). In addition, RNA folds into tertiary structures [12].

RNA secondary structure

Folding may begin by the formation or nucleation of some particularly stable secondary structure(s) [12]. The best known RNA structure, that of TRANSFER RNA, provides a wealth of information on nucleic acid structure in general, but also suggests how folding may occur. tRNA contains several secondary structure elements (Fig. N41) [12].

Bulges, bulge loops, and base mismatches. A single unpaired base may stack in the double helix or bulge out, but in either case the structure of the surrounding duplex will be slightly altered [4,12]. Bulges of more than one nucleotide, depending whether or not they stack in the helix, may cause RNA bending or even more distorted RNA duplexes [4]. Bulge loops appear in positions along a duplex where base-pairing interactions are weak because of base mismatches; the RNA helix may be distorted, and the unpaired bases may be free to interact and form a tertiary structure with another part of the RNA [12].

Cruciform, hairpin, and stem-loop structures. In the absence of complementary strands, at low nucleic acid concentrations, or under superhelical stress, cruciform or hairpin structures will often form from self-complementary DNA or RNA sequences. These are composed of a duplex stem and a terminal loop (Fig. N41).

Particular four-base loops (tetraloops) seem to be unusually stable and frequently occurring [4]. Three types, in order of percentage occurrence, are the GNRA, UNCG, and CUUG (where N = any base and R = purine). NMR structures of these loops show that the first and fourth base form a special base pair causing the loop to be closed by two bases with one or both of the two middle bases stacking onto the first base pair [4]. In addition, other specific hydrogen bonds have been observed, in one case between a base N7 and a sugar 2'-OH which provides additional stability to the interaction. The additional hydrogen bonding available in RNA from the 2'-OH may contribute to the enhanced stability of these loops compared with DNA loops of the same sequence, but this has not been firmly established. It has been suggested that loops such as these may be involved in the RNA folding process as stable secondary structure elements [4,12].

Tertiary interactions

Single strands. Many of the tertiary interactions important in RNA folding arise because of single-stranded regions found in bulges, loops or unfolded segments of RNA. Bases in single-stranded regions are often stacked and are accessible for interactions such as triple-base and PSEUDOKNOT type interactions with other parts of the RNA molecule [12].

Junctions. Regions connecting three or more stems are known as junctions; three-arm and four-arm junctions are illustrated in Fig. N41. These occur in tRNA, and in proposed structures of large RNA molecules as well as in the proposed structure of Holliday recombination intermediates in DNA [12] (*see* RECOMBINATION).

Pseudoknots. The interaction of hairpin loops with a single-stranded region in RNA forms a pseudoknot (Fig. N41) [4,12]. Pseudoknots are found in rRNA structures, and messenger RNA. In some cases they are thought to act as switches in translational control [12]. NMR and structure-mapping techniques have been useful in elucidating these RNA structures.

Base triples. RNA folding may be directed in some cases by the formation of triple base pairs (Fig. N40*a*). Three separate strands or a single strand of a loop may interact with a double-helical region to form a base triple or segment of triple helix. There are

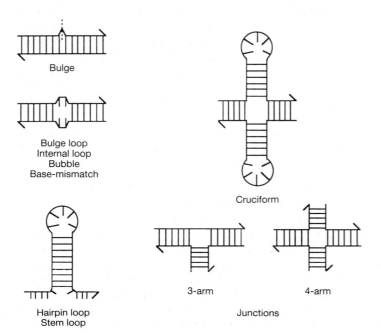

Bulge

Bulge loop
Internal loop
Bubble
Base-mismatch

Cruciform

Hairpin loop
Stem loop

3-arm   4-arm

Junctions

Pseudoknot

Fig. N41 Nomenclature and illustration of some RNA secondary and tertiary structural elements. Simple secondary structures include: bulge, bulge-loop or bubble, stem-loop or hairpin loop. Tertiary interactions include cruciform, three-arm and four-arm junction structures, and a simple pseudoknot shown as a schematic diagram and model [4,12].

a few examples of this in the structure of tRNA. RNA base triples have also been located in catalytic RNA structures such as the Group I introns (*see* RIBOZYMES) [12].

Other tertiary interactions. Other more complicated hydrogen bonding and stacking schemes occur in RNA [12,13]. For example, an 'axial interaction' occurs in the G-binding site of the self-cleaving *Tetrahymena* precursor RNA; this is a specific hydrogen bonding interaction between the guanine base which is inserted in the major groove perpendicular to the base pairs in a helical region [13]. The details of this interaction are not well understood, but structures such as this, as well as others that are currently unknown, are likely to be involved in shaping the tertiary structure and activity of folded RNA molecules [12,13].

M. CHURCHILL

1 Saenger, W. (1984) *Principles of Nucleic Acid Structure* (Springer-Verlag, New York).
2 Dickerson, R.E. et al. (1989) Definitions and nomenclature of nucleic acid structure parameters. *EMBO J.* **8**, 1–4.
3 Kennard, O. & Hunter, W.N. (1989) Oligonucleotide structure: a decade of results from single crystal X-ray diffraction studies. *Q. Rev. Biophys.* **22** (3), 327–379. Kennard, O. & Hunter, W.N. (1991) Single crystal X-ray diffraction studies of oligonucleotides and oligonucleotide–drug complexes. *Ang. Chem. Int. Ed. Engl.* **30**, 1254–1277.
4 Varani, G. & Tinoco, I. Jr. (1991) RNA structure and NMR spectroscopy. *Q. Rev. Biophys.* **24** (4), 479–532.
5 Travers, A.A. (1989) DNA conformation and protein binding. *Annu. Rev. Biochem.* **58**, 427–452. Travers, A.A. (1992) DNA conformation and configuration in protein–DNA complexes. *Curr. Opinion Struct. Biol.* **2**, 71–77.
6 Holbrook, S.R. et al. (1991) Crystal structure of an RNA double helix incorporating a track of non-Watson–Crick base pairs. *Nature* **353**, 579–581.
7 Jovin, T.M. (1991) Parallel-stranded DNA with trans-Crick–Watson base pairs. In *Nucleic Acids and Molecular Biology*, Vol. 5, 25–38 (Springer-Verlag, Heidelberg/New York/London).
8 Frank-Kamenetskii, M.D. (1990) Protonated DNA structures. In *Nucleic Acids and Molecular Biology*, Vol. 4, 1–8 (Springer-Verlag, Heidelberg/New York/London). Sklener, V. & Feigon, J. (1990). Formation of a stable triplex from a single DNA strand. *Nature* **345**, 836–838.
9 Sundquist, W.I. (1991) The structures of telomeric DNA. In *Nucleic Acids and Molecular Biology*, Vol. 5, 1–24 (Springer-Verlag, Heidelberg/New York/London).
10 Kang, C. et al. (1992) Crystal structure of four-stranded *Oxytricha* telomeric DNA. *Nature* **356**, 126–131.
11 Smith, F.W. & Feigon, J. (1992) Quadruplex structure of *Oxytricha* telomeric DNA oligonucleotides. *Nature* **356**, 164–168.
12 Tinoco, I., Jr. et al. (1990) RNA folding. In *Nucleic Acids and Molecular Biology*, Vol. 4, 205–226 (Springer-Verlag, Heidelberg/New York/London).
13 Yarus, M. et al. (1991) An axial binding site in the *Tetrahymena* precursor RNA. *J. Mol. Biol.* **222**, 995–1012.

nucleocapsid The complex of nucleic acid and protein coat in a virus particle. Some viruses may contain several nucleocapsids within an outer lipid envelope. *See*: ANIMAL VIRUSES; PLANT VIRUSES.

nucleocytoplasmic interactions The effect of the cytoplasm on nuclear gene expression and vice versa. It is studied by combining the nucleus of one cell with the cytoplasm of another by cell fusion or NUCLEAR TRANSPLANTATION. Before such experiments were carried out it was not clear whether the cytoplasm or the nucleus alone controlled development. Nuclear transplantation experiments showed that differences in gene expression are controlled by the cytoplasm even though the nucleus remains pluripotent. Somatic cell fusion has also been used to assess cytoplasmic regulation of gene activity. For example, human hepatocyte nuclei can be reprogrammed to express muscle-specific proteins when introduced into mouse muscle cells by cell fusion. The transfer of proteins between nucleus and cytoplasm can also be examined by these techniques. *See also*: SOMATIC CELL HYBRIDS.

nucleolar organizer Region of a chromosome associated with the NUCLEOLUS and corresponding to a cluster of RIBOSOMAL RNA genes.

nucleolus Non-membrane-bound dense, granular, and fibrillar spheroidal body within the nucleus, prominent in cells in INTERPHASE, but which disappears during MITOSIS, and within which most of a eukaryotic cell's RIBOSOMAL RNA is synthesized and RIBOSOMES assembled. The nucleolus forms around the nucleolar organizers — regions of chromosomes containing the genes for rRNA. *See also*: NUCLEAR MATRIX.

nucleoplasm The material inside the nucleus interior to the inner nuclear membrane and excluding the nucleolus.

nucleoplasmin MOLECULAR CHAPERONE involved in the assembly of nucleosomes (*see* CHROMATIN).

nucleoporins Glycoprotein peripheral components of the nuclear pore. *See*: NUCLEAR ENVELOPE.

nucleosides and nucleotides Nucleosides are composed of a purine or pyrimidine BASE covalently linked through the N9 (purine) or N1 (pyrimidine) to the C1 of the SUGAR D-ribose (or D-2-deoxyribose in the case of deoxyribonucleosides). Nucleotides (nucleoside phosphates) are made up of a nucleoside with the addition of up to three phosphate groups linked in series by phosphoanhydride bonds and linked to the sugar by a phosphoester bond (Fig. N42 and Table N7). The most common site of esterification is the 5′ position of the sugar ring; for example, the ribonucleoside triphosphate precursors of RNA are adenosine 5′-triphosphate (ATP), guanosine 5′-triphosphate (GTP), cytidine 5′-triphosphate (CTP), and uridine 5′-triphosphate (UTP); and the four deoxyribonucleotide precursors of DNA are deoxyadenosine 5′-triphosphate (dATP), deoxyguanosine 5′-triphosphate (dGTP), deoxycytidine 5′-triphosphate (dCTP), and thymidine 5′-triphosphate (TTP). Unless otherwise specified the abbreviations AMP, ADP, ATP, etc. represent the 5′- nucleotides.

Purine ribonucleotides (e.g. AMP, GMP) are synthesized *de novo* by a stepwise construction of the purine ring system on C1 of ribose 5-phosphate derived from 5′-phosphoribosyl-1-pyrophosphate (PRPP). In the major synthetic pathway, inosine monophosphate (IMP) is an intermediate. Pyrimidine ribonucleotides are derived from the coupling of orotic acid to a ribose 5-phosphate moiety derived from PRPP and subsequent decarboxylation to form uridine monophosphate; cytidine nucleotides are derived from uridine triphosphate. Deoxyribonucleotides are

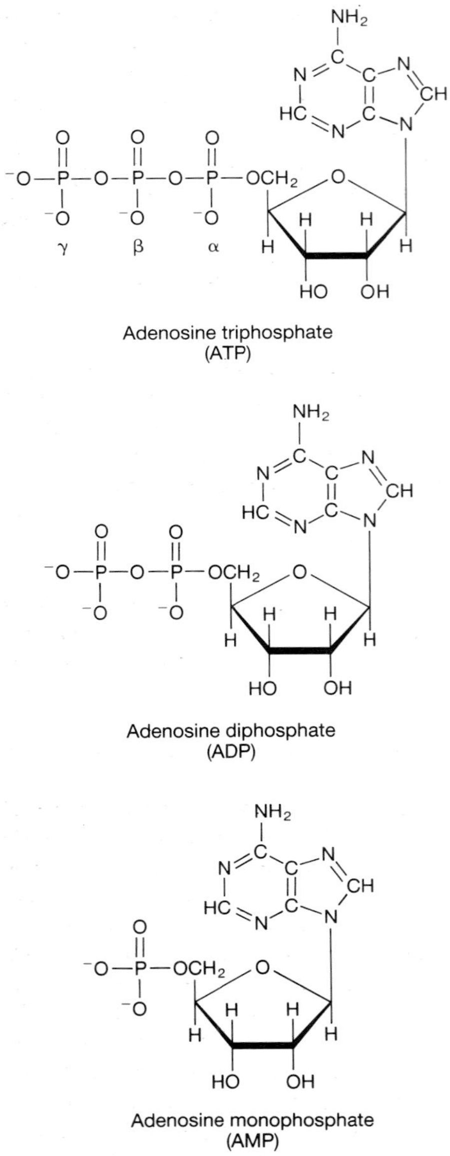

Adenosine triphosphate
(ATP)

Adenosine diphosphate
(ADP)

Adenosine monophosphate
(AMP)

Fig. N42 The ribonucleotides AMP, ADP, and ATP.

formed from the corresponding ribonucleotide by reduction of the 2′-OH on ribose catalysed by ribonucleotide reductase.

Nucleotides are involved in some way in most biological processes. Nucleoside triphosphates are the precursors for DNA and RNA synthesis, and nucleotides are activated intermediates in many reactions (*see* S-ADENOSYLMETHIONINE; SUGAR NUCLEO-TIDES). ATP is the universal energy currency of living cells. Most cellular processes are powered directly or indirectly by the hydrolysis of ATP to ADP or AMP. The Gibbs energy of hydrolysis of the phosphoanhydride bonds ($\Delta G^{\circ\prime} = -30.5$ kJ mol^{-1}) is more negative than that of ordinary phosphate esters. The active form of ATP is usually complexed in the cell with Mg^{2+} (MgATP) or Mn^{2+}.

The adenine nucleotide derivatives (NAD, NADP, FAD, FMN, CoA) are vital cofactors in many metabolic processes, with the

Table N7 Nomenclature for the nucleosides and nucleotides of the bases found in RNA and DNA

Base	Nucleoside	Nucleotide
Adenine	Adenosine	Adenosine monophosphate, (adenylic acid) (AMP) Adenosine diphosphate (ADP) Adenosine triphosphate (ATP)
	Deoxyadenosine	Deoxyadenosine monophosphate (dAMP), dADP, dATP
Guanine	Guanosine	Guanosine monophosphate (guanylic acid) (GMP), GDP, GTP
	Deoxyguanosine	dGMP, dGDP, dGTP
Cytosine	Cytidine	Cytidine monophosphate (cytidylic acid) (CMP), CDP, CTP
	Deoxycytidine	dCMP, dCDP, dCTP
Uracil	Uridine	Uridine monophosphate (uridylic acid) (UMP), UDP, UTP
Thymine	Thymidine*	Thymidine monophosphate (thymidylic acid) (TMP), TDP, TTP

*Thymidine, thymidine monophosphate, etc. usually refer to the deoxyribo-nucleoside and deoxyribonucleotide.

reduced forms of NAD and NADP (NADH and NADPH) being important sources of reducing power in cells. Guanine nucleotides are involved in the transduction of signals from many receptors (*see* G PROTEIN-COUPLED RECEPTORS; GTP-BINDING PROTEINS) and in PROTEIN SYNTHESIS. The cyclic nucleotides CYCLIC AMP and CYCLIC GMP act as second messengers (*see* SECOND MESSENGER PATHWAYS).

The importance of ribonucleotides as energy sources and cofactors in cellular processes may reflect their vital role in the evolution of life through the postulated RNA WORLD (*see also* MOLECULAR EVOLUTION).

nucleosome The repeating subunit in the organization of eukaryotic CHROMATIN. It is formed by ~200 bp of DNA double helix wrapped around a core particle which is an octamer composed of two copies each of HISTONES H2A, H2B, H3, and H4 (Fig. N43). Nucleosomes are connected by linker DNA which is, in most species, associated with histone H1, and associate with their neighbours to form a helical fibre of ~30 nm diameter. *See also:* PROTEIN–NUCLEIC ACID INTERACTIONS.

nucleotide(s) Constituents of nucleic acids (*see* DNA; NUCLEIC ACID STRUCTURE; RNA) which also have many other roles in all cells. *See:* NUCLEOSIDES AND NUCLEOTIDES.

nucleotide excision repair *See:* DNA REPAIR.

nucleotide pair *See:* BASE PAIR.

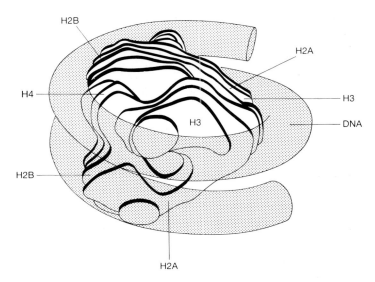

Fig. N43 Model of nucleosome core structure. It has also been found that the DNA is not wound smoothly around the protein core, but is sharply kinked (not shown here) at positions symmetrically disposed at a distance of about one and four double-helical turns in both directions from the nucleosomal pseudo-dyad axis. Richmond, T.J. et al. (1984) *Nature* **231**, 532–537.

nucleotide sequence   The order of nucleotide residues in a polynucleotide chain. In protein-coding genes the nucleotide sequence (of the EXONS only in eukaryotic genes) encodes the order of amino acid residues in the corresponding protein. *See*: DNA; GENE; GENETIC CODE; NUCLEIC ACID; NUCLEOSIDES AND NUCLEOTIDES; PROTEIN SYNTHESIS; RNA.

nucleotide substitution   BASE SUBSTITUTION.

nucleotide sugar   *See*: SUGAR NUCLEOTIDES.

nucleotide-gated ion channels   ION CHANNELS whose opening is regulated by nucleotides, including the cyclic nucleotides. *See*: OLFACTORY TRANSDUCTION; VISUAL TRANSDUCTION.

nucleotidyl hydrolases   *See*: DNA REPAIR.

nucleotidyltransferases   Enzymes that transfer a nucleoside monophosphate moiety from one compound (often a nucleoside triphosphate) to another. A large and varied group that includes the DNA POLYMERASES, RNA POLYMERASES, TERMINAL NUCLEOTIDYLTRANSFERASE, and the sugar phosphate nucleotidyltransferases (*see*: GLYCANS).

nucleus   (1) The organelle in eukaryotic cells that contains the CHROMOSOMES. It is large, often spherical, and is bounded by a double membrane (*see* NUCLEAR ENVELOPE). DNA REPLICATION, TRANSCRIPTION, RNA SPLICING, RNA PROCESSING, and most stages of RIBOSOME assembly take place within the nucleus. In INTERPHASE CELLS, a dense body, the NUCLEOLUS, is visible within the nucleus and is the site of rRNA synthesis and ribosome assembly. *See also*: CHROMOSOME STRUCTURE; MEIOSIS; MITOSIS; NUCLEAR MATRIX.
(2) Brain structure composed largely of neuronal cell bodies and dendrites.

nude mice   Strain of mice homozygous for the *nude* (*nu/nu*) mutation in which thymus development is lacking and which therefore do not produce mature T cells (*see* T CELL DEVELOPMENT). *nude* mice were used for thymic transplantation experiments which established the existence of MHC restriction of T cell responses (*see* MAJOR HISTOCOMPATIBILITY COMPLEX). The mutation also results in hairlessness, hence the name.

null allele   A nonfunctional ALLELE of a gene. It may arise either through deletion of the entire gene, or from a localized mutation of the gene that results in a completely inactive protein being produced from it.

numerical chromosome mutations   *See*: MUTATION; TRISOMY.

nurse cells   In *Drosophila* and other insects that undergo a full metamorphosis, female PRIMORDIAL GERM CELLS divide into 8 or 16 daughter cells, depending on species. One of these cells gives rise to the OOCYTE; the others become nurse cells. The cell divisions that give rise to the nurse cells and to the oocyte are incomplete and structures called ring canals remain between the cells and form intercellular channels between them, through which proteins and other factors can be transported. Nurse cells are highly POLYPLOID and suppply large amounts of RNA and protein to the growing oocyte. This investment from the nurse cells includes the components of the protein synthesis machinery and other factors to sustain early development of the embryo. Nurse cells also help determine the anterior–posterior polarity of the embryo. As the oocyte grows, the nurse cells gather at one end of the egg chamber and deposit the mRNA for a protein called *bicoid*, an important anterior determinant. *bicoid* mRNA is anchored at the end of the oocyte which is next to the nurse cells, and determines that this will become the anterior end of the oocyte. *See also*: DROSOPHILA DEVELOPMENT.

O

O-antigen Glycan side chain of the surface LIPOPOLYSACCHARIDE of Gram-negative bacteria, which varies from species to species and from strain to strain. *See*: BACTERIAL ENVELOPES.

O blood group *See*: ABO BLOOD GROUP SYSTEM.

O-linked oligosaccharide Carbohydrate chains attached to GLYCO-PROTEINS via the hydroxyl groups of amino-acid side chains (Fig. O1). When they contain *N*-acetyl-D-galactosamine (GalNAc) in α-linkage to serine or threonine they are also called mucin-type or Ser/Thr-linked oligosaccharides. Mucin-type oligosaccharides, like N-linked oligosaccharides, are added to glycoproteins in the endoplasmic reticulum and Golgi apparatus. O-linked oligosaccharides often have many of the same structural features that are found in N-linked oligosaccharides. A recently discovered type of O-linked glycosylation is the serine/threonine-linked *N*-acetyl-glucosamine (O-GlcNAc) found on cytoplasmic and nuclear proteins. At least six other monosaccharides, besides GalNAc and GlcNAc, can be linked directly to the hydroxyl groups of amino-acid side chains.

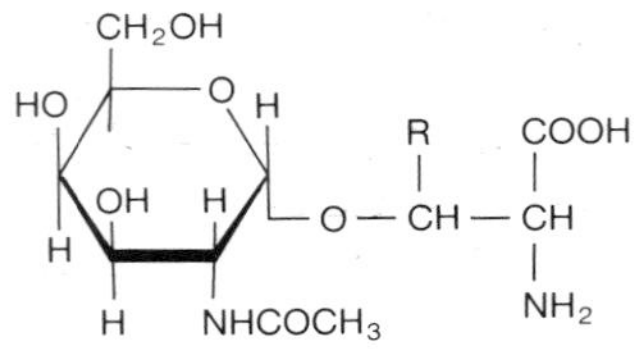

Fig. O1 O-linkage between *N*-acetyl-D-galactosamine and the hydroxyl group of an amino-acid side chain. Serine, R = H; threonine, R = CH$_3$.

obelin *See*: CALCIUM-BINDING PROTEINS.

occluding junction Tight junction. *See*: CELL JUNCTIONS.

ochre codon The TERMINATION CODON UAA. Any mutation resulting in the creation of a UAA codon is termed an ochre mutation. Such mutations can be suppressed by an ochre suppressor, which is a TRANSFER RNA with an altered anticodon such that it can bind and decode the UAA codon, allowing readthrough of the ochre mutation. *See also*: GENETIC CODE.

Oct-1 DNA-binding protein present in nuclei of many cell types that binds to the OCTAMER motif in immunoglobulin genes. It contains a POU domain (*see* HOMEOBOX GENES AND HOMEO-DOMAIN PROTEINS).

Oct-2 Lymphoid cell-specific TRANSCRIPTION FACTOR which binds the OCTAMER motif found in the heavy chain enhancer and V$_H$ gene promoter in the IMMUNOGLOBULIN GENES. It contains a POU domain (*see* HOMEOBOX GENES AND HOMEODOMAIN PROTEINS).

octamer-binding protein Protein that binds to the OCTAMER MOTIF.

octamer motif Consensus DNA sequence of eight bases found in the HEAVY CHAIN ENHANCER and the V$_H$ gene promoter. Similar motifs are found in other enhancers (e.g. in PAPOVAVIRUSES). *See*: IMMUNOGLOBULIN GENES.

octopine An OPINE, the condensation product of arginine and pyruvate, characteristic of *Agrobacterium tumefaciens* strains such as ACH5, B6S3, A6, 15955 (octopine strains). Also the name given to a family of opines all of which are based on pyruvate condensed with an amino acid (e.g. in octopinic acid, the amino acid is ornithine and in histopine, histidine).

octopine strains Strains of *Agrobacterium* which induce the formation of, and which can also utilize, opines of the OCTOPINE type.

octopine synthase (OS) Enzyme encoded by a gene in the T-region of octopine-type T$_I$ PLASMIDS which, in transformed plant cells, catalyses the condensation of arginine and pyruvate to OCTOPINE.

OEC OXYGEN EVOLUTION COMPLEX.

oestrogens STEROID HORMONES produced principally by the ovary, and which are involved in the reproductive cycle in females (Fig. O2). Oestrogens act via intracellular receptors of the STEROID RECEPTOR SUPERFAMILY.

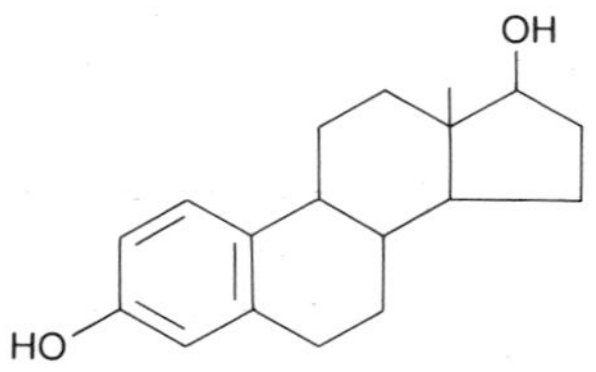

Fig. O2 β-Oestradiol, the main oestrogen secreted by the human ovary.

OFAGE Acronym for orthogonal field agarose gel ELECTRO-PHORESIS.

Ohno's law The hypothesis that X-linked genes in one mammalian species will be X-linked in others, as a result of the lethal effect of translocations of regions of the X-chromosome. *See:* X-CHROMOSOME INACTIVATION.

okadaic acid A fatty acid polyketide, which specifically inhibits protein phosphatases PP1 and PP2A (*see* PROTEIN PHOSPHATASES) at nanomolar concentrations. It is a potent TUMOUR PROMOTER. One of the causative agents of diarrhoetic seafood poisoning, it is produced by marine dinoflagellates which accumulate by filter feeding in the digestive glands of molluscs and marine sponges such as *Halichondria okadaii*, from which it was first extracted.

Cohen, P. et al. (1990) *Trends Biochem. Sci.* **15**, 98–102.

Okazaki fragment A short fragment of DNA generated on the lagging strand during DNA REPLICATION. Because DNA polymerase can only elongate a polynucleotide strand in a $5' \rightarrow 3'$ direction, a new DNA strand has to be synthesized discontinuously along one of the strands of the unwound duplex at the replication fork. Each Okazaki fragment is initiated as a short RNA primer and is ~1000–2000 nucleotides long in prokaryotes, and ~100–200 nucleotides long in eukaryotes. The gaps between fragments are subsequently filled in by DNA polymerase and joined by DNA ligase.

Ogawa, T. & Okazaki, T. (1980) *Annu. Rev. Biochem.* **49**, 421–457.

OKT series MONOCLONAL ANTIBODIES specific for the CD3 antigen (*see* T CELL RECEPTORS) and which are used to treat acute transplant rejection. They help to eliminate T cells by acting as lytic antibodies and also by opsonizing T cells for phagocytosis.

Olf-1 TRANSCRIPTION FACTOR specific to olfactory neurons. *See:* OLFACTORY TRANSDUCTION.

Olfactory signal transduction

THE human sense of smell involves the detection and discrimination of a very wide range of low concentrations (as low as 1 in 10^{12} parts) of airborne chemicals (odorants). In fish, the odorants are water soluble but the basic mechanism of detection appears extremely similar to that of mammalian olfaction. Invertebrates also make use of olfaction; insects, for example, detect very specific molecules (pheromones) at extreme dilution. The details of the mechanisms whereby any specific odorant interacts with its receptor and how these interactions generate neural signals have not yet been elucidated. However, considerable progress has been made suggesting that vertebrate and invertebrate olfaction, at least in part, are mediated by G PROTEIN-COUPLED RECEPTORS. In this respect, olfaction can be considered analogous to vision (*see* VISUAL TRANSDUCTION), to many types of hormone reception and neurotransmission, and to the mechanism of mating-factor reception in yeast, all of which involve G-protein linked receptor–effector coupling. In insect olfaction, odorants appear to stimulate a PHOSPHOLIPASE C [1]. This results in a transient increase in 1,4,5-inositol trisphosphate (IP$_3$). In vertebrates, there appear to be two second messengers that are generated in response to particular odorants: 3′,5′-cyclic adenosine monophosphate (cAMP) and IP$_3$ [1] (*see* SECOND MESSENGER PATHWAYS).

The olfactory epithelium

Mammalian olfactory reception and signal transduction take place in the olfactory neuroepithelium, which is located towards the top of the nasal cavity. Signal processing is carried out both in the olfactory bulb and in higher centres in the brain. The neuroepithelium is a heterogeneous tissue bathed in a thick layer of mucus, the secretion of numerous nasal glands. The mucus concentrates airborne odorants and contains secreted proteins that bind a range of odorants with affinities that reflect their efficacy in psychophysical experiments, for example pyrazine-binding protein [2,3]. These proteins may function in delivery of odorants to receptor cells or removal of specific odorants from their binding sites on the sensory neurons. Each half of the olfactory epithelium contains about 10^6 sensory olfactory neurons. The neurons are bipolar with a single dendrite that extends from the cell body to the surface of the olfactory epithelium, and an axon that crosses the cribriform plate to make a synaptic connection in the olfactory bulb. Several immotile cilia project from the club-like dendrite into the mucus thereby increasing the exposed surface area of the receptor neurons. These cilia are the site of odorant detection and of olfactory transduction. The neurons are unusual, not only because of their role in sensory reception, but also because they turn over throughout life, being replaced (every 4 weeks) by division and differentiation of stem cells located at the base of the neuroepithelium. The other cells of the neural epithelium may also have a more than passive role in olfaction. For example, the supporting columnar epithelial cells appear to be rich in enzymes that have been proposed to inactivate odorants after they have stimulated olfactory transduction [4].

Signal transduction

Early studies indicated that many odorants resulted in the production of cAMP. Olfactory cilia appeared to be rich in the enzyme ADENYLATE CYCLASE which produces cAMP from ATP (*see* SECOND MESSENGER PATHWAYS). Electrophysiological experiments indicated that the cilia also contained a cyclic nucleotide-gated ION CHANNEL that resembled that of vertebrate vision [5].

A major early breakthrough in the molecular characterization of the olfactory neuron was the characterization of a protein (olfactory marker protein, OMP) that is both specific for the olfactory neuron and present at very high levels [6]. The role of this protein in olfaction is still unknown. However, OMP has served as a marker of olfactory sensory neurons and has been extensively used to study neural degeneration and regeneration following surgical removal of the olfactory bulb. Bulbectomy results in the rapid and specific degeneration of the sensory neurons (reflected in a depletion of OMP and its mRNA) and has therefore proved a very powerful technique to study neuron-specific expression. Unilateral bulbectomy has demonstrated the olfactory neuron-specific expression of several proteins that appear to be linked to odorant-dependent production of cAMP [7]. These proteins include a G protein α subunit closely resembling, but distinct from, the $G_s\alpha$ which stimulates adenylate cyclase in response to adrenaline (epinephrine); an adenylate cyclase which appears to be activated by the G protein and is distinguished from the form found in other tissues by its very low basal activity; and a cyclic nucleotide-gated ion channel which is homologous to the light-activated channel of mammalian vision. All these proteins are highly enriched in the olfactory cilia.

Therefore it seems that odorant-induced stimulation of adenylate cyclase results from stimulation of membrane receptors related to those of vision and the β-adrenergic response. This results in the activation of the olfactory-specific G protein, G_{olf}, by promoting guanine nucleotide exchange at its binding site on the α subunit. In turn G_{olf} stimulates the olfactory-specific adenylate cyclase resulting in a rise in the cytoplasmic concentration of cAMP. The rise in concentration of cAMP opens the cyclic nucleotide-sensitive channel with depolarization of the olfactory neuron (Fig. O3). The interaction of an odorant molecule with a surface receptor molecule is highly amplified, as a single receptor can stimulate several G proteins; each cyclase that is activated catalyses the production of many molecules of cAMP; and the lifetime of the open channel allows passage of many ions. This amplification is critical, as studies have indicated that stimulation of a single receptor protein is sufficient for the generation of an ACTION POTENTIAL. Little is known about how other odorants stimulate the production of IP_3, but the sensitivity of this pathway to PERTUSSIS TOXIN suggests that it too is G-protein mediated.

Of equal importance to stimulation of signal transduction is the process of deactivation. Odorant removal from the mucus may be a role of the supporting epithelial cells (see above). The lifetime of

the active state of G proteins is controlled by the intrinsic GTPase activity of the α subunit, and as yet unidentified PHOSPHO-DIESTERASES probably hydrolyse the cyclic phosphodiester bond of cAMP to generate AMP. Rapid adaptation and/or desensitization to odorants is a feature of olfactory transduction. This process appears to involve the activation of a cAMP-dependent PROTEIN KINASE by odorants that stimulate cAMP production, and a PROTEIN KINASE C by odorants that stimulate IP_3 production [8].

Olfactory receptors

Perhaps the most interesting molecular components of olfactory transduction are the receptor proteins. These molecules are responsible for the actual recognition and discrimination of different odorants. About 30 distinct human anosmias (inability to smell a particular odorant) have been described, putting a lower limit on the number of receptor proteins of ~30.

The similarity of the proposed mechanisms of olfactory transduction to those of vision and hormone reception suggests that the receptors are likely to be related. Using the POLYMERASE CHAIN REACTION (PCR) to search for novel olfactory-specific G-protein linked receptors, a family of candidate proteins has been identified [9]. To do this, several degenerate PRIMERS for PCR were selected on the basis of sequence homology between known G protein-coupled receptors. All combinations of these primers were used to amplify many products from cDNA derived from olfactory tissue. The criterion proposed for any one of these PCR products to represent a receptor mRNA was that it should be uniform in size but heterogeneous in sequence. Appropriately sized products of the PCR reactions were analysed by digestion with RESTRICTION ENZYMES that have four-base recognition sequences and thus cut frequently. Digestion of products of similar

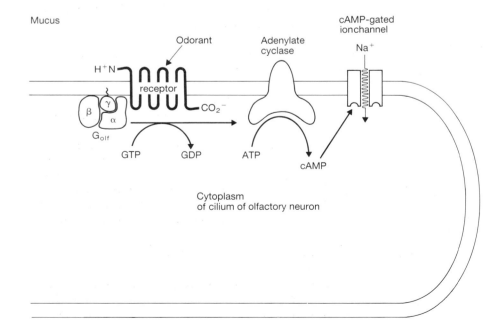

Fig. O3 The proposed pathway for cAMP-mediated olfactory transduction: after an odorant binds to its receptor in the plasma membrane of the sensory cilium, the receptor interacts with the heterotrimeric G protein, G_{olf}. This interaction catalyses guanine nucleotide exchange at its binding site on the α subunit of G_{olf}, thereby activating the protein. The activated G protein stimulates the membrane-associated adenylate cyclase to catalyse the conversion of ATP to cAMP. The rise in intracellular cAMP gates nonspecific ion channels in the plasma membrane of the cilium; this allows influx of Na^+ ions and results in a depolarization of the cell that is sufficient to trigger action potentials. Removal of odorant, and hydrolysis of GTP and cAMP, allow the cell to restore its resting ion balance.

size but different sequence with such restriction enzymes is likely to result in numerous different smaller fragments. The sum of the molecular weights of the digestion fragments will equal that of a homogeneous PCR product but exceed it if the PCR product is heterogeneous. In one case this was demonstrated, and allowed isolation of complete cDNAs from what seems to be a very large family of transcripts with related sequence.

The protein sequences deduced from the cDNA sequences have all the hallmarks of G protein-coupled receptors: seven membrane-spanning helices, N-terminal N-linked glycosylation sites, conserved extracellular cysteines that are probably disulphide linked, and other diagnostic sequence motifs. However, they seem to be significantly more closely related to one another than to other G-protein linked receptors. The proteins appear to be expressed in the olfactory epithelium but are not detectable in several other tissues. They are therefore extremely good candidate olfactory-receptor proteins. This receptor family is extremely large: estimates by SOUTHERN BLOTS of human chromosomal DNA suggest several hundred to several thousand different genes (representing up to 1% of the human genome).

In situ hybridization has been used to localize expression of individual receptor proteins and subfamilies of closely related receptor proteins [10,11]. There is a random distribution of olfactory neurons expressing specific receptor proteins in catfish epithelia [10]. Two levels of organization of olfactory receptor expression have been detected in mammals. There appear to be defined zones of the epithelium that limit the expression of particular subfamilies of receptors [11]. However, within a zone, expression appears to be random with up to 5% of olfactory neurons expressing members of a specific subfamily of receptors. Therefore, it is likely that discrimination of odours takes place because receptors with a defined ligand specificity are only present in a limited population of olfactory neurons. A major question in olfactory transduction is how discrimination is achieved at a molecular level. There are preliminary data that suggest one receptor may respond to several different odorants but the detailed ligand specificity of the olfactory receptors remains to be defined. It is also unclear whether the cloned receptors interact with G_{olf} or with a pertussis toxin-sensitive G protein that couples odorants to phospholipase C activation. Recently, mRNA for olfactory receptors was also detected in sperm, therefore it is possible that the functions of these proteins extend beyond their role in olfactory signal transduction.

N.J.P. RYBA

1 Boekhoff, I. et al. (1990) Rapid activation of alternative second messenger pathways in olfactory cilia from rats by different odorants. *EMBO J.* **9**, 2453–2458.
2 Bignetti, E. et al. (1985) Purification and characterisation of an odorant-binding protein from cow nasal tissue. *Eur. J. Biochem.* **149**, 227–231.
3 Pevsner, J. et al. (1988) Molecular cloning of odorant-binding protein: member of a ligand carrier family. *Science* **241**, 336–339.
4 Lazard, D. et al. (1991) Odorant signal termination by olfactory UDP glucuronosyl transferase. *Nature* **349**, 790–793.
5 Nakamura, T. & Gold, G.H. (1987) A cyclic nucleotide gated conductance in olfactory receptor cilia. *Nature* **325**, 442–444.
6 Margolis, F. (1985) Olfactory marker protein: from PAGE band to cDNA clone. *Trends Neurosci.* **8**, 542–546.
7 Reed, R.R. (1992) Signaling pathways in odorant detection. *Neuron* **8**, 205–209.
8 Boekhoff, I. & Breer, H. (1992) Termination of second messenger signaling in olfaction. *Proc. Natl. Acad. Sci. USA* **89**, 471–474.
9 Buck, L. & Axel, R. (1991) A novel multigene family may encode odorant receptors: a molecular basis for odor recognition. *Cell* **65**, 175–187.
10 Ngai, J. et al. (1993) Coding of olfactory information: topography of odorant receptor expression in the catfish olfactory epithelium. *Cell* **72**, 667–680.
11 Ressler, K.J. et al. (1993) A zonal organization of odorant receptor gene expression in the olfactory epithelium. *Cell* **73**, 597–609.

---

oligo(dG : dC) tailing A method of adding a single-stranded homopolymeric tail to the 3′ end of a DNA fragment. This technique is used to add complementary single-stranded ends onto blunt-ended double-stranded DNAs so that they may be joined together by complementary base pairing. Using the enzyme TERMINAL DEOXYNUCLEOTIDYLTRANSFERASE short lengths of deoxyguanosine (dG) residues are added to the 3′ end of one of the fragments to be joined. The same procedure is applied to the other DNA fragment in a separate reaction using deoxycytidine (dC) residues. The tailed fragments are then mixed to allow base-pairing between their complementary 3′ homopolymer tails. This technique is applied especially in the cloning of COMPLEMENTARY DNA into a VECTOR.

oligo(dT) cellulose A cellulose derivative used for affinity CHROMATOGRAPHY. Short chains of deoxythmidylic acid residues (dT) (normally up to ∼18) are bound to cellulose. The material can be used to purify selectively eukaryote mRNA, which has a 3′ poly(A) tail, from an RNA mixture, as nonpolyadenylated RNA (e.g. rRNA) does not bind to the affinity matrix.

oligodendrocytes *See*: GLIAL CELLS.

oligomycin Antibiotic which inhibits the F_0 channel in mitochondrial ATP SYNTHASE, thus inhibiting ATP synthesis.

oligonucleotide A short length of single-stranded polynucleotide chain usually <30 residues long.

oligonucleotide ligation assay A technique using the POLYMERASE CHAIN REACTION.

oligonucleotide-mediated mutagenesis A method of SITE-DIRECTED MUTAGENESIS. The gene to be mutated is cloned into a vector, the phage M13, which is then made single stranded. A synthetic OLIGONUCLEOTIDE homologous to a region of the cloned gene with the exception of a single nucleotide (that to be altered) is annealed to the single-stranded DNA. A second strand is synthesized *in vitro* using the oligonucleotide as a primer. This second strand thus contains the required mutation. The M13 phage containing the mutant gene is propagated and selected.

Hutchinson, C.A. et al. (1978) *J. Biol. Chem.* **253**, 6551.
Smith, M. (1985) *Annu. Rev. Genet.* **19**, 423–462.

oligonucleotide synthesizers Computer-controlled machines which assemble synthetic sections of DNA from the 3′ to the 5′ end by successively adding and coupling protected, activated deoxyribonucleotides (cyanoethylphosphoramidites) to a starting nucleotide anchored to the surface of a glass bead solid support. The efficiency of coupling at each stage is greater than 99.98% and oligonucleotides of length 100 are thus over 99% pure. Synthetic oligonucleotides are much used in modern molecular biology, particularly in OLIGONUCLEOTIDE-MEDIATED MUTA-GENESIS, in the POLYMERASE CHAIN REACTION for gene amplification, and in DNA CLONING, and they are the synthetic gene PROBES used to diagnose those diseases associated with mutated gene sequences.

oligosaccharide *See*: GLYCANS; GLYCOPROTEINS AND GLYCOSYL-ATION; SUGARS.

oligosaccharins Oligosaccharides, derived from fungal or plant cell walls, that exert physiological effects on higher plants at low (non-nutritional) concentrations. β-(1-3),(1-6)-Oligo-β-glucans produced from the cell walls of invading fungi by the action of plant endo-β-glucanases act as ELICITORS of PHYTOALEXIN production by higher plants as part of their defence strategy (*see* PLANT PATHOLOGY; PLANT WOUND RESPONSE). The most active elicitor is a heptasaccharide. Higher plant oligosaccharins include fragments, particularly nonasaccharides, derived from xyloglucans by the action of cellulase and which act as inhibitors of auxin- or acid-promoted growth at 10^{-7}–10^{-5} mol m^{-3}. Other xyloglucans at higher concentrations promote growth. Oligosaccharides derived from PECTINS (polymers rich in α(1-4)-linked D-galacturonic acid) act as auxin antagonists, stimulators of morphogenesis and lignification, and as inducers of proteinase inhibitor synthesis, phytoalexin production, ethylene synthesis, and the HYPERSENSITIVE RESPONSE to fungal infection. Such fragments may be derived from plant cell walls as a result of fungal attack and constitute a defence mechanism. *See also*: GLYCANS; PLANT CELL WALL MACROMOLECULES.

Aldington, S. et al. (1991) *Plant Cell Environ.* **14**, 625–636.

onc genes, onc region (1) The oncogenic genes of the T-region of the TI PLASMID and RI PLASMID of *Agrobacterium* strains, and which comprise the functions necessary to maintain the tumorous phenotype of the transformed plant cells. The main *onc* genes encode the synthesis of indoleacetic acid (an AUXIN) and isopentenyladenine (a CYTOKININ) in transformed cells.
(2) *See*: ONCOGENES.

Oncogenes

THE broadest definition of oncogenes includes any genetic entity that promotes tumour development [1,2]. Evidence for the existence of specific genes with malignant potential initially arose from studies of experimental tumour induction by oncogenic RETRO-VIRUSES. These viruses can be divided into two categories based on their tumorigenicity in animals and their ability to cause malignant TRANSFORMATION of susceptible target cells *in vitro*. The acutely transforming retroviruses induce tumours in infected

animals very rapidly (with latency periods of weeks) and are capable of causing cellular transformation *in vitro*. In contrast, the slowly transforming retroviruses require a longer latency period for tumour induction (months) and do not transform cultured cells. The genome of Rous sarcoma virus, an acutely transforming retrovirus that generates sarcomas in infected chickens, contains an additional RNA sequence of ~1500 nucleotides that is absent from the genomes of slowly transforming retroviruses. Genetic studies revealed that the extra sequence constitutes a single complementation group (denoted viral-*src* or v-*src*) which is essential for the malignant potential of Rous sarcoma virus. Other acutely transforming retroviruses were also found to harbour discrete sequences necessary for tumorigenesis. For example, the v-*myc* gene and the v-H-*ras* gene are required for the acute transforming activities of MC29 avian leukosis virus and Harvey murine sarcoma virus, respectively. To date, more than 20 distinct viral oncogenes (v-*onc* genes) have been identified within the genomes of different acutely transforming retroviruses.

Proto-oncogenes

In 1976 it was discovered that the genome of normal avian cells contains a sequence (designated cellular-*src* or c-*src*) that is nearly identical to v-*src* [3]. It soon became apparent that cellular homologues of v-*myc*, v-H-*ras*, and each of the other v-*onc* sequences are also present in vertebrate genomes. These observations confirmed the provocative notion that retroviral oncogenes have been derived from normal cellular counterparts by a process of retroviral TRANSDUCTION; during the replication of slowly transforming retroviruses rare recombinants with a dramatically increased malignant potential can emerge as the result of the acquisition of cellular (c-*onc*) sequences by the retroviral genome. Within their natural setting, the c-*onc* genes are presumed to mediate normal cellular functions that do not foment tumour development. However, on retroviral transduction these sequences acquire a novel malignant potential which is apparently responsible for the powerful tumorigenicity of acutely transforming retroviruses.

There are two general mechanisms by which transduced v-*onc* genes are rendered malignant. First, RNA transcription of v-*onc* can be influenced by regulatory sequences of the retroviral genome or by the loss of associated regulatory elements in the cellular genome; either factor might promote overexpression of v-*onc*, or inappropriate expression within a cell type in which c-*onc* transcription is normally repressed. Second, the v-*onc* gene product can be qualitatively altered as a consequence of retroviral transduction. For example, v-*onc* sequences often exhibit point mutations relative to their normal cellular homologue that encode malignant amino-acid substitutions in the viral oncoprotein.

Oncogenicity of slowly transforming retroviruses

Although the tumorigenicity of acutely transforming retroviruses can be attributed to the presence of transduced oncogenes, the more modest malignant potential of slowly transforming retroviruses, which lack v-*onc* sequences, is mediated in a different fashion. During productive retroviral infection, a double-stranded

DNA form of the replicating viral genome integrates into the host cellular genome, whereupon it is designated a provirus. Although the site of proviral integration within the cellular genome is largely random, in 1981 it was reported that >90% of B cell lymphomas induced by the slowly transforming avian leukosis virus (ALV) have the provirus inserted adjacent to the c-*myc* gene. This observation suggested that tumour induction by slowly transforming retroviruses results from the malignant activation of cellular *onc* sequences by retroviral integration at a proximal position in the host chromosome. Further support for this hypothesis soon emerged in studies of other slowly transforming retroviruses. For example, nearly 30% of breast carcinomas resulting from infection with mouse mammary tumour virus (MMTV) exhibit increased expression of the cellular *int-1* gene as a result of adjacent integration of an MMTV provirus. The long latency of tumour induction by slowly transforming retroviruses is likely to reflect the random nature of proviral integration; thus, numerous cycles of ALV replication would be required before retroviral insertion occurs within a chromosomal site appropriate for malignant activation of the c-*myc* gene.

Oncogenes and human cancer

Although oncogenic viruses are aetiological agents for some forms of human neoplasia, most human tumours show little evidence of viral involvement. Nevertheless, the study of retroviral tumour induction in animals has provided a conceptual background for the understanding of human malignancy. In particular, if cellular oncogenes can be malignantly activated by adjacent proviral integration, then it seemed feasible that these same genes might also be activated by other types of alterations that are not necessarily dependent on virus infection. A compelling demonstration of this idea emerged from gene transfer experiments conducted in 1981. Several research groups found that TRANSFECTION of genomic DNA derived from tumour cells can induce malignant transformation of recipient NIH 3T3 fibroblasts — a line of immortal but nonmalignant mouse cells. It soon became apparent that genomic DNAs from ~20% of human tumours, encompassing a wide spectrum of different malignancies, exhibit transforming activity in the NIH 3T3 transfection assay. Elaborate schemes were then developed to isolate the transforming agent from genomic tumour DNA. In this manner the NIH 3T3 transforming activity of each tumour could be attributed to a single genetic entity, and in most cases the transforming agent proved to be a member of the *ras* gene family — that is, either c-H-*ras*, c-K-*ras*, or N-*ras*. Moreover, each of the transforming *ras* genes identified by the NIH 3T3 transfection assay was found to harbour nucleotide point mutations that distinguished it from the nontransforming *ras* alleles present in normal genomic DNA (*see* GTP-BINDING PROTEINS). These point mutations generate amino-acid substitutions, often involving amino-acid residues 12 or 61, which severely alter the biochemical properties of *ras* proteins, presumably in a manner that promotes malignant cell transformation.

As two members of the *ras* gene family (c-H-*ras* and c-K-*ras*) are cellular homologues of retroviral oncogenes, the identification of altered cellular *ras* genes in human tumours provided convincing evidence to support a unifying theory of oncogenesis. According to this theory, there exist a number of genes that function in normal cellular development (i.e. the proto-oncogenes) which, if altered in an appropriate manner, can also serve to promote malignant development (Table O1). Moreover, malignant alteration of cellular proto-oncogenes can be achieved by a variety of distinct mechanisms, some of which involve viruses (e.g. retroviral transduction and retroviral integration) and some of which do not (e.g. point mutation, gene amplification, chromosomal TRANSLOCATION).

Oncogenes and chromosomal translocations

The existence of cellular proto-oncogenes had been presaged more than 20 years earlier by cytogenetic studies that revealed nonrandom chromosomal defects in cancer patients [4]. In 1960 the presence of an abnormal chromosome was reported in the malignant cells of nearly 90% of patients with CHRONIC MYELOID LEUKAEMIA (CML). Subsequent studies showed that this abnormality (the Philadelphia chromosome or Ph¹) was the product of a balanced reciprocal translocation involving chromosomes 9 and

Table O1 Some representative proto-oncogenes involved in human tumours

Proto-oncogene	Neoplasm(s)	Lesion
ABL	Chronic myelogenous leukaemia	Translocation
ERBB-1	Squamous cell carcinoma; astrocytoma	Amplification
ERBB-2 (NEU)	Adenocarcinoma of breast, ovary and stomach	Amplification
GIP	Carcinoma of ovary and adrenal gland	Point mutations
GSP	Adenoma of pituitary gland; carcinoma of thyroid	Point mutations
MYC	Burkitt's lymphoma / Carcinoma of lung, breast and cervix	Translocation / Amplification
L-MYC	Carcinoma of lung	Amplification
N-MYC	Neuroblastoma; small cell carcinoma of lung	Amplification
H-RAS	Carcinoma of colon, lung, and pancreas; melanoma	Point mutations
K-RAS	Acute myelogenous and lymphoblastic leukaemia; carcinoma of thyroid; melanoma	Point mutations
N-RAS	Carcinoma of genitourinary tract and thyroid; melanoma	Point mutations
RET	Carcinoma of thyroid	Rearrangement
ROS	Astrocytoma	?
K-SAM	Carcinoma of stomach	Amplification
SIS	Astrocytoma	?
SRC	Carcinoma of colon	?
TRK	Carcinoma of thyroid	Rearrangement

From [2].

22, designated t(9;22)(q34;q11) (*see* CHROMOSOME ABERRATIONS for explanation of terminology). In view of the strict association between the Ph^1 chromosome and CML, it was proposed that specific genes located near the translocation breakpoints of chromosome 9 and 22 are altered by the t(9;22)(q34;q11) in a manner that encourages the development of CML [5].

A second tumour-specific chromosomal abnormality, t(8;14)(q24;q32), was identified in tumour cell lines derived from patients with BURKITT'S LYMPHOMA. As cytogenetic techniques improved over the years it became apparent that most human tumour cells harbour chromosome defects, many of which are nonrandomly associated with particular types of neoplasms and thus are likely to serve as aetiological factors in the formation of those tumour types. Genetic evidence to support this supposition emerged from subsequent molecular studies of tumour-specific chromosomal abnormalities. For example, in CML patients the (9;22)(q34;q11) translocation generates a composite gene comprised of EXONS from the *bcr* locus on chromosome 22 and the c-*abl* gene on chromosome 9. The composite *bcr–abl* gene spans the chromosomal junction of t(9;22)(q34;q11) and encodes a FUSION PROTEIN with distinct biochemical properties which presumably promote CML development. It is noteworthy that c-*abl* is the cellular homologue of v-*abl*, a transduced oncogene harboured within the genome of the acutely transforming Abelson murine leukaemia virus (AbMLV).

The t(8;14)(q24;q32) of Burkitt's lymphoma also results in altered expression of a known cellular proto-oncogene. As a result of this translocation, the c-*myc* gene is transposed from its normal position on chromosome 8 into the immunoglobulin heavy chain locus on chromosome 14 (*see* IMMUNOGLOBULIN GENES). Thus studies of the (9;22)(q34;q11) and (8;14)(q24;q32) translocations provided further evidence that the cellular homologues of retroviral oncogenes can serve as important genetic determinants of human neoplasia. Moreover, the molecular analysis of other tumour-associated chromosome defects has implicated a number of additional genes in human tumorigenesis, many of which had not been recognized in previous animal studies [5].

Loss-of-function oncogenes: tumour suppressor genes

During tumour induction with acutely transforming retroviruses, the malignant character of transduced oncogenes (v-*onc*) is expressed in the presence of their normal cellular homologues (c-*onc*). This implies that v-*onc* sequences are genetically dominant over their cellular counterparts with respect to their ability to promote tumorigenesis. Likewise, many tumour-specific alterations of cellular proto-oncogenes that occur independently of viral infection (e.g., by somatic point mutation or chromosome rearrangement) are also dominant in nature. However, recent studies have demonstrated that, in addition to the proto-oncogenes that are malignantly activated by dominant gain-of-function mutations, there also exists a distinct class of genes which promote neoplasia upon loss of function. The activity of these genes was initially apparent upon analysis of somatic cell hybrids generated by fusing tumour cells with normal cells. When evaluated for several parameters of tumorigenicity, such

hybrids typically assume the nonmalignant phenotype of the normal fusion partner. However, these hybrids often give rise to tumorigenic revertants following the loss of specific chromosomes that were originally derived from the normal fusion partner. This phenomenon suggests that these chromosomes harbour genetic material that can suppress the malignant phenotype of the tumorigenic fusion partner (i.e. a TUMOUR SUPPRESSOR GENE).

Evidence for the existence of tumour suppressor genes also emerged from studies of RETINOBLASTOMA, a paediatric malignancy that occurs in two distinct forms. Patients with heritable retinoblastoma develop multiple tumours involving both eyes, and they transmit a gene for susceptibility to retinoblastoma (the *RB* gene) to half their offspring in a Mendelian fashion (*see* MENDELIAN INHERITANCE). In contrast, patients with sporadic retinoblastoma do not have a family history of the disease, and they only develop an isolated tumour in one eye. In order to explain these phenomena Alfred Knudson proposed that two genetic mutations are required for retinoblastoma formation. According to Knudson's hypothesis, the sporadic form of the disease arises on the rare occasion when both mutations occur somatically in a single retinal cell. However, patients with heritable retinoblastoma have already acquired one mutation by germ-line transmission from an affected parent, and thus are susceptible to tumour formation when a second mutation occurs somatically in any developing retinal cell. Inheritance of the germ-line mutation (i.e. the *RB* susceptibility allele) is presumably responsible for the familial association and multiple tumours characteristic of patients with heritable retinoblastoma.

The nature of the *RB* allele was suggested by cytogenetic studies which revealed interstitial deletions involving chromosomal band 13q14 in a significant proportion of retinoblastoma patients. Hence, the development of retinoblastoma seems to be associated with the loss of genetic material at 13q14. If so, then the two mutations required for retinoblastoma might represent the inactivation of both alleles of a single genetic locus at 13q14 — the *RB* susceptibility gene. Molecular genetic studies, including the isolation of the *RB* gene and its characterization in patients with either the sporadic or heritable disease, confirm that inactivation of both *RB* alleles is necessary for the formation of retinoblastoma. Hence, the normal *RB* gene serves as a tumour suppressor that contributes to oncogenesis after loss of function. Recent studies have led to the identification of distinct tumour suppressor genes that play a role in other malignancies, again as a result of the loss of gene function. These include the *WT1* and *NF1* loci implicated in WILMS' TUMOUR and NEUROFIBROMATOSIS, respectively.

p53

Loss or mutation of the p53 locus, another gene with tumour suppressor activity, is perhaps the most common genetic lesion observed in human neoplasia. Alterations of the p53 gene are frequently found in a number of different malignancies, including those responsible for much human mortality (e.g. breast, colon, and lung carcinomas). Moreover, germ-line transmission of mutant p53 alleles seems to be responsible for inheritance of Li-Fraumeni syndrome (LFS), a rare familial cancer condition.

Individuals afflicted with LFS are highly susceptible to a spectrum of different childhood and adult neoplasms. Hence, loss of p53 function can promote formation of diverse tumour types and clearly contributes to a significant proportion of human cancers.

Cellular proliferation and oncogenesis

Two major hypotheses have been advanced to explain the excessive cellular proliferation that drives tumour development. According to these, malignant cell growth results either from the abrogation of control mechanisms that normally limit cellular division (*see* CELL CYCLE) or from the failure of malignant cells to differentiate normally into mature noncycling progeny. It is not surprising, therefore, that the protein products of many proto-oncogenes participate in signal transduction pathways that regulate normal cell growth or differentiation. These pathways have a variety of extracellular, transmembrane, cytoplasmic, and nuclear components, any one of which might be subject to a malignant alteration that promotes tumour development. Accordingly, the products of proto-oncogenes include extracellular GROWTH FACTORS (e.g. c-*sis*), transmembrane GROWTH FACTOR RECEPTORS (c-*erbB*, c-*fms*), cytoplasmic regulatory proteins (c-*src* and the *ras* proteins), and nuclear factors that control RNA transcription (c-*myc*, c-*jun*, c-*fos*) or cell cycle progression (*PRAD1*). Further analysis of oncogenes and their protein products should reveal how the genetic lesions associated with malignancy engender the phenotypic features of tumour initiation and tumour progression [6].

R.J. BAER

See also: CHEMICAL CARCINOGENS AND CARCINOGENESIS; MUTATION; PAPOVAVIRUSES.

1 Cooper, G.M. (1990) *Oncogenes* (Jones and Bartlett, Boston).
2 Bishop, J.M. (1991) Molecular themes in oncogenesis. *Cell* **64**, 235–248.
3 Stehelin, D. et al. (1976) DNA related to the transforming gene(s) of avian sarcoma viruses is present in normal avian DNA. *Nature* **260**, 170–173.
4 Nowell, P.C. & Hungerford, D.A. (1960) Chromosome studies on normal and leukemic human leukocytes. *J. Natl Cancer Inst.* **25**, 85–109.
5 Rabbitts, T.H. (1991) Translocations, master genes and differences between the origins of acute and chronic leukemias. *Cell* **67**, 641–644.
6 Hunter, T. (1991) Cooperation between oncogenes. *Cell* **64**, 249–270.

oncogenesis The generation and development of a malignant tumour.

oncogenic Capable of causing a malignant tumour, applied usually to viruses.

oncomodulin *See*: CALCIUM-BINDING PROTEINS.

oncoprotein Protein encoded by an ONCOGENE.

oncoviruses *See*: RETROVIRUSES.

one gene–one enzyme hypothesis An idea first formulated by Beadle in 1945, now a cornerstone of molecular biology. It defined the unitary relationship between the elements of selection (GENES) and biochemical effectors (ENZYMES) and was originally based on mutational analysis of *Drosophila* eye colour and *Neurospora* biochemistry. It has since been extended into the one gene–one polypeptide hypothesis which broadens the concept to include all polypeptides, irrespective of function.

ONPG *o*-Nitrophenyl-β-D-galactopyranoside. An artificial substrate of the enzyme β-GALACTOSIDASE. When the substrate is cleaved, a yellow product (*o*-nitrophenol) is formed which is easily assayed spectrophotometrically at 420 nm. It is used as a histochemical stain in conjunction with a β-galactosidase REPORTER GENE in studies of development, location of gene expression, etc.

***ons* gene** Opine secretion gene. A gene in the T-region of the TI PLASMID that is involved in OPINE secretion in transformed plant cells. The 6a transcript of the ONC REGION.

ontogeny A synonym for embryonic development, the term ontogeny is most often associated with the statement, since discredited, that 'ontogeny recapitulates phylogeny' (Ernst Haeckel). By this, Haeckel referred to the observation that the early embryo of the higher vertebrates appears to go through a series of transitions whose morphology is similar to that of the embryos of primitive vertebrates. Thus the early mouse or chick embryo possesses primitive gill slits (pharyngeal or branchial arches), which disappear later in development.

oocyte A stage in the development of the female gamete during which the PRIMORDIAL GERM CELL develops into an egg. The oocyte is a large cell, possessing a large nucleus called the germinal vesicle, in which the chromosomes are arrested in MEIOSIS. The ooplasm comprises all non-nuclear cytoplasmic contents of the oocyte, including mRNA, ribosomes, mitochondria, yolk proteins, and other energy substrates and components of the cytoarchitecture. These can be either synthesized endogenously or transported into the oocyte after synthesis in other organs or accessory cells surrounding the ovary. *See also*: OOGENESIS; XENOPUS OOCYTE EXPRESSION SYSTEM.

oogenesis The process whereby descendants of diploid PRIMORDIAL GERM CELLS of the female (set aside early in development and contained within a specialized organ, the ovary) develop into eggs capable of fusing with haploid male gametes (sperm) and directing early development. Details vary between species but the process involves MEIOSIS to form a haploid egg and synthesis and assembly of a complex cytoarchitecture and organelles capable of directing early development.

In vertebrates, primordial germ cells migrate to the gonad and form oogonia, which undergo a period of mitotic proliferation and enter the first meiotic division as primary oocytes. In some animals, oogonia remain as a stockpile throughout life. In the human female however, all oogonia have entered meiosis by birth and there is no such reserve. In invertebrates, most mitotically active oogonia develop into nurse cells or other accessory cells which accumulate materials for transfer into the developing oocyte.

The primary oocyte is a large cell with a large nucleus (germinal vesicle) which enters a phase of growth and assembly of intracellular components. Primary oocytes remain arrested in prophase of the first meiosis for periods of from a few months to up to many years (in humans). DNA replicates, yielding two chromatids from each homologous paired chromosome. This is the growth phase of oogenesis, and primary oocytes acquire their external coats and cortical granules, and accumulate ribosomes, mRNA, mitochondria, tRNA, yolk, glycogen, and lipid. These components can be synthesized endogenously by the oocyte (in some species intense RNA synthesis is associated with decondensation of the still-paired chromosomes and formation of lateral loop 'lampbrush chromosomes'). Other components may be synthesized outside the oocyte in accessory cells of the ovary (follicle cells, NURSE CELLS) and transferred directly into the oocyte, or synthesized in other organs (e.g. liver), carried to the ovary by the bloodstream and transported into the oocyte by receptor-mediated ENDOCYTOSIS.

Maturation is the transformation of the oocyte into an egg which is capable of being fertilized. This process is triggered by a hormone from the soma, such as progesterone in frogs, or luteinizing hormone in mammals, and only occurs once the animal has become sexually mature. Oocytes are arrested at prophase I of meiosis, but at maturation the envelope surrounding the large oocyte nucleus breaks down and the chromosomes proceed to metaphase II of meiosis. Completion of this first meitoic division generates a secondary oocyte and a (much smaller) first polar body. Chromosomes recondense, the nuclear membrane breaks down and the replicated homologous chromosomes separate or are allocated to either the secondary oocyte or the first polar body. The second meiotic division is also asymmetric and produces the mature haploid ovum and a (smaller) second polar body. The stage at which the egg is released from the ovary (ovulation) and fertilized varies between species.

oogonia *See*: OOGENESIS.

ooplasm *See*: OOCYTE.

opa genes Genes encoding the opacity polypeptides of the BACTERIAL ENVELOPE of Neisseria gonorrhoea. *See*: BACTERIAL PATHOGENICITY.

opal codon The TERMINATION CODON UGA.

open circles A form of PLASMID circular DNA in which the supercoils have been released by the nicking of one of the strands of the duplex. *See also*: DNA SUPERCOILING.

open reading frame (ORF) A length of DNA or RNA sequence between an ATG (or AUG) translation start signal (INITIATION CODON) and a TERMINATION CODON, which can be potentially translated into a polypeptide sequence. The existence of an ORF in a region of DNA, however, does not constitute proof that the sequence is part of a GENE.

operator A region upstream of a bacterial gene (or operon) to which a REPRESSOR protein binds to modulate expression of that gene (or operon). *See*: BACTERIAL GENE EXPRESSION.

operon Unit of gene expression in bacteria in which a set of structural genes are transcribed as a single transcript under the control of a neighbouring regulatory gene. *See*: BACTERIAL GENE EXPRESSION; BACTERIAL GENE ORGANIZATION.

opiate receptors G PROTEIN-COUPLED RECEPTORS for opiate drugs, found in the central nervous system and on peripheral tissues. Their endogenous ligands are the opioid peptides. *See*: OPIOID PEPTIDES AND THEIR RECEPTORS.

opines Generic name for a diverse group of molecules synthesized in CROWN GALL cells. The genes for their synthesis in plants are derived from pathogenic bacteria of the genus *Agrobacterium*, which can also utilize the molecules as a sole source of carbon and nitrogen. With the exception of the AGROCINOPINES, all opines identified to date are imino acids formed by reactions on Schiff bases formed between amino acids and sugars or keto acids. *See*: AGROPINE; CUCUMOPINE; LEUCINOPINE; LYSOPINE; MANNOPINE; NOPALINE; OCTOPINE; SUCCINAMOPINE. *See also*: Ti PLASMID.

Opioid peptides and their receptors

THE mammalian opioid peptides were discovered and first isolated in 1975 as endogenous ligands of the opiate receptors in the central and peripheral nervous systems [1]. The three major classes of mammalian opioid peptides so far identified are the endorphins, the enkephalins, and the dynorphins, each class being liberated from a separate inactive precursor polypeptide by post-translational processing enzymes [2–4]. Opioid peptides are present in both brain and peripheral tissues and have been implicated in the control of pain, responses to stress, and numerous other functions (see below).

The polypeptide precursors of the endorphins, enkephalins, and dynorphins are the products of three distinct genes, prepro-opiomelanocortin (POMC), preproenkephalin A (PENK), and pre-proenkephalin B (more commonly known as preprodynorphin, PDYN), respectively. Each precursor contains a short 20–25 residue hydrophobic SIGNAL PEPTIDE at the N terminus which, when cleaved by a signal peptidase in the endoplasmic reticulum (*see* PROTEIN TRANSLOCATION), results in the liberation of the prohormone containing the biologically active opioid peptides (Fig. O4) [5]. Opioid peptide sequences are located primarily between pairs of dibasic amino acids (e.g. Lys-Arg or Lys-Lys) which allows them to be generated through enzymatic digestion of the precursor at these recognition sites by intracellular endopeptidases, which appear to be either aspartyl or serine proteinases [5,6]. The propeptide convertases PC1 (also known as PC3), PC2, PC4, PC5, furin, and PACE4 are members of a family of SERINE PROTEINASES homologous to the yeast Kex2 proteinase

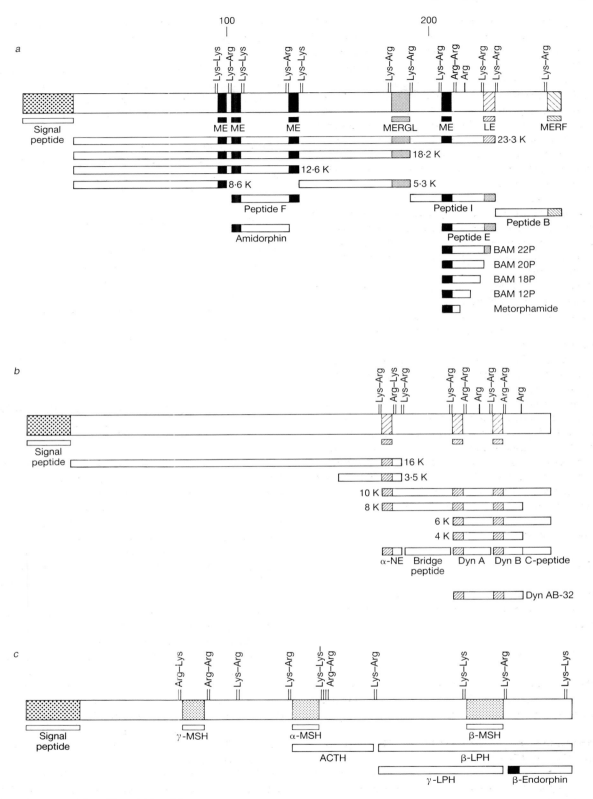

Fig. O4 Schematic representation of the opioid peptide precursors. *a*, Preproenkephalin A; *b*, preprodynorphin; *c*, prepromelanocortin. The locations of component peptides bounded by paired basic amino acid residues are indicated. Relative molecular mass $\times 10^{-3}$ of each peptide is given (e.g. 23.3K).

and structurally related to bacterial subtilisin. These endopeptidases appear to be definitively involved in the tissue- and precursor-specific processing of the opioid precursor proteins and neuropeptides [6–8]. Opioid peptides can also be generated through endoproteolytic cleavage at single, usually Arg, basic residues, this additional type of recognition site being most commonly observed in the prodynorphin precursor. Endopeptidase activity is then usually followed by carboxypeptidase removal of the remaining cleavage site basic residue from the C terminus of the molecule. A single enzyme appears to be responsible for this function although it has been given several names: carboxypeptidase E (CPE) or H (CPH) and enkephalin convertase [5].

The opioid peptides produced from each precursor all possess the same N-terminal enkephalin pentapeptide sequence, Tyr-Gly-Gly-Phe-Met (or Leu), which is essential for activation of the different types of opioid receptor. Once released from the cell, opioid peptides are then exposed to a variety of extracellular enzymes that can either completely inactivate each peptide or, by a modulatory action at single, not necessarily basic, amino acid recognition sites, cleave the parent molecule to yield a shorter fragment that may retain biological activity but demonstrates a quite different affinity and selectivity profile for the opioid receptors [4,5,9].

Opioid precursor processing is not uniform, with tissue specificity of processing pathways, mainly through differential expression of processing enzymes, an important determinant of the heterogeneity observed in the nature of the major end-products derived from each precursor [2–9]. This is exemplified in particular by some tissues displaying almost complete processing of the precursor to yield a high proportion of the bioactive opioid peptides whereas other tissues exhibit more incomplete processing and, consequently, demonstrate much lower precursor to opioid peptide ratios. Even when processing is complete, the end-products may differ between tissues as a result of having been produced through separate pathways. POST-TRANSLATIONAL MODIFICATIONS such as phosphorylation, N-acetylation, C-amidation, sulphation, and glycosylation also demonstrate tissue specificity although, when found to occur, the modified peptides are almost exclusively inactive at opioid receptors.

This entry outlines the three major mammalian opioid peptidergic systems and their receptors. Further information will be found in [1–31] and references therein.

Proenkephalin A

This is the physiological precursor for the [Met⁵]- and [Leu⁵]-enkephalins (Fig. O4*a*). The sequence of the complete 267 amino acid precursor (preproenkephalin) has been deduced from cloned cDNA. It has an M_r of 29 786 and consists of a 24 amino acid signal peptide at the N terminus which is cleaved to yield the PENK precursor molecule. PENK itself consists of a non-opioid N-terminal region termed synenkephalin with the sequences of the enkephalin-containing opioid peptides located in the remaining portion of the molecule (Fig. O4*a*) [3,10]. The human preproenkephalin gene contains four exons separated by three introns and has some homology with the POMC and prodynor-

phin precursor genes. It has been localized to the distal half of the long arm of chromosome 8 (q23 – q24). The rat gene is similar in organzation but contains only three exons. An mRNA of around 1.2 kb is produced from the human gene, of which 801 bases are translated into preproenkephalin [3,10].

Distribution

In the central nervous system (CNS), neuronal perikarya containing PENK mRNA have been shown by *in situ* HYBRIDIZATION studies to be widely distributed [11,12], being found at virtually all levels of the neuraxis in nuclei associated with mesencephalic, diencephalic, and telencephalic structures, in brainstem nuclei, and in the dorsal horn of the spinal cord. Enkephalin-containing neurons form both local circuits and long-term projections, with a particularly dense fibre-terminal network innervating the globus pallidus. While the predominant source of PENK is neuronal, measurable levels of the precursor mRNA have also been found in testis and ovary, and in cultured glial cells.

Processing

Complete proteolytic processing of preproenkephalin would yield four copies of [Met⁵]-enkephalin (ME) and one copy each of [Leu⁵]-enkephalin (LE) and the C-terminal extended forms of [Met⁵]-enkephalin-Arg⁶-Phe⁷ (MERF) and [Met⁵]-enkephalin-Arg⁶-Gly⁷-Leu⁸ (MERGL) (Fig. O4; Table O2). In bovine and

Table O2 Amino acid sequences of PENK-derived mammalian opioid peptides

Peptide	Amino acid sequence
[Met⁵]-enkephalin	Tyr-Gly-Gly-Phe-Met
[Leu⁵]-enkephalin	Tyr-Gly-Gly-Phe-Leu
MERF (heptapeptide)	Tyr-Gly-Gly-Phe-Met-Arg-Phe
(MERGL) (octapeptide)	Tyr-Gly-Gly-Phe-Met-Arg-Gly-Leu
Peptide E	Tyr-Gly-Gly-Phe-Met-Arg-Arg-Val-Gly-Arg-Pro-Glu-Trp-Trp-Met-Asp-Tyr-Gln-Lys-Arg-Tyr-Gly-Gly-Phe-Leu
BAM-22P*	Tyr-Gly-Gly-Phe-Met-Arg-Arg-Val-Gly-Arg-Pro-Glu-Trp-Trp-Met-Asp-Tyr-Gln-Lys-Arg-Tyr-Gly
Metorphamide	Tyr-Gly-Gly-Phe-Met-Arg-Arg-Val-NH₂
Peptide F	Tyr-Gly-Gly-Phe-Met-Lys-Lys-Met-Asp-Glu-Leu-Tyr-Pro-Leu-Glu-Val-Glu-Glu-Glu-Ala-Asn-Gly-Gly-Glu-Val-Leu-Gly-Lys-Arg-Tyr-Gly-Gly-Phe-Met
Amidorphan	Tyr-Gly-Gly-Phe-Met-Lys-Lys-Met-Asp-Glu-Leu-Tyr-Pro-Leu-Glu-Val-Glu-Glu-Glu-Ala-Asn-Gly-Gly-Glu-Val-Leu-NH₂

* BAM-12P and 20P represent the first 12 and 20 amino acid residues from the N terminus.

human brain, the predominant end-products of proteolytic processing of PENK appear to be ME, LE, MERF, and MERGL, which all exhibit a parallel distribution with the highest levels found in the globus pallidus followed by the caudate-putamen and hypothalamus, with low levels in the cortex and cerebellum [3,5].

In the bovine adrenal medulla, the situation is more complicated. Processing does not seem to be complete and although there is a relatively high abundance of end-products, larger molecular weight opioid peptide fragments, many with biological activity, are prevalent. These include peptide F (M_r 3800), which contains a copy of [Met^5]-enkephalin at both its N and C termini, peptide E (M_r 3200), which contains a single copy of [Met^5]- and [Leu^5]-enkephalin at its N and C termini respectively, and the BAM peptides 12P, 20P and 22P, and metorphamide (Table O2; Fig. O4) [3,5].

Candidate intracellular endopeptidases involved in the proteolytic processing of PENK include the propeptide convertase PC1 [7,8] and two other serine proteinases: an adrenal trypsin-like enzyme (ATLE), and IRCM-1, a POMC-processing enzyme from the pituitary. Both appear to have specificity for paired basic amino-acid residues and have been shown to process PENK-derived intermediates correctly. In addition, an endopeptidase has been detected in bovine adrenal chromaffin granules that cleaves either at pairs of basic residues or at single basic sites with smaller peptides of 20 amino acids or fewer being more efficiently processed. The larger active PENK-derived peptides may also be converted to the free enkephalin pentapeptides by an extracellular system of proteolytic enzymes that includes angiotensin-converting enzyme (ACE) and endopeptidase 24.15.

Degradation

PENK-derived peptides are degraded in a extracellular pathway involving predominantly neutral endopeptidase 24.11 (NEP, also known as enkephalinase) and aminopeptidase N (APN), with a possible involvement of a dipeptidylaminopeptidase (DAP) [9]. Peptides larger than ME, LE, and MERF do not appear to be substrates for NEP and APN.

The physiological significance of these enzymes in enkephalin metabolism has been demonstrated by the antinociceptive activity of inhibitors of the enzymes. Kelatorphan, a non-specific inhibitor of NEP, APN, and DAP, has pronounced antinociceptive effects against acute and chronic noxious stimuli, especially when they are associated with inflammation. Kelatorphan-mediated antinociception is reversed by naloxone and the δ-opioid receptor antagonist ICI 174,864, the latter observation being consistent with kelatorphan-induced elevated extracellular levels of endogenous enkephalins.

Regulation of gene expression

A variety of stimuli affect tissue-specific PENK gene expression [10,13], including membrane depolarization and the activation of hormone receptors. Receptor-mediated increases in PENK mRNA accumulation are predominantly associated with a rise in free CALCIUM within the cell. The release of Ca^{2+} from intracellular stores is mediated by INOSITOL PHOSPHATES, particularly IP_3, produced by a G-protein activated signal transduction pathway (*see* SECOND MESSENGER PATHWAYS). Although the mechanism by which Ca^{2+} regulates PENK gene expression is not well understood, activation of a CA^{2+}/CALMODULIN-DEPENDENT PROTEIN KINASE may be involved. Stimulation of PROTEIN KINASE C by DIACYLGLYCEROL, another product of the phosphoinositide pathways, has also been implicated in the regulation of PENK gene expression.

In bovine adrenal chromaffin granules, a marked increase in PENK mRNA levels has also been produced by a rise in intracellular cyclic AMP.

Prodynorphin

Dynorphin A was originally isolated from pituitary extracts and sequenced as a peptide of 17 amino acids contaning [Leu^5]-enkephalin at the N terminus. Simultaneously with the discovery of dynorphin A, α-neoendorphin, a decapeptide also containing [Leu^5]-enkephalin at the N terminus, was isolated and sequenced from hypothalamic extracts. Subsequently, several other [Leu^5]-enkephalin extended peptides were isolated from brain and pituitary (Table O3; Fig. O4*b*) [4].

The human prodynorphin (PDYN) gene was cloned from a hypothalamic cDNA library and has four exons separated by three introns with the preprodynorphin precursor encoded in exons 3 and 4. A similar structural organization has been observed in the rat. The gene is transcribed into an mRNA of around 2.4 kb, which is translated to give a precursor protein of 248 amino acids consisting of a signal peptide at the N terminus with the remaining portion of the molecule containing the active opioid peptides (Fig. O4*b*) [4].

Distribution

Immunocytochemical and/or *in situ* hybridization studies have demonstrated a widespread distribution of PDYN-synthesizing perikarya and neuronal projections containing PDYN-derived peptides at all levels of the neuraxis from the most rostral areas of the telencephalon to the brainstem and spinal cord [11,12]. In many regions there is considerable overlap between PDYN and PENK pathways although cells containing PENK mRNA are substantially more abundant and found in many more areas than those containing PDYN mRNA. Cells expressing PDYN mRNA are relatively abundant in several hypothalamic nuclei (supraoptic and paraventricular nuclei), basal ganglia (nucleus accumbens and caudate-putamen), limbic system (granular layer of the dentate gyrus), midbrain periaqueductal grey (PAG), pontine parabrachial nucleus (PBN), medullary nucleus tractus solitarius (NTS) and spinal trigeminal nucleus in the brainstem, and in the dorsal horn of the spinal cord. Fibres and terminals containing PDYN-derived peptides form short local connections in addition to longer, rostro-caudal projections. Regions receiving a particularly dense innervation include the globus pallidus, substantia nigra (pars reticulata), region CA_2/CA_3 and dentate gyrus of the hippocampus, as well as the neural lobe of the pituitary. In many instances, PDYN-derived peptides are co-localized with other

Table O3 Amino-acid sequences for PDYN- and POMC-derived mammalian opioid peptides

Peptide	Amino acid sequence
Prodynorphin-derived peptides	
Dynorphin A*	Tyr-Gly-Gly-Phe-Leu-Arg-Arg-Ile-Arg-Pro-Lys-Leu-Lys-Trp-Asp-Asn-Gln
α-neoendorphin	Tyr-Gly-Gly-Phe-Leu-Arg-Lys-Tyr-Pro-Lys
β-neoendorphin	Tyr-Gly-Gly-Phe-Leu-Arg-Lys-Tyr-Pro
Rimorphin (Dynorphin B)	Tyr-Gly-Gly-Phe-Leu-Arg-Arg-Gln-Phe-Lys-Val-Val-Thr
Leumorphin (Dynorphin B-29)	Tyr-Gly-Gly-Phe-Leu-Arg-Arg-Gln-Phe-Lys-Val-Val-Thr-Arg-Ser-Gln-Glu-Asp-Pro-Asn-Ala-Tyr-Tyr-Glu-Glu-Leu-Phe-Asp-Val
Dynorphin AB-32	Tyr-Gly-Gly-Phe-Leu-Arg-Arg-Ile-Arg-Pro-Lys-Leu-Lys-Trp-Asp-Asn-Gln-Lys-Arg-Tyr-Gly-Gly-Phe-Leu-Arg-Arg-Gln-Phe-Lys-Val-Val-Thr
POMC-derived peptides	
Human β-endorphin†	Tyr-Gly-Gly-Phe-Met-Thr-Ser-Glu-Lys-Ser-Gln-Thr-Pro-Leu-Val-Thr-Leu-Phe-Lys-Asn-Ala-Ile-Ile-Lys-Asn-Ala-Tyr-Lys-Lys-Gly-Glu
γ-endorphin	Tyr-Gly-Gly-Phe-Met-Thr-Ser-Glu-Lys-Ser-Gln-Thr-Pro-Leu-Val-Thr-Leu
α-endorphin	Tyr-Gly-Gly-Phe-Met-Thr-Ser-Glu-Lys-Ser-Gln-Thr-Pro-Leu-Val-Thr

* Dynorphin A (1–8) represents the first eight residues of Dynorphin A.
† In porcine β-endorphin Tyr 27 is replaced by His and Glu 31 is replaced by Gln.

neurotransmitters including monoamines, peptides, and amino acids, with the potential for acting as either co-transmitters or neuromodulators.

Processing

Post-translational proteolytic processing [4,5] of the prodynorphin precursor at all available dibasic recognition sites (i.e. Arg-Arg as well as Lys-Arg and Arg-Lys) would liberate three copies of LE. However, this rarely happens and neither the PDYN precursor nor the dynorphin opioid peptides are considered to serve as precursors for LE which is derived almost exclusively from the processing of PENK.

Intracellular endoproteolytic cleavage of PDYN at Lys-Arg residues, probably by a serine proteinase member of the pro-peptide family, most likely either PC1 or PC2, yields only the larger peptides including α-neoendorphin, bridge peptide, dynorphin A_{1-17}, and leumorphin, in addition to other C-terminally extended forms of [Leu^5]-enkephalin (Table O3;

Fig. O4). In contrast, additional end-products of PDYN processing — dynorphin A_{1-8}, rimorphin and dynorphin AB_{32} — are produced by extensive cleavage of the precursor at single Arg residues (Fig. O4). A THIOL PROTEINASE displaying just such a property and designated dynorphin-converting enzyme (DCE), has been identified and purified from rat brain and bovine pituitary secretory granules. DCE has been shown to cleave both Dyn A_{1-17} and leumorphin at a single Arg residue to produce Dyn A_{1-8} and rimorphin (or Dyn B), respectively. In most cases endopeptidase cleavage at dibasic recognition sites is followed by the rapid removal of the remaining basic residue by carboxypeptidase E (CPE). A notable exception to this, however, is the production of β-neoendorphin from α-neoendorphin by CPE removal of the C-terminal Lys, a step which appears to take place very slowly. Consequently it would appear that CPE is a rate-limiting enzyme in the production of β-neoendorphin. In addition to the smaller active peptides derived from the PDYN precursor molecule, larger molecular weight intermediates are present in relatively high abundance in certain tissues such as anterior pituitary (Fig. O4).

Pro-opioimelanocortin

The pro-opioimelanocortin (POMC) precursor is unique amongst the opioid peptide precursor molecules in that it contains only one copy of a sequence relating to an active opioid peptide — β-endorphin$_{1-31}$ (β-EP) (Table O3; Fig. O4c). POMC also contains the sequences of several non-opioid peptides: the stress hormones ADRENOCORTICOTROPIC HORMONE (ACTH) and α-, β- and γ-MELANOCYTE-STIMULATING HORMONE (MSH). The presence of such diverse peptide species within a common precursor, in combination with their distribution throughout the hypothalamo-pituitary–adrenal axis and in discrete CNS pathways associated with the modulation of nociception, has suggested that POMC may form a link between endogenous pain control and stress response systems [2].

The human POMC gene comprises 7665 bp with three exons separated by two introns. A similar organization is found in bovine, mouse, and rat genes. The human POMC gene is transcribed into an mRNA of around 1.2 kb which is translated into the 267 amino acid precursor protein [2,10,14].

Distribution

The POMC gene is predominantly expressed in the pituitary but also in a few brain areas, particularly the arcuate nucleus of the hypothalamus (ARH) and the medullary nucleus tractus solitarius (NTS) [11,12]. POMC-containing neurons in the ARH project to several nuclei of the limbic system and hypothalamus, thalamus and the mesencephalic structures of the periaqueductal grey and nuclei associated with the rostral ventromedial medulla (RVM). The principal POMC neuronal projections from the NTS are local either within the medulla oblongata or to the parabrachial nucleus (PBN) and spinal cord. POMC mRNA has been demonstrated in many peripheral tissues of the gastrointestinal tract, and reproductive and immune systems.

Processing

The endoproteolytic processing of the POMC precursor is the most extensively characterized of all the opioid prohormones [2,6,8,15]. The principal mammalian proteinases responsible for the cleavage of the inactive precursor molecule at paired basic amino acid residues are the propeptide convertases PC1 and PC2, which have been detected almost exclusively in neurons and within POMC-enriched areas of the CNS [6,8,15].

POMC-derived peptides are found in substantially higher quantities in the pituitary, in both the anterior and intermediate lobes, than in either the brain or adrenals. Tissue-specific processing of POMC in the anterior and neurointermediate lobes yields variable amounts of intermediate products in relation to the principal active end-products. In the anterior lobe, endoproteolytic processing of POMC is predominantly at Lys-Arg to produce the intermediates β-lipotropin, containing the opioid peptide β-endorphin$_{1-31}$ (β-EP) at the C terminus, and a protein (M_r 22 000) containing ACTH (Fig. O4). However, although β-EP and ACTH are both produced following subsequent cleavage at the remaining available Lys-Arg sites, the processing in the anterior lobe is much less complete than in the neurointermediate lobe. Anterior lobe ratios of β-lipotropin and β-endorphin of 2 : 1 have been detected.

The principal products of POMC processing in the rat adeno-hypophysis are β-lipotropin, β-EP, ACTH, and a peptide of M_r 16 000 (Fig. O4). Species differences in the amino-acid sequence of POMC can result in an additional Lys-Arg dibasic pair, at which site cleavage will produce the stress hormone β-MSH as well as β-EP in the primate anterior pituitary (Fig. O4). In addition, further endoproteolytic cleavage at the Lys-Lys dibasic pair in β-EP yields β-EP$_{1-27}$ (Table O3), a fragment that displays agonist properties in peripheral tissues but has also been reported to be an antagonist of β-EP function in the brain and spinal cord.

Carboxypeptidase E activity then reduces β-EP$_{1-27}$ to β-EP$_{1-26}$. Further post-translational processing within β-EP has been identified at the Leu^{17}-Phe^{18} dipeptide to produce γ-endorphin, a weak opiate agonist, with the possible additional removal of Leu^{17} to yield α-endorphin.

In the neurointermediate lobe, POMC is cleaved at all available dibasic amino acid sites (i.e. Lys-Lys as well as Lys-Arg), yielding significantly higher amounts of the smaller opioid, as well as non-opioid, peptides relative to the larger molecular weight intermediate proteins. In the case of the opioid peptide end-products, full processing of β-lipotropin in the neurointermediate lobe yields β-EP and, subsequently, produces higher amounts of β-EP$_{1-27}$ and β-EP$_{1-26}$ as well as relatively more α- and γ-endorphin. β-EP and its fragments then undergo further processing by N-acetylation with both the acetylated and non-acetylated forms of β-EP$_{1-27}$ and β-EP$_{1-26}$ considered to be the predominant endorphins in the neurointermediate lobe. Indeed, due to this more complete processing of POMC, only negligible amounts of β-EP$_{1-31}$ are detected in intermediate lobe melanotrophs. Both N-acetylated forms have only negligible activity at opioid receptors (Table O4).

POMC processing in the brain has many qualitative similarities to that observed in the pituitary, with more complete processing of the precursor in more rostral brain regions, including the ARH, and much lower precursor to opioid peptide ratios observed in caudal regions, including the NTS and spinal cord. In the ARH, processing of β-lipotrophin to β-EP is virtually complete with β-EP$_{1-27}$ and β-EP$_{1-26}$ as major end-products. These are mainly nonacetylated, whereas a much higher degree of post-translational N-acetylation occurs in the NTS and spinal cord.

Regulation of gene expression

In the anterior pituitary, POMC gene expression is markedly increased following exposure of animals to various types of stress including chronic footshock and cold water swim stress. Enhanced gene expression appears to be mediated principally by the hormone corticotropin-releasing hormone (CRH) which is released from the hypothalamus in response to stress and causes an increase in the rate of POMC gene transcription [10,14]. CRH-mediated enhancement of POMC transcription can be inhibited by voltage-sensitive calcium channel antagonists, suggesting that Ca^{2+} is involved in gene regulation. The increase in intracellular Ca^{2+} is probably a consequence of the stimulation of adenylate cyclase activity by CRH. POMC gene expression in the adenohypophysis is also under negative feedback control of adrenal steroids. Several TRANSCRIPTION FACTORS involved in the regulation of POMC gene expression have been identified.

Opioid receptors

Opiates and opioid peptides inhibit neuronal activity in both CNS and peripheral tissues through interaction with opioid receptors [16,17]. Three major types of opioid receptors — μ, δ, and κ — have been unambiguously distinguished by radioligand binding assays, *in vitro* pharmacological assays on isolated smooth muscle preparations, electrophysiological and neurochemical assays, and behavioural models [18,19]. A fourth type, the epsilon (ε) receptor, has been described, but the data are still widely regarded as being insufficient to support a separate existence from the μ-receptor. The receptors have been subdivided into different subtypes ($μ_1$ and $μ_2$, $δ_1$ and $δ_2$, etc.) but these divisions remain equivocal. The notion of an opioid receptor complex composed of distinct but physically interacting μ-, δ- and κ-binding sites has been put forward [20]. Receptors may be located on either presynaptic (mainly κ but also μ) or postsynaptic (μ and δ) membranes.

Activation of all opioid receptors leads to an inhibition of neurotransmitter release although the precise effect depends on the type of neurotransmitter and the CNS region involved [21]. The principal membrane electrical consequences of opioid receptor activation, which may represent underlying mechanisms for inhibition of neuronal activity, are either the μ- or δ-mediated increased outward potassium current, or a decrease in voltage-sensitive calcium conductance which can be mediated through μ, δ, and κ receptors [16].

Opioid receptors have also been shown to be linked to adenylate cyclase and phospholipase C second messenger systems [22]. However, while μ- and δ-mediated inhibition of adenylate cyclase has been clearly demonstrated, the evidence for a κ-receptor modulation of activity of either adenylate cyclase or phospholipase C is more controversial.

The overall physiological consequences of opioid receptor activation of second messenger systems remain to be resolved but the inhibition of either cAMP or IP_3 and diacylglycerol accumulation, and the subsequent effects on PROTEIN KINASE activity, do not seem to be related to the inhibition of Ca^{2+} and K^+ conductance [16].

Opioid μ, δ, and κ receptors all appear to be functionally linked to these different effector systems through direct interaction with a G protein (*see* G PROTEIN-COUPLED RECEPTORS; GTP-BINDING PROTEINS) [23]. In most cases, whether the effector is an ion channel or intracellular enzyme, the effects of opioid agonists can be blocked entirely by pretreatment with PERTUSSIS TOXIN (PTx). This implies that all three types of receptor are coupled to the effector system through the the pertussis-sensitive G_i and/or G_o proteins. However, any interaction between the κ-receptor and the inositol phosphate transduction pathway through phospholipase C would suggest coupling via the PTx-insensitive G_p protein. There is also some evidence that opioid receptors, in certain circumstances, may be coupled to G_s, leading to stimulation of adenylate cyclase and increased production of cAMP.

Sustained or repeated exposure to opioid peptides or nonpeptide opiates can lead to the development of tolerance and dependence. It has been proposed that the molecular mechanisms underlying these phenomena may include a functional uncoupling of the opioid receptor–G protein complex.

The first cloning of an opioid receptor, the δ receptor, has only recently been achieved, by EXPRESSION CLONING using a cDNA library from the neuroblastoma × glioma hybrid cell line NG108-15 [24,25]. The cDNA encodes a protein of 371 amino acids containing the predicted seven-transmembrane spanning hydrophobic domains associated with the G protein-coupled receptor superfamily. Subsequently, expression cloning of rat and mouse brain cDNA libraries has isolated clones encoding respectively the μ (398 amino acids) and κ (380 amino acids) opioid receptors [26,27].

Various other opioid-binding proteins have been isolated as possible opiate receptors, but none has so far been unequivocally identified with a pharmacologically identified receptor [28–30].

Receptor selectivities of opioid peptides

In general the basic structural requirements for an opioid peptide to be able to activate a receptor are a phenolic aromatic amino acid at position 1 and a nonphenolic aromatic amino acid at position 4 [19]. The importance of the N-terminal tyrosine in opioid peptides is clearly demonstrated by the lack of affinity or efficacy displayed by des-tyrosyl analogues and peptides in which the N-terminal tyrosine has been modified.

The activities of the various opioid peptides are illustrated in Table O4.

Table O4 Mammalian opioid precursor derived peptide affinities and selectivities for μ, δ and κ-binding sites

	Peptide selectivity ratio			Binding affinity (K_i, nM)		
	μ	δ	κ	μ	δ	κ
Proenkephalin-derived peptides						
[Leu^5]-enkephalin	19	1.2	8210	16	1.0	6841
[Met^5]-enkephalin	9.5	0.9	4442	11	1.0	4936
[Met^5]-enkephalyl-Arg-Gly-Leu	6.6	4.8	79	1.4	1.0	16
[Met^5]-enkephalyl-Arg-Phe	3.7	9.4	3	1.0	2.5	25
Metorphamide	0.12	2.7	0.25	1.0	23	2.1
BAM 12	0.16	1.4	0.41	1.0	8.8	2.6
BAM 22	0.10	0.66	0.17	1.0	6.6	1.7
Peptide E	0.53	1.7	1.1	1.0	3.2	2.1
Peptide F	25	29	78	1.0	1.2	3.1
Prodynorphin-derived peptides						
Dynorphin A	0.73	2.4	0.12	6.1	20	1.0
Dynorphin A (1–8)	3.8	5.0	1.3	2.9	3.8	1.0
Dynorphin B	0.68	2.9	0.12	5.7	24	1.0
α-neoendorphin	1.2	0.57	0.20	6.0	2.9	1.0
β-neoendorphin	6.9	2.1	1.2	5.8	1.8	1.0
POMC-derived peptides						
β-endorphin (human)	2.1	2.4	96	1.0	1.1	46
β-endorphin (pig)	1.9	1.3	26	1.5	1.0	20
β-endorphin (1–27) (pig)	3.0	2.4	185	1.3	1.0	77
N-acetyl-β-endorphin (human)	2800	3000	N.D.	1.0	1.1	—

Adapted from [17].

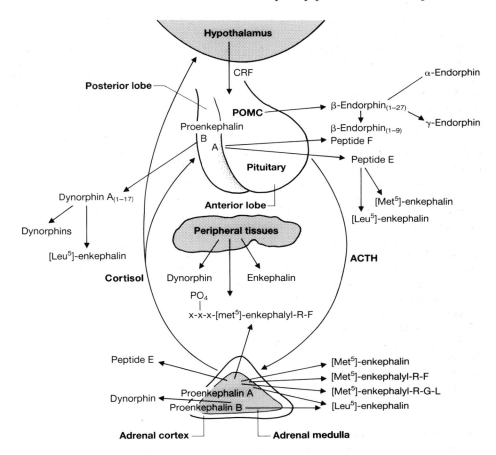

Fig. O5 Potential sources of circulating opioid peptides.

Possible roles of endogenous opioid peptides

Circulating opioid peptides (Fig. O5), particularly endorphins and enkephalins released from the adenohypophysis and adrenal medulla respectively, do not appear to have a direct physiological role. Rather, their main effect seems to be a hormonal modulatory influence on neuroendocrine control of stress and reproductive function. Endogenous opioid peptides have also been suggested to act as local regulators of paracrine and/or autocrine function in several peripheral organs and tissues, specifically those associated with the gastrointestinal tract, pancreas, kidneys and heart, and reproductive and immune systems (*see* [31] and references therein).

While the pituitary and adrenal medulla may be considered as the major peripheral sources of endogenous opioid peptides, the blood–brain barrier remains relatively resistant to penetration by such hydrophobic molecules. Opioid peptide modulation of CNS events, therefore, appears to be through localized release either from projection neurons to, or interneurons within, a specific area of the brain or spinal cord. Opioid peptides are contained in a widespread network of fibres and terminals, in many instances being co-localized with other neurotransmitters including monoamines, peptides, and amino acids, with the potential role of acting as either co-transmitters or neuromodulators.

Opioid peptides have been implicated in either the mediation or modulation of a multiplicity of CNS and peripheral functions including acute and chronic nociception and stress responses; regulation of the cardiovascular, respiratory, neuroendocrine, and immune systems; gastrointestinal motility and intestinal ion and gastric acid secretion; and thermoregulation. They may also exert an influence on many patterns of behaviour including food consumption, learning and memory, and mood and motivational processes [31].

Although few examples exist of the opioid peptidergic neuronal network exercising tonic control under normal conditions, the system seems to become activated under pathophysiological conditions. Involvement of opioid peptides either directly or indirectly in the pathophysiology of several common neurological (e.g. epilepsy, brain and spinal cord injury, Alzheimer's dementia), psychiatric (e.g. schizophrenia), and addictive (e.g. drug and alcohol addiction) disorders has been suggested [31].

However, with the possible exception of the potent anticonvulsant ($\mu + \kappa$) and neuroprotective (κ) properties of opioid agonists in the treatment of experimental epilepsy and cerebral ischaemia respectively, evidence for the involvement of opioid systems in many of these disorders is at best equivocal.

J.C. HUNTER

1 Hughes, J. et al. (1975) Identification of two related pentapeptides from the brain with potent opiate agonist activity. *Nature* **258**, 577–579.
2 Young, E. et al. (1992) Proopiomelanocortin biosynthesis, processing and secretion: functional implications. *Handbk Exp. Pharm.* (Herz, A., Ed.) **104**, Part I, 393–421.

3 Rossier, J. (1992) Biosynthesis of enkephalins and proenkephalin-derived peptides. *Handbk Exp. Pharm.* (Herz, A., Ed.) **104**, Part I, 423–447.

4 Day, R. et al. (1992) Prodynorphin biosynthesis and posttranslational processing. *Handbk Exp. Pharm.* (Herz, A., Ed.) **104**, Part I, 449–470.

5 Fricker, L.D. (1992) Opioid peptide processing enzymes. *Handbk Exp. Pharm.* (Herz, A., Ed.) **104**, Part I, 529–545.

6 Seidah, N.G. (1993) Mammalian paired basic amino acid convertases of prohormones and proproteins. *Annl N.Y. Acad. Sci.* **680**, 135–146.

7 Zhou, A. & Lindberg, I (1992) Purification and characterization of the prohormone convertase PC1 (PC3). *J. Biol. Chem.* **268**, 5615–5623.

8 Schafer, M.K-H. (1993) Gene expression of prohormone and proprotein convertases in the rat CNS: a comparative *in situ* hybridization analysis. *J. Neurosci.* **13**, 1258–1279.

9 Roques, B.P. et al. (1992) Peptidase inactivation of enkephalins: design of inhibitors and biochemical, pharmacological, and clinical applications. *Handbk Exp. Pharm.* (Herz, A., Ed.) **104**, Part I, 547–584.

10 Höllt, V. (1992) Regulation of opioid peptide gene expression. *Handbk Exp. Pharm.* (Herz, A., Ed.) **104**, Part I, 307–346.

11 Khatchaturian, H. et al. (1992) Anatomy and function of the endogenous opioid systems. *Handbk Exp. Pharm.* (Herz, A., Ed.) **104**, Part I, 471–497.

12 Masour, A. & Watson, S.J. (1992) Anatomical distribution of opioid receptors in mammalians: an overview. *Handbk Exp. Pharm.* (Herz, A., Ed.) **104**, Part I, 79–105.

13 Kley, N. & Leoffler, J.P. (1992) Molecular mechanisms in proenkephalin gene regulation. *Handbk Exp. Pharm.* (Herz, A., Ed.) **104**, Part I, 471–497.

14 Roberts, J.L. et al. (1992) Regulation of pituitary proopiomelanocortin gene expression. *Handbk Exp. Pharm.* (Herz, A., Ed.) **104**, Part I, 347–377.

15 Zhou, A. et al. (1993) The prohormone convertases PC1 and PC2 mediate distinct endoproteolytic cleavages in a strict temporal order during pro-opiomelanocortin biosynthetic processing. *J. Biol. Chem.* **268**, 1763–1769.

16 North, R.A. (1992) Opioid actions on membrane ion channels. *Handbk Exp. Pharm.* (Herz, A., Ed.) **104**, Part I, 773–797.

17 Corbett, A.D. et al. (1992) Selectivity of ligands for opioid receptors. *Handbk Exp. Pharm.* (Herz, A., Ed.) **104**, Part I, 645–679.

18 Chang, K-J. (1984) Opioid receptors: multiplicity and sequelae of ligand–receptor interactions. In *The Receptors* (Conn, P.M., Ed.), Vol. 1, 1–81 (Academic Press, New York).

19 Rees, D.C. & Hunter, J.C. (1990) Opioid receptors. *Comp. Med. Chem.* (Hansch, C. et al., Eds), Vol. 3, 805–846 (Pergamon, Oxford).

20 Rothman, R.B. et al. (1992) Allosteric coupling among opioid receptors: evidence for an opioid receptor complex. *Handbk Exp. Pharm.* (Herz, A., Ed.) **104**, Part I, 217–237.

21 Mulder, A.H. & Schoffelmeer, A.N.M. (1992) Multiple opioid receptors and presynaptic modulation of neurotransmitter release in the brain. *Handbk Exp. Pharm.* (Herz, A., Ed.) **104**, Part I, 125–144.

22 Childers, S.R. (1992) Opioid receptor-coupled second messenger systems. *Handbk Exp. Pharm.* (Herz, A., Ed.) **104**, Part I, 189–216.

23 Cox, B.M. (1992) Opioid receptor–G protein interactions: acute and chronic effects of opioids. *Handbk Exp. Pharm.* (Herz, A., Ed.) **104**, Part I, 145–188.

24 Kieffer, B.L. et al. (1992) The δ-opioid receptor: isolation of a cDNA by expression cloning and pharmacological characterisation. *Proc. Natl. Acad. Sci. USA* **89**, 12048–12052.

25 Evans, C.J. et al. (1992) Cloning of a δ opioid receptor by functional expression. *Science* **258**, 1952–1955.

26 Chen, Y. et al. (1993) Molecular cloning and functional expression of a μ-opioid receptor from rat brain. *Mol. Pharmacol.* **44**, 8–12.

27 Yasuda, K. et al. (1993) Cloning and functional comparison of κ and δ opioid receptors from mouse brain. *Proc. Natl. Acad. Sci. USA* **90**, 6736–6740.

28 Simon, E.J. & Gioannini, T.L. (1992) Opioid receptor multiplicity: isolation, purification, and chemical characterization of binding sites. *Handbk Exp. Pharm.* (Herz, A., Ed.) **104**, Part I, 3–26.

29 Goldstein, A. (1992) Expression cloning of cDNA encoding a putative opioid receptor. *Handbk Exp. Pharm.* (Herz, A., Ed.) **104**, Part I, 27–36.

30 Smith, A.P. et al. (1992) Characterization of opioid-binding proteins and other molecules related to opioid function. *Handbk Exp. Pharm.* (Herz, A., Ed.) **104**, Part I, 37–52.

31 Herz, A. (Ed.) (1992) *Handbk Exp. Pharm.* **104**, Part II, 3–823.

opsins The protein portions of vertebrate visual pigments such as RHODOPSIN and the CONE PIGMENTS. They are seven-helix membrane spanning proteins to which is bound the chromophore 11-*cis*-retinal via a Schiff base linkage to a specific lysine residue. Opsins with slightly differing amino-acid sequences yield visual pigments with maximal absorption at different wavelengths (*see* CONE PIGMENTS). *See*: VISUAL TRANSDUCTION. *See also*: BACTERIO-RHODOPSIN.

opsonization The process whereby a particle (often a bacterium) is rendered more attractive for ingestion by phagocytic white blood cells by becoming coated with specific ANTIBODY and/or COMPLEMENT. This process is one of the most important to the effective function of the humoral immune system as a defence against infection and is mediated by IgG and IgM antibodies and some complement components, especially C3b. *See*: ENDOCYTOSIS.

optical diffraction This is used with electron microscope images of regularly repeating objects to assess the quality of an image and the extent to which structural information at high RESOLUTION is contained within it. Such assessment can be both less subjective and more easily quantifiable when applied to the DIFFRACTION pattern of an object than to its direct image. A photographic negative from a microscope can be digitized and processed computationally by the techniques of image analysis to obtain a diffraction pattern from one or more areas within the photograph. Optical diffraction is an analogue technique, equivalent to these digital methods, which enables the observer to discriminate visually between diffraction patterns having more or less high resolution information, and thus to identify those regions of an image which are most highly ordered. This remains a more efficient method for the rapid identification of the location and boundaries of well preserved areas of an electron micrograph than its computational equivalent.

An optical diffractometer typically illuminates part of a micrograph by coherent light from a laser, under conditions for viewing the diffraction pattern and with facilities for both scanning the micrograph to illuminate successive regions and also masking an area of variable size and shape. *See also*: ELECTRON MICROSCOPY.

optical rotatory dispersion (ORD) The dispersion (or variation with wavelength) of optical rotatory power of a solution of a chiral molecule, having a handedness or ABSOLUTE CONFIGURATION.

optical rotatory power The rotation of a plane of polarized light by a solution of a chiral molecule.

ORD OPTICAL ROTATORY DISPERSION.

ORF OPEN READING FRAME.

organelle An intracellular membrane-bound structure in the eukaryotic cell, usually specialized for a particular function, for example CHLOROPLAST; ENDOPLASMIC RETICULUM; GOLGI APPARATUS; LYSOSOME; MITOCHONDRION; NUCLEUS. *See*: ANIMAL CELL; PLANT CELL.

organizer (1) Region of an embryo which determines an axis of development. The amphibian organizer was discovered by H. Mangold and H. Spemann in 1924, and consists of the MESODERM located at the dorsal lip of the BLASTOPORE. *See*: AMPHIBIAN DEVELOPMENT.
(2) NUCLEOLAR ORGANIZER.
(3) Microtubule organizing centres. *See*: MICROTUBULES.

origin of replication (ori) Specific site in DNA at which DNA REPLICATION is initiated. Usually designated *ori*. Bacterial and phage chromosomes have a single origin of replication; eukaryotic chromosomes have multiple origins. *See also*: ADENOVIRUS; ARS ELEMENT; CHROMOSOME STRUCTURE.

ornithine carbamoyltransferase (or transcarbamylase) deficiency Also known as ornithine transcarbamylase (OTC) deficiency. An X-LINKED DOMINANT condition due to deficiency of the urea cycle enzyme ornithine carbamoyltransferase (EC 2.1.3.3) and characterized by severe hyperammonaemia. It commonly presents at birth and the features, which include vomiting, irritability, seizures and coma, are frequently precipitated by protein-rich feeds.

ornithine cycle mutations Ornithine is a non-protein AMINO ACID synthesized from arginine, which has a role in the urea cycle, in the synthesis of proline, creatine, and polyamines, and has its own mitochondrial transport system. Diseases associated with hyperornithinaemia include gyrate atrophy (ornithine-oxoacid aminotransferase deficiency) and hyperornithinaemia/hyperammonaemia/homocitrullinaemia (due to deficiencies in mitochondrial transport). ORNITHINE CARBAMOYLTRANSFERASE DEFICIENCY belongs to the urea cycle mutations.

orthogonal field agarose gel electrophoresis (OFAGE) A type of pulsed-field gel ELECTROPHORESIS used for separating very large pieces of DNA in an agarose gel. The electrical field is switched between two sets of electrodes, each set at an angle to one another. This technique allows the resolution of whole chromosomes.

orthograde movement *See*: ANTEROGRADE MOVEMENT.

orthologous *See*: HOMOLOGY.

Orthomyxoviridae Family of enveloped RNA ANIMAL VIRUSES to which INFLUENZA VIRUS belongs. *See*: ANIMAL VIRUS DISEASES.

Oryza sativa Rice.

osmium tetroxide A fixative for ELECTRON MICROSCOPY.

osteoblast, osteoclast, osteocyte *See*: BONE DEVELOPMENT.

osteocalcin *See*: BONE GLA PROTEINS; CALCIUM-BINDING PROTEINS; EXTRACELLULAR MATRIX MOLECULES.

osteogenesis imperfecta A disease with a frequency of one in 10 000, characterized by bone fragility, often accompanied by other connective tissue abnormalities (blue sclerae, deafness, abnormal teeth). The phenotype varies greatly in severity from neonatal death to occasional fractures in adult life. It is inherited in AUTOSOMAL DOMINANT or RECESSIVE fashion, or may arise as a new mutation. It results from a defect in either the α1(1) or α1(2) collagen genes. *See also*: EXTRACELLULAR MATRIX MOLECULES.

osteoid *See*: BONE DEVELOPMENT.

osteonectin (SPARC, BM40) CALCIUM-BINDING PROTEIN first identified in bone, but which is also very abundant in EXTRACELLULAR MATRIX of many other tissues. Cells synthesizing it include osteoblasts, fibroblasts, and endothelial cells. It is a small GLYCOPROTEIN (M_r 43 000, polypeptide M_r 33 000), is phosphorylated, and binds several calcium ions including at least one at high affinity ($K_d = 10^{-6.5}$ M). Calcium binding induces a large conformational change. The sequence includes a very acidic N-terminal domain, but unlike the BONE GLA PROTEINS contains no γ-carboxyglutamic acid (Gla), has a cysteine-rich EGF-like domain (*see* GROWTH FACTORS), and a single 'EF hand' domain (*see* CALCIUM-BINDING PROTEINS) at the C terminus. The osteonectin homologue SC1 is an extracellular protein from brain.

osteopontin *See*: EXTRACELLULAR MATRIX MOLECULES.

OTC deficiency *See*: ORNITHINE CARBAMOYLTRANSFERASE DEFICIENCY.

otosclerosis An AUTOSOMAL DOMINANT condition causing conductive or mixed hearing loss in the second to fifth decades due to overgrowth of bone in the labyrinth of the ear. May be found histologically in up to 10% of individuals, but the PENETRANCE in affected pedigrees is low (40%) and the clinical signs may often remain undetected.

ouabain A steroid glycoside (Fig. O6) derived from the seeds of

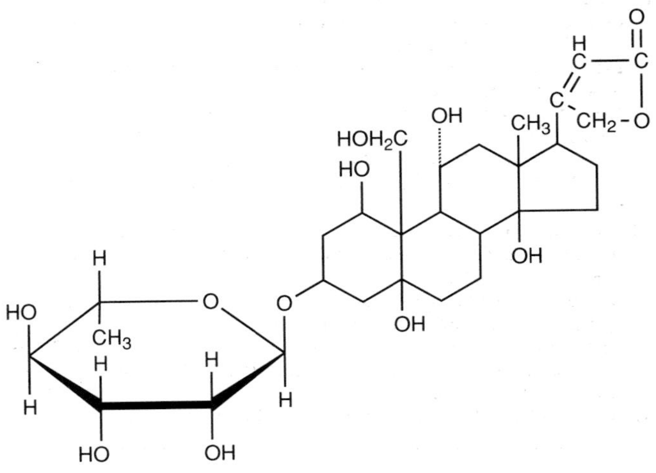

Fig. O6 Ouabain (β-L-rhamnosyl strophanthidin G).

Strophanthus gratus and also known as G-strophanthin. Ouabain and related steroids are known as cardiotonic steroids because of their stimulatory effects on the heart. Ouabain is a powerful inhibitor of the Na^+,K^+-ATPase pump (*see* MEMBRANE TRANSPORT). It inhibits Na^+/K^+ exchange by preventing dephosphorylation of the ATPase, and thus inactivating the ion pump. Ouabain contains a sugar residue and is therefore also known as a cardiac glycoside, although the sugar is not involved in inhibition.

ovalbumin Small protein (M_r 40 000) abundant in egg white and which is a nonenzymatic member of the SERPINS.

overdrive A DNA sequence of around 24 base pairs found near right border repeats of the TI PLASMID, and which appears to stimulate the initiation of T-strand production from the right border sequence.

overlapping genes Type of GENOME organization found in certain DNA BACTERIOPHAGES (such as ΦX174 and G4) and in other viruses in which parts of the genome can be read in more than one reading frame to give two, or rarely three different proteins.

oxidases OXIDOREDUCTASE enzymes in which O_2 is the electron acceptor.

β-oxidation The metabolic pathway in mitochondria and peroxisomes by which the coenzyme A (CoA) thioesters of fatty acids are progressively oxidized to acetyl-CoA, consisting of successive oxidation by flavin adenine dinucleotide, hydration, oxidation by NAD, and thiolysis by CoA. In mitochondria the reduced coenzymes are reoxidized by the ELECTRON TRANSPORT CHAIN with concomitant production of ATP, and the product, acetyl-CoA is further oxidized by the TRICARBOXYLIC ACID CYCLE.

oxidative phosphorylation The process by which the chemical energy released by the oxidation of reduced carbon compounds is used to drive the condensation of ADP and inorganic phosphate to form ATP. Oxidative reactions catalysed by the respiratory chain are coupled to ATP synthesis by transmembrane movements of protons (*see* ELECTRON TRANSPORT CHAIN). The arrangement of respiratory chain components in the mitochondrial inner membrane results in the extrusion of protons coupled to electron transfer, generating an electrochemical H^+ gradient. The return of protons down this gradient is channelled through the enzyme ATP SYNTHASE. The integrity of the membrane is essential for oxidative phosphorylation. Lipophilic weak acids and other compounds which make the membrane permeable to protons act as uncouplers of oxidative phosphorylation as ATP synthesis cannot be driven in the absence of a proton electrochemical gradient and the energy released by respiratory chain redox reactions is lost as heat. *See also*: CHEMIOSMOTIC THEORY.

oxidoreductases One of the main classes of enzymes (EC class 1) in the EC classification. They catalyse oxidoreduction reactions and include dehydrogenases, reductases, and oxidases.

oxygen evolution complex (OEC) An inappropriately named group of three peripheral THYLAKOID polypeptides (M_r 33 000, 23 000, and 16 000) which is associated with PHOTOSYSTEM II in chloroplasts and have an indirect role in oxygen evolution (*see* PHOTOSYNTHESIS). They are the products of the *psbO*, *psbP* and *psbQ* genes, respectively. Only the M_r 33 000 protein is known to occur in CYANOBACTERIA, and it stabilizes the Mn complex which is essential for water oxidation and is probably bound to interhelix loops of D1 and D2 on the luminal side of the membrane. The M_r 23 000 and 16 000 proteins increase the affinity of the water oxidation system for the essential ions Ca^{2+} and Cl^-.

oxyhaemoglobin The oxygenated form of HAEMOGLOBIN, in which one oxygen molecule is noncovalently bound to each of the four haem moieties via the iron atom.

oxytocin A polypeptide of nine amino acids (Fig. O7) released by the posterior pituitary gland. Its two known physiological roles are to stimulate uterine contraction during childbirth, and to cause milk ejection from the breast by contraction of the myoepithelial cells in response to suckling (let-down reflex). Its function in the male is unclear.

Cys — Tyr — Ile — Gln — Asn — Cys — Pro — Leu — Gln

Fig. O7 Oxytocin.

P

P The single-letter abbreviation for the AMINO ACID proline.

P blood group system BLOOD GROUP system consisting of a dominant P1 antigen (a determinant on a GLYCOLIPID), an allelic null antigen (P2), and rare NONALLELIC antigens. It is rarely of clinical importance except in the autoimmune disease paroxysmal cold haemoglobinuria, in which the Donath-Landsteiner antibody is an anti-P IgG. The structures of some antigens are known, borne variously on glycoproteins and/or glycosphingolipids. P^k is

Galα1,4Galβ1,4Glc-

in a ceramide trihexoside (*see* GLYCOLIPIDS) and P^l is

Galα1,4Galβ1,4GlcNAc-

in several types of molecule.

P element A TRANSPOSABLE GENETIC ELEMENT found in DROSOPHILA. It was discovered by investigators studying the phenomenon of hybrid dysgenesis. When males of certain strains, called P strains, are mated to females of another group of strains, called M, their offspring are often sterile. Offspring of the reciprocal cross are normal. This hybrid dysgenesis is caused by the activation of P element transposition in the GERM LINE of the hybrid offspring in the M cytoplasmic environment. The P element is 3 kb long and comprises a pair of 31-bp inverted repeats at its termini, which are required for integration, and a gene encoding transposase, the enzyme catalysing excision and integration. P elements are prevalent in wild strains of *Drosophila willistoni* but have rarely, if ever, been found in wild *D. melanogaster* strains until recently. It has been suggested that this transposon may have been horizontally transmitted from *D. willistoni* to *D. melanogaster* via parasitic mites. The technique of germ-line transformation in *Drosophila* (P ELEMENT-MEDIATED TRANSFORMATION) has been developed using P-element based VECTORS.

P element-mediated enhancer detection A technique for the identification of regulatory sequences within the DROSOPHILA genome which uses a P ELEMENT transposon containing the gene for *Escherichia coli* β-GALACTOSIDASE under the control of a weak constitutive promoter. On transformation into the *Drosophila* genome, if the element inserts near a transcriptional ENHANCER, this may stimulate the expression of the β-galactosidase gene in the same pattern of distribution as the endogenous gene normally regulated by the enhancer. The expression of β-galactosidase is easily detected using a histological stain. The method allows the *Drosophila* genome to be screened for genes expressed in particular patterns, tissues, or developmental stages.

P element-mediated transformation A technique for the stable integration of exogenous DNA into the GERM LINE of *Drosophila*. Vectors based on transposable P ELEMENTS containing sequences allowing integration together with a selectable marker gene are used to carry exogenous DNA. The P-element vector together with a helper plasmid, which provides a source of transposase, are injected into embryos at the syncytial blastoderm stage before pole cell formation. Adults deriving from injected embryos are bred, and individuals carrying integrated copies of the vector can be identified on the basis of the selectable marker, usually an eye colour gene. This technique has been of great importance for the *in vivo* analysis of gene function and regulation. *See also*: TRANSPOSABLE GENETIC ELEMENTS.

P face of lipid bilayer In FREEZE-FRACTURED membranes, the face representing the hydrophobic interior of the cytoplasmic half of the lipid bilayer.

P-glycoprotein transport system Membrane glycoprotein (M_r 170 000) encoded by the *MDR1* gene (in humans) which confers MULTIDRUG RESISTANCE on human cell lines and tumours, but whose normal biological function is unknown. The nucleotide sequence indicates six pairs of transmembrane domains and a cluster of potential N-linked glycosylation sites near the N terminus. It is an ATP-dependent active transporter and effects multidrug resistance by pumping hydrophobic drugs out of the cell. It also seems to form a chloride channel. It is a member of the ABC (ATP-binding cassette) superfamily of transporters which includes many bacterial transport systems, the putative peptide transporter from the MHC and the product of the cystic fibrosis transmembrane regulator (*see* CYSTIC FIBROSIS). *See also*: MEMBRANE TRANSPORT SYSTEMS.

P granules Granules in the egg of the nematode *Caenorhabditis elegans* which may represent CYTOPLASMIC DETERMINANTS with a role in the specification of GERM CELL fate. Using antibodies specific for P granules, it appears that they are initially spread evenly throughout the cytoplasm of the unfertilized egg. Just before the first cleavage, they move to the posterior end of the cell and are inherited by only one of the two daughter cells. This segregation is repeated at subsequent divisions such that the P granules end up in only those cells that will form the gametes. The molecular basis of their movement is unknown but it depends on some asymmetric property of the actin-based cytoskeleton.

P pilus *See*: BACTERIAL PATHOGENICITY.

P-site Site on ribosome to which the initiator aminoacyl-tRNA binds. Subsequently, elongation of the polypeptide chain takes place by repetitive cycles of transfer of the growing chain from peptidyl-tRNA bound in the P-site to incoming aminoacyl-tRNA bound in the A-site. *See:* PROTEIN SYNTHESIS; RIBOSOME.

P-type ATPases A family of ATPASES located in the membranes of eukaryotic and prokaryotic cells and which mediate the active transport of various ions across the membrane. They are distinct from the ATP SYNTHASE involved in OXIDATIVE PHOSPHORYLATION. They take their name from a common mechanism of action which involves phosphorylation of a conserved aspartate residue. Ion-dependent ATP hydrolysis and transient phosphorylation of the protein at this site changes the conformation of the protein, allowing ion transport across the membrane. The cycle is completed by dephosphorylation. In ATPases that transport two different ions in opposite directions across the membrane (e.g. Na^+, K^+-ATPase), the dephosphorylation reaction allows transport of the second ion. If exposed to an unusually steep and reversed gradient of the ion normally transported, P-type ATPases can act in reverse to generate ATP from ADP and inorganic phosphate. *See:* CA^{2+}-ATPASE; NA^+,K^+-ATPASE; H^+,K^+-ATPASE; K^+-ATPASE.

P1 phage A temperate BACTERIOPHAGE of *Escherichia coli*. When P1 infects *E. coli* it produces about 100 copies of itself but ~1% of the phage progeny are defective phage and carry a piece of the host's genomic DNA. The P1 phage can therefore be used to transfer random *E. coli* DNA fragments to genetically marked host strains and this process of TRANSDUCTION can be used to generate fine-structure GENETIC MAPS. More recently, P1 has been developed as a system for cloning large fragments (85–100 kb) of genomic DNA. The P1 phage has considerably greater insert capacity than either phage LAMBDA (10–20 kb) or COSMIDS (40–50 kb). YEAST ARTIFICIAL CHROMOSOMES (YACs) can accommodate larger inserts (several hundred kilobase) but an advantage of the P1 cloning system is that a bacterial host is used, and this can be a strain defective in restriction enzyme activity. Rearrangement or deletion of methylated genomic DNA inserts is therefore prevented.

Sternberg, N. (1990) *Proc. Natl. Acad. Sci. USA* **87**, 103–107.

P22 phage *See:* BACTERIOPHAGES; LAMBDOID PHAGES.

P40 A former name for interleukin-9 (IL-9).

p53 A DNA-binding TRANSCRIPTION FACTOR containing an acidic activation domain, which has been identified as a tumour suppressor protein. Apart from activating transcription from PROMOTERS with p53 binding sites, it has been reported to suppress transcription from various promoters lacking p53 binding sites and to inhibit helicase activity and DNA replication. Increased levels of p53 arrest cells in the G1 phase of the CELL CYCLE, and it has also been found to induce APOPTOSIS after DNA damage, to inhibit tumour cell growth, and to preserve genetic stability. Its binding by a variety of viral oncoproteins such as the

SV40 large T antigen (*see* PAPOVAVIRUSES) may be involved in mediating the oncogenic action of these viruses. TRANSGENIC MICE in which p53 function has been deleted have been produced, showing that it is not essential for normal development, but these animals do have an increased incidence of tumours. Studies on p53 in human tumours suggest that in tumours, cells lacking p53 function proliferate faster than those neoplastic cells containing one or two normal p53 alleles. *See also:* ONCOGENES; TUMOUR SUPPRESSOR GENES.

P450 *See:* CYTOCHROME P450.

P680 SPECIAL PAIR chlorophyll of PHOTOSYSTEM II reaction centre. *See also:* PHOTOSYNTHESIS.

P700 The SPECIAL PAIR of chlorophyll *a* molecules that forms the primary electron donor of the reaction centre of PHOTOSYSTEM I (*see* PHOTOSYNTHESIS). The apoprotein to which P700 is noncovalently bound is a heterodimer of closely related subunits called I and Ia or, preferably, A (M_r 83 000) and B (M_r 82 400), which are encoded by the *psaA* and *psaB* genes. The holoprotein contains one P700 per heterodimer and, in addition, about 100 light-harvesting chlorophyll *a* molecules (ANTENNA PIGMENTS).

P840, P870, P930 SPECIAL PAIR chlorophylls of the photosynthetic reaction centres of green sulphur bacteria, *Rhodobacter sphaeroides*, and *Rb. viridis* respectively. *See also:* PHOTOSYNTHESIS.

PA (1) PHASEIC ACID.
(2) PHYTOALEXIN.

pachytene A stage of MEIOSIS.

packaging The process whereby a newly replicated VIRUS or BACTERIOPHAGE genome is sequestered, or packaged, inside a protein shell (capsid), thus forming the mature virion. This process, also known as encapsidation, occurs towards the end of a productive virus or bacteriophage infection and is essential for propagation of the virus. *See also:* COSMIDS; LAMBDA; P1 PHAGE.

paclobutrazol 2*RS*,3*RS*-1-(4-Chlorophenyl)-4,4-dimethyl-2-(1,2,4-triazol-1-yl)pentan-3-ol (Fig. P1). M_r 293.5. A PLANT GROWTH RETARDANT. It causes severe dwarfing when applied to plants at $>10^{-6}$ M due predominantly to a reduction in their endogenous GIBBERELLIN content. The growth regulatory activity resides primarily with the 2*S*, 3*S*-enantiomer, which inhibits the

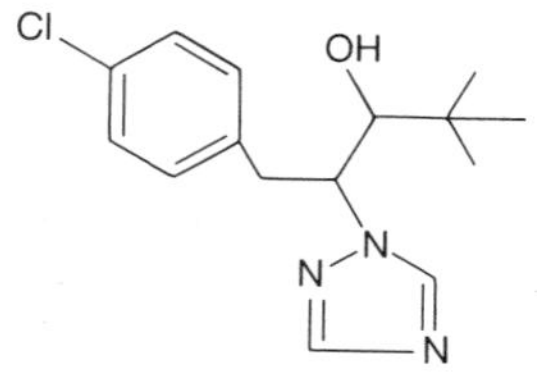

Fig. P1 Structure of paclobutrazol.

oxidation of *ent*-kaurene to *ent*-kaurenoic acid with an I$_{50}$ of $\sim 2 \times 10^{-8}$ M. The 2R,3R-enantiomer has some fungicidal activity as a result of blocking 14α-demethylation in the fungal sterol biosynthetic pathway. It is used commercially to restrict the growth of ornamental plants and fruit trees.

Hedden, P. & Graebe, J.E. (1985) *J. Plant Growth Regul.* **4**, 111–122.

PAGE Polyacrylamide gel ELECTROPHORESIS.

PAH POLYCYCLIC AROMATIC HYDROCARBON.

PAI Plasminogen activator inhibitor. *See:* SERPINS.

pair-distribution of a molecule The distribution of distances, *R*, between all atom pairs regardless of direction, each distance weighted by the product of the scattering lengths of the atoms in each pair to produce D(*R*). The pair-distribution function 'picture' is the PATTERSON FUNCTION 'picture' spun about its centre uniformly to all orientations (into and out of the paper), integrating to give a one-dimensional plot in *R*. *See:* Fig. N12 in NEUTRON SCATTERING AND DIFFRACTION.

pair-rule genes Class of SEGMENTATION GENES in which mutations give rise to defects in every other segment of the developing embryo. These genes are required to establish the 14 PARASEGMENTAL repeat units and are expressed in seven-stripe patterns at the blastoderm stage. The primary pair-rule genes — *hairy*, *even-skipped*, and *runt* — are required for the specification of the patterns of the other, secondary, genes. The characterized pair-rule genes all encode TRANSCRIPTION FACTORS and their activities are required for the patterning of the segment polarity and HOMEOTIC GENES. *See also:* DROSOPHILA DEVELOPMENT.

paired box A conserved sequence of 387 bp, encoding the paired domain, which was discovered in a set of *Drosophila* genes with sequence similarity to the pair-rule segmentation gene, *paired*. The protein products of these genes are likely to act as TRANSCRIPTION FACTORS and some also contain a homeodomain (*see* HOMEOBOX GENES AND HOMEODOMAIN PROTEINS). Paired box genes have been detected in a wide variety of species from nematode to humans. In the mouse, eight paired box (Pax) genes have been isolated, each expressed in a distinct spatiotemporal pattern during embryogenesis. Three Pax genes have been associated with developmental mutants. The Pax-1 gene corresponds to the genetic locus, *undulated*, which gives rise to vertebral defects. The Pax-3 gene is associated with the human Waadenburg syndromes which arise from cranial NEURAL CREST deficiencies, and Pax-6 corresponds to the mouse mutation *Small eye*, and the human locus responsible for the eye defect, aniridia.

PAL PHENYLALANINE AMMONIA-LYASE.

palindrome In double-stranded nucleic acids a palindromic sequence refers to a region in which the sequence on both strands is identical when read in an antiparallel direction (i.e. both strands read 5′ → 3′ or vice versa). This is illustrated by the recognition sequence for the RESTRICTION ENZYME *Eco*RI:

```
5′  |   3′
G A A| T T C
C T T |A A G
3′  |   5′
```

The upper strand, when read 5′-3′ (left to right) is GAATTC. The lower strand read 5′-3′ (right to left) is also GAATTC. The recognition sites for many type II restriction enzymes are of this kind. A palindromic sequence possesses an axis of dyad symmetry (or twofold rotational symmetry) as shown by the dotted line in the figure above. If the DNA fragment were rotated around this axis in the plane of the paper (i.e. clockwise or anticlockwise), so that the two strands change places, the sequence as read will be unchanged. The two 'halves' of a palindromic sequence, on either side of the axis of dyad symmetry, are refered to as INVERTED REPEATS because they consist of the same sequence in an inverse orientation. Inverted repeats have the potential to form intra-strand base pairs. This has consequences for the secondary structure of both DNA and RNA (*see* CRUCIFORM STRUCTURE).

palmitic acid *See:* FATTY ACIDS.

palmitoylation The post-translational addition of a palmitoyl group to a protein, by which it may be anchored to the plasma membrane. *See:* MEMBRANE ANCHORS.

pancreatic ribonuclease An ENDORIBONUCLEASE originally isolated from mammalian pancreatic juice (EC 3.1.27.5) and which cleaves RNA to 3′-phosphomononucleotides and 3′-phospho-oligonucleotides ending in Cp or Up, with 2′,3′-cyclic phosphate intermediates.

pancreatic trypsin inhibitor Non-serpin inhibitor of the proteolytic enzyme trypsin. *See:* PROTEINASE INHIBITORS. *See:* Fig. P74 in PROTEIN STRUCTURE for an illustration of the structure of bovine pancreatic trypsin inhibitor.

panhypopituitarism *See:* PITUITARY DWARFISM.

panoistic oocytes OOCYTES in which growth is autonomous, as the NURSE CELLS found in *Drosophila* and other insect species are absent. The growth of the oocyte is only assisted by the FOLLICLE CELLS, which provide yolk protein and metabolites.

pantothenic acid A vitamin, which is a cofactor of coenzyme A. It is very widely distributed in food; a deficiency state has not been identified with certainty in humans.

***pap* genes** Genes in *Escherichia coli* and other similar bacteria which encode components of the bacterial flagellum. *See:* BACTERIAL PATHOGENICITY.

papain THIOL PROTEINASE from papaya sap (EC 3.4.22.2), which cleaves preferentially on the carbonyl side of peptide bonds including Arg, Lys, and Phe-X-. It has been used to study ANTIBODY structure. Papain also has esterase, thiolesterase, transamidase, and transesterase activity.

PapD MOLECULAR CHAPERONE involved in assembly of P pili in *Escherichia coli*. *See:* BACTERIAL PATHOGENICITY.

paper chromatography *See*: CHROMATOGRAPHY.

papilloma Small tumour of epithelial cells which may be either benign (e.g. warts) or malignant, and which is caused by certain papilloma viruses (*see* PAPOVAVIRUSES).

papilloma virus *See*: PAPOVAVIRUSES.

Papovaviruses

THE family Papovaviridae encompasses the small nonenveloped spherical viruses found in a wide range of mammalian hosts [1]. They are double-stranded DNA viruses, some members of which — notably mouse polyoma virus, the simian virus SV40 and the papillomaviruses — have been intensively studied as models of 'eukaryotic' gene expression and tumorigenesis, because of their ability to TRANSFORM cells *in vitro*, and, in the case of papillomaviruses, their involvement in animal and human cancers [2].

The viral particles are skewed icosahedral in shape (45–55 nm in diameter), being composed of 72 protein units (capsomers) arranged to form a protein shell, or capsid, surrounding the viral minichromosome (see Fig. P3 below). The capsid is formed from one major virus-encoded protein and several minor species and, in polyomaviruses, the viral DNA is formed into nucleosomes around cellular histones H2, H3, and H4 (*see* CHROMATIN). The genome of these viruses is encoded in a single circular double-stranded DNA molecule, 5–8 kb in length.

The Papovaviridae comprises two genera: the polyomaviruses and the papillomaviruses. Although the pathology of each differs — infection by polyomaviruses is (with some exceptions) asymptomatic, whereas papillomaviruses are the cause of benign and malignant growths in the tissues of their host — the genome organization and the functions encoded therein have some common features (Fig. P2). Representative genomes from each genus have been sequenced and OPEN READING FRAMES (ORFs) and noncoding regulatory sequences identified. Each genome has three basic units: a noncoding region that contains the ORIGIN OF REPLICATION and ENHANCER/PROMOTER elements; a region from which nonstructural 'early' genes are transcribed following viral entry and uncoating; and the region in which the genes for the 'late', or structural proteins are located. Numbering of each genome begins at the origin of replication and proceeds, by convention, clockwise in the direction of TRANSCRIPTION for papillomaviruses and for early transcription in the case of polyomaviruses. In the latter, late transcripts are coded for on the opposite strand to early transcripts.

Polyomaviruses

Polyomaviruses have been isolated from a variety of hosts, including rodents, monkeys, and humans, and in general have little effect on the host species. In humans, the presence of JC virus in the brains of immunodeficient individuals has been associated with a condition known as progressive multifocal leukoencephalopathy, but the majority of the population carry a number of different viral strains with no ill effect. However, if mouse polyoma virus is introduced into newborn or immunocompromised rodents, or SV40 into hamsters, tumours develop at multiple sites. In addition, if virus is introduced into cells not PERMISSIVE for viral replication, transformation to a tumorigenic phenotype can occur.

The genome is divided into three functional units. Following entry of the virus into the host cell and uncoating of the genome in the nucleus, transcription of early RNA (by cellularly encoded RNA polymerase ll) occurs, which is differentially spliced (*see* RNA SPLICING) to yield either two or three mRNAs for the overlapping genes located on the early strand. Later in the viral life cycle transcription switches to producing late gene mRNAs, from the opposite strand. Viral regulation of these events is attributed to the product of the largest early gene, the large T antigen (LT), interacting with DNA sequences in the third unit, the noncoding region of the viral genome.

The noncoding region

There are some differences in the organization of the noncoding regions (ncr) of SV40 and mouse polyoma virus. The SV40 ncr contains a series of elements including repeats of 21 and 72 bp and a GC-rich palindromic sequence at the origin. LT binding sites, of the CONSENSUS SEQUENCE G[A/G]GGC, are also found in the ncr. Promotor sequences for the early and late transcriptional units and multiple enhancer elements have been identified. SV40 enhancer motifs bind a number of cellular factors present in HeLa cell extracts. The TPA responsive element (to which the AP1 transcription factor binds, *see* TRANSCRIPTION FACTORS) and OCTAMER-BINDING PROTEIN motifs are present, as are sites with which complexes termed AP2 to AP5 interact.

The mouse polyoma virus ncr differs in the array of enhancer sequences present, which interact with defined, murine cellular transcription factors such as PEBP 1 (or PEA 1), a member of the AP1 family. There are some homologies with sequences found in several other viral enhancers (e.g. ADENOVIRUS early region 1A and the bovine papillomavirus enhancers). Novel transcription factors binding to overlapping or adjacent sites have also been isolated (see [3]). The responsiveness of these elements varies depending on cell type and state of differentiation.

Replication, initiating from the origin and occurring bidirectionally, is dependent on functional LT, and in mouse polyoma virus, additionally involves enhancer elements. A cell-free SV40 replication system has been developed from which components essential to replication have been isolated to a high level of purity. Replication protein A, a complex of DNA polymerase α and primase, the catalytic subunit of PROTEIN PHOSPHATASE 2A, and LT are necessary to initiate DNA REPLICATION, and elongation requires PROLIFERATING CELL NUCLEAR ANTIGEN (PCNA), replication factor C, DNA topoisomerase 1 (*see* TOPOISOMERASE), and DNA POLYMERASE δ.

The early region

Polyomavirus early region organization follows one of two patterns. Most members of the family resemble SV40 (see Fig. P2), in encoding two major early gene products. Mouse and hamster

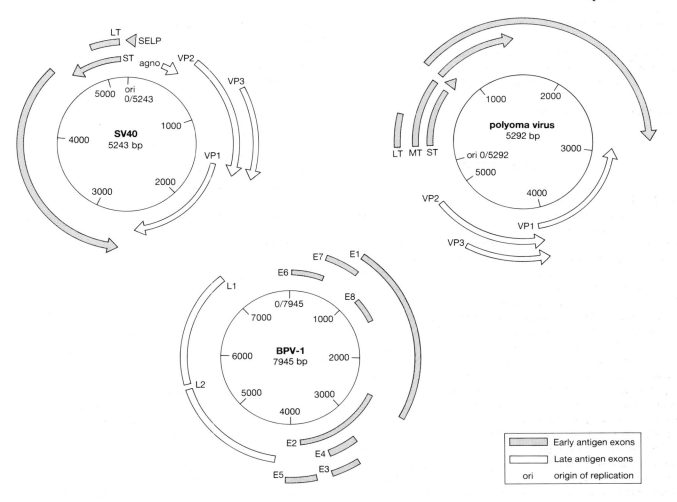

Fig. P2 Genomic organization of the papovaviruses. The circular genomic sequences, marked in base pairs (bp), of polyomaviruses (simian virus 40 (SV40) and mouse polyoma virus) and papillomaviruses (bovine papillomavirus, type-1 (BPV-1)) are shown. Potential protein-coding sequences are also indicated. The agnoprotein (agno) is a basic protein, expressed late in the viral life cycle of SV40, and SELP encodes a peptide of 23 amino acids.

polyomaviruses, however, encode a third gene product, middle T (MT) [4].

LT. The largest gene product of the early region of both SV40 and mouse polyoma virus, with an M_r of ~100 000 (100K), is known as the large T antigen (LT). There is a considerable degree of homology between the proteins from each source and they also share certain functional characteristics. SV40 LT is modified in a variety of ways post-translationally, including phosphorylation, glycosylation and ADP-ribosylation (*see* POST-TRANSLATIONAL MODIFICATION), and molecules have been isolated covalently bound to RNA. The protein contains a NUCLEAR LOCALIZATION SIGNAL (amino-acid sequence, in single-letter notation, PKK-KRKV) and is found predominantly in the nucleus of infected or transformed cells. There is also a minor population residing in a plasma membrane location. The nuclear form posesses DNA binding, ATPase and HELICASE enzymatic activities, and can interact directly with DNA polymerase α. It has a major role in viral DNA replication, binding as an oligomer at the origin of replication and initiating the process. It also may be involved in elongation of the replicating DNA strands. SV40 LT is capable of transcriptional activation and repression, acting to regulate early gene expression and activate late genes at the appropriate points in the viral life cycle. Cellular gene targets for these transcriptional regulatory activities have also been identified.

In addition to these activities, the protein is a potent transforming agent. Primary cells in culture can be transformed to a tumorigenic phenotype by the genes encoded in the early region. SV40 LT alone (or a combination of mouse polyoma virus LT and MT) is sufficient to achieve this. Regions of SV40 LT essential to this process, as defined by studies with mutant proteins, include sequences homologous with adenovirus early region 1A (E1A) and papillomavirus E7 gene products. In these proteins, this sequence has been shown to interact with the cellular TUMOUR SUPPRESSOR function encoded by the retinoblastoma gene (p105Rb). Additionally, a less well defined region is necessary for full transforming potential, which interacts with the tumour suppressor protein p53 [5].

Polyoma virus LT shares some functions with its SV40 counterpart. It probably has a similar role in replication; DNA binding and ATP-dependent helicase activity have been demonstrated, and it is also active in transcriptional activation/repression assays. It is less effective, however, in transformation assays and is capable only of promoting proliferation, or IMMORTALIZATION of primary tissue culture cells. Interaction with p105Rb has been shown, but no interaction with p53 has yet been demonstrated.

MT. Mouse and hamster polyomaviruses, but not SV40-type viruses, encode a 55K major early function. The hamster species is relatively poorly characterized, but mouse polyoma virus MT has been much studied. It is responsible for cellular alterations which lead to transformation of proliferating cells to a tumorigenic phenotype. MT is located in higher order structures in the cytoplasm, predominantly membrane and cytoskeletal in nature. In particular, a membrane-bound subpopulation associated with plasma membranes, essential to transformation, interacts with and activates the product of the cellular proto-oncogene c-*src* (*see* ONCOGENES). This complex is also directly associated with the 81K component of a phosphatidylinositol-3 kinase enzyme, which may also be found in association with activated platelet-derived growth factor receptor (*see* GROWTH FACTORS; GROWTH FACTOR RECEPTORS). In addition, MT interacts with the 60K, regulatory (A), and the 35K, catalytic (C), subunits of the cellular protein phosphatase 2A complex. The function of this interaction in transformation by MT is unknown.

The relevance of these interactions to the growth of the virus is similarly unclear. MT is necessary for lytic growth: mutations in the MT/ST unique region result in viruses that grow poorly. There is some correlation between this loss of function and correct post-translational modification of the major capsid protein, VP1, which results in a block in virion assembly.

ST. The small T gene product (23K in size) shares common sequence with the other early antigens of the polyomaviruses for most of its length; in mouse polyoma virus only the C-terminal four amino acids are unique to ST. It is dispensable for achieving most viral functions, but it appears to potentiate the actions of the other early antigens and may have a role in tumour formation *in vivo*. STs from both SV40 and polyoma virus interact with the A and C subunits of protein phosphatase 2A, which localizes the site of this interaction to a cysteine-repeat region common to polyoma virus MT and ST and SV40 ST.

The late region

This region encodes the three structural proteins of the virion. The major coat protein, VP1 (45K), is expressed at late times in the viral life cycle and is initially located in the nucleus. Later, it can be found in the cytoplasm of infected cells as well. In mouse polyoma virus it is modified post-translationally by phosphorylation and acetylation and has six isoelectric forms. It can spontaneously organize, *in vitro*, into aggregates of five molecules known as capsomeres, which, under the appropriate buffer conditions, can form capsids. Despite the capsomers being arranged in the capsid with axes of five- and sixfold symmetry (Fig. P3), only pentamers of VP1 are found.

The role of the minor coat proteins in the virion structure is less well defined. X-RAY CRYSTALLOGRAPHIC data have revealed that below the outer layer of VP1 capsomeres lies an electron-dense layer, probably composed of VP2 and 3 (35 and 23K, respectively). VP3 is entirely encoded within VP2 and so its contribution to viral infectivity is difficult to assess, although it is thought to be necessary for efficient viral reproduction. Both proteins accumulate in the nucleus and may be involved in transporting VP1 to this site. VP2 is modified post-translationally with myristic acid and loss of this modification reduces, but does not abolish, infectivity.

Papillomaviruses

It has not been possible to develop an *in vitro* lytic system for the study of papillomaviruses: virus and viral transcripts can only be obtained in very low amounts from infected epithelial tissues or papillomas. However, mRNAs isolated from these sources reveal overlapping and spliced transcripts, and promoters and POLYADENYLATION sites utilized vary, depending on the source and open reading frame (ORF) transcribed. Expression of the viral genes is tightly restricted by the level of differentiation of the cell in the epidermis. With the progression from undifferentiated basal layer cells to differentiated squamous cells, viral gene expression switches from early to late. Viral functions have mostly been analysed by cloning the ORFs into suitable vectors and expressing them in tissue culture. Association of some types of human papillomaviruses (HPVs) with cancerous conditions has stimulated a wealth of data on the functions encoded by these

Fig. P3 The packing arrangement of subunits in the papovavirus SV40. The crystal structure of SV40 revealed a new type of packing in which 72 pentamers of virion protein VP1 are arranged in a pattern of interlocking pentagons and hexagons. Twelve 5-coordinated pentamers lie on the 5-fold rotation axes of the icosahedron, each surrounded by five pentamers. Sixty 6-coordinated pentamers do not lie on any symmetry axes and each is surrounded by six pentamers. Based on [7].

viruses. However, bovine papillomavirus, type 1 (BPV-1) being more readily obtained than other members of the family, is commonly used as a prototype for studying viral functions. As with polyomaviruses, the genome has been divided into early and late regions (Fig. P2), although, unlike polyomaviruses, transcription of both regions occurs from a single strand.

Early region

The region containing early ORFs has been defined as the fragment of the genome (69% of the total) necessary for conferring transformation on cells in an *in vitro* assay. This region contains eight ORFs, five of which (E1, E2, E4, E6, and E7) appear to be found in other papillomavirus genomes, suggesting that these are essential functions.

E1. The E1 ORF gene is involved in episomal (extrachromosomal) replication of the viral genome. It bears some homology with SV40 LT in the regions involved in ATPase activity and nucleotide binding and its product can bind DNA, either alone, or complexed to E2 gene products.

E2. These polypeptides have the properties of transcriptional regulators [6]. They bind as dimers to the palindromic DNA sequence ACCGNNNNCGGT, which is found in several copies in the noncoding region of the virus. Simultaneous binding of E2 dimers to several of these sites results in a high level of transcriptional activation. There is also evidence for cooperation of E2 with cellular transcription factors such as AP1. Active E2 seems to repress promotor/enhancer activity for the ORFs E6 and E7. E2 is often found to be defective in material derived from HPV tumours.

E6, E7. The major proteins encoded by papillomaviruses that are relevant to transformation are transcribed from the E6 and E7 ORFs. These polypeptides are capable of transforming primary cells in a classical focus formation assay. The E6 protein, from HPV 16 and 18, can promote the breakdown of the tumour suppressor protein p53, in an ATP-dependent manner and utilizing the ubiquitin degradation pathway (*see* PROTEIN DEGRADATION). E7, which cooperates with the oncogene *ras* to transform primary cells, has been shown to interact with the underphosphorylated form of the p105Rb tumour suppressor gene. This interaction probably occurs through the N-terminal 37 amino acids, which share some homology with domains 1 and 2 of adenovirus E1A, also known to be involved with p105Rb binding. Therefore, transformation of cells by E6 and E7 may occur by a similar route to that of the DNA tumour viruses — adenovirus and SV40. E6 and E7 may also induce host chromosomal alterations.

E5. A further contribution to transformation, in some papillomaviruses, may come from the product of the E5 ORF. Studies on the 44 amino acid product of deer or bovine papillomavirus E5 ORFs have revealed a potential for the gene in inducing growth, cellular transformation, DNA synthesis, and activation of the platelet-derived growth factor receptor. Various mechanisms for the latter have been proposed, including direct binding to the receptor. Some homology with the β chain of platelet-derived growth factor (*see* GROWTH FACTORS) has been noted. The E5 protein has been isolated in association with a 16K protein thought to be important in cellular compartments in which processing of growth factor receptors occurs.

Late region

The late region has two large ORFs capable of coding for proteins of 55K (L1) and 50K (L2). L1 encodes the major structural capsid protein. The role of the minor capsid protein, L2, is not known.

N.S. KRAUZEWICZ

See also: ANIMAL VIRUSES; ANIMAL VIRUS DISEASE; EUKARYOTIC GENE EXPRESSION.

1 Salzman, N.P. & Howley, P.M. (Eds) (1987) *The Papovaviridae*, Vols 1 and 2 (Plenum, New York).
2 Doerfler, W. & Böhm, P. (Eds) (1992) *Malignant Transformation by DNA Viruses*, 1–85 (VCH, Cambridge).
3 Jones, N.C. et al. (1988) *Trans*-acting protein factors and the regulation of eukaryotic transcription: lessons from studies on DNA tumor viruses. *Genes Devel.* **2**, 267–281.
4 Pipas, J.M. (1992) Common and unique features of T antigens encoded by the polyomavirus group. *J. Virol.* **66**, 3979–3985.
5 Fanning, E. (1992) Simian virus 40 large T antigen: the puzzle, the pieces, and the emerging picture. *J. Virol.* **66**, 1289–1293.
6 McBride, A.A. et al. (1991) The papillomavirus E2 regulating protein. *J. Biol. Chem.* **266**, 18411–18414.
7 Liddington, R.C. et al. (1991) Structure of simian virus 40 at 3.8-Å resolution. *Nature* **354**, 278–284.

paracentric inversion *See*: CHROMOSOME ABERRATIONS.

paracrine Term applied to the action of local chemical mediators, which have their effects very close to the cells that secrete them.

paralogous *See*: HOMOLOGY.

paramylon A storage β1,3 glucan of the unicellular alga *Euglena gracilis*, formed by transfer from UDP-Glc, using a paramylon synthetase that is membrane-associated and solubilized by sodium deoxycholate. There is evidence for a GLYCOPROTEIN primer and the synthetase closely resembles CALLOSE synthetase in being stimulated by oligosaccharides of β1,3 and β1,4 glucans.

Paramyxoviridae Family of enveloped RNA ANIMAL VIRUSES comprising three genera — paramyxoviruses, morbiliviruses (measles, canine distemper), and pneumoviruses (respiratory syncytial virus, RSV). Virus particles are pleomorphic and ~150 nm in diameter, with surface spikes. The single-stranded RNA genome (M_r $5 \times 10^6 – 7 \times 10^6$) is usually (−) sense, but some particles contain (+)-sense genomes. *See*: ANIMAL VIRUS DISEASES; ANIMAL VIRUSES.

paranemin An INTERMEDIATE FILAMENT protein.

paraquat One of the low-potential ($E_m = -400$ mV) viologen dyes

(N,N′-dimethyl-γ,γ′-dipyridylium, also known as methyl violo-gen) that can act as an artificial electron acceptor for PHOTOSYS-TEM I and is closely related to DIQUAT. One-electron reduction yields a violet-coloured, auto-oxidizable free radical. Reaction with oxygen yields the superoxide anion in the first instance; this may form the basis of its herbicidal action which depends on light and thus active turnover of the electron transport system. *See also*: PHOTOSYNTHESIS.

Parascaris Nematode worm in which CHROMOSOME DIMINUTION occurs.

parasegments The initial metameres or repeat units in the DROSOPHILA embryo which are established at the blastoderm stage by the activities of the pair-rule class of SEGMENTATION GENES. These repeat units are out of phase with the later segmental units. *See*: DROSOPHILA DEVELOPMENT.

parasympathetic nervous system *See*: PERIPHERAL NERVOUS SYSTEM.

parathyroid hormone (PTH) A polypeptide hormone of 84 amino acids secreted by the parathyroid glands, located within the thyroid tissue. A major regulator of calcium homeostasis, it is released in response to falling plasma calcium concentration, and increases the absorption and mobilization of calcium through effects on the kidneys, intestine and bone.

paraxial mesoderm In vertebrate embryos, the MESODERM lying between the NOTOCHORD (which is the AXIAL MESODERM) and the INTERMEDIATE MESODERM. The paraxial mesoderm of the trunk gives rise to the SOMITES which in turn contribute cells to the dermis, skeletal muscles, and axial skeleton of the adult. In the head, the paraxial mesoderm contains fewer cells and its fate is less well understood; among other tissues, it probably contributes to some facial muscles. *See also*: AMPHIBIAN DEVELOPMENT; AVIAN DEVELOPMENT; MAMMALIAN DEVELOPMENT.

parenchyma In plants, a tissue composed of thin-walled, rela-tively undifferentiated cells, of various structure and function.

Parental genomic imprinting

IN its widest definition, genetic imprinting is considered to be any process in which EPIGENETIC information (i.e. information other than that encoded in the nucleotide sequence of genes) is intro-duced into chromosomes and is stably replicated together with the chromosome as cells divide. It may have a role in some aspects of CELL MEMORY in the development of multicellular organisms. Imprinting is carried over to progeny cells at cell division, can affect gene expression (often many cell generations later) and is reversible. Unlike heritable changes due to mutation or directed gene rearrangement (as in the IMMUNOGLOBULIN GENES), imprints can normally be removed from chromosomes without leaving behind any permanent alteration of the genetic material. Imprinting can be seen as a remote control switch; once turned it may have no effects for many cell generations, but eventually, gene expression and thus phenotype will depend on this initial switch.

Here, a particular class of imprints, those that mark the parental origin of genomes, chromosomes, and genes in mam-mals, will be considered [1]. Parentally imprinted genes are those whose expression depends on whether they are carried on a maternally or paternally derived chromosome. Some are expressed only from maternal chromosomes, others only from paternal chromosomes. In the past 5 years or so parental imprinting has progressed from an intriguing embryological observation to a firmly established biological principle, with far-reaching consequences for mammalian development and genetics, and human disease. Genes subject to parental imprint-ing have been identified and possible molecular mechanisms of imprinting suggested. This article focuses exclusively on parental imprinting of autosomes (*see* X-CHROMOSOME INACTIVATION for parental effects on the X-chromosome). Parental imprinting is also known in invertebrates and plants [2] but will not be discussed here.

Embryological evidence

Embryological investigations provided the first indication of the differential contribution of essential information for development by maternal and paternal chromosomes. In mammals, the pres-ence of both parental genomes in the zygote seems to be essential to produce a viable foetus. In the mouse, in which most work on parental imprinting has been done, monoparental diploid em-bryos cannot develop to term, irrespective of whether they carry two maternal or two paternal genomes, or of whether the ones with two maternal genomes are derived parthenogenetically or by nuclear transplantation [2,3]. They die soon after implantation at the latest.

The phenotype of parthenogenetic (or gynogenetic) and andro-genetic embryos is different from the earliest stages. Whereas parthenogenetic embryos develop in culture at a comparable rate to normal embryos at least to the blastocyst stage, only a variable proportion of androgenetic embryos reach that stage, in general more slowly. Following implantation, parthenogenetic and andro-genetic embryos have very different phenotypes indeed (Fig. P4). Parthenogenetic conceptuses can reach the 25-somite stage on day 10 of gestation with a relatively well-developed embryo proper, but their extraembryonic tissues are severely deficient, especially the trophoblast which is crucially important for placen-tation and hence nutrient transfer from the mother. Androgenetic embryos, by contrast, show very much the opposite phenotype: extraembryonic structures are well developed but the embryo itself is severely retarded and does not progress beyond the 4–6 somite stage.

In the mouse therefore, and probably in other mammals including humans, both parental genomes are necessary for development and they must provide the correct gene dosage at certain gene loci — which are termed imprinted loci. The

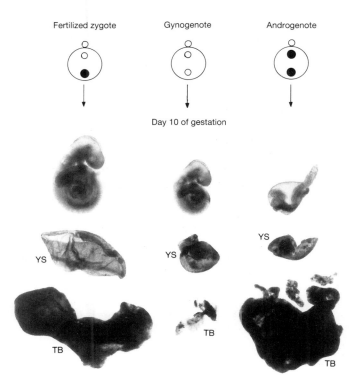

Fertilized zygote   Gynogenote   Androgenote

Day 10 of gestation

YS

YS

YS

YS

TB

TB

TB

TB

Fig. P4 The phenotypes of normal, gynogenetic, and androgenetic mouse embryos and extraembryonic tissues at 10 days. Gynogenetic and androgenetic embryos are obtained by pronuclear transplantation. TB, trophoblast; YS, yolk sac.

different phenotypes of the two types of experimental embryos indicate the cumulative actions of maternally or paternally imprinted genes.

Genetic evidence for imprinting

Uniparental disomy

The action of a small number of specific imprinted genes can be investigated genetically in embryos in which defined regions of one or other parental genome are DISOMIC (present in two copies), a condition known as uniparental disomy. Uniparental disomic embryos can be obtained by crossing mice heterozygous for particular chromosomal TRANSLOCATIONS. Because of NON-DISJUNCTION at meiosis some of the gametes will contain no copies of the translocation or the corresponding normal chromosome whereas others will contain both. Embryos arising from fusion of two such gametes will therefore contain two copies of a chromosome or part of a chromosome derived from the same parent. Uniparental disomic embryos have been obtained for most chromosomes and their development has been analysed [4].

Some of these embryos show altered phenotypes, indicating the presence of imprinted genes on the disomic segment. For example, mice paternally disomic for the proximal portion of chromosome 11 are larger than their normal littermates, whereas mice maternally disomic for the same chromosome are smaller. This suggests

that an imprinted gene or genes, whose difference in dosage causes the growth differences, is present on chromosome 11.

All imprinted regions identified so far determine phenotypes associated with growth and viability of the embryo, and in some cases perhaps, behaviour of neonates. The total number of parentally imprinted genes is unknown, but a minimum estimate is given by the number of identified imprinted chromosome segments (Fig. P5).

Imprinted genes

Four genes in the mouse have now been identified as undergoing parental imprinting; three of these have known functions so that one can now begin to interpret the observed phenotypes in the context of the expression or repression of these genes.

Mice inheriting a NULL ALLELE of the gene *Igf-2* for insulin-like growth factor II (*see* GROWTH FACTORS) from their father turn out to be considerably smaller than their littermates, whereas when the mutant allele is maternally inherited offspring show no growth deficiency [5] (Fig. P6). Consistent with this phenotype, *Igf-2* transcripts are produced from the paternal allele, but the maternal allele is almost completely repressed. The only exception to this imprinting is in the leptomeninges and choroid plexus in brain, where both alleles are expressed. This indicates the possibility that imprinting may be tissue- as well as stage-specific.

Additional evidence for maternal imprinting of *Igf-2* comes from the extremely low levels of *Igf-2* RNA in mouse embryos maternally disomic for the distal region of chromosome 7, where the *Igf-2* gene resides [6]. These embryos are smaller than their littermates and die *in utero* during the last third of the gestation period. *Igf-2* is expressed widely throughout the embryo, predominantly but not exclusively in extraembryonic tissues and in the mesodermal lineage, and is known to be an embryonic MITOGEN.

Near the *Igf-2* gene, inseparable by recombination and perhaps very close to it, is another imprinted gene, *H19* [7]. This gene, whose protein product is at present unknown, is imprinted in the opposite direction; it is expressed from the maternal chromosome but repressed on the paternal one. Its twofold overexpression in embryos disomic for maternal distal chromosome 7 could conceivably contribute to the lethality of these embryos mentioned above, as embryos transgenic for additional copies of the *H19* gene also die at around the same stage [8].

Tme (T maternal effect) is another imprinted locus. Deletions of the *Tme* locus such as T^{hp} or t^{wlub2} are lethal at late foetal stages when on the maternal chromosome but not when on the paternal chromosome. The *Tme* gene product is not known. The gene for the IGF-II receptor (*Igf-2r*) (the MANNOSE 6-PHOSPHATE RECEPTOR, *see also* LYSOSOMES; PROTEIN TARGETING) has been mapped to this region of mouse chromosome 17 and has been shown to be imprinted, with the maternal copy being the active one [9].

The *Snrpn* gene (Small nuclear ribonucleoprotein N) is on the middle part of chromosome 7 and is expressed from the paternal chromosome [10]. This gene is involved in the regulation of splicing and is predominantly expressed in the brain. Its possible absence in the Prader Willi syndrome (see below) may contribute to some of the symptoms of this disease.

The molecular mechanism of imprinting of these genes is at

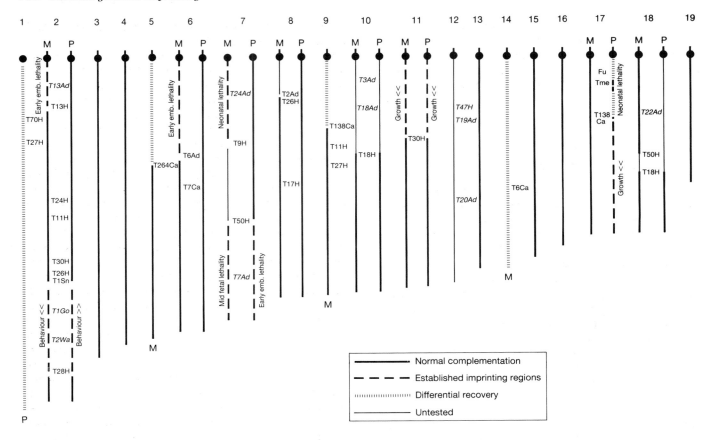

Fig. P5 Imprinted regions of the mouse genome. M, maternal disomy; P, paternal disomy. The phenotypes of maternal or paternal disomy are indicated. From [4].

present unknown (but see below for some possible mechanisms). Indeed it is not even known at precisely what level of gene regulation repression occurs. However, differences in DNA methylation and chromatin structure have recently been detected between parental alleles of some imprinted genes [3].

Chimaeric rescue

Cells from parthenogenetic or androgenetic embryos can be 'rescued' by incorporating them into CHIMAERIC EMBRYOS that also contain normal cells, and in which their later development and their effect on the chimaeric embryo can be tested. Allocation of cells to different lineages does not seem to be affected but selection against parthenogenetic (PG) and androgenetic (AG) cells occurs once lineages have been set apart [2,3]. PG cells are consistently lost from the extraembryonic tissues (in aggregation chimaeras) and in the foetus from mesodermal tissues, but are usually well represented in neuroectodermal tissues. In contrast, AG cells are selected against in neuroectodermal tissues but contribute substantially, perhaps to an even greater extent than normal cells, to mesodermal tissues such as skeletal muscle, heart muscle and the skeleton [11]. AG chimaeras that survive to term

(because they have a small proportion of AG cells) show severe skeletal abnormalities, partly due to hyperplasia of cartilage.

The basis of cell selection is at present unknown, but there could be a greater or reduced propensity to cell proliferation within a lineage as a result of overexpression or lack of AUTOCRINE or short range PARACRINE growth factors (such as IGF-II) or their receptors. It is certainly suggestive that PG cells are notoriously absent in the very tissues in which *Igf-2*, *Igf-2r* and

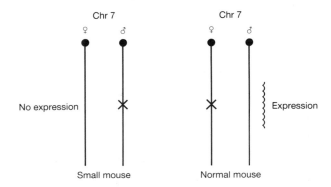

Fig. P6 Imprinting of the *Igf-2* gene. X, mutant allele. When the mutant allele of the *Igf-2* gene is paternally transmittted, offspring are small. When the same allele is maternally transmitted, offspring are of normal size. This is because *Igf-2* transcription is repressed on the maternal chromosome, and deficiency of the growth factor results in small size.

H19 are predominantly expressed. Chimaeric embryos have also been made from cells with maternal or paternal disomy of distal chromosome 7, on which *Igf-2* and *H19* reside [6]. Paternal disomics expressing a double dose of *Igf-2* cause an increase in size of the chimaeric embryo. This agrees with observations that AG chimaeras, which are disomic for the whole paternal genome, can be up to 50% larger than controls.

Imprinting of transgenes and possible molecular mechanisms

Some TRANSGENES in the mouse become parentally imprinted and such TRANSGENIC MICE therefore provide a model system for studying parental imprinting [12,13]. In some cases this imprinting seems to depend on where the transgene has become integrated into the host genome, in others the DNA sequence of the transgene seems more important. But it does not seem necessary for the transgene to become inserted into one of the known imprinted regions of the genome for imprinting to occur. For some transgenes, imprinting is reversible in the germ line; in others epigenetic modifications can persist and indeed accumulate over several generations.

One epigenetic modification identified in imprinted transgenes is DNA METHYLATION, a heritable modification that is known to be associated with effects on gene expression. Typically, the methylation pattern on the transgene reflects its mode of transmission: in most cases paternal transmission results in undermethylation (hypomethylation) whereas maternal transmission leads to hypermethylation (Fig. P7). In the offspring, the methylation differences are associated with expression or repression of the transgene, with hypermethylated transgenes being repressed. Differences in DNA methylation as well as chromatin structure have also been detected in some endogenous imprinted genes [3].

An important distinction among imprinted transgenes is

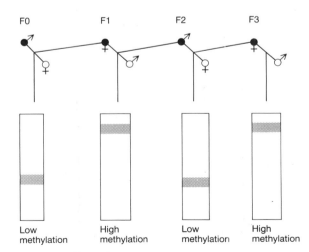

Fig. P7 Typical behaviour of an imprinted transgene. Paternal transmission of the transgene leads to low methylation of the transgene in the offspring, whereas maternal transmission leads to hypermethylation. The methylation pattern switches from generation to generation as the transgene is passed on from male to female to male.

whether imprinting is retained on a homozygous genetic background, as is the case for endogenous imprinted genes. Two transgenes in which this does occur may provide a closer approximation to endogenous imprinting [14,15]. DNA methylation does seem to be the primary imprinting signal for these transgenes. Methylation differences already exist in the gametes before fertilization and they persist during development. In two cases studied so far, the imprinted transgenes are unmethylated in foetal germ cells of both sexes irrespective of parental descent; in the female, the transgene then becomes methylated during the late stages of oogenesis, and the fully methylated copy is then kept methylated during preimplantation and postimplantation stages. Some endogenous sequences are also methylated during late oogenesis, and the oocyte contains an extraordinarily high level of methylase. In contrast, methylation of the transgene in mature spermatozoa is much lower and stays relatively low on the paternal chromosome following fertilization. A methylation imprint in the endogenous *Igf-2r* gene has recently been discovered that follows precisely the same pattern: it is methylated in the oocyte and remains so on the maternal chromosome but is undermethylated in sperm and the paternal chromosome [16].

A number of transgenes do not, however, retain imprinting on a homozygous background [12,13]. It is believed that their imprinting results from the action of genotype-specific modifier genes, different alleles of which are present in inbred strains of mice. The nature of the modifier genes that can affect epigenetic programming in mammals is at present unknown, although one has been chromosomally mapped [17]. There are, however, overt similarities between transgene modification in the mouse and POSITION EFFECT VARIEGATION in *Drosophila melanogaster*, and it is possible that the modifier genes work in a similar way to the enhancers and suppressors of position effect variegation. In the fly, a number of these genes have now been cloned, and some encode proteins that may be involved in the formation of HET-EROCHROMATIN chromosome domains (which can form stable epigenetic switches). A family of genes with homology to the *Drosophila* heterochromatin protein gene *HP1* has recently been identified in the mouse [18] and it will be important to see what kind of phenotype variant alleles of these genes can elicit.

Parental imprinting in human development and disease

Imprinting of the *Igf-2* and *H19* genes, and possibly others, is conserved in humans. Additional strong evidence for parental imprinting in the human genome comes from certain disease syndromes in which uniparental disomy has caused an aberrant phenotype [19,20]. PRADER–WILLI SYNDROME (PWS) and ANGEL-MAN SYNDROME (AS) can be the result of maternal and paternal disomy respectively of chromosome 15q11-13 [21]. The imprinted *Snrpn* gene is contained in this segment and may contribute to at least the PWS phenotype. Maternal disomy of human chromosome 7 has been found in association with growth retardation [19,20].

Beckwith–Wiedemann syndrome (BWS), a foetal overgrowth syndrome, can also be caused by disomy. It had been noticed that some sporadic BWS cases had a partial trisomy of chromosome 11p15.5 which was most often paternally derived. A number of

BWS cases with a paternal disomy of chromosome 11p15.5 have now been detected [22]. In humans, the *Igf-2* gene is contained in this chromosome segment, and as it is known to be maternally imprinted, the foetal overgrowth phenotype could conceivably be the result of *Igf-2* overexpression.

Increased or decreased dosage at imprinted genes because of disomic lineages can also contribute to predisposition to some cancer syndromes. For example, in Wilms' tumour and in rhabdomyosarcoma, allele losses of chromosome 11p frequently involve the maternal copy of that chromosome, so that the tumour is paternally disomic. Whether overexpressed imprinted growth factors (such as IGF-II) or epigenetic inactivation of TUMOUR SUPPRESSOR GENES contribute to tumorigenesis is at present not known [13,19].

In addition, an increasing number of MONOGENIC DISORDERS (usually dominant) are known to show parental effects in that maternal or paternal inheritance of the mutant gene has an effect on its penetrance or expressivity in the offspring [19,20].

Why imprinting?

Evolutionary biologists now believe that they have the answer to this question and their explanation is both elegant and consistent with most of the observations [23]. Consider in a placental mammal, a gene that acts in the foetus to acquire nutritional resources from the mother. Any growth factor gene that acts during intrauterine life, such as *Igf-2*, would qualify. Paternally derived alleles at this locus will tend to increase growth of the foetus, as this would maximize their chances of spreading through the population. By contrast, maternal alleles, although in principle subject to the same selective pressure, must also consider the burden for the mother of increased resource transfer, as this may compromise reproductive success for all the offspring produced by one mother. Over evolutionary time there will therefore be a trade off, a parental 'tug-of-war' over the most successful combination of expression levels of maternally and paternally derived alleles.

Such a theory predicts the behaviour of the genes for IGF-II and its receptor very accurately. Increased dosage of *Igf-2* can increase the size of the foetus. It should therefore be imprinted to be expressed when paternally inherited. The receptor, by contrast, presumably acts to bind IGF-II and thereby to decrease its availability for binding to the type I receptor, which is thought to be the main route of IGF-II growth factor activity. *Igf-2r* expression therefore acts negatively on the IGF-II axis, and is therefore predicted to be imprinted in the opposite direction, as it indeed is.

W. REIK

1 Surani, A. & Reik, W. (1992) Genomic imprinting in mouse and man. *Semin. Dev. Biol.* **3**, 73–160.
2 Solter, D. (1988) Differential imprinting and expression of maternal and paternal genomes. *Annu. Rev. Genet.* **22**, 127–146.
3 Surani, M.A. et al. (1993) The inheritance of germline-specific epigenetic modifications during development. *Phil. Trans. R. Soc. Lond.* **B339**, 165–172.
4 Cattanach, B.M. & Beechey, C.V. (1990) Autosomal and X-chromosome imprinting. *Development* (Suppl.) 63–72.
5 De Chiara, T.M. et al. (1991) Parental imprinting of the mouse insulin-like growth factor II gene. *Cell* **64**, 849–859.
6 Ferguson-Smith, A.C. et al. (1991) Embryological and molecular investi-
gations of parental imprinting on mouse chromosome 7. *Nature* **351**, 667–670.
7 Bartolomei, M.S. et al. (1991) Parental imprinting of the mouse H19 gene. *Nature* **351**, 153–155.
8 Brunkow, M.E. & Tilghman, S.M. (1991) Ectopic expression of the H19 gene in mice causes prenatal lethality. *Genes Devel.* **5**, 1092–1101.
9 Barlow, D.P. et al. (1991) The mouse insulin-like growth factor type-2 receptor is imprinted and closely linked to the Tme locus. *Nature* **349**, 84–87.
10 Cattawach, B.M. et al. (1992) A candidate mouse model for Prader Willi syndrome which shows an absence of Snrpn expression. *Nature Genet.* **2**, 270–274.
11 Barton, S.C. et al. (1991) Influence of paternally imprinted genes on development. *Development* **113**, 679–688.
12 Reik, W. et al. (1990) Imprinting by DNA methylation: from transgenes to endogenous gene sequences. *Development* (Suppl.) 99–106.
13 Sapienza, C. (1990) Sex-linked dosage-sensitive modifiers as imprinting genes. *Development* (Suppl.), 107–114.
14 Chaillet, J.R. et al. (1991) Parental-specific methylation of an imprinted transgene is established during gametogenesis and progressively changes during embryogenesis. *Cell* **66**, 77–84.
15 Sasaki, H. et al. (1991) Inherited type of allelic methylation variations in a mouse chromosome region where an integrated transgene shows methylation imprinting. *Development* **111**, 573–581.
16 Stöger, R. et al. (1993) Maternal-specific methylation of the imprinted mouse Igfr2r locus identifies the expressed locus as carrying the imprinting signal. *Cell* **73**, 61–72.
17 Engler, P. et al. (1991) A strain-specific modifier on mouse chromosome 4 controls the methylation of independent transgene loci. *Cell* **65**, 939–948.
18 Singh, P.B. et al. (1991) A sequence motif found in a *Drosophila* heterochromatin protein is conserved in animals and plants. *Nucleic Acids Res.* **19**, 789–794.
19 Reik, W (1989) Genomic imprinting and genetic disorders in man. *Trends Genet.* **5**, 331–336.
20 Hall, J.G. (1990) Genomic imprinting: Review and relevance to human disease. *Am. J. Hum. Genet.* **46**, 857–873.
21 Nicholls, R.D. et al. (1989) Genetic imprinting suggested by maternal heterodisomy in non-deletion Prader Willi syndrome. *Nature* **342**, 281–285.
22 Henry, I. et al. (1991) Uniparental paternal disomy in a genetic cancer-predisposing syndrome. *Nature* **351**, 665–666.
23 Moore, T. & Haig, D. (1991) Genomic imprinting in mammalian development: a parental tug-of-war. *Trends Genet.* **7**, 45–49.

---

parietal endoderm *See*: EXTRAEMBRYONIC MEMBRANES.

Parkinson's disease *See*: NEURODEGENERATIVE DISORDERS.

paroxysmal nocturnal haemoglobinuria *See*: GPI ANCHORS.

parsimony *See*: MOLECULAR PHYLOGENY.

parthenogenesis The process whereby eggs develop following activation in the absence of a sperm. In some species, comprised entirely of females, MEIOSIS is modified and results in a DIPLOID gamete that does not need fertilization for normal development. In certain insects, for example, the oocyte is activated by one of the polar bodies after the second meiotic division; and in one species of grasshopper, diploid ova are formed by two mitotic divisions without meiosis. Haploid parthenogenesis is also widespread, being used for example in the Hymenoptera (ants, wasps, and bees); fertilized (diploid) eggs develop into females while unfertilized (haploid) eggs develop into males.

In mammals, a number of exogenous stimuli besides normal fertilization, for example a rise in internal calcium ion concentra-

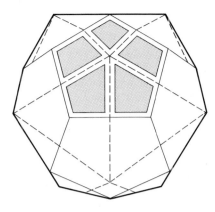

Fig. P8 Diagrammatic representation of the arrangement of subunits in the capsid of canine parvovirus.

from a single gene. VP1 and VP2 are alternatively spliced forms, and VP3 is derived by post-translational cleavage of the N-terminal 15–20 residues of VP2. The packing of the capsid subunits (Fig. P8) resembles that of picornaviruses, except that in canine parvovirus each capsomer is composed of a single protein subunit and not of three different proteins as in picornaviruses (*see* Fig. R15 in RHINOVIRUSES). *See*: ANIMAL VIRUSES.

Tsao, L. et al. (1991) *Science* **251**, 1456–1464.

passive transport Transport across a membrane that does not require expenditure of energy. *See*: MEMBRANE TRANSPORT SYSTEMS.

Patch clamp and its applications

PATCH clamp is an electrophysiological technique which enables ionic currents to be measured across some portion of the cell membrane in a variety of recording configurations. It was devised in the 1970s by Neher and Sakmann [1–4] who were jointly awarded the Nobel Prize for Physiology or Medicine in 1991 for inventing and initially applying the technique in cell biophysics and molecular neurobiology. Patch clamp allows the study of cells from many tissues, not just neurons and muscles, but also epithelial cells, cells of the immune system, and even plant cells. Membrane currents in cells of any size can be recorded, from the smallest lymphocytes of mammals, cellular organelles, and liposomes, with a diameter of a few micrometres, to the giant oocytes of lower vertebrates (diameter 1 mm). Cells can be studied in a variety of situations (e.g. after dissociation from tissue by gentle enzymatic treatment, in tissue culture, or within a thin tissue slice [5]), as well as at different stages of development.

Patch clamp allows membrane currents to be recorded with varying degrees of resolution. At one extreme, the currents flowing across the whole cell membrane can be measured, whereas at the other, the ionic current attributable to a single ION CHANNEL can be revealed. Single-channel recording represents the only procedure presently available in molecular biology whereby the conformational state of a single protein can be followed in real time, as patch clamp can reveal the stepwise flickerings of membrane current representing the 'open' and 'closed' states of a single ion channel in a patch of cell membrane.

The patch pipette

The recording micropipette used for patch clamp is probably the single most important element of the system. Patch pipettes are filled with a suitable solution and pressed onto the cell membrane to establish a giga-Ohm seal. One can then go on to establish a whole-cell or single-channel recording configuration. The type of glass tubing used in the manufacture of the pipette, the initial thickness of the tubing wall, the taper of the tip, and the size of the tip opening are all critical for the success of the procedure.

Borosilicate or hard glass is most commonly used to make the

tion induced by a calcium ionophore, can initiate completion of meiosis and allow normal development through to the somite stage. An ill-characterized defect in some parthenogenetic cells, concerning maternal imprinting of certain genetic loci whereby their function becomes deficient without the paternally-derived gene, causes susequent developmental failure (*see* PARENTAL GENOMIC IMPRINTING). Such maternal imprinting is particularly deleterious for embryonic mesodermal tissues and the extraembryonic placental tissues.

partial digest Usually refers to a digest of a DNA molecule by a RESTRICTION ENZYME in conditions such that only some of the restriction sites available to the enzyme are cleaved. The term may also be used to refer to incomplete digestion of a nucleic acid by a general ENDONUCLEASE.

partial dominance (intermediate dominance, semidominance) The situation in which the HETEROZYGOTE for a pair of ALLELES at a given locus has a different phenotype from either of the two HOMOZYGOUS states. Examples from human genetic disease include FAMILIAL HYPERCHOLESTEROLAEMIA, α_1-ANTITRYPSIN deficiency and the THALASSAEMIAS, in which the heterozygotes show a relatively mild but abnormal phenotype. Intermediate dominance and semidominance are used synonymously.

particle gun *See*: MICROBALLISTICS.

partition chromatography *See*: CHROMATOGRAPHY.

parvalbumin *See*: CALCIUM-BINDING PROTEINS.

Parvoviridae A family of nonenveloped DNA ANIMAL VIRUSES infecting vertebrates and invertebrates. The structure of canine parvovirus has been determined to 3.25 Å resolution. It is an icosahedral particle of outer diameter 180 Å, containing a single-stranded DNA of ~ 5000 bases, which encodes the capsid proteins and two nonstructural proteins involved in regulation of transcription. The capsid of 60 subunits is made up from three similar but not identical capsid proteins, which are all derived pipette. Its sealing properties are generally good, particularly if using thick-walled tubing; the thicker the wall the greater the thickness of the rim of the pipette tip and the better the seal. A thick rim also generates lower noise levels and may also contribute to seal stability. With thick-walled glass, however, for a given tip diameter the resistance of the electrode will be higher than with thin-wall glass, and this will contribute to increased series resistance when whole-cell recording.

Soda lime or soft glass is commercially available in the form of thin-wall haematocrit capillaries. It is used in the manufacture of low resistance pipettes. The sealing properties of thin-walled glass are generally not as good as thick walled and the noise level is higher, but owing to the low resistance the pipettes are good for whole-cell recording. Other types of speciality glass used include Kovar sealing glass and aluminosilicate glass — both hard glasses.

Membrane current recording configurations

There are two general configurations of patch clamp recording, one for recording whole-cell currents, and the other for single-channel recording. In turn, each configuration can have a number of variants. The steps followed in achieving the different recording modes are illustrated in Fig. P9.

Whole-cell patch clamp

To record the membrane current of the whole cell, a patch pipette filled with a buffered solution mimicking the intracellular medium is pressed onto the cell surface and gentle suction applied to form a seal of giga-Ohm resistance. The patch of membrane underneath the pipette can be removed by further suction to achieve electrical continuity with the inside of the cell membrane. The cell's membrane currents can then be measured in response (1) to VOLTAGE CLAMP (to reveal the cell's current–voltage relationship) or (2) to external application of a pharmacological agent while clamping the membrane voltage at a desired value. In either case it is the average current through many channels that is recorded. This recording configuration also allows the MEMBRANE POTENTIAL to be recorded under current clamp. An advantage of this configuration is that the low access resistance allows voltage clamping to be achieved rapidly (<0.5 ms); however, cells much bigger than 30–40 µm diameter may be difficult to voltage clamp sufficiently uniformly. Although pharmacological agents may be easily applied in the bathing solution, timed introduction of chemicals internally (e.g. SECOND MESSENGERS) requires elaborate perfusion devices.

An essential feature of whole-cell patch clamp is that the cell is dialysed to some extent by the solution in the patch pipette. This can be put to advantage whereby various chemicals can be introduced into the cytoplasm. A useful procedure is to fill the patch pipette with a fluorescent dye (e.g. lucifer yellow) to investigate the possibility that the patched cell may be dye coupled (i.e. electrically coupled) to neighbouring cells. However, as the internal environments of cells can be extremely diverse and cellular signalling may depend on complex molecular mechanisms, sometimes occurring in series, intracellular dialysis can

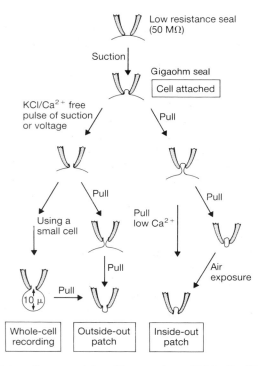

Fig. P9 Schematic representation of the procedures which lead to different patch clamp recording configurations — cell-attached, whole-cell recording, outside-out patch, and inside-out patch. The top frame is the configuration of a pipette in simple mechanical contact with a cell. Upon slight suction the seal between membrane and pipette increases in resistance by two to three orders of magnitude, forming a cell-attached patch. The improved seal allows a 10-fold reduction in background noise. This stage is the starting point for manipulations to isolate membrane patches in two different configurations — outside-out and inside-out. Alternatively, voltage clamp currents from whole cells can be recorded after disruption of the patch membrane if cells of sufficiently small diameter are used. The manipulations include withdrawal of the pipette from the cell (pull), short exposure of the pipette tip to air, and short pulses of suction or voltage applied to the pipette interior while cell attached. From [3].

cause serious problems in following a cell's functioning by whole-cell patch clamp. In particular, the phenomenon of wash-out has been noted, whereby a cell's response to an externally applied chemical is lost as dialysis progresses.

In order to prevent wash-out, the perforated patch recording mode has been introduced [6,7]. The patch pipette is partially filled with a weak concentration of antibiotic — most commonly nystatin or amphotericin B which are monovalent cation IONO-PHORES. Inclusion of the ionophore inside the patch pipette gradually opens perforations in the membrane beneath the pipette thus lowering the access resistance and ultimately allowing electrical contact to be established with the cell interior. In this way, whole-cell patch clamp mode can be achieved while preventing divalent ions and bigger molecules (such as second messengers) from leaving the cytoplasm.

Single-channel recording

Single-channel activity can be recorded when the patch pipette is initially pressed against the cell membrane surface to form a seal — the so-called cell-attached patch or on-cell patch clamp mode whereby the currents flowing across the patch of membrane beneath the pipette are recorded (Fig. P9). The main advantage of this mode is that the cytoplasm is undisturbed. However, there are several disadvantages: (1) any receptors that can be studied are those within the patch, so their stimulation by agonists requires dialysis of the patch pipette (alternatively, the agonists can be pre-loaded into the pipette); (2) the membrane potential cannot be measured unless an independent microelectrode is used; (3) the internal medium cannot be changed.

The two most common single-channel recording modes are those involving excised membrane patches, one in which the cytoplasmic side of the membrane is facing outward (inside-out patch) and the other where the external surface of the membrane faces outward (outside-out patch) (Fig. P9). In both cases, the voltage across the membrane patch can be controlled and a single or a few receptor/ion channels can be isolated and studied. Clearly, outside-out patches are ideal for investigating LIGAND-GATED ION CHANNELS, whereas inside-out patches are generally useful for elucidating membrane conductances activated by second messengers which can readily be applied during the recording by bath perfusion.

Applications of patch clamp recording

With the advent of single-channel recording techniques one can directly observe the activity of individual ion channels on a macromolecular level [8,9]. This technique has been used to probe the properties of voltage- and neurotransmitter-activated channels (*see* LIGAND-GATED ION CHANNELS; VOLTAGE-DEPENDENT ION CHANNELS), the channels that comprise electrical junctions (gap junctions) (*see* CELL JUNCTIONS) and ion channels modulated by intracellular second messengers. Features of the channel such as the amount of current it passes when open (single-channel conductance), the average amount of time the channel spends open or closed (single-channel kinetics), and its voltage- or ligand-gating characteristics can be measured. Some representative applications of patch clamp recording techniques in the investigation of cell membrane function are described below.

Voltage-dependent ion channels

A wide variety of voltage-dependent (or voltage-activated) ion channels have been investigated using single-channel techniques; they include channels which pass Na^+, K^+, and Ca^{2+} (Fig. P10). By combining the techniques of molecular biology and single-channel recording, information has been gathered at a molecular level about the operation of these channel types.

In the *Drosophila* mutant *Shaker*, for example, a rapidly inactivating voltage-gated K^+ channel termed the A-type channel is affected, leading to defects in neurotransmitter release at the neuromuscular junction and abnormal movements. Several alternative *Shaker* gene products generated by RNA SPLICING have been expressed in *Xenopus* oocytes (*see* XENOPUS OOCYTE EXPRESSION SYSTEM) and studied at the single-channel level. A comparison of the kinetics of activation and inactivation of ShA and ShB

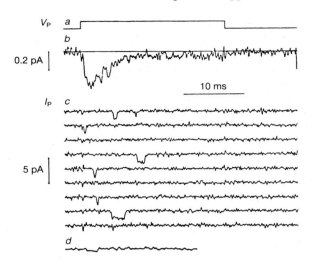

Fig. P10 Single Na^+ channel currents from cultured rat muscle cells recorded with the cell-attached patch clamp technique. *a*, Imposed membrane potential (V_p), held at -30 mV (with respect to the resting potential) and depolarized by 40 mV to $+10$ mV for ~23 ms at 1 s intervals. *b*, Average of a set of 300 current (I_p) records elicited by these pulses. The trace is very similar to the Na^+ current recorded (e.g. from the squid giant axon) by conventional two-electrode voltage clamp. *c*, Nine successive individual records from the above set. Square pulses of inward current (average size 1.6 pA) can be seen in most of the records; these correspond to the transient opening of individual channels by membrane depolarization. *d*, Record taken when two-thirds of the Na^+ in the pipette had been replaced with tetramethylammonium ions; the single-channel current is reduced accordingly. From [2].

showed that upon a DEPOLARIZING pulse, ShA channels opened at the beginning of the pulse, stayed open briefly (5–10 ms), then closed. ShB products likewise opened briefly and closed, but continued to open and close periodically for the duration of the pulse. A variation of ShB opened in long groups or bursts (20–30 ms) of openings. These gene products differed in their N- and C-terminal regions and thus demonstrated that the inactivation properties of the channel are, at least to some extent, controlled by these regions.

Ligand-gated ion channels

Ion channels activated by neurotransmitters have been widely studied by single-channel techniques (Fig. P11). The most direct way of investigating the properties of such channels is to include the transmitter of interest in the solution filling the patch pipette. On-cell patches can then be used to record channel activity. Likewise, an outside-out patch can be obtained and the transmitter of interest can be applied directly to the patch. These approaches have been used extensively to describe ion channels activated by the excitatory amino acid transmitter glutamate (*see* EXCITATORY AMINO ACID RECEPTORS). The differences, for example, between the properties of the *N*-methyl-D-aspartate (NMDA) and non-NMDA (AMPA, kainate, quisqualate, 4-AP) glutamate receptor-linked channels have been described. Both channel types exhibit multiple conductance states, display bursting behaviour, and certain of the channels rapidly desensitize. The conductance of the NMDA receptor-activated channel is generally larger

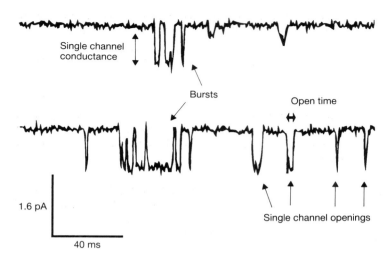

Single channel
conductance

Bursts

Open time

Single channel openings

1.6 pA

40 ms

Fig. P11 Single-channel records of glutamate-activated ion channels from retinal horizontal cells of the teleost fish, white bass (*Roccus chrysops*). The records were obtained from an inside-out patch, held at – 70 mV and activated by a brief application of 100 μM glutamate. Brief channel openings as well as groups or bursts of openings occurred in response to glutamate. The mean single-channel conductance is 13 pS and the mean open time is 8 ms.

than that of non-NMDA channels. Subunits of both NMDA and non-NMDA channels have been cloned. It is now possible to express individual subunits and combinations of subunits to determine the properties of the receptor and see how various combinations affect function.

Gap junction channels

Intracellular dialysis with lucifer yellow via patch pipette can reveal coupling of cells clamped in the whole-cell mode. However, cells can still be electrically coupled even if the gap junctions are too small to allow passage of the dye. The properties of the channels (connexons) that make up gap junctions or electrical synapses (*see* CELL JUNCTIONS) have also been studied by patch clamp [10]. Junctional conductance through individual channels has been studied in pairs of coupled cells in which both cells are whole-cell voltage clamped. In situations where the cells are only weakly coupled or are exposed to a compound which uncouples them, it is possible to observe single-channel events. The single-channel characteristics of cells such as clonal rat liver cells, pancreatic acinar cells, cardic myocytes, hepatocytes, and retinal horizontal cells, for example, have been investigated. In single retinal horizontal cells, it has been possible to study the single-channel conductance and gating characteristics of hemi-gap junction channels [11]. It was thus possible to show directly that the neuromodulator DOPAMINE suppressed the junctional current by activation of a cAMP-dependent PROTEIN KINASE. Although elevation of cGMP level also suppressed the junctional current, this effect was not mediated by the same kinase.

Modulation of ion channels by second messengers

A powerful use of single-channel recording techniques is in the study of POST-TRANSLATIONAL MODIFICATION of ion channels. The on-cell patch technique is particularly useful for demonstrating the existence of a second messenger that modulates channel activity, as it leaves the intracellular medium undisturbed while allowing the second messenger system to be activated by bath application of the neuromodulator [12]. The increased cytoplas-

mic level of second messenger can still affect the ion channels within the membrane patch under study. For instance, the action of the neurotransmitter 5-HYDROXYTRYPTAMINE (5-HT, serotonin) on a K+ channel (the S-channel) in sensory neurons of *Aplysia* has been studied in this way. Using the on-cell patch configuration, openings and closings of one to several channels were recorded over time. 5-HT applied to the cell was seen to reduce dramatically the number of openings of the S-channels and in fact silenced most of the activity in multichannel recordings. Because the patch of membrane was isolated from the exogenous 5-HT this could not be directly affecting the channels. Instead, a diffusible, intracellular second messenger was proposed to be at work. Subsequent studies showed this to be cAMP, the intracellular levels of which were raised by 5-HT activation of a 5-HT-dependent adenylate cyclase. Single channels recorded in inside-out patches were also closed when exposed to the catalytic subunit of PROTEIN KINASE A, suggesting that the modulated channel was likely to be phosphorylated. In this way, the modulation of many other types of K+ and Ca2+ channels has been shown to be coupled to G proteins either directly or through second messengers.

Patch clamp recording has also shown that not all channel modulation by second messengers involves phosphorylation. In the special case of vertebrate photoreceptors, for example, cGMP gates a cation conductance (the so-called transduction channel) in excised inside-out membrane patches in the absence of ATP or GTP, that is, without necessitating the presence of a kinase (*see* VISUAL TRANSDUCTION). Thus, it seems that these channels are being modulated by cGMP by some allosteric interaction and not through phosphorylation [13].

Secretion

Another important membrane function studied by patch clamp is secretion from cells by EXOCYTOSIS in response to electrical, hormonal or antigenic stimulation. The fusion of secretory vesicles with the plasma membrane leads to a rise in the overall membrane surface area and this can readily be detected as an increase in capacitance in the whole-cell recording mode [14]. As vesicles fuse with the plasma membrane one by one, the cell membrane capacitance increases in a stepwise manner. This technique has been used to investigate the relation between Ca^{2+} and secretion in mast cells, which secrete histamine (among other chemicals) as part of an inflammatory or immune response. Mast cells are a useful model as their surface area increases two to four times on stimulation. It has also been possible to correlate directly the rate of capacitance change (i.e. exocytosis) in adrenal chromaffin cells with the release of catecholamines, the latter being measured by an electrochemical detector [15]. A significant delay (5–100 ms) was found between the application of a depolarizing pulse (through the patch pipette) and occurrence of secretion, unlike the neuromuscular junction where the delay is <1 ms.

Another application of patch clamp methodology for the study of cellular secretion is in sniffer pipettes, where excised patches [16] of transmitter-sensitive (i.e. receptor-containing) membranes or whole cells [17] are patch clamped and brought near the secretory cell. Any chemical transmitter(s) released from the latter by stimulation may then be detected by the current induced in the membrane in the sniffer pipette.

M.B.A. DJAMGOZ

J.E.G. DOWNING

E.M. LASATER

See also: CYSTIC FIBROSIS.

1 Neher, E. et al. (1978) The extracellular patch clamp: A method for resolving currents through individual open channels in biological membranes. *Eur. J. Physiol.* **375**, 219–228.

2 Sigworth, F.J. & Neher, E. (1980) Single Na^+ channel currents observed in cultured rat muscle cells. *Nature* **287**, 447–449.

3 Hamill, O.P. et al. (1981) Improved patch clamp techniques for high-resolution current recording from cells and cell-free membrane patches. *Eur. J. Physiol.* **391**, 85–100.

4 Neher, E. & Sakmann, B. (1992) The patch clamp technique. *Sci. Am.* **266**, 28–35.

5 Edwards, F.A. et al. (1989) A thin slice preparation for patch clamp recordings from neurones of the mammalian central nervous system. *Eur. J. Physiol.* **414**, 600–612.

6 Horn, R. & Marty, A. (1988) Muscarinic activation of ionic currents measured by a new whole-cell recording method. *J. Gen. Physiol.* **92**, 145–159.

7 Rae, J.L. & Fernandez, J. (1991) Perforated patch recordings in physiology. *News Physiol. Sci.* **6**, 273–277.

8 Levitan, I.B. & Kaczmarek, L.K. (1991) *The Neuron* (Oxford University Press, Oxford).

9 Hille, B (1992) *Ionic Channels of Excitable Membranes*, 2nd edn (Sinauer, Sunderland, MA).

10 Spray, D.C. et al. (1991) Distinctive gap junction channel types connect WB cells, a clonal cell line derived from rat liver. *Am. J. Physiol.* **260**, C513–527.

11 DeVries, S.H. & Schwartz, E.A. (1992) Hemi-gap junction channels in solitary horizontal cells of the catfish retina. *J. Physiol., Lond.* **445**, 210–230.

12 Kaczmarek, L.K. & Levitan, I.B. (1987) *Neuromodulation* (Oxford University Press, Oxford).

13 Fesenko, E.E. et al. (1985) Induction by cyclic GMP of cationic conductance in plasma membrane of retinal rod outer segment. *Nature* **313**, 310–313.

14 Penner, R. & Neher, E. (1989) The patch-clamp technique in the study of secretion. *Trends Neurosci.* **12**, 159–163.

15 Chow, R.H. et al. (1992) Delay in vesicle fusion revealed by electrochemical monitoring of single secretory events in adrenal chromaffin cells. *Nature* **356**, 60–63.

16 Hume, R.I. et al. (1983) Acetylcholine release from growth cones detected with patches of acetylcholine receptor-rich membranes. *Nature* **305**, 632–634.

17 Tachibana, M. & Okada, T. (1991) Release of endogenous excitatory amino acids from ON-type bipolar cells isolated from the goldfish retina. *J. Neurosci.* **11**, 2199–2208.

patching Clustering of membrane proteins on the surface of animal cells in culture in response to binding with antibodies or lectins, which demonstrates the ability of membrane proteins to move laterally in the membrane.

pathogenesis-related proteins (PR proteins) General term to describe proteins responsible for the plant pathogenic capacity of fungi and bacteria and for proteins synthesized by the host plant in response to infection by such organisms and by viruses. PR proteins were initially identified as bands or spots on one- or two-dimensional polyacrylamide gel ELECTROPHORESIS of protein extracts of pathogens. The appearance of a novel band or spot during the time course of an infection defined the protein as pathogenesis related. Subsequent characterization of PR proteins has resulted in the identification and cloning of degradative enzymes (e.g. cutinases, proteases) that are crucial for successful pathogenesis. *See also*: PLANT PATHOLOGY.

pathovar A sub-classification of a single pathogenic species according to its host range. Just as the term cultivar describes a population of host plants with the same resistance characteristics, the term pathovar describes a population of the pathogen with shared virulence characteristics. *See also*: PLANT PATHOLOGY.

Pattern formation

IN the development of a multicellular organism, a single cell — the fertilized egg — gives rise to a multitude of specialized cell types arranged in a precise spatial pattern. How is this spatial pattern created? In part, the answer depends on cell movements that mould the shape of the body and adjust the positions of cells that have already adopted different characters. The heart of the matter, however, lies in the mechanisms by which specialized characters are assigned at the outset according to the cells' positions, so that each type of cell is generated at an appropriate location. In developmental biology, the term pattern formation refers especially to this latter process [1–3].

Cells at one end of the embryo must, for example, take on special characteristics appropriate to the head (in an animal) or the shoot (in a plant); cells at the other end must take on characteristics appropriate to a tail or a root. As each tissue develops, its cells must further diversify according to their positions to give ultimately a regular fine-grained pattern of differentiated cell types. In general, the final complex pattern is generated through a series of relatively simple steps: the overall large-scale organization of the body is established early, when the embryo is still small, and progressively finer levels of detail are embroidered on this basic plan as the body grows.

Whereas the final steps of pattern formation involve selection of a mode of overt cell DIFFERENTIATION — as photoreceptor cell, or cartilage cell, or smooth muscle cell, for example — the early steps involve specializations of cell character of a less obvious kind, which commit the cell to serve as a precursor of a certain range of cell types and to generate those cell types according to certain rules appropriate to a particular region of the body. An example is the formation of the primary GERM LAYERS of an animal embryo, in which the cells become specialized as ECTO-DERM, MESODERM, and ENDODERM, each with a distinctive, but still wide, range of possibilities of differentiation open to it. Another example is the specification of the main body axis, in which the cells of the future anterior, middle and posterior parts of the body (head, trunk and tail, or shoot, stem and root) acquire intrinsically different characters. The position-dependent character of a cell in this context is often referred to as its positional value, and cues that determine the cell's positional value are said to supply it with POSITIONAL INFORMATION. Cells with different positional values — for example, cells at different levels along the embryonic body axis — are not necessarily destined to differentiate into visibly different cell types, but they are none the less nonequivalent (*see* EQUIVALENCE GROUPS), in the sense that they are committed to generate structures (such as vertebrae) with different shapes and/or different fine-grained patterns.

The sequential elaboration of a more and more detailed pattern depends on cell memory: influences acting transiently at an early stage in development leave a lasting imprint on the character of a cell and its progeny. The later influences, rather than overriding the early positional information, build upon it and refine it. Thus the final choice of differentiated state is governed in a combinatorial fashion by items of positional information received at different stages in development.

The study of pattern formation centres on two broad questions: (1) how are cells supplied with positional information; and (2) how do the cells register and interpret this information? In particular, what are the biochemical cues that tell a cell where it is, and what molecular form does a cell's internal record of its positional value take?

Intracellular versus extracellular positional cues

In principle, cells can be supplied with positional information in two fundamentally different ways: by extracellular signals, and by intracellular signals. Extracellular signals organize a pattern by influencing the characters of cells that are already in existence. Intracellular signals organize a pattern within a single cell, and the intracellular pattern is then converted into a multicellular pattern by subdivision of that cell; this mode of pattern formation thus depends on asymmetric cell division in which, typically, localized CYTOPLASMIC DETERMINANTS in the parent cell become segregated into separate progeny cells, making them different already at the moment of their birth.

In animals, at least, examples of pattern formation through asymmetric cell division are largely confined to the very earliest stages of development, when the fertilized egg cleaves. In most species (mammals being an exception), the egg is an exceptionally large cell with easily detectable asymmetries, which determine

the orientation of the future anteroposterior and dorsoventral axes of the embryo (*see* AMPHIBIAN DEVELOPMENT). Localized cytoplasmic determinants are thought to be present in the egg, although for the most part they are poorly characterized. The mass of POLE GRANULES or pole plasm seen in many species (*see* DROSOPHILA DEVELOPMENT) is an important example: this material, initially concentrated at one end of the egg, becomes segregated into the future germ-cell lineage, and is thought to be a determinant of the difference between germ cells and somatic cells.

In insects, pattern formation by intracellular signalling is taken to an extreme. The egg undergoes repeated rounds of nuclear division without cell division, giving rise to a syncytial embryo in which many nuclei share a common cytoplasm. Eventually, this cytoplasm becomes partitioned into normal mononucleate cells; but by that stage (corresponding, in *Drosophila*, to about 6000 cells) the pattern of the body has already been sketched out in considerable detail. The process in *Drosophila* has been analysed at the molecular and genetic level so thoroughly as to have become one of the chief paradigms of pattern formation; most of the genes involved have been cloned (*see* DROSOPHILA DEVELOPMENT). By the technique of *in situ* HYBRIDIZATION, one can demonstrate the emergence of patterns of localized expression of these regulatory genes, through which the body plan becomes delineated — as in most developing systems — well before gross anatomical signs of differentiation are visible. Different sets of genes and gene products organize the main anteroposterior axis of the body, the specialized regions at the extreme ends of this axis, and the dorsoventral axis. The anteroposterior patterning mechanism in particular illustrates some important principles, relevant also to systems whose patterning depends on extracellular signals.

Morphogen gradients and pattern formation along the head-to-tail axis

The polarity of the *Drosophila* anteroposterior axis is defined initially by a gradient in the concentration of a signalling molecule that diffuses from a source at one end of the axis; the local concentration — declining from a maximum at one end of the egg to a minimum at the other — serves as an indicator of position and controls the spatial pattern of gene expression. A molecule that provides positional information in this way is called a MORPHOGEN. In the *Drosophila* egg, the morphogen is a protein, the product of the *bicoid* gene. The *bicoid* messenger RNA is deposited during oogenesis at the anterior end of the egg and remains anchored there, but the protein translated from it is free to diffuse, setting up a concentration gradient with a maximum at that end. The *bicoid* protein induces expression of other genes in a concentration-dependent manner: each regulated gene has its own characteristic threshold for response to the morphogen, and so becomes switched on in a specific spatial domain. Genes induced by the primary morphogen in their turn generate more localized spatial signals to govern finer details of patterning. Eventually, through a complex cascade of such interactions, the early embryo becomes organized into a series of segments, each with a similar internal pattern defined by the localized expression

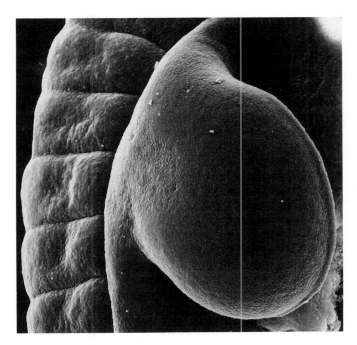

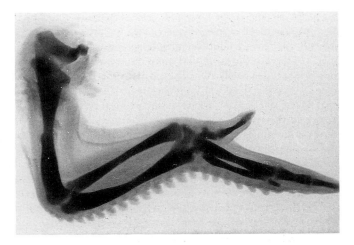

Fig. P13 Embryonic chick wing at 10 days of incubation. The internal tissues of the limb bud have differentiated to form a miniature version of the adult skeletal pattern.

Fig. P12 A chick wing bud, seen by scanning electron microscopy, at 4 days of incubation. The apical ectodermal ridge, which provides an essential signal for proximodistal outgrowth and patterning, can be seen at the distal edge of the tongue-shaped outgrowth. The skeletal elements have not yet begun to differentiate and the internal tissues of the bud appear almost uniform in conventional sections at this stage; but it can be shown that regional differences of HOM/Hox gene expression are already becoming established. Courtesy of Paul Martin; from [2].

of SEGMENTATION GENES, and each distinguished from other segments by the expression of a particular set of homeotic SELECTOR GENES (homeotic genes, HOM genes).

The role of HOM/*Hox* genes

The products of the segmentation genes and the HOM genes jointly define each cell's anteroposterior positional value in the insect embryo. They can be regarded as molecular 'address labels': the HOM gene products specify the district; the segmentation gene products specify, so to speak, the house number. A mutation that obliterates a HOM gene or makes it active in an inappropriate segment can cause a grotesque disturbance of the body plan — a HOMEOTIC MUTATION — giving rise, for example, to a fly that has legs instead of antennae growing out of its head, or one whole body segment converted into a duplicate of another.

HOM genes provide a unifying theme in animal pattern formation: they have been highly conserved in evolution, and seem to serve as specifiers of positional value along the anteroposterior axis throughout most of the Animal Kingdom, from nematodes to mammals (*see* HOMEOBOX GENES AND HOMEODOMAIN PROTEINS). The HOM genes of insects, and their homologues — often called *Hox* genes — in other species, can be recognized by two features: they contain a highly conserved homeobox sequence, involved in binding to DNA, and they occur in tightly linked clusters in the genome, within which the chro-

mosomal ordering of the individual HOM genes generally matches the spatial ordering of their expression domains in the body. In mammals, for example, there are four homologous clusters of *Hox* genes with these properties, all expressed in a regular series of domains along the anteroposterior axis of the body. (The pattern is most clearly visible in the embryonic central nervous system and especially in the hindbrain, which is segmented, rather like the body axis of *Drosophila*.) As in *Drosophila*, the products of the HOM/*Hox* genes seem to regulate development of the regions in which they are expressed, and artificially mutated expression patterns cause anatomical abnormalities.

It is not yet clear how cells in the early embryo of a mammal or other vertebrate are initially supplied with positional information to control HOM/*Hox* gene expression along the main body axis: the mechanism cannot be the same as in *Drosophila*, since the vertebrate embryo is not a syncytium, and positional signals have to pass from one cell to another.

Pattern formation in the vertebrate limb

Perhaps the best understood example of positional signalling and pattern formation in vertebrates is the development of the limbs, which has been studied mainly in the chick. The limbs arise from tongue-shaped buds that project from the flanks of the early embryo (Fig. P12). Each bud consists at first of undifferentiated embryonic connective tissue, or MESENCHYME, encased in a jacket of embryonic epidermis (ectoderm). The mesenchyme will differentiate to form the characteristic pattern of tissues of the skeleton (Fig. P13). Two signalling systems operate early in limb development to provide the mesenchyme cells with positional information along two axes — proximodistal (base to tip) and anteroposterior (thumb to little finger). (The specification of the third or dorsoventral axis is not well understood.) By grafting fragments of tissue between limb buds, it has been shown that the assignment of the proximodistal component of positional value depends on an influence from a specialized portion of the epidermis at the far end of the limb bud, called the apical ectodermal ridge. Mesenchyme cells in a region called the progress zone just beneath this ridge, while remaining apparently undifferentiated, take on progressively more and more distal positional values as they divide. Cells artificially removed from the influence of the ridge at an early stage will form proximal structures (e.g. upper arm); those removed at a late stage will form distal structures (digits). In the normal course of development, cell proliferation causes cells to overflow from the progress zone, in such a way that the most proximal part of the limb consists of cells whose ancestors spent the shortest time under the influence of the apical ectodermal ridge, while the most distal part of the limb consists of cells whose ancestors spent the longest time under its influence: it is the time spent under the influence of the ridge that dictates the positional value and hence the type of skeletal structure formed by the cells. The chemical nature of the signal from the ridge is not yet known.

The anteroposterior patterning of the limb bud is thought to be controlled by a morphogen gradient analogous to the *bicoid* gradient in *Drosophila*. The source of the morphogen appears to be a specialized patch of mesenchyme called the polarizing zone at the posterior margin of the bud. If a small amount of tissue from this region is grafted into the anterior side of a young limb bud, the cells on the anterior side of this host limb bud are diverted from their normal behaviour and a mirror-symmetric 'double-posterior' limb results: the anterior part of the hand, for example, consists of a mirror-image duplicate of the little-finger-to-thumb sequence of digits found in the posterior part (Fig. P14). Such experiments strongly suggest that the anteroposterior component of the positional value of the mesenchyme cells, and hence the type of skeletal structure that they form, is dictated by the graded concentration of a morphogen secreted by the cells of the polarizing zone. The action of a polarizing-zone graft can be mimicked by implanting an artificial source of retinoic acid (*see* RETINOIDS) on the anterior side of the limb bud, and this, coupled with the observation that retinoic acid is normally present in the limb bud, has suggested that retinoic acid may either be the natural morphogen here, or be a trigger for production of the morphogen.

What then is the molecular basis of positional value in this system? Remarkably, it has been found that at least two clusters of HOM/*Hox* genes are expressed in the limb bud — the members of one cluster in a series of spatial domains subdividing the limb along the anteroposterior axis, the members of another cluster in a series of domains subdividing it along the proximodistal axis. Moreover, there is evidence that retinoic acid can regulate HOM/*Hox* gene expression. All this suggests that the products of HOM/*Hox* genes may define the positional values of cells in the vertebrate limb bud, just as they appear to define positional values along the main head-to-tail axis of the body both in vertebrates and in invertebrates.

Pattern formation through cell contact interactions

In the examples of pattern formation discussed above, the major domains of the body are mapped out by mechanisms in which diffusible signalling molecules supply positional information over a broad tract of tissue (or morphogenetic field). In the later stages

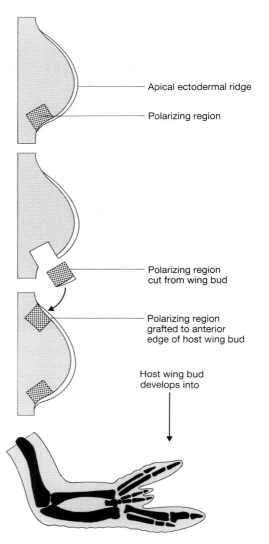

Apical ectodermal ridge

Polarizing region

Polarizing region cut from wing bud

Polarizing region grafted to anterior edge of host wing bud

Host wing bud develops into

Fig. P14 Experiment demonstrating the role of the polarizing region in controlling the 'thumb-to-little finger' sequence of digits along the anteroposterior axis of the chick wing bud. The polarizing region is thought to be the source of a morphogen whose local concentration defines which type of digit, if any, will form at a given site. From [2].

of development, during production of the ultimate fine-grained pattern of differentiated cell types, short-range signals transmitted directly via cell–cell contacts seem to be more important. This is most clearly seen in sensory epithelia such as the eye of *Drosophila*, where the various types of photoreceptor cells, lens cells and other supporting cell types are generated in a precise, repetitive mosaic. The patterning depends on signals whereby a cell that has embarked on a certain course of differentiation regulates the mode of differentiation of neighbouring cells that are in contact with it. Many of the genes coding for components of this type of signalling pathway in the *Drosophila* eye have been identified. Contact-mediated lateral inhibition represents an important special case: a cell that has begun to differentiate in a certain way inhibits its immediate neighbours from doing the same. This can give rise to a regular alternation of cell types. The products of the gene *Notch* and of other members of the set of so-called neurogenic genes are responsible for such lateral inhibition in *Drosophila*; they play a role not only in the eye and other sensory epithelia, but also in the development of mesodermal tissues. There are indications, especially from studies of the eye and the ear, that similar principles and homologous genes operate in vertebrates also.

Plants

Pattern formation in plants raises much the same questions as pattern formation in animals, but fewer answers are available. Recent studies, however, have applied to plant development the same methods of genetic analysis that have proved so successful in *Drosophila*, and these studies are beginning to reveal analogous mechanisms. For example, in flowering plants, the sequential pattern of elements in a flower — the sepals, petals, stamens and carpels — is specified by a set of highly conserved genes that define the differences between these appendages in much the same way that HOM genes define the differences between segments in an insect (*see* FLOWER DEVELOPMENT). Very little is known, however, about the morphogen gradients or other positional cues that control expression of these genes.

J.H. LEWIS

See also: INDUCTION.

1 Akam, M. & Gerhart, J. (Eds) (1991) Pattern formation and developmental mechanisms. *Curr. Opinion Genet. Devel.* **1**, 167–303.
2 Alberts, B. et al. (1989) *Molecular Biology of the Cell*, 3rd edn (Garland, New York).
3 Roberts, K. et al. (Eds) (1991) Molecular and cellular basis of pattern formation. *Development*, Suppl. 1.

Patterson function, Patterson synthesis A FOURIER SYNTHESIS which represents the vector distances between diffracting objects. It is calculated from the set of DIFFRACTION intensities alone and does not require their phases (*see* PHASE PROBLEM). The position and height of any peak in the map represent respectively the vector between two objects and the product of their two scattering powers (ATOMIC SCATTERING FACTOR and ATOMIC SCATTERING LENGTH for diffraction from X-rays and neutrons respectively). For molecules of only a few atoms, the peaks within a Patterson map can be directly interpreted to reveal atomic positions.

For macromolecules of N atoms there are up to N^2 vectors, all transposed to a common origin, such that they cannot be interpreted in terms of all the individual atoms. However, at very low RESOLUTION when the diffraction of, say, a tetrameric protein is effectively that arising from only four objects (the monomers), then the Patterson map may reveal the arrangement of the four subunits. In the method of MULTIPLE ISOMORPHOUS REPLACEMENT, a difference Patterson synthesis (calculated with coefficients $[F_{nat} - F_{deriv}]^2$) from the structure amplitudes of the native protein and HEAVY ATOM DERIVATIVE respectively) represents all the vectors between a few heavy atoms whose positions can thence be determined. *See*: PAIR-DISTRIBUTION; X-RAY CRYSTALLOGRAPHY.

Patterson, A.L. (1934) *Phys. Rev.* **46**, 372–376.
Rossmann, M.G. & Arnold, E. (1993) In *International Tables for X-ray Crystallography* (1992) Vol. B Reciprocal Space (Shmueli, U., Ed.) 230–263 (Kluwer Academic Publishers, Dordrecht).

PBL PERIPHERAL BLOOD LYMPHOCYTES.

pBR322 pBR322 (Fig. P15) is a widely used PLASMID vector for cloning DNA in *Escherichia coli*. pBR322 is itself a recombinant plasmid containing the DNA replication origin (ori) from the plasmid CoLE1, thus enabling it to be propagated in *E. coli*. In addition, it contains two antibiotic-resistance genes that confer resistance either to ampicillin (amp^R) or to tetracycline (tet^R). The antibiotic-resistance genes contain RESTRICTION SITES, so that foreign DNA can be inserted into these genes, an event that inactivates the antibiotic-resistance gene, thereby allowing identification of recombinants.

Bolivar, F. et al. (1977) *Gene* **2**, 95–113.

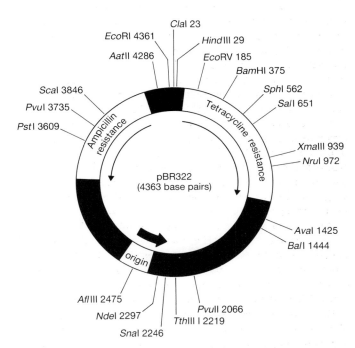

Fig. P15 Plasmid pBR322.

Pc The POLYCOMB locus of *Drosophila melanogaster*.

PC Phosphorylcholine, the immunodominant determinant (*see* HAPTEN) of the pneumococcal C-polysaccharide. BALB/c strain mice produce anti-PC antibodies of limited diversity, first revealed by the cross-reactive IDIOTYPES and subsequently confirmed by structural and molecular genetic studies. More than 80% of anti-PC antibodies from BALB/c mice with the immunoglobulin heavy chain HAPLOTYPE IgH^a share an idiotypic determinant, known as the T15 idiotype. *See also*: IMMUNOGLOBULIN STRUCTURE.

PC12 cells Cells originally cloned from a transplantable rat phaeo-chromocytoma (a tumour of the adrenal medulla). PC12 cells have many of the properties of adrenal medullary cells, for example being able to synthesize, store, and secrete CATECHOLA-MINES. They also respond to NERVE GROWTH FACTOR or transfection with the proto-oncogene *ras* by ceasing proliferation, becoming electrically excitable and extending neurites, thus providing a model *in vitro* system for the study of neuronal DIFFERENTIATION.

PCNA PROLIFERATING CELL NUCLEAR ANTIGEN. *See also*: PAPOVA-VIRUSES.

PCR POLYMERASE CHAIN REACTION.

PCR cycle (1) Photosynthetic carbon reduction cycle (*see* CALVIN CYCLE).
(2) A cycle of DNA replication in the POLYMERASE CHAIN REACTION.

PDGF Platelet-derived growth factor. *See*: GROWTH FACTORS.

PDYN Prodynorphin, a precursor for the opioid peptide dynorphin. *See*: OPIOID PEPTIDES AND THEIR RECEPTORS.

pea enation mosaic virus group A monotypic group named after the type virus. It is a multicomponent virus with isometric particles 28 nm in diameter which sediment as two components. The capsids are made up of a single protein species, M_r 22 000. The genome comprises two species of (+)-strand single-stranded RNA, the bottom component containing RNA 1 (M_r 1.7×10^6) and the middle component RNA 2 (M_r 1.3×10^6). There is no information on genome organization. *See also*: PLANT VIRUSES.

Francki, R.I.B. et al. (1985) In *Atlas of Plant Viruses*, Vol. 2, 39 (CRC Press, Boca Raton, FL).

peak D The psbA reaction centre polypeptide of PHOTOSYSTEM II. It is also known as the 32kD (32K) protein, the herbicide-binding protein (as it is the target of a number of herbicides), and the photogene 32 (because of increases in transcript levels observed on illumination). The protein appears to be very susceptible to damage during photosynthesis, necessitating a rapid turnover at high levels of light. *See*: PHOTOSYNTHESIS.

Pearson's syndrome *See*: MITOCHONDRIOPATHIES.

pectins A mixed group of highly complex polysaccharides found in the middle lamella and primary walls of higher plants and algae of the Characeae, and as gums. They are vital to the maintenance of adhesion between plant cells.

Pectinic acids are based on repeated sequences of α1,4-D-galactopyranosyluronic acid with α1,2-L-rhamnopyranosyl residues interspersed within them and usually some methyl ester content. Where the proportion of rhamnose is high, or residues of rhamnose and galacturonic acid alternate, rhamnogalacturonan is a better term, though it can be used of all pectinic acids. Most pectinic acids carry branches.

Pectic acids are linear polymers of α1,4D-galacturonic acid, prepared under partial degradation conditions so that other sugars and methyl esters are lost and the molecular weight may be much reduced. Industrial pectins, as used in foodstuffs, are generally of this type. Whole pectins represent the unfractionated pectic materials extracted from plant tissues under defined, mild conditions: they usually contain heterogeneous populations of pectinic acids, as well as ARABINOGALACTANS. Cold water-soluble pectins are those pectins extractable with cold water or cold solutions of chelating agents such as EDTA or sodium hexametaphosphate. Hot water-soluble pectins are those extracted, after removal of cold water-soluble pectins, by hot water or hot solutions of chelating agents. Residual pectins cannot be extracted without the use of degradative methods.

Pectinic acids can be subdivided into three types. Type I is strongly acidic with a low methyl ester content, contains short side-chains attached to galacturonic acid and has occasional rhamnosyl residues in the backbone. Type II is of larger molecular size and contains all the features of type I pectinic acids, with, additionally much larger branches called 'neutral blocks'. These neutral blocks have cores of galactan, with peripheral arabinosyl residues and arabinan sequences. Type II pectinic acids also contain methyl ether groups and have some substitution of their rhamnosyl residues, probably by neutral blocks. Twelve or more different sugars can be present in such a pectinic acid. Type O pectinic acid is known only from biosynthetic studies (see below). It has a higher methyl ester content than type I pectinic acid and is less acidic. It lacks neutral blocks, but probably contains both rhamnose and short branches. The short branches of pectins vary with their source, but commonly contain xylose (glycosidically linked to galacturonan), galactose, fucose, glucuronic acid and 4-O methyl glucuronic acid.

Pectinic acids are unstable under most conditions. Above pH 8.0 they may undergo β-elimination at esterified residues of galacturonic acid with cleavage of the backbone. Below pH 6.0 arabinofuranosides hydrolyse, releasing arabinose or oligosaccharides. At neutrality, nonesterified carboxyl groups of galacturonic acid slowly hydrolyse the glycosidic link of the residues to which they belong. However, parts of pectinic acids resist degradation, so making their full structural analysis extremely difficult.

Most pectinic acids are water soluble and assume a flexible, extended configuration in solution. In some, self-association occurs. Metal ions, such as calcium and heavy metals, bind strongly to pectinic acids, by interaction with both hydroxyl and carboxyl groups and cause large conformational changes and precipitation. Much of the calcium in plant cell walls is sequestered by pectins.

Nothing is known of the initiation and priming of pectinic acid synthesis. The galacturonan backbone is synthesized from UDP-GalA by a membrane-bound complex containing an unknown number of glycosyl transferases. Methyl ester groups are added from S-ADENOSYLMETHIONINE during or after polymerization, so that the first product is a type O pectinic acid. It is suggested that short side chains are added simultaneously with the methyl ester groups and that rhamnose is also present at this stage. The type O pectinic acid rapidly converts to types I and II, by partial removal of methyl ester and addition of neutral blocks respectively. Pectic arabinogalactan is the precursor of the neutral blocks, which are probably generated by trans-glycosylation. Splicing of galacturonan by rhamnogalacturonan to yield larger molecules cannot be excluded. Pectin synthesis is, therefore, extremely complicated and must involve many glycosyl transferases and the presence of several kinds of recognition sites within the glycans. *See also*: GLYCANS.

PEG POLYETHYLENE GLYCOL.

penetrance A figure (usually given as a percentage) that expresses the probability that an abnormal gene will exert at least one of its phenotypic effects in an individual inheriting that gene. If the penetrance is less than 100% it is described as incomplete penetrance: this may result from variation between unrelated affected individuals in the precise genetic defect, and/or interaction with other genetic or environmental modifiers. Easily confused with EXPRESSIVITY, a term that embodies the variability in severity and clinical presentation between different manifesting carriers of the same genetic disease.

penicillin A β-lactam antibiotic produced by *Penicillium* species, and active against Gram-positive bacteria. It was discovered by Alexander Fleming in 1929, after he noticed that cultures of *Staphylococcus aureus* had been killed by the fungus *Penicillium notatum* which had contaminated the plates. In the 1940s, penicillin was isolated by Florey and Chain and colleagues. A range of different naturally occurring penicillins has since been discovered and semi-synthetic derivatives produced. Their common structural feature is the β-lactam ring, and the variable 'R' group (Fig. P16), which differs between various types.

Penicillins exert their bacteriocidal action by preventing formation of peptide (pentaglycine) cross-linkages during biosynthesis of the PEPTIDOGLYCAN cell wall (*see* BACTERIAL ENVELOPES). The structure of penicillin closely resembles that of the natural substrate (D-Ala–D-Ala–) of the transpeptidase enzyme which catalyses the cross-linking, and penicillin thus acts as a substrate analogue inhibitor of the enzyme. The inhibition is irreversible. Because the bacteria are unable to synthesize a structurally sound cell wall, they become susceptible to osmotic shock and lyse. Penicillins are only effective on actively growing bacteria (*see* PENICILLIN SELECTION). The various penicillins have different patterns of specificity for particular bacterial species.

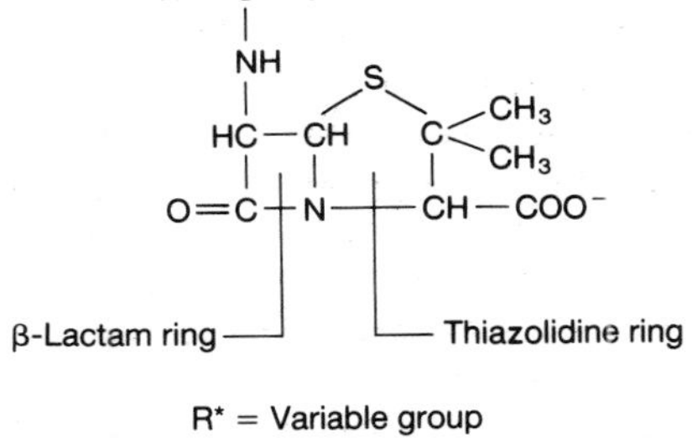

R* = Variable group

Fig. P16 Penicillin.

The β-lactam ring of penicillins is cleaved by β-LACTAMASES (e.g. penicillinases), thus rendering the molecule nonfunctional. Penicillin resistance in bacteria is due to the possession of these β-lactamase enzymes.

penicillin selection A widely used enrichment procedure for the isolation of AUXOTROPHIC mutants of *Escherichia coli*. The method is based on the fact that PENICILLIN will kill only actively growing bacterial cells by interfering with the synthesis of the cell wall, causing cell lysis. Following MUTAGENESIS the bacteria are incubated in a defined medium containing penicillin, but lacking the nutrient for which the auxotrophy is sought. For example, if selecting for histidine auxotrophs the medium would lack histidine. Within an hour or so >99% of all the non-histidine requiring PROTOTROPHS will have lysed whereas the histidine-requiring auxotrophs, which are unable to grow as a consequence of their deficiency, will survive in the presence of the penicillin.

PENK Proenkephalin A, a precursor for the enkephalins. *See*: OPIOID PEPTIDES AND THEIR RECEPTORS.

pentose phosphate pathway An alternative pathway of glucose degradation to the use of glycolysis and the tricarboxylic acid cycle, the first step of which is the enzymatic oxidation of glucose 6-phosphate by glucose-6-phosphate dehydrogenase. The pathway has three main activities depending on the tissue and the metabolic state. The primary purpose is to generate reducing power in the form of NADPH. The second is to convert hexoses to pentoses required in the synthesis of nucleic acids. The third is the complete oxidative degradation of pentoses by converting them to hexose which can enter the glycolytic pathway. By far the commonest defect associated with the pathway is GLUCOSE-6-PHOSPHATE DEHYDROGENASE DEFICIENCY; defects of other enzymes are rarely associated with clinical disease.

pentraxins A strongly conserved family of plasma proteins, usually pentamers having a symmetric ring of identical subunits. The human pentraxins are the prototype acute phase protein C-REACTIVE PROTEIN (CRP), and SERUM AMYLOID P COMPONENT (SAP). Pentraxins have specific calcium-dependent ligand binding and can activate COMPLEMENT. Their precise function remains unclear, as no polymorphism or deficiency of either protein has been reported, but they may have a nonspecific role in host defence. Human CRP is not glycosylated, but human SAP has a single typical complex biantennary GLYCAN side chain per subunit, which lacks the microheterogeneity characteristic of GLYCOPROTEINS.

pepsin ACID PROTEINASE from gastric juice which cleaves proteins preferentially on the carboxyl side of Phe or Leu residues (EC 3.4.23.1). It is formed from an inactive precursor, pepsinogen, by a cleavage of a single peptide bond.

peptidases Enzymes that hydrolyse the peptide linkage between amino acids in a POLYPEPTIDE. In the EC classification, the term peptidase is reserved for those enzymes that cleave either single residues or dipeptides at the termini of a polypeptide (aminopeptidases and carboxypeptidases) or which degrade small peptides (EC class 3.4). Endopeptidases that cleave internal peptide bonds in proteins are classified as PROTEINASES.

peptide bond The covalent amide linkage between amino acid residues in a polypeptide chain, which is formed by linking the α-carboxyl carbon of one residue with the α-amino nitrogen of another. *See*: PROTEIN STRUCTURE.

peptide neurotransmitters A large number of bioactive peptides are secreted by neurons (*see e.g.* NEUROPEPTIDES; OPIOID PEPTIDES AND THEIR RECEPTORS; PERIPHERAL NERVOUS SYSTEM), many as co-transmitters with one of the classical NEUROTRANSMITTERS such as noradrenaline. A few have been shown to have classical transmitter activity (e.g. substance P in spinal cord neurons involved in nociception, and the LHRH-like peptide that mediates synaptic transmission in sympathetic ganglia in the frog), but many others are thought to act as neurohormones or neuromodulators, acting in a PARACRINE fashion to modulate the actions of other transmitters.

peptide synthesis Peptides of up to ~ 150 amino acids can be synthesized by automated solid-phase methods (Merrifield synthesis). The required C-terminal amino acid, with its amino group blocked with a *tert*-butyloxylcarbonyl (*t*-BOC) group to avoid unwanted reactions, is attached to an insoluble matrix such as polystyrene beads through its carboxyl group and the *t*-BOC protecting group then removed by dilute acid. Amino acids are then added stepwise to the immobilized amino acid in the required sequence. The incoming amino acid has its α-amino group blocked by *t*-BOC. Protecting groups are also added to potentially reactive side chains. Its carboxyl group is activated by reaction with dicyclohexylcarbodiimide (DCC) and reaction of the free amino group of the immobilized amino acid or peptide with the activated carboxyl group results in formation of a peptide bond and release of dicyclohexylurea, leaving the growing peptide attached to the matrix. The *t*-BOC protecting group is then removed. The next amino acid is then added, and so on. The matrix is washed after each reaction to remove unwanted reagents and products. The completed peptide is released from the matrix by reaction with hydrogen fluoride which does not attack peptide bonds.

peptide transporters Proposed products of the *TAP* genes, which are involved in the transport of the small peptides produced by processing of viral antigens in the cytoplasm into the ENDOPLASMIC RETICULUM for onward transport to the cell surface in a complex with class I MHC MOLECULES. *See*: ANTIGEN PROCESSING AND PRESENTATION.

peptidoglycan A general term for a family of GLYCANS found in cell walls of bacteria, CYANOBACTERIA and actinomycetes: in Gram-negative organisms they form part of the inner wall (*see* BACTERIAL ENVELOPES). The glycan consists of repeating disaccharide units, forming chains of indeterminate length, cross-linked by non-protein peptide, so that, overall, it forms a giant cagework that envelops the organism and restricts its osmotic swelling. The glycan is more highly conserved than the peptide. Typically it is of the form

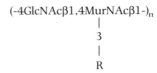

where glucosamine is always present and usually *N*-acetylated. The acyl muramic acid is commonly *N*-acetyl and less commonly of other types (e.g. *N*-glycolyl in mycobacteria) and normally of gluco-configuration (rarely manno-). In the methanogens, *N*-acetyl talosaminuronic replaces acyl muramic acid. R represents peptide. Synthesis of peptidoglycan involves: (1) the assembly of the part of the peptide on UDP-MurNAc, which occurs by the addition of individual amino acids and dimers of D-alanine, without involvement of tRNA; (2) the transfer of UDP-MurNAc-peptide to undecaprenyl phosphate and the addition of *N*-acetyl glucosamine from UDP-GlcNAc; (3) the addition of ammonia and (in many species) the addition of further amino acids from aminoacyl-tRNA to form a branch; and (4) the addition of the fully aminoacylated disaccharide to existing peptidoglycan. These are considerable interspecific variations in the amino-acid sequences of the peptides (though they are constant with a species) and transfers from tRNA may not always be involved: where they are, the tRNA is peculiar to this process and is not involved in protein synthesis. Non-protein amino acids are commonly found in the peptide. Glutamic acid is always the second amino acid and D-alanine occurs as the fourth residue. L-alanine is usually the first residue, although serine or glycine may replace it. Branching can occur at the glutamyl residue, which is always linked at its carboxyl group and can be variously substituted.

Cross-linking of the peptide occurs by transpeptidation at the third amino acid (usually lysine or *meso*-diaminopimelic acid; rarely others), so that the branch added at stage 3 forms the cross-link. It precedes addition of the disaccharide to glycan. Variations in the pattern or frequency of cross-linking form chemotypes of microbial species. Several antibiotics inhibit peptidoglycan synthesis. BACITRACIN prevents the conversion of undecaprenyl pyrophosphate to undecaprenyl phosphate, so inhibiting the undecaprenol phosphate cycle. Vancomycin prevents the transfer of peptidyl disaccharide from undecaprenyl pyrophosphate onto growing proteoglycan. Cycloserine and O-carbamyl serine prevent the dipeptide of D-alanine from forming, and PENICILLINS inhibit transpeptidation reactions and hence, cell wall assembly.

peptidyl prolyl *cis-trans* isomerases (PPIases) Abundant and universal proteins which catalyse the isomerization of X-prolyl peptide bonds and can accelerate the folding of proline-containing proteins (*see* POST-TRANSLATIONAL MODIFICATION OF PROTEINS; PROTEIN FOLDING). They can be divided into two unrelated groups — the CYCLOPHILINS and the FK500-binding proteins (FKBP). Both bind to immunosuppressive drugs — CYCLOSPORIN and FK500 respectively — but immunosuppression does not seem to be due to inhibition of the PPIase activity. The cyclophilin-related *ninaA* membrane protein in the eye of *Drosophila* is required for folding and stability of RHODOPSIN.

peptidyl transferase The enzyme complex associated with the RIBOSOME which catalyses peptide bond formation during the elongation phase of PROTEIN SYNTHESIS. The nascent peptide in the P-site of the ribosome is transferred to the amino group of the incoming aminoacyl-tRNA in the A-site of the ribosome. The peptidyl transferase complex is also involved in the termination of protein synthesis, catalysing the hydrolysis of the tRNA–peptide bond by transferring the C-terminal residue of the nascent polypeptide chain from the P-site tRNA onto a water molecule.

Studies in bacteria suggest that the enzyme complex is an integral part of the 50S ribosomal subunit. Ribosome reconstitution experiments indicate that the ribosomal proteins L6, L11, and L16 are critical to peptidyl transferase activity. Proteins L2, L5, L15, L18, L25 and L27 have also been implicated in peptidyl transferase activity, as also has 23S rRNA.

Some bacterial protein synthesis inhibitors (e.g. CHLORAMPHENICOL and LINCOMYCIN) appear to exert their effects through inhibiting the action of peptidyl transferase. CYCLOHEXIMIDE inhibits peptidyl transferase activity in the eukaryotic ribosome.

peptidyl-tRNA tRNA to which a polypeptide chain is attached. *See*: PROTEIN SYNTHESIS.

per The *period* gene of *Drosophila melanogaster*. *See*: BIOLOGICAL CLOCKS.

perdurance The persistence of a gene's product after the removal of the gene from a cell. In CLONAL ANALYSIS in DROSOPHILA, using X-ray induced MITOTIC RECOMBINATION, homozygous mutant clones induced late in development may not exhibit a mutant phenotype. This may reflect a lack of requirement for the gene at this stage, or alternatively, may be due to the persistence of sufficient wild-type gene product, derived from the heterozygous parental cell, to rescue the mutant progeny cells.

perforin (cytolysin) Pore-forming protein synthesized by immunocompetent CYTOTOXIC T CELLS. It is contained in cytoplasmic granules which are released in a directed fashion when the T cell is activated by encounter with a target cell. After release the perforins polymerize in the membrane of the target cell, forming ion-permeable channels.

perikaryon *See*: NEURON.

perineurium The sheath surrounding each bundle of nerve fibres in a peripheral nerve. The outermost layers are composed of densely packed longitudinally arranged collagen fibres, with some fibroblasts and macrophages. The deeper layers have a pronounced basement membrane.

***period* (*per*)** Genetic locus in *Drosophila* involved in the determination of biological rhythms. *See*: BIOLOGICAL CLOCKS.

peripheral blood lymphocytes (PBL) Mature immunocompetent circulating B CELLS and T CELLS. *See*: B CELL DEVELOPMENT; T CELL DEVELOPMENT.

peripheral membrane protein Membrane protein attached to the face of a membrane by electrostatic and hydrogen bond interactions and which is easily detached from the membrane by, for example, a change in pH. *Cf.* INTEGRAL MEMBRANE PROTEIN. *See*: MEMBRANE STRUCTURE.

The peripheral nervous system

THE peripheral nervous system consists of all neurons that are not confined within the brain and spinal cord. The major divisions of the peripheral nervous system are the somatic (or somatomotor), the sensory and the autonomic nervous systems, each defined according to anatomical and functional principles. The autonomic nervous system is subdivided into sympathetic, parasympathetic and enteric nervous systems, each with defined neural pathways and different embryological derivations from neural crest tissue. However, with our increasing knowledge of intrinsic innervation, from both a structural and functional basis, it is apparent that there are autonomic neurons and ganglia that are neither strictly parasympathetic nor sympathetic, for example pelvic ganglia and intracardiac ganglia. All the divisions of the peripheral nervous system have a distinctive neuronal chemical coding and pharmacology [1–3] (Fig. P17), and many features survive in primary tissue culture, a technique that has proven useful for electrophysiological and molecular biological studies that have advanced our knowledge of neuronal function.

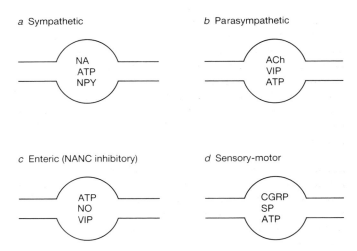

a Sympathetic

NA
ATP
NPY

b Parasympathetic

ACh
VIP
ATP

c Enteric (NANC inhibitory)

ATP
NO
VIP

d Sensory-motor

CGRP
SP
ATP

Fig. P17 Schematic diagram of the principal cotransmitters in autonomic and sensory nerves. From terminal varicosities of *a*, sympathetic; *b*, parasympathetic; *c*, enteric nonadrenergic, noncholinergic (NANC) inhibitory; *d*, sensory-motor nerves, various transmitter substances can be released together. ACh, Acetylcholine; ATP, adenosine-5′-triphosphate; CGRP, calcitonin-gene related peptide; NA, noradrenaline; NO, free radicals of nitric oxide; NPY, neuropeptide Y; SP, substance P; VIP, vasoactive intestinal polypeptide.

The somatic motor nervous system

The somatic motor (or efferent) nervous system consists of those motor neurons that have their cell bodies in the nuclei of the ventral horn of the spinal cord, and that innervate skeletal muscles (*see* MUSCLE; MYOGENESIS). Truncal and peripheral muscles are innervated by axons leaving the spinal cord via the ventral root, passing through the dorsal and ventral rami communicantes. Centrally, these neurons receive inputs from corticospinal pyramidal tract motor neurons. Skeletal muscles of the head and neck are innervated by neurons that have their cell bodies in pontine and medullary motor nuclei of cranial nerves V, VII, XI and XII, and that course with these cranial nerves as well as the glossopharyngeal (IX) and vagus (X) nerves. The central input to these neurons comes from neurons of the corticobulbar pyramidal tract.

Somatic motor neurons transmit to skeletal muscle with ACETYLCHOLINE, which acts on NICOTINIC RECEPTORS in the sarcolemma. Although there is evidence that ATP is released from somatic motor neurons, and that it can modulate (either *per se* or through its breakdown product ADENOSINE) the release and action of acetylcholine, there is no evidence that it acts as a true neurotransmitter in this division of the peripheral nervous system [4,5].

Some somatic neurons may contain NEUROPEPTIDES, in particular opioid peptides such as β-endorphin (*see* OPIOID PEPTIDES). Calcitonin gene-related peptide (CGRP) has been found at motor end-plates, but in contrast to other divisions of the peripheral nervous system, neuropeptides are not well represented in somatic nerves.

The somatic sensory nervous system

The somatic sensory (or afferent) nervous system of the torso and limbs consists of neurons that have their cell body in a dorsal root ganglion, from which a central process extends through the dorsal root to terminate in the substantia gelatinosa of the dorsal horn of the spinal cord. As they enter the spinal cord the fibres may send short ascending or descending collaterals that also terminate in the substantia gelatinosa.

The peripheral branch of the sensory neuron ends, in the organ that it innervates, as a bare nerve ending or a specialized structure (e.g. muscle spindles, heat and touch receptors). The afferent system for the face and head mostly has its cell bodies in the semilunar ganglion of the trigeminal nerve (V), but some are in the geniculate ganglion of the facial nerve (VII), and the superior ganglia of the glossopharyngeal (IX) and vagus (X) nerves. However, proprioceptive neurons innervating joints, tendons, and muscles of the face and head have their cell bodies in the mesencephalic trigeminal tract in the upper pons and mesencephalon.

Primary sensory neurons with their cell bodies in the dorsal root ganglia also function physiologically or pathophysiologically as motor neurons. In the wheal and flare formation of Lewis's triple response the vasodilatation is caused by afferent impulses passing along collaterals of the sensory nerves (the so-called axon reflex) which terminate in close proximity to cutaneous blood vessels. The neurotransmitter that causes this vasodilatation is CGRP and/or substance P, and in a subpopulation of nerves, possibly ATP [4–6]. These two peptides occur mainly, but not exclusively, in sensory nerves.

The sympathetic nervous system

The sympathetic nervous system is basically a two-neuron efferent system with the cell bodies of the preganglionic neurons situated in the lateral grey columns of the spinal cord from T1 to L3, having a single axon that leaves the cord via the ventral root. The sympathetic nerves separate from the somatic motor nerves, and form the white rami communicantes that lead to the paravertebral ganglia. These form two sympathetic chains running along each side of the spinal cord from the level of C1 down to S3. The preganglionic nerves either synapse with postganglionic nerve cells in the paravertebral ganglia, or pass through without synapsing to collateral prevertebral ganglia, or they enter the sympathetic chain and ascend or descend several segments before synapsing. The prevertebral ganglia are located relatively remotely from the spinal cord, in perivascular plexuses sited at the origin of arteries. The most prominent of these collateral ganglia are the coeliac, superior mesenteric, and inferior mesenteric ganglia. The preganglionic neurons that synapse in the paravertebral chain are very short, whereas the splanchnic nerve, formed from the preganglionic neurons that pass through without synapsing, and subsequently pass through the diaphragm to synapse in the coeliac, superior mesenteric and inferior mesenteric ganglia, are long.

The postganglionic fibres emerge from the sympathetic chain as the grey rami communicantes, and tend to run periarterially. The head receives its postganglionic sympathetic innervation from the superior cervical ganglion, which lies at the top of the sympathetic chain, above the middle and inferior cervical ganglia. Thoracic viscera receive their postganglionic sympathetic innervation predominantly from the sympathetic chain ganglia at levels C1 to T5, abdominal viscera from T5 to T12 via the coeliac ganglion, with the distal colon and genito-urinary tract receiving innervation from the chains at levels from L1 to S3 via the inferior mesenteric ganglion. Postganglionic nerves from the cervical ganglia innervate the heart, thyroid gland, and upper limb sweat glands and blood vessels.

The predominant neurotransmitter in preganglionic sympathetic nerves is acetylcholine, which mediates fast nicotinic transmission in sympathetic ganglia [7]. However, these preganglionic neurons also contain other neurotransmitters such as GLUTAMATE, γ-aminobutyrate (GABA), or a variety of neuropeptides [8,9] that contribute to synaptic transmission. Postganglionic sympathetic neurons predominantly transmit with varying proportions of ATP and NORADRENALINE, and many populations also contain neuropeptide Y (NPY) [2]. Some postganglionic sympathetic neurons maintain a cholinergic phenotype, and many of these also contain vasoactive intestinal polypeptide (VIP) [9].

The parasympathetic nervous system

This division, like the sympathetic system, is a two-neuron

system. Preganglionic nerves originate from the midbrain and medullary centres of cranial nerves III, VII, IX and X, and from the lateral grey horns of the spinal cord at S2 to S4. The parasympathetic cranial nerves III, VII and IX synapse in the cranial ganglia, namely the ciliary, pterygopalatine, submandibular and otic ganglia. The vagus nerve and the parasympathetic sacral outflow synapse in local ganglia (sometimes called terminal ganglia), which are admural or intramural to visceral organs. From these local ganglia axons project to target tissues such as visceral smooth muscle, cardiac muscle, and secretory cells.

Preganglionic parasympathetic neurons are predominantly cholinergic, although they may also contain neuropeptides such as VIP, substance P and enkephalin (*see* OPIOID PEPTIDES). Postganglionic parasympathetic neurons are also characteristically cholinergic, but in various subpopulations also contain ATP [5] and neuropeptides such as NPY, VIP, substance P, and enkephalin [9].

The enteric nervous system and local ganglia

The enteric nervous system represents a highly developed system of local ganglia and their interconnections that innervates the gastrointestinal tract. Local ganglia are found closely apposed to, or embedded in, the walls of visceral organs such as the airways, heart, liver, pancreas, ureter, urinary bladder, gall bladder and some blood vessels. Each organ may possess several local ganglia, and their size may vary from microganglia containing two or three neurons to much larger ganglia containing hundreds of neurons. The function, architecture and neural connections of the majority of local ganglia are not well documented, and although they may receive preganglionic parasympathetic input they may also be innervated by postganglionic sympathetic nerves. In addition they may contain sensory neurons and interneurons.

The enteric nervous system contains two major ganglionated plexuses, from which are derived several nonganglionic plexuses. The myenteric plexus lies between the outer longitudinal muscle layer and the inner circular muscle layer, containing a system of interconnected ganglia stretching from the oesophagus down as far as the internal anal sphincter. Along its length its precise morphology varies from region to region. Neurons project into the longitudinal and circular smooth muscle layers, to the nonganglionic deep muscular plexus of the circular muscle, and to the submucosal plexus. Like the myenteric plexus, the submucosal plexus extends the length of the gut, and is situated in submucosa between the circular muscle layer and the muscularis mucosae. It sends projections to the circular muscle, myenteric plexus, muscularis mucosae, and intestinal epithelial cells. The enteric nervous system contains sensory neurons that have their cell bodies in the ganglionated plexuses. The ganglia also contain interneurons, and are the source of motor neurons supplying the smooth muscles and epithelial cells.

Neurons in the enteric nervous system altogether contain almost every conceivable transmitter substance [10]. Motor neurons transmit with acetylcholine, noradrenaline, ATP, VIP, substance P, and free radicals of NITRIC OXIDE (NO) [5,11,12]. Enteric neurons contain various combinations of neuropeptides, and

descriptions of their projections are available [10]. There are no definitive markers for sensory or motor neurons, or interneurons in the enteric nervous system, but some substances such as enkephalin, GABA and 5-HYDROXYTRYPTAMINE (serotonin) do not act as neuromuscular transmitters, and are probably confined to interneurons [13]. Other local ganglia also contain a whole gamut of transmitter substances, including acetylcholine, ATP and neuropeptides [6,9]. As in the enteric nervous system, there are no definitive markers, but excitatory motor activity is mediated via acetylcholine and substance P whereas nonadrenergic, noncholinergic inhibitory neurons utilize variable combinations of ATP, VIP and NO.

Peripheral neurons *in vitro*

Primary tissue culture of peripheral neurons has been extensively used in experiments to determine transmitter or neurochemical content, receptor expression, electrophysiological properties, and biochemical propensities of such cells. Discrete ganglia, for example sympathetic paravertebral and prevertebral ganglia, dorsal root ganglia and myenteric ganglia can be dissected out and grown as explants *in vitro*. For local ganglia that lie embedded within a muscular wall, and that are not distinct enough to be dissected and grown as explants, dissociated cell culture preparations can be made that allow easy visualization of cells free from surrounding tissue. The main advantages of working in cell culture are that identifiable cells are directly accessible for study, and that the environment of the cells and experimental conditions can readily be controlled and modified [14].

Continuous cell lines are not available from peripheral neurons, unlike central neurons; however, the phaeochromocytoma cell line PC12 derived from the adrenal medulla has been used as a model of peripheral neurons, principally to study aspects of neurochemical storage and secretion. This is because adrenal medullary CHROMAFFIN CELLS are embryologically derived from the neural crest, and because these cells lie postsynaptic to sympathetic preganglionic neurons, and may be regarded as modified neurons themselves.

The localization and co-localization of neuropeptides and neurogenic amines have been studied in, for example, dissociated ganglia from the myenteric plexus, heart, trachea and urinary bladder, and explants of sympathetic chain, superior cervical ganglia, myenteric plexus and dorsal root ganglia. Cultures have also been used extensively for electrophysiological characterization of neurons and trophic interactions. Expression of receptors by cultured neurons has been examined in several ways: pharmacologically, autoradiographically and by examining the expression of specific messenger RNAs using *in situ* HYBRIDIZATION [15,16].

C.H.V. HOYLE
G. BURNSTOCK

See also: NEURON; SYNAPTIC TRANSMISSION.

1 Burnstock, G. (1986) The changing face of autonomic neurotransmission. (The First von Euler Lecture in Physiology.) *Acta Physiol. Scand.* **126**, 67–91.
2 Burnstock, G. (1990) Cotransmission. The Fifth Heymans Lecture — Gent, February 17, 1990. *Arch. Int. Pharmacodyn. Ther.* **304**, 7–33.

3 Burnstock, G. & Hoyle, C.H.V. (Eds) (1992) *The Autonomic Nervous System. Vol. I: Autonomic Neuroeffector Mechanisms* (Harwood Academic Publishers, Chur).

4 Hoyle, C.H.V. & Burnstock, G. (1991) ATP receptors and their physiological roles. In *Adenosine in the Nervous System* (Stone, T.W. , Ed.) 43–76 (Academic Press, London).

5 Hoyle, C.H.V. (1992) Transmission: purines. In *The Autonomic Nervous System. Vol. I: Autonomic Neuroeffector Mechanisms* (Burnstock, G. & Hoyle, C.H.V., Eds) 367–407 (Harwood Academic Publishers, Chur).

6 Dockray, G.J. (1992) Transmission: peptides. In *The Autonomic Nervous System. Vol. I: Autonomic Neuroeffector Mechanisms* (Burnstock, G. & Hoyle, C.H.V., Eds) 409–464 (Harwood Academic Publishers, Chur).

7 Buckley, N.J. & Caulfield, M. (1992) Transmission: acetylcholine. In *The Autonomic Nervous System. Vol. I: Autonomic Neuroeffector Mechanisms* (Burnstock, G. & Hoyle, C.H.V., Eds) 257–323 (Harwood Academic Publishers, Chur).

8 Fillenz, M. (1992) Transmission: noradrenaline. In *The Autonomic Nervous System. Vol. I: Autonomic Neuroeffector Mechanisms* (Burnstock, G. & Hoyle, C.H.V., Eds) 323–366 (Harwood Academic Publishers, Chur).

9 Morris, J.L. & Gibbins, I.L. (1992) Co-transmission and neuromodulation. In *The Autonomic Nervous System. Vol. I: Autonomic Neuroeffector Mechanisms* (Burnstock, G. & Hoyle, C.H.V., Eds) 33–119 (Harwood Academic Publishers, Chur).

10 Furness, J.B. & Costa, M. (1987) *The Enteric Nervous System* (Churchill Livingstone, Edinburgh).

11 Hoyle, C.H.V. & Burnstock, G. (1989) Neuromuscular transmission in the gastrointestinal tract. In *Handbook of Physiology, Section 6: The Gastrointestinal System, Vol I: Motility and Circulation* (Wood, J.D., Ed.) 435–464 (American Physiological Society, Bethesda, MD).

12 Sanders, K.M. & Ward, S. (1992) Nitric oxide as a mediator of nonadrenergic noncholinergic neurotransmission. *Am. J. Physiol.* G379–G392.

13 Hills, J.M. & Jessen, K.R. (1992) Transmission: γ- aminobutyric acid (GABA), 5-hydroxytryptamine (5-HT) and dopamine. In *The Autonomic Nervous System. Vol. I: Austonomic Neuroeffector Mechanisms* (Burnstock, G. & Hoyle, C.H.V., Eds) 465–508 (Harwood Academic Publishers, Chur).

14 Hassall, C.J.S. et al. (1989) The use of cell and tissue culture techniques in the study of regulatory peptides. In *Regulatory Peptides* (Polak, J.M., Ed.) 113–136 (Birkhüser, Basel).

15 Furuyama, T. et al. (1992) Co-expression of glycine receptor α subunit and GABA$_A$ receptor γ subunit mRNA in the rat dorsal root ganglion cells. *Mol. Brain Res.* **12**, 335–338.

16 Smith, K.E. et al. (1992) Differential regulation of muscarinic and nicotinic cholinergic receptors and their mRNAs in cultured sympathetic neurons. *Mol. Brain Res.* **12**, 121–129.

peripherin An INTERMEDIATE FILAMENT protein.

periplasmic binding proteins *See*: BACTERIAL CHEMOTAXIS; BACTERIAL ENVELOPE.

periplasmic space In yeast cells and bacteria, the area between the cytoplasmic membrane (i.e.plasma or inner membrane) and the cell wall. *See*: BACTERIAL ENVELOPE.

permeases Bacterial membrane transport proteins that carry small molecules across the cell membrane, for example lactose permease, encoded by gene Y of the *Escherichia coli* LAC OPERON. Lactose permease is an integral membrane protein (M_r 47 000) with 12 putative transmembrane helices. It acts as a symport for lactose by coupling its transport into the cell with the transport of a proton down the proton gradient across the bacterial cell membrane. Other permeases use the proton gradient to transport other sugars and amino acids into the cell. *See also*: BACTERIAL ENVELOPES; MEMBRANE TRANSPORT SYSTEMS.

permissive cell A cell which supports the replication of a given virus and undergoes lysis as a result. In a nonpermissive cell such a productive infection does not occur; instead the cell undergoes an abortive infection, or, in some cases, will become a stably TRANSFORMED CELL. Permissive cells can be used in PLAQUE ASSAY tests to titrate the number of infectious particles in a sample.

permissive signal A signal necessary for full DIFFERENTIATION of a developing tissue but which does not influence the developmental FATE of the cells. *Cf.* INDUCTION; INSTRUCTIVE SIGNAL.

peroxidase An oxidoreductase catalysing reactions of the general type: donor + H$_2$O$_2$ = oxidized donor + H$_2$O (EC 1.11.1). *See also*: HORSERADISH PEROXIDASE; LACTOPEROXIDASE.

peroxisomes Intracellular organelles of eukaryotic cells of up to ~0.5 μm in diameter surrounded by a single membrane. In hepatocytes they contain D-amino acid oxidase, urate oxidase, and catalase amongst other enzymes. In all cells peroxisomes contain one or more enzymes that use molecular oxygen to remove hydrogen atoms from organic substrates:

$$RH_2 + O_2 \rightarrow R + H_2O_2$$

Peroxisomal catalase utilizes the H$_2$O$_2$ to oxidize a variety of substrates including phenols:

$$H_2O_2 + R'H_2 \rightarrow R' + 2H_2O$$

Peroxisomes are diverse organelles and in different cells of the same organism can contain different sets of enzymes. In germinating plant seeds a type of peroxisome known as a glyoxysome is present. This contains the enzymes of the glyoxylate cycle which are capable of net conversion to glucose of acetyl CoA from fatty acid degradation. It is not clear whether new peroxisomes always derive from pre-existing peroxisomes or whether they can form from the ENDOPLASMIC RETICULUM. Enzymes in the lumen of the peroxisome are synthesized in the cytosol and then targeted to the organelle. The C-terminal sequence SKL (single-letter amino-acid notation) seems to be necessary and sufficient for targeting proteins to the lumen of the peroxisome.

pertussis *See*: BORDETELLA PERTUSSIS; PERTUSSIS TOXIN.

pertussis toxin A protein toxin secreted by the bacterium *Bordetella pertussis*, the causative agent of whooping cough. Entry of the toxin into a target cell leads to failure of inactivation of ADENYLATE CYCLASE activity. This effect is caused by the toxin-catalysed ADP-RIBOSYLATION of the α subunit of the inhibitory G protein G$_i$, which prevents its dissociation into α and βγ subunits (*see* GTP-BINDING PROTEINS). The G$_i$ α subunit is thus rendered unable to interact with receptors and thus does not inhibit adenylate cyclase activity on receptor activation.

PEST sequence Amino-acid sequence (PEST — Pro (P)-Glu (E)-Ser (S)-Thr (T)) that appears to label some proteins for degradation. *See*: PROTEIN DEGRADATION AND TURNOVER.

PET POSITRON EMISSION TOMOGRAPHY.

***pet* genes** (1) Photosynthetic electron transport genes. A series of genes coding for components of the electron transport system of thylakoid membranes of oxygenic photosynthetic organisms. In algae and higher plants the genes for CYTOCHROME *f* (*petA*), CYTOCHROME *b*-563 (*petB*), and the 17K subunit IV of the CYTO-CHROME *bf* COMPLEX (*petD*) are in the CHLOROPLAST GENOME, whereas those for the RIESKE IRON–SULPHUR PROTEIN (*petC*), PLASTOCYANIN (*petE*), and FERREDOXIN (*petF*) are in the nucleus. *See also*: CYANOBACTERIA; PHOTOSYNTHESIS.
(2) Nuclear genes encoding mitochondrial proteins in yeast, which give rise to *petite* mutants in SACCHAROMYCES CEREVISIAE.

petiole The stalk-like part of a leaf between its lamina and its attachment to the stem.

***petite* mutant** *See*: SACCHAROMYCES CEREVISIAE.

Petunia Ornamental flowering plant that has become a model organism for plant genetics, studies of FLOWER DEVELOPMENT, and PLANT GENETIC ENGINEERING. *See also*: TRANSPOSABLE GE-NETIC ELEMENTS: PLANTS.

PFK PHOSPHOFRUCTOKINASE.

Pfr The form of PHYTOCHROME that absorbs maximally at around 730 nm.

PFU PLAQUE-FORMING UNIT.

PHA Phytohaemagglutinin. *See*: LECTINS.

phaeochromocytoma cells Cells of a paraganglionoma of the adrenal medulla or urinary bladder. They generally secrete adrenaline and noradrenaline, and exhibit a strong chromaffin reaction. A line of phaeochromocytoma cells, P12, is widely used in studies of secretion.

phaeophytin (PHAEO (PHEO)) An early degradation product of CHLOROPHYLL in which the central Mg^{2+} is replaced by $2H^+$. Two phaeophytin *a* molecules (derived from chlorophyll *a*) are found in each reaction centre of PHOTOSYSTEM II, and two BACTERIOPHAEOPHYTIN molecules per bacterial reaction centre; these act as important intermediates in the rapid photoreduction of Q_A. *See also*: PHOTOSYNTHESIS.

phage BACTERIOPHAGE.

phage immunity The inability of a lysogenic bacterium carrying a particular strain of temperate phage as a PROPHAGE to be infected with the same strain of phage. *See*: BACTERIOPHAGES; LAMBDA.

phage induction The activation of a PROPHAGE thereby inducing entry into the LYTIC cycle of phage infection. *See*: BACTERIO-PHAGE; INDUCTION (short entry); LAMBDA; LYSOGEN.

phagocyte A cell specialized to carry out phagocytosis (*see* ENDOCYTOSIS). In mammals, phagocytes are chiefly MACRO-PHAGES, MONOCYTES and NEUTROPHILS.

phagocytic vacuole (phagosome) Membrane-bounded vesicle, or vacuole, derived from the plasma membrane and surrounding particulate ligands internalized by phagocytosis. Often used to describe the first vesicle formed during phagocytosis as opposed to PHAGO-LYSOSOMES which are formed by the fusion of phago-cytic vacuoles with LYSOSOMES. *See*: ENDOCYTOSIS.

phagocytosis Process through which cells internalize large, op-sonized particles (e.g. bacteria, yeast, red blood cells, and certain intracellular parasites) in large membrane-bounded vesicles. *See*: ENDOCYTOSIS.

phagolysosome Vesicles in which the digestion of ligands inter-nalized by phagocytosis occurs. Formed by the intracellular fusion of PHAGOCYTIC VACUOLES, or phagosomes, with LYSOSOMES. *See*: ENDOCYTOSIS.

phagosome PHAGOCYTIC VACUOLE. *See also*: ENDOCYTOSIS.

phakosin An INTERMEDIATE FILAMENT protein.

phalloidin Highly toxic alkaloid from the toadstool *Amanita phal-loides*, the Death Cap (also the source of the AMANITIN group of toxins). It binds to ACTIN filaments and stabilizes them, preventing their depolymerization. As it binds so stably to actin filaments it is also used as a conjugate with fluorescent dyes to stain actin filaments and to visualize the organization of the MICROFILAMENT components of the cytoskeleton in the cell.

pharmacogenetics The study of genetic factors affecting re-sponse to pharmacological agents. Examples of genetically deter-mined adverse effects include malignant hyperthermia after exposure to anaesthetics, and haemolysis in association with GLUCOSE-6-PHOSPHATE DEHYDROGENASE DEFICIENCY in response to a variety of drugs. *See also*: PORPHYRIA; PRIMAQUINE SENSITIV-ITY.

phase problem A problem that arises in DIFFRACTION analysis and crystallography where the observable diffraction pattern allows measurement of only the amplitudes of diffraction and not their relative phases. In order to reconstruct an image this set of phases must be known. The phase problem is the evaluation of these unmeasurable quantities. In microscopy, an image is formed by optical or electron lenses focused upon the diffracting object, whereas for hard X-rays or neutrons no lenses are avail-able and the other methods of evaluating phases must be used.

Phase information can be derived from several sources:
1 The intensity distributions within a diffraction pattern (DIRECT METHODS, used for small molecules).
2 The diffraction pattern of a closely related structure (ISOMOR-PHOUS DERIVATIVE; MOLECULAR REPLACEMENT; MULTIPLE ISO-MORPHOUS REPLACEMENT, methods often applicable to macro-molecules).

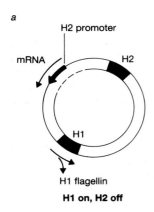

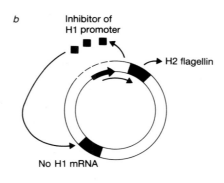

H1 on, H2 off | **H1 off, H2 on**

Fig. P18 The mechanism of phase variation in *Salmonella*.

3 Known structural relationships within the crystalline entity (NON-CRYSTALLOGRAPHIC SYMMETRY or MOLECULAR AVERAGING, useful for viruses and oligomeric proteins).
4 Knowledge of the differential scattering power of regions within the crystal (SOLVENT FLATTENING or CONTRAST VARIATION).
These methods may be used in combination and to improve trial phases, but all have limited applicability for macromolecules, which diffract to less than atomic RESOLUTION.

phase variation The phenomenon seen in populations of some pathogenic bacteria which undergo a reversible, and apparently random, change in phenotype. It usually refers to the random switching by some strains of the bacterium *Salmonella typhimurium* from production of one type of the flagellar protein flagellin to another type. Certain strains of *S. typhimurium* have two genes which encode two distinct types of flagellin, H1 and H2 respectively. A given cell expresses only one of these genes (and hence only one type of flagellin). If, however, bacteria expressing one type of flagellin are grown in culture for many generations, some bacteria expressing the other type of flagellin will appear.

The expression of the H1 and H2 genes is determined by the orientation of a 970-bp sequence present in the H2 PROMOTER. In one orientation, the H2 gene is transcribed producing a POLYCIS-TRONIC mRNA which encodes H2 flagellin, a transcriptional inhibitor of the H1 flagellin gene and a factor required for inversion of the promoter element. In the opposite orientation, H2 transcription is switched off; no H2 flagellin or H1 transcriptional inhibitor is therefore synthesized, allowing the H1-type flagellin gene to be expressed (Fig. P18). Switching between the two genetic states occurs via a nonregulated site-specific inversion of the promoter element. *See also*: BACTERIAL PATHOGENICITY.

phaseic acid (PA) (−)-3-Methyl-5-{8-[1*R*,5*R*-dimethyl-8*S*-hydroxy-3-oxo-6-oxa-bicyclo-(3,2,1)-octane]}-2*Z*,4*E*-pentadienoic acid (Fig. P19). M_r 280.3. It is a metabolite of the plant hormone ABSCISIC ACID. It is thought to form spontaneously from the true metabolite, 8′-hydroxyabscisic acid (previously known as metabolite C), which is formed by a CYTOCHROME P450-mediated oxidation of abscisic acid. It will apparently inhibit the induction of α-amylase synthesis in barley aleurone by GIBBERELLIN, but has no activity in most bioassays for abscisic acid. It is further

metabolized by reduction to DIHYDROPHASEIC ACID or by the formation of glucosyl conjugates. *See also*: PLANT HORMONES.

Milborrow, B.V. (1969) *Chem. Commun.* 966–967.

Fig. P19 Structure of phaseic acid.

phaseolotoxin *See*: PLANT PATHOLOGY.

phasmid A type of VECTOR composed of a linearized PLASMID, or repeats of it, flanked at each end by segments of phage LAMBDA containing all the genes required for lytic function. Phasmid recombinants are packaged *in vitro* before infection.

phenocopy A PHENOTYPE that closely resembles a (genetic) disease state, but has arisen by a completely different (environmentally determined) mechanism.

phenocritical period CONDITIONAL MUTATIONS may be used to give an indication of when in development a particular gene is required. For example, for a TEMPERATURE-SENSITIVE MUTATION, the beginning of the phenocritical (or temperature-sensitive) period is defined as the latest time during development at which a shift from the restrictive to permissive temperature results in a mutant phenotype. The end of the period is indicated by the earliest time when a shift from permissive to restrictive temperature gives wild-type development. Timings of requirement for gene function obtained using the phenocritical period should be treated with caution.

phenotype Physical traits displayed by an individual. These include both visible traits such as size, structure, colour, etc, and also invisible traits, such as the ability to synthesize a particular enzyme. The phenotype is determined both by the organism's GENOTYPE and by its environment.

phenotypic variability Variability in the clinical features of a particular genetic condition. This may be due to admixture of several different disease loci, interaction of the disease gene with other genetic or environmental modifiers, PLEIOTROPIC effects of the abnormal gene, or unknown factors. EXPRESSIVITY is an alternative term when referring to a single disease gene.

phenylalanine (Phe, F) An aromatic AMINO ACID.

phenylalanine ammonia-lyase (PAL) Enzyme (EC.4.3.1.5) catalysing the deamination of L-phenylalanine to cinnamic acid (Fig. P20); the elimination of the pro-S proton from C-3 of phenylalanine is antiperiplanar to produce *trans*-cinnamic acid.

PAL is a tetramer, the precise composition of which varies in different plants and tissues. In parsley, the relative molecular mass (M_r) is 330 000 and subunit M_r is 83 000. Multiple forms exist or can be induced in many tissues; in bean, there are at least four isoforms. The enzyme from leaves of monocotyledons and some other plants also catalyses the deamination of L-tyrosine to *trans*-4-hydroxycinnamic acid (*p*-coumaric acid). PAL is important in catalysing the first committed stage in the biosynthesis of many plant phenolics, including lignin, flavonoids and chlorogenic acid. It is induced by many environmental signals, especially light, tissue damage and pathogen attack.

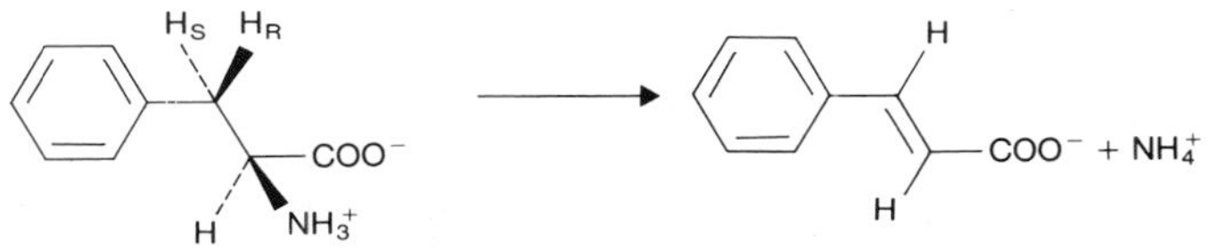

Fig. P20 The deamination of L-phenylalanine to cinnamic acid catalysed by phenylalanine ammonia-lyase.

phenylketonuria (PKU) An AUTOSOMAL RECESSIVE condition occurring in around one in 12 000 live births in North European populations. It is characterized by fair hair, pale skin, microcephaly and severe mental retardation. The disease is caused by a deficiency in the liver enzyme phenylalanine hydroxylase (phenylalanine 4-monooxygenase, EC 1.14.16.1), which converts phenylalanine to tyrosine. Strict control of dietary phenylalanine in early infancy lowers the plasma concentrations of phenylalanine and prevents mental retardation. There is now widespread neonatal screening for PKU through the Guthrie blood spot test. This has virtually eliminated PKU as a major cause of mental retardation.

phenylpropanoids *See*: PLANT CELL WALL MACROMOLECULES.

pheochromocytoma *See*: PHAEOCHROMOCYTOMA.

pheophytin *See*: PHAEOPHYTIN.

Philadelphia chromosome (Ph¹) An abnormal chromosome 22 found in the malignant cells of nearly 90% of patients with CHRONIC MYELOID LEUKEMIA (CML). It is the product of a balanced reciprocal translocation (*see* CHROMOSOME ABERRATIONS) involving chromosomes 9 and 22, designated *t*(9;22) (q34;q11). *See also*: ONCOGENES.

ΦX174 Single-stranded DNA BACTERIOPHAGE with a genome of 5375 bases, which was the first genome to be completely sequenced, by F. Sanger and colleagues in 1977.

phloem The main tissue that conducts organic solutions in vascular plants. It consists of SIEVE ELEMENTS, companion cells, PARENCHYMA and SCLERENCHYMA.

phorbol esters Polycyclic alcohols derived from croton oil which activate PROTEIN KINASE C as a result of their resemblance to the second messenger DIACYLGLYCEROL (*see* SECOND MESSENGER PATHWAYS). Although not carcinogenic by themselves, they act as TUMOUR PROMOTERS through their persistent activation of protein kinase C and thus of signalling pathways involved in cell proliferation.

Phormidium laminosum Filamentous moderately thermophilic CYANOBACTERIUM, which does not fix nitrogen. From Section III of the classification by Rippka et al. The first oxygen-evolving PHOTOSYSTEM II preparations were obtained from *P. laminosum*.

Rippka, R. et al. (1979) *J. Gen. Microbiol.* **111**, 1–61.

phosphatases Enzymes (EC 3.1.3) that hydrolyse phosphoric monoester bonds, resulting in the removal of a phosphate group. They include the PROTEIN PHOSPHATASES which are involved in regulating the activity of numerous proteins (*see* PROTEIN PHOSPHORYLATION), the acid and alkaline phosphatases of wide specificity, numerous sugar phosphatases and the polynucleotide phosphatases that remove phosphate groups from the ends of a polynucleotide chain. Phosphatases are used in RECOMBINANT DNA TECHNOLOGY to remove phosphate groups from the 5′ ends of a cleaved CLONING VECTOR to prevent its recircularization during DNA ligation.

phosphatides PHOSPHOLIPIDS.

phosphatidylcholine (lecithin), phosphatidylethanolamine, phosphatidylserine The major bilayer-forming PHOSPHOLIPIDS in biological membranes. *See*: MEMBRANE STRUCTURE.

phosphatidylinositol A glycerophospholipid (*see* Fig. P22 below) found in biological membranes and which, through the phosphorylated form, PHOSPHATIDYLINOSITOL 4-5-BISPHOSPHATE, is the precursor to the second messengers inositol 1,4,6-trisphosphate and diacylglycerol. *See*: MEMBRANE STRUCTURE; PHOSPHOINOSITIDES; PHOSPHOLIPIDS; SECOND MESSENGER PATHWAYS.

phosphatidylinositol 4,5-bisphosphate (PtdIns(4,5)P$_2$, PIP$_2$) A PHOSPHOINOSITIDE present in all eukaryotic cells, particularly in the plasma membrane (Fig. P21). It is the major substrate for receptor-stimulated PHOSPHOINOSITIDASE C, with the consequent formation of inositol 1,4,6-trisphosphate and diacylglycerol, and probably also for receptor-stimulated INOSITOL PHOSPHOLIPID 3-KINASE. *See*: SECOND MESSENGER PATHWAYS.

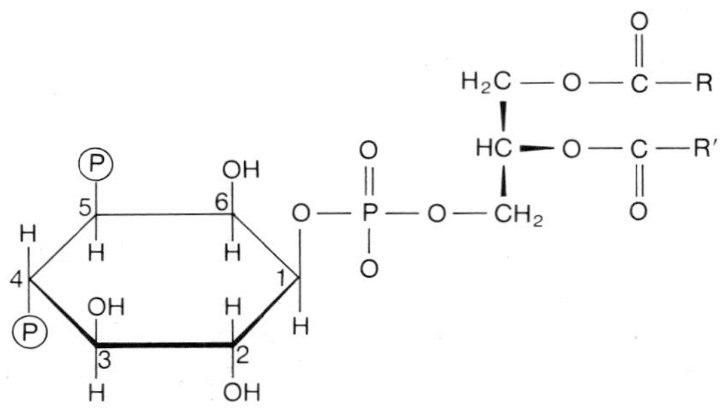

Fig. P21 Phosphatidylinositol 4,5-bisphosphate.

phosphatidylinositol-3 kinase INOSITOL PHOSPHOLIPID 3-KINASE.

phosphatidylserine *See*: MEMBRANE STRUCTURE; PHOSPHOLIPIDS.

phosphodiester bond Covalent linkage

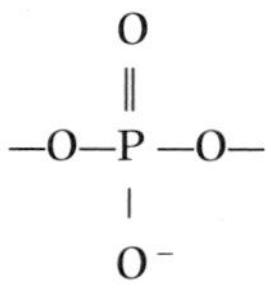

which is, for example, the linkage between nucleotide residues in a polynucleotide chain. *See*: DNA; NUCLEIC ACID STRUCTURE; RNA.

phosphodiesterases Enzymes that hydrolyse phosphodiester (P—O) bonds. EC 3.1.4.

phosphofructokinase Transferase enzyme catalysing the reaction:

ATP + D-fructose 6-phosphate = ADP + D-fructose 1,6-bisphosphate

Phosphofructokinase is an allosteric enzyme, only becoming activated when aggregated into oligomers. The active mammalian muscle enzyme is composed of four subunits of total M_r 360 000. The crystal structures of the enzymes from *Bacillus stearothermophilus* and *Escherichia coli* have been determined (Plate 5*d*). *See*: ALLOSTERIC EFFECT; PROTEIN STRUCTURE.

phosphoinositidase C Generic name for eukaryotic PHOSPHOLIPASES C that are specific for the PHOSPHOINOSITIDES PtdIns(4,5)P$_2$, PtdIns4P, and PtdIns. They exist as multiple structurally related subfamilies (β, γ, δ) and their activity is controlled

by cell-surface receptors either through G proteins (β) or as a result of receptor tyrosine phosphorylation (γ). *See*: SECOND MESSENGER PATHWAYS.

phosphoinositides (inositol phospholipids) A family of membrane glycerophospholipids, the most abundant of which is phosphatidylinositol (PtdIns) (Fig. P22): this is 1,2-diacyl-*sn*-glycerol-3-phospho-1D-*myo*inositol. PtdIns forms a few per cent of the membrane lipids of eukaryotic cells, and these cells also contain much smaller quantities of several PtdIns-derived lipids with monoester phosphate groups on the inositol ring: these are phosphatidylinositol 4-phosphate (PtdIns4P), PHOSPHATIDYL-INOSITOL 4,5-BISPHOSPHATE (PtdIns(4,5)P$_2$), phosphatidylinositol 3-phosphate (PtdIns3P), phosphatidylinositol 3,4-bisphosphate (PtdIns(3,4)P$_2$), and phosphatidylinositol 3,4,5-trisphosphate (PtdIns(3,4,5)P$_3$. PtdIns(4,5)P$_2$ is the substrate of the receptor-activated PHOSPHOINOSITIDASES C and probably also of the receptor-controlled INOSITOL PHOSPHOLIPID 3-KINASE(S): whether these enzymes catalyse reactions with PtdIns or PtdIns4P remains uncertain. Complex PtdIns-glycan structures serve as covalently attached membrane anchors for a variety of cell-surface proteins (*see* GPI ANCHORS). *See also*: MEMBRANE STRUCTURE; SECOND MESSENGER PATHWAYS.

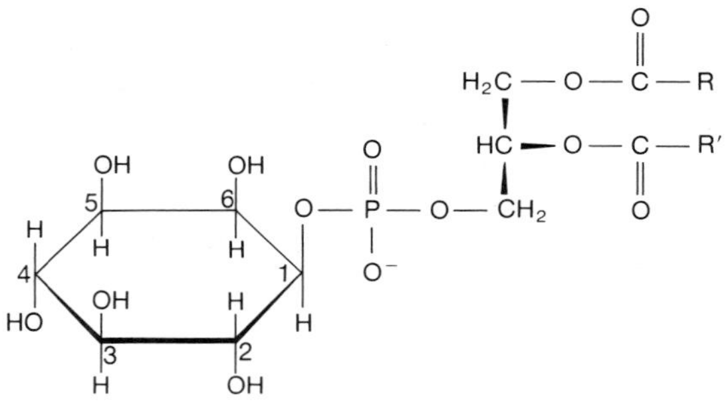

Fig. P22 Phosphatidylinositol.

phospholipase(s) Enzymes that cleave PHOSPHOLIPIDS by the hydrolysis of C—O or P—O bonds. They are classified into phospholipases A$_1$, A$_2$ (which are carboxylic ester hydrolases), C, and D (which are PHOSPHODIESTERASES), according to the bonds hydrolysed (Fig. P23). Phospholipases are present in large

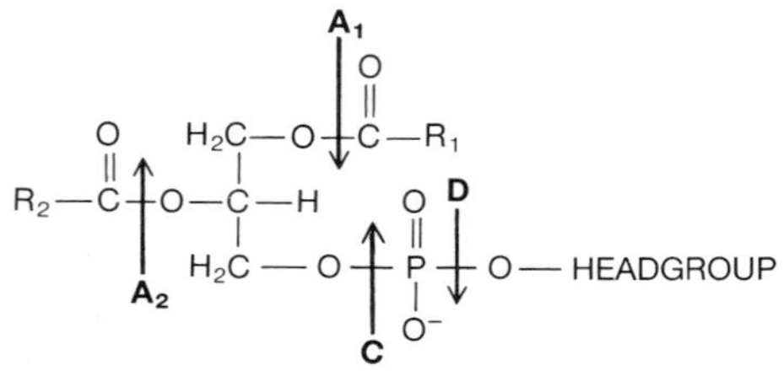

Fig. P23 The sites of action of the different classes of phospholipases. R$_1$ and R$_2$ are fatty acyl chains.

amounts in bacterial secretions, digestive secretions and snake venoms, and are also involved in the intracellular generation of lipid second messengers from membrane phospholipids (*see* ARACHIDONIC ACID; PHOSPHOLIPASE C; PLANT WOUND RESPONSES; SECOND MESSENGER PATHWAYS).

phospholipase C Generic term for enzymes that cleave either the glycerol–phosphate bond in glycerophospholipids or the sphingosine–phosphate bond in sphingophospholipids. Bacterial phospholipases C, some of broad lipid specificity and some with narrower specificities (e.g for sphingomyelin or for phosphatidyl-inositol and the GPI ANCHORS of cell-surface proteins), are widely used as biological and analytical tools. Although the receptor-controlled PHOSPHOINOSITIDASES C are the best characterized eukaryotic phospholipases C, there is clear evidence for other regulated phospholipases C with different lipid substrates (e.g. phosphatidylcholine or sphingomyelin). *See:* SECOND MESSENGER PATHWAYS.

phospholipid(s) A class of amphipathic phosphorus-containing LIPIDS derived from glycerol phosphate or the long chain amino alcohol sphingosine, which are essential constituents of biological membranes (Table P1). The glycerol-based phospholipids (phos-phoglycerides) are composed of a hydrophobic tail of two fatty acid molecules esterified through their carboxyl groups to C1 and C2 of glycerol 3-phosphate to form phosphatidate, and a hydro-philic polar headgroup bound by an ester linkage to the phos-phate moiety (Fig. P24). In the SPHINGOLIPID sphingomyelin the hydrophobic unit comprises the hydrocarbon chain of sphingo-sine and a fatty acid esterified to the amino group of sphingosine (*see* Fig. S16*a*). In naturally occurring phosphoglycerides one of the fatty acids is often saturated and one unsaturated, and the

Table P1 The main phospholipids of eukaryotic membranes

Glycerol phosphate derivatives (phosphoglycerides)
Phosphatidylcholine (lecithin)
Phosphatidylethanolamine
Phosphatidylserine
Phosphatidylinositol
Phosphatidylglycerol (plant cells)
Diphosphatidylglycerol (cardiolipin) (plant cells, mitochondria)

Sphingosine derivatives
Sphingomyelin (animal cells)
Phytosphingosine-containing plant phospholipids

fatty acids are usually even-numbered, 16- and 18-carbon fatty acids being most common (e.g. palmitate, C_{16}, saturated, and oleate, C_{18}, unsaturated), although chains of 14 to 24 carbons are found. The phospholipid phosphatidylglycerol is found in plant cell membranes and diphosphatidylglycerol (cardiolipin) (Fig. P24*d*) is common in plant cell membranes and is found in animal tissues mainly in mitochondrial membranes in the heart. The PHOSPHOINOSITIDES (e.g. phosphatidylinositol and PHOSPHATI-DYLINOSITOL 4,5-BISPHOSPHATE, which may also be classified as GLYCOLIPIDS) are precursors of the second messengers INOSITOL 1,4,5-TRISPHOSPHATE and DIACYLGLYCEROL (*see* SECOND MESSEN-GER PATHWAYS). In eukaryotic cells, phospholipids are mainly synthesized on the cytoplasmic face of the ENDOPLASMIC RETICU-LUM (*see* MEMBRANE LIPIDS). *See also:* MEMBRANE STRUCTURE; MEMBRANE SYNTHESIS.

phospholipid bilayer *See:* MEMBRANE STRUCTURE.

Fig. P24 *a*, Phosphatidylserine;
b, phosphatidylethanolamine;
c, phosphatidylcholine; *d*, diphosphatidylglycerol.

phospholipid exchange proteins (phospholipid transfer proteins) Soluble cytosolic proteins that have been shown to mediate the exchange of phospholipids between different membranes *in vitro* and are believed to be involved in the import of lipids into mitochondrial, perioxisome, and chloroplast membranes from the endo plasmic reticulum and other cellular membranes *in vivo*.

phosphoprotein Protein with a phosphoryl group covalently attached to an amino acid side chain (usually serine, threonine, or tyrosine). *See*: PROTEIN KINASES; PROTEIN PHOSPHORYLATION.

phosphorylase Refers to glycogen phosphorylase (EC 2.4.1.1) which degrades GLYCOGEN from the nonreducing end to glucose-1-phosphate in the reaction:

$$(1,4\text{-}\alpha\text{-}D\text{-Glycosyl})_n + P_i = (1,4,\text{-}\alpha\text{-}D\text{-glycosyl})_{n-1} + \alpha\text{-}D\text{-glucose 1-phosphate}$$

See also: PHOSPHORYLASE KINASE.

phosphorylase kinase A Ca^{2+}-dependent serine/threonine-specific PROTEIN KINASE which phosphorylates glycogen phosphorylase. Phosphorylase kinase was the first protein kinase to be discovered. It seems to be highly specific for phosphorylase, although it also phosphorylates and inactivates glycogen synthase in cell-free assays. It contains four types of subunit — α (M_r 140 000), β (M_r 130 000), γ (M_r 45 000), and δ (M_r 17 000) — with the native enzyme being $\alpha_4\beta_4\gamma_4\delta_4$. The γ subunit is the catalytic subunit, which is constitutively active when isolated. The δ subunit is calmodulin (*see* CALCIUM-BINDING PROTEINS) and phosphorylase kinase is unusual among Ca^{2+}-dependent protein kinases in that calmodulin remains tightly bound even in the absence of Ca^{2+}. The α and β subunits are related to each other, and their detailed functions are not known, although they exert additional regulatory functions on $\gamma\delta$. Phosphorylation of the α and/or β subunits by CYCLIC AMP-DEPENDENT PROTEIN KINASE increases the sensitivity to Ca^{2+} and the maximal velocity. The α subunit exists as tissue-specific ISOFORMS, and the form found in fast-twitch MUSCLE fibres conveys the ability to be further activated by exogenous calmodulin or TROPONIN C in addition to the endogenous δ subunit.

Heilmeyer, L.G. (1991) *Biochim. Biophys. Acta* **1094**, 168–174.

phosphorylation The covalent addition of a phosphoryl group to a compound using a donor of high phosphate transfer potential, often ATP. The enzymes catalysing phosphorylation are known generally as kinases or phosphotransferases. The phosphorylation of many proteins by PROTEIN KINASES alters their activity and is an important cellular control mechanism (*see e.g.* CELL CYCLE; GROWTH FACTOR RECEPTORS; PROTEIN PHOSPHORYLATION; PROTEIN KINASE C; SECOND MESSENGER PATHWAYS). *See also*: END-LABELLING; OXIDATIVE PHOSPHORYLATION; PHOSPHATASES.

phosphorylcholine (PC) Small molecule used as a defined hapten in the study of immunoglobulin antigen-binding sites. It is also an intermediate in the synthesis of the PHOSPHOLIPID phosphatidylcholine. *See*: IMMUNOGLOBULIN STRUCTURE; PC.

phosphoserine, phosphothreonine, phosphotyrosine *See*: PROTEIN PHOSPHORYLATION.

phosphotransferases A large group of enzymes that transfer a phosphoryl group from one substrate to another or from one position to another in the same molecule (EC 2.7). *See*: KINASES; PROTEIN KINASES; PROTEIN PHOSPHORYLATION.

photo-affinity labelling A technique which makes use of the capacity of certain compounds to make covalent bonds after UV irradiation with molecules to which they are noncovalently bound at a specific binding site (e.g. a receptor). Photo-affinity labels can thus be used as markers for the molecules with which they can be induced to bind. For example, the auxin analogue azoidoindole-3-acetic acid can be used to label auxin-binding proteins in plants. To aid detection, photo-affinity labels can be conjugated with radioactive or fluorescent tags. Labelling may, however, be of low specificity due to the highly reactive species generated by irradiation.

photogene A gene whose transcript levels are increased in response to increased levels of illumination. This may be due to either increased levels of TRANSCRIPTION or modulation of RNA stability. Photogene 32 is a light-regulated gene (known to be controlled by PHYTOCHROME in peas and mustard) which is in the CHLOROPLAST GENOME and encodes the Q_B-binding subunit (herbicide-binding protein; D1) of the PHOTOSYSTEM II reaction centre. *See also*: PHOTOSYNTHESIS.

photolyases Enzymes that catalyse the light-dependent repair of cyclobutane PYRIMIDINE DIMERS in DNA. The enzymes studied from bacteria and fungi are of M_r ranging from 50 000 to 70 000. *See*: DNA REPAIR.

photomorphogenesis Developmental responses induced by light treatments that are neither directional nor periodic. This term is normally restricted to plant and fungal phenomena. Photomorphogenesis does not include PHOTOTROPISM, PHOTOPERIODISM or phenomena associated with PHOTOSYNTHESIS. The photoreceptors for photomorphogenic phenomena are PHYTOCHROME and CRYPTOCHROME.

photoperiodism The response of an organism to the relative length of light and dark periods, for example the flowering of many plants, which is triggered by the lengthening and shortening of the days as the seasons change.

photophosphorylation The formation of ATP during PHOTOSYNTHESIS.

photoreactivation *See*: DNA REPAIR.

photoreactive yellow protein *See*: PROTEIN STRUCTURE for illustration of structure (Fig. P82).

photoreceptor A light-absorbing molecule which can initiate a specific photochemical reaction leading to a biophysical, biochemical, physiological, developmental or behavioural change in an organism. Known photoreceptors include: BACTERIOCHLOROPHYLL, BACTERIORHODOPSIN, CHLOROPHYLL, CRYPTOCHROME, PHYTOCHROME, PROTOCHLOROPHYLL, RHODOPSIN, and certain CAROTENOID, BILIPROTEIN, FLAVOPROTEIN and HAEM pigments.

photoregulation Regulation by light of biophysical, biochemical, physiological, developmental or behavioural processes.

photorespiration Light-induced oxygen uptake and CO_2 release which is a concomitant of the glycolate cycle in plant glyoxysomes. *See also*: PEROXISOMES; PHOTOSYNTHESIS.

Photosynthesis

PHOTOSYNTHESIS is in essence a light-induced redox process in which carbon dioxide is reduced to some readily metabolizable storage compound by an external reductant. The nature of the reductant provides a primary classification of the different types of photosynthetic organism. In particular, the ability to use the universal donor, water, distinguishes the oxygenic organisms, that is plants (including the various groups of algae), CYANOBACTERIA, and PROCHLOROPHYTES, from the prokaryotic organisms that are conventionally referred to as photosynthetic bacteria (see [1] for an overview of photosynthesis).

In all types of photosynthesis, the central piece of molecular machinery is the PHOTOSYNTHETIC REACTION CENTRE, in which the primary redox process takes place. This process is a light-induced transfer of an electron over a substantial distance from the primary donor (a special pair of CHLOROPHYLL molecules) to an acceptor, which is either a QUINONE or an IRON–SULPHUR CENTRE. *In vivo* the transfer is from one side of a coupling membrane to the other, which helps to stabilize the initial charge separation against a wasteful, spontaneous back reaction. Oxygenic organisms are further distinguished by having two types of reaction centre — PHOTOSYSTEMS I and II — whereas photosynthetic bacteria manage with only one.

The coupling membrane of photosynthetic organisms can take many different forms, but in all cases it separates two distinct aqueous phases between which a PROTONMOTIVE FORCE is developed in the light. The protonmotive force drives the synthesis of ATP (adenosine triphosphate) by the transmembrane enzyme ATP SYNTHASE. In photosynthetic bacteria the coupling membrane is formed from vesicular or tubular invaginations of the cell membrane which give rise to small vesicles known as chromatophores when the cells are broken. The chromatophores are inverted relative to the intact cell, so that what was originally the external (periplasmic) side of the membrane has become the interior of the chromatophore. In cyanobacteria the thylakoid membranes seem to be more clearly distinct from the plasma membrane. In eukaryotic cells thylakoid membranes form extensive vesicle systems within the CHLOROPLASTS and, in higher plants, are usually differentiated into stacks of appressed disks called grana and the unappressed stroma lamellae connecting different granal stacks.

Photosynthetic bacteria have been favourite objects of experimental investigation because their photosynthetic apparatus is in many ways simpler than that of higher plants. The purple nonsulphur bacteria (e.g. *Rhodobacter sphaeroides*, *Rb. viridis*, and *Rhodospirillum rubrum*) form a very versatile group that can utilize a variety of simple organic compounds or H_2 gas as reductants for CO_2. Under aerobic conditions, however, photosynthetic pigment production is suppressed and the cells grow solely by using the substrate for respiration. The related group of purple sulphur bacteria (e.g. *Chromatium vinosum*) are strict anaerobes that can only grow photosynthetically. These organisms utilize simple sulphur compounds such as sulphide or thiosulphate as reductants. Although there are many similarities between the purple nonsulphur bacteria and the green sliding bacterium *Chloroflexus aurantiacus*, the strictly anaerobic green sulphur bacteria, such as *Chlorobium*, have some distinctive properties and seem to be related to the recently discovered *Heliobacterium chlorum*.

There are two distinct phases to the overall photosynthetic reaction, although regulatory processes tie them closely together. The primary energy conversion involves assemblies of pigment molecules, reaction centres and electron transport components organized within a membrane according to CHEMIOSMOTIC principles. The 'light reactions' lead to the production of ATP and NAD(P)H (reduced nicotinamide dinucleotide (phosphate)) and involve a complex and structurally highly ordered molecular apparatus. The reduction of CO_2, on the other hand, involves metabolic reactions that require no direct input of light and take place in aqueous phases of the cell.

Light-harvesting mechanisms

An important feature of photosynthetic organisms is their ability to photosynthesize effectively under a wide variety of light intensities. This requires the differentiation of the pigments into two main types, those directly concerned with the photochemical reaction, and those which act only as light-absorbing molecules (antenna or light-harvesting pigments), organized in such a way that the excitation energy (or 'exciton') can be efficiently transferred to a reaction centre [2]. The transfer must occur extremely rapidly (less than 1 ns) if the energy is not to be dissipated by fluorescence or as heat before it can be used in the reaction centre, so the antenna pigments must have fixed spatial relationships to one another. This is achieved by binding to specific light-harvesting proteins, which often have repetitive structural features. The simplest are probably the LH1 and LH2 proteins of purple bacteria which are made up of repeating units of a dimer of small, related, BACTERIOCHLOROPHYLL-binding subunits (α, β). A very elaborate, but quite well understood, type of structure is that of the PHYCOBILISOMES of cyanobacteria and red algae (rhodophytes). These are large particulate structures that are easily visible attached to the outer surface of the thylakoid membranes in electron micrographs. The pigments are open-chain tetrapyrroles (PHYCOBILINS) covalently linked to α and β protein subunits to form PHYCOBILIPROTEINS (phycoerythrin, phycocyanin and allophycocyanin) which form repeating units in the phycobilisome held together by colourless linker polypeptides.

Less well understood is the organization of the phycobilins in cryptophytes (members of the Cryptophyta, a group of flagellate eukaryotic algae occurring in fresh and coastal waters), where they are located inside the thylakoid lumen. In green plants, related families of nuclear *cab* genes code for chlorophyll proteins which when linked to their prosthetic groups — chlorophylls *a* and *b* — make up the light-harvesting complexes of photosystems II (LHCII) and I (LHCI) (see Fig. P26 below). It is not yet clear whether cyanobacteria, which have no chlorophyll *b*, possess any light-harvesting chlorophyll protein complex analogous to LHCI. The organization of light-harvesting chlorophyll complexes in those algae that have chlorophyll *c* rather than chlorophyll *b* (e.g. brown seaweeds) is also incompletely understood.

Photosynthetic reaction centres

Purple bacteria

The solution of the X-ray crystal structure (*see* X-RAY CRYSTAL-LOGRAPHY) of the reaction centres from two purple bacteria, *Rhodobacter viridis* and *Rb. sphaeroides*, has provided a basic understanding of the relation between structure and function which is likely to be applicable to all photosynthetic reaction centres [3]. The reaction centre is a protein complex embedded in the membrane to which are bound the pigments and cofactors involved in light-induced electron transfer across the membrane. To a first approximation the protein components may be considered to act as a scaffold holding together, in the correct relative positions, the pigments and cofactors. These consist of a special pair of bacteriochlorophyll molecules forming the primary electron donor, two accessory bacteriochlorophylls, two BACTE-RIOPHAEOPHYTINS, two quinone molecules (Q_A and Q_B) and a single nonhaem iron atom (Fig. P25). The most striking feature of the arrangement is the two-fold symmetry between pairs of molecules about an axis joining the special pair at one side of the

structure and the iron atom near the other. The accessory chlorophylls and phaeophytins appear to form two pathways for the electron through delocalized π-electron rings, from the special pair chlorophylls to Q_A and Q_B respectively. This transfer could be extremely rapid, as is necessary if the initial charge separation is to occur with a high quantum yield. Functional evidence, however, shows that under normal circumstances only the Q_A side is active. The nearly symmetric arrangement is maintained by the protein components, for the pigment-binding polypeptides are a pair of homologous subunits (L and M), each of which forms five membrane-spanning helices.

The symmetry is broken by the H subunit, which contains a single membrane-spanning helix and a globular domain on the cytoplasmic side of the membrane and a four-haem CYTOCHROME subunit on the periplasmic side, and which is present in *Rb. viridis* and many other purple bacteria but not in *Rb. sphaeroides*.

Photosystem II

Although the full three-dimensional structures of other reaction centres are not known, there is clearly a close similarity between those of photosystem II (Fig. P26) and the purple bacteria (without the cytochrome subunit). There is evidence for a similar arrangement of pigments and cofactors, and the photosystem II D1 and D2 polypeptides are at least topologically and functionally analogous to the L and M subunits respectively. There are several important differences, however, mainly connected with the unique ability of photosystem II to photo-oxidize water. The special pair of chlorophylls (P680) has a very strongly oxidizing redox potential (>1 V), the Mn_4 centre involved in water oxidation is probably bound to interhelix loops of D1 or D2 on the luminal side of the membrane, and the redox potentials involved are high enough for a tyrosine residue of D1 to act as an electron transfer intermediate ('Z') between the Mn centre and P680. On the other hand, the acceptor side arrangement of primary and

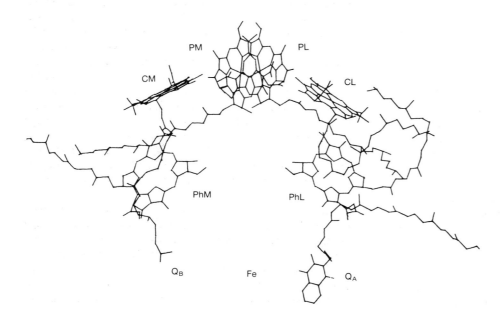

Fig. P25 Spatial arrangement of the cofactors in the photosynthetic reaction centre from *Rhodobacter viridis*. PL, PM, bacteriochlorophyll molecules that form the special pair and are associated with the L and M protein subunits, respectively; CL, CM, the 'accessory' bacteriochlorophyll molecules; PhL, PhM, bacteriophaeophytin molecules. The right-hand branch, associated mainly with subunit L, is the active arm. *In vivo* the membrane would be horizontal and at right angles to the plane of the paper. After [6].

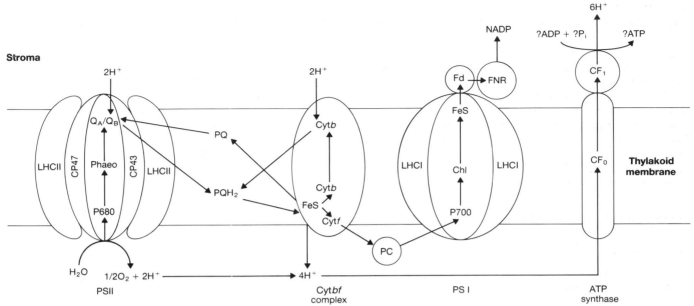

Fig. P26 The electron transport system of chloroplasts drawn to indicate its proton-translocating properties and its association with protein complexes and chlorophyll proteins within the thylakoid membrane. PSI, PSII, photosystems I and II respectively; Cyt, cytochrome; P680, P700, the reaction centre special pair chlorophyll in PSI and PSII respectively; Phaeo, phaeophytin; Chl, chlorophyll *a*; Q_A, Q_B, tightly and loosely bound plastoquinone molecules associated with the D2 and D1 polypeptides, respectively, of PSII; PQ, plastoquinone; FeS, iron–sulphur centres; PC, plastocyanin; Fd, ferredoxin; FNR, ferredoxin-NADP$^+$ oxidoreductase; CP43, CP47, light-harvesting chlorophyll *a* proteins tightly associated with the reaction centre of PSII; LHCI, LHCII, light-harvesting chlorophyll *a/b* proteins associated with PSI and PSII respectively; CF$_0$, CF$_1$, components of the proton-translocating ATP synthase. ?ADP, ?ATP signify that the chemiosmotic stoichiometries relating electron transport to phosphorylation are not known with certainty.

secondary quinones (Q_A and Q_B) is almost identical. Closely associated with the D1D2 dimer is the enigmatic cytochrome *b*-559, which has no counterpart in purple bacteria.

Green sulphur bacteria and photosystem I

The reaction centres of green sulphur bacteria show some similarity to those of purple bacteria in the nature of the special pair and the presence of a cytochrome subunit on the donor side, but the other pigments and cofactors show more similarity with photosystem I, especially the presence of iron–sulphur centres, including a polypeptide of relative molecular mass (M_r) 9000 that contains two 4Fe–4S centres (subunit C in photosystem I). The main protein subunits are much larger (70–80K) than in purple bacteria, but in both green bacteria and photosystem I they occur as a heterodimer of homologous polypeptides. These observations suggest that all photosynthetic reaction centres are built on similar principles, with an almost symmetric arrangement on a two-fold axis and a series of pigment intermediates that potentially form a double pathway for the electron across the membrane.

Electron transport

For effective photosynthesis, reaction centres must be built into electron transport systems that stabilize the initial charge separa-tion and lead to the reduction of NAD(P)$^+$ and synthesis of ATP. The simplest type occurs in purple bacteria where the electrons from ubiquinol, which has been reduced by the reaction centre, are allowed to flow back in a controlled fashion, down the thermodynamic gradient, to the oxidized special pair. The key intermediates in this process are the cytochrome bc_1 complex and at least one other cytochrome. The electron transfer process is coupled to the transfer of protons from the cytoplasmic side to the periplasmic side of the membrane by a redox-loop mechanism combined with an additional protonmotive 'Q-cycle' in the bc_1 complex. Overall there is a cyclic flow of electrons around the reaction centre leading to CYCLIC PHOSPHORYLATION by chemi-osmotic processes. The external reductant can donate electrons into the ubiquinol pool or to a cytochrome, but ubiquinol is not a strong enough reductant to reduce the NAD$^+$ required for CO_2 fixation directly. This is achieved by so-called 'reversed' electron transport in which some of the energy liberated by the cyclic process is used to drive the electrons 'uphill' from the quinol to NAD$^+$.

The iron–sulphur centres of green sulphur bacteria and pho-tosystem I, on the other hand, have redox potentials low enough for direct reduction of NAD(P)$^+$ (via ferredoxin and ferredoxin-NADP$^+$ reductase). This leads to a noncyclic process in which the overall reaction is light-driven electron transfer from the external reductant to the pyridine nucleotide coupled to noncyclic phos-phorylation. In addition, cyclic phosphorylation (around photosystem I in the case of oxygenic organisms) may occur to provide extra ATP if required.

Electron transport in oxygenic photosynthesis

A fundamental difference between photosynthetic bacteria and oxygenic organisms is the occurrence of two types of reaction centre — photosystem I and photosystem II — in the membranes of the latter (Fig. P26). These are linked in series in a manner usually referred to as the 'Z-scheme' because the electron transport system can be written out with the two light reactions represented by two arms of the Z; photosystem II spans a considerably more oxidizing range of redox potentials than photosystem I, although the 'dark' electron transfer from Q_B on the acceptor side of photosystem II to P700 on the donor side of photosystem I remains thermodynamically downhill and coupled to proton transfer and phosphorylation. The components involved in this intersystem electron transfer are the plastoquinone pool (PQ), the cytochrome *bf* complex (which is homologous with the cytochrome bc_1 complex of purple bacteria) and the copper-containing protein, plastocyanin (PC) [4]. There is a remarkable segregation of protein complexes between grana and stroma lamellae, with photosystem II being concentrated in the appressed membranes of grana stacks, photosystem I and the ATP SYNTHASE in stroma lamellae and the unappressed parts of the grana membranes, and the cytochrome *bf* complex in all parts.

Photodissipation

An important corollary of the ability of plants to photosynthesize effectively at low light intensities is the phenomenon of light saturation. At high light intensities there is a tendency for the antenna systems to deliver excitons to the reaction centres faster than the electron transport system or, ultimately, the carbon reduction enzymes, can utilize the trapped energy. Plants thus possess mechanisms for photodissipation (the controlled conversion of light energy into heat), in order to avoid destructive photochemical side reactions [5]. The need for this is particularly acute in photosystem II which generates oxidants powerful enough to cause oxidation and bleaching of the bulk chlorophyll. More than one molecular process may be involved but the major type seems to be one in which the physical state of chlorophyll is modified (presumably by the binding protein) in such a way that the excited state relaxes very rapidly to the ground state with the generation of heat. This may occur either in the reaction centre or in the antenna system. In the latter case it is connected with the conversion of violaxanthin into zeaxanthin (part of the 'xanthophyll cycle'). In photosystem I, P700$^+$ is a much weaker oxidant (as are the special pairs in photosynthetic bacteria) and is capable of dissipating excitation energy. Not infrequently the capacity of the photodissipative processes is overloaded, for example by bright sunshine on a frosty morning. The last ditch defence involves the phenomenon of photoinhibition in which the plant responds by a destruction of part of the photosystem II reaction centre, especially the D1 polypeptide (the product of photogene-32) [7]. There is evidence for an elaborate cycling of photosystem II units and component polypeptides between the grana stacks and stroma lamellae, as part of the repair process.

Carbon dioxide fixation

In order to fix CO_2 a metabolic cycle is necessary to regenerate the acceptor molecule, which in nearly all photosynthetic organisms is ribulose bisphosphate. The carboxylation reaction gives rise to 3-phosphoglyceric acid which is reduced by the reversal of glycolytic reactions to triosephosphate. This is the primary product; part of it is used for synthesis of the preferred storage compound, depending on the organism, and part of it is recycled, essentially by a reversal of the reactions of the pentose phosphate pathway, to yield ribulose 5-phosphate which in turn is phosphorylated by phosphoribulokinase to give ribulose bisphosphate. The key enzymes of the cycle (the reductive pentose phosphate or CALVIN CYCLE) are phosphoribulokinase, ribulose bisphosphate carboxylase, glyceraldehyde phosphate dehydrogenase, and the two phosphatases for fructose-1,6-bisphosphate and sedoheptulose-1,7-bisphosphate. These are subject to a remarkable system of metabolic control to ensure that although the biosynthetic pathway operates in the light, a respiratory process can operate in the dark without wasteful metabolic cycling. Similarly, the ATP synthase enzyme is tightly controlled to ensure that ATP is not drained away by hydrolytic activity at night. Although an elaborate series of ALLOSTERIC EFFECTS is involved, the key process signalling the transition from dark to light is reduction of each of these enzymes (except the carboxylase itself) by the thiol/disulphide protein THIOREDOXIN. As soon as the photosynthetic electron transport system starts to turn over, ferredoxin is reduced and in turn reduces thioredoxin. This reaction is catalysed by thioredoxin reductase, which is incompletely characterized but known to contain an active disulphide group and a 4Fe–4S centre.

Ribulose bisphosphate carboxylase/oxygenase (RUBISCO) has a low turnover number and is probably the most abundant protein on earth. The oxygenase activity, in which O_2 substitutes for CO_2 giving one molecule of phosphoglycollate and one of phosphoglycerate, appears to be a wasteful side reaction that occurs in all organisms, although there is some variation in the ratio of carboxylase to oxygenase activity. In green plants there is a metabolic pathway in which some of the carbon of phosphoglycollate is recovered as phosphoglycerate. This is responsible for the phenomenon of photorespiration (light-induced oxygen uptake and CO_2 release) which results in many plants having a relatively high compensation point (the light intensity at which the rates of photosynthesis and respiration just balance).

There are two kinds of exception to the rule that all CO_2 fixation goes by RUBISCO. The first is that of the green sulphur bacteria in which a reversed citric cycle driven by reduced ferredoxin replaces the reductive pentose phosphate cycle. The carboxylase enzymes are pyruvate synthase and α-ketoglutarate synthase, in which acetyl-CoA and succinyl-CoA are reductively carboxylated by ferredoxin. It is notable that these organisms are strict anaerobes in which one would not expect to find an oxidative citric cycle and also that the reductive carboxylations are involved in assimilatory reactions in a number of other groups of microorganism.

The second is not properly an exception but a subtle and complicated way in which a whole group of higher plants, the C4 plants (especially maize, sugar cane and some other tropical grasses) have avoided the effects of the oxygenase reaction of RUBISCO shown by the much wider group of C3 plants. The activity of RUBISCO yields two molecules of the C3 acid phosphoglycerate and occurs in the C4 plants as well, but in the latter group a primary carboxylation into oxaloacetic acid (C4) is catalysed by phosphoenolpyruvate (PEP) carboxylase, which has a much higher affinity for CO_2 than RUBISCO. The C4 syndrome is associated with a different leaf structure (Kranz anatomy). RUBISCO is confined to the large 'Kranz cells' surrounding the vascular bundles, and is shielded from the atmosphere by the mesophyll cells to which PEP carboxylase is confined. The mesophyll cells are in direct contact with atmospheric CO_2 via the stomata in the epidermises. The initial fixation products, malate or aspartate, are transported to the bundle sheath cells where they are decarboxylated for refixation of the CO_2 by RUBISCO. C3 acids, such as alanine, move in the opposite direction to provide the acceptor molecule for PEP carboxylase. All this acts, in effect, as a CO_2 concentrating device for the benefit of RUBISCO, and C4 plants have very low compensation points, but it is not achieved without cost. C4 plants have a higher quantum requirement, because they need more ATP to fix one molecule of CO_2, but they can achieve higher rates of photosynthesis at high light intensities and can better afford to keep their stomata closed in the heat of the day in order to avoid desiccation.

D.S. BENDALL

See also: ELECTRON TRANSPORT CHAIN.

1 Gregory, R.P.F. (1989) *Biochemistry of Photosynthesis*. 3rd edn (Wiley, Chichester).
2 Cramer, W.A. & Knaff, D.B. (1990) *Energy Transduction in Biological Membranes* (Springer-Verlag, New York).
3 Nitschke, W. & Rutherford, A.W. (1991) Photosynthetic reaction centres: variations on a common structural theme? *Trends Biochem. Sci.* **16**, 241–245.
4 Cramer, W.A. et al. (1991) Electron transport between photosystem I and photosystem II. *Curr. Topics Bioenergetics*, **16**, 179–222.
5 Bendall, D.S. (1989) *Plants Today* **2**, 188–192.
6 Deisenhofer, J. et al. (1984) X-ray structure analysis of a membrane protein complex. *J. Mol. Biol.* **180**, 385–398.
7 Barber, J. & Anderson, B. (1992) Too much of a good thing: light can be bad for photosynthesis. *Trends Biochem. Sci.* **17**, 61–66.

photosynthetic carbon reduction (PCR) cycle *See*: CALVIN CYCLE.

photosynthetic reaction centre *See*: PHOTOSYNTHESIS; REACTION CENTRE.

photosystem I The assembly of light-harvesting pigment–protein complexes and REACTION CENTRE in thylakoid membranes of plants and CYANOBACTERIA that catalyses the light-dependent electron transfer from PLASTOCYANIN (or CYTOCHROME C-552 in some algae and cyanobacteria) to FERREDOXIN. Polypeptide subunits are named alphabetically in accordance with the gene nomenclature *psaX*. The holocomplex contains a heterodimer of large hydrophobic polypeptides that bind most of the pigments and cofactors (subunits A and B), some small hydrophobic subunits (I, J, and K), and several hydrophilic subunits that are involved in interaction with reaction partners: C (the M_r 9000 [4Fe–4S] protein that acts as terminal electron acceptor of the photosystem and electron donor to ferredoxin), D (ferredoxin-binding subunit, M_r 18 000, formerly known as II), E (M_r 10 000), F (plastocyanin binding, M_r 17 000), G (M_r 11 000), and H (M_r 10 000). In addition, the eukaryotic photosystem contains at least three polypeptides of the chlorophyll *a/b* complex — LIGHT-HARVESTING COMPLEX I. *See also*: PHOTOSYNTHESIS.

photosystem II The assembly of light-harvesting pigment–protein complexes and REACTION CENTRE in thylakoid membranes of plants and CYANOBACTERIA that catalyses the light-dependent electron transfer from water to PLASTOQUINONE. The reaction centre consists of a heterodimer of homologous polypeptides (D1 and D2, M_r 32 000, products of the *psbA* and *psbD* genes) that bind the special pair of chlorophyll *a* molecules (P680) and other pigments and cofactors and, closely associated with it, CYTOCHROME B-559 (*psbE* and *F* genes), and the light-harvesting chlorophyll *a* proteins CP47 (*psbC*) and CP43 (*psbD*); all these are chloroplast encoded. The structure of the D1D2 heterodimer is thought to be closely similar to that of the LM heterodimer of purple bacteria. At least 12 additional photosystem II polypeptides have been described, some integral, some peripheral, some chloroplast encoded, some nuclear encoded. The best known are three peripheral, nuclear-encoded polypeptides associated with the luminal surface of the thylakoid that are sometimes referred to, not very appropriately, as the OXYGEN EVOLUTION COMPLEX (OEC). In addition, there are various light-harvesting pigment–protein complexes, depending on the taxonomic group. *See also*: PHOTOSYNTHESIS.

phototransduction The conversion of light energy incident on a receptor into an appropriate cellular response. *See*: BACTERIO-RHODOPSIN; BLUE LIGHT RESPONSES; PHYTOCHROME; VISUAL TRANSDUCTION.

phototropism Movement of plant organs by differential growth, in response to light. Coleoptiles and most stems grow towards light (positively phototropic); most roots are insensitive to light, but some are negatively phototropic. The process is sensitized by blue light.

phragmoplast An array of MICROTUBULES which forms late in MITOSIS in cells of higher plants. Golgi vessels (*see* GOLGI APPARATUS) coalesce upon it to assemble many small membrane-bounded vesicles into a large, flat vesicle that divides the cell into two. *See also*: CYTOKINESIS.

***phy* gene(s)** Genes encoding the apoprotein portion of the plant photoreceptor protein family, PHYTOCHROME.

phycobilins A group of open-chain tetrapyrrole pigments (bile pigments) that act as the chromophores of PHYCOBILIPROTEINS. The chromophores are covalently linked to cysteine residues of the protein through thioether bonds. Appropriate treatment of allophycocyanin and phycocyanin yields phycocyanobilin, and phycoerythrin yields the red pigment phycoerythrobilin. R-phycocyanin yields both phycocyanobilin and phycoerythrobilin, whereas R-phycoerythrin yields phycoerythrobilin and a third pigment, phycourobilin.

phycobiliproteins A group of proteins that contain covalently-linked phycobilins as chromophores and act as light-harvesting assemblies associated with photosystem II in red algae (Rhodophyta), Cryptophyta and cyanobacteria. They also occur in CYANELLES. The protein contains closely related α and β subunits of M_r 17 000–22 000. The main types are PHYCOERYTHRIN, PHYCOCYANIN and ALLOPHYCOCYANIN. *See also*: PHOTOSYNTHESIS; PHYCOBILISOME.

phycobilisome A highly organized structure occurring in red algae and CYANOBACTERIA, and containing PHYCOBILIPROTEINS arranged in such a way that the energy of absorbed light quanta is efficiently funnelled towards the centre of the structure and then to the REACTION CENTRE of PHOTOSYSTEM II by being passed from pigments absorbing at shorter wavelength to those absorbing at longer wavelength. Within the phycobilisome the phycobiliproteins are arranged in rod-like structures which contain stacks of disks, each made up of a pair of $α_3β_3$ trimers. Additional colourless linker polypeptides are present. *See also*: PHOTOSYNTHESIS.

phycocyanin A PHYCOBILIPROTEIN ($λ_{max}$ ~620 nm) in which the chromophore is phycocyanobilin, an open-chain tetrapyrrole, and which occurs in Rhodophyta, Cryptophyta, and CYANOBACTERIA (where it is the major pigment responsible for the blue colour).

phycoerythrin A PHYCOBILIPROTEIN ($λ_{max}$ 640–670 nm) in which the chromophore is phycoerythrobilin. It is the major pigment of red algae, but also occurs in Cryptophyta and some CYANOBACTERIA.

phylloplane Leaf surface.

phylogenetic method *See*: ALIGNMENT.

phylogeny The evolutionary history and line of descent of a species or other taxonomic group. In molecular biology, the evolutionary history of a protein or nucleic acid. *See*: MOLECULAR EVOLUTION; PHYLOGENY.

Physarum polycephalum A member of the ACELLULAR SLIME MOULDS which forms a multinucleate motile plasmodium from which stalked sporangia arise. *Physarum* has been particularly studied for the role of cytoskeletal components and associated proteins in cytoplasmic streaming. *See*: ACTIN-BINDING PROTEINS; RNA EDITING.

physical map/mapping *See*: GENE MAPPING; HUMAN GENE MAPPING; RESTRICTION MAPPING.

phytin Mixture of calcium and magnesium salts of phytic acid, the hexaphosphate ester of myoinositol. Occurs widely as phosphate storage compound, especially in cereal grains.

phytoalexins Low molecular weight antimicrobial compounds that are synthesized by and accumulate in plants after their exposure to microorganisms. Phytoalexin accumulation is almost always associated with a previous HYPERSENSITIVE RESPONSE, and there is evidence that they accumulate preferentially in the resulting dead cells. Phytoalexins have been extensively characterized from the Leguminosae, Solanaceae, and other plant families and include a wide spectrum of chemical structures: isoflavonoids, stilbenes, terpenes, etc. *See also*: PLANT PATHOLOGY; PLANT WOUND RESPONSES.

Phytochrome

PHYTOCHROME (Pr, Pfr) is the collective name given to a family of PHOTORECEPTOR proteins unique to plants and of vital importance in sensing variations in the natural light environment and inducing appropriate modifications of metabolism and development in response.

Phytochrome is unique among biological photoreceptors in that its action is crucially dependent upon its property of photochromicity. Photochromic substances undergo a change in their absorption spectrum on absorption of light. Phytochrome exists in two interconvertible states. One form, Pr, has an absorption maximum of around 660 nm in the red (600–700 nm) region of the spectrum; the other, Pfr, which is considered to be the biologically active form, absorbs maximally at around 730 nm in the far-red (700–760) region. Absorption of light by either form results in photoconversion to the other form.

In continuous, broad-band light, phytochrome consequently 'cycles' between Pr and Pfr, the equilibrium proportions of Pr and Pfr being determined by the relative proportions of red and far-red light in the actinic beam. This photochromic behaviour allows phytochrome to detect the relative balance of red and far-red radiation in the natural environment, thus endowing it with the capacity to detect the presence and proximity of neighbouring vegetation. It is also involved in the photoregulation of many other developmental and biochemical processes.

Genes that code for the members of the phytochrome family have been characterized in several species, and the chemistry of the phytochrome APOPROTEINS and their CHROMOPHORES has been studied for some years. The molecular mechanism of action of phytochrome is, however, still a matter of speculation.

Physiology

Classical physiological studies in the 1940s and 1950s, at the USDA Laboratory in Beltsville, Maryland, pioneered by Sterling Hendricks and Harry Borthwick, led to the concept of a photochromic photoreceptor existing in two photointerconvertible forms, absorbing, respectively, red (around 660 nm) and far-red

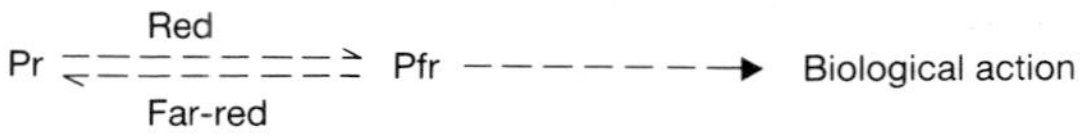

Pr $\underset{\text{Far-red}}{\overset{\text{Red}}{\rightleftharpoons}}$ Pfr $----\rightarrow$ Biological action

Fig. P27 Phytochrome exists in two interconvertible forms; Pr has maximum absorption at 665 nm and is photoconverted to Pfr, which has maximum absorption at 730 nm and is photoconverted to Pr. Pfr is considered to be the biologically active form.

(around 730 nm) radiation. The conclusions were based partly on extensive and accurate action spectra, but mainly on conceptually simple, but crucially significant, experiments which showed that the developmental effects induced by brief exposure to red light could be negated, or reversed, by an immediate subsequent brief exposure to far-red light. This led to the scheme shown in Fig. P27 [1], which still forms the basis of current thought concerning the action of phytochrome.

For many years it was considered that only a single phytochrome existed, but it was very difficult to reconcile this view with the many different ways in which light regulates growth and development. Red/far-red reversibility, the classic criterion of phytochrome involvement, has been shown for developmental responses at all stages of the plant life cycle, including the photoregulation of seed and spore germination, stem extension, leaf expansion, photosynthetic and FLAVONOID pigment formation, enzyme synthesis, and the expression of many light-regulated genes [2]. Red/far-red reversible responses also show that phytochrome participates in the perception of photoperiodic signals important in the induction of flowering and bud dormancy. In mature light-grown plants, phytochrome has been shown by other criteria (i.e., by manipulation of the ratio of red to far-red light) to be very important in the natural environment, where it senses the presence and proximity of neighbouring and shading vegetation and mediates the shade avoidance strategy through which plants compete for the available radiant energy. Light reflected from or transmitted by green vegetion is depleted in red and relatively enriched in far-red radiation, and this spectral signal of vegetation proximity is detected by phytochrome through a change in the steady-state proportions of Pr and Pfr [3].

Types and terminology

The inability to reconcile multiple modes of phytochrome response with the concept of a single photoreceptor has been resolved in recent years by immunochemical and recombinant DNA investigations, which have shown that the phytochromes are coded for by a family of related, but distinct, genes. All higher plants that have been studied have revealed the existence of a family of *phy* genes, although lower plants, such as mosses and ferns, appear to have only a single *phy* gene.

The terminology of phytochrome is currently confusing; although several molecular forms may be recognized at the gene level, only two can be recognized immunochemically, and distinction between even these two forms at the physiological level is controversial.

One form, Type I, accumulates in dark-grown tissues, and has been best characterized chemically, especially that isolated from oat (*Avena*) and pea (*Pisum*). Type I is synthesized as Pr, in which state it is remarkably stable to proteolytic degradation. However, the Pfr form of Type I is very unstable; in oats it has a half-life at 25 °C of around 30 min. Type I phytochrome is thus described as light labile.

The synthesis of Type I is rapidly downregulated by Pfr, which acts both at the level of TRANSCRIPTION and on messenger RNA stability or processing. Thus, upon continuous irradiation, the synthesis of Type I phytochrome is strongly inhibited, and existing Type I is rapidly degraded; eventually a steady state is reached, usually representing about 3% of the original level of phytochrome present in the dark.

Immunological characterization with highly specific MONOCLONAL ANTIBODIES has shown that most of this residual phytochrome in light-treated plants is immunochemically distinct from Type I. Consequently, the term Type II has been coined for the pool of phytochrome that predominates in light-grown tissues. Type II Pfr is thought to be about as stable as Type II Pr, and there is no indication of downregulation of Type II synthesis. Consequently, Type II phytochrome is described as being light stable. Immunochemical evidence suggests that Type II is present in both dark-grown and light-grown tissues, and is synthesized constitutively. These distinctions are supported by as yet incomplete chemical studies which indicate minor differences in relative molecular mass (M_r), amino-acid composition and spectral properties between Types I and II.

Molecular structure

Phytochromes consist of an apoprotein composed of a single polypeptide chain. This polypeptide contains two main domains, with further important subdomains in the N-terminal portion. A linear tetrapyrrole (bilin) chromophore is attached to the N-terminal domain.

The apoprotein

The amino-acid sequences of Type I phytochrome apoprotein from oat, rye, rice, maize, pea and courgette, and of phytochrome from light-grown peas and *Arabidopsis thaliana* have been derived from cloned nucleotide sequences. All the sequences have some regions of high HOMOLOGY but others of low homology. Most research has concentrated on oat Type I phytochrome (Fig. P28). Four *phyA* genes are expressed in oats, but the derived amino-acid sequences are >98% identical; this small diversity may be related to ploidy. The oat Type I molecule consists of 1128 amino acids and has an M_r of 124 900 (124.9K).

The mean HYDROPATHY INDEX for Type I phytochrome is >1.06, consistent with a cytosolic protein that does not span cellular membranes. Oat Type I phytochrome exists in solution as a stable dimer, having an apparent M_r of 253K. A value of 5.6 nm has been derived for the Stokes radius of the dimer yielding a frictional ratio of 1.37, indicating the extent to which the molecule diverges from a sphere, which would have a frictional ratio of 1.0. Thus, the dimer has an elongated, ellipsoidal shape.

a

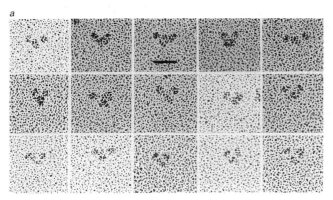

b

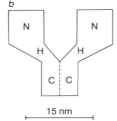

15 nm

Fig. P28 *a*, Electron micrographs of phytochrome molecules from *Avena sativa* after rotary shadowing. Bar = 24 nm. *b*, Schematic of the phytochrome polypeptide depicting approximate dimensions and arrangement of the structural domains, based on electron micrographs. N, N-terminal domain; C, C-terminal domain; H, hinge region. Adapted from [9].

A phytochrome monomer consists of up to four domains. Proteolysis of Pfr by endogenous proteases yields two main polypeptides, a 74K fragment released from the N terminus and a 55K C-terminal fragment. The spectral properties of the 74K fragment, which includes the chromophore, are indistinguishable from those of the native 124K molecule, and it migrates as a globular spherical polypeptide on gels. Thus, about 60% of the molecule represents a discrete globular domain which, apart from the continuity of the polypeptide chain, lacks stable linkages with the rest of the molecule.

The 55K C-terminal domain elutes as a 160K species on gel filtration, apparently behaving as a non-spherical dimer of polypeptides derived from the C-terminal end of the molecule. This evidence is consistent with a model in which dimerization is a function of the C-terminal end of the monomer, allowing the chromophoric N-terminal portions to be relatively free.

Using small angle X-ray scattering, Furuya [4] proposed a model for the dimeric structure of pea Type I phytochrome that visualizes the chromophoric domain as disk-like but not involved in dimerization, which is a property of the 55K C-terminal domain.

Further proteolysis reveals the presence of important subdomains at the N terminus. Incubation of Pr in crude extracts allows endogenous proteases to excise a 6–10K fragment to yield a remaining N-terminal domain of 114–118K. Loss of the 6–10K segment causes a shift of Pfr absorption maximum from 730 to 722 nm, an increase in the rate of dark reversion of Pfr to Pr, an increase in reactivity of the chromophore towards the oxidant tetranitromethane, and a change in CIRCULAR DICHROISM spectrum.

Sequence comparisons and monoclonal antibody studies of proteolytic fragments have shown conserved regions close to the chromophore attachment site and in the C-terminal portion. An EPITOPE located in the C-terminal portion is identical in all phytochromes examined from a range of cereals, dicots, ferns and algae, indicating conservation of surface structure. The amino-acid compositional data for Type I from oat, rye, courgette and pea indicate that 46–50% of the amino acids are polar in nature, within the range expected for soluble proteins. Phytochrome has no prosthetic group other than the chromophore, but may be both glycosylated and phosphorylated.

Chromophore attachment

The site, or sites, of chromophore attachment have been studied by amino-acid sequencing of chromopeptides. Type I oat phytochrome bears only a single chromophore per monomer attached to a cysteine residue as a thioether. CHOU–FASSMAN ANALYSIS predicts that the chromophore is attached at a β-TURN in a mildly hydrophilic region between two strongly hydrophobic segments, the more extensive of which may form a cavity within which the chromophore is housed. Other techniques also indicate that the chromophore is buried within the molecule and is not exposed to the hydrophilic external medium. HYDROPATHY PROFILE analysis indicates that the N-terminal 6–10K polypeptide segment important for the maintenance of stable chromophore–apoprotein interactions is largely hydrophilic; this segment may shield the chromophore from the external medium.

The chromophore

The phytochrome chromophore is an open chain TETRAPYRROLE, a bilin, and is closely related to that of PHYCOCYANIN; on this basis, the appropriate name for the chromophore is phytochromobilin. Phycocyanin is an algal pigment consisting of an apoprotein with several bilin chromophores, but unlike phytochrome, it is not photochromic. The photochromicity of phytochrome must, therefore, be the result of the particular interrelationship between its chromophore and its apoprotein.

Chromophore cleavage has not been successful in characterizing phytochromobilin. *In situ* chromic acid oxidation releases the individual pyrrole units with their side substitutions intact, and from the nature of the products a model for phytochromobilin has been proposed which differs from phycocyanobilin only by the substitution of a methyl for a vinyl group in ring D (Fig. P29) [5]. The very small difference between phytochromobilin and phycocyanobilin clearly cannot account for the photochromicity of phytochrome. Studies of chromopeptides have been useful, and have revealed that the chromophore is attached via a thioether linkage between ring A and cysteine. The structure of the Pr chromophore is now reasonably certain, at least for Type I phytochrome, but that of Pfr is still not fully resolved. Most

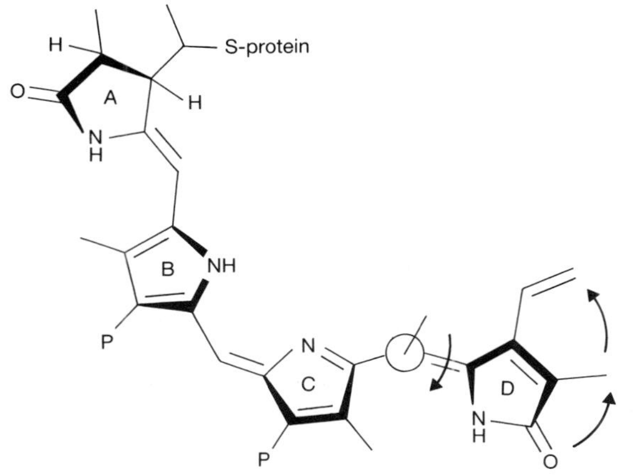

Fig. P29 The phytochrome chromophore, Pr-phytochromobilin. P, propionyl group. Redrawn from [10].

evidence fits the view that a Z–E isomerization occurs between rings C and D, so that Pfr has a more extended conformation than Pr. Stabilization of the isomeric structures of the Pr and Pfr phytochromobilin must be dependent on interactions with the polypeptide backbone.

Genes and gene regulation

Phytochrome is better understood at the molecular genetic than at the chemical level. Phytochrome is encoded in the nucleus by a small family of genes: at least three (*phyA, phyB, phyC*) and up to five *phy* genes having been identified in *A. thaliana*. The members of the family are differentially expressed in relation to developmental and environmental (e.g. light) signals, and the gene products may have different roles in regulating plant development. The first phytochrome gene to be cloned and characterized was that for the Type I phytochrome of oats. More recently, three phytochrome-encoding sequences from *A. thaliana* have been identified and characterized [6]. These are termed *phyA, phyB,* and *phyC*; the existence of two further genes is suspected. The *phyA* gene is thought to represent Type I *Arabidopsis* phytochrome and *phyB* to represent Type II phytochrome. The *phyA, phyB,* and *phyC* genes show no more than 60–70% homology, indicating substantial divergence and relatively ancient branching in the evolution of this photoreceptor family.

Regulation of phytochrome gene expression is complex. Using transcript-specific hybridization PROBES, the patterns of expression of the *phyA, phyB* and *phyC* genes in *A. thaliana* have been determined [6]. The level of *phyA* mRNA was high in dark-grown seedlings, and was only slightly reduced by a red light pulse, but was markedly depressed by several hours exposure to white light; neither *phyB* nor *phyC* transcript levels were markedly affected by light. Two types of transcript from the *phyA* gene were observed, the lower M_r form being strongly downregulated by light whereas the larger M_r form was only clearly present in plants treated with white light. This pattern resembles the expression of the pea *phyA* (Type I) homologue, where the production of different transcripts

is due to multiple start sites within a complex PROMOTER. The existence of upstream READING FRAMES in the pea Type I and in the *A. thaliana phyA, phyB* and *phyC* 5′ leader regions implies the potential for complex transcriptional regulation.

Mechanism of phytochrome action

The molecular mechanism of phytochrome is frustratingly elusive. Phytochrome can certainly act rapidly, the most rapid response being effects on electrical potentials, which occur within seconds; the downregulation of Type I gene transcription is also induced within seconds. Many phenomena regulated by phytochrome have characteristics consistent with action at membranes, but as yet the biochemical evidence indicates that phytochrome is cytosolic and is unlikely to interact directly with cellular membranes. On the other hand, the biochemical evidence relates only to Type I phytochrome.

The strongest evidence for a membrane location of active phytochrome comes from work with the filamentous red alga *Mougeotia*, whose strap-like chloroplast rotates under the influence of red and far-red light [7]. Polarized light treatments have shown that the photoreceptors responsible for the perception of the light signals must be organized on a relatively stable structure in such a way that their chromophores are aligned parallel to each other. This dichroic orientation of phytochrome molecules is also seen in a few other cases, and strongly suggests that, in such plants, at least some of the cellular phytochrome must be associated with membranes or other stable structures.

All attempts to date to establish cell-free phytochrome-mediated responses that may be amenable to biochemical investigation have proved either fruitless or nonreproducible. It is at least possible that the different species of phytochrome, although they possess closely similar photochemical properties, may have different molecular mechanisms of action.

Intense effort is being invested in characterization of the phytochrome gene family from several sources. To study the role of individual phytochromes, transgenic plants that overexpress, or underexpress, individual members of the gene family are being produced [8]. Physiological and biochemical analysis of transgenic plants, created using both sense and ANTISENSE gene constructs, will be of value in determining the roles of the individual phytochromes. Transient expression studies are also being performed in studies of the regulation of the expression of the phytochrome genes. Similar approaches are being used with point-mutated and deleted phytochrome gene constructs to probe those regions of the molecule that are required for biological activity. Expression of phytochrome genes in microorganisms, such as *Escherichia coli*, yeast or baculovirus, should eventually yield sufficient amounts of the individual apoproteins to study their assembly with the chromophore. There are hopes that this route will eventually produce sufficient intact phytochrome for crystallization and subsequent X-ray analysis.

H. SMITH

See also: PLANT HORMONES.

1 Borthwick, H.A., Hendricks, S.B. et al. (1952) A reversible photoreaction controlling seed germination. *Proc. Natl. Acad. Sci. USA* **38**, 662–666.

2 Kendrick, R.E. & Kronenberg, G.H.M. (Eds) (1986) *Photomorphogenesis in Plants* (Martinus Nijhoff, Dordrecht).

3 Smith, H. (1982) Light quality, photoperception and plant strategy. *A. Rev. Pl. Physiol.* **33**, 481–518.

4 Furuya, M. (1989) Molecular properties and biogenesis of phytochrome I and II. *Adv. Biophys.* **25**, 133–167.

5 Rüdiger, W. (1986) The chromophore. In *Photomorphogenesis in Plants* (Kendrick, R.E. & Kronenberg, G.H.M., Eds) (Martinus Nijhoff, Dordrecht).

6 Sharrock, R. & Quail, P.H. (1989) Novel phytochrome sequences in *A. thaliana*: structure, evolution, and differential expression of a plant regulatory photoreceptor family. *Genes Dev.* **3**, 1745–1757.

7 Haupt, W. (1983) Photoperception and movement. *Phil. Trans. R. Soc.* **B303**, 467–478.

8 Keller, J. et al. (1989) Expression of a functional monocotyledonous phytochrome in transgenic tobacco. *EMBO J.* **8**, 1005–1012.

9 Jones, A.M. & Erickson, H.P. (1989) Phytochrome structure. *Photochem. Photobiol.* **49**, 479–483.

10 Rüiger, W. & Thümmler, F. (1991) Phytochrome, the visual pigment of plants. *Angew. Chem. Int. Ed.* **30**, 1216–1228.

phytochromobilin *See*: PHYTOCHROME.

phytoglycogen A branching α-GLUCAN of some plants, which is similar to AMYLOPECTIN, but is more highly branched (at about every 12 residues) and can be larger. It is abundant in 'waxy' mutants of maize.

phytohaemagglutinin *See*: LECTINS.

phytohormone *See*: PLANT HORMONES.

phytol Long chain isoprenoid alcohol that is esterified to the protochlorophyllide *a* ring system in chlorophyll.

Phytophthora infestans A HEMIBIOTROPHIC oomycete pathogen of solanaceous crops, especially potato, on which it causes late blight. *P. infestans* was the causal agent of the great Irish potato famine in the 19th century. It does well in cold damp periods, quickly destroying foliage. The hyphae survive winter months within tubers, and spread both sexually and asexually through rain-borne spores of various types (oospores, zoospores, sporangia). *See also*: PLANT PATHOLOGY.

phytoreovirus group Named from the Greek *phytos*, plant, and reovirus. One of the two genera of plant viruses in the family REOVIRIDAE, the other being the FIJIVIRUS GROUP. The isometric particles have a 70 nm diameter outer shell made up of four polypeptide species (M_r 130 000, 96 000, 36 000 and 35 000) surrounding an inner core 59 nm in diameter comprising three polypeptide species (M_r 160 000, 118 000 and 58 000). The particles contain 12 species of double-stranded RNA with M_r values ranging from 3.0×10^6 to 0.3×10^6. Gene functions are not known for most of the RNA species. These are considered to be insect viruses becoming adapted to plants. *See also*: PLANT VIRUSES.

Matthews, R.E.F. (1982) *Intervirology* **17**, 85.

phytotropin A group of inhibitors of polar AUXIN transport, so called because of their effect on tropic responses. They include NAPHTHYLPHTHALAMIC ACID (NPA). 2,3,5-Triiodobenzoic acid (TIBA, Fig. P30), is often also termed a phytotropin although it does not meet the structural criteria of the definition. Phytotropins inhibit auxin transport noncompetitively by blocking its efflux from the basal regions of cells. They are thought to bind to the auxin efflux carrier, but not at the auxin-binding site. Phytotropins and the MORPHACTINS have been valuable experimental tools in studies of auxin transport. *See also*: PLANT HORMONES.

Kateckar, G.F. & Geissler, A.E. (1980) *Plant Physiol.* **66**, 1190–1195.

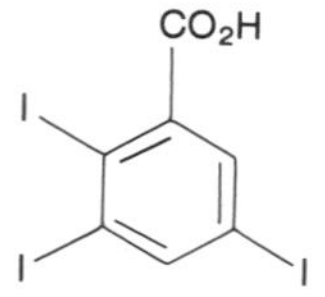

Fig. P30 Structure of 2,3,5-triiodobenzoic acid.

Picornaviridae Family of small nonenveloped RNA viruses to which POLIOVIRUS, RHINOVIRUSES, and FOOT-AND-MOUTH DISEASE VIRUS belong. The family contains four genera: Aphthovirus (e.g. foot-and-mouth disease virus), Cardiovirus (e.g. encephalomyocarditis virus), Enterovirus (e.g. poliovirus), and Rhinovirus (the common cold viruses). Crystal structures of members of all genera have been solved. All have icosahedral particles ($T = 1$) with a capsid composed of 60 copies of a capsomer of three types of virion protein (VP1, VP2, and VP3) associated with VP4. The capsid polypeptides and viral enzymes are derived by cleavage of a large precursor polypeptide (a POLYPROTEIN), the single translation product of virus mRNA, which is cleaved during synthesis into three precursor proteins (P1, P2, and P3). The genome is a single-stranded (+)-sense RNA of around 7500 nucleotides with a VPg protein bound to the 5′ end through a uridine residue, and a 3′ poly(A) tract. Cardioviruses and aphthoviruses contain a poly(C) tract near the 5′ end. Picornavirus replication takes place in the cytoplasm via a (−)-strand intermediate which acts as a template for (+)-strand synthesis. Both (−)- and (+)-sense RNAs have VPg attached to the 5′ end through a short stretch of U residues, and VPg-U is thought to act as a primer for RNA synthesis. (+)-Sense RNA may be used as mRNA after cleavage of the VPg or be packaged into virions. Assembly of viral capsids is by a sequence of aggregation and cleavage *in situ* of precursor proteins (*see* POLIOVIRUS). *See also*: Plate 7c–e; ANIMAL VIRUSES; ANIMAL VIRUS DISEASE.

pilin Protein subunit of bacterial PILUS. *See also*: BACTERIAL PATHOGENICITY.

pilus (*pl.* pili) Proteinaceous filament which protrudes from the surface of some bacterial cells, and which enables the bacterium to attach to other cells or surfaces. The F pilus (or sex pilus) is involved in bacterial CONJUGATION and is only found on 'male' (donor) cells containing the F PLASMID. If there is no F pilus, mating does not take place. The F pilus is also the receptor for the filamentous single-stranded DNA BACTERIOPHAGE M13, the virus attaching to the tip of the F pilus. Pili also act to target bacteria to specific receptor sites on host cell membranes. For example, *Neisseria gonorrhoeae* uses pili to attach to cells in the human genitourinary tract, allowing the bacteria to colonize these cells and cause the disease gonorrhoea. For P pili and the mechanism of antigenic variation involving pilus protein (pilin) in *Neisseria* see BACTERIAL PATHOGENICITY. *See also*: PLASMIDS.

pin proteins *See*: PLANT WOUND RESPONSES.

pinF Plant-inducible locus associated with the virulence (*vir*) region of the TI PLASMID. Sometimes referred to as the *virF* gene.

pinocytic vesicle Membrane-bounded vesicles responsible for fluid phase ENDOCYTOSIS (pinocytosis). Clathrin-coated vesicles account for the bulk of fluid phase endocytosis in certain cell types such as baby hamster kidney (BHK) cells. However, vesicles without morphologically visible coats may account for some fluid phase endocytosis in other cells. The term 'pinocytic vesicles' has also been used to describe vesicles which are now known as ENDOSOMES. *See also*: COATED PITS AND VESICLES.

pinocytosis Alternative term for fluid phase ENDOCYTOSIS.

PIP₂ PHOSPHATIDYLINOSITOL 4,5-BISPHOSPHATE. *See also*: SECOND MESSENGER PATHWAYS.

pisatin An isoflavonoid PHYTOALEXIN produced by several species of peas. *See*: PLANT PATHOLOGY.

Pit-1 TRANSCRIPTION FACTOR regulating expression of the prolactin and growth factor genes in the pituitary. *See also*: HOMEOBOX GENES AND HOMEODOMAIN PROTEINS.

pituitary dwarfism A rare disease which is usually sporadic and of unknown cause, although two Mendelian forms are known. Proportionate short stature (resulting from failure to secrete GROWTH HORMONE) is the cardinal feature, and is sometimes accompanied by deficiency of other pituitary hormones (particularly gonadotrophins and adrenocorticotrophic hormone). Growth may be improved by regular injections of growth hormone.

pituitary gland An aggregation of specialized secretory cells at the base of the brain, attached by a stalk to the HYPOTHALAMUS. It consists of two main lobes: the anterior lobe consists of endocrine cells which, under the influence of hypothalamic peptides, secrete protein hormones which regulate the function of the thyroid, the gonads, the adrenal cortex, and other endocrine glands; the posterior (neural) lobe is a reservoir for antidiuretic hormone (vasopressin) and oxytocin. The smaller intermediate lobe secretes α-melanocyte-stimulating hormone. Anterior pituitary hormones include ADRENOCORTICOTROPIN; FOLLICLE-STIMULATING HORMONE; GROWTH HORMONE; LUTEINIZING HORMONE; THYROID-STIMULATING HORMONE.

PKR RNA-ACTIVATED PROTEIN KINASE. *See also*: INTERFERONS.

PKU PHENYLKETONURIA.

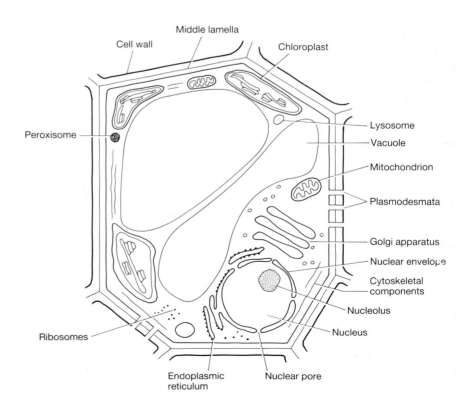

Fig. P31 Generalized diagram of a higher plant cell from photosynthetic tissue.

placode A localized thickening of the surface epithelium of the embryo, usually circular, which gives rise to some of the facial organs, such as the lens, the ear, and the nose. At least some of the placodes are thought to result from INDUCTION by underlying tissues.

plakoglobin Protein of M_r 82 000, localized to intermediate and desmosome junctions. May interact directly with the cytoplasmic domains of cadherins (*see* CELL ADHESION MOLECULES). It is related to another intermediate junction component, β-CATENIN, and to the product of the *Drosophila* SEGMENT POLARITY GENE *armadillo. See also*: CELL JUNCTIONS; DROSOPHILA DEVELOPMENT.

plant cell The cells of green algae and higher plants are highly diverse in size and structure. The common features found in many plant cells are illustrated in Fig. P31.

plant cell signalling Far less is known about the molecular mechanisms of signal transduction in plant cells than in animal cells. PLANT HORMONES and other plant signal molecules (such as the microbial elicitors of phytoalexin synthesis and wound responses) may act at specific receptors (e.g. auxin-binding proteins, glucan elicitor binding proteins). Although cyclic AMP occurs in plants, there is no evidence for its action as a classical second messenger. However, an inositol phospholipid second messenger pathway similar to those in animal cells and resulting in changes in cytoplasmic free CALCIUM concentrations is thought to operate in some PHYTOCHROME-mediated responses to light and in the cellular responses to some PLANT HORMONES. A heterotrimeric G protein (*see* GTP-BINDING PROTEINS) has been identified as part of the signal transduction mechanism for blue-light induced responses such as suppression of stem elongation and regulation of stomatal opening. *See also*: PLANT PATHOLOGY; PLANT WOUND RESPONSES.

alkali (24% potassium hydroxide) under conditions which prevent degradation by oxidation with atmospheric oxygen. The residual material is composed mainly of polymers of glucose and is termed α-cellulose. It also contains small but varying amounts of the sugars (xylose, mannose and galactose) making up the hemicelluloses of the wall. It constitutes the fibrils of the wall.

Cellulose is defined as an extended chain (degree of polymerization, 8000–12 000) of β(1,4) glucosyl units (a β-glucan). The conformation of the chains is stabilized by inter- and intrachain hydrogen bonds. Cellulose is the main polymer of the microfibrils (3.5 nm diameter) each of which consists of about 36 glucan chains held together in ordered arrays by intermolecular hydrogen bonds. Microfibrils (Fig. P32) make up the fibrillar phase of the wall and contribute to its tensile strength. The cellulose microfibrils are spun out by mobile enzyme complexes at the outer surface of the PLASMA MEMBRANE. As the microfibrils are laid down with a definite initial orientation that can change during the growth of the wall, it is very probable that their orientation is controlled by CYTOSKELETAL elements in the cytoplasm which therefore have some influence on the movement of the enzyme complex.

Pectic substances

Pectic polysaccharides are classified as water-soluble polysaccharides extracted from the wall by warm water and a chelating agent for calcium at a pH of about 4.5 which prevents degradation by transelimination.

The basic structure of the polymers is that of a core consisting of a chain of α(1,4)-linked D-galacturonosyl units interspersed with occasional L-rhamnosyl units linked α(1,2) in the chain. A variable number of the carboxyls of the chain can be esterified by methyl groups and the hydroxyls of the uronic acids are randomly esterified by acetate at O-2 and O-3. Attached to the main

Plant cell wall macromolecules

THE plant cell wall is a composite of two phases comprising a fibrillar component embedded in a matrix. The bulk of the material of the cell wall is polysaccharide; CELLULOSE constitutes the fibrils, and hemicellulose and pectic substances the matrix. In addition to the polysaccharides, at various stages of wall development protein, lignin, lipids and waxes are laid down.

Polysaccharides

These are classified empirically on the basis of methods for their extraction and solubility, with additional chemical analyses to establish individual polysaccharides and various criteria for their purity [1].

Cellulose

Cellulose is designated as the polysaccharide left after extraction of delignified, lipid-extracted wall material, with boiling strong

Fig. P32 Freeze-etch sections through a young plant cell wall to show the cellulose microfibrils *in situ*.

chain are short side chains composed of L-fucose, D-xylose and D-galactose or even single units of these SUGARS.

Some of the core polygalacturonosyl chains have highly branched arabinogalactan side substituents of large molecular weight which may also occur as separate neutral polymers. The arabinogalactans consist of an outer shell of L-arabinofuranosyl (arabinose) radicals linked α(1,3) and α(1,5) attached to a core main chain of β(1,4) galactosyl units which can also carry some branches of galactose linked β(1,3) and β(1,6).

This complex mixture of acidic and neutral polysaccharides is metabolically modified during growth and development. It is one of the principal components, together with water, of the matrix of the young wall and the variable nature of its composition is related to its most striking property of forming reversible gels with water. It influences the water distribution in the wall of the young growing cell, especially in the matrix, and hence it affects the relationship between the matrix polysaccharides and the microfibrils. This in turn influences the texture and mechanical properties of the wall. These are important features during growth of the young wall.

Hemicelluloses

The hemicellulose polysaccharides are extracted from the wall by treatment with hot alkali, after removal of pectic substances and lignin. They form the bulk of the matrix of the mature wall.

In young walls an important component of the hemicelluloses is xyloglucan. This polysaccharide is composed of a β(1,4) glucan backbone with side chains of xylose attached α(1,6). Some of these carry a β(1,2) galactose unit and sometimes this galactose residue is substituted by a terminal L-fucosyl radical linked α(1,2). The xyloglucan forms, by means of hydrogen bonding, a linkage between the glucans of the microfibrils and the hemicelluloses of the matrix. Alterations in the matrix–fibril interaction bring about changes in the important stress transmission characteristics between these two phases and this has a very important effect on the properties of the wall. Any compound that forms bridges between the cellulose of the fibrils and the polymers of the matrix, even though present in relatively small amounts, can have such a major influence upon the properties of the wall composite.

As the wall develops, the principal polysaccharides deposited in the matrix are xylans in angiosperms and glucomannans in gymnosperms. Xylans are large molecules with a degree of polymerization of about 150–200. Attached to the β(1,4) xylosyl backbones are short terminal side chains of 4-O-methyl-D-glucuronic acid (methyl-ester and free acid units) joined by α(1,2) bonds. These are probably distributed randomly along the chain in the ratio of approximately 1 glucuronosyl radical to 10 xylosyl radicals. *In vivo*, about 50% of the xylose groups are acetylated at C-3 and C-2. The grasses have large amounts of xylans in the hemicellulose fraction and these have an additional side chain of L-arabinofuranose attached by α(1,3) links. Arabino-(4-O-methylglucurono)xylans also occur in gymnosperms but in much smaller amounts.

Glucomannans and galactoglucomannans form the bulk of the hemicelluloses of gymnosperms (ratio of galactose : glucose : mannose, around 1 : 1 : 3). These polysaccharides consist of chains (degree of polymerization about 100) of randomly arranged D-glucose and D-mannose radicals joined by β(1,4) links. The D-galactose residues are attached to the main chain as terminal branches linked α(1,6). *In situ*, the mannose units of the polymer are acetylated at C-2 or C-3.

The extended chains of the hemicellulose molecules are arranged in the matrix in a parallel orientation to that of the microfibrils and a proportion of them are closely applied to the microfibrils by hydrogen bonds so that they can act as keying substances between the bulk hemicelluloses of the matrix and the fibrils. These are difficult to separate from the cellulose when the wall is extracted with hot alkali and their presence accounts for the xylose, mannose and galactose found in α-cellulose preparations.

Synthesis and deposition of wall polysaccharides

It is probable that the enzyme complexes for the synthesis of hemicelluloses, pectins and cellulose from nucleoside diphosphate sugars are constructed, at least in part, on the ENDOPLASMIC RETICULUM and then transferred to the GOLGI APPARATUS where the bulk of the synthesis takes place.

The complex for cellulose synthesis is not normally functional within the Golgi apparatus and is transported to active sites at the plasma membrane. Hemicelluloses and pectic substances are, however, formed within the cisternae and vesicles of the Golgi apparatus and they are transported within the vesicles to the plasma membrane, where fusion occurs and the polysaccharides are packed into the wall [2].

Lignin

Lignin is an aromatic, insoluble, high molecular weight constituent of the cell wall, and is derived by the enzymatic dehydrogenation and subsequent polymerization of coumaryl, coniferyl, and sinapyl alcohols. The guaiacyl type (3-methoxy-4-hydroxy) is found in gymnosperms and the syringyl-guaiacyl type (3,5-dimethoxy-4-hydroxy) in dicotyledons. Cinnamyl alcohols are the immediate building units of lignin and these are transported to the wall where oxidation to mesomeric phenoxy radicals occurs, catalysed by an isoenzyme of peroxidase. Rapid polymerization of these free radicals takes place to give a three-dimensional covalently linked network with some covalent bonds being formed between the polysaccharides and the lignin.

The lignin penetrates the wall from the outside (middle lamella and primary wall) inwards during the early stages of secondary thickening. It is a hydrophobic filler material which replaces water in the cell wall and finally encrusts the microfibrils and the matrix polysaccharides; this ensures a good stress transfer between the components and also ensures that the layers and components of the wall will not slip with respect to one another. The system becomes one in which the linear polysaccharide polymers are enclosed in a cross-linked polymer cage. At this stage the wall has great tensile strength because of the presence of the microfibrils, and a rigid structure because of the lignified matrix.

Other bridging linkages between wall polymers

Covalent linkages between the polymers of the wall can also be established by the action of wall peroxidase on the tyrosine residues of proteins that are present in the wall. Oxidation produces an isodityrosine link between the polypeptides. Ferulic acid esters occur on the arabinose residues of the acidic matrix polysaccharides in the cell walls of some herbaceous plants and grasses. Esters on two separate polymers may be oxidized to give phenoxy free radicals and these can then form a diferulic ester linkage which would join them together. This provides a mechanism for possible linkages between polysaccharide–polysaccharide and polysaccharide–lignin.

The formation of these covalent cross-linkages between the polymers of the primary wall is believed by some workers to limit plant cell wall extensibility and have some part in the mechanisms for the control of cell growth and extension.

Proteins

Most of the proteins found in the cell wall are GLYCOPROTEINS. Some are enzymes such as isoenzymes of peroxidase, phosphatases and amylases, and others are structural proteins and glycoproteins. These latter are usually hydroxyproline-rich and glycine-rich proteins, and in addition to having a structural function, some are also LECTINS.

Skins and protective coverings of the wall

The walls of the epidermal cells of leaves and many other aerial organs of plants are covered with a protective film of wax overlying a skin of cutin which penetrates into the pectic substances of the outer layers of the primary wall. These substances are spread over the whole surface of the leaf or other organ and they serve to prevent excess water loss and act as an aid against mechanical injury. The waxes are a complex mixture containing long-chain alkanes, alcohols, ketones and fatty acids. The alcohols and fatty acids occur in the form of long chain esters and are also found uncombined. Some hydroxy fatty acids occur and these are present as polymeric ester-linked compounds known as estolides. Cutin is a complex mixture of polymeric fatty acids mainly interconnected by ester, peroxide, and ether linkages, probably as a three-dimensional network. Hydroxy fatty acids predominate in the mixture so that the polymers are mainly estolides. Present in the cuticular layer are tannins and phenylpropanoids.

A suberin complex is present as a component of periderm (outer bark) walls. It is made up of suberin, which is an aliphatic polyester, and polyphenylpropanoids, which are lignin-like and make up a large proportion of the complex. It may partially or completely enclose the cell and it is formed as scars in wounded tissue. Cork cambium produces many suberized cells in a periderm. This periderm — cork — is light, buoyant, and impervious to decay.

D.H. NORTHCOTE

See also: GLYCANS.

1 Northcote, D.H. (1972) Chemistry of the plant cell wall. *Annu. Rev. Plant Physiol.* 23, 113–132.
2 Northcote, D.H. (1989) Control of plant cell wall biogenesis. *ACS Symp. Ser.* 399, 1–15 (American Chemical Society, Washington, DC).

plant developmental mutants *See:* FLOWER DEVELOPMENT; PLANT EMBRYONIC DEVELOPMENT.

plant embryonic development The genetic basis of plant embryonic development has been most extensively studied in DICOTYLEDONS such as ARABIDOPSIS, and *Antirrhinum* (Fig. P33). From the study of mutant lines it is estimated that some 50 genes (out of an estimated total of 4000 essential genes) are involved in embryonic pattern formation in *Arabidopsis*. The main axes — apical–basal (A–B, shoot tip to root) and periphery–centre (radial) — are set up separately and independently of each other in the embryo. The axes are established early in embryogenesis and elaborated on in later stages. Pattern formation generates the basic features of the organisation, whereas morphogenetic activity such as localized cell division or changes in cell shape brings about the structure of the seedling.

Genetic studies indicate that the first asymmetric division of the zygote distributes positional determinants required for A–B polarity, and that some system of POSITIONAL INFORMATION is involved in determining cell fate in the early embryo. The A–B axis seems to be partitioned into three main regions, apical, central, and basal, and is thought to be established by the octant-stage embryo. In *Arabidopsis* and other species (e.g. *Antirrhinum*), mutations have been discovered that specifically delete different parts of the A–B axis (see Table P2).

Table P2 Genes involved in embryonic development in *Arabidopsis*

Class	Genes	Phenotype of mutant seedling
Apical–basal pattern deletion	*gnom*	Deletions of terminal regions (cotyledons and radical) leaving a 'hypocotyl' which does not, however, resemble the seedling hypocotyl in structure
	gurke	Deletion of apical region (cotyledons)
	fackel	Deletion of central region (hypocotyl)
	monopteros	Hypocotyl, radicle, and root meristem deleted
	shoot meristemless	Seedlings look normal but lack a shoot meristem and therefore cannot produce a shoot
Radial pattern defect	*keule*	Epidermal cells are abnormal. The seedling is rough surfaced, often elongated, reduced cotyledons
	knolle	Epidermal cell layer is not distinguishable from other cells. Seedling is round or tuber shaped

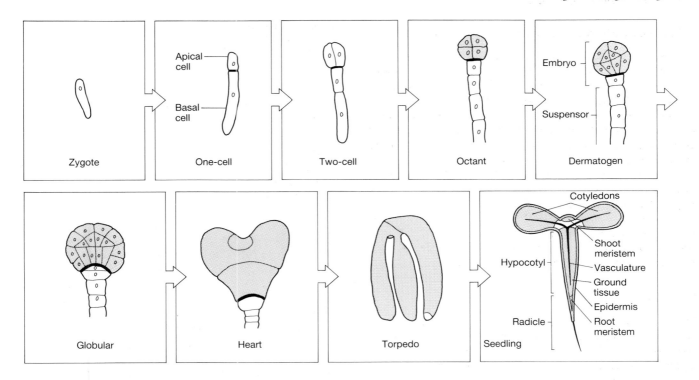

Fig. P33 Stages in development of the *Arabidopsis* embryo. The bold line marks the boundary between descendants of the apical cell and those of the basal cell. 'One-cell' and 'two-cell' stages refer to the number of cells in the embryo proper — the apical cell and its descendants. The *Arabidopsis* seedling consists of two embryonic leaves (cotyledons), an embryonic stem (hypocotyl), and an embryonic root (radicle). Two MERISTEMS (shoot and root) have been set aside. From Weigle, 1993.

The genes involved in setting up the A–B axis seem to act in a hierarchical fashion at different stages of development. Of the genes identified so far, *GNOM* acts earliest: certain *gnom* mutations perturb development before the first division of the zygote. In these mutants the zygote does not elongate as much as the wild-type and divides more symmetrically, followed by divisions in random planes, rather than in the fixed planes of cleavage of the wild type.

GNOM and *SHOOT MERISTEMLESS* are also required for the regeneration of root and shoot meristems in culture (under hormonal influence) but *MONOPTEROS* is not.

Mutations affecting the radial axis have also been obtained in *Arabidopsis*, in which the epidermis is lacking.

Mayer, U. et al. (1991) *Nature* **353**, 402–407.
Weigel, D. (1993) *Curr. Biol.* **3**, 443–445.

Plant genetic engineering

THE genetic engineering of plants, in the modern sense, is the development and application of the technology for plant genetic transformation through the direct manipulation of the plant genome and plant gene expression by the introduction of novel DNA. There are now several instances in which agronomically useful genes from a variety of sources have been genetically engineered into plants and it is likely that, in the near future, such transformed or transgenic plants will be used in crop production for the first time. Whether or not there will be subsequent widespread use of genetically engineered plants depends on consolidating the technology for transformation, extending the range of species that can be regenerated in culture, and identifying and isolating useful genes.

Plant transformation

Production of a phenotypically transformed plant depends on the ability to introduce DNA into the plant genome so that it is expressed at an appropriate level and in the required cell types. In order to ensure that all the cells of the plant will contain the altered genome, for most applications it is also necessary to be able to regenerate the individual transformed cells into fully differentiating plants and to select for or identify the transformed cells in a background of nontransformed cells.

The most widely used methods of transformation employ a derivative of the tumour inducing (Ti) plasmid from the bacterium *Agrobacterium tumefaciens* as a DNA VECTOR (*see* TI PLASMID). In the natural situation, following infection with this pathogenic soil bacterium, part of the Ti plasmid — the T-DNA — is transferred into the nuclear genome of the plant. The parts of the T-DNA essential for tumorigenesis are a border sequence and the genes controlling synthesis and sensitivity to hormones and synthesis of secondary metabolites in the plant cell. However, of these, only the border sequence is necessary for transformation of the plant

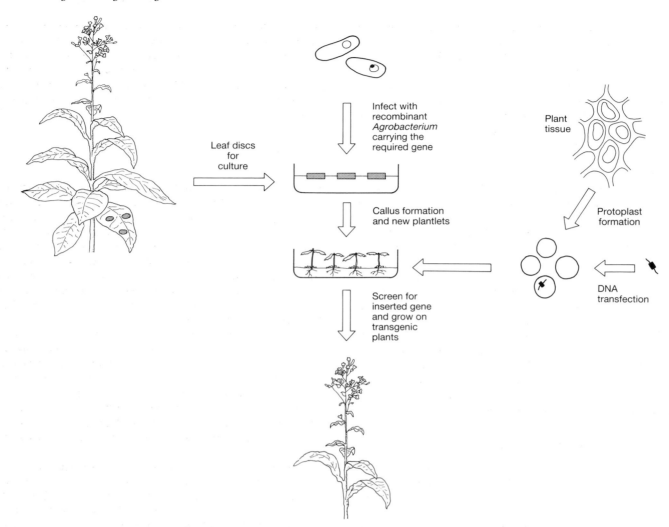

Fig. P34 Strategies of gene transfer into plants. In one route, agrobacteria carrying Ti plasmids containing the gene to be transferred are used to infect leaf disks in culture. Plants are regenerated from this transformed tissue. In the other, protoplasts derived from cultured plant tissue are transformed with the required gene and then grown up into individual plants.

genome [1]: the other elements of the T-DNA have been replaced in most Ti plasmid vectors with the new gene being engineered into the plant genome and with a dominant marker gene for selection of the transformed cells. A commonly used marker gene encodes resistance to kanamycin although other drug- or herbicide-resistance genes have been used.

Transformation is normally carried out by co-cultivation of the plant tissue to be transformed together with *A. tumefaciens* carrying the modified T-DNA (Fig. P34). The plant tissue is often an explant from leaves, stems or other organs. After an initial co-cultivation with *A. tumefaciens*, the tissue is subsequently cultured without the bacterium under conditions that will select for the phenotype of the marker gene in the T-DNA, and in the presence of hormones to promote shoot and root development from the cells at the wounded surface of the explant. The regenerated plants and their progeny are then screened for the phenotype of the novel gene.

This procedure works best with, and was devised for use with, Solanaceous plants tobacco, tomato and potato. It has also been used successfully with plants from other families including *Arabidopsis thaliana* (a small plant with a simple genome which is widely used in genetic studies), *Brassica napus* (oilseed rape) and cotton. In principle, plants of any family could be transformed by *A. tumefaciens* but in practice the range is restricted owing to the difficulty of regenerating many species in culture, including most of the cereals.

In one alternative transformation procedure there is no need for a regeneration step: seeds of *Arabidopsis thaliana* imbibed in the presence of *A. tumefaciens* germinated and produced progeny transformed with the T-DNA. Unfortunately, this procedure has not been successful with other species [2].

Direct transfer of DNA

Direct DNA transfer into protoplasts (plant cells from which the wall has been removed) avoids the need to transfer DNA into the *Agrobacterium* vector plasmids. Uptake of DNA is promoted by ELECTROPORATION or treatment of the cells with polyethylene glycol and/or calcium (*see* TRANSFECTION) and the DNA becomes integrated into the plant genome. The transformed protoplasts are then regenerated in culture. This technique is thus most successful with Solanaceae and other species that are easily regenerated from protoplasts, although there are reports of the method being applied to some more recalcitrant species, including rice and maize [3].

Other methods of direct transfer of DNA overcome the problems of regeneration by introducing DNA into plant MERISTEMS or germ-line tissue, rather than dedifferentiated protoplasts or wound-induced CALLUS, as in the methods described above. There are reports of successful transformation by injection of DNA into the floral primordium or even into pollen precursor cells in the anther.

The most bizarre method of plant cell transformation uses a modified gun to fire DNA-coated tungsten particles into the target tissue (*see* MICROBALLISTICS). There are now second-generation particle guns in which the particles, which may also be of gold, are propelled electrostatically or by pressurized gas. The major achievement of the particle gun so far is the transformation of soybean [4], but as meristematic tissue is regenerable from most if not all plants, the method may well succeed even with wheat and barley, which until now have resisted all attempts at transformation.

Genes for engineering into plants

The genes that are engineered into plants — the transgenes — comprise PROMOTER and transcriptional terminator sequences which control the expression of the associated coding sequence (*see* EUKARYOTIC GENE EXPRESSION; EUKARYOTIC GENE STRUCTURE; TRANSCRIPTION). Cell-specific promoters have been used for some applications, but if the coding sequence is to be expressed constitutively throughout the plant, a frequent choice is the 35S RNA promoter from cauliflower mosaic virus (*see* CAULIMOVIRUS GROUP).

The coding sequence in the transgene may originate from any type of organism as illustrated in the examples described below.

Herbicide resistance

Herbicides are widely used for weed control, particularly in intensive crop production systems. It is hoped that use of herbicide-resistant crop plants will allow a lower treatment of herbicide to be used more effectively than now, with benefits to the environment and the efficiency of crop husbandry. The herbicide can be applied as a low dose during growth of the herbicide-resistant plants, whereas the current practice is to apply high levels of herbicide to purge the field of weeds before the crop germinates. Herbicides act by inhibiting enzymes of plant metabolism. In plants engineered to resist herbicides the effect has been achieved in several different ways: (1) by overexpression of the target enzyme of the herbicide; (2) by expression of a herbicide-insensitive homologue of the target enzyme isolated

either from a mutant plant or from bacteria; or (3) by expression of a bacterial enzyme that inactivates the herbicide [5]. Resistance has been achieved to several different herbicides including glyphosate (by overexpression), sulphonylurea compounds (by expression of herbicide-insensitive enzymes) and phosphinothricin and bromoxynil compounds (by expression of inactivating enzymes).

Virus resistance

Virus-resistant plants of several species have been engineered by transgenic expression of viral coat protein genes. Coat protein-mediated resistance has been achieved to potato viruses X, Y, and leafroll in potato, tomato mosaic virus in tomato and in sugar beet to the rhizomania virus. The other examples include many types of positive-strand RNA viruses and tomato spotted wilt virus, which has a double-stranded genome. The resistance probably results from perturbation of the balance in the infected transgenic plant between coat protein production and virus multiplication [6]. The production of excess coat protein may 'mop up' viral RNA before it has been able to replicate.

Other types of transgenic resistance to virus infection are based on expression of viral sequences. For example, resistance to GEMINIVIRUSES has been engineered by transformation with a defective-interfering form of the viral genome (*see* DEFECTIVE-INTERFERING VIRUS) or with a viral gene oriented in the Ti plasmid vector so that it is expressed in the plant in the ANTISENSE rather than the sense orientation. Transgenic expression of viral SATELLITE RNA has also been used to engineer resistance to certain RNA viruses.

The defective-interfering and satellite RNA strategies may both act by decoy mechanisms, with the product of the transgene competing with the intact genome for the viral replicase (*see* PLANT VIRUSES). The antisense RNA is thought to form a duplex with and thereby destabilize or inactivate the mRNA function of the sense RNA from the infecting virus. There are also recent reports of virus resistance through expression of parts of the viral replicase gene, also with positive-strand RNA viruses.

Insect resistance

Plants have been engineered to insect resistance by transfer of genes from a bacterium and from cowpea (*Vigna unguiculata*). The bacterium *Bacillus thuringiensis* produces an insecticidal protein endotoxin (*see* δ-ENDOTOXIN). To achieve high level expression of the endotoxin gene in plants it was necessary to mutate the coding sequence without affecting the protein sequence but so that the pattern of CODON USAGE followed that of a plant rather than a bacterial gene. The insecticidal gene from cowpea encodes an inhibitor of the proteolytic enzyme TRYPSIN and is thought to act by inactivating the digestive system of the insect so that it starves [7] (Fig. P35).

Male sterility

Genetically male-sterile plants (*see* MITOCHONDRIAL GENOMES: PLANTS) are desirable for the production of hybrid seed as they

Fig. P35 Insect resistance in tobacco plant (*Nicotiana*) as a result of expression of the cowpea trypsin inhibitor gene. *a*, Control; *b*, expressing the trypsin inhibitor gene. Courtesy Paul Boseley, Agricultural Genetics Company.

avoid the need for expensive and laborious removal of anthers from flowers to prevent self-fertilization. In a novel transgenic method of generating functionally male-sterile plants, the development of the pollen cells was ablated specifically by expression of fungal or bacterial RIBONUCLEASE transgenes fused to a pollen-specific promoter from the plant [8].

Flower colour

Changes in flower colour have been produced in several ways. The most direct involves the transfer between plant species of genes involved in pigment formation. There are also examples in which the production of pigment-forming enzymes was reduced or eliminated through the use of antisense RNA technology, as described above for virus resistance [9].

Introduction of additional genes for enzymes in the pigment-forming pathway modifies flower colour, but by reducing rather than increasing the amount of pigment production. The mechanism of this phenomenon, called co-suppression, is not understood [10], but is thought to involve reversible modification of the introduced DNA.

Fruit quality

The modification of fruit quality in tomatoes is one of the best examples of antisense techniques in plant genetic engineering [11]. Genetically engineered tomatoes have been produced in which there is reduced production of the ethylene-forming enzyme or of polygalacturonase. Ethylene is the ripening hormone of tomato and polygalacturonase is involved in softening cell walls in the fruit during ripening. The reduced production of these components gives fruit which may be stored unripe until treated with ethylene or which does not soften and rot as readily as natural fruit.

Future developments

These examples illustrate the ways in which the properties of plants may be modified by genetic engineering: gene expression may be increased or reduced, manipulated temporally by the use of cell-specific promoters or changed qualitatively by the transfer of genes between unrelated plants or even from different types of organism. The use of novel transformation methods will allow this technology to be applied to most types of crop plant and the scope of applications will expand with the range of available genes [12]. In this context, the use of gene isolation methods based on GENOME WALKING or TRANSPOSON TAGGING will be particularly important: through these techniques it is becoming possible to isolate genes which control the basic stages in plant development with which it will become possible to manipulate the growth and form of crop plants.

The major unknowns for plant genetic engineering concern the stability of the transgenes: it is not yet known to what extent maintenance or expression of the transgene is stable in the genome or whether the cosuppression effect will apply generally. There is also an economic factor: it may not always be possible to apply the advanced technology of plant transformation, particularly in developing countries. In these situations, there may be potential for the use of viral vectors which promote expression and systemic spread of novel genes without transformation of the plant genome following infection of the plant. Both DNA and RNA viruses are suitable for this purpose.

D. BAULCOMBE

See also: GENETIC ENGINEERING.

1 Binns, A.N. (1988) Cell biology of *Agrobacterium* infection and transformation of plants. *Annu. Rev. Microbiol.* **42**, 575–606.
2 Feldmann, K.A. & Marks, M.D. (1987) *Agrobacterium*-mediated transformation of germinating seeds of *Arabidopsis thaliana*: A non-tissue culture approach. *Mol. Gen. Genet.* **208**, 1–9.
3 Potrykus, I. (1991) Gene transfer to plants: Assessment of published approaches and results. *Annu. Rev. Plant Physiol. Plant Mol. Biol.* **42**, 205–225.
4 Christou, P. et al. (1990) Soybean genetic engineering — commercial production of transgenic plants. *Trends Biotechnol.* **8**, 145–151.
5 Quinn, J.P. (1990) Evolving strategies for the genetic engineering of herbicide resistance in plants. *Biotech. Adv.* **8**, 321–333.
6 Beachy, R.N. et al. (1990) Coat protein-mediated resistance against virus infection. *Annu. Rev. Phytopathol.* **28**, 451–474.
7 Gatehouse, A.M.R. et al. (1991) Novel insect resistance using protease inhibitor genes. In *Molecular Approaches to Crop Improvement* (Dennis, E.S. & Llewellyn, D.J., Eds) 63–77 (Springer-Verlag, Wien/New York).
8 Mariani, C. et al. (1992) A chimaeric ribonuclease-inhibitor gene restores fertility to male sterile plants. *Nature* **357**, 384–387.
9 van der Krol, A.R. et al. (1988) An anti-sense chalcone synthase gene in transgenic plants inhibits flower pigmentation. *Nature* **333**, 866–869.
10 Napoli, C. et al. (1990) Introduction of a chimeric chalcone synthase gene into petunia results in reversible co-suppression of homologous genes in trans. *Plant Cell* **2**, 279–289.
11 Kramer, M. et al. (1989) Progress towards the genetic engineering of tomato fruit softening. *Trends Biotechnol.* **7**, 191–194.
12 Fraley, R.T. (1992) Sustaining food supply. *Trends Biotechnol.* **7**, 40–43.

plant growth retardants A group of synthetic plant growth regulators (*see* PLANT HORMONES) that reduce plant stature by inhibiting cell expansion in the subapical MERISTEM without affecting the pattern of development. Most plant growth retardants have been shown to function primarily by inhibiting the biosynthesis of the GIBBERELLIN hormones. Three groups of retardants, which affect different steps in the gibberellin biosynthetic pathway, are recognized: quarternary ammonium or phosphonium salts, such as CCC and CHLORPHONIUM CHLORIDE inhibit *ent*-kaurene oxidase; N-containing hetercyclic compounds, such as ANCYMIDOL, PACLOBUTRAZOL, UNICONAZOLE and TETCYCLACIS, inhibit *ent*-kaurene oxidase; and the ACYLCYCLOHEXANEDIONES inhibit GA_{20} 3β-hydroxylase. Other metabolic pathways, for example that for sterol biosynthesis, may be secondary targets for some retardants.

Cathey, H.M. (1964) *Annu. Rev. Plant Physiol.* **15**, 271–302.

plant growth substances *See:* PLANT HORMONES.

Plant hormones

PLANT hormones, also known as phytohormones or plant growth substances, are naturally occurring substances which, in low concentrations, influence plant physiological processes [1]. In contrast to animal hormones, they do not always act at a distance from their site of synthesis, which is often not localized in particular tissues. However, there are examples where hormones move within the plant from a site of production to a site of action. The term 'plant growth regulator', although also used for plant hormones, encompasses all compounds, natural and synthetic, which, when applied to plants, evoke a physiological response.

Five groups of plant hormones are generally recognized: AUXINS, GIBBERELLINS, CYTOKININS, ABSCISIC ACID, and ethylene. These differ in their chemical nature and in their physiological function, although their functions overlap to a certain degree. They are ubiquitous in higher plants. The function of further, widely distributed groups of active compounds, for example BRASSINOSTEROIDS and JASMONIC ACID, are less clearly defined and they are not yet fully accepted as hormones. There are, in addition, numerous compounds of limited distribution with some physiological activity, but whose function is unclear.

Auxins

INDOLE-3-ACETIC ACID (IAA) first isolated from a higher plant in 1946, is the most common auxin and is ubiquitous. Other naturally occurring auxins with more restricted distribution are 4-chloroindole-3-acetic acid and phenylacetic acid. There are also many synthetic auxins, some of which have agronomic applications as growth regulators and herbicides.

The biosynthesis of IAA in plants is still unclear. Three routes from tryptophan have been proposed [2]; two major pathways via tryptamine or indolepyruvic acid, which are converted to indoleacetaldehyde and hence to IAA. In brassicas, indole-3-acetaldoxime is an intermediate. There is some evidence that L-tryptophan must be first converted to the D enantiomer in a gibberellin-stimulated step. Recently the intermediacy of tryptophan was questioned and indole was proposed as the direct precursor [3]. Metabolism of IAA is by esterification with inositol or glucose, amide formation with glutamic or aspartic acids, peroxidative decarboxylation or oxidation without decarboxylation. Hydrolysis of IAA conjugates is a potential source of free auxin.

IAA levels are increased at infection sites of plants transformed with the bacteria *Agrobacterium tumefaciens* or *Pseudomonas savastanoi*. Two genes, *iaaM* and *iaaH*, coding for tryptophan monooxygenase and indoleacetamide hydrolase, respectively, are transferred from the bacteria as part of the T-DNA (*see* TI PLASMID).

Auxins are known to promote cell enlargement and division, apical dominance (repression of outgrowth of lateral buds), root initiation, differentiation of vascular tissue, and ethylene biosynthesis, and to mediate tropistic responses. They inhibit leaf and fruit abscission and are involved in flower and fruit development. The mechanism for stimulating cell elongation has been studied in detail. The most rapid observable effect is the HYPERPOLARIZATION of the plasma membrane, maximal within two minutes, due to the stimulation of H^+-ATPases (*see* ATPASES; MEMBRANE TRANSPORT SYSTEMS), resulting in H^+ secretion and acidification of the extracellular space. There is correlative evidence associating this process with early growth, although the mechanism is unclear.

An auxin-binding protein of M_r 22 000 has been purified from membranes of *Zea mays* (maize) and cloned. The mature protein contains a KDEL sequence at the C terminus which results in its retention in the lumen of the ENDOPLASMIC RETICULUM, where most of the protein is located. However, antibodies to the protein inhibit H^+-ATPase activity and hyperpolarization indicating that the protein may function as an auxin receptor on the plasma membrane.

Auxins alter the expression of several genes of unknown function [4]. The effect is rapid (5–30 min), specific to auxin and correlated with cell division or elongation, the genes being differentially expressed in auxin-responsive tissues. An auxin-responsive element has been identified within the PROMOTER of a gene expressed during the division of tobacco mesophyll protoplasts.

As well as moving in the vascular systems and diffusing into cells in common with the other plant hormones, auxins are unique in undergoing polarized transport. Their basipetal transport from shoot tips, which is associated with apical dominance, is thought to be mediated by specific influx and efflux carriers on the plasma membrane. Inhibitors of auxin transport such as 2,3,5-TRIIODOBENZOIC ACID, the MORPHACTINS, and PHYTOTROPINS such as NAPHTHYLPHTHALAMIC ACID, inhibit the efflux carriers and cause loss of apical dominance.

Gibberellins

Since their discovery as secondary metabolites of the phytopathogenic fungus GIBBERELLA FUJIKUROI (the structure of the major component, GA_3, (*see* GA_n) was determined in 1956) more than 90 fully characterized GIBBERELLINS (GAs) have been detected in higher plants and fungi. They are found in greatest concentration in immature seeds, but are present in all growing tissues. Gibberellin A_1 is the compound active in regulation of stem extension in many, if not all, species. Related compounds, ANTHERIDIOGENS, induce antheridia production in gametophytes of some ferns.

Gibberellins are tetracyclic diterpenoid carboxylic acids and are biosynthesized from mevalonic acid via the hydrocarbon *ent*-kaurene [5]. The two-step cyclization of geranylgeranyl pyrophosphate by *ent*-kaurene synthetase, a soluble enzyme complex, is the first committed step to GAs. It is inhibited by a group of quaternary ammonium or phosphonium growth retardants, for example CHLORMEQUAT CHLORIDE. *ent*-Kaurene is oxidized by microsomal NADPH-dependent monoxygenases to GA_{12} which is the source of GAs by way of further oxidation to all GAs by soluble 2-oxoglutarate-dependent dioxygenases. Oxidation of *ent*-kaurene is inhibited by N-containing heterocyclic growth retardants, including PACLOBUTRAZOL, TETCYCLACIS, and UNICONAZOLE. The soluble oxidases are inhibited by ACYLCYCLOHEXANEDIONE growth retardants. Deactivation of GAs is by 2β-hydroxylation and further oxidation at carbon-2. Glucosyl ethers and esters are the most common conjugates.

Although GAs induce or stimulate many physiological processes, including stem extension and leaf expansion, fruit set and development (often in the absence of fertilization), seed germination and maleness in dioecious flowers, details of its mode of action are known only for germinating cereal grains. In the cereal aleurone layer GA induces the synthesis and secretion into the endosperm of hydrolytic enzymes, for example α-amylase, that break down the macromolecular reserves for use by the embryo [6]. Gibberellin increases the rate of expression of genes for several of these enzymes, an effect that is abolished in the presence of abscisic acid (ABA). Conserved base sequences which confer GA-inducible, ABA-repressible expression have been recognized in the promoter region of some of these genes, and potential factors that interact with these sequences have been identified. Gibberellin receptor proteins have not been identified, although they may be present in the plasma membrane.

Abscisic acid

ABSCISIC ACID (ABA), which is the only member of its class, was originally isolated as a growth retardant. It is produced in roots and mature leaves, particularly in response to water deficiency, and also in developing seeds.

As a sesquiterpene, ABA is produced from mevalonic acid. Although in the fungi CERCOSPORA ROSICOLA and *C. cruenta* it is formed directly from farnesyl pyrophosphate, in higher plants the generally accepted biosynthetic pathway involves oxidative cleavage of a CAROTENOID [7]. The conversion of the xanthophyll, 9′-*cis*-neoxanthin, to *cis*-XANTHOXIN is suggested to be the rate limiting step and to be promoted by water deficiency. *cis*-Xanthoxin is oxidized to ABA-aldehyde and then to ABA by constitutive cytosolic enzymes, the final step being catalysed by a molybdenum-containing aldehyde oxidase. Metabolism of ABA is via 8′-oxidation to PHASEIC ACID (PA), and subsequent reduction of the ketone to give both DIHYDROPHASEIC ACID (DPA) epimers. Glucosyl ester and ether conjugates of ABA and its metabolites are also produced.

Abscisic acid affects a range of physiological processes; major functions are in the adaptation of plants to changes in their environment. In response to water shortage and to soil flooding ABA induces stomatal closure and, in some cases, the synthesis of proteins that may protect tissues from the effects of desiccation. There is some evidence that ABA acts as a message from roots, in which water deficit or excess in the soil is first perceived, to

leaves. In developing seeds, ABA prevents precocious germination and induces the production of proteins, including some that protect the embryo during desiccation. In germinating cereal grains ABA counteracts the induction by GA of hydrolytic enzymes in the aleurone layer. It is also involved in the response of plants to wounding, possibly in conjunction with jasmonic acid (*see* PLANT WOUND RESPONSES).

Two responses to ABA, involving changes in ion fluxes and in gene expression, are recognized [8]. The decrease in turgor of stomatal guard cells induced by ABA is due principally to an efflux of K^+ ions, resulting either from a direct effect of ABA on K^+ channels or mediated by Ca^{2+} ions, the cytoplasmic concentration of which increases in guard cells in response to ABA. The rate of expression of numerous genes for proteins that are induced by desiccation, cold, wounding and other stresses is increased in the presence of ABA, as is that of genes encoding storage proteins or proteins conferring desiccation tolerance which are expressed during the middle to late stages of embryo development. ABA-response elements, such as the consensus sequence CACGTG, have been recognized in the promotor regions of several ABA-induced genes, including wheat *Em* (early methionine-labelled), rice *Rab* (Responsive to ABA) and cotton *LEA* (late embryo ABA) genes. DNA-binding proteins belonging to the leucine zipper class (*see* PROTEIN–NUCLEIC ACID INTERACTIONS; TRANSCRIPTION FACTORS) and identified in wheat and rice nuclear extracts bind to the consensus sequence and promote expression. The mechanism of signal transduction for ABA is still unknown and there is no convincing evidence for a receptor.

Cytokinins

Originally discovered from their ability to promote cell division in tissue cultures, cytokinins are N^6-substituted adenine, adenosine and adenosine 5′-phosphate derivatives. They occur naturally as the free compounds and as components of certain tRNAs. The first cytokinin found, KINETIN (6-furfurylaminopurine), was a product of autoclaved DNA. Natural cytokinins include N^6(Δ-2-isopentenyl)adenine, *trans*-zeatin (*see* ZEATIN), and DIHYDRO-ZEATIN, and their N^9-ribosides and N^9-riboside-5′-monophosphates (*see also* BENZYLADENINE). They are produced predominantly in root tips and developing seeds.

Although cytokinins may be formed by the hydrolysis of tRNAs, their biosynthesis in higher plants is more likely to be initiated by the condensation of dimethylallyldiphosphate with adenosine monophosphate, catalysed by isopentenyl transferase [9]. The resulting isopentenyladenine ribotide is then *trans* hydroxylated to zeatin ribotide. Reversible dephosphorylation and deribosylation yield the free base. All three forms, base, riboside and ribotide, which are rapidly interconverted, are biologically active. Zeatin is reduced to dihydrozeatin, which is more stable and biologically active than zeatin, by a soluble, NADPH-dependent reductase that is specific for the free base. Cytokinins may be N-glucosidated at the 3, 7 or 9 positions to form inactive conjugates; the 7 and 9-glucosides are formed irreversibly. In contrast O-glycosylation of zeatin and dihydrozeatin is reversible and the products are considered storage forms. Apart from conjugate formation, regulation of cytokinin levels commonly involves side-chain cleavage as a result of cytokinin oxidase activity.

The *ipt* (formerly *tmr*) gene in the Ti plasmid of *Agrobacterium tumefaciens* codes for an isopentenyl transferase and causes elevated cytokinin levels when expressed in infected plants. The *tzs* gene, which is close to the *vir* (virulence) region of the Ti plasmid, also codes for an isopentenyl transferase.

In the presence of auxin, cytokinins induce cell division in tissue culture and in crown gall tumours and, when present at high levels relative to auxin, will promote shoot initiation in these tissues. Cytokinins stimulate leaf expansion by cell enlargement, promote chloroplast development and delay leaf senescence. At the molecular level cytokinin action is poorly understood. Proteins that bind cytokinin have been identified, but their functions are unknown.

Ethylene

The physiological activity of this gas was first noted through the propensity of smoke to synchronize flowering and fruit ripening, and of coal gas to induce leaf abscission. First found to be released naturally by ripening fruit, it is produced by most plant tissues, particularly in response to stress and in maturing and senescing organs.

Ethylene is produced from methionine via (−)-S-ADENOSYL-METHIONINE (SAM) and ACC (1-aminocyclopropane-1-carboxylic acid) [10]. Sulphur is conserved in this process by recycling 5′-methylthioadenosine. The conversion of SAM to ACC is catalysed by a pyridoxal phosphate-requiring aminotransferase (ACC SYNTHASE), which is the key regulatory enzyme for ethylene biosynthesis and is induced by numerous stimuli, including wounding, infection, and hormones, particularly IAA. There is a MULTIGENE FAMILY for ACC synthase, the members of which are differentially regulated in a tissue- and inducer-specific manner [11]. Ethylene is formed from ACC through the action of ACC OXIDASE (ethylene-forming enzyme), a soluble enzyme with a requirement for O_2, Fe^{2+} and ascorbate. The ACC oxidase gene has been cloned from ripening tomato fruit and the enzyme shown to belong to a class of dioxygenases. Expression of ANTISENSE RNA to ACC oxidase or ACC synthase in transgenic tomato plants suppresses ethylene production and fruit ripening (*see* PLANT GENETIC ENGINEERING). ACC is also converted to ethylene in transgenic plants containing ACC deaminase of fungal or bacterial origin. Ethylene is metabolized by oxidation to CO_2 or to ethylene oxide and ethylene glycol. *N*-malonylation of ACC, catalysed by ACC *N*-malonyltransferase, may serve to regulate the level of this intermediate. The process is probably not readily reversible.

Ethylene evokes a wide range of physiological responses (*see* ETHYLENE). There is compelling evidence that it is involved in ripening of climacteric fruit (which show a burst of respiration), and in leaf and fruit abscission in dicotyledonous species, flower senescence, stem extension of aquatic plants, gas space (aerenchyma) development in roots, and epinastic curvature in flooded plants. Responses of plants to wounding and pathogen attack are to some extent mediated by ethylene (*see* PLANT PATHOLOGY; PLANT WOUND RESPONSES). Many responses involve increased gene expression. For example, expression of polygalacturonase and cellulase is induced in ripening fruit and in leaf abscission layers, as is that of chitinases, β-1,3-glucanases and hydroxyproline-rich proteins in ethylene-treated or infected leaves. Membrane-bound, highly lipophilic binding proteins for ethylene have been isolated, but not yet identified as receptors.

Other growth regulators

Many naturally occurring compounds, other than members of the main hormone classes described above, cause physiological changes when applied to plants. Although it is unclear whether most of these compounds function as natural growth regulators a strong case can be made in certain instances. The brassinosteroids form a very large group of compounds, including BRASSINOLIDE, the first member to be isolated, from rape pollen, and one of the most active. Although the biosynthesis of brassinolide is not yet known, it is assumed to originate from common plant sterols by a series of hydroxylation and alkylation reactions [12]. The B-ring lactone may be formed by a Baeyer–Villiger type oxidation of a 6-keto compound, many examples of which occur naturally. The function of brassinosteroids is unclear. They promote growth in certain systems, usually by cell enlargement, when applied at very low (≤nM) concentrations. Although many of their effects mimic those of auxins, GA or cytokinin, there are some unique responses, including that utilized in the characteristic bean second internode bioassay where a combination of cell enlargement and division causes swelling, bending, and splitting of the stem.

Jasmonic acid and its methyl ester are formed in a series of reactions initiated by the action of lipoxygenase on linolenic acid [13]. Although they induce senescence and inhibit growth when applied to plants, it is unclear if these effects reflect their normal function. In common with water deficit, wounding and increased nitrogen availability they stimulate transcription of storage proteins in soybean leaves and may function to direct nitrogen redistribution. Methyl jasmonate is volatile and acts as an interplant signal. It is released on wounding or pathogen attack and induces the synthesis of PROTEINASE INHIBITORS in leaves of the same or neigbouring plants (*see*: PLANT WOUND RESPONSES).

P. HEDDEN

See also: PLANT CELL WALL MACROMOLECULES.

1 Davies, P.J. (1987) *Plant Hormones and their Role in Plant Growth and Development* (Martinus Nijhoff, Dordrecht).

2 Reinecke, D.M. & Bandurski, R.S. (1987) Auxin biosynthesis and metabolism. In *Plant Hormones and their Role in Plant Growth and Development* (Davies, P.J., Ed.) 24 (Martinus Nijhoff, Dordrecht).

3 Wright, A.D. et al. (1991) Indole-3-acetic acid biosynthesis in the mutant maize orange pericarp, a tryptophan auxotroph. *Science* **254**, 998–1000.

4 Key, J.L. (1989) Modulation of gene expression by auxin. *BioEssays* **11**, 52–57.

5 Graebe, J.E. (1987) Gibberellin biosynthesis and control. *Annu. Rev. Plant Physiol.* **38**, 419–465.

6 Jones, R.L. & Jacobsen, J.V. (1991) Regulation of synthesis and transport of secreted proteins in cereal aleurones. *Int. Rev. Cytol.* **126**, 49–88.

7 Parry, A.D. & Horgan, R. (1991) Carotenoids and abscisic acid (ABA) biosynthesis in higher plants. *Physiol. Pl.* **82**, 320–326.

8 Heatherington, A.M. & Quatrano, R.S (1991) Mechanism of action of abscisic acid at the cellular level. *New Phytol.* **119**, 9–32.

9 McGaw, B.A. (1987) Cytokinin biosynthesis and metabolism. In *Plant*

Hormones and Their Role in Plant Growth and Development (Davies, P.J., Ed.) 76 (Martinus Nijhoff, Dordrecht).

10　Yang, S.F. & Hofmann, N.E. (1984) Ethylene biosynthesis and its regulation in higher plants. *Annu. Rev. Plant Physiol.* **35**, 155–189.

11　Rottmann, W.H. et al. (1991) 1-Aminocyclopropane-1-carboxylate synthase in tomato is encoded by a multigene family whose transcription is induced during fruit and floral senescence. *J. Mol. Biol.* **222**, 937–961.

12　Mandava N.B. (1988) Plant growth-promoting brassinosteroids. *Annu. Rev. Plant Physiol.* **39**, 23–52.

13　Anderson, J.M. (1989) Membrane-derived fatty acids as precursors to second messengers. In *Second Messengers in Plant Growth and Development. Plant Biology, Vol. 6* (Boss, W.F. & Morré, D.J., Eds) 181 (Liss, New York).

---

plant lectins *See*: LECTINS.

plant mitochondrial genome *See*: MITOCHONDRIAL GENOMES: PLANT.

Plant pathology

DISEASE in plants is caused by a variety of biological agents, including viruses, bacteria and fungi, and also by abiotic factors such as nutrient deficiencies which cause physiological disorders. The PLANT VIRUSES are dealt with elsewhere in this volume; this article discusses some aspects of fungal and bacterial diseases of plants.

Fungi and bacteria cause a wide range of economically significant plant diseases which each year account for the loss of many millions of tonnes of food and plant products. In temperate climates fungi are the most serious disease-causing microbial pathogens, whereas bacteria present a greater problem in tropical and semi-tropical regions, where warm, humid conditions promote bacterial growth.

Plant pathogenic fungi

Plant pathogenic fungi constitute an extensive and diverse group with representative genera found within five taxonomic classes and two divisions of the kingdom Fungi (Table P3).

In addition to strict taxonomic classification, plant–fungal host–pathogen interactions are often described according to the mode of nutrition of the parasite. Biotrophic pathogens are obligate parasites which require a living plant for survival. They successfully evade detection by their hosts and actively penetrate the plant cuticle by the formation of appressoria. Nutritional benefit is derived via specialized structures known as haustoria, which invaginate the plasma membrane of the host cell. Biotrophic pathogens of major agricultural importance include the mildews (e.g. *Bremia* spp.) and rusts (*Puccinia* spp.).

Necrotrophic pathogens are SAPROPHYTES whose life strategy involves killing the host plant. This may occur via the production of host-specific, or host-nonspecific toxins. Important necrotrophs include the southern corn leaf blight fungus (*Helminthosporium maydis*), which devastated the North American corn crop in 1970 (*see* MITOCHONDRIAL GENOMES: PLANT), the wilts (e.g. *Verticillium* spp.), and damping-off fungi (e.g. *Pythium* spp.).

Table P3 Representative fungal pathogens and the diseases they cause

Classification	Fungus	Disease
Myxomycota	*Plasmodiophora*	*Brassica* club-root
Eumycota		
Mastigomycotina	*Bremia*	Downy mildew
	Peronospora	Mildew
	Phytophthora	Potato late blight
	Pythium	Damping-off
	Synchytrium endobioticum	Warts
Ascomycotina	*Erysiphe*	Mildew
	Giberella	Rots
	Leptosphaeria	Cankers
	Nectria	Galls
	Ceratocystis ulmi	Dutch elm vascular wilt (Dutch elm disease)
	Verticillium	Wilt
Deuteromycotina	*Alternaria*	Leaf spots and blights
	Fusarium	Wilt
	Helminthosporium	Leaf blight
Basidiomycotina	*Melampsora*	Rusts
	Puccinia	Rusts
	Ustilago	Smuts
	Stereum	Wood rot

Hemibiotrophs initiate infection as biotrophs and then progress to saprophytic growth during the latter stages of infection. The devastating Irish potato famine of the 1840s was caused by the hemibiotrophic pathogen *Phytophthora infestans*.

Plant pathogenic bacteria

Table P4 shows the main classes of bacterial pathogens and the diseases they cause. The genera *Agrobacterium*, *Erwinia*, *Pseudomonas*, and *Xanthomonas* contain the majority of economically significant phytopathogenic bacteria.

Distinctions on the basis of nutrition are not so readily applied to bacterial plant pathogens, which unlike fungi, rarely have exacting nutritional requirements. Nevertheless, the crown-gall pathogen, *A. tumefaciens*, and *P. syringae* pv. *phaseolicola*, causal agent of halo blight of beans, live and multiply within host tissue and can therefore be considered biotrophs. Similarly, because of their ability to kill cells in advance of colonization, soft-rotting bacteria such as *Erwinia* can be considered necrotrophs.

Impact of molecular genetics on plant pathology

Application of the tools and techniques of modern molecular genetics to the study of plant pathology has revolutionized our understanding of plant–pathogen interactions. Until the 1970s few studies attempted to examine the genetic basis of pathogenicity. Classical plant pathology was primarily concerned with the interaction between pathogen and host at the level of host defence, or with the effects of the environment on the progress of disease. In addition, the majority of genetical studies involved

Table P4 Representative bacterial pathogens and the diseases they cause

Bacterial pathogen	Disease
Gram-negative	
Agrobacterium	Hypertrophy (galls)
Erwinia	Soft rot (especially of tubers), necrosis, vascular wilt
Pseudomonas solanacearum	Vascular wilt
P. syringae pv. *glycinea*	Leaf spot
P. syringae pv. *phaseolicola*	Halo blight
P. syringae pv. *savastoni*	Canker
P. syringae pv. *syringae*	Leaf spot
Xanthomonas	Leaf spot, necrosis
Gram-positive	
Rhodococcus	Hypertrophy
Curtobacterium	Vascular wilt
Clavibacter	Vascular wilt
Streptomyces	Scab
Mycoplasma-like organisms (MLOs)	Yellowing and stunting
Spiroplasma	Yellowing and stunting

pv., Pathovar–a race of the pathogen which produces a particular disease on a particular host.

fungi, as gene exchange systems were not available for bacteria. While classical genetics and plant breeding studies remain indispensable to the understanding of plant–pathogen interactions, the advent of RECOMBINANT DNA TECHNOLOGY has greatly enhanced the scope of experimental approaches [1].

There have been many difficulties associated with the application of recombinant DNA technology to fungi. The most limiting has been the inability to transform many fungi with DNA. Even when gene TRANSFORMATION of fungal protoplasts has been achieved, the number of transformants rarely exceeds 10^4 per microgram of DNA. However, a number of technological advances, such as ELECTROPORATION, have made gene manipulation in a wide range of fungi now possible [2].

The study of Gram-negative bacterial pathogens has benefitted enormously from molecular genetics, primarily because molecular techniques developed for *Escherichia coli* and other medically important bacteria (*see e.g.* DNA CLONING; GENETIC ENGINEERING; VECTORS) can be applied either directly, or with little modification, to the study of pathogenicity (*see e.g.* TI PLASMID). The Gram-positive plant pathogens have proven less amenable to genetic analysis; nevertheless advances have been made in understanding the genetics of pathogenicity of these bacteria [3]. With the exception of the spiroplasmas, mycoplasma-like organisms (MLOs) cannot be cultured *in vitro* and have therefore received little attention.

Resistance of plants to pathogens

The meaning of some of the terms used by plant pathologists to describe plant–pathogen interactions is not immediately obvious and can be confusing. The following paragraphs should therefore be read carefully! (see [4]).

Plants differ in their ability to resist particular pathogens. Some plants never succumb to disease, even under conditions conducive to disease development, whereas others show varying degrees of resistance and still others are susceptible to infection under a range of environmental conditions.

Two contrasting categories of plant resistance are frequently encountered. The first is host resistance, which commonly occurs between plants and their pathogens and is one in which the pathogen and host are INCOMPATIBLE (see below). Such resistance is governed by one (monogenic) or a few genes (oligogenic), and the plant defends itself by defence mechanisms activated in response to attempted infection. This type of resistance is absolute under most environmental conditions, but a small number of mutations (possibly only one) in the pathogen may produce a new race able to infect the previously resistant variety [5].

The second category of plant resistance is termed nonhost resistance and relies upon a variety of general (POLYGENIC) defence mechanisms to repel pathogen attack. Nonhost resistance is the most common type of resistance in diverse natural ecosystems, where plants resist most species of pathogen they encounter. Nonhost resistance is likely to result from nutritional or ecological incompatibility between host and pathogen, or as a result of pre-existing host barriers such as the cutins and suberins of the plant CUTICLE. Such constitutive resistance is passive, because it does not depend upon a physiological response by the host to infection.

Nonhost resistance may also be active. In some instances a nonpathogen may overcome pre-existing barriers, yet be recognized by the plant and effectively restricted by induced, or secondary defences, such as the hypersensitive response (HR) (see below).

Host resistance and the hypersensitive response

The ability of plant pathogens to cause disease in only a small number of plants with which they may come in contact has always been of interest to plant pathologists, particularly because of the need to breed disease-resistant crop plants.

Detailed studies on the interaction between flax (*Linum usitatissimum*) and one of its fungal pathogens, *Melampsora lini*, concerning the simultaneous inheritance of fungal virulence and host resistance led to the gene-for-gene hypothesis governing host specificity [6]. This hypothesis postulates that corresponding genes for resistance (*R* genes) and avirulence (*A* genes) exist in the host and fungus, respectively. Resistance is associated with the genetically incompatible interaction of *R : A* genes, and susceptibility with the genetically compatible interaction of *R : a, r : A*, or *r : a* genes. An incompatible interaction leads to resistance through induction of the hypersensitive response (HR) which is characterized by membrane DEPOLARIZATION and rapid death of host cells at the site of infection which effectively localizes the fungus to necrotic flecks of tissue. Associated with hypersensitive cell death is the activation of secondary responses in neighbouring cells which lead to the accumulation of PHYTOALEXINS, PROTEINASE INHIBITORS, and other antimicrobial compounds (*see* PLANT WOUND RESPONSES).

Subsequent work in other plant–pathogen systems, particularly lettuce downy mildew (*Bremia lactucae*), has shown that gene-for-gene interactions are widespread [7]. In general, pathogen avirulence and host resistance are dominant. Linkage between host resistance genes is common, but linkage between avirulence genes is rarely observed. Gene-for-gene resistance, which is also known as vertical resistance by plant breeders, has been popular in breeding programmes owing to the ease with which individual genes can be transferred to high-yielding cultivars. However, a single mutation in the fungus (e.g. *Aa* to *aa*) is sufficient for virulence and therefore sole reliance upon vertical resistance for plant protection is extremely unwise. Horizontal resistance, which is polygenic and not based on gene-for-gene determinants, is more difficult to introduce into cultivars, but is also more likely to resist being overcome by novel virulence genes in the pathogen [5].

Difficulties associated with molecular genetic analysis of fungi have meant that investigations on the molecular basis of host specificity have progressed more rapidly in bacteria. Initial studies involved *Pseudomonas syringae* pv. *glycinea* and its host, soybean (*Glycine max*). By transferring clones from a genomic library (*see* DNA LIBRARIES) from a race which gave an incompatible (avirulent) reaction on soybean to other races of the pathogen and assaying the transconjugants for their reaction on differential tester plant cultivars, a COSMID clone was identified which determined the ability to cause disease in certain soybean cultivars. Transfer of this clone, containing the *avrA* gene, to other races of the pathogen caused them to develop an avirulent reaction on cultivars upon which they had previously developed a virulent reaction [8].

Demonstration of avirulence genes constitutes a major success of molecular genetic analysis of bacterial pathogens and a number of avirulence genes have now been identified from both *P. syringae* races and *Xanthomonas campestris* races [9]. Cloning and sequencing of avirulence genes has shown that some avirulence gene products have sequences in common, such as the *avrB* and *avrC* genes from *P. syringae* pv. *glycinea*, but generally avirulence genes encode dissimilar products with no sequence homology to known proteins [10,11]. The mode of action of the avirulence genes is not certain, but the product of the *avrD* gene of *P. syringae* pv. *tomato*, which codes for a small cytoplasmic protein (M_r 34 000), does not appear to be the ELICITOR recognized by the plant resistance gene. Instead, AvrD is thought to catalyse the formation of a unique low molecular weight elicitor molecule which interacts with the product of the plant resistance gene (termed the receptor), to trigger the hypersensitive response [3].

Circumstantial evidence indicates that the products of *avrB* and *avrC* from *P. syringae* pv. *glycinea* may function in a similar manner, and recent work shows that outer membrane protein porins may be involved in deployment of the elicitor [12] (*see* BACTERIAL ENVELOPE).

Host–pathogen recognition

In fungi, much work has focused upon the possible biochemical mechanisms underlying the recognition of pathogenic fungi by their hosts. Three theories have been advanced to explain the specific interaction of resistance and avirulence genes [7,13]. The specific-elicitor specific-receptor model postulates that resistance gene products are plant membrane-bound receptors (possibly LECTINS) which specifically recognize constitutive fungal elicitor molecules and transduce the signal to the nucleus. According to this theory, the ligands for the resistance gene products are not avirulence gene products, but GLYCOPROTEIN elicitors modified by avirulence gene-encoded glycosyl transferases. The toxic dimer model argues that a heterodimer, consisting of the products of the resistance and avirulence genes, is directly toxic to cells, leading to HR cell collapse and its sequelae. Lastly, the ion-channel defence model incorporates the first two theories by arguing that the resistance gene products are ion-channel-forming membrane-bound receptors which either recognize the avirulence gene products (constitutive or diffusible peptide elicitors), or recognize downstream metabolites of the avirulence gene products. Recognition leads to the opening of ion channels which disrupt ionic homeostasis and lead to cell death.

A number of fungal elicitors have been purified and a range of compounds, including the fungal wall oligomer CHITOSAN, the oomycete membrane lipid ARACHIDONIC ACID, and peptides from *Cladosporium fulvum*, have been shown to trigger the HR [14]. However, these elicitors are nonspecific and induce the HR irrespective of host genotype. It is possible that they may be responsible for induction of nonhost resistance, but they are unlikely to be avirulence gene products.

One specific elicitor has been isolated, a peptide of 27 amino acids from *C. fulvum*. This peptide induces the HR on resistant but not susceptible hosts, and the gene encoding the peptide (*avr9*) has been cloned from a genomic library of *C. fulvum* [15]. Isolation of the *avr9* gene and other elicitors which interact specifically with plant receptors should facilitate identification and cloning of plant resistance genes. The cloning and characterization of a plant resistance gene–pathogen avirulence gene pair will provide a more complete understanding of the molecular basis of recognitional specificity and is an area of research poised for tremendous growth.

Fungal pathogenicity determinants

For a fungus successfully to parasitize a plant it must penetrate the cuticle and gain access to the cells below this impervious barrier. Some fungi, such as *Nectria haematococca*, secrete the enzyme cutinase and directly penetrate the cuticle, whereas others gain access through stomatal pores or damaged tissue. Once inside the leaf, biochemical evidence shows that other enzymes, particularly pectinases, are major pathogenicity determinants.

The role of pathogenicity genes in overcoming nonhost resistance was elegantly illustrated by insertion of the cutinase gene from the pea pathogen *Fusarium solani* f.sp. *pisi* into an isolate of a wound-infecting *Mycosphaerella* sp., which upon acquisition of the ability to hydrolyse cutin was able to infect unwounded plants [16]. Genetic studies of the involvement of extracellular enzymes in pathogenicity are now progressing and there is considerable effort directed toward development and use of genetic strategies to identify other basic pathogenicity determinants in fungi [2].

Not all pathogenicity genes involve aggressive enzyme attack. The ability of bean rust to cause disease depends upon the ability of its germinating spores to recognize the architecture of bean stomatal guard cells specifically and thereby gain access to the intercellular leaf spaces. Pathogenicity may also result from fungal genes which allow the pathogen to tolerate induced secondary defences (such as the HR), rather than to overcome passive primary ones. For example, it has been established that pisatin demethylase, a CYTOCHROME P450 detoxifying enzyme that acts on the pea phytoalexin pisatin, is a pathogenicity determinant of the pea pathogen *Nectria haematococca*. Phytoalexin detoxification is an area of very active molecular research [17].

Fungi also produce a range of toxins which are likely to contribute toward pathogenicity. Fungal toxins constitute a diverse range of low molecular weight secondary metabolites which, on the basis of their spectrum of activity, can be divided into either host-specific or -nonspecific toxins. While their contribution to virulence appears certain, genetic studies with Tox⁻ mutants have yet to be performed.

Bacterial pathogenicity determinants

Plant pathogenic bacteria produce a range of extracellular compounds, including toxins, extracellular degradative enzymes, extracellular polysaccharide (EPS), and plant growth-promoting substances, which are either established or suspected contributors to pathogenicity. Many of these factors have been examined from a molecular perspective and Tables P5 and P6 list known virulence factors.

Several bacterial pathogens produce EPS which is considered to contribute to virulence. EPS produced by the tomato wilt pathogen *P. solanacearum* is thought to plug xylem vessels and facilitate dispersal of the bacterium throughout the plant vascular system. Growth-promoting substances such as cytokinins and auxins (*see* PLANT HORMONES) are produced by a number of bacteria, of which the most extensively studied are those produced by *A. tumefaciens* and *P. syringae* pv. *savastoni*. Mutational analysis has shown that genes encoding cytokinins and auxins are required for normal pathogenicity [1].

Erwinia spp. capable of causing soft rot produce an extensive array of extracellular pectinases which can degrade plant cell constituents. Pectate lyase (*pel*) genes from *E. chrysanthemi* have been cloned by expression of an *E. chrysanthemi* cosmid library in *E. coli* and between three and five *pel* genes have been found in all strains. Determination of the precise contribution of *pel* genes to pathogenicity has been accomplished by the construction of mutants lacking individual gene products by MARKER EXCHANGE MUTAGENESIS [1].

Toxins have often been considered major determinants of pathogenicity in plant pathogenic pseudomonads. TRANSPOSON MUTAGENESIS with Tn5 has been used to generate Tox⁻ mutants in a range of *P. syringae* pathovars and in *P. tolaasii* [18,19], and the contribution of particular toxins to pathogenesis has been examined [18]. Toxin production has in all instances been shown to contribute toward virulence, but only in *P. tolaasii*, the causal organism of brown blotch disease of mushrooms, has the toxin (tolaasin) been shown to be a major determinant of pathogenicity [19]. The structure of most pseudomonad toxins is now known and a number of genes involved in the biosynthesis, resistance,

Table P5 Virulence factors with a suspected role in bacterial pathogenicity

Virulence factor	Organism	Compound
Extracellular degradative enzymes	*A. tumefaciens* *E. chrysanthemi*	Polygalacturonase Pectin methylesterase Galacturonidase Pectate lyase Protease Glucanase
	E. carotovora	Pectate lyase Polygalacturonase Protease
	P. solanacearum *P. syringae* pv. *lachrymans* *X. campestris*	Polygalacturonase Pectate lyase Polygalacturonase Pectate lyase Glucanase Protease Amylase
	P. solanacearum	Glucanase Phospholipases Hemicellulases Amylases
Extracellular polysaccharide	*P. solanacearum* *E. stewartii* *E. amylovora* *X. campestris*	
Plant growth substances	*A. tumefaciens*	IAA Cytokinin
	P. syringae pv. *savastoni* *P. amygdali*	Cytokinin IAA Cytokinin
	Rhizobium spp. *C. fasciens* *E. herbicola*	Cytokinin Cytokinin Auxin
Other substances	*P. syringae*	Ice nucleation factor

A, Agrobacterium; C, Corynebacterium; E, Erwinia; P, Pseudomonas; X, Xanthomonas. IAA, indole acetic acid. pv., pathovar — a race of the pathogen which produces a particular disease on a particular host.

and export of these diverse toxins are currently the subject of intensive investigation [18,19].

Pathogenicity genes of bacteria encoding unknown products

The study of bacterial pathogenicity factors has recently been extended to pathogenicity genes encoding unknown products. By screening large numbers of mutants for their ability to cause disease when inoculated onto a host plant a number of novel plant interaction phenotypes have been identified; the most interesting of these are *hrp* ('harp') genes. Hypersensitive *response* and *pathogenicity* (*hrp*) genes were initially identified in *P. syringae* pv. *phaseolicola* [20], but similar genes have now been found in

Table P6 Host nonspecific toxins produced by plant pathogenic pseudomonads

Class/structure	Producing organism	Toxin	Mechanism or site of action
Lipodepsinonapeptide	*Pseudomonas syringae* pv. *syringae*	Syringomycin Syringotoxin Syringostatin	Plasma membrane
Lipodepsipeptide	*P. tolaasii* *P. reactans*	Tolaasin WLIP	Plasma membrane (ion channel formation)
Polyketide	*P. syringae* pv. *atropurpurea* *P. syringae* pv. *tomato* *P. syringae* pv. *glycinea* *P. syringae* pv. *morsprunorum* *P. syringae* pv. *maculicola*	Coronatine	?
β-Lactam	*P. syringae* pv. *tabaci* *P. syringae* pv. *coronafaciens* *P. syringae* pv. *garcae*	Tabtoxin	Glutamine synthetase
Sulpho-diaminophosphinyl peptide	*P. syringae* pv. *phaseolicola*	Phaseolotoxin	Ornithine transcarbamoylase
Hemithioketal	*P. syringae* pv. *tagetis*	Tagetitoxin	Chloroplast RNA polymerase
Enol-ether amino acid	*P. andropogonis*	Rhizobitoxine	Sulphur metabolism Ethylene synthesis

WLIP, white line inducing principle.

other plant pathogenic pseudomonads, xanthomonads and also in a range of plant pathogenic *Erwinia* spp. [21].

Inactivation of *hrp* genes has shown that they control the ability of phytopathogenic bacteria to cause disease on susceptible plants and to elicit the HR on resistant plants [21]. SATURATION MUTAGENESIS of the *hrp* genes from *P. syringae* pv. *phaseolicola* with a reporter transposon, Tn3-spice, revealed seven complementation groups over a 22 kbp region of the chromosome, with an additional locus, *hrpM*, located elsewhere [21]. Detailed analysis of the *hrp* cluster from *P. syringae* pv. *syringae* has shown the presence of at least eight transcriptional units with organization similar to that reported in *P. syringae* pv. *phaseolicola* [22]. Little is known about how the *hrp* genes function at the biochemical level during pathogenesis, but some progress has been made in understanding *hrp* regulation. A regulatory gene, *hrpS*, has been identified which shares homology with the NtrC subfamily of bacterial regulatory proteins [21]. In addition, the *ntrA* gene of *P. syringae* pv. *phaseolicola*, encoding the σ^{56} factor (*see* TRANSCRIPTION), is reported to be required for the expression of several *hrp* loci.

Plant-inducible bacterial genes

Complex signalling is known to occur in the *Rhizobium*–plant and *Agrobacterium*–plant interactions (*see* NODULATION and TI PLASMID respectively) and recent studies show that other plant–pathogen interactions may involve equally sophisticated signal exchange. Studies using reporter gene fusions to *hrp* genes have shown that *hrp* OPERONS are specifically induced by plant signal molecules and show enhanced expression during compatible reactions in which disease develops and in incompatible interactions in which

the HR develops [21,22]. The *syrB* gene of *P. syringae* pv. *syringae*, which is required for production of syringomycin (Table P6), is also regulated by plant metabolites, particularly phenolic compounds [18], and plant-inducible PROMOTERS have also been detected in *E. chrysanthemi* and *X. campestris* pv. *campestris* [1]. Recent advances in recombinant DNA technology provide a variety of tools and techniques for the study of signal exchange between microorganism and plant and it is likely that this area, which holds the key to understanding plant–microbe interactions, will receive increasing attention.

P.B. RAINEY

K.B. BECKMAN

1 Daniels, M.J. et al. (1988) Molecular genetics of pathogenicity in phytopathogenic bacteria. *Annu. Rev. Phytopathol.* **26**, 285–312.

2 Hargreaves, J. & Turner, G. (1992) Gene transformation in plant pathogenic fungi. In *Molecular Plant Pathology*, Vol. 1 (Gurr, S.J. et al., Eds) 79–97 (Oxford University Press, Oxford).

3 Hennecke, H. & Verma, D.P.S. (Eds) (1991) *Advances in Molecular Genetics of Plant–Microbe Interactions*, Vol. 1 (Kluwer Academic Publishers, Dordrecht).

4 Federation of British Plant Pathologists (1973) *A Guide to the Terms in Plant Pathology. Phytopathol. Pap.* **17** (Commonwealth Mycological Institute, Kew).

5 Agrios, G.N. (1988) *Plant Pathology*, 3rd edn (Academic Press, San Diego).

6 Flor, H.H. (1955) Host–parasite interactions in flax rust — its genetics and other implications. *Phytopathology* **45**, 680–685.

7 Keen, N.T. (1982) Specific recognition in gene-for-gene host-parasite systems. *Adv. Plant Pathol.* **1**, 35–82.

8 Staskawicz, B.J. et al. (1984) Cloned avirulence gene of *Pseudomonas syringae* pv. *glycinea* determines race-specific incompatibility on *Glycine max* (L.) Merr. *Proc. Natl. Acad. Sci. USA* **81**, 6024–6028.

9 Keen, N.T. & Staskawicz, B. (1988) Host range determinants in plant pathogens and symbionts. *Annu. Rev. Microbiol.* **42**, 421–440.

10 Tamaki, S.J. et al. (1991) Sequence domains required for the activation of

avirulence genes *avrB* and *avrC* from *Pseudomonas syringae* pv. *glycinea*. *J. Bacteriol.* **173**, 301–307.

11 Ronald, P.C. (1992) The cloned avirulence gene *avrPto* induces disease resistance in tomato cultivars containing the *Pto* resistance gene. *J. Bacteriol.* **174**, 1604–1611.

12 Li, T.-H. et al. (1992) Phenotypic expression of the *Pseudomonas syringae* pv. *syringae* 61 *hrp/hrm* gene cluster in *Escherichia coli* MC4100 requires a functional porin. *J. Bacteriol.* **174**, 1742–1749.

13 Gabriel, D.W. & Rolfe, B.G. (1990) Working models of specific recognition in plant–microbe interactions. *Annu. Rev. Phytopathol.* **28**, 365–391.

14 Dixon, R. (1986) The phytoalexin response: elicitation, signalling and control of host gene expression. *Biol. Rev.* **61**, 239–291.

15 Van Kam, J.A.L. et al. (1991) Cloning and characterization of complementary DNA avirulence gen *avr9* of the fungal pathogen *Cladosporium fulvum*, causal agent of tomato leaf mold. *Mol. Plant–Microbe Interact.* **4**, 52–59.

16 Dickman, M.B. et al. (1989) Insertion of cutinase gene into a wound pathogen enables it to infect intact host tissue. *Nature* **342**, 448–451.

17 Van Etten, H.D. et al. (1989) Phytoalexin detoxification: Importance for pathogenicity and practical implications. *Annu. Rev. Phytopathol.* **27**, 143–164.

18 Gross, D.C. (1991) Molecular and genetic analysis of toxin production by pathovars of *Pseudomonas syringae*. *Annu. Rev. Phytopathol.* **29**, 247–278.

19 Rainey, P.B. et al. (1992) Biology of *Pseudomonas tolaasii*, cause of brown blotch disease of the cultivated mushroom. *Adv. Plant Pathol.* **8**, 95–117.

20 Lindgren, P.B. et al. (1986) Gene cluster of *Pseudomonas syringae* pv. *phaseolicola* controls pathogenicity of bean plants and hypersensitivity on nonhost plants. *J. Bacteriol.* **168**, 512–522.

21 Willis, D.K. et al. (1991) *hrp* genes of phytopathogenic bacteria. *Mol. Plant–Microbe Interact.* **4**, 132–138.

22 Xiao, Y. et al. (1992) Organization and environmental regulation of the *Pseudomonas syringae* pv. *syringae* 61 *hrp* cluster. *J. Bacteriol.* **174**, 1734–1741.

plant reovirus *See*: PHYTOREOVIRUS GROUP.

plant rhabdovirus group A group of the family RHABDOVIRIDAE resembling the animal and insect members of this family in many respects. They have the characteristic bullet-shaped or bacilliform particles with a lipid outer coat in which are embedded glycoproteins. Inside this coat is a matrix protein layer which encloses the ribonucleoprotein core comprising (–)-strand RNA and several proteins. The genomic RNA of sonchus yellow net virus is transcribed to five mRNAs for structural proteins. These structural genes are arranged on the genome in the same order as those of animal rhabdoviruses but there are differences between plant and animal rhabdoviruses in nonstructural genes. As with PHYTOREOVIRUSES plant rhabdoviruses are considered to be insect viruses becoming adapted to plants. *See also*: PLANT VIRUSES.

Heaton, L.A. et al. (1989) *Proc. Natl. Acad. Sci. USA* **86**, 8665–8668.

Plant viruses

THERE are more than 650 viruses that infect plants. They cause a range of diseases, the most common natural symptom being mosaics or mottles of dark green, light green and sometimes yellow; the pattern of the mosaic is usually regulated by the distribution of veins in the leaves. Other symptoms include necrosis, either as spots (lesions) or spreading along veins (streaks), deformation or outgrowths, and general yellowing or reddening of the leaves frequently caused by blockage of the vascular tissue.

Some viruses have wide host ranges; alfalfa mosaic virus (AlMV, Fig. P36*d*), for example, can infect more than 430 species in 50 plant families. Others have narrow host ranges, sometimes limited to just one species; barley yellow mosaic virus, for example, will only infect barley (*Hordeum vulgare*).

Plant viruses are transmitted naturally in a variety of ways which are usually specific for any one virus. To infect a plant the virus has to be introduced through both the cuticle and the cellulose cell wall, which requires mechanical damage. Some viruses, such as tobacco mosaic tobamovirus (TMV, Fig. P36*h*) (*see* TOBAMOVIRUSES), move from plant to plant through broken leaf hairs, the result of leaves rubbing together. Most viruses, however, have a biological vector which may be a sucking or piercing insect (aphids, leafhoppers), a nematode or a fungus (various lower fungi). The interactions between the virus and its vector are often very specific.

Viruses are classified on the basis of the composition and organization of their GENOME together with other characters such as particle morphology. There are currently 34 groups of plant viruses (Fig. P37; *see also individual entries for each group*). Further details of individual viruses and groups of viruses can be found in [1–4].

Structure

Most plant viruses have very simple particles, with the nucleic acid genome being encapsidated in a coat (capsid) of protein subunits of no more than three types. These simple particles come in three basic shapes.

1 Isometric (spherical) particles are usually constructed on basic ICOSAHEDRAL SYMMETRY principles with some variation in detail to allow for specific connections to be made between subunits of the coat. The most common form is composed of 180 subunits, but there are some examples of 60 subunits, and of 420 subunits.

2 Bacilliform particles (e.g. AlMV) can be considered as being based on icosahedra cut in two with a tubular portion being built between the two halves.

3 Many plant viruses have rod-shaped particles; some are short and rigid, others long and flexuous. The protein subunits of rod-shaped viruses are arranged in a helix with the nucleic acid winding from one end of the rod to the other embedded in a groove within the subunits (Fig. P38).

Particles are stabilized by two main forms of bonding: (1) ionic interactions between the nucleic acid and basic groups on the protein; and (2) protein–protein interactions which may be hydrophobic, or involve a pH-sensitive carboxylate–carboxylate interaction, or be mediated by divalent cations. All the rod-shaped particles and many spherical particles are stabilized by both protein–protein and protein–nucleic acid interactions. For some of the spherical and bacilliform particles protein–nucleic acid interactions predominate whereas for others stabilization is almost totally by protein–protein bonds; in the latter case 'empty' particles not containing nucleic acid are found.

Some plant viruses have more complex particles; the plant RHABDOVIRUS and TOMATO SPOTTED WILT VIRUS groups have particles enclosed in a lipid coat.

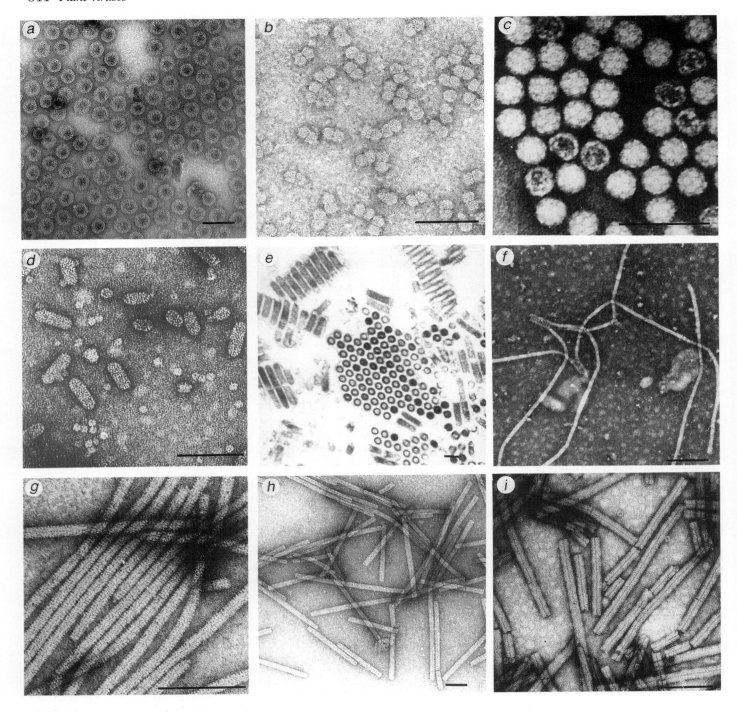

Fig. P36 Virus particle morphology. *a*, Cauliflower mosaic virus; *b*, maize streak geminivirus; *c*, turnip yellow mosaic tymovirus; *d*, alfalfa mosaic virus; *e*, eggplant mottled crinkle rhabdovirus; *f*, beet yellows closterovirus; *g*, potato virus X potexvirus; *h*, tobacco mosaic tobamovirus; *i*, tobacco rattle tobravirus. Bar = 100 nm.

Viral genomes

The majority of plant viruses, about 75%, have (+)-strand (messenger-sense) single-stranded (ss) RNA as their genome.

There are, however, viruses which have (–)-strand ssRNA (complementary to (+)-strand RNA), double-stranded (ds) RNA, or ss- or dsDNA (see Fig. P37).

In some viruses, the genome needed for infection is divided between two or more separate molecules (segmented genomes), which in some cases may be encapsidated in separate particles (multicomponent viruses). Examples of such multicomponent viruses with (+)-strand RNA are TOBRAVIRUSES (two components, rod-shaped particles), HORDEIVIRUSES (three components, rod-shaped particles), COMOVIRUSES (two components, spherical

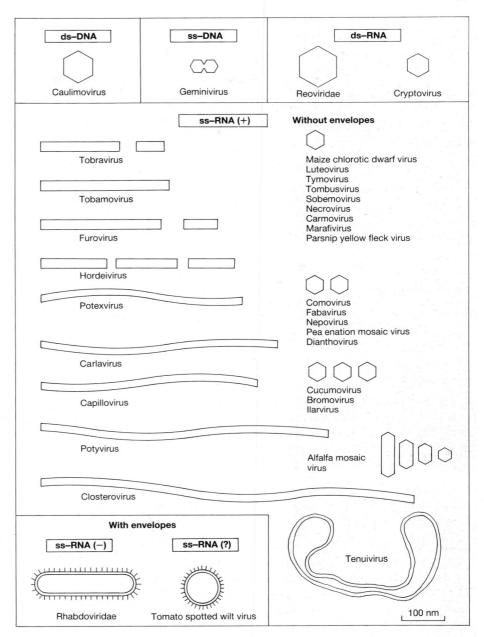

Fig. P37 Groups of plant viruses. The diagram shows the shape of the particle and the genome type of the 34 recognized groups. Courtesy R.I.B. Francki.

particles), BROMOVIRUSES (three components, spherical particles) and AlMV (three components, bacilliform particles).

CRYPTOVIRUSES carry two dsRNA species in each particle, the plant REOVIRUSES have either 10 (FIJIVIRUS GROUP) or 12 (PHYTOREOVIRUS GROUP) different dsRNAs per particle and TOMATO SPOTTED WILT has three species of (–)-strand RNA per particle. B group GEMINIVIRUSES have two species of circular ssDNA encapsidated in separate particles.

The genomes of (+)-strand RNA viruses range in size from about 4 kilobases (kb) (e.g. SOBEMOVIRUSES) to about 20 kb (citrus tristeza CLOSTEROVIRUS). The 5′ end of the RNA genome can be in the form of a CAP structure (m⁷G5′ppp5′...) (e.g TMV), a ppA (e.g. tobacco necrosis NECROVIRUS), or a VPg, which is a small protein linked through a phosphodiester bond from an amino

acid to the 5′ end of the nucleic acid (e.g. comoviruses). The 3′ end of the RNA genome can be a POLY(A) TAIL (e.g. comoviruses), an unphosphorylated nucleotide (X-OH), or can be in the form of a TRANSFER RNA-like structure that will accept amino acids; the 3′ end of the RNA of the TMV type strain accepts histidine, those of TMV cowpea strain and of TYMOVIRUSES accept valine, and bromoviruses, CUCUMOVIRUSES and hordeiviruses accept tyrosine.

Most viruses encode at least four, and often more, proteins [5]. These are the virus capsid protein(s), some proteins (usually two) involved in viral replication, and a protein involved in cell-to-cell movement of the virus; other viral proteins may include a helper protein that facilitates transmission by vector, and proteins of unknown functions.

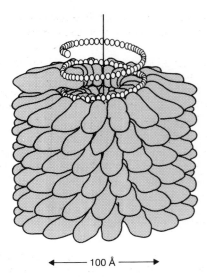

←——— 100 Å ———→

Fig. P38 Diagrammatic structure of the TMV virus showing the arrangement of nucleic acid and protein subunits. The protein subunits form a tight helical array with 16⅓ units per turn, and the RNA is packed in between at a radius of about 40 Å from the helix axis. There are three nucleotides per protein subunit [8].

Cycle of infection

In a typical infection cycle, virus is acquired from an infected plant by a vector, the interaction of the virus with the vector being determined by the viral coat protein and in many cases also by a transmission helper factor. The vector introduces the virus into a cell of another plant where the nucleic acid genome is freed from the coat — the uncoating process.

In many (+)-strand RNA viruses uncoating is effected by ribosomes attaching to the 5′ end of the RNA which is revealed on relaxation of the particle structure. As the ribosomes translocate along the RNA during translation they remove the coat protein subunits [6].

This primary round of translation produces the viral polymerases which replicate the viral genome. That is followed by secondary translation in which large quantities of coat protein are produced and the new viral genomes are encapsidated.

An early translation product is the cell-to-cell movement protein which interacts with PLASMODESMATA to enable infection units, which may not be complete particles, to spread to adjacent cells. When the infection reaches the vascular tissue, complete virus particles enter the phloem SIEVE TUBES and are transported to other parts of the plant. Cell-to-cell spread from the sieve tubes leads to full systemic infection of the plant [7].

Genome replication

(+)-strand RNA viruses

The replication of (+)-strand RNA viruses is via a (−)-strand RNA synthesized by an RNA-DEPENDENT RNA POLYMERASE. One or more of the proteins in the replication complex is encoded by the virus. By comparing amino-acid sequences of proteins from a wide range of prokaryotes and eukaryotes, sequence motifs characteristic of RNA polymerases and RNA HELICASES (RNA unwinding proteins) have been identified in viral proteins.

Two basic types of viral RNA polymerase have been recognized. Possession of one or the other correlates with other viral features, and on this basis both animal and plant (+)-strand RNA viruses can be divided into two supergroups — the SINDBIS VIRUS-like group which has SUBGENOMIC MRNAS and does not have a VPg, and the PICORNAVIRUS-like group which uses the POLYPROTEIN strategy of translation and has VPgs on its RNAs.

The detailed molecular biology of RNA → RNA replication is, however, poorly understood. The sequences at the 3′ and 5′ ends of the (+) and (−) strands are considered to be important in recognition by the polymerase complex. Since many subgenomic RNAs are transcribed from a longer (−)-strand there must also be internal recognition sites for initiation of transcription.

Other RNA viruses

The particles of viruses with dsRNA genomes, and most probably also those with (−)-strand RNA genomes, contain RNA-dependent RNA polymerase activity and both these types of viral genome replicate RNA → RNA.

DNA viruses

The CAULIMOVIRUSES, which have dsDNA genomes, replicate through an RNA intermediate. The viral DNA forms a minichromosome in the nucleus of the host cell and is transcribed to give an RNA of slightly more than genome length, that is, with a TERMINAL REDUNDANCY. This RNA is the template for the formation of dsDNA in the cytoplasm using virus-encoded REVERSE TRANSCRIPTASE, as in the RETROVIRUSES.

The dsDNA geminiviruses replicate in the nucleus. It is thought that either they encode their own DNA-DEPENDENT DNA POLYMERASE or, more likely, that they contribute protein subunits to the host polymerase which enables it to overcome the normal restriction of only one round of DNA replication per cell division.

Translation of (+)-strand RNA viruses

There are two distinct strategies for translating the genetic information on the (+)-strand RNA genome into protein. These have to overcome the limitation that eukaryotic ribosomes will only initiate translation near the 5′ end of an RNA so that protein-coding sequences to the 3′ side of the first OPEN READING FRAME would not be translated.

In the polyprotein strategy followed for example by COMOVIRUSES and POTYVIRUSES, all the genetic information is encoded in one continuous open reading frame which is translated to give a large polyprotein. This is then cleaved to produce the functional proteins.

In the subgenomic strategy on the other hand, individual CISTRONS are made available for translation in various ways. In some cases the genome is divided into monocistronic pieces, which can each act as an mRNA (see, for example, some genes of the ALFALFA MOSAIC VIRUS GROUP). In cases where one of the

Table P7 Abbreviations for some plant viruses

AbMV	Abutilon mosaic virus
AlMV, AMV	Alfalfa mosaic virus
ACLV	Apple chlorotic leafspot virus
ACMV	African cassava mosaic virus
ASGV	Apple stem grooving virus
BCTV	Beet curly top virus
BGMV	Bean golden mosaic virus
BBWV	Broad bean wilt virus
BMV	Brome mosaic virus
BSMV	Barley stripe mosaic virus
BWYV	Beet western yellows virus
BYDV	Barley yellow dwarf virus
CaMV, CMV	Cauliflower mosaic virus
CarLV	Carnation latent virus
CarMV	Carnation mottle virus
CarRSV	Carnation ringspot virus
CBMV	Common bean mosaic virus
CLV	Cassava latent virus
CPMV	Cowpea mosaic virus
CPSMV	Cowpea severe mosaic virus
CuMV	Cucumber mosaic virus
FDV	Fiji disease virus (sugar cane)
GFV	Grapevine fanleaf virus
KLV	Kalanchoe latent virus
LMV	Lettuce mosaic virus
LNYV	Lettuce necrotic yellow virus
LTSV	Lucerne transient streak virus
MCDV	Maize chlorotic dwarf virus
MDV	Maize dwarf virus
MRFV	Maize rayado fino virus
MSV	Maize streak virus
PLRV	Potato leaf roll virus
PNRSV	Prunus necrotic ringspot virus
PVX	Potato X virus
PVY	Potato Y virus
PYDV	Potato yellow dwarf virus
RDV	Rice dwarf virus
SBMV	Southern bean mosaic virus
SCMV	Sugar cane mosaic virus
SMV	Soybean mosaic virus
SNMV	Solanum nudiflorum mottle virus
SYNV	Sonchus yellow net virus
TAV	Tomato aspermy virus
TBSV	Tomato bushy stunt virus
TBV	Tulip breaking virus
TBRV	Tobacco black ring virus
TCV	Turnip crinkle virus
TEV	Tobacco etch virus
TGMV	Tomato golden mosaic virus
TMV	Tobacco mosais virus
TNV	Tobacco necrosic virus
TRSV	Tobacco ringspot virus
TRV	Tobacco rattle virus
TSWV	Tomato spotted wilt virus
TTV	Tobacco tumour virus
TYLCV	Tomato yellow leaf curl virus
TYMV	Turnip yellow mosaic virus
TYTV	Tomato yellow top virus
VCV	Vicia cryptic virus
VMV	Velvet mottle virus
WDV	Wheat dwarf virus
WMV	Wheat mosaic virus
WSV	Wheat streak virus
WCCV	White clover cryptic virus

genome segments contains more than one cistron, an additional copy of the 3′ cistron is present in the genome and acts as the mRNA (*see e.g.* ALFALFA MOSAIC VIRUS GROUP; BROMOVIRUS GROUP). This copy is known as subgenomic and is not needed for infection. In other cases, there are multiple start sites for transcription on the genomic RNA so that various subgenomic mRNAs with different 5′ ends but the same 3′ end are produced from the (−)-strand replication intermediate (as in TMV, *see* TOBAMOVIRUS GROUP). In some cases (e.g. TMV) the first cistron on a viral mRNA also has a leaky TERMINATION CODON which can be suppressed by a host tRNA (SUPPRESSOR tRNA), allowing some of the ribosomes to read through into the next cistron.

R. HULL

See also: ANIMAL VIRUSES; PROTEIN SYNTHESIS; TRANSCRIPTION; VIROIDS.

1 Murant, A.F. & Harrison, B.D. (Eds) (1970 to date) *Descriptions of Plant Viruses* (Commonwealth Mycological Institute/Association of Applied Biologists, Wellesbourne).
2 *The Plant Viruses* (various volumes, 1985 to date) (Plenum Press, New York).
3 Hull, R. et al. (1989) *Virology: Dictionary and Directory of Animal, Bacterial and Plant Viruses* (Macmillan, London).
4 Francki, R.I.B. et al. (Eds) (1991) Classification and Nomenclature of Viruses. *Arch. Virol.* (Suppl. 2) (Springer, Vienna).
5 Goldbach, R. & Wellink, J. (1988) Evolution of plus-strand viruses. *Intervirology* **29**, 260–267.
6 Wilson, T.M.A. (1988) Structural interactions between plant RNA viruses and cells. *Oxford Surv. Plant Mol. Cell. Biol.* **5**, 89–144.
7 Hull, R. (1989) The movement of viruses in plants. *Annu. Rev. Phytopathol.* **27**, 213–240.
8 Klug, A. & Caspar, D.L.D. (1960) *Adv. Virus Res.* **7**, 225–325.

Plant wound responses

A WOUND response can be induced in a plant by physical injury or through the effects of invasion by pests and pathogens. The response is known to occur both locally at the site of damage and systemically, in distant regions of the plant. The wound response encompasses a wide range of events, from changes in the structural organization of the cell wall and changes in the activity of enzymes already synthesized, through to changes in protein composition brought about by wound-induced activation of specific genes [1].

Injury undoubtedly causes cellular damage, and certain features of a wound response reflect the harmful effects of that damage. However, it is also clear that many of the responses to wounding are defensive and as such lead to local and systemic changes of direct benefit to the plant. In this context, wound inducible genes, such as those encoding the serine proteinase inhibitor (pin) proteins [2,3] have been used in transgenic applications leading to improved resistance to pests (*see* PLANT GENETIC ENGINEERING) [4]. Similarly, the gene regulatory elements responding to a wound stimulus can also be used separately in transgenic applications to trigger the expression of other broad-spectrum defence genes at the injury or invasion site.

Since plant development, unlike that of animals, is plastic, environmental stimuli directly influence developmental programming. Events that occur naturally during development are often very similar to those induced by environmental stimuli [1]. Injury can be regarded as such a stimulus, and wound-induced changes in one tissue or organ of a plant can be identical to those that are developmentally regulated in another. An example is the regulation of *pin* genes in potato plants. Expression of *pin2* is developmentally regulated, as the gene product is abundant in potato tubers — specialized storage organs — but levels are negligible elsewhere until the plant is damaged, when pin protein accumulates throughout the aerial tissue [5]. The stimulus in the wound response is injury, but injury leads to events that constitute molecular signals. Certain signals are local to the damaged cells but as the response is also systemic, there must be mechanisms for long-range cell–cell communication and the means to coordinate the local and systemic events that give rise to the total response of the plant.

Events at the local site of injury

Events within the cell wall

Two of the most rapid wound-induced events known involve changes to the architecture of the cell wall. One is the synthesis of CALLOSE, a β1,3 GLUCAN polysaccharide which is synthesized by a membrane-bound enzyme complex located at the cell surface. Loss of cellular integrity triggers callose synthase activity [6]. Callose synthesis can be stimulated *in vitro* by a number of agents, of which calcium, polyamines and FATTY ACID derivatives may reflect endogenous activators [6]. The rapidity of callose induction and the inability of plasma membrane vesicles from the same tissue to synthesize CELLULOSE (a β1,4 glucan) *in vitro*, suggests that the same enzyme complex might be responsible for the synthesis of both polysaccharides, the catalytic specificity then being dependent on prevailing conditions at the cell surface.

In addition to changes in polysaccharide composition, the organization of wall components can also be altered. Wound-induced insolubilization of cell wall proteins occurs *in situ* at the damage site and is thought to arise from protein cross-linking mediated by hydrogen peroxide (H_2O_2) [7].

Both these changes in cell wall architecture are wound-inducible but equally, both can be developmentally regulated. It is important to note that changes such as these do not reflect changes in gene activation, emphasizing the role of post-transcriptional and indeed, post-translational events in the wounded plant.

A third change in the cell wall during the wound response arises indirectly, from the known effects of cell wall fragments on wound-inducible genes such as those encoding pin proteins. Fragments of PECTIN, a complex heteropolysaccharide consisting of an α1,4 polygalacturonide backbone, induce *pin* gene activation when applied *in vitro* to plant tissue [8]. It is possible that oligogalacturonides also act as signals in the whole plant during a wound response. Hydrolases responsible for pectin fragmentation may be located and activated in the wall. Alternatively, the enzymes may be in the vacuole, and on loss of cellular compartmentation together with loss of turgor, may be drawn into the cell wall and mix with the extracellular polysaccharides.

Membrane-related events

Damage to cellular membranes also leads to membrane DEPOLARIZATION in the surviving cells and a consequent change in ion fluxes. Concurrent with these events, degradation of membrane lipids liberates free fatty acids which in turn can also cause a broad range of effects [9]. Notably, the products of deacylation of membrane PHOSPHOLIPIDS and galactolipids (*see* GLYCOLIPIDS) inhibit mitochondrial electron transport (*see* ELECTRON TRANSPORT CHAIN), uncouple OXIDATIVE PHOSPHORYLATION, and provide fatty acid substrates for α-oxidation, which is maintained for several hours before carbohydrate is once more used as the energy source. A form of wound-induced respiration is then maintained, reaching a maximum 2–3 days following injury and involving newly synthesized *b*-type CYTOCHROMES [10].

The C_{18} unsaturated fatty acids linoleic and linolenic acid can also act as substrates in the production of important signalling molecules, including JASMONIC ACID and its methyl ester, cucurbic acid, tuberonic acid, and traumatic acid [9]. When applied *in vitro*, jasmonic acid and its precursors (linolenic acid and octadecenoic intermediates) can induce the accumulation of *pin* transcripts [11].

Jasmonates may therefore be endogenous activators of wound-induced genes. The 13-lipoxygenase enzymes and phospholipase A_2 (*see* PHOSPHOLIPASE) may prove to be key regulatory enzymes in the signalling cascade. In some experimental systems lipoxygenases are wound induced and in others, jasmonate has been shown to induce lipoxygenase. As yet however, no causal relationship has been established between lipid degradation products and the enzymes responsible for their production and the induction of wound-induced genes.

Similarly, the precise role of oxidative reactions such as the release of superoxide anions, lipid peroxidation, and the regulation of oxidative damage by GLUTATHIONE remains unknown, although there is good evidence from a wide range of experimental systems that they are involved in plant wound responses [12,13]. Likewise, wounding is known to trigger a transient local accumulation of the plant growth regulator ethylene (*see* PLANT HORMONES), as evidenced by measurements of product, enzyme activities, and changes in steady-state levels of messenger RNAs corresponding to 1-aminocyclopropane carboxylic acid synthase, and ethylene-forming enzyme [14–16]. But again, the chain of events linking ethylene to an overall wound response has not yet been defined for any experimental system.

Induction of cell division

In addition to the rapid events described above, damage can induce changes in cell division in adjacent unwounded cells; this response is dependent on the morphogenetic potential of the damaged tissue [10]. Three principal responses have been observed:

1 For most leaf tissue and tissues of monocot species, no cell division is induced; rather, increased rates of wall lignification are triggered in the adjacent cells to form a protective layer (*see* PLANT CELL WALL MACROMOLECULES).

2 In other systems the wound stimulus gives rise to irregular cell divisions and the proliferation of CALLUS, which can eventually differentiate to repair the damaged organ.

3 The third response, typical of potato tuber tissue, involves the induction of regular patterns of cell division, in which the MITOTIC SPINDLES of the PARENCHYMA orientate perpendicular to the wound surface, leading to the formation of a new layer of cells parallel to and protecting the damage site.

Changes in gene expression patterns and gene activation

Local wound-induced gene expression has been studied in a number of ways [1]. The construction of cDNA libraries from the mRNA of wounded tissue (*see* DNA CLONING; DNA LIBRARIES), followed by differential screening using PROBES generated from unwounded tissue, has led to the identification of wound-inducible transcripts. This approach has been applied to the wound response of a variety of monocot and dicot species and a range of organs and tissues. The expression of wound-inducible genes has been analysed with respect to timing and cell specificity, and in some cases, the wound-induced genes have been isolated and promoter activity studied in transgenic plants.

An alternative approach is the study of a particular gene whose expression can be predicted to change in response to wounding. The prediction may arise from a knowledge of the cell biology of wounding, or from known changes in enzyme activities. For example, PHENYLALANINE AMMONIA-LYASE (PAL) catalyses the first reaction in the biosynthesis of phenylpropanoids from L-phenylalanine. This is a key step in the biosynthesis of lignin as well as of plant defence-related products such as some PHYTO-ALEXINS, and as such is highly regulated [17]. A MULTIGENE FAMILY encodes different PAL ISOENZYMES, and promoter activity of each different gene is regulated by developmental and environmental signals [18]. As well as the PAL genes, those encoding other enzymes of phenylpropanoid metabolism such as CHALCONE SYNTHASE, as well as those encoding hydroxyproline-rich GLYCO-PROTEINS (structural components of the cell wall) have been studied in relation to local effects of wounding [18].

The systemic wound response

The plant as a whole also responds to wound damage. This systemic response of cells distant from the injury site has been studied mainly in relation to foliar damage — the effects of leaf wounding on tissues elsewhere in the plant. This area of study was initiated by the observation that pin proteins accumulated in the wounded leaf of a potato or tomato plant and also in distant unwounded leaves [2]. The systemic response can be prevented by removal of the damaged leaf within a set time span, implying that some sort of signal(s) is emitted from the wound site and is responsible for the long-range effects. Systemic accumulation of pin proteins has also been observed after wounding of tubers and roots in potato plants, suggesting that damage to any tissue will lead to distant effects [5].

The effect mainly studied involves the two serine proteinase inhibitors (pin1 and pin2), and changes in pin activity, pin protein, *pin* transcripts, and *pin* gene promoter activity have been followed [1]. However, it is unlikely that the systemic response is restricted to activation of these genes and as work progresses, more systemically responding genes are being identified.

Cell signalling in the wound response

Loss of cellular integrity through damage leads to an array of signalling events induced in the intact cells adjacent to the injury, and in cells at progressively greater distances from the injury site. For the former cell population, there is mainly a direct effect in which signals are recognized and then transduced within an intracellular pathway culminating in wound-induced changes, whether at the transcriptional or post-transcriptional level. In parallel (or in consequence) long-range signalling events occur and these must involve passage through a transport conduit away from the local site to other regions of the plant. In turn, the distant cells that receive these systemic signals will be involved in another chain of causal events linking recognition to intracellular effects [19].

Cell signalling in the wound response therefore involves three separate categories of event: (1) those in adjacent living cells; (2) those involving a long-distance conduit; and (3) those involving the distant cells responsive to the systemic signal(s). Within the lamina of a wounded leaflet it is clear that signals can diffuse quite readily throughout the APOPLAST. It is equally clear that for wider access and effects outside that region — into the petiole, stem, other leaves, and root system — use of an established long-distance transport conduit is a necessity [20].

Within this theoretical framework, there are currently a great many contenders for the causal agent in the systemic wound response, and interest in systemic signalling has largely overshadowed local signalling, even though it is the local events that set up the systemic response. Two classes of candidate exist — mobile chemical signals and physical signals.

Most of the evidence for chemical signals relies on a bioassay in which compounds are applied through the cut surface of a stem of an excised plant, and *pin* gene expression is assayed in the leaves. Application of cell wall fragments, ABSCISIC ACID (ABA), jasmonic acid, or a peptide called systemin leads to elevated levels of *pin* gene products [19–21]. Whole plant experiments, the use of ABA-deficient mutants of potato [22] and of phenotypic mutants of tomato expressing a prosystemin ANTISENSE gene [23], indicate a role at some level for both ABA and systemin in the systemic wound response, but not necessarily as the causal systemic signal(s). They could equally well be involved in amplification of the response at the distant sites. AUXIN may also be involved, but to repress the response rather than as an activator [24]. No evidence exists to date for cell wall fragments or jasmonate as long-distance mobile signals, and there are indications that exogenous fragments applied to a wound site on a leafblade do not exit from the lamina [25]. As yet, it is unclear which, if any or all, of these compounds act as local signals, whether local to the injury site or within the transduction chain of systemically responding distant cells.

Cell wall fragments cause membrane depolarization [26]. Their effect and the effect of wounding can both be inhibited by application of aspirin and more specific ion transport inhibitors such as fusicoccin [27]. It is possible that rapid changes in ion fluxes at the surface membranes of adjacent cells establish longer range electrical events and these events constitute one potential mechanism for a physical system of systemic signalling. Certainly, systemic electrical signals can be correlated with wounding by physical injury or heat stimulus [28].

Hydraulic events constitute an alternative physical mechanism for long-distance signalling [29] and may prove to be particularly significant in many responses as plants exist under high levels of water tension, with turgor regulating many aspects of development and defence, and with direct consequences on signalling pathways through modulation of stretch-activated ion channels.

It is probable that the relative roles of chemical and physical signals will change with the developmental state of the plant, particularly given the developmental regulation of cell junction number and the decrease in SYMPLASTIC continuity as the maturity of the plant and differentiation of cell types increases.

D.J. BOWLES

See also: PLANT PATHOLOGY.

1 Bowles, D.J. (1990) Defence-related proteins in higher plants. *Annu. Rev. Biochem.* **59**, 873–907.
2 Green, T.R. & Ryan, C.A. (1972) Wound-induced proteinase inhibitor in plant leaves: a possible defence mechanism against insects. *Science* **175**, 776–777.
3 Ryan, C.A. (1990) Proteinase inhibitors in plants: genes for improving defences against insects and pathogens. *Annu. Rev. Phytopathol.* **28**, 425–449.
4 Hilder, V.A. et al. (1987) A novel mechanism of insect resistance engineered into tobacco. *Nature* **330**, 160–163.
5 Pena-Cortes, H. et al. (1988) Systemic induction of proteinase inhibitor 2 gene expression in potato plants by wounding. *Planta* **174**, 84–89.
6 Kauss, H. (1990) Role of the plasma membrane in host–pathogen interactions. In *The Plant Plasma Membrane* (Larsson, C. & Moller, I.M., Eds) 321–350 (Springer-Verlag, Heidelberg).
7 Bradley, D.J. et al. (1992) Elicitor and wound-induced oxidative cross-linking of a proline-rich plant cell wall protein: a novel rapid defence response. *Cell* **70**, 21–30.
8 Bishop, P. et al. (1981) Isolation and characterization of proteinase inhibitor inducing factor activity from tomato leaves. Identity of poly- and oligogalacturonide fragments. *J. Biol. Chem.* **259**, 13172–13177.
9 Anderson, J.M. (1989) Membrane-derived fatty acids as precursors to second messenger. In *Second Messengers in Growth & Development* (Boss, W.F. et al., Eds) 181–212 (Liss, New York).
10 Kahl, G. (1982) Molecular biology of wound healing. In *Molecular Biology of Plant Tumours* (Academic Press, New York).
11 Farmer, E.E. & Ryan, C.A. (1992) Octadecanoid precursors of jasmonic acid activate the synthesis of wound-inducible proteinase inhibitors. *Plant Cell* **4**, 129–134.
12 Alscher, R.G. (1989) Biosynthesis and anti-oxidant function of glutathione in plants. *Physiol. Plant.* **77**, 457–464.
13 Sutherland, M.W. (1991) The generation of oxygen free radicals during host plant responses to infection. *Physiol. Mol. Plant Path.* **39**, 79–94.
14 Van der Straeten et al. (1990) Cloning and sequence of two different cDNAs encoding 1-aminocyclopropane-1-carboxylate synthase in tomato. *Proc. Natl. Acad. Sci. USA* **87**, 4859–4863.
15 Smith, C.J.S. et al. (1986) Rapid appearance of an mRNA correlated with ethylene synthesis encoding a protein of molecular weight 35 000. *Planta* **168**, 94–100.
16 Hamilton, A.J. et al. (1991) Identification of a tomato gene for the ethylene-forming enzyme by expression in yeast. *Proc. Natl. Acad. Sci. USA* **88**, 7434–7437.
17 Hahlbrock, K. & Scheel, D. (1989) *Annu. Rev. Plant Physiol. Plant Mol. Biol.* **40**, 347–359.
18 Dixon, R.A. & Lamb, C.J. (1990) *Annu. Rev. Plant Physiol. Plant Mol. Biol.* **41**, 339–367.
19 Bowles, D.J. (1991) The wound response of plants. *Curr. Biol.* **1**, 165–167.
20 Bowles, D.J. (1993) Local and systemic signals in the wound response. In *Semin. Cell Biol.* (Academic Press).
21 Ryan, C.A. (1992) The search for proteinase inhibitor inducing factor. *Plant Mol. Biol.* **19**, 123–133.
22 Pena-Cortez, H. et al. (1989) Abscisic acid is involved in the wound induced expression of the proteinase inhibitor II gene in potato and tomato. *Proc. Natl. Acad. Sci. USA* **86**, 9851–9855.
23 McGurl, B. et al. (1992) Structure, expression and antisense inhibition of the systemin precursor gene. *Science* **255**, 1570–1573.
24 Thornburg, R.W. Li, X. (1990) Auxin levels decline in tobacco foliage following wounding. *Plant Physiol.* **93**, 500–504.
25 Baydoun, E.A-H. & Fry, S.C. (1985) The immobility of pectic substances in injured tomato leaves and its bearing on the identity of the wound hormone. *Planta* **165**, 269–276.
26 Thain, J.F. et al. (1990) Oligosaccharides that induce proteinase inhibitor activity in tomato plants cause depolarization of tomato leaf cells. *Plant Cell Environ.* **13**, 569–574.
27 Doherty, H.M. & Bowles, D.J. (1990) The role of pH and ion transport in oligosaccharide-induced proteinase inhibitor accumulation in tomato plants. *Plant Cell Environ.* **13**, 851–855.
28 Wildon, D.C. et al. (1992) Electrical signalling and systemic proteinase inhibitor induction in the wounded plant. *Nature* **360**, 62–65.
29 Malone, M. & Stankovic, B. (1991) Surface potentials and hydraulic signals in wheat leaves following localized wounding by heat. *Plant Cell Environ.* **14**, 431–436. Malone, M, et al. (1994) The relationship between wound-induced proteinase inhibitors and hydraulic signals in tomato seedlings. *Pl. Cell Environ.* **17**, 81–87.

plaque, plaque assay Clear zones produced by the lytic infection by a BACTERIOPHAGE (e.g. bacteriophage LAMBDA) on a lawn of bacteria. If a suspension of a large number of bacteria is infected with a smaller amount of phage, and plated on agar, the uninfected cells produce a dense, translucent 'lawn' on bacteria. However, an infected cell is lysed, releasing progeny phage which then infect adjacent cells. This process ultimately leads to small clear zones of lysed cells in the bacterial lawn. Each plaque is the result of an initial infection by a single bacteriophage. The number of plaques can therefore be used to calculate the titre of phage. *See also*: TURBID PLAQUE.

plaque-forming units (PFU) The number of BACTERIOPHAGE particles present in a given suspension. The number of PLAQUES formed on a lawn of bacteria is inversely proportional to the dilution of bacteriophage suspension added. Therefore, the 'titre' of a bacteriophage-containing solution is defined in terms of 'plaque-forming units' (PFU). If each phage particle in the suspension produces a plaque then the efficiency of plating (EOP) is said to be 1.

plaque purification The BACTERIOPHAGE particles in a given PLAQUE are descended from a single initial infectious phage. Therefore, a plaque constitutes a source of a pure phage line — a fact that allows one to plaque-purify recombinant bacteriophage constructs made *in vitro*.

plasma cell Terminally differentiated antibody-producing cell that differentiates from a B cell after antigenic stimulation. *See*: ANTIBODIES; B CELL DEVELOPMENT; IMMUNOGLOBULIN GENES.

Plasma lipoproteins and their receptors

INSOLUBLE LIPIDS are transported in plasma as soluble lipoprotein particles that comprise a core of hydrophobic cholesteryl esters and triglyceride surrounded by a surface monolayer of amphipathic PHOSPHOLIPIDS with which free CHOLESTEROL and the apolipoproteins are associated. The apoproteins serve as cofactors for enzymes involved in metabolism of the lipid core and as recognition factors that allow both secretion and regulated uptake of lipoproteins. Lipoproteins are characterized by their density; this is determined by their size which dictates the relative amount of lipid and protein [1].

Metabolism

When large triglyceride-rich lipoproteins, very low density lipoprotein (VLDL) from the liver and CHYLOMICRONS (CM) from the gut, enter the plasma their triglyceride core is hydrolysed by lipoprotein lipase on the external capillary endothelium to release fatty acids. Triglyceride-depleted chylomicron remnants are rapidly cleared by receptor-mediated uptake into the liver; a proportion of VLDL remnants are cleared but some are further modified in plasma by lipoprotein lipase, hepatic lipase and the cholesteryl ester transfer protein (CETP), eventually accumulating as low density lipoproteins (LDL), the major carrier of cholesterol in plasma [1]. LDL and some VLDL remnants are taken up by LDL-receptor-mediated ENDOCYTOSIS, a process that is transcriptionally regulated to maintain intracellular cholesterol homeostasis; some is also cleared by ill-defined nonspecific mechanisms. LDL is susceptible to cell-mediated oxidative damage, resulting in modified LDL that is taken up avidly by macrophages and other cells expressing the scavenger receptor (see later); this process may contribute to atherogenesis (the deposition of fatty material on the walls of arteries) [2].

During hydrolysis of the triglyceride core, excess surface components are generated from VLDL and chylomicrons which are precursors for high-density lipoproteins (HDL). Initially discoidal, these nascent particles of phospholipid, free cholesterol and apolipoproteins are rapidly converted to spherical HDL by the enzyme lecithin : cholesterol acyltransferase (LCAT). HDL is thought to protect against premature coronary heart disease by participating in reverse cholesterol transport, whereby cholesterol is continually transferred from peripheral tissues to HDL, esterified by LCAT and then exchanged by means of CETP for triglyceride in lipoproteins containing apolipoprotein B. This excess cholesterol can then be delivered to the liver for LDL-receptor-mediated uptake and excreted via the bile [1].

Apolipoprotein B

The apolipoproteins are of two types, those that remain an integral part of the lipoprotein with which they are secreted and those that are more loosely associated and can exchange between particles. The properties of the apolipoproteins are summarized in Table P8.

The nonexchangeable apoproteins include apoB100 from the liver and apoB48 found in chylomicrons derived from the intestine [3]. ApoB100, with 4536 amino acid residues, is one of the longest single polypeptide chains known; apoB48 is identical to its first 2152 N-terminal residues. Both proteins are the product of a single gene that spans 49 kilobases (kb) and comprises 29 exons; with 7572 bases, exon 26 of the *apoB* gene is one of the longest exons known. In liver and intestine, a single primary messenger RNA transcript is produced and spliced, but in the human intestine a single base in codon 2153 is edited to produce a new stop codon (Fig. P39) (*see also* RNA EDITING). Editing occurs post-transcriptionally by means of a cytidine deaminase, an apparently simple enzyme with no known requirement for low molecular weight cofactors or an additional energy source and with no detectable RNA component. The sequence requirements for editing *in vitro* reside in a 55-nucleotide or less fragment of *apoB* mRNA, and SITE-DIRECTED MUTAGENESIS has shown that an 11-base region starting five bases 3' to the edited base is essential, but that bases immediately adjacent have little influence. The physiological significance of *apoB* mRNA editing in the intestine is not clear, but may lie in the observation that apoB48, while retaining all the structural determinants for lipoprotein assembly and secretion, lacks the LDL receptor-binding domain that resides in the C-terminal region of apoB100. Editing seems to be subject to hormonal and developmental regulation [3].

As the size of the plasma pool of LDL is an important risk factor for coronary heart disease, strenuous efforts have been made to identify factors that regulate apoB synthesis and secretion. Surprisingly, under conditions in which secretion of VLDL and apoB vary widely, *apoB* gene transcription and the level of mRNA remains constant. The details are unclear, but any regulation is post-translational in that a proportion of newly synthesized apoB is degraded before it can assemble with lipids to form VLDL [3].

Mutations in the *apoB* gene cause the inherited disorder hypobetalipoproteinaemia, in which truncated forms of apoB are synthesized because new stop codons have been introduced. Familial defective abetalipoproteinaemia is an inherited disorder that results in HYPERCHOLESTEROLAEMIA because a single mutation in codon 3500 of the *apoB* gene causes a defect in the binding site for the LDL receptor and reduces uptake of plasma LDL [4].

The soluble apolipoprotein gene family

Newly secreted VLDL and chylomicrons also contain the soluble apoproteins of which apoAI, -AII and -AIV, apoE, and apoCI, -CII and -CIII are the most abundant. These apoproteins are gradually lost as triglyceride hydrolysis proceeds until LDL is formed in which apoB is the sole protein [1]. The soluble apoproteins are all members of a MULTIGENE FAMILY that has evolved by gene duplication from an ancestor most closely related to apoCI, the smallest apolipoprotein (*see* MOLECULAR EVOLUTION). Gene structure has been strongly conserved in that each gene comprises four exons, with introns in the 5' untranslated leader region and in the coding regions for the SIGNAL PEPTIDE and the mature protein. Variation between genes occurs in the size of the fourth exon. Most of these genes are found in two gene clusters (Fig.

Table P8 Plasma apolipoproteins

Apolipoprotein	Amino-acid residues in mature protein	Gene	Chromosomal localization (human)	Lipoprotein	Function
apoAI	243	1.9 kb, 4 exons	*apoAI–CIII–AIV* gene cluster 11q33	HDL, chylomicrons	Activates LCAT; ligand for HDL receptor
apoAII	77	1.34 kb, 4 exons	1p21	HDL	None known
apoAIV	376	2.6 kb, 3 exons	*apoAI–CIII–AIV* gene cluster	Chylomicrons, HDL	Synthesis and secretion of chylomicrons
apoB100	4536	43 kb, 29 exons	2p23	VLDL, LDL	Secretion of lipoproteins containing endogenous triglyceride; ligand for LDL receptor
apoB48	2152	43 kb, 29 exons	2p23	Chylomicrons	Secretion of chylomicrons from the gut
apoCI	57	4.4 kb, 4 exons	*apoE–CI–CII* gene cluster 19q13.1	HDL, VLDL, chylomicrons	? Activates LCAT
apoCII	79	3.32 kb, 4 exons	*apoE–CI–CII* gene cluster	HDL, VLDL, chylomicrons	Cofactor for lipoprotein lipase
apoCIII	79	3.1 kb, 4 exons	*apoAI–CIII–AIV* gene cluster	HDL, VLDL, chylomicrons	? Regulates/inhibits receptor-mediated uptake of apoB-containing lipoproteins
apoE	299	3.6 kb, 4 exons	*apoE–CI–CII* gene cluster	HDL, VLDL, chylomicrons	Specific ligand for uptake of remnant lipoproteins by LDL receptor and chylomicron remnant receptor
apo(a)	Highly polymorphic: due to different numbers (~15–35) of an 114-amino acid repeated unit		6q27	Lp(a)	None known

HDL, high-density lipoprotein; LCAT, lecithin–cholesterol acyltransferase; LDL, low-density lipoprotein; VLDL, very-low-density lipoprotein.

P40) with the exception of the *apoAII* gene, found on chromosome 2. The apolipoprotein genes are expressed in the liver, the intestine or both, with the exception of *apoE* which is more widely expressed, notably in regenerating nerve fibres and macrophages loaded with cholesterol. Tissue-specific expression of apolipoprotein genes involves a complexity of TRANSCRIPTION FACTORS, including HNF-1α, C/EBP and a novel member of the STEROID RECEPTOR family. Each member of a gene cluster is transcribed independently [6].

ApoAI–CIII–AIV gene cluster

ApoAI is the major component of HDL, a lipoprotein with a protective role against coronary heart disease. In a search for

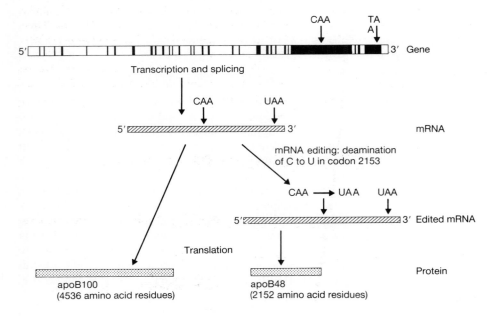

Fig. P39 Synthesis of apoB100 and apoB48 in liver and intestine.

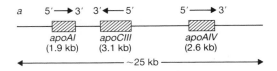

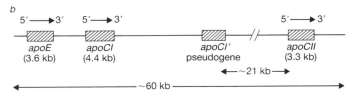

Fig. P40 Organization of human apolipoprotein genes. *a*, The *apoAI–CIII–AIV* gene cluster on chromosome 11; *b*, the *apoE–CI–CII* cluster on chromosome 19.

genetic determinants of HDL concentration, association between polymorphic markers in the *apoAI–CIII–AIV* gene cluster and plasma lipid levels has demonstrated that variation at this locus influences plasma triglyceride levels. This is probably due to *apoCIII* as TRANSGENIC MICE overexpressing human *apoCIII* are hypertriglyceridaemic [7]. Variation at the *apoAI* locus may also affect HDL levels; several rare polymorphic forms of apoAI are found in individuals with low HDL. Mutations in *apoAI* can cause a rare form of familial systemic AMYLOIDOSIS [8].

ApoE–CI–CII gene cluster

Inheritance of different alleles of the POLYMORPHIC *apoE* gene is responsible for 10% of the variation in plasma cholesterol in most populations. Individuals HOMOZYGOUS for one variant, *apoE2*, can develop type III dysbetalipoproteinaemia if an additional genetic or environmental factor is present. Some much rarer alleles of *apoE* produce dominant expression of this disorder in heterozygous individuals. ApoE is a ligand for the LDL receptor and its effects on plasma cholesterol are probably mediated by differences in the affinity of the LDL receptor for lipoproteins carrying variant apoE proteins [9]. The factors that regulate *apoE* gene transcription have been investigated extensively by the expression of gene constructs in transgenic mice and involve complex interactions between factors that bind elements in the 5′ promoter

region, in the first intron and in 3′ regions many kilobases distant from the structural gene [10].

ApoCII is an essential cofactor for lipoprotein lipase and mutations in the *apoCII* gene are one of the causes of the recessive inherited disorder familial hyperchylomicronaemia (type I hyperlipidaemia) [11]. The function of apoCI is less clear, but transgenic mice expressing human apoCI are hypertriglyceridaemic [10], possibly because apoCI displaces apoE from lipoprotein particles.

Lipoprotein(a)

Lipoprotein(a) (Lp(a)) [12] comprises a particle of LDL with apolipoprotein(a) associated by disulphide linkage to apoB. A high concentration of Lp(a) in plasma seems to be a major independent risk factor for coronary heart disease. The structure of apo(a) predicted from its cDNA shows remarkable homology to PLASMINOGEN (Fig. P41). The *apo(a)* gene is one of the most polymorphic loci known, with >20 alleles differing in the number of identical repeated kringle IV domains. The concentration of Lp(a) in plasma varies over a 100-fold range and is under strong, probably more than 90%, genetic control. Alleles with fewer kringle IV domains tend to give rise to high concentrations of Lp(a) in plasma. Apo(a) is synthesized in the liver of man and other primates, but has not been found in any other species apart from the hedgehog. Neither apo(a) nor Lp(a) has any known function and thus apo(a) may be a vestigial protein.

Lipoprotein-modifying enzymes

The genes for the enzymes and receptors involved in lipoprotein metabolism are described in Table P9. Some inherited disorders of plasma lipoproteins are due to deficiencies in the plasma enzymes involved in their metabolism. Mutations in the gene for lipoprotein lipase give rise to the recessive disorder familial hyperchylomicronaemia, in which triglyceride-rich lipoproteins accumulate in plasma because lipolysis is impaired. Several different mutations have been identified and these have helped to delineate functional domains in the enzyme. One mutation, Gly 188→Glu, has been identified in patients of diverse origins, suggesting that it arose in a distant common ancestor. Hepatic lipase deficiency due to genetic defects has been described but has less marked effects on lipoprotein metabolism and does not invariably cause hyperlipidaemia.

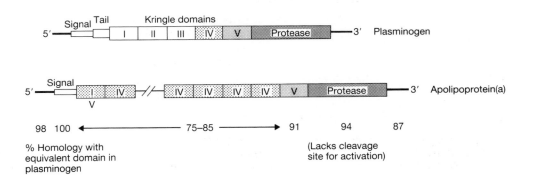

Fig. P41 Sequence similarities between the mRNAs for plasminogen and apolipoprotein(a).

Table P9 Enzymes and receptors involved in lipoprotein metabolism

Enzyme or receptor	Amino-acid residues in mature protein	Gene	Chromosomal localization (human)	Physiological function
Lipoprotein lipase	475	30 kb, 10 exons	8p22	Hydrolysis of triglyceride in VLDL and chylomicrons Requires apoCII as cofactor
Hepatic lipase	476	60 kb, 9 exons	15q21	Hydrolysis of triglyceride in small VLDL, IDL and HDL
Lecithin–cholesterol acyl transferase (LCAT)	416	4.2 kb, 6 exons	16q22.1	Esterification of cholesterol to cholesteryl esters Requires apoAI as cofactor
Cholesteryl ester transfer protein (CETP)	476	25 kb, 16 exons	16q21	Exchange and transfer of cholesteryl esters and triglyceride between lipoproteins; may exist as complex with apoD and LCAT
LDL receptor	731	45 kb, 18 exons	19p13.1	Specific cellular uptake of LDL and remnants of apoB-containing lipoproteins
HDL receptor	M_r 110 000–120 000	Not known	Not known	Mobilization of intracellular cholesterol Removal of cholesterol from HDL in liver
Scavenger receptor (acetyl-LDL receptor)	Type I 453 Type II 349	Not known	Not known	Uptake of modified LDL by macrophages

HDL, high-density lipoprotein; IDL, intermediate-density lipoprotein; LDL, low-density lipoprotein; VLDL, very-low-density lipoprotein.

Mutations in the gene for LCAT cause the rare recessive familial disorder of LCAT deficiency, caused by accumulation of unesterified cholesterol in plasma lipoproteins and leading to proteinuria, haemolytic anaemia, and diffuse corneal opacities [13]. Mutations in the LCAT gene that are less deleterious to enzyme function are now known to cause the milder inherited deficiency disorder fish-eye disease, so named because of the characteristic corneal opacities.

The genes for lipoprotein lipase and hepatic lipase are closely related and share some homology with that for LCAT. However CETP is apparently unrelated to any of the other apoproteins or enzymes involved in lipoprotein metabolism. CETP does not seem to be essential for health as its absence in individuals with inherited defects in the CETP gene is associated with longevity [8].

Lipoprotein receptors

The LDL receptor

The structural gene for the LDL receptor spans ~45 kb on human chromosome 19 and comprises 18 exons that delineate distinct structural and functional domains of the protein (Fig. P42) [14]. The N-terminal extracellular ligand-binding domain comprises seven cysteine-rich repeats with similarity to domains found in a number of proteins in the COMPLEMENT cascade. The second domain shows extensive homology with the EPIDERMAL GROWTH FACTOR (EGF) precursor and itself contains several internal repeats; this domain is important for recycling of receptors from the endosomal compartment to the cell surface (*see* COATED PITS AND VESICLES; ENDOCYTOSIS). The third domain is a threonine- and serine-rich region that is the site of attachment of the majority of the O-linked sugars (*see* GLYCOPROTEINS) on the receptor protein. The fourth is a membrane-spanning domain of 22 amino acids and the fifth domain is the C-terminal cytoplasmic tail that

contains the sequence Asn-Pro-X-Tyr (NPXY), an internalization signal that directs the LDL receptor to clathrin-coated pits on the cell surface. The arrangement of the LDL receptor gene supports the view that genes for multifunctional proteins have evolved by shuffling exons derived from other functionally unrelated proteins.

Confirmation of the function of each domain has come from identification of the effect of specific mutations in the LDL receptor gene of patients with familial hypercholesterolaemia (FH), a codominant inherited disorder of lipoprotein metabolism that is associated with a markedly increased risk of premature coronary heart disease. Many different mutations in the receptor gene cause FH, precluding simple DNA-based diagnosis. In addition to studies on naturally occurring mutations, detailed analysis by site-directed mutagenesis has revealed that the repeats in the binding domain are not equivalent; repeats 4 and 5 only are required for binding lipoproteins in which apoE is the ligand whereas additional adjacent repeats are required to bind LDL.

Transcription of the LDL receptor gene is regulated by intracellular sterol flux, although it is not yet known whether cholesterol or an oxidized derivative such as 25-OH-cholesterol is the active agent. Sterol-mediated regulation has been localized to an element of 8–10 nucleotides in the PROMOTER region of the receptor gene (sterol response element, SRE-1); this element is also found in the promoter regions of the genes for hydroxylmethylglutaryl-CoA reductase (HMG-CoA reductase) and HMG-CoA synthase, two key enzymes in the biosynthesis of cholesterol. Sterol-mediated regulation of the HMG-CoA reductase gene involves an element that overlaps with but is not identical to the SRE-1 sequence. Site-directed mutagenesis of the LDL receptor gene promoter has suggested that transcription is promoted by a protein factor that binds to the SRE-1, and that the function of this protein is modulated by a second sterol-sensitive protein. None of the

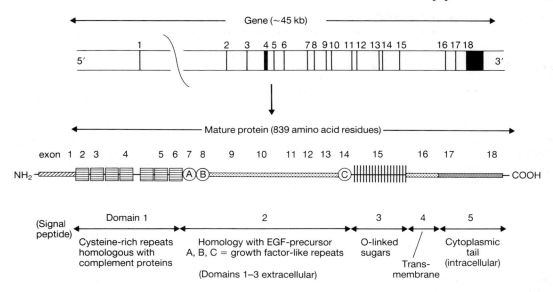

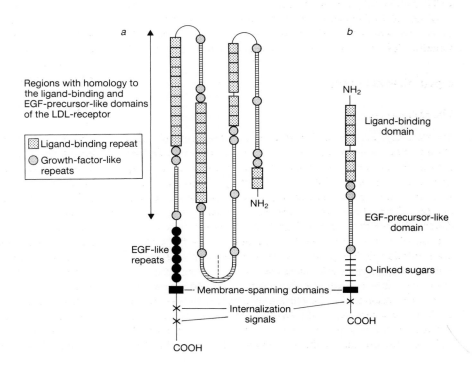

Fig. P42 Domain structure of the LDL receptor and its gene.

putative transcription factors have yet been identified unequivocally. LDL receptor gene transcription may also be regulated by other factors, for example INSULIN, but it is not clear whether these modulate transcription directly or through an increased demand for cholesterol by promoting cell growth.

LDL-receptor-related protein

The LDL-receptor-related protein (LRP) was discovered when cDNA libraries were screened with a probe to the cysteine-rich repeat common to the LDL receptor and complement proteins [15]. Its structure (Fig. P43a) suggested that it might function as a cell-surface receptor, and the independent observation that it bound apoE suggested that it might be the elusive chylomicron remnant receptor. Although it has subsequently been shown to have some of the desired properties, it has also been found to be identical to the hepatic α_2-macroglobulin receptor that binds protease–α_2-macroglobulin complexes. This suggests that LRP might have wide specificity for the uptake and clearance of large unwanted proteins from the circulation. Binding of chylomicron remnants to LRP is greatly enhanced in the presence of lipoprotein lipase, but with no requirement for enzyme activity.

Fig. P43 Similarity between the predicted structures of the LDL-receptor-related protein (LRP) (*a*) and the LDL receptor (*b*).

The scavenger receptor

Two closely related species of cDNA for the scavenger receptor have been cloned, originally from a bovine lung cDNA library but also from other species [16]. The mature protein comprises a dimer or trimer of polypeptide chains, but it is not known whether homodimers or heterodimers are present in cells. Each cDNA can produce an active scavenger receptor when expressed independently in heterologous cells. Receptors comprised of either type of monomeric unit have in common a terminal cytoplasmic tail domain, a membrane-spanning domain, a helical coiled coil domain and a collagen-like domain. The type I receptor also has an N-terminal cysteine-rich domain that is absent from the type II receptor; this domain is unlikely to be functionally important unless it marginally affects specificity. The identity and structure of the genes have not yet been described and no inherited disorders of scavenger receptor function are known.

A.K. SOUTAR

1 Havel, R.J. & Kane, J.B. (1989) Structure and metabolism of plasma lipoproteins. In: *The Metabolic Basis of Inherited Disease*, 6th edn, Vol. II (Scriver, C.R. et al., Eds) 1129–1138 and following chapters (McGraw–Hill, New York).
2 Witzum, J.L. & Steinberg, D. (1991) Role of oxidised low density lipoprotein in atherogenesis. *J. Clin. Invest.* **88**, 1785–1792.
3 Scott, J. (1990) Regulation of the biosynthesis of apolipoprotein B$_{100}$ and apolipoprotein B$_{48}$. *Curr. Opinion Lipidol.* **1**, 96–103.
4 Farese, J.R. et al. (1992) Apolipoprotein B gene mutations affecting cholesterol levels. *J. Int. Med.* **231**, 643–652.
5 Chan, L. & Li, W–H. (1991) Apolipoprotein variation among different species. *Curr. Opinion Lipidol.* **2**, 96–103.
6 Zannis, V.I. et al. (1992) Molecular biology of the human apolipoprotein genes: gene regulation and structure/function relationship. *Curr. Opinion Lipidol.* **3**, 96–113.
7 Breslow, J.L. (1992) The genetic basis of lipoprotein disorders. *J. Int. Med.* **231**, 627–631.
8 Tall, A.L. (1992) Metabolic and genetic control of HDL cholesterol levels. *J. Int. Med.* **231**, 661–668.
9 Mahley, R.W. et al. (1990) Apolipoprotein E: genetic variants provide insights into its structure and function. *Curr. Opinion Lipidol.* **1**, 87–95.
10 Taylor, J.M. et al. (1991) Expression of the human apolipoprotein E/apolipoprotein C1 gene locus in transgenic mice. *Curr. Opinion Lipidol.* **2**, 73–80.
11 Lalouel, J–M. et al. (1992) Lipoprotein lipase and hepatic triglyceride lipase: molecular and genetic aspects. *Curr. Opinion Lipidol.* **3**, 86–95.
12 Utermann, G. (1990) Lipoprotein(a): a genetic risk factor for premature coronary heart disease. *Curr. Opinion Lipidol.* **1**, 404–410.
13 Assmann, G. et al. (1991) Lecithin: cholesterol acyltransferase deficiency and fish-eye disease. *Curr. Opinion Lipidol.* **2**, 110–117.
14 Soutar, A.K. & Knight, B.L. (1990) Structure and regulation of the low density lipoprotein (LDL)–receptor and its gene. *Br. Med. Bull.* **46**, 891–916.
15 Brown, M. et al. (1991) The low–density lipoprotein receptor–related protein: double agent or decoy? *Curr. Opinion Lipidol.* **2**, 65–72.
16 Freeman, M. et al. (1990) An ancient, highly conserved family of cysteine-rich protein domains revealed by cloning I and type II murine macrophage scavenger receptors. *Proc. Natl. Acad. Sci. USA* **87**, 8810–8814.

plasma membrane (plasmalemma) Membrane bounding the surface of all living cells, forming a semipermeable barrier between the cytoplasm and the outside environment. It is composed of a phospholipid bilayer in which are embedded proteins serving the functions of enzymes, transporters, ion channels, ion pumps, receptors for hormones and growth factors, etc. It regulates the entry and exit of most solutes and ions, few substances being able to diffuse through unaided. It is part of the functional ENDOMEMBRANE SYSTEM of the eukaryotic cell. *See*: BACTERIAL ENVELOPE; ENDOCYTOSIS; EXOCYTOSIS; MEMBRANE STRUCTURE; MEMBRANE TRANSPORT SYSTEMS.

plasma proteins Proteins of the fluid component of blood. Almost half consists of SERUM ALBUMIN, which maintains plasma osmotic pressure and binds many substances. Other important components include the IMMUNOGLOBULINS, COMPLEMENT, COAGULATION FACTORS, α$_1$-ANTITRYPSIN and specific transport proteins (e.g. HAPTOGLOBIN, LIPOPROTEINS, and TRANSFERRIN).

plasma thromboplastin component Factor X. *See*: BLOOD COAGULATION AND ITS DISORDERS.

plasmacytoma Isolated bone tumour due to a malignant proliferation of PLASMA CELLS. Evolution to multiple MYELOMA may be observed.

plasmalemma PLASMA MEMBRANE.

plasmalemmasome In plant cells, a membranous structure formed between the PLASMALEMMA and the cell wall. It consists of tubules, cisternae and vesicles. Lomasomes and paramural bodies are similar structures.

Plasmaviridae *See*: BACTERIOPHAGES.

Plasmid(s)

PLASMIDS are extrachromosomal genetic elements composed of DNA or RNA and found in both eukaryotic and prokaryotic cells. They carry genetic information, are not part of a CHROMOSOME, but can propagate themselves autonomously in cells. Mostly circular molecules, although linear plasmids have also been found, plasmids range in size from those that encode no more than 10 genes to those that encode several hundred.

The distinction between plasmids and viruses is sometimes difficult to make as many viruses exist for long periods in a dormant, non-infectious state in cells, propagating themselves according to the cellular growth rate. In contrast to viruses however, plasmids do not form infectious particles.

Types of plasmid

A representative selection of types of plasmid is given in Table P10. A wide variety of different genes may be carried by plasmids and these are sometimes used to name them [1,2]. Thus there are virulence plasmids, which encode toxins or other proteins that increase the virulence of bacteria, metabolic plasmids (including degradative plasmids) encoding a variety of catabolic or other enzymes, colicin plasmids (COL PLASMIDS), which encode antibacterial proteins called COLICINS, drug-resistance plasmids (R FACTORS), which confer resistance to antibiotics on their bacterial

Table P10 Examples of bacterial plasmids

Plasmids	Original host	Relative molecular mass	Copy no.	Incompatibility group	Conjugation	Characteristics
F	*Escherichia coli* K-12	63×10^6	1–2	FI	+ (F pili)	Integrates to form Hfr strains and can form F′ plasmids
R1	*Salmonella paratyphi*	58×10^6	1–2	FII	+ (F pili)	Drug-resistance plasmid, encoding resistance to ampicillin, chloramphenicol, fusidic acid, kanamycin, streptomycin, and sulphonamide
RP1	*Pseudomonas aeruginosa*	38×10^6	1–2	PI	+	Wide host range, drug-resistance plasmid encoding resistance to ampicillin, kanamycin, and tetracycline
ColV, I-K94	*Escherichia coli*	85×10^6	1–2	FI	+ (F pili)	Col and virulence plasmid encoding colicins V and I as well as COMPLEMENT resistance which increases bacterial virulence
ColE1-K30	*Escherichia coli*	4.3×10^6	15	?	–	Col plasmid specifying colicin E1
pBR322		2.9×10^6	40	?	–	Cloning vector. Encodes resistance to ampicillin and tetracycline. Made by combining parts of three other plasmids (R1, pSC101 and pMB1)
TOL (pWWO)	*Pseudomonas putida*	78×10^6	?	?	+	Degradative plasmid (carries genes for toluene catabolism)
Ti plasmids	*Agrobacterium tumefaciens*	90×10^6–160×10^6	1–2	?	+	Cause crown gall disease in plants; modified versions widely used as vectors in plant genetic engineering

hosts, and tumour-inducing plasmids (*see* TI PLASMID) which can be transferred from bacteria to plant cells to induce the formation of crown gall tumours.

Many of these important traits specified by plasmids are encoded by transposons, genetic elements which can transfer copies of themselves to other replicons with which they have little, if any, sequence homology (*see* TRANSPOSABLE GENETIC ELEMENTS).

Structure and function

The terminology of plasmid structure and function derives mainly from work on bacterial plasmids, especially those from *Escherichia coli*. This entry therefore concentrates on the essential features of bacterial plasmids as the same terminology has been adopted by researchers studying plasmids from other sources. The first plasmid to be discovered was the fertility factor, F FACTOR (F plasmid), in *E. coli* strain K-12, which confers the ability to conjugate.

The next best-studied group of plasmids are those found in fungi, particularly the 2μM PLASMID (2μ plasmid) from *Saccharomyces cerevisiae* (baker's yeast), also called the 2μm circle. The 2μm circle is one of the few nuclear plasmids found in fungi. Plasmids have also been discovered in plant and fungal mitochondria (*see* MITOCHONDRIAL GENOMES: LOWER EUKARYOTE; MITOCHONDRIAL GENOMES: PLANT).

Structure

Most plasmids are circular DNA molecules — sometimes referred to as covalently closed circles. The double helices of the DNA often have additional twists; they are SUPERCOILED or superhelical (Fig P44). Superhelical turns are introduced into the molecule by GYRASES. A nick in one of the two polynucleotide strands of a covalently closed circular plasmid produces an open circular molecule; the molecule loses its superhelical twists and is said to become 'relaxed'.

Replication

The number of plasmids per cell (or per bacterial chromosome) varies with different plasmids and the particular copy number is characteristic of the plasmid. Plasmids coexist stably with their host cells implying that they replicate very efficiently and that, at

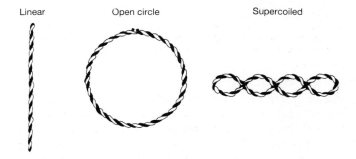

Linear   Open circle   Supercoiled

Fig. P44 Molecular forms of plasmid DNA.

least for plasmids with a low copy number, there are mechanisms for segregating newly replicated copies into daughter cells. The coupling of the rate of replication of plasmids to the growth rate of their host cells appears to be brought about largely by regulating the initiation of new rounds of plasmid replication — the more rapidly the host cell grows, the more frequently new rounds are initiated. Once initiated, rounds of replication are completed at about the same rate at both high and low cell growth rates. The frequency of initiation also determines plasmid copy number, which can vary between one and about one thousand per cell.

Plasmids are examples of REPLICONS. These were originally defined [3] as 'units of DNA capable of independent replication which set up specific systems of signals allowing, or preventing, their own replication'.

Plasmid replication and its control involves the products of both plasmid and chromosomal genes. Many of the steps in plasmid replication are similar to those in chromosomal replication, although there are also specific differences. Some plasmids replicate unidirectionally, that is, replication begins at a specific point, the origin, and the replication fork moves around the molecule in a single direction. Others replicate bidirectionally, that is, two replication forks are generated at an origin and they move in opposite directions around the chromosome (the circular chromosomes of several bacteria are also replicated bidirectionally) (*see* DNA REPLICATION). Yet other plasmids are replicated by means of a ROLLING CIRCLE MECHANISM.

Bacterial plasmids that can replicate in a wide range of bacterial genera are referred to as wide host range (or promiscuous) plasmids. Some plasmids of this type can replicate in many, if not all, genera of Gram-negative bacteria. Curing agents, for example an ACRIDINE DYE like acridine orange, are substances which interfere with plasmid replication, leading to the loss of plasmids from host cells.

Plasmid incompatibility

Different plasmids that can stably coexist in the same cell line are said to be compatible; those that cannot are termed incompatible.

Incompatible plasmids cannot be stably inherited; as the host cell replicates, daughter cells are produced which contain only one of the two kinds of plasmid. Bacterial plasmids can be classified into about 30 incompatibility groups. Plasmids from the same group cannot stably coexist in the same cell line unless a strong selection is applied for coexistence.

Incompatibility is believed to be a consequence of the mechanisms that control the initiation of plasmid replication or of the mechanisms controlling the partitioning of daughter plasmids at cell division.

Conjugation and plasmid transmissibility

Many plasmids transfer themselves from one host cell to another by CONJUGATION. This process involves plasmid-encoded proteins which are involved in the attachment of the host cell to another cell and the transfer of a plasmid copy to it. This mode of transmission is in contrast to two other means of plasmid transfer which can be demonstrated in the laboratory, namely TRANSFORMATION and TRANSDUCTION. Transformation refers to the uptake of DNA by cells from solution. Transduction is the transmission of DNA by BACTERIOPHAGES. Phages sometimes transfer pieces of DNA, including plasmids, either instead of, or as well as, their own chromosomes (*see* LAMBDA).

Transmission by bacterial conjugation, for example the transmission of the F plasmid from *E. coli*, involves sex pili (sometimes called fimbriae) which are hollow tubes composed of protein subunits called pilin, encoded by the plasmid, that project from the bacterial cell. The mechanism of conjugation in *E. coli* is not well understood, but pili are able to retract and this may be a part of the mechanism which draws two bacteria together (Fig. P45). At high ratios of donor to recipient, some of the recipients are killed — a phenomemon known as lethal zygosis.

Some conjugative plasmids (plasmids that induce conjugation) encode the constitutive production of sex pili, which are correlated with conjugative ability; each cell containing the plasmid has at least one pilus. Other plasmids regulate the production of sex pili and of other proteins involved in conjugation so that their

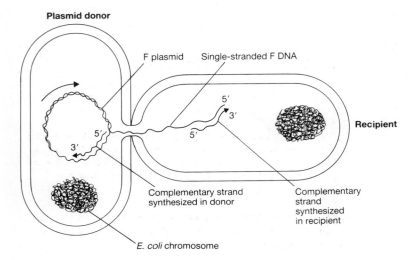

Fig. P45 Bacterial conjugation. The F plasmid found in *E. coli* transfers a single-stranded DNA copy of itself. In the recipient, a DNA strand complementary to the transferred strand is synthesized to reform double-stranded DNA.

synthesis is repressed in most cells. Sometimes, the repression mechanism of this latter type can inhibit the formation of the conjugative mechanism and pilus formation by plasmids of the former type, when both types are in the same cell. The inhibitory plasmids are said to bring about fertility inhibition.

Some plasmids can prevent the entry of certain other plasmids into a cell by the phenomenon of surface exclusion. Thus, an *E. coli* strain which is F$^+$ (i.e. it carries an F plasmid) will be a poor recipient for the transfer of F from another strain. The effect of surface exclusion can be overcome by starving the recipients of nutrients, when they become F$^-$ PHENOCOPIES.

Plasmid-mediated gene transfer

The F plasmid was first recognized because of its ability to transfer chromosomal genes from one *E. coli* strain to another. The ability of F and of other plasmids to transfer genes in this way is a reflection of their ability to integrate into the bacterial chromosome — they are able through RECOMBINATION to become physically part of the chromosome. When the F plasmid integrates into the chromosome of the host bacterium, an HFR STRAIN is formed, so called because it is now able to transfer chromosomal genes at *High f*requency in contrast to an F$^+$ strain. Plasmids and other genetic elements which can exist in either integrated or autonomous forms are called episomes. The integration of the F plasmid into the chromosome involves special DNA sequences, present on both plasmid and chromosome, called insertion sequences (*see* TRANSPOSABLE GENETIC ELEMENTS).

An Hfr donor transfers a part of the bacterial chromosome during conjugation. Only rarely is a copy of the entire chromosome transferred, an event which takes at least 100 minutes. Excision of an integrated F plasmid from the bacterial chromosome can give rise to F′ (F-prime) plasmids in which a piece of the bacterial chromosome is attached to the F plasmid DNA.

Conjugative plasmids can also transfer non-conjugative plasmids, a process known as plasmid mobilization.

Plasmid vectors

Plasmids are extensively used as VECTORS to clone DNA molecules (*see* DNA CLONING; GENETIC ENGINEERING). Most plasmid vectors are made by taking DNA from a variety of replicons (plasmids, bacteriophage chromosomes and bacterial chromosomes) and joining it together (using RESTRICTION ENZYMES and DNA LIGASE) to form a plasmid which has an ORIGIN OF REPLICATION, a selectable MARKER (usually an antibiotic-resistance gene) and a PROMOTER for expressing genes of interest in the required host cell (*see* EXPRESSION VECTORS). The Ti plasmids are widely used as vectors in PLANT GENETIC ENGINEERING as they can integrate part of their DNA into the recipient plant cell.

Plasmid vectors can also be made which can replicate in two or more unrelated cells such as yeast and bacteria, or bacteria and animal cells. These are called SHUTTLE VECTORS.

K.G. HARDY

1 Hardy, K.G. (1986) *Bacterial Plasmids*. 2nd edn (Van Nostrand Reinhold, Wokingham, UK).
2 Hardy, K.G. (1993) *Plasmids: a Practical Approach*. 2nd edn (Oxford University Press, Oxford).

3 Jacob, F. et al. (1963) On the regulation of DNA replication in bacteria. *Cold Spring Harbor Symp. Quant. Biol.* **28**, 29–348.

2 µm plasmid A circular DNA PLASMID of 6318 bp present in most strains of the yeast SACCHAROMYCES CEREVISIAE at between 50 and 100 copies per haploid genome. A plasmid of similar size is also found in *Zygosaccharomyces* species. The 2 µm plasmid contains two 599-bp perfect INVERTED REPEATS (IVR1 and IVR2), which separate unique regions of 2774 bp and 2346 bp respectively (Fig. P46). Recombination between the two inverted repeat sequences, mediated by the plasmid-encoded *FLP* gene product, generates two forms of the plasmid (A and B) which exist in equal proportions in most strains. The plasmid is located in the nucleus and is packaged into NUCLEOSOMES typical of chromosomal DNA. Four plasmid-borne genes are required for its stable maintenance and propagation; *REP1* and *REP2* which encode proteins that interact with the *REP3* DNA sequence to regulate *FLP* gene expression, and with the origin of DNA replication (ARS). The 2 µm plasmid has been widely used in the development of multicopy recombinant YEp plasmids for the genetic manipulation of *S. cerevisiae*.

Murray, J.A.H. (1987) *Mol. Microbiol.* **1**, 1–4.

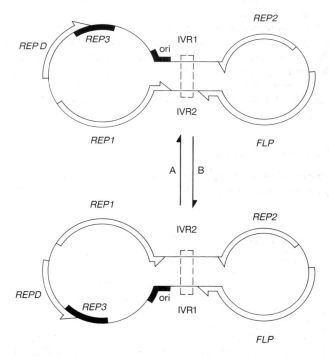

Fig. P46 The 2 µm plasmid of *Saccharomyces cerevisiae*. The forms A and B are shown. Also indicated are the plasmid-borne genes *REP1*, *REP2*, *REP3*, and *REPD* together with the *FLP* gene. The origin of DNA replication (ori) and the two inverted repeat sequences (IVR1 and IVR2) are also shown.

plasmid maintenance sequence Sequence on a plasmid, other than the ORIGIN OF REPLICATION, that is required for maintenance in a host cell.

plasmid rescue The extraction of a recombinant plasmid from a host other than *Escherichia coli* and introduction into *E. coli* by TRANSFORMATION for further study.

plasmid vectors *See*: COLE1; COSMID; EXPRESSION VECTOR; GENETIC ENGINEERING; PBR322; PLANT GENETIC ENGINEERING; 2μM PLASMID; TI PLASMID; VECTOR.

plasmin, plasminogen Plasmin is a SERINE PROTEINASE (EC 3.4.21.7) found in blood plasma, produced by specific cleavage of the proenzyme plasminogen by plasminogen activators (*see* TISSUE PLASMINOGEN ACTIVATOR; UROKINASE). In blood plasma it degrades FIBRIN into soluble products. *See*: BLOOD COAGULATION AND ITS DISORDERS; SERPINS.

plasminogen activator *See*: TISSUE PLASMINOGEN ACTIVATOR; UROKINASE.

plasminogen activator inhibitor (PAI) *See*: SERPINS.

plasmodesmata (*sing.* plasmodesma) Fine cytoplasmic channels through the walls of plant cells, usually between two cells and connecting their protoplasts. They are lined by PLASMALEMMA.

plasmodium Multinucleate mass of acellular motile protoplasm formed during the life cycle of ACELLULAR SLIME MOULDS.

Plasmodium Genus of mosquito-borne sporozoan protozoans that infect red blood cells and cause malaria. Four species, *P. falciparum*, *P. vivax*, *P. ovale* and *P. malariae* are pathogenic in man, the first-named being the major cause of mortality. The heterozygous state for certain genetic disorders (e.g. SICKLE CELL ANAEMIA and THALASSAEMIA) provides increased resistance to malaria.

plasmogamy Fusion of the cytoplasms of two or more cells prior to nuclear fusion.

plastids A family of cytoplasmic organelles of plant cells which possess their own genomes and which includes PROPLASTIDS, CHLOROPLASTS, ETIOPLASTS, LEUCOPLASTS, and CHROMOPLASTS. There is no counterpart in animal cells. *See also*: CHLOROPLAST GENOME.

plastocyanin A small blue protein (M_r 10 500), containing a single copper atom per molecule, which is located in the thylakoid lumen of chloroplasts and mediates electron transfer between CYTOCHROME F and PHOTOSYSTEM I in the photosynthetic electron transport system of oxygenic organisms. In plants and algae its reactivity is dominated by its highly acidic nature, but the protein from CYANOBACTERIA is weakly acidic or basic. All plastocyanins have a very similar structure of the Greek key β-barrel type (*see* PROTEIN STRUCTURE for illustration). In plants it is nuclear encoded (*petE* gene). Some algae and cyanobacteria can replace it by cytochrome c_6 when grown under copper deficient conditions. *See also*: PHOTOSYNTHESIS.

plastome The genome of any plastid. *See*: CHLOROPLAST GENOME.

plastoquinone A substituted benzoquinone which is very hydrophobic by virtue of a long isoprenoid side chain and which occurs in the thylakoid membranes of all oxygenic photosynthetic organisms (*see* PHOTOSYNTHESIS). The bulk of the active plastoquinone occurs as a 'pool' able to diffuse within the lipid phase of the membrane and catalyse electron transfer between PHOTOSYSTEM II and the CYTOCHROME BF COMPLEX and simultaneous transfer of protons from stroma to lumen. Q_A in photosystem II is a tightly bound form of plastoquinone which normally undergoes only one-electron reduction to the semiquinone. A molecule of plastoquinone from the pool can be bound reversibly to the Q_B site and to quinone-binding sites of the cytochrome *bf* complex. *See also*: PHOTOSYNTHESIS.

platanetin 6-(3,3-Dimethylallyl)-3,5,7,8-tetrahydroxyflavone (Fig. P47). It is isolated from bud scales of *Platanus acerifolia*, a species of plane tree. A potent uncoupler of OXIDATIVE PHOSPHORYLATION.

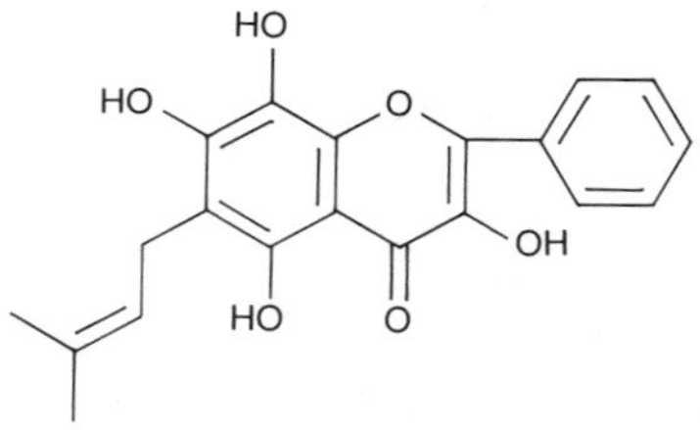

Fig. P47 Platanetin.

platelet A small anuclear cell of the blood, produced by budding from MEGAKARYOCYTES in the bone marrow. The main function of platelets is to maintain HAEMOSTASIS, which they achieve by adhesion to damaged surfaces (*see* VON WILLEBRAND DISEASE FACTOR), shape change, self-aggregation, and activation of the blood coagulation pathway. *See*: BLOOD COAGULATION AND ITS DISORDERS.

platelet-derived growth factor *See*: GROWTH FACTORS.

β-pleated sheet Descriptive name for β-sheet secondary structure in proteins where alternating Cα atoms and side chains are above and below the plane of the sheet. This occurs whether the β-strands are parallel or antiparallel. *See*: PROTEIN STRUCTURE.

plectin An INTERMEDIATE FILAMENT protein.

pleiotropy The case in which a mutation in a single gene produces a multiplicity of different effects. For example, the *wingless* mutation in birds not only affects wings, but also leads to the absence of the metanephric kidney and of the air sacs in the lungs. Thus, *wingless* seems to be concerned with a range of processes rather than displaying a one gene–one character relationship.

ploidy The number of copies of the haploid genome in each cell. Haploid organisms have a single copy, diploid organisms have two copies, while POLYPLOID organisms have more than two copies (e.g. tetraploid, four copies). Prokaryotes are usually haploid and eukaryotes usually diploid or polyploid. In eukaryotes, the diploid state is maintained in somatic cells by chromosome duplication during MITOSIS. In the GERM CELLS however, MEIOSIS leads to the formation of haploid GAMETES.

pluripotency The potential of a cell or tissue to develop into more than one type of cell or tissue if placed in an appropriate environment. The process of development can be regarded as a progressive restriction in potency from the totipotent fertilized egg to unipotent TERMINALLY DIFFERENTIATED cells. If a cell or tissue can adopt more than one fate, it is defined as pluripotent, for example, the pluripotent haematopoietic stem cell of bone marrow which is committed to form blood cells, but not yet committed as to the precise type (*see* HAEMATOPOIESIS).

PMA The PHORBOL ESTER 4-phorbol-12-myristate 13-acetate.

podophyllotoxin *See*: LIGNANS.

Podoviridae *See*: BACTERIOPHAGES.

point group A set of non-translational SYMMETRY elements passing through a single point. There are 32 three-dimensional crystallographic point groups consistent with a space-filling lattice, thus defining the 32 crystal classes. However, only 11 of these contain solely axes of rotation and are thus compatible with chiral molecules such as proteins and nucleic acids, which are comprised of L-amino acids and D-nucleotides respectively. The other point groups all contain either mirror planes of reflection and/or centres of inversion, and so can occur only in crystals of racemic mixtures or centrosymmetric structures.

Crystallographic point groups are identified by the symmetry of the DIFFRACTION pattern; for the 11 enantiomorphic point groups this is unambiguous, but additional information is needed for the unique identification of the remaining 21 point groups.

International Tables for X-ray Crystallography (1992) Vol. A Space-Group Symmetry (Hahn, T., Ed.) (Kluwer Academic Publishers, Dordrecht).
Stout, G.H. & Jensen, L.H. (1989) *X-ray Determination: A Practical Guide* (John Wiley, New York).

point mutation A single base change in a DNA strand, for example a G residue altered to a T. Such a MUTATION may alter the identity of the CODON in which it lies thereby creating a MISSENSE MUTATION or NONSENSE MUTATION.

poison sequence Certain DNA sequences that cannot be cloned in *Escherichia coli* as they cause cell death.

polar coordinate model Model proposed to explain the way in which information specifying a cell's position (and hence state of differentiation) in the adult is organized during development. It suggests that this 'POSITIONAL INFORMATION' occupies a two-dimensional grid and that component cells are assigned coordi-nates ('positional values') along the two axes (proximo-distal and circumferential in the case of limbs). This model can account for the types of patterns resulting from grafts during development or REGENERATION. *See also*: INTERCALARY REGENERATION; PATTERN FORMATION.

polar direction A direction in a crystal which is unique: the action of the SYMMETRY elements does not produce any direction equivalent to it. Thus a polar POINT GROUP contains one or more polar directions, for example the DYAD AXIS in point group p2, whereas a nonpolar point group contains no polar directions, for example point group p32 where the orthogonal triad and dyad axes act upon other so that no direction is unique.

polar fibres *See*: MITOSIS.

polar granules Specialized organelles, localized to the posterior pole of insect eggs, which are required for the formation of the germ-line pole cells. Transplantation of posterior pole cytoplasm containing polar granules to other positions in the egg can lead to the ectopic specification of germ-line cells. This experiment provided strong support for the concept of CYTOPLASMIC DETERMINANTS already localized in the egg before fertilization. Several MATERNAL EFFECT genes of the 'grandchildless' class in *Drosophila* are known to be required for the formation of polar granules and the protein product of the gene *vasa* is localized to the granules.

polar lobe *See*: MOLLUSC DEVELOPMENT.

polar side chains The AMINO ACIDS with polar side chains are serine, threonine, cysteine, asparagine, glutamine, histidine, tyrosine, and tryptophan (see Table A5 and Fig. A22). All these side chains can form HYDROGEN BONDS. Cysteine makes only occasional hydrogen bonds, through its S atom. If polar side chains are internal within a protein molecule they usually form hydrogen bonds with each other or with the polypeptide backbone.

polarity (1) Of egg or embryo. Asymmetry about an embryonic axis revealed by any regional difference in cytoarchitecture or developmental potential within an egg or embryo (e.g. animal–vegetal polarity which can take the form of an asymmetric distribution of yolk and organelles, etc. between animal and vegetal poles). *See also*: AMPHIBIAN DEVELOPMENT; ANTERIOR-POSTERIOR AXIS; DORSO-VENTRAL AXIS; DROSOPHILA DEVELOPMENT; MATERNAL CONTROL OF DEVELOPMENT; ZONE OF POLARIZING ACTIVITY.
(2) Of a cell. *See*: EPITHELIAL POLARITY.
(3) Of microfilaments and microtubules. *See*: MICROFILAMENTS; MICROTUBULES.

poliovirus Virus of the genus Enterovirus of the family PICORNA-VIRIDAE. It is a small icosahedral nonenveloped RNA virus (22–30 nm in diameter). Each capsid contains a (+)-sense single-stranded RNA ($M_r \sim 2.5 \times 10^6$) of 7500 nucleotides with a 5′ VPg protein cap and a 3′ poly(A) tail. The virus mRNA is translated into a POLYPROTEIN which is cleaved to produce four capsid

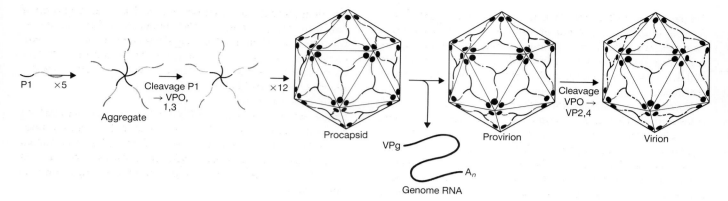

Fig. P48 Poliovirus capsid proteins are derived by sequential cleavage from the precursor protein P1. Five copies of P1 aggregate to form a pentamer followed by cleavage of P1 into P0 (precursor to VP2 and VP4), VP1, and VP3, and aggregation of 12 copies of the pentamer to form a procapsid. Genomic RNA is incorporated into the procapsid to form the provirion. *In situ* cleavage of VP0 to VP2 and VP4 produces the mature poliovirus particle.

proteins (VP1, VP2, VP3 and VP4) and the viral RNA polymerase, VPg protein and the viral protease. Assembly of the virion takes place in a series of steps in which precursor proteins aggregate and are cleaved *in situ* (Fig. P48) (*see also* Plate 7*c*, *d*). *See*: ANIMAL VIRUSES; ANIMAL VIRUS DISEASES.

poly(A) tail A homopolymer of adenosine monophosphate of between 40 and 200 nucleotide residues which is added in the nucleus to the 3′ end of most eukaryotic mRNAs after TRANSCRIPTION. The poly(A) tail itself is not encoded in the DNA but the site of its attachment is signalled by a POLYADENYLATION site. *See*: RNA PROCESSING.

polyacrylamide gel electrophoresis (PAGE) *See*: ELECTROPHORESIS.

polyadenylation The post-transcriptional addition in the nucleus of a sequence of adenine nucleotides (the POLY(a) TAIL) to the 3′ end of most eukaryotic mRNAs. The site of polyadenylation is signalled by a specific sequence in the RNA. *See*: RNA PROCESSING.

polyadenylation site mutant A mutation of the consensus sequence required for addition of poly(A) to the 3′ end of mature mRNA (*see* RNA PROCESSING) and which results in premature mRNA degradation. A mutation of this type in the α_2-globin gene is an important cause of HAEMOGLOBIN H disease in the Middle East.

polyamines (Pas) In plant physiology, Pas refer principally to the diamine putrescine (Put), the triamine spermidine (Spd), and the tetra-amine spermine (Spm), which have growth regulating activities at concentrations of $\sim$1 mol m^{-3}. Pas are claimed to be ubiquitous in plants and derived from arginine or ornithine via decarboxylase and propylaminotransferase activities. Pas are essential for normal growth and may also exert regulatory effects on

cell division, leaf senescence, and ETHYLENE production and action. The role of Pas in plant development is frequently probed using α-difluoromethylornithine (DFMO) and α-difluoromethyl-arginine (DFMA) which block ornithine or arginine decarboxylases and thus inhibit Pa biosynthesis. Polyamines are detected in nmol per gram fresh weight amounts by fluorescence of dansylated derivatives following high performance liquid CHROMATOGRAPHY or thin layer chromatography separation of perchloric acid extracts.

polyamine site Modulatory site on the NMDA receptor where the binding of polyamines such as spermidine, spermine and putrescine enhances the binding of channel-blocking ligands such as MK-801 and TCP, suggesting that this site potentiates the action of NMDA receptor agonists. *See*: EXCITATORY AMINO ACID RECEPTORS.

polycentric chromosome Chromosome with multiple CENTROMERES.

polycistronic Applied to mRNA transcribed from bacterial operons which encodes several different polypeptides that are then translated sequentially from the mRNA. *See*: BACTERIAL GENE EXPRESSION; PROTEIN SYNTHESIS.

polyclone The founding group of cells whose descendants form a developmental COMPARTMENT in *Drosophila*.

Polycomb (Pc) locus Genetic locus in *Drosophila melanogaster*. The *Pc* gene was originally identified as a mutation which gave rise to additional sexcomb bristles on the legs of male flies. More extensive examination of various alleles of *Polycomb* has shown that it is responsible for maintaining the orderly expression of genes in the ANTENNAPEDIA and BITHORAX COMPLEXES (ANT-C and BX-C). Genes in the ANT-C and BX-C are thought to be crucial to the determination of the identities of segments and imaginal disks, in that inappropriate expression of these genes causes homeotic transformations of adult structures (*see* HOMEOTIC GENES). These genes are initially expressed in the blastoderm in particular regions under the control of the SEGMENTATION GENES, which are transiently expressed TRANSCRIPTION FACTORS (*see* DROSOPHILA DEVELOPMENT). In *Polycomb* mutants, the pattern of ANT-C and BX-C gene expression in the blastoderm is initially normal, but around the time of GERM BAND extension, when the protein products of the segmentation genes cease to be expressed, it becomes erratic: in the most extreme *Pc* alleles, all of the ANT-C and BX-C genes are turned on in all segments.

These observations have led to the hypothesis that *Pc* participates in the maintenance, as opposed to the initiation, of gene expression. Very little is known about how this kind of gene regulation works, although the molecular characterization of the *Pc* gene and its protein product has been illuminating. *Pc* protein binds to POLYTENE CHROMOSOMES at more than 60 different sites, including many of the gene loci which it has been shown to regulate genetically. In addition, *Polycomb* protein will bind ectopic copies of a 4-kb regulatory DNA sequence from the ANT-C. Polycomb protein does not appear to bind DNA directly but it has been suggested that it may participate in a heterochromatin-like complex.

polycyclic aromatic hydrocarbons (PAH) Compounds of hydrogen and carbon arranged in multiple, unsaturated (aromatic) rings. They are widespread in nature and arise by heating or burning organic matter. They include many carcinogens: benzo[a]pyrene and 7,12-dimethylbenz[a]anthracene are well-known examples. PAHs often do not react directly with DNA, but are carcinogenic because they are converted into highly reactive derivatives by metabolic inactivation. After entry into the body they are broken down and converted into hydrophilic derivatives by microsomal mixed function oxidases (CYTOCHROMES P450) in the liver by steps that include oxidation, hydroxylation, and conjugation with water-soluble substances such as glucuronic acid. Some intermediates in the detoxification process are unstable and react with DNA to form mutagenic adducts. *See*: CHEMICAL CARCINOGENS AND CARCINOGENESIS; DNA REPAIR.

polycystic kidney disease A genetically heterogeneous group of conditions associated with cyst formation in the kidneys, leading to progressive renal failure. The commonest type, AUTOSOMAL DOMINANT polycystic kidney disease type I is also associated with cysts in the liver and berry aneurysms of the cerebral arteries; the defect is located on human chromosome 16.

polycythaemia An abnormally increased red cell mass. It usually develops in adult life either as a primary abnormality (polycythaemia rubra vera), or secondary to other disease. Rare genetic causes include HAEMOGLOBIN variants and methaemoglobinaemia (*see* HAEMOGLOBIN M).

Polydnaviridae Family of enveloped insect viruses. Each particle contains many double-stranded DNAs of variable M_r (total M_r $80 \times 10^6 - 170 \times 10^6$).

polydystrophy A condition characterized by the presence of many congenital abnormalities of the connective tissues.

polyethylene glycol (PEG) Polymer used in the form M_r 7000–9000 for a range of procedures in molecular and cell biology, including precipitation of viruses and nucleic acids, induction of cell fusion, and (together with Ca^{2+}) facilitation of the TRANSFORMATION of plant and animal cells (*see* GENETIC ENGINEERING; PLANT GENETIC ENGINEERING; SOMATIC CELL HYBRIDS; TRANSFECTION). It is also used as an additive in CRYSTALLIZATION in wider size range M_r 400–20 000.

polygalacturonase Enzyme (EC 3.2.1.15) from plant cells which is responsible for degradation of cell wall matrix carbohydrates (*see* PLANT CELL WALL MACROMOLECULES). It catalyses the random hydrolysis of 1,4-α-D-galacturosiduronic linkages in pectate and other GALACTURONANS. It is also present in other organisms (e.g. fungi and bacteria).

polygenic Determined by several genes; applied to characters such as height which although having a measure of heritability, are continuously variable in the population and do not show a pattern of inheritance typical of a single-gene simple Mendelian trait. Such characters are presumed to be determined by the additive effects of several different genes usually also with an environmental contribution. Examples include height, intelligence, plasma lipid levels, and cleft lip. *See*: CONTINUOUS VARIATION; MENDELIAN INHERITANCE.

polyimmunoglobulin receptor (polyIg receptor) A receptor for polymeric immunoglobulin molecules, such as polyIgA, present on the basolateral surface of epithelial cells of exocrine glands (e.g. of the gut mucosa and salivary glands). After binding of immunoglobulin the receptor–ligand complex undergoes ENDOCYTOSIS and is transported from one face of the epithelium to the other by TRANSCYTOSIS. During this process, the ligand-binding portion of the receptor is cleaved proteolytically to form the SECRETORY COMPONENT which remains associated with the polyIg when it is secreted. The polyIg receptor is a 773-amino acid polypeptide with a single membrane-spanning region, and is a member of the IMMUNOGLOBULIN SUPERFAMILY, the extracellular portion comprising five immunoglobulin-like domains.

polylinker A synthetic DNA fragment containing multiple RESTRICTION SITES. Polylinkers are inserted into VECTORS in order to provide a range of cloning sites — restriction sites at which DNA with compatible cohesive ends can be inserted.

polymerase(s) Nucleotidyltransferases which catalyse the template-dependent synthesis of polynucleotide chains from ribonucleoside or deoxyribonucleoside 5'-triphosphates. They direct the sequential addition of nucleotide residues to one end (the 3' end) of a polynucleotide by catalysing the formation of a phosphodiester bond between the free 3'OH of the sugar moiety of the terminal nucleotide residue and the innermost P atom (5'P) of a free nucleoside triphosphate, forming the covalently linked sugar–phosphate backbone of nucleic acids. The different types of template-dependent polymerase fall into four main classes: DNA polymerases (DNA-dependent DNA polymerases, EC 2.7.7.7) which catalyse DNA REPLICATION; RNA polymerases (DNA-dependent RNA polymerases, EC 2.7.7.6) which catalyse the TRANSCRIPTION of DNA into RNA; RNA replicases (RNA-dependent RNA polymerases, EC 2.7.7.48) which catalyse replication of viral RNA genomes; and reverse transcriptase (RNA-dependent DNA polymerase, EC 2.7.7.49) which catalyses the synthesis of RETROVIRUS DNA provirus from viral genomic RNA. *See*: DNA; DNA POLYMERASES; POLYMERASE CHAIN REACTION; REVERSE TRANSCRIPTASE; RNA; RNA POLYMERASES.

Polymerase chain reaction

THE polymerase chain reaction (PCR) is an *in vitro* method for amplifying specific DNA sequences. Starting with trace amounts of a particular nucleic acid sequence from any source, PCR enzymatically generates millions or billions of exact copies, thereby making genetic analysis of tiny samples a relatively simple process. PCR was invented by Kary Mullis of Cetus Corporation in 1983, and is now a widely used technique in molecular biology, with direct applications in the fields of genetic research, medical diagnosis, and forensic science.

PCR involves a repetitive series of temperature cycles with each cycle comprising three stages: denaturation of the template DNA at >91°C to separate the strands of the target molecule, then cooling to ~50°C to allow annealing to the template of single-stranded oligonucleotide PRIMERS which are specifically designed to flank the region of DNA of interest, and finally, extension of the primers by DNA polymerase at about 72°C (Fig. P49) [1]. Thus, one cycle of PCR doubles the number of target DNA molecules, since the newly synthesized strands can themselves act as templates in the next round of amplification (Fig. P50). This logarithmic growth in product implies that 20 PCR cycles could theoretically result in a million-fold amplification of a DNA fragment whose length is defined by the 5′ ends of the primers. In practice however, the efficiency of the amplification reaction is inversely proportional to the length of the segment amplified: most reactions are designed to amplify segments of less than 2 kb in size, but amplification of fragments of up to 12 kb long has been achieved. Furthermore, the exponential generation of product is attenuated in the later PCR cycles owing to the 'plateau

effect' in which either the reactants become exhausted or the presence of more than 0.3–1 pmol of specific product inhibits subsequent polymerization reactions. In addition, highly degraded samples can display a lag phase before the exponential phase: if no intact template molecules exist, the two primers can be sequentially extended from smaller fragments of overlapping sequence during the first few PCR cycles by a process known as 'jumping PCR', until the 3′ ends of the extended primers overlap and enable exponential amplification to ensue [2].

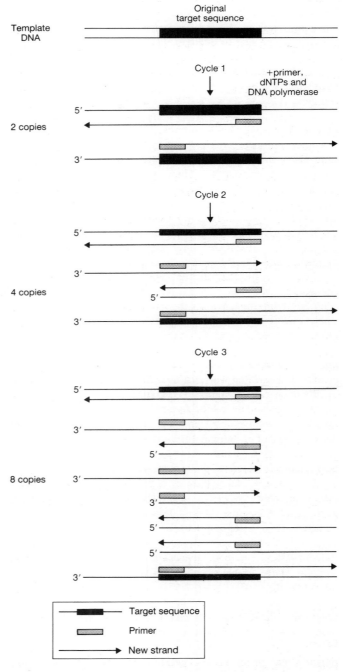

Fig. P50 Each temperature cycle results in a net doubling of template molecules to be amplified in the next cycle of the reaction.

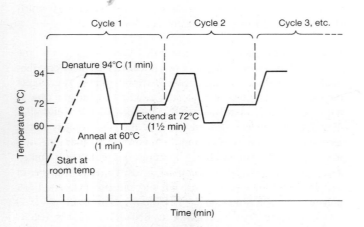

Fig. P49 Each cycle of the PCR reaction comprises three stages at different temperatures: denaturation, annealing of the primers, and the extension reaction.

Typically, the PCR reaction is performed in a volume of 50–100 µl overlain with oil to prevent evaporation during the temperature cycling. The reaction mix comprises about 20 nmol of each of the four deoxynucleoside triphosphates (dATP, dCTP, dGTP, dTTP), 10–100 pmol of primers, magnesium salts and buffers, DNA polymerase, and the target DNA sample.

The first experiments with PCR utilized the KLENOW FRAGMENT of *Escherichia coli* DNA polymerase I to extend from the 3′ end of the annealed primers by incorporation of free deoxynucleotides present in the reaction mix. However, fresh enzyme has to be laboriously added after each cycle because it is irreversibly inactivated by the denaturation step. The enzymes which are now universally used are thermostable, such as the Taq polymerase isolated from the extreme thermophilic bacterium *Thermus aquaticus*, and can withstand the repeated exposures to high temperature required for strand separation in a typical PCR reaction. Hence Taq polymerase is added once only before temperature cycling and the process is automated by the use of microcomputer-controlled 'intelligent heating blocks' or water baths. Each primer pair has a reaction optimum for maximum target yield with minimal production of spurious fragments caused by adventitious false priming at non-target sites. These conditions are determined empirically by altering the magnesium, primer, and enzyme concentrations and adjusting the temperature cycling parameters. Ideally, between-primer sequence complementarity should be avoided to prevent formation of the 'primer-dimer' artefact, whereby a primer concatenate is amplified in preference to the desired target sequence. Computer programs have been described which aid in primer design [3].

One of the main advantages of PCR is the exquisite sensitivity of the technique. However this sensitivity is also its major drawback, making it highly susceptible to contamination. Contamination can come from other samples which are being co-processed, from the operator, the environment, or from PCR products which have been amplified previously. Use of appropriate positive and negative controls highlights whether a contamination problem exists, and the chances of it occurring can be minimized by adopting a carefully thought-out experimental regime. Physical separation of sample preparation and the setting-up of PCR reactions from the subsequent handling and analysis of the amplification products is essential. Ideally, they should be carried out in different laboratories, but where space is at a premium separate positive displacement hoods may suffice. Separate sets of pipettes, preferably positive displacement, should be used for sample preparation, setting up the PCR reaction, and for postamplification manipulations. Operators should wear gloves at all times and replace them frequently. Under extreme circumstances, it has even been necessary for operators to wear face masks and hair nets to prevent DNA contained in their skin flakes from contaminating the process.

Research applications

Genomic cloning

Before the invention of PCR, many of the preparative and analytical processes used in nucleic acid research were labour intensive and time consuming. For instance, in order to compare a well-characterized gene with an uncloned mutant form, it was necessary to first construct a genomic library of the mutant organism (*see* DNA LIBRARIES), then isolate the required clone by HYBRIDIZATION. Only then could the mutation be characterized by DNA SEQUENCING. With PCR however, genes can be synthesized *in vitro*: large quantities of DNA template can be amplified directly from trace amounts of genomic DNA, using suitable primers designed from predetermined sequence data from the native gene. The DNA synthesized from PCR can then be sequenced directly. Alternatively, by incorporating suitable RESTRICTION SITES into the primer sequences, the PCR product can be cloned into an appropriately cleaved VECTOR to facilitate further manipulations.

Strategies also exist for amplifying and sequencing DNA segments whose flanking DNA sequences are unknown, by a process known as reverse PCR (Fig. P51). Such an approach has potential use in CHROMOSOME WALKING and involves cutting the DNA template with a RESTRICTION ENZYME at positions external to the region to be characterized. The fragment is then circularized by ligation and amplified using primers which are complementary to the known segment but are positioned with their 3′ ends pointing in opposite directions (Fig. P51). This results in amplification of the unknown sequences which have been previously joined in the ligation stage.

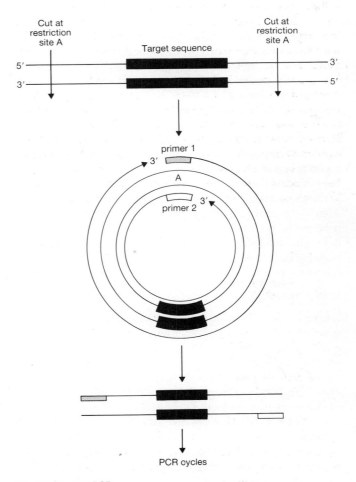

Fig. P51 Reverse PCR.

Generating template for sequencing

Modified PCR techniques have proved ideal for generating DNA template for sequencing and have been directly applied to bacterial colonies and phage plaques, thereby obviating any DNA preparation [4]. One approach is asymmetric PCR, in which unequal amounts of the two primers are used. During the first 20 or so cycles double-stranded DNA is produced in an exponential fashion but beyond this the primer at lower concentration becomes depleted and subsequent cycles generate predominantly single-stranded DNA by extension from the nonlimiting primer. This single-stranded DNA accumulates linearly and is more amenable to sequencing by the Sanger method than the corresponding double-stranded molecule [5].

Single-stranded DNA template for sequencing can also be generated by strand-specific elution from a solid support: one of the PCR primers is BIOTINYLATED at the 5′ end, and, following symmetric amplification, the PCR product is immobilized by specific binding of the biotin group to streptavidin-coated magnetic beads. The nonbiotinylated strand is removed by alkali treatment and the immobilized strand can then be sequenced directly. This approach is amenable to automation as the bound DNA template can be moved to the side of the reaction vessel by a magnet during washing stages.

RNA analysis

RNA molecules can also be analysed with PCR by first converting them to cDNA by reverse transcription using the enzyme REVERSE TRANSCRIPTASE. The resulting DNA transcript can then be amplified by the normal PCR process. With this approach, even rare mRNA molecules can be detected and monitored. If only part of the amino-acid sequence of a protein is known, these data can be used to synthesize degenerate primers which can amplify a segment of cDNA. This in turn can then be used as a probe to screen a cDNA library for the desired full-length clone. Alternatively, if only one end of a sequence is known, the cDNA generated using the known primer can be tailed with a homopolymer (poly(G)) sequence to which an 'anchor' primer that includes a poly(C) stretch can bind and act as the 'unknown' PCR primer. This strategy has been used to amplify T CELL RECEPTOR transcripts, the 5′ ends of which encode the variable (and therefore unknown) region of the receptor (*see* T CELL RECEPTOR GENES).

In vitro mutagenesis

Studies on the effect of altering DNA sequence with regard to its structure and function is another area of application for PCR, by exploiting the ability of the oligonucleotide primers to bind to the template DNA even when they are not an exact sequence complement. Hence, a product can be rapidly generated into which has been incorporated a base substitution, deletion, or addition within the sequence defined by the primers. The function of this modified form can then be compared with the native molecule. This is especially valuable in the study of PROTEIN–NUCLEIC ACID INTERACTIONS, such as in transcription complexes (*see* CONTROL OF EUKARYOTIC TRANSCRIPTION; TRANSCRIPTION; TRANSCRIPTION FACTORS) [6].

Forensic applications

Forensic investigations often involve the analysis of biological material deposited at the scene of a crime such as hairs, blood, saliva, and semen stains. The technique of DNA profiling has revolutionized forensic biology (*see* DNA TYPING: FORENSIC APPLICATIONS) but requires large amounts (>100 ng) of DNA which must also be relatively undegraded. The PCR reaction is not limited by these constraints and a number of approaches have been developed. The first to be used in casework was the DOT BLOT analysis of the highly polymorphic DQα gene of the human leukocyte antigen (HLA) complex (*see* MAJOR HISTOCOMPATIBILITY COMPLEX). This test detects and distinguishes between six of the eight possible DQα alleles using allele-specific oligonucleotide (ASO) probes. In essence this method comprises amplification, with BIOTINYLATED primers, of a 242-bp fragment which spans the region of sequence variation. This is then characterized using a reversed dot-blot format in which the PROBE and not the amplified DNA is immobilized on a nylon membrane. Hybridization is detected by a colour reaction mediated by attachment of a streptavidin/HORSERADISH PEROXIDASE conjugate to the biotinylated PCR product.

Another forensic PCR test is amplification of loci which vary in length according to the number of repeated sequences they contain (VARIABLE NUMBER TANDEM REPEAT LOCI (VNTR loci)). Analysis simply involves resolving and visualizing the products of PCR on agarose or acrylamide gels (*see* ELECTROPHORESIS). This is comparable to the original DNA fingerprinting technique (*see* DNA TYPING) and is likely to be more informative than the dot-blot method.

An alternative to sizing VNTR loci is to assay their patterns of variant repeat units, by utilizing PCR primers that selectively anneal within the tandem array of repeat sequences. This process is called minisatellite variant repeat (MVR) mapping and generates a highly informative digital DNA profile from as little as 10 ng DNA [7].

A more sensitive approach is to amplify and analyse the highly variable noncoding region of mitochondrial DNA, by either direct sequencing or dot-blot analysis [8]. Human cells contain 1000–10 000 mitochondrial DNA copies, so mitochondrial amplification is ideal for the analysis of highly degraded samples.

Molecular palaeontology

PCR technology has already founded new scientific fields of molecular archaeology and palaeontology with direct sequencing being applied to the analysis of preserved remains from, for example, a 7000-year-old human brain and even a 20-million year-old plant fossil [9], thereby providing new insights into evolutionary history (*see also* MOLECULAR EVOLUTION; MOLECULAR PHYLOGENY).

Medical uses

Medical applications of PCR are broadly in the areas of research, genetic counselling, and clinical investigations. The first paper published on PCR, in 1985, described its use in detecting the sickle-cell mutation [1]. Tests have subsequently been developed for other haemoglobinopathies including β-thalassaemia (*see* HAE-MOGLOBIN AND ITS DISORDERS), as well as for other genetic diseases such as PHENYLKETONURIA, DUCHENNE MUSCULAR DYS-TROPHY [10], and CYSTIC FIBROSIS. PCR is ideally suited to PRENATAL DIAGNOSIS since it is rapid and only trace amounts of foetal DNA are required for amplification. Sexing embryos generated by *in vitro* fertilization is now also possible by taking a single cell from a blastomere comprising 6–10 cells, and amplifying a Y chromosome-specific repetitive element. Such a diagnostic procedure is invaluable for couples at risk of transmitting diseases such as X-linked mental retardation.

PCR is ideal for the detection of pathogens because it is possible to amplify a specific sequence present at low concentration within a complex DNA mixture. With regard to detection of viral infections, a test for HIV-1 (*see* IMMUNODEFICIENCY VIRUSES) has been developed which can detect the presence of the virus earlier than any discernible antibody response in the infected person. Likewise, PCR has been used in the detection of herpes, hepatitis, and cytomegaloviruses, as well as the human papilloma viruses which are associated with cervical cancer.

Alternative amplification methods

Variations on PCR include the use of a chimaeric primer which includes the phage T7 RNA polymerase recognition sequences so that after multiple PCR cycles the signal can be further boosted by generating RNA transcripts from the PCR products.

A transcription-based amplification system (TAS) has also been developed [11]. Each cycle of TAS involves two stages: cDNA synthesis and RNA transcription. cDNA synthesis is performed using an RNA template, reverse transcriptase and a chimaeric primer which is complementary to the RNA sequence and also contains an RNA polymerase binding site. Up to 100 copies of RNA are generated and passed on to the RNA transcription step which increases the copy number of the template for cDNA synthesis in the next cycle of TAS. Thus, unlike PCR, few cycles are required to generate a million copies of the target molecule.

Ligation chain reaction (LCR) is another PCR alternative [12]. A thermostable LIGASE is used to specifically link two adjacent oligonucleotides which hybridize to a complementary target with perfect base-pairing at the junction. These oligonucleotide dimers can be exponentially amplified by repeated thermal cycling in the presence of a second set of adjacent primers complementary to the first. This approach enables single base pair differences to be detected at the ligation junction and the entire process can be automated.

Another variation is oligonucleotide ligation assay (OLA) which uses only one set of primers and results in a linear increase in product per cycle. Product capture and detection can be automated by attaching biotin to one primer pair and a suitable reporter group to the other [13].

K.M. SULLIVAN

1 Saiki, R.K. et al. (1985) Enzymatic amplification of β-globin genomic sequences and restriction site analysis for diagnosis of sickle cell anaemia. *Science* **230**, 1350–1354.
2 Paabo, S. et al. (1989) Ancient DNA: extraction, characterization, molecular cloning, and enzymatic amplification. *Proc. Natl. Acad. Sci. USA* **86**, 1939–1943.
3 Rychlik, W. et al. (1990) Optimization of the annealing temperature for DNA amplification in vitro. *Nucleic Acids Res.* **18**, 6409–6412.
4 Gussow, D. & Clackson, T. (1989) Direct clone characterization from plaques and colonies by the polymerase chain reaction. *Nucleic Acids Res.* **17**, 4000.
5 Gyllensten, U.B. & Erlich, H.A. (1988) Generation of single-stranded DNA by the polymerase chain reaction and its application to direct sequencing of the HLA DQa locus. *Proc. Natl. Acad. Sci. USA* **85**, 7652–7656.
6 Higuchi, R. et al. (1988) A general method of in vitro preparation and specific mutagenesis of DNA fragments: study of protein and DNA interactions. *Nucleic Acids Res.* **16**, 7351–7367.
7 Jeffreys, A.J. et al. (1991) Minisatellite repeat coding as a digital approach to DNA typing. *Nature* **354**, 204–209.
8 Stoneking, M. et al. (1991) Population variation of human mtDNA control region sequences detected by enzymatic amplification and sequence-specific oligonucleotide probes. *Am. J. Hum. Genet.* **48**, 370–382.
9 Golenberg, E.M. et al. (1990) Chloroplast DNA sequence from a Miocene *Magnolia* species. *Nature* **344**, 656–658.
10 Chamberlain, J. et al. (1988) Deletion screening of the Duchenne muscular dystrophy locus via multiplex DNA amplification. *Nucleic Acids Res.* **16**, 11141–11156.
11 Kwoh, D.Y. et al (1989) Transcription-based amplification system and detection of amplified human immunodifficiency virus type I with a bead-based sandwich hybridization format. *Proc. Natl. Acad. Sci. USA* **86**, 1173–1177.
12 Barany, F. (1991) Genetic disease detection and DNA amplification using cloned thermostable ligase. *Proc. Natl. Acad. Sci. USA* **88**, 189–193.
13 Landegren, U. et al. (1988) DNA diagnostics — molecular techniques and automation. *Science* **242**, 229–237.

polymorphism The existence of a character in two or more variant forms in a population and where the least common form is present in more than 1% of individuals. *See*: BALANCED POLYMORPHISM; DNA TYPING; GENETIC POLYMORPHISM; HUMAN GENE MAPPING; HYPERVARIABLE DNA; RFLP; VARIABLE NUMBER TANDEM REPEATS.

polynucleotide chain Chain of nucleoside 5'-monophosphate residues linked through phosphodiester bonds from the 3'-OH of one sugar to the 5'-OH of the next. An RNA chain is composed of ribonucleotides, a DNA strand of deoxyribonucleotides. *See also*: NUCLEIC ACID STRUCTURE.

polynucleotide kinase Transferase enzyme that transfers a phosphoryl group from ATP to 5' terminal hydroxyl groups on DNA and RNA chains pretreated with phosphatase to remove the 5' phosphate groups. Polynucleotide kinase purified from *Escherichia coli* infected with T4 phage is generally used. Its main use is to end label nucleic acid fragments with radioactive phosphorus (^{32}P) derived from γ-^{32}P-ATP. *See*: END-LABELLING.

polyoma virus *See*: PAPOVAVIRUSES.

polypeptide A chain of amino acid residues linked through peptide bonds between the α-carboxyl carbon of one amino acid residue and the α-nitrogen of the next. *See*: PROTEIN STRUCTURE.

polypeptide backbone In a protein, the chain of regularly repeating

from which the side chains (R) project. *See*: PROTEIN STRUCTURE.

polyphyletic Composed of several independent lines of descent; applied to a taxonomic group. *Cf.* MONOPHYLETIC.

polyploid Organism in which the somatic cells contain more than two sets of a haploid genome, as in TRIPLOIDS ($x = 3$), or TETRA-PLOIDS ($x = 4$). Many plants, especially crop plants, are polyploids, for example, bread wheat (*Triticum aestivum*), which is hexaploid ($x = 6$). Naturally occurring polyploids such as bread wheat generally contain more than one type of haploid genome and have derived from interspecies crosses. Some animals are polyploids.

polyposis coli, genetic A heterogeneous disorder, the commonest of which (familial adenomatous polyposis) is AUTOSOMAL DOMINANT with a frequency of one in 10 000. Multiple adenomatous polyps develop in the gastrointestinal tract from the second decade onwards; prophylactic total colorectomy is indicated because of the high chance of malignant transformation in one or more polyps. The genetic defect is located on human chromosome 5 and may also be implicated in sporadic tumours of the colon. *See also*: TUMOUR SUPPRESSOR GENES.

polyproline helix Type of helical SECONDARY STRUCTURE formed by poly-L-proline. It is less tightly coiled than the normal α-helix and lacks the hydrogen bonds between successive turns of the helix. Polyproline helices are stabilized by steric repulsion between the rings of the proline residues. Polyproline-type helices are found in procollagen, which has a repeat unit of -Gly-Pro-Pro-. *See*: EXTRACELLULAR MATRIX MOLECULES.

polyprotein A protein which is processed after synthesis into multiple copies of the same peptide or into different polypeptides or proteins. Many eukaryotic polypeptide growth hormones and other small peptide signalling molecules are synthesized from precursors of this type. The precursor to epidermal growth factor (EGF), for example, is a transmembrane protein of 1168 amino acids, from which the 53 amino acid EGF is cleaved. The α mating factor in the yeast *Saccharomyces* is produced from a polyprotein containing multiple copies of the sequence. The polyprotein product of the pro-opiomelanocortin gene can be processed in various ways to yield ADRENOCORTICOTROPIN, β-ENDORPHIN, and melanocyte-stimulating hormones (*see* EUKARYOTIC GENE STRUCTURE; OPIOID PEPTIDES AND THEIR RECEPTORS). Cleavage of polyproteins usually occurs in secretory vesicles at basic residues. Polyproteins are also characteristic of some groups of viruses (*see* ANIMAL VIRUSES; NEPOVIRUS GROUP; PLANT VIRUSES).

polyribosome, polysome Messenger RNA that is being translated simultaneously by several RIBOSOMES. *See*: PROTEIN SYNTHESIS.

polysaccharide Any polymeric SUGAR, either as a pure saccharide or as a glycoconjugate (*see* GLYCANS).

polyspermy The situation when more than one sperm FERTILIZES an egg and contributes its DNA to the formation of a CLEAVAGE NUCLEUS. This is incompatible with normal development and therefore 'blocks to polyspermy' have evolved. 'Fast' blocks include the restriction of sperm entry to one channel in the egg which becomes congested with cortical granule contents and hence impenetrable as soon as a sperm has entered, and the rapid (but reversible) sperm-induced membrane DEPOLARIZATION which occurs in many eggs. Additional more persistent 'slow' blocks also exist and in the majority of eggs these occur as a result of modifications to egg coverings induced by cortical granule contents. *See also*: CORTICAL REACTION.

polytene chromosomes Giant chromosomes found in secretory tissue (studied in the salivary glands of *Drosophila* and other insects especially), formed by many rounds of chromosome replication without separation. As a consequence the chromosomes elongate and thicken. *See*: DROSOPHILA.

polyubiquitination The addition of multiple UBIQUITIN molecules to a protein. *See*: POST-TRANSLATIONAL MODIFICATION; PROTEIN DEGRADATION.

POMC Pro-opiomelanocortin, the polypeptide precursor for β-endorphin and ADRENOCORTICOTROPIC HORMONE. *See*: OPIOID PEPTIDES AND THEIR RECEPTORS.

Pompe disease An AUTOSOMAL RECESSIVE, severe GLYCOGEN storage disease due to deficiency of acid α-glucosidase, which catalyses the breakdown of glycogen to glucose in LYSOSOMES (type II glycogenosis). Heart failure and/or hypotonia usually lead to death in the first year (classical Pompe disease), but milder forms of the enzyme deficiency compatible with survival into adulthood are known. *See also*: GLYCOGENOSES.

pons The region of the brainstem connecting the midbrain and the medulla oblongata (*see* CENTRAL NERVOUS SYSTEM).

population genetics The study of the distribution and dynamics of genetic variation among individuals in free-living populations. Fundamental to most aspects of population genetics is the concept of allele frequency (GENE FREQUENCY), the prevalence of variation at specific loci within the population. Population genetics makes use of genetic markers which are selectively neutral in order to obtain an unbiased sample of the genetic variation present at all loci. Methods frequently employed in population genetics include ALLOENZYME electrophoresis, quantitative genetics, natural phenotypic variation, and DNA-based methods such as RFLPs and variation in mitochondrial DNA (*see* DNA TYPING). Population genetic studies are often important parts of larger studies of animal and plant ecology, distribution, and evolution.

Cooke, F. & Buckley, P.A. (1987) *Avian Genetics: A Population and Ecological Approach* (Academic Press, London).

porins Transmembrane proteins in the outer membrane of Gram-negative bacteria which mediate the passive transport of ions and small molecules. Similar proteins are found in the outer mitochondrial membrane. *See*: BACTERIAL ENVELOPES; MEMBRANE TRANSPORT SYSTEMS; MITOCHONDRIAL BIOGENESIS.

porphyria A group of disorders of PORPHYRIN metabolism. Divided into hepatic and erythropoietic types according to the site of the enzyme defect; clinical features vary widely. The commonest conditions in the United Kingdom are acute intermittent porphyria (AUTOSOMAL DOMINANT with low PENETRANCE), associated with gastrointestinal and neuropsychiatric disturbances often precipitated by drugs; and porphyria cutanea tarda, commonly related to excessive alcohol intake.

porphyrin The generic name for a group of pyrrole ring-containing organic compounds widely distributed in plants and animals, which serve as cofactors in many proteins. The porphyrin structure contains a central 16-membered ring formed from 12 carbon and four nitrogen atoms, contributed by four pyrrole rings. The four nitrogen atoms on the inside of the ring together bind a metal atom. The best-known porphyrins are HAEM in which the metal is iron, and CHLOROPHYLL in plants, in which it is magnesium.

position effect The effect on gene expression of a change in the location of a particular gene on the chromosome.

positional cloning A strategy for isolating a gene whose gene product is completely unknown which starts from a knowledge of its position on the chromosome. *See*: CHROMOSOME JUMPING; CHROMOSOME WALKING; HUMAN GENE MAPPING.

positional information The MORPHOGEN in a MORPHOGENETIC FIELD imparts positional information to the cells comprising that field. From this information, the cell is believed to derive information defining its position within the field in respect to the antero-posterior (rostro-caudal), proximo-distal, and dorso-ventral axes of the embryo. A corollary of this is that if one experimentally alters the value of the positional information of a cell, its FATE should be changed accordingly. The most famous example of the concept of positional information is the 'French Flag' model of L. Wolpert. In this model the French and American flags are considered as fields whose x and y axes define positions on the surface of each flag. The flags are initially white, but take on their natural colours during 'flag development'. The argument goes that a piece of French flag transplanted from the bottom right-hand corner (region destined to produce *red flag*) to the top left of the developing American flag (region destined to produce *stars*) will give rise to blue flag material (the colour of the equivalent region of the French flag). The transplant can only give rise to colours of the French flag, because of its genetic origin, but it takes its positional information from where it has been transplanted in the developing American flag. *See*: PATTERN FORMATION.

positive control Type of regulation of gene expression (generally used in connection with INDUCIBLE bacterial operons) in which a gene (or operon) is expressed only when a specific activator protein(s) binds to the promoter. *Cf.* NEGATIVE CONTROL.

positive strand RNA viruses Single-stranded RNA viruses in which the genome is equivalent to the mRNA. *See*: ANIMAL VIRUSES; PLANT VIRUSES.

positron emission tomography (PET) Positron emission tomography combines the principles of computerized tomography and radioisotope imaging to monitor changes in blood flow to different regions of the body. It is used particularly to study the localization of neuronal activity in the brain by studying changes in blood flow. Positron-emitting isotopes are injected or inhaled, and an X-ray source and detector are rotated around the head. Differences in radiodensity that are registered reflect the tissue distribution of the positron-emitting isotope.

post-mitotic neuron A differentiated NEURON, which does not divide further.

post-translational import Proteins destined for one of the various compartments of the eukaryotic cell may be targeted to their destination from the cytoplasm either during translation (co-translational import) or after it (post-translational import). Examples of the latter are import into the nucleus, mitochondria and peroxisomes. As transport into the nucleus occurs through a large nuclear pore, nuclear proteins can fold and assemble before import. For the other organelles, the protein may be preserved in a partially unfolded state by binding to a chaperone protein (*see* MOLECULAR CHAPERONES) before meeting its target membrane and being imported. In contrast, import into the endoplasmic reticulum (ER) is usually co-translational; however, some small proteins may enter the ER after translation. *See also*: PROTEIN TRANSLOCATION.

Post-translational modification of proteins

TEXTBOOKS represent proteins chemically as a series of amino acid residues linked by peptide bonds and, in discussing protein biosynthesis, they emphasize the process by which such a linear polypeptide is assembled (*see* PROTEIN STRUCTURE; PROTEIN SYNTHESIS). But in most cases, the formation of a functional protein requires chemical events distinct from, and subsequent to, the sequential construction of the POLYPEPTIDE CHAIN. Such chemical processes are described as post-translational modifications [1–7]. They may be modifications which introduce new functional groups into a protein, providing it with a chemical property absent from the side chains of the 20 genetically encoded amino acids. Or they may be reversible modifications which act to switch a protein between two distinct functional states. They may be modifications of the N-terminal free amino group, or of specific side chains, or even specific proteolytic cleavage of the polypeptide representing the genetically encoded translation product. They may be the products of specific enzyme activity, or arise from non-enzymatic processes. They may occur so early in the life of an individual protein molecule as to be more properly described as co-translational, rather than post-translational modifications, or they may mark late events leading to the protein's final degradation and turnover (*see* PROTEIN DEGRADATION). They may take place in the cytoplasm, or in a subsequent subcellular compartment, or extracellularly. It is clearly possible to classify and discuss post-translational modifications in terms of any one of these distinctions. The comprehensive listings in [1,2] use a classification by residue modified. Post-translational modifications will be discussed here in terms of their location in the cell and their place in a protein's 'life history'.

Modifications in the cytoplasm: synthesis, targeting and turnover

Although many of the most widely recognized post-translational modifications are characteristic of secretory or cell-surface proteins, most proteins, whatever their ultimate cellular destination, undergo some modification. For proteins synthesized completely within the cytoplasm, the earliest and most widespread are removal or modification of the N-terminal residue. Where the residue following the initiator methionine (Met) is relatively small, there is the strong possibility that the N-terminal Met will be removed cotranslationally by a specific methionine aminopeptidase. Whether or not this process has occurred, many cytoplasmic proteins are subject to N-acetylation on the N-terminal α-amino group; again the pattern of residues near the N-terminus of the protein determines whether this modification occurs. The functions of these modifications seem to relate to the rate of protein turnover; differences in sequence and modification at the N-terminus seem to correlate with propensity to degradation by specific pathways (see below); most stable cytoplasmic proteins are N-acetylated.

A further N-terminal modification found in proteins synthesized in the cytoplasm is N-acylation by a long-chain FATTY ACID — myristic acid. Such N-terminal myristoylation serves to anchor a cytoplasmic protein to the cytoplasmic face of the PLASMA MEMBRANE. Another modification leading to membrane attachment is acylation of an —SH group of a cysteine (Cys) residue, a reaction which seems to be limited to Cys residues close to the C-terminus of proteins. *Ras* proteins (*see* GTP-BINDING PROTEINS) and others are subject to this modification, which involves either a fatty acyl group (palmitoylation) or a prenyl (farnesyl or geranylgeranyl) residue. Many other modifications of side chains in cytoplasmic proteins are known, and several of these are known to have regulatory functions (see below). But for many, no clear function is yet established. A large number of protein methylases and demethylases are known to catalyse transfer of methyl groups from S-ADENOSYLMETHIONINE to the side chains of lysyl, arginyl, glutamyl, and aspartyl residues, and to remove them by hydrolysis. The existence of enzymes catalysing both methylation and demethylation suggests some regulatory role for these modifications, but the only well-established case is the role of carboxyl-group methylation in the adaptation phenomenon of BACTERIAL CHEMOTAXIS.

A final cytoplasmic modification of considerable significance is ubiquitinylation which acts as a marker, committing proteins to degradation (*see* PROTEIN DEGRADATION). Studies on the mechanism of an ATP-dependent proteolytic system from reticulocytes led to the discovery of a major pathway for selective protein degradation, and it was soon found that a heat-stable small polypeptide was essential for the action of this system. This small polypeptide (76 residues) is UBIQUITIN. It is transferred to substrate proteins to form isopeptide bonds between the C-terminal glycine of ubiquitin and ε-amino groups of lysyl residues in the substrate; further ubiquitin molecules can then be conjugated to Lys 48 of the bound ubiquitin. The basis of the specificity in selection of protein substrates for ubiquitinylation is clearly a key feature determining the relative turnover rates of proteins; to some extent it is determined by the N-terminal residue of the substrate protein (the N-end rule), but several parallel ubiquitinylation processes with different specificities may be operational (and, of course, there are other pathways for protein degradation).

Intracellular reversible modifications modulating protein function

The paradigm of a post-translational modification with a regulatory role is phosphorylation and its reversal, dephosphorylation (*see also* PROTEIN PHOSPHORYLATION). From the time of the initial demonstration of this modification and its role in regulating the activities of enzymes involved in GLYCOGEN degradation, it has provided the leading example of the complexity of regulation that can be achieved by post-translational modification. PROTEIN KINASES leading to phosphorylation of serine (Ser) and threonine (Thr) side chains were the first to be defined; many such kinases are known, each with distinct regulatory mechanisms and with substrate specificities ranging from highly selective to very broad. Among the commonest regulators of protein kinase activity are CYCLIC AMP and the combination of Ca^{2+} ions with the small protein calmodulin (*see* CALCIUM; CALCIUM-BINDING PROTEINS). The phosphorylation of Ser and Thr residues effected by protein kinases can be reversed by PROTEIN PHOSPHATASES and again many enzymes are known in this class. Further proteins act as inhibitors of protein phosphatases or kinases. Exquisite sensitivity and complexity of regulation can be achieved by cascades or networks of protein kinases, phosphatases and their inhibitors each with different regulators and specificities, acting on a target protein whose activity is modulated by conversion between the phosphorylated and dephosphorylated state. Target proteins include not only metabolic enzymes but structural proteins and proteins involved in cell signalling and the CELL CYCLE. A less well-characterized type of protein phosphorylation occurs on side chains of tyrosyl residues. Many cell-surface receptors and viral transforming proteins have tyrosine-specific protein kinase activity (*see* GROWTH-FACTOR RECEPTORS; ONCOGENES). These proteins frequently modify residues within the same polypeptide (autophosphorylation) but other substrates have been less easy to define and the functional significance of tyrosyl phosphorylation remains incompletely understood.

Up to 10% of cytosolic proteins are subject to phosphorylation; other regulatory modifications are known, but none are so widespread or so well characterized. The net charge on nuclear proteins such as HISTONES can be modified by acetylation, phosphorylation and by poly-ADP-RIBOSYLATION.

Acetylation of lysyl side chains of proteins is a widespread modification whose function is poorly characterized. In the case of histones, acetylation seems to modulate histone–DNA interaction

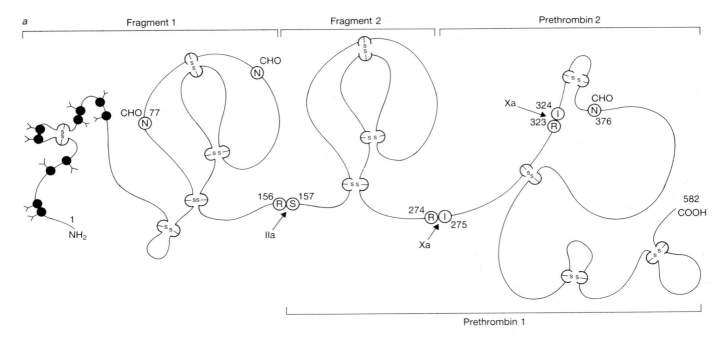

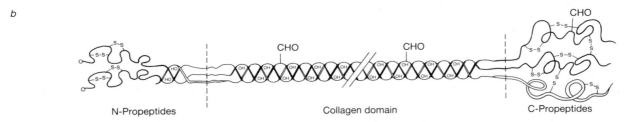

Fig. P52 Multiple post-translational modifications in extracellular proteins. *a*, Schematic diagram of prothrombin. Disulphide bonds (-SS-) and sites of N-linked glycosylation (CHO) are indicated. Note also the γ-carboxyglutamyl (Gla) residues in the N-terminal portion of the molecule ( ). The primary points of cleavage by thrombin (IIa) and by Factor Xa are indicated by arrows. The active thrombin is formed by Xa cleavage which generates first prethrombin 2 (inactive) noncovalently associated with fragments 1 and 2 and then cleaves within prethrombin 2 to generate active thrombin. This last cleavage is analogous to the conversion of chymotrypsinogen to chymotrypsin. The active part is the cleaved prethrombin 2, but it depends for its activity on the covalent disulphide bond holding the cleaved fragments together (275–323 + 324–582) and on the noncovalent association with fragments 1 and 2. R, Arg; S, Ser; T, Thr. *b*, Schematic diagram of a type I procollagen molecule. Disulphide bonds and sites of hydroxylation and glycosylation are indicated. Dashed lines indicate the sites of specific proteolysis to yield the mature collagen molecule.

and hence to influence the packing and overall structure of CHROMATIN; a crude correlation exists between extent of acetylation and activity in gene transcription.

One highly specific modification with a regulatory function is the nucleotidylation of bacterial glutamine synthetase; this key enzyme of nitrogen metabolism is regulated by adenylylation of a specific tyrosyl residue and this process is controlled by another protein which is itself regulated by a reversible uridylylation.

Early co- and post-translational modifications in the secretory pathway

Secreted and cell-surface proteins of eukaryotes are subject to a range of characteristic post-translational modifications (see Fig. P52), some of which occur while the protein is still undergoing translation, translocation into the lumen of the ENDOPLASMIC RETICULUM (ER) and folding into the biologically active conformation (*see* PROTEIN TRANSLOCATION). Indeed some of these modifications are essential for attaining this conformation. Thus the presence of intra- and intermolecular disulphide bonds is highly characteristic of secreted proteins and extracellular domains of cell-surface proteins, and the formation of these bonds is essential for folding and progression through the secretory pathway. The process is catalysed by the enzyme protein disulphide isomerase, and may involve GSSG (oxidized glutathione) as oxidant.

The characteristic prolyl-4-hydroxylation of collagens (and collagenous sequences in other proteins) stabilizes the triple-helical conformation and is also essential for initial folding (*see* EXTRACELLULAR MATRIX MOLECULES); when this process is inhibited, newly synthesized procollagen polypeptides assemble into disulphide-linked trimers, but do not fold into the characteristic triple-helical conformation. The hydroxylase responsible for this modification employs O_2 and α-ketoglutarate as substrates and Fe^{2+} and ascorbate (vitamin C) as cofactors (failure of collagen hydroxylation is the major pathology of the vitamin C deficiency disease scurvy). Interestingly, prolyl-4-hydroxylase is a heterotetrameric species, two subunits of which are identical to protein disulphide isomerase. These enzymes are soluble resident proteins of the ER lumen.

Two other modifications are common to nearly all proteins entering the secretory pathway, and function on nascent polypeptide chains. The first is the proteolytic cleavage of the targeting SIGNAL PEPTIDE (see below and PROTEIN TARGETING; PROTEIN TRANSLOCATION). The second is N-glycosylation of asparagine (Asn) residues located in the specific tripeptide sequence Asn-Xaa-Thr/Ser (see GLYCANS; GLYCOPROTEINS AND GLYCOSYLATION). A preformed oligosaccharide is transferred from the lipid carrier dolichyl pyrophosphate to the amide group of the Asn residue; the enzyme responsible is an integral protein of the ER membrane and comprises several subunits. Not all Asn residues located in the defined primary sequence unit (sequon) are N-glycosylated, and some show partial occupancy; this implies either that other primary sequence features determine specificity, or that the modification can be blocked sterically, suggesting that protein folding competes with glycosylation. The functions of glycosylation remain incompletely defined; glycosylation can affect a protein's solubility and physical properties, its interactions with other proteins and with cell-surface receptors, its resistance to proteases, its immunogenicity and its lifetime. In many cases these properties are defined by modifications of the initial oligosaccharyl side chain which occur later in the secretory pathway (see below).

In contrast to signal peptide cleavage and glycosylation, some significant modifications are restricted to a very few proteins. γ-Carboxylation of glutamyl residues introduces a novel functionality into proteins, the dibasic acid, and hence confers a characteristic high affinity for Ca^{2+} ions. This modification was first discovered in proteins of the blood clotting pathway (see BLOOD COAGULATION AND ITS DISORDERS), but is also found in proteins of bone and other systems. The carboxylase responsible is an integral protein of the ER and the reaction involves CO_2, molecular oxygen and uses vitamin K as a cofactor. Only groups near the N terminus of PROTHROMBIN and other blood-clotting proteins are γ-carboxylated, and the specificity seems to be determined by the PROSEQUENCE of these proteins, which is subsequently cleaved (see below).

Later modifications in the secretory pathway

Subsequent to folding and initial modification in the ER, secreted proteins can undergo further modifications at various stages in the secretory pathway. The modifications include further proteolytic events such as the generation of mature yeast α mating factor from its precursor by the *KEX2* gene product, and the various cleavages of hormone precursors, such as proinsulin, which occur in vertebrate secretory vesicles and granules (see EXOCYTOSIS; INSULIN AND ITS RECEPTOR; PROTEIN SECRETION).

There is also further glycosylation, including O-glycosylation of Ser and Thr side chains, which occurs in the GOLGI APPARATUS, and extensive modification of N-linked glycan chains, which begins with the removal of terminal glucose residues in the ER and continues with a wide variety of modifications in the Golgi to generate the striking range of complex biantennary and triantennary N-glycan chains, some of which contain terminal fucose or sialic acid groups. Enzymes responsible for these modifications have characteristic locations in subcompartments of the secretory pathway and the modifications can be used to trace progress along the pathway. The nature of these modifications is characteristic both of the protein substrate and of the cell in which it is expressed, a fact of considerable significance for the expression of high value recombinant proteins (see GENETIC ENGINEERING). In many cases an individual protein displays a great heterogeneity in its processed glycan side chains even at an individual site in the protein; after many years work it remains unclear whether this reflects a rather relaxed modification with little need for stringent control, or a highly sophisticated system designed to generate a subtle level of variety of glycoforms.

Some glycoprotein processing in the secretory pathway generates targeting signals such as the formation of the specific mannose 6-phosphate residue which acts as a signal for targeting to the LYSOSOME, or the construction of the phosphatidylinositol glycan structure which acts as a membrane anchor for a variety of cell-surface proteins (see GPI ANCHOR). Other modifications late in the secretory pathway include the conversion of terminal -X-Gly sequences in some hormones to -X-NH_2 amides, the sulphation of tyrosyl residues which is quite common, especially in neuropeptides and hormones, and the distinctive iodination of tyrosyl residues in THYROGLOBULIN which generates the precursors of THYROID HORMONES.

Specific proteolysis as a post-translational modification

Many biologically active proteins are generated by specific post-translational PROTEOLYSIS from larger initial translation products. These proteolytic events tend either to be comparatively early processes associated with the initial synthesis and targeting of the protein, or later events which generate the active product at a specific time or place in response to a signal. The early events, in which the proteases act on proteins which are not yet fully folded, include the co-translational cleavage of signal sequences from eukaryotic secretory proteins as they are translocated into the ER lumen; the enzyme responsible is an integral protein and has not been extensively characterized. Translocation into other organelles, specifically mitochondria and chloroplasts, is also associated with cleavage of an N-terminal signal; in these cases the nature of the sequences constituting a translocation signal is not fully defined, but they have a propensity to form a positively charged amphipathic α-HELIX. These translocations are post-translational events and the full-length translation product is translocated in a partially unfolded state and is processed by a metalloproteinase in the organellar matrix (see MITOCHONDRIAL BIOGENESIS; PROTEIN TRANSLOCATION).

The later proteolytic processes modify fully folded proteins and are usually associated with some sort of regulation, either of secretion or of biological activity. Some of these proprotein conversions occur within the secretory pathway (see above), but the majority are extracellular. Many viral proteins and some hormone precursors are generated as polyproteins which are cleaved by a multitude of proteolytic events into the biologically active end products. In other cases, activation of a precursor depends on a single proteolytic event. The most familiar of these 'activating' proteolyses are the ZYMOGEN conversions involved in the activation of digestive enzymes and the enzymes of the blood-clotting cascade (see Fig. P52). But there are many examples of this widespread form of irreversible enzyme regulation which contrasts with the reversible regulation achieved by phosphorylation/dephosphorylation.

Nonenzymatic modifications

All the modifications discussed so far have been physiological processes catalysed by specific enzymes. But some chemical modification, especially of long-lived proteins, is nonenzymatic and apparently nonphysiological. Thus many proteins which turn over slowly, such as haemoglobin, red cell membrane proteins, albumin, collagen and myelin undergo a nonenzymatic reaction between glucose and free amino groups to form so-called advanced glycosylation end products; this process is enhanced by high blood glucose concentrations and is used is a marker for blood glucose control in diabetes. Other nonenzymatic modifications indicative of protein 'ageing' include deamidation of Asn and glutamine (Gln) side chains and oxidation of various residues, particularly the formation of methionine sulphoxides. This modification is known to inactivate a number of proteins and is thought to account for the pronounced loss of α_1-antitrypsin activity (*see* SERPINS) in the airways of cigarette smokers, which may potentiate the onset of emphysema in this group.

R.B. FREEDMAN

1 Uy, R. & Wold, F. (1977) *Science* **198**, 890–896.
2 Wold, F. (1981) *Annu. Rev. Biochem.* **50**, 783–814.
3 Freedman, R.B. & Hawkins, H.C. (Eds) (1980) *The Enzymology of Post-translational Modification of Proteins*, Vol. 1 (Academic Press, London).
4 Freedman, R.B. & Hawkins, H.C. (Eds) (1985) *The Enzymology of Post-translational Modification of Proteins*, Vol. 2 (Academic Press, London).
5 Wold, F. & Moldave, K. (Eds) (1984) *Methods in Enzymology* **106** (Academic Press, New York).
6 Wold, F. & Moldave, K. (Eds) (1984) *Methods in Enzymology* **107** (Academic Press, New York).
7 Harding, J.J. & Crabbe, M.J.C. (Eds) (1991) *Post-Translational Modification of Proteins* (CRC Press, Baton Rouge) 262 pp.

posterior group genes A group of MATERNAL EFFECT genes in *Drosophila*, mutations in which result in the lack of abdominal segments. The majority of these genes, with the exception of *nanos* and *pumilio*, are also required for formation of the pole plasm and for pole cell development. Posterior group genes direct the localization and stepwise assembly of the POLAR GRANULES, whose formation is a prerequisite for the localization of the posterior determinant which specifies abdominal development. The gene *nanos* encodes the posterior determinant. Its mRNA is localized at the posterior pole and *nanos* product acts in the abdominal region to inhibit the translation of *hunchback* maternal RNA. *See also*: DROSOPHILA DEVELOPMENT.

postsynaptic Applied to neuron, receptors, etc. on the receiving side of a synapse. *See*: SYNAPTIC TRANSMISSION.

postsynaptic density Area of dense cytoskeletal material under the plasma membrane of the postsynaptic neuron at a chemical synapse. *See*: NEURONAL CYTOSKELETON; SYNAPTIC TRANSMISSION.

potassium K^+. *See*: MEMBRANE POTENTIAL; NA^+-CA^{2+} EXCHANGER; NA^+, K^+-ATPASE; POTASSIUM CHANNELS; VOLTAGE-GATED ION CHANNELS.

potassium channels ION CHANNELS that conduct potassium ions (K^+). Both voltage-gated and unregulated potassium channels are known. Three classes of voltage-gated potassium channels have been cloned:
1 The *Shaker* class (protein with six transmembrane helices and a presumed conducting pore region P).
2 Inwardly rectifying channels such as the potassium channel from heart atrium (two transmembrane helices and a P region).
3 The minK channel which is a single-pass membrane protein.
See: VOLTAGE-GATED ION CHANNELS.

The potassium leak channel is a non-gated channel found in the plasma membrane of mammalian and other cells through which K^+ flows out of the cell down its concentration gradient, and which, together with the NA^+, K^+-ATPASE, is involved in setting up the MEMBRANE POTENTIAL.

potency Of embryonic cells, the total fates of a cell or tissue region which can be achieved by manipulating the environment of the cell or tissue.

potexvirus group Sigla from the type member, potato virus X. A group of plant viruses with flexuous rod-shaped particles, 470–580 nm long and 13 nm wide. The coat protein subunits are arranged in a helix around the genomic (+)-sense RNA. The genome organization is shown in Fig. P53. The 5' gene reads through a weak termination codon to give the P166 and P191 which have sequence homologies with RNA replicases. The 3' terminal genes are translated from subgenomic mRNAs. The 3' terminal gene encodes the coat protein but the functions of the other two are unknown. *See also*: PLANT VIRUSES.

Milne, R.G. (Ed.) (1988) *The Plant Viruses*, Vol. 4 (Plenum, New York).

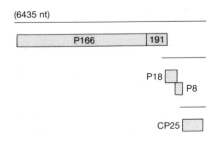

Fig. P53 Genome organization of potato virus X (PVX). The lines represent the RNA species and the boxes the proteins (P) with M_r given as $\times 10^{-3}$; nt, nucleotides.

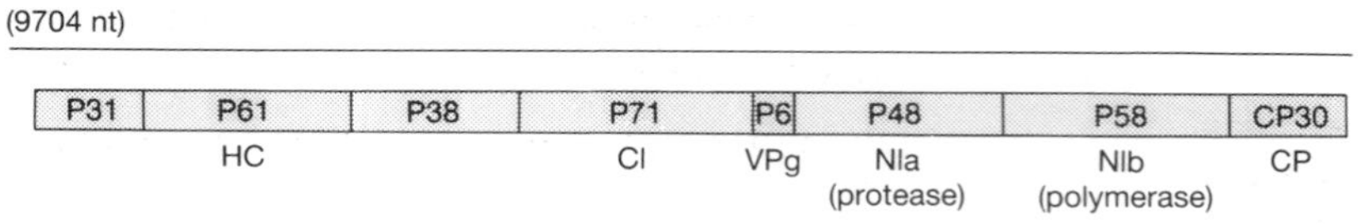

(9704 nt)

P31	P61	P38	P71	P6	P48	P58	CP30
	HC		CI	VPg	NIa (protease)	NIb (polymerase)	CP

Fig. P54 Genome organization of potato virus Y (PVY). The lines represent the RNA species and the boxes the proteins (P) with M_r given as $\times 10^{-3}$; nt, nucleotides.

potyvirus group Sigla from the type member, potato virus Y. A group of plant viruses with flexuous rod-shaped particles, 680–900 nm long and 11 nm diameter. The coat protein subunits (M_r 32 000–36 000) are arranged in a helix around the genomic (+)-sense RNA. Genome organization is shown in Fig. P54. The RNA is translated to give a POLYPROTEIN which is processed into the following products: P31, function unknown; P61, aphid transmission helper component; P38, function unknown; P71, cytoplasmic inclusion protein function unknown; P6 linked covalently to the 5′ of the viral RNA (VPg); P48, nuclear inclusion protein a which is a protease; P58, nuclear inclusion protein b which is considered to be the RNA polymerase; CP30, the viral coat protein. *See also*: PLANT VIRUSES.

Milne, R.G. (Ed.) (1988) *The Plant Viruses*, Vol. 4 (Plenum, New York).

POU domain This protein domain, which occurs in a family of DNA-binding proteins, is named after the founding members of the family — P (Pit-1), O (Oct-1/2), U (Unc-86). It consists of two parts. The POU-specific domain is a conserved domain of 76 amino acids which is separated by a spacer of variable length from a POU homeodomain of 60 amino acids which exhibits homology to the classic homoedomain. The POU-specific domain appears to be required for high affinity site-specific binding as well as for DNA-dependent protein–protein interactions. The POU homeodomain is critical for DNA binding. Several members of the large POU domain gene family affect the differentiation of specific cell types in development, for example, *pit-1* is required for the appearance and proliferation of several of the distinct cell types in the anterior pituitary. *See also*: HOMEOBOX GENES AND HOMEODOMAIN PROTEINS; TRANSCRIPTION FACTORS.

Poxviridae Family of large enveloped double-stranded DNA animal viruses which includes VACCINIA VIRUS (cowpox), monkey pox viruses, and the human variola virus (the agent of smallpox which was exterminated from human populations in 1976). It comprises the subfamilies Chordopoxvirinae and Entomopoxvirinae. The ovoid virus particles (length around 300 µm) have a complex structure, with a nucleoprotein capsid containing a linear genome (120–300 kb) surrounded by a lipid envelope containing virus-encoded glycoprotein spikes. The large poxvirus genome contains around 200 genes. As well as encoding virion proteins, the genome encodes many proteins involved in virus replication and transcription such as DNA POLYMERASE, the enzyme THYMIDINE KINASE which catalyses a step in the synthesis of deoxyribonucleotides required for DNA replication, a DNA-dependent RNA POLYMERASE, and RNA PROCESSING enzymes. Unlike other DNA viruses, poxviruses replicate entirely in the cytoplasm, establishing discrete centres of viral synthesis; uniquely amongst DNA viruses, the virion contains a virus-encoded RNA polymerase, which after uncoating of the viral

particle transcribes a set of EARLY GENES to initiate infection. The two strands of the vaccinia virus genome are joined covalently by phosphodiester bonds and form a single-stranded circle on denaturation. The mode of genome replication is still not completely determined. *See also*: ANIMAL VIRUSES.

ppGpp and pppGpp Guanosine polyphosphates synthesized by the bacterium *Escherichia coli* when starved of amino acids. *See*: MAGIC SPOT; STRINGENT REPONSE.

Pr The form of PHYTOCHROME that absorbs maximally at around 660 nm.

PR Progesterone receptor. *See*: STEROID RECEPTOR SUPERFAMILY.

PR proteins PATHOGENESIS-RELATED PROTEINS. *See also*: PLANT PATHOLOGY.

Prader–Willi syndrome Syndrome in which short stature is associated with obesity and mild mental retardation. It is associated with an absence of paternal chromosome region 15q11q13. *See*: CONTIGUOUS GENE SYNDROMES; PARENTAL GENOMIC IMPRINTING.

pre-B cell B CELL precursor in which the immunoglobulin heavy chain genes have been rearranged and an IgM heavy chain expressed but in which the light chain genes have not yet been rearranged. *See*: B CELL DEVELOPMENT; IMMUNOGLOBULIN GENES.

pre-mRNA The initial RNA transcript from a protein-coding eukaryotic gene before the introns are removed by RNA SPLICING to generate mRNA.

pre-T cells (triple positive cells) Bone-marrow derived cells found in the thymus and committed to the T CELL lineage. They are more mature than pro-T cells but have not yet finished the process of cell differentiation. Pre-T cells have begun to rearrange their T CELL RECEPTOR GENES and they express both CD4 and CD8 accessory molecules. When these cells first express a T cell receptor–CD3 complex at the cell surface they are sometimes called triple positive cells. Thymocytes remain in this category until they have completed thymic education — both positive and negative selection — and become ready for export as mature T cells. *See*: T CELL DEVELOPMENT.

prealbumin Former name for TRANSTHYRETIN.

preclimacteric Stages in fruit maturation preceding CLIMACTERIC rise in respiration, associated with the biochemical changes during ripening. Difficult to identify in fruit with no pronounced climacteric, such as citrus fruit.

precursor RNA processing proteins *See*: PRP PROTEINS.

preimplantation stage embryo All mammalian embryos obtain nutrients and discard waste via the maternal circulation. Contact with maternal blood vessels is made by embryonic invasion of the uterine epithelium during a process called implantation. At stages previous to this, mammalian embryos are called preimplantation embryos. They develop free in the reproductive tract during their progress down the oviduct into the uterus. The length of this stage and the number of cleavage cycles undergone varies between species. Early development is controlled exclusively by the maternal genome but at some stage during cleavage (depending on species) the embryonic genome becomes activated. The latter stages of preimplantation development involve morphological change into a blastocyst and differentiation of two cell lineages, the inner cell mass which forms embryo and EXTRAEMBRYONIC MEMBRANES, and the trophectoderm which forms extraembryonic membranes. *See also*: MAMMALIAN DEVELOPMENT.

prenatal diagnosis The detection of a genetic disease before birth. It is usually applied when a couple are known to be at risk of having a child with a serious genetic disorder which can be identified before birth.

Prenatal diagnosis is carried out by either direct inspection of the foetus for malformation, analysis of amniotic fluid (amniocentesis), chorion villus sampling, or, most recently, direct sampling of fertilized ova after *in vitro* fertilization. Chorion villus sampling involves removal of a small amount of the chorionic tissue which surrounds the foetus between 9 and 13 weeks gestation and which is of foetal rather than maternal origin.

Amniotic fluid or chorionic tissue can be used for chromosome analysis, to detect microscopically detectable CHROMOSOME ABERRATIONS or for the detection of a variety of enzyme deficiencies by biochemical analysis. They can also be used, as can cells removed from fertilized eggs, for analysis of DNA directly for genetic defects.

There are a variety of approaches to the analysis of foetal DNA for prenatal diagnosis. Total cellular DNA can be studied or the particular region that contains the gene of interest may be amplified by the POLYMERASE CHAIN REACTION (PCR). If the mutation causing the genetic defect is known, it can be identified in the extracted DNA using oligonucleotide PROBES for the mutated sequence, or, if it alters a restriction site, its presence can be detected directly by changes in the pattern of DNA fragments found after treatment with the appropriate restriction enzymes (*see* RESTRICTION ANALYSIS).

If the molecular defect is not known it may be possible to determine whether the foetus is affected using restriction fragment length polymorphism (RFLP) linkage analysis (*see* HUMAN GENE MAPPING). In this case a family study is carried out to determine whether a particular polymorphism is marking the parental chromosomes that carry the defective gene. If it is, the presence of the marker in the foetal DNA is an indication that the foetus has also received the mutant gene from its parents.

Prenatal diagnosis using foetal DNA analysis has been used widely for the detection of the THALASSAEMIAS and other disorders of human haemoglobin (*see* HAEMOGLOBIN AND ITS DISORDERS), haemophilia (*see* BLOOD COAGULATION AND ITS DISORDERS), CYSTIC FIBROSIS and muscular dystrophy (*see* DUCHENNE MUSCULAR DYSTROPHY).

Weatherall, D.J. (1991) *The New Genetics and Clinical Practice*, 3rd edn (Oxford University Press, Oxford).

prepattern Hypothesis formulated by Stern in 1954 to explain the effects of specific mutations on the pattern of bristle formation in insects. He proposed that the pattern is built in two stages: first, a prepattern of distribution of MORPHOGEN is established; second, the bristle pattern is formed by cells responding to singularities in the prepattern. In MOSAIC ANALYSIS, patches of cells carrying mutations in genes responsible for the prepattern might be expected to have long-range effects. Mutations in genes responsible for the competence to respond to the prepattern should be CELL-AUTONOMOUS. In the prepattern hypothesis the complexity of the final pattern is cryptically present in the prepattern and this contrasts with the POSITIONAL INFORMATION hypothesis where complexity can be generated by differential cell response to a simple morphogenetic gradient. *See also*: BRISTLE PATTERN.

preprophase band In plant cells, the ring of MICROTUBULES around the nucleus. It forms just before MITOSIS and marks the plane in which the future spindle equator will lie and hence the plane of cell division.

preprotein A protein which still contains the SIGNAL SEQUENCE that directs its insertion into or through membranes. *See*: PROTEIN TRANSLOCATION.

presynaptic Applied to neuron, receptors, etc. on the transmitting side of a synapse. *See*: SYNAPTIC TRANSMISSION.

presynaptic facilitation The enhancement of the presynaptic activity of a cell by external stimuli (i.e. neuromodulators) acting on the presynaptic terminal. Stimuli capable of inducing facilitation cause an enhanced influx of Ca^{2+} into the presynaptic terminal which leads to an increase in the amount of neurotransmitter released in response to the arrival of an electrical impulse. *See also*: SYNAPTIC TRANSMISSION.

presynaptic inhibition The reduction of the presynaptic activity of a cell by external stimuli (i.e. NEUROMODULATORS) acting on the presynaptic terminal. This effect is mediated by a synaptically mediated decrease in the Ca^{2+} influx into the terminal, or by an increased conductance to Cl^-, which leads to a decrease in the height of the ACTION POTENTIAL in the presynaptic terminal and a consequent inhibition or decrease in the amount of neurotransmitter released. *See also*: SYNAPTIC TRANSMISSION.

presynaptic web Area of cytoskeletal material under the plasma membrane of the axon terminal. *See*: NEURONAL CYTOSKELETON.

Pribnow box Short conserved sequence found in the PROMOTERS of many bacterial genes and centred approximately 10 base pairs upstream of the transcription start site. It has the CONSENSUS SEQUENCE TATAAT and is required for the correct positioning of

RNA polymerase at the transcription start site. It is also known as the – 10 site. *See*: BACTERIAL GENE EXPRESSION; TRANSCRIPTION.

primaquine An antimalarial drug used to eliminate hepatic forms of *Plasmodium vivax* and *P. ovale*. When given to individuals with GLUCOSE-6-PHOSPHATE DEHYDROGENASE DEFICIENCY it may cause severe haemolysis.

primary cell culture Cultured cells derived directly from normal tissues rather than from tumours or from CELL LINES. Primary cultures are usually from skin or embryonic tissues and are predominantly FIBROBLASTS. Unlike cell lines, primary cell cultures will only undergo a limited number of cell divisions.

primary immune response The IMMUNE RESPONSE that occurs following the first exposure to an ANTIGEN. The rate of development and magnitude of this response depends upon the immunogenicity of the antigen, but it is usually slower and smaller than the response following subsequent exposure to the same antigen (the SECONDARY IMMUNE RESPONSE). In the primary response there is usually a lag period of several days between the introduction of the antigen and the appearance of specific ANTIBODY or T lymphocytes (T CELLS) in the blood. Serum antibody is of low TITRE and usually of low AFFINITY, with a relative predominance of IgM. *See also*: B CELL DEVELOPMENT; T CELL DEVELOPMENT.

primary induction, primary embryonic induction *See*: NEURAL INDUCTION.

primary lysosome Historically it was supposed that newly formed LYSOSOMES derived directly from the GOLGI APPARATUS as small vesicles of ~0.1 μm in diameter containing ACID PHOSPHATASE but no endocytosed material, and which were thus called primary lysosomes. The model largely derived from the observation in neutrophils of azurophilic granules which can fuse with phagosomes to form phagolysosomes. This model, for which the evidence has always been weak and mainly morphological, is now disputed and the origin of mature lysosomes is thought to be more complex.

primary nondisjunction *See*: NONDISJUNCTION.

primary oocyte *See*: OOGENESIS.

primary structure The AMINO-ACID SEQUENCE of a protein or the NUCLEOTIDE sequence of DNA or RNA.

primary transcript The RNA produced by TRANSCRIPTION of a gene, and which often must undergo further processing to yield a functional mRNA or other structural RNA. *See*: RNA PROCESSING; RNA SPLICING; RIBOSOMAL RNA.

primase RNA POLYMERASE that synthesizes the short RNA primers for DNA REPLICATION using DNA as a template. The enzyme in *Escherichia coli* is of M_r 60 000 and synthesizes a short stretch of ~5 nucleotides.

primed-burst paradigm Pattern of stimulation mimicking natural hipppocampal THETA RHYTHM. This pattern of stimulation (a single priming pulse followed, after about 150 ms, by a 100 Hz burst of two or more stimuli) is effective in inducing LONG-TERM POTENTIATION.

primer (1) In DNA REPLICATION a short stretch of RNA that is synthesized by a specialized RNA polymerase (PRIMASE) using the DNA as a template and which is then elongated by DNA polymerase. The primer is subsequently excised and the gap filled in by a DNA polymerase.
(2) Short oligonucleotide of defined sequence which is annealed to the DNA template to initiate the POLYMERASE CHAIN REACTION.

primitive lattice, primitive unit cell A description of particular SYMMETRY arrangements found in crystals having only one LATTICE point in each UNIT CELL. In contrast, a non-primitive cell or lattice has an additional lattice point at the centre of the unit cell or at the centres of one (or all three) pair(s) of opposite faces of the unit cell. These arrangements, which are referred to as centred lattices or centred unit cells, can be recognized by the pattern of SYSTEMATIC ABSENCES seen in the DIFFRACTION pattern from the crystal. *See*: X-RAY CRYSTALLOGRAPHY.

primitive streak A groove that forms along the longitudinal axis of the bilaminar germ disk early in the development of mammalian and avian embryos. During GASTRULATION, where the two-layer embryo becomes a three-layer disk, cells of the epiblast (primitive ECTODERM) migrate toward the midline of the germ disk where they detach from the epiblast and move beneath it, forming the invagination of the primitive streak. Some of the epiblast cells displace the cells of the underlying hypoblast (primitive ENDODERM), while others migrate between the two layers to form the MESODERM. Soon after gastrulation commences, HENSEN'S NODE, which initially lies at the end of the primitive streak, regresses caudally, laying down the notochord. *See also*: AVIAN DEVELOPMENT; MAMMALIAN DEVELOPMENT.

primordial germ cells (PGC) Small diploid precursor cells of sperm and eggs. It is possible to follow the life histories of primordial germ cells in a number of species, using their high level of alkaline phosphatase enzyme activity as a marker, and certain common features of their development have been discovered. They usually arise as a small number of cells in the posterior of the organism, such as the posterior pole of the *Drosophila* blastoderm, or at the posterior end of the vertebrate PRIMITIVE STREAK. They are highly motile and prolific. Shortly after they are first identifiable, they begin to migrate towards the developing gonads, dividing frequently as they go. Eventually, large numbers reach the gonad, become stationary, and commmence MEIOSIS. It is not until they reach the developing gonads that male and female primordial germ cells become distinguishable.

primordium A structure or region which is the precursor of an adult structure or organ. The German word ANLAGE is often used as a synonym.

primosome Multisubunit assembly including the PRIMASE which forms on DNA and initiates the synthesis of RNA PRIMERS for DNA REPLICATION.

prion A hypothetical proteinaceous entity lacking nucleic acid, which has been proposed as the causal agent of TRANSMISSIBLE SPONGIFORM ENCEPHALOPATHIES.

pro-T cells (thymic progenitor cells, triple negative cells) Bone marrow stem cells that first start to colonize the foetal thymus of the mouse at around day 11 of gestation. From the time they arrive, these cells undergo a period of cell division, massively expanding the thymocyte pool. Pro-T cells lack most of the surface CD ANTIGENS associated with more mature T cells, including CD4, CD8 and CD3. The only real evidence that these are cells committed to the T cell lineage is that they are found in the thymus. Once isolated, pro-T cells cannot be distinguished from pro-B cells. *See*: B CELL DEVELOPMENT; T CELL DEVELOPMENT.

probe (1) Any molecule that specifically binds to a nucleic acid sequence or to a protein that is being searched for, and which can be labelled so that the required targets can then be detected. Probes for nucleic acid sequences are radiolabelled or chemically tagged oligonucleotides of complementary sequence to the DNA or RNA sequence being sought. Specific MONOCLONAL ANTIBODIES are used as probes to detect proteins. *See*: DNA CLONING; HUMAN GENE MAPPING; HYBRIDIZATION; SOUTHERN BLOTTING. (2) A reporter group whose physicochemical properties are a function of its environment and which can be attached to a macromolecule and monitored spectroscopically.

processed pseudogene PESUDOGENE which possesses a 3′ poly(A) tract, and lacks introns and promoter and other regulatory sequences, and which is therefore thought to be derived from REVERSE TRANSCRIPTION of an mRNA.

processing *See*: ANTIGEN PROCESSING AND PRESENTATION; POST-TRANSLATIONAL MODIFICATION OF PROTEINS; RNA PROCESSING.

processivity The capacity of DNA POLYMERASE to synthesize a stretch of polynucleotide chain before disengaging from the nucleic acid template.

Prochloron The first PROCHLOROPHYTE genus identified. *Prochloron didemni* grows as a unicellular ectosymbiont, usually in the cloacal cavity of didemnid ascidians.

Prochlorophyta, prochlorophytes Photosynthetic prokaryotes containing chlorophyll *a* and chlorophyll *b* but lacking PHYCOBILIPROTEINS, and therefore resembling plant CHLOROPLASTS. The first identified was *Prochloron*; two free-living types, including *Prochlorothrix*, have been found also. It is not yet fully clear whether they represent the descendants of the organisms that formed chloroplasts by endosymbiosis, or a cyanobacterial lineage that has independently acquired chlorophyll *b* and lost phycobiliproteins. *See also*: CYANOBACTERIA.

Prochlorothrix The second PROCHLOROPHYTE genus identified. It grows as a free-living, filamentous organism in fresh water.

procollagen The precursor of the extracellular matrix protein collagen. *See*: EXTRACELLULAR MATRIX MOLECULES; POST-TRANSLATIONAL MODIFICATION OF PROTEINS.

proconvertin A SERINE PROTEINASE involved in the blood clotting pathway.

productive rearrangement Rearrangement occurring within IMMUNOGLOBULIN GENES or T CELL RECEPTOR GENES which yields a rearranged gene from which a functional product is synthesized. *See*: B CELL DEVELOPMENT; GENE REARRANGEMENT; T CELL DEVELOPMENT.

prodynorphin Polypeptide precursor to the opioid peptide dynorphin and other opioid peptides. *See*: OPIOID PEPTIDES AND THEIR RECEPTORS.

proenkephalin A Polypeptide precursor to the enkephalins and other opioid peptides. *See*: OPIOID PEPTIDES AND THEIR RECEPTORS.

proenzyme The inactive precursor of an enzyme, which is activated by specific proteolytic cleavage (*see* POST-TRANSLATIONAL MODIFICATION OF PROTEINS).

profilin *See*: ACTIN-BINDING PROTEINS.

progenitor cell Undifferentiated cell whose lineal descendants differentiate along the appropriate pathway to produce a fully differentiated phenotype. All differentiated cells have, by definition, a progenitor cell type, for example neuroblasts for neurons (*see* NEUROBLAST DIFFERENTIATION) and GERM CELLS for gamete cells.

progesterone STEROID HORMONE involved in the female reproductive cycle. It is produced mainly by the ovary (corpus luteum) and the placenta and acts on the endometrium of the womb to prepare it for implantation. Continued secretion is essential for maintenance of a pregnancy to term. It has an antiovulatory effect if given at certain times during the menstrual cycle and is an ingredient of contraceptive pills. It acts via intracellular receptors of the STEROID RECEPTOR SUPERFAMILY.

programmed cell death In many developing organisms, both vertebrate and invertebrate, cells die in particular positions at particular times as part of the normal morphogenetic process. The histological picture is one of nuclear collapse and cell condensation (APOPTOSIS). Death may arise as a result of failure to acquire TROPHIC FACTORS necessary for continuing survival, as in the developing vertebrate nervous system, where neuronal death appears to be initiated by the expression of a set of appropriate genes. In the developing nematode, *Caenorhabditis elegans*, individual identified cells die in specific sublineages, mostly in the nervous system, and death may be prevented by mutations in several genes of the *ced* (cell death abnormal) family, whose function is under active investigation. *See also*: CAENORHABDITIS DEVELOPMENT; T CELL DEVELOPMENT.

progress zone In the developing limb, a population of mesodermal cells immediately subjacent to the APICAL ECTODERMAL RIDGE of the vertebrate limb bud. These cells have high mitotic activity, and leave the progress zone from its proximal side as growth of the limb proceeds. Positional identities are imparted to the cells as they leave, in a proximal-to-distal temporal sequence. Thus, the first cells to leave the progress zone are specified with proximal fates, while cells that remain in the progress zone and undergo further divisions are specified with more distal fates. The maintenance of the progress zone is dependent (through unknown mechanisms) on the continuing presence of the apical ectodermal ridge. *See also*: LIMB DEVELOPMENT; PATTERN FORMATION.

proinsulin The precursor protein from which insulin is produced. *See*: INSULIN AND ITS RECEPTOR.

projection neuron Neurons with long axons that 'project' from one area of the brain to another. Examples are some dopamine-containing neurons which have their cell bodies in the ventral tegmental area of the substantia nigra and their axon terminals in the frontal lobe of the cerebral cortex.

prokaryotes Mostly single-celled microorganisms, with a simple internal organization, lacking organelles such as mitochondria, chloroplasts, endoplasmic reticulum, and Golgi apparatus, and in which the DNA (usually in the form of a single circular molecule) is not enclosed in a nucleus. Prokaryotes also lack a CYTOSKELE-TON. In those prokaryotes that possess a cell wall, the composition and organization of the wall is quite different from that of the walls of fungi or plants (*see* BACTERIAL ENVELOPE). Because of the lack of a nuclear membrane, gene TRANSCRIPTION and translation can occur simultaneously (*see* BACTERIAL GENE EXPRESSION). Prokaryotic genomes also differ considerably in respect of gene structure and organization from those of the nuclear genomes of EUKARYOTES. There is little repetitive DNA and with very rare exceptions, prokaryote genes lack INTRONS. In many cases, functionally related genes are arranged contiguously in a single transcription unit (operon) under the control of a single regulatory gene (*see* BACTERIAL GENE ORGANIZATION). Prokaryotes are generally considered to include the true bacteria (EUBACTERIA, including CYANOBACTERIA and PROCHLOROPHYTES), the ARCHAE-BACTERIA, and simple unicellular microorganisms of doubtful affinity such as rickettsias, mycoplasmas, spiroplasmas, and chlamydiae. Evolutionary relationships within the prokaryotes and between prokaryotes and eukaryotes are currently the matter of much debate. *See also*: BACTERIAL CHEMOTAXIS; BACTERIAL PATHOGENICITY; BACTERIAL PROTEIN EXPORT.

prolactin Protein hormone produced by the anterior pituitary under the influence of the hypothalamus. Its main effects are on the mammary gland, where it stimulates tissue proliferation and control of milk secretion, and the corpus luteum, where it is involved in the maturation of ovarian follicles. It also has general anabolic effects on metabolism.

prolamellar body A paracrystalline array of lipid and protein that occurs in the plastids (ETIOPLASTS) of leaves of plants grown in the dark. Few recognizable THYLAKOID membranes are formed in the dark, but upon illumination they develop from the prolamellar body which disappears.

prolamine A general term for small globular proteins insoluble or sparingly soluble in water but soluble in alcohol.

proliferating cell nuclear antigen (PCNA) *See*: PAPOVAVIRUSES.

proline (Pro, P) A secondary amino acid with the aliphatic side chain bonded to both the carboxyl C and the amino N to form a cyclic structure. *See*: AMINO ACIDS.

prometaphase The stage of MITOSIS or MEIOSIS that follows PROPHASE and precedes METAPHASE. During prometaphase, the already condensed CHROMOSOMES interact with the MITOTIC SPINDLE and become both oriented and positioned. Their orientation assures that the SISTER CHROMATIDS of each chromosome are attached to opposite poles of the spindle. Their positioning moves them to the plane equidistant from the two spindle poles, which is called the METAPHASE PLATE. The importance of prometaphase is that during this time the chromosomes become arranged in such a way that the simple separation of sister chromatids and their migration apart during ANAPHASE will almost always lead to an accurate segregation of the genes before cell division.

promiscuous DNA DNA sequences present in organelles other than those in which they are normally expressed. The promiscuous sequences need not themselves be expressed. They represent the results of transposition, rather than ancestral sequences, but the mechanism of transposition is not known. *See*: MITOCHON-DRIAL GENOMES: PLANT.

promoter The region of a gene at which initiation and rate of TRANSCRIPTION are controlled. It contains the site at which RNA polymerase binds and also sites for the binding of regulatory proteins (e.g. REPRESSORS; TRANSCRIPTION FACTORS). Most promoters in both prokaryotic and eukaryotic genes are located in nontranscribed DNA upstream of the start site of transcription; an exception is eukaryotic genes that are transcribed by RNA polymerase III, which have internal promoters. *See*: BACTERIAL GENE EXPRESSION; BACTERIAL GENE ORGANIZATION; EUKARYOTIC GENE EXPRESSION; EUKARYOTIC GENE STRUCTURE.

35S promoter The PROMOTER controlling transcription of the 35S RNA in cauliflower mosiac virus (*see* CAULIMOVIRUS GROUP), which is used as a promoter in EXPRESSION VECTORS introduced into transgenic plants when expression of the introduced gene throughout the plant is required. *See*: PLANT GENETIC ENGINEER-ING.

pronucleus The male or female gamete nucleus present in the fertilized egg before nuclear fusion.

proofreading (1) In DNA REPLICATION, the capacity of the DNA POLYMERASES involved in the DNA chain elongation to excise a mispaired base by virtue of their $3' \rightarrow 5'$ exonuclease activity. RNA polymerases do not possess exonuclease activity; RNA synthesis is therefore more error-prone than DNA synthesis. (2) In PROTEIN SYNTHESIS, the ability of many aminoacyl-tRNA synthetases to prevent mischarging of tRNA with the wrong amino acid (*see* GENETIC CODE; TRANSFER RNA). Amino acids that are wrongly activated by the synthetase because of their structural resemblance to the correct species are hydrolysed by the synthetase before they can be transferred to tRNA. A second proofreading stage occurs at the ribosome: the time lag between binding of aminoacyl-tRNAs and peptide bond formation allows time for the dissociation of most incorrectly bound aminoacyl-tRNAs.

pro-opiomelanocortin (POMC) Polypeptide precursor to β-endorphin and other non-opioid peptides (adrenocorticotropic hormone, β-lipotropin, and melanocyte-stimulating hormone). *See:* OPIOID PEPTIDES AND THEIR RECEPTORS.

properdin Alternative name for FACTOR P, a protein involved in the alternative pathway of COMPLEMENT fixation.

prophage A BACTERIOPHAGE genome integrated into the chromosome of its bacterial host cell. Phages that can integrate into the host genome are known as temperate phages, and cells carrying such prophages are known as LYSOGENS. Prophages can remain in the chromosome for many cell generations until some stimulus leads to their induction. The phage genome then excises from the chromosome and enters the LYTIC CYCLE of phage replication.

prophase The stage of MITOSIS or MEIOSIS during which CHROMOSOMES condense. Its beginning is difficult to define, because initial chromosome condensation may begin immediately after DNA REPLICATION, but there is usually a gap, called G2, between the completion of DNA replication (S phase) and the onset of microscopically visible condensation, the operational definition of prophase onset (*see* CELL CYCLE). Prophase ends with the onset of the interaction between the chromosomes and the MITOTIC SPINDLE. Prophase is important because most chromosomes are long relative to the diameter of a cell, often exceeding it by more than 1000-fold. The condensation process results in a compaction of the chromosome into a form that can be managed by the mechanical action of the spindle, assuring that each daughter cell will receive one copy of every gene. The mechanisms of condensation are poorly understood, but they include the supercoiling of pre-existing coils of CHROMATIN. The prophase that precedes meiosis is unusually long and complicated because it is during this time that homologous chromosomes pair and RECOMBINATION occurs. During prophase, there are also major changes in the organization of the cytoplasm.

proplastid A small, unpigmented and undifferentiated form of PLASTID that occurs in meristematic tissues. Proplastids develop into other forms of plastid, such as CHLOROPLASTS or ETIOPLASTS in leaves, as the cell differentiates.

prosequence Part of a protein that assists in its folding but which is removed from the mature protein molecule. *See:* MOLECULAR CHAPERONES; POST-TRANSLATIONAL MODIFICATION OF PROTEINS; PROTEIN FOLDING.

prostacyclins, prostaglandins *See:* EICOSANOIDS.

prostanoids Compounds derived from prostanoic acid, such as the prostaglandins. *See:* EICOSANOIDS.

prosthetic group Nonprotein component of an enzyme, involved in the reaction mechanism (*see* MECHANISMS OF ENZYME CATALYSIS). Prosthetic groups may be organic (e.g. pyridoxal phosphate or flavin mononucleotide), inorganic (e.g. Fe, Cu, or Zn), or both, for example HAEM.

proteases Enzymes that degrade proteins by hydrolysing peptide bonds between amino acid residues. Also known as PROTEINASES.

protease nexin *See:* SERPINS.

proteasome A nonlysosomal multicatalytic proteinase complex found in eukaryotic cells. Similar proteolytic complexes are also found in prokaryotes. Proteasomes are thought to associate with other proteins to form the 26S proteinase complex that catalyses ubiquitin-dependent protein degradation. Individual proteasomes are cylindrical particles of $M_r \sim 700\,000$, each composed of 15–20 proteins. They have a broad substrate specificity, degrading proteins to small peptides in an ATP-dependent reaction. Proteasomes are distributed throughout the cytosol and are also found in the nucleus of many cell types. *See:* PROTEIN DEGRADATION AND TURNOVER.

protein Large organic molecule containing carbon, oxygen, hydrogen, and nitrogen, as well as small amounts of other elements such as sulphur, and with relative molecular mass (M_r) ranging from a few thousands to several millions. The term was introduced by J.J. Berzelius in 1838 and was derived from the Greek *proteios*, 'of the first rank'. Thousands of different proteins are present in a cell, the synthesis of each type of protein being directed by a specific GENE. Proteins serve many vital functions both within cells and as extracellular molecules. They make up much of the cellular structure; as ENZYMES they catalyse biochemical reactions; they regulate gene expression (*see e.g.* TRANSCRIPTION; TRANSCRIPTION FACTORS); they regulate the movement of ions and other solutes into and out of the cell (*see* MEMBRANE TRANSPORT SYSTEMS); and they act as receptors for hormones and other signalling molecules.

Proteins are made up of AMINO ACID residues covalently linked together by PEPTIDE BONDS into a linear, unbranched POLYPEPTIDE chain, the order of the amino acids in any particular protein being specified by the order of nucleotides in the gene that specifies that protein (*see* GENETIC CODE; PROTEIN STRUCTURE; PROTEIN SYNTHESIS). After synthesis, the polypeptide chain folds to adopt a particular conformation unique to each protein and which is determined by the amino-acid sequence of the chain (*see*

MOLECULAR CHAPERONES; PROTEIN FOLDING). Many proteins also undergo further POST-TRANSLATION MODIFICATION before they become fully functional. Many proteins are composed of two or more identical or different subunits, each subunit being composed of a single polypeptide. Such multisubunit proteins are known as oligomers — homo-oligomers (homomers) are composed of identical subunits, hetero-oligomers (heteromers) of different subunits. The ability of proteins to associate with each other and with other macromolecules to form covalently and noncovalently bonded complexes allows the SELF-ASSEMBLY of larger structures such as viral capsids (*see* ANIMAL VIRUSES; BACTERIOPHAGES; PLANT VIRUSES) and cytoskeletal components (*see* INTERMEDIATE FILAMENTS; MICROFILAMENTS; MICROTUBULES).

The properties and function of any particular protein are largely determined by the physical and chemical properties of its topologically and chemically varied surface. The great range of possible surface configurations of proteins makes possible their important function as specific recognition molecules, capable of binding a wide variety of ligands (*see e.g.* ANTIBODIES; CELL ADHESION MOLECULES; EXTRACELLULAR MAXTRIX MOLECULES; IMMUNOGLOBULIN STRUCTURE; INTEGRINS; RECEPTORS and cross references therein; T CELL RECEPTORS). Proteins secreted from cells also act as intercellular signalling molecules (e.g. some hormones and GROWTH FACTORS). *See also:* ALLOSTERIC EFFECT; HAEMOGLOBINS; MECHANISMS OF ENZYME CATALYSIS; MEMBRANE STRUCTURE; MEMBRANE TRANSPORT SYSTEMS; MOLECULAR EVOLUTION: SEQUENCES AND STRUCTURES; PROTEIN DEGRADATION AND TURNOVER; PROTEIN ENGINEERING; PROTEIN–NUCLEIC ACID INTERACTIONS; PROTEIN SECRETION; PROTEIN TARGETING; PROTEIN TRANSLOCATION.

protein A Protein of M_r 42 000 from the bacterium *Staphylococcus aureus* which binds to IgG (*see* ANTIBODIES) from a wide range of species including human, rabbit, donkey, pig, and guinea-pig. It is used as a secondary reagent in IMMUNOELECTRON MICROSCOPY, light microscopic immunocytochemistry and also other immunological and biological techniques.

Goding, J.W. (1978) *J. Immunol. Meth.* **20**, 241–253.

protein C *See:* BLOOD COAGULATION AND ITS DISORDERS.

protein C inhibitor *See:* SERPINS.

protein crystallography *See:* ELECTRON CRYSTALLOGRAPHY; X-RAY CRYSTALLOGRAPHY.

Protein degradation and turnover

PROTEINS in eukaryotic cells are continuously turning over, with the half-life of individual proteins varying from a few minutes to weeks[1]. Among the proteins with the shortest half-lives are key regulatory proteins and the products of cellular ONCOGENES. The rapid turnover of such proteins allows a rapid response to changes in cellular environment. There are several distinct pathways of protein degradation, the activity of which can be influenced by nutritional and hormonal conditions. In addition, rates of degradation of certain proteins depend on the cellular concentrations of specific metabolites. Characteristics of protein degradation as well as many of the components of the degradative pathways are similar in different types of eukaryotic cells [1,2].

Protein structure and rates of degradation

Rates of degradation are probably influenced by several factors and different factors may be important for different proteins. General correlations have been suggested between various physicochemical properties of proteins and their intracellular half-lives, but such correlations have not held up well when large numbers of proteins have been examined. Analysis of primary structure (amino-acid sequence) of proteins shows that many of those with short half-lives contain one or more PEST regions — regions rich in the amino acids proline (P), glutamic acid (E), serine (S) and threonine (T) — whereas proteins with long half-lives do not[1]. A role for the N-terminal amino acid (the so-called N-end rule) has also been implicated in the selection of proteins as substrates for the ubiquitin system of degradation (see below).

Binding of ligands (substrates, products of metabolic pathways etc.) can influence the half-life of individual enzymes, as can COVALENT MODIFICATION reactions [3] such as oxidation or PHOSPHORYLATION, and conjugation with UBIQUITIN. It is likely that factors such as these are involved in the regulation of protein turnover in addition to the control of the proteolytic machinery of the cell.

Pathways of protein degradation

The pathways of intracellular protein degradation can be split into two groups — lysosomal and nonlysosomal. Nonlysosomal pathways include both ubiquitin-dependent and ubiquitin-independent mechanisms. The detailed sequence of steps in the degradation pathway has not yet been elucidated for any individual cellular protein although in many cases it has been possible to establish the primary pathway. For example, the proportion of protein degradation by LYSOSOMES, both for bulk cellular protein and for individual proteins, has been estimated using inhibitors of lysosomal proteolysis. Such studies show that in general proteins with relatively short half-lives are degraded by nonlysosomal mechanisms. Relatively long-lived proteins seem to be degraded within the lysosomes and the increase in protein degradation occurring during starvation can usually be attributed to lysosomal proteolysis [2,4].

The ubiquitin-dependent pathway is believed to be important for the ATP-dependent elimination of abnormal proteins during exposure of cells to HEAT SHOCK and other stresses. The involvement of the ubiquitin pathway is indicated by a decrease in rates of degradation of abnormal proteins in mammalian cells containing a temperature-sensitive ubiquitin-activating enzyme, and by genetic analysis in yeast [5]. Ubiquitin-dependent proteolysis may also be important for degradation of certain regulatory proteins such as CYCLINS which have a role in the control of the CELL CYCLE [6].

Lysosomal degradation

There are several pathways for the uptake of cellular proteins into lysosomes for degradation by lysosomal proteinases (e.g. cathepsins B, H, L, and D) [4]. Microautophagy, which may be the main pathway of uptake in well-nourished cells, involves internalization of protein-containing vesicles at the lysosome surface. Macroautophagy, which increases during starvation, involves the fusion of lysosomes with autophagic vacuoles which form when membranes derived from the smooth ENDOPLASMIC RETICULUM engulf cytoplasmic proteins. During extended periods of starvation and in cultured cells deprived of serum, direct uptake of selected proteins from the cytosol can occur. The mechanism of this latter process apparently involves the recognition of specific sequence motifs in proteins destined for lysosomal uptake [4]. These motifs, exemplified by KFERQ (single-letter amino-acid notation), are bound by a peptide recognition protein, prp73. This is a constitutively expressed member of the Hsp70 family (*see* HEAT SHOCK), which increases in concentration in response to starvation.

Ubiquitin-dependent protein degradation

Ubiquitin is a small (M_r ~8500) and highly conserved protein which has been found conjugated to free amino groups of proteins through its C-terminal glycine residue. Such conjugation may have a variety of functions, including the marking of proteins for intracellular proteolysis. There are several different ubiquitin genes, one of which is a heat-shock gene.

The ubiquitin pathway for protein degradation has been elucidated largely by studies with reticulocyte lysates [7] and with yeast [5]. It is a complex multicomponent system (Fig. P55). The

first step in the pathway involves the ATP-dependent activation of ubiquitin by the ubiquitin-activating enzyme, E_1, a protein of M_r 100 000. Ubiquitin becomes activated by formation of a thioester between its C-terminal glycine and a cysteinyl residue of E_1.

The second step in the pathway involves the transfer of activated ubiquitin to a ubiquitin carrier protein, E_2. There are many different E_2 species, at least two of which seem to be involved in protein breakdown [5]. Other variants of E_2 may have a role in the conjugation of ubiquitin to proteins for purposes unrelated to protein turnover.

The third step involves transfer of ubiquitin from E_2 to the protein substrate and usually requires a ubiquitin protein ligase, E_3. There are two types of E_3 molecule, each specific for different amino acid residues at the N termini of substrate proteins [8]. The bond formed during conjugation of ubiquitin to a protein is an isopeptide bond between the C-terminal glycine of ubiquitin and an ε-amino group of a lysine residue of the other protein.

From the structure of a few ubiquitin–protein conjugates, it seems that the signal for degradation is multiubiquitination — the conjugation of many ubiquitin molecules attached as a linear or branched polyubiquitin chain [8]. Ubiquitin-conjugated proteins destined for proteolytic breakdown are thought to be degraded by an ATP-dependent 26S proteinase complex but the lysosomes could also be involved.

Ubiquitin-independent pathways of nonlysosomal proteolysis

Other pathways of nonlysosomal protein breakdown are poorly defined although there is good evidence for their existence. A large number of intracellular proteinases have been identified but some are likely to be involved in specific protein processing rather than general protein degradation. In yeast, for example, there are more than 40 peptidases, of which only seven have been found within vacuoles (lysosomes). One abundant proteinase complex which probably plays a major role in both ubiquitin-independent and ubiquitin-dependent pathways of nonlysosomal proteolysis in yeast as well as higher eukaryotic cells is the multicatalytic proteinase or proteasome [2,9,10]. Each complex of M_r 700 000 seems to contain many different types of subunit and multiple distinct catalytic centres. Several of the genes specifying proteasome components are essential for cell proliferation. The proteasome catalyses ubiquitin-independent protein degradation but several reports suggest that it can also associate with as yet poorly characterized components to form the 26S proteinase which degrades ubiquitin–protein conjugates. The proteasome therefore appears to be involved in ubiquitin-dependent protein degradation and has also been implicated in the processing of antigens for presentation by the MHC class I pathway [11]. Proteasomes are found both in the nucleus and in the cytoplasm [12].

A.J. RIVETT

1 Rechsteiner, M. et al. (1987) Protein structure and intracellular stability. *Trends Biochem. Sci.* **12**, 390–394.
2 Rivett, A.J. (1990) Eukaryotic protein degradation. *Curr. Opinion Cell Biol.* **2**, 1043–1049.
3 Stadtman, E.R. (1990) Covalent modification reactions are marking steps in protein turnover. *Biochemistry* **29**, 6323–6331.

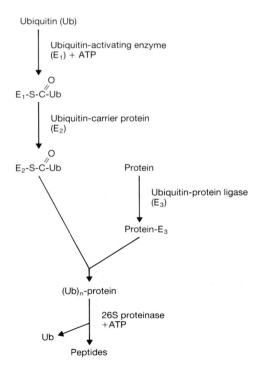

Fig. P55 The ubiquitin pathway for intracellular protein degradation.

4 Dice, J.F. (1990) Peptide sequences target cytosolic proteins for lysosomal proteolysis. *Trends Biochem. Sci.* **15**, 305–309.
5 Jentsch, S. et al. (1991) Genetic analysis of the ubiquitin system. *Biochim. Biophys. Acta* **1089**, 127–139.
6 Rechsteiner, M. (1991) Natural substrates of the ubiquitin proteolytic pathway. *Cell* **66**, 615–618.
7 Hershko, A. (1991) The ubiquitin pathway for protein degradation. *Trends Biochem. Sci.* **16**, 265–268.
8 Finlay, D. & Chau, V. (1991) Ubiquitination. *Annu. Rev. Cell Biol.* **7**, 25–69.
9 Orlowski, M. (1990) The multicatalytic proteinase complex, a major extralysosomal proteolytic system. *Biochemistry* **29**, 10289–10297.
10 Rivett, A.J. (1993) Proteasomes: multicatalytic proteinase complexes. *Biochem. J.* **291**, 1–10.
11 Goldberg, A.L. & Rock, K.L. (1992) Proteolysis, proteasomes and antigen presentation. *Nature* **357**, 375–379.
12 Rivett, A.J. & Knecht, E. (1993) Proteasome location. *Curr. Biol.* **3**, 127–129.

protein disulphide isomerase Enzyme (EC 5.3.4.1) involved in the isomerization of disulphide bonds in proteins during folding. It is present in the ENDOPLASMIC RETICULUM, but not the cytosol, of eukaryotic cells. The mammalian enzyme is a homodimer of subunit M_r 57 000, each of which contains duplications of domains with strong homology to THIOREDOXIN. *See*: POST-TRANSLATIONAL MODIFICATION OF PROTEINS.

Protein engineering

PROTEIN engineering covers the deliberate design and production of proteins with structures and properties that are, it is hoped, better than those of natural proteins.

Production of engineered proteins

The properties of proteins, including their three-dimensional form, derive ultimately from their amino-acid sequence (*see* PROTEIN STRUCTURE). The techniques of protein engineering are therefore concerned with changing the amino-acid sequence of a protein; this can be achieved in several ways.

Chemical synthesis and modification

Direct chemical synthesis of small proteins is achieved by repetitively coupling an activated derivative of the required amino acid to the end of a growing POLYPEPTIDE attached to a solid support (*see* PEPTIDE SYNTHESIS).

Computer-controlled wet chemistry enables sequences of both natural and unnatural amino acids to be made. Unnatural amino acids, for example the conformationally restricted 2,2-dimethylglycine, have the property of forcing the polypeptide to adopt a particular SECONDARY STRUCTURE, such as a 3_{10}-HELIX, and can lead to more stably folded new materials than are available from natural amino acids (*see* PROTEIN FOLDING).

Changing the charge of an already synthesized protein by reacting surface lysine residues with an acid anhydride produces proteins that are much more stable when injected as drugs into the circulatory system, as they are rendered resistant to attack by proteolytic enzymes. A modified engineered TISSUE PLASMINOGEN ACTIVATOR, which is used to dissolve blood clots in the treatment of thrombosis, has been produced in this way.

Chemical synthesis is presently the preferred route to construct new and very stable protein-based machines for NANOTECHNOLOGY.

Production from chemically synthesized or modified DNA sequences

Chemical synthesis of GENE fragments is the basis of the second, more usual means of production of engineered proteins (*see* OLIGONUCLEOTIDE SYNTHESIZERS; SYNTHETIC DNA). Bacterial and mammalian cells will translate the nucleotide sequence of any DNA into a linear sequence of amino acids making up a polypeptide chain (*see* GENETIC CODE; GENETIC ENGINEERING; PROTEIN SYNTHESIS). Completely synthetic genes of lengths up to 100 or more nucleotides can be made specifying any desired polypeptide sequence. Synthetic gene segments can also be joined to parts of natural genes to create hybrid genes coding for newly designed protein sequences. Alternatively, the synthetic fragment can be used to replace or extend a part of a natural gene by OLIGONUCLEOTIDE MISMATCH MUTAGENESIS or the POLYMERASE CHAIN REACTION (PCR). The final step is to incorporate the synthetic gene into a PLASMID of high copy number to direct the synthesis of the designed polypeptide chain in either cultured eukaryotic cells or in bacteria (*see* EXPRESSION VECTOR; GENETIC ENGINEERING). These new materials can be very cheap: the better bacterial systems produce one-half of their total protein as the engineered, synthetic protein.

One problem with the genetic route to new proteins is that there are not yet efficient ways of incorporating nonbiological amino acids. Natural enzymes also often contain prosthetic groups which are essential for the catalytic activity. Organic conductors and transition state metal chelators can easily be inserted into short, chemically synthesized peptides, and the recent development of a protein ligase which specifically links a cyanoethyl ester-activated peptide to a much larger section of protein which has been made by the genetic route (in the same manner as DNA LIGASE links together DNA segments) is opening the way to much more versatile synthetic engineered proteins than can be constructed from the 20 or so natural amino acids.

Folding of engineered proteins

The useful properties of proteins — as ENZYMES, HORMONES, ION CHANNELS and membrane pores, ANTIBODIES or as drug delivery systems — depend upon the polypeptide chain adopting a unique three-dimensional folded conformation — the tertiary structure of the protein (*see* PROTEIN FOLDING; PROTEIN STRUCTURE). The unique folding is responsible for the protein being very selective in the small molecules it will recognize.

Sadly, the code which relates the fold of a natural protein to its amino-acid sequence is unknown and while polypeptide chains of almost any amino-acid sequence can be made, it is rare for unnatural sequences to fold to a unique three-dimensional structure. Thus the present state-of-the-art in protein engineering avoids the folding problem by making only relatively small

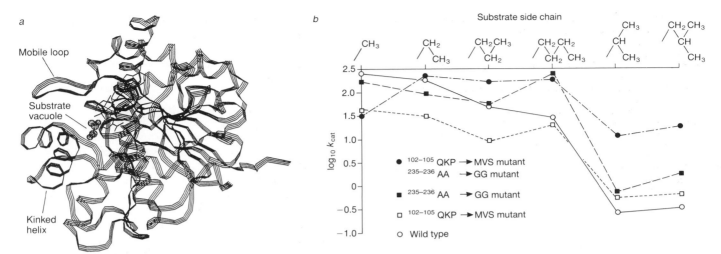

Fig. P56 *a*, Ribbon diagram of the enzyme lactate dehydrogenase from *Bacillus stearothermophilus*. The substrate, pyruvate, is shown as small dotted circles. *b*, Changes that were made in the wild-type enzyme to convert it into a form that could accommodate large branched-chain substrates, such as α-ketoisocaproate. To accommodate the larger substrate, the inside surface of the mobile loop (residues 102–105) was made more mobile by changing a proline to a serine, and also made more hydrophobic (MVS). In addition, the flanking helix was also made more mobile by the incorporation of glycine residues (GG), which destabilize the helix. Neither the helix mutant nor the loop mutant alone were satisfactory. Only when both changes were made was the novel α-ketoisocaproate dehydrogenase produced. *a* J.J. Holbrook; *b*, Adapted from [3].

changes in surface regions of existing large and thermally stable, self-folding protein frameworks. This is protein redesign distinct from protein design *de novo* which would make no reference to the fold of known natural proteins. Design of folds of secondary structure that are unknown in nature has not yet been achieved. Folds that mimic natural ones but use amino-acid sequences that are not direct copies of natural sequences are, however, starting to appear — for example the synthetic four-helix bundle FELIX [1].

Another way to avoid having to design a self-folding protein framework is to link together two or more globular DOMAINS each of which carries a useful biological function. This is supramolecular protein engineering and succeeds because the N- and C-ends of protein chains are usually on the solvent surface of globular proteins and may be joined without interfering with the folding of each domain. Linkage of the required domains is achieved by linking together the appropriate DNA sequences and translating them in cultured cells.

This route has been used to join the enzymatic domain (the B chain) of the protein tissue plasminogen activator to an anti-fibrin antibody. In this way the activator is targeted to blood clots, which are largely composed of fibrin (*see* BLOOD COAGULATION AND ITS DISORDERS). There it activates the precursor (PLASMINO-GEN) of the fibrin-hydrolysing enzyme PLASMIN, which is trapped in the clots. Plasminogen is therefore activated only at the clot, which it then dissolves, and does not lead to a general hydrolysis of circulating FIBRINOGEN and subsequent bleeding.

Design tools

Like other forms of engineering, the design and proving phases of protein engineering use very powerful three-dimensional computer graphics to display the structures of existing proteins which have been determined by protein X-RAY CRYSTALLOGRAPHY and by NMR SPECTROSCOPY and to display the engineered proteins whose structures are being optimized by energy minimization and molecular dynamics.

Enzyme engineering

Enzymes can be engineered to have enhanced properties. Natural selection normally optimizes enzymes for action on natural chemicals under natural conditions. There is, therefore, greatest scope for protein engineering to improve enzymes directed against unnatural substrates or for use under unnatural conditions.

A particularly unnatural condition is in the household washing machine, where enzyme-containing washing powders have to act effectively. For this purpose the proteolytic bacterial enzyme alcalase has been engineered to be (1) stable at 70°C, (2) stable at very alkaline pH, (3) resistant to denaturation by detergent, and (4) resistant to oxidation by the chlorine in the bleach. This latter was achieved by removing the amino acid sensitive to oxidation — methionine.

Engineered enzymes are also used to produce required STEREOISOMERS of a compound for medical use. Stereoisomers often occur in chemicals containing an asymmetric carbon atom. Many modern drugs have such asymmetric centres and the different isomers can have very different effects. Since the trage-dies that resulted when the stereoisomer of a drug that was active against morning sickness turned out to impair foetal develop-ment, the pharmaceutical industry is increasingly introducing drugs which are single compounds, not the naturally occurring mixtures of two stereoisomers. Highly selective enzymes that can discriminate between the two stereoisomers of a compound are often used in the production process, and their usefulness is being extended by protein engineering. For example, the enzyme lactate dehydrogenase normally reduces $CH_3.C=O.COOH$ (pyruvic acid) to $CH_3.H^+COH.COOH$ (the L-lactic acid isomer) but has been rationally engineered (Fig. P56) to specifically transform substrates with very large side chains such as $-CH_2.CH.(CH_3)_2$ as required for modern single-compound drug production (*see* BIOTRANSFORMATION and BIODEGRADATION).

Engineering new catalytic properties into enzymes is still rare as the protein designer must construct the new substrate-binding site without altering the precise arrangement in space of the catalytic residues. A less rational but simpler approach to engineering new enzymes is to raise an antibody to a transition state analogue of the substrate. Any device which stabilizes the transition state of a chemical reaction will catalyse that reaction (*see* MECHANISMS OF ENZYME CATALYSIS). Such antibodies therefore often have weak catalytic activity; they are named ABZYMES (*Antibody-enzymes*). The weak activity can be improved by using protein engineering to insert extra catalytic amino acids into the antibody molecule.

A unique achievement is an abzyme catalysing the Diels–Alder chemical reaction. No natural enzyme is known to catalyse this carbon–carbon bond-forming reaction.

Engineered antibodies

Protein engineering has been exploited in the design of engineered antibodies [2] for medical use in humans. Highly specific MONOCLONAL ANTIBODIES have many applications in medicine. But most have to be made via the mouse HYBRIDOMA route; and the human body tends to react against these mouse (or other animal) antibodies, neutralizing their activity and clearing them rapidly from the body. 'Humanized' antibodies, which escape neutralization by the human immune system, can be produced by protein engineering. DNA encoding the specific antigen-binding site on the mouse antibody can be inserted by genetic engineering into the 'framework' DNA derived from a suitable human antibody gene.

This engineered gene can then be inserted into an appropriate cultured cell to produce the mouse–human hybrid antibody. As most of the surface amino-acid residues of the engineered protein are derived from a human antibody the immune system does not neutralize it. The recently developed ability to make in bacteria large quantities of antibodies from a synthetic gene that codes for a 20-residue peptide which links the heavy and light chain COMPLEMENTARITY DETERMINING REGIONS (CDRs) now avoids the expense and limitation of antibody production in cultured animal cells. For example, engineered hybrid antitumour antibodies that regress tumours by selectively binding to malignant but not normal cells have been produced in this way.

J.J. HOLBROOK

1 Richardson, J.S. & Richardson, D.C. (1990) The *de novo* synthesis of proteins. In *Proteins: Form and Function* (Bradshaw, R.A. & Purton, M., Eds) 173–182 (Elsevier Trends Journals, Cambridge). This book also contains a general introduction to site-directed mutagenesis, protein engineering, bacterial proteinase optimization and the problems of protein folding and protein design.
2 Winter, G. & Milstein, C. (1991) Man-made antibodies. *Nature* **349**, 293.
3 Wilks, H.M. et al. (1990) Designs for a broad substrate specificity keto acid dehydrogenase. *Biochemistry*, **29**, 8587–8591.

protein evolution *See*: MOLECULAR EVOLUTION: SEQUENCES AND STRUCTURES.

protein export *See*: BACTERIAL PROTEIN EXPORT; PROTEIN SECRETION.

protein fingerprint A pattern of spots, characteristic of the protein concerned, which is produced by visualization of fragments separated by two-dimensional ELECTROPHORESIS of a digest of the purified protein by a specific cleavage reagent such as the proteinase trypsin.

Protein folding

THE biological activity of a protein depends on its folding into a highly organized three-dimensional structure under physiological conditions. The central questions of protein folding are:

1 Why does it occur? What are the forces that make a well-defined conformation thermodynamically favourable?

2 How does it occur? What mechanisms permit rapid and efficient self-organization?

3 How is the 3-D structure encoded by the amino-acid sequence? These issues are, of course, intimately related and the goal of applying the principles of folding to, for example, predict structures from sequence or to design sequences compatible with particular structures (and functions?) (*see* PROTEIN ENGINEERING) is likely to require a broad understanding of each. At present there is still some way to go towards this objective but folding is the subject of intensive study.

Although the folding of all proteins must, of course, obey the same physicochemical laws, the balance of forces involved is likely to be very different in determining the structures of, for example, water-soluble proteins compared with those that traverse membranes. To date the vast majority of studies have focused on globular proteins in aqueous solution and these are considered in this discussion.

The native state

The conformation of a protein under conditions where it is biologically active is called its native structure (*see* PROTEIN STRUCTURE). Such structures can be determined in detail by X-RAY CRYSTALLOGRAPHY and NMR SPECTROSCOPY. Globular proteins are organized into compact structural DOMAINS, larger proteins comprising several such domains with varying degrees of interaction between them. Although these domains may depend on their mutual interaction for stability, they can be regarded as quasi-independent folding units. There are obviously wide differences in the details of the 3-D structures of different proteins but it now seems that there may be only a limited number of gross chain topologies compatible with forming a stable structural domain. It is quite commonly observed that proteins with no obvious sequence homology fold into similar structures; for example, the IMMUNOGLOBULIN FOLD is not only found in a huge range of related proteins in the IMMUNOGLOBULIN SUPERFAMILY but is also closely related to folds observed in proteins as disparate as CRYSTALLINS (eye-lens proteins), PLASTOCYANINS (photosynthetic electron carriers) and some viral coat proteins. Thus the folding code is clearly degenerate.

Globular domain structures are characterized by various general features. Side chains are partitioned (but only approximately) according to hydrophobicity, with nonpolar groups packed predominantly into a very compact, water-excluding core while surfaces are formed mainly from charged and polar groups. Some polar groups inevitably are removed from solvent and these generally form intramolecular hydrogen bonds; in particular, a substantial proportion of backbone amide —NH and —CO groups are generally involved in regular hydrogen bonded arrays, defining the secondary structure of the protein. The folding problem is essentially one of understanding how all these features are optimally achieved for a given sequence.

The native structure of a protein can be broken down under various conditions, for example at extremes of temperature or pH or in the presence of high concentrations of chemical agents such as urea. This process, called denaturation, leads to states which are not biologically active and are typically much less ordered than the native. Denaturation is often reversible, demonstrating that the native state is thermodynamically stable, relative to unfolded forms, under physiological conditions. In many cases the native structure may actually be the most stable possible conformation but it is hard to be sure that there are not others which would be more stable, if only they were kinetically accessible. In one case, a type of protease inhibitor called a SERPIN, the native state has actually been shown to be metastable, converting slowly to an alternative structured form that is not biologically active.

Thermodynamics of folding

The key to understanding why proteins fold is to probe the thermodynamics of the process; this can be achieved using various techniques. MICROCALORIMETRY provides a direct route to estimating the enthalpy and entropy of unfolding at the midpoint of the thermal transition and extrapolation is possible to estimate their values, and therefore also the overall free energy of folding under physiological conditions. Alternatively, various spectroscopic probes can be used to monitor denaturation as a function of temperature or denaturant concentration and quantitative analysis of the denaturation curves can then provide estimates of thermodynamic parameters.

The huge loss in conformational entropy consequent upon folding represents a major thermodynamic obstacle. To overcome this, the noncovalent interactions formed within the protein globule and in the surrounding water must be optimized. The balance of free-energy contributions is exquisite and a subset of native interactions may not be sufficient to ensure stability; thus proteins often fold in an all-or-none manner. The stability of the native structure under a given set of conditions depends, therefore, essentially on its free energy relative to that of a disordered state. Denatured states have often been considered to approximate to a random coil but it has now become clear that in some proteins, at least, they may actually have extensive residual structure. The molten globule model describes many such states, characterized by substantial compactness and a high residual secondary structure content. The nature of such partially folded states is currently the subject of considerable interest. In practice, the stability of native states is generally fairly small, in the region of 10–50 kJ mol^{-1}, and the balance of interactions which leads to this net free energy remains incompletely understood.

The sequestration of nonpolar side chains in the protein core appears to be accompanied by a favourable free energy change and this 'hydrophobic effect', analogous in some ways to the immiscibility of oil and water, certainly provides a major driving force for folding. The essence of the effect is simply that noncovalent interactions (including HYDROGEN BONDS in the water and VAN DER WAALS FORCES, especially in the compact protein core) are better optimized overall when nonpolar groups and water molecules are segregated than when they interact directly. However, complications arise because compensatory changes in the structure of water take place when exposed nonpolar groups are solvated, leading to large, temperature-dependent enthalpy and entropy changes. As a result, the physicochemical principles remain controversial in detail.

The hydrophobic effect alone cannot account for the stability of folded structures and other factors must also be important. Hydrogen bonding may well be a major one: in an unfolded state the potential of both backbone and side-chain polar groups will be satisfied by interaction with water, whereas in the folded state many of these will be replaced by intramolecular interactions within the protein (e.g. in SECONDARY STRUCTURE) and increased water–water interaction. The net consequence of this will depend on how optimal the lengths and geometries of the different sets of hydrogen bonds are. This remains a matter of some controversy but there is some evidence for a favourable overall contribution to the free energy of folding.

Some of these general principles of protein stability have been elucidated by comparison of the thermodynamics of unfolding with the dissolution of model compounds; further insight comes from correlations between aspects of the 3-D structures of different proteins, such as the amount of nonpolar surface area buried, and their folding thermodynamics. Protein engineering now affords a powerful method for analysing individual contributions, for example by studying the effects of deletion of specific groups (Fig. P57). These may be interpretable quite directly in terms of the contribution of interactions made by the group in question to the overall stability, although care needs then to be taken to ensure that the structure is not otherwise altered by the change in sequence.

Accounting quantitatively for the thermodynamics of folding remains severely problematic but various simplified approaches to estimating the feasibility of folding into a particular structure have been developed. Some use empirical methods to estimate explicitly the free energy of folding while others are simply based on statistical analysis of patterns of residues observed in proteins of known structure. For example, one can assess whether a sequence is likely to be compatible with a particular domain fold by examining factors such as the likely secondary structural preferences of regions of the sequence, the distribution of polar and nonpolar side chains between the core and surface and how

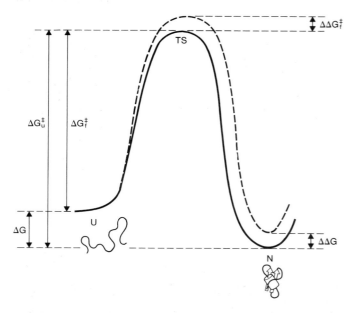

Fig. P57 Thermodynamics of folding of a model protein. The full line is a two-dimensional representation of the conformational free energy surface for the wild-type protein. The states indicated are the unfolded (U), native (N) and transition states (TS). In this simple case, folding is assumed to be wholly cooperative, with no metastable intermediates. On the left are indicated the free energies of the overall transition (ΔG) and of the activation barriers to folding ($\Delta G_f^{\ddagger}$) and unfolding ($\Delta G_u^{\ddagger}$). The dotted line shows the possible result of a protein engineering experiment in which a specific group has been deleted from the protein. The change in the free energy of folding ($\Delta\Delta G$) is a measure of the contribution to the stability of the native structure from the interactions of this group. The effect of the mutation on the kinetics of folding is reflected in the increase in its activation free energy ($\Delta\Delta G_f^{\ddagger}$). In this example the effect is virtually the same on the free energies of the transition state and the native state, with the result that the unfolding rates are unaltered ($\Delta\Delta G_u^{\ddagger}$ ∼0). This would imply that the interactions of this particular group are essentially fully native-like in the transition state.

compact the resulting core would be. In this way, the key determinants of the thermodynamic feasibility of folding are, at least implicitly, taken into account.

Folding pathways

The number of hypothetically possible conformations of a protein is so large that it could not fold in a reasonable time by a process of purely random conformational searching. Instead there must be a mechanism that channels molecules towards the native structure via partially organized intermediates. Crucial questions, which remain the subject of controversy, are: (1) to what extent this implies the existence of a unique folding pathway; (2) whether the need for a kinetically efficient pathway is itself an important determinant of the folded structure; and (3) what general features characterize the sequence of events in a folding process. Although general answers are available to none of these questions, considerable progress is currently being made.

Classical studies of refolding kinetics used spectroscopic probes, such as absorption, fluorescence or CIRCULAR DICHROISM (CD) in conjunction with rapid mixing in a stopped-flow system. Except

in rather small peptides, the timescale of overall refolding is seldom faster than around 100 ms which is conveniently suited to this methodology. However, while these methods remain the basic tools of the trade, on their own they can provide only very limited mechanistic information. The problem is that since folding intermediates are so ephemeral, powerful but intrinsically slow structural probes, such as NMR, can only be applied indirectly.

Two experiments that can provide more detailed, residue-specific data are illustrated in Fig. P58. The first monitors the sequence in which the disulphide bridges in a protein form during refolding from a state in which they have all been reversibly broken. Since such a cross-link will only form in cooperation with the stabilization of folded structure, this provides a reflection of the folding pathway. For a few small proteins, particularly the BOVINE PANCREATIC TRYPSIN INHIBITOR, it has been possible to map out a pathway in considerable detail.

The other experiment uses hydrogen exchange labelling to distinguish between amides protected, usually through specific hydrogen bonds, and those still exposed to solvent at different stages during refolding. In some ways the two methods are analogous in that they monitor indirectly formation of structural elements that afford stability to the individual interactions observed.

Protein engineering is applicable to the study of intermediates and transition states on a folding pathway in much the same way as to the study of native and denatured states (Fig. P57). In this case, free energy perturbations are deduced from changes in the kinetics of folding or unfolding processes, using simple transition state theory. This method promises to be very powerful. For example, in combination with a simple fluorescence probe, it has already been used to provide a remarkably detailed structural picture of an intermediate and the late transition state in the refolding of a small enzyme, barnase. In conjunction with a wider range of structural probes, still further resolution of folding pathways should be possible.

The characteristic cooperativity of folding suggests that isolated structural elements might not be expected to be stable. This led to the idea that a very early 'nucleation' event might be the rate-limiting step in folding. In kinetic studies, however, the transition state of folding appears to be a highly ordered state, with many features of the native structure. Thus initiation of folding seems to be a surprisingly facile process but it is still not wholly clear how it happens. Two models that have been widely discussed are illustrated in Fig. P59 and described below:

1 The framework model. This emphasizes the crucial role of interactions between residues adjacent in the sequence at this early stage, proposing that elements of the secondary structural framework form rapidly and essentially independently before being organized more slowly into a globular structure. Various lines of experimental evidence have been cited in support of this model: for example, NMR studies of protein fragments have revealed that some isolated sequences do tend to fold into native-like secondary structural elements, including discrete helices and reverse turns. Stopped-flow CD studies have lent further credence to the model, demonstrating that substantial secondary structure typically forms within 1 ms, while the overall tertiary structure formation follows much more slowly.

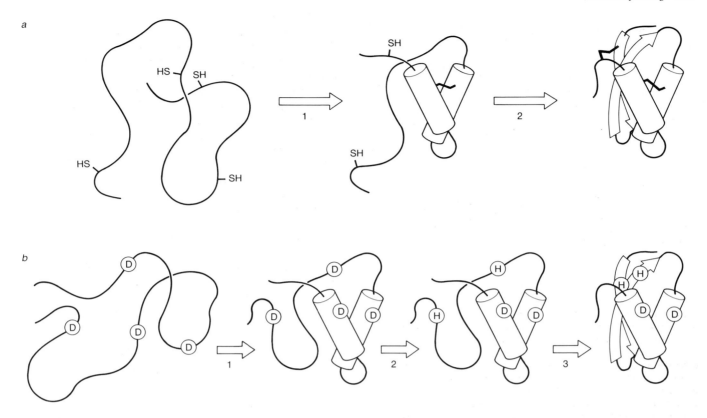

Fig. P58 Experiments for probing folding pathways. *a*, The disulphide cross-links of a protein are reduced to free thiols (—SH), causing it to unfold. (1) Refolding is initiated by adjusting the redox state. The sequence in which the disulphides (shown by heavy angled lines) re-form is a reflection of the folding pathway, since they only become stable when there are sufficient cooperating noncovalent interactions. (2) Ultimately the native state will be restored. The reaction can, however, be stopped at various points by blocking remaining free thiols; analytical methods can then be used to determine the distribution of disulphides which have formed at that stage. By doing this after varying times, the pathway can be elucidated. *b*, A protein is unfolded by action of a chemical denaturant in D_2O; this causes exchangeable hydrogens — for example, backbone amide NHs — to be replaced by deuterons (D). (1) Folding is initiated by dilution with excess H_2O; at this stage the pH is chosen such that these deuterons will not exchange back. (2) After a chosen interval, the pH is raised so that exchange is very fast and any deuterons that remain accessible to solvent are replaced by hydrogens (H); elements of folded structure of sufficient stability will, however, protect amides involved. (3) After this short pulse the pH is again dropped and folding allowed to proceed to completion. Amides that are buried in elements of stable structure are now protected from solvent exchange and the pattern of labelling that occurred during refolding can be assayed by NMR spectroscopy. By varying the time allowed before labelling the folding pathway can be mapped out.

2 The hydrophobic collapse model. This is based on the idea that the large thermodynamic driving force for exclusion of water from around nonpolar groups will tend to cause them to cluster together in any way that would lead to a substantial reduction in accessible surface area. Thus an early event in folding might be collapse to a fairly condensed but not necessarily very specific structure which would then reorganize more slowly to form the final folded conformation. This model would still predict the formation of secondary structure but in this case as a consequence of the collapse, in order to optimize hydrogen bonding and side chain packing in the condensed state. Experimental investigation of this possibility is less straightforward since convenient probes of refolding do not directly measure hydrophobic collapse. Some results, based on detecting changes in residue mobility and average inter-residue distances, suggest that the main increase in overall packing density actually tends to occur as an intermediate stage, following a much more rapid initial secondary structure formation; this led to the idea of a condensed state with a high secondary structure content — a 'molten globule' — as a general type of intermediate in folding pathways. However, much faster clustering of some nonpolar residues has been demonstrated by fluorescence methods, so the timescale on which hydrophobic interactions become important remains a complex issue.

At present there are insufficient data to be sure of the general validity of either of these models and the truth may indeed incorporate elements of both. Now that information on the behaviour of individual residues is becoming obtainable, a clearer view may emerge. Protein engineering methods and hydrogen exchange labelling are now enabling partially structured states to be characterized in unprecedented detail, providing fascinating insight into the hierarchical organization of the native state. Much remains to be sorted out, however; complications have been observed, such as parallel alternative folding pathways and the formation of non-native structure which has subsequently to rearrange to form the 'correct' set of interactions, although it is not clear how general and how mechanistically important these

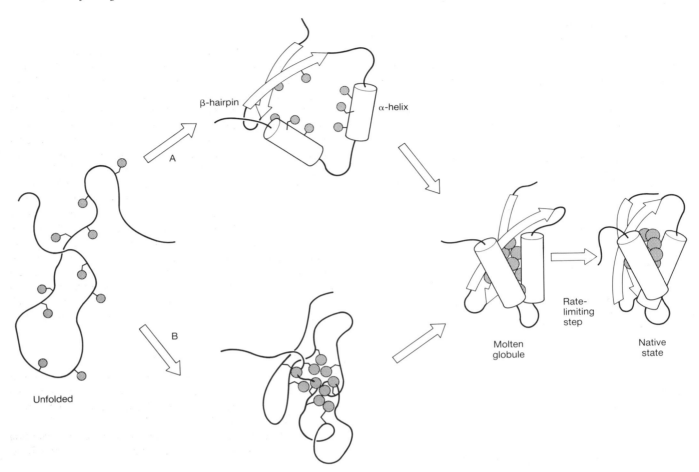

Fig. P59 Models for the folding mechanism of a hypothetical single-domain protein. Hydrophobic groups which ultimately form the core of the native structure are represented by the filled circles (●). Starting from the unfolded state, which has no persistent ordered structure, alternative ways in which folding might be initiated are considered. Route A corresponds to the framework model, in which secondary structural elements (in this case two helices and a β-hairpin) first form independently, driven by interactions between residues close together in the sequence. Subsequently these elements tend to coalesce with clustering of hydrophobic groups to form a relatively compact intermediate or molten globule (MG). An alternative route

(B) supposes that the initiation event is a hydrophobic collapse, driven by the favourable free energy of excluding nonpolar groups from water. The clustering of these groups then prompts the backbone to become organized into secondary structure, to optimize both hydrogen bonding and stereochemistry, leading again to a molten globule-type folding intermediate. At this stage side chains are still relatively mobile and nonspecifically packed. The final step involves completion, adjustment and locking of the structure to form the highly ordered, almost solid-like core of the final native state. This late event appears typically to be rate limiting.

features are. It is also true that a great deal happens too fast to be resolved using standard stopped-flow kinetic methods, so that new technical advances are needed if we are to be able to probe the very early stages of organization.

Folding *in vivo*

In recent years the concept of spontaneous self-organization of protein structures has undergone something of a revolution (*see* SELF-ASSEMBLY). Although many proteins will apparently renature without assistance *in vitro*, others do not do so efficiently and the problem often becomes severe in the case of proteins that assemble into oligomeric structures. It is now clear that correct folding is not left to chance in a cell, but that mechanisms exist to help ensure that the correct native structure is achieved.

MOLECULAR CHAPERONES are a range of proteins which exist in all cellular compartments where folding occurs, apparently to assist in folding and assembly processes. This is a quite complex, often ATP-driven operation, which is not yet understood in detail. Chaperones are not specific to particular proteins but appear to be able to interact with any polypeptide which has as yet very little folded structure, thereby protecting it from unproductive processes such as aggregation. Whether folding is actually promoted while the polypeptide is still bound to the chaperone is a matter of controversy. There are also cellular enzymes that appear to catalyse other events constituting significant barriers to folding — x-proline peptide bond isomerization (*cis–trans* isomerization) and disulphide interchange. None of these aids to folding actually determines the native structure, however; they simply ensure that it is achieved successfully. The information encoding the 3-D structure is contained within a protein's own amino-acid sequence. Thus, while *in vitro* renaturation may not, in some cases, be a good model for the process of folding as it happens in the cell, its study remains a valid approach to teasing out the details of how the native structure is determined by the sequence.

An excellent series of reviews, covering most aspects of folding is to be found in [1] and other useful reviews in [2–4].

P.A. EVANS

1 Creighton, T.E. & Kim, P.S. (Eds) (1991) Folding and binding. *Curr. Opinion Struct. Biol.* **1**, 3–60.
2 Kim, P.S. & Baldwin, R.L. (1990) Intermediates in the folding reactions of small proteins. *Annu. Rev. Biochem.* **59**, 631–666.
3 Creighton, T.E. (1990) Protein folding. *Biochem. J.* **270**, 1–16.
4 Matouschek, A. & Fersht, A.R. (1991) Protein engineering in analysis of protein folding pathways and stability. *Meth. Enzymol.* **202**, 82–112.

---

protein G A monomeric protein (M_r 63 000) from human group G streptococcus. It possess two or three antibody-binding sites and binds IgGs (*see* ANTIBODIES) from a wide variety of species. Compared with PROTEIN A it binds better to rat, mouse and goat IgGs and is a useful secondary reagent in IMMUNOELECTRON MICROSCOPY.

Björk, L. & Kronvall, G. (1984) *J. Immunol.* **133**, 969–974.

protein import *See*: MITOCHONDRIAL BIOGENESIS; PROTEIN TRANSLOCATION.

Protein kinase(s)

THE protein kinases are enzymes that catalyse PROTEIN PHOSPHO-RYLATION by transferring the terminal phosphate from ATP to a side chain of the protein:

$$\text{Protein} + \text{ATP} \rightarrow \text{Phosphoprotein} + \text{ADP}$$

Protein phosphorylation is normally a reversible process, but conversion of the phosphoprotein back to the unmodified protein is catalysed by PROTEIN PHOSPHATASES which bring about a distinct reaction, a hydrolysis not involving ATP or ADP. In eukaryotic cells, protein kinases serve at least three general functions:

1 They are the major agencies through which extracellular signal molecules acting at cell-surface RECEPTORS produce their intracellular effects (*see* CYCLIC AMP-DEPENDENT PROTEIN KINASE; GROWTH FACTOR RECEPTORS; PROTEIN KINASE C; SECOND MESSENGER PATHWAYS).

2 They are the major agencies through which events that occur discontinuously in the CELL CYCLE (e.g. DNA SYNTHESIS; MITOSIS) are initiated and timed.

3 They also seem in some cases to protect the cell against toxic changes in metabolites, a form of stress response (*see* AMP-ACTIVATED PROTEIN KINASE).

The introduction of the negatively charged phosphate group usually causes a marked change in function of the target protein (*see* PROTEIN PHOSPHORYLATION).

Structure

All protein kinases for which the amino-acid sequence has been determined consist of a catalytic domain of M_r ~35 000 coupled to regulatory regions or domains. The catalytic and regulatory domains may be on the same subunit (e.g. MYOSIN LIGHT CHAIN KINASE; CA²⁺/CALMODULIN-DEPENDENT PROTEIN KINASE II) or on separate subunits (e.g. CYCLIC AMP-DEPENDENT PROTEIN KINASE; PHOSPHORYLASE KINASE). In all cases the catalytic domains (of which about 200 have now been sequenced) show clear sequence similarity [1], and all are probably therefore derived from a common ancestor. The tertiary structure of one member of this large family (the catalytic subunit of cAMP-dependent protein kinase) has been determined by X-RAY CRYSTALLOGRAPHY [2]. It consists of a small N-terminal domain and a larger C-terminal domain (Fig. P60). The active site is in the cleft between the two domains, with MgATP binding to residues in the N-terminal domain and the protein substrate (represented in the crystal structure by a nonphosphorylatable peptide inhibitor) forming interactions with the C-terminal domain. It is likely that all protein kinase catalytic domains are folded in a similar manner.

Specificity

The phosphate groups inserted by protein kinases are esterified on the hydroxyl moieties of serine, threonine, or tyrosine side chains. A different class of protein kinases phosphorylates histidine residues, but the function of these is poorly understood. Around 200 protein kinases have now been sequenced, counting ISOFORMS and homologues in different species. About two-thirds

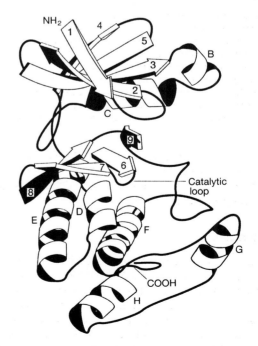

Fig. P60 Diagram of the conserved catalytic core of the catalytic subunit of cAMP-dependent protein kinase. Adapted from [2].

Table P11 Phosphorylation site motifs

Protein kinase	Phosphorylation site motif
Cyclic AMP-dependent protein kinase	**RR**X<u>S</u>
Cyclic GMP-dependent protein kinase	**R/K**X<u>S</u>
Protein kinase C	<u>S</u>X**K/R**
Calmodulin-dependent protein kinase II	**R**XX<u>S</u>
Casein kinase II	<u>S</u>XX**E**
Casein kinase I	**S(p)**XX<u>S</u>
Glycogen synthase kinase-3	<u>S</u>XXX**S(p)**
p34/cdc2	<u>S</u>**P**X**K/R**

In each case the most common motif is shown, although minor variations in these motifs do occur. The phosphorylation site is shown as <u>S</u> (serine), although most of these protein kinases will also phosphorylate threonine at this position. Residues critical in kinase recognition are shown in bold type: **S(p)** refers to phosphoserine.

of these are specific for serine or threonine residues, with the remaining third being specific for tyrosine residues (*see* GROWTH FACTOR RECEPTORS). Both classes are members of the same MULTIGENE FAMILY, although they fall into distinct subfamilies by sequence analysis [1]. It seems likely that they evolved from an ancestral nonspecific type, and indeed a third class of protein kinase has recently been described which phosphorylates both tyrosine and serine/threonine residues. This group, exemplified by the *wee1* gene product of the yeast SCHIZOSACCHAROMYCES POMBE [3], is difficult to distinguish from the serine/threonine-specific subfamily on the basis of amino-acid sequence alone.

Many protein kinases will rapidly phosphorylate small synthetic peptides (10–16 residues) based on the amino-acid sequences around sites phosphorylated on the target proteins. In these cases it seems that the tertiary structure of the target protein is not important for recognition, although it is clearly essential that the target site is exposed on the surface of the protein. A few protein kinases (e.g. phosphorylase kinase) do not efficiently phosphorylate short peptides, and in these cases it seems that the secondary/tertiary structure of the target is important for recognition.

Phosphorylation studies with protein targets and synthetic peptides have shown that most protein kinases recognize a simple amino-acid sequence motif which involves a neighbouring basic or acidic residue [4]. A few of these motifs are shown in Table P11, although it should be realized that these cannot be rigorously applied. For example, the commonest motif for recognition by cAMP-dependent protein kinase is **RRXS**, but physiological substrates are known where the motif used is **KRXXS**, or **RRS** (single-letter amino-acid notation). It is now known that the adjacent basic residues (**RR** or **KK**) interact with four glutamate side chains, each on a different loop of the catalytic subunit of the kinase. These loops must be sufficiently flexible to allow different spacings between the double basic residue motif and the phosphorylated serine/threonine.

Regulation

Protein kinases are regulated in a number of different ways:

Constitutively active

Some protein kinases seem to be intrinsically active and have no known form of regulation. This may of course merely indicate that the mechanism of regulation has not been found. However, in some cases the target sites also appear to be constitutively phosphorylated in cells, for example those phosphorylated by casein kinase II. Phosphate at these sites may be inserted during or immediately after synthesis of the target protein, and may not turn over rapidly. Regulation of these phosphorylation events, if it occurs, must be by modulation of protein phosphatases.

Second-messenger dependent

An important class of protein kinases are activated by the binding of second messengers. This class includes cAMP-dependent protein kinase (protein kinase A), cGMP-dependent protein kinase, and protein kinase C. There are also several protein kinases which are activated by CALCIUM and CALMODULIN, for example Ca^{2+}/calmodulin-dependent protein kinase II, phosphorylase kinase, myosin light chain kinase, elongation factor 2 kinase. The second-messenger-dependent protein kinases mediate most, if not all, intracellular responses to these second messengers. In the absence of the second messenger, the catalytic domain is maintained in an inactive state by binding to an inhibitor region which may be on a separate subunit (e.g. cAMP-dependent protein kinase) or on the same subunit (most others). This inhibitory region may contain an actual phosphorylation site sequence, in which case it is phosphorylated when the second messenger binds. This is known as AUTOPHOSPHORYLATION. In other cases the inhibitory region resembles a phosphorylation site but does not have a phosphorylatable residue, in which case it is known as a PSEUDOSUBSTRATE SITE. Examples of some autophosphorylation and pseudosubstrate sites are shown in Table P12.

Extracellular-signal activated (receptor protein kinase)

In these cases the protein kinase catalytic domain is present as the cytoplasmic domain of a transmembrane receptor in which the external domain binds an extracellular messenger. This binding activates the internal protein kinase, probably by causing aggregation of receptors. Most well-studied examples are receptors for polypeptide hormones or growth factors in which the protein kinase is tyrosine specific, such as the receptors for insulin, epidermal growth factor and platelet-derived growth factor (*see* GROWTH FACTOR RECEPTORS; INSULIN AND ITS RECEPTOR). However, a few recent examples have been found where the protein kinase domain is more closely related to serine/threonine kinases, for example the transforming growth factor-β receptor (TGFβ receptor).

Protein-kinase regulated

Some protein kinases are themselves regulated by phosphorylation by other protein kinases. A PROTEIN KINASE CASCADE is a sequence of protein kinases in which one phosphorylates and activates the next. Although few well characterized examples of

Table P12 Comparison of the sequence of a phosphorylation site on a representative target protein, and autophosphorylation/pseudosubstrate sites on the protein kinase itself

Protein kinase	Phosphorylation/ pseudosubstrate site	Location and type of site
cAMP-dependent	RTK RSNSV	Phosphorylase kinase β subunit, P
	GRR RRGAI	RI subunit of protein kinase, PS
	RFD RRVSV	RII subunit of protein kinase, AP
Myosin light chain	KKRP QRATS	Smooth muscle light chain, P
	RRKWQKTGH	Smooth muscle light chain kinase, PS
	KKAK RRAA AEGGS	Skeletal muscle light chain, P
	KRRWKKNF I AVSA	Skeletal muscle light chain kinase, PS
Ca^{2+}/calmodulin-dependent II	QA TRQA SI	Synapsin 1 site 2, P
	CMHRQE TV	α isoform of kinase, AP
Protein kinase C	I VRK RTLR RLL	EGF receptor, P
	FARK GALR EKN	α/β isoform of protein kinase, PS

Single-letter amino acid notation. The phosphorylation/pseudosubstrate site is shown in bold type; basic residues implicated in site recognition are underlined. AP, autophosphorylation site; EGF, epidermal growth factor; P, phosphorylation site; PS, pseudosubstrate site.

this phenomenon have been described, the number is increasing. Examples include activation of phosphorylase kinase by cAMP-dependent protein kinase, activation of AMP-activated protein kinase by its kinase kinase, and the four component cascade:

MAP kinase kinase kinase → MAP kinase kinase
→ MAP kinase → S6 kinase

There are also examples in which phosphorylation inactivates the protein kinase, for example phosphorylation of the protein kinase c-*src* on tyrosine 527 by c-*src* kinase, and phosphorylation of p34/cdc2 on tyrosine 14 and threonine 15, by the *wee1* protein kinase and another unidentified kinase (*see* CELL CYCLE).

D.G. HARDIE

1 Lindberg, R.A. et al. (1992) Dual-specificity protein kinases: will any hydroxyl do? *Trends Biochem. Sci.* **17**, 114–119.
2 Knighton, D.R. et al. (1991) Crystal structure of the catalytic subunit of cAMP-dependent protein kinase. *Science* **253**, 407–414, see also 414–420.
3 Featherstone, C. & Russell, P. (1991) Fission yeast p107^{wee1} mitotic inhibitor is a tyrosine/serine kinase. *Nature* **349**, 808–811.
4 Kemp, B.E. & Pearson, R.B. (1990) Protein kinase recognition sequence motifs. *Trends Biochem. Sci.* **15**, 342–346.

protein kinase A CYCLIC AMP-DEPENDENT PROTEIN KINASE.

Protein kinase C

PROTEIN kinase C (PKC) encompasses a family of PROTEIN KINASES that are both structurally and functionally related [1–3]. These enzymes catalyse the transfer of phosphate from ATP to serine or threonine side chains in a wide variety of protein substrates. To date members of this kinase subfamily have been identified in mammals [1–3], *Drosophila* [4], *Xenopus* [5], nematode [6], and yeast [7], but not in prokaryotes. PKC is considered to act as the intracellular 'receptor' for the second messenger DIACYLGLYCEROL (*see* SECOND MESSENGER PATHWAYS). Many cellular agonists stimulate the production of diacylglycerol through the activation of PHOSPHOLIPASES C [8] (or D) (Table P13) and as such PKC is likely to have a key role in the response to these agonists. This is emphasized by the fact that biologically active PHORBOL ESTERS can directly mimic the effect of diacylglycerol [9].

Structure

PKC proteins are monomeric and range in size from 592 to 737 amino acids. Seven mammalian PKC genes have so far been identified. These are classified according to the Greek alphabet as α to η [1–3, 10]. Differential messenger RNA splicing (*see* RNA SPLICING) within the coding region has been clearly documented for the β gene giving rise to two distinct products $β_1$ and $β_2$. A truncated form of PKC-ε has also been described and is defined as $ε_2$. The genes for PKC-α, -β and -γ have been mapped to human chromosome 17 at q22-q24, chromosome 16 at p12-q11.1 and chromosome 19 at q13.2-q13.4 respectively.

The predicted amino-acid sequences for the PKCs contain a series of highly conserved regions (C_1–C_4 in Fig. P61) interspersed with variable regions (V_0–V_5) which show little overall similarity. Between the V_1 and C_1 regions in all these enzymes is a PSEUDO-SUBSTRATE SITE that has been shown to play an essential part in suppressing the activity of the HOLOENZYME. The C_1 region itself confers the diacylglycerol/phorbol ester binding activity of the protein. This DOMAIN is made up of a cysteine-rich tandem repeat (except for PKC-ζ which contains only one cysteine-rich unit); the presence of either of the cysteine-rich repeats has been shown to confer phorbol ester binding.

The C_2 domain seems to be responsible for the Ca^{2+}-dependent

Table P13 Agonists stimulating inositol lipid breakdown

Agonist	Cell type
Carbachol	Astrocytoma cells
LTD-4	Basophilic leukaemic cells
5-Methyltryptamine, 5-Hydroxytryptamine (5HT, serotonin)	Blowfly salivary glands
Adrenaline, 5HT	Corneal epithelium
Bradykinin, platelet-derived growth factor	Fibroblasts
Thyroid-releasing hormone	GH3
Vasopressin, angiotensin II	Glomerulosa cells
Vasopressin, adrenaline, angiotensin, EGF	Liver
Carbachol, oxytocin	Myometrium
P2 (purinergic)	Neutrophil/monocyte progenitors
fMet-Leu-Phe	Neutrophils
Caerulein, carbachol, substance P, cholecystokinin	Pancreatic acinar cells
Thrombin	Platelets
fMet-Leu-Phe	Polymorphonuclear leukocytes
Acetylcholine, thrombin, endothelin	Smooth muscle
Vasopressin	WRK-1 cells

A selection of agonists acting on a variety of cell types serves to illustrate the broad spectrum of agents that elicit inositol lipid breakdown.

interaction of PKC with membranes. This is inferred from the fact that PKC-α, -β and -γ can (nonproductively) bind phospholipid bilayers in the presence of Ca^{2+} while PKC-ϵ and -δ, which have no C_2 domain, show no such Ca^{2+}-dependent interaction. It is likely therefore, but unproven, that PKC-ζ and -η behave like PKC-ϵ and -δ in this respect.

The catalytic portion of the molecule is comprised of the C_3–V_5 regions. The V_3 region seems to behave like a hinge between the regulatory (V_0–V_3) and catalytic domains. This region becomes exposed upon activation and is the primary site of proteolysis by calpain and trypsin. Limited proteolysis by either of these proteases leads to the generation of a constitutively active kinase domain fragment (C_3–V_5), which is frequently referred to as PKM catalytic fragment. The predicted sequences for this C-terminal region are homologous to all other known serine-, threonine- and tyrosine-specific protein kinases, defining this as the catalytic domain.

Effectors

As isolated, the PKCs in their holoenzyme forms show very little kinase activity towards most substrates (it should be noted however that protamine sulphate is unusual in that it can serve as a substrate in the absence of PKC activators). For optimal activity, diacylglycerol (or phorbol esters) and phospholipids are required and in the cases of PKC-α, -β_1, -β_2 and -γ (which contain the C_2 domain), Ca^{2+} is also required for most substrates.

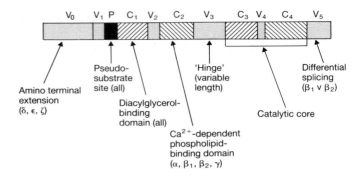

Fig. P61 Domain organization of the PKC family.

Certain fatty acids including ARACHIDONIC ACID have also been shown to stimulate activity of certain PKCs *in vitro*, but it is not clear whether this is of physiological significance.

Activity

The typical PKC substrate, lysine-rich HISTONE, is often used to study activity, but is a very poor substrate for PKC-ϵ and -δ [3]. By contrast, all the PKC enzymes studied to date will phosphorylate serine-containing synthetic peptides based upon their own and indeed each others' pseudosubstrate sites. These all contain basic residues both N-terminal and C-terminal to the target serine. Thus all the PKCs appear to show a specificity towards substrate sequences containing basic residues.

The majority of PKC enzymes have not been compared using any of the few physiological substrates defined so far. On the basis of studies with synthetic peptide substrates, however, it can be concluded that there are differences in primary structural requirements between different enzymes.

Physiological substrates

Although a number of potential physiological substrates for PKC have been described, few have been defined rigorously. Of those that have, a number are membrane associated (e.g. the receptor for epidermal growth factor (EGF) (*see* GROWTH FACTOR RECEPTORS), c-Src (*see* ONCOGENES), and the T CELL RECEPTOR) and this is consistent with the membrane association of activated (i.e. diacylglycerol/phospholipid-bound) PKC. In the case of the EGF receptor, phosphorylation is associated with a decrease in receptor affinity for EGF and a small reduction in receptor tyrosine kinase activity. In the case of other cell surface substrates there is circumstantial evidence for an increased rate of internalization following PKC-dependent phosphorylation. Certain substrates such as the MARCKS protein (*m*yristylated *a*lanine-rich *C*-kinase *s*ubstrate, also known as 80K), appear to be phosphorylated in a PKC-dependent fashion even in a soluble state; it is possible that such substrates simply diffuse to the active, membrane-bound enzyme. However, MYOSIN light chain which is itself associated with the CYTOSKELETON appears to be phosphorylated at a specific site by PKC and it has been suggested that a proteolysed form of PKC can be produced following membrane-associated activation (*see* 'downregulation' below).

By contrast, although *in vitro* PKC can phosphorylate the RIBOSOMAL PROTEIN S6, and activation of PKC *in vivo* leads to S6 phosphorylation, it is now clear that S6 phosphorylation is regulated through a KINASE CASCADE; this is likely to be one of many such kinase cascades.

Assessing activation

Several strategies have been used to assess the role of agonist-induced PKC activation in eliciting cellular responses. Three general strategies are routinely employed. The first is characterized by assessment of the phosphorylation state of the MARCKS protein (a PKC-specific substrate); if increased phosphorylation is observed, it is inferred that PKC has become activated. The strategy is thus to assess what seems to be a specific consequence of PKC activation. A second approach is to determine the location of PKC itself. If cells are extracted in the presence of Ca^{2+}-chelators PKC shows a predominantly cytoplasmic distribution. However, on activation *in vivo* the active complex formed in the presence of diacylglycerol proves to be relatively stable to extraction; that is, the activated PKC is stabilized in its membrane-associated form. As such a so-called 'translocation' assay can be used to define any increase in perceived membrane-associated PKC, this type of analysis is in reality not a 'translocation' assay but a 'membrane stabilization' assay.

The third strategy used to assess the role of PKC in a response involves the phenomenon of downregulation. Chronic exposure of cells to the nonmetabolizable phorbol esters leads to chronic activation of PKC with a concomitant increase in its rate of proteolysis, apparently through initial cleavage in the V_3 domain. In many instances, but not all, this leads to a severe net loss of PKC (the steady-state level may reduce to <5%). Under such conditions, washing and rechallenging cells with phorbol esters no longer elicits typical responses — that pathway is inoperative. Under these conditions it is possible to test whether cellular responses to other agonists (i.e. non-phorbol esters) are also blocked. If a response is impaired or lost in downregulated cells it indicates a role for PKC in that response.

P.J. PARKER

See also: CALCIUM-BINDING PROTEINS; G PROTEIN-COUPLED RECEPTORS; GROWTH FACTOR RECEPTORS.

1 Nishizuka, Y. (1984) The role of protein kinase C in cell surface signal transduction and tumour promotion. *Nature* **308**, 693–695.
2 Nishizuka, Y. (1988) The molecular heterogeneity of protein kinase C and its implications for cellular regulation. *Nature* **334**, 661–665.
3 Parker, P.J. et al. (1989) Protein kinase C — a family affair. *Mol. Cell. Endocr.* **65**, 1–11.
4 Rosenthal, A. et al. (1987) Structure and nucleotide sequence of a *Drosophila melanogaster* protein kinase C gene. *EMBO J.* **6**, 433–441.
5 Chen, K.-h. et al. (1988–89) Molecular cloning and sequence analysis of two distinct types of *Xenopus laevis* protein kinase C. *Second Messengers and Phosphoproteins* **12**, 251–260.
6 Tabuse, Y. et al. (1989) Mutations in a protein kinase C homolog confer phorbol ester resistance on *Caenorhabditis elegans*. *Science* **243**, 1713–1716.
7 Levin, D.E. et al. (1990) A candidate protein kinase C gene, PKC1, is required for the *S. cerevisiae* cell cycle. *Cell* **62**, 213–224.
8 Meldrum, E. et al. (1991) The PtdIns-PLC superfamily and signal transduction. *Biochim. Biophys. Acta* **1092**, 49–71.
9 Castagna, M. et al. (1982) Direct activation of calcium-activated, phospholipid-dependent protein kinase by tumour-promoting phorbol esters. *J. Biol. Chem.* **257**, 7847–7851.
10 Osada, S. et al. (1990) A phorbol ester receptor/protein kinase, nPKCη, a new member of the protein kinase C family predominantly expressed in lung and skin. *J. Biol. Chem.* **265**, 22434–22440.

---

protein kinase cascade A chain of two or more PROTEIN KINASES in which one phosphorylates and activates the other, and this process is repeated until the end of the chain is reached. The classic example is the activation of PHOSPHORYLASE KINASE by CYCLIC AMP-DEPENDENT PROTEIN KINASE, but longer cascades (e.g. RAF → MAP KINASE KINASE → MAP KINASE → ribosomal subunit kinase p90^{rsk}) have been reported. As each protein kinase acts catalytically, a considerable amplification of the initial signal can be achieved. *See also*: PROTEIN PHOSPHORYLATION.

Protein–nucleic acid interactions

THE interaction of proteins with both RNA and DNA is fundamental to the process and regulation of gene expression and DNA REPLICATION. In consequence there has been a considerable incentive to understand the nature of such interactions. A whole range of molecular genetic techniques have been used to identify and study biologically relevant protein–nucleic acid complexes. However, the application of the biophysical techniques of X-RAY CRYSTALLOGRAPHY and NMR SPECTROSCOPY has been necessary in order to elucidate the precise three-dimensional structures of both the isolated components and the details of their interaction.

In recent years, using these structural techniques, remarkable progress has been made towards understanding the structural and chemical basis of the interactions between sequence-specific DNA-binding proteins and their binding sites. In contrast we are only just beginning to gain an insight into the nature of protein–RNA interactions.

Nucleic acid recognition

What does the recognition of nucleic acids involve? Like all molecular interactions protein and nucleic acid molecules will interact if the free energy of their interaction, under the conditions in question, is such that the complex is more stable than the free components. In other words the interaction between the two molecules must be sufficiently strong to outweigh the energy and entropy cost in bringing them together and excluding solvent from the interface. Since, in aqueous solution, intermolecular forces are weak, the combined effect of many individual interactions is required for the formation of a stable complex. Hence interacting macromolecules have highly complementary surfaces so as to maximize the surface area of interaction between them and hence the number of potential intermolecular interactions. This can be thought of as shape recognition. The nature and arrangement of the individual intermolecular interactions can be thought of as chemical recognition. So when proteins and nucleic acids interact, the complementary shapes of the molecules allow the detailed chemical interactions to occur.

Nucleic acid-binding proteins themselves can be divided into two classes: (1) those that bind highly specifically to a particular binding site; and (2) those that bind nonspecifically with little reference to the base sequence. For the first class it is critical that the protein discriminates between closely related binding sites. This is typified by the binding of TRANSCRIPTION FACTORS to their gene PROMOTERS (*see* BACTERIAL GENE EXPRESSION; EUKARYOTIC GENE EXPRESSION; TRANSCRIPTION) and the charging of TRANSFER RNA (*see also* GENETIC CODE). In the one case the transcription factor must bind to only one, or a few specific sites within a continuous stream of nonspecific DNA. In the other case the tRNA synthetase must discriminate between 20 tRNAs with closely related three-dimensional structures.

Proteins that process and package nucleic acids (e.g. HISTONES in eukaryotes and protein HU in the bacterium *Escherichia coli*), illustrate the second class of nucleic acid-binding proteins. Specific and nonspecific nucleic acid binding is discussed in terms of both shape and chemical recognition.

What is being recognized

DNA structure

Early FIBRE DIFFRACTION studies of DNA revealed that the double helix could adopt discrete conformations depending upon the degree of hydration. Two models for the extreme conformations (differing in their helical parameters, *see* NUCLEIC ACID STRUCTURE) were termed A- and B-forms. It is now generally accepted that both the conformation and rigidity of DNA are determined by local base stacking and that the regular A- and B-forms are not adequate to describe DNA in solution. For any one stretch of helix the width and depth of the major and minor grooves, the displacement of base pairs from the helix axis, the helical periodicity and the global bend of the DNA are determined by the sequence of bases and consequently the structure and flexibility of the double helix is continuously variable. Proteins binding to DNA can access the base pairs in either the major or minor grooves. Since the structural variation and deformability of the double helix affects the accessibility and position of hydrogen bonding groups in the major and minor grooves it is an essential part of protein–nucleic acid recognition. In DNA close to the B-form structure the major groove can accommodate protein secondary structure more easily than the minor groove. In A-like DNA the converse is true. Furthermore, in the major groove the pattern of hydrogen bond donors and acceptors is unique for each base pair whereas in the minor groove it is not possible to distinguish between AT and TA base pairs nor between GC and CG base pairs. Consequently the major groove is better suited to making sequence-specific interactions.

RNA structure

RNA differs from DNA in that it has an additional 2′ hydroxyl on the ribose (*see* NUCLEIC ACID STRUCTURE: RNA). This affects the conformation of the backbone such that double-stranded RNA has helical parameters close to those of A-form DNA. However, more importantly, most RNAs occur as intramolecularly folded single molecules and hence show a significantly greater structural variety than DNA. Many of the possible structural elements in RNA are illustrated in the crystal structure of tRNA (*see* NUCLEIC ACID STRUCTURE; TRANSFER RNA). These include bubbles, bulges, triple helices, pseudoknots and stem-loop structures. It is clear therefore that proteins which recognize RNA are likely to have as wide a structural variation as the RNAs to which they bind.

Shape recognition: nucleic acid-binding proteins

Much of our knowledge of DNA recognition derives from structural analyses of the DNA-binding domains found within transcription factors. A number of different classes of DNA-binding motifs have emerged (Fig. P62 and [1,2]). Each class is typified by sequence and structural conservation. Whereas the conservation of sequence is rather weak for the helix-turn-helix motifs, it is particularly evident in the zinc-binding motifs which have a highly conserved structural framework (see below).

Each of the known DNA-binding motifs has been duplicated during evolution and has subsequently diverged within the individual protein environments so as to bind to different sites in DNA. Comparison of the different DNA-binding motifs and how they have evolved to bind to different sequences is very revealing of the features important for DNA recognition.

At the time of writing structures of six DNA-binding motifs have been solved. Four of these have rigid DNA-binding secondary structures that interact in the major groove of DNA (Fig. P62). In three of these (the helix-turn-helix motif, the zinc-finger motif and the steroid hormone receptor motif), this interaction is mediated by a short α-helix which protrudes from the surface of the protein. In the fourth motif with a rigid secondary structure that binds in the major groove (termed ribbon-helix-helix motif) a two-stranded antiparallel β-sheet performs a role similar to that of the α-helix. (*See* PROTEIN STRUCTURE for an explanation of the terms used to describe protein structure.) Recognition, however, is not always mediated by rigid secondary structural elements. Two known DNA-binding motifs (the basic leucine zipper motif and the HU β-strand motif) have 'arms' that are flexible in the absence of DNA, but are thought to become ordered on binding, wrapping round the double helix. In the basic leucine zipper motif α-helices are induced in the major groove and in the HU motif β-strands bind in the minor groove.

Chemical recognition: the intermolecular interface

The same physical rules that determine protein and nucleic acid structure govern their specific and nonspecific interactions. The forces involved include HYDROGEN BONDS, VAN DER WAALS FORCES (or 'London dispersion'), HYDROPHOBIC INTERACTIONS, global electrostatic interaction and local SALT BRIDGE interactions.

Considering protein–DNA interactions first, it was anticipated that the comparatively uniform phosphate backbone of the DNA would be important for nonspecific contacts and that the pattern of hydrogen bond donors and acceptors in the major groove

a

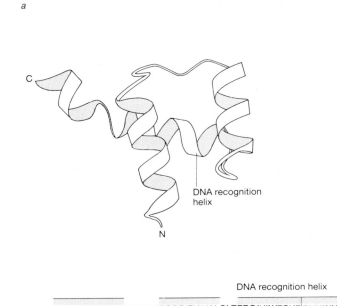

DNA recognition helix

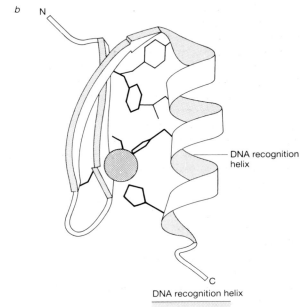

b

DNA recognition helix

DNA recognition helix

RKRGR**Q**TYTRY**Q**TLELEKE**F**HFNRYLTRRRRIEIAHAL**C**LTERQIKI**WF**QN**R**RN**K**W**K**KEN

DNA recognition helix

DRP**YS**C**DHPGC**DKAF**VR**NHD**L**IRH**K**KS**H**QEKA

c

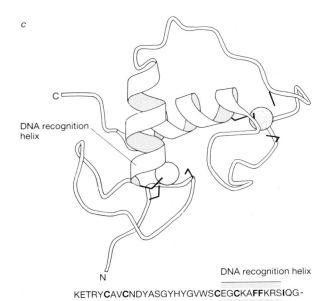

DNA recognition helix

DNA recognition helix

KETRY**C**AV**C**NDYASGYHYGVWS**C**EG**C**KA**FF**KRSIQG-

-HND**Y**M**C**PATNQ**C**TIDKNRRKS**C**QA**C**RL**R**K**C**Y EVG**M**MKG

d

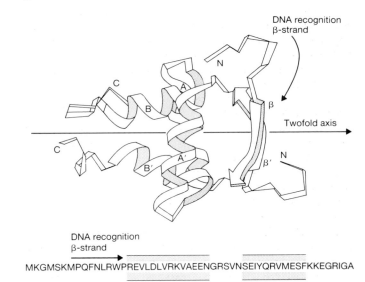

DNA recognition β-strand

Twofold axis

DNA recognition β-strand

MKGMSKMPQFNLRW**P**REVLDLVRKVAEENGRSVNSEIYQRVMESFKKEGRIGA

Fig. P62 DNA-binding domains. Schematic representation of the three-dimensional structure of: *a*, helix-turn-helix motif from the antennapedia protein from *Drosophila* (from [5]); *b*, zinc-finger motif from the yeast transcription factor SWI5 (from [6]); *c*, double-loop-zinc-helix motif from the human oestrogen receptor; *d*, the dimeric ribbon-helix-helix motif from the prokaryotic Arc repressor (from [7]). The amino-acid sequence of each domain is shown below the structure. In each motif characteristic amino-acid residues are shown in bold. The structures illustrated were solved using NMR spectroscopy. Structures of related proteins have been determined using X-ray crystallography.

would be important for specific binding. Furthermore it was also speculated that there would be some kind of amino acid–nucleotide 'recognition code' analogous to the genetic code.

The high resolution protein–DNA structures have now revealed a three-dimensional network of interactions that ties the two molecules together. As expected, contacts involving the phosphate backbone of the DNA seem to orient the protein in such a way that specific contacts can be made in the major groove. However, it has become clear that no simple amino acid–base recognition code exists, although certain common themes have emerged. In particular, arginine and glutamine residues have longish side chains that are able to make bidentate contacts to individual bases. Indeed, the most commonly occurring interaction is of arginine with the N7 and O6 of guanine. Glutamine can interact similarly with the N7 and N6 of adenine. However, both these amino acids are seen to make a variety of other contacts to bases. Finally the bulky 5′-methyl group of thymine is suited to making van der Waals contacts with the methyls of several amino acids, although it may also have an important negative role in sterically preventing incorrect binding.

Although many of these observations were largely anticipated, the crystal structures of protein–DNA complexes have revealed a number of surprises. The first is the role of water in mediating amino acid–base contacts; the second is that proteins can recognize the local structure of DNA through the spatial pattern of the backbone phosphates. The third is that long side chains may be buttressed by interacting with other side chains of the backbone of the protein. Fourthly, the zinc-finger motif illustrates the role of a zinc ion in contributing to a phosphate–histidine interaction. Finally, partial nonspecific binding in the complex of the glucocorticoid receptor DNA-binding domain has revealed that amino acid–base contacts can contribute to nonspecific binding.

Chemical recognition of RNA has been observed in the structures of tRNA synthetases with their cognate tRNAs (*see* GENETIC CODE). Like the protein–DNA interactions a complex network of hydrogen bonds is involved, but in this case interactions are made both with regions that are double helical and with unpaired bases in the anticodon loop which bind in discrete pockets in the protein.

The structures of known DNA- and RNA-binding proteins and their interactions with nucleic acids are described below. More detailed descriptions may be found in [1–4] and references therein.

Recognition of DNA

The helix-turn-helix motif

Our understanding of the structural basis of DNA recognition originated with studies of the helix-turn-helix (HTH) motifs observed originally in the early 1980s in prokaryotic and bacteriophage transcriptional regulators and now more recently in the homeotic proteins in eukaryotes (*see* HOMEOBOX GENES AND HOMEODOMAIN PROTEINS). At the time of writing, three-dimensional structures have been determined for 10 protein–DNA complexes. These have revealed a common structural motif and similar mode of binding to DNA. The minimal 20-residue HTH motif consists of two helices packed together at between 90° and 120° with a tight turn between them (Fig. P62*a*). It is not, in general, a stably folded structure when isolated from the rest of the protein and is packed against a variety of different supporting structures in the different HTH proteins. The residues in the turn and on the inside surface of the two-helix 'elbow' are partially

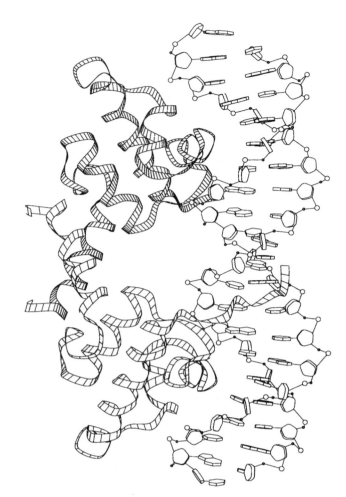

Fig. P63 Dimer of λ repressor binding to DNA.

conserved. The prokaryotic HTH proteins (e.g. λ repressor, *see* LAMBDA) bind to DNA as symmetric dimers and are arranged such that the two HTH motifs interact with successive turns of the DNA double helix (Fig. P63). The C-terminal helix of the protein (also known as the recognition helix) lies in the major groove.

While the orientation of the HTH motif in the major groove is broadly similar in all these complexes, there is considerable variation in the precise angle of the protein with respect to the major groove. In the eukaryotic HTH motifs in particular, the recognition helix is substantially longer, which results in a significantly different orientation. Furthermore, whereas the prokaryotic HTH proteins all bind to DNA as dimers the eukaryotic proteins are monomeric.

The zinc-finger motif

In 1985 a sequence motif of 30 amino acids (termed 'zinc finger') was observed in the *Xenopus* transcription factor TFIIIA where nine such motifs are arranged sequentially and direct sequence-specific DNA binding. This motif is defined by four metal ligands

a

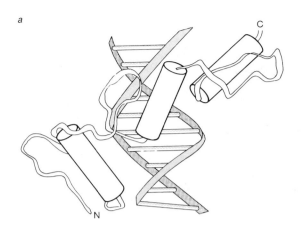

b

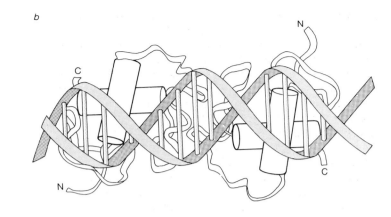

Fig. P64 Protein–DNA complexes. Schematic representation of the crystal structure of: *a*, the DNA-binding domain of the mouse transcription factor Zif268, containing three zinc-finger motifs, complexed to its recognition sequence; *b*, a dimer of the DNA-binding domain of the rat glucocorticoid receptor complexed to its palindromic recognition sequence.

(arranged: Cys-X$_{(2-5)}$-Cys-X$_{(12-13)}$-His-X$_{(2-5)}$-His) and three conserved hydrophobic residues. Analysis using NMR SPECTROSCOPY revealed that each 30-residue motif folds to form a discrete structural unit consisting of an irregular β-sheet packed against a short helix, C-terminal to the sheet. A zinc ion is tetrahedrally ligated by two histidines at the C-terminal end of the helix and two cysteines within the sheet (Fig. P62*b*). The three conserved hydrophobic residues form a core to the domain. It is now clear that this motif is used extremely widely in eukaryotic DNA-binding proteins from yeast to man. Indeed, to date, more than 200 different cDNA sequences have been found to encode zinc-finger motifs, amounting to more than 1200 individual zinc fingers. The modular nature of the zinc-finger motif, with variable numbers of fingers, or recognition units, arranged tandemly, may account for the widespread occurrence of these motifs.

The crystal structure of three zinc-finger domains from the mouse protein Zif268 complexed with 11 bp of DNA illustrates this modular mode of binding: each finger makes equivalent contacts in the DNA major groove with no interaction between adjacent fingers. The binding site is 3–4 bp to which hydrogen bonds are made by residues toward the N terminus of the α-helix (Fig. P64*a*).

The steroid receptor zinc-binding motif

The steroid and thyroid hormone receptors are ligand-activated transcription factors containing a highly conserved domain of around 70 amino acids which directs sequence-specific DNA binding. This independently folded domain contains two zinc-binding sequence motifs, which loosely resemble the zinc finger motif. The structure determined by NMR spectroscopy shows that the two zinc finger-like motifs fold to form a single structural domain and are thus quite distinct from the independently folded zinc fingers. The structure consists of two helices oriented per-

pendicularly to each other. A zinc ion, coordinated by four conserved cysteines, holds the base of a loop at the N terminus of each helix. This domain has been described as a 'double loop-zinc-helix' (Fig. P62*c*).

The crystal structure of this domain from the glucocorticoid receptor complexed to its 15-bp binding site shows that, like the HTH proteins, the receptor DNA-binding domains bind to DNA as dimers with one of the two helices in each monomer lying in the major groove of the DNA and buttressed by the second helix (Fig. P64*b*) (*see also* Plate 6*b*). This structure therefore shares features of the helix-turn-helix motif, although the orientation of the buttressing helix is reversed.

The ribbon-helix-helix motif

The extensive use of the HTH for specific DNA recognition in prokaryotes meant that there was considerable surprise when it was found that the prokaryotic MetJ, Arc and Mnt repressors interact with DNA in quite a different fashion. In the MetJ repressor a two-stranded antiparallel β-sheet, formed between a symmetric dimer of the protein, interacts with the DNA in the major groove (Fig. P62*d*). Since two helices are important for the stability of the structure, this has been termed a ribbon-helix-helix motif. Formation of a tetrameric complex on binding to DNA results in the two protein dimers binding in the major groove on opposite faces of the DNA double helix.

The HU/IHF motif

In prokaryotes, DNA is packaged in a complex with the dimeric protein HU. The structure of this protein dimer has been solved in the absence of DNA (Fig. P65*a*). It consists of a core of four α-helices from which two partially disordered β-ribbons protrude. These β-ribbons are thought to wrap around the DNA double helix.

Intriguingly, while HU binds nonspecifically to DNA, a closely related protein, the *E. coli* IHF protein (integration host factor), binds highly specifically. This illustrates how specific intermolecular interactions can be built onto a standard nonspecific structure. Chemical protection studies indicate that IHF binds in the

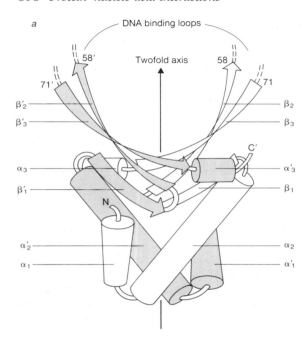

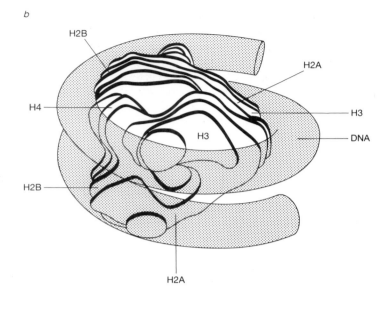

Fig. P65 DNA packaging proteins. Schematic representation of the crystal structure of: *a*, the HU dimer from *Bacillus stearothermophilus* [7]; *b*, the nucleosome core particle in which 145 bp DNA wrap around the histone octamer (the DNA is not wrapped smoothly around the core but is kinked in four places, not shown).

minor groove of the DNA. This is the only known sequence-specific DNA-binding motif that binds in the minor groove; as the minor groove is rather narrow in B-form DNA it is likely that this motif will be extremely sensitive to variations in the local DNA conformation.

The TATA-box binding factor

The crystal structures of complexes of the eukaryotic general transcription factor TATA-box binding factor (TBF, TFIIDτ) with its cognate TATA box DNA from yeast and *Arabidopsis* have been solved. The protein is made up of a curved 10-stranded anti-parallel β-sheet which forms the DNA-binding surface, surmounted by four α-helices. TBF binds DNA in the minor groove, with the long axis of the protein almost parallel with the DNA axis, forcing the DNA to bend sharply and partly unwind at the TATA element [8].

The basic leucine zipper (bZip) motif

In 1988 a number of transcription factors were found to share a common sequence pattern of leucine residues repeated at intervals of seven residues. This motif mediates the dimerization of these proteins, through the formation of an α-helical coiled coil structure. A basic region adjacent to the 'zipper' directs sequence-specific DNA binding. At the time of writing, the structure of this region is not known, but there is evidence that on binding to

DNA, it adopts an α-helical conformation. Two related models for the protein–DNA interaction have been suggested: the 'scissors grip' and 'induced helical fork' models. In these models, the intertwined helices in the zipper continue as individual helices into the basic region and grip the DNA in the major grooves on either side of the double helix.

Other DNA-binding domains

Although we now understand how the DNA-binding domains described above interact with DNA, there are an increasing number of DNA-binding domains identified by sequence homology and DELETION MUTAGENESIS. The structures of several of these have been solved very recently. These include the DNA-binding domain of the yeast GAL4 family of transcriptional activators, and the papilloma E2 protein (*see* PAPOVAVIRUSES). These structures reveal further variation in how to achieve complementarity of shape, yet the principles of specific recognition are the same.

Restriction enzymes

The prokaryotic restriction enzymes cleave DNA at short target sites of specific sequence. All restriction enzymes show specificity of cleavage, but whereas some also bind specifically to their target sites others show no particular binding specificity. Structures have been determined for both *Eco*R1 and *Eco*RV complexed with their recognition elements. It is clear that these structures are considerably more complex than the transcription factor domains. Both crystal structures reveal that the protein binding causes distortions in the phosphate backbone of the DNA. The crystal structure of a dimer of *Eco*RI bound to its palindromic recognition element reveals that the protein induces a pronounced kink in the DNA phosphate backbone at the point of cleavage. This kink causes the DNA to unwind by ~28° and to open both the major and minor grooves by about 3.5Å. A 'four-barrelled' α-helix motif (two helices from each monomer) interacts in the widened major groove while two extended arms wrap around the DNA double helix. One of these consists of an antiparallel β-sheet and the other of α-helix and extended chain. This second arm is joined to the four-barrelled α-helix motif by a region of extended chain which lies in the major groove of the DNA.

The structure of *Eco*RV is rather different. The main body of the protein lies over the minor groove of the DNA. Two recognition loops wrap around the double helix and interact in the major grooves on the opposite face of the DNA.

Although structures for only two restriction enzymes are available it is clear that they are structurally quite distinct, yet they both cause significant distortion of the DNA on binding and partially encircle the DNA. These contacts are more extensive than those seen in the transcription factor DNA-binding domains and may be important to compensate for the energy cost of distorting the DNA structure.

Nonspecific DNA-binding proteins

All the DNA-binding domains described so far bind comparatively strongly to specific DNA sequences. In addition, they have weaker but significant affinity for DNA of noncognate sequence. This property almost certainly enhances the rate of finding their cognate elements in a stream of nonspecific DNA. However, nonspecific DNA binding is important in its own right for many biological processes. These include the processing and packaging of DNA.

DNA processing enzymes. The crystal structure of the DEOXY-RIBONUCLEASE DNase I complexed with an octamer DNA duplex shows that most contacts are made to the DNA in the minor groove. As a consequence the groove is widened by 3Å and a bend of greater than 20° is made in the helix axis. This requirement for a particular minor groove width is consistent with the low cutting rates observed in regions which have a characteristically narrow minor groove. In contrast to the DNA, the protein structure is essentially unchanged on DNA binding, reflecting the rather rigid αββα structure. The backbone at the bend in the DNA is sandwiched between the ends of the two β-sheets.

DNA polymerase is one enzyme for which a lack of sequence specificity is most critical for its function in copying DNA (*see* DNA REPLICATION). The structure of the KLENOW FRAGMENT of *E. coli* DNA polymerase I (lacking the 5′ to 3′ exonuclease activity) has been determined in the absence of DNA. A cleft within the polymerase is thought to enclose the DNA duplex with flexible regions of the protein moving so as to surround the DNA completely. Low resolution studies of other polymerases have revealed a similar architecture.

DNA packaging. The prokaryotic packaging protein HU has been discussed above since homologous proteins bind to specific sequences.

In eukaryotes, DNA is tightly packaged into a dense material called CHROMATIN. This is made up of a repeating unit called a nucleosome which is packed to form a solenoid-like structure. Remarkably, in 1984 (before any other protein–DNA complex had been solved) the structure of the nucleosome was solved to 7 Å resolution. The nucleosome core consists of a disk containing two copies of each of four proteins (histones) around which 145 bp of DNA is wrapped in 1.75 turns. At this resolution few details of the structure are resolved, but it is clear that the histone proteins are in close contact with the phosphate backbone of the DNA double helix (Fig. P65*b*). The tight circular path of the DNA requires that it be somewhat distorted and as the nucleotide sequence affects the deformability of the double helix it was predicted, and has now been demonstrated, that nucleosomes exhibit some preference in their placement on the DNA.

DNA binding by small peptides. Many of the DNA-binding proteins described above possess, in addition to their well-characterized DNA-binding domain, short proline-rich peptide sequences (e.g. SPKK (single-letter AMINO-ACID NOTATION) repeats in the termini of histone proteins). It has been suggested that these may contribute to DNA binding by interacting in the minor groove, essentially nonspecifically, but with a preference for a narrow minor groove and hence AT-rich regions. Binding studies support this proposal, although further direct evidence has not been forthcoming.

Recognition of RNA

Our understanding of how proteins interact with RNA is much more limited than that of protein–DNA interactions [3,4]. To date the only crystal structures containing proteins complexed with RNA that have been solved are two tRNA synthetases, of different classes, complexed with their cognate tRNAs, and the helical virus TOBACCO MOSAIC VIRUS. However, since it is now possible to prepare large quantities of synthetic RNA, it is to be hoped that rapid progress will be made in this area.

FIBRE DIFFRACTION studies of helical viruses have shown details of the interaction between the coat protein of tobacco mosaic virus and its single-stranded RNA, in an extended chain with three nucleotides per protein monomer [9]. The interaction is both specific, for the recognition and initiation of virus assembly, and general, for the elongation and packaging of all possible sequences.

tRNA synthetases

The 20 tRNA synthetases are used for charging the appropriate tRNAs with their corresponding amino-acid residues. They have been divided into two classes on the basis of sequence homologies (*see* GENETIC CODE). The structures of a number of these enzymes and tRNAs have been determined in isolation. More recently the structure of one enzyme from each class has been determined in complex with its cognate tRNA [5]. These are the class I *E. coli* glutaminyl-tRNA synthetase and the class II yeast aspartyl-tRNA synthetase (Fig. P66*a*). Remarkably, the two protein structures are very different and interact with their cognate RNAs in a very

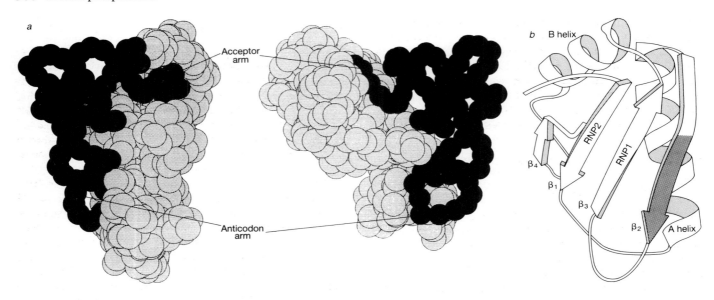

Fig. P66 RNA-binding proteins. Schematic representation of the crystal structure of: *a*, two tRNA-synthetase–tRNA complexes, the glutaminyl-tRNA synthetase–tRNA^{Gln} complex from *E. coli* (left) and the yeast aspartyl-tRNA synthetase–tRNA^{Asp} complex (right) (courtesy D. Moras); *b*, the RNA-binding domain of the human U1A small nuclear ribonucleoprotein (from [3]).

of the RNA loop. The loops between the β-sheet and α-helical regions form a pair of 'basic jaws' that are believed to sandwich the phosphates in the 5′ stem.

J.W.R. SCHWABE

D. RHODES

See also: STEROID RECEPTOR SUPERFAMILY.

1 Steitz, T. (1990) Structural studies of protein–nucleic acid interaction: the sources of sequence-specific binding. *Q. Rev. Biophys.* **23**, 205–280.
2 Harrison, S.C. (1991) A structural taxonomy of DNA-binding domains. *Nature* **353**, 715–719.
3 Nagai, K. (1992) RNA–protein interactions. *Curr. Opinion Struct. Biol.* **2**, 131–137.
4 Moras, D. (1992) Aminoacyl-tRNA synthetases. *Curr. Opinion Struct. Biol.* **2**, 138–142.
5 Harrison, S.C.A. & Aggarwal, A.K. (1990) DNA recognition by proteins with the helix-turn-helix motif. *Annu Rev. Biochem.* **59**, 933–969.
6 Neuhaus, D. et al. (1992) Solution structures of two zinc fingers domains from SW15, obtained using two-dimensional ^{1}H NMR spectroscopy: a zinc finger structure with a third strand of β-sheet. *J. Mol. Biol.* **228**, 637–651.
7 Phillips, S.E.V. (1991) Specific β-sheet interactions. *Curr. Opinion Struct. Biol.* **1**, 89–103.
8 Kim, Y. et al. (1993) *Nature* **365**, 512–520; Kim, J.L. et al. (1993) *Nature* **365**, 520–527.
9 Namba, K. et al. (1989) Visualization of protein–nucleic acid interactions in a virus. Refined structure of intact tobacco mosaic virus at 2.9Å resolution by X-ray fiber diffraction. *J. Mol. Biol.* **208**, 307–325.

different manner, yet both perform the same function. The two proteins are large and make extensive contacts with their respective tRNAs. However, the glutaminyl and aspartyl synthetases recognize opposite faces of the tRNA and so appear to exhibit mirror image conformations, although both proteins contact the anticodon stem and loop as well as the acceptor stem. In both complexes the protein conformation is essentially unchanged on binding tRNA, whereas the tRNA itself is substantially distorted on binding to the synthetase. In particular the anticodon loop is distorted so as to interact with the protein.

Double-helical RNA has a structure like that of A-form DNA with a wide shallow minor groove and a deeper narrower major groove. Thus it is reasonable that RNA-binding proteins would interact in the minor groove. Indeed this was observed in the structure of the class I synthetase in complex with its cognate tRNA. However, the class II synthetase interacts in the major groove. Such access to the major groove can be gained at the ends of regions of duplex RNA.

The ribonucleoprotein (RNA-binding) domain

Although the two tRNA synthetases are the only proteins whose structure in complex with RNA has been solved, a model based on both biochemical data and protein structure has been suggested for the complex between the RNA-binding domain of U1A (a small nuclear ribonucleoprotein involved in RNA splicing) and its cognate RNA stem-loop structure. The domain has a rigid structure with two helices packed against a four-stranded β-sheet (Fig. P66*b*). It is thought that the RNA binds to the surface of the β-sheet. The right-handed twist of the β-sheet matches the twist

Protein phosphatases

PROTEIN phosphatases are enzymes that reverse the actions of PROTEIN KINASES by cleaving phosphate from serine, threonine, and/or tyrosine residues in proteins in the neutral pH range. They are structurally and functionally distinct from the acid and alkaline phosphatases. The cellular roles of protein phosphatases are many and diverse, and are listed for particular phosphatases below. The protein phosphatases are divided into three groups according to catalytic function: (1) protein phosphatases which dephosphorylate serine and threonine residues; (2) protein phosphatases which dephosphorylate tyrosine residues; (3) protein phosphatases which dephosphorylate serine, threonine and tyrosine residues. Structurally, the protein serine/threonine phosphatases comprise two GENE FAMILIES both of which are distinct from the gene family comprising the protein tyrosine phosphatases and serine/threonine/tyrosine phosphatases [1–8].

Protein serine/threonine phosphatases (PPs)

Activity of the catalytic subunits

Four major types of protein phosphatase catalytic subunits which dephosphorylate serine and threonine residues have been identified in the cytosol of eukaryotic cells by protein chemistry and enzymology. They are termed protein phosphatases 1, 2A, 2B, and 2C (PP1, PP2A, PP2B and PP2C, the human genome symbols being *PPP1*, *PPP2*, *PPP3* and *PPM1* respectively). Although they have overlapping activities *in vitro*, they can be distinguished by their action on PHOSPHORYLASE KINASE and by their sensitivity to certain activators and inhibitors.

PP1 is potently inhibited by the thermostable proteins INHIBITOR 1 and INHIBITOR 2 and dephosphorylates the β subunit of phosphorylase kinase specifically, whereas type 2 protein phosphatases are unaffected by the inhibitor proteins and dephosphorylate the α subunit of phosphorylase kinase preferentially. The three type 2 protein phosphatases can be distinguished by their requirement for divalent cations and their response to the tumour promoter OKADAIC ACID and the toxin MYCROCYSTIN LR. PP2A has no absolute requirement for divalent cations, whereas PP2B is a Ca^{2+}-dependent, calmodulin-stimulated enzyme and PP2C is dependent on Mg^{2+}. PP2A is inhibited by subnanomolar concentrations of okadaic acid and microcystin, whereas PP2B is more than 1000-fold less sensitive and PP2C is resistant to these toxins. PP1 is also extremely sensitive to okadaic acid and microcystin although the K_i value for okadaic acid is higher than that observed with PP2A.

Other protein phosphatase catalytic subunits have been defined by molecular cloning (see below) but the nature of their substrate specificities and sensitivity to inhibitors is largely unknown.

Structures of the catalytic subunits

The amino-acid sequences of PP1, PP2A, PP2B and PP2C catalytic subunits have been elucidated by cDNA cloning (see DNA CLONING). Mammalian PP1 (M_r 37 000–38 000 (37–38K)) and PP2A (M_r 36K) are about 50% identical, whereas PP2B (M_r 58–59K) is ~40% identical to both PP1 and PP2A in the catalytic domain (Fig. P67). PP2B also possesses a long regulatory C-terminal extension. In contrast, PP2C (M_r 42–43K) is unrelated in the catalytic domain. The protein serine/threonine phosphatases therefore comprise two distinct gene families.

The catalytic subunits of PP1, PP2A and PP2B have been cloned from mammals, the fruitfly *Drosophila*, yeast, the mould *Aspergillus* (PP1 only) and plants (PP1 and PP2A only), but do not appear to be present in bacteria. The structures of the catalytic subunits of PP1 and PP2A have remained remarkably constant through evolution and are the most conserved of all known enzymes. Bacteriophage LAMBDA encodes a related phosphatase (221 amino acids) which dephosphorylates serine residues. The three sequences GDXHG, GDXVDRG and RGNHE (single-letter notation; X may be any amino acid) are conserved in all members of the PP1/PP2A/PP2B family including that from phage λ.

ISOFORMS of most PPs are present in eukaryotes. In addition, a number of other catalytic subunits have been identified in the PP1/PP2A/PP2B family which cannot readily be classified as isoforms. These include: PP4 (mammals); PPX, PPV and PPY (*Drosophila*); PPZ1, PPZ2, SIT4 and PPH3 (*Saccharomyces cerevisiae*). The detection of 11, 15 and 12 PP genes in mammals, *Drosophila* and yeasts respectively, predicts that more members of this family are likely to be uncovered in higher eukaryotes.

Protein phosphatase 1 (PP1)

Human gene symbol *PPP1*, also designated AMD (ATP, Mg^{2+}-dependent) phosphatase, PP1 dephosphorylates many proteins *in vitro*, including GLYCOGEN PHOSPHORYLASE and GLYCOGEN SYNTHASE, which it also dephosphorylates *in vivo*. It is found in both nucleus and cytoplasm of all eukaryotic cells investigated and has pleiotropic actions, being involved in the regulation of glycogen metabolism, MUSCLE contractility, CALCIUM ion channels, PROTEIN SYNTHESIS, and cell division.

Fig. P67 Diagrammatic alignment of protein serine/threonine phosphatase catalytic subunits. Solid segments indicate the conserved regions. The open segments indicate the variable regions. The number of residues in the mammalian isoforms is indicated. PP2B has two inserts in the conserved regions of seven and six residues respectively. CaM, Calmodulin-binding domain; I, region involved in suppression of activity in the absence of Ca^{2+} and calmodulin; R, region that interacts with the 19K regulatory subunit. Regions interacting with PP1 and PP2A regulatory subunits have not been defined.

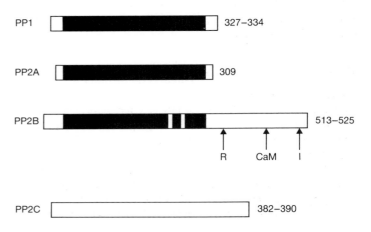

The catalytic subunit of PP1 may be complexed to one of a number of different regulatory subunits. Three isoforms of the catalytic subunit (>90% identity) are known in mammals — α, β and γ. PP1γ is alternatively spliced at the C terminus. The genes are widely dispersed in the human genome with *PP1CA* (PP1α) at chromosome 11q13, *PP1CB* (PP1β) on chromosome 2p23, and *PP1CC* (PP1γ) on chromosome 12q24.

Regulatory subunits are involved in targeting the catalytic subunit to particular locations and/or modifying its substrate specificity or activity. The R_G subunit (M_r 124K; 1109 amino acids) has been cloned from rabbit skeletal muscle and targets the catalytic subunit to glycogen particles and probably also to the sarcoplasmic reticulum. It plays a part in the hormonal control of glycogen metabolism. Phosphorylation of the R_G subunit in response to INSULIN results in activation of PP1 with ensuing dephosphorylation and inactivation of phosphorylase and dephosphorylation and activation of glycogen synthase. Phosphorylation of the R_G subunit at a different site in response to ADRENALINE causes release of the catalytic subunit into the cytosol, where it is inhibited by phosphorylated inhibitor 1. The R_G subunit gene is expressed in skeletal muscle, diaphragm and cardiac muscle, but not in brain, liver, kidney and lung. Different regulatory subunits target the catalytic subunit of PP1 to myofibrils in skeletal muscle and smooth muscle, where the enzyme is likely to be involved in the dephosphorylation of MYOSIN light chains.

PP1 dephosphorylates a regulatory subunit (phospholamban) of the sarcoplasmic reticulum Ca^{2+}-ATPase, suppressing Ca^{2+} uptake into the sarcoplasmic reticulum of cardiac muscle. It regulates protein synthesis by dephosphorylating the initiation factor eIF2 and ribosomal protein S6. Mutants of *Drosophila*, *Aspergillus* and *Schizosaccharomyces pombe* that are defective in PP1 catalytic subunit are blocked in MITOSIS, being unable to complete chromosome separation at anaphase. It is not clear whether this form of PP1 is bound to a regulatory subunit, but a likely candidate in *Schiz. pombe* is sds22$^+$ (M_r 32K) which has 11 leucine-rich internal repeats of 22 residues.

PP1 can be purified from the cytosol as an inactive form complexed to inhibitor 2 (F_C and PP1 I). Incubation with Mg^{2+} and ATP leads to activation of PP1 through phosphorylation of threonine 72 on inhibitor 2 by glycogen synthase kinase 3 (also termed F_A).

Protein phosphatase 2A (PP2A)

Human gene symbol *PP2*, also designated PCS (polycation-stimulated) phosphatase, PP2A has been found in the nucleus and cytoplasm of all eukaryotic cells investigated. The catalytic subunit is bound to a regulatory subunit of 65K (the A subunit or PR65) and usually also a second regulatory subunit which may be either 55K (B subunit or PR55), 54K (the B' subunit), or 72K. Two isoforms of the catalytic subunit (α and β, ~98% amino acid identity), two isoforms of the A subunit (87% identity) and two isoforms of the B subunit (calculated from the sequence to be 52K) (86% identity) have been cloned from mammals. The α catalytic subunit PPP2CA has been mapped to chromosome 5q23-31 and the β isoform PPP2CB to chromosome 8p11.2-12. The A subunits have 15 imperfect repeating units each consisting of 39 amino acids with a high content of regularly arranged leucine/isoleucine residues. The catalytic subunit, the A regulatory subunit, and the α isoform of the mammalian B subunit are probably present in all cells while the β isoform of the B subunit may be specific to neurons.

PP2A dephosphorylates enzymes involved in the regulation of glycolysis (6-phosphofructo-2-kinase and pyruvate kinase), glycogenolysis (phosphorylase kinase α subunit), gluconeogenesis (fructose-2,6-bisphosphatase and pyruvate kinase), amino-acid degradation (phenylalanine hydroxylase), lipid metabolism (acetyl-CoA carboxylase), CATECHOLAMINE synthesis (tyrosine hydroxylase) and protein synthesis (elongation factor 2). The catalytic subunit has also been identified as a negative regulator of the dephosphorylation and activation of p34^{cdc2} protein kinase in *Xenopus* and *Schiz. pombe* and therefore as a suppressor of the G2 to M transition of the CELL CYCLE. Disruption of CDC55, the *S. cerevisiae* homologue of the B regulatory subunit, leads to a delay in septation and/or cell separation. A complex of the A subunit and the catalytic subunit can associate with the tumour antigens of small DNA tumour viruses (PAPOVAVIRUSES). Certain protein kinases in signal transduction pathways, such as MAP KINASE and ribosomal protein S6 kinases, are inhibited by PP2A. However, no mechanisms for regulating PP2A activity *in vivo* have been defined.

Protein phosphatase 2B (PP2B)

Human gene symbol *PPP3*, also designated calcineurin CNA, PP2B is a Ca^{2+}-dependent, calmodulin-stimulated protein phosphatase found in the cytosol of all mammalian tissues but particularly abundant in brain where it comprises up to 1% of total protein. PP2B is a heterodimer of two subunits — a 58–59K catalytic subunit (A) (see above) and a 19K regulatory (B) subunit, which binds Ca^{2+}. In the presence of Ca^{2+}, calmodulin binds reversibly to the catalytic subunit converting it to a more active trimeric enzyme.

Isoforms (α or 1, β or 2 and γ) of the catalytic subunit are encoded by at least three genes and the diversity is further increased by alternative splicing (*see* RNA SPLICING) of the primary RNA transcripts. The two α isoforms differ by an insert of 10 amino acids near the C terminus, as do the β2 and β3 isoforms. The β1 isoform has an insert of 54 amino acids near the N terminus and a different C terminus to β2 and β3. A possible functional difference between these isoforms may result from the fact that the C terminus is involved in the suppression of activity in the absence of Ca^{2+} and calmodulin. All PP2Bβ isoforms contain 11 consecutive proline residues near the N terminus (amino acids 11–22). The gene for the α isoform has been mapped to chromosome 4 and that for the β isoform to chromosome 10. Brain contains the largest amounts of PP2Bα and β messenger RNA, although detectable levels are present in most other tissues. The γ isoform is specifically expressed in testis. PP2Bα, β and γ isoforms show ~80% amino acid identity to each other.

The regulatory B subunit has been cloned from mammals and *S. cerevisiae*. It has '4 EF hand' Ca^{2+}-binding domains (*see* CALCIUM-BINDING PROTEINS), indicating that it contains high affinity Ca^{2+} binding sites. It has 35% sequence identity with calmodulin and

29% identity with TROPONIN C. The N-terminal glycine is myristoylated which may explain why it is partly associated with the particulate fraction as well as the cytosol in brain.

The specificity of PP2B is much more restricted than PP1, PP2A and PP2C, and the most effective substrates so far identified are proteins that regulate other protein kinases and phosphatases. PP2B dephosphorylates inhibitor 1 and its isoform DARPP-32, the regulatory subunit of PROTEIN KINASE A, the α subunit of phosphorylase kinase, calmodulin-dependent cAMP phosphodiesterase, G substrate (a specific substrate for cyclic GMP-dependent protein kinase present in the cytosol of cerebellar Purkinje cells) and GAP43(a growth-associated protein involved in axon growth during nerve development and regeneration). The physiological roles of PP2B are not clearly elucidated, but a major role may be to allow extracellular signals that act via Ca^{2+} to attenuate those that act through CYCLIC AMP (*see* SECOND MESSENGER PATHWAYS). PP2B may be involved in the regulation of ion channels in both neuronal and non-neuronal cells. In lymphocytes, immunosuppressant drugs such as CYCLOSPORIN A and FK506 exert their pharmacological effects through PP2B, by forming a complex with cytoplasmic drug receptors which then binds to and specifically inhibits PP2B. In *S. cerevisae*, PP2B is involved in the recovery of cells from cell cycle arrest induced by mating pheromone.

Protein phosphatase 4 (PP4)

Human gene symbol *PP4*, also designated PPX. This protein phosphatase is found in all mammalian tissues. The catalytic subunit (307 amino acids) shows 65% identity to PP2A and 45% to PP1 and is probably complexed to regulatory subunit(s). Although distributed throughout the cytoplasm and the nucleus, PP4 localizes intensely to CENTROSOMES.

SIT4/phosphatase V

SIT4 is a protein phosphatase (311 amino acids) identified in *S. cerevisae* through a mutation (*sit4*) that restored transcription to the *HIS4* gene in the absence of *trans*-acting DNA binding factors (GCN4, BAS1 and BAS2) which are normally required for HIS4 transcription. It affects the transcription of a number of genes. SIT4 shows 55% amino acid identity to *S. cerevisae* PP2A catalytic subunit and 42% to PP1. It is required in late G1 of the cell cycle for progression into S phase. Protein phosphatase V is the *Drosophila* homologue of SIT4.

Protein phosphatase 2C (PP2C)

Human gene symbol *PPM1*. This monomeric enzyme is found in the cytosol of mammalian cells, turkey gizzard, *Drosophila* and yeast. Unlike PP1, PP2A and PP2B, the catalytic subunit of PP2C is not known to interact with any other proteins. Two isoforms of mammalian PP2C have been cloned, PP2Cα (or 1) from rat, rabbit and man (M_r 42K) and PP2Cβ (or 2) from rat (M_r 43K). The two isoforms show 76% amino acid identity to each other but no sequence similarities to the catalytic domains of PP1, PP2A and PP2Bs.

PP2C is dependent on Mg^{2+} or Mn^{2+} for activity (10–20 mM Mg^{2+} is optimal). However, it is unlikely that changes in the concentration of Mg^{2+} ions regulate its activity *in vivo*. PP2C dephosphorylates many proteins *in vitro* but, in contrast to PP1 and PP2A, none of its physiological roles are defined. A potential role may be in the regulation of CHOLESTEROL biosynthesis, as PP2C possesses high activity against hydroxymethylglutaryl-CoA reductase kinase, which inactivates HMG-CoA reductase, the rate-limiting enzyme of this pathway.

Protein tyrosine phosphatases (PTPs)

PTPs are a family of intracellular and integral membrane phosphatases which dephosphorylate tyrosine residues in proteins. PTPs have been identified in mammals, *Drosophila* and *Schiz. pombe* and are implicated in the control of normal and neoplastic growth and proliferation. They have also been found encoded by PLASMIDS in bacteria of the genus *Yersinia*, where they are implicated in pathogenicity.

Structure

PTPs may be divided into nonreceptor (PTPN) and receptor (PTPR) forms.

Nonreceptor PTPs. These are found in the cytosol and comprise a catalytic domain which may be either followed or preceded by a regulatory domain. PTP1B (M_r 50K) from human placenta, PTP1 (M_r 50K) from rat brain and TC-PTP (M_r 48K) from human T CELLS (which is 65% identical in amino-acid sequence to PTP1B) are made up of a catalytic domain followed by a regulatory domain, in which there is a hydrophobic section near the C terminus which may be responsible for localizing the PTPs to the particulate fractions. PTP1B purified from the placenta has lost this site, because it is truncated by 11K compared with the structure deduced from the cDNA, probably as an artefact of isolation.

Mammalian nonreceptor PTPs with long N-terminal regulatory regions preceding the catalytic domain include: PTP1C (M_r 69K), which has two adjacent copies of *src* homology region 2-like domains (*see* SH2 DOMAIN; TYROSINE KINASES); MEG (M_r 106K) and PTPH1 (M_r 104K), which have sequence similarities to the cytoskeleton-associated proteins band 4.1, EZRIN and talin (*see* ACTIN-BINDING PROTEINS); STEP, a neurone-specific PTP (M_r 46K); and MEG2 (M_r 68K), which has sequence similarities in the regulatory region to retinaldehyde-binding protein (Fig. P68).

Bacteria of the genus *Yersinia*, the causative agent of bubonic plague and a range of other gastrointestinal syndromes which may be fatal, contain plasmids which encode 50K PTPs, where the catalytic domain is in the C-terminal half. A 62K PTP identified in *Schiz. pombe* has a similar structure.

Receptor PTPs. These comprise an N-terminal extracellular domain, a single transmembrane domain and an intracellular variable region which is followed by a phosphatase catalytic region (Fig. P68). Most receptor PTPs have two catalytic domains separated by a more variable 58-residue section, but PTP1β and

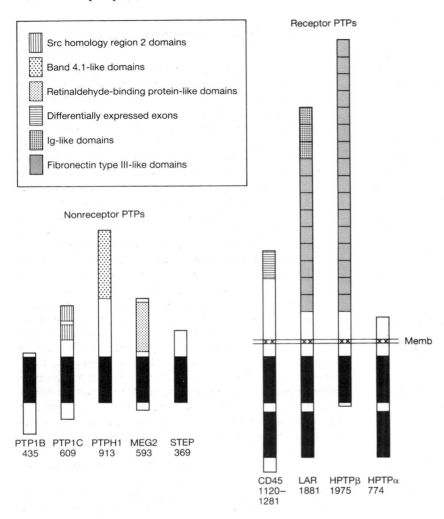

Fig. P68 Diagrammatic representation of human intracellular and transmembrane protein tyrosine phosphatases. A single example of each type of structural organization is presented with the number of amino acids in that molecule. Other members in the same class are given in the text. Solid segments indicate tyrosine phosphatase catalytic domains, cross-hatched segments show membrane-spanning regions and open boxes represent segments with no known sequence similarities. Known sequence similarities of other regions are given in the key.

Drosophila PTP10D have only one catalytic domain. The extracellular region varies in structure and at least nine distinct forms of receptor PTPs have been isolated from eukaryotes. They can be classified into four major structural types.

1 Type 1 is CD45 (leukocyte common antigen), which has a heavily glycosylated external segment and a cysteine-rich region.
2 Type 2 structures have between one and three tandem IMMUNOGLOBULIN-LIKE DOMAINS linked to two to nine FIBRONECTIN TYPE III REPEATS, resembling the structure of neural adhesion molecules (*see* CELL ADHESION MOLECULES). LAR (LEUKOCYTE COMMON ANTIGEN RELATED PROTEIN), human RPTPmu (1432 amino acids) and the *Drosophila* enzymes DLAR (1997 amino acids) and DPTP (1439 amino acids) are in this class, differing from each other in the number of each type of repeat.
3 Type 3 structures contain only multiple fibronectin type III domains. They include human PTPβ and the neuron-specific *Drosophila* DPTP99A (1231 and 1061 amino acids) and DPTP10D (1631 and 1558 amino acids) which are alternatively spliced in the extracellular domain.
4 Type 4 molecules have small external domains (27–123 residues). This class comprises human receptor-PTPα (HPTPα), HLRP

(793 amino acids), HPTPε (681 amino acids) and PTP18. It is likely that the extracellular segment of most if not all receptor PTPs is glycosylated.

The catalytic domain(s) of all PTPs is characterized by an essential cysteine residue located within the consensus sequence [I/V]HCXAGXXR[S/T]G. However, in the second catalytic domain of HPTPγ, PTP18 and DPTP99A the cysteine is replaced by aspartic acid. At least 19 distinct genes encoding PTPs have been identified in humans, and a report of 27 in the protochordate *Styela plicata* (a sea squirt) suggests that the PTP family in mammals may be very much larger. PTP genes are widely dispersed throughout the human genome with the following map positions: PTPN1 (PTP1B), 20q13; PTPN2 (TC-PTP), 18p11; PTPN6 (PTP1C), 12p12-13; PTPRC (CD45), 1q31-q32; PTPRF (LAR) 1p33-p32; PTPRLI (RPTP mu), 18p11.2; PTPRL2 (PTPα), 20p13; PTPRG (PTPγ), 3p21.

Function

In vitro, PTPs dephosphorylate the autophosphorylated insulin and EGF receptors and many artificial substrates phosphorylated on tyrosine residues (*see* GROWTH FACTOR RECEPTORS). The activity of PTPs can be inhibited by orthovanadate, molybdate and zinc. However, the specific function of each PTP is largely unknown. From studies of overexpression of PTPs it has been suggested that nonreceptor PTPs may participate in MITOGENIC signalling pathways. The structures of PTPH1 and MEG suggest that they may dephosphorylate proteins localized at the junction between the membrane and the cytoskeleton.

Receptor type I PTPs have been implicated in T-cell signal transduction pathways, where they are involved in coupling receptors to the generation of the second messenger inositol trisphosphate (*see* CD45; SECOND MESSENGER PATHWAYS; T-CELL ACTIVATION). Receptor type II and III PTPs resemble molecules that are involved in HOMOPHILIC and HETEROPHILIC interactions respectively, raising the possibility that they may influence cell–cell interactions during development and morphogenesis. The intracellular location of the *Drosophila* enzymes DLAR, DPTP99A and DPTP10A suggests that they may be involved in neuron outgrowth and guidance.

Protein serine/threonine/tyrosine phosphatases (PT/SPs)

Phosphatases which dephosphorylate serine and threonine residues as well as tyrosine residues (PT/SPs) are distantly related in sequence to the PTPs. They have been found in mammals, *Xenopus*, *Drosophila* and yeasts, where they are essential for cell division, and in VACCINIA VIRUS, where they may be involved in pathogenicity.

Vaccinia virus encodes a 20K protein distantly related to PTPs (~20% amino acid identity). It possesses the sequence (I/V)HCXAGXXR(S/T)G containing the essential cysteine residue characterizing the PTP family, but when expressed in bacteria it dephosphorylates substrates containing phosphoserine as well as phosphotyrosine. Activity is blocked by vanadate but not by okadaic acid. This PT/SP may have a role in viral replication.

Cdc25 (*see* CELL CYCLE) shows 21% identity to vaccinia virus phosphatase in a stretch of 92 amino acids near the C terminus with conservation of the essential cysteine in the sequence XHCXXXXXRXX. Cdc25 has been shown to dephosphorylate tyrosine and threonine residues *in vitro*, and *in vivo* causes the activation of p34^{cdc2} kinase at the G2 to M phase of the cell cycle.

P.T.W. COHEN

See also: CALCIUM; PROTEIN PHOSPHORYLATION.

1 Cohen, P. (1989) The structure and regulation of protein phosphatases. *Annu. Rev. Biochem.* **58**, 453–508.
2 Cohen, P. & Cohen, P.T.W. (1989) Protein phosphatases come of age. *J. Biol. Chem.* **246**, 21435–21438.
3 Cohen, P.T.W. et al. (1990) Protein serine/threonine phosphatases: an expanding family. *FEBS Lett.* **268**, 355–359.
4 Shenolikar, S. & Nairn, A. (1991) Protein phosphatases: recent progress. *Adv. Second Messenger Phosphoprotein Res.* **23**, 1–121.
5 Tonks, N.K. & Charbonneau, H. (1989) Protein tyrosine dephosphorylation and signal transduction. *Trends Biochem. Sci.* **14**, 497–500.
6 Hunter, T. (1989) Protein tyrosine phosphatases: the other side of the coin. *Cell*, **58**, 1013–1016.
7 Tonks, N.K. (1990) Protein phosphatases: key players in the regulation of cell function. *Curr. Opinion Cell Biol.* **2**, 1114–1124.
8 Fischer, E.H. et al. (1991) Protein tyrosine phosphatases: a diverse family of intracellular and transmembrane enzymes. *Science* **253**, 401–406.

Protein phosphorylation

PROTEIN phosphorylation is a regulatory covalent modification of a protein, in which the phosphate group is transferred from ATP to a side chain of the protein, catalysed by a PROTEIN KINASE. Protein dephosphorylation is the reverse process, a hydrolysis of a phosphate group from the side chain of a protein, catalysed by a PROTEIN PHOSPHATASE (Fig. P69). In addition to regulatory protein phosphorylation events, some enzymes use covalently bound phosphate as catalytic intermediates, but this differs in that the phosphate group is both inserted and removed by the enzyme itself and turns over during each cycle of catalysis.

Given the prevailing concentrations for ATP, ADP and phosphate in the cell, the equilibrium of a protein kinase reaction lies well in favour of protein phosphorylation, and that for a protein phosphatase reaction lies well in favour of dephosphorylation. By controlling the rate of the protein kinase and/or the protein phosphatase reaction, a phosphorylation–dephosphorylation cycle is therefore well placed to act as a molecular switch or trigger [1]. Along with the GTP ↔ GDP cycle of GTP-BINDING PROTEINS, phosphorylation–dephosphorylation seems to represent the major regulatory mechanism of eukaryotic cells which switches activity from one state to another, rather than merely exerting a homeostatic control. As protein kinases and protein phosphatases act catalytically, a phosphorylation–dephosphorylation cycle is also an effective way of amplifying a small initial signal. A particularly large amplification is possible is two or more protein kinases are arranged in series — a system known as a PROTEIN KINASE CASCADE.

There are at least three general functions for phosphorylation–dephosphorylation cycles in eukaryotic cells:

1 They are the major mechanism by which extracellular signal molecules acting at cell surface RECEPTORS produce their ultimate intracellular effects (*see* PROTEIN KINASES; SECOND MESSENGER PATHWAYS).

2 They are the major mechanism through which events which occur discontinuously in the CELL CYCLE are initiated and timed.

3 They also seem in some cases to protect the cell against toxic changes in metabolites, a form of stress response (*see* AMP-ACTIVATED PROTEIN KINASE).

Regulatory protein phosphorylation normally occurs on serine or

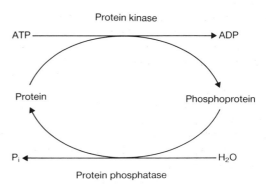

Fig. P69 The phosphorylation–dephosphorylation cycle.

Table P14 Examples of proteins regulated by phosphorylation–dephosphorylation

Class of protein	Protein	Protein kinase	Effect of phosphorylation
Metabolic enzyme	Phosphorylase	Phosphorylase kinase	$\downarrow K_a$ for activator, AMP
	L-pyruvate kinase	cAMP-dependent PK	$\uparrow K_m$ for substrate, $\uparrow\downarrow$ sensitivity to allosteric effectors
	Hormone-sensitive lipase	cAMP-dependent PK	$\uparrow V_{max}$
	Tyrosine hydroxylase	cAMP-dependent PK	$\uparrow V_{max}$, $\downarrow K_m$ for cofactor
	Glycogen synthase	Glycogen synthase kinase-3	$\uparrow K_a$ for activator, glucose 6-phosphate
	Glycogen synthase	Calmodulin-dependent PKII	$\uparrow K_a$ for activator, glucose 6-phosphate
	Acetyl-CoA carboxylase	AMP-activated PK	$\downarrow V_{max}$, $\uparrow K_a$ for activator, citrate
	HMG-CoA reductase	AMP-activated PK	$\downarrow V_{max}$
	Hormone-sensitive lipase	AMP-activated PK	Prevents phosphorylation by cAMP-dependent PK
Protein kinase	Phosphorylase kinase	cAMP-dependent PK	$\uparrow V_{max}$, $\uparrow$ sensitivity to Ca^{2+}
	MAP kinase	MAP kinase kinase	$\uparrow V_{max}$
	p34/cdc2	wee1 protein kinase?	$\downarrow V_{max}$
	AMP-activated PK	AMP-activated PK kinase	$\uparrow V_{max}$
Receptor	β_2-adrenergic receptor	β_2-adrenergic receptor kinase	Homologous desensitization
	β_2-adrenergic receptor	cAMP-dependent PK	Heterologous desensitization
Motor protein	Smooth muscle myosin light chain	Smooth muscle MLCK	Initiates contraction
	Skeletal muscle myosin light chain	Skeletal muscle MLCK	$\uparrow$ Force of contraction
	Troponin C	cAMP-dependent PK	$\uparrow$ Rate of dissociation of Ca^{2+}
Cytoskeletal	Lamin B	p34/cdc2, MAP kinase	Disaggregation of nuclear lamina
Ion channel	L-type, voltage-gated Ca^{2+} channel	cAMP-dependent PK	$\uparrow Ca^{2+}$ entry
Ion pump regulator	Phospholamban	cAMP-dependent PK	$\uparrow Ca^{2+}$ uptake by sarcoplasmic reticulum
Transcription factor	CREB	cAMP-dependent PK	Activates transcription
	c-Jun	MAP kinase	Activates transcription

This list is not comprehensive; examples have been chosen to reflect the variety of protein type regulated. MLCK, myosin light chain kinase; PK, protein kinase.

threonine residues (*see* PROTEIN KINASES) but it can occur on tyrosine (*see* GROWTH FACTOR RECEPTORS; TYROSINE KINASES) and histidine residues (*see* BACTERIAL CHEMOTAXIS). Introduction or removal of one or more negatively charged phosphate groups on the protein often causes a marked change in function. The proteins regulated by phosphorylation–dephosphorylation fall into several functional classes. Table P14 lists representative examples of these classes, and summarizes the varied effects of phosphorylation on protein function.

In a few cases, protein phosphorylation may act by direct steric blocking of a binding site or catalytic site, for example the inhibition of the cell cycle-regulating protein kinase, cdc2, by phosphorylation at tyrosine 14, which is in the ATP-binding site. More commmonly, at least in eukaryotes, protein phosphorylation produces a conformational change which affects binding or catalysis indirectly. The best understood example is the dimeric enzyme GLYCOGEN PHOSPHORYLASE, which has been studied in both phosphorylated and dephosphorylated forms by X-RAY CRYSTALLOGRAPHY [2]. In this case the phosphorylation site is at serine 14 in the N-terminal region. Phosphorylation at this site weakens a loose intrasubunit association of the N-terminal region, and the phosphate group forms interactions with arginine residues on both the same and the other subunit. This pulls the two subunits together at one point, causing a relative rotation of the subunits, and the effect is transmitted to the active site by a subtle conformational change within each subunit.

D.G. HARDIE

1 Hunter, T. (1987) A thousand and one protein kinases. *Cell* **50**, 823–829.
2 Barford, D. et al. (1991) Structural mechanism for glycogen phosphorylase control by phosphorylation and by AMP. *J. Mol. Biol.* **218**, 233–260.

protein S *See*: BLOOD COAGULATION AND ITS DISORDERS.

Protein secretion

THE process by which proteins move along the secretory pathway to their final destinations in the PLASMA MEMBRANE or external medium is known as protein secretion. The fundamental difficulty faced by secretory proteins and those membrane proteins which straddle a cellular membrane (integral membrane proteins), is how the hydrophilic stretches of amino acids present in most of these proteins can be transferred across membranes whose major constituents are hydrophobic lipids (*see* MEMBRANE STRUCTURE; PROTEIN TRANSLOCATION). Two distinct types of translocational machinery have evolved to solve this problem. In one, a ribonucleoprotein complex acts as an adaptor whose role is to ensure that the N termini of nascent secretory and membrane proteins engage the membrane-bound translocational machinery. This design allows a large variety of proteins to use the same translocational machinery, and it is probable that the majority of secretory and membrane proteins in eukaryotes and prokaryotes utilize this pathway. The second mechanism involves the partic-

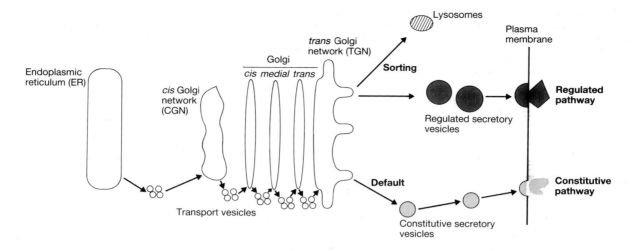

Fig. P70 The secretory pathway in eukaryotic cells.

ipation of membrane-bound translocators. These are protein complexes which include integral membrane proteins and whose function is to transport specific proteins across cellular membranes.

Each type of translocational machinery is found in both prokaryotes and eukaryotes, and a comparison of the corresponding types reveals a similar molecular logic underlying each process, a feature which points to a common evolutionary origin. However, many biochemical processes require the presence of membranes, so that the larger size of most eukaryotic cells and the consequent reduction in their surface area : volume ratios has been accompanied by the development of complex membrane-bound internal compartments. In prokaryotic cells such as bacteria, the translocational machinery is exclusively located in the plasma membrane (*see* BACTERIAL ENVELOPE; BACTERIAL PROTEIN EXPORT). In eukaryotic cells this is generally not the case and the ENDOPLASMIC RETICULUM (ER) membrane provides the appropriate site.

Protein secretion in eukaryotic cells

Secretion through the endoplasmic reticulum

The pioneering studies of George Palade and co-workers some 30 years ago established the general intracellular route followed by secretory proteins on their exodus to the cell surface. He studied the synthesis and movement of the digestive enzyme precursor, chymotrypsinogen, in the acinar cells of the rat pancreas. These cells were known to accumulate this and other enzyme precursor proteins in membrane-bound storage compartments called zymogen granules, before the fusion of these granules with the plasma membrane, and the liberation of their contents after the cell had received an appropriate stimulus.

Palade established that the translation of this secretory protein began in the cytosol along with the majority of other proteins (*see* PROTEIN SYNTHESIS), but that newly synthesized chymotrypsino-

gen soon became associated with the ER before moving on to the GOLGI APPARATUS, and finally, to the zymogen granules. This model of secretory protein movement has been refined over the years and extended to include integral membrane proteins and to cater for secretory and membrane proteins that are continuously transported to the plasma membrane (constitutive secretion), and also those proteins like chymotrypsinogen whose secretion is regulated (regulated secretion).

The current model, shown schematically in Fig. P70, owes much to the use of cell-free protein translation and translocation systems which allow the fine detail of the process of protein secretion to be studied *in vitro*. Evidence that similar pathways operate *in vivo* has come from the discovery in the yeast, SACCHAROMYCES CEREVISIAE, of various mutants (*sec* mutants) which are blocked at different stages of secretion. The early steps in the pathway are shown in Fig. P71. Translation begins in the cytosol and the emergence of the N-terminal amino acids from the large subunit of the ribosome is the decisive early event which leads to the subsequent segregation of the protein into the secretory pathway. This N-terminal (usually) section, called the SIGNAL SEQUENCE or signal peptide, is recognized by a ribonucleoprotein particle comprising a 7S RNA and six polypeptide chains, called the SIGNAL RECOGNITION PARTICLE (SRP). SRP binds to the signal peptide and slows down translation. This translational retardation is relaxed when the complexed SRP encounters the DOCKING PROTEIN (also called the SRP receptor), a heterodimeric protein embedded in the ER membrane via its larger subunit. Although the interaction between SRP and its receptor is quickly followed by the dissociation of SRP from both the nascent protein and its receptor, the signal peptide is transferred into the ER membrane. As translation proceeds, the emerging nascent polypeptide chain is vectorially discharged across the ER membrane into the ER lumen. The translocation complex falls apart during the passage of integral membrane proteins as specific stretches of hydrophobic acids called STOP-TRANSFER SEQUENCES are inserted into the membrane. This results in the anchorage of the protein in the membrane at that point, although protein translocation can in certain circumstances be reinitiated by amino-acid sequences distal to the stop-transfer sequence. It is the relative arrangement

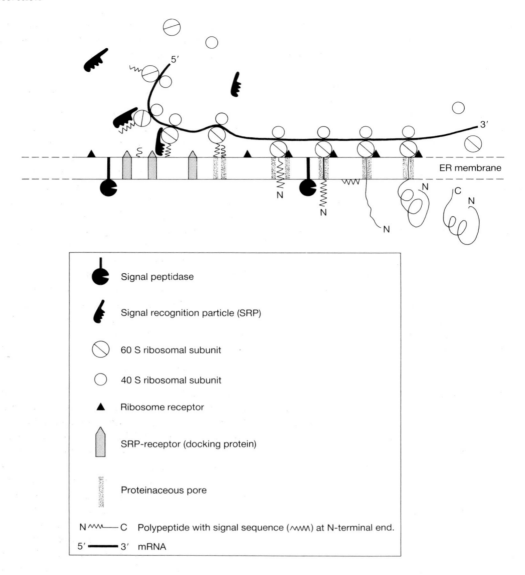

Fig. P71 Translocation of a polypeptide across the endoplasmic reticulum.

and number of stop-transfer and cleavable or noncleavable signal sequences within the amino-acid sequence that are ultimately responsible for the final orientation of the membrane protein in the membrane (*see* PROTEIN TRANSLOCATION).

The translocation process above is depicted as a CO-TRANSLATIONAL process, that is, translocation only occurs during on-going translation. This tight coupling of the two processes is believed to occur *in vivo*, although under certain circumstances (in yeast) it can be relaxed. During translocation, the attachment of the polysome to the ER membrane is believed to be reinforced by the binding of the ribosome directly to a ribosome receptor protein present in the ER membrane and consolidated by the presence of new nascent peptides arising from further ribosome recruitment. The exact physical environment around the polypeptide as it traverses the membrane is not known although a

pore lined by protein is the favoured model. This proteinaceous pore is thought to be a dynamic structure whose individual protein subunits may change according to the protein being translocated. Although ATP is required for this process to occur *in vitro*, its direct involvement *in vivo* remains unclear; in contrast to the situation in bacteria (see below) an energized ER membrane is not required.

The emergence of the polypeptide into the ER lumen is followed by several events. First, the signal peptide is usually cleaved by a membrane-associated complex of six polypeptides called SIGNAL PEPTIDASE; there are, however, many integral membrane proteins and a few secretory proteins which do not undergo cleavage. Second, folding and, in some cases, assembly of polypeptides into distinctive tertiary and quaternary structures occurs in the ER (*see* PROTEIN FOLDING). Often these events are accompanied by covalent changes to the polypeptide, including disulphide bond formation and the addition of sugars (*see* GLYCO-PROTEINS AND GLYCOSYLATION; POST-TRANSLATIONAL MODIFICA-

TION). There is some evidence pointing to the existence of proteins in the ER which prevent the misfolding of nascent polypeptide chains: such proteins have been called MOLECULAR CHAPERONES. It appears that the further onward movement of proteins along the secretory pathway depends on their attaining the correct three-dimensional structure, and that exit from the ER acts as a type of quality control preventing misfolded proteins or incompletely assembled oligomeric proteins from reaching the cell surface (*see e.g.* ANTIGEN PROCESSING AND PRESENTATION).

Following these processing events, the proteins are packaged into vesicles in a specialized region of the ER, and, after budding off, these vesicles move to, and fuse with membranes of the *cis* Golgi network (CGN, also known as the SALVAGE COMPARTMENT or 'intermediate compartment'). Vesicular shuttles are also responsible for transporting the protein between the various compartments (*cis*, *medial* and *trans*) of the Golgi apparatus, finally delivering their contents to the TRANS GOLGI NETWORK (TGN). Biochemical and morphological analysis of cell-free reconstituted transport systems from animal cells have identified some of the steps and molecules involved in the budding and fusion processes. It is now known that a family of coat proteins known as COPS are exclusively involved in the budding process (*see* COATED PITS AND VESICLES; GOLGI APPARATUS) whereas a tetrameric protein called the *N*-ethyl maleimide-sensitive fusion protein (NSF) along with three helper SNAPS (α, β, and γ) are implicated in the fusion event. Common to both events is the participation of different families of small GTP-BINDING PROTEINS. Yeast mutants encoding many of these proteins have now been isolated (e.g. *sec18* encodes NSF, *sec17* encodes α-SNAP) and the phenotype of such mutants is consistent with the roles of the encoded proteins deduced *in vitro*.

During the transit from ER to TGN, further translational modifications may be performed on the proteins by a battery of enzymes which reside in specific compartments of the organelles. Sorting of proteins to their final destinations occurs in the TGN. A variety of possible fates are available to proteins in the TGN, including targeting to the LYSOSOMES, the SECRETORY GRANULES and the plasma membrane. It is generally believed that secretory and membrane proteins of the constitutive pathway do not require any sorting signals to guide them to the plasma membrane. The constitutive pathway is accordingly termed the default pathway. In contrast, sorting signals are required for all other proteins which enter the secretory pathway (*see* GOLGI APPARATUS; LYSOSOMES; PROTEIN TARGETING) and in some cases receptor proteins present in the TGN have been identified and shown to provide the specificity of recognition needed for sorting. As a result of the various recognition processes, secretory and membrane proteins destined for the secretory granules are sorted into vesicles coated with a protein called clathrin, whereas secretory and membrane proteins of the constitutive pathway are packaged by default into vesicles which have a different protein coat, and are shuttled to the plasma membrane (*see* COATED PITS AND VESICLES). The fusion of these vesicles with the plasma membrane leads to the release of secretory proteins into the surrounding medium (*see* EXOCYTOSIS), whereas membrane proteins are retained in the plasma membrane. In some cell types, notably epithelial and neuronal cells, where regions of the plasma membrane contain specific subsets of proteins and are functionally

distinct, exocytosis is polarized, with different populations of vesicles being targeted to specific regions of the plasma membrane (*see* EPITHELIAL POLARITY).

Secretion directly via the plasma membrane

Most known secretory and membrane proteins utilize the pathway described above. However it is now clear that some secretory proteins do not enter the secretory pathway but are directly exported through the plasma membrane. Although their mechanism of exit is not yet clear, the participation of membrane transporters like the product of the *S. cerevisiae STE6* gene is a possibility. The STE6 protein is a single-chain integral membrane protein, and is responsible for the export of the mating pheromone a, which is a short peptide. This transporter shares significant regions of amino-acid sequence homology with other eukaryotic transporters (e.g. the P GLYCOPROTEIN and the cystic fibrosis transmembrane conductance regulator protein, *see* CYSTIC FIBROSIS) and with a family of bacterial ATP-driven protein translocators involved in protein secretion (see below). This probably reflects an underlying similarity in structure and function between the various transporters.

Protein secretion in bacteria

In bacteria, the site of protein entry into the secretory pathway is the plasma membrane which in Gram-negative bacteria is called the inner membrane. Subsequent destinations include the external medium, the periplasmic space, or the outer membrane (Gram-negative bacteria) (*see* BACTERIAL ENVELOPE). It is the convention to regard passage into, or through, the plasma membrane as the topological equivalent of entry into the ER in eukaryotic cells. The finding that many bacterial secretory proteins have cleavable signal sequences indicated that there might be mechanistic similarities between the processes in prokaryotes and eukaryotes. Indeed, it has been shown that eukaryotic proteins can be secreted from bacteria and *vice versa*. However, it seems that bacteria have several secretory pathways which, although they often share certain steps, also feature a number of redundant components. Genetic dissection of the components involved in bacterial protein secretion has identified a number of genes whose products are thought to contribute to the translocation machinery in the plasma membrane. These genes are called *Sec* genes and include *SecA*, *SecY*, *SecE*, *SecD*, and *SecF*. In one secretory pathway it is believed that nascent polypeptides are prevented from misfolding in the cytoplasm by any of a number of molecular chaperones, including SecB, GroEL, GroES, DnaK, and possibly, trigger factor (*see* HEAT SHOCK; MOLECULAR CHAPERONES). The stabilized polypeptides may then be transferred to the membrane-bound translocational machinery after first interacting with SecA. In another pathway, more comparable to the major one in eukaryotes, nascent chains are bound by a ribonucleoprotein complex which has a functional resemblance to the eukaryotic SRP even though only a 4.5S RNA and a protein M_r 48 000 have so far been identified. This complex is eventually transferred to the complex formed by SecA, Y, E, D, and F for further processing. Both these pathways require an energized cytoplasmic membrane for translocation to occur, although a direct role for ATP *in vivo* remains unclear. Translocation can occur post-translationally. The choice of pathway seems to be dictated by the nature of the secretory protein and the exact composition of the secretory apparatus available in a particular cell. Thus, when the usual pathway taken by a particular protein is unavailable, as a result of mutation for example, the other pathway can often be used.

The final pathway available in bacteria involves ATP-driven protein translocator complexes which reside in the plasma membrane. Several of these transporters are known and the best studied is that responsible for the secretion of haemolysin B by *Escherichia coli*. This transporter is a homodimer which forms a translocation pore in the membrane. Cytoplasmic domains of these subunits bind to ATP, the hydrolysis of which is thought to provide the energy for translocation. In other bacterial transporter proteins the domains providing the pore and the ATP-binding site are on separate proteins.

Conclusions

Protein secretion is vital to the viability of all living cells. There are striking similarities in the process in eukaryotes and prokaryotes but also differences. Although the major secretory route in eukaryotic cells is far more complex than that in prokaryotes, involving transit through a number of discrete membrane-bounded compartments, the critical events which promote protein insertion through a membrane are similar. Thus, prokaryotic and eukaryotic signal sequences can be interchangeable, and the major player in eukaryotic secretion, SRP, also has a bacterial homologue.

In both eukaryotes and prokaryotes, certain proteins can bypass the major secretory routes because dedicated protein translocation complexes facilitate their transit through the cytoplasmic (prokaryotic) or plasma (eukaryotic) membrane. The energy for this process is thought to be ATP, and since the Sec secretion pathway in bacteria also uses ATP, at least *in vitro*, a case has been made for the Sec pathway being a more promiscuous version of the ATP-driven transporters. Thus more than one type of system can coexist in the same membrane in prokaryotes, and this also seems to be the case in eukaryotes where it has been shown recently that cytosolically generated viral peptides are transported into the ER lumen by protein translocators (*see* ANTIGEN PROCESSING AND PRESENTATION).

There are, unsurprisingly, differences between secretion in eukaryotic and prokaryotic cells. Secretion via the Sec pathway requires an energized cytoplasmic membrane, which is not the case in SRP-mediated transport in eukaryotes. It is also clear that in bacteria, translocation through the Sec pathway is not obligatorily coupled to translation, and often translocation begins some time after translational initiation, and, for some proteins, the process can be completely post-translational. In eukaryotes it is believed that the two processes are usually tightly coupled.

Further information on this topic will be found in [1–4].

A. COLMAN

1 Rothman, J.E. & Orci, L. (1992) Molecular dissection of the secretory pathway. *Nature* **355**, 409–415.

2 Huttner, W. & Tooze, S. (1989) Biosynthetic protein transport in the secretory pathway. *Curr. Opinion Cell Biol.* **1**, 648–654.

3 Pfeffer, S. & Rothman, J.E. (1987) Biosynthetic protein transport and sorting by the endoplasmic reticulum and Golgi. *Annu. Rev. Biochem.* **56**, 829–852.

4 Nunnari, J. & Walker, P. (1992) Protein targeting to and translocation across the membrane of the endoplasmic reticulum. *Curr. Opinion Cell Biol.* **4**, 573–580.

protein sequence The order of amino acids in a protein chain, conventionally listed starting from the N terminus.

protein sequencing The AMINO-ACID SEQUENCE of a PROTEIN can either be determined directly by chemical analysis of the purified protein or can be deduced from the NUCLEOTIDE SEQUENCE of the corresponding GENE. Even with modern fully automated methods, complete direct sequencing of a moderately sized protein is far more laborious and time-consuming than the determination of the corresponding nucleotide sequence from a cloned gene, and this indirect method is now the main way of obtaining amino-acid sequences of proteins (*see* DNA SEQUENCING). Where small amounts of a purified protein whose gene is unknown are available, direct sequencing may be used to determine a short N-terminal sequence. This is then used to construct a nucleic acid PROBE for isolating the corresponding gene, from which the complete amino-acid sequence of the protein can be determined. Limited direct sequencing is also used to compare the sequence of the mature protein with that deduced from the DNA sequence, which represents the unprocessed protein.

The amino-acid sequence of a peptide of up to 50 amino acids can be determined by the technique of Edman degradation (named after its inventor Pehr Edman). The N-terminal amino acid is labelled on the α-amino group by phenylisothiocyanate and released from the peptide as a cyclic derivative (a phenyl-thiohydantoin amino acid) under mildly acidic conditions which do not degrade the rest of the peptide chain, and identified by CHROMATOGRAPHY. This cycle is repeated on the shortened peptide chain until the peptide has been completely sequenced. Protein sequencing by Edman degradation is now a completely automated procedure, the instrument being known as a sequenator. Sequenators incorporating gas-liquid or pulsed-liquid chromatography to identify the released amino acid derivatives can analyse from 200 to 1 pmole quantities of protein. A sequenator can perform up to 3.5 cycles of Edman degradation per hour so that even with automation, the effort involved in directly sequencing a protein of some thousands of amino acids is still considerable.

Peptides for sequencing by Edman degradation are produced from larger proteins by specific cleavage by CYANOGEN BROMIDE or the enzymes TRYPSIN (tryptic peptides) or CHYMOTRYPSIN, and separated by chromatography. Each of these reagents cleaves the protein at different amino-acid residues. By comparing the sequences of peptides produced by trypsin with those of peptides produced by chymotrypsin for example, chymotryptic peptides that overlap two tryptic peptides can be identified, thus allowing ordering of the tryptic peptides into a complete sequence.

The positions of DISULPHIDE BONDS can be determined by specific cleavage of the protein into peptides under conditions in which the disulphide bonds remain intact. The mixture is electrophoresed on paper to separate the peptides and the sheet exposed to perform acid vapour which cleaves disulphide bonds. Subsequent electrophoresis in a perpendicular direction results in a diagonal line of the peptides not affected by this treatment, with the cysteic acid-containing peptides, which have a different electrophoretic mobility from that of their parent disulphide-linked peptides, lying off the diagonal. Sequencing of the cysteic acid peptides can then identify the position of the disulphide bond.

Protein structure

CHEMICALLY, protein molecules are long polymers of amino-acid residues typically containing several thousand atoms. They are linear polymers, composed of a polypeptide backbone with a side chain attached at each residue. Ions, small organic ligands and even water molecules are also integral parts of many protein structures. Biochemically and physiologically, proteins have a variety of roles in life processes: there are structural proteins (ranging from viral coat proteins to epidermal keratin); catalytic proteins (the enzymes (*see* MECHANISMS OF ENZYME CATALYSIS)); transport and storage proteins (e.g. HAEMOGLOBIN, MYOGLOBIN and FERRITIN); regulatory proteins including hormones and the proteins that regulate gene TRANSCRIPTION; and the recognition proteins of the immune system (IMMUNOGLOBULINS and T CELL RECEPTORS).

The polypeptide chains of proteins show a great variety of folding patterns, with a number of common underlying features. These include the explicit recurrence of conformations such as α-helices and β-sheets, and common properties of folding such as dense packing of hydrophobic residues in protein interiors.

Over a thousand protein structures are now known. Most were determined by X-RAY CRYSTALLOGRAPHY or NEUTRON SCATTERING AND DIFFRACTION, some by NMR SPECTROSCOPY, a few by FIBRE DIFFRACTION or ELECTRON MICROSCOPY. From these we have derived our understanding both of the functions of individual proteins, in terms of physical–organic chemical explanations of catalytic activity of enzymes, and of the general principles of protein architecture and folding. The Protein Data Bank at Brookhaven National Laboratory, in the USA, archives and distributes protein structures (*see also* DATABASES).

Under physiological conditions of solvent and temperature, proteins fold spontaneously into a 'native' three-dimensional conformation; that is, the AMINO-ACID SEQUENCE dictates the three-dimensional structure (*see* PROTEIN FOLDING). The amino-acid sequence of a protein is known as its primary structure and this in turn dictates its secondary structure (the regions that form helices and sheets) and tertiary structure (the spatial relationships and interactions between the helices and sheets). For oligomeric proteins, the quaternary structure is the packing together of subunits. From the point of view of the logic of life, the folding of proteins is the point at which nature makes the leap from the one-dimensional world of genetic information stored in the GENETIC CODE to the three-dimensional world we inhabit [1,2].

How is the native state determined? Any possible folding of the chain places different residues in proximity. The interactions of the side chains and backbone, with one another and with the solvent, determine the energy of the conformation. Proteins have evolved so that one folding pattern produces a set of interactions that is significantly more favourable than all others. This is the native state. If we could calculate sufficiently accurately the energies and entropies of different conformations, and if we could computationally examine a large enough set of possible conformations to be sure of including the correct one, we should be able to predict each protein's structure from its amino-acid sequence on the basis of *a priori* principles. So far this is not possible.

The smallest folding units of proteins are known as domains. A domain is a portion of a polypeptide chain of a protein — often but not always a contiguous region of the chain — that forms a compact structural unit. Domains are the smallest subsets of a protein that are credibly stable on their own. Indeed, examples are known in which enzymes that are composed of several polypeptide chains in one organism are single polypeptide chains in another; the regions corresponding to the separate proteins in the former constitute individual domains in the latter. Similar domains are also seen to recur in different proteins, not always with the same function (*see* MOLECULAR EVOLUTION: SEQUENCES AND STRUCTURES).

Properties of the side chains

Protein structures are determined by the sequences of the amino-acid side chains appearing along the polypeptide chain. As each side chain is chosen from a set of 20 (*see* AMINO ACIDS; GENETIC CODE), a protein is in principle a message written in an alphabet of 20 letters. However, residues are occasionally subject to POST-TRANSLATIONAL MODIFICATIONS, or to the introduction of an unusual amino acid — selenocysteine — at the time of synthesis (in glutathione peroxidase and a very few other examples).

The 20 side chains vary in physicochemical properties. The smallest, glycine, consists of only a hydrogen atom; one of the largest, phenylalanine, contains a benzene ring. Some side chains, of which phenylalanine is an example, contain only hydrocarbon groups, and are electrically neutral. Because of the thermodynamically unfavourable interaction of hydrocarbons with water, they are called 'hydrophobic' residues. The congregation of hydrophobic residues in protein interiors, predicted by W.J. Kauzmann before the first protein structures were determined, is an important contribution to protein stability. Some side chains are polar. Asparagine and glutamine contain amide residues; serine, threonine and tyrosine contain hydroxyl groups. Polar side chains can participate in HYDROGEN BONDING. Other side chains are charged: aspartic acid and glutamic acid contain carboxyl groups and are negatively charged; lysine and arginine contain amino and guanidinium groups, respectively, and are positively charged. The polar or charged atoms appear at or near the distal ends of these side chains. The proximal atoms are nonpolar. This facilitates formation of SALT BRIDGES between charged residues.

Backbone conformations

Common to all protein structures is the linear polypeptide chain,

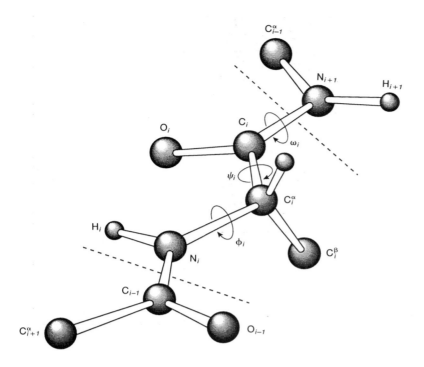

Fig. P72 The conformation of a polypeptide chain can be specified approximately by the values of the angles of internal rotation in the backbone. Each residue has three such angles: ψ is a rotation around the N–Cα bond, ϕ is a rotation around the Cα–C bond, and ω is a rotation around the peptide bond itself. Usually $\omega = 180°$, the *trans* conformation. This drawing shows a dipeptide in the fully extended conformation, $\psi = \phi = \omega = 180°$.

shown in Fig. P72. The PEPTIDE BOND itself tends to be planar, with the *trans* conformation energetically more favourable — and far more common — than the *cis* conformation. Rotation is permitted around the N–Cα and Cα–C single bonds of all residues except proline. The sequence of ϕ, ψ and ω angles of all residues in a protein defines the backbone conformation. (However, small variations in the bond lengths and angles of the backbone atoms of the residues make it impossible to reconstruct a backbone conformation accurately from a list of conformational angles and a library of standard residues.)

Steric interactions limit the allowed values of conformational angles. For ω, observed values cluster tightly around $180°$ (*trans*) and, rarely, $0°$ (*cis*). *Cis*-peptides are energetically less unfavourable adjacent to prolines than adjacent to other residues, and *cis*-prolines account for most *cis*-peptides in proteins. For ϕ and ω angles, the allowed ranges tend to fall into regions shown in a Sasisekharan–Ramachandran plot: a graph in which each residue corresponds to a point with abscissa ϕ and ordinate ω (see Fig. P73). Broken lines in the figure delimit sterically allowed regions. The conformations of most amino acids fall into either the right-handed α-helix (α_R) or β-regions (see Figs P77 and P78). By symmetry, glycine residues can also populate the left-handed α-helix (α_L) region.

Figure P73 shows the typical distribution of residue conformations in a well-determined protein structure, ribonuclease A. Most residues fall in or near the allowed regions, but the folding forces a few into energetically unfavourable states.

Side-chain conformation

Most side chains have one or more degrees of conformational freedom. For example, serine and threonine can have different rotational isomers (rotamers) around the Cα–Cβ bond. Arginine has five angles of internal rotation. Side-chain conformations are specified by angles χ_i, where χ_1 refers to the rotation around the Cβ–Cγ bond, and successive χ angles to successive bonds, proceeding out along the side chain. Different amino acids have preferred side chain conformations, which tend to correspond to staggered conformations around the bonds. Sometimes particular local interactions can be important; for example, the hydroxyl groups of serine and threonine can form hydrogen bonds to the backbone, giving them effectively hydrophobic rather than polar character.

Interactions that stabilize proteins and bind ligands

Two important general principles are:

1 Proteins are only marginally stable, and achieve stability only in narrow ranges of conditions of solvent and temperature. Outside these boundaries proteins lose their definite compact structure and form denatured states with disordered backbone conformations, and few if any specific interactions among the residues. The free energy of stabilization of proteins is typically only ~40 kJ mole^{-1}.

2 Folding is a global process: the entire protein — or at least an entire domain — is necessary for stability.

Protein structures depend on a variety of chemical forces for their stability and for their affinity and specificity for ligands.

Covalent and coordinate chemical bonds

Many proteins contain covalent chemical bonds in addition to those of the polypeptide backbone and the side chains. DISULPHIDE BRIDGES between cysteine residues are quite common.

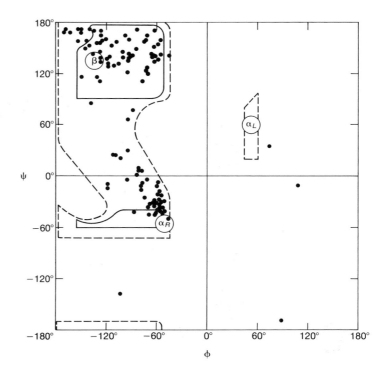

Fig. P73 Sasisekharan–Ramachandran plot of the regions of sterically permitted values of ψ and ϕ for a *trans* dipeptide. Solid lines enclose regions estimated to include the maximum tolerable limits of steric strain. The diagram indicates three standard conformations: the right-handed α-helix (α_R); the β-sheet (β); and the left-handed α-helix (α_L). The α_L region is occupied mainly by glycine residues. The dots correspond to the conformations of the residues in a well-defined, high-resolution protein structure, ribonuclease A (7RSA). Most but not all of the points lie within the 'allowed' regions.

Figure P74 shows the small protein bovine pancreatic trypsin inhibitor, which contains three disulphide bridges. Disulphide bridges can also link different polypeptide chains, as in INSULIN and immunoglobulins.

In CYTOCHROME C, covalent chemical bonds link the HAEM group to the protein: a histidine and a methionine are bonded to the iron atom of the haem, and two cysteine residues are bonded to carbon atoms of the haem. In contrast, in globins only the iron atom of the haem group is bound to the protein (*see* HAEMOGLOBIN AND ITS DISORDERS).

The binding of metal ions to proteins typically involves coordi-nate chemical bonds. Figure P75 shows the copper-binding site of PLASTOCYANIN (*see also* Plate 4*c*).

Hydrogen bonding

In the unfolded state, polar atoms of proteins make hydrogen bonds to water. In the folded state, the hydrogen-bonding potential of atoms buried in the interior of the protein — including of course the main-chain polar atoms of the peptide groups — must be satisfied. (The energy of sacrificing a hydrogen bond is of the same order of magnitude as the energy of stabilization of the

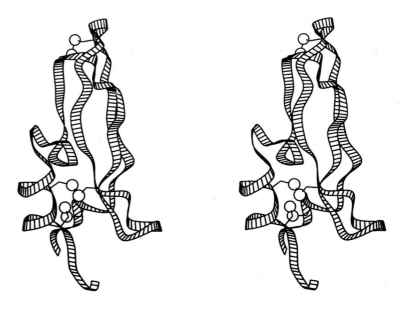

Fig. P74 Stereopair diagrams of bovine pancreatic trypsin inhibitor, a small protein (58 residues) containing three disulphide bonds (circles).

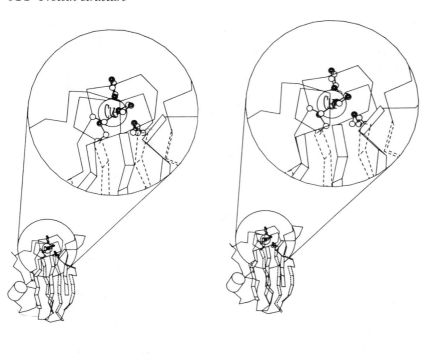

Fig. P75 Stereopair depictions of the copper-binding site of poplar leaf plastocyanin. Plastocyanin is a typical double β-sheet sandwich structure: it contains two β-sheets packed face to face, with the strands in both sheets running in a predominantly up or down direction. The copper ion is ligated by four residues — two histidines, a cysteine and a methionine. Two of these residues are in one of the sheets and a third in a loop linking two strands of the same sheet.

a

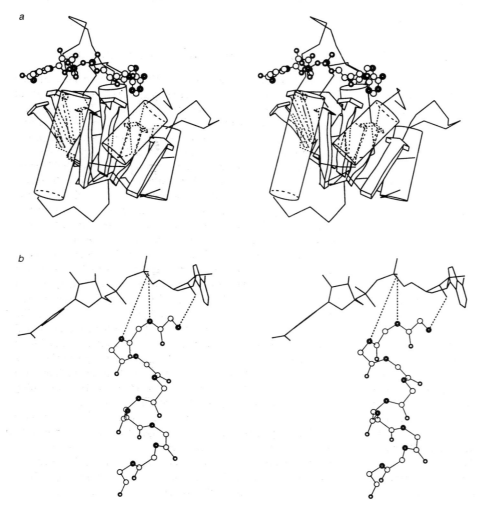

b

Fig. P76 Stereopairs of *a*, the nucleotide-binding domain (depicted in ball-and-stick convention) of horse liver alcohol dehydrogenase. *b*, Detail of hydrogen bonding between helix and cofactor.

native state of the protein.) The formation of helices and sheets is a common way of forming hydrogen bonds to main-chain atoms.

Hydrogen bonds also contribute to the binding of cofactors and substrates. Figure P76 shows the binding of nicotinamide adenine dinucleotide (NAD) to a domain of horse liver alcohol dehydrogenase. NH groups protruding from the ends of two of the helices form hydrogen bonds to phosphate groups of the cofactor.

The hydrophobic effect, van der Waals forces and packing

It is thermodynamically unfavourable for nonpolar atoms of a protein to be exposed to water. The interior of a globular protein contains hydrophobic residues sequestered away from water. (Charged residues are fairly rigorously excluded from protein interiors. However, protein interiors do not contain exclusively nonpolar atoms; the main chain, containing peptide groups, must pass through the interior, and some polar side chains are also buried. Buried polar atoms are generally involved in hydrogen bonds. Conversely, about half the residues on the surface of a protein are neutral and nonpolar.)

The sequestering of uncharged and nonpolar side chains into protein interiors provides another common mechanism of binding ligands. The haem group of the globins is largely nonpolar but contains two propionic acid groups at one edge. As might be expected, the nonpolar part occupies a crevice in the interior of the protein, whereas the charged acidic groups protrude into the solvent.

The packing of atoms in protein interiors contributes in two ways to the stability of the structure. In addition to the exclusion of nonpolar residues from contact with water, there is the force of attraction between protein atoms themselves. The strength of VAN DER WAALS forces depends on the distance between the atoms involved. At long distances the forces are attractive. The nearer the atoms, the stronger the force, until the atoms are actually 'in contact', when the forces become repulsive and strong. To maximize the total cohesive force, therefore, as many atoms as possible must be brought as close together as possible.

It is the requirement for a dense packing that imposes structure on the protein interior. Interior interfaces between helices and sheets show good packing between surfaces of complementary shape. This requirement for complementarity of internal surfaces biases the relative positions and orientations of packed helices and sheets, which in turn influences the overall geometry of observed folding patterns.

Complementarity of opposing surfaces is also responsible for the specificity with which many enzymes bind their substrates. CHYMOTRYPSIN, TRYPSIN and ELASTASE are three SERINE PRO-TEASES with different specificities: chymotrypsin cleaves peptide bonds adjacent to large flat nonpolar side chains such as phenylalanine, trypsin cleaves peptide bonds adjacent to long, positively charged side chains (arginine and lysine) and elastase cleaves peptide bonds next to small nonpolar side chains. Each protease contains a crevice near its catalytic site into which the side chain adjacent to the scissile bond must insert. The size, shape, and charge distribution of the atoms that line the pocket create a surface complementary to the side chain to be selected.

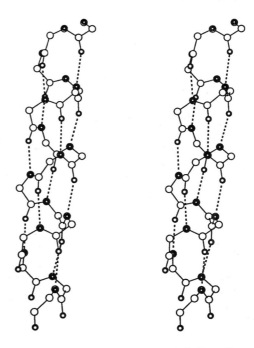

Fig. P77 Stereopair of a standard right-handed α-helix. Dotted lines depict hydrogen bonds.

Common architectural features of protein structures

Most proteins contain α-helices and/or β-sheets. In some fibrous proteins virtually all of the residues belong to one of these types of structure: wool contains α-helices; silk β-sheets. Typical globular proteins contain regions of helix or sheet, connected by loops. Helices and sheets are well known 'standard' structural pieces that form components of protein conformations: Fig. P77 shows a standard α-helix and Fig. P78 shows an idealized parallel β-sheet (β-sheets can be formed with parallel or antiparallel strands — compare Figs P78 and P79 — or with mixed strand directions). Are there any other recurrent structural patterns? Can we analyse protein architecture by making a list of standard pieces and another list of the way in which they are put together?

Certain combinations of helices and strands of sheet have been identified as supersecondary structures. Figure P79 shows a β-sheet from the LECTIN concanavalin A and a hairpin turn created by two successive antiparallel strands of β-sheet and the loop between them (see Fig. P83c for the entire structure). The corresponding structure with antiparallel helices appears in ROP (repressor of primer), a dimeric 'four-helix bundle' consisting of two chains, each of which forms two packed antiparallel helices (Fig. P80). Figure P76b shows a typical β-α-β unit containing two successive antiparallel strands of β-sheet, with the intervening section of chain forming a helix packed against them.

In only rare cases would such structures be stable on their own, but they form the basis of complete protein structures, or even of domains. Thus ROP (Fig. P80) contains two pairs of helices. The nucleotide-binding domain of alcohol dehydrogenase contains multiple β-α-β units assembled side-by-side (Fig. P76a). In fact this unit consists of two pairs of β-α-β-α-β-α units.

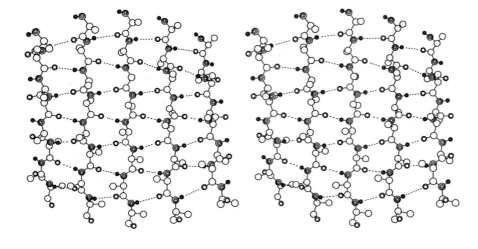

Fig. P78 Stereopair of an ideal parallel β-sheet. Dotted lines depict hydrogen bonds.

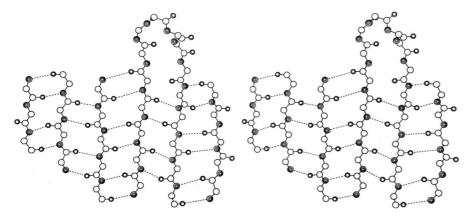

Fig. P79 Stereopair of an antiparallel β-sheet from the lectin concanavalin A from jackbean. The hairpin turn between the second and third strands from the right includes the loop linking two successive strands of sheet that follow each other in the chain and are hydrogen bonded to each other in the structure. Dotted lines depict hydrogen bonds.

In contrast, Fig. P81 shows triose phosphate isomerase, a barrel structure created by eight β-α units closed into a cylinder with the strands forming a closed parallel sheet, and with the α-helices packed against the outside of the sheet. The same folding pattern has been found in over 20 enzymes with different functions.

In proteins containing primarily β-sheets, two sheets often pack face to face (like the sheets on a bed in fact), in one of two typical relative orientations: strands nearly parallel or strands nearly perpendicular. Plastocyanin (Fig. P75) is an example of parallel strand orientation. The strands may alternate in direction — from N to C terminus — indeed, they must, but if the directionality of the chain is ignored the strands are almost parallel. Photoreactive yellow protein (Fig. P82) illustrates the alternative double β-sheet packing, with the strands nearly perpendicular.

Without wanting to oversimplify the extent to which recurrent structural patterns account for the folds of complex proteins, Fig. P83 shows a selection of structures, chosen by the criterion that the common structural features discussed here are relatively easy to spot. In some cases the secondary structures are distinguished by schematic representations — helices shown as cylinders and strands of sheet as large arrows or as polygonal ribbons — but in others the reader is given the opportunity to parse the structure for themselves (*see also* Plates 3–5).

Conformational changes in proteins

The figures reproduced here show only snapshots of particular conformations of proteins. Many proteins undergo conformational changes as part of their mechanisms of action, often in

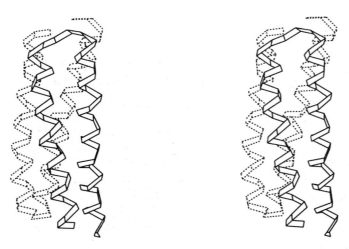

Fig. P80 Stereopair of ROP (repressor of primer), a dimeric four-helix bundle.

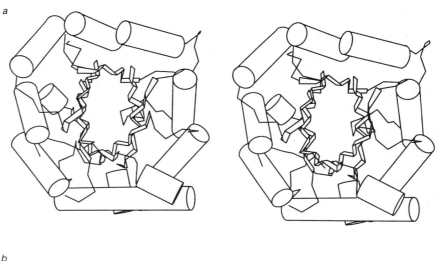

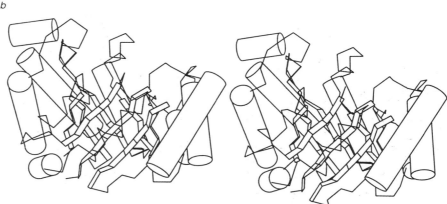

Fig. P81 Stereopairs of the β-barrel of chicken triose phosphate isomerase (TIM). This is constructed from eight β-α units with the strands closed into a roughly cylindrical β-sheet, with the α-helices packed outside. Cylinders, α-helices; folded ribbon, β-sheet. *a*, View into the barrel from the C termini of the strands; *b*, view perpendicular to axis of the barrel. Selected helices in *b* are not represented by cylinders because they would obscure the sheet.

response to the binding of ligands. An enzyme may have one state in which a cleft is 'open' to receive a substrate and/or cofactor, and a 'closed' state in which it engulfs the substrate and/or cofactor after binding. In the more complex cases of ALLOSTERIC proteins, the conformational change is transmitted between subunits to produce cooperativity of binding.

Figure P84 shows the open and closed forms of the enzyme citrate synthase. The open form contains only the substrate, citrate, but the closed form contains both the substrate and the cofactor, coenzyme A.

Adult human haemoglobin consists of four subunits, two α chains and two β chains. There are two allosteric states of

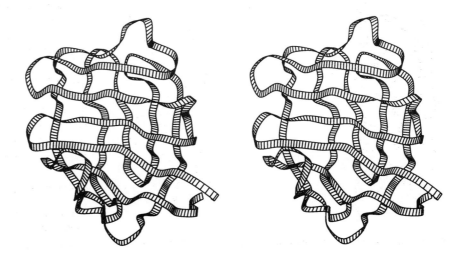

Fig. P82 Stereopair diagram of structure of photoreactive yellow protein, showing orthogonal packing of two sheets.

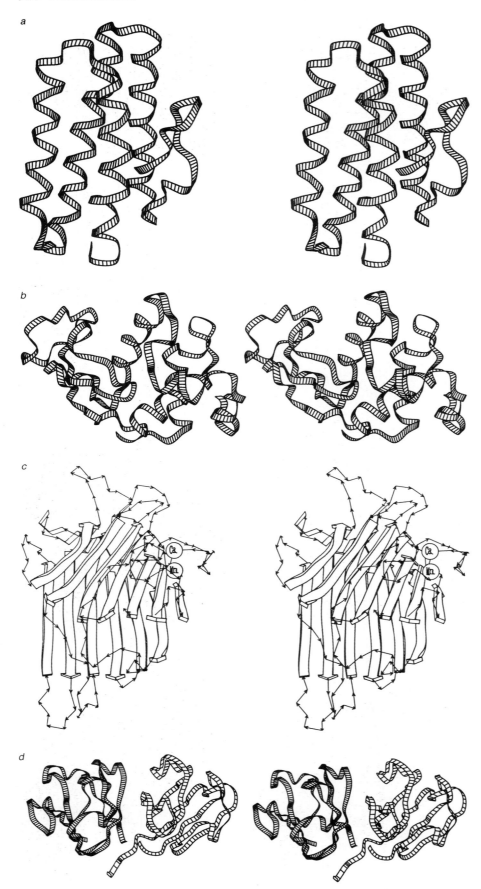

Fig. P83 A selection of folding patterns, all depicted as stereopairs, illustrating the conformations described in the text as recurring structural patterns. *a*, Haemerythrin; *b*, hen egg white lysozyme; *c*, concanavalin A. *d*, γ-crystallin.

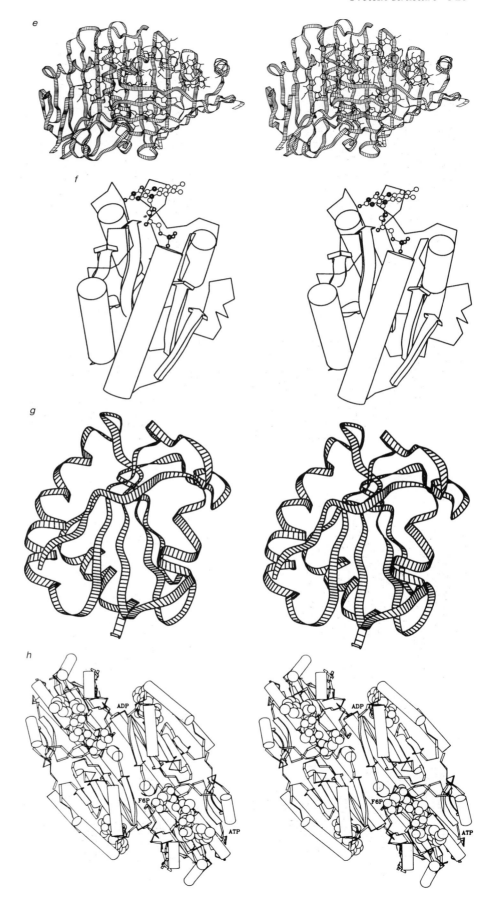

Fig. P83 *Continued e*, bacteriochlorophyll-*a* protein; *f*, flavodoxin; *g*, thioredoxin; *h*, phosphofructokinase, R state, showing substrate fructose 6-phosphate (F6P), cofactor ATP and effector ADP. The molecule is a tetramer, of which only two subunits are shown (*see also* Plate 5*d*).

a

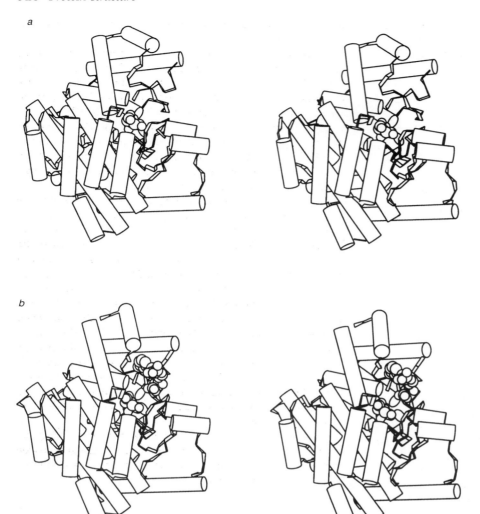

b

Fig. P84 Conformational change in citrate synthase. *a*, Stereopair of open form, containing only substrate; *b*, stereopair of closed form, containing both substrate and coenzyme A. The substrate binds in a cleft between two domains, and after both the substrate and coenzyme are bound the domains close around them.

assembly of the subunits—the oxy and deoxy quaternary structures—in which the subunits are packed together in different ways. Concomitant with the change in quaternary structure are changes in tertiary structure producing changes in the shapes of the surfaces of the subunits, altering the relative energies of the alternative packings. Figure P85 shows one aspect of the allosteric changes in haemoglobin. This figure shows the α_1–β_1 dimer in oxy and deoxy forms, superposed on the α_1–β_1

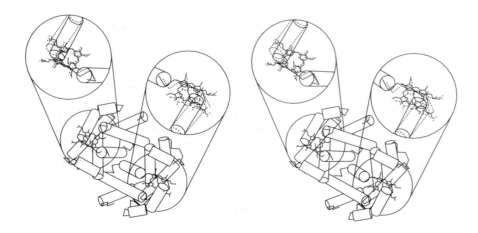

Fig. P85 Stereopair of the α_1–β_1 dimer of haemoglobin in oxy (solid lines) and deoxy (broken lines in blow-up) forms.

interface. In the blown-up regions only the F-helix, FG-corner, G-helix and haem group are shown. In this frame of reference there is a small shift in the haem groups, and a shift and conformational change in the FG-corners.

A.M. LESK

1 Branden, C. & Tooze, J. (1991) *Introduction to Protein Structure* (Garland Publishing, New York).
2 Lesk, A.M. (1991) *Protein Architecture: A Practical Approach* (Oxford University Press, Oxford).

protein structure determination *See*: CIRCULAR DICHROISM; ELECTRON CRYSTALLOGRAPHY; ELECTRON MICROSCOPY; ESR/EPR SPECTROSCOPY; EXAFS; IMAGE ANALYSIS; NEUTRON SCATTERING AND DIFFRACTION; NMR SPECTROSCOPY; X-RAY CRYSTALLOGRAPHY.

Protein synthesis

THE instructions for making PROTEINS in living cells are encoded in the nucleotide sequence of nucleic acids (DNA, or for some viruses, RNA; *see* GENETIC CODE). Protein synthesis itself is the final step in the decoding and expression of the protein-coding information stored in nucleic acids and is the process by which the amino acid chain of a protein (a POLYPEPTIDE CHAIN) is assembled in the correct sequence. The sequence of amino acids in a polypeptide chain is determined by the sequence of nucleotides in a MESSENGER RNA (mRNA), which is itself a copy of the nucleotide sequence of the corresponding GENE (*see* TRANSCRIPTION). The correspondence of amino acid sequence to nucleotide sequence follows the rules of the GENETIC CODE in which each triplet of three consecutive nucleotides (a codon) in the mRNA, encodes a particular amino acid. Decoding of mRNA to produce a polypeptide chain is also termed translation. A full account of the process of protein synthesis is given in [1].

Translation occurs on subcellular particles called RIBOSOMES. Each ribosome is made up of two nonidentical subunits ('large' and 'small') each of which contains one or more RIBOSOMAL RNA (rRNA) molecules and 20–50 different RIBOSOMAL PROTEINS. Several ribosomes may simultaneously translate the same mRNA

molecule; such groups of ribosomes are referred to as polyribosomes or polysomes.

In eukaryotic cells, translation occurs outside the nucleus, primarily in the cytoplasm; proteins encoded by mitochondrial or chloroplast DNA are, however, translated in these organelles (*see* CHLOROPLAST GENOME; MITOCHONDRIAL GENOMES). The essential mechanism of translation is extremely similar in all cases but there are important differences in detail, especially between eukaryotic cytoplasmic translation on the one hand, and translation in prokaryotes and in cellular organelles on the other. Henceforward, 'eukaryotic' will refer to translation in the cytoplasmic compartment of eukaryotic cells.

Proteins are synthesized starting with their N termini (*see* PROTEIN STRUCTURE), corresponding to translation of the mRNA in the 5' to 3' direction. Translation of mRNA may be conveniently divided into three stages (Fig. P86): initiation, where the correct site on the mRNA for commencing translation is identified and binding of the ribosome to the mRNA occurs; elongation, during which the coding sequence of the mRNA directs the synthesis of the polypeptide chain; and termination, which occurs when the ribosome encounters a stop or termination codon signalling the end of the coding sequence of the mRNA which results in release of the completed polypeptide chain and the ribosome from the mRNA. During the elongation stage, TRANSFER RNAS (tRNAs) carrying the appropriate amino acid recognize the codons in mRNA by means of anticodon : codon interactions and thus deliver amino acids for addition to the growing peptide chain in the correct order.

Many bacterial mRNAs contain more than one coding sequence (cistron) (i.e. they encode more than one polypeptide chain) and are termed polycistronic (*see* BACTERIAL GENE EXPRESSION; BACTERIAL GENE ORGANIZATION).

Initiation

Translation commences at an initiation codon. This is generally AUG although other closely related codons such as GUG may also be used, especially in bacteria. Since AUG encodes the amino acid methionine, this is the first amino acid incorporated (even when non-AUG codons are employed). In bacteria the methionine is modified to *N*-formylmethionine. AUG is the only codon to code

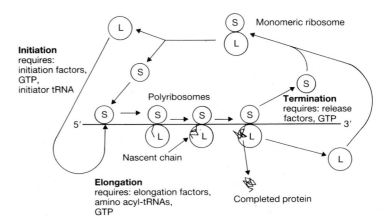

Fig. P86 The ribosome cycle. A generalized scheme for translation in bacteria and eukaryotes is shown. Specific features of the bacterial system (e.g. Shine–Dalgarno sequence, polycistronic mRNAs) and of the eukaryotic system (5' cap, scanning) are not shown. S, small ribosomal subunit; L, large ribosomal subunit.

for methionine, and different tRNAs exist for methionine as the initial amino acid (initiator tRNA) and for methionine in internal positions within polypeptide chains. Initiation is mediated by proteins termed initiation factors (IFs in bacteria, eIFs (eukaryotic IFs) in eukaryotes).

As the AUG codon is ambiguous (it can indicate either the start of translation of the mRNA or merely the location of methionine residues within proteins) mechanisms must exist to distinguish between these functions. In bacteria, the initiator AUG ('start codon') is distinguished from internal AUGs on the basis of an interaction between complementary sequences in the rRNA of the small ribosomal subunit (16S rRNA) and a purine-rich sequence immediately upstream of the start codon (the SHINE–DALGARNO SEQUENCE) in the mRNA.

In eukaryotes no such interaction occurs: current evidence generally favours the Kozak scanning model for translational initiation in eukaryotes. All eukaryotic cellular cytoplasmic mRNAs have at their 5′ end a cap consisting of 7-methylguanosine triphosphate linked to the first nucleotide of the mRNA itself by a 5′ : 5′-phosphodiester bond. Several eukaryotic initiation factors (eIF-4A, eIF-4B, eIF-4E and eIF-4F) can interact directly or indirectly with the cap (and are therefore termed cap-binding proteins). They are believed to mediate the unwinding of regions of secondary structure within the 5′-LEADER region (5′-untranslated region or 5′-UTR) of the mRNA which interfere with initiation. eIF-4A has ATP-dependent RNA HELICASE activity. This unwinding facilitates the binding of the ribosome to the mRNA and its subsequent scanning along the mRNA in a 5′ → 3′ direction to locate the start AUG codon: translation in eukaryotes generally starts at the first AUG from the 5′-end. However, this is not always the case and the context (the nucleotide sequence) around the AUG plays an important, although largely unexplained, part in its recognition by the ribosome as the actual initiation codon.

Cytoplasmic RNAs in eukaryotic cells are generally associated with mRNA-binding proteins in the form of messenger ribonucleoprotein complexes (mRNPs). A specific protein (M_r ~70 000) is associated with the poly(A) tail of the mRNA. mRNA-binding proteins have an important role in regulating the translation of mRNA molecules, as they can modulate the ease with which mRNAs can be translated, probably by altering their secondary structure.

Some viral mRNAs, particularly those of picornaviruses, depart from the mechanism of initiation described above, having evolved different mechanisms to favour the translation of their uncapped mRNAs at the expense of host mRNAs (*see* PICORNAVIRUS; PLANT VIRUSES).

The other major process in initiation is the attachment of the (formyl)methionyl-initiator tRNA to the ribosome, at the peptidyl- or P-site. This is mediated by another initiation factor (IF-2 in bacteria, eIF-2 in eukaryotes), which forms a ternary complex with the tRNA and GTP. eIF-2 is an important control point in the translational control of protein synthesis in eukaryotes. It is inactivated through phosphorylation by PROTEIN KINASES which are themselves regulated by haem (e.g. in reticulocyte lysates) or by double-stranded RNA (this is the protein kinase induced by INTERFERON). In all cases the attachment of the small ribosomal subunit precedes that of the large one. The assembly of the complete 70S (bacterial) or 80S (eukaryotic) initiation complex allows commencement of the next stage of translation — elongation.

Elongation

This process is essentially identical in all organisms (Fig. P87). Immediately after initiation, the ribosomal P-site is occupied by the initiator methionyl-tRNA and the next codon is aligned with the vacant ribosomal A (aminoacyl)-site. Entry of the correct cognate AMINOACYL-TRNA (whose anticodon matches the codon in the A-site) is mediated by an elongation factor (EF-Tu in bacteria and eEF-1 in eukaryotes, associated as a ternary complex with GTP). Formation of the PEPTIDE BOND between the methionine moiety of methionyl-tRNA and the amino acid carried by the incoming aminoacyl-tRNA then follows, catalysed by the peptidyltransferase activity associated with the large ribosomal subunit. The GTP is also hydrolysed to GDP and P_i. The second elongation factor (EF-G in bacteria and eEF-2 in eukaryotes, both of which also bind GTP) then mediates the translocation step in which the spent tRNA leaves the P-site, the peptidyl-tRNA moves from the A- to the P-site and the ribosome moves by the equivalent of one codon (three nucleotides) relative to the mRNA to align the next codon with the A-site. This step is also associated with GTP hydrolysis. Peptide-chain elongation consists of repetitive cycles of this elongation process, the nascent chain being extended by one amino acid residue at each cycle.

The effect of DIPHTHERIA TOXIN on eukaryotic protein synthesis is due to its ability to inactivate eEF-2 by ADP-RIBOSYLATION.

Aminoacylation of tRNA

Attachment of amino acids to tRNA serves both to 'activate' them energetically for incorporation into the polypeptide chain and to provide a mechanism for inserting the correct amino acid corresponding to a particular codon in the mRNA. Amino acids attach to the 3′ end of tRNA in an aminoacylation reaction catalysed by enzymes termed AMINOACYL-TRNA SYNTHETASES. Although only 20 amino acids are commonly found in proteins, the existence of 61 sense codons (i.e. codons specifying amino acids) in the genetic code requires the existence of more than 20 tRNAs. (A different tRNA is not necessarily required for every codon owing to the operation of WOBBLE in codon : anticodon interactions.) However, only one aminoacyl-tRNA synthetase exists for each amino acid, so each enzyme must correctly recognize two or more distinct tRNA molecules. The rules governing recognition of the correct tRNA by aminoacyl-tRNA synthetases are not yet completely understood. The importance of charging tRNA with the correct amino acid is underlined by the proofreading mechanisms utilized by certain aminoacyl-tRNA synthetases.

Termination

Termination occurs when the translating ribosome encounters a termination or stop codon (UAA, UAG or UGA). Since no tRNA exists to decode such codons, elongation ceases. The termination process involves release of the now complete polypeptide chain, the final tRNA and the ribosomal subunits, which are then free to

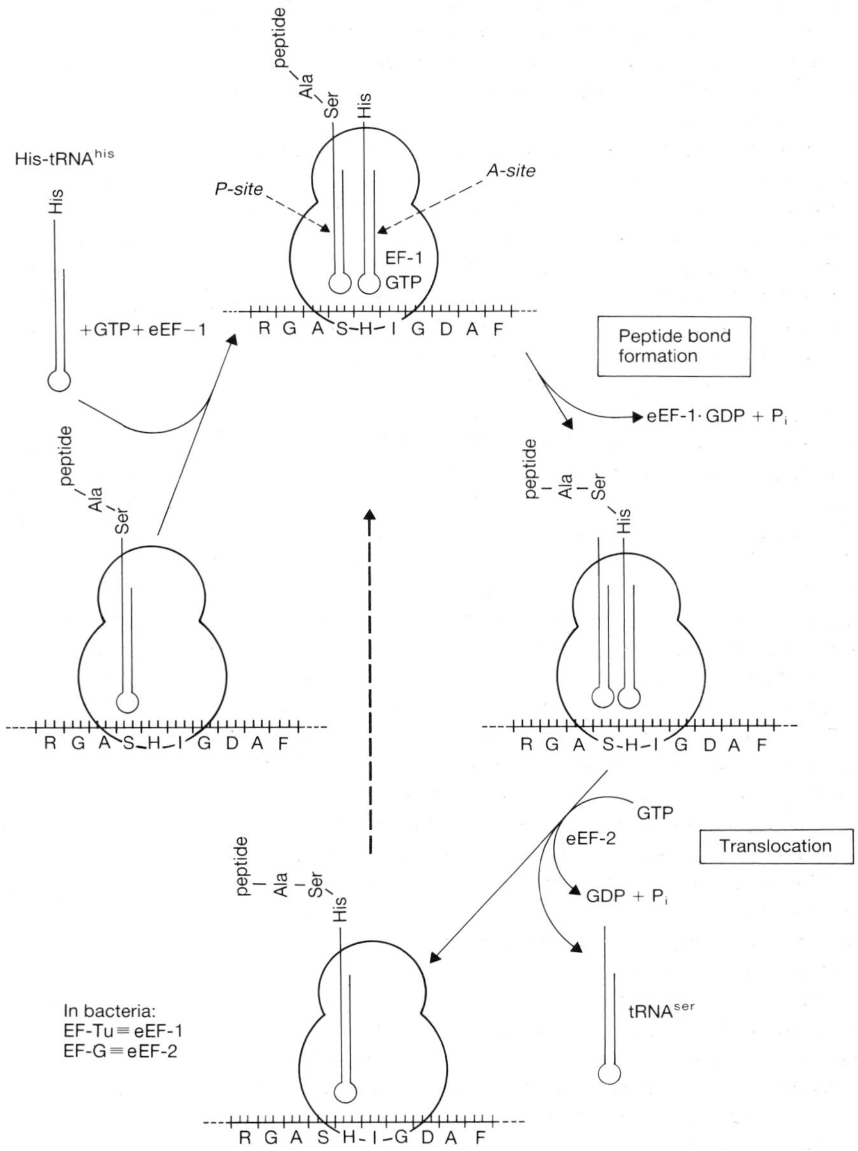

Fig. P87 Elongation. A single cycle of elongation is shown: the broken arrow shows the entry of the ribosome into the next cycle. The eukaryotic process is shown here: the corresponding bacterial factors are indicated in the bottom left-hand corner. The codons in mRNA are given in the single-letter amino-acid notation. Only the A- and P-sites on the ribosome are shown.

In bacteria:
EF-Tu ≡ eEF-1
EF-G ≡ eEF-2

participate once more in mRNA translation: this process requires proteins termed release factors and in bacteria, also EF-G.

Translational fidelity

Fidelity refers to the accuracy of translation. Errors might arise at initiation, elongation or termination. At elongation, the process of ensuring correct recognition of the codon by the anticodon of the right aminoacyl-tRNA is a compromise between accuracy and speed, and error rates of approximately 1 per 1000 residues have been estimated. Low levels of minor errors (infrequent replacement of one amino acid by another in elongation) can apparently be tolerated by cells.

At termination, misreading is a consequence of readthrough past a stop codon and generally arises from the existence of SUPPRESSOR TRNA which either inserts an amino acid at particu-

lar stop codons or causes FRAMESHIFTING. Readthrough would be expected to lead to synthesis of extended, structurally and functionally aberrant proteins, but is a required part of the synthesis of certain proteins, for example certain viral proteins in eukaryotes (*see* ANIMAL VIRUSES; PLANT VIRUSES).

Translation of mRNA from a different species in intact cells or cell-free systems (*see* IN VITRO TRANSLATION) such as the reticulocyte lysate can give rise to errors due to premature termination as a consequence of marked differences in CODON BIAS between the two species.

Post-translational events

The polypeptide chain released from the ribosome is not necessarily the final functional form of the protein, and it may undergo post-translational modification(s) (e.g. limited proteolysis, glycosylation, phosphorylation), assembly into a larger multisubunit protein (or other macromolecular assemblies) or translocation to other sites in the cell (e.g. to organelles, or through the secretory pathway). *See*: POST-TRANSLATIONAL MODIFICATION OF PROTEINS; PROTEIN PHOSPHORYLATION; PROTEIN STRUCTURE; PROTEIN TRANSLOCATION.

Antibiotic inhibitors of translation

A number of antibiotics and other agents inhibit mRNA translation usually by interacting with ribosomes and impairing specific steps in the process. These compounds include CHLORAMPHENICOL, CYCLOHEXIMIDE, ERYTHROMYCIN, PUROMYCIN, STREPTOMYCIN and TETRACYCLINE. Such agents have proven extremely useful as (for example) antibacterial agents owing to the selectivity for bacterial (70S) ribosomes rather than eukaryotic (80S) ribosomes, or to their selective entry into bacterial cells. These compounds have also been very important in investigating the structures and mechanisms of components of translation, in particular of ribosomes themselves.

C.G. PROUD

See also: CONTROL OF EUKARYOTIC GENE EXPRESSION; PROTEIN ENGINEERING; RNA.

1 Lewin, B. (1990) *Genes IV*, Chs 1–7 (Oxford University Press, Oxford).

Protein targeting

PROTEIN targeting is the process whereby different proteins are sorted and directed to different destinations within the cell. In prokaryotes, these destinations may be the cytoplasm, the inner or outer membranes, the intermembrane space (periplasm) or secretion from the cell (*see* BACTERIAL ENVELOPE; BACTERIAL PROTEIN EXPORT). Eukaryotic cells, however, are more complicated, containing many different organelles and compartments. Apart from a few proteins that are encoded and synthesized *in situ* in mitochondria and chloroplasts, all cellular proteins, whatever their final destination, are encoded in nuclear DNA and synthesized on the translational apparatus in the cytosol (*see* PROTEIN SYNTHESIS). Therefore the amino-acid sequence of a protein must contain, in some form, the information required to target it to its correct destination.

The 'signals' that specify protein targeting take various forms. The simplest is a precise sequence of amino acids conserved among all proteins sharing a particular destination. More usual is a short region of the protein in which general physical characteristics, such as arrangements of charge and hydrophobicity, define the sorting signal, but in which the amino-acid sequence varies within these constraints; this type of sequence is exemplified by the SIGNAL SEQUENCES of bacterial and eukaryotic secretory proteins. Finally, some targeting signals may only be formed after the protein has folded, taking the form of a surface 'patch' which comprises amino-acid residues from distant parts of the protein's primary sequence.

The different types of targeting signal are interpreted in different ways. In some cases, short linear signals are recognized while the protein is still being translated and is emerging from the ribosome. In other cases, translation is completed before the targeting process; if this involves a subsequent crossing of a membrane, chaperone proteins (*see* MOLECULAR CHAPERONES) may be required to preserve the protein in a conformation competent for membrane translocation. In either case the signal sequence may be cleaved from the protein after it has served its purpose. Targeting 'patches', in contrast, can only be formed after a protein has been translated and folded. Finally, some targeting signals function indirectly, by specifying a particular form of POST-TRANSLATIONAL MODIFICATION to the protein. It is this modified form that is then recognized and targeted correctly.

Some of the signals and cellular machinery involved in protein targeting are well understood, but many remain to be discovered. The pathways followed by newly synthesized proteins are depicted in Fig. P88, and the principal targeting processes in a eukaryotic cell are summarized below. They exemplify each of these different forms of targeting signal and recognition process. More information about the various processes and the relevant references can be found elsewhere in the encyclopedia (*see* PROTEIN TRANSLOCATION for mechanisms of crossing membranes; and BACTERIAL PROTEIN EXPORT, EPITHELIAL POLARITY, GOLGI APPARATUS, LYSOSOMES, MITOCHONDRIAL BIOGENESIS, and PROTEIN SECRETION for more detailed information on particular systems).

Protein targeting to, and from, the endoplasmic reticulum

The earliest sorting decision for eukaryotic proteins depends on the presence or absence of a signal sequence targeting the protein to the ENDOPLASMIC RETICULUM (ER). Proteins containing a signal sequence are recognized by a cytoplasmic ribonucleoprotein particle — the SIGNAL RECOGNITION PARTICLE (SRP). As the signal sequence is usually encoded at the N-terminal end of the protein, it emerges from the ribosome while translation is still in progress. At this point it is recognized by SRP, and the complex of messenger RNA (mRNA), ribosome, nascent protein and SRP attaches via the DOCKING PROTEIN to the membrane of the ER (*see* Fig. P89 in PROTEIN TRANSLOCATION ACROSS MEMBRANES). This process occurs with nearly all transmembrane and secretory proteins. The signal sequence involved has a core of hydrophobic amino acids, but otherwise there is no strict conservation between signal sequences of different proteins (*see also* BACTERIAL PROTEIN EXPORT). After targeting has been accomplished, and the N-terminal portion has entered the lumen of the ER, the sequence is usually cleaved from the polypeptide chain by SIGNAL PEPTIDASE.

The next sorting decision involves discrimination between secretory and membrane proteins. As both undergo protein translocation across the ER membrane, this discrimination is simply achieved by the presence of a STOP-TRANSFER SEQUENCE in membrane proteins, which results in their permanent attachment to membranes.

Once within the ER, a third level of protein targeting of membrane and secretory proteins takes place, namely transport

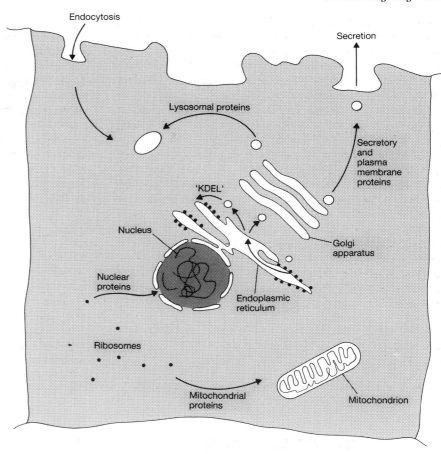

Fig. P88 The various pathways followed by newly synthesized proteins in a eukaryotic cell.

to the different membrane compartments of the cell. A small group of soluble proteins remain in the ER. These possess the targeting signal KDEL (Lys-Asp-Glu-Leu) or a close variant. This relatively well-conserved sequence acts as a RETENTION SIGNAL for the ER; in fact these proteins leave the ER by VESICLE-MEDIATED TRANSPORT along with secretory proteins, and are then selectively recognized and returned to the ER. The rest of the soluble proteins are transported to the GOLGI APPARATUS where proteins destined for the LYSOSOMES are selectively modified with mannose 6-phosphate. The signal for this modification is a 'patch' formed by amino-acid residues dispersed along the protein's sequence which come together when the protein folds. The second stage of targeting of lysosomal proteins is binding by the mannose 6-phosphate receptor which diverts these proteins into vesicles leading to the lysosomes. Soluble proteins which enter the ER but lack either of these signals are secreted, via the Golgi complex, from the cell (*see* PROTEIN SECRETION).

Membrane proteins that enter the ER may also be destined for one of several different compartments: the ER itself, the Golgi complex, lysosomes or the cell surface. Those proteins which lack signals for retention in the ER or Golgi complex, or diversion to lysosomes, reach the cell surface. Some potential targeting signals for these intracellular destinations have been identified, but there are no specific amino-acid sequences conserved among all the membrane proteins resident within a particular compartment (*see* EPITHELIAL POLARITY). Once on the cell surface, some membrane proteins can re-enter the cell via coated pits (*see* COATED PITS AND VESICLES). The signal for targeting to coated pits is again not a precise sequence of amino acids, but often includes a tyrosine or other aromatic residue close to the cytoplasmic side of the membrane.

As these different membrane compartments are connected by vesicle-mediated transport, another targeting process takes place, namely that each vesicle leaving one compartment is precisely destined for another. This may involve different Rab proteins (*see* GTP-BINDING PROTEINS) on the cytoplasmic side of each different type of vesicle.

A separate mechanism exists for directing certain proteins that lack ER signal sequences to the cytoplasmic side of membranes. Such proteins may be modified with myristate, palmitate, isoprenoid derivatives or methyl groups, allowing them to associate with the hydrophobic lipid bilayer of the membrane.

Protein targeting from the cytoplasm

Distinct mechanisms exist for targeting proteins to each of the other organelles in the eukaryotic cell. In each case targeting occurs after translation of the protein is completed. Proteins destined for the nucleus contain one or more regions of basic amino acids — the NUCLEAR IMPORT SIGNAL — which direct these proteins to and through large structures in the NUCLEAR ENVELOPE known as nuclear pores.

MITOCHONDRIAL BIOGENESIS is more complicated in that the mitochondrion contains four separate destinations, the outer membrane, the intermembrane space, the inner membrane and the matrix. Mitochondrial proteins must enter one or more membranes after their synthesis; to allow this, their folding is controlled by chaperone proteins both inside and outside the mitochondrion. The signals for mitochondrial targeting, and for subsequent sorting within the mitochondria, again have general characteristics in common but no precisely conserved sequence; like ER signal sequences, they are often cleaved from the protein after targeting has taken place. In plants, CHLOROPLAST proteins are targeted by mechanisms broadly similar to those of mitochondria, with the additional complication that proteins destined for the inside of the THYLAKOIDS must cross three separate membranes.

PEROXISOMES also acquire their proteins after translation. The targeting signals in this case are not subsequently removed from the protein. For one class of peroxisomal proteins, a sequence of three amino acids at the C terminus acts as the targeting signal.

J. ARMSTRONG

See also: MEMBRANE STRUCTURE; PROTEIN STRUCTURE; PROTEIN SYNTHESIS.

1 Glick, B. & Schatz, G. (1991) Import of proteins into mitochondria. *Annu. Rev. Genet.* **25**, 21–44.
2 Pelham, H. (1989) Control of protein exit from the endoplasmic reticulum. *Annu. Rev. Cell Biol.* **5**, 1–23.

Protein translocation across membranes

MANY newly synthesized proteins of eukaryotic cells must cross the hydrophobic phospholipid bilayer of membranes while moving from their sites of synthesis in the cytoplasm to their final destinations in subcellular organelles such as MITOCHONDRIA, NUCLEUS, ENDOPLASMIC RETICULUM (ER) or PEROXISOMES where they perform their function. Prokaryotes also have to export some proteins over the cytoplasmic membrane to the BACTERIAL ENVELOPE (*see also* BACTERIAL PROTEIN EXPORT). As proteins are hydrophilic, mechanisms have evolved which not only target them to the appropriate membrane but overcome the energy barrier of insertion into membranes and translocation across them.

In general, specific short sequences within the newly synthesized polypeptide chain are responsible for targeting and translocation. Targeting sequences have been identified by the use of genetic engineering to create chimaeric proteins in which a candidate targeting sequence is attached to a cytoplasmic protein, which is then found to be rerouted to a non-cytoplasmic location.

Proteins destined for secretion or for the endomembrane systems of the cell must all enter the ER as the first stage of their journey. Proteins destined for mitochondria and chloroplasts are exported directly into those organelles from the cytoplasm (*see* PROTEIN TARGETING for a general discussion of the various pathways).

Newly synthesized proteins need to be in loosely folded or unfolded conformations to translocate. Many proteins start to translocate across the ER membrane co-translationally while they are still attached to the ribosome, and thus do not fold until they reach the lumen. Some proteins destined for the ER or for mitochondria bind in unfolded conformation to MOLECULAR CHAPERONE proteins such as the HEAT-SHOCK protein of M_r 70 000 (Hsp70) [1], shortly after synthesis, and are not released until they are close to the membrane. Concomitant ATP hydrolysis is required for release of these protein precursors. GroELS, trigger factor and SecB are molecular chaperones that aid translocation of some newly synthesized proteins into the periplasm of bacteria (*see* BACTERIAL ENVELOPE; BACTERIAL PROTEIN EXPORT).

In vitro systems

Much of our understanding of the mechanism of translocation has been obtained from studies of translation and translocation *in vitro*. Messenger RNA (mRNA) for the protein under study is obtained by transcription of cloned DNA. The mRNA is then added to a supernatant from a high-speed centrifugation (about $100\,000g_{av}$) of cell lysates generated from yeast, reticulocytes, wheat-germ or mammalian cell lines and a radioactively labelled protein is synthesized. If membrane vesicles from the appropriate subcellular organelle are present, the newly synthesized protein may translocate into the interior of the vesicle. Translocation can be detected as a loss in molecular mass of the protein as the topogenic (targeting) sequence is subsequently cleaved or by the protection that the translocated protein acquires against added proteases.

The endoplasmic reticulum

Proteins that enter the secretory pathway to the cell surface through the GOLGI APPARATUS (*see* PROTEIN SECRETION) or through the endosomal system to LYSOSOMES, initially translocate across the ER membrane into the lumen. Proteins destined for the plasma membrane or endosomal membranes only partly translocate and become integrated into the membrane bilayer. Most secretory and several membrane proteins are initially synthesized with a SIGNAL SEQUENCE 15–30 residues long at their N-terminal ends. This is cleaved off after translocation into the lumen of the ER by SIGNAL PEPTIDASE. Signal peptides have three functions: (1) to keep the polypeptide unfolded; (2) to interact with SIGNAL RECOGNITION PARTICLE (SRP); and (3) to interact with the phospholipid bilayer and the proteinaceous translocation complex of the membrane.

Large proteins are threaded across the ER membrane while they are being synthesized and are still attached to the ribosome (*see* VECTORIAL DISCHARGE). The ribonucleoprotein SRP first binds the signal sequence as it emerges from the ribosome when the nascent chain is about 80 residues long, and acts on the ribosome to halt or slow elongation until contact is made with the SRP-receptor (the DOCKING PROTEIN) in the ER membrane (Fig. P89) [2]. Interaction with SRP prevents nascent chains from folding up or aggregating in the cytoplasm before reaching the ER membrane.

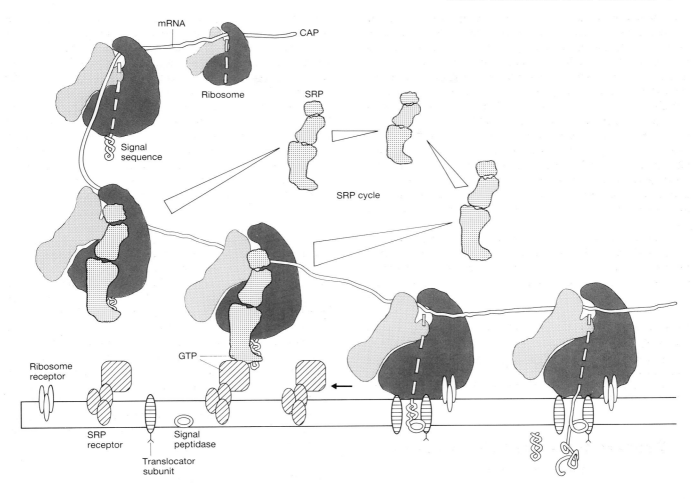

Fig. P89 Targeting by the signal recognition particle (SRP). SRP binds to the complex of nascent chain and ribosome as the signal sequence emerges from the ribosome, halting elongation of the nascent chain until contact is made with the SRP receptor (docking protein) in the endoplasmic reticulum membrane, which allows elongation to continue. GTP is required for successive steps in which SRP is released from the ribosome, the nascent chain translocates through a translocation complex in the membrane, and SRP is released from its receptor and recycles. The signal sequence is cleaved by signal peptidase in the lumen.

Binding of GTP to the arrested complex of SRP, nascent chain and ribosome enhances attachment to docking protein, and catalyses release of the ribosome and the nascent chain from SRP, thus ensuring unidirectionality of the translocation process, resumption of polypeptide elongation, and insertion of the signal sequence into the membrane. GTP hydrolysis is required for release of SRP from docking protein, and a subsequent round of interaction with ribosome and nascent chain. While translocation is taking place the ribosome remains attached to the ER membrane by an integral ribosome receptor protein, giving rise to the characteristic studded appearance of the rough ER under the electron microscope. The hydrophobic signal sequence then inserts spontaneously into the phospholipid bilayer, looping across so that the positively charged 'n-region' of the sequence remains

cytoplasmic. The signal sequence and the following polypeptide move laterally into a complex of proteins through which translocation of the nascent chain occurs. In their original signal hypothesis, Blobel and Dobberstein [3] envisaged that arrival of the signal sequence at the membrane triggered assembly of subunits into the translocation complex, making an aqueous tunnel through which the polypeptide passed, which disassembled after translocation. The protein subunits comprising the complex have not been rigorously identified, but candidates include: (1) a M_r 35 000 (30K) glycoprotein (signal sequence receptor α, SSRα) which has been shown to cross-link to photoreactive diaziridine groups placed within the translocating nascent chain; (2) a 22K (SSRβ) integral membrane which associates with SSRα; (3) homologues of Sec61; (4) Sec62 and (5) Sec63 of yeast; and (6) a 43K integral membrane protein which has been shown to cross-link to a photoreactive chemically synthesized signal sequences. The ATPase BIP (immunoglobulin heavy-chain binding protein) may rest on the luminal side of the complex, receiving polypeptides as they translocate and acting as a molecular chaperone for folding and oligomerization.

Oligosaccharides are covalently attached to some proteins during translocation, catalysed by an enzymatic complex consisting of membrane-bound 66K and 63K glycoproteins and a 48K protein, and in the lumen of the ER an endoprotease (signal peptidase) cleaves off N-terminal signal sequences, allowing proteins to fold to their native form. A few secreted proteins such as ovalbumin, and some membrane proteins, retain their signal sequences uncleaved in the mature protein.

Some protein precursors such as pre-promelittin and antibacterial proteins secreted by frog skin are so small (70–80 residues) that they are fully synthesized before the signal sequence completely emerges from the ribosome. In this instance SRP has no role, and the precursors instead bind to Hsp70, and translocate post-translationally when they are close to the membrane. If TERMINATION CODONS are deliberately removed, and providing ATP/GTP is available, larger proteins will also translocate in the absence of translation as the nascent chain remains attached to the ribosome. Although larger, yeast α mating factor translocates *in vitro* into yeast MICROSOMES post-translationally, indicating that SRP may not be absolutely essential in yeast.

Yeast *sec* mutants

Probable members of the translocation pathway across the ER in yeast (*Saccharomyces cerevisiae*) have been identified using TEMPERATURE-SENSITIVE MUTANT strains that fail to secrete proteins such as acid phosphatase, invertase or α mating factor at the nonpermissive temperature. These are the Sec proteins. Strains mutant for the *sec* genes *61*, *62* and *63* build up secretory precursors in the cytoplasm. They exhibit synthetic lethality — where two defective or partially functioning proteins exaggerate the deficiency in translocation — and are thus thought to act in tandem and be present as a complex, probably the translocation complex. Sec61p is a polytopic membrane protein (see below) of 51K, and is similar in structure to the SecY protein of *Escherichia coli*; Sec62p and 63p are also integral membrane proteins.

Membrane protein assembly

Proteins are anchored into membranes by stretches of about 19 consecutive hydrophobic amino acids, and may contain one (monotopic), two, or more (polytopic) transmembrane regions. The topography of many membrane proteins, which is established upon insertion at the ER (even though they are later moved to the cell surface), depends on the number and type of signal sequences, transmembrane regions and the charge distribution in the immediate vicinity either side of the transmembrane sequences.

Membrane proteins are classified into three types. Class 1 (e.g. G-protein encoded by the VESICULAR STOMATITIS VIRUS) are targeted and inserted via SRP by a signal sequence that is subsequently cleaved, and the N-terminal end of the protein translocates until a second hydrophobic sequence — the STOP-TRANSFER SEQUENCE — is reached, which stops transfer. The stop-transfer sequence then moves from the translocation complex into the phospholipid bilayer, leaving the C-terminal region cytoplasmic. Class 2 and 3 proteins have uncleaved signal sequences which interact with SRP and the translocation complex, but also double as stop-transfer sequences. Class 2 proteins (e.g. asialoglycoprotein receptor) have more basic residues on the N-terminal side of the stop-transfer sequence, and end up with the N-terminal end cytoplasmic and the C-terminal side luminal.

Class 3 proteins have more basic residues on the C-terminal side of the stop-transfer sequence, and end up in the opposite orientation. Polytopic proteins such as opsin (the protein component of RHODOPSIN) have a number of signal/stop transfer sequences and each domain is inserted one after the other after targeting by SRP. The characterstic '7 membrane-spanning domain' topography of rhodopsin and numerous other related receptor proteins (*see* G PROTEIN-COUPLED RECEPTORS) is determined by the charge distribution either side of the most N-terminal signal/stop transfer sequence.

Protein translocation in bacteria

About 25% of proteins synthesized in the bacterial cytoplasm are targeted to the inner membrane, periplasm, outer membrane (*see also* BACTERIAL ENVELOPE; BACTERIAL PROTEIN EXPORT), or are secreted. Many of these proteins are synthesized with N-terminal signal sequences similar in structure to their eukaryotic counterparts. They are cleaved in the periplasm by the membrane protein leader (signal) peptidase I, a 36K protease that is similar in function but not structure to eukaryotic signal peptidase.

One protein, the 50-residue coat protein encoded by the bacteriophage M13, is synthesized with a 23-residue signal sequence, and is integrated into the inner membrane after translation by a mechanism that follows Wickner's membrane trigger hypothesis [4]. This suggested that the role of the signal sequence is to allow the precursor to expose hydrophobic regions on the surface of the protein so that membrane insertion is spontaneous. The insertion and cleavage of M13 coat protein requires only the phospholipid components of the membrane, the electrochemical potential across the membrane and leader peptidase.

Other proteins are translocated by mechanisms that involve components identified in bacterial *sec* mutants. Pro-OmpA, the precursor of an outer membrane protein, is stabilized in an unfolded conformation first by binding to the product of *secB*, a soluble tetrameric protein of M_r 64 000, suggested to be the prokaryotic equivalent of SRP. The complex then binds to a peripheral membrane ATPase, SecA. ATP hydrolysis by SecA stimulated by the presence of the polytopic integral membrane protein SecY and acidic phospholipid components of the membrane, catalyses release and subsequent translocation of pro-OmpA [5] (Fig. P90). SecA and SecY are close to the translocating polypeptide, and, together with SecE, form the translocase.

Secretion of *E. coli* haemolysin, a 107K protein, is mediated by a 27-residue C-terminal targeting sequence. Secretion depends on the products of three genes, HylB, C and D. HylB couples the energy of ATP hydrolysis to translocation. HylB is partly homologous to yeast STE6, a 1290-residue ATP-driven transport system responsible for export of α mating factor directly from the cytoplasm, MDR, the MULTIDRUG RESISTANCE protein that pumps drugs out of mammalian cells, and CFTR, a transport protein defective in CYSTIC FIBROSIS. Similar proteins, known as TAP1 and TAP2, in antigen-presenting cells are responsible for the translocation of antigenic peptides through the ER membrane to the lumen where they form a complex with class I MHC proteins (*see* ANTIGEN PROCESSING AND PRESENTATION). The product of *hylC* is a fatty-acyl carrier protein.

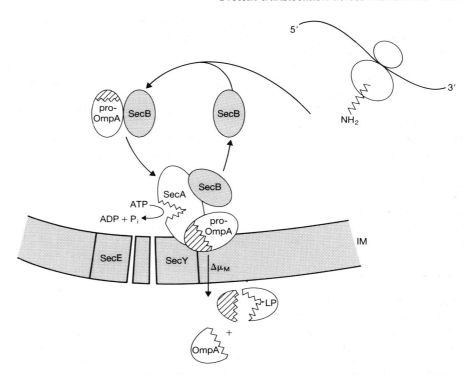

Fig. P90 A model for translocation of pro-OmpA through the bacterial inner membrane. SecB binds to pro-OmpA, keeping it in an open conformation. Near the membrane, pro-OmpA binds to a peripheral ATPase, SecA, which hydrolyses ATP as it hands on pro-OmpA to SecY in the translocation complex. After translocation, which requires the electrochemical potential gradient ($\Delta_{\mu,M}$) across the membrane, the signal sequence is cleaved by leader peptidase I (LP) in the periplasm.

Translocation into mitochondria

Proteins are imported from their site of synthesis in the cytoplasm to one of four locations in mitochondria; the matrix, the intermembrane space, the outer membrane or the inner membrane (*see also* MITOCHONDRIAL BIOGENESIS). Many mitochondrial proteins are synthesized as precursors, which are imported post-translationally with the aid of nucleotide triphosphate hydrolysis and the electrochemical potential across the inner membrane. Many precursors of matrix proteins such as malate dehydrogenase contain N-terminal presequences (transit sequences) of 10–70 residues which are cleaved off after import. The inner membrane ATP/ADP carrier protein (the adenine nucleotide translocator) in contrast possesses internal transit sequences which are not cleaved, and apocytochrome *c* translocates spontaneously through the outer membrane and is not proteolytically processed.

The essential portion of a transit sequence contains relatively high numbers of positively charged (lysines and arginines) and hydroxy (serines and threonines) amino acids and hydrophobic residues grouped together. These sequences are capable of adopting α helices in apolar solvents, detergents or acidic phospholipids such as cardiolipin, which are predominant in mitochondrial membranes. More detailed structural analyses reveal a HELIX-TURN-HELIX model, with the helix closest to the cleavage site being more stable. The helices are amphiphilic, with one side of the helix hydrophilic and positively charged and the other hydrophobic. A transit sequence may interact with and disrupt phospholipid bilayers by adopting an orientation in which its positive surface lies on the surface and interacts with the acidic groups of phospholipids while the hydrophobic face is buried in the apolar interior of the bilayer.

Import into the matrix is thought to proceed across both inner and outer membranes where they are closely apposed. Premature folding in the cytoplasm is prevented by cytoplasmic chaperones such as Hsp70. Receptors on the outer membrane then bind to the precursor proteins before translocation. A 19K protein in *Neurospora* promotes import of many, but not all, mitochondrial precursors, and a 72K receptor recognizes the internal transit sequences of the ATP/ADP carrier protein. After capture by receptors, the transit sequence may then insert its C-terminal amphiphilic helix into the outer membrane bilayer, and diffuse rapidly to the import sites. Components of the mitochondrial translocation complex have been identified by cross-linking to a chimaeric precursor which became stuck in a transmembrane orientation by virtue of a tail consisting of trypsin inhibitor. Two separate protein channels in the two membranes line up at the translocation site. In yeast, subunits of the outer complex include ISP42 and ISP6, and the inner channel includes ISP45/Mpilp and MA56. In the matrix, transit peptides are cleaved by a divalent cation-dependent processing peptidase [6].

A mitochondrial Hsp70 in the matrix binds transiently to protein precursors after translocation. Binding is disrupted by ATP and, as with BiP in the ER lumen, ATP hydrolysis may provide the driving force for transmembrane translocation. A second mitochondrial heat-shock protein, Hsc60, also binds precursors and releases them on ATP hydrolysis. Its sequence is similar to that of GroEL in *E. coli*.

Proteins such as cytochrome c_1 of the bc_1 complex, which is assembled on the outer surface of the inner membrane, and cytochrome b_2, a soluble protein in the intermembrane space, have complex transit sequences. An N-terminal amphiphilic sequence brings about import into the matrix, and is then cleaved to expose an additional stretch of 20 hydrophobic residues, preceded by some basic residues, similar in structure to the ER signal sequences. These precursors never enter the matrix completely as translocation is interrupted by the hydrophobic sequence, which acts as a stop-transfer signal.

Import into chloroplasts and the nucleus

Transit sequences of chloroplast precursors are similar in structure to mitochondrial transit sequences, but contain fewer positively charged residues and more serines and threonines. Membrane translocation is post-translational and requires ATP hydrolysis. Thylakoid proteins are imported first into the stroma by virtue of their cleavable transit sequences. There they bind first to chaperonin 60, and then to an Hsp70 homologue. Insertion into the thylakoid membrane requires a PROTONMOTIVE FORCE.

The nuclear envelope is perforated by pore complexes (nuclear pores, *see* NUCLEAR ENVELOPE) through which mRNA, transfer RNA and ribosomal subunits are exported and proteins are imported. Proteins of M_r <40 000 diffuse readily, whereas import of larger proteins involves recognition of nuclear targeting sequences which cause the pores to widen. Nuclear targeting sequences such as those of ADENOVIRUS E1a protein, and the large T antigens of PAPOVAVIRUSES, contain several adjacent positively charged residues while that of the molecular chaperone nucleoplasmin consists of two clusters of basic charges. Import is post-translational, requiring Ca^{2+} and ATP. Transport through the nuclear pore may thus be similar to transport of other proteins through translocation complexes in their respective membranes. It is only in the nucleus, however, that translocation may be viewed under the electron microscope.

B.M. AUSTEN

See also: MEMBRANE STRUCTURE.

1 Chirico, W.J. et al. (1988) 70K Heat shock related proteins stimulate protein translocation into microsomes. *Nature* **332**, 44–47.
2 Walter, P. & Blobel, G. (1981) Translocation of proteins across the endoplasmic reticulum III. Signal recognition particle (SRP) causes dignaldependant and site specific arrest of chain elongation that is released by microsomal membranes. *J. Cell Biol.* **91**, 557–561.
3 Blobel, G. & Dobberstein, B. (1975) Transfer of proteins across membranes. Presence of proteolytically processed and unprocessed nascent immunoglobulin light chains on membrane bound complexes. *J. Cell Biol.* **67**, 835–851.
4 Wickner, W. (1979) The assembly of proteins into biological membranes; the membrane trigger hypothesis. *Annu. Rev. Biochem.* **48**, 23–45.
5 Saier, M.H. et al. (1989) Insertion of proteins into bacterial membranes; mechanism, characteristics and comparisions with eukaryotic processes. *Microbiol. Rev.* **53**, 333–366.
6 Hartl, F.-U. et al. (1989) Mitochondrial protein import. *Biochem. Biophys. Res. Commun.* **988**, 1–45.

protein tyrosine kinase PROTEIN KINASE that phosphorylates protein substrates on tyrosine side chains. *See:* GROWTH FACTOR RECEPTORS; ONCOGENES.

proteinases The EC preferred name for the proteolytic enzymes, also known as proteases or endopeptidases, that hydrolyse internal peptide bonds in proteins and polypeptides (*see* PROTEIN STRUCTURE). General degradative proteinases are nonspecific in

their activity and digest proteins to peptides (*see* LYSOSOMES; PROTEIN DEGRADATION AND TURNOVER). Other proteinases are more specific in their activity, cleaving a particular protein at one or several predetermined points (*see* POST-TRANSLATIONAL MODIFICATION OF PROTEINS). Proteinases are classified separately from PEPTIDASES, which split individual amino acid residues or dipeptides off one or other end of a polypeptide chain or which act only on small peptides (e.g. dipeptides). The proteinases are further divided on the basis of their catalytic mechanism: important groups include the ACID PROTEINASES, the SERINE PROTEINASES, the THIOL PROTEINASES, the CARBOXYL PROTEINASES, and the METALLOPROTEINASES. *See also:* MECHANISMS OF ENZYME CATALYSIS.

proteinase inhibitors Proteins that specifically inhibit PROTEINASES by binding to them. The SERPINS constitute one large family of SERINE PROTEINASE inhibitors. Other unrelated proteinase inhibitors include bovine pancreatic trypsin inhibitor (BPTI), a small protein (M_r 6000) which forms a stable and long-lived complex with trypsin, binding to its active site. *See also:* PROTEIN STRUCTURE.

proteoglycan Type of glycoconjugate in which one or more GLYCOSAMINOGLYCAN chains are covalently linked to POLYPEPTIDE. Other types of glycoprotein glycan are usually present as well. Proteoglycans can form very large aggregates, for example the proteoglycan aggregates from cartilage (total M_r ~2 million) in which proteoglycan monomers arise at regular intervals on opposite sides of a long filament of HYALURONIC ACID (Fig. P91). *See also:* EXTRACELLULAR MATRIX MOLECULES; GLYCANS.

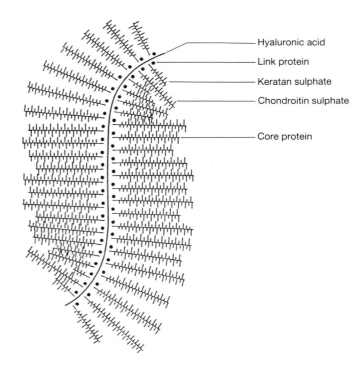

Hyaluronic acid
Link protein
Keratan sulphate
Chondroitin sulphate
Core protein

Fig. P91 Schematic diagram of a proteoglycan aggregate from cartilage.

proteolipid Conjugates of protein and lipid which are distinguishable from lipoproteins by their insolubility in water and solubility in organic solvents.

proteolysis The degradation of proteins by hydrolysis of the peptide bonds of the polypeptide backbone. Proteolysis *in vivo* is mediated by PROTEINASES and PEPTIDASES.

prothorax The most anterior thoracic segment, T1, in insects.

prothrombin (Factor II) The penultimate factor in the coagulation cascade. It is activated to THROMBIN by proteolytic cleavage by Factor Xa (*see* Fig. P52 in POST-TRANSLATIONAL MODIFICATION OF PROTEINS). Thrombin cleaves fibrinogen to form FIBRIN; it also activates Factors V, VIII, XIII, protein C, and the platelet release reaction. Prothrombin deficiency is very rare. The prothrombin time is a commonly used coagulation test. *See also:* BLOOD COAGULATION AND ITS DISORDERS.

protochlorophyll, protochlorophyllide Yellow pigment, a precursor to CHLOROPHYLL, which is produced by immature chloroplasts (etioplasts) and which consists of the Mg-protoporphyrin moiety not yet esterified to PHYTOL.

protofilament *See:* MICROTUBULES.

protomer Structural unit of the viral capsid of the PICORNAVIRIDAE, comprising virion proteins derived by proteolytic cleavage from a single precursor POLYPROTEIN.

proton pumps Proteins in the membranes of bacterial and eukaryotic cells that actively transport protons (H^+) across the membrane. *See:* ATP SYNTHASE; BACTERIORHODOPSIN; H^+,K^+-ATPASE; MEMBRANE TRANSPORT SYSTEMS; V-TYPE-ATPASE.

protonema The filamentous early stage of the GAMETOPHYTE of mosses. Also applied to similar stages of the prothallus of some ferns and to the filamentous precursor of a more complex body or a microscopic generation in some algae, especially Rhodophyta (red algae) and Phaeophyta (brown algae).

protonmotive force The proton electrochemical difference across a membrane expressed in millivolts; the force mediating energy transduction in OXIDATIVE PHOSPHORYLATION and PHOTOPHOSPHORYLATION as described by the CHEMIOSMOTIC THEORY. Protonmotive force (Δp) is equal to:

$$\Delta\mu_{H^+}/F = \Delta\psi - (2.3RT/F)\Delta pH$$

where $\Delta\mu_{H^+}$ is the proton electrochemical difference across the membrane, F is the Faraday constant, $\Delta\psi$ is the difference in electrical potential, R is the gas constant, T is the absolute temperature and ΔpH is the difference in pH.

proto-oncogene A term coined by M.J. Bishop in 1981, to describe the cellular homologue of a viral gene implicated in neoplastic transformation (an ONCOGENE). Proto-oncogenes serve normal cellular functions, but may contribute to neoplasia when their expression is altered by point mutation, translocation, or amplification. Examples of such genes are *ras*, *myc* and *abl*, prefixed with c- to distinguish them from their viral counterparts (prefixed v-).

protoplast Naked cell prepared by enzymatic (or sometimes mechanical) removal of the cell wall. Protoplasts are easily prepared from leaf mesophyll and other tissues of many plant species and also from other organisms with cell walls.

protoplast fusion The fusion of PROTOPLASTS (plant cells with walls removed), induced by POLYETHYLENE GLYCOL or electric fields, for the purposes of hybrid cell production. *See:* PLANT GENETIC ENGINEERING.

protostomes A major group of bilaterally symmetric animals (i.e. excluding protozoa, sponges, and coelenterates) which together with the DEUTEROSTOMES represent two main evolutionary lines displaying distinct and characteristic basic plans of development. Protostomes (including nematodes, flatworms, annelids, and molluscs) share the following characteristics: (1) mosaic (or determinative) form of development (i.e. where cell fate is fixed early in development); (2) SPIRAL CLEAVAGE; (3) mouth forms at or near the site of the BLASTOPORE.

prototroph Bacterial or fungal strain displaying the minimum WILD TYPE nutritional requirements for the strain. *Cf.* AUXOTROPH.

protozoa A heterogeneous group of wall-less nonphotosynthetic unicellular eukaryotes, thought to be POLYPHYLETIC in origin. Members of this group possess many features of interest to molecular biology including ANTIGENIC VARIATION, nonstandard GENETIC CODES, RIBOZYMES, RNA EDITING, and TRANS-SPLICING. Amoeboid protozoa have also been intensively studied in regard to CELL MOTILITY and cytoskeletal function. *See also:* MITOCHONDRIAL GENOMES; LOWER EUKARYOTE.

provirus A viral genome integrated into the genome of its host cell. In this state it can be transmitted by a cell to its progeny cells, and when integrated into the genome of a germ cell, from generation to generation. Proviruses can excise from the genome and resume a productive viral life cycle either spontaneously or in response to a variety of treatments. Animal viruses that can enter a provirus state include the RETROVIRUSES and the PAPOVAVIRUSES.

proximo-distal axis Longitudinal axis along a limb, from point of attachment to body to limb tip.

PrP, PrPc A neuronal membrane protein which in modified form (PrPSc) is a component of the fibrils found in the brains of animals with scrapie and other TRANSMISSIBLE SPONGIFORM ENCEPHALOPATHIES.

PRP proteins Precursor RNA processing proteins from the yeast *Saccharomyces cerevisiae*, which are involved in assembly of the spliceosome and RNA SPLICING. Some (e.g. PRP2, PRP16, PRP22, PRP5) contain an amino-acid motif — the DEAD or DEAH BOX — together with other conserved motifs, and are members of a larger family of RNA-dependent ATPases and ATP-dependent RNA HELICASES. Others are structural components of the U snRNPs.

psa genes, psb genes Genes for components of PHOTOSYSTEM I and PHOTOSYSTEM II respectively.

pseudoalleles Genes which behave as though they were ALLELES at the same gene LOCUS, but which can be separated by a recombination event.

pseudoautosomal region Region shared by mammalian X- and Y-chromosomes.

pseudocleavage CORTICAL CONTRACTIONS that develop as normal cleavage furrows but then resorb, for example before first cleavage in embryos of *Caenorhabditis elegans*, or in experimental fragments of eggs and embryos. The term is also applied to situations of abnormal cleavage where anucleate fragments of eggs or embryos are formed.

pseudodominance *See*: PARTIAL DOMINANCE.

pseudogene DNA sequence possessing close sequence similarity to a functional GENE but not expressed as a functional protein owing to the presence of deleterious MUTATIONS. Mutations may block initiation of TRANSCRIPTION, prevent correct RNA SPLICING or introduce premature TERMINATION CODONS. Most pseudogenes are presumed to have arisen as the result of a DNA duplication event followed by the accumulation of mutations and sequence divergence. Another class of pseudogenes (processed pseudogenes) lack INTRONS and resemble DNA copies of the functional gene's mRNA. Processed pseudogenes may be located on a different chromosome from the functional gene, possibly as the result of a DNA INTEGRATION event following REVERSE TRANSCRIPTION of the mRNA.

pseudo-Hurler polydystrophy *See*: LYSOSOMES; POLYDYSTROPHY.

pseudoknot *See*: NUCLEIC ACID STRUCTURE.

Pseudomonas A bacterial genus belonging to the family Pseudomonadaceae and encompassing a vast and heterogeneous group of organisms. Pseudomonads are unicellular, Gram-negative, rod-shaped bacteria, which are motile by polar flagella. They are respiratory chemo-organotrophs and strictly aerobic. Some are capable of utilizing nitrate or other nitrogenous oxides as terminal electron acceptors, while others are facultative chemo-organotrophs and able to oxidize H_2. The genus includes many microorganisms which are important in the microbiology of soils and waters, in plant pathology, human and veterinary medicine, in industrial microbiology, and bioremediation. Pseudomonads possess extreme metabolic versatility, with some species capable of utilizing more than 100 different carbon compounds (*see* PLASMIDS). Many are capable of degrading recalcitrant carbon compounds, including synthetic organics and pesticides. Pseudomonads are found in an extensive range of different habitats and are the dominant bacterial group found in soils and marine environments. Several, including the type species *Ps. aeruginosa*, are opportunistic pathogens, but only a few are true pathogens. A number can live at low temperatures and can cause spoilage of meat and dairy products. Their widespread occurrence combined with their nutritional versatility renders them prime participants in mineralization of organic matter and in nitrogen cycling. *See*: PLANT PATHOLOGY.

Palleroni, N.J. (1984) *Pseudomonas*. In *Bergey's Manual of Systematic Bacteriology* (Krieg, N.R. & Holt, J.G., Eds) 141–199 (Williams and Wilkins, Baltimore).

pseudosubstrate site Inhibitory sequences within regulatory subunits/domains of PROTEIN KINASES. Such sites retain primary sequence recognition motifs appropriate for the kinases' substrates but in place of a phosphorylatable amino acid (e.g. serine), have an amino acid with no side chain hydroxyl (e.g. alanine). *See also*: PROTEIN KINASE C.

PSI, PSII PHOTOSYSTEM I and PHOTOSYSTEM II respectively. *See also*: PHOTOSYNTHESIS.

psoralen Compound which intercalates into DNA and RNA double helices and which is used as a cross-linking reagent to investigate nucleic acid structure. On irradiation with ultraviolet light at 365 nm it forms a covalent link between pyrimidine residues on opposite strands aligned in a particular helical geometry, or between bases on closely juxtaposed single strands in macromolecular assemblies. The derivative AMT (4'-aminomethyl-4,5',8-trimethylpsoralen) has been used, for example, to investigate RNA–RNA base pairing in the SPLICEOSOME.

PTC Plasma thromboplastin component. Now known as Factor X. *See*: BLOOD COAGULATION AND ITS DISORDERS.

PtdIns Phosphatidylinositol. *See*: LIPIDS; PHOSPHOINOSITIDES.

PtdIns(4,5)P$_2$ PHOSPHATIDYLINOSITOL 4,5-BISPHOSPHATE. *See also*: SECOND MESSENGER PATHWAYS.

PTH (1) PARATHYROID HORMONE.
(2) Phenylisothiocyanate, a reagent used in PROTEIN SEQUENCING.

PTI Pancreatic trypsin inhibitor. *See*: PROTEINASE INHIBITORS.

Ptk2 cells An epithelial CELL LINE derived from the kidney of adult male potoroo (rat kangaroo).

puc operon An OPERON of purple bacteria containing the genes for LIGHT-HARVESTING COMPLEX II (*pucA*, *pucB* and *pucE*).

puf operon An OPERON of purple bacteria containing the genes for LIGHT-HARVESTING COMPLEX I (LHI) (*pufA* and *pufB*), and the L and M subunits of the REACTION CENTRE (*pufL* and *pufM*). In addition *pufQ* is involved in regulation of bacteriochlorophyll synthesis. *See also*: PHOTOSYNTHESIS.

puff Distended region of a POLYTENE CHROMOSOME, visible under the light microscope, at which transcription is taking place.

***puh* operon** An OPERON of purple bacteria expressing the gene for the H subunit of the REACTION CENTRE (*puhA*).

pulse-chase Experimental technique in which cells are very briefly labelled (the pulse) with a radioactively labelled precursor of a particular molecule or pathway under study, followed by incubation with non-labelled precursor (the chase). The fate of the radioactive label in the cell can then be detected.

pulsed-field gel electrophoresis *See*: ELECTROPHORESIS; ORTHO-GONAL FIELD AGAROSE GEL ELECTROPHORESIS.

purine The generic name for one class of organic nitrogenous BASE, based on the purine ring, of which adenine and guanine are found in DNA and RNA. A large number of other purines occur in nature, including hypoxanthine, xanthine, uric acid, caffeine, and theobromine.

purinergic neuron Neuron secreting adenosine or ATP as a NEURO-TRANSMITTER. *See*: PERIPHERAL NERVOUS SYSTEM; PURINERGIC SYSTEMS.

purinergic systems Those systems that use purine nucleosides or nucleotides in signal transmission in the nervous system, in the immune system, and in some epithelial and endothelial cells. In practice, adenine is the natural purine involved and the natural primary messengers are adenosine and adenine nucleotides. In 1972, ATP was proposed to be released as a NEUROTRANSMITTER from the terminals of certain purinergic neurons in the auto-nomic nervous system [1] and also as a co-transmitter, especially with NORADRENALINE (norepinephrine) or ACETYLCHOLINE. The receptors involved in these systems are known as purinoceptors.

In 1978, purinoceptors were classified into P_1 and P_2 classes [2]. P_1 receptors respond highly preferentially to adenosine and AMP; P_2 receptors to ATP and ADP; the P_1 types alone are generally antagonized by methylxanthines (e.g. caffeine). ATP may be broken down by exonucleotidases to adenosine, or it may survive and act at P_2 receptors. Adenosine is the active form in the PRESYNAPTIC INHIBITION of the release of certain neurotrans-mitters. Otherwise adenosine appears to function as a PARACRINE messenger, that is, acting not within synapses but by diffusion to a distant target, as in the vascular and immune systems and on some smooth muscles. Adenosine receptors have been divided pharmacologically thus far into A_1, A_2, and A_3 subclasses of the P_1 class and subtypes within these classes have been proposed [3].

ATP can act as a synaptic transmitter both in the PERIPHERAL NERVOUS SYSTEM and in the brain and spinal cord or, at some sites, as a paracrine messenger. In the vascular system, aggregat-ing platelets secrete ATP and ADP, which act on vascular endo-thelial cells to lead to the vasodilatory release of NITRIC OXIDE. The effects of P_2 receptors are important on smooth and cardiac muscles, on vascular endothelial cells, on epithelia of the lung and gastrointestinal tract and the kidney, on hepatocytes, on fibroblasts, on astrocytes, and on some exocrine glands, as well as on mast cells, lymphocytes, and other immune system cells. While ATP is the most potent natural agonist for many P_2 receptors, for the P_{2T} receptor of platelets only ADP is an agonist, and for the P_{2U} receptor UTP (not known to be released) is more potent. An important distinction is made between the P_{2X} and P_{2Y} classes of purinoceptors. P_{2X} receptors are transmitter-gated ion channels (*see* RECEPTORS), whereas P_{2Y} receptors are G PROTEIN-COUPLED RECEPTORS. The P_{2Y} receptors exist in numer-ous subtypes [3].

The primary structures of members of some of these classes have been elucidated by DNA cloning: for all A_1 [4] and A_2 [5] and further adenosine receptors, and for P_{2Y} receptors [6,7].

1 Burnstock, G. (1972) *Pharmacol. Rev.* **24**, 504.
2 Burnstock, G. (1978) In *Cell Membrane Receptors for Drugs and Hormones: A Multidisciplinary Approach* (Raven Press, New York).
3 Dubyak, G.R. & Feder, J.S. (Eds) (1990) *Annls NY Acad. Sci.* **603**.
4 Libert, F. et al. (1989) *Science* **244**, 569–572.
5 Libert, F. et al. (1991) *EMBO J.* **10**, 1677–1682.
6 Webb, T.E. et al. (1993) *FEBS Lett.* **324**, 219–225.
7 Lustig, K.D. et al. (1993) *Proc. Natl. Acad. Sci. USA* **90**, 5113–5117.

purinoceptors Receptors for purines (usually adenosine or ATP) present on neurons and other cells. They include LIGAND-GATED ION CHANNELS (P2X) and G PROTEIN-COUPLED RECEPTORS (P1, P2Y). *See*: PURINERGIC SYSTEMS.

puromycin An antibiotic from *Streptomyces* sp. which blocks bacterial PROTEIN SYNTHESIS by causing premature termination of nascent polypeptide chains. It does so because it structurally resembles the 3′ end of an aminoacylated tRNA (Fig. P92). Thus, it competes with aminoacyl-tRNAs for the A-site on the 50S ribosomal subunit. The nascent polypeptide (in the P-site) is transferred to puromycin by the enzyme complex peptidyl trans-ferase. Further peptide bond formation cannot occur, as the peptidyl–puromycin bond cannot be cleaved. Therefore, as the

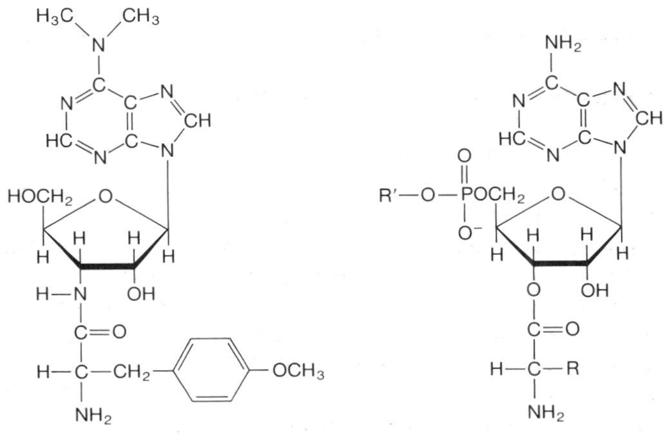

Fig. P92 Puromycin (*a*) compared with the aminoacyl terminus of an aminoacyl-tRNA (*b*). R, amino acid side chain; R′, the remainder of the tRNA molecule.

nascent polypeptide cannot be transferred to the next aminoacyl-tRNA, protein synthesis is prematurely terminated.

purple bacteria A group of photoautotrophic bacteria such as RHODOPSEUDOMONAS, which contain bacteriochlorophyll and the purple protein BACTERIORHODOPSIN.

purple membrane *See*: BACTERIORHODOPSIN; HALOBACTERIUM.

***pvt, pvt-like* locus** The position on chromosome 15 in mice (*pvt* locus) and chromosome 8, band 24, in humans (pvt-like locus) where the chromosomal TRANSLOCATION breakpoints occur involving the c-*myc* proto-oncogene from chromosomes 15 or 8 respectively. An equivalent *pvt* locus in rats is also a target for RETROVIRUS insertion causing T-cell LYMPHOMAS.

pyrimidine The generic name for one class of organic nitrogenous BASE, based on the pyrimidine ring, of which cytosine and uracil are found in RNA and cytosine and thymine in DNA. Other natural pyrimidines include isouramil in *Vicia faba* which can be toxic to individuals with GLUCOSE-6-PHOSPHATE DEHYDROGENASE DEFICIENCY.

pyrimidine dimer Lesion in DNA in which two adjacent pyrimidines on the same strand are covalently joined together by a photochemical reaction. Cyclobutane pyrimidine dimers are the photoproducts produced in highest yield by ultraviolet irradiation of DNA at wavelengths 220–300 nm. These dimers are formed by rearrangement of the 5,6 double bonds of the adjacent pyrimidines to form a cyclobutane ring. Another type of pyrimidine dimer is the (6–4) photoproduct, which is formed in UV-irradiated DNA with about one-third the yield of cyclobutane dimers. Both classes of photoproduct are toxic and mutagenic and are subject to nucleotide excision repair. In some organisms, there are additional specialized repair enzymes (photolyases and DNA glycosylases) for the cyclobutane type of dimer. *See*: DNA REPAIR.

Q

Q The single-letter abbreviation for the AMINO ACID glutamine.

Q, Qa In the mouse, a chromosomal region mapping within the TLa complex on chromosome 17 (*see* MAJOR HISTOCOMPATIBILITY COMPLEX); also, the proteins encoded by those loci. Qa antigens are structurally similar to the class I MHC MOLECULES. However, Qa antigens are less polymorphic than other class I products, and are expressed only on haematopoietic cells. Their function is unknown.

Q-banding Quinacrine banding. *See*: CHROMOSOME BANDING.

Q-cycle *See*: ELECTRON TRANSPORT CHAIN; PHOTOSYNTHESIS.

Q_B protein The D1 subunit or HERBICIDE-BINDING PROTEIN of PHOTOSYSTEM II. It contains the binding site for a PLASTOQUI-NONE molecule (Q_B) which acts as the secondary quinone acceptor and is able to exchange with the plastoquinone pool. *See also*: PHOTOSYNTHESIS.

Qβ replicase RNA-dependent RNA POLYMERASE from the BACTE-RIOPHAGE Qβ, which is often used as the replicase in *in vitro* experiments on the evolution of RNA sequences.

quantitative variation CONTINUOUS VARIATION.

quasi-equivalent positions A type of subunit arrangement that can occur in the protein coat of icosahedral viruses containing more than 60 subunits, which can be arranged with three units to each of the 20 icosahedral faces. In most larger viruses, having a TRIANGULATION NUMBER (T) greater than one, there are additional axes related to the 12 icosahedral pentad axes (which are each surrounded by five subunits). These ten (T-1) additional axes are surrounded by six subunits in an arrangement with strict threefold SYMMETRY but quasi-sixfold symmetry. The closely similar interactions between the subunits in pentamers and hexamers are described as being quasi-equivalent.

The theory of quasi-equivalence proposed by Caspar and Klug long before any detailed virus structures were known, suggested that there would be pentamers and hexamers respectively at the five-coordinated and six-coordinated positions on a capsid. A notable exception has since been found in the case of the PAPOVAVIRUSES SV40 and polyoma (*see* Fig. A822) only, where the subunits are always grouped in pentamers, even in the hexavalent positions; such that the T = 7 capsid has pentamers at both the 60 positions, which are six-coordinated, and the 12 pentamers on the icosahedral axes, which are five-coordinated.

Caspar, D.L.D. & Klug, A. (1962) *Cold Spring Harb Symp. Quant. Biol.* **27**, 1–24.
Klug, A. (1983) *Nature* **303**, 378–379.
Liddington, R.C. et al. (1991) *Nature* **354**, 278–284.

quasispecies RNAs, such as viral genomes, that differ slightly from each in nucleotide sequence and have arisen from a 'parental' molecule by small mutations. The term is used, for example, for the collection of variants of HIV that arise during an individual infection (*see* IMMUNODEFICIENCY VIRUSES) and of the variant nucleic acid molecules produced in *in vitro* experiments in evolution.

quaternary structure The level of structural organization in oligomeric proteins (i.e. those composed of more than one subunit) represented by the number and arrangement of the subunits and the interactions between them. *See*: ALLOSTERIC EFFECT; COOPERATIVITY; PROTEIN STRUCTURE.

quinacrine banding (Q-banding) *See*: CHROMOSOME BANDING.

quinic acid Hexahydro-1,3,4,5-tetrahydroxybenzoic acid (Fig. Q1). D(−)-quinic acid occurs in many plants, especially cinchona bark, coffee beans and tobacco leaves; and more widely as its 5-caffeoyl ester (chlorogenic acid). It is synthesized by the reduction of 3-dehydroquinic acid in the shikimate pathway.

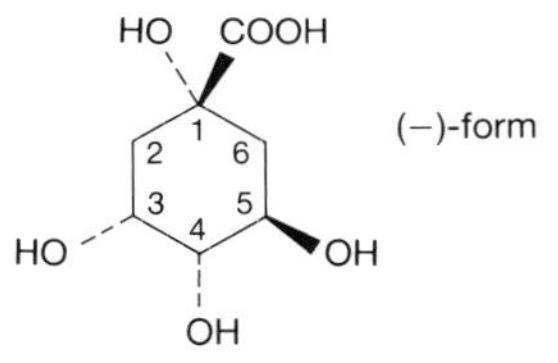

Fig. Q1 Quinic acid.

quisqualate Relatively nonselective agonist for EXCITATORY AMINO ACID RECEPTORS (Fig. Q2). It activates both non-NMDA and metabotropic receptors.

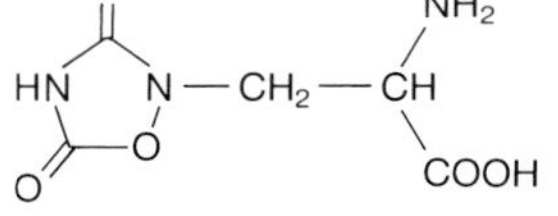

Fig. Q2 Quisqualate.

R

R The single-letter abbreviation for the AMINO ACID arginine.

R-factor The residual index R, a measure used in X-RAY CRYSTALLOGRAPHY to show the agreement between two sets of structure amplitudes (F) or intensities (I) which might be observed and calculated, crystal-1 and crystal-2, for example:

$$R = \frac{\sum_{hkl} |F_{obs} - F_{calc}|}{\sum_{hkl} F_{obs}}$$

or

$$R = \frac{\sum_{hkl} |I_1 - I_2|}{\sum_{hkl} I_1}$$

where the summation is over all reflections. The R-factor is often calculated for particular regions of RECIPROCAL SPACE, either for bands of angular RESOLUTION or for planes of successive zones. It has been shown that the expected values of the R-factor for comparing a set of observed structure amplitudes with those calculated from a model using the correct kind and number of atoms but placed at random in the unit cell, are 83% for CENTRIC and 59% for ACENTRIC reflections respectively (*see* WILSON STATISTICS).

Merging R-factors show the agreement on merging partial data sets of diffraction intensities, possibly from different crystals. Symmetry R-factors show the agreement between intensities or amplitudes of reflections expected to be equivalent to each other, by reason of the crystal SYMMETRY. Refinement R-factors monitor the progress of structure refinement and indicate the final quality of a molecular model and the diffraction data upon which it is based. For macromolecules these are in the region of 15–20%, whereas for most small molecules they are under 5%.

Free R-factors, now increasingly used as unbiased indicators of the quality of a derived model, are calculated from a small random subset of the data which has been completely excluded from the modelling and refinement process, thus providing cross-validation of the fit between model and data. Large differences between the refinement R-factor and the free R-factor indicate that the model has been misfitted or overfitted and that its information content is less than might otherwise be thought.

Wilson, A.J.C. (1950) *Acta Cryst.* **3**, 397–398.
Brünger, A.T. (1992) *Nature* **355**, 472–474.

R factor, R plasmid (drug-resistance factor/plasmid) Small, autonomously replicating PLASMID that confers resistance to certain antibacterial substances on the host bacterium. A single type of R factor may carry the genes required for resistance to a number of different drugs and can be transmitted across generic boundaries. R100, the first R factor discovered (in strains of *Shigella* in Japan in 1957) confers host resistance to streptomycin, tetracycline, sulfonamide, chloramphenicol, and mercury and has been found in *Shigella* spp., *Escherichia coli* and *Salmonella* spp. in many parts of the world. R factors are particularly and dangerously endemic among bacterial populations in hospitals where they reduce the efficacy of antibacterial therapy. *See also*: TRANSPOSABLE GENETIC ELEMENTS.

Davies, J. & Smith, D.I. (1978) *Annu. Rev. Microbiol.* **32**, 469–518.

R gene *See*: RESISTANCE GENES.

R-loop mapping The physical mapping of INTRONS using DNA–RNA HYBRIDIZATION in conjunction with electron microscopy. A mature RNA species is hybridized to single-stranded DNA containing the gene from which it was transcribed. The RNA molecules anneal to complementary regions within the DNA and loop out any intervening, noncomplementary DNA sequences in the form of distinct R-loops which can be subsequently visualized by electron microscopy. R-loops correspond to introns that are spliced out during the maturation of RNA (Fig. R1) (*see* RNA SPLICING).

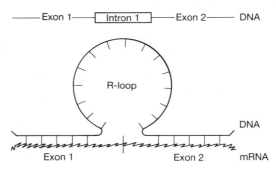

Fig. R1 R-loops correspond to introns that are spliced out during RNA processing and maturation.

R-ras *See*: GTP-BINDING PROTEINS.

R-state The high affinity or 'reactive' form of an ALLOSTERIC protein.

R1-DNA, R2-DNA Larger and smaller respectively of two plasmid-like DNAs from the mitochondria of South American maize plants of the RU GROUP. They are homologous with each other. *See*: MITOCHONDRIAL GENOMES: PLANT.

Rab proteins (*Ypt* proteins, *Sec4* protein) A family of small GTP-BINDING PROTEINS involved in vesicle-mediated transport during EXOCYTOSIS and ENDOCYTOSIS. In yeast, they are known as *ypt* proteins, or in one case *sec4* protein. They are distantly related to the *ras* oncogene product but are not themselves ONCOGENES. Different members of the family are located on the cytoplasmic surface of different membranes; for example Rab5 is found on early ENDOSOMES and on the plasma membrane, and Rab7 is found on late endosomes. The proteins are thought to be required for targeting each different type of transport vesicle to its particular cellular destination. After fusion of the vesicle to its target membrane, the *rab* protein may be released from the membrane to the cytoplasm, from where it can recycle for another round of vesicle-mediated transport.

Rac protein A small GTP-BINDING PROTEIN.

races of pathogens A race of a bacterial or fungal pathogen is genetically defined by its virulence against a host plant. Virulence is due to the possession of recessive avirulence (*A*) genes, which are complementary to dominant resistance (*R*) genes in the host. In a host plant such as potato, which possesses a defined number (11) of specific resistance genes (R_1–R_{11}) against the pathogen *Phytophthora infestans*, the races of the pathogen are defined in terms of their possession of avirulence genes (e.g. A_1–A_{11}). For instance, race 4,7 of *P. infestans* is virulent on a host possessing either or both of the resistance genes R4 and R7. The same race is avirulent on a host possessing any of the other resistance genes. *See also*: PLANT PATHOLOGY.

Racimo de Uva *See*: RU GROUP.

RAD, *rad* genes *See*: DNA REPAIR.

radial glial cells GLIAL CELLS that span the full radial thickness of the developing vertebrate NEURAL TUBE, and that provide guides for migrating neurons after the final mitosis of their NEUROBLAST precursors in the ventricular or subventricular zones. After the period of neuronal guidance, the radial glial cells develop further into astrocytes or oligodendrocytes. Migrating neurons have been observed *in vitro* in close contact with the glial cell surface, and an adhesion molecule, astrotactin (M_r 100 000), has been suggested to mediate the interaction in the developing cerebellum.

radial spoke A component of eukaryotic cilia and flagella. *See*: MICROTUBULES.

radiation chimaeras Animals which have been treated with a sublethal dose of X-irradiation sufficient to destroy the haematopoietic stem cells of the bone marrow, and which are then transfused with bone marrow cells from another animal of different MHC genotype. This reconstitutes an immune system in which the lymphoid T cell progenitors are of a different MHC type from the thymus in which they will develop. Such animals are used to study the role of the MHC in immune responses. *See*: MAJOR HISTOCOMPATIBILITY COMPLEX; T CELL DEVELOPMENT.

radiation damage The integrity of biological molecules is degraded by the ionizing radiation used in their study by techniques such as ELECTRON MICROSCOPY and X-RAY DIFFRACTION. The highest RESOLUTION structural detail is lost first when damage occurs; severe damage means that only a part of a structural data set is obtainable from a single sample or crystal.

Absorption of the incident energy leads to the formation of ions and free radicals, and breakage of some covalent bonds. Susceptibility to radiation damage appears quite variable between molecules, and between different crystalline forms of the same molecule. It depends on the total accumulated dose, and also the dose rate (as once free radicals are present, the molecular damage they cause depends on their natural diffusion rate and thus upon real time). Damage is reduced by making the sample as cold as possible (techniques of ELECTRON CRYOMICROSCOPY and cryo-crystallography both minimize this effect) and by collecting data as quickly as possible.

For X-rays, the fraction of the incident energy absorbed by a sample decreases with the wavelength such that the severity of radiation damage can be significantly reduced by using shorter wavelengths as are now available from SYNCHROTRON RADIATION. The energy of thermal neutrons is sufficiently low that they do not damage molecules studied by neutron diffraction (*see* NEUTRON SCATTERING AND DIFFRACTION).

Arndt, U.W. (1984) *J. Appl. Cryst.* **17**, 118–119.
Henderson, R. (1990) *Proc. R. Soc. Lond.* **B240**, 6–8.
Helliwell, J.R. (1992) *Macromolecular Crystallography with Synchrotron Radiation* (Cambridge University Press).

radioimmunoassay (RIA) Any method for detecting or quantitating ANTIGENS or ANTIBODIES utilizing radiolabelled reactants. RIA can be used to detect very small quantities of antigens or antibodies, even in complex mixtures. The methodology is relatively straightforward but requires that a sample of the material to be tested is available in a pure form and can be labelled with a radioactive isotope. A number of variations in the procedure are possible depending on whether labelled antigen or antibody is used. 'Sandwich' assays use two different antibodies.

radioisotope Radioactive form of an element (Table R1). Radioisotopes are used widely in molecular biology to label marker and detector molecules such as antibodies and oligonucleotide probes. The radioisotope of phosphorus in particular (^{32}P) is used to label

Table R1 Some commonly used radioisotopes

Isotope	Half-life
^{3}H	12.26 years
^{14}C	5730 years
^{22}Na	2.62 years
^{32}P	14.28 days
^{35}S	87.9 days
^{42}K	12.36 h
^{45}Ca	163 days
^{59}Fe	45.6 days
^{125}I	60.2 days
^{203}Hg	46.9 days

nucleic acids for a variety of applications (*see e.g.* DNA SEQUENC-ING; DNA TYPING; END-LABELLING; FOOTPRINTING; HYBRIDIZATION; PROBE). Molecules labelled with radioisotopes are detected by AUTORADIOGRAPHY.

radioligand A radioactively labelled ligand used to locate receptors or measure their binding parameters (*see* SCATCHARD PLOT).

radius of gyration For a single particle this quantity, defined in terms of classical mechanics, represents a kind of averaged dimension of the molecule or aggregate. The radius of gyration (R_g) can be determined by SMALL-ANGLE SCATTERING studies using X-rays or neutrons. For particles whose size is known (e.g. from relative molecular mass and density), R_g gives information about the particle shape. *See also*: NEUTRON SCATTERING AND DIFFRACTION.

raf A potential ONCOGENE encoding a serine/threonine PROTEIN KINASE which is involved in intracellular signal transduction of a wide range of stimuli inducing growth and/or differentiation. It can be activated by binding to an activated *ras* protein, and in turn phosphorylates and activates the protein kinase MAP KINASE KINASE. *See also*: PROTEIN KINASE CASCADE; RAS.

RAF The ribonucleoprotein-associated fold present in some RIBOSOMAL PROTEINS.

RAG-1, RAG-2 Recombination activator genes involved in the activation of GENE REARRANGEMENT in the IMMUNOGLOBULIN GENES. *See*: RECOMBINASE.

Ral protein A small GTP-BINDING PROTEIN.

Ramachandran plot A representation of the backbone torsion angles φ and ψ for a protein or polypeptide. *See*: PROTEIN STRUCTURE.

Rana Genus of frog. *See*: AMPHIBIAN DEVELOPMENT.

random coil Protein conformation found in proteins such as elastin (*see* EXTRACELLULAR MATRIX MOLECULES) in which the polypeptide backbone remains unfolded and can adopt a variety of partially extended random conformations, thus allowing it to stretch and recoil like elastic.

random conical tilt Image reconstruction technique in electron microscopy in which specimens are viewed in two orientations — untilted and tilted — and the images correlated.

Radermacher, M.J. (1988) *Electron Microsc. Techn.* **9** 359–394.

random X inactivation *See*: X-CHROMOSOME INACTIVATION.

Rap protein A small GTP-BINDING PROTEIN.

RAR RETINOID RECEPTOR.

ras Any of a family of cellular ONCOGENES encoding small GTP-binding proteins. *ras* genes have been found mutated in a wide range of human tumours. The *ras* protein (Ras) is a central component in intracellular signalling pathways involved in the transduction of stimuli that induce growth and/or differentiation. In mammalian cells it is activated by guanine nucleotide releasing factors and in the active state binds and activates the serine/threonine PROTEIN KINASE encoded by the *raf* proto-oncogene. The mutant Ras proteins found in tumour cells have reduced GTPase activity and therefore remain activated, resulting in an increased level of growth stimulatory signal. Ras is known to be involved in similar pathways in many eukaryotic organisms, including *Saccharomyces cerevisiae*, *Schizosaccharomyces pombe*, *Drosophila* and *Caenorhabditis*. *See*: GTP-BINDING PROTEINS; RAF.

rasGAP Protein that stimulates the GTPase activity of *ras* proteins. *See*: GTP-BINDING PROTEINS.

RB Right border. Sequences around the right-most limit of the T-region of a TI PLASMID. It contains a copy of the border repeat sequence and associated sequences (e.g. OVERDRIVE) which may enhance T-strand initiation.

RB1 The gene for familial RETINOBLASTOMA which is a TUMOUR SUPPRESSOR GENE.

rbc genes Genes encoding the large subunit (*rbcL*) and the small subunit (*rbcS*) of RIBULOSE BISPHOSPHATE CARBOXYLASE/OXYENASE (RUBISCO). In most eukaryotic cells *rbcL* is chloroplast encoded and *rbcS* is in the nuclear genome. In prokaryotic cells the two genes are cotranscribed as part of an operon. *rbcL* is invariably found in the CHLOROPLAST GENOME in one or two apparently identical copies and the number of copies of *rbcL* per cell is therefore very large because of the presence of multiple copies of the genome in each chloroplast. In higher plants and green algae *rbcS* occurs as a small, slightly heterogeneous, gene family (2–10 copies) in the nuclear genome. In several other algal groups it has been found in the chloroplast genome, closely linked to RBCL.

RBCL, RBCS Large and small subunits respectively of RIBULOSE BISPHOSPHATE CARBOXYLASE/OXYGENASE.

RBP RETINOL-BINDING PROTEIN.

RCA *Ricinus communis* agglutinin. *See*: LECTINS.

rDNA RIBOSOMAL DNA. Sometimes also used as an abbreviation for RECOMBINANT DNA.

rdt A transposable genetic element found in maize. *See*: TRANSPOSABLE GENETIC ELEMENTS: PLANTS.

re-annealing The reconstitution of a double-stranded nucleic acid from two complementary strands.

reaction centre The central process in PHOTOSYNTHESIS is light-induced electron transfer across a membrane against the electro-chemical gradient. This occurs in a protein complex referred to as a photosynthetic reaction centre. There are at least four different types, but all contain a special pair of CHLOROPHYLL molecules, near one side of the membrane, that acts as the primary electron donor and a QUINONE (in purple bacteria and PHOTOSYSTEM II) or IRON–SULPHUR CENTRE (PHOTOSYSTEM I and green sulphur bacteria), on the opposite side, that acts as acceptor; these pigments and cofactors are associated with a heterodimer of closely similar polypeptides.

reaction–diffusion models of pattern formation Models involving chemical reactions and spatial diffusion with the symmetry-breaking properties required for simulating embryological PATTERN FORMATION. The prototype was suggested by A. Turing in 1952 to explain repeating (periodic) patterns: two substances are initially distributed together homogeneously; one inhibits the production of, and diffuses more readily than, the other; the latter promotes the production both of itself and its antagonist, and with time will distribute in a series of concentration peaks placed at regular intervals. Variants of this model, for example using two substances whose synthesis is mutually exclusive locally but stimulatory at a distance, have since been proposed and analysed by computer simulation. Elegant though these models are, there is as yet no experimental evidence that periodic patterns in embryos are generated by reaction–diffusion processes.

Turing, A.M. (1952) *Phil. Trans. Roy. Soc. B* **237**, 37–72.

reading frame Any one of three ways a NUCLEOTIDE SEQUENCE can be read as a series of triplets. Messenger RNAs generally contain only one translatable reading frame which is dictated by the position of the INITIATION CODON. *See also:* FRAMESHIFT; GENETIC CODE; OPEN READING FRAME; PROTEIN SYNTHESIS.

readthrough The TRANSLATION of an mRNA beyond the normal TERMINATION CODON resulting in an extended readthrough POLYPEPTIDE. Alternatively, the term can be used to describe the TRANSCRIPTION of a DNA template past the natural transcription termination site resulting in an extended RNA transcript.

rearranging genes The only known vertebrate GENES that undergo permanent and directed rearrangement at the DNA level in somatic cells in order to be expressed are those encoding the immunoglobulins and the T cell receptors (*see* ANTIBODY; IMMUNOGLOBULIN GENES; T CELL RECEPTORS; T CELL RECEPTOR GENES). Immunoglobulins and T cell receptors recognize many different foreign molecules by way of the highly variable region at the N termini of their constituent polypeptide chains. This variability is in large part achieved by the segmented nature of their genes which consist of multiple V, D, and J segments that are brought together by GENE REARRANGEMENT (V-(D)-J recombination) to assemble a functional gene. Directed gene rearrangements in other organisms include those at the YEAST MATING-TYPE LOCUS, ANTIGENIC VARIATION in African trypanosomes, and PHASE VARIATION in bacteria.

reassociation kinetics Second-order kinetics describing the reassociation of denatured double-stranded nucleic acid molecules over time (*see* COT ANALYSIS; HYBRIDIZATION).

recA Gene in the bacterium *Escherichia coli* which encodes a protein involved in DNA RECOMBINATION and DNA REPAIR. *E. coli* mutants in which the *recA* gene has been inactivated are deficient in homology-dependent RECOMBINATION processes, TRANSDUCTION, and SOS REPAIR mechanisms. The *recA* gene product (RecA) catalyses the exchange of homologous strands during recombination and triggers proteolytic activity in SOS repair proteins such as LexA. *RecA* mutants are useful for the propagation of unstable RECOMBINANT DNA molecules and for introducing site-directed changes in DNA by oligonucleotide primer mismatch (*see* OLIGONUCLEOTIDE-MEDIATED MUTAGENESIS).

RecBCD complex Protein complex with exonuclease, endonuclease, and DNA helicase activity, which is involved in RECOMBINATION in *Escherichia coli*.

Receptor

A RECEPTOR can be defined as a structure (generally a protein) located on or in a cell, which specifically recognizes a binding molecule — a ligand — and thereby initiates either a specific biological response (the wide sense of receptor) *or* the transduction of a signal (the narrow sense of receptor). In the original proposal of the receptor concept in 1905 by the British physiologists T.R. Elliot and J.N. Langley the only type of ligand considered was an administered drug and the response was the first step leading to the observed effect of the drug. However, the essential two elements of (1) specificity of binding, and (2) consequent information flow were much earlier recognized by Paul Ehrlich in Germany who summarized the concept as '*Corpora non agunt nisi fixata*' — there is no drug action without binding. When SYNAPTIC TRANSMISSION via neurotransmitters was recognized later, and also the actions of hormones, the term receptor was naturally extended to include the structures which recognize those two classes of natural ligands, with the assumption that many drug receptors are in fact cellular receptors for native transmitters or hormones. The initial definition given above embraces all these types of ligand. It is important to note in this context that a drug may act at a site which is also the receptor for a particular transmitter molecule without necessarily exerting an identical action: more accurately, the ligand acts at a selective binding site within the receptor macromolecule. The possibility of different types of binding sites on one molecule is implicit in the full concept of a receptor. One drug may bind differently at the same location and be an antagonist of the transmitter action, and a third class of drug may bind at a different site on the same receptor and modulate the action of the natural transmitter. This extension of the concept of a receptor to allow for several types of binding site on one receptor macromolecule, and for the cases where there are modulatory sites on the receptor, is exemplified in the entries on GABA AND GLYCINE RECEPTORS and EXCITATORY AMINO ACID RECEPTORS.

It is now no longer thought productive to define receptors in terms of the interaction with exogenous drug molecules. Drugs may produce specific biological effects by action at various types of targets besides receptors for transmitters and hormones (e.g. at enzymes, the mitotic apparatus, nucleic acids, and, for oxidants, at lipids). The term 'drug' is too general for use now in the definition of a receptor. It is more logical to define the range of receptors for natural response-evoking ligands and then to enquire at which of those a given drug acts. In the wide definition of 'a receptor' derived thus, the response produced is not confined to signal transmission, and the definition encompasses receptors for such differing processes as the uptake of nutrients (*see* PLASMA LIPOPROTEINS AND THEIR RECEPTORS), cell adhesion (*see* CELL ADHESION MOLECULES; INTEGRINS), cell aggregation, and many other types of activity. Internalization after recognition (e.g. of antigen, *see* ANTIGEN PROCESSING AND PRESENTATION) by ENDOCYTOSIS (*see also* COATED PITS AND VESICLES) is a common type of receptor response not linked *per se* to signal transmission.

In the narrow sense definition, on the other hand, the term receptor means 'signal-transducing receptor' and the native ligands for the receptor activation are molecules which specifically carry information to a cell or between compartments within a cell (*see* SECOND MESSENGER PATHWAYS). These receptors may therefore be:

1 In the plasma membrane, as for the receptors for neurotransmitters, growth factors, morphogens, sensory stimulants, and most circulating hormones (*see* GROWTH FACTOR RECEPTORS; OLFACTORY TRANSDUCTION; VISUAL TRANSDUCTION).

2 In an organelle membrane (e.g. those receptors where the transduction process is the release of Ca^{2+} from an intracellular store).

3 In cytosolic solution, as for the receptors for steroid hormones, thyroid hormones, and retinoids which belong to a particular superfamily of transcription factors (*see* STEROID RECEPTORS).

Receptor actions in the immune system may, in different cases, come under the wide or narrow definition, so that recognition of an antigen may be linked to signal transduction and intracellular activation (*see* B CELL DEVELOPMENT; T CELL ACTIVATION; T CELL RECEPTORS; LYMPHOKINES).

The mere recognition of a ligand is insufficient for a structure to be termed a receptor; structures which bind a ligand without evoking a native biological response are termed 'acceptors'. Examples are a binding site for xenon (and cyclopropane) in the myoglobin structure [1], and the membrane acceptor sites for certain polypeptide toxins.

All signalling pathways in eukaryotes involve one or more of the signal-transducing receptors (which, where the context is obvious, are generally referred to simply as 'receptors'). The major exception is that of the electrical synapse and the related gap junction (*see* CELL JUNCTIONS). For the signal-transducing receptors, the receptor protein must contain a device for linkage to an effector system (*see* CALCIUM; G PROTEIN-COUPLED RECEPTORS; GROWTH FACTOR RECEPTORS; OLFACTORY TRANSDUCTION; SECOND MESSENGER PATHWAYS; VISUAL TRANSDUCTION; VOLTAGE-GATED ION CHANNELS). In the simplest case, this device is contained within the oligomeric transmembrane structure of

the LIGAND-GATED ION CHANNEL type of receptor (*see* NICOTINIC RECEPTOR) [2]. In all others, some additional binding site must be present on the receptor for the protein which contributes the first stage of the effector system; this may be a site which binds to a GTP-binding protein or to a specific PROTEIN KINASE or to a protein substrate for the receptor's own kinase activity (*see* RECEPTOR SUPERFAMILIES). Another class of such effector binding sites is found on the cytoplasmic receptors for steroid hormones, etc. These receptors are single polypeptides which contain a zinc finger or related domain for binding to a specific response element in the promoters of the genes whose expression they regulate (*see* PROTEIN–NUCLEIC ACID INTERACTIONS). This domain is distinct from the domain which binds the receptor-activating hormone. The transduction function of hormone binding to these receptors is therefore quite distinct in its nature from that of the other classes of signal-transducing receptors.

The intrinsic molecular linkage between the binding of the activated ligand and the coupling to an effector system requires that the binding induces a conformational change in the structure of the receptor protein. For the case of the G PROTEIN-COUPLED RECEPTORS, the receptor conformational change is relayed to specific domains in the three intracellular loops deduced for the polypeptide topography in the cell membrane. The first effector step is then tight binding of the trimeric G protein and the exchange of its bound GDP for GTP (*see* GTP-BINDING PROTEINS), leading on via enzymes and ion channels to the activated G protein-dependent pathways (*see* SECOND MESSENGER PATHWAYS). In the case of the ligand-gated ion channels, the receptor conformational change is relayed from the agonist binding site to the channel domain of the receptor itself (*see* NICOTINIC RECEPTORS). In the case of the eukaryotic opsin photoreceptors, the ligand is replaced by a photon and the primary conformational change it induces is the isomerization of 11-*cis*-retinal to all-*trans*-retinal, that is, a change in a covalently bound cofactor of the receptor. With the steroid hormone receptor superfamily, the conformational change induced causes the dissociation of the receptor from its masking *hsp* protein and the formation of its specific DNA-binding site. In other receptor classes, a functional enzyme catalytic site is induced.

E.A. BARNARD

1 Schoenborn, B.P. et al. (1965) Binding of xenon to sperm whale myoglobin. *Nature* **207**, 28–30.
2 Burgen, A. & Barnard, E.A.B. (Eds) (1992) *Receptor Subunits and Complexes* (Cambridge University Press, Cambridge).

receptor autoradiography The use of selective radioactively labelled ligands to visualize the distributions of RECEPTORS and binding sites in tissue, often brain slices. It is used in combination with histological and immunohistochemical methods to identify specific regions of the brain.

receptor binding *See:* SCATCHARD PLOT.

receptor desensitization Shift of a LIGAND-GATED ION CHANNEL from an activated state to an inactive state in the continued presence of agonist. The desensitized conformation of the receptor (i.e. agonist present) may differ from the closed conformation of the receptor (i.e. no agonist present).

receptor (or channel) expression When the DNA(s) encoding a receptor have been cloned, validation of these as giving the full protein structure needed for a functional receptor (or channel) must be demonstrated by one of the receptor expression techniques. The cell-free IN VITRO TRANSLATION systems used for water-soluble nonreceptor protein products are in general unable to demonstrate receptor or channel activity. Three levels of activity measurement are in use for recombinant receptor or channel expression. The first is the demonstration of the binding of radioligands specific for the receptor or channel, and determination of the corresponding affinities. This analysis is performed using transient TRANSFECTION of the cDNA (inserted into a suitable EXPRESSION VECTOR) into a cultured cell line, isolation of the membrane fraction, and binding and ligand displacement studies (*see* SCATCHARD PLOT). The second is the direct measurement of channel activity; this is applicable to the transmitter-gated ion channels (*see* RECEPTORS) and to the VOLTAGE-GATED ION CHANNELS. This activity of recombinant proteins is most commonly demonstrated in the XENOPUS OOCYTE EXPRESSION SYSTEM. For such receptors studied therein, the opening of the channel by appropriate agonists and the pharmacology of receptor or channel activation or block are diagnostic. The third approach is that needed for functional activity in other receptor classes. It requires the coupling of the expressed receptor to a transduction or SECOND MESSENGER PATHWAY in a host cell. The latter can be the *Xenopus* oocyte, using electrophysiological recording of one of its currents when induced by that coupling, or transfected cultured cells recorded under PATCH CLAMPING. The latter cells are also used for an alternative form of this approach in which second messenger changes induced when an agonist is applied to the implanted receptors are measured biochemically, for example, agonist-induced inhibition of cAMP formation. The latter general approach is required for receptors of the growth factor, types, and is often also used for recombinant G PROTEIN-COUPLED RECEPTORS.

receptor-mediated endocytosis ENDOCYTOSIS of extracellular ligands in membrane-bounded vesicles following the binding of ligand to specific cell-surface RECEPTORS. Ligand binding at the cell surface increases the efficiency of endocytosis compared with nonadsorptive fluid phase endocytosis. Ligands internalized by receptor-mediated endocytosis include nutrients (e.g. low density lipoprotein and transferrin), GROWTH FACTORS (e.g. epidermal growth factor and platelet-derived growth factor), polypeptide hormones (e.g. INSULIN), molecules to be cleared from the extracellular medium (e.g. α_2-macroglobulin–protease complexes and secreted lysosomal hydrolases) and opportunistic ligands such as viruses and toxins. Uptake usually occurs through clathrin-coated pits and coated vesicles (*see* COATED PITS AND VESICLES), and frequently the receptors involved are able to interact with components of the clathrin coat to facilitate their uptake. The term can also apply to other forms of endocytosis involving receptor–ligand complexes, including phagocytosis, but generally refers to endocytosis through coated pits.

receptor nomenclature Many cell-surface receptors for hormones, neurotransmitters, etc. are present in several different forms in the body. Receptors used to be defined solely on the basis of their pharmacology — that is, their ability to be inhibited by different sets of antagonists and activated by different sets of agonists. Receptor subtypes were receptors that were activated by the same endogenous LIGAND but nevertheless exhibited a different pharmacology. The impact of molecular cloning techniques has been to expand the number of receptor subtypes and RECEPTOR SUBUNITS such that molecular species are being identified before a defined characteristic pharmacology is known. In the case of the multisubunit receptors (e.g. EXCITATORY AMINO ACID RECEPTORS; GABA AND GLYCINE RECEPTORS) the situation is extremely complicated. In principle, large numbers of permutations and combinations of different subunits are possible, each combination with a potentially different pharmacology. A committee of the International Union of Pharmacological Societies (IUPHAR) has been set up to make specific recommendations regarding receptor nomenclature. An annually updated guide to receptor nomenclature is published.

Watson, S.P. & Girdlestone, D. (1993) Receptor nomenclature supplement. *Trends. Pharmacol. Sci.*

receptor potential A change in membrane potential in a sensory receptor cell in response to a stimulus, analogous to a POSTSYNAPTIC POTENTIAL.

receptor reconstitution Transmembrane RECEPTORS can be extracted from the cell membrane by solubilization in a detergent (e.g. Triton X-100). Often the solubilized proteins are then purified by affinity CHROMATOGRAPHY or similar procedures. The soluble purified protein may then be reconstituted into lipid vesicles by detergent removal and addition of phospholipid. Either lipids extracted from the original tissue or exogenous lipids such as cholesterol and azolectin may be used. Incorporation of the protein(s) of interest into lipid vesicles provides a convenient system in which to investigate receptor function and allosteric modulation with radioisotope flux studies and/or electrophysiology.

receptor recycling *See*: COATED PITS AND VESICLES; ENDOCYTOSIS.

receptor solubilization Use of detergent to extract membrane-spanning RECEPTORS from the cell membrane as a prelude to purification and biochemical analysis. *See*: RECEPTOR RECONSTITUTION.

receptor subunit The individual polypeptides that make up any RECEPTOR, but in particular the subunits of the neurotransmitter receptors of the LIGAND-GATED ION CHANNEL SUPERFAMILY. These receptors are hetero-oligomeric complexes of several subunits. The amino-acid sequences of a large number of ionotropic receptor subunits are now known and they share several characteristic features including four putative transmembrane regions.

receptor superfamilies Protein families composed of cell-surface and other RECEPTORS related by amino-acid sequence and/or overall structure. The superfamilies so far identified of the signal-transducing receptors are:

1 The G PROTEIN-COUPLED RECEPTORS with seven transmembrane domains (e.g. α- and β-adrenergic receptors, dopamine D1 and D2 receptors, and hundreds more).

2 The tyrosine kinase-containing receptors (e.g. insulin receptor, epidermal growth factor receptor, etc.; *see* GROWTH FACTOR RECEPTORS), or receptors that link to a tyrosine kinase (e.g. growth hormone receptor) or guanylate cyclase-containing receptors (natriuretic peptide receptors). All subunits in this superfamily contain a single transmembrane domain (or exceptionally, a membrane anchor).

3 The LIGAND-GATED ION CHANNEL SUPERFAMILY of neurotransmitter receptors.

4 The intracellular STEROID RECEPTOR SUPERFAMILY.

Barnard, E.A. (1992) *Trends Biochem. Sci.* **17**, 368–374.

receptor tyrosine kinases Protein TYROSINE KINASES that form part of cell-surface receptors (*see* GROWTH FACTOR RECEPTORS).

receptosome Early endocytic compartment involved in receptor-mediated ENDOCYTOSIS. Term now superseded by ENDOSOME.

Willingham, M.C. & Pastan, I. (1980) *Cell* **21**, 647–677.

recessive allele/trait ALLELE or trait in diploid organisms which is phenotypically expressed in the HOMOZYGOUS state but is masked by its DOMINANT counterpart in the HETEROZYGOUS state. Usually, the dominant allele produces a functional product while the product of the recessive allele is inactive or absent. *See:* MENDELIAN INHERITANCE.

recessive lethal Mutant ALLELE which is lethal when in the HOMOZYGOUS state. The majority of recessive lethals represent the loss of a vital cellular function.

recessive mutation A mutation which is manifest only in the HOMOZYGOUS or HEMIZYGOUS (as mutations on the X-chromosome in males) state; in HETEROZYGOTES the normal ALLELE masks the recessive allele. Recessive mutations tend to cause inactivation or absence of the gene product.

reciprocal cross Two crosses between the same pair of genotypes or phenotypes in which the source of the gametes (e.g. male or female) is reversed in one cross.

reciprocal lattice *See:* RECIPROCAL SPACE.

reciprocal recombination Genetic RECOMBINATION in which there is an equal and balanced exchange of material between the two DNAs undergoing recombination.

reciprocal space A concept which is useful for understanding the geometry of DIFFRACTION by crystals. In normal Cartesian geometry a crystal can be regarded as being a three-dimensional array of UNIT CELLS, all having identical contents and being equally spaced along three axes (*see* SYMMETRY). These axes define a lattice of points in real space, used to describe crystalline faces and morphology and as a reference origin to define the coordinates of atoms within the crystal unit cell.

The diffraction pattern from a crystal can be shown by BRAGG'S LAW ($\lambda = 2d\sin\theta$) to be an array of points each having a diffraction vector **S** (where $\mathbf{S} = 2\sin\theta/\lambda = 1/d$) whose magnitude is reciprocal to the interplanar spacing (d) of the planes giving rise to a particular diffraction peak, and whose direction is perpendicular to that set of planes of spacing d. The set of diffraction peaks from a crystalline lattice is itself a lattice, with axial dimensions inversely proportional to the real axes, and axial directions orthogonal to those of the corresponding planes in the real lattice. There is thus a reciprocal relationship between the lattice of the diffraction pattern and that of the crystal such that the former is often referred to as the reciprocal lattice and the diffraction pattern is said to be in reciprocal space.

The volume of the unit cell in real space (V) is the inverse of that of the reciprocal space unit cell ($V^* = 1/V$). Reciprocal cell dimensions ($\mathbf{a}^*, \mathbf{b}^*, \mathbf{c}^*$) are vectors related to those in real space ($\mathbf{a}, \mathbf{b}, \mathbf{c}$) by the following vector equations, in which K is the reciprocal space constant:

$$\mathbf{a}^* = K\frac{(\mathbf{b} \wedge \mathbf{c})}{V} \quad \mathbf{b}^* = K\frac{(\mathbf{c} \wedge \mathbf{a})}{V} \quad \mathbf{c}^* = K\frac{(\mathbf{a} \wedge \mathbf{b})}{V}$$

where $V = \mathbf{a} \cdot (\mathbf{b} \wedge \mathbf{c})$

$$\mathbf{a} = K\frac{(\mathbf{b}^* \wedge \mathbf{c}^*)}{V^*} \quad \mathbf{b} = K\frac{(\mathbf{c}^* \wedge \mathbf{a}^*)}{V^*} \quad \mathbf{c} = K\frac{(\mathbf{a}^* \wedge \mathbf{b}^*)}{V^*}$$

where $V^* = \mathbf{a}^* \cdot (\mathbf{b}^* \wedge \mathbf{c}^*)$

When $K = \lambda$ the reciprocal lattice is dimensionless; when $K = 1$ the reciprocal lattice has dimensions of reciprocal Ångstroms (Å^{-1}). Both conventions are found useful for different purposes.

The coordinates of lattice points in reciprocal space are the MILLER INDICES (hkl) denoting diffraction points. The diffraction vector $\mathbf{S}_{hkl}$ is conveniently calculated as the vector sum: $\mathbf{S}_{hkl} = h\mathbf{a}^* + k\mathbf{b}^* + l\mathbf{c}^*$. The coordinates of lattice points in real space are the ZONE AXIS symbols [U V W] relating to the interplanar spacing of zones in the diffraction pattern. *See:* X-RAY CRYSTALLOGRAPHY.

International Tables for X-ray Crystallography (1992) Vol. B Reciprocal Space (Shmueli, U. Ed.) (Kluwer Academic Publishers, Dordrecht).
Stout, G.H. & Jensen, L.H. (1989) *X-ray Determination: a Practical Guide* (John Wiley, New York).

reciprocal translocation *See:* CHROMOSOME ABERRATIONS.

reckless phenotype Phenotypic instability in *Escherichia coli recA* mutants caused by the degradation of excessive amounts of DNA by the product of the *recBCD* genes — EXONUCLEASE V. Control of this enzyme is somehow deficient in *recA* mutants of *E. coli*.

recombinant (1) Organism, cell, or virus which has undergone genetic RECOMBINATION with the exchange of genetic material between two DNAs. *See also:* MEIOSIS; MENDELIAN INHERITANCE. (2) A RECOMBINANT DNA; the term recombinant protein is often used for proteins produced from a recombinant DNA in a hetero-

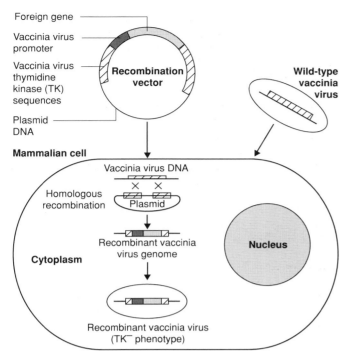

Fig. R2 Introduction of a foreign gene into vaccinia virus by homologous recombination. Vaccinia has a large double-stranded genome of around 187 000 kb and is thus difficult to manipulate *in vitro*. The required gene is therefore first inserted into a plasmid containing a vaccinia virus promoter and flanking sequences homologous to vaccinia DNA inessential for replication. Infection of plasmid-containing cells with vaccinia leads to a low level of homologous recombination between plasmid and virus with the transfer of the vaccinia promoter and foreign gene to the vaccinia virus genome. The foreign gene is usually inserted into the viral *tk* (thymidine kinase) gene, and recombinant viruses identified by their tk⁻ phenotype. Unlike most DNA viruses, vaccinia and other poxviruses replicate in the cytoplasm of infected cells.

logous host, for example, human insulin produced in a bacterium.

recombinant DNA (rDNA, rtDNA) A composite DNA containing sequences from different sources, for example a human protein-coding sequence inserted into a plasmid or bacteriophage VECTOR. Sometimes called chimaeric DNA. In a general way, the term has now come to be used for any DNA that is the result of manipulation *in vitro*. See: DNA CLONING; GENETIC ENGINEERING; PLANT GENETIC ENGINEERING; TRANSGENIC TECHNOLOGIES.

recombinant DNA techniques/technology An umbrella term for a range of techniques for isolating, analysing and manipulating DNA *in vitro*. See: DNA CLONING; DNA SEQUENCING; GENETIC ENGINEERING; HUMAN GENE MAPPING; HYBRIDIZATION; PLANT GENETIC ENGINEERING; RECOMBINANT DNA; RESTRICTION ANALYSIS; RESTRICTION ENZYMES; RESTRICTION MAPPING; TRANSGENIC TECHNOLOGIES.

recombinant vaccines Recombinant DNA technology is now being applied to the development of new VACCINES, in particular to replace existing expensive or less effective vaccines, and to try to develop vaccines against the human immunodeficiency virus (HIV) responsible for AIDS, and diseases such as malaria for which conventional vaccines are unavailable. Recombinant vaccines are of two main types:

1 Subunit vaccines. Vaccines comprising the required antigenic protein component which has been made by expression *in vitro* in bacterial or animal cells. These do not contain live or killed bacteria or viruses. An effective vaccine against HEPATITIS B VIRUS has been developed using hepatitis B surface antigen made in yeast cells.

2 Live or killed vaccines. These are based on a viral or bacterial carrier into which genes for the required antigens have been transplanted (Fig. R2). Effective live vaccines based on a VACCINIA VIRUS carrier containing novel antigenic determinants have been developed (e.g. against rabies and rinderpest in animals).

Because of its long history of clinical use, its robustness and its ability to replicate in a wide range of host cells, vaccinia virus is one of the main vectors used in developing recombinant vaccines against a variety of disease agents for human and animal use (although live vaccinia can provoke severe reactions in a minority of people). An oral live recombinant vaccinia vaccine incorporating the rabies virus glycoprotein has been successful in controlling rabies in the wild fox population in parts of western Europe. Experimental recombinant vaccinia vaccines include those being developed against HIV and malaria. Other microorganisms being considered as carriers include vaccine strains of poliovirus and *Mycobacterium*.

recombinant vaccinia virus *See*: RECOMBINANT VACCINES.

Recombinase

IN the immune system, somatic DNA recombination plays a key role in generating functional antigen receptor genes [1]. This recombination is essential for the activation and diversification of both immunoglobulin and T cell receptor (TCR) genes (*see* IMMUNOGLOBULIN GENES; T CELL RECEPTOR GENES). The enzymatic machinery responsible for assembling V, (D), and J segments is called V-(D)-J recombinase.

Recombination signal sequences

Recombination takes place at specific sites adjacent to the coding sequence of the germ-line V, D and J gene segments. Joining takes place between two recombination signal sequences (RSSs) which are each composed of a heptamer and a nonamer. The heptamer and nonamer sequences are highly conserved, with consensus sequences CACTGTG and GGTTTTTGT respectively [2]. Joining takes place when one RSS consists of a heptamer and a nonamer separated by a spacer of 12 base pairs (bp) and the other consists of a heptamer and a nonamer separated by a 23-bp spacer (Fig. R3). Joining never occurs between two RSSs with the same spacer length. This is the so-called 12-23 rule. It has been shown that two pairs of RSSs are sufficient to cause the V-(D)-J type of recombination, when the 12-23 rule is satisfied [3].

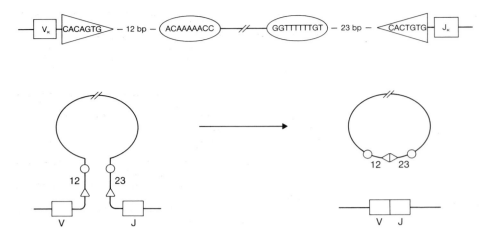

Fig. R3 Recombination signal sequences (RSSs) for V-(D)-J joining. Two blocks of sequences are highly conserved by V-(D)-J joining sites. They are a palindromic heptamer, CACTGTG, and an A/T-rich nonamer, GGT T T T TGT, or their complementary sequences. Spacer lengths separating the heptamer and nonamer are also conserved, which are either 12 bp or 23 bp. Joining occurs between two sets of the heptamer and nonamer, when one set contains the 12-bp spacer and the other contains the 23-bp spacer. Joining usually occurs by a deletion mechanism excising the intervening DNA between the two RSSs. The looped-out structure is cleaved into circular DNA which contains two RSSs joined in head-to-head fashion.

Joining occurs by a DNA deletion mechanism when the two RSSs are arranged in a head-to-head fashion, 7mer–9mer...9mer–7mer. Circular DNA containing a SIGNAL JOINT has been isolated as a reciprocal recombination product of V-(D)-J joining. Joining occurs by DNA inversion when two RSSs are oriented in the same direction, 7mer–9mer...7mer–9mer.

Although the heptamer and nonamer sequences are generally well conserved, substantial deviation is found in many naturally occurring RSSs. Mutation analysis with an artificial substrate has revealed that only a few nucleotides in the consensus sequences are essential [4]. In the heptamer, three nucleotides adjacent to the recombination site have to be conserved, whereas others can be changed without losing the substrate activity.

Enzymatic activity

At least three activities have been postulated for the V-(D)-J recombinase complex: a DNA binding activity, an endonuclease activity and a DNA ligase activity. The DNA binding activity would recognize recombination signal sequences and bring two gene segments together in the proper orientation to satisfy the 12-23 rule. The endonucleolytic activity would then cleave DNA at the recombination site. Finally, the ligase activity would join two pairs of cleaved DNA ends covalently to complete the recombination reaction.

Two more activities may be associated with the recombinase to diversify the junctional sequences of immunoglobulin and T cell receptor genes; they are an EXONUCLEASE activity and a TERMINAL DEOXYNUCLEOTIDYLTRANSFERASE (TdT) activity [5] (*see* N REGION DIVERSITY).

Expression

Expression of the recombinase is strictly regulated in a tissue-specific manner. It is found only in lymphocytes at the early stage in their differentiation. Furthermore, V-(D)-J joining occurs sequentially in a time-ordered fashion. In B cells, for example, D and J are joined before V and D. As soon as the μ polypeptide is produced, further rearrangement is terminated in the heavy-chain gene and light-chain gene rearrangement is initiated.

As immunoglobulin and T cell receptor genes follow the same joining rule, it seems likely that a common recombinase is responsible for V-(D)-J joining in both B and T cells. It is not well understood, however, why functional rearrangement of immunoglobulin genes occurs only in B cells, whereas T cell receptor genes rearrange only in T cells.

Candidate proteins

Several nuclear proteins have been reported as candidates for recombinase components. Some are endonucleolytic activities which introduce double-strand breaks at sites near the RSS [6].

DNA-binding proteins have been isolated in pre-B cells using a gel mobility shift assay (*see* BAND RETARDATION). A nuclear protein that interacts with the 23-bp RSS has also been isolated [7]. This protein contains a sequence related to the integrase motif. A cDNA clone coding for an HMG1-related DNA-binding protein has isolated with 12-bp RSS probes [8]. Using RFLP, genetic linkage was suggested for this gene and *RAG*.

RAG genes

Recombination activator genes, *RAG1* and *RAG2*, have been isolated [9] by their ability to convert a fibroblast cell line into a cell line with the ability to rearrange the immunoglobulin genes. *RAG1* and *RAG2* are sufficient to activate V-(D)-J joining in recombination-negative cells [10]. It is not clear yet whether these genes code for the V-(D)-J recombinase itself. The *RAG2* gene may be responsible for the GENE CONVERSION that occurs in chicken immunoglobulin genes (*see* GENERATION OF DIVERSITY). Selective expression of the *RAG2* gene has been demonstrated in chicken B cells in which immunoglobulin gene conversion was actively taking place [11]. However, the issue is still controversial, and is yet to be studied further.

Another genetic locus which controls V-(D)-J joining is CB17*scid* [12], which has been mapped to mouse chromosome 16. A defect in the locus causes SEVERE COMBINED IMMUNODEFICIENCY. In the *scid* mouse, both B cell and T cell maturation are blocked owing to failure of DJ joining. An extensive deletion is found at the DJ joint in the *scid* immunoglobulin and T cell receptor genes, indicating that an exonuclease function is somehow deregulated.

V gene replacement

The V-(D)-J recombinase is also responsible for V gene replacement. This is a correction mechanism to rescue out-of-frame V(D)J structures. As the recombinase cannot regulate V-(D)-J joining to preserve the reading frame of the genetic code, a productive (i.e. in frame) recombinant can only be generated in one event out of three. V gene replacement removes the V gene portion of an assembled V(D)J structure, and replaces it with a new sequence of the germ-line V gene. A heptamer-like sequence embedded in the V coding sequence is used as a truncated RSS, and recombines with the 3′-RSS of the germ-line V gene. This variant type of V-(D)-J joining between one complete RSS and a heptamer alone can also be found in some chromosomal abnormalities in lymphocytic tumours [13,14]. Examples include TRANSLOCATION t(14;18)(q32;q21) by the *bcl-2* PROTO-ONCOGENE in human follicular lymphomas, and t(1;14)(p32;q11) by *tal-1* in T cell acute lymphoblastic leukaemias (*see* CHROMOSOME ABERRATIONS).

Isotype (class) switching

Besides V-(D)-J joining, immunoglobulin heavy-chain genes undergo a different type of DNA rearrangement, isotype switching, at later stages of B cell development. The switch recombinase replaces the heavy-chain constant (C_H) gene closest to the VDJ structure with the other C_H gene further downstream (i.e. to the 3′ side). Intramolecular DNA deletion is the most likely mechanism for the class switch recombination, which excises the looped-out intervening DNA [15]. Unlike V-(D)-J joining, switch recombination is an antigen-dependent process. Some LYMPHOKINES are known to direct isotype switching in a class-specific manner at the level of DNA recombination. For example, interleukin-4 (IL-4) induces the switch from μ to γ1 and then to ε in a successive manner [16]. A cytokine — transforming growth factor β (TGFβ) — directs the switch from μ to α. Recombination usually takes place in switch (S) regions which are rich in repetitive sequences. The switch recombinase seems to be distinct from the V-(D)-J recombinase, because the heptamer–nonamer motif for V-(D)-J joining is not found in S regions.

H. SAKANO

1 Tonegawa, S. (1983) Somatic generation of antibody diversity. *Nature* **302**, 575–581.
2 Sakano, H. et al. (1979) Sequences at the somatic recombination sites of immunoglobulin light-chain genes. *Nature* **280**, 288–294.
3 Akira, S. et al. (1979) Two pairs of recombination signals are sufficient to cause immunoglobulin V-(D)-J joining. *Science* **238**, 1134–1138.
4 Hesse, J.E. et al. (1989) V(D)J recombination: a functional definition of the joining signals. *Genes Dev.* **3**, 1053–1061.
5 Alt, F. et al. (1982) Organization and reorganization of immunoglobulin genes in A-MuLV-transformed cells: Rearrangement of heavy but not light chain genes. *Proc. Natl. Acad. Sci. USA* **78**, 4118–4122.
6 Kataoka, T. et al. (1984) Isolation and characterization of endonuclease J: a sequence-specific endonuclease cleaving immunoglobulin genes. *Nucleic Acids Res.* **12**, 5995–6010.
7 Matsunami, J. et al. (1989) A protein binding to the $J_κ$ recombination sequence of immunoglobulin genes contains a sequence related to the integrase motif. *Nature* **342**, 934–937.
8 Shirakata, M. et al. (1991) HMG1-related DNA-binding protein isolated with V-(D)-J recombination signal probes. *Mol. Cell. Biol.* **11**, 4528–4536.
9 Shatz, D. & Baltimore, D. (1988) Stable expression of immunoglobulin gene V(D)J recombinase activity by gene transfer into 3T3 fibroblasts. *Cell* **53**, 107–115.
10 Ottenger, M.A. et al. (1980) RAG-1 and RAG-2, adjacent genes that synergistically activate V(D)J recombination. *Science* **248**, 1517–1523.
11 Carlson, L.M. et al. (1991) Selective expression of RAG-2 in chicken B cells undergoing immunoglobulin gene conversion. *Cell* **64**, 201–208.
12 Bosma, G.O. et al. (1983) A severe combined immunodeficiency mutation in the mouse. *Nature* **301**, 527–530.
13 Showe, L.C. & Croce, O.M. (1987) The role of chromosomal translocations in B- and T-cell neoplasia. *Annu. Rev. Immunol.* **5**, 253–277.
14 Boehm, T. & Rabbitts, T.H. (1989) A chromosomal basis of lymphoid malignancy in man. *Eur. J. Biochem.* **185**, 1–17.
15 Honjo, T. & Kataoka, T. (1988) Organisation of immunoglobulin heavy chain genes and allelic deletion model. *Proc. Natl. Acad. Sci. USA* **75** 2140–2144.
16 Yoshida, K. et al. (1990) Immunoglobulin switch circular DNA in the mouse infected with *Nippostrongylus brasiliensis*: evidence for successive class switching for μ to ε via γ1. *Proc. Natl. Acad. Sci. USA* **87**, 7829–7833.

Recombination

THE term 'genetic recombination' was first used to describe the apparently random exchanges (crossing-over) of parts of two homologous chromosomes that were seen during MEIOSIS. It was first observed in the X-chromosome of the fruit fly, *Drosophila melanogaster* (*see* DROSOPHILA). These exchanges can be detected by the production of recombinant progeny if the chromosomes involved carry different genetic information (i.e. different alleles at particular loci) but it is certain that crossovers and exchange of chromosomal material occur even when the two chromosomes are seemingly identical. Since the recognition of DNA as the genetic material, the term recombination has come to cover a variety of events in which new genetic combinations result from the physical interaction of two DNA molecules.

The process of recombination underlies much of the science of genetics (*see e.g.* HUMAN GENE MAPPING; LINKAGE MAP; MENDELIAN INHERITANCE). The discoveries of recombination in bacteria, especially in *Escherichia coli*, and in its BACTERIOPHAGES, were key events in the development of molecular biology, enabling the structure of prokaryotic genes to be analysed on a fine scale and opening up the mechanism of recombination itself to genetic and, later, biochemical dissection.

Recombination events are frequent — perhaps one crossover for every 5 kb in some yeast chromosomes — and universal, although there are instances in which recombination does not occur, as in meiosis in the male fruit fly. Recombination is considered to have played an important part in evolution, particularly in sexually reproducing organisms, by creating a wider range of phenotypes for natural selection to operate upon than would have been achieved by MUTATION alone. It has also come to be a necessary stage in the replication of the DNA of certain bacteriophages, notably phage T4.

Despite the fundamental importance of recombination in biology its molecular mechanisms are still only partially understood. Progress towards an adequate description of recombination in bacteria has been rapid in recent years, however, mainly as the result of the study of recombination-defective mutants. One consequence of these studies has been to show that recombination plays an important part in the repair of damaged DNA (*see* DNA REPAIR) and may well have evolved as a by-product of a mode of repair rather than in its own right. The importance of recombination processes in the cell is illustrated by the discovery that for some stages at least, there exist two, and possibly more, enzymes able to catalyse the same processes.

General, or homologous, recombination will be described first: this depends upon a substantial degree of homology, the longer the better, between the DNA sequences involved. The enzymes which act in general recombination show no sequence specificity. Homologous recombination must be distinguished from certain specialized types of recombination and some chromosomal rearrangements (internal recombination) in which the DNAs have only short homologous sequences and rely upon enzymes specific for the limited class of exchanges they are able to catalyse. This type of recombination is often known generally as site-specific recombination. These two types of recombination are thought to involve similar physical mechanisms and are discussed in more detail below.

A third kind of mechanism is however necessary to explain the rare events which are attributed to so-called illegitimate recombination. It is likely that these recombinants arise during DNA REPLICATION due to the switching of the replication complex from one template to a second template, perhaps as a result of damage in the first template. The two templates and the daughter strand are processed by replication and repair enzymes to give a complete chromosome. This is known as the COPY CHOICE MECHANISM. If the switch does not occur in the correct register the final product either lacks a sequence or has an extra sequence at the point of crossing-over and so differs from legitimate recombination in which the sequence at the crossover is preserved exactly.

Recombination also occurs between RNA genomes and may be as frequent in certain eukaryotic virus families, for example the coronaviruses, as in bacterial DNA viruses [1]. In this case the preferred model is a copy-choice mechanism, but one in which the transfer is often at a point of homology. However, this area of research has been much less well studied than DNA recombination and will not be discussed further.

General recombination

The problem

The process of general recombination can be formally described in simple terms. If two homologous chromosomes each carrying two genetic markers (i.e. two loci with different alleles that can be distinguished phenotypically) are brought together in the same cell, they can undergo a process of DNA exchange such that the resulting chromosomes carry combinations of alleles that differ from both parental chromosomes (Fig. R4).

But this simplicity conceals great complexity at the molecular level. At least one of the two DNA molecules must be activated for

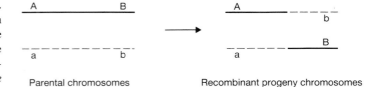

Fig. R4 Formation of recombinant chromosomes by homologous recombination between two chromosomes carrying different ALLELES, *A* or *a*, *B* or *b*, at two different genetic loci.

the process of recombination to be initiated; the two molecules must be brought together within a region of homology for synapsis to occur; one or more DNA strands must be broken and some form of pairing with the other molecule take place; the complex must be resolved so that the exchange is complete; and any remaining breaks or nicks repaired so that the final outcome is continuous complete DNA molecules. And all this must happen despite the topological difficulties imposed by the double helical nature of the two interacting molecules.

A molecular model

The first convincing model mechanism for recombination was based on the assumption that breaks in corresponding strands in the two helices could lead to pairing of the free ends with the complementary strand on the other DNA helix (Fig. R5*a*) [2]. Further exchange of strands between the two molecules leading to movement of the junction (branch migration) strengthens the interaction (Fig. R5*b*). This structure can isomerize by rotation of the molecules about one another to give a structure that has come to be called a Holliday junction or Holliday structure (Fig. R5*c*). If new cuts are made to resolve this structure there are two possibilities. If they are made at the s sites as indicated on the diagram two recombinant chromosomes are formed (the 'splice' recombinant) with a HETERODUPLEX region Bb between A and C. But if the cuts are made at the p sites there is no recombination between A and C. A 'patch' recombinant is formed with a heteroduplex region Bb within otherwise parental chromosomes.

Mismatch repair and gene conversion

In both of the cases above a segment of hybrid DNA is formed in which one strand is derived from one parent and the complementary strand from the other. If the two strands in the hybrid segment are genetically identical (i.e. of exactly complementary DNA sequence), resolution will proceed with no further change in the DNA sequence. However, if the DNA strands are genetically different (i.e. not exactly complementary) the region is said to be heteroduplex and could be a target for a MISMATCH REPAIR system. Mismatch repair normally occurs in newly replicated DNA to correct errors of replication, but bacteria are known to recognize mismatches in heteroduplex DNA introduced as such into cells. If the hybrid DNA occurs within a coding sequence, mismatch repair can convert one allele into the other. For example, in the case illustrated above, the mismatch Bb could be

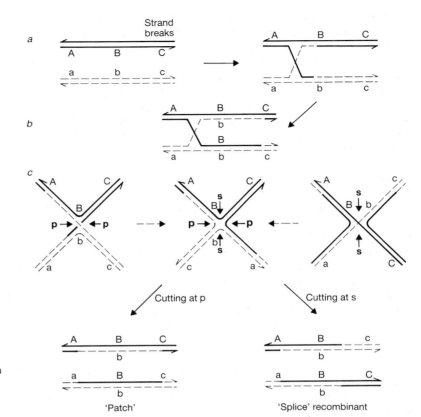

Fig. R5 The formation and resolution of crossovers in homologous recombination. *a*, Homologous duplex DNA molecules pair (synapsis). Breaks in corresponding strands in the duplexes allow the crossover and exchange of DNA strands, with the free ends pairing with the opposite duplex. *b*, Further strand exchange and movement of the crossover (branch migration) stabilizes the recombination intermediate (the Holliday structure), with the formation of hybrid DNA (Bb). *c*, The four-way DNA junction can be resolved by cleavage to give two alternative outcomes. Cutting at s and rejoining gives 'splice' recombination and two recombinant molecules; cutting at p and rejoining gives 'patch' recombination and two parental molecules. In both cases a stretch of heteroduplex DNA (Bb) is formed between the initial crossover point and the point at which the crossover was resolved.

corrected to either BB or bb. This could account for the phenomenon of GENE CONVERSION, which is frequently observed in fungal crosses, in which a single act of recombination generates progeny that do not segregate in the expected Mendelian fashion (*see* MENDELIAN INHERITANCE). The non-Mendelian ratios are most easily detected in crosses involving closely linked markers, that is, where a recombination event occurring between the loci is likely to generate hybrid DNA covering the coding region of one or other locus.

Another instance where mismatch repair may occur is in bacterial TRANSFORMATION (e.g. of *Streptococcus pneumoniae*). In transformation, only one strand of the donor DNA enters the cell and undergoes recombination with the bacterial chromosome to give patch-type recombinants. Thus, for all the genetic markers involved in the cross heteroduplex DNA is necessarily created. For any particular gene, certain alleles are found to transform wild-type strains of *S. pneumoniae* relatively inefficiently compared with other alleles. However, all alleles of a gene transform with equal efficiency strains in which mismatch correction is defective, suggesting that the failure of some alleles to transform in wild-type strains is due to their elimination by mismatch correction.

Remaining questions

Many modifications of the Holliday model have been proposed, but its essentials, the creation of hybrid DNA and the resolution of the resulting Holliday structures, apply in a wide range of systems

and form the basis for interpreting biochemical studies of recombination. The main problem is still the nature of the initiating event, and one proposal is that one of the chromosomes suffers a double-stranded break. If this is so, it would explain why transformation of the yeast *S. cerevisiae* with a plasmid linearized by a double-stranded break (e.g. cut by a restriction nuclease) in a region of homology with a region of chromosomal DNA leads to efficient recombination between plasmid and chromosome. This application of homologous recombination is now much used as a method for inducing targeted GENE DISRUPTION.

Although these molecular models were first suggested to account for phenomena most easily studied in fungal systems the best experimental evidence for them has come from the study of bacterial mutants. Biochemical studies of these mutants have identified a number of the proteins involved in bacterial recombination.

Bacterial recombination proteins

In *recA* mutants of *E. coli*, in which the RecA protein is absent or inactive, recombination is virtually abolished. These strains are also highly sensitive to ultraviolet light and X-irradiation because recombination is not available as a form of repair and they also cannot use an inducible set of repair processes known as SOS REPAIR. In the wild type, synthesis of the RecA protein in UV-irradiated cells is greatly enhanced and it can constitute as much as 4% of the newly synthesized protein.

The RecA protein has proved to be remarkably versatile, given its small size (352 amino acids): it serves both as the synapsis-generating protein and as the key link in SOS induction. It can form a complex with ATP and single-stranded DNA (ssDNA), in which the protein polymerizes to form a helical cage enfolding and stretching the ssDNA to 1.5 times the length it would have in the B-form of double-stranded DNA (dsDNA). The helix is right-handed with six RecA molecules per turn and three nucleotides complexed to each RecA monomer. It is probable that in the cell the formation of this complex is assisted by the removal of secondary DNA structure by ssDNA binding protein (SSB protein). The RecA complex can interact with dsDNA and when a region of homology of sufficient length (probably 30–50 nucleotides) is found, strand transfer is initiated (Fig. R6). Strand transfer starts from the 3′ end of the ssDNA and leads to the displacement of the homologous strand with substantial levels of ATP hydrolysis, but since as many hydrogen bonds are made as are broken, most of this hydrolysis may serve to displace RecA monomers from the newly formed dsDNA by altering the affinity of the protein for DNA. The region of initial matching of sequences must be long, between three or four helical turns in B-DNA, but once strand transfer has been initiated, mismatches, damaged DNA (e.g. PYRIMIDINE DIMERS), and short regions of nonhomology can be accommodated. Four-stranded intermediates have also been demonstrated *in vitro* and it is likely that they can be formed *in vivo*. Tensions due to supercoiling would, *in vivo*, be relieved by DNA topoisomerases. Thus RecA can catalyse the formation of Holliday structures, the key recombination intermediate [3].

In *E. coli* the most recent evidence suggests that once strand transfer is initiated the process is continued and regulated by either RecG protein or by a complex of RuvA and RuvB proteins. These proteins were also identified by means of mutants. Mutants in either *ruvA* or *ruvB* are partially defective for recombination and more sensitive than normal to DNA damage. This combination of defects again suggests that these proteins are involved in repair by recombination. Strand transfer catalysed by the RuvAB complex is quicker than by RecA and requires less ATP hydrolysis.

Resolution, the cutting of the DNA chains at the Holliday junctions so as to separate the interacting molecules without loss of continuity, was expected to require enzymes with high specificity for these junctions. Although phage-encoded enzymes with the required properties have been found, the *E. coli* enzyme has eluded detection until recently. It is now known to be the product of the *ruvC* gene. RuvC, a protein of M_r 19 000, cleaves only four-stranded Holliday junctions and does so in such a way as to yield splice or patch products equally. The two cuts are made symmetrically across the junction of the four stranded intermediate. The formation and resolution of recombination intermediates has now been demonstrated *in vitro* to require only RecA and RuvC proteins [4].

Thus the pathway from initial substrate to final resolution can be summarized as follows: RecA protein initiates synapsis between ssDNA and dsDNA and initiates strand exchange and branch migration. The displaced strand must be broken and itself interact to form a four-strand Holliday junction, again with the aid of RecA. Branch migration is continued at an increased pace by the RuvAB complex (or RecG protein) until at some point the junction becomes the substrate for RuvC proteins and the complex is resolved by cleavage of the crossover strands leaving nicks in both chromosomes which are sealed by the action of DNA LIGASE. During this process and subsequently, other enzymes — DNA polymerases, topoisomerases, and mismatch repair enzymes act to facilitate the process and modify the final product.

Two problems remain. It is still a matter for dispute whether Holliday junctions can isomerize as suggested in the model, because of the complex rotations required. The generation of ssDNA with a free 3′ end, which may be a necessary step in the formation of the synapse, is still unresolved, although the formation of Holliday junctions between a plasmid and a plasmid with a short gap in one strand has been demonstrated *in vitro*.

This latter function may be the role of the RecBCD complex. Mutations in the *recB* and *recC* genes cause a drop to 1–10% of normal levels of recombination and a greatly increased sensitivity to DNA damaging agents. Such mutants are, however, on the borderline of viability; often less than 50% of the cells in a culture form colonies on transfer to solid media. The RecBCD protein complex is large (M_r ~330 000) and is remarkably versatile. In the presence of ATP it can be: (1) an EXONUCLEASE, active against both ss- and dsDNA; (2) a dsDNA ENDONUCLEASE; and (3) a DNA helicase. Detailed analysis of these activities has suggested a model for its role in recombination [5]. It is proposed that the complex binds to the free ends of dsDNA, unwinds it, a process which as with other helicases is dependent upon ATP hydrolysis, and generates single-strand tails. These can reassociate or be complexed by SSB protein. When the complex passes the sequence 5′ GCTGGTGG 3′ (the CHI SEQUENCE), it is activated to become an endonuclease and cut that strand 4–6 nucleotides on the 3′ side. Helicase action continues, generating ssDNA with a free 3′ end which could serve for the nucleation of the RecA complex.

The chi sequence is found to occur well above its expected frequency and at a level sufficient for one to be present within every 5 kb of *E. coli* DNA so that recombination could be started

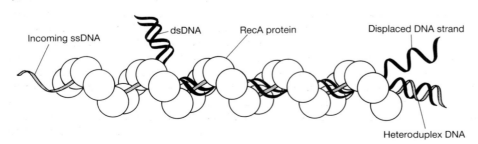

Incoming ssDNA · dsDNA · RecA protein · Displaced DNA strand · Heteroduplex DNA

Fig. R6 Role of RecA protein in recombination. The incoming single DNA strand is shaded. Adapted from [3].

in every region of the chromosome. RecBCD-like activities have been found in other bacteria but not all recognize the chi sequence. The combined action of RecA, RecBCD and SSB proteins in forming joint molecules has been demonstrated *in vitro* [6].

The *recB*, *recC* and *recD* mutants have revealed the true complexity of the battery of genes involved in the various aspects of recombination. Suppressor mutations, *sbcB* and *sbcC* in combination, restore recombination proficiency and UV resistance to *recB* and *recC* mutants. The *sbcB* mutation causes the loss of function of exonuclease I which is active on ssDNA, a finding which emphasizes the role of ssDNA in recombination. The multiply mutant *recBC sbcBC* strains can then be screened for mutations which again reduce recombination proficiency. To date no fewer than seven more genes, *recF*, *recJ*, *recN*, *recO*, *recP*, *recQ*, and *recR*, as well as the *ruvA*, *B*, and *C* and *recG* genes referred to above, have been identified and the gene products partially characterized. The activities discovered include an ssDNA binding protein (RecF), an ssDNA exonuclease (RecJ), and a dsDNA helicase (RecQ). These gene products can perform all the processes necessary for recombination normally carried out by the RecBCD complex and may also be involved in the residual formation of recombinants normally observed in *recB* and *recC* strains. It is very likely that some DNA substrates require processing beyond the capacities of the enzymes of the normal pathway and that the activities of the different proteins overlap one another. However it is clear that genes with other functions in normal cells (e.g. the UvrD helicase) may also be involved as is shown by the increased sensitivity of mutants to DNA damage. Many of these genes are under the control of the LexA repressor (see below) and are thought to be brought into action when the cell contains a high level of damaged DNA.

SOS repair

When wild-type *E. coli* DNA is damaged, and especially when the replication complex reaches a lesion that prevents further synthesis (e.g. an abasic lesion or a pyrimidine dimer), a number of genes involved in DNA repair, UV mutagenesis and inhibition of cell division (to prevent cell division before repair) are induced. These include *recA* itself, *uvrA*, and *uvrB* (involved in EXCISION REPAIR), some of the other *rec* genes listed above, the *ruvAB* operon, the *umcDC* operon (involved in UV mutagenesis), and *sfiA*, a gene encoding a protein that inhibits filamentation. These genes collectively contribute to the process known as SOS repair and are normally repressed by the LEXA repressor which binds to sequences called LexA boxes in their PROMOTERS. The SOS pathway is induced when an activated form of RecA, RecA*, binds to LexA in such a way that a bond in LexA becomes liable to spontaneous hydrolysis. The progressive fission of LexA repressor molecules lifts repression and transcription is initiated, especially for the genes like *recA* which possess strong promoters, thus accelerating the proteolysis of LexA. This coprotease action of RecA* also causes INDUCTION in lysogens by fission of the λ repressor (*see* LAMBDA). It is also responsible for the activation of UmuD protein to its active form UmuD′, which when complexed with UmuC causes DNA polymerase III to synthesize past lesions

which have hitherto blocked DNA synthesis. This action, too, requires the presence of RecA and, *in vitro*, SSB protein. This trans-lesion synthesis may itself be damaging in that it can be mutagenic (hence the name SOS repair) if the replacement nucleotides are incorrect.

All bacteria so far examined have a homologue of RecA so similar in structure and size that the gene when cloned into an appropriate vector will complement *recA* NULL MUTANTS of *E. coli* in all its functions (except for homologues from the more remotely related genera which cannot supply the coprotease function). One of the many genes of *S. cerevisiae* involved in meiotic recombination, *RAD51*, encodes a protein with some sequence homology with RecA and mutants in *RAD51* are similarly defective in both recombination and repair. However, in eukaryotic cells not only are the systems more complex and involve more proteins, but mitotic and meiotic cells utilize different sets of proteins for these functions.

Proteins homologous to RecBCD are also widespread among bacteria but no eukaryotic counterpart has yet been found. A variety of proteins active against Holliday junctions have been detected in cell-free extracts from eukaryotic cells.

Specialized or site-specific recombination

In addition to general recombination a number of forms of recombination have evolved which do not require long stretches of DNA homology. These include: chromosomal rearrangements (internal recombination), as in PHASE VARIATION in *Salmonella typhimurium*, the inversion of the FLP sequence in the replication of the 2μm PLASMID of *S. cerevisiae*, and the rearrangement of vertebrate immunoglobulin and T cell receptor genes (*see* GENE REARRANGEMENT); transposition of mobile genetic elements such as transposons (*see* TRANSPOSABLE GENETIC ELEMENTS); and the INTEGRATION of bacteriophage chromosomes with host chromosomes in the formation of lysogens. These events are catalysed by specific proteins (recombinases — including integrases and transposases) which recognize short DNA sequences which become the crossover regions (*see* RECOMBINASE). Despite the apparently large difference in sequence and the absence of any involvement of the proteins of homologous recombination, the Holliday model is nevertheless followed, in outline at least. Breakage of the strands leads to strand transfer and the creation of Holliday structures, but the resolution is highly directional and specific.

The best studied example of site-specific recombination is the integration of the chromosome of phage λ with the *E. coli* chromosome, which is a necessary step in the establishment of the lysogenic state for this as for many other bacteriophages (*see* LAMBDA). Integration occurs at specific points, the *att* site, on both chromosomes and requires only one phage-encoded protein — integrase (the recombinase). This system has lent itself to experimental dissection because the integration of a small form of the phage into a small plasmid carrying the bacterial *att* site (*attB*) can be easily followed *in vitro*. The requirements for efficient integration prove remarkably simple: the phage DNA must be present as a supercoil and only two proteins are essential — integrase and a host-encoded, histone-like protein, integration host factor (IHF). The sequence requirements are also small. Both chromosomes must contain an identical palindromic sequence of 15 bp embedded in sequences which are similar but not identical. This 31-bp core is all that is required in the target site *attB*, but this core in *attλ* is embedded asymmetrically in a sequence of ~200 bp which provides binding sites for IHF and integrase. Three molecules of IHF introduce sharp bends (~140°) into the *att* sequence. Integrase has two distinct DNA-binding sites; three molecules of integrase bind so that with IHF they make a protein core around which the λ DNA is tightly wrapped. The second integrase binding site is used to bind to the ends of the 15 bp essential sequence in both the phage and bacterial sites, thus achieving synapsis and the correct alignment of the two sites.

Nicks are then introduced into the DNA with the attachment of the integrase to the 3′ phosphate of the nicked DNA strand through covalent phosphotyrosine bonds. Strand transfer takes place to create a Holliday structure and this is resolved by nicking of the other strand and joining to give the recombinant. This process does not require ATP because the product is less supercoiled than the substrates. It is highly directional — the breaks must be introduced first into two particular strands — so that the process always leads to integration. The reverse process, excision, which occurs on phage INDUCTION, requires in addition to integrase another phage-encoded protein and a different host factor [7].

Integrase is one of a family of closely related proteins, including the FLP recombinase, which catalyse site-specific recombinations of this type. It is also related to the type I topoisomerases. Transposition similarly involves the formation of tightly organized complexes in which the DNA is highly constrained so that resolution of the Holliday junction is specific and yields only one product.

K.A. STACEY

1 Lai, M.M.C. (1992) RNA recombination in animal and plant viruses. *Microbiol. Rev.* **56**, 61–79.
2 Holliday, R. (1964) A mechanism for gene conversion in fungi. *Genet. Res.* **5**, 282–304.
3 West, S.C. (1992) Enzymes and molecular mechanisms of genetic recombination. *Annu. Rev. Biochem.* **61**, 603–640.
4 Dunderdale, H.J. (1991) Formation and resolution of recombination intermediates by *E. coli* RecA and RuvC proteins. *Nature* **354**, 506–510.
5 Smith, G.R. (1987) Mechanism and control of homologous recombination in *Escherichia coli*. *Annu. Rev. Genet.* **21**, 179–201.
6 Roman, L.J. et al. (1991) Rec BCD-dependent joint molecule formation promoted by *Escherichia coli* RecA and SSB proteins. *Proc. Natl. Acad. Sci. USA* **88**, 3367–3371.
7 Landy, A. (1989) Dynamic, structural and regulatory aspects of λ site-specific recombination. *Annu. Rev. Biochem.* **58**, 913–949.

recombination-deficient mutant An organism, typically *Escherichia coli*, in which one or more enzymes required for RECOMBINATION have been inactivated by mutation. Recombination-deficient strains of *E. coli* are useful for the propagation of VECTORS carrying unstable heterologous DNA.

recombination fraction A measure of the probability of a RECOMBINATION event occurring between two linked loci. The recombination fraction (ψ) gives the expected percentage of recombinant genotypes (with respect to the parental genotypes) in a given mating. ψ is also a measure of the GENETIC DISTANCE between the loci, as the distance affects the probability of a recombination event between them. For example, a recombination fraction of 0.1 represents a genetic distance of 10 cM, and an expected frequency of recombinants of 10%. The value of ψ can vary between 0 (i.e. no recombination and complete linkage) to 0.5 (i.e. completely random recombination and therefore no linkage).

recombination signal sequence (RSS) Sequence in immunoglobulin and T cell receptor genes at which somatic recombination takes place. *See*: GENE REARRANGEMENT; RECOMBINASE.

recoverin *See*: CALCIUM-BINDING PROTEINS.

redox potential The tendency of a redox couple (a mixture of oxidized and reduced species) to donate electrons. The standard redox potential $E^{o'}$ (also known as the midpoint potential, E_m) of a compound is the electrical potential relative to the hydrogen electrode when the oxidized and reduced forms are both at a concentration of 1 M at pH 7, 25 °C. Couples with more negative redox potential are more strongly reducing (e.g. NADH/NAD, standard redox potential – 320 mV); those with positive redox potential are more strongly oxidizing (e.g. H_2O/O_2, standard redox potential + 816 mV).

reductases Name given to many OXIDOREDUCTASE enzymes in which the electron acceptor is not molecular oxygen.

refinement, crystallographic least squares *See*: LEAST SQUARES METHOD.

refinement of macromolecular structures Molecular structures are refined from initial models, so as to obtain the best agreement between the observed data and that calculated from the current model, by variation of the model parameters.

For NMR data (*see* NMR SPECTROSCOPY), calculations of distance geometry yield a family of solutions which are consistent with the observed spectra. These structures are displayed by superposition of atomic coordinates — highlighting those parts of a molecule where the spectra do not define a conformation. For crystallographic data, a single set of atomic coordinates is refined with additional parameters to indicate the extent of conformational variability: occupancy factors reflect alternative sites (e.g. side chain conformations in proteins); TEMPERATURE FACTORS reflect the extent of disorder which may be either static or dynamic.

For crystal structure analyses, refinement of macromolecular structures presents more difficulties than that of small molecules where DIFFRACTION data extend to far higher RESOLUTION (typically 0.8 Å or less). In the case of small molecules the number of data observables greatly exceeds the number of refinable parameters such that refinement procedures are mathematically well determined and usually converge rapidly. In contrast, for macromolecules the ratio of observations to parameters is of the order of unity, or less, such that additional information is necessary for refinement to converge. Detailed knowledge of amino acid, nucleotide and other molecular structures (derived in turn from many crystallographic analyses of small molecules) can be applied here to define the expected geometry within a macromolecule; for example, bond lengths and angles, planarity of aromatic rings and peptide bonds, and chirality of tetrahedral asymmetric carbon atoms.

The early protein crystal structures were refined so as to agree exactly with the rigid constraints of stereochemistry. This sometimes produced large regions of discrepancy between the observed and calculated electron density maps, while maintaining precise standardized geometry. More recently, increased computational facilities allow the geometrical information to be applied as RESTRAINTS against which the model is refined to agree as closely as is consistent with known errors in the observed data, but with a penalty term in the refinement for departures from them. The refinement algorithms incorporate weighting schemes whereby the relative importance given to the restraints from geometrical factors (energy term) and to those from the diffraction data (X-ray term) can be varied. Thus, if the energy terms are given overwhelming weight the refinement becomes constrained rather than restrained, whereas if the X-ray terms are given overwhelming weight it becomes essentially a free refinement. For most protein crystals this would produce unreasonably poor stereochemistry but in a few cases of extremely well diffracting protein crystals unrestrained refinement has been successful. These refined protein models show that there are real variations in geometry (e.g. peptide bond torsion angles vary by a few degrees from exact planarity; proline residues are occasionally *cis* though usually *trans*) such that a restrained refinement is more realistic than one which is constrained.

Current algorithms for least squares refinement of macromolecules include within a common formalism many components of the energy term, and of the observational term, which for crystallography is the X-ray diffraction data and for NMR is the derived set of interatomic distances and torsion angles. For both cases, the radius of convergence of the refinement is increased by use of SIMULATED ANNEALING to avoid the problems associated with local minima in the function being refined. *See*: X-RAY CRYSTALLOGRAPHY.

Brunger, A.T. (1991) *Curr. Opinion Struct. Biol.* **1**, 1016–1022.
Jensen, L.H. (1985) *Meth. Enzymol.* **115**, 227–234.

reflections A term often used in X-RAY CRYSTALLOGRAPHY for the DIFFRACTION intensities. This arises from the analogy of diffraction to reflection from a series of planes, as expressed in the derivation of BRAGG'S LAW.

regeneration Some capacity to repair damaged cells, tissues, or organs exists in all organisms, but the mechanism and degree of regeneration achieved vary considerably. Plants alone have the ability to regenerate a whole new organism from a single differentiated cell, but lower animals, for example sponges and coelenterates, have considerable powers of regeneration. Such lower animals regenerate by MORPHALLAXIS: tissues remaining after injury are respecified and reorganized to replace the missing parts

with a minimum of cell division. Higher organisms capable of regenerating parts of the body produce a regeneration BLASTEMA, an aggregate of proliferating undifferentiated cells endowed with the ability to DIFFERENTIATE into the appropriate parts of the regenerate (EPIMORPHOSIS). The cells constituting the blastema may derive from undifferentiated cells already present in the animal (for example in flatworms), or from previously differentiated cells present locally at the site of injury (for example in amphibian limb regeneration).

Regeneration of the adult urodele (salamander) limb is a favoured model for the experimental investigation of regeneration and PATTERN FORMATION in vertebrates. The blastema cells arise from a population of cells proximal to the amputation site, and transplantation experiments have shown that muscle cells and dermal fibroblasts contribute a large proportion of the blastemal cells. Neurons invade the blastema soon after amputation, and a wide variety of experiments have shown that the initiation of proper limb regeneration depends on the presence of nerves in the regenerate. Regeneration fails, for example, following limb denervation, and requires the presence of a threshold number of nerve fibres to take place. One candidate molecule responsible for these effects is glial growth factor which is present in the blastema and can stimulate local mitosis. The apical epidermal cap that forms as the epidermis covers the wound site also stimulates the division of subjacent cells, perhaps in a manner analogous to the APICAL ECTODERMAL RIDGE during normal LIMB DEVELOPMENT.

Patterns of regeneration may be modelled using relatively simple rules. In the regenerating insect and amphibian limbs, for example, the results of a wide variety of experimental perturbations are predicted successfully by the POLAR COORDINATE MODEL, according to which the cells of the regenerate are specified by their positions within a system of polar coordinates. Each cell has a circumferential value specifying the anterior–posterior axis, and a radial value specifying the proximo-distal axis. Exposure of the amphibian limb blastema to a solution of retinoic acid (*see* RETINOIDS) causes the subsequent regenerated structures to adopt a more proximal pattern; regeneration at the wrist level, for example, can be made to produce the entire complement of limb structures. Whether endogenous retinoids play a role during limb regeneration is unclear.

regional specification The process whereby cells in different positions in the embryo are assigned and become COMMITTED to different fates. Understanding how regional specification arises is a central problem of developmental biology. *See*: PATTERN FORMATION.

regulation In embryos, the ability of a portion of a MORPHOGENETIC FIELD to reconstitute the entire field. At early stages of development, when the entire embryo is a single field, regulation defines the ability of parts of some embryos to reconstitute the entire embryo. In 2-cell-stage amphibians, for example, separation of the two BLASTOMERES can allow each blastomere to produce a whole embryo.

regulatory gene Any GENE involved in regulating the expression of another. Some regulatory genes (e.g. those encoding REPRES-

SORS and TRANSCRIPTION FACTORS) produce proteins that bind to specific DNA sequences in the PROMOTER or other regulatory regions of the genes they regulate. The regulatory protein interacts with the transcriptional machinery to induce, enhance or repress gene expression. Other regulatory genes, identified previously by classical genetic studies, may represent the regulatory DNA sequences (e.g. promoter regions) themselves. *See also*: BACTERIAL GENE EXPRESSION; EUKARYOTIC GENE EXPRESSION; TRANSCRIPTION.

Reichert's membrane The BASEMENT MEMBRANE of the murine parietal endoderm (*see* EXTRAEMBRYONIC MEMBRANES). During mouse development, at the egg cylinder stage of development, cells derived from the primitive endoderm migrate to cover the entire inner surface of the mural trophectoderm (the PARIETAL ENDODERM) and produce a typical basement membrane containing laminin, type IV collagen, and entactin.

relative molecular mass (M_r) The mass of a molecule relative to 1/12 of the mass of an atom of $^{12}_{6}C$ (12 atomic mass units on the atomic mass scale). The term is used synonymously with molecular weight but is not identical as it represents a ratio, whereas molecular weight is equivalent to molecular mass, and is expressed in units of mass. M_r can be calculated by summing the relative atomic masses of the atoms in the molecule or, for some macromolecules, can be determined directly by ultracentrifugation techniques or by comparison with the behaviour of compounds of known M_r on gel permeation chromatography or electrophoresis. Molecular weights are often expressed in daltons (1 dalton = 1 atomic mass unit) or kilodaltons (= 1000 daltons).

relaxed conditions *See*: NUCLEIC ACID HYBRIDIZATION.

relaxed mutant An *Escherichia coli* mutant unable to mount the STRINGENT RESPONSE to amino-acid starvation or other forms of nutrient deprivation. Most mutations map to two loci corresponding to the *relA* gene and the gene encoding the RIBOSOMAL PROTEIN L11. *relA* mutants are unable to synthesize guanosine tetraphosphate (ppGpp), the effector molecule of the stringent response. RIBOSOMES of L11 gene mutants are unable to 'idle' when an uncharged TRANSFER RNA occupies the ribosomal A site (*see* PROTEIN SYNTHESIS).

release channel *See*: CALCIUM.

release factor In PROTEIN SYNTHESIS a protein that recognizes the TERMINATION CODON, catalyses the release of the complete polypeptide from the last tRNA and expels tRNA and mRNA from the ribosome. *Escherichia coli* possesses two release factors — RF1, which recognizes the UAA and UAG termination codons, and RF2, which recognizes the UAA and UGA termination codons. Only one release factor, eRF, has been found in eukaryotes. Little is known about the function of the mammalian eRF, although unlike prokaryotic RFs, it requires the hydrolysis of GTP for activity.

releasing factor Generally refers to a hormone that controls endocrine function by causing or inhibiting the release of a hormone from its site of storage. Most of the hormones known as releasing factors are small peptides secreted by neurons of the medial hypothalamus which act on the pituitary gland.

renaturation The reassociation of a double-stranded nucleic acid from its two constituent strands (*see* DNA; HYBRIDIZATION), or the refolding of a previously unfolded protein into its native state (*see* PROTEIN FOLDING; PROTEIN STRUCTURE).

renaturation analysis *See*: COT ANALYSIS; HYBRIDIZATION.

renin An ACID PROTEINASE (EC 3.4.23.15) which cleaves the peptide donor ANGIOTENSINOGEN to give angiotensin I. It is formed from prorenin by proteolytic cleavage in plasma and kidney.

Reoviridae Family of nonenveloped RNA viruses with members infecting plants or animals. The double-stranded RNA genome is divided between 10 to 12 segments of linear RNA (total M_r $12 \times 10^6 - 20 \times 10^6$) packaged within a single virion. *See*: ANIMAL VIRUSES; PHYTOREOVIRUS GROUP; PLANT VIRUSES.

repertoire The total number of different antigen specificities present in the lymphocyte (T CELL and B CELL) population. The theoretical repertoire of antigen specificities generated during the development of T cells and B cells may be of the order of 10^{10} different receptors. It is therefore substantially higher than the number of lymphocytes available at a given time. The actual repertoire is strongly selected during lymphocyte development to produce a diverse and self-tolerant population of mature lymphocytes (*see* B CELL DEVELOPMENT; T CELL DEVELOPMENT). The primary repertoire is that diversity arising directly out of the process of gene rearrangement during lymphocyte development.

repetitive DNA DNA sequences that occur more than once in a haploid genome. Reassociation kinetics (*see* COT ANALYSIS) define two forms of repetitive DNA in eukaryotes, which have subsequently been isolated and analysed. The fast component consists of short sequences TANDEMLY REPEATED thousands of times, for example sequences found at chromosome telomeres (*see* CHROMOSOME STRUCTURE) and the VARIABLE NUMBER TANDEM REPEATS (VNTRs). The intermediate component consists of longer sequences dispersed throughout the genome and which are repeated on average several hundred times although some (e.g. the human ALU SEQUENCES) are present in up to 10^6 copies. The content of higher eukaryotic genomes is often >50% repetitive DNA (*see* GENOME ORGANIZATION).

replacement vector VECTOR derived from phage LAMBDA, in which genes not essential for LYTIC infection are replaced with the DNA to be cloned.

replica plating A method of repeatedly transferring cells from a single master plate, typically an agar plate, onto different test media such that individual colonies retain their original spatial relationship to one another. Colonies with a selectable phenotype (usually growth or no growth) as a result of MUTATION or TRANSFORMATION can then be readily detected by comparing the replica plates with the original master plate. In the original versions of the technique, transfer was by a velvet pad pressed onto the surface of the plate; filter paper is now used.

replica *See*: ELECTRON MICROSCOPY.

replicase RNA-dependent RNA POLYMERASE that replicates the genomes of RNA viruses. *See*: ANIMAL VIRUSES; BACTERIOPHAGES; PLANT VIRUSES.

replication *See*: DNA REPLICATION.

replication bubble (replication eye) Electron micrographic appearance of double-stranded DNA undergoing replication at specific internal replication origins. Each bubble, or eye, is formed by one (unidirectional) or two (bidirectional) replication forks growing away from the origin (Fig. R7). *See*: DNA REPLICATION.

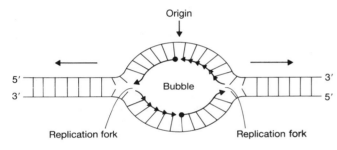

Fig. R7 The bidirectional growth of a replication bubble.

replication origin *See*: DNA REPLICATION; ORIGIN OF REPLICATION.

replication protein A Multisubunit protein complex from eukaryotic cells that binds to single-stranded DNA and is thought to have a role in unwinding DNA for DNA REPLICATION. It is composed of three polypeptides (M_r 70 000, 34 000, and 13 000).

replicative form (RF) The double-stranded form adopted by the circular genome of small, single-stranded DNA BACTERIOPHAGES, such as M13 and ΦX174, upon infection of a host bacterium. The double-stranded genome is able to serve as a template for RNA TRANSCRIPTION and DNA REPLICATION using host-derived mechanisms.

replicon A region or unit of a CHROMOSOME served by a single ORIGIN OF REPLICATION. *See*: CHROMOSOME STRUCTURE; DNA REPLICATION.

reporter gene A coding sequence attached to heterologous PROMOTER or ENHANCER elements and whose product is easily and quantifiably assayed when the construct is introduced into tissues or cells of the same origin as the regulatory elements. Reporter genes commonly used in the study of eukaryotic gene expression

include bacterial genes encoding β-GALACTOSIDASE (*lacZ*), CHLORAMPHENICOL ACETYLTRANSFERASE (*cat*) and β-GLU-CURONIDASE (*GUS*).

reporter transposon A specialized transposon (*see* TRANSPOSABLE GENETIC ELEMENTS) which contains, in addition to an antibiotic-resistance gene, a promoterless REPORTER GENE encoding a gene product which can be readily detected and assayed. Because the reporter gene has no promoter, it can only be transcribed from adjacent host promoters and this provides valuable information on the location, direction, and regulation of host transcription signals.

repression The prevention of transcription from a gene, resulting in its non-expression. This may be achieved by a variety of mechanisms. *See*: BACTERIAL GENE EXPRESSION; CHROMATIN; DNA METHYLATION; EUKARYOTIC GENE EXPRESSION; LAC OPERON; LAMBDA; TRANSCRIPTION; TRANSCRIPTION FACTORS.

repressor The product of a REGULATORY GENE which binds to a particular site in the DNA or RNA of a structural gene to prevent its expression. Repressors form the switch in many feedback control loops in bacteria where their DNA-binding capability can be allosterically regulated by a substrate for (an inducer) or a product of (a corepressor) the metabolic pathway controlled by the OPERON whose expression they regulate. The classic examples of repressors are the LacI protein which negatively regulates the expression of the LAC OPERON in *Escherichia coli* and the cI protein of bacteriophage LAMBDA which suppresses gene expression in the lysogenic state. *See*: BACTERIAL GENE EXPRESSION; LAC REPRESSOR; LAMBDA REPRESSOR; PROTEIN–NUCLEIC ACID INTERACTIONS.

repressor of primer (ROP) *See*: PROTEIN STRUCTURE for an illustration of the structure (Fig. P80).

repulsion The situation in which two linked genetic traits *A* and *B* (alleles *A*, *a* and *B*, *b* respectively; *A* and *B* dominant over *a* and *b*) are heterozygous, and each chromosome carries one dominant and one recessive allele (genotype *Ab*/*aB*). The opposite of coupling (genotype *AB*/*ab*). Equal proportions of the two genotypes are expected, unless the traits are in LINKAGE DISEQUILIBRIUM.

RER Rough ENDOPLASMIC RETICULUM.

RES RETICULOENDOTHELIAL SYSTEM.

residual bodies Dense bodies, observed morphologically, which contain ACID HYDROLASES and materials resistant to enzymatic degradation. They are thought to derive from LYSOSOMES, but their fate is unknown.

resistance genes The R genes of plants, which are involved in resistance to a specific strain of pathogen in gene-for-gene resistance (*see* PLANT PATHOLOGY). The first R gene to be cloned, a tomato gene for resistance to *Pseudomonas syringae*, is a PROTEIN KINASE.

resolution (1) The final separation into individual molecules of DNAs undergoing RECOMBINATION, which occurs by branch migration after strand exchange has taken place.

(2) In crystallographic analyses, the resolution of an ELECTRON DENSITY MAP indicates the extent of the DIFFRACTION data included in its calculation, and thus the amount of detail to be seen within the map. Traditionally, a map calculated at 3Å (0.3 nm) refers to a map which includes all data out to 3Å (whereas physicists would expect atoms separated by 2.2Å to be just resolved by such a map). Higher resolution refers to finer details such that a resolution of 10Å is much lower than that of 2Å.

The resolution (*d*) of a diffraction intensity is inversely related to its angle of diffraction (θ) by BRAGG'S LAW:

$$\lambda = 2d \sin (\theta)$$

for wavelength λ. The diffraction data can be envisaged as points in a regular three-dimensional lattice in RECIPROCAL SPACE, where the distance of each point from the origin is the reciprocal of its resolution, such that the number of intensities measurable out to a given resolution equals the number of reciprocal LATTICE points within a sphere of that radius. Thus, the number of intensities increases in proportion to the third power of the resolution; namely, doubling the resolution increases the number of intensities by a factor of 8.

The highest angle STRUCTURE FACTORS relate features of closest *d* spacing; resolutions of 6Å, 3Å, and 1.5Å may be considered as being low, medium, and high resolutions respectively for macromolecules and often new structures are analysed in these successive stages. In a 6Å map the molecular boundary can be delineated from the solvent, and the positions and lengths of α-helices (but not the direction of the polypeptide backbone) can be seen; in a 3Å map the course of a polypeptide chain can be traced, and the positions of many side chains located, but the orientation of peptide bonds may be unclear until 2.5Å, where clear carbonyl groups define the direction of the polypeptide chain; in a 1.5Å map all atoms except hydrogen would be visible and aromatic groups have a central hole in the electron density.

However, the effective resolution of an electron density map may be less than that expected if there are errors in the data — either in the observed STRUCTURE AMPLITUDES or in the derived phases — or if the crystal itself is disordered. For many large molecules, the higher resolution intensities are often much weaker than the others, being correspondingly harder to measure accurately. See: ELECTRON MICROSCOPY; X-RAY CRYSTALLOGRAPHY.

resolvase ENDONUCLEASE involved in the resolution into two separate DNA molecules of the four-way DNA junction (HOLLIDAY STRUCTURE) formed during RECOMBINATION.

respiratory chain The respiratory chain of MITOCHONDRIA transfers electrons from organic reductants, particularly NADH and succinate, to molecular oxygen (O_2). The eukaryotic respiratory chain consists of a series of alternating hydrogen and electron carriers embedded in or associated with the mitochondrial inner membrane. Electrons are transferred from NADH to ubiquinone, succinate to ubiquinone, ubiquinol to cytochrome *c* and cyto-chrome *c* to O_2 in reactions catalysed by multisubunit enzyme complexes that can be purified from membrane preparations. Prokaryotic respiratory chains involve similar components, associated with the cell membrane. See: ELECTRON TRANSPORT CHAIN.

restorer of fertility *See*: RF.

restraint An ideal value of a quantity (e.g. bond length or angle) to which a model is restrained to be compatible, and refined to approach more closely. *Cf.* constraint, a fixed value imposed. *See*: MOLECULAR DYNAMICS; NMR SPECTROSCOPY; REFINEMENT OF MACROMOLECULAR STRUCTURES.

restriction *See*: CLONAL RESTRICTION; MHC RESTRICTION; RESTRICTION ENZYMES; RESTRICTION AND MODIFICATION.

restriction analysis Analysis of DNA by treatment with specific RESTRICTION ENZYMES and separation of the resulting fragments by size by gel ELECTROPHORESIS. The pattern of fragments produced is then usually selectively probed with a sequence-specific PROBE to detect fragments belonging to a particular gene or other sequence of interest. The patterns of restriction fragments produced can be used to detect the presence of mutations (e.g. for comparing normal and disease genes) or to compare DNAs from different individuals or species for forensic, ecological and evolutionary studies (*see* DNA TYPING; MOLECULAR EVOLUTION; PRENATAL DIAGNOSIS), or to construct physical maps of DNA (*see* RESTRICTION MAPPING; RFLP; HUMAN GENE MAPPING).

restriction and modification Bacterial defence mechanism requiring the action of METHYLASES and RESTRICTION ENZYMES to protect the cell from invading DNA (such as PLASMID or BACTERIOPHAGE DNA). Each restriction enzyme recognizes and cleaves at or near a specific, short, and often PALINDROMIC DNA sequence. The same sequence in the host bacterium's DNA is recognized and protected by an endogenous methylase that modifies it by adding methyl groups to adenine or cytosine residues, thus preventing the restriction enzyme from binding and cleaving host DNA at that site. Each bacterial species has different modification and restriction activities. Any invading DNA from a different host type will be modified at different sites and can be degraded by the defending host's restriction enzymes. *See also*: DNA METHYLATION.

restriction endonuclease *See*: RESTRICTION ENZYMES.

restriction endoribonuclease *See*: RIBOZYMES.

Restriction enzymes

RESTRICTION enzymes are ENDONUCLEASES that cleave DNA in response to a recognition site on the DNA [1]. The recognition site (restriction site) consists of a specific sequence of nucleotides in the DNA duplex, typically 4–8 base pairs (bp) long. These endonucleases are found in many species of bacteria, where they function in RESTRICTION AND MODIFICATION SYSTEMS (R/M systems) [2]. The bacterial R/M system is analogous to an immune system, in that it enables the bacterium to distinguish its own DNA from foreign DNA and to eliminate the latter. The discovery of restriction–modification systems followed the observation more than 30 years ago that a single cycle of phage growth in a particular bacterial host could alter the host range of the progeny phage; they had become 'restricted' in their host range as a result of 'modification' in the original host. Restriction–modification is achieved by a combination of two enzyme activities: a modification methyltransferase and a restriction endonuclease. The former transfers methyl groups from S-ADENOSYLMETHIONINE to specific bases within the recognition sequence, generally one in each strand: if one strand is already methylated, the second strand remains a substrate (*see* DNA METHYLATION). The latter cleaves the DNA but only if the recognition site is not methylated in either strand. Methylation of just one strand blocks restriction activity, so the bacterial DNA is protected from the endonuclease even after its semiconservative replication. However, DNA that is foreign to the cell carrying the R/M system will lack the appropriate methylation and, if such DNA enters the cell, it is likely to be cleaved by the restriction enzyme.

Types of restriction enzyme

R/M systems have been classified into three types, named I, II and III, on the basis of their genetics and enzymology [2]. In type I systems, the recognition of the DNA sequence and both enzyme activities are all carried out by a single oligomeric protein. The type I proteins contain three different subunits: one to recognize the DNA sequence; a second to methylate the DNA at the recognition site; and a third to cleave the DNA. The latter reaction is induced by the recognition site being unmethylated in both strands but it does not occur at that site: instead, the DNA is cleaved at random positions, often several kilobases away from the recognition site. S-adenosyl methionine is required as a cofactor for both the methyltransferase and endonuclease activities of type I systems and the latter activity also needs both Mg^{2+} ions and ATP.

In type II systems, restriction and modification are carried out by separate proteins which independently recognize the same DNA sequence. Type II modification enzymes are generally monomeric proteins and their restriction enzymes are usually dimers of identical subunits [1]. Within each type II system, there is no homology between the restriction and modification enzymes. The modification enzymes from different type II systems are often homologous to each other, but the restriction enzymes are a diverse group of proteins [2]. In contrast to type I systems, type II endonucleases require only Mg^{2+} ions as a cofactor for DNA cleavage, and they cleave the DNA at specific phosphodiester bonds within (or adjacent to) the recognition site.

The structures of two type II endonucleases, *Eco*RI and *Eco*RV, have been determined by X-RAY CRYSTALLOGRAPHY: the two structures have little in common and the way in which they interact with their recognition sequences is completely different (*see* PROTEIN–NUCLEIC ACID INTERACTIONS) [3,4]. In both cases, the parts of the enzymes that interact with the DNA have structures that are unlike any of the standard protein motifs for DNA recognition, but these structural elements still contact the bases in the target sequence by networks of hydrogen bonds. For *Eco*RI, these interactions result in the enzyme binding more strongly to its recognition sequence than to anywhere else on the DNA, although part of the interaction energy is used to distort the structure of the DNA prior to catalysis. In contrast, the *Eco*RV restriction enzyme binds all DNA sequences, including the recognition site, with equal affinity but only the specific complex carries out the DNA cleavage reaction. The DNA in the specific complex of *Eco*RV at its recognition site is severely distorted while that in its nonspecific complex has a regular structure. This in turn results in only the specific complex being able to bind the Mg^{2+} ions that are needed for the catalytic reaction.

Type III systems, like type I, use a single oligomeric protein for both activities but they contain only two sorts of subunits: one to recognize the target sequence on the DNA and carry out the methyl transfer reaction; a second to cleave the DNA. Type III enzymes cleave the DNA 25–27 bp away from the recognition site.

Type II restriction enzymes and their applications

Type II restriction enzymes have extensive applications throughout molecular biology, in the dissection of DNA and in the construction of recombinant DNA molecules; without them the spectacular advances in molecular genetics over the past two decades could not have taken place (*see e.g.* DNA CLONING; DNA SEQUENCING; DNA TYPING; GENE MAPPING; GENETIC ENGINEERING; RECOMBINANT DNA; RESTRICTION ANALYSIS; RFLP). These applications depend on the specificity of the enzymes for their recognition sites on DNA: that is, their ability to discriminate between the recognition sequence and any other DNA sequence. Under standard reaction conditions, the reaction rate at the recognition sequence can be more than a million times faster than that at DNA sequences just one base pair different. (But certain special reaction conditions can lead to increased activity at sites one base pair different from the recognition site: this is sometimes known as 'star' activity.)

Moreover, type II enzymes, in contrast to types I and III, cleave the DNA at fixed locations relative to their recognition sites. Hence, under appropriate conditions, the only products observed from a reaction of a type II restriction enzyme on a DNA substrate will be a defined set of DNA fragments (restriction fragments), as a result of the DNA being cut at each copy of the recognition site and nowhere else. The individual fragments can be separated from each other on the basis of size by ELECTROPHORESIS through either agarose or polyacrylamide. The construction of a RESTRICTION MAP then involves determining the position and the orientation of each fragment in the DNA substrate.

More than 2300 type II restriction endonucleases have been identified to date from a wide range of bacterial species and, between them, these enzymes cleave DNA at 230 different DNA sequences [5]. In many cases, two or more restriction enzymes from different bacteria recognize the same DNA sequence: the enzymes are then called isoschizomers. Some examples of recognition sites for type II restriction enzymes are given in Table R20. The majority of restriction enzymes recognize unique nucleotide sequences, 4–8 bp long, at symmetric sites where both strands have the same 5′–3′ sequence. Of the examples in Table R2, *Alu*I, *Eco*RI, *Eco*RV, *Kpn*I, and *Not*I all have sites of this sort. But in some cases, the site is degenerate, in that one or more positions in the sequence can be occupied by different bases: for example, *Hae*II requires a purine as the first base in its recognition sequence but this can be either A or G (Table R2). In other cases, the site is 'hyphenated' in that it contains two series of specified bases separated by one or more unspecified base(s): namely the sites for *Hinf*I and *Bgl*I in Table R2.

The number of sites of a particular restriction enzyme on a DNA molecule is determined by the precise sequence of that DNA. However, the statistical frequency of a given sequence in random DNA is $1/4^N$, where N is the number of base pairs in the target sequence. Hence, the statistical probability of a 4-bp recognition site, such as that for *Alu*I, is one per 256 bp. In reality, the average size of restriction fragments in an *Alu*I digest may be close to 256 bp, but this average will include some fragments that are much larger than 256 bp and others that are much smaller. The probability of finding an 8-bp sequence, such as a *Not*I site, is much lower than that for a 4-bp sequence; only one per 65 536 bp. Consequently, *Not*I produces much larger DNA fragments than *Alu*I. Moreover, CG sequences in higher eukaryotes are often methylated; some restriction enzymes are inhibited by this methylation and these cleave eukaryotic DNA less frequently than might be anticipated.

Type II enzymes cleave both strands of the DNA, but the enzymes vary in the positions at which they cut the DNA relative to their recognition sites. Some cut each strand in the 5′ half of the recognition sequence whereas others act in the 3′ half; the former produce double-stranded DNA fragments with single-strand extensions at the 5′ ends whereas the latter leave single-strand extensions at the 3′ ends (e.g. *Eco*RI and *Not*I for 5′ extensions, *Kpn*I and *Hae*II for 3′ extensions; Table R2). Others cut both strands at the centre of the site to yield blunt-ended fragments (e.g. *Alu*I and *Eco*RV). There also exists a subset of type II restriction enzymes, known as type IIs, that recognize asymmetric DNA sequences and cleave both strands of the DNA at fixed distances away from the recognition site (e.g. *Mbo*II in Table R2).

All type II restriction enzymes cleave phosphodiester bonds to leave 5′ phosphates and 3′ hydroxyl groups. The terminus can therefore be acted on by DNA LIGASE to join two or more restriction fragments together. In particular, two restriction fragments with the same single-strand extensions ('sticky ends') will base pair with each other spontaneously and can then be joined covalently by DNA ligase. This useful property was exploited to make the first recombinant DNAs (*see* GENETIC ENGINEERING).

S.E. HALFORD

1 Bennett, S.P. & Halford, S.E. (1989) Recognition of DNA by type II restriction enzymes. *Curr. Topics Cell. Regul.* **30**, 57–104.
2 Wilson, G.G. & Murray, N.E. (1991) Restriction and modification systems. *Annu. Rev. Genet.* **25**, 585–627.
3 Rosenberg, J.M. (1991) Structure and function of restriction endonucleases. *Curr. Opinion Struct. Biol.* **1**, 104–113.
4 Winkler, F.K. (1992) Structure and function of restriction endonucleases. *Curr. Opinion Struct. Biol.* **2**, 93–99.
5 Roberts, R.J. & Macellis, D. (1992) Restriction enzymes and their isoschizomers. *Nucleic Acids Res.* **20**, 2167–2180.
6 Smith, H.O. & Nathans, D. (1973) A suggested nomenclature for bacterial host modification and restriction systems and their enzymes. *J. Mol. Biol.* **81**, 419–423.

Table R2 Recognition sites for some type II restriction enzymes

Microorganism	Enzyme*	Sequence†
Arthrobacter luteus	*Alu*I	A G↓C T T C↑G A
Escherichia coli RY13	*Eco*RI	G↓A A T T C C T T A A↑G
Escherichia coli J62[pGL74]	*Eco*RV	G A T↓A T C C T A↑T A G
Klebsiella pneumoniae	*Kpn*I	G G T A C↓C C↑C A T G G
Nocardia otitidis-caviarum	*Not*I	G C↓G G C C G C C G C C G G↑C G
Haemophilus aegyptius	*Hae*II	R G C G C↓Y Y↑C G C G R
Haemophilus influenzae	*Hinf*I	G↓A N T C C T N A↑G
Bacillus globigii	*Bgl*I	G C C N N N N↓N G G C C G G N↑N N N N C C G
Moraxella bovis	*Mbo*II	G A A G A N N N N N N N N↓N C T T C T N N N N N N N↑N N

* Restriction enzymes are named after the *Genus* and *species* from which they are isolated [6].
† The sequences are written 5′–3′ in the top strand and 3′–5′ in the bottom strand. The phosphodiester bonds cleaved by each enzyme are marked ↓ and ↑ in top and bottom strands respectively. R and Y signify a purine (A or G) pyrimidine (C or T) base respectively, and N any base.

restriction fragment Fragment of DNA produced by cleavage of a larger piece of DNA by RESTRICTION ENZYMES.

restriction fragment length polymorphism *See*: RFLP.

restriction map, restriction mapping A linear physical map of a region of DNA, drawn to scale and showing the relative positions of the target sites of various RESTRICTION ENZYMES (Fig. R8; Table R3). A restriction map is prepared as a preliminary to more detailed sequence analysis. It is typically constructed by digesting the DNA with several different pairs of restriction enzymes. For each pair, the DNA is digested with each enzyme individually (a single digest) and with both enzymes (a double digest). The products of each digestion are separated by agarose gel ELECTROPHORESIS and the DNA fragment sizes estimated by comparison against DNA size standards. The restriction map is deduced from the permutations of restriction site positions that would give rise to the observed fragment sizes.

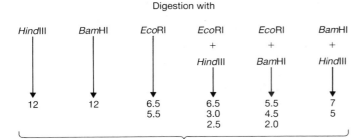

Digestion with

*Hind*III	*Bam*HI	*Eco*RI	*Eco*RI + *Hind*III	*Eco*RI + *Bam*HI	*Bam*HI + *Hind*III
12	12	6.5	6.5	5.5	7
			5.5	3.0	4.5	5
				2.5	2.0	

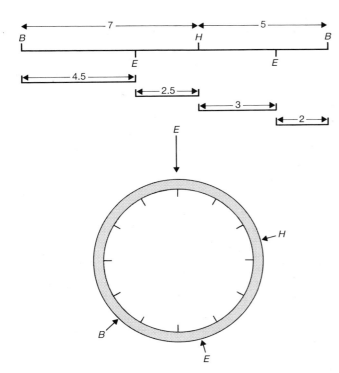

Fig. R8 Construction of a restriction endonuclease map of a circular DNA molecule of 12kb. The DNA is digested with the endonucleases indicated and the size (in kb) of the products is determined by agarose gel electrophoresis. The positions of the recognition sites for the three enzymes can be deduced from the sizes of the fragments.

Table R3 Symbols for cleavage sites of commonly used restriction enzymes

Restriction endonuclease	Symbol	Restriction endonuclease	Symbol
*Alu*I	Al	*Kpn*I	K
*Ava*I	A	*Mbo*I	M1
*Bam*HI	B	*Mbc*II	M2
*Bcl*I	Bc	*Msp*I	M
*Bgl*I	Bg	*Pst*I	P
*Bgl*II	B2	*Pvu*I	Pu
*Cla*I	C	*Pvu*II	Pv
*Dde*I	D	*Sac*I	Sc
*Eco*RI	E	*Sal*I	S
*Eco*RII	E2	*Sau*3A	S3
*Eco*RV	E5	*Sma*I	Sm
*Hae*II	He	*Sph*I	Sp
*Hae*III	Ha	*Sst*I	Ss
*Hinc*II	Hc	*Taq*I	T
*Hind*III	H	*Xba*I	X
Hinf I	Hf	*Xho*I	Xh
*Hpa*I	Hl	*Xma*III	Xm
*Hph*I	Hp		

proteins possess retention signals for their target compartment, although in many cases these signals have not yet been precisely identified. The clearest example is the sequence KDEL (ADEL or HDEL in some organisms) at the C terminus of soluble ER proteins.

reticulocyte lysate *See*: IN VITRO TRANSLATION.

reticuloendothelial system (RES) A collective term for the non-lymphoid cells which are involved in the nonspecific clearance of foreign material or partly degraded or abnormal proteins from the circulation and which is generally taken to include the endothelial cells of the liver and other organs, and the reticular cells in lymphoid organs.

retina In vertebrates, the innermost layer of the eyeball, containing the cells required for the reception and transmission of visual stimuli. A layer of pigmented epithelium separates the retina from the underlying tissues of the eyeball. Above this lies the photosensitive layer of rod and cone cells, which make connections with subsequent layers of bipolar and horizontal neurons, and the multipolar ganglionic neurons, whose axons form the optic nerve. *See*: EYE DEVELOPMENT; VISUAL TRANSDUCTION.

retinal The chromophore of the visual pigments RHODOPSIN and the CONE PIGMENTS, and of BACTERIORHODOPSIN. It is a derivative of the isoprenoid RETINOL.

retinitis pigmentosa (RP) Genetically determined disorder characterized by night blindness, constriction of the visual fields, and changes in the fundus of the retina. The disease can be X-linked (6%), AUTOSOMAL DOMINANT (10%), or autosomal recessive (84%). The mutations causing some forms of autosomal recessive RP are located in the RHODOPSIN gene on human chromosome 3.

restriction point A transition point between S-phase and MITOSIS in the CELL CYCLE of eukaryotic cells.

ret A potential ONCOGENE encoding a receptor protein with tyrosine kinase activity (*see* GROWTH FACTOR RECEPTORS).

retention signal One class of sorting signal involved in the processes of PROTEIN TARGETING: mutation or removal of such a signal allows the protein to be transported onward to another cellular compartment. Those proteins that enter the ENDOPLASMIC RETICULUM (ER) may subsequently travel by vesicle-mediated transport via the GOLGI APPARATUS to the cell surface. If the final destination of the protein is one of these intracellular compartments, however, it must be diverted from this pathway and retained within either the ER or Golgi apparatus. Therefore such

retinoblastoma A malignant tumour arising from the cells of the retina. There are two patterns of occurrence. In one group there is an onset in early childhood, tumours in both eyes, a propensity to multiple tumours and a high risk of developing other cancers. The other group is comprised of patients who have disease in one eye only, in whom the tumour develops later in life, and who do not seem to be at risk of having a second malignancy.

The first group appear to have inherited a GENE which makes them more likely to develop a retinoblastoma; the other group probably have an acquired MUTATION of this gene. The gene involved has been assigned to chromosome 13 (13q14) and has been designated the *RB* locus. It is thought that the development of a retinoblastoma involves two mutational events at the *RB* loci. Hereditary cases are thought to carry a germ-line mutation which inactivates the *RB* locus on one of the pair of chromosomes 13, and which is present in all somatic and germ cells; tumour formation only occurs, however, when a second, and in this case somatic, mutation occurs in a retinal cell at the *RB* locus on the other homologous chromosome. Because the inactivation of the normal ALLELE at the *RB* locus leads to tumour formation, the normal allele is termed a TUMOUR SUPPRESSOR GENE.

In sporadic cases, which are extremely rare, it is assumed that mutations at the *RB* locus on both chromosomes occur as chance events in the same retinal cell.

In hereditary cases, there are several ways in which the normal allele of the *RB* gene might be inactivated in somatic cells. One involves mitotic NONDISJUNCTION, whereby the chromosome carrying the normal allele is lost because of an error during cell division in the retinal cell. It could also result from a similar event with subsequent reduplication of the chromosome carrying the mutant *RB* allele or from mitotic RECOMBINATION between the mutant *RB* allele and the CENTROMERE resulting in HOMOZYGOSITY of the mutant allele along with the distal part of the chromosome. A similar effect would be produced by a spontaneous point mutation of the normal allele. Many of these predictions have now been confirmed by RFLP analysis using markers flanking the *RB* locus in DNA obtained from retinoblastomas. The product of the *RB* gene is a phosphoprotein of M_r 110 000 with DNA binding activity. A large number of different mutations of the *RB* locus have been defined in families with familial retinoblastoma. *See also*: ONCOGENES.

Weinberg, R.A. (1989) *Blood* **74**, 529–532.

retinoic acid *See*: RETINOIDS.

retinoid(s) Analogues or metabolites of vitamin A (RETINOL). They include retinal, the chromophore of the visual pigments RHODOPSIN, CONE PIGMENTS, and BACTERIORHODOPSIN, and metabolites of retinol with a presumed importance in development.

Their importance in development was first recognized by the deleterious effects of vitamin A deficiency on mammalian embryos. At the other extreme, large doses of vitamin A were found to be teratogenic, causing a wide range of abnormalities such as craniofacial malformations, and neural tube, limb, and heart defects. The type and severity of the defect depends on when, during gestation, the embryo is exposed to the retinoid. If the

Fig. R9 Retinoic acid.

embryo is treated during the critical period of development of a certain tissue, that tissue will be significantly affected. More recently, retinoids have been shown to have a role in PATTERN FORMATION during development. In the developing limb of the chick embryo, digit pattern is thought to be defined by an area of cells — the ZONE OF POLARIZING ACTIVITY (ZPA) at the posterior margin of the limb bud. If another ZPA is transplanted to the anterior margin of the limb bud, mirror-image digit duplications result. It has been suggested that the ZPA specifies digit pattern by setting up a morphogenetic gradient across the limb bud, so that with two ZPAs, one on either side of the limb bud, there are apposed MORPHOGENETIC FIELDS. Retinoic acid (Fig. R9) (a metabolite of vitamin A) applied locally to the anterior margin of the limb bud also results in digit duplication. This, and the fact that a concentration gradient of endogenous retinoic acid has been demonstrated across the limb bud, imply that this retinoid may be the MORPHOGEN in question. Exogenous retinoids have also been shown to change pattern formation in regenerating amphibian limbs, whereby more is regenerated than was originally missing. Mouse embryos exposed to retinoic acid also show a change in pattern formation in the form of a shift in vertebra type and homeobox gene expression along the embryonic axis. All tissues where retinoids have been shown to affect development contain retinoid-binding proteins and nuclear RETINOID RECEPTORS, providing further evidence for a physiological role of retinoids in normal development.

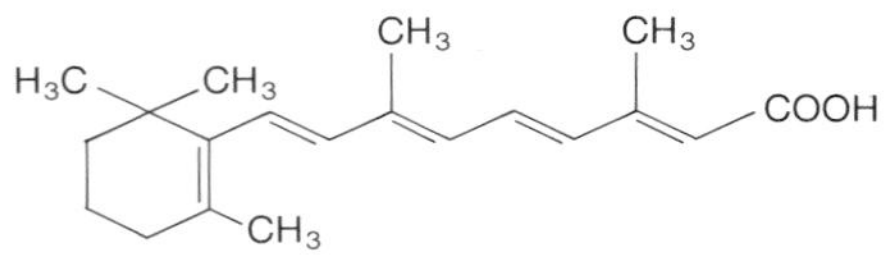

Retinoic acid also induces the differentiation of certain cells, such as EMBRYONAL CARCINOMA cells, in culture.

Marshall, H. et al. (1992) *Nature* **360**, 737–741.

retinoid receptors (RAR) Intranuclear DNA-binding proteins belonging to the STEROID RECEPTOR SUPERFAMILY which act as receptors for retinoic acid (*see* RETINOIDS). They bind to DNA as heterodimers with a related protein, the RETINOID X RECEPTOR, and when complexed with their specific ligand act as TRANSCRIPTION FACTORS.

retinoid X receptor (RXR) Protein of the STEROID RECEPTOR SUPERFAMILY which forms a DNA-binding heterodimer with some other members of the family that acts as a TRANSCRIPTION FACTOR in the presence of the specific ligand.

retinol Vitamin A (Fig. R10), an isoprenoid whose derivatives include retinal, the chromophore of the visual pigment RHODOPSIN, and retinoic acid, which has been implicated in developmental processes (*see* RETINOIDS). Retinol phosphate is involved in the shuttling of glycosyl residues across the membranes of the GOLGI APPARATUS during glycan assembly (*see*

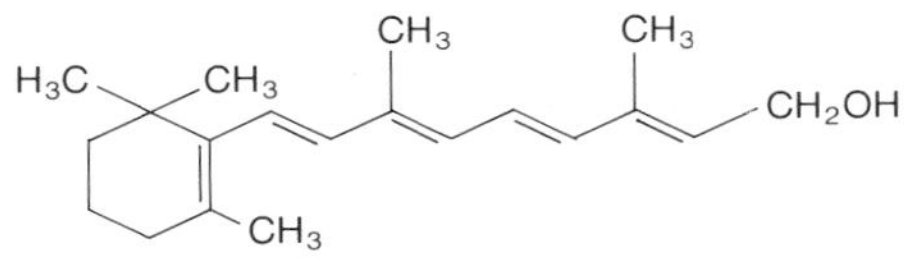

Fig. R10 Retinol.

GLYCANS; GLYCOPROTEINS AND GLYCOSYLATION). The phosphoryl glycosides are less stable than those of DOLICHOL.

retinol-binding protein (RBP) A plasma protein of 182 amino acids which is synthesized by hepatocytes and which transports RETINOL (vitamin A) from its storage site in the liver to the tissues where it is required. Each protein binds one molecule of retinol inside a β-BARREL core. The complex of retinol-binding protein and retinol is itself tightly bound by the plasma protein TRANS-THYRETIN which prevents its loss by the kidney. Retinol-binding protein has a strong structural, and weak sequence, homology with the protein β-LACTOGLOBULIN from milk. The structure is illustrated in Plate 3c.

Newcomer, M.E. et al. (1984) *EMBO. J.* **3**, 1451–1454.
Sawyer, L. (1987) *Nature* **327**, 659.

retraction fibre *See*: CELL MOTILITY.

retrograde chromatolysis A change seen in NEURONS in which neuronal processes have been damaged. There is a breakdown of rough ENDOPLASMIC RETICULUM and its disappearance from the central portion of the cell, and a displacement of the nucleus from its usual position towards the cell periphery. The effects of damage begin near the site of the lesion and extend back to the cell body.

retrograde movement/transport The movement of material towards the cell body from the more distal areas of a neuron. *See*: MICROTUBULE-BASED MOTORS.

retroposon, retrotransposon Mobile genetic element that transposes via an RNA intermediate (retrotransposition), for example the yeast TY ELEMENT. *See*: TRANSPOSABLE GENETIC ELEMENTS.

retroviral vectors The genomes of RETROVIRUSES have been adapted as highly efficient VECTORS for introducing novel genes into animal cells. The demonstration that retroviral ONCOGENES (v-*onc*) are altered forms of normal cellular genes illustrated the potential of retroviruses as genetic vectors. Moreover, the acutely transforming retroviruses are often replication-defective because the v-*onc* genes have replaced one or more of the genes essential for viral replication. Such defective oncogenic retroviruses can only be propagated in the presence of a wild-type 'helper' virus, which supplies the missing gene products in *trans*. Imitation of these phenomena *in vitro* and exploitation of other molecular facets of the retroviral life cycle has produced current retroviral vector technology in which retroviruses have been adapted as vehicles for the delivery and expression of cloned genes in a wide variety of cells, for both experimental and therapeutic purposes.

Retroviral infection represents the most efficient method of gene transfer available compared with other physical methods (*see* TRANSFECTION). Infection with high titre retroviruses can potentially lead to modification of nearly every cell in the target population. Genes carried in retroviral vectors are transferred as single copies without (generally) any adverse side effects on the recipient cell and, unlike other transfer techniques, are integrated in a predictable stable configuration, usually in transcriptionally active chromosomal regions. The chromosomal site of integration is, however, still random.

The production of recombinant retroviral vectors carrying a gene of interest is carried out in two distinct stages.

1 A recombinant vector DNA genome carrying the gene to be introduced is constructed *in vitro* using RECOMBINANT DNA TECHNIQUES. This construct must include DNA sequences needed for the efficient expression of the inserted genes; these may be provided by the viral long terminal repeats (LTRs) and must also include relevant splicing signals. In addition, appropriate safety modifications are made: genes essential for viral replication are deleted so that the transferred genome alone cannot direct the formation of infectious virions inside the infected cell — the recombinant genome is made replication defective.

However, the vector must retain sequences that allow it to be packaged into infectious virions. These include the regions of the retroviral genome required in *cis* for its efficient incorporation into a retroviral particle — the packaging signal (Psi), the tRNA primer binding site (– PBS), the 3′ regulatory sequences for reverse transcriptase (+ PBS), and the LTRs, which contain the sequences required for genomic RNA association, reverse transcriptase and integrase functions, as well as directing expression of the virion RNA to be packaged into the viral particle (*see* RETROVIRUSES).

2 Once constructed, the vector DNA is introduced into retroviral packaging cells by transfection. These cells provide the missing proteins required in *trans* to incorporate the vector RNA transcripts into viral particles with the required host range [1].

Gene delivery via retroviruses may be used for various purposes. One is simply to mark a target cell population for easy identification. Integration of the provirus into the genome may also lead to novel mutations which can be more easily located by virtue of their linkage to the retroviral genome.

More commonly, retroviral vectors are used to deliver a new gene to a specific cell type, where its expression is driven from sequence elements included in the vector and the effects of expression can be monitored. If the expressed gene encodes a correct version of a defective gene, its transfer into diseased cells may even allow correction of the genetic deficiency, either on a cellular level or in the animal as a whole. Such approaches may eventually lead to GENE THERAPY for specific monogenic disorders [2].

1 Vile, R.G. (1991) In *Methods in Molecular Biology*, Vol. 8 (Collins, M., Ed.) (Humana Press, NJ).
2 Friedmann, T. (1989) *Science* **244**, 1275–1281.

Retroviruses

THE family Retroviridae are enveloped single-stranded RNA viruses infecting animals (*see also* ANIMAL VIRUSES). They are unique among animal RNA viruses in that their multiplication involves the synthesis of a DNA copy of the RNA which is then integrated into the genome of the infected cell.

The family consists of three groups: the spumaviruses (or foamy viruses) such as the human foamy virus (HFV); the lentiviruses, such as the human immunodeficiency virus (HIV) types 1 and 2, as well as visna virus of sheep; and the oncoviruses (although not all viruses within this group are oncogenic) [1]. The oncoviruses are further subdivided into groups A, B, C and D on the basis of particle morphology, as seen under the electron microscope during viral maturation.

1 Small A-type particles represent the immature particles of the B- and D-type viruses seen in the cytoplasm of infected cells. These particles are not infectious.

2 B-type particles bud as mature virions from the plasma membrane by the enveloping of intracytoplasmic A-type particles. At the membrane they possess a toroidal core of ~75 nm, from which long GLYCOPROTEIN spikes project. After budding, B-type particles contain an eccentrically located, electron-dense core. The prototype B-type virus is mouse mammary tumour virus (MMTV).

3 No intracytoplasmic particles can be observed in cells infected by C-type viruses. Instead, mature particles bud directly from the cell surface via a crescent 'C'-shaped condensation which then closes on itself and is enclosed by the plasma membrane. Envelope glycoprotein spikes may be visible, along with a uniformly electron-dense core. Budding may occur from the surface plasma membrane or directly into intracellular vacuoles. The C-type viruses are the most commonly studied and include many of the avian and murine leukaemia viruses. Bovine leukaemia virus (BLV), and the human T-cell leukaemia viruses types I and II (HTLV-I/II) are similarly classified as C-type particles because of the morphology of their budding from the cell surface. However, they also have a regular hexagonal morphology and more complex genome structures than the prototypic C-type viruses such as the murine leukaemia viruses (MLV).

4 D-type particles resemble B-type particles in that they show as ring-like structures in the infected cell cytoplasm, which bud from the cell surface, but the virions incorporate short surface glycoprotein spikes. The electron-dense cores are also eccentrically located within the particles. Mason Pfizer monkey virus (MPMV) is the prototype D-type virus.

Life cycle

Retroviruses are defined by the way in which they replicate their genetic material: during replication the RNA is converted into DNA, the reverse (retro-) of the normal direction of flow of genetic information (Fig. R11). Following infection of the cell a double-stranded molecule of DNA is generated from the two molecules of RNA which are carried in the viral particle by the molecular process known as REVERSE TRANSCRIPTION. The DNA form becomes covalently integrated in the host cell genome as a provirus, from which viral RNAs are expressed with the aid of cellular and/or viral factors. The expressed viral RNAs are packaged into particles and released as infectious virions [2].

The retroviral genome

The retrovirus particle is composed of two identical RNA molecules. Each genome is a positive sense, single-stranded RNA molecule (*see* ANIMAL VIRUSES), which is capped at the 5′ end and polyadenylated at the 3′ tail (*see* RNA PROCESSING). The prototype C-type oncoviral RNA genome (e.g. MLV) contains three OPEN READING FRAMES called *gag*, *pol* and *env*, bounded by regions that contain signals essential for expression of the viral genes (Fig. R12). At the 5′ end of the RNA genome is the R region which is duplicated at the 3′ end of the genome. The U5 region contains the polyadenylation signals for viral transcripts and separates R from a short stretch of noncoding RNA, which includes the transfer RNA (tRNA) primer binding site (−PBS). This site is complementary to the 3′ terminal 16–19 nucleotides of a host cell-derived tRNA molecule which is used as a primer for DNA synthesis in reverse transcription. The *gag* region encodes the structural proteins of the viral capsid. The *pol* region encodes a viral proteinase as well as the proteins for genome processing, including reverse transcriptase, ribonuclease H and endonuclease enzymatic activities. The *env* region specifies the glycoproteins of the viral envelope (Figs R12 and R13). 3′ to the coding sequences lie further regulatory sequences; the +PBS sequence involved in reverse transcription, followed by the U3 region which contains sequences that will act as the PROMOTER for transcription of the proviral genome, and a copy of R.

The more complex genomes of HTLV-I/II, the lentiviruses and the spumaviruses carry additional open reading frames which encode regulatory proteins involved in the control of genome expression [3,4].

Virion structure and infection of the cell

The diploid virus particle contains the two RNA strands complexed with *gag* proteins, viral enzymes (*pol* gene products) and host tRNA molecules within a 'core' structure of *gag* proteins. Surrounding and protecting this capsid is a lipid bilayer, derived from host cell membranes and containing viral envelope proteins, generating a viral particle of about 100–150 nm across (Fig. R13). The *env* proteins bind to the cellular receptor for the virus and the particle probably enters the host cell via receptor-mediated ENDOCYTOSIS and/or membrane fusion [5].

Reverse transcription

After the outer envelope is shed, the viral RNA is copied into DNA by reverse transcription. This is catalysed by the reverse transcriptase enzyme encoded by the *pol* region and uses the host cell tRNA packaged into the virion as a primer for DNA synthesis. Reverse transcription involves both the RNA molecules and there are at least two 'jumps' of template by the reverse transcriptase. The first occurs between the two RNAs (intermolecular) and the second occurs from one end to the other of one RNA molecule (intramolecular). In this way the RNA genome is converted into

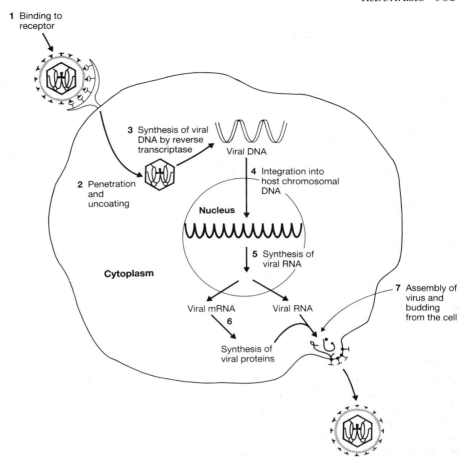

Fig. R11 The retroviral life cycle.

the more complex DNA genome of Fig. R14. This sequence of reactions is described in more detail in [6].

Integration

The double-stranded linear DNA produced by reverse transcription — the provirus — may, or may not, have to be circularized in the nucleus before integration into the host cell genome. The provirus now has two identical repeats at either end, known as the long terminal repeats (LTR). Each LTR consists of a U3 R U5 cassette and represents an intact regulatory region, containing all the transcriptional regulatory sequences required for expression of the viral DNA genome (Fig. R14). The junction between the two

joined LTR sequences produces the site recognized by a *pol* product — the integrase protein — which catalyses integration, such that the provirus is always joined to host DNA two base pairs (bp) from the ends of the LTRs [7]. A duplication of cellular sequences is seen at the ends of both LTRs, reminiscent of the integration pattern of TRANSPOSABLE GENETIC ELEMENTS. Integration is thought to occur essentially at random within the target cell genome.

Expression of the provirus

TRANSCRIPTION, RNA SPLICING and TRANSLATION of the integrated viral DNA is mediated by host cell proteins. Variously

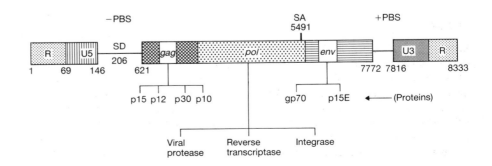

Fig. R12 The RNA genome of MLV as found in the viral particle. SD, donor splice site; SA, splice acceptor; – PBS, tRNA-binding site; + PBS, (+) strand DNA priming site.

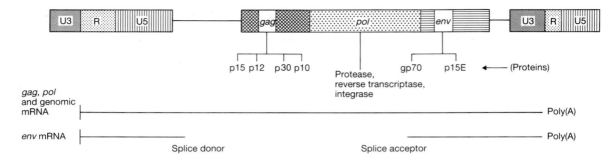

Fig. R14 Proviral DNA genome of MLV integrated into the host cell DNA.

spliced transcripts are generated. In the case of the human retroviruses HIV-1/2 and HTLV-I/II viral proteins are also used to regulate gene expression. The interplay between cellular and viral factors is important in the control of virus latency and the temporal sequence in which viral genes are expressed [8,9].

Classical C-type viruses use a singly spliced messenger RNA (mRNA) to express the envelope proteins. In Moloney murine leukaemia virus (MoMLV), a splice donor signal downstream (i.e. to the 3′ side) of the 5′ LTR is used in conjunction with a major 3′ splice acceptor site upstream (i.e. to the 5′ side) of the *env* sequences [1]. Full length transcripts of viral RNA (initiated at the start of the R region in the 5′ LTR and terminated at a polyadenylation site at the end of the R region in the 3′ LTR) serve both as mRNA for *gag* and *pol* proteins and as genomic RNA for packaging into particles (see Fig. R14). The splicing patterns of HIV-1/2 and HTLV-I/II are more complex, reflecting the greater number of encoded proteins and points of control of viral gene expression [9].

Packaging of viral components

gag and *pol* mRNAs are translated to yield a POLYPROTEIN precursor which is subsequently cleaved, by a viral proteinase encoded at the start of the *pol* region, into the mature proteins (see Fig. R12). The *env* gene is also usually translated as a polyprotein precursor which is cleaved to give two proteins

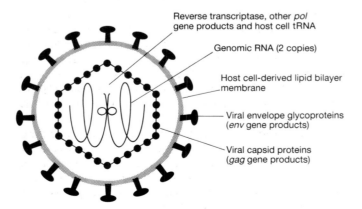

Fig. R13 Structure of a retroviral particle.

that are glycosylated by cellular enzymes. Full length RNA genomic transcripts (see Figs R11 and R12) are packaged into viral particles, following recruitment of the viral structural proteins. Packaging into virion particles requires an RNA signal known as the Psi sequence, which lies just 3′ of the 5′ LTR and therefore downstream of the donor splice site. This ensures that any spliced messages will not include the Psi site, and so will not be packaged. Hence, only full length genomic transcripts become packaged into virions [10].

Transmission

Retroviruses can be transmitted horizontally and vertically (*see* ANIMAL VIRUS DISEASE). Efficient infectious transmission of retroviruses requires the expression on the target cell of receptors which specifically recognize the viral envelope proteins, although viruses may use receptor-independent, nonspecific routes of entry at low efficiency. In addition, the target cell type must be able to support all stages of the replication cycle after virus has bound and penetrated. Vertical transmission occurs when the viral genome becomes integrated in the germ line of the host. The provirus will then be passed from generation to generation as though it were a cellular gene. Hence endogenous proviruses become established which frequently lie latent, but which can become activated when the host is exposed to appropriate agents.

Pathogenesis

Retroviruses have been linked to a wide range of diseases, including anaemia, neurological disorders, immune suppression, and malignancy.

Oncogenic transformation

The oncoviruses (often called the RNA tumour viruses) have been subdivided into two groups of pathogens, namely the acutely transforming and slow transforming retroviruses (Table R4).

Acutely transforming retroviruses. This group can transform cultured cells (*see* TRANSFORMED CELLS) and can cause disease rapidly in susceptible animals. These viruses usually carry an ONCOGENE (v-*onc*) within the viral genome, which is directly responsible for their tumorigenicity, and which is different in each type of virus. The viral oncogenes have been derived from cellular genes that the viruses have acquired, probably as a result

Table R4 Examples of acute- and slow-transforming retroviruses and some of the cellular genes they activate oncogenically by retroviral transduction and insertional mutagenesis respectively

Acutely transforming		Slow-transforming	
Retrovirus	Transduced cellular proto-oncogene	Retrovirus	Cellular gene(s) activated by insertional mutagenesis
Rous sarcoma virus (RSV)	c-*src*-1	Avian leukosis virus (ALV)	c-*myc*, *erb*-B, c-*myb*, Ha-*ras*
Abelson murine leukaemia virus (AMLV)	c-*abl*	Gibbon ape leukaemia virus	interleukin-2
Avian erythroblastosis virus (AEV)	c-*erb*-1	Mouse mammary tumour virus (MMTV)	*int*-1, *int*-2, *int*-3, *int*-4
Rat Harvey murine sarcoma virus (MSV)	c-Ha-*ras*-1	Murine leukaemia virus (MLV)	*lck*, *pim*-1, *evi*-1
Rat Kirsten murine sarcoma virus	c-Ki-*ras*-2	Spleen focus-forming virus (SFFV)	*spi*-1
Chicken MC29	c-*myc*	Feline leukaemia virus (FeLV)	c-*myc*
Simian sarcoma virus (SSV)	c-*sis*	Friend murine leukaemia virus	p53, Ha-*ras*, *fms*

The acquisition of the c-*onc* gene by the wild-type retrovirus is usually accompanied by mutation to the coding sequence of the gene to convert it to a viral oncogene (−*onc*). Transduction of c-*onc* also usually removes an essential part of the viral genome so the resulting recombinant, acutely transforming retrovirus cannot replicate autonomously. Therefore, for propagation, coinfection with a nondefective helper virus is required to provide the viral structural proteins in *trans*. RSV is the exception because the *src* oncogene has been transduced into the viral genome outside of the essential viral structural genes, *gag*, *pol* and *env*. RSV is therefore an acutely transforming, replication-competent retrovirus.

of the inclusion of cellular RNA within a viral particle. Subsequent recombination between viral and cellular RNA during reverse transcription leads to the incorporation of the cellular sequences into the viral genome and delivery of this novel unit into the host cell DNA [11]. If the transduced gene normally has a central role in control of cellular growth and differentiation, the changes in coding sequence and/or control of expression that it undergoes on incorporation into the viral genome can render it oncogenic. Such cellular PROTO-ONCOGENES (c-*onc*) may become oncogenic by being placed under novel, virally determined transcriptional control (both quantitatively and temporally), and/or by sustaining critical mutations to the coding sequence [12]. However, full cellular transformation usually requires the expression of v-*onc* in conjunction with other genetic and epigenetic changes within the target cell.

Slow transforming retroviruses. This group does not contain a 'classical' oncogene. The mechanism of transformation is believed rather to involve the insertion of provirus near, or in, the coding region of a cellular proto-oncogene, called insertional mutagenesis. The strong promoter and enhancer sequences within the viral LTRs can exert transcriptional effects from distances of up to several kilobase pairs from the proto-oncogene. The normal regulation of expression of the cellular gene is disrupted, and overexpression or inappropriately timed expression can contribute to transformation.

A hypothesis of receptor-mediated leukaemogenesis has also been proposed, whereby virus binding to cell surface molecules may induce a physiological response, such as a growth signal, which induces a target cell population to proliferate. This increases the chances of subsequent genetic alterations within the cell pool, some of which may lead to transformation.

HTLV-I is also a slow-transforming virus, causally associated with adult T-cell leukaemia (ATL), but it probably promotes T-cell transformation by a different pathway involving virally encoded

regulatory proteins, especially p40^{tax}, which transactivate expression of cellular proto-oncogenes. HIV-1 and 2 have also been implicated in both the direct and indirect promotion of various types of malignancy (such as Kaposi's sarcoma) which present much more frequently in AIDS patients than in the general population. However, the direct role of HIV in malignant transformation remains doubtful as many patients who are immunosuppressed as a result of other infections or treatments (e.g. transplant recipients) also develop tumours at increased rates.

Immunosuppression

The D-type viruses are not aetiologically associated with malignancy, although MPMV was initially associated with a mammary tumour in a rhesus monkey. D-type viruses cause immune suppression in simian primates but by an unknown mechanism. Immune suppression is also a feature of infection by the lentiviruses (e.g. HIV and SIV, *see* IMMUNODEFICIENCY VIRUSES) and variant strains of feline leukaemia virus (FeLV). In infection with HIV and FeLV large amounts of unintegrated proviral DNA have been observed, which may be related to the pathogenesis.

The lentiviruses, including HIV-1/2 and visna virus of sheep, are associated with slow progressive disease leading to immune suppression and neurological disorders. HIV is the widely recognized causative agent of the acquired immunodeficiency disease syndrome (AIDS) [13], although this has been disputed [14]. The immunodeficiency viruses are covered in more detail elsewhere in the encyclopedia.

The spumaviruses are highly cytopathic to cells in culture, producing the characteristic highly vacuolated cytoplasm which gives the viruses their name, but there is little evidence of pathogenicity *in vivo*. Simian foamy virus has caused transient immune suppression in rabbits and there is a single report of virus isolation from patients with thyroiditis.

Other disease associations

Retroviruses have been linked to other diseases, especially neurological disorders. HTLV-I, for example, is associated with tropical spastic paraparesis, a condition similar in some respects to multiple sclerosis. The mechanisms by which retroviruses promote these, and other, pathogenic effects are not yet fully understood, but it seems likely that the discovery of new retroviruses will be accompanied by new insights into disease.

R. VILE

1 Weiss, R.A. et al. (1985) *RNA Tumor Viruses* (Cold Spring Harbor Laboratory, New York).
2 Varmus, H. (1988) Retroviruses. *Science* **240**, 1427–1435.
3 Schulz, T.F. & Weber, J. (1990) The human T-lymphotropic viruses types 1 and 2. In *AIDS and the New Viruses* (Weiss, R.A. & Dalgleish, A.G., Eds) 125–162 (Academic Press, New York).
4 Haseltine, W.A. (1990) Molecular biology of HIV-1. In *AIDS and the New Viruses* (Weiss, R.A. & Dalgleish, A.G., Eds) 11–40 (Academic Press, New York).
5 Hunter, E. & Swanstrom, R. (1990) Retrovirus envelope glycoproteins. In *Curr. Topics Microbiol. Immunol.* **157**, 187–253.
6 Panganiban, A.T. & Fiore, D. (1988) Ordered interstrand and intrastrand DNA transfer during reverse transcription. *Science* **241**, 1064–1069.
7 Brown, P.O. (1990) Integration of retroviral DNA. In *Curr. Topics Microbiol. Immunol.* **157**, 19–48.
8 Cullen, B.R. & Greene, W.C. (1989) Regulatory pathways governing HIV-1 replication. *Cell* **58**, 423–426.
9 Greene, W.C. et al. (1989) HIV-1, HTLV-1 and normal T cell growth: transcriptional strategies and surprises. *Immunol. Today* **10**, 272–278.
10 Linial, M.L. & Miller, A.D. (1990) Retroviral RNA packaging: sequence requirements and implications. In *Curr. Topics Microbiol. Immunol.* **157**, 125–152.
11 Bishop, J.M. (1988) The molecular genetics of cancer. *Leukaemia* **2**, 199–208.
12 Weinberg, R.A. (1989) Oncogenes, antioncogenes and the molecular basis of multistep carcinogenesis. *Cancer Res.* **49**, 3713–3721.
13 Moss, A.R. (1990) Clinical epidemiology of AIDS and HIV infection: what do we expect in the second decade of the epidemic. In *AIDS and the New Viruses* (Weiss, R.A. & Dalgleish, A.G., Eds) 1–10 (Academic Press, New York).
14 Duesberg, P.H. (1987) Retroviruses as carcinogens and pathogens: expectations and reality. *Cancer Res.* **47**, 1199–1220.

reversal potential *See*: EQUILIBRIUM POTENTIAL.

reverse genetics A strategy for studying gene structure and function which has been made possible by the development of techniques of SITE-DIRECTED MUTAGENESIS which enable precisely predetermined mutations to be introduced into cloned DNAs *in vitro*. The mutated DNA is then introduced into a cultured cell or organism, or used in a cell-free system, to study the effects of the mutation.

reverse PCR POLYMERASE CHAIN REACTION technique for selectively amplifying DNA sequences whose flanking regions are unknown.

reverse transcriptase An RNA-dependent DNA POLYMERASE (EC 2.7.7.49) which synthesizes DNA on an RNA template during the life cycle of RETROVIRUSES and other retrotransposable elements (*see* TRANSPOSABLE GENETIC ELEMENTS). The enzyme is part of the viral particle. Cellular reverse transcriptases are suspected but not yet identified. Reverse transcriptase also has DNA-dependent DNA polymerase and RIBONUCLEASE H (RNaseH) activities. Purified reverse transcriptase is used *in vitro* to generate COMPLEMENTARY DNA from polyadenylated mRNAs for use in GENETIC ENGINEERING and other recombinant DNA applications. The crystal structure of reverse transcriptase from the human IMMUNODEFICIENCY VIRUS HIV-1 has been determined. *See also*: CAULIMOVIRUS; PLANT VIRUSES.

Kohlstaedt, L.A. et al. (1992) *Science* **256**, 1783–1790.

reverse transcription The synthesis of DNA on an RNA template, which occurs in the life-cycle of RETROVIRUSES, and possibly in some cellular phenomena (*see* PROCESSED PSEUDOGENES; TRANSPOSABLE GENETIC ELEMENTS). It is mediated by the enzyme REVERSE TRANSCRIPTASE.

reversion The change from a mutant phenotype to the original WILD TYPE phenotype, which is usually due to BACK MUTATION.

revertant An organism in which a particular mutant phenotype has been reverted to WILD TYPE by a second MUTATION that either reverses the original mutation or COMPLEMENTS it. *See also*: SUPPRESSOR MUTATIONS.

Reye syndrome *See*: MITOCHONDRIOPATHIES.

Rf Restorer of fertility. Nuclear genes restoring male fertility to maize lines showing cytoplasmic male sterility. Different types of cytoplasmic male sterility are suppressed by different *Rf* genes. *See*: MITOCHONDRIAL GENOMES: PLANT.

RF REPLICATIVE FORM.

RFLP Restriction fragment length polymorphism. A DNA POLYMORPHISM which results from a loss or creation of a site at which a particular RESTRICTION ENZYME cuts. DNA carrying the different 'allelic' forms will give different sizes of DNA fragments on digestion with the appropriate restriction enzyme. Where a RFLP is closely linked to the defective allele at a disease locus (e.g. in HUNTINGTON'S CHOREA and SICKLE CELL ANAEMIA) it can sometimes be used to detect the presence of the defective gene in a CARRIER or affected foetus even if the disease locus itself has not yet been accurately mapped or cloned (*see* PRENATAL DIAGNOSIS). Polymorphic restriction sites provide useful markers for building up physical and genetic maps (*see* HUMAN GENE MAPPING).

RGD sequence The amino-acid sequence Arg-Gly-Asp, which has been identified as a recognition motif in proteins that bind INTEGRINS. *See also*: EXTRACELLULAR MATRIX MOLECULES.

Rh blood group system The Rh (Rhesus) antigen D is clinically the most important blood group antigen after the ABO antigens because the majority of D-negative people make anti-D (Rhesus factor) following transfusion of D-positive red cells. Anti-D was the commonest cause of severe haemolytic disease of the newborn until the passive administration of anti-D immunoglobulin (often called Rh immunoglobulin) to D-negative mothers was introduced to suppress the primary Rh immunization of the D-negative mother by a D-positive foetus. Rh is a complex system, limited to red cells and their precursors: 46 serological determinants are recognized. The main antigens appear to be controlled by at least two, possibly three, closely linked loci located on the short arm of chromosome 1 in humans. Rh antigens are integral membrane proteins which require lipid for their expression; they are attached to the red cell cytoskeleton and their absence compromises red cell integrity. Cells lacking Rh antigens have abnormal shape and decreased survival. The D antigen is associated with a nonglycosylated membrane polypeptide, M_r of ~30 000, and with polypeptides of M_r ~45 000–100 000 which are glycosylated.

rhabdomyosarcoma Tumour of striated muscle.

Rhabdoviridae Family of enveloped RNA viruses which infect animals and plants. Particles are bullet-shaped or round-ended rods $130–380 \times 50–95$ nm, with surface projections. They contain a ($-$)-sense single-stranded RNA (M_r ~4.0×10^6). This is transcribed into several ($+$)-sense RNAs corresponding to the five virion proteins. Rhabdoviruses include VESICULAR STOMATITIS VIRUS and rabies. *See*: ANIMAL VIRUSES; PLANT RHABDOVIRUS GROUP.

rhamnose *See*: SUGARS.

Rhesus blood group *See*: RH BLOOD GROUP SYSTEM.

rheumatoid factor ANTIBODIES of the IgM class found in patients with rheumatoid arthritis. They are directed against antigenic determinants on IgG antibodies and form IgM–IgG–complement complexes which are deposited in joints.

rhinoviruses Genus of PICORNAVIRIDAE. Like other picornaviruses, the icosahedral virion is composed of a ($+$)-sense single-stranded RNA enclosed in a capsid composed of 60 each of four different protein subunits (VP1, VP2, VP3, and VP4) (Fig. R15), which are derived by proteolytic cleavage of a polyprotein pre-

cursor (*see* Fig. P48 in POLIOVIRUS). Parts of VP1, VP2, and VP3 form the surface of the virion whereas VP4 is internal and binds RNA. Rhinoviruses are responsible for around 50% of cases of the common cold. More than 90 different serotypes of rhinoviruses exist and this has precluded the development of a successful vaccine. The receptor for the major group of rhinoviruses is the intercellular adhesion molecule ICAM-1 (*see* CELL ADHESION MOLECULES). Binding takes place in a cleft on the virus surface which surrounds the VP1 pentamer. The crystal structure of a human virus complexed with its receptor has been determined. *See*: ANIMAL VIRUSES.

Rossmann, M.G. et al. (1985) *Nature* **317**, 145–153.
Olson, N.H. et al. (1993) *Proc. Natl. Acad. Sci. USA* **90**, 507–511.

rhizobacteria Bacteria which inhabit the RHIZOSPHERE.

rhizobia A group of soil bacteria with the ability to induce nitrogen-fixing nodules on the roots of legumes. All are in the α-group of Gram-negative bacteria but within that group they are scattered widely taxonomically and comprise several different genera (*Rhizobium, Azorhizobium, Bradyrhizobium*). *See*: NITROGEN FIXATION; NODULATION.

rhizobitoxine *See*: PLANT PATHOLOGY.

Rhizobium A genus of nitrogen-fixing bacteria. *See*: NITROGEN FIXATION; REGULATION OF GENE EXPRESSION; NODULATION; RHIZOBIA.

rhizoplane The root surface.

rhizosphere The narrow zone of soil surrounding living plant roots which is directly influenced by the activity of the roots.

rho protein, rho factor (ρ) (1) An ancillary protein required for the efficient termination of TRANSCRIPTION by prokaryotic RNA polymerase at rho-dependent terminator sites. *See also*: BACTERIAL GENE EXPRESSION.
(2) *ras*-related GTP-BINDING PROTEINS.

Rhodobacter Genus of purple nonsulphur photosynthetic bacteria, also known as *Rhodopseudomonas. Rhodobacter* is the name currently preferred. Various species have been widely used to study the molecular biology of PHOTOSYNTHESIS in the simpler system afforded by bacteria. The REACTION CENTRE from *Rhodobacter viridis* was the first integral membrane protein to have its high-resolution molecular structure determined by X-ray crystallography. *Rhodobacter capsulata* has been used to study the function of individual amino-acid residues in the reaction centre complex by SITE-DIRECTED MUTAGENESIS. *Rhodobacter sphaeroides* has been the most extensively used for functional studies. The reaction centre has been crystallized and its structure determined to high resolution; it differs from that of *Rhodobacter viridis* in lacking a tightly bound CYTOCHROME subunit.

Deisenhofer, J. et al. (1985) *Nature* **318**, 618–624.

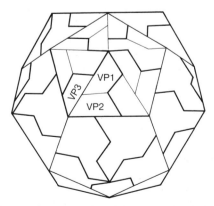

Fig. R15 Diagrammatic representation of the packing of subunits in the icosahedral capsid of rhinovirus HRV14. The thickly outlined unit corresponds to the VP1, VP3, VP0 protomer identified from assembly experiments.

Rhodococcus A genus of Gram-postive nocardioform bacteria. Rhodococci are rod to branched filamentous shaped bacteria which are nonmotile, aerobic, and chemo-organotrophic. Rhodococci are widely distributed in nature and are frequently isolated from soil, freshwater and marine habitats, and from the guts of blood-sucking arthropods. Some species are pathogenic for man and animals and one, *R. fascians*, causes leaf galls in many plants and fasciation in sweet peas. *See*: PLANT PATHOLOGY.

Goodfellow, M. (1984) *Rhodococcus*. In *Bergey's Manual of Systematic Bacteriology* (Sneath, P.H.A. et al., Eds) 1472–1481 (Williams and Wilkins, Baltimore).

Rhodopseudomonas Old name for RHODOBACTER.

rhodopsin Purple protein photoreceptor pigment found in the disk membrane of the rods of the vertebrate retina, which is composed of a protein opsin ($M_r \sim 40\,000$) complexed to the chromophore 11-*cis*-retinal. Isomerization of 11-*cis*-retinal to all-*trans*-retinal by a photon of light (photobleaching) produces a conformational change in the protein, which produces a signal that is transduced through SECOND MESSENGER PATHWAYS to activate an ion channel in the rod outer membrane (*see* VISUAL TRANSDUCTION). Rhodopsin is a member of the seven transmembrane domain family of G PROTEIN-COUPLED RECEPTORS (Fig. R16). *See also*: BACTERIORHODOPSIN.

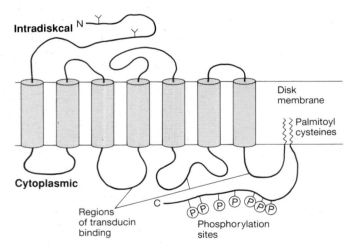

Fig. R16 Model for the topography of rhodopsin in the disk membrane. 11-*cis*-retinal (not shown) is attached by its aldehyde group to the ε-amino group of a lysine residue within the lipid bilayer, forming a protonated Schiff base. It lies within a pocket in the protein with its long axis parallel to the plane of the membrane. Excitation by light causes the isomerization of 11-*cis*-retinal to all-*trans*-retinal which is accompanied by a marked conformational change in the retinal. Photoexcited rhodopsin activates transducin. The Schiff base linkage in rhodopsin then becomes deprotonated and is hydrolysed to yield opsin and 11-*trans*-retinal which dissociates from the protein, which therefore becomes colourless (photobleaching). Rhodopsin is reformed by recombination with 11-*cis*-retinal regenerated in the dark from all-*trans*-retinal. Deactivation of photoexcited rhodopsin also involves phosphorylation by rhodopsin kinase, a serine/threonine PROTEIN KINASE, at C-terminal sites. Phosphorylated rhodopsin then binds the inhibitory protein arrestin, which blocks further binding of transducin.

Rhodospirillum rubrum Purple nonsulphur photosynthetic bacterium widely used to study PHOTOSYNTHESIS.

rhombomeres Neural segments which occur during the development of the vertebrate HINDBRAIN. First described by Orr in 1889, the seven rhombomeres represent the only part of the higher vertebrate central nervous system that is overtly segmented. A specific relationship exists between the site of origin of the cranial nerves and individual rhombomeres. Thus the glossopharyngeal nerve (IX) exits from rhombomere 6, the facial (VII) from 4, and the trigeminal (V) from 2. It is also known that rhombomeres 3 and 5 do not contribute to the cranial NEURAL CREST, a phenomenon associated with ensuring that the migrating neural crest cells find their way to the correct branchial arch. The Hox genes (*see* HOMEOBOX GENES AND HOMEODOMAIN PROTEINS) are known to have their anterior limits of expression at specific rhombomere boundaries. The rhombomere is also an example of a vertebrate lineage restriction unit: once the rhombomere identities are set up, cells cannot migrate across the border into an adjacent rhombomere. *See also*: SEGMENTATION; VERTEBRATE NEURAL DEVELOPMENT.

Ri plasmid Root-inducing plasmid from the bacterium *Agrobacterium rhizogenes*, the agent of HAIRY ROOT DISEASE, and analogous in many respects to the TI PLASMID.

RIA RADIOIMMUNOASSAY.

ribonuclease(s) (RNases) Enzymes involved in the processing or degradation of RNA molecules by hydrolytic cleavage of phosphodiester bonds. Like other NUCLEASES, ribonucleases may cleave an RNA molecule internally (ENDORIBONUCLEASES), or at or near one end (EXORIBONUCLEASES). The degree of specificity for the cleavage site varies between RNases. RNases involved in the general turnover of RNA tend to be less specific than those involved in particular maturation processes, such as tRNA processing (*see* RIBONUCLEASE H; RIBONUCLEASE P) or RNA SPLICING. Unlike the DNA-cutting RESTRICTION ENZYMES, relatively specific RNases do not have multibase sequence requirements; rather they are directed by a conjunction of primary and secondary structures within the RNA molecule. A wide range of RNases are used in studies of RNA structure and RNA–protein interactions. *See also*: BARNASE; MUNG BEAN NUCLEASE; PANCREATIC RIBONUCLEASE; RIBOZYMES; RNA RESTRICTION ENDONUCLEASE; S1 NUCLEASE.

ribonuclease A (RNase A) Mammalian RIBONUCLEASE responsible for routine degradation of cellular RNA. *See*: Fig. P73 in PROTEIN STRUCTURE for illustration of structure.

ribonuclease H (RNase H) RIBONUCLEASE activity which is part of the REVERSE TRANSCRIPTASE molecule. The RNase H activity of retroviral reverse transcriptases is a 3′ or 5′ exonuclease that degrades the viral RNA template after it has been copied into DNA. That associated with cellular reverse transcriptases has endonuclease activity.

ribonuclease L (RNase L) RIBONUCLEASE that is activated after treatment of cells with INTERFERON and which degrades viral and cellular RNAs.

ribonuclease P (RNase P) A ubiquitous ENDORIBONUCLEASE responsible for generating the mature 5′ end of TRANSFER RNA molecules. The enzyme is very specific, using the highly conserved tertiary structure of tRNA as a substrate identifier. RNase P contains a 377-nucleotide RNA component in addition to a protein of M_r 13 800, and the RNA is required for catalytic activity. The RNA portion of the eubacterial enzyme (e.g. that from *Escherichia coli*) can function as a RIBOZYME.

ribonuclease S1 S1 NUCLEASE.

ribonucleic acid *See*: RNA.

ribonucleoprotein A noncovalent complex of RNA and protein which is the natural state of RNA *in vivo*.

ribonucleoprotein-associated fold (RAF) Also called the Rossmann fold, a protein fold associated with RIBOSOMAL PROTEINS that bind rRNA, and other nucleotide-binding proteins such as the NAD⁺-DEPENDENT DEHYDROGENASES.

ribonucleoprotein particle (RNP) A RIBONUCLEOPROTEIN complex that behaves as a discrete particle under the electron microscope or during centrifugal separation. SEDIMENTATION VALUES of RNPs vary between 15 and >200S. The RNA fraction of any particular RNP is presumed to dictate the population of proteins that associate with it to form the particle. The most abundant RNP is the RIBOSOME, containing three different RNAs and >70 different proteins. The large protein fraction of RNPs containing eukaryotic mRNA (mRNPs), also called informosomes, consists almost entirely of two types of protein, M_r 50 000 and 70 000. Also associated with mRNPs are prosomes, a class of small RNPs made up from a variable combination of >20 proteins and one of several pRNAs. *See also*: RNA SPLICING; SIGNAL RECOGNITION PARTICLE.

ribophorin Ribophorins I and II are integral ENDOPLASMIC RETICULUM membrane GLYCOPROTEINS which form part of the ribosome-binding and PROTEIN TRANSLOCATION machinery involved in cotranslational insertion of nascent polypeptides into or through the ER membrane. *See also*: PROTEIN SECRETION; PROTEIN TARGETING.

riboprobe EXPRESSION VECTOR systems for the preparation *in vitro* of radioactively labelled single-stranded RNA PROBES or microgram quantities of defined RNA transcripts from cloned DNA inserts. They usually exploit PROMOTERS from either the SP6 or T7 bacteriophage.

ribose The type of SUGAR found in RNA. DNA contains the 2-deoxy form (deoxyribose).

ribosomal DNA (rDNA) Eukaryotic DNA consisting of a cluster of

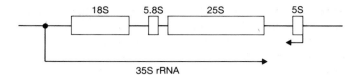

Fig. R17 Organization of the rDNA repeat on chromosome XII of the yeast *Saccharomyces cerevisiae*.

40 to several hundred copies of a repeat containing the tandemly arranged genes for the 5.8S, 18S, 25S, and, in lower eukaryotes, 5S RIBOSOMAL RNA (Fig. R17). Ribosomal DNA corresponds to the NUCLEOLAR ORGANIZER found at the core of the NUCLEOLUS and is often found in extrachromosomal DNA in a form that can be rapidly amplified. *See also*: EUKARYOTIC GENE STRUCTURE; RNA.

ribosomal genes Genes that encode the structural components of the RIBOSOME, that is, the genes for ribosomal proteins and rRNAs.

ribosomal proteins (r-proteins, R-proteins) Abundant, low molecular weight (M_r <35 000), mostly basic proteins intrinsic to the structure and function of the RIBOSOME. The prokaryotic ribosome contains ~52 ribosomal proteins whereas the eukaryotic cytoplasmic ribosome contains ~75. Proteins from the small ribosomal subunit have the prefix 'S' (e.g. S10), those from the large subunit have the prefix 'L' (e.g. L23). Many ribosomal proteins bind directly to the rRNA backbone at sites with specific primary and/or secondary structures. A number are substrates for specific PROTEIN KINASES and PROTEIN PHOSPHATASES (e.g. eukaryotic ribosomal protein S6), that may allow for the fine tuning of PROTEIN SYNTHESIS. A small subset of ribosomal proteins are acidic and can be exchanged on and off the ribosome. Ribosomal protein genes are found in only one or two copies per haploid genome. Their expression is coordinately regulated by various means to produce equimolar amounts of protein that are roughly stoichiometric with rRNA concentrations. The crystal structures of several ribosomal proteins have been solved including those of proteins L6, L9, L7/L12, L30, and S5 and S6. Initially work focused on those which were readily obtainable but with the advent of cloning techniques there is an emphasis on those of greatest interest. Several of the ribosomal proteins that form a complex with the RNA have a similar motif, with an α-helix and a sheet of three β-strands, also seen in the structure of the RNA-binding domain of the small nuclear RNA-binding protein U1 from ribonucleoprotein A. This topology looks like becoming another widely occurring fold — the ribonucleoprotein-associated fold or ROSSMANN FOLD.

ribosomal RNA (rRNA) The major structural component of the RIBOSOME, consisting of three evolutionarily conserved RNA species (see Table R5) that together make up ~80% by mass of the total cellular RNA. The rRNAs have complex secondary and tertiary structures (*see* Fig. R29 in RNA) which create the frame around which RIBOSOMAL PROTEINS can assemble. Prokaryotic rRNAs are first transcribed as a single large precursor RNA which

Table R5 The rRNA composition of large and small ribosomal subunits from prokaryotes and eukaryotes.

Organism	rRNA species in	
	Small subunit	Large subunit
Escherichia coli	16S	23S
		5S
Saccharomyces cerevisiae	18S	25S
		5.8S
		5S
Rat	18S	28S
		5.8S
		5S

is then cleaved to release the individual species. Eukaryotic large and small subunit rRNAs are transcribed by RNA polymerase I and are similarly products of a precursor maturation process with the exception of 5S rRNA, which is transcribed independently yet coordinately with other rRNAs by RNA polymerase III. *See also*: EUKARYOTIC GENE STRUCTURE; RIBOSOMAL DNA; rRNA GENE AMPLIFICATION.

ribosome Large RIBONUCLEOPROTEIN particle which is present in many copies in all cells and which is the site of PROTEIN SYNTHESIS. All ribosomes consist of two subunits of unequal size, the large and small subunit, whose size and composition differ between prokaryotic and eukaryotic cell, although the overall architecture is similar. Ribosomes from the bacterium *Escherichia coli* have a sedimentation coefficient of 70S, a total M_r of ~2 700 000, and a diameter of ~200 Å. They are composed of a large subunit of 50S and a small subunit of 30S. The 50S subunit is made up of 34 different proteins (RIBOSOMAL PROTEINS) termed L1–L34 and the RIBOSOMAL RNAS 23S and 5S rRNA. The 30S subunit contains 21 ribosomal proteins (S1–S21) and a 16S rRNA.

Eukaryotic ribosomes, which occur in the cytoplasm, and in many cells are found clustered at the cytoplasmic face of the ENDOPLASMIC RETICULUM, have a sedimentation coefficient of 80S and M_r 4 200 000. They are composed of a large subunit of 60S and a small subunit of 40S. The large subunit contains three rRNAs (5S, 28S, and an rRNA unique to eukaryotes, 5.8S rRNA), and 50 proteins, L1–L50. The small subunit contains proteins S1–S33 and an 18S rRNA.

Each ribosomal subunit has been shown to be able to self-assemble *in vitro* from its constituent proteins and RNA. The rRNA makes up almost two-thirds of the ribosome mass, and is essential for ribosome structure and function. rRNAs are folded by internal base-pairing into complex three-dimensional conformations (*see* RNA). Ribosomal architecture (Fig. R18) has been studied by electron microscopy, including immunoelectron microscopy, which provides information on surface features, by neutron scattering and diffraction, and by cross-linking studies, both of which provide information on the location of ribosomal proteins relative to each other (Fig. R19). The sequences of all the rRNAs have been determined, and FOOTPRINTING and primer extension studies have identified possible protein binding sites on

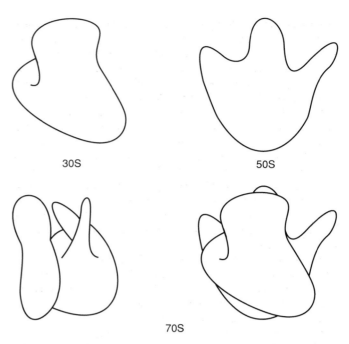

30S   50S

70S

Fig. R18 Proposed architecture of the *E. coli* 70S ribosome and its subunits. The mRNA binding site and the 3′ end of the 16S RNA are located on a plateform between the upper and lower parts of the 30S subunit. The two tRNA binding sites are situated in the cleft between this platform and the upper third of the subunit. On the 50S subunit, the peptidyltransferase site is thought to be located in the valley between two of the protuberances. The finger-like projection contains the GTPase site that powers the movements of tRNAs and mRNA. Adapted from Lake, J.A. (1985) *Annu. Rev. Biochem.* **54**, 507–530.

rRNA, enabling integrated models of the 30S subunit to be constructed. Crystal structures of several ribosomal proteins — L6, L7, L9, L12, and L30, and S5 and S6 — have so far been determined.

Binding sites for mRNA and aminoacyl-tRNAs are located on the 30S subunit. Prokaryotic mRNA initially binds to the ribosome by base-pairing via the SHINE–DALGARNO REGION with the extreme 3′ end of the 16S RNA. There are two binding sites for charged tRNAs carrying an amino acid, the A-site at which the incoming aminoacyl-tRNA binds, and the P-site to which the

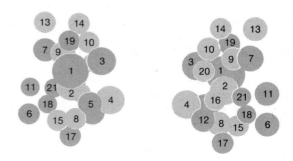

Fig. R19 The arrangement of proteins in the 30S subunit of *E. coli*. Left, cytoplasmic face; right, subunit interface side. From Moore, P.B. (1988) *Nature* **331**, 223–227.

tRNA translocates after the growing polypeptide chain has been transferred to it (*see* PROTEIN SYNTHESIS). Streptomycin, an antibiotic which interferes with the fidelity of translation by prokaryotic, but not eukaryotic ribosomes, interacts with the 16Sr RNA, as does spectinomycin, an antibiotic which inhibits the translocation of peptidyl-tRNA from the A-site to the P-site.

The peptidyl transferase activity that catalyses the formation of peptide bonds is located on the 50S subunit as is the site of GTPase activity which is essential for the movement of mRNA and tRNAs on the ribosome (*see* GTP-BINDING PROTEINS; PROTEIN SYNTHESIS). The exact site of peptidyl transferase activity has not yet been determined, but by analogy with RIBOZYMES it is possible that enzymatic activity is due to the RNA rather than to protein. The exit site by which the growing polypeptide chain leaves the ribosome is positioned on the opposite side of the 50S subunit to the GTPase site.

In eukaryotes, ribosomes are assembled in the NUCLEOLUS from rRNAs transcribed in the nucleolus and ribosomal proteins imported from the cytoplasm. Assembled ribosomes are then exported from the nucleus to the cytoplasm.

rRNAs show highly conserved secondary structure and their sequences have evolved relatively slowly. They are thus much used to study the evolutionary relationships of distantly related groups such as prokaryotes and eukaryotes, and the various groups within the prokaryotes (*see* MOLECULAR PHYLOGENY).

Hill, W.E. et al. (Eds) (1990) *The Ribosome: Structure, Function and Evolution* (American Society for Microbiology, Washington DC).
Noller, H.F. (1991) *Annu. Rev. Biochem.* **60**, 191–227.

ribosome-binding site The site in prokaryotic mRNA which binds to the ribosome. *See:* BACTERIAL GENE EXPRESSION; PROTEIN SYNTHESIS; SHINE–DALGARNO SEQUENCE.

ribosylation *See:* ADP-RIBOSYLATION.

Ribozymes

RIBOZYMES are catalytic RNA molecules that can promote specific biochemical reactions without the need for ancillary proteins. The RNA-catalysed reactions can either be intramolecular (autocatalytic), for example self-splicing or self-cleaving, or intermolecular, using other RNA molecules as substrates and involving multiple turnovers of the ribozyme. This latter case is an example of a true enzymatic reaction as the catalyst (the ribozyme) is recovered unchanged after each reaction and can thus catalyse many reactions.

There are four major classes of ribozyme: (1) and (2) the self-splicing group I and group II introns; (3) the RNA moiety of the ubiquitous enzyme ribonuclease P (RNase P); and (4) the self-cleaving RNAs from viroids and satellite virus RNAs (Figs R20–R23). Only RNase P RNA is thought to be naturally enzymatic, but most of the other ribozymes can be made to have enzymatic activity. Because other cellular processes have essential RNA components (e.g. guide RNAs in RNA EDITING, ribosomal RNAs (rRNAs) in translation (*see* PROTEIN SYNTHESIS), and small

nuclear RNAs (snRNAs) in nuclear messenger RNA (mRNA) splicing (*see* RNA SPLICING), it is likely that other catalytic RNAs will be discovered.

Ribozymes promote cleavage, and ligation in some cases, of specific phosphodiester bonds at rates $\sim 10^{10}$-fold higher than the uncatalysed reaction. The rates are generally lower than similar reactions catalysed by protein enzymes, but this seems to be due to a rate-limiting step for product release rather than a rate-limiting chemical step. Typical turnover numbers are in the range 0.5–2.0 molecules per minute. All the reactions proceed through a displacement reaction that involves the nucleophilic attack of a ribose hydroxyl or of water. The ribozyme probably promotes the reaction by making a proton from the attacking nucleophile more labile or by stabilizing the transition state. At least some ribozymes are thought to be metalloenzymes. The reaction products of group I and group II introns and of RNase P RNA have 5′ phosphates and 3′ hydroxyls. The reaction products of the self-cleaving RNAs have 5′ hydroxyls and 2′,3′-cyclic phosphates. All the ribozymes require a divalent cation (generally magnesium). As with protein enzymes, the reaction specificity and the catalytic activity are conferred by the secondary and tertiary structure of the RNA.

Molecular manipulation of the various ribozymes is yielding 'designer ribozymes' that can be used, for example, as RNA RESTRICTION ENZYMES for molecular biological analyses. Ribozymes are also being studied as possible prophylatic or therapeutic agents [1]. Because the replication of some subviral agents such as HEPATITIS DELTA VIRUS and certain plant viruses probably proceeds through a ribozyme-catalysed step, the ribozyme itself is a potential target for therapeutic agents that disrupt its activity. The observation that RNA can serve both as a replication template and as a catalyst has also fuelled speculations about RNA as the primordial self-replicating molecule (*see* MOLECULAR EVOLUTION; RNA WORLD).

Group I introns

Introns are noncoding sequence elements that interrupt the coding sequences (exons) in many eukaryotic genes. They are removed from the primary RNA transcript by a precise cleavage-ligation reaction, called splicing. There are four classes of introns, each with a distinct splicing pathway — group I, group II, transfer RNA (tRNA), and nuclear mRNA (*see* RNA SPLICING). So far only group I and group II introns are known to have ribozyme activity.

Group I introns are characterized by four conserved sequence elements and by the ability to be folded into a highly characteristic secondary structure (Fig. R20 and reviewed in [2]). Some of these introns are self-splicing *in vitro* and it seems likely that all contain, or once contained, some catalytic activity. The best studied group I intron is that contained within the rRNA precursor from the protist *Tetrahymena thermophila*, but all those characterized have the same splicing pathway.

The intron promotes a two-step transesterification reaction in which the 3′ hydroxyl of a guanosine cofactor makes a nucleophilic attack at the phosphodiester bond at the 5′ splice site (exon–intron boundary) to form a free 5′ exon and an intron–3′ exon intermediate containing a phosphodiester bond to the co-

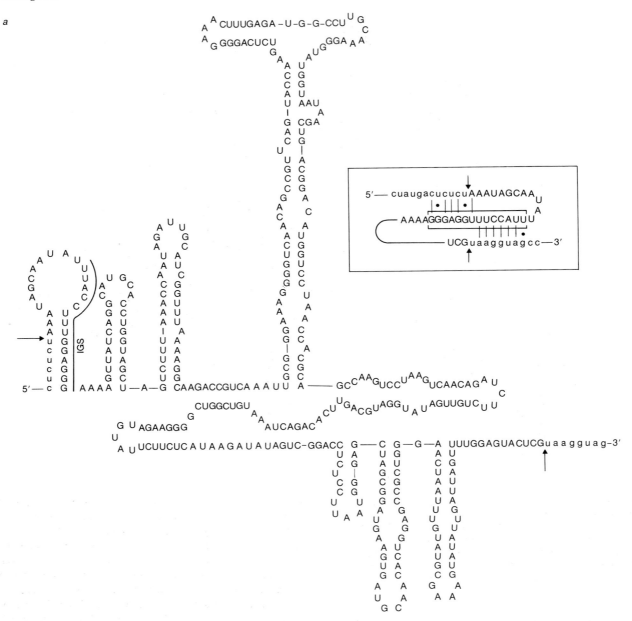

Fig. R20 Proposed RNA secondary structure for the group I rRNA intron from *Tetrahymena thermophila*, one of the four main classes of ribozyme. The other classes are illustrated in Figs R21–R23. The structures are all based on phylogenetic comparisons and computer modelling, but in most cases experimental evidence has been obtained for the interactions. Arrows indicate 5′ and 3′ splice sites. The internal guide sequence (IGS) is overlined. The insert shows the IGS (boxed) and the base-paired exon sequences (lower case). Adapted from [6].

factor guanosine. In the second step, the free 3′ hydroxyl of the 5′ exon makes a nucleophilic attack at the 3′ splice site to form the ligated exons and release the intron with the guanosine cofactor. The net number of phosphodiester bonds is unchanged and no

external source of energy is required. Under the appropriate conditions, the reverse reaction has been demonstrated.

The exons are held in precise alignment for the phosphotransfer reactions by specific interactions with the intron that involve elements of an internal guide sequence (IGS), which positions the 5′ exon and to a lesser extent the 3′ exon. In addition the intron contains a specific binding site for the guanosine cofactor. The released *Tetrahymena* intron can undergo a further intramolecular reaction in which the free 3′ hydroxyl of the intron makes a nucleophilic attack at the phosphodiester bond either 15 or 19 nucleotides from the 5′ end of the intron to form a circular molecule and release a short oligomer (15 or 19 nucleotides long) containing the noncoded guanosine.

A shortened form of the *Tetrahymena* intron (L-19) lacks a

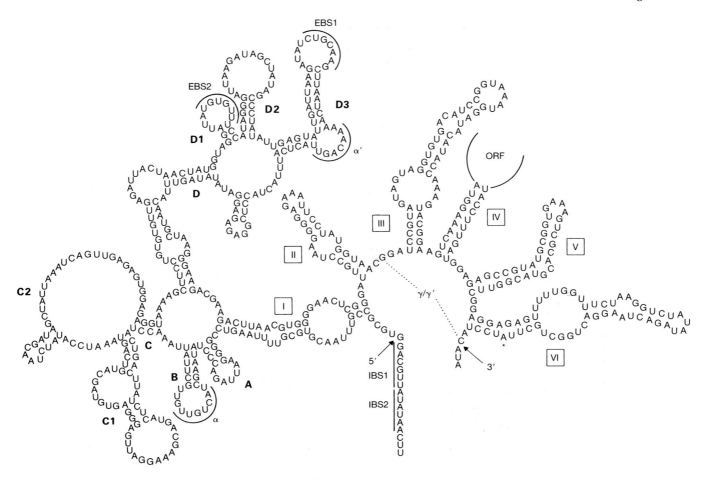

Fig. R21 The group II intron structure of the cytochrome oxidase subunit I RNA from the filamentous fungus *Podospora anserina*. Domains of core secondary structure are numbered I–VI. Domain I is composed of subdomains labelled A, B, C, and D, which are further subdivided as indicated. EBS and IBS represent exon and intron binding sites respectively. Other tertiary interactions are designated α, α', γ and γ'. *, The adenosine involved in the splicing reaction. Arrows indicate the 5' and 3' splice sites. From [7].

substrate for any intramolecular reactions but retains catalytic activity, being able to catalyse multiple reactions on added substrates. L-19 is thus a true enzyme. It can function as a PHOS-PHOTRANSFERASE and ACID PHOSPHATASE, as a sequence-specific ENDORIBONUCLEASE (Fig. R24), as a NUCLEOTIDYLTRANSFERASE, and as an RNA LIGASE. It also has some activity as a sequence-specific endonuclease acting on single-stranded DNA. All these activities can be thought of as variations of the forward and reverse reactions in self-splicing. The endoribonuclease specificity can be altered by changing the nucleotide sequence within the IGS.

Group II introns

Group II introns are characterized by conserved sequence elements and by the ability to be folded into a characteristic

secondary structure that is distinct from that of group I introns (Fig. R21 and reviewed in [3]). Some are self-splicing *in vitro*. Like group I introns, group II introns are spliced by a two-step transesterification reaction. However, the reaction is initiated by the nucleophilic attack of the 2' hydroxyl of an adenosine residue, contained within the intron sequence, at the phosphodiester bond at the 5' splice site; the 5' exon is freed and an intron-3' exon, branched, circular intermediate is formed called a lariat. The lariat contains 3',5'- and 2',5'-phosphodiester bonds at the branch site. In the second step, the 3' hydroxyl of the 5' exon makes a nucleophilic attack at the 3' splice site to form the ligated exons and release the intron, which is still in the form of a lariat. The net number of phosphodiester bonds is unchanged and no external energy source or cofactor is required. The reverse reaction has been shown. The self-splicing group II introns can catalyse intermolecular (*trans*-splicing) reactions, but enzymatic (turnover) activity has not yet been demonstrated.

Ribonuclease P RNA

Ribonuclease P (RNase P) is a ubiquitous endoribonuclease that removes the 5'-terminal ends of precursor tRNA (pre-tRNA) as one of several steps involved in tRNA maturation. It consists of an RNA–protein complex. Experiments *in vitro* have shown that the

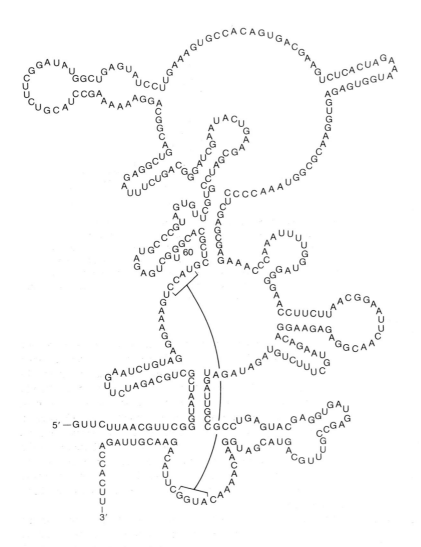

Fig. R22 Secondary structure model for Rnase P RNA from the eubacterium *Bacillus subtilis*. Some potential tertiary interactions are shown. From [4].

RNA moiety of RNase P from EUBACTERIA functions as a ribozyme (Fig. R22 and reviewed in [4]). It requires higher monovalent and divalent cation concentrations than the HOLO-ENZYME, but the kinetics of the reactions are similar. Thus, the small and basically charged RNase P protein is probably only involved in shielding electrostatic repulsions between the negatively charged substrate and enzyme. The protein has no detectable enzymatic activity.

Eubacterial RNase P RNA is the only ribozyme known to act as an enzyme (i.e. to catalyse multiple reactions) *in vivo*. The one-step reaction involves the hydrolysis of the phosphodiester bond at the cleavage site by water. The products have 5′ phosphates and 3′ hydroxyls. No energy source or cofactors are required. The ribozyme recognizes the coaxial stacking of the tRNA acceptor stem and the TΨC arm (*see* TRANSFER RNA). Ribozyme activity has not been demonstrated for RNase P RNA from EUKARYOTES, ARCHAEBACTERIA and cellular organelles (mitochondria and chloroplasts), which cannot be folded into the same RNA secondary structure as that from eubacteria.

Self-cleaving RNAs

A number of small RNAs pathogenic to plants and animals undergo a site-specific autocatalysed cleavage reaction *in vitro* [5]. These catalytic RNAs are classified into four main groups: (1) the VIROIDS (independently replicating, circular, rod-shaped plant pathogens); (2) VIRUSOIDS (encapsidated viroid-like satellite RNAs, which require a helper virus for replication); (3) linear satellite RNAs (similar to virusoids but encapsidated as linear molecules); and (4) HEPATITIS DELTA VIRUS (an animal pathogen that requires a helper virus for replication). These RNAs probably replicate by a ROLLING CIRCLE mechanism. The multimeric linear molecules that are produced are thought to be processed into unit-size progeny by the self-cleavage reaction *in vivo*. In addition, the RNA transcript of the newt *Notophthalmus viridescens* satellite II DNA (not directly related to the pathogenic satellite RNAs) undergoes a self-cleavage reaction *in vitro*. The most common structural motif contains 13 conserved nucleotides that are present within a secondary structure domain of ~30 nucleotides, which is known

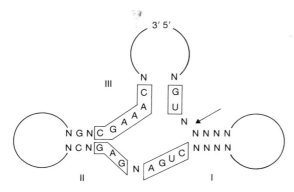

Fig. R23 Hammerhead structure from the self-cleaving RNAs. Conserved nucleotide sequences are shown boxed, nonconserved nucleotides are designated by an N, structural domains are indicated by I, II, and III, and the cleavage site is shown by the arrow. From [5].

as a hammerhead (Fig. R23). This domain is necessary and sufficient for catalytic activity *in vitro*. The self-cleaving domain derived from the genomic and antigenomic RNA from hepatitis delta virus consists of an 85-nucleotide region that can be folded into a PSEUDOKNOT-like structure. The minus strand of tobacco ringspot virus satellite RNA contains a catalytic domain of 50

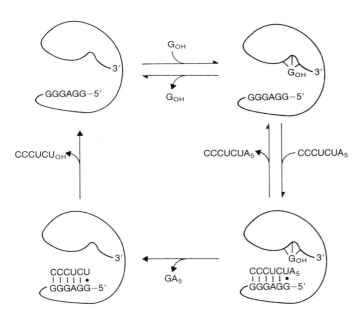

Fig. R24 The *Tetrahymena* L-19 as an endoribonuclease. The element of the IGS that binds the 5′ exon sequence is shown as 5′-GGAGGG-3′. This element can bind a substrate molecule containing the sequence 5′-CCCUCU-3′. In the scheme shown, an oligonucleotide with the sequence 5′-CCCUCUAAAAA-3′ binds to the IGS sequence. The 3′ hydroxyl of a noncovalently bound guanosine cofactor then makes a nucleophilic attack at the phosphodiester bond between U6 and A7, and a short oligonucleotide is released with the guanosine cofactor covalently attached at the 5′ end (GAAAAA). The cleaved substrate is released and L-19 is ready for another reaction. Substrate and guanosine cofactor bind the L-19 independently; the order of binding is shown arbitrarily. Larger RNAs will also be cleaved if they contain the consensus sequence. The IGS can be altered to change the substrate specificity. From [8].

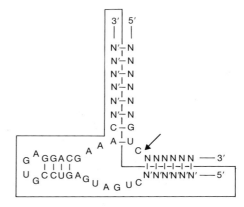

Fig. R25 The self-cleaving hammerhead domain as a restriction endoribonuclease. The enzyme is shown as the boxed sequence, and N′ signifies residues that can be changed to alter the specificity for substrate. The substrate contains the sequence $(N)_6GUC(N)_6$, where N can be any nucleotide and GU are conserved nucleotides; the C at the cleavage site (arrow) is not conserved *in vivo*, but it improves the cleavage reaction *in vitro*. See text for further details. Adapted from [9].

nucleotides, known as a HAIRPIN structure, and a 14-nucleotide substrate domain. The catalytic and substrate domains base-pair together to form two short intramolecular stems. Another self-cleaving RNA has been identified from the mitochondrial plasmid transcibe of *Neurospora* (VS RNA). The catalytic motif is unknown but it appears to be unrelated to the other self-cleaving RNAs. All the self-cleaving RNAs undergo a one-step reaction that involves the nucleophilic attack of the 2′ hydroxyl on the ribose ring at the phosphodiester bond to be cleaved to form products with 5′ hydroxyls and 2′,3′-cyclic phosphates. No cofactors or energy source are required. In some cases the reverse reaction has been demonstrated.

The intramolecular autocatalytic cleavage reaction can be converted into an intermolecular enzymatic reaction by providing the enzymatic activity and substrate (the sequence to be cleaved) as two separate RNA molecules that hybridize to form the catalytically active structure (*see* HYBRIDIZATION). Each enzyme unit remains intact after the reaction and consequently is able to cleave multiple units of substrate. This ribozyme can also be engineered to make a highly specific restriction endoribonuclease (Fig. R25).

N.K. TANNER

See also: MECHANISMS OF ENZYME CATALYSIS.

1 Cech, T.R. (1988) Ribozymes and their medical implications. *J. Am. Med. Assoc.* **260**, 3030–3034.
2 Cech, T.R. (1990) Self-splicing of group I introns. *Annu. Rev. Biochem.* **59**, 543–568.
3 Michel, F. et al. (1989) Comparative and functional anatomy of group II catalytic introns — a review. *Gene* **82**, 5–30.
4 Pace, N.R. & Smith, D. (1990) Ribonuclease P: function and variation. *J. Biol. Chem.* **265**, 3587–3590.
5 Symons, R.H. (1989) Self-cleavage of RNA in the replication of small pathogens of plants and animals. *Trends Biochem. Sci.* **14**, 445–450.
6 Burke, J.M. et al. (1987) Structural conventions for group I introns. *Nucleic Acids Res.* **15**, 7217–7221.
7 Schmidt, U. et al. (1990) Self-splicing of the mobile group II intron of the

filamentous fungus *Podospora anserina* (COI II) in vitro. *EMBO J.* **9**, 2289–2298.

8 Herschlag, D. & Cech, T.R. (1990) DNA cleavage catalyzed by the ribozyme from *Tetrahymena*. *Nature* **344**, 405–409.

9 Haseloff, J. & Gerlach, W.L. (1988) Simple RNA enzymes with new and highly specific endoribonuclease activities. *Nature* **334**, 585–591.

ribulose *See:* SUGARS.

ribulose bisphosphate *See:* CALVIN CYCLE; SUGARS.

D-ribulose-1,5-bisphosphate carboxylase/oxygenase (RUBISCO)

The carboxylase (EC 4.1.1.39) responsible for virtually all photosynthetic CO_2 fixation; the most abundant protein in leaves and probably of the biosphere. It catalyses the reaction

$$\text{D-ribulose-1,5-bisphosphate} + CO_2 \rightarrow 2 \; \text{3-phospho-D-glycerate}$$

In most photosynthetic organisms, both eukaryotic and prokaryotic, it is a protein of M_r 560 000 made up of eight identical large subunits (LSU, $M_r \sim 53\,000$), which contain the catalytic sites, and eight identical small subunits (SSU, $M_r \sim 14\,900$) of unknown function. The enzyme from the bacterium *Rhodospirillum rubrum*, however, is a homodimer of large subunits, and some other members of the Rhodospirillaceae (including *Rhodobacter*) contain two types of enzyme: type I, which is the normal L_8S_8 form, and type II, a smaller homo-oligomer of large subunits. The crystal structure of the dimer from *R. rubrum* has been determined and that of a plant L_8S_8 enzyme at 3 Å resolution. The L subunits of bacterial and plant RUBISCO are of similar structure, each containing three distinct domains: N-terminal, a central barrel, and a small helical C-terminal domain. L subunits are arranged in pairs with the catalytic sites at the interface between two L subunits. The L_8S_8 molecule is keg-shaped with pairs of L subunits surrounding a central axis and tetramers of S subunits at each end (Fig. R26). In all plant species the large subunit is encoded within the CHLOROPLAST GENOME; in some species the small subunits are encoded in the nuclear genome and imported into the chloroplast (*see* PROTEIN TRANSLOCATION). Assembly of RUBISCO requires chaperone proteins (*see* MOLECULAR CHAPERONES). All forms of the enzyme have an apparently wasteful oxygenase activity associated with the same active centre. *See also:* CALVIN CYCLE; PHOTOSYNTHESIS.

ricin A LECTIN from castor oil plant *Ricinus communis*, which is highly toxic to animal cells. It consists of two subunits, one of which is the lectin and binds to the cell surface, enabling the other, toxic, subunit to enter the cell.

Rieske iron–sulphur centre/protein A high-potential [2Fe–2S] centre originally described in terms of its characteristic EPR spectrum in mitochondria by Rieske where it occurs in the cytochrome bc_1 complex, and which also occurs in thylakoid membranes in the cytochrome bf complex. The protein of chloroplasts (M_r 20 000) is nuclear encoded and has its redox centre on the luminal side of the membrane near the C terminus. There is probably a single membrane-spanning helix near the N terminus. *See:* ESR/EPR SPECTROSCOPY; ELECTRON TRANSPORT CHAIN; IRON-SULPHUR PROTEINS; PHOTOSYNTHESIS.

rifampicin Synthetic analogue of the RIFAMYCIN group of antibiotics.

rifamycin Quinone antibiotic produced by *Streptomyces mediterranei* which inhibits prokaryotic TRANSCRIPTION by binding to the β subunit of RNA polymerase and preventing the formation of phosphodiester bonds (Fig. R27).

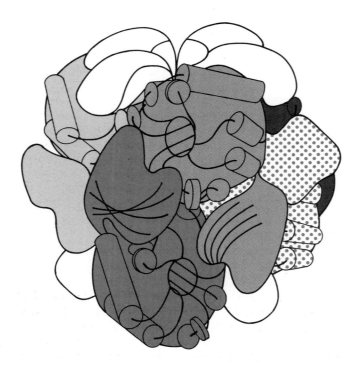

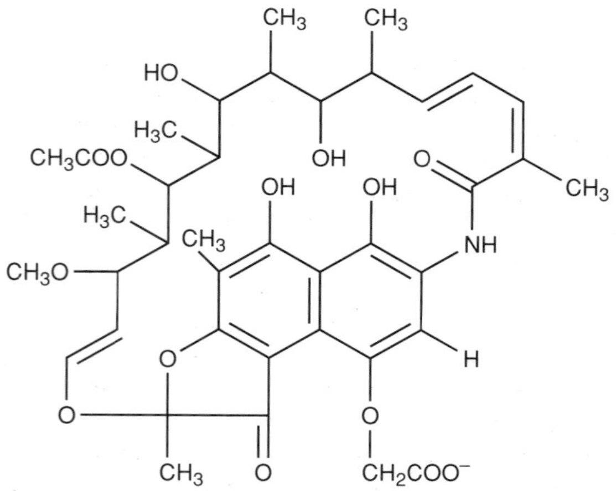

Fig. R26 Schematic drawing of RUBISCO from tobacco chloroplasts, to illustrate the spatial arrangement of subunits. The large subunits are variously shaded, the small subunits are unshaded. The small helical domains are omitted for clarity. Adapted from Chapman, M.S. et al. (1987) *Nature* **329**, 354–356.

Fig. R27 Structure of rifamycin.

rishitin A sesquiterpenoid PHYTOALEXIN of the Solanaceae, which has been especially well studied in potato tubers. Rishitin, which is synthesized via the acetate–mevalonate pathway, was the first phytoalexin identified. Its role in inhibiting fungal growth in potato is uncertain, as rishitin is not found in potato leaves where resistance is effectively expressed.

RNA (ribonucleic acid)

THE nucleic acid RNA is a polymer of monoribonucleotides which is found in all cells and in some viruses. Both prokaryotic and eukaryotic cells contain messenger RNA (mRNA), transfer RNA (tRNA) and ribosomal RNA (rRNA), all of which are involved in various ways in the accurate conversion of genetic information (in the form of DNA) into protein (*see* PROTEIN SYNTHESIS). This process in eukaryotic cells also involves two additional classes of RNA — heterogeneous nuclear (or pre-messenger) RNA (hnRNA or pre-mRNA) and small nuclear RNAs (snRNA).

Apart from its role in the translation of genetic information, RNA is also part of the SIGNAL RECOGNITION PARTICLE, which is involved in the translocation of proteins into the endoplasmic reticulum (*see* PROTEIN TRANSLOCATION), and part of the bacterial enzyme RIBONUCLEASE P. Also, RNA PRIMERS are essential in the replication of DNA (*see* DNA REPLICATION), some RNA molecules are involved in the regulation of gene expression or plasmid copy number (*see* ANTISENSE RNA), and some RNAs have catalytic function (*see* RIBOZYMES).

The genomes of some viruses are composed of RNA (*see* ANIMAL VIRUSES; BACTERIOPHAGES; PLANT VIRUSES; RETROVIRUSES), where it functions as the primary carrier of genetic information.

All cellular RNAs are synthesized by the process of TRANSCRIPTION, in which a complementary RNA copy of a DNA template is made. In the replication of many RNA viruses, however, RNA itself is the template for the synthesis of new viral RNA. Most classes of RNA molecule (with the exception of tRNA) occur naturally as complexes with specific proteins, forming RIBONUCLEOPROTEIN PARTICLES (RNP).

Structure

All RNAs have a similar basic chemical structure (*see also* NUCLEIC ACID STRUCTURE). Each monoribonucleotide unit consists of a base glycosidically linked to a phosphorylated ribose sugar (*see* NUCLEOSIDES AND NUCLEOTIDES). These units are covalently linked through 3′,5′-phosphodiester bonds into an unbranched single-stranded chain (Fig. R28). The four major bases in RNA are the purines adenine and guanine, and the pyrimidines cytosine and uracil (A, G, C, and U, respectively). The thymine nucleoside (T) in DNA is transcribed into U in RNA. Minor bases occur in some RNAs, generally as a result of covalent post-transcriptional modifications. For example, tRNA contains a number of minor bases including *N,N*-dimethylguanine, dihydrouracil, ^6N-iso-pentenyladenine and 4-thiouracil [1]. Also, methylation of a small proportion of A residues (to form methyl^{6}A) occurs in some mammalian mRNAs.

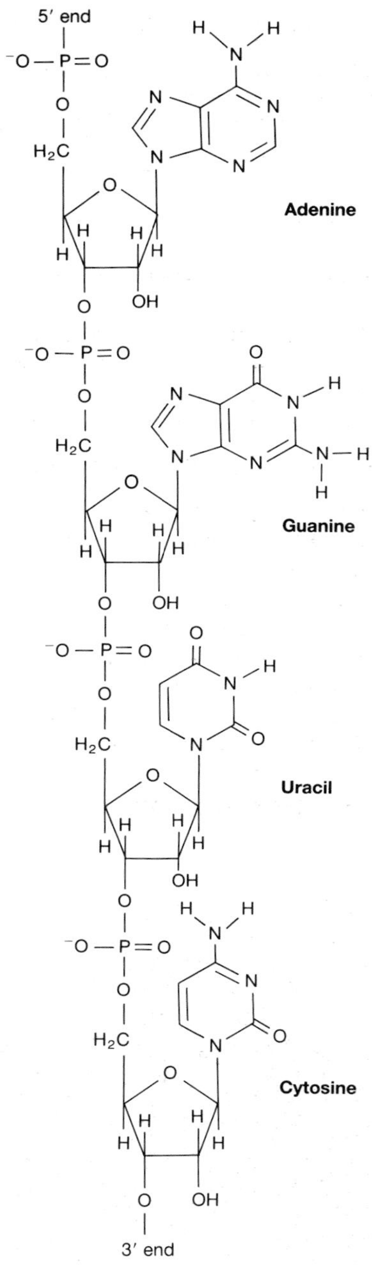

Fig. R28 Chemical structure of RNA.

Regions of complementarity within a single-stranded RNA sequence can base-pair with one another to generate double-stranded regions of SECONDARY STRUCTURE (hairpin loops). U base-pairs with A, and G with C, via hydrogen bonding, and base-stacking interactions can contribute to the thermodynamic stability of these double-stranded regions. Guidelines for the estimation of the thermodynamic stability of RNA secondary structures [2], and computer programs for the prediction of RNA secondary structure formation [3] are available. Further folding can take place to form complex TERTIARY STRUCTURES. The cloverleaf structure of the tRNA molecule is possibly the best defined RNA secondary structure, and the tertiary structure of tRNA has now been resolved by X-ray crystallography [4] (*see* TRANSFER RNA). In some cases single-stranded loops at the head of the double-stranded stem within the hairpin can base-pair with sequences outside of the hairpin to form more complex structures called PSEUDOKNOTS.

Messenger RNA

Messenger RNA is a temporary complementary copy of the SENSE STRAND (anticoding strand) of protein-coding DNA. In eukaryotes, this major intermediate in gene expression is transcribed from protein-coding genes by RNA polymerase II (*see* GENE; RNA POLYMERASE; TRANSCRIPTION;). It is usually transcribed as a relatively long pre-mRNA (also called primary transcript or hnRNA) which is then processed, still within the nucleus, to remove INTRONS (*see* RNA SPLICING). Further post-transcriptional modifications are made to most eukaryotic mRNAs to add a 5'-cap structure and a 3'-poly(A) tail (*see* RNA PROCESSING). In addition, there are now a number of examples of non-templated nucleotides being added post-transcriptionally within existing mRNA transcripts (*see* RNA EDITING). The mature mRNA molecule is then transported to the cytoplasm where it is translated into protein on the ribosome (*see* PROTEIN SYNTHESIS).

Messenger RNAs range in length from a few hundred to many thousands of nucleotides depending largely upon the size of the protein they encode. An mRNA contains a region that specifies the protein sequence (the protein-coding region), flanked on either side by untranslated regions (5' and 3' untranslated regions). The coding region represents a sequence of CODONS, each of three nucleotides. The sequence of codons specifies the amino-acid sequence of the polypeptide chain, with each codon representing a particular amino acid (*see* GENETIC CODE).

The primary and secondary structure of the 5' untranslated region strongly influences the frequency with which ribosomes initiate translation on an mRNA. The function of the 3' untranslated region is unknown, but in some cases it affects the stability of the mRNA.

As described above, the synthesis and translation of eukaryotic mRNAs encoded by nuclear DNA are separated into different compartments, being carried out in the nucleus and in the cytoplasm, respectively. (Mitochondrial and chloroplast genes are transcribed and translated within the organelles — *see* CHLOROPLAST GENOME; MITOCHONDRIAL GENOMES.) Prokaryotes on the other hand lack a nuclear compartment, and because their protein-coding genes lack introns, and their mRNAs are not generally subject to post-transcriptional modification, synthesis and translation of prokaryotic mRNAs can occur simultaneously, with translation beginning at the free end of the mRNA as it is being transcribed from the DNA. Because of this close linkage of transcription and translation in prokaryotes, events at the translational level can directly influence transcription, as in the phenomenon of attenuation [5] (*see* BACTERIAL GENE EXPRESSION).

In prokaryotes the synthesis of many functionally related proteins is co-regulated by grouping their coding sequences into an operon, the whole of which is transcribed using a single promoter into a continous polycistronic mRNA encoding more than one polypeptide chain. Almost all eukaryotic mRNAs on the other hand are monocistronic (i.e. they encode a single polypeptide chain) and each eukaryotic protein-coding gene generally has its own PROMOTER region.

These fundamental differences have important implications for the regulation of gene expression in prokaryotes and eukaryotes (*see* BACTERIAL GENE EXPRESSION; BACTERIAL GENE ORGANIZATION; EUKARYOTIC GENE EXPRESSION; EUKARYOTIC GENE STRUCTURE).

Ribosomal RNA

Ribosomal RNA is a major structural component of the RIBOSOME. In the eukaryotic nucleolus, it is copied by RNA polymerase I from ribosomal DNA as a single transcript (45S). This is then processed into the large (25–28S, about 5000 nucleotides) and 5.8S rRNAs (about 160 nucleotides), which are assembled into the large ribosome subunit, and the small rRNA (18S, about 2000 nucleotides) (Fig. R29) which is incorporated into the small ribosome subunit [1]. The large subunit also contains 5S rRNA (about 120 nucleotides) which is transcribed by RNA polymerase III in the nucleus. The 45S rRNA gene and the 5S rRNA gene are present in multiple copies in nuclear DNA.

In prokaryotes, the 16S and 23S rRNAs correspond to the 18S and 28S rRNAs found in the small and large ribosomal subunits of the mammalian cells, respectively. During the initiation of translation in *Escherichia coli*, a conserved sequence near the 3' end of the 16S rRNA interacts with the SHINE–DALGARNO SEQUENCE at the 5' end of the mRNA, thus helping to locate the small ribosomal subunit close to the initiation codon [6] (*see* PROTEIN SYNTHESIS).

Transfer RNA

Transfer RNAs are small RNAs (70–80 nucleotides) which have a key role in translation (*see also* GENETIC CODE; TRANSFER RNA). They act as 'adaptor' molecules, each tRNA molecule having a 3-base ANTICODON complementary to one (or two) of the codons, and also having a site at which the appropriate amino acid is attached. An amino acid is added to its cognate tRNA by an aminoacyl-tRNA synthetase to form the 'charged' tRNA molecule. During translation, the anticodon on the charged tRNA pairs with the codon on the mRNA within the ribosome, and the amino acid is then transferred from the tRNA to the growing polypeptide chain. Accuracy in the mRNA codon : tRNA anticodon interaction, and in the recognition of amino acids and tRNAs by aminoacyl-tRNA synthetases is central to the insertion of the appropriate amino acid into the polypeptide chain during translation. tRNAs in eukaryotes are transcribed by RNA polymerase III, and they contain a number of minor bases which comprise less than 5% of the total base content [1]. (*See* TRANSFER RNA for the structure of a tRNA molecule.) Most tRNA genes are present in multiple copies in nuclear DNA.

a

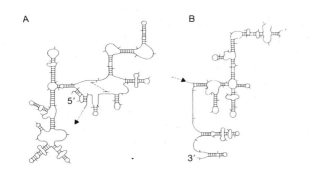

b

c

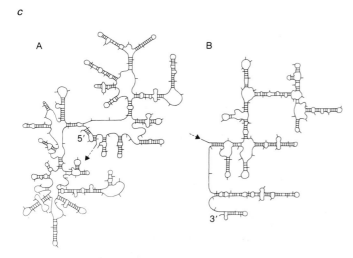

Fig. R29 Comparison of proposed secondary structures of the smaller ribosomal RNA. *a*, 12S rRNA from human mitochondrion; *b*, 16S RNA from *Escherichia coli* (prokaryote); *c*, 18S rRNA from *Xenopus laevis* (eukaryote). Each structure has been separated (at the arrows) into two parts (A and B) to illustrate the similarities between RNAs from different sources. Adapted from [1].

Small nuclear RNAs

Both the accurate removal of introns during splicing, and the post-transcriptional addition of a 3′-poly(A) tail during the expression of most eukaryotic genes, require specific snRNA molecules. These snRNA sequences, which exist complexed with specific proteins in small nuclear riboprotein particles (snRNPs), recognize specific sequences at the cleavage sites at intron/exon junctions, the lariat formation sequences within introns, or polyadenylation signals close to the 3′ end of most eukaryotic mRNAs (*see also* RNA SPLICING; SPLICEOSOMES).

RNA genomes

It has been suggested that DNA genomes have evolved from primordial RNA genomes [7] (*see* MOLECULAR EVOLUTION). Some viruses have RNA genomes. RNA tumour viruses and other retroviruses have single-stranded RNA genomes which are converted into DNA by the enzyme REVERSE TRANSCRIPTASE followed by integration of the DNA 'transcript' into the genome of the host eukaryotic cell (*see* RETROVIRUSES). Other RNA viruses such as TOBACCO MOSAIC VIRUS and POLIOVIRUS have single-stranded RNA genomes which are replicated by an RNA-dependent RNA polymerase. Single-stranded RNA viruses are divisible into (+)-strand and (−)-strand RNA viruses on the basis of whether the genome itself, or a complementary RNA copy, is translated to produce the viral proteins. A third class of RNA viruses, including the reoviruses, has a double-stranded RNA genome (*see* ANIMAL VIRUSES; PLANT VIRUSES).

A.J.P. BROWN

1 Adams, R.L.P. et al. (Eds) (1986) *The Biochemistry of the Nucleic Acids*, 10th revised edn (Chapman & Hall, London).
2 Tinoco, I. et al. (1973) Improved estimation of secondary structure in ribonucleic acids. *Nature* **246**, 40–41.
3 Zuker, M. & Stiegler, P. (1981) Optimal computer folding of large RNA sequences using thermodynamics and auxiliary information. *Nucleic Acids Res.* **9**, 133–148.
4 Quigley, G.J. & Rich, A. (1976) Structural domains of transfer RNA molecules. *Science* **194**, 796–806.
5 Bertrand, K. & Yanofsky, C. (1976) Regulation of transcription termination in the leader region of the tryptophan operon of *Escherichia coli* involves tryptophan or its metabolic product. *J. Mol. Biol.* **103**, 339–349.
6 Shine, J. & Dalgarno, L. (1974) The 3′-terminal sequence of *Escherichia coli* 16S ribosomal RNA: complementarity to nonsense triplets and ribosome binding sites. *Proc. Natl. Acad. Sci. USA* **71**, 1342–1346.
7 Joyce, G.F. (1989) RNA evolution and the origins of life. *Nature* **338**, 217–224.

RNA-activated protein kinase (PKR) Serine/threonine PROTEIN KINASE induced by INTERFERON whose activity is dependent on the presence of double-stranded RNA. It phosphorylates the α subunit of eukaryotic initiation factor eIF-2 thus inhibiting PROTEIN SYNTHESIS. *See*: INTERFERON.

RNA capping *See*: RNA PROCESSING.

RNA-dependent DNA polymerase REVERSE TRANSCRIPTASE. *See* RETROVIRUSES.

RNA-dependent RNA polymerase Replicase. *See*: ANIMAL VIRUSES; BACTERIOPHAGES; PLANT VIRUSES; RNA POLYMERASES.

RNA editing

RNA editing is defined as a process responsible for any differences between the final sequence of a MESSENGER RNA (mRNA) and its genetically determined template, with the exclusion of RNA SPLIC-ING and TRANSFER RNA (tRNA) modification. The first example of RNA editing was observed during the study of the *coxII* transcript in the mitochondria of trypanosomes [1]. It has since been found in organellar transcripts of other organisms, as in *Physarum polycephalum* mitochondria, higher plant mitochondria and chloroplasts, as well as in unique nuclear transcripts in mammals. By extension, phenomena such as the POLYADENYLATION of vertebrate mitochondrial mRNAs to create stop codons, and G (guanosine) insertion in paramyxoviruses can also be considered as RNA editing processes. Table R6 summarizes the different kinds of editing.

Trypanosome mitochondria

In trypanosomes [2], RNA editing involves the specific addition and deletion of Us (uridine) in protein-coding sequences and has been shown to correct FRAMESHIFTS, to generate INITIATION CODONS and even to create entire OPEN READING FRAMES as in the case of the pan-edited *coxIII* RNA of *Trypanosoma brucei* where the reading frame is obtained after the insertion of 558 Us and the deletion of 40 Us encoded by the DNA. RNA editing in trypanosomal mitochondria is post-transcriptional and must proceed unidirectionally from 3′ to 5′ as many partially edited transcripts can be found, always containing an edited 3′ end and an unedited 5′ end.

In spite of the large number of U deletions/insertions, after RNA editing the deduced protein sequences are almost or completely identical between different trypanosomes. Editing seems to be a key event in the regulation of translation (PROTEIN SYNTHESIS). Us are also inserted in 5′ and 3′ noncoding regions. Studies of editing patterns at different stages of the biological cycle of trypanosomes suggest that transcripts are edited in a developmental stage- and transcript-specific manner.

Template information for the specific insertion/deletion of Us is provided by so-called guide RNAs (gRNA). These short RNAs (60–80 nucleotides) are complementary to the edited RNA sequences provided that G : U pairs are allowed. These gRNAs have been shown to be encoded in maxicircles as well as in the minicircles that constitute the trypanosomal mitochondrial genome (*see* MITOCHONDRIAL GENOMES: LOWER EUKARYOTES). gRNAs also contain a nonencoded poly(U) tail of variable length (5–24 Us).

The precise mechanism of mRNA editing in trypanosomes is still unknown. Several models have been proposed in which gRNAs play an important part. The models involve the pairing of the 5′ end of the gRNA (called the anchor segment) with the mRNA and the 3′ to 5′ insertion of Us derived from the poly(U) tail at mismatch positions until a fully complementary gRNA–mRNA duplex has been formed. The exact mechanism of U insertion/deletion is not yet known but may be accomplished either by separate cleavage and ligation events or by a trans-esterification process (Fig. R30).

Higher plant organelles

RNA editing by C to U conversion has been found in mitochondrial transcripts of higher plants [3]. Up to now, all transcripts analysed contain editing sites and this is true for all the plants studied, which are all ANGIOSPERMS. Editing sites tend to be conserved between different plants. More than 300 editing sites have been found, most of them in coding regions. Almost all are conversions of C to U with a few inverse U to C conversions. Some editing events create start or stop codons. A few editing sites have been found in noncoding regions. The vast majority of the editing sites are at the first or second position of the codons, almost always leading to a change in the amino acid encoded. The modification of the predicted protein sequence after RNA editing ranges from <1% to >10%. These modifications have been verified in the direct protein sequence of wheat ATP9. The 'edited' deduced protein sequences are more similar to their homologous non-plant sequences than are the corresponding genomic sequences.

The mRNAs of a number of genes are present in the mitochondria as a pool of partially edited transcripts [4]. Comparison of the editing state in unprocessed, processed and polysomal RNAs (i.e.

Table R6 Different kinds of RNA editing

Type of editing	Co- or post-transcriptional	Mechanism	Information	Transcripts of
U insertions/deletions	Post-	Cleavage/ligation or transesterification	External guide RNAs	Trypanosome mitochondria
C insertions	Post-(?)	?	?	*P. polycephalum* mitochondria
C→U conversion	Post-	Sequence-specific cytidine deaminase	Primary RNA sequence	Mammalian apo-B100 mRNA
A→G conversion	Post-	?	?	Mammalian brain mRNA
C→U and U→C conversions	Post-	?	?	Higher plant mitochondria
C→U conversions	Post-(?)	?	?	Higher plant chloroplasts
A→I conversions	Post-	dsRNA specific	dsRNA	Eukaryotic RNAs
G insertion	Co-	Stuttering of the polymerase	Template sequence	Paramyxoviruses
A addition	Post-	Poly(A) polymerase	?	Vertebrate mitochondria

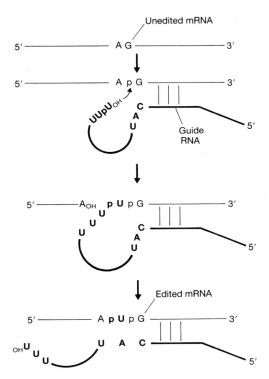

Fig. R30 A possible mechanism for the insertion of uridine into mitochondrial mRNAs in trypanosomes. The guide RNA (sequence in bold) pairs with the unedited mRNA and uridines from the poly(U) tail are inserted into the mRNA at mismatch positions in a two-step transesterification reaction.

RNAs being translated) provides evidence that RNA editing is a post-transcriptional phenomenon and that the final proteins are identical. In contrast to the situation in trypanosomal mitochondria, there is no evident polarity for RNA editing in plant mitochondria and the editing mechanism is unknown.

RNA editing by C to U conversion has also been described in transcripts of the genes coding for *rpl2* and *psbl* from maize and tobacco chloroplasts respectively. These two conversions create an AUG initiation codon from an ACG threonine codon at positions which are conserved as initiation codons in other plants. No partially edited transcript has been described and it seems that RNA editing in chloroplasts, albeit similar to editing in plant mitochondria, might be less extensive.

Mammalian apolipoprotein B mRNA

A unique RNA editing site occurs in the 14-kilobase (kb) RNA coding for the mammalian plasma lipoprotein apolipoprotein B100 (apoB100) (*see* PLASMA LIPOPROTEINS AND THEIR RECEPTORS [5]). This editing event (a C to U conversion at position 6666 of the transcript) changes a glutamine (CAA) codon into a stop (UAA) codon, leading to the synthesis of the short apolipoprotein B48. This process is developmentally regulated, as apoB100 is found in liver whereas apoB48 is present in the intestine in humans.

An *in vitro* system for this specific editing has been developed that enables the mechanism of the C to U conversion to be analysed and the sequence requirements to be determined. The reaction is carried out by a site-specific cytidine deaminase. A 27S proteinaceous macromolecular editing complex (the editosome) has been identified and a protein of M_r 40 000 was also found to bind specifically to apoB100 mRNA around the editing site. There is experimental evidence that the editing enzyme does not contain RNA. The sequences surrounding the edited C are important for the specificity of recognition of the editing site. Sequences 3′ to the editing site are necessary and sufficient to support site-specific editing and editosome assembly, whereas sequences upstream (i.e. to the 5′ side) of the editing site serve to enhance these functions.

Other cases

In the slime mould *Physarum polycephalum* the transcript coding for the mitochondrial α subunit of ATP SYNTHASE differs from its template gene by the insertion of 54 Cs at a relatively regular spacing [6]. These insertions correct 54 frameshifts (all − 1). Two other protein-coding genes (*atp9* and *cob*) and the genes for rRNAs are also edited by C insertion. The mechanism of this process is unknown.

Another type of RNA editing has been discovered by comparison of the sequences of the genes and transcripts coding for the GluR2 subunit of the glutamate receptor ion channels in the mammalian central nervous system (*see* EXCITATORY AMINO ACID RECEPTORS). At a single position in homologous segments of the various transcripts, a glutamine codon (CAG) in the gene is modified in the mRNA into an arginine (CGG) by an A to G transition. Not all transcripts are edited with the same efficiency. The modification from glutamine to arginine profoundly affects the properties of ion flux in these channels.

In *Xenopus* oocytes, an unwinding/modifying activity modifies double-stranded RNAs (dsRNAs) by converting numerous As into Is (inosines) by the action of a dsRNA-specific adenosine deaminase. This post-transcriptional process is present in a number of eukaryotic cells. Its biological role is not clear and it has been postulated that it may serve as signal for the degradation of dsRNAs, possibly as part of a defence mechanism against viruses and VIROIDS.

The co-transcriptional addition of Gs has been described in paramyxoviruses, a class of negative-strand RNA viruses (*see* ANIMAL VIRUSES). During transcription of the P gene of these viruses, one or more Gs are inserted at a specific purine-rich site of the transcript, leading to the synthesis of two (or three) different proteins with a common N terminus but a different C terminus. The mechanism of this particular type of editing is a co-transcriptional stuttering of the viral polymerase, that is, the polymerase pauses and the nascent transcript slips, inducing the insertion of a G at the slippage positions.

Another form of RNA processing that was reported earlier may be, in retrospect, classified as RNA editing. Genetic information in vertebrate mitochondria is highly compacted (*see* MITOCHONDRIAL GENOMES: ANIMAL) and most of the termination codons are truncated (to T or TA of the TAA codon). A full stop codon is created immediately following transcription, by polyadenylation which provides the missing As. The activity responsible is likely to be a terminal adenosine transferase.

Evolution

The various editing processes described above are found in a wide range of organisms and genetic systems and display a wide range of putative mechanisms. It is therefore unlikely that they all share the same evolutionary origin. However, the fact that most of the different RNA editing processes occur in organelles may indicate that their origin is linked with the prokaryotic endosymbiont involved in the creation of the ancestor of eukaryotic cells (*see* ENDOSYMBIONT HYPOTHESIS). It is especially interesting to note that the plant organelles (mitochondria and chloroplasts) share the same type of RNA editing: C to U conversion. This could be an indication of a common origin for these two RNA editing events. Another interesting observation is that RNA editing mainly involves pyrimidines: C/U insertions, and C to U and U to C conversions. It has been speculated that RNA editing has been developed to incorporate pyrimidines into purine-rich RNAs, which were likely to be the primitive constituents of the so-called RNA WORLD. It is very difficult to estimate the selective advantage of maintaining RNA editing in any of the cases described above. It has often been suggested that these processes could have appeared as a result of GENETIC DRIFT and that they bring an advantage to the genetic system that maintains RNA editing by adding a new means of regulating gene expression.

J.M. GRIENENBERGER

See also: GENETIC CODE; TRANSCRIPTION.

1 Benne, R. et al. (1986) Major transcript of the frameshift *coxII* from trypanosome mitochodria contains four nucleotides that are not encoded in the DNA. *Cell*, **46**, 819–826.
2 Stuart, K. (1991) RNA editing in trypanosomatid mitochondria. *Annu. Rev. Microbiol.* **45**, 327–344.
3 Gray, M. W. et al. (1992) Transcription, processing and editing in plant mitochondria. *Annu. Rev. Plant Physiol. Plant Mol. Biol.* **43**, 145–175.
4 Gualberto, J. M. et al. (1991) Expression of the wheat mitochondrial *nad3 rps12* transcription unit: correlation between editing and mRNA maturation. *Plant Cell* **3**, 1109–1120.
5 Powell, L. N. et al. (1987) A novel form of tissue-specific RNA processing produces apolipiprotein-B48 in intestine. *Cell* **50**, 831–840.
6 Cattaneo, R. (1991) Different types of messenger RNA editing. *Annu. Rev. Genet.* **25**, 71–88.

RNA helicase A HELICASE which unwinds double-stranded RNA. Proteins with known ATP-dependent RNA helicase activity include the murine INITIATION FACTOR eIF4A and human p68, members of the DEAD BOX family of proteins.

RNA ligase Enzyme that catalyses the ATP-dependent formation of a phosphodiester bond between the 3′ end of one RNA molecule and the 5′ end of another. RNA ligase activity is required for the splicing of eukaryotic precursor tRNAs.

RNA localization The localized expression of certain proteins in somatic cells and of certain developmental proteins in the egg (*see* RNA, LOCALIZED MATERNAL) is mediated by localization of mRNA to the site of synthesis, rather than by PROTEIN TARGETING. In some cases mRNA localization has been observed to be by movement of ribonucleoprotein particles containing the mRNA along cytoskeletal tracks.

Wilhelm, J.E. & Vale, R.D. (1993) *J. Cell Biol.* **123**, 269–274.

RNA, localized maternal Many classical experiments led to the postulation of maternal CYTOPLASMIC DETERMINANTS which are localized in the egg and which are responsible for the specification of the embryonic axes. Ultraviolet irradiation and biochemical studies suggested that RNA might be a component of these determinants. More recently, in *Drosophila* and *Xenopus*, specific mRNAs have been shown to be localized within the unfertilized egg. In *Drosophila*, the mRNA for the anterior determinant, *bicoid*, is localized at the anterior pole. Localized to the posterior pole are: the RNA for the posterior determinant, *nanos*, *oskar* mRNA, which has roles in both germ cell determination and in the localization of *nanos*, and also cyclin B mRNA. In *Xenopus* several localized maternal transcripts have been reported, although none has yet been attributed a developmental function. Vg-1, a member of the TGF-β family of GROWTH FACTORS, and Xcat-2, which shows sequence similarity to *nanos*, are localized to the vegetal pole and mRNAs An1–An3 are restricted to the animal hemisphere. An2 encodes the α chain of mitochondrial ATP SYNTHASE and An3 appears to encode an RNA HELICASE.

RNA mapping *See*: S1 MAPPING.

RNA-mediated recombination The RECOMBINATION between a reverse transcript and its homologous chromosomal allele. It has been shown to occur in the yeast *Saccharomyces cerevisiae*, where it requires the reverse transcription function of the yeast TY ELEMENT (a RETROTRANSPOSON). RNA-mediated recombination has been proposed as a mechanism for the removal of introns from a gene (by substitution of the gene with a cDNA copy lacking introns) and for the GENE CONVERSION of members of MULTIGENE FAMILIES which is thought to maintain the sequence similarity between duplicated sequences.

RNA plasmids Single- and double-stranded RNA molecules found in the S and RU maize cytoplasmic lines. They appear to replicate by an RNA-DEPENDENT RNA POLYMERASE, and it has been suggested that they are of viral origin. They are not homologous with the rest of the mitochondrial genome. *See*: MITOCHONDRIAL GENOMES: PLANT; RU GROUP.

RNA polymerases Enzymes which catalyse the synthesis of RNA. DNA-dependent RNA polymerases catalyse the synthesis of an RNA complementary in sequence to a DNA template (*see* TRANSCRIPTION). Prokaryotic organisms have a single multisubunit 'core-enzyme' whose activity is modulated by ancillary factors (sigma, σ; rho, ρ) that help determine which set of genes is transcribed and where the polymerase terminates (*see* TRANSCRIPTION). RNA polymerases encoded by BACTERIOPHAGES are less complex, having simple phage-specific sequence and cofactor requirements. RNA polymerases isolated from *Escherichia coli* infected by phage SP6 or T7 are used to synthesize large amounts of RNA from cloned DNA in *in vitro* transcription reactions. The crystal structure of RNA polymerase from bacteriophage T7 has been determined. Eukaryotes possess three functionally different multisubunit polymerases, each responsible for transcribing a different class of genes. RNA polymerase I transcribes all rRNAs except 5S rRNA (in higher eukaryotes). RNA polymerase II transcribes all other RNA except small RNAs such as tRNA and 5S rRNA which are transcribed by RNA polymerase III (*see* EUKARYOTIC GENE EXPRESSION; EUKARYOTIC GENE STRUCTURE).

Virus-encoded REPLICASES and TRANSCRIPTASES are RNA-dependent RNA polymerases involved in replicating and transcribing the RNA genomes of RNA viruses and bacteriophages (*see* ANIMAL VIRUSES; BACTERIOPHAGES; PLANT VIRUSES). Some transcriptases (e.g. of the VESICULAR STOMATITIS VIRUS) are multifunctional enzymes involved in both transcription and replication.

PRIMASES are specialized DNA-dependent RNA polymerases which in some systems, synthesize the short RNA primers for DNA REPLICATION.

Sousa, R. et al. (1993) *Nature* **364**, 593–599.

RNA primer *See*: DNA REPLICATION; PRIMER.

RNA processing Many RNAs must undergo post-transcriptional processing after TRANSCRIPTION is complete in order to yield a mature RNA. Transcripts from RIBOSOMAL RNA (rRNA) and TRANSFER RNA (tRNA) genes in both prokaryotes and eukaryotes undergo post-transcriptional processing to release mature rRNA and tRNA molecules. Prokaryotic messenger RNA (mRNA) undergoes little further processing after transcription, and translation can begin even before transcription is completed. Eukaryotic nuclear RNA, on the other hand, is subject to considerable post-transcriptional processing.

Three major kinds of processing events have been identified.
1 Eukaryotic RNAs transcribed by RNA polymerase II (i.e. most protein-coding RNAs) are modified before they leave the nucleus by the addition of a cap structure at the 5′ end which plays a major part in translation, facilitating attachment to the ribosome (*see* PROTEIN SYNTHESIS). The cap is an added 5′ terminal G, methylated on the 7-position of the purine base and linked to the initiating nucleotide (usually A or G) by an unusual 5′–5′ triphosphate linkage — the reverse orientation to that normally found in RNA. A nuclear enzyme, guanylyl transferase, catalyses condensation between GTP and the 5′ terminal phosphate of nuclear mRNA (Fig. R31). The new G residue is methylated by the enzyme guanine-7-methyltransferase and the next two residues may also be modified by methylation(s).

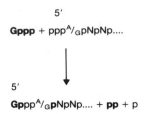

Fig. R31 Cap formation.

2 The majority of class II transcripts (with the exception of histone mRNAs) have a poly(A) tail, 40–100 residues long, added to their 3′ ends in the nucleus (polyadenylation). This involves both cleavage of the primary transcript at the 3′ end of the mRNA sequence together with a coupled polyadenylation reaction. Polyadenylation is important for the stability and translatability of mRNA and is intimately involved in the termination of transcription. Mammalian polyadenylation signals are now well defined: they contain two elements, the AAUAAA sequence 20–30 nucleotides before and a diffuse GU-rich sequence immediately after the 3′ end of the mRNA.
3 Transcripts of all three gene classes are usually subject to endo- or exonucleolytic cleavage followed, in the case of introns in class II and some other genes, by exon splicing (*see* RNA SPLICING) to form the functional mature species. 5S RNA is an exception, being synthesized directly in the mature form. *See also*: RIBONUCLEASE P.

RNA replicase *See*: REPLICASE; RNA POLYMERASES; RNA REPLICATION.

RNA replication All cellular RNA is synthesized by TRANSCRIPTION catalysed by RNA polymerases using DNA as a template. Most viruses containing RNA as their genetic material replicate their genomes by RNA-dependent RNA polymerases (viral replicases) (*see e.g.* INFLUENZA VIRUS; POLIOVIRUS). There are various strategies for RNA replication depending on the virus and whether the RNA genome is single or double stranded (*see* ANIMAL VIRUSES; PLANT VIRUSES). The exceptions are the animal RETROVIRUSES which first copy their RNA genome into DNA using the enzyme REVERSE TRANSCRIPTASE.

RNA restriction enzyme Enzyme capable of recognizing and cleaving at a specific site within an RNA chain. *See*: RIBOZYME.

RNA sequencing RNA is sequenced indirectly by first making a cDNA copy using reverse transcriptase (*see* COMPLEMENTARY DNA). This copy can then be rapidly sequenced using one of the standard DNA SEQUENCING methods.

RNA splicing

RNA splicing is the process that takes place in EUKARYOTIC nuclei in which INTRONS are removed from the primary transcripts (*see* TRANSCRIPTION) of protein-coding genes and EXONS ligated to form functional MESSENGER RNA. Some other types of eukaryotic RNA, particularly TRANSFER RNA, are also processed in a similar way and there are a few instances of RNA splicing known from prokaryotes.

In 1977 the totally unexpected finding of split genes in an ADENOVIRUS [1] overthrew the assumption, gained from work on bacterial and phage genes, of a strict one-to-one relationship between the nucleotide sequence of a gene and amino-acid sequence of the protein it encodes. Later that year, the same phenomenon was shown in eukaryotic nuclear genes, where protein-coding regions of genes, known as exons, were split or interrupted by noncoding regions or introns, sometimes known as intervening sequences (IVS). For general reviews of the topic see [2–5].

Nuclear mRNA splicing

Most nuclear mRNA precursors (pre-mRNAs, sometimes known as heterogeneous nuclear RNA or hnRNA) in higher eukaryotes contain multiple introns which must be precisely excised by RNA splicing prior to transport over the nuclear membrane into the cytoplasm for translation (*see* PROTEIN SYNTHESIS). Pre-mRNA splicing occurs by a two-step mechanism [6,7].

In the first step cleavage occurs at the 5′ splice site (the splice donor site). The guanosine residue at the 5′ end of the intron is covalently linked through a 2′–5′ phosphodiester bond to an adenine residue, typically 20–50 nucleotides upstream from the 3′ splice site (the splice acceptor site), within a recognition element designated the branch point sequence (BPS). The cleavage and branch formation reactions have not been resolved kinetically and are thought to occur simultaneously.

The second step involves cleavage at the 3′ splice site and ligation of the exons. The intron is released as a lariat RNA (from Spanish for lasso) containing the 2′–5′ linkage at the branch point. The phosphate moieties at both the 5′ and 3′ splice sites are conserved in the products. Both steps may be *trans*-esterification reactions, where a hydroxyl group reacts with a phosphodiester bond displacing a hydroxyl group while forming a new phosphodiester bond. In addition to the conserved sequences at the 5′ (AG::GUAAGU) and 3′ (Y$_n$CAG::G; Y is either T or C) splice sites, the BPS and flanking exon sequences participate in splice site recognition. In yeast the BPS is highly conserved and is known as the TACTAAC box, whereas in higher eukaryotes it is more difficult to identify the BPS solely on the basis of DNA sequence (YNYRAY; R is either A or G).

The splicing reaction takes place within spliceosomes [8] which are high molecular weight (50–60S) particles that contain pre-mRNA, the highly abundant (usually >10^5 per cell) small nuclear ribonucleoprotein particles (snRNPs, sometimes known as snurps) U1, U2, U4/U6 and U5, and non-snRNP splicing factors. The snRNAs found in the snRNPs have distinctive properties of a trimethyl cap structure at the 5′ end of the RNA, an internal sequence AUUUUUG and are found associated with a common set of core polypeptides recognized by poly- or MONOCLONAL ANTIBODIES of the Sm type. Spliceosomes can be assembled *in vitro* and have allowed the functions of some spliceosome components to be determined. The U1 snRNP binds to the 5′ splice site, whereas the U2 snRNP binds, together with the U2 auxiliary factor (U2AF) and ATP, to the BPS. In both cases binding involves the formation of complementary base pairs between snRNA and pre-mRNA. U5 may associate with the 3′ splice site (Fig. R32).

Alternative splicing

Some pre-mRNAs are alternatively spliced in different cell types or at different times during development. Regulated alternative splicing can lead to the production of different proteins from a

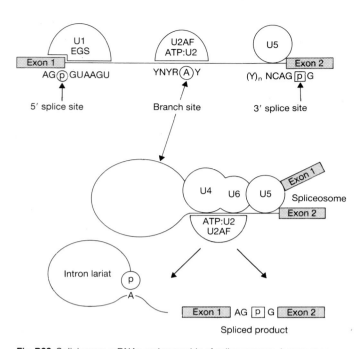

Fig. R32 Splicing pre-mRNAs and assembly of spliceosomes. A precursor mRNA is drawn with the intervening sequence or intron located between the 5′ and 3′ splice sites. The intron is flanked by exons 1 and 2. Consensus sequences at the splice sites and branch site are shown (Y, pyrimidine; R, purine; N, any base; p, phosphate). Addition of this RNA to a nuclear extract results in recognition of the 5′ splice site by U1 snRNP. The external guide sequence (EGS) forms complementary base pairs with the consensus sequence at the 5′ splice site. It is likely that U5 snRNP binds to the 3′ splice site at this stage. Further incubation results in the formation of the spliceosome containing at least U2, U4, U5 and U6 snRNAs and multiple proteins including U2AF (intermediate). The two RNAs that are intermediates in the splicing reaction, the 5′ exon and the intron lariat linked to the 3′ exon, are only found in the spliceosome. The products of the reaction, the intron lariat and the spliced product, are shown on the bottom line. The fate of the phosphate residues can be seen by following the circle or square. The *trans*-esterification reaction that occurs in the spliceosome is similar to that for Group II introns in figure (self-splicing).

single pre-mRNA (*see e.g.* IMMUNOGLOBULIN GENES) or the splicing can function as an on–off switch during development. Virtually every imaginable pattern of alternative splicing has been reported but the mechanisms for selection of 5′ and 3′ splice sites are unclear. The majority of pre-mRNAs in higher eukaryotes contain multiple introns with many cryptic 5′ and 3′ splice sites scattered through both introns and exons. Cryptic sites look like splice sites but are not normally used. The splicing machinery must discriminate between normal and cryptic splice sites and avoid exon skipping by selecting 5′ and 3′ splice sites that are within the same intron. The magnitude of the problem is illustrated by the DYSTROPHIN pre-mRNA that contains 65 exons and 2 million nucleotides of RNA with one intron at least 200 000 nucleotides long. Differences in the activities or amounts of general splicing factors such as SF2 can have a dramatic effect on the selection of different 5′ splice sites. Furthermore, splice site selection can be negatively regulated by repressor proteins that bind specifically to recognition sequences in pre-mRNA and thus

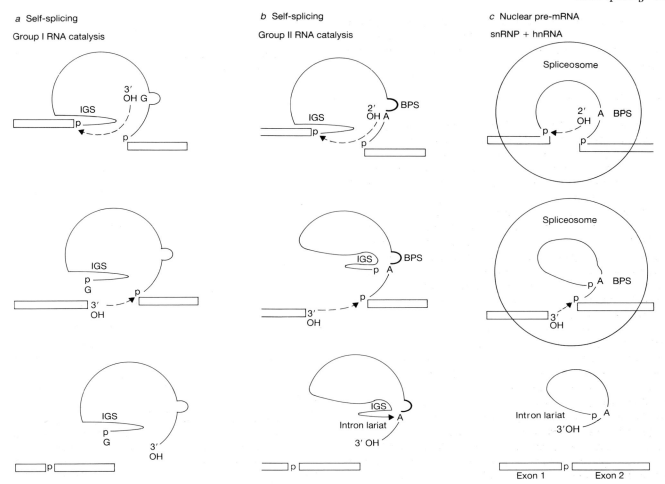

Fig. R33 Comparisons of self-splicing and nuclear pre-mRNA splicing mechanisms. *a*, *b*, Splicing mechanisms of self-splicing introns of the group I and group II type; and *c*, the splicing mechanism for pre-mRNAs (sometimes known as hnRNA). All the proposed mechanisms for RNA splicing shown occur by a two-step *trans*-esterification mechanism that results in retention of the phosphate groups at the splice sites. RNA catalysis for the group I and II type introns is catalysed by RNA structures within the intron (the internal guide sequence at the 5′ splice site and a site marked by a bold semicircle, which is the branch point sequence (BPS) in group II introns). Group I splicing uses a guanosine cofactor (G), whereas type II splicing uses an adenosine (A) residue within the intron. The results of the group I and group II reaction are thus different; group I reactions result in the release of a linear intron whereas the group II product is a lariat RNA. Nuclear pre-mRNA splicing occurs within a multicomponent complex, the spliceosome, which promotes the splicing reaction (see Fig. R32).

influence the usage of 5′ and 3′ splice sites and result in alternative splicing. A typical example of tissue-specific alternative splicing in the rat α-tropomyosin gene is shown in Fig. R33.

Self-splicing

1981 brought another remarkable discovery, that of self-splicing introns in the RIBOSOMAL RNA (rRNA) of the protist *Tetrahymena* which led to a whole new interest in catalytic RNA and the role of RNA in evolution [9] (*see* RIBOZYMES). Self-splicing introns have been found in fungal mitochondria (*see* MITOCHONDRIAL GENOMES), protist nuclei, plant chloroplasts (*see* CHLOROPLAST GENOMES), CYANOBACTERIA [10] and eubacterial BACTERIOPHAGES [11,12]. RNA precursors of *Tetrahymena* rRNA and other similar introns self-splice *in vitro* using a guanosine cofactor and magnesium ions in the absence of protein in an RNA-catalysed reaction whose rate can be increased by the binding of specific proteins.

Self-splicing RNAs can be assigned to two groups on the basis of conserved consensus sequences and closely associated defined secondary structures called the internal guide sequence (IGS) (*see* GUIDE SEQUENCE). In group I introns, such as that of *Tetrahymena* rRNA, self-splicing occurs in the presence of a guanosine cofactor in two steps, both of which are *trans*-esterifications that do not require the input of energy. In group II introns, self-splicing occurs in the absence of a nucleotide cofactor as the 2′ hydroxyl group at the branch site participates in the first *trans*-esterification reaction to produce a lariat RNA that is subsequently found in the excised intron.

Mechanistically, group I and II self-splicing reactions are very similar to pre-mRNA splicing (Fig. R34). As no energy input is required, group I and II introns have been shown to undergo reverse splicing *in vitro*, that is, both reactions are reversible. The

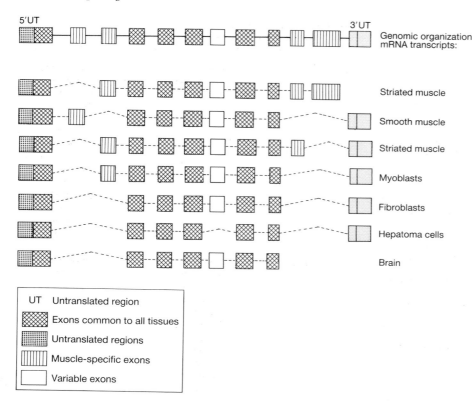

Genomic organization
mRNA transcripts:

Striated muscle

Smooth muscle

Striated muscle

Myoblasts

Fibroblasts

Hepatoma cells

Brain

UT Untranslated region
Exons common to all tissues
Untranslated regions
Muscle-specific exons
Variable exons

Fig. R34 Tissue-specific alternative splicing of the hnRNA product of the rat α-tropomyosin gene. The genomic organization is shown on the top line. Boxes represent exons, including those encoding the 5′ and 3′ untranslated (UT) regions, and the thick line represents the intervening sequences. The organization of the mRNA transcripts present in different tissue types is shown below. The dashed lines indicate the regions removed to produce each mRNA. A region of hnRNA that functions as an exon in one tissue may be part of an intron in another. Cross-hatched boxes, exons common to all tissues; striped boxes, muscle-specific exons; plain boxes, exons with variable usage.

internal guide sequences of group I introns are essential for directing accurate splicing but can be removed physically from the intron and as discrete RNA species function as external guide sequences (EGS). This is functionally analogous to the role of U1 snRNA as an external guide sequence in pre-mRNA splicing. External guide RNAs are used in other RNA-catalysed events in the cell such as the addition of telomeres to chromosomes (*see* CHROMOSOME STRUCTURE and RNA EDITING).

RNA editing

The second major jolt to the assumption that the nucleotide sequence of every mRNA would be a simple copy of the sequence of its DNA template occurred with the finding that individual nucleotides, generally U, are post-transcriptionally inserted, deleted or altered in sequences by RNA EDITING [13]. Extensive RNA editing is found in mitochondrial pre-mRNAs in the parasitic protozoa *Trypanosoma*, *Leishmania* and *Crithidia*, and has also been found in fungal and plant mitochondria, and in a single example of a mammalian nuclear gene (*see* RNA EDITING). The precise editing mechanism is still unknown but appears to have similarities to reactions occurring in RNA self-splicing.

Trans-splicing

In trypanosomes (*Trypanosoma*) a common 39-nucleotide LEADER SEQUENCE is added to virtually all mRNAs post-transcriptionally, and in nematodes (*Caenorhabditis elegans* and *Ascaris lumbricoides*) a common 22-nucleotide spliced leader (SL) sequence is spliced onto the 5′ end of 10–20% of mRNAs [14]. In trypanosomes, *trans*-splicing may serve the dual role of providing a mechanism for processing multicistronic pre-mRNAs as well as providing a unique mechanism for RNA cap addition (*see* RNA PROCESSING). *Trans*-splicing has been shown to occur *in vitro* via a Y-shaped intermediate and to require some snRNPs (e.g. U2 and U4/U6) but not others (e.g. U1). SL may activate their own 5′ splice sites (self-splice) as nearly identical secondary structure exist for the SL RNAs in trypanosomes and nematodes, thus avoiding the need for U1 snRNA as an external guide sequence. *Trans*-splicing also occurs in chloroplasts (*see* CHLOROPLAST GENOME).

RNA splicing in prokaryotes

As transcription and translation are linked in prokaryotes the scope for extensive post-transcriptional modification of RNA is limited. However, self-splicing group I introns requiring a guanosine cofactor have been found in prokaryotes, first in the T4 bacteriophage that infects *Escherichia coli* [11] and later in the SPO1 bacteriophage of *Bacillus subtilis* [12]. Group I introns have also been found in the tRNA^{Leu} genes of eight cyanobacterial species [10]. These introns are extremely similar in sequence, structure and location to the tRNA^{Leu} intron of plant chloroplasts suggesting a phylogenetic link between bacteria and eukaryotic organelles.

E.J. MELLOR

1 Berget, S.M. et al. (1977) Spliced segments at the 5′ terminus of adenovirus 2 late mRNA. *Proc. Natl. Acad. Sci. USA* **74**, 3171–3175; Chow, L.T.

et al. (1977) An amazing sequence arrangement at the 5′ ends of adenovirus 2 messenger RNA. *Cell* **12**, 1–8.

2 Darnell, J. et al. (1990) *Molecular cell biology*, 2nd edn (Scientific American Books, San Francisco, CA).

3 Watson, J.D. et al. (1987) *The Molecular Biology of the Gene* 4th edn (Benjamin/Cummings, San Diego, CA).

4 Rio, D.C. (1992) RNA processing. *Curr. Opinion Cell Biol.* **4**, 444–452.

5 Woolford, J. L. Jr, & Peebles, C.L. (1992) RNA splicing in lower eukaryotes. *Curr Opinion Genet. Devel.* **2**, 712–719.

6 Green, M.R. (1986) Pre-mRNA splicing. *Annu. Rev. Genet.* **20**, 671–708

7 Ruby, S.W. & Abelson, J. (1987) An early hierarchic role of U1 small nuclear ribonucleoprotein in spliceosome assembly. *Science* **242**, 1028–1035.

8 Guthrie, C. & Patterson, B. (1988) Spliceosomal snRNAs. *Annu. Rev. Genet.* **22**, 387–419.

9 Cech, T.R. et al. (1981) *In vitro* splicing of the ribosomal RNA precursor of Tetrahymena: involvement of a guanosine nucleotide in the excision of the intervening sequence. *Cell* **27**, 487–496.

10 Xu, M.-Q et al. (1990) Bacterial origin of a chloroplast intron: conserved self-splicing group I introns in cyanobacteria. *Science* **250**, 1566–1570; Kuhsel, M.G. et al. (1990) An ancient group I intron shared by eubacteria and chloroplasts. *Science* **250**, 1570–1573.

11 Belfort, M. (1990) Phage T4 introns: self-splicing and mobility. *Annu. Rev. Genet.* **24**, 363–385.

12 Goodrich-Blair, H. et al. (1991) Self-splicing group I intron in the DNA polymerase gene of *Bacillus subtilis* bacteriophage SPO1. *Cell* **63**, 417–424.

13 Cech, T.R. (1991) RNA editing: the world's smallest intron. *Cell* **64**, 667–669.

14 Sharp, P.A. (1987) *Trans*-splicing: variation on a familiar theme. *Cell* **50**, 147–148.

RNA stability The relative longevity or resistance to degradation of RNA molecules in the cell. The rate at which a single RNA species is degraded is usually expressed as a half-life — the time taken for half the molecules in a finite pool of a particular RNA species to be degraded by endogenous RIBONUCLEASES. Prokaryotic mRNAs are rapidly turned over; the more stable species among them having half-lives of a few minutes. Eukaryotic mRNAs have a broader stability range that varies from several minutes for unstable species to several hours for the more stable species. Different RNA species possesses factors that increase stability (such as 5′ cap structure, 3′ poly(A) tract and secondary structures), or decrease stability (for example, internal ribonuclease cleavage sites). *See also*: BACTERIAL GENE EXPRESSION; RNA PROCESSING.

RNA synthesis during early development During early development the maternally derived RNA decays and zygotic transcription initiates. The stage of development at which the newly synthesized embryonic gene products become dominant varies across species and between genes. In mammals the maternal message decays rapidly before the onset of major zygotic transcription but in other species there is considerable overlap. The major embryonic transcript class is high-complexity nuclear RNA, the vast majority of which never accumulates outside the nuclear compartment. Over 10^4 TRANSCRIPTION UNITS appear to be utilized in early sea urchin embryos.

RNA tumour viruses *See*: RETROVIRUSES.

RNA viruses Viruses containing RNA as the genetic material (Table R7).

Table R7 RNA virus families

	Bacteriophages	Plant virus groups	Animal viruses
ds		Rhabdoviridae	Arenaviridae
		Tomato spotted wilt virus	Bunyaviridae
		Tenuvirus	Caliciviridae
			Coronaviridae
			Filoviridae
			Flaviviridae
			Nodaviridae
			Paramyxoviridae
			Picornaviridae
			Retroviridae
			Rhabdoviridae
			Tetraviridae
			Togaviridae
			Toroviridae
ss	Leviviridae	Alfalfa mosaic virus	Birnaviridae
		Bromovirus	Reoviridae
		Capillovirus	
		Carlavirus	
		Carmovirus	
		Closterovirus	
		Comovirus	
		Cryptovirus	
		Cucumovirus	
		Dianthovirus	
		Fabavirus	
		Furovirus	
		Hordeivirus	
		Ilarvirus	
		Luteovirus	
		Maize chlorotic dwarf virus	
		Marafivirus	
		Necrovirus	
		Nepovirus	
		Parsnip yellow fleck virus	
		Pea enation mosaic virus	
		Potexvirus	
		Potyvirus	
		Reoviridae	
		Sobemovirus	
		Tobravirus	
		Tobamovirus	
		Tombusvirus	
		Tymovirus	

ds, double-stranded; ss, single-stranded.

RNA world A proposed stage of precellular and early cellular evolution in which the genetic replicators and metabolic catalysts ('enzymes') and cofactors would have been RNA and RIBONUCLEOTIDE based, and in which proteins and DNA had not yet evolved. A role for RNA as both replicator and enzyme would resolve the paradox raised by the present day vital role of proteins in all cellular processes. Proteins cannot direct their own replication, so the first replicators must have been RNA; however, in present-day cells the conversion of the genetic information in RNA into proteins itself requires proteins. The existence of a self-replicating system dependent on RNA for all its functions resolves the paradox and allows for the gradual evolution of proteins and their eventual takeover as metabolic catalysts. *See:* MOLECULAR EVOLU-TION; RIBOZYMES.

Gesteland, R.F. & Atkins, J.F. (Eds) *The RNA World* (Cold Spring Harbor Laboratory Press, New York).

RNase, RNAse RIBONUCLEASE.

RNase A *See:* RIBONUCLEASE A.

RNase H RIBONUCLEASE H. *See:* REVERSE TRANSCRIPTASE.

RNase L RIBONUCLEASE L. *See:* INTERFERONS.

RNase P RIBONUCLEASE P. *See also:* RIBOZYMES.

RNasin™ A commercially available proteinaceous, RIBONU-CLEASE inhibitor that inactivates a variety of ribonucleases, particularly mammalian RIBONUCLEASE A. RNasin is used to minimize RNA degradation in reactions such as IN VITRO TRANSCRIPTION, IN VITRO TRANSLATION, and the synthesis of cDNA.

RNP Abbreviation used for RIBONUCLEOPROTEIN or RIBONUCLEO-PROTEIN PARTICLE.

Robertsonian translocation Translocation involving two ACRO-CENTRIC chromosomes in which the two long arms are joined into one very long chromosome and the remaining tiny fragment is usually lost.

rods (1) Photoreceptor cells in the vertebrate retina that are responsible for nocturnal (monochromatic) vision. They contain the visual pigment RHODOPSIN with a maximal light absorption at wavelength 496 nm. *See:* VISUAL TRANSDUCTION.
(2) Rod-shaped viruses. *See:* ANIMAL VIRUSES; BACTERIOPHAGES; PLANT VIRUSES.

***roi* locus** Root-inducing locus. *See:* IPT GENE.

rolling circle replication A mode of DNA replication found for some circular DNA molecules including certain BACTERIOPHAGE genomes (e.g. phage LAMBDA) and also found in *Xenopus* oocytes during amplification of extrachromosomal RIBOSOMAL DNA. DNA synthesis initiates at a single origin from which a sole replication fork proceeds around the template indefinitely. As the fork revolves, the newly synthesized strand displaces the previously synthesized strand from the template, producing a characteristic tail. After a number of revolutions (*n*), the displaced strand will contain *n* contiguous single-stranded copies of a sequence complementary to the template DNA (R35). The displaced strand may then be made double-stranded and processed into single or multimeric copies of the original DNA. A similar process is thought to occur in the replication of some circular RNAs, for example those of VIROIDS.

Root effect An exceptionally strong alkaline BOHR EFFECT shown by many bony fish haemoglobins. *See:* HAEMOGLOBINS.

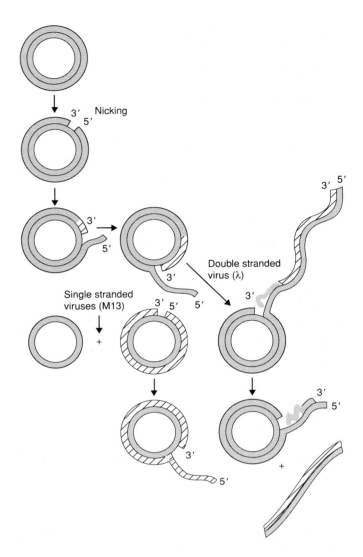

Fig. R35 Schematic of rolling circle replication in single-stranded and double-stranded viruses. From Singer, M. & Berg. P. (1991) *Genes and Genomes* (University Science Books, CA/Blackwell Scientific Publications, Oxford).

root nodule *See:* NODULATION.

***ros* locus** Chromosomal locus in *Agrobacterium tumefaciens* which affects plasmid-encoded virulence (*vir*) gene expression.

Rossmann fold A mononucleotide-binding motif found in many protein structures. M.G. Rossmann first recognized the frequent occurrence in nucleotide-binding proteins of this α/β structure where a parallel β-sheet is surrounded on both sides by α-helices. With more examples becoming known, a wider variation is seen in this motif such that only the central part of the original motif can now be regarded as essential for nucleotide binding.

Branden, C.-I. & Tooze, J. (1991) *Introduction to Protein Structure* (Garland, New York).

Rossmann, M.G. et al. (1975) . In *The Enzymes*, Vol. XI (Boyer, P.D., Ed.) 61–102 (Academic Press, New York).

rostro-caudal axis The ANTERIOR–POSTERIOR AXIS.

Rot curve The equivalent of a Cot curve when RNA and DNA are being hybridized. *See*: COT ANALYSIS; HYBRIDIZATION.

rotary shadowing *See*: ELECTRON MICROSCOPY.

rotation axes, rotational symmetry *See*: SYMMETRY; SYSTEMATIC ABSENCES.

rotation function *See*: MOLECULAR REPLACEMENT.

rough endoplasmic reticulum (RER) *See*: ENDOPLASMIC RETICULUM.

Rous sarcoma virus (RSV) A RETROVIRUS and the first virus identified as a tumour virus, discovered by Peyton Rous in 1911. It causes sarcomas in chickens and may be spread by either horizontal or vertical transmission. The first viral ONCOGENE to be characterized molecularly was the v-*src* gene of RSV. Its characterization led to the identification of the normal homologous proto-oncogene (c-*src*) in vertebrate cells.

Rous, P. (1911) *J. Exp. Med.* **13**, 397–411.

RP RETINITIS PIGMENTOSA.

***rpl* genes, *rps* genes** Genes for polypeptides of the large and small subunits respectively of the RIBOSOME.

rRNA *See*: RIBOSOMAL RNA; RNA.

rRNA gene amplification The RIBOSOMAL RNA genes are preferentially amplified at a certain developmental stage in several species. Among the best studied are the amplification of rDNA during OOGENESIS in *Xenopus*, and the preferential replication of DNA fragments carrying rRNA genes concomitant with the formation of the transcriptionally active MACRONUCLEUS in the protist *Tetrahymena*. In *Tetrahymena* up to 10^4 copies of these fragments can be found packaged within multiple NUCLEOLI. Each fragment contains two copies of the rRNA TRANSCRIPTION UNIT together with a region of nontranscribed DNA. *See also*: GENE AMPLIFICATION.

RS recombination A RECOMBINATION event that can occur in the mouse immunoglobulin κ light chain locus in which the RS (recombining sequence) element located 25 kb downstream of the C_κ gene can recombine with sequences upstream of the C_κ gene thereby deleting the C gene (*see* IMMUNOGLOBULIN GENES). The majority of B cells expressing their λ light chain genes carry an RS recombination in at least one of their κ alleles. The biological significance of this is not fully understood, but it may ensure that no secondary functional rearrangements occur on the κ locus in λ-expressing B cells. A sequence with homologous function in the human κ locus is termed the KAPPA DELETING ELEMENT.

RSS RECOMBINATION SIGNAL SEQUENCE. *See*: GENE REARRANGEMENT; RECOMBINASE.

RSV ROUS SARCOMA VIRUS. *See*: RETROVIRUSES.

rtDNA RECOMBINANT DNA.

RU group A group of MALE FERTILE maize cytoplasmic lines (also designated S*), indigenous to South America. The group is named after one of the races within it, Racimo de Uva. The mitochondria contain plasmid-like elements similar to those in some MALE STERILE lines. *See*: MITOCHONDRIAL GENOMES: PLANT.

RUBISCO, RuBPCase RIBULOSE BISPHOSPHATE CARBOXYLASE/OXYGENASE.

ruffling Cells such as FIBROBLASTS crawling over a surface extend thin cytoplasmic protrusions (microspikes) and sheet-like processes (lamellipodia) from their forward or advancing edge. During cell movement some of these dynamic surface extensions adhere to the ground while others fold back and sweep backwards over the cell surface. This membrane activity is known as ruffling, and can be induced by some GROWTH FACTORS in certain cell types. Some ruffles may cause the formation of large (0.2–1.0 μm) ENDOSOMES by a process known as macropinocytosis. *See*: CELL MOTILITY.

run-on/run-off transcription Run-on transcription is a method devised to analyse TRANSCRIPTION in eukaryotes qualitatively and quantitatively. Intact nuclei are rapidly isolated and then incubated with radioactive nucleoside triphosphates (NTPs). RNA polymerase molecules that were actively transcribing before the nuclei were isolated, continue, or 'run-on', in the presence of the radioactive NTPs to yield labelled transcripts. These, when isolated, separated, and detected radiologically give a faithful representation of the *in vivo* transcription profile.

Run-off transcription describes the *in vitro* synthesis of RNA using, typically, the bacteriophage SP6 or T7 RNA polymerase systems. DNA carrying the template sequence for the required transcript is cloned in front of a T7 or SP6 RNA polymerase-specific PROMOTER in a PLASMID designed for this function (e.g. Bluescript). The resulting recombinant plasmid is linearized directly 3′ of the cloned sequence and then incubated with the RNA polymerase of choice and NTPs. The RNA polymerase will commence transcription at the promoter, continue through the template and run off the end to yield a full-length transcript.

***ruv* genes, Ruv proteins** *See*: DNA REPAIR; RECOMBINATION.

RXR RETINOID X RECEPTOR.

ryanodine receptor Calcium channel identified by binding of the plant alkaloid ryanodine. *See*: CALCIUM.

S

s SEDIMENTATION COEFFICIENT, which is generally measured in SVEDBURG UNITS.

S (1) SVEDBERG UNIT.
(2) The single-letter abbreviation for the AMINO ACID serine.

S cytoplasm Source of S-type cytoplasmic male sterility (cmsS) in maize. Distinguished from other types by the nuclear genes that restore fertility. *See*: MITOCHONDRIAL GENOMES: PLANT.

S-factor Plasmid-like element associated with cytoplasmic male sterility. *See*: MITOCHONDRIAL GENOMES: PLANT.

S-locus The locus determining gametophytic SELF-INCOMPATIBILITY in flowering plants.

S-phase The phase of the eukaryotic CELL CYCLE in which DNA synthesis occurs.

S protein Vitronectin. *See*: COMPLEMENT; EXTRACELLULAR MATRIX MOLECULES; SERPINS.

S—S linkage *See*: DISULPHIDE BOND.

S–S switching *See*: CLASS SWITCHING; GENE REARRANGEMENT; IMMUNOGLOBULIN GENES.

S-value The magnitude of the SEDIMENTATION COEFFICIENT of a particle or macromolecule measured in the ultracentrifuge. *See*: SVEDBERG UNIT.

S1-DNA, S2-DNA Larger and smaller, respectively, of two plasmid-like DNAs found in the mitochondria of male-sterile maize plants of the cmsS group. Copies are also found integrated into the mitochondrial genome, and partial copies are found integrated in the mitochondrial genome of male-fertile plants. *See*: MITOCHONDRIAL GENOMES: PLANT.

S1 mapping A technique that provides information on the sequence features present in mRNA molecules. A radioactively labelled single-stranded DNA fragment prepared from a cloned gene is annealed to its mRNA to form a double-stranded nucleic acid. S1 NUCLEASE (which preferentially degrades single-stranded DNA) is then used to degrade the noncomplementary single-stranded regions of the RNA–DNA hybrid. The size of the remaining DNA fragment(s) is then determined by polyacrylamide gel ELECTROPHORESIS and AUTORADIOGRAPHY. S1 mapping can be used to define the number and approximate location of INTRONS within a cloned gene, to define the 5′ end of the mRNA, and to identify the extent of any heterogeneity at the 5′ end of an mRNA.

Berk, A.J. & Sharp, P.A. (1977) *Cell* **12**, 721–732.

S1 nuclease An ENDONUCLEASE (EC 3.1.30.1) from the mould *Aspergillus oryzae* that preferentially degrades single-stranded DNA or RNA. It has found many applications in molecular biology, but most commonly in the S1 MAPPING of mRNA molecules. It can also be used to remove the single-stranded portions of the STICKY ENDS generated by some RESTRICTION ENZYMES to produce blunt-ended DNA, and to remove noncomplementary single-stranded DNA overhangs when constructing COMPLEMENTARY DNA. S1 nuclease does however often degrade DNA within a double-stranded region to produce ragged ends, and it is therefore often replaced by the milder MUNG BEAN NUCLEASE. S1 nuclease has also been used to demonstrate the presence of short interspersed repeated sequences in genomic DNA, and to transform genomic DNA from the Z-DNA to the B-DNA conformation associated with gene activation (*see* DNA; NUCLEIC ACID STRUCTURE).

S100 α and β *See*: CALCIUM-BINDING PROTEINS.

Saccharomyces cerevisiae

THE yeast *Saccharomyces cerevisiae*, long exploited by man for its ability to ferment glucose to ethanol (in brewing) and carbon dioxide (in baking), has in the last decade achieved a new significance as a model organism for studies in eukaryotic cell and molecular biology [1,2]. Many aspects of higher eukaryotic cell structure and function are found in this unicellular microbe and can be probed with powerful molecular genetic tools. Much of our current understanding about eukaryotic gene structure and function, secretion, and the cell cycle and its regulation comes from this simplest of eukaryotes.

Although the use of yeast in brewing was first recorded as long ago as 6000 BC, it was not until 1680 and the development of the microscope by Antoni van Leeuwenhoek that yeast's characteristic budding morphology was first described. In 1832 it was first recognized as a fungus (by Persoon and Fries) and brewing yeast was named *Saccharomyces cerevisiae* by Meyer some five years later. By the end of the 19th century it was being used in the first *in vitro* biochemical studies of fermentation and metabolism by Buchner and Ehrlich. Genetic studies were pioneered by Winge, Lindegren, and their colleagues in the 1930s and 1940s, prompted by their discovery that both haploid and diploid forms of *S. cerevisiae* could be stably maintained and that the MENDELIAN INHERITANCE of genes during sexual reproduction could be easily monitored following sporulation of the diploid. The description of a method for genetically transforming *S. cerevisiae* with PLASMID DNA in 1978 finally paved the way for the development and application of the full range of recombinant DNA techniques. REVERSE GENETICS strategies are easily implemented in yeast and have been used to study all aspects of yeast biology. Throughout this genetic revolution *S. cerevisiae* has also remained a favoured organism for more classical physiological and biochemical studies.

Structure and life cycle

S. cerevisiae is a member of the ASCOMYCOTINA. The genus *Saccharomyces* consists of some nine species characterized by ovoid or ellipsoidal cells (Fig. S1*a*), the formation of ascospores contained within a sac-like ascus (Fig. S1*b*) and efficient conversion of sugars to alcohol. The cells of *S. cerevisiae* can be up to 20 μm long but, unstained, yield little internal detail under the light microscope (Fig. S1*a*). Internal detail is distinguishable in electron micrographs of thin-sectioned cells (Fig. S1*c*) with the main recognizable features being the nucleus, one or more mitochondria, a large vacuole and a comparatively thick cell wall.

Under ideal nutritional conditions yeast cells divide once every 90–120 minutes with new cells developing as buds which separate from the mother cell after receiving a full complement of chromosomes. These daughter cells can then generate new cells by budding. Scanning electron micrographs of mother cells (Fig. S1) demonstrate the raised 'bud scar' which identifies the site at which the daughter cell separated. Each mother can produce more than 10 daughter cells. Although superficially, budding is one of the more peculiar features of the yeast life cycle, proteins involved in the budding process have close homologues in higher eukaryotic cells suggesting that budding may be akin to the general process of polarization of cell structure along a defined axis.

S. cerevisiae cells can be stably maintained as either HAPLOIDS (chromosome number 16) or DIPLOIDS (chromosome number 32). Haploid cells are one of two mating types designated *a* and *a* respectively. Diploid cells are formed by the fusion of haploid cells of opposite mating type and will grow and divide as such unless exposed to nutritional deficiency whereupon the diploid cell sporulates to produce four haploid ascospores [3]. The mating type of *S. cerevisiae* is determined by a single locus, the *MAT* locus, the *MATa* allele specifying the α-mating type and the *MATa* specifying the a-mating type. In the wild, haploid *S. cerevisiae* strains are homothallic, being able to switch their mating type at a very high frequency, whereas most cultivated laboratory strains are heterothallic, that is they do not switch their mating type, as

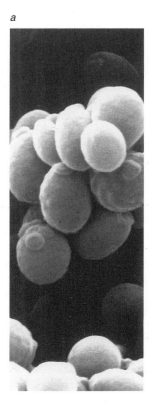
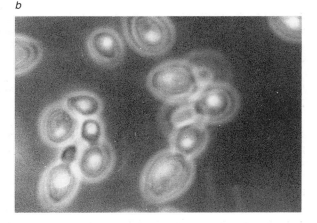
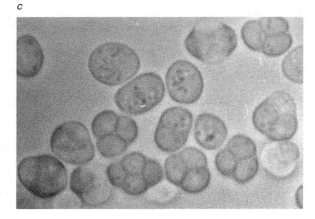

Fig. S1 Different views of the budding yeast *Saccharomyces cerevisiae*. *a*, Scanning electron micrograph showing newly formed buds; *b*, as seen under the electron microscope; *c*, yeast asci containing haploid ascospores.

a consequence of a mutation in the *HO* gene (*see* YEAST MATING-TYPE LOCUS).

Genome and chromosomes

Haploid cells of *S. cerevisiae* contain ~1.65×10^7 bp of DNA, a genome some three times larger than that of the bacterium *Escherichia coli*, but almost 200 times smaller than the human genome. Eighty-five to ninety per cent of the DNA is located in the nucleus where it is partitioned into 16 chromosomes which range in size from 240 kb to 1.6 Mb [4]. Of the remaining DNA, 10–15% is found in the mitochondria as a 75-kb circular molecule (mtDNA, *see* MITOCHONDRIAL GENOMES: LOWER EUKARYOTES) and 3–5% is found as a 6.3-kb circular DNA plasmid called the 2 μM PLASMID. There are other genetic elements in *S. cerevisiae*, the most intensively studied being the 'killer' trait encoded by a family of double-stranded RNA plasmids.

Unlike higher eukaryotes, the *S. cerevisiae* genome contains little REPETITIVE DNA with some 90% of the nuclear genome being single-copy sequence. There is no evidence for families of highly repeated simple-sequence DNA except in the telomere-associated regions (immediately subterminal regions, *see* CHROMOSOME STRUCTURE). There are however three different TRANSPOSABLE GENETIC ELEMENTS — Ty, sigma and tau — each of which occurs in about 20–50 copies per haploid genome and which are probably analogous to the MODERATELY REPETITIVE INTERSPERSED SEQUENCES which characterize higher eukaryotic genomes (*see* GENOME ORGANIZATION). A number of genes are present in multiple copies in the *S. cerevisiae* nuclear genome including 360 tRNA genes, and 100–120 rRNA genes. Many of the genes that encode abundant proteins, for example glycolytic enzymes, ribosomal proteins, and histones, are present in two and sometimes three copies.

Yeast chromosomal DNA, like that of all other eukaryotic cells, is complexed with the four main histones — H2A, H2B, H3 and H4 — to form a CHROMATIN fibre, yet apparently lacks histone H1 — the histone important for compaction of higher eukaryotic chromatin during mitosis. Yeast chromatin also contains a number of acidic nonhistone-like proteins of undefined function. *S. cerevisiae* chromatin is organized into nucleosomal subunits which, as in mammalian cells, each contain 146 bp of DNA although the nucleosome linker DNA is somewhat shorter (20 bp in yeast, 40–60 bp in mammalian chromatin). The 2 μm plasmid is also present as chromatin in the nucleus. Metaphase chromosomes of *S. cerevisiae* cannot be visualized, either because they are too small or because the lack of histone H1 prevents their effective compaction.

The major functional elements of yeast chromosomes, namely centromeres (CEN), telomeres (TEL), and origins of replication (ARS, autonomously replicating sequences), have also been defined at the molecular level for many of the 16 chromosomes (*see* CHROMOSOME STRUCTURE). These elements have been used to construct artificial chromosomes (YACs or yeast artificial chromosomes) although these are less mitotically stable than their natural counterparts suggesting that there are additional important *cis*-acting chromosomal elements yet to be defined. The various components of *S. cerevisiae* chromosomes have also been widely exploited in the development of a broad range of CLONING VECTORS.

The entire DNA sequence of chromosome III has now been determined by a consortium of European laboratories [5]. At 315 kb it is one of the smallest chromosomes of this simple eukaryote but represents the first reported sequence of an entire chromosome from any organization.

Gene organization and expression

Protein-coding genes of *S. cerevisiae*, just like their mammalian counterparts, are transcribed in the nucleus by RNA POLYMERASE II (polII) and its ancillary TRANSCRIPTION FACTORS [6]. The PROMOTERS of these genes contain several conserved *cis*-acting elements including the TATA box, which binds the general transcription factor TFIID, and the CCAAT box, which binds the heteromeric HAP2/3/4 complex (*see* EUKARYOTIC GENE EXPRESSION; EUKARYOTIC GENE STRUCTURE; TRANSCRIPTION FACTORS). Further upstream (i.e. to the 5′ side) of these sequences are other *cis* elements implicated in promoter regulation — the UPSTREAM ACTIVATOR SEQUENCES (UAS) which bind gene-specific transcriptional activators (and which resemble in many respects ENHANCERS) and UPSTREAM REPRESSOR SEQUENCES (URS), which bind negatively controlling protein factors. The high degree of conservation of structure and function of eukaryotic transcription factors is highlighted by the demonstration that both general factors (e.g. TFIID) and specific transcriptional activators (e.g. Gal4, Gcn4) from *S. cerevisiae* can bind to mammalian promoters and participate in the transcriptional mechanism. Although sequences 3′ to the coding region of a transcribed gene are important for correct termination of polII transcription the precise nature and function of these *cis*-acting sequences remains unclear for *S. cerevisiae*. All polII-generated transcripts are capped post-transcriptionally at their 5′ end with 7-methylguanosine (m^7G) and are polyadenylated at their 3′ end (*see* RNA PROCESSING), the latter even being the case for histone mRNAs which are not generally polyadenylated in mammalian cells.

Few *S. cerevisiae* polII-transcribed genes have INTRONS, the principal exception being a small group of genes encoding ribosomal proteins. These introns contain the conserved 5′ and 3′ splice sites. The branch point sequence (5′ UACUAAC 3′) is highly conserved within *S. cerevisiae* introns, and is significantly less degenerate than the analogous mammalian sequence. The introns are spliced by two transesterification reactions mediated, as in higher eukaryotes, by spliceosomes comprised of a number of snRNPs [7] (*see* RNA SPLICING). Introns also occur in several tRNA genes (transcribed by RNA polymerase III), in which the typically short intron (15–40 nucleotides) is always located immediately adjacent to the ANTICODON of the tRNA. These introns are spliced out in a two-step reaction involving a tRNA ENDONUCLEASE and a tRNA LIGASE. Other genes transcribed by polIII are those encoding 5S rRNA and a number of other small RNAs (e.g. snRNAs).

RNA polymerase I (polI) transcribes the multicopy ribosomal DNA of yeast which encodes the 5.8S, 18S, and 25S rRNAs; they are initially synthesized as a precursor rRNA (35S) which is post-transcriptionally cleaved into the mature rRNAs.

S. cerevisiae mRNAs are translated by an 80S ribosome-mediated mechanism essentially similar to that in all eukaryotic cells [8] (*see* PROTEIN SYNTHESIS), with the exception that *S. cerevisiae* (and other fungi) require a third elongation factor (EF-3) in addition to the conserved elongation factor 1 (EF-1) and elongation factor 2 (EF-2). The function of EF-3 in protein synthesis has yet to be defined. *S. cerevisiae* mRNAs require a total of 46 tRNA species for their translation with 20 of these being abundant, reflecting the strong CODON BIAS found in many yeast mRNAs.

Once a protein is synthesized in *S. cerevisiae* it may be targeted to one of the following cellular compartments depending on the nature of the targeting signal encoded by the protein: mitochondria, nucleus, peroxisomes, endoplasmic reticulum (ER), or periplasmic space [9] (*see* MITOCHONDRIAL BIOGENESIS; PROTEIN SECRETION; PROTEIN TARGETING; PROTEIN TRANSLOCATION). Proteins translocated across the ER membrane enter the secretory pathway, being subsequently transported to the GOLGI APPARATUS and then, via secretory vesicles, to the cell surface. Many of the steps involved in secretion have been defined by analysing *sec* (secretion-defective) mutants. Proteins entering this pathway may also be diverted to the VACUOLE, a lysosome-like structure containing several hydrolytic enzymes. Although *S. cerevisiae* seems to have a system analogous to the mammalian signal recognition particle (SRP) for linking protein translation with its translocation across the ER membrane (i.e. CO-TRANSLATIONAL translocation) there is also an SRP-independent mechanism. Some proteins are also retained within the lumen of the ER, particularly those implicated in protein folding, such as BIP and PROTEIN DISULPHIDE ISOMERASE. The nature of the C-terminal ER RETENTION SIGNAL for such proteins in *S. cerevisiae* (-HDEL) is similar to the mammalian signal (KDEL).

Much of our understanding of how cytoplasmically synthesized proteins can be targeted to the mitochrondria comes from studies with *S. cerevisiae*, particularly the importance of mitochondrial membrane receptors and the role of heat shock protein-related chaperonins (*see* HEAT SHOCK; MITOCHONDRIAL BIOGENESIS; MOLECULAR CHAPERONES).

Many *S. cerevisiae* proteins undergo POST-TRANSLATIONAL MODIFICATION (e.g. acylation, glycosylation and phosphorylation) in which various nonprotein chemical groups are attached. Generally the carbohydrate group added to *S. cerevisiae* GLYCOPROTEINS is less variable than those found in mammalian cells and is especially rich in mannose residues. Both O-linked and N-linked glycosylation occur. Studies with *S. cerevisiae* have also demonstrated the importance of protein phosphorylation in regulating diverse cellular activities including transcription, translation, and conjugation.

Cell division and its regulation

Cell division in *S. cerevisiae*, as in all eukaryotes, involves the participation of spindles upon which the chromosomes line up before the regular segregation process (*see* MITOSIS). Attachment of the chromosome to the spindle fibres is by the centromere. *S. cerevisiae* has no recognizable equivalent of the mammalian KINETOCHORE. Following chromosome segregation and nuclear cleavage the cell division cycle terminates with the division of the cytoplasm of the mother cell to produce the daughter bud (cytokinesis). The CELL CYCLE of *S. cerevisiae* consists of the sequential M-G1-S-G2 phases observed in other eukaryotes, with DNA synthesis being limited to the S phase [10]. Cell division in *S. cerevisiae* differs from other eukaryotic cells in having a closed mitosis — the nuclear membrane does not break down before spindle formation.

Numerous mutants with conditional defects in the cell division cycle (*cdc* mutants) have been used to define the steps required for normal progression through the cycle. Several components of the regulatory machinery have been identified including the *CDC28* gene product, a PROTEIN KINASE which controls the 'start' of the cell cycle (the transition from G1 to S). *CDC* gene products have also been implicated in DNA replication (e.g. *CDC9* encodes a DNA LIGASE) and SPINDLE POLE BODY duplication (e.g. *CDC31*). Several genes for CYCLIN, which is implicated in eukaryotic cell cycle control, have been described. Studies with *S. cerevisiae* have contributed significantly to our understanding of how eukaryotic cells become committed to a cell cycle and how this cycle is executed.

Gene manipulation

A major attraction of *S. cerevisiae* to researchers in all fields of cell and molecular biology is the ease with which it can be genetically manipulated by either random mutagenesis or SITE-DIRECTED MUTAGENESIS. Furthermore, the ability to maintain *S. cerevisiae* as a haploid organism has facilitated the isolation and phenotypic characterization of RECESSIVE MUTATIONS and their subsequent genetic analysis. This is achieved by mating the haploid mutant strain with another genetically marked haploid, but of opposite mating type, to generate a diploid heterozygous at the locus of interest. The inheritance of the mutant gene is then followed by TETRAD ANALYSIS after the heterozygous diploid has been induced to undergo meiosis (i.e. sporulation). For nuclear single-gene mutations two of the meiotic products (spores) will carry the mutant gene and the remaining two the wild-type gene, giving a 2 : 2 segregation pattern. Deviation from this pattern can indicate any one of a number of genetic phenomena in the diploid — GENE CONVERSION, ANEUPLOIDY or POLYGENIC CONTROL of the genetic trait being examined. A ratio of 4 : 0 indicates a NON-MENDELIAN genetic trait; the respiration-deficient PETITE MUTATIONS, some of which have genetic defects in the mitochondrial genome, show either a 4 petite : 0 wild type inheritance pattern (a suppressive *petite*) or a 0 petite : 4 wild type pattern (a neutral *petite*) on tetrad analysis.

Genetic mapping techniques that exploit tetrad analysis have been used to assign the chromosomal linkage of well over 600 gene mutations thereby generating a complex GENETIC MAP of all 16 chromosomes of *S. cerevisiae* [11]. A PHYSICAL MAP of the *S. cerevisiae* genome has also been constructed on the basis of the location of sites for the rare-cutting RESTRICTION ENZYMES *Not*I and *Sfi*I.

The easy application of recombinant DNA techniques in the genetic manipulation of *S. cerevisiae* has further enhanced its utility as a model eukaryote [12]. After the discovery of a recombinant plasmid-based transformation system in 1978, a vast repertoire of recombinant plasmids has been developed for introducing cloned DNA sequences into *S. cerevisiae*. Different classes of plasmid provide for different modes of maintenance of the cloned DNA sequence in the cell (Table S1). Transformants are simply selected by the ability of a plasmid-borne gene (e.g. *URA3* encoding orotidine-5′-phosphate decarboxylase) to complement a host genetic deficiency (e.g. *ura3*, conferring a uracil requirement on the host). A series of YAC vectors have been developed which can replicate in *S. cerevisiae* with up to 1000-kb DNA inserts and these have proved of immense value in mammalian genome cloning and mapping projects (*see* HUMAN GENE MAPPING).

The large number of well-defined yeast mutants with easily recognizable phenotypes has meant that the corresponding wild-type genes can be directly cloned by complementation of the host mutation by a plasmid-borne sequence; well over 500 complete *S. cerevisiae* gene sequences (some 10% of the total predicted number of genes in the *S. cerevisiae* genome) are now available in the major DNA sequence databases (e.g. EMBL, GenBank; *see* DATABASES AND INFORMATION HANDLING IN MOLECULAR BIOLOGY).

The discovery that cloned *S. cerevisiae* genes, introduced into *S. cerevisiae* by transformation on a YIp plasmid (see Table S1), recombine with their chromosomal homologues has led to the development of reverse genetic methods for generating specific gene mutations in the genome. These methods have now been refined to such an extent that any desired mutational change can be made in a cloned gene *in vitro* (*see* SITE-DIRECTED MUTAGENESIS) and the modified form of that gene then integrated into its original chromosomal location, by HOMOLOGOUS RECOMBINATION, precisely replacing the endogenous wild-type allele. Such a gene replacement strategy can be carried out efficiently because, unlike in higher eukaryotic cells, integration via a nonhomologous recombination mechanism does not seem to occur in *S. cerevisiae*.

Biotechnological applications

The brewing and baking industries represent the oldest and possibly the largest of all biotechnology-related industries [13]. In beer production *S. cerevisiae* is used primarily in top fermentations of wort to produce the beers and ales traditionally consumed in the UK. The natural tendency of *S. cerevisiae* to flocculate, that is to form large cellular aggregates, during fermentation has helped the brewer but is little understood in mechanistic or genetic terms. In addition to producing ethanol and carbon dioxide, a number of higher alcohols (fusel alcohols) are synthesized by *S.*

cerevisiae. Most strains are highly tolerant to ethanol facilitating their use in the production of ethanol for other uses (e.g. 'gasohol'). One drawback of *S. cerevisiae* as a fermentative organism is its inability to utilize polysaccharides such as starch and cellulose although these metabolic 'defects' can be overcome by genetic manipulation, by introducing the genes encoding the necessary amylolytic enzymes.

More than 2 million tonnes of *S. cerevisiae* are produced commercially each year worldwide, with much of the yeast biomass being used in the human and animal food industries.

More recently *S. cerevisiae* has found a new biotechnological outlet as a host for the commercial-scale expression of RECOMBINANT PROTEINS of pharmaceutical value [14]. This is primarily because of our detailed understanding of gene expression in *S. cerevisiae* and its wide range of plasmid VECTORS. *S. cerevisiae* is also recognized by regulatory authorities such as the US Food and Drug Administration (FDA) as a GRAS (Generally Regarded As Safe) host for such exploitation, and by the fact that many of the post-translational processes typical of a mammalian cell are found in this simple eukaryote (*cf.* the bacterium *Escherichia coli*; *see* GENETIC ENGINEERING). Various strategies have been developed for expressing recombinant proteins. These include constitutive, high-level intracellular expression using endogenous promoters from genes such as *PGK1* and *ADH1*; regulated expression using either the *GAL1* promoter (regulated by choice of carbon source) or the *CUP1* promoter (regulated by heavy metals); and high level extracellular synthesis by coupling an appropriate signal sequence to the N terminus of the recombinant protein. The signal peptide of the *S. cerevisiae* α mating factor (MFα) (pre-pro) precursor has been widely used to convert recombinant proteins into a secretory form. A wide range of biopharmaceutical products expressed in *S. cerevisiae* have now either been approved for sale or are in clinical trials, with the hepatitis B vaccine Recombivax of Merck, Sharp and Dohme (*see* RECOMBINANT VACCINES) representing one of the first recombinant products available worldwide.

M.F. TUITE

1 Strathern, J.N. et al. (Eds) (1982) *Molecular Biology of the Yeast Saccharomyces: Life Cycle and Inheritance* (Cold Spring Harbor, New York).
2 Strathern, J.N. et al. (1982) *Molecular Biology of the Yeast Saccharomyces: Metabolism and Gene Expression* (Cold Spring Harbor, New York).
3 Herskowitz, I. (1988) Life cycle of the budding yeast *Saccharomyces cerevisiae*. *Microbiol. Rev.* **52**, 538–552.
4 Newlon, C.S. (1989) In *The Yeasts*, Vol. 3 (Rose, A.H. & Harrison, J.S., Eds), 57–116 (Academic Press, New York).
5 Struhl, K. (1989) Molecular mechanisms of transcriptional regulation in yeast. *Annu. Rev. Biochem.* **58**, 1051–1080.
6 Ruby, S.W. & Abelson, J. (1991) Pre-mRNA splicing in yeast. *Trends Genet.* **7**, 79–85.
7 Tuite, M.F. (1989) In *The Yeasts*, Vol. 3 (Rose, A.H. & Harrison, J.S., Eds), 161–204 (Academic Press, New York).
8 Sheckman, R. (1985) Protein localization and membrane traffic in yeast. *Annu. Rev. Cell Biol.* **1**, 115–143.
9 Wheals, A.E. (1987) In *The Yeasts*, Vol. 1 (Rose, A.H. & Harrison, J.S., Eds), 283–290 (Academic Press, New York).
10 Mortimer, R.K. et al. (1992) Genetic map of *Saccharomyces cerevisiae*, edition 11. *Yeast* **8**, 817–902.
11 Guthrie, C. & Fink, G.R. (Eds) (1991) *Guide to Yeast Genetics and Molecular Biology* (Academic Press, New York).
12 Oliver, S.G. et al. (1992) The complete DNA yeast chromosome III. *Nature* **357**, 38–46.
13 Oliver, S.G. (1991) In *Biotechnology Handbooks 4* — Saccharomyces (Tuite, M.F. & Oliver, S.G.), 213–248 (Plenum, New York).

Table S1 The classes of plasmid used for transformation of the yeast *Saccharomyces cerevisiae*

Plasmid type	Mode of maintenance	Key elements*
YIp (Integrative)	Integrated: single or multicopy	S
YEp (Episomal)	Autonomous: multicopy	S + ORI (2μ)
YRp (Replicative)	Autonomous; multicopy	S + ARS
YCp (Centromeric)	Autonomous: single copy	S + ARS + CEN
YLp (Linear)	Autonomous: single copy	S + ARS + CEN + TEL

* ARS, autonomously replicating sequence; CEN, centromere; ORI (2μ), origin of DNA replication from the 2μ plasmid; S, selectable marker; TEL, telomere.

14 Barr, P.J. et al. (Eds) (1989) *Yeast Genetic Engineering* (Butterworths, Boston).

SAF Scrapie-associated fibrils. *See*: TRANSMISSIBLE SPONGIFORM ENCEPHALOPATHIES.

salicylic acid 2-Hydroxybenzoic acid. M_r 138.12. Its role as an endogenous regulator in plants was first recognized as the trigger of thermogenesis in the *Arum* spadix. It is also involved in plant defence and is an inducer of PATHOGENESIS-RELATED PROTEINS. *See*: PLANT WOUND RESPONSES.

Salmonella typhimurium Gram-negative bacterium of the family Enterobacteriaceae. This species is motile with peritrichous flagella and is a facultative anaerobe. The molecular genetics of *S. typhimurium* have been intensively studied as an alternative to *Escherichia coli* as a model organism. *See also*: BACTERIAL ENVELOPE; BACTERIAL GENE EXPRESSION; BACTERIAL GENE STRUCTURE; O-ANTIGEN; PHASE VARIATION.

salt bridge (electrostatic bond) Noncovalent bond formed when a charged group attracts an oppositely charged group. Salt bridges may be formed between enzymes and their substrates, and between the subunits of a multimeric protein (*see e.g.* HAEMOGLOBINS) or in virus CAPSIDS. Salt bridges are also known as ion pairs, ionic bonds, or salt linkages.

saltatory movement (of organelles) Abrupt, seemingly random, jerking movements of some organelles.

salvage compartment Name originally given to that part of the *cis* Golgi network of eukaryotic cells involved in salvaging escaped ENDOPLASMIC RETICULUM proteins and returning them to the ER. It is now known that proteins can also be salvaged from the Golgi stack. *See also*: GOLGI APPARATUS.

Warren, G. (1987) *Nature* **327**, 17–18.
Pelham, H.R.B. et al. (1992) *Trends Cell Biol.* **2**, 183–185.

SAM S-ADENOSYLMETHIONINE.

sam *See*: ONCOGENES.

Sandhoff disease One of the severe AUTOSOMAL RECESSIVE GM_2-gangliosidoses. It results from deficiency of the β subunit of the enzyme HEXOSAMINIDASE, which is a component of both isoenzyme forms (A and B). This causes a build-up of ganglioside GM_2 and other GLYCOPLIPIDS of neuronal sphingolipid metabolism. The clinical features resemble the commoner TAY–SACHS DISEASE. Both prenatal and carrier state diagnosis is possible by enzyme analysis.

Sanger method A widely used method of DNA SEQUENCING.

Sanger reaction A method for determining the N-terminal amino acid residue of a protein, originally developed by Frederick Sanger. The N-terminal residue can be labelled with 1-fluoro-2,4-kinitrobenzene, which reacts with the NH_2 group to form a coloured dinitrophenyl derivative of the peptide. The bond formed is stable under conditions that hydrolyse peptide bonds, so treatment with 6 M hydrochloric acid yields only the derivatized N-terminal residue, which can then be identified by high performance liquid CHROMATOGRAPHY. Dansyl chloride can also be used as the derivatizing agent as it requires much less material.

Sanger, F. & Tuppy, H. (1951) **49**, 463–490.

SAP SERUM AMYLOID P COMPONENT.

sarcalumenin *See*: CALCIUM-BINDING PROTEINS.

sarcoma A malignant tumour of connective tissue or other tissue of mesodermal origin, for example osteosarcoma (malignant tumour of bone). The demonstration by Peyton Rous in 1911 that avian sarcoma could be transmitted by a cell-free extract led eventually to the identification of the RETROVIRUS, Rous sarcoma virus, the first tumour-causing virus to be recognized.

sarcomere In striated MUSCLE the morphological and functional repeating unit, visible under the electron microscope, which forms a single contractile unit.

sarcoplasmic reticulum (SR) An extensive membrane-bounded compartment in skeletal MUSCLE, which is a store of intracellular Ca^{2+} (*see* CALCIUM). On stimulation of muscle, calcium channels in the SR open to release calcium, which is involved in causing muscle contraction. This calcium release channel has homology to the inositol trisphosphate receptor which acts as a calcium channel to release calcium from intracellular stores in many types of cells, but its opening is controlled by voltage changes transmitted via the T-tubules (*see* CALCIUM). The calcium stores of the SR are replenished by the action of Ca^{2+}-ATPase in the SR membrane, which acts as an ion pump (*see* MEMBRANE TRANSPORT SYSTEMS) to transport Ca^{2+} from the cytosol into the SR lumen. *See*: MUSCLE.

Sasisekharan–Ramachandran plot *See*: PROTEIN STRUCTURE.

satellite cell *See*: MYOGENESIS.

satellite DNA Highly repeated sequences (present in at least 10^4 copies per genome) that are arranged tandemly into clusters. In many species, these clusters are of sufficiently different base composition from that of the bulk of genomic DNA that DNA containing them into a distinct band on isopycnic collects DENSITY GRADIENT CENTRIFUGATION. They are often found in abundance around CENTROMERES (*see* CHROMOSOME STRUCTURE). *See also*: HYBRIDIZATION; SATELLITE RNA.

satellite RNA (1) RNA arising from the transcription of SATELLITE DNA.
(2) Episomal component of plant RNA viruses dependent on the virus for replication and other functions necessary for spread in infected plants. Satellites are not required by the helper virus and do not share extensive sequence homology with it. They can be either circular or linear. Satellite viruses encode their own coat

proteins; satellite RNAs are encapsidated by the helper virus. Also known as virus satellites or virus satellite RNA. *See also*: PLANT VIRUSES; VIROIDS; VIRUSOIDS.

satellite sequence *See*: SATELLITE DNA; SATELLITE RNA.

satellite virus *See*: SATELLITE RNA.

saturated fatty acids FATTY ACIDS which contain no C=C or C≡C bonds.

saturation mutagenesis The introduction of single base-pair substitution MUTATIONS at most if not all positions in a defined nucleotide sequence. This technique enables the mapping of important functional regions, such as transcription initiation sites in a PROMOTER, protein binding sites, etc. Similarly, the introduction of MISSENSE MUTATIONS individually at each codon in a structural gene can reveal important regions or motifs within the encoded protein. Mutants can be generated by IN VITRO MUTAGENESIS by, for example, incubation of cloned DNA with hydroxylamine, which promotes A → T substitutions in the DNA sequence. The target DNA may be an entire bacterial chromosome, or DNA cloned in a suitable vector.

SAUR Acronym for *Small Auxin Up RNAs*. They were isolated as cDNA clones from excised soybean cotyledons treated with AUXIN. They comprise a family of highly homologous clustered genes, whose transcription is activated by auxin, reaching maximum rates within 5 min of auxin treatment. The production of the RNAs, which are about 550 nucleotides long, correlates with growth redistribution during the gravitropic response of soybean hypocotyls.

McClure, B.A. & Guilfoyle, T.J. (1987) *Plant Mol. Biol.* **9**, 611–623.

SAXS Acronym for small-angle X-ray scattering. *See*: SMALL-ANGLE SCATTERING.

SBMA X-linked spinal and bulbar muscular atrophy. *See*: KENNEDY'S DISEASE.

scaffold attachment regions *See*: NUCLEAR MATRIX.

scanning electron microscopy (SEM) *See*: ELECTRON MICROSCOPY.

scanning mutagenesis Also known as linker scanning, this IN VITRO MUTAGENESIS strategy enables mutations to be targeted within a specific region of a cloned DNA sequence (e.g. a PROMOTER). It involves the generation of both 5′ and 3′ deletions of the target DNA sequence and their subsequent ligation via a synthetic oligonucleotide linker.

McKnight, S.L. & Kingsbury, R.L. (1982) *Science* **217**, 316–324.

scanning tunnelling microscopy Electron microscopic technique which can achieve atomic-scale imaging. It has been used to study double-helical DNA.

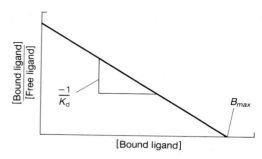

Fig. S2 Scatchard plot.

Driscoll, R.J. et al. (1990), *Nature* **346**, 294–296.

Scatchard plot A method for determining binding parameters from data on the binding of a radioactively labelled ligand (a radioligand) to a RECEPTOR (or other binding protein). The plot (introduced by G. Scatchard in 1949) for binding to soluble proteins is shown in Fig. S2. It is equally applicable to membrane-bound receptors and to isolated receptors. It represents a plot of r/c versus r for different ligand concentrations, where r = the ratio of moles ligand bound per mole protein (often expressed as amount of ligand bound per milligram protein) and c = concentration of unbound ligand. It requires:

1 The attainment of equilibrium in the binding reaction at a saturable set of sites.
2 An accurate knowledge of the specific radioactivity of the ligand sample used and of its radio-purity.
3 Measurement of the amount of radioactivity bound at a series of ligand concentrations (preferably 10 or more) covering the entire range over which the specific sites present can bind it.
4 Measurement of the protein content of the sample.

The amount of ligand bound (plotted on the x-axis) is calculated per milligram of sample protein. A linear Scatchard plot is obtained if there is only a single class of binding site present and if there is no COOPERATIVITY of binding at the sites on one receptor complex. The intercept on the x-axis is the value of B_{max}, that is, the number of the binding sites per milligram of protein present. Where a purified receptor protein is studied, the B_{max} is an indicator of its purity. For example the NICOTINIC RECEPTOR has two high-affinity sites per receptor molecule for acetylcholine (or for competitive antagonists and for its neurotoxins) and dividing by the known molecular weight of the pentameric protein predicts a B_{max} of 8000 nanomoles per milligram protein for a pure specimen. The slope of a linear Scatchard plot is the negative of the affinity, which is the reciprocal of K_d, the equilibrium binding (dissociation) constant for the ligand–receptor site interaction.

Curved Scatchard plots denote (if experimental artefacts are excluded) cooperativity in the binding (convex upwards, positive cooperativity; concave downwards, negative cooperativity), or they denote (concave downwards) multiple affinities, due to two or more classes of binding sites. It is usual now to fit binding data and derive Scatchard plots by one of the standard computer programs for ligand binding analysis. These will determine iteratively the best-fit values of B_{max} and K_d and whether the data show the best statistical fit to a one-site, two-site, or three-site population, giving the corresponding parameters for each of these.

scatter factor HEPATOCYTE GROWTH FACTOR.

scattering factor *See*: ATOMIC SCATTERING FACTOR.

scattering intensity distribution($S(Q)$) In a solution-scattering experiment in SMALL-ANGLE SCATTERING, when scattering from a complete molecule is considered, constructive and destructive wave interference occurs for scattering from atom to atom. In this case the scattering is not uniform in all directions defined by the complete solid angle 4π steradians (*cf.* SCATTERING PARAMETER). The scattering (S) from one molecule can still be expressed as an effective impenetrable scattering cross-section, but this now has to be composed of scattering in different small elements of solid angle in the different scatter directions defined by Q. Thus the scattered intensity is expressed as a molecular scattering cross-section per unit solid angle in direction Q. This can be calculated from measured scatter 'spectra' knowing the concentration of the molecules in the specimen and calibrating with a spectrum from a known concentration of a standard specimen. $S(Q)$ for molecules in solution is the intensity to be expected by averaging taking into account the ideally random distribution of molecular orientations. *See*: NEUTRON SCATTERING AND DIFFRACTION.

scattering length *See*: ATOMIC SCATTERING LENGTH.

scattering-length density (ϱ) Macromolecules consist of many atoms and, from the point of view of scattering and diffraction studies may be considered to be a distribution of scattering-length density. The value of the scattering-length density at any point in the molecule is the sum of the atomic scattering lengths in a small volume element surrounding that point divided by the volume element. This terminology is used in NEUTRON SCATTERING AND DIFFRACTION; the equivalent for X-RAY DIFFRACTION is ELECTRON DENSITY.

scattering parameter (Q) In a SMALL-ANGLE SCATTERING experiment the molecules ideally are in random orientation in solution such that the scattering is symmetric about the direction of the incident beam. The scattering parameter (Q) thus varies only with the angle of scattering (2θ) from the incident beam (of wavelength λ) and is defined as:

$$Q = 4\pi \, (\sin \theta)/\lambda$$

scattering vector (S) A vector which describes the geometry of the incident and diffracted beams. It is defined by the difference between the unit vectors describing the incident (s_0) and scattered beam (s) of wavelength λ by:

$$\mathbf{S} = 2\pi\lambda \, (\mathbf{s_0} - \mathbf{s})$$

See: NEUTRON SCATTERING AND DIFFRACTION.

scavenger receptor Receptor on macrophages for damaged low-

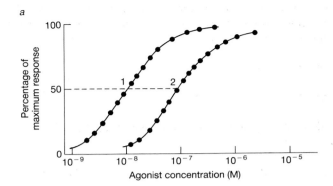

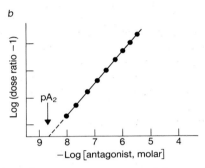

Fig. S3 *a*, Dose–response curves in the absence (1) and the presence (2) of a competitive antagonist. *b*, Schild plot derived from dose–response curves obtained at a series of increasing antagonist concentrations.

density lipoproteins. *See*: PLASMA LIPOPROTEINS AND THEIR RECEPTORS.

Schild plot A method of comparing the potency of competitive antagonists on the responses to an agonist drug. It is applicable to pharmacological measurements, for example on an isolated tissue preparation or *in vivo*, or to current responses from intracellular or PATCH-CLAMP recording. Dose–response curves (strictly concentration–response curves), in which the percentage of maximum response is plotted against log (agonist concentration) should be sigmoid and displaced towards the high agonist–concentration side by the added antagonist, parallel to the curve obtained without antagonist and with no depression of the maximum response (Fig. S3*a*). The latter is a requirement for competitive behaviour. Then, a fixed level of the response (e.g. 50% of the maximum) is compared on the curve obtained with the agonist alone and a curve with antagonist added: the ratio of those two agonist concentrations is the dose ratio (strictly concentration ratio) for that antagonist level. In the Schild plot, log (dose ratio – 1) is plotted against – log [antagonist] (Fig. S3*b*).

The slope of the Schild plot should be 1 if the antagonist is truly competitive with the agonist and if COOPERATIVITY and receptor heterogeneity are absent. This requires perfection of all experimental conditions and is rarely attained exactly. The intercept of the line on the *x*-axis (Fig. S3*b*) is termed the pA$_2$ value, this being the negative log of that antagonist concentration which would produce a twofold shift in the agonist dose–response curve.

pA_2 is ideally equal to pK_B, where K_B is the concentration of the antagonist required to occupy 50% of the receptors. Again, in the ideal case this equals the K_d value obtained for the antagonist from RADIOLIGAND binding, using it either itself radioactively labelled or as a competitor for a labelled agonist.

The Schild method is a null one, in that no assumptions are required about the coupling between the agonist binding reaction and the observed response.

Rang, H.P. & Dale, M.M. (1990) *Pharmacology*, Ch. 1 (Churchill-Livingstone, Edinburgh).

Schizophrenia

SCHIZOPHRENIA is a common disorder with a lifetime risk approaching 1%. As a psychiatric disease, schizophrenia has no objective physiological or biochemical diagnostic tests. In the past this has meant that the frequency and scope of diagnosis has varied according to social or individual whim, although it has improved with the increasing application of standardized diagnostic criteria. The disease is usually first diagnosed during early adulthood. Key symptoms include hallucinations (particularly 'hearing voices'), delusions, and abnormal experiences, such as the feeling that control of one's thoughts has been lost, perhaps to some outside agency. The patient loses empathy with others, becoming withdrawn and demonstrating inappropriate or blunted mood. Within this general syndrome several subtypes have been described including paranoid, residual, or disorganized schizophrenia; these reflect differences in the relative patterns of symptoms but have not proved to be of much value in understanding their relationships to changes in the brain or to different aetiological factors.

Of more interest in this respect is the distinction between two syndromes in schizophrenia. Patients having primarily positive symptoms (delusions, hallucinations, incongruous affect) have been described as type I, while those with negative symptoms (withdrawal, loss of drive, flattened affect) have the type II syndrome. There is still some dispute as to whether these subtypes relate to distinct disease processes or whether, at the other extreme, they are different expressions of a single disorder. It is the type I syndrome that best responds to classical antipsychotic drugs, which are far less effective at ameliorating negative symptoms. Conversely, the syndrome of negative symptoms has occasionally (if inconsistently) been associated with identifiable abnormalities of the brain.

Neuropathology and neurochemistry

Certainly there are structural differences in the central nervous system in schizophrenia [1]. Schizophrenic patients tend to have smaller brains with larger ventricular volumes, reflecting relative deficits of neurons. Recent post-mortem morphometric studies, as well as modern imaging techniques in living patients (notably magnetic resonance imaging), strongly indicate that regions of the medial temporal lobe (hippocampus, amygdala and hippocampal gyrus) are particularly affected, with diminished numbers and/or disorganization of groups of neurons. These neuronal deficits are often more apparent in the left hemisphere. As well as the morphological indications of neuronal deficits in temporal lobe structures, there are also neurochemical correlates of these changes, particularly in the hippocampus. Thus a marker for GABAergic neurons (i.e. neurons producing GABA (γ-aminobutyric acid)) in this region is diminished in schizophrenia, as is the neuropeptide CHOLECYSTOKININ, also found in GABA-containing neurons. Deficits of the kainate subtype of EXCITATORY AMINO-ACID RECEPTORS have also been reported and, paralleling the neuropathology, these various neurochemical abnormalities often appear to be greater in the left hemisphere.

A few neuropathological investigations, as well as imaging studies of neural activity by measurement of blood flow or energy metabolism, sometimes in combination with neuropsychological tasks, have indicated the frontal cortex as another site of dysfunction in schizophrenia. The relative reduction in frontal cortical activity has been found to correlate with negative symptoms of the disease. Neurochemical changes are not easy to integrate with these observations; there have been some reports consistent with an increase of glutamatergic innervation in the frontal cortex.

Pharmacology

The predominant neurochemical hypothesis for the past two decades has, however, been the 'dopamine hypothesis' of schizophrenia: that a hyperactivity of DOPAMINE neurotransmission occurs in the disease. The evidence for this is mainly circumstantial and derives from the observations that dopamine agonists and drugs increasing synaptic dopamine (e.g. amphetamine) can induce a psychosis that is indistinguishable from acute paranoid schizophrenia. Further support comes from the drugs used to treat schizophrenia [2]; while these ANTIPSYCHOTIC DRUGS (Table S2) have effects on receptors to a range of neurotransmitters, it is only their ability to block the D_2 subtype of DOPAMINE RECEPTORS that correlates with their clinical efficacy. Post-mortem neurochemical studies have provided little direct evidence for the dopamine hypothesis, and the increase in dopamine D_2 receptor density that has been observed in brain tissue from schizophrenic patients is usually ascribed not to the disease

Table S2 Common antipsychotic drugs

Chemical class	Examples
Phenothiazines	Chlorpromazine Fluphenazine Thioridazine
Thioxanthenes	Clopenthixol Flupenthixol
Butyrophenones	Haloperidol
Diphenylbutyl piperidines	Pimozide
Dibenzazepines	Clozapine
Substituted benzamides	Sulpiride Remoxipride

process but to an effect of long-term treatment with dopamine antagonist drugs. Certainly animal studies have shown that chronic administration of these drugs can increase both numbers of, and messenger RNA for, D_2 receptors. A further problem is that D_2 receptors are found primarily in the striatum of the brain, a region that is mainly concerned with the control of movement and not particularly implicated in the complex behaviours that make up schizophrenia. The only indication of a dopamine dysfunction in schizophrenia is that of an asymmetric increase in the (left hemisphere) amygdala, although it seems likely that this might be a secondary effect of other neuronal abnormalities observed in the temporal lobe (see above).

Recently a new perspective in which to consider antipsychotic drug action has been provided by the discovery, based on the identification of their mRNAs, of other dopamine receptors. Thus the D_3 site may be important in mediating antipsychotic drug action as its antagonist pharmacology resembles that of the D_2 receptor and it is expressed in a relatively greater proportion outside the striatum. The D_4 receptor also resembles the D_2 and D_3 molecules with ~40% homology and, as well as being expressed in regions that include the amygdala and frontal cortex, D_4 may have a greater affinity for clozapine, a drug that is unusual in its notable efficacy in patients who do not respond to other antipsychotics. While we have yet to see whether D_3 and D_4 receptors are important in human brain function, the potential of drugs acting selectively at these sites in the treatment of schizophrenia is being actively investigated.

Aetiology and genetics

The symptoms seen in schizophrenia can occur in various other diseases of the brain including EPILEPSY and HUNTINGTON'S CHOREA as well as some metabolic disorders. Along with the wide variation in symptoms shown by schizophrenic patients, this might suggest schizophrenia to be the product of several disease processes. Certainly many aetiological factors, both genetic and environmental, have been proposed. Of the possible nongenetic causes, several pre- or perinatal events are postulated to be responsible for the development of the disease. Any abnormalities of the brain in schizophrenia seem to be established before the onset of the symptoms of the disease; imaging studies provide no indication of a continuing atrophic process and in autopsy samples there is an absence of the usual markers of previous neuro-degeneration. Along with histological findings suggestive of abnormal neuronal migration, these studies implicate an insult to the brain during development as a causative factor. Obstetric complications and viral infection of the mother or foetus *in utero* have variously been proposed to contribute to the later emergence of schizophrenia.

The disease is partly genetically transmitted, as demonstrated by a CONCORDANCE of up to 40–50% between monozygotic twins, although the level of concordance is very sensitive to the diagnostic criteria used. This may indicate that environmental factors determine whether or not the disease appears in individuals with a genetic predisposition to schizophrenia, although an alternative explanation is that either environmental or simple genetic aetiologies are separately responsible in different cases. Using families

in which a clear hereditary transmission is present, searches have been undertaken for genes of major effect [3]. These have usually been initiated following single observations of families in which chromosomal abnormalities have been associated with syndromes that included schizophrenic symptoms. An initial report apparently linking schizophrenia with markers on chromosome 5 in seven Icelandic and British families was not subsequently reproduced in a range of other European and American pedigrees, and the original report is now considered to be a false positive result. Linkage to chromosome 11, hypothesized following the report of a large pedigree in which a balanced 1:11 chromosomal translocation segregated with schizophrenia [4], has proven more positive, but can still only account for a small proportion of hereditary disease. Association with genes on the 'autosomal' part of the sex chromosomes has also been postulated, with some supporting evidence.

The dopamine D_2 receptor has attracted some interest as a candidate gene, particularly as it is encoded on chromosome 11. So far the investigations of such candidate genes as possible sites of genetic abnormality in schizophrenia have developed from naive hypotheses that have usually yielded negative or inconsistent data in simpler biochemical studies. This reflects the lack of evidence for a consistent, and primary, neurochemical abnormality in the disease.

Genetic studies in schizophrenia are complicated by the difficulties of a circumscribed diagnosis; many pedigrees in which schizophrenia is prominent also contain individuals with AFFECTIVE DISORDERS and other psychiatric diagnoses. Results are often sensitive to whether or not such cases are included as 'schizophrenic', although the concept of psychosis as a continuum, in which manic-depressive and schizoaffective diseases are not genetically differentiated from schizophrenia, has proven useful in some analyses.

G.P. REYNOLDS

1 Roberts, G.W. (1990) Schizophrenia: the cellular biology of a functional psychosis. *Trends Neurosci.* **13**, 207–211.
2 Reynolds, G.P. (1992) Developments in the drug treatment of schizophrenia. *Trends Pharmacol. Sci.* **13**, 116–121.
3 Ciaranello, R.D. & Ciaranello, A.L. (1991) Genetics of major psychiatric disorders. *Annu. Rev. Med.* **42**, 151–158.
4 St Clair, D. et al. (1990) Association within a family of a balanced autosomal translocation with major mental illness. *Lancet* **336**, 13–16.

Schizosaccharomyces pombe Fission yeast. Unicellular eukaryotic microorganism of the ASCOMYCOTINA which has been intensively studied in connection with the mating type locus and the genetics of control of the CELL CYCLE. Unlike the budding yeast, SACCHAROMYCES CEREVISIAE, to which it is only distantly related, it divides by transverse division of the elongated cylindrical cells. GENETIC NOMENCLATURE in *Schiz. pombe* is quite distinct from that of *S. cerevisiae*. See: YEAST MATING-TYPE LOCUS.

Schwann cells GLIAL CELLS which form the myelin sheath around the axons of neurons in the PERIPHERAL NERVOUS SYSTEM.

SCID, *scid* mice *See:* SEVERE COMBINED IMMUNODEFICIENCY.

sclerenchyma Plant tissue composed of hard-walled cells.

scrapie *See*: TRANSMISSIBLE SPONGIFORM ENCEPHALOPATHIES.

scyrps SMALL CYTOPLASMIC RIBONUCLEOPROTEIN PARTICLES.

SDP SHORT DAY PLANT.

SDS The detergent SODIUM DODECYL SULPHATE.

SDS-PAGE Sodium dodecyl sulphate-polyacrylamide gel electrophoresis. *See*: ELECTROPHORESIS.

sea urchin development *See*: ECHINODERM DEVELOPMENT.

Sec mutants Mutants of either ESCHERICHIA COLI or SACCHAROMYCES CEREVISIAE showing a defect in secretion, for example *secB* (*E. coli*), *secl8* (*S. cerevisiae*). *See*: PROTEIN SECRETION; PROTEIN TRANSLOCATION.

***sec4* protein** RAB PROTEINS. *See also*: GTP-BINDING PROTEINS.

Secale cereale Rye.

Second messenger pathways

CELLS are sensitive to an extraordinary number of chemical signals: these control cell and tissue behaviour and integrate body function. Such signals may originate close to the responding cells (e.g. neurotransmitters, prostaglandins), be distributed throughout multicellular bodies (e.g. classical hormones), or travel between organisms (e.g. odours). The 'target cells' for each signal are the cells that bear RECEPTORS for that agent. Each extracellular stimulus acts at a unique receptor (or at multiple types of receptor), but onward transmission of information into cells is channelled through a limited number of signalling pathways built into the plasma membrane. As a result, one signalling pathway may mediate the effects of many extracellular agents on many intracellular responses in many target cells. The principles of the use of this economical style of information transmission by the cells of multicellular organisms are summarized in Fig. S4. This article will concentrate exclusively on cellular signalling in animal cells. Signal transduction in plants is less well understood; the classical cyclic AMP second messenger pathway (see below) appears to be absent, but there is evidence for the inositol phospholipid signalling pathway (*see* PLANT CELL SIGNALLING; PLANT HORMONES).

Recognition of extracellular stimuli by receptors

Ligand-specific receptor proteins in the cell-surface membrane recognize extracellular stimuli and discriminate between closely related molecules. Striking tissue-to-tissue variations in the pharmacology of responses to many agents provided the first evidence that a single stimulus often controls cells through more than one

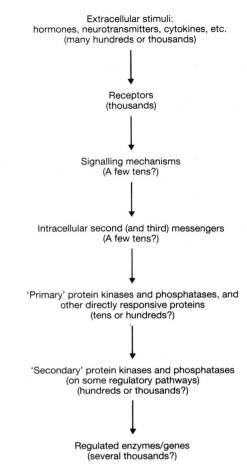

Fig. S4 A schematic summary of the manner in which cells use a limited number of signal transduction pathways to transmit to the cell interior the diverse regulatory information that is brought by a very large number of stimuli which act at an even larger number of receptors. This summary also gives an approximate idea of the magnitude of the genetic resources that cells invest in machinery for the receipt and interpretation of extracellular information.

type of receptor (e.g. MUSCARINIC and NICOTINIC RECEPTORS for accetylcholine; H_1 and H_2 receptors for histamine), and the recent molecular cloning of receptor proteins has revealed a far greater receptor diversity than originally suspected.

Thus the genome of every cell encodes many more receptors than the number of different chemical signals that any individual cell ever has to recognize: there are probably a few thousand receptor genes, representing a few per cent of the genome. A small subset of these genes is expressed in a differentiated cell at a particular time: this allows each cell to respond only to the limited range of stimuli for which it is a target, and each cell's receptor repertoire is constantly undergoing environmental regulation rather than remaining static. For example, the events by which ANTIGENS initiate proliferation and differentiation of lymphocytes into mature immune effector cells involve correct sequential expression of a series of LYMPHOKINE receptors (*see also* B CELL DEVELOPMENT; T CELL DEVELOPMENT). This selective control of receptor gene expression is still little understood.

Onward transmission of receptor signals to the cell interior: general principles

The genetic load of encoding so many receptors would be multiplied if each receptor had a unique way of transmitting its message into the cell. Fortunately, this is not the case (see Fig. S4). Instead, the biochemical machinery of the plasma membrane translates the multifarious incoming information into a simpler set of ionic, electrical, and chemical signals that control intracellular events. Table S3 summarizes some of the mechanisms involved [1], and lists some receptors that use these and other signalling mechanisms. These mechanisms are well described in [2,3] and knowledge in this field is annually updated in [4].

Table S3 Examples of the signal transmission mechanisms used by receptors

Each is summarized as: Extracellular stimulus → Receptor → Effector protein → Messenger(s) → Messenger target(s) (For further details, see text)

Receptors with intrinsic protein tyrosine kinase activity (RTKs)

Growth factors, etc. → RTK activation → P-Tyr motifs → SH2 proteins → SH2 protein targets

Hepatocyte growth factor (HGF) (≡ Scatter factor)	Platelet-derived growth factor (PDGF)	Insulin-like growth factor 1 (IGF-1)
Epidermal growth factor (EGF)	Macrophage colony-stimulating factor (M-CSF)	Nerve growth factor (NGF)
Insulin		
Haemopoietic stem cell factor (SCF) (≡ *Steel* factor)		

Receptors with intrinsic guanylate cyclase activity (RGCs)

Stimulus → RGC activation↑ → [cGMP]↑ → cGMP-activated protein kinase↑ or cGMP-regulated ion channel

Atrial natriuretic peptide	*Escherichia coli* enterotoxin receptor (gut)

7-span receptors (7SpR) that activate adenylate cyclase (AC)

Stimulus → 7SpR → G_s → AC activity↑ → [cAMP]↑ → cAMP-dependent protein kinase activity↑

Histamine (H_2 receptor)	Adrenergic (β_1 and β_2 receptors)	Adrenocorticotropic hormone (ACTH)
Glucagon	Thyroid stimulating hormone (TSH)	Growth hormone-releasing hormone (GHRH)
Prostacyclin	Vasopressin (V_2 receptor)	Receptors mediating olfaction (a large family)
Some taste receptors?		

7-span receptors that inhibit adenylate cyclase (AC)

Stimulus → 7SpR → G_i → AG activity↓ → [cAMP]↓ → cAMP-dependent protein kinase activity↓

Adrenergic (α_2 receptor)	Prostaglandin E_1 (some tissues)	Opioid peptides (δ receptor)
Adenosine (A_1 receptor)	Acetylcholine (muscarinic receptors — two subtypes)	

7-span receptors that activate a cyclic nucleotide phosphodiesterase (PDE)

Stimulus → 7SpR → G_t → cGMP-PDE activity↑ → [cGMP]↑ → cGMP-regulated ion channel activity↑

Rhodopsin (vertebrates: four forms, in retinal rod cells and in three types of colour-selective retinal cone cells)
Some taste receptors?

7-span receptor that opens a hyperpolarizing K⁺ channel

Acetylcholine → 7SpR → G_i → K⁺ channels open → Cell hyperpolarization

Acetylcholine (muscarinic receptor — one subtype)

7-span receptors that activate phosphoinositidase C (PIC)

Stimulus → 7SpR → G_q → PIC activity↑ → PtdIns(4,5)P_2 hydrolysis → $\begin{cases} (Ins(1,4,5)P_3\!\uparrow \to \text{cytosol } [Ca^{2+}]\!\uparrow \\ ([1,2\text{-diacylglycerol}]\!\uparrow \to \text{PKC activation}\!\uparrow \end{cases}$

Adrenergic (α_1 receptor)	Histamine (H_1 receptor)	Acetylcholine (muscarinic receptors, two subtypes)
Vasopressin (V_1 receptor)	Substance P/other tachykinins	Rhodopsin (invertebrates, e.g. *Drosophila*)
Thromboxane A_2	Endothelins (three receptors)	Thyrotropin releasing hormone (TRH)
Some taste receptors?	Glutamate ('metabotropic' receptor)	Platelet activating factor (PAF or AGEPC)

Some receptor proteins incorporate both (1) an outward-facing recognition site through which they respond to an extracellular stimulus; and (2) a mechanism for the onward transmission of the received stimulus to the cell interior; these receptors probably comprise a minority of the cell's total repertoire. Most receptors of this general design so far identified fall into one of two major families — the LIGAND-GATED ION CHANNELS, which are discussed elsewhere, and receptor tyrosine kinases (RTKs, see below and Table S3). A smaller group comprises the ligand-stimulated guanylate cyclases (Table S3 and [5]), and recent studies have shown that other receptors include plasma membrane protein tyrosine phosphatases (e.g. the CD45 molecule of lymphocytes) (*see* PROTEIN PHOSPHATASES) or protein serine-threonine kinases (*see* PROTEIN KINASES).

Many more cell-surface receptors signal through relay systems in which three components act in sequence:

receptor → G protein → effector protein

G proteins are guanine-nucleotide dependent coupling proteins (*see* GTP-BINDING PROTEINS; G PROTEIN-COUPLED RECEPTORS) which are abbreviated individually to G_*, where the subscript identifies a particular G protein. The effector protein is usually an enzyme or an ion channel (Table S3 and [6]). The hundreds or thousands of receptors that function in this way seem all to be members of a single structural family of polypeptides, typified by the rhodopsins (the photoreceptors of the eye, *see* VISUAL TRANS-DUCTION). These '7-span' receptors are embedded within the plasma membrane, with their polypeptide chains traversing the membrane seven times. The sites through which they respond to extracellular stimuli are exposed to the external medium, and their coupling to G proteins, and thence to effectors, is achieved primarily through the intracellular polypeptide loop that links the fifth and sixth transmembrane domains (*see* G PROTEIN-COUPLED RECEPTORS).

The G proteins through which these receptors pass on their information comprise a smaller evolutionary family, with probably a few tens of members [7–9]. The effector proteins that the G proteins activate or inhibit are structurally and functionally far more heterogeneous than the 7-span receptors or G proteins (Table S3).

Two major evolutionary achievements expressed by the 7-span receptor and G-protein families are : (1) they permit multifarious extracellular signals to channel their cellular control information through only a limited number of effector mechanisms; and (2) they facilitate the evolutionary emergence of multiple receptors for single stimuli, with different receptors often using different signalling mechanisms (e.g. the various muscarinic acetylcholine receptors, see Table S3) [6–9].

The second messenger concept

In the mid-1950s Earl Sutherland's laboratory discovered that CATECHOLAMINES such as adrenaline (when they act through β-ADRENERGIC RECEPTORS) transmit information to the cell interior by stimulating the synthesis of adenosine 3′,5′-cyclic monophosphate (CYCLIC AMP, cAMP) by the enzyme ADENYLATE CYCLASE (see Table S3). cAMP serves as an intracellular nucleotide second messenger which, at micromolar concentrations,

activates a CYCLIC AMP-DEPENDENT PROTEIN KINASE that phosphorylates, and thus activates or inactivates, key enzymes of hormone-regulated metabolic pathways: these include muscle glycogen phosphorylase and the hormone-sensitive lipase of adipose tissue (*see also* PROTEIN PHOSPHORYLATION). As a result of this discovery, Sutherland recognized that cAMP was likely to be the first of a substantial number of intracellular second messengers which (1) would be produced in cells in response to activation of receptors; and (2) would mediate the onward transmission of the information sensed by cell-surface receptors in a form capable of regulating intracellular events. Figure S4 is a generalized modern statement of this idea of information flow into cells. Although Sutherland's original concept remains secure, subsequent progress has shown that the intracellular messengers formed by cells in response to activation of receptors are remarkably diverse: some are freely diffusing water-soluble molecules like cAMP, but others remain bound in membranes either as lipophilic molecules such as 1,2-DIACYLGLYCEROL or as modified amino-acid sequences (e.g. phosphotyrosine motifs) within membrane proteins (see below).

Receptors with intrinsic or associated protein tyrosine activity

The best understood receptor proteins that exhibit an intrinsic enzyme activity are the receptor tyrosine kinases (RTKs) (Table S3 gives examples from mammals, but much key evidence on RTK function has come from *Drosophila* and *Caenorhabditis*). Each RTK is a transmembrane protein in which the extracellular portion includes a domain that binds an extracellular stimulus and the intracellular segment includes a protein tyrosine kinase domain that catalyses the phosphorylation on tyrosine (Tyr) residues both of its own cytoplasmic domain and of other substrate protein(s). In most RTKs, typified by the epidermal growth factor (EGF) receptor, the stimulus recognition and kinase elements are towards the opposite ends of a single polypeptide chain that spans the membrane, but the insulin and insulin-like growth factor (IGF-I) receptors exist as disulphide-bridged $\alpha_2\beta_2$ tetrameric proteins that include two ligand-binding and two kinase sites (*see* GROWTH FACTOR RECEPTORS; INSULIN AND ITS RECEPTOR).

Stimulation of an RTK by its cognate ligand activates its tyrosine kinase activity, causing phosphorylation of the cytoplasmic domain of the RTK and then phosphorylation of other protein substrates. The initial receptor phosphorylation is an interchain event in which one polypeptide chain of a receptor dimer phosphorylates the other. In the insulin and IGF-I receptors this occurs within a pre-existing receptor oligomer, whereas dimerization of other receptors and the consequent interchain phosphokinase activity is driven by ligand binding. Once a receptor is phosphorylated, its kinase site opens up so as to give access for the phosphorylation of other proteins.

The tyrosine-phosphorylated receptors and other tyrosine-phosphorylated proteins then associate with a number of target proteins which interpret to the cell interior the signals received by the receptors. These proteins that physically associate with some or all activated RTKs include:

1 RasGAP, a GTPase-activating protein which modulates the

activity of the Ras proteins (a trio of small GTP-dependent switch proteins encoded by c-*ras* proto-oncogenes; *see* GTP-BINDING PROTEINS).

2 Raf-1, a serine–threonine protein kinase encoded by the c-*raf* proto-oncogene.

3 The cytoplasmic protein tyrosine kinase Src which is encoded by the c-*src* proto-oncogene.

4 The phospholipase PHOSPHOINOSITIDASE Cγ (PICγ) and INOSITOL LIPID 3-KINASE (or phosphatidylinositol 3-kinase (PI3K)), which respectively hydrolyse and phosphorylate inositol-containing phospholipids (*see* PHOSPHOINOSITIDES).

5 Shc protein, the product of the *shc* gene [10–14].

Of these, RasGAP, c-Src, PICγ, the M_r 80 000 (80K) regulatory subunit of PI3K, and Shc all contain one or more copies of a shared polypeptide sequence motif, known as the Src homology domain type 2 (SH2 DOMAIN) which spans ~100 amino-acid residues. SH2 domains mediate interactions between activated, and thus tyrosine-phosphorylated, RTKs and SH2-containing effector proteins. The different variants of SH2 domains recognize particular tyrosine-phosphorylated motifs: for example, the sequence Y^PMDMS (single-letter amino-acid notation) activates PI3K. Thus PICγ, PI3K, and RasGAP, and presumably the many other SH2-containing proteins that are now being identified by cloning studies, interact with RTKs at different tyrosine-phosphorylated sites. These tyrosine-phosphorylated sequence motifs generated as a consequence of receptor activities can therefore be thought of as receptor-tethered intracellular second messengers [12].

It remains to be fully explained how, downstream of the interactions of SH2-containing proteins with phosphotyrosine-containing sequence motifs that are generated as a result of RTK activation, cells respond with long-term changes such as proliferation or the controlled expression of differentiated characteristics. However, it seems that the chain of events that links receptor activation to these gene-regulatory events involves ~10 proteins, with key roles for Ras, Raf-1, the Mitogen-Activated Protein kinases (MAP KINASES, protein kinases capable of phosphorylating both serine–threonine and tyrosine residues on target proteins) and a wide spectrum of gene-regulatory proteins. Figure S5 depicts the likely sequence of components in one such pathway [12–14].

Seven-span receptors that interact with G proteins to control the formation of intracellular second messengers

A substantial number of the many 7-span receptor species at the cell surface (*see* RECEPTORS) are now known to control cells by influencing adenylate cyclase activity; some stimulate cAMP formation and others inhibit it (e.g. Table S3 and [15]). Thus most or all of the extracellular information delivered to these diverse receptors is summarized for the cell interior simply as a rise or fall in the intracellular concentration of cAMP. The first stage of this simplification is a division of receptors into a stimulatory group and an inhibitory group, each of which communicates with adenylate cyclase through a different G protein: these stimulatory and inhibitory G proteins are known as G_s and G_i respectively. Although it is clear that activation of these αβγ heterotrimeric G proteins by stimulated receptors provokes their

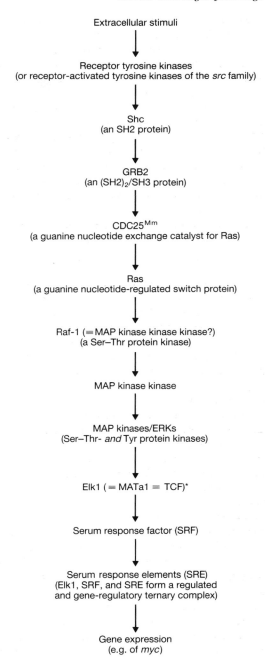

Fig. S5 Possible sequence of events in the transmission of signals from at least some receptor tyrosine kinases at the cell surface to the control of (some) genes. See text for details.

dissociation into free α subunits and βγ complexes, with the quantities of the free $α_s$ and $α_i$ within the membrane serving as the main regulators of adenylate cyclase activity, the molecular details of the stimulatory/inhibitory interplay between G_s, G_i, and adenylate cyclase remain a matter of some argument [7,9].

Mutations of G_i and G_s can disable the normal activation of adenylate cyclase by receptors, so rendering cells refractory to normal control by stimuli that signal through the regulation of adenylate cyclase. Moreover, in the relatively few cells that respond to elevation of cytosolic [cAMP] by proliferating, mutations that either permanently inactivate G_i or constitutively activate G_s are pro-proliferative, meaning that these mutant G proteins can function as oncogenes that contribute one of the multiple steps needed to convert normal cells to malignancy [7–10].

A second large family of 7-span receptors feeds its input to the cell through a very different signalling system based on the hydrolysis of the minor membrane phospholipid PHOSPHATIDYL-INOSITOL 4,5-BISPHOSPHATE (PtdIns(4,5)P_2) by phosphoinositidase C (see Table T78) [16–18]. Although this signalling system was also discovered in the 1950s, it only became understood in the 1980s. By recruiting different isoenzymes of PIC, it can be activated either through RTKs (via PICγ, see above) or by 7-span receptors coupled to G proteins (via the β isoenzymes of PIC).

As with adenylate cyclase, control of PIC activity by 7-span receptors is indirect, with G proteins serving as intermediaries. Since the coupling of some receptors (e.g. that for the neutrophil chemotactic peptide fMet-Leu-Phe) to PIC is abolished by treatment of cells with pertussis toxin treatment, while activation of PIC by other receptors (e.g. the V_1 vasopressin receptor) survives this treatment, it seems likely that at least two G proteins can couple receptors to PIC activation. The pertussis-resistant G protein, which was recently identified as G_Q (also known as G_{11}), activates the β isoenzymes of PIC, but the pertussis-sensitive G protein and the PIC isoenzyme that it activates remain to be defined.

PIC-catalysed hydrolysis of PtdIns(4,5)P_2 yields two products, water-soluble inositol 1,4,5-trisphosphate (Ins(1,4,5)P_3), which diffuses into the cell interior, and lipid-soluble 1,2-diacylglycerol (1,2-DAG), which remains associated with the cytoplasmic surface of the cell membrane: each has a unique role as an intracellular messenger. Within cells there is a membrane compartment, probably a portion of the ENDOPLASMIC RETICULUM, into which Ca^{2+} is continuously pumped by an ATP-driven pump, so holding the cytoplasmic Ca^{2+} concentration of 'resting' cells at $\sim 0.1\ \mu M$ (*see* CALCIUM). When receptors cause a rise in the intracellular concentration of Ins(1,4,5)P_3 this compound binds to receptors on the membrane enclosing this Ca^{2+} store and triggers a rapid release of Ca^{2+} into the cytoplasm (*see also* CALCIUM). The Ins(1,4,5)P_3-sensitive receptor proteins responsible for this Ca^{2+} mobilization are a small family of ligand-gated Ca^{2+} channels consisting of oligomers (possibly tetramers) of polypeptide subunits of M_r 250 000 (250K): they probably function in a manner basically similar to the neurotransmitter-regulated ion channels of the plasma membrane. As a result of cell stimulation, the cytoplasmic Ca^{2+} concentration often rises briefly to $>0.5\ \mu M$ within seconds after the application of a stimulus. Ins(1,4,5)P_3 is inactivated both by dephosphorylation and by entry into a complex series of metabolic pathways that interconvert a number of previously unknown inositol polyphosphate isomers, the biological functions of which are still uncertain [17]. Downstream of the Ins(1,4,5)P_3-stimulated mobilization of Ca^{2+}, cells often exhibit temporally and spatially complex changes in cytosolic Ca^{2+} concentration, most notably repetitive spikes of $[Ca^{2+}]$ whose frequency is regulated by the agonist concentration and travelling waves of $[Ca^{2+}]$ that traverse the cell

from an initiating site that is thought to be an Ins(1,4,5)P_3-responsive Ca^{2+} store [17]. 1,2-Diacylglycerol, the other messenger molecule formed from PtdIns(4,5)P_2, activates one or more of a family of isoenzymatic protein kinases (the PROTEIN KINASES C, PKC) that can also be activated by PHORBOL ESTER tumour promoters [18]. The latter compounds were originally of interest because they enhanced the tumour yield in skin treated with carcinogens, but they were later found partially to mimic many effects of cell stimulation (such as platelet aggregation and lymphocyte proliferation), indicating that they act by subverting a normal cellular signalling process [18].

Some of the intracellular targets of Ca^{2+} and PKC are known: for example, Ca^{2+} (in combination with the intracellular Ca^{2+} receptor protein, calmodulin; *see* CALCIUM-BINDING PROTEINS) activates glycogen breakdown, smooth muscle contraction, and the pumping of excess Ca^{2+} from cells; and PKC controls a plasma membrane ion channel that exchanges intracellular K^+ ions for extracellular Na^+ and thus controls intracellular pH. However, we do not yet fully understand either the mechanisms by which these two signals often synergize in the activation of cells nor those by which they control longer term cellular responses.

The future

Past research has identified, and given us a partial understanding of, some of the biochemical mechanisms by which cells respond to extracellular controls. Several are mentioned in Table S3, some of which are discussed above, and yet others are either only just emerging or are yet to be discovered. There is evidence, including the recent identification by cloning of many 'orphan' 7-span receptors and G proteins that await assignment of function [6,9], that yet more signalling reactions will turn out to be under G protein control. Candidates include a phospholipase A_2 (which liberates free arachidonate and thus initiates EICOSANOID synthesis) and additional C-type and D-type phospholipases (which hydrolyse, respectively, the glycerol-phosphate bonds of phospholipids and the bonds between phosphate and a headgroup such as choline, *see* PHOSPHOLIPASE) [19]. The latter enzymes liberate 1,2-DAG, ceramide (*N*-acylsphingosine), or phosphatidate (*see* LIPIDS), the cellular regulatory functions of which are yet to be fully defined, from their glycerophospholipid and phosphosphingolipid substrates.

R.H. MICHELL

1 Michell, R.H. (1988) *Trends Pharmacol. Sci.* (April), centrefold.
2 Barritt, G.J. (1992) *Communication within Animal Cells* (Oxford University Press).
3 *Trends Biochem. Sci.* (1992): issue on 'Signal transduction: crosstalk' **17**, 367–443.
4 *Curr. Opinion Cell Biol* (1989–1992): annual review issues on 'Cell Regulation' **1**, 157–235; **2**, 165–237; **3**, 169–234; **4**, 141–273.
5 Thompson, D.K. & Garbers, D.L. (1990) Guanylyl cyclase in cell signalling. *Curr. Opinion Cell Biol.* **2**, 206–211.
6 Iismaa, T.P. & Shine, J.G. (1992) G protein-coupled receptors. *Curr. Opinion Cell. Biol.* **4**, 195–202.
7 Bourne, H.R. et al. (1991) The GTPase superfamily: conserved structure and molecular mechanism. *Nature* **349**, 117–127.
8 Spiegel, A.M. (1992) G proteins in cellular control. *Curr. Opinion Cell Biol.* **4**, 203–211.

9 Hepler, J.R. & Gilman, A.G. (1992) G proteins. *Trends Biochem. Sci.* **17**, 383–387.

10 Pazin, M.J. & Williams, L.T. (1992) Triggering signaling cascades by receptor tyrosine kinases. *Trends Biochem. Sci.* **17**, 374–378.

11 Cantley, L.C. et al. (1991) Oncogenes and signal transduction. *Cell* **64**, 281–302.

12 Pawson, T. & Gish, G.D. (1992) SH2 and SH3 domains: from structure to function. *Cell* **71**, 359–362.

13 French, P.J. et al. (1993) *Life Sci. Adv. — Oncology* **12**, 57–84.

14 McCormick, F. (1993) How receptors turn Ras on. *Nature* **363**, 15–16.

15 Bentley, J.K. & Beavo, J.A. (1992) Regulation and function of cyclic nucleotides. *Curr. Opinion Cell Biol.* **4**, 233–240.

16 Michell, R.H. et al. (Eds) (1989) *Inositol Lipids and Cellular Signalling* (Academic Press, London).

17 Berridge, M.J. (1993) Inositol trisphosphate and calcium signalling. *Nature* **361**, 315–325.

18 Asaoka, Y. et al. (1992) Protein kinase C, calcium and phospholipid degradation. *Trends Biochem. Sci.* **17**, 414–417.

19 Liscovitch, M. (1992) Crosstalk amongst multiple signal-activated phospholipases. *Trends Biochem. Sci.* **17**, 393–399.

secondary immune response The immune response that occurs following a second or repeated exposure to an ANTIGEN. In the humoral response, specific IgG ANTIBODIES of high affinity appear rapidly and at high titre in the serum, being produced from an already expanded population of specific MEMORY CELLS (*see* AFFINITY MATURATION; HYPERMUTATION; B CELL DEVELOPMENT). T cell responses are also enhanced as the result of activation of memory T cells.

secondary lymphoid organs/tissues Lymphoid tissues in which antigen recognition by mature T and B cells, and proliferation of antigen-activated cells takes place, such as lymph nodes and spleen. They are distinguished from primary lymphoid tissues such as bone marrow and thymus in which lymphocyte development takes place.

secondary lysosomes Intracellular organelles containing ACID HYDROLASES and endocytosed materials undergoing digestion. The origin of LYSOSOMES is no longer thought to be by formation of a primary lysosome which becomes a secondary lysosome after fusion with an endocytic vesicle and thus the term is becoming redundant. Recent evidence suggests the presence of a pre-lysosomal compartment between the *trans* Golgi network and the mature lysosome with complex membrane traffic pathways linking these organelles and ENDOSOMES.

secondary structure Level of structural organization in proteins described by the folding of the polypeptide chain into structural motifs such as α-helices and β-sheets which involve HYDROGEN BONDING of backbone atoms. Secondary structure is also formed in nucleic acids, especially in single-stranded RNAs, by internal base-pairing. *See*: NUCLEIC ACID STRUCTURE; PROTEIN STRUCTURE.

secretin A polypeptide hormone of 27 amino acids secreted by the S cells of the mucosa of the upper intestine. Acid instilled into the duodenum stimulates its release; its main physiological effects are to cause pancreatic secretion of bicarbonate and enzymes, and to inhibit GASTRIN release from the stomach. The latter is the basis of the secretin test for pancreatic gastrinoma in Zollinger–Ellison syndrome, in which injection of secretin does not lead to reduction in gastrin output.

secretion *See*: BACTERIAL PROTEIN EXPORT; EXOCYTOSIS; PROTEIN SECRETION.

secretion vector An EXPRESSION VECTOR containing, in addition to the promoter, a secretion SIGNAL SEQUENCE followed by an in-frame cloning site. These vectors can direct the secretion of proteins expressed from structural genes inserted in the correct reading frame at the cloning site. Examples include the MFα (mating factor α)-based secretion vectors for *Saccharomyces cerevisiae*, and the OmpA-based secretion vectors for *Escherichia coli*. *See also*: BACTERIAL PROTEIN EXPORT; SECRETION.

secretor gene An allelic gene system determining whether or not antigens of the ABO BLOOD GROUP SYSTEM are secreted by the salivary glands and intestine. The first autosomal linkage to be discovered in humans (in 1951) was between the secretor locus and the LUTHERAN BLOOD GROUP.

secretory cells In the most general sense, any cell that produces and secretes a product. It is used especially of those cells highly specialized for the production and secretion of digestive enzymes, hormones, neurotransmitters, and neuromodulators etc. *See*: EXOCYTOSIS; PROTEIN SECRETION; SYNAPTIC TRANSMISSION.

secretory component (SC) Accessory molecule bound to secreted, multimeric forms of IgA ((IgA)$_2$) and IgM ((IgM)$_5$) and which mediates their transport across epithelial membranes. SC is the immunoglobulin-binding domain of a larger receptor, the POLY-IMMUNOGLOBULIN RECEPTOR (poly-Ig receptor), expressed on the basal membrane of epithelial cells. The SC subunit is produced by degradation of the poly-Ig receptor during poly-Ig movement across the membrane.

secretory IgA Dimeric form of IgA consisting of two molecules of IgA ($\alpha_2 L_2$) linked by a J CHAIN and with SECRETORY COMPONENT attached. Secretory IgA is the most abundant immunoglobulin in mammalian secretions. *See also*: ANTIBODY.

secretory proteins Proteins such as digestive enzymes, hormones, and antibodies that are released from the cell after synthesis. They are almost invariably GLYCOPROTEINS and, in eukaryotic cells, undergo glycosylation during their passage from the cytoplasm through the endoplasmic reticulum and Golgi apparatus to the plasma membrane and release. *See*: BACTERIAL PROTEIN EXPORT; EXOCYTOSIS; PROTEIN SECRETION.

secretory vesicles Membrane-bounded vesicles filled with product for secretion. *See*: EXOCYTOSIS; PROTEIN SECRETION; SYNAPTIC TRANSMISSION.

sedimentation coefficient, sedimentation constant (s) The sedimentation coefficient of a particle in solution in a gravitational field is defined as the velocity of sedimentation (v, in cm s^{-1}) divided by the centrifugal field (in cm s^{-2}). Sedimentation coefficients are generally expressed as SVEDBERG UNITS (S) (10^{-13} seconds). As the sedimentation coefficient is dependent on the mass of the particle it can be used to estimate molecular mass provided that an independent measure of the diffusion coefficient is also available and that the partial specific volume is known: if proteins are regarded as spherical, s is proportional to mass approximately as $m^{2/3}$. *See also*: SEDIMENTATION EQUILIBRIUM.

sedimentation equilibrium When a purified protein sample is subjected to ultracentrifugation at a relatively low speed, sedimentation is counterbalanced by diffusion and a smooth gradient of protein concentration develops. The molar mass of the protein can then be calculated from the dependence of concentration on distance from the rotation axis. This method of determining the mass of a protein can be carried out on native undenatured proteins and thus provide accurate molecular weight values for multimeric proteins, and does not require the diffusion coefficient to be known.

seed storage proteins Families of simple proteins abundant in plant seeds (Table S4). Improvement of the nutritional value of

Table S4 Sources of some seed storage proteins

Protein	Plant species
Arachin	*Arachis hypogaea* (groundnut or peanut); 13S storage globulins
Avenin	*Avena sativa* (oats); storage globulins
Cocosin	*Cocos nucifera* (coconut); 11S storage globulins
Conarchin	*Arachis hypogaea*; 7–8S storage globulins
Concocosin	*Cocos nucifera*; 7S storage globulins
Conglutin	*Lupinus albus* (white lupin); 11S storage globulins
Conglycinin	*Glycine max* (soybean); 7S storage globulins
Convicine	*Vicia faba* (broad bean)
Crambin	*Crambe abyssinica*
Cruciferin	*Brassica napus* (rapeseed); 12S storage globulins
Cucurbitin	*Curcurbitaceae*; 11S storage globulins
Edestin	*Cannabis sativa* (hemp); 13S storage globulins
Excelsin	*Bertholletia excelsa* (Brazil nut)
Gliadin	*Triticum* (wheat); storage prolamines
Gluten	A complex formed from gliadins and glutenins of wheat
Glutenin	*Triticum*; storage prolamines
Glycinin	*Glycine max*; 11S storage globulins
Helianthin	*Helianthus annuus* (sunflower); 11S storage globulins
Hordein	*Hordeum* spp. (barley); storage prolamines
Kafirin	*Sorghum* (sorghum or milo)
Legumin	Originally from *Pisum sativum* (pea); now more generally applied to families of 11S storage globulins from legumes
Napin	*Brassica napus*
Oryzin	*Oryza sativa* (rice)
Pennisetin	*Pennisetum* sp. (millet)
Phaseolin	*Phaseolus vulgaris* (French bean); 7S storage globulins
Psophocarpin	*Psophocarpus tetragonolobus* (winged bean)
Secalin	*Secale cereale* (rye)
Vicilin	Originally from *Pisum sativum* (pea); now more generally applied to families of 7S storage globulins from legumes
Vicine	*Vicia faba*
Zein	*Zea mays* (maize); storage prolamines

Shotwell, M.A. & Larkins, B.A. (1989). In *The Biochemistry of Plants* (Stumpf, P.K. & Conn, E.E., Eds) Vol. 15, Ch. 7, 297–345 (Academic Press, New York).

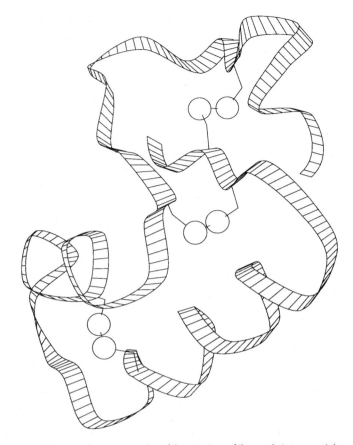

Fig. S6 Schematic representation of the structure of the seed storage protein crambin. The circles represent disulphide bonds.

seed proteins (some of which are low or lacking in certain amino acids essential to the human diet) is one of the goals of PLANT GENETIC ENGINEERING. The structure of crambin is shown in Fig. S6.

segment Basic morphological unit of both the invertebrate and vertebrate body plan along the anterior–posterior axis. *See*: DROSOPHILA DEVELOPMENT; SEGMENTATION; VERTEBRATE NEURAL DEVELOPMENT.

segment polarity genes Mutations in this class of genes in *Drosophila* give rise to embryos with pattern defects in portions of every segment. Their role is to provide POSITIONAL INFORMATION across the segment. The 14 genes in this group have a wide variety of expression patterns, from uniform expression to a single thin stripe per segment, and their products fall into a diverse array of functional types. Some (e.g. *engrailed* and *gooseberry*) encode TRANSCRIPTION FACTORS, others (e.g. *fused* and *shaggy*) encode PROTEIN KINASES, the *wingless* gene produces a secreted product with homology to a vertebrate ONCOGENE *int-1*, and the product of the *armadillo* gene is likely to be associated with CELL JUNCTIONS. How this set of genes function is unknown. One suggestion is that one gene (e.g. *wingless*) might produce a graded MORPHO-

GEN and that the other genes are involved in the establishment of this gradient and in its interpretation to generate diverse cell states across the segment.

segmentation Segments are the basic morphological units of both the vertebrate and the invertebrate body plan. Among invertebrates, annelids and arthropods have segmented body plans but form segments in very different ways. Among arthropods, the process has been best studied in the long germ band insect, *Drosophila*. Here cells are allocated to segments on the basis of their position within the blastoderm through the action of the SEGMENTATION GENES. The segments form simultaneously along the length of the embryo. In short germ band insects segments are formed sequentially from a growth zone. In annelids, represented by studies in the leech, cells are allocated to segments on the basis of an indeterminant lineage generated by a set of teloblast stem cells. The annelids and arthropods are believed to derive from a common segmented precursor. The identification of homologues of *Drosophila* segmentation genes in other invertebrates should clarify the relationships between segmentation processes that appear highly diverse. Segmental identity, although also specified early in development, is determined independently of the segmentation process.

In insects, the nervous system, the epidermis, and the MESODERM are segmented, but the only tissues displaying overt segmentation in vertebrates are the HINDBRAIN and the SOMITES. The degree of segmentation in the vertebrate head is currently unclear, despite the obvious segmentation associated with the developing cranial nerves. The importance of segmentation in evolution lies in its provision of the basic building blocks for body structures. Organisms with a complex body plan are thought to have evolved from ancestors with repeated identical segmental units. During evolution, these simple segments subsequently differentiated from each other to give the characteristics of the different genera.

Theories of vertebrate segmentation were common at the end of the 19th century, on account of the contemporary interest in the theory of evolution. Recent molecular genetic studies have reawakened interest in this subject, as many of the genes responsible for segmentation in *Drosophila* have homologues in vertebrate species. The most notable success of the molecular genetic approach has been to show that the homologues of many of the HOMEOTIC GENES of *Drosophila* — the vertebrate HOX GENES — are expressed in vertebrates before the onset of any morphologically apparent segmentation. Two areas of particular interest are the segmental development of the brain, and that of the segmented mesoderm (somites) and vertebral column. In each case it appears that particular classes of Hox genes are expressed in discrete domains, consistent with a role in specifying identity along the rostro-caudal axis. Specific Hox genes are expressed in the developing segments of the hindbrain (RHOMBOMERES), and in specific regions of the developing trunk. In many cases, the rostral limit of Hox gene expression coincides with a particular morphological boundary in the long axis of the developing embryo; this expression limit may be important in imparting POSITIONAL INFORMATION to developing tissues, each region of tissue expressing a particular 'code' of Hox genes. Segmentation may have evolved as a strategy for developing regional differences along the long axis of the body with greater efficiency. *See also*: ANNELID DEVELOPMENT; DROSOPHILA DEVELOPMENT; HOMEOBOX GENES AND HOMEODOMAIN PROTEINS; VERTEBRATE NEURAL DEVELOPMENT.

segmentation genes In 1980 Nüsslein-Volhard and Wieschaus identified a set of zygotic genes which are required for the establishment of correct segmentation of the *Drosophila* embryo. These genes were divided into three classes according to their mutant phenotypes: GAP GENES, PAIR-RULE GENES, and SEGMENT POLARITY GENES. The molecular genetic analysis of these genes has provided the basis for our current understanding of the processes of PATTERN FORMATION in early embryogenesis in *Drosophila*. *See*: DROSOPHILA DEVELOPMENT.

segmented genomes Viral genomes composed of two or more separate molecules of nucleic acid. Segmented genomes occur in both PLANT and ANIMAL VIRUSES.

segregation The separation of the pair of ALLELES at any genetic LOCUS during the formation of gametes, first described by Gregor Mendel (*see* MENDELIAN INHERITANCE). Generally the segregation of alleles at one locus is quite independent of the segregation of alleles at any other (Mendel's First Law), unless the loci concerned are very close together on the chromosome. In this case certain combinations of alleles (i.e. those carried on the same chromosomes) will be observed to be inherited together (*see* LINKAGE).

Mendel, G. (1866) Versuche über Pflanzenhybriden. *Verh. naturf. Ver. Brünn* **4**, 3–44. Royal Hort. Soc. (1901) translation reprinted by Harvard University Press (1948).

selectins A family of mammalian carbohydrate-binding adhesion molecules found on leukocytes and endothelium. They include the leukocyte homing receptor (LAM-1, L-selectin), the endothelial leukocyte adhesion molecule (ELAM-1, E-selectin), and CD62 (P-selectin) on platelets and endothelial cells (Table S5). Selectins are membrane glycoproteins with an N-terminal extracellular domain of 117–120 amino acids homologous to some C-TYPE ANIMAL LECTINS. The N-terminal domain is followed by an EGF-like motif of 34–40 amino acids and numbers of 62-amino acid CONSENSUS REPEATS (which are present in many proteins regulating complement action) (Fig. S7). The selectins mediate interactions between leukocytes and endothelium in lymphocyte recirculation, inflammation, and blood clotting. Binding is calcium-dependent.

Springer, T.A. (1990) *Nature* **346**, 425–433.
Lawrence, M.B. & Springer, T.A. (1991) *Cell* **65**, 859–873.

selection The implementation of conditions that enable the discrimination of cells displaying a required PHENOTYPE, for example the growth of bacteria in medium containing antibiotics to select for cells containing antibiotic resistance genes.

selective immunoglobulin deficiency DYSGAMMAGLOBULINAEMIA.

Table S5 The classes of selectins

Selectin	Previous names	Expressed on	Activity	Possible ligands
E-selectin	ELAM-1	Endothelial cells	Synthesized in response to inflammatory agents. Mediates adhesion of neutrophils, monocytes, and some lymphocytes	Sialylated, fucosylated polylactosamines that contain sialyl Lewis x antigen (see LEWIS BLOOD GROUP)
L-selectin	Leukocyte homing receptor, gp90^{mcl}, LAM-1, LECAM-1, LEC-CAM-1	Leukocytes	Mediates binding of lymphocytes to endothelium during lymphocyte recirculation and neutrophil migration at sites of inflammation	Fucosylated glycans that include Lewis x antigen
P-selectin	CD62, GMP-140, PADGEM	Platelets and endothelial cells	Associated with granules. Brought to cell surface after stimulation by thrombogenic agents. Mediates binding to neutrophils and monocytes at site of injury	?

selector gene The concept of selector genes in *Drosophila* derived from the discovery of lineage COMPARTMENTS together with the observation that in certain HOMEOTIC MUTATIONS the transformed areas corresponded exactly with compartments. It was proposed that selector genes would be stably expressed in all the cells of a particular compartment and would direct the cells within the compartment to follow a specific developmental fate. One of the functions of a selector gene would be to control cell–cell affinities such that cells from different compartments do not mix. The homeotic genes and other homeobox genes such as ENGRAILED fulfil, to some extent, the selector gene criteria. *See also*: HOMEOBOX GENES AND HOMEODOMAIN PROTEINS.

self-assembling epitope carrier High molecular weight protein complexes carrying one or more peptide EPITOPES on their outer surface in their native conformation, which have been used in the development of vaccines. An epitope from a pathogenic virus, for example, can be incorporated into a nonpathogenic viral particle by inserting the sequence encoding the epitope into the genome of the nonpathogenic virus. The protein carrying the desired epitope is synthesized and incorporated into the nonpathogenic viral particle, which thus acts as a self-assembling carrier for the antigen. The concept has also been applied to the development of an AIDS vaccine. By fusing the gene encoding the HIV-1 coat protein to the TyA gene of the yeast transposon Ty1, a TyA-HIV fusion protein can be produced in yeast which self-assembles into a spherical retroviral-like capsid (Ty-VLP) with the HIV proteins on the outer surface.

Valenzuela, P. et al. (1985) *Bio/Technology* **3**, 323–325.

self-assembly The principle of self-assembly states that all the information required for the assembly of a multicomponent structure resides solely within the components themselves; it is the specific binding properties of these components that directs the assembly process. Self-assembly is widespread in biological systems — in the folding of polypeptide chains into their final conformation, and in the assembly of viral particles, multimeric proteins and nucleoprotein particles, for example. The idea of self-assembly arose from the discovery that infectious particles of

Fig. S7 Selectins and their ligands.

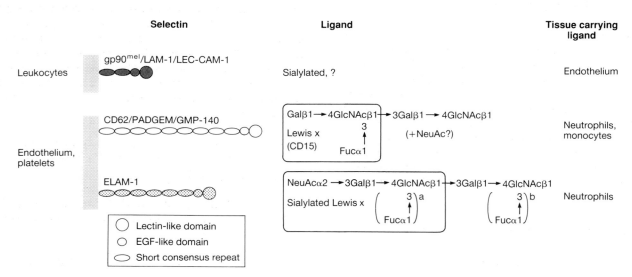

TOBACCO MOSAIC VIRUS could be reconstituted from dissociated protein subunits and purified viral RNA, and from subsequent observations that many isolated denatured proteins can refold spontaneously into functional structures after the denaturing agent is removed.

As applied to proteins and protein-containing structures, the principle of self-assembly holds that the information for the structure, and hence the function, of a protein is contained entirely within the aminoacyl sequence(s) of the polypeptide chain(s) of which it is composed. Each polypeptide chain interacts with itself as it is synthesized to assume a folded conformation (*see* PROTEIN FOLDING; PROTEIN STRUCTURE) of lower free energy. Once folded, the protein may be able to bind to other macromolecules, such as other proteins and nucleic acids, in a highly specific manner so that complex oligomeric structures can self-assemble from their components. Self assembly is a spontaneous process requiring no input of energy nor any steric information extrinsic to the polypeptide chains. In many cases, however, self-assembly has to be assisted by pre-existing proteins termed MOLECULAR CHAPERONES, which inhibit kinetically favoured incorrect assembly pathways that would produce nonfunctional structures; some chaperones also use ATP hydrolysis to carry out their function. Nonenveloped virions are believed to form by a self-assembly process (*see* TOBACCO MOSAIC VIRUS) but in most cases there is little detailed information on the processes involved. *See also*: PROTEIN FOLDING; PROTEIN STRUCTURE; PROTEIN TRANSLOCATION.

self-cleaving RNA *See*: RIBOZYMES.

selfish DNA A term coined by Francis Crick in 1979. Such DNA is defined by its ability to spread by making additional copies of itself within the genome, without making a specific contribution to the phenotype. Examples of apparently selfish DNA include RETROPOSONS (which in the human are represented by the ALU and LINE families of repeated sequences among others), and SIMPLE SEQUENCE DNA. *See*: TRANSPOSABLE GENETIC ELEMENTS.

Doolittle, W.F. & Sapienza, C. (1979) *Nature* **284**, 601–603.
Orgel, L.E. & Crick, F.H.C. (1979) *Nature* **254**, 604–607.

self-incompatibility A system that prevents inbreeding in flowering plants. It is of two types, gametophytic and sporophytic. The former is determined by a single nuclear locus with many alleles (S-alleles). The pollen genotype (which is haploid) is expressed during pollen tube development and if its S-allele is the same as one of those in the style tissue, tube growth ceases in the style (cessation of growth is often associated with CALLOSE deposition, but is probably mediated in some species by RIBONUCLEASES). In the sporophytic system, pollen type is determined by the male parent, probably mediated through materials deposited in the outer layers of the pollen. Tube growth is inhibited on the stigma surface.

self-splicing introns *See*: RIBOZYMES; RNA SPLICING.

SEM Scanning ELECTRON MICROSCOPY.

semiconservative replication The generation of daughter DNA duplexes each of which contains one complete strand from the parental DNA duplex and one newly synthesized strand. This is the usual mode of replication of cellular double-stranded DNA and is known as 'semiconservative' as only one strand of the parental DNA duplex is passed to each daughter cell. *See*: DNA REPLICATION.

Meselson, M. & Stahl, F.W. (1958) *Proc. Natl. Acad. Sci. USA* **44**, 671–682.

Semliki Forest virus (SFV) Small enveloped alphavirus (group A arbovirus) of the Togavirus family (Togaviridae) (*see* ANIMAL VIRUSES), named after the Semliki forest in Uganda where it was first isolated. It is closely related to Sindbis virus. The virions are 75-nm diameter spherical particles with a total M_r of 46×10^6. Each particle comprises a 25-nm diameter icosahedral ($T = 4$) nucleocapsid, containing a single 42S (+)-sense RNA and 240 copies of the M_r 30 000 capsid protein, surrounded by a lipid bilayer membrane containing 240 copies of the viral membrane, or spike, glycoprotein complex. The RNA genome encodes four structural and three nonstructural proteins. The nonstructural proteins include RNA-dependent RNA polymerase and endoprotease activities, together with inhibitors of host cell transcription and translation. The structural proteins, which are synthesized from a 26S SUBGENOMIC MRNA, are the capsid protein and the spike glycoprotein complexes, each of which contains three heterodimers of the P62 (M_r 62 000) and 49K E1 glycoproteins (M_r 49 000). Progeny virions are generated by budding through the plasma membrane of an infected cell. During virus assembly the P62 protein is proteolytically cleaved to yield the E2 (M_r 52 000) and E3 (M_r 10 000) polypeptides which, together with E1, form the mature spike glycoprotein complex responsible for receptor binding and membrane fusion.

SFV was the first mammalian virus for which the mechanism of entry into cells was understood in detail. Virions bind to cell-surface receptors and are internalized by receptor-mediated ENDOCYTOSIS in clathrin-coated vesicles (*see* COATED PITS AND VESICLES). The low pH environment in ENDOSOMES triggers a conformational change in the spike glycoprotein complex which leads to fusion of the viral membrane with the endosome membrane, thereby injecting the viral nucleocapsid into the target cell cytosol where replication occurs. Together with VESICULAR STOMATITIS VIRUS and INFLUENZA VIRUS, SFV has been used extensively in studies of membrane biosynthesis and transport.

As with related alphaviruses, SFV can replicate in arthropod vectors such as mosquitoes and vertebrate hosts. Laboratory strains infect a wide range of tissue culture cells of mammalian and insect origin.

Simons, K. et al. (1982) *Sci. Am.* **246**, 58–66.
Kielian, M. & Helenius, A. (1986) In *The Togaviridae and Flaviviridae* (Schlesinger, S. & Schlesinger, M.J., Eds) 91–119 (Plenum, New York).
Choi, H-K. et al. (1991) *Nature* **354**, 37–43.
Simons, K. & Warren, G. (1984) *Adv. Protein Chem.* **36**, 79–132.

Sendai virus Paramyxovirus which induces membrane fusion and which is used to induce cell fusion in the formation of SOMATIC CELL HYBRIDS.

senile dementia *See*: NEURODEGENERATIVE DISORDERS.

sense strand The DNA strand that is transcribed into mRNA.

sensory neuron A NEURON that acts as a SENSORY RECEPTOR itself, or which synapses with a sensory receptor cell and carries sensory signals from the periphery to the central nervous system. *See*: OLFACTORY TRANSDUCTION; VISUAL TRANSDUCTION.

sensory receptor Electrically active cell which responds to a sensory stimulus (e.g. light, touch, movement) by producing a change in membrane potential (receptor potential) which, depending on the type of receptor, is either converted into an ACTION POTENTIAL, or transmitted across a synapse to generate an action potential in an associated neuron. *See e.g.*: OLFACTORY TRANSDUCTION; VISUAL TRANSDUCTION.

sensory rhodopsin Membrane protein of the BACTERIO-RHODOPSIN family from *Halobacterium* spp. which acts a light receptor for phototaxis.

septate junction These are cell–cell junctions of invertebrate (insect) species. Pleated septate junctions are found in insect epithelial tissues, and sometimes spuriously called septate 'desmosomes'. They display a regular gap of 15–20 nm across which run septa, giving a ladder-like appearance. They can be pleated, giving a honeycomb effect. Freeze-fracture analysis reveals rows of intramembrane particles spaced 10–20 nm apart. Their function appears to be mainly adhesive rather than occluding and they often lie in circumferential belts around the lateral surfaces of epithelial cells.

Continuous or smooth septate junctions (zonula continua) are found circumferentially localized around many cells of endodermal origin. These also have a gap of 15–20 nm but have no clear septa. After tracer impregnation, however, septa can be seen, and also 'rods' between the septa. Freeze-fracture analysis shows continuous ridges, which may correspond to the septa. These also appear to be primarily adhesive in function. *See also*: CELL JUNCTIONS.

Lane, N.J. (1982) In *Insect Ultrastructure* (King, R.C. & Akai, H., Eds) Vol. 1, 402–433 (Plenum, New York, 1982).

sequenator Machine in which automated PROTEIN SEQUENCING is carried out.

sequence The order of amino acid residues in a protein chain or the order of nucleotide residues in a nucleic acid. *See*: ALIGNMENT; DATABASES AND INFORMATION HANDLING IN MOLECULAR BIOLOGY; DNA; DNA SEQUENCING; MOLECULAR EVOLUTION; MOLECULAR EVOLUTION: SEQUENCES AND STRUCTURES; PROTEIN STRUCTURE; PROTEIN SEQUENCING; RNA.

sequential model The Koshland–Némethy–Filmer model of cooperativity in proteins. *See*: ALLOSTERIC EFFECT; COOPERATIVITY.

SER Smooth ENDOPLASMIC RETICULUM.

serine (Ser, S) A small AMINO ACID with a hydrophilic side chain.

serine proteinases, serine proteases A large and ubiquitous group of PROTEINASES in which a histidine and a serine residue at the active site are involved in catalysis (EC 3.4.21). The serine proteinases include the proteinases of the blood clotting and fibrinolytic pathways (*see* BLOOD COAGULATION AND ITS DISORDERS), and some COMPLEMENT components as well as general proteolytic enzymes such as TRYPSIN and CHYMOTRYPSIN (Table S6). The activity of these proteinases needs to be carefully controlled and in many cases an inactive precursor is formed first which is activated by specific cleavage (*see* POST-TRANSLATIONAL MODIFICATION OF PROTEINS). They are inhibited by the SERPIN family of proteins, many of which are also activated by proteolytic cleavage. Many eukaryotic serine proteinases are members of a large SUPERFAMILY of proteins — the serine proteinase superfamily — which also includes nonenzymatic proteins such as HAPTOGLOBIN and apolipoprotein Lp(a) (*see* PLASMA LIPOPROTEINS AND THEIR RECEPTORS) which have lost their enzymatic function during evolution. Microbial serine proteinases, such as subtilisin from *Bacillus subtilis*, seem to represent a separate line of descent from the eukaryotic serine proteinase superfamily. *See*: MECHANISMS OF ENZYME CATALYSIS; MOLECULAR EVOLUTION: SEQUENCES AND STRUCTURES.

serotonin An alternative name for 5-HYDROXYTRYPTAMINE.

serotonin receptors 5HT RECEPTORS.

serotype A strain of bacterium or virus which can be identified immunologically by specific antibodies by virtue of its antigenic difference from other strains.

Table S6 Some members of the eukaryotic serine proteinase superfamily

Simple proteinases
Cathepsin G
Chymotrypsin
Elastase
Kallikrein
Trypsin

Fibrinolytic system
Plasmin
Plasminogen activators (urokinase and tissue plasminogen activator)

Blood clotting pathway
Coagulation factors VII, IX, X, XI, XII, protein S
Prothrombin
Proconvertin

Complement components
C1 (C1r and C1s)
C2
Factor B
Factor D
Factor I

Nonenzymatic
Haptoglobin
Lp(a)

Serpins

THE serpins are a well-defined family of SERINE PROTEINASE INHIBITORS with a highly ordered tertiary structure formed by some 400 residues. They differ quite markedly from the other families of serine proteinase inhibitors which are all much smaller proteins that have independently evolved a common, fixed, reactive centre conformation. The serpins, however, have a mobile reactive centre that can adopt varying conformations, and one puzzle has been why such a relatively complex protein has developed into the predominant family of serine proteinase inhibitors in higher organisms and, in particular, in human plasma.

The reason for this evolutionary success probably lies in the complexity of the serpins and their consequent ability to adapt to different functional needs. Examples of this are the modifications of the framework structure to allow the modulation of inhibitory activity, which occurs with the activation of the plasma inhibitor ANTITHROMBIN by HEPARIN or the locking of the plasminogen activator inhibitor PAI-1 into an inactive latent state. Other modifications have involved the complete loss of inhibitory activity as in the egg white protein OVALBUMIN (Fig. S8), in the plasma hormone carriers thyroxine- and cortisol-binding globulins, and in the peptide-donor ANGIOTENSINOGEN.

The serpins evolved some 300 million years ago and are now widely distributed in eukaryotes including plants. An everyday example from plants is the Z4 inhibitor from barley; this serpin survives the brewing process to end up as the main protein in beer, and is said to contribute to its characteristic taste. Serpins have not been found in prokaryotes but numbers of them have been identified in vaccinia and other viruses where they probably function as inhibitors of the proteolytic steps involved in the host's immune response. Examples are the inhibition of the interleukin-1 converting enzyme by vaccinia virus and the inhibition of the complement-activating C1-esterase by the rabbit myxoma virus.

Structure and function

Some 40 members of the family are now known; they are all strongly homologous and their sequences can be confidently aligned in terms of their deduced secondary and tertiary structure. The archetype of the family is α_1-antitrypsin, also known as α_1-proteinase inhibitor. It has the typical structure of a serpin with 30% composed of helices and 40% formed by three β-sheets. One of these sheets, the five- or six-stranded A sheet, is a dominant feature of the molecule with the reactive centre peptide loop extending from its middle (fourth) strand (Plate 4*d* and Fig. S9 (ovalbumin)).

There are two main areas of variation within the serpins; these occur in the free N terminus of the molecule and in the peptide loop, situated near the C terminus of the molecule, that contains the reactive centre. The reactive centre functions as an ideal substrate, with a central residue that matches the proteolytic specificity of the target proteinase (e.g. an arginine for a thrombin-like proteinase, or an alanine or methionine for an elastase). It is useful for comparisons between serpins to denote

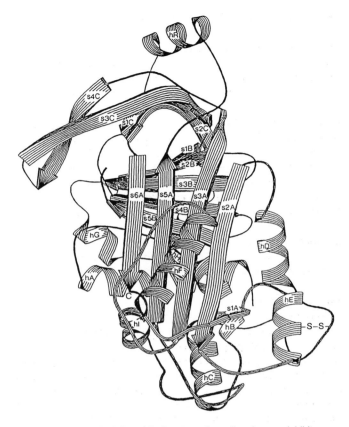

Fig. S8 Schematic depiction of the intact serpin ovalbumin, a noninhibitory serpin.

the reactive centre residue as P_1 with residues N terminal to it as P_2, P_3 etc. and C-terminal as P'_1, P'_2 etc. With this notation the reactive centre can be described as lying on a peptide loop that extends from P_{17} at the end of strand 5 of the A sheet to P'_4 at the commencement of the strand 1 of the C sheet (Fig. S9*b*).

The structure of the reactive centre loop in the active inhibitory form of the serpins is not known. By deduction, it is likely to have a conformation similar to the canonical form independently evolved in each of the smaller serine proteinase inhibitors. These all have a tightly constrained peptide loop of some 12 residues. This loop fits the active-site cleft of the cognate serine proteinases, with the P_1 residue of the inhibitor positioned close to the catalytic centre of the proteinase and its flanking residues interacting with subsites in the cleft. The end effect is a tight complex of proteinase and inhibitor in which the active site is constrained in the catalytic (Michaelis) form.

The serpins form similarly tight complexes with their cognate proteinases, and in plasma these complexes are removed from the circulation and catabolized by the liver. The serpin–proteinase complex can, however, slowly dissociate to give the intact proteinase and inhibitor, or it can be rapidly dissociated with detergent to give the covalently linked acyl–enzyme complex. These findings, with other evidence, suggest that the serpin–proteinase conformation is closer to the tetrahedral complex rather than the Michaelis intermediate formed by the small inhibitors. Recent

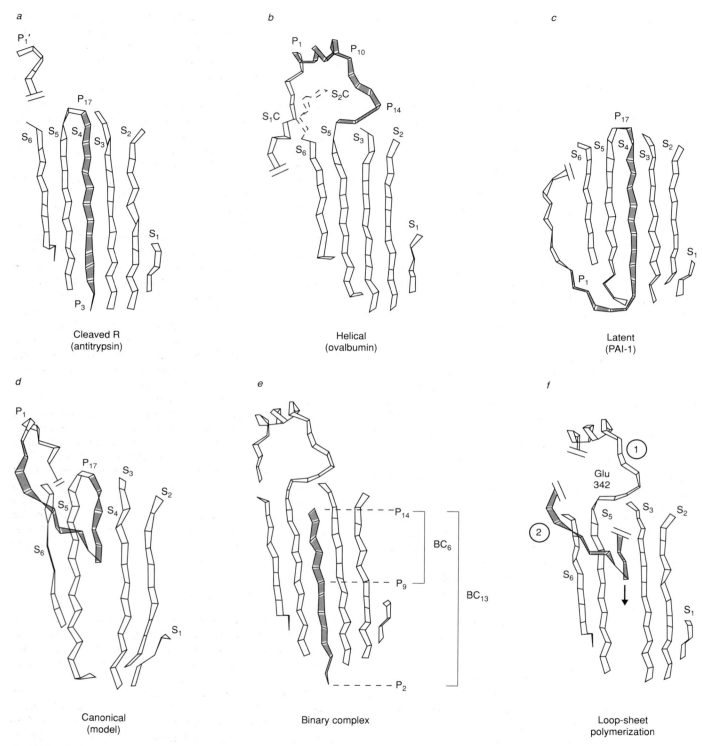

Fig. S9 The A sheet of serpins showing alternative conformations of the reactive centre peptide loop (shaded). The mobility of the reactive centre can readily be envisaged commencing with the helical form *b*. *a*, Cleavage at the reactive centre P_1 leads to the incorporation of the cleaved peptide to give the R form. *c*, Opening of the A sheet by mild denaturation results in an induced insertion of the loop into the A sheet, seen in an extreme form in latent PAI-1. *d*, The active inhibitory form is likely to adopt the partially inserted canonical conformation. *e*, Insertion of the loop is competed for by free peptide homologues, shown here as the 6- and 13-residue peptides BC6 and BC13 respectively. *f*, In a similar way, insertion of the loop of molecule 1 is subject to competition from the loop of molecule 2, leading to loop–sheet polymerization, as occurs at high temperatures or spontaneously in the Z mutant (Glu342Lys) of antitrypsin. P_1, P_{10}, P_{14}, etc. are positions of residues in the reactive centre; S_1, S_2, S_3, etc. denote the individual β-strands of the A sheet.

evidence suggests that the uniqueness of the serpins is not in the conformation at the time of docking with proteinase but in the accompanying changes that result in a locking of the complex.

Conformations of the reactive centre loop

The ability of the reactive centre loop of the serpins to adopt different conformations is confirmed by three different structures. The first crystallographic structure of a serpin was that of α_1-antitrypsin, modified by cleavage at its P_1 reactive centre (Fig. S9a). The susceptibility of the serpins to such proteolytic cleavage, not only at the reactive centre but also at the adjacent P_8–P'_5 residues, had long been realized and it was deduced from this that the reactive centre is sited on an exposed peptide loop. The unexpected finding with the structure of antitrypsin was that this cleaved loop was completely incorporated back into the A sheet with a consequent separation of the P_1 methionine from the P'_1 serine by 70 Å. As predicted, this shift from an exposed loop to a strand stabilized by sequential hydrogen bonds is accompanied by an increase in stability of the molecule. This property of undergoing a major conformational change on loop cleavage is characteristic of the inhibitory serpins and has been lost in most of the noninhibitors such as ovalbumin and angiotensinogen. The structure of intact ovalbumin (Fig. S9b) showed that its reactive centre loop was completely excluded from the A sheet and in an exposed position on the molecule as a three-turn helix extended on two peptide stalks. There is apparent mobility of the helix on these stalks but even so the helix would need to open in order to form an active inhibitory conformation. Examination of the structure however indicated that the N-terminal stalk of the helix was in a position to re-enter the A sheet but that this was blocked by mutations in the hinge origin of the stalk (P_{14}–P_{10}). This hinge region is consistently occupied by small side-chain residues, principally alanines in inhibitory serpins but larger polar residues in noninhibitors. The conclusion from the structure of ovalbumin is that partial re-entry of the reactive centre loop into the A sheet is a prerequisite for the inhibitory activity of the serpins. Thus in ovalbumin, the helical form represents the extreme of the extended form of the loop whereas the cleaved R form of antitrypsin represents the other extreme of complete reincorporation of the loop into the A sheet.

Further support for the concept of mobility of the loop comes from the demonstration that in the inhibitory serpins overinsertion of the loop can be induced to give a locked inactive conformation. Incubation of antithrombin at 60°C or exposure to 2 M guanidinium chloride results in a fraction with a profound change in properties. There is a complete loss of inhibitory activity, a change in stability that matches that of the S–R change, and a loss of the vulnerability of the inhibitor to proteolytic loop cleavage. These changes in properties remain constant even after removal by dialysis of the denaturing agent, but the inhibitor can subsequently be converted back to normal properties and function by exposure to, and then dialysis from, 8 M urea. This induced conformation of the serpins is denoted as the L state as its properties are identical to those previously observed in an unusual latent form of the plasminogen activator inhibitor PAI-1.

PAI-1 has a key function in the control of fibrinolysis (*see* BLOOD COAGULATION AND ITS DISORDERS). It has long been known to exist *in vitro* in a stable latent form and its existence in the circulation in an active form is believed to depend on its forming a complex with another plasma protein, VITRONECTIN. In latent PAI-1 (Fig. S9c), there is total incorporation of the loop into the A sheet, with the linked release of strand 1 of the C sheet around the β-sheet, allowing for the overall movement of the reactive centre P_1–P'_1 from one pole of the molecule to the other. This movement is spontaneous and complete in PAI-1, but in the L state of the other serpins there is evidence to suggest that it involves somewhat less than complete insertion, probably to P_7 rather than to P_3. PAI-1 must therefore have a conformational adaptation that favours total incorporation of the loop into the A sheet.

Another structure which supports the concept of mobility of the loop is that of the annealed binary complex shown in Fig. S9e. The binary complexes are formed by the incubation of free peptide homologues of the reactive centre loop with inhibitory serpins. The homologue peptide anneals into the position of strand 4 of the A sheet and prevents the entry of the molecular only reactive centre loop back into the sheet. The binary complex formed in this way has the physical properties of R antitrypsin but in other ways resembles ovalbumin, with a complete loss of inhibitory activity but substrate-like vulnerability to cleavage at the reactive centre. Peptides of varying lengths can be annealed into the molecule and comparisons of the physical properties of these, together with studies of dysfunctional variants of plasma inhibitors, provide clues as to the active inhibitory conformation of the serpins.

Current knowledge of the conformation of the reactive centre loop in the active inhibitor is based on deductions from these observations and experimental results rather than from direct structural data. These suggest that inhibition occurs as a two-stage process of docking and locking. A working hypothesis is that the active inhibitory form of the serpins has a loop conformation that extends and moves into the A sheet when docking with the proteinase occurs. This is followed by a locking of the complex with further movement of the loop beyond that of the conformation in Fig. S9d, in which the loop is modelled on the canonical conformation of the small inhibitors.

Plasma and tissue serpins

Some 10% of the total plasma proteins are proteinase inhibitors, almost all of them serpins. The relatively well-defined specificity of individual serpins suits them for their task as specific inhibitors of the inflammatory cascades of the plasma. It is evident that as the serine proteinases that initiate these cascades have evolved, so there has been a matching evolution of the cognate serpins. For example THROMBIN, the final proteinase of the coagulation pathway, is inhibited by antithrombin; C1-esterase, the initiator of the COMPLEMENT cascade, is inhibited by C1-inhibitor; and the fibrinolytic pathway is controlled similarly at two levels — by PAI-1, the inhibitor of plasminogen activator, and by antiplasmin, the inhibitor of PLASMIN. This balanced system of checks and counterchecks to proteolysis enables these critical cascades in the blood to be controlled. The situation has even more inbuilt checks in that closely related serpins may have opposing effects; antithrombin, for example, functions as an inhibitor of coagulation, whereas the Protein-C inhibitor acts as a procoagulant.

The serpins' function is to maintain the balance within and between the inflammatory cascades. It is illustrative to consider the conditions under which these balances break down. One such situation is in the shock syndrome that accompanies serious injuries or septicaemia. Here there is massive consumption of the plasma proteinase inhibitors which then allows uncontrolled proteolysis with activation of coagulation, fibrinolysis, the complement and kinin cascades to give the usually fatal condition of disseminated intravascular coagulation. The same consequences follow bites by some species of rattlesnake whose venom contains METALLOPROTEINASES which specifically cleave the reactive centre loops of serpin. Although serpins of most mammals are vulnerable to loop cleavage by rattlesnake metalloproteinases this is not so for the snake's natural prey, the opossum; this has evolved a loop structure resistant to cleavage — a nice illustration of the innate diversity of the structure of the loop and its consequent ability to evolve rapidly to meet survival needs.

The fine balances between proteinases and inhibitors also occur at tissue as well as plasma level. The control of cell proliferation in growing tissue depends on the resorption of the adjacent extracellular matrix. This is achieved by developing a thin shell of proteolytic activity around each developing cell. A number of proteinases are involved but a critical contribution is that of plasminogen and the activating proteinases of the fibrinolytic pathway: TISSUE PLASMINOGEN ACTIVATOR (tPA) and UROKINASE. To balance these, a series of inhibitors is secreted at the cell surface, the key ones being the serpins PAI-1 and PAI-2 along with the specific cell surface proteinase inhibitor — protease nexin. The action of these inhibitors is confined to the vicinity of the cell surface by specific binding mechanisms but as well as this, other mechanisms have evolved to confine the sphere of inhibition. A good example is the recently identified elastase inhibitor of the monocyte. This has a cysteine at its P_1 reactive centre which needs to be in the reduced form to maintain inhibitory activity. As a consequence the monocyte elastase inhibitor has only a brief period of activity once it is released from the reducing environment of the cell into the extracellular space where it is readily oxidized.

In order to localize inhibitory activity within the circulation, a number of the plasma and tissue serpins have developed anionic subsites on the molecule that allow them to bind to the heparan side chains of the GLYCOCALYX of endothelial and other cells. These sites are usually referred to as 'heparin-binding sites' as they bind the sulphated oligosaccharide HEPARIN. It is necessary, however, to emphasize that heparin as such is a pharmaceutical product obtained by hydrolysis of animal tissue extract and is not a physiological compound of human plasma. Thus the 'heparin-binding' serpins — antithrombin, heparin cofactor II, PAI-1, protein-C inhibitor, and protease nexin — originally evolved their anionic sites in order to allow them to bind to tissues, and in particular to the endothelial cells that line the microcirculation. The activation of proteinase inhibition that notably takes place, with antithrombin and heparin cofactor II, on binding to heparin/heparans is almost certainly a subsequent evolutionary development which has been refined to give what is now an elegant switching mechanism.

The prime heparin-binding site on the serpins is formed by a positive external face on the D helix. This specifically binds to a pentasaccharide motif present in the heparans of the endothelial cells and in heparin. The binding of the isolated pentasaccharide itself to antithrombin induces a conformational change with an opening of the A sheet and a partial reinsertion of the reactive centre loop. This gives a change in loop conformation with consequent increased inhibitory activity. As well as the conformational change, there are two other contributions to the heparin activation of inhibition of thrombin by antithrombin. The longer heparins/heparans extend from the high-affinity binding site on the D helix of antithrombin around to the reactive centre pole of the molecule, giving first a neutralization of anionic sites on antithrombin and then a bridging with anionic exosites on thrombin to give a stable ternary complex.

An interesting further refinement occurs in heparin cofactor II. This inhibitor is activated by dermatan sulphate (*see* EXTRACELLULAR MATRIX MOLECULES), which binds to the D helix, as in antithrombin, but this binding in addition displaces the N-terminal tail of the molecule. The tail of heparin cofactor II has a sequence similar to that of the leech anticoagulant hirudin which specifically binds to the active centre cleft of thrombin. Thus heparin cofactor II forms an especially tight complex with thrombin stabilized by its N-terminal, as well as reactive centre, interaction.

Molecular pathology of plasma serpins

The plasma serpins are subject to the same random mutations as all proteins and careful screening has identified a large number of mutants, some resulting in nonexpression of the serpin and others in the production of a dysfunctional protein. Antithrombin in particular has an apparently high frequency of variants and as many as one in 350 people may have a dysfunctional allelic product although only a fraction of these result in thromboembolic disease. The high observed frequency is due to the ready detectability of mutations of arginyl residues in antithrombin as these often involve the heparin-binding site. The frequent mutations at arginine reflect the presence of CpG 'hot-spots' coding for arginines and the ease of detection of these mutants in antithrombin gives an index of the overall frequency of hot-spot mutations. Mutations of antithrombin at its heparin-binding site usually cause only minor problems in a heterozygote. Severe familial thromboembolic disease is likely to occur when mutations cause a complete loss of inhibitory activity, as in mutations affecting the P_1 reactive centre residue or the critical P_{14}–P_{10} hinge region of the loop.

By far the most frequent of the serpin variants are the deficiency variants of α_1-antitrypsin with one in 10 Northern Europeans being a carrier of either the severe Z deficiency variant or the milder S variant. Homozygotes for the Z variant are likely to develop the degenerative lung disease emphysema in early adult life and in childhood are at risk of developing a fatal liver cirrhosis. Biochemically, the abnormality is of interest as it affects the secretion of α_1-antitrypsin rather than its synthesis. Homozygotes for the Z mutation accumulate α_1-antitrypsin in the endoplasmic reticulum of their hepatocytes. The blockage of processing results in a plasma deficiency of α_1-antitrypsin and consequently a failure to protect the lung tissue against the elastase released by white blood cells. The molecular basis for the failure in secretion of Z antitrypsin has recently been shown to be due to a mechanism unique to the serpins, of intermolecular loop–sheet polymerization (Fig. S9*f*). The mutation in Z antitrypsin results in the replacement of a conserved glutamic acid at position 342 by a lysine. This is the P_{17} site at the base of the hinge of the reactive centre loop. The effect of the replacement of the P_{17} glutamate by lysine is to open the A sheet and hence favour the insertion of the loop from another Z antitrypsin molecule to give end-to-end polymerization. This results in the production of tangles of fibrils that distort the liver cells with consequent cell death and eventually the development of liver cirrhosis.

The Z variant is the most frequent of the mutants of antitrypsin but perhaps the most interesting mutation was that detected in a child from Pittsburgh with a severe bleeding disorder. This was shown to be due to a variant antitrypsin in which the P_1 methionine had been replaced by an arginine. The consequence was that the Pittsburgh antitrypsin had lost the ability to inhibit elastase and changed its function to become a highly active inhibitor of thrombin. This new inhibitory activity against thrombin was independent of heparin binding and as antitrypsin is present in 10-fold greater concentration than antithrombin the end effect was an excessive inhibition of thrombin with a consequent failure of coagulation. The mutation is understandably rare and only two cases have been recorded. Its significance, however, was to provide firm evidence for the placement of the reactive centre of the serpins and it underlined the key contribution of the P_1 residue in defining inhibitory specificity. Furthermore, this experiment of nature, in which a single amino-acid change caused a complete change in function, provided a preview of prospects which are now being realized — the ability to design and engineer serpins with specified inhibitory activities.

R.W. CARRELL

See also: PROTEIN STRUCTURE.

1 Bode, W. & Huber, R. (1991) Ligand binding: proteinase–inhibitor interactions. *Curr. Opin. Struct. Biol.* **1**, 45–52.
2 Carrell, R.W. & Boswell, D.R. (1986) Serpins: the superfamily of plasma serine proteinase inhibitors. In *Proteinase Inhibitors* (Barrett, A.J. & Salvesen, G., Eds) 405–419 (Elsevier Biomedical Press, Amsterdam).
3 Huber, R. & Carrell, R.W. (1989) Implications of three-dimensional structure of α_1-antitrypsin for structure and function of serpins. *Biochemistry* **28**, 8951–8966.
4 Carrell, R.W. & Evans, D.L.L. (1992) Serpins: mobile conformations in a family of proteinase inhibitors. *Curr. Opin. Struct. Biol.* **2**, 438–446.

serum amyloid A An ACUTE PHASE PROTEIN.

serum amyloid P component (SAP) The major DNA- and CHROMATIN-binding glycoprotein of blood plasma. It is classed as a PENTRAXIN, and can act as a LECTIN, recognizing the cyclic 4,6-pyruvate acetal of β-D-galactose, as well as several polyanions. It is a common constituent of tissue AMYLOID deposits. The subunit molecular conformation of SAP resembles that of lectins from leguminous plants, even though there is little sequence homology and a different mode of carbohydrate binding.

serum response element (SRE) Regulatory DNA sequence found in the promoters of immediate early genes such as *fos*, which is activated in response to binding of the serum response factor (SRF).

sevenless (sev) A *Drosophila* gene involved in EYE DEVELOPMENT.

severe combined immunodeficiency (SCID) A profound state of IMMUNODEFICIENCY characterized by a deficiency of both cell-mediated and humoral immunity and which may be caused by a variety of primary defects.

In humans it is a diagnostic category of primary immunodeficiency states in which both cellular and humoral immunity are profoundly impaired. Three main forms of SCID have been well described in humans: (1) the classical form which involves genetic lesions affecting both T cell and B cell development (e.g. ADENOSINE DEAMINASE DEFICIENCY); (2) an X-linked form resulting in defective T cell development (since B cell function requires T cell help this condition also results in severe combined immunodeficiency); and (3) an inability to express class II MHC MOLECULES on the surface of lymphocytes and other ANTIGEN-PRESENTING CELLS (type II BARE LYMPHOCYTE SYNDROME) which results in an inability to present antigens to HELPER T CELLS. Apart from these three major categories, there are numerous other rarer defects which cause SCID but as yet are not fully understood.

In mice, the state of severe combined immunodeficiency caused by the *scid* mutation (on chromosome 16) when homozygous is thought to result from an inability to rearrange immunoglobulin or T cell receptor genes correctly. When T and B cells reach the stage of development at which this rearrangement normally occurs (*see* B CELL DEVELOPMENT; T CELL DEVELOPMENT) the defect in recombination leads to aberrant cuts and rejoins in the coding segments of the receptor genes; this culminates in a great reduction in the production of functional receptors. Hence, *scid* mice have greatly reduced or virtually undetectable levels of T and B cells. The primary defect is not specific to T and B cells but can also affect other myeloid cell lineages and fibroblasts, which exhibit unusual sensitivity to ionizing radiation. It has been suggested that the *scid* mutation leads to an inability to repair DNA damage induced by ionizing radiation (*see* DNA REPAIR) or in the process of gene rearrangement. About 15% of *scid* mice have detectable levels of B and T cells suggesting that the defect is partially leaky.

severin *See*: ACTIN-BINDING PROTEINS.

sex chromosomes Chromosomes that form nonhomologous pairs in one of the sexes, for example the X- and Y-chromosomes in humans, the W and Z in birds, in which the female is the heterogametic sex, and the X-chromosome in *Drosophila*, and which are involved in determining sex. *See*: SEX DETERMINATION; SEX LETHAL; X-CHROMOSOME; Y-CHROMOSOME.

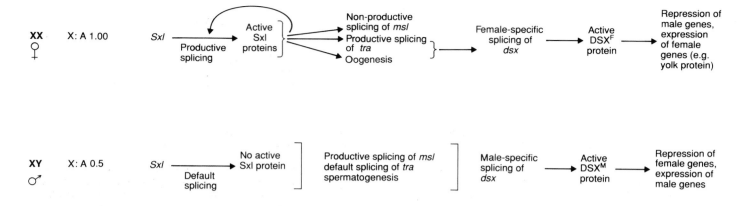

Fig. S10 Regulatory pathway of somatic sexual determination in *Drosophila melanogaster*. From Hodgkin, 1990.

sex determination The genetic and molecular basis of sex determination is best understood in three types of organism: the fruit-fly *Drosophila melanogaster*, the nematode worm *Caenorhabditis elegans*, and mammals. In *Drosophila* and mammals there are two sexes, male and female, whereas in *Caenorhabditis elegans* the two sexes are male and hermaphrodite (in other nematodes they are male and female). In all these organisms, somatic sexual phenotype is determined chromosomally, and is fixed early in development. Many other strategies of sex determination exist however: in hymenopterans females arise from fertilized eggs and are diploid, whereas males arise from unfertilized eggs and are haploid; in some reptiles (e.g. turtles and crocodilians) the sex of the hatchling is influenced by the prevailing temperature during a critical period of incubation; and some fishes and invertebrates can change sex in adult life.

The sexual phenotype of the germ line is determined chromosomally in *Caenorhabditis*, is influenced by interactions with the somatic gonad in *Drosophila*, and in mammals is dependent on the sex of the somatic gonad.

In *Drosophila*, *Caenorhabditis* and mammals, sex is determined by genes located on the SEX CHROMOSOMES, which unlike the AUTOSOMES differ between the sexes. In both *Drosophila* and *Caenorhabditis*, sex is determined by the ratio of the number of X-chromosomes to the number of sets of autosomes, although at the molecular level the mechanisms of sex determination differ between the two organisms. A quite different system of sex determination operates in mammals. The two sexes, male and female, have the genetic constitution XY and XX respectively, with the Y-chromosome being actively male-determining.

Genetic studies have identified three aspects of sex determination which are regulated by largely independent mechanisms, although the same genes may be involved in more than one of these processes.

1 Somatic sexual development involves the differentiation of sexually dimorphic characters in the soma, including the differentiation of reproductive structures.

2 Germ-line sexual development determines the type of gamete (sperm or ovum) produced.

3 In the organisms discussed here, the unequal numbers of X-chromosomes in the different sexes require mechanisms of dosage compensation to equalize total gene expression from the X-chromosome sets.

Sex determination by X : autosome ratio. In both *Drosophila* and *Caenorhabditis*, in one sex there are two sets of autosomes and two X-chromosomes (2X; 2A, ratio 1), whereas in the other there are two sets of autosomes and one X-chromosome (X; 2A, ratio 0.5). *Drosophila* has two sexes, male and female. In some *Drosophila* species there is no counterpart to the X-chromosome and males are XO, but in *D. melanogaster* males are XY as there is a small Y-chromosome, which is, however, not male-determining. *Caenorhabditis elegans* has two sexes, male (XO) and hermaphrodite (XX). In hermaphrodites, a limited number of sperm are produced early in development, the remainder of the germ cells developing into oocytes. In other nematode species, XX worms are female, producing only oocytes.

The X : autosome ratio is the primary sex-determining signal. In *Drosophila*, its target is the SEX-LETHAL gene on the X-chromosome, which acts as a master regulatory gene controlling not only sexual phenotype but also dosage compensation. In *Drosophila*, the X : A ratio is read as a transcriptional signal dependent on the concentration of active heterodimers between the X-linked helix-loop-helix DNA-binding proteins (Sis-a and Sis-b) and the autosomally encoded *daughterless* helix-loop-helix gene product. In females, early transcriptional activation produces an *Sx1* product which then autoregulates *Sx1* to maintain the gene in a functional state. This feedback acts at the level of RNA SPLICING to yield female-specific mRNAs with an open reading frame of 350 amino acids containing an RNA recognition motif. In male flies, male-specific splicing of *Sx1* early in development produces an RNA containing a stop codon which results in a truncated nonfunctional protein. *Sx1* acts by initiating a cascade of alternative splicing events by regulating the splicing of the TRANSFORMER gene (Fig. S10). In female flies, the active *Sx1* product directs the productive splicing of *tra* transcripts (*see* TRANSFORMER GENES); active *tra* protein is therefore produced in female but not in male flies.

The ultimate target for female-specific splicing by the TRANSFORMER gene products is DOUBLE-SEX (*dsx*) RNA, which shows sex-specific differences in splicing, but which produces a func-

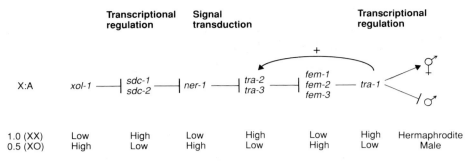

	Transcriptional regulation	Signal transduction		Transcriptional regulation			

X:A							
1.0 (XX)	Low	High	Low	High	Low	High	Hermaphrodite
0.5 (XO)	High	Low	High	Low	High	Low	Male

Fig. S11 Regulatory pathway of somatic sexual determination in *Caenorhabditis elegans*. In XO (male) animals high levels of *xol-1* gene product depress *sdc* action, and allow high levels of *her-1* gene activity. This inhibits the action of *tra-2* and *tra-3* genes, resulting in high levels of expression of the *fem* genes and a low level of *tra-1* expression. In XX (hermaphrodite) animals a low level of *xol-1* expression on the other hand results in high levels of *sdc* gene products. These depress the activity of the *her-1* gene, allowing high levels of *tra-2* and *tra-3* activity, which negatively regulates *fem* gene activity, and thus allows high levels of *tra-1* gene activity and subsequent hermaphrodite development. From Kuwabara & Kimble, 1992.

tional protein in both sexes — DSX^M in males, DSX^F in females, both of which act as transcriptional regulators. They possess a common DNA-binding domain related to the well known ZINC FINGER motif. Genetic studies suggest that, directly or indirectly, DSX^M represses female genes in males, whereas DSX^F (together with *intersex*) represses male genes in females, possibly by differential activity on the same genes. For example, one target of the DSX proteins is the ENHANCER that directs sex- and tissue-specific expression of the *Yolk protein (Yp)* genes, which are expressed only in females. At this enhancer, DSX^M and DSX^F can bind to three sites, binding of DSX^M repressing transcription whereas binding of DSX^F activates transcription.

In *Caenorhabditis*, multiple short noncoding DNA elements on the X-chromosome seem to be involved in producing the X : autosome ratio signal, but their mode of action is not yet known. The targets of the X : A ratio are the X-linked *xol* and *sdc* genes, with *xol* regulating *sdc* activity. In XX (hermaphrodite) animals the level of *xol-1* is low and the *sdc* genes are active whereas in XO (male) animals, *xol-1* is a negative regulator of *sdc* gene action.

Somatic sexual phenotype in *Caenorhabditis* is determined by a hierarchy of autosomal genes activated by the X-linked *sdc* and *xol* genes (Fig. S11) and involving the feminization (FEM) and TRANSFORMER GENES. Several components are contributed maternally (*sdc-1, fem-1, fem-2, fem-3*, and *tra-3*) and initial commitment to one sex occurs early in embryonic development. The last-acting gene in this cascade is the transformer gene *tra-1* whose activity is necessary and sufficient to determine all aspects of female somatic sexual phenotype. This set of genes is also involved in determining germ-line sexual phenotype (see below).

sdc-1 has been identified as a zinc finger protein and may therefore bind to DNA and regulate gene transcription. The *her-1* and *fem-3* gene products are novel proteins whose activity is unknown. The *her-1* protein may be secreted. *tra-2* encodes a membrane protein, possibly a receptor, and is present in both sexes. *tra-1* encodes a zinc finger protein that is presumed to be a transcriptional regulator. CELL–CELL INTERACTIONS are of importance in the early events of sex determination. Neither *sdc-1* nor *her-1* acts in a CELL-AUTONOMOUS manner, whereas *tra-1* acts

cell-autonomously. Thus *sdc-1* and *her-1* may act to regulate the production of a signal and *tra-1* would be part of the receiving mechanism.

In the determination of germ-line sex in *Drosophila*, active *Sx1* is required for oogenesis but not for spermatogenesis. The *tra-2* gene, which acts in female somatic sex determination has a role in males in spermatogenesis, although not apparently in the determination of germ-line sexual phenotype.

In *Caenorhabditis elegans*, germ-line sex is determined by the same genes that control somatic sexual phenotype, but their detailed interactions are different. In hermaphrodites, additional genes that permit the initial phase of spermatogenesis are involved.

Mammals. In mammals development of the female sexual phenotype appears to be a basic default pathway, which is switched into a male-specific pathway early in embryonic development by the expression of a specific gene on the Y-chromosome, the testis-determining gene (*SRY* in humans, *Sry* in mice). In mammals, commitment of the gonad to become testis is the primary determinant of maleness. In the absence of such commitment, development as a female ensues.

SRY (the sex-determining region of the Y) was first identified in humans from studies of XX men who are male because one of their X-chromosomes bears a small translocated region of the Y-chromosome including *SRY*. *SRY*, and its murine homologue *Sry*, have homologies with both the Mc mating-type protein from fission yeast (*Schizosaccharomyces*) and nonhistone HMG proteins. The protein product of *Sry* has been shown to bind DNA. *Sry* (and by inference human *SRY*) was confirmed in the mouse as the testis-determining gene by its introduction as a TRANSGENE into otherwise XX mice, who developed as males, by its male-specific expression in the gonadal somatic cells responsible for testis determination, and its absence in XY mice that developed as females. Detailed examination of the DNA of humans with various genetically based abnormalities in sexual development is also consistent with *SRY* being the testis-determining gene. In particular, the occurrence of XY females has been associated with novel point mutations in the *SRY* gene.

Although *Sry* has been identified as a critical switch in the pathway of male sexual development, little is known of other genes in the pathway responsible for the interpretation of the *Sry* signal and the maintenance of male development. At a physiological level, the further development of male reproductive structures and the degeneration of female structures requires the steroid hormone TESTOSTERONE, which is produced by the male-committed gonad and which acts at intracellular receptors of the STEROID RECEPTOR SUPERFAMILY to activate genes involved in male differentiation. Studies of testicular-feminization syndrome in XY humans and mice show that XY individuals bearing an active *Sry* gene can develop as phenotypic females (although they still have testes) as a result of the insensitivity of their embryonic tissues to testosterone, due to an X-linked mutation in the testosterone receptor.

Dosage compensation. In animals in which the different sexes carry different numbers of X-chromosomes, gene expression from the single X in one set must be made equal to total gene expression from the two X-chromosomes in the other set. The X-chromosomes in all species are large chromosomes carrying an appreciable proportion of the total genome (around 20% in *Drosophila*) as well as the small number of genes involved specifically in sex determination. In *Drosophila*, both X-chromosomes are transcribed in females, and to compensate, there is an increased rate of transcription from the single X-chromosome in males, so that the level of X-related transcripts is the same in both sexes. Increased gene expression is via a twofold increase in transcription, in male flies only, of sex-linked loci under the control of ENHANCER elements on the X-chromosome. The *trans*-acting factors involved include the *MSL* set of genes (including *mle* (maleless), *msl-1, 2-*, and *-3* (male-specific lethal)), which encode TRANSCRIPTION FACTORS.

In *Caenorhabditis*, equalization of transcripts may be by down-regulation of transcription in XX animals. Genes affecting dosage compensation (the dosage compensation dumpies or DCD genes *dpy-21, -26, -27, -28*) have been found, mutations in which are lethal in XX but not XO animals.

Dosage compensation in female mammals is achieved by random inactivation of one X-chromosome in somatic cells (*see* X-CHROMOSOME INACTIVATION). In birds, however, the male is ZZ and the female ZW but there is no evidence for dosage compensation for the Z-chromosome. *See also*: KLINEFELTER SYNDROME; TURNER SYNDROME.

Hodgkin, J. (1990) *Nature* **344**, 721–728.
Kuwabara, P.E. & Kimble, J. (1992) *Trends Genet.* **8**, 164–168.
Coschigano, K.T. & Wensink, P.C. (1993) *Genes Dev.* **7**, 42–54.
Sinclair, A.H. et al. (1990) *Nature* **346**, 240–245.
Gubbay, J. et al. (1990) *Nature* **346**, 245–250.
Goodfellow, P.N. & Lovell-Badge, R. (1993) *Annu. Rev. Genet.* **27**, 71–92.

sex determining region of the Y (SRY, sry) The male sex-determining gene, located on the Y-chromosome. *See*: SEX DETERMINATION.

Sex-lethal (Sxl) The activity of the *Sxl* gene in *Drosophila* controls dosage compensation and both somatic and germ-line sex. The X : A(utosome) ratio provides the primary signal to switch the *Sxl* gene into either functional (female) or nonfunctional (male) mode. The X : A ratio is read as a transcriptional signal dependent on the concentration of active heterodimers between the X-LINKED helix-loop-helix DNA-binding proteins (Sis-a and Sis-b) and the autosomally encoded *daughterless* helix-loop-helix gene product. In females, early transcriptional activation produces *Sxl* product which then autoregulates *Sxl* to maintain the gene in a functional state. This feedback acts at the level of splicing to yield female-specific mRNAs with an open reading frame of 350 amino acids containing an RNA recognition motif. Only nonfunctional transcripts are produced in males. *Sxl* acts to initiate a cascade of alternative splicing events by regulating the splicing of the TRANS-FORMER gene. *See also*: SEX DETERMINATION.

sex linked Describes a genetically determined disease or condition seen overwhelmingly in one sex. In the case of X-linked conditions in humans, the disease is due to a RECESSIVE allele carried on an X-chromosome and is usually only seen in males; females heterozygous for the allele are symptomless carriers.

SH2 domain *src* homology domain 2. A protein domain first identified in the product of the *src* cellular ONCOGENE, a TYROSINE PROTEIN KINASE, and which has since been discovered in numerous other intracellular proteins involved in signal transduction. It is a domain of ~100 amino acids which binds certain phosphotyrosine-containing sequences and is thought to be involved in protein–protein interactions in signal transduction. *See*: SECOND MESSENGER PATHWAYS.

SH3 domain *src* homology domain 3. A protein domain of ~50 amino acid residues present in the product of the *src* cellular ONCOGENE, a tyrosine PROTEIN KINASE, and some other proteins involved in signal transduction, and also in a number of cytoskeletal proteins, such as α-spectrin (*see* ACTIN-BINDING PROTEINS). It is thought to be involved in protein–protein interactions. *See also*: ACTIN-BINDING PROTEINS; INSULIN AND ITS RECEPTOR.

shadowing The directional deposition of a fine layer of electron-dense metal on specimens for ELECTRON MICROSCOPY. It adds bulk to linear macromolecules such as DNA and delineates surface features of larger structures.

Shaker *Drosophila* mutant with a genetic defect in a class of potassium ion channels. *See*: VOLTAGE-GATED ION CHANNELS.

shear *See*: NUCLEIC ACID STRUCTURE.

shearing The mechanical (i.e. nonenzymatic) cleavage of DNA molecules. Genomic DNA can be sheared by passage through a small-bore hypodermic needle. This approach has been used to generate DNA fragments for subcloning into VECTORS to produce random genomic libraries (*see* DNA CLONING; DNA LIBRARIES).

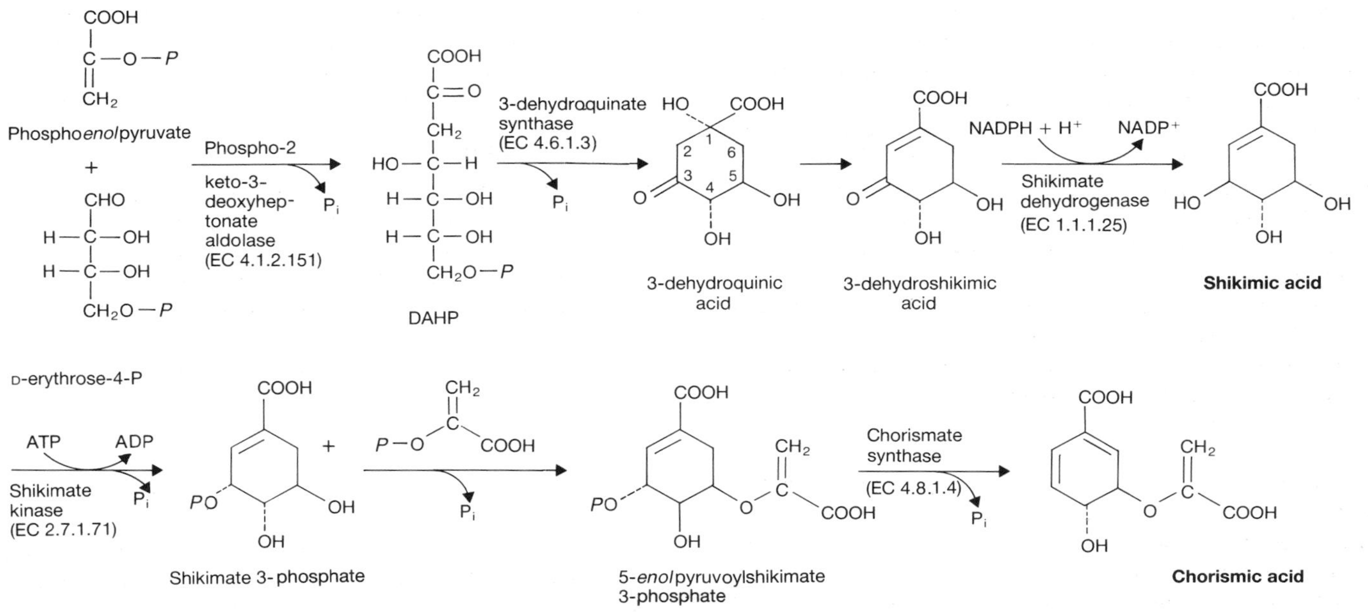

Fig. S12 Shikimic acid pathway.

β-sheet (beta-sheet) An element of protein secondary structure. *See*: PROTEIN STRUCTURE.

***shi* locus** Shoot-inducing locus in the Ti PLASMID. *See*: IAAH/M GENES.

Shigella Genus of nonmotile bacteria of the Enterobacteriaceae which includes *Shigella dysenteriae* (*Sh. shigae*) and *Sh. flexneri* which cause acute bacillary dysentery. *Sh. dysenteriae* type 1 (*Sh. shigae*) produces an exotoxin. *See also*: BACTERIAL PATHOGENICITY.

shikimic acid pathway Pathway of aromatic AMINO ACID biosynthesis in bacteria, fungi, and plants, leading to phenylalanine, tyrosines, and tryptophan (Fig. S12). It involves two stages: the conversion of (1) phosphoenolpyruvate and erythrose 4-phosphate to chorismate, with shikimic acid as intermediate, and (2) chorismate to phenylalanine and tyrosine or tryptophan. The two stages are regulated by feedback inhibition of isoenzymes of the first reaction in each stage, by each of the amino-acid products.

Shine–Dalgarno region/sequence A purine-rich region of between 6 and 8 bp just upstream from the AUG translation INITIATION CODON in prokaryotic mRNA. In *Escherichia coli*, it has the following sequence: 5′ AGGAGG 3′ and lies around seven bases 5′ to the AUG initiation codon. This sequence acts as a ribosome-binding site and is important for the efficient initiation of translation. It was originally discovered in *E. coli* by John Shine and Larry Dalgarno in 1974 because of its complementarity to the 3′ end of the small (16S) rRNA present in the 30S ribosomal subunit. The initial positioning of mRNA on the 30S subunit requires base pairing between the 16S rRNA and the Shine–Dalgarno sequence. *See*: BACTERIAL GENE EXPRESSION; PROTEIN SYNTHESIS.

Shine, J. & Dalgarno, L. (1974) *Proc. Natl. Acad. Sci. USA*, **71**, 1342–1346.

short consensus repeat Amino-acid sequence 65–70 residues long, including four conserved cysteine residues that form intra-chain disulphide bonds, folding the sequence into a knot. Repeats of this sequence are found in some COMPLEMENT RECEPTORS, in PROTEIN H, and in the SELECTIN family of adhesion molecules.

short day plant (SDP) Plant that is induced to flower by exposure to a regime in which the nights (i.e. periods of darkness) are longer than a certain critical value.

shotgun cloning A cloning method in which total genomic DNA is randomly sheared mechanically or digested with restriction enzymes and the fragments ligated into a suitably prepared CLONING VECTOR. It is used to prepare genomic libraries (*see* DNA LIBRARIES) and to subclone restriction fragments of DNA from one plasmid to another. Subsequent screening is required to identify the desired recombinant. This approach to DNA cloning negates the need for purification of the desired DNA fragment, and is particularly appropriate when subcloning a known target DNA sequence from a large cloned DNA fragment. *See*: DNA CLONING.

shuttle vector Autonomously replicating VECTOR (usually a circular PLASMID) which contains ORIGINS OF REPLICATION and selectable MARKERS permitting it to be taken up by, and maintained in, two different host organisms, for example *Escherichia coli* and *Saccharomyces cerevisiae*.

sialic acid *See*: N-ACETYLNEURAMINIC ACID; NEURAMINIC ACID.

sickle cell anaemia A common haemoglobinopathy in people of African descent, less frequent in other tropical races, caused by HOMOZYGOSITY for production of SICKLE CELL HAEMOGLOBIN. The abnormal haemoglobin polymerizes into fibres in response to reduced oxygen tension, causing the red cells to deform to a characteristic 'sickle' shape, leading to blockage of capillaries and haemolysis. The disease is characterized by chronic haemolysis and acute 'crises' due to sickling of red cells (*see* HAEMOGLOBIN AND ITS DISORDERS).

Sickle cell trait is the HETEROZYGOUS state for the gene for sickle cell anaemia. The condition is asymptomatic except in conditions of extreme hypoxia, and is protective against malaria. Its main clinical importance is that offspring of two sickle cell trait carriers have a one in four risk of being homozygous for the gene and thus of suffering from sickle cell anaemia.

sickle cell haemoglobin HAEMOGLOBIN S.

side chains (1) Of an amino acid or polypeptide backbone, the group attached to the αC atom of the amino acid. They can be classified into polar, nonpolar, charged, and aromatic side chains (*see* AMINO ACIDS) and confer particular chemical and physical reactivities on the amino acid within a protein structure. *See also*: PROTEIN STRUCTURE.

(2) Of a glycoprotein, the oligosaccharide groups attached to the polypeptide backbone. *See*: GLYCANS; GLYCOPROTEINS AND GLYCOSYLATION.

siderophore Low molecular weight, high affinity, virtually iron (FeIII)-specific ligand which is synthesized by microorganisms in response to iron limitation (Fig. S13). *See*: BACTERIAL PATHOGENICITY.

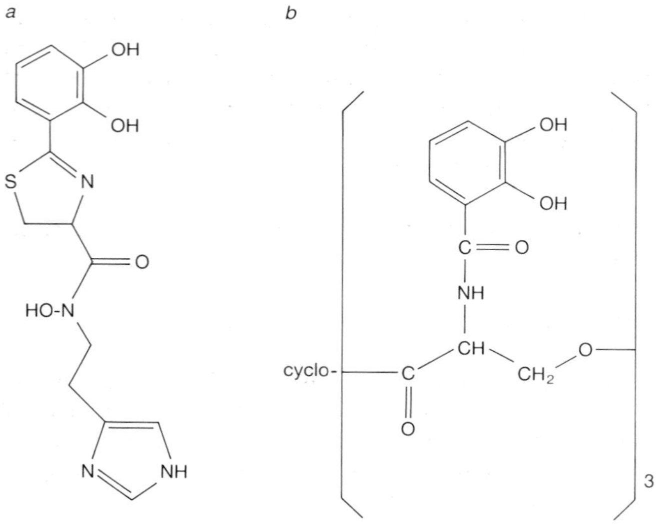

Fig. S13 The siderophores anguibactin (*a*) from *Vibrio anguillarum* and enterobactin (*b*) from *Escherichia coli*.

sieve plate The perforated region of the wall between two SIEVE TUBE elements, and consisting of one or more sieve areas with their characteristic pores. Also used for similar structures which join two trumpet hyphae end to end in complex brown algae.

sieve tube A series of cells connected end to end forming the conducting elements of the PHLOEM of angiosperms. The cells bear SIEVE PLATES through which substances are translocated along the tube.

sigla A name formed from letters or other characters taken from the words in a compound term. Virus group names are often sigla, for example, TOBAMOVIRUS GROUP from 'tobacco mosaic virus group'.

sigma (σ) (1) Sigma factor, sigma regulatory subunit. A dissociable subunit of bacterial RNA POLYMERASE which is responsible for ensuring that the polymerase starts TRANSCRIPTION from the correct site in a gene. Sigma factor increases the affinity of RNA polymerase for the PROMOTER and decreases its affinity for DNA in general. A number of different sigma factors exist, each typically recognizing a different consensus sequence, and thus directing polymerase interaction to a different range of promoters. For example, σ^{70} (M_r 70 000) is the one usually associated with the polymerase and recognizes the classical -35, -10 consensus sequences (*see* BACTERIAL GENE EXPRESSION; TRANSCRIPTION); σ^{54} recognizes a consensus sequence in the promoters of NITROGEN FIXATION GENES and σ^{32} recognizes the *Escherichia coli* HEAT SHOCK element. The controlled production of different sigma factors is also involved in the developmental switch from vegetative growth to endospore production in *Bacillus* (*see* BACTERIAL DIFFERENTIATION).

(2) A TRANSPOSABLE GENETIC ELEMENT found in SACCHAROMYCES CEREVISIAE.

Losick, R. & Pero, J. (1981) *Cell*, **25**, 582–584.

signal II An EPR signal induced in the light and originating in PHOTOSYSTEM II which has the characteristics of a free radical with additional fine structure. Different kinetic forms have been described, II_{slow}, II_{fast} and $II_{very\ fast}$, according to their rates of relaxation. The signals originate in the oxidized forms of two intermediates on the oxidizing side of photosystem II. Normally the immediate electron donor to P680 is 'Z', which is the side chain of Tyr 161 of subunit D1 and responsible for signal II_{vf}. The corresponding residue on subunit D2 is Tyr 160 which can also be oxidized by P680 (donor 'D') and gives rise to signal II_s, but it is only very slowly rereduced. Signal II_f is a modified form of II_{vf} produced when Mn is removed from the water-oxidizing enzyme. *See also*: ESR/EPR SPECTROSCOPY; PHOTOSYNTHESIS.

signal joint During GENE REARRANGEMENT in the IMMUNOGLOBULIN GENES and T CELL RECEPTOR GENES the sites of recombination of the V, D and J gene segments are determined by heptamer and nonamer recognition sequences which abut the coding regions to be joined. During rearrangement of the V(D)J gene segments the heptameric signal sequences are fused to each other giving rise to a signal joint which does not form part of the final coding sequence of the assembled gene. The crossover site in the signal joint is precisely at the heptamer borders.

signal peptidase Integral membrane enzyme of the ENDOPLASMIC RETICULUM membrane which cleaves N-terminal SIGNAL SEQUENCES on proteins as they emerge into the lumen of the endoplasmic reticulum. The enzyme solubilized with neutral detergents requires phospholipid to be active. Its mechanism is unknown, and as a proteinase it is in a class of its own as none of the typical group-specific reagents inhibit it. The purified complex from hen oviduct consists of a protein of M_r 19 000 and a differentially glycosylated form of the same protein (M_r 22 000/ 24 000), and from canine pancreas, four protein subunits of M_r 25 000, 21 000, 18 000, and 12 000 as well as a glycosylated form (M_r 22 000/23 000). The 21 000 and 18 000 subunits are homologous to the Sec11 protein in yeast; in the *sec11* strain uncleaved secreted proteins accumulate at the nonpermissive temperature. *See also*: PROTEIN TARGETING; PROTEIN TRANSLOCATION.

Shelness, G.S. & Blobel, G. (1990) *J. Biol. Chem.* **265**, 9512–9519.

signal sequences, signal peptides Short sequences that direct newly synthesized secretory or membrane proteins to and through membranes of the ENDOPLASMIC RETICULUM, or from the cytoplasm to the periplasm across the inner membrane of bacteria (*see* BACTERIAL ENVELOPE; BACTERIAL PROTEIN EXPORT), or from the matrix of MITOCHONDRIA into the inner space, or from the stroma of CHLOROPLASTS into the thylakoid. They are often, but not universally, in an N-terminal location and are cleaved off by SIGNAL PEPTIDASES after the protein has crossed the membrane. Signal sequences are not homologous, but contain three common structural features — a hydrophobic core, known as the h-region, comprising at least eight uncharged residues flanked by a polar basic region (n-region) on the N-terminal side, and a hydrophilic region (c-region) of about six residues terminating at a small uncharged residue. This residue contributes the carboxyl group of the peptide bond that is cleaved by signal peptidase. The signal sequence is also known as the leader peptide or translated leader sequence. Fusing such a sequence (e.g. from the YEAST MATING FACTOR MFα) to a gene that is to be expressed in a heterologous host ensures secretion of the recombinant protein from the cells (*see* GENETIC ENGINEERING). *See also*: PROTEIN SECRETION; PROTEIN TARGETING; PROTEIN TRANSLOCATION.

von Heijne, G. (1985) *J. Mol. Biol.* **184**, 99–105.

signal recognition particle (SRP) A rod-shaped RIBONUCLEOPROTEIN containing a 7SL RNA of 300 nucleotides and six protein subunits of M_r 72 000, 68 000, 54 000, 19 000, 14 000, and 9 000. It targets proteins containing SIGNAL SEQUENCES to the ENDOPLASMIC RETICULUM membrane. SRP binds signal sequences in nascent chains as they emerge from the ribosome and halts or slows elongation until contact is made with DOCKING PROTEIN in the membrane. The 19K subunit mediates binding of the 54K subunit to the 7SL RNA backbone, the 14/9K heterodimer mediates elongation arrest and the 68/72K dimer binds to docking protein. The 54K subunit possesses a consensus GTP-binding domain in its N-terminal section, and a methionine-rich C-terminal domain which folds into several amphiphilic sequences and binds the signal sequence. GTP binding and/or hydrolysis may be required to allow the signal sequence to leave SRP and insert into the membrane before translocation. SRP is isolated by affinity chromatography of salt extracts of mammalian MICROSOMES, and homologous components are found in yeast and bacteria. *See also*: PROTEIN TARGETING; PROTEIN TRANSLOCATION.

Bernstein, H.D et al. (1989) *Nature* **340**, 482–486.

signal transduction The process by which the information contained in an extracellular physical or chemical signal (e.g. a hormone or GROWTH FACTOR) is received at the cell by the activation of specific RECEPTORS and conveyed across the plasma membrane, and along an intracellular chain of signalling molecules, to stimulate the appropriate cellular response. *See*: CELL-SURFACE RECEPTORS; OLFACTORY TRANSDUCTION; SECOND MESSENGER PATHWAYS and cross-references therein; VISUAL TRANSDUCTION.

signalling receptors Those RECEPTORS whose activation by their cognate ligand or physical stimulus (e.g. light) leads to the transduction of a signal and the activation of the intracellular biochemical response pathways. *See*: CELL-SURFACE RECEPTORS; SECOND MESSENGER PATHWAYS and cross-references therein.

silencer Name given to presumed protein factors that negatively control expression of eukaryotic genes, and also to the DNA sequences involved in such repression.

silent allele NULL ALLELE.

silent mutation A change in a DNA sequence which does not result in a detectable phenotypic change. Mutations in STRUCTURAL GENES are silent if they do not change the amino acid inserted or if they result in substitution of a residue that does not affect protein function. Silent mutations can also be located outside protein-coding regions. *See*: MOLECULAR EVOLUTION; MOLECULAR PHYLOGENY; MUTATION.

silica gel A desiccant used for the storage of hygroscopic chemicals.

simian virus 40 (SV40) *See*: PAPOVAVIRUSES.

simple-sequence DNA DNA composed of tandemly repeated short noncoding sequences, which occurs within the genomes of most eukaryotes, often in thousands of copies. *See*: GENOME ORGANIZATION; HIGHLY REPETITIVE DNA; SATELLITE DNA; VARIABLE NUMBER TANDEM REPEATS.

simulated annealing A method used in the REFINEMENT OF MACROMOLECULAR STRUCTURES, which has a much larger radius of convergence than conventional refinement procedures, wherein a model system is adjusted by minimizing the difference between calculated and observed values. Traditional methods (usually based on a LEAST SQUARES METHOD) cannot distinguish a local minimum from a global minimum in the function being minimized, as they have no means of overcoming a barrier between minima. Simulated annealing can cross these barriers by taking energetically 'uphill' steps in the target function E_{total} with a probability $\exp(-E_{total}/k_bT)$ using an effective temperature T as a control parameter, where k_b is the Boltzmann constant. Typically the system is slowly cooled — so restricting access to successively smaller regions of phase space until a configuration is obtained which is closer to the global minimum than the starting configuration. The method derives its name from the analogy with the physical annealing of an amorphous solid (local order, high energy) to a crystalline solid (long-range order, lower energy) by first liquefying the solid, then cooling or annealing it slowly so as to avoid the system becoming trapped in another metastable state.

Simulated annealing has become an invaluable tool for the rapid refinement of molecular structures derived from either crystallography or NMR SPECTROSCOPY; it is incorporated into various computer programs for refinement, the most widely known of which is now probably XPLOR written by Axel Brünger.

Brünger, A.T. (1988) *J. Mol. Biol.* **203**, 803–816.
Brünger, A.T. (1991) *Annu. Rev. Phys. Chem.* **42**, 197–223.

Sindbis virus A single-stranded (+)-sense RNA virus of the alphavirus genus of the TOGAVIRIDAE. It produces a lytic infection in most cell types but can enter a persistent state in post-mitotic neurons.

SINEs RETROTRANSPOSON-like sequences found in vertebrate genomes as *Short INterspersed DNA sequence Elements*, also known as short interspersed sequences (*cf.* LINE). They range in length from 130 to 300 bp but in contrast to SIMPLE-SEQUENCE DNA, SINEs are generally not tandemly repeated but are present as isolated dispersed elements. They are found in many thousands of copies per genome. The most abundant family of SINEs in mammals is that of the ALU SEQUENCES which are present at about 10^6 copies per human genome.

Singer, M.F. (1982) *Cell* **28**, 433–434.

single-channel recording *See*: PATCH CLAMP.

single-copy DNA/sequence A specific DNA sequence found in only one or very few copies per genome. The term may refer to a typical protein-coding gene with a single locus per haploid genome, or an episomal PLASMID present in one copy per genome. *See*: GENOME ORGANIZATION.

single gene disorder MONOGENIC DISORDER.

single isomorphous replacement (SIR) A method for solving the crystallographic PHASE PROBLEM by using crystals of only a single HEAVY ATOM DERIVATIVE which is isomorphous to the crystal under study. The SIR method is a special case of the more commonly used MULTIPLE ISOMORPHOUS REPLACEMENT (MIR) method. With SIR there remains a twofold ambiguity in the determined phase, which is solved either by the use of MIR or by combining the use of SIR with observations of ANOMALOUS SCATTERING. *See*: X-RAY CRYSTALLOGRAPHY.

single-strand binding protein *See*: SSB PROTEIN.

Siphoviridae *See*: BACTERIOPHAGES.

SIR SINGLE ISOMORPHOUS REPLACEMENT.

SIROAS Acronym for SINGLE ISOMORPHOUS REPLACEMENT with optimized anomalous scattering.

sis Gene encoding the B chain of platelet-derived growth factor (*see* GROWTH FACTORS), and which is a potential ONCOGENE. A mutant version of *sis* is the oncogene in the RETROVIRUS simian sarcoma virus.

sister cells The two cells produced by mitotic division of the parental cell.

sister chromatids A pair of genetically identical pieces of double-stranded DNA and associated proteins, produced by DNA replication of a eukaryotic CHROMOSOME. In MITOSIS, sister chromatids are attached to one another as they condense during PROPHASE and become organized during PROMETAPHASE. They separate from one another at the onset of ANAPHASE.

sister chromatid exchange (intrachromosomal recombination) Crossing-over between SISTER CHROMATIDS (*see* RECOMBINATION). This may occur either within the bivalent during MEIOSIS or between the duplicated chromatids of somatic cells.

site-directed mutagenesis Any of various techniques by which defined mutations can be made *in vitro* in a cloned DNA. These techniques enable cloned genes to be mutated at will and the functions of particular DNA sequences to be elucidated by the effects of introducing the mutated gene into a cultured cell or organism. This 'reverse genetics' approach has provided a powerful method for dissecting regulatory regions of genes and for altering genes for reintroduction as TRANSGENES. *See*: CASSETTE MUTAGENESIS; DELETION MUTAGENESIS; OLIGONUCLEOTIDE-MEDIATED MUTAGENESIS.

site-specific recombination RECOMBINATION between two DNA sequences which occurs only at a specific site. It is guided by an enzyme that recognizes a short nucleotide sequence and initiates recombination at a specific site within it. Unlike general recombination, extensive base pairing between the two molecules need not be involved. Site-specific recombination was first detected in bacteriophage LAMBDA, where its ability to INTEGRATE into the *Escherichia coli* chromosome is mediated by the *att* sites. In this case the enzyme brings lambda and chromosomal *att* sites together, then initiates the DNA cutting and religation reactions. This type of recombination is thought to be commonly employed by DNA sequences that move in and out of a host genome (e.g. RETROVIRUSES; TRANSPOSABLE GENETIC ELEMENTS).

sitting drop A technique used in the CRYSTALLIZATION of macro-molecules.

situs inversus *See*: SYMMETRY AND ASYMMETRY IN ANIMAL DE-VELOPMENT.

SIV Simian IMMUNODEFICIENCY VIRUSES.

skeletal muscle Striated muscle. MUSCLE attached to the skeleton and which is under voluntary motor control. It is composed of transversely striped fibres, and the contractile apparatus is highly organized into myofibrils.

SLE SYSTEMIC LUPUS ERYTHEMATOSUS.

sliding filament model Model mechanism for contraction of skeletal muscle. *See*: MUSCLE.

slimes Most fungal slimes are elaborated GLYCANS of fungal walls, particularly specialized mannans, which form extended capsules. Plant slimes are a heterogenous group of glycans, many of which are either PECTINIC ACIDS, ARABINOGALACTANS, AMY-LOIDS, or GALACTOGLUCOMANNANS. Some have specific functions such as lubricating root caps. All are water soluble and give solutions of a slimy or slippery texture (*see also* GUMS). Most animal slimes are highly hydrated GLYCOPROTEINS.

slime moulds Heterogeneous group of eukaryotic microorganisms comprising the ACELLULAR SLIME MOULDS such as *Physarum polycephalum* and the CELLULAR SLIME MOULDS such as *Dictyostelium discoideum*. *See also*: DICTYOSTELIUM DEVELOPMENT.

slot blot A method of detecting and quantitating a particular DNA or RNA sequence by HYBRIDIZATION to a homologous PROBE in a similar manner to a DOT BLOT, excepting that the apparatus used concentrates samples onto a nitrocellulose or nylon membrane in an array of thin rectangular slots.

slow virus Viruses that produce disease only many years after infection (*see e.g.* the lentiviruses in RETROVIRUSES).

small-angle scattering Low-angle scattering of neutrons or X-rays is a technique which can indicate the overall size and shape of a particle in solution. As the particles are not orientated, the scattering data represent the spherically averaged transform of an individual particle. Comparing the data with those expected from trial models determines the particle dimensions and possibly also the distribution of individual components within it, if these have distinguishable scattering densities. Nucleic acids scatter more strongly than proteins because of the phosphate groups present in the former, and thus complexes between proteins and nucleic acids can be usefully studied in solution. Such studies have been applied, for example, to virus particles, where it was shown that in addition to the regularly ordered proteins forming the spherical capsid, part of the polypeptide chain extends far into the central cavity of the virus. *See also*: DIFFRACTION; NEUTRON SCATTERING AND DIFFRACTION.

small cytoplasmic ribonucleoprotein particles Any of a number of small RIBONUCLEOPROTEIN particles found in the cytoplasm, such as the SIGNAL RECOGNITION PARTICLE.

small Hsp Small heat shock proteins. Any of a variety of small proteins (M_r of less than 60 000) that are induced by HEAT SHOCK in eukaryotic cells.

small nuclear ribonucleoprotein particles (snRNPs, snurps) Complexes of SMALL NUCLEAR RNAs (snRNAs) with proteins. Several snRNPs (the Sm class) have a vital role in RNA SPLICING, with different snRNPs recognizing the left and right splice junctions. For example snRNP U1 contains an RNA (U1) of 165 nucleotides and at least 11 proteins (U1-specific proteins A and C, and the common U snRNP proteins B′, B, D1, D2, D3, E, F, and G). One of its functions is recognition of the 5′ splice site. In yeast, all the snRNPs are found in the SPLICEOSOME and are responsible for its assembly. snRNPs are often found to be AUTOANTIGENS associated with the autoimmune disease SYSTEMIC LUPUS ERYTHEMATOSUS. Antibodies from patients suffering from this disease selectively precipitate snRNPs from cells of humans and other animals (but not from yeast or plants) suggesting a high degree of evolutionary conservation in their structure.

small nuclear RNA (snRNA) A group of intranuclear RNA molecules ranging from 90 to 220 nucleotides long. ELECTROPHORESIS defines six discrete snRNA species, designated U1 to U6. They are abundant and have long half-lives, but unlike other stable RNAs (e.g. rRNA and tRNA), snRNAs are not directly involved in cytoplasmic protein synthesis. All snRNAs described to date are complexed with proteins to form SMALL NUCLEAR RIBONUCLEO-PROTEIN PARTICLES (snRNPs). Some snRNAs seem to have a role in RNA SPLICING and RNA PROCESSING. snRNA U3 has been implicated in the maturation of the 28S rRNA in the nucleolus, whereas U1 is involved in RNA splicing in the nucleoplasm, its role being the recognition and bringing together of adjacent splice junctions (*see also* SPLICEOSOME).

small ribosomal RNA 5S RIBOSOMAL RNA.

small T antigen Protein produced by PAPOVAVIRUSES which may have a role in tumour formation.

Smith degradation A chemical procedure for the structural analysis of glycans in which 1,2 diols are oxidized to yield aldehydes, by means of periodic acid. Free formaldehyde is released where 1,2,3 triols are present. Reduction of insoluble aldehyde with borohydride produces substrates for further cycles of degradation. Differential oxidation can be achieved by the careful control of temperature and periodate concentration, facilitating discrimination between flexible and inflexible diols and conformation isomers of the latter.

The periodate-Schiff staining procedure of histochemistry is related to the Smith degradation and a form of the degradation itself can be used to distinguish between sialic acids variously acylated at C7, 8 and 9, in histological sections.

smooth endoplasmic reticulum (SER) Regions of ENDOPLASMIC RETICULUM which do not carry RIBOSOMES on their cytoplasmic surface.

smooth muscle MUSCLE tissue found in the walls of arterial vessels, the alimentary canal, and within internal organs. It differs from skeletal and cardiac muscle in internal structure, being composed of uninucleate spindle-shaped cells.

snake venom phosphodiesterase EXONUCLEASE (EC 3.1.15.1) which has been used in the sequence determination of RNA. It digests single-stranded RNA or DNA from the 3′ end releasing nucleoside 5′-monophosphates.

SNAPs (soluble NSF attachment proteins) Peripheral membrane proteins which bind NSF to the SNAP RECEPTORS. There are three forms: α (M_r 35 000), β (M_r 36 000), and γ (M_r 39 000), one of which, α-SNAP, is homologous to Sec17p, one of the proteins needed for transport from the ENDOPLASMIC RETICULUM to the GOLGI APPARATUS in the yeast SACCHAROMYCES CEREVISIAE. *See also*: PROTEIN SECRETION.

Clary, D.O. et al. (1990) *Cell* **61**, 709–721.

SNAP receptors (SNAREs) Receptors that bind in a MgATP-regulated fashion to NSF/SNAPs (*see* PROTEIN SECRETION; PROTEIN TARGETING) and are found on both transport vesicles (v-SNAREs) and target membranes (t-SNAREs). The complex of SNAREs, NSF, and SNAPs binds the vesicle to the membrane with which it is to fuse. Specificity is provided by different v- and t-SNARE pairs at each of the vesicle-mediated steps in the cell including those involved in regulated secretion. At the synapse, the v-SNARE on synaptic vesicles is synaptobrevin, whereas the t-SNARE on the presynaptic membrane is syntaxin.

Söllner, T. et al. (1993) *Nature* **362**, 318–324.
Warren, G. (1993) *Nature* **362**, 297–298.

SNAREs *See*: SNAP RECEPTORS.

snRNA SMALL NUCLEAR RNA.

snRNPs, snurps SMALL NUCLEAR RIBONUCLEOPROTEIN PARTICLES.

sobemovirus group Sigla from the type member southern bean mosaic virus. A group of single component plant viruses with isometric particles 30 nm diameter. The capsid structure is icosahedral ($t = 3$), the coat protein subunits having a M_r of 30 000. The genome is (+)-sense RNA which encodes four proteins (Fig. S14). The 5′ gene is thought to frameshift to the next OPEN READING FRAME to give two products which have sequence homology to RNA REPLICASES. The coat protein is translated from

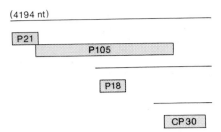

Fig. S14 Genome organization of southern bean mosaic virus (SBMV). The lines represent the RNA species and the boxes the proteins (P) with M_r given as $\times 10^{-3}$; nt, nucleotides.

a subgenomic mRNA from the 3′ end. The other protein of unknown function is also translated from a subgenomic mRNA. *See also*: PLANT VIRUSES.

Hull, R. (1988) In *The Plant Viruses*, Vol. 3 (Koenig, R., Ed.) 113–146 (Plenum, New York).

SOD SUPEROXIDE DISMUTASE.

sodium For related entries *see also* NA⁺.

sodium–calcium exchanger An antiport system in many eukaryotic cells which actively exports Ca⁺ ions out of the cell using energy derived from the Na⁺ gradient across the membrane. One Ca^{2+} ion is exported for every three Na⁺ ions that enter. *See also*: MEMBRANE TRANSPORT SYSTEMS.

sodium–calcium exchanger, sodium–calcium, potassium exchanger MEMBRANE TRANSPORT SYSTEMS involved in CALCIUM transport in various cells. *See*: NA⁺–CA²⁺ EXCHANGER; NA⁺–CA²⁺, K⁺ EXCHANGER.

sodium channels ION CHANNELS that allow the passage of sodium (Na⁺) ions. *See*: VOLTAGE-GATED ION CHANNELS.

sodium dodecyl sulphate (SDS) Amphipathic detergent (CH_3 $(CH_2)_{10}CH_2SO_4Na$) which is used to disrupt noncovalent interactions in proteins and membranes (i.e. denaturation). It is used particularly in the extraction (solubilization) of membrane proteins. ELECTROPHORESIS in the presence of SDS (SDS-PAGE) is widely used to separate and size proteins.

sodium–hydrogen exchanger An antiport system in mammalian and other eukaryotic cells, which is driven by the gradient of Na⁺ across the plasma membrane and which is involved in movement of H⁺ out of the cell and cytoplasmic pH regulation. *See also*: MEMBRANE TRANSPORT SYSTEMS.

sodium pump Commonly refers to the ubiquitous NA⁺, K⁺-ATPASE ion pump of eukaryotic cell plasma membranes, which maintains a low level of intracellular sodium by transporting Na⁺ out of the cytoplasm against its concentration gradient. *See*: MEMBRANE TRANSPORT SYSTEMS.

Solanum tuberosum The potato, a widely used host for PLANT GENETIC ENGINEERING experiments.

solenoid A higher-order structure proposed for CHROMATIN, formed by further coiling of a strand of NUCLEOSOMES.

soluble NSF attachment proteins *See*: SNAPS.

solution hybridization *See*: COT ANALYSIS; HYBRIDIZATION.

solvent flattening An approach towards improving the phases used in X-RAY CRYSTALLOGRAPHY for the calculation of initial ELECTRON DENSITY MAPS, to facilitate the interpretation of an unknown structure. A trial map often contains contiguous volumes of higher density (from which the molecular envelope can be determined) separated by regions of lower density appropriate to the solvent. The electron density of solvent regions is expected to be constant such that where these regions show variation in density this can be attributed to errors in either the measured STRUCTURE AMPLITUDES or the present set of trial phases. Construction of a modified electron density map containing the original values inside the molecular envelope and an appropriate constant value elsewhere (for the flattened solvent) enables a new set of phases to be calculated by means of a FOURIER TRANSFORM. Combination of these new phases and the previous trial phases, with suitable weighting of the two phase sets, can be expected to yield an electron density map which is more readily interpretable. The procedure is often used iteratively. Its success depends on many factors including the validity of the molecular envelope and the relative volumes occupied by protein and solvent.

Wang, B.C. (1985) In *Methods in Enzymology*, Vol. 115 (Diffraction Methods for Biological Macromolecules) (Wyckoff, H. et al., Eds) 90–112 (Academic Press, New York).
Leslie, A.G.W. (1987) *Acta Cryst.* **A43**, 134–136.

solvent scattering-length density Sum of the scattering lengths of all the atoms in unit volume of the solvent. *See also*: NEUTRON SCATTERING AND DIFFRACTION.

solvent structure The environment of biological molecules has an important influence on their structure and stability. Hydrophobic proteins found in membranes are surrounded by lipids; hydrophilic molecules are stabilized by hydrogen bonds to the solvent. Detailed structural information from X-RAY CRYSTALLOGRAPHY has revealed a shell of tightly bound water molecules occupying specific positions near the surface of protein molecules. Typically, the next shell of ordered water molecules is less tightly bound, with sites apparently occupied only partially. Interpretation of an ELECTRON DENSITY MAP during a crystal structure determination involves identifying atoms of both protein and solvent. An accurate model for the solvent, and its contribution to the DIFFRACTION, enables a more accurate diffraction pattern to be calculated. Thence, by minimization of the differences between observed and calculated diffraction, the model for the molecular structure can be successively refined to reveal more significant details.

Apart from water, specific ions from the solvent are bound to many protein structures. For nucleic acids, bound polycations are often necessary for their crystallization. Specifically bound water molecules aid the stability of mismatches in duplex DNA, as seen in NUCLEIC ACID STRUCTURE, and ordered water molecules within A/T-rich B-DNA form a spine of hydration in the narrow groove. *See also*: PROTEIN STRUCTURE.

Schulz, G.E. & Schirmer, R.H. (1979) *Principles of Protein Structure* (Springer-Verlag, Heidelberg).
Saenger, W. (1984) *Principles of Nucleic Acid Structure* (Springer-Verlag, Heidelberg).

somaclonal variation The genetic variation that arises in a population of cultured plant cells.

somatic cell Body cell, any cell other than the gametes or their germ cell precursors.

somatic cell genetics The study and manipulation of the genome of somatic cells as opposed to that of germ-line cells. *See also*: SOMATIC CELL HYBRIDS.

Somatic cell hybrids

FUSION of two or more cells or cell fragments produces cell hybrids which have been used widely in biological research ever since the first methods using inactivated Sendai virus preparations to produce controlled fusion were published [1]. Fusion takes place at the plasma membrane which is temporarily breached before re-forming as a single fluid lipid bilayer around the new cell hybrid. Membrane fusion occurs spontaneously as part of many physiological and pathological processes (e.g. fertilization of egg by sperm and muscle fibre formation in normal development, and multinucleate giant cell formation in certain viral infections). In the laboratory, however, the frequency of fusion is enhanced by the use of agents which perturb membranes by physical or (bio)chemical means. The most frequently used fusogen currently is POLYETHYLENE GLYCOL (PEG) which allows the surface membranes of neighbouring cells in dense aqueous suspension to become sufficiently closely juxtaposed for fusion to take place. Frequency of cell fusion can be controlled by adjusting cell density and the ratio of different parent cells. To ensure that hybrid cells are efficiently recovered it is essential to have positive selection protocols for the desired fusion product and, if necessary, against the parental components which might outgrow the early, often slow-growing and unstable, hybrid cell. The HAT selection system is one of the best known, oldest, and still most effective [2]. *De novo* nucleotide synthesis is inhibited using AMINOPTERIN (A); therefore the cell is forced to make nucleotides by scavenger synthesis using the enzymes hypoxanthine phosphoribosyltransferase (HPRT), thymidine kinase (TK) and adenine phosphoribosyltransferase (APRT), for which hypoxanthine (H) and thymidine (T) are supplied from the medium. Given one parental cell which is HPRT$^-$ and one which is HPRT$^+$ and using a second selective marker in a reciprocal fashion, only hybrid cells will be selected (Fig. S15). Other

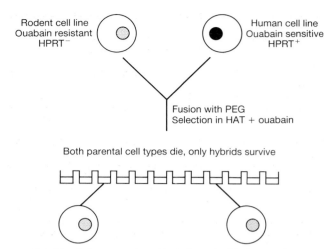

Rodent cell line
Ouabain resistant
HPRT$^-$

Human cell line
Ouabain sensitive
HPRT$^+$

Fusion with PEG
Selection in HAT + ouabain

Both parental cell types die, only hybrids survive

Hybrid cell clones, produced by independent fusion events

Fig. S15 Production of cell hybrids.

selection systems can be imposed, now that cellular TRANSFECTION with active genes is possible, by the insertion of drug resistance marker genes into one of the parental lines. Following the stage of actual fusion the two nuclei remain separate in a heterokaryon until synchronous mitosis allows both sets of chromosomes to aggregate together on a single metaphase plate. The products of this cell division are the founders of the hybrid cell clone which may well subsequently evolve and change as inevitable selection for improved growth capacity is imposed.

Details of fusion techniques and related cell culture and selection details are described in [3] and [4].

Different types of cell hybrids and their uses

In the laboratory most fusions are carried out between somatic as opposed to germ cells. The fusion of two whole cells gives rise to somatic cell hybrids of varying genetic stability. The degree of stability depends on the species of cells undergoing fusion, their tissue of origin, and the state of their respective chromosome complements. Intraspecific hybrids, often between an established cell line and a transiently cultured or cell type isolated *de novo*, are generally more stable than interspecies hybrids. The instability of the latter type of hybrid cell has been utilized in human gene mapping which makes use of the parasexual segregation of human chromosomes from rodent–human hybrid cells to identify chromosomes carrying particular genes (*see* HUMAN GENE MAPPING for this use of somatic cell hybrids). A variety of established mouse cell lines are the rodent cells most frequently used to make human–rodent cell hybrids, but Chinese hamster and rat cell lines are also employed. Why the chromosomes of one species are lost in preference to another's is not fully understood, but it is probably related to the relative rates of chromosome replication and cell division. As a rule of thumb the chromosomes lost are from the species with the higher $2n$ chromosome number. Thus hybrids of hamster and mouse ($2n = 20$ and $2n = 40$ respectively) lose mouse chromosomes preferentially and are produced for mouse gene mapping. Chromosome behaviour following fusion

can also be influenced by the CELL CYCLE phase of partner cells at the time of fusion. Fusions between cells at different stages of the cell cycle have been used to study control factors and interactions.

Fusion of limited lifespan fibroblasts from patients with a variety of similar biochemical abnormalities has been used to define COMPLEMENTATION GROUPS and hence independent mutant genes implicated in, for example, LYSOSOMAL STORAGE DISEASES [5] and in XERODERMA PIGMENTOSUM involving different DNA REPAIR functions [6].

Functional heterogeneity in other important circumstances has been uncovered by cell fusion. It was something of a surprise to many that when two different malignant cell types are fused in intraspecific crosses so that the products are relatively stable chromosomally, the resulting hybrids are mostly nontumorigenic in isogenic hosts. However, if the hybrids are allowed to grow for some time, some subclones with specific chromosome loss revert to malignancy. This work led to the suggestion that TUMOUR SUPPRESSOR GENES exist and even permitted the identification of chromosomes that carry them [7].

Whole-cell fusions between cells of similar ontogenic origin are generally more efficient than those between cells of widely different derivation. This may be explained in part by the phenomenon, much studied in the early days of cell hybrids, of extinction and re-expression of tissue-specific genes [8,9]. In general, however, the control networks are highly complex and it has not proved possible to identify the key control elements in most cases in whole-cell fusions.

In contrast, fusion of cells belonging to the same CELL LINEAGE has been widely exploited to produce stable HYBRIDOMAS which continue to express tissue-specific products. Established rodent myeloma cell lines fused with freshly isolated lymphocytes from lymphoid organs of immunized animals produce a high frequency of B CELL hybrids which continue to secrete the immunoglobulin product of a single B lymphocyte (*see* MONOCLONAL ANTIBODIES) whose specificity can be defined by screening [10]. This technology has been of enormous utility in research, in clinical analysis and therapy as well as commercially, particularly in the area of diagnostics. Similarly, T CELL hybrids have been used to study function using fusion of established T cell lines as a means of clonal isolation and expansion of single T cells under conditions which permit them to maintain much of their functional capacity [11].

Cell fusion can be used for immortalizing critical cellular functions and components which would otherwise be lost because we do not yet know how to grow a number of different cell types in permanent culture [12]. It can also be used to isolate and subsequently define by gene mapping key rearranged chromosomes with a critical role in malignancy and disease [13].

The techniques of cell fusion can be used to combine fragments of cells in order to dissect the interaction of different components. Thus cells enucleated with the aid of CYTOCHALASIN B, termed cytoplasts, can be fused with whole cells or with the isolated nuclei of different donor cells to study nucleo-cytoplasmic interactions, for example between mitochondrial and nuclear genes. Transient products called cybrids can be produced by fusing different cytoplasts [3].

Microcell fusion

Nuclei isolated by cytochalasin B treatment are enveloped in a fragment of plasma membrane and can participate in fusion just like whole cells. For many studies it is useful to fuse only a fraction of the nuclear contents — for example a single chromosome. This can be attempted by using the microcell fusion technique in which micronucleus formation is elicited by prolonged exposure of the donor cells to COLCEMID and enucleation is carried out with the aid of cytochalasin B. Partially purified micronuclei are fused with the required recipient cells [3,4]. More recently this technique has been most widely used in the search for tumour suppressor genes on specific chromosomes and chromosome fragments which can be made selectable by transfection of drug resistance markers, for example.

Single chromosomes and selected chromosome fragments can also be introduced into cells by chromosome-mediated gene transfer [4,14], which utilizes the uptake of isolated MITOTIC CHROMOSOMES into recipient cells. Calcium chloride precipitation or more recently LIPOFECTION is used for such CHROMATIN transfer by techniques which are very similar to DNA transfer technology [15].

Radiation hybrids

Cell hybrids containing only small fragments of a single, specific, usually human, chromosome, mainly for gene mapping purposes (*see* HUMAN GENE MAPPING), are produced by a recently resurrected technique. A hamster cell hybrid bearing a single human chromosome is subjected to a calibrated, but lethal dose of irradiation, leading to random chromosome fragmentation. The irradiated cell is rescued by fusion with a normal rodent cell. Small fragments of the human chromosome are retained in the new hybrid clones. Gene order and distance can be deduced by statistical analysis.

As the genome has become more and more easily manipulated it has become possible to engineer the uptake of small DNA fragments and specific genes into cells. Many new genome transfer techniques have been devised, whose origins can be traced back to cell fusion methods. The protocols used for the production of transgenic animals, especially those involving transformed embryonic stem cells (*see* TRANSGENIC TECHNOLOGIES), the current attempts to engineer the uptake of large genomic constructs in the form of YEAST ARTIFICIAL CHROMOSOME clones, and some of the methods contemplated for GENE THERAPY all derive to some extent from the evolving technology of somatic cell hybrids.

V. VAN HEYNINGEN

1 Harris H. & Watkins, J.F. (1965) Hybrid cells derived from mouse and man: artificial heterokaryons of mammalian cells from different species. *Nature* **205**, 640–646.
2 Szybalski, W. et al. (1962) Genetic studies with human cell lines. *Natl. Cancer Inst. Mongr.* **7**, 75–79.
3 Shay, J.W. (Ed.) (1982) *Techniques in Somatic Cell Genetics* (Plenum, New York).
4 Goodfellow, P.N. et al. (1988) Techniques for mammalian genome transfer. In *Genome Analysis — A Practical Approach* (Davies, K.E., Ed.) 1–18 (IRL Press, Oxford).
5 Galjaard, H. et al. (1975) Genetic heterogeneity in GM1 gangliosidosis. *Nature* **257**, 60–62.
6 De Weerd-Kastelein, E.A. et al. (1972) Genetic heterogeneity of xeroderma pigmentosum demonstrated by somatic cell hybridization. *Nature New Biol.* **238**, 80–83.
7 Harris, H. (1988) The analysis of malignancy by cell fusion: the position in 1988. *Cancer Res.* **48**, 3302–3306.
8 Peterson, J.A. & Weiss, M.C. (1972) Expression of differentiated functions in hepatoma cell hybrids: induction of mouse albumin production in rat hepatoma–mouse fibroblast hybrids. *Proc. Natl. Acad. Sci. USA* **60**, 1282–1287.
9 Gourdeau, H. & Fournier, R.E.K. (1990) Genetic analysis of mammalian cell differentiation. *Annu. Rev. Cell Biol.* **6**, 69–94.
10 Kohler, G & Milstein, C. (1975) Continuous cultures of fused cells secreting antibody of predefined specificity. *Nature* **256**, 495–497.
11 Kappler, J. et al. (1982) Antigen presentation by Ia$^+$ B cell hybridomas to H-2-restricted T cell hybridomas. *Proc. Natl. Acad. Sci. USA* **79**, 3604–3607.
12 Geurts van Kessel, A.H.M. et al. (1981) Characterisation of a complex Philadelphia translocation (1p-;9q + ;22q-) by gene mapping. *Hum. Genet.* **58**, 162–165.
13 van Heyningen, V. et al. (1985) Molecular analysis of chromosome 11 deletions in aniridia-Wilms tumor syndrome. *Proc. Natl. Acad. Sci. USA* **82**, 8592–8596.
14 Klobutcher, L.A. & Ruddle, F.H. (1981) Chromosome mediated gene transfer. *Annu. Rev. Biochem.* **50**, 533–554.
15 Murray, E.J. (Ed.) (1991) Gene transfer and expression protocols. In *Methods in Molecular Biology 7* (Humana Press, Clifton NJ).

somatic hypermutation The high rate of POINT MUTATION occurring in the V regions of IMMUNOGLOBULIN GENES in B cells after antigenic stimulation. It is 10^3 to 10^4 times higher than the rate of spontaneous mutation in other mammalian genes. Somatic hypermutation is responsible for the AFFINITY MATURATION of humoral immune responses, in which ANTIBODIES of higher affinity emerge as the response proceeds and dominate the late stage of primary responses and the secondary and tertiary responses. The mutations are found to cluster in the V region exons and flanking sequences of both heavy and light chains, in the HYPERVARIABLE REGIONS contributing to the antigen-binding site. This is thought to be a result of the selection of high affinity antibody-producing cells by antigen rather than restriction of mutation events to those regions. Somatic hypermutation takes place in activated B cells in germinal centres of secondary lymphoid tissues. Its mechanism and cause are not known, and it does not seem to occur in T cells. *See also:* B CELL DEVELOPMENT; IMMUNOGLOBULIN STRUCTURE.

somatic mesoderm In vertebrates, the LATERAL PLATE MESODERM splits into two layers separated by a cavity known as the intra-embryonic coelom (the future pleural, pericardial, and peritoneal cavities). The somatic mesoderm comprises all the lateral plate mesoderm other than that surrounding the gut tube (SPLANCHNIC MESODERM).

somatic mutation Any mutation that occurs in a somatic cell, as opposed to a germ-line cell. Somatic mutations are passed on to the progeny of the affected cell (e.g. in the development of a tumour) which form a clone of mutant cells within the body. Somatic mutations in animals are not inherited by the offspring of the parent in which they occur, but in plants, where there is no defined germ line, they theoretically can be if they occur in a cell that subsequently gives rise to a germ cell. In the IMMUNOGLO-BULIN GENES in B CELLS the variability of the V regions is enhanced by somatic mutations (*see* SOMATIC HYPERMUTATION).

somatic recombination RECOMBINATION that occurs in SOMATIC CELLS. *See*: GENE REARRANGEMENT; MITOTIC RECOMBINATION.

somatomedins Polypeptide growth hormones produced by the liver under the influence of GROWTH HORMONE. Somatomedin-1 is now known as insulin-like growth factor I (*see* GROWTH FACTORS).

somatomotor nervous system *See*: PERIPHERAL NERVOUS SYSTEM.

somatopleure The dorsal sheet of LATERAL PLATE MESODERM that forms the outer wall of the COELOM. The mesoderm of the limb bud in higher vertebrates (apart from the skeletal muscles and dermis) is derived from this layer.

somatosensory system *See*: PERIPHERAL NERVOUS SYSTEM.

somatostatin A cyclic polypeptide composed of 14 amino acids; also called growth hormone release inhibiting hormone. Widely distributed in the brain, notably the hypothalamus; also secreted in the pancreatic D cells and gut. Inhibits the release of a wide variety of hormones from the pituitary gland and the gastrointestinal tract.

somatotropin GROWTH HORMONE.

somite An epithelial rosette of cells derived from the PARAXIAL MESODERM, forming a segmental series along the long axis of vertebrate embryos. The cells of a somite give rise to the sclerotome (forming vertebrae), the myotome (forming muscle), and the dermotome (forming dermis). Shortly after the somite is formed, its rostral half is invaded by cells of the NEURAL CREST, which contribute to the dorsal root and autonomic ganglia, the segmental ganglia of the PERIPHERAL NERVOUS SYSTEM. Following somite differentiation, the cells of the sclerotome contribute to the vertebrae, and it is for this reason that the somite was originally referred to as the 'proto-vertebra'. One major area of contention surrounding the somite arises from the fact that the segmental register of vertebrae is out of phase with that of the somites by half a segment. To explain this, 19th century embryologists, notably Remak, proposed that the somite undergoes a resegmentation. Under this model, the rostral half of the somite becomes the caudal half of the vertebra. Recent studies of the phenomenon have so far failed to give a clear answer on this issue, although the evidence tends to favour it.

Somites develop as epithelial rosettes of cells lying either side of the NEURAL TUBE in the vertebrate embryo. These bud off from the unsegmented paraxial mesoderm during early development. Somites form progressively in a rostro-caudal direction. The mechanism of somite formation is not well understood, but studies on the role of CELL ADHESION MOLECULES such as N-cadherin and NCAM show that these are expressed in the paraxial mesoderm before segmentation, and that N-cadherin is

crucial in maintaining close contact between cells of the newly formed somite epithelium. While it is likely that adhesion molecules are important in somite formation, experiments involving heat-shock treatment of embryos, and others using local applications of retinoic acid (*see* RETINOIDS), have shown that further factors, such as the timing of the CELL CYCLE, are also involved.

son-of-sevenless (sos) *Drosophila* gene involved in EYE DEVELOPMENT. It is part of a signal transduction pathway and encodes a guanine nucleotide releasing protein that interacts with a *Drosophila* Ras protein (*see also* GTP-BINDING PROTEINS).

sorcin *See*: CALCIUM-BINDING PROTEINS.

Sorghum bicolor An important cereal crop of the Americas, Africa, India and China. It is also known as great millet, Guinea corn, kaffir corn or dari. Cytoplasmic male sterility in *Sorghum* has been much studied. *See*: MITOCHONDRIAL GENOMES: PLANT.

sos *Drosophila* gene SON-OF-SEVENLESS, involved in EYE DEVELOPMENT.

SOS pathway/repair/response DNA REPAIR mechanisms in bacteria which are induced as a part of the cellular response to DNA damage, for example by UV irradiation, or following inhibition of DNA REPLICATION. The SOS response ensures a greater cellular capability to repair damaged DNA. The damaging agent induces the production of the RecA protein (*see* RECOMBINATION) which triggers proteolytic activity and cleavage of the REPRESSOR protein LexA. This in turn induces all the OPERONS to which the LexA repressor was bound, including genes encoding the UvrA and UvrB proteins, which are involved in DNA excision repair.

Walker, G.C. (1984) *Microbiol. Rev.* **48**, 60–93.

Southern analysis, Southern blotting The technique developed by Edward Southern in 1975 in which denatured DNA is transferred from the agarose gels in which fragments have been separated by ELECTROPHORESIS to a nitrocellulose or nylon membrane laid over the gel, before HYBRIDIZATION with a complementary nucleic acid PROBE (*see* Fig. H19 in HYBRIDIZATION). This step is required as hybridization to the gel itself is very inefficient. The protocol originally involved the use of paper towels to draw buffer through the agarose gel by capillary action (capillary blotting). This method has now been largely superseded by either ELECTROBLOTTING or VACUUM BLOTTING procedures which are more rapid. The technique is ubiquitous in molecular genetics and its numerous applications usually revolve around the identification of a particular DNA sequence within a mixture of RESTRICTION FRAGMENTS, for example to determine the presence, position, and number of copies of a gene in the genome. It is also an integral technique in DNA TYPING. *See also*: NORTHERN BLOTTING; WESTERN BLOTTING.

Southern, E.M. (1975) *J. Mol. Biol.* **98**, 503–517.

southern corn leaf blight Disease of maize (*Zea mays*) caused by the fungus *Bipolaris* (*Helminthosporium maydis*) which has caused massive crop failure in the past when monocultures of a particular self-sterile maize variety proved exceptionally susceptible to the toxin produced by a strain of the fungus. *See*: MITOCHONDRIAL GENOMES: PLANT.

Sp1 A general eukaryotic TRANSCRIPTION FACTOR involved in the initiation of transcription by RNA polymerase II. It contains three zinc fingers which are involved in its binding to DNA. *See*: PROTEIN–NUCLEIC ACID INTERACTIONS.

Kadonaga, J.T. et al. (1987) *Cell* **51**, 1079–1090.

SP6 polymerase An RNA POLYMERASE derived from the SP6 BACTERIOPHAGE of *Salmonella typhimurium*, and which is highly specific for the SP6 viral PROMOTER. By placing a structural gene downstream of the SP6 promoter in a RECOMBINANT DNA, SP6 RNA polymerase can be used to produce large quantities of RNA transcript from this DNA. This system is widely used for IN VITRO TRANSCRIPTION.

space group One of the set of 230 possible three-dimensional groups of SYMMETRY elements consistent with a space-filling lattice; only 65 of these (from 11 of the 32 possible crystallographic POINT GROUPS) are compatible with enantiomorphic molecules such as proteins and nucleic acids. Recognition of the correct symmetry and its utilization potentially enables great simplification of crystal structure determination. For a known point group, the choice of space group is determined by the presence or otherwise in the DIFFRACTION pattern of SYSTEMATIC ABSENCES which arise as a direct consequence of translational symmetry within the crystal UNIT CELL. This may be due to additional lattice points in the crystal lattice or the translational symmetry elements associated with each lattice point; only screw axes of rotation are relevant for enantiomorphic molecules.

International Tables for X-ray Crystallography (1992) Vol. A Space-Group Symmetry (Hahn, T., Ed.) (Kluwer Academic Publishers, Dordrecht).
Stout, G.H. & Jensen, L.H. (1989) *X-ray Determination: A Practical Guide* (John Wiley, New York).

space lattice *See*: SYMMETRY.

spacer DNA DNA separating one gene from the next in a co-transcribed gene cluster (transcribed spacer), or one gene cluster from the next (untranscribed spacer), as, for example, in the rRNA genes of eukaryotes. The 18S, 28S, and 5S rRNAs of higher eukaryotes are encoded by repeated TRANSCRIPTION UNITS each containing an 18S, a 28S, and a 5S rRNA coding sequence, separated by spacer DNA. Transcription produces a single RNA transcript encoding the 18S, 28S, and 5S rRNAs and their spacers. Subsequent RNA PROCESSING releases the separate rRNAs. Each transcription unit is also separated from the next by a length of DNA called a spacer DNA, which is not transcribed.

spacer RNA The transcription product of SPACER DNA. In higher eukaryotes, the three different RIBOSOMAL RNA species are produced as a single transcript. The rRNA coding sequences are separated by regions of spacer RNA which contain the EXONUCLEASE cleavage sites which are cut to generate the mature rRNAs.

SPARC Osteonectin. *See*: CALCIUM-BINDING PROTEINS.

SPEC1 *See*: CALCIUM-BINDING PROTEINS.

special pair All photosynthetic REACTION CENTRES seem to contain a special pair of CHLOROPHYLL molecules (chlorophyll *a* in oxygenic organisms, bacteriochlorophyll *a* in photosynthetic bacteria) over which an exciton can be delocalized and which acts as the primary electron donor. These pigments are named according to the wavelength maximum of the long wave absorption band in the resting, reduced state: P700 (PHOTOSYSTEM I), P680 (PHOTOSYSTEM II), P870 (*Rhodobacter sphaeroides*), P930 (*Rb. viridis*), P840 (green sulphur bacteria), and differ in the redox potentials that they span. *See also*: PHOTOSYNTHESIS.

specialized transduction *See*: TRANSDUCTION.

species phylogeny *See*: HOMOLOGY.

specific activity The specific radioactivity of a labelled species measures the relative abundance of radioactive molecules in a labelled sample as units of radioactivity per unit mass.

specification The state of COMMITMENT of cells to generate a particular PHENOTYPE that is present when the cells are isolated from the embryo. By this definition, the specification of cells from a region is not necessarily the same as the normal developmental FATE of those cells, although specification is often loosely defined in this latter way. Thus, the presumptive neural plate of an amphibian blastula will differentiate into epidermis when cultured *in vitro*, and these cells are therefore specified as epidermal at this developmental stage. In the embryo, neuralization of these cells, so that they develop into neurons, depends upon their receiving an inductive signal from the adjacent mesoderm.

specificity (1) Of enzyme for substrate, *see* ENZYMES; MECHANISMS OF ENZYME CATALYSIS; PROTEIN STRUCTURE.
(2) Of base pairing, *see* BASE PAIR; DNA; HYBRIDIZATION; NUCLEIC ACID; RNA.
(3) Of antigen receptors for antigen, *see* ANTIBODIES; IMMUNOGLOBULIN STRUCTURE; T CELL RECEPTORS.
(4) In respect of LONG-TERM POTENTIATION, the fact that synaptic potentiation is confined to the conditioned (tetanized) pathway and is not observed at synapses in a nonconditioned pathway terminating on the same group of neurons. This can be explained by the need for the presynaptic terminals to be stimulated and release the ligand for the postsynaptic receptors which induce long-term potentiation; no long-term potentiation can therefore be induced in an inactive pathway.

spectrometry, spectroscopy *See*: CIRCULAR DICHROISM; ESR/EPR SPECTROSCOPY; EXAFS; NMR SPECTROSCOPY.

spectrin ACTIN-BINDING PROTEIN which forms part of the red cell membrane skeleton. In other cells it is known as fodrin. *See also*: CALCIUM-BINDING PROTEINS; MEMBRANE STRUCTURE.

Spemann organizer *See*: AMPHIBIAN DEVELOPMENT; INDUCTION; ORGANIZER.

sperm–egg interaction The interaction between sperm and egg is the first step on the pathway to FERTILIZATION. In mammals this interaction begins when sperm attach themselves loosely and then bind tightly to receptors on the surface of the ZONA PELLU-CIDA, the egg's outer coat. In mice, the sperm receptor molecule has been identified as a GLYCOPROTEIN, ZP-3, consisting of some 400 amino acids, from which branched oligosaccharide side chains extend. The oligosaccharides are of several different sorts and not all ZP-3 molecules have the same sequence of sugar chains. A particular set of these sugar side chains — the O-linked oligosaccharides — are responsible for binding to the egg during fertilization.

sperm maturation The final phase of differentiation of sperm from haploid spermatids (*see* SPERMATOGENESIS). A mature sperm consists of a head, where the nucleus and acrosome are located, a long flagellum for a tail, and the midpiece, which consists of a large number of mitochondria located under the base of the head. During maturation from spermatids, which are small round haploid cells with a normal cytoplasm, the microtubule array which forms the sperm tail polymerizes and extends, the Golgi apparatus secretes enzymes into a vesicle which will evolve into the acrosome, the chromatin in the nucleus becomes highly condensed, and all excess cytoplasm (except for the mitochondria), including ribosomes, endoplasmic reticulum, and other organelles, is stripped away. *See also*: ACROSOME REACTION.

spermatocyte, spermatid *See*: SPERMATOGENESIS.

spermatogenesis The production of the mature male gamete. In mammals, spermatogenesis starts at puberty in the seminiferous tubules of the testes. The PRIMORDIAL GERM CELLS reach the gonad during embryonic development and divide to form type A1 spermatogonia. At puberty the A1 cells divide mitotically, adjacent to the tubule basal lamina, nourished by the adjacent Sertoli cells of the tubular epithelium. A1 division produces more type A1 cells as well as the type A2 spermatogonia; the latter divide via several intermediate stages to produce the primary spermatocytes that enter MEIOSIS. The first meiotic division yields a pair of secondary spermatocytes; these complete the second meiotic division to form the haploid spermatids, by now positioned close to the lumen of the seminiferous tubule, where they differentiate into mature sperm cells (*see* MATURATION OF SPERM).

Gene transcription during spermatogenesis occurs mainly during meiotic prophase, when specific RNA transcripts needed for the generation of viable sperm are expressed. In many species the gene for protamine, a protein that replaces the nuclear histones in sperm, is transcribed; translation into protein takes place later, at the spermatid stage, when other genes are also transcribed.

spherocytosis *See*: HEREDITARY SPHEROCYTOSIS.

spheroplast Fungal (e.g. yeast) or bacterial cell from which the cell wall has been removed. The removal of a fungal cell wall by gentle enzymatic degradation with lytic enzymes produces spherical membrane-bound structures enclosing the cell contents. Such spheroplasts can remain viable for extended periods if maintained in an osmotically stabilized buffer, containing, for example, 1.2 M sorbitol. Spheroplasts are formally distinguished from PROTOPLASTS in that they are devoid of any cell wall material. Spheroplasts suspended in osmotically buffered nutrient agar can regenerate their cell walls. Preparation of spheroplasts is usually necessary for efficient genetic TRANSFORMATION of fungi by exogenous DNA or for gentle cell disruption.

Eddy, A.A. & Williamson, D.H. (1957) *Nature* **179**, 1252–1253.

sphingolipids MEMBRANE LIPIDS that contain the long-chain amino alcohol sphingosine in place of glycerol (Fig. S16). They include the phospholipid sphingomyelin and many GLYCOLIPIDS, such as the ceramides and gangliosides. Nervous tissue is particularly rich in sphingolipids. Phospholipids containing phyto-sphingosine, a saturated C_{18} dihydroxysphingosine with the additional secondary alcohol group adjacent to that in sphingosine, have been found in plant tissues.

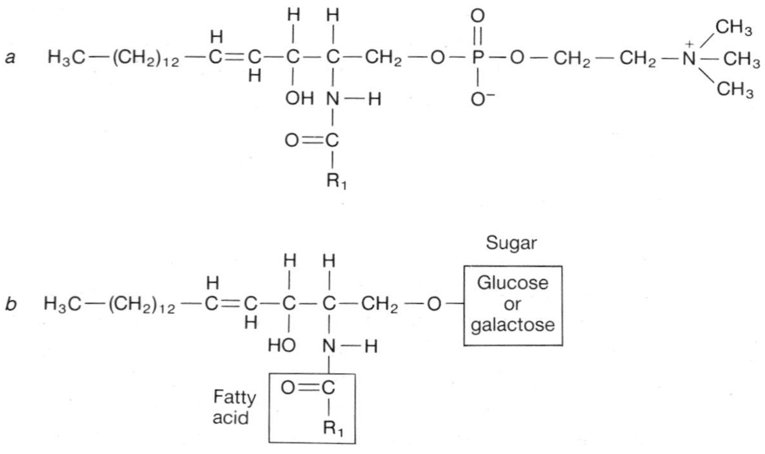

Fig. S16 Sphingolipids. *a*, The phospholipid sphingomyelin; *b*, ceramide, a precursor for glycosphingolipids.

sphingomyelin A PHOSPHOLIPID containing sphingosine instead of glycerol, a single fatty acid, and a phosphorylcholine headgroup. *See*: SPHINGOLIPIDS.

spin labelling The construction of a molecule containing a group (the spin label) such as an N–O (nitroxide) group, which contains an unpaired electron and which can be detected by ESR/EPR SPECTROSCOPY. The technique has been used to study the motion of membrane lipids.

spinach *Spinacia olearacea*, the leaves of which are the commonest source of chloroplasts for the experimental study of PHOTOSYNTHESIS. Not to be confused with other leafy vegetables which are sometimes loosely referred to as spinach (e.g. *Beta vulgaris* (spinach beet, perpetual spinach, Swiss chard) and *Tetragonia tetragonioides*, New Zealand spinach).

spinal cord Part of the vertebrate central nervous system through which sensory and motor signals are relayed between periphery and brain (Fig. S17). Certain simple reflexes are mediated at the level of the spinal cord only. The spinal cord has a segmental organization with sets of spinal nerves serving different areas of the body along the antero-posterior axis. Sensory information enters the spinal cord via the dorsal root ganglia; motor information leaves the spinal cord via the ventral roots. Sensory signals are carried within the spinal cord to brain nuclei by ascending pathways (the ventral and lateral spinothalamic tracts (pain, temperature), the dorsal columns (touch, pressure, vibration), the dorsolateral tracts, and the dorsal and ventral

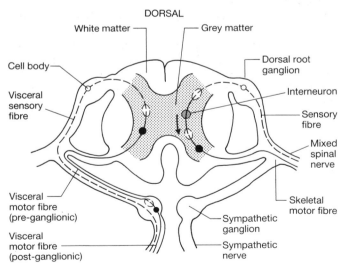

Fig. S17 A simplified schematic view of the connections between spinal cord and peripheral nervous system. Only one set of neurons is shown on each side for clarity and simple reflex circuits are illustrated. Sensory information from the viscera and body surface enters the spinal cord via the dorsal root ganglia. Apart from the simple reflex circuits, incoming nerve fibres synapse on ascending neuronal tracts which carry information to nuclei of the mid- and hindbrain, from which it is relayed to the cerebral cortex. Descending motor information leaves the spinal cord through the ventral roots.

spinocerebellar tracts. Motor signals are relayed from the brain to peripheral tissues in a separate set of pathways. Much interest is at present focused on protein TROPHIC FACTORS released by damaged neurons and GLIAL CELLS and their role in possible repair processes within the spinal cord.

spindle *See*: MITOTIC SPINDLE.

spindle fibre A cluster of MICROTUBULES and associated proteins in the MITOTIC SPINDLE that forms a bundle of sufficient size to be visible in the light microscope. There are spindle fibres that connect the spindle poles with the KINETOCHORES on each chromosome ('chromosomal' or 'kinetochore fibres'). Chromosomal fibres can both pull and push on the chromosomes to which they are attached, so they are a significant part of the mitotic machinery. Other spindle fibres are readily seen in the region between the chromosomes as they separate during anaphase (the 'interzone fibres'). Some of the interzone fibres appear to run from one spindle pole to the other, so they have been called 'interpolar fibres' or 'pole-to-pole fibres'. Electron microscopy shows that these are composed of microtubules that are generally shorter than the pole-to-pole distance. Such microtubules emanate from each centrosome and interdigitate near the spindle's equator. In the region where they overlap, they are bunched together by a dense matrix, and intermicrotubule bridges are often seen. The short microtubules may therefore be connected mechanically to make a fibre that is indeed an interpolar fibre, even though its components do not run the whole distance. *See also*: MITOSIS.

spindle pole body Structures found at the ends of the MITOTIC SPINDLES of many fungi. They are built into a large pore in the NUCLEAR ENVELOPE and serve the role of a CENTROSOME, organizing the cell's MICROTUBULES. *See also*: MITOSIS.

spines *See*: DENDRITES.

spinobulbar muscular atrophy *See*: KENNEDY'S DISEASE.

spinocerebellar ataxia type I A hereditary neurodegenerative disease characterized by degeneration of the cerebellum, spinal cord, and brainstem, which becomes manifest only in adulthood. The defect involved is an expansion of a trinucleotide repeat (GAG) in a gene located on chromosome 6. Affected patients possess between 41 and 81 repeats compared with the 25–36 present in unaffected individuals. Other genetic diseases in which trincleotide expansion is involved are HUNTINGTON'S CHOREA, KENNEDY'S DISEASE, and MYOTONIC DYSTROPHY.

Orr, H.T. (1993) *Nature Genet.* **4**, 221–226.

spiral cleavage A form of CLEAVAGE pattern characteristic of many invertebrate PROTOSTOMES (including annelids, molluscs, flatworms, echiuroids, and nematodes) It is associated with a MOSAIC (determined) form of development where the fate of individual cells (BLASTOMERES) is fixed early in development and may be dependent on asymmetric distribution of cytoplasmic components. The pattern of cleavage is invariant and the origin and fate of individual blastomeres can be identified and followed, generating a complete cell lineage for an individual organism. The first two cleavage planes are vertical to the animal–vegetal axis and perpendicular to one another. The third and subsequent cleavage planes occur at an angle (other than 90° or 180°) to the animal–vegetal axis. Division is often asymmetric. The direction of cleavage is referred to as dextral (or right-handed) if cleavage planes are orientated in a clockwise direction when viewed from the animal pole and sinistral (or left handed) if orientated anti-clockwise. Dextral and sinistral cleavages alternate and hence the cells of one tier come to lie over the cleavage furrows of the tier below. Cells of the vegetal tier (often but not invariable larger) are called macromeres whereas cells of the animal tier are known as micromeres.

spiroplasmas Small helical prokaryotic parasitic microorganisms that grow intracellularly and cause various diseases in plants. *See:* PLANT PATHOLOGY.

splanchnic mesoderm MESODERM surrounding the gut tube which gives rise to the smooth muscle of the gut wall. Otherwise called visceral mesoderm.

splanchnopleure The ventral sheet of LATERAL PLATE MESODERM that forms the inner wall of the COELOM.

splice site The sequences within an intron at which RNA SPLICING is initiated. Introns are always flanked by 5′ GT and 3′ AG dinucleotides, which together with the 25–30 nucleotides around them define the exon–intron splice sites.

Mount, S.M. (1982) *Nucleic Acids Res.* **10**, 459–472.

spliceosomes Large multisubunit nuclear RIBONUCLEOPROTEIN particles found in eukaryotes (including yeast) as 40–60S particles, and which associate with pre-mRNA and catalyse RNA SPLICING. Spliceosomes assemble on the intron to be spliced before the intitial cleavage of the 5′ splice site. The formation of the spliceosome relies on base pairing between RNA molecules and on RNA–protein and protein–protein interactions. Spliceosomes contain at least five distinct SMALL NUCLEAR RIBONUCLEOPROTEINS (snRNPs). In both yeast and mammals spliceosomes are composed of three ribonucleoprotein subunits, the U1 and U2 snRNPs, and the (U4/U6, U5) tri-snRNP, as well as other protein factors.

Green, M.R. (1991) *Annu. Rev. Cell. Biol.* **7**, 559–599.

splicing *See:* RNA SPLICING. Can also refer to DNA splicing which occurs as a result of somatic recombination, for example in the immunoglobulin loci, where DNA is cut at specific sites and rejoined to form a functional gene (*see* GENE REARRANGEMENT). *See also:* TRANS-SPLICING.

splicing mutation Any mutation affecting gene expression by affecting correct RNA SPLICING. Splicing mutations may be due to mutations at intron–exon boundaries which alter splice sites.

Spm (En) Suppressor mutator (enhancer), a transposable genetic element found in maize. *See:* TRANSPOSABLE GENETIC ELEMENTS: PLANTS.

spo genes Mutants of the bacterium *Bacillus subtilis* defective in sporulation have led to the identification of *spo* genes. These are divided into classes according to the stage in sporulation blocked by the mutation, for example, *spo0* mutants fail to exit from the vegetative phase into the sporulation pathway. The *spo0* mutations have identified a phosphorelay system of signal transduction that regulates the activity of a TRANSCRIPTION FACTOR Spo0A. Later acting *spo* genes regulate the differentiation of the mother cell–forespore complex. SpoIIID encodes a DNA-binding protein involved in compartment-specific transcriptional regulation and required for the gene rearrangement of the *spoIVCB/spoIIIC* genes to generate a mother cell-specific σ factor. *See also:* BACTERIAL DIFFERENTIATION.

sponge cell aggregation The species-specific reaggregation of dissociated cells from marine sponges has provided a model system for cell–cell recognition processes. Reaggregation of *Microciona* is dependent on Ca^{2+} and requires a large PROTEOGLYCAN-like aggregation factor (MAF) together with a receptor located in the plasma membrane.

spore Specialized propagative cell produced by bacteria, fungi, protozoa, and cryptogams (algae, mosses and liverworts, ferns). Many microorganisms produce spores in response to adverse environmental conditions. They are usually tough, thick-walled cells, with impermeable outer walls and a marked decrease in metabolic activity. Some bacteria, such as *Bacillus* species produce an intracellular endospore; in fungi and other multicellular organisms spores are produced in specialized structures as part of a complex life cycle (*see e.g.* NEUROSPORA CRASSA; SACCHAROMYCES CEREVISIAE). *See also:* BACTERIAL DIFFERENTIATION.

sporophyte The spore-bearing generation which alternates with the GAMETOPHYTE in the life cycles of land plants. It is usually DIPLOID giving rise to spores by MEIOSIS. Some algae also have sporophyte generations, but their spores are not necessarily derived meiotically.

spot desmosome Desmosome. *See:* CELL JUNCTIONS.

spumaviruses *See:* RETROVIRUSES.

SR (1) SARCOPLASMIC RETICULUM.
(2) SYNCHOTRON RADIATION.

src Cellular gene encoding a tyrosine PROTEIN KINASE, which when mutated (as in the v-*src* gene of Rous sarcoma virus) acts as an ONCOGENE. *See also:* RETROVIRUSES; SH2 DOMAIN; SH3 DOMAIN.

Src homology domain *See:* SH2 DOMAIN; SH3 DOMAIN.

SRE SERUM RESPONSE ELEMENT; STEROL RESPONSE ELEMENT.

SRP SIGNAL RECOGNITION PARTICLE.

SRP receptor DOCKING PROTEIN.

SRY, sry The sex-determining region of the Y, the male sex-determining gene located on the Y-chromosome of human and mouse respectively. *See:* SEX DETERMINATION.

SSB protein Single-stranded DNA binding proteins (also known as helix-stabilizing proteins) have a vital role in DNA REPLICATION. They bind most single-stranded DNA regions in a cell in a cooperative manner along the nucleic acid backbone, leaving the bases accessible for base pairing and promoting the conversion of single- to double-stranded DNA. The behaviour of SSB proteins strongly destabilizes the DNA double helix, and so promotes the helix opening process, which in turn prepares a straightened template for DNA POLYMERASE. SSB proteins are also important in DNA REPAIR and RECOMBINATION. The binding of these proteins greatly speeds the rejoining of separated complementary single-stranded DNA and similarly promotes recombination between homologous sequences.

Sancar, A. et al. (1981) *Proc. Natl. Acad. Sci. USA* **78**, 4274–4278.

SSU Small subunit of RIBULOSE BISPHOSPHATE CARBOXYLASE/OXYGENASE.

SSV1 group *See:* BACTERIOPHAGES.

ST Small T antigen. *See:* PAPOVAVIRUSES.

star activity A property of some RESTRICTION ENZYMES that results in a change in their cleavage site specificity under certain nonoptimal reaction conditions. The enzyme *Eco*RI for example usually recognizes and cuts double-stranded DNA at the following sequence:

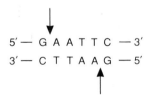

By changing the salt or pH conditions in this reaction, *Eco*RI star activity can be exhibited so that the enzyme now recognize and cuts at the sequence:

Other enzymes that exhibit star activity include *Bam*HI, *Xba*I, *Sal*I, and *Pst*I.

Polisky, B. et al. (1975) *Proc. Natl. Acad. Sci. USA* **72**, 3310–3314.

starch A storage GLUCAN of many plants, containing a mixture of AMYLOSE, AMYLOPECTIN and, sometimes PHYTOGLYCOGEN. Amylose and amylopectin are synthesized together in four stages.
1 A primer precursor bearing short glucan sequences is formed by transfer of glucose from UDP-Glc (not ADP-Glc) to a membrane-associated GLYCOPROTEIN by means of 'transferase I', which has no requirement for Mg^{2+} or Mn^{2+} and has moderate heat resistance. The product is not degraded by β-amylase.
2 A less stable 'transferase II' then adds further glucosyl residues from UDP-Glc or ADP-Glc, in the presence of Mg^{2+} or Mn^{2+} to generate long chains of an α-1,4 glucan primer that are labile to β-amylase. ADP-Glc is the better glucosyl donor in particulate preparations of transferase II, unless Glc-1-phosphate is added, and is the only effective donor if the enzyme is solubilized: it is, therefore, presumed to be the physiological donor.
3 The primer chains are further extended by 'starch synthetase', or synthase, a generally membrane-bound enzyme that can be solubilized and sometimes occurs in a soluble form *in vivo*. Both UDP-Glc and ADP-Glc seem to be able to act as glucosyl donors *in vitro*, although many plants seem to contain and use one rather than the other. Soluble forms of the synthase tend to transfer most effectively from ADP-Glc, though the reverse has been described in barley seeds. Some synthesis of glucan by reversal of starch phosphorylase might occur in a few plants, but there is little evidence for it.
4 A branching enzyme then transfers short blocks of glycan from the reducing terminals of the α1,4 glucan chains to positions a few residues before the (new) terminal positions and joins them by ways of α1,6 links. Chain extension then occurs again.

PHYTOGLYCOGEN is considered to be formed in a manner exactly analogous to AMYLOPECTIN. AMYLOSE synthesis hardly involves branching enzyme, but it may act as a precursor for amylopectin. As with GLYCOGEN, the degradation of starch in plants is by phosphorolysis, by way of starch phosphorylase, to glucose 1-phosphate rather than by hydrolysis to glucose by an amylase (α-glucosidase). Debranching is by way of a specific hydrolase.

start codon INITIATION CODON. *See:* GENETIC CODE; PROTEIN SYNTHESIS.

stationary phase The period in a cell culture which arises when one or more essential factors become growth limiting. This results in the end of the exponential growth or log phase and a cessation in the increase of the total cell number in the culture. If this period is extended, there can be a marked decrease in cell viability. Removal of the constraints to growth, for example by the addition of the limiting nutrient(s), can restore the cells to an actively growing state after a lag period.

statocyte A cell that contains STATOLITHS or other means of detecting the direction of gravity.

statolith A granule which can settle in a cell in response to gravity and is thought to be responsible for the perception of gravity. In plants they are usually starch grains, but can also be vesicles containing barium sulphate.

Steel (*Sl*) Mouse COAT COLOUR GENE encoding the ligand for the receptor tyrosine PROTEIN KINASE Kit (encoded by the coat colour gene *W*).

stem cell Founder cell of embryonic or other cell lineage. Most cells in a lineage DIFFERENTIATE as they undergo successive cell divisions, becoming more differentiated with each generation. In contrast, stem cells are unique in a lineage in that they divide to produce two daughter cells with different fates: one is another stem cell identical to the mother cell, and the other is a lineage progenitor cell which will divide to produce more differentiated cells. It is the ability to undergo both self-renewal and differentiation that is characteristic of stem cells. In adult mammals, they occur in most tissue systems, for example the bone marrow, where they give rise to all blood cells, and muscle (satellite cells) (*see* HAEMATOPOIESIS; MYOGENESIS).

stem cell factor Protein on the surface of bone marrow stromal cells which binds to a receptor encoded by the *kit* proto-oncogene on developing B CELLS, stimulating their proliferation.

STEM Scanning transmission ELECTRON MICROSCOPY.

stem-loop Secondary structure formed in single-stranded nucleic acid comprising a base-paired 'stem' surmounted by a single-stranded 'loop'. *See*: NUCLEIC ACID STRUCTURE; RIBOZYMES; RNA; TRANSFER RNA.

stereo pair (stereo views) Two images representing the appearance of an object as seen by the left and right eyes respectively. Stereoscopic drawings are easily computed for a given separation between the eyes and distance from object to viewer. They may be viewed through special glasses; when printed side by side many people can fuse the two pictures unaided into a single three-dimensional image. *See* illustrations in IMMUNOGLOBULIN STRUCTURE; PROTEIN STRUCTURE.

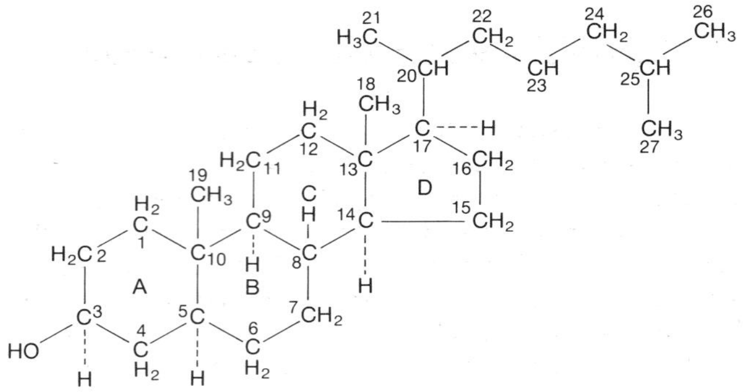

Fig. S18 The steroid cholestanol showing the conventional numbering of the ring system.

stereocilia (1) Large microvilli projecting from the surface of the hair cells in the inner ear, which contain mechanically gated ion channels that are responsible for producing a RECEPTOR POTENTIAL in response to movement of the cells in response to sound vibrations, and displacements due to gravity and acceleration. They are stiffened internally by ordered longitudinal arrays of actin filaments (*see* MICROFILAMENTS).
(2) Similar structures that develop at the animal pole of echinoderm embryos (*see* ECHINODERM DEVELOPMENT).

stereology A body of methods which allow three-dimensional quantities to be obtained from measurements made on two-dimensional sections. Modern stereology enables the quantitation of number, length, surface and volume of biological structures. *See*: IMMUNOELECTRON MICROSCOPY.

Gundersen, H.J.G. (1986) *J. Microsc.* **143**, 3–45.

steroids, steroid hormones Steroids comprise a large group of lipids whose structure is based on a fused reduced ring system (Fig. S18). In mammals, steroids produced by the adrenal cortex and gonads (Fig. S19) act as hormones (*see individual entries*). Steroids and related compounds (e.g. RETINOIDS) act at intracellular receptors (*see* STEROID RECEPTOR SUPERFAMILY).

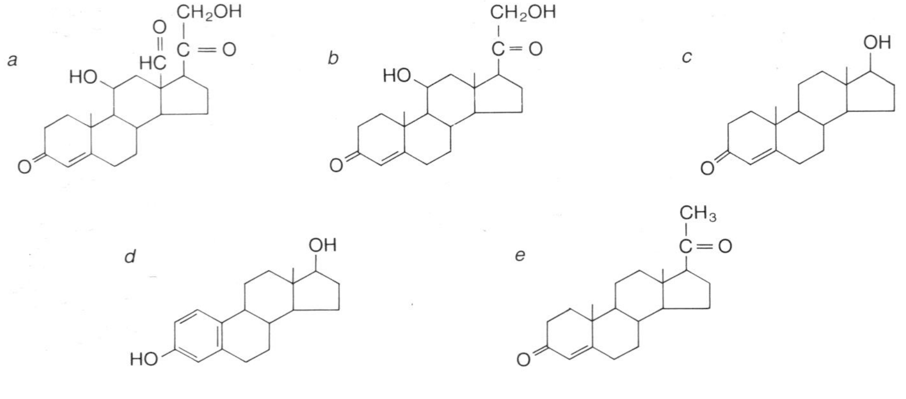

Fig. S19 Steroid hormones. *a*, The MINERALOCORTICOID aldosterone; *b*, the CORTICOSTEROID corticosterone; *c*, TESTOSTERONE; *d*, the OESTROGEN β-oestradiol; *e*, PROGESTERONE.

Steroid receptor superfamily

THE principal mode of action of STEROID HORMONES is to modulate the TRANSCRIPTION of target genes in the nucleus, leading to alterations in levels of mRNA synthesis. Receptors for steroid hormones are high affinity ligand-binding proteins ($K_d \sim 10^{-10}$M) which exist in low concentrations (2×10^4–6×10^4 molecules) in cells. Their biochemical characterization and purification initially led to the cloning of cDNAs encoding the GLUCOCORTICOID and OESTROGEN receptors. Since then a number of highly homologous proteins have been identified, which comprise a superfamily of nuclear receptors (Table S7). Some of these proteins represent receptors for other steroid hormones (androgens, PROGESTERONE, MINERALOCORTICOIDS, VITAMIN D), whereas other members of the family bind structurally unrelated ligands such as THYROID HORMONE and retinoic acid (*see* RETINOIDS). A number of these proteins have been designated 'orphan' receptors, as no cognate ligand has yet been identified. Some receptors (GR, ER, AR, MR, see Table S7 for abbreviations) are encoded by a single gene whereas others (TR, RAR, RXR, PPAR) are encoded by two or more genes. Receptor-related proteins have been isolated from the fruit fly *Drosophila*, the nematode *Caenorhabditis elegans* and many higher eukaryotes, but not from yeast, suggesting that this family of proteins evolved at the metazoan stage.

The nuclear receptors can be aligned on the basis of primary amino-acid sequence homology to delineate distinct domains (Fig. S20). The N-terminal region (A/B domain) varies in length and composition and is poorly conserved between different receptors. A central cysteine-rich region (C domain) exhibits the greatest homology, and is followed by the C-terminal region (D/E/F domains) which also contain some conserved sequence motifs. These receptors also share a common mode of action as ligand-inducible TRANSCRIPTION FACTORS. Following synthesis, the receptors translocate to the nucleus (*see* PROTEIN TARGETING)

Table S7 Members of the steroid receptor superfamily

Hormone receptors	Orphan receptors	*Drosophila* receptor-related proteins
Glucocorticoid (GR)	Peroxisome proliferator (PPAR)	Ecdysone receptor (EcR)
Androgen (AR)		Embryonic gonad (EGON)
Mineralocorticoid (MR)	Apo A1 regulatory protein(ARP-1)	
Progesterone (PR)	erb-A related (EAR-1, EAR-2)	Fushi tarazu factor 1 (FTZ-F1)
Oestrogen (ER)		
Vitamin D (VDR)	Hepatocyte nuclear factor (HNF-4)	Knirps (KNI)
Thyroid hormone (TR)		Knirps related (KNRL)
Retinoic acid (RAR)	Steroidogenic factor (SF-1)	Seven-up (SVP)
Retinoid X (RXR)		Tailless (TLL)
	Nerve growth factor inducible (NGFI-B)	Ultraspiracle (USP)
	Oestrogen receptor related (ERR-1, ERR-2)	
	Chicken ovalbumin upstream (COUP)	

Nuclear receptors in higher eukaryotes comprise hormone receptors whose ligand is known and orphan receptors with no known ligand. A number of receptor related proteins have been isolated from *Drosophila* and the ligand has been identified in one case (EcR).

and bind with high affinity to specific regulatory DNA sequences or response elements, usually located in the PROMOTER region of target genes. The presence of ligand results in activation or repression of target gene transcription (Fig. S21). Deletion and mutational analyses have allowed various receptor functions to be localized to these domains.

Subcellular location

Following synthesis, some receptor proteins (GR, PR) remain predominantly cytosolic and form high molecular weight com-

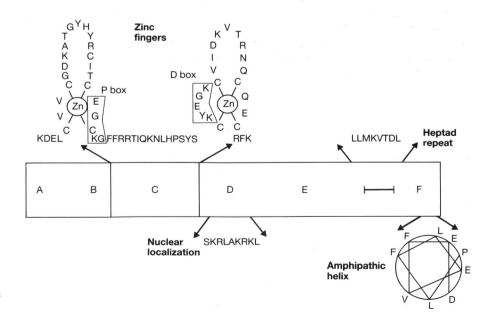

Fig. S20 The organization of nuclear receptors into six domains (A–F) is shown. Conserved amino acid sequence motifs including the zinc fingers, nuclear localization signal, a heptad repeat involved in dimerization and a putative amphipathic helix mediating transactivation, are depicted.

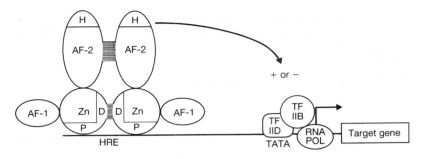

Fig. S21 An outline of the mechanism of nuclear receptor action. The P box region of receptors mediates interaction with regulatory DNA sequences or hormone response elements (HRE). Dimerization is mediated by the D box as well as sequences in the C-terminal region. The activity of the transcription initiation complex is modulated in a ligand-independent (AF-1) and ligand-dependent (AF-2) manner, leading to target gene activation or repression.

plexes with HEAT SHOCK proteins (Hsp90 and Hsp70). The presence of ligand leads to the dissociation of receptor from this complex followed by translocation of the receptor–ligand complex to the nucleus. In contrast, other receptors (TR, RAR), do not form complexes with Hsp90 and are constitutively nuclear. For both classes of receptor, transport into the nucleus is dependent on a conserved motif of basic residues in the D domain which acts as a nuclear localization signal, analogous to that identified in the SV40 T antigen (*see* PAPOVAVIRUSES: PROTEIN TRANSLOCATION) (*see* Plate 6*b*). This energy-dependent nuclear translocation is counterbalanced by passive efflux into the cytosol, such that receptors shuttle between the two compartments.

Target gene recognition

The invariant cysteine residues within the C domain coordinate zinc to form two finger-like motifs which direct sequence-specific DNA binding (*see* PROTEIN–NUCLEIC ACID INTERACTIONS) (*see* Plate 6*b*). A large number of response elements in target genes, mediating transcription regulation, have been identified, enabling consensus binding sites to be deduced. On the basis of their DNA binding characteristics, the nuclear receptors can be subdivided into three groups:

1 Almost all steroid receptors (GR, PR, AR, MR, ER), favour binding as homodimers to hexanucleotide motifs (AGAACA or AGGTCA), arranged in an inverted repeat configuration.

2 A second group, comprising TR, VDR, RAR, and PPAR, bind preferentially as heterodimers to response elements consisting of AGGTCA half sites arranged as direct repeats. In each case, they form heterodimers with a common partner — the retinoid X receptor (RXR), which is another member of this superfamily. Specificity is conferred by the fact that a given heterodimer will preferentially recognize a direct repeat with a particular spacing, for example 3 bp for VDR, 4 bp for TR, and 2 or 5 bp for RAR. RXR may act as a 'booster' protein in these contexts, enhancing the DNA binding and transcriptional activation properties of its partner.

3 Some orphan receptors such as NGFI-B and SF-1 are capable of binding to DNA as monomers and recognize a single AGGTCA motif in the appropriate context.

In all cases the receptor–DNA interaction is mediated by a region called the P box at the base of the first finger (Fig. S20), with different amino acids dictating response element recognition. Thus, receptors such as GR, PR, MR, and AR which bind to AGAACA response elements (Fig. S22), contain the peptide se-

quence CGSCKV (single-letter notation) within the P box, whereas TR, VDR, RAR, PPAR and NGFI-B, which recognize the hexanucleotide AGGTCA, contain the residues EGCKG in this region.

The ability of receptors to form homo- and heterodimers is mediated by at least two protein–protein interactions between adjacent monomers. First, another region within the zinc finger domain, called the D box (Fig. S20), stabilizes homo- and heterodimer formation on response elements. Second, heptad motifs consisting of hydrophobic amino acids at the first, fifth, and eighth positions have been identified in the C-terminal D/E/F domain. These sequences may be capable of forming a dimerization interface analogous to the LEUCINE ZIPPER motif in other transcription factors such as FOS and JUN. Mutational analyses indicate that the last heptad (Fig. S20) is required for homodimer formation in the oestrogen receptor and for heterodimer formation in the thyroid hormone and retinoic acid receptors.

Future studies are likely to increase the repertoire of known interactions between individual receptors as well as the configurations of response elements. This multiplicity represents a mechanism for generating diversity, enabling nuclear receptors to regulate a large number of target genes in a temporally regulated and tissue-specific manner.

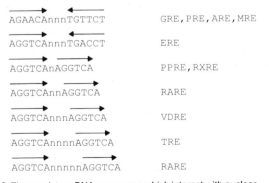

Fig. S22 The regulatory DNA sequences which interact with nuclear receptors have been analysed to derive consensus response elements. An arrow signifies a half-site (AGGTCA or AGAACA) on the coding or noncoding strand which binds a receptor monomer. Intervening non-conserved nucleotides are denoted n. ARE, androgen response element; ERE, oestrogen response element; GRE, glucocorticoid response element; MRE, mineralocorticoid response element; PPRE, peroxisome proliferator response element; PRE, progesterone response element; RARE, retinoic acid response element; RXRE, retinoid X response element; TRE, thyroid hormone response element; VDRE, vitamin D response element.

Ligand binding

For most nuclear receptors, the residues involved in ligand binding are widely distributed across the D/E/F domains. These amino acids are thought to form a pocket that accommodates the ligand, which is usually hydrophobic in nature. Specific cysteine residues have been shown to be important for ligand binding in GR and ER but determinants in other receptors have not yet been mapped. The binding of ligand is associated with marked conformational changes within the D/E/F domains, as has been shown by CIRCULAR DICHROISM spectroscopy for TR or altered sensitivity to enzymatic digestion for PR. For some receptors (GR, PR) it also leads to dissociation from heat shock proteins. However, the hallmark of nuclear receptors is that ligand binding mediates activation or repression of target gene transcription.

Activation and repression of transcription

It is thought that receptors activate transcription by enhancing the activity of the transcription initiation complex. Although there is evidence for a direct interaction between receptors (e.g. PR, ER) and general transcription factors such as TFIIB (*see* TFIIA–H), other 'adaptor' or 'co-activator' proteins may also be involved in this process. Nuclear receptors contain two types of activation function. The first (AF-2) is contained in the D/E/F domains and is dependent on the presence of ligand. A putative amphipathic α-helical sequence motif in the F domain (Fig. S20) which is conserved in a number of nuclear receptors may mediate this. Mutations in this region markedly impair transactivation by the oestrogen and thyroid hormone receptors. The A/B domain encodes the second activation function (AF-1), which is ligand-independent. Some receptors (GR), exhibit strong AF-1 activity whereas in others it is weak (ER) or absent (VDR). The genes for some receptors (TR, RAR, RXR, PPAR) are alternatively spliced to generate a number of protein isoforms with divergent sequences within the A/B domain. These isoforms exhibit differing AF-1 activity, which also varies depending on the nature of the target gene as well as cell type.

In contrast, transcriptional repression by nuclear receptors is less well understood. Some receptors (TR, RAR) that are constitutively nuclear are able to bind response elements in the absence of ligand. This interaction is associated with an inhibition of basal promoter activity or 'transcriptional silencing' and the sequences which mediate this also map broadly to the D/E/F domain. Nuclear receptors are also capable of repressing transcription in a ligand-dependent manner following interaction with specific target gene promoter sequences. These negative response elements are poorly defined and may differ in configuration from the positive response elements described above. Current models suggest that nuclear receptors inhibit gene transcription by blocking the action of positive transcription factors. They may either interfere with the transcription factors' ability to bind DNA or block their activity via protein–protein interactions.

A third type of interaction is exemplified by the interaction of GR with the transcription factor AP-1 on a composite response element which exhibits dual binding specificities. This response element responds both positively and negatively to GR in different cell types, reflecting either a synergistic interaction between GR and c-Jun homodimers or mutual antagonism between GR and Fos–Jun complexes.

Post-translational modification

Another factor which modulates the transcriptional activity of nuclear receptors is PHOSPHORYLATION. The glucocorticoid receptor undergoes phosphorylation at several serine residues within the A/B domain which probably contribute to the AF-1 activity of this receptor. The progesterone receptor is phosphorylated on serine and threonine residues at several stages in its pathway of action. Some residues are phosphorylated in the basal state, with others following the binding of ligand and following interaction with DNA (mediated by DNA-dependent protein kinase, DNA-PK).

This mechanism may also allow some receptors to be activated in the absence of ligand. DOPAMINE, acting through its cell-surface receptor, activates PR, ER, VDR, TR, and COUP by phosphorylation. The oestrogen receptor is known to undergo tyrosine phosphorylation and can be activated by epidermal growth factor — which is coupled to a cell-surface receptor with tyrosine kinase activity (*see* GROWTH FACTORS; GROWTH FACTOR RECEPTORS). Future studies may provide further examples of 'cross talk' between cell-surface receptor signalling cascades and nuclear receptor action.

Nuclear receptor mutations in disease

The ONCOGENE v-*erbA*, found in the avian erythroblastosis virus, is highly homologous to the thyroid hormone receptor but transcriptionally inactive owing to a deletion of the putative amphipathic helix at the C terminus. This mutant protein blocks wild-type thyroid hormone and retinoic acid receptor action, with arrested proliferation and differentiation of erythroid progenitor cells leading to erythroleukaemia.

Human acute promyelocytic leukaemia is associated with a characteristic balanced chromosomal TRANSLOCATION t(15;17) (q22;q21) (*see* CHROMOSOME ABERRATIONS). This translocation results in a chimaeric transcript, derived from the retinoic acid receptor (RARα) gene fused to a novel gene (PML), whose function is unknown. The predicted fusion protein may interfere with normal pathways of PML action, blocking myeloid differentiation. Treatment with all-*trans*-retinoic acid can overcome this block and result in disease remission.

Nuclear receptor mutations also form the basis of a number of inherited human diseases. Androgen insensitivity syndrome is an X-LINKED disorder of male sexual development of variable severity, ranging from a female phenotype (testicular feminization) to undervirilization or infertility. It is associated with a variety of point mutations throughout the AR gene leading to the synthesis of truncated or transcriptionally defective mutant proteins. An adult-onset form of motor neuron disease (KENNEDY'S DISEASE), with spinal muscular atrophy and androgen insensitivity, is characterized by an increase in the number of polymorphic trinucleotide CAG repeats in the coding region of the androgen receptor gene. The mutation leads to an expansion in the polyglutamine sequence motif within the A/B domain, but its effect on androgen receptor function remains unclear. Other hormone insensitivity syndromes include glucocorticoid resistance, thyroid hormone resistance, and vitamin D-resistant rickets. These disorders are all due to nuclear receptor defects and characterized by target organ refractoriness to hormone action in the presence of supraphysiological concentrations of ligand. Vitamin D resistance is inherited as an AUTOSOMAL RECESSIVE trait and associated with homozygosity for mutations in the zinc-finger domain which abolish DNA binding. In contrast, dominantly inherited thyroid hormone resistance is characterized by heterozygosity for mutations restricted to the D/E/F domains. The mutant proteins are transcriptionally impaired yet capable of inhibiting wild-type receptor action in a dominant negative manner analogous to v-erbA.

V.K.K. CHATTERJEE

1 Evans, R.M. (1988) The steroid and thyroid hormone receptor superfamily. *Science* **240**, 889–895.
2 Parker, M.G. (Ed.) (1991) *Nuclear Hormone Receptors* (Academic Press, New York).
3 Leid, M. et al. (1992) Multiplicity generates diversity in the retinoic acid signalling pathways. *Trends Biochem. Sci.* **25**, 427–433.
4 Lazaar, M.A. (1993) Thyroid hormone receptors: multiple forms, multiple possibilities. *Endocr. Rev.* **30**, 184–193.
5 Truss, M. & Beato, M. (1993) Steroid hormone receptors: interaction with deoxyribonucleic acid and transcription factors. *Endocr. Rev.* **30**, 459–479.

sterols STEROIDS with 8 to 10 carbon atoms in the side chain and an alcoholic hydroxyl group at position 3 (Fig. S23). They are important constituents of membranes. The sterol cholesterol is transported in the blood in the form of lipoprotein particles (*see* PLASMA LIPOPROTEINS AND THEIR RECEPTORS).

Fig. S23 *a*, Cholesterol, an abundant sterol of animal membranes; *b*, ergosterol, a sterol of yeast membranes.

sterol response element (SRE) Sequence element within the PROMOTER of the LDL receptor gene, which mediates regulation of gene expression by sterols. It is also found in genes for enzymes involved in CHOLESTEROL biosynthesis. *See*: PLASMA LIPOPROTEINS AND THEIR RECEPTORS.

sticky ends (cohesive ends) The single-stranded DNA termini protruding from fragments of double-stranded DNA following digestion with a RESTRICTION ENZYME that produces a staggered cut. These termini can have either the 5'-phosphate protruding (e.g. cut by *Eco*RI) or the 3'-OH protruding (e.g. cut by *Pst*I). Complementary sticky ends can pair by standard BASE PAIRING and can be covalently joined by the enzyme DNA LIGASE. DNA molecules cut with restriction enzymes that have different recognition sites, but generate the same sticky ends, for example:

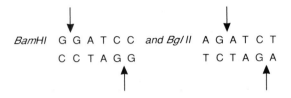

*Bam*HI G G A T C C and *Bgl* II A G A T C T
　　　　　　C C T A G G　　　　　　　T C T A G A

can also be rejoined covalently in this manner. *Cf.* BLUNT ENDS.

stomata (*sing.* stoma) Pores in the epidermis of leaves and other organs usually of aperture varied by two GUARD CELLS.

stomatal complex The stomatal GUARD CELLS in plant leaves and any neighbouring epidermal cells which help to control the size of the aperture.

stomate Having stomata.

stop codon TERMINATION CODON. *See also*: GENETIC CODE; PROTEIN SYNTHESIS.

stop-transfer sequence A region of uncharged, hydrophobic residues in a polypeptide which, as a transmembrane region, integrates a protein into a phospholipid bilayer. It is thought that initial insertion of some membrane proteins into the bilayer occurs via the SIGNAL SEQUENCE and SIGNAL RECEPTOR PARTICLE (SRP), and translocation of the nascent chain ensues until the hydrophobic stop-transfer sequence reaches the translocation complex. At this point, translocation stops, and the stop-transfer sequence moves laterally from the complex into the bilayer. Some stop-transfer sequences also double as signal sequences, which insert via SRP into the phospholipid bilayer, are not cleaved but remain as the transmembrane region. *See also*: MEMBRANE STRUCTURE; PROTEIN TARGETING; PROTEIN TRANSLOCATION.

Blobel, G. (1980) *Proc Natl. Acad. Sci. USA* **77**, 1496–1500.

β-strand (beta-strand) One of the basic conformational structures in proteins, a short stretch of polypeptide chain in an almost fully extended conformation. *See*: PROTEIN STRUCTURE.

strand displacement A type of DNA REPLICATION in which one DNA strand only is replicated, the other parental strand being displaced from the template. This type of replication is found in mitochondria, plasmids and some phages.

strand separation (1) Separation of the strands of the DNA double helix that occurs during DNA REPLICATION.
(2) *In vitro* separation of the two strands of a DNA duplex by heating (at 95°C) or by exposure to high alkali conditions. A necessary step in many molecular biological procedures (e.g. DNA SEQUENCING; HYBRIDIZATION).

streptavidin A tetrameric protein from the prokaryote *Streptomyces avidinii* that binds to BIOTIN. It is used in techniques ranging from nucleic acid HYBRIDIZATION to IMMUNOELECTRON MICROSCOPY as a component of detector complexes to locate molecules that have been BIOTINYLATED.

streptolysin O Enzyme from *Streptococcus* spp. used to permeabilize animal cells.

Streptomyces A genus of Gram-positive mycelial- and spore-forming prokaryotes which commonly inhabit soil. They produce more than half of all known antibiotics, and have a genome around three times the size of that of *Escherichia coli*.

streptomycin A basic aminoglycoside antibiotic (Fig. S24) produced by *Streptomyces griseus* and active against Gram-positive and Gram-negative bacteria. It inhibits bacterial but not eukaryotic PROTEIN SYNTHESIS, by interfering with the binding of formylmethionyl-tRNA to the ribosome, thus inhibiting initiation. It also causes misreading of the mRNA template. Sensitivity of the ribosome to streptomycin is determined by ribosomal protein S12 in the 30S subunit.

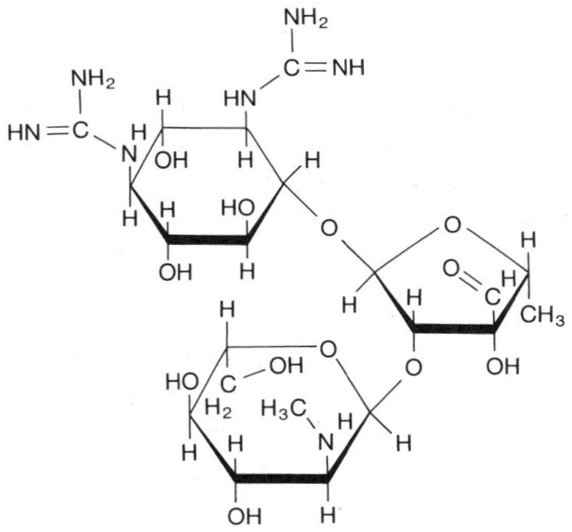

Fig. S24 Streptomycin.

stress fibres Bundles of actin filaments (MICROFILAMENTS) lying parallel with the substrate surface, seen in cultured mammalian cells. *See*: CELL MOTILITY.

stress responses, stress-induced proteins *See*: COLD SHOCK; HEAT SHOCK.

stretch receptor Specialized mechanoreceptor neuron in muscle which detects muscle tension as a result of the deformation of its dendrites caused by muscle fibre contraction and relaxation. In vertebrates stretch receptors are organized into muscle spindles.

striated muscle MUSCLE attached to the skeleton and which is under voluntary control. It has a transversely striped appearance under the microscope. Each muscle fibre is composed of fused muscle cells containing a highly organized contractile apparatus. *See*: MUSCLE. *See also*: MYOGENESIS.

stringency The HYBRIDIZATION conditions determining the degree of annealing of two nucleic acids. Hybridization under stringent conditions requires a perfect or near perfect sequence match. Hybridization under relaxed conditions allows hybridization between sequences with less than 100% identity, and is used, for example, to search for HOMOLOGOUS sequences in another species. Greater stringency can be achieved by reducing the salt concentration or increasing the temperature of the hybridization.

stringent control The strict control of a factor (e.g. PLASMID copy number) to a limited quantity per genome, usually one. The alternative is RELAXED CONTROL (as seen with most multicopy plasmids).

stringent response The reversible shutdown of a wide range of cellular activities in bacterial cells exposed to poor growth conditions, and caused by the absence of sufficient amino acids to sustain protein synthesis. This mechanism enables the cell to survive until nutritional conditions improve. As a result of this response, the production of mRNA, rRNA, and tRNA drops to about 5–10% of levels under normal conditions. The reponse is mediated by the polyphosphates ppGpp and pppGpp, which are produced in conditions of amino-acid starvation, and which, by a complex pathway, regulate the transcription of these genes.

stroma The aqueous phase of CHLOROPLASTS, analogous to the matrix of MITOCHONDRIA, in which the THYLAKOID membranes are embedded and which contains many enzymes and other proteins, especially those of the carbon reduction cycle (CALVIN CYCLE). The term 'stroma lamellae' refers to thylakoid membranes that are not stacked to form grana, so that their external surfaces are fully accessible to the stroma. *See also*: PHOTOSYNTHESIS.

structural gene A DNA sequence that can be transcribed by RNA polymerase to produce a protein-coding mRNA, a tRNA, or an rRNA. *Cf.* REGULATORY GENE.

structure amplitude A quantity representing the magnitude of a STRUCTURE FACTOR but not its phase (*see* PHASE PROBLEM). The structure amplitude equals the square root of the corresponding DIFFRACTION intensity from a crystal. *See*: X-RAY CRYSTALLOGRAPHY.

structure factor The FOURIER TRANSFORM of the ELECTRON DENSITY at all points within the ASYMMETRIC UNIT of a crystal is the set of structure factors in RECIPROCAL SPACE. These vectors determine the DIFFRACTION pattern of a crystal such that the set of measurable diffraction intensities is the set of squares of the magnitudes of the structure factors. These magnitudes, or STRUCTURE AMPLITUDES, can thus be derived from the observed intensities, whereas the crystallographic PHASE PROBLEM means that the phase of the structure factor is generally only calculable indirectly and from additional information. *See*: X-RAY CRYSTALLOGRAPHY.

Coppens, P. (1993) The Structure Factor. In *International Tables for Crystallography* Vol. B Reciprocal Space (Shmueli, U., Ed.) 10–22 (Kluwer Academic Publishers, Dordrecht).

Stuhrmann plot A particular form of graph relating the reciprocal of solvent density to the square of the RADIUS OF GYRATION measured from the SMALL-ANGLE SCATTERING by an object immersed in solvents of differing densities. At the origin of the plot its slope indicates whether the higher or lower scattering regions of the particle are more towards the outside of the particle, while its curvature at the origin depends on departures from symmetry in the positions of regions of different scattering within the particle. *See also*: NEUTRON SCATTERING AND DIFFRACTION; X-RAY DIFFRACTION.

Styela *See*: ASCIDIAN DEVELOPMENT.

subclone DNA clone generated by the transfer of part of a cloned DNA fragment from one recombinant plasmid to another.

subgenomic RNA In some (+)-strand RNA viruses, an mRNA containing the 3′ gene(s) which are unavailable for translation from genomic RNA. It is synthesized on a partial (–)-strand template which is produced after replication of the genomic RNA. *See*: ANIMAL VIRUSES; PLANT VIRUSES.

sublimons Rearranged forms of plant mitochondrial DNA present at low levels. It is suggested that they may suddenly be amplified and become the predominant genomic form. *See*: MITOCHONDRIAL GENOMES: PLANT.

Small, I.D. et al. (1987) *EMBO J.* **6**, 865–869.

submitochondrial particles MITOCHONDRIA disrupted by ultrasonication form inverted inner membrane vesicles that have been widely used to study OXIDATIVE PHOSPHORYLATION and other reactions. The external surface of these submitochondrial particles is directly accessible to substrates, including NADH, that are not transported across the inner membrane.

Darley-Usmar, V.M. et al. (Eds) (1987) *Mitochondria: A Practical Approach* (IRL Press, Oxford).

substance P A bioactive peptide secreted from neurons in both gut and central nervous system. It is able to cause contraction of smooth muscle. In the central nervous system it is present in small-diameter sensory axons concerned with NOCICEPTION ('pain endings'), and may be involved in some way in the perception of pain. It acts at G PROTEIN-COUPLED RECEPTORS.

substantia nigra Structure within the midbrain which contains cell bodies of dopamine-containing neurons.

substrate guidance *See*: HAPTOTAXIS.

subunit vaccines Vaccines composed of purified antigens (e.g. the hepatitis B virus surface antigen) rather than killed or attenuated microorganisms. Protein antigens for such vaccines are now readily produced by GENETIC ENGINEERING. *See also*: RECOMBINANT VACCINES.

succinamopine An OPINE which is formed from the conjugation of asparagine and 2-oxoglutaric acid. It exists in D,L- and L,L-forms. Succinamopine-type strains of *Agrobacterium tumefaciens* (e.g. EU6, AT181 or T10/37) can utilize both forms. Strain Bo542, however, can only catabolize the L,L- form.

sucrose density gradient *See*: DENSITY GRADIENT CENTRIFUGATION.

sugars A large group of low molecular weight CARBOHYDRATES, generally considered as including monosaccharides, monosaccharide derivatives such as amino sugars, and oligosaccharides (di-, trisaccharides, etc.). They are water-soluble unbranched hydroxyaldehydes or hydroxyketones (or derivatives thereof) and most have a sweet taste.

1 Monosaccharides. The smallest molecules generally considered as carbohydrates are the 3-carbon sugars glyceraldehyde and dihydroxyacetone (Fig. S25). Aldehyde sugars are known as aldoses, ketone sugars as ketoses (Table S8). Monosaccharides are also classified by the number of carbon atoms. Aldose sugars are named generically trioses (three C atoms), tetroses (four C), pentoses (five C), hexoses (six C), heptoses (seven C). Generic

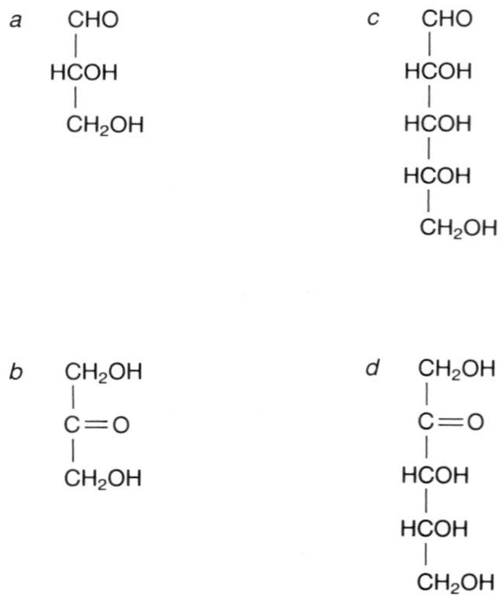

Fig. S25 The common features of monosaccharides. *a*, D-glyceraldehyde (an aldose); *b*, D-dihydroxyacetone (a ketose); *c*, D-ribose (an aldose); *d*, D-ribulose (a ketose). The carbon skeleton is unbranched and each carbon atom except one bears a hydroxyl group. One carbon atom bears a carbonyl oxygen; the carbonyl oxygen may reside on a terminal carbon atom (aldoses) or on a centrally placed carbon atom (ketoses). In the common ketoses, the carbonyl oxygen resides on the C2 atom. The molecules are shown in linear form for convenience.

Table S8 Some naturally occurring aldoses and ketoses

No. of carbon atoms	Aldoses	Ketoses
5	D-Ribose	D-Ribulose
	L-Arabinose	
	D-Xylose	
6	D-Glucose	D-Fructose
	D-Mannose	D-Sorbose
	D- and L-Galactose	

names for the ketoses are formed by the insertion of 'ul' (e.g. pentulose, heptulose).

Monosaccharides are optically active (i.e. they can rotate a beam of plane-polarized light either to the left, *levo*rotatory, or to the right, *dextro*rotatory), and are thus sometimes prefixed with *d*- (or (+)) or *l*- (or (–)) to indicate the sign of rotation. However, the optical sign gives no indication of the molecular configuration at the various centres of asymmetry in the molecule. The convention of nomenclature using the prefixes D- and L- indicates the configuration around the centre of asymmetry most distant from

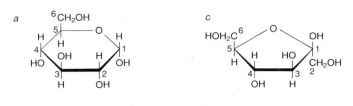

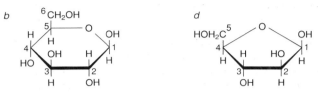

Fig. S27 *a*, α-D-Glucopyranose; *b*, β-D-glucopyranose; *c*, β-D-fructofuranose; *d*, β-D-arabinofuranose.

the carbonyl atom end of the molecule (Fig. S26). A structure of given atomic composition (e.g. $C_6H_{12}O_6$) can exist as numerous STEREOISOMERS.

A further complication of nomenclature in monosaccharides arises from their cyclic structure. In glucose, for example, the carbonyl oxygen can form an oxygen bridge between carbons 1 and 5 and so D-glucose can exist in two stereoisomeric forms α- and β-, depending on the configuration of the hydroxyl group on C1 (Fig. S27). When a six-membered ring is formed, the structure is termed a pyranose; when a five-membered ring is formed (e.g. in fructose and arabinose) the structure is termed a furanose. The structural formulae for glucose and other monosaccharides can also be written to indicate the molecular conformation (Fig. S28).

Only glucose and fructose are naturally abundant as free monosaccharides. Other sugars are common as units of oligosaccharides or polysaccharides or other glycans (e.g. L-arabinose, D- and L-galactose, D-mannose, and D-xylose, *see* GLYCANS; GLYCOPROTEINS), and in nucleotides and nucleic acids (D-ribose, in RNA). Common abbreviations for the sugars found in glycans are given in Table S9.

Monosaccharides participate in many metabolic reactions in the form of sugar phosphates: glucose 6-phosphate, glucose 1-phosphate, fructose 6-phosphate, fructose 1,6-bisphosphate, and ribulose 1,5-bisphosphate are important metabolic intermediates (*see* CALVIN CYCLE; GLYCOLYSIS; PHOTOSYNTHESIS). As precursors for GLYCAN synthesis they are generally in the form of SUGAR NUCLEOTIDES.

a   CHO   1
    |
    HCOH   2
    |
    HOCH   3
    |
    HCOH   4
    |
    HCOH   5
    |
    CH₂OH   6

b   CH₂OH   1
    |
    C=O   2
    |
    HOCH   3
    |
    HCOH   4
    |
    HCOH   5
    |
    CH₂OH   6

c   CHO
    |
    HCOH
    |
    HOCH
    |
    HOCH
    |
    CH₂OH

Fig. S26 *a*, D (+)-Glucose; *b*, D (–)-fructose; *c*, L-arabinose. The molecules are shown in the linear form for convenience.

Fig. S28 The 'chair' conformation of glucose.

Table S9 Abbreviations for commonly occurring sugars

Ara	Arabinose
Fuc	Fucose
Gal	Galactose
GalNAc	*N*-acetylgalactosamine
Glc	Glucose
GlcA	Glucuronic acid
GlcNAc	*N*-acetyglucosamine
Man	Mannose
Mur	Muramic acid
MurNAc	*N*-acetylmuramic acid
NeuAc, NAN	*N*-acetylneuraminic acid (sialic acid)
Rib	Ribose
Ru	Ribulose
Sia	Sialic acid
Xyl	Xylose

2 Monosaccharide derivatives. The most important and commonly encountered are:

Carboxylic acids formed by oxidation of monosaccharides. There are three main groups: (1) aldehyde oxidized to carboxyl, for example gluconic acid (from glucose); (2) primary hydroxyl group remote from the aldehyde oxidized to carboxyl (uronic acids), for example glucuronic acid (Fig. S29) derived from glucose, and L-iduronic acid (*see* GLYCOSAMINOGLYCANS); (3) dicarboxylic acids such as saccharic acid.

Sugar alcohols formed from monosaccharides by reduction, for example sorbitol and mannitol (Fig. S30). Carbocyclic alcohols such as INOSITOL are related.

Amino sugars, in which a hydroxyl group is replaced by an amino group, for example glucosamine (Fig. S31) and galactosamine. Such sugars may be further modified, for example N-ACETYLGLUCOSAMINE, a common residue in many polymers (*see* GLYCANS; GLYCOPROTEINS AND GLYCOSYLATION). Sulphuryl and phosphoryl groups may also be added (*see* keratan sulphate in GLYCOSAMINOGLYCANS).

Deoxy sugars, in which a hydroxyl group is replaced by a hydrogen, for example 2-deoxyribose, the sugar found in DNA, L-rhamnose (6-deoxy-L-mannose), and L-fucose (6-deoxy-L-galactose).

Sialic acids, which are derivatives of NEURAMINIC ACID and are found in many tissues as constituents of proteins, lipids, and polysaccharides. *See*: GLYCANS; GLYCOLIPIDS; GLYCOPROTEINS AND GLYCOSYLATION; LEWIS BLOOD GROUP; SIALIC ACID.

Muramic acid (Fig. S32), a complex *N*-acetylamino sugar acid

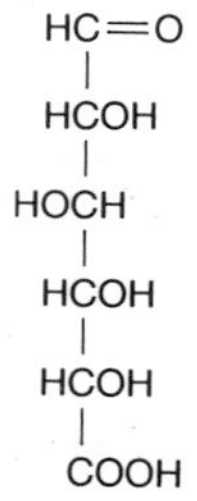

Fig. S29 D-glucuronic acid.

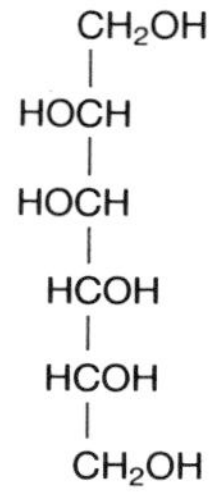

Fig. S30 D-mannitol.

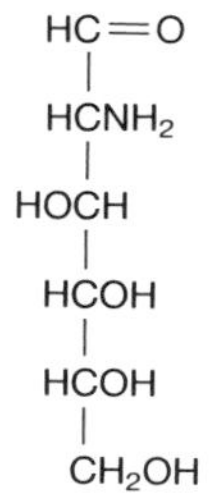

Fig. S31 D-glucosamine.

derivative which is a constituent of the peptidoglycans of bacterial cell walls (*see* BACTERIAL ENVELOPES).

3 Oligosaccharides. Monosaccharides are linked together to form oligosaccharide chains by a glycosidic linkage between the anomeric (carbonyl) carbon atom of one and (usually) an alcoholic hydroxyl of another (in sucrose, a disaccharide composed of glucose and fructose units, the glycosidic linkage is between the two anomeric carbons). The linkage may be in either the α or β configuration (Fig. S33).

sugar nucleotides (nucleotide sugars) Glycosidic phosphoryl esters of nucleoside diphosphates or (more rarely) nucleoside monophosphates (*see* NUCLEOSIDES AND NUCLEOTIDES). They are common sugar donors for glycosyl transfers and the major intermediates for interconnecting sugar skeletons for biosynthesis of glycans. They are synthesized by reactions of the form either

$$\text{sugar } 1\text{-P} + \text{NTP} \rightarrow \text{NDP-sugar} + \text{PP}$$

or

$$\text{sugar} + \text{NTP} \rightarrow \text{NMP-sugar} + \text{PP}$$

Several classes of sugar nucleotide are known. UDP sugars are of

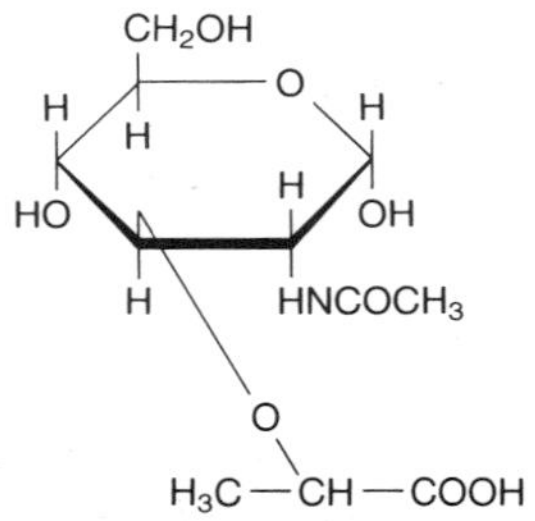

Fig. S32 Muramic acid.

a CH₂OH

b CH₂OH CH₂OH

c CH₂OH CH₂OH

Fig. S33 *a*, Sucrose, α-D-glucopyranosyl-β-D-fructofuranoside; *b*, cellobiose (β1,4 glycosidic linkage); *c*, maltose (α1,4 glycosidic linkage). The numerals refer to the carbon atoms linked by the glycosidic bond. The asterisks in sucrose denote the anomeric carbons.

wide occurrence, with most being derived from UDP-Glc, though UDP-amino sugars have a separate origin. TDP sugars are common in prokaryotes and a few are known in plants; CDP sugars are almost, but not quite wholly, confined to prokaryotes as are CDP sugar alcohols. GDP sugars are widespread and tend to derive from GDPMan while ADP sugars occur (rarely) in prokaryote and plants. Sialic acids and KDO (3-oxo-3-deoxymanno-octulosonic acid, of bacteria, etc.) occur as CMP sugar nucleotides. In eukaryotes, sugar nucleotides are usually synthesized in the cytoplasm by membrane-associated enzymes and are interconverted by a range of epimerases, decarboxylases etc., which are more firmly membrane-bound. The generation of 4-oxo-intermediates is critical in sugar nucleotide interconversions and the flexibility of ene-diol rearrangements makes for great biosynthetic versatility. CMP sialic acid synthesis occurs in eukaryotic nuclei.

Kochetkov, N.K. & Shibaev, V.N. (1973) *Adv. Carb. Chem. Biochem.* **28**, 307–400.
Stoddart, R.W. (1984) *Biosynthesis of Polysaccharides*, 27–57 (Croom Helm, London).

sugar transporters *See*: MEMBRANE TRANSPORT SYSTEMS.

suicide substrate Mutant DNA sequences used to study DNA RECOMBINATION which mimic a normal site-specific recombination sequence such as *att* (*see* LAMBDA). They contain the common core region where recombination is thought to take place and the nicking of one of the strands within this core sequence gives a substrate that halts the recombination reaction at an intermediate stage. Work using suicide substrates established that the exchange of single strands in the recombination event occurs sequentially rather than simultaneously.

sulphatases Enzymes that hydrolyse sulphuric esters, removing a sulphate group. EC 3.1.6. *See*: ARYL SULPHATASE; HUNTER SYNDROME; LYSOSOMES.

sulpholipids Sulphur-containing GLYCOLIPIDS such as 6-sulpho-6-deoxy-α-glucosyl diacylglycerol which is found in small amounts in chloroplast thylakoid membranes (6-deoxy glucose is also known as quinovose).

superantigens Antigens that bind to and stimulate all T CELLS that carry receptors containing a given β-chain variable element, regardless of the type of D_β, J_β or of the α chain (*see* T CELL RECEPTORS; T CELL RECEPTOR GENES). This is in contrast to the normal mode of antigen binding to the T cell receptor in which the antigen-combining site and therefore the antigen specificity is determined by the amino-acid sequence of both the α and β chains. The most likely explanation for the phenomenon is that superantigen binding occurs outside the classically defined antigen-binding site. Superantigens can stimulate a high proportion of T cells (some V_β families are represented in up to 20% of T cells); they are the most powerful MITOGENS known. Superantigens, in common with normal antigens, induce a proliferative response in their target T cell population when they are associated with class II MHC MOLECULES (*see* MAJOR HISTOCOMPATIBILITY COMPLEX). Superantigens, however, bind outside the groove in the MHC molecule where peptides for presentation to T cells are usually nested and they do not show MHC restriction. An artificial division classifies superantigens as endogenous, such as the self superantigens of the mouse MLS SYSTEM and exogenous, such as the staphylococcal enterotoxins. Mls and other endogenous superantigens are now believed to be encoded by RETROVIRUSES.

supercoiling, supertwisting A conformational state of double-helical DNA, first discovered in circular DNA molecules, in which the double helix is further twisted into a more compact structure (Fig. S34). Naturally occurring circular DNA molecules are negatively supercoiled (i.e. with a left-handed twist) which facilitates unwinding and strand separation of the DNA helix (*see* DNA REPLICATION; TRANSCRIPTION). The number of turns of super-

Fig. S34 Supercoiled DNA.

helix is called the writhing number (*W*). Relaxation of a super-coiled structure is mediated by type I TOPOISOMERASES which make single-strand breaks in the helix. The bacterial enzyme DNA GYRASE, a type II topoisomerase, introduces negative supercoils into DNA. *See also*: LINKING NUMBER; PLASMIDS; TWIST.

superfamily A group of proteins that contain amino-acid sequences with significant similarities to sequences in other proteins, implying that these regions of the proteins have similar folding patterns and are related in evolution. The term superfamily is normally used where the sequences are <50% identical and the term family for sequences with >50% identity. In a protein these segments may be variable in number and be associated with other superfamily domains. The terms are also applied to the genes encoding these proteins. *See*: GROWTH FACTOR RECEPTORS; HAEMOGLOBINS; IMMUNOGLOBULIN SUPERFAMILY; MOLECULAR EVOLUTION; SERINE PROTEINASES.

supergene family A commonly used but inappropriate term, as the genes within the family are not 'super'; the term SUPERFAMILY or MULTIGENE FAMILY should be used.

superhelix, superhelical *See*: SUPERCOILING.

superoxide dismutase (SOD) Ubiquitous enzyme (EC 1.15.1.1) which catalyses the dismutation of superoxide radicals:

$$O_2^{\cdot-} + O_2^{\cdot-} + 2H^+ = O_2 + H_2O_2$$

It is a metalloprotein, the eukaryotic enzymes containing both copper and zinc (Fig. S35), whereas those from most prokaryotes contain maganese or iron. In humans, a familial variant of AMYOTROPHIC LATERAL SCLEROSIS has been linked to defects in cytosolic superoxide dismutase (encoded by the *SOD-1* gene).

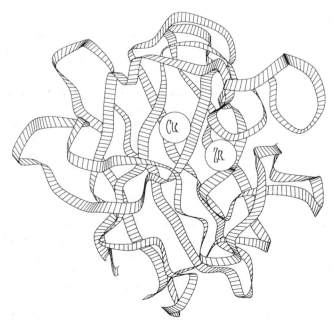

Fig. S35 Superoxide dismutase.

supersecondary structure *See*: PROTEIN STRUCTURE.

suppressor factor (TsF) Factor proposed to be present in cell-free supernatants from lymphocyte cultures which have been reported to suppress an immune response in an antigen-specific manner. Many other mediators (such as PROSTAGLANDINS) will nonspecifically downregulate immune responses; however, the term suppressor factor is generally reserved for antigen-specific factors. Most interactions are reported to show some genetic restriction, either to immunoglobulin heavy-chain type or to certain MHC genes (I-J in the mouse, *see* MAJOR HISTOCOMPATIBILITY COMPLEX). The significance of this is unknown. TsF is believed to operate through a nonspecific third-party cell, called the acceptor cell. To date no putative suppressor factors have been cloned.

suppressor mutation Any mutation that results in reversion of a mutant PHENOTYPE to a WILD-TYPE phenotype. Intragenic suppressor mutations usually consist of a second compensating mutation that restores either the original coding sequence or the correct READING FRAME. Extragenic suppressor mutations occur in genes encoding macromolecules involved in translation (e.g. elongation factors, tRNAs, tRNA synthetases, ribosomal proteins) with the majority being mutant tRNAs (*see* SUPPRESSOR TRNA). This latter class tends to suppress NONSENSE MUTATIONS.

suppressor T cells (T$_S$ cells) An ill-defined class of T CELLS that are presumed to act to downregulate immune responses. Suppressor cell activity has variously been reported as being antigen-specific, IDIOTYPE-specific and even nonspecific. Suppressor activity may even in some cases be explicable as a lack of T cell help. Complex networks of cell interactions have been deduced: thus, suppressor inducer cells act on suppressor T cells to cause the release of SUPPRESSOR FACTORS. These in turn seem to require a third cell type to mediate suppression. The classical suppressor T cell, characterized in cell transfer experiments, carries the CD8 surface marker (*see* CD antigens).

suppressor tRNA A TRANSFER RNA with one or more base changes in its anticodon which makes it recognize a different codon than that corresponding to the amino acid it carries. Nonsense suppressor tRNAs translate one or other of the three TERMINATION CODONS, inserting an amino acid at that position. Nonsense mutations can therefore be suppressed by such mutant tRNAs. The measurement of termination codon readthrough by suppressor tRNAs, and the cellular consequences of the expression of a suppressor tRNA, have been used to identify factors affecting the accuracy of translation. *See also*: GENETIC CODE.

Capecchi, M.R. & Gussin, G.N. (1965) *Science* **149**, 417–422.

SV40 Simian virus 40. *See*: PAPOVAVIRUSES.

Svedberg unit (S) Unit in which the SEDIMENTATION COEFFICIENT of a particle or macromolecule in the ultracentrifuge is generally expressed. It is named after T. Svedberg who developed the technique of ultracentrifugation and its use for separating proteins and estimating their molecular weights. Svedberg's basic equation for the molecular weight of a protein is

$$M = \frac{RTs}{D(1 - \bar{v}\rho)}$$

where M is the molecular weight, D the diffusion coefficient (cm^2 s^{-1}), T the absolute temperature, R the gas constant, $\bar{v}$ the partial specific volume of the protein, and ρ the density of the solvent. s, the sedimentation coefficient, is the rate of sedimentation in a unit field and has the dimensions of time (seconds). It usually lies between 1×10^{-13} and 200×10^{-13} seconds. 10^{-13} seconds is the Svedberg unit. The Svedberg unit is used in the names for the ribosomal subunits (e.g. 30S and 50S in *Escherichia coli*) and the RIBOSOMAL RNAS (e.g. 5S, 16S and 23S rRNA in bacteria).

swainsonine An inhibitor of mannosidases that interferes with glycoprotein processing. *See*: GLYCANS; GLYCOPROTEINS AND GLYCOSYLATION.

Swiss 3T3 cells *See*: 3T3 CELLS.

switch recombination, switch region, switch sequence *See*: CLASS SWITCHING; GENE REARRANGEMENT; IMMUNOGLOBULIN GENES.

switch variant A cell or CELL LINE that has been generated by immunoglobulin heavy-chain CLASS SWITCHING (*see* GENE REARRANGEMENT; IMMUNOGLOBULIN GENES). Although a small number of cell lines have been shown to undergo immunoglobulin class switching at high frequency *in vitro* (a process that can be affected by MITOGENS or INTERLEUKINS), most cell lines only give rise to switch variants at low frequency. Although such variants can arise by gene deletion through switch recombination, many *in vitro* generated switch variants are found to harbour gene deletions in which one end-point of the deletion is not within an S region.

***sym* plasmid** 'Symbiosis' PLASMIDS in some strains of rhizobia that carry genes involved in determining host range and the early steps of infection of legume roots. *See*: NODULATION.

symmetry Symmetry in molecular aggregates is an inevitable consequence of the repeated use of a limited number of specific chemical interactions. Polypeptides show symmetry at the secondary level of PROTEIN STRUCTURE where α-helices and β-sheets are two different arrangements whereby each successive amino-acid residue is positioned to make an identical hydrogen bond with another residue. The double helix of nucleic acids is a further example of symmetry at the secondary structure level (*see* DNA; NUCLEIC ACID STRUCTURE). Oligomeric proteins often have symmetric quaternary structures; most tetrameric proteins have three mutually orthogonal twofold (or DYAD) rotation axes by which one subunit is related in turn to each of the three other subunits (*see e.g.* HAEMOGLOBINS). Viral capsids have units arranged with helical or icosahedral symmetry (*see* ANIMAL VIRUSES; PAPOVAVIRUSES; PICORNAVIRUSES; PLANT VIRUSES); the

limited coding capacity of viral genomes requires multiple copies of one, or a small number of, coat protein molecule(s) to form a capsid around the nucleic acid. The requirement that these molecules assemble correctly implies specific recognition, which in turn, implies symmetry and the repeated use of specific patterns of chemical bonding between the protein molecules. Large spherical viruses with more than 60 coat protein subunits exhibit QUASI-EQUIVALENT bonding with small departures from perfect symmetry. Regular arrays of separate molecules are said to be crystalline — whether in two-dimensional sheets or three-dimensional crystals. Through the study of crystals and their regular planar faces the possible symmetric arrangements were defined.

The complete mathematical description of all the possible symmetry operations in crystals was a major achievement of the 19th century, being then based only on observations of crystal morphology — long before crystal structure determination became possible. A crystal is a regular array of identical 'building blocks' stacked in three dimensions so as to fill completely the crystal volume. The symmetry of such an arrangement is most easily visualized by considering that of the periodic LATTICE first and then the symmetry elements within the repeating units.

A space lattice is an array of points each having exactly the same environment in an identical orientation to every other point. The UNIT CELL of a crystal is a parallelopiped-shaped block containing a complete unit of repeat in each direction; the whole volume of the crystal is built up by regular assembly of unit cells translated by the cell dimension (a, b, c) along each of its three edges (parallel to the x, y, z-coordinate axes respectively) which are related by angles α, β, γ (between the axes yz, zx, xy respectively). In most crystals there are symmetry elements within the unit cell, such that this contains more than one copy of the basic repeated object or ASYMMETRIC UNIT, which is related to all other identical objects within the unit cell by the operation of these symmetry elements and to the contents of all other unit cells by the lattice translation vectors **a**, **b**, **c**.

The presence of one or more rotation axes of symmetry requires definite relationships between the lengths of the unit cell axes and the angles between them. A pattern with an n-fold axis of rotational symmetry is invariant when rotated by $360/n°$ about the axis. The requirement that a crystal unit cell be replicated indefinitely in three dimensions implies that only rotation axes where $n = 2, 3, 4,$ or 6 are consistent with a space-filling crystalline lattice. This concept can be visualized most easily by considering that when tiling a wall with tiles which are regular rectangles, triangles, squares, or hexagons, there are no gaps between the tiles, whereas with regular pentagons, heptagons, or octagons, there are always spaces in between.

The seven crystal systems are classified according to the highest degree of rotational symmetry present; taking these together with the possibilities of having more than one lattice point per unit cell gives the 14 Bravais lattices, whose characteristics are tabulated (Table S10). There are seven primitive lattices and seven centred lattices, which give rise to SYSTEMATIC ABSENCES in the diffraction pattern.

The symmetry elements within an asymmetric unit may be mirror planes of reflection or centres of inversion or axes of

Table S10 The seven crystal systems with unit cells, rotational symmetry, and lattice types

Crystal system	No. of independent unit cell parameters	Relationships between unit cell parameters		Order of rotational symmetry	Lattice types*
		Edges a b c	Angles α β γ		
Triclinic	6	$a \neq b \neq c$	$\alpha \neq \beta \neq \gamma \neq 90°$	1	P
Monoclinic	4	$a \neq b \neq c$	$\alpha = \gamma = 90°$ $\beta > 90°$	2	P C
Orthorhombic	3	$a \neq b \neq c$	$\alpha = \beta = \gamma = 90°$	$2^{\dagger}$	P C F I
Tetragonal	2	$a = b \neq c$	$\alpha = \beta = \gamma = 90°$	4	P I
Trigonal	2	$a = b \neq c$	$\alpha = \beta = 90°$ $\gamma = 120°$	3	P
Hexagonal	2	$a = b \neq c$	$\alpha = \beta = 90°$ $\gamma = 120°$	6	P
Cubic	1	$a = b = c$	$\alpha = \beta = \gamma = 90°$	$3^{\ddagger}$	P F I

* For nomenclature of lattice symbols *see* SYSTEMATIC ABSENCES.

† Orthorhombic system has three orthogonal dyad axes (twofold axes).

‡ Cubic system has four triad axes along all directions of the four body diagonals of a cube.

rotation or a combination of these such as an inverse axis of rotation. Reflection and inversion operations always change the hand of an enantiomorphic object such that biological macromolecules can only crystallize in the absence of these symmetry elements. Thus the only symmetry elements considered further here are the rotation axes. Combining these with the unit cell of the seven crystal systems generates the set of possible crystallographic POINT GROUPS, in which all symmetry elements pass through a single point, or origin, which is thus unaffected by any of their operations. Combining these point group symmetries with the possible translational symmetry elements within the unit cell generates the SPACE GROUPS, which define all possible arrangements of molecules which are consistent with an infinitely repeating lattice. *See also*: ZONE.

International Union of Crystallography (1992) *International Tables for X-ray Crystallography*, Vol. A, Space-group symmetry (Hahn, T., Ed.) (Kluwer, Dordrecht).

Stout, G.H. & Jensen, L.H. (1989) *X-ray Structure Determination: A Practical Guide* (Wiley, New York).

symmetry and asymmetry in animal development In vertebrates, the basic body plan follows a roughly bilateral arrangement, with a plane of symmetry aligned along the head–tail axis. But this bilateral symmetry is not absolute. Most obviously, many organ systems occupy characteristic positions on the left or right of the body (e.g. heart, pancreas, stomach, liver). But there also exist other, more subtle, manifestations of left–right asymmetry, including the laterality of brain functions in higher vertebrates and behavioural features such as left- or right-handedness. In lower animals, where the body plan is bilateral (e.g. arthropods), there is no obviously left–right asymmetry, suggesting that left–right differences were acquired relatively recently in evolution. In agreement with this, the earliest fossils of apparently asymmetric animals belong to the groups Soluta, Cincta, Cornuta, and Mitrata (dating from between the Lower Cambrian and Upper Carboniferous periods, 550–320 million years ago).

During development of higher vertebrate embryos, the earliest visible left–right asymmetry is seen in HENSEN'S NODE, within which the primitive pit and primitive groove display a left-sided slant. However, experimental evidence from the chick embryo suggests that there is covert left–right symmetry even before the appearance of the PRIMITIVE STREAK, because rotations of the HYPOBLAST to the left or to the right produce secondary embryonic axes at different, characteristic frequencies. At later stages of development, the left–right asymmetry becomes more obvious; the head of the embryo always starts to turn to the right (so that the right ear faces in the original dorsal direction); the primitive heart tube shows a characteristic bending to the left at very early stages.

Many invertebrates display symmetries other than bilateral, and some have no symmetry at all. Thus, many echinoderms are radially symmetric and certain molluscs (especially the gastropods) have helical (spiral) symmetry. It is thought that in at least some of these animals, the handedness of the spiral symmetry is inherited in part by EPIGENETIC mechanisms, perhaps by mechanisms analogous to cortical inheritance of the polarity of cilia in protozoans.

To date, only one gene has been found to be involved directly in setting up left–right asymmetry in vertebrates: the *iv* locus of the mouse. In *iv/iv* homozygotes (situs inversus mutant), the probability of any one organ system being on the left or the right is 50%. This implies that the role of the *iv* gene product is to introduce left–right information into an otherwise stochastic system. The gene has not yet been cloned and the molecular bases of left–right asymmetry are still mysterious.

sympathetic nervous system *See*: PERIPHERAL NERVOUS SYSTEM.

symplast The living protoplasmic component of the plant body considered as a continuous system of CYTOPLASTS connected by PLASMODESMATA.

symport MEMBRANE TRANSPORT SYSTEM in which two different substances are cotransported in the same direction.

synapse A specialized junction between two nerve cells or between nerve and muscle cell, across which signals are transmitted. At a chemical synapse the plasma membrane of the axon terminal of the transmitting neuron and that of the receiving cell are separated by a small gap. Chemical NEUROTRANSMITTERS are released from the axon terminal, diffuse across this synaptic cleft, and stimulate specific receptors on the postsynaptic membrane. At electrical synapses an electrical signal carried by ions is transmitted directly from one cell to another via gap junctions (*see* CELL JUNCTIONS). *See also*: SYNAPTIC TRANSMISSION.

synapse elimination Late stage of development of the vertebrate nervous system, when neurons retract certain synaptic terminals while retaining others. This does not involve cell death, and is thought to be an adjustment mechanism ensuring proper quantitative (and possibly qualitative) matching of synaptic connections between pre- and postsynaptic populations of cells. First described at the mammalian neuromuscular junction, it has since been detected in many areas of the central nervous system, for example at the Purkinje cells of the cerebellum. The mechanism is thought to involve competition between presynaptic neurites for survival-promoting molecules released by the postsynaptic cell, but there is no direct evidence for this. *See*: VERTEBRATE NEURAL DEVELOPMENT.

synapsins *See*: ACTIN-BINDING PROTEINS.

synapsis *See*: MEIOSIS.

synaptic cleft Small gap of 20–30 nm between the plasma membranes of the transmitting and receiving neurons at a chemical synapse. *See*: SYNAPTIC TRANSMISSION.

synaptic density Dense area immediately underneath the plasma membranes of presynaptic and postsynaptic terminals at a synapse, and which consist of a meshwork of cytoskeletal elements. *See*: NEURONAL CYTOSKELETON; SYNAPTIC TRANSMISSION.

synaptic folds Infolding of the plasma membranes of a muscle cell and its associated axon terminal at a neuromuscular synapse. *See*: MOTOR END-PLATE.

synaptic plasticity Changes in the efficiency of SYNAPTIC TRANSMISSION evoked by changes in the neuronal environment. This phenomenon is thought to be the molecular mechanism underlying learning and memory and to be involved in the maturation of the nervous system. EXCITATORY AMINO ACID RECEPTORS are believed to be involved in synaptic plasticity. *See also* LONG-TERM POTENTIATION.

Synaptic transmission

SYNAPTIC transmission describes the process by which a neuron communicates a signal to another cell at a defined site — the synapse. This usually occurs by the transmitting neuron releasing a chemical transmitter (NEUROTRANSMITTER) from a terminal release site, but in some cases there may be a direct electrical connection between the sending neuron and the receiving cell — an electrical synapse.

Neurons can influence the behaviour and transmitting proper-ties of other neurons in ways other than direct synaptic transmission, by, for example, the release of substances that act on neurons distant from the site of release. This type of action is known as neuromodulation to distinguish it from synaptic transmission.

Chemical synapses

Chemical transmission in the CENTRAL NERVOUS SYSTEM (CNS) and in the peripheral ganglia of the autonomic nervous system (*see* PERIPHERAL NERVOUS SYSTEM) from neuron to neuron takes place across well-defined synapses (Fig. S36*a*, *c*). In the somatomotor system, chemical transmission from nerve fibres to skeletal muscle cells also occurs across highly specialized synapses — the neuromuscular junctions (Fig. S36*b*); however, neuromuscular junctions in autonomic effectors such as smooth muscle are comparatively unspecialized (Fig. S36*d*).

The cell types to which neurons transmit are more diverse in the autonomic nervous system than in the CNS, passing signals to other nerve cells, smooth muscle and cardiac muscle cells, secretory cells of endocrine and exocrine glands, interstitial cells of Cajal and epithelial cells in the gastrointestinal tract, epithelial cells lining the airways, and even making transient contact with mobile cells such as mast cells. In contrast, in the CNS synapses are restricted to neurons, and in the somatic system the motor neurons synapse only with skeletal muscle fibres.

In its simplest form, a chemical synapse is characterized at the electron microscopic level by an accumulation of SYNAPTIC VESICLES containing neurotransmitter substances close to the presynaptic membrane, which appears thicker and denser than the remaining plasmalemma. This presynaptic region is apposed to the target cell, separated by a synaptic cleft of 20–30 nm. The postsynaptic membrane of the target cell also appears to be thicker and more electron dense than the general plasmalemma (Fig. S37).

The synaptic vesicles have various morphologies that may partially depend on the fixation conditions, but may also indicate their transmitter content. The three most common types of vesicle found in central and peripheral nerves are: small clear vesicles, often circular or elliptical, 20–40 nm in diameter which contain ACETYLCHOLINE; small granular vesicles 50–90 nm in diameter, with an electron-dense core, containing monoamines such as NORADRENALINE (norepinephrine) or 5-HYDROXYTRYPTAMINE (serotonin); and large granular vesicles, 120–150 nm in diameter, which contain NEUROPEPTIDES, but may also be associated with monoamines. Other transmitter substances such as ATP, glutamate, γ-aminobutyrate (GABA) and glycine, proteins such as chromogranin and serotonin-binding protein, and enzymes such as dopamine β-hydroxylase are also contained within certain synaptic vesicles.

As far as the transmission of excitability is concerned, chemical synapses are strictly unidirectional. However, substances do pass from the postsynaptic cell to the presynaptic neuron, and this mechanism has been exploited in tracer studies determining the projection of neurons. A label such as HORSERADISH PEROXIDASE is applied to a discrete area of an organ, and after a suitable period of time and subsequent processing the label can be found

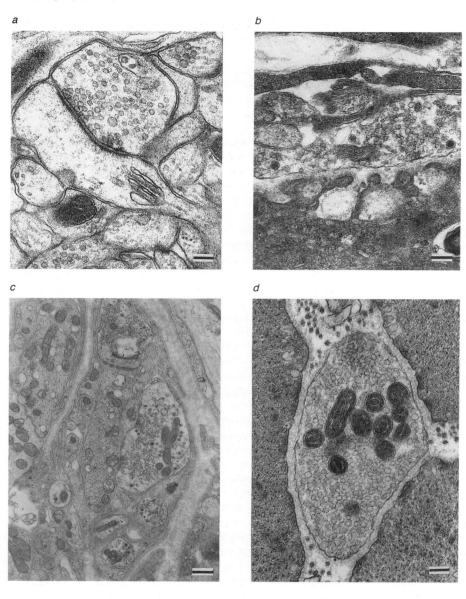

Fig. S36 Electron micrographs of synapses in the central and peripheral nervous systems. *a*, Rat cerebral cortex: the nerve terminal synapsing with a dendritic spine is full of small clear synaptic vesicles. Note the increased electron density of the pre- and postsynaptic membranes. *b*, Rat gastrocnemius neuromuscular junction: the membrane of the muscle (lower region of micrograph) is thrown up into folds at the motor end-plate where the nerve terminal contains numerous small clear synaptic vesicles, a few large opaque vesicles, and a mitochondrion. *c*, Cat inferior mesenteric ganglion: in the centre of this low-power micrograph a dendrite can be seen to make synaptic contacts with two nerve terminals (one above, one on the right), both of which contain numerous small clear, and fewer large opaque vesicles, and some mitochondria. Note the pre- and postsynaptic membrane densities similar to those seen in the central synapse in *a*. *d*, Rat urinary bladder: a terminal varicosity packed with small clear vesicles, and containing several mitochondria, is surrounded by three smooth muscle cells that are separated from one another by a few collagen fibrils. The neuromuscular distance varies from 30 to 60 nm, and neither the neural nor muscular cell membranes show synaptic specializations (the darkenings on the muscle cell membranes are not related to the nerve terminal, but are sites of insertion of actin filaments). Also particularly note the lack of the folding of the muscle membrane that is seen in *b*. Scale bars, 100 nm in *a, b* and *d*, and 200 nm in *c*. *a, b, c* and *d* courtesy of J. Parnavelas, G. Vrbova, D. Tomlinson, and G. Gabella, respectively.

in neurons separated by one or more synapses some distance away, having been transported retrogradely.

Electrical synapses

Gap junctions representing electrical synapses have comparatively fewer morphological features (*see* CELL JUNCTIONS). Here there is a narrow gap of 2–4 nm between the two cells, which have clusters of channel-forming proteins spanning their plasmalemma in register with one another. Thus the cytosol of the two cells is connected, providing a low resistance pathway for the spread of electrical current, and also for the diffusion of low molecular weight molecules. Gap junctions can pass electrical current in both directions, but some have been found that rectify, and allow current to pass preferentially in one direction only. In mammals this type of electrical synapse has been found in the

CNS in regions such as the trigeminal mesencephalic nucleus, vestibular nucleus, and inferior olivary nucleus.

Chemical synaptic transmission

For synaptic transmission to occur, an ACTION POTENTIAL generated in the presynaptic neuron invades the presynaptic nerve terminal. During the DEPOLARIZING phase of the action potential voltage-dependent calcium channels open, allowing Ca^{2+} to enter the terminal and elevate the cytosolic concentration of free Ca^{2+} (*see*: CALCIUM; VOLTAGE-DEPENDENT ION CHANNELS). This induces EXOCYTOSIS of the transmitter from its vesicular store into the synaptic cleft. From here the transmitter diffuses to its receptor on the postsynaptic cell where, once bound, this chemical signal is transduced into a cellular response. The molecular mechanism of signal transduction varies widely and is entirely

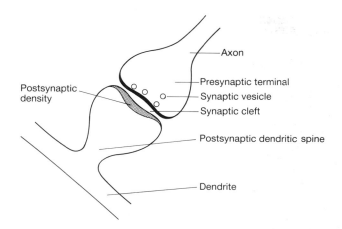

Fig. S37 Diagrammatic representation of a 'typical' chemical synapse between neurons.

dependent on how the receptor is coupled to a SECOND MESSENGER PATHWAY. Most postsynaptic receptors are coupled either directly or indirectly to ion channels, and regulate the ionic permeability of the cell membrane through electrocoupling, causing substantial changes in the MEMBRANE POTENTIAL of the cell when they are activated. Other receptors are pharmacocoupled so that when they are activated, intracellular levels of Ca^{2+} can increase and the cell respond, without any great change in membrane potential [1].

The electrical response that can be recorded in a postsynaptic cell as a result of either spontaneous or evoked stimulation of a presynaptic neuron is called a postsynaptic potential. If the postsynaptic potential change is a depolarization, with the membrane potential becoming more positive, and tending to generate action potentials it is called an excitatory postsynaptic potential (EPSP). On the other hand, a HYPERPOLARIZING change, making the membrane potential more negative, and suppressing excitation is called an inhibitory postsynaptic potential (IPSP). In the case of skeletal muscle the electrical response is termed an end-plate potential (EPP), because the transmitter is released at the highly specialized MOTOR END-PLATE. End-plate potentials are implicitly excitatory responses: the equivalent of IPSPs are not found in skeletal muscle. Miniature end-plate potentials (MEPPs) can be recorded in skeletal muscle, and are small-amplitude transient depolarizations due to the spontaneous release of transmitter (acetylcholine) from motor neuron terminals.

The equivalent responses in smooth muscle cells are called junction potentials rather than synaptic potentials, because the neuromuscular junction is relatively unspecialized, lacking the rigidly fixed relationship found between presynaptic and postsynaptic cells, and lacking apparent specialization of the postjunctional region of the smooth muscle cell membrane. Both excitatory junction potentials (EJPs) and inhibitory junction potentials (IJPs) can be recorded in smooth muscle cells.

Excitatory and inhibitory postsynaptic potentials

Synaptic potentials were recorded in the CNS in the pioneering microelectrode studies of Eccles and his colleagues who recorded directly from the cell bodies of motor neurons in the spinal cord while stimulating the Ia afferent nerve fibres from agonistic and antagonistic skeletal muscles (see [2,3]). The EPSPs are graded according to the stimulus intensity, and if large enough will cause voltage-dependent sodium channels to open in the postsynaptic cell membrane and generate action potentials. Their reversal potential (*see* EQUILIBRIUM POTENTIAL) is around 0 mV owing to a dominant Na^+ influx and concomitant K^+ efflux. Fast IPSPs (f-IPSPs) can be recorded in motor neurons when the Ia afferents of the antagonistic muscle are stimulated. These are short-latency transient hyperpolarizations, graded in amplitude, with a reversal potential of ~ -80 mV, which in this type of neuron is close to both the potassium and chloride equilibrium potentials. During this synaptic potential the dominant ion flux is that of Cl^-; if the intracellular concentration of Cl^- is increased by microinjection, the reversal potential becomes more positive. The neurotransmitter responsible for f-IPSPs in the spinal cord is probably glycine, acting through receptor-operated ion channels (*see* GABA AND GLYCINE RECEPTORS).

These types of synaptic potentials have been well studied in peripheral ganglia, which are generally more accessible for experimental manipulation than neurons in the CNS. In many ganglia, stimulation of the fibre tracts evokes synaptic potentials that can be recorded with a microelectrode. Fast excitatory postsynaptic potentials (f-EPSPs) are transient depolarizations with a latency of the order of tens of milliseconds, a duration usually less than 100 ms, and an amplitude that is graded according to the intensity of the stimulus, of the order of millivolts (Fig. S38). If the level of depolarization during the EPSP is sufficient to activate voltage-dependent sodium channels action potentials will be generated (Fig. S38).

In the vast majority of cases the f-EPSP is mediated via acetylcholine acting on NICOTINIC RECEPTORS which are ligand-gated ion channels. This ion channel has a greater permeability to Ca^{2+} than Na^+, but under physiological conditions with an extracellular concentration of Na^+ ~ 100 times that of Ca^{2+}, more sodium than calcium ions flow through the channel. The channel is also permeable to K^+, but the reversal potential of ~ -10 mV for the ion channel is substantially more positive than the resting membrane potential, and is not closer to the Na^+ equilibrium potential of roughly $+50$ mV because of this simultaneous increase in permeability for K^+ [4–6]. An example of a noncholinergic f-EPSP is found in the rat superior cervical ganglion, where the EPSP is mediated by GABA [7].

A slow hyperpolarization, that is a slow IPSP (s-IPSP), follows the nicotinic EPSP in many ganglia, and this may be followed by a slow depolarization (s-EPSP). Most often IPSPs are studied in the presence of hexamethonium to block the nicotinic EPSP because, with the possible exception of a small population of cells in myenteric ganglia [8], neurons in peripheral ganglia have not been found to have only inhibitory responses. The s-IPSP usually has a latency in the range of 100–400 ms, and is usually mediated via acetylcholine acting on MUSCARINIC RECEPTORS (*see also* G PROTEIN-COUPLED RECEPTORS). Like the EPSP its amplitude depends on the strength of the applied stimulus.

The mechanisms underlying s-IPSPs have not been fully eluci-

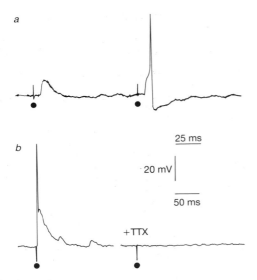

a

25 ms

20 mV

50 ms

b

+TTX

Fig. S38 Synaptic potentials recorded in postganglionic cells of neonatal (14-day-old) rat paratracheal ganglia (an autonomic ganglion). Presynaptic nerves were stimulated by a single pulse of electrical field stimulation (●), evoking after a short latency (*a*) an excitatory postsynaptic potential (EPSP) which, when increased, evoked an EPSP of sufficient amplitude to evoke an action potential. In *b*, stimulation of the presynaptic nerve evoked an EPSP and action potential that after a 5-min incubation with tetrodotoxin (TTX, 1 μM) were abolished, indicating that the response is indeed synaptic. The small transients coincident with the stimuli are stimulus artefacts. Figure courtesy of Fiona M. Reekie.

dated, and are often obscured by the difficulty of separating them from s-EPSPs that may arise almost simultaneously; however, in many ganglia the slow hyperpolarization is associated with a decreased membrane resistance due to an increased K^+ conductance. In this case the muscarinic receptor is coupled to a potassium channel which when open allows K^+ to leave the cells since the potassium equilibrium potential (~ -90 mV) is more negative than the resting membrane potential (~ -65 mV). In contrast, in the rat superior cervical ganglion the slow hyperpolarization is associated with an increase in membrane resistance, and the IPSP reverses close to the chloride equilibrium potential, and therefore involves a closing of chloride channels. In peripheral neurons and smooth muscle cells the chloride equilibrium potential is more positive (~ -25 mV) than the resting membrane potential: closing chloride channels will prevent Cl^- leaving the cell, and thus cause a relative hyperpolarization. The net result of opening K^+ channels or closing Cl^- channels is broadly the same in this type of cell, resulting in hyperpolarization and decreased excitability.

Slow EPSPs recorded in mammalian and amphibian peripheral ganglia are usually mediated via muscarinic receptors, but in some bullfrog ganglionic cells and in the guinea-pig inferior mesenteric ganglion, the neuropeptides luteinizing hormone releasing hormone (LHRH) and substance P, respectively, have been proposed as mediators of s-EPSPs [9,10]. The amplitude of s-EPSPs, is dependent upon the strength of the applied stimulus, and decreases when the membrane potential is hyperpolarized, and increases when it is depolarized. There is also an increase in

the membrane resistance, so the overall picture is consistent with there being a suppression of K^+ conductance. The tonic potassium current that is suppressed is called the M-current [11], and is voltage dependent. Inhibition of the M-current results in increased excitability of the cell. The muscarinic receptor that mediates this inhibition and the s-EPSP is coupled to a G-protein system that regulates the potassium ion channel [12].

End-plate potentials and miniature end-plate potentials

The end-plate potential of skeletal muscle is in many ways similar to a fast EPSP. Evoked EPPs are transient depolarizations with a short (milliseconds) latency of onset, graded amplitude that can caused action potential discharge if the threshold of activation of voltage-dependent sodium channels is reached, and a duration of the order of tens of milliseconds. They are due to the release of acetylcholine from motor neuron terminals at the motor end-plate acting on postsynaptic nicotinic receptors that cause an increase in Na^+ and K^+ conductance. The EPP reverses at around -15 mV because the increase in Na^+ conductance is much greater than that of K^+ (if there were no simultaneous increase in K^+ conductance the reversal potential would be nearer the sodium equilibrium potential of $+50$ mV).

Miniature end-plate potentials are recorded with microelectrodes inserted into the muscle cell close to the end-plate, and are observed as low-amplitude (~ 0.5 mV as opposed to several mV) depolarizations with a similar time-course to EPPs. They are due to the spontaneous release of a small number of quanta of acetylcholine from the motor end-plate [13], whereas the EPP is due to roughly a hundred quanta released together as a result of an action potential invading the nerve terminal.

Excitatory and inhibitory junction potentials

Both these types of synaptic potentials are recorded in smooth muscles of autonomic neuroeffectors. In many types of blood vessel, the vas deferens and urinary bladder EJPs are purinergic and are due to the release of ATP from the prejunctional neuron acting on postjunctional P_{2X} purinoceptors to evoke an increase in cation conductance. Cholinergic EJPs are typically found in non-sphincteric muscle of the gastrointestinal tract, and adrenergic EJPs are usually found in sphincteric muscles of the gastrointestinal tract, and accessory muscles such as the rat anococcygeus [14–16]. In the vas deferens and blood vessels, where noradrenaline is a cotransmitter with ATP in sympathetic nerves, the adrenergic response is a slow depolarization or a pharmacocoupled response rather than a fast junction potential [16,17]. Similarly, in the urinary bladder, acetylcholine causes a relatively slow depolarization rather than the junction potential that is due to ATP.

The transient depolarization of EJPs in the gut has a longer time course, often 0.5–1.0 s, and a longer latency, usually above 50 ms, than is seen for EPSPs. The EJP amplitude is graded according to the strength of the applied stimulus, and as for EPSPs, if the depolarization is great enough, action potentials will be evoked, this time due to the activation of voltage-dependent Ca^{2+} channels in the smooth muscle membrane. However, in the vas deferens and some blood vessels the latency of the EJP is very short, partly because the ion channel concerned is gated by the P_2 purinoceptor [5,15]. The reversal potential of EJPs is always more positive than the resting membrane potential, involving increases in Na^+ and Ca^{2+} conductance, or in the case of the rat anococcygeus muscle, Cl^- conductance.

Inhibitory junction potentials evoked by a single stimulus applied to the prejunctional neuron are transient hyperpolarizations, with a latency ~100 ms, an amplitude up to ~30 mV, but usually ~10 mV, and a duration from 0.8–4.0 s, depending on the tissue in which they are recorded. As for all other types of synaptic potential their amplitude is graded according to the stimulus strength. The transmitter for IJPs is largely unknown, but in some places such as the guinea-pig taenia coli it is likely to be ATP or a related purine. In all cases where an IJP can be evoked by a single stimulus the response is due to a calcium-dependent increase in potassium conductance. Reversal potentials for IJPs are close to the potassium equilibrium potential (~ -90 mV), and they are inhibited by potassium-channel blockers [14,15].

Vasoactive intestinal polypeptide (VIP) is a potent smooth muscle relaxant in gastrointestinal smooth muscle, but is unlikely to be responsible for the IJP because it has no electrical activity comparable to that of the IJP. The free radical of NITRIC OXIDE acts as an inhibitory neuromuscular transmitter in some smooth muscles, particularly sphincteric and circular muscle [18]; its effects are mediated via activation of cytosolic GUANYLATE CYCLASE. The time course of nitrergic transmission is slow compared with the single IJP, so nitric oxide may mediate relatively late slow hyperpolarizations, but does not mediate the faster hyperpolarization that involves the opening of calcium-dependent K^+ channels.

C.H.V. HOYLE

See also: LONG-TERM POTENTIATION; NEURON; NEURONAL CYTOSKELETON.

1 Van Breman, C. & Saida, K. (1989) Cellular mechanisms regulating $[Ca^{2+}]_i$ in smooth muscle. *Annu. Rev. Physiol.* **51**, 315–329.
2 Coombs, J.S. et al. (1955) The specific ionic conductances and the ionic movements across the motoneuronal membrane that produce the inhibitory postsynaptic potential. *J. Physiol., Lond.* **130**, 326–373.
3 Coombs, J.S. et al. (1955) Excitatory synaptic action in motoneurons. *J. Physiol., Lond.* **130**, 374–395.
4 Karczmar, A.G. et al. (Eds) (1986) *Autonomic and Enteric Ganglia: Transmission and its Pharmacology* (Plenum, New York).
5 Benham, C.D. (1992) Signal transduction mechanisms. In *The Autonomic Nervous System, Vol. I: Autonomic Neuroeffector Mechanisms* (Burnstock, G. & Hoyle, C.H.V., Eds) 215–256 (Harwood Academic Publishers, Chur).
6 Buckley, N.J. & Caulfield, M. (1992) Transmission: acetylcholine. In *The Autonomic Nervous System, Vol. I: Autonomic Neuroeffector Mechanisms* (Burnstock, G. & Hoyle, C.H.V., Eds) 257–322 (Harwood Academic Publishers, Chur).
7 Eugéne, D. (1987) Fast non-cholinergic depolarizing postsynaptic potentials in neurons of rat superior cervical ganglia. *Neurosci. Lett.* **78**, 51–56.
8 Hodgkiss, J.P. & Lees, G.M. (1984) Slow intracellular potentials in AH-neurons of the myenteric plexus evoked by repetitive activation of synaptic inputs. *Neuroscience* **11**, 255–261.
9 Dockray, G.J. (1992) Transmission: peptides. In *The Autonomic Nervous System, Vol. I: Autonomic Neuroeffector Mechanisms* (Burnstock, G. & Hoyle, C.H.V., Eds) 409–464 (Harwood Academic Publishers, Chur).
10 Morris, J.L. & Gibbins, I.L. (1992) Co-transmission and neuromodulation. In *The Autonomic Nervous System, Vol. I: Autonomic Neuroeffector Mecha-nisms* (Burnstock, G. & Hoyle, C.H.V., Eds) 33–119 (Harwood Academic Publishers, Chur).
11 Brown, D.A. & Adams, P.R. (1980) Muscarinic suppression of a novel voltage-sensitive K^+-current in a vertebrate neurone. *Nature*, **283**, 673–676.
12 Brown, D.A. (1990) G-proteins and potassium currents in neurons. *Annu. Rev. Physiol.* **52**, 215–242.
13 Del Castillo, J. & Katz, B. (1954) Quantal components of the endplate potential. *J. Physiol., Lond.* **124**, 560–73.
14 Hoyle, C.H.V. & Burnstock, G. (1989) Neuromuscular transmission in the gastrointestinal tract. In *Handbook of Physiology, Section 6: The Gastrointestinal System, Vol I: Motility and Circulation* (Wood, J.D., Ed.) 435–464 (American Physiological Society, Bethesda, MD).
15 Brock, J.A. & Cunnane, T.C. (1992) Electrophysiology of neuroeffector transmission in smooth muscle. In *The Autonomic Nervous System, Vol. I: Autonomic Neuroeffector Mechanisms* (Burnstock, G. & Hoyle, C.H.V., Eds) 121–214 (Harwood Academic Publishers, Chur).
16 Hoyle, C.H.V. (1992) Transmission: purines. In *The Autonomic Nervous System, Vol. I: Autonomic Neuroeffector Mechanisms* (Burnstock, G. & Hoyle, C.H.V., Eds) 367–407 (Harwood Academic Publishers, Chur).
17 Hoyle, C.H.V. & Burnstock, G. (1991) ATP receptors and their physiological roles. In *Adenosine in the Nervous System* (Stone, T.W., Ed.) 43–76 (Academic Press, London).
18 Sanders, K.M. & Ward, S. (1992) Nitric oxide as a mediator of nonadrenergic noncholinergic neurotransmission. *Am. J. Physiol.* **262**, G379–G392.

synaptic vesicles Small vesicles containing NEUROTRANSMITTER which are present at axon terminals. *See*: EXOCYTOSIS; SYNAPTIC TRANSMISSION.

synaptobrevin Protein present on the SYNAPTIC VESICLE membrane which acts as a SNAP RECEPTOR.

synaptosome A preparation consisting of pinched-off axon terminals and the associated postsynaptic membrane, formed when brain tissue is homogenized. *See*: SYNAPTIC TRANSMISSION.

synaptotagmin Protein present in SYNAPTIC VESICLE membranes in neuron axonal terminals. It is known to bind Ca^{2+}, and may be the calcium sensor for EXOCYTOSIS.

synchrotron radiation (SR) High-energy physics research into subatomic structure led to the construction of particle accelerators in which synchrotron radiation was an unwanted product, in that its use was parasitic on machines optimized for high-energy physics — but much biological data was obtained from these early machines (at sites such as Hamburg, Stanford, and Cornell). However, synchrotron radiation soon became recognized as an important research tool in biology, chemistry, and physics, and dedicated synchrotron sources were built with machine parameters optimized for this use. The first such dedicated high-energy source was the Daresbury Synchrotron Radiation Source (SRS) which opened in 1981. Since then, the biological applications of synchrotron radiation have increasingly become a major part of the scientific programme of this and other synchrotron radiation sources. All modern synchrotron radiation sources are now storage rings designed to give continuous beams with stable positions and radiation of specific properties produced by use of magnetic insertion devices such as wigglers and undulators.

Synchrotron radiation provides an extremely powerful source of electromagnetic radiation over a wide range of wavelengths from X-rays to infrared. Its main advantages for studying macro-molecules are: a high flux (photons per second) that is well collimated, giving an almost parallel beam of high brightness (flux per unit solid angle) from a small source, which gives high brilliance (brightness per unit area); the wavelength is tunable over a wide range and for broad or very fine bandwidth. Other properties of the radiation are its polarization, defined time structure, and calculable spectra.

Synchrotron radiation has proved invaluable in many X-RAY CRYSTALLOGRAPHY applications for crystals which are weakly diffracting, have large unit cells, or suffer severe RADIATION DAMAGE. Rapid data collection has enabled time-resolved studies of enzymatic activity.

SMALL-ANGLE SCATTERING studies of particles in solution and FIBRE DIFFRACTION analyses of MUSCLE fibres have been other successful applications to DIFFRACTION. In spectroscopy, the use of synchrotron radiation has opened up the whole area of EXAFS for studying metalloproteins.

Helliwell, J.R. (1992) *Macromolecular Crystallography with Synchrotron Radiation* (Cambridge University Press).

syncytial blastoderm Stage in insect embryogenesis when the cleavage nuclei lie around the periphery of the egg but prior to their separation by cell membranes. *See*: DROSOPHILA DEVELOPMENT.

syndromes *See*: ANGELMAN SYNDROME; ACQUIRED IMMUNE DEFICIENCY SYNDROME; BECKWITH–WIEDEMANN SYNDROME; BLOOM'S SYNDROME; COCKAYNE'S SYNDROME; CONTIGUOUS GENE SYNDROMES; CRIGLER–NAJJAR SYNDROME; DIGEORGE SYNDROME; DOWN'S SYNDROME; EDWARD'S SYNDROME; EHLERS–DANLOS SYNDROME; FRAGILE X SYNDROME; HUNTER SYNDROME; KALLMANN SYNDROME; KARTAGENER'S SYNDROME; KLINEFELTER SYNDROME; LESCH–NYHAN SYNDROME; LI–FRAUMENI SYNDROME; MARFAN SYNDROME; MARTIN BELL SYNDROME; MCLEOD SYNDROME; PRADER–WILLI SYNDROME; TURNER SYNDROME; WAGR SYNDROME; WISKOTT–ALDRICH SYNDROME.

Synechococcus Unicellular rod-shaped CYANOBACTERIA from Section I of the classification of Rippka et al. The genus includes those designated *Anacystis nidulans*.

Rippka, R. et al. (1979) *J. Gen. Microbiol.* **111**, 1–61.

Synechocystis Unicellular CYANOBACTERIA from Section I of the classification of Rippka et al. They differ from *Synechococcus* in the pattern of cell division and cell shape, being coccoid rather than rod-shaped. *Synechocystis* PCC6803 is widely used in genetic analysis, being transformable and able to grow photoheterotrophically in the absence of a functional PHOTOSYSTEM II.

Rippka, R. et al. (1979) *J. Gen. Microbiol.* **111**, 1–61.

synemin An INTERMEDIATE FILAMENT protein.

syntaxin A SNAP RECEPTOR protein on the presynaptic membrane of NEURONS.

syntenic genes Genes located on the same chromosome.

synteny Correspondence in gene order between the chromosomes of different species.

synthetases An alternative name for enzymes of EC class 6, the LIGASES.

synthetic lethal A chromosome carrying a LETHAL MUTATION, which is derived from a normal chromosome by CROSSING-OVER.

Synthetic Theory of Evolution At the beginning of the 20th century biologists were divided by a conceptual gap. Some were interested in exploring the programme of evolutionary biology established in detail by Charles Darwin (*see* DARWINIAN EVOLUTION), while others were following the theories of particulate inheritance established by Gregor Mendel (*see* MENDELIAN INHERITANCE). Each group championed different, incompatible, evolutionary theories. The gap was bridged in the 1930s and 1940s with the development of the 'Synthetic Theory of Evolution' or 'Neodarwinism', so-called because it brought together Darwinian NATURAL SELECTION and Mendelian genetics in complete agreement. This came about in part through the development of theories in POLYGENIC INHERITANCE and POPULATION GENETICS, and in part through the realization that evolutionary explanations (ultimate causation) and genetic explanations (proximate causation) were complementary, and not mutually exclusive.

Mayr, E. & Provine, W. B. (1980) *The Evolutionary Synthesis: Perspectives on the Unification of Biology* (Harvard University Press, London).

syringomycin, syringotoxin, syringostatin Toxins produced by *Pseudomonas syringae* strains. *See*: PLANT PATHOLOGY.

systematic absences Classes of DIFFRACTION intensities from crystals which are systematically absent, always having zero intensity as a direct consequence of the presence of additional lattice points in the crystal LATTICE or translational SYMMETRY elements (such as screw axes of rotation) associated with each lattice point.

A primitive lattice has no general systematic absences; centred lattices have one or more additional lattice points each with an identical environment to that at the UNIT CELL origin. The additional lattice points may be at the centre of the unit cell (body-centred lattice) or of one pair of its principal faces (C face-centred lattice) or all three pairs of its principal faces (all face-centred lattice). With centred lattices, half of the expected diffraction intensities are always absent (and three-quarters are absent for the all face-centred lattices). The extra one (or three) lattice point(s) reduce(s) the volume of the ASYMMETRIC UNIT to one half (or one quarter) of that expected from the dimensions and symmetry elements of the unit cell — thereby reducing the magnitude of the task of structure determination (Table S11).

A screw axis of rotational symmetry generates systematic absences only along the direction parallel to the axis. For an axis of n-fold rotational symmetry (where n can be 2, 3, 4, 6) the screw translation is a fraction m/n of the unit cell repeat in that direction (where m is any integer from 1 to $(n-1)$; the cases $m = 0$ and $m = n$

Table S11 Systematic absences from lattices

Lattice type	Lattice symbol	For every atom at position (x, y, z) expressed in fractional coordinates there is an identical atom at position:	Diffraction intensity *h k l* is systematically absent unless indices *h k l* satisfy conditions:
Primitive	P	Not applicable	No condition
Body-centred	I	$(x + 1/2, y + 1/2, z + 1/2)$	$h + k + l = 2n$
A face-centred*	A*	$(x, y + 1/2, z + 1/2)$	$k + l = 2n$
B face-centred*	B*	$(x + 1/2, y, z + 1/2)$	$l + h = 2n$
C face-centred	C	$(x + 1/2, y + 1/2, z)$	$h + k = 2n$
All face-centred	F	$(x, y + 1/2, z + 1/2)$ and $(x + 1/2, y, z + 1/2)$ and $(x + 1/2, y + 1/2, z)$	$k + l = 2n$ and $l + h = 2n$ and $h + k = 2n$ — i.e. *hkl* all even or all odd.

*These are nonconventional settings rarely found in the literature but are useful to illustrate the symmetric nature of the positions of extra atoms and their resulting systematic absences; with these it is clearer that an F-lattice is essentially a P-lattice with the three extra lattice points of lattice types A and B and C.

are identical, both corresponding to a pure rotation axis with zero translation component (Table S12).

International Union of Crystallography (1992) *International Tables for X-ray Crystallography, Vol. A* (Reidel, Dordrecht).

Stout, G.H. & Jensen, L.H. (1989) *X-ray Determination: A Practical Guide* (Wiley, New York).

systemic lupus erythematosis (SLE) A chronic, remitting, AUTO-IMMUNE disease characterized by glomerulonephritis, arthritis, and skin rashes, due to deposition of IMMUNE COMPLEXES. It affects women predominantly (10 : 1 male) with an overall incidence of one in 700 (women aged between 20 and 60). Autoantibodies to DNA, nucleoproteins, SMALL NUCLEAR RIBONU-CLEOPROTEINS, and other antigens are present. The presence of antibodies against native double-stranded DNA is diagnostic.

systemin Peptide that induces a systemic wound response when applied to a plant stem. *See*: PLANT WOUND RESPONSES.

Table S12 Axial systematic absences from screw axes of rotation

Axis type	Axis symbol		With a screw axis parallel to the z-axis, for each angle of rotation there is a translation expressed as a fraction of the cell repeat **c**		Diffraction intensity *0 0 l* is systematically absent unless:
Screw dyad	2_1		+ 1/2		$l = 2n$
Screw triad	3_1	3_2	+ 1/3	+ 2/3	$l = 3n$
Screw tetrad	4_1	4_3	+ 1/4	+ 3/4	$l = 4n$
	4_2		+ 2/4		$l = 2n$
Screw hexad	6_1	6_5	+ 1/6	+ 5/6	$l = 6n$
	6_2	6_4	+ 2/6	+ 4/6	$l = 3n$
	6_3		+ 3/6		$l = 2n$

For a screw axis parallel to the x or y axis, the reflection condition applies to reflections of classes *h 0 0* and *0 k 0* respectively and the test is for *h* or *k* to be a certain multiple of integer *n*.

T

2,4,5-T 2,4,5-trichlorophenoxyacetic acid (Fig. T1). M_r 255.5. A highly active synthetic AUXIN, which acts as a herbicide at high concentration (*see also* 2,4-D). It achieved some notoriety after its use as a defoliant in Vietnam. Its use in agriculture has been discontinued because of possible contamination with 2,3,7,8-tetrachlorodibenzo-*p*-dioxin, a highly toxic by-product of its manufacture.

Hamner, C.L. & Tukey, H.B. (1944) *Science* **100**, 154–155.

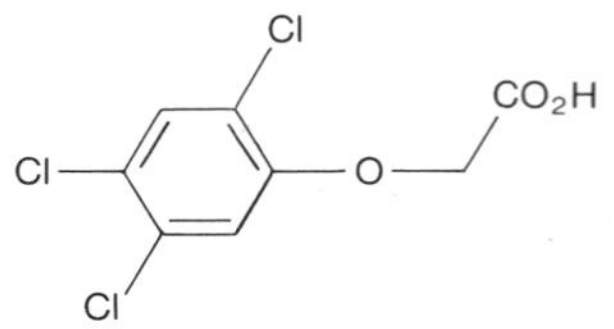

Fig. T1 Structure of 2,4,5-T (2,4,5-trichlorophenoxyacetic acid).

T (1) The pyrimidine BASE thymine.
(2) The single-letter abbreviation for the AMINO ACID threonine.

T The mouse mutation BRACHYURY.

T_m The melting or TRANSITION TEMPERATURE of double-stranded DNA or RNA. *See:* HYBRIDIZATION.

T antigen (1) Protein produced by PAPOVAVIRUSES and involved in transformation.
(2) Precursor to the M and N blood group antigens. *See:* MN BLOOD GROUP.

T cell Class of LYMPHOCYTE which mediates immune recognition and effects CELL-MEDIATED IMMUNE RESPONSES. In general, T cells recognize protein antigens after they have been processed into peptide fragments and become associated with MHC MOLECULES (*see* ANTIGEN PRESENTATION; ANTIGEN PROCESSING; MAJOR HISTOCOMPATIBILITY COMPLEX) on the surfaces of antigen-presenting or target cells. T cells expressing an αβ T CELL RECEPTOR constitute the majority of the peripheral T cells circulating in the blood; those expressing γδ receptors are found commonly in epithelial tissues, and their precise role is unknown. All T cells express their antigen-specific receptor at the cell surface as a complex with the CD3 antigen (*see* T CELL RECEPTOR). T cells can be classified by their function: regulatory T cells include both HELPER T CELLS and SUPPRESSOR T CELLS, whereas the effector population consists of CYTOTOXIC T CELLS. *See also:* B CELL; T CELL DEVELOPMENT; T CELL RECEPTOR GENES.

T cell activation In the course of a normal immune response, the binding of ligand (normally the antigenic complex of peptide and MHC MOLECULE) to the T CELL RECEPTOR complex (TCR–CD3) on the surface of a T CELL initiates intracellular changes, usually leading to the proliferation of the T cell concerned and the production of LYMPHOKINES. The transduction of the signal through the cell membrane requires the presence of TCR–CD3 and the associated co-receptors CD4 or CD8. Activation entails the stimulation of at least two second-messenger generating pathways: (1) two PROTEIN KINASES (pp56^{lck} and pp59^{fyn}) associated with CD4 and TCR polypeptides respectively, phosphorylate the η chain of the CD3 complex (*see* T CELL RECEPTORS); (2) they also phosphorylate and activate PHOSPHOLIPASE C, which then generates the second messengers diacylglycerol and inositol 1,4,5-trisphosphate (*see* SECOND MESSENGER PATHWAYS). These processes lead to mobilization of intracellular Ca^{2+}, the influx of extracellular Ca^{2+} (*see* CALCIUM), and activation of PROTEIN KINASE C.

T cell development

THE process of T cell development or differentiation describes the maturational events that begin with pluripotential bone marrow stem cells and end with a diverse population of specialized functional T CELLS. All T cells are derived from a common precursor cell pool and they recognize and respond to antigens using the same system of receptors (*see* T CELL RECEPTORS; T CELL RECEPTOR GENES). T cell development and differentiation occurs in the THYMUS. Mature T cells (*see* CYTOTOXIC T CELL; HELPER T CELL; SUPPRESSOR T CELL) leaving the thymus bear antigen receptors (T cell receptors, TCR) that recognize antigen in the form of peptide fragments complexed with MHC MOLECULES — the products of the MAJOR HISTOCOMPATIBILITY COMPLEX (MHC) (*see* ANTIGEN PROCESSING AND PRESENTATION). In the thymus these pivotal cells of the immune system are both positively selected for self MHC recognition and negatively selected for receptors that are reactive to other self antigens. This paradoxical requirement, sometimes called thymic education, ensures that mature T cells express an effective, self-tolerant and diverse repertoire of receptors that can coordinate immune function to the benefit of the individual. T cell development is extensively reviewed with full references in [1].

The great majority of circulating human T cells express receptors of the αβ type but there is a distinct population of T cells which express a different form of the T cell receptor, the γδ receptor (*see* T CELL RECEPTOR). Although γδ T cells seem normally to differentiate in the thymus like αβ T cells, not all of them

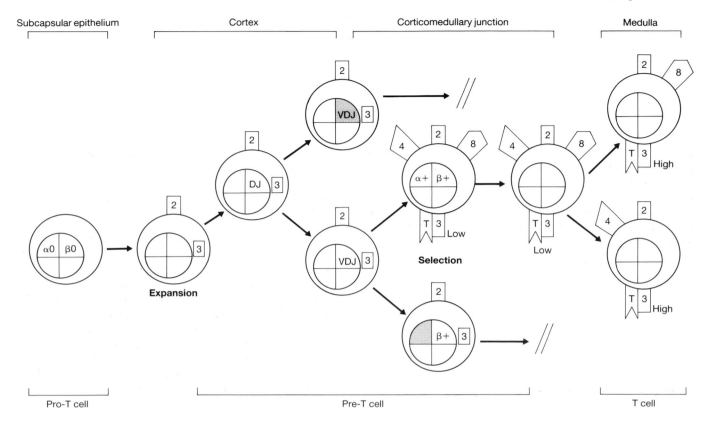

Subcapsular epithelium   **Cortex**   **Corticomedullary junction**   **Medulla**

Expansion

Selection

Pro-T cell   Pre-T cell   T cell

Fig. T2 A generalized scheme for T cell differentiation in the thymus. Pro-T cells undergo massive proliferation in the cortex of the thymus and begin the process of T cell receptor gene rearrangement. Shading indicates nonproductive rearrangement which leads to cell death. The expression and location of various CD markers (arabic numerals) is indicated as well as the level of expression of the T cell receptor at the cell surface. As differentiation proceeds, the cells appear concentrated at the locations indicated along the top of the figure.

have an absolute requirement for intrathymic development and they are not restricted to recognition of antigen in conjunction with MHC molecules. The summary of T cell development that follows refers largely to αβ T cells.

Within the thymus, four main structural regions are readily distinguishable (see Fig. T2). The subcapsular region, which is the most exterior, is the arrival point for bone marrow-derived stem cells. These cells move into the cortex where they go through a stage of rapid proliferation. The corticomedullary junction is the most likely physical site for the process of selection. Here, the thymocytes (as the developing T cells are called) can be seen partly internalized by so-called thymic nurse cells. The thymic medulla contains the most mature thymocytes, which are phenotypically indistinguishable from mature T cells.

Stages in T cell development

Bone marrow stem cells arrive at the foetal thymus in waves of colonization. In the mouse, the first wave occurs at about day 11

of foetal development. This early colonization is vital to the proper functioning of the mature immune system. Thus, newborn mice in which the thymus is removed (thymectomy) will have no mature αβ PERIPHERAL T CELLS throughout their adult life. In contrast, thymectomy in the adult has no great effect on immune function.

The cells that enter the thymus have no distinctive characteristics that mark them apart from prospective B lymphocytes (B CELLS); they do not carry the αβ T cell surface marker proteins CD4 or CD8 and have therefore been referred to as double negative cells. Because these cells also do not express the TCR–CD3 functional receptor complex (*see* T CELL RECEPTOR) they are sometimes called triple negative cells. It seems likely that the thymic environment provides the signals for commitment to the T cell lineage. Although thymic hormones have been used to substitute for the thymus at various parts of the differentiation pathway, no *in vitro* substitute for the organ has yet been developed.

The immature thymocytes, or pro-T cells, give rise to all the subsets of the T cell lineage and account for about 5% of thymocytes in the neonate. Colonization of the thymus is followed by a period of rapid expansion in the thymocyte population. The CD2 antigen (*see* CD ANTIGENS) is one of the first markers that these early thymocytes express that distinguishes them as cells of the T lineage: messenger RNA for CD2 can be detected in the foetal thymus as early as day 13 of gestation in the mouse. As the cells mature they go on to express the accessory (co-receptor) molecules for MHC–T cell receptor binding — CD4 and CD8 —

and are known as double positive cells. Triple positive cells are also found, which express low levels of TCR–CD3 in addition to CD4 and CD8.

Receptor gene rearrangement

The key event in the process of T cell differentiation is the rearrangement of the T CELL RECEPTOR GENES and their expression as a functional TCR–CD3 complex at the cell surface (*see also* GENE REARRANGEMENT). In the foetal mouse thymus, this begins on day 14 of gestation, when the locus encoding the β chain of the receptor rearranges. T cell receptor gene rearrangement is complete by day 17. Unlike B cells, T cells do not undergo SOMATIC MUTATION in the receptor genes in the course of development (*see* B CELL DEVELOPMENT). This means that the primary or potential T cell repertoire is completely formed by day 17, and explains, at least in part, why adult thymectomy is not harmful to the individual.

Positive and negative selection

The diversity of the T cell repertoire (*see* GENERATION OF DIVERSITY; T CELL RECEPTOR; T CELL RECEPTOR GENES) arises from the variety of combinations of variable (V), J and D (β chain only) gene elements together with the random addition and deletion of nucleotides at the junctions (*see* N-REGION DIVERSITY). Because the receptors function as recognition elements for antigens and there is a selective advantage in being able to recognize as many foreign antigens as possible, it is clear that there will be pressure to make the repertoire as diverse as possible. However, it is equally clear that the emergence of self-reactive clones of T cells is a potentially lethal event. The randomly generated primary repertoire must be selected to remove thymocytes which have failed to rearrange their T cell receptor genes productively, as well as those with autoreactive and potentially harmful receptor specificities.

Positive selection

Two major events shape the T cell repertoire of antigen-binding receptors. The first round of T cell culling is called positive selection; positive selection eliminates all those cells which cannot bind to class I or class II MHC products within the thymus and ensures that all mature T cells will be able to interact with self MHC. Many workers have analysed the mechanisms which underlie MHC restriction. Studies with RADIATION CHIMAERAS showed that it was the MHC type of the thymus in which the T cells were developing and not that expressed by the cells themselves, that imprinted restriction. Thus, the most likely candidates for cells involved in positive selection are the cells of the radio-resistant thymic epithelium.

Positive selection has recently been clearly demonstrated using mice transgenic for already rearranged α and β T cell receptor genes [2]. Because the TRANSGENES are already rearranged, rearrangement of the endogenous receptor genes is largely suppressed and the majority of thymocytes express this single receptor combination. Large numbers of mature T cells can be found in transgenic mice with the same MHC HAPLOTYPE as that from which the original T cell receptor genes were cloned. In contrast, transgenic animals with different MHC haplotypes fail to positively select and as a consequence have few mature T cells.

The class of the MHC molecule to which the original T cell receptor was restricted has a powerful influence on the phenotype of the mature T cell population in the transgenic mice. Thus, if the original T cell clone from which the receptor was isolated is CD4$^+$ and MHC class II restricted, then the T cell population of the transgenic animal is skewed towards expression of CD4.

Positive selection occurs in the cortex of the thymus and acts on immature triple positive cells. The selection process involves at least three molecules: T cell receptor, MHC molecules and a co-receptor molecule — either CD4 or CD8. Mature T cells bearing αβ receptors restricted to interactions with class I MHC molecules carry the co-receptor molecule CD8 whereas those with class II-restricted receptors carry the co-receptor molecule CD4. These mature cells are therefore sometimes known as single positive T cells. The co-receptors CD8 and CD4 bind to class I and class II MHC molecules respectively to stabilize the cross-linking of the T cell and the cell bearing the MHC–antigen complex. CD8 cells are mainly destined to become cytotoxic T cells, which can recognize antigen in conjunction with ubiquitous class I MHC molecules on any type of cell, whereas CD4 cells will become helper T cells, whose interactions are restricted to cells bearing class II MHC molecules — mainly professional antigen-presenting cells and other T cells.

Negative selection

The population of thymocytes which remains after positive selection will almost certainly contain cells with autoreactive receptor combinations. Negative selection (also called clonal deletion) is an important pathway that prevents these cells from reaching the periphery with pathogenic consequences. The best evidence for negative selection comes from studies of V$_\beta$ gene usage and SUPERANTIGENS. Superantigens bind to families of T cells that bear receptors encoded by particular V$_\beta$ genes. If the superantigen is present during T cell development (i.e. when it is a self-antigen), all V$_\beta$ receptors are deleted from the mature repertoire. Deletion occurs at the triple positive stage and most probably occurs after engagement with ligands expressed on thymus cortical epithelium. The outcome of the overall selective process is a population of mature, self-tolerant T cells.

Resolving the paradox

The contradictions inherent in positive and negative selection have provoked much debate. Two models have emerged to explain the apparent paradox. The altered ligand hypothesis suggests that MHC molecules within the thymus are essentially different from those elsewhere, perhaps because they bind a different set of peptides. These peptides would allow the delivery of unique signals to the emerging T cell population. The second main hypothesis, the affinity hypothesis, proposes that the cells that form the functional repertoire are selected to fall within a window of receptor affinities. Those cells with low or no affinity for MHC molecules are disposed of (positive selection), and high affinity clones are subsquently deleted (negative selection), leaving only those cells that interact weakly with self.

R.A. LAKE

1 *Immunol. Today* (1991) **12**(2), 65–92.
2 Kiselow, P. et al. (1988) Positive selection of antigen-specific T cells in thymus by restricting MHC molecules. *Nature* **335**, 730–733.

T-cell growth factor (TCGF) Interleukin-2. *See*: IL-2.

T cell receptor(s)

THE T cell receptor is the structurally variable antigen-specific receptor present on the surface of T lymphocytes (T CELLS). T cell receptors, which are essential to the IMMUNE RESPONSE, differ in several ways from the cell-surface IMMUNOGLOBULINS which serve as antigen receptors on B lymphocytes (B CELLS):

1 Immunoglobulins are made up of four polypeptide chains (two heavy chains and two light chains) and possess two antigen recognition sites; T cell receptors are dimers and possess only one (Fig. T3).

2 Immunoglobulins are found either as membrane-bound receptors or in a secreted form (antibodies); T-cell receptors only exist as membrane-bound surface receptors.

3 Immunoglobulins recognize soluble and native antigens; the T cell receptors only recognize protein antigens once they have been processed and converted into small peptides and bound to MHC MOLECULES at the surface of antigen-presenting cells (e.g. macrophages, monocytes, B cells and dendritic cells) (*see* ANTIGEN PROCESSING AND PRESENTATION; MAJOR HISTOCOMPATIBILITY COMPLEX) (Fig. T4). This phenomenon is known as MHC-restricted recognition or MHC restriction.

The 'education' of T cells to recognize the MHC molecules of their own body (i.e. self MHC) takes place in the thymus (*see* T CELL DEVELOPMENT). The conventional T cell receptor, or αβTCR, is expressed on most T lymphocytes and consists of two disulphide-linked chains (α and β), noncovalently associated with the CD3 proteins to form the functional TCR–CD3 complex at the cell surface (Fig. T5). Recombinant DNA technology has led to the identification of another type of T cell receptor, the γδTCR, which comprises two chains, γ and δ, also associated with the CD3 proteins, at the cell surface of a subset of T cells. T cells express either the αβTCR or the γδTCR.

αβ T cells

The αβ T cell receptors are expressed at the surface of T cells that also express the CD ANTIGENS CD3 and either CD4 or CD8. These αβ cells represent 90–99% of the mature peripheral T cells in humans. The CD3 CD4 αβ T cells recognize antigens as peptides bound to class II MHC molecules and are, for the most part, HELPER T CELLS. The CD3 CD8 αβ T cells recognize antigens as

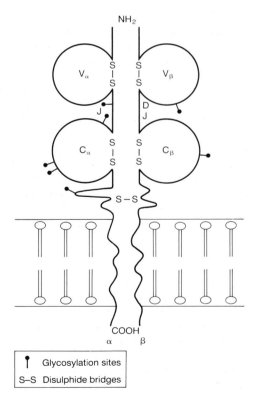

Fig. T3 Schematic representation of the αβ T cell receptor. V, Variable region; C, constant region; J, joining segment; D, diversity segment. Modified from [5].

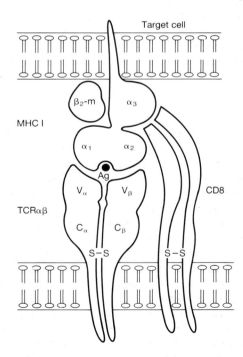

Fig. T4 Schematic representation of the αβ T cell receptor of a CD3 CD8 cytotoxic T cell recognizing the antigenic peptide bound to a class I MHC molecule. β2-m, β2-microglobulin; α1, α2, α3, domains of the MHC class I α chain. From [5].

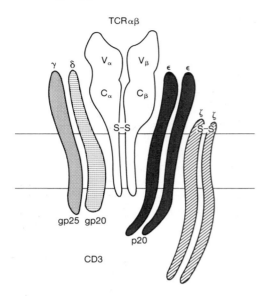

TCRαβ

Fig. T5 Schematic representation of the human αβTCR–CD3 complex. See text for explanation. Modified from [5].

peptides bound to class I MHC molecules and are mostly CYTO-TOXIC T CELLS. CD4 and CD8 are molecules that interact with a nonpolymorphic region of class II and class I MHC molecules, respectively.

The ability of αβ T cells to recognize a potentially infinite array of antigen–MHC combinations resides in the variable N-terminal portions of the TCR α and β chains. TCR α and β chains are subject to ALLELIC EXCLUSION so that a unique αβ receptor type is expressed at the surface of a given T cell clone (*see* T CELL DEVELOPMENT; T CELL RECEPTOR GENES).

γδ T cells

The γδ T cell receptors (see Table T1 below) are mainly expressed at the surface of T cells that express CD3 but lack CD4 or CD8. There are also smaller subsets of CD3 CD4 γδ and CD3 CD8 γδ T cells. The γδ T cells represent 1–10% of peripheral T cells (mean 3%) and 0.2–0.9% of the THYMOCYTES in humans (*see* T CELL DEVELOPMENT). Like the α and β chains, the γ and δ chains have clonally unique structures and are subject to allelic exclusion. The γδ T cells can recognize a heterogeneous array of ligands, as, for example, HEAT SHOCK proteins from either bacteria or autologous cells, whether presented or not by classical or nonclassical MHC molecules. The function of these γδ cells could be to eliminate body cells that are under stress, for example as a result of infection or transformation to malignancy. The function of the γδ T cell receptors, the nature of their cognate antigens and the context in which these antigens are recognized are the object of extensive studies. The representation of γδ T cells in different tissues and their abundance relative to αβ cells vary considerably between vertebrates.

Structure of the T cell receptor chains

The α and β chains of the αβTCR, and the γ and δ chains of the

γδTCR possess an N-terminal extracellular region which comprises a variable domain of 102–119 amino acids and a constant domain of 83–113 amino acids, a transmembrane region of 20–24 residues, and a small cytoplasmic tail of 4–12 amino acids. Each variable and constant domain has a structure similar to that of the immunoglobulin domain and is characterized by a series of multistranded antiparallel β-sheet bilayers and an intrachain disulphide loop of 63–69 amino acids (*see* IMMUNOGLOBULIN STRUCTURE; IMMUNOGLOBULIN SUPERFAMILY). Each chain contains a cysteine proximal to the transmembrane region which is involved in the formation of an α–β and γ–δ interchain disulphide bond, respectively. The human TCR γ_2 chain, encoded by the constant region gene TRGC2 (*see* T CELL RECEPTOR GENES), is an exception in that it has no cysteine in that region and is nondisulphide-linked to the δ chain (*see* TCRγ). For the nomenclature of the human TCRγ genes, see [1].

All the clonotypic α, β, γ, and δ chains and the associated CD3 proteins possess a hydrophobic and probably helicoidal membrane-spanning region which has the unusual feature of containing charged amino acids: one or two positive charges for the TCR α,β, γ, and δ chains, and one negative charge for the CD3 proteins, which probably interact to stabilize the TCR–CD3 complex (Fig. T6).

Synthesis

As for immunoglobulins, the constant domain of the TCR chains is encoded by a constant region gene (C) whereas the variable domain is encoded by the joining together of noncontiguous DNA segments: a variable (V) gene and a joining (J) segment for the α and γ chains; a V gene, a diversity (D) segment and a J segment for the β and δ chains (*see* GENE REARRANGEMENT; IMMUNOGLOBULIN GENES; RECOMBINASE; T CELL RECEPTOR GENES).

Owing to the localization of the TCRδ locus within the TCRα locus (*see* T CELL RECEPTOR GENES), some V_α genes may be joined to D_δ and J_δ segments and expressed with the unique C_δ gene in δ chains, some V_δ genes may be joined to J_α segments and expressed with the unique C_α gene in α chains.

The murine and human TCRβ locus (*see* T CELL RECEPTOR GENES) contains two isotypic constant region genes, TRBC1 and TRBC2, that encode proteins that differ by only four (murine) or six (human) amino acid residues. Any of the V_β genes may be expressed with one or the other constant region gene. Expression of a particular C_β gene does not correlate with function or specificity.

The human TCRγ locus (*see* T CELL RECEPTOR GENES) also contains two isotypic constant region genes TRGC1 and TRGC2. The respective $C_\gamma1$ and $C_\gamma2$ encoded domains differ by two replacements in 163 amino acids, and an insertion of 16 or 32 amino acids, which characterize two allelic forms of $C_\gamma2$. One of the two amino acid replacements affects the cysteine involved in the interchain disulphide bridge, so that the γ1 chain is disulphide linked to the δ chain whereas the γ2 chain is not. Allelic forms of $C_\gamma1$ have been described, which differ by three amino acids in their extracellular region.

The constant domain of the murine γ chain is encoded by one of three isotypic C genes (C1, C2 and C4), two of them associated with a unique V gene, the third one with four V genes (*see* T CELL

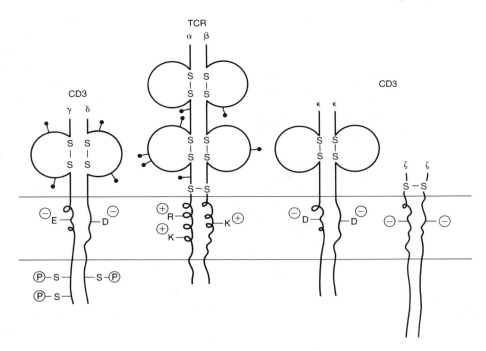

Fig. T6 Representation of the αβTCR–CD3 complex showing the positively and negatively charged amino acids (single-letter notation) of the transmembrane region as well as the intracytoplasmic phosphorylation sites of the CD3γ, CD3δ and CD3η chains. Modified from [5].

RECEPTOR GENES). Murine C$_\gamma$1 and C$_\gamma$2 differ by only six replacements in 290 amino acids, and an insertion of five amino acids in C$_\gamma$1. C$_\gamma$4 differs significantly in sequence from the other C$_\gamma$ genes (including the PSEUDOGENE C$_\gamma$3) with about 66% overall amino acid identity. In addition the C$_\gamma$4 sequence contains an insertion of 23 amino acids (compared with C$_\gamma$2).

Glycosylation of TCR chains

The T cell receptor chains are GLYCOPROTEINS. The human TCRα chain (M_r 43 000–49 000 (43–49K)) contains five N-linked complex oligosaccharides attached to a polypeptide backbone of 32K. The human TCRβ chain (38–44K) contains one high-mannose and one complex N-linked glycan side chain attached to a polypeptide backbone of 34K. The murine TCRα chain (40–50K) contains four N-linked oligosaccharides of the complex type attached to a polypeptide backbone of 28K. The murine TCRβ chain (40–50K) contains two or three high-mannose glycans attached to a polypeptide of 32K.

There are three types of human γ chains depending on the constant region domain (Fig. T7): the γ1 chains (40K) whose constant region encoded by the TRGC1 gene is disulphide-linked to the δ chain and has three N-glycosylation sites, the γ2(2×) chains (40 and 44K, which represent two different degrees of glycosylation) encoded by the allelic TRGC2 gene with duplication of exon 2 (and therefore 16 amino acids longer than γ1), and the γ2(3×) chains (55K) encoded by the allelic TRGC2 gene with triplication of exon 2 (and therefore 32 amino acids longer than γ1) [2]. The γ2(2x) and γ2(3x) chains are nondisulphide linked to the δ chain and have four or five N-glycosylation sites, respectively.

The murine TCRγ, when glycosylated, is a 35K glycoprotein with a polypeptide backbone of 32K, which is disulphide linked to a 45K TCRδ chain. The polypeptide backbone of the TCRδ glycoprotein is 31K. Glycosylation of the murine γ chains depends

on the constant domains. Indeed the C$_\gamma$1 domain contains a single site for N-linked glycosylation whereas the C$_\gamma$2 domain contains none. However the C$_\gamma$1 site is absent in some strains. The C$_\gamma$4 domain also contains a single site for N-glycosylation.

Diversity and repertoire of the αβ T cell receptors

The diversity of T cell receptor chains depends on two mechanisms: combinatorial diversity which is a consequence of the number of V, D and J segments (*see* T CELL RECEPTOR GENES), and N-REGION DIVERSITY which creates an extensive and clonal somatic diversity at the V-J and V-D-J junctions (*see* GENERATION OF DIVERSITY). The combinatorial association of α and β chains, and of γ and δ chains and the junctional diversity each affect the specificity of the T cell receptor and therefore play a part in the development of the T cell REPERTOIRE. The functional αβTCR repertoire in an individual is selected from a population of cells containing randomly rearranged α and β genes, by a combination of positive and negative selective events that occur during intrathymic maturation (*see* T CELL DEVELOPMENT).

Both helper and cytotoxic αβ cells draw upon the same pool of α and β variable genes in the production of their antigen receptors. Fifty-three V$_\beta$ genes belonging to 22 subgroups and 40 V$_\alpha$ genes belonging to 29 subgroups have been identified so far in humans. Both V$_\alpha$ and V$_\beta$ contribute to the antigen-binding site. There is no simple and general correlation between usage of any V$_\alpha$ or V$_\beta$ genes and phenotype, function, specificity, or MHC restriction. However, in some systems, receptors encoded by particular V$_\alpha$ and V$_\beta$ genes may be selected by particular antigen–MHC configurations.

Diversity and repertoire of the γδ T cell receptors

The combinatorial diversity of the γ and δ chains is more limited

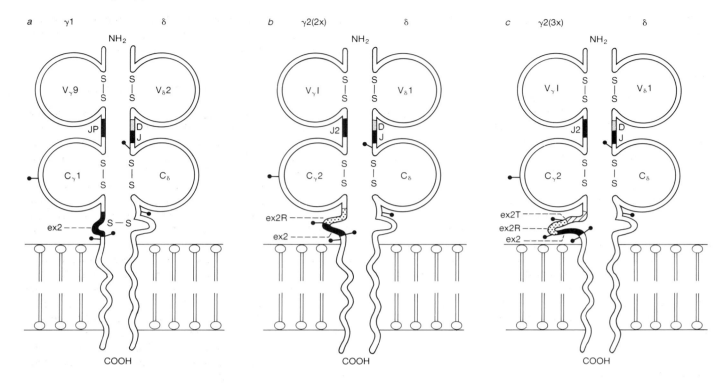

Fig. T7 Schematic representation of the three types of human γδ T cell receptors. There are three types of human γδTCR, distinguished by their γ chains. *a*, The γ1δ receptor, in which the 40K γ1 chain is disulphide linked to the δ chain; *b*, γ2(2 ×)δ and *c*, γ2(3 ×)δ, in which the 40 or 44K γ2(2 ×) chain and the 55K γ2(3 ×) chain are characterized by a duplication or triplication of exon 2 respectively, and are nondisulphide linked to the δ chain. From [2,3].

present only early during foetal thymic development. DEC and r-IEL arise from these foetal thymic precursors. Unlike DEC, i-IEL can arise in the adult in the absence of thymic influence. In the human, epithelial tissues contain γδ T cells in significantly lower numbers than in the mouse. Representation of the γδ T cells in

than that of the α and β chains. There are eight functional V_γ genes and five J segments for the human TRG locus, and possibly eight V_δ genes, and three D and three J segments for the human TRD locus [3] (*see* T CELL RECEPTOR GENES).

This potential diversity is restricted by the preferential usage of some V genes or D and J segments. In humans, most of the peripheral γδ T cells express a preferential V_γ and V_δ rearrangement (Table T1). In the mouse, there is a tight correlation between TCR V_γ usage and tissue localization (Table T1). In adult mice, Thy-1 marrow-derived dendritic cells resident in the skin (Thy-1 dendritic epidermal cells, DEC) are γδ T cells expressing a monomorphic receptor. Murine reproductive intraepithelial (r-IEL) lymphocytes express a monomorphic γ chain associated with the same δ chain. Both DEC and r-IEL receptors contain no N-diversity, which suggests that they might have a role in immunological surveillance for stress-induced self-antigen rather than for foreign antigens. Murine intestinal intraepithelial (i-IEL) cells express a monomorphic γ chain and use a variety of V_δ genes. The i-IEL cells, which are CD8 γδ, express the CD8α but not the CD8β (Lyt-3) molecules. TCR gene utilization is more complex in the murine lung although a subset of resident pulmonary lymphocytes (RPL) express a monomorphic δ chain (BID cells).

Murine epithelia are seeded by waves of thymic precursors

Table T1 Types of γδ T cells and their receptor diversity

Cells	Gene usage	N-region diversity	Repertoire
Human			
Peripheral γδ cells	Preferentially $V_\gamma 9$-J_γP-$C_\gamma 1$ $V_\delta 2$-$D_\delta 3$-$J_\delta 1$-C_δ	+	Diverse
Mouse			
Thy-1 dendritic epidermal cells (DEC)	$V_\gamma 5\{3\}$-$J_\gamma 1$-$C_\gamma 1$ $V_\delta 1$-$D_\delta 2$-$J_\delta 2$-C_δ	None	Monomorphic
Reproductive intraepithelial cells (r-IEL)	$V_\gamma 6\{4\}$-$J_\gamma 1$-$C_\gamma 1$ $V_\delta 1$-$D_\delta 2$-$J_\delta 2$-C_δ	None	Monomorphic
Intestinal intraepithelial cells (i-IEL)	$V_\gamma 7\{5\}$-$J_\gamma 1$-$C_\gamma 1$ various δ genes, $V_\delta 4$ most prevalent	+	Diverse
Resident pulmonary cells (RPL)	Various	+	Diverse
BID subset of RPL	Various γ genes $V_\delta 7$-$D_\delta 2$-$J_\delta 1$-C_δ	None	Restricted

The nomenclature of the murine V genes is not yet decided. The 'A' nomenclature in [4] is used here. The alternative numbering ('B') is given in { }.

adult human skin is not significantly different from that in peripheral blood.

The TCR–CD3 complex

In addition to the αβ or γδ heterodimers which are structurally unique for each clone of T cells, the T cell receptor–CD3 complex comprises additional nonpolymorphic chains, the CD3γ, δ, ε, ζ and η proteins, which are identical in all T cells. The CD3 proteins are responsible for coupling TCR occupancy to intracellular signal transduction pathways (*see* SECOND MESSENGER PATHWAYS) that result in the events comprising T CELL ACTIVATION. Each CD3 chain contains an N-terminal extracellular domain, a transmembrane segment and a cytoplasmic domain. The transmembrane regions of the CD3 proteins have a predicted α-helix configuration and contain a negatively charged amino acid (aspartic acid for the CD3δ and ε chains, glutamic acid for the CD3γ chain). The cytoplasmic regions of the CD3 chains are longer than those of the TCR α, β, γ, and δ chains and presumably have an important role in the interaction with cytoplasmic components that are directly involved in the transduction of the antigen-binding signal.

The CD3γ, δ, and ε proteins

The CD3γ (25K in human, 21K in mouse) and CD3δ (20K in human, 25K in mouse) are glycoproteins bearing N-linked oligosaccharide side chains (two for human CD3γ and δ, one for mouse γ and three for mouse δ), and whose protein sizes are ~16K. CD3ε (20K in human, 25K in mouse) is not glycosylated. All these chains are likely to contain intrachain disulphide bonds and are members of the immunoglobulin superfamily. The three CD3γ, δ, and ε chains probably arose by GENE DUPLICATION. The genes encoding these three chains are found within a region 50–300 kilobases (kb) long on human chromosome 11q23 and mouse chromosome 9. The highly homologous CD3γ and δ genes lie within 1.5 kb of each other. The CD3ε cytoplasmic tail contains an exceptionally large number of basic amino acids followed by a short stretch of prolines, also found in the cytoplasmic tail of the CD antigen CD2. Activation of T cells results in phosphorylation of one or two serine residues on the CD3γ chain and of one serine residue on the CD3ε chain.

The CD3 ζ and η chains

The zeta (ζ) chain is a 16K nonglycosylated protein with no sequence or structural homology to the other CD3 or TCR chains. In contrast to the other chains, ζ has a very short extracellular domain of 6–9 amino acids, with the vast majority of the ζ protein (113 amino acids) existing as the cytoplasmic domain. The transmembrane region has a negatively charged amino acid. The ζ chain is encoded by a gene found on human and mouse chromosome 1. In the majority of receptors, ζ exists as a disulphide-linked 32K (ζζ) homodimer. Five to twenty per cent of ζ seems to be disulphide-linked to a 21K nonglycosylated protein designated as CD3-eta (CD3η) or p21, in a 37K (ζη) heterodimer. Cells

possess both ζζ homodimers and ζη heterodimers with a ratio of homodimers to heterodimers ranging from 5 : 1 to 10 : 1. Activation of T cells results in tyrosine phosphorylation of CD3η.

Biosynthesis and assembly of the TCR–CD3 complex

A T cell cannot recognize antigen unless the TCR–CD3 complex is correctly assembled and efficiently transported and expressed at the cell surface. The genes encoding the TCR–CD3 proteins are expressed sequentially during the final stages of T cell maturation. CD3γ, CD3δ and CD3ε are synthesized by the earliest recognizable thymocytes but the proteins remain inside the cell. Assembly of the TCR components takes place within the ENDOPLASMIC RETICULUM (ER) and begins soon after biosynthesis. Within the endoplasmic reticulum there is a transient, noncovalent association with a 26K nonglycosylated protein, the CD3ω or TRAP (T cell receptor associated protein). Upon further maturation, T cells begin to express TCRβ chains intracellularly. After synthesis of the TCRα chain, there is assembly of the full receptor within the ER, and processing of the N-linked side chains of TCR and CD3 chains in the GOLGI APPARATUS. The TCR–CD3 complex is then transported to the plasma membrane (*see* PROTEIN TARGETING). Incomplete TCRs are directed from the Golgi to the LYSOSOMES where they are rapidly degraded, or are retained for long periods within the ER.

M.-P. LEFRANC

1 Lefranc, M.-P. & Rabbitts, T.H. (1990) A nomenclature to fit the organization of the human T cell receptor γ and δ genes. *Res. Immunol.* **141**, 615–618.
2 Lefranc, M.-P. & Rabbitts, T.H. (1989) The human T-cell receptor γ (TRG) genes. *Trends Biochem. Sci* **14**, 214–218.
3 Lefranc, M.-P. & Rabbitts, T.H. (1990) Genetic organization of the human T cell gamma and delta loci. *Res. Immunol.* **141**, 565–577.
4 Lefranc, M.-P. (1990) The mouse T cell receptor gamma genes. *Res. Immunol.* **141**, 693–695.
5 Lefranc, M.-P. (1990) Organization of the human T-cell receptor genes. *Eur. Cytokine Network* **1**, 121–130.

T cell receptor genes

THE T cell receptor genes encode the CLONOTYPIC structures of the T cell receptor complex which serves as the specific antigen receptor on the surface of T CELLS (*see* T CELL RECEPTORS). Four T cell receptor (TCR) loci are known, designated α, β, γ, and δ (or TRA, TRB, TRG, and TRD in humans). These give rise to the α, β, γ, and δ chains respectively. α and β chains combine to form the αβ T cell receptor (αβTCR) and γ and δ chains combine to form the γδTCR [1,2,3].

In human and mouse the TCR loci reside at three distinct chromosomal sites, and are located in humans on chromosome 14q11 (TCRα and TCRδ), chromosome 7p15 (TCRγ), and chromosome 7q35 (TCRβ). In a similar fashion to the IMMUNOGLOBULIN GENES, each locus comprises a number of variable (V) genes, diversity (D) segments (only known in β and δ genes), a number of joining (J) elements and one or two (TCRβ and TCRγ loci) constant region (C) genes. The δ locus is contained within the α locus. The

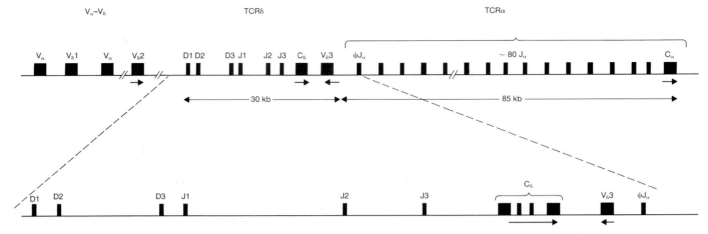

Fig. T8 Organization of the human T cell receptor α (TRA) and δ (TRD) loci. TRD is embedded in TRA; three D and three J segments precede the single C_δ gene located 85 kb upstream of the single C_α gene. The $V_\delta 3$ gene is located 3 kb 3′ of C_δ in an inverted orientation. Rearrangement between the δRec sequence (not shown on figure), located upstream of $D_\delta 1$ and the ψJ_α sequence results in the deletion of the D_δ and J_δ segments and the C_δ gene. This deletion would precede the V_α–J_α rearrangement. Forty V_α genes belonging to 29 subgroups and 75 J_α segments have so far been identified. The organization of the mouse TCRα and δ loci is similar but there are only two D and two J segments. Three murine δRec sequences have been identified. From [5].

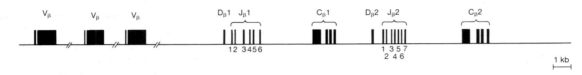

Fig. T9 Organization of the human T cell receptor β (TRB) locus. There are two C region genes, each associated with a group of J_β segments (six $J_\beta 1$ and seven $J_\beta 2$) and one D segment ($D_\beta 1$ or $D_\beta 2$). Fifty-three V_β genes belonging to 22 subgroups have so far been identified. The organization of the mouse TCRβ locus is similar but there are only six functional $J_\beta 2$ segments. From [5].

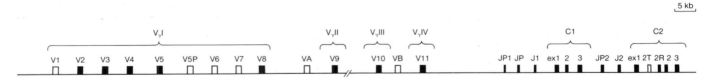

Fig. T10 Organization of the human T cell receptor γ (TRG, TCRG) locus. There are two C genes, five J segments and in most cases 14 V genes, which belong to four subgroups. Nine V_γ genes, five of them functional, and four PSEUDOGENES belong to subgroup I, whereas subgroups II, III and IV each consists of a single V gene — V9, V10, and V11. The pseudogenes VA and VB do not belong to any of these subgroups. An allelic variation in the number of VI genes from seven to ten can be found, and results from deletion of the V4 and V5 genes or insertion of an additional gene — V3P — between V3 and V4. From [6].

organization of the human α and δ loci (Fig. T8), the human β locus (Fig. T9), and the human and murine γ loci (Figs T10, T11) are illustrated.

In general V genes are located 5′ to C elements; however, in the TCRβ and TCRδ loci, V genes are known to reside 3′ of the C genes and thus show reversed transcriptional polarity (Fig. T12). The orientation of the V elements with respect to the C genes determines the mechanisms of GENE REARRANGEMENT; deletional rearrangement occurs for V genes located 5′ of C, whereas rearrangement by inversion is used for V genes located 3′ of the C region. All known J elements in the TCR gene complexes tend to occur in clusters. The tandem arrangement of the TCRδ/TCRα locus and the mouse TCRγ locus (*see* Figs T8, T11) requires particularly tight control over V gene usage in V-(D)-J rearrange-

ments. The chromosomal structure of germ-line TCRα/δ, TCRγ and TCRβ loci is formally very similar (Fig. T12). Indeed their structure suggests that they have arisen by gene duplication events involving individual elements and/or whole gene clusters with subsequent divergence (*see* IMMUNOGLOBULIN SUPERFAMILY; MOLECULAR EVOLUTION).

The coordinated regulation of TCR gene rearrangement and expression ensures that two general constraints on antigen receptor surface expression are met in T cells. First, only one allele of each TCR locus is expressed at the cell surface, a process referred to as ALLELIC EXCLUSION; this ensures that an individual T cell only produces receptors of a single antigen specificity. Second, each T cell expresses only one of the two classes of heterodimeric TCRs — αβ or γδ.

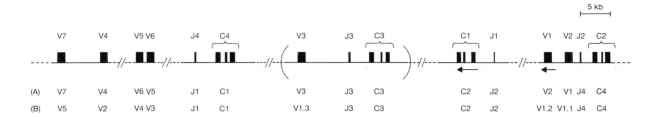

Fig. T11 Organization of the murine γ T cell receptor locus. Of the three functional C genes, C$_\gamma$2 and C$_\gamma$4 are associated with a unique J segment and V genes, respectively V2 and V1 (nomenclature 'A' [7]) belonging to the same subgroup, and are in an inverted orientation on the chromosome. C$_\gamma$1 is associated with a single J segment and with four V genes belonging to four different subgroups. C$_\gamma$3 is a pseudogene which is deleted together with the associated J3 segment and V3 gene in several strains.

The TCRα/δ locus

The TCRα locus contains within it the TCRδ locus (Fig. T8). There is one C-region (C$_\alpha$) gene which has four exons, of which the last is entirely noncoding. The J$_\alpha$ region is unusually large and extends over about 75 kilobases (kb) in humans and encompasses an estimated 75–100 J$_\alpha$ elements; the ψJ$_\alpha$ element (the DELTA-DELETING ELEMENT), which is used in a site-specific deletion event which removes the TCRδ locus, is located at the 5' end of the J$_\alpha$ region. The V$_\alpha$ genes (~40) are located at an unknown distance — at least 100 kb — 5' of the J$_\alpha$ region and are interspersed with Vδ genes. They are subdivided into several families on the basis of the protein sequence of their gene products. A T cell enhancer/silencer element is located 3' of the C$_\alpha$ gene, which activates TCRα expression in αβ T cells but silences it in γδ T and non-T cells. Enhancer elements in the J–C intron have not been defined.

The TCRδ locus is located within the TCRα locus. Its C region is encoded in four exons. In humans, three D and three J regions

are located within ~30 kb upstream of C$_\delta$. A single V$_\delta$ gene is found with inverted transcriptional polarity downstream of C$_\delta$ and at the 5' end of the J$_\alpha$ region. The majority of V$_\delta$ genes are located 5' of the DJC$_\delta$ region and interspersed with V$_\alpha$ genes; their usage is markedly biased during γδTCR formation in foetal thymic ontogeny (*see* T CELL RECEPTORS). There is a T cell-specific enhancer in the J$_\delta$3–C$_\delta$ intron.

The TCRβ locus

The TCRβ locus is arranged as a duplicated complex of D, J, and C regions on human chromosome 7q34. Its two C genes are very similar, each being encoded in four exons, and are thought to be functionally indistinguishable. A single D element is associated with several J elements, located upstream of each C gene. The entire DJC duplex is contained within 18 kb (in humans). The V$_\beta$ complex comprises ~50–100 genes all located within 600 kb upstream of D$_\beta$1. The distance between D$_\beta$1 and the most downstream V$_\beta$ element is less than 80 kb. In the mouse, a single V$_\beta$ element is located 3' of the DJC cluster. The V$_\beta$ genes fall into several families on the basis of protein homology, and are organized in similar fashion in human and mouse. A T cell-specific ENHANCER is located 3' of the DJC duplex, but none has yet been found in the J–C intron regions.

The TCRγ locus

The TCRγ locus is arranged as a duplicated JC complex on human chromosome 7p15. Upstream of the C$_\gamma$1 gene, encoded in three exons, are located three J$_\gamma$ segments, whereas the downstream C$_\gamma$2 region is preceded by two J$_\gamma$ segments, all encompassed within 40 kb. In humans, two HAPLOTYPES have been described for the TCRγ locus. In allele I, the C$_\gamma$2 gene contains a duplicated second exon, whereas the C$_\gamma$2 gene in allele II carries a triplicated second exon. The 14 known human V$_\gamma$ genes (eight are functional) fall into four subgroups, encoded within 100 kb of DNA upstream of the JC$_\gamma$ duplex; the distance between the most downstream V$_\gamma$ gene and the most upstream J$_\gamma$1 segment is 16 kb. In the mouse, the organization of the TCR$_\gamma$ locus is different in that several VJC complexes are arranged in tandem (Fig. T11). A T cell-specific enhancer has been localized in the mouse downstream of one VJC cluster.

In contrast to C$_\gamma$1, the second exon in the C$_\gamma$2 gene does not encode the third cysteine residue characteristic of all other TCR C regions, and which is involved in interchain bonding in the TCR heterodimer (*see* T CELL RECEPTORS). Therefore, depending on the usage of C$_\gamma$ regions in a particular cell, either disulphide-linked

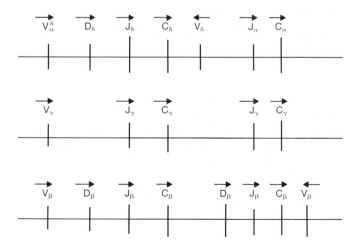

Fig. T12 Comparison of the four known human TCR gene complexes, not drawn to scale. The transcriptional polarity of genes is indicated by arrows above the gene segment. The figure highlights the overall similarity of the three separate chromosomal loci.

($C_\gamma 1$) or noncovalently associated ($C_\gamma 2$) $\gamma\delta$TCR heterodimers are formed. However, the two intrachain immmunoglobulin-like folds formed utilize the universally conserved cysteine residue pairs encoded by V_γ and C_γ regions.

TCR gene rearrangement

Rearrangement of the TCR loci to produce functional genes encoding the receptor polypeptide chains takes place during T CELL DEVELOPMENT in the thymus. Rearrangement seems to be initiated (as suggested for immunoglobulin gene rearrangement) by prior transcription of V segments. There is ample evidence that the same V(D)J RECOMBINASE complex that rearranges immuno-globulin loci also effects rearrangement of TCR genes.

Studies in the ontogeny of V(D)J rearrangements in the TCR loci have shown that TCRδ rearrangements are among the earliest events in T cell differentiation in foetal life, although considerable overlap in the timing of TCR rearrangement exists, and indeed, T cells have been described displaying both TCRδ and TCRα rearrangements alongside TCRβ and TCRγ rearrangements. However, only αβ and γδ rearrangements are expressed together and the αβ and γδ cells may represent separate and distinct lineages.

D–J rearrangements generally precede V–DJ rearrangements in the TCRβ locus, whereas V–D, D–J and V–J joining seem to occur simultaneously in the TCRδ locus. In the TCRα/δ locus, utilization of interspersed V_δ and V_α occurs with little apparent overlap. In the majority of gene rearrangements — those producing γδ receptors and the later wave of αβ receptors — the selection of V, J, or D segments seems to be random.

Selection of V genes for rearrangement does, however, occur in some cases, and seems to be one mechanism by which the specificity of antigen receptors (and thus their possible function) can be regulated. In the mouse, for instance, foetal thymocytes carrying the γδTCR receptor show strictly age-dependent waves of V_γ and V_δ usage. It is thus possible that ontogenetically controlled V gene usage is one way of generating functionally distinct subsets of T cells (*see* T CELL RECEPTORS).

Site-directed mutagenesis has shown that the antigen-binding pocket of αβ TCR polypeptides is mainly encoded in the junctional region between the V and C segments, and that contact with MHC molecules and co-receptor molecules is mediated by both V and C segments. Contributors to the antigen-binding region are D and J segments, as well as non-germ-line encoded nucleotides (N REGION) sandwiched between them in the process of V-(D)-J recombination, presumably by the action of the enzyme TERMI-NAL DEOXYNUCLEOTIDYLTRANSFERASE (TdT). The contribution to junctional diversity by N nucleotides varies from one TCR gene to the other and also varies with the stage in T-cell ontogeny. In foetal mouse thymocytes expressing γδTCR, the junctional diversity is low, because only one D element is used and the two N regions (between V-D and D-J junctions) are very short, as most cells do not express TdT. In the adult mouse thymus, many γδ cells carry TCRδ genes with greatly increased junctional diversity, as up to three D segments together with four corresponding N regions (some encoding four or more amino acids) join V and J region sequences in rearranged alleles.

TCR gene expression

The expression of rearranged TCR alleles is regulated by a complex interplay of T cell-specific transcriptional ENHANCER and SILENCER elements (*see* EUKARYOTIC GENE EXPRESSION), and such *cis*-acting regulatory sequences have been identified in all four TCR loci. For instance the human TCRα enhancer is only active in αβ cell lines but inactivated in γδ cells and non-T cells by a nearby silencer element.

Furthermore, whereas TCRβ messenger RNA can be found in T cells bearing αβ or γδ receptors, αβ T cells generally do not express TCRγ mRNA, even though they might carry a function-ally rearranged VJCγ allele. This transcriptional silencing of functionally rearranged alleles seems to be one of the mech-anisms ensuring that T cells only express one type of receptor complex on their surface.

Aberrant rearrangements

The V-(D)-J recombinase, utilized in physiological DNA rear-rangements in antigen receptor loci, is also implicated in the formation of interlocus rearrangements. This can lead to cytoge-netically visible CHROMOSOMAL ABERRATIONS such as inversions and translocations [4]. These aberrant rearrangements occur among antigen receptor genes, forming chimaeric receptors such as between TCRβ and TCRγ, and TCRα and the immunoglobulin heavy chain locus in humans. Their functional significance is unclear. Alternatively, TCR genes are found repositioned in new chromosomal domains not encoding antigen receptor genes in T cell tumours; these changes disrupt the integrity of TRANSCRIP-TION UNITS, operationally defined as T cell ONCOGENES, which become transcriptionally deregulated either under the influence of TCR ENHANCER elements or by interference with the colinear-ity of their PROMOTER sequences. TCRα and δ genes are more frequently involved in these chromosomal abnormalities than TCRβ. No molecular cloning has yet been reported implicating the TCRγ locus in any such tumour-specific abnormality.

T. BOEHM

1 Abbas, A.K. et al. (1991) *Cellular and Molecular Immunology* (Saunders, Philadelphia).
2 Davis, M.M. & Bjorkman, P.J. (1988) T-cell antigen receptor genes and T-cell recognition. *Nature* **334**, 395–401.
3 Boehm, T. & Rabbitts, T.H. (1989) The human T cell receptor genes are targets for chromosomal abnormalities in T cell tumours. *FASEB J.* **3**, 2344–2359.
4 Rabbitts, T.H. & Boehm, T. (1991) Structural and functional chimerism results from chromosomal translocation in lymphoid tumors. *Adv. Immunol.* **50**, 119–146.
5 Lefranc, M.-P. (1989) Organisation des genes des recepteurs αβ et γδ des lymphocytes T humains. *Med. Sci.* **5**, 754.
6 Lefranc, M.-P. et al. (1989) Molecular mapping of the human T cell receptor gamma (TRG) genes and linkage of the variable and constant regions. *Eur. J. Immunol.* **19**, 989.
7 Lefranc, M.-P. (1990) The mouse T cell receptor gamma genes. *Res. Immu-nol.* **141**, 693–695.

t-complex The t-complex of mouse chromosome 17 covers ~30 Mb. Chromosome 17 in certain mice carries two INVERSIONS within the region when compared to the wild type, and RECOM-

BINATION between the wild type and these t-chromosomes is thereby suppressed. Markers along the t-chromosome are inherited as a unit, the t-haplotype, and are transmitted through the sperm of heterozygotes more frequently than the wild-type homologue. When homozygous, 'distorter' genes within the complex cause male sterility, a disadvantage that has resulted in the presence within the t-haplotype of recessive mutations causing lethal defects of early development. Examples include failures of implantation, blastocyst formation, and ectodermal and mesodermal differentiation. The mutations BRACHYURY (*T*) and *Fused* map within the complex, as does the MAJOR HISTOCOMPATIBILITY COMPLEX, H-2. The well-advanced genetics of the region makes genes within the complex good candidates for further study at the molecular level.

T-complex The T-strand of the TI PLASMID of *Agrobacterium tumefaciens* coated with virE2 protein and with the virD2 protein covalently bound to the 5′ end. This complex is transferred by the bacterium to the plant cell where NUCLEAR IMPORT SIGNALS in the proteins promote transfer into the plant cell nucleus.

T cytoplasm Texas cytoplasm, source of T-type cytoplasmic male sterility (cmsT) in maize. Distinguished from other types by the nuclear genes that restore fertility. It confers susceptibility to southern corn leaf blight (caused by the fungus *Bipolaris* (*Helminthosporium*) *maydis*), and the widespread use of cmsT in maize breeding in the USA greatly contributed to a disastrous epidemic of southern corn leaf blight in the early 1970s. The race T of the fungus produces a toxin, T toxin, which causes cmsT mitochondria to become uncoupled and leaky. *See*: MITOCHONDRIAL GENOMES: PLANT.

T-DNA DNA derived from the T-REGION of the TI PLASMID or RI PLASMID, and which is present in transformed plants. In the case of bipartite T-regions, one or both portions may be present independently and at differing copy numbers.

T-even phage A group of virulent BACTERIOPHAGES (T2, T4, T6) of similar size which can infect *Escherichia coli*. The T-even phages have characteristic anisometric (i.e. elongated) head morphology.

T lymphocyte *See*: T CELL.

T-odd phage A group of virulent BACTERIOPHAGES (T1, T3, T5, T7) with similar sized genomes which infect *Escherichia coli*. The T-odd phages have characteristic isometric head morphology.

T-region The portion of a TI PLASMID or RI PLASMID that is bounded by border repeat sequences and which is transferred to plant cells.

T-state The low affinity or unreactive form of an ALLOSTERIC protein.

T-strand Single-stranded DNA derived from the T-region of the TI PLASMID of *Agrobacterium tumefaciens*. In the form of a T-complex it has the virD2 protein covalently attached to the 5′ end and is coated by the virE2 gene product, a protein that binds nonspecifically to single-stranded DNA. This is probably the form of the DNA that is transferred to plant cells.

T tubules TRANSVERSE TUBULES.

T1–T11 antigens *See*: CD ANTIGENS.

T2 phage A T-EVEN virulent BACTERIOPHAGE of *Escherichia coli*. It is very similar to T4 phage and can recombine with the T4 phage in a mixed infection.

T3 complex Now largely obsolete designation for the CD3 complex of polypeptides associated with the antigen-specific CLONOTYPIC receptor in the T CELL RECEPTOR complex. *See also*: CD ANTIGENS.

T4 DNA ligase A DNA LIGASE purified from the bacterium *Escherichia coli* infected with T4 PHAGE. The enzyme is widely used for the *in vitro* ligation of DNA fragments with COHESIVE ENDS (*see* DNA CLONING). The protein is a monomer of M_r 68 000 and requires Mg^{2+} and ATP to catalyse the formation of phosphodiester bonds between adjacent 5′-P and 3′-OH termini.

T4 phage A T-EVEN virulent BACTERIOPHAGE of *Escherichia coli* which has a very large double-stranded linear DNA genome of ~173 kb. The genome encodes ~135 different gene products of which ~30% are required for phage assembly, the remainder encoding metabolic enzymes. Bacteriophage T4 has played an important part in the development of molecular genetics, particularly through the studies of Seymour Benzer in the 1950s and 60s on the T4 *rII* locus, mutations in which prevent PLAQUE formation. Detailed genetic mapping studies of the *rII* locus provided the first insights into the internal organization of a GENE. One unusual feature of the T4 genome is that it contains no cytosine residues, but rather has a modified base, 5-hydroxymethyl cytosine (HMC), which pairs with guanine in place of cytosine (*see* BASES). Furthermore, HMC has glucose-like sugars covalently linked to it forming glucosylated HMC.

Benzer, S. (1961) *Proc. Natl. Acad. Sci. USA* **47**, 403–415.

T4 polynucleotide kinase *See*: POLYNUCLEOTIDE KINASE.

T4 RNA ligase An enzyme purified from *Escherichia coli* infected with T4 PHAGE which can catalyse the formation of a phosphodiester bond between 5′-P and 3′-OH groups on both single-stranded DNA and single-stranded RNA molecules. It is therefore distinct from T4 DNA LIGASE. It requires ATP as a cofactor and Mg^{2+}.

T7 phage A T-ODD PHAGE of *Escherichia coli*. The 39 936-bp linear double-stranded DNA genome is encased in a viral head with a very short tail. There are 55 genes on the T7 genome including some whose reading frames overlap (*see* OVERLAPPING GENES). PROMOTER sequences from the T7 phage genome are widely exploited in VECTORS for *in vitro* TRANSCRIPTION (*see* EXPRESSION VECTORS).

T7 RNA polymerase A DNA-dependent RNA POLYMERASE encoded by the T7 PHAGE with a high specificity for the PROMOTER sequences of T7 genes. It is a single polypeptide of M_r 98 000 and is widely used to synthesize mRNA transcripts *in vitro* utilizing EXPRESSION VECTORS carrying a suitable T7 promoter.

Ta1 *See:* TRANSPOSABLE GENETIC ELEMENTS.

tabtoxin *See:* PLANT PATHOLOGY.

TACTAAC box A highly conserved CONSENSUS SEQUENCE present in all INTRONS of mRNAs of the yeast SACCHAROMYCES CEREVISIAE. As the RNA sequence UACUAAC it is located near the 3′ end of introns and is required for lariat formation during RNA SPLICING. The A residue in bold type forms an unusual 2′–5′ PHOSPHODIESTER BOND with the free 5′-P of the cleaved intron.

TAF TBF-associated factor. Denotes any of the TRANSCRIPTION FACTORS that associate with the TATA BOX BINDING FACTOR (TBF) to form the transcription factor TFIID.

tagetitoxin *See:* PLANT PATHOLOGY.

TAL TYROSINE AMMONIA-LYASE.

talin *See:* ACTIN-BINDING PROTEINS.

Tam Transposable genetic element from *Antirrhinum majus*. *See:* TRANSPOSABLE GENETIC ELEMENTS: PLANTS.

tandem duplication A pair of adjacent, usually short, DNA sequences lying in series within a much larger DNA sequence.

tandem repeats A series of identical DNA sequences lying adjacent to each other in the same orientation, within a much larger DNA sequence. GENES may also be tandemly repeated (e.g. the genes encoding 5S rRNA are usually present as single or multiple clusters of tandem repeats, *see* RIBOSOMAL RNA).

TAP genes Genes encoding the peptide transporter that transports antigenic peptides from the cytoplasm into the endoplasmic reticulum during ANTIGEN PROCESSING AND PRESENTATION. *TAP-1* and *TAP-2* are located in the MHC III region of the MAJOR HISTOCOMPATIBILITY COMPLEX.

Taq polymerase Thermostable DNA POLYMERASE from the bacterium *Thermus aquaticum*, used in the POLYMERASE CHAIN REACTION.

targeting *See:* GENE TARGETING; PROTEIN TARGETING.

taste blindness The inability to taste the chemical phenylthiocarbonate (PTC); individuals who cannot taste PTC are HOMOZYGOUS for an AUTOSOMAL RECESSIVE gene.

TATA box A DNA sequence motif with the CONSENSUS SEQUENCE TATA(A/T)A(A/T). The TATA box is usually found ~20–30 bp upstream of the start site of TRANSCRIPTION in most eukaryotic PROMOTERS recognized by RNA polymerase II. In *Saccharomyces cerevisiae* promoters its position relative to the RNA start site is much more variable, being anywhere between 40–110 bp upstream of the start site. The TATA box shows sequence identity with a similar promoter element found in prokaryotic gene promoters, the so-called – 10 consensus sequence TATA. A general TRANSCRIPTION FACTOR, TFIID, binds to the eukaryotic TATA box as part of a general eukaryotic transcription complex which is required for RNA polymerase II to bind and initiate TRANSCRIPTION. The TFIID–TATA box interaction may be important for the correct positioning of RNA polymerase II with respect to the RNA start site. Gene promoters lacking the TATA consensus sequence are rarely found. *See also:* EUKARYOTIC GENE EXPRESSION.

TATA box binding factor/protein (TBF, TBP, TFIIDτ) Protein required for the initiation of TRANSCRIPTION from most eukaryotic promoters. It is a subunit of TFIID, the TRANSCRIPTION FACTOR that binds to the TATA BOX in promoters for genes transcribed by RNA polymerase II and which mediates binding of other transcription factors to form the basal transcriptional complex (*see* EUKARYOTIC GENE EXPRESSION). TBF is also required for transcription from genes transcribed by polymerase III, which do not have a TATA box, where it is associated with TFIIB and TFIIIC.

tau proteins A class of microtubule-associated proteins found in neurons. They are found also in the paired helical filaments in ALZHEIMER'S DISEASE. *See:* NEURONAL CYTOSKELETON.

taxis Directional movement of a cell towards a source of attractant. Two subtypes of taxis are distinguished as being of potential biological importance: CHEMOTAXIS — directional attraction of a cell or cell process by a diffusible substance; and HAPTOTAXIS — directional attraction of a cell or cell process by a surface-bound factor.

taxol Drug (Fig. T13) isolated from the Pacific yew (*Taxus brevifolia*) which hyperstabilizes MICROTUBULES and inhibits their function, and which is being developed as an anticancer drug.

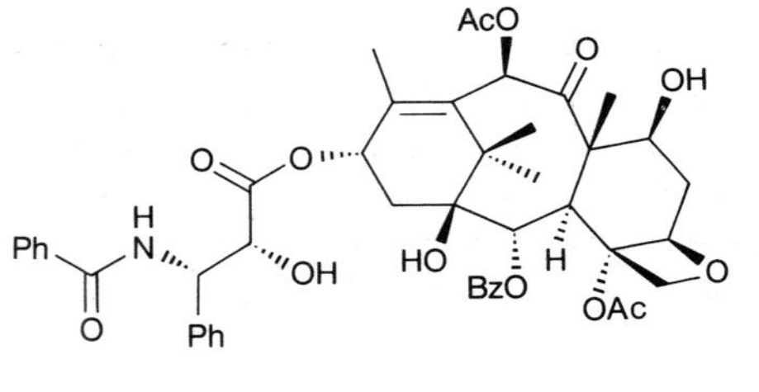

Fig. T13 Taxol.

Tay–Sachs disease One of the AUTOSOMAL RECESSIVE G_{M2}-gangliosidoses. It is caused by deficiency of the α subunit of the enzyme HEXOSAMINIDASE, a component of the type A ISOENZYME. This results in build-up of GANGLIOSIDE G_{M1}, an intermediate in neuronal SPHINGOLIPID metabolism and other GLYCOLIPIDS. In the classical disease, symptoms are rapidly progressive from six months of age, including seizures, blindness, and psychomotor regression, with death usually within three years; later onset forms occur. It is particularly common (1 : 2000) amongst Ashkenazi Jews, in whom screening for the carrier state is worthwhile.

TBF, TBP TATA BOX BINDING FACTOR/PROTEIN.

TBSV TOMATO BUSHY STUNT VIRUS.

TC10, TC21 *See*: GTP-BINDING PROTEINS.

TCR The T CELL RECEPTOR, the antigen-specific CLONOTYPIC component of the T cell receptor complex.

TCRα, TCRβ, TCRδ, TCRγ *See*: T CELL RECEPTORS; T CELL RECEPTOR GENES.

TDP The nucleotide thymidine diphosphate, normally denoting the deoxyribonucleotide. *See*: NUCLEOSIDES AND NUCLEOTIDES.

Tectoviridae *See*: BACTERIOPHAGES.

teichoic acid Family of GLYCANS of bacterial cell walls in which sugar alcohols, alternating sugar alcohols and sugars, or a sugar alcohol glycosidically linked to a sugar forms a polymer with phosphate bridges, that is, of one of the repeating forms -A-P-A-P-, -A-P-B-P-A-P-B-, or -AB-P-AB-P-. Further substitution with sugars (via glycosides) and D-alanine to give side chains commonly occurs. Synthesis involves undecaprenyl phosphate (*see* UNDECA-PRENOL AND DOLICHOL CYCLES). *See also*: BACTERIAL ENVELOPES.

teichuronic acid A family of GLYCANS from the walls of Gram-positive bacteria which are linear, alternating copolymers of uronic acid with hexose or *N*-acetyl hexosamine. They are synthesized in response to severe phosphate depletion, provided that the level of magnesium ions at the cell surface remains low.

telocentric Chromosome with a terminal CENTROMERE.

telomerase A RIBONUCLEOPROTEIN enzyme with REVERSE TRAN-SCRIPTASE activity that synthesizes the G-rich strands of telomeres (the structures sealing the ends of chromosomes). The enzyme RNA contains a sequence that acts as a template for synthesis of the tandem repeats which are added to the protruding 3′ end of the G-rich strand of the telomere. *See*: CHROMOSOME STRUCTURE.

Blackburn, E.H. (1991) *Nature* **350**, 569–572.

telomere Structure formed of multiple repeating DNA sequences found at the ends of eukaryotic chromosomes and which serves to seal the ends. *See*: CHROMOSOME STRUCTURE.

telophase The stage of MITOSIS or MEIOSIS following ANAPHASE during which the two, already segregated, sets of CHROMOSOMES decondense and return to their normal INTERPHASE condition. Telophase begins with the reformation of the NUCLEAR ENVELOPE and the onset of CHROMATIN decondensation. Its end is difficult to define, because the nuclear processes of interphase, such as RNA synthesis, begin gradually. During telophase the cytoplasms of the two daughter cells also return to their interphase condition, including the reformation of an interphase array of MICROTU-BULES and MICROFILAMENTS and the reformation of the GOLGI APPARATUS. Processes associated with the interphase plasma membrane, like PINOCYTOSIS and ruffling, also recommence.

TEM TRANSMISSION ELECTRON MICROSCOPY.

temperate cycle Life cycle of BACTERIOPHAGE in which the virus enters the PROPHAGE state, does not replicate and does not cause lysis of the infected cell. *See e.g.* LAMBDA; LYSOGEN.

temperate phage BACTERIOPHAGE capable of entering either a LYTIC or TEMPERATE CYCLE in contrast to VIRULENT phages which are only capable of lytic growth. Temperate phage produce TURBID PLAQUES rather than the clear plaques produced by virulent phages on a lawn of host bacteria. Bacteriophage LAMBDA is an example of a temperate phage.

temperature factor Common name for the Debye–Waller factor which describes the thermal motion of atoms within a crystal. This motion causes a decrease in the diffracted intensities by a factor:

$$\exp[-B(\sin^2\theta/\lambda^2]$$

where θ is the Bragg angle, λ is the wavelength and B the Debye–Waller or temperature factor. This is related to the mean square total displacements $\bar{U}^2$ by $B = 8\pi^2\bar{U}^2/3$ such that root mean square (r.m.s.) displacements of 0.5 Å and 1.0 Å correspond to B-factors of about 6 and 27 respectively.

For most macromolecules, the temperature factor — or B-factor — is the only refinable parameter describing the state of order or disorder in crystals. Data are usually insufficient to distinguish between static and dynamic disorder, except for small structures of only a few atoms studied at more than one temperature. Some crystals of macromolecules are quite disordered at high RESOLUTION; this is reflected by the magnitude of the diffracted intensities decaying very rapidly with resolution and it can be quantified by an overall B-factor for the diffraction data set relating it to a standard model.

A refined crystallographic model includes the atomic coordinates of every non-hydrogen atom and also an atomic B-factor, though these are sometimes averaged within an amino acid side chain, or along regions of the polypeptide chain. Typically, amino acid side chain atoms have higher B-factors than the backbone atoms, and any mobile or flexible regions of a polypeptide will be refined with very high B-factors. For crystals of small organic molecules, individual atom anisotropic B-factors can be refined to show directional variation in the localization of atoms.

The term temperature factor is thus used by crystallographers in several distinct ways; the nonspecific term B-factor may be less misleading in many situations where the disorder within a crystal is of unknown origin. Similarly, use of the terms 'hotter' and 'colder' to describe regions within a molecule having higher and lower atomic B-factors is a shorthand abstraction with no physical reality.

Atomic thermal motions do increase with temperature by laws of classical thermodynamics, a fact exploited in the SIMULATED ANNEALING method used in the REFINEMENT OF MACROMOLECULAR STRUCTURES where molecular dynamics calculations are performed at an elevated temperature, enabling high energy barriers to be crossed before slow cooling to identify the molecular conformation with minimum energy. *See*: X-RAY CRYSTALLOGRAPHY.

Stout, G.H. & Jensen, L.H. (1989) *X-ray Structure Determination: A Practical Guide* (Wiley, New York).

temperature-sensitive (*ts*) mutants A class of CONDITIONAL MUTANTS which express their mutant PHENOTYPE only at a higher temperature (the restrictive or nonpermissive temperature), yet show a WILD-TYPE phenotype at a lower temperature (the permissive temperature). At the nonpermissive temperature the product of the mutant gene is believed to be nonfunctional as a consequence of heat-induced conformational changes in its structure. Temperature-sensitive mutants are valuable for studying the function of essential genes in macromolecular syntheses and in the CELL CYCLE.

template A POLYNUCLEOTIDE chain (either RNA or DNA) that is utilized during either DNA REPLICATION or TRANSCRIPTION as a guide to the synthesis of a second polynucleotide chain with a complementary base sequence. For example, one of the two strands of DNA within a gene acts as a template for mRNA synthesis by RNA polymerase. The DNA strand which acts as the template in transcription is sometimes referred to as the template strand and is therefore complementary in base sequence to the mRNA.

tenascin *See*: EXTRACELLULAR MATRIX MOLECULES.

tensin *See*: ACTIN-BINDING PROTEINS.

tenuin Protein associated with intermediate junctions. *See*: CELL JUNCTIONS.

tenuivirus group Name from Latin *tenuis*, thin or fine, referring to the virus particles. Group of plant viruses with filamentous particles, 8 nm in diameter and of varying length, occasionally branched. The particles contain four or possibly five species of single-stranded RNA some of which appear to be of (−)-sense. The genome organization has not been determined. *See also*: PLANT VIRUSES.

Toriyama, S. (1986) *Microbiol. Sci.* **3**, 347–351.

teratocarcinoma Malignant tumour comprising undifferentiated STEM CELLS and cells (derived from them) of several adult tissue types, for example gut epithelium, muscle, cartilage, and bone.

The undifferentiated cells grow *in vitro*, being known as embryonal carcinoma cells (EC cells). Teratocarcinomas may arise spontaneously, for example from PRIMORDIAL GERM CELLS or OOCYTES in certain strains of mice, or may be produced experimentally by grafting early mouse embryos to suitable nonuterine sites (such as the testis or kidney capsule) in genetically identical hosts.

teratogen Any substance that can cause abnormal embryonic development if present in the environment of the early embryo. Examples are the rubella virus and the drug thalidomide.

teratology The branch of embryology concerned with the study of 'monsters' (Gk, *teras*), or abnormalities of development, although the term is used more commonly to describe experiments studying the effects of noxious substances on embryonic development. The immunologist Peter Medawar pronounced: 'The classification and investigation of abnormal embryos ... are the subject matter of 'teratology', a word that suggests pretensions to the stature of a science (a designation not really deserved) ... , but teratology has not — as had at one time been hoped — thrown a flood of light upon developmental processes, and it has not helped us very notably in the interpretation of normal development. Teratology is more deeply in debt to embryology than the other way around.'

terminal deoxynucleotidyltransferase (TdT, terminal transferase) Enzyme (EC 2.7.7.31) which catalyses the non-templated addition of nucleotides to the 3′ end of a DNA strand. It is responsible for generating additional diversity at the joints of V, D, and J segments in immunoglobulin heavy chain genes and T cell receptor genes (*see* GENERATION OF DIVERSITY; N-REGION DIVERSITY). Once the DNA is cut during the recombination process an exonuclease activity is able to remove several base pairs which TdT then replaces with random nucleotides as its polymerase activity is independent of a nucleic acid template. TdT activity is found in cells of the early B and T cell lineages (*see* B CELL DEVELOPMENT; T CELL DEVELOPMENT) and is no longer found in cells that are expressing functional immunoglobulins or T cell receptors.

It is also used *in vitro* in recombinant DNA work to add HOMOPOLYMER TAILS to DNA fragments to aid cloning.

terminal differentiation The end state of a pathway of DIFFERENTIATION. For example, in erythropoiesis it is the red blood cell, in B CELL DEVELOPMENT the antibody-secreting PLASMA CELL, and in neuroblast differentiation the fully functional post-mitotic NEURON.

terminal group genes A class of genes in *Drosophila* which, when mutant, give rise to embryos lacking the unsegmented terminal regions, the acron and telson. The specification of the termini involves a signal transduction pathway that leads from a signal generated in the somatic follicle cells to the response in the oocyte. The gene *torso-like* is required in a subpopulation of follicle cells for the production, during oogenesis, of a localized ligand which may be anchored in the VITELLINE MEMBRANE. The

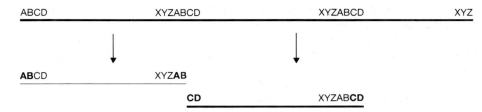

Fig. T14 The generation of terminal redundancy following cleavage of concatamers of phage genomes. The terminal redundancies are in bold type.

receptor is the product of the *torso* gene and is a receptor tyrosine kinase (*see* GROWTH FACTOR RECEPTORS) present uniformly in the plasma membrane of the egg. After fertilization, the ligand is activated and induces the local activation of the *torso* kinase at the termini. This leads to a signal transduction cascade involving *l(1)pole hole*, the *Drosophila* homologue of the vertebrate proto-oncogene D-*raf* which encodes a serine/threonine PROTEIN KINASE. The end product of the pathway is the transcriptional activation, at the termini, of the GAP GENES *huckebein* and *tailless*. *See also*: DROSOPHILA DEVELOPMENT.

terminal redundancy The duplication of a portion of a BACTERIOPHAGE genome at either end. Such terminal redundancy is generated as a consequence of the cleavage of the CONCATAMERS of phage genomes into lengths that are greater than that of a single genome, thereby duplicating the first segment of the genome to be packaged at the end of the packaged genome (Fig. T14).

terminal transferase *See*: TERMINAL DEOXYNUCLEOTIDYLTRANSFERASE.

terminal web A specialized cytoskeletal structure underlying the apical plasma membrane in epithelial cells of the intestinal brush border. It consists of a network of SPECTRIN overlying a layer of INTERMEDIATE FILAMENTS. The bundles of actin filaments that form the core of microvilli extend into the terminal web and are linked together by spectrin. *See*: CELL MOTILITY.

terminase *See*: LAMBDA.

termination (1) In TRANSCRIPTION, a TERMINATOR signal within the DNA sequence 3′ to the coding sequence of the gene is recognized by RNA polymerase, followed by completion of the final PHOSPHODIESTER BOND. The transcription complex then dissociates from the DNA. *See also*: BACTERIAL GENE EXPRESSION; EUKARYOTIC GENE EXPRESSION.
(2) In translation (*see* PROTEIN SYNTHESIS) termination is signalled by the arrival of a TERMINATION CODON in an mRNA at the ribosomal A site. A RELEASE FACTOR then binds to the ribosome and catalyses the cleavage of the polypeptide from the peptidyl-tRNA located at the ribosomal A site. The ribosome then dissociates from the mRNA.

termination codon Any of the CODONS UAA, UGA, UAG, which signal the TERMINATION of translation. They are also called stop codons or nonsense codons. *See also*: GENETIC CODE; NONSENSE MUTATION; PROTEIN SYNTHESIS.

terminator A DNA sequence located 3′ to the coding region of a

GENE which causes RNA polymerase to terminate TRANSCRIPTION of the gene and to dissociate from the DNA. In prokaryotic cells, the terminator sequence is usually located within the transcribed region and in some cases requires an additional factor, the RHO PROTEIN, to facilitate termination. Prokaryotic terminators are classified as being either rho-dependent or rho-independent. Unlike promoter sequences, there appears to be no CONSENSUS SEQUENCE associated with terminators in either prokaryotic or eukaryotic genes.

terpenes Generic name for substances composed of the 5-carbon isoprenoid unit (Fig. T15). Derivatives of terpenes or terpenes conjugated with other compounds form a large number of substances found in plants including essential oils such as citral, geraniol, camphor, and menthane, many pigments (e.g. CAROTENOIDS), plant hormones (e.g. GIBBERELLINS), and the vitamin A (*see* RETINOL) and squalene of animals. Nomenclature is based on a C_{10} unit: thus monoterpenes (C_{10}); sesquiterpenes (C_{15}); diterpenes (C_{20}); triterpenes (C_{30}); tetraterpenes (C_{40}); polyterpenes (> C_{40}).

Fig. T15 *a*, Isopentenyl pyrophosphate; *b*, geranyl pyrophosphate, the basic monoterpene unit.

tertiary structure The overall three-dimensional conformation of a biopolymer. For proteins this involves side chain interactions and packing of secondary structure motifs. For nucleic acids this may be the packing of STEM-LOOPS or SUPERCOILING of double helices. *See*: NUCLEIC ACID STRUCTURE; PROTEIN STRUCTURE.

testis-determining region Region on the Y-chromosome involved in male sex determination, and which contains the *SRY* gene (humans) or *Sry* gene (mouse).

testosterone A male sex hormone, a STEROID HORMONE produced by the testis and which is required for maturation and function of the testes and development of secondary sexual characteristics. It acts at intracellular receptors of the STEROID RECEPTOR SUPERFAMILY.

tetanic stimulation High-frequency stimulation (typically of 100 Hz and between 20 and 100 stimuli) applied to an afferent pathway in order to induce LONG-TERM POTENTIATION.

tetanus toxin Protein neurotoxin produced by the bacterium *Clostridium tetani*. In the activated form it is a METALLOPROTEIN-ASE which blocks NEUROTRANSMITTER release by cleaving the SNAP RECEPTOR protein synaptobrevin-2, an integral membrane protein of small synaptic vesicles. Tetanus toxin is produced as a single polypeptide chain which is cleaved to generate two chains linked by a single disulphide bond. The heavy chain (M_r 100 000) binds specifically to neuronal cells and allows entry of the light chain (M_r 50 000) which after reduction of the disulphide bond prevents neurotransmitter release by cleaving synaptobrevin-2.

Schiavo, G. et al. (1992) *Nature* **359**, 832–835.

tetcyclacis 5-(4-Chlorophenyl)-3,4,5,9,10-pentaaza-tetracyclo-5,4,1, $O^{2.6}, O^{8.11}$-dodeca-3,9-diene (formerly BAS 106.W) (Fig. T16). M_r 273.5. A PLANT GROWTH RETARDANT. It inhibits the oxidation of *ent*-kaurene to *ent*-kaurenoic acid in the GIBBERELLIN biosynthetic pathway. It is also reported to inhibit phytosterol biosynthesis resulting in reduced cell division and to block ABSCISIC ACID metabolism (hydroxylation at C8′) conferring enhanced stress tolerance, but these are relatively minor effects. It has been used to produce more compact rice seedlings with improved rooting and tillering characteristics, but this application has been discontinued.

Rademacher, W. et al. (1984) *Biochemical Aspects of Synthetic and Naturally Occurring Plant Growth Regulators* (Menhennett, R. & Lawrence, D.K., Eds) 1–11 (British Plant Growth Regulation Group, Wantage, UK).

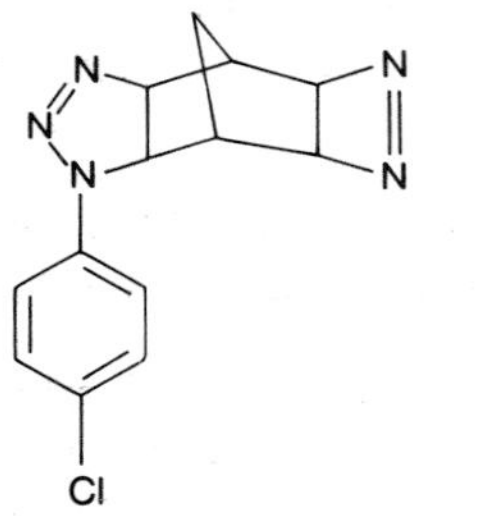

Fig. T16 Structure of tetcyclacis.

tetracyclines A group of antibiotics produced by *Streptomyces* species, and their related semisynthetic derivatives. They inhibit both Gram-positive and Gram-negative bacteria, and rickettsiae, and are also used as growth-promoting antibiotics. They inhibit bacterial PROTEIN SYNTHESIS by inhibiting the peptidyl transferase activity of the prokaryotic 50S ribosomal subunit. Clinically important tetracyclines include aureomycin and other chlortetracyclines, terramycin, and tetracycline (Fig. T17).

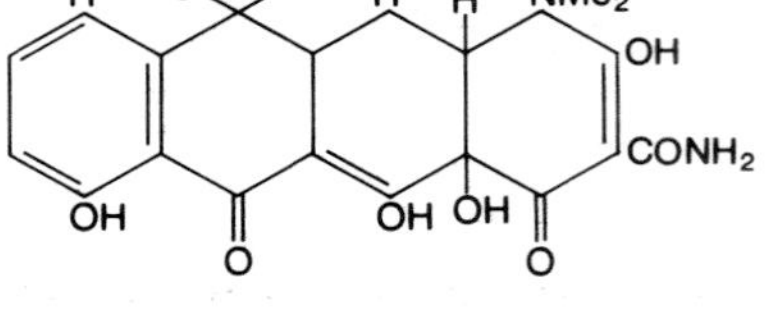

Fig. T17 Tetracycline.

tetrad analysis A method of genetic analysis that can be applied to ascomycete fungi. *See:* NEUROSPORA CRASSA.

Δ^9-tetrahydrocannabinol The psychoactive ingredient of marijuana. G PROTEIN-COUPLED cannabinoid receptors have been found in brain and in peripheral tissues (spleen).

Matsuda, L.A. et al. (1990) *Nature* **356**, 561–564.
Munro, S. et al. (1993) *Nature* **365**, 61–65.

Tetrahymena A genus of ciliated PROTOZOAN (class Ciliatea). Each cell contains two types of nuclei: a hyperpolyploid MACRONUCLEUS which controls the growth and development of the organism, and one or two diploid MICRONUCLEI which are involved only in sexual reproduction. About 10% of the DNA sequences present in the micronuclei are also found in the macronucleus. Self-splicing introns were first discovered in studies on ribosomal biogenesis in *Tetrahymena thermophila* (*see* RIBOZYMES).

tetraploid Organism or cell whose somatic nuclei contain four copies of each of the CHROMOSOMES (autotetraploid, AAAA) or which have arisen by hybridization between related species followed by production of unreduced gametes (allotetraploids, AABB). Tetraploids may also be designated as 4x where x is the original haploid number of chromosomes. Because tetraploids have an even number of chromosomes they can undergo regular MEIOSIS. Stable tetraploid species are common in plants.

Tetraviridae Family of insect RNA viruses. *See:* ANIMAL VIRUSES.

tetrodotoxin (TTX) Neurotoxin from the ovaries of puffer fish which acts by selectively blocking voltage-gated sodium channels (*see* VOLTAGE-GATED ION CHANNELS).

Texas red™ A sulphonyl chloride derivative of sulphorhodamine whose emission spectrum (peak ~620 nm) shows very little overlap with FLUORESCEIN and so is widely used with fluorescein in two-colour fluorescence microscopy.

TF TRANSCRIPTION FACTOR. The groups of general transcription factors which form the transcription initiation complexes required for basal levels of transcription from eukaryotic promoters transcribed by RNA POLYMERASES I, II, and III, are termed TFI, TFII, and TFIII respectively. *See also:* EUKARYOTIC GENE EXPRESSION; TRANSCRIPTION.

TFI TRANSCRIPTION FACTORS forming the basal transcription initiation complex on promoters of genes transcribed by RNA polymerase I (*see* EUKARYOTIC GENE EXPRESSION; EUKARYOTIC GENE STRUCTURE).

TFIIA—H A group of TRANSCRIPTION FACTORS which form the transcription initiation complex required for basal TRANSCRIPTION from promoters of genes transcribed by RNA polymerase II (Table T2). *In vitro* studies have shown that additional factors are also required for the formation of an initiation complex responsive to TRANSCRIPTIONAL ACTIVATORS.

Table T2 Transcription factors forming the basal transcription complex at promoters of genes transcribed by RNA polymerase II.

Transcription factor	Molecular details	Probable function
TFIIA	Monomer	Stabilizes TFIID complex on TATA box
TFIIB	Monomer, N-terminal region required for binding to TFIID : DNA complex	Binds to TFIID : DNA complex and provides binding site for TFIIF : polymerase complex
TFIID	Multisubunit protein containing the TATA-box binding subunit (TFIIDτ)	Binds to TATA box to initiate assembly of the transcriptional initiation complex
TFIIE	Heterotetramer of two different subunits (M_r α, 57 000; β, 34 000). α contains a zinc finger motif and leucine zipper motif	Enters preinitiation complex after TFIIF : polymerase. Involved in polymerase binding to DNA
TFIIF	Heterodimer (M_r α, 74 000; β, 30 000)	Binds to RNA polymerase II and may be involved in polymerase binding to the TFIID : DNA complex
TFIIH	Monomer	Enters preinitiation complex after polymerase, TFIIF, and TFIIE

TFIID A multisubunit TRANSCRIPTION FACTOR which binds to the TATA BOX in eukaryotic PROMOTERS transcribed by RNA polymerase II (*see* EUKARYOTIC GENE EXPRESSION; EUKARYOTIC GENE STRUCTURE) and which plays a central role in the formation of the initiation complex for basal transcription on such promoters. In higher eukaryotes, it comprises the TATA-box binding subunit (TFIIDτ) and at least seven associated proteins. It is present in all eukaryotic cells and is highly conserved in terms of both structure and function; for example, mammalian TFIID will function in the lower eukaryote SACCHAROMYCES CEREVISIAE. Bound TFIID binds the polymerase–TFIIF complex.

TFIII A group of eukaryotic TRANSCRIPTION FACTORS that form the transcriptional complex required for basal TRANSCRIPTION from promoters of genes transcribed by RNA polymerase III.

TFIIIA A TRANSCRIPTION FACTOR necessary for the TRANSCRIPTION of the 5S rRNA genes by RNA polymerase III although it is apparently not required for the transcription of tRNA genes by the same polymerase. TFIIIA has an M_r of 37 000 and binds directly to DNA via zinc fingers at a site internal to the transcribed region (*see* PROTEIN–NUCLEIC ACID INTERACTIONS).

TGF Transforming growth factor. *See:* GROWTH FACTORS.

TGN TRANS GOLGI NETWORK.

thalamus Large, paired ovoid masses of grey matter in the brain, located in the diencephalon, below the lateral ventricles, which contain nuclei that are the site of termination of many sensory pathways.

thalassaemia Severe heritable anaemias of two types — α-thalassaemia and β-thalassaemia — which are a serious health problem in certain populations around the Mediterranean and in the Third World. The diseases are caused by mutations in the genes which encode HAEMOGLOBIN. In α-thalassaemia the α-globin genes on chromosome 16 are affected and there is a deficiency of α-globin, whereas in β-thalassaemia there is a deficiency of β-globin as a result of defects in the β-globin gene cluster on chromosome 11. In both types the synthesis of normal haemoglobin is reduced, causing severe anaemia in homozygotes, and the excess globin chains precipitate out within the red cells. Other symptoms include hepatosplenomegaly and bone malformation. The diseases are clinically heterogeneous, reflecting the involvement of a wide variety of mutations within the α- and β-globin gene clusters. These mutations are either gross deletions of genes and control elements, generally leading to a severe phenotype, or more subtle regulatory mutations affecting transcription or translation of the proteins. β-Thalassaemia is sometimes ameliorated by the continued expression of the foetal γ-chains (*see* HEREDITARY PERSISTENCE OF FOETAL HAEMOGLOBIN); otherwise treatment is by blood transfusion. Both forms of thalassaemia can be detected by PRENATAL DIAGNOSIS. *See also:* HAEMOGLOBIN AND ITS DISORDERS.

thaumatin Protein isolated from berries of the tropical plant *Thaumatococcus danielli* which is intensely sweet to the taste.

theta (θ) Notation for the BRAGG ANGLE in DIFFRACTION studies, where 2θ equals the angle of diffraction, the angle between incident and diffracted rays. *See:* X-RAY CRYSTALLOGRAPHY.

theta replication, theta ring Mode of replication in some circular DNAs such as plasmids and bacterial chromosomes, where bidirectional replication starting from a unique origin of replication displaces the non-template DNA strand and leads to a structure resembling the Greek letter θ as viewed in the electron microscope (Fig. T18).

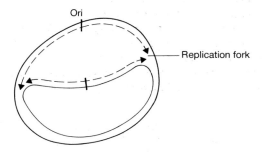

Fig. T18 Theta replication.

theta rhythm One of the natural firing rhythms (4–12 Hz) of the hippocampal EEG rhythm recorded *in situ*. This firing pattern is mimicked by primed-burst stimulation, which very effectively induces LONG-TERM POTENTIATION.

thick filament, thin filament *See*: MUSCLE.

thin layer chromatography *See*: CHROMATOGRAPHY.

thioredoxins Small proteins of M_r 12 000 that are involved in certain redox processes by virtue of a disulphide group that can be reversibly oxidized and reduced. Chloroplasts contain two thioredoxins that are involved in light regulation of carbon fixation; thioredoxin *f* activates by reduction certain enzymes of the reductive pentose phosphate cycle (CALVIN CYCLE), including fructose bisphosphatase, sedoheptulose bisphosphatase, phosphoribulokinase, and NADP-linked glyceraldehyde phosphate dehydrogenase, wherease thioredoxin *m* deactivates glucose-6-phosphate dehydrogenase. Thioredoxin is reduced in the light by FERREDOXIN in a reaction catalysed by FERREDOXIN-THIOREDOXIN REDUCTASE. *See also*: PHOTOSYNTHESIS.

thorax Anatomical region of adult insects which bears the legs and wings. There are three thoracic segments, T1–T3, each of which bears a pair of legs. T2 also bears a pair of wings, while T3 bears the halteres (*see* IMAGINAL DISKS). In the larva, the segment corresponding to T1 is termed the prothorax, T2 the mesothorax, and T3 the metathorax.

threonine (Thr, T) An AMINO ACID with an aliphatic hydroxyl side chain, and which has two centres of asymmetry.

thrombin SERINE PROTEINASE (EC 3.4.21.5) (M_r 39 000) found in blood plasma which cleaves FIBRINOGEN to FIBRIN in the blood coagulation cascade in a calcium-dependent reaction (*see* BLOOD COAGULATION AND ITS DISORDERS). It cleaves specifically after arginine residues. Thrombin is synthesized as an inactive precursor — PROTHROMBIN — which is activated by proteolytic cleavage by Factor Xa (*see* Fig. P52 in POST-TRANSLATIONAL MODIFICATION OF PROTEINS). Thrombin is inhibited by ANTITHROMBIN, a member of the SERPIN family of proteinase inhibitors.

thrombocyte PLATELET.

thrombomodulin *See*: BLOOD COAGULATION AND ITS DISORDERS.

thrombospondin Extracellular adhesion molecule (*see* EXTRACELLULAR MATRIX MOLECULES) which is a homotrimeric GLYCOPROTEIN (M_r 140 000). The protein has several high-affinity calcium-binding sites ($K_d \sim 10^{-7}$ M); they fall in a region of the protein that has eight repeats very rich in aspartic acid and asparagine residues, resembling 'EF hands' but without the flanking helices (*see* CALCIUM-BINDING PROTEINS).

thromboxanes *See*: EICOSANOIDS.

Thy-1 A DIFFERENTIATION ANTIGEN present on the surface of T CELLS and other haemopoietic cells, but also expressed on neu-

rons, fibroblasts, and vascular endothelial cells. Structurally related to the IMMUNOGLOBULINS and considered to be a member of the IMMUNOGLOBULIN SUPERFAMILY.

thylakoid The 'flattened sac'-like structures that form the coupling membranes of CHLOROPLASTS (*see* PHOTOSYNTHESIS). The thylakoid membrane separates the stroma aqueous phase from an internal aqueous phase or lumen; the latter becomes markedly acidified upon illumination. Thylakoid membranes occur in many different forms. In chloroplasts of most higher plants they are differentiated into the stacked disks called grana, and the stroma lamellae which are fully accessible to the stroma. The light-harvesting and electron transport systems of photosynthesis are closely associated with the thylakoid membranes.

thymic progenitor cells PRO-T CELLS. *See*: T CELL DEVELOPMENT.

thymidine The nucleoside containing thymine. *See*: NUCLEOSIDES AND NUCLEOTIDES.

thymidine kinase (TK) An enzyme (EC 2.7.1.21) which catalyses the reaction:

ATP + thymidine = ADP + thymidine 5′-monophosphate

a step in the synthesis of thymidine triphosphate (dTTP), an important precursor of DNA synthesis. It allows the salvage of THYMIDINE produced by the breakdown of DNA. Thymidine kinase is unique in that it is usually active in dividing tissues but shows very low activity in nongrowing tissues. Fungi, including SACCHAROMYCES CEREVISIAE, apparently lack TK activity. The *tk* gene from herpes simplex virus (HSV) has been widely used as a SELECTABLE MARKER in the TRANSFORMATION of mammalian cells. Mutant cells which lack *tk* (i.e. tk⁻) will grow on 5-BROMODEOXYURIDINE, a compound which can only be incorporated (and therefore kill cells) if a cell is tk⁺. Tk⁻ cells that have been transformed to tk⁺ by transformation or TRANSFECTION with DNA carrying the HSV *tk* gene can be selected for by growth on HAT MEDIUM which contains hypoxanthine, aminopterin, and thymidine.

thymine (T) A pyrimidine, one of the four nitrogenous BASES present in DNA. In DNA it pairs with adenine (*see* BASE PAIR). *See also*: NUCLEOSIDES AND NUCLEOTIDES.

thymine dimer Structure formed between two adjacent thymines in DNA under the action of ultraviolet light. In normal cells thymine dimers can be excised from DNA and the damage repaired (*see* DNA REPAIR) but the cells of people suffering from the genetic defect XERODERMA PIGMENTOSUM are deficient in the ability to carry out this repair and such people are as a result exceptionally sensitive to sunlight.

thymocyte Bone-marrow derived cell maturing within the thymus into a T CELL. *See*: T CELL DEVELOPMENT.

thymoma Tumour originating in the thymus, usually benign, associated with symptoms of myasthenia.

thymosin *See*: ACTIN-BINDING PROTEINS.

thyroglobulin Iodine-containing glycoprotein which is the main store of iodine in the thyroid gland.

thyroid hormones Iodine-containing protein hormones produced by the thyroid gland. The major hormones are thyroxine (T_4) (*see* Fig. T19) and triiodothyronine (T_3); they have a pleiotropic stimulatory effect on metabolism and/or growth in many organs, exerted at the level of gene expression. Receptors for these hormones belong to the STEROID RECEPTOR SUPERFAMILY in which the ligand–receptor complex acts as a transcriptional activator. Both deficiencies and overproduction of thyroid hormones are common and clinically important. Human thyrotoxicosis (overproduction of thyroid hormone) is usually AUTOIMMUNE in origin (Graves' disease) and is associated with the MHC (HLA) antigens B8 and DR3. Several rare hereditary defects of thyroid hormone synthesis have been described, including Pendred syndrome, associated with deafness.

thyroid-stimulating hormone (thyrotropin, TSH) Glycoprotein hormone produced by the anterior pituitary, comprising an α-chain (92 amino acids) and a β-chain (112 amino acids). It stimulates production of thyroid hormones (e.g. thyroxine) from the thyroid and fatty acid release from fat cells.

thyroxine L-3,5,3′,5′-tetraiodothyronine (Fig. T19). One of the principal THYROID HORMONES, also called T_4.

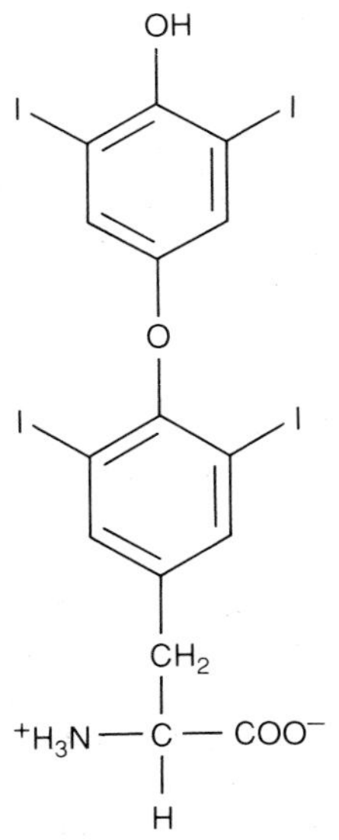

Fig. T19 Thyroxine.

Ti Former abbreviation for the antigen-specific αβ component of the T CELL RECEPTOR complex, now becoming superseded by designations for individual molecules within the complex.

Ti plasmid

THE Ti plasmid is the tumour-inducing PLASMID of the plant-pathogenic bacterium *Agrobacterium tumefaciens*, which causes the disease crown gall, a plant tumour [1]. Part of the plasmid is routinely transferred into the plant cell genome during infection, and so the Ti plasmid is now widely used, in modified form, as a VECTOR for introducing novel genes into plants (*see* PLANT GENETIC ENGINEERING).

Agrobacterium and crown gall

Agrobacterium tumefaciens is a Gram-negative, coliform, soil-borne plant-pathogenic bacterium which causes crown gall on a wide range of mainly dicotyledonous plants. Infection, normally at a wound site near soil level (at the crown of the plant), is followed by the proliferation of cells to form a gall (crown gall). The gall cells can be cultured *in vitro* in the absence of the inducing bacteria, and display a unique altered phenotype in that their continued growth in culture is not dependent on the presence of PLANT GROWTH HORMONES (e.g. AUXINS and/or CYTOKININS) added to the medium.

In addition to becoming autotrophic for growth hormones (hormone autotrophy) the 'transformed' plant cells also overproduce and secrete one or more unique, small molecules generically referred to as OPINES [2]. The majority of these are imino acids formed by reactions on Schiff bases formed between amino acids and sugars or ketoacids. The precursors for these opines are all small molecules, freely available as normal metabolic intermediates in plant cells. The particular families of opines synthesized are characteristic of the inducing strain of *A. tumefaciens* and the inducing strain can also utilize the specific opines secreted by the gall cells as a sole source of carbon and nitrogen.

As a result, the OPINE CONCEPT has been postulated which maintains that the pathogen induces opine production in the host cells to produce an environment for growth and propagation of the pathogen. A scheme for classifying virulent strains has been developed which is based on the type of opine utilized by specific strains and synthesized in gall cells produced as a result of transformation by that strain. Strains are designated for example, nopaline-type or octopine-type.

Biological function of the Ti plasmid

In 1977, it was demonstrated that the transformation phenomenon was related to the transfer of a defined segment of DNA (the T-DNA, see below) from a specific region (T-region) of a large plasmid, the Ti plasmid, carried by virulent strains of the bacterium. These plasmids are of the order of 200 kilobase pairs (kbp) in size and contain the genes that determine tumorigenicity, genes involved in the transfer of plasmid DNA into the host genome, the genes encoding enzymes of opine catabolism, and genes for viral exclusion functions and plasmid maintenance, among others.

In all cases examined to date, at least one of the opines synthesized in gall tissue also functions as an inducer for the conjugal transfer of the Ti plasmid from the pathogen to suitable recipient bacterial strains, and this capability is again strain and opine specific. In general, in the presence of the opine, conjugal transfer frequencies increase 1000-fold.

Plasmids are common in wild isolates of *Agrobacterium* and vary in size from 5 kbp to more than 400 kbp, but only some of these carry the necessary genes to confer tumorigenicity on the bacterium. In some strains, however, some of the opine catabolic functions are found on plasmids other than the Ti plasmid. The molecular mechanisms of crown gall tumorigenesis are reviewed in [3].

T-DNA

The T-region, the portion of the Ti plasmid from which the T-DNA is derived, is delimited by imperfect DIRECT REPEATS (border repeat sequences) of 25 base pairs and may be one continuous region or may be divided into two (T_L and T_R) regions. In strains C58 or T37, which are characterized by their metabolism of nopaline, the T region is monopartite and some 23 kbp long. In strains ACH5 and 15955 which are characterized by their metabolism of octopine, the T-region is bipartite: T_L is 13 kbp long whereas T_R is 8 kbp long. In the octopine Ti plasmids, the T_L-region contains the 'oncogenic' (*onc*) functions and the gene for OCTOPINE SYNTHASE whereas the T_R-region contains the genes for the synthesis of the mannityl opines (MANNOPINE and MANNOPINIC ACID and their corresponding lactonized forms, AGROPINE and agropinic acid).

In the large, monopartite T-region of nopaline Ti plasmids, the *onc* genes and NOPALINE SYNTHASE occupy the right-hand half of the 23 kbp. Immediately to the left of this region lies the gene for another opine group, AGROCINOPINES A and B. The remaining five or six genes on the nopaline T-region have, as yet, no known function.

onc functions

The *onc* functions, carried by the T-region, are responsible for the hormone-independent growth of transformed plant cells. Two genes (*iaaH* and *iaaM*) encode enzymes that convert tryptophan to indoleacetamide and finally to indoleacetic acid (an auxin). These gene loci are known as the *shi* (shoot-inducing) or *tms* (tumour morphology shooty) loci because, when mutated or removed, the tumour phenotype is shoot proliferation resulting from the high cytokinin, low auxin hormonal imbalance.

One gene (*ipt*) is responsible for the synthesis of the cytokinin isopentenyl adenine. This locus is known as the *roi* (root-inducing) or *tmr* (tumour morphology rooty) locus because, when mutated or removed, the tumour phenotype is root proliferation resulting from the high auxin, low cytokinin hormone imbalance.

There are two other genes within the *onc* region as it is currently defined. One of these (the *tml* locus) affects tumour size (tumour morphology large) in some species (e.g. *Kalanchoe*). However, at least one strain of *Agrobacterium* (AT181) lacks this gene altogether and so its importance in pathogenicity is not clear. The final gene (*ons*) encodes a product which appears to be involved in opine secretion.

Transfer of T-DNA to the plant genome

The plasmid sequences to be transferred to plants are generated as a single-stranded molecule (T-strand) starting at the right border sequence (5′ end) and terminating at the left border sequence (3′ end) (Fig. T20) [4,5]. The plasmid-encoded virD2 protein (see virulence genes below) is covalently bound to the 5′ end of the T-strand and the DNA molecule is coated with a protein that nonspecifically binds single-stranded DNA, the product of the plasmid *virE2* gene. Together these comprise the T-complex. Initiation and termination of T-strand production is also affected by the sequences around the border repeat. In particular, a sequence known as 'OVERDRIVE' found associated with the right border repeat has a significant effect in stimulating T-strand production.

A key feature of the transformation efficiency of the *Agrobacterium* transformation system appears to be the presence of two NUCLEAR IMPORT SIGNALS (KRXR in single-letter amino-acid notation) located near the C terminus of the virD2 protein. The *virE2* gene product, of which there may be as many as 600 molecules per T-complex (for a 20 kb T-strand), also contains signals that can assist in transfer of the T-complex across the nuclear pore. These signals ensure that the T-strand is transferred in a polar fashion into the plant cell nucleus with high efficiency.

Both Ti plasmid and bacterial chromosomal functions are involved in the mechanism of pathogenicity. A number of chromosomal genes (*chv* genes) have been identified that play a part in the interaction of the bacterium with plant cells. These genes encode a range of products associated with cell wall or membrane modifications or modulate VIRULENCE GENE (*vir* gene) expression in some way. The *ros* chromosomal locus also affects the host–pathogen interaction apparently also by modulating the expression of some of the *vir* genes.

Virulence (*vir*) genes

The *vir* genes themselves, some of which have been described already, span about 30 kbp located counter-clockwise of the T-region. They encode the products necessary for the transfer of the plasmid sequences (T-DNA) to the plant cell and are activated by exposure to wounded or metabolically active plant cells. This activation is triggered by a group of small molecules produced by such plant cells. The first of these to be identified was acetosyringone (3′,5′-dimethoxy-4′-hydroxyacetophenone). Since then, many related molecules have been identified which have similar, albeit weaker, effects.

The *vir* genes are activated through a signal transduction system based on protein phosphorylation and similar in principle to many other bacterial signal transduction systems [6]. The sensor/transmitter is the transmembrane virA protein, and the receiver/regulator is the virG protein, which is activated by a kinase activity associated with the virA protein. The virA protein, which, in association with the *chvE* gene product, detects the signal molecules, is structurally related to a wide range of protein histidine kinases involved in such diverse bacterial systems as chemotaxis (*see* BACTERIAL CHEMOTAXIS), sporulation, nitrogen metabolism regulation, hexose phosphate uptake, phosphoglyc-

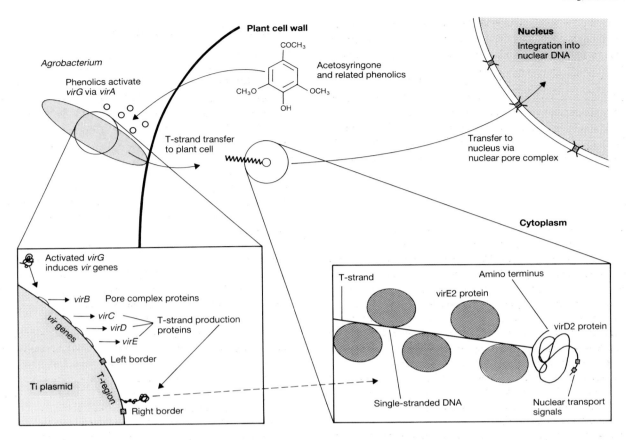

Fig. T20 Summary diagram of *Agrobacterium* infection and gene transfer.

erate transport and F plasmid expression and aerobic metabolism. The *virG* gene product, once activated, switches on the remaining *vir* genes (*virB*, *virC*, *virD*, *virE*) and, in some cases, genes which do not appear to be absolutely required for expression of the virulent phenotype but are none the less induced by the same system (*tzs*, *pinF*).

The 11 products of the *virB* OPERON most probably encode the proteins needed to form the conjugation-like pore in the bacterial surface which allows transfer of the T-DNA complex into the plant cell.

Those products of the *virC*, *virD* and *virE* operons whose functions are known are responsible for the release and packaging of the single-stranded derivative of the T-region (the T-strand). Single-strand nicks are made at or near the third base pair in the border repeat and the bottom strand is unwound from the double-stranded molecule until the left border is reached where the process terminates and the T-strand is released.

One other gene found in the *vir* region of some, but not all Ti plasmids, is the *tzs* gene which results in transribosylzeatin production in the bacterium which can have positive or deleterious effects on the interaction between the two organisms, depending on the plant species involved.

Ti plasmids as vectors for genetic engineering

The genes contained within the T-region, between the two border repeats, have no apparent effect on the transfer mechanism, only on the expression of the disease phenotype in plant cells. As a result, the size of the T-region can be artificially altered: reduced to a few thousand base pairs or increased to 50 kbp or more without seriously affecting the transfer efficiency.

Genes from other sources can therefore be spliced into this region and be automatically transferred into the plant cell and integrated into the nuclear DNA. They replace the *onc* genes, which are unwanted in this context. Modified versions of the Ti plasmid, either contained within *Agrobacterium* or as isolated DNA, are therefore used widely as vectors in plant genetic engineering.

Because the action of the *vir* genes on the T-region is mediated by soluble protein products, it was found that the *vir* genes were still effective in mediating transfer of the T-region into plant cell DNA if they were located on a separate plasmid. A binary vector system has therefore been developed in which a 'helper' Ti plasmid lacking a T-region is used to drive the transfer of an isolated, reconstructed T-region which can contain a 'foreign' gene to be transferred and which is carried on a separate plasmid. The advantage of such systems is that they have allowed the construction of modern vectors with selection marker genes contained between a right and left border sequence without having to manipulate the very large Ti plasmid directly.

A.G. HEPBURN

1 Braun, A.C. (1978) Plant tumors. *Biochim. Biophys. Acta* **516**, 167.
2 Tempé, J. & Petit, A. (1983) La piste des opines. In *Molecular Genetics of the Bacteria-Plant Interaction* (Puhler, A., ed.) 14ff (Springer-Verlag, New York).
3 Kado, C.I. (1991) Molecular mechanisms of crown gall tumorigenesis. *Crit. Rev. Plant Sci.* **10**, 1
4 Zambryski, P. (1988) Basic processes underlying *Agrobacterium*-mediated DNA transfer to plant cells. *Annu. Rev. Genet.* **22**, 1.
5 Zambryski, P.C. (1992) Chronicles from the Agrobacterium–plant cell DNA transfer story. *Annu. Rev. Plant Physiol. Plant Mol. Biol.* **43**, 465–490.
6 Albright, L.M. et al. (1989) Prokaryotic signal transduction mediated by sensor and regulator protein pairs. *Annu. Rev. Genet.* **23**, 311.

TIBA 2,3,5-Triiodobenzoic acid. *See:* PHYTOTROPIN.

tight junction *See:* CELL JUNCTIONS.

TIM TRIOSE PHOSPHATE ISOMERASE.

TIM barrel Protein domain with eight parallel β-strands surrounded by eight α-helices. This was first seen in the enzyme TRIOSE PHOSPHATE ISOMERASE but many examples are now known. *See:* MECHANISMS OF ENZYME CATALYSIS; MOLECULAR EVOLUTION; PROTEIN STRUCTURE (Fig. P81).

TIP Tumour-inducing principle. Early term used to describe the factors causing the CROWN GALL disease of plants. Now known to be the transfer of genes from the infecting bacterium (*Agrobacterium tumefaciens*) to the plant cell.

tissue plasminogen activator (tPA, TPA) SERINE PROTEINASE synthesized by a variety of tissues which cleaves plasminogen to plasmin and hence stimulates fibrinolysis. It is involved in the control of plasmin activity at cell surfaces in developing tissues and in blood coagulation (*see* BLOOD COAGULATION AND ITS DISORDERS). Human recombinant TPA is used clinically as a thrombolytic agent. *See:* SERPINS for inhibitors; *see also* UROKINASE.

tissue-specific gene/promoter Gene that is expressed only in a particular cell type, or in a particular tissue distribution. The PROMOTERS that direct such specific gene expression are often known as tissue-specific promoters. *See:* EUKARYOTIC GENE EXPRESSION; TRANSCRIPTION FACTORS.

titin *See:* ACTIN-BINDING PROTEINS; MUSCLE.

TL, TLa In the mouse, a complex of genetic loci mapping telomeric to the H-2 complex (*see* MAJOR HISTOCOMPATIBILITY COMPLEX) and spanning ~1 centimorgan of DNA, and which encodes the MHC class I Qa and TL lymphoid differentiation antigens. Similar in structure to MHC class I antigens, the Qa and TLa gene products are less polymorphic than transplantation antigens and are expressed selectively: Qa antigens are expressed only on haematopoietic cells, and TL antigens only on thymocytes and some leukaemic cells. Their functions are unknown.

T$_L$-DNA, T$_R$-DNA The left- and right-most segments respectively of the bipartite T-regions found in octopine TI PLASMIDS and the supervirulent Ti plasmid of strain Bo542. T$_L$ normally contains the ONC GENES. T$_R$ normally contains genes for OPINE synthesis.

TLC Thin layer chromatography. *See:* CHROMATOGRAPHY.

tml locus Tumour morphology large locus of the TI PLASMID of *Agrobacterium tumefaciens*. The gene product is of unknown function and appears to affect tumour size on only some plant species.

TMP The nucleotide thymidine monophosphate, usually denoting the deoxyribonucleotide. *See also:* NUCLEOSIDES AND NUCLEOTIDES.

tmr locus Tumour morphology rooty locus of the T-region of the TI PLASMID of *Agrobacterium tumefaciens*. It is the site of the gene (*ipt* locus) which encodes CYTOKININ synthesis and is only expressed in transformed plant cells.

tms locus Tumour morphology shooty locus. The two genes (*iaaH, iaaM*) that encode AUXIN synthesis in the T-region of TI PLASMIDS and are only expresssed in transformed plant cells.

TMV Tobacco mosaic virus. *See:* PLANT VIRUSES; TOBAMOVIRUSES.

Tn An abbreviation for bacterial transposons and used as a prefix to identify a specific transposon, for example, TnA, Tn7, Tn3, etc. (*see* TRANSPOSABLE GENETIC ELEMENTS).

TNFα (cachectin), TNFβ (lymphotoxin, LT) *See:* TUMOUR NECROSIS FACTOR.

Tnt 1 A transposable genetic element found in tobacco. *See:* TRANSPOSABLE GENETIC ELEMENTS: PLANTS.

tobacco mosaic virus (TMV) A rod-shaped plant RNA virus that causes mosaic symptoms on tobacco (*Nicotiana*), tomato, and a wide range of other plants (see Figs P36h and P38 in PLANT VIRUSES). It is of historical importance as the first virus to be recognized as a new type of submicroscopic disease-causing agent, (by Ivanovski in 1892), and as the virus in which SELF-ASSEMBLY was first demonstrated. It has been the subject of much work on the molecular biology of virus assembly and of the development of X-RAY DIFFRACTION techniques for the intact virus (*see* FIBRE DIFFRACTION) and the circular subassembly of the coat protein, the disk (Fig. T21). The X-RAY CRYSTALLOGRAPHY analysis of the protein disk was a landmark proving the technical feasibility of all subsequent virus crystallography (Fig. T22). *See:* PLANT VIRUSES; TOBAMOVIRUS GROUP for illustration of genome arrangement, (Fig. T23).

Crowther, R.A. & Amos, L.A. (1971) *J. Mol. Biol.* **60**, 123–130.
Champness, J.N. et al. (1976) *Nature* **259**, 20–24.
Bloomer, A.C. et al. (1978) *Nature* **276**, 362–368.

tobamovirus group Sigla from TOBACCO MOSAIC VIRUS, the type member. Group of viruses with rigid rod-shaped particles, 300 nm long and 18 nm in diameter. The coat protein subunits of M_r 17 000 are arranged in a helix around the genomic (+)-sense RNA (*see* Fig. P38 in PLANT VIRUSES). The genome organization is shown in Fig. T23. The 5′ gene reads through a weak termina-

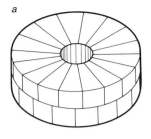

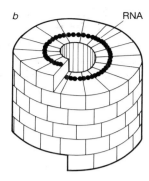

Fig. T21 Packing of subunits in the tobacco mosaic virus protein disk which self-assembles from capsid proteins, and in the virus. *a*, The disk contains 34 subunits in two rings of seven. *b*, The virus capsid is composed of helically arranged subunits with 16⅓ subunits per turn of the helix. The RNA present in the virus particle is sandwiched between the subunits as shown. In the transition from a 17-fold disk to a 16⅓-fold helix, subunits slide over each other by about 10Å.

tion codon to give two products which have sequence homology to RNA replicases. The P30 is involved in cell-to-cell movement of the virus in the plant and is translated from a subgenomic mRNA. The 3′ gene product is the coat protein also translated from a subgenomic mRNA. *See also*: PLANT VIRUSES.

van Regenmortel, M.H.V. & Fraenkel-Conrat, H., Eds (1986) *The Plant Viruses* Vol 2, Chapters 1–8 (Plenum, New York).

tobravirus group Sigla from the type member, tobacco rattle virus. Group of MULTICOMPONENT VIRUSES with rigid rod-shaped particles of two lengths. The long particles are 180–215 nm in length and contain RNA 1. The short particles are 46–114 nm in length and contain RNA 2. Both particles are 21–23 nm in diameter and have capsids of a single coat protein species (M_r 22 000) arranged in a helix. The genomic RNAs are (+)-sense, the genome organization being shown in Fig. T24. The 5′ gene of RNA 1 reads through a weak termination codon to give two products which have sequence homology to RNA replicases. The other two genes are translated from subgenomic mRNAs. The P29 is involved in cell-to-cell spread but the function of the other protein is unknown. The 5′ gene of RNA 2 is the coat protein. The rest of RNA 2 varies between strains, the PSG strain having no other genes whereas other strains have one or more genes of unknown function. *See also*: PLANT VIRUSES.

Harrison, B.D. & Robinson, D.J. (1986) In *The Plant Viruses*, Vol. 2 (van Regenmortel, M.H.V. & Fraenkel-Conrat, H., Eds) 339 (Plenum, New York).

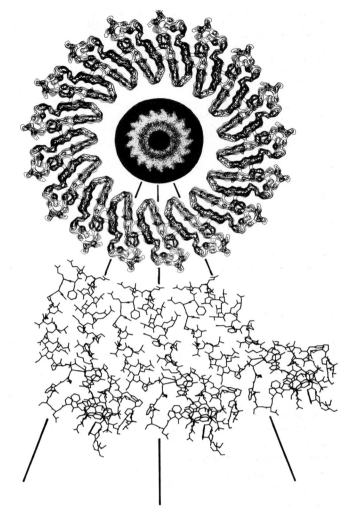

Fig. T22 The protein disk of tobacco mosaic virus viewed from above at successive stages of resolution. At the centre is a rotationally filtered electron microscope image at about 22Å resolution. Around it is a slice through the 5Å resolution electron density map of the disk obtained by X-ray analysis. On the outside is part of the atomic model built from the 2.8Å map. From Bloomer et al., 1978.

TOCSY *See*: NMR SPECTROSCOPY.

Togaviridae Family of enveloped RNA ANIMAL VIRUSES with spherical particles 40–70 nm in diameter. An icosahedral nucleocapsid contains a single molecule of (+)-sense single-stranded

(6395 nt)

P126	P183

| P30 |

| CP17 |

Fig. T23 structure of genome organization of tobacco mosaic virus (TMV). The lines represent the RNA species and the boxes the proteins (P) with M_r given as $\times 10^{-3}$; nt, nucleotides.

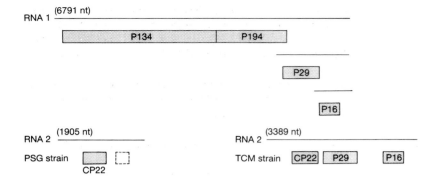

RNA 1 (6791 nt)

P134 P194

P29

P16

RNA 2 (1905 nt)

PSG strain

CP22

RNA 2 (3389 nt)

TCM strain CP22 P29 P16

Fig. T24 Genome organization of tobacco rattle virus (TRV). The lines represent the RNA species and the boxes the proteins (P) with M_r given as $\times 10^{-3}$; nt, nucleotides.

RNA (M_r 4×10^6). They include Sindbis virus and SEMLIKI FOREST VIRUS.

tolaasin *See*: LIPODEPSIPEPTIDE; PLANT PATHOLOGY.

tolerance A state of immunological nonresponsiveness induced to selected antigen(s) by previous exposure to the same antigen or antigenic determinants, usually in modified form or dose. Tolerance to self-MHC products and other self antigens occurs, at least in part, by deletion of self-reactive T cells during development in

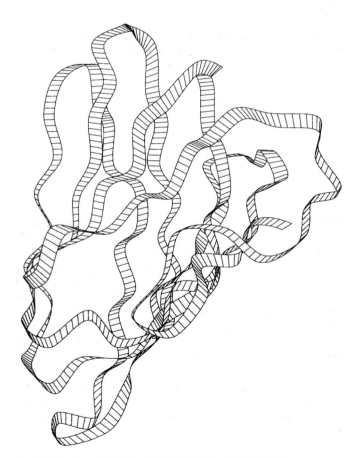

Fig. T25 Structure of capsid protein of tomato bushy stunt virus.

the thymus (*see* T CELL DEVELOPMENT). Breakdown of this clonal deletion process is one way in which AUTOIMMUNITY arises.

tomato bushy stunt virus (TBSV) The type virus of the TOMBUS-VIRUS GROUP (*see* Fig. T25 and Plate 7*a*).

tomato spotted wilt virus group Named after the type member. Group of plant viruses with spherical particles, 80–110 nm in diameter and comprising a lipid envelope surrounding a ribonucleoprotein core. The particles contain four protein species: the nucleocapsid protein (M_r 27 000), two membrane glycoproteins (M_r 58 000 and 78 000), and a large protein (M_r 120 000). The genome comprises three linear single-stranded RNAs, S (3.1 kb), M (5.4 kb), and L (8.2 kb) which are complexed with N proteins and form circular nucleocapsids. S RNA has ambisense strategy; this and other properties show that TSWV has many similarities to animal BUNYAVIRUSES. *See also*: PLANT VIRUSES.

Kormelink, R. et al. (1991) *Virology* **181**, 459–468.

tombusvirus group Sigla from tomato bushy stunt virus (TBSV), the type virus. Group of viruses with isometric particles, 30 nm in diameter (Plate 7*a*). The 180 coat protein subunits (M_r 41 000) are arranged in icosahedral symmetry around the genomic (+)-strand RNA. The crystal structure of the capsid protein from TBSV has been determined (Fig. T25). The protein has three domains: the N-terminal one interacts with RNA inside the capsid; the second surface domain covers the spherical capsid surface whereas the third protruding domain associates in dimers to give surface projections. The genome organization for cucumber necrosis virus, the only member so far sequenced, is shown in Fig. T26. The 5′ gene reads through a weak termination codon to give two products which have sequence homology to RNA polymerases. The coat protein is translated from a subgenomic mRNA as are the 3′ proteins, the functions of which are not known. *See also*: PLANT VIRUSES.

Martelli, G.P. et al. (1988) In *The Plant Viruses*, Vol. 3 (Koenig, R., Ed.) 13–72 (Plenum, New York).

tonofilament *See*: KERATIN FILAMENTS.

tonoplast The single cytoplasmic membrane around a VACUOLE in a plant cell.

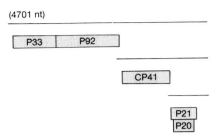

(4701 nt)

Fig. T26 Genome organization of cucumber necrosis virus (CNV). The lines represent the RNA species and the boxes the proteins (P) with M_r given as $\times 10^{-3}$; nt, nucleotides.

topoisomerases A class of enzymes (EC 5.99.1) found in both prokaryotes and eukaryotes which catalyse the interconversion of different topological isomers of DNA, thereby causing a change in the LINKING NUMBER. This involves the transient breakage of one or both strands of DNA and can result in the removal of negative or positive SUPERCOILS from DNA or the introduction of negative supercoils. There are two classes of topoisomerases: type I, which cleave only one of the two DNA strands and change the linking number in steps of one; and type II, which cleave both strands and change the linking number in steps of two. Type II topoisomerases are also known as DNA GYRASES. The cleavage sites generated by topoisomerases usually consist of a 5′-OH group and a 3′-P group, the latter being linked to a tyrosine residue in the enzyme. *See:* DNA; NUCLEIC ACID STRUCTURE.

Wang, J. (1985) *Annu. Rev. Biochem.* **54**, 665–699.

Toroviridae Family of enveloped RNA ANIMAL VIRUSES causing enteric infections in man and other mammals. The particles are elongated, 35–170 nm long, or bend into an incomplete torus. The single-stranded (+)-sense RNA genome ($M_r \sim 6.5 \times 10^6$) has a poly(A) tract at the 3′ end. The virus particle contains a major nucleocapsid phosphoprotein, two envelope proteins and a glycoprotein 'spike' protein. Four subgenomic mRNAs are produced in infected cells.

Torpedo californica The electric eel, whose electric organ is a rich source of NICOTINIC RECEPTORS.

totipotency The potency of a cell describes its potential for development into a range of possible FATES when placed in the appropriate environment. A totipotent cell can develop into all possible fates when implanted into the various regions of the embryo. If there is some restriction of fate, but the cell is not determined as to one fate alone, it is described as pluripotent.

toxins Toxic proteins and other compounds produced by living organisms. *See:* BACTERIAL PATHOGENICITY; BOTULINUM TOXIN; CHOLERA TOXIN; CYANOBACTERIA; DIPHTHERIA TOXIN; δ-ENDO-TOXIN; ENTEROTOXINS; PERTUSSIS TOXIN; PLANT PATHOLOGY; TETRODOTOXIN; α-TOXINS.

α-toxins Polypeptides such as α-bungarotoxin found in the venom of *Bungarus* snakes, which bind with very high specificity to nicotinic acetylcholine receptors (*see* NICOTINIC RECEPTORS). When conjugated to radioactive, fluorescent, or other markers they have been utilized as acetylcholine-specific markers.

toxoid A protein toxin that has been inactivated by heat or chemical treatment but still retains its antigenicity and can therefore be used as a vaccine.

TPA (1) 12-O-Tetradecanol phorbol-13 acetate. *See:* PHORBOL ESTER.
(2) TISSUE PLASMINOGEN ACTIVATOR.

TR Thyroid hormone receptor. *See:* STEROID RECEPTOR SUPERFAMILY.

tracker dye A chemical dye added to protein or nucleic acid samples prior to ELECTROPHORESIS of the samples through either acrylamide or agarose. The dye allows the experimenter to monitor the rate of electrophoretic migration of the sample. Commonly used dyes are BROMOPHENOL BLUE and XYLENE CYANOL FF. Both dyes comigrate with small double-stranded nucleic acids (bromophenol blue, 12–100 bp; xylene cyanol FF, 45–460 bp), the equivalent size depending on the electrophoretic conditions employed (e.g. concentration of agarose or acrylamide, buffer components used, etc.).

***trans*-ACPD** *trans*-1RS,3RS-*cis*-1-aminocyclopentyl-1,3-dicarboxylate (Fig. T27). Selective agonist for one subtype of EXCITATORY AMINO ACID RECEPTOR — the METABOTROPIC RECEPTOR.

Fig. T27 *trans*-ACPD.

***trans*-ACPD receptor** G protein-coupled EXCITATORY AMINO ACID RECEPTOR. Activated by glutamate and by the agonists quisqualate, ibotenate, and TRANS-ACPD, but not by AMPA. There are several subtypes of this receptor but the pharmacology and structures have not yet been characterized. ACPD receptors are linked to a variety of intracellular biochemical cascades. The regulation of phosphoinositide metabolism via a G protein and the subsequent regulation of intracellular Ca^{2+} levels may be important in LONG-TERM POTENTIATION. *See also:* G PROTEIN-COUPLED RECEPTORS; SECOND MESSENGER PATHWAYS.

***trans*-acting** A term used to describe a gene which produces a product that influences the expression of another gene elsewhere in the genome. *Cf.* CIS-ACTING. *See:* EUKARYOTIC GENE EXPRESSION; TRANSCRIPTION FACTORS.

***trans* Golgi network (TGN)** A tubulovesicular compartment found on the *trans* side of the Golgi stack and involved in sorting of transported proteins and lipids to their correct destinations including lysosomes, secretory granules and the plasma membrane. Also known as the *boulevard périphérique*, the *trans* tubular network and the *trans* Golgi reticulum. *See also*: GOLGI APPARATUS.

Griffiths, G. & Simons, K. (1986) *Science* **234**, 438–443.

trans-splicing The joining of two separate pieces of RNA to produce a complete RNA, which occurs in the expression of some trypanosome genes and plant chloroplast genes. *See*: ANTIGENIC VARIATION; CHLOROPLAST GENOMES; RNA SPLICING.

transcript RNA produced by TRANSCRIPTION from a DNA template by RNA polymerase.

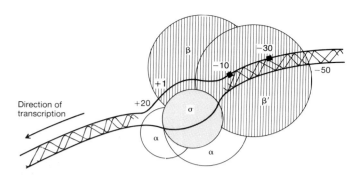

Fig. T28 Open complex formation after binding of RNA polymerase to the promoter. The open complex formed as a result of the interaction between at least five subunits of the RNA polymerase and the prokaryotic promoter. The DNA is melted over a region of ~30 bp including the Pribnow box and the direction of transcription is indicated with an arrow.

Transcription

TRANSCRIPTION is the term used to describe the synthesis of RNA from a DNA template. In order to be expressed, genetic information encoded in DNA has first to be transcribed into a complementary copy of RNA, which is then either used directly, as in the case of structural RNAs (e.g. transfer RNAs (tRNA) and ribosomal RNAs (rRNA)) or, as messenger RNA (mRNA), is translated into protein (*see* PROTEIN SYNTHESIS).

A stretch of DNA that is transcribed as a single continuous RNA strand, a transcript, is called a transcription unit. A unit of transcription may contain one or more sequences encoding polypeptides (translational open reading frames (ORF) or cistrons). In prokaryotes POLYCISTRONIC mRNAs are common. In eukaryotes, MONOCISTRONIC mRNAs are the general rule, but some transcription units encode more than one polypeptide as a consequence of alternative transcriptional start sites and/or alternative pathways of RNA SPLICING or other types of post-transcriptional RNA PROCESSING.

Transcription is the principal point at which gene expression is controlled in both prokaryotes and eukaryotes. The existence of an unstable 'messenger' RNA that acted as an intermediate between DNA and protein synthesis was first predicted by Jacob and Monod in 1961 [1]. Transcription is catalysed by a high molecular weight (M_r >500 000) complex of several proteins known as DNA-dependent RNA polymerase.

For general reviews and references on the subject see [2–5].

Transcription in prokaryotes

In prokaryotes there is one type of RNA polymerase (containing subunits $\alpha_2\beta\beta'\sigma$ which binds to DNA at specific sequences called PROMOTERS near the beginning of each transcription unit. The dissociable sigma factor (σ) of the RNA polymerase HOLOENZYME is concerned with the recognition of promoter sequences and is only associated with polymerase during promoter recognition, the formation of an open promoter complex (Fig. T28) and initiation of transcription. The usual σ factor (σ^{70}) can be replaced with different σ subunits which give the polymerase specificity for different promoters (*see e.g.* HEAT SHOCK; NITROGEN FIXATION

GENES). An asymmetric interaction involving conserved regions of the promoter centred at position – 10 (TATAAT, sometimes known as the Pribnow box) and – 35 (TTGACA) with respect to the site of RNA initiation orients the polymerase on the DNA SENSE STRAND in a position to initiate transcription.

Binding of RNA polymerase at promoters is accompanied by substantial conformational changes in both the DNA and protein resulting in local opening of the DNA duplex to expose hydrogen-bonding groups on each strand. The initiating ribonucleoside triphosphate (rNTP) (*see* NUCLEOSIDES AND NUCLEOTIDES) pairs with its complementary deoxynucleotide on the sense strand of the DNA. The next rNTP pairs with the next DNA base and the RNA polymerase catalyses the formation of a phosphodiester bond with the 3′OH of the initiating rNTP with elimination of pyrophosphate PP$_i$. During elongation the DNA strands remain open or dissociated over a length of ~17 base pairs (bp), in a 'transcription bubble' that moves along the helix with the polymerase. With the formation of each successive phosphodiester bond, the polymerase moves along the DNA sense strand from 3′ to 5′. Because nucleotide pairing is antiparallel, the RNA strand grows in the 5′ to 3′ direction at a rate of 30 to 60 nucleotides per second by the addition of NTP monomers at the 3′ end. The DNA helix re-forms behind the enzyme, so that the 5′ end of the RNA is released as a free single strand. Transcription occurs *de novo* and does not require DNA or RNA PRIMERS.

Termination of transcription in prokaryotes is signalled by a hairpin and loop structure formed in the transcribed RNA. Some structures are formed by sequence alone and others are modulated by a protein factor known as rho (ρ). The specific bypassing of transcription termination signals, ANTITERMINATION, is used in bacteria and their phages (e.g. LAMBDA) as a powerful means of temporally controlling gene expression.

The initiation of transcription in prokaryotes is commonly controlled by the binding of regulatory proteins (known as activator or repressor proteins) to specific DNA sequences (e.g. OPERATOR regions) that overlap the promoter and thus occlude polymerase binding. The binding of the activator or repressor proteins to DNA is often regulated by a small co-effector molecule such as an amino acid or a sugar that alters the conformation of the protein to allow or prevent DNA binding (*see* ALLOSTERY).

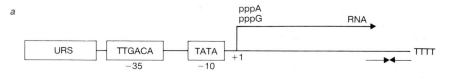

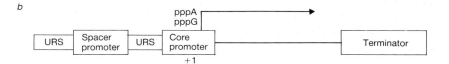

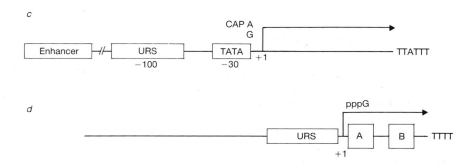

Fig. T29 Representation of the four main classes of promoters and terminators for gene transcription. *a*, Prokaryotic genes; *b*, eukaryotic rDNA; *c*, eukaryotic protein-coding genes; *d*, eukaryotic tRNA genes. General transcription factors interact with the core promoters or TATA sequences and sequence-specific DNA-binding proteins interact with the upstream regulatory sequence (URS) and with enhancer sequences to modulate transcriptional activity. The hairpin-loop structure at the end of prokaryotic operons (*a*) is represented by converging arrows. Other sequences associated with termination of transcription in different classes of genes are shown. The structures found at the 5′ ends of transcribed RNAs are shown. The boxes A and B within the coding region of tRNA genes are internal promoters recognized by polymerase III.

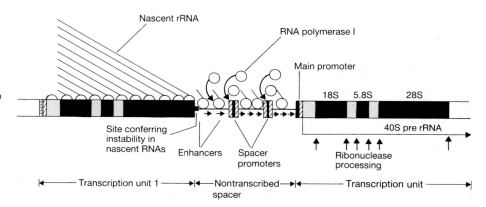

Transcription in eukaryotes

The process of transcription elongation in eukaryotes is similar to that described for prokaryotes, but there are differences in the mechanisms used to control the initiation and termination of transcription (Fig. T29).

There are three RNA polymerases in eukaryotic cells which are responsible for transcribing distinct classes of genes (*see* EUKARYOTIC GENE STRUCTURE) and which can be distinguished on the basis of their sensitivity to the fungal toxin α-AMANITIN. Polymerase I transcribes the rRNA genes (except 5S RNA) and is totally resistant to amanitin; polymerase II transcribes genes coding for proteins (mRNA) and a class of genes coding for small nuclear RNAs (snRNA) that are largely involved in RNA process-ing and is highly sensitive to amanitin (inhibited at $0.1\ \mu g\ ml^{-1}$); and polymerase III transcribes genes that code for a number of small RNAs including tRNA, 5S RNA, and 7SL RNA and shows intermediate sensitivity to amanitin (inhibited at $10\text{–}500\ \mu g\ ml^{-1}$). The physically small class III genes are often found reiterated and clustered at hundreds of copies per haploid genome and contain promoters located within the transcribed region. Many hundreds of copies of the tandemly linked class I rRNA genes are separated from each other by spacer DNA containing spacer promoters, ENHANCERS and a termination site immediately upstream from the next main promoter (Fig. T30).

The 5′ and 3′ sequences flanking the transcribed region largely control transcription initiation and termination. In class II genes, signals in the 5′ flanking region specify the RNA start site, fix the

Fig. T30 Structure of rDNA from *Xenopus* transcribed by polymerase I. Two transcription units separated by one nontranscribed spacer are shown. Each transcription unit is transcribed into a 40S pre-rRNA (solid and stippled boxes) which is processed by endo- and exoribonucleases into the mature 18S, 5.8S, and 28S rRNAs (solid boxes). The transcription initiates at a site close to the main promoter and initiation is enhanced by sequences (enhancers and spacer promoters) in the nontranscribed spacer which function to attract RNA polymerase I (open circles). At the junction of transcription unit I and the nontranscribed spacer there is a sequence that confers instability in the nascent RNA and functions to terminate transcription. Not to scale.

maximum rate of transcription initiation and regulate transcription in response to environmental stimuli or a developmental programme. Unlike the prokaryotic RNA polymerase, purified eukaryotic RNA polymerases cannot by themselves initiate transcription specifically or accurately at promoters. Accurate initiation requires the addition of multiple general TRANSCRIPTION FACTORS of the general classes TFI, TFII and TFIII (e.g. TFIIA, B, D, and E together with RNA polymerase II) (Fig. T31). These proteins are often sufficient for basal level transcription from some promoters. TFIID is the only one of these class II factors that interacts directly with the DNA at the general CONSENSUS SEQUENCE TATAAA (known as the TATA box) found in most eukaryotic promoters and which functions to determine efficient transcription initiation 25 to 30 bases downstream (i.e. to the 3′ side). The TATA binding protein (TBP, a component of TFIID) is required for transcription of all three classes of eukaryotic gene. Class I and III genes do not contain a TATA box and TBP is found associated with other transcription factor complexes, for example TFIIIB and TFIIIC during transcription initiation at class III promoters [6,7]. Short sequences known as upstream promoter elements, at which specific DNA–protein complexes form, are found up to 100 base pairs upstream (i.e. to the 5′ side) of the RNA initiation site and may aid TFIID binding to the TATA box and thus contribute to basal level transcription.

Sequences that enhance the basal rate of transcription are known as ENHANCERS. These elements may also mediate specific regulatory effects, being active only under certain physiological conditions. Enhancers can function at a distance (several kilobases) from the start point of transcription, and are often found in the 5′ flanking regions of genes but have also been found in the transcribed region and the 3′ flanking region. Enhancers are the sites at which regulatory proteins whose function is to activate or repress transcription may bind. In genes regulated by the same stimuli, the same transcription factor is likely to bind to a conserved DNA sequence in each promoter [8].

Chromatin and transcription

DNA in eukaryotes is not naked but in the form of a DNA–protein (histone) complex known as CHROMATIN. The structure of chromatin, particularly the position of nucleosomes, can influence transcription of genes and it is with chromatin, rather than naked DNA, that transcription factors regulating gene expression must interact. The basic nucleosome structure of chromatin is maintained during transcription but there are differences between active genes (and genes which have the potential to be transcribed) and bulk inactive chromatin. In many genes the DNA becomes generally sensitive or, at specific sites, hypersensitive, to attack by DNase I or micrococcal nuclease. Hypersensitive sites are often found close to enhancer elements and may reflect the presence of specific DNA–protein complexes at the enhancer [9].

In vitro transcription systems

Bacterial cell-free DNA-directed coupled transcription and translation allows the *in vitro* expression of any gene contained on a bacterial plasmid or a bacteriophage genome provided the rele-

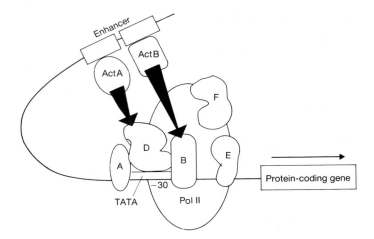

Fig. T31 Formation of a preinitiation complex at the promoter of a gene transcribed by RNA polymerase II (Pol II). The general transcription factors TFIIA, B, E, and F together with the polymerase associate with TFIID bound to the TATA box of the promoter. The two activator proteins ActA and ActB, bound to the enhancer, are shown functioning to enhance the rate and extent of preinitiation complex formation around the TATA box.

vant bacterial promoter signals are present. Specific RNA transcripts from cloned cDNA inserts can be prepared using the promoter sequences and purified RNA polymerases of the bacteriophages T7 or SP6 (*see* EXPRESSION VECTORS). This RNA can be modified by capping the 5′ end *in vitro* and is then a suitable substrate for translation in a mammalian or plant extract. *In vitro* transcription systems from eukaryotes have been developed for all three polymerases which recognize a pre-formed DNA protein complex at gene promoters, not naked DNA. RNA is synthesized on a truncated template using the so-called RUN-OFF assay and the process shows a high degree of fidelity.

Reverse transcription

DNA can be synthesized from an RNA template by the enzyme REVERSE TRANSCRIPTASE, in a process requiring a primer such as tRNA (*see* PLANT VIRUSES; RETROVIRUSES).

E.J. MELLOR

See also: ANIMAL VIRUSES; BACTERIAL GENE EXPRESSION; EUKARYOTIC GENE EXPRESSION.

1 Jacob, F. & Monod, J. (1961) Genetic regulatory mechanisms in the synthesis of proteins. *J. Mol. Biol.* **3**, 318–356.
2 Darnell, J. et al. (1990) *Molecular Cell Biology* 2nd edn (Scientific American Books, San Francisco, CA).
3 Watson, J.D. et al. (1987) *The Molecular Biology of the Gene*, 4th edn (Benjamin/Cummings, Menlo Park, CA).
4 Laskey, R. & Scott, M.P. (Eds) (1993) Gene expression and differentiation. *Curr. Opinion Genet, Dev.* **3**.
5 Bloom, K & Green, M. (Eds) (1992) Nucleus and gene expression. *Curr. Opinion Cell Biol.* **4**, 377–567.
6 Schultz, M.C. et al. (1992) Variants of the TATA-binding protein can distinguish subsets of RNA polymerase I, II and III promoters. *Cell* **69**, 697–702.
7 Cormack, B.P. & Struhl, K. (1992) The TATA-binding protein is required for transcription by all three nuclear RNA polymerases in yeast cells. *Cell* **69**, 685–696.

8 Zavel, L. & Reinberg, D. (1992) Advances in RNA polymerase II transcription. *Curr. Opinion Cell Biol.* **4**, 488–495.

9 Svaren, J. & Horz, W. (1993) Histones, nucleosomes and transcription. *Curr. Opinion Genet. Dev.* **3**, 219–225.

transcription-based amplification system A technique using the POLYMERASE CHAIN REACTION.

transcription complex Complex of RNA POLYMERASE and TRANSCRIPTION FACTORS which is assembled on the PROMOTER region of a gene to initiate TRANSCRIPTION. *See also*: TFI, TFII, TFIII.

Transcription factors

THE regulatory regions of eukaryotic genes contain specific short DNA sequences which control the TRANSCRIPTION of that particular gene (*see* EUKARYOTIC GENE EXPRESSION; EUKARYOTIC GENE STRUCTURE). Such sequences act by binding particular regulatory proteins known as transcription factors which interact with each other and the RNA POLYMERASE enzyme itself in order to modulate transcription. The nature of the transcription factors which bind to a specific gene is critical in determining its pattern of transcription. Thus if the gene binds predominantly transcription factors which are active in all cell types it will be expressed in all cells whereas if it binds factors which are synthesized or are active in only one cell type, the gene will be expressed in a cell type-specific manner.

The critical role of transcription factors in the process of transcription has led to their intensive study [1–3]. Such studies reveal that these proteins have a modular structure in which distinct regions of the protein mediate each of its different functions (Fig. T32). Thus in the case of the glucocorticoid receptor, a transcription factor which mediates the induction of specific genes in response to glucocorticoids (*see* STEROID RECEP-

TORS), different regions of the protein mediate its ability to bind to DNA, its ability to activate gene transcription following such binding), and its ability to bind glucocorticoid hormones. An extreme example of such modularity is provided by the herpes simplex virus virion protein VP16. This protein possesses a region capable of activating transcription but lacks any DNA-binding activity and must therefore form a complex with the cellular DNA-binding protein Oct-1, in order that DNA binding and consequent transcriptional activation can occur. Hence in this case, DNA binding and transcriptional activation domains are located on different proteins.

In most transcription factors however, specific regions of the same protein mediate its ability to bind to DNA and its ability to activate transcription by interacting with other factors following binding. The various structural elements in these factors which can mediate either DNA binding or transcriptional activation will be discussed in turn and are summarized in Table T3.

DNA-binding domains

A number of different motifs which can mediate DNA binding have been identified in different transcription factors and have been used to classify these factors into families with a common related DNA-binding domain (*see also* PROTEIN–NUCLEIC ACID INTERACTIONS) [4,5].

The zinc finger

One of the first transcription factors to be purified and cloned was TFIIIA, which has a critical role in the transcription of the 5S ribosomal RNA genes by RNA polymerase III. The DNA-binding region of this factor contains nine repeats of a 30-amino acid sequence of the form Tyr/Phe-X-Cys-X-Cys-X_{24}-Cys-X_3-Phe-X_5-Leu-X_2-His-X_{34}-His-X_5 where X is a variable amino acid. Each of these repeats therefore contains two invariant pairs of cysteine and histidine residues which coordinate a single atom of zinc. This results in a finger-like structure (Fig. T33a) in which the conserved phenylalanine and leucine residues and several basic residues in the finger project from the surface of the protein [6,7].

Multiple examples of this zinc finger motif have subsequently been identified in a number of transcription factors for genes transcribed by RNA polymerase II including Sp1, the *Drosophila* Krüppel protein (*see* DROSOPHILA DEVELOPMENT), the yeast ADRI protein and many others. Interestingly, a single mutation in one of the zinc finger motifs of the Krüppel protein, in which a cysteine is replaced by a serine that cannot bind zinc, results in a mutant fly whose appearance is exactly identical to that produced by complete deletion of the gene. Hence the ability to bind zinc is essential for DNA-binding activity and therefore for the functioning of the protein as a transcription factor.

A similar zinc-binding motif is also found in the DNA-binding regions of members of the steroid/thyroid hormone receptor family which bind specific steroid hormones and either activate or repress gene expression by binding to specific DNA sequences in target genes [6]. In this case however, the binding region consists of two fingers, each of which contains four cysteine residues coordinating the zinc atom rather than two each of

a   GCN4

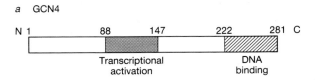

b   Glucocorticoid receptor

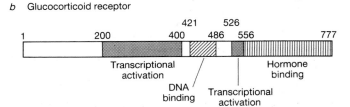

Fig. T32 Domain structure of the yeast GCN4 transcription factor (*a*) and the mammalian glucocorticoid receptor (*b*). Note the distinct domains which are active in DNA binding or transcriptional activation. Numbers are amino-acid positions.

Table T3 Transcription factor domains

Domain	Role	Factors containing domain	Comments
Homeobox	DNA binding	Numerous *Drosophila* homeotic genes; related genes in other organisms	DNA binding mediated via helix-turn-helix motif
Cysteine-histidine zinc finger	DNA binding	TFIIIA, Krüppel, Sp1, etc.	Multiple copies of finger motif
Multi-cysteine zinc finger	DNA binding	Steroid/thyroid hormone receptor family	Single pairs of fingers, related motifs in adenovirus E1A and yeast GAL4, etc.
Basic element	DNA binding	C/EBP, Fos, Jun, GCN4	Often found in association with leucine zipper
Leucine zipper	Protein dimerization	C/EBP, Fos, Jun, GCN4, Myc	Mediates dimerization which is essential for DNA binding by adjacent domain
Helix-loop-helix	Protein dimerization	Myc, *Drosophila* daughterless; MyoD, E12, E47	Mediates dimerization which is essential for DNA binding by adjacent domain
Amphipathic acidic α-helix	Gene activation	Yeast GCN4 and GAL4, steroid/thyroid receptors	Probably interacts directly with TFIID
Glutamine-rich region	Gene activation	SP1	Related regions in Oct-1, Oct-2, AP2, etc.
Proline-rich region	Gene activation	CTF/NF1	Related regions in AP2, Jun, Oct-2

cysteine and histidine and it also lacks the conserved phenylalanine and leucine residues found in the other type of finger (Fig. T33*b*). In addition the two-finger element is present only once in each receptor, as opposed to the multiple fingers ranging from 2 to 37 found in genes encoding cysteine–histidine fingers. The two types of finger may not therefore be evolutionarily related.

The existence of related DNA-binding regions in the steroid receptors which bind to distinct but related DNA sequences has allowed detailed study of features in the protein which are important for sequence-specific DNA binding. The oestrogen receptor and the glucocorticoid receptor bind to related but distinct sequences in their target genes. Alteration of two amino acids in the N-terminal finger of the oestrogen receptor to those found in the same region of the glucocorticoid receptor results in

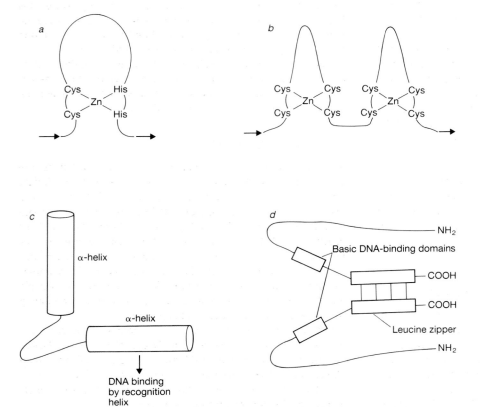

Fig. T33 DNA binding domains. *a*, The cysteine–histidine zinc finger. *b*, The multicysteine zinc finger. *c*, The helix-turn-helix motif. *d*, The leucine zipper and adjacent basic DNA-binding domain following dimerization.

a receptor which can bind to the DNA sequence normally bound by the oestrogen receptor and hence switch on oestrogen-responsive genes even though the other 775 amino acids of the protein are derived from the glucocorticoid receptor. Hence these two amino acids are critical in determining the precise DNA sequence that is bound by the factor.

Although best analysed in the steroid/thyroid receptor gene family, similar multicysteine fingers are found in other transcription factors such as the yeast transcription factor GAL4 and the ADENOVIRUS E1A protein.

The helix-turn-helix motif

A number of HOMEOTIC GENES which dramatically affect the development of the fruit fly *Drosophila melanogaster* have been described (*see also* DROSOPHILA DEVELOPMENT). These genes contain a common highly conserved region of 60 amino acids which is known as the homeobox or homeodomain and which mediates the DNA-binding ability of these proteins [8] (*see* HOMEOBOX GENES AND HOMEODOMAIN PROTEINS). Structure predictions for this region indicated that it could form a helix-turn-helix motif in which an α-helical region is followed by a β-turn and then another α-helical region (Fig. T33*c*). This structure, originally predicted on the basis of homology to bacterial REPRESSOR proteins, has now been directly confirmed by NMR spectroscopic studies of the *Antennapedia* protein homeodomain from *Drosophila* (*see* NMR SPECTROSCOPY).

In the complex of bacterial repressor and DNA, the first protein helix lies across the major groove of the DNA whereas the second helix, known as the recognition helix, lies partly within the major groove and makes sequence-specific contact with the DNA (*see* PROTEIN–NUCLEIC ACID INTERACTIONS).

In various homeodomain proteins, for example the products of the genes *paired (prd)* and *fushi-tarazu (ftz)*, there is a similar role for the second helix in sequence-specific DNA recognition by the homeodomain; a mutation at residue 9 of this helix which results in substitution of the serine present in the *prd* protein by the glutamine present in the *ftz* protein confers on the *prd* protein the ability to bind to binding sites for *ftz* protein. Similar dramatic effects of other alterations at this position have also been demonstrated.

Following its original identification in *Drosophila* homeotic genes, similar homeodomains containing the helix-turn-helix motif have also been identified in the yeast mating type transcriptional regulatory proteins, in a variety of amphibian and mammalian transcription factors and in plants such as *Antirrhinum*.

The leucine zipper and the basic DNA-binding domain

Another element found in several transcription factors such as the liver-specific transcription factor C/EBP, the yeast factor GCN4 and the PROTO-ONCOGENE proteins Fos and Jun is the leucine zipper [9]. In this structure, leucine residues occur every seven amino acids in an α-helical structure such that the leucines occur every two turns on the same side of the helix.

Rather than acting directly as a DNA-binding motif however,

the zipper facilitates the dimerization of the protein by interdigitation of two leucine-containing helices on different molecules (Fig. T33*d*). Dimerization results in the correct protein structure for DNA binding by the adjacent region which in C/EBP, Fos and Jun is highly basic, distinct from those discussed so far, and which interacts directly with the acidic DNA.

The Fos and Jun proteins both bind to sequences in DNA known as AP-1 sites, which mediate gene induction following PHORBOL ESTER treatment. But whereas the Jun protein can bind specifically to this sequence as a homodimer, the Fos protein can only do so after formation of a heterodimer with Jun. This difference is directly due to a difference in the leucine zipper motif of the two proteins which prevents Fos homodimer formation. If the Fos leucine zipper region is replaced by that of Jun the chimaeric protein can now bind to DNA through the basic region of Fos. The requirement for dimer formation of these proteins before DNA binding thus introduces another potential regulatory point in the control of gene expression [10].

Although originally identified in leucine zipper-containing proteins, the basic DNA-binding domain has also been identified by homology comparisons in a number of other transcriptional regulatory proteins including two proteins, E12 and E47, that bind to the immunoglobulin enhancer (*see* IMMUNOGLOBULIN GENES), the muscle regulatory protein MyoD1 (*see* MYOGENESIS), and the *Drosophila* daughterless protein. In these cases, the basic domain is associated with an adjacent region that can form a helix-loop-helix structure in which two amphipathic helices (containing all the charged amino acids on one side of the helix) are separated by an intervening nonhelical loop. Although originally thought to be the DNA-binding domain of these proteins, this helix-loop-helix motif is now believed to have a similar role to the leucine zipper in mediating protein dimerization and facilitating DNA binding by the adjacent basic DNA-binding motif.

Hence both the leucine zipper and the helix-loop-helix motif are critical in allowing dimerization of the factors which contain them. This dimerization is critical for subsequent DNA binding mediated by the basic DNA-binding domain.

Other DNA-binding motifs

Although most DNA-binding domains analysed so far fall into the three classes described above, not all do. Thus the DNA-binding domains of the transcription factors AP2, CTF/NF1 and SRF are distinct from the known motifs and from each other. As more factors are studied, DNA-binding motifs similar to those of these proteins are likely to be identified and they will become founder members of new families of DNA-binding motifs. Indeed this process is already underway; the similarity between the mammalian factor HNF-3 and the *Drosophila* fork head factor for example has led to the identification of a new family of proteins containing the so called fork head DNA-binding motif.

It is clear therefore that a number of different structures exist that can mediate sequence-specific DNA binding. Several of these are common to a number of different transcription factors, with differences in the precise amino-acid sequence of the motif in each factor controlling the precise DNA sequence which it binds and hence the target genes for the factor.

Activation domains

In some cases, the DNA-bound transcription factor simply acts as an inhibitory molecule blocking the binding to DNA of a positively acting factor and thereby repressing transcription (*see* EUKARY-OTIC GENE EXPRESSION). In most cases however, following DNA binding the factor interacts with other factors or the RNA polymerase itself to stimulate transcription. Such interaction is dependent on specific activation domains in the molecule which are normally distinct from the region which mediates DNA binding [11].

When the activating regions of a number of transcription factors were compared, it was found that, although they had no amino-acid sequence homology, they all possessed a high proportion of acidic amino acids [12]. These acidic amino acids are arranged in such a way that they form an amphipathic α-helix in which all the negative changes are displayed along one surface of the helix. This structure is essential for transcriptional activation; a peptide which can form an acidic amphipathic helix can activate transcription when linked to the DNA-binding domain of the yeast transcription factor Gal4 whereas the same amino acids placed in a random order cannot.

Although this acidic activation domain is common to a number of transcription factors isolated from organisms as diverse as yeast and man, other nonacidic activation domains have also been described. The activation domain of the transcription factor Sp1 contains a glutamine-rich region, and that of CTF/NF1 is very rich in proline. Similar glutamine- or proline-rich regions have been found in other transcription factors which indicates that activation domains of these types are not confined to a single protein.

It is likely that the various activation domains act by interacting with other protein factors to facilitate transcription. Although this may occur by direct interaction with the RNA polymerase itself, it seems more likely, at least in the case of the acidic activation domain, that its effect is mediated either through the TATA-box binding factor TFIID or through another transcription factor, TFIIB [13]. Both these factors are components of the basal transcriptional complex which forms on the gene promoter before transcription and which interacts with RNA polymerase II itself (*see* EUKARYOTIC GENE EXPRESSION). By stimulating the assembly or stability of this complex, activation domains can therefore act to enhance the rate of transcription. Specific regions of transcription factors can activate transcription following DNA binding, by interacting with another bound factor and facilitating the assembly of a stable transcriptional complex.

D.S. LATCHMAN

1 Latchman, D.S. (1991) *Eukaryotic Transcription Factors* (Academic Press, London).
2 Johnson, P.F. & McKnight, S.L. (1989) Eukaryotic transcriptional regulatory proteins. *Annu. Rev. Biochem.* **58**, 799–839.
3 Faiset, S. & Meyer, S. (1992) Compilation of vertebrate-encoded transcription factors. *Nucleic Acids Res.* **20**, 3–26.
4 Struhl, K. (1989) Helix-turn-helix, zinc finger and leucine zipper motifs for eukaryotic transcriptional regulatory proteins. *Trends Biochem. Sci.* **14**, 137–140.
5 Schleif, R. (1988) DNA binding by proteins. *Science* **241**, 1182–1187.
6 Evans, R.M. & Hollenberg, S.M. (1988) Zinc fingers, guilt by association. *Cell* **52**, 1–3.
7 Klug, A. & Rhodes, D. (1987) Zinc fingers: a novel protein motif for nucleic acid recognition. *Trends Biochem. Sci.* **12**, 464–469.
8 Affolter, M. et al. (1990) Homeodomain proteins and the regulation of gene expression. *Curr. Opinion Cell Biol.* **2**, 485–495.
9 Abel, T. & Maniatis, T. (1989) Action of leucine zippers. *Nature* **341**, 24–25.
10 Jones, N. (1990) Transcriptional regulation by dimerization, two sides to an incestuous relationship. *Cell* **61**, 9–11.
11 Ptashne, M. (1988) How eukaryotic transcriptional activators work. *Nature* **335**, 683–689.
12 Sigler, P.B. (1988) Acid blobs and negative noodles. *Nature* **333**, 210–212.
13 Sharp, P.A. (1991) TFIIB or not TFIIB? *Nature* **351**, 16–18.

---

transcription initiation site The site upstream from the translation start site of a gene where RNA polymerase begins to transcribe the DNA. In both prokaryotes and eukaryotes the DNA sequence immediately upstream of the transcription start site is often characterized by the presence of a TATA BOX. This sequence is localized about 15–30 nucleotides upstream of the actual transcription start site. *See*: BACTERIAL GENE EXPRESSION; BACTERIAL GENE ORGANIZATION; EUKARYOTIC GENE EXPRESSION; EUKARYOTIC GENE STRUCTURE; TRANSCRIPTION.

transcription unit A sequence of DNA that is transcribed from a single PROMOTER. In eukaryotes, transcription units are usually single GENES; in bacteria and other prokaryotes, several genes may be grouped together to form a single transcription unit under the control of a promoter — an OPERON. *See*: BACTERIAL GENE EXPRESSION.

transcriptional activators, transcription activating proteins *See*: EUKARYOTIC GENE EXPRESSION; TRANSCRIPTION; TRANSCRIPTION FACTORS.

transcriptional terminator *See*: TERMINATOR; TRANSCRIPTION.

transcytosis Transfer of macromolecules across epithelial and endothelial cell monolayers by means of membrane-bounded vesicles. Transcytosis is an essential function of epithelia. It mediates the absorption of digestive products from the gut, the secretion of antibodies that bathe and protect mucosal surfaces, the resorption functions of the kidney, and milk production in the mammary gland. In polarized epithelia (*see* EPITHELIAL POLARITY), transcytosis is mediated through the constitutive ENDOCYTIC PATHWAY and occurs either from the apical (luminal) domain to the basolateral (serosal) domain or vice versa.

Vectorial transport of molecules is mediated by specific receptors which are targeted to the appropriate plasma membrane domains by signals contained in the amino-acid sequences of their cytoplasmic domains (*see* PROTEIN TARGETING). Differences in the environmental conditions across the monolayer facilitate the transfer by promoting association or dissociation of the ligand and receptor. For example, in neonatal rats passive immunity is established by the transcytosis of maternal immunoglobulin across gut epithelia. In the low pH conditions of the gut, immunoglobulin binds to receptors on the apical membrane of duodenal epithelial cells. The immunoglobulin–receptor complexes are internalized through coated pits, delivered to apical endosomes, and subsequently sorted into vesicles (transcytotic vesicles) which deliver them, either directly or via the basolateral endosome system, to the basolateral membrane. In the neutral pH conditions of the basolateral surface the immunoglobulin dissociates from the receptor. In contrast, IgA is captured from the blood by IgA receptors at the sinusoidal membrane of, for example hepatocytes, the complex is transported via the endosomal system to the bile cannaliculus (apical membrane) and released by proteolytic cleavage of the IgA receptor.

Transcytosis also occurs in endothelial cells where blood constituents are transported to the interstitial space through endothelial cells (extravascularization). In this situation selective transcytosis contributes to the integrity of endothelial boundaries such as the blood–brain barrier. In contrast to epithelial cells, endothelial cells seem to have two different forms of transcytosis. They undergo constitutive coated vesicle-mediated endocytosis which may mediate the transcytosis of certain components, as described above for epithelial cells. In addition, endothelial cells contain numerous noncoated invaginations of the plasma membrane, termed plasmalemmal vesicles, which seem to mediate selective transport of macromolecules directly across the cell without the involvement of the endocytic pathway. *See also*: ENDOCYTOSIS.

Milici, A.J. et al. (1987) *J. Cell Biol.* **105**, 2603–2612.
Hopkins, C.R. (1991) *Cell*, **66**, 827–829.
Rodman, J.S. et al. (1990) *Curr. Opinion Cell Biol.* **2**, 664–672.
Simons, K. & Fuller, S. (1985) *Annu. Rev. Cell Biol.* **1**, 243–288.

transcytotic vesicle Membrane-bounded vesicles involved in TRANSCYTOSIS. Presumed to form from apical or basolateral endosomes.

transdetermination A change in COMMITMENT of IMAGINAL DISK cells. It is demonstrable by serial transplantation of imaginal discs in adult flies (when the disk cells continue to proliferate) followed by grafting into metamorphosing larvae, when disk cells will occasionally show a change of fate, for example from antenna to leg. The transdetermined state is relatively stable, being heritable through many cell divisions, and occurs preferentially in certain directions (e.g. from genital fate to leg fate, but not vice versa).

transdifferentiation A change in COMMITMENT of differentiated cultured cells. Cells of one PHENOTYPE will sometimes transform spontaneously to a new phenotype which is stably inherited in subsequent cell divisions. *See also*: DIFFERENTIATION.

transducin (G$_t$, T) G protein in retinal rod cells that couples signal transduction from photo-excited RHODOPSIN to the cellular response. Transduction is activated by excited rhodopsin and in turn activates a CYCLIC GMP PHOSPHODIESTERASE. *See*: GTP-BINDING PROTEINS; SECOND MESSENGER PATHWAYS; VISUAL TRANSDUCTION.

transducing phage *See*: TRANSDUCTION.

transduction (1) A genetic transfer process in bacteria, originally discovered in 1952 by Norton Zinder and Joshua Lederberg, in which a portion of a bacterial chromosome is transferred to a new host bacterium by becoming incorporated into a BACTERIOPHAGE particle. Phages that can mediate transduction are known as transducing phages. The chromosomal DNA transduced from the donor bacterium can recombine with the host bacterium's chromosome. Two forms of transduction exist: generalized transduction and specialized transduction. In generalized transduction an essentially random region of bacterial chromosome is packaged into the head of the bacteriophage particle usually in place of part or all of the bacteriophage genome. The P1 PHAGE is an example of a bacteriophage capable of mediating generalized transduction, being able to transfer up to 90 kb chromosomal DNA at any one time. In specialized transduction, only a small but specific portion of the donor chromosome is transferred and usually as part of the viral genome. The bacteriophage LAMBDA is able to mediate specialized transduction as a consequence of escaping the host chromosome following a lysogenic infection and carrying part of the flanking chromosome region with it. As integration of the lambda genome occurs at a specific site in the *Escherichia coli* chromosome (between the *gal* and *bio* genes) only one region of the *E. coli* chromosome can be so transferred. Lambda can transduce no more than 5 kb of chromosomal DNA at any one time. Transduction can also be exploited as a GENETIC MAPPING tool for *E. coli* by, for example, looking at the frequency of cotransduction of two genes during generalized transduction. (2) SIGNAL TRANSDUCTION.

Transfection

TRANSFECTION is the genetic modification of cultured animal cells by the uptake of DNA from the culture medium. The DNA to be transfected is usually in the form of recombinant PLASMIDS or other types of DNA VECTOR containing the genes of interest (*see* GENETIC ENGINEERING; RECOMBINANT DNA). Transfection is analogous to the process of TRANSFORMATION whereby bacterial cells can take up exogenously added DNA. Transfection of DNA into plant cells may also be achieved by first removing the cell wall to produce a PROTOPLAST (*see* GENETIC ENGINEERING; PLANT GENETIC ENGINEERING).

Genes can be introduced into cells by transfection either transiently or stably. In transient gene transfer, the expression of the recombinant gene product can be detected within the animal cell from 12 hours up to 80 hours post-transfection, and generally the introduced DNA does not become integrated into the cell's chromosomes. In other cases the DNA becomes permanently integrated into the cell's chromosomes to form stably transfected cell lines.

As the spontaneous entry of DNA into the animal cell and its subsequent expression in the nucleus is very inefficient, several methods have been developed to facilitate the process, including the use of polycations such as DEAE–dextran [1,2], calcium phosphate coprecipitation [3], electroporation [4], lipofection (LIPOSOME fusion) [5], RETROVIRUS VECTORS [6] and microinjection [7]. The most widely used methods for introducing DNA into animal cells are the calcium phosphate- or DEAE-dextran-mediated transfections.

DEAE-dextran- and calcium phosphate-mediated transfection

The transient transfection of animal cells in culture can be achieved using the DEAE (*Di*ethyl*amino*ethyl)-dextran method [1,2], where uptake of DNA by the cell is significantly improved following reaction of the DNA with the polycation. This technique is not, however, suitable for the transfection of certain cell lines (owing to the toxicity of the polymer) or for establishing stable cell lines. Alternatively, DNA transfection can be achieved by presenting larger amounts of the nucleic acid to the cell as a calcium phosphate–DNA coprecipitate [3]. Although the exact mechanism by which each of these techniques improves transfection is unclear, they presumably act by facilitating ENDOCYTOSIS of the DNA into the cytoplasm of the cell, from where it is transferred to the nucleus.

The efficiency of expression of plasmid DNA transfected by these methods can be increased by additional treatments such as glycerol shock and/or CHLOROQUINE treatment [9] following transfection. Treatment of cells with sodium butyrate has also been shown to increase the transcriptional activity of plasmids containing the SV40 enhancer region [9].

Electroporation

Electroporation [4] provides a way of transfecting animal cells generally with greater efficiencies than the chemical methods described above. A brief, high voltage electrical pulse is applied to the cells, which alters the conformation of pores within the plasma membrane, allowing the DNA to pass directly through these pores into the cytoplasm. The two important pulse parameters that need to be optimized for each cell type are the field strength and the time constant, as both cell survival and transfection efficiency are related to the intensity of the field (field strength = voltage/distance between the electrodes) and to the length of the pulse (time constant).

Lipofection

Lipofection [5] is a lipid-mediated DNA transfection procedure, in which artificial membrane vesicles (liposomes) prepared from cationic lipids form stable complexes with DNA, and then fuse with the plasma membrane, ensuring high efficiency transfer of the DNA into the cell. The efficiency of this procedure varies with the cell line under study, but can be more efficient than the calcium phosphate or DEAE-dextran methods described above.

Other techniques

Other techniques such as the direct microinjection of DNA [7] into the nucleus of animal cells, and protoplast fusion [8], which involves the direct transfer of DNA to mammalian cells by cell fusion with bacterial SPHEROPLASTS (protoplasts) harbouring the recombinant plasmid, may be used in cases where the transfection techniques described above have proved unsuccessful. Similarly, the use of a RETROVIRUS VECTOR system [6] allows for the efficient conversion of the vector DNA into highly transmissible virus capable of infecting many cell types.

Transient and stable transfection

All the techniques described above have been used successfully to establish transiently transfected cell lines, and are equally suitable for the establishment of stably transfected cell lines, with the exception of the DEAE–dextran method. For transient expression, cells are harvested 48–60 hours post-transfection and assayed for expression of the recombinant DNA by detection of RNA or the gene product. For stable transfections, the cells are placed in selective medium 18–24 hours post-transfection, and maintained in selective culture for 2–3 weeks, until individual colonies can be cloned and propagated for assay.

The selection of transfected cell lines depends on detecting a phenotypic change associated with the transfer of the DNA to the recipient cell. If the gene of interest confers a selectable phenotype, the use of a recipient cell line lacking expression of that gene enables selection of transfected cells in which the defect has been complemented. For example, the dihydrofolate reductase gene (DHFR) is necessary for the biosynthesis of purines. Cells lacking the gene, such as DHFR-deficient CHO (Chinese hamster ovary) cell lines, will not grow on medium lacking purines. The DHFR gene can therefore be used as a dominant marker to select and amplify transfected genes in such cells growing in medium lacking purines. If, however, the gene of interest does not convey a selectable phenotype, the technique of cotransfection must be used. Cotransfection involves using a selectable plasmid marker gene (e.g. encoding antibiotic resistance) in addition to the DNA of interest. This allows the selection of cells which have integrated the DNA by screening for expression of the selectable marker. These clones can then be rescreened for expression of the desired gene.

The use of either transiently or stably tranformed cell lines will depend on the aims of a particular investigation. Transient expression experiments, for example, are usually carried out to determine which of the various regulatory elements of a gene should be included in a vector for expression of that gene within a particular cell type (promoter analysis), whereas stable transfections are used when high amounts of the recombinant DNA product are required.

Irrespective of the method used to establish a transfected cell line, the efficiency of transient or stable tranfection will depend on the cell line under study. Certain cell lines are more suitable for the construction of transient compared with stable cell lines (e.g. COS-7 versus CHO). Cell lines may be more efficiently transformed by one particular transfection technique, and the ability of some cell lines to assimilate DNA can vary widely. The optimal quantity of DNA for transfection also has to be determined, as it may depend on the copy number of the transfecting plasmid within a particular cell type. Therefore, the DNA transfection techniques described above must be optimized to suit the particular cell type, in order to maximize the uptake and expression of exogenously added DNA in either a transient or stable fashion.

L.H. NAYLOR

See also: XENOPUS OOCYTE EXPRESSION SYSTEM.

1 McCutchan, J.H. & Pagano, J.S. (1968) *J. Natl. Cancer Inst.* **41**, 351–357.
2 Kawai, S. & Nishizawa, M. (1984) *Mol. Cell. Biol.* **4**, 1172–1174.
3 Graham, F.L. & Van der Eb, A.J. (1973) *Virology* **52**, 456–467.
4 Neumann, E. et al. (1982) Gene transfer into mouse lyoma cells by electroporation in high electric fields. *EMBO J.* **7**, 841–845.
5 Felgner, P.L. et al. (1987) *Proc. Natl. Acad. Sci. USA* **84**, 7413–7417.
6 Cepko, C.L. et al. (1984) Construction and application of a highly transmissible murine retrovirus shuttle vector. *Cell* **37**, 1053–1062.
7 Capecchi, M.R. (1980) High efficiency transformation by direct microinjection of DNA into cultured cells. *Cell* **22**, 479–488.
8 Schaffner, W. (1980) *Proc. Natl. Acad. Sci. USA* **77**, 2163–2167.
9 Gorman C. (1985) In *DNA Cloning: A Practical Approach* (Glover, D.M., Ed.) (IRL Press, Oxford).

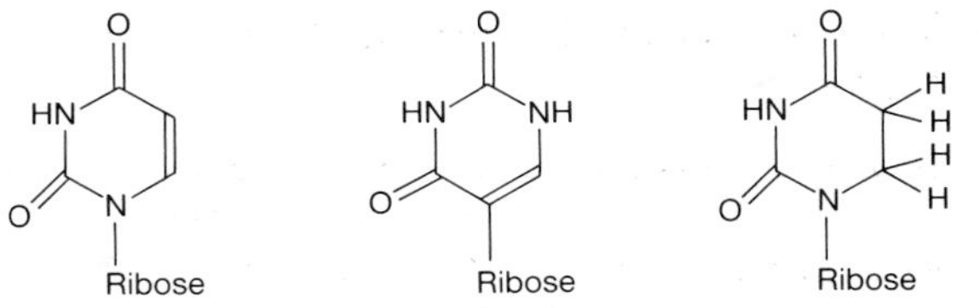

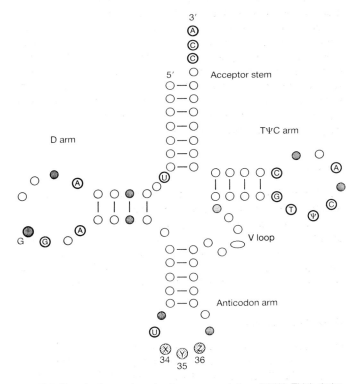

Fig. T34 Modified bases in tRNA are produced by modification of the four usual bases: A, U, C, G. Examples of modified nucleosides used in tRNA and derived from uridine are shown. Pseudouridine (ψU) has an interchange of N and C atoms. The saturation of the double bond to form dihydrouridine abolishes the aromatic character of the ring.

Transfer ribonucleic acid (tRNA)

TRANSFER RNA (tRNA) is a family of small nucleic acids that mediate the translation of the nucleic acid code into the amino-acid sequence of a protein (*see* GENETIC CODE; PROTEIN SYNTHESIS).

Trying to understand how the nucleotide sequence of a messenger RNA could be translated into an amino acid led to the hypothesis, formulated by Francis Crick, that an adaptor molecule was needed which matched the CODON in mRNA to its particular amino acid [1]. Small ribonucleic acids were isolated, with two major properties: they could recognize a codon in mRNA and could also able to bind the amino acid corresponding to that codon. The main function of tRNAs is to carry amino acids to the RIBOSOMES and to incorporate the correct amino acid into the nascent protein chain. The pool of tRNAs can be subdivided in 20 subsets, each one specific for a given amino acid. tRNAs that specify the same amino acid are named isoacceptors.

To fulfil their functions tRNAs interact with many proteins and nucleic acids. Some partners, like ELONGATION FACTORS or ribosomal sites, are common to all tRNAs, implying common recognition patterns. Others, like AMINOACYL-TRNA SYNTHETASES — the enzymes that attach amino acids to tRNAs — have to discriminate among isoacceptor groups, to charge a tRNA with its corresponding amino acid.

Structure

Primary structure

tRNAs are between 74 and 90 ribonucleotides long. Most of the bases are adenine (A), uracil (U), guanine (G), and cytosine (C), but up to 10% of the bases are modified during tRNA maturation (Fig. T34) (*see* RNA PROCESSING). tRNA sequences show a very high degree of conservation, the principal feature being the terminal CCA sequence which is present in all tRNAs [2].

Secondary structure

A Watson–Crick base-pairing analysis shows that all tRNAs present intramolecular complementary regions, leading to a cloverleaf secondary structure diagram [3], with four major arms, composed of base-paired stems and unpaired loops (Fig. T35). The acceptor arm contains the 3′ and 5′ ends of the molecule. The free 2′ or 3′ hydroxyl group of the terminal adenine at the 3′-terminal CCA is the primary site of aminoacylation. The anticodon loop contains the ANTICODON base triplet at positions 34–36. The D and TψC arms are named respectively after the dihydrouridine, which is always present, and the three strictly conserved nucleotides. The extra arm, consisting of 3–5 bases for Class 1 tRNAs or 13–21 bases for Class 2, is the most variable region of the molecule (V loop).

Fig. T35 Cloverleaf secondary structure representation of tRNA. Thick circles represent invariant bases. Hatched circles indicate semi-invariant (constant purine or pyrimidine) or specially modified bases. Position 37 adjacent to the anticodon (34–36) is almost always modified. Shaded grey circles indicate the anticodon nucleotides X, Y, Z.

Tertiary structure

High resolution crystallographic studies revealed that tRNAs have a common L-shape structure, composed of two nearly perpendicular limbs of ~70 Å long and 20 Å thick (Fig. T36a) [4–7]. The core of the structure is formed by the TψC and the D arms which make tertiary base hydrogen bonds. The TψC and acceptor arms form a continous RNA helix, as do the D and anticodon arms. Most of the conserved residues are responsible for the stability of the structure, supporting the general character of the known crystal structures. Of special importance is the conformation of the anticodon loop. The coding triplet (34–36) and the adjacent hypermodified base 37 stack over the anticodon stem and form a continous helix. The loop conformation is stabilized by a hydrogen bond between the imino proton of the invariant U33 and the phosphate group of residue 36 in what is known as a U-turn.

Stereochemistry

Most bases involved in the cloverleaf secondary structure follow the regular Watson–Crick base-pairing scheme. Yet some nonstandard hydrogen bonds occur, mainly between constant residues involved in tertiary structure interactions (*see* BASE PAIR).

Figure T36b shows some typical triplets at the hinge of the L and the interactions made by A15 and T54 with U48 and A58 respectively, in tRNA^{Asp}. In the structures of free tRNA^{Phe} and tRNA^{Asp} more than 70 bases are stacked. The base stacking pattern is close to the one observed in A-RNA (*see* NUCLEIC ACID STRUCTURE) and plays an important part in structure stabilization. The importance of base stacking and nonstandard hydrogen-bonding pattern is illustrated by the fact that the acceptor and the TψC arm make a continuous helix, although bases 7 and 49 are not linked by a sugar–phosphate backbone.

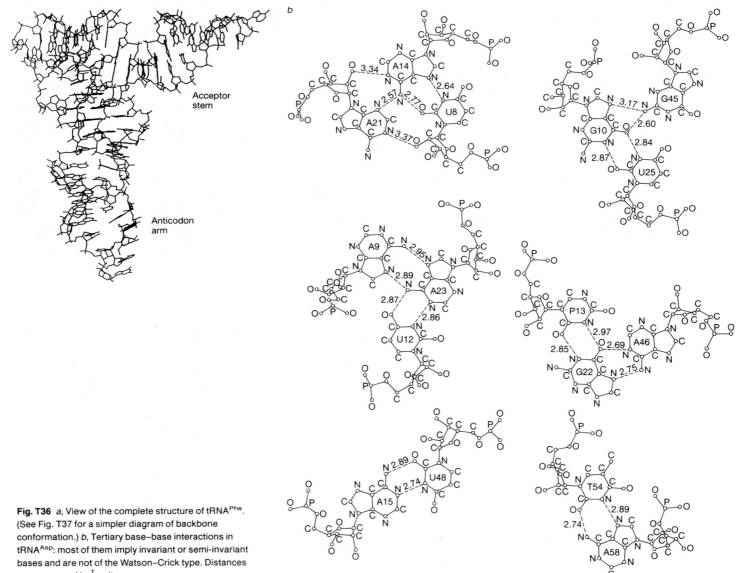

Fig. T36 *a*, View of the complete structure of tRNA^{Phe}. (See Fig. T37 for a simpler diagram of backbone conformation.) *b*, Tertiary base–base interactions in tRNA^{Asp}: most of them imply invariant or semi-invariant bases and are not of the Watson–Crick type. Distances are expressed in Å units.

Dynamical aspects

The core of the molecule is rigid whereas the extremities of the arms are very flexible and can adapt themselves to another molecule. Figure T37 displays the conformation of the phosphate backbones of tRNA^{Asp} as a free molecule (*a*) or complexed with its cognate aminoacyl-tRNA synthetase (*b*). The hinge region shows no major changes whereas the anticodon loop conformation is totally different [8,9].

Function

The protein synthesis cycle

Amino acid–anticodon correspondence. To translate genes into proteins, a correspondence between a codon and an amino acid must exist. This is realized when tRNAs are aminoacylated with their corresponding amino acid by their cognate aminoacyl-tRNA synthetases which have to discriminate among different species of tRNAs. Correct charging is therefore essential to the delivery of the correct amino acid at the right position in the polypeptide chain. Mischarging (i.e. the aminoacylation of a tRNA with an amino acid which does not correspond to its anticodon) will automatically result in the wrong amino acid being incorporated into the polypeptide chain (*see* GENETIC CODE).

Deciphering the mRNA. This crucial step of protein synthesis requires a perfect decipherment of the codons in mRNA by the anticodons in tRNAs. In this three-stage process, tRNAs interact with various cell components which have to recognize features common to all tRNAs.

Initiation. The translation of an mRNA into a protein is initiated when Met-tRNA^{iMet}, the initiator tRNA, is positioned on the P site in the 30S subunit of the ribosome bound to mRNA. Then the 50S subunit completes the ribosome and an elongator aminoacyl-tRNA enters the A site.

Elongation. Aminoacyl elongator tRNAs, which form ternary complexes with elongation factors and GTP, bind to the acceptor site of the ribosome, where they interact with the mRNA. The polypeptide chain is then transferred from the peptidyl-tRNA in the P site to the aminoacyl-tRNA located in the A site. This is followed by translocation of one codon along the mRNA and the release of a free tRNA (Fig. T38).

Termination. Translation stops when a nonsense codon, for which no tRNA exists, is reached (*see* PROTEIN SYNTHESIS).

Translation and aminoacylation

The aminoacyl-tRNA. When tRNAs are aminoacylated by their cognate synthetases the correspondence between codons and amino acids is established. A free hydroxyl group of the terminal ribose is esterified by the carboxyl group of the amino acid. Class I synthetases (*see* GENETIC CODE) transfer the amino acid to the 2′ OH of the ribose, whereas Class II enzymes link the amino acid to the 3′ OH group of the sugar [10,11]. Aminoacylation is a two-step reaction. The activation of the amino acid produces an aminoacyl-adenylate which is then transferred to the tRNA molecule.

Specificity of aminoacylation. The accuracy of this reaction is central for the fidelity of protein synthesis, because once a tRNA is aminoacylated, the activated amino acid will be incorporated in the growing protein chain with no further control. Aminoacyl-tRNA synthetases, kinetic parameters and tRNAs control the accuracy of the reaction.

tRNA identity. tRNAs have a small number of determinants as proven by identity swap experiments. Few examples are well documented: tRNA^{Ser} has been changed to tRNA^{Leu} *in vivo*. Subsequently, experiments showed that the anticodon is an important determinant for many tRNAs (e.g. tRNA^{Asp}, tRNA^{Met}) but in some cases (tRNA^{Ala}) identity can rely on a single base pair (G3–U70) [12].

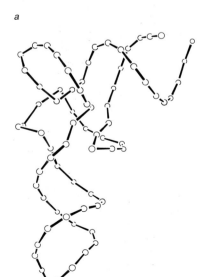

a   *b*

Fig. T37 Schematic views of the sugar–phosphate backbone of the free (*a*) and complexed (*b*) yeast tRNA^{Asp}. The least-square superposition of the 30 phosphate atoms forming the core of the molecule shows a large deviation in the anticodon loop (G30–U40); the largest displacement is of the order of 20 Å for the phosphate U35.

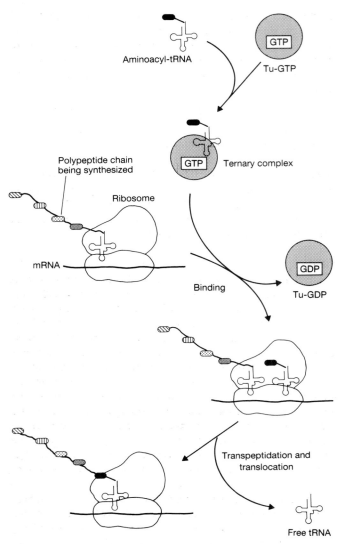

Fig. T38 Role of tRNA in protein synthesis. Tu-GTP/GDP is the complex of elongation factor and guanine nucleotide. *See* PROTEIN SYNTHESIS for more details.

tRNA decodes mRNA by base pairing

Codon–anticodon interactions. Experiments designed to decipher the genetic code, with *in vitro* protein synthesis systems, provide an indirect proof of tRNA–mRNA interactions (*see* GENETIC CODE).

Wobble. The anticodon recognizes its complementary codon on the mRNA through base pairing. The second and third anticodon bases follow the usual Watson–Crick base-pairing scheme. However, on the first position, nonstandard base pairs such as G : U occur. The wobble hypothesis was the first stereochemical explanation for these interactions and their relationship with code degeneracy [13,14].

Interaction model. Binding studies between tRNAs and synthetic trinucleotides confirmed that codon anticodon interaction occurs through base pairing [15]. Additional support came from the crystal structure of yeast tRNA^{Asp} which suggests a model for codon–anticodon recognition. Owing to the peculiar anticodon of tRNA^{Asp}, in the crystal tRNA molecules form dimers by direct association of their anticodons. Each tRNA can thus be seen as an mRNA for the other molecule. Three Watson–Crick base pairs are formed which stack over the anticodon stem in a helical fashion [16].

Recognition involves only codon–anticodon interactions. This can be demonstrated *in vitro* where cysteinyl-tRNA^{Cys} can be chemically converted to alanyl-tRNA^{Cys}. Added to a cell-free system, it can be shown that alanine is then incorporated into a polypeptide instead of cysteine. *In vivo*, the fact that nonsense mutations can be suppressed by mutations in the anticodon of the tRNA constitutes another proof.

Additional functions

Although the major role of tRNA is to act as adaptor during protein synthesis, other functions have been detected [17]. The process of ATTENUATION, a bacterial mechanism for regulating gene expression, involves tRNAs. Aminoacyl-tRNAs act as aminoacyl donors in other synthesis processes. Aminoacyl-phosphatidylglycerol synthesis or interpeptide bridge formation in various bacterial mureins are examples (*see* BACTERIAL ENVELOPE). In these transfer reactions the reactive group of the amino acid is its carboxy group. Some tRNAs, such as tRNA^{Trp} or tRNA^{Pro}, function as PRIMERS for some viral reverse transcriptases (*see* RETROVIRUSES). tRNA-like structures — RNA molecules which can react with one or more tRNA-specific enzymes — are known in viral RNA genomes. Modelling studies on plant viral tRNA-like structures indicate that they contain an RNA PSEUDOKNOT and that the resulting tertiary structure mimics the L-shape structure of tRNAs. tRNA-like structures have a role in the translational regulation of some bacterial genes (*his* operon, *thrS* gene, *metG* gene) [18].

D. MORAS

A. POTERSZMAN

See also: RNA.

1 Crick, F.H.C. (1957) *Biochem. Soc. Symp.* 14–25.
2 Sprinzl, M. (1989) Compilation of tRNA sequences and sequences of tRNA genes. *Nar 231*, **17**, r1–r172.
3 Holley, R.W. et al. (1965) Structure of a ribonucleic acid. *Science* **147**, 1462–1465.
4 Schimmel, R. et al. (Eds) (1979) *Transfer RNA: Structure, Properties and Recognition* (Cold Spring Harbor Laboratory, New York).
5 Moras, D. et al. (1980) Crystal structure of tRNA^{Asp}. *Nature* **288**, 669–674.
6 Saenger, W. (1988) *Landolt-Bornstein Numerical Data and Functional Relationships in Science and Technology.* New Series Group VII, Vol. 1, *Nucleic Acids*, Subvol. B, *Crystallographic and Structural Data II*, 1–29 (Springer-Verlag, Berlin/Heidelberg).
7 Basavappa, R. (1991) The 3Å structure of yeast initiator tRNA: functional implications in initiator/elongator discrimination. *EMBO J.* **10**, 3105–3111.
8 Rould, M. et al. (1989) Structure of E. coli glutaminyl-tRNA synthetase complexed with ATP at 2.8 Å resolution. *Science* **246**, 1135–1142.
9 Ruff, M. et al. (1991) Class II aminoacyl-tRNA: crystal structure of yeast aspartyl-tRNA synthetase complexed tRNA^{Asp}. *Science* **252**, 1682–1689.
10 Eriani G. et al. (1990) Partition of tRNA synthetases into two classes based on mutually exclusive sets of sequence motifs. *Nature* **347**, 203–206.
11 Moras, D. (1992) Structural and functional relationships between aminoacyl-tRNA synthetases. *Trends Biochem. Sci.* **17**, 159–164.

12 Normanly, J. & Abelson, J. (1989) tRNA identity. *Annu. Rev. Biochem.* **58**, 1029–1049.

13 Crick, F.H.C. (1966) Codon–anticodon pairing: the wobble hypothesis. *J. Mol. Biol.* **19**, 548–555.

14 Fuller, W. & Hodgson, A. (1967) Conformation of the anticodon loop in tRNA. *Nature* **215**, 817–821.

15 Uhlenbeck, O.C. (1972) Complementary oligonucleotide binding to transfer RNA. *J. Mol. Biol.* **65**, 25–41.

16 Moras, D. et al. (1986) Anticodon–anticodon interaction induces conformational change in tRNA: yeast tRNA^{Asp}, a model for tRNA–mRNA recognition. *Proc. Natl. Acad. Sci. USA* **83**, 932–936.

17 Schimmel, R. et al. (Eds) (1980) *Transfer RNA: Biological Aspects* (Cold Spring Harbor Laboratory, New York).

18 Mans, R. et al. (1991) tRNA-like structures: structure, function and evolutionary significance. *Eur. J. Biochem.* **210**, 303–324.

transferases One of the main classes of enzymes (EC class 2) in the EC classification. They transfer a group (e.g. methyl, glycosyl, acetyl, etc.) from one compound to another.

transferrin Iron transport protein of M_r 77 000 (apotransferrin) present in blood plasma. Each apotransferrin molecule binds two ferric ions to form transferrin. It is taken up by cells by receptor-mediated ENDOCYTOSIS via specific TRANSFERRIN RECEPTORS. *See also:* COATED PITS AND VESICLES.

transferrin receptor Type II homodimeric transmembrane GLYCOPROTEIN with disulphide-linked subunits of M_r 95 000, responsible for the binding and uptake of TRANSFERRIN. Transferrin receptors on the cell surface bind extracellular iron-laden transferrin and become internalized into clathrin-coated vesicles and delivered to ENDOSOMES (*see* COATED PITS AND VESICLES; ENDOCYTOSIS). Here the low pH causes the bound iron to dissociate from the transferrin; however, the transferrin receptor remains occupied. The receptor then recycles back to the plasma membrane with the apotransferrin still bound, where, at neutral pH, the apotransferrin is released.

transformation (1) Genetic: the introduction of one or more exogenous DNA molecules into a bacterial cell and their expression. The process was first discovered in pathogenic strains of *Diplococcus* (now *Streptococcus*) *pneumoniae* by Fred Griffith in 1928, a discovery that eventually led to the identification of DNA as the genetic material — Griffith's 'transforming principle'. In some bacterial species (e.g. *Bacillus* and *Haemophilus* species) transformation occurs naturally although the cells must be in a 'competent' state to allow uptake of DNA. In other bacterial species (e.g. *Escherichia coli*) the competent state, although not naturally occurring, can be artificially induced in the laboratory. The details of the process of transformation are still largely undefined but it does involve the reversible binding of double-stranded DNA to receptors on the bacterial cell surface in an energy-dependent manner, followed by the irreversible uptake of the DNA into the cell. In naturally occurring transformation systems the introduced DNA is usually integrated into the host genome. The use of transformation to introduce plasmid DNA into bacterial or yeast cells in which a competent-like state has been induced, coupled with the ability to detect transformants by the acquisition of a SELECTABLE MARKER gene, has been essential for the development of RECOMBINANT DNA TECHNOLOGY and GENETIC ENGINEERING. Transformation can also be used in the genetic mapping of bacterial genomes. The term is also now more loosely used to denote the introduction of genetic material to cells other than bacteria, for example plant cells (*see* PLANT GENETIC ENGINEERING).

Cohen, S.N. et al. (1972) *Proc. Natl. Acad. Sci. USA* **69**, 2110–2114.

(2) Neoplastic: the conversion of a normal cell into a tumour cell. *See:* ONCOGENES; TRANSFORMED CELL.

transformed cell Cultured cell which has acquired certain properties of a tumour cell, such as the capacity for unlimited proliferation and the capacity for anchorage-independent growth in culture, but which does not necessarily cause tumours when transplanted into an animal. *See also:* ONCOGENES.

transformer (*tra*) gene (1) In *Drosophila*. These genes act in the sexual differentiation pathway. The *tra-1* gene is transcribed in both sexes, but produces a functional product only in females due to sex-specific splicing regulated by the SEX LETHAL (*Sxl*) gene product. The predicted *tra* protein is highly basic, with a large number of interspersed arginine-serine dipeptides, a feature common to several genes involved in RNA SPLICING. The *tra-2* gene acts in conjunction with *tra-1* to control the expression of the *doublesex* locus. There is a requirement for *tra-2* in the female soma and also in the male germ line for spermatogenesis. The *tra-2* gene produces multiple transcripts, some of which are exclusive to the male germ line. The *tra-2* proteins contain an arginine-serine domain and an RNA recognition motif. The expression of *tra-2* in the soma is not sex specific but the *tra-2* protein is only active in combination with the female-specific *tra-1* product.

(2) In *Caenorhabditis*. These genes are required for the hermaphrodite/female sexual differentiation pathway in both germ line and somatic tissues. *tra-1* has two transcripts which are similarly expressed in both XX and XO animals. The predicted *tra-1* products contain zinc-finger DNA binding motifs and are likely to function as TRANSCRIPTION FACTORS. Restriction of their activity to females may involve interaction with FEM GENE products. The *tra-2* gene is required for the regulation of the *fem* genes and is likely to encode an integral membrane protein involved in the transduction of an extracellular signal. It is expressed in both sexes but in males it may be inactivated by the secreted product of the *her-1* gene.

transforming growth factor (TGF) *See:* GROWTH FACTORS.

transforming oncogenes ONCOGENES that can be identified by their capacity for neoplastic TRANSFORMATION of NIH 3T3 fibroblasts, a cell line of immortal but nonmalignant mouse cells. They generally encode an abnormal form of some protein normally involved in pathways of cell growth or differentiation, in contrast to TUMOUR SUPPRESSOR genes for which neoplastic transformation is the result of loss of the gene and its normal function.

transgene A GENE that has been transferred from one species to another by GENETIC ENGINEERING. *See also*: PLANT GENETIC ENGINEERING; TRANSGENIC TECHNOLOGIES.

transgenic Applied to a plant or animal into which a gene from another species (the transgene) has been introduced. The term is sometimes used more broadly to describe any organism whose genome has been altered by *in vitro* manipulation of the early embryo or fertilized egg to induce, for example, specific GENE KNOCKOUT. *See*: GENETIC ENGINEERING; P ELEMENT-MEDIATED TRANSFORMATION; PLANT GENETIC ENGINEERING; TRANSGENIC TECHNOLOGIES.

transgenic mice Numerous strains of laboratory mice carrying an introduced TRANSGENE, or in which an endogenous gene has been specifically inactivated (GENE KNOCKOUT) have been produced by TRANSGENIC TECHNOLOGY. They are of use in studies of the regulation of gene expression (*see* FUSION GENES), gene rearrangement in IMMUNOGLOBULIN GENES and T CELL RECEPTOR GENES, oncogenesis, and in the study of development.

Transgenic technologies

TRANSGENESIS is a term used to describe the artificial introduction of new genetic material into the germ line of both animals and plants. As such, it is a form of genetic manipulation and, as our abilities improve, the definition of transgenesis has become broadened to include not only the introduction of foreign DNA into the germ line but also designer gene modifications which to date usually involve the insertion of new extraneous DNA but do not necessarily need to do so. An area which should also logically be included is 'somatic transgenesis' or somatic GENE THERAPY, where cells or organs in whole organisms are genetically modified but not necessarily the germ line.

Three main routes to transgenesis in mammals have been described (Fig. T39). These are:

1 Integration of retroviral vectors into an early embryo.
2 Injection of DNA into the PRONUCLEUS of a newly fertilized egg.
3 The incorporation of genetically manipulated EMBRYONIC STEM CELLS into an early embryo.

This entry outlines techniques of transgenesis in mammals, particularly as applied to mice; genetic modification of plants and of bacteria is dealt with elsewhere (*see* EXPRESSION VECTORS;

GENETIC ENGINEERING; PLANT GENETIC ENGINEERING; TI PLASMID). Numerous other methods are used in other multicellular organisms (*see e.g.* P ELEMENT TRANSFORMATION for *Drosophila*); and some (e.g. MICROBALLISTIC techniques) may become important in mammals.

Retroviral infection

If PREIMPLANTATION EMBRYOS are exposed to RETROVIRUS, then a proportion of the embryonic cells will stably integrate proviral sequences into their genome, usually as one copy per cell. In 1976 it was first shown that adult mice derived from embryos exposed to infection with Moloney murine leukaemia virus (MoMLV) could transmit integrated proviral sequences through the germ line [1]. Unfortunately, preimplantation mouse embryos are not permissive for MoMLV expression from the LTR promoter (*see* RETROVIRUSES) so that genes driven by this promoter are not expressed in the embryo. Upon integration, the provirus is subject to *de novo* methylation [2], effectively shutting off proviral TRANSGENE expression even in cell lineages derived from the original infected cells. This problem has been circumvented somewhat in present RETROVIRAL VECTORS by the use of an internal promoter to give expression to the transgene carried by the vector. Such a promoter can be aimed at providing either ubiquitous expression of a transgene (e.g. the herpes simplex virus thymidine kinase promoter [3]), or more cell-specific expression (e.g. the β-globin promoter to give expression in haematopoietic tissues [4]).

One of the limits to using retroviruses to generate transgenic mice is the packaging limits of the virus which restrict the insert size to ~9 kb. Any insert much larger than this and the viral RNA cannot be packaged into the viral capsid.

It should also be noted that depending on the developmental stage at which the retroviral infection occurs, and the number of embryonic cells infected, the resulting offspring are MOSAIC for the proviral insert and indeed different somatic cells might contain proviruses integrated at different genomic location. It is not until the provirus is transmitted through the germ line that an animal is obtained that can be truly referred to as transgenic.

Pronuclear injection

Until recently, the main route for generating transgenic mammals was through injection of DNA into the pronucleus of a fertilized egg. This DNA integrated into the genome and could be passed

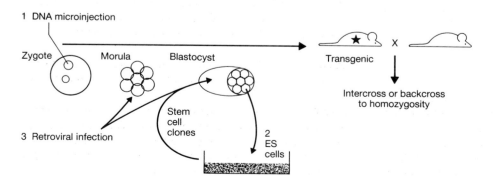

Fig. T39 The three major routes to transgenesis in mammals. ES cells, embryonic stem cells.

through the germ line as stable genetic information. In this way transgenic mice, rats, pigs, cows, and sheep have been produced. This method has the advantage that large fragments of DNA (up to 250 kb) can be injected into the pronucleus, thus allowing complete genes with associated regulatory sequences to integrate into the genome.

The transgene is often found in a multicopy, head-to-tail concatameric array thought to result from homologous recombination between injected DNA molecules before integration. The site of integration in the genome is often associated with rearrangements and duplications, making it difficult to clone flanking sequences adjacent to a transgene.

The site of integration of the transgene in the genome can affect its expression so that founder animals with different integration sites often show widely different levels of expression. Nevertheless, appropriate temporal and spatial expression of transgenes has been observed in several instances irrespective of chromosomal location. It has been found that for the β-globin gene, appropriate tissue-specific expression can be almost guaranteed at any genomic location if a sufficiently large piece (38 kb) of adjacent sequence is included containing DNase hypersensitive sites both 5′ and 3′ of the gene [5]. These flanking sequences have been termed LOCUS CONTROL REGIONS.

Other factors have also been shown to affect transgenic expression including the strain of mouse used [6]. In one experiment, C57/BL6 × SJL hybrid mice were found to be more likely than C57/BL6 inbred mice to express a growth hormone transgene. In addition, the presence of heterologous introns within the coding region can also enhance the expression of some transgenes in mice [7].

Embryonic stem cells (ES cells)

The use of stem cells — either pluripotent embryonic stem cells to repopulate the whole embryo and germ line, or other stem cells (e.g. from bone marrow) to repopulate a specific somatic cellular compartment is now becoming more important.

These cells, which may be maintained in tissue culture, lend themselves to specific genetic manipulation and selection *in vitro* before reconstitution of the embryo and thus allow screening and/or selection of infrequent genetic changes. One particularly significant application is the use of HOMOLOGOUS RECOMBINATION that occurs between the endogenous genome of the cell and incoming DNA which is homologous to the endogenous locus and also carries novel or altered sequences [8]. The repair processes of the cells are able to replace the endogenous gene with the incoming DNA construct. This method may be used both for GENE KNOCKOUT and for more subtle genetic modification.

Future directions

The zygote microinjection route to transgenesis is well established for many species but needs to be made more efficient in species other than the mouse. There needs to be an increase both in the efficiency of production of transgenics and also in the expression of the transgenes. Promising new strategies are being developed and the rules underlying the successful expression of transgenes

are becoming better understood. Much of the fundamental work will remain in mice, but rats are particularly attractive animal models for physiological and pharmacological research. Embryonic stem cell methods are needed in rats. Larger animals may be particularly useful in some studies, and the domestic farm animals are the best options, with the pig being an attractive large animal model.

Improvements are to be expected in homologous recombination techniques and methods to utilize SITE-SPECIFIC RECOMBINATION.

M. EVANS

See also: GENE TARGETING; GENETIC ENGINEERING; HUMAN GENE MAPPING; MAMMALIAN DEVELOPMENT; RECOMBINANT VACCINES.

1 Jaenisch, R. (1976) Germline integration and mendelian transmission of the exogenous moloney leukemia virus. *Proc. Natl. Acad. Sci. USA* **73**, 1260–1264.
2 Jahner, D. et al. (1982) *De novo* methylation and expression of retroviral genomes during mouse embryogenesis. *Nature* **298**, 623–626.
3 Stewart, C.L. et al. (1987) Expression of retroviral vectors in transgenic mice obtained by embryo infection. *EMBO J.* **6**, 383–388.
4 Soriano, P. et al. (1986) Tissue-specific and ectopic expression of genes introduced into transgenic mice by retroviruses. *Science* **234**, 1409–1413.
5 Grosveld, F. et al. (1987) Position independent, high level expression of the human β-globin gene in transgenic mice. *Cell* **51**, 975–985.
6 Brinster, R.L. et al. (1985) Factors affecting the efficiency of introducing foreign DNA into mice by microinjecting eggs. *Proc. Natl. Acad. Sci. USA* **82**, 4438–4442.
7 Palmiter, R.D. et al. (1991) Heterologous introns can enhance expression of transgenes in mice. *Proc. Natl. Acad. Sci. USA* **88**, 478–482.
8 Bradley, A. (1991) Modifying the mammalian genome by gene targeting. *Curr. Opinion Biotechnol.* **2**, 823–829.

transient expression When animal cells in culture are TRANSFECTED with PLASMID or other DNA, only a very small proportion of that DNA becomes integrated into the GENOME of the transfected cell. However, ~50% of the cells will take up the DNA into their nuclei where it will persist for several days without integrating into the chromosomes. Genes carried by the unintegrated DNA will be expressed and regulated in line with the endogenous genes for up to 72 hours after transfection. Such transient gene expression can be used to identify key regulatory elements in PROMOTERS of cloned genes fused to a suitable REPORTER GENE (e.g. that for CHLORAMPHENICOL ACETYLTRANSFERASE).

transilluminator An item of equipment used to visualize fluorescent ETHIDIUM BROMIDE–DNA complexes. It generates ultraviolet light (suitable wavelengths being either 254 nm or 302 nm) from below the ethidium bromide-stained sample (e.g. an agarose gel), the UV light passing through an excitation filter which blocks out visible light thereby enhancing the detectable fluorescence.

transit peptide Peptide sequence that targets proteins to the chloroplast. *See*: PROTEIN TARGETING; PROTEIN TRANSLOCATION.

transition A single BASE PAIR change in a DNA sequence which results in an alteration from a purine–pyrimidine base pair (e.g. A–T) to another (e.g. G–C) (Fig. T40). *Cf.* TRANSVERSION. Transitions represent the major class of point mutation induced by a variety of mutagens. *See also*: MUTATION.

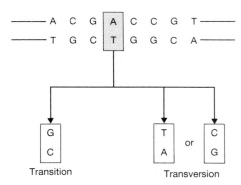

Fig. T40 The generation of transitions and transversions by single base-pair changes in DNA.

transition state The highest-energy species and the most unstable species in a chemical reaction pathway, in which chemical bonds are in the process of being formed and broken. Enzymes catalyse biochemical reactions by stabilizing the transition state and thus favouring reaction. The complementarity of the enzyme active site to the structure of the transition state leads to an increase in binding energy as the structure of the substrate and the conformation of the enzyme changes to that of the transition state in the course of the reaction and produces a reaction pathway in which the Gibbs free energy of the transition state is lower than in the reaction in the absence of the enzyme. The Gibbs energy of activation of the reaction is therefore decreased and the rate of the reaction increased. The term 'induced fit' describes the structural changes that occur on enzyme–substrate binding. *See:* MECHANISMS OF ENZYME CATALYSIS.

transition state analogue A compound that is a structural mimic of the TRANSITION STATE of a biochemical reaction. Such compounds bind more strongly to the enzyme active site than does the ground state of the normal substrate, indicating that the complementarity of an enzyme active site is greatest for the transition state. For lysozyme, for example, which catalyses the hydrolysis of β(1,4)-linked units of *N*-acetylglucosamine (NAG), one transition state analogue is a lactone derivative of *N*-acetylglucosamine in which the pyranose ring is in the half-chair conformation. This binds 100-fold more tightly to lysozyme than do the normal pyranosides in the full-chair conformation in NAG polymers. Unlike the natural transition state, transition state analogues form stable complexes with the enzyme and thus inhibit the reaction.

transition temperature (T_m) (1) Of double-stranded nucleic acids, the temperature at which the two strands separate. *See:* HYBRIDIZATION.
(2) Of membrane lipids, the temperature at which the fatty acyl chains of membrane lipids change from a rigid ordered state, in which all C—C bonds are in the *trans* conformation, to a fluid disordered state in which some are in the *gauche* configuration, introducing bends into the chain. The actual transition temperature for any lipid depends on the degree of saturation of the fatty acyl chain and its length — the higher the degree of saturation and the longer the chain the higher the transition temperature.

transitional endoplasmic reticulum A specialized region of the ENDOPLASMIC RETICULUM (ER) which funnels all newly synthesized proteins assembled in the ER to the GOLGI APPARATUS. There is evidence to suggest that it detaches periodically to form the *cis* Golgi network.

Palade, G. (1975) *Science* **189**, 347–358.

translation The step in PROTEIN SYNTHESIS at which the genetic information encoded in mRNA is used to synthesize a polypeptide chain.

translation function *See:* MOLECULAR REPLACEMENT.

translational control Control of gene expression which is imposed at the level of PROTEIN SYNTHESIS rather than at the earlier stages of TRANSCRIPTION. Examples include the translation of maternal mRNAs deposited in oocytes only at a particular stage in development after fertilization.

translational symmetry *See:* SYMMETRY.

translocation (1) Chromosomal: the transfer of a segment of one CHROMOSOME to another nonhomologous chromosome (*see* CHROMOSOME ABERRATIONS). There are two types of chromosomal translocation: a nonreciprocal interstitial translocation which involves a one-way transfer; and a more common reciprocal translocation which involves a two-way exchange between the two chromosomes. Translocations can result in gross rearrangement of chromosomes, altering both their size and the relative position of the CENTROMERE. Several human genetic diseases are associated with chromosomal translocations (e.g. in some individuals with DOWN'S SYNDROME). Translocation may also activate proto-oncogenes (e.g. in BURKITT'S LYMPHOMA in humans, *see* ONCOGENES).
(2) The movement of a polypeptide across a biological membrane. *See:* PROTEIN TRANSLOCATION.
(3) The reaction which mediates the movement of a ribosome along an mRNA during PROTEIN SYNTHESIS.
(4) The movement of sugars through the phloem of a vascular plant.

translocation mapping Mapping genes in relation to breakpoints in chromosome TRANSLOCATIONS. *See:* HUMAN GENE MAPPING.

transmembrane domain (TM) A region of an integral membrane protein that spans the membrane and serves to anchor the protein in the membrane (*see* MEMBRANE STRUCTURE). In most cases, putative transmembrane domains have been identified from the HYDROPATHY PLOTS of the amino-acid sequence derived from cloned cDNA as short sequences of ~20 hydrophobic amino acids. Such regions might adopt an α-helical conformation. Many membrane proteins contain several transmembrane domains (*see* G PROTEIN-COUPLED RECEPTORS; LIGAND-GATED ION CHANNEL

SUPERFAMILY). *See also*: PROTEIN TRANSLOCATION; STOP-TRANSFER SEQUENCES.

transmembrane protein A protein that spans the whole width of a membrane. Transmembrane proteins are anchored in the membrane by one or more hydrophobic TRANSMEMBRANE DOMAINS which are thought to form α-helices. *See*: MEMBRANE STRUCTURE; PROTEIN TRANSLOCATION.

transmethylases *See*: METHYLTRANSFERASES.

Transmissible spongiform encephalopathies

THE transmissible spongiform encephalopathies (TSEs) are degenerative diseases of the central nervous system which naturally affect humans (Creutzfeldt–Jakob disease (CJD), Gerstmann–Straussler syndrome (GSS), kuru), sheep and goats (scrapie), cattle (bovine spongiform encephalopathy (BSE)), mink (transmissible mink encephalopathy (TME)), and mule deer, elk and antelope (chronic wasting disease). Spongiform encephalopathies have also been diagnosed in captive species of zoo antelope and in domestic cats.

These disorders can be transmitted by inoculation and the incubation period of the experimental disease is chiefly determined by dose, route of infection, host and donor species and genotype, and strain of pathogen. Even so, the relative importance of host genetics and infection in the spread and epidemiology of the naturally occurring diseases of humans and animals remains controversial. Genetic loci have been identified in sheep (*Sip*) and mice (*Sinc*) which determine the incubation time of experimentally induced disease, and the incidence of related diseases in humans (familial CJD and GSS) is also genetically determined. Transmission to susceptible animals may occur naturally by ingestion of diseased tissues or feed (as in BSE) but the transmissible factor or pathogen has yet to be fully characterized. Progress has been hindered by the time and difficulties of measuring infectivity (which requires animal bioassay), the lack of an immune response to infection and inadequate *in vitro* systems for studying the replication of the agent.

Clinical symptoms are variable but can include mental deterioration and incoordination of limb movement in humans. In domestic animals abnormal behaviour, impaired gait, itching, and apprehension are frequently observed. These neurological signs occur late in the pathogenesis of the disease and may take months or years to develop.

Membrane-bound cysts (vacuoles) in nerve tissue, an increase in the number and size of astroglial cells, and the accumulation of scrapie (or TSE)-associated fibrils (SAF) in brain are diagnostic features of these neurodegenerative diseases (Fig. T41). SAF are an aggregated, protease-resistant form (PrPSc) of a neuronal membrane protein, PrPc (sometimes called PrP). They copurify with high titres of infectivity on subcellular tissue fractionation. Much of the molecular biology of scrapie and related diseases has focused on this protein [1–4].

a

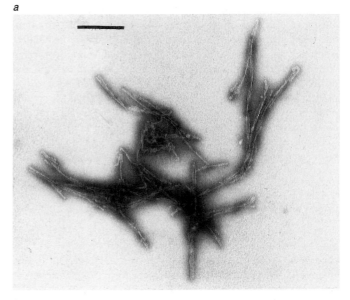

b

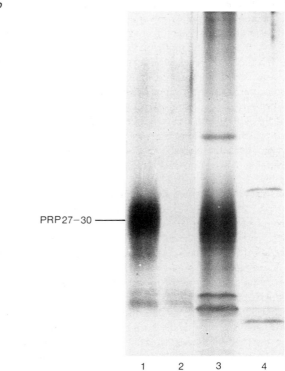

PRP27–30 ——

1   2   3   4

Fig. T41 Scrapie-associated fibrils (SAF) and PrPc protein. *a*, SAF from scrapie-affected mouse brain. Scale bar, 200 nm. *b*, SDS-polyacrylamide gel electrophoresis and silver staining of proteins extracted from (1) hamster scrapie brain, (2) normal hamster brain, (3) cow BSE brain, (4) normal cow brain. Proteases are used in the preparation of these fractions and PrP27-30 is clearly seen as a disease-specific product.

What is the TSE agent?

In the 1950s, infectivity from extracts of scrapie-affected sheep was shown to pass through filters with pores small enough to retain all but VIRUSES. The infectious factor does have some other attributes of a conventional virus, such as phenotypic variation, but what sets scrapie apart from other diseases is the relative resistance of this virus-like activity to inactivation by normal virucidal procedures, such as boiling or exposure to ionizing and ultraviolet radiation. This may indicate that, unlike a virus, a nucleic acid template is not required for replication of the scrapie phenotypes and that a protein may be the sole or an integral part of the infectious particle.

Two novel structures for the infectious particle of the TSEs have been put forward to explain the biological and physicochemical properties of infectivity: a prion and a virino.

Prions and virinos

A prion is a 'proteinaceous infectious particle which resists inactivation by procedures which modify nucleic acids' and has come to be overused as a general word for the scrapie-like agents. One version of a prion structure has some form of a protein as the sole component of the pathogen [5].

A virino is a composite structure of host protein and a host-independent molecule (possibly a small nucleic acid) which determines the strain of scrapie [6].

Which, if either, of these model structures is correct remains to be seen. Much current research is focused on PrPc in some form or other — as the prion protein, as the virino coat protein or as a cellular receptor for a more conventional pathogen.

In the prion model, the infectious agent or prion is a modified form of PrPc. The protein is encoded in one EXON by a single-copy host gene, so this gene structure precludes the modification of PrPc by an alternative splicing mechanism (*see* RNA SPLICING). The conversion of PrPc into the modified form (PrPSc) is proposed to occur by either direct interaction of PrPc with PrPSc (the 'infectious agent'), or in the case of sporadic or familial disease (for example, GSS or CJD or natural scrapie) by a stochastic mechanism — a rare, chance event which sparks off the self-replicating, catalytic conversion of PrPc to PrPSc.

Studies with infected cell cultures have shown that this conversion is a post-translational event which takes place following the processing of PrPc in the *trans* Golgi network (*see* GOLGI APPARATUS) [1,3], although its exact nature is unknown. No disease-specific covalent modification of PrPc has been defined, suggesting that the only difference between PrPSc and PrPc might be conformational. However, other possibilities not yet excluded by experimental evidence include the synthesis of different PrPc polypeptides encoded by a post-transcriptionally edited PrPc messenger RNA or as the result of a ribosomal FRAMESHIFT similar to that observed in RETROVIRUSES.

A great many of the physicochemical and epidemiological features of these diseases can be explained by this simple one-component model. But as yet the mechanism by which different strains of agent produce different phenotypes in the same strain of mouse (which encodes a single PrPc amino-acid sequence) is more difficult to envisage. One suggestion has been that strains of scrapie are encoded by multiple, metastable conformations of PrPc.

The proposed structure of the virino provides a more conventional explanation for strain variation. In this two-component model, a second molecule (possibly a small nucleic acid) binds to host protein (PrPc?), and this protein coat provides protection from the host immune system and degradative processes. In this virino complex, it is the second molecule, not PrPc, that determines the strain of the infectious particle and the phenotype of the infection.

Recently, prions and virinos have been unified [5]: prions have been called apo-prions, virinos renamed holo-prions and the second component of the virino has been dubbed a co-prion, but, whatever the nomenclature, it is clear that the application of molecular biological techniques to the structure and function of PrPc and the spongiform encephalopathies has revolutionized our thinking and brought us close to a molecular understanding of these enigmatic diseases.

The PrPc protein

The structure of PrPc (Fig. T42) is virtually constant in mammalian species, including humans, mouse, and cow, and there may be an homologous gene in lower organisms. It is a GLYCOPROTEIN of M_r 33 000–35 000 (33–35K) which is anchored to the cell plasma membrane by a phosphatidylinositol glycolipid (*see* GPI ANCHOR) attached to its C-terminal amino acid. This normal isoform of the protein is completely degraded by proteases under conditions which leave a 27–30K protease-resistant core of the SAF isoform (PrP27-30). PrP27-30 lacks the first 67 amino acids of the mature PrPc protein (Fig. T42).

The PrPc protein is clearly implicated as a key factor in determining host susceptibility and survival time. Mutations in and around the PrPc gene are linked to the alleles of *Sip* in sheep and *Sinc* in mice, and to the incidence of familial CJD and GSS in humans [3]. These mutations map to several domains of the protein and current research is aimed at understanding the molecular mechanism by which these amino-acid sequence changes might make the difference between the death or survival of animals exposed to scrapie or BSE.

Transgenic mouse studies have also shown that the effects of the *Sinc* alleles in mice, such as relative susceptibility and survival times following scrapie exposure, can be drastically altered by expression and overexpression of PrPc genes from other species. For example, overexpression of the hamster PrPc gene in TRANSGENIC mice shortened their survival time when injected intracerebrally with the Sc237 strain of hamster scrapie to 50–150 days (depending on the level of expression) compared with >700 days in non-transgenic littermates. But the survival times of the transgenic mice when challenged with a mouse-adapted scrapie isolate were similar to or even longer than their non-transgenic littermates [7]. Hence, PrPc may be a product of the *Sinc* or host equivalent locus. However, it has also been pointed out that a protein (or nucleic acid) can be encoded by an open reading frame on the DNA strand opposite in sense to that encoding the PrPc protein sequence and that this putative molecule (an anti-PrPc) may have some role in controlling the timing and incidence of disease [8].

Two main lines of evidence therefore indicate a central role for PrPc in the development of the scrapie-like diseases.

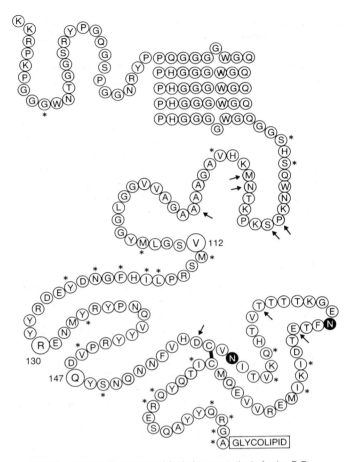

Fig. T42 The amino-acid sequence (single-letter notation) of ovine PrPc protein. This sequence was deduced from the nucleotide sequence of PCR amplified genomic DNA from sheep [15], and is associated with relatively short scrapie incubation times. Protein polymorphisms linked to the alleles of *Sip*, the ovine scrapie incubation time gene, are shown numbered within large circles; variable amino acids in other species which are linked to survival time or incidence of disease are arrowed; variation in the number of octapeptide repetitive sequences has also been linked to familial CJD (25); asterisks mark amino-acid positions which are variable but not linked to disease in a comparison of PrPc proteins from other species; asparagine (N) residues in black circles are potential glycosylation sites. The glycolipid attachment site is inferred from other PrPc proteins. Adapted from [1] by W. Goldmann.

1 Modified forms of PrPc (SAF/PrPSc) accumulate in the brain and some peripheral tissues (spleen, lymph nodes) during the development of scrapie and related diseases. This accumulation parallels increases in the titre of the infectious agent in these tissues, and some workers have claimed co-purification of infectivity with a fraction of M_r 30 000 on SDS–polyacrylamide gel ELECTROPHORESIS [9] and gel filtration CHROMATOGRAPHY. On the other hand, titre and PrPSc have recently been shown to dissociate in a strain-specific way following treatment of infected animals with a polyene antibiotic, amphotericin B [10]. Further work is needed to determine whether PrPc is strictly required for expression of infectivity.

2 There are linkages of the PrPc gene to the alleles of *Sinc* in *Sinc* congenic mouse strains, *Sip* in sheep and to the incidence of the human familial diseases. This linkage is supported by transgenic animal studies. Transgenic mouse studies show that PrPc influences host susceptibility to disease, and in some cases expression of a GSS-linked variant of PrPc (where murine PrPc codon 101 is changed to read leucine instead of proline) results in a spontaneous neurodegenerative disease with the histopathological and clinical features of TSE [11]. The SAF isoform of PrPc has not been detected in this case and further transmission data are not yet available, although it remains possible that infectious particles have been produced *de novo* in the brains of these transgenic mice.

Is PrPc a vital protein?

What is the normal function of PrPc and how is it affected by disease? Its location on the cell surface and other biochemical evidence has provoked speculation that the protein acts as a cell-surface receptor, a glial cell growth factor, an acetylcholine receptor inducer, a cell adhesion molecule or (on lymphocytes) as a lymphoid cell activator. However, the protein's function may be nonessential or shared by other molecules because, even though it is widely expressed in embryonic [12] and adult tissues [13], the deletion of the murine PrPc gene from embryonic stem cells by homologous recombination does not prevent these cells developing into fertile chimaeric mice whose offspring, PrPc null mutants, appear normal [14]. In the near future, challenge of these mice with strains of different infectivity will show whether other molecules can substitute for PrPc during the development of disease, and will provide a test of its putative role as the precursor of the infectious particle — a crucial test of the prion and virino models of the infectious particle.

J. HOPE

1 Hope, J. & Baybutt, H. (1991) The key role of the membrane protein PrP in scrapie-like diseases. *Semin. Neurosci.* **3**, 165–171.

2 Prusiner, S.B. (1991) Molecular biology of prion diseases. *Science* **252**, 1515–1522.

3 Chesebro, B. (Ed.) (1991) Transmissible spongiform encephalopathies: scrapie, BSE and related disorders. *Curr. Topics Microbiol. Immunol.* **172** (Springer-Verlag, Berlin/Heidelberg).

4 Bock, G. & Marsh, J. (Eds) (1988) Novel infectious agents and the central nervous system. *Ciba Foundation Symp.* **135** (Wiley, Chichester).

5 Weissmann, C. (1991) A 'unified theory' of prion propagation. *Nature* **352**, 679–683.

6 Dickinson, A.G. & Outram, G.W. (1988) Genetic aspects of the unconventional virus infections: the basis of the virino hypothesis. In *Ciba Foundation Symp.* **135**, 63–83 (Wiley, Chichester).

7 Prusiner, S.B. et al. (1990) Transgenetic studies implicate interactions between the homologous PrP isoforms in scrapie prion replication. *Cell* **63**, 673–686.

8 Goldgaber, D. (1991) Anticipating the anti-prion protein? *Nature* **351**, 106.

9 Brown, P. et al. (1990) Conservation of infectivity in purified fibrillary extracts of scrapie-infected hamster brain after sequential enzymatic digestion or polyacrylamide gel electrophoresis. *Proc. Natl. Acad. Sci. USA* **87**, 7240–7244.

10 Xi, Y.G. et al. (1992) Amphotericin B treatment dissociates in vivo replication of the scrapie agent from PrP accumulation. *Nature* **356**, 598–601.

11 Hsiao, K. et al. (1990) Spontaneous neurodegeneration in transgenic mice with mutant prion protein. *Science* **250**, 1587–1590.

12 Manson, J. et al. (1992) The prion protein gene: a role in mouse embryogenesis? *Development* **15**, 117–122.

13 Manson, J. et al. (1992) Expression of the PrP gene in the brain of Sinc congenic mice and its relationship to the development of scrapie. *Neurodegeneration* **1**, 45–52.

14 Bueler, H. et al. (1992) Normal development and behaviour of mice lacking the neuronal cell-surface PrP protein. *Nature* **356**, 577–582.

15 Goldmann, W. et al. (1991) Different scrapie-associated fibril proteins (PrP) are encoded by lines of sheep selected for different alleles of the Sip gene. *J. Gen. Virol.* **72**, 2411–2417.

---

transmission electron microscopy (TEM) *See*: ELECTRON MICROSCOPY.

transneuronal degeneration Degeneration of a neuron following removal of its inputs or targets but not involving direct damage to the cell.

transplacement The precise replacement of a GENE at its normal chromosomal location with a modified form of the same gene. The replacement gene is usually introduced on a plasmid vector. Integration into the chromosome is achieved by a double RECOMBINATION event. This methodology was first developed with diploid strains of the yeast SACCHAROMYCES CEREVISIAE.

transplantation antigen Any one of the ubiquitously expressed MHC class I protein products: H-2K, D, and L in the mouse, and HLA-A, B, and C in human (*see* MAJOR HISTOCOMPATIBILITY COMPLEX; MHC MOLECULES). The term refers to the capacity of these ubiquitously expressed glycoproteins to serve as recognition elements for ANTIBODIES and effector T CELLS in the rejection of ALLOGRAFTS.

transport systems *See*: INTRACELLULAR TRANSPORT; MEMBRANE TRANSPORT SYSTEMS.

transport vesicles Small, spherical, membrane-bounded compartments which ferry proteins and many lipids between membrane-bounded compartments in eukaryotic cells. *See*: ENDOPLASMIC RETICULUM; GOLGI APPARATUS; LYSOSOMES; PROTEIN SECRETION.

transporters MEMBRANE TRANSPORT SYSTEMS. The term is used in particular for those systems that use the energy contained in an ion gradient across the plasma membrane to transport ions or other solutes (e.g. sugar transporters, bacterial permeases). It may also refer to the peptide transporters that transfer peptides from cytoplasm into the ENDOPLASMIC RETICULUM (*see* ANTIGEN PROCESSING AND PRESENTATION).

Transposable genetic elements

TRANSPOSABLE elements are DNA sequences that can move (transpose) from one place to another in the GENOME of a cell. The first transposable elements to be recognized were the *Activator/ Dissociation (Ac/Ds)* elements of maize (*Zea mays*). These were identified by Barbara McClintock as being responsible for unstable MUTATIONS in this species [1] (*see* TRANSPOSABLE GENETIC ELEMENTS: PLANTS). Since then, transposable elements have been found in a wide range of organisms, both prokaryotic and eukaryotic. Reviews discussing the best studied of these elements can be found in [2].

Transposable elements in the genome are characterized by being flanked by direct repeats of a short sequence of DNA that has been duplicated during transposition and is called a target-site duplication. Virtually all transposable elements, whatever their type and mechanism of transposition, make such duplications at their site of insertion (target site). In some cases the number of bases duplicated is constant, in others it may vary with each transposition event. Most transposable elements have inverted repeat sequences at their termini. These terminal inverted repeats may be anything from a few bases to a few hundred bases long and in many cases they are known to be necessary for transposition.

Prokaryotic transposable elements

Prokaryotic transposable elements have been studied most extensively in *Escherichia coli* and other Gram-negative bacteria but are also present in Gram-positive species. They are generally called insertion sequences (IS) (Fig. T43*a*) if they are less than about 2 kilobases (kb) long, or transposons (Tn) (Fig. T43*b, c*) if they are longer. BACTERIOPHAGES such as Mu (for 'mutator', *see* MU PHAGE) and D108 which replicate by transposition make up a third type of transposable element. Elements of each type encode at least one polypeptide required for their own transposition and this is called a transposase. Transposons often include genes coding for functions unrelated to transposition and most transposons that have been studied in detail carry antibiotic-resistance genes (*see* PLASMIDS; R PLASMIDS).

Transposons can be divided into two classes according to their structure. Elements of one class, known as compound or composite transposons, have copies of an IS element at each end, usually in inverted orientation (Fig. T43*b*). Transposition of composite transposons requires transposases coded by one of their terminal IS elements. Transposons of the second class, such as Tn*3*, have terminal inverted repeats of about 30 base pairs (bp) and do not contain sequences from IS elements (Fig. T43*c*).

Mechanisms of transposition

Transposition is either conservative or replicative although some elements such as bacteriophage Mu and the insertion sequence IS*1* can transpose both conservatively and replicatively. During replicative transposition one copy of the transposing element remains at the donor site and another is inserted at the target site. Tn*3* transposes replicatively. It encodes two proteins required for transposition — a transposase that initiates transposition and a RESOLVASE that performs SITE-SPECIFIC RECOMBINATION between the two copies of the element produced during transposition.

An element transposing conservatively is excised from one site and inserted at the other. This mechanism is used by the composite transposon Tn*10*. It codes for a transposase that makes double-stranded cuts at the ends of a transposing element to excise it from the donor molecule and staggered single-stranded cuts at the target site to allow it to insert.

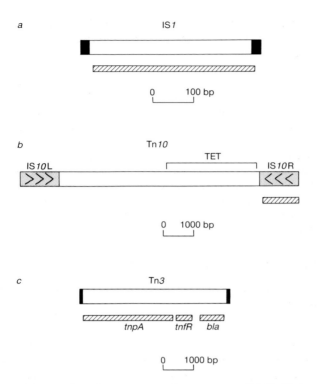

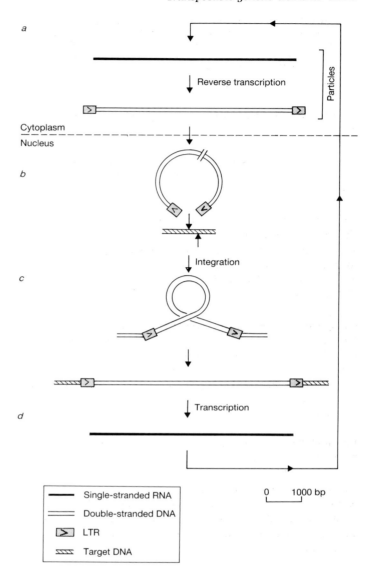

Fig. T43 Bacterial insertion sequences and transposons. *a*, IS*1*. The filled regions at the ends indicate the terminal inverted repeats. The hatched line indicates the region coding for a protein required for transposition. *b*, The composite transposon Tn*10*. The stippled regions indicate inverted copies of the insertion sequence IS*10* that lie at its ends. Each copy of IS*10* has short terminal inverted repeats. The hatched line indicates a regions of the right-hand copy of IS*10* that codes for a function required for transposition of Tn*10*. The left-hand copy of IS*10* is mutant and cannot provide this function. The region of Tn*10* responsible for resistance to tetracycline (TET) is indicated. *c*, The transposon Tn*3*. The filled regions at the ends indicate the terminal inverted repeats. The hatched lines indicate sequences coding for the following: *tnpA* and *tnpR*, proteins required for transposition; *bla*, the enzyme β-lactamase responsible for ampicillin resistance.

Eukaryotic transposable elements

Eukaryotic elements can also be classified according to their structure and mechanism of transposition. The primary distinction is between elements that transpose via an RNA intermediate and elements that transpose directly from DNA to DNA.

Retrotransposons

Elements that transpose via an RNA intermediate are often referred to as retrotransposons and their most characteristic feature is that they encode polypeptides that are believed to have REVERSE TRANSCRIPTASE activity. There are two types of retrotransposon: some resemble the integrated proviral DNA of a RETROVIRUS in that they have long direct repeat sequences, LONG TERMINAL REPEATS (LTRs), at each end. *Copia* elements of the fruit fly (*Drosophila melanogaster*), TY ELEMENTS of yeast (*Saccharomyces cerevisiae*), *Ta1* elements of the plant *Arabidopsis thaliana* and *IAP* of mice are elements of this type. The similarity between these

Fig. T44 Transposition of the retrotransposon *Ty1* of *S. cerevisiae*. *a*, Full-length *Ty1* RNA is packaged into virus-like particles made up of *Ty1*-encoded protein. These include a reverse transcriptase that copies this RNA into linear double-stranded DNA with LTRs at each end. *b*, The linear DNA enters the nucleus together with *Ty1*-encoded proteins. It associates with target chromosomal DNA that is cleaved with a *Ty1*-encoded nuclease. *c*, Integration takes place when the 3′ hydroxyl ends of *Ty1* DNA are joined to the 5′ phosphates of the target. *d*, The cycle is completed by transcription of *Ty1* RNA from the integrated DNA.

retrotransposons and proviruses extends to their coding capacity. They contain sequences related to the *gag* and *pol* genes of a retrovirus, suggesting that they transpose by a mechanism related to a retroviral life cycle. This has been confirmed for *Ty1* elements (Fig. T44).

Retrotransposons of the second type have no terminal repeats. They also code for *gag*- and *pol*-like polypeptides and transpose by reverse transcription of RNA intermediates but do so by a mechanism that differs from that of retrovirus-like elements. LINE

(L1) elements in mammalian genomes, *I* and *jockey* elements of *Drosophila* and *Tad* elements of the mould *Neurospora crassa* are of this type.

Transposition by reverse transcription is a replicative process and does not require excision of an element from a donor site.

Other eukaryotic elements

Elements that transpose without recourse to an RNA intermediate have inverted repeats at their termini. In most cases these repeats are 10–40 bp long. *Ac/Ds* and *Spm/En* elements in *Z. mays*, *Tam* elements in *Antirrhinum majus* and *P* elements in *D. melanogaster* (Fig. T45) are of this type. Although these elements are believed to transpose conservatively, excising from one site before inserting at another, they increase in copy number as a result of transposition. *Ac/Ds* elements do so by transposing from a replicated region of the genome to an adjacent site that has yet to replicate. *P* elements increase in number because they excise to leave a chromosomal break that can be repaired using the SISTER CHROMATID or HOMOLOGOUS CHROMOSOME as a template for DNA synthesis. The sequence of the donor site can be restored complete with its *P* element if the sister chromatid is used.

Mutations due to insertions of elements of this type are often unstable and revert to an apparently wild-type state under conditions that allow transposition. This is due to complete or partial excision of the element concerned. In plants this can result in striking variegation in the appearance of petals, leaves, seeds and other tissues, which has led to the identification of several new elements.

Transposable elements as mutagens

Transposable elements are an important source of spontaneous mutations and must have influenced the ways in which genes and genomes have evolved. They can inactivate genes by inserting within them and can cause gross chromosomal rearrangements either directly through the activity of their transposases or indirectly as a result of recombination between copies of an element scattered around the genome. Transposable elements that excise often do so imprecisely and may produce ALLELES coding for altered gene products if the number of bases added or deleted is a multiple of three (*see* MUTATION).

Transposable elements themselves may evolve in unusual ways. If they were inherited like other DNA sequences then copies of an element in one species would be more like copies in

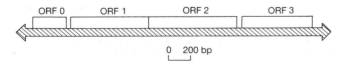

```
    ORF 0        ORF 1           ORF 2              ORF 3
```

```
    0    200 bp
```

Fig. T45 The structure of a complete *P* element of *D. melanogaster*. The arrowheads indicate terminal inverted repeats. The open boxes indicate four open reading frames, ORF 0–3, that are joined to code for the single polypeptide required for *P* element transposition. The first three open reading frames can be joined to code for a polypeptide that regulates transposition. Redrawn from [3].

closely related species than copies in more distant species. This is not always the case, suggesting that transposable elements are occasionally transmitted horizontally from one species to another.

Transposable elements as experimental tools

In addition to their intrinsic biological importance, transposable elements have become useful experimental tools. They can be used as probes to identify RECOMBINANT DNAs carrying genes into which they have inserted, a procedure known as TRANSPOSON TAGGING. Some elements can act as VECTORS for genetic transformation experiments. *P* elements of *D. melanogaster* have been particularly useful in this respect and have made a major contribution to understanding the molecular biology of this species.

D.J. FINNEGAN

See also: MITOCHONDRIAL GENOMES: LOWER EUKARYOTE; P ELEMENT MEDIATED TRANSFORMATION.

1 McClintock, B. (1957) *Cold Spring Harbor Symp. Quant. Biol.* **16**, 13–47.
2 Berg, D. & Howe, M.M. (1989) *Mobile DNA* (American Society of Microbiology, Washington, DC).
3 Finnegan, D.J. (1990) Transposable elements and DNA transposition in eukaryotes. *Curr. Opinion Cell Biol.* **2**, 475.

Transposable genetic elements: plants

TRANSPOSABLE genetic elements were first proposed in maize (*Zea mays*) by Barbara McClintock in 1948. She termed them controlling elements because of their effect on gene expression during plant development [1]. The transposable elements identified by McClintock were of the conservative type (*see* TRANSPOSABLE GENETIC ELEMENTS), which transpose by excising from one site and inserting into another. Subsequently, both transposons that move conservatively and those that move replicatively have been isolated from plants (see Tables T4 and T5).

Transposons cause insertion MUTATIONS in the genes into which they transpose (Fig. T46). Conservative elements subsequently excise during further transposition and, in the process, often restore the function of the host gene, which can be observed phenotypically as REVERSION. Somatic reversion may occur at a high frequency and, if the host gene encodes a CELL-AUTONOMOUS phenotype, may be seen as variegation or phenotypic instability within an individual. The activity of transposable elements in maize was first studied through their effects on pigmentation and storage product accumulation in the corn cob. Variegation in the colour of individual kernels can be caused by transposon insertion in a pigment gene and subsequent somatic excision at various times during development. The variegated phenotype is a measure of a conservative transposon's ability to excise, and has been used to study transposon function. As well as the effect on the locus into which the element transposes, its presence can lead to chromosomal rearrangements, such as deletions, inversions, and chromosomal breaks which may affect neighbouring loci.

Table T4 Common plant transposable elements that move by excision

| Element | | | |
Autonomous	Non autonomous	Plant species	Target site duplication
AC	DS	Maize	8 bp
Spm (En)	dSpm (I)	Maize	3 bp
Dt	rdt	Maize	9 bp
*Mu-Ml	Mn	Maize	9 bp
Tam1	Tam2, Tam4?	Snapdragon	3 bp
Tam3		Snapdragon	8 bp (5 bp?)
	dTph1	Petunia	8 bp

*May also transpose replicatively

Table T5 Common plant transposable elements that move replicatively

Element	Element type	Plant species
Tnt 1	Retroviral-like	Tobacco
Tnt 1	Retroviral-like	*Arabidopsis*
Bs1	Retroviral-like	Maize
Cin4	Nonviral retrotransposon	Maize

In plants, conservative transposons tend to form families (as seen for P ELEMENTS of *Drosophila*) consisting of autonomous (regulator) and nonautonomous (receptor) members. Autonomous members encode specific protein(s) (transposase) required for transposition of family members and also have the specific terminal repeats required for transposition. Nonautonomous elements have the terminal repeats but do not encode functional transposase and are therefore dependent on transposase encoded by autonomous elements for their movement. Nonautonomous elements often arise from autonomous ones by internal mutation. The best characterized families are the Ac/Ds and Spm/dSpm (also known as En/I) families of maize.

The autonomous Ac elements (~4.6 kb) encode a single protein (transposase) while Spm (~8.3 kb) encodes two functionally distinct proteins. The Ac transposase and one of the Spm proteins (tnpA) bind to subterminal repeats to bring the ends of the element together during transposition. The second Spm protein (tnpD) may serve the cutting function during excision. A protein not encoded by Ac may serve this function for Ac. Both Ac and Ds cause target site duplications of 8 bp on insertion (*see* TRANSPOSABLE GENETIC ELEMENTS) which is not usually lost on excision of the element. Members of the Spm family cause 3 bp target site duplications.

There is a total of around 35 Ac and Ds elements in the maize genome and many different kinds of Ds elements occur; they include elements of different lengths in which different internal regions have been deleted or altered, and complex elements containing multiple copies of Ds sequences.

The activity of plant transposons is strongly correlated with the degree of methylation of their DNA. DNA METHYLATION may repress expression of the transposase genes, thus suppressing transposition. Methylation may also affect the affinity of transposase for its target DNA motifs. Methylation of specific regions of

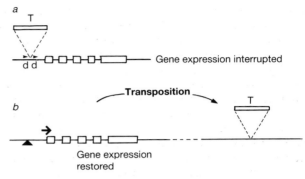

Fig. T46 Illustration of the phenotypic consequences of a transposon insertion into a gene required for pigment biosynthesis in snapdragon flowers. *a*, When the transposon (T) is inserted at the locus (in this case in the promoter region of the gene) expression is prevented and so no colour is formed. d, Direct duplication of target sequence generated by insertion. *b*, When the transposon excises, gene expression (shown by arrows), and consequently pigment production, are restored. Somatic excision can occur at a high frequency as shown by revertant pigment-producing sectors in the otherwise colourless flower. The earlier in development of the flower that the excision occurs, the larger the coloured sector, as all cells derived from the revertant cell carry the restored functional gene. On self-fertilization the majority of plants show the somatically unstable phenotype, but excision in the germ line gives rise to some fully pigmented revertant progeny.

transposon DNA may vary during plant development and this is correlated with changing frequencies of transposition.

Plant transposons have been used very successfully to identify new genes by TRANSPOSON TAGGING, and to study their function, especially in the two species where transposons are best characterized — maize and snapdragon (*Antirrhinum majus*). The Ac/Ds and Spm/dSpm systems from maize have been introduced into other plant species (including *Arabidopsis*, tobacco, potato, tomato, *Petunia*, carrot) by plant transformation (*see* PLANT GENETIC ENGINEERING) to establish heterologous transposon tagging systems.

C. MARTIN

1 McClintock, B. (1948) Mutable loci in maize. *Carnegie Inst. Wash. Ybk* **47**, 155–169.
2 Fedoroff, N.V. (1983) Controlling elements in maize. In *Mobile Genetic Elements* (Shapiro, J., Ed.) 1–63 (Academic Press, New York).
3 Peterson, P.A. (1987) Mobile elements in plants. *Crit. Rev. Plant Sci.* **6**, 105–208.
4 Gierl, A. et al. (1989) Maize transposable elements. *Annu. Rev. Genet.* **23**, 71–85.

transposase Enzyme that mediates transposition by TRANSPOSABLE GENETIC ELEMENTS. In bacteria, transposases are usually encoded by the insertion sequence (IS) elements of transposons. Transposases seem to have two roles in transposition: the cleavage of the target DNA sequence and recognition of the ends of the transposons. The synthesis of transposases is under the tight control of a REPRESSOR which in turn controls the frequency of transposition.

transposition The process whereby a transposon or other TRANSPOSABLE GENETIC ELEMENT moves to a new location either within the same chromosome or on a new chromosome. Replicative transposition involves the transfer of a copy of the transposon, leaving the parental transposon in its original chromosomal location, a process initiated by one or more TRANSPOSASE enzymes. Conservative transposition involves the excision of the transposon from one location and its integration into another.

Replicative transposition in the bacterial transposon Tn3 is a two-step process. In the first step the transposon forms a CO-INTEGRATE at the target site in the host chromosome where it is then replicated. In the second step, the co-integrate so formed is resolved by a RESOLVASE, an enzyme encoded by the transposon, to generate a single copy of Tn3 integrated at the target site. In contrast, the transposon Tn10 seems to transpose by excision from its original site and integration into its new one without any replication (conservative transposition).

The yeast transposon Ty1 (*see* TY ELEMENT) is transposed by a mechanism very different from that found in bacteria as it seems to require an RNA intermediate (*see* RETROTRANSPOSON).

transposon Genetic element that can transpose to a different position in the genome. *See*: TRANSPOSABLE GENETIC ELEMENT; TRANSPOSITION.

transposon mutagenesis A technique for generating mutants (in both bacteria and eukaryotes) by nonspecific insertion of mobile genetic elements (transposons), (*see* TRANSPOSABLE GENETIC ELEMENTS) into the DNA required to be mutated. This powerful technique not only causes insertion mutations which disrupt gene function (*see* MUTATION), but, because of the presence of an identifiable marker — the transposon — facilitates identification and cloning of the disrupted gene. Transposon mutagenesis is typically accomplished by conjugal transfer or transfection of a PLASMID vector carrying a transposon to a recipient cell in which the plasmid will not replicate. Once the transposon-loaded vector is introduced into a recipient cell, the transposon inserts a copy of itself into the recipient DNA. Because the plasmid cannot replicate it is soon diluted out and lost. Such a transposon delivery vector is known as a suicide plasmid, because it 'commits suicide' once introduced into the recipient. Potential mutants are isolated by their ability to grow in the presence of a particular antibiotic to which resistance has been conferred by the transposon. A specific detection or screening assay is then required to identify mutants.

Simon, R. et al. (1983) *Bio/Technology* **1**, 784–791.

transposon tagging A method of cloning genes for which no HYBRIDIZATION probe exists. Essentially, a DNA LIBRARY is prepared from the genomic DNA of a mutant strain which is known to have the target gene inactivated by a TRANSPOSABLE GENETIC ELEMENT (transposon). The gene library is then screened with the transposon as the hybridization probe to identify clones carrying the transposon and surrounding DNA sequences — which must represent the target gene sequences. This strategy has been exploited in the isolation of genes from *Drosophila* and plants.

transthyretin (TTR) A tetrameric protein (monomer M_r 55 000; 127 amino acids) found in plasma and cerebrospinal fluid. It transports thyroxine and, indirectly, retinol via a tight complex with RETINOL-BINDING PROTEIN. Mutant forms of transthyretin are associated with familial AMYLOIDOSIS.

Terry, C.J. et al. (1993) *EMBO J.* **12**, 735–741.

transverse tubules Infoldings of the plasma membrane which make contact with the sarcoplasmic reticulum in skeletal muscle cells and which may be involved in transfer of signals from the cell surface to the contractile apparatus. *See*: CALCIUM; MUSCLE.

transversion A single base pair change in a DNA sequence which results in a purine–pyrimidine base pair (e.g. A–T) being replaced by a pyrimidine–purine base pair (e.g. T–A or C–G) (see Fig. T40 in TRANSITION.) *See also*: MUTATION.

traumatic acid *See*: PLANT WOUND RESPONSES.

treadmilling The flow of actin subunits through an actin filament in which subunits are assembling at the plus end at an identical rate to the rate of disassembly at the minus end. The length of the filament remains the same although there is a net flow of subunits through it. *See*: MICROFILAMENTS.

triacylglycerols Esters of FATTY ACIDS with the trihydroxyl alcohol glycerol (Fig. T47), also known as triglycerides or neutral fats. They are abundant in tissues as storage lipids, as in mammalian adipose tissue and subcutaneous fat, and as oils in seeds. In animals, the fatty acids most commonly found in triacylglycerols are palmitic, stearic, and oleic acids. The variety of triacylglycerols in plant seeds is much greater than in animal tissues with more than 300 different fatty acids having been identified in seed oils. Whether triacylglycerols are liquid or solid at room temperature depends on the types of fatty acid they contain and the position on glycerol at which a particular fatty acid is esterified. Triacylglycerols are stored in lipid bodies or oil droplets surrounded by a single layer membrane. They are synthesized in the ENDOPLASMIC RETICULUM (ER), where they are thought to accumulate between the leaflets of the ER membrane, eventually being released as an oil droplet.

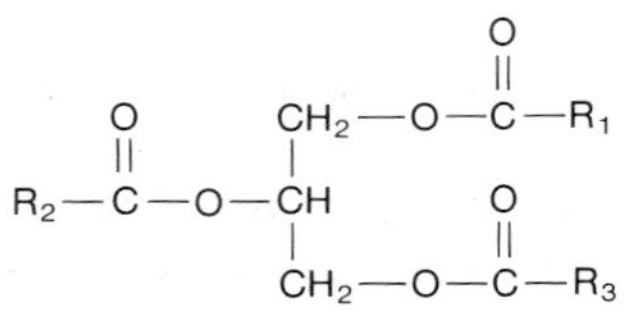

$$
\begin{array}{l}
O \\
\| \\
CH_2-O-C-R_1 \\
| \\
R_2-C-O-CH O \\
| \| \\
CH_2-O-C-R_3
\end{array}
$$

Fig. T47 General structure of a triacylglycerol.

tricarboxylic acid cycle (TCA, Krebs cycle, citric acid cycle) The cyclic pathway by which the 2-carbon acetyl groups of acetyl-CoA are oxidized to carbon dioxide and water. The cycle was discovered by Hans Krebs and involves production and consumption of citrate, hence the synonyms Krebs cycle and citric acid cycle. Its primary function is the production of NADH and reduced flavin adenine dinucleotide ($FADH_2$), whose re-oxidation is energetically coupled to the production of ATP by OXIDATIVE PHOSPHORYLATION. A major secondary function is to interconvert the carbon skeletons of important compounds in intermediary metabolism. The cycle occurs in the cytoplasm of bacteria or in the matrix of MITOCHONDRIA.

2,3,5-triiodobenzoic acid *See:* PHYTOTROPIN.

trinucleotide repeats DNA sequence composed of repeated trinucleotides, found in certain genes. A change in the number of repeats is the presumed cause of various genetic diseases. *See:* FRAGILE X SYNDROME; HUNTINGTON'S DISEASE; KENNEDY'S DISEASE; MYOTONIC DYSTROPHY.

triose phosphate isomerase (TIM) A dimeric enzyme (E.C. 5.3.1.1) which catalyses the intramolecular oxidoreduction reaction:

D-glyceraldehyde 3-phosphate = D-dihydroxyacetone phosphate

In this enzyme catalytic activity has evolved to reach about the maximum possible. It contains a characteristic structure, the TIM barrel (*see* Fig. P81 in PROTEIN STRUCTURE). *See:* MECHANISMS OF ENZYME CATALYSIS; MOLECULAR EVOLUTION: SEQUENCES AND STRUCTURES.

triple negative cells PRO-T CELLS. *See:* T CELL DEVELOPMENT.

triple positive cells PRE-T CELLS. *See:* T CELL DEVELOPMENT.

triploid Organism or single cell whose somatic nuclei contain three copies of a basic haploid chromosome set. Triploids may be designated as $3x$, where x is the original number of chromosomes. Triploids generally undergo highly irregular MEIOSIS.

triskelion Soluble form of CLATHRIN, consisting of three copies of clathrin heavy chain and three copies of clathrin light chain. Triskelions (Fig. T48) can self-assemble *in vitro* into a polyhedral lattice identical to that found on clathrin-coated vesicles (*see* COATED PITS AND VESICLES).

trisomy A chromosome additional to the normal complement (i.e. $2N+1$) so that in each somatic nucleus one particular chromosome is represented three times rather than twice.

trisomy 18 (Edward's syndrome) TRISOMY of chromosome 18 is one of the few trisomies of AUTOSOMES in which the affected infant reaches term. The frequency of occurrence is 1 : 8000 births with an increasing incidence with increasing maternal age. The features of the syndrome include characteristic facial changes, simple fingerprints, congenital heart defects and renal

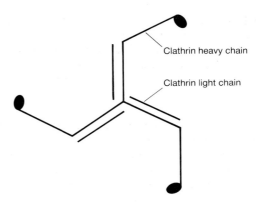

Fig. T48 Model of a triskelion.

abnormalities. Affected females seem more likely to survive to term than males; around 10 females are born with trisomy 18 to every 3 males. The life expectancy of infants with trisomy 18 is limited to a few months.

trisomy 21 *See:* DOWN'S SYNDROME.

Triticum aestivum Bread wheat.

Triton X-100 Nonionic detergent (polyethylene glycol *p*-isooctylphenyl ether) widely used for solubilizing biological membrane components. Its large micelle size can make its removal (e.g. by dialysis) difficult, however.

trk Potential ONCOGENE which encodes the high affinity receptor for NERVE GROWTH FACTOR. *See also:* GROWTH FACTOR RECEPTORS.

***trn* genes** Genes encoding TRANSFER RNAS (usually of plants), often identified by the potential of the transcript to form cloverleaf secondary structures, with certain conserved residues.

tRNA TRANSFER RNA.

trophectoderm One of the two differentiated cell types of the mammalian BLASTOCYST, trophectoderm is a monolayer of typical epithelium which forms the outer layer of the blastocyst surrounding the inner cell mass. During postimplantation development, cells of mural trophectoderm (which form the walls of the blastocyst and are not in contact with inner cell mass) cease to divide and become large and polyploid (primary trophoblastic giant cells). The remaining polar trophectoderm cells remain diploid and proliferate rapidly while in proximity to the inner cell mass. In the absence of this influence these cells can become polyploid and either replace the mural trophoblastic giant cells or penetrate the egg cylinder forming extraembryonic ectoderm which develops into the chorion, or penetrate into the endometrium to form the placenta. Some of these cells and cells of the chorion become polyploid (secondary giant cells). *See also:* EXTRAEMBRYONIC MEMBRANES; MAMMALIAN DEVELOPMENT.

trophic factors Substances produced by one cell type that maintain the survival of another cell type. Trophic interactions have been described most frequently in the nervous system, between interconnected neuronal populations, and are mediated by trophic factors (molecules) such as CILIARY NEUROTROPHIC FACTOR, NERVE GROWTH FACTOR, and NEUROTROPHIN-3.

trophoblast Those parts of the TROPHECTODERM that cease to divide and become POLYPLOID (giant cells). Primary trophoblastic giant cells arise from those mural and polar trophectoderm cells which lose contact with the INNER CELL MASS as they migrate around the BLASTOCOEL cavity. Together with parietal endoderm these cells form the parietal yolk sac. Secondary giant cells arise from those cells of polar trophectoderm which lose contact with the inner cell mass by penetrating the uterine endometrium to form part of the chorioallantoic placenta (in the mouse). *See also:* EXTRAEMBRYONIC MEMBRANES.

tropoelastin Monomer of the EXTRACELLULAR MATRIX MOLECULE elastin.

tropomyosin *See:* ACTIN-BINDING PROTEINS; MUSCLE.

troponin A complex of three polypeptide chains — troponin C (M_r 18 000), troponin I (M_r 24 000), and troponin T (M_r 37 000) — which is involved in the regulation of striated MUSCLE contraction by CALCIUM. It is bound to tropomyosin (*see* ACTIN-BINDING PROTEINS in the thin filament of muscle (via the elongated troponin T subunit) at a rate of one troponin complex per tropomyosin molecule. Troponin I binds to actin and troponin C is the CALCIUM-BINDING PROTEIN of the complex. Troponin C consists of two calcium-binding domains linked by an extended α-helix. Each domain contains two Ca^{2+}-binding sites; low affinity sites in the N-terminal domain, high affinity sites in the C-terminal domain. In resting muscle the high affinity sites are occupied by either Ca^{2+} or Mg^{2+}. When the cytosolic Ca^{2+} is raised, the low affinity sites are also occupied leading to a conformational change in the troponin complex that is transmitted to tropomyosin, changing its position on the actin filament and in some way facilitating or increasing the effectiveness of actin interaction with the S1 head of MYOSIN.

Trp Abbreviation for the AMINO ACID tryptophan.

trypanosomes A group of parasitic PROTOZOA, typified by the genus *Trypanosoma* (class Mastigophora) that cause severe illness in both animals and humans (e.g. African sleeping sickness (*T. brucei*), and the South American Chagas' disease, *T. cruzi*). African trypanosomes (e.g. *T. brucei*) are able to evade the human IMMUNE RESPONSE by rapidly changing the nature of their coat protein antigen — the so-called variant surface glycoproteins (VSGs) (*see* ANTIGENIC VARIATION).

trypsin A SERINE PROTEINASE digestive enzyme (EC 3.4.21.4) secreted in the pancreatic juice. It is synthesized as an inactive precursor, trypsinogen, from which the active enzyme is cleaved autocatalytically or by the action of enteropeptidase by scission at

a single peptide bond, liberating the hexapeptide Val-$(Asp)_4$-Lys from the N-terminal end (Fig. T49). Trypsin preferentially cleaves peptide bonds involving the carboxyl group of Arg or Lys. It is used in protein analysis to produce smaller 'tryptic peptides' for sequencing.

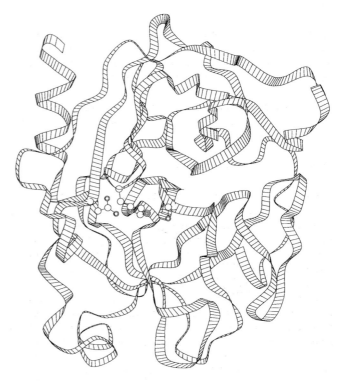

Fig. T49 Diagram of the structure of the enzyme trypsin, showing the active site.

tryptic peptides PEPTIDES produced from a longer polypeptide chain by hydrolysis with the enzyme TRYPSIN, which specifically cleaves on the carboxyl side of lysine and arginine residues.

tryptophan (Trp, W) An aromatic AMINO ACID.

***ts* mutants, *ts* mutations** *See:* TEMPERATURE-SENSITIVE MUTANTS.

TSE TRANSMISSIBLE SPONGIFORM ENCEPHALOPATHIES.

TSH THYROID-STIMULATING HORMONE.

TTP Thymidine (deoxythymidine) 5′-triphosphate, a nucleotide precursor for DNA synthesis. *See:* NUCLEOSIDES AND NUCLEOTIDES.

TTR The plasma protein TRANSTHYRETIN.

TTX The neurotoxin TETRODOTOXIN.

tuberonic acid *See:* PLANT WOUND RESPONSES.

tubulin Protein composed of two subunits α and β (M_r ~50 000 each) which forms the protein subunit of MICROTUBULES. It is present in almost all eukaryotic cells and is particularly abundant in the brain (10–20% of soluble protein). The α- and β-tubulins are encoded by multigene families (e.g. *Drosophila* has four α-tubulin genes and four β-tubulin genes). *See also*: NEURONAL CYTOSKELETON.

tumorigenesis The complete process which, starting from the initial first event(s) in one cell, followed by one or more further steps, leads to clonal expansion of a (pre)malignant cell population, eventually manifesting itself as a tumour.

tumour antigen Cell-surface antigens that appear on CANCER CELLS but not on normal cells from the same source, such as carcinoembryonic antigen.

tumour cells *See*: CANCER CELLS.

tumour necrosis factor (TNF, cachectin, lymphotoxin, LT) A CYTOKINE with many actions, including the mediation of inflammatory responses through its vasodilatory effects, stimulation of tumour cell proliferation accompanied by a cytotoxic action, stimulation of T-cell mediated immunity to tumour cells, and an ill-defined action leading to wasting in cancer patients (cachexia, hence alternative name cachectin). TNF also stimulates ANGIOGENESIS but has a destructive action on tumour vasculature. TNF binds to two distinct cell surface receptors, one expressed on a wide variety of cells (and probably mediating the cytotoxic action on tumour cells), the other expressed primarily on lymphoid and myeloid cells. Early hopes that therapy of cancer patients with TNF might have useful clinical results was dampened by toxic effects of the treatment; indeed, TNF may promote tumour growth and invasiveness in certain human cancers. Attempts are now being made to engineer the molecule to increase selective toxicity for tumour cells. Two nonallelic forms of TNF — TNFα and TNFβ — are preferentially expressed in different cell types.

TNFα is produced by macrophages, eosinophils and natural killer (NK) cells and is encoded by a gene on human chromosome 6p23. The mature protein of 157 amino acids has an M_r of 17 000 and has no potential glycosylation sites, but does possess cysteine residues capable of forming disulphide bonds. Human and mouse TNFα show 86% homology at the amino-acid level and are cytostatic or cytotoxic for a variety of tumour cell types both *in vitro* and *in vivo*. Normal tissues are also affected by TNFα and it is intimately associated with IL-1 in inflammatory responses (*see* LYMPHOKINES). TNFα shares 30% homology with TNFβ and both increase phagocytic activity of neutrophils, and have important roles in cell interactions and autoimmune pathology.

LTNFβ is produced exclusively by human T CELLS. It is a GLYCOPROTEIN of 171 amino acids (M_r 25 000) with a single N-glycosylation site at residue 62. The gene is located on human chromosome 6p23 some 1200 base pairs from the TNFα gene. Despite the close proximity of the two genes they are preferentially expressed in different tissues, probably due to differences in the PROMOTER regions. Both TNFα and TNFβ bind to a common receptor, and show similar antitumour effects *in vivo* and *in vitro*, although not all cells which bind TNFs are susceptible to their toxic effects. TNFβ increases the phagocytic activity of neutrophils (in synergy with IFNγ), and in common with the α form its production is enhanced by IL-2. *See*: LYMPHOKINES.

tumour promoter Compound that increases the effectiveness of a carcinogen but is not itself carcinogenic, for example croton oil and PHORBOL ESTERS.

Tumour suppressor genes

TUMOUR suppressor genes (TSGs) are a class of genes believed to be involved in different aspects of normal control of cellular growth and division. Their common characteristic is that it is their inactivation, usually by genetic means, which contributes to tumour development. This contrasts with the other main class of genes involved in neoplasia — the dominantly acting oncogenes (*see* ONCOGENES), whose activation, again by various genetic means, also leads to a malignant phenotype. To emphasize this opposition, and to reflect the possibility that TSGs directly antagonize the action of oncogenes, TSGs are sometimes referred to as anti-oncogenes. TSGs must usually be inactivated at both ALLELES for manifestation of their oncogenic effect and are thus also referred to as recessive oncogenes. However, there are examples of phenotypic effects at the cellular level in the heterozygous condition, making this term inaccurate.

The first evidence for the existence of TSGs came from the unexpected observation that the fusion of normal cells with tumour cells resulted in hybrids which were nontumorigenic. However, with the loss of chromosomes, which occurs naturally when hybrid cells are grown in culture, the tumorigenic phenotype was sometimes restored. Furthermore, the introduction of normal chromosomes into tumorigenic hybrids can cause reversal to the nontumorigenic phenotype. Thus expression of genes from the normal parent suppresses tumorigenicity [1].

As the normal functioning of these genes suppresses tumorigenicity, regions of chromosomes which are lost in tumours are likely to carry these TSGs. Karyotype analysis of a wide variety of tumour types has shown that chromosomal deletions occur with a high frequency and in a tumour-specific manner, and thus point to the likely location of TSGs [2]. RFLP analysis, which allows comparison of a patient's normal and tumour genotype, has confirmed the prevalence of allelic loss inferred from cytogenetic analysis. This approach has allowed sufficient localization of TSGs to highlight regions within chromosomes warranting further intensive study, and this has led to the isolation of several TSGs.

The study of ALLELOTYPES, besides identifying regions of allelic loss, has also been very important in revealing the underlying mechanism of tumour suppressor gene activation [3]. Heritable RETINOBLASTOMA, for example, is a dominant condition which predisposes to retinal tumours at an early age. It turns out that it is the normal allele, from the unaffected parent, which is lost in the tumour (Fig. T50). This loss of heterozygosity reveals an

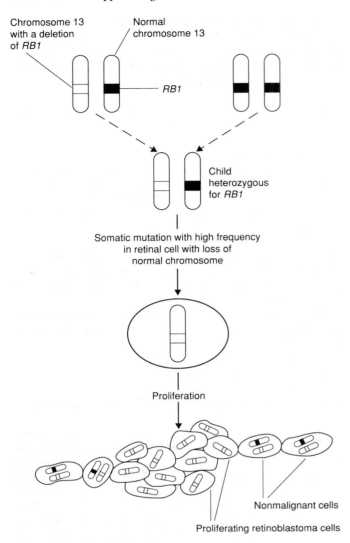

Chromosome 13 with a deletion of *RB1*

Normal chromosome 13

RB1

Child heterozygous for *RB1*

Somatic mutation with high frequency in retinal cell with loss of normal chromosome

Proliferation

Nonmalignant cells

Proliferating retinoblastoma cells

Fig. T50 Schematic representation of the inheritance of retinoblastoma and the appearance of the tumour after allelic loss.

underlying mutation in the remaining chromosome. In the case of inherited tumours such as retinoblastoma, the mutant allele is inherited and the normal allele is lost by some somatic mechanism. This second event occurs with a high frequency, explaining why the disease is inherited as a dominant trait, although at a cellular level the locus involved is recessive. For the 'sporadic' cases of these tumours which arise purely somatically, inactivation of TSGs requires genetic damage to both alleles at a tumour suppressor gene locus. This is frequently mutation, or submicroscopic deletion, on one chromosome, and deletion or whole chromosome loss of the other homologous chromosome.

Several genes have now been isolated which to a greater or lesser extent fulfil the criteria for a tumour suppressor gene [4]. These are: (1) localization to a region of a chromosome showing consistent allelic loss; (2) genetic damage of the remaining allele (although inactivation may be by epigenetic means); and (3) growth inhibition or reduction in tumorigenicity on reintroduc-

tion of the gene into the tumour type in which it is inactivated. The gene isolated from a region deleted in retinoblastoma cells — the *RB1* gene — most adequately meets these criteria, but other candidate TSGs which fulfil some of the criteria are *WT1*, a gene isolated from a region (11p13) occasionally deleted in inherited and somatically arising WILMS' TUMOUR, *NF1*, a gene thought to be involved in the development of NEUROFIBROMATOSIS, and *P53*, a gene which seems to be inactivated in a wide variety of tumours. More recently, several genes involved in the development of colon cancer have been isolated, namely *DCC*, *MCC*, and *APC* (*FAP*), with some of the characteristics of TSGs.

TSGs have only been isolated and cloned relatively recently and so we do not yet know their normal biochemical function and the reasons why their inactivation contributes to tumour development. The most extensively studied TSGs are *RB1* and *P53*. Both code for nuclear PHOSPHOPROTEINS which are involved in CELL CYCLE regulation. The transforming proteins of several DNA tumour viruses can form complexes with the Rb protein and p53 protein and this interaction is necessary for the transformation process [5] (*see e.g.* ADENOVIRUS; PAPOVAVIRUSES). Mice in which the *P53* gene has been inactivated by GENE KNOCKOUT develop normally although they show marked tumour susceptibility at an early age. This observation, initially surprising, is explained by the suggestion that p53 is required to inhibit DNA synthesis following DNA damage (e.g. by radiation) thus allowing time for effective DNA REPAIR. Cells with mutant p53 do not respond to DNA damage in this way and pass through the cell cycle, perpetuating potentially oncogenic lesions.

The *WT1* gene product is a protein with four zinc fingers and therefore likely to be a TRANSCRIPTION FACTOR. The *NF1* gene product seems to be a GTPase-activating protein for the GTP-BINDING PROTEIN p21^{ras}. The TSGs involved in the development of colon carcinoma are the least well characterized; the DCC protein sequence has considerable homology to neural CELL ADHESION MOLECULES such as NCAM, whereas the APC protein is a large protein (2843 amino acids) with coiled coil conformation in its N-terminal region.

TSGs are inactivated in a wide variety of tumours [6]. For example *RB* inactivation is seen not just in retinoblastoma, but in many other tumour types including breast tumours, bladder carcinoma, and lung tumours. Similarly, *P53* is inactivated in probably an even wider range of tumours. When a tumour suppressor gene is inactivated in a particular type of tumour, its inactivation can be detected in samples from almost all patients, and where it is not detected this may be because of the detection strategy rather than the fact that these tumours are exceptional. This situation contrasts with that for the transforming oncogenes which are activated in only a small proportion of patients with a given tumour type. This is probably a reflection of the roles of these genes in tumour progression — TSGs being more often involved in the early stages of tumour development.

In the case of inherited predispositions to certain cancers the isolation of the relevant genes allows the use of methods of DNA analysis (*see e.g.* RESTRICTION ANALYSIS; PCR) to detect members of a family who have inherited a mutant gene. Affected individuals can be monitored closely so that therapeutic intervention will be possible at an early stage of the disease.

One goal of cancer research has been to identify differences between normal and tumour cells which can be exploited for therapeutic strategies. TSGs and their products represent possible targets [7].

P.H. RABBITTS

See also: CONTIGUOUS GENE SYNDROME.

1 Harris, H. (1988) The analysis of malignancy by cell fusion. *Cancer Res.* **48**, 3302–3306.
2 Anderson, M.C.M. & Spandidos, D.A. (1990) Onco-suppressor genes. In *Molecular Biology of Cancer Genes* (Sluyser, M., Ed.) (Ellis Horwood, London).
3 Cavenee, W.K. et al. (1983) Expression of recessive alleles by chromosomal mechanisms in retinoblastoma. *Nature* **305**, 779–784.
4 Marshall, C.J. (1991) Tumour suppressor genes. *Cell* **64**, 313–326.
5 Green, M.R. (1989) When the products of oncogene and anti-oncogenes meet. *Cell* **56**, 1–3.
6 Ponder, B.A.J. (1988) Gene losses in human tumours. *Nature* **335**, 400–403.
7 Sager, R. (1989) Tumour suppressor genes. The puzzle and the promise. *Science* **246**, 1406–1412.

tunicamycin An antibiotic which inhibits N-glycosylation by inhibiting the enzyme N-acetylglucosamine transferase (Fig. T51).

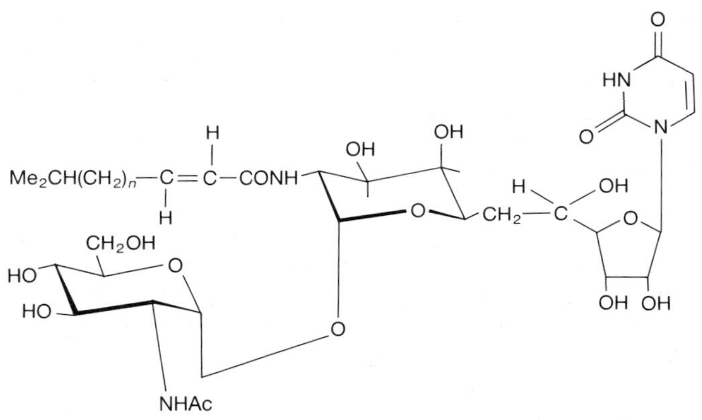

Fig. T51 Tunicamycin.

tunicate development *See:* ASCIDIAN DEVELOPMENT.

turbid plaques Discrete cloudy areas produced in a lawn of bacteria infected by a TEMPERATE PHAGE (*see also* BACTERIOPHAGES). These areas represent foci of infection in which some progeny phages of the original infected cell enter the prophage state in their bacterial host cells, which are therefore enabled to multiply, producing minicolonies in the middle of a clear area of lysed bacteria.

β-turn A secondary PROTEIN STRUCTURE motif connecting two β-strands.

Turner syndrome Syndrome characterized by a phenotypic female with gonadal dysgenesis, short stature, and sexual immaturity. These women have primary amenorrhoea and are infertile.

More than 50% have only one X-chromosome (chromosomal constitution 45,X), ~17% have an ISOCHROMOSOME of the long arm of the X-chromosome, 16% are mosaics, and 10% have deletions of the short arm on one of their X-chromosomes. The overall incidence is one in 2500 female births.

turnover number The number of moles of substrate that are reacted per mole of enzyme per unit time.

12–23 rule The rule governing the joining of V, (D) and J segments to make the immunoglobulin (and T cell receptor) variable region coding sequence. Two segments are joined together only if they are flanked by recombination signal sequences containing a spacer of 12 base pairs and 23 ± 1 base pairs respectively. *See:* GENE REARRANGEMENT; RECOMBINASE.

twist (1) One form of puckering found in five-membered SUGAR rings, where three atoms are coplanar with one atom lying to each side of this plane. This contrasts with the envelope form in which four atoms are coplanar.
(2) In supercoiled DNA the number of double-helical twists or turns, which is related to the LINKING NUMBER. *See also:* SUPERCOILING.

twitchin *See:* IMMUNOGLOBULIN SUPERFAMILY.

two-dimensional gel electrophoresis *See:* ELECTROPHORESIS.

Ty element The major class of TRANSPOSABLE GENETIC ELEMENT in the yeast SACCHAROMYCES CEREVISIAE (*T*ransposon *y*east). The most widely studied member of this class, Ty1, is ~6000 bp long and is flanked by DIRECT REPEATS of 334 bp, the so-called delta elements (δ). The integrated Ty1 element is flanked by a 5-bp repeat of host DNA sequence and is present at between 25–35 copies per haploid genome. Ty1 elements are efficiently transcribed by RNA polymerase II from a PROMOTER within the left δ element to generate an abundant 5.6 kb mRNA. This RNA encodes two major polypeptides, TyA and TyB, the latter being further cleaved into at least three proteins (Fig. T52). TyB is initially synthesized as a fusion with TyA via a RIBOSOMAL FRAMESHIFT. The Ty-encoded polypeptides show significant homology to analogous proteins from RETROVIRUSES. Ty1 transposes via an RNA intermediate in a process analogous to the retroviral life cycle. A 5.6-kb RNA transcript is reverse transcribed in a virus-like particle which consists mainly of the structural protein encoded by the TyA gene.

tymovirus group Sigla of turnip yellow mosaic, the type member. Group of viruses with isometric particles, 29 nm in diameter which sediment as two components, one nucleoprotein and the other nucleic acid free. The 180 coat protein subunits (M_r 20 000) are arranged in icosahedral symmetry around the (+)-strand genomic RNA, appearing in clusters of five or six at the pentamer and trimer vertices respectively. The genome organization is shown in Fig. T53. The 5′ P69 has unknown function. The P206 reads through a weak termination codon to give the P221, both of which have sequence homology to RNA polymerases. The coat

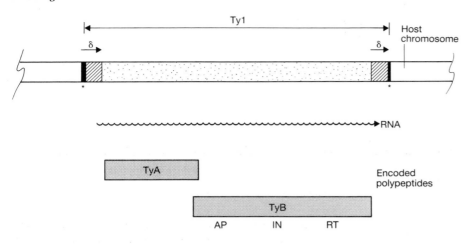

Fig. T52 The structure of the yeast transposable element Ty1 and its encoded products.

protein at the 3′ end is translated from a subgenomic mRNA. *See also*: PLANT VIRUSES.

Hirth, L. & Givord, L. (1988) In *The Plant Viruses*, Vol. 3 (Koenig,R., Ed.) 163–212 (Plenum, New York).

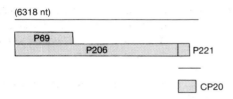

Fig. T53 Genome organization of turnip yellow mosaic virus (TYMV). The lines represent the RNA species and the boxes the proteins (P) with M_r given as $\times 10^{-3}$; nt, nucleotides.

tyrosinaemia Heritable metabolic disorder characterized by increased blood levels of tyrosine and urinary excretion of acetic and lactic acids. The symptoms include cirrhosis of the liver and renal problems. There are several types of tyrosinaemia all of which are rare and inherited as AUTOSOMAL RECESSIVE traits. One type is caused by a deficiency of fumaryl acetoacetate hydrolase and is characterized by the presence of vitamin D-resistant rickets in addition to hepatic and renal involvement. The fumaryl acetoacetate hydrolase gene has been localized to the distal portion of the long arm of chromosome 15 (15q23–25). Another form is caused by the deficiency of hepatic tyrosine aminotransferase (TAT) and is characterized by eye involvement: the TAT gene has been localized to the long arm of chromosome

16 (16q22.1). A third type of tyrosinaemia is caused by the deficiency of *p*-hydroxyphenylpyruvic acid oxidase and shows involvement of the central nervous system. The symptoms of tyrosinaemia can be alleviated by the adoption of a diet low in tyrosine.

tyrosine (Tyr, Y) An aromatic AMINO ACID.

tyrosine ammonia-lyase (TAL) Enzyme that catalyses the deamination of L-tyrosine to *trans-p*-coumaric acid (4-hydroxycinnamic acid) (Fig. T54). It is probably identical with PHENYLALANINE AMMONIA-LYASE, especially in leaves of MONOCOTYLEDONS.

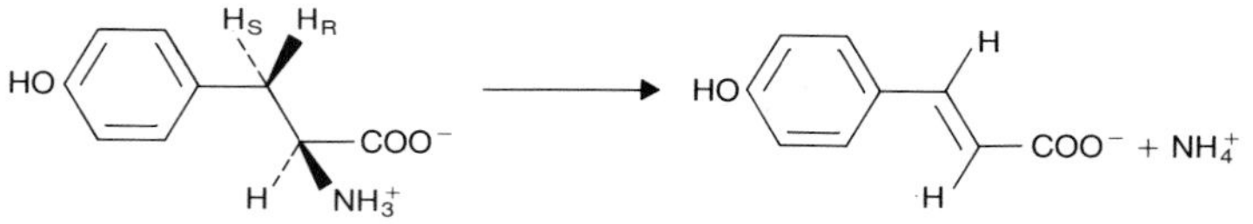

Fig. T54 The deamination of tyrosine by tyrosine ammonia-lyase.

tyrosine kinase, tyrosine protein kinase Protein tyrosine kinase. Any PROTEIN KINASE that phosphorylates proteins on tyrosine side chains. The best-studied examples of tyrosine kinases are the proto-oncogene product Src (*see* ONCOGENES) and the GROWTH FACTOR RECEPTORS. *See also*: PROTEIN PHOSPHORYLATION.

***tzs* gene** Gene for the synthesis of transribosylzeatin in nopaline strains of *Agrobacterium*. The gene is located on the TI PLASMID near the virulence (*vir*) region and is activated with the *vir* genes.

U

U The pyrimidine BASE uracil, which replaces thymine in RNA.

U1, U2, etc. SMALL NUCLEAR RIBONUCLEOPROTEINS involved in RNA SPLICING.

UAA One of the three codons that specify termination of translation. Sometimes referred to as the 'ochre' termination codon. *See*: GENETIC CODE; PROTEIN SYNTHESIS.

UAG One of the three codons that specify termination of translation. Sometimes referred to as the 'amber' termination codon. *See*: GENETIC CODE; PROTEIN SYNTHESIS.

UAS UPSTREAM ACTIVATOR SEQUENCE.

ubiquitin A small highly conserved protein (M_r ~8500, 76 amino acids) (Fig. U1) found in all eukaryotes, and with a role in PROTEIN DEGRADATION AND TURNOVER. It is also found attached to HISTONES. *See also*: POST-TRANSLATIONAL MODIFICATION.

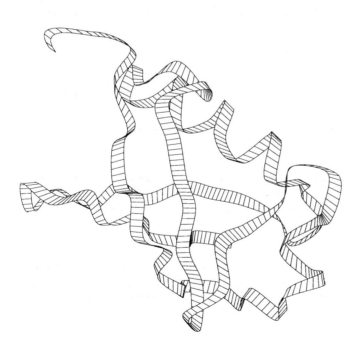

Fig. U1 Structure of ubiquitin.

ubiquitination The covalent addition of UBIQUITIN to a protein. *See*: POST-TRANSLATIONAL MODIFICATION; PROTEIN DEGRADATION AND TURNOVER.

UDP The ribonucleotide uridine diphosphate. *See*: NUCLEOSIDES AND NUCLEOTIDES.

UDP-sugars *See*: SUGAR NUCLEOTIDES.

UGA One of the three codons that specify termination of translation. Sometimes referred to as the 'opal' termination codon. In certain contexts it can be translated as a 'sense' codon; for example in some MITOCHONDRIAL GENOMES it is translated as a tryptophan codon, and in some bacterial and mammalian mRNAs it is used as the codon for the rare amino acid selenocysteine. *See*: GENETIC CODE; PROTEIN SYNTHESIS.

ultracentrifugation techniques Techniques for cell fractionation and the separation and sizing of macromolecules which depend on the differential sedimentation properties of particles of different sizes, shapes, and densities in the ultracentrifuge. *See*: CELL FRACTIONATION; DENSITY GRADIENT CENTRIFUGATION; SEDIMENTATION COEFFICIENT; SVEDBERG UNIT.

ultrastructure Structure of a cell below the level that can be resolved in the light microscope.

ultraviolet irradiation of eggs and embryos Ultraviolet microbeam irradiation provides an effective surgical tool for the ABLATION of cells and has been used extensively in *Caenorhabditis elegans* and *Drosophila* to establish FATE MAPS and to study the developmental potential of the remaining cells. UV irradiation has also contributed to studies on maternal CYTOPLASMIC DETERMINANTS. In *Xenopus*, UV irradiation of the vegetal pole leads to disruption of the dorso-ventral axis. This effect was originally thought to be due to destruction of an axial determinant but it was later found that embryos could be rescued after irradiation by rotation. The UV irradiation may damage the subcortical microtubule array which is required for the subcortical rotation necessary to establish the dorso-ventral axis. In insects, UV irradiation appears to damage the anterior organizing activity and can result in the formation of embryos with double abdomens. *See*: AMPHIBIAN DEVELOPMENT; CAENORHABDITIS DEVELOPMENT; DROSOPHILA DEVELOPMENT.

ultraviolet radiation (UV radiation, UV light) Electromagnetic radiation of wavelengths between those of the violet end of the visible light spectrum (around 380 nm) and X-rays (10 nm). Although invisible to the human eye, it can be captured on photographic film. The ultraviolet spectrum is subdivided by wavelength into A (400–320 nm), B (320–280 nm) and C (280–10 nm) bands. Much UV radiation, and virtually all the shorter wavelengths (UV-C) is absorbed by the stratospheric ozone layer. The UV-B wavelengths are the most potentially damaging to life. Ultraviolet light is a potent MUTAGEN because nucleic acids (both DNA and RNA) absorb light in the UV region with the pyrimidine bases thymine and cytosine being especially reactive. UV light causes the hydration of cytosine by the insertion of a water molecule ring at a C=C bond, whereas it disrupts the C=C bonds in thymine and will generate thymine dimers (*see* PYRIMIDINE DIMERS). Such thymine dimerization may be the primary mutagenic effect produced by UV. *See*: CHEMICAL CARCINOGENS AND CARCINOGENESIS; DNA REPAIR.

UMP The ribonucleotide uridine monophosphate *See*: NUCLEOSIDES AND NUCLEOTIDES.

***umu* genes** *See*: DNA REPAIR.

uncoating ATPase Abundant cytoplasmic protein (M_r 71 000) which catalyses the ATP-dependent uncoating of clathrin-coated vesicles *in vitro* (*see* COATED PITS AND VESICLES). The uncoating ATPase is a member of the 70K HEAT SHOCK protein family (Hsp70).

uncoordinated (*unc*) mutants Over a hundred loci affecting movement have been identified in the nematode *Caenorhabditis elegans*. Many of these loci encode structural and regulatory components of the contractile apparatus; for example *unc-54* is the structural gene for the major MYOSIN heavy chain, myoB. Other loci affect developmental processes, for example, *unc-86* encodes a TRANSCRIPTION FACTOR required for the regulation of cell fate in specific neuronal lineages. *See*: CAENORHABDITIS DEVELOPMENT; CAENORHABDITIS NEURAL DEVELOPMENT.

undecaprenol An isoprenoid of prokaryotes, involved in glycosyl transfers as either undecaprenyl phosphate or pyrophosphate (*see* GLYCANS; UNDECAPRENOL AND DOLICHOL CYCLES) (Fig. U2). In both cases the glycosides are primarily acting as sugar acceptors for transfers from SUGAR NUCLEOTIDES and are donors onto growing glycan. They also act as shuttles of glycosyl residues across the surface membranes of prokaryotes. The unsaturation of the α-isoprene unit makes both the phosphoryl and pyrophosphoryl glycosides less stable than their dolichyl counterparts (*see* DOLICHOL).

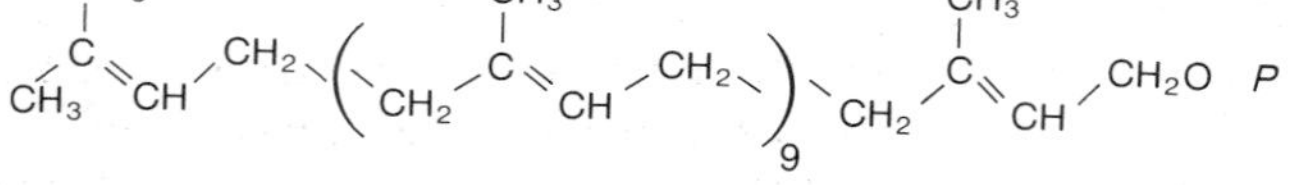

Fig. U2 Undecaprenol.

undecaprenol and dolichol cycles In prokaryotes, glycosyl transfers via the lipid UNDECAPRENOL are involved in the synthesis of repeating structures in TEICHOIC ACIDS, PEPTIDOGLYCAN, the O-antigen chains of LIPOPOLYSACCHARIDES and some capsular polysaccharides (*see* BACTERIAL ENVELOPES). DOLICHOL-dependent transfers have a similar role in assembling so-called glucosaminoglycans of some archaebacterial glycoproteins and bacterial cellulose.

In eukaryotes, dolichol-dependent transfers produce single small GLYCANS, which either remain as such or act as primers for dolichol-independent synthesis of larger repeating or nonrepeating glycans (*see* GLYCOPROTEINS AND GLYCOSYLATION).

In all cases there is an isoprenyl phosphate or pyrophosphate cycle involved. There are three general forms (Fig. U3 (*a–c*)). The simplest (Fig. U3*a*) only occurs with some types of teichoic acid synthesis. It involves the addition of sugar from a sugar nucleotide to undecaprenyl phosphate to yield a nucleoside diphosphate and an undecaprenyl phosphoryl sugar, to which a phosphoryl sugar alcohol unit is then added directly from (e.g.) CDP-glycerol. The whole is then transferred to a growing teichoic acid chain, releasing undecaprenyl phosphate and extending a glycan of the form (-AB-P-AB-P-)$_n$. The cycle does not involve a pyrophosphoryl link, nor does it release free phosphate: in principle it is reversible

The second type of cycle (Fig. U3*b*) involves an initial transfer of sugar phosphate from sugar nucleotide to undecaprenyl phosphate to yield an undecaprenyl pyrophosphoryl sugar and a nucleoside monophosphate. If this acts directly as a donor of phosphoryl sugar to an acceptor chain, a teichoic acid of the form (-A-P-A-P)$_n$ results and undecaprenyl phosphate recycles. Alternatively, an additional phosphoryl sugar alcohol phosphoryl unit may be added first so that the final transfer gives a teichoic acid chain of the type (-A-P-B-P-A-P-B-P-)$_n$. Such cyles involve the formation of a pyrophosphoryl link, but do not release free phosphate; they are potentially reversible.

The third type of cycle occurs with both undecaprenyl and dolichyl isoprenes and involves both the formation of a pyrophosphoryl link and the release of free phosphate. Such cycles are irreversible (Fig. U3*c*).

There are two subtypes of the third type of cycle: the first (and simpler) only involves the linkage of sugar to isoprene via pyrophosphate, the second involves sugars linked both via pyrophosphate and phosphate (Fig. U4). Cycles for peptidoglycan and lipopolysaccharide synthesis are of the former type and those for bacterial polysaccharide (*see* CELLULOSE) and eukaryotic glycan synthesis (*see* DOLICHOL) of the latter. Details of the synthesis of bacterial glycoprotein glycan are still uncertain.

undulator A magnetic insertion device inside a high-energy particle activator. *See*: SYNCHROTRON RADIATION.

undulin *See*: EXTRACELLULAR MATRIX MOLECULES.

unequal crossing-over A nonreciprocal RECOMBINATION event. This type of recombination between arrays of similar repeated sequences may be responsible for the expansion and contraction of such arrays.

uniconazole (*E*)-1-(4-chlorophenyl)-4, 4-dimethyl-2-(1,2,4-triazol-1-yl)-1-penten-3-ol (also known as S-3307 and XE-1109) (Fig. U5). M_r 291.5. A PLANT GROWTH RETARDANT. It inhibits the biosynthesis of GIBBERELLINS at *ent*-kaurene oxidase. The

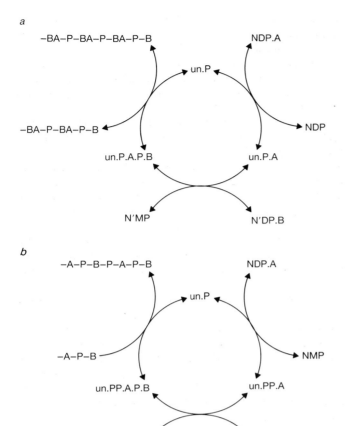

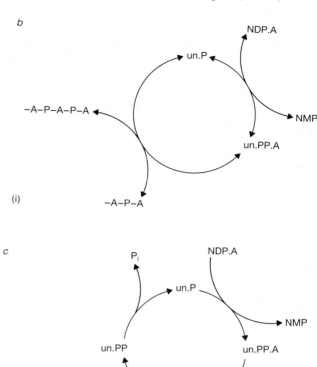

Fig. U3 Types of undecaprenol cycles involved in glycosyl transfers in the synthesis of prokaryotic teichoic acids (*a,b*), peptidoglycans and some other carbohydrate polymers.

commercial product is enriched in the more active *S*-enantiomer.

Izumi, K. et al. (1985) *Plant Cell Physiol.* **26**, 821–827.

unidentified reading frame (URF) A potential open READING FRAME in a DNA sequence whose encoded product is unknown.

uniparental disomy *See*: PARENTAL GENOMIC IMPRINTING.

uniparental inheritance Inheritance determined by one parent only, usually through organellar DNA. But *see also* PARENTAL GENOMIC IMPRINTING.

uniport MEMBRANE TRANSPORT SYSTEM in which one substance is transported in one direction.

unique DNA *See*: SINGLE COPY DNA.

unit cell A volume element which is the basic block from which any crystal is built. It is a parallelopiped in shape, with edges of lengths a, b, c which equal the repeat distances in three dimen-

sions of the regular crystalline stacking. Depending on the absence or presence of SYMMETRY elements within the crystal, the unit cell contains one or more ASYMMETRIC UNITS, each having one copy of the basic repeating motif of the crystal (though this motif may correspond to more than one molecule). By convention, the coordinate system of unit cells is chosen to reflect the full symmetry. *See*: X-RAY CRYSTALLOGRAPHY.

universal genetic code *See*: GENETIC CODE.

unsaturated fatty acids FATTY ACIDS with one or more C=C or C≡C bonds in the hydrocarbon chain.

3′ untranslated region (3′ UTR) the region between the TERMI-NATION CODON of an mRNA molecule and its extreme 3′ end, a region not usually translated by ribosomes. In eukaryotic mRNAs it encompasses the POLY(A) TAIL. It is usually between 50 and 200 nucleotides long, although some mammalian mRNAs have 3′ UTRs up to 1000 nucleotides long, and is believed to have a role in both translation initiation (*see* PROTEIN SYNTHESIS) and in determining the stability of an mRNA molecule. Some transiently unstable mRNAs (e.g. that for the PROTO-ONCOGENE c-*myc*), contain an AU-rich sequence (5′ AUUUA 3′) in their 3′ UTRs.

5′ untranslated region (5′UTR) The region between the 5′ end of

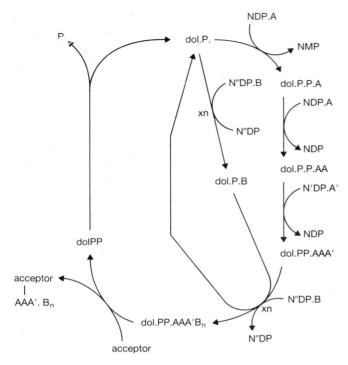

Fig. U4 Dolichol cycle involved in glycosyl transfers in the synthesis of prokaryotic polysaccharides and eukaryotic glycans.

an mRNA molecule and the AUG CODON used to initiate translation, which is important in directing ribosomes to the correct initiation site. In prokaryotic mRNAs this region contains the SHINE–DALGARNO SEQUENCE — the ribosome-binding site. The equivalent regions in eukaryotic mRNAs lack a ribosome-binding site, but are usually short (<100 nucleotides) and devoid of secondary structure, a characteristic that seems to facilitate the scanning mechanism of initiation (*see* PROTEIN SYNTHESIS). There are several examples of the 5′ UTR being used to regulate translation (e.g. HEAT SHOCK protein synthesis).

untranslated regions (3′UTR, 5′UTR) The regions located both 5′ and 3′ of the coding region of an mRNA which, although required for translation of the mRNA, do not define any part of the structure of the encoded polypeptide. *See:* BACTERIAL GENE EXPRESSION; EUKARYOTIC GENE EXPRESSION; EUKARYOTIC GENE STRUCTURE; GENE.

upregulation Of receptors, an increase occurring *in vivo* or in culture in the number of molecules of a given receptor type at a

particular location. This is usually due to an increase in TRANSCRIPTION of the gene(s) encoding the receptor protein, but it may also be due to stabilization of the mRNA involved or of the assembled protein product, or to a decrease in the removal of the receptor (e.g. by inhibition of its internalization by ENDOCYTOSIS). Such a change is induced by various treatments, for different receptors. Thus, upregulation often occurs experimentally upon chronic treatment with an antagonist. *In vivo* it occurs with many receptors when agonist availability is depressed, for example, by the loss of the afferent nerve supply or by chronic presynaptic block of transmitter release. Other cases can occur, however; nicotine treatment of rats, for example, upregulates the numbers of neuronal NICOTINIC RECEPTORS but not their affinity.

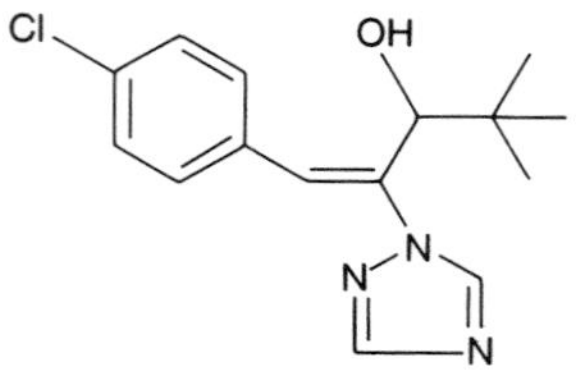

Schwartz, R.D. & Keller, K.J. (1983) *Science* **220**, 214–216.

upstream Located on the 5′ side of a coding sequence, start site of transcription, etc. in DNA.

upstream activator sequence (UAS) Generic term for ENHANCER-like elements located within PROMOTER regions of genes transcribed by RNA polymerase II in the yeast SACCHAROMYCES CEREVISIAE. Like mammalian enhancers, UASs are *cis*-acting elements which can function in either orientation with respect to the transcription start site in the promoter and can mediate their effect at variable distances from that start site. Their primary role is regulatory: UASs bind one or more TRANSCRIPTION FACTORS which activate expression of the gene associated with the UAS. The most intensively studied UAS is that located in the promoters of *GAL* genes of *S. cerevisiae*. In this case the UAS (UAS$_G$) has been functionally defined as a 17-bp PALINDROMIC SEQUENCE which is repeated two or more times in the promoters of each of the genes subject to galactose regulation (namely *GAL1*, *GAL2*, *GAL7* and *GAL10*). The UAS$_G$ is the binding site for the Gal4 transcription factor (M_r 100 000) which binds as a dimer. Virtually every gene of *S. cerevisiae* examined contains a UAS between 1400 and 100 nucleotides upstream of the TATA BOX and in general, deletion of a gene's UAS abolishes its transcription. *See also*: EUKARYOTIC GENE EXPRESSION.

Guarente, L. (1984) *Cell* **36**, 799–800.

upstream regulatory sequences (URS) Sequences to the 5′ side of the core PROMOTER region of a eukaryotic gene, to which additional TRANSCRIPTION FACTORS bind.

uracil (U) A pyrimidine, one of the four nitrogenous BASES present in RNA and which is also the base in the uridine nucleotides involved in GLYCAN synthesis and other metabolic reactions. In RNA it pairs with adenine (*see* BASE PAIR). *See also*: NUCLEOSIDES AND NUCLEOTIDES.

uracil-DNA glycosylase Enzyme that catalyses the release of uracil residues from DNA. Its physiological role is to correct deaminated cytosine residues in DNA. Uracil-DNA glycosylases are ubiquitous; they are found in bacteria, yeast and mammalian cells, and are also encoded by herpes viruses and EPSTEIN–BARR VIRUS. In *Escherichia coli* the enzyme is the product of the *ung*

Fig. U5 Structure of uniconazole.

gene. The genes for uracil-DNA glycosylases in various organisms encode proteins of M_r ranging from 25 000 to 35 000, and display a high degree of amino-acid sequence conservation. In *E. coli*, the enzyme can destroy DNA strands which are highly substituted with uracil in place of thymine, and this is the basis of one popular system for SITE-DIRECTED MUTAGENESIS of cloned DNA segments. *See:* DNA REPAIR.

uranyl acetate Used as a stain in ELECTRON MICROSCOPY and as a heavy atom derivative in X-RAY CRYSTALLOGRAPHY.

urea cycle mutations *See:* CARBAMOYL-PHOSPHATE SYNTHETASE DEFICIENCY; CITRULLINAEMIA; ORNITHINE CARBAMOYLTRANS-FERASE DEFICIENCY.

ureides Compounds containing the ureido group, -NH.CO.NH$_2$. The most important plant ureides are allantoic acid (2,2-diureidoacetic acid), allantoin (allantoic lactam), and L-citrulline (5-ureido-2-aminopentanoic acid). They are important as transport and storage forms of nitrogen in certain families: allantoin and allantoic acid in the Aceraceae, Boraginaceae, Hippocastanaceae and Platanaceae; citrulline in the Betulaceae and Juglandaceae.

URF UNIDENTIFIED READING FRAMES.

uridine The nucleoside containing the BASE uracil. *See:* NUCLEOSIDES AND NUCLEOTIDES.

urogastrone Epidermal growth factor. *See:* GROWTH FACTORS.

urokinase (uPA) Plasminogen activator synthesized by human kidney and other cells, originally isolated from urine. It is a SERINE PROTEINASE (EC 3.4.21.31) which cleaves plasminogen to plasmin and hence stimulates fibrinolysis. It is involved in the control of plasmin activity at cell surfaces in developing tissues. *See also:* TISSUE PLASMINOGEN ACTIVATOR. *See:* SERPINS for inhibitors.

URS UPSTREAM REGULATORY SEQUENCES.

usnic acid Compound with antibacterial activity, isolated from lichens (Fig. U6). The (*R*)-form and (*S*)-form are found in *Usnea* species, (*S*)-form only in *Letharia vulpina* and *Ramolina tingitana*.

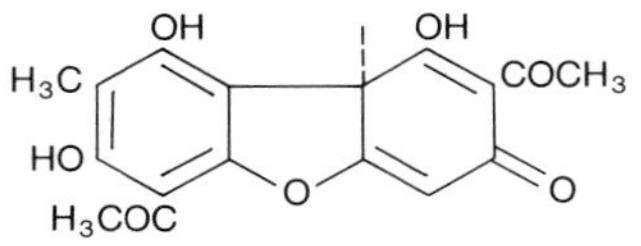

Fig. U6 Usnic acid.

UTP Uridine 5′-triphosphate, a nucleotide precursor for RNA synthesis. *See:* NUCLEOSIDES AND NUCLEOTIDES.

3′ UTR, 5′ UTR 3′, and 5′ UNTRANSLATED REGIONS of an mRNA.

uvomorulin A CELL ADHESION MOLECULE of the cadherin family, which is involved in cell–cell binding in epithelial and other tissues, and in embryonic development. *See:* EPITHELIAL POLARITY.

***uvr* genes** *See:* DNA REPAIR.

V

V The single-letter abbreviation for the AMINO ACID valine.

V$_H$, V$_\kappa$, V$_\lambda$ The V REGIONS of the immunoglobulin heavy chain, kappa chain and lambda chain (light chains) respectively (*see* IMMUNOGLOBULIN GENES).

V$_m$ The UNIT CELL volume per molecular weight for a protein crystal. *See*: MATTHEWS NUMBER.

V-(D)-J joining/recombination The process of directed somatic DNA rearrangement in B cells and T cells that leads to the formation of a functional immunoglobulin or T cell receptor gene. *See*: GENE REARRANGEMENT.

V gene Variable (V) gene segment which is assembled to form part of a complete IMMUNOGLOBULIN or T CELL RECEPTOR variable (V) region EXON. *See*: GENE REARRANGEMENT; IMMUNOGLOBULIN GENES; T CELL RECEPTOR GENES.

V region Variable region. Part of ANTIBODY or T CELL RECEPTOR which varies in sequence and confers antigen specificity on the molecule. *See also*: GENE REARRANGEMENT; GENERATION OF DIVERSITY; IMMUNOGLOBULIN STRUCTURE; T CELL RECEPTOR GENES.

V-type ATPases Vacuolar proton ATPASES. Protein complexes found in the membranes of the vacuoles of yeasts and other fungi and which actively transport protons from the cytoplasm into the vacuole through the coupled hydrolysis of ATP. Similar proton-transporting ATPases are found in other eukaryotic cells in the membranes of clathrin-coated vesicles, endocytic and exocytic vesicles, ENDOSOMES, GOLGI APPARATUS, LYSOSOMES, and SECRETORY GRANULES, where they are involved in acidification of these organelles. They are multisubunit enzymes with total $M_r > 400\,000$. ATPase activity and proton-transport activity are located in different subunits. *See*: COATED PITS AND VESICLES; ENDOCYTOSIS; EXOCYTOSIS.

Mellman, I. et al. (1986) *Annu. Rev. Biochem.* **55**, 663–700.
Adachi, I. et al. (1990) *J. Biol. Chem.* **265**, 967–973.

vaccine A preparation of a viral, bacterial or other pathogenic agent or of their isolated ANTIGENS which can be administered prophylactically to induce immunity. Conventional types of vaccine include:
1 Killed vaccines: preparations of the killed or inactivated viral or bacterial disease agent.
2 Live attenuated vaccines: strains of the disease agent that are nonpathogenic but which still induce specific immunity.

3 Toxoids: inactivated bacterial toxins.
Recombinant DNA techniques are being used in the development of vaccines for pathogens for which no vaccines yet exist, or for which existing vaccines are relatively ineffective or difficult to administer. Vaccines under development include those against protists such as the malaria agent *Plasmodium*, against multicellular parasites such as *Schistosoma*, and against viruses such as the human IMMUNODEFICIENCY VIRUS. *See also*: RECOMBINANT VACCINES.

vaccinia virus ANIMAL VIRUS (family Poxviridae), the cause of cowpox, a sporadic disease of cows and man, probably spread from an unknown wild host. It is related to the smallpox virus, and inoculation with vaccinia (first used by Edward Jenner in 1796) protects against smallpox (hence the term 'vaccination'). A worldwide vaccination programme by the World Health Organization using a vaccinia vaccine led to the official eradication of smallpox in 1977. Vaccinia virus is one of the main carriers used in the development of genetically engineered vaccines. *See also*: RECOMBINANT VACCINES.

vacuolar proton ATPases *See*: V-TYPE ATPASES.

vacuole Membrane-bounded vesicle of eukaryotic cells. In animal cells the term is usually applied to large vesicles (>100 nm in diameter) to which functions cannot be definitely ascribed; it may refer to LYSOSOMES, ENDOSOMES, SECRETORY VESICLES, phagocytic organelles or other membrane-bounded organelles. In yeast, the vacuole is the degradative compartment corresponding to the lysosome. In many protozoa, specialized phagocytic organelles give rise to a digestive organelle often termed the 'digestive vacuole'. In plant and fungal cells the vacuole is a large central membrane-bounded space containing hydrolytic enzymes. In plant cells it has a space-filling function and is important in maintaining cell turgor. It corresponds functionally to the lysosome of animal cells.

valine (Val, V) A neutral AMINO ACID with an aliphatic side chain.

valinomycin Antibiotic from *Streptomyces* sp. (Fig. V1) that acts as a mobile ion carrier or IONOPHORE, transporting K$^+$ across membranes.

VAMP A synonym for synaptobrevin. *See*: SNAP RECEPTORS.

vancomycin Glycopeptide antibiotic from *Streptomyces* sp. which inhibits bacterial cell wall synthesis by inhibiting PEPTIDOGLYCAN synthesis (*see* BACTERIAL ENVELOPES).

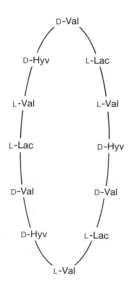

Fig. V1 Valinomycin, a cyclic molecule made up of D-valine (D-Val), L-lactate (L-Lac), D-hydroxyvalerate (D-Hyv) and L-valine (L-Val).

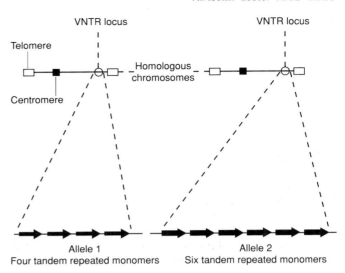

Fig. V2 Allelic variation at VNTR loci consists of different numbers of the same short DNA sequence (represented by arrows) arranged in a tandem (continuous) array. These tandemly repeated sequences tend to be clustered towards chromosomal telomeres (*see* CHROMOSOME STRUCTURE).

van der Waals force A weak nonspecific attractive force which occurs between two atoms. It is strongest when the atoms are 3–4 Å apart; at closer distances repulsive forces come into play. It has a bond energy of ~4.18 kJ mol^{-1}. Multiple van der Waals forces are important in interactions between two complementary surfaces (e.g. in antigen–antibody recognition, *see* IMMUNOGLOBULIN STRUCTURE).

vapour diffusion A technique used in CRYSTALLIZATION of macromolecules.

variability plot Comparison of the amino-acid sequences of variable (V) regions of IMMUNOGLOBULINS and T CELL RECEPTORS shows regions that are rather conserved (*see* FRAMEWORK REGIONS) and regions with a high degree of variability between individual molecules (*see* HYPERVARIABLE REGIONS). The latter define the COMPLEMENTARITY DETERMINING REGIONS and thus the antigen specificity of the immune receptors. Numerical values for the variability can be calculated by dividing the number of different amino acids which occur at a given position by the frequency of the most common amino acid at that position. Variability plots for both the heavy and light chain V regions of antibody molecules show that some residues are invariant (variability of 1). In contrast complementarity determining regions can easily be recognized by their high variability values.

variable gene activity theory of development Considerable evidence indicates that, with a few exceptions, cells exhibiting diverse states of DIFFERENTIATION contain equivalent GENOMES. Thus, the majority of development proceeds through the differential regulation of gene activity rather than by alterations in the gene complement.

variable number tandem repeats (VNTRs) TANDEMLY REPEATED DNA sequences which show variation within populations in the number of monomer units found together at specific loci. Monomer lengths vary from 11 to 60 bp in VNTR loci which are used as a source of RFLPs for gene mapping. One class of 'hypervariable' VNTR, the minisatellite, is the basis of DNA TYPING methods (Fig. V2). In most cases, no functional significance to the length variation is known, and only rarely are these variable DNA sequences present within coding genes.

Nakamura, Y. et al. (1987) *Science* **235**, 1616–1622.

variable region (V region) Region of ANTIBODY or T CELL RECEPTOR which varies in sequence and structure from molecule to molecule and confers antigen specificity on the molecule. *See also*: COMPLEMENTARITY DETERMINING REGIONS; GENE REARRANGEMENT; GENERATION OF DIVERSITY; IMMUNOGLOBULIN STRUCTURE; T CELL RECEPTOR GENES.

variant Burkitt's lymphoma Most Burkitt's lymphomas, about 90%, carry the TRANSLOCATION t(8;14)(q24;q32); the variant Burkitt's lymphomas carry either of the specific translocations t(2;8)(p12;q24) or t(8;22)(q24;q21). There are ~5% of each variant form.

variant surface glycoprotein (VSG) The variable surface antigen of African trypanosomes. See: ANTIGENIC VARIATION.

varicella–zoster virus DNA virus of the family HERPESVIRIDAE (*see also* ANIMAL VIRUSES) which causes chickenpox in humans on primary infection and becomes latent in dorsal root spinal ganglia. On reactivation, the virus descends the axon and reinfects the epidermis causing shingles; this usually occurs only once or not at all in a lifetime.

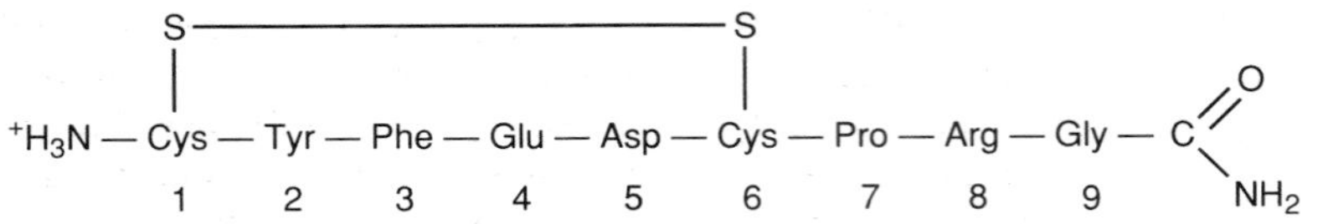

Fig. V3 Arginine vasopressin.

vasculogenesis *See*: ANGIOGENESIS.

vasoactive intestinal peptide (VIP) Peptide of 28 amino acids produced by axon terminals in the PERIPHERAL NERVOUS SYSTEM.

vasopressin Also called antidiuretic hormone (ADH). A nonapeptide hormone produced by the posterior lobe of the pituitary gland. Arginine vasopressin (Fig. V3) is the natural antidiuretic hormone of man and mammals other than the pig. It increases the permeability of the distal convoluted tubule of the kidney and collecting ducts to water by a cyclic AMP-dependent mechanism and hence leads to the production of more concentrated urine.

VCAM-1 *See*: CELL ADHESION MOLECULES.

VDR Vitamin D receptor. *See*: STEROID RECEPTOR SUPERFAMILY.

vector (1) A small carrier DNA molecule into which a DNA sequence can be inserted for introduction into a new host cell where it will be replicated, and in some cases expressed. Vectors can be derived from PLASMIDS, BACTERIOPHAGES or PLANT or ANIMAL VIRUSES. They must contain an ORIGIN OF REPLICATION recognized by the proposed host cell and in the case of EXPRESSION VECTORS, promoter and other regulatory regions recognized by the new host cell. A vector containing a 'foreign' gene is introduced into a cell by TRANSFORMATION, TRANSFECTION, or by making use of viral entry mechanisms. There are now many specialized vectors tailored for various applications. *See also*: BOVINE PAPILLOMA VIRUS; CLONING VECTOR; CO-INTEGRATING PLASMID; COSMID; EXPRESSION VECTOR; GENETIC ENGINEERING; INTEGRATION ·VECTOR; LAMBDA; PAPOVAVIRUSES; PHASMID; PLANT GENETIC ENGINEERING; 2µM PLASMID; RETROVIRAL VECTORS; SECRETION VECTOR; SHUTTLE VECTOR; TI PLASMID; YEAST ARTIFICIAL CHROMOSOME.
(2) Mathematical entity possessing both magnitude and direction. Vectors (**a**) are often represented by their three orthogonal components (a_x a_y a_z). For two vectors **x** **y** of length $|x|$, $|y|$ with angle θ between them, scalar multiplication $r = x \cdot y$ is a directionless number of magnitude $|x|$ $|y|$ cos θ, and vector multiplication is itself a vector in the direction perpendicular to both **x** and **y** with magnitude $|r| = |x|$ $|y|$ sin θ.
(3) Insect or other agent that carries a virus or other disease agent (e.g. parasitic protozoa such as those causing MALARIA) from one host to another.

vectorial discharge The threading of the extended nascent polypeptide chain across the ENDOPLASMIC RETICULUM membrane from the cytoplasm to the lumen (*see* PROTEIN TRANSLOCATION), while still attached to the ribosome. Discharge may take place while elongation from the ribosome is still occurring (co-translational import), but translocation does not depend on ongoing translation, and the two events can be uncoupled providing the nascent chain remains attached to the ribosomes and ATP and GTP are available. *See also*: PROTEIN TARGETING.

vectorial labelling Directed labelling of membrane components which can distinguish the two faces of the membrane.

vegetal pole In amphibian and other asymmetric eggs, the yolk-rich pole of the cleaving early zygote. The diametrically opposite pole, the animal pole, is relatively low in yolk concentration, and the zygote nucleus is usually displaced towards the animal pole.

ventral root *See*: SPINAL CORD.

ventricles Cavities within the brain which are filled with cerebrospinal fluid and lined with ependymal cells, a type of GLIAL CELL.

Vertebrate neural development

THE vertebrate nervous system arises from the neural epithelium of the early embryo, a simple sheet of cells that eventually produces many hundreds or even thousands of different cell types. Functional arrays of NEURONS are assembled into clusters (nuclei) or layers, in stereotyped positions where they will receive appropriate afferent inputs and from which they can extend their axons to synapse with appropriate target cells. A rough estimate gives the number of synapses in the human brain as 10^{15}. This article is concerned with the formation of the ground pattern of the nervous system, up to the onset of neural activity and the regressive processes, such as APOPTOSIS (cell death) and axon elimination, that later reinforce and refine the initial pattern of connectivity. The questions are how does a part of the embryonic ECTODERM become COMMITTED to a neural fate; how do cells in the neural epithelium know their position; how is their positional information interpreted into cell type-specific DIFFERENTIATION; and how do growing axons navigate towards and recognize their correct synaptic targets? Until recently, analysis of early developmental stages has been hampered by the large number of precursor cells in the neural epithelium and our inability to distinguish between them. New molecular and antigenic markers now enable developmental biologists to visualize and perturb processes occurring within this 'white sheet' of cells.

The nervous system has its origins during GASTRULATION when a region of the dorsal ectoderm becomes organized as the NEURAL PLATE. During neurulation the neural plate folds into a tube

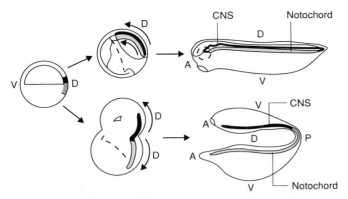

Fig. V4 Diagram showing the location of midline mesodermal (hatched) and neural (filled) structures in normal (upper) and exogastrulated (lower) *Xenopus* embryos. In normal development, involution of marginal zone cells (left) brings the mesoderm anteriorly beneath the dorsal ectoderm (centre), which has extended in the same direction from noninvoluting cells of the marginal zone. During exogastrulation, these two cell populations extend in opposite directions (arrows), producing separate endomesodermal mass and ectodermal sac that are connected together posteriorly (right). The later expression of neural markers in the ectodermal sac of the exogastrula indicates that neural induction does not depend on radial signals from underlying mesoderm. Adapted from [2].

(NEURAL TUBE) whose anterior end rapidly expands into distinct vesicles representing forebrain, midbrain and hindbrain. The hindbrain region in particular is then further partitioned to form a series of true neural segments — RHOMBOMERES [1]. These early morphological features of the neural tube dictate the overall plan of the CENTRAL NERVOUS SYSTEM (CNS) and predict its later regional specializations.

Commitment to neural fate

The commitment of ectodermal cells to a neural fate is controlled by intercellular signals that emanate from the Spemann organizer, the dorsal blastopore lip of amphibian embryos or Hensen's node of amniote embryos, which forms at the posterior end of the axis (*see* AMPHIBIAN DEVELOPMENT; AVIAN DEVELOPMENT). A subpopulation of these cells, involuting and extending anteriorly within the embryo to form the dorsal MESODERM, passes inducing signals radially to the overlying ectoderm (*see* INDUCTION). Noninvoluting cells of the organizer region extend anteriorly to form the notoplate (presumptive floor plate) at the dorsal midline. Neural inducing signals may also emanate from these cells, spreading tangentially through the plane of the ectoderm; substantial neural differentiation occurs within dorsal ectoderm in experimental situations where mesoderm is prevented from involuting. For example, *Xenopus* exogastrulae (Fig. V4) express two pan-neural markers, NCAM (*see* CELL ADHESION MOLECULES) and a neurofilament-like protein NF3 (*see* NEURONAL CYTOSKELETON), and certain classes of spinal cord neuron differentiate in the dorsal wall of the empty ectodermal sac [2].

Major elements of antero-posterior pattern are also imposed on the neural epithelium by signals spreading in the plane of the ectoderm. Thus, dorsal ectoderm explanted from early *Xenopus* gastrulae expresses position-specific neural genes (e.g. *engrailed-2*,

Krox-20, Xlhbox-1 and *-6*) at appropriate 'axial' levels provided that the ectoderm remains in contact at its posterior margin with dorsal mesoderm [3]. Tangential signals from the dorsal mesoderm are sufficient for patterned expression; the same result is obtained with ventral ectoderm, where these HOMEOBOX GENES are not normally expressed. The most anterior of the genes expressed in explanted gastrula ectoderm is *engrailed-2*, a marker of the posterior midbrain. This, and regions posterior to it, constitute the epichordal neural plate whose morphogenesis appears to depend on convergence and extension movements of both involuting and noninvoluting cells of the dorsal midline that substantially elongate the axis and bring signalling cells within short range of their targets. Anterior to the epichordal neural plate is the forebrain region, markers of which have not been detected in exogastrulae; forebrain and eye induction may require vertical signals from the most rostral (earliest involuting) mesoderm that forms the prechordal plate.

Recent studies of mesoderm induction [4] have shown that the ACTIVIN pathway may be linked to neural induction. Animal cap tissue isolated from blastulae injected with a truncated activin receptor (which acts as a dominant negative mutation) spontaneously expresses the pan-neural marker NCAM without any contact with mesoderm cells. If activin acts normally to maintain the epidermal pathway in ectoderm cells, then the neural inducer could be an activin antagonist.

Hox genes

The emergence of regional patterns along the anteroposterior axis of the nervous system appears to depend on the expression of position-specifying genes. Perhaps the single most important discovery in recent years has been that a set of genes that specify positional value along the anteroposterior axis of *Drosophila* is conserved in vertebrates. These homeotic selector genes (HOM genes) (*see* DROSOPHILA DEVELOPMENT) are the master control genes that coordinate the regulators of all processes involved in the development of structures appropriate to axial position. Consistent with their serving a similar function, the vertebrate homologues of these genes, known as Hox [5], also have a clustered organization in which the relative position of each gene in the cluster reflects its boundary of expression along the anteroposterior axis (*see* HOMEOBOX GENES AND HOMEODOMAIN PROTEINS). Duplications during evolution of the vertebrate genome have increased the number of Hox genes such that mammals possess four copies of genes that are represented singly in *Drosophila*. Divergence between these paralogous genes would be expected to increase the resolution of pattern control. Hox genes are expressed in overlapping domains along the neuraxis of the early embryo, those at the 3′ ends of the clusters being expressed most anteriorly, in the hindbrain. There is a striking correspondence between the expression boundaries of these genes and the interfaces between rhombomeres. The overlapping distribution of Hox transcripts in the hindbrain region suggests the possibility that their proteins may operate in a combinatorial or hierarchical manner to control the phenotypic specialization of rhombomeres. Accumulating evidence supports this view; loss-of-function mutations of rostrally expressed Hox genes in mice may result in

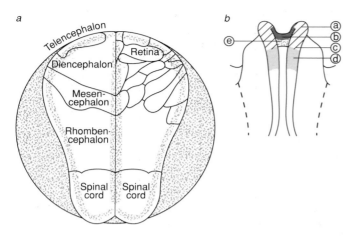

Fig. V5 Fate map of the amphibian (*a*) and chick (*b*) neural plates. The rostral-most region gives rise to the infundibulum and hypothalamus, whereas the telenchephalon derives from more posterior and lateral regions. a, Presumptive area of the nasal ectoderm; b, neural ridge containing the hypophyseal placode; c, presumptive region containing the olfactory placode; d, presumptive territory of the ectoderm of upper beak; e, presumptive territory of the diencephalon (i.e. infundibulum and hypothalamus). *a* Adapted from [16]; *b* adapted from [17].

severe malformation or absence of segmentally derived neural structures [6]. Although functional studies are still few, the conclusion seems inescapable that homeobox genes direct pattern formation, establishing the plan of the neuraxis.

Other homeobox genes

Hox genes are not expressed anterior to the hindbrain. But more rostral levels of the axis, the mesencephalon, diencephalon, and telencephalon, display spatially restricted expression of other transcriptional control genes which have homeoboxes divergent from the Hox type. For these genes too, there is remarkable conservation between flies and vertebrates. Homologues of the GAP GENES *orthodenticle (otd)* and *empty spiracles (ems)*, that may function as homeotic selector genes in the specification of particular head segments in *Drosophila*, have recently been cloned from mice and found to be expressed in overlapping domains that encompass the entire rostral neuraxis with the exception of the ventral diencephalon. As with the Hox genes, the nested expression of the *otd* and *ems* homologues suggests that their proteins may be employed in a combinatorial manner to specify the identity of mesencephalic, telencephalic and dorsal diencephalic regions.

Fate maps of both amphibian and chick embryos show that the ventral diencephalon, whose derivatives include the hypothalamus and hypophysis, develops at the most rostral extremity of the neural plate (Fig. V5). The telencephalon, initially posterior and lateral to the terminal region, reaches the front later in development through the pronounced overgrowth of the dorsolateral rim of the neural plate. Specification of the ventral diencephalon may be under the control of a further homeobox gene, *dlx*, a homologue of *Drosophila Distal-less*, which is expressed at the rostral tip

of the amphibian neural plate and later in the hypothalamic (and striatal) regions of the mouse forebrain.

Dorsoventral patterning

The nervous system exhibits a characteristic dorsoventral pattern, particularly prominent in the spinal cord, where different neuronal types differentiate with respect to their position down the dorsoventral axis. Motor neurons develop in the ventral third, relay neurons develop in the middle third, smaller interneurons develop in the dorsal third. The most dorsal region, represented by the neural folds that mark the transition from neural to epidermal ectoderm, produces the migratory NEURAL CREST cells that give rise to the glia and the majority of neurons in the PERIPHERAL NERVOUS SYSTEM. Specification of cell fate within the ventral region of the cord appears to depend on signals emanating from the NOTOCHORD, whose cells are the direct descendants of midline mesoderm cells involved in neural induction.

Experiments on avian embryos have demonstrated that both floor plate differentiation and motor neuron differentiation depend on signals emanating from the ventral midline [7]. Removal of the notochord from beneath the neural plate results in a normal sized spinal cord from which these ventral cell types are absent. Dorsal cell types and dorsal-specific surface antigens appear in their place. Implantation of a supernumerary notochord alongside and in contact with the lateral neural plate results in an additional floor plate at its point of contact and the formation of an additional column of motor neurons in the dorsal cord. Thus it appears that neural induction leaves the neural plate with a 'dorsal' fate, awaiting further signals from the midline mesoderm that endow proximal neural plate tissue with a ventral fate.

A number of developmental control genes containing the *Drosophila paired*-type box (*see* PAIRED BOX) have recently been cloned which are expressed in sharply defined dorsoventral domains in the neural tube [8]. These members of the *Pax* gene family are expressed during the period when the neural plate is labile and are profoundly affected by notochord removal or addition, with altered domains of expression presaging the repatterning of cell types that eventually follow such manipulations.

Cell lineages

That notochord and floor plate are capable of switching dorsal neural tube cells to fates characteristic of ventral tube indicates both the multipotentiality of spinal neuroepithelial cells and the operation of environmental factors in the selection of cell phenotype. A conditional role of the local environment on cell fate determination appears to preponderate in the CNS as a whole. Defining the set of phenotypes to which a single cell can give rise (its potential), or normally does give rise (its fate), is an essential step towards understanding the mechanisms whereby diversity is generated, although lineage analyses alone cannot reveal how cell fate decisions are made, whether by interactive mechanisms or by cell-autonomous, lineage-dependent means. Stereotyped or limited cell movements could result in determinate-looking cell lineages arising from multipotent precursors. Until recently, no technique was available for marking unambiguously a single cell and its clonal descendants in higher vertebrates, where neural precursor cells are small and often inaccessible. Now two powerful methods have been developed: iontophoretic microinjection of intensely fluorescent vital dyes (e.g lysinated rhodamine dextran) and retrovirus-mediated gene transfer (*see* RETROVIRAL VECTORS). Each satisfies only a subset of the requirements of an ideal lineage tracer, and few of these properties are shared, but in the one system (the dorsal root ganglion) so far analysed by both methods, the results are the same: a single premigratory neural crest cell, whose descendants migrate into and populate a ganglion, can give rise to a broad variety of both neuronal and GLIAL CELL types [9,10].

Development involves a series of decisions during which the developmental potential of a cell is progressively restricted. The classic embryological test for commitment to a single fate, or narrow range of fates, is to move a cell to a different environment; if uncommitted, a cell would be expected to give rise to a different phenotype. A promising system for the analysis of commitment mechanisms is the mammalian neocortex. Forming in the dorsal telencephalic wall, the cortex is a laminated structure in which the neurons of each layer not only share characteristic morphologies and projection patterns but are born during the same short period of development. The laminar architecture is built up progressively, in an inside-out sequence, so that cells contributing to later-formed, superficial laminae are born later and migrate through the earlier-formed laminae on their migration route from the ventricular surface. Heterochronic transplantation has shown that cells in the ventricular layer remain multipotent up to a point early in their last cell cycle [11]; if presumptive layer 6 precursors are moved to the ventricular zone of an older animal where layer 2 is being generated, the transplanted cells populate layer 2 in the host and form neuronal types characteristic of that laminar position (Fig. V6). Temporally regulated environmental cues appear to be able to regulate the choice of fate. If, however, the cells are moved just before their final mitosis they migrate out only as far as their presumptive lamina of origin and differentiate accordingly. Thus, there appears to be a critical period during the last cell cycle when commitment to the laminar identity typical of birthday takes place. Thus, cells are assigned a laminar identity whilst still in the ventricular zone, some time before commencing their migration out to that layer. Intriguingly, only the radial extent of cell migration seems to be fixed during this period; recombinant retroviral marking studies suggest that tangential migration is widespread and variable in extent [12], and heterotopic grafting experiments have shown that area-specific fates are decided much later in cortical development [13].

Axon guidance

An aspect of neuronal specification that has received a great deal of attention while revealing few of its secrets is the guidance of axons; their growing tips (growth cones) may navigate over considerable distances through varied terrains to find their targets, often with great accuracy. Axons of certain systems may simply utilize the guidance structure presented by the surfaces of pre-existing axons. The following examples underscore the im-

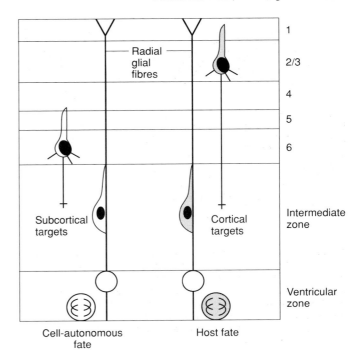

Fig. V6 Cell fate in the cerebral cortex. The diagram shows the two possible outcomes of transplanting presumptive layer 6 neurons into an older host where layer 2 is being generated. If fate is cell-autonomous, the transplanted cells would migrate to layer 6 and project subcortically. If fate is decided by local factors, the cells would migrate to layer 2 and project intracortically. Both outcomes are possible depending on where the cells are in the cell cycle at the time of transplantation. Adapted from [18].

portance of selective fasciculation in vertebrate nervous systems and reveal parallels with invertebrate systems where some of the guidance cues concerned are already being described at the molecular level (*see* INSECT NEURAL DEVELOPMENT; CAENORHABDITIS NEURAL DEVELOPMENT). Along the amphibian visual pathway, between retinal ganglion cell and the contralateral optic tectum, the growth cones of retinal ganglion cells establish their midline crossing point in the ventral diencephalon (the optic chiasm) by following a pathway provided by an earlier-formed commissural axon tract [14]. In the early embryonic brains of lower vertebrates, where the first neurons to differentiate are very few and can be individually identified, a framework of orthogonally arranged axon tracts provides a simple scaffold for many later developing and more complex projections.

But what guides the growth cones of the earliest axons in such systems? There is little evidence for pre-aligned substratum guides, such as channels of intercellular space, that could influence the navigation of the pioneers. Rather, evidence exists for the operation of several other types of mechanism, acting singly or in concert. Vectors may be provided by patterned expression of growth-promoting molecules on neuroepithelial cell surfaces (e.g. NCAM) or in their extracellular matrix (e.g. laminin) (*see* CELL-MATRIX INTERACTIONS; EXTRACELLULAR MATRIX MOLECULES). Pathway specificity could be conferred by temporally regulated expression of these ligands or their receptors. Alternatively, soluble molecules secreted by target cells can form concentration gradients towards which growth cones respond by chemotaxis; examples of systems where growth cone chemotaxis has been implicated by *in vitro* studies include the peripheral trigeminal/whisker field system, the commissural axon/floor plate interaction in the spinal cord, and the corticopontine projection. A third mechanism, involving the expression of repulsive molecules on cell surfaces, provides both no-go areas to axons in general (e.g. the caudal sclerotome and regions of developing cartilage) and more selective inhibition. Glial cells at the mammalian optic chiasm, for example, through which growth cones from the nasal retina pass on their way to the contralateral superior colliculus, express a surface molecule that prevents ingress by growth cones from the temporal region of the retina and forces their deflection back onto the ipsilateral optic tract. This mechanism is not entirely selective, however, as some later arriving temporal axons do enter the contralateral superior colliculus. These projection errors appear to arise only once the chiasm is already crowded with axons and are removed later in development, sharpening the initial segregation of left and right visual fields to opposite sides of the brain.

The establishment of topographically ordered neural connections has also been most intensively studied in the visual system. The mechanisms whereby precise connections become established at their appropriate sites are thought to involve the targeting of retinal axons by position-specifying molecules that are known to be distributed in gradients across both the retina and optic tectum. Although *in vitro* experiments lead us to suppose that such molecules might steer the growth cones of retinal axons precisely to their target cells, recent evidence suggests that this is probably not the case. Retinal axons in the developing rat have a diffuse and imprecise projection over the superior colliculus, appearing to be unresponsive to guidance cues once within their target field. Instead, it seems that later forming side branches show the earliest topographic specificity and it may be that these are induced and/or stabilized by position-specifying molecules [15].

This article can cover only a limited number of topics in what is a rapidly advancing field. Our understanding of the ontogeny of neural pattern is still in its infancy, but the manifest power of modern tools, from molecular biology especially, heralds a period of explosive growth.

A.G.S. LUMSDEN

See also: MAMMALIAN DEVELOPMENT; NEURONAL CYTOSKELETON; PATTERN FORMATION.

1 Lumsden, A. (1990) The cellular basis of segmentation in the developing hindbrain. *Trends Neurosci.* **13**, 329–335.
2 Ruiz i Altaba, A. (1992) Planar and vertical signals in the induction and patterning of the *Xenopus* nervous system. *Development* **116**, 67–80.
3 Doniach, T. et al. (1992) Planar induction of anteroposterior pattern in the developing central nervous system of *Xenopus laevis*. *Science* **257**, 542–545.
4 Hemmati-Brivanlou, A. & Melton, D.A. (1992) A truncated activin receptor inhibits mesoderm induction and formation of axial structures in *Xenopus* embryos. *Nature* **359**, 609–614.
5 McGinnis, W. & Krumlauf, R. (1992) Homeobox genes and axial patterning. *Cell* **68**, 283–302.
6 Lufkin, T. et al. (1991) Disruption of the Hox 1.6 homeobox gene results in defects in a region corresponding to its rostral domain of expression. *Cell* **66**, 1105–1119.
7 Yamada, T. et al. (1991) Control of cell pattern in the developing nervous system: polarizing activity of the floor plate and notochord. *Cell* **64**, 635–647.
8 Goulding, M. (1992) Paired box genes in vertebrate neurogenesis. *Semin. Neurosci.* **4**, 327–335.
9 Bronner-Fraser, M. & Fraser, S.E. (1989) Developmental potential of avian trunk neural crest cells *in situ*. *Neuron* **3**, 755–766.
10 Frank, E. & Sanes, J.R. (1991) Lineage of neurons and glia in chick dorsal root ganglia: analysis *in vivo* with a recombinant retrovirus. *Development* **111**, 895–908.
11 McConnell, S.K. & Kaznowski, C.E. (1991) Cell cycle dependence of laminar determination in developing cerebral cortex. *Science* **255**, 434–440.
12 Walsh, C. & Cepko, C. (1992) Widespread dispersion of neuronal clones across functional regions of the cerebral cortex. *Science* **255**, 434–440.
13 Schlagger, B.L. & O'Leary, D.D.M. (1991) Potential of visual cortex to develop an array of functional units unique to somatosensory cortex. *Science* **252**, 1556–1560.
14 Taylor, J. (1992) The development of the retinofugal projection within the early formed scaffold of axon tracts in the vertebrate CNS. *Semin. Neurosci.* **4**, 357–363.
15 Simon, D.K. & O'Leary, D.D.M. (1992) Development of topographic order in the mammalian retinocollicular projection. *J. Neurosci.* **12**, 1212–1232.
16 Eagleston, G.W. & Harris, W.A. (1990) Mapping of the presumptive brain region in the neural plate of Xenopus laevis. *J. Neurobiol.* **21**, 427–440.
17 LeDouarin, N. (1986) Cephalic placodes and neurogenesis. *Trends Neurosci.* **9**, 175–180.
18 McConnell, S.K. (1992) The control of neuronal identity in the developing cerebral cortex. *Curr. Opinion Neurobiol.* **2**, 23.

vertical transmission Transmission of a virus or other infectious agent from parent (usually mother) to offspring (*cf.* HORIZONTAL TRANSMISSION). It may result from any interaction between mother and offspring both before and during birth and during the immediate postnatal period. Some viruses are TERATOGENIC when passed by the maternal route: rubella virus (Togaviridae) causes severe defects to the heart, eyes and ears of the foetus if contracted through the mother during the first three months of pregnancy. Some RETROVIRUSES are transmitted through many generations as PROVIRUSES. *See also*: ANIMAL VIRUS DISEASE; ANIMAL VIRUSES.

very low density lipoprotein (VLDL) *See*: PLASMA LIPOPROTEINS AND THEIR RECEPTORS.

vesicle Small sac of membrane in the cytoplasm of eukaryotic cells in which proteins, other secretory products, and membrane are transported between compartments of the endomembrane system and from the site of synthesis to the plasma membrane for secretion. Long-lasting vesicles filled with material awaiting secretion are often known as granules. *See*: COATED PITS AND VESICLES; EXOCYTOSIS; GOLGI APPARATUS; PROTEIN SECRETION; PROTEIN TRANSLOCATION; SYNAPTIC TRANSMISSION; VESICLE-MEDIATED TRANSPORT.

vesicle-mediated transport Means by which proteins and many lipids are ferried between physically distinct membrane-bounded compartments in eukaryotic cells. A cytoplasmic coat (*see* COATED PITS AND VESICLES) is required to effect the budding of a vesicle from one compartment and both docking and fusion mechanisms are needed to effect specific delivery to the next compartment on the pathway. *See*: GOLGI APPARATUS; PROTEIN SECRETION; SNAPs; SNAP RECEPTORS.

vesicular stomatitis virus (VSV) Enveloped virus of the family RHABDOVIRIDAE, genus Vesiculovirus (*see* ANIMAL VIRUSES). It is closely related to the human viral pathogens Cocal, Chandipura, and Piry and more distantly to rabies virus. The particles are ~50 nm in diameter and 150 nm long. In fixed samples they are frequently observed as bullet-shaped structures, but naturally may be bacilliform. VSV comprises two major serotypes, 'Indiana' and 'New Jersey', which show considerable genetic diversity. The virions contain one molecule of (–)-sense single-stranded RNA. For the Indiana strain this genome contains 11 162 nucleotides and encodes five polypeptides. The polypeptides are designated (3′ → 5′): N, nucleocapsid protein; NS, nonstructural protein; M, matrix protein; G, glycoprotein; and L, large proteins. The N protein is the major structural protein of the virus and encapsidates the genomic viral RNA; L and NS are found associated with the viral nucleocapsid particle and together function as the viral RNA-dependent RNA polymerase. G is the only integral membrane protein of the virus and forms the externally projecting homotrimeric spike glycoprotein complexes required for receptor binding and membrane fusion. The M protein is associated with the inner face of the viral membrane.

VSV causes oral lesions ('soremouth') in cattle and also infects horses, pigs and humans. In the laboratory, VSV replicates in a wide range of cells, including insect, fish, and most animal cell lines. Owing to its extensive range of host cells, ability to generate conditional lethal mutations, and capacity to switch off host cell protein synthesis, VSV has been used extensively in studies of membrane biosynthesis and transport.

Emerson, S.U. (1985) In *Virology* (Fields, B.N. et al., Eds) (Raven Press, New York).
Doms, R.W. et al. (1988) *J. Cell Biol.* **107**, 89–99.

V$_H$ subgroup On the basis of amino-acid sequence homologies, immunoglobulin heavy chain variable genes (V$_H$ genes) are divided into V$_H$ subgroups. There are three subgroups (VHI–III) in mice and humans. As increasing DNA sequence information has become available, the V$_H$ genes have been further divided into V$_H$ gene families. There are nine murine V$_H$ gene families and six human V$_H$ families, members within a family displaying 80% or more DNA sequence similarity. *See*: IMMUNOGLOBULIN GENES.

villin *See*: ACTIN-BINDING PROTEINS.

vimentin An INTERMEDIATE FILAMENT protein.

vinblastine, vincristine Anticancer plant alkaloids isolated from the Madagascar rosy periwinkle *Catharanthus (Vinca) roseus*. Vinblastine and vincristine are now used in clinical practice.

vinculin *See*: ACTIN-BINDING PROTEINS; CELL JUNCTIONS.

VIP Vasoactive intestinal peptide. *See*: PERIPHERAL NERVOUS SYSTEM.

vir genes The virulence genes carried on TI PLASMIDS and RI PLASMIDS of *Agrobacterium* sp. They encode the signal transduction, transcriptional activation and T-strand production and transfer functions. *virA* is part of the *Agrobacterium* signal trans-

duction system for T-strand transfer. It encodes the sensor/transmitter protein which detects the small phenolic molecules that induce the virulence (*vir*) genes. The *virB* operon encodes up to 11 proteins which are largely associated with the membrane of *Agrobacterium tumefaciens*. These may be the pore complex proteins which effect T-strand transfer to plant cells. The *virC* operon encodes proteins involved in T-strand production. virC1 binds to the OVERDRIVE sequence. The *virD* operon encodes proteins involved in T-strand production. VirD1 has a GYRASE activity and is required for relaxing the Ti plasmid DNA. virD2 is a site-specific single-strand cutting ENDONUCLEASE which nicks the border repeat sequences. virD2 also becomes covalently attached to the 5′ end of the T-strand and may, through the action of two NUCLEAR TARGETING SIGNALS near its C terminus, direct the T-strand to the plant cell nucleus. The *virE* operon encodes proteins involved in T-strand production. The virE2 protein binds nonspecifically to single-stranded DNA and coats the T-strand complex. *virF* is the *pinF* gene of the Ti plasmid. It is not apparently required for virulence but is induced with the other virulence (*vir*) genes. *virG* encodes part of the *Agrobacterium* signal transduction system which stimulates the production of the T-strand. The VirG protein is the receiver/regulator protein which functions as the transcription activator of the *vir* region. It is activated by a PROTEIN KINASE activity associated with the *virA* gene product which detects the signal molecules produced in wounded plant cells.

vir mutant The abbreviation has been applied to both *viridis* and *virescens* mutants of higher plants. These are both mutations resulting in CHLOROPHYLL deficiency, but have distinct properties.

viral envelope Lipid and protein bilayer outer covering of some virus particles, which is derived from the host cell membrane during budding of the virus from the cell. Virus-encoded glycoproteins are embedded in the lipid bilayer. *See*: ANIMAL VIRUSES; IMMUNODEFICIENCY VIRUSES; INFLUENZA VIRUS; RETROVIRUSES; SEMLIKI FOREST VIRUS; VESICULAR STOMATITIS VIRUS. Families of enveloped viruses are listed in Table V1.

Table V1 Families of enveloped viruses

Plant virus groups	Animal viruses
Tomato spotted wilt virus	Arenaviridae
	Baculoviridae
	Bunyaviridae
	Coronaviridae
	Filoviridae
	Flaviviridae
	Hepadnaviridae
	Herpesviridae
	Orthomyxoviridae
	Paramyxoviridae
	Polydnaviridae
	Poxviridae
	Retroviridae
	Rhabdoviridae
	Togaviridae

viral oncogene *See*: ONCOGENES.

virescens mutant Higher plant mutants in which pigmentation has been affected in such a way that the seedlings first appear as pigment deficient but gradually intensify in colour and may eventually look virtually normal. Such mutants are known for barley, maize, cotton, peanuts, and soybeans. In most cases greening is retarded at lower temperatures and this is associated with impaired assembly of chloroplast 70S ribosomes.

virgin B cell, virgin T cell Immunocompetent lymphocytes that have acquired antigen specificity but have not yet encountered antigen. *See*: B CELL DEVELOPMENT; T CELL DEVELOPMENT.

viridis mutant A group of chlorophyll-deficient mutants of barley. The mutants *vir-h, -q, -n* and *-zb* have been shown to be deficient in PHOTOSYSTEM I, and *vir-n* lacks subunits A and B and two smaller subunits. Other mutants lack the PHOTOSYSTEM II reaction centre (*vir-c, -e, -zd* and *-m*).

virino Hypothetical disease entity containing both protein and nucleic acid that has been postulated as the infectious agent in the TRANSMISSIBLE SPONGIFORM ENCEPHALOPATHIES.

virion A complete virus particle. *See*: ANIMAL VIRUSES; PLANT VIRUSES; VIRUS STRUCTURE.

Viroids

VIROIDS are unencapsidated, small, single-stranded, circular RNA molecules which infect plants (Table V2) [1]. In many cases they cause disease symptoms and are economically important; potato spindle tuber viroid (PSTVd), for example, causes tuber elongation and yield loss in potatoes and tomatoes, and citrus exocortis viroid (CEVd) causes dwarfing and bark scaling in citrus. In other cases they are latent, at least in some hosts, such as *Columnea* latent viroid which is latent in *Columnea* species but causes symptoms in tomato.

Classification

At least 13 distinct viroids have been characterized as well as

Table V2 Some plant viroids and their abbreviations

ASBVd	Avocado sunblotch viroid
CCCVd	Coconut cadang cadang viroid
CCMVd	Chrysanthemum chlorotic mottle viroid
CEVd	Citrus exocortis viroid
CPFVd	Cucumber pale fruit viroid
CSVd	Chrysanthemum stunt viroid
HSVd	Hop stunt viroid
PSTVd	Potato spindle tuber viroid
TASVd	Tomato apical stunt viroid
TPMVd	Tomato plant macho viroid

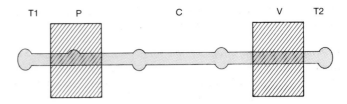

Fig. V7 Stylized predicted secondary structure of a viroid with the suggested domains identified.

many sequence variants of some. They contain between 240 and 375 nucleotides. The circular RNA undergoes extensive intramolecular BASE PAIRING resulting in a rod-like structure. From the RNA sequences two major groups of viroids have been recognized [2], one based on PSTVd and having a high G + C content, and the other based on avocado sunblotch viroid and having a low G + C content.

Five sequence DOMAINS have been identified in the molecules of members of the PSTVd group (Fig. V7). The C (central) domain is the most highly conserved between viroids; the sequence of the core region of this domain has been proposed as a criterion for classification of PSTVd viroids into two groups, one based on PSTVd and the other on apple scar skin viroid. The P (pathogenicity) domain is involved, at least in PSTVd, in modulating the severity of symptoms in tomato *cv*. Rutgers. The V (variable) domain is the most variable region, in which there can be <50% homology between otherwise closely related viroids. The T (terminal) domains are possibly involved in the recognition of host enzymes responsible for replication.

However, recent evidence points to extensive recombination between equivalent domains of different viroids [3]. This raises questions about classifications based on domain sequences and also points to how viroids could evolve.

Replication

Viroids are found in the nuclei of infected plants and replicate by way of an oligomeric negative-strand RNA intermediate to give oligomeric positive-strand RNA; this is suggestive of a ROLLING CIRCLE mechanism of replication. They do not encode any proteins and it is thought that (+) to (−) strand replication involves host DNA-dependent RNA polymerase II, and (−) to (+) strand synthesis DNA-dependent RNA polymerase I. A specific sequence of avocado sunblotch viroid has been identified as being involved in self-cleavage (*see* RIBOZYMES); how the oligomeric molecules of other viroids are processed to the monomer circular form is still uncertain.

R. HULL

See also: PLANT VIRUSES.

1 Diener, T.O. (Ed.) (1987) *The Viroids* (Plenum, New York).
2 Koltunow, A.M. & Rezainan, M.A. (1989) A scheme for viroid classification. *Intervirology* **30**, 194–201.
3 Rezaian, M.A. (1990) Australian grapevine viroid — evidence for extensive recombination between viroids. *Nucleic Acids Res.* **18**, 1813–1818.

virulence genes Genes carried by pathogenic microorganisms that determine their ability to cause disease.

virulent Term applied generally to any bacterium or virus that causes disease. Applied to BACTERIOPHAGE lambda or other TEMPERATE PHAGES it refers to a mutant phage that is able to propagate in a bacterial host LYSOGENIC for the same or related wild-type phage.

virus *See*: ANIMAL VIRUSES; BACTERIOPHAGES; PLANT VIRUSES and individual virus groups and viruses listed in Table V3.

virus satellite *See*: SATELLITE RNA.

virus vectors VECTORS derived from viral genomes. *See*: BACULO-VIRUS VECTORS; DNA CLONING; EXPRESSION VECTORS; GENETIC ENGINEERING; LAMBDA; RETROVIRAL VECTORS; VACCINIA VIRUS.

Table V3 Virus groups and individual viruses covered in the encyclopedia.

Bacteriophages	Plant virus groups	Animal viruses
Lambda	Alfalfa mosaic virus	Adenovirus
Lambdoid phages	Bromovirus	Arenaviridae
M13 phage	Capillovirus	Baculoviridae
Mu phage	Carlavirus	Birnaviridae
P1 phage	Carmovirus	Bunyaviridae
P22 phage	Caulimovirus	Caliciviridae
Q_β phage	Closterovirus	Cardioviruses
T-even phage	Comovirus	Coronaviridae
T-odd phage	Cryptovirus	Corticoviridae
T2 phage	Cucumovirus	Cystoviridae
T4 phage	Dianthovirus	Epstein–Barr virus
T7 phage	Fabavirus	Enteroviruses
	Fijivirus	Filoviridae
	Furovirus	Flaviviridae
	Geminivirus	Foot-and-mouth
	Hordeivirus	disease virus
	Ilarvirus	Hepadnaviridae
	Luteovirus	Hepatitis viruses
	Machlovirus	Herpesviridae
	Marafivirus	Immunodeficiency viruses
	Necrovirus	Influenza virus
	Nepovirus	Inoviridae
	Phytoreovirus	Iridoviridae
	Plant rhabdovirus	Orthomyxoviridae
	Potexvirus	Papovaviruses
	Potyvirus	Paramyxoviridae
	Sobemovirus	Parvoviridae
	Tenuivirus	Picornaviridae
	Tobamovirus	Poliovirus
	Tobravirus	Polydnaviridae
	Tomato spotted wilt group	Poxviridae
	Tombusvirus	Reoviridae
	Tymovirus	Retroviruses
		Rhabdoviridae
		Rhinoviruses
		Semliki Forest virus
		Tetraviridae
		Togaviridae
		Toroviridae
		Vaccinia virus
		Vesicular stomatitis virus

virusoid (satellite RNA, virus satellite) Encapsidated plant pathogenic RNAs ranging in size from 320 to 400 nucleotides long and which require co-infection with a specific virus, a helper virus, for replication and encapsidation. Originally discovered as a group of satellite RNAs derived from the SOBEMOVIRUS plant virus group, they were called virusoids because they shared with the VIROIDS a circular RNA configuration and a compact RNA structure. The term is now often applied generally to both the linear and circular SATELLITE RNAs of plant viruses. *See also*: PLANT VIRUSES.

visceral endoderm *See*: EXTRAEMBRYONIC MEMBRANES.

visceral sensory system *See*: PERIPHERAL NERVOUS SYSTEM.

visinin *See*: CALCIUM-BINDING PROTEINS.

Visna virus *See*: RETROVIRUSES.

visual pigments *See*: CONE PIGMENTS; OPSINS; RHODOPSIN.

Visual transduction

PHOTORECEPTORS are our interface with the visual world about us. There are two main classes in the vertebrate eye: the sensitive rods, responsible for nocturnal vision (also known as scotopic vision after the Greek *skotos*, dark), and the less sensitive cones, responsible for diurnal or photopic vision. Rod vision is monochromatic in mammals, whereas the trichromacy of colour vision arises from the existence of three different photopigment molecules in red-, green-, and blue- sensitive cones. All these different photoreceptors transduce light into an electrical signal in essentially the same way, so much of what follows refers to a generic vertebrate photoreceptor.

Photoreceptors must convert light energy into an electrical signal with high gain; the minute energy of a single photon ($\sim 4 \times 10^{-19}$ J) obstructs the passage of about 10^6 Na$^+$ ions in a dark-adapted mammalian rod. Photoreceptors must also adapt, or reduce this high gain as the ambient light level rises. As photoreceptors adapt, the response speeds up and, as in many physical instruments, there is a trade-off between sensitivity and temporal resolution.

The general structure of photoreceptors is shown in Fig. V8. The outer segments of both rods and cones contain a stack of membrane-bound disks, in which the light-sensitive photopigment is located. The pigment is formed by the conjugation of 11-*cis* retinal (derived from vitamin A, or all-*trans* retinol) with the integral membrane protein opsin. The process of phototransduction begins with the absorption of a photon of light by the photopigment, called RHODOPSIN in rods; no very satisfactory names exist for the three cone pigments. The wavelengths of maximum absorption of light by human photoreceptors, determined by microspectrophotometry of intact outer segments [1], are: rods, 496 nm, in the blue-green part of the spectrum; blue cones, 419 nm; green cones, 530 nm; red cones, 558 nm.

Photoreceptors in neonatal mammals, including humans, are

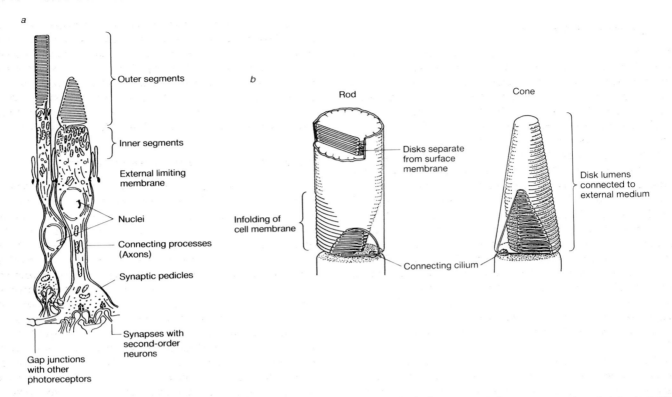

Fig. V8 Diagram of the structure of vertebrate rods and cones. The characteristic differences in morphology between rods and cones shown here are observed in many species, although the reason why a rod-like outer segment is associated with a high sensitivity has not been established. *a*, Main features of a rod (left) and a cone. The outer segment contains about 2000 membranous disks, and is attached to the inner segment by a short stalk containing the 9 + 2 arrangement of MICROTUBULES characteristic of a motile cilium. The synaptic pedicle communicates with second-order cells (horizontal and bipolar cells). *b*, Principal elements of the outer segments of rods and cones. Disks are formed in both types of receptor by an out-pouching of membrane at the base of the outer segment.

insensitive compared with their adult counterparts. The main cause is a deficiency in the supply of 11-*cis* retinal in the neonatal retina, because sensitivity is raised to adult levels by supplying exogenous 11-*cis* retinal [2]. The deficiency may arise because the concentration of retinoids, which in other systems have been shown to be an important morphological signal (*see* RETINOIC ACID), must be kept low during development.

Structure of the light-sensitive molecule

The amino-acid sequences of the four opsin molecules involved in human vision (rhodopsin and the red-, green- and blue-sensitive cone opsins) have now been determined [3]. The opsin family are integral membrane proteins, consisting of around 360 amino acids and $M_r \sim 39\,000$. All members contain seven probable transmembrane α-helical domains, and are members of the wider class of G PROTEIN-COUPLED RECEPTORS which include the muscarinic receptor and the β-adrenergic receptor.

The light-sensitive chromophore, 11-*cis* retinal, is linked to a lysine residue in the middle of the seventh helix via a protonated Schiff base (*see* RHODOPSIN for the structure of the molecule). The entire 11-*cis* retinal molecule lies within the hydrophobic part of the membrane, in a hydrophobic pocket formed by the transmembrane α-helical domains. The absorption spectrum of native 11-*cis* retinal peaks in the ultraviolet, and the shift of this spectrum into the visible range is achieved partly by protonation of the Schiff base nitrogen. Further 'fine-tuning' of the absorption to produce the observed spectra in the different photoreceptors is produced by differences between the four opsin molecules, but the physical basis of this important process is not as yet understood.

The genes encoding the red and green pigments are located in humans in a head-to-tail tandem array on the X chromosome and show strong sequence homology, as expected for the products of a GENE DUPLICATION event in the recent past [3]. Expression of both these genes is controlled by a single small region (<0.6 kb) located about 3 kb upstream of the red gene transcription initiation site [4]. There is much less homology between rhodopsin, the blue cone pigment and the red/green pigment, showing that the genes for these pigments diverged at a much earlier date, probably before the vertebrate radiation. Analysis of cloned genes from colour-blind individuals has demonstrated that the defects responsible for COLOUR BLINDNESS and other abnormalities of colour vision are due to MUTATIONS in the genes for the various cone opsins (*see* CONE PIGMENTS). The commonest colour vision defects are various forms of red or green colour blindness or anomalous colour vision, and are almost invariably seen in males because of the location of the genes for the corresponding pigments on the X-chromosome.

The light-sensitive cascade

Absorption of a photon of light by 11-*cis* retinal causes an extremely rapid isomerization to the all-*trans* configuration. This conformational change in the interior of the opsin molecule alters its cytoplasmic configuration and initiates the sequence of events outlined in Fig. V9. Rhodopsin is free to diffuse laterally within the plane of the membrane, and a photoisomerized molecule can therefore encounter and activate a large number (about 500) of TRANSDUCIN molecules. Transducin is a member of the G-protein family (*see* GTP-BINDING PROTEINS), and like other G proteins is a peripheral membrane protein consisting of three subunits — α, β, and γ. Interaction with an isomerized rhodopsin causes release of the GDP bound to the transducin α subunit (termed α_t or T_α) and uptake of a GTP in its place. In the GTP-bound state the α subunit dissociates from the β and γ subunits and is then free to interact with and activate the cyclic GMP phosphodiesterase (PDE). The interaction between α_t and PDE removes two inhibitory subunits from the PDE; no further gain is introduced at this stage, as the ratio between active PDE and active α_t seems to be less than 1 : 1. Activation of the PDE causes a rapid drop in the intracellular level of cGMP, by cleavage of the cyclic 3′,5′ diester bond to form GMP [5].

It has been known for many years that the ultimate effect of light is to close cation-selective channels in the outer segment membrane, but the link between the drop in cGMP concentration and channel closure remained obscure [6]. Some evidence favoured an alternative hypothesis in which a release of Ca^{2+} by light would close the channels. The problem was resolved by elegant experiments in which ionic channels were recorded by the PATCH-CLAMP method in isolated patches of outer segment membrane [7]. The channels were found to be opened by cGMP, with at least three molecules interacting cooperatively to cause channel opening. No other cofactor is required to open channels, and (contrary to the calcium hypothesis) calcium ions are not effective in closing channels. The cDNA sequence of the cGMP-gated channel has now been established, and shows some homology with voltage-gated channels [8]. The channel appears to be composed of a number (probably four) of identical subunits.

Ionic currents in the outer segment membrane

From the discussion above it is clear that phototransduction generates a high gain by the cascading of a series of stages of lower gain, as in a photomultiplier tube. The end-point of the pathway is the suppression of Na^+ influx through the outer segment membrane, which in turn produces a HYPERPOLARIZATION of the photoreceptor MEMBRANE POTENTIAL because potassium efflux through K^+-selective channels in the inner segment membrane continues in the light [6]. This hyperpolarization causes a suppression of the release of NEUROTRANSMITTER from the synaptic pedicle which communicates with second-order neurons (horizontal and bipolar cells) in the retina (Fig. V8*a*). It is only further on in the visual pathway — at the level of amacrine and ganglion cells in the inner retina — that ACTION POTENTIALS are generated. The action potentials are transmitted to the brain along the ganglion cell axons which make up the optic nerve.

The major current carrier through the light-sensitive channel in normal conditions is Na^+, but the channel is not particularly selective for Na^+ — in fact it discriminates poorly between monovalent cations, and Na^+ carries the majority of the light-sensitive current simply because it is the most abundant external cation [6]. About 10% of the light-sensitive current is carried by Ca^{2+}, and this calcium influx plays an important part in the process of light adaptation (see below). Recent studies of ion fluxes through channels in isolated patches of membrane under voltage clamp have shown that the energy barrier profile of the channel consists of two barriers near the external membrane surfaces, with a single energy well located centrally [9].

Inactivation of the light-sensitive cascade

The activity of the pathway shown in Fig. V9 is terminated at several points. Activated rhodopsin is phosphorylated at several cytoplasmic sites near its C terminus by a specific rhodopsin kinase (*see* PROTEIN KINASES; PROTEIN PHOSPHORYLATION). The phosphorylation, which does not of itself terminate the activity of rhodopsin, is the signal for the binding of an arrestor molecule (variously called arrestin, 48K protein and S-antigen) which 'caps' the active rhodopsin and prevents interaction with further transducin by steric hindrance [10]. Rhodopsin subsequently decays through several states, ending in the release of all-*trans* retinal, but these states do not appear to be of much importance for visual transduction. The all-*trans* retinal diffuses from the cell after release and is reconverted to the 11-*cis* form by a retinal isomerase located in the pigment epithelium layer behind the photoreceptor layer; it then re-enters the cell where it recombines spontaneously with opsin to regenerate rhodopsin. The activity of

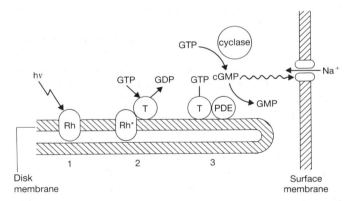

Fig. V9 Schematic diagram of activation of the light-sensitive cascade. In step 1 the 11-*cis* retinal chromophore of a rhodopsin molecule, Rh, absorbs a photon of light and is converted to all-*trans* retinal, forcing the rhodopsin molecule into the active state, Rh˙. In step 2 Rh˙ interacts with transducin (T), converting it into the active form by catalysing release of GDP bound to the α subunit of T (α_t) and subsequent uptake of GTP. About 500 such interactions occur per Rh˙ in rods, but probably fewer in cones. In step 3 active α_t activates cGMP phosphodiesterase (PDE) by removing two inhibitory γ subunits, thus catalysing the hydrolysis of cGMP to GMP. Removal of cGMP allows the light-sensitive channels to close. The diagram does not show the inactivation of the various steps in the pathway. The interaction of calcium with the pathway is shown in Fig. V10.

transducin is terminated by its intrinsic GTPase activity, which recent evidence shows is greatly enhanced by interaction with the inhibitory γ subunit of PDE [11].

Finally, the recovery of cGMP to its dark level is speeded by an ingenious mechanism (Fig. V10) which relies on the fall of intracellular free calcium concentration caused by light (see below). In darkness the GUANYLATE CYCLASE is partially inhibited by combination with a calcium-dependent protein called recoverin [12], and is released from inhibition when the calcium level falls after a flash of light. The recovery of cGMP after a flash is therefore speeded by the drop in calcium.

The same mechanism explains many of the phenomena observed during light adaptation. A dim steady light speeds the rate constant of degradation of cGMP, by the action on the PDE outlined in Fig. V9, and also speeds the rate constant of synthesis of cGMP by the indirect action shown in Fig. V10, mediated by the effect of a fall in intracellular Ca^{2+} concentration on the guanylate cyclase. The effect of a flash of light, when given in the light-adapted state, is therefore to cause a fall in cGMP concentration which is both smaller in size and more rapid than in darkness, because of the dual action of light on both the rate constant α of synthesis of cGMP and on the rate constant β of its breakdown (see Fig. V10). Both the reduction in gain and the speeding of the light-sensitive pathway, which correspond to the psychophysical observations that sensitivity is reduced and temporal resolution improved in the light-adapted state, can be explained by the mechanism outlined in Fig. V10 (reviewed in [6]).

Recent evidence shows that calcium also prolongs the activity of the PDE [13], although it has not yet been established at which point in the pathway shown in Fig. V9 the action is taking place. The fall in calcium in the light-adapted state therefore speeds the turn-off of the PDE, an effect which is in the same direction as the calcium regulation of the guanylate cyclase. The concentration of cGMP — and consequently of open light-sensitive channels — is therefore regulated in push–pull fashion by calcium ions, by an action both on the guanylate cyclase responsible for synthesizing cGMP and on the PDE responsible for breaking it down.

Calcium movements in the outer segment

From the preceding section it is clear that CALCIUM plays a crucial role in the process of light adaptation. The only route for calcium entry is through the light-sensitive channels themselves [6,14], and the free calcium concentration in the outer segment cytoplasm consequently falls when the channels are closed by light. The fall is rapid (~100 ms in mammalian rods) because calcium is extruded from the outer segment by an unusual and powerful calcium transport mechanism in which four Na^+ ions enter the outer segment in exchange for one Ca^{2+} and one K^+ ion (see Fig. V10). Calcium transport is therefore energized by both the potassium and the sodium gradients [15]. This Na : Ca,K transporter is probably a more powerful descendant of the Na : Ca exchange which is widespread in the membranes of other cells (*see* MEMBRANE TRANSPORT). The extra power of the Na : Ca,K exchange may be required in the outer segment because the large Na^+ influx through light-sensitive channels reduces the Na^+ gradient below that which would enable a Na : Ca exchange to reach the low calcium levels (<100 nM) necessary to disinhibit the guanylate cyclase [12] and initiate the process of light adaptation (see Fig. V10). Both the Na : Ca exchange and the Na : Ca,K exchange are currently under study from both a molecular and a functional point of view, and in the near future a better understanding of these two important calcium transport mechanisms and of their relation to one another will no doubt emerge.

P.A. MCNAUGHTON

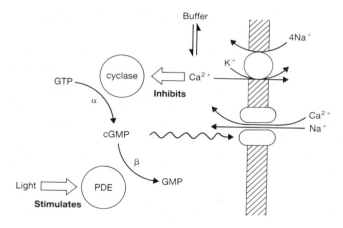

Fig. V10 Mechanisms controlling intracellular calcium in photoreceptor outer segments, and interaction of calcium with the light-sensitive pathway. Calcium carries about 10% of the current through the light-sensitive channels, which are the only route for calcium entry. Within the outer segment much of the calcium is rapidly and reversibly bound to buffer systems [14]. Calcium is pumped from the outer segment by an exchange of stoichiometry $4\,Na^+ : 1\,Ca^{2+}, 1\,K^+$. Calcium has two principal actions on the light response: it inhibits the guanylate cyclase via the protein recoverin [12], and it inhibits the turn-off of the phosphodiesterase [13]. The effect of the fall of calcium occurring after a flash of light is therefore to speed recovery both by promoting the resynthesis of cGMP and by hastening the turn-off of the light-sensitive cascade.

1 Dartnall, H.J.A. et al. (1983) Human visual pigments: microspectrophotometric results from the eyes of seven persons. *Proc. R. Soc.* **B220**, 115–130.
2 Ratto, G.M. et al. (1991) Development of the light response in neonatal mammalian rods. *Nature* **351**, 654–657.
3 Nathans, J. et al. (1986) Molecular genetics of colour vision: the genes encoding blue, green and red pigments. *Science* **232**, 193–202.
4 Wang, Y. et al. (1992) A locus control region adjacent to the human red and green visual pigment genes. *Neuron* **9**, 429–440.
5 Stryer, L. (1986) Cyclic GMP cascade of vision. *Annu. Rev. Neurosci.* **9**, 87–119.
6 McNaughton, P.A. (1991) Light responses of vertebrate photoreceptors. *Physiol. Rev.* **70**, 847–883.
7 Fesenko, E.E. et al. (1985) Induction by cyclic GMP of cationic conductance in the plasma membrane of the retinal rod outer segment. *Nature* **313**, 310–313.
8 Kaupp, U.B. et al. (1991) The cyclic nucleotide-gated channels of vertebrate photoreceptors and olfactory epithelium. *Trends Neurosci.* **14**, 150–157.
9 Zimmermann, A.L. & Baylor, D.A. (1992) Cation interactions within the cyclic GMP-activated channel of retinal rods from the tiger salamander. *J. Physiol, Lond.* **449**, 759–783.
10 Kuhn, H. et al. (1984) Light-induced binding of 48-kDa protein to

photoreceptor membranes is highly enhanced by phosphorylation of rhodopsin. *FEBS Lett.* **176**, 473–478.

11 Arshavsky, V.Y. & Bownds, M.D. (1992) Regulation of deactivation of photoreceptor G-protein by its target enzyme and cGMP. *Nature* **357**, 416–417.

12 Dizhoor, A.M. et al. (1991) Recoverin: a calcium-sensitive activator of retinal rod guanylate cyclase. *Science* **251**, 915–918.

13 Kawamura, S. & Murakami, M. (1991) Calcium-dependent regulation of cyclic GMP phosphodiesterase by a protein from frog retinal rods. *Nature* **349**, 420–423.

14 Lagnado, L. et al. (1992) Calcium homeostasis in the outer segments of retinal rods from the tiger salamander. *J. Physiol., Lond.* **455**, 111–142.

15 Cervetto, L. et al. (1989) Extrusion of calcium from rod outer segments is driven by both sodium and potassium gradients. *Nature* **337**, 740–743.

vitamin D Family of fat-soluble vitamins derived from cholesterol (in mammals) or other sterols and required for normal bone growth. A deficiency causes rickets in children. It acts at intracellular receptors of the STEROID RECEPTOR SUPERFAMILY.

vitamin D binding protein *See*: ACTIN-BINDING PROTEINS.

vitelline membrane (1) In vertebrates, the GLYCOPROTEIN membrane that invests the egg immediately superficial to the plasma membrane. In mammals it is a relatively thick, translucent layer known as the ZONA PELLUCIDA, which mediates sperm–egg recognition via its constituent glycoproteins.
(2) In invertebrates, the covering over the embryo lying immediately external to the plasma membrane of the egg.

vitellogenesis In fish, amphibians, reptiles, and birds the YOLK PROTEINS deposited in growing oocytes are derived from a precursor protein, vitellogenin, which is synthesized under oestrogen control in the liver of mature females. Oestrogen induces vitellogenin at both the transcriptional and translational levels. In *Xenopus*, in response to hormone vitellogenin mRNA rises from undetectable levels to constitute roughly half the total cellular mRNA. Oestrogen also specifically stabilizes the vitellogenin mRNA and its half-life rises from 16 hours to 3 weeks. In insects the yolk proteins are synthesized mainly in the fat body and subsequently secreted into the haemolymph to be taken up by the growing OOCYTE. The synthesis of yolk proteins is under the control of juvenile hormone, ECDYSONE, and a brain-derived neurosecretory hormone.

vitellogenin *See*: VITELLOGENESIS; YOLK PROTEINS.

vitronectin Extracellular matrix protein (M_r 72 000), also present in blood plasma where it it is known to bind the SERPIN plasminogen activator inhibitor PAI-1. It is also involved in scavenging terminal complexes of COMPLEMENT components. *See*: EXTRACELLULAR MATRIX MOLECULES.

VLDL Very low density lipoprotein. *See*: PLASMA LIPOPROTEINS AND THEIR RECEPTORS.

VNTR locus *See*: DNA TYPING; VARIABLE NUMBER TANDEM REPEATS.

voltage clamp An important electrophysiological technique that enables the MEMBRANE POTENTIAL to be maintained at a chosen fixed value, and to be switched from one value to another. Thus, the nature and time course of the current(s) flowing across the membrane at particular values of the membrane potential can be determined to reveal the voltage-activated CONDUCTANCE(S). The technique is also used when testing a cell's current response to an applied agonist without obscuring the effect by any change in membrane potential. It its classic form, voltage clamp is achieved by two electrodes: one to record the membrane potential (to compare with the 'command' voltage) and the other to inject the current required to maintain the membrane potential at the chosen level. This method is, however, only possible with large cells. Low-resistance patch electrodes enable voltage clamping of small cells and excised patches of membrane (*see* PATCH CLAMP).

Jones, S.W. (1990) In *Neuromethods*, Vol. 14 (Boulton, A.A. et al., Eds) 143–192 (Humana Press, Clifton, NJ).

voltage-dependent blockade A block of ION CHANNEL activity which is dependent on the cell membrane potential. For example, at membrane potentials more negative than -35 mV Mg^{2+} ions are drawn into the NMDA receptor-operated ion channel (*see* EXCITATORY AMINO ACID RECEPTORS). This causes a blockade of the channel, preventing entry of extracellular Na^+ or Ca^{2+} and limiting the contribution of NMDA receptors to SYNAPTIC TRANSMISSION. At depolarized membrane potentials Mg^{2+} ions are expelled from the channel. This allows the entry of Na^+ and Ca^{2+} through the channel and allows NMDA receptor activation to contribute to synaptic transmission.

Voltage-gated ion channels

ION channels are pore-forming proteins which span the plasma membrane of cells (*see* MEMBRANE STRUCTURE). As the lipid bilayer is electrically nonconducting, ion channels are the major route for ion movement into or out of the cell. Most ion channels are 'gated', in that they only open in response to a specific stimulus, which for voltage-gated channels, is a change in the electrical potential across the cell membrane (compare LIGAND-GATED ION CHANNELS). In the resting state, the potential inside the cell is normally found to be negative (typically around -80 mV) in comparison to the outside. Making the inside of the cell less negative (i.e. depolarization) usually causes voltage-gated ion channels to open. Depolarization *in vivo* is brought about by a variety of means, for example by neurotransmitter action at ligand-gated ion channels. Experimentally, cells can be depolarized by using PATCH CLAMP and VOLTAGE CLAMP techniques. Another key feature of voltage-gated ion channels is their selectivity for different ion species; the main types found are selective for either potassium, sodium, or calcium.

The precise mechanisms by which voltage-gated ion channels open in response to depolarization are not known; however it seems clear that the channel protein undergoes a conformational change which opens the pore, so allowing the passage of ions. This process is known as activation. After a period of a few

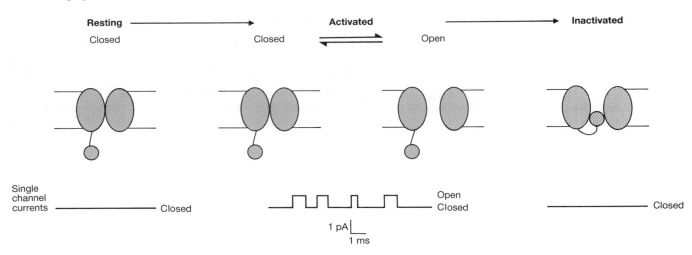

Fig. V11 A schematic representation of voltage gating of ion channels. Under resting conditions, channels are closed in the resting state such that ions cannot pass through. Voltage-dependent activation of channels allows them to alternate rapidly between activated closed and activated open states. While open, the channel is permeable to ions and currents can be observed (lower trace). Inactivation is proposed to occur by movement of a ball and chain motif (see text). Channels do not conduct ions in the inactivated state.

milliseconds the channel closes again. If the stimulus persists, such channel openings can occur repeatedly (Fig. V11). Eventually, even though the cell is still depolarized, channel activity may cease and the channel is then unable to open again until after the cell membrane potential returns to normal (inactivation). As well as being controlled by membrane potential, voltage-gated ion channels can also be blocked by drugs and toxins and modulated by intracellular mediators which can phosphorylate the channel [1].

Voltage-gated ion channels are of great importance in cellular physiology. In neurons, for example, sodium and potassium channels are involved in the generation and propagation of an impulse along the fibre. Following an initial depolarization (for example as a result of neurotransmitter action (*see* SYNAPTIC TRANSMISSION)), sodium channels open transiently, causing a further depolarization and the generation of an ACTION POTENTIAL. The further depolarization in turn causes the eventual opening of potassium channels which bring the membrane potential back towards the resting level and terminates the action potential. Voltage-gated calcium channels are also important physiologically. For example, depolarization at nerve terminals as the result of the arrival of a nerve impulse causes calcium channels at the terminal to open; the calcium that then enters the terminal leads to release of neurotransmitters (*see* EXOCYTOSIS) which in turn trigger post-synaptic events at the synapse (*see* SYNAPTIC TRANSMISSION).

Molecular structure

Early studies on the primary structure of ion channel proteins (of sodium and calcium channels) involved first purifying the pro-

teins from membranes using a labelled ligand. A partial amino-acid sequence was then obtained, and from this oligonucleotide probes were constructed with which the specific DNA for that particular ion channel could be isolated from a cDNA library (*see* DNA CLONING; DNA LIBRARIES). The complete amino-acid sequences of the channel proteins were then obtained by sequencing the DNAs, and the topology of the protein in the membrane deduced from the primary structure.

In the case of potassium channels, it proved impossible to purify the proteins directly, so an alternative strategy involving the *Shaker* mutant of the fruit fly *Drosophila* was used. The mutant is known as *Shaker* because flies shake their legs when anaesthetized with ether. This is as a result of a defect in a potassium channel protein encoded by the *Shaker* gene. Because of the detailed knowledge of the genetic map of *Drosophila*, the technique of CHROMOSOME WALKING could be used to isolate the *Shaker* locus without any prior knowledge of the protein sequence [2]. The *Shaker* locus was found to contain a region homologous to parts of the sodium and calcium channels. To confirm that the locus coded for the potassium channels, *Shaker* cDNA clones were transcribed *in vitro* to mRNA which was then translated in *Xenopus* oocytes (*see* XENOPUS OOCYTE EXPRESSION SYSTEM). It was then confirmed by electrophysiological examination of the oocytes that potassium channels had indeed been expressed.

cDNAs for other sodium, calcium, and potassium voltage-gated ion channels from mammalian tissues have since been isolated and sequenced using probes derived from the first clones. The technique of expression cloning (*see* XENOPUS OOCYTE EXPRESSION SYSTEM) has also been used to identify cDNAs encoding ion channels. This technique is based on screening cRNAs (derived from a cDNA library of a suitable tissue such as brain) by the electrophysiological analysis of channel currents in oocytes injected with these cRNAs.

Of the three channels (potassium, sodium, and calcium) the potassium channel protein is the smallest and simplest in structure. Most potassium channel sequences so far obtained contain six hydrophobic membrane-spanning sequences (known as S1 to S6, see Fig. V12) which are predicted to span the membrane. Both the N and C termini are located on the intracellular side of

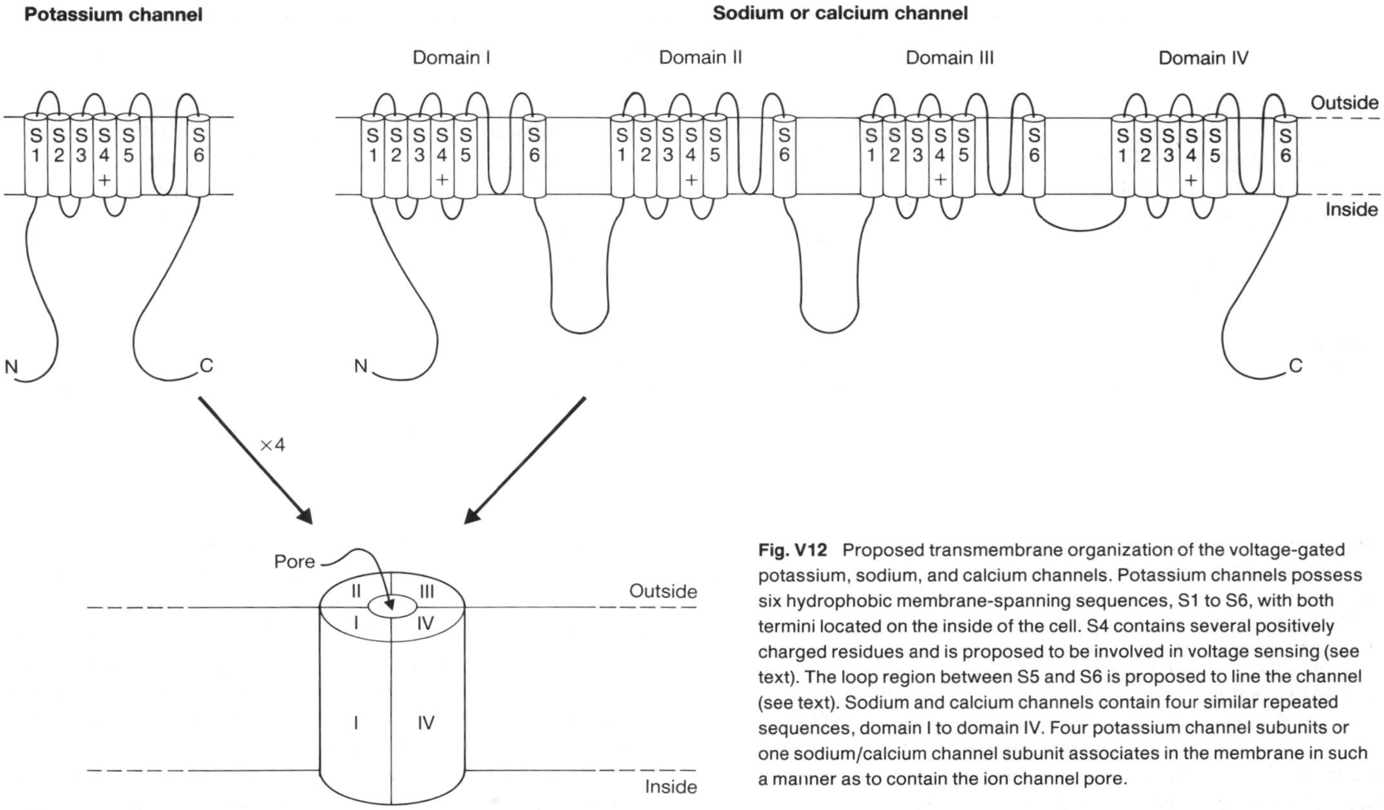

Potassium channel

Sodium or calcium channel

Fig. V12 Proposed transmembrane organization of the voltage-gated potassium, sodium, and calcium channels. Potassium channels possess six hydrophobic membrane-spanning sequences, S1 to S6, with both termini located on the inside of the cell. S4 contains several positively charged residues and is proposed to be involved in voltage sensing (see text). The loop region between S5 and S6 is proposed to line the channel (see text). Sodium and calcium channels contain four similar repeated sequences, domain I to domain IV. Four potassium channel subunits or one sodium/calcium channel subunit associates in the membrane in such a manner as to contain the ion channel pore.

the membrane. The model is consistent with HYDROPATHY PLOTS, the lack of a SIGNAL SEQUENCE, the intracellular location of probable phosphorylation sites, and the extracellular location of probable glycosylation sites (*see* GLYCOPROTEINS AND GLYCOSYLATION; PROTEIN TRANSLOCATION). Both sodium and calcium channels are around four times larger than potassium channels. These larger channels possess four similar motifs, each containing six membrane-spanning regions (Fig. V12). Such a structure is adequate to form the complete channel; for the smaller potassium channel, probably four subunits are required. There is considerable homology between these three channel types and it is likely that they all function in a similar manner. Work on the structure–function relationships of voltage-gated ion channels will be discussed below using potassium channels as examples, but the mechanisms outlined probably apply more widely. The channel subunits described above are sufficient to form the ion channel pore. However, there are also other subunits associated with these channels. Examples of these are the α_2, β, δ, and λ subunits associated with the calcium channel, the β_1 and β_2 subunits associated with the sodium channel and a β subunit associated with the potassium channel. Although these subunits may modulate function of the ion channel subunit, little is understood of the mechanisms and they will not be discussed further here.

Potassium channels: relating structure to function

Much of the functional study of ion channels has involved electrophysiological analysis of potassium channel currents using the voltage clamp technique in oocytes that have been injected with cRNA. Using this method in combination with site-directed mutagenesis, it is possible to begin to establish functional roles for precise regions of channel structure. Described below are some of the principal aspects of molecular function of voltage gated-ion channels, illustrated with particular reference to potassium channels.

A crucial aspect of voltage-gated ion channel function is the ability to sense changes in membrane potential. A comparison of all voltage-gated ion channel amino-acid sequences shows that the region S4 (Fig. V12) always contains a unique series of several positively charged arginine or lysine residues separated by two or three other residues (usually hydrophobic in nature). Changes in membrane potential are believed to force these positively charged amino acids to move in the electric field across the membrane (perhaps by rotation of a helix). Depolarization, for example, would cause the S4 region to move towards the outside of the cell. This would then allow a structural change to occur, opening the pore to the passage of ions. For potassium channels, support for this theory has been provided in experiments in which mutant *Shaker* potassium channels were expressed in *Xenopus* oocytes [3]. Mutations in which the positively charged amino acids of S4 were individually replaced by the neutral amino acid glutamine caused changes in the degree of activation at any particular membrane potential. On the other hand, other properties such as the selectivity of the channel for potassium were not changed in these mutants, suggesting that S4 is not concerned with channel properties other than voltage sensing.

Similar results for the voltage-sensing role of the S4 region have also been found for sodium channels, and it is also expected that the calcium channel S4 region has a similar function.

Molecular mechanisms underlying channel inactivation have also been studied by these techniques. One proposed mechanism for inactivation is based on the so-called ball and chain model. For this, inactivation is suggested to occur by movement of a part of the intracellular portion of the ion channel (the 'ball') in such a manner that it swings on a 'chain' in order to block the inner mouth of the ion channel pore (Fig. V11). This channel block by the ball is thought to be caused by exposure by channel activation of a previously hidden binding site for the ball close to the inner opening of the pore. Evidence for this model in potassium channels is as follows. Deletions and point mutations over the first N-terminal 19 amino acids prevented or dramatically slowed inactivation of *Shaker* potassium channels expressed in oocytes, implicating this region as the ball [4]. Consistent with this, inactivation was restored in *Shaker* deletion mutants by addition of a synthetic peptide corresponding to the first 20 amino acids [5]. Deletions over the next 60 or more amino acids resulted in speeding up of inactivation whereas insertions slowed inactivation, implicating this region as the chain [4] (which could move faster if the chain is shorter). The ball contains positively charged amino acids, and so it is likely that the receptor to which it binds is negatively charged. One such binding region has been located on the intracellular loop sequence between S4 and S5; mutations in this region resulted in changes in the nature of inactivation [6]. The ball and chain model is almost certainly an oversimplification, and does not seem to be the only mechanism for inactivation. A slower inactivation process also occurs possibly through movement of the channel mouth; point mutations near the extracellular mouth of the channel can perturb this slow inactivation component.

The amino-acid residues lining the ion channel pore are structurally and functionally important because they define the ionic selectivity of the channel and its CONDUCTANCE, and also form sites to which drugs and toxins can bind and block the channel. Several studies have suggested that the pore may be lined by the loop region between S5 and S6, also known as the H5 region. Charybdotoxin is a polypeptide of 37 amino acids found in scorpion venom which inhibits various kinds of potassium channels, including *Shaker* channels, by blocking the pore on the external side. The toxin binding sites have been located on the H5 region by making point mutations in this region, injecting the mRNA derived from the mutant DNA into oocytes and measuring the block of potassium currents by the toxin [7]. The location of the toxin binding sites suggests that the H5 region lines the pore near the outer mouth of the channel. Similar results have been obtained with the snake venom toxin dendrotoxin [8] (a polypeptide of 60 amino acids) and with tetraethylammonium (TEA) applied externally [9]. Interestingly, the H5 region probably extends much further into the channel because a block caused by application of TEA from the intracellular side is prevented by appropriate mutation of the H5 region.

Further support for the idea that the H5 region lines the pore comes from experiments studying the ionic selectivity of the channel after mutations of this region [10]. Mutant channels were less selective: channel conductance to rubidium and ammonium ions was increased relative to potassium.

The H5 region has also been implicated in lining the pore of other types of potassium channels such as two rat brain potassium channels, NGK2 and DRK1 [11]. These two channels differ in their single-channel conductance and in their sensitivity to block by TEA. The conductance of NGK2 channels is around three times higher than for DRK1 channels; NGK2 channels are more sensitive to externally applied TEA, whereas DRK1 channels are more sensitive to internal TEA. A chimaeric mutant of these two channels has been made in which the H5 region of the NGK2 channel was transferred to the DRK1 channel. This resulted in channels which possessed conductance and TEA sensitivity characteristics of the NGK2 channel. This is again as would be expected if the H5 region is responsible for lining the pore of these potassium channels.

Two further potassium channel types have been discovered with transmembrane topologies different from that of *Shaker*-like channels. One type is predicted to possess one hydrophobic membrane-spanning sequence, probably with the N terminus on the outside and the C terminus on the inside of the cell [12]. The other type of potassium channel is proposed to possess two membrane-spanning sequences with both termini on the inside [13]. Interestingly, the loop sequence between these two membrane-spanning regions bears a high degree of homology with the H5 region described above. Little is currently understood of the functioning of these channels and it is not yet known whether sodium and calcium channel homologues of these channel types exist.

Here we have attempted to outline current understanding of the relationship between structure and function of voltage-gated ion channels with particular reference to potassium channels. A combination of the techniques of molecular biology and electrophysiology should continue to provide useful tools for the future understanding of ion channels.

D. WRAY

G.G. WILSON

1 Wilson, G.G. et al. (1993) Modulation by cAMP of a cloned rat brain potassium channel expressed in oocytes. *J. Physiol., Lond.* **467**, 167.

2 Timpe, L.C. et al. (1988) Expression of functional potassium channels from Shaker cDNA in Xenopus oocytes. *Nature* **331**, 143–145.

3 Papazian, D.M. et al. (1991) Alteration of voltage-dependence of Shaker potassium channel by mutations in the S4 sequence. *Nature* **349**, 305–310.

4 Hoshi, T. et al. (1990) Biophysical and molecular mechanisms of Shaker potassium channel inactivation. *Science* **250**, 533–538.

5 Zagotta, W.N. et al. (1990) Restoration of inactivation in mutants of Shaker potassium channels by a peptide derived from ShB. *Science* **250**, 568–571.

6 Isacoff, E.Y. et al. (1991) Putative receptor for the cytoplasmic inactivation gate in the Shaker K+ channel. *Nature* **353**, 86–90.

7 MacKinnon, R. & Miller, C. (1989) Mutant potassium channels with altered binding of charybdotoxin, a pore-blocking peptide inhibitor. *Science* **245**, 1382–1385.

8 Hurst, R.S. et al. (1991) Identification of amino acid residues involved in dendrotoxin block of rat voltage-dependent potassium channels. *Mol. Pharmacol.* **40**, 572–576.

9 MacKinnon, R. & Yellen, G. (1990) Mutations affecting blockade and ion permeation in voltage-activated K+ channels. *Science* **250**, 276–279.

10 Yool, A.J. & Schwarz, T.L. (1991) Alteration of ionic selectivity of a K+ channel by mutation of the H5 region. *Nature* **349**, 700–704.

11 Hartmann, H.A. et al. (1991) Exchange of conduction pathways between two related K$^+$ channels. *Science* **251**, 942–944.

12 Takumi, T. et al. (1988) Cloning of a membrane protein that induces a slow voltage-gated potassium current. *Science* **242**, 1042–1045.

13 Ho, K. et al. (1993) Cloning and expression of an inwardly rectifying ATP-regulated potassium channel. *Nature* **362**, 31–38.

von Gierke's disease A genetic disorder of GLYCOGEN storage which results from the deficiency of glucose 6-phosphatase.

von Hippel–Lindau disease Very rare heritable disease with variable symptoms, of which the common diagnostic criteria are retinal angiomas (tumours composed of blood or lymph vessels) and cerebellar hemiangioblastomas. Symptoms are apparent by the third decade of life. The disease is inherited as an AUTOSOMAL DOMINANT trait with complete PENETRANCE and the gene responsible has been localized to the short arm of chromosome 3. The affected tissues are all derived from the NEUROECTODERM, suggesting that the gene may be involved in the regulation of development of these tissues.

von Recklinghausen's disease *See*: NEUROFIBROMATOSIS.

von Willebrand disease factor A multimeric glycoprotein present in plasma, platelets and subendothelial connective tissue. It is necessary for the adhesion of platelets to regions of vascular damage, and it stabilizes the coagulation factor Factor VIII (referred to as FVIII : WF in this context). Significant deficiency (von Willebrand disease) affects one in 8000 individuals and causes a bleeding disorder with features of both platelet dysfunction and haemophilia; autosomal dominant and recessive forms occur. *See also*: BLOOD COAGULATION AND ITS DISORDERS.

VPg protein Small protein of around 22 amino acids covalently attached to the 5′ end of the (+)-sense genomic RNA of certain viruses. *See*: ANIMAL VIRUSES; NEPOVIRUSES; PICORNAVIRIDAE; PLANT VIRUSES.

VSG VARIANT SURFACE GLYCOPROTEIN.

VSV VESICULAR STOMATITIS VIRUS.

vulva formation *See*: CAENORHABDITIS DEVELOPMENT.

W

W The single-letter abbreviation for the AMINO ACID tryptophan.

WAGR syndrome A CONTIGUOUS GENE SYNDROME characterized by the association of WILMS' TUMOUR (a childhood nephroblastoma) with *Aniridia* (failure to form an iris), a variety of Genitourinary abnormalities, and mental *Retardation*.

Waldenstrom's macroglobulinaemia *See*: IMMUNOCYTOMA.

Watson–Crick base pairing *See*: BASE PAIR.

waxes *See*: PLANT CELL WALL MACROMOLECULES.

wee1 Gene from the yeast *Schizosaccharomyces pombe* involved in control of the CELL CYCLE which encodes a PROTEIN KINASE that can phosphorylate proteins on serine, threonine, or tyrosine residues.

western blotting A method for detecting one (or more) specific proteins in a complex protein mixture such as a cell extract, which is named by analogy with Southern hybridization to detect DNA sequences, and northern blotting to detect RNAs. The procedure involves fractionating the protein mixture by denaturing SDS-polyacrylamide gel ELECTROPHORESIS and transferring and immobilizing the mixture onto a solid membrane of either nitrocellulose or nylon by ELECTROBLOTTING. The loaded membrane is then incubated with an ANTIBODY raised against the protein of interest. The antibody–antigen complex so formed on the membrane can then be detected by a procedure which involves the application of a second antibody, raised against the first antibody, and to which an enzyme (e.g. HORSERADISH PEROXIDASE) has been covalently linked. The insoluble reaction product generated by enzyme action can then be used to indicate the position of the target protein on the membrane. The sensitivity of detection can be increased by amplification of the signal using either the BIOTIN–STREPTAVIDIN system or by CHEMILUMINESCENCE detection.

Towbin, H. et al. (1979) *Proc. Natl. Acad. Sci. USA* **76**, 4350–4354.

wheat germ translation system A cell-free extract prepared from wheat germ which is able to support IN VITRO TRANSLATION of a wide variety of natural and synthetic mRNAs. These lysates show only low levels of translation of endogenous mRNAs thereby avoiding the necessity of removing such mRNAs with MICROCOCCAL NUCLEASE.

wheat-germ agglutinin A LECTIN found in wheat germ.

wiggler A magnetic insertion device. *See*: SYNCHOTRON RADIATION.

wild type For any given LOCUS, the ALLELE most frequently found in natural populations is designated the wild-type allele. All other alleles are therefore non-wild type. The term can also be used to describe a phenotype.

Wilms' tumour A childhood nephroblastoma (solid tumour of the kidney) affecting one in 10 000 children, usually appearing within the first five years of life. A susceptibility to Wilms' tumour is associated with inheritance of defects in several different genes, including the TUMOUR SUPPRESSOR GENE *WT-1* which maps to chromosome 11p13. Tumours arise from mesenchymal STEM CELLS that would normally differentiate into parts of the nephron. Around 5–10% also contain ectopic tissues such as bone and cartilage. Wilms' tumour also appears as one of the manifestations of the WAGR syndrome — a CONTIGUOUS GENE SYNDROME. The *WT-1* gene has been cloned and encodes a zinc finger protein which is presumed to be a TRANSCRIPTION FACTOR.

Wilson plot A plot used to determine the absolute scale factor for a set of DIFFRACTION intensities; this must be known, for example, to contour an ELECTRON DENSITY MAP of a crystal in defined units (e.g. electrons per $\mathring{A}^3$). The plot shows the variation with angle of diffraction (i.e. with RESOLUTION) of the ratio of mean diffracted intensity to the sum of squares of the ATOMIC SCATTERING FACTORS (calculable from a known molecular weight and atomic composition) per ASYMMETRIC UNIT. For a structure consistent with WILSON STATISTICS, the overall scale factor and TEMPERATURE FACTOR of the diffraction data can be determined from the slope and intercept of this linear plot. *See*: X-RAY CRYSTALLOGRAPHY.

Stout, G.H. & Jensen, L.H. (1986) *X-ray Structure Determination: A Practical Guide* (Wiley, New York).
Wilson, A.J.C. (1942) *Nature* **150**, 151–152.

Wilson statistics Relationships between the distribution of DIFFRACTION intensities expected from a crystal and its SYMMETRY and UNIT CELL contents were studied by A.J.C. Wilson in the 1940s. He showed that, for a structure of equal atoms, distributed uniformly in a crystal, the average intensity equals the sum of squares of the ATOMIC SCATTERING FACTORS; that is, the average intensity depends only upon the contents of the unit cell and not upon the atomic position. However, the distribution of individual intensities does depend upon the atomic positions; in a noncentrosymmetric crystal the intensities cluster more closely around their mean value than for a centrosymmetric crystal —

which has relatively more reflections which are very weak and more reflections which are very strong. One consequence of this clustering is that the expected value of the residual index (R-FACTOR) comparing observed STRUCTURE AMPLITUDES with those calculated from a model using the correct kind and number of atoms but placed at random in the unit cell, is lower for ACENTRIC reflections (59%) than for CENTRIC reflections (83%). Axes of rotational symmetry also influence an intensity distribution; the best known examples being the icosahedral spikes, of very strong intensities, along the directions of fivefold axes of isometric viruses.

Most crystals of small organic molecules obey Wilson statistics reasonably well, whereas inorganic crystals often have very unequal atoms, and macromolecules all have a very nonuniform distribution of atoms with solvent channels between molecules. In the macromolecular case, Wilson statistics can be regarded as a null hypothesis where atoms are uniformly placed throughout the cell. Structure determination and phasing procedures are then judged by the extent to which they generate a distribution of intensities having a higher LIKELIHOOD of observing the experimental intensities than those arising from the Wilson intensity distribution. *See*: WILSON PLOT; X-RAY CRYSTALLOGRAPHY.

Wilson, A.J.C. (1950) *Acta Cryst.* **3**, 397–398.
Stout, G.H. & Jensen, J.H. (1989) *X-ray Structure Determination: A Practical Guide* (Wiley, New York).

Wilson's disease A rare AUTOSOMAL RECESSIVE disease of copper metabolism (frequency 1 : 100 000). Defects in biliary excretion and incorporation of copper into CAERULOPLASMIN lead to tissue build-up of copper (which is highly toxic), notably in the liver, brain, and kidneys. Cirrhosis, progressive neurological degeneration, and renal tubular defects result. Copper deposits in the cornea are characteristic (Kayser–Fleischer ring). Treatment is with the copper chelator penicillamine.

wilt diseases *See*: PLANT PATHOLOGY.

Wiskott–Aldrich syndrome An X-LINKED immunodeficiency characterized by eczema, thrombocytopenia (reduced numbers of platelets), and susceptibility to bacterial infections. The disease maps to band p11.1 in the short arm of the X-chromosome.

There is a deficiency of the CD ANTIGEN CD43 and other membrane glycoproteins as a result of glycosylation defects, but the significance of this is unclear. *See also*: X-CHROMOSOME INACTIVATION.

WLIP White line inducing principle. A toxin produced by some plant pathogenic *Pseudomonas* spp. *See*: PLANT PATHOLOGY.

wobble Freedom in the pairing of the third base of the codon to the first base of the anticodon which allows the anticodons of some TRANSFER RNAS to pair with several codons (*see* GENETIC CODE). Allowed pairings are given in Table W1.

Crick, F.H.C. (1966) *J. Mol. Biol.* **19**, 548–555.

Wolman disease A rare AUTOSOMAL RECESSIVE disease caused by deficiency of lysosomal acid lipase, an enzyme involved in lipid hydrolysis. Lipid-engorged lysosomes can be demonstrated in many tissues; severe gastrointestinal and neurological symptoms usually result in death by six months of age. A less severe defect leads to cholesteryl ester storage disease. *See also*: LYSOSOMES.

***wox* gene** A nuclear gene (in the case of green plants) coding for a peripheral membrane polypeptide (M_r 33 000) associated with PHOTOSYSTEM II. The polypeptide is not strictly essential for the oxidation of water *in vitro*, but stabilizes the Mn complex involved. The gene is now known as *psbO*, but other names previously used include *psb1* and *oee1* (oxygen-evolving enhancer).

writhe, writhing number (*W*) The number of turns of superhelix in a supercoiled DNA molecule. *See*: SUPERCOILING.

Table W1 Allowed pairings at the third base of the codon according to the wobble hypothesis

First base of anticodon	Third base of codon
C	G
A	U
U	A or G
G	U or C
I	U, C, or A

X

X-chromosome The larger of the two sex chromosomes in mammals, and in some other animals, such as insects (*see* DROSO-PHILA). In mammals, females carry two X-chromosomes in their diploid cells, males an X and a Y. *See also*: SEX DETERMINATION; X-CHROMOSOME INACTIVATION.

X-chromosome inactivation

IN somatic cells of normal female mammals one of the two X-chromosomes is transcriptionally inactive, and fails to respond to signals eliciting transcription of the other, active, X-chromosome. As a result normal XX females and XY males effectively have equal doses of X-linked gene products (for reviews see [1–4]). Thus, X-chromosome inactivation (XCI) is a system of dosage compensation, with an end result comparable with that seen in dosage compensation systems in *Drosophila melanogaster* and other insects, but with a different underlying mechanism (*see* SEX DETERMINATION). In individuals with abnormal numbers of X-chromosomes, for example, XO, XXY, XXX, or XXXX, only a single X-chromosome remains active. Thus, the system should be considered as maintaining dosage compensation through activity

of a single X-chromosome. In triploids either one or two X-chromosomes may be active, and in tetraploids two X-chromosomes remain active, suggesting that a ratio of AUTOSOME sets to X-chromosomes of 2A : 1X may be involved in determining the number of active X-chromosomes. X-chromosome inactivation is found in both placental mammals and marsupials, and in a rudimentary form in monotremes [5], but is not known to occur in any other group.

In placental mammals X-chromosome inactivation has been studied mainly in human and mouse. At the beginning of embryonic development in the mouse both X-chromosomes are in the active state, and their differentiation into active and inactive forms occurs first in the late blastocyst in the TROPHECTODERM and primitive ENDODERM cell lineages, which give rise to the endodermal component of the EXTRAEMBRYONIC MEMBRANES. Somewhat later, inactivation occurs in the primitive ECTODERM cell lineage which gives rise to the embryo proper. This first appearance of the inactive X-chromosome (Xi) is known as the initiation of X-inactivation. Once initiation has occurred the same X-chromosome remains inactive in the mitotic descendants of each cell, throughout the life of the animal, giving rise to clones of cells with the same X-chromosome active (Xa). If the female is HETEROZYGOUS for a marker gene this gives rise to variegation,

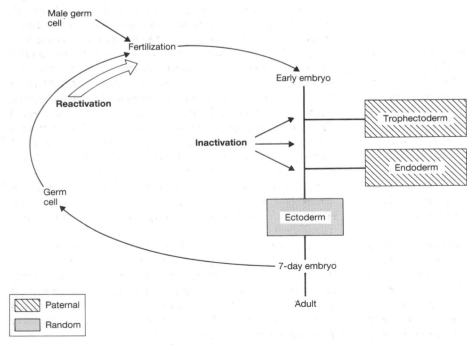

Fig. X1 Cycle of X-chromosome activity in the mammalian female. The imprinted type of paternal X-chromosome inactivation occurs in the trophectoderm and primitive endoderm cell lineages. Reactivation occurs in the oocyte at the approach to meiosis, and presumably persists after fertilization to reactivate the inactive X-chromosome derived from the male germ cell.

1134

detectable visibly, histochemically, or in other ways, according to the type of marker. An exception to the constancy of activity state occurs in the germ cells, where reactivation of the Xi in females occurs at about the time of the onset of MEIOSIS, so that oocytes have both X-chromosomes active. In male germ cells, by contrast, the single X-chromosome becomes inactive at the late spermatogonial stage (Fig. X1).

Properties of the inactive X-chromosome

The CHROMATIN of the Xi is in a condensed state, and can be recognized during interphase by the formation of the sex chromatin body, which lies adjacent to the nuclear membrane. The active X-chromosome is decondensed and lies more centrally in the nucleus. The presence or absence of sex chromatin is not, however, a reliable indicator of X inactivation, as in some cells sex chromatin is not visible although other evidence indicates that an Xi is indeed present.

Another feature of the Xi is its late replication of DNA. Replication begins later than in the Xa or in autosomes, and the inactive chromosome can be distinguished by labelling with [³H]thymidine or bromodeoxyuridine (BrdU).

Yet a third feature concerns methylation of cytosine bases (*see* DNA METHYLATION). Although there is no difference in overall methylation of Xa and Xi, genes on the Xi show a distinct pattern of methylation. Cytosines in CpG islands in 5′ promoter regions are heavily methylated on the Xi, and little methylated on the Xa [6,7].

A further property of the Xi involves differential sites of hypersensitivity to digestion with deoxyribonuclease (DNase), indicating altered binding of proteins to DNA and different packaging around nucleosomes [8].

At present the causal relations among these various properties are not understood, and this is discussed further below.

Ohno's Law

Ohno [9] argued that, as a result of the effectively haploid GENE DOSAGE of the X-chromosome, TRANSLOCATION of genes between the X-chromosome and autosomes would lead to disruption of regulatory processes and inviability, and hence any such translocations which occurred during evolution would not be viable. He therefore postulated that genes X-linked in one mammalian species would be X-linked in all. This has become known as Ohno's Law and has been well borne out, with no known exceptions so far among placental mammals. By comparing mouse and human the X-chromosome can be divided into five segments within each of which the order of genes has been conserved, but which have been rearranged in evolution with respect to each other [3].

Genes on the long arm (Xq) of the present day human X-chromosome are also X-linked in marsupials, but those on the human short arm (Xp) are autosomal. In a monotreme, the platypus, human Xq genes are again X-linked, even though, as judged by the synchronous replication of the major part of the monotreme X-chromosome, only a small segment is inactivated [5]. Human Xp genes are autosomal as in marsupials [10]. Thus,

during evolution a segment corresponding to the present human Xp has been gained from autosomes in the line leading to placental mammals, presumably while X-inactivation itself was also undergoing evolution [5,10].

X-chromosome activity in germ cells

X-chromosome activity in germ cells differs from that in somatic cells. Oocytes have two Xa, whereas in the male germ cell the single X-chromosome becomes inactive. Correct X-chromosome dosage is important to the survival and function of germ cells. In chromosomally XO females, the rate of loss of oocytes from the ovary is abnormally high, leading to a shortened reproductive life in mice, and complete sterility in humans. In male mice with supernumerary X-chromosomes (e.g. XX, XXY), germ cells die at the spermatogonial stage. One possible explanation of these phenomena is that germ cells require different amounts of X-linked gene products, two doses being lethal to a male germ cell, but important for a female germ cell. Another possibility is that failure of pairing, and presence of unpaired sites at meiosis, as for example, in XO oocytes, is detrimental to a germ cell [4]. In oocytes this mechanism may be important, but in XXY males death of germ cells occurs too early to be due to pairing failure.

Imprinting

In placental mammals, in the embryo proper, either the maternal or paternal X-chromosome may be inactivated at random in different cells, leading to the variegated effect in heterozygotes already mentioned. In marsupials, however, there is nonrandom inactivation, with the paternal X preferentially inactivated in all cells. This is termed imprinting (*see* PARENTAL GENOMIC IMPRINTING) with the maternally and paternally derived X-chromosomes bearing an imprint of their parental origin. Similar imprinting is seen in the trophectoderm and primitive endoderm cell lineages of mice and rats, again with the paternal X-chromosome preferentially inactivated [2,4] (see Fig. X1). Imprinting also seems to occur in human extraembryonic tissues, which preferentially express the maternal X-chromosome, although this point is not as well established as in the mouse. In tissues with paternal X inactivation there is of course no variegation in heterozygotes; rather, the phenotype is that determined by the maternal X chromosome. Imprinting appears not to be an essential feature of X inactivation in those tissues in which it occurs; in embryos with inappropriate parental origin of X-chromosomes single X activity is still maintained. In chromosomally X^{P}O embryos the single X remains active, whereas in parthenogenetic X^{m}X^m embryos, inactivation of one X^m occurs in both the embryo and the extraembryonic cell lineages, although perhaps at less than normal frequency [3,4].

Imprinting is of course not limited to the X-chromosome, but also occurs widely among autosomes, with differential expression of maternally and paternally inherited genes or chromosome segments. Presumably X-chromosome imprinting shares its mechanism with autosomal imprinting, but this mechanism is not known. Differential methylation of DNA has been suggested, but this point has not yet been substantiated.

Nonrandom X inactivation resulting from cell selection

Although X inactivation in the embryo of placental mammals is typically random, in certain abnormal situations the observed result, either in the whole animal or in certain tissues, may in fact be nonrandom. One type of abnormality which frequently leads to nonrandom X-chromosome expression involves CHROMOSOME ABERRATIONS. In human females heterozygous for X-chromosome deletions the abnormal X-chromosome is typically inactive in all cells. Conversely, in females heterozygous for X-autosome translocations, the normal X-chromosome is in most cases inactive in all or nearly all cells. A single mouse X-autosome translocation, T(X;16)16H, also results in heterozygotes having the normal X inactive in all cells, and in this case it has been possible to show that this is the result of cell selection.

At the time of initiation, random inactivation of either X-chromosome occurs. Those cells with the translocated X-chromosome inactive are genetically unbalanced, owing to spreading of inactivation into attached autosomal material, and failure of inactivation of one X-chromosome segment (see below), and they die (reviewed in [1]). The resulting embryo thus has the normal X-chromosome as the Xi in all its surviving cells. It is inferred that cell selection is the explanation also for the numerous human cases.

Cell selection is also thought to be the explanation for nonrandom X-chromosome inactivation in certain tissues in heterozygotes for some diseases. In human heterozygotes for the X-linked deficiency of the enzyme hypoxanthine phosphoribosyl transferase (HPRT), all blood cells have the normal X-chromosome active, whereas in fibroblasts there is a typical mixture of cells with one or other X-chromosome active. Similarly, in the X-linked WISKOTT–ALDRICH SYNDROME all B and T lymphocytes in heterozygotes have the normal X-chromosome active [11].

Applications of X inactivation

The variegation produced in females heterozygous for suitable marker genes has been valuable in studying clonality of tumours and normal tissues. Because the initial event occurs early in embryogeny, and remains highly stable thereafter, the size, shape and number of patches in any tissue in the adult depend on the manner of clonal growth of cells. The size of the primordial cell pool in the mouse liver and other tissues, and the monoclonality of liver tumours, has been shown by histochemical staining in mice heterozygous for deficiencies of the X-linked enzymes ornithine transcarbamylase (OTC) and glucose 6-phosphate dehydrogenase (G6PD) (reviewed in [12]).

Nonrandom inactivation due to cell selection also has applications. Using either protein polymorphisms or the differential methylation of the inactive X-chromosome combined with restriction fragment length polymorphisms (RFLPS), one can detect nonrandom inactivation in particular tissues of females heterozygous for X-linked diseases, and this can give information as to the cell type in which the disease process acts. Nonrandom inactivation can also be used for diagnosing whether a female is heterozygous in diseases such as the WISKOTT–ALDRICH syndrome. An example is provided by the heterozygotes for Wiskott–Aldrich syndrome in whom nonrandom inactivation is seen in B and T lymphocytes [11].

Mechanism of X inactivation

X-inactivation centre

Information relevant to the mechanism of X-chromosome inactivation has come not only from aspects of the phenomenon already mentioned but also from study of mouse X-autosome translocations. In all of these except T(X;16)16H random or near random inactivation of one or other X-chromosome is still retained. In such translocations only one of the two X-chromosome segments undergoes inactivation, whereas the other remains active in all cells. Inactivation spreads from the inactive segment into the attached autosomal material.

The interpretation is that there is an inactivation centre on the X-chromosome. At the time of initiation of X inactivation a single centre becomes blocked, and the X-chromosome with this blocked centre becomes the Xa. Inactivation centres on all other X-chromosomes initiate inactivation which spreads along the chromosome in both directions. Segments lacking an inactivation centre remain active (Fig. X2).

There has recently been considerable progress in identifying the inactivation centre. Its location has been mapped quite accurately in both man and mouse from study of translocations and deletions. A human gene termed *XIST* has been found which maps to the region of the centre and has the unique property of being expressed from the Xi but not the Xa [13]. A homologous gene in the mouse, *Xist*, also maps to the region of the centre and shows a similar unique pattern of expression from the Xi [14,15]. The combination of location and unique properties of *Xist* makes it a strong candidate for a role in the inactivation centre but its function is not yet known. No protein product has yet been identified. In the male *Xist* is expressed only in the testis, suggesting a role also in the inactivation of the single X-chromosome in male germ cells [16]. The factor *Xce* in the mouse, different alleles of which affect the probability of inactivation of a particular X chromosome, also maps to the region of the X-inactivation centre.

Spreading and maintenance of inactivation

In X-autosome translocations inactivation may spread into the attached autosomal material for a considerable distance, over 20 centimorgans (cM) of recombination distance or some chromosome bands of physical distance, but appears not to spread to the end of the autosome. This suggests that there may be some factor in the X-chromosome which promotes spread, or some inhibitory factor in autosomes. Once spreading has occurred inactivation is maintained with a remarkably high level of fidelity in the embryo and adult of eutherian mammals, but in the mouse at least, the stability is not quite complete. In mouse females with nonrandom X-chromosome inactivation due to translocation T(X;16)16H, reactivation of the inactive allele has been found at a low frequency both for a true X-linked gene, that for OTC, and for attached autosomal coat colour genes [3]. In marsupials and in eutherian extraembryonic membranes stability is less complete,

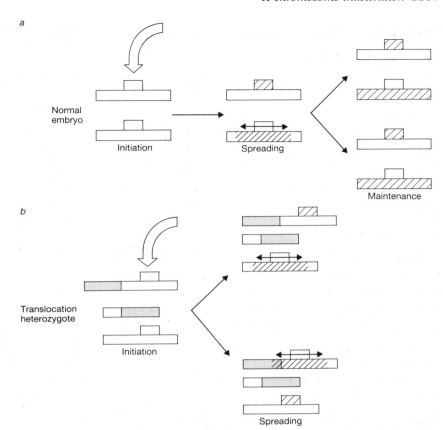

Fig. X2 Stages in initiation and maintenance of X-chromosome inactivation. The inactivation centre is shown as a box. Active X-chromosome material is white, inactive material is hatched and autosomal material is stippled. *a*, Normal female. A trans-acting factor (large arrow) blocks one inactivation centre, and the unblocked centre initiates spreading. *b*, Female heterozygous for an X-autosome translocation. Only one of the two segments into which the X is broken carries an X-inactivation centre. The inactivating signal can spread from this centre, but the segment without the centre remains active. From [13].

and in appropriate conditions reactivation can occur in cultured cells, or in marsupials even *in vivo*.

For the mechanism of spreading the existence of 'way stations' or booster elements along the X-chromosome which promote the spread has been suggested [1,6,7]. It is not known if genes respond individually to the spreading signal or in blocks. In an attempt to elucidate this, searches have been made for genes that escape inactivation. Several have now been found on the human X-chromosome, for example *UBE1*, *RPS4*, *ZFX*. It is not known whether they escape the original spreading of inactivation or whether stabilization fails so that they are reactivated. By contrast the mouse homologues of these genes are inactivated normally [17,18]. Detailed study of these genes may provide insight into the mechanism of spreading or of stabilization.

The nature of the spreading signal is unknown. Methylation of critical sites in DNA of the inactive X-chromosome has been suggested, but is now thought to be part of stabilization rather than spreading. The evidence for this is that differential methylation is not found on the marsupial inactive X-chromosome, and that the DNA of the Xi of extraembryonic membranes of mice is not modified [2,4,6]. In both cases the inactivation is of the imprinted type, with the paternal X-chromosome affected, and in both cases is also less stable. Thus, methylation may stabilize the inactive state, and there may be some interaction between imprinting and stabilization of inactivation.

Detailed studies of methylation have shown that in the human *PGK1* gene (for an isoenzyme of phosphoglycerokinase) almost all sites in CpG islands of the promoter region are methylated. Treatment with the demethylating agent 5-AZACYTIDINE leads to loss of methylation and to reactivation of the gene, and the pattern of results shows that it is possible to envisage a stable self-replicating state in which methylation inhibits transcription and transcription inhibits methylation [6].

The property of late replication of the inactive X-chromosome is also thought to be important in the maintenance of inactivation [2,6,7]. As with methylation it is possible to envisage a feedback situation in which late replication prevents transcription, and transcription is required for early replication [6]. There remain the questions as to means by which differential methylation or late replication are first brought about. Suggestions for this involve interaction of DNA with proteins. Studies of the DNase sensitivity of the human *PGK1* promoter region showed that the active gene had several FOOTPRINTS resulting from protein binding, whereas the inactive gene had none [8]. The results suggest that two NUCLEOSOMES were bound to the promoter region of the inactive *PGK1* gene. A further suggestion for the original spreading signal in X inactivation concerns a change in state of the chromatin fibre, brought about by scaffold attachment proteins [6,7]. The hypothesis is that binding of these proteins to DNA, at scaffold attachment regions (SARs) (*see* NUCLEAR MATRIX), would lead to reeling in of the DNA fibre, which would be thrown into loops, bringing neighbouring SARs into apposition and thus condensing the fibre. This changed state of the fibre would then result in differential access to methylase and also to asynchronous replication. Recent advances, including the cloning of the *Xist* gene, as well as a detailed knowledge of the structure of X-linked genes should lead to rapid strides in future in the elucidation of the primary events in X inactivation.

M.F. LYON

1 Gartler, S.M. & Riggs, A.D. (1983) Mammalian X-chromosome inactivation. *Annu. Rev. Genet.* **17**, 155–190.
2 Grant, S.G. & Chapman, V.M. (1988) Mechanisms of X-chromosome regulation. *Annu. Rev. Genet.* **22**, 199–233.
3 Lyon, M.F. (1988) The William Allan memorial award address: X-chromosome inactivation and the location and expression of X-linked genes. *Am. J. Hum. Genet.* **42**, 8–16.
4 Lyon, M.F. (1989) X-chromosome inactivation as a system of gene dosage compensation to regulate gene expression. *Prog. Nucleic Acid Res. Mol. Biol.* **36**, 119–130.
5 Graves, J.A.M. & Watson, J.M. (1991) Mammalian sex chromosomes: evolution of organization and function. *Chromosoma* **101**, 63–68.
6 Riggs, A.D. (1990) Marsupials and mechanisms of X-chromosome inactivation. *Aust. J. Zool.* **37**, 419–441.
7 Riggs, A.D. (1990) DNA methylation and late replication probably aid cell memory, and type I DNA reeling could aid chromosome folding and enhancer function. *Phil. Trans. R. Soc. Lond.* **B326**, 285–297.
8 Pfeifer, G.P. & Riggs, A.D. (1991) Chromatin differences between active and inactive X chromosomes revealed by genomic footprinting of permeabilized cells using DNase I and ligation-mediated PCR. *Genes Dev.* **5**, 1102–1113.
9 Ohno, S. (1967) *Sex Chromosomes and Sex-linked Genes* (Springer-Verlag, Berlin).
10 Watson, J.M. et al. (1991) Sex chromosome evolution: platypus gene mapping suggests that part of the human X chromosome was originally autosomal. *Proc. Natl. Acad. Sci. U.S.A.* **88**, 11256–11260.
11 Greer, W.L. et al. (1989) X-chromosome inactivation in the Wiskott–Aldrich syndrome: a marker for detection of the carrier state and identification of cell lineages expressing the gene defect. *Genomics* **4**, 60–67.
12 Lyon, M.F. (1988) Clones and X-chromosomes. *J. Pathol.* **155**, 97–99.
13 Lyon, M.F. (1991) The quest for the X-inactivation centre. *Trends Genet.* **7**, 69–70.
14 Borsani, G. et al. (1991) Characterization of a murine gene expressed from the inactive X chromosome. *Nature* **351**, 325–329.
15 Brockdorff, N. et al. (1991) Conservation of position and exclusive expression of mouse *Xist* from the inactive X chromosome. *Nature* **351**, 329–331.
16 Richler, C. et al. (1992) X inactivation in mammalian testis is correlated with inactive X-specific transcription. *Nature Genet.* **2**, 192–195.
17 Ashworth, A. et al. (1991) X-chromosome inactivation may explain the difference in viability of XO humans and mice. *Nature* **351**, 406–408.
18 Kay, G.F. et al. (1991) A candidate spermatogenesis gene on the mouse Y chromosome is homologous to ubiquitin-activating enzyme E1. *Nature* **354**, 486–489.

X-fragile chromosome syndrome *See* FRAGILE X SYNDROME.

X-linked agammaglobulinaemia (XLA) Very rare severe immune deficiency inherited as an X-LINKED recessive trait. Affected boys lack circulating B lymphocytes (B CELLS) and ANTIBODIES and are unusually prone to bacterial infections, especially of the respiratory tract and intestine. The disease becomes manifest at around six months of age as the levels of maternally derived immunoglobulins decrease. Treatment is by immunoglobulin replacement therapy. The disease gene is located on the proximal long arm of the X-chromosome and is thought to encode an activity involved in the development of B cells, as pre-B cells are present in the bone marrow (*see* B CELL DEVELOPMENT). A gene for XLA has been identified which apparently encodes a member of the *src* family of protein TYROSINE KINASES.

Vetrie, D. et al. (1993) *Nature* **361**, 226–233.

X-linked disorders Disorders caused by a mutant (usually recessive) allele located on an X-chromosome. X-linked diseases typically are manifest only in males whereas female heterozygotes are symptomless carriers. *See e.g.*: DUCHENNE MUSCULAR DYSTROPHY; HAEMOPHILIA A.

X-ray crystallography

ELUCIDATION of molecular structures by X-ray crystallography has contributed greatly to the transformation, over the past 70 years, of our understanding of the processes occurring in living cells. In particular, X-ray crystallography has revealed the structures of proteins and nucleic acids and their constitutent monomers, and also of drugs, hormones and vitamins. Knowledge of these three-dimensional structures has had a profound impact upon the whole of biology. The structural detail is at a higher resolution than is available by ELECTRON MICROSCOPY and allows atomic models to be built. (Fig. X3), for much larger molecules than can yet be studied by NMR spectroscopy.

In recent years, the successes of macromolecular crystallography have been even more dramatic, depending on technical advances which have reduced the timescale for solving the structure of a new protein, in favourable cases, to a few months, rather than the several years previously needed. The process of structure analysis comprises a sequence of operations, and there are now several standard procedures but, for macromolecules, the process is seldom routine. This entry will consider the theoretical basis of crystallography, the historical foundations upon which modern analyses depend, the recent innovations underpinning the rapid progress, the practical implementation of a crystal structure analysis, the assessment of the derived structural results, and finally, the future prospects of structure determination by this and other techniques. Further information will be found in [1–8].

Theoretical basis

X-RAY DIFFRACTION from individual molecules is far too weak to be measurable, whereas diffraction from a crystal concentrates the scattering into specific directions (REFLECTIONS) with zero scattering elsewhere. The signal is further enhanced by the number of molecules in the crystal — typically 10^{15} or more. The total diffracted intensity from organic molecules (wherein all atom types scatter similarly) depends on the volume of the crystal itself, whereas the number of reflections is directly proportional to the volume of the crystal UNIT CELL, the basic repeating unit of which the crystal is comprised. Thus, for crystals of a given size, the average intensity decreases as the unit cell size increases, so that, for small molecules, diffracted intensities are, on average, stronger and easier to measure than for macromolecules — from which they are both weaker and far more numerous.

X-rays are scattered by the electrons surrounding atoms and, in the absence of resonance effects such as ANOMALOUS SCATTERING, the scattering from an element of volume (dv) at position **r** scatters in proportion to the electron density at that point ρ (**r**).

This scattered wave has a phase shift, relative to that scattered from the origin, which depends critically on both the position vector **r** and the direction ($\hat{\mathbf{k}}_0$, $\hat{\mathbf{k}}$) of the incident and scattered wave vectors. This phase shift, which is the essence of diffraction, is given by

$$2\pi\mathbf{r}\cdot(\mathbf{k} - \mathbf{k_0})/\lambda = 2\pi\mathbf{r}\cdot\mathbf{S}$$

where

$$\mathbf{S} = (\mathbf{k} - \mathbf{k_0})/\lambda$$

and

$$|\mathbf{S}| = 2s = 2\sin\theta/\lambda$$

where **S** is the scattering vector at angle θ for X-rays of wavelength λ. These equations are closely related to BRAGG'S LAW.

The dependence of diffraction on structure is seen from this relationship between electron density ρ (**r**) at each position r and the phase shift $2\pi\mathbf{r}\cdot\mathbf{S}$. On adding the scattering from all the elements within an object of volume V, the total scattering is given by the integral

$$F(\mathbf{S}) = \int_V \rho\,(\mathbf{r})\exp2\pi i\,\mathbf{r}\cdot\mathbf{S}\,dv \qquad (1)$$

This integral is equivalent to a summation of the scattering distributions from all the atoms within the object. For every atom this sum includes an ATOMIC SCATTERING FACTOR (f_j) which has been tabulated for every atom type; the atomic position ($\mathbf{r}_j = x_j$, y_j, z_j for the jth atom); and a TEMPERATURE FACTOR (B_j) which represents the mobility of the atom.

$$F(\mathbf{S}) = \Sigma_j f_j \exp\,(-B_j s^2)\exp\,2\pi i\,(\mathbf{r}\cdot\mathbf{S})$$

The periodic character of a crystal makes the scattering factor F(**S**) a discrete function with maxima at integral values of **S**(h,k,l) where (h,k,l) can be considered as points in the RECIPROCAL LATTICE. This discrete function F(h,k,l), whose values are the STRUCTURE FACTORS, is given by

$$F(h,k,l) = \Sigma_j f_j \exp\,(-B_j s^2)\exp\,2\pi i\,(hx_j + ky_j + lz_j)$$
$$= |F(h,k,l)|\,\exp\,(i\varphi) \qquad (2)$$

where the sum is over all atoms in the unit cell. Structure factors are COMPLEX NUMBERS having both an amplitude $|F(h,k,l)|$ and a phase φ, whereas the actual measured quantity is the diffracted intensity I(h,k,l), which is proportional to the square of the STRUCTURE AMPLITUDE. The phase angle is thus lost in the measurement, giving rise to the PHASE PROBLEM of diffraction.

$$I(h,k,l) = k\,|F(h,k,l)|^2 \qquad (3)$$

The proportionality constant (k) between the intensity and the square of the structure amplitude depends on several calculable experimental factors: intensity of the incident X-ray beam; crystal volume; and geometry of the diffraction measurement. However, an empirical scale factor is usually used to place the data on the absolute scale.

Every atom j is thus seen, by the precise relationships of equations (2) and (3), to contribute to each diffracted intensity I(h,k,l); this contrasts with the situation in many spectroscopic

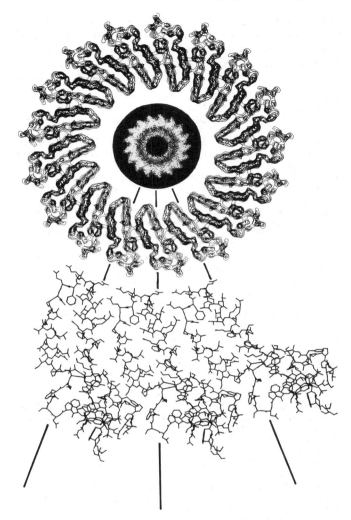

Fig. X3 Stages in the structure analysis of the protein disk of tobacco mosaic virus (TMV). The central portion shows an electron micrograph of the TMV disk after image processing to enhance the 17-fold rotational symmetry. Surrounding this is a section through the 5 Å electron density map, which shows two antiparallel α-helices, joined by a hairpin bend at the inner part of each protein molecule, and a more complex density towards the outer perimeter of the protein disk. The lower section shows the corresponding part of the atomic model built from the 2.8 Å electron density map. From [11].

techniques wherein specific observations can be separately assigned to specific structural features. In most cases, there are far more intensity observations than the number of atomic parameters (usually x, y, z, B for each atom). For crystals of small molecules the overdetermination (ratio of observations to structural parameters) is considerable, but for protein crystals this is reduced by the exponential decay with scattering angle, which results from atomic mobility and disorder within the crystal, which imposes an upper limit on the maximum RESOLUTION at which diffraction is observable. For diffraction measurements from most small-molecule crystals this limit is governed by the X-ray wavelength, but very few proteins diffract as well as this, and with their larger unit cells the number of intensities to be measured is very much greater.

The problem of crystal structure analysis is to recover the structure from the measured intensities. The nonlinearity of the equations, coupled with the distributed nature of the observations precludes the direct solution of the system of equations. However, once an initial trial model is known, these equations are a sound basis for refining the atomic parameters of the model. Recovery of the molecular structure, or electron density ρ (**r**), of the crystal unit cell from the structure factors F(*h*,*k*,*l*) is possible because (1) shows that the structure factors are the FOURIER TRANSFORM of the electron density. The Fourier transform function has the property that it can be mathematically inverted by means of an inverse transform; thus the electron density is also the Fourier transform of the set of structure factors:

$$\rho(\mathbf{r}) = \rho(x,y,z) = \Sigma_h \Sigma_k \Sigma_l \, F(h,k,l) \exp \{ -2\pi i(hx_j + ky_j + lz_j) \} \quad (4)$$

For a crystal, where the scattering vector **S** is discrete, rather than continuous, the inverse Fourier integral reduces to the threefold summation over the indices *h*, *k*, *l* (sometimes referred to as MILLER INDICES) shown in (4). The summation ranges are infinite in principle but, in practice, include only the sphere of observation where $s < s_{max}$, corresponding to a resolution limit of $d_{min} = 0.5/s_{max}$.

Calculation of the electron density from (4) requires the set of structure factors, whereas only the structure amplitudes can be measured. Recovery of the structure, from which the atomic model is derived by interpretation of the ELECTRON DENSITY MAP, thus reduces to the phase problem of crystallography, that is the evaluation of the phases, or initial estimates of them, which are not recorded in the diffraction pattern. Several methods of overcoming the phase problem are considered below. Derivation of an atomic model from the contoured electron density map of a macromolecule depends on a wealth of accumulated chemical detail, much of which was itself derived from crystal structure analyses of small molecules. The success of these earlier studies laid the foundations on which larger structural analyses all depend.

Historical foundations

Crystal structures of amino acids and nucleotides first revealed their detailed atomic structure and geometry. However, before the specialized field of macromolecular crystallography started to show structural detail, a variety of molecules of intrinsic biological interest in themselves, and not simply as units of a large linear polymer, were crystallized and their three-dimensional structures solved by X-ray analysis. Hormones, drugs and vitamins all come within this category and provide classic examples of the power of X-ray crystallography to determine complex biological molecules. Not until 1945 did the structure of cholesteryl iodide unequivocally display the connectivity of atoms in a steroid. Even the chemical formula of the β-lactam antibiotic PENICILLIN remained uncertain until its crystal structure was solved almost 20 years after Fleming's discovery of the drug. The formula of the cobalt-containing vitamin B_{12}, a planar molecule related to but distinct from the porphyrins, was determined from the electron density

map calculated in 1955 in the laboratory of Dorothy Crowfoot Hodgkin, who also played a leading role in the determination of the two structures mentioned previously.

Crystal structure analysis of individual amino acids revealed the geometry of these units and the peptide bond by which they are linked together in proteins. These results not only confirmed the planarity of the peptide linkage, as deduced from resonance theory, but also yielded accurate molecular dimensions, which in turn led Linus Pauling to predict that certain configurations permitting the formation of hydrogen bonds along the polypeptide chain would be especially stable. Two of these configurations, the right-handed α-helix and the β-sheet, are now recognized as universal features of PROTEIN STRUCTURE, having been found in most proteins whose conformation is known. Knowledge of these secondary structure motifs is now used to assist the interpretation of a macromolecular electron density map. It is usually also necessary to know the primary structure (i.e. the nucleotide or amino acid sequence) of the biopolymer. When sequences had to be determined from the protein itself rather than from the corresponding gene sequence the need for a sequence was a major limitation on structure determinations. When partial or no sequence data have been available, attempts to deduce this from the electron density map have had only moderate success.

Before the advent of modern computers, Fourier syntheses were calculated by hand and great reliance was placed on interpreting projections of a crystal structure. Suitable projections can, and did, give useful information about a small structure, but any projection of a macromolecular structure is almost entirely uninterpretable in terms of atomic structure. Thus, the successful determination of protein structures had to await the development of the early computers.

Recent innovations

Over the past 10 years or so, every step of a crystallographic analysis has been enormously facilitated by technical developments. DNA CLONING and protein expression systems have revolutionized our ability to obtain the molecule of interest in sufficient purity and quantity for crystallization. SITE-DIRECTED MUTAGENESIS enables molecules designed with a particular modification to be produced. Rapid DNA SEQUENCING reveals the primary structure of the polypeptide or polynucleotide. Protein crystals can usually now be frozen in liquid nitrogen while preserving their high-resolution detail, and thus becoming far less subject to RADIATION DAMAGE.

The X-ray beam from a SYNCHROTRON RADIATION SOURCE has an intensity several orders of magnitude above that yet attainable from a laboratory source, and allows selection of one or more particular wavelengths — whether to optimize the ANOMALOUS SCATTERING resonance effect or to minimize the absorption by, and thus damage caused to, the crystal. Use of a broad spectrum of wavelengths, rather than a monochromatic beam, again increases by several orders of magnitude the rate at which intensities can be measured in Laue diffraction. Electronic area detectors using high-speed, position-sensitive counters provide for simultaneous measurement of many reflections, just as were originally recorded on X-ray film, while IMAGE PLATE cameras with readout by laser scanning act as if a reusable film, without any wet processing or chemical fog.

Greater computing power and memory enable the storage before data reduction of many digitized diffraction patterns, for analysis which optimizes over the complete data set the fitting of reflection profiles in two or three dimensions, and the tracking of any parameter shifts during the experiment. Computer graphics workstations enormously facilitate the display of electron density maps and, with highly developed software routines, their interpretation into the coordinates of a folded polypeptide chain and the visualization of a complex three-dimensional structure, as shown by many figures in this volume (*see e.g.* Plates 3–7). The RESTRAINED refinement of a macromolecule is now routinely feasible but is still very demanding of computing resources. Finally, the growing power of structural databases and motif recognition algorithms facilitates our understanding of key aspects of molecular functions and relationships; with new structures now being determined at an ever-increasing rate, it is especially necessary to present effectively all the information arising from crystallographic analyses.

Practical implementation of a crystal structure analysis

Crystallization

This is often the rate-limiting stage of the whole analysis. For a new molecule, many trials of different conditions are usually needed. CRYSTALLIZATION techniques and practice still remain something of an art but, increasingly, rational approaches to the problem are being successfully applied. Where material is scarce, trials are conducted on a micro scale using only 1 µg of protein for each condition, but a successful crystal structure determination usually requires 10 mg or more of the macromolecule. When crystals cannot be grown from an intact molecule, they can often be grown from a fragment of it, say by deleting a terminal peptide whose conformation is disordered, or from a single domain: for example, myosin heads do form crystals whereas intact molecules give fibres; Fab fragments of antibodies form crystals far more readily than does the Fc fragment with its flexible linkage to two Fab domains.

The first crystals obtained may be too small to diffract well or too disordered to diffract at high enough resolution to provide an atomic model. Crystal quality can often be improved by exploring conditions close to those which gave the first crystals, by variation of pH, choice of precipitant (e.g. $(NH_4)_2SO_4$, PO_4^-, polyethylene glycol, MPD), buffer, temperature and protein concentration. Sometimes it may be necessary to add a specific ligand, or improve the homogeneity of the protein sample. Crystallization experiments in microgravity conditions inside a space shuttle have claimed some success, but unwanted convection can be greatly reduced by the use of gels for slowing the process of crystallization within an earthbound laboratory.

Unlike crystals of small molecules, those of proteins or nucleic acids rapidly dry out, becoming badly disordered unless drying is prevented. Protein crystals are usually mounted inside a thin-walled glass capillary together with a drop of their mother liquor. The capillary is then mounted directly on a camera. However, the crystal may tend to slip and both the glass and the liquid also contribute to the overall diffraction. These factors may be reduced by means of specially flattened or tapered capillaries, or mounting in agarose gels, and by minimizing the volume of liquid in the capillary. Recently, the use of frozen crystals has allowed the capillary to be dispensed with, as the crystal is surrounded by a film of cryo-solvent held in a loop of fine fibre.

Characterization

Characterization of new crystals involves determining both their UNIT CELL and crystallographic POINT GROUP (from which is learnt the SYMMETRY of the molecular packing within the crystal) and also the resolution limit to which diffraction intensities can be measured, which itself depends on the regularity of the packing and the extent of any static or dynamic disorder within the crystals. Occasionally, crystals are found to be twinned, with interpenetrating packing arrangements. The presence of either significant disorder or twinning, or both, normally prevents further analysis, and so crystallization of another crystal form is necessary. Full SPACE GROUP determination was traditionally done by measuring the intensities of entire planes or ZONES within reciprocal space using a precession camera, but the long exposures this requires — coupled with its inefficiency of measurement for a given dose of irradiation — have led many workers now to defer this step until after complete data acquisition to obtain the maximum data from what might be a single precious crystal.

Data measurement

Measurement of all diffraction intensities out to the desired resolution is subject to various constraints. The required accuracy of measurement depends on the use to be made of the intensities: errors in data from a native protein will affect the final atomic model; errors in intensities from a derivative, or from its corresponding native protein, determine the precision of any isomorphous or anomalous differences to be used in phase determination, which imposes stringent requirements on the accuracy if the difference is to have useful phasing power. Multiple measurements, of two or more asymmetric units of the diffraction pattern, increase the precision of the merged data but, if the crystals suffer severe radiation damage, this places an upper limit on the time available for observing the diffraction patterns and thus on the precision attainable. Data measured from different crystals may not merge as well as batches of data from the same crystal. The point group of the crystal symmetry determines which fraction of the spherical volume in reciprocal space contains the necessary minimum of one asymmetric unit. The unit cell dimensions determine how many diffracted intensities are included within this volume, to a given resolution. The sizes of the crystal and of the X-ray beam, together with the angular factors of crystal mosaicity and X-ray collimation and wavelength dispersion all influence the proximity of neighbouring intensities and thus the extent to which they begin to overlap at higher resolution. The aperture of the detector to be used, and its distance from the crystal, determine the number of diffraction orders measurable simultaneously and their separation distance. All these considerations require that the optimum experimental strategy must be separately designed in each case.

With a laboratory X-ray source and copper tube, the wavelength is fixed at 1.54 Å (CuK$_\alpha$), although for small-molecule crystals a molybdenum tube is advantageously used, giving a wavelength of 0.71 Å (MoK$_\alpha$). Electronic area detectors have fast readouts but relatively small apertures, while the newer image plate cameras require a few minutes to scan each image, but have much larger apertures allowing images of up to 2000 pixels in both dimensions. This is similar to earlier photographic images obtained on X-ray film, which has higher background noise levels. Images record that part of the three-dimensional diffraction pattern obtained by successive small rotations of the mounted crystal about a known camera axis (Fig. X4*a*). To obtain a complete dataset, it is often necessary to collect some further images by rotating about an axis parallel to a different direction within the crystal. A typical dataset might include a total angular rotation of between 30° and 180°, subdivided into successive images each of angular width between 0.1° (for an electronic detector) and 1.0° (for image plate or film), requiring a total exposure time of between several hours and a few days. Some intensities are fully recorded on a single image, others are partially recorded on two or more successive images.

With monochromatic synchrotron radiation the exposure time for each image may be as short as a few seconds, although the next image must be delayed by a few minutes when using an image plate to enable the scanner to read out the previous image. Even faster data collection can be achieved through use of Laue diffraction with a spectrum of wavelengths and a stationary crystal (Fig. X4*b*); in favourable cases a complete diffraction pattern can be recorded as a single image. This permits time-resolved experiments for the study of enzyme kinetics. Optimized anomalous dispersion data can provide phase information when the crystal contains atoms with measurable ANOMALOUS SCATTERING: at least three complete datasets, each collected at a carefully chosen wavelength, are required. The anomalous scattering from sulphur atoms is sufficient only for very small proteins; normally the presence of a metal atom bound to the molecule is necessary.

Data reduction

Data reduction from a set of images eventually produces a set of reflections with indices (h, k, l) reduced to the asymmetric unit, and each having an observed intensity (I) with the standard deviation of that measurement (σ_I). The process involves several successive steps which are becoming increasingly automated: determination or verification of the crystal orientation and unit cell parameters; also of the beam and detector positions and characteristics; this enables indexing of all the reflections followed by refinement of all parameters; estimation of the X-ray background between the Bragg reflections; determination from the stronger intensities of the profiles of all diffraction spots; integration of pixels contributing to each reflection, by direct summation or by profile-fitting, with due allowance for pixel variances in each case, to give a background-subtracted intensity. If this is

done separately for each image, the output datafile may contain both fully recorded and partially recorded reflections. When images are assembled so that successive slices through a diffraction maxima can be analysed together, peak profiles can be fitted in three dimensions, with narrow angular images, or the pixels of two or more wide slices are simply added so that intensities partially recorded on each slice can then be treated as if fully recorded on the overall slice.

Intensities from each image must be placed upon a common scale. The relative scale factors are rarely constant but can be refined to allow empirically for: changes in the beam intensity or illuminated volume of the crystal; change of detector sensitivity; use of different crystals, radiation damage and increasing overall temperature factor, differential absorption of the primary and diffracted beams as they pass through the crystal in different directions; wavelength-dependent factors in the case of Laue diffraction. Scaling parameters are refined by LEAST SQUARES minimization of the residual differences.

Finally, the scaled data are merged and analysed statistically. Within a dataset of many thousands of intensities there may inevitably be a few outliers, which can safely be rejected at this point. However, if there are many deviant measurements, there is something seriously wrong with the dataset. Careful study of the statistical analyses should provide clues to what has gone wrong. The internal consistency of multiple measurements of the same I(h,k,l), is monitored by the R-FACTOR, a residual index traditionally used by crystallographers. The adequacy of a dataset also depends on its percentage completeness and the cumulative distribution of intensities against their standard deviations.

Subsequent stages in a structure determination use, not the intensities, but the structure factors related to them as in (3). Any intensity that has been determined as being a negative quantity (by experimental error after background substration and other corrections) thus requires special consideration before taking its square root. The usual procedure is to truncate the intensity distribution, paying heed to the standard deviation of each of the measured intensities, to constrain all intensities to be of positive definitive.

Phase determination

For small molecules, phase determination almost invariably relies on the so-called DIRECT METHODS based on statistical relationships between intensities, or occasionally on the heavy-atom method, but these are not generally applicable to macromolecules. Any heavy atom, or cluster of atoms, has a scattering factor which is too small a fraction of the total scattering from the protein to indicate the phase angle for the whole molecule. However, an extension of this method known as isomorphous replacement is very widely used. This requires datasets from crystals of both the native protein and the protein plus one or more atoms whose atomic coordinates are known. Figure X5 shows how these two sets of amplitudes determine the protein phase except for a twofold ambiguity, which is resolved by use of further heavy-atom derivatives (MULTIPLE ISOMORPHOUS REPLACEMENT), the anomalous scattering contribution of the heavy atom, or both of these. Graphical methods were used originally to

a

b

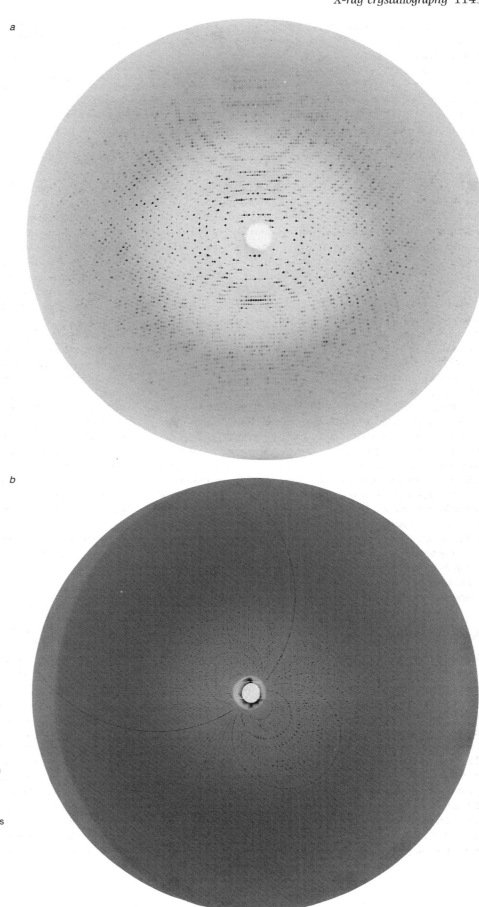

Fig. X4 X-ray diffraction patterns from a crystal of the disk of tobacco mosaic virus (TMV). This complex of 34 molecules of the viral coat protein (molecular mass 600 000 daltons) forms orthorhombic crystals with unit cell dimensions a = 228 Å, b = 224 Å, and c = 174 Å. *a*, Rotation photograph obtained with monochromatic X-rays (λ = 0.88 Å) by rotating the crystal through 0.6° about a horizontal symmetry axis. *b*, Laue diffraction pattern obtained with a broad wavelength spectrum from a stationary crystal with its symmetry axes tilted about both the horizontal and vertical axes of the image by approximately 10°.

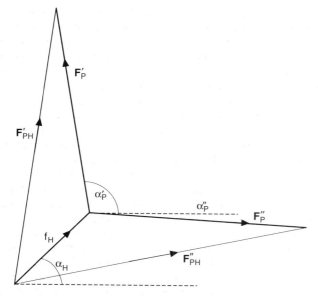

Fig. X5 The phase triangle as used in the isomorphous replacement method. A heavy atom occupying a known position in the unit cell has a calculated scattering vector (f_H) with phase angle α_H. The observed scattering amplitudes of the native protein (F_P) and the protein plus heavy atom (F_{PH}) allow only two possible values for the protein phases α'_P and α''_P.

determine the most probable phase, α_P, but computers are now invariably used, and due allowance is made for experimental errors in the observed amplitudes, in order to determine the best phase, which is the centroid of the probability distribution for all possible phases. The best phase can be shown to be that producing minimum error in an electron density map. The limitation on this method is the preparation of suitable heavy-atom isomorphous derivatives, which often proves as difficult as obtaining the crystals of the molecule in the first place, or lack of isomorphism between the native and derivative crystals such that phases can be calculated only at low resolution, often 5 Å or below.

The large number of relatively weak intensities from a protein crystal prevents the widespread use of direct methods of phase determination as employed routinely for small molecules. Methods utilizing phase relationships between three or more reflections have not yet been used successfully for protein phases *ab initio*, except in a few very favourable special cases, but they are applicable for extending phases to higher resolution. A related method utilizes restraints imposed on the phases whenever there is more than one copy of a molecule within the asymmetric unit. This commonly occurs with those many proteins which exist as oligomers and so crystallize with several identical molecules in an asymmetric unit. If the relative position and orientation of the multiple copies of a molecule are known, and these molecules are often related by NON-CRYSTALLOGRAPHIC SYMMETRY, an initial set of phases can be improved by an iterative procedure: first averaging the electron density map over multiple copies of the molecules, then back-transforming the averaged map to obtain calculated amplitudes and phases, and lastly using these phases with the observed amplitudes to produce an improved map. This cyclic refinement procedure proved invaluable in the crystallo-

graphic analyses of viruses and has produced reliable phases starting from the approximate phases determined from a single isomorphous derivative.

Although *ab initio* phase determination by direct methods has succeeded in only a few special cases, there is now considerable interest in developing the application of this more widely through rigorous testing of possible phases, ranked according to their LIKELIHOOD, and constraining maps to fit, in a MAXIMUM ENTROPY sense, the statistical expectations of density distribution and all other known properties of macromolecular electron density maps such as positivity and atomicity, solvent content within the crystal — where the fitting is optimized. The early analysis of intensity distributions gave rise to WILSON STATISTICS, in which the atoms are assumed both to be all of equal scattering power and also to be uniformly distributed throughout the unit cell. Extensions of this theory are being developed to make allowance for the density expected in solvent regions, which will be lower than the mean electron density of the protein volume.

A third class of methods, known as MOLECULAR REPLACEMENT, involves searching the PATTERSON MAP of a crystal (representing a map of all the interatomic vectors, with the cluster of short vectors around the origin including both intra- and intermolecular vectors), for features matching those of a known molecular fragment or a closely related molecule. The atomic coordinates of this are used to calculate the Patterson map expected when the known molecule is placed in a large cell, where all short Patterson vectors will be intramolecular. The ROTATION FUNCTION expresses the overlap between the observed and calculated Patterson maps, at all possible relative orientations. It has a maximum at the orientation representing that rotation of the known fragment necessary to bring it into the same orientation as it has in the new crystal. Having first oriented the known fragment, it must be positioned within the unit cell of the crystal. This can be done either by use of a TRANSLATION FUNCTION or by searching all possible positions and monitoring the R-factor between the observed structure amplitudes and those calculated for the model at each trial position. When the known molecule can be correctly oriented and placed within the unit cell, a set of structure factors can be calculated as in (2). Combining these phases with the observed structure amplitudes from the crystal allows an initial map of the crystal to be calculated.

A fourth method of phase determination is applicable only when the new crystal is isomorphous to one of known structure. The known phases can be used directly with the new amplitudes to calculate a map using amplitudes F_{obs} from the new crystal and phases α_{calc} from the known structure, but the difference from the parent structure will be shown more clearly by a difference Fourier where the amplitudes are ($F_{obs} - F_{calc}$). This method can be seen to be a special case of the molecular replacement method, wherein the molecular orientation and translation are the same in both the known and unknown crystals, which also have the same unit cell. It is a very useful method for studying the binding of a series of ligands which can be diffused into crystals without disrupting the molecular packing.

The quality of a set of phases is assessed by the nonuniformity of the phase probability distributions. The figure of merit (*m*) represents the distance of the centroid of this distribution from its origin; for each reflection (*hkl*) a vector to the centroid, when expressed in polar coordinates, has amplitude (*m*) and phase angle (α_{best}) representing the best phase of that structure factor $\mathbf{F}_{hkl}$. Phase angles can generally have any value between 0 and 2π radians (or 0 and 360°) and a uniform distribution of phase angles is to be expected over the reflections within a dataset. There may be classes of reflections, however, for which the phases are constrained by reason of the crystal SYMMETRY. For any centro-symmetric projection of the unit cell, the phases of reflections in the ZONE normal to the projection axis (i.e. a two-, four-, or six-fold rotation axis with another such axis orthogonal to it) are constrained to be either 0 or π if the centre of symmetry is at the origin, else to be $\pi/2$ or $3\pi/2$, with no other values being allowed. These centric reflections clearly have a bimodal phase probability distribution quite distinct from that of the general acentric reflections. Centric reflections usually have a higher figure of merit for their phases than do acentric ones; their amplitudes also have different distributions as is reflected by higher expectation value of the R-factor for centric reflections (*see:* WILSON STATISTICS).

Improving the trial phases

Improving the trial phases may be quite unnecessary if the first electron density map computed has clear indications of a continuous backbone with side chains protruding in specific directions at regular intervals along it. However, as increasingly difficult problems are being studied, the first map is often not readily interpretable. This may result from many factors (e.g. disordered crystals, poorly isomorphous derivatives, imprecise anomalous differences, or use of a molecular replacement model which is not sufficiently similar to the unknown molecule), and is usually indicated by the quality of the phasing statistics. In practice, the improvement is usually achieved by means of density modification algorithms in real space whilst the effectiveness of the process is assessed by monitoring in reciprocal space the improvement in phasing statistics.

If the molecular envelope is visible, from regions of lower density solvent separating contiguous higher density regions of macromolecule, where the relative volumes of these two regions can be defined by the molecular weight per asymmetric unit, then the solvent density can be assumed to be featureless and an additional constraint placed upon the phases by the process of SOLVENT FLATTENING. Whenever there are two or more copies per asymmetric unit of a molecule or subunit, and these can be assumed to have the same detailed structure (which is likely in all parts of the molecule except those involved in intermolecular interactions) then these subunit densities can be constrained to be their average values by algorithms which impose non-crystallographic symmetry upon the electron density map. This averaging will directly enhance the interpretability of the map and, when this map is Fourier transformed to give the calculated structure amplitudes and phases, it will also give symmetry-averaged phases which can be combined in an iterative procedure with the initial phase angles to give improved phases. Sometimes a fragment of the molecule may be known, either from another molecule, correctly oriented and positioned in the new unit cell, or from interpretation of part of the new structure;

in such cases the atoms of the molecular fragment are used to calculate partial structure factors. Comparison of the amplitudes of these with the observed structure amplitudes, relative to the size ratios of the fragment and the intact molecule, indicates how much reliance can be placed upon the phase angles from this partial fragment when combined with the initial phases.

All of these sources of additional phasing power yield combined phases which, when used in a Fourier transform with the observed amplitudes, will give an improved electron density map. Special measures are necessary to avoid the risks of bias arising from inappropriate phase combinations which might perpetuate an incorrectly recognised fragment, a real departure from symmetry within an oligomer, or a wrongly assigned solvent boundary; all of these errors would degrade the final model, but would normally be detected during the refinement process.

Accurate phases are the key to a clearly interpretable map: classical studies have shown that errors in phases have a far more deleterious effect than errors in amplitudes: for example, an electron density map of a small organic molecule computed from calculated amplitudes with random phases is uninterpretable whereas one from randomised amplitudes with calculated phases is indeed far from correct but it does allow the atomic positions and a fragment of the molecule to be identified. With macromolecules, where data rarely extend to atomic resolution, it is often more efficient to incorporate all prior knowledge of the structure at this stage in order to facilitate the subsequent step of interpretation and fitting of a model — but judgement is necessary to avoid over-reliance upon any trial phases which are incapable of sufficient improvement by means of these statistical methods.

Fitting a model to a map

This is the most exciting stage of the whole structure determination, when one first sees details of the molecular conformation and hints about its mode of action. Techniques have evolved rapidly such that the earlier physical models are now rarely used — whether wire components and screws or push-fit plastic parts. Three-dimensional interactive computer graphics systems facilitate every stage: tracing the backbone chain of continuous unbranched polymers, and marking regions of ambiguity; placing C_α atoms at the appropriate separation along this backbone and close to the branching density of sidechains; recognizing secondary structural elements such as α-helices and β-sheets which can be built directly before their side chains are added later when the primary sequence and the density backbone have been placed correctly in register with each other; making connecting loops between these elements which may be clearly visible or highly disordered.

These connecting loops can now increasingly be built by examination of a library of pentapeptide fragments: fitting the first residue onto the last of those residues already built and selecting, from the library of those conformations known to exist in other structures, the best fit to the density map. This step is repeated at each position along the loop, with the overlap giving confirmation of the choice made or indication of other alternatives. The complete model may be built in this way or, if there is too much uncertainty in interpreting the electron density map, the partial model fragment may be used for phase improvement as described above — resulting in a new map wherein parts of the still unknown fragment should now have become clearer. The iteration cycle of these steps as necessary may also include refinement of the model, at its present level of completeness.

Model refinement

Refining the model ensures its optimal fit to all of the available data, which include both the experimentally derived structure amplitudes (the X-ray term) and the accumulated stereochemical knowledge about bond lengths and angles, planar groups and torsion angles, together with the stabilizing effects on a molecule of hydrogen bonds and other non-covalent interactions (the empirical energy term). REFINEMENT OF MACROMOLECULAR STRUCTURES is achieved by a simultaneous minimization of the residuals in these two terms, appropriately weighted according to the resolution and precision of the X-ray data. The major variable parameters are the positional coordinates (x_j, y_j, z_j) of every atom (j), and the corresponding temperature factors (B_j). If the energy term dominates the overall residual, the refinement is essentially rigidly constrained to ideal stereochemistry irrespective of its agreement with the measured intensities, while if the X-ray term is too dominant the model will have bad stereochemistry and the refinement may not converge, unless it is at very high resolution. With intermediate weighting, between these two extremes, the refined model is restrained to have good stereochemistry while also fitting the X-ray data as well as is possible.

Refinement algorithms are becoming increasingly powerful but can only move the model towards a local minimum in the energy function, not a global minimum, unless they include some means of overcoming the barriers separating these minima. This is achieved computationally by means of SIMULATED ANNEALING coupled with MOLECULAR DYNAMICS; the manual equivalent is to rebuild part of the model, which may anyway still be necessary. It is usual to monitor progress (in reciprocal space) of the refinement by close inspection (in real space) of the current model superimposed both upon the difference map (calculated by means of a Fourier transform using amplitude coefficients $(F_{obs} - F_{calc})$ and phases α_{calc}) and also upon the observed density (computed from F_{obs} and α_{calc}) in order to check whether further manual intervention is needed to rebuild any parts of the model. The use of calculated phases in computing the observed map could introduce bias into this map. Any unwanted effects of this are checked by use of an 'omit map' for each region of the asymmetric unit in turn: all atoms in that region are excluded from the summation of (2) such that they do not contribute to the phase angle α_{calc}. The density in this region of the structure is now independent of the current model for the region.

At the end of a successful refinement, the difference map will show no significant density and the omit map will agree with the final atomic model. This often includes not only the atoms of the protein itself but also many molecules of bound water, wherever there is evidence of hydrogen bonding from the solvent atoms to the protein. If too many solvent oxygen atoms are included, the refinement R-FACTOR can become inappropriately low, but this

can be checked for by monitoring also the free R-factor as a means of cross-validation of any apparent improvement to the model.

Description of the structure

The structure can be described mathematically by a list of parameters (as now routinely deposited in the Brookhaven Protein Databank, PDB), pictorially by a series of diagrams, and verbally by analysing the salient features. All these have advantages for particular purposes, but can also mislead if taken in isolation or in an unintended context. Crystallographic structures are normally each depicted as having one precise conformation, in contrast to a structure determined by NMR SPECTROSCOPY where the normal representation is of a galaxy of possible conformations. Here the identity of ill-defined molecular regions is emphasized; this information is also available from a crystal structure, in the atomic temperature factors — whose significance, however, is less obvious to anyone looking only at a schematic diagram of the molecule. Contacts between molecules within the crystal lattice may also influence the conformation of certain regions of the molecular surface, but the residues involved in these interactions are rarely marked, either in depictions of the molecule or in lists of atomic coordinates, once these become separated from the crystallographic details of unit cell and symmetry, whence the interactions can be deduced. A detailed drawing of a crystallographically determined structure may thus not contain the whole truth; whether or not it also contains nothing but the truth has to be assessed in the light of all the evidence.

Assessing a crystal structure

Published pictures of macromolecules determined from X-ray analysis of crystals present an air of certainty, but their validity must be checked before detailed interpretations are made therefrom: an erroneous picture might well look equally convincing. The early structures were not fully refined, by which process errors of chain tracing, primary sequence or torsion angles would normally be detected, whereas new structures now are both invariably refined and also subjected to an increasing number of consistency checks against the existing structural database. The main criteria for judging any crystal structure are its resolution and extent of refinement.

The quoted nominal resolution of a crystal structure analysis is the minimum Bragg spacing $(d_{min} = \frac{1}{2}s_{max})$ included in the dataset; the effective resolution may be less than this if mobility or disorder reduce the extent of measurable diffraction. The accuracy of locating atomic positions is far higher though than the limit of point resolution, such that at 2 Å resolution standard deviations in atomic positions are of the order of 0.2 Å, while neighbouring atoms are linked by covalent bonds with standard deviations of the order of 0.02 Å in bond lengths. The R-factor of a well-refined structure is usually below 0.20; the free R-factor will be higher than this, but if too high could be indicative of overfitting or misfitting of the model.

The extent of any atomic mobility can be judged by the

β-values, or temperature factors, which are often higher for surface residues than for those in the core of a molecule. Surface residues involved in crystal lattice contacts may have β-values far lower than those of neighbouring residues; these low values will not be appropriate for an isolated molecule in solution. Some molecules undergo conformational changes at a certain pH, or in the presence of particular ligands, such that the conditions under which the crystals were grown become relevant to which structure is being observed. Caution is thus needed to avoid inappropriate use of some details, but other aspects of macromolecular crystal structures have provided an enormous wealth of information leading directly to major biological insights: the future promises still greater achievements for the techniques of crystallography.

Future prospects

The scope of biological X-ray crystallography already covers a wide spectrum from detailed chemical mechanisms of catalysis to the architecture and assembly pathway of a complete virus particle. In conjunction with other techniques, especially electron microscopy, the structural results available have provided many of the foundations of contemporary molecular biology.

Future prospects include the use of even more powerful synchrotron radiation sources now being developed, and very intense microfocus X-ray sources for individual laboratories. Many more proteins will be successfully crystallized for the first time from quite small sample preparations; other crystals will be improved to diffract at yet higher resolutions. All subsequent steps of a crystal structure determination will become easier for non-expert practitioners, as programs include more internal diagnostic tests. Small protein structures will be refined to a degree comparable with that already achieved for small molecules, while lower resolution studies with less well-ordered specimens continue to give useful information about the structure and dynamics of a wide variety of systems of biological importance. Advances in electron microscopy and diffraction techniques for two-dimensional crystals will yield atomic models for many molecules unable to form three-dimensional crystals. Ever smaller crystals will yield to X-ray analysis. The range of molecular sizes accessible to NMR spectroscopy will steadily increase to give more atomic structures determined by the two techniques — NMR and X-rays — but often studied in different environments.

The avalanche of new structures will continue to provide fresh insights into the relationships between molecules, and greater understanding of the dependence of function on the underlying structures, whether these are static or dynamic. These ideas will more effectively guide the design of modified molecules, for applications in medicine, agriculture and many industrial processes and products.

A.C. BLOOMER

1 Glusker, J.P. (Ed.) (1981) *Structural Crystallography in Chemistry and Biology*. Benchmark Papers in Physical Chemistry and Chemical Physics, Vol. 4 (Hutchinson Ross, Stroudsberg, PA).
2 Blundell, T.L. & Johnson, L.N. (Eds) (1976) *Protein Crystallography* (Academic Press, London).
3 Heliwell, J.R. (1992) *Macromolecular Crystallography* (Cambridge University Press).
4. Wyckhoff, H.W., Hirs, C.H.W. & Timasheff, S.N. (1985) Diffraction Methods for Biological Macromolecules, *Meth. Enzymol.*, Vols 114 & 115 (Academic Press, New York).
5 Hahn, T. (1992) *International Tables for X-ray Crystallography, Vol. A*, Space-group Symmetry (Kluwer Academic, Dordrecht).
6 Shmueli, U. (1993) *International Tables for X-ray Crystallography, Vol. B*, Reciprocal Space (Kluwer Academic, Dordrecht).
7 Wilson, A.J.C. (1992) *International Tables for X-ray Crystallography, Vol. C*, Mathematical, Physical and Chemical Tables (Kluwer Academic, Dordrecht).
8 Branden C. & Tooze, J. (1991) Introduction to Protein Structure (Garland, New York).
9 Bloomer, A.C. et al. (1978) *Nature* **276**, 362–368.

X-ray diffraction X-ray wavelengths (10^{-2}–10 nm) are similar to the dimensions of biological molecules and the separation of their constituent atoms. Thus X-ray analysis has been used extensively for the elucidation of molecular structures, contributing greatly to our understanding of biochemical reactions. X-ray microscopy has recently become possible with the development of X-ray lenses, but only for the longer X-ray wavelengths. The power of X-ray analysis for studying macromolecules has been through the study not of images but of DIFFRACTION patterns.

X-rays are scattered by the electrons surrounding an atom, such that the strength of diffraction varies with atomic number — making hydrogen atoms difficult to detect using X-rays (in contrast to their ease of detection using NEUTRON SCATTERING AND DIFFRACTION). Absorption of X-rays by biological specimens is low, so that a specimen thickness of well over 1 μm is required for X-ray scattering to be significant: thus even with macromolecules, two-dimensional crystals (e.g. bilayers of lipids and proteins) scatter too weakly for study by X-rays, although they can be studied by ELECTRON DIFFRACTION.

The intensities of an X-ray diffraction pattern can be measured, but not the associated phases necessary for computation of an image. These must be determined by ancillary means; this being the PHASE PROBLEM inherent in crystallography and related techniques.

Globular proteins and other molecules which form three-dimensional crystals allow a direct determination of their structure by X-RAY CRYSTALLOGRAPHY. Such structural analyses, requiring measurement of a three-dimensional diffraction pattern, are potentially at atomic resolution, if the crystals are sufficiently well ordered, although the largest structures are often determined in several stages of increasing resolution. Fibrous molecules do not crystallize but may form ordered fibres naturally, or be induced to do so, for study by FIBRE DIFFRACTION analysis where the observed diffraction pattern is rotationally averaged about the axis of the fibre. This gives two-dimensional data which allow an indirect approach to structure determination at medium resolution by model-building studies to compare the observed pattern with that expected from a proposed trial structure, and was the method used to determine the double-helical structure of DNA. Complexes which are neither crystalline nor form ordered fibres may be studied in solution by SMALL-ANGLE SCATTERING. Here the diffraction pattern is spherically averaged and so this method can only give one-dimensional data — and hence low resolution structural detail — the basic shape and dimensions of a particle.

XANES X-ray absorption near edge structure. *See*: EXAFS.

***xantha* (*xan*) mutants** A group of yellow barley mutants thought to be affected in CHLOROPHYLL biosynthesis.

Xanthomonas A bacterial genus belonging to the family Pseudomonadaceae which is closely related to the genus PSEUDOMONAS. Xanthomonads are yellow pigmented, aerobic, motile, rod-shaped bacteria, and all are plant pathogens. Five species are currently recognized, of which one, *X. campestris*, is divided into more than 100 pathovars which are distinguishable solely on the basis of plant host range. Many xanthomonads produce high molecular weight acidic polysaccharides which, in addition to their importance as virulence factors (*see* PLANT PATHOLOGY), are widely used in industry (xanthan gum).

Bradbury, J.F. (1984) *Xanthomonas*. In *Bergey's Manual of Systematic Bacteriology* (Krieg, N.R. & Holt, J.G., Eds) 199–210 (Williams and Wilkins, Baltimore).

xanthoxin 5-[1*S*,2*R*-epoxy-4*S*-hydroxy-2,6,6-trimethylcyclohexyl]-3-methyl-2*Z*,4*E*-pentadienal (Fig. X6). M_r 250.3. A biosynthetic precursor of the plant hormone ABSCISIC ACID, to which it is converted in two oxidative steps. It is derived by oxidative cleavage of the CAROTENOID 9'-*cis*-neoxanthin. Its formation is thought to be a regulatory step for abscisic acid biosynthesis. *See also*: PLANT HORMONES.

Parry, A.D. & Horgan, R. (1991) *Physiol. Plant.* **82**, 320–326.

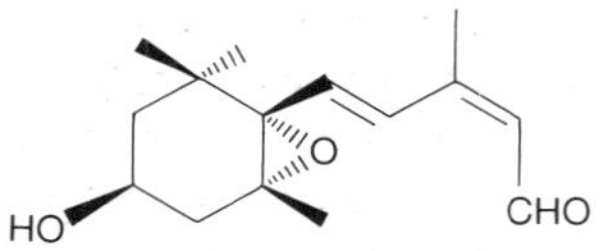

Fig. X6 Structure of xanthoxin.

xenogeneic Pertaining to another species. Applied, for example, to ANTIBODIES raised in one species by immunization with an antigen from another. Such antibodies are also termed heterologous.

xenograft A graft of tissue from one species to another.

Xenopus laevis The African clawed toad or frog (Order Anura (Salientia), family Pipidae), much used in experimental developmental biology because of the robustness of its eggs and embryos, and the relative ease of experimental surgical manipulations on the developing embryo. Its large OOCYTES are also used for the expression of injected RNAs and thus the expression cloning of genes (*see* XENOPUS OOCYTE EXPRESSION SYSTEMS). *See also*: AMPHIBIAN DEVELOPMENT; INDUCTION; PATTERN FORMATION.

Xenopus oocyte expression system

OOCYTES are unfertilized eggs which can, in principle, perform all the biochemical steps required to produce a fully functional protein — from translation to insertion of protein into plasma membranes — when injected with messenger RNA. In particular, the large oocytes of the clawed frog *Xenopus laevis* are convenient to work with and extremely efficient at performing this role for 'foreign' mRNAs introduced into them, and have become indispensable in studies of the molecular basis of function of proteins of many origins. They have been particularly useful in the isolation and study of ion channels and neurotransmitter receptors from the nervous system, but even plant and viral proteins can be expressed in this system. Oocytes can also be used to study expression and processing of secretory proteins.

Xenopus oocytes are non-uniform in structure with two distinct regions — the yellowish vegetal hemisphere and the dark brown animal hemisphere (*see* AMPHIBIAN DEVELOPMENT) — and *in situ* are surrounded by several layers of tissue (Fig. X7). Microvilli and ENDOPLASMIC RETICULUM are more abundant in the animal hemisphere and this is where the oocyte's phosphatidylinositol SECOND MESSENGER system is located. Of some importance in evaluating the oocyte's endogenous membrane properties [1], the follicle cells also possess receptors and ion channels and can contribute to the oocyte's overall electrophysiological responses. The endogenous characteristics of oocytes show considerable seasonal and batch variability and need to be monitored periodically during expression studies.

Injection of mRNAs

There are numerous protocols for the extraction of RNA, depend-

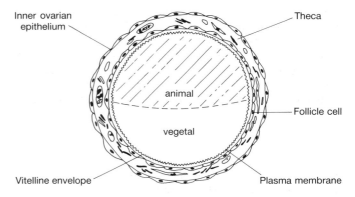

Fig. X7 Schematic diagram of morphological features of a stage V/VI oocyte of *Xenopus laevis* and surrounding tissues. The follicular oocyte is surrounded by a number of cellular and noncellular layers. The vitelline membrane is a noncellular fibrous layer which remains after collagenase treatment. This is surrounded by a monolayer of some 10^4 follicle cells with macrovilli forming gap junctions with the microvilli of the oocyte. The theca is a fibrous connective tissue layer containing blood vessels, nerve fibres, smooth muscle, and fibroblasts. The inner ovarian epithelium is a continuation of the ovary wall. From [14].

ing on the origin of the biological material. An extraction procedure for whole tissue homogenates is the guanidinium thiocyanate method which allows a wide range of poly(A^+) mRNA sizes to be selected and purified on an oligo(dT)-cellulose column. However, with the increasing demand for the simplification of extraction and purification procedures, there are now many commercial kits which facilitate the process and give relatively good yields. Following isolation, the mRNA is usually fractionated, most commonly on a sucrose density gradient (*see* DENSITY GRADIENT CENTRIFUGATION). This gives information on the relative size of the mRNA encoding the protein (also indicating whether subunits encoded by different sizes of mRNA are involved) and can increase expression by concentrating the relevant mRNA. Each oocyte is injected into the cytoplasm with 40–60 nl of poly(A^+) mRNA or cRNA (mRNA derived from complementary DNA) at a concentration of $\sim 1\,\mu g\,\mu l^{-1}$. Alternatively, cDNA (10 nl at a concentration of $0.1\,\mu g\,\mu l^{-1}$) may be injected directly into the nucleus. Injected oocytes are incubated at constant temperature (20°C) under sterile conditions whereupon expression can occur over 24 h to >10 days.

Detection of expressed proteins

Proteins expressed in *Xenopus* oocytes may be detected and characterized in several different ways depending on the nature of the protein. Electrophysiological recording can be used if the expressed protein can generate an electrical signal, as in the case of receptors, ion channels, and electrogenic ion pumps and transporters [2]. Two main techniques are available:

1 Two-electrode VOLTAGE CLAMP. This is the most widely used means of recording from oocytes, whereby one electrode records the MEMBRANE POTENTIAL while the other injects current. The large size of the oocyte (>1 mm diameter) facilitates impalement with two electrodes although the inherently large capacitance ($\sim$300 nF) limits the clamping speed.

2 Patch clamp. This has been used mostly in structure–function studies of expressed proteins, especially receptors or ion channels, whereby information about single channel kinetics can be provided (*see* PATCH CLAMP).

Alternative techniques for the screening of expressed proteins rely on radiotracers or binding of labelled ligands or toxins (e.g. α-BUNGAROTOXIN to NICOTINIC RECEPTORS; OUABAIN to the α-subunit of NA$^+$,K$^+$-ATPASE). When the expressed protein is an enzyme, biochemical assays of membrane homogenates may be used. A screening technique of particular relevance to phosphatidylinositol-linked receptors expressed in *Xenopus* oocytes involves fluorometric measurement of intracellularly released Ca^{2+}.

Some recent expression studies

The application of RECOMBINANT DNA TECHNOLOGY, including *in vitro* SITE-DIRECTED MUTAGENESIS studies, has resulted in many studies characterizing the function of specific domains or subunits of protein expressed in *Xenopus* oocytes. Some representative examples of the different strategies adopted and the types of proteins expressed are summarized below.

Voltage-dependent ion channels

Ion channels from various tissues have been expressed and studied after injection of mRNAs or cRNAs (Fig. X8). Cloning of the voltage-sensitive Na$^+$ channel (*see* VOLTAGE-DEPENDENT ION CHANNELS) suggested that it contains four homologous repeats, each consisting of six putative transmembrane segments. It was proposed from the largely positive residue composition of one of these segments (S4) that this area is involved in the depolarization-induced activation of the channel (i.e. this is the site of voltage sensitivity). A similar situation probably exists in voltage-dependent K$^+$ and Ca^{2+} channels. To test this hypothesis, point mutations were introduced into the region of S4, thereby either neutralizing the positive residues or replacing them with negative ones [3]. On expressing the mutant clone in *Xenopus* oocytes, the net effect of reducing the positive charges in S4 was indeed a reduction in the voltage dependence of activation. Deletions at the linkage between repeats III and IV reduced the rate of inactivation of the channel, implying that this region is important for channel closure. The ionic selectivity 'filter' of the channels seems to be due to the presence of clusters of negatively charged amino acid residues in short segments deep in the channel protein between S5 and S6 in each of the four repeats [3].

Neurotransmitter receptors coupled directly to ion channels

Differential expression of subunits. Neurotransmitter receptors coupled directly to ion channels are essentially ligand-gated ion

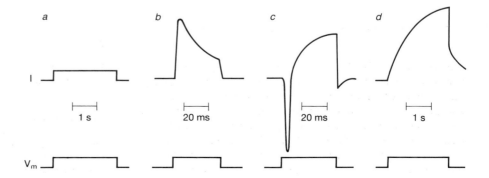

Fig. X8 Qualitative illustration of different types of voltage-dependent ion channel activity expressed in *Xenopus* oocytes by injection of mRNA from different tissues. *a*, The ohmic current of an uninjected (control) oocyte. *b–d*, Membrane currents with different kinetics recorded from different oocytes injected with (*b*) intermediate-sized mRNA from brain; *c*, large-sized mRNA from brain; *d*, small-sized mRNA from uterus, for comparison. I, Membrane current; V_m, membrane voltage. From [15].

channels. The functional role of the various subunits has been investigated by injection of cRNAs encoding one or more of the subunits of a specific receptor/channel. The GABA$_A$ receptor, for example, is composed of up to four different polypeptide subunits (α, β, γ, δ) some of which have several variants (*see* GABA AND GLYCINE RECEPTORS). Expression of either α or β separately in oocytes results in the formation of a functional channel with characteristics similar to those of the native receptor [3]. This suggests that both α and β subunits can substitute for each other to form a functional homo-oligomeric channel complex, although receptors formed by the combination of α and β subunits desensitized more rapidly, showed greater outward rectification and displayed smaller single-channel conductances than those assembled from α and γ subunits.

GABA$_A$ receptors are also the target for various chemicals including barbiturates, benzodiazepines (e.g. diazepam), steroid anaesthetics, and alcohol. Benzodiazepine enhancement of GABA responses was obtained with either of the γ subunit variants, γ2S and γ2L, whereas barbiturate (e.g. pentobarbital) action did not require any specific subunit. Ethanol enhancement of GABA action was seen only if the expressed receptor contained the γ2L subunit (in addition to γ and β subunits). On the other hand, the ethanol sensitivity of the GABA$_A$ receptor complex depended not only on the presence of the γ2L subunit but also on the phosphorylation state of this subunit.

In contrast to the GABA$_A$ receptor, work with muscle nicotinic acetylcholine receptors (nAChRs) indicates that of the four subunits (α, β, γ, δ) most are necessary to form a functional channel. Only when α and δ were separately omitted did the receptor still show some, albeit weak, activity. *Xenopus* oocytes have also played a crucial part in expanding our knowledge of EXCITATORY AMINO ACID RECEPTOR subtypes. Through a combination of molecular cloning and expression studies, at least two families of subunits for the non-NMDA class of glutamate receptor have been identified [4]:

1 GluR1–GluR4 which code for AMPA-sensitive glutamate receptors with four transmembrane domains and may exist in the central nervous system as heteromers or homomers.

2 GluR5–Glu7 and KA1, which have only $\sim$30% amino-acid sequence identity relative to the GluR1–4 genes, and code for kainate-preferring receptors. NMDA receptor genes have been isolated by 'expression cloning' in *Xenopus* oocytes (see below), and it has been possible to gain an insight into the contribution of individual subunits to channel properties.

Co-translation of mRNAs from different sources. The functional role of different subunits in receptor/ion channels can be studied by co-translating mRNAs from different tissues where the analogous proteins have different properties. Expression of mouse–*Torpedo* 'hybrid' nicotinic acetylcholine receptors in *Xenopus* oocytes by injecting a mixture of muscle mRNAs encoding each subunit, enabled a detailed characterization of subunit contribution to overall properties of the receptor. By comparing the channel conductances of different combinations, the γ and δ subunits were found to influence channel open time, while the β subunit shortened it. Voltage dependency was found to reside in the β and δ subunits.

Cat muscle nAChRs desensitize more slowly than analogous receptors from *Torpedo* electric organ when each is expressed separately in *Xenopus* oocytes. To investigate the molecular basis of this difference, hybrid cat–*Torpedo* muscle nAChRs were expressed, and the difference tracked down to the γ subunit. On the other hand, in neuronal nAChRs, β subunits were found to determine the time course of desensitization (Fig. X9) [5].

Invertebrate neurotransmitter receptors. There are considerable pharmacological differences in receptors to the same endogenous ligand in vertebrates and invertebrates. Determination of the molecular bases of this dissimilarity would be particularly useful to the pharmaceutical and agrochemical industries in designing novel and more specific chemical pesticides and insecticides.

Amino-acid receptors have been expressed in *Xenopus* oocytes from muscle mRNA of the locust (*Schistocerca gregaria*) [6]. Several differences were found between these amino-acid receptors and analogous ones from vertebrates. In particular, although the locust GABA receptor gated Cl$^-$ and was inhibited by picrotoxin,

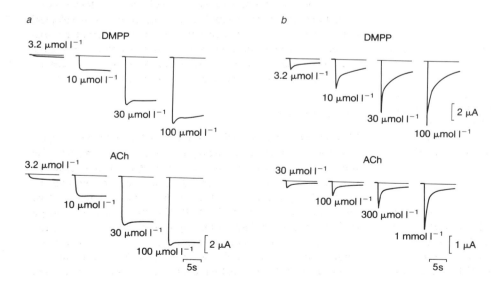

Fig. X9 Subunit dependence of desensitization characteristics of rat neuronal nicotinic ACh receptors expressed in *Xenopus* oocytes. *a*, Effect of 1,1-dimethyl-4- piperazine (DMPP) and acetylcholine (ACh) on receptors composed of $\alpha_3\beta_4$ subunits. *b*, Effect of DMPP and ACh on receptors composed of $\alpha_3\beta_2$ subunits. Individual membrane currents elicited by DMPP (top row) and ACh (bottom row) on voltage-clamped oocytes are shown (holding potential = – 80 mV). Modified from [5].

it was insensitive to both baclofen and bicuculline, indicating that it is distinct from the vertebrate GABA$_A$ and GABA$_B$ receptors. A GABA receptor subunit from the mollusc *Lymnaea stagnalis* has recently been sequenced. The polypeptide has ~50% sequence homology to the vertebrate GABA$_A$ receptor β subunit and when expressed forms functional GABA-activated Cl$^-$ channels which are blocked by bicuculline. A glutamate receptor from *Drosophila melanogaster* has also been cloned. The subunit, designated DGlu-II, shows 37–38% homology to the rat glutamate receptor subunits GluR1, 4 and 5. Invertebrate nAChR receptors have also been studied (*see* NICOTINIC RECEPTORS).

Receptors linked to second messengers

An early example of an expression study using the intracellular signalling mechanisms of *Xenopus* oocytes was the cloning of the acetylcholine MUSCARINIC RECEPTOR (mAChR) and its expression in oocytes from rat brain mRNA [7]. The role of the inositol phosphate second messenger system in receptor activity was shown by pre-labelling the cells with [^{3}H] inositol and then monitoring the increased formation of labelled inositol 1,4,5-trisphosphate (InsP$_3$) following a brief exposure of oocytes to acetylcholine. Additionally, direct injection of InsP$_3$ into oocytes elicited oscillatory currents with identical onset, amplitude, and waveform to those induced by acetylcholine. Incubation of mRNA-injected oocytes with PERTUSSIS TOXIN suppressed the response, indicating that a GTP-BINDING PROTEIN was also involved. The clone had a 30% sequence identity with the β-ADRENERGIC RECEPTOR, one of the first indications of the now large family of G PROTEIN-COUPLED RECEPTORS.

Melatonin receptors have been expressed in *Xenopus* oocytes from mRNA from pars tuberalis (PT) of the sheep pituitary, and have been shown to *inhibit* second messenger turnover [8]. A functional melatonin receptor was produced which seemed to be coupled via G proteins to a signal transduction system with the characteristics of an InsP$_3$/Ca^{2+} second messenger system. This shows a lack of fidelity in G-protein coupling as in native PT cells melatonin receptors couple to the second messenger cyclic AMP.

Gap junction proteins

Xenopus oocytes can also be used to investigate gap junction proteins (connexins) (*see* CELL JUNCTIONS). Pairs of oocytes injected with *Xenopus* embryonic cRNA and placed side by side became coupled by voltage-sensitive gap junctions [9]. Oocyte pairs constructed with one oocyte expressing the voltage-independent heart gap junction and the other expressing the endogenous voltage-dependent connexon [10], suggest that the gating of the two half-junctions are independent of each other.

Ion pump subunits

The ion pump Na$^+$,K$^+$-ATPase consists of two subunits — α and β. The larger α subunit (M_r 84 000–120 000) is thought to be catalytic, while the smaller β subunit (M_r 40 000–50 000) is a glycoprotein probably involved in the transport and/or the correct assembly of the α subunit to form the complete ATPase complex.

A functional ATPase can be obtained in *Xenopus* oocytes only by co-injection of both α and β subunit mRNAs. In some cases, the introduced α subunit can couple with an endogenous β subunit or with the β subunit from H$^+$,K$^+$-ATPase. Both α and β subunits occur in several isoforms and screening of human genomic libraries with α subunit probes has revealed a family of homologous genes. The various α and β subunit forms are expressed in a tissue-specific manner suggesting functional specialization of the isoforms with the corresponding Na$^+$,K$^+$-ATPases having differing sensitivity to cardiac glycosides such as ouabain. A possible basis for differences in ouabain sensitivity has been provided by expression in *Xenopus* oocytes of a ouabain-resistant receptor from mRNA with a mutation of a cysteine residue deep in the transmembrane M1 region [11].

Neurotransmitter uptake systems

An important feature of the control of synaptic transmission is the re-uptake of neurotransmitter by the presynaptic neuron. The *Xenopus* oocyte is a particularly good system for pharmacological characterization of neurotransmitter transporters because of: (1) the presence of a large transmembrane Na$^+$ gradient; (2) their large size, which results in large easily detectable neurotransmitter accumulation; and (3) lack of significant endogenous uptake.

The importance of understanding neurotransmitter uptake systems is reflected in the variety expressed to date, including those for dopamine, noradrenaline, serotonin (5HT), glutamate, and GABA. Transporters expressed in oocytes generally possess all the physiological attributes displayed *in vivo*. Such fidelity of translation in the oocyte has impelled detailed study of the transporters. Examples of such work include the effect of specific inhibitors such as cocaine on the dopamine transporter or the age and region-dependent distribution of brain transporters [12].

Both the GABA transporter and cocaine-sensitive dopamine transporter have recently been cloned in *Xenopus* oocytes [12]. The cDNA clones (designated GAT-1 and DAT1 respectively) encode proteins with 67% sequence homology, 12 putative transmembrane segments, and four potential glycosylation sites. Both proteins are thought to exist as single subunits, but they may be a multimer of identical subunits. Evidence from pharmacological data and gene cloning indicates that the neurotransmitter transporters as a whole constitute a MULTIGENE FAMILY. They show little homology to other non-neurotransmitter transporters, GABA and glycine receptors, or anion exchangers/antiporters.

Expression cloning

This cloning method uses the 'universal' ability of *Xenopus* oocytes as a translation system to identify the gene encoding a particular protein. It was first used to clone a subtype of the serotonin receptor — the 5HT$_{1c}$ receptor [13]. Since then, many different neuroactive proteins have been isolated and identified. Two approaches may be used (Fig. X10). The 'hybridization' strategy enables cloning of large transcripts that cannot be inserted correctly into phage vectors, but relies on nonexpression of clones which may occur naturally owing to the fickleness of different oocyte batches. Thus, a further 'control' RNA batch is required to

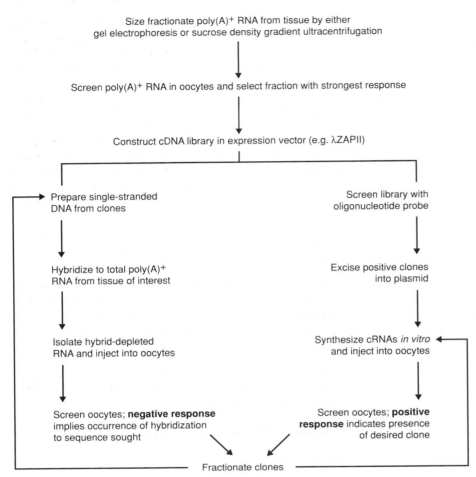

Size fractionate poly(A)+ RNA from tissue by either gel electrophoresis or sucrose density gradient ultracentrifugation

Screen poly(A)+ RNA in oocytes and select fraction with strongest response

Construct cDNA library in expression vector (e.g. λZAPII)

Prepare single-stranded DNA from clones

Screen library with oligonucleotide probe

Hybridize to total poly(A)+ RNA from tissue of interest

Excise positive clones into plasmid

Isolate hybrid-depleted RNA and inject into oocytes

Synthesize cRNAs *in vitro* and inject into oocytes

Screen oocytes; **negative response** implies occurrence of hybridization to sequence sought

Screen oocytes; **positive response** indicates presence of desired clone

Fractionate clones

Fig. X10 Two different strategies for expression cloning in *Xenopus* oocytes.

monitor expression in the oocytes. The *in vitro* RNA synthesis method has the advantage of not requiring large quantities of RNA and identifies only a functional, full length cloned sequence. But it does require some previous information on the sequence of the protein expected, in order to construct the oligonucleotide probe initially used for screening the library.

Expression cloning has successfully identified and isolated the genes encoding the various glutamate receptor subtypes, as well as those for peptide receptors and ion channels. However, expression cloning does not guarantee success. It is a lengthy procedure and *Xenopus* oocytes do not express all proteins with the same efficiency. A further problem arises when proteins are comprised of multiple subunits. Although a 'functional' protein may be expressed from one subunit, the additional subunits may have regulatory roles and confer additional properties on the protein. It would be a much more arduous task to expression clone several proteins simultaneously and the successful clonings to date have been for proteins which are functional in homomeric form. Nevertheless, expression cloning has frequently succeeded where more conventional approaches have failed.

S.P. FRASER

M.B.A. DJAMGOZ

1 Fraser, S.P. & Djamgoz, M.B.A. (1992) In *Current Aspects of the Neurosciences* (Osborne, N.N., Ed.) 267–315 (Macmillan, Basingstoke).

2 Fraser, S.P. et al. (1993) Electrophysiology of *Xenopus* oocytes: An expression system in molecular neurobiology. In *Electrophysiology, A Practical Approach* (Wallis, D.I., Ed.) 65–86 (Oxford University Press).

3 Stühmer, W. (1991) Structure–function studies of voltage-gated ion channels. *Annu. Rev. Biophys. Chem.* **20**, 65–78.

4 Hennebery, R.C. (1992) Cloning of the genes for excitatory amno acid receptors. *BioEssays* **14**, 465–471.

5 Cachelin, A.B. & Jaggi, R. (1991) β Subunits determine the time course of desensitization in rat α3 neuronal nicotinic acetylcholine receptors. *Eur. J. Physiol.* **419**, 579–582.

6 Fraser, S.P. et al. (1990) Amino acid receptors from insect muscle: electrophysiological characterization in *Xenopus* oocytes following expression by injections of mRNA. *Mol. Brain Res.* **8**, 331–341.

7 Kubo, T. et al. (1986) Cloning, sequencing and expression of complementary DNA encoding the muscarinic acetylcholine receptor. *Nature* **323**, 411–416.

8 Fraser, S.P. et al. (1991) Melatonin receptor mRNA expression in *Xenopus* oocytes: Inhibition of G-protein activated response. *Neurosci. Lett.* **124**, 242–245.

9 Ebihara, L. et al. (1989) Cloning and expression of a *Xenopus* embryonic gap junction protein. *Science* **243**, 1194–1195.

10 Swenson, K.I. et al. (1989) Formation of gap junctions by expression of connnexins in *Xenopus* oocyte pairs. *Cell* **57**, 145–155.

11 Canessa, C.M. et al. (1992) Mutation of a cystein in the first transmembrane segment of Na,K-ATPase α subunit confers ouabain resistance. *EMBO J.* **11**, 1681–1687.

12 Amara, S.G. & Kuhar, M.J. (1993) Neurotransmitter transporters: recent progress. *Annu. Rev. Neurosci.* **16**, 73–93.

13 Lubbert, H. et al. (1987) cDNA cloning of a serotonin 5-HT$_{1c}$ receptor by electrophysiological assays of mRNA-injected *Xenopus* oocytes. *Proc. Natl. Acad. Sci. USA* **84**, 4332–4336.

14 Snutch, T.P. (1988) The use of *Xenopus* oocytes to probe synaptic transmission. *Trends Neurosci.* **11**, 250–256.

15 Levitan, I.B. & Kaczmarek, L.K. (1991) *The Neuron, Cell and Molecular Biology* (Oxford University Press).

xeroderma pigmentosum (XP) AUTOSOMAL RECESSIVE trait with an incidence of one in 100 000 births. Affected individuals are abnormally sensitive to sunlight and show an increased incidence of skin carcinomas and corneal scarring. The basic defect is an inability to repair UV-induced damage to DNA owing to the loss or mutation of one of the several genes involved in DNA REPAIR. PRENATAL DIAGNOSIS is possible.

Xg^a blood group The Xg^a antigen is encoded by an X-linked gene, located at Xp22.3-7pter. Gene frequencies, XG^a 0.659 and XG 0.341, made XG the most used marker in linkage studies of X-borne conditions and in investigations of patients with sex-chromosome aneuploidy before the introduction of DNA PROBES. The Xg^a antigen is red cell specific, carried by a sialoglycoprotein of M_r ~22 000–29 000. Anti-Xg^a is not clinically significant. A genetical relationship is postulated between XG and the gene controlling the red cell 12E7 quantitative polymorphism. 12E7 antigen, a 32 500 M_r sialoglycoprotein encoded by a PSEUDO-AUTOSOMAL LOCUS *MIC2*, is not polymorphic on other tissues.

XLA X-LINKED AGAMMAGLOBULINAEMIA.

XO Condition in which one X-chromosome is absent in somatic cells of phenotypic females (also symbolized as 45,X). *See*: TURNER SYNDROME.

XP XERODERMA PIGMENTOSUM.

XPLOR Widely used computer software for the REFINEMENT OF MACROMOLECULAR STRUCTURES which uses an approach known as SIMULATED ANNEALING.

XXY trisomy *See*: KLINEFELTER'S SYNDROME.

xylans Heterogeneous group of GLYCANS of bacteria, algae and land plants which contain D-xylose as at least the major sugar: most are heteropolymers. Xylans of red algae contain irregular sequences of D-xylopyranosyl residues in β1,3 and β1,4 linkage, while a homopolymeric β1,3 xylan is known in the green alga *Caulerpa filiformis*. Little is known of their biosynthesis, though UDP-Xyl is the putative xylosyl donor: several transferases may be involved. Xylans of terrestrial plants are linear polymers of β1,4-linked D-xylopyranosyl residues, variously and irregularly (but not randomly) substituted by short side chains containing L-arabinose, D-galactose, D-xylose and 4-OMeD-glucuronic acid. The xylosyl residues may have acetyl substituents at C2 or C3.

The main xylan chains wind into triple helices from which the side chains project. The nature of the side chains is species-restricted and their synthesis must involve several glycosyl transferases and sugar nucleotides. UDP-Xyl and a single transferase appear to suffice to produce the main chain. Methyl ether groups are derived from *S*-adenosylmethionine and are added after polymerization. Nothing is known of initiation, though a GLYCO-PROTEIN primer is the likely initiator. Xylans are a major constituent of HEMICELLULOSE.

Aspinall, G.O. (1970) *Polysaccharides*, 103–115 (Pergamon Press, Oxford).

Stoddart, R.W. (1984) *The Biosynthesis of Polysaccharides* 236–239 (Croom-Helm, London).

xylem The chief water-conducting tissue of vascular plants. It consists of vessels and/or tracheids which are filled with water when functional, and fibres, PARENCHYMA and ray cells. It also serves a supporting function, especially the secondary xylem which is the wood of dicotyledons and conifers.

xylene cyanol FF Dark blue/black dye of M_r 554.6 (Fig. X11), which is used as a molecular weight marker in both agarose and polyacrylamide gel ELECTROPHORESIS.

xylogenesis Formation and maturation of xylem.

xylose A pentose SUGAR.

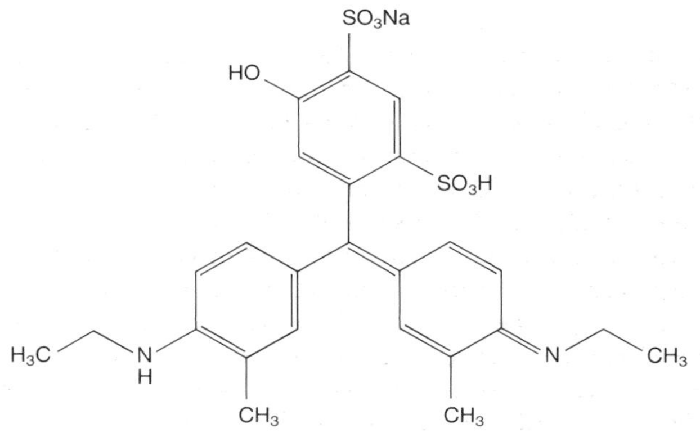

Fig. X11 Xylene cyanol FF.

Y

Y The single-letter abbreviation for the AMINO ACID tyrosine.

Y-box factors DNA-binding proteins, for example YB-1 from human cells, which bind to the CCAAT-containing Y-box in human MHC class II genes.

Y-chromosome The smaller of the two sex chromosomes in mammals. It is carried only by males, who have one X- and one Y-chromosome in their somatic cells. It bears little HOMOLOGY to the X-chromosome and carries few genes. It carries genes essential for male sex determination, including the testis-determining gene. *See*: SEX DETERMINATION; Y LINKAGE.

Y linkage Specific pattern of genetic transmission observed for gene loci located on the Y sex chromosome. Such loci are only found in males and show male-to-male transmission only. Genes on the Y-chromosome do not have an allelic copy within the cell, so there are no DOMINANT or RECESSIVE Y-linked traits. However, loci at the tip of the short arm of the Y-chromosome do have homologous copies on the short arm of the X-chromosome and behave as autosomal traits. This PSEUDOAUTOSOMAL REGION is involved in the pairing of the sex chromosomes during MEIOSIS. Y-linked traits are passed directly from father to son; an affected father will have affected sons and normal daughters. There are no clinically important Y-linked disorders. An example of a Y-linked trait is that of hairy ear rims, a phenomenon common in parts of India.

YAC YEAST ARTIFICIAL CHROMOSOME.

yeast Member of a unicellular group of ascomycete fungi. *See*: SACCHAROMYCES CEREVISIAE; SCHIZOSACCHAROMYCES POMBE; YEAST MATING-TYPE LOCUS.

yeast artificial chromosome (YAC) A vector constructed from components of yeast chromosomes that permit its maintenance and replication *in vivo*. These include a centromere, telomeres, and origins of replication (ARS ELEMENTS) (*see* CHROMOSOME STRUCTURE). Very long lengths of DNA (up to 1000 kb) can be cloned and maintained in a YAC (*see* DNA LIBRARIES).

yeast development *See*: SACCHAROMYCES CEREVISIAE; SCHIZO-SACCHAROMYCES POMBE.

yeast mating factors Protein pheromones produced by complementary mating types of the yeasts *Saccharomyces cerevisiae* and *Schizosaccaromyces pombe*, and which induce fusion of gametes of opposite mating types. They are denoted **a** and α in *S. cerevisiae*,

and P and M in *Schiz. pombe*. *See also*: YEAST MATING-TYPE LOCUS.

Yeast mating-type locus

SEXUAL differentiation in yeast is directed by the mating-type genes. The existence of two complementary mating types allows haploid cells of opposite mating type to form a diploid zygote by conjugation, a process equivalent to gamete fusion. A third cell type, heterozygous for both mating types, allows MEIOSIS to occur before sporulation, thereby regenerating the haploid state. So-called homothallic strains of the yeasts *Saccharomyces cerevisiae* and *Schizosaccharomyces pombe* are self-fertile by switching back and forth from one mating type to the other in a clonal haploid cell population. They do so by DNA rearrangements in which information from either of two silent loci is mobilized and moved to the mating-type locus where it is expressed [1–5].

Mating-type information in yeasts is studied for at least three reasons:

1 As an example of the directed transfer of DNA from one genetic locus to another.

2 As an example of differential gene expression and silencing mediated by additional sequences lying outside the mating-type loci.

3 As an example of regulatory cascades of differentiation which are controlled by the mating-type genes and by additional environmental cues.

Yeasts are unicellular fungi. They fill an ecological niche of disconnected, carbohydrate-rich microenvironments. From the diploid state, and in response to nutritional depletion (especially for nitrogen), they sporulate after meiosis — usually by formation of four-spored asci. The spores are dormant and more resistant than vegetative cells, and can germinate when transferred to a new medium that supports growth. As they are haploid, zygotes must be formed before a new round of spores can be generated.

The two species of yeast best analysed genetically are the prototype budding yeast *S. cerevisiae* and the fission yeast *Schiz. pombe*. Both yeasts tend to be self-fertile (homothallic) when isolated from nature, although heterothallic strains, which are stable for mating type, are usually preferred for laboratory experiments. The ability of homothallic strains to switch mating types allows zygote formation and sporulation within a clonal culture derived from a single haploid spore.

Generally speaking, the two yeasts (which are not closely related) employ quite comparable systems for mating-type control and interconversion. But at a more detailed level, the mechanistic

Table Y1 Nomenclature and definition of terms

	Saccharomyces cerevisiae	*Schizosaccharomyces pombe*
Gene symbols and conventions	Three-letter capitals for dominant wild-type alleles	Three-letter lower-case symbols throughout
Mating types	**a**, α	*P* (plus), *M* (minus)
Active mating-type locus (chromosome)	*MAT* (III, of 16)	*mat1* (II, of 3)
Silent storage loci (distance from active locus)	*HML*-α (180 kb) *HMR*-**a** (120 kb)	*mat2-P* (15 kb) *mat3-M* (30 kb)
Cassette length, differential core sequence	**a**: 642 bp α: 747 bp	*P*: 1113 bp *M*: 1127 bp
Homology boxes for initiation and resolution	*Z* (239 bp) *X* (704 bp)	*H1* (59 bp) *H2* (135 bp)
Mating-type functions	*MCM1* (= *PRTF*, *GRM*), a transcription factor interacting with *MAT* proteins	*Pc* (118 aa): early *P* function, specifies *P* cells
	MCM1 alone activates **a**-specific promoters	*Pm* (= *Pi*) (159 aa; pheromone-inducible): late *P* function for induction of *mei3* and meiosis
	MCM1 + α1 activates α-specific promoters	*Mc* (181 aa): early *M* function, specifies *M* cells
	MCM1 + α2 represses **a**-specific promoters	*Mm* (= *Mi*) (42 aa): late *M* function for induction of *mei3* and meiosis
	α1 repressed by **a**1-α2	*pat1* (= *ran1*): protein kinase suppressing all sexual functions during growth, modulated during conjugation by unknown factors, inactivated by *mei3* before meiosis
	a1-α2 represses haploid-specific promoters	
	a2 no function	
	RME1 (repressed by **a**1-α2) represses sporulation-specific promoters	

and regulatory differences become so profound that a common origin of homothallism (if it ever existed) has become much disguised by divergent evolution. As the conventional jargon and gene symbols are quite different for the two yeasts, species-specific terms have been avoided throughout this text and are listed in Table Y1. The life cycles, too, are differently organized (Fig. Y1). The comparable features of both yeasts will first be described, and then the specific differences.

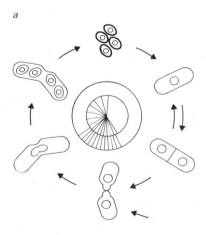

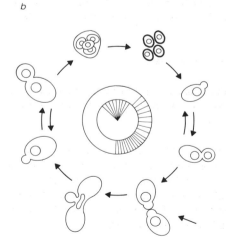

Fig. Y1 Life cycles of fission (*a*) and budding (*b*) yeast. *a*, In the haplontic life cycle of fission yeast, pheromone production and response are induced by nutritional depletion, and meiosis occurs in zygotes. The natural life cycle is haploid throughout, limiting the diploid phase to the zygotic fusion nucleus, which enters meiosis immediately. Gamete differentiation is coupled to nutritional depletion (the same conditions that also further sporulation). Diploid cell lines in fission yeast can be selected in the laboratory. Mating-type switching continues in diploid cells, even if these are heterozygous for mating type, and the switching pattern is directed by each target cassette individually. *b*, In the haplo-diplontic life cycle of budding yeast, pheromone production and response are possible throughout the haploid phase, and only meiosis in diploid cells is induced by starvation. Hatched sectors: nitrogen starvation (inner circle), which is crucial during the induction phase, but no longer necessary during spore formation; pheromone production and response (outer circle).

Common features

Cells of opposite mating type carry DNA sequences of different informational content (mating-type cassettes) at the active mating-type locus (see Fig. Y2). In addition, complementary mating-type cassettes are located at two transcriptionally silent storage loci, from which copies can be retrieved for transposition in the course of mating-type interconversion.

The three cassette loci are located on the same chromosome, all oriented in the same direction. The cassette at the active mating-type locus gives rise to two divergent transcripts starting from an internal PROMOTER region. The same promoters are also present at the storage loci, but they are kept inactive by external SILENCER signals flanking these loci. The gene products of the mating-type cassettes are regulatory proteins, which cooperate with additional TRANSCRIPTION FACTORS and direct the processes of both zygote formation (conjugation) and meiosis.

If there is only one active mating-type locus, as in haploid cells, this can result in differentiation of gamete-like cells, able to fuse with one another. The gametes communicate by secreting their own mating pheromone (Table Y1) and responding to the complementary one. These factors resemble peptide hormones of higher organisms, and are encoded elsewhere in the genome. Meiosis requires simultaneous activity in the same cell of both the different mating-type cassettes — the usual hallmark of the diploid state.

Mating-type switching

The switching process is a unidirectional GENE CONVERSION involving a limited segment of nonhomologous DNA, where both initiation and resolution of the RECOMBINATION event are guided by flanking boxes of homologous DNA, present at all three cassette loci (Fig. Y2). Additional external sequences (of still unknown significance, and omitted in the figure) are shared by only two of the three loci.

The initiation site at the target cassette (the active mating-type locus) is activated by a double-strand cut, close to the boundary between the core segment and the adjacent homology box. The exterior 3'-end may then act as a primer for copy synthesis, using the appropriate silent cassette as a template. The newly synthesized strand carrying switched mating-type information is rejoined to the target locus within the limits of the other homology box (the resolution end), and synthesis in the opposite direction, now using the extra strand as a template, finishes the switching event. The donor cassette remains unchanged during the process, but the target cassette has received new information — usually for the opposite mating type.

Mating-type switching occurs shortly before or during S-phase of chromosomal DNA replication in the CELL CYCLE. The cuts that initiate switching are not generated at random but follow a tightly controlled pattern in cell division pedigrees. Equilibrium between the two mating types is approached after just a few rounds of cell division.

Fission yeast

The fission yeast cell cycle (Fig. Y1a) has a well-defined mitotic phase and leads to the formation of two equivalent sister cells.

Only one of two sister cells receives a newly switched mating-type cassette at each switching event. Conjugation of sister cells is, in fact, the most frequently observed way of forming an ascus in homothallic *Schiz. pombe* strains. The pattern of mating-type switching is composed of two successive stem cell lineages for each mating type. These lineages result from different states of chromosomal imprinting at individual cassettes rather than from physiological asymmetry at the cellular level.

The potential target cassettes exist in two states: virgin (unswitchable) and switchable. When an unswitchable parent cassette is replicated, only one of the daughter cassettes will be switchable. When a switchable cassette is replicated, only one of the daughter cassettes is switched (and becomes virgin again); the other one will still be switchable. If the non-switched daughter cassette is fully conserved from the parent cell without being used as a template in any strand, no DNA sequence needs to be discarded in return for the insertion of a new cassette.

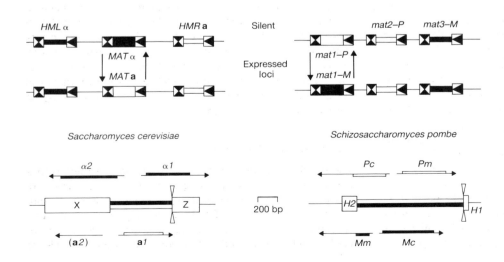

Saccharomyces cerevisiae

Schizosaccharomyces pombe

200 bp

Fig. Y2 Comparative configurations of mating-type genes in budding and fission yeast. *a*, Order and orientation of the various cassettes involved in mating-type switching (not drawn to scale). The constant but silent storage loci are indicated above the diagram, the expressed and switching cassettes between the alternate configurations. This drawing emphasises the two sets of flanking homology boxes common to all three loci in each yeast. *b*, Approximate dimensions of the expressed mating-type cassettes, including the divergent transcripts and corresponding translation frames for both mating types. The location of the switch-associated double-stranded break is also indicated at the initiation site. Mating-type switching is stopped by a/α heterozygosis in *S. cerevisiae*, but remains unaffected by *P/M* heterozygosis in *Schiz. pombe*.

The switchable state of target cassettes is correlated with double-strand breaks at the initiation site. These breaks persist throughout most of the cell cycle and are not lethal, even in the absence of donor cassettes. The ends are protected by bound proteins and appear to be sealed before replication of unswitched cassettes. Such features are reminiscent of specialized TOPO-ISOMERASE action. The essential region conferring target site specificity extends remarkably far away from the break point, some 200 base pairs (bp) into the cassette and outside as well. A restriction-type endonuclease for this process has not been found and may not in fact be present.

Switch-deficient mutants have been isolated in some 10 different *swi* genes. By and large they fall in three classes, affecting the generation of the double-strand break, its utilization for switch initiation, or the final resolution of switching intermediates. Mutants of the third class, in particular, are prone to faulty rearrangements by aberrant resolution events. All the *swi* mutants are somewhat leaky, and only triple mutants spanning all three classes have a really tight phenotype. A few are pleiotropic, also affecting general recombination and/or radiation sensitivity. In addition, two double mutants of the early class are lethal, defining a more general, essential function. All this is indicative of a highly integrated mechanism making use of the various *swi* gene products. Interestingly, the common partner in both lethal double-mutants has recently been shown to be a single missense mutation for DNA polymerase [6], further strengthening the functional link between mating-type switching and DNA replication in fission yeast.

All three loci carrying mating-type cassettes are closely linked in *Schiz. pombe*. Some 15 kilobases (kb) of DNA containing at least one essential gene separate the active mating-type locus from the nearest silent cassette, and another 15 kb lacking any other marker fill in between the two silent cassettes. The entire region is low in general recombination, except for a hotspot around the double-strand break. The domain around the silent cassettes, in particular, is completely devoid of meiotic crossing-over. At least three gene functions contribute to this blockage, which is reminiscent of heterochromatinization (*see* HETEROCHROMATIN; X-CHROMOSOME INACTIVATION). Enforced suppression of recombination in this area may be a safeguard against excessive rearrangements between the partially repetitive sequences. Occasionally, however, rearrangements still occur and the most frequently used heterothallic strains are the result of deletions or duplications. Some other deletions are lethal.

The haploid life cycle of fission yeast is based on the effective shutdown of all sexual activities during vegetative growth and reinforced by the tight coregulation of conjugation and meiosis. Both mating-type cassettes must be expressed for gamete differentiation, and deletions of the active mating-type locus are sterile. Both mating-type cassettes encode an early function and a late one. The early functions are needed for production of the respective mating pheromone and receptivity towards the complementary mating factor. The two late functions combined are specifically needed for meiosis. As one of these late functions is only induced by the pheromone response system, meiosis itself depends upon pheromone activation in *Schiz. pombe*. The structural genes for the pheromones and corresponding receptors, as well as specific inducers of meiosis, are the most prominent targets for being controlled by the mating-type functions, but these genes are located elsewhere in the genome.

The suppression of sexual activities during vegetative growth is mediated by a multifunctional PROTEIN KINASE, the activity of which is modulated first by nutritritional deprivation and initial mating-type expression and again in response to pheromone reception. Its complete inactivation is necessary for the induction of meiosis, and the ultimate inactivation factor is specifically induced by the two late mating-type functions in combination. This cascade normally ensures that haploid cells depleted for nitrogen first differentiate as gametes, then conjugate, and only thereafter undergo meiosis and sporulation.

Budding yeast characteristics

The budding mode of cell division in *S. cerevisiae* results in asymmetric distribution of many cell components and also in differently timed cell cycles in the two descendants, of which only the new bud is termed a daughter cell and the other one is termed the mother cell. As a new mitotic spindle is already formed at the beginning of bud emergence, the nuclear division cycle, too, is very much distorted in comparison with other organisms.

The life cycle includes a regular succession of haploid and diploid stages during vegetative growth. All haploid cells are potential gametes, producing their own pheromone (α or **a**) and being responsive to the complementary mating factor. As soon as both kinds are present in the same culture, the vegetative cell division cycle is transiently stopped by G1 arrest, zygotes are formed by conjugation, and vegetative growth is resumed by budding at the diploid level. The zygote nucleus is heterozygous for mating type. This condition defines a third recognizable cell type, which no longer responds to either pheromone. Further mating-type switching is stopped as well. When such cells are starved of nitrogen, they regenerate the haploid state after meiosis and sporulation.

Newly switched cells of this yeast always arise in pairs of the same mating type. This suggests that the resident mating-type cassette is discarded whenever it is replaced by a new one and that this happens before replication. Moreover, the switching pattern is dominated by the cellular asymmetry of the budding process. Only mother cells are switchable, which means that both descendants after the next budding can have switched mating type. New buds, however, are always unswitchable before they become 'experienced' by giving birth to at least one bud of their own. In consequence of these switching rules, conjugation of cousin cells is the most frequently observed class of zygote formation in homothallic strains of *S. cerevisiae*.

The *HO* locus and its function

The switchable state in budding yeast is also correlated with a double-strand break at the target cassette. These breaks are transient, and are introduced by a sequence-specific ENDO-NUCLEASE, the product of the *HO* gene. The effective target sequence spans about 20 bp. The ends are progressively degraded, and the cut is lethal if no donor cassette is available for sealing of the gap by switching. The cuts are also lethal in mutant strains deficient in double-strand break repair genes. The standard laboratory strains of *S. cerevisiae* are heterothallic because they cannot make the cut, due to a simple *ho* mutation. All the rest of their originally homothallic back-up system is still potentially intact.

All the *swi* mutations affecting the frequency of mating-type switching in budding yeast are either linked to or affect the regulation of the *HO* gene, which is one of the most rigorously controlled genes in *S. cerevisiae* [7]. This gene is only expressed when three conditions are met. The cell must be haploid (i.e. not heterozygous for mating type), it must be a mother cell in the budding cycle, and it must be in late G1 phase of the nuclear division cycle. Three sets of control boxes specific for any one of these parameters are scattered over some 1.4 kb of DNA upstream of the *HO* coding region, and expression is regulated in a combinatorial fashion by positive and negative transcription factors.

HO expression in haploid cells depends on simultaneous activity of two positive factors, which only coincide in mother cells. One of these is specific for late G1 phase and depends on a prominent cell cycle regulator, which at mitosis acts as a CYCLIN-associated protein kinase, but in G1 phase contributes to another function still waiting for a mechanism. The G1/S boundary is passed soon after anaphase in mother cells but only much later in the buds, which have a considerably prolonged cell cycle — mostly due to a long G1 period of cellular growth to reach the size of a mother cell. The second transcription factor is unstable and synthesized periodically. It is only allowed access to the nucleus during late anaphase. Although its original concentration in mother and daughter nuclei is very similar, this factor is not retained in the daughter nuclei long enough to be present at late G1 when it would be needed [8].

The *in vivo* patterns of *HO* expression and mating-type switching are tightly correlated. Conversely, if *HO* is artificially brought under different control mechanisms, the timing of double-strand cuts changes accordingly so that G2 cells and/or buds can be induced to switch effectively.

The three cassette-bearing loci in *S. cerevisiae* are scattered over an entire chromosome, the active locus lying in the middle and the two silent ones close to the telomeres (*see* CHROMOSOME STRUCTURE). The silent cassettes are not only suppressed transcriptionally but are also rendered inaccessible to the *HO* endonuclease. In mutants that allow expression of the storage loci, all three cassettes become available for cutting and can now serve as target sites for mating-type switching.

Mating-type functions

As in *Schiz. pombe*, divergent transcripts are encoded by the active mating-type cassettes in *S. cerevisiae*, but only three of the four different transcripts are functionally significant. The mating-type gene products act on four sets of other genes: α-specific genes and **a**-specific genes expressed in the respective mating type, haploid-specific genes expressed in both mating types, and sporulation-specific genes expressed in **a**/α cells induced for sporulation. Deletion of the active mating-type locus does not induce sterility but allows mating as a stable **a** cell, as **a**-specific genes are constitutively expressed in the absence of α information.

The α cassette encodes two functions: α2 is a repressor of **a**-specific genes, and α1 is a positive regulatory factor for α-specific genes, which leads to mating as an α cell. The **a** cassette encodes a single important function: **a**1. This function is dispensable in haploid **a** cells, but in **a**/α cells it associates with α2 to form a repressor of both α1 and all the other haploid-specific genes, such as *HO*. Above all, a general inhibitor of sporulation-specific genes is also among the haploid-specific functions repressed by **a**1-α2. Therefore, **a**/α cells can no longer mate with any mating type, nor can they continue to switch their mating-type cassettes, but sporulation-specific genes become expressible.

Current trends and perspectives

There are still many unresolved problems in this mating-type system, which provides an intricate and hopefully tractable example of differentiation at the unicellular level.

The directionality of mating-type switching, in particular, cannot yet be explained in any detail. How is it that the resident cassette is usually replaced by a copy of the opposite mating type rather than by completely homologous information? In *S. cerevisiae*, at least, each mating type seems to prefer a particular silent locus as a donor — irrespective of what information is actually carried there.

The silencing mechanism for the storage loci is also not yet fully resolved, and the detailed interactions of the transcriptional regulatory proteins encoded by the locus with their target genes remain to be worked out.

The chain of pheromone-activated signal transduction is still subject to most active research [9]. Membrane-spanning receptor proteins bind the corresponding pheromone at their exterior domain. This leads to interaction of the intracellular domain with G PROTEINS and other factors, eventually resulting in the activation of various pheromone-inducible promoters.

Last but not least, the regulatory coupling of meiosis (and also conjugation in fission yeast) to nutritional depletion is yet another field of uncharted opportunities [10]. CYCLIC AMP and *ras*-related functions (*see* GTP-BINDING PROTEINS; ONCOGENES) are implicated in both yeasts, but their specific roles seem to be quite different.

R. EGEL

1 Herskowitz, I. (1989) A regulatory hierarchy for cell specialization in yeast. *Nature* **342**, 749–757.

2 Gutz, H. & Schmidt, H. (1990) The genetic basis of homothallism and heterothallism in *Saccharomyces cerevisiae* and *Schizosaccharomyces pombe*. *Semin. Devel. Biol.* **1**, 169–176.

3 Egel, R. et al. (1990) Sexual differentiation in fission yeast. *Trends Genet.* **6**, 369–373.

4 Klar, A.J.S. (1992) Developmental choices in mating-type interconversion in fission yeast. *Trends Genet.* **8**, 208–213.

5 Haber, J.E. (1992) Mating-type gene switching in *Saccharomyces cerevisiae*. *Trends Genet.* **8**, 446–452.

6 Singh, J. & Klar, A.J.S. (1993) DNA polymerase-α is essential for mating-type switching in fission yeast. *Nature* **361**, 272–273.

7 Nasmyth, K. & Shore, D. (1987) Transcriptional regulation in the yeast life cycle. *Science* **237**, 1162–1170.

8 Tebb, G. et al. (1993) Swi5 instability may be necessary but is not sufficient for asymmetric *HO* expression in yeast. *Genes Dev.* **7**, 517–528.

9 Marsh. L. et al. (1991) Signal transduction during pheromone response in yeast. *Annu. Rev. Cell Biol.* **7**, 699–728.

10 Nielsen, O. (1993) Signal transduction during mating and meiosis in *S. pombe. Trends Cell Biol.* **3**, 60–65.

yolk General term embracing the major storage products of the egg which are used as energy sources during embryonic development (*see* OOGENESIS). Eggs vary in the amount of yolk they contain and it is virtually absent from eggs of mammals, where embryonic nutrients are derived from the mother. The amount and distribution of yolk in an egg can influence the types of CLEAVAGE it undergoes. The proteins making up the yolk of yolky eggs are commonly lipophosphoproteins and are synthesized either in accessory cells of the ovary or in the liver. In amphibia, vitellogenin is the lipophosphoprotein synthesized by the liver. It is transported to the ovary, taken up into oocytes by receptor-mediated ENDOCYTOSIS, converted into the high-phosphate protein phosvitin, and the high-lipid protein lipovitellin, and packaged into large crystalline structures (yolk platelets) for storage. During early embryonic development these energy sources are mobilized by the breaking down of the yolk proteins by enzymes such as phosphoprotein phosphatases and the proteolytic cathepsins.

ypt proteins RAB PROTEINS. *See also*: GTP-BINDING PROTEINS.

Z

Z (1) The single-letter abbreviation for the AMINO ACIDS glutamine or glutamic acid.

(2) In PHOTOSYSTEM II the side chain of Tyr 161 (Y_z) of the D1 subunit which acts as the redox intermediate between the Mn complex of the water-oxidizing enzyme and P680. It was originally identified by its EPR signal (signal II_{vf}). *See also*: ESR/EPR SPECTROSCOPY; PHOTOSYNTHESIS; SIGNAL II.

Z Atomic number.

Z disc *See*: MUSCLE.

Z-DNA *See*: DNA; NUCLEIC ACID STRUCTURE.

Z scheme A term introduced by Hill and commonly used to describe the photosynthetic electron transport system of oxygenic organisms, in which two photosystems (represented by the two arms of the 'Z') are connected together in series (Fig. Z1). It derives from a display of the scheme on a normal redox potential diagram in which the O_2 electrode is positive and the H_2 electrode negative. Many prefer to display it in the opposite orientation, so that the light reactions are seen to drive the electrons upwards on the paper. Turning the scheme on its side, as a compromise, produces a 'Z'. *See also*: PHOTOSYNTHESIS.

Hill, R. (1965) *Essays in Biochemistry* **1**, 121–151.

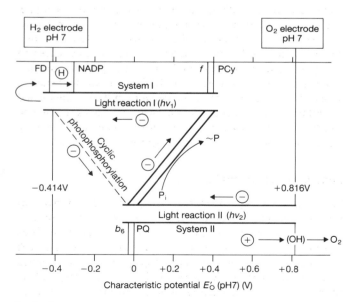

Fig. Z1 The original Z scheme. Redrawn from Hill (1965).

Z4 inhibitor *See*: SERPINS.

Zea mays Maize (corn, sweetcorn).

zeatin (Z) 6-[4-Hydroxy-3-methylbut-2E-enylamino]purine (Fig. Z2). M_r 219.2. A naturally occurring CYTOKININ first isolated and characterized from immature seeds of *Zea mays* (maize). Zeatin and its N^9-riboside and -ribotide, each of which has cytokinin activity, are widely distributed in plants. They are produced in root tips and young immature seeds, in which they are present at $10–1000\ \text{pmol}\,\text{g}^{-1}$ fresh weight, and are transported in the xylem from roots to shoots. Zeatin is produced from ISOPENTENYLADENINE by *trans*-hydroxylation and is metabolized by cleavage of the side chain to form adenine, and catalysed by cytokinin oxidase, or by the formation of O-, N^7- and N^9-glucosides. The N-glucosides are formed irreversibly, whereas the O-glucoside can be hydrolysed to zeatin and is probably a storage form. Zeatin riboside and other cytokinin ribosides are also components of tRNA, although this is probably not an important source.

Letham, D.S. (1963) *Life Sci.*, 569–573.

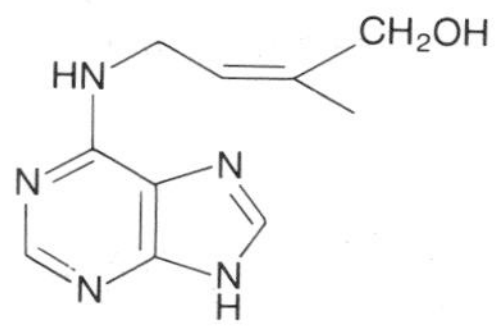

Fig. Z2 Structure of zeatin.

Zebrafish development

THE zebrafish (*Brachydanio rerio*), a native of the Indian subcontinent, has long been a popular denizen of tropical fish tanks in hotel lobbies and hallways. In the past 10 years, it has successfully made the transition to the research laboratory to become one of the major vertebrate species used in the study of various aspects of developmental biology and has joined the elite group of model developmental systems [1,2] (*see* AMPHIBIAN DEVELOPMENT; AVIAN DEVELOPMENT; CAENORHABDITIS DEVELOPMENT; DROSOPHILA DEVELOPMENT; MAMMALIAN DEVELOPMENT).

What are the advantages offered by this hardy little teleost? The primary requirement of any laboratory animal is that it is easy both to look after and to breed from. The zebrafish is both. Thirty to fifty adult fish will live happily in a five gallon tank at 28.5°C, fed regularly on pellets and the occasional gourmet bloodworm. Under these conditions, the female will lay a batch of up to 300 eggs about once a week. Conveniently, the females lay at dawn, a time easily regulated by an artificial 14 hour light/10 hour dark laboratory lighting system. Eggs, fertilized by male fish, may be siphoned from the bottom of the tank, or removed from spawning chambers placed inside the breeding tank. Artificial fertilization can also be achieved by gently squeezing eggs from the females and fertilizing with sperm similarly obtained from the males. This method is important for genetic studies, as will be described later.

Fertilized eggs, incubated at 28.5°C, begin gastrulation within 5–6 hours, and develop into a recognizable fish by 24 hours post-fertilization (hpf). Throughout this time, the embryo is transparent, making it very easy to follow the principal developmental events through a microscope. Indeed, the transparency enables every cell in the embryo to be seen in the whole intact animal and allows a full appreciation of the true beauty of the developing embryo.

What follows is a three-part description of the use of zebrafish in current developmental research. First, the sequence of embryonic development will be described in more detail; we will then consider several of the major areas of experimental analysis and finally, a brief discussion will be given on the hurdles still to be overcome in this system, and on the likely directions of future research.

Zebrafish development

The major stages of development are depicted in Fig. Z3. Upon fertilization, the embryo consists of a single dome of cytoplasm sitting atop the yolk mass. Within an hour, this cap of cytoplasm cleaves, but the yolk cell does not; subsequent cleavages produce a mound of blastomeres perched above the yolk. The early

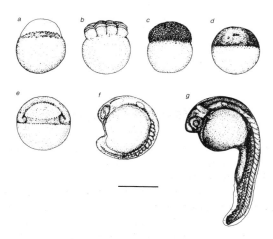

Fig. Z3 Stages in the development of the zebrafish embryo. The drawings are all left side views and are scaled identically. *a*, 0.5 hpf, before the first cleavage; *b*, 16-cell stage which is reached at 1.5 hpf; *c*, 3 hpf, the midblastula; *d*, 4 hpf, the late blastula; *e*, gastrulation the shield stage, reached at 6 hpf; *f*, the 14-somite stage reached at 16 hpf; *g*, the 24 hpf embryo. From [6].

cleavages are usually in predictable positions: cleavages 1 to 5 are all vertical, with the first horizontal cleavage producing the 64-cell stage 2 hours after fertilization. The midblastula stage is reached at 3 hours, with the blastomeres remaining high above the yolk cell. By this time, the blastula can be divided into a single outer layer of flattened cells, the outer enveloping layer, a deep layer (DEL) of more rounded cells, and a syncytial cell layer located above the yolk. Cell movement begins during the blastula stage, when cells of the DEL begin to jostle, and the mound of blastomeres begins to flatten to produce first an oblong (3.7 hpf), and then a spherical embryo (4 hpf). This thinning of the blastoderm reflects the beginning of a migration of blastula cells downwards over the margins of the yolk cell, a process known as epiboly. By 5.3 hpf the leading edge of the blastoderm has reached halfway down the yolk cell (50% epiboly) and an additional set of cell movements initiates the process of gastrulation. At the margin of the blastoderm, the DEL cells begin to turn inwards, a process known as involution, migrate towards the future dorsal side of the embryo (convergence), and spread rostrocaudally (extension).

These movements — involution, convergence, and extension — are formally equivalent to the gastrulation movements described in detail in *Xenopus* (*see* AMPHIBIAN DEVELOPMENT). A difference, however, is that involution in the zebrafish appears to initiate simultaneously, or nearly so, all around the periphery of the blastoderm, whereas in *Xenopus*, this movement occurs first in a localized position, the blastopore lip. By 5.7 hpf the involution process gives the appearance of a ring on the periphery of the blastoderm: this is the germ ring stage. This is followed by the shield stage, at which the convergent movement of gastrulating cells to the dorsal position has created a local thickening of the germ ring termed the embryonic shield (6 hpf). At the shield stage, the embryonic axes can be visualized for the first time: the shield is dorsal and the animal pole marks rostral.

Gastrulation continues for several hours, and the completion of epiboly is reached when the yolk cell is completely covered with a thin layer of cells. Within an hour of closure of the yolk plug (9.6 hpf), the tailbud is visible and the first SOMITE forms, initiating the rapid regionalization of the MESODERM. Specific neural structures such as the optic vesicles (11.5 hpf), otic placodes and hindbrain segments (16 hpf) and otic vesicle (19 hpf) soon become visible. Body movements, in the form of random twitches, initiate at about 18 hpf, as neural circuitry in the peripheral and central nervous systems becomes connected. Somites continue to be added at a rate of about one pair every 30 minutes so that by 24 hours, 30 of these segmental structures are present.

Principal experimental progress

In recent years experimentation using zebrafish embryos has been very fruitful in several fields of developmental research. Here we will briefly describe some of the progress in three different areas of study. First we will discuss how early cell movements, cell lineage, and fate mapping have been examined by following fluorescently labelled cells in living embryos. Secondly we outline how analysis of developmental mutations has led to several important insights into how specific gene defects can affect development, and finally we show how the relative simplicity of the nervous system has allowed detailed examination of the early development of neurons and their processes.

Analysis of cell lineage and movement

As the developing zebrafish embryo is transparent, cells injected with a fluorescent dye can be observed under the microscope in living embryos against an optically clear background. Cells injected early in development may be observed over long periods of time as they migrate and divide within the embryo. The drawback of prolonged observation is that excitation of the fluorescent dye can be phototoxic to the labelled cell. To overcome this, labelled cells are observed using very little fluorescent excitation through a signal intensification camera. The camera image is then passed through a computer image enhancement program and labelled cells are viewed on a monitor screen. This technology enables the same cells to be repeatedly viewed during development without any detriment to their viability and health.

The ability to follow cells through development allows detailed LINEAGE MAPS of single cells to be constructed (*see also* CAENORHABDITIS DEVELOPMENT). This is a considerably more difficult task when using other lineage markers (such as horseradish peroxidase or retroviruses) for which there is no way of examining the labelled cells at any point between the time of injection and fixation. For this reason the transparent fish embryo is possibly the best vertebrate model system for analysis of early cell movements and divisions in the embryo.

These cell labelling methods have established that clonally related cells injected at cleavage or blastula stages populate a wide range of tissues within the developing embryo [2]. The degree of scattering of clonally related cells also highlights the extensive cell movements which occur during epiboly, involution, convergence and extension. In contrast to the indeterminate fate of cells injected during early blastula stages, the progeny of cells present during gastrula stages are restricted to particular tissues. However, even though cells may be restricted to form a single tissue, they may still form different cell types within the one tissue. Therefore, CELL FATE is not determined by a cell lineage-dependent mechanism in the zebrafish embryo. These results focus our attention on the roles of cell position and CELL–CELL INTERACTIONS in controlling important developmental decisions. To study such phenomena, embryologists have traditionally taken one of two approaches: cell transplantation to examine responses and behaviours of cells in ectopic positions within the embryo; and genetics, whereby a search is made for MUTATIONS which affect early development. The latter approach can only be made on systems which are amenable to genetic analysis, and the zebrafish has many of the necessary attributes.

Genetic analysis of developmental mechanisms

The zebrafish lends itself to genetic studies for a number of reasons; the generation time is relatively short (only 2 or 3 months) and females generate large numbers of eggs which develop rapidly into transparent embryos easily analysed in large numbers. In addition, techniques have been developed which further simplify the search for developmentally interesting mutations (Fig. Z4). George Streisinger, in the 1970s, at the University of Oregon, pioneered the zebrafish as a system in which to study developmental genetics (see [1,2] for reviews). He discovered that embryos would develop essentially normally as haploids, that is, with only half the normal amount of genetic material. Haploid embryos are generated by fertilizing eggs squeezed from the female with sperm inactivated by exposure to ultraviolet (UV) light. Such UV-sperm are capable of activating the egg and initiating development, but fail to donate their genetic material, leaving the developing embryo with only the haploid female set of chromosomes. Haploid fish go through the early stages of development and die soon after hatching. Using haploid analysis, any RECESSIVE mutation present in the female HETEROZYGOTE will be revealed in 50% of the offspring in a single generation.

This method can also be used for identifying mutations induced in the female gametes using chemical or gamma-irradiation mutagenesis. This is achieved by mutagenizing embryos at a stage when 3–4 primordial germ cells have been determined. After exposure to the mutagen, the embryos are grown up and a percentage of their offspring (derived from the germ cells present in the mutagenized embryo) will carry a mutation which is then revealed by fertilization with UV-inactivated sperm. Some eggs carrying the mutation must also be fertilized with normal sperm to produce a breeding stock of fish carrying the mutation.

Two methods have also been established for creating HOMOZYGOUS diploids. The first relies on the fact that mature eggs are held in the second meiotic division (*see* MEIOSIS) and are released to complete this division by fertilization. Treatment of eggs with hydrostatic pressure immediately after fertilization with UV-sperm prevents formation of the second polar body and leaves the egg containing a diploid set of chromosomes, entirely derived from the mother. The method does not produce complete homozygosity, however, because genetic crossing-over (RECOMBINATION) occurs before the first meiotic division. The further a gene is from the CENTROMERE, the more likely it is that a crossover event will lead to that gene being heterozygous in the pressure-treated progeny. Because of this, the 'early pressure' method can be useful for mapping genes with respect to their location along the chromosome.

The second method for making homozygous diploids creates truly homozygous fish and relies on heat shocking eggs during the first cell cycle, to prevent the first mitotic division. Heat shock is less useful than 'early pressure', as it generates large numbers of nongenetic abnormalities, but it is useful for producing fertile ISOGENIC lines of fish which can be used as a stock for mutagenesis studies. Such isogenic clonal lines, free of lethal mutations, are obtained as parthenogenetic offspring of single homozygous females originally produced by heat shock.

Several mutations that affect genes of interest to developmental biologists have now been isolated. Of these, we will describe two that have now been well characterized. The first, a mutant named *spadetail* (*spt-1*) [3], produces a phenotype in which the mesoderm of the trunk is severely depleted as a result of abnormal migration of mesodermal precursor cells towards the tail. The depletion in the trunk region leads to disruption in somite

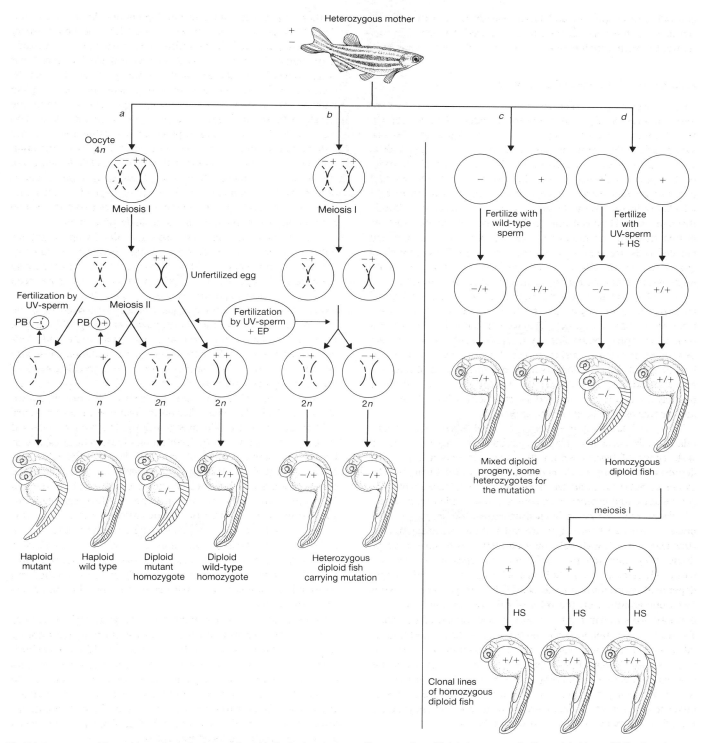

Fig. Z4 A summary of the main genetic strategies mentioned in the text. Starting from a wild-type mother, heterozygous for a recessive mutation, four procedural routes are indicated. *a*, Creation of haploids or, if early pressure (EP) is used to prevent second polar body (PB) formation, the generation of homozygous diploids. Additionally, at a frequency dependent upon the distance of the mutant gene from the centromere, recombination will result in the generation of fish heterozygous for the mutated locus (*b*). Route *c* shows normal development generating heterozygotes for the maintenance of recessive genes. Route *d* illustrates the generation of homozygous diploid fish using UV-irradiated sperm and heat shock (HS) during the first cell cycle. The chromosomal content of eggs in panels *c* and *d* has been simplified for clarity. See text for more details. Adapted from [2].

...mation in this region. Ho and Kane were able to assay whether the abnormal accumulation of cells in the tail was due to a CELL-AUTONOMOUS effect in the mesodermal cells by cotransplanting wild-type and mutant mesodermal precursor cells (each labelled with a different fluorescent dye) into the same host embryo, and then following the cells using the image enhancement techniques described above. Independent of the genotype of the host embryo, the mixture of transplanted cells always sorts out, with wild-type cells migrating normally to the trunk and mutant cells migrating abnormally into the tail. Thus the *spadetail* mutation acts cell autonomously in mesodermal precursor cells, altering their abilities to migrate normally throughout the body.

In a second embryonic lethal mutation, *cyclops*, the ventral midline of the entire central nervous system is absent [4]. Within the spinal cord the defect is very specific — just a single line of ventral floorplate cells is missing. Again, elegant MOSAIC ANALYSIS from cell transplantation experiments has been used to investigate the nature of the mutation. In this case, wild-type cells transplanted into mutant embryos formed normal floorplate, thereby rescuing the mutant phenotype. What is more, the presence of floorplate formed from wild-type cells actually induced mutant cells also to form floorplate. This indicates that the mutation is not due to an inability of ventral midline cells to form floorplate cells in *cyclops* embryos; rather it is more likely that, in *cyclops*, these cells lack the ability to respond to a floorplate induction signal (it is postulated that lateral induction of floorplate by wild-type cells in mutant embryos acts via a separate, secondary signal).

Developmental neurobiology

More work has been done on the development of the nervous system of the zebrafish than on any other aspect of its development (see [5]) and we will only briefly describe some of the advantages of the system.

It was known for many years that at least one pair of cells was uniquely identifiable in lower vertebrates — this was the Mauthner cell, a reticulospinal neuron involved in generating escape responses. In recent years it has become apparent that single cells and very small clusters of cells can be identified throughout the developing fish nervous system. Each group of cells can be identified on the basis of its location and axonal projection pattern. The small number of such neurons means that the nervous system is sufficiently simple to study the development and axonal outgrowth of single cells in great detail. Added to this, the small size and transparency of the embryo allow neuronal development to be examined in the whole animal. The remarkably simple embryonic nervous system of the zebrafish appears to share many features with developing invertebrate nervous systems, and indeed, techniques initially used to study invertebrate neural development, such as laser ablation of cells and single cell transplantation, are now successfully being used to further analyse early zebrafish neural development.

Future prospects

The zebrafish has emerged as an important vertebrate model system for studying developmental mechanisms for many different reasons: ease of breeding; hardiness; rapid generation time; small size and transparency of embryos; simplicity of the early nervous system; ability to create mutant stocks; and cell marking, tracing, and transplantation techniques. Genetics will continue to play an important part in the future and should provide an increasing number of mutants affecting a wide range of developing tissues and organs as mutational screening becomes more sophisticated and specific.

Molecular analysis of zebrafish development is currently a very active field of research though it is true to say that, at present, the fish lags behind *Xenopus* and the mouse in this area of enquiry. Molecular genetics is one field in which the zebrafish has great potential and indeed there is currently a great deal of work aimed at finding and isolating developmental regulatory genes. The most common approach is to look for zebrafish homologues of genes already identified and isolated in insects or in other vertebrates. Using this approach, a large number of zebrafish developmental genes, including HOMEOBOX GENES, and *pax* genes (genes containing the PAIRED BOX) and *Wnt*-related genes, have now been isolated. The relatively new approach of performing *in situ* HYBRIDIZATIONS on whole embryos has been used to great effect in fish embryos, allowing very detailed analysis of the regional expression patterns of these potentially important patterning genes. It is hoped that the experimental potential of the zebrafish embryo, already exploited in other fields of research, will also prove useful in trying to examine the functions of these, and other, important developmental control genes. Furthermore, several recent reports have shown that it is now feasible to insert DNA into the fish genome, and although this technology is still in its embryonic stages, it surely points the way towards a future in which INSERTION MUTAGENESIS will be another tool available to the researcher working with the zebrafish embryo.

N. HOLDER
S.W. WILSON

See also: VERTEBRATE NEURAL DEVELOPMENT.

1 Kimmel, C.B. & Warga, R.M. (1988) Cell lineage and developmental potential of cells in the zebrafish embryo. *Trends Genet.* **4**, 68–74.
2 Kimmel, C.B. (1989) Genetics and early development of zebrafish. *Trends Genet.* **5**, 283–288.
3 Ho, R.K. & Kane, D.A. (1990) Cell-autonomous action of zebrafish spt-1 mutation in specific mesodermal precursors. *Nature* **348**, 728–730.
4 Hatta, K. et al. (1991) The *cyclops* mutation blocks specification of the floorplate of the zebrafish central nervous system. *Nature* **350**, 339–341.
5 Eisen, J.S. (1991) Developmental neurobiology of the zebrafish. *J. Neurosci.* **11**, 311–317.
6 Kimmel, C.B. et al. (1992) Lineage specification during early embryonic development of the zebrafish. In *Cell–Cell Interactions in Early Development* (Gerhart, J., Ed.) 203–226 (Wiley & Sons, New York).

Zidovudine AZR.

zinc-finger motif/protein DNA-binding protein secondary structure motif found in many TRANSCRIPTION FACTORS in which a Zn atom is bound to the base of a loop of polypeptide chain. *See*: PROTEIN–NUCLEIC ACID INTERACTIONS (*see* Fig. P62); TRANSCRIPTION FACTORS (*see* Fig. T33).

zipper, zippering *See*: (1) ENDOCYTOSIS; (2) LEUCINE ZIPPER.